Biology Today

Contributing Consultants

Biology Today

Del Mar, California

Communications Research Machines,
Inc., Del Mar, California 92014. Library
of Congress Catalog Card Number:
72-176334 Manufactured in the United
States of America
Second Printing

Preface

Like religion or politics or group therapy, science is a peculiarly human endeavor that seeks newer and better solutions to the problems plaguing us all. The horizons of science continue to expand as it becomes ever more responsive to social priorities of the day.

The preeminence of the physical sciences in the 1940s was triggered in part by the sense of urgency to end a horrifying world war. The technological means to do so had emerged out of the work of the Rutherfords, the Bohrs, the Einsteins of physics. After the war officially came to a close, many physical scientists found that they could apply their knowledge to biological problems. In exploring the mysteries of atoms, physicists reinforced the interrelatedness of the whole of scientific, and human, endeavor. Their work also provided incredible perceptions for the biological sciences. Before, studies of all living things had floated disconnectedly above the physical and chemical substrate of life, and the biologists were left to map the manifest diversity of it. Armed with perceptions at the molecular level, biology—traditionally a science of careful and elaborate observation—is now finding solutions to problems of immediate concern.

Many of these advances in biology directly or indirectly involve information processing. Man has access to most of the information in the biosphere. If earth is to be preserved as a habitat for humans and other living things, then humans must better examine, codify, interpret, and apply this information. Watson and Crick's insight into the structure of DNA was a dramatic step forward in our understanding of the basic units of biological information. When we discover how these and other units are rearranged against randomness—how they form an amoeba, a potato, or a man, or by extrapolation, quantum mechanics or a Bach concerto—then we will know better how to utilize this information in problem-solving.

Virology, cancer research, immunology, applied genetics—these are a few of the common denominators in the existing equation of social priorities for biological inquiry. Yet, unhappily, we have come to realize that economic and technological man, reckless of biological balance, threatens the existence of all living things, himself included. Constructive change can only come through intelligent commitment—a commitment that requires an understanding of and reverence for the basic mechanisms of life. The goal of this book is to help convey that essential understanding.

John H Painter Jr.

Publisher
Del Mar, California

Contents

IV
The Continuity of Life
204–311

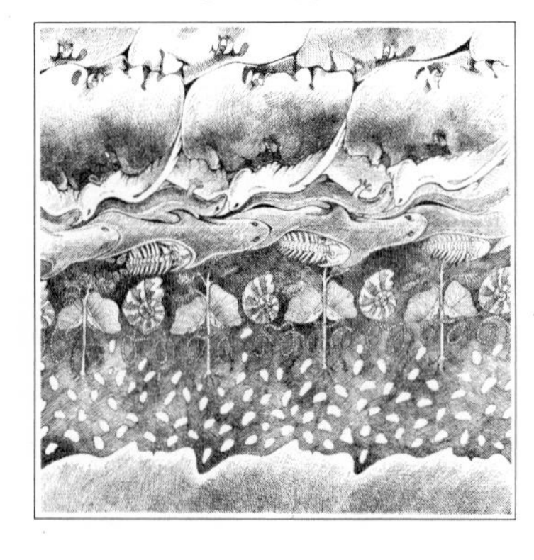

VI
Biological Behavior
558–631

VIII
The Human Organism
792–947

What Is Life?

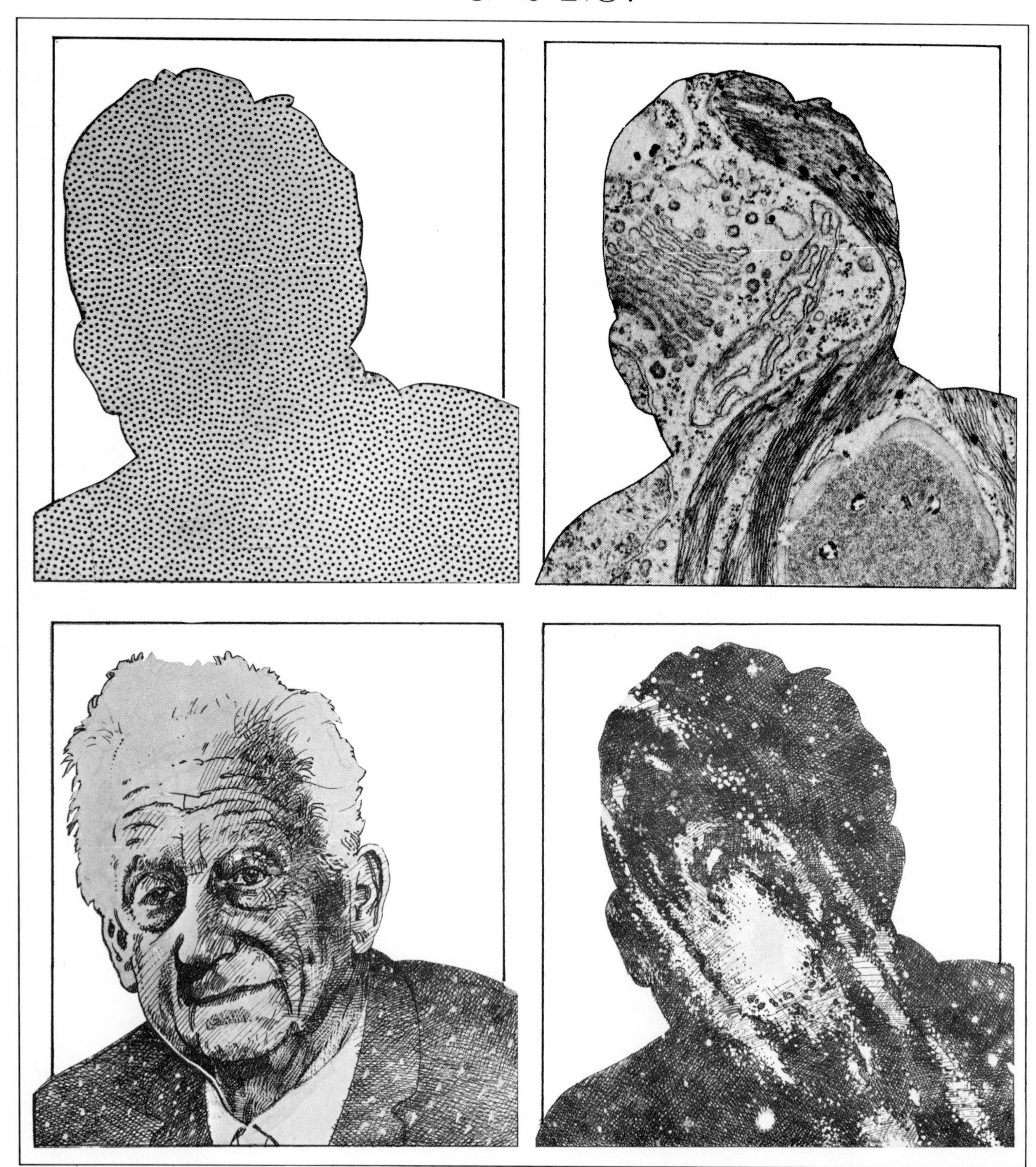

Albert Szent-Györgyi, who considers himself to be an American, was born into a Hungarian family of prominent scientists 77 years ago. He was unwillingly involved in his early life in the two World Wars and in the political intrigues of the big powers engaged in these wars. Throughout his life, he has been quietly searching for the principles on which nature and all of life are organized. In 1937 Szent-Györgyi's search resulted in the isolation of vitamin C, for which he was awarded the Nobel Prize. His protest against man's "idiocy," which he believes is evident in the irrational pursuit of war and politics that characterizes our Western culture, has been capsulized in his two short books, The Crazy Ape *and* What's Next. The Crazy Ape *warns youth against the gerontocracy that rules the world and urges mankind to take advantage of technological skills in order to create a psychologically and socially progressive world where humanistic values are paramount. To him, research is "not a systematic occupation but an intuitive artistic vocation."*

"What is life?" The question has been asked innumerable times but has been answered to the satisfaction of few. Science is based on the experience that nature gives intelligent answers to intelligent questions. To senseless questions, nature gives senseless answers—or no answers at all. If nature has never provided an answer to this question, perhaps something is wrong with the question.

The question is wrong indeed. It has no sense, for life in itself does

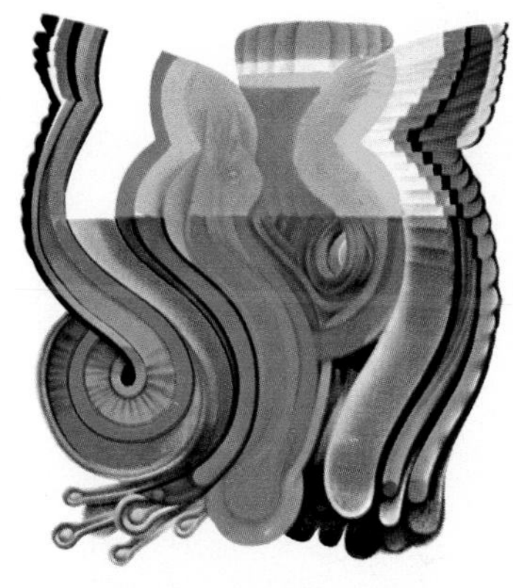

not exist. No one has seen or measured life. Life is always linked to material systems; what man sees and measures are living systems of matter. Life is not a thing to be studied; rather, "being alive" is a quality of some physical systems.

A look at the living world reveals an incredible variety of shapes, sizes, forms, and colors. There seems to be an infinite variability among living systems. How can man approach such complexity? How can he ask intelligent questions?

One key to an intelligent approach may be the simple fact that things can be put together in two different ways: randomly or meaningfully. Things put together in random fashion form a senseless heap. Nine persons selected at random and placed together probably will form nothing more

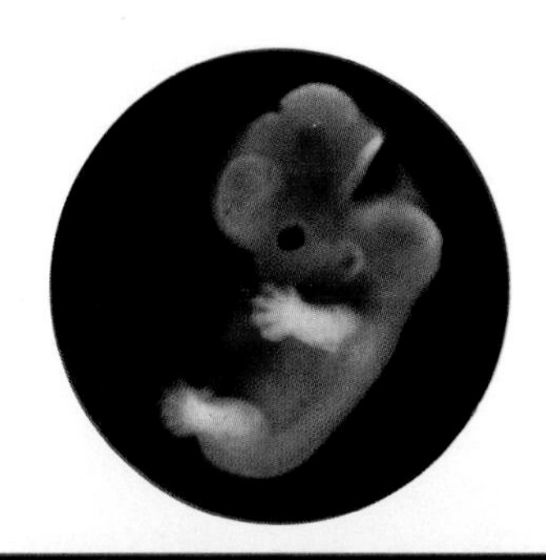

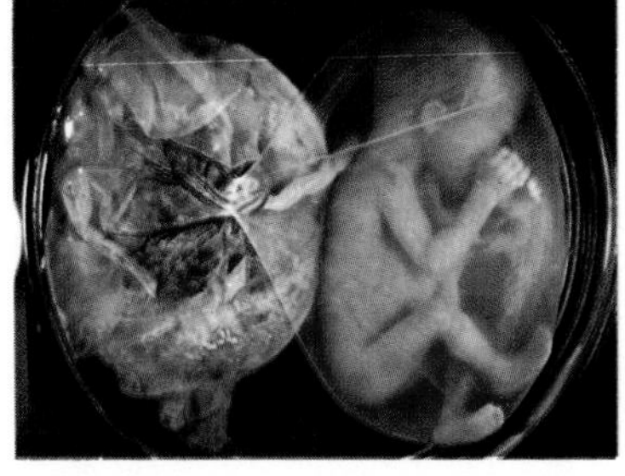

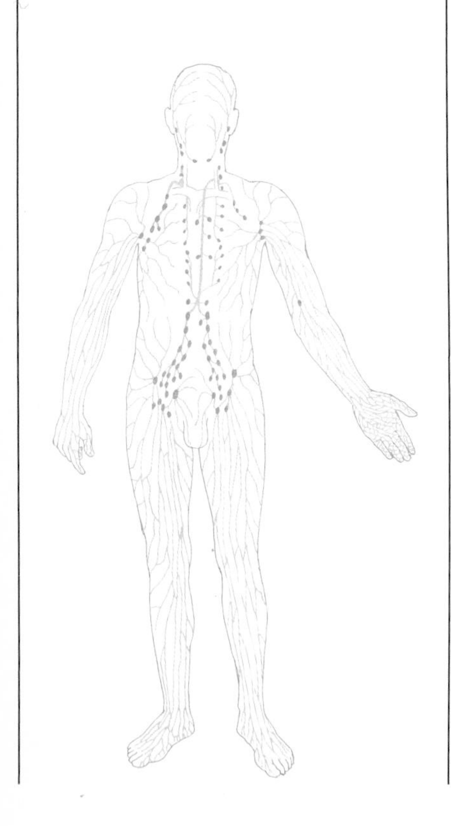

than a slightly puzzled collection of nine individuals. Nine persons selected and combined in a meaningful fashion may form a championship baseball team. The whole in this case is more than the sum of its parts – it is what is called organization.

If an atomic nucleus is combined with electrons, an atom is formed. This atom is something entirely new, quite different from electrons or nuclei alone. When atoms are combined, molecules are formed. Again, a new thing is generated with strikingly different qualities. Smaller molecules – say, amino acids – may be combined to form a "macromolecule" – perhaps a protein. This macromolecule has a number of amazing qualities. It demonstrates self-organization – the ability to create more complex, higher structures. It may act as an enzyme to speed up a particular

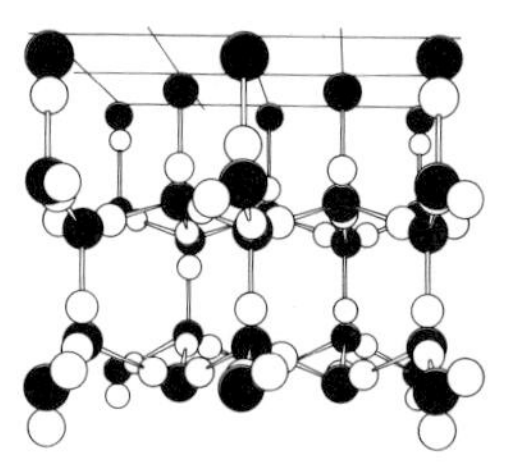

chemical reaction, or it may act as an antibody to neutralize the effects of some other specific protein molecule. Proteins can be created in a literally inexhaustible variety of forms, each with its own qualities.

Macromolecules may be combined to form small "organelles," such as mitochondria or muscle fibrils. When they are combined, the result is a cell – the unit of life, the miracle of creation – capable of reproduction and of independent existence.

The more complex the system, the more complex its qualities. Organs may be built from cells; from organs may come an individual organism, such as a human being. Individuals in turn may be combined

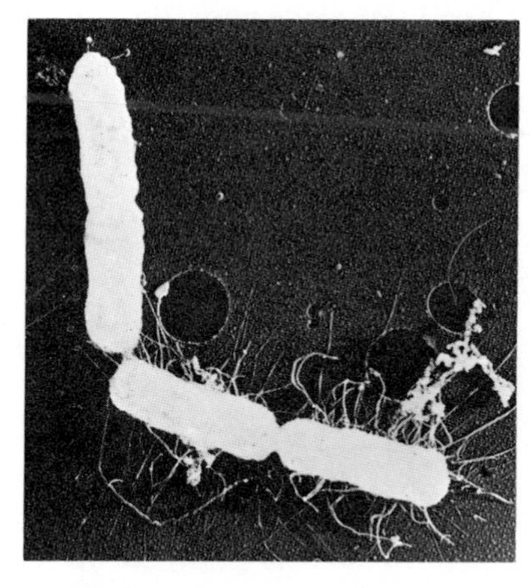

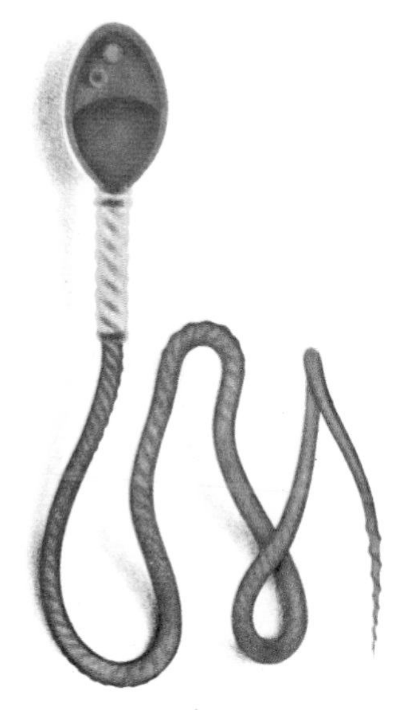

to form societies or populations, which again have their own rules. At each level of complexity are new qualities not present in the simpler levels. The study of each level yields new information for the biologist.

The history of biology has been marked by a penetration into ever smaller dimensions. In the sixteenth century, Vesalius was dependent on his unaided eyesight for his study of the human body. In the following century, the optical microscope led to the discovery of many new details of structure. Marcello Malpighi observed the capillary vessels that complete the cycle of blood circulation and showed that even such tiny insects as the silkworm have an intricate internal structure. Anton van Leeuwenhoek described blood cells and the compound eyes of insects. Robert Hooke described the cellular structure of plants.

As microscopes were improved, more and more details of structure were described. By the nineteenth century, it was becoming clear that all complex organisms are composed of semi-independent units called cells. The major structural features of cells were established. Bacteria were discovered and studied.

In this century, the electron microscope has taken the scientist down to molecular dimensions, and he has learned to observe with x-rays as well as with visible light. Organic chemistry was established in the nineteenth century, and by the beginning of this century, it was clear that this approach could be applied to the study of living systems. Biologists have had to learn

a new anatomy—the anatomy of molecules. Chemists and physicists have penetrated the atom, first finding the elementary particles and then moving still deeper into the realm of wave mechanics. The discovery of the wave properties of the electron has given a deep insight into the nature of biological reactions.

As scientists attempt to understand a living system, they move down from dimension to

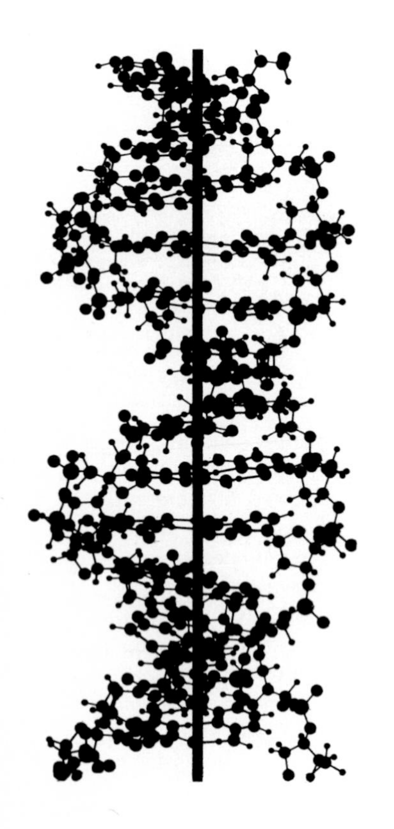

dimension, from one level of complexity to the next lower level. I followed this course in my own studies. I moved from anatomy to the study of tissues, then to electron microscopy and chemistry, and finally to quantum mechanics. This downward journey through the scale of dimensions has its irony, for in my search for the secret of life, I ended up with atoms and electrons, which have no life at all. Somewhere along the line, life has run out through my fingers. So, in my old age, I am now retracing my steps, trying to fight my way back toward the cell.

I have concluded that life is not linked to any particular unit; it is the expression of the harmonious collaboration of all. As I descended through the levels of complexity, I studied simpler units and found myself speaking more and more in the language of chemistry and physics.

J. F. Danielli has shown that the subcellular organs of various cells are interchangeable. They can be transferred from one cell to another, much as organs can be transplanted from one human individual to another. The parts of the cell have no individuality. The

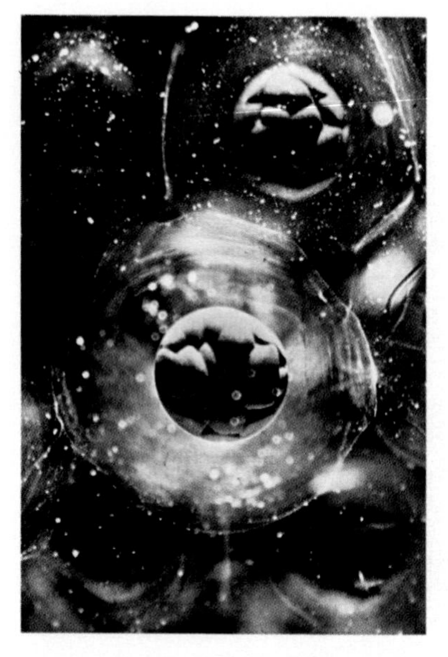

quality of individuality resides in the higher organization—in the cell or the individual.

No one yet knows the higher principle that holds a cell together. Perhaps the answer will be found in irreversible thermodynamics. The good working order of a living cell may correspond to a stable state with a high probability of occurrence. Perhaps some new principle—as yet undiscovered—keeps the cell together. Living systems do not only maintain their good working order but they all tend to improve it, to make the working structure more complex. When the fundamental principle that holds the cell together is found, perhaps we

will then also understand what brought together the first living system and understand what drives living systems toward self-perfection.

Scientists know today that rather complex molecules—amino acids, nucleic bases, even macromolecules—can under certain conditions be built without intervention of living systems. They are still seeking the principle that brought them together for the first time and that makes these systems improve themselves by building more complex structures, capable of more complex functions.

Biology is a very young science. It has called itself a separate science for only some eight decades. No one

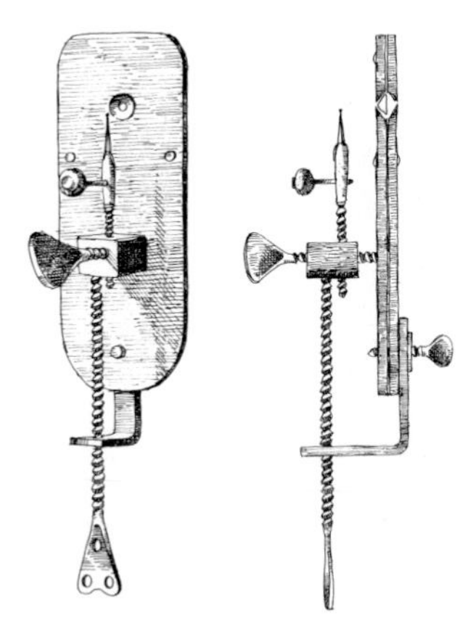

can expect it to find the answers to all questions. The most important questions are yet unanswered—perhaps unasked. What biologists can do—what they are doing at present—is to ask questions that seem answerable with present techniques. They ask questions about structure and function, from the nature of consciousness down to the behavior of electrons, hoping that some day all of this detailed knowledge will come together in a deeper understanding.

Perhaps some day they will find a new way of looking at things. The best scientists, with the aid of giant computers, cannot yet fully explain the behavior of three electrons moving within an atom. Yet those three electrons—even dozens of electrons—know exactly what to do and never miss. In the essence, nature may be far simpler than is believed.

To see the solutions, scientists must preserve a certain naïveté, a

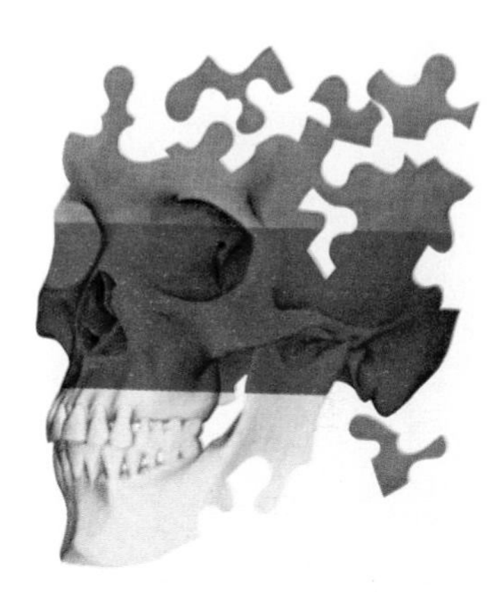

childish simplicity of the mind, an ability to recognize a miracle when they see it every day. The solution may be far closer than it seems. It was a hundred years ago that H. P. Bowditch, one of the first American physiologists, showed that after a frog's heart has been stopped for a while, its first beats are rather weak. The heart gradually regains its original strength, with the record of the heartbeat rising like a series of

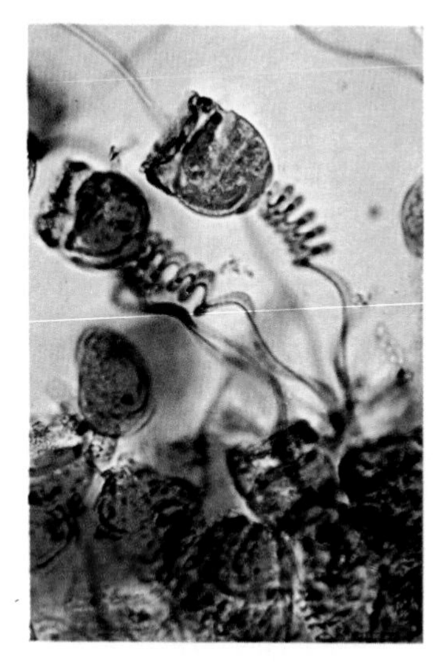

stair steps. Bowditch called this phenomenon "the staircase."

One might expect the heart to be stronger after a rest. Yet what is observed here can be considered a general quality of living systems. Life generates life: rest or inactivity causes life to fade away. Muscles weaken if they are not used; they become stronger if exercised regularly.

This principle is one of self-organization, one of the most striking differences between a living system and a nonliving one. A machine is worn out by usage. A living system is worn out by inactivity. Living systems are able to organize and improve themselves.

S. J. Hajdu and I tried to discover the mechanism that produced the Bowditch staircase. We found that potassium leaks out from the heart fibers into the surrounding fluid

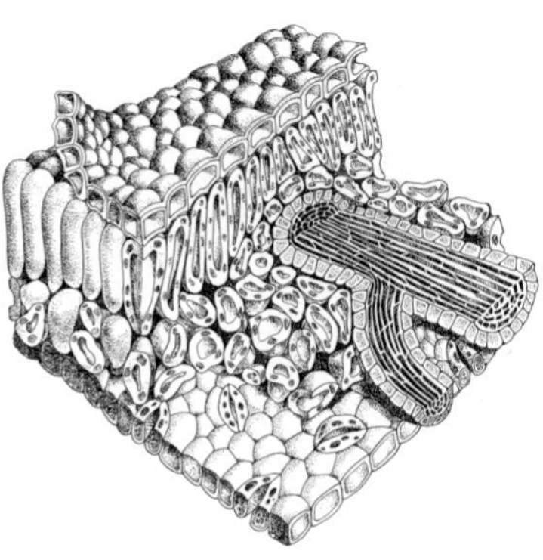

when the heart is not beating. When the heart resumes its activity, the potassium is pumped back into the heart fibers. The change in strength of the heartbeats is caused by this change in the distribution of potassium. The movement of the potassium back into the fibers, against the potassium concentration gradient, increases the amount of organization or order in the system. The entropy of the system is decreased. Ernst Schrödinger in 1944 suggested that the ability to decrease entropy is the most characteristic feature of living systems. In a living system, the decrease of entropy leads to further decrease of entropy, to greater order. The increase of

entropy leads to further increase of entropy, the maximum state of entropy being death.

Further pursuit of these thoughts would lead into abstract speculation. I would like to point out, however, that these abstruse questions are not at all far from the sickbeds of suffering patients. One of our most important drugs is digitalis, which is used to stimulate failing hearts. Hajdu and I showed that digitalis helps the heart to pump potassium back into its fibers and to retain it. The inability of a heart to maintain its potassium concentration may be one cause of its failure.

Living systems are clearly different from nonliving systems. There must therefore be a difference in the way these two kinds of systems are thought about. Physics is undoubtedly the most basic science. In a way, biology is only an applied science, applying physics as a tool for the analysis of living systems. However, there are distinctions between physical and biological events.

Suppose a process, left to itself, is likely to occur 999 times in one way for each time that it occurs in another. The physicist concerns himself primarily with the first way. Physics is the science of the probable.

Biology is the science of the improbable. On principle, biological reactions must be improbable. If man's cells worked only through the probable reactions, they would soon run down. In order to regulate itself, a cell must use improbable reactions and must make them take place through very specific tricks. The cell may make use of just that one way in a thousand that the physicist ignores.

The cell finds a way to make the reaction go at just the right moment and at the desired rate. The reaction may be improbable, but if it is thermodynamically possible, the cell will find a way to use it.

Physically, all of us—you and I—are improbable. The probability of atoms happening to come together in the complex structure that makes up my body is so tiny that it is practically equal to zero.

Another difference between the physical and biological approaches to the study of a reaction lies in the matter of isolation. The physicist is apt to attempt to isolate the reaction he wishes to study. In biology, single reactions are rarely encountered. Most biological reactions are parts of

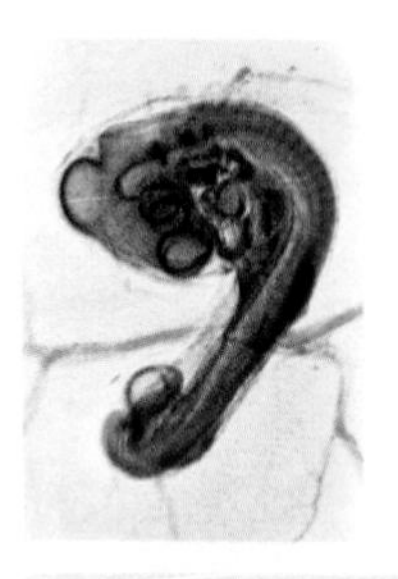

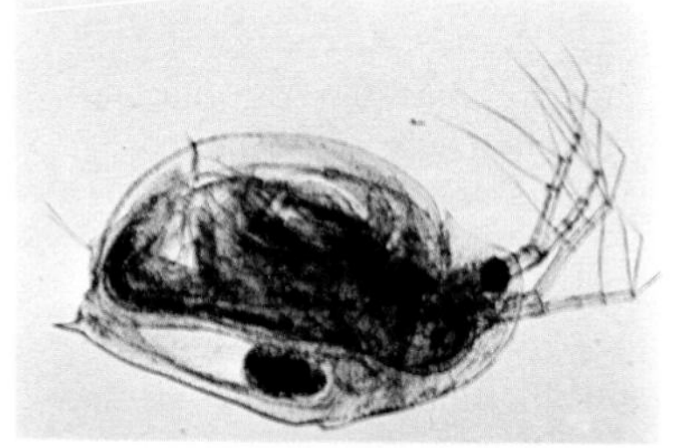

complex chains. They cannot be fully understood except as members of the chain, or even as parts of an entire living system — the living biological entity.

For example, one of the most important biological reactions is the "electron flow" that underlies photosynthesis and biological oxidation. These processes generate the energy that keeps living systems going. In these reactions, electrons "flow" from molecule to molecule. If an electron moves from molecule A to molecule B, it leaves a positive electric charge on molecule A. This charge tends to pull the electron back toward A. Electrons could not move against such a strong electrostatic attraction. However, if molecule A is a member of a chain and simultaneously receives an electron from a third molecule, its positive charge will be neutralized. There will then be nothing to pull the electron back toward A. Thus, where electrons cannot move from A to B in an isolated system, they can "flow" without difficulty if A and B are members of a chain.

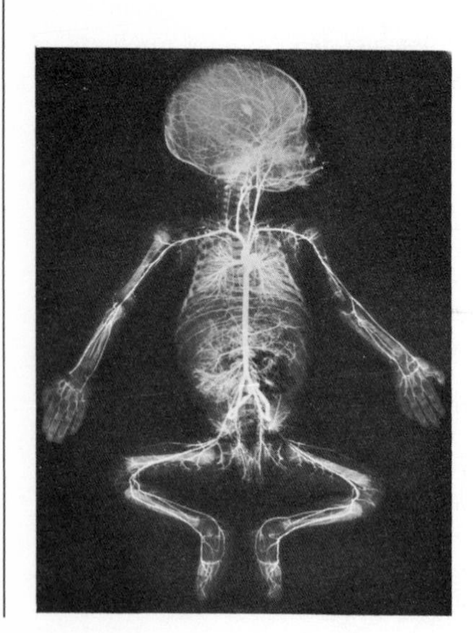

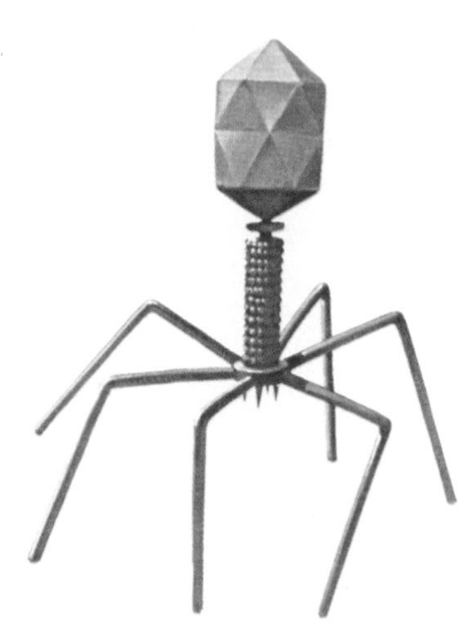

Like many biological researchers, I have often worked for long periods, using all the tricks of chemistry, wave mechanics, and mathematics to understand a certain reaction. In the end, I have found that the cell carries out this reaction in the only way that it could be accomplished. In my long research career, one of the greatest mysteries to me was the way in which a living cell—without the aid of computers or even a brain—could find this single path to the necessary result.

In one way, the discoveries of genetics have made our understanding of evolution even more difficult. A cell does not directly alter the molecules in evolution. Instead, it alters the code of the nucleic acid in its chromosomes. Then all of the descendants of that cell make the appropriate changes in the molecules involved in the reaction. In most cases, a number of genes must be altered to accomplish a meaningful change in a chemical reaction. If all of these genes were not changed simultaneously, only confusion would result.

According to present ideas, this change in the nucleic acid is accomplished through random variation. The nature of protein molecules formed in the cell is determined by the code of the nucleic acid. The protein cannot alter the nucleic acid. If I were trying to pass a biology examination, I would vigorously support this theory. Yet in my mind I have never been able to accept fully the idea

that adaptation and the harmonious building of those complex biological systems, involving simultaneous changes in thousands of genes, are the results of molecular accidents.

The feeding of babies, for example, involves very complex reflexes. These reflexes require extremely complex mechanisms, both in the baby and in the mother, which must be tuned to one another. Similar mechanisms are involved in the sexual functions of male and female animals. These mechanisms must be tuned precisely to one another in order to achieve successful copulation. Thousands of genes must be involved in the coding of these mechanisms. The probability that all of these genes should have changed together through random variation is practically zero, even considering that millions or billions of years may have been available for the changes.

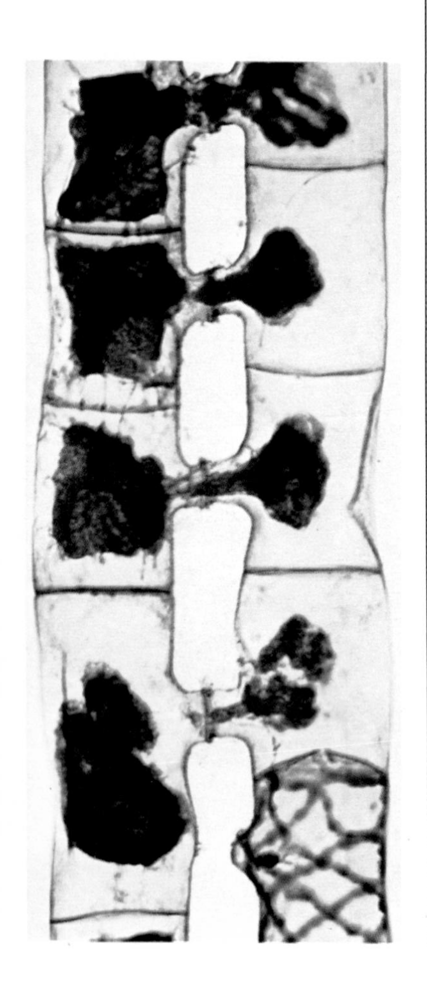

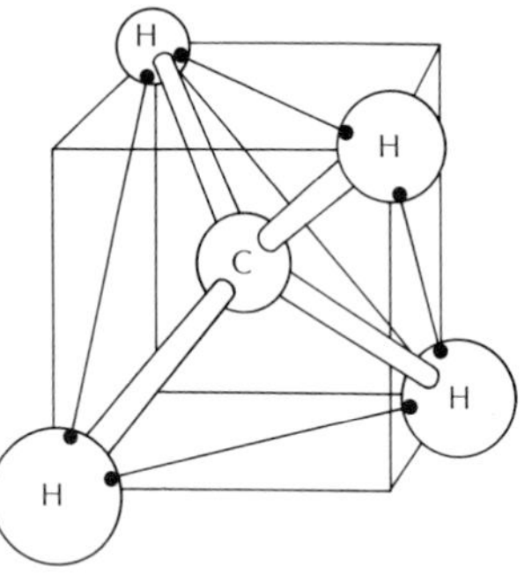

I have always been seeking some higher organizing principle that is leading the living system toward improvement and adaptation. I know this is biological heresy. It may be ignorance as well. Yet I think often of my student days, when we biologists knew practically nothing. There was then no quantum theory, no atomic nucleus, and no double helix. We knew only a little about a few amino acids and sugars. All the same, we felt obliged to explain life. If someone ventured to call our knowledge inadequate, we scornfully dismissed him as a "vitalist."

Today also we feel compelled to explain everything in terms of our

present knowledge. Identical twins are often exactly alike in the smallest details of physical appearance, indicating that the instructions for building this entire structure must have been encoded in the genetic materials that they share. All the same, I have the greatest difficulty in imagining that the extremely complex structure of the central nervous system could be totally described in the genetic codes. Thousands of nerve fibers grow for long distances in order to find the nerve cell with which they can make a meaningful junction. Surely the nucleic acid did not contain a blueprint of this entire network. Rather, it must have contained instructions that gave the nerve fiber the "wisdom" to search for and locate the only nerve cell with which it could make a meaningful connection. Perhaps this guiding principle also is related to the way in which the first living system came together.

I do not think that the extremely complex speech center of the human brain, involving a network

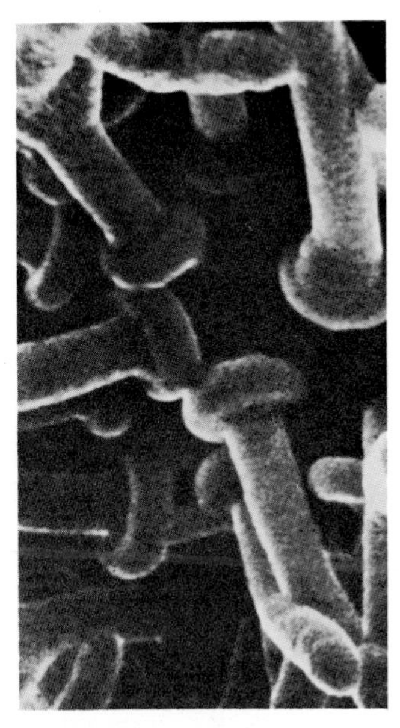

formed by thousands of nerve cells and fibers, was created by random mutations that happened to improve the chances of survival of individuals. I must believe that man built a speech center when he had something to say, and he developed the structure of this center to higher complexity as he had more and more to say. I cannot accept the notion that this capacity arose through random alterations, relying on the survival of the fittest. I believe that some principle must have guided the development toward the kind of speech center that was needed.

Walter B. Cannon, the greatest of American physiologists, often spoke of the "wisdom of the body." I doubt whether he could have given a more scientific definition of this "wisdom." He probably had in mind some guiding principle, driving life toward harmonious function, toward self-improvement.

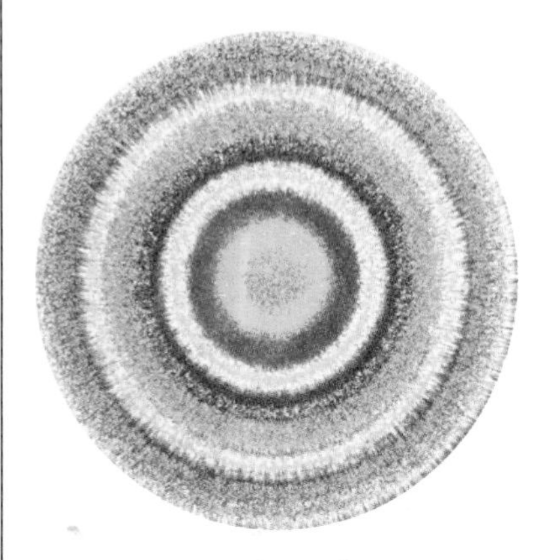

Life is a wondrous phenomenon. I can only hope that some day man will achieve a deeper insight into its nature and its guiding principles and will be able to express them in more exact terms. It is this mysterious quality of life that makes biology the most fascinating of sciences. To express the marvels of nature in the language of science is one of man's noblest endeavors. I see no reason to expect the completion of that task within the near future.

—Albert Szent-Györgyi
Woods Hole, Massachusetts

Biology Today

1

The Study of Life

Life is unified in two ways. In the first place, organisms are dependent upon one another for the chemical substances which are essential to their existence, and each type of organism plays a part in maintaining the entire world of life as a going concern. In the second place, all organisms are related to one another through a long line of evolutionary descent. A billion years ago or more, the first organisms appeared upon the earth. All the evidence indicates that they were simpler than even the humblest of the organisms with which we may become acquainted by viewing them through the most powerful microscopes. Through the long aeons that have passed since that time, the descendants of those organisms have developed through evolution to become the myriad of living forms both great and small which populate the earth today.

—Clarence W. Young (1938)

1

Vitalists and Mechanists

Biology is often defined as "the study of life" — but biology has never succeeded in proposing a simple definition of life. In fact, biology today is tending toward the view that there is no sharp dividing line between living and nonliving systems.

Francis H. C. Crick, who shared the 1962 Nobel Prize for his work in discovering the molecular structure of deoxyribonucleic acid (DNA), has said: "It is notoriously difficult to define the word 'living.' In many cases we all know whether something is alive or dead. You are alive; cats and dogs are alive; whereas a rock or a pane of glass is dead. But the word 'dead' is a bad one, because it half implies that the object was once alive and is now dead. It is interesting that there is no *simple* word for something that is not alive and never has been" (Crick, 1966).

In the past, biologists argued that living things are alive because they contain a "vital force." Few modern biologists, if any, would count themselves as vitalists. Crick continues: "Vitalism implies that there is some special force directing the growth or behavior of living systems which cannot be understood by our ordinary notions of physics and chemistry. Exact knowledge is the enemy of vitalism."

Crick is an outstanding example of the school of biological thought that has largely replaced vitalism. This new group of biologists believe that living as well as nonliving systems can be explained by physical and chemical laws. Because this view implies that biology eventually can be reduced to statements about physics and chemistry, its adherents often are called reductionists or mechanists.

Although most modern biologists hold that living systems operate according to basic laws of chemistry and physics, they also recognize that living systems are distinguished by their incredible complexity. In fact, Crick points out, "It is essentially true that *all* very highly ordered complex objects are biological." Complexity and organization are the features most characteristic of living systems. They are so complex and highly organized that in many ways they seem to have properties quite different from those of simpler, nonliving systems.

Not many years ago, cynics were fond of pointing out that a man is made up of nothing more than 97 cents' worth of chemicals. Since then the price of the chemicals has risen, but the statement can be criticized for other reasons as well. Obviously, a human being is not a random heap of simple chemicals, any more than a painting by Picasso or Van Gogh is a random splash of a few pennies' worth of paint. The unique properties of a living system are due primarily to the organization of its chemical components.

When chemists began to study the properties of matter, they quickly realized that the substances in living systems are quite different from those in the nonliving world. The substances that make up the earth, sea, and air are relatively stable. Water can be frozen to ice or boiled to vapor, but it can always be returned to its original liquid form by reversing the procedure that changed it. Metals, salts, and other minerals also can be melted, even vaporized, but will recrystallize in their original forms when cooled. On the other hand, most of the substances making up living systems are very unstable. When materials such as wood and sugar are heated, they char and burn. These processes are not readily reversible; the ashes and smoke that result cannot be restored to the original material by cooling or by any simple chemical or physical process.

In 1807 a Swedish chemist, Jöns Jakob Berzelius, suggested that the

Figure 1.1 (left). Illustration depicting the costume worn by physicians during the Marseilles plague in 1720.

Figure 1.2 (right). A modern model of the cell.

substances derived from living systems be called *organic* substances; all other chemicals he called *inorganic* substances. Although organic substances can be easily converted into inorganic ones (through burning, for example), Berzelius found it impossible to create organic substances from inorganic ones. He suggested that only living systems have the power to produce organic substances.

Alchemists and early chemists had long recognized that all matter is made up of a limited number of simple substances called *elements*. The organic substances proved to be composed almost entirely of only four elements: carbon, oxygen, hydrogen, and nitrogen—elements that also are found in inorganic substances. In fact, analysis of organic substances revealed that they contain no unique material substances; any element that is found in organic substances also is found in inorganic matter. Most early biologists therefore concluded that the unique properties of living systems are caused by some "vital force" that can be imparted to matter only by another living system.

VITALISM AND SPONTANEOUS GENERATION

From the early Greek philosophers through the natural philosophers of the seventeenth century, most educated or observant men believed that living organisms are generated spontaneously from nonliving matter. Frogs and insects emerge from the mud after a spring rain. Maggots or worms arise spontaneously in decaying meat. In 1668 this belief was challenged through a series of brilliant experiments performed by Francesco Redi, an Italian poet, physician, and naturalist. Redi's experiments still are cited as examples of careful and valid scientific investigation (Interleaf 1.1). Redi showed that worms arise in decaying meat only if adult insects are able to lay their eggs in the meat. Other researchers soon showed that aphids, fleas, lice, and the larvae in plant galls are produced only from eggs laid by adult animals.

For a short time, the theory of spontaneous generation of living creatures was discredited. Then in 1677 Anton van Leeuwenhoek, using hand-ground lenses, discovered organisms of microscopic size. Microscopic examination of any sample of broth or water revealed a teeming population of microorganisms. These microorganisms—bacteria, yeasts, and protozoans—are so much smaller and simpler in appearance than other organisms that their spontaneous generation seemed quite plausible.

For two centuries, controversy raged about the spontaneous origin of microorganisms. Some experimenters showed that filtering, boiling, or chemically treating samples of broth or water killed the microorganisms. Other experimenters, however, showed that the microorganisms reappeared in many cases despite the most careful attempts to avoid contamination by living microorganisms.

Finally, in the early nineteenth century, Lazzaro Spallanzani and Theodor Schwann showed that organisms do not appear in a broth that has been

Figure 1.3 (right). Anton van Leeuwenhoek's hand-ground lenses.

Figure 1.4 (lower left). Lazzaro Spallanzani.

Figure 1.5 (lower right). Theodor Schwann.

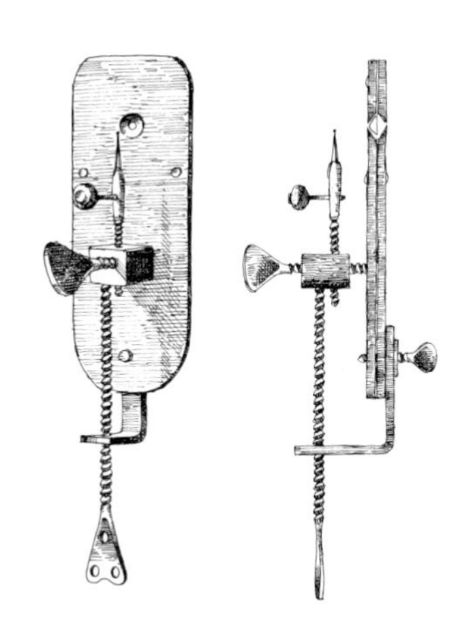

Figure 1.6. Francesco Redi.

Interleaf 1.1

FRANCESCO REDI'S EXPERIMENTS ON SPONTANEOUS GENERATION OF INSECTS

Francesco Redi (1621–1697) of Italy was a distinguished scholar, philologist, physician, and poet, as well as a naturalist of wide interests. As a scientist, he is best known for a series of experiments in which he showed quite conclusively that maggots are not spontaneously generated in decaying meat. Redi set out in 1668 to test this hypothesis experimentally. At the time, there was little clear understanding among scientists of the importance of experimentation or of the best way in which to carry out experiments. Redi's work is of importance not only for his conclusions but because he designed his experiments so carefully that there could be little doubt of the correctness of his conclusions. The quotations from Redi's report that follow are taken from the English translation by Bigelow (1909).

Redi began with a simple observation to test the truth of the ancient belief that "the putrescence of a dead body, or the filth of any sort of decayed matter engenders worms." He obtained three dead snakes ("the kind called eels of Aesulapius") and put them in an open box to decay. "Not long afterwards I saw that they were covered with worms of a conical shape and apparently without legs. These worms were intent on devouring the meat, increasing meanwhile in size, and from day to day I observed that they likewise increased in number; but although of the same shape, they differed in size, having been born on different days."

After the meat had been consumed by the worms, leaving only the bones, the worms disappeared, apparently escaping from the box. To discover what happened to the worms, Redi repeated the experiment, this time carefully sealing all holes through which the worms might escape. After three days, when the decaying meat became covered with worms, Redi noted this time that there were two different kinds of worms alike in form, but one kind large and white and the other kind smaller and pink. When the meat was gone, the worms tried to escape from the box but were unable to do so. "On the nineteenth day of the same month some of the worms ceased all movements, as if they were asleep, and appeared to shrink and gradually to assume a shape like an egg. On the twentieth day all the worms had assumed the egg shape, and had taken on a golden white color, turning to red, which some darkened, becoming almost black. At this point the red, as well as the black ones, changed from soft to hard, resembling somewhat those chrysalids formed by caterpillars, silkworms, and similar insects."

Redi separated the red and black egg-shape objects and put them in glass vessels sealed with paper. After eight days, each of the red objects broke open "and from each came forth a fly of grey color, torpid and dull, misshapen as if half finished, with closed wings; but after a few minutes they commenced to unfold and to expand in exact proportion to the tiny body which also in the meantime had acquired symmetry in all its parts. Then the whole creature, as if made anew, having lost its grey color, took on a most brilliant and vivid green; and the whole body had expanded and grown so that it seemed incredible that it could ever have been contained in the small shell." The black objects broke open after 14 days "to produce certain large black flies striped with white, having a hairy abdomen, of the kind that we see daily buzzing about the butcher's stalls."

Redi concluded from these observations that the worms are the immature forms of flies and "began to believe that all worms found in meat were derived directly from the droppings of flies, and not from the putrefaction of the meat." He remembered seeing both the green and the large black flies hovering over the meat before the worms appeared. To test this new hypothesis, Redi set up another experiment, for "belief would be in vain without the confirmation of experiment."

He prepared four large, wide-mouth flasks, each containing a different kind of meat: snake, fish, eels, and milk-fed veal. Each of the flasks was carefully closed and sealed. Then Redi prepared an identical set of flasks, but he left this set open. Within a few days, the meat in the open flasks became covered with worms, and flies were seen freely entering and leaving the open flasks. However, "in the closed flasks I did not see a worm, though many days had passed since the dead flesh had been put in them. Outside on the paper cover there was now and then a deposit, or a maggot that eagerly sought some crevice by which to enter and obtain nourishment. Meanwhile the different things placed in the flasks had become putrid."

This experiment is noteworthy because it represents one of the earliest examples of the deliberate use of a control group in a biological experiment. Redi prepared two identical sets of flasks; one set was open and the other set was sealed. By using various kinds of meat, Redi was able to test the effect of another variable. His results showed that the appearance of worms occurred only in open flasks but was not dependent upon the kind of meat used. Redi repeated his experiment many times, using different kinds of vessels and different kinds of meat and keeping the vessels under different weather conditions at different seasons of the year. He even tried burying pieces of meat underground. The results were always consistent with his hypothesis: worms appeared in decaying meat only if adult flies were able to place their droppings on the meat. He even tried using dead maggots and flies as the meat in the flasks, but he observed the same results.

Although these experiments might seem to offer indisputable evidence that worms cannot arise in dead flesh of any kind unless flies are allowed to make deposits in the flesh, Redi recognized another possible interpretation of his observations. In every case, the sealing of the meat to prevent the entry of flies had also prevented the free entry and circulation of air. It could be argued that only the lack of fresh air kept maggots from arising spontaneously in the sealed meat samples.

Therefore, Redi set up another experiment. He put samples of meat and fish in a large vase closed with a fine veil through which air could circulate freely, but with holes too small for flies to penetrate. He put the vase inside a framework covered with the same kind of netting. "I never saw any worms in the meat, though many were to be seen moving about on the net-covered frame. These, attracted by the odor of the meat, succeeded at last in penetrating the fine meshes and would have entered the vase had I not speedily removed them." Redi noticed that some flies left deposits ("fly specks") on the netting, whereas others left live worms. "I noted that some left six or seven [live worms] at a time there and others dropped them in the air before reaching the net. Perhaps these were of the same breed mentioned by Scaliger, in whose hand, by a lucky accident, a large fly deposited some small worms, whence he drew the conclusion that all flies bring forth live worms directly and not eggs. But what I have already said on the subject proves how much this learned man was in error. It is true that some kinds of flies bring forth live worms and some others eggs, as I have proved by experiment."

Redi's report is a model of good scientific procedure in several respects. Not only are his experiments cleverly designed to provide unambiguous tests of his hypotheses, but he describes exactly what he did and what results he observed. He separates his interpretations and beliefs from these factual accounts of the experiments. Science has been built upon this sort of reporting; each investigator is expected to describe his work and observations meticulously and factually, without letting his own ideas or beliefs color the facts. Later researchers may disagree with his interpretations, but their own theories must be consistent with his experimental observations.

With the hindsight of a few centuries of experience, it is easy to see how unrecognized assumptions and beliefs crept into Redi's supposedly factual accounts of his observations. It is not nearly as easy to recognize the assumptions and beliefs that underlie contemporary "factual" accounts. However, the cumulative progress of science would be impossible if it were necessary for each scientist to distrust the reports of those who disagree with him and to repeat all of their experiments to be sure that they did not "fudge" the results. To permit other scientists to look for other possible interpretations of his results, the modern biologist is careful to include in his report a great many details that Redi did not think worth reporting. For example, he would probably include information on the weather conditions, the species names of the organisms used or observed, the exact sizes of the flasks and of the openings in the netting, and the exact numbers and sizes of the worms and flies that were observed. He would be sure to note whether the flasks were observed continuously and to explain just how he determined, for example, that *all* of the red objects broke open to release green flies. Because of this exact reporting, other scientists are able to think of possible alternative interpretations of the experiments without as much need to perform a new experiment in order to test each one.

Figure 1.7 (left). Pierre Berthelot at work in his laboratory.

Figure 1.8 (right). Friedrich Wöhler.

heated to the temperature of boiling water for nearly an hour—*if* all air reaching the broth has been similarly heated. The proponents of spontaneous generation could argue only that the heated air was somehow damaged and inadequate for the support of life. Schwann countered this argument, however, by showing that heated air is suitable for breathing.

The evidence against spontaneous generation of life was consistent with the prevailing theory of vitalism. This theory holds that living systems can arise from nonliving matter only through the intervention of an already living organism to impart a vital force to the matter. In short, each organism can arise only as the offspring of a parent organism similar to itself.

But how did all the various kinds, or species, of organisms originate? If the ancestors of modern organisms were not generated spontaneously—if each organism has always arisen from parents like itself—it seems necessary to suppose either that life has always existed in its present forms or that there was a supernatural creation of the many different species at some time in the past. Those who disproved spontaneous generation were unable to present a convincing scientific explanation of the origin of life.

THE RISE OF MECHANISM

Even while the theories of vitalism were being developed, they received a damaging blow from an unexpected quarter. In 1828 a German chemist, Friedrich Wöhler, was studying an inorganic substance now called ammonium cyanate. Wöhler analyzed the crystals that are produced when this substance is heated. He was startled to find that the crystals were urea, the major solid component of mammalian urine and a substance definitely considered to be organic.

Other chemists attempted to create organic substances from inorganic materials, and reports of their successes soon began to accumulate. All doubt was removed by the late 1850s, when Pierre Berthelot succeeded in producing such organic substances as alcohols, methane, acetylene, and

Figure 1.9. Svante August Arrhenius.

benzene from inorganic chemicals. It was clear that at least the simpler organic substances can be produced without the addition of a vital force.

Vitalism, however, was not completely discredited. Many biologists still felt that the more complex organic substances—particularly the proteins, which play important roles in living systems—could not be synthesized outside a living organism. Nevertheless, the weight of scientific opinion began to shift toward a mechanistic view of life. This approach holds that organic substances possess no special vital force but can be described in terms of the same chemical and physical laws that apply to inorganic substances. Organic substances, then, are distinguished only by the complexity of their organization.

In 1858 Charles Darwin and Alfred Russel Wallace simultaneously published the theory of evolution through natural selection, which each man had developed independently. This theory proposed a mechanism to explain how the vast array of organisms currently inhabiting the earth could have developed from simpler organisms that lived in the past. In each generation, those individuals possessing variations that helped them to survive or to reproduce would be likely to produce more offspring than those with unfavorable variations. If it is assumed that the offspring inherit characteristics similar to those of their parents, this natural selection of favorable variations, carried on over many thousands of generations, could lead to the eventual development of all modern organisms from a single, simple ancestral living system.

As this theory gained acceptance, the problem of the origin of life became both simpler in form and more difficult to solve. It was no longer necessary to account for the independent origin of many species but only to explain how the first, simple organism arose. Yet at almost the same time, the experiments of Louis Pasteur proved beyond reasonable doubt that even microorganisms cannot arise spontaneously (Interleaf 1.2).

Most scientists felt that it was futile to inquire into the origin of life. It was even suggested that life, like matter, has always existed. A Swedish chemist, Svante August Arrhenius, proposed the idea that life-bearing particles are scattered through space. They fall on all planets and germinate to produce organisms wherever conditions are favorable. Arrhenius' theory, however, does not account for the origin of life; it merely pushes the origin of the first living system further away in time and space.

By the late nineteenth century, most biologists had accepted a mechanistic view of living systems and had rejected the theories of vitalism and spontaneous generation. Yet as biologist George Wald (1954) comments: "Most modern biologists, having reviewed with satisfaction the downfall of the spontaneous generation hypothesis, yet unwilling to accept the alternative belief in special creation, are left with nothing." The question of the origin of life faded into the background as biologists applied themselves to the more rewarding task of investigating the mechanisms by which living systems operate. Application of the techniques and theories of the physical sciences to the study of organisms proved remarkably fruitful in terms of new understanding.

The mechanistic view of life was widely accepted by biologists long before experimenters actually succeeded in creating complex organic molecules from inorganic substances. In fact, it was not until 1969 that biochemists succeeded in synthesizing one of the complex protein molecules that plays an active chemical role in living organisms (Hirschmann, *et al.*,

Figure 1.10. A caricature of Louis Pasteur.

Interleaf 1.2

LOUIS PASTEUR'S EXPERIMENTS ON SPONTANEOUS GENERATION OF MICROORGANISMS

The controversy about spontaneous origin of life raged for a long time after the brilliant experiments of Redi (Interleaf 1.1). Although Redi demonstrated that maggots arise in decaying meat only if flies lay their eggs or deposit live young there, he continued to believe in the spontaneous generation of other kinds of insects. By the early eighteenth century, other researchers had demonstrated that insects of many different kinds arise only as offspring of adult insects. However, the microorganisms discovered by Anton van Leeuwenhoek in 1677 seemed to be a different case. Attempts to carry out experiments modeled after those of Redi on the origin of microorganisms led to conflicting results.

The major figures in a great debate that went on near the middle of the eighteenth century were John Needham, an English Catholic priest and naturalist, and Lazzaro Spallanzani, an Italian abbot and biologist. Needham, with the help of the Comte de Buffon, attempted to apply Redi's experimental methods to the study of the origin of microorganisms. He boiled mutton broth, thus killing all the microorganisms present in it. He placed the sterile broth in a well-sealed flask and left it for a few days. When he opened the flask, he found the broth swarming with microorganisms. He obtained similar results when he repeated the experiment with a variety of organic solutions—water in which various animal and vegetable substances had been soaked. Thus, Needham concluded that microorganisms do arise spontaneously.

Spallanzani realized that it would be difficult to be sure that no microorganisms enter the broth after it has been boiled. The adult microorganisms were so tiny that they could barely be seen in the microscopes of the time, and it seemed likely to Spallanzani that the eggs or other reproductive forms of these creatures might even be undetectable under the microscope. He repeated Needham's experiments, boiling the liquid for half an hour and then placing it in loosely corked flasks. After eight days, he examined the liquids and found them filled with microorganisms. Spallanzani next prepared five flasks containing a liquid prepared by soaking seeds in water. He left one flask open and sealed the other four completely by melting the glass and closing the openings. He then boiled the liquid in the flasks, varying the length of boiling for the four closed flasks. After two days, he found the open flask swarming with microorganisms. A closed flask boiled for half a minute contained only the smaller kinds of microorganisms, and the other closed flasks boiled for one to two minutes contained only extremely minute organisms. In further experiments, he found that boiling a sealed flask for 30 to 45 minutes is sufficient to ensure that no new microorganisms will appear in the flask so long as it remains sealed.

Other theorists objected to Spallanzani's experiments on the grounds that heating the air in the sealed flasks makes it unfit to support life. Theodor Schwann demonstrated that no microorganisms arise in broth to which only heated air is allowed access. Furthermore, he demonstrated that animals can live in such previously heated air. The work of Spallanzani and Schwann might seem to have disproved the possibility of spontaneous generation, but many scientists were unconvinced. When others tried to repeat these experiments, they often observed the generation of microorganisms. Although it could be argued that some leak in the flasks or some flaw in the procedure had permitted microorganisms to get in from the outside, such arguments were unlikely to convince the researcher who believed that he had done his work carefully. A prominent advocate of spontaneous generation was the naturalist Félix Archimède Pouchet, who published a large book in 1859 describing the process of spontaneous generation and the conditions under which it occurs.

Related to the debate over spontaneous generation was another conflict about the nature of fermentation and decay. The most common view was that set forth by the noted German chemist Justus von Liebig, who argued that decay and fermentation are peculiar processes that occur only in organic matter as the last stages of the process of death. On the other hand, Spallanzani and Schwann found that no decay occurs in the microorganism-free broths that have been boiled in sealed flasks. Therefore, they concluded that decay is a process carried out by living microorganisms. Schwann went on to show that yeast is composed of microorganisms and that the presence of live yeast organisms is necessary for the alcoholic fermentation of sugar.

Figure 1.11. Louis Pasteur in his laboratory.

Both of these debates were to be resolved essentially as a result of the work of one man, Louis Pasteur. A physical chemist, Pasteur began his career with research on the optical properties of certain crystals that are produced in the course of fermentation. He then became interested in the process of fermentation itself and in 1857 published a paper on the souring of milk. He was able to isolate the substance that produces lactic acid, which in turn makes the milk sour. The substance proved to be a ferment, and it was later shown to be a mass of bacteria. Pasteur then turned to the study of the alcoholic fermentation of sugar as it occurs in the process of wine making. He found that the fermentation is dependent upon the presence of certain yeasts or molds that live on the skins of ripened grapes. Pasteur's study of the defects of some wines convinced him that a number of different organisms, including some kinds of bacteria, participate in the fermentation process and that these organisms must be present in the proper proportions if a good wine is to be formed. Pasteur (1866) concluded that both fermentation and decay are the results of activity carried out by microorganisms that reach foods and dead flesh largely in the form of tiny airborne germs.

To confirm his hypothesis, Pasteur filtered air by drawing it through guncotton. He dissolved the guncotton and showed that a residue of microscopic spherical and rod-shape objects remained. He also drew air that had already been filtered through guncotton and showed that in this case no microscopic residue remains when the guncotton is dissolved. Thus, he confirmed that the air is full of microscopic objects that could be the germs of his hypothesis.

Still the supporters of spontaneous generation were no more convinced by Pasteur's experiments than they had been by those of Spallanzani and Schwann. They argued that Pasteur had merely shown the air to contain microscopic objects but that he had provided no evidence that these objects are alive or can become alive. They still argued that the creation of microorganisms in decaying organic material occurs spontaneously and that decay is caused by a vital force leaving the dying matter to enter the newly created microorganisms.

In answer to these objections, Pasteur conducted another series of experiments that were similar in design to the original experiments of Redi. He placed fermentable substances in flasks, then drew the neck of the flask out into a long, narrow S shape. Air can enter and leave the flask through the narrow opening, but any particles in the air are likely to be trapped on the walls of the long, curving neck. The flasks and their contents were heated to the temperature of boiling water for some time, then left sitting in still air. No fermentation was observed in the flasks. However, if the necks were broken off so that atmospheric dust and other small particles could enter with the air, the contents began to ferment within a few hours and microorganisms abounded in the liquid (Pasteur, 1861). Although this experiment did not immediately convince all of the proponents of spontaneous generation that they were wrong, it did carry great weight with the majority of the scientific community. Within 20 years after the publication of Pasteur's work, the idea of spontaneous generation had essentially been abandoned.

Pasteur's work carried great impact not only because of the elegance of his experimental technique but also because of his success in confirming other implications of his germ theory. Pasteur's work on fermentation led to important practical improvements in many industrial processes, and his application of the germ theory to medical practices had dramatic results. Furthermore, as biologists learned more about microorganisms, these little creatures began to seem more and more similar to larger organisms, and the idea that they might form spontaneously began to seem quite peculiar.

Figure 1.12. Linus Pauling (left) in Oslo to receive the 1962 Nobel Prize.

1969). In 1970 other researchers succeeded in synthesizing a gene, one tiny part of the complex DNA molecule that carries hereditary information in a living organism (Agarwal, *et al.*, 1970). It may be many years before confirmation of the mechanistic view is obtained through the complete synthesis of a simple living creature from inorganic materials. However, few biologists today doubt that such a synthesis is theoretically possible.

THE ORIGIN OF LIFE

By the middle of the twentieth century, sufficient progress had been made in the understanding of the chemical and physical mechanisms of life to permit biologists to make some progress toward a theory about the origin of life. Extrapolation of the evolutionary process backward in time leads to the hypothesis of a single species of living systems—probably quite simple in nature—that were the ancestors of all later life on earth. It is possible to suppose that these first living systems were one of a great variety of complex physiochemical systems—most of which were not alive—that had been generated by the physical and chemical processes on the primitive earth.

Darwin anticipated this view in a letter written in 1871 to J. D. Hooker, the botanist who studied and classified the plant samples Darwin collected during the voyage of the *Beagle*. "It is often said," wrote Darwin, "that all the conditions for the first production of a living organism are now present, which could ever have been present. But if (and oh what a big if) we could conceive in some warm little pond with all sorts of ammonic and phosphoric salts,—light, heat, electricity &c. present, that a protein compound was chemically formed, ready to undergo still more complex changes, at the present day such matter would be instantly devoured, or absorbed, which would not have been the case before living creatures were formed."

In essence, the modern view is that spontaneous generation of simple organisms may have been possible under the very different chemical and physical conditions that are thought to have existed on the primitive earth (Chapter 32). From a variety of evidence, scientists now believe that the earth, the other planets, and the sun were all formed about 5 billion years ago from a cloud of cosmic dust at low temperatures. The oldest rocks yet found to contain apparent remains of living organisms were formed about 3.1 billion years ago. Because the time between the formation of the earth and the appearance of living systems was probably only 1 or 2 billion years or less, it has been suggested that the organic substances from which life formed were a part of the dust cloud that condensed into the earth (Robinson, 1966). Arrhenius' hypothesis that life may have come in the form of spores or seeds from elsewhere in space is not totally disproven either. One major reason for the careful sterilization of unmanned rockets to the moon and to other planets has been the hope of eventually testing Arrhenius' theory by looking for such spores on planets where life has not existed.

MODERN MECHANISM AND REDUCTIONISM

Over the past two centuries, the mechanistic approach to biology has steadily gained adherents as it has accumulated an impressive record of successes in explaining living systems. It is now generally accepted that both living and nonliving systems are governed by the same physical and chemical principles.

Indeed, the most spectacular advances in biology during this century have come through mechanistic and reductionistic approaches. Important

Figure 1.13. Apollo 12 lunar sample seen under polarized light.

Figure 1.14. Two prominent modern biologists. Jonas Salk (above) developed the Salk polio vaccine. François Jacob (below) is seen receiving the 1965 Nobel Prize.

advances in molecular biology and molecular genetics have come as the most modern understandings of chemistry and physics were applied to the study of living cells. In more and more cases, it is becoming possible to explain some characteristic of an organism in terms of the physiological systems within the organism, to explain these systems in terms of cellular interactions, and to explain the cellular interactions in terms of chemical and physical processes. On the other hand, there has not been notable success in moving toward new generalizations at higher levels of organization. That is, biologists are still struggling to delineate the general principles that govern the functioning of populations and of the ecosystem.

Because of the successes of the reductionistic approach, many scientists have concluded that the goal of biological study should be the eventual explanation of living systems entirely in terms of chemistry and physics, with everything fully measured and quantified at the molecular level. On the other hand, another large group of biologists argues that living systems are unique, and they may be expected to obey certain principles at higher levels of organization that cannot readily be reduced to simpler physical laws. Although it might be theoretically possible to explain the outcome of a presidential election in terms of a complete biochemical understanding of the American population, there is great practical benefit in having principles of behavioral, political, and social sciences to apply to such studies. Furthermore, the statistical processes involved in interactions of large numbers of cells or individual organisms may make complete reductionism impossible. Thus, it may be possible to explain in terms of physiological mechanisms why an individual is likely to behave in certain ways, but it may still be necessary to have general principles of social behavior to explain the particular distributions of behavior observed in a population of interacting individuals.

Advance in any science requires an interplay between theory and experiment, between imaginative guesswork and careful testing, between broad generalizations and exact measurements. Any good experiment is designed because the experimenter wants to test some hypothesis that he has created. Even simple observation of nature must be guided by some preconceptions about what is sufficiently significant to be observed. On the other hand, even the most brilliant theoretical scheme is of little scientific interest unless it can be tested by experimentation or observation in the real world.

Biology today is a complex, diverse, and incomplete field of knowledge. Many different groups of researchers and theorists have explored different aspects of living systems. They have used different assumptions about how to ask their questions, different vocabularies, and very different experimental techniques. As a result, biological knowledge seems somewhat like a huge jigsaw puzzle with a great many missing pieces. Here and there, segments of the picture can be fitted together very nicely. The broad relationships between these segments are fairly clear, but there still are many gaps and scattered loose pieces whose proper position seems very uncertain.

In the chapters that follow, the broad picture has been outlined and the details filled in here and there. It is important to keep in mind that behind almost every sentence stating some apparently simple fact may lie a history of experimentation and debate as complex as the controversy about spontaneous generation. What seems today to be a safe generalization may tomorrow be proven by some critical experiments to have been a faulty interpretation of the available evidence. What seems today to be a minor

Figure 1.15. The mapping of molecules.

exception to some general rule may prove tomorrow to be the key to an important new understanding of living systems.

From the level of molecular interactions on a time scale of seconds or minutes to the level of population interactions on a time scale of millions of years, the biologist is involved in the study of the most complex systems known to exist in the universe. The understanding accumulated in the past few centuries is staggering in its scope and variety, but the amount yet to be learned about life is far greater. Very few of the chapters in this book represent summaries of aspects of living systems that are thoroughly understood. Although most modern biologists are mechanists and reductionists, it will be a very long time before an introductory biology book can be written in which all important facts about even a single living system are set forth clearly and simply as the logical result of chemical and physical laws.

FURTHER READING

Asimov (1964) provides a brief and very readable summary of the history of biology. For more detailed information on the development of the scientific study of living systems, see Commoner (1966), Crick (1966), Crombie (1959), Dampier (1958), Eiseley (1969), Hall (1954), Rook (1963), Simpson (1969), and Singer (1959). More general discussions of the nature of science will be found in books by Butterfield (1957), Conant (1951), Nagel (1961), and Nash (1963).

2
The Variety of Life

The instructor stepped up to the podium, shuffled his notes, and began his lecture: "Without grass, life as we know it would be impossible." The resulting explosion of laughter from the students was somewhat disconcerting, for he had been about to launch into a very serious discussion of ecological food pyramids. Communication gaps such as this one occur with unfortunate regularity in daily life, but they usually are cleared up by further conversation. In scientific circles—where thousands of individuals must communicate via the written word in monthly periodicals and must rely upon the reports of others for reliable information—the problem becomes more severe. It may be vitally important to know exactly what kind of organism is meant by a name such as "grass" in a scientific report.

In attempting to categorize the incredible variety of living things that exist in the world, biologists have developed a complex system of classification and terminology. It has been said that a first-year biology student must learn more new words than does a student in an introductory language course. This statement may be a slight exaggeration, but it does not seem so to many students, who wonder why things cannot be said in "plain English." Present systems of nomenclature and classification were worked out slowly as it became necessary to deviate from the "everyday" language of other kinds of writings.

The present system of classification was developed in the eighteenth century. Although the system has been revised extensively since then, the basic principles and rules of *biosystematics* (biological classification and nomenclature) have remained essentially unchanged since 1758.

PRINCIPLES OF CLASSIFICATION

Biological classification systems are hierarchical. That is, an individual is assigned to one small set of very similar individuals, and that set is grouped with other sets to form a larger set of individuals with some similarities, and so on. Hierarchical classification schemes are common in many areas of life and are built into the everyday language. The hierarchy used in biological classification has several levels of sets, with a general name given to each level, or *taxon*. A set at one level in the hierarchy is made up of one or more sets of the level below it. For example, an order contains one or more families, and each family contains one or more genera. Thus, the kingdom is the most inclusive taxon, and the species is the least inclusive (most specific) taxon. Two individuals classified as members of the same species are very similar to one another in many ways, whereas two individuals classified as members of the same kingdom may share only a few common characteristics. Additional levels of classification—such as suborders, superfamilies, and so on—are added to the hierarchy in cases where it seems important to emphasize certain similarities or differences.

The modern system of classification, or *taxonomy*, is based upon a system first worked out in the eighteenth century by Carolus Linnaeus. Each species is assigned a two-part, or *binomial*, name. The first part of the full name is that of the genus, and the second part of the name is that of the species. Linnaeus wrote in Latin—the common scientific language of his age, much as English is today—and formal systematic names are still written in Latin.

The basic unit of classification is the species. All higher taxa represent abstractions—groupings created in the human mind and not corresponding to any actual aggregations in nature. Species also are abstractions, but they are more closely related to natural groupings than are higher taxa. For most kinds of organisms, the individual is a unit that has clear physical significance. Individuals live together in populations, which can be defined relatively clearly for sexually reproducing organisms. A *population* consists of those individuals living in the same area who actually or potentially are related to one another. In other words, any two individuals who can mate with one another are members of the same population, as are their parents, offspring, and other ancestors and descendants. A species can be defined simply as the group including all populations that could interbreed if they lived in the same area.

Each species is named by genus and species, and an elaborate set of rules exists to ensure that no two species will be assigned the same name. The name of the human species, for example, is *Homo sapiens*—a name originally assigned by Linnaeus. There are no other species of the genus *Homo* living today, but some fossil remains are thought to represent other species of *Homo*. This genus is grouped in the family Hominidae (hominids) with some other genera of fossil creatures similar to man but not similar enough to be considered part of the genus *Homo*. The family Hominidae is part of the superfamily Hominoidea (hominoids), which includes the great apes as well as the various manlike genera. The superfamily Hominoidea is part of the suborder Anthropoidea (anthropoids), which includes all of the tailless apes that walk erect or semierect. This suborder is part of the order Primates. Among other characteristics that distinguish the primates are flat nails instead of claws, forward facing eyes, and large brains. This order is part of the class Mammalia (mammals), which includes all animals that suckle their young. The Mammalia are grouped in the phylum Chordata (chordates), which includes all organisms having a vertebral column or a dorsal, tubular nerve cord. Finally, the Chordata are part of the kingdom Animalia (animals).

At each taxon, the systematic classification indicates a range of other organisms that share certain characteristics with the species being classified. Certain characteristics have been chosen as the basis for classification, whereas others have been largely ignored. Thus, organisms of similar size or color are not likely to be grouped together at any level of this system. Originally, groupings were based upon characteristics that seemed most basic or "natural," but the modern system is based upon an attempt to indicate evolutionary relationships.

THE CONSTANCY OF SPECIES

Today, the system of classification is based chiefly upon theories of evolution. Species grouped in the same genera are considered to be "more closely related" than are two species in different genera. Two species grouped in the same genus are thought to have shared a common ancestral species in the relatively recent past. All species grouped in a single family presumably shared common ancestors in the more distant past. Because theories of the evolutionary history of organisms are continually being revised and debated, the higher levels of taxonomy are also in a state of continual revision. At any given time, several different schemes of classification are likely to be proposed and defended by various biologists. To some extent, this continual revision and debate counteracts the advantages of a uniform system of nomenclature. However, the rules by which names are proposed and adopted ensure that the meaning of any particular name will be clear to those who read a paper in which it is used, although a considerable amount of research in the voluminous literature of taxonomy may be necessary to determine the exact meaning of a particular item.

It is more difficult to define species in organisms that reproduce asexually. In fact, there are a number of difficulties related to a precise definition of species. However, it is now generally agreed that populations are significant, natural groupings of individuals, and that in most cases species can be defined as meaningful, unambiguous groupings, where members of one species are reproductively isolated from those of another species. Higher taxa are much more arbitrary and reflect current views of significant functional and structural similarities. The system of classification used in this book represents a compromise among some of the newer proposals that seem to be gaining favor among biologists. This system reflects current theories about the evolutionary relationships among organisms, but it is only one of many similar schemes currently in use. The evolutionary nature of taxonomic systems can be illustrated through a brief consideration of the changes that have occurred in the groupings of the most general taxon, the kingdoms.

HOW MANY KINGDOMS?

Early students of life were aware of little more than the higher plants and animals. It was natural for them to divide all organisms into two great realms, or kingdoms, of life—the immobile plants that gather nutrients through roots and the active, food-ingesting animals. Such a major division was recognized by Aristotle and became part of the formal system of classification in the work of Linnaeus.

As early as 1860, various authors began to suggest that the single-cell organisms should be regarded as a separate kingdom—neither plant nor animal—and the names Protista and Protoctista were suggested for this kingdom. In some schemes, the simpler multicellular organisms—such as fungi and algae—were included in this third kingdom. However, through the early twentieth century, many biologists continued to think only in terms of two kingdoms—plants and animals—with some uncertainties of

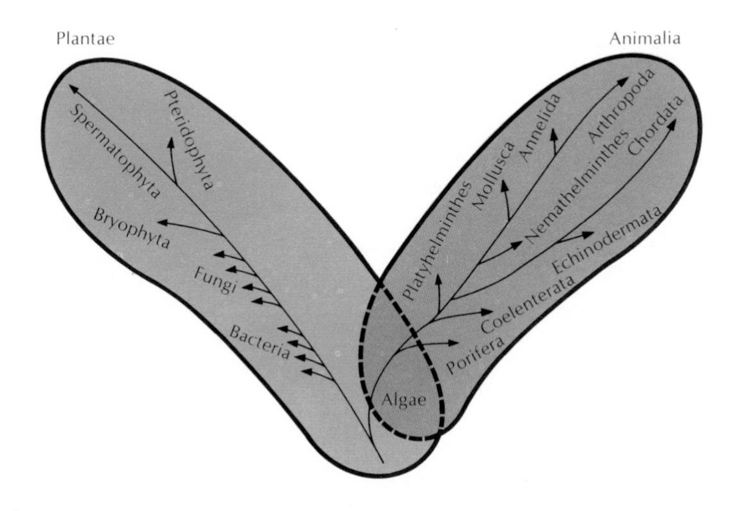

Figure 2.1. The two-kingdom system of classification.

classification among simpler organisms. In evolutionary terms, this method of classification implies that plant and animal lines diverged very early in the history of life (Figure 2.1).

As more was learned about microorganisms, simple multicellular organisms, and evolutionary history, the ranking of the Protista as a separate kingdom became more and more common. But even as the three-kingdom system came into common use, evidence was accumulating that other major divisions should be made in the grouping of organisms. Studies of single-cell organisms revealed two very different kinds of cell structure. *Procaryotic* organisms – bacteria and blue-green algae – lack a cell nucleus and are fundamentally different from other cells in many important ways (Chapter 8). The distinction between procaryotic cells and the *eucaryotic* (nucleated) cells of all other organisms was shown to represent very fundamental and extensive differences in basic organization.

Further study of fungi has revealed that these organisms depend upon a supply of organic molecules in their environment as animals do, yet they feed by absorbing substances through cell walls much as plants do. The complexity of fungi makes it unreasonable to group them with the Protista, yet it seems clear that they represent an independent line of evolution that diverged from plants and animals early in the history of life.

On the basis of the most recent information about evolutionary relationships, R. H. Whittaker (1969) has proposed a five-kingdom system that recognizes the fungi as a separate group of equal importance, with plants and animals as evolutionary lines of multicellular organisms that have developed from single-cell organisms (Figure 2.2). Whittaker's system appears to be gaining favor among biologists, and a modified version of his system follows (Whittaker, 1969).

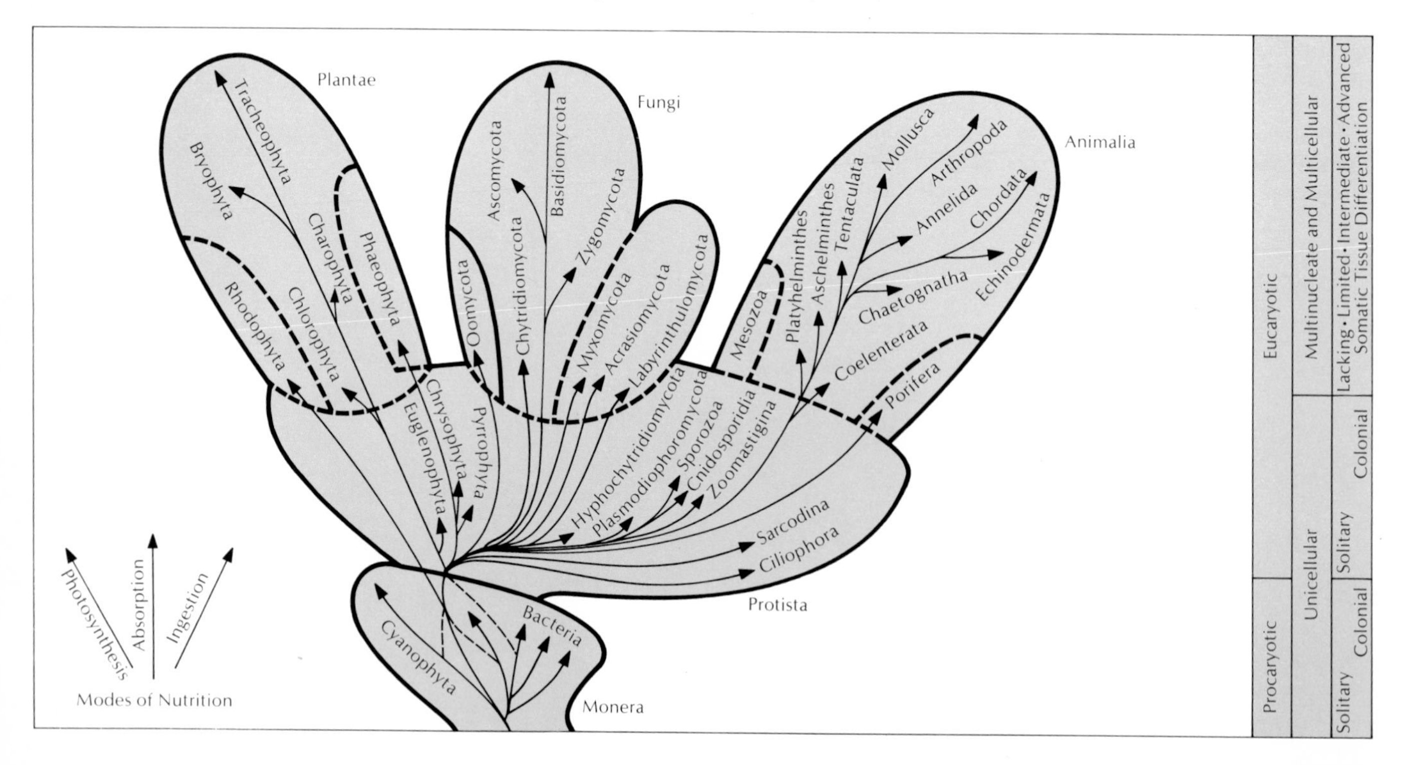

Figure 2.2. Whittaker's five-kingdom system of classification.

KINGDOM MONERA [procaryotes]
- Phylum Schizophyta, or Schizomycetes [bacteria]
- Phylum Cyanophyta, or Myxophyta [blue-green algae]

KINGDOM PROTISTA [protists]
- Phylum Euglenophyta [euglenophytes]
- Phylum Chrysophyta [golden algae and diatoms]
- Phylum Xanthophyta [yellow-green algae]
- Phylum Pyrrophyta [dinoflagellates and cryptomonads]
- Phylum Hyphochytridiomycota [hyphochytrids]
- Phylum Plasmodiophoromycota [plasmodiophores]
- Phylum Sporozoa [sporozoans]
- Phylum Cnidosporidia [cnidosporidians]
- Phylum Zoomastigina [animal flagellates]
- Phylum Sarcodina [rhizopods]
- Phylum Ciliophora [ciliates and suctorians]

KINGDOM PLANTAE [plants]
- Phylum Rhodophyta [red algae]
- Phylum Phaeophyta [brown algae]
- Phylum Chlorophyta [green algae]
- Phylum Charophyta [stoneworts]
- Phylum Bryophyta [liverworts, hornworts, and mosses]
- Phylum Psilophyta [psilophytes]
- Phylum Lycopodophyta [club mosses]
- Phylum Arthrophyta [horsetails]
- Phylum Pterophyta [ferns]
- Phylum Cycadophyta [cycads]
- Phylum Coniferophyta [conifers]
- Phylum Anthophyta [flowering plants]

KINGDOM FUNGI [fungi]
- Phylum Myxomycophyta [slime molds]
- Phylum Eumycophyta [true fungi]

KINGDOM ANIMALIA [animals]
- Phylum Mesozoa [mesozoans]
- Phylum Porifera [sponges]
- Phylum Archaeocyatha [extinct organisms]
- Phylum Cnidaria [coelenterates]
- Phylum Ctenophora [comb jellies]
- Phylum Platyhelminthes [flatworms]
- Phylum Nemertea [ribbon worms]
- Phylum Acanthocephala [spiny-headed worms]
- Phylum Aschelminthes [pseudocoelomate worms]
- Phylum Entoprocta [pseudocoelomate polyzoans]
- Phylum Bryozoa [sea mosses, or moss animals]
- Phylum Brachiopoda [brachiopods, or lampshells]
- Phylum Phoronida [phoronid worms]
- Phylum Mollusca [molluscs]
- Phylum Sipunculoidea [peanut worms]
- Phylum Echiuroidea [spoon worms]
- Phylum Annelida [segmented worms]
- Phylum Arthropoda [arthropods]
- Phylum Brachiata [beard worms]
- Phylum Chaetognatha [arrow worms]
- Phylum Echinodermata [echinoderms]
- Phylum Hemichordata [acorn worms]
- Phylum Chordata [chordates]

Monera

Figure 2.3 (above). Blue-green algae *Anabaena*, representing the phylum Cyanophyta, and green algae on a pond.

Figure 2.4 (below). Colonial blue-green algae. Some of the blue-green algae may function as nitrogen-fixers in aquatic ecosystems and thus play a role in the global nitrogen cycle.

Protista

Figure 2.5 (opposite left). Diatoms. Note the variety in form among the types represented. These organisms are of major economic importance because they form the base of the oceanic food chain and thus support much of the marine flora and fauna.

Figure 2.6 (opposite right). Close-up photograph of a radial diatom.

Whittaker regards plants, fungi, and animals as three groups of organisms that, through evolution, have come to specialize in three different modes of nutrition: photosynthesis, absorption, and ingestion. Protista include a variety of diverging lines of evolution, all composed of organisms that share a eucaryotic, single-cell level of organization. Monera are the simplest known organisms and are presumed to be similar to the early organisms from which the other kingdoms evolved.

Further revisions of kingdoms may result as more is learned about the early evolution of life. Whatever system is accepted, however, it should be emphasized that systems of classification represent categories superimposed on nature by the human mind in its search for order, and any system may be expected to fit nature somewhat imperfectly. The classification and naming of organisms in itself does not represent significant knowledge about the living world. A classification system simply provides a means of summarizing existing knowledge and provides a stimulus for new researches and theories. It is a good idea to keep in mind that learning the name of an organism tells one little about the thing itself, although it may explain something about the theories of the person who invented the name.

Various groups of organisms are considered in detail elsewhere in this book, but a brief pictorial survey of the five kingdoms will help to clarify the nature of the classification system and to illustrate the diversity of life.

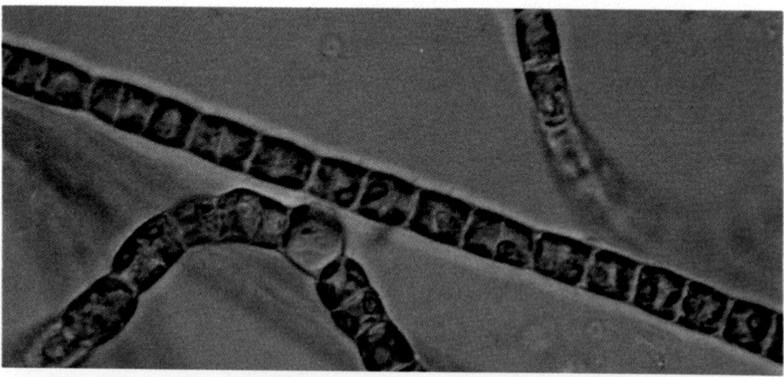

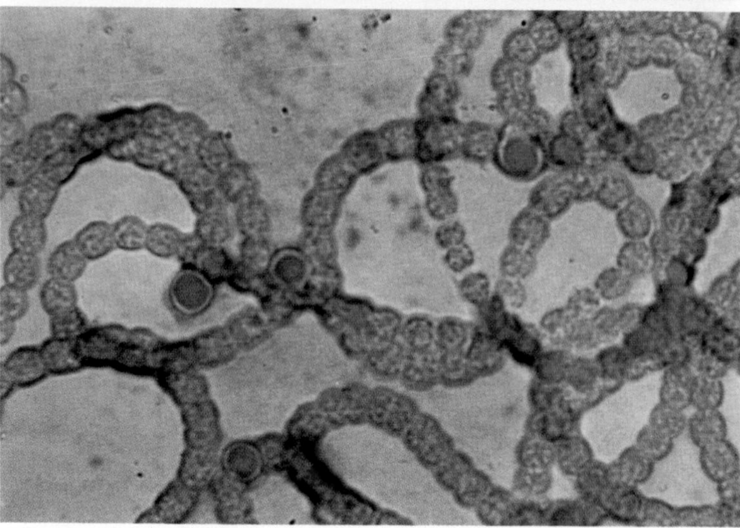

Figure 2.7 (middle). A radiolarian test or shell. Radiolarians are marine amoebas with silica shells. (Eric Gravé)

Figure 2.8 (middle left). Close-up photograph of the extended pseudopod of the shelled amoeba *Arcella* (Phylum Protozoa). (Eric Gravé)

Figure 2.9 (middle right). *Paramecium aurelia*, representing the ciliate class of protozoans. (Eric Gravé)

Figure 2.10 (lower left). A representative marine ciliate. Note the winglike extensions of the pellicle (surface membrane), which function as flotation devices. (Eric Gravé)

Figure 2.11 (lower right). *Euplotes*, an advanced ciliate. The ciliophora are structurally the most complex of the protozoa. (Eric Gravé)

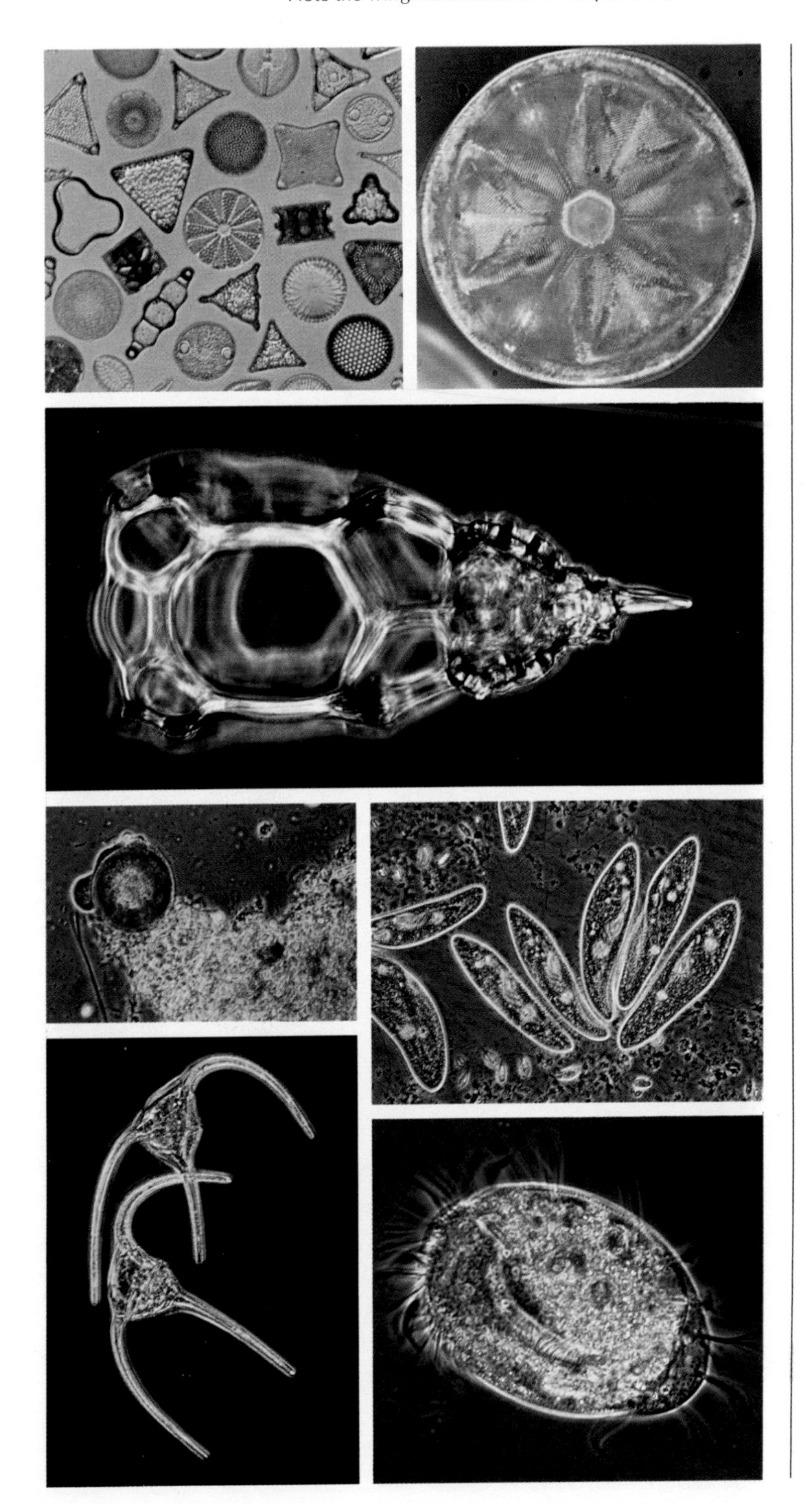

Fungi

Figure 2.12 (above). Photograph of the plasmodium, or vegetative body, of a slime mold (Kingdom Fungi). Most slime molds are free-living saprophytes, feeding on dead organic matter such as wood or leaf litter on the forest floor.

Figure 2.13 (middle). Close-up photograph of the sporangia, or fruiting bodies, that develop from the slime mold plasmodium during the reproductive period.

Figure 2.14 (lower left). Bracket fungi growing on the trunk of a tree. This organism feeds on the dead bark tissue of the tree.

Figure 2.15 (lower right). The basidium, or club-shape fruiting body, commonly called a mushroom. These fungi act as saprophytes in breaking down dead organic matter to obtain their energy.

Plantae

Figure 2.16 (opposite left). A desmid, one of the green algae (Phylum Chlorophyta) in the Kingdom Plantae. *(Eric Gravé)*

Figure 2.17 (upper middle). Close-up photograph of strands of *Spirogyra*, a common fresh-water green alga. (Eric Gravé)

Figure 2.18 (upper right). Close-up photograph of a thallus (sheetlike) liverwort (Phylum Bryophyta). Note the globular, fingerlike reproductive units.

Figure 2.19 (middle left). *Lycopodium*, a club moss. Note the terminal spore sacs.

Figure 2.20 (center). Venus flytrap. This insectivorous plant is a native of marsh regions, and it supplements its nitrogen intake with insect protein.

Figure 2.21 (middle right). *Welwitschia marabilis*, a unique desert plant indigenous to the Kalahari Desert in southwest Africa.

Figure 2.22 (lower left). Snow plant, of the alpine zone.

Figure 2.23 (lower middle). Passion flower.

Figure 2.24 (lower right). Elephant-foot tree, a member of the palm family and native to Mexico.

Animalia

Figure 2.25 (upper left). A simple vase-shape sponge. These organisms are filter-feeders, straining out planktonic organisms from the water.

Figure 2.26 (upper right). The medusa, or jellyfish, form of a coelenterate.

Figure 2.27 (middle left). *Obelia* polyp. Note the ring of tentacles surrounding the mouth.

Figure 2.28 (center). Coral polyps, the well-known reef-building coelenterates.

Figure 2.29 (middle right). Sea pens, colonial coelenterate polyps that feed extended as shown, but retreat into a central stalk when disturbed.

Figure 2.30 (lower left). The honeycomb worm, a marine tube-building annelid worm.

Figure 2.31 (lower middle). The white-lined nudibranch, a gastropod member of the phylum Mollusca. Note the fingerlike gills on its back.

Figure 2.32 (lower right). Close-up photograph of the Great Scallop, a representative molluscan bivalve.

Figure 2.33 (above). Octopus, a cephalopod mollusc without an external shell covering.

Figure 2.34 (below). The mystery snail, another gastropod mollusc.

Arthropod members include:

Figure 2.35 (upper left). Spiny lobster, a crustacean.

Figure 2.36 (upper middle). The crustacean shrimp.

Figure 2.37 (upper right). A wolf spider, an arachnid.

Figure 2.38 (middle left). The ox beetle, an insect.

Figure 2.39 (middle right). The harvest ant, an insect.

Figure 2.40 (lower left). A woolly aphid, also an insect.

Chordate members include:

Figure 2.41 (lower middle). The reef-dwelling lion fish, a true bony fish.

Figure 2.42 (lower right). The amphibian tree frog.

Figure 2.43 (upper left). A tokay gecko, a reptile.

Figure 2.44 (upper right). The alligator, also a reptile.

Figure 2.45 (middle left). Rhinoceros viper, a reptile.

Figure 2.46 (center). Avocet, a shore bird.

Figure 2.47 (middle right). A blue-crowned pigeon.

Figure 2.48 (lower left). Tasmanian gray kangaroo, a marsupial mammal.

Figure 2.49 (lower middle). Opposum, also a marsupial.

Figure 2.50 (lower right). A white-footed deer mouse, a placental mammal.

Figure 2.51 (above). A hippopotamus, an even-toed hooved mammal (an artiodactyl).

Figure 2.52 (middle) Snow leopard, a carnivorous mammal.

Figure 2.53 (lower left). Hyrax, a small, herbivorous mammal that is related to the elephant.

Figure 2.54 (lower right). An orangutan, a primate.

FURTHER READING

For more detailed discussion of the topics of this chapter, see Unit Seven of this book, particularly Chapters 32, 36, 37, and 38.

Dickinson (1967) describes the life and work of Linnaeus. Good general discussions of biological classification are given by Blackwelder (1967), Copeland (1956), Mayr (1942, 1969), and Savory (1963). More specific discussions of plant classification are given by Bell (1967), Bell and Woodcock (1968), Davis and Heywood (1963), and Solbrig (1970). Animal classification is discussed by Blackwelder (1963), Mayr (1963, 1969), Mayr, *et al.* (1953), and Simpson (1961). For detailed discussion of the rules by which species and higher taxa are named, see works by Smith and Williams (1970), and Stoll, *et al.* (1964). Romer (1968) gives a good overview of the diversity of life and its evolutionary significance.

For further discussions of the nature of species, see works by Dobzhansky (1937), Mayr (1957), and Simpson (1951). Alston and Turner (1963) discuss biochemical contributions to problems of classification, and Sokal (1966) describes a recent quantitative approach to classification.

II
The Physical Basis of Life

If the scheme of philosophy which we now rear on the scientific advances of Einstein, Bohr, Rutherford and others is doomed to fall in the next thirty years, it is not to be laid to their charge that we have gone astray. Like the systems of Euclid, of Ptolemy, of Newton, which have served their turn, so the systems of Einstein and Heisenberg may give way to some fuller realization of the world. But in each revolution of scientific thought new words are set to the old music, and that which has gone before is not destroyed but refocussed. Amid all our faulty attempts at expression the kernel of scientific truth steadily grows; and of this truth it may be said—The more it changes, the more it remains the same.

—Eddington (1927)

3
Atoms and Bonds

The study of chemistry has become essential for an understanding of life sciences. Without the dramatic applications of chemical principles, biology today would still be a descriptive science supporting a good deal of theory but lacking experimental foundation. The physical sciences have given a firm scientific basis to biology. In turn, the synthesis of disciplines called biochemistry has led to vast improvements in medicine. For example, the developments of anesthesia, antibiotics, and other drug therapy have been crucial steps in the successful application of biology to the practice of medicine. The doctor has come a very long way from the days when he dosed his patient with fox fangs and bled him with leeches. Without an application of chemistry to biology and thus to medicine, life would still be nasty, brutish, and short.

Biological processes can be understood in terms of their chemical behavior and their adherence to the same physical laws that govern inanimate systems. Certain biological processes such as human memory are so complex that they have not yet been described in chemical terms, but there is every expectation that they will be so explained. Much of the excitement of biological research stems from the ever-present question: How can molecules comprised of only a few elements be constructed so as to carry out the function of such an intricate organ as the human brain? As biology has been reinforced with chemistry, it has become obvious that, at the molecular level, all living organisms are remarkably alike.

The study of the interaction of light with atoms and molecules has led to an understanding of how plants transfer energy from the sun to highly ordered molecules. The concept of chemical charge can explain how an organism converts the stored energy of fats and carbohydrates into useful work. Bonding between atoms provides the basis for the three-dimensional structure of complex arrays of molecules. Thus, chemistry is central to biology because the chemical structure and function of molecules are interdependent. It is the sum of the function of all the molecules in a living cell that determines the development and proliferation of that cell and distinguishes it from its neighbors.

ATOMIC THEORY

As early as the fifth century B.C., the idea that all things are made up of tiny atoms was suggested by the Greek philosopher Leucippus and his student Democritus. For centuries there was little support for such a view, but in the early years of the nineteenth century John Dalton, an English chemist, found that he could best explain his experimental results by using the idea that the various substances are composed of minute "atoms." His data indicated that the atoms combined with each other in a number of fixed ratios to form compounds.

From Dalton's theory comes the basis of modern chemistry. Before plunging into a discussion of that chemistry—particularly in relation to the living world—one other aspect of Dalton's theory should be made clear.

Neither Democritus nor Dalton ever saw an atom. Nor has anyone else, for although the term "atom" is used in describing chemical reactions and living processes, the atom is a *model*. It is a model in the sense that it is a representation of reality—as good a representation as can be developed from available information—but nevertheless an abstraction.

Dalton's was a very simple atomic model. As more information has been gathered, the atomic model has become increasingly complex in

Figure 3.1a (left). Dalton's table of atomic weights that he derived in 1803.

Figure 3.1b (right). John Dalton, an English chemist and physicist, who presented the first clear statement of atomic theory.

ELEMENTS

	wt		wt
Hydrogen	1	Strontian	46
Azote	5	Barytes	68
Carbon	54	Iron	50
Oxygen	7	Zinc	56
Phosphorus	9	Copper	56
Sulphur	13	Lead	90
Magnesia	20	Silver	190
Lime	24	Gold	190
Soda	28	Platina	190
Potash	42	Mercury	167

detail and comprehensive in its ability to explain actual phenomena. In the following discussions, a model of the atom is used that is complex enough to account for most important chemical properties, yet one considerably simplified from the complete modern understandings of atomic structures.

Dalton postulated that all substances are made up of small, indivisible particles called atoms and that each element contains only a single kind of atom. Atoms can combine to form molecules, which are the basic particles of chemical compounds. Thus, Dalton pictured the molecule of carbon dioxide (then called "fixed air") as a combination of an atom of carbon and two atoms of oxygen. Chemists later devised a shorthand notation for describing atoms and molecules, using the first one or two letters of the names of the elements. In many cases, the abbreviation is based upon the Latin or German name of the element. For example, "Na," the abbreviation for sodium, is derived from the Latin term "natrium." Subscript numerals are used to indicate the numbers of each kind of atom in a molecule; thus, carbon dioxide is written as CO_2.

Dalton knew from chemical analyses that about 3 grams of carbon combine with each 8 grams of oxygen when carbon dioxide is formed. Because each molecule of CO_2 contains one carbon atom and two oxygen atoms, any sample of CO_2 must contain twice as many oxygen atoms as it does carbon atoms. Therefore, if a certain number of carbon atoms weighs 3 grams, the same number of oxygen atoms must weigh 4 grams. In other words, each oxygen atom is 4/3 as heavy as a carbon atom.

In this fashion, Dalton was able to deduce the relative weights of the atoms of the 30 or so elements known at the time. The hydrogen atom

Figure 3.2 (left). A model of the nuclear atom. The classic model of the atom is a dense nucleus of neutrons and protons surrounded by electrons. Shown here is an oxygen atom with 8 protons (orange), 8 neutrons (yellow), and 8 electrons.

Figure 3.3 (right). The electron cloud representation of the atom. The negatively charged electrons are distributed diffusely around a positively charged nucleus. Note that the atom does not have a sharp boundary.

proved to be the lightest, and therefore Dalton arbitrarily set its weight equal to 1. Some of Dalton's relative *atomic weights* later proved to be erroneous because he had made incorrect assumptions about the number of atoms in certain molecules. For example, he assumed that the formula for water is HO; later studies showed that it is in fact H_2O. Within a few decades after Dalton's work, however, complete tables of atomic weights were available. For various reasons, the values of atomic weights were later calculated by giving oxygen a weight of 16.00. On this scale, hydrogen has an atomic weight of 1.008 instead of exactly 1.00. Recently, the standard has been revised again, but the values of the atomic weights have been changed only slightly by these redefinitions.

The Nuclear Atom

Only in the most simplified models can an atom be regarded as a simple particle. It is clear from many kinds of evidence that almost all of the mass of an atom is concentrated in a tiny fraction of total volume. This *nucleus* carries a positive electrical charge; the remaining volume of the atom has an equal negative electrical charge but very little mass.

The nucleus contains two kinds of particles: protons and neutrons. The *proton* carries the smallest amount of positive electrical charge yet observed, which can thus be called one unit of positive charge (+1). The *neutron* has nearly the same mass as the proton but carries no electrical charge.

The region surrounding the nucleus is occupied by particles called electrons. The *electron* carries an electrical charge equal in strength to that of

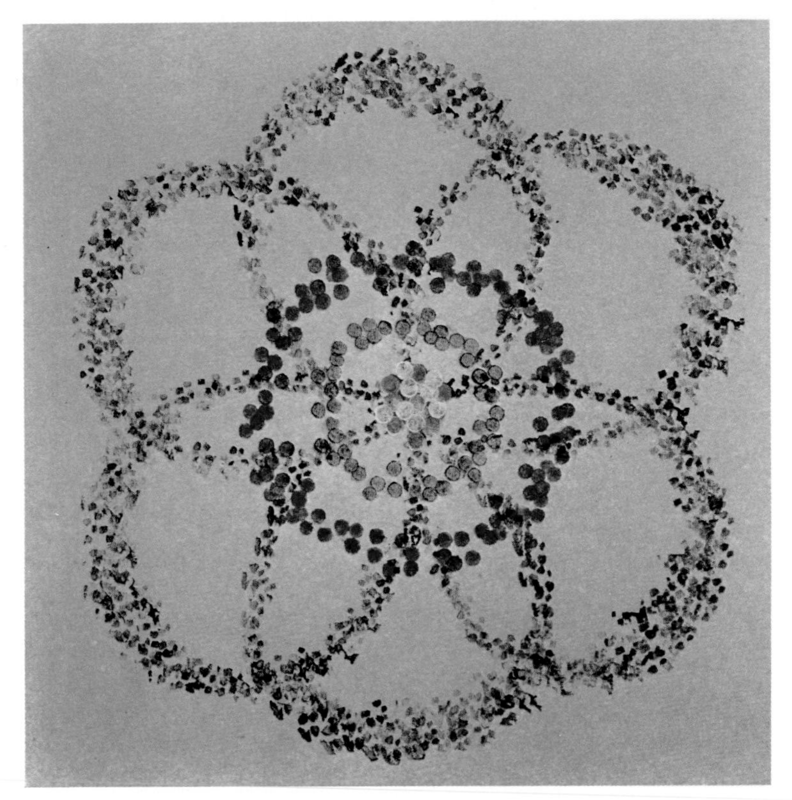

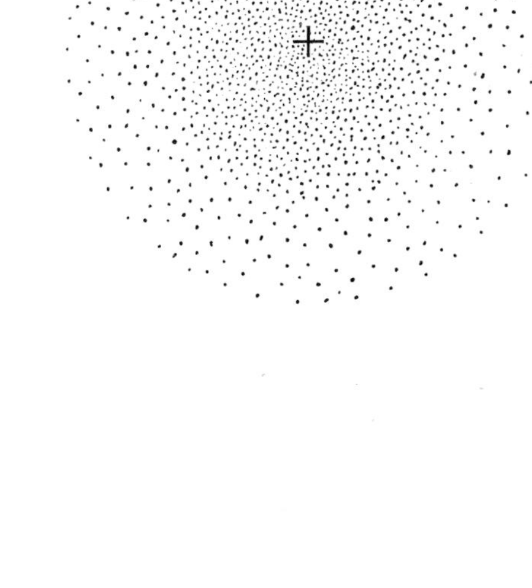

Figure 3.4 (above). The hydrogen molecule. *D* refers to the distance between the two nuclei in H_2.

Figure 3.5 (below). The isotopes of hydrogen.

the proton but of negative sign—in other words, one unit of negative charge (−1). The mass of an electron is only 1/1840 of the mass of a proton or neutron.

Other subatomic particles have been observed to exist briefly during high-energy nuclear reactions, such as those that occur in atomic bomb explosions and in the sun. In order to explain the observed properties of nuclear reactions, physicists are currently developing models in which the proton and neutron are themselves regarded as composed of yet smaller units. However, a simplified model that pictures the atom as composed of the three basic particles—protons, neutrons, and electrons—is sufficient to explain chemical reactions.

The lightest and simplest atom is that of hydrogen. Its nucleus is composed of a single proton, and a single electron occupies the space around the nucleus. The diameter of the nucleus is about 10^{-13} centimeter (0.0000000000001 cm). The diameter of the entire atom is more difficult to define because it has no sharp boundary. It is possible, however, to measure the distance between the nuclei of two atoms joined together in a molecule. Half of this distance may be considered as a convenient representation of the radius of the atom. Such measurements indicate that the diameter of the hydrogen atom is about 10^{-8} cm.

As a way of visualizing the spatial relationships within the atom, suppose that it were enlarged to the size of a football stadium. The nucleus could be represented by a gum drop resting on the center of the field. The electron may be represented by a fly buzzing about the stadium, usually to be found somewhere inside the stadium but occasionally moving outside its boundaries as well. The picture serves as a reminder that most of the volume of an atom is empty space; it is also important to recall that nearly all of the mass of the atom is concentrated in the nucleus.

The hydrogen atom is distinguished from all other elements by the single proton in its nucleus. An atom of helium contains two protons in its nucleus; an atom of lithium contains three protons. In each case, the positive electrical charge of the protons is balanced by the negative charge of an equal number of electrons in the space surrounding the nucleus. Each element is made up of atoms whose nuclei contain a particular number of protons; this number is called the *atomic number* of the element. Carbon, with six protons in its nucleus, has an atomic number of 6, and the atomic number of oxygen is 8.

Although all atoms of a particular element have the same number of protons in the nucleus, not all have the same mass. About 0.016 percent of all hydrogen atoms, for example, have a mass about twice as great as that of a proton. These atoms contain one proton and one neutron in the nucleus. Such an atom is called hydrogen-2 (^2H) and is said to have a mass number of 2. The *mass number* is equal to the sum of the number of protons and the number of neutrons in the nucleus. The two kinds of hydrogen atoms are called *isotopes* of hydrogen. All of the isotopes of a particular element have very similar chemical properties; they differ only in the number of neutrons in the nucleus of the atom.

Most chemical elements are made up of mixtures of isotopes. Oxygen, for example, has three stable isotopes. About 99.8 percent of natural oxygen is composed of oxygen-16 (with 8 protons and 8 neutrons in the nucleus). The remaining atoms are those of oxygen-17 (with 8 protons and 9 neutrons) and oxygen-18 (with 8 protons and 10 neutrons).

Although all three of these oxygen isotopes are stable, isotopes of other

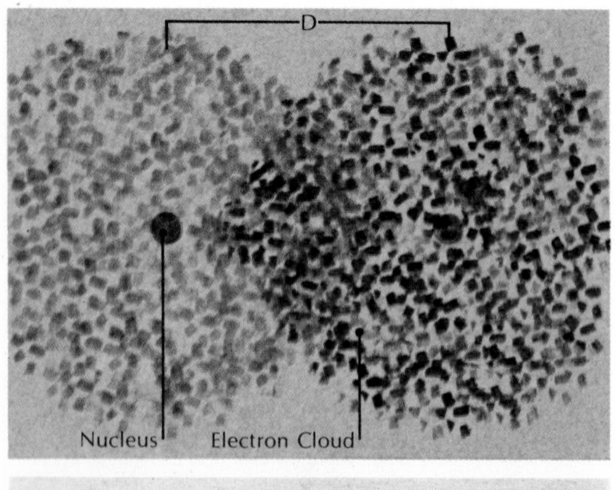

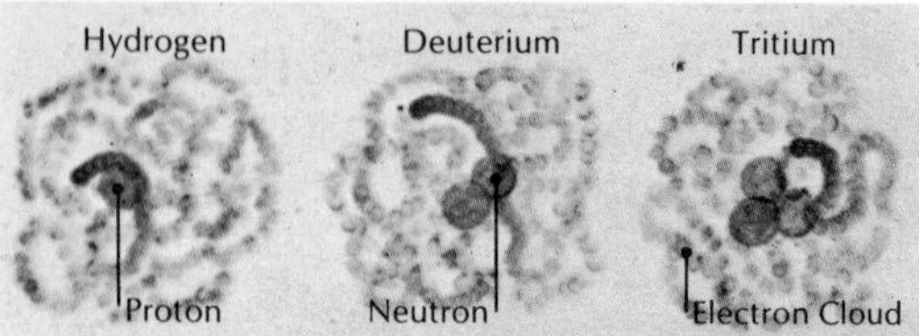

elements do have unstable nuclei. These isotopes are said to be radioactive, and their nuclei undergo radioactive decomposition with the characteristic emission of such high-energy fragments as γ-rays (very high-energy light) or β-particles (high-energy electrons). Each radioactive isotope decomposes distinctively at a characteristic rate, the number of decomposition products depending upon the number of unstable nuclei present. With an appropriate detector, scientists can determine the concentration and nature of a particular radioactive isotope by observing the quantity and quality of its decomposition products.

The use of unusual isotopes of elements has had a sweeping impact in biology. Radioisotopes of carbon and hydrogen, discovered only a few decades ago, have allowed biologists to describe just how plants convert carbon dioxide, water, and sunlight into sugar and how humans convert that sugar back into usable energy, water, and carbon dioxide. Radioisotopes of sulfur and phosphorus were employed to prove that the carrier of genetic information was deoxyribonucleic acid (DNA). Today, radioisotopes such as iron, iodine, chromium, and cobalt are used routinely in

Table 3.1
Radioisotopes That Are Useful to Biologists

ISOTOPE	Atomic Number	Natural Abundance	Halflife	Mode of Decay
H^3	1	synthetic	12.3 years	$\beta-$
C^{14}	6	synthetic	5730 years	$\beta-$
O^{18}	8	.204 %	stable	none
P^{32}	15	synthetic	14.3 days	$\beta-$
I^{131}	.53	synthetic	8.1 days	$\beta-$

medicine, and many other isotopes are frequently used in the laboratory to uncover a vast array of biological processes. Table 3.1 shows the properties of a number of isotopes useful to biologists.

An atom is electrically neutral because the positive charge of the protons in the nucleus is balanced by the negative charge of an equal number of electrons around the nucleus. However, electrons can be removed from or added to an atom, giving it a net electrical charge. Such an electrically charged atom is called an *ion.* (Positively charged ions are called *cations;* negatively charged ions are *anions.*) For example, the electron may be removed from an atom of hydrogen, leaving only the nucleus with a charge of 1+. Chemists use the symbol H^+ to represent this hydrogen ion. Because there is an electrical force of attraction between the proton and the electron in the hydrogen atom, energy is required to remove the electron and form the hydrogen ion.

The amount of energy needed to remove an electron from the atom is called the *first ionization energy* (I_1) of the element; the energy needed to remove a second electron is called the *second ionization energy* (I_2), and so on. Table 3.2 shows the ionization energies for some of the lighter elements. Note the variations in the ionization energies of the elements; some elements lose electrons and form ions much more easily than do others. The first ionization energy may be regarded as a measure of the strength with which the atom holds its "loosest" electron. This property determines the chemical behavior of the elements.

The energies are measured in kilocalories per mole—a mole is 6.02 ×

Figure 3.6. The first ionization potentials for the first three periods of the periodic table.

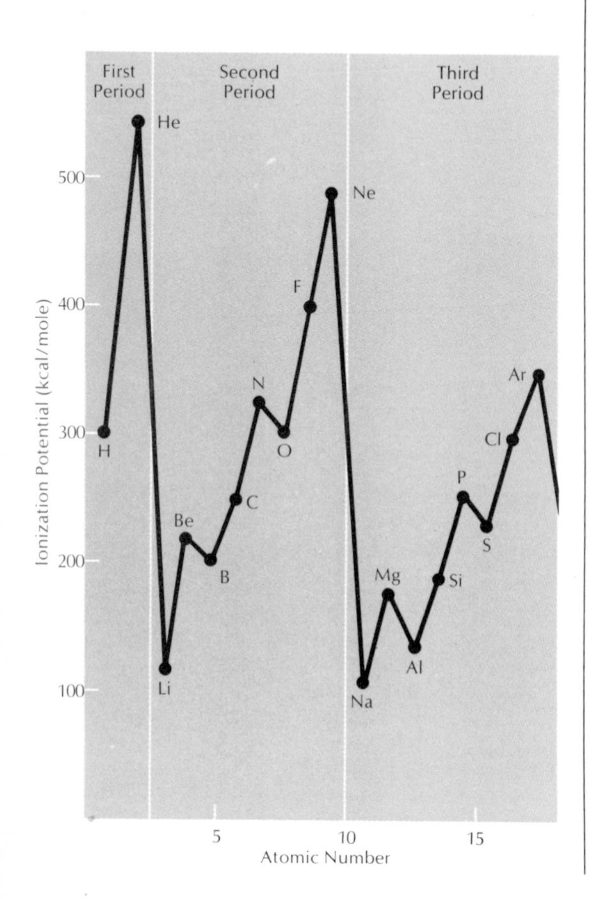

10^{23} atoms, a number calculated such that the weight of a mole of atoms equals the atomic weight in grams. The same holds true for molecules—that is, 6.02×10^{23} molecules has the same molecular weight in grams. When discussing biological reactions, scientists speak of fractions of a mole because the amount of materials involved are often millionths or billionths

Table 3.2
Ionization Energies of Some Elements (kcal/mole)

ELEMENT	Chemical Symbol	Atomic Number	I_1	I_2	I_3	I_4
Hydrogen	H	1	313.5			
Helium	He	2	566.9	1254		
Lithium	Li	3	124.3	1744	2823	
Beryllium	Be	4	214.9	419.9	3548	5020
Boron	B	5	191.3	580.0	874.5	5980
Carbon	C	6	259.6	562.2	1104	1487
Nitrogen	N	7	335.1	682.8	1094	1786
Oxygen	O	8	314.0	810.6	1267	1785
Fluorine	F	9	401.8	806.7	1445	2012
Neon	Ne	10	497.2	947.2	1500	2241
Sodium	Na	11	118.5	1091	1652	2280
Magnesium	Mg	12	176.3	346.6	1848	2521
Aluminum	Al	13	138.0	434.1	655.9	2767
Silicon	Si	14	187.9	376.8	771.7	1041
Phosphorus	P	15	254	453.2	695.5	1184
Sulfur	S	16	238.9	540	807	1091
Chlorine	Cl	17	300.0	548.9	920.2	1230
Argon	Ar	18	363.4	637.0	943.3	1379
Potassium	K	19	100.1	733.6	1100	1405
Calcium	Ca	20	140.9	273.8	1181	1550

Source: B. H. Mahan, *University Chemistry*, 2nd ed. (Reading, Mass.: Addison-Wesley, 1969), p. 444.

of a mole (micromoles or nanomoles, respectively). Another unit of energy is the kilocalorie. A kilocalorie is 1,000 calories and is often replaced in discussion of food energy by the synonymous term "Calorie" (note the capital C). One calorie (small c) is defined as the amount of energy required to raise the temperature of 1 gram of water by 1° Centigrade. A kilocalorie is the amount of energy required to keep a 100 watt light bulb lit for 40 seconds. Human beings require about 2,000 Calories of food energy per day to sustain normal activity—slightly less than the continuous burning of a 100 watt bulb.

Electrons and Chemical Properties

When the first ionization energies of Table 3.2 are plotted on a graph (Figure 3.6), a regular pattern is visible. The first ionization energies of the elements helium, neon, argon, krypton, and xenon are much higher than those of the neighboring elements, whereas the first ionization energies of lithium, sodium, potassium, rubidium, and cesium are extremely low.

These regularities correspond to chemical properties of the elements. Lithium, sodium, potassium, rubidium, and cesium belong to a group of elements called *alkali metals.* They form metallic solids and react readily with many other elements, releasing large amounts of energy in most of these reactions. The alkali metals all tend to form positive ions, such as Li^+,

Figure 3.7. The relative atomic sizes of the inert gases.

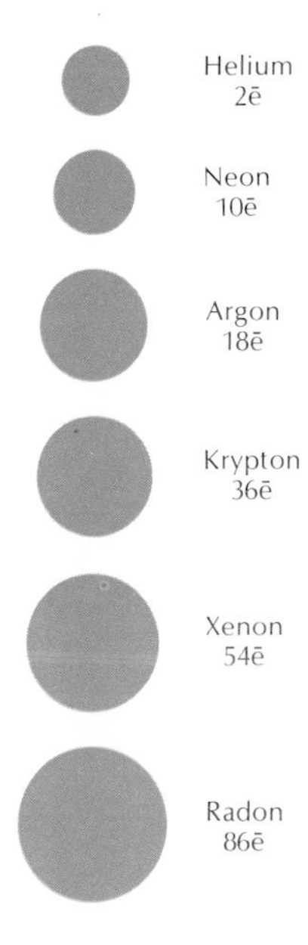

Na^+, and K^+. These properties appear to be due to the ease with which an electron can be removed from the alkali atom.

Helium, neon, argon, krypton, and xenon are strikingly different from the alkali metals in their chemical properties. These elements belong to a group called the *inert gases*, and they are distinguished by their almost total lack of chemical activity. Under normal conditions, the inert gases do not take part in any chemical reactions. Again, these properties are consistent with the fact that it is extremely difficult to remove an electron from the atom of an inert gas.

The elements that have an atomic number one greater than the alkali metals form a group called the *alkaline-earth metals:* beryllium, magnesium, calcium, strontium, barium, and radium. These elements are also chemically active, but the atom of an alkaline-earth metal tends to lose two electrons in chemical reactions, forming an ion such as Be^{++}, Mg^{++}, or Sr^{++}. The tendency to lose two electrons can be readily explained on the basis of the information in Table 3.1. Lithium, for example, easily loses its first electron, but a much greater amount of energy is needed to remove a second electron. Beryllium, on the other hand, loses its first and second electrons relatively easily but strongly resists the removal of a third electron.

Both the lithium ion (Li^+) and the beryllium ion (Be^{++}) are left with two electrons, the same number of electrons as are present in the neutral atom of the inert gas helium. In fact, there is a general tendency for atoms to gain or lose electrons in such a number that the remaining electron structure will be similar to that of an inert gas. It appears that the electron structure of the inert gases is an extremely stable one. The *halogens* (fluorine, chlorine, bromine, and iodine) hold their electrons almost as tightly as do the inert gases. However, these elements tend to gain an electron in chemical reactions, forming ions such as F^- and Cl^- and thus attaining electron populations similar to those of inert gases.

The electron populations of the inert gases are 2, 10, 18, 36, 54, and 86 — a series that shows a certain elusive regularity. (Note the differences between successive numbers in the series.) The regularity is best shown in the *periodic table* (Figure 3.8), in which the elements are listed in order of increasing atomic number. Each vertical column of the table includes elements of similar chemical properties.

The periodic table was developed in 1869 by the Russian chemist Dmitri Mendeléev on the basis of the 63 elements then known. During the following decades, chemists discovered many new elements, guided in their search by the blank spaces in Mendeléev's table. Mendeléev had arranged the elements in order of increasing atomic weight, although he had to make a few exceptions to this rule in order to keep elements of similar chemical properties in the same column. Although the table proved extremely useful in organizing the elements, chemists had no theory to account for the regularities that it revealed.

In the early years of the twentieth century, a number of new discoveries helped to make sense of the table. The discovery of the subatomic particles (electrons and protons) led to Ernest Rutherford's model of the nuclear atom and to Henry Moseley's measurements of atomic numbers. With this information, it became clear that the regularities in the periodic table reflect the tendency for atoms to achieve certain stable electron populations. However, there was still no explanation for the tendency to seek these particular electron populations.

The clues that led toward an explanation came — as so often happens in

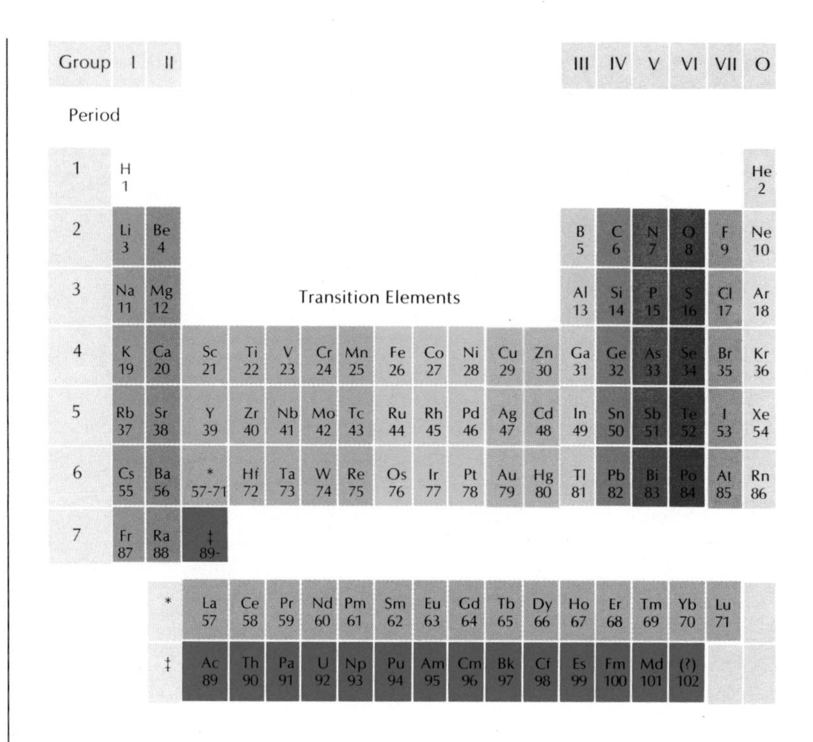

Figure 3.8 (above). Periodic table of the elements, first developed by Dmitri Mendeléev in 1869.

science—from studies that seemed completely unrelated. The German physicist Max Planck concluded from his studies of energy radiated by heated objects that electromagnetic energy (such as light or radio waves) is not continuous but, like matter, comes in small, discrete packages. Planck called these packages of energy *quanta*, and he showed that the energy of a single quantum of radiation, or photon, is related to the frequency of the radiation by the simple equation

$$E = h\nu \tag{3.1}$$

where E = energy of a single quantum,
h = a number called Planck's constant,
and ν(nu) = the frequency of the radiation.

Planck's quantum theory, published in 1900, represented such an extreme departure from the accepted laws of physics that physicists were hesitant to accept it. It proved so useful, however, that its publication is today considered to mark the dividing line between classical and modern physics.

One of the most impressive successes of the quantum theory was its application to the structure of the atom. It had been known for half a century that gaseous elements produce radiation of certain characteristic frequencies when the gas is burned or when an electrical discharge is passed through it. A solid heated to high temperatures produces radiation of a wide range of frequencies, the frequency of greatest radiation depending upon the temperature of the solid. When light from such an incandescent solid is analyzed in the device called a spectroscope (Figure 3.10), a continuous spectrum is produced. In contrast, the radiation from a gaseous element produces a series of bright lines; each element produces its own character-

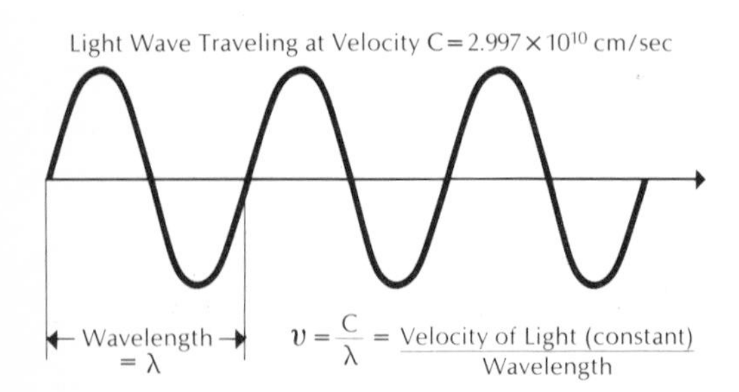

Figure 3.9 (below). The energy of a light wave, or photon, is given by the relative E = $h\nu$ (Planck's constant). As ν increases, the energy of the photon increases. Note that the wavelength (λ) becomes smaller as the energy increases. Thus, x-rays or ultraviolet rays that have short wavelengths are more energetic than visible or infrared light rays, which have longer wavelengths.

istic line spectrum. These lines are not randomly placed but occur in series with very regular patterns. Table 3.3, for example, shows the line spectrum of hydrogen gas. There are two groups of lines—one in the visible part of

Table 3.3
Lines in the Spectrum of Hydrogen

LINE	Frequency (cycles/second)	Wavelength (centimeters)	Energy of Photons (kcal/mole)
Visible Group			
1	4.57×10^{14}	6.57×10^{-5}	43.6
2	6.17×10^{14}	4.86×10^{-5}	58.8
3	6.91×10^{14}	4.34×10^{-5}	65.9
4	7.31×10^{14}	4.10×10^{-5}	69.7
5	7.55×10^{14}	3.97×10^{-5}	72.1
6	7.71×10^{14}	3.89×10^{-5}	73.6
Ultraviolet Group			
1	24.65×10^{14}	1.22×10^{-5}	235.2
2	29.22×10^{14}	1.03×10^{-5}	278.8
3	30.81×10^{14}	0.97×10^{-5}	294.0
4	31.49×10^{14}	0.95×10^{-5}	301.1
5	31.96×10^{14}	0.94×10^{-5}	304.9

Source: Adapted from Chemical Education Material Study, *Chemistry: An Experimental Science* (San Francisco: Freeman, 1963), p. 255.

the spectrum, the other in the ultraviolet part. Each group shows a regular decrease in the spacing between lines with increasing frequency.

Although physicists had developed mathematical expressions to summarize the regularities in line spectra, they had no explanation for their existence. One of Rutherford's students, Niels Bohr, felt that spectra must be related to the structure of the nuclear atom proposed by Rutherford. Bohr felt that a combination of Planck's quantum theory and Rutherford's

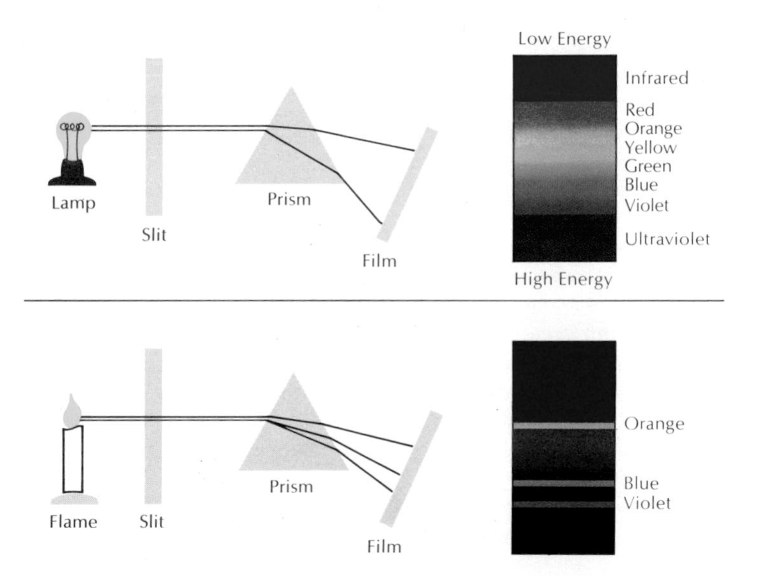

Figure 3.10. A continuous spectrum (upper portion) is generated by the hot filament of an incandescent bulb. A line spectrum (lower portion) is produced by the excitation of electronic energy levels in the atoms of an element heated by a flame. (*From* General Chemistry, *by Linus Pauling. Third edition. W. H. Freeman and Company.* © *1970*)

Figure 3.11 (above). Origin of the line spectra of elements. Electrons in excited energy levels fall into levels of lower energy. During this process, they emit a photon of light whose energy is equal to the energy difference between the initial and final energy level of the electron.

Figure 3.12 (below). Diagram showing the ultraviolet energy level of hydrogen. When an electron excited to the third energy level in hydrogen drops to the second energy level, it emits a photon equal in energy to the difference between the two levels.

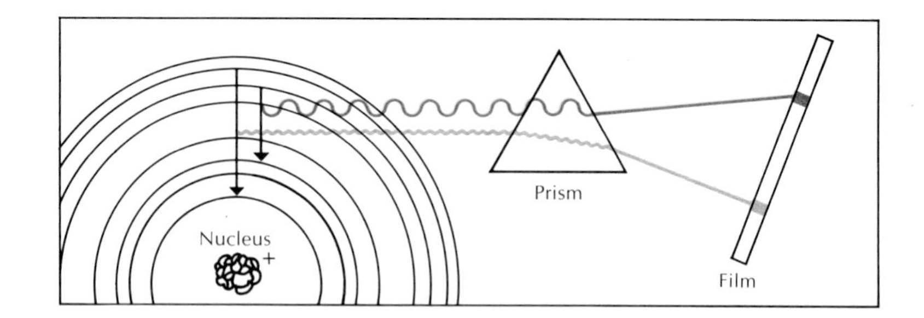

nuclear atom model might produce a better explanation of the line spectra. Bohr's quantized model of atomic structure did indeed prove useful and became the basis of modern models.

Using Planck's equation (Formula 3.1), one can translate the frequency of the lines in the hydrogen spectrum into terms of energy. The righthand column of Table 3.3 shows the energy per mole of *photons* (quanta of light) for each of the hydrogen spectral lines. Bohr suggested that an electron does not move randomly about the nucleus but can exist only at certain levels of energy. When an electron moves from one energy level to a lower one, a photon is produced with energy just equal to the difference between the two energy levels. When the hydrogen is heated or subjected to electrical discharge, electrons are *excited*—that is, they obtain energy and move to the higher energy levels. As they drop back to the more stable, lower energy levels, electrons emit photons of just those frequencies corresponding to the energy differences between the possible levels.

Because the lines in the ultraviolet group are produced by photons of greater energy than those producing the visible lines, it seems logical to suppose that the ultraviolet lines are produced by electrons falling from higher energy levels to the lowest possible energy level. Using this assumption and assigning an arbitrary energy value of 0.0 kcal/mole to the lowest energy level, one can obtain the set of energy levels shown in Figure 3.12. The series of lines in the ultraviolet group does not end with the sixth line but continues with an apparently infinite number of lines spaced ever more closely. The limit approached by the lines of highest energy is 313.5 kcal/mole. Presumably, this number corresponds to the amount of energy released as a free electron is captured by the atom and moves into the lowest possible energy level.

What would happen if an electron moved from the third to the second energy level? The energy of the electron would drop from 278.8 to 235.2 kcal/mole, and thus a photon with energy of 43.6 kcal/mole should be emitted. Such photons are emitted, producing the first line of the visible group. Similarly, all of the visible lines can be explained by movements of electrons from higher levels to the second energy level. Similar series of lines are formed in the infrared (low-energy) photons, just as would be predicted for electrons falling to the third, fourth, and higher levels.

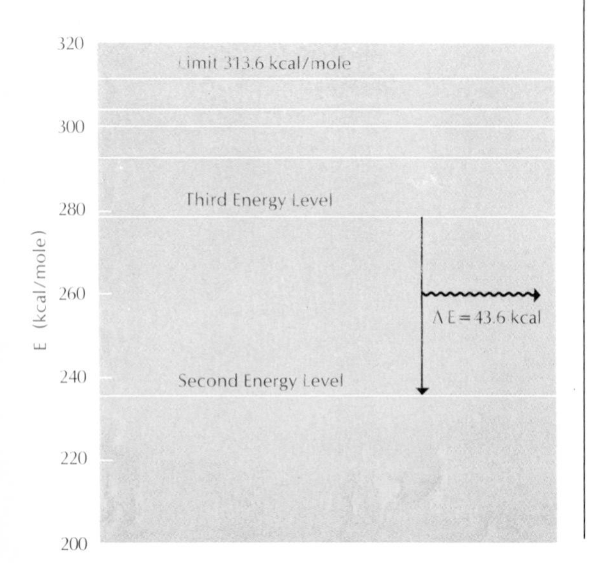

This model indicates that 313.5 kcal/mole of energy are released as a free electron falls into the first energy level. Therefore, 313.5 kcal/mole of energy should be needed to remove an electron from the atom, and this number is indeed the ionization energy of hydrogen (Table 3.2).

Bohr's model provided an explanation of the relationship between the various groups of lines in the hydrogen spectrum. It was based, however, upon the assumption (unjustified by the laws of physics developed through study of macroscopic objects) that electrons could exist only at certain discrete energy levels within the atom. Furthermore, Bohr's model did not provide quantitatively correct explanations of the spectra of elements other than hydrogen. For these reasons, physicists were reluctant to accept Bohr's model. The contradictions and problems were eventually resolved in the theories of *quantum mechanics*, developed during the 1920s.

In the modern model of atomic structure, the motion of each electron in an atom is described by an *orbital*, a description of the space in which the electron is most likely to be found. According to the theories of quantum mechanics, it is not possible to describe an exact path for an electron. All

Figure 3.13. The Bohr model of a hydrogen atom. This model postulates an atom in which an electron travels in one of several orbits. When the electron falls from a higher energy orbit to a lower energy level, it emits light of a certain wavelength. Each group of arrows represents energy transitions that give rise to spectral lines. These energy transitions are shown in the spectral lines below.

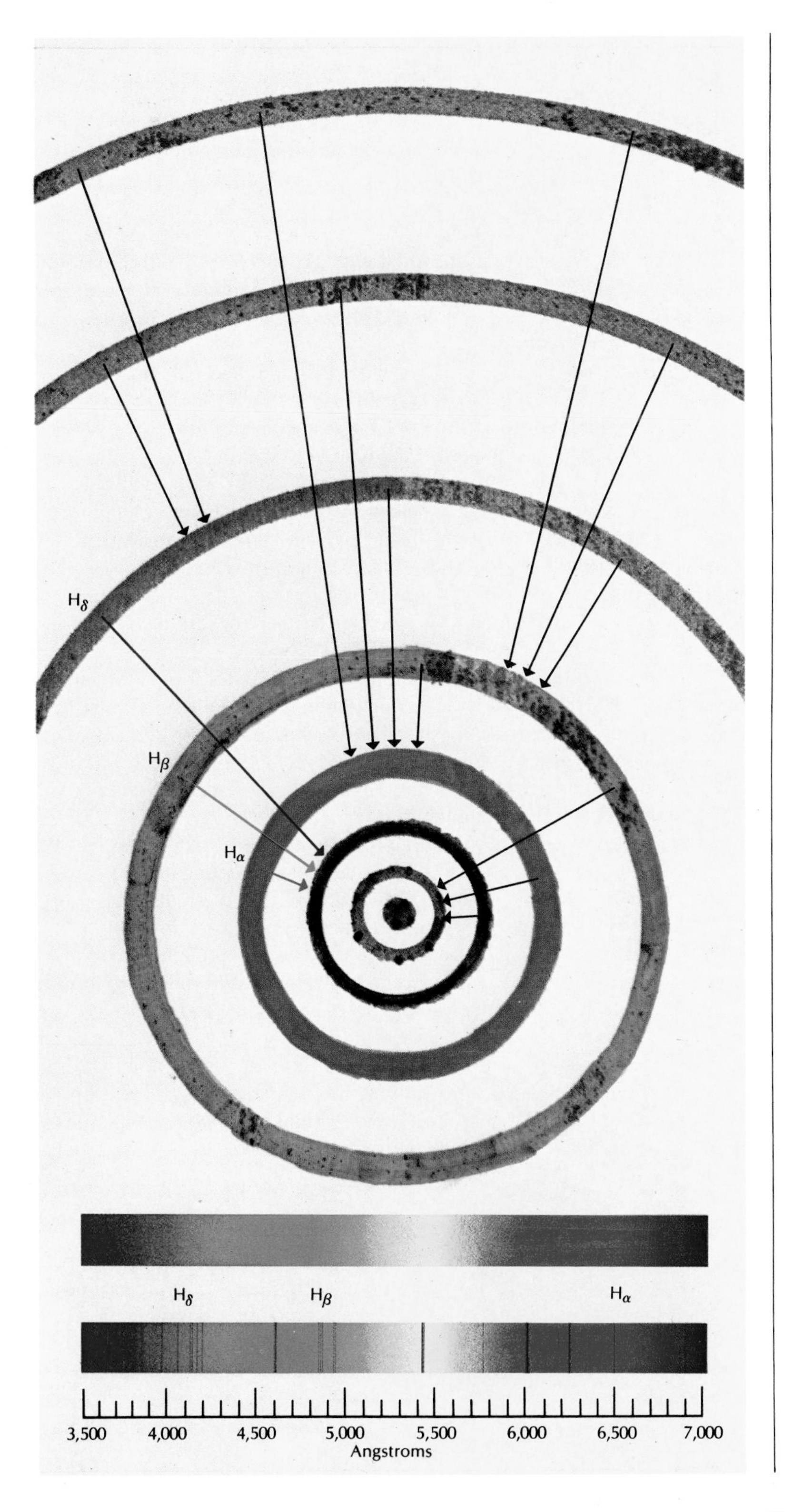

Figure 3.14. Cross section of an atom. Although there is a probability that the electrons of an atom may be found anywhere in space, a sphere may be defined so that the probability of finding the electrons inside is nearly 100 percent.

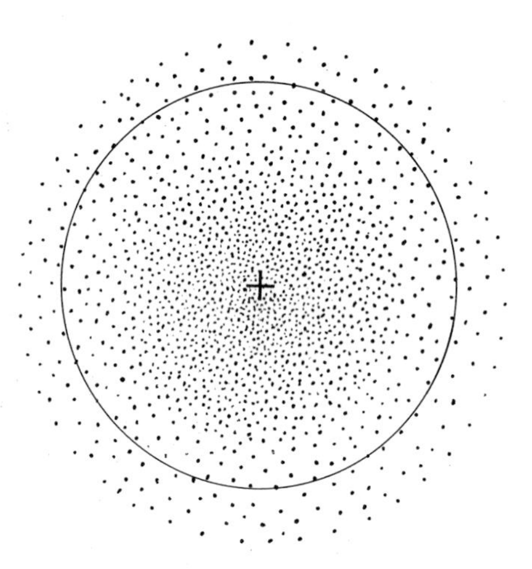

that can be done is to predict the probability of finding the electron at any given point in space around the nucleus. Each pair of electrons in an atom moves in a distinct orbital of a particular energy level, and the shapes of the orbitals help to explain the positions of atoms within a molecule.

Because the energy levels of some orbitals are nearly identical, it is possible to describe most of the chemical properties of elements in terms of groups of orbitals, or *electron shells*. Each of the inert gases has a full outer electron shell, and other elements tend to gain or lose electrons in order to achieve a full outer shell. The first shell can hold only 2 electrons; the inert gas helium, for example, fills this shell with its 2 electrons. The second shell can hold 8 electrons; the 10 electrons of neon fill both the first and second shells. The third shell also can hold 8 electrons; the 18 electrons of argon fill the first, second, and third shells.

In the next row of the periodic table, potassium and calcium show the properties to be expected if they contain, respectively, 1 and 2 electrons in the fourth shell. However, the next 10 elements—scandium through zinc—behave chemically as if they also have only 2 electrons in the outer shell. Apparently, after 2 electrons have been placed in the fourth shell, the third shell is able to accept another 10 electrons. The 10 elements (zinc through scandium) are called transition metals and are somewhat similar in their chemical and physical properties. After the third shell has been expanded to 18 electrons, the filling of the fourth shell resumes; the properties of gallium show that it has 3 electrons in the fourth shell. The fourth shell is completed with 8 electrons in the krypton atom.

The number of electrons in each shell for each atom is indicated in the periodic table. The elements of atomic numbers 57 through 71 differ only in the number of electrons in the third from the outermost shell. Because chemical properties are determined primarily by the outer electrons, these elements are very similar chemically—so similar, in fact, that it is difficult to separate them by chemical methods.

This picture of electron structure provides a model that accounts for the regularities summarized in the periodic table. Each element in a particular *column* has a similar population of electrons in its outer shells; each tends, through the gain or loss of electrons, to achieve a full outer shell.

The ionization energies shown in Table 3.2 can be explained in terms of this model. The ionization energy represents the energy needed to take an outer electron from its normal energy level and move it to a level so high that it is free from the nucleus. In the lithium atom, it is relatively easy to remove the single electron from the second shell, but a much greater amount of energy is needed to remove the remaining electrons from the lower energy level of the first shell. The beryllium atom has two electrons in the second shell that can be removed relatively easily.

CHEMICAL BONDS

The helium atom, with two electrons filling its first shell, is very stable. The hydrogen atom has only one electron in this shell and is much more reactive chemically. It readily becomes bonded to other atoms to form molecules. For example, hydrogen gas at normal temperatures is made up not of individual hydrogen atoms but of diatomic (two-atom) molecules composed of two hydrogen atoms bound together. Chemists symbolize this molecule as H_2.

The formation of this bond can be understood by visualizing what happens as two hydrogen atoms approach each other. When the atoms are

Figure 3.15. Formation of the chemical bond. From right to left, A depicts two free atoms that are moving in space. In B, the atoms are starting to interact, and the energy of the pair is decreasing as they share their electrons. The bond is fully formed in C, and the energy of the pair is minimized at its most stable internuclear distance, r_0. In D, any attempt to push the atoms closer together causes the energy of the system to rise rapidly because of internuclear repulsion. This most stable distance, r_6, is the band length.

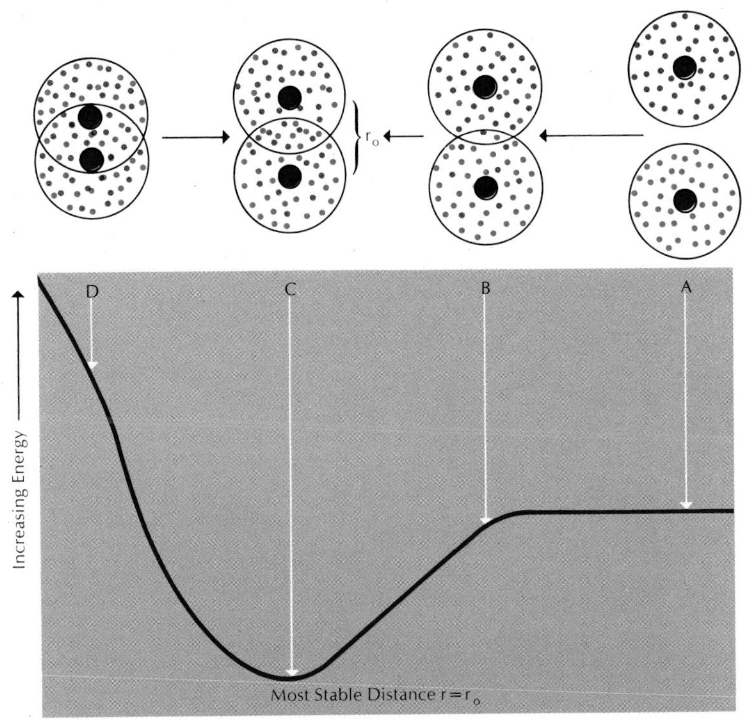

close together, there is electrical attraction between the nucleus of each atom and the electron of the other. There is also electrical repulsion between the two nuclei and between the two electrons. The attractive forces will tend to hold the two atoms together, and the repulsive forces will tend to push them apart. Because it is known that the two atoms do tend to remain together at some characteristic internuclear distance, the attractive forces must be stronger than the repulsive forces.

Each hydrogen atom has a "vacancy" in its outer shell. When the two atoms are close together, the unfilled orbitals overlap. In the region of overlap, each electron can be shared by both nuclei. Because each electron is attracted to both nuclei and is able to move within a stable orbital, there are strong attractive forces holding the two atoms together. In fact, it appears that the two electrons correlate their movements in such a way that they stay far apart from each other, minimizing the repulsive forces between the two electrons and helping to hold the molecule together.

In the H — H bond, electrons are shared equally by the two nuclei. In this way, each atom has approximated the stable electron population of a full outermost shell. Such a bond, in which electrons are shared between nuclei, is called a *covalent bond.* Electron-dot formulae are used to symbolize the electrons of the outermost shell involved in chemical bonds. Each hydrogen atom, with its single electron, may be represented as H ·, and the hydrogen molecule with its shared pair of electrons may be written as H:H.

A helium atom has two electrons filling its outer energy level (He:). If two helium atoms approach each other, little overlap of orbitals can occur because each outer shell is fully occupied. The helium atoms therefore do not form bonds but remain as independent atoms.

The oxygen atom has six electrons in its outer shell, with room for two

Figure 3.16. Types of chemical bonds. Shown here in the water molecule are covalent bonds, which bind atoms through shared electrons; hydrogen bonds, which are strong electrical links; and van der Waals forces, which are weaker electrical links between the oxygen nucleus and electrons of an oxygen atom.

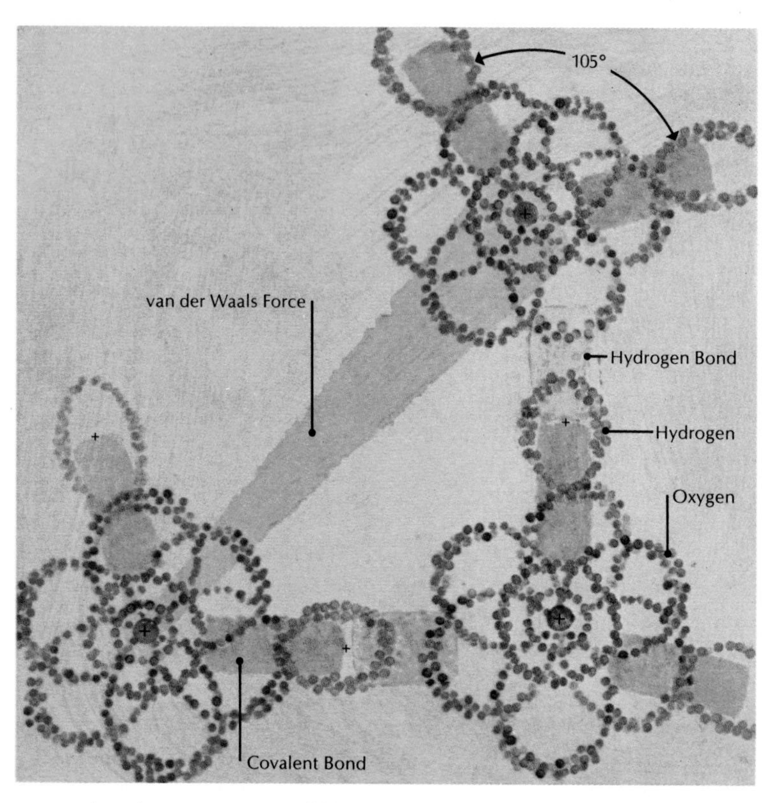

more. Therefore, oxygen would be expected to form covalent bonds, sharing two electrons with other atoms. For example, each of the unpaired electrons could be shared with a hydrogen atom, forming a molecule that could be represented as H—O—H or as H_2O, the water molecule.

$$:\ddot{\underset{\cdot}{O}}: + H\cdot + H\cdot \rightarrow :\ddot{O}:H \atop \quad\quad\quad\quad\quad\quad\quad\quad\quad\quad H \tag{3.2}$$

The oxygen nucleus has a charge of 8^+, whereas the hydrogen nucleus has a charge of only 1^+. Thus, the oxygen nucleus exerts a much stronger attractive force on the shared electrons than does the hydrogen nucleus, and the shared electrons are most likely to be found nearer the oxygen nucleus. As a result, the portion of the molecule near the oxygen nucleus has an excess of negative charge, whereas the portions near the hydrogen nuclei have an excess of positive charge. A bond such as the O—H bond, in which the two nuclei do not equally share the electrons, is a polar covalent bond and is said to have partial ionic character. In a completely *ionic bond*, the bonding electrons are not shared at all but are held entirely by one of the nuclei; the two atoms are held together solely by the electrical attraction between the cation (the atom that has lost an electron) and the anion (the atom that has gained an electron).

Such ionic bonds are formed in table salt (sodium chloride, NaCl). Within a crystal of salt, there is a regular latticelike arrangement of sodium (Na^+) and chloride (Cl^-) ions, with each ion attracted to neighboring ions of the

Figure 3.17. Sodium chloride (NaCl) — a cubic ionic crystal. The brown spheres are sodium ions and the gray spheres chloride ions.

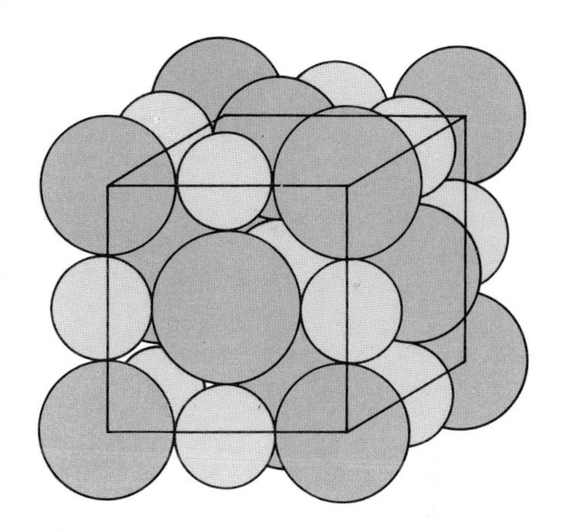

opposite charge. Because no covalent bonds are formed, there is really no molecule of NaCl. Table salt dissolves readily in water because the Na^+ and Cl^- ions separate readily under the electrical attractions of the positive and negative ends of the water molecule.

Oxygen, like hydrogen, forms a diatomic gas (O_2 molecules). In this case, two pairs of electrons must be shared between the two oxygen atoms:

$$:\dot{\underset{\cdot}{O}}: + :\dot{\underset{\cdot}{O}}: \rightarrow :\dot{\underset{\cdot}{O}}::\dot{\underset{\cdot}{O}}: \quad (3.3)$$

Such a *double bond* is stronger than a single bond (where only one pair of electrons is shared) and draws the two nuclei closer together. In a simple diagram, the double bond may be represented by the symbol O=O.

Nitrogen has only five electrons in its second shell and therefore can share three pairs of electrons with atoms such as hydrogen, forming the molecule of ammonia (NH_3).

$$:\dot{\underset{\cdot}{N}}\cdot + 3H\cdot \rightarrow :\overset{H}{\underset{H}{\ddot{\underset{\cdot\cdot}{N}}}}:H \quad (3.4)$$

Like hydrogen and oxygen, nitrogen exists as a diatomic gas (N_2). Here, the two atoms share three pairs of electrons, forming a triple bond that may be represented as N:::N or as N≡N.

Organic Chemistry

Carbon has four electrons in its outer shell and therefore can share four pairs of electrons with hydrogen atoms to form methane (CH_4).

$$\cdot\dot{\underset{\cdot}{C}}\cdot + 4H\cdot \rightarrow H:\overset{H}{\underset{H}{\ddot{\underset{\cdot\cdot}{C}}}}:H \quad (3.5)$$

The importance of carbon to the chemistry of living systems arises from its exceptional tendency to form strong, stable covalent bonds with other carbon atoms. Thus, long chains of carbon atoms may be formed, with various other atoms bonded to the remaining unpaired electrons of the carbon atoms. Because the carbon chains can branch or form rings, millions of carbon compounds can be formed.

Other elements such as silicon or boron form strongly bonded chains in the elementary state, but carbon is unique in that it forms strong chains even when various other atoms are bonded to the remaining unpaired electrons of the carbon atoms. When chains of carbon atoms branch or reunite to form rings, additional structures are available. With the endless possibilities of bonding among carbon atoms, it is not surprising that the study of carbon compounds is an entire field in itself, a field known as organic chemistry. The name denotes the central importance of carbon-containing molecules (or organic compounds) to living organisms. Much of the recent intense search for extraterrestrial life has focused on looking for organic molecules on the various planets and in specimens of rock and dust retrieved from the moon. The existence of past or present life would be

Figure 3.18. The tetrahedral structure of methane. Note that the tetrahedral arrangement of electron pairs minimizes interelectronic repulsion.

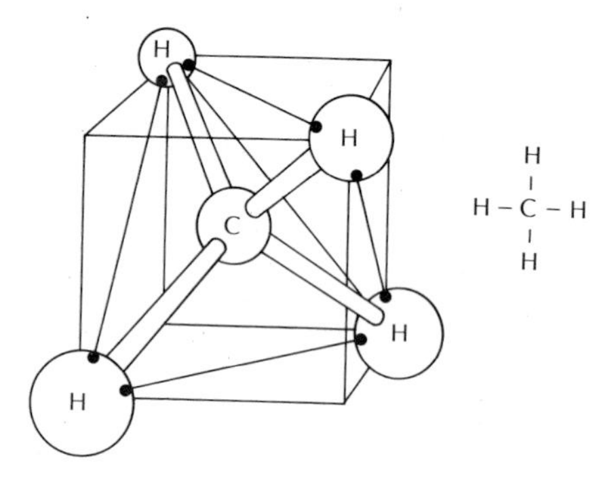

indicated if scientists found organic compounds. Thus far, the search has proved fruitless.

A most important feature of organic chemistry is the carbon atom bound to four other atoms (commonly termed tetravalent carbon). The covalent carbon-hydrogen bonds in such a compound as methane are called *single bonds*. In such tetravalent carbon compounds, the four bonds of the central atom are directed toward the vertices of a regular tetrahedron. Each bond forms an angle of 109.5° with its neighbors. This angle is the maximum separation of bonds attainable with four substituents bound to a central atom, and it permits great bond strength together with the least unfavorable interactions between nonbonded atoms. The remarkable strength of diamonds is accounted for by the fact that they consist of a three-dimensional lattice of tetrahedral carbon atoms.

Carbon can enter into multiple bond formation with itself or other atoms. In a compound such as ethylene or formaldehyde, two electrons from a carbon atom are shared with two from a second atom to make a four-electron bond, called a *double bond*. When three electrons from carbon are shared with three from a second atom, as in the case of acetylene or hydrogen cyanide, a six-electron bond, or *triple bond*, is formed.

```
H       H      H
 \     /        \
  C=C            C=O
 /     \        /
H       H      H
 Ethylene      Formaldehyde

H—C≡C—H        H—C≡N
 Acetylene     Hydrogen Cyanide
```

All the above compounds are used extensively in the manufacture of plastics. In addition, acetylene is used as a fuel in welding; hydrogen cyanide is a well-known poison; and formaldehyde, due to its usefulness as a preservative, is widely used for embalming.

Although stronger than a single bond, a double bond is not fully twice as strong; similarly, a triple bond is not 3/2 as strong as a double bond. Thus, organic molecules containing multiple bonds often tend to undergo reactions in which single bonds are created at the expense of destroying double or triple bonds. Many reactions involving carbon compounds in living systems depend on the reactivity of a few of these multibonded functional groups of atoms, such as the >C=O bond (carbonyl group) depicted in formaldehyde. Another rather special type of multiple bonding in organic chemistry is exemplified by benzene. This compound and others with related structures are called aromatic substances—presumably because many that are volatile have a pleasing aroma. The additional stability of these compounds is a unique property of certain organic ring compounds containing alternating double bonds. Many such aromatic structures are found in molecules essential for living organisms. Like the substituents attached to double and triple bonds, the substituents attached to aromatic rings lie in a single plane.

Organic molecules typically contain tens or hundreds of atoms. Compounds containing only carbon and hydrogen are called *hydrocarbons*.

Gasoline is a mixture of hydrocarbons, each containing about eight carbon atoms. Paraffin is a hydrocarbon containing about 20 carbon atoms. Polyethylene is a hydrocarbon containing thousands of carbon atoms linked together in a simple chain (Figure 3.19).

Polyethylene can be considered as composed of the basic—CH_2—unit repeated over and over again. Molecules that are built up by joining together simpler units (*monomers*) are called *polymers*. Polyethylene is a simple polymer because it contains the same monomer repeated over and over again (a homopolymer). It is possible to construct polymers in which a number of different monomers are joined together. Such polymers, called heteropolymers, have interesting properties and play important roles in living systems. Both proteins and nucleic acids are heteropolymers.

Thus, carbon forms the backbone of most organic molecules, and the strongly bonded chains of carbon atoms can be considered a framework to which other atoms are bonded. The other three major elements of organic compounds—hydrogen, oxygen, and nitrogen—provide a range of bonding capabilities, forming one, two, and three covalent bonds, respectively. With only these four elements, it is possible to build an almost infinite range of molecules with almost any desired structure or chemical property. From these possibilities, the variety of living systems is constructed.

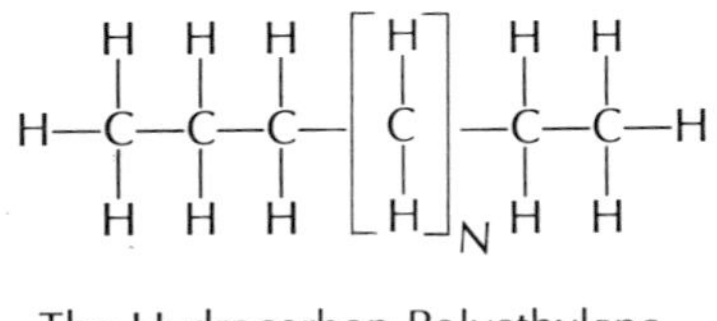

Other Reactive Groups

There is a bewildering variety of organic compounds. Some degree of simplicity can be obtained because there are a few important *reactive groups* of atoms that appear in many different kinds of molecules. Hydrocarbon molecules in which one or more hydrogen atoms are replaced by hydroxyl (—OH) groups are called *alcohols*. More than one group may replace hydrogen atoms in a hydrocarbon chain. Crystalline sugar consists of a hydrocarbon chain with alcohol groups and an aldehyde or carbonyl group replacing hydrogens attached to each carbon in the chain (Figure 3.20). The sensation of sweetness is associated with the aldehyde or carbonyl group; sugars taste sweet and perfumes have a pleasant smell because of the presence of one of these groups.

Like many biochemicals, sugars can be polymerized to form long chains. The sugar chains, or polysaccharides, include such substances as starch and cellulose. Sugars can react with other kinds of molecules to form nucleosides (a sugar combined with a nitrogenous base), nucleotides

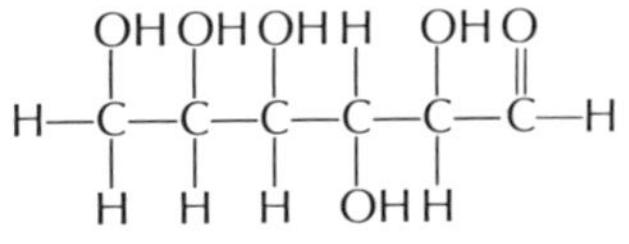

Figure 3.19 (above). The structural formula of polymeric polyethylene. Each molecule is composed of thousands of "monomer" CH_2 units.

$$H-\underset{H}{\overset{H}{C}}-\underset{H}{\overset{H}{C}}-\underset{H}{\overset{H}{C}}-\left[\underset{H}{\overset{H}{C}}\right]_N-\underset{H}{\overset{H}{C}}-\underset{H}{\overset{H}{C}}-H$$

The Hydrocarbon Polyethylene

Figure 3.20. Space-filling model (lower left) and structural formula (lower right) of glucose, a crystalline sugar.

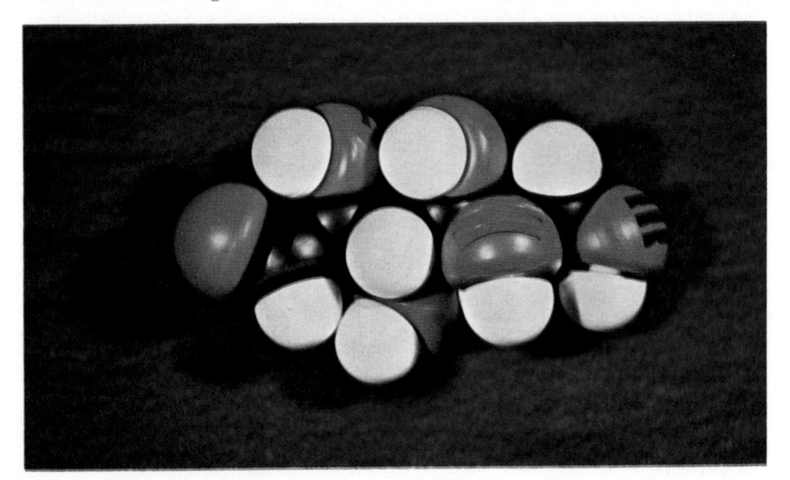

$$H-\underset{H}{\overset{OH}{C}}-\underset{H}{\overset{OH}{C}}-\underset{H}{\overset{OH}{C}}-\underset{OH}{\overset{H}{C}}-\underset{H}{\overset{OH}{C}}-\overset{O}{\overset{\|}{C}}-H$$

The Sugar Glucose

Figure 3.21. Structural formula (upper left) and space-filling model (upper right) of formic acid, the simplest organic acid.

Figure 3.22. Space-filling model (middle right) and structural formula (lower right) of lauric acid, a fatty acid.

$$H-\overset{\overset{\displaystyle O}{\|}}{C}-OH$$

Formic Acid

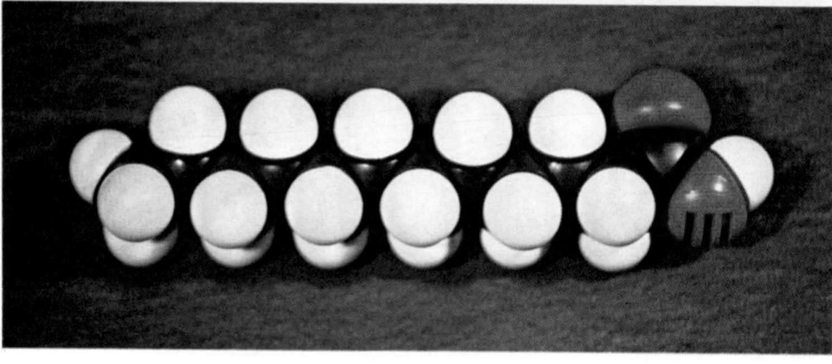

$$H-\underset{H}{\overset{H}{\overset{|}{\underset{|}{C}}}}-\underset{H}{\overset{H}{\overset{|}{\underset{|}{C}}}}-\underset{H}{\overset{H}{\overset{|}{\underset{|}{C}}}}-\underset{H}{\overset{H}{\overset{|}{\underset{|}{C}}}}-\underset{H}{\overset{H}{\overset{|}{\underset{|}{C}}}}-\underset{H}{\overset{H}{\overset{|}{\underset{|}{C}}}}-\underset{H}{\overset{H}{\overset{|}{\underset{|}{C}}}}-\underset{H}{\overset{H}{\overset{|}{\underset{|}{C}}}}-\underset{H}{\overset{H}{\overset{|}{\underset{|}{C}}}}-\underset{H}{\overset{H}{\overset{|}{\underset{|}{C}}}}-\underset{H}{\overset{H}{\overset{|}{\underset{|}{C}}}}-\overset{\overset{\displaystyle O}{\|}}{C}\diagdown OH$$

Lauric Acid—A Fatty Acid

(sugar, nitrogenous base, and phosphoric acid), glycoproteins (sugar polymer and protein), and other complex associations.

The carboxyl group (—COOH) is characteristic of the *organic acids*. The simplest organic acid is formic acid (Figure 3.21), which causes ant bites to be irritating. Another simple organic acid, acetic acid, gives vinegar its sour taste. The *fatty acids* are long hydrocarbon chains with a carboxyl group on the end (Figure 3.22).

An acid may react with an alcohol, eliminating a water molecule and forming an *ester* (Figure 3.23). This reaction is somewhat similar to the acid-base neutralization reaction that occurs in inorganic chemistry. However, alcohols do not act like bases; they do not ionize in water to give hydroxide ions. In fact, alcohols are nonelectrolytes.

One of the principal constituents of cell membranes is a *lipid* formed by the reactions of fatty acids, such as stearic acid, with an alcohol called glyc-

Figure 3.23. Space-filling model (upper left) and structural formula (upper right) of the ester ethyl acetate. The water molecule to the right of the ester is the other product of the reaction of ethyl alcohol and acetic acid, which formed the ester.

Figure 3.24 (middle right). Structural formula of a lipid molecule.

Figure 3.25. Space-filling models of amino acids (lower left) and structural formula (lower right) of the amino acid alanine. Amino acids may exist in two forms, L and D, which are related to each other by a mirror plane. Only L amino acids (the left model) are found in nature.

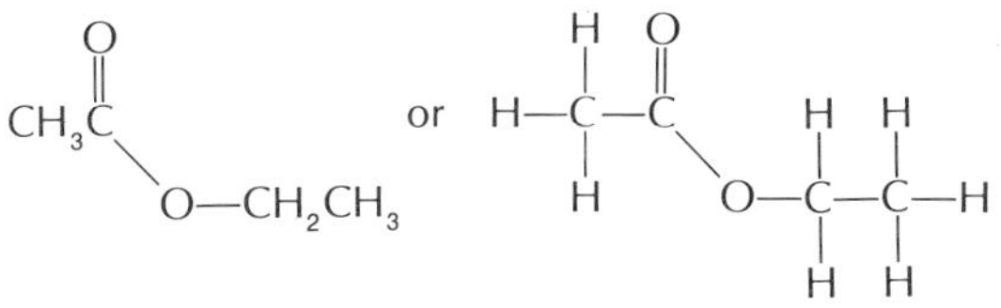

$CH_3C(=O)—O—CH_2CH_3$ or $H—C(H)(H)—C(=O)—O—C(H)(H)—C(H)(H)—H$

The Ester Ethyl Acetate

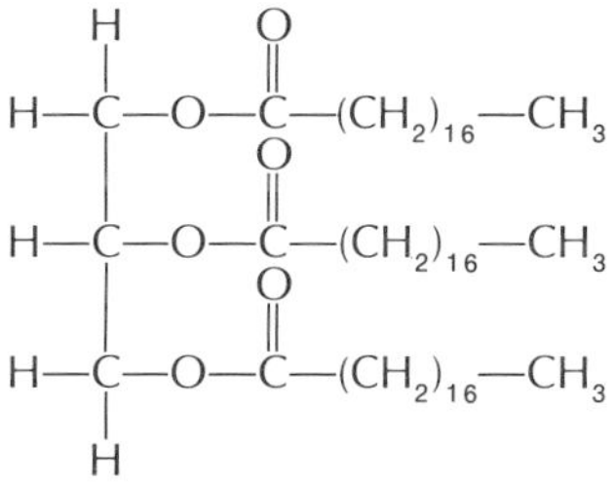

$H—C(H)—O—C(=O)—(CH_2)_{16}—CH_3$
$H—C—O—C(=O)—(CH_2)_{16}—CH_3$
$H—C(H)—O—C(=O)—(CH_2)_{16}—CH_3$

A Lipid Molecule

$(H)(H)N—C(R)(H)—C(=O)—OH$

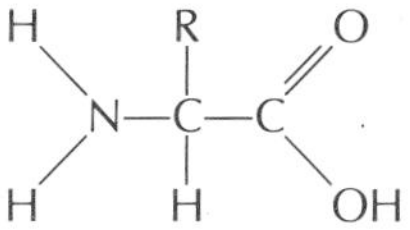

An Amino Acid

erol (Figure 3.24). Lipids are composed of an organic alcohol linked to organic acids through ester bonds.

The *amino acids* are among the most important small organic molecules. In these molecules, two reactive groups are found – a carboxyl group on one end and a basic configuration, the amino group ($—NH_2$), on the next carbon in the chain. There are about 20 commonly occurring amino acids in nature. All of them have the basic structure shown in Figure 3.25, in which R may represent any of various reactive groups.

Electronegativity and Hydrogen Bonding

The electron-pulling ability of a nucleus involved in a covalent bond is called the *electronegativity* of that atom. The relative electronegativities of a few elements will serve as an example, where the symbol > means greater than and ≈ means equivalent to:

$$F > O > N \approx Cl > Br > C \approx S > P \approx H \quad (3.6)$$

The greater the electronegativity, the greater the tendency for the nucleus to draw the shared electron pair. The greater the difference in the electronegativity of two atoms, the more ionic will be the character of a bond between them. A bond of partial ionic character is often called a polar

Figure 3.26. Crystallographic model (above) and space-filling model (below) of a water molecule, showing the bond angle and length.

van der Waals Radius of Hydrogen = 1.2 A
σ+
Covalent Bond Length = 0.965 A
104.5°
2σ−
van der Waals Radius of Oxygen = 1.4A
σ+
Direction of Dipole Moment

bond because it possesses positively and negatively charged poles. It can be seen from the ordering given above that hydrogen fluoride (HF) is more polar than hydrogen chloride (HCl), which is more polar than hydrogen bromide (HBr). Water (H_2O) is more polar than ammonia (NH_3). The H—C bonds in methane are not very polar because the electronegativities of hydrogen and carbon are not very different.

There is an entire spectrum of bonds of varying polarity, ranging from completely covalent bonds such as H — H at one extreme to completely ionic bonds such as Na^+Cl^- at the other extreme. Most bonds in organic molecules are near the covalent end of this spectrum. These strong covalent bonds give organic materials much of their stability and structure.

The existence of polar bonds makes possible another form of bonding called *hydrogen bonding.* To emphasize the polar nature of the water molecule, its electron-dot formula can be written as

$$\begin{array}{l} 2\delta^- \\ :\ddot{O}:H\delta^+ \\ \ \ddot{H} \\ \ \ \delta^+ \end{array} \tag{3.7}$$

In this representation, the symbol δ represents a fraction of a unit charge.

If another water molecule approaches, there is a weak attraction between the positive hydrogen ends of one water molecule and the negative oxygen end of the other. This attraction creates something like a weak ionic bond between the two molecules. It is this type of bonding that produces the rigid structure of ice, and similar bonds give some semicrystalline structure to liquid water. Such bonds are called hydrogen bonds.

In organic molecules, hydrogen bonds occur most often when a polar bond containing a hydrogen atom occurs near a polar bond containing either an oxygen atom or a nitrogen atom. The hydrogen bonds play important roles in determining the three-dimensional structures of protein and nucleic acid molecules.

Water

All the complex substances in living organisms make up only a small fraction of the weight of living tissues. The principal component is water, which accounts for more than 75 percent of the weight of most tissues. Some exceptions are hair, horn, solid bone, spores, seeds, and so on, all of which are metabolically relatively inert. When spores (dormant bacteria) are transformed into cells showing active metabolism, an increase in water content always takes place.

To a chemist, water is an unusual substance possessing remarkable properties. It is a compound of extreme stability. It is a remarkable solvent. It does not mix with most organic substances but is strongly attracted by most inorganic substances, including itself. When frozen into a solid, it expands (unlike most other substances, which contract upon solidification). But because it is such an abundant substance and because of its paramount importance for living systems, water is often taken for granted without explicit discussion of its properties.

All of water's oddities can be understood in terms of its molecular structure (Figure 3.26). In the water molecule, there are eight valence electrons. Four are involved in covalent bonds between the oxygen atom and

Figure 3.27 (above). The tetrahedral structure of water.

Figure 3.28 (below). Water molecules arranged in ice form. In the model shown, the molecules of ice have been pulled apart to show their configuration more clearly, whereas in an actual structure, the molecules would be closer together. In a similar model showing water in liquid form, the molecules would be more loosely organized, and they would be further apart and joined by more hydrogen bonds.

the hydrogen atoms. The remaining four electrons are in nonbonding orbitals on the oxygen atom. The shape of the water molecules is an isosceles triangle, and the H—O—H bond angle is approximately 105°. The electronegativity of the oxygen atom makes the O—H bonds polar, with a net positive character at the hydrogen atoms and a net negative character at the oxygen atom. If a tetrahedron is described with the oxygen atom at its center and the hydrogen atoms at two of its corners, then the regions of highest density of the nonbonding electrons are directed toward the remaining two corners of the tetrahedron (Figure 3.27). This symmetry minimizes the electrical repulsions between the eight valence electrons.

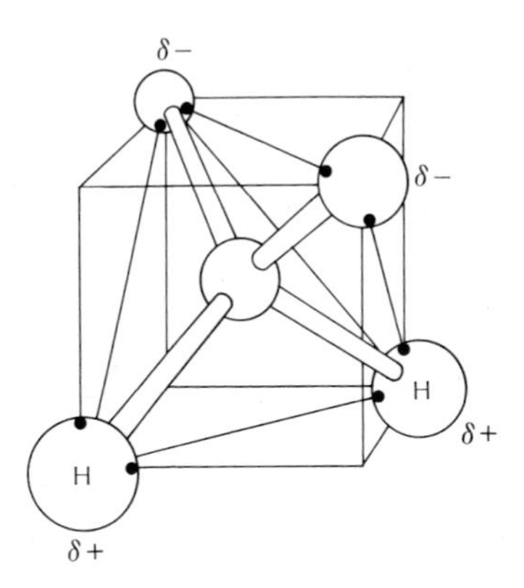

The water molecule is thus an electrically polar structure. In a group of water molecules clustered together, a positively charged region on one molecule tends to be attracted toward a negative region on another molecule. There are two positive regions on each molecule—the hydrogen atoms—and two regions with negative character—the nonbonding electron densities. Thus, each water molecule can have four nearest neighbors. Each of the nonbonding pairs of electrons attracts a positive hydrogen atom on a neighboring water molecule, and each of the hydrogens attracts the oxygen ends of neighboring molecules. Thus, each oxygen atom is at the center of a tetrahedron of four other oxygen atoms. Equivalently, water molecules can be regarded as spheres, each with four nearest neighbors (Figure 3.28). This ordered structure represents the molecular arrangement in ice.

When ice melts, the higher *coordinated* structure of the crystal breaks in many places, but not all of the hydrogen bonds are broken. The molecules tend to pack closer together so as to fill up some of the vacant space. This packing results in the greater density of liquid water. When ice is converted to water at 0° C, only about 15 percent of the hydrogen bonds are broken. Cold water contains interconnected groups of water molecules whose structures are based on the same pattern as in ice, only with some broken bonds. As the temperature continues to rise, increasing thermal motion of the molecules tends to make the liquid expand again. The aggregates of water molecules are in dynamic equilibrium, constantly breaking up and re-forming, so that there are no permanent crystalline structures. As the temperature rises, more of the hydrogen bonds are broken, but a considerable amount of this *crystalline* structure remains even at 100° C.

It requires a remarkably large amount of energy to melt ice, to raise the temperature of liquid water, or to vaporize water. All of these phenomena (heat of fusion, heat capacity, and heat of vaporization, respectively) can be explained in terms of the highly ordered structure of water and are essential in maintaining a relatively constant temperature for living systems.

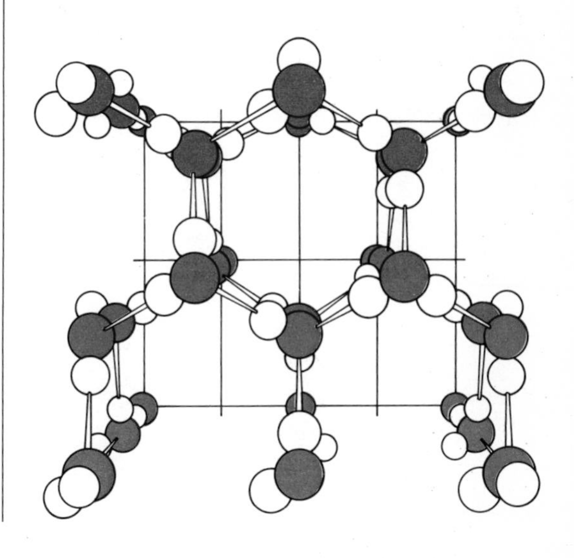

When the surface area of a liquid is increased, molecules that were formerly in the interior, surrounded by and attracted to neighboring molecules, must be brought to the surface. Work must then be done against the attractive forces operating between molecules in the interior. These forces are exactly those against which work must be done in vaporizing the liquid. In vaporization, however, the molecules are removed completely from the intermolecular forces, whereas for molecules on the surface the intermolecular forces remain except in the direction pointing away from the liquid phase. The result is that, at the surface, a molecule is in a state of higher energy than when it was in the interior. In accord with the tendency of all systems to attain a state of minimum energy, the surface of the fluid will

Figure 3.29. A water skipper feeding on water. This insect is able to walk on the surface of the pond because the water has such a high surface tension.

assume a shape that minimizes the surface area, so as to maximize the number of molecules in the interior and minimize the *surface energy* of the system. Any tendency to deform the surface from this minimum area will be counteracted by the tendency of the molecules to remain in the interior.

It is therefore not surprising that substances with high heats of vaporization have high surface tensions. Water has the highest surface tension of any known liquid, with the possible exception of certain metals in the liquid state and some molten salts. This high surface tension accounts for the ability of water to rise in the capillary spaces of soil and plants. The surface tension is responsible for the ability of water skippers and other insects to walk on the surface of a pond.

Water is an excellent solvent for any polar substance, whether ionic or covalent. In the case of ionic substances (salts), the electrical attractions operating between ions in the crystalline state are overcome by electrical attractions between the ions and the oppositely charged moieties of the water molecules. For example, solid sodium chloride can be dissolved in water, even though the crystal itself is an extremely stable configuration, because the process of dissolution lowers the energy of the system still further. The sodium ions will be surrounded by water molecules with their oxygen atoms oriented toward the sodium ion; the chloride ion will be surrounded by water molecules with their hydrogen atoms oriented toward the ion. If this system was not more stable than the initial state consisting of solid NaCl and water, then dissolution would not occur.

Numerous nonionic compounds also dissolve in water. Notable examples are molecules containing the polar amino, hydroxyl, carboxyl, and keto groups, all of which can form hydrogen bonds with the water molecules. On the other hand, substances containing large hydrocarbon moieties and other nonpolar groups are generally only slightly soluble, if at all, in water. Such substances tend to be concentrated in droplets that coalesce at the surface in films because they are not attracted by the water molecules.

Figure 3.30. Salt dissolving in water. Polar water molecules cluster around each sodium and chloride ion as the salt dissolves.

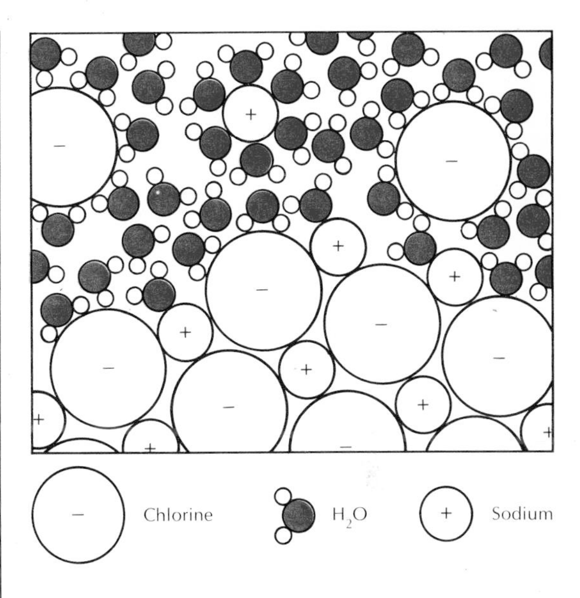

Bond Energies

Table 3.4 shows *average bond energies* for some of the bonds that are important in organic compounds. The average bond energy represents the approximate amount of energy needed to break that bond in any compound in which it occurs. The actual bond energy in a particular molecule will vary somewhat from these average values, depending upon the other atoms that are attached to the atoms involved in the bond. The values in this table provide a reasonable estimate of the strength of various bonds involved in organic molecules and of the energy that can be released as such a bond is formed.

Table 3.4
Average Bond Energies

BOND	Average Bond Energy (kcal/mole)
C≡C	199.6
C=O	178
C=C	145.8
H—F	134.6
Si—F	129.3
O—H	110.6
C—F	about 110
H—H	104.2
H—Cl	103.2
C—H	98.7
N—H	93.4
Si—O	88.2
H—Br	87.5
C—O	85
C—C	82.6
S—H	81.1
C—Cl	80
C—N	80
C—S	62.0
Si—Si	42.2
O—O	33.2

Source: Linus Pauling, *The Nature of the Chemical Bond* (Ithaca, N.Y.: Cornell, 1960), p. 85; B. H. Mahan, *University Chemistry*, 2nd ed. (Reading, Mass.: Addison-Wesley, 1969), p. 458.

IONS, ACIDS, AND BASES

Because most biochemical reactions take place in solution, the study of the nature of molecules in the dissolved state is essential. When ionic substances are dissolved in water, the polar water molecules cluster around the ions, separating them and yielding a solution of positively charged and negatively charged ions scattered throughout the water molecules. The substance being dissolved is called the *solute*, and the substance in which it is dissolved is called the *solvent*. If an electric voltage is applied across the solution, the positive ions move toward the negative terminal, and the negative ions move toward the positive terminal. This flow of ions produces an

Figure 3.31. Experimental apparatus used to determine the conductivity of a solution.

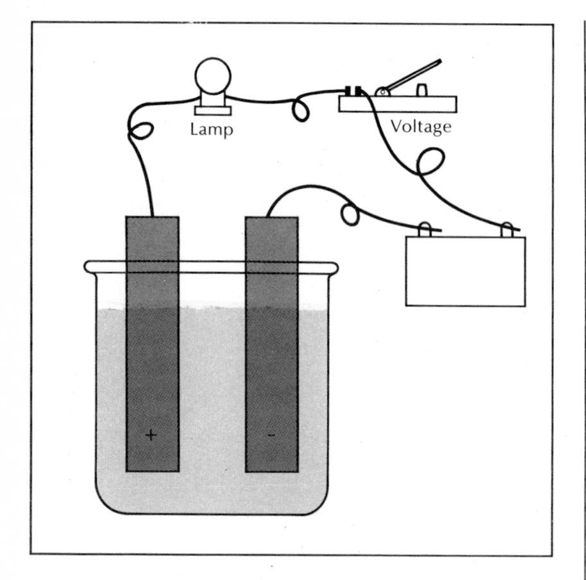

electric current within the solution, and the solution conducts an electric current. The magnitude of the current is proportional to the number of ions in the solution. The simple apparatus pictured in Figure 3.31 can be used to determine the current flow and hence the number of ions. If the bulb burns brightly, the solute is a *strong electrolyte;* if the bulb burns dimly, the solute is a *weak electrolyte;* if the bulb does not light up at all, the solute is a *nonelectrolyte.*

Compounds that are nearly completely ionic, such as NaCl, NaOH, $CaCl_2$, and KBr, dissolve in water to become strong electrolytes. Compounds with polar bonds, such as $HgCl_2$, $PbBr_2$, and $CuCl_2$, are weak electrolytes. Compounds with nonpolar covalent bonds, such as CH_4, N_2, and most biomolecules, are nonelectrolytes. It is difficult, if not impossible, to know from the chemical formula whether a compound is a strong or weak electrolyte or a nonelectrolyte. One must either be knowledgeable in chemistry or perform a test such as that shown in Figure 3.31.

Two particular ions are of great importance in chemistry—the hydrogen ion (H^+) and the hydroxide ion (OH^-). A compound that releases hydrogen ions when dissolved in water is called an *acid,* and one that releases hydroxide ions or accepts hydrogen ions is called a *base.* Acids and bases are especially important because many chemical reactions are accelerated in acidic or basic solutions. Acids are characterized by their ability to dissolve many metals with the simultaneous release of hydrogen gas; by a characteristic sour taste (the sour taste of lemons and vinegar is due to the presence of weak acids), and by the color changes they produce in certain dyes (such as litmus paper). Bases taste bitter, feel slippery (soap is slippery because it contains small amounts of base), and change the colors of the indicator dyes in a fashion opposite to that of acids.

Some of the more common acids are nitric acid (HNO_3), hydrochloric acid (HCl), sulfuric acid (H_2SO_4), and phosphoric acid (H_3PO_4). Some common inorganic bases are sodium hydroxide (NaOH), calcium hydroxide ($Ca[OH_2]$), and ammonium hydroxide (NH_4OH).

If mixed in the proper proportions, acids and bases exactly neutralize each other to give a solution that has the properties of neither an acid nor a base. The products of this *neutralization* reaction are an ionic compound, or *salt,* and water. The following equations are examples of neutralization reactions:

$$NaOH + HCl \rightarrow NaCl + H_2O \tag{3.8}$$

$$Ca(OH)_2 + 2\,HNO_3 \rightarrow Ca(NO_3)_2 + 2\,H_2O \tag{3.9}$$

$$3\,KOH + H_3PO_4 \rightarrow K_3PO_4 + 3\,H_2O \tag{3.10}$$

The concentration of hydrogen ions in solution is so important to most chemical reactions that chemists use a special notation when discussing it. They have devised a *pH scale,* which varies between 1 and 14, to specify the acidic or basic character of a solution (Interleaf 3.1). A pH of 7.0 means that the solution is perfectly neutral—that is, the concentrations of H^+ and OH^- in the solution are equal and the solution is neither acidic nor basic. A pH lower than 7.0 means that the solution is acidic; the lower the pH, the more H^+ ions are free in solution and the more acidic the solution is. A pH greater than 7.0 means that the solution is basic; the greater the

Interleaf 3.1

THE pH SCALE

The pH is defined as being equal to the negative logarithm (to the base 10) of the hydrogen ion concentration in moles per liter of solution:

$$pH + -\log_{10} [H+] = +\log_{10} \frac{1}{[H+]}$$

Analysis shows that there is 10^{-7} gram (g) of hydrogen ions present in a liter of pure water. For hydrogen ions, 10^{-7} g is equal to 10^{-7} moles (because H has an atomic weight of 1). Therefore, the pH of pure water is $\log_{10} \frac{1}{.0000001}$ of 7. Pure water is chemically neutral, that is to say it is neither acid nor base, and the concentration of hydrogen ions must equal the concentration of hydroxide ions (the only base that can be present in pure water).

Table 3.5
pH Values for Various Solutions

SOLUTION	pH
Pure gastric juice	about 0.9
Orange juice	2.6–4.4
Vinegar	3.0
Grapefruit juice	3.2
Tomato juice	4.3
Urine	4.8–7.5
Saliva	6.6
Milk	6.6–6.9
Distilled water	7.0
Intestinal juice	7.0–8.0
Blood serum	7.4
Tears	7.4
Pancreatic juice	7.5–8.0
Egg white	8.0
Sea water	8.0

Source: E. S. West and W. R. Todd, *Textbook of Biochemistry* (New York: Macmillan, 1951), p. 49.

pH, the fewer H^+ ions and the more basic the solution. The pH of a number of solutions is given in Table 3.5.

VALENCE AND OXIDATION NUMBER

The electrons in the outer shell that can be shared with other atoms are called the *valence electrons*. The *valence* of an element is a number representing the number of electron pairs that can be shared in covalent bonds or the number of electrons that are gained or lost in the formation of ions. Thus, hydrogen has a valence of 1, oxygen has a valence of 2, nitrogen has a valence of 3, and carbon has a valence of 4. The concept of valence was part of early attempts to explain the regularities summarized in the periodic table and has largely been abandoned since more complete explanations of bonding capability have arisen.

However, a related concept—*oxidation number*—is of considerable importance in modern chemistry. For simple *atoms* and *ions*, the oxidation number is equal to the net charge on the atom or ion. Thus, the oxidation number of H^+ is +1, that of Be^{++} is +2, that of Cl^- is −1, and that of He is 0. In *molecules*, however, the oxidation number is assigned on a more arbitrary basis: each shared pair of electrons is assigned to the nucleus that attracts it most strongly. The oxidation number is then calculated as the charge that the atom would bear if the electrons were actually completely held by those nuclei. For example, in H_2O the shared electrons of the O—H bonds are attracted more strongly by the oxygen than by the hydrogen. Thus, all the shared electrons are assigned to the oxygen atom, giving it a net charge of −2. (Its own eight electrons are balanced in charge by the eight protons in its nucleus; the two electrons contributed by the hydrogen atoms provide an excess of negative charge.) The hydrogen atoms each lose one electron, giving each hydrogen atom a net charge of +1. Thus, in H_2O hydrogen is assigned an oxidation number of +1 and oxygen is assigned an oxidation number of −2. The assignment of electrons in calculating oxidation numbers is an arbitrary mathematical operation and does not represent the actual distribution of electrons in the molecule. In a bond where electrons are shared equally, the shared electron pair is split between the two atoms; thus, in H_2 the oxidation number of hydrogen is 0.

In the O—H bond, oxygen attracts the electrons more strongly than does hydrogen (Formula 3.6); thus, the electron pairs in H_2O are assigned to the oxygen atom, giving oxidation numbers of +1 for hydrogen and −2 for oxygen. In the covalent molecular compound HCl, chlorine attracts the electrons more strongly than does hydrogen; the electron pair is assigned to the chlorine atom, giving oxidation numbers of +1 for hydrogen and −1 for chlorine.

In many reactions, the oxidation numbers of elements are changed. For example, water can be formed from hydrogen and oxygen gases:

$$2\,H_2 + O_2 \rightarrow 2\,H_2O + 136.6\ \text{kcal/mole}\ O_2 \qquad (3.11)$$

In this reaction, the oxidation number of hydrogen is changed from 0 (in H_2) to +1 (in H_2O). In such a case where the oxidation number is increased, the element is said to have been *oxidized*. This process of oxidation may be described as a loss of electrons; the molecule that accepts the electrons (in this case, oxygen) is called the *oxidizing agent*, or *oxidant*. In the same reaction, the oxidation number of oxygen is changed from 0 (in O_2) to -2 (in H_2O). Such a decrease in oxidation number is described by saying that the oxygen has been *reduced*. Reduction is a gain of electrons; the molecule that donates the electrons (in this case, hydrogen) is called the *reducing agent*, or *reductant*.

A reaction in which oxidation numbers are changed is called an *oxidation-reduction reaction*, or *redox reaction*. In any such reaction, a reductant donates electrons and is oxidized, while an oxidant accepts the electrons and is reduced. In many cases, the electrons are not actually transferred from one molecule to the other but are shared in a polar covalent bond. Nevertheless, it is useful to think of such reactions as processes of electron transfer in which a certain amount of energy is absorbed or released as the electrons are transferred. This way of looking at oxidation-reduction reactions helps to clarify many of the chemical processes involved in living systems.

FURTHER READING

The Chemical Education Material Study (1963) presents an exceptionally clear introduction to basic ideas of chemistry. An excellent paperback treatment of many of the topics of this chapter is given by Herz (1963). A particularly current and clear college chemistry text is the one by Mahan (1969). Pauling (1960) clarifies modern ideas about bonding and molecular structure

4
Chemistry of Life

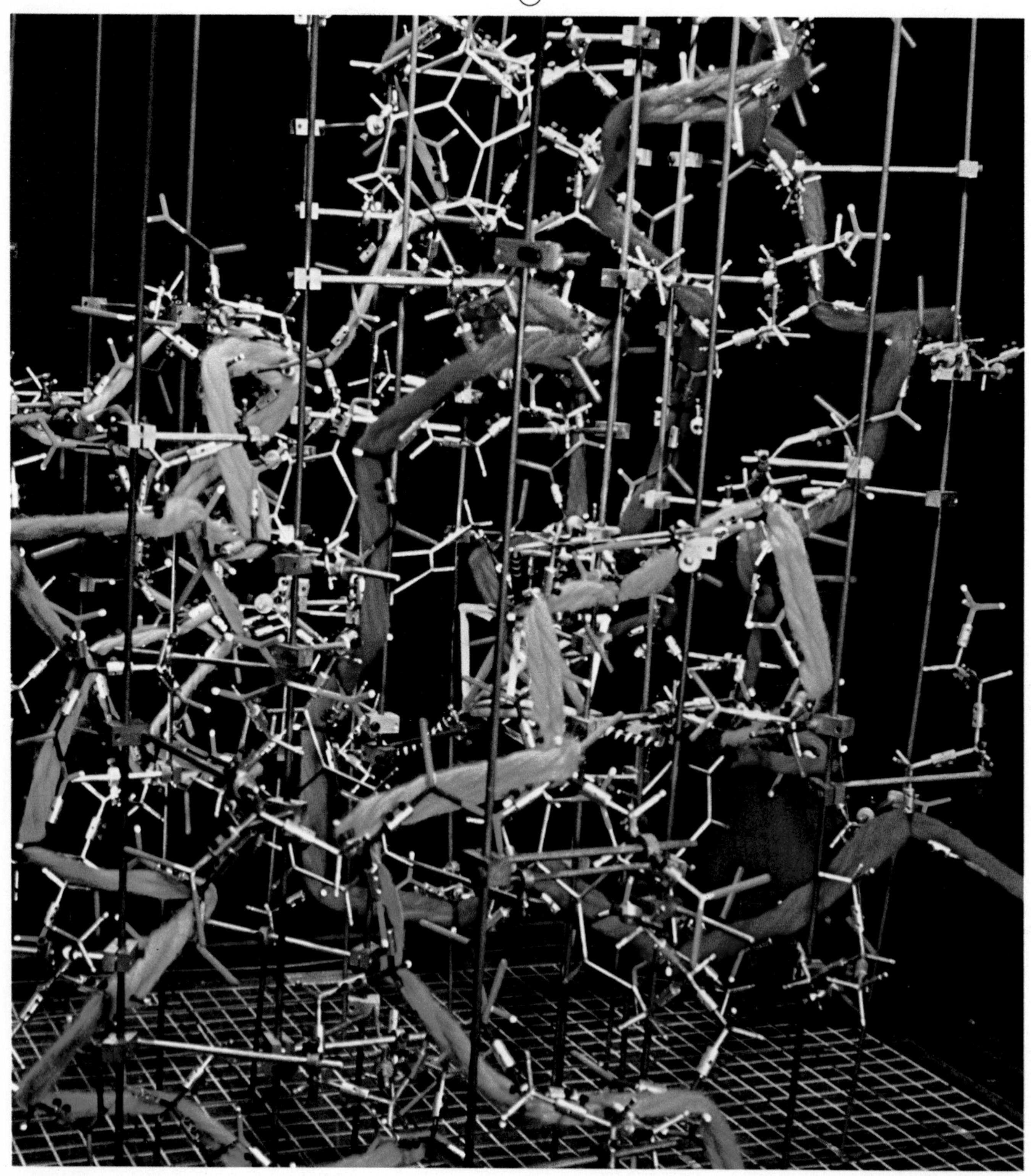

Energy transfers within living systems are accomplished through changes involving matter—that is, through chemical changes. These chemical reactions are of almost endless variety, taking place in structures of all degrees of complexity. Throughout biology, such reactions are encountered repeatedly, for they underlie most of the phenomena observed on a macroscopic scale.

When a chemical reaction occurs, bonds within the original molecules (the reactants) are broken, and new bonds are formed to create molecules of the products. One result of these changes is the release or the absorption of energy. For example, consider the chemical reaction that occurs when coal is burned. Coal is composed primarily of carbon, which combines with oxygen gas in the air to form the gas carbon dioxide. The reaction can be summarized as follows:

$$C + O_2 \rightarrow CO_2 + 94.0 \text{ kcal/mole} \tag{4.1}$$

In this reaction, the double bond in the oxygen molecule is broken, and two new double bonds are formed in the carbon dioxide molecule (which could be written as $O{=}C{=}O$). The formation of the $C{=}O$ bonds releases far more energy than is consumed in breaking the $O{=}O$ bond; in other words, the products are at a lower energy state than the reactants. Thus, for each mole of carbon that is burned, a mole of CO_2 is formed, and 94.0 kilocalories of energy are released. When coal is burned, this energy is given off as heat and light.

Each molecule has a certain energy content, or *heat content, H.* The heat released during the reaction represents the difference between the heat content of the reactants and that of the products. Thus, chemists symbolize the *heat of reaction* as ΔH (delta H is the difference between the two values of H) and define it as the heat content of the products minus the heat content of the reactants. When coal is burned, the reactants have a greater heat content than the products; thus, ΔH for this reaction has a negative sign. In general, a negative ΔH indicates a reaction releasing heat, an *exothermic* reaction. The reaction is customarily written as follows:

$$C + O_2 \rightarrow CO_2 \quad \Delta H = -94.0 \text{ kcal/mole} \tag{4.2}$$

Once a fire has been started, coal continues to burn until all the available carbon or oxygen is consumed. Just as electrons tend to move spontaneously to the lowest available energy level (releasing photons of light energy as they do so), molecules tend to move to the lowest available energy state (releasing heat energy as they do so).

However, coal can be stored in the presence of oxygen for years without noticeable change. The reaction occurs only when coal is heated to a high temperature (the flash point) to get the reaction started. Why must energy be supplied to begin the reaction? The molecular model provides a simple explanation.

THE COLLISION THEORY

The chemical reaction can occur only when a molecule of oxygen and an atom of carbon approach each other closely. In fact, it seems logical to suppose that chemical reactions involve collisions between molecules, atoms, or ions of the reactants. This *collision theory* provides a simple picture

Figure 4.1 (above). Enthalpy changes in the combustion of coal in the presence of oxygen to form carbon dioxide.

Figure 4.2 (below). The collision theory.

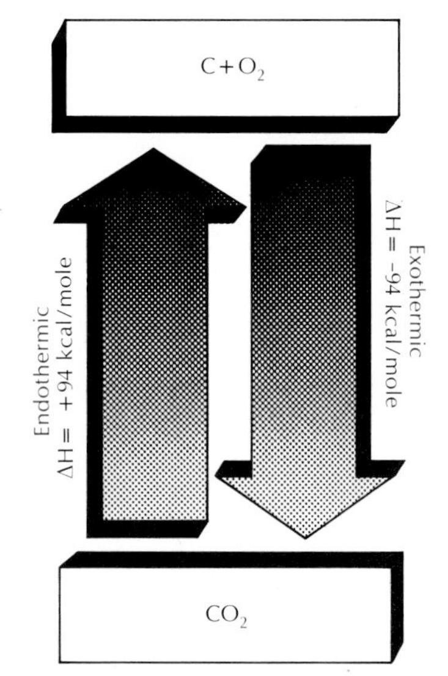

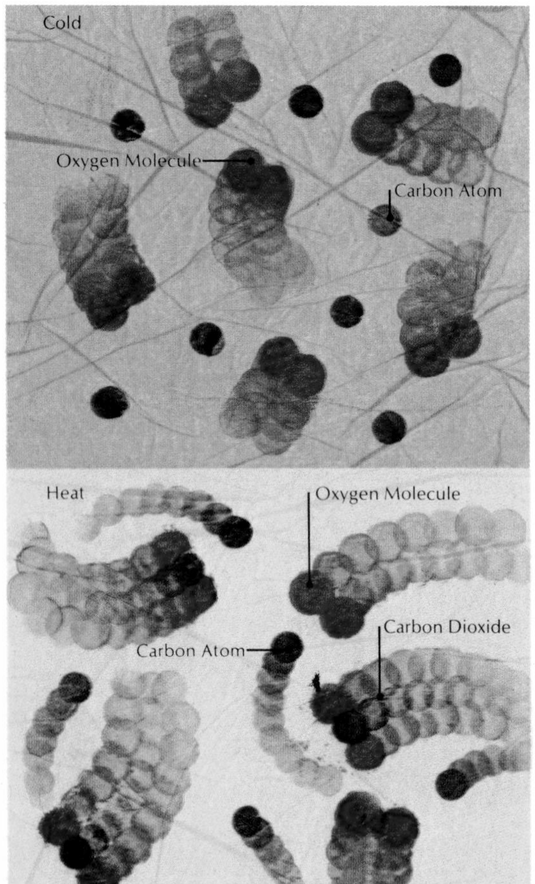

of the mechanism for the reaction. An oxygen molecule collides with a carbon atom. The force of the collision depends on the speed of the molecules involved, and speed is a function of temperature. At a low temperature, the molecule moves slowly, and the energy of the collision is not sufficient to break the double bond in the oxygen molecule. The carbon atom and the oxygen molecule simply bounce apart; no chemical change occurs. At a high temperature, however, the molecules are moving rapidly and collision occurs with enough energy to break the O — O bond. For an instant, some sort of intermediate grouping of the atoms exists. This *activated complex* has a high heat content and is unstable. It quickly readjusts into the CO_2 molecule as the two C — O bonds form, releasing energy.

The energy relationships during the reaction can be graphed as in Figure 4.4. The high energy content of the activated complex forms a barrier to the progress of the reaction. Until energy is available to form the complex, the reaction cannot occur. Once the reaction has begun, the energy released is sufficient to push other molecules across the energy barrier and to keep the reaction going; once coal has ignited, the fire keeps burning without further heat supplied to it.

The collision theory implies that chemical reactions occur more rapidly if the concentrations of the reactants are increased, because collisions then occur more frequently—as is the case for most chemical reactions. The theory also implies that reactions proceed more rapidly at higher temperatures, because molecules move more rapidly at elevated temperatures, tend to collide more frequently, and more of the collisions are energetic enough to pass the energy barrier of the activated complex. These anticipated effects seem to occur in most chemical reactions.

Chemical reactions are reversible. If a molecule of CO_2 acquires enough energy during a collision with another molecule, it can form the activated complex, which can break apart to form carbon and oxygen. However, the energy barrier to be crossed in this direction is much higher (moving from right to left in Figure 4.4). More-energetic collisions are needed to cross the barrier in this direction; thus, at any given temperature, the rate at which this reverse reaction occurs is lower than the rate at which the forward reaction occurs. Yet unless the temperature is so low that no CO_2 molecules have enough energy to cross the barrier, a small amount of CO_2 will continually be breaking down to form C and O_2.

Equilibrium

As the reaction proceeds and the reactants are used up, the frequency of collisions between C and O_2 decreases. At the same time, as more CO_2 is formed, the frequency of collisions leading to the reverse reaction increases. An *equilibrium* is reached when the rates of the forward and reverse reactions are exactly equal. In most cases, however, the amounts of products and reactants are not equal when equilibrium is reached. At equilibrium, no further macroscopic changes are detectable; the amounts of C, O_2, and CO_2 in the system remain constant, despite the fact that individual molecules are still undergoing the forward and reverse reactions.

It is essential to separate the two distinct features controlling the course of chemical reactions. On the one hand, the *energy charge* accompanying a reaction dictates whether or not the formation of products is a favorable process. On the other hand, the *activation energy*—the energy required to push the reactants to the stage of the activated complex—dictates how fast the reaction will occur. The magnitude of the energy change is indicative of

Figure 4.3 (above). Burning of nitrogen in the air to produce nitrous oxides. This type of reaction is one of those responsible for the production of photochemical smog from the exhaust products of internal combustion engines. Only the lower reaction goes to completion because it is the only one of the three reactions with sufficient energy and the proper geometry to form the activated complex that goes to products. Increasing the temperature and pressure of the reaction causes more collisions to take place and thus increases the likelihood of product formation.

Figure 4.4 (below). A reaction coordinate diagram for the burning of coal in air (ΔE = energy of activation). Energy must be applied initially so the reactants will gain sufficient energy ($\Delta E^*_{FORWARD}$) to reach the activated state. Once this energy is supplied (as in the form of a lighted match), it is sufficient to cause the reaction to continue until equilibrium is reached. In this case, almost all of the reactants will end up as CO_2 because the products have lower energy. If the reaction were run backward, the $\Delta E^*_{REVERSE}$ would be much greater than the $\Delta E^*_{FORWARD}$.

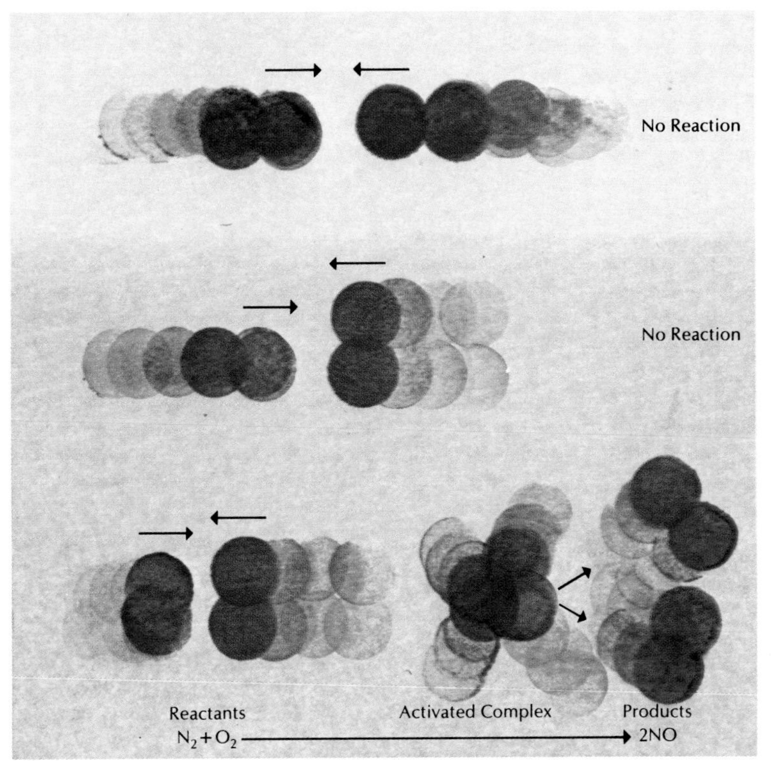

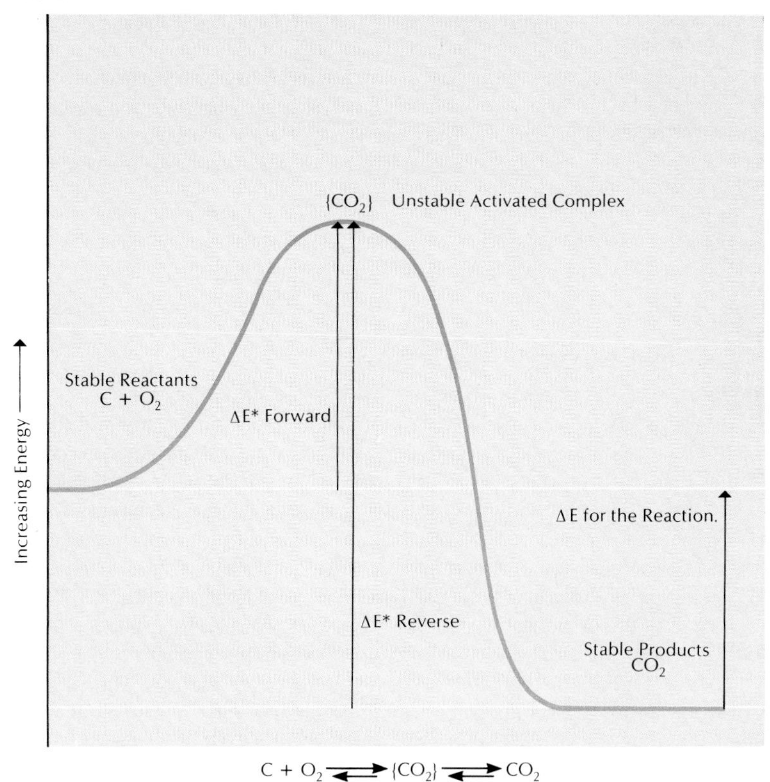

Figure 4.5. Action of a catalyst portrayed graphically. A catalyst (ΔE_c^*) lowers the energy barrier over which the reactants (A and B) must pass in order to form products (AB). The amount of energy released in the overall reaction is not altered by the catalyst, nor is the final equilibrium, which is governed solely by the difference in energy between reactants and products. The catalyzed reaction reaches equilibrium more rapidly, because on the average more molecules have enough energy to get over the barrier in a given period of time than in an uncatalyzed reaction.

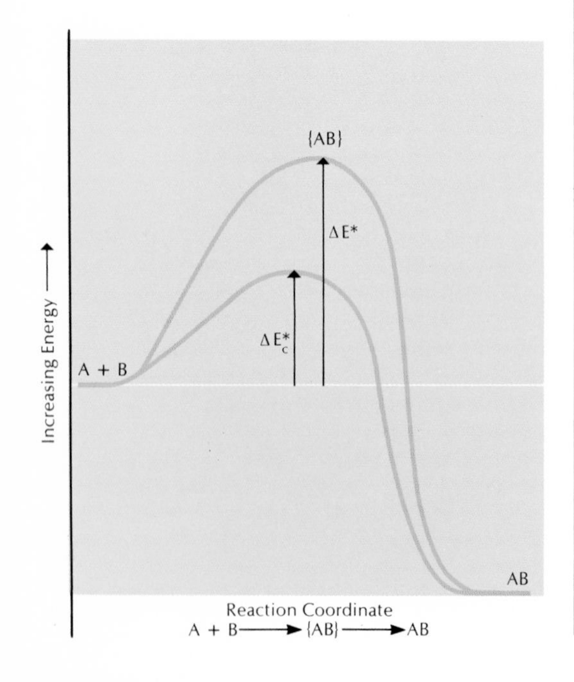

what the equilibrium concentration of reactants and products will be, and the amount of activation energy required controls the rate of reaction—how long it takes to reach this equilibrium.

Most chemical reactions occur slowly at moderate temperatures, because few collisions are energetic enough to produce the activated complex. Although the final equilibrium may heavily favor the products over the reactants, at low temperatures it may take hours, days, even months to reach equilibrium. Yet most living systems operate at temperatures of less than 100° F—relatively low temperatures in chemical terms. How does the organism manage to carry out the vast number of reactions needed for life processes at rapid rates?

Catalysts

A reaction can be speeded up if the concentrations of the reactants are increased; organisms use many different mechanisms to bring the reactants together in high concentrations. For example, reactants may migrate near each other and be assembled on cell membrane surfaces. An even more favorable increase in reaction rate can be obtained if the energy barrier to the reaction is lowered; then more collisions will be successful in forming the products. For most chemical reactions, a lower energy barrier can be produced by the addition of a suitable *catalyst*, a substance that is not consumed in the reaction but provides an alternate mechanism for the reaction. In the presence of the molecules of the catalyst, a sequence of activated complexes can be formed to carry out the reaction, with each of the new activated complexes having a lower heat content than the uncatalyzed complex (Figure 4.5). Although the rate of reaction along the old path is not altered, a much higher rate of reaction is made possible along the new path. Because the energy barriers for the forward and reverse reactions are lowered by the same amount, the catalyst speeds up both the forward and reverse reactions equally. Thus, the same equilibrium ratio of products to reactants is reached, but it is reached more rapidly than in the uncatalyzed reaction.

ENZYMES

Almost every one of the myriad chemical reactions that occur in living systems is catalyzed by an *enzyme*, a complex, three-dimensional protein. Each enzyme catalyzes only a particular chemical reaction but is an extremely efficient catalyst for that reaction.

Because of their specificity, enzymes play an important role in determining which reactions are carried out in the living organism. Suppose that a certain compound *A* can be converted in the living system either to compound *B* or to compound *C*. Both reactions, $A \rightarrow B$ and $A \rightarrow C$, proceed very slowly at the temperature of the living system. If there were present an enzyme E_{AB} that catalyzed the reaction $A \rightarrow B$ and no corresponding enzyme E_{AC} were present to speed up reaction $A \rightarrow C$, then *A* would be converted almost entirely to *B* and very little of *C* would be formed.

If both of the enzymes, E_{AB} and E_{AC}, could be created by the living system but each was created only under certain conditions, then the production of *B* or *C* could be regulated to meet varying circumstances. The organism would be able to produce E_{AB} only under those conditions where *B* was needed and to produce E_{AC} only when *C* was needed. The creation of these enzymes occurs in chemical reactions regulated by still other enzymes. The

Figure 4.6 (above). Generalized schematic diagram showing an enzyme catalyzing the conversion of a complementary shaped substrate molecule to products. The enzyme molecule remains unchanged during the reaction.

Figure 4.7 (below). Competitive inhibition of enzymatic catalysis by a molecule that reversibly binds at the catalytic site of the enzyme, thereby competing with the substrate molecule.

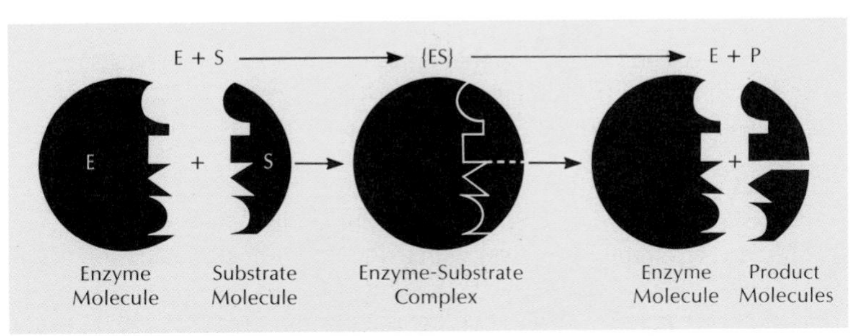

information needed to direct the formation of this complex network of enzymes comes ultimately from the genetic mechanism.

It is thought that the reactant molecule, the *substrate* of the enzyme, attaches to a particular *active site* on the enzyme in such a position that the molecule is activated to react with another molecule. After the reaction occurs, the newly formed product molecule leaves the site, making room for another substrate molecule. Apparently, the substrate molecule fits into the active site of the enzyme, rather in the manner of a key fitting into a particular lock (Figure 4.6).

Because the fit must be exact, an enzyme is highly specific and, in most cases, accepts only a few kinds of structurally similar substrate molecules. Nevertheless, because one enzyme molecule can service substrate molecules over and over again, only a few enzyme molecules are needed to catalyze the reaction of many millions of substrate molecules. The shape of the enzyme molecule probably changes slightly as it attaches to the substrate molecule. This slight change is thought to make the substrate molecule somewhat unstable and thus hasten the reaction.

In some cases, a particular molecule may resemble the substrate sufficiently closely to become attracted to the active site but be sufficiently different that no reaction occurs. When such a molecule occupies the active site, the enzyme is blocked, or *inhibited*. No substrate molecules can then attach to the active site, and the inhibited enzyme ceases to catalyze the reaction. If the inhibiting molecule becomes covalently bonded to the active site or permanently alters it, the enzyme is *irreversibly inhibited*.

If the inhibiting molecule binds reversibly to the active site, then those enzyme molecules bound to inhibitors are no longer free to catalyze the conversion of substrate to product, and the overall rate of reaction diminishes. An example of this phenomenon is provided by the familiar human protein hemoglobin, which catalyzes the transportation of oxygen from the living tissues to those cells where the oxygen is used. This process is inhibited by carbon monoxide, which binds to the active site of hemoglobin some 200 times as avidly as oxygen. Because the atmosphere is about one-fifth oxygen, it would require a concentration of only 0.1 percent carbon monoxide to inhibit fully one-half of all oxygen transport. Such a process, in which the substrate and the inhibitor compete reversibly for the same active site, is called *competitive inhibition*. That this process is reversible is illustrated by the practice of Tokyo traffic policemen who are subjected to very high concentrations of carbon monoxide. They periodically are relieved of duty and revive themselves by breathing pure oxygen.

Inhibition of selected reactions could provide a means of controlling the chemical mechanism of an organism. Many common drugs, such as penicillin, are selective inhibitors, inhibiting some of the important reactions in

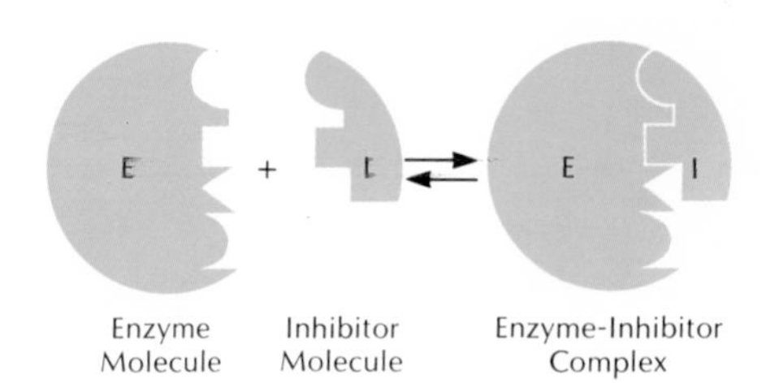

Figure 4.8 (above). J. Willard Gibbs, who derived the second law of thermodynamics.

Figure 4.9a (below). Diagram illustrating the principle of entropy. Two metal blocks are placed in an insulated enclosure that prevents heat from entering or leaving the system of the two blocks. Initially, the block on the left is hot, and the one on the right is cool. After a time, however, the heat flows into the cool block and the temperature becomes uniform throughout the two blocks, an example of the increasing entropy of the system. At this point, it is impossible for the system to return to its original state without the expenditure of some energy from an outside source.

the infecting organism but not inhibiting important reactions in the host organism. The search for a drug that cures cancer is essentially a search for some reaction occurring in the tumor cells that could be inhibited by a molecule that would not inhibit any important reactions in normal cells.

AVAILABLE ENERGY AND ENTROPY

The energy available for useful work is usually discussed in terms of the second law of thermodynamics, which states that energy-releasing processes occur to increase entropy. Entropy (the Greek *trope* means to turn or change) is a broad measure of the amount of disorder or randomness of a system. As the disorder of a system increases, the entropy increases; completely random distribution of all the molecules of a system would represent a condition of maximum entropy. The production of energy in a system is generally accompanied by the loss of a portion of that energy to the surroundings. The lost energy serves to heat the surroundings, but it is not available for useful work within the system. When the random motion of molecules is increased, this wasted heat serves to increase the entropy of the surroundings. A number of familiar processes are marked by an increase in entropy—heat flows from warm objects to cool ones, compressed gas when released escapes to the atmosphere, and corroding metal washes away in the rain ultimately to be diluted in the sea. None of the reverse of these spontaneous processes will occur unless energy is expended; all of them are marked by an increase in randomness—an increase in entropy.

Living systems are marked by their high degree of order. Plants maintain this low entropy content by consuming the energy of the sun to achieve photosynthesis—a process that involves the conversion of numerous small molecules in the surroundings to highly organized energy-rich macromolecules within the plant (Chapter 5). On the other hand, animals maintain their low entropy content at the expense of increasing the entropy of their environment. Man, for example, consumes the highly organized molecules produced by plants or lower animals and excretes a number of very much smaller, less organized molecules—primarily carbon dioxide—that are energy poor.

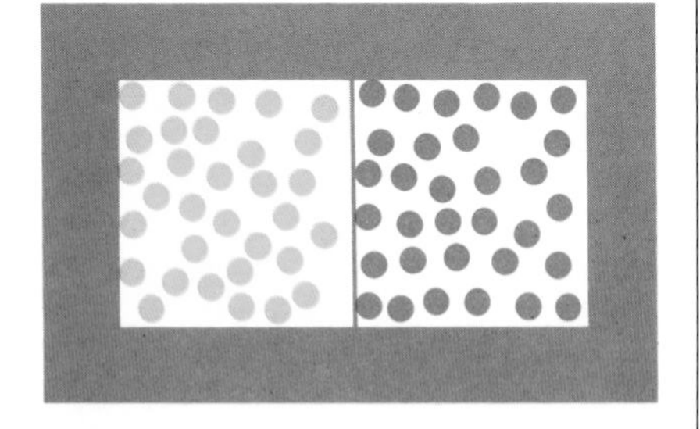

When discussing the chemical reactions that occur in living systems, biologists prefer to define entropy as that portion of the energy change from converting reactants to products that is unavailable for useful work. Chemists find it useful to define a new expression of energy change that takes into account both the change in heat content (ΔH) and the entropy change. This expression is the *free energy* change (ΔG) and is a measure of the energy that is actually available for useful work. A reaction with a negative ΔG (release of free energy) is called an *exergonic* reaction; one with a positive ΔG (consumption of free energy) is called *endergonic*.

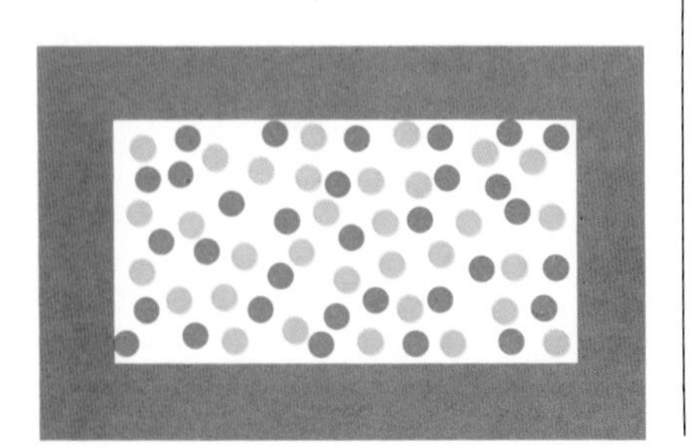

METABOLISM: TRANSFER OF ENERGY

The formation of CO_2 from C and O_2 is a strongly exergonic reaction. The C=O bonds in CO_2 form readily with the release of large amounts of energy; conversely, these bonds can be broken only if large amounts of energy are supplied. In general, if a bond is formed only by the input of energy (an endergonic reaction), energy will be released when it is broken (an exergonic reaction). *High-energy bonds*—those that release at least 5 kcal/mole when broken—are of great importance in the transfer of energy within the living system. High-energy bonds involving phosphorus and

Figure 4.9b. An interpretation of the principle of entropy. Fuels that provide heat — the stars, the sun, fossil fuels, and so on — all result in an increase in the randomness of molecular motions, or entropy, which has also been described as the "general running down" of the universe.

Figure 4.10. A model of coupled reactions. Living cells must carry out many reactions that are energetically unfavorable. The reaction C→ D is an example of an endergonic reaction requiring the expenditure of free energy for its completion. Reactions of this type are coupled to exergonic reactions such as A→B. This coupling is done through enzymatic cofactors that store and transfer the energy obtained in the first reaction (A→B) for use in the reaction C→D. In many instances, the cell utilizes ATP as the cofactor for coupling endergonic and exergonic reactions.

oxygen play major roles in metabolic processes of all living organisms because energy is shuttled from exergonic reactions to endergonic ones through this mechanism of energy storage.

Pyrophosphoric acid ($H_4P_2O_7$) can be formed from orthophosphoric acid (H_3PO_4) in the following reaction:

$$2H_3PO_4 \rightarrow H_4P_2O_7 + H_2O \qquad (4.3)$$

or

$$\mathrm{HO{-}\overset{\overset{\displaystyle O}{\|}}{\underset{\underset{\displaystyle OH}{|}}{P}}{-}OH + HO{-}\overset{\overset{\displaystyle O}{\|}}{\underset{\underset{\displaystyle OH}{|}}{P}}{-}OH \rightarrow HO{-}\overset{\overset{\displaystyle O}{\|}}{\underset{\underset{\displaystyle OH}{|}}{P}}{-}O{-}\overset{\overset{\displaystyle O}{\|}}{\underset{\underset{\displaystyle OH}{|}}{P}}{-}OH + H{-}O{-}H} \qquad (4.4)$$

This reaction is endergonic ($\Delta H = +6.1$ kcal/mole). The pyrophosphate linkage (P—O—P) is formed only through the input of energy. The reverse reaction is exergonic and proceeds spontaneously with the release of energy. In this reverse reaction, pyrophosphoric acid is broken down in a reaction with water (called *hydrolysis*) to form orthophosphoric acid and energy:

$$H_4P_2O_7 + H_2O \rightarrow 2\,H_3PO_4 \quad \Delta H = -6.1 \text{ kcal/mole} \qquad (4.5)$$

The chemist who wishes to prepare pyrophosphoric acid simply heats orthophosphoric acid, thus supplying the energy needed to carry out the endergonic reaction. At the higher temperature, the equilibrium is shifted toward the pyrophosphoric acid, although a considerable amount of orthophosphoric acid remains when equilibrium is reached. However, an organism operates at a relatively low temperature. Can it possibly create a compound such as pyrophosphoric acid?

An enzyme will not shift the equilibrium (as does the heating); it will merely increase the rate at which the equilibrium is reached. In this case, an enzyme would allow pyrophosphoric acid to be more rapidly hydrolyzed at low temperatures. The organism solves this sort of problem by transferring energy in high-energy bonds from exergonic reactions to carry

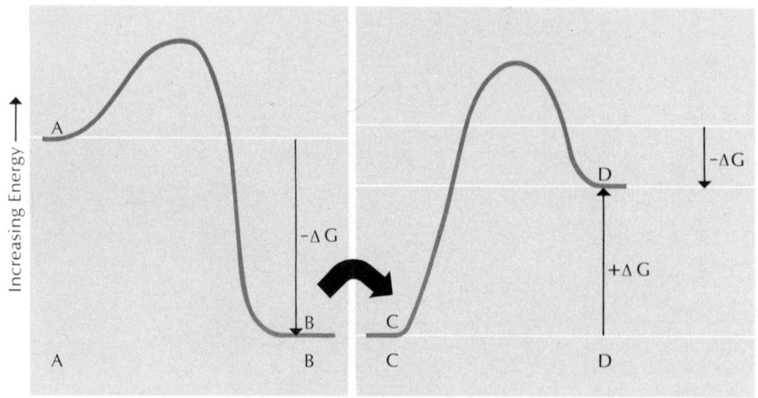

Figure 4.11 (above). The structural formula of adenosine triphosphate (ATP).

Figure 4.12 (below). The hydrolysis of adenosine triphosphate to adenosine diphosphate and inorganic phosphate.

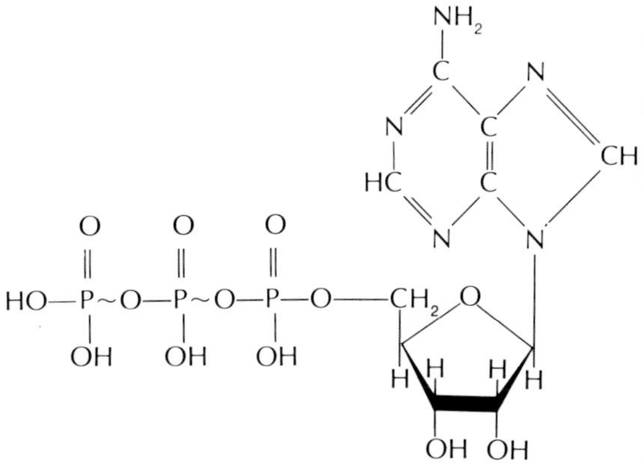

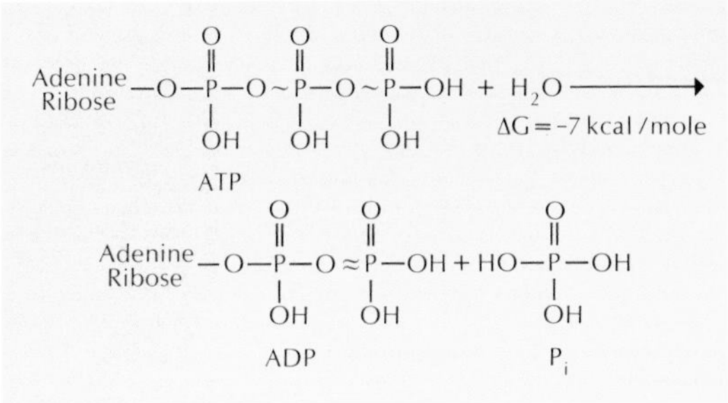

out endergonic reactions. The molecules that can shuttle energy in high-energy bonds are called *cofactors*. An exergonic and an endergonic reaction coupled by cofactors are called *coupled reactions*.

Figure 4.11 shows the structural formula of a molecule of a substance called adenosine triphosphate (ATP). Note that it has two pyrophosphate linkages. These high-energy bonds, like those in pyrophosphoric acid, can easily be broken by hydrolysis (the addition of water). An example of the hydrolysis of ATP is shown in Figure 4.12; the reaction can be summarized as follows:

$$ATP + H_2O \rightarrow ADP + H_3PO_4 \quad \Delta G = \text{about } -7 \text{ kcal/mole} \qquad (4.6)$$

As would be expected, the formation of ADP (adenosine diphosphate) from ATP is quite exergonic. Conversely, the formation of ATP from ADP and H_3PO_4 (the reverse reaction) is quite endergonic.

In the living system, energy that is released as ATP is being hydrolyzed to form ADP can be used as a source of energy for such endergonic reactions as the formation of pyrophosphoric acid. Suitable enzymes are required to carry out both reactions at moderate temperatures, and most strongly energetic reactions in living systems involve at least four groups of molecules: reactants, enzymes, cofactors, and products.

A similar coupling of exergonic and endergonic reactions is involved in the formation of the ATP itself. The following reaction, involving the hydrolysis of phosphoenolpyruvic acid (another phosphorus-containing compound), is even more exergonic than the conversion of ATP to ADP:

$$CH_2COH_2PO_3COOH + H_2O \rightarrow CH_3COCOOH + H_3PO_4 \qquad (4.7)$$

In the organism, the following reaction takes place:

$$\text{phosphoenolpyruvic acid} + ADP \rightarrow \text{pyruvic acid} + H_3PO_4 \qquad (4.8)$$

This reaction can be regarded as the coupling of the two reactions that have already been discussed:

$$\text{phosphoenolpyruvic acid} + H_2O \rightarrow \text{pyruvic acid} + H_3PO_4 \qquad (4.9)$$

and

$$ADP + H_3PO_4 \rightarrow ATP + H_2O \qquad (4.10)$$

$$\text{phosphoenolpyruvic acid} + ADP + H_2O + H_3PO_4 \rightarrow \text{pyruvic acid} + ATP + H_3PO_4 + H_2O \qquad (4.11)$$

The amount of H_2O and of H_3PO_4 is the same before and after the reaction; therefore, the overall reaction can be summarized as:

$$\text{phosphoenolpyruvic acid} + ADP \rightarrow \text{pyruvic acid} + ATP \qquad (4.12)$$

In the living system, this reaction is facilitated by a particular enzyme that catalyzes the transfer of the phosphate group (PO_4) from the phosphoenolpyruvic acid to ADP.

The first reaction (Formula 4.9) is strongly exergonic; the second (Formula 4.10) is strongly endergonic. However, the total reaction is

Adenosine Triphosphate (ATP)

Figure 4.13. An example of a machine-punched code on paper tape. The tape bears the message: "We hold these Truths to be self-evident, that all Men are created equal. . . ."

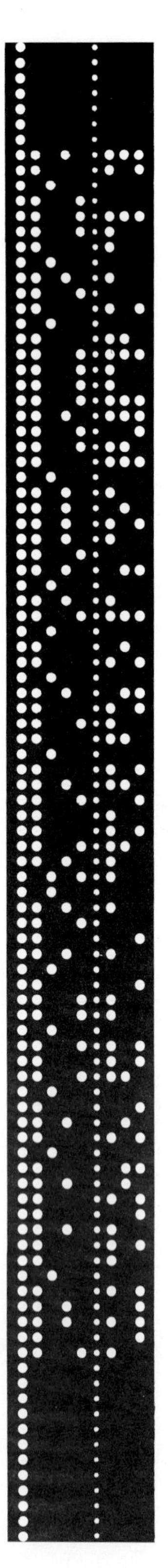

moderately exergonic, because the positive ΔG of the second reaction is more than balanced by the negative ΔG of the first reaction. Thus, the energy obtained by hydrolyzing phosphoenolpyruvic acid is used to convert ADP to ATP. Energy has been stored in the high-energy pyrophosphate linkage of the ATP. This energy can now be used to carry out some other endergonic reaction, which can be coupled with the hydrolysis of ATP.

ATP is the negotiable currency of the energy exchange in the organism. Because the breakage of the pyrophosphate linkage in ATP is so exergonic, when the appropriate catalysts are present, the energy stored in the ATP can be spent for many purposes. The use of the energy is accomplished by coupling the breakdown of the ATP with an endergonic reaction. Because most of the reactions involved in growth—that is, in building the complex molecules of the living system—are endergonic, this energy source is very important to the system. It is noteworthy that, under the conditions of the living cell, ATP is a relatively stable compound, requiring enzymatic catalysis in order to undergo rapid reactions. It should be emphasized that the stability of a compound is reflected by its reluctance to decompose and is not determined by the energy change occurring on decomposition.

GENETICS: TRANSFER OF INFORMATION

What sort of system exists to transfer the information about this complex chemical mechanism to the next generation of organisms? What would be the requirements of such an information system? An exploration of these questions must begin with a look at the nature of information itself.

Information, in essence, is a message—a readable message. The message may be in various forms and may be read in various ways. It may be stored in one form and delivered in another. Nevertheless, certain characteristics are indispensable. Information implies structure. Consider a message composed of paragraphs, which are made up of sentences, words, and letters. Understanding of the message is made possible by recognition of the structure of the letters.

But not all messages are written. In spoken messages, the structure of the sounds is important. Mathematical equations, musical notation, sculpture, a passage of orchestral music—each of these forms of information has a structure. Alter the structure and the message may be altered; alter the structure sufficiently and the message may be lost entirely.

Messages set forth in a linear form, such as writing, have an important characteristic in their sequence. RUN does not carry the same message as does URN, and RNU is nonsense in English. Furthermore, the message must be read in a particular direction; RUN read backward becomes NUR, meaningless in English but meaningful in German. Other message forms, such as music, are also linear. Beethoven's Fifth Symphony played backward does not carry the message that Beethoven intended; if a tape recording of the music is cut up and spliced in a new sequence, an interesting tape collage may result, but the message will again be altered.

How could these characteristics be used to design a system to carry the information (the "blueprint," so to speak) needed for the construction of a molecule of the enzyme E_{AB}? Because the information is to be used in reactions among molecules, it is reasonable to suppose that the information itself will be stored in some sort of molecular structure.

The kind of molecule needed to store this information depends upon the complexity of the information to be stored. The structure of a skyscraper requires a more elaborate set of blueprints than does the structure of a homemade coffee table. Similarly, the complexity of the blueprint for enzyme E_{AB} depends upon the complexity of the enzyme molecule itself.

Enzyme E_{AB}, like every other enzyme, is a protein, a particular type of heteropolymer. A blueprint for the construction of a homopolymer could be relatively simple. It must specify the monomer to be used, the way in which the monomers are to be linked together, and the number of monomers in the polymer. Because the monomers are identical, sequence is irrelevant. However, for a heteropolymer such as an enzyme, the blueprint must also specify the sequence in which the various monomers are to occur in the chain.

Because the protein to be constructed is a linear molecule, the blueprint can be readily encoded in another linear molecule. The length of the molecule must be sufficient to include all of the information needed to specify fully the structure of the protein; because the protein is a macromolecule, the message must also be stored in a macromolecule. Finally, the message—when read in the proper direction—should specify the sequence of monomers in the protein chain.

From this reasoning, it seems that the information should be stored in a linear macromolecule. This information molecule must be a heteropolymer, for a homopolymeric message could not carry the information needed to construct the proper sequence of different monomers in the heteropolymeric protein. Proteins contain some 20 different monomers, but the information molecule would not necessarily have to contain this many different monomers. Just as 26 letters can be used to encode many different words, so groups of a few different monomers can be used to encode information about the 20 monomers found in proteins.

The message molecule used by organisms has just the characteristics deduced above. It is a heteropolymer called *ribonucleic acid* (RNA), a linear macromolecule made up of four different kinds of monomers. In certain viruses (RNA viruses), RNA is the ultimate repository of genetic information. In other viruses and all other organisms, RNA is the working genetic information—in a sense it is the copy of the blueprint taken to the construction site rather than the master copy kept in the office. The master copy is deoxyribonucleic acid (DNA). DNA is also a linear polymer composed of four different kinds of monomers, differing slightly in structure but similar to the same four kinds as in RNA.

Biochemists have succeeded in determining the basic structure of the nucleic acids, they have built some simple information molecules from scratch, and they are beginning to develop the ability to design and to build RNA and DNA molecules that will carry any desired bit of genetic information.

Direct observation of a molecule is sometimes quite valuable, particularly with such powerful techniques as x-ray diffraction. However, certain kinds of information can be obtained only by taking apart a substance such as RNA. The first problem is to obtain a sample of pure RNA that can be studied. In one common approach, the molecule is hydrolyzed; as in the case of the pyrophosphate linkage in ATP, many complex biochemicals can be broken apart through reaction with water. Hydrolysis may be carried out under mild conditions with the aid of an enzyme, or it may be done under harsher conditions in the presence of hot acid or alkali.

The pieces are separated and studied in the hope of understanding their structure. It may be necessary to separate these pieces still further before a simple molecule is obtained that can be recognized by its properties. Then the biochemist must figure out how the pieces were linked together in the original molecule and how that molecule functions. This sort of endeavor, carried out over the past 120 years with increasingly sophisticated techniques, has yielded an enormous amount of information about the chemistry of living systems. In particular, it has permitted rapid progress toward an understanding of the transfer of information within the organism. This understanding begins with a closer look at the structure of the proteins, which are constructed according to the information stored in DNA and RNA.

Proteins

The monomers that make up the protein polymer are *amino acids*. The simplest amino acid, glycine, consists of a carbon atom bonded to a carboxyl group (COOH), an amine group (NH_2), and two hydrogen atoms:

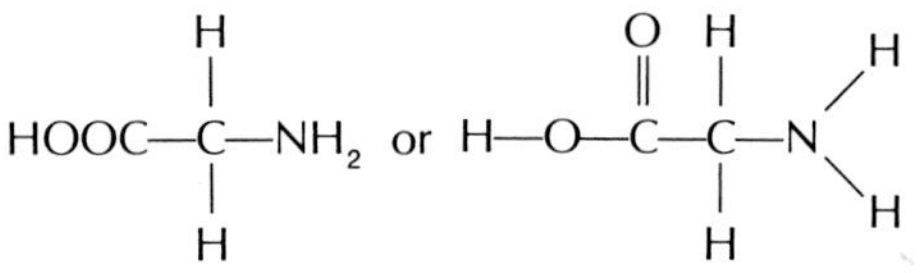

$$\begin{array}{ccc} & \mathrm{H} & \\ & | & \\ \mathrm{HOOC{-}} & \mathrm{C} & \mathrm{{-}NH_2} \\ & | & \\ & \mathrm{H} & \end{array} \quad \text{or} \quad \begin{array}{ccccc} & \mathrm{O} & \mathrm{H} & & \mathrm{H} \\ & \| & | & \diagup & \\ \mathrm{H{-}O{-}} & \mathrm{C} & \mathrm{{-}C{-}} & \mathrm{N} & \\ & & | & \diagdown & \\ & & \mathrm{H} & & \mathrm{H} \end{array}$$

The other 20 or so amino acids are similar, but one of the hydrogen atoms attached to the central (alpha) carbon is replaced by a more complex group. In alanine, for example, one of the hydrogens is replaced by a methyl (CH_3) group:

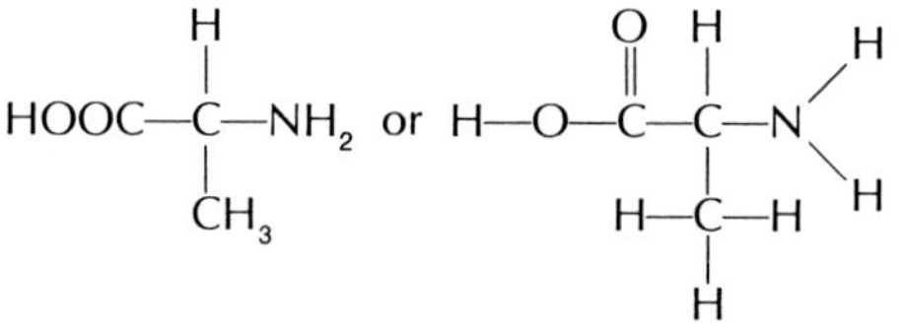

$$\begin{array}{ccc} & \mathrm{H} & \\ & | & \\ \mathrm{HOOC{-}} & \mathrm{C} & \mathrm{{-}NH_2} \\ & | & \\ & \mathrm{CH_3} & \end{array} \quad \text{or} \quad \begin{array}{ccccc} & \mathrm{O} & \mathrm{H} & & \mathrm{H} \\ & \| & | & \diagup & \\ \mathrm{H{-}O{-}} & \mathrm{C} & \mathrm{{-}C{-}} & \mathrm{N} & \\ & & | & \diagdown & \\ & & \mathrm{H{-}C{-}H} & & \mathrm{H} \\ & & | & & \\ & & \mathrm{H} & & \end{array}$$

An amino acid can form linkages to other amino acids at either end. The reaction proceeds through several intermediate stages, but the net result is to free a water molecule and to attach the nitrogen in the amine group of one amino acid directly to the carbon in the carboxyl group of the other amino acid. As an example, the combination of two glycine molecules could occur in this fashion:

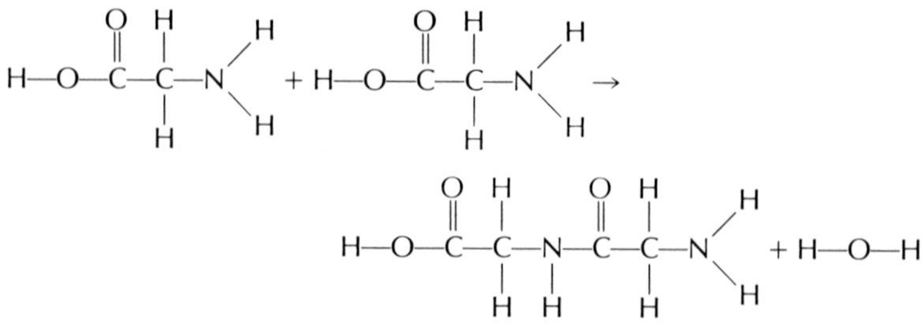

$$\begin{array}{ccccc} & \mathrm{O} & \mathrm{H} & & \mathrm{H} \\ & \| & | & \diagup & \\ \mathrm{H{-}O{-}} & \mathrm{C} & \mathrm{{-}C{-}} & \mathrm{N} & \\ & & | & \diagdown & \\ & & \mathrm{H} & & \mathrm{H} \end{array} + \begin{array}{ccccc} & \mathrm{O} & \mathrm{H} & & \mathrm{H} \\ & \| & | & \diagup & \\ \mathrm{H{-}O{-}} & \mathrm{C} & \mathrm{{-}C{-}} & \mathrm{N} & \\ & & | & \diagdown & \\ & & \mathrm{H} & & \mathrm{H} \end{array} \rightarrow$$

$$\begin{array}{cccccccc} & \mathrm{O} & \mathrm{H} & & \mathrm{O} & \mathrm{H} & & \mathrm{H} \\ & \| & | & & \| & | & \diagup & \\ \mathrm{H{-}O{-}} & \mathrm{C} & \mathrm{{-}C{-}} & \mathrm{N{-}} & \mathrm{C} & \mathrm{{-}C{-}} & \mathrm{N} & \\ & & | & | & & | & \diagdown & \\ & & \mathrm{H} & \mathrm{H} & & \mathrm{H} & & \mathrm{H} \end{array} + \mathrm{H{-}O{-}H}$$

The linkage formed shown as:

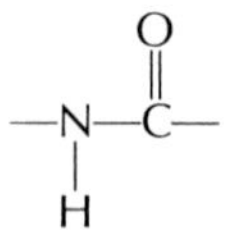

$$\begin{array}{ccc} & & \mathrm{O} \\ & & \| \\ \mathrm{{-}N} & \mathrm{{-}} & \mathrm{C{-}} \\ | & & \\ \mathrm{H} & & \end{array}$$

is called a peptide linkage, and the resulting compound is called a *peptide*. In this case, because there are two amino acids in the peptide, it is a dipeptide. However, more amino acids could be added to either end of the chain. In fact, *polypeptides* exist with many thousands of amino acids joined together. One of the smallest of the proteins is salmine, a polypeptide containing 58 amino acids joined by peptide linkages.

The many proteins needed by an organism are synthesized from the amino acids. Some amino acids can be synthesized directly by the organism from simple nitrogen-containing inorganic compounds. Animal organisms must obtain other amino acids by breaking apart protein molecules in food. The human organism, for example, requires a supply of 8 *essential amino acids* in the proteins of the diet; the other 12 amino acids used in the synthesis of proteins can be built by the human organism if sufficient nitrogen is supplied in the diet. The dietary requirements vary from species to species. For example, humans can synthesize glycine, whereas young chicks must have this amino acid supplied in their diets. Rats can synthesize glycine but must be supplied histidine, which is not an essential amino acid for humans. Thus, an organism not only requires a certain amount of protein in its diet but requires certain kinds of protein that supply its essential amino acids—the amino acids that it cannot synthesize.

Each of the amino acids except glycine exists in two isomeric forms, which are identical in chemical properties but can be distinguished by optical properties. The alpha carbon atom of the amino acid is bonded to four different groups of atoms (except in glycine, where two of the bonds are to hydrogen atoms). The spatial arrangement of these bonds can be represented by placing the alpha carbon in the center of a tetrahedron with the four groups at the corners (Figure 4.14). Note that there are two possible configurations of the four groups. These two arrays are the mirror image of one another and are called optical isomers. They possess a kind of "handedness" and cannot be superimposed upon one another. In this three-dimensional representation, it is clear that the *D* isomer cannot be converted into the *L* isomer by any simple rotation of the molecule; bonds must be broken and re-formed to convert one isomer into the other. All of the amino acids found in the proteins of living systems are of the *L* form, and living systems are unable to make use of *D* isomers in building proteins. There is no adequate chemical explanation for this preference; proteins made of *D* amino acids can be synthesized in the laboratory. Apparently, the information

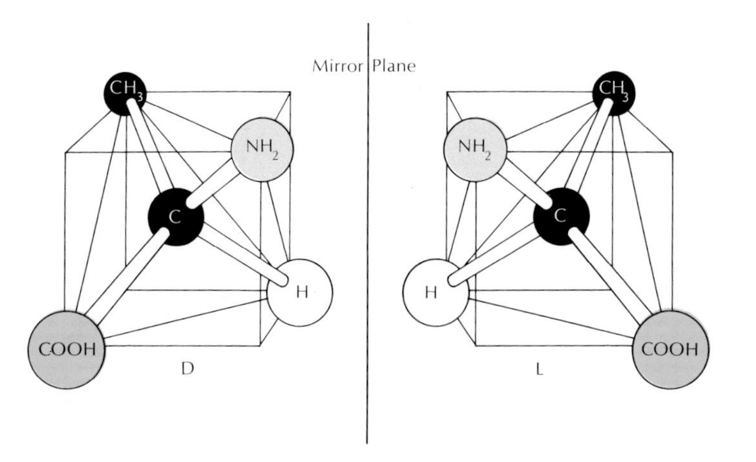

Figure 4.14. Optical isomers of the amino acid alanine. Only the L-form is found in nature.

Figure 4.15. Schematic model of a right-handed α-helix. There are three amino acids in one hydrogen-bonded loop. Note that all $-CO$ and $-NH$ groups form hydrogen bonds.

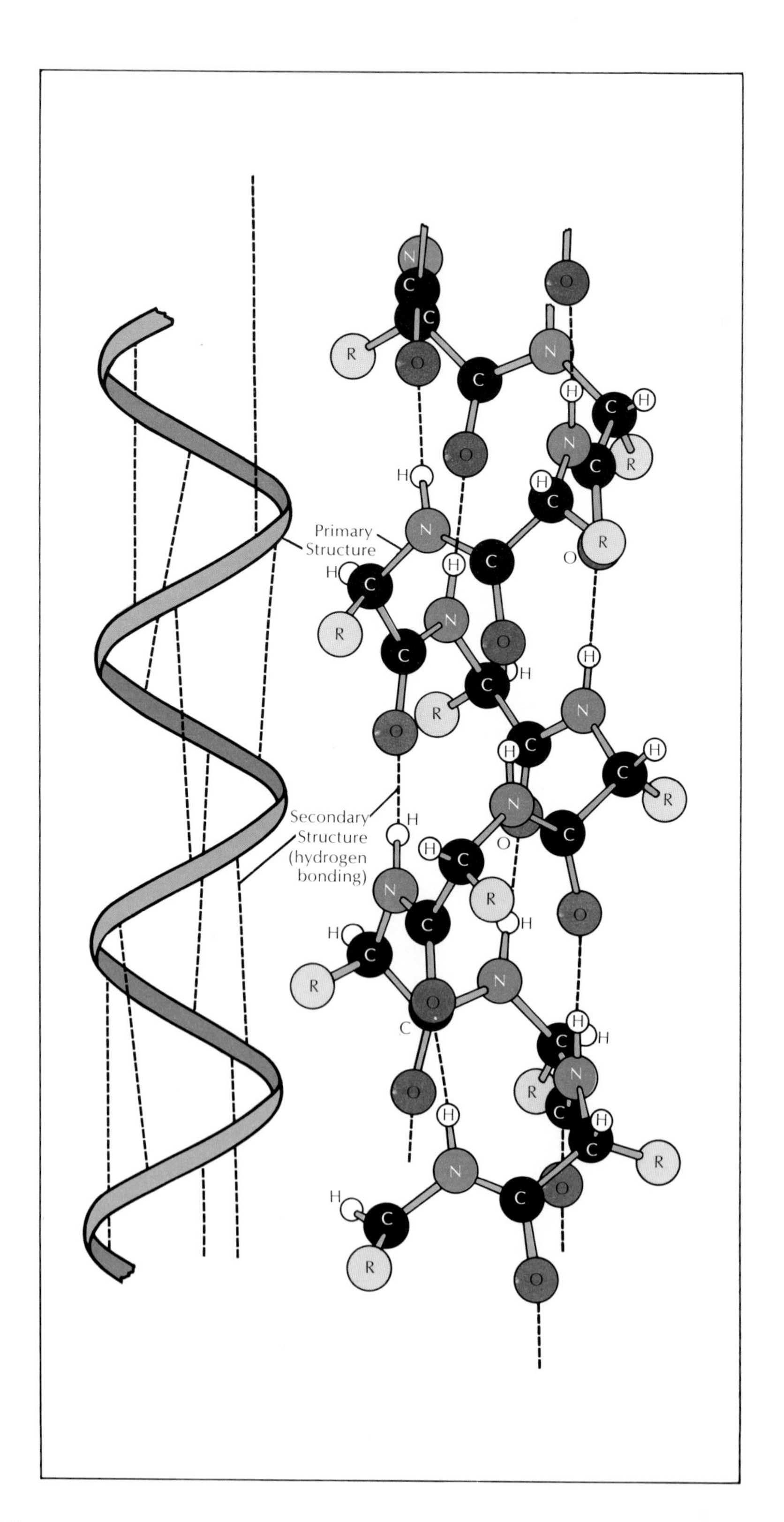

Figure 4.16. Schematic model of a protein molecule, comprised primarily of α-helical structures.

encoded in the nucleic acids specifies the use of *L* amino acids, rather than the *D* form, for protein synthesis.

Thus far, a protein molecule has been described as a long, straight polypeptide chain. In many cases, however, the chain winds into a coil of one sort or another. This coil may be a rather random twisting, or it may be a very regular winding, resulting in a structure like that of a helical spring. One such helical structure that has been identified is the right-handed alpha helix (Figure 4.15). In some proteins—called *globular* proteins—the coiled structure is bent, folded (often several times), and twisted until the overall shape of the molecule is spherical or ellipsoidal. Those proteins with a more-or-less linear structure are called *fibrous* proteins. Most enzymes are globular proteins. Most of the proteins making up structural elements of organisms are fibrous.

The globular protein molecule is held in its twisted position by bonds other than those in the polypeptide backbone (Figure 4.15). The coils of the alpha helix are held together by hydrogen bonds (shown as H — O).

Still other bonds account for the folding and bending of the coil. In the portion of a protein molecule shown in Figure 4.16, a *disulfide bridge* (—S—S—) has formed between two amino acids, both cysteine, some

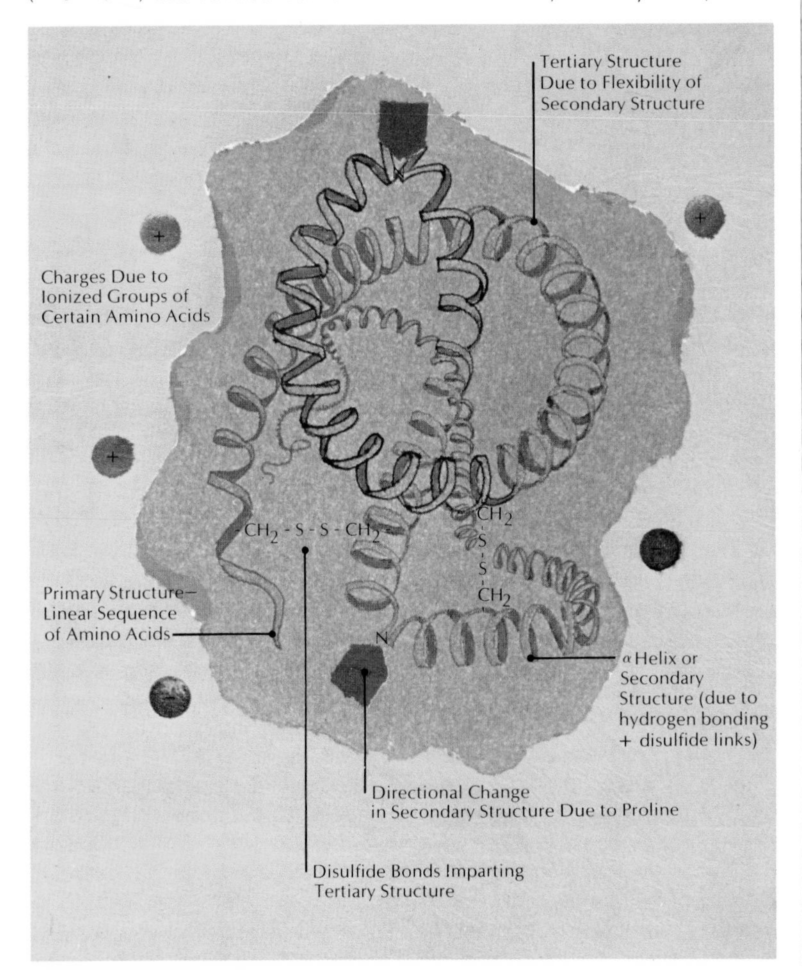

Figure 4.17 (above). Structural formula for a section of ribonucleic acid (RNA). One of each of the nucleotide bases found in RNA is shown.

Figure 4.18 (below). Nomenclature for the chemical subunits comprising the building blocks of RNA—the ribonucleotide bases.

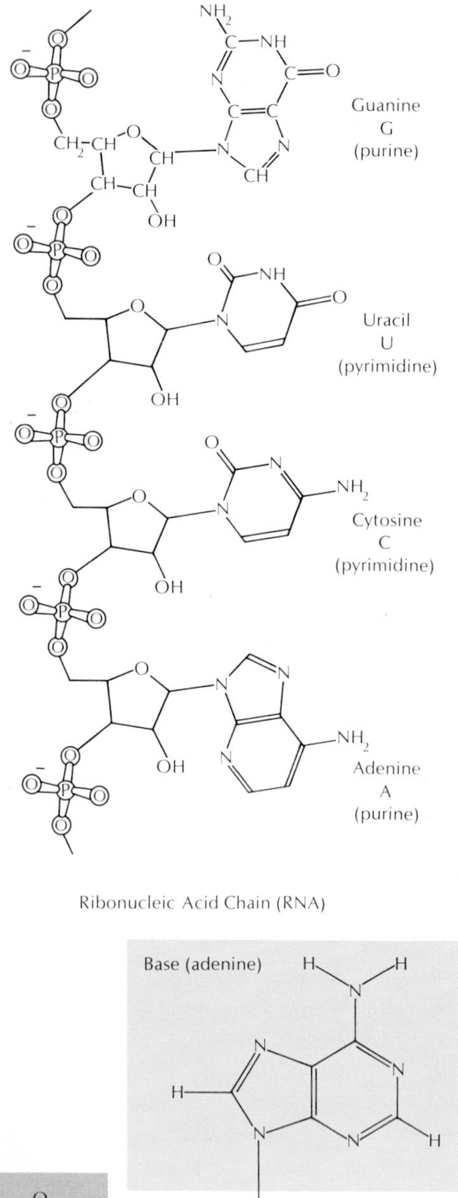

Guanine G (purine)

Uracil U (pyrimidine)

Cytosine C (pyrimidine)

Adenine A (purine)

Ribonucleic Acid Chain (RNA)

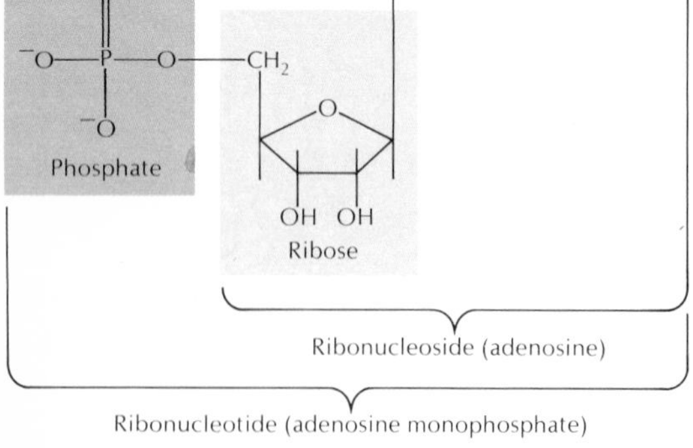

Base (adenine)

Phosphate

Ribose

Ribonucleoside (adenosine)

Ribonucleotide (adenosine monophosphate)

distance apart on the chain. Cysteine has a CH_2SH group attached to the alpha carbon. The hydrogen atoms have been lost, and a covalent bond has been formed between the sulfur atoms of the two cysteine molecules. The disulfide bridge holds this portion of the protein molecule in a looped shape. Breakage of such bonds may greatly change the shape and therefore the properties of a protein molecule, even though the polypeptide backbone itself remains intact. Thus, the biochemical properties of many proteins may be altered by conditions not extreme enough to break apart the polypeptide chains. For example, many proteins are denatured (so altered in structure that they lose their useful properties) by slight heating, gentle stirring, or treatment with weak acids.

Many of the globular protein molecules are made up of several independent polypeptide chains associated by mutual attractive forces or joined by disulfide bonds. Hemoglobin is made up of four such chains or subunits. In some of the fibrous protein molecules, independent polypeptide chains are similarly bonded together and, in some cases, even twisted together into multiple coils or helices.

The amount of information needed to describe the twisting, folding, and cross-linkages in large protein molecules is tremendous. It would seem that the RNA molecule would have to be many times as large as the protein molecule whose "blueprint" it contains. However, biochemists now think that this information need not be supplied separately. When a sequence of amino acids is strung together in the proper order, the twisting and folding seem to occur automatically as a result of the nonbonded interactions between regions of the chain. Thus, the complex structure of the protein is a result of the sequence of amino acids in its backbone and need not be specified separately.

RNA

The information molecule, RNA, is also a polymer, but it is not a protein. RNA contains four kinds of *nucleotide* monomers (Figure 4.17). Each of the nucleotides is made up of a sugar, a phosphate group, and an organic base. In each of the four nucleotides the sugar is ribose. They are therefore called *ribonucleotides*, and the polymeric structure is known as ribonucleic acid (RNA). The phosphate group is also identical in the four ribonucleotides, which differ only in the base attached to each. The four bases are adenine, guanine, uracil, and cytosine (Figure 4.18). Because each of these bases contains nitrogen, they are sometimes called the nitrogenous bases.

The polynucleotide RNA is formed by linkages between the sugar of one nucleotide and the phosphate group of the next. Thus, the backbone of the polymer is a series that could be represented as—sugar—phosphate—sugar—phosphate—. The bases extend off to the side of this backbone.

Just as there are many proteins, each with its own sequence of amino acids, so are there many RNAs, each with its own sequence of nucleotides. This multiplicity makes it possible for each RNA to carry the instructions for the building of a particular protein. However, there are 20 different amino acids used in building proteins and only 4 nucleotides available in RNA to encode the sequence of amino acids for building the protein. If each nucleotide represented one amino acid in the code, RNA could not carry complete and unambiguous instructions for the construction of proteins.

However, four different symbols are quite adequate for spelling out any message if the symbols are grouped to form words. If the symbols *A*, *G*, *U*, and *C* are used to represent the four nucleotides, 16 different "two-letter

Figure 4.19 (above). The chemical structures for the ribose sugars and nucleic acid bases found in RNA and DNA.

Figure 4.20 (below). Schematic model of the double-helix configuration of the DNA molecule.

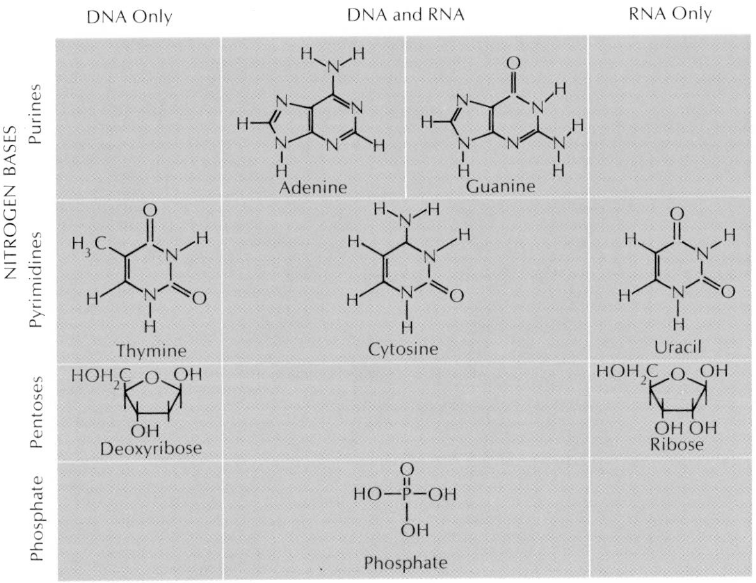

NITROGEN BASES /	DNA Only	DNA and RNA	RNA Only
Purines		Adenine; Guanine	
Pyrimidines	Thymine	Cytosine	Uracil
Pentoses	Deoxyribose		Ribose
Phosphate		Phosphate	

words" can be written: *AA, GA, CA, UA, AG, GG, CG, UG, AU, GU, CU, UU, AC, GC, CC,* and *UC*. A language using only two-symbol words could not provide a unique word for each of the 20 amino acids. However, a little work with pencil and paper shows that these 4 symbols can make up 64 different three-letter words. This number is more than sufficient to provide a different word for each amino acid. In fact, many of the amino acids could be represented by two or more different words (synonyms, if the analogy to language is continued).

Recently, many lines of research have shown that RNA does use a "three-letter-word" code (the so-called *genetic code*) to represent the amino acids. The three-nucleotide groups are called *codons*, which are discussed in detail in Chapter 15. These codons are arranged in sequence along the RNA molecule without spaces or overlap. The message must be read by looking at the first three nucleotides, then the next three, and so on.

In building a protein, a process of *translation* occurs in which the sequence of ribonucleotide triplet codes is expressed as a linear sequence of the corresponding amino acids in the protein molecule. There must exist a mechanism that determines where to start reading the RNA message, recognizes each codon in sequence, adds the appropriate amino acid to the growing polypeptide chain, and ends the chain at the proper point. The deciphering of the details of this mechanism has been one of the major successes of biology in recent years (Chapter 15).

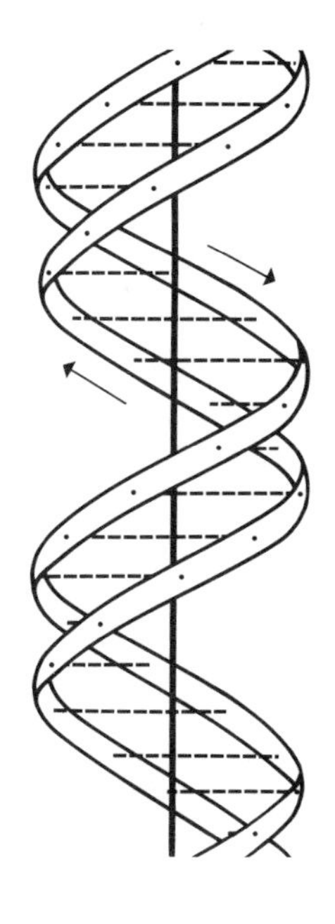

DNA

In most organisms, RNA is only a working copy of the basic genetic information, which is stored in the molecules of DNA (deoxyribonucleic acid). DNA is a polynucleotide very similar to RNA. Each of the nucleotides of DNA is also composed of a phosphate group, a sugar, and a base. In DNA the sugar is slightly different from that in RNA in that it lacks one oxygen atom (Figure 4.19). This sugar is called deoxyribose, from which is derived the name deoxyribonucleic acid. The bases found in the deoxyribonucleotides are adenine, guanine, cytosine, and thymine. Except for the presence

Figure 4.21. Complementary hydrogen bonding of purine and pyrimidine base pairs in DNA. Note the triple bond between cytosine and guanine and the double bond between adenine and cytosine.

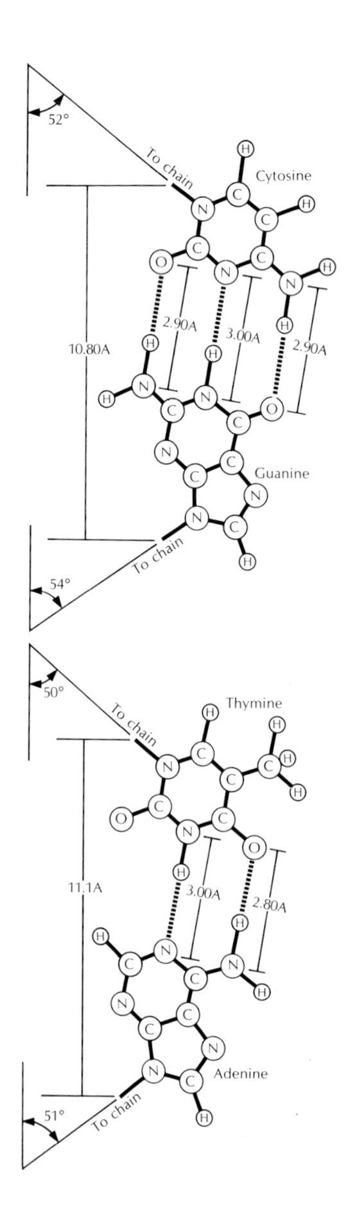

of thymine instead of uracil, these bases are the same as those found in the ribonucleotides. As in RNA, the structure of the polynucleotide DNA consists of a phosphate-sugar backbone, with the bases attached along the side. However, for DNA there is one very important complication to this simple picture.

DNA must specify the sequence of ribonucleotides in RNA, which in turn specify the sequence of amino acids in proteins. What specifies the sequence of deoxyribonucleotides in DNA? The answer is other DNA. If a cell divides, the two new daughter cells must have the same genetic information that was present in the parent cell. Therefore, if DNA is the fundamental genetic material, the daughter cells must have the same kind of DNA that was present in the parent. The macromolecular structure of DNA must lend itself to the guidance of its own replication as well as the guidance of RNA synthesis.

An early clue to the nature of the DNA structure came from work by Erwin Chargaff (1955) and others who were trying to determine the base composition of DNA samples from various sources. They found, as expected, that the composition varies from species to species—bacterial DNA differs in its proportions of the four bases from human DNA, which in turn differs from the DNA of cattle. For each species, however, the proportion of adenine is always equal to that of thymine, and the proportion of guanine is always equal to that of cytosine. Another clue to the DNA structure came from x-ray diffraction measurements made by M. H. F. Wilkins, Rosalind Franklin, and others (Wilkins, Stokes, and Wilson, 1953). These measurements suggested that the DNA molecule is coiled into a helical structure.

The structure of DNA, which has since been confirmed by many kinds of evidence, was proposed by James Watson and Francis Crick (1953)—the famous double helix (Figure 4.20). According to the Watson-Crick model, discussed further in Chapter 15, the DNA molecule is composed of two polynucleotide chains intertwined in a helical spiral. The two chains are connected by hydrogen bonds between their bases; each base on one chain is weakly bound to a base on the other chain. The two sugar-phosphate chains run in opposite directions; that is, the sequence of atoms goes one way in one chain and the opposite way in the other chain.

In order for the chains to fit together in the structure indicated by the x-ray diffraction measurements, Watson and Crick proposed that a large base group on one chain must always be paired with a small base group on the other chain. In fact, because of the shapes and hydrogen bonding of the base groups, only two pairings are possible: adenine with thymine, and guanine with cytosine (Figure 4.21). Thus, wherever adenine appears in one chain, it must be paired with thymine on the other chain, and guanine on one chain must always be paired with cytosine on the other. The pairs of bases are joined by hydrogen bonds.

The two strands of the DNA molecule are not identical but are oriented in opposite directions. The bases on the one strand are complementary to the bases on the other. Suppose that the two strands could be separated, breaking the weak hydrogen bonds that hold the strands together but keeping the sugar-phosphate backbones of the two chains intact. If the environment contained deoxyribonucleotides and if a mechanism existed to put them together to form new polynucleotide strands, each of the parent strands of DNA could build upon itself a new companion strand. Because of the base pairing restrictions, each of the new companion strands would be complementary to the strand on which it was built. Therefore, the final

Figure 4.22 (above). Complementary bonding of specific purine-pyrimidine pairs results in the production of complementary DNA strands.

Figure 4.23 (middle). The DNA molecule in the process of replication according to the Watson-Crick model.

Figure 4.24 (below). Transcription of a DNA template code to an RNA molecule. The enzyme RNA polymerase attaches to the DNA molecule, opening up a short section of the double helix for transcription. As RNA polymerase moves along the DNA template, the growing RNA strand peels off and attaches to a ribosome, while hydrogen bonds re-form the complementary DNA strands. *Printed by permission from J. D. Watson,* Molecular Biology of the Gene, *© 1970, J. D. Watson*

result would be two complementary DNA strands, each identical to the parent strand (Figure 4.22).

Actually, the mechanism is far more complex than this simple description. For example, to separate the two strands without breaking the sugar-phosphate backbones is impossible because of the intertwining of the strands. In the living system, the strands unwind gradually as complementary strands are built, and the backbones of the strands are temporarily broken and then rejoined to permit the separation of the two strands. Some organisms contain single-stranded DNA. In these organisms, the original strand presumably synthesizes a short-lived complementary strand, which then synthesizes a duplicate of the original DNA.

The Watson-Crick model fulfills the requirement for self-duplication of the information stored in the DNA molecule. No translation is involved in the DNA self-duplication; it is simply a process of *replication*. The property of mutation can be easily explained by this model. If for any reason an error should occur in the copying, that error would be passed on in future replications. Suppose, for example, that a small segment of the strand were reversed during the unwinding process. When this segment built a complementary strand, the new strand would match the reversed sequence, not the original sequence. When this new DNA molecule, in turn, separated and formed new complementary strands, the error (termed a mutation) would continue to be copied on every subsequent replication of the DNA.

The process of creating RNA from the information on the DNA molecule is similar to that involved in duplicating the DNA. The new strand being built is composed of ribonucleotides instead of deoxyribonucleotides, and the base uracil (instead of thymine) must be paired with the base adenine. An exact replication is not carried out, yet the copying is on a one-for-one basis, unlike the translation involved in creating a protein from the RNA molecule. The creation of RNA from DNA is called *transcription*. Because the RNA molecule is usually single stranded, only one of the DNA strands is used for transcription of the RNA. It is not always the same strand of DNA, and the mechanism by which one or the other is chosen is an area of active research. It is this phase of transcription that appears to determine how protein synthesis responds to changing environmental conditions.

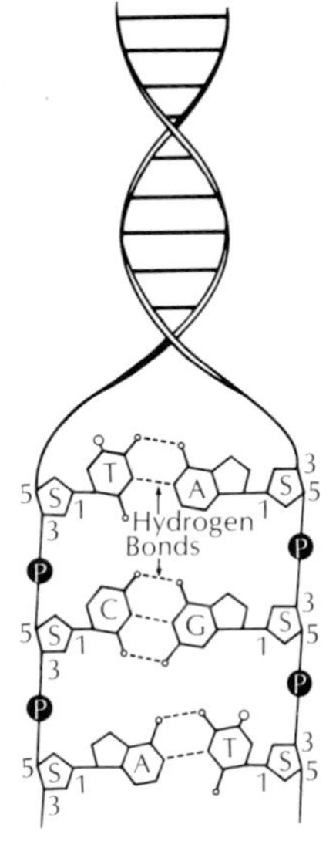

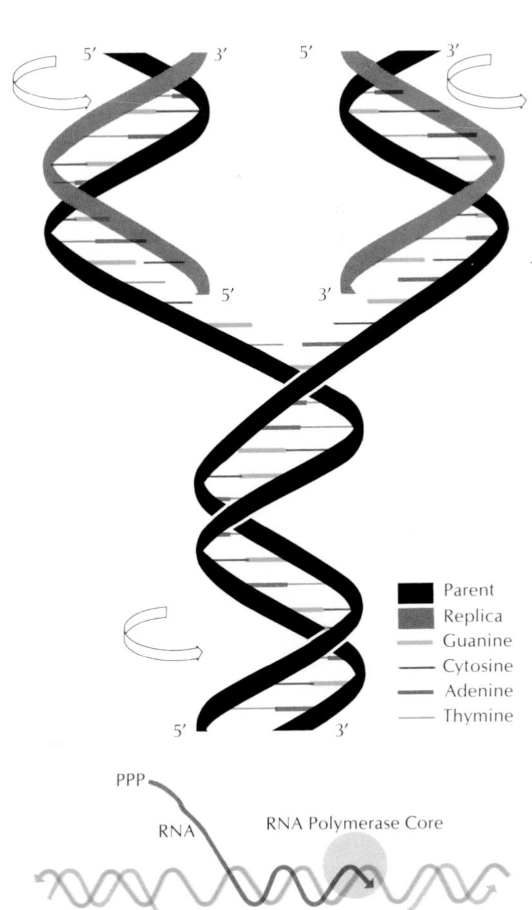

LIVING CHEMICAL SYSTEMS

Although living systems contain atoms of only a few elements, the unusual properties of the carbon atom permit the formation of an incredible variety of complex molecules. The atoms and molecules themselves clearly are not alive. Yet a combination of certain complex molecules, involved in an intricate network of chemical reactions, can possess all of those properties that are ascribed to living things. Biochemists have made a striking case for the view that living things are merely very complicated mechanisms obeying the same laws of chemistry and physics as nonliving systems.

FURTHER READING

Baker and Allen (1965) discuss the chemistry of living systems. A clear and well-illustrated introduction to the chemistry of proteins is given by Dickerson and Geis (1970). Watson (1968) presents a lively and controversial account of the development of the Watson-Crick model of DNA structure.

Scientific American articles pertinent to this chapter include those by Changeux (1965), Crick, (1954, 1962), and Holley (1966).

5
Photosynthesis

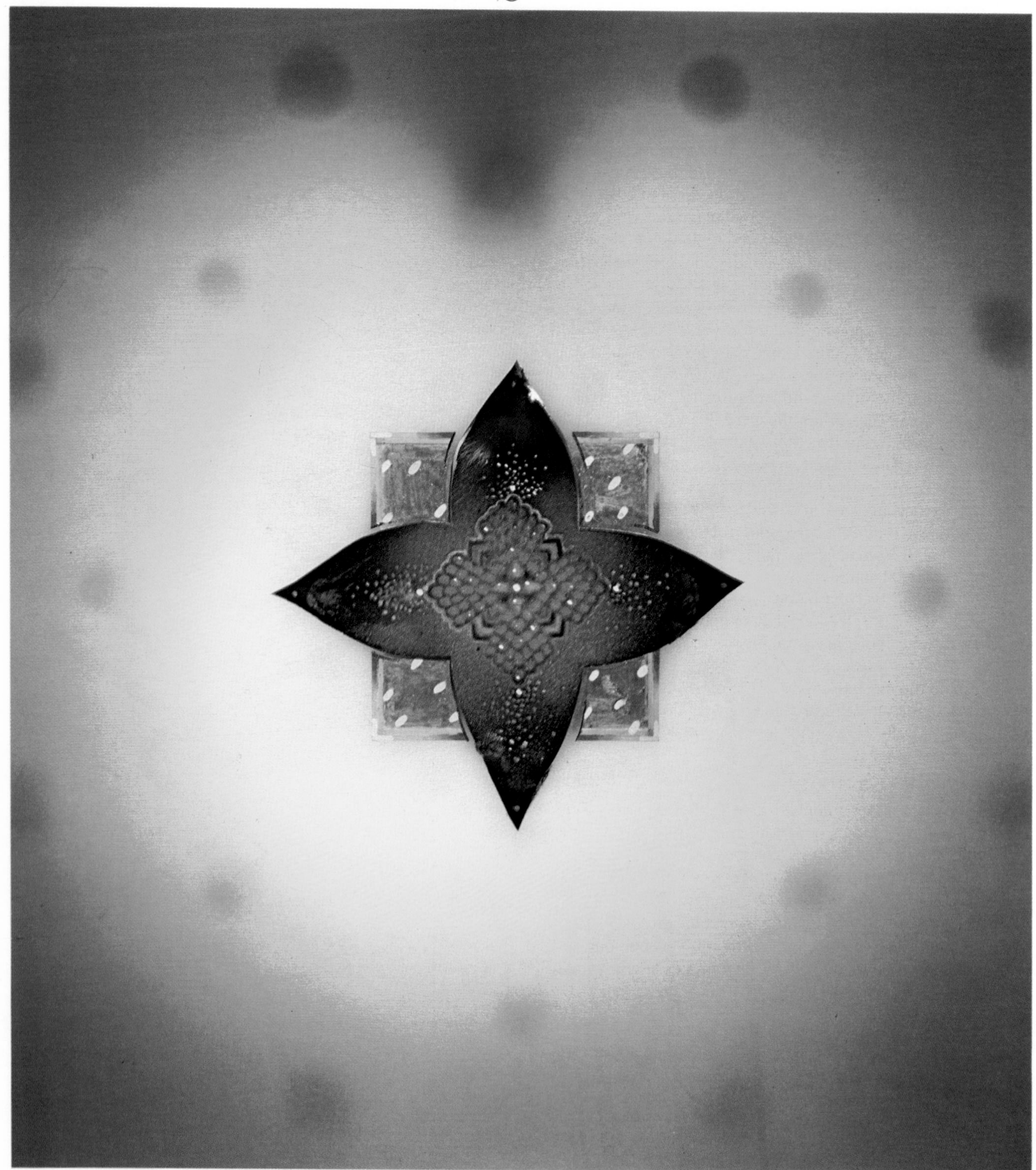

All biological processes dissipate energy. They convert free energy—energy available to do mechanical, chemical, or biological work—into heat. Living systems are therefore dependent upon the steady supply, transfer, and storage of free energy.

Today, the widespread occurrence of life on earth, forming an almost continuous biosphere, is made possible only because certain organisms have the ability to use photons of solar energy for the synthesis of reduced molecules containing high-energy bonds that can store chemical energy. This conversion of radiant energy to chemical energy occurs through the process called *photosynthesis*; it occurs in all green plants, in the blue-green algae, and in some bacteria. The photosynthetic process is important not only as a source of energy for the biosphere but as the source of essentially all organic molecules in the biosphere. As a vital by-product, photosynthesis produces the oxygen molecules of the atmosphere. Clearly, this process is of primary importance to the existence of life on earth, and biologists have devoted a great deal of attention to its study for the past two centuries (Interleaf 5.1).

THE PHOTOSYNTHETIC MECHANISM

In the photosynthetic process, plants use light energy to convert two extremely stable, low-energy molecules (carbon dioxide and water) into an unstable, energy-rich system consisting of organic matter and free oxygen. This energy-rich system fuels almost all other life processes. Most of the organic molecules created through photosynthesis are sugars, which plants can polymerize to form polysaccharides such as starch and cellulose. The simple sugar *glucose* (Figure 5.2) can be taken as typical of the organic molecules produced by photosynthesis.

The photosynthetic formation of glucose can be represented by the following chemical formula:

$$6\,CO_2 + 6\,H_2O \rightarrow \{CH_2O\}_6 + 6\,O_2 \quad \Delta G = +686\ \text{kcal/mole} \quad (5.1)$$

A generalized formula for photosynthesis of a single carbon unit can be written as follows:

$$CO_2 + H_2O \rightarrow \{CH_2O\} + O_2 \quad \Delta G = +114\ \text{kcal/mole} \quad (5.2)$$

This reaction is strongly endergonic. In the absence of catalysts, the reverse reaction (oxidation of carbohydrate to form carbon dioxide and water, with release of thermal energy) occurs only at high temperatures, as in the burning of sugar, wood, or paper. The plant is able to carry out this process at room temperature through an extremely complex system of catalyzed reactions whose sum is the overall reaction of Formula 5.2.

In photosynthesis, energy is supplied by light quanta (photons) whose energy content depends upon the frequency or wavelength of the light. In red light, each mole of photons provides about 40 kcal of energy. To provide the 114 kcal needed to combine a mole of CO_2 with a mole of H_2O, *at least* 3 moles of photons are needed. In other words, the reduction of a single atom of carbon (in a molecule of CO_2) by a single molecule of H_2O requires energy from at least 3 photons. There is an insignificantly small probability that 3 photons would happen to strike just as the 2 molecules

Figure 5.1. Energy cycle of the biosphere. Photosynthetic organisms utilize light energy from the sun to synthesize large carbohydrate molecules from simpler compounds. Oxygen is produced as a by-product. Respirating organisms degrade the large carbohydrate molecules synthesized by plants, utilizing the energy obtained to sustain life functions and produce water and carbon dioxide as waste products. Because energy is constantly being lost as heat, the cycle requires constant input of light energy for its continued functioning.

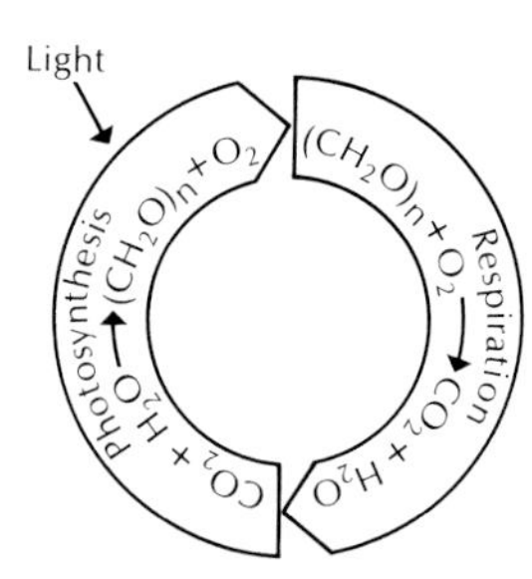

THE "DISCOVERY" OF PHOTOSYNTHESIS

It was recognized long ago that plants provide the ultimate source of food for nearly all organisms on earth. Until the seventeenth century, however, it was assumed that plants create their tissues from materials extracted out of the soil. About 1648 a Flemish alchemist, Jan van Helmont, conducted an experiment to test this theory. He grew a tree in a tub of soil, adding nothing but water for five years. In that time, the tree gained 164 pounds, but the soil weighed only 2 ounces less than it had at the beginning of the experiment. Helmont considered these results as proof of his belief that water is the basic substance of which plants (indeed, all matter, according to Helmont) are made. Some 80 years later, the English naturalist Stephan Hales published the results of his studies of plants, from which he concluded that air is also an important source of nutrients for plant growth.

In the late eighteenth century—sometimes called the age of pneumochemistry (from the Greek word for breath)—chemists were busy identifying and studying the various kinds of air (now called gases). In 1771 the most noted of the English pneumochemists, the Reverend Joseph Priestley, showed that plants "improve air" that has been "damaged" by the breathing of animals or the burning of candles. Some eight years later Jan Ingenhousz, a Dutch physician at the Imperial Court of Vienna spending a summer of research in England, found that this improvement is caused by a rapid chemical reaction induced by light in the green-colored tissues (leaves and stalks) of plants. The improvement of the air was described as "dephlogistication" until Antoine Lavoisier showed that this process could better be described as "enrichment in oxygen."

In 1782 a Swiss pastor, Jean Senebier, discovered that the reaction now called photosynthesis occurs only in the presence of "fixed air" (now called carbon dioxide); in 1804 another learned citizen of Geneva, Théodore de Saussure, showed that water also is necessary for the reaction to take place. In 1796 Ingenhousz proclaimed that the process of photosynthesis is the main if not the only source of all organic matter in plants and thus, indirectly, also in animals feeding on plants. The early guesses of Helmont and Hales proved to be correct; plants create their tissues from carbon dioxide and water through the process of photosynthesis. In this process, oxygen is released into the atmosphere.

Some 50 years after the discovery of the chemical nature of photosynthesis, Julius Robert Mayer (a doctor of medicine famed as codiscoverer of the first law of thermodynamics) first recognized that photosynthesis converts light energy into chemical energy. Thus, by 1845 it was recognized that photosynthesizing plants represent the biggest chemical factory on the surface of the earth, as well as the ultimate power station of the biosphere.

Jan van Helmont

Joseph Priestly

Julius Robert Mayer

Figure 5.2 (above). The structural formula of glucose.

Figure 5.3 (below). Experimental evidence proving that the origin of liberated oxygen in photosynthesis is from the photolysis of water molecules. Isotope tracer experiments using oxygen-18 are shown above. Red type represents radioactive oxygen.

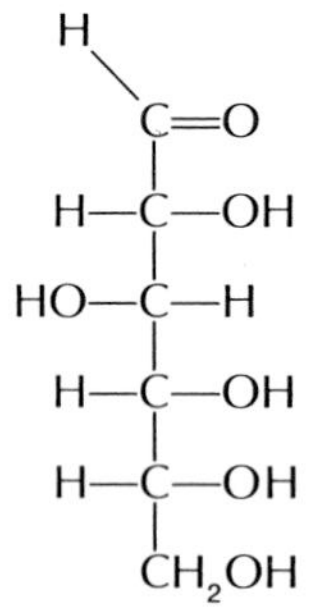

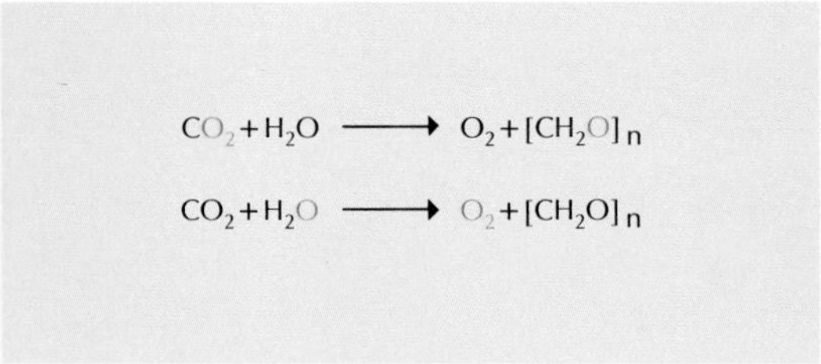

H
C=O
H—C—OH
HO—C—H
H—C—OH
H—C—OH
CH_2OH

$CO_2 + H_2O \longrightarrow O_2 + [CH_2O]n$

$CO_2 + H_2O \longrightarrow O_2 + [CH_2O]n$

collided; the organism must possess some mechanism for gathering energy from photons and providing it at the proper time and place to convert the simple molecules to sugar.

Precise measurements indicate that, under the most favorable conditions (dim light and sufficient—but not too much—carbon dioxide), *8* photons are needed by the plant cell to consume one molecule of CO_2 and to liberate one molecule of O_2. Three photons contain enough energy to carry out the reaction; the plant uses 8 photons. Therefore, the plant is able to use the light energy with an efficiency of about 35 percent (3/8), a much higher efficiency than has been achieved in the laboratory using artificial photosynthetic systems with visible light.

Originally, it was assumed that the photosynthetic process involves the removal of oxygen from carbon dioxide, followed by hydration of the remaining carbon:

$$\text{first,} \quad CO_2 \rightarrow C + O_2 \tag{5.3}$$

$$\text{then,} \ C + H_2O \rightarrow \{CH_2O\} \tag{5.4}$$

According to this mechanism, the oxygen molecules are formed from oxygen atoms in the CO_2 molecules. It is possible to test this hypothesis by using isotopes of oxygen—CO_2 can be prepared with oxygen-18 rather than normal oxygen-16. When this heavy CO_2 is used in photosynthesis, the oxygen gas that is liberated is not made up of oxygen-18. However, if H_2O prepared with oxygen-18 is used in photosynthesis, the oxygen gas liberated is entirely made up of oxygen-18. It appears that the oxygen gas is formed from oxygen atoms in the H_2O. The mechanism of photosynthesis therefore must involve the *photolysis*, or splitting of water, followed by the reduction, or hydrogenation, of CO_2:

$$\text{first,} \quad 2\,H_2O \rightarrow O_2 + 4\,H \tag{5.5}$$

$$\text{then,} \ 4\,H + CO_2 \rightarrow \{CH_4O_2\} \rightarrow \{CH_2O\} + H_2O \tag{5.6}$$

The second part of this mechanism involves the formation of an unstable intermediate (CH_4O_2), which readily loses a water molecule to form the carbohydrate unit (CH_2O). The water thus formed is not recycled back to the first reaction but must be liberated as a waste product or used elsewhere in the plant. If this recycling did not occur, the use of heavy oxygen in H_2O ought to result in only part of the liberated oxygen being heavy.

This model of the photosynthetic mechanism indicates that *4* hydrogen atoms must be transferred from water molecules to a single molecule of CO_2 to carry out photosynthesis. Because 8 photons are needed to carry out the process, it appears that 2 photons are needed for the transfer of each hydrogen atom. Not all this energy is stored in the final product molecules; some is lost at each step in the process.

The modern tendency is to describe photosynthesis as an oxidation-reduction reaction in which 4 hydrogen atoms from water are transferred to CO_2. Energetically, this transfer is an "uphill" process.

STAGES OF PHOTOSYNTHESIS

In the presence of adequate amounts of CO_2 and H_2O, it would be expected that the rate of photosynthesis—measured by the rate of production of

Figure 5.4 (above). Light saturation in photosynthesis. The photosynthetic rate is proportional to light intensity, but only up to a saturation point at which the rate remains constant regardless of light intensity.

Figure 5.5 (below). The light and dark stage reactions of photosynthesis. At low light intensity, the overall reaction rate is limited by the light-dependent production of intermediate compounds. At precisely the saturation point of light intensity, the light-dependent stage is producing intermediate compounds at the maximum pace at which the dark stage enzymatic reactions can complete the reaction series. If more light is supplied, there is no change in the overall rate because the total reaction series cannot proceed faster than the dark stage rate.

oxygen—would be proportional to the intensity of light available (the rate at which photons are being made available to carry out the reaction). Experiments show that this proportionality is valid only up to a certain point, beyond which the rate of photosynthesis remains constant regardless of further increases in light intensity (Figure 5.4). F. F. Blackman, an English plant physiologist, suggested in 1905 that this "light saturation" occurs because photosynthesis is a two-stage process.

In the *light-limited* (or *photochemical*) *stage*, light energy is used to produce high-energy (chemically reactive) molecules; the rate of this process is nearly proportional to the light intensity. The light-limited stage is the energy storage phase, which includes the photochemical process. In the *dark* (or *enzymatic*) *stage*, the unstable intermediate molecules are converted into stable final products (oxygen and sugar) through reactions not involving light energy. The dark reactions to reduce CO_2 to sugar are catalyzed, as are all metabolic reactions, by enzymes. Cofactors also play important roles in the dark reactions.

Because enzyme molecules are very large, only a fairly small number of them can be present in the cell. Moreover, each enzyme molecule can transform only a limited number of substrate molecules each second. The enzymatic stage therefore has a maximum rate imposed by the maximum rate at which the enzyme can process substrate molecules. Many reactions and many enzymes are involved in the enzymatic stage, so that the slowest enzyme determines the ceiling rate of the entire stage. The maximum rate

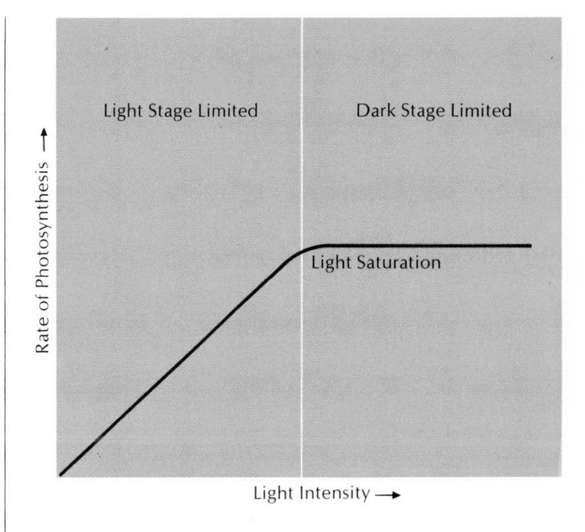

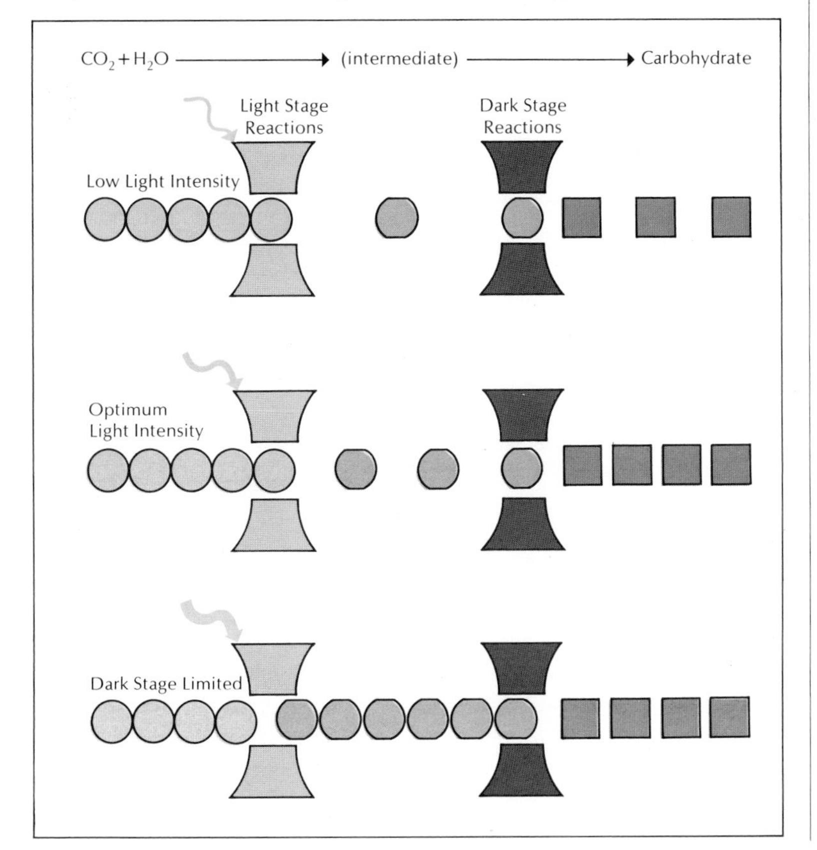

Figure 5.6. The two-step relationships of the light reactions of photosynthesis.

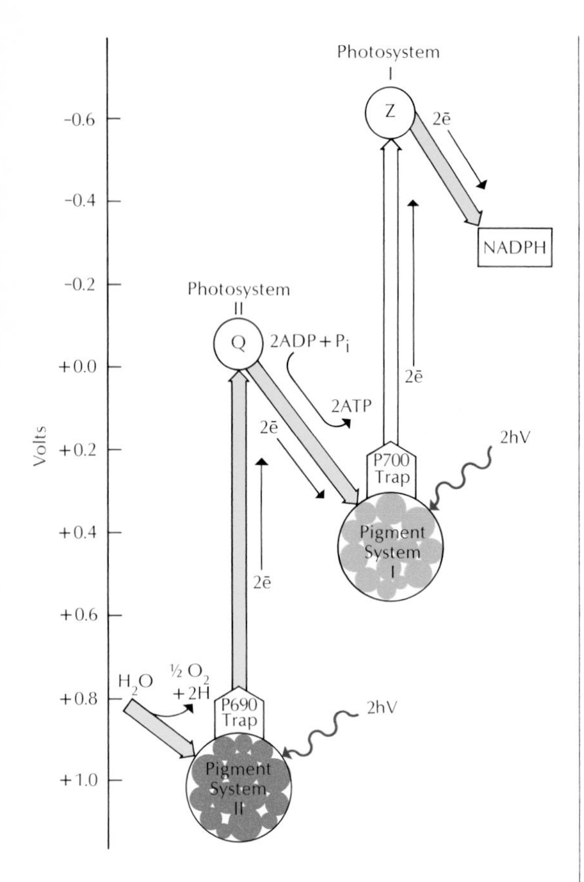

of this stage is achieved when all of the molecules of an enzyme involved in the process are loaded with substrate molecules and are engaged in their transformation.

Once the enzymatic stage has reached its maximum rate, further increases in the rate of the photochemical stage will simply cause an accumulation of the energy-rich intermediate molecules. These intermediates are unstable and spontaneously return to their original forms if they are not used immediately for conversion into stable end products. The instability of the intermediate products is confirmed by the fact that production of oxygen ceases almost instantly when the light supply is cut off.

The Light-Limited Stage

Because eight photons and four hydrogen atoms are needed to process each molecule of CO_2, it appears that each hydrogen atom is pushed "uphill" energetically in two steps, each involving a push by one photon of energy. This energetic pushing, and the trapping of the energized hydrogen, is the heart of the light-limited stage. However, many details of the two-step process in this stage remain obscure. In particular, its precise physical nature is not understood, nor is it known how the two steps are coordinated so effectively that the overall photosynthetic process can proceed with very little loss of absorbed energy quanta. Furthermore, the details of the enzymatic splitting of water to provide the hydrogen that is pushed remain unclear. In spite of the gaps, an outline of the process can be given from what is known.

Each of the two energy pushes involves a separate set of reactions. These sets are known as photochemical reaction II (PCII) and photochemical reaction I (PCI). (The Roman numerals were assigned in order of the detailed study of the steps, not in order of their occurrence in the photosynthetic process.) Hydrogen atoms are removed from water (leaving the oxygen in some unstable intermediate form) and pushed up the energy hill in PCII (Figure 5.6). The hydrogen atoms then travel down a series of redox reactions in which part of the energy is captured for the photosynthetic phosphorylation of ADP to ATP (Chapter 3). The hydrogen atoms are then given another push to an even higher energy level by PCI. Finally, the hydrogen atoms pass into another cofactor sequence, which results in the reduction of the cofactor NADP (nicotinamide adenine dinucleotide phosphate) to its reduced state, NADPH.

The net result of the light-limited stage, then, is the transfer of hydrogen atoms from water molecules to the high-energy compound NADPH, with the production of ATP molecules. Oxygen is released as a by-product of this activity. In the dark stage, NADPH acts as a reductant for the conversion of CO_2 to carbohydrate, and ATP provides chemical energy for this conversion. The nature of the intermediate molecule from which the hydrogen atoms are removed at the beginning of the light-limited stage is not yet completely known.

The Dark Stage

Several sequences of enzymatic or dark reactions are involved in the photosynthetic process. The cofactor sequence occurring between PCII and PCI and that occurring after PCI have been discussed as part of the light-limited stage. Other enzymatic reactions are involved in the formation of an intermediate molecule from the water molecule and the conversion of the intermediate to O_2 after the removal of the hydrogen atoms at the beginning of PCII. The heart of the dark stage, and by far the best understood part of the photosynthesis, is the sequence of enzymatic reactions by which CO_2 is reduced to carbohydrate.

THE REDUCTION OF CO_2

The dark reactions involved in the reduction of CO_2 utilize the energy and reducing power trapped in the light-limited stage. Three processes are involved in this sequence of reactions: (1) *incorporation* of CO_2 into a carbon dioxide acceptor; (2) *reduction* of this complex using the ATP and NADPH produced in the light-limited stage; and (3) *transformation* of the reduced complex (which is probably a molecule containing three carbon atoms) into a six-carbon sugar such as glucose or fructose.

1. Carbon-dioxide incorporation. If the product of CO_2 incorporation is a three-carbon compound, the acceptor would be expected to be a two-carbon compound. Yet no suitable compound of this type has been identified in photosynthetic systems; it appears that two-carbon compounds are chemically too active to be tolerated by cells in substantial quantities. This puzzle was solved by Melvin Calvin and A. A. Benson, who discovered that a five-carbon sugar (a pentose, $C_5H_{10}O_5$) can bind CO_2 and then split into two three-carbon molecules in two closely associated enzymatic steps.

The five-carbon sugar involved is ribulose, and the actual CO_2 acceptor is the diphosphate of this sugar, ribulose diphosphate (RuDP). This molecule combines with CO_2 to form an unstable six-carbon compound ($\{C_6\}$), which splits through hydrolysis into two molecules of phosphoglyceric acid (PGA). The reactions may be abbreviated as follows (complete structural formulae are shown in Figure 5.7):

$$\text{RuDP} + CO_2 \rightarrow \{C_6\} \tag{5.7}$$

$$\{C_6\} + H_2O \rightarrow 2\ \text{PGA} \tag{5.8}$$

2. Reduction of the intermediate. The PGA is next reduced to a triose (three-carbon sugar) by the high-energy products resulting from the light-limited stage. In this reduction, the carbon atoms gain the appropriate number of hydrogen atoms and sufficient energy to bring them to the sugar level ($C_3H_6O_3$).

In this process, PGA first gains a high-energy phosphate group to become the much more energetic molecule diphosphoglyceric acid (DPGA).

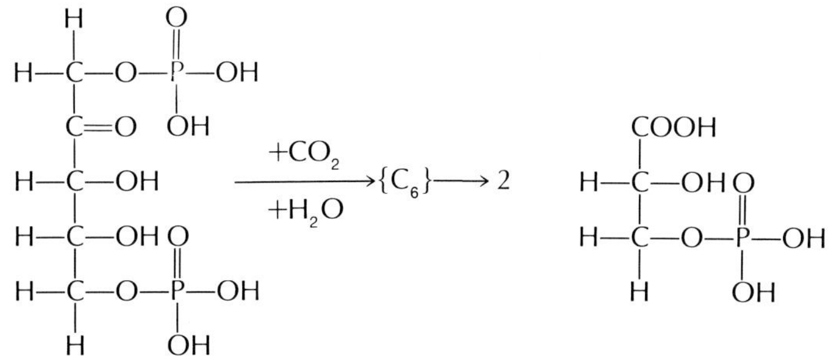

```
   H      O
   |      ||
H—C—O—P—OH
   |      |
   C=O   OH                                          COOH
   |           +CO2                                   |
H—C—OH      ————————→{C6}———→ 2          H—C—OH O
   |           +H2O                                   |       ||
H—C—OH O                                    H—C—O—P—OH
   |      ||                                          |       |
H—C—O—P—OH                                    H      OH
   |      |
   H     OH
```

Ribulose Diphosphate (RuDP) — 3-Phosphoglyceric Acid (PGA)

Figure 5.7. The structural formulae of ribulose diphosphate (RuDP) and 3-phosphoglyceric acid (PGA).

Figure 5.8 (above). The structural formulae of the reduction of the intermediate.

Figure 5.9 (below). An experiment illustrating the coupling of the light and dark reactions in photosynthesis.

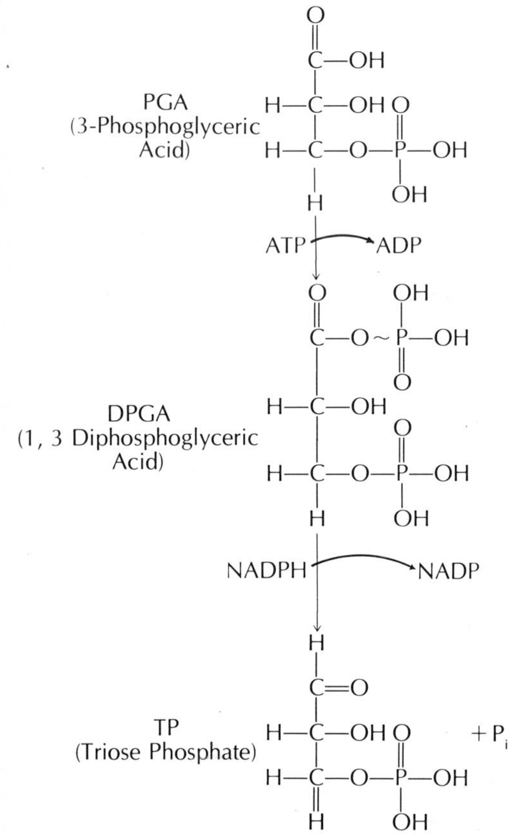

PGA (3-Phosphoglyceric Acid)

ATP → ADP

DPGA (1, 3 Diphosphoglyceric Acid)

NADPH → NADP

TP (Triose Phosphate) $+ P_i$

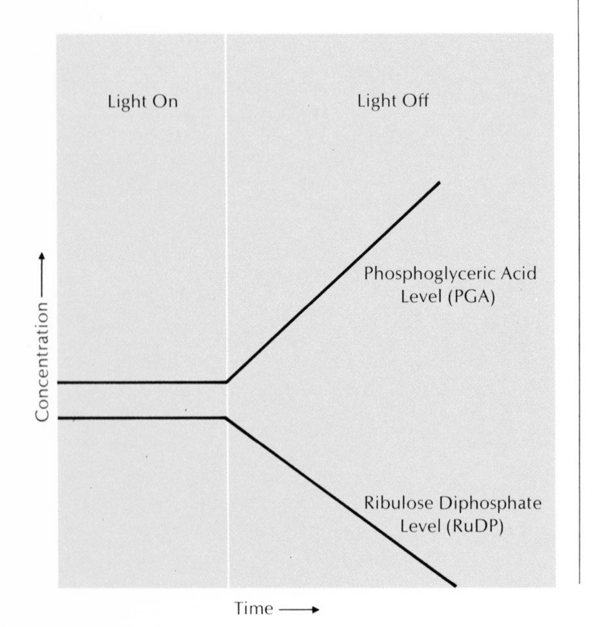

DPGA is then reduced to triose phosphate (TP) as NADPH is oxidized back to NADP (see Figure 5.8 for complete structural formulae):

$$\text{PGA} + \text{ATP} \rightarrow \text{DPGA} + \text{ADP} \quad (5.9)$$

$$\text{DPGA} + \text{NADPH} \rightarrow \text{TP} + \text{NADP} + P_i \quad (5.10)$$

Because both of the energy-rich compounds used in this step (ATP and NADPH) are produced in the light-limited stage and because ADP, P_i, and NADP are returned to the light-limited stage, the two stages of photosynthesis are intimately coupled molecularly and energetically.

If the mechanism proposed for these two steps is valid, turning off the light that falls on an actively photosynthesizing cell should lead to an immediate increase in concentration of PGA and decrease in concentration of RuDP. These changes should occur because reactions 5.7 and 5.8 can continue in the dark, whereas reactions 5.9 and 5.10 are dependent upon a supply of ATP and NADPH produced in the light-limited stage. As soon as the light is turned off, the supply of energy-rich compounds will be rapidly used up, and the transformation of PGA to TP will slow down, while production of PGA from RuDP continues. This expectation was confirmed experimentally (Figure 5.9).

3. *Transformation of the carbohydrate.* Finally, a sequence of enzymatic reactions combines two triose molecules to form a single six-carbon sugar (hexose). First, the TP is partially converted to another three-carbon sugar phosphate, dihydroxyacetone phosphate (DHAP). An equilibrium is reached with about 60 percent TP and 40 percent DHAP. Next, one molecule of TP and one molecule of DHAP combine to form a molecule of fructose diphosphate (FDP), which loses a phosphate group to form fructose monophosphate (FMP). The FMP is partially converted to glucose monophosphate (GMP), which loses a phosphate group to form glucose. To form 1 glucose from 6 CO_2 molecules, a total of 18 ATP and 12 NADPH will be required.

$$\text{TP} \rightleftarrows \text{DHAP} \quad (5.11)$$

$$\text{TP} + \text{DHAP} \rightarrow \text{FDP} \quad (5.12)$$

$$\text{FDP} \rightarrow \text{FMP} + P_i \quad (5.13)$$

$$\text{FMP} \rightleftarrows \text{GMP} \quad (5.14)$$

$$\text{GMP} \rightarrow \text{glucose} + P_i \quad (5.15)$$

In fact, only 1/6 of the triose phosphate is converted to glucose. The other 5/6 of the TP goes through another complex sequence of reactions to produce ribulose diphosphate (RuDP), which can then act as the CO_2 acceptor to keep the cycle going. The net result of the sequence is the conversion of 10 three-carbon sugar molecules into 6 five-carbon sugars. A major role in this sequence is played by the cofactor *thiamine*, which aids in moving two-carbon units from one sugar to another. Thiamine is required in the human diet and is also known as vitamin B_1.

The complete cycle of carbon compounds involved in the dark stage includes some 13 reaction steps that are catalyzed by at least 11 separate enzymes. Melvin Calvin received the Nobel Prize in 1961 for his work in

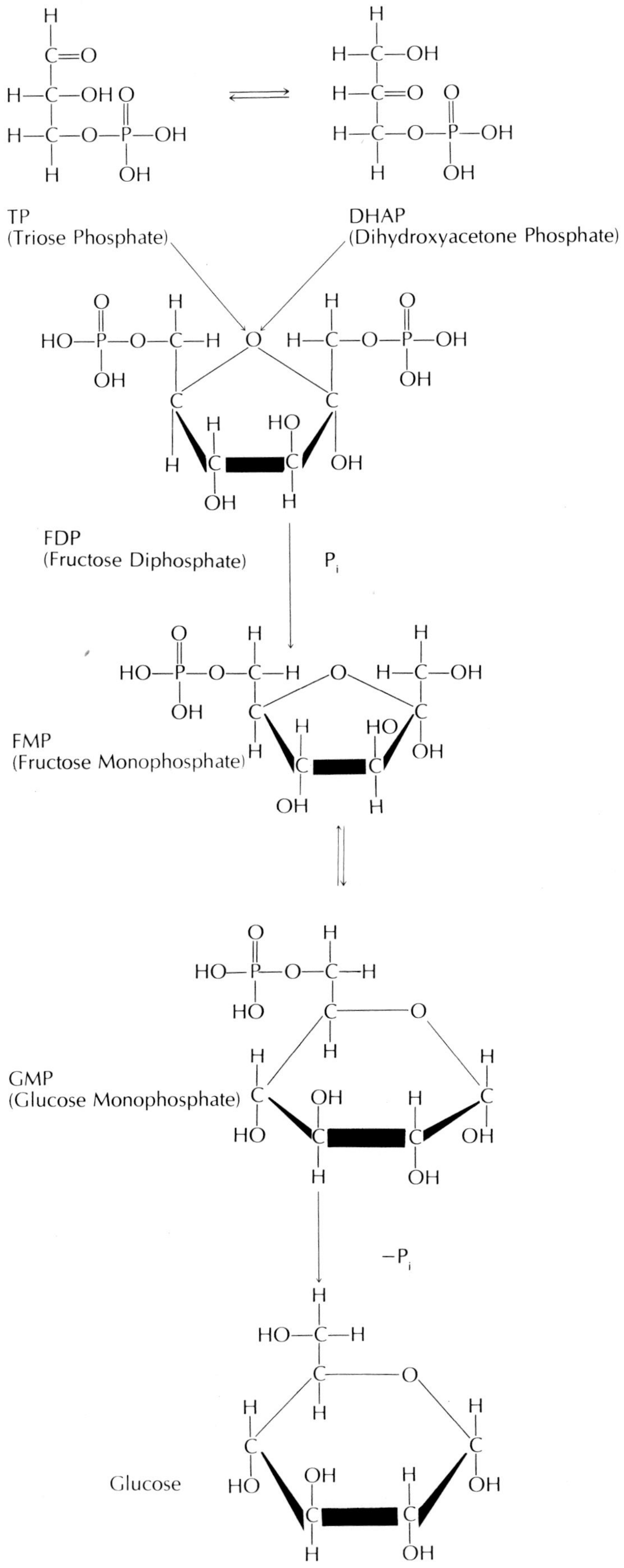

TP
(Triose Phosphate)

DHAP
(Dihydroxyacetone Phosphate)

FDP
(Fructose Diphosphate)

P_i

FMP
(Fructose Monophosphate)

GMP
(Glucose Monophosphate)

$-P_i$

Glucose

Figure 5.10. Structural formulae for the transformation of the carbohydrate.

Figure 5.11 (above). The photosynthetic formation of glucose from CO_2 via the Calvin cycle. Inputs are shaded in brown; products are in yellow. Clockwise from the left, the abbreviations designate the following terms: PGA = 3-phosphoglyceric acid; TP = glyceraldehyde 3-phosphate; DHAP = dihydroxyacetone phosphate; FDP = fructose 1,6-diphosphate; FMP = fructose 6-phosphate; GMP = glucose 6-phosphate; X5P = xylulose 5-phosphate; E4P = erythrose 4-phosphate; SDP = sedoheptulose 1,7-diphosphate; S7P = sedoheptulose 7-phosphate; R5P = ribose 5-phosphate; Ru5P = ribulose 5-phosphate; RuDP = ribulose 1,5-diphosphate.

Figure 5.12 (below). Absorption spectra of three types of photosynthetic organisms, showing the regions of the spectrum that can be used for photosynthesis. Regions of intense absorption correspond to the absorption maxima of the respective forms of chlorophyll contained by the organisms.

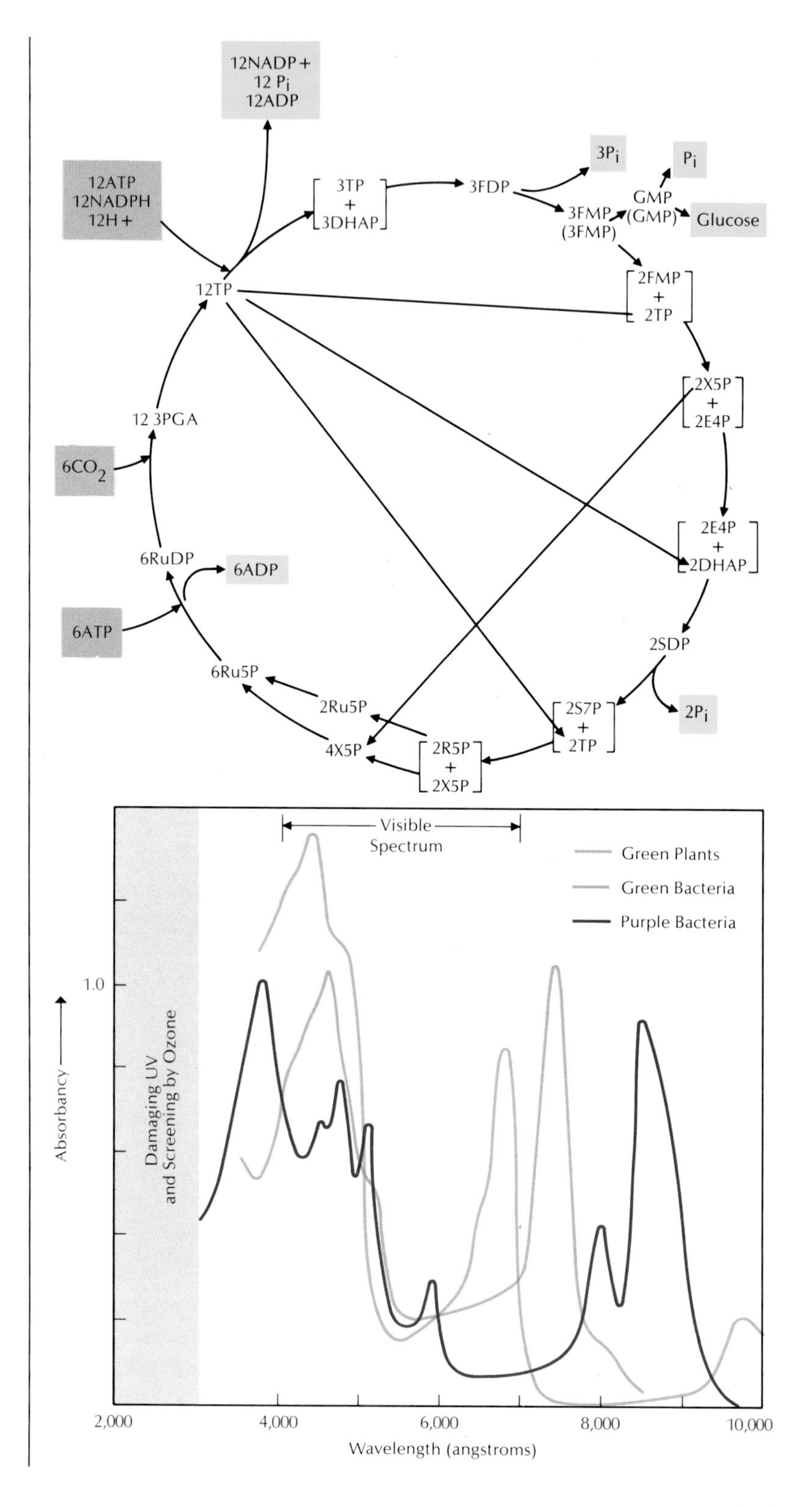

clarifying the details of this cycle, which is now known as the *Calvin cycle* (Figure 5.11). Each trip around this cycle consumes 1 molecule of CO_2 and 4 hydrogen atoms and produces 1/6 of a glucose molecule.

THE LIBERATION OF OXYGEN

The details of the enzymatic reaction sequence leading from H_2O to O_2 are all but unknown. In the steps of this sequence, hydrogen atoms must be removed from the water molecule and sent into the photochemical reaction PCII. To liberate 1 oxygen molecule, 4 hydrogen atoms must be removed. This process cannot happen in a single step; rather, some intermediates must be involved, but nothing is known about their nature. It is clear, however, that these reactions are enzymatic rather than photochemical. Interestingly, the reverse process in respiration, where molecular oxygen is taken into the reaction, is also poorly understood.

Energy-Trapping Systems

The reactions of the dark stage are not very different from enzymatic reactions involved in other processes of cellular metabolism. The photochemical reactions of the light-limited stage, however, are unlike any other reactions occurring in living systems. A better understanding of these photochemical reactions is therefore of great interest to biochemists and biophysicists.

It appears from various experiments that the two photochemical reaction steps, PCII and PCI, occur in two separate reaction centers and are supplied with energy from two separate photon-trapping systems. These two energy-trapping systems absorb light of slightly different frequencies. A biologist can thus study the two reaction steps separately because light of a frequency is absorbed solely or chiefly by one of the systems. For example, in green plants and algae, far-red light (wavelengths greater than 700 millimicrons) seems to be absorbed preferentially in system I, whereas light at about 650 millimicrons is absorbed preferentially in system II. In red algae, the situation is even more striking, for green light is absorbed preferentially in system II, whereas red light is absorbed preferentially in system I.

In recent years, many studies have been made of the properties of the two systems, the reactions that each carries out, and the cooperation between the two systems. It even has been possible to separate, at least partially, the proteins that make up the two systems.

THE PHOTOSYNTHETIC UNIT

The reactions of the photochemical phase are made possible by *pigments*, unusual catalysts that are able to absorb photons. Because these molecules absorb light of particular frequencies, the color of the visible light passing through the photosynthetic cell is altered.

The involvement of a photon in a chemical reaction is a two-step process. First, the photon is absorbed by a pigment molecule; the absorbed energy raises one or more of the electrons in the pigment molecule to an energy-rich, excited state. In the excited or oxidized state, the pigment molecule becomes a catalyst, but unlike other catalysts it is able to contribute to the catalyzed reaction all or part of its stored energy. It can therefore catalyze endergonic, or uphill, reactions such as the transfer of hydrogen atoms in photosynthesis, something no ordinary catalyst can do. Thus, a chlorophyll molecule, excited by absorption of red light, can contribute

Figure 5.13 (above). Computed percentage of absorption of the different wavelengths of visible light by the chlorophyll, carotenoid, and phycocyanin pigments of Chroococcus, a blue-green alga. In the red region, absorption is due mainly to chlorophyll, and in the orange and yellow regions, it is due mainly to phycocyanin.

Figure 5.14 (below). Schematic portrayal of electron resonance in three-dimensional photosynthetic units.

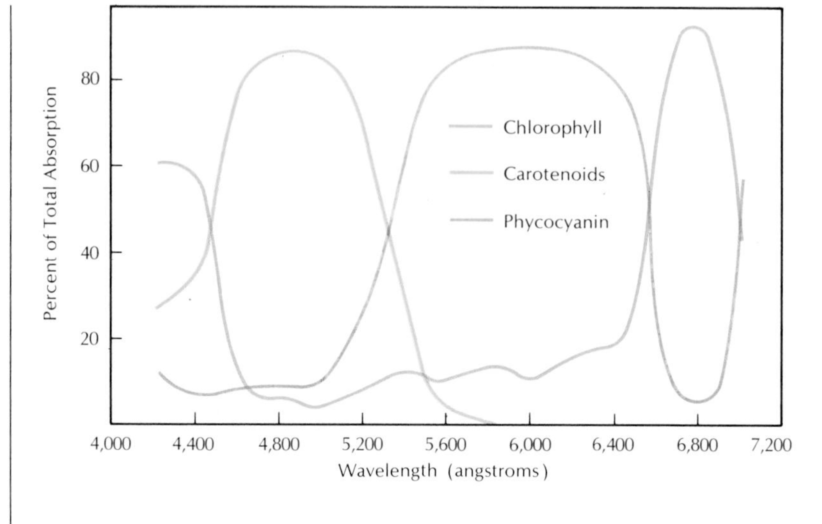

about 40 kcal/mole of energy to the reaction that it catalyzes. Its electrons then drop back to the low-energy state, and the chlorophyll molecule is ready to absorb another photon.

Not all pigment molecules present in a photosynthesizing plant cell function directly as catalysts. This function may be the responsibility of only the small proportion that are associated with certain enzymatic reaction centers. Pigment molecules are small in comparison to enzyme molecules; the molecular weight (sum of the atomic weights of all atoms in the molecule) of chlorophyll is about 1,000, whereas that of a typical enzyme is about 100,000 to 1,000,000. Thus, the cell has room for several hundred pigment molecules for each enzyme molecule.

The cell has a need for all these pigment molecules in order to capture a supply of photons. Good organic pigments, such as chlorophyll, are relatively efficient in the capture of light as compared to other molecules. Even so, a molecule of such a pigment, exposed to direct sunlight at midday, will not absorb more than about 10 photons per second. On the other hand, an ordinary enzyme molecule may transform substrate molecules at rates of tens of thousands of substrate molecules per second. Thus, the cell can use hundreds or even thousands of pigment molecules to gather enough light energy for use in a single enzymatic reaction center. The problem is how the cell gets these photons where they are needed—a process that has received a great deal of study.

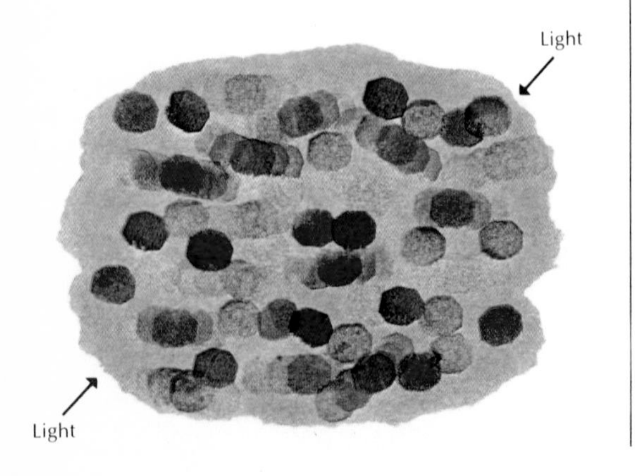

It appears that pigment molecules are organized in the living organism into *photosynthetic units*. These units are three-dimensional bodies, or perhaps two-dimensional islands, within which pigment molecules can exchange energy by a process of *electron resonance:* In effect, the excitation energy is passed from molecule to molecule. Ultimately, the energy can be passed to the single enzymatic reaction center in the unit. This photophysical step of photon capture and energy transfer can be regarded as the first step of the photochemical process.

The mechanism of this migration of excitation energy has been studied in crystals and in concentrated solutions of pigments. However, the actual transfer mechanism in the living cell is far from understood. It is not known, for example, whether there are separate units for the two primary photochemical systems, PCII and PCI. There is evidence that one type of unit contains about 300 chlorophyll molecules per enzymatic reaction center, but whether this unit serves PCII, PCI, or both has not yet been discovered.

Much of the information about the units has come from studies of photosynthesis during very brief flashes of light. If the flash is intense enough, each unit captures at least one photon and thus sets its enzymatic reaction center to work. The intensity of the flashes can be increased until the production of oxygen ceases to increase, indicating that "flash saturation" (each unit being busy) has been reached. At flash saturation, 1 oxygen molecule is produced for about each 2,400 chlorophyll molecules. Because 8 photons are needed to liberate 1 oxygen molecule, 8 units must be involved. Thus, it appears that each unit contains about 300 (2,000/8) pigment molecules.

The shortest dark interval between flashes that will permit a full oxygen yield on each flash is about 0.1 second at room temperature. Apparently, this amount of time is needed for the rate-limiting enzyme to process the supply of substrate molecules that it receives as a result of a single flash. The average time required for the rate-limiting enzyme to process a single substrate molecule is thought to be about 0.01 second.

The Pigments

Complete understanding of the photophysical and photochemical processes in photosynthesis has been delayed by the fact that each photosynthetic organism contains an assortment of pigments. Only one pigment is found in all plants; this pigment is chlorophyll *a* (Figure 5.15), which is yellowish-green in solution. It seems reasonable to suppose that this pigment alone takes part in the direct transfer of energy to the enzymatic reaction center, whereas all of the other pigments serve only as components of the photon-gathering apparatus. In fact, a large proportion of the chlorophyll *a* itself must be assigned to this accessory role, for each photosynthetic unit contains some 300 molecules of this pigment.

The function of the other pigments is unknown. Many plants have a second chlorophyll (blue-green chlorophyll *b* in green land plants and green algae, and brownish chlorophyll *c* in brown algae and diatoms). All plants have an assortment of yellow or orange pigments called carotenoids, and the red and blue-green algae also possess an assortment of orange-red and blue pigments of the phycobilin group.

It seems that all these pigments contribute photons with different efficiencies to the systems PCII and PCI, but the division between the two systems is not simple. Phycobilins are located primarily, but not exclusively, in

Figure 5.15 (above). The structural formula of chlorophyll. In chlorophyll *a*, $X = —CH_3$. In chlorophyll *b*, $X = —CHO$.

Figure 5.16 (below). Structural formula of β-carotene, a pigment occurring in plants.

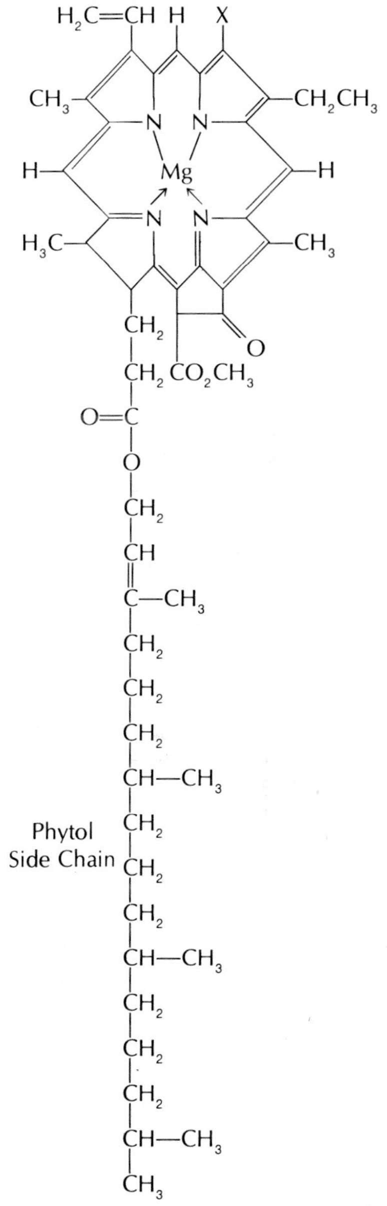

Phytol Side Chain

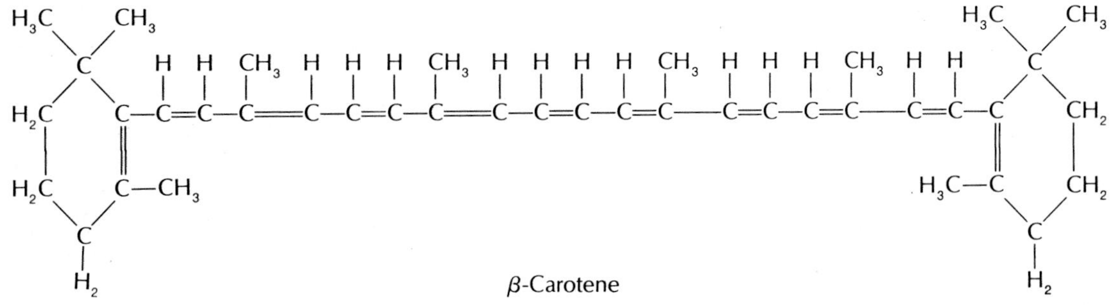

β-Carotene

Figure 5.17 (above). Absorption bands of chlorophylls *a* and *b*. Both absorb in the red region of the spectrum and hence are colored green.

Figure 5.18 (below). The absorption spectrum of chlorophyll compared with the light energy emitted from the sun. Note that the maximum energy comes through at about 4,500 A in the blue region. Chlorophyll usually absorbs about 6,500 A in the red region.

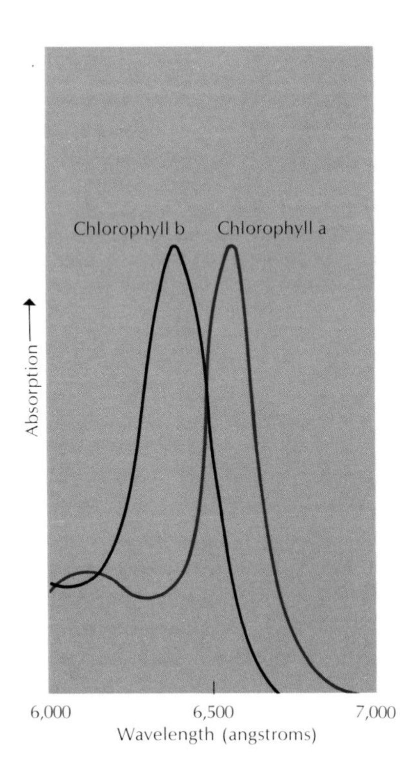

PCII, as is chlorophyll *b*. Chlorophyll *a* is located primarily, but not exclusively, in PCI in red algae, but is about equally abundant in PCII and PCI in green plants and green algae. To further complicate the situation, it appears that plants contain more than one form of each pigment; these forms differ somewhat in the frequencies of light they absorb (Figure 5.17). For example, there are at least three and perhaps more forms of chlorophyll *a* in green plants. Yet when these different forms are removed from the organism and analyzed, they prove to be chemically identical. The differences in absorption spectra must arise from differences in the ways that the molecules are arranged in the cell.

It has been suggested that the function of the accessory pigments is to gather photons of frequencies not absorbed by chlorophyll and to pass the captured energy on to the chlorophyll for use in the enzymatic reactions. Much remains to be learned, however, about the photosynthetic pigments and their exact role in the photochemical process.

The chlorophylls, carotenoids, and phycobilins appear to function primarily in capturing and transferring light energy. Other pigments that are present in very small amounts appear to act directly in the redox reactions of the photochemical process. One pigment, known as P700 because its absorption peak is at 700 millimicrons, apparently acts as the "trap" in PCI, collecting the energy gathered by the photosynthetic unit and acting as the reductant (electron donor) for the second uphill push of the hydrogen atoms to NADP. One or two molecules of P700 are present for each photosynthetic unit of PCI. P700 is apparently another form of chlorophyll *a*. It has not been isolated outside of the living system. Another pigment, P690, has been identified recently as the trap that collects energy from the photosynthetic units of PCII.

Compounds called cytochromes (iron-containing proteins) have been shown to play a role as cofactors between the two photochemical reac-

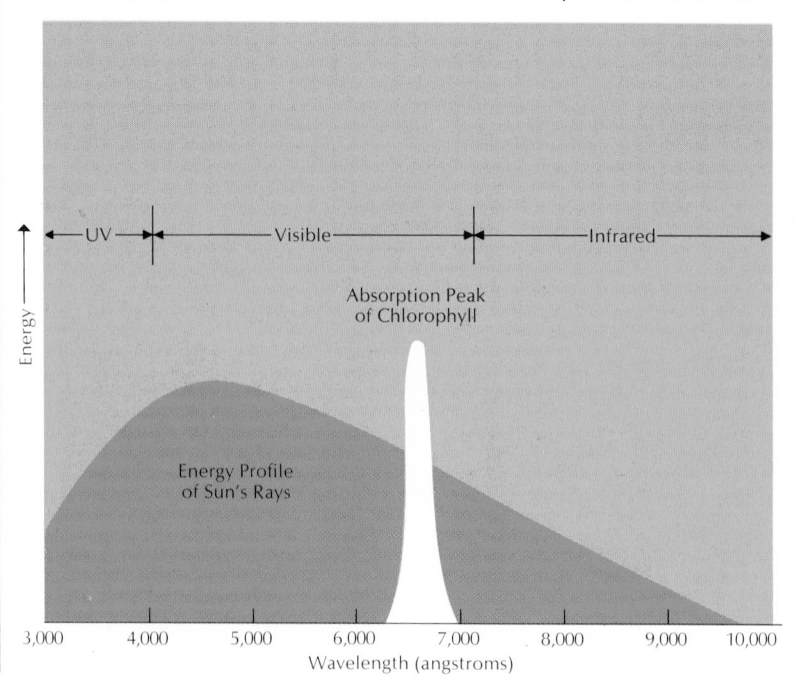

Figure 5.19 (above). Action spectrum of photosynthesis of *Chlorella*.

Figure 5.20 (below). Structural formula of a prosthetic group of a phycocyanin.

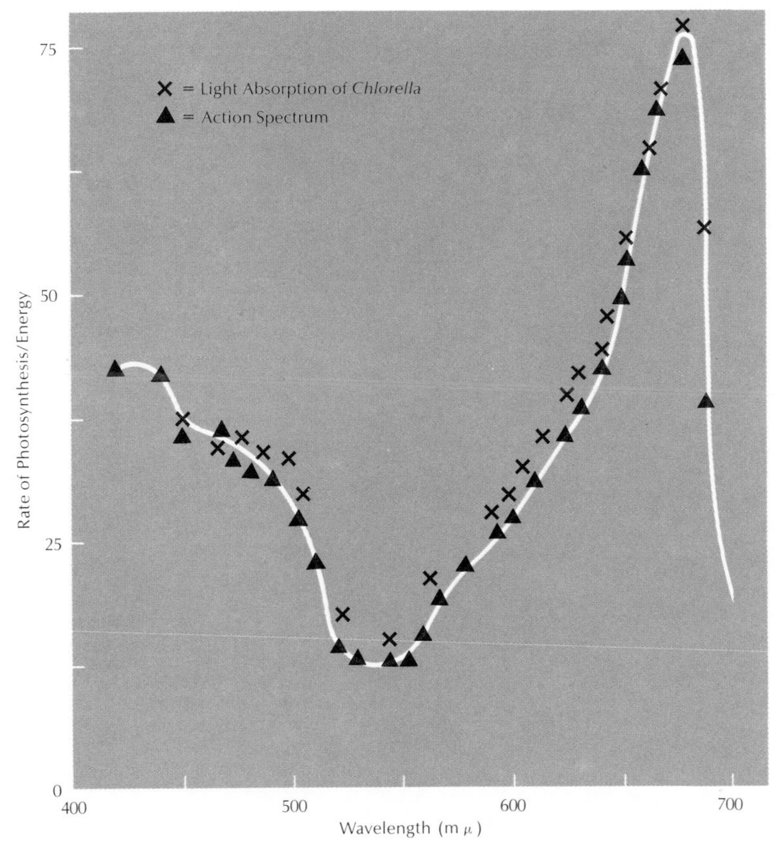

tions. Two kinds of cytochrome take part in redox reactions producing ATP in cellular respiration. At least two kinds of cytochrome, differing in their oxidation potentials, are present in photosynthetic cells. One is a form called cytochrome *b*; the other is called cytochrome *f*. It appears that cytochrome *b* may function as the oxidizing agent in PCII, accepting electrons from an as yet unknown intermediate in the $H_2O \rightarrow O_2$ sequence. Cytochrome *f* may act as an electron donor for PCI, passing electrons on to P700. Reduced cytochrome *b* and oxidized cytochrome *f* then react, probably through a sequence of intermediates, to restore each to its original

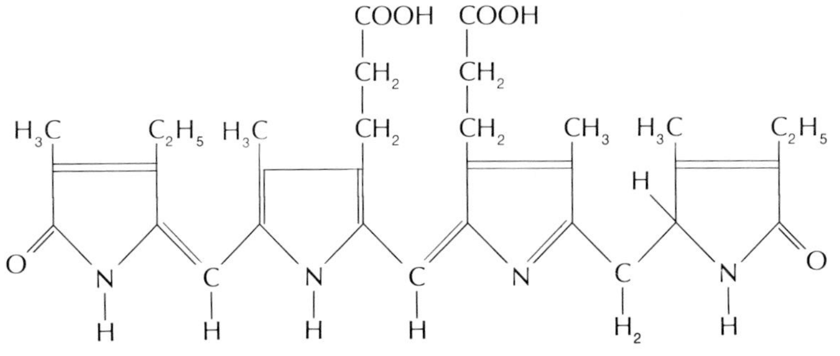

Figure 5.21. A collage of leaves, showing the abundance of pigments and forms found in nature. In addition to the green chlorophylls, there are two main accessary pigments—the carotenoids (yellows to reds) and the phycobilins (reds to blues). These pigments provide a full complement for light absorption in the visible spectrum and help to increase the efficiency of photosynthesis. Color also plays an important role in insect attraction, pollination, and seed dispersal via the fruit.

Figure 5.22. Electron micrograph of a thin section of cells of the purple sulfur bacterium *Chromatium sp.* strain D. (× 72,000)

state. This reaction releases about 9 kcal/mole of energy, which is used to synthesize ATP. Pigments called plastoquinone and plastocyanin appear to act as intermediate cofactor pairs in this reaction sequence.

VARIATIONS OF THE PHOTOSYNTHETIC REACTION

Just as a chemist may learn about the structure of a molecule by taking it apart and examining the pieces, he may learn about a reaction by blocking or altering various steps in the sequence and observing the results. Progress toward an understanding of photosynthesis was impeded for a long time because this approach seemed impossible: photosynthesis appeared to be an all-or-nothing process not subject to partial alterations. It was therefore an important landmark when, in 1937, the English plant biochemist Robert Hill discovered that the oxygen-evolving photochemical processes of photosynthesis can be made to operate without the carbon-dioxide-reducing dark reactions if a substitute oxidant such as a ferric compound is supplied instead of CO_2.

Studies of this *Hill reaction* have played an increasingly important role in recent research into the chemical mechanisms of photosynthesis. The Hill reaction indicates that PCII and the associated dark reactions that form free O_2 molecules can function by themselves without the CO_2-transforming dark reactions. The substitute oxidant can be provided before or after PCI. In other words, the Hill reaction may involve only one photochemical step or may involve both of them. A weaker oxidant is used if PCI is to be eliminated, because only the first upward push of the hydrogen atoms is to be accomplished.

Biologists have prepared samples of material from photosynthesizing cells that are apparently enriched in one or the other of the photochemical systems. These samples show different capacities for carrying out the Hill reaction. Such studies have helped to clarify parts of the mechanism involved in the photochemical steps and the intermediate dark reactions.

Another variation of the photosynthetic process occurs in pigmented

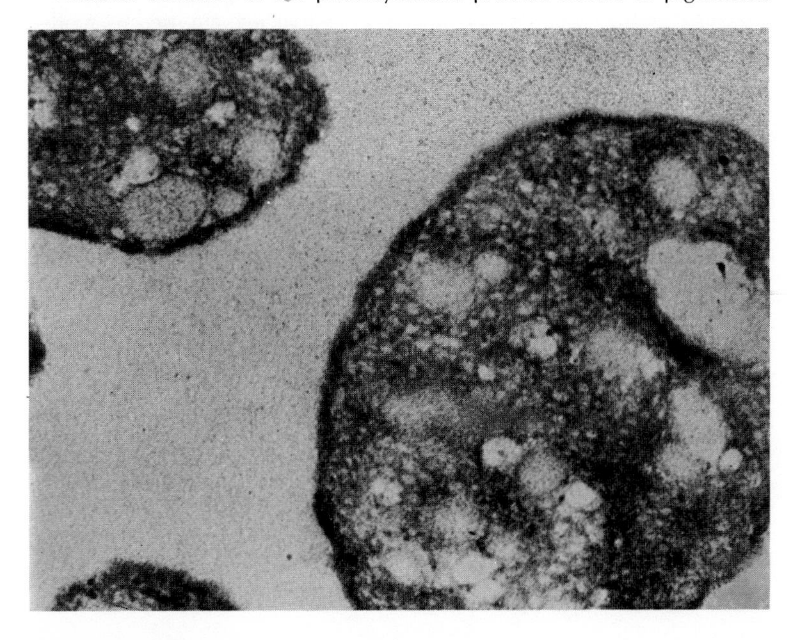

Figure 5.23. Fossil plants. These fossil ferns were found in Illinois and are approximately 300 million years old. They were a dominant plant form during the Upper Carboniferous period.

bacteria. These organisms are unable to oxidize water, but they can synthesize sugar in light if they are supplied with reductants such as hydrogen sulfide (H_2S), free hydrogen, or some organic hydrogen donors. These reductants are oxidized to produce various materials (such as sulfur in the case of H_2S), but no oxygen is liberated.

Most of the pigmented or purple bacteria contain as their main pigment bacteriochlorophyll, which is very similar to chlorophyll but absorbs light of long wavelengths at the very end of the visible spectrum and even in the near infrared. Thus, bacterial photosynthesis can proceed in infrared light. The bacteria store much less energy than do the green plants. Whether this low efficiency is due to the low-energy content of the far-red photons is not yet known. The bacterial photosynthesis also requires eight quanta for each molecule of CO_2 reduced and thus apparently also involves two photochemical steps, although the energy stored in the process is quite small.

Other bacteria utilize the process of *chemosynthesis*, in which the energy needed to reduce carbon dioxide is supplied by exothermic chemical reactions rather than by capture of photons. The hydrogen bacteria, for example, synthesize carbohydrates using energy derived from oxidizing hydrogen gas to form water.

EVOLUTION OF PHOTOSYNTHESIS

Because every living system today uses ATP as a medium of energy exchange, it seems likely that the earliest organisms also used ATP or a similar molecule. Molecules of ATP may have been present in the primeval "organic soup" and could have been used by the early organisms. However, the supply of randomly created ATP could not have supported life for long. Early in the history of life on earth, successful organisms must have been those with some mechanism that would convert chemical energy in the other organic molecules into the common coinage of ATP, which could be utilized throughout the living system.

Mechanisms such as fermentation and glycolysis, which convert sugars to waste products such as alcohols or organic acids, are used by modern bacteria that survive in reducing environments. These mechanisms are also used by other organisms for temporary energy supply when oxygen is not available. Similar though simpler mechanisms were probably developed by the early organisms.

Eventually, the supply of organic molecules suitable for these reactions must also have been depleted. The supply of new organic molecules created by random processes probably decreased over the first billion years of the earth's history, not only because early organisms consumed the molecules at an increasing rate but because the amount of ultraviolet light available at the surface decreased.

At first, most of the ultraviolet light in the solar radiation reached the earth's surface, where it could provide energy for organic syntheses. However, the light molecules of hydrogen gas must have gradually escaped from the earth's gravitational pull and been lost to space. After most of the free hydrogen gas was gone from the atmosphere, free oxygen and ozone (O_3 or $O{-}O{=}O$, an unstable and highly reactive molecule) would begin to form in the upper atmosphere as the ultraviolet light broke down molecules of water vapor. As long as free hydrogen was available, the water molecules would have been quickly re-formed by combination of hydrogen and oxygen. The layer of oxygen and ozone absorbed much of the ultraviolet light before it reached the surface, thus probably significantly decreasing the rate at which organic molecules were synthesized.

The diminution of ultraviolet radiation did have some advantages for living systems. Complex organic molecules decompose readily when struck by the high-energy photons of ultraviolet light. In fact, the existence of modern land organisms is made possible by the shielding effects of the oxygen and ozone in the atmosphere. As long as the unshielded ultraviolet radiation reached the surface, even the simplest organisms would have been forced to remain several meters below the water surface. Thus, the appearance of the oxygen-ozone layer in the upper atmosphere may have been the event that made possible the development of the first organisms.

As traces of free oxygen began to reach the lower atmosphere, the rate of synthesis of organic molecules must have decreased still further. With the supply of organic molecules in the environment dwindling, organisms apparently were forced to develop other sources of chemical energy and precursor molecules.

Some scientists have argued that chemosynthesis was the earliest mode developed by living systems for building organic molecules and for obtaining free energy from the environment, with bacterial photosynthesis developing later in evolution, to be followed by green-plant photosynthesis when more efficient pigments and enzymes had been developed. In this view, the production of oxygen would have been a late stage in the evolution of photosynthesis, coming long after the ability to synthesize carbohydrates was developed. Others feel that chemosynthesis and bacterial photosynthesis are special adaptations developed late in evolution to permit bacteria to survive in environments that could not be utilized by normal photosynthetic organisms.

It is becoming clear that many of the reactions that make up the sequences in the photosynthetic process are similar or identical to reactions that occur in other metabolic processes. Even the photosynthetic pigments may have existed in biological systems before the entire photosynthetic mechanism was developed. Thus, the evolution of the first photosynthetic organism may not have required the development of a complex mechanism from nothing but simply may have required the recombining of existing mechanisms in a new way.

No matter how the first photosynthetic organisms arose, there can be no doubt that they had a profound effect upon the earth. Carbon dioxide, which had been accumulating as a waste product of fermentation processes, now became a useful source of carbon for building new organic molecules. Oxygen, given off in increasing quantities as photosynthetic organisms became more common, became a major constituent of the atmosphere. The old reducing atmosphere was replaced by the present oxidizing one, and it was no longer possible for organic molecules to be synthesized spontaneously. From the time of the development of photosynthesis, all organisms were dependent upon the photosynthetic organisms for their supply of organic molecules and of chemical energy.

FURTHER READING

The most useful recent discussions of photosynthesis are those by Fogg (1968), Levine (1969), and Rabinowitch and Govindjee (1969). Of largely historical interest—as an indication of how ideas have changed in the past few decades—are books and articles by Arnon (1960), Bassham (1962), Bassham and Calvin (1957), Kamen (1963), Rabinowitch (1948), Rabinowitch and Govindjee (1965), and Wald (1959).

6

Respiration

Through the process of photosynthesis, plants are able to convert solar energy into chemical energy and to synthesize all the organic molecules they require from such simple inorganic compounds as carbon dioxide, water, and ammonia. The plants are the major group of *autotrophic* ("self-feeding") organisms. The photosynthetic and chemosynthetic bacteria are also autotrophs. All other organisms are *heterotrophic*—they must obtain organic molecules as foodstuffs from the environment.

The plant photosynthesizes sugar, and from this substance it can derive practically everything else it needs for life: polysaccharides—such as starch and cellulose—proteins, nucleotides, nucleic acids, coenzymes, and lipids. When an animal eats a plant, all of these organic molecules are brought to the metabolic machinery of the animal. The animal organism, however, cannot simply use these molecules to construct its own tissues. The plant substance must first be demolished, the chemical energy converted to usable form, monomers salvaged, and new macromolecules characteristic of animal systems built up. In most heterotrophs, the breakdown of foodstuffs is accomplished in an oxidation process called *respiration*, whose overall chemical equation is the reverse of the overall photosynthetic process:

$$\{CH_2O\} + O_2 \rightarrow CO_2 + H_2O \qquad \Delta G = -114 \text{ kcal/mole} \tag{6.1}$$

This chemical process is sometimes called cell respiration to avoid confusion with breathing and other physical methods by which oxygen is brought into large organisms. Plants also carry out respiration, breaking down some of the glucose they have photosynthesized.

Why convert CO_2 to carbohydrates in the first place, if they are only to be broken down again later? The carbohydrates and other macromolecules play many important roles in the structure and function of the organism. Furthermore, the carbohydrates provide a good form for long-term energy storage. The sugars are an ideal medium in which to transfer energy from one organism to another or, in polymerized forms, to store energy within the organism for long periods of time. In the process of respiration, energy is transferred from carbohydrates and other food molecules to ATP, which can be used by the cell to perform cell work. Carbon dioxide and water are liberated as waste products and are cycled back to the photosynthetic organisms to be used again as building blocks for organic macromolecules.

The *metabolism* of a cell or of an entire organism is the sum of all the chemical processes that the cell or organism is capable of performing. The metabolic pathway from which most plants and animals derive energy for cell work involves the breakdown of the six-carbon sugar glucose. Laboratory measurements of the energy released by the complete oxidation of one mole of glucose (180 grams) are made using a device known as a bomb calorimeter. As the name of the device implies, it can measure large amounts of energy release in a short period of time. Such a measurement shows that the complete oxidation of glucose to CO_2 and H_2O releases 686 kcal. The same amount of energy is released when a cell oxidizes one mole of glucose.

However, there is one important difference: the cell releases and captures the energy in small steps instead of one great explosive process. Energy that is released in small increments is far more useful than an explosive,

Figure 6.1 (left). Schematic diagram illustrating the principal modes of nutrition. Autotrophic organisms (above) utilize materials from the physical world by structuring them into organic molecules with either light energy via ATP (photosynthesis) or chemical bond energy (chemosynthesis). Heterotrophs (below) make use of preexisting organic molecules that are acquired through an engulfing or swallowing process (holotrophism), through absorption (saprotrophism), or through absorption from a living host (parasitism).

Nutrients thus acquired are chemically processed and the necessary energy extracted.

Figure 6.2 (right). A bomb calorimeter used in laboratory measurements of energy released during the oxidation of glucose.

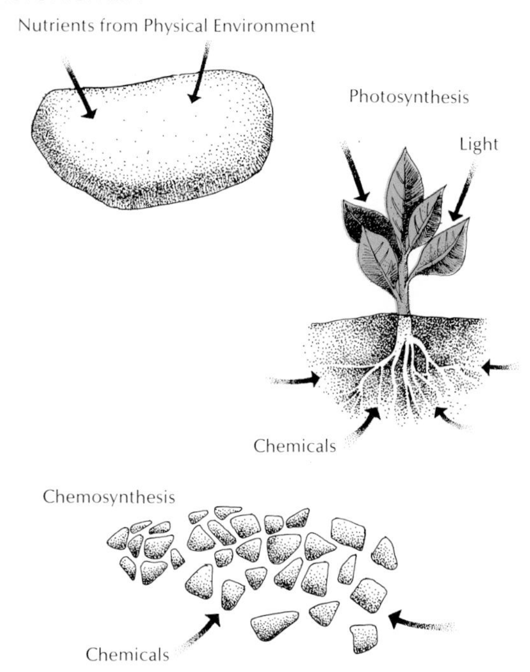

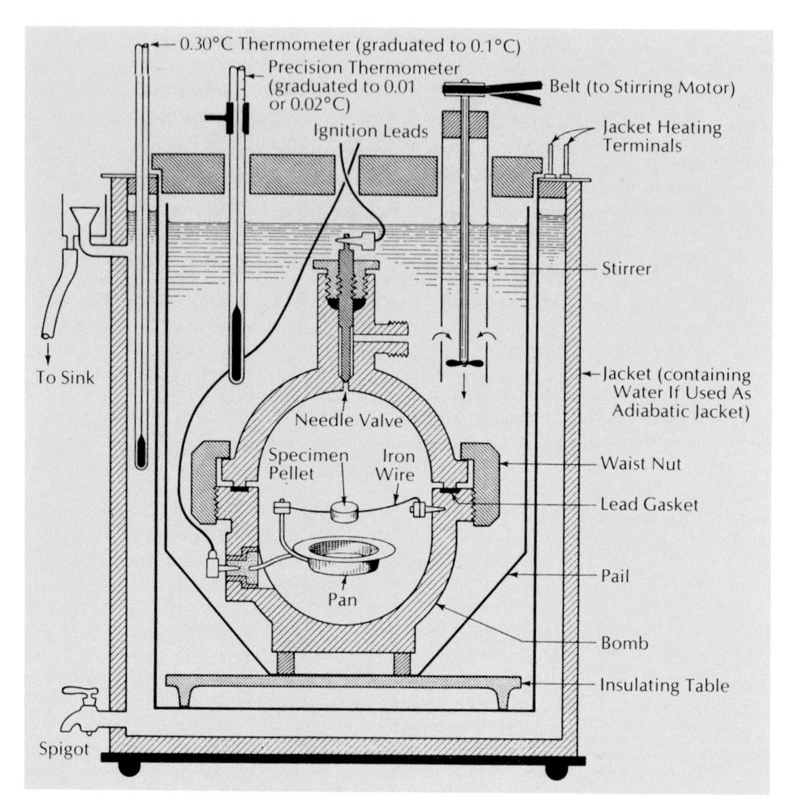

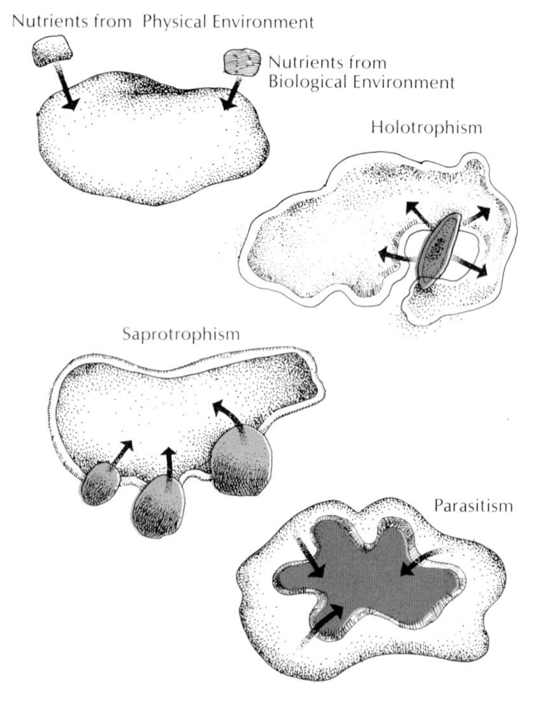

uncontrollable release. In a gasoline engine, small amounts of fuel are burned under controlled conditions to push the pistons and to transform energy fairly efficiently into power for the automobile wheels. Far more usable energy is obtained than could be derived from the sudden and explosive burning of several gallons of gasoline.

The first phase in the breakdown of organic molecules with release of energy is accomplished in processes known as fermentation and glycolysis, processes that do not require free oxygen and stop before the organic molecules are completely broken down to CO_2 and H_2O. Heterotrophic organisms living in reducing or anaerobic environments must use these processes as their only source of chemical energy. Other organisms make use of these processes as an emergency source of energy when oxygen is temporarily unavailable. These processes are not as efficient as respiration in capturing free energy during the breakdown of organic molecules.

THE EMBDEN-MEYERHOF PATHWAY

The sequence of reactions involved in the metabolism of glucose (Figure 6.3) is called the Embden-Meyerhof pathway in honor of the two men who did the definitive research on this sequence in the 1920s and 1930s. This pathway begins with the input of a simple six-carbon sugar molecule, glucose, and ends with either the production of two molecules of a three-carbon acid, lactic acid, or with the production of ethanol and CO_2, depending on the cell in which the process occurs. In the sequence of reactions, the stable glucose molecule is first energized and made unstable; then it is partially dismantled or degraded, liberating a small part of the chemical

Figure 6.3. The Embden-Meyerhof pathway of glucose metabolism. Note that the first 7 steps consist primarily of endergonic reactions that require ATP. In steps 7 – 10, partial oxidation of the three-carbon sugar takes place with simultaneous production of ATP.

energy in the glucose and partially oxidizing it. A description of the detailed steps is useful as an example of the kinds of mechanisms involved in metabolic pathways.

In the *steps 1 – 3* of the pathway, glucose is converted to fructose diphosphate by the addition of two high-energy phosphate groups from ATP molecules. The net result of these three steps is to convert the glucose to a higher-energy, unstable intermediate, while converting two molecules of ATP to ADP. Thus, the organism has invested two ATP "coins" of energy to prepare the glucose molecule for breakdown.

In *step 4*, the unstable six-carbon molecule is split into two three-carbon units, phosphoglyceraldehyde and dihydroxyacetone phosphate. These two compounds exist in equilibrium with each other (*step 5*).

In *step 6*, the phosphoglyceraldehyde gains a phosphate group from inorganic phosphate (P_i) and loses two hydrogen atoms to the coenzyme nicotinamide adenine dinucleotide (NAD). As the phosphoglyceraldehyde is oxidized to diphosphoglycerate, the NAD is reduced to NADH. As in all redox reactions, energy is transferred from the reductant (phosphoglyceraldehyde) to the reduced product (NADH). As the phosphogly-

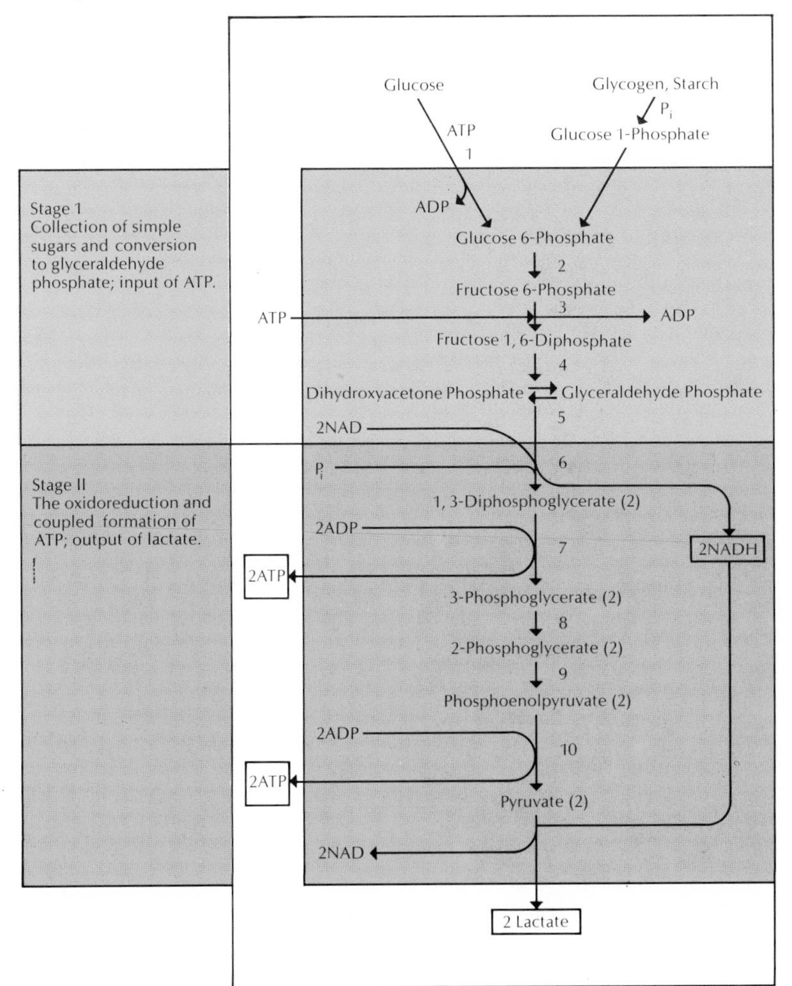

Figure 6.4. Alcoholic fermentation involves the metabolism of the six-carbon sugar glucose (top) to 2 two-carbon ethanol molecules. The six-carbon molecule is split when fructose is fragmented into 2 molecules of glyceraldehyde phosphate. Each of these molecules is further metabolized through four more intermediate compounds before ethanol (ethyl alcohol) is produced.

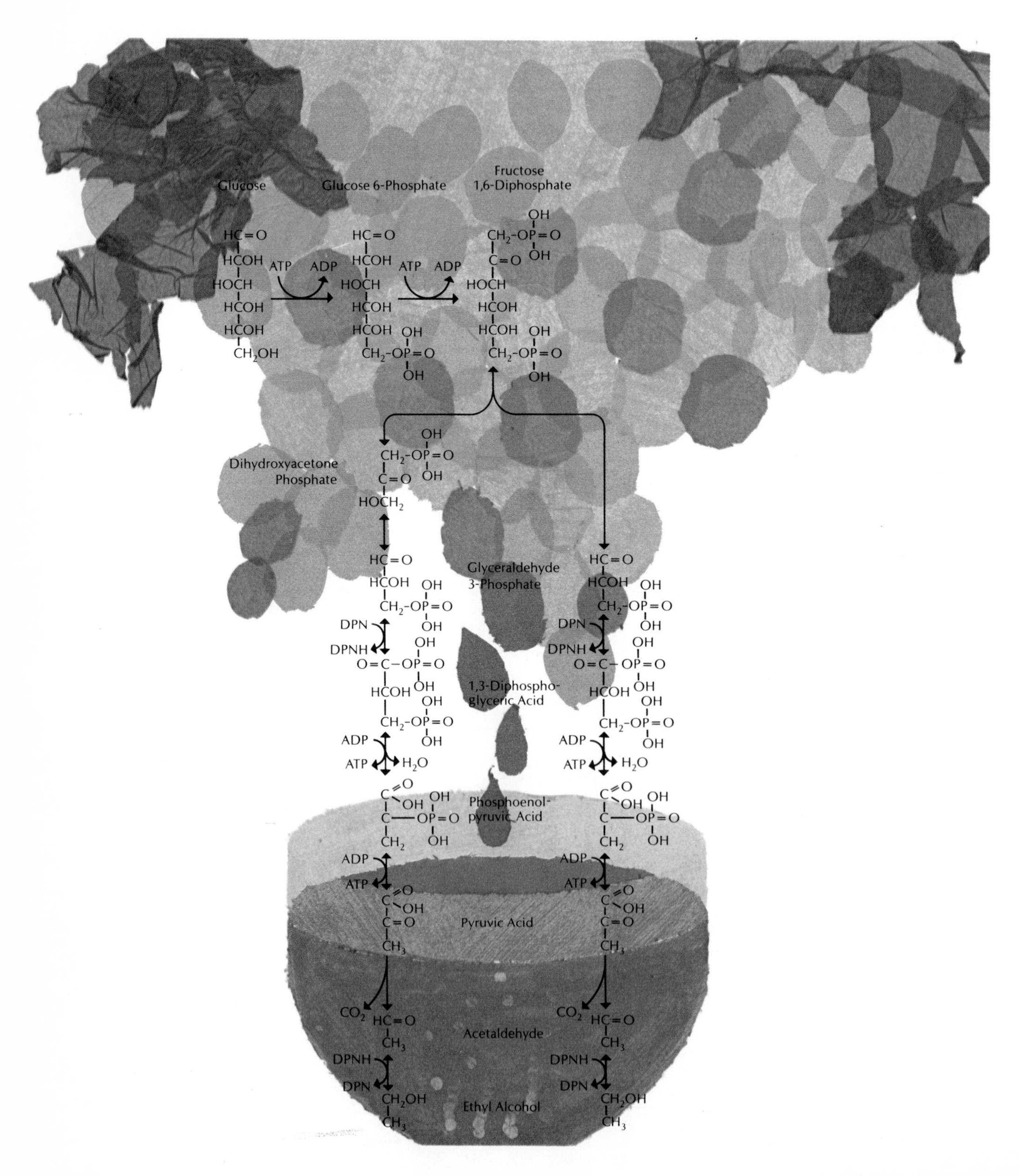

ceraldehyde is used up in step 6, the dihydroxyacetone phosphate is converted to phosphoglyceraldehyde by step 5. Thus, eventually two molecules of phosphoglyceraldehyde will enter step 6 for each molecule of glucose that entered step 1.

In *step 7*, a phosphate group is transferred from diphosphoglycerate to ADP, forming phosphoglycerate and ATP. In this step, some energy is transferred from the diphosphoglycerate to the ATP. Because two molecules of ATP are produced in this step for each molecule of glucose entering the pathway, this step restores the ATP that was invested in steps 1 and 3. Such a reaction, in which ADP is converted to ATP by the simple transfer of a phosphate group from an organic molecule, is called substrate-level phosphorylation.

In *steps 8* and *9*, the bonds of the organic molecule are rearranged with the loss of a water molecule, forming phosphoenolpyruvate. This molecule is unstable enough to carry out another phosphorylation of ADP (*step 10*), forming pyruvate and ATP. As one glucose molecule moves through the sequence, two molecules of ATP are used in steps 1 and 3, but four molecules of ATP are produced in steps 7 and 10. Thus, the net result for the organism is the production of two molecules of ATP that can be used to supply chemical energy elsewhere for useful work.

These ten steps of the Embden-Meyerhof pathway occur in both fermentation and glycolysis. In alcoholic fermentation, which is carried out by organisms such as yeast, CO_2 is released from pyruvate, forming acetaldehyde. Acetaldehyde is reduced at the expense of one NADH to form ethanol. The overall reaction for alcoholic fermentation may be written as

$$\text{Glucose} + 2\ P_i + 2\ \text{ADP} \rightarrow 2\ \text{ethanol} + 2\ CO_2 + 2\ \text{ATP} \qquad (6.2)$$
$$\Delta G = -36\ \text{kcal/mole}$$

The breakdown of the glucose to ethanol and CO_2 releases about 56 kcal/mole; about 20 kcal/mole of this energy is stored in the ATP molecules for use elsewhere in the organism; the balance of the energy is accounted for by an increase in entropy.

Some microorganisms carry out fermentations of amino acids; others use sugars as energy sources but produce waste products other than alcohols. The types of organic molecules fermented and the waste products emitted are used as a means of classification for many bacteria and other microorganisms. In every case, the fermentation process is a relatively inefficient means of retrieving chemical energy from organic molecules.

Glycolysis is a particular extension of the Embden-Meyerhof pathway in which pyruvate is reduced to lactic acid using NADH. Some bacteria utilize this process; soured milk is caused by bacterial production of lactic acid. It also occurs in muscles forced to contract under anaerobic conditions. The process of glycolysis has been thoroughly studied because it occurs as the first step in glucose breakdown in the respiration of most animal species. The overall reaction of anaerobic glycolysis is:

$$\text{glucose} + 2\ P_i + 2\ \text{ADP} \rightarrow 2\ \text{lactic acid} + 2\ \text{ATP} \qquad (6.3)$$
$$\Delta G = -32\ \text{kcal/mole}$$

The breakdown of glucose to lactic acid releases about 52 kcal/mole, of which about 20 kcal/mole is stored in the high-energy linkages of ATP.

Figure 6.5 (above). The structural formula of glycogen.

Figure 6.6 (below). During intense muscular activity, animals often supplement the energy from aerobic respiration by another pathway, anaerobic glycolysis. This process results in the accumulation of toxic lactic acid, and it builds up an oxygen debt. During the resting period, faster breathing occurs, which repays this debt and causes the complete oxidation of the lactic acid. In mammals, the lactic acid is carried to the liver, returned to the muscle, and transformed into glycogen.

Branch

Branch Point

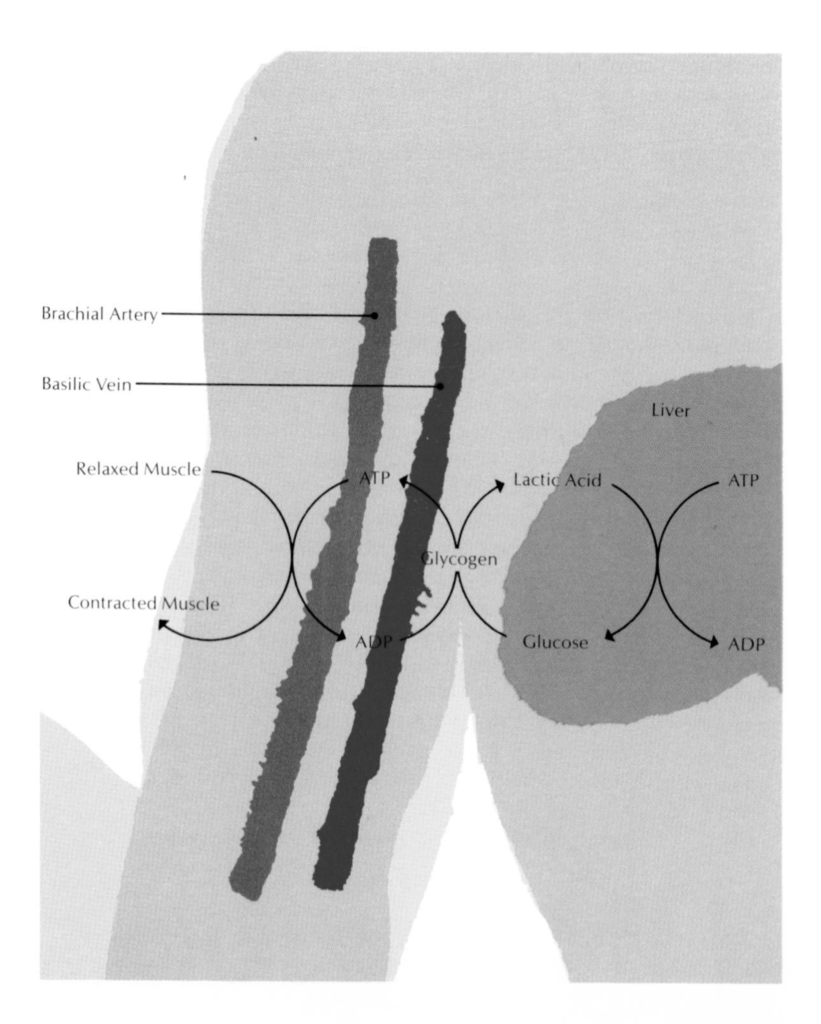

Figure 6.7 (right). Schematic diagram illustrating the sequential action of the glycolytic chain of enzymes. The product of one reaction becomes the substrate of the next enzyme. Thus, the chain of enzymes can be considered a type of disassembly line that takes in glucose at one end and turns out lactic acid at the other. (Refer to the key for specific enzymes and intermediate compounds in the series.)

Figure 6.8 (left). The redox reactions of the coenzyme nicotinamide adenine dinucleotide (NAD). NAD, the oxidized form, is a hydrogen acceptor. When combined with two hydrogens, it becomes $NADH_2$, the reduced form. The release of hydrogen converts it back to NAD and decreases the free energy of the molecule; acceptance of hydrogens increases its free energy.

Much of the chemical energy of glucose remains stored in lactic acid.

In the muscles of animals, glucose is stored in the form of *glycogen*, a complex polymer of glucose (Figure 6.5). When a muscle contracts and relaxes rapidly, glycogen disappears and lactic acid accumulates. When the concentration of lactic acid becomes high enough, the muscle ceases to contract. Energy to power contraction is obtained in the form of ATP through glycolysis of the glycogen, but lactic acid accumulates as a toxic waste product. When the blood circulation returns to such a muscle, the lactic acid diffuses into the blood and is transported to the liver, where glucose is resynthesized from the lactic acid. This process involves the expenditure of energy by the liver cells.

Only a small percentage of the energy released in the reaction is captured by ATP, and a great deal of chemical energy remains stored in the organic molecules discarded as waste products. In organisms capable of respiration, fermentation is usually carried out as a first step in the breakdown of organic molecules. The pyruvate of the fermentation is then oxidized further to release CO_2 and H_2O.

Each step in the pathway is catalyzed by a specific enzyme. The reaction sequence may be regarded as a chain of enzymes, each of which converts its substrate into a form suitable to act as substrate to the next enzyme (Figure 6.7). This chain of enzymes can be regarded as a "disassembly line" that takes in glucose at one end and turns out lactic acid at the other.

The NAD^+-NADH coenzyme pair is found in all organisms. Just as the ATP-ADP pair plays a universal role in transferring phosphate groups and

Nicotinamide

Adenine

NAD

Ribose

Diphosphate

Ribose

$-2(H) \rightleftharpoons +2(H)$

NADH

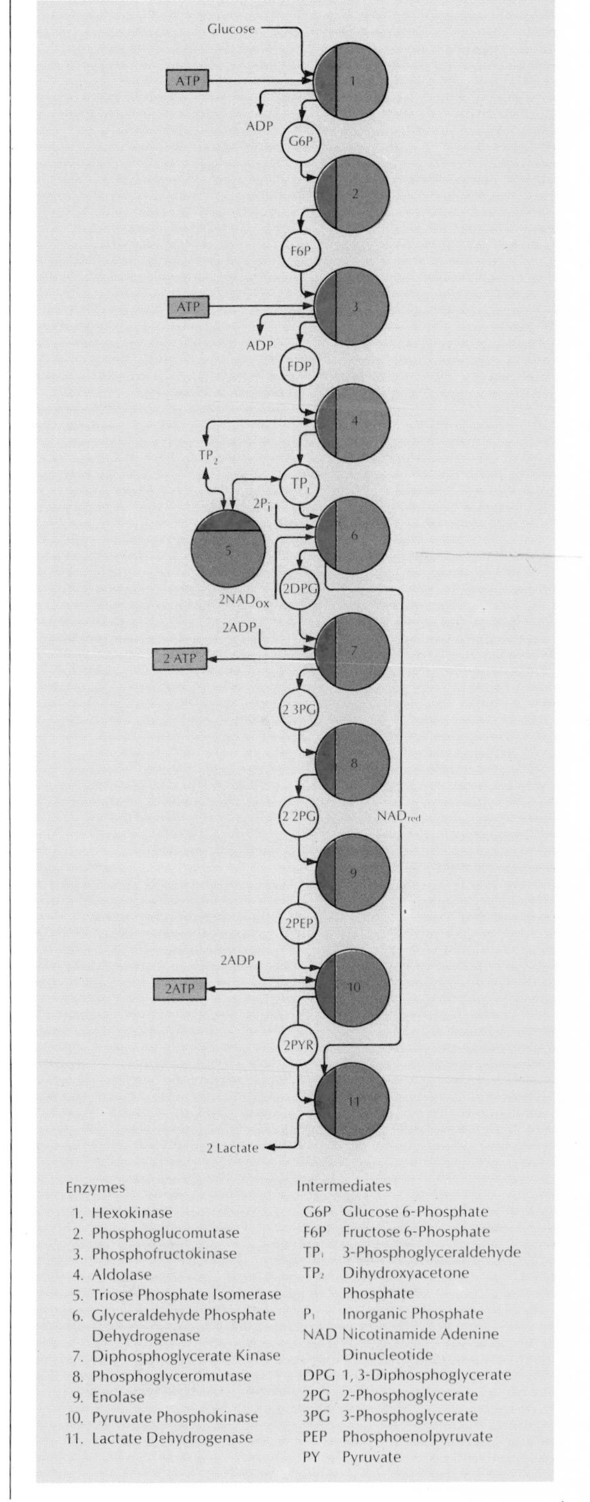

chemical energy in all living systems, the NAD-NADH pair plays a universal role in transferring electrons (or hydrogen atoms). Notice that in both glycolysis and alcoholic fermentation, reduced NADH must be oxidized back to NAD, using as its final electron acceptor pyruvate or acetaldehyde. If this impotent carrier were not reoxidized, the metabolic machinery would soon be halted because the continued dissimilation of glucose requires a constant supply of NAD (step 6). Because there are only very small quantities of this coenzyme in the cell, it must constantly shuttle back and forth in a cyclic fashion between oxidized and reduced states. As soon as NADH donates its electrons to an acceptor, it is ready to accept electrons from step 6 or other oxidative reactions of cellular metabolism. NAD^+ was formerly known as DPN (diphosphopyridine nucleotide), and this older name may be found in much of the older literature on metabolism. NAD^+ is a complex molecule, containing a group called nicotinamide, which acts as the electron donor and receptor. Nicotinamide is one of the B vitamins, and its absence from the diets of humans and other higher animals causes the disease pellagra.

THE KREBS CITRIC ACID CYCLE

In aerobic organisms, the pyruvate produced as the end product of glycolysis is degraded to CO_2 in a reaction sequence called the Krebs citric acid cycle. During this cycle, hydrogen is removed by coenzyme teams and moved through the oxidative phosphorylation sequence, where it ultimately reduces oxygen to water.

The oxygen-requiring continuation of the Embden-Meyerhof pathway was postulated in 1937 in a simplified form by Hans Adolf Krebs and W. A. Johnson. In aerobic organisms, the pyruvate produced in glycolysis moves directly into this reaction sequence. Compounds derived from the breakdown of lipids, fatty acids, and amino acids can also enter this sequence at various points along the way.

During glycolysis, glucose is partially oxidized and split into two three-carbon molecules. In the Krebs cycle—the heart of the metabolic proc-

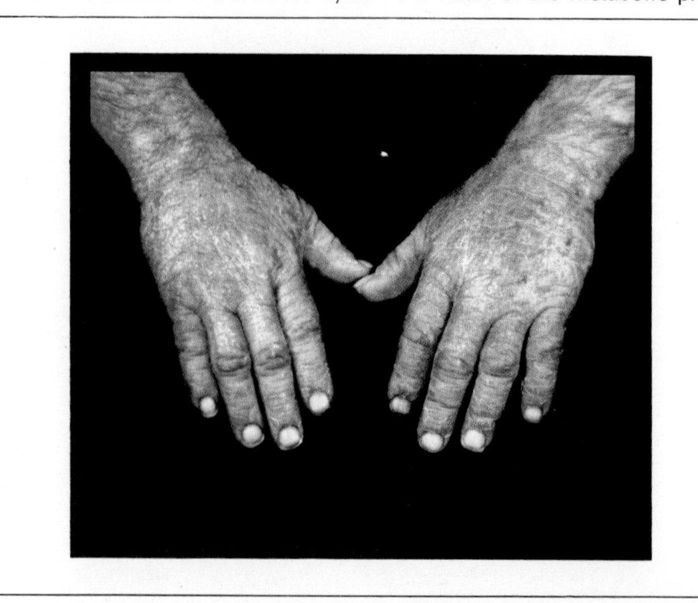

Figure 6.9. Dietary absence of nicotinamide, one of the B vitamins, can cause the disease pellagra. The secondary effects of this disorder are shown in this photograph of scaly dermatitis over the back of the hands and wrists.

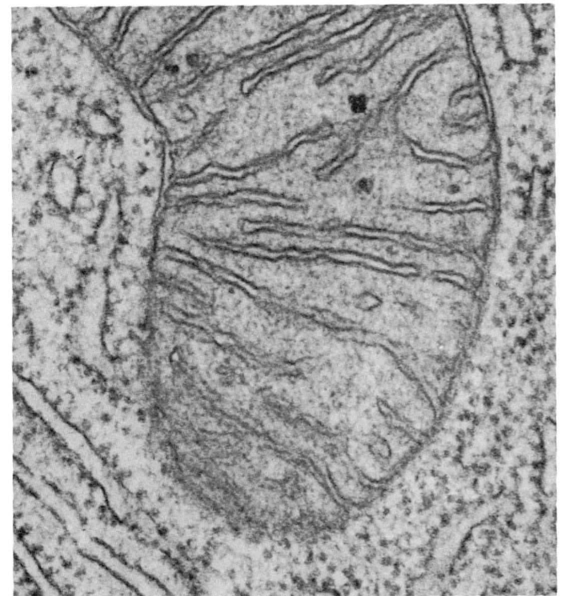

Figure 1 (×77,500)

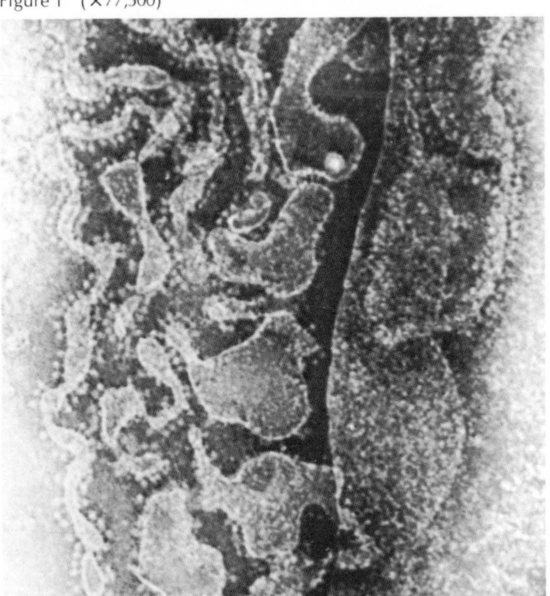

Figure 2 (×136,700)

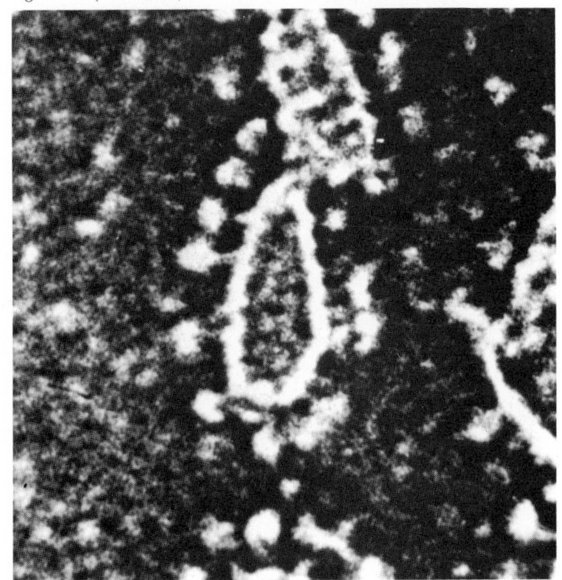

Figure 3 (×623,300)

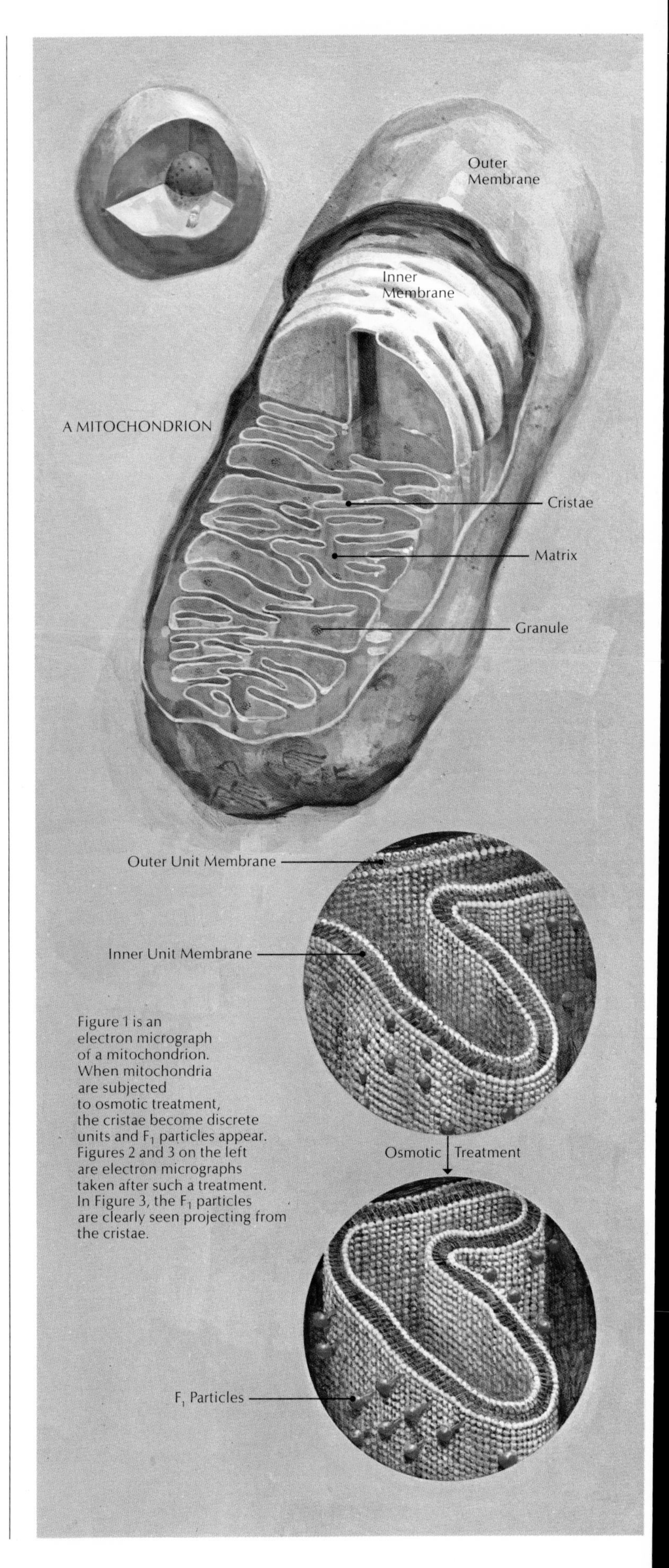

Figure 1 is an electron micrograph of a mitochondrion. When mitochondria are subjected to osmotic treatment, the cristae become discrete units and F_1 particles appear. Figures 2 and 3 on the left are electron micrographs taken after such a treatment. In Figure 3, the F_1 particles are clearly seen projecting from the cristae.

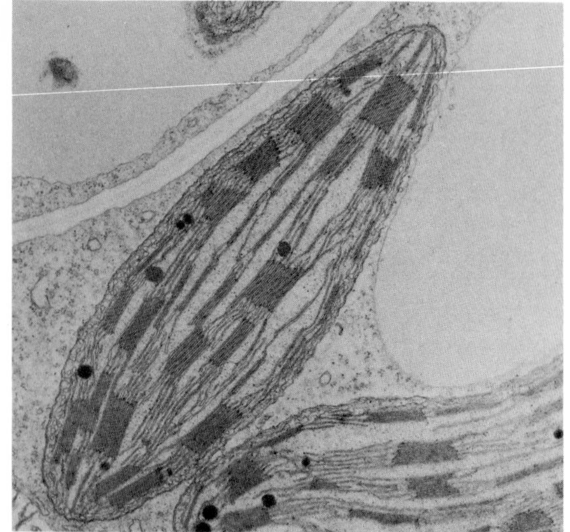
Figure 4 (×10,700)

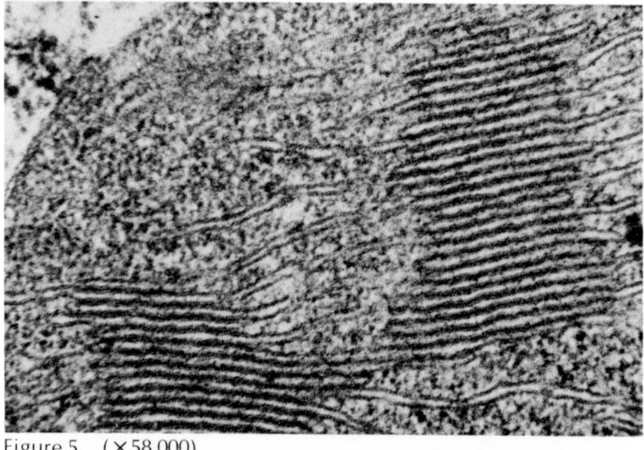
Figure 5 (×58,000)

Figure 6 (×310,500)

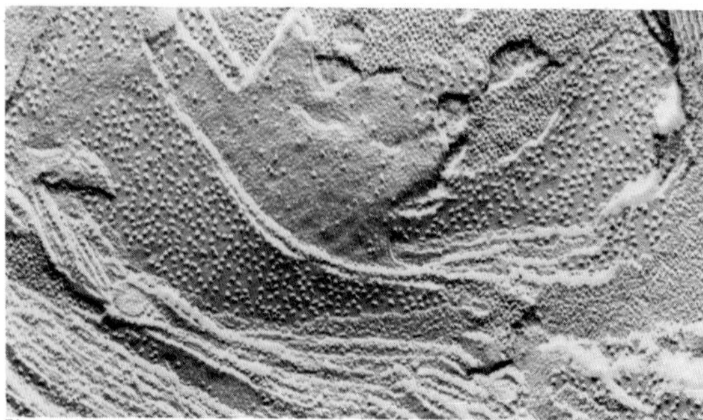
Figure 7 (×38,500)

A series of electron micrographs showing increased magnification of chloroplast structure (Figure 4). Figure 5 shows two grana stacks and a portion of the outer membrane. Figure 6 shows an enlarged region of a granum. In this figure, the staining techniques are reversed from Figure 5 so that the loculus is black (L), and the inner membrane of the thylakoid is white. Figure 7 is an electron micrograph taken by the freeze-etch technique showing the ultrastructure of thylakoids. The different subunit arrays may correspond to Photosystems I and II and the ATP coupling system.

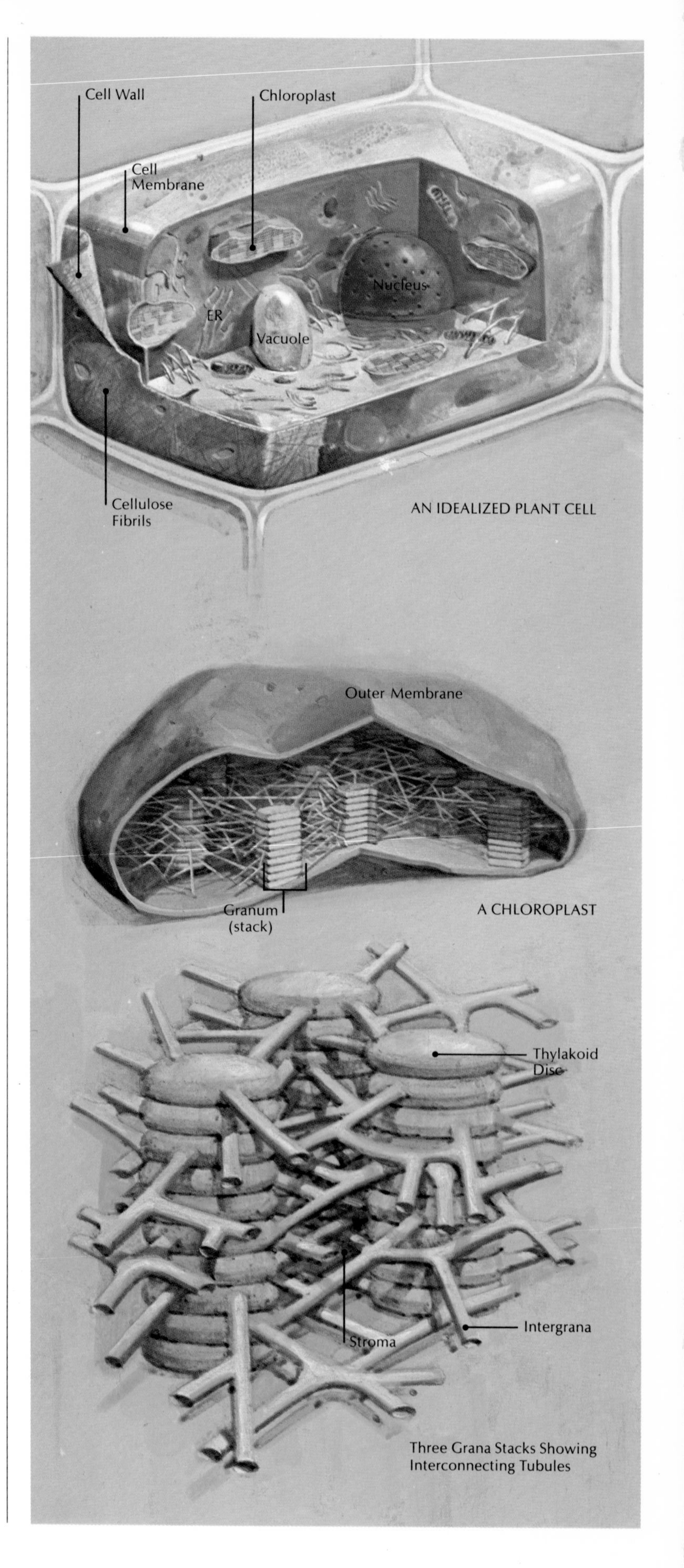

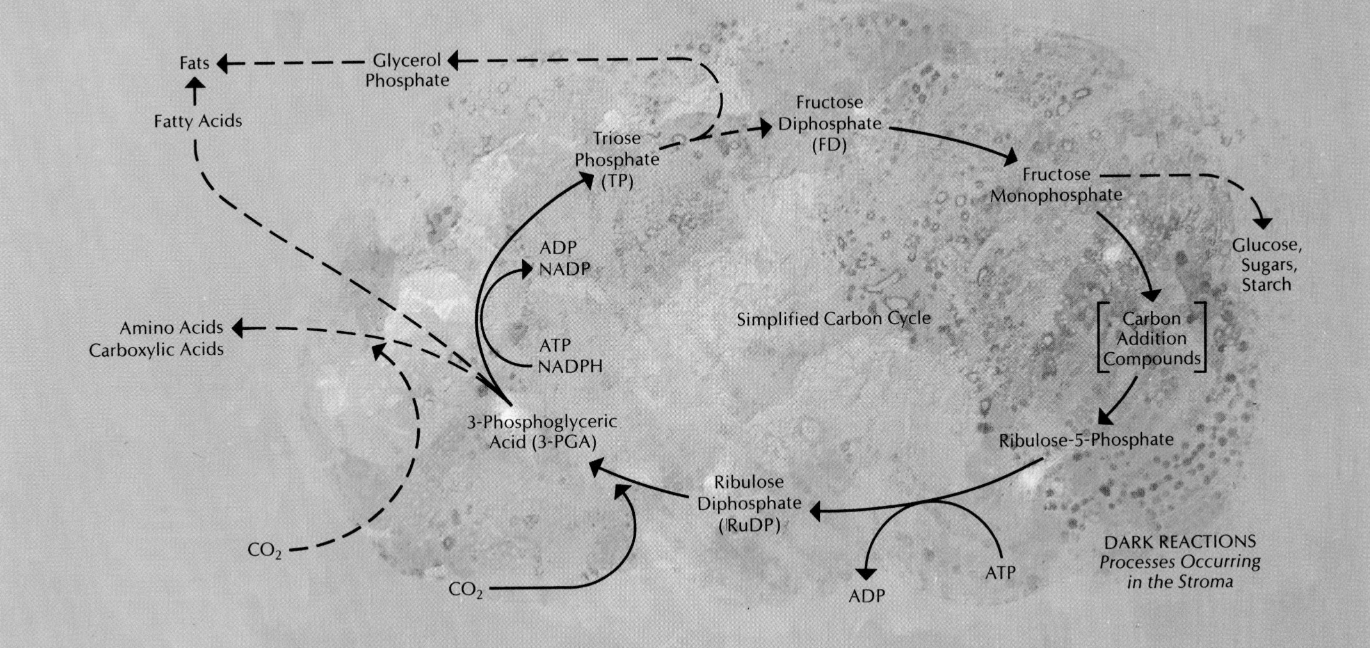

Fats
Glycerol Phosphate
Fatty Acids
Fructose Diphosphate (FD)
Triose Phosphate (TP)
Fructose Monophosphate
Glucose, Sugars, Starch
ADP
NADP
ATP
NADPH
Amino Acids
Carboxylic Acids
Simplified Carbon Cycle
Carbon Addition Compounds
3-Phosphoglyceric Acid (3-PGA)
Ribulose-5-Phosphate
Ribulose Diphosphate (RuDP)
CO_2
CO_2
ATP
ADP
DARK REACTIONS
Processes Occurring in the Stroma

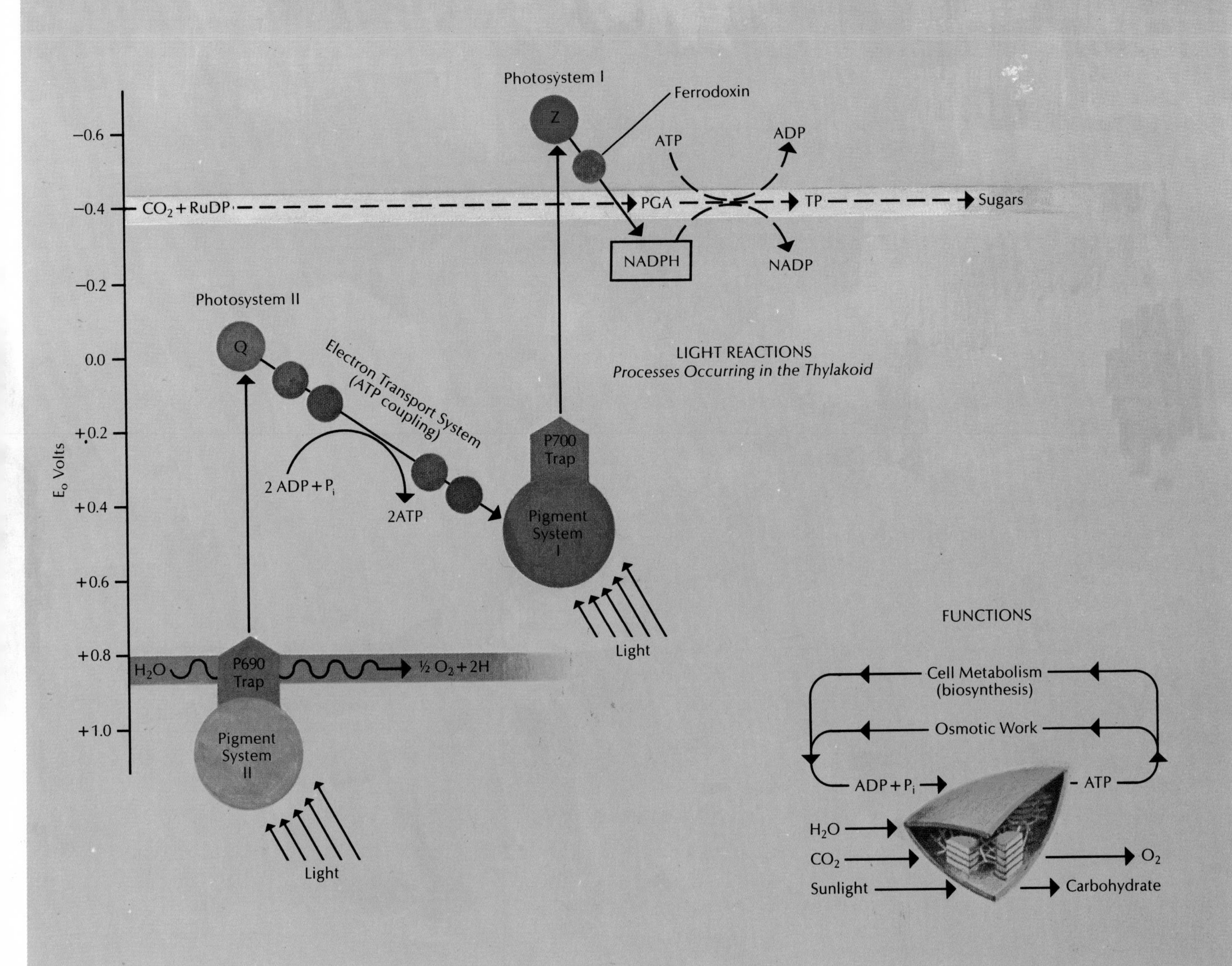

Photosystem I
Ferrodoxin
Z
ATP
ADP
−0.6
−0.4
$CO_2 + RuDP$
PGA
TP
Sugars
NADPH
NADP
−0.2
Photosystem II
Q
0.0
Electron Transport System (ATP coupling)
LIGHT REACTIONS
Processes Occurring in the Thylakoid
+0.2
P700 Trap
E_o Volts
2 ADP + P_i
+0.4
2ATP
Pigment System I
+0.6
Light
FUNCTIONS
+0.8
H_2O
P690 Trap
½ O_2 + 2H
Cell Metabolism (biosynthesis)
+1.0
Osmotic Work
Pigment System II
ADP + P_i
ATP
H_2O
CO_2
O_2
Sunlight
Carbohydrate
Light

D

Interleaf 6.1

STRUCTURE AND FUNCTION

Important biochemical concepts often become lost in what appears to be a series of dry reactions depicted by circles, arrows, and strangely named substances. These concepts can become more meaningful by the introduction of a major approach in modern biology—that of structure and function. By looking at structural parts, biologists can better understand the functioning of systems and can then use this knowledge of components to understand how these systems interact to perform specific tasks. With the use of modern techniques, biologists have probed the subcellular world and elucidated much of the mystery of larger biosystems.

This concept is not new; it has its basis in all inquisitive thought. It has been expressed in invention, in primitive man's understanding of natural forces, and in the work of poets and architects. During the Middle Ages, man viewed the universe as an interaction between the macro and microcosm. Today, through a more sophisticated use of technology, man has further developed his understanding of larger systems (behavioral, organismic, ecological) and noted parallels in the elucidation of smaller systems (molecular pathways, organelles, cells). These systems do not mirror one another—each has its own unique aspects—but there is a continual interaction of architecture and function between the two. An understanding of this interrelationship ultimately expands the way we view our universe.

The panels that follow will graphically show the relationship between the biochemical pathways described in Chapters 5 and 6 and the structural units that carry out these processes. These structural units are two main organelles that have extremely important roles in plants and animals; they are the mitochondrion and the chloroplast.

Both mitochondria and chloroplasts are energy units. Their efficiency is increased by the inward convolutions of their membranes. This arrangement is common in living systems because it increases the surface area to volume ratio and thereby maximizes the space available for reactions to occur. Plants use sunlight, CO_2, and water as fuels to produce the energy molecule ATP and the electrons necessary for sugar formation. Animals, on the other hand, depend on oxygen and the metabolic breakdown of food for ATP synthesis. Plants also have mitochondria, but 30 times as much ATP is produced in the chloroplasts. Much of what follows is condensed and abbreviated, but it is hoped that the combination of this brief introduction and the detailed graphics will give the reader an appreciation of the concept of structure and function.

The Mitochondrion

The use of the electron microscope has revealed much of the structural mystery of subcellular systems. Although there is a limitation in dealing with processed and fixed material, much can be learned from reading electron micrographs. They provide a tangible base from which structural details can be studied. Electron micrographs of mitochondria show a highly complex ultrastructure. A mitochondrion is basically a double-membrane system with an outer membrane that envelops a highly convoluted inner membrane. Both membranes have what is termed the unit membrane structure; that is, they are composed of a double layer of lipids with interspersed protein molecules. These membranes also provide the basic building blocks upon which more ornate structures are added for specific functions. The main regions of the mitochondrion are labeled in panel B.

Further details of structure can be seen after a treatment that causes the mitochondria to swell. New particles, previously within the membrane, now cover the cristae surface with a polygonal head attached to the membrane by a stem (Figure 3). These particles have been variously termed F_1, or elementary, particles. They appear regularly dispersed along the cristae, with some estimates of up to 10,000 per mitochondrion. They are an important link in understanding the energy role of the organelle.

The functions of mitochondria are shown in panel E. The metabolic breakdown of food pumps intermediaries through a complex series of enzymatic reactions

Continued on H

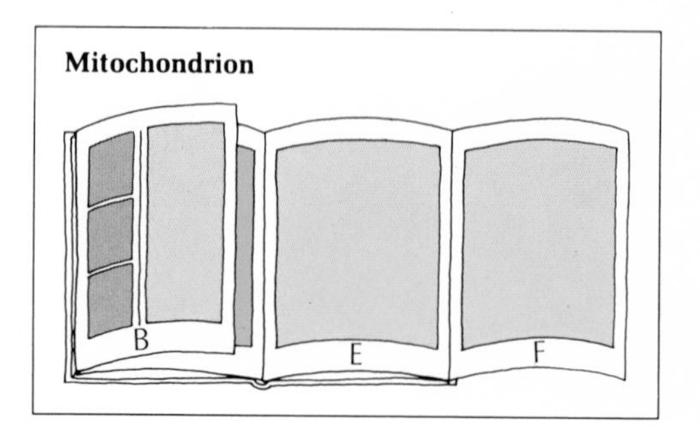

Outer Membrane
Carbon Cycle
NADP
NADPH
TP
3-PGA
Inner Membrane
ATP Coupling System
Photosystem I
Loculus
O_2
H_2O
Photosystem II
Carotenoid
Chlorophyll *b*
Chlorophyll a

G

Outer Membrane
Inner Membrane
Cristae
NAD
FP
b
c
a
a_3
F_1
KREBS CYCLE
Matrix
Lipid
Protein
Unit Membrane

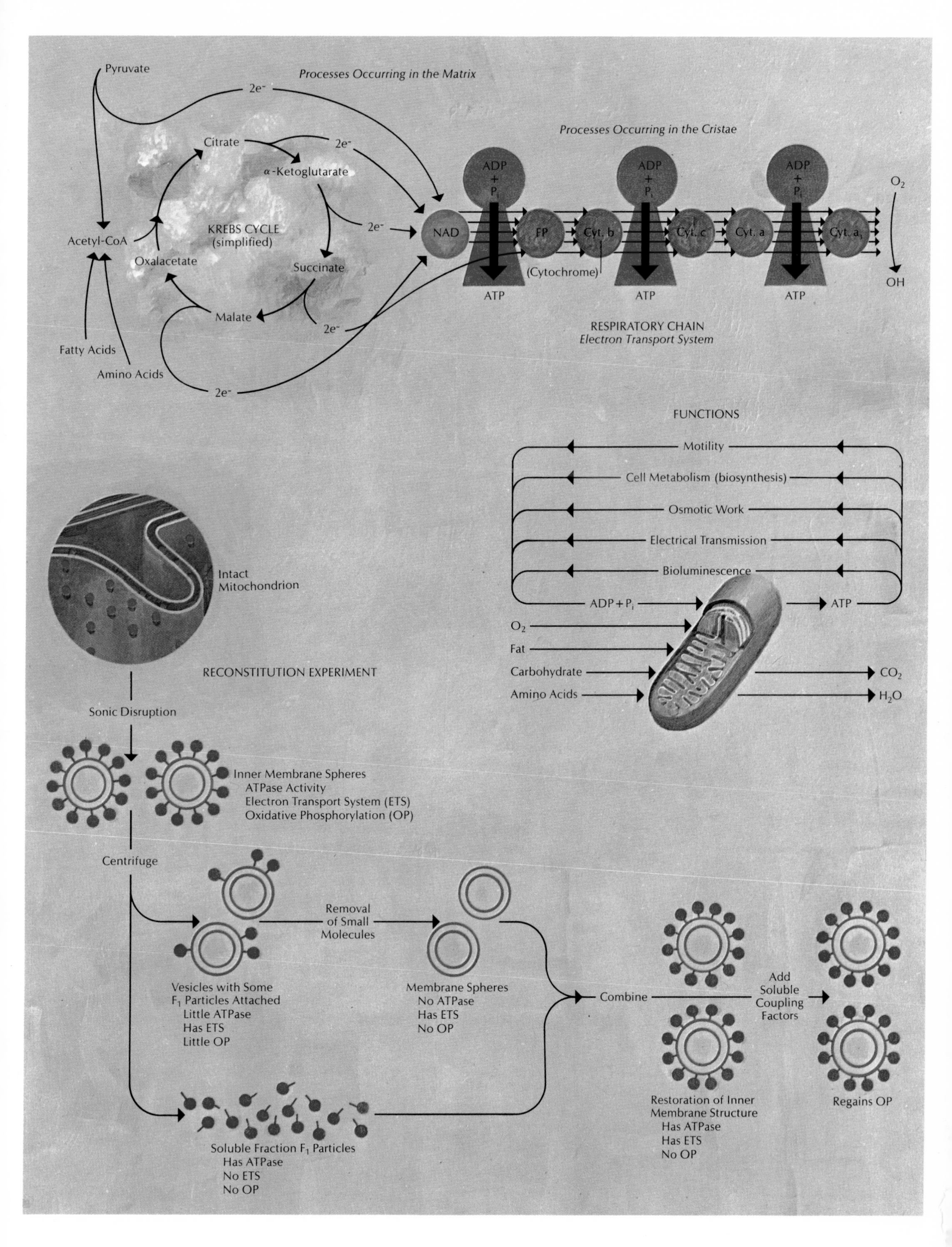
Processes Occurring in the Matrix
Pyruvate
2e⁻
Citrate
α-Ketoglutarate
2e⁻
2e⁻
KREBS CYCLE
(simplified)
Acetyl-CoA
Oxalacetate
Succinate
Malate
2e⁻
Fatty Acids
Amino Acids
2e⁻
Processes Occurring in the Cristae
ADP + P_i
ADP + P_i
ADP + P_i
O_2
NAD
FP
Cyt. b
Cyt. c
Cyt. a
Cyt. a_3
(Cytochrome)
ATP
ATP
ATP
OH
RESPIRATORY CHAIN
Electron Transport System
FUNCTIONS
Motility
Cell Metabolism (biosynthesis)
Osmotic Work
Electrical Transmission
Bioluminescence
ADP + P_i
ATP
O_2
Fat
Carbohydrate
Amino Acids
CO_2
H_2O
Intact
Mitochondrion
RECONSTITUTION EXPERIMENT
Sonic Disruption
Inner Membrane Spheres
ATPase Activity
Electron Transport System (ETS)
Oxidative Phosphorylation (OP)
Centrifuge
Removal
of Small
Molecules
Vesicles with Some
F_1 Particles Attached
Little ATPase
Has ETS
Little OP
Membrane Spheres
No ATPase
Has ETS
No OP
Combine
Add
Soluble
Coupling
Factors
Soluble Fraction F_1 Particles
Has ATPase
No ETS
No OP
Restoration of Inner
Membrane Structure
Has ATPase
Has ETS
No OP
Regains OP

known as the Krebs cycle; this action in turn provides electrons for the respiratory chain, where ATP is formed through the process of oxidative phosphorylation.

Extensive research has elucidated the complexities of these reactions and helped to relate them to the formation of ATP. Cristae increase in number with an increase in oxidative activity of the mitochondrion. By adding enzymes that digest proteins in the membrane, researchers have clarified the role of enzymes in ATP production. And experimenters using the technique of sonic oscillation (which breaks the organelle into smaller particles) have seen still smaller membrane spheres capable of oxidative phosphorylation. These small fragments have intact F_1 particles that contain a special enzyme, ATPase, involved in coupling oxidation and phosphorylation. A flow diagram of these reconstitution experiments is shown in panel E. Other work has shown that the enzymes involved in the Krebs cycle are localized in the matrix. The work of many investigators has laid the groundwork of a reconstitution of structure and function. Panel F shows an interpretation of various components of mitochondrial structure (relative sizes of the molecules are not to scale). Lipids, membrane proteins, cytochromes, and F_1 particles are all depicted interacting with the Krebs cycle. This hypothetical model shows some of the dynamic aspects of this once invisible organelle.

The Chloroplast

Electron micrographs have also revealed a highly complex ultrastructure for the chloroplast. This organelle has basically the same double-membrane system as the mitochondrion. The convolutions of the inner membrane make up a vast network of stacked membranous discs, known as grana, that are interconnected by tubules (panel C). The individual discs that make up the grana are called thylakoids. Within each chloroplast are the basic materials of the photosynthetic machinery.

Two main series of operations take place in the chloroplast. The *light reactions* are dependent upon the sun's energy, whereas the *dark reactions,* the pathways of carbohydrate metabolism, are light independent. Panel D shows a simplified version of this carbon cycle necessary for sugar formation. Two photosystems with pigments (chlorophylls, carotenoids, and so on), carrier and acceptor molecules, and electron transport chains carry out the light reactions. The basic plan postulated for this operation is also shown in panel D. The reactions are sequential and initially depend upon the action of light to excite the pigment molecules. Photosystem II, via electron resonance, traps electrons in a special pigment (P690); this action provides the energy to split water and, with an acceptor molecule Q, to raise the free energy of the system. This energy then "flows" down an electron transport chain, providing the energy to form ATP. Photosystem I is also excited by light with a similar array of harvesting pigments. The electrons in this system are thought to provide the energy, via carrier Z, to form NADPH. The enzyme ferrodoxin catalyzes this photoreduction. NADPH is then utilized in the dark reactions. These complicated reactions imply that a highly organized molecular structure exists to carry out these functions. Chemical analyses indicate that the pigment chlorophyll is one of the main components of the chloroplast. These molecules can be seen as having a structure-function role analogous to structural lipid molecules.

Special electron microscopy freeze-etch-cleavage techniques (Chapter 10) reveal a repeating array of smaller particles, termed quantasomes, within the thylakoids. Reconstitution experiments show that two specific units within this quantasome array may correspond to Photosystems I and II, whereas other units act as coupling factors for ATP formation. Much research is being done to elucidate the nature of these particles, but many details are still unknown and several theories are being proposed to explain the complicated relationships of the photosynthetic apparatus. One such hypothetical model is shown in panel G. It depicts the interaction of chloroplast structure and function—the beauty of light and molecules working together in this vital life process.

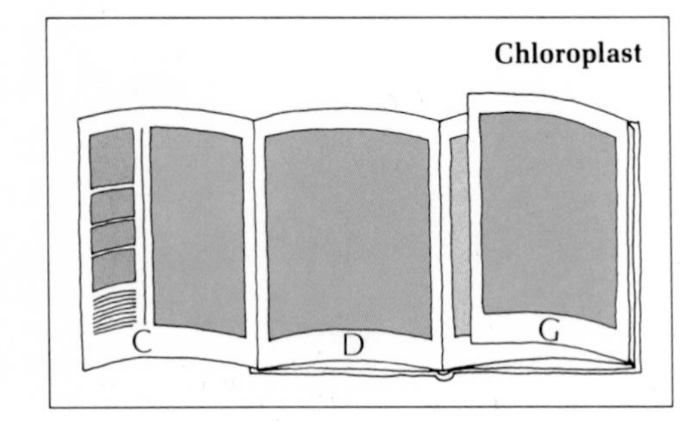

Figure 6.10. The Krebs citric acid cycle and respiratory chain phosphorylation. Major intermediate compounds, each requiring a specific enzyme to continue the reaction series, are shown in the Krebs cycle. Note the function of NAD in transporting high-energy electron (H) to the respiratory chain, where it is passed along to various cofactors and ultimately reduces oxygen and forms metabolic water.

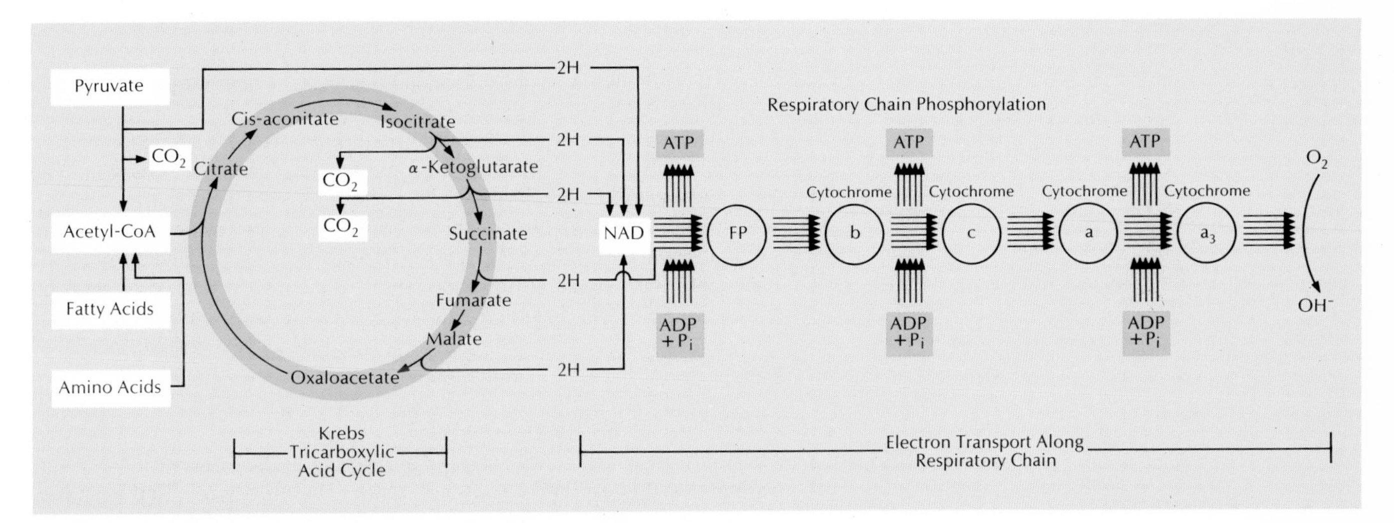

esses—the oxidation process is completed. Eventually, all of the hydrogen atoms are removed from the carbon compounds and are used to reduce cofactor molecules.

The process of aerobic glycolysis can be represented by the following overall reaction:

$$\text{glucose} + 2\ \text{NAD} + 2\ \text{ADP} + 2\ P_i \rightarrow 2\ \text{pyruvate} + 2\ \text{NADH} + 2\ \text{ATP} \quad (6.4)$$

The two molecules of NADH enter the oxidative phosphorylation sequence directly

$$2\ \text{NADH} + 6\ P_i + 6\ \text{ADP} + O_2 \rightarrow 2\ \text{NAD} + 2\ H_2O + 6\ \text{ATP} \quad (6.5)$$

The two molecules of pyruvate pass through the Krebs cycle and oxidative phosphorylation with the following overall reaction:

$$2\ \text{pyruvate} + 5\ O_2 + 30\ \text{ADP} + 30\ P_i \rightarrow 6\ CO_2 + 4\ H_2O + 30\ \text{ATP} \quad (6.6)$$

The sum of reactions 6.4, 6.5, and 6.6 yields the overall equation for respiration

$$\text{glucose} + 6\ O_2 + 38\ \text{ADP} + 38\ P_i \rightarrow 6\ CO_2 + 6\ H_2O + 38\ \text{ATP} \quad (6.7)$$

For every molecule of glucose that is completely oxidized in respiration, 38 molecules of ATP are produced. The free energy of oxidation of glucose is about −686 kcal/mole. Because each mole of ATP stores about 7 kcal of energy, the respiration of a mole of glucose stores about 266 kcal in the form of 38 moles of ATP. The efficiency of respiration is thus about 42 percent (266/686); the other 58 percent of the chemical energy in the glucose is lost.

The complete reaction sequence of the Krebs cycle is shown in Figure 6.10. Before it can enter the main cycle, the pyruvate is oxidized to a two-carbon compound, acetic acid. The acetic acid is bonded to a coenzyme molecule called coenzyme A (CoA), which acts as a carrier of acetyl

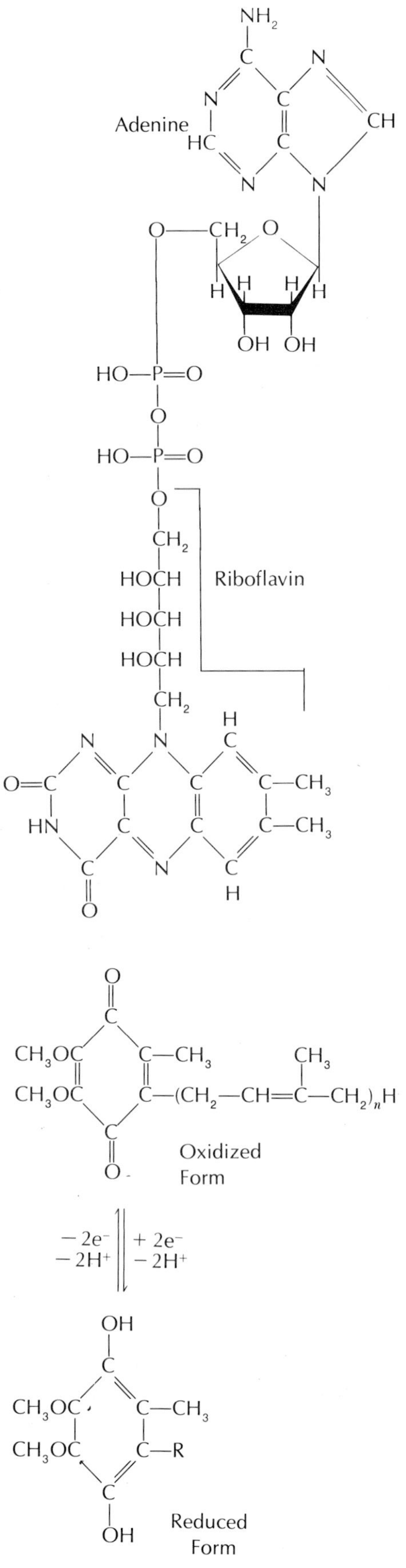

Adenine

Riboflavin

Oxidized Form

$-2e^-$ $-2H^+$ | $+2e^-$ $-2H^+$

Reduced Form

Figure 6.11. The flavin adenine dinucleotide (FAD) is shown above and coenzyme Q (CoQ) is shown below. CoQ_6 is found in a few microorganisms, and CoQ_{10} is present in numerous mammals.

groups, much as ATP acts as a carrier of phosphate groups. During this step, the third carbon atom of the pyruvate is oxidized to CO_2. The oxidant for this reaction is NAD, which is reduced to NADH. The NADH enters the oxidative phosphorylation sequence.

The acetyl-CoA complex is also formed in the breakdown of amino acids, fatty acids, and lipids. Acetyl-CoA enters the Krebs cycle by donating its acetyl group to a four-carbon acid to form the six-carbon *citric acid*, for which the cycle is named. At three more steps in the cycle, reduction of nicotinamide cofactors (NAD or NADP) occurs. The reduced cofactors move into the oxidative phosphorylation sequence. In two other steps of the cycle, ATP is produced directly by substrate-level phosphorylation.

Many of the intermediates formed in the Krebs cycle can serve as precursors for other reaction pathways.

OXIDATIVE PHOSPHORYLATION

Two turns of the Krebs cycle completes the oxidation of glucose to CO_2, yet no more ATP is available to do cell work. The energy released from the glucose is captured by the compounds NADH and $FADH_2$. In order to convert this energy into energy stored in ATP, these reduced compounds are gradually oxidized by a series of electron transfer reactions at the same time ADP is phosphorylated to ATP. This process is known as oxidative phosphorylation.

During oxidative phosphorylation, enzymes called *dehydrogenases* transfer hydrogen to various cofactors such as nicotinamide nucleotides (NAD^+), flavin nucleotides (FAD), quinones (coenzyme Q), and the iron-containing pigments cytochromes. The final step of oxidative phosphorylation requires oxygen as a hydrogen acceptor, thus forming H_2O.

The engineer designing a gasoline engine will be extremely pleased with himself if he succeeds in designing an engine that turns 25 percent of the chemical energy of the gasoline into useful work. The reactions of the Krebs cycle and of oxidative phosphorylation add 34 ATP molecules to the 2 formed during glycolysis, so that a total of 36 ATP molecules are formed during the complete oxidation of glucose to CO_2 and H_2O. This formation represents 360 kcal/mole captured in ATP compared to 686 kcal/mole available in glucose for an efficiency of 54 percent.

REGULATION OF RESPIRATION

In most organisms, there is usually an abundant supply of glucose and oxygen. What, then, prevents the organism from breaking down all of its supply of glucose through respiration? This total consumption would be most undesirable, for energy stored in ATP is lost rapidly through the breakdown of ATP to ADP and P_i unless the ATP is used rapidly in coupled reactions.

The rate at which the Krebs cycle reactions occur appears to be limited chiefly by the ratio of ATP to ADP in the organism. When a great deal of work is being done, ATP is converted to ADP and the ADP concentration rises. With an abundance of ADP available, the reactions of the Krebs cycle proceed rapidly, turning the ADP back into ATP. If little work is being done, the ATP concentration rises and the ADP concentration decreases. With little ADP available, the reactions of the Krebs cycle cannot proceed very rapidly. This feedback mechanism quite precisely regulates the rate of glucose oxidation to match the amount of work being done.

If the supply of oxygen to a microorganism or cell is limited, the final

Figure 6.12. ATP-ADP feedback system. The diagram illustrates how ATP-ADP concentrations in organisms regulate the rate of respiration. When a great deal of work is being done, ATP is converted to ADP and the ADP concentration rises. With an abundance of ADP available, the reactions of the Krebs cycle can proceed rapidly, converting ADP back into ATP. If little work is being done, the ATP concentration rises and the ADP concentration decreases. With a diminished ADP supply, the reactions of the Krebs cycle cannot proceed very rapidly. Thus, the ATP-ADP feedback system precisely regulates the rate of glucose oxidation to match the amount of work being done.

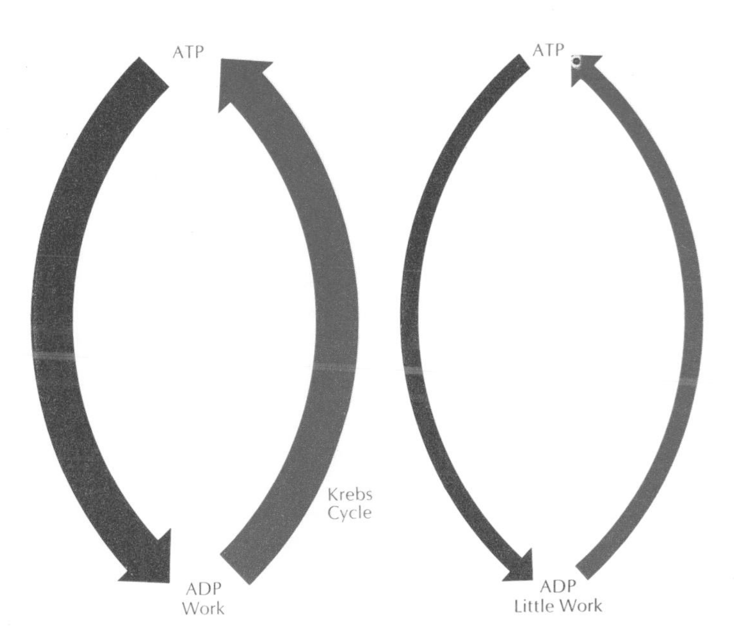

step of oxidative phosphorylation cannot proceed very rapidly. As each enzyme in that sequence becomes so overloaded with substrate molecules that it cannot pass along to the next step, NADH begins to accumulate and the supply of NAD dwindles, slowing down the reactions of the Krebs cycle so that pyruvate begins to accumulate. This reaction supplies enough NAD to keep the glycolytic process going for some time. Thus, the organism can continue to produce small amounts of ATP without oxygen, but at the cost of accumulating the toxic waste product, lactic acid. Some microorganisms are able to eliminate this substance from the cell and thus to survive indefinitely in anaerobic conditions. In human muscle cells, on the other hand, the accumulation of lactic acid eventually interferes with the process of muscle contraction, and after a period of extreme exertion, the muscles may cramp.

INTERMEDIARY METABOLISM

Until now, cellular metabolism has been discussed from the standpoint of the breakdown of food molecules, primarily glucose, through a series of discrete steps in which glucose is completely oxidized to CO_2 and H_2O. The ultimate goal of this breakdown is the generation of ATP, which can then be used by the cell to drive unfavorable chemical reactions. The process whereby the molecules are broken down is called *catabolism.* The other side of metabolism, that of biosynthesis, is called *anabolism.* Both catabolism and anabolism occur simultaneously within the cell. Catabolic processes also generate other compounds of importance to the cell.

The ATP produced in catabolic reactions is used to drive energetically unfavorable reactions, such as the assembly of macromolecules. Their direct synthesis involves only a limited number of precursors, such as amino acids, nucleotides, and sugars. These small molecules are not the only ones present in cells; there are a wealth of small molecules involved in cellular metabolism. As in glycolysis or the Krebs cycle, all molecules are made sequentially and specifically in a number of discrete steps, each one being catalyzed by a different enzyme. Each step produces an *intermediate*, and the ordered breakdown or synthesis is called *intermediary metabolism.*

Nucleic acids are highly ordered molecules and use directly the high-energy triphosphates of adenosine, guanosine, thymine, and cytosine for their synthesis. These precursors cannot directly arise by hydrolysis of nucleic acids, for this process yields only the monophosphates. These precursors must first be resynthesized to the triphosphate level, requiring the expenditure of 2 ATP equivalents. Alternatively, the bases and sugars may be synthesized separately and then linked together. The new synthesis of amino acids includes using portions of glysine, aspartic acid, formic acid, and nitrogen atoms donated from glutamine. Energy is also required for this process. For instance, the new construction of ATP from ribose and the components of the purine ring requires the expenditure of 8 ATPs.

The monomers of carbohydrates are simple sugars. Carbohydrates, such as glycogen, starch, and cellulose, are composed of repeating units of glucose linked together in different ways to form these macromolecules. Some carbohydrates may be used only as storage products—food reserves to be used during periods of starvation—whereas others are used in building structures such as cell walls. The surface factors of red blood cells, which determine blood type, are carbohydrates, and lipids are also known to contain carbohydrates.

The most common form of fats are the triglycerides. They are synthesized by the sequential combination of glycol, which is easily synthesized from a glycolytic intermediate, with three fatty acids. The fatty acids themselves are synthesized from the repeated addition of acetyl-CoA, a 2-carbon unit from the Krebs cycle, onto a small precursor molecule consisting of CO_2 and acetyl-CoA. The fatty acids are hooked by an ester linkage onto the glycerol.

The degradation of these compounds occurs by the hydrolysis of the fatty acids from the glycerol, followed by the oxidation of the fatty acids—2 carbons at a time—in the form of acetyl-CoA units, which may pass directly into the Krebs cycle. Glycerol may enter the glycolytic pathway.

This brief outline of cellular metabolism is not complete, but it should serve to illustrate several points. All macromolecules serving as food must first be broken down, at least to the level of monomers. These monomers may serve either as sources of energy by further oxidation or they may be constructed into new and unique macromolecules typical of that cell. All metabolic conversions occur in discrete steps, each step being catalyzed by a different enzyme. Some but not all of these steps require input of ATP. The cell is thus able to maintain a balance between catabolic and anabolic processes.

RESPIRATION AND PHOTOSYNTHESIS

The processes of respiration and photosynthesis are coupled by the flow of carbon, oxygen, and hydrogen through the biosphere. CO_2 and H_2O are consumed in photosynthesis, forming organic molecules and releasing O_2. In respiration, the organic molecules are oxidized to CO_2 and H_2O as O_2 is consumed. Each process depends upon the other for its supply of raw materials. In plants, both processes occur in the same organism. Some of the organic molecules formed in photosynthesis are broken down by respiration to provide ATP for the energy needs of the organism. Heterotrophic organisms are capable only of respiration and are dependent upon the photosynthetic organisms for a supply of organic molecules and O_2.

In the flow of energy through the biosphere, these two processes also play major roles. Radiant energy from the sun is partially converted to chemical energy through photosynthesis; the balance of the radiant energy is lost. The stable molecules formed in photosynthesis are broken down in respiration, with part of the chemical energy being stored in the short-term ATP storage and the balance being lost. The chemical energy of ATP is then used to do various kinds of useful work. However, all of the energy eventually is converted to heat as it passes through various processes. One of the most important forms of work done is the synthesis of the various special organic molecules needed for growth and reproduction of the organism.

FURTHER READING

Good general introductions to metabolic processes are given in books by Conn and Stumpf (1966) and Lehninger (1970). Lehninger (1961) briefly summarizes the energy transfer processes of photosynthesis and respiration. The initial description of the Krebs citric acid cycle was given by Krebs (1950), and more complete descriptions and discussions are given by Krebs and Kornberg (1957). Lehninger (1964) discusses the respiration processes in relation to the structure of the mitochondrion, which will be discussed in Chapter 7 of this book.

Further Readings for Chapter 5 are also relevant for this chapter.

Organization

One of the insights that has been most helpful to biologists is the so-called Cell Theory. It is difficult to say precisely what that phrase encompasses. In part, this is because the cellular nature of living systems was revealed not in a dramatic single resolution but rather through a series of clarifications that were developed throughout most of the mid-nineteenth century.

The basic postulates of the Cell Theory are, first, that all organisms are composed of subunits resembling one another in the possession of a certain set of organelles and a boundary; second, that these entities arise only through the division of pre-existing cells. . . .

The establishment of the Cell Theory, though attended by far less ceremony than the nearly contemporaneous Darwinian Revolution, has made it possible for biologists to deal coherently with what would otherwise be a hopeless welter of diversity. The doctrine of evolution offered explanations for the enormous breadth of the living spectrum; the cell theory offered hopeful assurances that these variations, despite their extent, had a theme, that the theme was the cellular organization of living systems, and that one might hope to comprehend some of the basic whys of life without inspecting an infinite series of special cases. That expectation has been amply fulfilled. . . .

—Donald Kennedy (1965)

7
Eucaryotic Cells

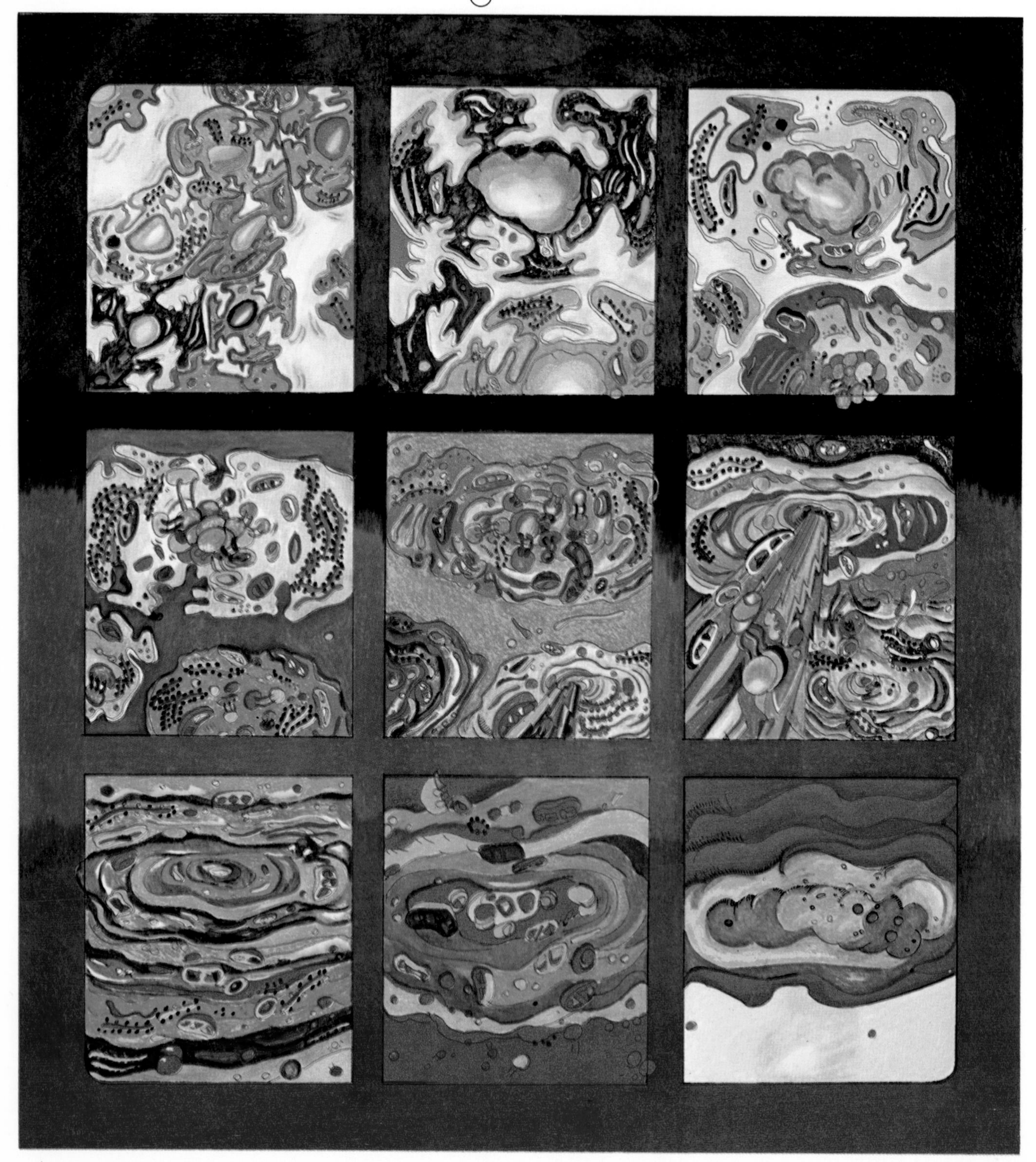

In 1969 thousands of Americans became fascinated by the microorganisms that Theodor Rosebury described in *Life on Man*. This delightful book was widely reviewed and soon after publication was prominent on the best-seller list. It sparked magazine and newspaper articles, and its author was a widely sought guest on radio and television talk shows. In contrast, the discoveries that Rosebury describes received little attention at the time they were made. Even when the cell theory—one of the great scientific steps in understanding the organization of living systems—was set forth in 1838–1839, it received no coverage in the newspapers, which were filled with articles on cross-Atlantic ship travel, international treaties, and the Opium War.

DEVELOPMENT OF THE CELL THEORY

Basically, the cell theory states that all higher organisms are composed of combinations of simpler subunits called cells. This deceptively simple idea proved to be one of the major keys to the recognition of underlying structural and functional similarities among the great variety of organisms that were being described and classified.

By the seventeenth century, scientists had made many observations concerning the structures and functions of living organisms, but the power of observation was limited by man's eyesight. The invention and use of the microscope gave man the power to observe smaller and smaller objects. The use of the microscope resolved many previously unanswerable questions. For example, it was known that the heart pumped the blood into the arteries and that the veins somehow collected the blood and returned it to the heart, but it was not until the microscope revealed the tiny capillaries that the connections between arteries and veins and the true nature of the circulatory system became established.

One of the earliest microscopists was Robert Hooke. In 1662 he was appointed curator of experiments for the Royal Society of London. This distinguished body of scientists met weekly, and it was Hooke's job to set up an experimental demonstration for their observation at each meeting. Many of his demonstrations were simply observations of familiar objects under the microscope. One such demonstration was made using a thin slice of cork (Figure 7.1). Hooke observed that the cork appeared to be composed of many little boxes, or cells, lined up end to end. For this observation, Hooke is credited with the discovery of the *cell*.

Among the great early microscopists was Anton van Leeuwenhoek, who owned a drapery shop in the Dutch town of Delft. In his spare time, Leeuwenhoek built hundreds of tiny, single-lens microscopes and used them to make remarkably accurate observations. Although Hooke and others developed more powerful compound microscopes, Leeuwenhoek managed to see more than anyone with his delicately ground lenses, many no larger than a pinhead. Many of his drawings show cellular structures in plant and animal tissues, but Leeuwenhoek limited himself to reports of what he saw and offered no interpretations. Biologists were interested in his descriptions of "animalcules" ("little animals"), which he first saw in 1675 and thereafter observed in great variety. Leeuwenhoek also described the small objects in blood later called red blood cells. Although microscopists continued to describe cellular structures throughout the eighteenth century, they failed to suspect that a better understanding of organisms could emerge from careful comparison of microscopic structures.

Figure 7.1 (below). The microscope Robert Hooke used to observe the microscopic structure of cork.

Figure 7.2 Anton van Leeuwenhoek (far right) ground hundreds of fine lenses in order to observe sperm cells, yeasts, and bacteria. One of his early microscopes is shown at right.

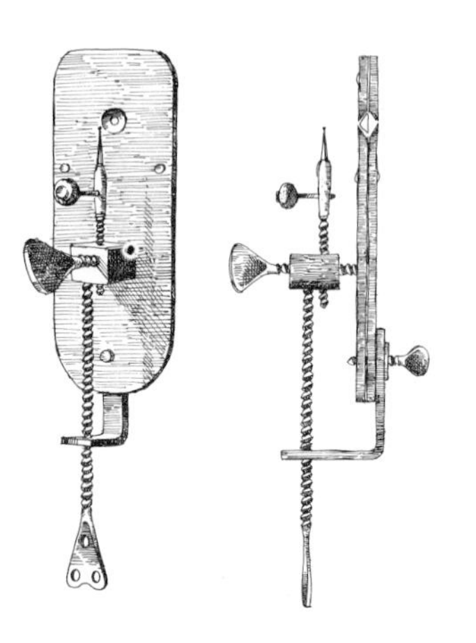

About the same time, microscopists were discovering that the interior of the cell is not the simple fluid originally described. Leeuwenhoek had described small dark objects within some cells, and other investigators began to report a confusing array of "inclusions" in various kinds of cells.

Among the biological writings of the early nineteenth century, there are precursors of the cell theory. The German natural philosopher Lorenz Oken elaborated a complex theory in which all of nature was regarded as reflecting the ideal characteristics of man. His extensive writings include one passage that seems to be a brilliant forecast of the cell theory. He states that "all organic beings originate from and consist of vesicles or cells."

Another early cell theory was set forth in 1824 by the French physiologist René Joachim Henri Dutrochet. He concluded from his microscopic studies that plants are composed entirely of cells and that plant growth occurs both through increase in the volume of cells and through the addition of new cells. Dutrochet then turned to the study of animal tissues and concluded that they too are composed of fluid-filled cells. He theorized that various plant and animal tissues are of different natures only because they contain different fluids in their cells.

With the benefit of hindsight, it is easy to see that Oken and Dutrochet were on the right track. However, as Canadian physician William Osler once commented, "In science, the credit goes to the man who convinces the world, not to the man to whom the idea first occurs." Neither Oken nor Dutrochet convinced the scientific world of the importance or the universality of cells.

That task was successfully accomplished by two young biologists—the botanist Matthias Jakob Schleiden and the zoologist Theodor Schwann, who in 1839 joined forces to become the "public relations" men for the cell theory. Schleiden had begun his career as a lawyer but became so depressed over his lack of success in that profession that he attempted suicide. Upon his recovery, Schleiden turned to the study of plants. He denounced the systematic collections of various species made by most botanists as so much "hay" and devoted his own efforts to careful microscopic analysis of plant structures. Schleiden (1838) argued that all higher plants "are aggregates of fully individualized, independent, separate beings, namely the cells themselves."

During a visit to the University of Berlin, Schleiden enthusiastically described plant cells and their nuclei to his friend Theodor Schwann, a me-

Figure 7.3 (left). Lorenz Oken was among the first theorists to set forth an early version of the cell theory.

Figure 7.4 (middle). Matthias Jacob Schleiden, a German botanist whose ideas were instrumental in the development of the cell theory.

Figure 7.5 (right). Theodor Schwann, a German zoologist and a colleague of Schleiden's.

thodical young physiologist and researcher. On the basis of his own independent study, largely with animal tissues, Schwann (1839) greatly extended Schleiden's conclusions. Schwann pointed out that animal tissues also are universally composed of cells. He wrote a book setting forth the idea that all organisms—from oak trees and tigers to men—are composed of individual cells. The fertilized egg from which an animal grows—whether the large egg of a bird, the small egg of a frog or a fish, or the microscopic ovum of a mammal—is a single cell, with a surrounding membrane and a nucleus much like those of any cell found in animal tissues. The development of an animal occurs, said Schwann, through the creation of new cells. He concluded that animals and plants are composed entirely of cells and of substances produced by cells and that the cells, to some extent, are independent living units, although they are subordinate to the entire organism.

By the fortieth anniversary of the publication of Schwann's book, the cell theory was so well established that an international ceremony was held in Schwann's honor. Tribute was also given to Schleiden for his important contribution to the recognition of the universal importance of cells in plants. Thus, by the late nineteenth century, Schleiden and Schwann were being credited as the "fathers" of the cell theory. The publications of Schleiden and Schwann were followed by rapid progress in the understanding of plant and animal organization.

The cell is recognized today as the basic subunit of any living system. A single cell is a clearly defined unit, bounded by a membrane that separates it from other cells or from the outside environment. The definition of the cell as a biological unit, however, has more basis than merely the existence of a physical boundary. A cell contains all of the genetic information, all of the translational molecules, and all of the enzymes that are essential to the life of that cell. In short, a cell is the simplest unit that can exist as an independent living system.

THE STRUCTURE OF CELLS

To most of the early investigators—particularly those specializing in the study of plant tissues—the cell appeared to be a fluid-filled wall or bladder with a granular or dense nucleus in the fluid. In later studies, attention shifted from the cell wall to the material inside the cell.

Gradually, biologists found that most animal cells are filled largely with protoplasm, a viscous, granular, constantly moving fluid inside the cell.

Figure 7.6. Early staining techniques were often crude and inefficient. The use of stains, however, helped in the microscopic examination of cell nuclei and other large organelles. These specimens of various marine animals were made in the middle of the nineteenth century using carmine dye, port wine, and other primitive staining media.

Furthermore, they discovered that the boundary of the living animal cell is a very thin membrane; the highly visible walls of plant cells are nonliving structures that lie outside the membrane.

With the further development of microscopes and staining techniques, more details of cellular structure were discovered. The term "protoplasm," with its implications of a homogeneous living substance, has now been almost entirely abandoned. The terms "nucleoplasm" (for the material of the nucleus) and "cytoplasm" (for the material inside the cell but outside the nucleus) are still used. Still, even these terms are decreasingly appropriate in view of the modern recognition that they refer to collections of distinct molecular structures rather than to simple and uniform substances.

The cells of plants, animals, and most microorganisms are similar in their basic structure. These *eucaryotic* ("having a true nucleus") cells are surrounded by a cell membrane, or plasma membrane, and contain a nucleus surrounded by a nuclear membrane.

Within the cytoplasm of the eucaryotic cell are a number of *organelles*, various specialized structures that perform particular functions and contain

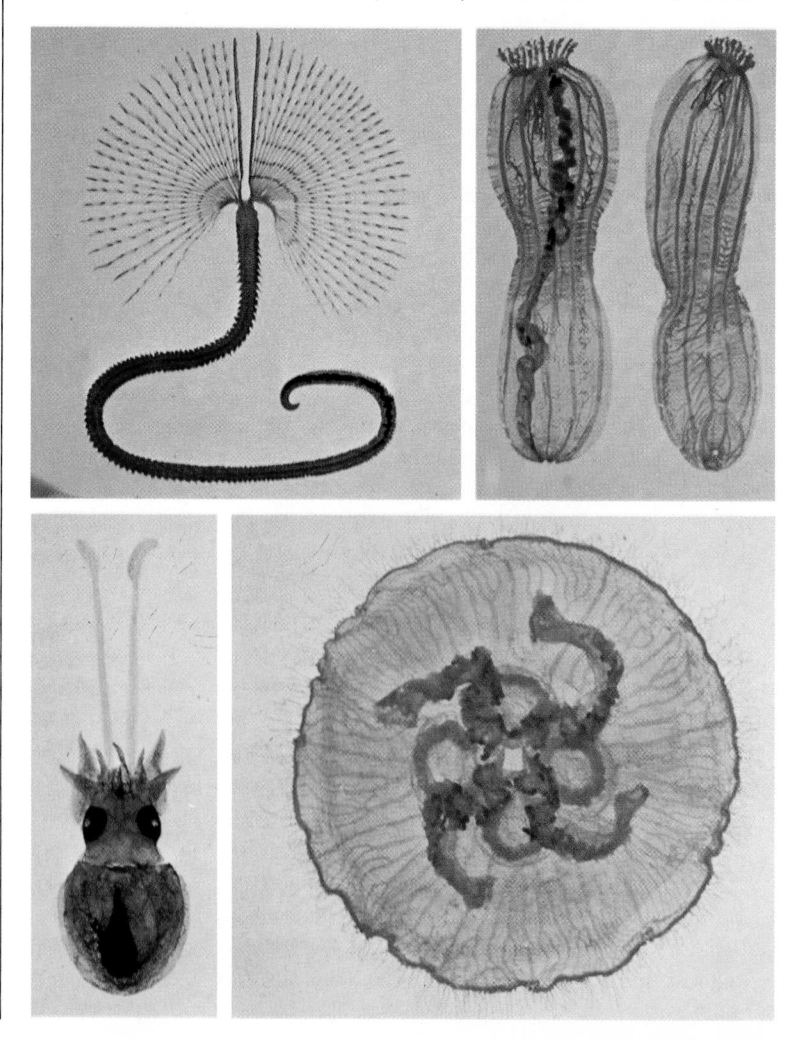

Figure 7.7. A size comparison chart of different types of cells. The ostrich egg and the bird eggs are reduced in size by one-half.

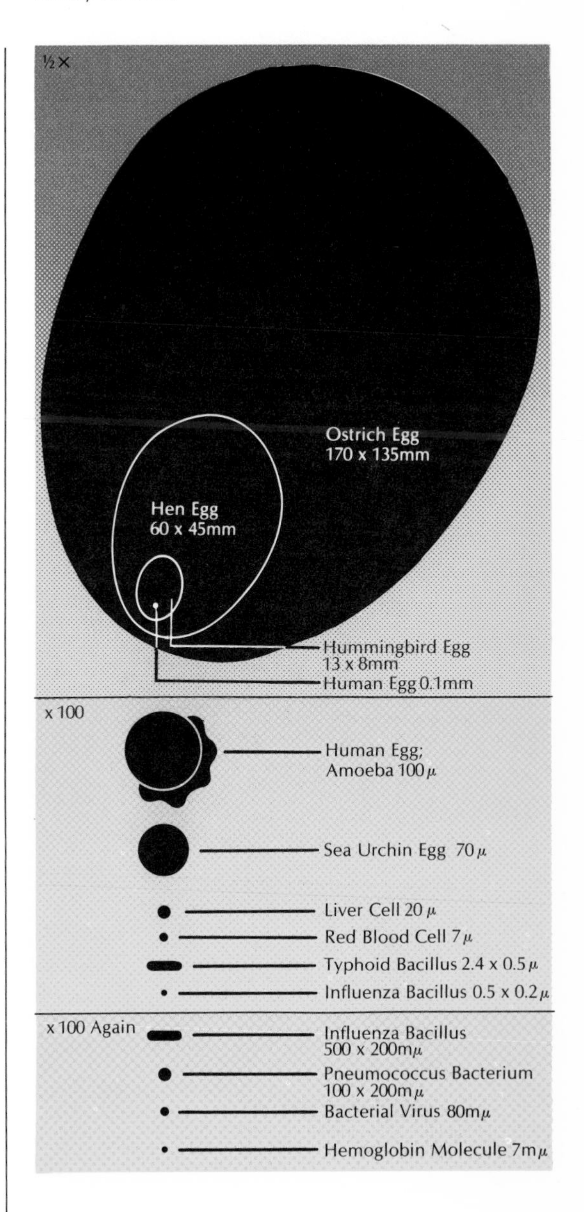

specialized membranes. Although there are many variations on the basic theme of eucaryotic cell structure—for example, kinds of cells in which one or more parts are developed in greater numbers or complexity—the basic structural pattern of eucaryotic cells is remarkably constant throughout the realm of life.

OBSERVATION OF CELLULAR COMPONENTS

What am I, Life? A thing of watery salt
Held in cohesion by unresting cells . . .

These lines by John Masefield convey a sense of the dynamic, continuous activity in living cells. Nevertheless, much of what has been learned about cells has been gained from the study of static structure, and static representations of cell structure are mere snapshots—frozen moments picked out of a miniature, tumultuous maelstrom of unending biochemical activity. If the descriptions, photographs, and diagrams of this chapter create an image of cells as complex but motionless pieces of molecular machinery, perhaps Masefield's words will help to stress their active, ever-changing nature.

Eucaryotic cells vary greatly in size. Some single cells, such as eggs and certain protozoans, are large enough that they can be seen with the unaided eye. Most eucaryotic cells, however, are about ten times smaller than the smallest object visible to the naked eye, but they can be seen with the aid of a microscope. Some of the larger organelles within eucaryotic cells can be distinguished with a light microscope, but their structural details can be detected only with the electron microscope (Interleaf 7.1). Many cellular components are too small to be seen except by electron microscopy.

It is difficult to develop a feeling for the sizes of submicroscopic cellular components. In examining micrographs, one must keep in mind that large and striking structures, such as the nucleolus, may be only 0.00005 inch in diameter. Cellular features are so small that it is awkward to express their size in fractions of an inch. Therefore, sizes of submicroscopic features are expressed in appropriately small units, *microns* and *angstroms*. One micron (μ) is equal to 10^{-6} meter (0.000001 meter, or 0.0001 centimeter), or about 0.00004 inch. One angstrom (A) is equal to 10^{-10} meter, 10^{-4} micron, or about 0.000000004 inch. The *millimicron* ($m\mu$), equal to 10^{-3} micron or 10 angstroms, is also used occasionally. Most eucaryotic cells are approximately 10 to 30 microns in diameter.

The human eye sees most things by reflected light—the light that bounces off the objects being observed. An object under a microscope, however, is normally observed by transmitted light—the light that has passed through the object. To be seen with a microscope, the object must be optically dense—it must absorb some of the light that passes through it. If it absorbs light of some wavelengths (colors) more than others, it appears to have the color of the wavelengths that are not absorbed. To be observed in the electron microscope, objects must absorb (or, more accurately, scatter) electrons.

Until recently, observation of a cell or subcellular structure was possible only after a lengthy treatment. The cell is killed, chemically fixed so that its constituents do not decompose, and its water then replaced with some solvent as a preparation for the addition of an embedding matrix. This matrix—usually a wax or plastic—holds the cell and its components rigidly in position while the cell is sliced into minutely thin sections. The sections then are stained with colored dyes (for light microscopy) or are stained with

Figure 7.8. A comparison between an electron micrograph of an onion root cell (above) and a light micrograph of the same type of cell (below).

Interleaf 7.1
LIGHT AND ELECTRON MICROSCOPY

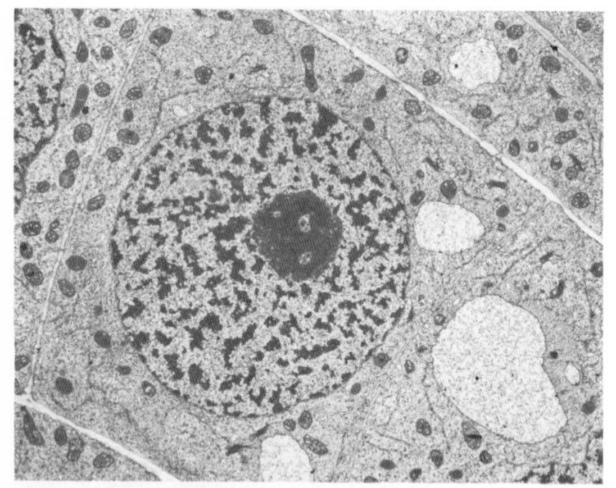

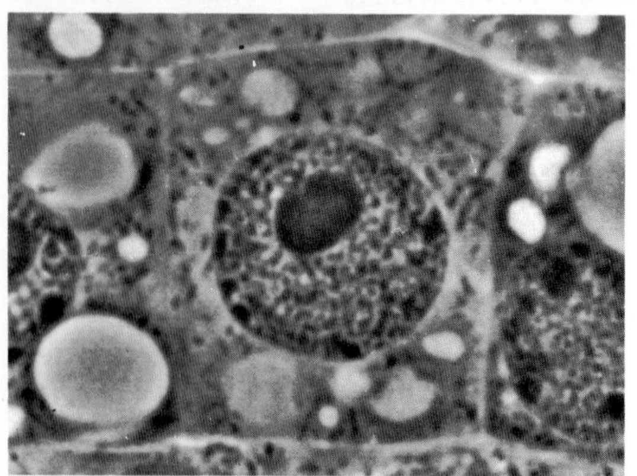

The ability to see details of small structures is limited by the *resolution* of the optical system being used. Resolution refers to the minimum distance between two points at which they can be distinguished; any two points separated by a distance smaller than the resolution will be seen as a single point. Figure 7.8 shows two sections through cells of an onion root. Both micrographs have a linear magnification of X 1,000 (that is, the distance between two points in the image is 1,000 times as great as the distance between the corresponding points in the actual specimen). One micrograph was taken with a light microscope, and the other was taken with an electron microscope. The image obtained with the light microscope is blurry and indistinct, whereas much finer detail can be seen in the electron micrograph. The finer resolution of the electron micrograph is of obvious value in studying cellular structure. In fact, much greater magnifications can be reached with clarity of detail.

The resolution of a light microscope is limited both by the wavelength of light being used and by certain characteristics of the microscope itself. The most significant limitation is due to the diffraction of light—the bending of light waves as they pass around the edge of an object. Diffraction causes the edge of an image to be blurred, and it is impossible to distinguish two point images very close together. In a light microscope, the diffraction becomes greater (and the resolution becomes poorer) as the wavelength of light is increased, as the optical density of the material between the object and the lens is reduced, and as the aperture (angle through which light is admitted) of the lens is reduced. The best resolution is achieved by using a dense oil between the object and the lens and by using violet light of short wavelengths. Even with an ideal microscope, the best resolution that could be achieved under such conditions is about 1,700 A, or 0.17μ. In fact, the best light microscopes have a resolution of about 0.25μ with white light.

Other effects that once posed limitations to the resolution of light microscopes have been overcome through the design of lenses that compensate for these effects. *Chromatic aberration* results when a lens focuses light of different colors (wavelengths) at slightly different distances from the lens. This effect severely hampered the observations of early microscopists, who saw each small object surrounded by rings of various colors. Achromatic lenses, which eliminate chromatic aberration through the use of two kinds of glass that counteract each other, were introduced about 1830. *Spherical aberration* causes light passing through the center of the lens to be focused in a different plane from light passing through the edges of the lens. This effect also was soon minimized by proper lens design.

Figure 7.9 shows simplified diagrams of light and electron microscopes. In order to emphasize the similarities between the two instruments, the light microscope is inverted, and the dimensions and details of the two instruments have been somewhat distorted. In the light microscope, light from a hot filament (or from the sun) is passed through a condenser lens to produce a parallel beam of light. This beam passes through the specimen and is then focused by the objective lens. An eyepiece lens is used to magnify the image produced by the objective lens.

Light microscopes reached essentially their theoretical limits of resolution late in the nineteenth century, when instruments with oil immersion lenses and condenser lenses became generally available. Even the use of shorter wavelength ultraviolet light would only improve the resolution by a factor of about 2 to about 0.1 micron.

The electron microscope is based on the fact that a beam of electrons has wave properties with very short wavelengths. The first experimental electron microscopes were built in the early 1930s, and commercial models became available in 1939. The electrons are drawn from a hot filament by an electric field. The beam of electrons is focused by magnetic fields, which are produced by electromagnets. A visible image is produced when the electrons strike a coated screen, whose molecules emit visible light when struck by electrons. In most electron microscopes, this screen swings out of the way so that the electrons can fall directly onto photographic film and produce a micrograph. Because electrons are scattered by gas molecules, clear

Figure 7.9. A diagram comparing similar components in a light microscope (left) and an electron microscope (right). Shown below is a scanning electron micrograph of the compound eye of the fly.

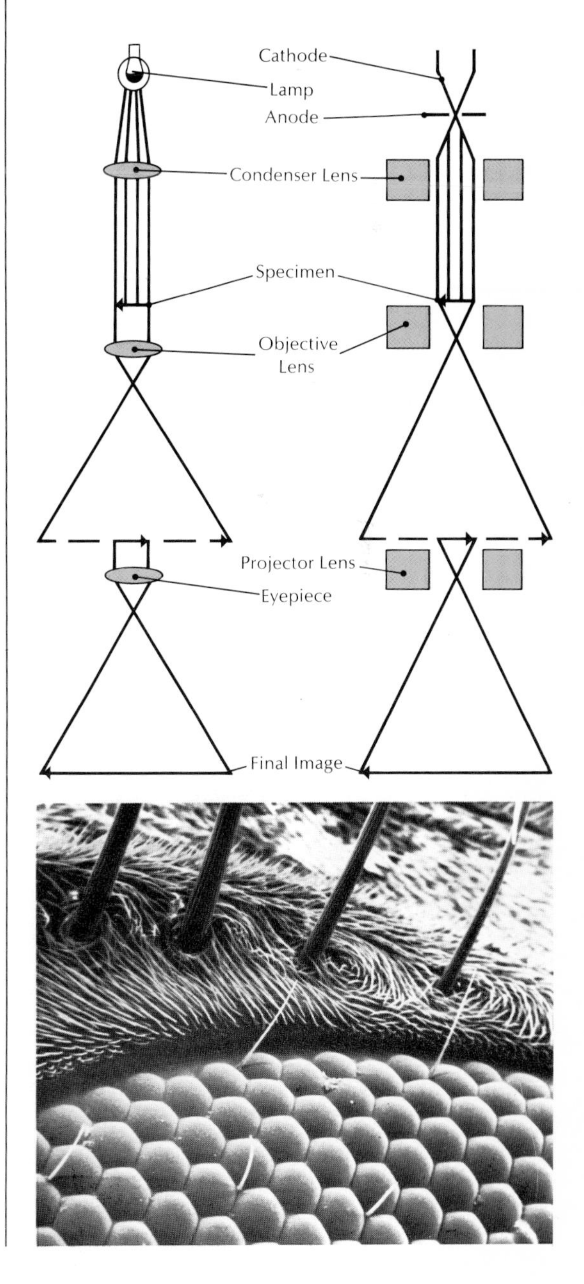

images are formed only if a vacuum is maintained within the electron microscope. Because electrons are scattered so easily, the specimen must be very thin—a few hundred angstroms or less for most biological specimens.

In the light microscope, contrast between dark and light areas of the image results chiefly from absorption of light by parts of the specimen. In the electron microscope, dark areas result where parts of the specimen scatter electrons out of the beam so that they are not focused onto the image. With a specimen thicker than a few hundred angstroms, most of the electrons are scattered and a uniformly dark image results. With a very thin specimen, most of the electrons pass through the specimen except where they are scattered by the heavy atoms of a metal stain.

In the light microscope, the image is focused by moving the glass lenses. In the electron microscope, the focal length of the magnetic lenses is changed by altering the current flowing through the electromagnets. In general, however, the path of the electron beam through the electron microscope is similar to that of the light beam through the light microscope.

The wavelength of the electron beam used in the typical electron microscope is about 0.05 A. With an ideal instrument design, it would theoretically be possible to achieve resolutions of about this order in electron microscopy. At the present state of the art, however, the resolution of the electron microscope is limited primarily by factors other than diffraction. The equivalent of chromatic aberration exists in the electron microscope if the electron beam contains electrons of varying velocities, but this effect has largely been eliminated by use of very stable voltage supplies to draw the electrons from the filament. However, spherical aberration is a very serious limitation in the electron microscope because the magnetic field in the lenses differs greatly in strength from the edge to the center of the lens. In order to minimize spherical aberration, it is necessary to use a very narrow aperture, which permits the image to be formed only from the part of the electron beam that passes through the center of the lenses. This narrow aperture greatly increases the diffraction effects, so that the best resolution achieved by current electron microscopes is around 1 to 2 A. Further improvements in instrument design should make possible a tenfold increase in the resolution and make possible the visualization of molecular structures.

A major problem in electron microscopy has been the development of techniques for specimen preparation. New devices were developed to cut extremely thin sections without seriously distorting the structures. Methods were devised to support these thin slices on metal grids covered with fine films of carbon or collodion. Materials were found for embedding, fixing, and staining the specimens. Because the heavy-metal stains that must be used to make biological specimens visible under the electron microscope are largely toxic to living organisms, because living organisms are too thick to be transparent to an electron beam, and because specimens must be completely dehydrated before being placed in the vacuum of the microscope, it is impossible to observe living cells with the electron microscope. Therefore, there is no direct way to determine how much the *ultrastructure* (small details of structure visible only by electron microscopy) is altered by the techniques of preparation.

Various techniques of specimen preparation are used for electron microscopy. *Electron stains* are electron-scattering (or electron-dense) materials that combine preferentially with certain parts of the biological structure. Like the stains of light microscopy, electron stains darken certain parts of the structure in the image. *Negative staining* involves the use of a general film of electron-dense material that is pushed aside by the biological molecules, producing a negative or light image of the biological structure. *Heavy-metal shadowing* is accomplished by spraying a fine film of electron-dense metal over the specimen at an angle, so that heavy deposits are built up on one side of the structures and a transparent shadow is left on the other side. In the recently developed *scanning electron microscope*, a beam of electrons is scattered from the surface of a thicker specimen, making possible micrographs that show three-dimensional structures.

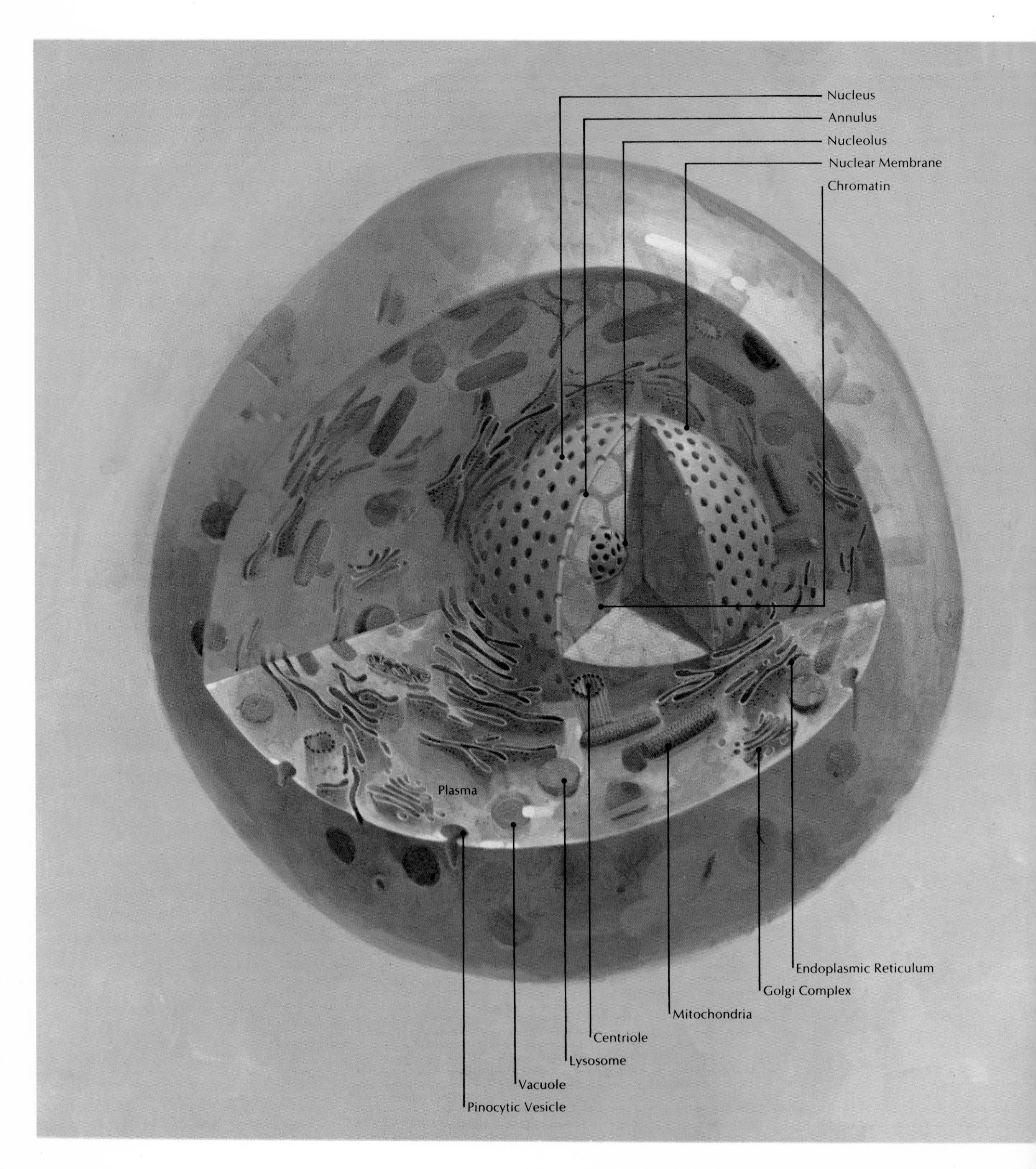
Nucleus
Annulus
Nucleolus
Nuclear Membrane
Chromatin
Plasma
Endoplasmic Reticulum
Golgi Complex
Mitochondria
Centriole
Lysosome
Vacuole
Pinocytic Vesicle

Figure 7.10. A three-dimensional drawing of a eucaryotic animal cell showing its organelles suspended in the cytoplasm (opposite page). Shown at upper left on this page is the nucleolus, which contains RNA and protein and is the site of rRNA synthesis. At the upper right is a Golgi apparatus, a stack of folded membranes that packages and transports materials to be secreted. The mitochondria (lower left) are the sites of cellular respiration. The endoplasmic reticulum (ER) at the lower right forms a complex system of cisternae, or sacs, throughout the cell. Rough ER is studded with ribosomes and is extensive in protein-secreting cells. Smooth ER can be seen in the lower righthand corner.

Nucleolus

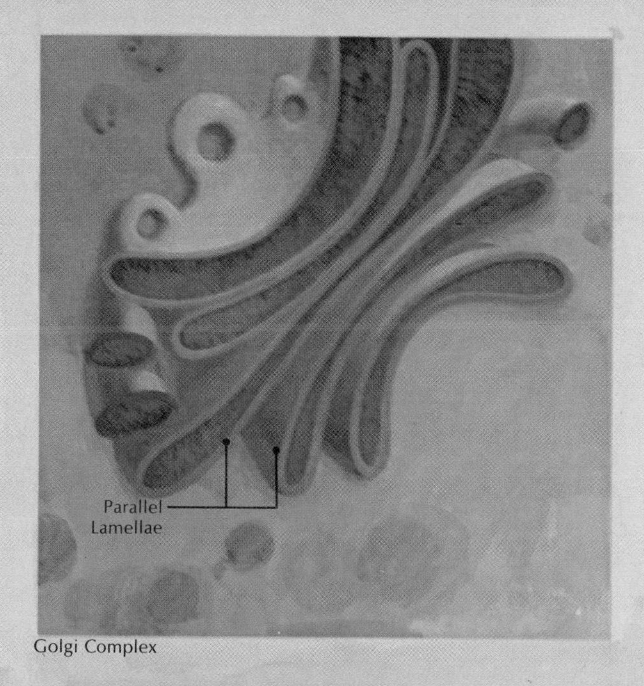

Golgi Complex

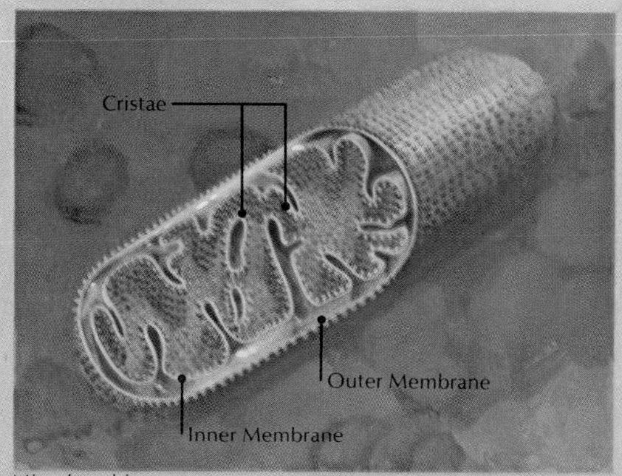

Mitochondria

Ribosome

Endoplasmic Reticulum

Figure 7.11a. Electron micrograph and diagram of a cell nucleus, the most prominent feature of a eucaryotic cell. Several dense nucleoli and scattered chromatin material can also be seen. (× 4,000)

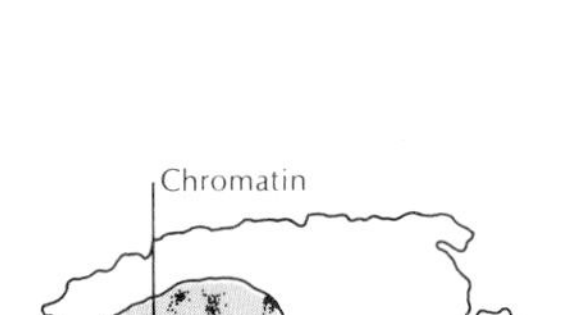

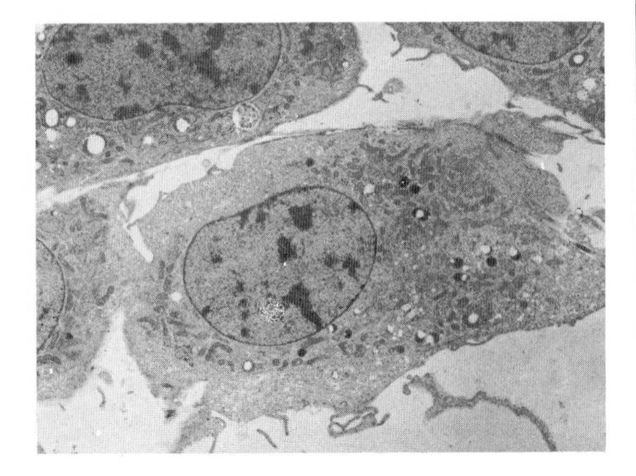

electron-dense metals (for electron microscopy). Some dyes are general or nonspecific stains; others specifically stain certain cellular constituents.

Biology is the study of life, but the student of cell structure usually observes only dead, preserved materials. He must infer how they looked when they were alive. He does not see the actual materials of the cell but only the staining materials. Furthermore, he is forced to deduce a dynamic, continually changing picture of cellular activity from a series of instantaneous, static "snapshots" of the structure.

Special phase contrast and interference microscopes have been developed to enhance the visibility of structures within living cells. With these instruments, it is possible to observe processes in living cells, processes that formerly had been deduced only from the study of a great many sections of fixed and stained cells. However, no means of observing living cells in the electron microscope is yet available.

THE CELL NUCLEUS

Generally, the *nucleus* is the most prominent feature of a eucaryotic cell (Figure 7.11). The contents of a nucleus are separated from the surrounding cytoplasm by a nuclear membrane, which is clearly visible in electron micrographs (Figure 7.12). This membrane is double; it appears to consist of two membranes, each about 70 A thick, separated by a distance of about 150 to 200 A.

Closely spaced around the nuclear membrane are small, round structures called *annuli*, which are formed where the inner and outer membranes come together to form a much thinner membrane. These annuli sometimes are called nuclear pores, but this name is misleading because they are not open channels between the nucleoplasm and the cytoplasm. Annuli are thought to be selective barriers that permit the passage of macromolecules such as RNA but, at the same time, prevent the free exchange of ions between nucleoplasm and cytoplasm. The precise function of the membrane in the annuli is not known.

Early staining techniques for light microscopy revealed what appeared to be granular material scattered through the nucleus. This *chromatin* seemed to gather into threadlike bodies, or *chromosomes*, just before cell division. Biochemical analyses show that the chromatin and chromosomes consist of DNA in close association with RNA and protein. The electron microscope reveals that the granular-appearing chromatin in nondividing cells consists of the chromosomes in unwound form. During cell division, the chromosomes coil and condense to such an extent that their threadlike structure becomes visible under the light microscope. Electron micrographs of the highly compacted chromosomes of dividing cells have not revealed the detailed structure of chromosomal material. In thin sections, the chromosomes appear as densely packed aggregations of tangled fibers. Even with special techniques for isolation of individual chromosomes, the chromosome still resembles disorganized, packed masses of yarn.

Within the nucleus are one or more large bodies, the *nucleoli* (Figure 7.13). In most kinds of cells, the nucleus contains one or two nucleoli, but exceptions are numerous. The nucleolus contains large amounts of RNA and protein and is now known to be the site of the synthesis of ribosomal RNA. A small piece of chromosomal DNA, the *nucleolar organizer*, lies within the nucleolus and apparently carries information that directs the formation of the nucleolus itself and of ribosomal RNA.

Figure 7.11b (above). Electron micrograph view of the nuclear region of an onion root tip cell made by the freeze-etch preparation technique. The relatively new freeze-etching technique involves a splitting of membranes, which allows extremely detailed examination of membrane faces. (× 27,000)

Figure 7.12 (middle). Electron micrograph showing the double nuclear membrane. (× 26,000)

Figure 7.13 (below). A large central nucleolus is clearly visible in this electron micrograph and diagram. Also visible are large quantities of darkly stained chromatin material. (× 14,000)

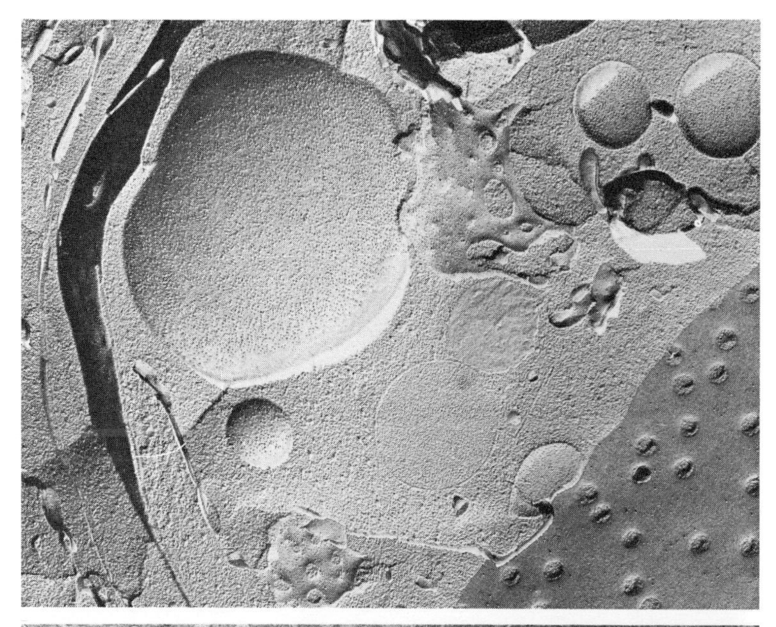

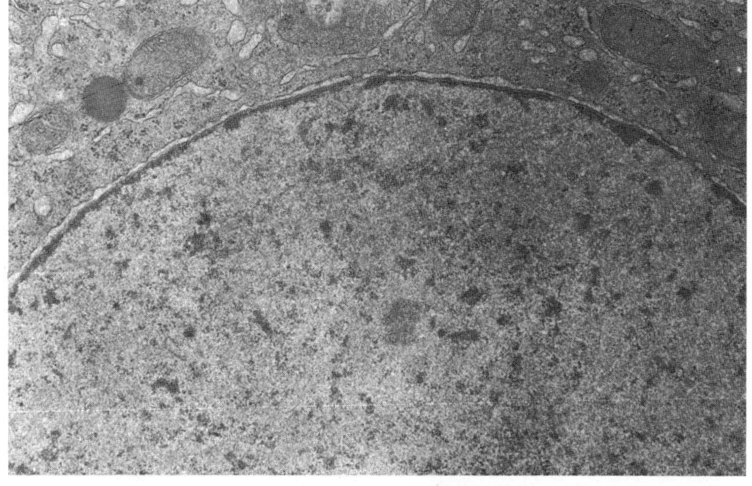

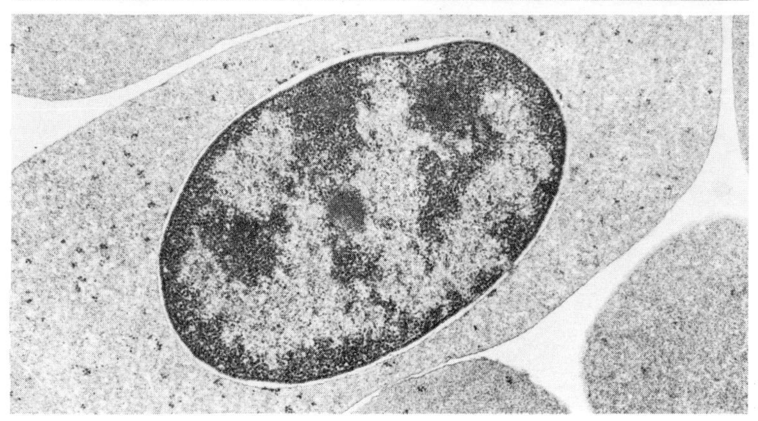

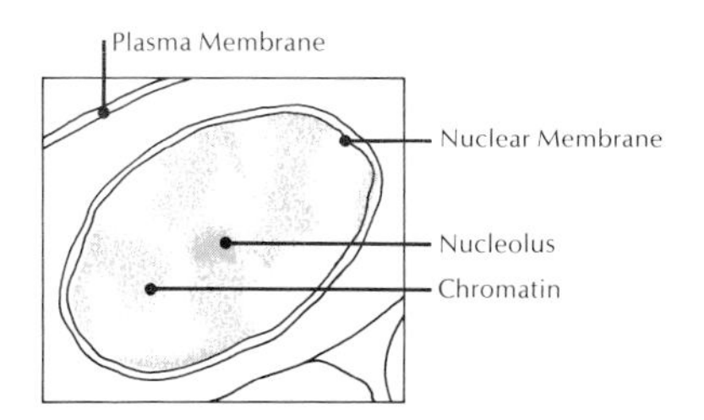

Although nucleoli vary in size among different types of cells and among different organisms, they are usually large enough to be discerned easily with the light microscope. Nucleoli are larger and denser in cells carrying on active protein synthesis. They are smaller and less dense in metabolically inactive cells. Electron micrographs show that the nucleolus is made up of two kinds of granules. One kind is about 150 A in diameter; the other, about 75 A. These granules are thought to be the precursors of ribosomes. In some nucleoli, the granules are arranged into fibers called *nucleolonemata*. No membrane is visible around the nucleolus, and the nucleoplasmic material appears to extend in various zones within the nucleolus.

CYTOPLASM

The cytoplasm of the eucaryotic cell consists of all the materials outside the nuclear membrane, including the outer cytoplasmic membrane of the cell. Some cells—such as amoebae and some egg cells—contain a relatively large amount of cytoplasm and relatively small nuclei. Others, such as thymus cells, are almost all nucleus with very little cytoplasm. Depending upon the functions of certain cells, the cytoplasm can contain such specialized products as hemoglobin, starch, or yolk. The cytoplasm contains the organelles and also bodies of inactive or structureless materials—such as droplets of lipid or starch—called *inclusions*. The types and numbers of organelles and inclusions in the cytoplasm vary with the activity and type of cell. Despite the variations from cell to cell, certain general features that hold true for most eucaryotic cells can be described.

Plasma Membrane

Every cell is surrounded by a plasma membrane, or cell membrane. The existence of this membrane was postulated long ago because cytoplasm flows

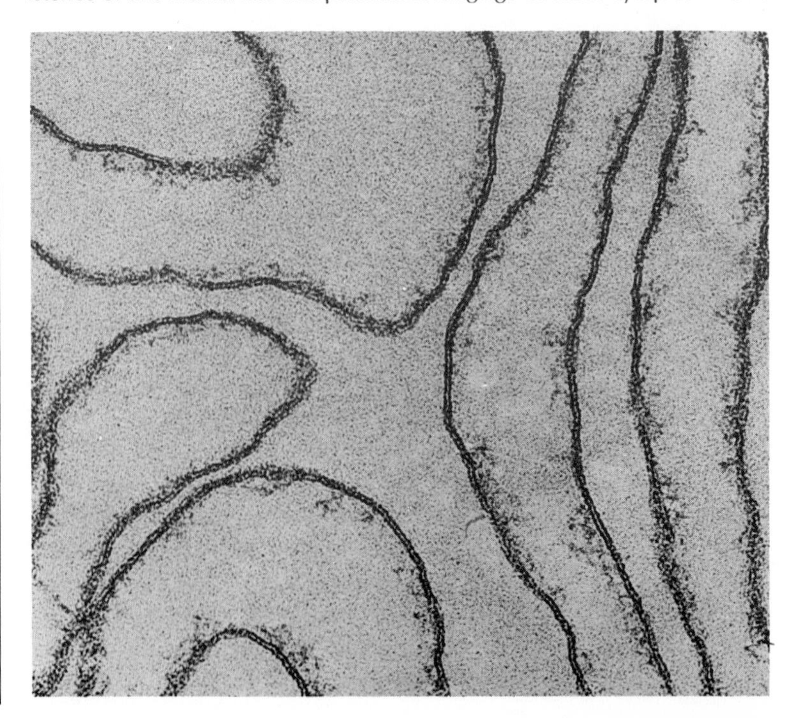

Figure 7.14. Electron micrograph of a plasma membrane. Note the two dark lines separated by a light line, together making up the unit membrane. (× 500,000)

out of the cell if the cell surface is punctured or torn. With the light microscope, the membrane can be seen—if at all—only as a very thin edge. The electron microscope, however, shows the plasma membrane as a dense line about 90 A thick. In high-resolution micrographs of favorably sectioned and stained material, the cell membrane appears as two 30 A, electron-dense lines separated by an electron-transparent space about 30 A thick. The cell membrane is composed largely of proteins and lipids, but the structural arrangement of these components within the membrane is still not known. Current theories of membrane structure and function are discussed in Chapter 10.

The plasma membrane separates the highly unstable and chemically reactive molecules of the cell's interior from the external environment. The membrane regulates the flow of materials into and out of the cell with exquisite precision and efficiency. Substances such as water freely diffuse across the cell membrane; water molecules move toward the side of the membrane where the water concentration is lowest. The movement of ions, however, depends upon their size, charge, and chemical nature.

In many kinds of cells, the membrane is folded to form tiny projections called *microvilli* (Figure 7.15). These structures are most abundant in cells that specialize in the absorption of substances from the external environment—for example, intestinal cells—and they apparently serve to increase the surface area and absorptive capability of the cell. In some specialized cells, such as nerve cells or light-receptor cells in the eye, the plasma membrane is complexly folded to produce special structures related to the specific functions of these cells. Cellular structures called *desmosomes* seem to provide a form of tight connection between adjacent cells in multicellular organisms.

The plasma membrane is far more than a simple envelope surrounding the living protoplasm. In fact, it is a complex structure that plays an active role in the life processes of the "unresting cell."

Figure 7.15. Electron micrograph of microvilli.

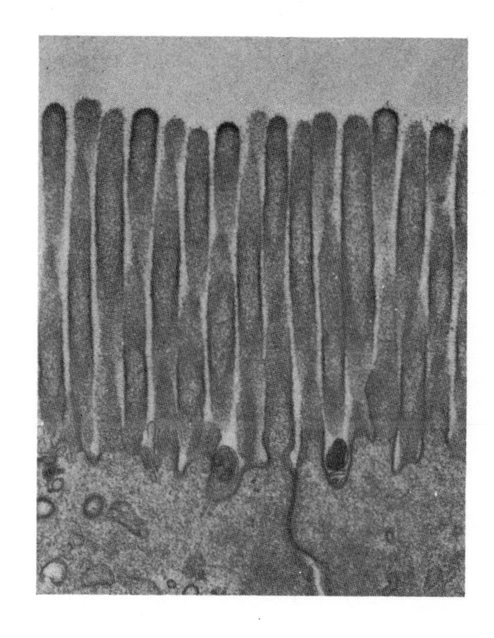

Ribosomes

Most of the RNA found in the cytoplasm is associated with protein in distinct particles called ribosomes. The ribosomes catalyze construction of proteins and are most numerous in cells that are actively synthesizing protein. Studies indicate that a ribosome consists of two separate subunits, each containing ribosomal RNA (rRNA). These two parts can be seen in some high-resolution electron micrographs. The larger subunit provides binding sites for transfer RNA (tRNA); the smaller subunit binds to a molecule of messenger RNA (mRNA). The message that specifies the amino acid sequence is read from the mRNA, and the molecules of tRNA insert their attached amino acids into the growing polypeptide chain in accordance with that sequence.

The formation of ribosomes begins in the nucleolus, where molecules of rRNA are transcribed from part of the DNA of the nucleolar organizer. The RNA molecule is split into two unequal parts, which acquire protein to form two unequal particles (ribosomal precursors). These particles move independently into the cytoplasm, where two units (one of each kind) combine to form a ribosome. A complete ribosome is composed of about 60 percent RNA and 40 percent protein.

Ribosomes, which are about 170 A in diameter, may be found scattered randomly through the cytoplasm, but they are often found in clusters or

Figure 7.16. Rosettes of ribosomes of the endoplasmic reticulum. (× 54,000)

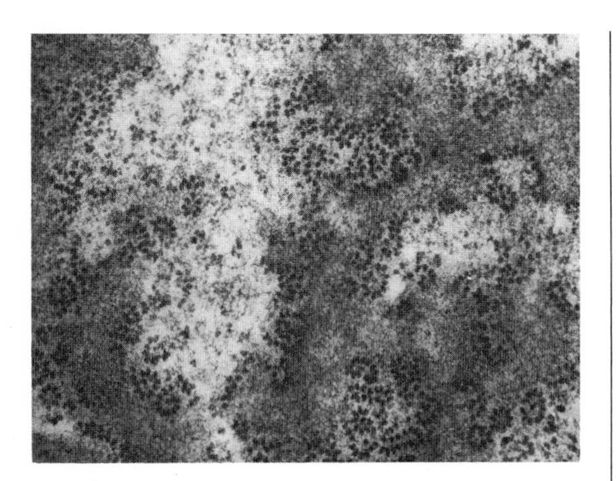

rows (*polysomes*), which contain several ribosomes attached to a single mRNA molecule. In eucaryotic cells engaged in synthesis and secretion of proteins, a large proportion of the cell's ribosomes are found attached to membranes within the cytoplasm.

Endoplasmic Reticulum

Before the advent of electron microscopy, biochemists variously described the cytoplasm as a watery mixture of organic molecules and soluble salts, as a complex mixture of solutes and gels, or—in more sophisticated theories—as "a complex polyphase colloidal system." Light microscopists, on the other hand, tended to emphasize the presence of organelles and of regions that were revealed by staining. The microscopists found considerable evidence to suggest the presence of regular submicroscopic structures within the cytoplasm. The electron microscope has confirmed that the cytoplasm of most cells contains an elaborate system of internal membranes.

The *endoplasmic reticulum* consists of membrane sheets folded through the cytoplasm, forming a complex system of tubules, vesicles, and sacs (cisternae). The membrane of the endoplasmic reticulum (ER) has a unit structure similar to that of the plasma membrane and nuclear membrane, and—in some places—the membranes of the ER may be continuous with the plasma and nuclear membranes.

Some parts of the ER membranes (*rough ER* or granular ER) are studded with ribosomes. Other parts of the membrane are smooth, with no ribosomes attached. Rough ER is particularly extensive in protein-secreting cells. Proteins synthesized on ribosomes on the ER are released into the cisternae of the ER. They then move into another part of the cell to be transported to their destination. Cells that produce and secrete nonprotein substances (for example, hormone-producing cells of the testicles and the adrenal gland) often contain large numbers of thin tubules of smooth ER. The relationship of smooth ER to hormone synthesis is not yet understood.

Golgi Apparatus

While experimenting with staining techniques on nerve cells, the Italian physician Camillo Golgi noticed the presence within the cells of a complex of vesicles (Figure 7.17). Similar structures were observed in nerve and secretory cells from many kinds of animals. For many years, the Golgi apparatus, as these vesicles came to be called, was the subject of considerable controversy. Because it cannot be seen in living cells, there was some reason to suspect that the structure might be an artifact of staining.

Electron micrographs confirmed the existence of the Golgi apparatus and provided some details of its structure. Its form in animal cells is highly variable, but in all cases it is composed of membranes similar in appearance to smooth ER. These membranes are folded into vesicles.

A common form of Golgi apparatus is the *dictyosome*, which consists of a stack of cisternae surrounded by a netlike halo of tubules and small spherical vesicles. Dictyosomes are found in many animal cells and are present in plant cells.

One function of the Golgi apparatus is to package and transport materials to be secreted to the exterior of the cell. Substances to be excreted accumulate in the vesicles of the Golgi apparatus. These vesicles enlarge, separate from the Golgi apparatus, and move to the plasma membrane. The membrane of the vesicle fuses with the plasma membrane, and the con-

Figure 7.17 (above). Electron micrograph showing a cross section of a Golgi apparatus, which is comprised of a stack of cisternae or vesicles. Note also the dictyosome, a stack of well-defined vesicles typically surrounded by a halo of smaller branching tubules. (× 50,000)

Figure 7.18 (below). Electron micrograph and diagram illustrating the variety of shapes of mitochondria. Each mitochondrion is surrounded by a double membrane and has the internal arrangement of folded membranes called cristae. (× 36,000)

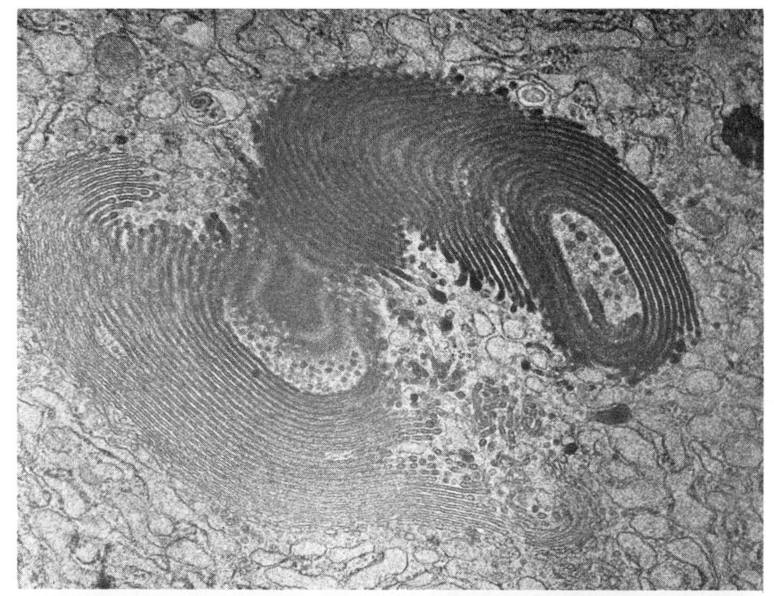

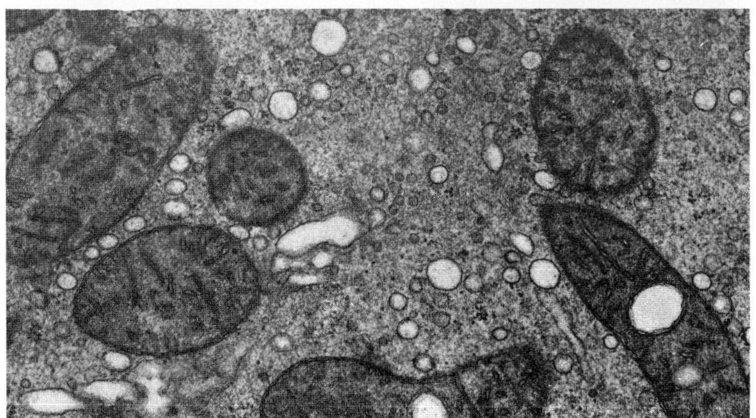

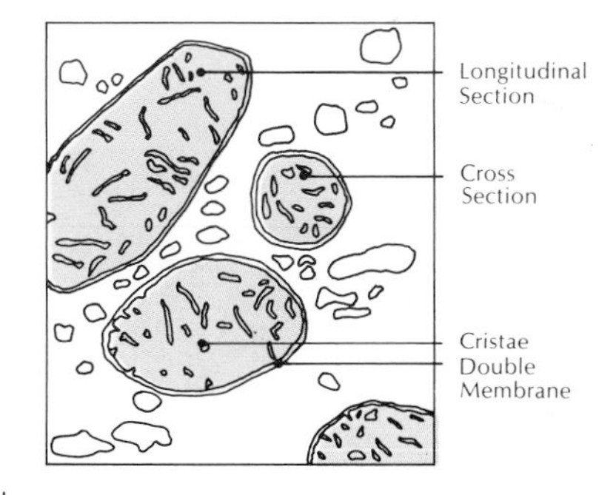

tents of the vesicle are discharged to the exterior of the cell. This process has been observed in the secretion of plant cell walls and in the secretion of enzymes and other substances by animal cells.

Mitochondria

Mitochondria are small, generally oval organelles found in nearly all eucaryotic cells, but they are absent from procaryotic cells (Figure 7.18). These structures serve as cellular "power plants." Cells that have high energy requirements—muscle cells, for example—contain many large mitochondria, whereas cells with low energy requirements have smaller mitochondria in sparser numbers. It is in the mitochondrion that cellular respiration takes place, with liberation of CO_2 and H_2O and phosphorylation of ADP to ATP (Chapter 6). Only a few types of highly specialized eucaryotic cells, such as mammalian red blood cells, lack mitochondria.

Mitochondria are not uniformly distributed through the cytoplasm of most cells but are often found "where the action is." For example, there are

Figure 7.19. Electron micrograph of lysosomes. (× 50,000)

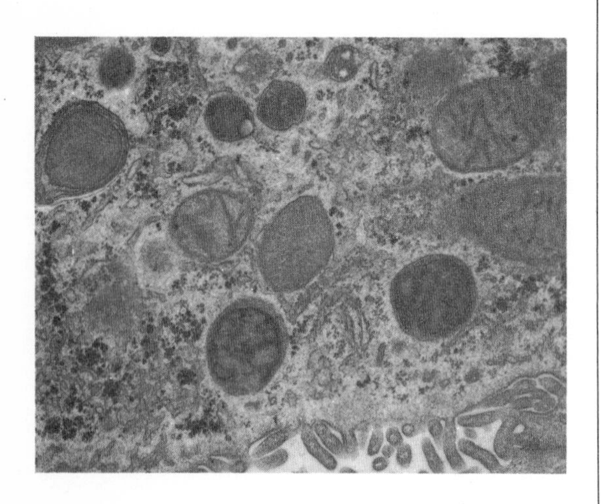

large accumulations of mitochondria near the basal membrane of kidney tubule cells, where a great deal of energy is used in transporting ions across the cell membrane.

Most mitochondria are about 0.5μ in diameter and from 0.5 to 7.0μ in length. They range in number from a single mitochondrion per cell in one kind of eucaryotic alga to hundreds of thousands of mitochondria per cell in some amoebae. The mitochondrion is surrounded by a double membrane. The fluid-filled space between the membranes may be 60 to more than 200 A wide. In most mitochondria, the inner membrane is enfolded to form *cristae*—sheets or tubules that extend across the interior of the mitochondrion. The material within the inner membrane, the *matrix*, often contains fibers, granules, or droplets.

The surfaces of the cristae in contact with the matrix are covered with small particles that are about 80 A in diameter and are attached to the cristae by stalks. There are more than 10,000 of these particles in each mitochondrion. These stalked particles are thought to contain the enzymes that catalyze electron-transfer reactions of oxidative phosphorylation. The enzymes presumably are arranged and grouped in such a way as to lead the intermediate molecules sequentially from one reaction step to the next. The enzymes of the Krebs cycle (Chapter 6), on the other hand, are located within the fluid matrix.

The glycolytic reactions and other steps that lead to the formation of acetyl–coenzyme A occur in the cytoplasm. The acetyl–coenzyme A complex is taken into the mitochondrion, where the Krebs cycle reactions occur in the matrix. As electrons are removed in the oxidation of the carbon compound, they are passed to the enzyme sequences in the particles of the cristae membranes, where the electrons eventually reduce oxygen to form water. The ATP formed during phosphorylation is transferred back to the outside of the mitochondrion, where it may participate as an energy source in various metabolic reactions (Lehninger, 1964).

Recent experiments have shown that mitochondria contain their own DNA, which is different in size and base composition from chromosomal DNA. Mitochondria can replicate independently of the nucleus, through a process similar to binary fission. The DNA of the mitochondrion is in the form of a circular strand.

Lysosomes

Although approximately the same size as mitochondria, lysosomes are organelles of very different structure, function, and origin. They are sacs of hydrolytic enzymes enclosed within a single unit membrane that isolates the enzymes from the rest of the cytoplasm (Figure 7.19). They are believed to form by the pinching off of sacs from the Golgi apparatus (Novikoff, *et al.*, 1964). The hydrolytic enzymes within a lysosome vary somewhat from cell to cell but typically include enzymes that hydrolyze proteins and nucleic acids.

Lysosomes, with their very simple structure, were not detected until the 1950s, and their function is still not certain. They have not been proven to exist in plant cells. They apparently act as disposal units of the cytoplasm, for within them are found the remains or fragments of mitochondria, ingested food particles and microorganisms, worn-out red blood cells, and any other debris that may have become incorporated within the cytoplasm. Lysosomes are most abundant in cells that specialize as scavengers within

Figure 7.20. Chloroplast in a leaf cell. The stroma lamellae, or large thylakoids, can be seen extending between grana from one end of the chloroplast to the other, with small thylakoids interspersed between them. (× 35,000)

the multicellular organism and in cells that participate in the breakdown of other cells (as in the absorption of the tail structure of a tadpole as it develops into a frog).

Once a lysosome has performed its function, it is expelled from the cell by the same mechanism that moves secretory materials out through the plasma membrane. In a dying cell, the membrane of the lysosome is broken down, and the hydrolytic enzymes are released into the cytoplasm—leading to irreversible changes and destruction of macromolecules, which in turn leads to the death of the cell. Malfunctions of the lysosomes are apparently involved in a number of human diseases, including cancer, and the discharge of enzymes from the lysosomes into the cytoplasm may produce cell damage or death.

Plastids

Plastids, which are present only in plant cells, are similar to mitochondria in several respects. They have a double membrane, as well as a system of internal membranes; they contain their own DNA and ribosomes and therefore may be able to reproduce independently of the nucleus; their DNA differs from that of the chromosomes in significant ways. Plastids, however, show a greater size range than do mitochondria. Whereas mitochondria are involved in making free energy available from the chemical energy of carbohydrates, most plastids are involved in the reverse process—the capture of solar energy with which carbohydrates are synthesized.

Plastids are of two general types: *chromoplasts* (which contain pigments) and *leucoplasts* (which are colorless). *Chloroplasts*, which contain the green pigment chlorophyll, are the best known of the chromoplasts (Figure 7.20). It is in the chloroplasts that photosynthesis takes place. Other

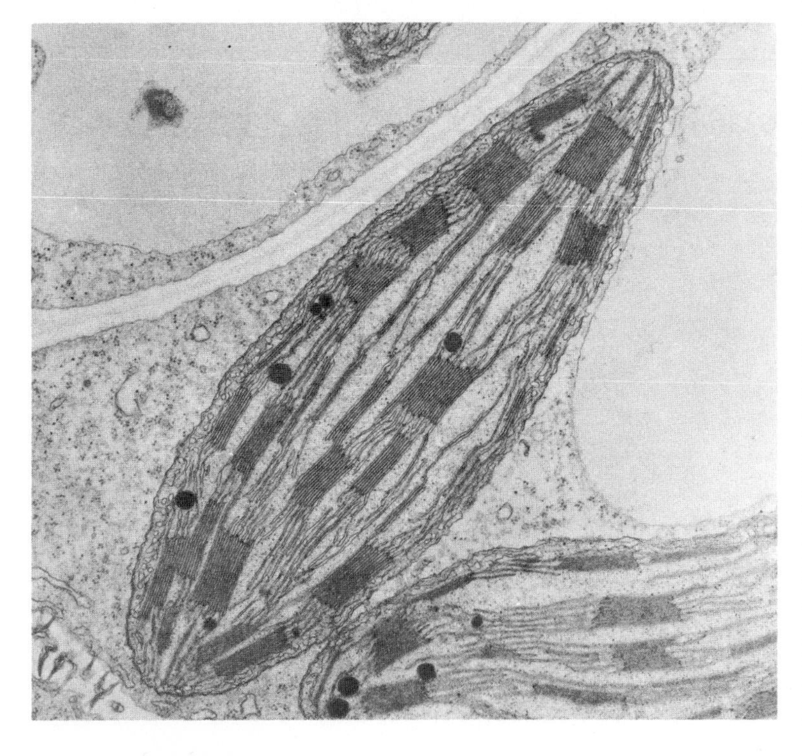

Figure 7.21 (above). Electron micrograph of a leucoplast, a structure specialized for storing polysaccharides. (× 86,000)

Figure 7.22. Electron micrograph of a chloroplast showing grana in portion of a developing leaf cell (middle). The stacks of membranes, or grana, in chloroplasts contain chlorophyll. (× 70,000). Shown below is a diagram of the structure of the grana.

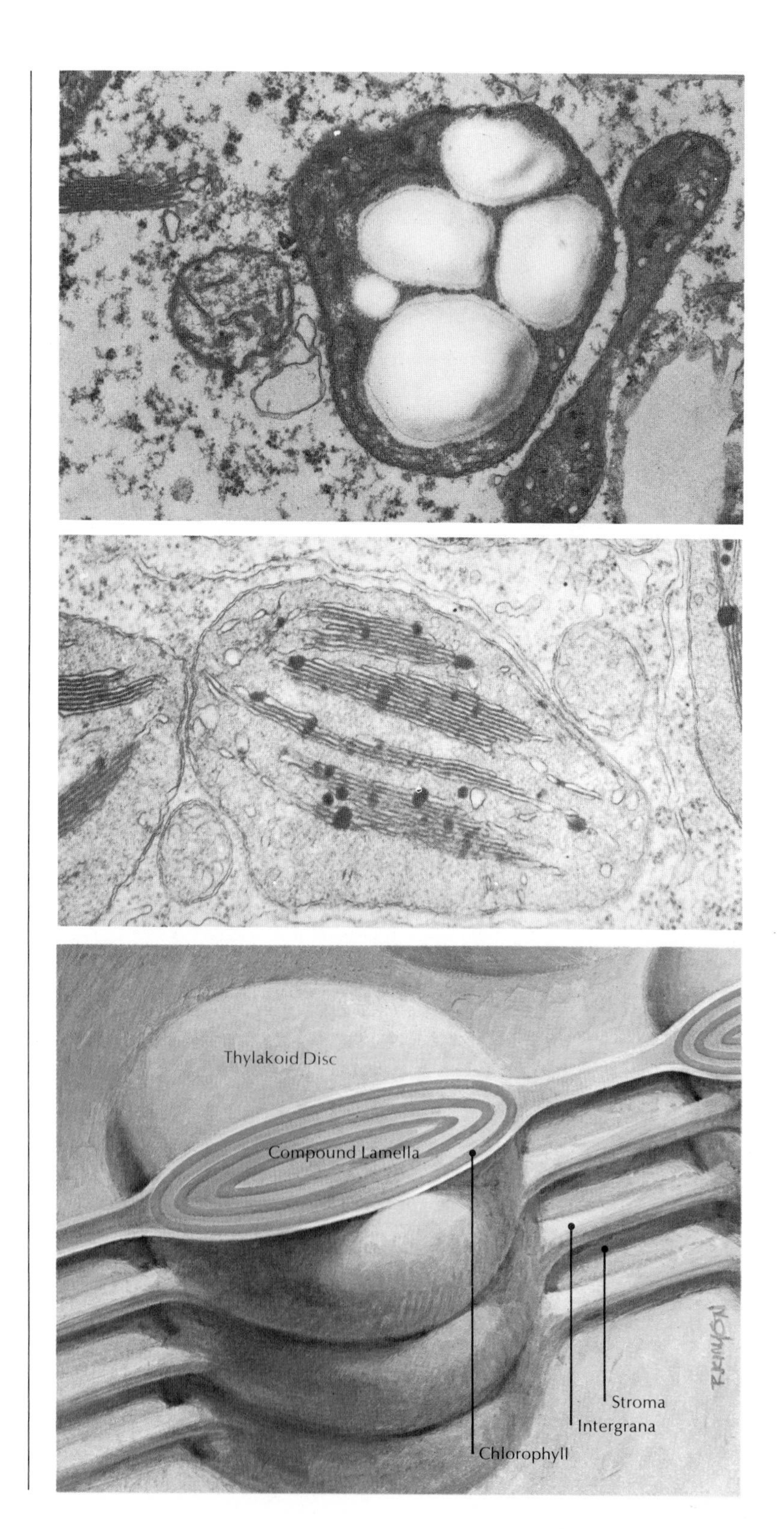

chromoplasts contain other kinds of pigments and often are brilliantly colored. The yellow color of carrots, the red color of tomatoes, and—in some cases—the various colors of flowers are due to such chromoplasts.

Leucoplasts are the sites for conversion of glucose to starch and to lipids or proteins; these products then are stored in the leucoplasts (Figure 7.21). The most conspicuous group of leucoplasts is the starch-storing *amyloplasts*, which are found in many fruits and vegetables. The whiteness of potatoes is due to the presence of abundant amyloplasts.

Because of the great importance of photosynthesis in the carbon cycle and energy flow of the biosphere, chloroplasts have been thoroughly studied. In higher plants, mosses, ferns, and some algae, chloroplasts are disc-shape bodies about 2 to 4μ in diameter and 1μ or less in thickness. In many algae, however, chloroplasts assume more elaborate shapes: stars, spirals, perforated sheets, and so on. One of the algae most frequently used in photosynthesis research is *Chlorella*, which has a single, cup-shape chloroplast occupying the bulk of the cell.

The outer membrane of the chloroplast is similar in structure to the plasma membrane of the cell, whereas the inner membrane is a complex system made up of flattened sacs called *thylakoids*. In higher plants, small thylakoids are stacked one upon the other to form a unit called a *granum*; these grana are interspersed with larger thylakoids. The material surrounding grana and thylakoids is the *stroma*, which contains dissolved salts, enzymes, more widely spaced membranes, ribosomes involved in chloroplast protein synthesis, and the DNA of the chloroplast. In red algae, the thylakoids are tightly spaced and scattered through the stroma; in brown algae, they are grouped into small numbers of layers. In some cases, the thylakoids between grana form tubular structures. Within the thylakoids and grana, the pigment molecules are stacked in regularly arranged units that facilitate the photosynthetic reactions.

Chloroplasts generally are not present in the cells of plants grown in the dark. Present instead are much smaller bodies called *proplastids*, which develop into chloroplasts if the plant is placed in the light. Many proplastids contain an elaborate structure called a prolamellar body, which develops into grana in the presence of light. Chlorophyll is synthesized from a precursor called protochlorophyll as proplastids develop into chloroplasts.

The term "proplastid" also is used for small plastids that lack any elaborate internal structure. Such simple proplastids are thought to be the parent structure from which all different plastid types develop. These simple proplastids reproduce by binary fission, as do the mature chloroplasts of simple plants. Until recently, it was thought that mature chloroplasts of higher plants are incapable of division, but mounting evidence from electron microscopy indicates that chloroplast division is probably a common process in higher plants as well.

Centrioles

Eucaryotic animal cells contain two inconspicuous pairs of bodies called centrioles. These organelles are also present in the sperm cells of algae, mosses, fungi, ferns, and some gymnosperms, but they are missing from other plant cells. Each centriole is a cylinder about 0.2μ in diameter, composed of nine parallel triplets of hollow, cylindrical *microtubules*. The two members of a centriole pair normally differ in length and are at right angles to each other. One is about 0.5μ in length, the other usually about 0.2μ long.

Figure 7.23. Electron micrograph and diagram of the structure of centrioles. The centriole consists of nine parallel triplets of microtubules. A similar pattern of microtubule arrangement is found in the structure of cilia and flagella. (× 130,000)

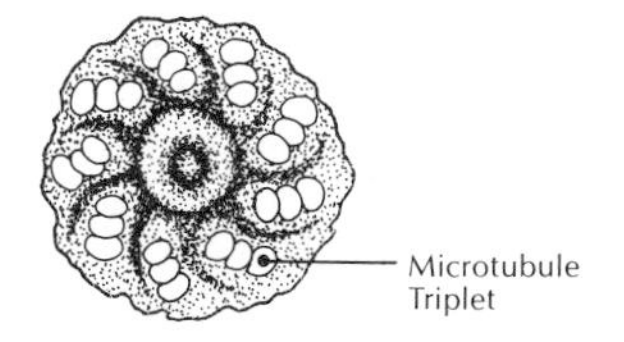

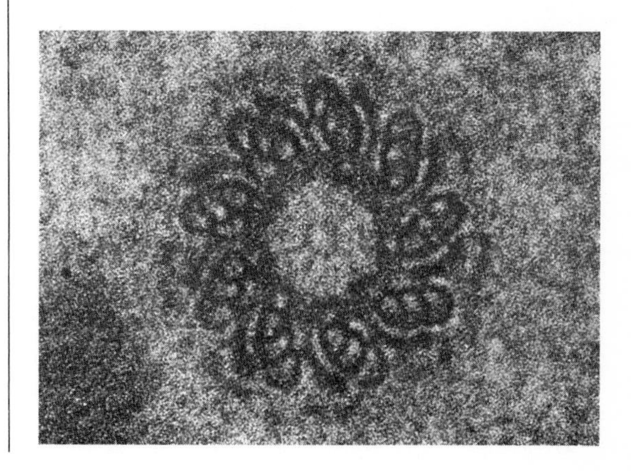

Figure 7.24. Electron micrograph of microtubules in cross section. The dark area to the right is chromatin. (× 35,000)

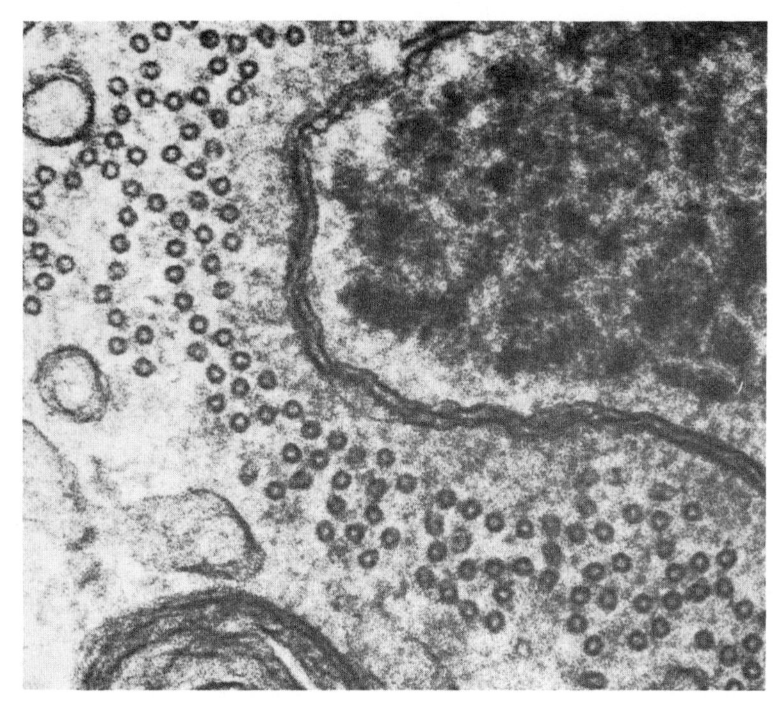

The size and structure of centriole pairs are remarkably constant throughout all the various eucaryotic cells that contain them.

Microtubules are common in the cytoplasm of most eucaryotic cells. In some animal cells, they are oriented in regular patterns in the cytoplasm and appear to channel the flow of substances through the cytoplasm. In the cells of higher plants, microtubules are particularly abundant in the cytoplasm near the plasma membrane and are also found extending along streams of cytoplasm that move through the cell interior. In some cells, microtubules apparently serve as a "cytoskeleton" that gives structural integrity to the cell. Microtubules are a prominent feature of nerve cells and fibers, but their function is unknown. Recent findings suggest that microtubules are involved in phenomena of cell movement, including the so-called amoeboid motion characteristic of amoebae, slime molds, and certain blood and tissue cells. It has been suggested that the regular shape of annuli, or nuclear pores, in the nuclear membrane may indicate that the pore is formed by a circular array of microtubules.

Each microtubule in turn is made up of 13 *filaments* arranged in a circular pattern (Ledbetter and Porter, 1964). The filaments are cylinders about 40 to 60 A in diameter. Other sorts of filaments are found free in the cytoplasm of many kinds of cells, where they act as structural supports and may serve other functions. The filaments are composed of proteins.

Cilia and Flagella

Cilia and flagella are hairlike appendages of cells. They extend from the plasma membrane to the exterior and are bounded by an outfolding of the membrane. "Cilium" and "flagellum" are relative terms. "Flagellum" generally is used for longer structures and "cilium" for shorter ones, but the two kinds of appendages have identical microstructures. Cilia are com-

Figure 7.25 (above). Electron micrograph of a sample of butterfly sperm with cross-sectional views of their flagella. Note the arrangement of microtubules within one flagellum—nine parallel ducts of microtubules with two single central fibrils. (× 65,000)

Figure 7.26 (below). Electron micrograph showing large vacuoles. Starch storage plastids can be seen surrounding the nucleus, with dictyosomes, mitochondria, and ER in the cytoplasm. Chloroplasts, not seen in this cell, are also common features of plant cells. (× 6,000)

monly about 10 or 20μ long, and flagella can be as much as thousands of microns long (in the sperm of some insects, for example). However, all cilia and flagella are about 0.2μ in diameter.

Most cilia and flagella are capable of motion. Because of their length, flagella usually move in an undulating fashion, whereas cilia move with simple, oarlike strokes. Their activity propels the cell to which they are attached or moves things past a stationary cell. Some kinds of cells have hundreds of cilia, some have only a few, and many have none at all. Most flagellated cells have only a single flagellum, but in algae and fungi the flagella usually occur in pairs. Many unicellular organisms move by means of cilia or flagella. A sperm cell is propelled by a single, long flagellum. The meeting of sperm and egg is further facilitated in many organisms by the motion of cilia on cells that line the female reproductive tract. In lungs, cilia move foreign particles such as dust and soot out of the respiratory tract.

A flagellum is an extension of one of the two centrioles of the flagellated cell. In a ciliated cell, many extra centrioles may be formed and may serve as *basal bodies* from which cilia develop. Two of the microtubules of each centriolar triplet extend the full length of the flagellum or cilium. One of each pair of microtubules possesses enzyme molecules that are essential to the motility of the appendage. An additional pair of microtubules extend up the center of the flagellum or cilium. If this central pair of microtubules is missing, the appendage is nonmotile. The arrangement of microtubules in the flagellum or cilium invariably forms the characteristic array called the *9 + 2 pattern.*

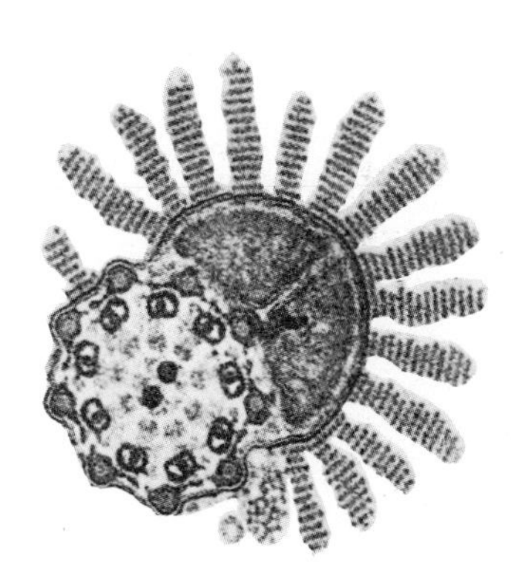

Vacuoles

Vacuoles are membrane-surrounded spaces found within all kinds of cells (Figure 7.26). They vary in size more than any other organelle. In fact, vacuoles might be considered to lie in a vaguely defined position between true organelles and simple inclusions. Some vacuoles play an active role in cell processes, whereas others serve merely as storage depots.

The largest vacuoles appear in plant cells, where they may make up most of the cell's volume. Vacuoles in plant cells are filled with fluid or "cell sap" under a pressure that helps to maintain the shape of the cell and to give the plant a rigid structure. A plant wilts because the amount of water available is insufficient to maintain the fluid pressure of the vacuoles. This loss of pressure leads to cell collapse.

In mature plant cells, a single vacuole may occupy 90 percent or more of the volume of the cell. The cytoplasm, nucleus, and plastids of such cells are pressed against the cell wall by the large central vacuole. Although the cells at the tips of roots and shoots as well as certain specialized cells in the plant body lack large central vacuoles, almost all of these cells have collections of small vacuoles.

Water is the major component of the fluid in large vacuoles of plant cells. Dissolved in this water are salts, sugars, pigments, and other substances. The red color of many flowers is due to pigments concentrated in the vacuoles of the flower petal cells. In citrus fruits, the contents of the vacuoles are quite acidic, giving these fruits their characteristic sour taste. In some cases, the fluid in the vacuole is so acidic that the cytoplasm would be severely damaged if it were exposed to the vacuolar contents.

Recent evidence suggests that the vacuolar membrane in many plant cells is closely associated with the ER. Some electron micrographs show the

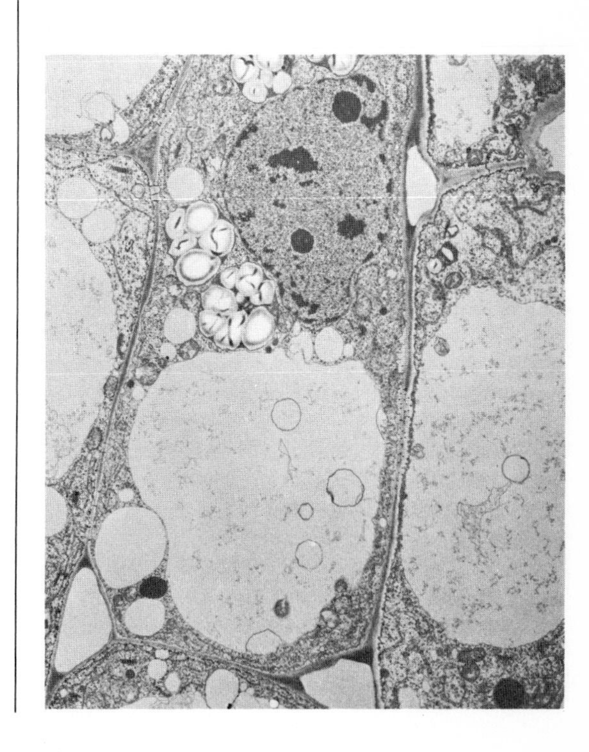

Figure 7.27. Scanning electron micrograph of chitin, an extracellular polysaccharide.

ER attached to the vacuolar membrane and continuous with it. However, ribosomes have never been observed attached to the vacuolar membrane.

The origin, development, and function of vacuoles in young plant cells are not clearly understood. At first, vacuoles are inconspicuous and, in many cases, are invisible in the light microscope. As the cells mature, many small vacuoles appear, forming larger vacuoles by joining together.

Vacuoles in microorganisms and animal cells exhibit great variability in function and in size. Fresh-water protozoans such as amoebae and paramecia possess *contractile vacuoles*, which may be quite intricate and constantly changing in shape. Because the water outside the cell is at a higher concentration than the water within the cytoplasm, these small, fresh-water organisms experience a constant diffusion of water through the plasma membrane into the cytoplasm. Contractile vacuoles accumulate the excess water and periodically pump it out through a pore that they form in the cell membrane. The membrane is immediately repaired after the vacuole completes its contraction. Without contractile vacuoles, these fresh-water organisms would burst from the accumulated influx of water from the exterior. Other kinds of vacuoles in animal cells play a relatively passive role and might better be considered to be inclusions.

CYTOPLASMIC INCLUSIONS

In the rather arbitrary category of cytoplasmic inclusions are grouped all particles, droplets, storage granules, and other substances that are relatively inert with respect to the metabolic activities of the cell. They vary in size from glycogen granules (about 150 to 300 A in diameter) to crystals of various sorts that are visible at low powers of the light microscope. In the intermediate size range are lipid droplets, yolk granules, pigment granules, virus inclusions (often in crystalline form), and other crystals. Crystals of the organic base guanine are found in the surface cells of fishes, amphibians, and lizards and in the light-reflecting cells of the eyes of many nocturnal animals. Cells that contain guanine crystals impart a silvery luster to the tissues they form. Calcium oxalate crystals are common in plant cells. They may be so numerous that sucking on the stalk of the plant (for example, "dumb cane," or *Dieffenbachia*) makes one unable to talk. The crystals become lodged in the mouth and throat, causing these tissues to swell.

EXTRACELLULAR STRUCTURES

Many types of cells are embedded in a matrix of material produced by the cells themselves. Bone cells are interspersed within a matrix formed chiefly of crystals of an inorganic compound, hydroxyapatite. Ligaments and tendons (the "gristle" of meat) derive their toughness from the substance collagen. The hard external skeleton of insects and other arthropods is composed chiefly of chitin, a polysaccharide that surrounds the cells that synthesize it. Although produced by the cells, these materials are deposited outside the plasma membrane.

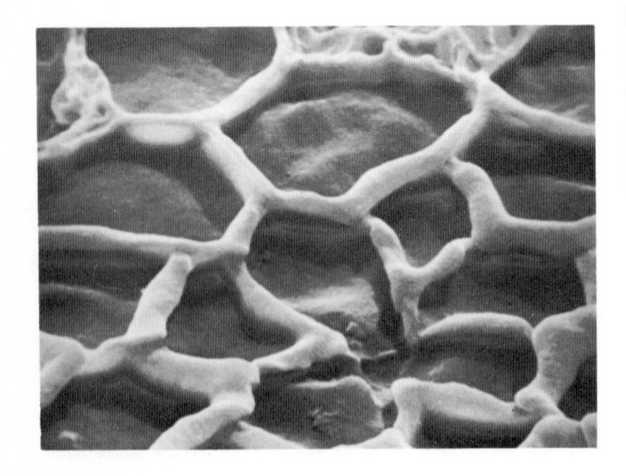

Cell walls outside the cell membrane are an important part of the structure of most plant cells. Many aspects of a plant, including its general form as well as its mode of cell division and growth, are determined by the nature of the cell walls.

A young plant cell undergoing division and elongation is surrounded by a single, thin, elastic *primary wall*, about 1 to 3μ thick (Figure 7.28). This primary wall increases greatly in area as the cell grows. In many cells,

Figure 7.28 (left). Electron micrograph showing a fairly thin cell wall of a plant cell. (× 6,000)

Figure 7.29 (right). Microfibrils in the primary cell wall. (× 2,800)

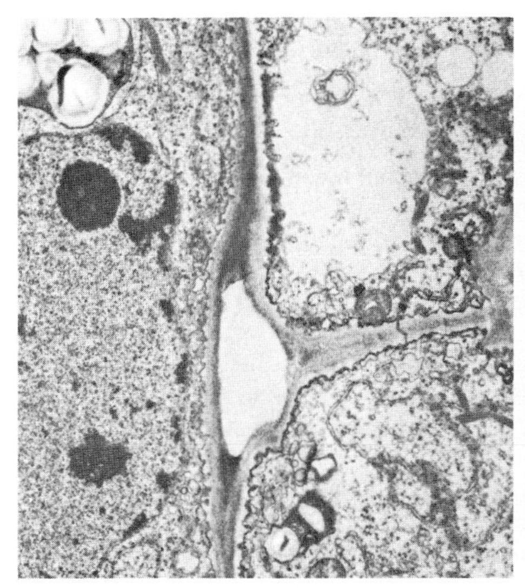

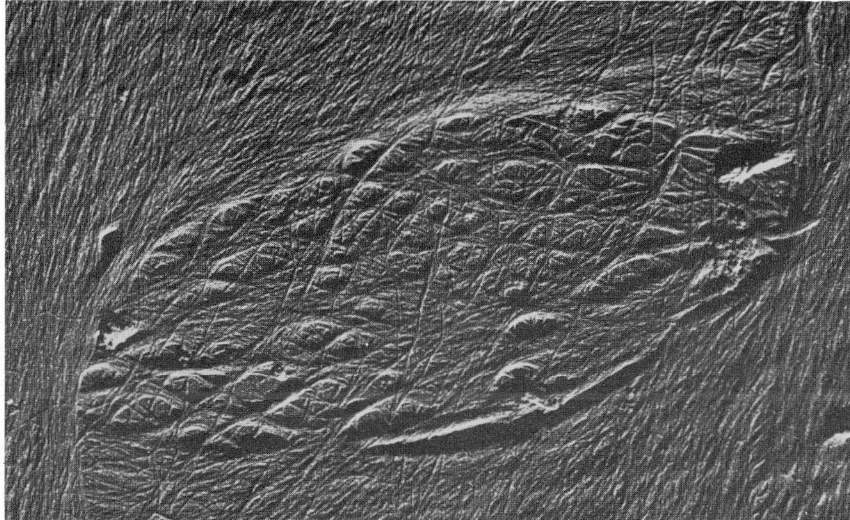

when growth ceases, a rigid *secondary wall* about 5 to 10μ thick forms between the cell membrane and the primary wall. Between the primary walls of adjacent cells is a layer called the *middle lamella*, which serves as a matrix to hold the cells together. Through the walls of adjacent cells, there may be passages called *plasmodesmata*, which connect the cells' cytoplasms.

Electron micrographs reveal both primary and secondary walls as a series of layers, with each layer made up of *microfibrils* embedded within a matrix (Figure 7.29). Microfibrils are long chains of cellulose, which is a polymer of glucose. The amount of cellulose varies greatly in different kinds of plant cell walls. Some cells, such as the hair cells of the cotton seed (from which commercial cotton is obtained), have walls of almost pure cellulose. Others, such as the cells at the growing tip of a root, have little cellulose.

Microfibrils can be made up of polymers formed from units other than glucose. Such noncellulose microfibrils are less common than cellulose microfibrils in the walls of most land plant cells, but in some algae they make up the entire cell wall. Many fungi have cell walls made up of chitinous microfibrils.

The matrix between the microfibrils consists of nonfibrous macromolecules that provide flexibility while holding the microfibrils together. A substance called lignin binds the microfibrils together in many types of plant cell walls. In balsa wood, where lignin is absent, the plant tissues usually are brittle. Whereas lignin is present only in some types of plant cell walls, materials called hemicelluloses and pectic substances are present in the matrix of all plant cell walls. Both of these substances are polymers of sugar units; the exact molecular nature of the two substances is unclear. The pectic substances are soluble in hot water and appear to be the chief components of the middle lamellae. Vegetables are easier to chew after cooking because hot water dissolves the pectin of the middle lamellae, allowing the cells to separate. Changes in the cell walls of fruits during ripening also are related to modification of the chemical nature of the middle lamellae.

Pectin forms a thick solution in water and, in the presence of acids and sugars, "sets" to form a gel. Fruits that have a high pectin content can easily be prepared into jams or jellies. Ripe fruit or fruit poor in pectin (such as strawberries) can be made into jellies by the addition of commercial pectin preparations.

The middle lamellae and primary walls of young plant cells are composed primarily of cellulose, hemicellulose, and pectic substances. All these compounds are macromolecular carbohydrates. As the cells mature, lignin may be added to the matrix of the walls. Lignin is not a carbohydrate, but its detailed molecular structure is not definitely known. Lignin never is found alone in cell walls but is always associated with cellulose. Lignin is most abundant in the secondary wall, particularly in woody plants, but is also found in the primary wall of mature plant cells.

Among other components found in plant cell walls are waxes that provide an impermeable surface for leaves, stems, and fruits. Cell walls of many grasses and plants such as the horsetails contain silica—the major component of sand and glass. Various proteins, which probably play a role in the synthesis and growth of the cell wall, have also been detected.

Current investigations of the process of cell-wall formation have revealed that the cellulose microfibrils form near to, but outside of, the cell membrane in a pattern that is determined by the cytoplasm of the cell. Microtubules just inside the cell membrane apparently play a role in the orientation of microfibrils but are not directly involved in this synthesis.

There is evidence that the Golgi apparatus is involved in the synthesis of the matrix materials of the walls and the middle lamellae. Vesicles pinched off from the Golgi apparatus apparently move to the cell membrane, fuse with the membrane, and deposit their contents in the cell wall. The ER plays a role in cell-wall synthesis in some cells, but its role is far less understood.

In many mature plant cells, particularly in the xylem—or wood—of trees, strong and rigid cell walls of dead cells provide great strength and rigidity. However, not all plant cells have rigid walls. Most plant cells are surrounded only by primary walls or by thin secondary walls, which are not strong enough to retain their shape without the additional rigidity provided by pressure of the cytoplasm and the large central vacuole. The firmness of most nonwoody plant tissues depends upon the balance between the external wall pressure and the countering internal pressure.

In organs and tissues of multicellular animals, cells are held together by various intercellular substances. Hyaluronic acid (a polymer made up of sugar units combined with proteins) is a jellylike material that binds together many animal cells. Some bacteria secrete an enzyme (hyaluronidase) that dissolves this substance and assists the bacteria in penetrating animal tissues. A similar enzyme is secreted by sperm cells, permitting them to penetrate the coat of jellylike substances that surrounds an egg cell.

Skin cells are mounted on a basal lamina made up of layers of fibers of collagen embedded in a matrix. Collagen fibers provide rigidity and strength. Fibers of an elastic protein called elastin are abundant in flexible tissues such as skin and the walls of large arteries.

The process of aging in multicellular animals is intimately involved with changes in intercellular materials. As aging proceeds, more and more collagen fibers are formed between cells, cross-linkages appear between individual fibrils, and the elastin fibers become thicker and less flexible. Thus, the skin becomes less pliant, the joints stiffen, and the muscle tissues become tougher and stringier. In effect, the processes that bind cells together merely continue to form ever more rigid connections between cells until the rigidity of structure impedes the functioning of the organism. Similar processes occur in plants but are of less hindrance to the organism because the plant needs little mobility.

STRUCTURE AND FUNCTION

As new techniques permit increasingly close investigation of cell structure, more and more parts of the cell are found to be precisely and intricately organized at the molecular level. Specific cellular ultrastructures (structures so small that they are revealed only by electron microscopy) now are known to guide the myriad chemical reactions that must occur in coordinated fashion to maintain cells as living systems.

One noted modern cell biologist recalls that he was told in his introductory biology course that although much remained to be learned about the cell, it was certain that all important biochemical reactions occur in water solutions within the cytoplasm. The use of electron microscopy has drastically changed this view. "Now," he says, "with only slight exaggeration, I can state that it is certain that most important biochemical reactions occur on the surfaces of membranes or within other specialized structures."

Some reactions are known to occur in solution within the cytoplasm, but many reactions do appear to be carried out on membrane surfaces. The study of such reactions poses a great challenge for biochemists.

Although the emphasis of this chapter has been upon cell structure, the topics of cell structure and cell function cannot be separated. The boundaries between the study of cell structure or physiology and the study of cell function or biochemistry are largely disappearing in modern cell biology, or cytology. Biologists studying cells are beginning to approach the goal of explaining living organisms in terms of the chemical processes and physical structures within the cell. The classification of various cellular structures into neat, static categories—as has been done in this chapter—is useful for analysis. However, the living cell—an unresting cell—is constantly changing and carrying out biochemical reactions. Not all cells or cellular structures can readily fit into such neat categories, because these categories fail to represent the dynamic and continuous nature of the cell interior. A listing of cell parts no more completely describes the living cell than does a listing of organs describe a living human. For a more complete picture of the unresting cell, the structural picture of this chapter must be combined with the information about processes summarized in Unit Two and in following chapters of this book.

FURTHER READING

For further details of the history of the cell theory, see books by Hughes (1959), Nordenskiöld (1960), Singer (1959), and Taylor (1963).

There are a great many books available on the structure and function of eucaryotic cells. One of the most complete sources of information (although now somewhat out-of-date) is the six-volume collection edited by Brachet and Mirsky (1959–1964). Among the other useful introductions to cell biology are books by DuPraw (1968), Fawcett (1966), Gerard (1961), Kennedy (1965), Mercer (1962), Stern and Nanney (1965), and Swanson (1969). Buvant (1969), Frey-Wyssling and Mühlethaler (1965), and Jensen (1964) offer more detailed treatment of plant cells.

Further information about electron microscopy and excellent collections of micrographs of cellular structures may be found in books by Jensen and Park (1967) and Ledbetter and Porter (1970).

8
Procaryotic Cells

Such seemingly diverse phenomena as the multicolored formations in springs and ponds in Yellowstone National Park, colored mainly by blue-green algae, and an epidemic of typhoid fever, brought on by disease-causing bacteria, actually have much in common. The blue-green algae bacteria and mycoplasmas are classified as procaryotic cells—a fact that distinguishes these organisms from all other living things.

PROCARYOTES AND EUCARYOTES

Although procaryotes are capable of independent existence, they lack many of the features characteristic of eucaryotic cells. Procaryotes lack nuclear and intracellular membranes, and the hereditary material (DNA) is not segregated from the rest of the cellular contents by a nuclear membrane, although it may tend to gather in certain areas.

The procaryotic cells of bacteria, blue-green algae, and mycoplasmas are the smallest and simplest organisms, but each procaryotic cell is equipped with the necessary biochemical machinery to maintain itself and to divide. Although some procaryotes attach themselves into groups of cells, they do not form cooperative units in which one cell depends upon others for its survival.

The nucleated eucaryotic cells—which include most of the cells observed by the microscopists who developed the cell theory—are larger and more complex than procaryotic cells. A single eucaryotic cell—like a procaryotic cell—may exist independently of other cells as a complete organism. Other kinds of eucaryotes exist in cooperative units of a few dozen cells, forming a colonial organism or a simple animal or plant. The higher animals and plants are made up of many trillions of interdependent eucaryotic cells.

The structural differences between the most complex procaryotic cell and the simplest eucaryotic cell are far more significant than the differences between cells from an oak tree and from a human. Plants and animals are diverging lines of organisms descended from some common eucaryotic ancestor, and their cells are constructed along the same general eucaryotic patterns. In many external features, a green algal cell and a blue-green algal cell appear to be similar, and, until recently, these groups were lumped together in most classification schemes. In cellular structure, however, it is now clear that the green algal cell is more similar to a human cell than to the procaryotic blue-green algal cell.

Procaryotic and eucaryotic cells do have common features. Both are limited at the outer boundary by cell membranes of similar thickness and gross structure, although the chemistry of the membrane differs for the two groups. Both use DNA as the macromolecule that carries hereditary information. Both use RNA as the macromolecule that carries information from the DNA to the ribosomes, which are the sites of protein synthesis. And there is much evidence that both procaryotic and eucaryotic cells use essentially the same genetic code.

There are other similarities between the two kinds of cells. Most of what scientists have deduced about genetic control mechanisms comes from studies of procaryotic cells and viruses. The fact that these findings also seem applicable to eucaryotic cells attests to the unity of life in a most compelling way.

Eucaryotic cells are generally regarded as more complex descendants of ancestral procaryotes (Margulis, 1970). It is unlikely that the eucaryotes

Figure 8.1. Electron micrograph (× 40,000) and diagram of a "typical" procaryotic cell.

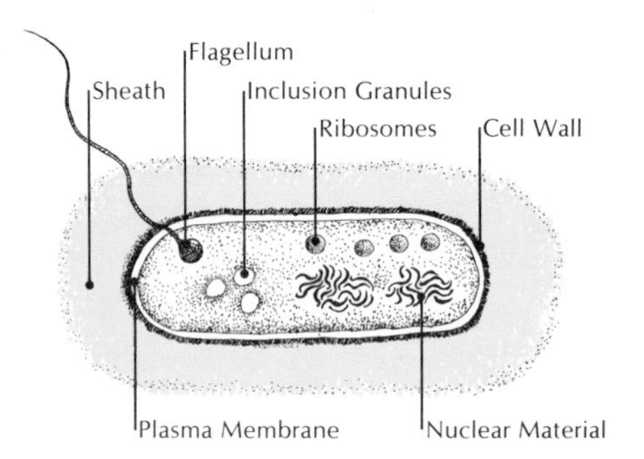

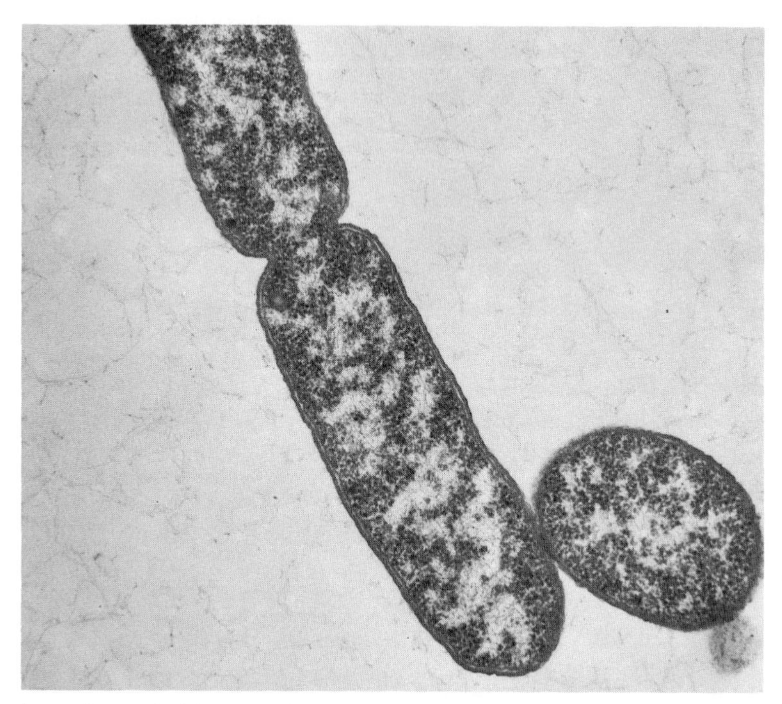

have descended from modern types of procaryotes, but it does seem that both have descended from a common ancestor.

BACTERIA

The bacteria are a group of procaryotic microorganisms that are so minute they can barely be seen with the light microscope. Although bacterial cells may be found in attached groups, they are never organized into cooperative, multicellular organisms in which one cell is dependent upon another for its survival. Most bacterial cells are from 2 to 5μ in length, although a few kinds are as long as 100μ and some other kinds as short as 0.2μ.

Very little of bacterial structure is revealed by the light microscope. Based upon their overall shapes, bacteria have been divided into three general groups: the rodlike *bacilli* (singular, bacillus), the spherical *cocci* (singular, coccus), and the corkscrew-shape *spirilla* (singular, spirillum). Other classifications of bacteria are based upon the ways that they are stained and the changes that they produce in the environment. Some types of bacteria possess long whiplike extensions called *flagella*. The rapid lashing of the flagellum propels the organism through the medium.

The light microscope did reveal some details of the processes by which bacteria reproduce. All bacteria reproduce through a process of division called *binary fission*, a continuous action typical of procaryotic cells (Figure 8.3). A bacterium simply grows larger for a time, then divides across its middle to form two equal daughter cells. The daughter cells then repeat this process of growth and division. The time between bacterial cell divisions can be as short as 20 minutes.

Certain stains demonstrate the presence of nucleic acids and protein in the larger bacteria. Staining reveals that the nucleic acids contained in the dividing bacterial cell are apportioned about equally between the two

Figure 8.2 (above). The three major groups of bacteria based upon their overall shapes. At the upper left are both rodlike *bacilli* and spherical *cocci* bacteria. At the upper middle are predominately bacilli, and at the lower middle are the corkscrew-shape *spirilla*. The diagrams of bacteria denote both shape and growth pattern—that is, chains (streptococcus), small groupings, or irregular clusters (staphylococcus).

Figure 8.3. (lower left). Electron micrograph of a dividing bacterium, *Bacillus subtilis*. Note the two nuclear areas in the two daughters and the cell wall that is forming between the two. In the diagram of binary fission at lower right, note the lack of an elaborate mitotic apparatus, as is observed in eucaryotic cells. (× 23,000)

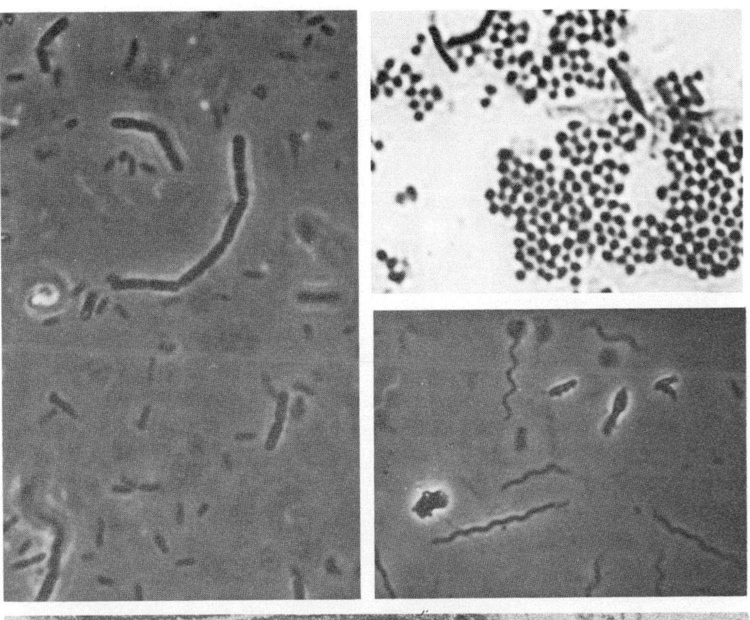

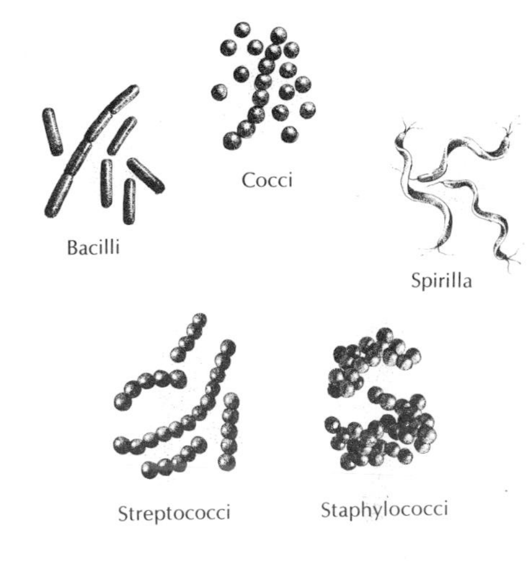

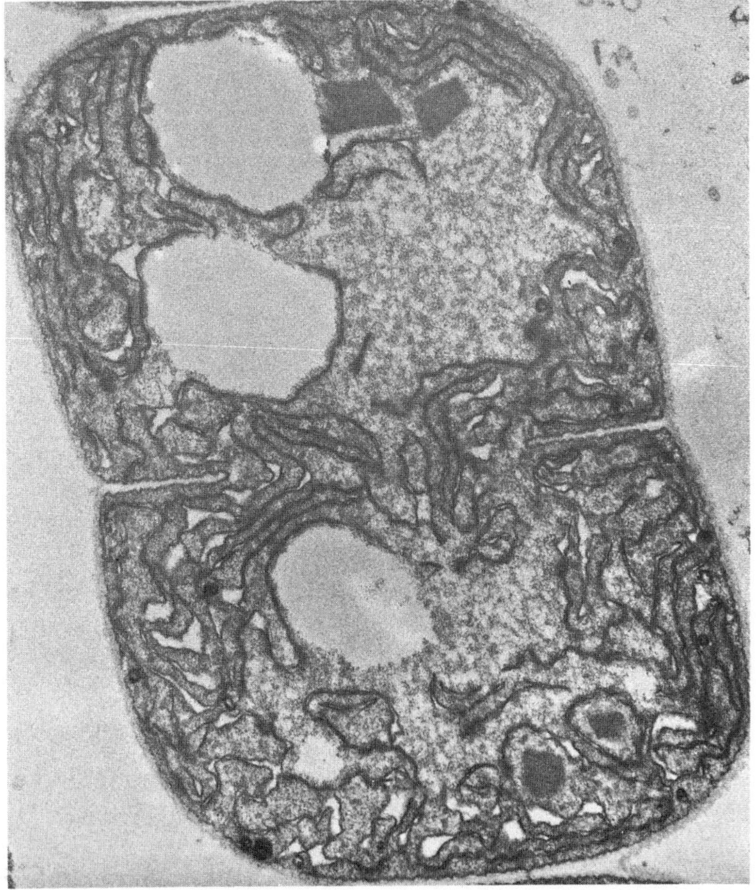

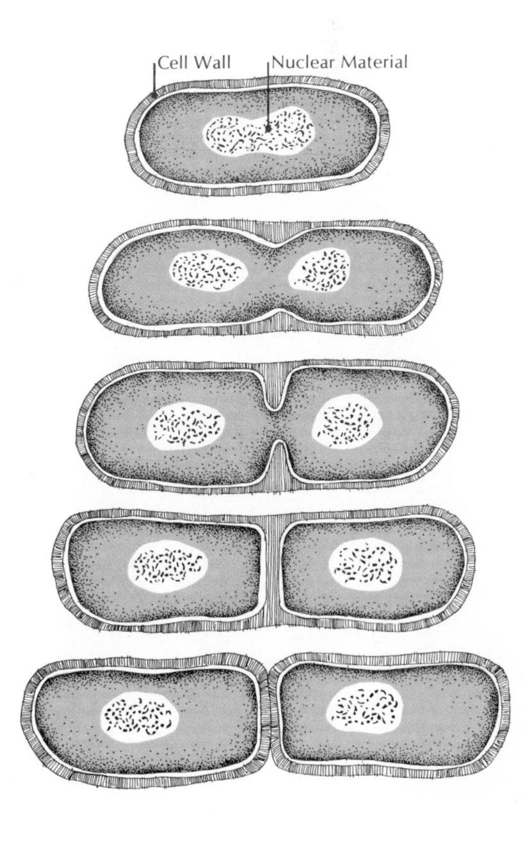

Figure 8.4. An artist's interpretation of a swarm of rapidly dividing bacteria.

Figure 8.5. Electron micrograph of a sporulating bacillus (left). The spore is the large, dark oval at one end of the cell and is surrounded by a clearly visible spore coat. The opaque areas at the other end of the cell are not vacuoles but are areas containing fatty material. In the diagram of spore formation at right, the spore coat is forming around the essential nucleic acids of the bacterium. The remaining portion of the cell disintegrates. Thus, the spore is a "resting stage" in which the vital genetic information of the cell is stored until favorable conditions for normal function are restored.

daughter cells. However, neither chromosomes nor the other regular structures and processes associated with eucaryotic cell division are visible with the light microscope in a dividing bacterial cell. Bacteria divide very rapidly. If nutrients and space were available for continued division at the 20-minute intervals observed in laboratory cultures, a bacterium could produce offspring of mass greater than the mass of the earth in less than two days.

Some bacteria also reproduce by *budding,* a process in which a much smaller daughter cell is divided from the parent bacterium. Although the daughter cell that buds off receives a very small portion of the cytoplasm of the parent cell, the DNA is divided approximately equally between the parent and the daughter cells.

Spore formation, on the other hand, is not a reproductive process, and it is characteristic of only certain species of bacteria (Figure 8.5). When environmental conditions are unfavorable for survival, the bacterial cell of these species forms a virtually impermeable membrane—a *spore coat*—within its cytoplasm. This coat surrounds the nucleic acids and a small part of the cytoplasm. After the spore coat is formed, the remaining parts of the cell that surround the spore disintegrate. Thus, a spore represents a "resting stage" in which the vital parts of the cell are preserved in a quiescent state until favorable conditions for normal metabolism are restored.

Spores can withstand a temperature of 100° C for several hours in a slightly alkaline solution, whereas bacterial cells are killed almost instantly at this temperature. The existence of these resistant spores played an important role in the controversy over spontaneous generation. Those investigators who happened to use slightly acidic boiling water to sterilize their equipment and nutrients found no bacteria arising afterward. Others who happened to use slightly alkaline solutions did not completely destroy the spores and consequently observed bacteria that seemed to be generated spontaneously.

Exposure to a higher temperature—about 120° C for a few minutes—destroys most bacterial spores. Water cannot be heated to such a temperature

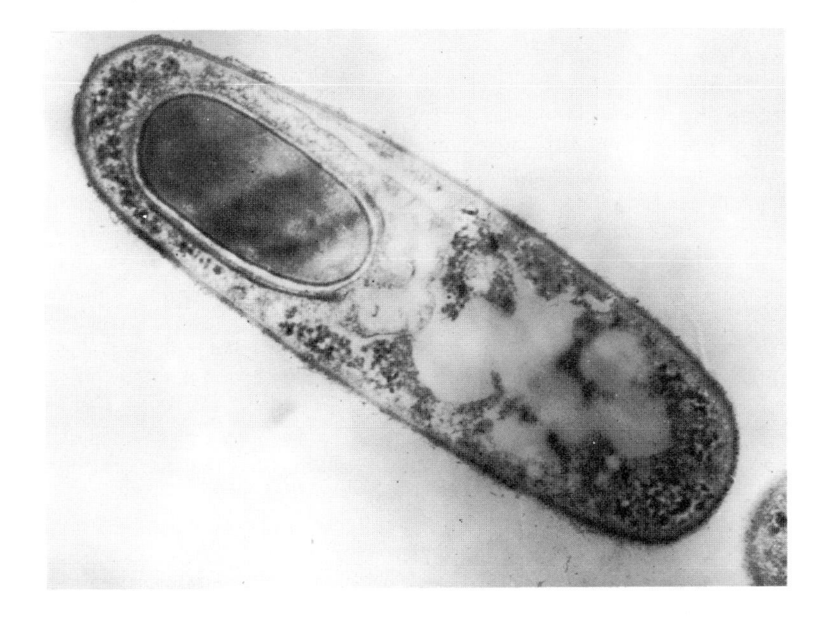

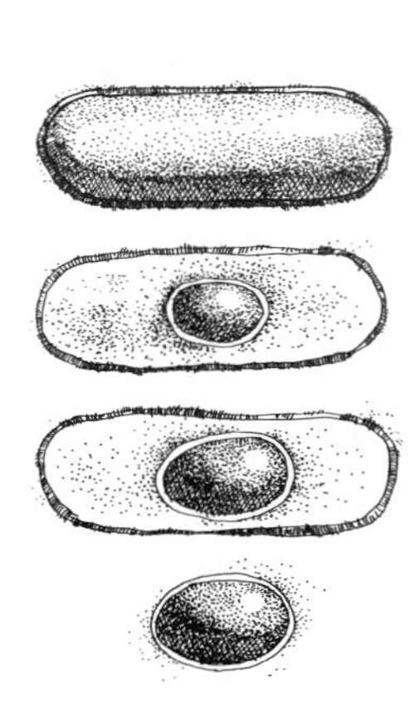

Figure 8.6. Electron micrograph of conjugation in the colon bacteria *Escherichia coli*. In this primitive type of sexual reproduction, genetic material is transferred from a "donor cell" to "a recipient cell," which will later divide by fission. Its daughter cells will possess genetic information from two different parent cells, thus increasing the genetic variation within the population. (× 49,000)

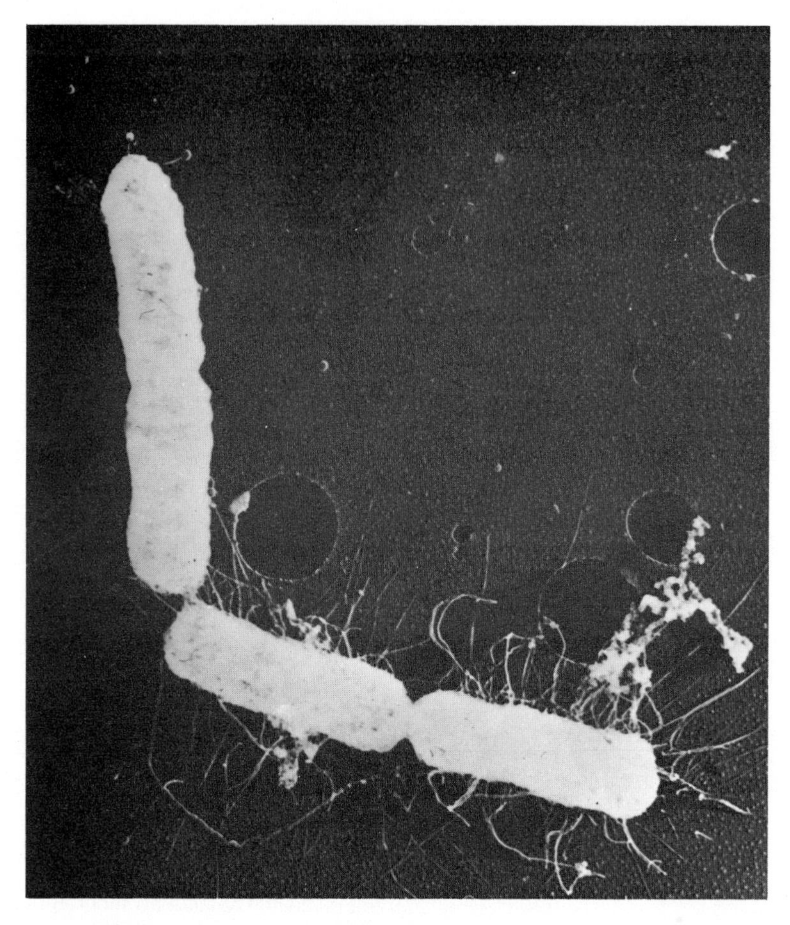

at normal atmospheric pressure, but medical and laboratory equipment can be sterilized effectively in a pressurized device called an autoclave, in which water boils at about 120° C.

The electron microscope reveals that the genetic material—fine filaments of DNA—is clustered in a nuclear area of the bacterial cell but is not surrounded by a nuclear membrane. The DNA, however, does appear to be intimately associated with the membrane surrounding the bacterium. Bacterial cytoplasm contains various organelles, but some kinds of organelles found in eucaryotic cells are missing in bacterial cells. The bacterium is enclosed in a cell membrane that is surrounded by a cell wall. In some species, the cell wall is enveloped by a protective sheath, or *capsule*. The bacterial cytoplasm contains many ribosomes, which appear in the electron microscope as small dots. The ribosomes are composed of proteins and the nucleic acid RNA. They are often found in clusters called *polyribosomes*, which are the sites where bacterial proteins are synthesized.

In 1947 E. L. Tatum and his student Joshua Lederberg discovered that some species of bacteria can engage in a form of sexual reproduction called *conjugation* (Figure 8.6). A fine bridge is formed between the mating cells, and some DNA from the donor cell moves through this bridge into the recipient cell. The recipient cell then reproduces by the usual binary fission. Some of its daughter cells have parts of the DNA from the donor cell in place of parts of the original DNA of the recipient cell. Thus, the daughter cells possess genetic information that is a combination of the information from two different parent cells. Such *genetic recombination* within a population creates much greater variation than could arise through random mutations.

Bacteria obtain energy through the catabolism of organic molecules in the environment. Bacteria are responsible for much of the decay of the remains of dead plants and animals, ultimately breaking down the complex organic molecules of the dead organisms into carbon dioxide, ammonia, and water that can be recycled through the biosphere. *Aerobic bacteria* use oxygen and carry out respiratory processes similar to those of eucaryotic organisms. *Anaerobic bacteria* do not require oxygen but make use of processes such as fermentation and glycolysis to obtain their energy.

More than 15,000 species of bacteria are known, many of direct importance to human life. Of particular interest in relation to theories about early life on earth are the *autotrophic bacteria*, which manufacture their own carbon-containing compounds from CO_2. The *photosynthetic bacteria* use bacteriochlorophyll to capture solar energy; the *chemosynthetic bacteria* derive energy from the oxidation of compounds such as ammonia (NH_3), nitrites (compounds containing the ion NO_2^-), or hydrogen sulfide (H_2S). Far more common are the *heterotrophic bacteria*, which derive their energy from the oxidation of organic molecules and are thus dependent on other organisms for their food supply.

Bacteria as Disease-Causing Organisms

Early interest in bacteria was centered around the harmful disease-causing organisms. In the late nineteenth century, Louis Pasteur, who is often considered the father of bacteriology, developed the idea that bacteria can cause disease—an idea now known as the germ theory of disease.

The work of Casimir Davaine of France in the 1860s showed that the blood of cattle dying from a disease called anthrax contained large numbers of microscopic, rodlike bodies, which he called bacteridia (the Greek *bakteria* means rod or staff). When Davaine injected a healthy animal with the smallest amount of blood he could prepare from a diseased animal, the animal that received the injection soon developed anthrax. In Germany a few years later, C. J. Eberth showed that bacteria can be filtered from the blood and that anthrax will not develop in a healthy animal injected with filtered, bacteria-free blood from a diseased animal.

In an attempt to halt an outbreak of anthrax among the cattle of his district, a German country doctor named Robert Koch developed methods for the study of bacteria (Figure 8.7). Koch found that anthrax bacteria can be grown in laboratory containers if the bacteria are supplied with the proper nutrients. He found a number of culture media in which the bacteria will thrive. Blood serum (the clear, yellowish fluid that remains after blood clots) proved to be a particularly suitable medium for anthrax bacteria. Koch also devised an incubator to maintain his bacterial cultures at temperatures similar to those of the fluids inside the body.

Koch showed that bacteria growing in the centers of his cultures were separated from one another and grew slowly. Near the surface of the culture, however, where oxygen from the air was abundant, he observed that the bacteria grew longer and joined end-to-end in long threads. Wherever these bacterial threads contacted the air, they were transformed into

Figure 8.7. Robert Koch (above), a German country doctor, developed many of the bacteriological techniques still in use today; early micrograph (middle) of anthrax bacillus; and stained anthrax bacteria (below).

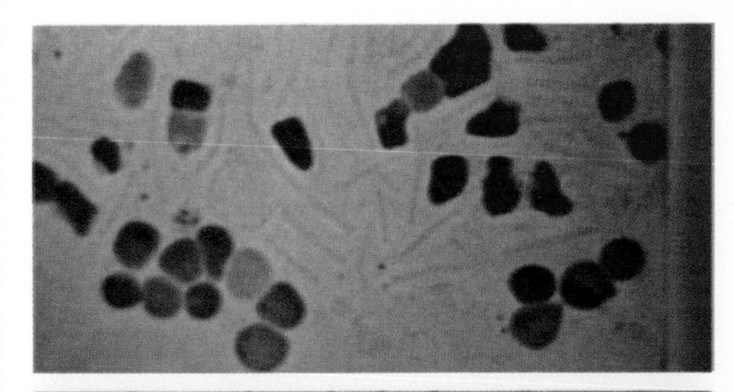

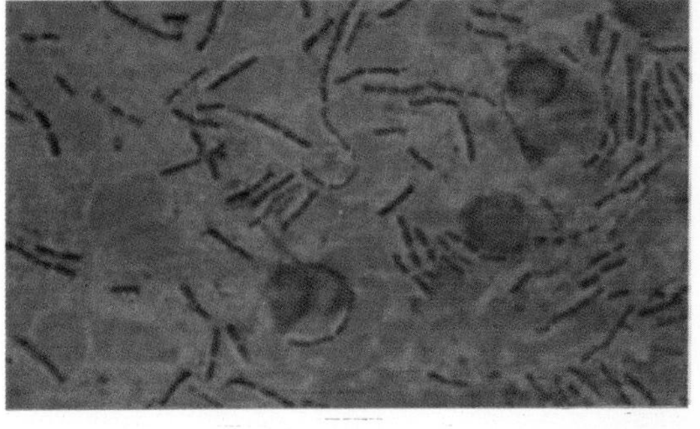

Figure 8.8. Photograph of a bacteria culture of *Clostridium botulinus* grown on an egg yolk medium. Botulus toxin in exceedingly small amounts can produce the deadly food poisoning called botulism.

minute, nongrowing, round bodies, which Koch called spores. The bacterial spores changed back again into bacterial cells when they were transferred to a fresh culture medium. Koch showed that bacterial spores are extremely resistant to damage by drying, heating, or chemical treatment.

In the 1870s, when Koch was beginning his work on bacteria, the newly established German synthetic dye industry was turning out many new dyes. Koch found several dyes that can be used to stain bacteria, increasing their visibility under the light microscope. He also developed a method for the preparation of pure cultures of a single type of bacteria. The fluid that contains the bacteria is repeatedly diluted until each drop of the final dilution contains only a few microorganisms. With a small dropper or hypodermic needle, minute amounts of this fluid are spotted onto the surface of a transparent, jellylike nutrient medium. At each spot where a single bacterium has been deposited and has reproduced, a colony of that particular type of bacteria is formed. This procedure is known as *cloning*. The techniques of bacterial staining and solid-medium culturing developed by Koch have not been greatly changed over the years, and similar techniques are still used by bacteriologists.

The bacteria most familiar to the layman are those that cause diseases in man, in his domestic animals, and in his cultivated plants. Among the diseases caused by bacterial invasion of the human body are typhoid fever, cholera, plague, dysentery, scarlet fever, diphtheria, tuberculosis, and wound infections such as gangrene.

Bacterial invasion of the human body produces the disease state in a number of ways. In some cases, the invading bacteria become so numerous that they successfully compete with the host cells for nutrients and oxygen. In other cases, the bacteria produce a poison, or *toxin*, that disrupts the normal processes of the host. The deadly food poisoning called botulism is caused by an extremely lethal toxin. It has been estimated that as little as 3.5×10^{-7} grams of botulinus toxin is sufficient to kill a human. From this information, one can calculate that a little more than one kilogram of the pure toxin would be sufficient to eradicate the population of the earth.

But, even from the prejudiced human viewpoint, bacteria do far more good than harm. Bacteria play key roles in the production of buttermilk,

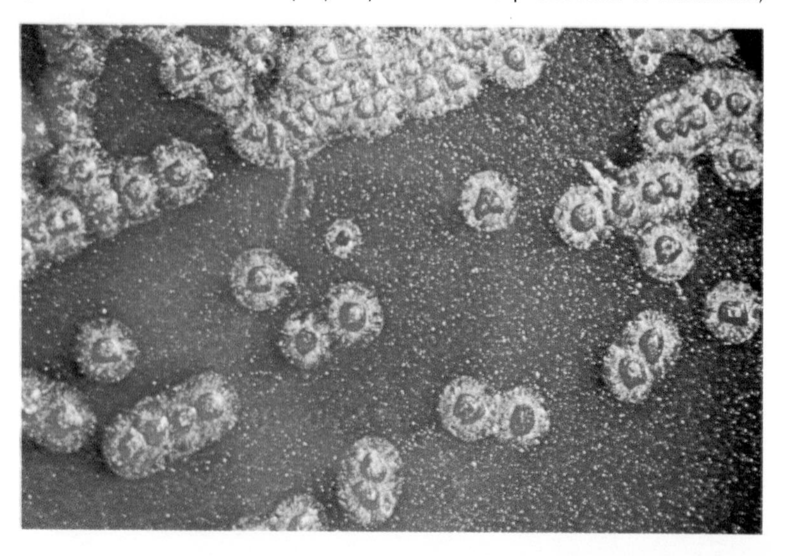

Figure 8.9a (above) Photograph of the blue-green alga *Nostoc*. (Courtesy Carolina Biological Supply Company)

Figure 8.9b (below). Filamentous forms of blue-green algae. The upper filament represents the species *Spirulina versicolor*, and the lower two filaments belong to the genus Arthrospira. In flamingos, Arthrospira directly contributes to the birds' pink color; the pigment comes from carotenes in the blue-green algae that make up part of the birds' diet.

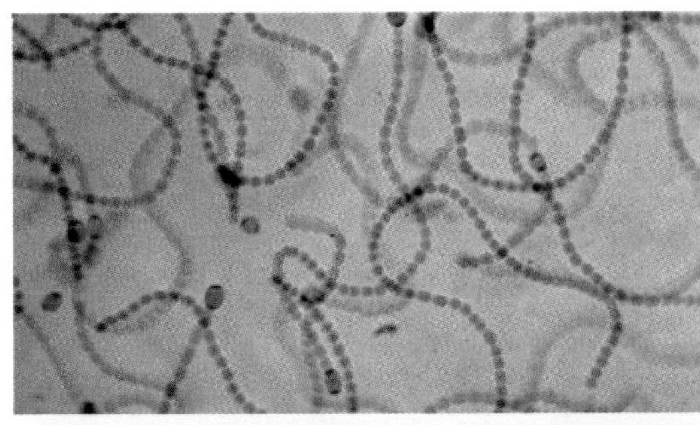

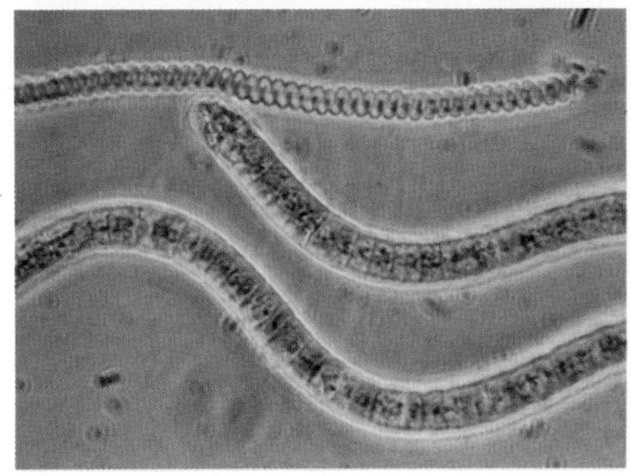

cheese, vinegar, and other foodstuffs. They also play key roles in modern sewage disposal plants. Bacteria are used in various industrial procedures, such as removing hairs in the preparation of leather. In promoting the decay of dead organisms, bacteria form a vital link in the carbon and nitrogen cycles of the biosphere. The bacteria help to break down the complex organic molecules and to restore carbon and nitrogen to the ecosystem in the form of simple inorganic molecules that can be used by plants. The human intestinal tract contains a bacterial population essential for normal health. Among other things, these intestinal bacteria synthesize vitamin K, which is required by the human organism for normal blood clotting.

Several methods are used to keep harmful or decay-causing bacteria out of human foodstuffs. Because the metabolic activity of bacterial cells is enzymatically regulated, it is extremely slow or completely stopped at low temperatures. Refrigeration, therefore, is an effective method of slowing food decay. Drying is an effective means of food preservation because many bacterial cells are destroyed by extreme dehydration and because their enzymes are inactive in the absence of water. Salting is effective in food preservation for the same reason.

Various organic chemicals that kill or inhibit bacteria are useful in food preservation. Many of these antibacterial chemicals are produced as waste products of fermentation by various microorganisms. Thus, wine is preserved by alcohol, sauerkraut by lactic acid, and cheese by lactic and propionic acids. In each case, the inhibitory chemicals are formed during the fermentation process that produces the particular foodstuff. Other antibacterial chemicals are artificially added to most modern packaged foods.

Sterilization by heating in a pressure cooker or autoclave destroys bacterial cells and spores and their enzymes. Home-canned foods can be dangerous because simple boiling at atmospheric pressure does not destroy spores, particularly in alkaline foods.

BLUE-GREEN ALGAE

Blue-green algae represent the other prominent group of procaryotic organisms. As in the bacterial cell, the DNA of a blue-green algal cell is localized in a nuclear area or "nucleoid," but no nuclear membrane surrounds it. Reproduction occurs only through binary fission. All of the 1,500 or so known species of this group are photosynthetic. All species of blue-green algae contain the photosynthetic pigments chlorophyll *a* and phycocyanin (a blue pigment). Many species do contain additional accessory pigments of various colors, but none contains other forms of chlorophyll. The pigments are arranged on infoldings of the cytoplasm, making the algal cell interior appear more highly structured than the bacterial cell interior.

The blue-green algal cell is surrounded by a cell wall of cellulose. In many species, the outer portion of the cell wall becomes covered with a slimy substance, which sometimes is present as a very thick layer. This slime layer apparently protects the cell from dehydration and facilitates intercellular interaction and filament formation.

Some species form filaments in which a number of cells are joined together. In a few species, the filaments show slow, twisting movements, but the mechanism by which this motion is accomplished is unknown. Many species can live in moist earth, surviving long dry spells by forming resistant *resting spores*. Some species survive at temperatures as high as 70° C and are found growing in hot springs. The multicolored formations in springs

Figure 8.10 (above left). Diagram of the general structure of a blue-green alga. Note that the nuclear material, composed of DNA fibrils, is localized in a nuclear region but remains unbounded by a nuclear membrane. Shown at the upper right is an electron micrograph of the blue-green alga *Anabaena*. Clearly visible are the cell wall and the layers of photosynthetic membranes. (× 6,800)

Figure 8.11 (below). Life cycle of the pleuropneumonialike organism (PPLO) *Mycoplasma laidlarvii*. Note the double-track system of reproduction. Large cells may reproduce with or without the formation of elementary bodies. (From "The Smallest Living Cells" by H. J. Morowitz and M. E. Tourtellotte. © 1962 by Scientific American, Inc.)

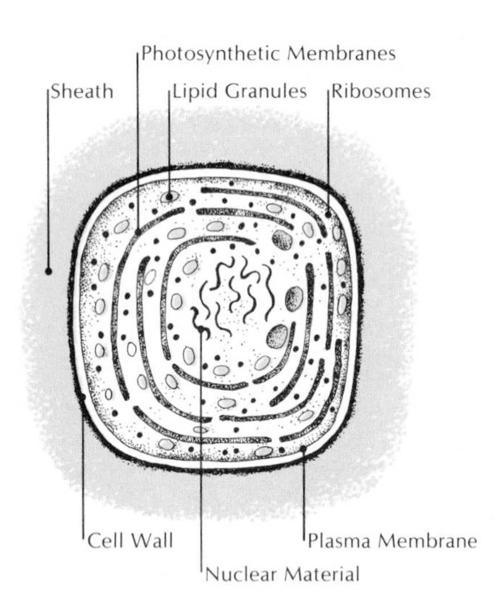

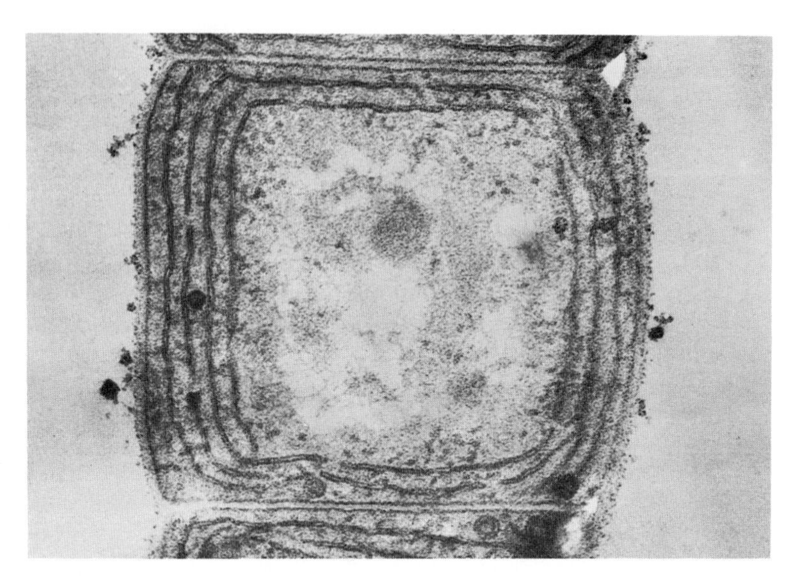

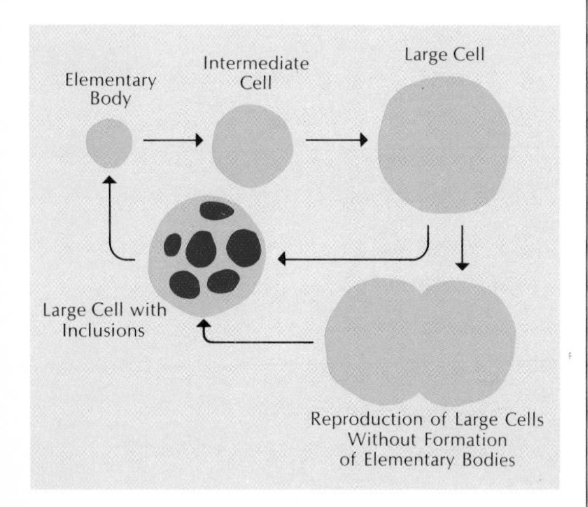

and ponds in Yellowstone National Park are colored mainly by various species of blue-green algae containing a wide variety of accessory pigments.

The first photosynthetic organisms may have been similar to modern photosynthetic bacteria or blue-green algae. Some of these early organisms, incapable of adapting to the increasing concentration of free oxygen produced in the atmosphere by photosynthetic activities, may have developed the ability to invade other cells and to derive nourishment from the material of the host cell's cytoplasm. Eventually, according to this hypothesis, these invaders lost most of their cellular components and the ability to survive independently. At the same time, the host cells evolved a dependency on the symbionts as a source of carbohydrates formed through photosynthesis. A true mutualistic relation, or *symbiosis*, was the result—the ancestor of the modern photosynthetic eucaryotic cell whose photosynthetic pigments are grouped in chloroplasts.

MYCOPLASMAS

The smallest known cells belong to a group called the pleuropneumonialike organisms (PPLO), or mycoplasmas. The existence of extraordinarily minute organisms responsible for pleuropneumonia, a highly contagious disease of cattle, was suggested by Louis Pasteur. However, early investigators found that these organisms could not be trapped in porcelain filters that remove bacterial cells from blood. In 1898 E. I. E. Nocard and P. P. E. Roux succeeded in growing pleuropneumonia organisms in a complex medium, demonstrating that they are not dependent on larger cells for their survival. It was not until 1931 that W. J. Elford developed a set of special filters that revealed the pleuropneumonia organisms to be about 0.13 to 0.15μ in diameter, about a tenth the size of a typical bacterium.

About 30 different species of PPLO have been identified, ranging in diameter from 0.1 to 0.25μ (Figure 8.11). One species causes a form of human pneumonia. Harold Morowitz and Mark Tourtellotte (1962) studied one of the smallest of the PPLO, an organism that normally lives free rather than invading larger organisms. This species forms *elementary bodies* that are spherical and about 0.1μ in diameter. During its life cycle of a few days, the organism becomes larger and forms a cell about 1μ in diameter. The large cells may divide to form other large cells, or within themselves they may develop small bodies, which are released as elementary bodies.

Morowitz and Tourtellotte also studied another PPLO that causes a respiratory disease in poultry. This organism is found as a cell about 0.25μ in diameter and does not appear to pass through a life cycle involving changes in size. Chemical analysis shows that the cell contains both DNA and RNA. The DNA is in the form of double helices, and the RNA is present chiefly as ribosomes. Various proteins are found in the cytoplasm of this PPLO cell. The cell contains lipids similar to those found in animal cells and has a flexible membrane about 100 A in thickness, also similar to the plasma membrane of animal cells. Although the electron microscope reveals little of the internal structure of the cell, Morowitz and Tourtellotte were able to propose a model based upon their chemical studies (Figure 8.12).

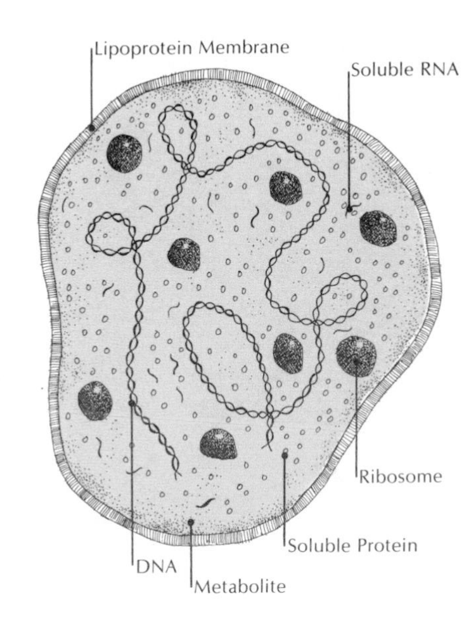

Figure 8.12. A model diagram illustrating the organization of *Mycoplasma gallisepticum*. Chemical analysis has revealed the presence of DNA in the form of a double helix and RNA in ribosomal form. In addition, the cell contains lipids similar to those found in animal cells and has a flexible membrane similar to the plasma membrane of animal cells.

Mycoplasmas, as these minute organisms are now called, may not be the smallest cells. If smaller organisms exist, they may have escaped detection by methods now in use. Yet no cell could be very much smaller than the mycoplasma. Because cell membranes are about 100 A (0.01μ) in thickness, it does not seem possible for a cell to be less than about 0.025μ in diameter, or about one-tenth the size of the mycoplasma. In order to contain the protein molecules necessary for metabolic reactions, Morowitz and Tourtellotte estimated that a cell would need a diameter of at least 0.04μ, or little less than half the size of the smallest mycoplasmas.

The procaryotic cells represent the simplest and smallest organisms yet known to be capable of independent survival in suitable environments. Even the smallest mycoplasmic cell is far more complex than the very simple DNA-enzyme system postulated to have been the first step in the evolution of living systems. The procaryotes may be descendants of organisms developed relatively early in evolution (but long after the first living systems). Or they may be descendants of organisms developed much later in evolution through simplification of more complex organisms. In any case, they represent the smallest and simplest living systems known to exist in the modern world. Mycoplasmas cannot be much larger than the smallest possible cell. If smaller and simpler living systems do exist, they cannot be based upon the common cellular structure that pervades all of life.

FURTHER READING

More information about bacteria and blue-green algae will be found in books by Gunsalus and Stanier (1960–1964), Jacob and Wollman (1961), Simon (1963), Sistrom (1969), Stanier, Doudoroff, and Adelberg (1970), and Thimann (1963); also articles by Braude (1964), Cairns (1966), Delbrück and Delbrück (1948), Echlin (1966), Hotchkiss and Weiss (1956), Wollman and Jacob (1956), and Wood (1951).

9
Viruses

Viruses have three important characteristics. First, they are very small. As many as 20,000 viruses could be fitted comfortably inside a small bacterial cell. Some large protein molecules are larger than some of the smaller virus particles. Second, viruses are structurally simple. In fact, there is little room for complexity. Simple viruses are nothing more than DNA or RNA packed inside a protective protein coating. Many viruses in purified form can crystallize just as do pure proteins. Third, viruses cannot reproduce outside of cells. When free, a virus is inert. It has none of the characteristics of a living system. It is no more alive than is the material inside a bottle of organic chemicals on a laboratory shelf. When the virus invades a cell, however, it is able to take over the metabolic machinery and to direct the cell's machinery to produce new virus particles instead of carrying out the normal cellular processes. In most cases, this takeover results in the death of the cell, which is ruptured as the newly made virus particles escape into the environment.

The debate about whether viruses are living organisms did not end with the discovery of their chemical nature and structure. In light of the characteristics described above, the question remains: Are viruses alive or not? Attempts to answer this question have not proven very fruitful, and biologists now accept the viruses as unique kinds of biological systems—neither living nor nonliving but as a sort of bridge between the two categories.

Every cell—even the simplest mycoplasma—has both DNA and RNA, ribosomes, thousands of enzymes, and an outer cell membrane that regulates the entry and exit of water, salts, foods, gases, and other components involved in metabolism. Such a cell is truly alive and independently self-sufficient. In contrast, a virus is an incomplete biological entity—a stripped-down structure usually containing only one kind of nucleic acid—either DNA or RNA, either single- or double-stranded—and a protein coat. Viruses are tiny packages of genetic information—or "genetic vectors."

The nucleic acid of the virus carries genetic information that directs the takeover of a host cell and the production of new viruses. Like the genetic information of a cell, the viral nucleic acid is subject to mutational changes. The protein coat of the virus serves two major functions. First, it protects the viral nucleic acid from enzymes such as DNase and RNase, which are abundant in organisms and would destroy bare nucleic acid molecules. Second, by showing a specific affinity for the surfaces of certain kinds of host cells, the protein coat permits the virus to attach itself only to appropriate host cells and not to other kinds of cells within which it could not reproduce. In other words, the protein coat determines the *host range* of the virus.

The simplicity and specificity of a virus make it a remarkable research tool for molecular biologists. It can be regarded as a naturally occurring hypodermic that can inject nucleic acid into a particular kind of cell. Used in this way, viruses have revealed a great deal of information about the genetic mechanisms of cells.

Viruses exist in a wide range of sizes, structures, and types. They vary in size from poliovirus (about 100 A in diameter) to rickettsia and other large disease-causing viruses (about 0.5μ in diameter). Most viruses are spherical, but some are bricklike, rodlike, cubic, or irregular in shape (Figure 9.2). Some contain DNA, whereas others contain RNA; a few contain both kinds of nucleic acids. Some viruses have a coat made of a single protein, whereas others have coats made of many proteins. Some viruses contain lipid and

Figure 9.1 (above). A T-2 *E. coli* bacteriophage osmotically shocked. The long strands are DNA. (× 50,000)

Figure 9.2. Examples of the shapes and sizes of viruses. Influenza virus (middle left); T-5 (middle right); and tobacco mosaic virus particles (below).

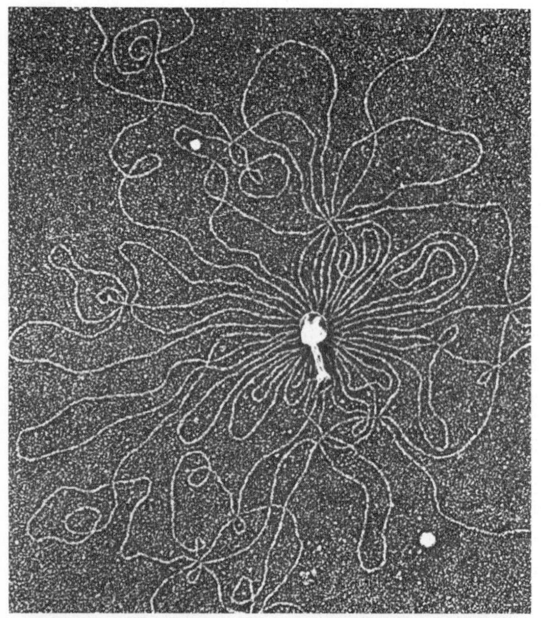

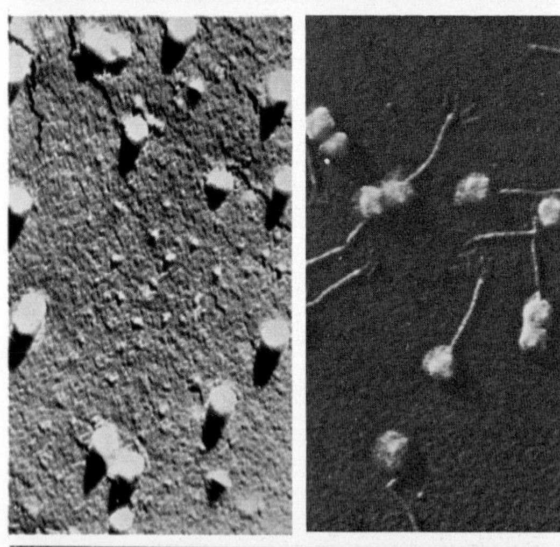

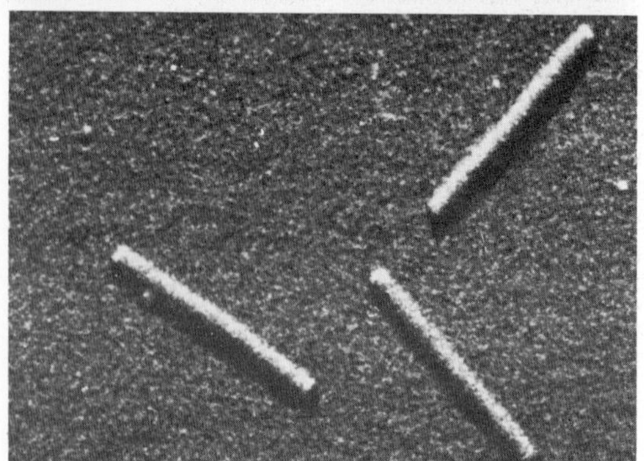

carbohydrate components in addition to their nucleic acid and protein. Cells of nearly every type of organism—from bacteria to plants, insects, and animals—are susceptible to attack by some type of virus.

TMV

The tobacco mosaic virus (TMV) is typical of the rod-shape viruses that attack plant cells (Figure 9.3). It is composed of a coiled molecule of RNA, surrounded by a protein coat, or *capsid*, made up of many identical protein subunits, or *capsomeres*. The entire particle, containing nucleic acid and capsid, is known as a *virion*.

In the mid-1950s, Heinz Fraenkel-Conrat and his colleagues separated TMV into its RNA core and protein overcoat components. These separated parts were then recombined to form infectious viral particles. This study showed that the virion is formed through a process of self-assembly. Under proper conditions of acidity, temperature, and salt concentration, capsomeres spontaneously gather around the viral RNA core to form a rod-shape virion. In another part of this study, it was demonstrated that both the RNA core and the protein coat must be present in order to produce fully infective virus particles (Fraenkel-Conrat and Williams, 1955).

In 1957 groups in both Berkeley and Germany independently reported that the RNA core of TMV, stripped of its protein coat, is still slightly infectious if rubbed into tobacco plant leaves. Typical lesions (wounds) develop on the leaves, and the infected cells produce normal virions containing both RNA and protein (Figure 9.4). The viral RNA is less than 1 percent as infectious as the normal TMV virion, but these experiments show that the RNA molecule alone contains all the genetic information necessary to infect a cell and to cause production of complete new virions. This study corroborated earlier virus and cell experiments that had demonstrated that genetic information is carried by nucleic acids alone.

The single-stranded RNA molecule of TMV contains about 7,000 nucleotides—enough to code about five to ten viral proteins. Obviously, one of these proteins must be the capsid protein of the virion. There are several different strains of TMV, each of which makes a slightly different capsid protein. After cell infection, the new virions that are formed always have capsid proteins similar to those of the parental virus, regardless of the genetic make-up of the host cell. Therefore, the information that specifies the amino acid sequence of the capsid protein must be stored in the viral RNA, not in the genetic molecules of the host cell.

The nucleic acid core of the TMV particle consists of a single strand of RNA. In order to reproduce itself, this RNA strand must serve as a template for the transcription of a complementary RNA strand. Such a process of RNA replication does not normally occur in cells and is apparently accomplished by an RNA replicase enzyme that must be one of the other proteins coded by the viral RNA. Thus, the first step after the viral RNA enters the cell must be production of the RNA replicase through the use of the cell's ribosomes to translate the viral RNA. After the replicase is formed, the viral RNA can be duplicated, and production of capsid and other proteins that assist in the takeover of the cellular machinery proceeds more rapidly.

As the capsid protein molecules are formed, they spontaneously fold into the uniquely shaped capsomeres. These capsomeres tend to aggregate into disc-shape units, which are aggregated to form the rodlike capsid around an RNA molecule. In the absence of viral RNA, capsomeres can be

Figure 9.3. Sequence of steps in the assembly of TMV from its RNA and protein subunits. A typical TMV has a rod shape and is composed of a coiled molecule of RNA (yellow) surrounded by a protein coat, or capsid, made up of many identical protein subunits, or capsomeres (blue). The entire particle is known as a virion.

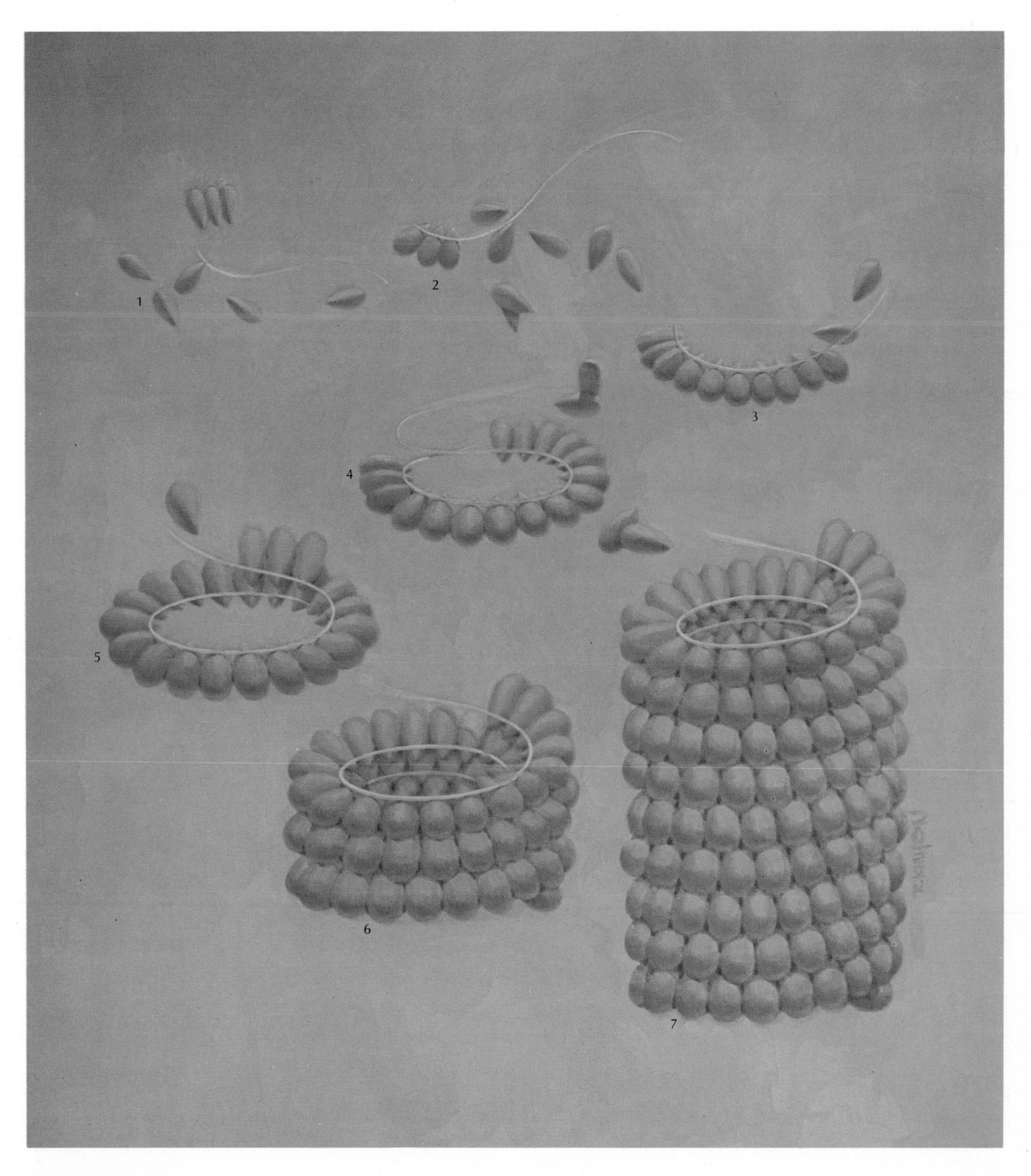

Figure 9.4 (above). A bottle containing TMV crystals frozen by W. M. Stanley in 1935. When rubbed into a tobacco plant, the crystals still cause infection.

Figure 9.5 (below). Diagram and electron micrograph of a T-even bacteriophage. (× 420,000)

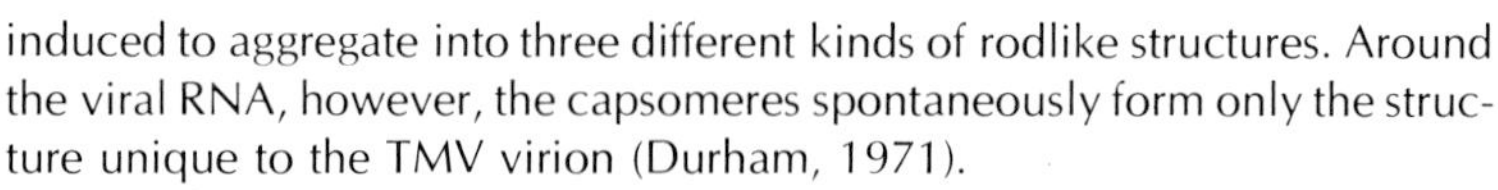

induced to aggregate into three different kinds of rodlike structures. Around the viral RNA, however, the capsomeres spontaneously form only the structure unique to the TMV virion (Durham, 1971).

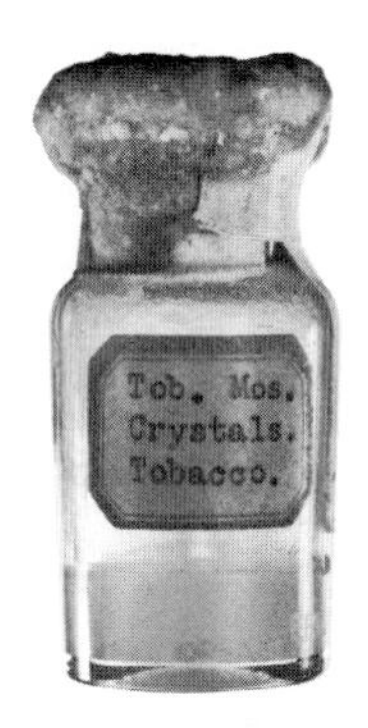

A final confirmation of the role of viral RNA in capsid protein synthesis came from studies in which mutations were induced chemically. A change of a single base among the 7,000 bases in the viral RNA molecule may cause a corresponding change of a single amino acid in the capsid protein. Many of these mutations profoundly affect the biological behavior and infectivity of the mutant viruses, suggesting that the folding of the capsid protein can be altered by the change of a single amino acid in its sequence.

BACTERIOPHAGE

In 1915 an English bacteriologist, Frederick William Twort, was dismayed to find that a number of his bacterial colonies were dying. The cells of the bacteria ruptured, leaving nothing but a foggy liquid. He discovered that this liquid infected other bacterial cells, even after filtration. Thus, Twort announced the discovery of viruses that infect bacteria. This discovery was

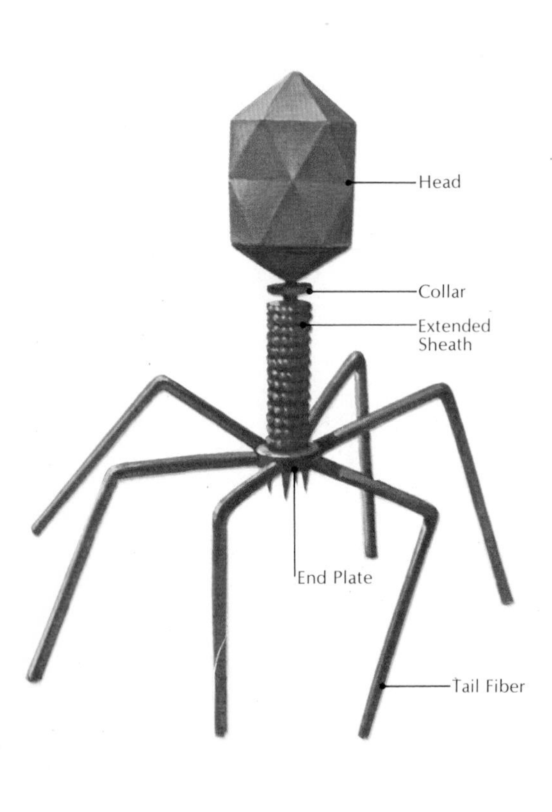

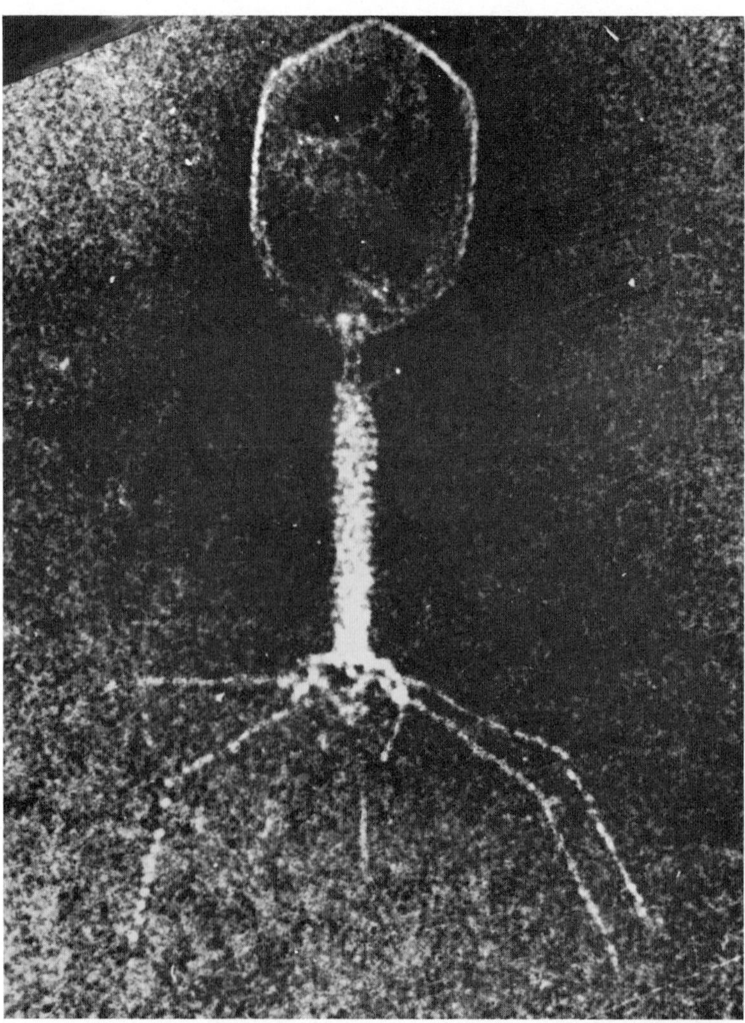

Figure 9.6 (above). T-4 bacteriophages adsorbed on an *E. coli* B cell wall. Phages are connected to the bacterial cell surfaces by short tail fibers, which will eventually penetrate the cell wall.

Figure 9.7 (below). Electron micrograph of a T-4 infected *E. coli*. The bacteriophage protein remains as an empty "ghost" attached to the outer surface of the cell wall. (× 55,000)

independently repeated two years later by the Canadian bacteriologist Félix Hubert d'Hérelle, who named the viruses *bacteriophage* (bacteria eaters). Although Twort emerged the victor in the subsequent dispute about who should get credit for first discovering bacterial viruses, d'Hérelle's name for the viruses is still used.

The DNA-containing bacteriophage—often called phage by biologists who work with them—have been of enormous help in understanding gene action. One family of phage that infect the bacterium *Escherichia coli*—a common subject for laboratory studies—is generally designated by strain as T-1, T-2, T-3, and so on. The T-even strains (T-2, T-4, T-6, and so on) are *virulent* particles that kill the host bacterial cell. Expression of the DNA in T-even bacteriophage now is better understood than the gene action of any other virus or of any organism.

A typical particle of a T-even phage has a large hexagonal head, which contains its genetic material—double-stranded DNA (Figure 9.5). The DNA must be tightly coiled within the head, for it is a long molecule with a molecular weight of about 120 million—long enough to carry information for the production of more than 100 average-size proteins. The protein coat of the head is composed of subunits, each with a molecular weight of about 80,000. Associated with the DNA inside the head is an internal protein, which probably helps to hold the DNA in a tightly coiled position. At the base of the head is a protein collar with a tail assembly attached. The intricate phage tail consists of a sheath of 144 contractile protein subunits wrapped around a hollow protein core. The top of the core-sheath complex is attached to the collar region of the head. The bottom of this complex is attached to a flat, hexagonal base plate, which has six spikes protruding from it. Extending from these spikes are long, kinked tail fibers made up of several different proteins.

This elaborate structure acts as a microsyringe to inject its viral DNA through the tough polysaccharide cell wall of the bacterial host cell. To infect a bacterial cell, a T-even phage must make contact with it (Figure 9.6). The phage, with no mechanism for moving itself, is carried along in the random thermal motion of molecules and other submicroscopic particles. Eventually, the phage happens to bump into a host cell that possesses appropriate protein receptor sites on its surface. The tail fibers of the phage "recognize" the receptor sites and attach themselves by weak, noncovalent bonds. Inappropriate species of bacteria do not possess suitable receptor sites, and the phage does not attach itself to such a cell.

Although the attachment process is reversible, what immediately follows is not. An enzyme produced by the attached phage digests a small hole in the tough bacterial cell wall. The contractile sheath of the phage tail then contracts (possibly using energy provided by hydrolysis of bound ATP), the head of the phage collapses, and the DNA contents of the head are extruded through the tail into the cytoplasm of the host cell. Only the viral DNA and a small amount of its internal protein enter the bacterial cell. Most of the phage protein remains as an empty "ghost" on the outer surface of the cell wall (Figure 9.7).

Before injection of the viral DNA, the biochemical activities of the *E. coli* cell are governed by its own chromosome, a single, long, circular molecule of DNA. After injection, the bacterial cell contains an additional chromosome—the viral DNA. The consequences of this extra chromosome are disastrous for the bacterial cell. In less than 30 minutes, the injected

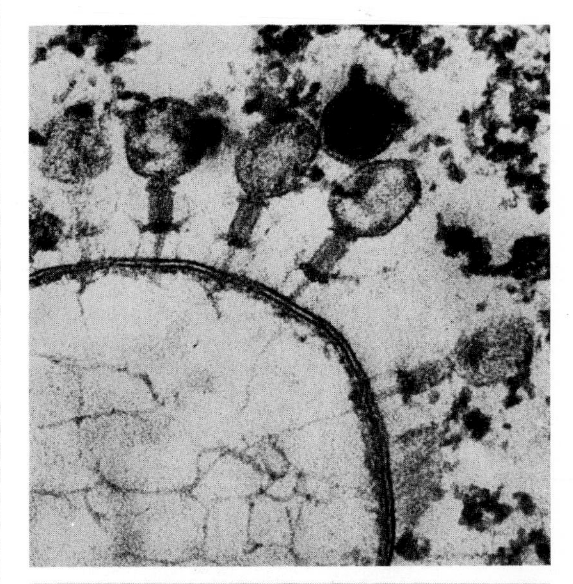

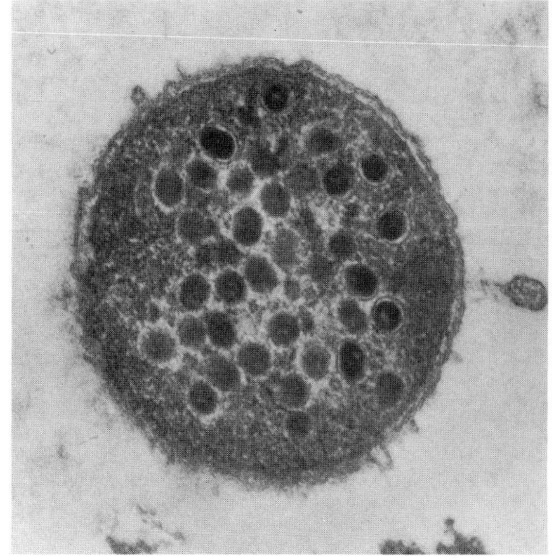

Figure 9.8. Life cycle of the double-stranded T-2 virus.
Printed by permission from J. D. Watson, Molecular Biology of the Gene, © *1965, J. D. Watson*

viral DNA takes over the direction of the cellular mechanism and directs the production of 100 or more new viruses. The release of the newly produced viruses is brought about by the disruption and death of the host cell (Figure 9.8).

The sequence of events during production of new phage particles is quite well known. Immediately after the viral DNA enters the cell, it interacts with the host cell's RNA polymerase and ribonucleotide precursors to make early viral mRNA. From this mRNA, the ribosomes and enzymes of the host cell produce early viral enzymes and other proteins that are necessary for replication of the viral DNA.

The viral DNA differs from the bacterial DNA in that it has a base called 5-hydroxy-methylcytosine in place of normal cytosine. Glucose molecules are attached to this unusual base (Figure 9.9). Among the early viral enzymes produced are DNases that attack the bacterial DNA but do not degrade the viral DNA. Apparently, the modified cytosine of the viral DNA protects it from the action of these enzymes. Other early viral enzymes include RNases that help break down the RNA of the host cell, and enzymes that convert cytosine to 5-hydroxy-methylcytosine. The latter process prevents the replication of host cell DNA and promotes the replication of viral

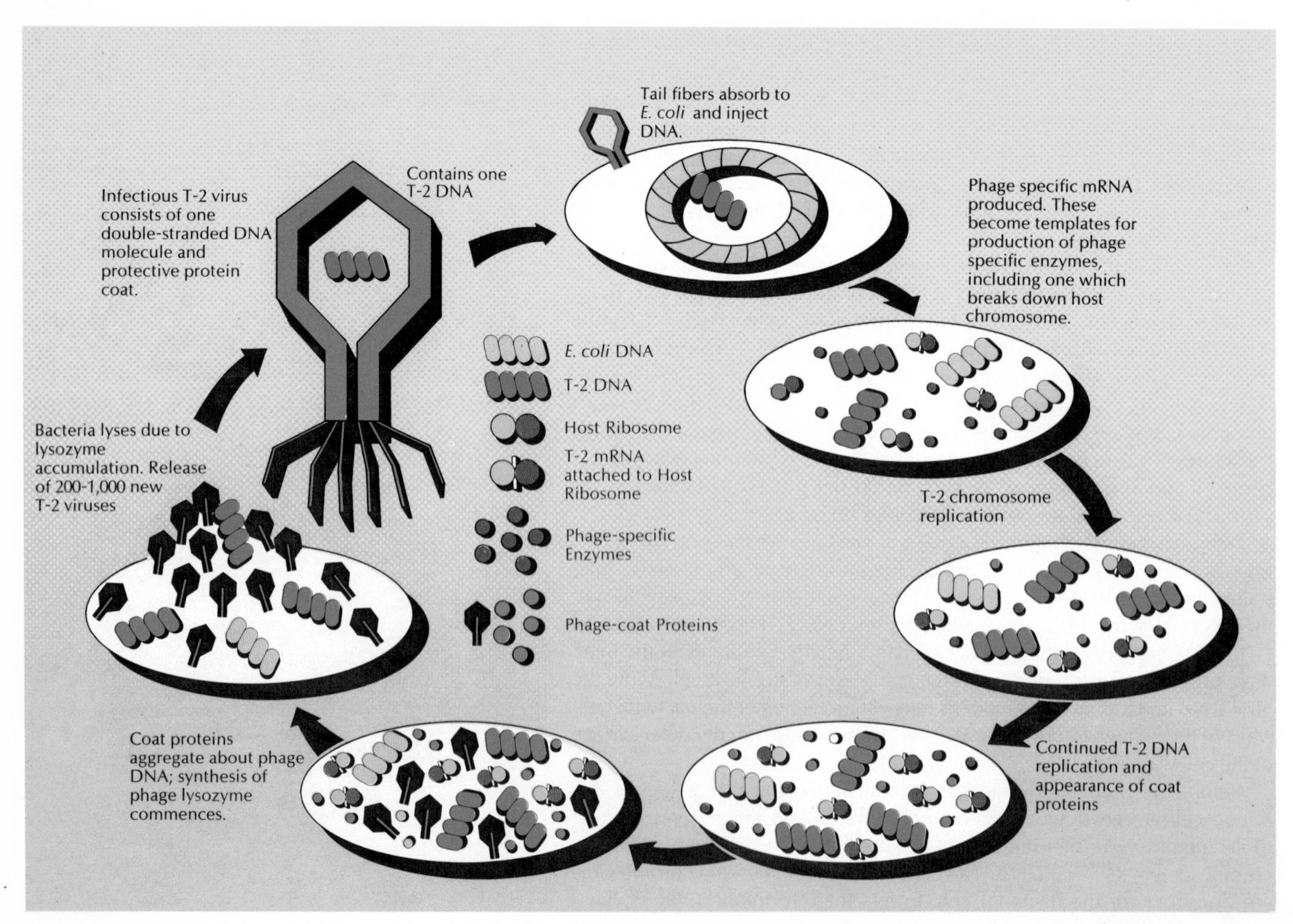

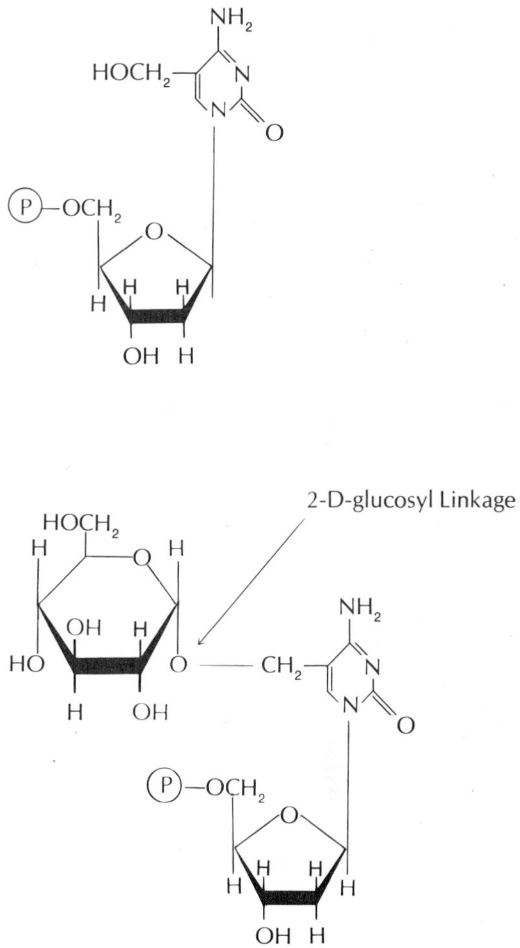

Figure 9.9. Structural formulae showing the base 5-hydroxy-methylcytosine (above) and the same base with a glucose molecule attached (below). Viral DNA contains this base instead of the normal cytosine.

DNA. The DNA of the host cell is broken down by the DNase into its nucleotide subunits, which then are assembled by another early enzyme (a specific DNA polymerase) and produce about 100 copies of the viral DNA.

Although the viral DNA contains specifications for about 100 viral proteins, only a few of these are produced in the early stage of viral infection. The various parts of the viral DNA are activated to produce mRNA in a regular time sequence. Thus, various viral proteins are produced as needed in an assembly-line sequence, culminating in the production and release of complete phage particles.

Many control processes are needed to achieve this time sequence. For example, the host RNA polymerase has a control protein (sigma factor) that causes the polymerase to attach to certain sites on the host DNA. One of the proteins produced by the viral DNA inactivates the sigma factor of host polymerase molecules. Apparently, these sigma factors are replaced by new control proteins that are synthesized according to information in the viral DNA. The viral sigma factors then combine with and modify the host RNA polymerase so that the polymerase will activate synthesis of mRNA from parts of the viral DNA that become active late in the viral replication sequence.

Other viral enzymes guide the production of new tRNA molecules at appropriate times. It even appears that some viral proteins modify host ribosomes to make them suitable for viral protein synthesis.

Finally, as the synthesis of copies of viral DNA is completed, the late viral proteins are produced. These include internal protein and the proteins needed to build the head and tail assemblies of the 100 or more new viruses. The process of phage assembly was studied by R. S. Edgar and his colleagues, who demonstrated that this complex structure is produced through a process of self-assembly (Wood and Edgar, 1967).

The studies were carried out by using many phage mutants, each strain being defective in one essential viral protein. Each of these mutant strains is unable to complete the production of phage particles. Because a specific protein that is necessary for the next step is missing, the assembly of phage particles comes to a halt at some intermediate stage.

Suppose, for example, that two defective mutant phage strains are unable to complete the production of the head unit of the phage particle. Cultures of strain A produce incomplete head structures (H_A), whereas cultures of strain B produce different incomplete head structures (H_B). When extracts of the two strains are mixed together, however, complete phage head units are formed.

Apparently, H_A and H_B are precursors of the normal head structure:

$$\text{Precursor I} \xrightarrow{+\text{ Protein I}} \text{Precursor II} \xrightarrow{+\text{ Protein II}} \text{Head Structure}$$

One mutant strain lacks protein I and therefore halts assembly with precursor I. The other strain lacks protein II and halts production with precursor II. When extracts of the two strains are mixed, both proteins are present and normal head assembly can be completed.

Is H_A precursor I or precursor II? Does strain A lack protein I or protein II? In some cases, these questions can be answered by examination with the electron microscope—one precursor may be more similar to a completed head structure than the other. If the phage assembly sequence cannot be determined by electron microscopy, it can be determined by experimental tests. H_A can be isolated from strain A and added to extracts of strain B. If H_A is precursor I, strain B will possess protein I and lack protein II. Therefore, the H_A will be converted to H_B, but assembly will stop there. On the other hand, if H_A is precursor II, strain B will lack protein I and possess protein II. Therefore, the H_A will be converted into complete head structures.

After many such experiments, Edgar and his colleagues determined the sequence in which the phage particle is assembled. They discovered not a single linear assembly line but several subassembly lines converging to the final phage particle. Different proteins, produced by different parts of the viral DNA, control each assembly step.

Most of the information of the viral DNA codes for structural proteins, but apparently some of the information leads to production of proteins that act as catalysts (enzymes) for the assembly process. There is a head subassembly line, a tail subassembly line, and still another subassembly line for tail fibers. When completed heads encounter tails that are complete except for tail fibers, the two join together spontaneously. Note that this joining occurs only when each of the separate parts is completed. Finally, when the head-tail assemblies encounter completed tail fibers, the fibers attach to the base-plate spikes to complete the formation of fully infectious phage particles.

When the new phage particles are completed, the last step of viral infection, *cell lysis*, takes place. A very late portion of the viral DNA codes for the production of the enzyme lysozyme, which breaks down the rigid cell wall of the bacterium. With the bursting of the bacterial wall, hundreds of new viruses are released into the medium, where they eventually collide with new host cells to trigger a new cycle of infection and viral replication.

ANIMAL VIRUSES

Most viruses that attack human and other animal cells reproduce in a fashion similar to the phages, producing virus particles with an exposed protein coat surrounding a DNA or RNA core. With some viruses, however, the protein coat is enclosed by a membrane derived from the host cell. The cell surface buds off to form a lipid-protein coating over the nucleic acid–protein virus particle. Among the viruses with such membrane coatings are the DNA-containing herpes virus and the RNA-containing influenza virus.

Some of the more complex viruses contain their own enzymes, such as RNA polymerase, nucleotide hydrolases, and so forth. These enzymes assist the viral nucleic acid in the rapid takeover of the host cell. Some viruses, such as the influenza virus, contain the enzyme neuraminidase in their membranous envelopes. Apparently, this enzyme plays a role in the budding of the viruses from the surface of an infected cell.

Animal viruses do not possess an injection apparatus like that of the T-even phages, but their capsid proteins do recognize specific receptor sites on the host cell membrane. It appears that the virus particle (or a portion of it) passes directly through the cell membrane into the cytoplasm (Figure 9.10). Encapsulated viruses apparently fuse their viral membrane with the host cell membrane, thus delivering the internal viral nucleic acid and internal proteins into the cytoplasm of the host cell.

After entry into the cytoplasm, the viral particle is soon stripped of capsid proteins by cell enzymes, resulting in the release of the viral nucleic acid. In the case of DNA viruses, the viral DNA acts directly as a template

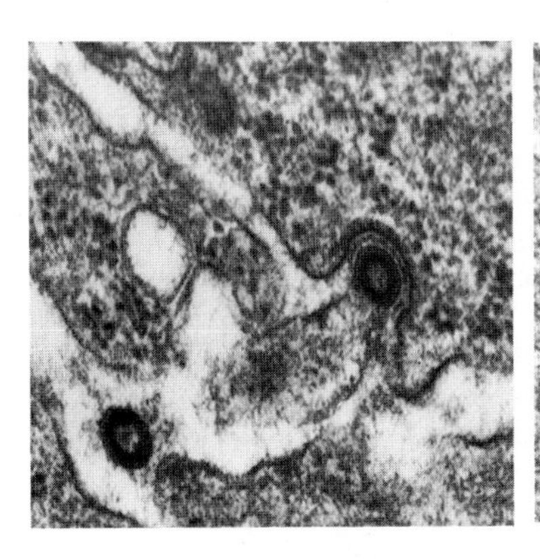

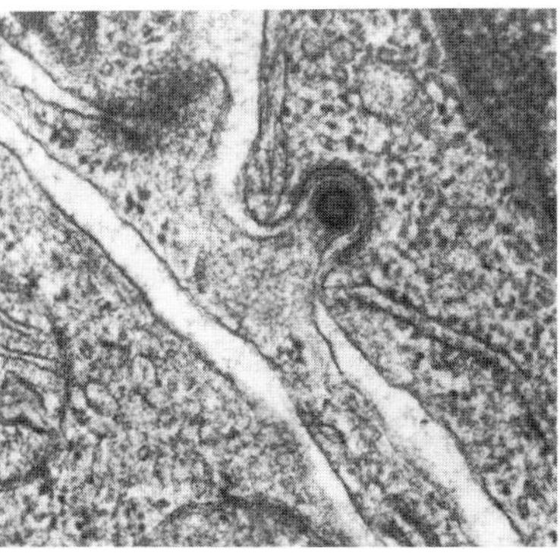

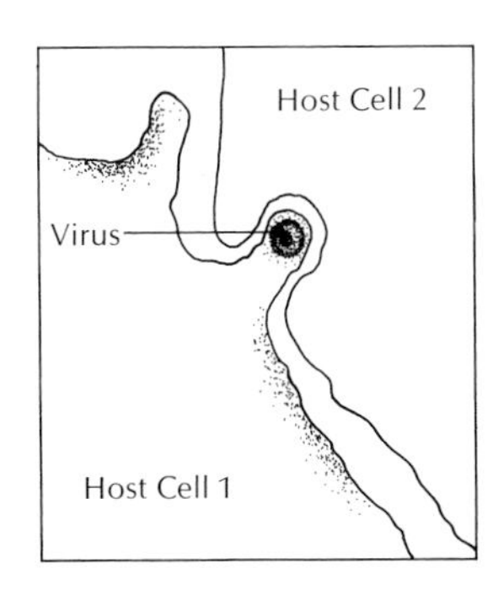

Figure 9.10. Electron micrographs and a diagram showing the direct cell to cell transfer of a Rittner virus.

(in most cases with host RNA polymerase) for the transcription of viral mRNA (which guides the formation of viral proteins) and as a template for the replication of viral DNA.

RNA viruses replicate in much the same way, but in most cases the viral RNA acts as a messenger RNA that codes directly for viral proteins by association with host ribosomes. As with TMV, one of the proteins produced by viral RNA is an RNA replicase, which then catalyzes the replication of the viral RNA. The RNA replicase makes complementary strands of the viral RNA and then makes duplicates of the original strands on the complementary strands.

Recent experiments with certain RNA cancer viruses indicate that these viruses produce a DNA polymerase that brings about formation of viral DNA by using the viral RNA as a template. The viral DNA formed inside the host cell then serves as the template for the transcription of new viral RNA. The discovery of DNA synthesis guided by an RNA template forced the modification of the so-called central dogma of molecular biology, which asserted that genetic information flows only from DNA to RNA to protein. (F. H. C. Crick, however, has pointed out that he was careful not to exclude the possibility of RNA-directed DNA synthesis from his original statement of the central dogma.) Now that biologists have begun to look for RNA-directed DNA synthesis and RNA replication in normal cells, they are finding that even in cells not infected by viruses the picture is not as simple as indicated by the simplified form of the central dogma. Enzymes capable of guiding DNA synthesis from RNA templates apparently do exist in normal mammalian cells, where they may play a role in processes such as differentiation or memory (Scolnick, *et al.*, 1971). The existence of these enzymes in normal cells is not yet fully confirmed, and the significance of RNA-directed DNA synthesis will probably not become clear until its role and extent have been more fully studied.

Both RNA and DNA viruses act as packets of genetic information. The details of the information vary in different viruses, but the overall message they bring into infected cells is similar. In effect, the viral nucleic acid tells the cell: "Make the proteins that I specify. Use some of these proteins to make many copies of my nucleic acid. Wrap these new viral genes in new viral capsid protein to assemble complete new viral particles."

If this cycle of virus reproduction and cell destruction were to continue indefinitely, no animal could survive a viral infection because all its cells would soon be destroyed. In fact, the organism possesses various defense mechanisms that destroy many of the virus particles before they can infect other cells. In addition, each virus attacks only certain kinds of cells. For example, common-cold viruses grow mainly in the cells lining the respiratory passages, viruses causing intestinal disorders grow mainly in gut cells, and hepatitis virus invades chiefly liver cells. This specificity is partly due to the fact that the virus particle attaches only to cells possessing appropriate receptor sites on their membranes. Thus, most viral diseases attack only certain tissues of the organism.

The destruction of cells that are susceptible to viral infection may be prevented by the immune responses of the animal body. The immune system, stimulated by the presence of the foreign virus particles, produces molecules called antibodies, which combine with the viral coat protein. The antibodies adhere firmly to the viral particle and prevent it from attaching to or penetrating cells. Antibodies that neutralize a particular kind of foreign particle are produced in great quantities after that kind of particle enters the body. The immune system can produce antibodies much more quickly in response to any future invasion of the same kind of particles (Chapter 22). Thus, one attack by a particular virus (or microorganism) confers greater or lesser immunity to ill effects from future attacks by the same type of particle.

Immunity to viruses can last for varying periods of time. Most childhood virus diseases (measles, mumps, chickenpox) usually induce lifelong immunity. These diseases *are* childhood diseases because most people suffer an attack during childhood and thereafter are immune.

Some common-cold viruses and other respiratory viruses, such as influenza, induce weaker immunity. These viruses can reinfect the body after a period of months or years. This weak immunity may be partly explained by the fact that antibody concentrations are much lower in nose and throat fluids than in the bloodstream. In addition, influenza viruses are extremely versatile infecting agents because they are able to withstand drastic mutation of the parts of the viral RNA that code for membrane protein without losing their infectivity.

Every few years, a new virulent mutant of influenza virus arises in the human population. This mutant has a membrane protein that does not combine strongly with the antibodies previously produced. Therefore, the antibodies already present in the body do not immediately neutralize an invasion of the new mutant virus strain. If the influenza mutant is almost totally resistant to existing antibodies, a pandemic sweeps the world. Such a pandemic of influenza has occurred about every ten years, and as the human population becomes larger and denser, the likelihood of more frequent influenza pandemics increases.

Many viral diseases can be prevented by vaccination. Edward Jenner's vaccination against smallpox involved the injection of cowpox viruses into the human body. These viruses are so similar to smallpox viruses that they cause the production of antibodies that combat both kinds of viruses. The cowpox virus, however, causes relatively little damage in the human body. The antibodies produced to combat the cowpox infection are effective enough to handle most future invasions of either cowpox or smallpox viruses.

In dealing with most viral diseases, biologists have not been lucky enough to find a closely related, relatively harmless virus that can be used as a natural vaccine. Instead—as Louis Pasteur did in producing rabies

Figure 9.11. An artist's interpretation of a pandemic. Every few years, a new mutant strain of an influenza virus arises in the human population. If this virulent mutant is resistant to antibodies, a pandemic occurs. Such a pandemic has been occurring about every 10 years.

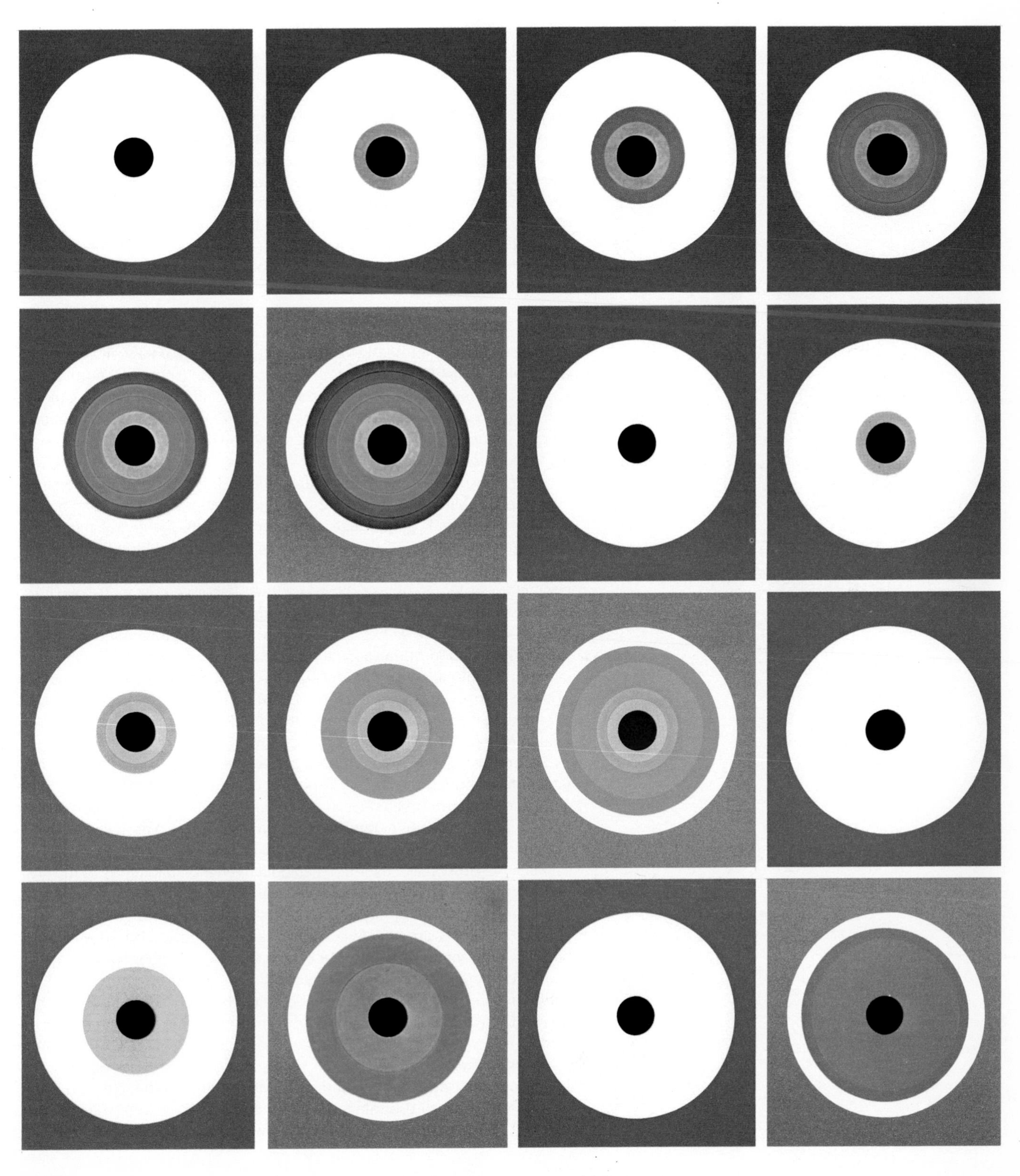

vaccine—they have been forced to produce forms of the disease virus with reduced infectivity. Biologists can produce this type of vaccine by physically or chemically treating the normal virus or by searching for mutant strains with the desired properties.

Vaccines have now been developed for many of the human viral diseases, but vaccination against influenza viruses is not yet completely successful. Because influenza viruses mutate so drastically, a new vaccine must be developed for each major new strain of virus. Often, an influenza pandemic is well under way before a specific vaccine can be prepared in quantity. Vaccination does little or no good after the body has already been infected by viruses. Furthermore, influenza vaccines are sometimes ineffective, or they may produce relatively severe illnesses in a significant proportion of the population.

INTERFERON

It has been known for nearly 40 years that when an animal cell is infected with a virus, it becomes resistant to further infection (*superinfection*) by the same or different viruses. This phenomenon is called *interference*. Culture fluid taken from infected cell cultures can confer resistance upon uninfected cell cultures. The active principle was shown to be a protein and was named *interferon*.

Interferon probably plays a natural role in the body's defense against viral infection. It appears soon (12 to 18 hours) after infection, long before the immune system can produce significant amounts of antiviral antibodies (about 5 days). The cells that are first infected may be killed by the virus, but they will produce and release interferon, which is transported via the circulatory system to other cells, which then become resistant to infection.

Research on interferon has centered upon two questions. First, what is the mechanism by which cells are stimulated to produce interferon? Not only viral infection but also bacterial *endotoxins* and synthetic substances can induce animal cells to produce and release interferon. The most potent of these *inducers* has turned out to be double-stranded RNA (dsRNA). Under the right conditions, minute amounts of dsRNA can confer resistance to viral infection.

Second, how does interferon cause cells to become resistant to viral infection? Although the answer is not yet clear, most of the evidence suggests that interferon-treated cells can distinguish viral mRNA from cellular mRNA; the cells refuse to translate the viral message into protein, but translation of cellular messages is unimpaired.

Because viruses take over the metabolic (genetic) machinery of the host cell, until now it has been impossible to find a drug that interrupts the viral replication cycle without also disturbing the normal expression of cellular genes. Antibiotics are effective against bacterial (but not viral) infections because they interfere with aspects of bacterial metabolism that are not employed by animal cells. Thus, penicillin or sulfa drugs are effective as antibiotics because they block synthesis, respectively, of the polysaccharide cell wall and of certain coenzymes—processes that are not required by animal cells.

Interferon-treated cells seem to be resistant to most types of virus, yet interferon does not seem to interfere with normal cellular metabolism. Original attempts to use interferon as an antiviral drug were fraught with difficulty. First, interferon is one of the most active proteins known; con-

Figure 9.12. The life cycle of a lysogenic bacterial virus. *Printed by permission from J. D. Watson,* Molecular Biology of the Gene, *© 1965, J. D. Watson.*

versely, the amounts of activity that can be isolated represent extremely minute amounts of protein. Second, interferon's action is *species-specific*—it is active only on cells from the same species as the cells that produce it. It is not feasible to use humans as a source of commercial interferon. Third, when interferon is injected into the bloodstream, it is cleared from the circulatory system fairly rapidly. In spite of these difficulties, interferon has been used to confer resistance upon humans to some viruses—for example, the cold-causing *rhinoviruses*. Interferon also causes resorption of the fetus in pregnant mammals; it thus has possibilities for use in solving overpopulation problems.

Inducers of interferon, notably dsRNA, have been used to stimulate production of interferon by the organism itself. These drugs also pose problems; they are moderately toxic and induce fever. Nevertheless, inducers of interferon have provided remarkable protection against viral infection and have caused regression of established infections. Moreover, there is ample evidence that some forms of cancer are correlated with the occurrence of viral chromosomes in the cancer cells, and interferon has been shown to cause regression of viral tumors. The discovery of interferon thus has great potential in the prevention and cure of viral disease.

LYSOGEN

Certain other aspects of virus activity are of enormous significance for medicine and of value for the study of gene action. In many cases, viral DNA will break into segments during viral reproduction. These segments may

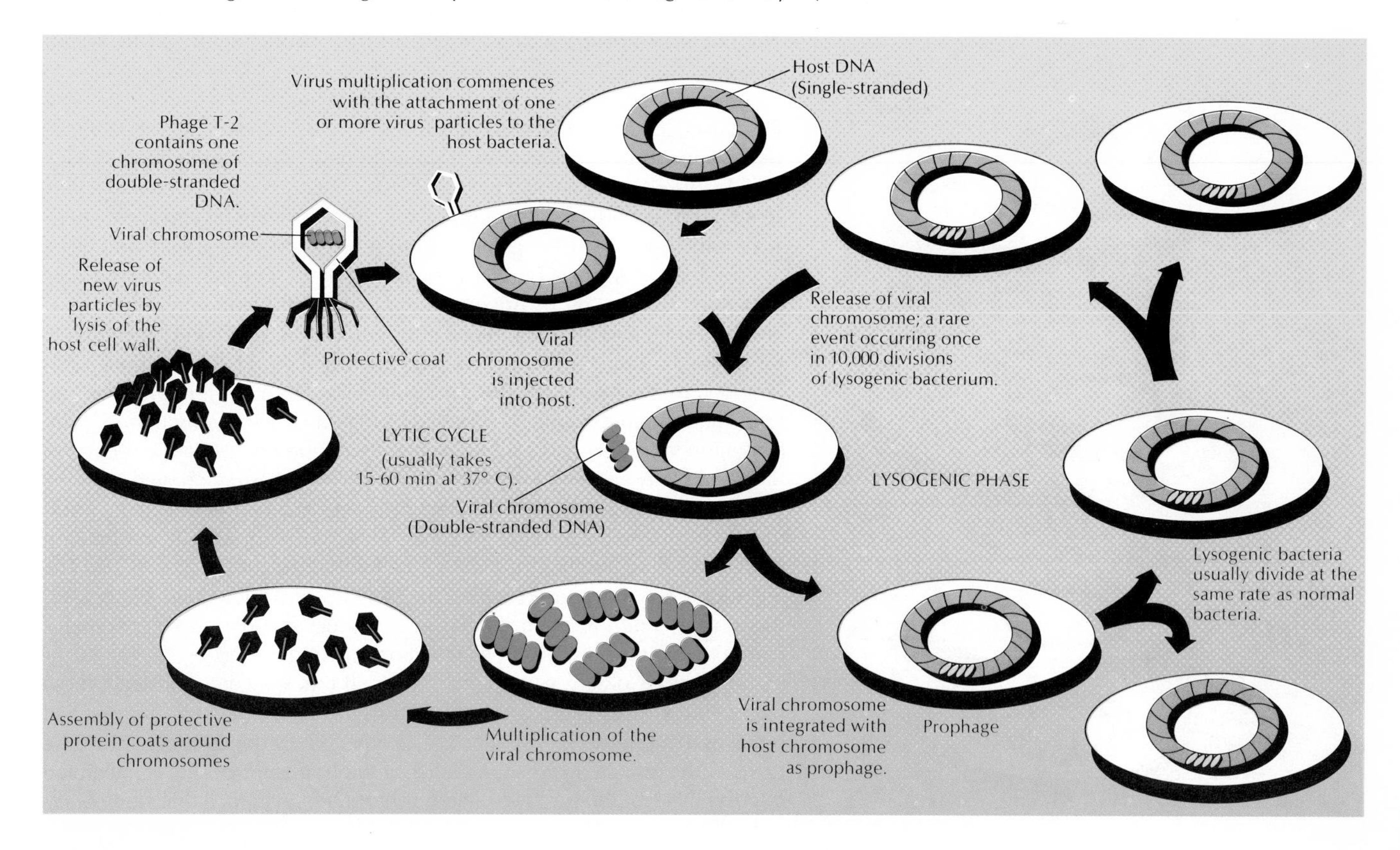

Figure 9.13. The process of transduction by a bacteriophage.

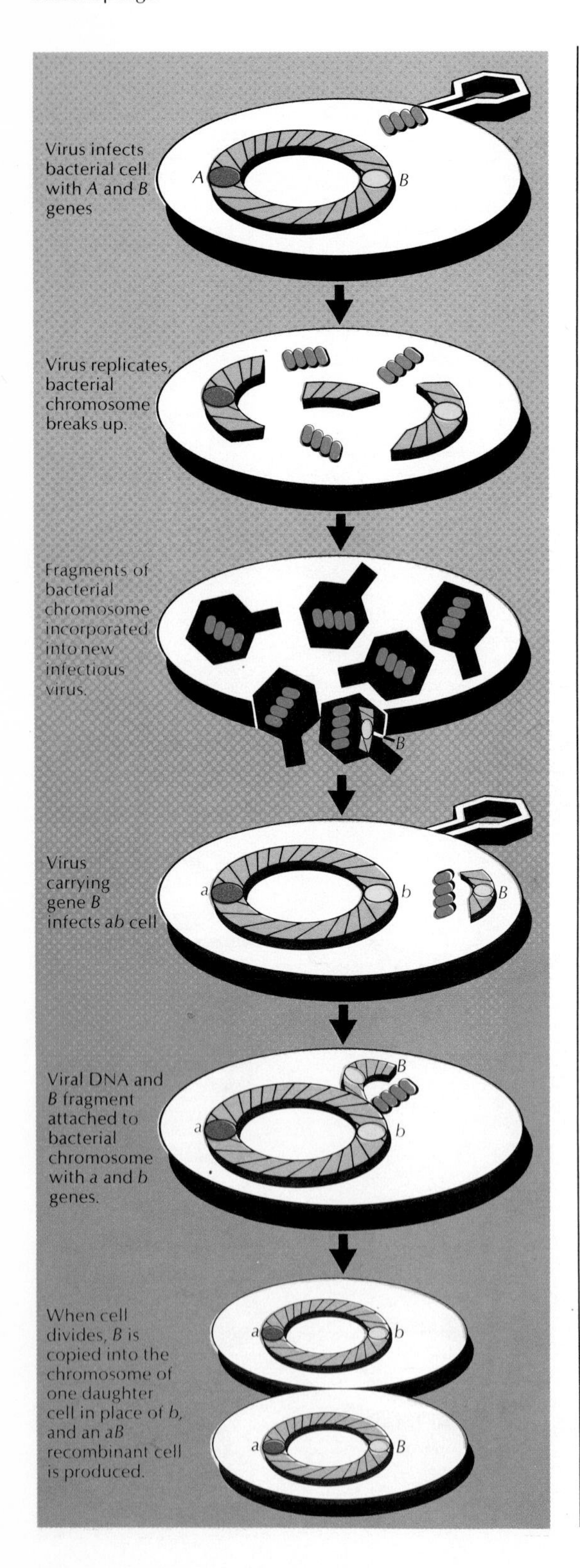
Virus infects bacterial cell with A and B genes
A
B
Virus replicates, bacterial chromosome breaks up.
Fragments of bacterial chromosome incorporated into new infectious virus.
B
Virus carrying gene B infects ab cell
a
b
B
Viral DNA and B fragment attached to bacterial chromosome with a and b genes.
B
a
b
When cell divides, B is copied into the chromosome of one daughter cell in place of b, and an aB recombinant cell is produced.
a
b
a
B

then recombine to form new genes or gene sequences. If a single cell is simultaneously infected with more than one virus, important new combinations of genetic material can occur from the various recombining DNA segments.

Some very small viruses—called *satellite viruses*—can reproduce only when they infect a cell that also is infected by a larger virus. Apparently, the satellite virus does not carry all the genes needed for its reproduction but must use some proteins coded by the other virus' genetic information. Sometimes the satellite virus is so small that its nucleic acid probably carries only a single gene coding for its own coat protein.

In bacteria infected with certain phage, such as phage lambda, a most unusual phenomenon occasionally takes place. These phage normally reproduce their DNA as small loops, wrap the DNA in a protein coat, and then burst the cell. In a certain proportion of the infected bacteria, however, the viral DNA does not replicate. Instead, its small loop of DNA attaches itself to the large loop of bacterial DNA. Nucleases then break both DNA loops at the point of their attachment, and the broken ends are joined to form one large loop that contains both bacterial and viral DNA. Thus, the viral DNA becomes an integral part of the host cell chromosome and is replicated every time the host cell divides.

This integrated state, or *lysogen*, may persist for thousands of generations, with each pair of daughter cells carrying the viral DNA within the chromosomal loop. Information in the integrated viral DNA produces protein, which prevents independent replication of the viral DNA. This protein also prevents replication of any new viral DNA entering the lysogenic cell. The bacterial cell thus gains immunity from further infection by related viruses.

On rare occasions, a lysogenic bacterium suddenly separates the viral DNA from the integrated chromosome. The released viral DNA then begins to replicate in its usual fashion, producing new viruses and killing the host cell. In this way, the lysogenic viral infection becomes activated into a virulent viral infection. Agents such as x-rays, ultraviolet light, and some chemicals can cause most lysogenic cells to release their integrated viral DNA.

In some cases, the viral DNA comes out of the integrated chromosome imperfectly. A part of the original viral DNA is left behind on the bacterial chromosome, and an equal piece of original bacterial DNA is included in the released viral DNA loop. After this abnormal viral DNA multiplies, it is wrapped in a protein coat just as normal viral DNA would be. When this virus infects another cell, the viral DNA may be missing so much of its information that it is unable to replicate and kill the host cell. The section of bacterial DNA carried along with the viral DNA may recombine with the DNA of the new host cell. Thus, the virus acts as a carrier of genetic information from one bacterium to another. This transfer of genetic information from one cell to another by a virus is called *transduction* (Figure 9.13). The technique has been used widely in genetics and molecular biology for many years and has been crucial in a number of important experiments clarifying the mechanisms of gene action.

Some kinds of viruses do not kill the cell that they enter. Instead, these *oncogenic* viruses cause a *transformation* of the animal cell, and the cell behaves in a fashion very similar to a cancer cell. Evidence suggests that the DNA of the oncogenic virus, like that of the lysogenic phage, is integrated into the DNA of the host cell. There is mounting evidence that study of recombinations of viral DNA, satellite viruses, and lysogenic viruses may provide important clues to the processes that cause cancer and other human diseases.

SYNTHESIS OF VIRUSES

Because the genetic information carried by a virus is relatively simple, viruses were logical choices for the first attempt at synthesis of a biological system. Viral nucleic acids were the first major components of biological systems to be synthesized in the laboratory. Viral RNA was replicated many thousandfold, using viral RNA as a template for the action of purified viral replicase acting on RNA precursors. When these newly synthesized RNA molecules were exposed to bacterial cells stripped of their rigid cell walls, the viral RNA replicated in the cells and killed them, producing thousands of viruses complete with protein coats (thus demonstrating the accuracy of the synthesized RNA information). In this experiment, the viral RNA was replicated outside of a living cell, but an initial supply of viral RNA was used as a template as well as a supply of purified viral RNA replicase. Thus, this experiment is a long way from the complete synthesis of viral RNA from simple precursors.

A similar replication of viral DNA outside the cell was later accomplished. In this case, DNA precursors, DNA polymerase, and a ligase enzyme (to close the newly synthesized DNA loops) were used to produce fully infectious synthetic DNA from a viral DNA, which acted as a template.

ORIGIN OF VIRUSES

Because viruses are unable to replicate without using the mechanism of a cell, they could not have existed as primitive living systems that evolved before the existence of cells. Despite the simplicity of their structure, some scientists believe that viruses are relative latecomers in the evolutionary story, developing from cells that became parasitic on other cells and eventually lost most of their cytoplasmic machinery.

The simplicity of viruses has been of great value to scientists in their attempts to understand basic principles of living systems. In the interactions of viral systems with almost every kind of cell, biologists have found many clues to the basic nature of gene action. It is ironic that much of current knowledge about genetic mechanisms of living systems has come through the study of viral systems, which cannot fully be classified as living. Yet scientists often find that study of the unusual or odd phenomenon leads to important insights into the nature of the normal world.

FURTHER READING

Vaccination and immunity are discussed in greater detail in Chapter 22. Among the many books on viruses, particularly good introductions to the field are those by Fraenkel-Conrat (1962), Luria and Darnell (1968), and Stanley and Valens (1961). For more detailed information on the molecular biology of viruses, see books by Burnet and Stanley (1956–1959), Fenner (1968), and Fraenkel-Conrat (1968).

Bacteriophage are described in greater detail by Hayes (1969) and Stent (1963). For further information about oncogenic viruses, see the books by Gross (1961) and Harris (1964).

10
Membranes

Cell biologists of the late nineteenth century thought that the cell membrane, or plasma membrane, was simply an invisibly thin film that held together the living contents of the cell. Information gained from biochemical studies and electron microscopy has greatly changed this view of the membrane. It is now known that every cell—procaryotic or eucaryotic—is surrounded by a membrane about 100 A thick. This membrane has a regular molecular structure and plays an active role in the life of the cell. Within the eucaryotic cell, similar membranes form the boundaries of the organelles and provide sites for many enzymatic reactions required by the cell.

The plasma membrane surrounding the cell serves as a passive barrier that holds together the contents of the cell and protects them from the conditions of the external environment. Yet a living cell must continually interact with its environment, obtaining metabolites and discarding waste products. The plasma membrane serves as an active envelope that regulates this vital flow of materials to and from the cell interior. Within the eucaryotic cell, membranes play a similar role in maintaining the integrity and specialized conditions of each compartment of the cell, while simultaneously regulating and facilitating the necessary interchanges among compartments.

Many of the membranes inside the eucaryotic cell play another important role. Membranes of the endoplasmic reticulum (ER), Golgi apparatus, mitochondria, and plastids—and even parts of the plasma membrane itself—provide structural sites for the regular organization of multienzyme complexes (Chapter 4). Substrate molecules are passed from one enzyme to another along the membrane in highly organized and efficient biochemical reaction sequences.

Protein synthesis is carried out on membrane-associated ribosomes, which apparently carry out this process while floating freely in the cytoplasm. Membranes, however, also appear to be important in protein synthesis, for ribosomes tend to cluster along the nuclear membrane, the rough ER, and the plasma membrane. In procaryotic cells, which lack the ER and a nuclear membrane, some of the ribosomes are distributed along the inner surface of the plasma membrane. There is growing evidence that the process of DNA replication normally occurs at a site on a membrane, with the long DNA molecule moving past a stationary enzyme site as it is replicated.

One of the better-studied, membrane-mediated processes is photosynthesis (Chapter 5). Enzyme and cofactor molecules involved in photosynthesis are arranged in a precise pattern within the complex membrane system of the chloroplast—sequentially transferring electrons and substrate molecules from one step in the process to the next, much like a factory assembly line. The photosynthetic pigments, enzymes, and cofactors are arranged side-by-side on these membrane stacks to facilitate an efficient assembly-line process. When the appropriate molecules involved in photosynthesis are mixed in a test tube, without the ordering provided by membranes, the photosynthetic process is weak and inefficient.

Biochemists have had to develop new techniques and theories for the study of these membrane-mediated reactions, for traditional chemical techniques and theories deal largely with the study of reactions in solutions. A cell's membranes are only a few macromolecules thick. Clearly, an understanding of the properties of membranes will be attained only through a full knowledge of membrane molecular structure. Although much has been learned about membrane structure, many important questions have been

Figure 10.1 (above). Phospholipid structure. Phospholipids are complex molecules built around the three-carbon alcohol glycerol. An organic base is attached to one of glycerol's three carbon atoms and fatty acids to the remaining two carbon atoms. A variety of bases and fatty acids are found in membrane phospholipids. In the symbol used here to represent a phospholipid molecule, the polar head is represented by the circle, the hydrocarbon tails by the two sticks.

Figure 10.2a (middle). Schematic diagram of a micelle in water. Phospholipid molecules form various stable configurations in different environments. In water, the molecules are arranged with the hydrophobic tails shielded from the water by the hydrophilic ends.

Figure 10.2b (below). Schematic diagram of a micelle in oil. The hydrophilic ends are interacting with each other (and with any water that happens to be in the environment), and the fatty acid ends are interacting with the surrounding oil.

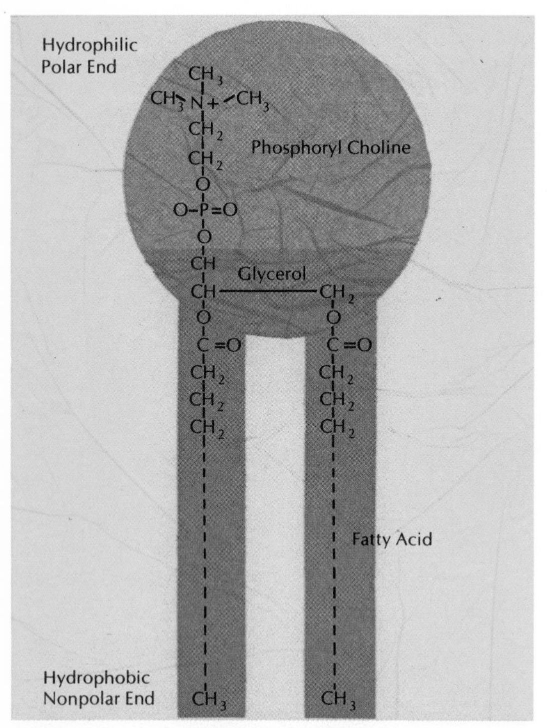

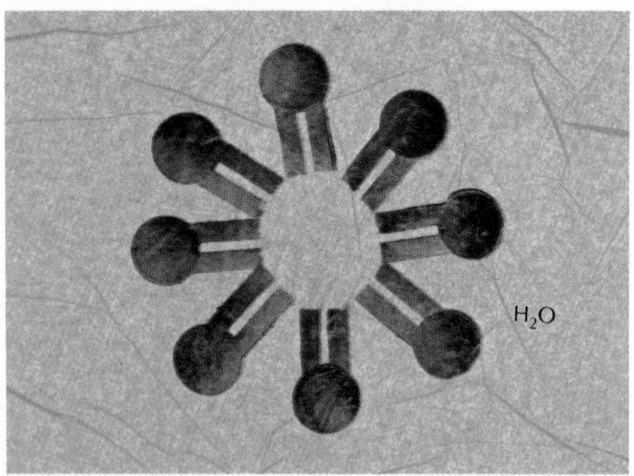

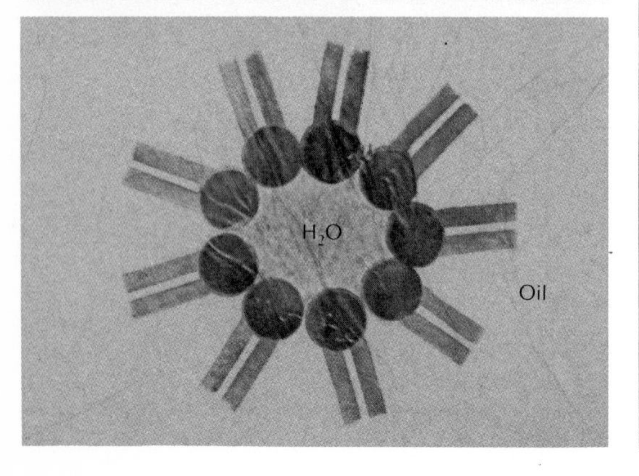

answered only tentatively or not at all. Membrane structure is an area of active research at the present time, and almost every month brings new experimental findings.

COMPOSITION OF MEMBRANES

Nonpolar molecules, soluble in fatty or oily solvents, pass relatively freely through cell membranes. Polar molecules, soluble in water but insoluble in fatty solvents, move much less freely through the membranes. For these reasons, it was suggested early in this century that the plasma membrane consists of a thin layer of lipids. But more recent chemical analyses have shown that cell membranes are composed of lipids, proteins, and carbohydrates. All three components contribute to membrane properties. The lipids make the membrane relatively impermeable to ions and polar molecules. Long-chain protein molecules apparently give the membrane strength, elasticity, and the ability to expand and contract. Other proteins have enzymatic functions, including active transport of molecules across the membrane and metabolic reactions carried out on the membrane surface. The carbohydrates play an important role in chemical interactions between the cell and its surroundings. Thus, the plasma membrane is far more than a simple envelope; it is a complex molecular system that performs many intricate chemical and physical processes essential to the survival of the cell.

Membrane *lipids* are mainly of a class of compounds called phospholipids. The structure of a *phospholipid* is built around the three-carbon alcohol, glycerol. A polar phosphate group is attached to one of glycerol's three carbon atoms, and nonpolar fatty acids are attached to the other two carbon atoms (Figure 10.1). Because of the attached polar group, this end of the phospholipid molecule is *hydrophilic*—that is, it has a strong attraction for water molecules. The fatty acid groups, however, are nonpolar, making the other end of the phospholipid molecule *hydrophobic*—that is, it is more strongly attracted to other nonpolar molecules than to water molecules.

Small droplets called *micelles* are formed when a phospholipid is added to a water solution. The arrangement of individual phospholipid molecules within a droplet has been determined by thermodynamic calculations and by various physical measurements. These studies reveal that the most stable arrangement assumed by most phospholipid molecules is that in which the hydrophobic ends of the molecules are shielded from the surrounding water and the hydrophilic ends are in contact with the water (Figure 10.2a).

If any fat-soluble, nonpolar compounds are present in the water solution, they tend to collect within the micelle at the hydrophobic ends of the lipid molecules. Soaps and detergents also are molecules with hydrophobic and hydrophilic ends. They act in a fashion similar to the phospholipids in that they isolate small amounts of oily substances within micelles, which move freely through a water solution. Thus, an insoluble oily film can be converted into a large number of micelles, which are easily washed away by water. Phospholipid molecules are oriented in the opposite direction if a micelle is formed in a nonpolar solvent such as oil. In this case, polar substances tend to collect within the micelle (Figure 10.2b).

Phospholipids are one of three major classes of lipids; the other two are triglycerides and steroids. A *triglyceride* molecule consists of glycerol in which a fatty acid group is attached to each of the three carbon atoms. Triglycerides serve as a major form of long-term energy storage in animal cells but appear to play no part in membrane structures. A *steroid* is a complex alcohol containing a four-ring carbon structure. Cholesterol is a steroid found in the membranes of many kinds of animal cells. In various glandular cells, cholesterol is converted into steroid hormones—substances that are circulated through the organism, altering the rates of metabolic processes in many kinds of cells. The role of cholesterol in the many other kinds of animal cells where it is found is not yet known, but the rigid, planar cholesterol molecules may add strength to the lipid portion of the membrane.

Proteins are the second major type of molecule in cell membranes. Because of technical problems in isolation and purification of membrane proteins, relatively little is known about the types and properties of protein molecules in the membrane. A number of nonmembrane proteins have been thoroughly investigated, and their three-dimensional arrangements, or conformations, fall into four major categories (Figure 10.4). Although the conformational states of most membrane proteins have not been determined, there is evidence that they share some of the conformations of the better-studied proteins. Membrane proteins also may assume certain unusual conformations as a result of lipid-protein interactions within the membrane.

The *carbohydrate groups* of the plasma membrane may be linked either to proteins, forming complexes called *glycoproteins*, or to lipids, forming *glycolipids*. Both glycoproteins and glycolipids appear to confer unusual properties on the cell surface. Parts of these molecules coat the outer surface of a cell, forming chemical configurations that make the surface of that cell unique—different from one type of cell to another, different from one

Figure 10.3 (above). A negative-contrast micrograph of a lipid micelle. (× 74,000)

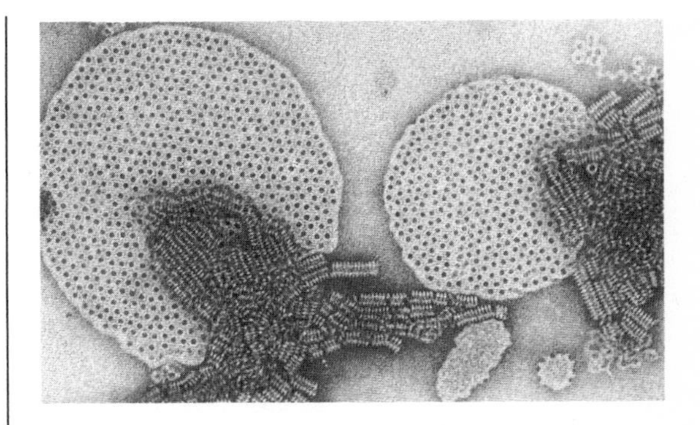

Figure 10.4 (below). Protein structure. The linear structure of amino acids in a protein is referred to as its *primary* structure. Interchain disulfide bridges and hydrogen bonds determine its *secondary* structure. The three-dimensional structure is referred to as the *tertiary* structure of the protein.

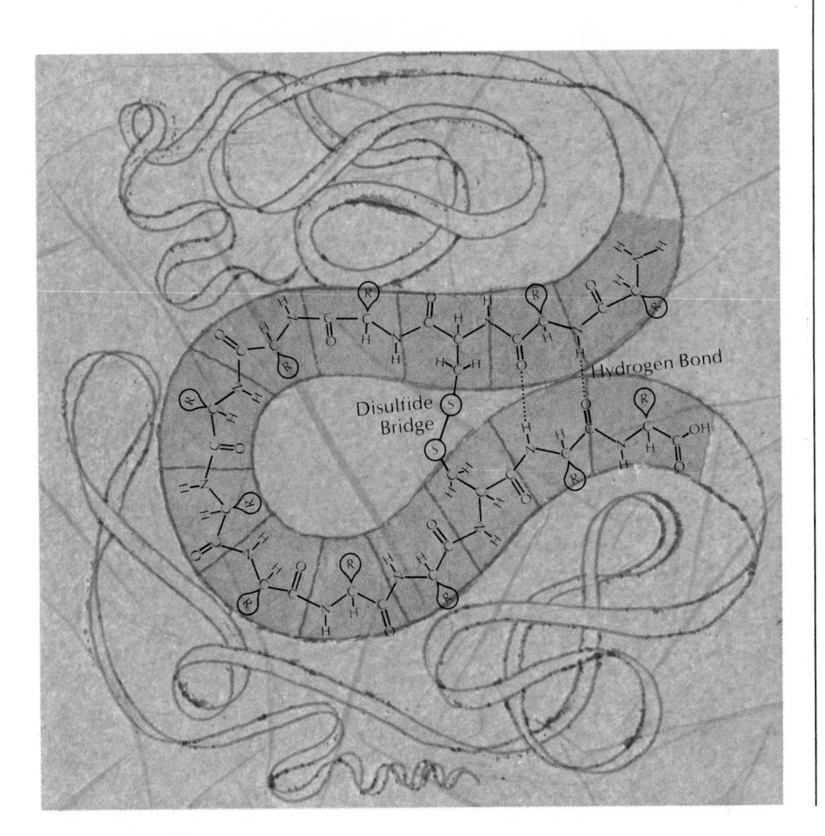

Figure 10.5. Lipid monolayer. When applied to a water surface, extracted membrane phospholipids will spread out to form a layer one molecule thick (monolayer). This lipid monolayer can be compressed to its minimum surface area, this area measured, and the results compared with calculated surface areas of the original cell membrane. These data suggest that the possible configuration of lipids in the cell membrane is a bilayer.

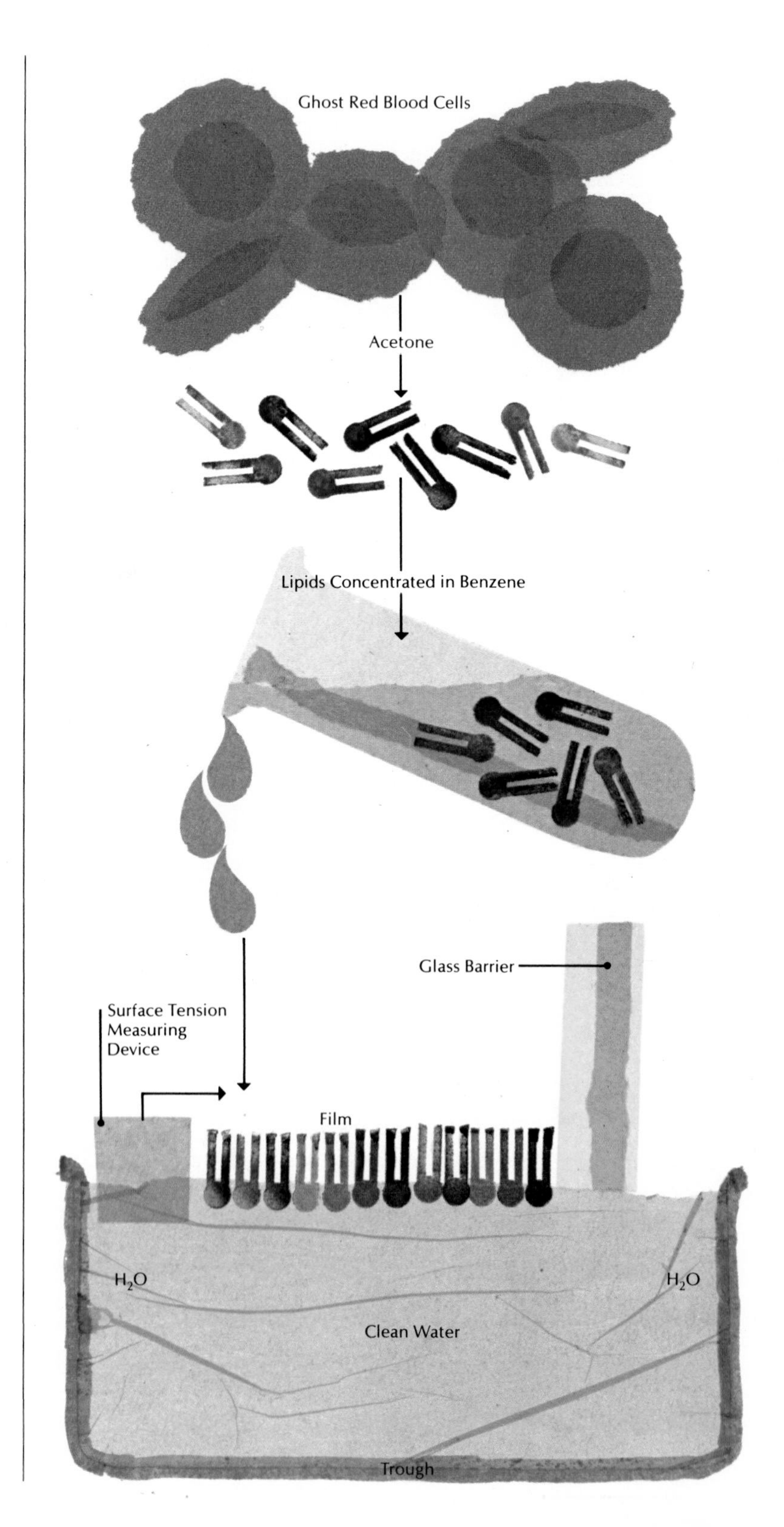

Figure 10.6. Red blood cell ghost membranes seen in an electron micrograph of a thin section of several red blood cell membranes. The triple-layer structure is seen most clearly when the section cuts directly across a membrane. This triple-layer configuration is represented by two highly visible electron-dense lines, separated by a lightly stained space. Each of the three layers in this structure is about 20 to 30 A in thickness.

species to another, and even different from one individual organism to another. These surface chemical configurations are intimately involved in interactions among cells. Cells react in certain ways to other cells bearing surface groups similar to their own; they react in quite different ways to cells bearing different groups. Surface groups that are unique to an individual organism are responsible for the rejection reactions that often follow organ transplantations.

Electron microscopy shows that most of the internal membranes have a structure similar to that of the plasma membrane. This observation and the available chemical data suggest that the internal membranes are composed of similar lipids, proteins, and carbohydrates, although the ratios of these substances probably vary from one kind of membrane to another and from one kind of cell to another.

Molecular Structure of Membranes

When a polar-nonpolar substance such as phospholipid is placed on a water surface, it spreads out to form a layer only one molecule thick (Figure 10.5). Within this monomolecular layer, or *monolayer*, the molecules are oriented with their polar, hydrophilic ends in the water and their nonpolar, hydrophobic ends pointing into the air. In a device called a film balance, a monolayer can be formed on the surface of a shallow trough of water and then compressed to align all of the molecules in the layer. The study of lipid monolayers with this device offered the first clues about the possible structure of cell membranes.

The mammalian red blood cell has been used for many membrane studies. The red blood cell is a highly specialized cell that lacks a nucleus and has a relatively simple internal structure. When these cells are placed in distilled water, the water floods and the cell contents escape, leaving so-

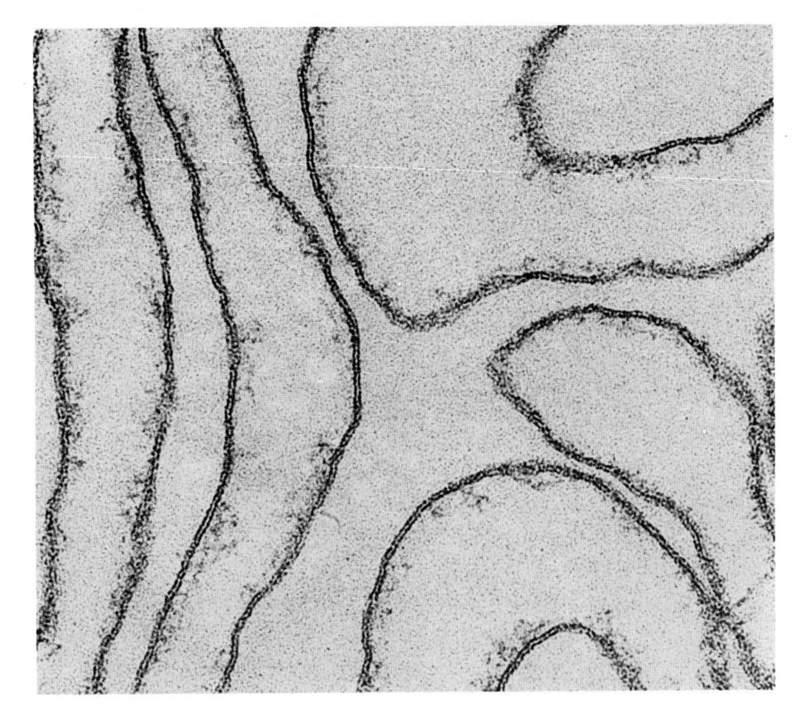

Figure 10.7. Three possible structures for a lipid bilayer.

called red cell ghosts, which consist of the plasma membranes. (Analysis of quantities of such ghosts provided the best chemical information about plasma membrane composition.)

In a classic experiment, the lipids from ghosts were dissolved in acetone, concentrated by evaporation, and redissolved in a small amount of benzene. The benzene-lipid solution then was used to form a monolayer in a film balance; after the benzene evaporated, a lipid monolayer remained. The area of this lipid monolayer was compared with the calculated surface area of the original cells. The data are consistent with the theory that the lipid in the cell membrane is evenly distributed in a bimolecular layer.

At least three structures could be imagined for such a lipid bilayer (Figure 10.7). Because the cell membrane is bounded on both sides by water solutions, the early researchers concluded that the most likely structure for the lipid bilayer in the cell membrane is one with the hydrophilic ends facing outward and the hydrophobic ends together in the center of the bilayer.

J. F. Danielli and his coworkers (1935) measured the surface tension of cell membranes and found it to be much lower than the measured surface tension of lipid monolayers and bilayers. They suggested that the outer surfaces of the lipid bilayer are coated with protein molecules. The polypeptide chains of the proteins, lying at right angles to the fatty acid chains of the lipids, add strength to the membrane but reduce its surface tension. Nonpolar groups on the protein chains extend into the lipid bilayer toward the hydrophobic ends of the lipid molecules. Polar groups on the protein chains are directed in the opposite direction toward the water solution. These specific attractions hold the membrane structure together without the formation of covalent bonds among the various molecules.

During the 1950s, techniques were developed for the study of cell membranes with the electron microscope. The membrane, which is far too thin to be visible with even the most powerful light microscope, could be observed for the first time. Using a special stain and very high magnification, electron microscopists observed the membrane as a triple-layer structure, formed by two electron-dense lines separated by a lightly stained space (Figure 10.8). Each of the three layers in this structure is about 20 to 30 A in thickness. Because this structure was found to be characteristic of all cell membranes, J. D. Robertson (1959) suggested that it be called the unit membrane configuration. The structures observed are consistent with a model similar to the Danielli model of cell membrane structure. Until

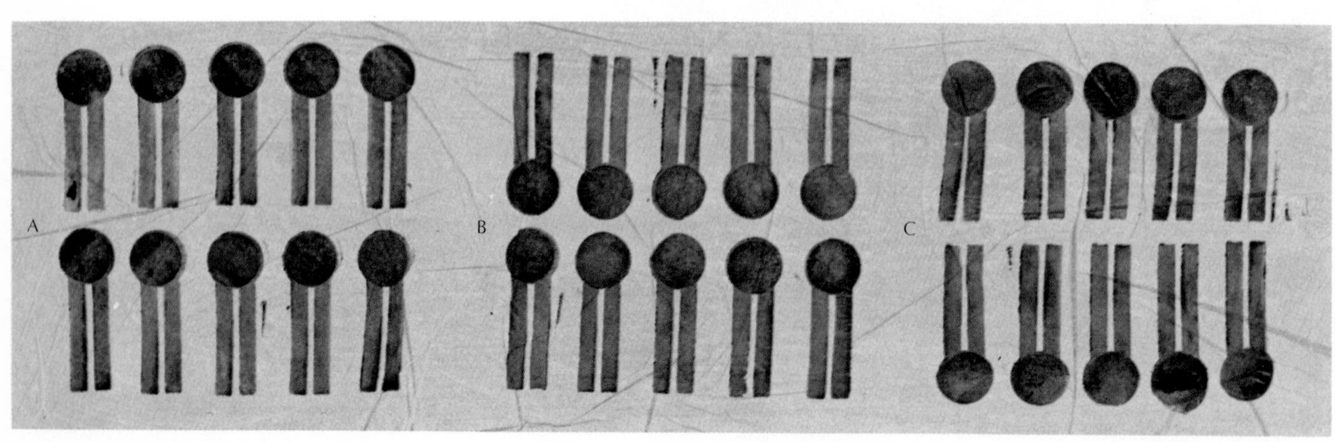

Figure 10.8. Triple layer, or unit membrane, configuration. Electron micrograph analysis indicates membrane structure is triple-layered. This structure is consistent in all cell membranes. Variations visible in electron micrographs indicate that multiple lipid bilayers may exist in some parts of the membrane. A possible arrangement of these lipid bilayers is illustrated.

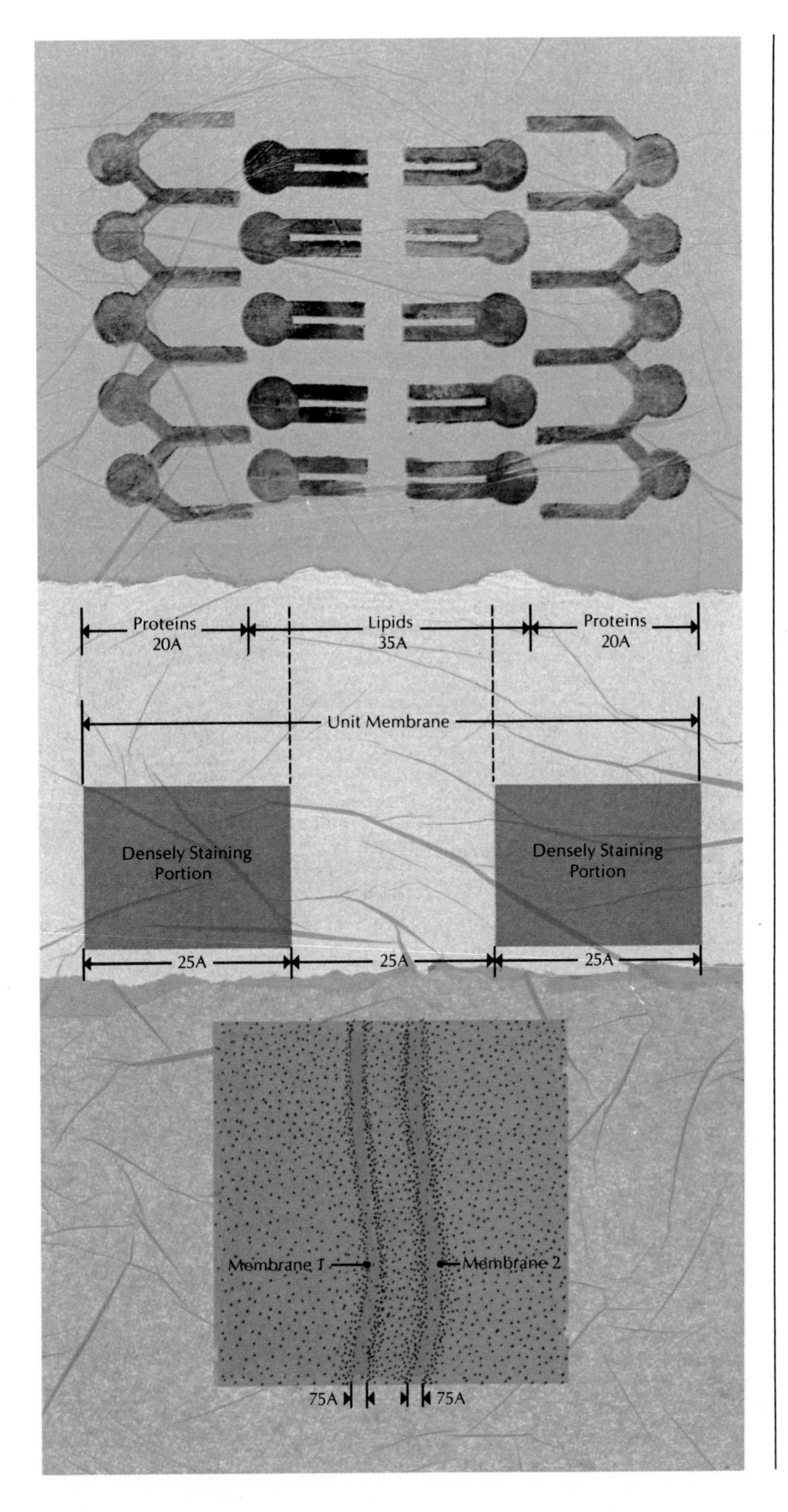

Figure 10.9 (right). The lipid micelle–protein configuration is an alternative to the Danielli model but is also consistent with experimental data.

Figure 10.10 (below). Indirect evidence suggests that pores lined with polar groups may exist in the membrane. If present, they would facilitate the passage of ions and polar molecules. However, pore structures are too small to be visible in electron micrographs, and their existence remains a matter of conjecture.

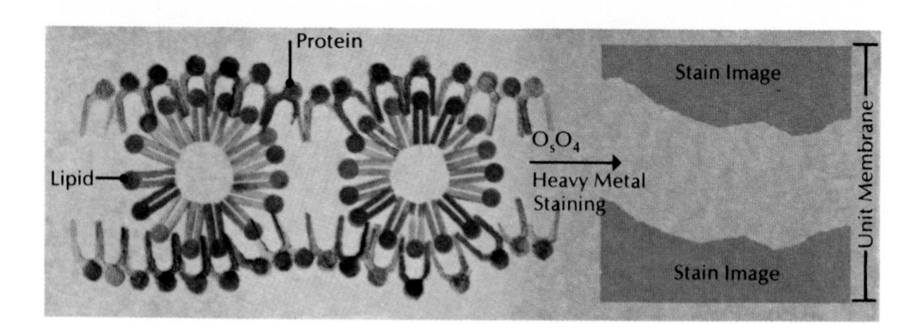

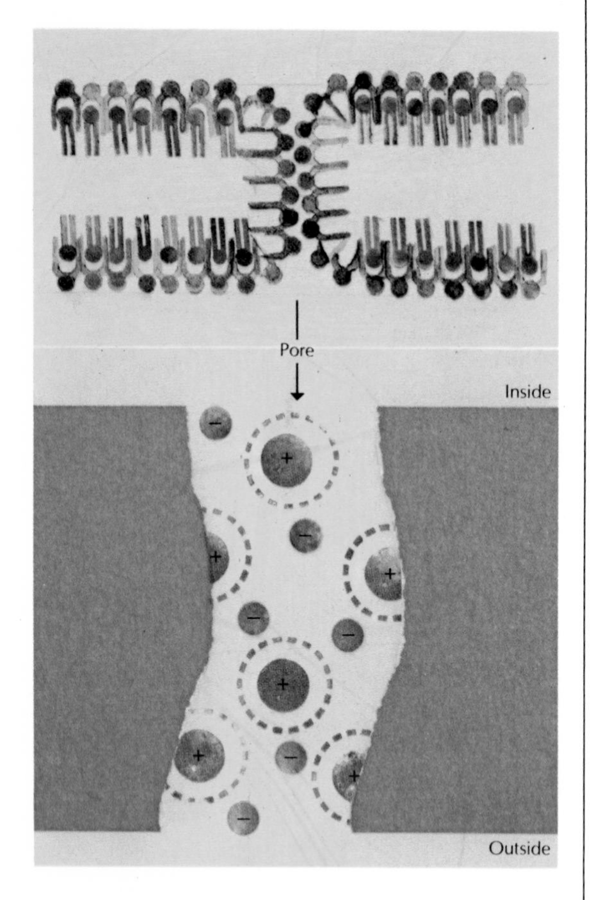

recently, this Danielli model of membrane structure has been widely accepted (Davson and Danielli, 1943). It remains one possible structure for cell membranes, but recent research has emphasized other possible structural arrangements.

Alternatives to the Danielli model that are also consistent with the experimental data have been proposed. The triple-layer image could be produced by a structural arrangement of lipid molecules such as that shown in Figure 10.9. Even though the lipid molecules are arranged in micelles, the image seen in a cross-sectional electron micrograph would be difficult to distinguish from that of a lipid bilayer.

Although the unit membrane structure is remarkably constant in biological membranes, there are some variations visible in electron micrographs. Variations in spacing between the dark bands suggest that multiple lipid bilayers may exist in some parts of the membrane (Figure 10.10). Indirect evidence suggests that pores lined with polar groups may exist in the membrane. Ions and small polar molecules would pass more readily through these pores. Such structures are too small to be visible in electron micrographs, and their existence remains a matter of conjecture. There also is evidence that the nature of the proteins and lipids varies from place to place over the membrane, probably creating a mosaic pattern of varying properties.

The position of the glycoproteins and glycolipids within the membrane structure is unknown. Most of the sugar groups of these molecules are along the cell surface, but there is some evidence that other parts of the molecules may extend into the lipid portion of the membrane.

FREEZE-CLEAVAGE STUDIES

Recently, interesting new discoveries about cell membranes have been made with a technique of electron microscopy called *freeze-cleavage preparation*. Instead of chemically fixing and staining the cell, biologists use a process of rapid freezing. This technique greatly reduces the possibility of rearrangement of the membrane components during preparation. The sample is then sectioned in a very cold, evacuated chamber. Instead of cutting a smooth section, a blade cleaves the sample along natural lines of weakness.The sample is then coated with a heavy metal to make the surface visible in the electron microscope.

The process is diagrammed in Figure 10.11a as if it were being carried out on a single ghost embedded in ice. In practice, the procedure involves the cleavage of hundreds of randomly oriented ghosts by a single fracture. Close inspection of the membrane surface revealed in these electron micrographs shows that the surface is covered with globular units spaced at regular intervals (Figure 10.11b). Further studies have shown that these units are entirely absent from some membranes—for example, the membrane sheaths surrounding some nerve cells—and are extremely abundant in others—such as the membranes of red blood cells.

At first, it was assumed that such micrographs show the inner or outer surface of the membrane. However, D. Branton suggested that the cleavage plane passes down the middle of the membrane. This hypothesis has been confirmed by the addition of an etching step to the freeze-cleavage process (Figure 10.12b). In a photomicrograph produced with the freeze-etching technique, two surfaces can be seen—one produced by cleaving and one by etching (Figure 10.12a). The cleaved surface is covered by the globular

Figure 10.11a (upper left). The effect of the freeze-cleavage technique on cell membranes. This technique involves the rapid freezing of the cells, presumably to prevent alteration of membrane components during preparation. The cells are then sectioned in a cold, evacuated chamber. Samples are prepared for the electron microscope by coating them with a heavy metal.

Figure 10.11b (upper right). Electron micrograph showing a platinum replica of a red blood cell membrane prepared by freeze-cleaving. Note the textured appearance of the surface illustrated. Analysis of the membrane surface revealed in these electron micrographs shows that the surface is covered with globular units spaced at regular intervals.

Figure 10.12a (lower left). Electron micrograph of a platinum replica of a red blood cell ghost membrane prepared by the freeze-cleavage-etching technique. Comparative analysis of the freeze-cleavage surface and the etched surface is possible.

Figure 10.12b (lower right). The freeze-cleavage technique coupled with the etching process. With the addition of the etching process, both the cleavage surface and the outer surface layer of the membrane are exposed and available for comparative analysis.

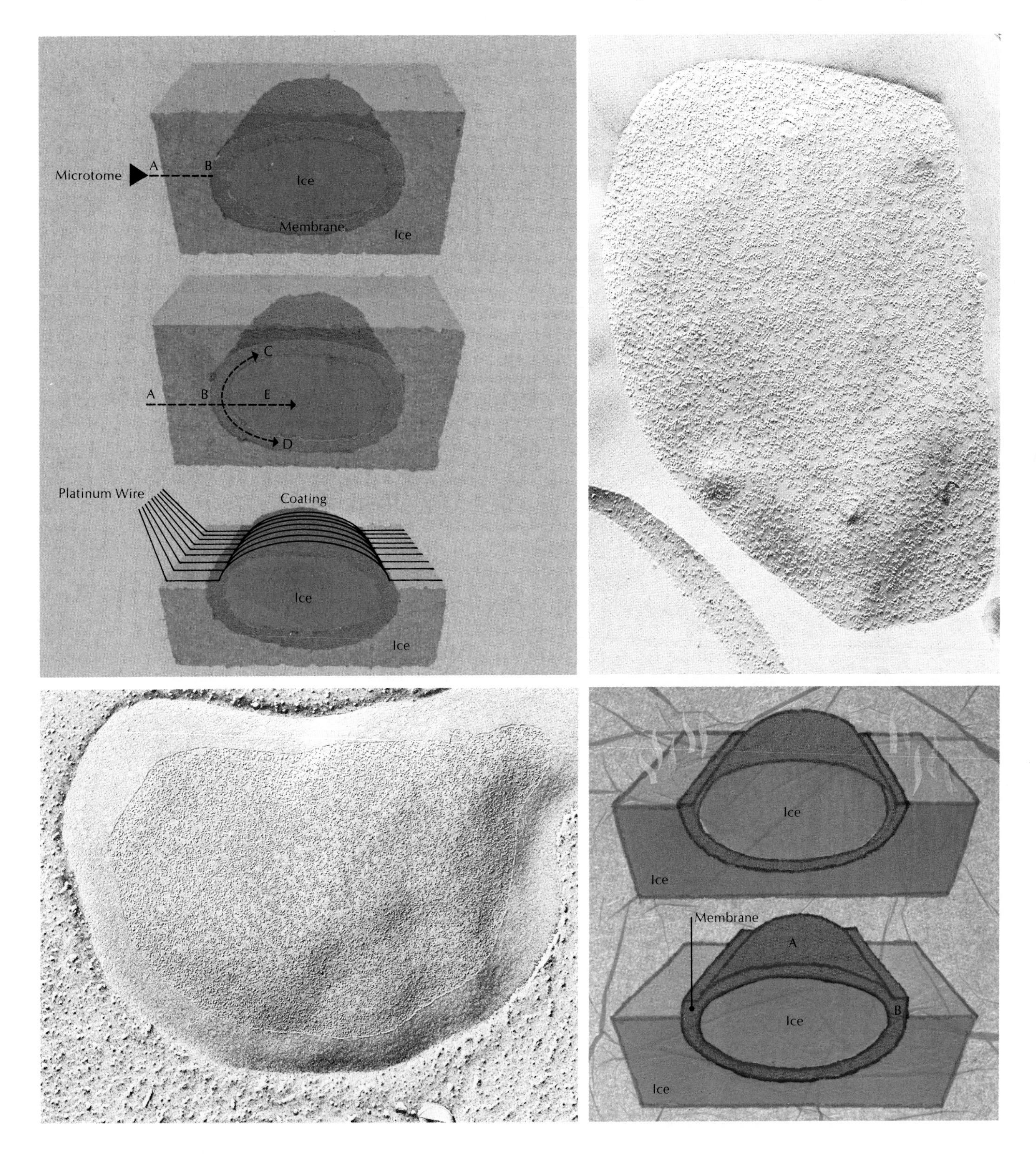

Figure 10.13. Lipid micelle–protein globule model. This diagram depicts the globule as a bilayer arrangement of lipids modified into a micelle configuration surrounding protein molecules. Glycoproteins may be anchored in the membrane by associating with these globule units.

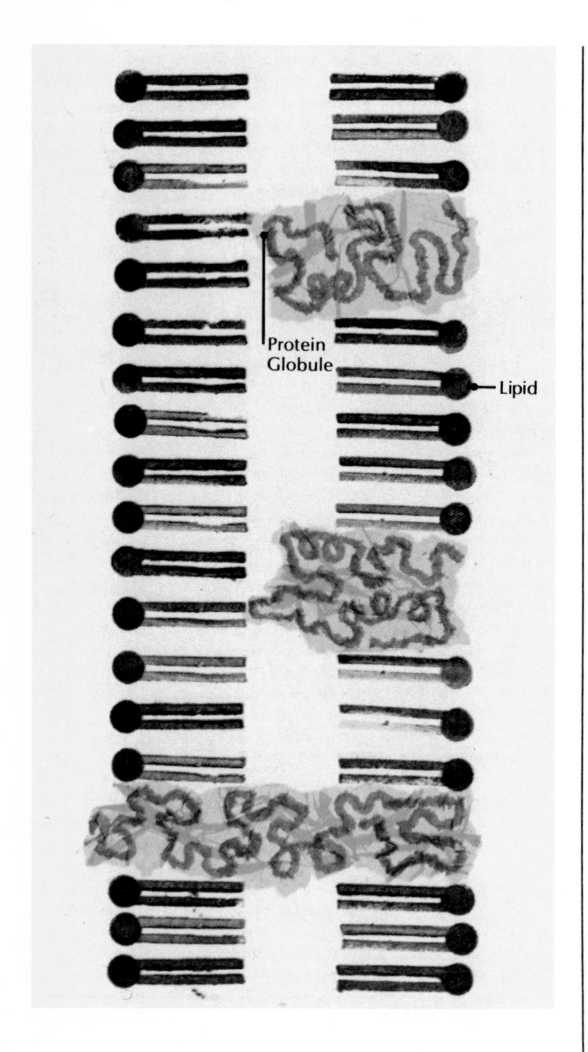

particles, whereas the etched surface is relatively smooth. If the etched surface is the true outer surface of the membrane, the cleaved surface must lie at some level within the membrane. Because the experiment was done with ghosts, the globules cannot be internal structures of the cell.

Further confirmation that the globules lie inside the membrane was obtained by labeling the outside surfaces of the red blood cells with marker proteins. After the freeze-etching procedure, these marker proteins are observed on the relatively smooth, etched surface.

The nature of the globules is not yet known. One possible model of membrane structure incorporating these new findings is shown in Figure 10.13. The globules are indicated, but their molecular structure is unknown. Tentative evidence suggests that part of the globular unit may be composed of protein molecules surrounded by a lipid micelle. The bilayer arrangement of lipids is retained as the basic membrane structure in this model, although some scientists feel that the globules are evidence against the bilayer model. However, the cleavage through the middle of the membrane is similar to the cleavage that occurs when artificial lipid bilayers are subjected to the freeze-cleavage process. The variation in numbers of globules within various kinds of membranes suggests that these globules are specialized structures that are present only in membranes that carry out complex functions. The simpler membranes, such as those of the myelin sheath wrapped around some nerve cells, may be similar in structure to the simple bilayer model. The model shown in Figure 10.13 suggests that parts of the glycoproteins are anchored in the membrane in association with the globular units, but this suggestion also is a tentative hypothesis based upon slight evidence.

TRANSPORT OF MOLECULES ACROSS MEMBRANES

Membranes serve as barriers that separate different compartments within the cell and also separate the cell from its external environment. They are selective barriers, transporting needed substances into cells and unwanted substances or secretions out of them. The concentrations of ions and molecules within a cell are maintained at levels suitable for the processes of life, and the cell's volume is regulated by the amount of material kept within it. Because of their obvious importance to living systems, the mechanisms by which substances are transported across membranes have been under intensive study.

Molecules passing through a membrane may move by one of three basic mechanisms: diffusion, facilitated transport, or active transport. If the concentrations of any particular substance are unequal on the two sides of the membrane, more molecules will strike the membrane on the side of higher concentration than will strike it on the other side. Thus, the overall flow of a substance to which the membrane is permeable will be from the side of higher concentration toward the side of lower concentration, until eventually an equilibrium is reached and the two concentrations separated by the membrane are equal. Transport by diffusion is always *passive*—that is, it requires no expenditure of energy.

Most biological membranes are semipermeable—that is, permeable to certain molecules and not to others. All, however, are somewhat permeable to water. Because the cell cytoplasm enclosed by the membrane is a highly concentrated solution of various molecules (the membrane is impermeable to most of them), the concentration of water molecules inside the cell is lower than the concentration of water molecules in pure water. For this reason, most animal cells placed in water burst because water diffuses into the cell in response to the difference in concentration. Diffusion of water or other solvent molecules across a semipermeable membrane is known as *osmosis* (Figure 10.14). Plant cells, unlike animal cells, have rigid cellulose walls outside their membranes and if immersed in water can withstand the osmotic pressure without bursting. Conversely, higher vertebrate cells shrink in size if placed in sea water, which is about three times saltier than blood. Water leaves these cells because it is more concentrated inside than out. Thus, a fresh-water fish dies of thirst if placed in the ocean.

Nonpolar inorganic molecules (such as oxygen and carbon dioxide gases) and lipidlike substances (such as hydrocarbon anesthetic drugs) can move across membranes at rapid rates and without any apparent selectivity. These nonpolar molecules probably penetrate the membrane directly at any randomly selected site, and their transport is diffusion. On the other hand, the great selectivity with which particular ions and polar molecules

Figure 10.14 (left). A demonstration of osmosis in which a thistle tube filled with a colored sugar solution is immersed in water. The two solutions are separated by a semipermeable membrane. Water moves along its concentration gradient (as the membrane is impermeable to sugar) until an equilibrium stage is reached. This inward pressure raises the sugar solution in the tube.

Figure 10.15 (right). The upper portion of the figure compares diffusion and osmosis. In diffusion, a permeable solute or molecule such as water is able to freely diffuse across the membrane following a concentration gradient. In osmosis, a pressure is built up due to the presence of nonpermeable solutes, resulting in the flow of water across the membrane. The lower portion of the figure shows the effect of solute molecule size on osmotic pressure. A large molecule, such as a protein, inhibits the movement of smaller permeable molecules through the pores, causing a large difference in osmotic pressure across the membrane (inside relative to outside). A smaller solute is able to pass through the pore, resulting in a smaller difference in osmotic pressure. Small molecules (on the right) pass through easily, resulting in no osmotic pressure.

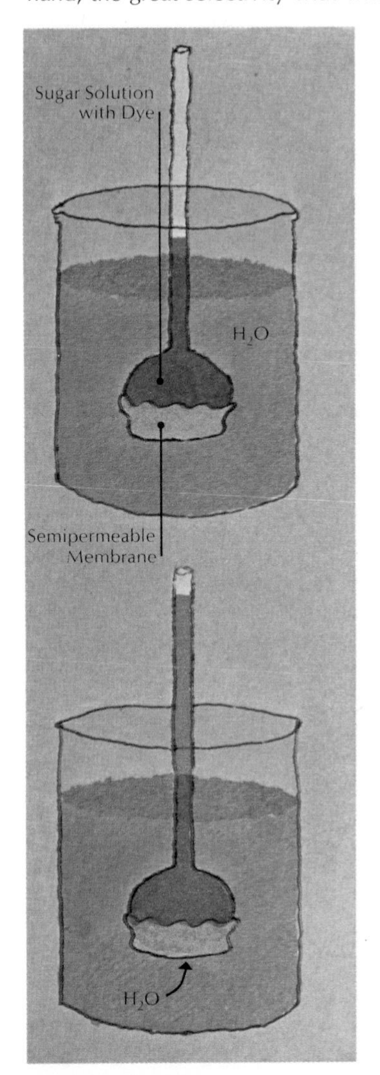

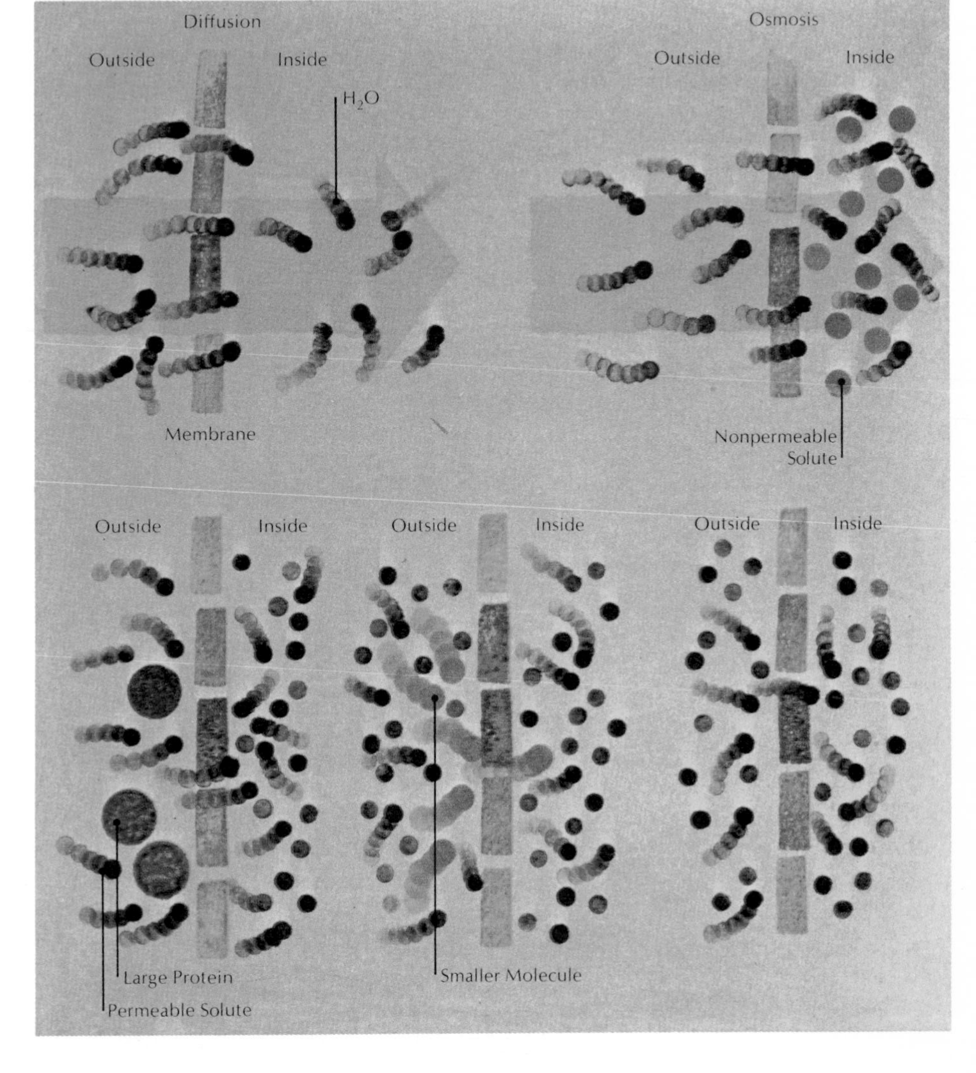

are moved across membranes and the rates at which transport of these molecules occurs are consistent with the view that there are relatively few sites on membranes specialized for the transport of ions and polar molecules.

The lipid layer of membranes behaves as if it has pores, which allow free passage of small polar molecules and through which ions pass in a controlled manner. However, most polar molecules do not diffuse across membranes except as mediated by *carriers*, which are specific for various types of molecules needed by the cell. Such carriers are not visible with the electron microscope, but their existence may readily be demonstrated by physiologists. Recently, biochemists have isolated membrane proteins that appear to be able to function as carriers; like enzymes, these proteins bind the molecules to be transported. It is postulated that molecules can move through the lipid portion of the membrane to deliver their bound substrates to the other side. Transport across a membrane is *passive* if it occurs only from a region of high concentration to one of low concentration and therefore requires no expenditure of energy. If carriers are utilized in passive transport, such transport is said to be facilitated.

Some substances, however, are moved across the membrane toward the side of higher concentration in a direction opposite that expected for passive transport. Some mechanism of *active transport*, utilizing chemical energy to move the molecules against the natural direction of diffusion, must exist. A mechanical model of such a mechanism is shown in Figure 10.16.

The molecular mechanisms responsible for active transport are currently being studied intensively. One mechanism that has been shown to operate for several cases of active transport of sugars involves the immediate phosphorylation by ATP of the sugar as it arrives inside the cell. The resulting sugar phosphate is not bound by the carrier protein, so that the free sugar concentration inside the cell is kept very low. The transport system responds to this difference in concentration by bringing more sugars into the cell. Thus, in a sense, this active transport is really passive.

ENDOCYTOSIS

Another transport mechanism is involved in the passage of very large molecules (such as proteins) or multimolecular particles (such as viruses and bacteria) into a cell. This process of *endocytosis* involves the infolding of the membrane to form a small vesicle within the cytoplasm. This vesicle contains materials trapped from the exterior of the cell. The plasma membrane then fuses across the opening, and the vesicle detaches from the membrane and moves into the cytoplasm.

This process occurs on a relatively large scale in amoeboid cells that feed upon bacteria and other relatively large particles. In such cases, the process is called *phagocytosis* (cell eating) and can readily be observed with the light microscope. The amoeboid cell flows around the particle until it has completely engulfed the particle. The cell membrane then fuses, and the vesicle containing the particle moves into the cytoplasm, where digestive enzymes are introduced into the vesicle. The process of phagocytosis is used by amoeboid cells (such as some white blood cells) in animal bodies; these cells remove many of the bacteria, viruses, and other foreign particles that find their way into the intercellular spaces of the body.

It has also been observed that some cells take minute droplets of extracellular fluid into the cytoplasm. Although the droplets are so tiny that they

Figure 10.16. A mechanical model of active and facilitated transport. Parts A and B are two hypothesized mechanisms for active transport. As the solute is being transported against a concentration gradient (more solute molecules on the inside than outside), energy in the form of ATP breakdown is required. In part A, a mobile carrier-solute complex moves the solute into the cell. In part B, a series of fixed carriers relay the solute into the cell. Parts 1, 2, and 3 are hypothesized mechanisms for facilitated transport. In this system, solute movement is along a concentration gradient (diffusion — no energy requirement) but polar molecules require carriers because they do not readily diffuse across the membrane. Part 1 shows a carrier protein within the membrane in a globular conformation. An exposed terminal peptide complexes with the solute and, through a conformation shift followed by a dissociation of the complex, releases the solute into the cytoplasm. Part 2 shows a mobile carrier analogous to that in the active transport model. Part 3 depicts a carrier model in the form of a rotating mechanical device. Transportation of the solute involves a complexing and 180° rotation of the carrier into the cell, where the solute diffuses into the cytoplasm.

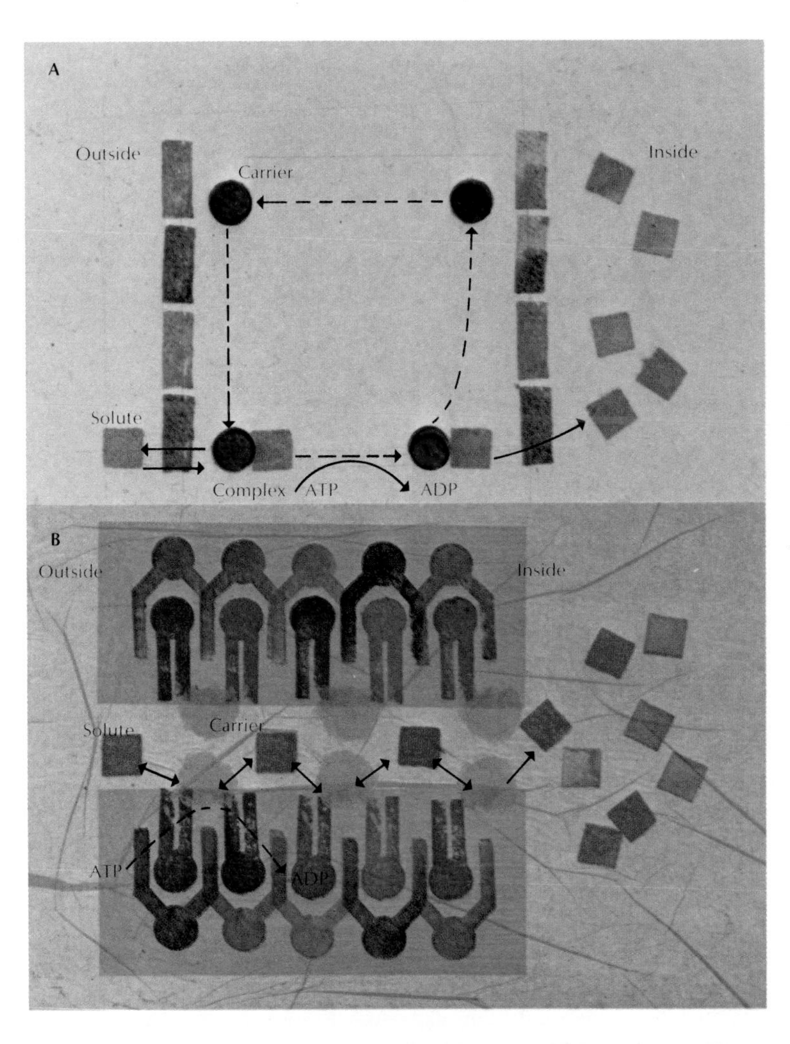

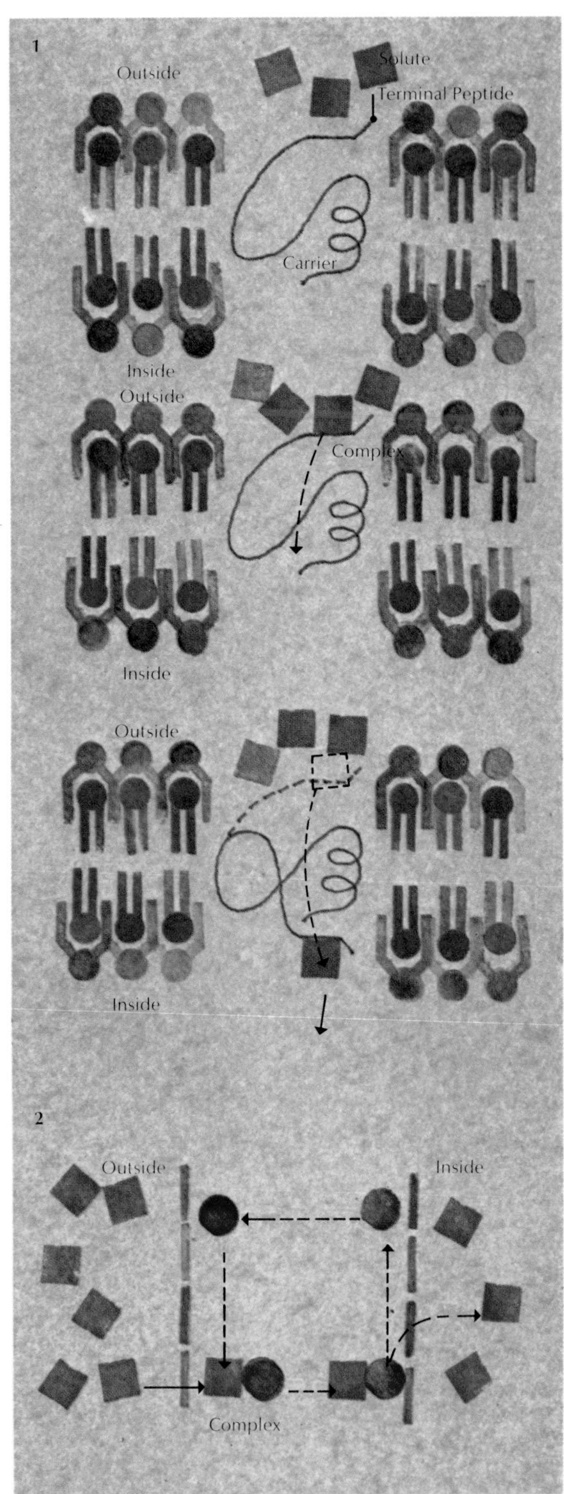

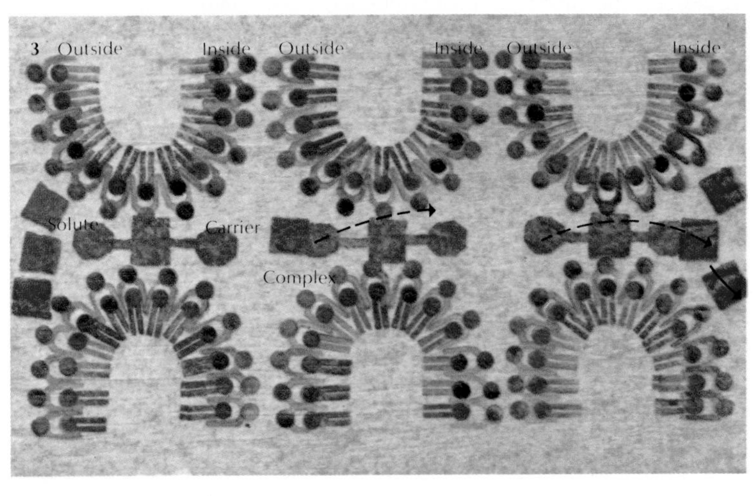

Figure 10.17 (above). Phagocytosis in the amoeba *Chaos chaos*. To capture its prey, the amoeba extends pseudopodia to encircle and engulf it. The cell membrane then fuses, and the vacuole containing the captured prey moves into the cytoplasm, where digestion takes place. (*Eric Gravé*)

Figure 10.18 (below). Electron micrograph of capillary endothelium. The invaginated membranes are forming pinocytic vesicles.

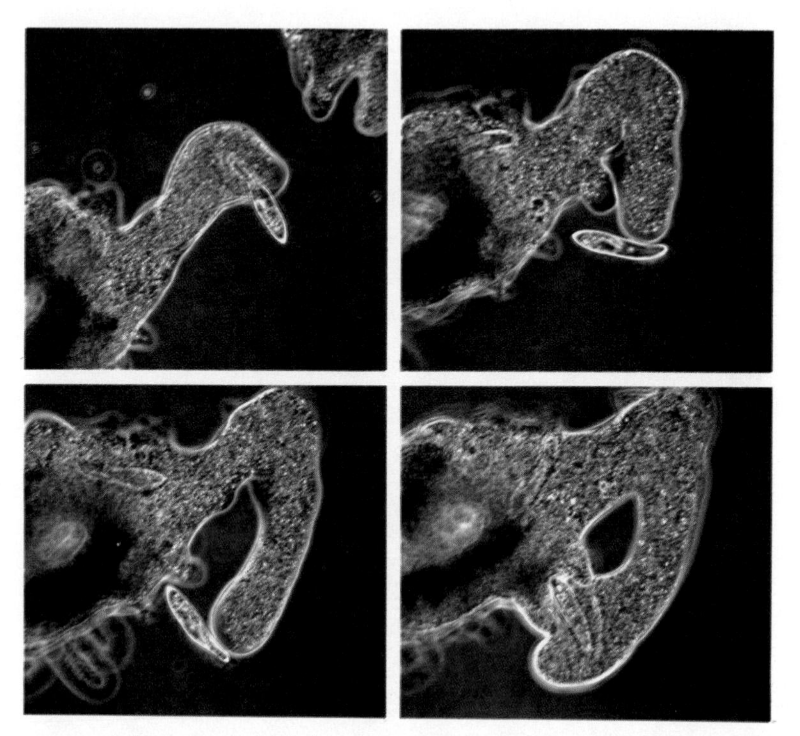

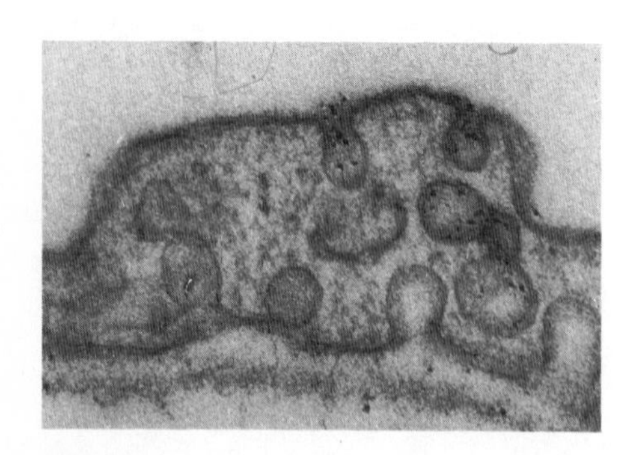

can barely be seen in the light microscope, what could be seen suggested that the mechanism of this process, called *pinocytosis* (cell drinking), is similar to that of phagocytosis. Electron micrographs have confirmed that pinocytosis does involve the infolding of minute vesicles and the subsequent pinching off of those vesicles into the cytoplasm.

Part of the membrane buckles in by some unknown mechanism, entrapping a portion of the external solution that may contain some large molecules. The pinched-in membrane forms a cuplike structure, which continues to deepen until the edges fuse to form a closed vesicle. The vesicle detaches from the membrane and moves into the interior of the cytoplasm with its contents. The large molecules may be broken down by enzymes to units small enough to pass through the vesicle membrane into the cytoplasm, or the vesicle membrane may be broken down to permit the large molecules to enter directly into the cytoplasm.

A similar process in reverse is used by certain cells as a way to excrete protein. Proteins produced for secretion in gland cells are packaged in vesicles by the Golgi complex. These vesicles move to the plasma membrane and pass through a series of stages similar to pinocytosis, but in reverse. The proteins thus are released outside the plasma membrane.

Processes of endocytosis presumably involve the use of ATP energy to rearrange the membrane conformation. There is evidence that the process may be triggered by the presence of protein molecules in the solution outside the cell, but the triggering and infolding mechanisms remain unknown.

MEMBRANES: THE NEXT CHALLENGE FOR MOLECULAR BIOLOGY

Within the cell, membranes function as parts of chemical factories—as floor space for the organized arrangement of systems of enzymes. They act as phase separators, creating and maintaining various volumes of different chemical compositions. The cell membranes of nerve fibers act to transmit electrochemical signals.

In each of these situations, the membranes react with great sensitivity to physical and chemical factors in the environment. This sensitivity reaches its greatest development in the sensory receptor cells, which utilize membrane structures to convert various forms of energy from the external environment into nerve signals. These mechanisms will constitute the principal challenge for the next phase of molecular biology.

Current ignorance of receptor mechanisms and membrane structure and synthesis is comparable to the ignorance of the molecular basis of genetics 30 or 40 years ago, when biologists knew that there are genes, that the genes are located on chromosomes, and that the genes are arranged in a linear order. They also knew that the chromosomes contain proteins and nucleic acids, but for several decades biologists assumed that the proteins represent information storage and that the nucleic acids—with their apparently limited and monotonous structure—represent a structural backbone. With respect to membranes, there now exists a similar degree of uncertainty as to the relative roles of protein and lipid: Which one is primarily involved in structural support and which one is primarily involved in the sensitive reactions of the membrane?

Just as the deciphering of the molecular mechanism of the linear gene represented the major biological breakthrough of the first part of this century, the understanding of the molecular mechanisms of two-dimensional membranes may represent the next step for molecular biology.

FURTHER READING

The general references on eucaryotic cell structure listed in Further Reading for Chapter 7 will provide more information on membrane structure. Also of interest are articles by Danielli and Davson (1935), Dippell (1962), Fawcett (1958), Hokin and Hokin (1965), Holter (1961), Robertson (1959, 1960, 1962), and Solomon (1960). Bangham (1971) provides an interesting example of the current debate over various models of membrane structure. Korn (1966) offers a summary of objections to Robertson's unit membrane theory.

11
Specialized Cells and Tissues

Although unicellular organisms are extremely small, they exist in a great variety of shapes and structures. There are thousands of species of procaryotes, unicellular algae and fungi, and protozoans, or protists. Each species differs from the others in structural details, metabolic processes, or life cycle. Autotrophic organisms require only inorganic nutrients and synthesize their own organic materials. Heterotrophic organisms survive in a wide variety of environments in which they can obtain various organic materials as food. Some organisms are encased in hard shells or slimy coatings. Others utilize flagella or amoeboid extensions for motility. Some are attached to fixed surfaces. Yet each unicellular organism must possess a full complement of structures or organelles if it is to survive independently.

Most plants, animals, and fungi larger than microscopic size are multicellular organisms, each made up of thousands, millions, or trillions of individual cells functioning cooperatively to maintain integrated processes throughout the entire organism. Therefore, most cells in such organisms are highly specialized. Various parts of the "typical" eucaryotic cell structure are modified, missing, or increased in size and number to adapt that particular cell to the performance of specific functions in the organism. These individual cells have lost their ability to survive independently (except in laboratory cultures, where they can be supplied with the nutrients and environment normally provided by other cells). Their structures have become so specialized that they must depend upon other cells in the organism for essential life processes.

The line that separates unicellular from multicellular organisms is not always sharp. Many kinds of organisms lie somewhere between fully independent single cells and fully interdependent cells of a multicellular organism. These intermediate cell forms are of particular interest to biologists. By studying them, biologists hope to learn what kinds of processes have led to the evolution of multicellular organisms.

Some unicellular organisms join together to form filaments or bodies, but the cells in these groupings retain their individuality. Any cell separated from the group can survive independently. Among some algae and fungi, the fusion of two or more cells forms a single, large, multinucleate cell—such organisms are called *coenocytes* (shared cells). The slime molds, or *Myxomycophyta*, are of particular interest because they can exist in various parts of their life cycle as coenocytic bodies, as multicellular structures with cellulose cell walls, and as independent cells (Figure 11.1).

Some protists, such as *Pleodorina* and *Volvox*, form colonies composed of large numbers of cells joined together. The individual cells of *Volvox* are linked by strands of cytoplasm (Figure 11.2). Most cells in a *Volvox* colony are flagellated, chloroplast-containing cells. The vegetative cells of *Volvox* are embedded in a matrix of extracellular material and are incapable of individual motility or of reproduction. Only a few reproductive cells in one part of the colony retain the capability to reproduce by either sexual or asexual processes. *Volvox* appears to be a very primitive form of multicellular organism with the simplest kind of cell specialization.

Of greater complexity are the sponges, which are composed of groups of cells with a variety of specialized functions: food-gathering cells, food-transporting cells, skeleton-making cells, maintenance and repair cells, reproductive cells, and cells that attach the entire creature to the surface on which it lives (Figure 11.3). The different kinds of cells in a colonial protist

Figure 11.1 (left). An adult slime mold with developing sporangia, or fruiting bodies. *(Courtesy Carolina Biological Supply Company)*

Figure 11.2 (right). *Volvox*, a colonial protistan, consisting of many flagellated cells. Cells are arranged in a single layer, with each cell in direct contact with the external environment. Located in the interior of the sphere are daughter colonies, produced by the fusion of gametes from specialized reproductive cells. *(Courtesy Carolina Biological Supply Company)*

Figure 11.3 (below). Photograph of a sponge taken at Cayman Island, British West Indies. Hidden in the body wall (diagram) are hundreds of tiny porelike openings, through which water is drawn into the internal cavity of the animal. Particulate matter in the water is trapped by the flagellated collar cells, and water passes on through the animal.

or a sponge are not completely interdependent. Any one kind of cell can become modified to perform the function of another. Because the cells retain some degree of independence, these organisms are regarded as simple or primitive multicellular organisms.

The corals, jellyfish, sea anemones, and other coelenterates represent a more complex or advanced form of multicellularity. They exhibit cell specialization of a fairly high degree—such as sensory cells of various types, a primitive sort of nerve cell, muscle cells, and the highly specialized stinging cells called nematocysts. In addition, the coelenterates show a simple form of tissue organization (Figure 11.4). The outer surface of the organism is covered with a layer of specialized cells that form an *ectodermal* (skin) tissue. The inner surfaces are lined with an *endodermal* tissue composed of different specialized cells.

In the plant kingdom, organisms also can be arranged in a sequence of increasing cell specialization and tissue development. Between the groups of unicellular algae and the simple multicellular plants are a variety of intermediate forms that can be regarded either as colonies of unicellular organisms or as simple multicellular organisms.

Among the coelenterates are other organisms that might be regarded as colonies of specialized multicellular individuals or as primitive examples of organisms with relatively independent organs. In the Portuguese man-of-war, for example, there are individual organs specialized to provide services of locomotion, flotation, feeding, and reproduction for the entire colony or single organism (Figure 11.5). Most of the more complex animals possess a body organization in which tissues make up a great variety of dif-

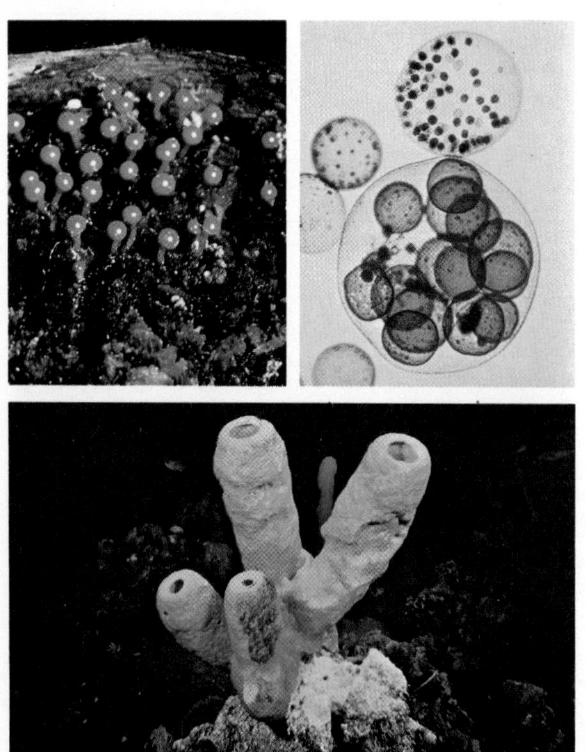

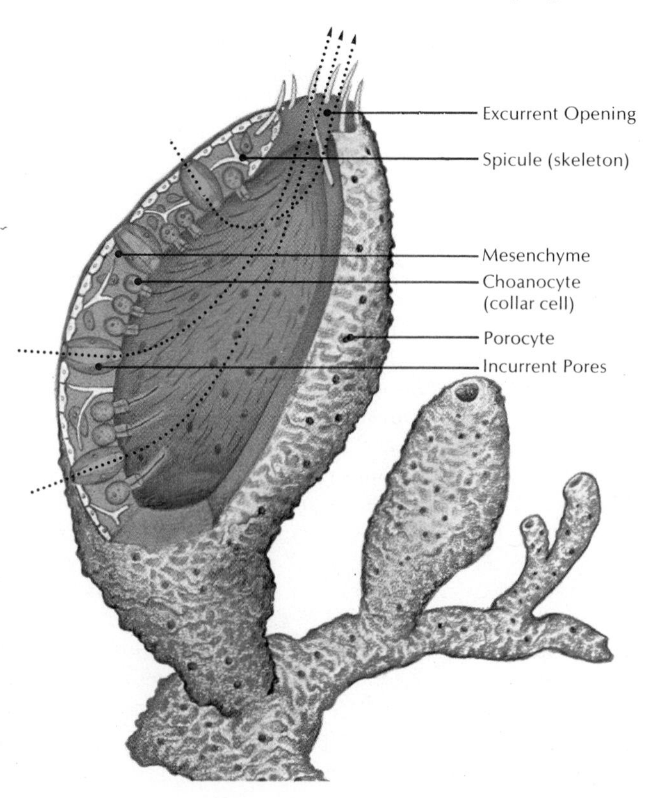

Figure 11.4 (above). The coelenterates show a simple form of tissue organization. The outer surface is covered with ectodermal cells, and endodermal cells line the inner surface.

Figure 11.5 (lower left). Portuguese man-of-war. This coelenterate colony is made up of hundreds of individuals in four distinct classes based on function: feeding polyps, protective or defensive polyps, float polyps, and reproductive polyps.

Figure 11.6 (lower right). Simple, cuboidal epithelium in the lining of a kidney tubule. Such cells are bound together by a thin layer of intercellular cement and an underlying basement membrane composed of collagen fibers embedded in a matrix. *(Courtesy Carolina Biological Supply Company)*

ferent organs, each specialized for a particular task in the life of the organism. In complex plants, on the other hand, there are relatively few organ structures, but many functions are carried out by specialized tissues.

Although the intermediate kinds of organisms are of great interest because they offer clues about the ways that cell specialization, tissues, and organs might have evolved from simpler organisms, it will be useful here to examine some of the specialized cells and tissues of more "advanced" multicellular plants and animals. Such an organism is a collection of cells of many different structural types and functions. Cells of similar type and function are organized into *tissues*. Various tissues may be coordinated to form an *organ* that carries out a more complex yet unified function. A survey of some major types of plant and animal tissues reveals some of the ways in which cell specialization makes multicellularity possible.

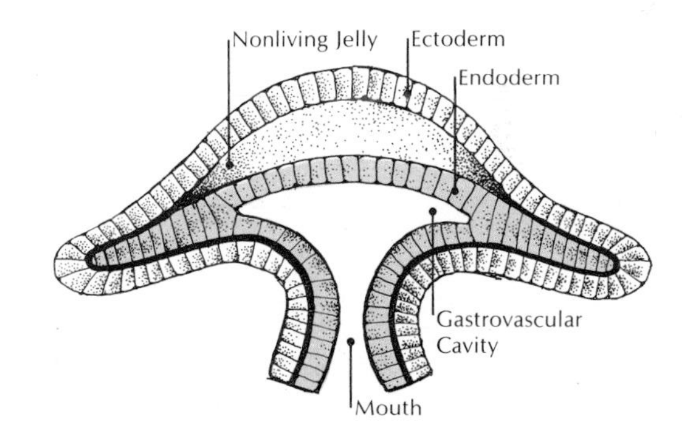

ANIMAL EPITHELIAL TISSUE

The epithelial tissue of an animal is an aggregation of cells that cover surfaces and thus "contain" the organism (Figure 11.6). Epithelial cells must adhere tightly to one another to form a continuous sheet. If they are damaged, they must replace themselves rapidly in order to maintain the integrity of the surface. The ways in which epithelial cells are attached to one another serve as good examples of the various forms of cell-cell attachment found throughout multicellular organisms.

Between adjacent cells is a thin layer of *intercellular cement*, whose major component is a mucopolysaccharide secreted by the epithelial cells themselves. The layer of epithelial cells rests upon a basal lamina, which

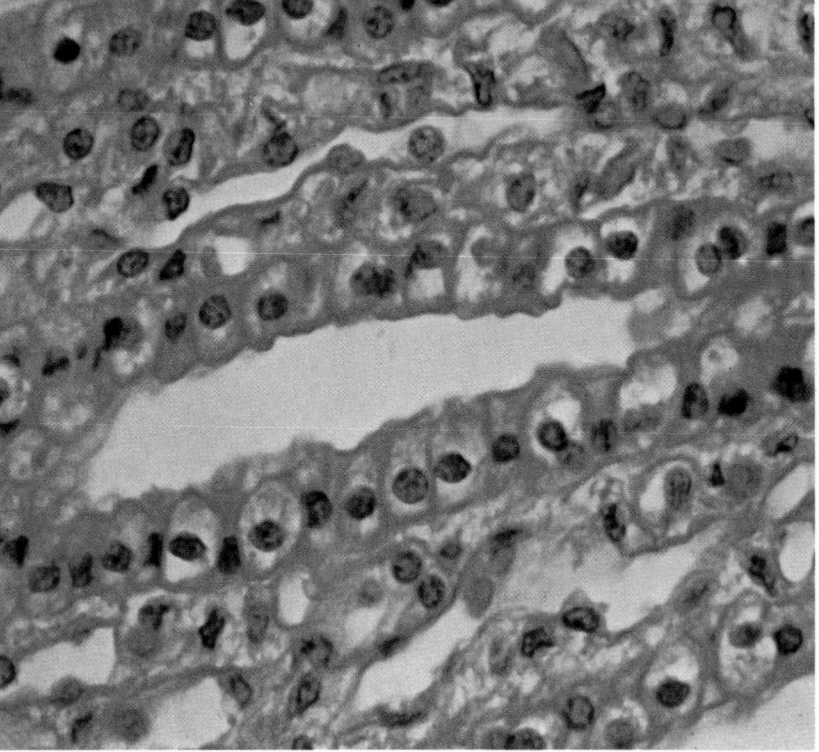

Figure 11.7. Photograph of the bioluminescent feathered sea star. Bioluminescence in this organism produces the golden glow effect and is due to cells that are able to convert chemical energy into light. The emission of light is dependent upon nervous stimulation of specialized cells in light-producing organs.

consists of collagen fibers embedded in a matrix. The collagen and the matrix also are secreted by the epithelial cells. The collagen fibers add strength to the tissue.

Many mechanical attachments exist between cells. The simplest of these attachments are *interdigitations*, or tongue-and-groove associations, of the membranes of adjacent cells. Because interdigitations increase the area of surface contact between cells, the amount of intercellular cement holding the cells together is also increased. Other forms of mechanical attachment are provided by specialized structures such as *desmosomes*, which are thought to act as attachments that anchor cells.

The skin that covers the outer surface of an animal is a good example of exterior epithelial tissue (Montagna, 1965). It serves primarily to hold the organism together and to protect it from the external environment, but its properties vary from place to place to meet particular needs of different parts of the organism.

Many epithelial cells are specialized to secrete substances onto the surface of the organism. One basic kind of secretion is *mucus*, a mucopolysaccharide that forms a protective covering over the outer cell surface. Certain cells in the skins of many organisms secrete poisons of various sorts; these cells apparently have evolved from mucus-secreting cells. Epithelial cells lining the interior of the digestive tract secrete enzymes that break down molecules of ingested food so that the material can be absorbed by other cells of the intestinal epithelium.

Sensory reception is another kind of specialization of epithelial tissues. Sensory receptor cells are triggered by various changes in the external or internal environment. Some sensory receptor cells are modified ciliated cells in which the cilia form the actual receptor organelles that respond to the stimulus from the environment. The modified cilia are instrumental in translating this stimulus into an electrochemical change that can be transmitted to the central nervous system (Chapter 26).

Some epithelial cells are specialized as pigment-bearing cells, which either protect the organism by absorbing harmful radiation or conceal it in

Figure 11.8 (left). Allium leaf epidermis. Note the thick, dark cell walls and the relatively large central vacuoles of these cells. Epidermal cells such as these typically secrete a waxy cuticle to prevent dehydration and to protect the plant from mechanical injury. Also visible in this photograph are numerous stomata-guard cell complexes, which serve to regulate plant transpiration. *(Courtesy Carolina Biological Supply Company)*

Figure 11.9 (right). Photograph of a partial section through a woody stem. The periderm layer consists of cork cells and their products, which comprise bark.

one way or another from predators (protective coloration). Bioluminescent cells, which are capable of converting chemical energy into light, are another example of epithelial specialization involving pigments (Figure 11.7).

Other forms of specialization exist within epithelial tissues, but these examples give some idea of the complexity of functions that epithelial cells serve and of the corresponding range of specialized cell structures that exist.

PLANT SURFACE TISSUE

Like the epithelial tissue of an animal, the surface tissue of a plant serves as a protective covering for the outer surface of the organism. *Epidermal tissue* covers the surfaces of roots, stems, and leaves in most plants (Figure 11.8). Like the skin of an animal, epidermal tissue is made up of a thin layer (usually one cell in thickness) of flattened, interdigitating cells. Most epidermal plant cells have thickened outer walls, relatively large central vacuoles, and a relatively small amount of cytoplasm. Many of the epidermal cells on the aerial parts of the plant secrete cuticle, a water-resistant, waxy substance that forms a surface layer, protecting the plant from dehydration and invasion by parasites.

The epidermal cells of the roots have no cuticle covering but are specialized to absorb water. Some of the root epidermal cells form long, hairlike extensions into the soil. Some of the epidermal cells of the aerial part of the plant may be specialized to form spines, hairs, or glands, all of which play roles in the protection or functioning of the plant.

In adult trees, the epidermis is replaced by another tissue, the *periderm*, which is composed of cork cells (Figure 11.9). Cork cells secrete a waterproof coating of suberin and then die, so that the surface of the periderm is a thick layer of hollow, water- and injury-resistant cork cells. The periderm forms the familiar bark of a tree.

ANIMAL SUPPORTIVE TISSUE

The body of an animal is supported by a number of different forms of tissue. Skeletal tissues such as chitin, cartilage, and bone provide a more-or-less

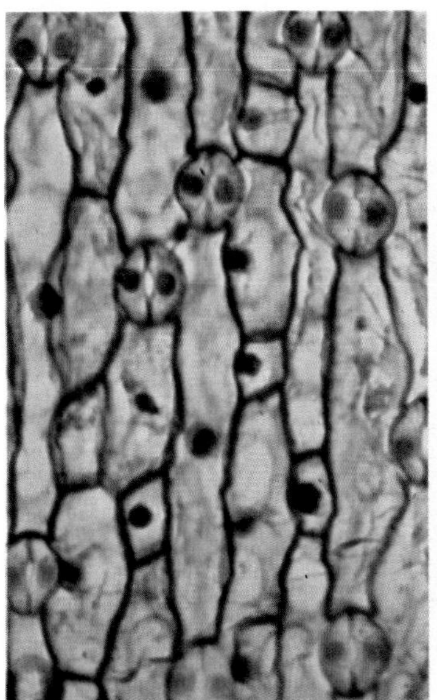

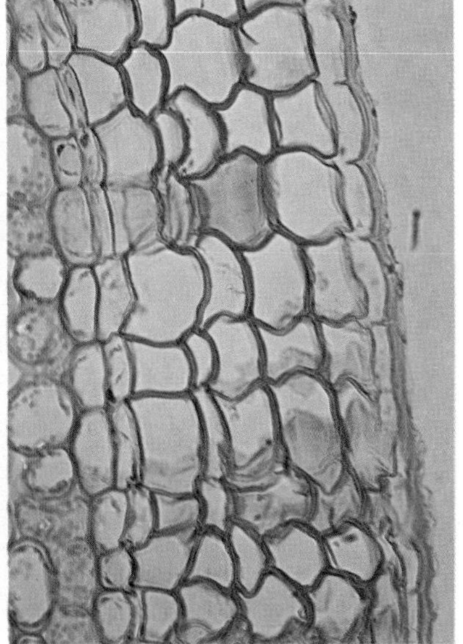

Figure 11.10 (left). Human bone cross section. (*Courtesy Carolina Biological Supply Company*)

Figure 11.11 (upper right). Hyaline cartilage. Note the clear, almost transparent matrix in which are embedded lacunae, or cell spaces. Each lacuna contains one or more chondryocytes, or cartilage cells. (*Courtesy Carolina Biological Supply Company*)

Figure 11.12 (below). Fibroelastic connective tissue. The most conspicuous components of this tissue type are the large number of threadlike fibers, some of them tough and strong (collagen) and others elastic and flexible (elastin). (*Courtesy Carolina Biological Supply Company*)

rigid framework that gives structural support and provides a system of levers by which the organism can move parts of its body. The cells of skeletal tissues are specialized in the secretion of particular kinds of extracellular materials (Figure 11.10).

Structurally, connective tissues are made up of rather generalized cells, called *fibroblasts*, which synthesize extracellular fibers such as collagen and elastin (Figure 11.12). These fibers bind tissues together, provide support for the tissues and for the organs formed from them, and join skeletal members to each other and to the muscles that move them. These

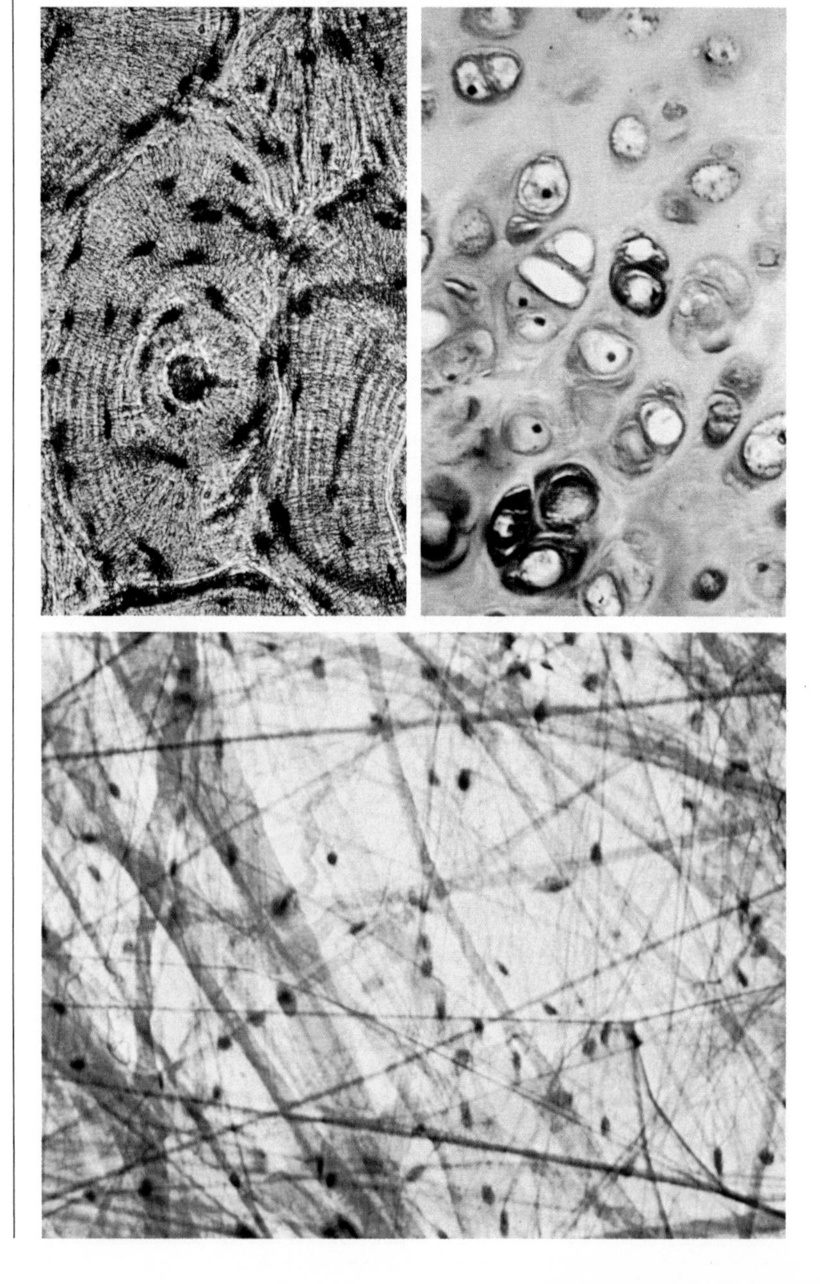

connective fibers are an extremely important part of the structure of an animal. Collagen, in fact, is the most abundant single protein in the human body and the most common protein in the entire animal kingdom.

PLANT MERISTEMATIC AND FUNDAMENTAL TISSUES

The body of a plant contains a number of regions of nonspecialized cells that retain the capability of division to produce a continuous supply of cells that can specialize to form new tissues. The undifferentiated tissues that are sites of active cell division are called *meristems* (Figure 11.13). Although there is great variation in characteristics among meristematic cells, most tend to have thin walls and to be small, closely packed, and filled largely with cytoplasm. Meristematic tissues form the growing tips of roots and shoots. In many plants, there are layers of meristematic tissue near the surface of branches and stems. These lateral meristems enable the plant body to continue to grow thicker throughout its life. The existence of meristematic tissues that continue to produce new cells (and therefore new tissues and organs) throughout the life of a plant is one of the major differences between plants and animals. In most higher animals, the vast majority of tissues and organs are formed early in life, and most of the cells of the body lose the ability to divide.

Most of the plant body is made up of *fundamental tissues*, each of which is composed largely of a single kind of specialized cell. *Sclerenchyma* is a tissue composed of cells that secrete thick cell walls rich in lignin and then die. Thus, the mature sclerenchymatic tissue is composed of a network of lignified walls with minute pores that were once the cell interiors. Some sclerenchymatic cells become elongated into fibers such as those of flax and hemp. Sclerenchyma serves primarily as a supportive tissue for the plant body. *Collenchyma* is another supportive tissue with thickened cell walls, but the collenchymatic cells remain alive through most of the plant's life span.

Much of the body of a plant—particularly of the lower plants—is made up of a tissue composed of living, nonspecialized, thin-wall, large-vacuole cells that may occasionally begin meristematic activity or cell specialization to form other tissues. This *parenchymatic* tissue gives support and rigidity to the plant body and serves as a site for storage of nutrients and water.

Figure 11.13. Photograph of a longitudinal section through a shoot apex, showing leaf buds and the apical meristem region. Meristematic tissue is characterized by its ability to undergo repeated mitotic divisions, thereby accounting for plant growth.

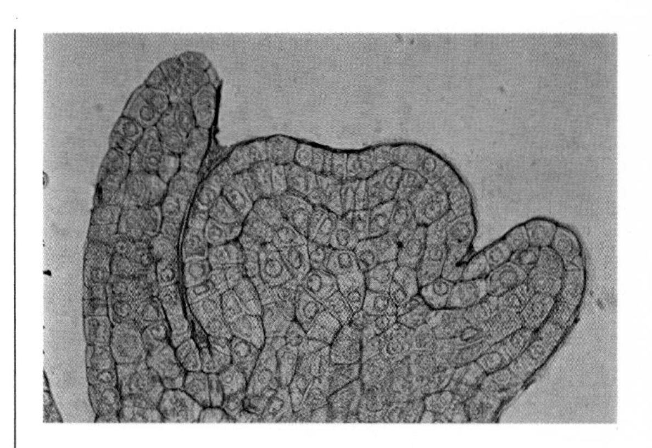

ANIMAL NERVOUS TISSUE

As an animal embryo develops, some epithelial cells specialize to form nervous tissue. Comparison of more primitive organisms suggests that nervous tissue evolved through further specialization of epithelial tissue. Nervous tissue, however, has acquired such a distinctive and specialized function that it is regarded as a tissue in its own right. The individual cells of nervous tissue serve to integrate the organism's activities at all levels.

Nervous tissue is made up of two types of cells, *neurons* and *interstitial cells*, both of which are characterized by fairly extensive fiberlike projections. Neurons create the electrochemical signal called the *nerve impulse* and transmit this impulse to other cells—neurons, muscle cells, or gland cells. Interstitial cells play a supportive role—binding neurons together, forming insulating coverings for the nerve fibers, or providing the neurons with certain necessary nutrients.

Neurons represent an extreme example of cellular specialization. By the time a vertebrate animal is born, nearly all its nerve cells have been formed.

Neurons retain their nuclei, and their DNA continues to code for the production of enzymes and other proteins needed in various neural activities. Other portions of the DNA message are somehow irreversibly "switched off," so that a fully specialized neuron is incapable of mitosis and therefore cannot divide (Chapter 13). If a nerve cell dies, it can never be replaced. An individual human acquires his greatest number of neurons within the first few years of his life. Thereafter, the number of neurons in the body decreases steadily. Fortunately, the neurons of the central nervous system are well protected from physical and metabolic damage by the skull and vertebral column and by a peculiar physiological barrier called the blood-brain barrier, so that neuron degeneration probably does not become a signifcant phenomenon until late in life. Aging processes that damage the heart and the blood vessels can cause the death of neurons as a secondary effect because neurons require a constant supply of oxygen and glucose, and interference with these supplies brings about neuron death. The death of significant numbers of neurons may bring about progressive senility or paralysis.

A typical neuron consists of a cell body and one or more cytoplasmic extensions (Figure 11.14). The cell body contains the nucleus, mitochondria, Golgi apparatus, ribosomes, smooth and rough ER, and clusters of microtubules, which often appear in the light microscope as cytoplasmic fibers called *neurofibrils*. Also, there are structures called *Nissl bodies*, which are layers of rough ER cisternae.

The extensions from the cell body are of two types: the *axon*, which is relatively long; and the *dendrite*, which is relatively short and branched. Most neurons have multiple dendrites and a single axon. In most neurons, the axon extends from one side of the cell, and dendrites extend from many parts of the cell body. In sensory neurons, the cell body is located near the center of the axon, and the dendrites are attached to the end of the axon where impulses originate (for example, at sensory receptors in the skin). The transmitting end of the axon bears a number of fine branches called the *axonal arborization*.

The axons of many neurons are wrapped in a *myelin sheath*, another example of cell specialization. The myelin sheath is composed of the membranes of interstitial cells and is wrapped around the axons to form several concentric layers. The cells that wrap around peripheral nerve fibers—that is, nerve fibers outside of the brain and spinal cord—are called *Schwann cells* (because they were first described by Theodor Schwann). The cells that wrap around axons within the central nervous system (brain and spinal cord) are called *oligodendrocytes*. The axon, with its surrounding sheath, is called a nerve fiber.

Some axons are exceedingly long. For example, a nerve terminating in a blood vessel in the foot of a giraffe has its cell body and dendrites in the spinal cord. Its axon is a single, continuous fiber, perhaps eight or nine feet in length, extending from the spinal cord to the end of the foot. Thousands of Schwann cells are required to wrap a myelin sheath around such a long nerve fiber. Between each pair of successive Schwann cells is a gap called a *node of Ranvier*. Such nodes are not as conspicuous between the oligodendrocytes of the central nervous system.

ANIMAL MUSCLE TISSUE

Muscle cells are specialized for the function of contraction. The simplest muscle cells are found in the coelenterates, where certain epithelial cells

Figure 11.14. Diagram of a typical neuron.

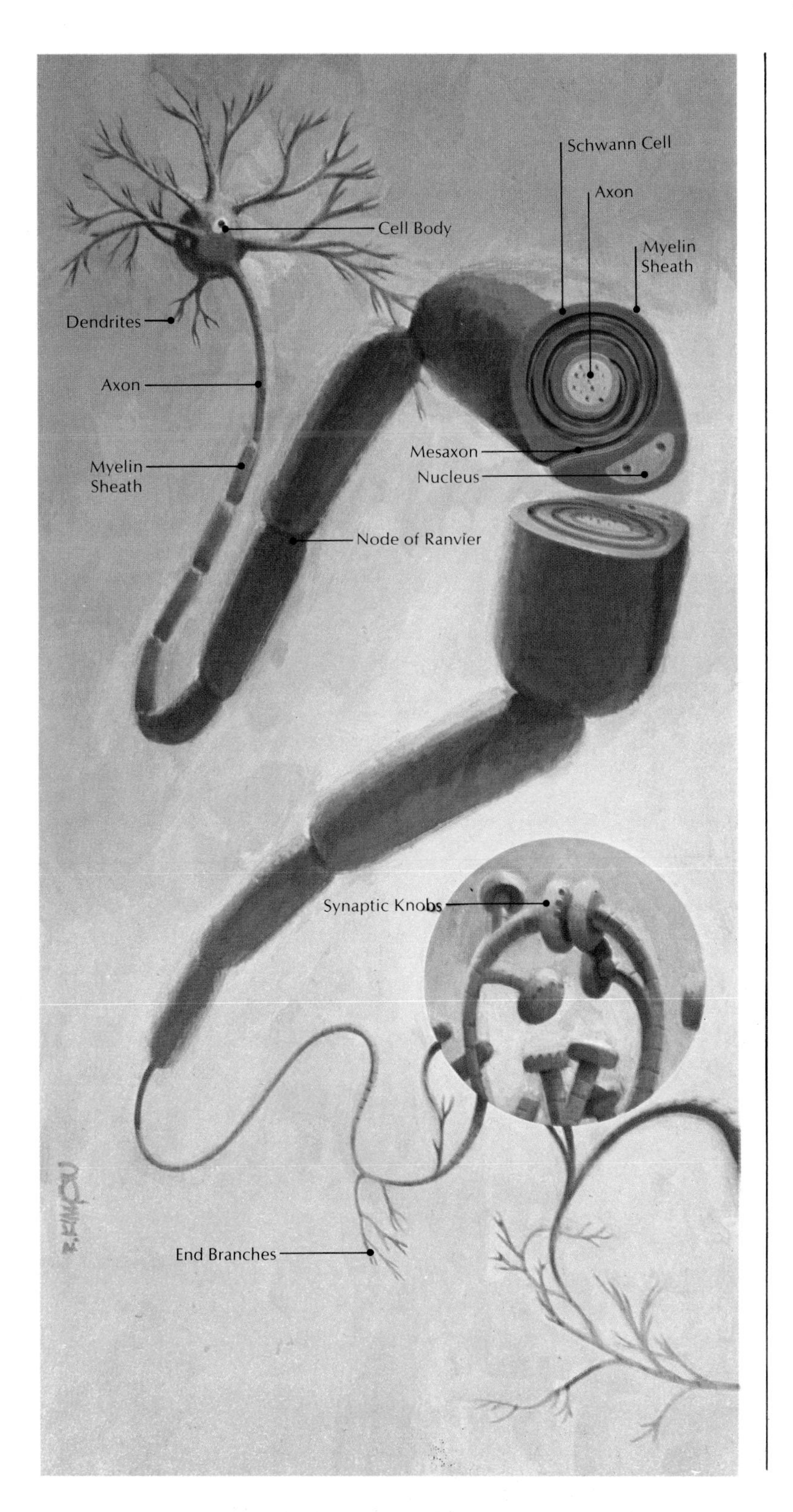

Figure 11.15 (above). Skeletal muscle. Cross striations (alternate dark and light bands), resulting from myofilament arrangement, may be seen in the longitudinal sections. *(Courtesy Carolina Biological Supply Company)*

Figure 11.16 (below). Electron micrograph and diagram of skeletal muscle. Note the dark A-bands that alternate with the light I-bands. In the center of each A-band is a light H-band, and in the center of each I-band is a thin, dark Z-line. Also visible is the sarcoplasmic reticulum.

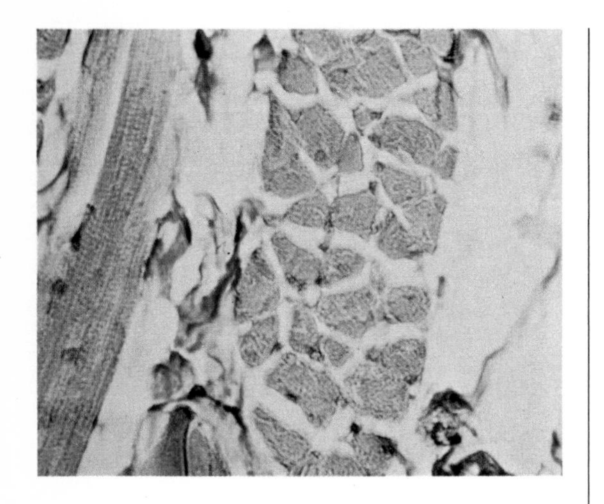

have two extensions from the cell base running parallel to the surface of the organism. Within these extensions are fibrils that respond to nervous stimulation by contracting.

Three different types of muscle tissue are found in higher animals: skeletal, cardiac, and smooth muscle. Skeletal and cardiac muscle cells are characterized by transverse lines, or *striations*, which are visible in the light microscope. Smooth muscle lacks these striations.

The specializations of muscle cells are so extreme and of such fundamental importance to animal organisms that they merit rather detailed consideration.

Skeletal Muscle Cells

The cells of skeletal muscle appear as fibers in the light microscope. A single fiber measures about 100 μ in diameter and a few millimeters to a few centimeters in length. Each fiber shows a banded pattern of transverse striations with dark bands, or *A-bands*, alternating with light bands, or *I-bands* (Figure 11.15). In the center of each I-band is a thin, dark line (*Z-line*), and in the center of each A-band is a light *H-band*. The repeating unit of this pattern is called a *sarcomere*. Each fiber is a single cell with several nuclei surrounded by a thin but tough cell membrane called the *sarcolemma*. When a nerve impulse triggers the muscle cell, the cell contracts. Biochemical analyses show that about 20 percent of the weight of a muscle cell is protein; the balance is water and dissolved substances.

The light microscope reveals each cell to be composed of a number of *myofibrils* about 1μ in diameter. The myofibrils also are striated; in fact, the striations of the fiber are due to the aligned striations of its myofibrils. Between the myofibrils in the cell are mitochondria and a complex membrane structure called the *sarcoplasmic reticulum*. The nuclei lie near the edge of the cell, adjacent to the sarcolemma.

Electron micrographs of myofibrils reveal that they are made up of still smaller *myofilaments*. There are two kinds of myofilaments in each myofibril; the thicker filaments are about 100 A in diameter and 1.5μ in length; the thinner filaments are about 50 A in diameter and about 2μ in length.

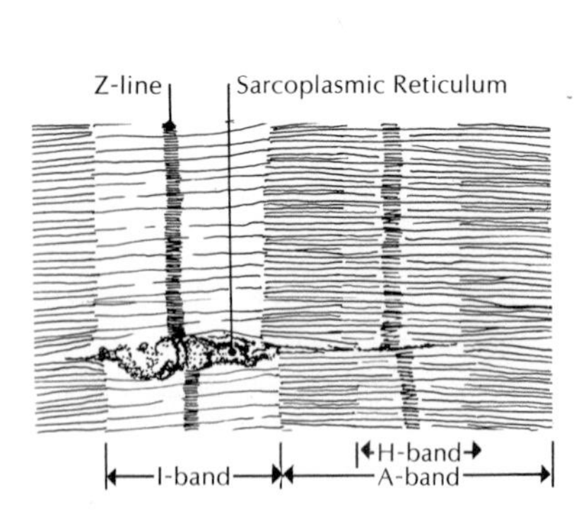

Figure 11.17 (above). Human cardiac muscle. Note the conspicuous cross striations and intercalated discs (opaque cross bands) characteristic of this type of muscle tissue. *(Courtesy Carolina Biological Supply Company)*

Figure 11.18 (below). Smooth muscle. Note the obvious lack of cross striations and the large, centrally placed nuclei, typical of this type of muscle tissue. *(Courtesy Carolina Biological Supply Company)*

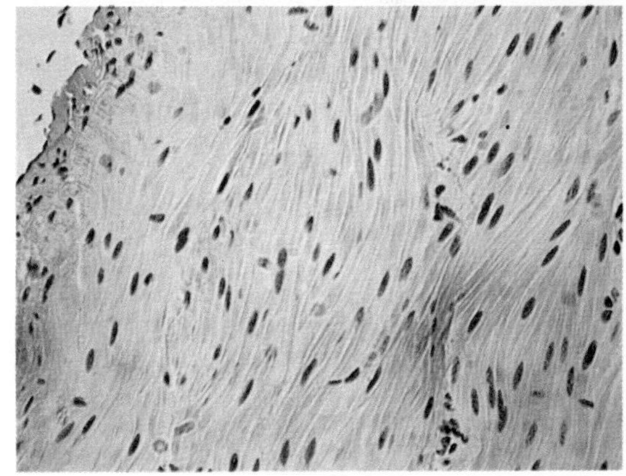

The dark A-band is made up of both thick and thin filaments; the light I-band is made up of thin filaments alone, whereas the light H-band is made up only of thick filaments. The Z-line is a narrow zone of very dense material not arranged in filaments (Figure 11.16). In a section cut across the A-band, the thick and thin filaments can be seen to be arranged in a regular pattern.

Electron micrographs at high magnification show that the thick and thin filaments are linked by cross-bridges, which extend from the thick filaments at intervals of about 60 A. These cross-bridges probably play an important role in the process of contraction (Chapter 25).

The protein in the myofibrils is composed chiefly of myosin and actin, with smaller amounts of tropomyosin. *Myosin* makes up about one-half of the protein in the myofibril. When myosin is extracted from the myofibril, this protein forms filaments about 0.2μ in length and 100 A in diameter. These filaments, similar in appearance to the thick filaments of the myofibrils, have numerous side projections that appear similar in spacing to the cross-bridges of the myofibrils (H. E. Huxley, 1965). The hypothesis that the thick filaments are composed of myosin is confirmed by the observation that the A-bands disappear from a myofibril when the myosin is extracted from it.

The other major protein of myofibrils, *actin*, forms a globular molecule in pure solution. When placed in a solution with salt and ATP concentrations similar to those of the muscle cell, actin forms long fibers. The assumption that actin forms the thin filaments is confirmed by the observation that the structures of the I-band disappear when actin is extracted from the muscle cell.

Less than 3 percent of the protein in the myofibril is *tropomyosin*. This protein has a molecular weight much smaller than those of actin and myosin; it is thought to make up the structure of the Z-lines. Under the electron microscope, crystals of tropomyosin have structures similar to that of the Z-line (H. E. Huxley, 1965). The sarcoplasmic reticulum, a modified type of ER, forms a regular structure related to the myofibrils. Tubules of this structure lie along the surface of the A-band region, with small, saclike structures in the I-band region.

Cardiac Muscle Cells

Cardiac muscle, found only in vertebrate animals, forms the bulk of the heart. It also may extend a short distance along the walls of the large arteries that emerge from the heart. A modified type of cardiac muscle constitutes the heart's so-called neuromuscular tissue, which functions as an internal impulse-conducting system within the organ.

Cardiac muscle is intermediate in some respects between skeletal muscle and smooth muscle. It is striated like skeletal muscle but is not under voluntary control.

Smooth Muscle Cells

Smooth muscle tissue contracts or relaxes very slowly. In vertebrates, smooth muscle is found in internal organs such as the stomach, intestines, and blood vessels, and its contraction is involuntary.

This muscle tissue is called smooth because it lacks the striated pattern formed by the orderly array of thick and thin filaments in skeletal and cardiac muscle cells. Smooth muscle cells are about 10μ in diameter, tapering

Figure 11.19 (below). Photograph of a cross section through a Ranunculus root. Note the cytological difference between the heavy-wall xylem cells and the small, thin-wall phloem cells.

with a single, centrally located nucleus. Smooth muscle cells prepared for electron microscopy appear to contain only thin (actin) filaments. Recent results with different techniques of fixation, however, show that smooth muscles may also have a thick type of filament. The "true" ultrastructure of smooth muscle remains to be discovered.

PLANT VASCULAR TISSUE

Just as nervous and muscle tissues are characteristic of higher animals and essential to their way of life, so vascular, or conductive, tissue is characteristic of higher plants and plays a major role in the functioning of those organisms. Some of the cells of vascular tissues are specialized to serve as tubes through which fluids can be moved from one part of the plant body to another. The two major kinds of vascular tissues, xylem and phloem, are both complex tissues made up of a number of different specialized cells.

Xylem forms a conducting system that extends from the roots to the tips of all the shoots, leaves, and other appendages of the plant body (Figure 11.19). Through this tissue, water and dissolved nutrients and minerals move from the soil to all the cells of the plant. The fluids move through hollow tubes that are formed by specialized cells called *tracheids* and *vessel elements*, which form thick cell walls and then die to leave these walls as hollow tubes. Thick-wall fibers, formed by the death of sclerenchyma cells, help to support the xylem tissue. The only living cells in mature xylem are parenchyma cells scattered between the thick-wall, dead cells. Xylem is the wood of the plant, and it provides support for the plant body as well as transport of fluids.

Phloem forms another conductive system that extends through much of the plant body. In the phloem, the organic materials produced through photosynthesis or cell secretion are moved from one part of the plant body

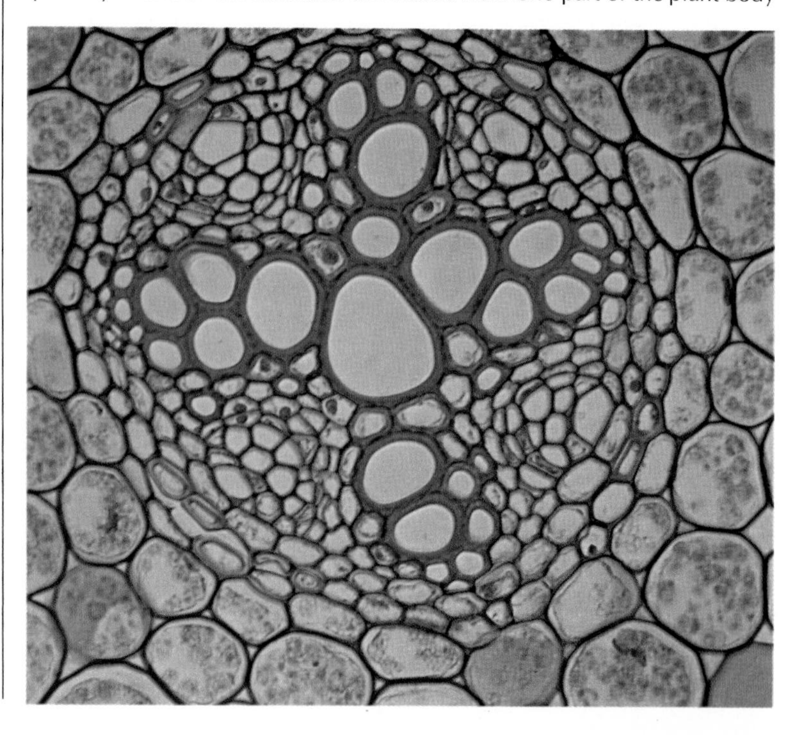

to another. The fluids move through specialized *sieve cells*, which have relatively thin walls and which retain their cytoplasm at maturity, although the nucleus disintegrates. Closely associated with the sieve cells are companion cells, which retain both cytoplasm and nucleus at maturity. Like the xylem, the phloem also contains sclerenchymatic and parenchymatic cells, which provide support and nutrient storage.

OTHER SPECIALIZED CELLS

The examples discussed in this chapter by no means exhaust the catalog of specialized cells of higher animals and plants. The body of an average human contains about 100 trillion (10^{14}) cells of many thousands of different specialized types. Some cells are specialized for the synthesis of various necessary substances. Others are specialized for the transport of materials from one solution to another. Still others are specialized to play various roles in the integration or the reproduction of the organism. Some of these other specialized cells in animals will be discussed in Chapter 16 and in Unit Five. The process of differentiation, through which cells become specialized, is discussed in Chapters 17 and 18.

Only a cell biologist or a histologist (student of tissues) is apt to be familiar with the structural details of all the many different types of specialized cells or to be fluent in the use of the many names coined to describe specialized structures. It is important, however, to realize that the basic structure of the eucaryotic cell is modified in many ways in multicellular organisms, thus producing highly specialized cell structures with highly specialized functions. Each specialized cell, in turn, is a complex mechanism of highly ordered macromolecules. The molecular approach to biology has not stolen the wonder or beauty from the contemplation of life; rather, it has increased the sense of awe at the precision with which living systems are constructed and operate.

FURTHER READING

For further information on tissues and specialized cells in higher animals, see books by Arey (1963), Bloom and Fawcett (1962), and Patt and Patt (1969). Similar information about higher plants will be found in books by Eames and MacDaniels (1947), Esau (1965), and Jensen (1964). For more general discussions of specialized cell structures, see books by Brachet and Mirsky (1961), DeRobertis, Nowinski, and Saez (1965), Loewy and Siekevitz (1963), Stern and Nanney (1965), and Swanson (1969).

Among the *Scientific American* articles that discuss various specialized cells and tissues are those by Allen (1962), Comroe (1966), H. E. Huxley (1958, 1965), Kennedy (1967), Miller, Ratliff, and Hartline (1961), D. S. Smith (1965), H. W. Smith (1953), Speirs (1964), and Wurtman and Axelrod (1965).

IV
The Continuity of Life

"Genetics" is a twentieth-century word and a twentieth-century science.
The word was coined by an English biologist, William Bateson, to designate that branch of biology which deals with the underlying causes of inherited resemblances and differences between individuals, and hence with the evolution of all living things.
Asserting that "the essential process by which the likeness of the parent is transmitted to the offspring . . . is as utterly mysterious to us as a flash of lightning to a savage," Bateson in 1902 exhorted his fellow biologists to engage more actively in the experimental study of heredity. He promised them that "an exact determination of the laws of heredity will probably work more change in man's outlook on the world, and in his power over nature, than any other advance in natural knowledge that can be clearly foreseen."
. . . In the sixty-some years that have intervened since Bateson's quoted remarks appeared in the *Journal of the Royal Horticultural Society,* the process by which the genes (units of hereditary material) are transmitted from parent to child has become so well understood that man, alone among species, now possesses the ability to control his own evolution.

—George and Muriel Beadle (1966)

12
Mendelian Genetics

Gregor Johann Mendel is usually described as an obscure Moravian monk. According to the traditional legend, the research results of this cloistered priest long held the key to inheritance and evolution. He had published his account of the laws of heredity in 1866, but only in the proceedings of an obscure natural history society. A third of a century later, after Mendel's death, three botanists chanced upon his paper and—the legend has it—all the pieces of the evolutionary puzzle fell into place.

Almost every line in this legend is misleading, if not plainly false. A different picture emerges from an examination of Mendel's life and researches in their historical context.

Two hours' drive north of Vienna is the town of Brno, Czechoslovakia. The little town of Hynčice, where Mendel was born in 1822, lies to the northeast a few hours from Brno. The whole area was then part of Austria. Mendel's hometown was called Heizendorf and lay in the province of Silesia. Brno—then Brünn—was in the province of Moravia. The Mendel family had lived in Heizendorf since the late seventeenth century. Like many others in the district, they were of German descent, and Mendel spoke German as his first language.

His parents named him Johann, but on his admission as a novice to the Augustinian monastery at Brünn in 1843 he was renamed Gregor. Mendel's sharp mind already had been evident to his teachers, both in school and at the University Philosophical Institute of Olmütz. Because he was not a particularly churchy young man, his joining the monastery was as much a step in his academic career as a response to a devout calling. Many monastery members were full-time teachers, either at the Philosophical Institute in Brünn or at the local Gymnasium (preparatory school). From time to time others would leave for a few years to serve as professors in universities. Mendel himself was a substitute teacher of science, first at the Gymnasium in Znaim and then, until his election as abbot in 1868, at the Technical High School in Brünn. In 1850, at the request of the high school, Mendel took an examination to qualify as a regular science teacher. He was essentially self-taught in the natural sciences, but his colleagues apparently convinced him that he was ready to take this difficult examination, which normally followed several years of university study. Although Mendel performed surprisingly well on the examination, his inadequate training was obvious, and the examiners refused to qualify him as a teacher but complimented him for his industry and talent. To provide Mendel with the education he needed, the monastery sent him in 1851 to the University of Vienna for two years of studies in science and mathematics.

On his return to the monastery, Mendel not only undertook his arduous plant-hybridization experiments but made full use of the excellent monastery library to keep abreast of developments in other fields. He took several trips abroad, traveling to Italy more than once and in 1862 joining a group that visited the Industrial Exhibition in London.

In addition to being an alert, intellectual, and energetic man of the world, Mendel was a highly practical person. He had spent his childhood on a farm, and later he was a founding member of the Moravian and Silesian Agricultural Society, winning awards for the development of new fruit and vegetable varieties. He was active in the fire brigade in Heizendorf and was elected chairman of a bank in Brünn. Stocky, kind, and quietly genial, the monk was dearly loved by his pupils and fellow clerics. Yet there was an underlying intensity and nervousness about him. Like many college students, he wrote bad poetry and got sick from worry over examinations. Later, as abbot, he fought a stubborn but futile resistance to new taxes on monasteries. There are reports of his cigar consumption hitting 20 a day and his pulse 120 a minute in these closing years, before he finally fell victim to Bright's disease in 1884.

MENDEL'S PAPER

Mendel's fame rests upon a single paper. He read it to the Brünn Natural Science Society at two evening meetings, on February 8 and March 8, 1865. The next year it appeared in the fourth volume of the proceedings (Mendel, 1866; Křiženecký, 1965; Stern and Sherwood, 1966).

The core of Mendel's paper comprises three major generalizations: the principle of dominance, the principle of segregation, and the principle of independent assortment, or recombination. These principles may be summarized as follows: (1) When parents differ in one characteristic, their hybrid offspring resemble one of the parents, not a blend of the two characters—the principle of *dominance*. (2) When a hybrid reproduces, its reproductive cells are of two kinds—half transmitting the dominant character of one parent, and the other half transmitting the recessive character of the other parent—the principle of *segregation*. (3) When parents differ in two or more pairs of characters, each pair shows dominance and segregation independently of the other pairs, so that all possible combinations of the various pairs occur in their chance frequencies in the reproductive cells of the hybrid—the principle of *independent assortment*, or *recombination*.

These fundamental and important principles were not the work of a naïve priest puttering with some peas to pass the time between morning and evening prayers. Mendel had received a comprehensive scientific education at the University of Vienna. After returning to Brünn, he crossed many species of plants, undertook microscopic investigations, kept bees and mice, studied the effects on plants of environmental changes, and statistically examined various meteorological data.

Two crucial features of Mendel's researches on heredity can be traced to his training at the University of Vienna. First, he analyzed his results with a mathematical sophistication rare among the naturalists and horticulturalists of his day. Second, his interpretation of his observations emphasizes the reproductive cells as the link between parents and offspring. Mendel's professors at Vienna had included Christian Doppler (of Doppler-effect fame) and Andreas von Ettinghausen—two physicists with a penchant for applied mathematics. Mendel studied botany with the eminent Franz Unger, whose contributions to cell theory included the identification of the male reproductive cells of mosses. Furthermore, Unger was a speculative theorist who, in a book published in 1852, denied the constancy of species and suggested in rather vague terms that the plant kingdom developed through natural processes. The pantheistic tone of Unger's book provoked the Catholic press to call for his dismissal. The students responded with a petition for his retention. The effect of this affair on Mendel is unknown, but the mutability of species and the history of the plant kingdom were hot topics at the time, and Mendel possibly undertook his investigation of heredity in the hope of shedding some light on the issues. Within a few months of his return from Vienna, Mendel had established some 34 pure strains of peas, obviously planning for some kind of hybridization experiments. The experi-

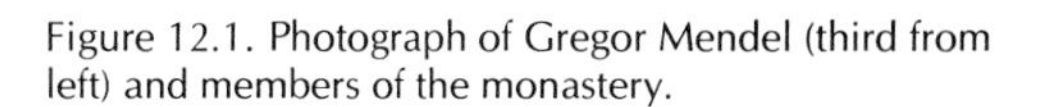

Figure 12.1. Photograph of Gregor Mendel (third from left) and members of the monastery.

ments reported in Mendel's famous paper were begun in 1856 and virtually completed by 1863.

Mendel was not the first person to cross plants, nor was he the first hybridizer trying to discover the mechanism of inheritance. Previous investigators who had crossed species or varieties had noted how the dominance of some characters produces a uniformity in the first-hybrid generation. They also had noticed the greater variability of the generation produced by breeding the first-generation hybrids among themselves. However, no one before Mendel had studied large numbers of single-character crosses with the intention of developing a general theory that would describe the statistical distribution of single characters among successive hybrid generations.

Just how Mendel reached his theoretical conclusions will always remain obscure. The core of his interpretation is the assumption that the hereditary elements or factors present in a hybrid are not irreversibly blended together but can reappear separately in the next generation. It is far from clear what first led Mendel to this assumption. There is agreement that most of his experimental results are too good to be true. That is, the close fit between his recorded results and the frequencies expected from theoretical calculations is extremely improbable. Both repetitions of Mendel's experiments and modern statistical analysis make it very unlikely that Mendel could have observed results so close to his theoretical expectations. This evidence does not necessarily imply deliberate dishonesty; the improbably close fit is more likely due to the unconscious bias in scoring plants or to termination of counts when ratios were close to the predicted number. In any case, it appears that many if not all of Mendel's experiments were performed to confirm his hunches about what the laws of inheritance must be, not to discover those laws.

Mendel's initial thinking may have gone something like this: Often it makes no difference to the results of a cross whether a male with the character being investigated is bred with a female lacking it or vice versa. Parents, then, transmit equal and equivalent contributions to their offspring. Because the offspring receives equal contributions from both parents, the reproductive cell of each parent must contribute half of the normal adult

Figure 12.2. Mendel's microscope.

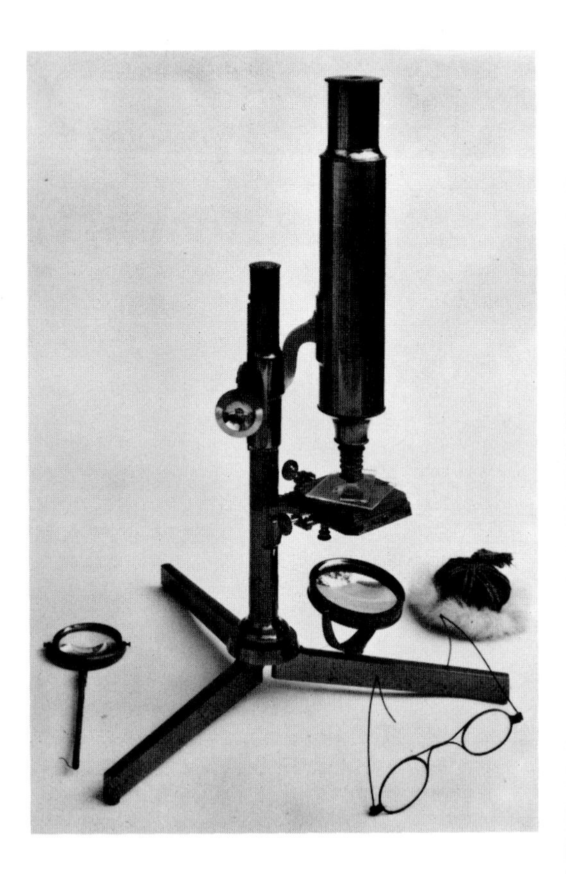

hereditary factors for each character. Without such halving in the reproductive cells, there would be an indefinite build-up of the hereditary material. When the half-doses of hereditary factors meet in a fertilized egg, they must either mix (like red and blue ink) or remain effectively discrete (like red and blue billiard balls). The ability of characters to skip a generation and reappear in the next indicates that there is no irreversible mixing. There are thus discrete and permanent hereditary factors responsible for visible characters. The frequency of various characters in any generation is determined solely by the chances for each of the possible combinations of hereditary factors in fertilization.

There is no way of knowing whether Mendel reasoned his way to his basic principles by the route described above. It is easy to see, however, that the explanation Mendel gave for his crucial experiments follows directly from these few premises. Further description of the experiments will be facilitated somewhat by using terminology developed long after Mendel's paper was published. The original parental generation is designated the P generation; the offspring of P make up the F_1 (first filial) generation; the offspring of F_1 are the F_2 generation; and so on.

MENDEL'S EXPERIMENTS

Many observable characters of pea plants are influenced by heredity—characters such as the length and color of the stem; the size and form of leaves; the position, color, and size of flowers; the form and size of seeds; and the color of seed coats and seed contents. After preliminary work, Mendel chose for further study the seven pairs of characters outlined below.

1. *Characters related to the form of ripe seeds.* The seeds may be *round* (or roundish, with only shallow surface depressions), or they may be irregularly angular and deeply *wrinkled.*
2. *Characters related to the color of seed contents.* The contents of ripe seeds may be *yellow* to orange or may have a more-or-less intense *green* tint.
3. *Characters related to the color of the seed coat.* The seed coat may be *white* (such a seed produces a plant with white flowers), or it may be *gray* to brown, with or without violet spots (such a seed produces a plant with reddish-violet flowers).
4. *Characters related to the form of ripe pods.* The ripe pods may be *inflated* with a smooth surface, or they may be deeply *constricted* between the seeds and more-or-less wrinkled.
5. *Characters related to the color of unripe pods.* The unripe pods may be light to dark *green*, or they may be vividly *yellow.*
6. *Characters related to the position of flowers.* The flowers may be *axial* (distributed along the main stem), or they may be *terminal* (bunched at the top of the stem).
7. *Characters related to the length of the stem.* Stem length varies greatly in different strains of pea plants, but Mendel chose one pure-breeding strain with *short* stems (9 to 18 inches) and another with *long* stems (6 to 7 feet).

For each pair of characters, Mendel obtained two pure-breeding strains, differing only in that single pair of characters. He chose these characters for study because in each pair the two contrasting characters are always distinct; no intermediate characters appear in crosses between the two pure-breeding strains.

Mendel began his famous experiments by cross-pollinating plants for each pair of strains differing in a single character. For example, he dusted flowers of plants from the pure-breeding, round-seed strain with pollen from plants of the pure-breeding, wrinkled-seed strain. Other wrinkled-seed plants were fertilized with pollen from round-seed plants. Similar

Figure 12.3a (above). A monohybrid cross between pea plants differing in seed types. One parent is homozygous dominant for round seeds (RR), and the other parent is homozygous recessive for wrinkled seeds (rr). The phenotype of the F_1 offspring is round, but note that the genotype is Rr, or heterozygous. If two of these round F_1 plants are mated, their offspring (F_2) will show a 3 to 1 phenotypic ratio and a 1:2:1 genotypic ratio.

Figure 12.3b (below). The same F_1 cross—diagrammed in a Punnett square—shows in more detail the origin of the 3 round to 1 wrinkled seed phenotypic ratio and the underlying 1:2:1 genotypic ratio—that is, 1 homozygous round (RR), 2 heterozygous round (Rr), and 1 homozygous wrinkled (rr).

cross-fertilizations were performed for each pair of pure-breeding strains. In each case, the pure-breeding plants are the P generation for the crossing experiment.

Mendel carefully collected the seeds that were produced on the P-generation plants, recorded their characters, planted them, and recorded the characters of the resulting adult plants. These seeds and the plants into which they develop are the individuals of the F_1 generation. Mendel called them hybrids, but modern geneticists prefer to use this term only for individuals produced by cross-breeding between two different species. In each cross, Mendel found that all the F_1 individuals resemble one of the contrasting parental characters. For example, all the F_1 seeds collected after the cross of wrinkled-seed and round-seed plants show the round-seed character. Thus, Mendel called round seeds a *dominant* character and wrinkled seeds a *recessive* character. In a cross between pure-breeding strains with dominant and recessive characters, all the F_1 individuals show the dominant character. The recessive character is not expressed in the F_1 generation, but it must be present in some form because it is passed on to the F_2 generation. Mendel found the following characters to be dominant: round seeds, yellow seed contents, gray seed coats, inflated pods, green pods, axial flower distribution, and long stems.

The characters related to shape of seed and to color of seed contents are observed as the seeds are collected from the parental plants. The other characters can be observed in the offspring only after the seeds have been planted and have developed into mature plants in the year after the cross was made. Thus, Mendel's experiments required a great deal of time as well as careful record keeping.

The F_2 generation for each cross was produced by self-fertilization of the F_1 plants. In each case, some of the F_2 individuals showed the recessive character that had disappeared in the F_1 generation. For each cross, the ratio of individuals showing the dominant character to individuals showing the recessive character averaged about 3:1. None of the F_2 individuals showed characters intermediate between the characters of the original parental strains.

For example, the cross between round-seed and wrinkled-seed parents produced 253 F_1 seeds, all of them round. When the plants grown from these seeds were self-fertilized the next year, 7,324 seeds were collected. Of these F_2 seeds, 5,474 were round and 1,850 were wrinkled, a ratio of 2.96 to 1. Table 12.1 shows the characters observed in the F_2 generations of the various crosses. The ratios of dominant to recessive varied from 2.82:1

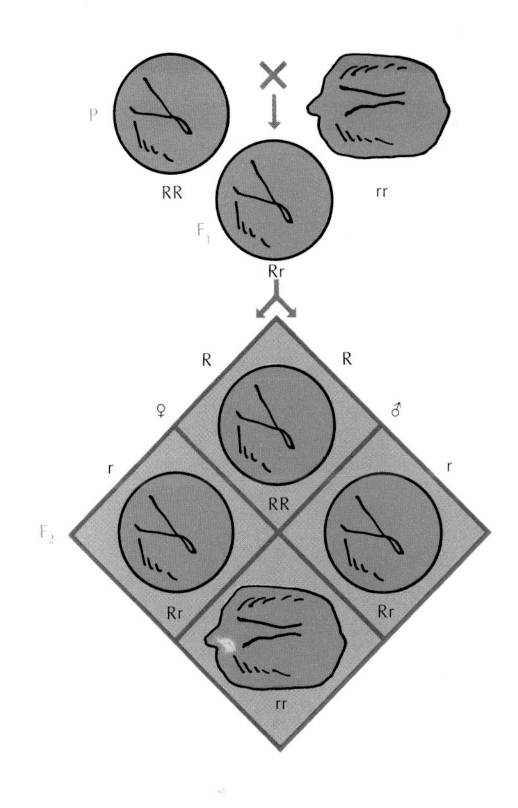

		Male Gametes R	Male Gametes r
Female Gametes	R	RR (round seed)	Rr (round seed)
Female Gametes	r	Rr (round seed)	rr (wrinkled seed)

Table 12.1
Ratios of Characters in F_2 Generations of Mendel's Seven Crossing Experiments

Dominant Character: Character	Number of F_2 Individuals	Recessive Character: Character	Number of F_2 Individuals	RATIO
Round seed	5,474	Wrinkled seed	1,850	2.96:1
Yellow seed contents	6,022	Green seed contents	2,001	3.01:1
Gray seed coats	705	White seed coats	224	3.15:1
Inflated pods	882	Constricted pods	299	2.95:1
Green pods	428	Yellow pods	152	2.82:1
Axial flowers	651	Terminal flowers	207	2.14:1
Long stems	787	Short stems	277	2.84:1

to 3.15:1. Like any statistician, Mendel concluded that these ratios really represent a constant 3:1 ratio. To put the matter less flippantly, any statistically minded inquirer will not try to explain the exact ratio actually obtained in a particular experiment but will try to discover why similar experiments always yield a ratio of about 3:1. That is exactly what Mendel did.

The results expected from various crosses can readily be predicted according to Mendelian principles by use of a simple checkerboard diagram. Each column represents a different kind of male gamete, or reproductive cell. Each row represents a different kind of female gamete. Each box within the diagram represents one possible kind of zygote that can result from fusion of gametes. If there are four boxes in the diagram, then one-quarter, or 25 percent, of the offspring may be expected to be of the kind described by that box.

In the characters related to seed form and color of seed contents, Mendel found that most pods contained seeds of both characters. Similarly, the plants grown from seeds collected from a single pod usually show both characters studied in the other crosses. The ratio between dominant and recessive characters in seeds from a single pod or even in seeds from a single plant may be quite far from the average 3:1. For example, one of the F_1 plants in the first cross yielded 43 round F_2 seeds and only 2 wrinkled seeds. The 3:1 ratio is obtained only as an average of the results of many individual crosses.

The next year an F_3 generation was produced by self-fertilization of the F_2 plants. The F_2 individuals that showed recessive characters produced only F_3 individuals showing recessive characters. However, the F_2 individuals that showed dominant characters proved to be of two kinds. One-third of these dominant-character F_2 individuals produced only dominant-character F_3 individuals. The other two-thirds of the dominant-character F_2 individuals produced F_3 individuals with both dominant and recessive characters in the ratio 3:1. In other words, all of the recessive-character and one-third of the dominant-character F_2 individuals proved to be pure-breeding. The other two-thirds of the dominant-character F_2 individuals showed the same pattern of offspring as the F_1 generation had shown.

For example, all the wrinkled F_2 seeds grew into plants that yielded only wrinkled seeds when self-fertilized. Of 565 round F_2 seeds, 193 grew into plants that yielded only round seeds when self-fertilized. The other 372 round F_2 seeds grew into plants that yielded round and wrinkled seeds in a ratio of about 3:1 when self-fertilized.

The distribution of characters in the F_3 generation makes it clear that the 3:1 ratio of dominant to recessive characters in the F_2 generation actually reflects a 2:1:1 ratio. Among the offspring of the F_1 hybrids, one-half are F_2 hybrids showing the dominant character, one-quarter are pure-breeding dominants, and one-quarter are pure-breeding recessives. At the time he wrote his paper, Mendel had carried most of his crosses through four or five generations and found that the 2:1:1 ratio continues to appear among the offspring of hybrids in each generation, whereas the pure-breeding lines continue to produce offspring like their parents.

Mendel explained his observations by assuming that each seed contains two hereditary factors, or elements, affecting a particular character. One of these factors is obtained from each parent. For example, in the characters related to seed form, let *R* represent a factor for the round-seed character and *r* represent a factor for the wrinkled-seed character. The pure-breeding plants of the P generation can be represented as *RR* and *rr*. Cross-breeding

Figure 12.4a (above). A dihybrid cross, illustrating independent assortment, between a round, yellow pea plant (RRYY) and a wrinkled, green pea plant (rryy) carried through the F_2 generation. Those members of the F_1 generation are completely heterozygous (RrYy) for both traits and, when mated, yield 16 possible genetic combinations in four classes with the following phenotypic distribution: 9 yellow round, 3 yellow wrinkled, 3 green round, and 1 green wrinkled.

Figure 12.4b (below). Only the dihybrid cross is outlined here, but the distribution of F_2 genotypes is illustrated.

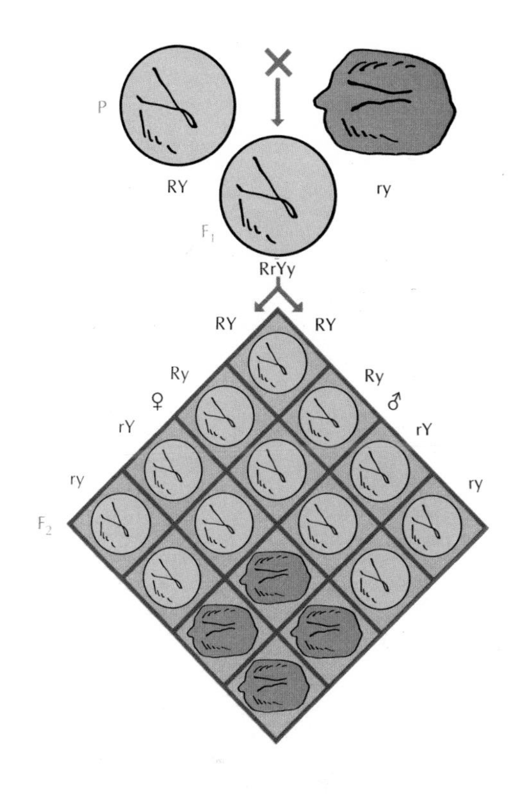

produces an F_1 seed that obtains one factor from each parent and thus is *Rr*. The seed becomes round because the dominant *R* is expressed despite the presence of the recessive *r*.

When the *Rr* seed grows into a plant, it will produce, in equal numbers, reproductive cells carrying *R* and cells carrying *r* factors. Four combinations are thus equally likely to be found among the F_2 seeds produced by self-fertilization: *RR, rR, Rr*, and *rr*. Thus, the F_2 generation will consist of pure-breeding dominants, hybrids, and pure-breeding recessives in the ratio 1:2:1, just as Mendel had found experimentally. Because half of the F_2 individuals possess exactly the same *Rr* factors that the F_1 generation possessed, further self-fertilizations of these individuals should produce the same ratios of offspring. Mendel's hypotheses can best be visualized through the simple checkerboard diagram that was developed by later workers in this field.

Next, Mendel carried out crosses involving two or more pairs of characters. In one experiment, for example, pure-breeding plants with round yellow seeds were crossed with pure-breeding plants with wrinkled green seeds. The plants of the P generation may be represented as *RRYY* and *rryy*. All of the reproductive cells from one strain carry the factors *RY* and all of those from the other strain carry *ry*. Thus, all the F_1 plants will possess the factors *RrYy*. As expected, all the F_1 seeds were round and yellow. If the factors are assorted randomly in the production of reproductive cells (assuming, however, that each reproductive cell gets only one of each kind of factor), four kinds of reproductive cells are possible: *RY, Ry, rY*, and *ry*. The checkerboard diagram shows that the F_2 individuals will have four different combinations of characters in the ratio 9 : 3 : 3 : 1. Mendel's experimental results confirmed this prediction.

Crosses involving more than 2 independently assorting factor pairs can be diagrammed in the same way, but the situation rapidly becomes more complex. With 3 pairs of factors, the predicted ratio of distinguishable kinds of individuals in the F_2 generation is 27 : 9 : 9 : 9 : 3 : 3 : 3 : 1. With 4 factor pairs, there are 16 possible kinds of individuals, and with 5 factor pairs the number of kinds is 32. With only 23 different, independently assorting pairs of factors, the offspring of a cross between multihybrid individuals could be of about 8 million different observable types. It is easy to see why Mendel was impressed with the ability of his simple hereditary factors to account for variability in nature.

In fact, his theoretical principles could account for the results Mendel obtained in numerous experiments. He counted over 10,000 plants in the 8

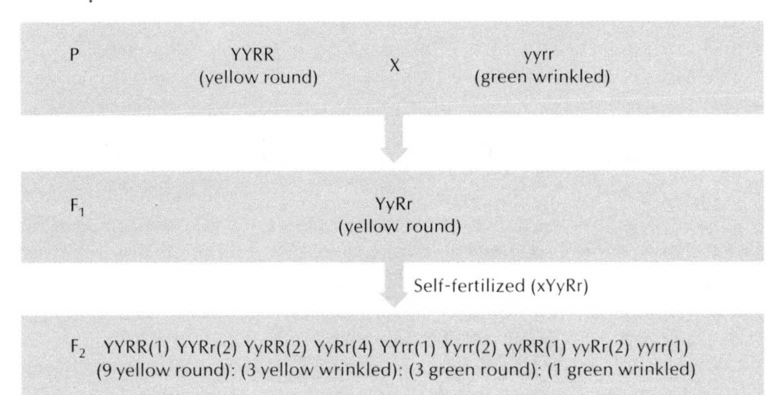

years he gathered data for his paper. Unfortunately, those biologists of his time who did read Mendel's paper were skeptical of his results and theories, and Mendel was unable to convince them to repeat his laborious experimental work in order to check his conclusions. Mendel's powerful principles, however, gave rise in the early years of this century to many of the basic ideas of *genetics*, as the study of inheritance then came to be called.

Other useful terms were introduced early in this century. Individuals showing the same observable characters are said to be of the same *phenotype;* those possessing the same set of hereditary factors, or *genes*, are said to be of the same *genotype*. Pea plants of different genotypes (*RR* and *Rr*) both display the same round-seed phenotype. An individual whose genotype comprises identical factors (*RR* or *rr*) is called *homozygous*. An individual with the hybrid genotype *Rr* is called *heterozygous*. The word "gene" was not introduced until 1909 and has meant various things at various times, but one of its original uses was for the factors within cells that produce the characters studied by Mendel.

REACTIONS TO MENDEL'S PAPER

Mendel did not live to take part in modern genetics. In his own lifetime, no one showed even a slight understanding of his work. This failure was a bitter disappointment to Mendel. Contrary to legend, his paper was discussed at those two evening meetings of the Brünn society, but no one really understood the significance of his conclusions. In a letter written the next year, Mendel commented sadly: "I encountered, as was to be expected, divided opinion; however, as far as I know, no one undertook to repeat the experiments."

On publication the following year, the paper quickly became widely available. Of the 115 copies of the journal sent to subscribers, 8 copies went to Berlin, 4 to the United States, and a couple to the Royal Society and the Linnean Society in England. In addition, 40 reprints of Mendel's article were ordered and presumably distributed to biologists around the world. A standard bibliography on plant hybridization published in 1881 mentions Mendel's paper several times.

A number of scientists must have read Mendel's paper, but for several reasons they were not impressed. First, no agreement existed over the details of animal and plant reproduction. Darwin, for example, believed that more than one pollen grain is required for a single fertilization in plants. Second, anyone glancing at the title of Mendel's paper and then skimming over its contents could well get the erroneous impression that Mendel intended the conclusions to apply only to pea plants. Mendel did discuss the results of some crosses among bean plants, which generally supported his results with peas. However, the beans also presented some new phenomena that Mendel had tentatively explained by making some further assumptions about the combined effects of several factors influencing the same character. Third, mathematics—however elementary—was seldom used in discussions of plant breeding. Naturalists would have been baffled by the formulae, and those scientists who could have followed Mendel's mathematical arguments were unlikely to bother reading an article on pea hybrids. Fourth, there was no immediate follow-up from Mendel—no further articles on his research, much less a book. When another article by Mendel finally did appear in 1870, the news was bad. In experiments with hawkweed (*Heiracium*), Mendel found that both the F_1 and the F_2 generations of a hybrid cross showed only the dominant character. The recessive character failed to reappear in later generations, a result that was consistent with older theories of blending inheritance rather than with Mendel's theory of discrete hereditary factors. It is now known that hawkweed usually reproduces asexually and that many of Mendel's supposed crosses of hawkweed plants were not crosses at all.

With hindsight, one might bemoan Mendel's bad luck in choosing hawkweed plants for his later experiments. However, it would be more realistic to marvel at his good luck in choosing pea plants for his initial experiments. Modern knowledge of plant genetics reveals that few plants could be as suitable as peas for demonstrating simple genetic properties. Even with peas, Mendel was fortunate, for many genes are linked together on chromosomes and therefore do not assort independently. The chance of anyone happening to choose seven independently inherited characters in peas is only 1 in 163.

Mendel made a concerted effort to explain his discoveries to the famous Swiss botanist and cell theorist Karl von Nägeli. Mendel sent his paper to Nägeli and followed it with ten letters, but to no avail. Nägeli's replies are lost, but he never accepted Mendel's major generalizations. This rejection may have been partly due to Nägeli's own experience with hawkweed. At any rate, all Nägeli did for Mendel was to urge him to concentrate on hawkweed and therefore to lose confidence in his earlier conclusions.

Nägeli was an evolutionist, but Mendel's letters to him make only passing or indirect reference to the problem of the origin of species. The nature of Mendel's thought on this question is difficult to establish. A copy of the German translation of Darwin's *On the Origin of Species*, annotated in Mendel's handwriting, has been found among Mendel's belongings, so it is clear that Mendel carefully studied Darwin's theories. However, the markings do not indicate how far Mendel agreed with Darwin's proposals. Mendel's classic paper points out that any conclusions about inheritance will have important implications for hypotheses about evolution, but that is all he says. Some passages in Mendel's paper seem to allude to Darwin's book, but they may be a response to an 1849 book written by the German plant hybridizer Carl Friedrich von Gärtner.

It is often stated that Darwin would have recognized the significance of Mendel's paper if he had read it. Confidence in such speculations is always misplaced and especially so in this case. Darwin may well have seen an early reference to Mendel's paper, but he did not pursue it. Had he done so, he probably would have made little sense of Mendel's arguments. In the first place, Darwin had difficulty with even simple equations. Second, Mendel's conclusions were in opposition to a longstanding conviction of Darwin's—drawn from the study of domesticated animals and plants—that the variability in any species is due to the disturbing effects of external conditions on the reproductive system. Third, by the time Mendel's paper appeared, Darwin had developed his own highly comprehensive theory of inheritance, "pangenesis," a theory he constructed to explain a far wider range of phenomena than Mendel had considered. Darwin, then, would probably have tried to explain Mendel's results in terms of his own broader theoretical scheme.

Hindsight suggests that Mendel would have had better luck in interesting Darwin's cousin Francis Galton. Galton, a keen Darwinian, loved

Figure 12.5. Erich von Tschermak, one of the men who rediscovered Mendel's work around the turn of the century.

mathematics and was already applying it to the study of inheritance. In his later work, Galton came close to postulating the existence of independently segregating hereditary factors. He and Darwin together might have discussed Mendel's work and tried to incorporate Mendel's principles into a genetical theory of natural selection. However, there is no evidence that Galton ever read Mendel's paper.

A man less reticent than Mendel would have sent reprints and letters to men like Darwin, Galton, and other evolutionists. But that was not Mendel's way. Part of his reticence may have been due to prudence rather than to shyness. It may have been risky for a priest and teacher to take a stand on such issues. Mendel surely had memories of the attacks on his former professor Unger. Yet, on the other hand, Mendel was never one to dodge opposition.

The neglect of Mendel's work was a personal tragedy. His tragic flaw was his modesty. The truly effective theorist is humble before facts but just a little arrogant before his fellow scientists. Mendel's admirers often say that he alone founded the entire science of genetics—a sincere compliment to a deserving hero, but also an ironic misstatement of the historical facts. Mendel's whole life is a demonstration of the fact that it takes more than one man—however talented—to initiate a new field of science. Genetics was not established as a science until Mendel's principles were rediscovered by a number of other scientists around 1900, and there was a further delay of 30 years before Mendelian inheritance and Darwinian natural selection were coupled finally into a widely acceptable evolutionary theory.

REDISCOVERY OF MENDEL'S WORK

According to the usual account, Mendel's principles were independently rediscovered about 1900 by three men—Hugo De Vries, Carl Correns, and Erich von Tschermak. Each of these men found at the last minute before publication of his own paper that he had been scooped 40 years earlier. The republication of Mendelian theories supposedly shed light upon the nature of heredity and evolution and led to rapid and unified progress in the field of genetics.

The timing of this rediscovery was no chance affair. In the 40 years following Mendel's publication, a series of fundamental innovations in biological theory had prompted a general search for universal laws governing the inheritance of single characters. By 1900 the cell theory was generally established, the nature and role of male and female sex cells in fertilization had been clarified, and nuclear fusion had been identified as the crucial event in fertilization. Several diverse hypotheses had proposed the existence of subunits within the cell nucleus, and some theorists had suggested that these submicroscopic particles are involved in the transmission of hereditary characters. Many biologists were aware of the central role of chromosomes in cell division, and some had speculated that the chromosomes might be involved in heredity. In short, although Mendelian principles were far from being obvious consequences of the basic biological theories of 1900, these theories were compatible with Mendelian principles, whereas the prevailing theories of 1866 had not been.

Furthermore, the biologists studying heredity at the turn of the century were far more ready to accept mathematical concepts than had been the plant breeders of Mendel's day. Statistical analyses were being used in a variety of biological researches, and it is not surprising that a number of investigators independently realized that a better understanding of heredity might come through statistical analysis of the offspring characters in crossing experiments.

However, rediscovery of Mendel's principles did not provide instant clarification of the problem of heredity. Although some investigators using Mendelian theories did make rapid progress toward an understanding of the nature of the gene, many biologists remained unconvinced of the validity of Mendel's principles for more than a decade after their rediscovery. Like Nägeli with his hawkweed, many biologists working with organisms other than the pea plant knew from firsthand experience that Mendel's principles in their simple form could account for only a few of the phenomena of heredity.

Nor did the rediscovery do anything to settle disputes among evolutionary theorists. Far from it. For various reasons, there were numerous rivals to the Darwinian theory of natural selection by 1900. Many of the Mendelians became vigorous opponents of Darwinian selectionists. De Vries' mutation theory was one attempt to explain the origin of new species by a mechanism other than natural selection.

Only in the 1930s did geneticists reach sufficient understanding of the distribution of genes in a population over many generations to permit a satisfactory synthesis of the theories of evolution and genetics. Thus, it was nearly two decades after the rediscovery before Mendelian principles of inheritance were generally accepted (in a modified form) and more than three decades before those principles were used to provide a convincing explanation of the mechanism of evolution.

Before turning to the developments in genetics research after 1900, it will be useful to examine the nuclear processes involved in cell division, many of which were first described during the period between Mendel's original publication and its later rediscovery.

FURTHER READING

Moore (1963) in his first 18 chapters gives a brief and clear account of the development of genetics from Mendel through the early 1960s. Olby (1966) presents a full account of the hereditary theories of Darwin and other nineteenth-century biologists.

Stern and Sherwood (1966) and Voeller (1968) furnish useful collections of important historical papers in English translations. Interesting and contrasting early interpretations of Mendelian principles are given by Bateson (1913) and Morgan, *et al.* (1915).

13
Cell Division

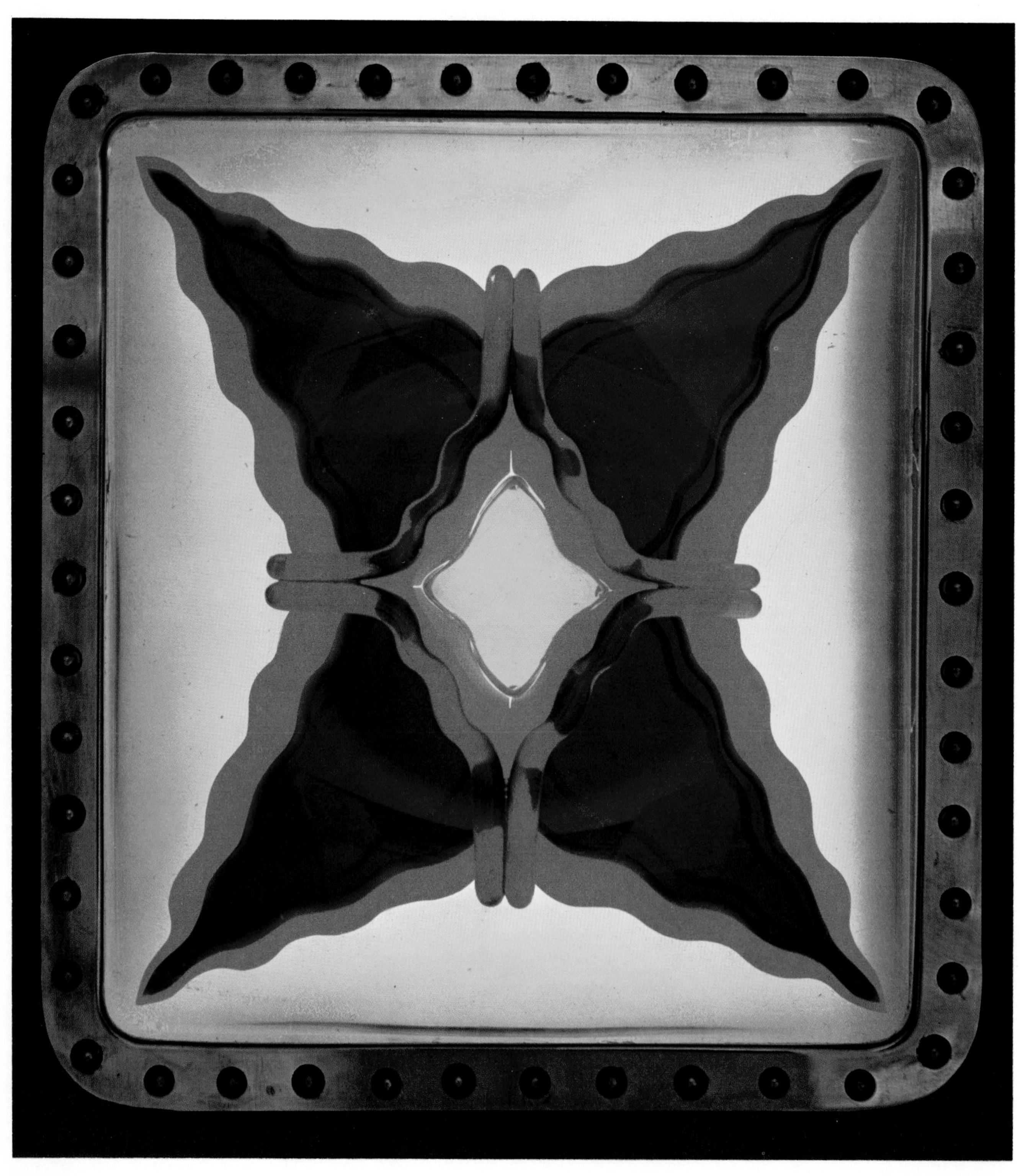

In 1866 Mendel wrote of hypothetical hereditary factors carried by the pollen cell and the female germinal cell. Two factors affecting each trait are combined when pollen and germinal cells fuse during fertilization. The new plant that develops shows characters related to the pairs of factors that it has received. When it produces pollen and germinal cells, it contributes one factor for each trait to each of these cells. With the knowledge of cells and of reproductive processes available at the time, Mendel could not go much further in proposing a physical mechanism of heredity.

During the years following Mendel, cytologists made tremendous progress in filling out the details of cell theory. Against the new background of knowledge about cell and organism reproduction, Mendel's principles took on physical significance for their twentieth-century rediscoverers. Many twentieth-century geneticists shifted their attention toward an attempt to locate the hereditary factors within the cell and to understand how the factors are carried and assorted. This physical understanding proved crucial in the further extension of Mendel's laws to accommodate a number of observations that are not compatible with those laws as Mendel stated them.

The most detailed nineteenth-century descriptions of division in animal cells were made by the German anatomist Walther Flemming, who used the newly developed oil-immersion microscope lens. He showed that the threadlike structures split lengthwise and that half of each thread moves into each of the daughter nuclei (Figure 13.2). Flemming introduced the term *mitosis* (the Greek *mitos* means thread) to describe this orderly process of nucleus and cell division. He called the dark staining material within the nucleus "chromatin." The name "chromosomes" for the threadlike bodies composed of chromatin was suggested a few years later by another researcher.

During the 1880s, Flemming and other microscopists worked out most of the details of the mitosis process. These experimenters discovered that every cell of an organism contains the same number of chromosomes. This number is reduced during the formation of egg and sperm cells; fusion of sperm and egg nuclei during fertilization restores the full chromosome number in the offspring. The process by which chromosome number is halved during formation of sperm and egg cells was described in detail. This special kind of cell division is now called *meiosis* (the Greek *meioûn* means to lessen).

HAPLOIDY AND DIPLOIDY

Chromosomes are present in the nucleus of every eucaryotic cell. In a multicellular organism, every cell (except sperm and egg cells) contains exactly the same number of chromosomes as any other cell in the organism. Reproduction of most unicellular organisms and multiplication of cells within a multicellular organism are accomplished through the process of cell division called *mitosis*. In this process, each daughter cell receives a set of chromosomes that matches the set present in the parent cell. The special process of division called *meiosis* results in a halving of the chromosome number in the daughter cells. Such a process is necessary for sexually reproducing organisms if each generation is to have the same number of chromosomes in its cells as the parents had.

The processes of mitosis and meiosis do not occur in procaryotic cells. These cells have neither a nucleus nor distinct, threadlike chromosomes.

Figure 13.1. Early drawings of the mitosis process.

Figure 13.2. The Flemming bodies shown in this micrograph were named after the German anatomist Walther Flemming, who introduced the term "mitosis."

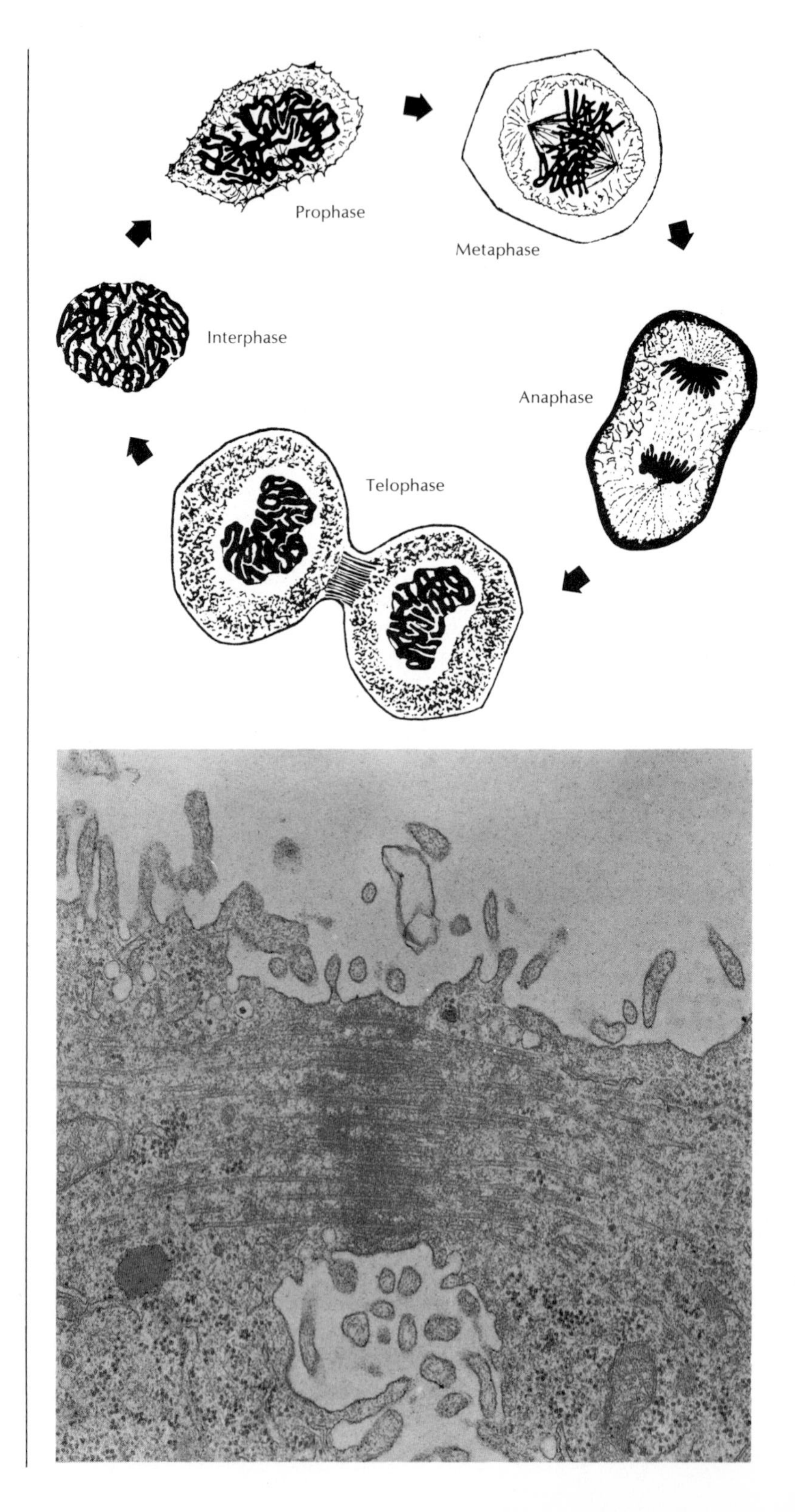

The procaryotic equivalent of a chromosome is a single molecule of DNA. This molecule is duplicated before cell division, and during fission one copy moves into each daughter cell. The mechanism by which the copies are properly divided between the daughter cells is not known.

Most eucaryotic cells contain a number of chromosomes that can be distinguished microscopically on the basis of size and shape. In 1900 it was thought that all chromosomes within a nucleus are equivalent, but cytologists soon showed that the chromosomes are of distinct kinds and that most cells contain a pair of each kind of chromosome. Meiosis results in the formation of sperm and egg cells with only a single chromosome of each kind. A cell with only one chromosome of each kind is called a *haploid* cell, and the number of chromosomes that it contains is called the *haploid number* (n) for the species. The zygote, or fertilized egg cell, is formed by fusion of sperm and egg cells and thus contains $2n$ chromosomes—a pair of each kind. This condition is called *diploidy*. In most higher plants and animals, body cells are diploid throughout the life cycle. Haploid sex cells are produced by meiotic divisions of certain specialized body cells. Diploidy is restored after fertilization.

Other patterns are found in the life cycles of many simpler plants and animals and of most protists. For example, the cells of many organisms are haploid through much of the life cycle. The fusion of haploid sex cells forms a diploid zygote, which quickly undergoes meiosis to restore the haploid condition in the new organism. In unicellular organisms reproducing asexually, the diploid condition may be produced briefly by a duplication of the chromosomes just before cell division. Nevertheless, the division involved in reproduction restores the haploid number of chromosomes in the offspring cells.

MITOSIS

The process of eucaryotic cell reproduction involves a complex mechanism that ensures the proper duplication and division of the chromosomes. For descriptive purposes, the process of mitosis has been divided into five stages, or phases (Figure 13.3). In living cells, mitosis proceeds smoothly and without sharp changes from one stage to the next; the stages are thus artificial descriptive aids, not inherent discontinuities in the process.

The nondividing phase in which no chromosomes are visible with the light microscope is called *interphase*. Under the light microscope, the chromatin appears granular or in the form of very thin, randomly coiling threads. Early investigators called this phase the resting stage, but subsequent research has shown that it is, in fact, a period of active preparation for the coming cell division.

Under the light microscope, the first visible signs of mitosis occur during *prophase*, as the chromosomes become distinctly visible. The chromatin appears to gather into long, thin, highly twisted chromosomes, each made up of two coiled filaments called *chromatids*, which lie side by side and are joined at a point called the *centromere*. Early in prophase, the cell becomes rounder, and the cytoplasm becomes denser and more viscous. As prophase progresses, the chromatids become shorter and thicker, the centromeres become more obvious, and the chromosomes move toward the nuclear membrane. The *spindle*—a structure made up of thin filaments stretching across the cell between two poles—begins to form. The nuclear membrane disintegrates, and the chromosomes move toward the center of the

Figure 13.3. Shown here is a sequence of light microscope photographs of mitotic stages in the division of an onion root cell. Mitosis is a continuous, dynamic process, and this sequence must be viewed in that context. *(Courtesy Carolina Biological Supply Company)*

1. Interphase. Note the darkly stained nucleus, containing what appears to be granular chromatin. Often called the "resting stage", this is a period of high molecular activity within the cell nucleus.
2. Prophase. Chromosomes are quite distinct and, as depicted in the drawing, are composed of coiled, threadlike filaments called chromatids.
3. Early metaphase. Chromosomes are beginning to become oriented on the equatorial plate with their "arms" pointing toward the poles.
4. Late metaphase. Chromosomes have become noticeably oriented on the equatorial plate.

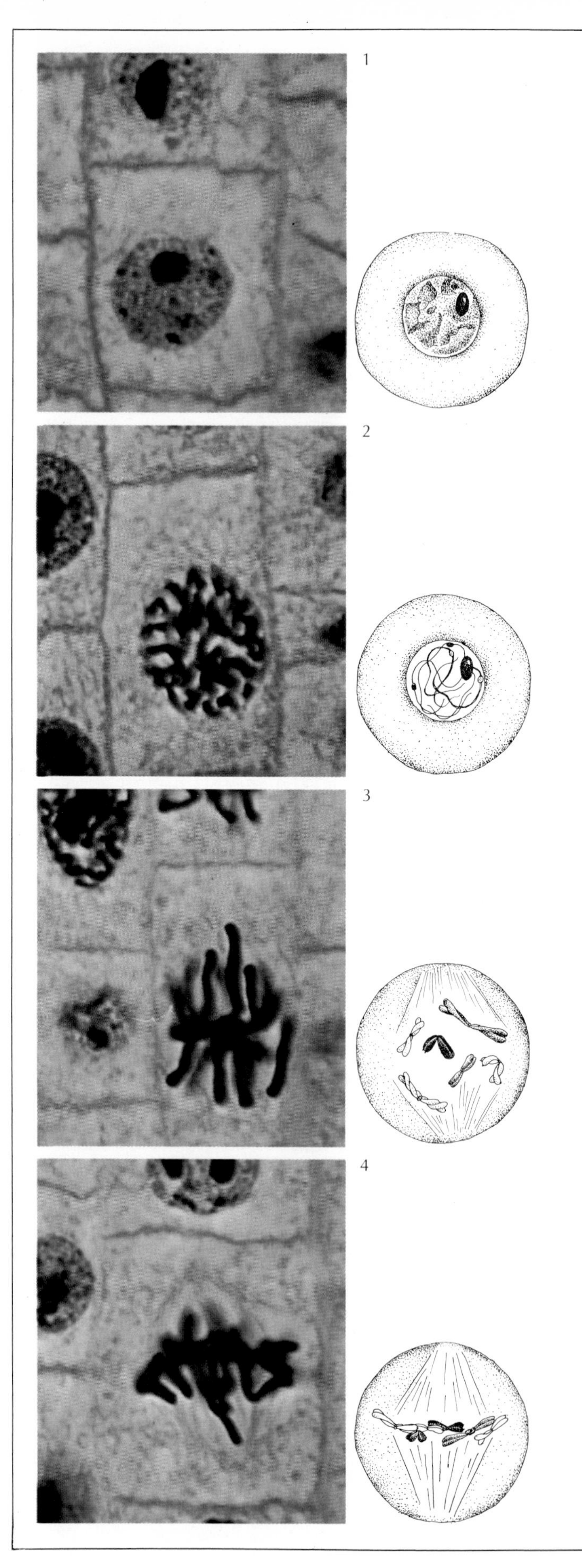

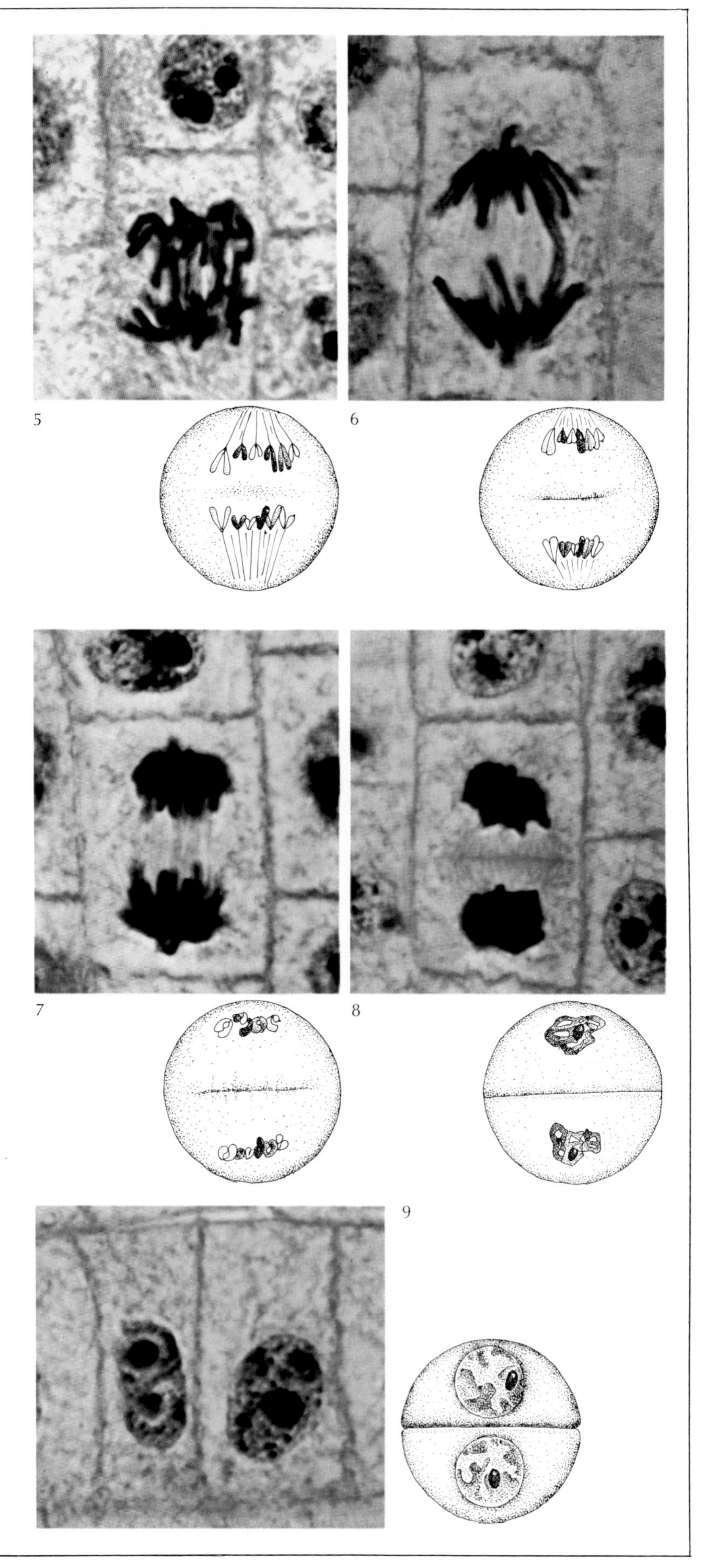

5. Early anaphase. Chromatids are beginning to separate, with division of the centromere joining the two chromatids of each chromosome.
6. Late anaphase. Separation of chromatids continues with movement occurring toward the poles. The spindle filaments attached to each chromatid appear to be pulling them from the equatorial plate.
7. Early telophase. The clumping of daughter chromatids at the poles can be seen. Spindle filaments joining chromatids across the equatorial plate are also visible. These spindle filaments apparently pull the chromatids apart during their movement across the cell.
8. Late telophase. Cytoplasmic division is visible in this photograph, with a cell plate forming along the equatorial plate. Individual chromosomes have lost their definition, and the formation of daughter nuclei is in progress.
9. Interphase (daughter cells). Two new identical daughter cells have formed. These are conspicuous from surrounding cells by their reduced size.

Figure 13.4. Phase-microscope photograph of isolated mitotic apparatuses from sea urchin eggs magnified 1,000 diameters. In the center of the photograph is a mitotic apparatus composed of a spindle with asters at either end. The light area in the center of the asters represents the mitotic poles. (× 450)

cell, where they arrange themselves on a plane perpendicular to the spindle axis about midway between the two poles. When the chromosomes have become noticeably oriented on the *equatorial plate* (the imaginary plane described above), the cell is said to be in *metaphase*. By metaphase, some of the filaments of the spindle have become attached to the centromeres of the chromosomes.

The next phase, or *anaphase*, begins with the separation of the chromatids. The centromere appears to split in two, freeing the two chromatids that make up each chromosome. The chromatids move to opposite poles of the spindle, where they gather into compact groups. During this movement, the chromatids look as if they are being dragged across the cell by the spindle filaments attached to the centromere. The filaments between the two groups of chromatids, or daughter chromosomes, look as if they were being stretched.

The gathering of the daughter chromosomes at the two poles marks the beginning of *telophase*, during which the chromosomes become longer and thinner, nuclear membranes form around the two groups of chromosomes, and finally the cytoplasm is divided to form two daughter cells, each containing one of the newly formed nuclei. The two daughter cells now return to the stage of interphase.

Mitosis is a continuous, dynamic process. The cycle of events has no beginning or end. Interphase offers a convenient starting place for description because the cell structure seems simplest in this stage and because most cells spend about two-thirds of their lives in interphase.

The Spindle

The spindle was so named because it reminded early investigators of the old-fashioned spindle used in making thread by hand. Its shape is similar to two cones joined together at their bases. The poles are the vertices of the cones, and the chromosomes at metaphase typically lie where the bases of the two cones make contact. Short chromosomes may lie entirely within the equatorial plate at metaphase; longer chromosomes usually have only their centromeres on the plate, with the arms extending into the cones or oriented more-or-less at random.

The structure of the spindle is not yet well understood. In many kinds of cells, filaments or fibers radiate into the cytoplasm from the poles, forming *asters*. The spindle, the asters, and the centrioles (in those cells that possess them) make up the *mitotic apparatus*. The mitotic apparatus has been isolated from some kinds of cells during division (Figure 13.4). Analysis of the isolated apparatus shows that the main component is a protein. Water, RNA, and ATP are also present in significant amounts (Mazia, 1955). As much as 15 percent of the protein in the cytoplasm may be utilized in the formation of the mitotic apparatus. The organization of the protein into fibers probably involves the formation of disulfide bridges that cross-link the various protein molecules. Microscopic studies suggest that the centromeres—and, in animal cells, the centrioles—play a role in the formation of the spindle fibers. It is not yet known whether the RNA plays a role in mitosis.

There is still considerable debate about how the chromosomes are separated in the spindle. The protein fibers of the spindle appear in electron micrographs as microtubules. Visual observation in the light microscope suggests that the spindle fibers contract between the chromosomes and

Figure 13.5. Micrographs of the various stages of mitosis in fixed and stained pollen grains (microspores) of a spiderwort plant (*Tradescantia*). (a) Pollen grain of a spiderwort plant (*Tradescantia*). (b) A cell with one interphase nucleus, containing many chromatin strands but no distinguishable nucleus. (c) A flattened cell with a nucleus in prophase. Thick, intertangled chromosomal threads are clearly visible. (d) Late prophase with six doubled chromosomes. Each chromosome can be clearly seen to be composed of two chromatids. (e) Late anaphase with two groups of chromosomes clustered at the poles of the spindle (not visible). (f) Late telophase with nuclear membranes formed around the two sets of daughter chromosomes. Note the disparity in size between the two nuclei (see text for explanation).

the poles and then elongate between the pairs of chromosomes, with the fibers pushing or pulling the chromosomes apart. There is some evidence that ATP may be necessary for the contraction and elongation of the spindle fibers. Electron micrographs confirm that spindle fibers attach to the chromosomes at the centromeres. It has been postulated that the fibers leading from centromere to pole begin to be formed at the centromere and extend toward the pole as they grow. Some investigators have suggested that the fibers contract with expenditure of ATP energy, perhaps in a process analogous to the contraction of myofibrils in muscle cells. Others have suggested that the fibers between centromere and pole decrease in length by loss of protein molecules, whereas those between centromeres grow longer by addition of protein molecules. None of the many hypotheses under consideration has been confirmed experimentally, and the mechanism of chromosome movement during anaphase remains a mystery.

Chromosomal Changes During Mitosis

Figure 13.5 shows the various stages of mitosis observed in fixed and stained pollen grains (microspores) of a spiderwort plant (*Tradescantia*). In Figure 13.5a, the cell contains one interphase nucleus with many chromatin strands but no distinguishable chromosomes. Figure 13.5b shows a cell that has been somewhat flattened to make the structures more visible. The nucleus of this cell is in early prophase, with thick chromosomal threads that are still long, contorted, and intertangled. Figure 13.5c is a nucleus in late prophase with six separate chromosomes, each clearly double. Figure 13.5d shows a metaphase cell in a view looking toward one pole of the spindle, showing the six double chromosomes lying on the equatorial plate. Each chromosome can be clearly seen to be composed of two chromatids. The spindle itself is not made visible by the staining procedures used in preparing these cells. Figure 13.5e shows a cell in late anaphase with the two groups of chromosomes clustered at the poles of the spindle. Because of restricted space, the spindle is rather short in this cell. Figure 13.5f shows a pollen grain in late telophase, with nuclear membranes

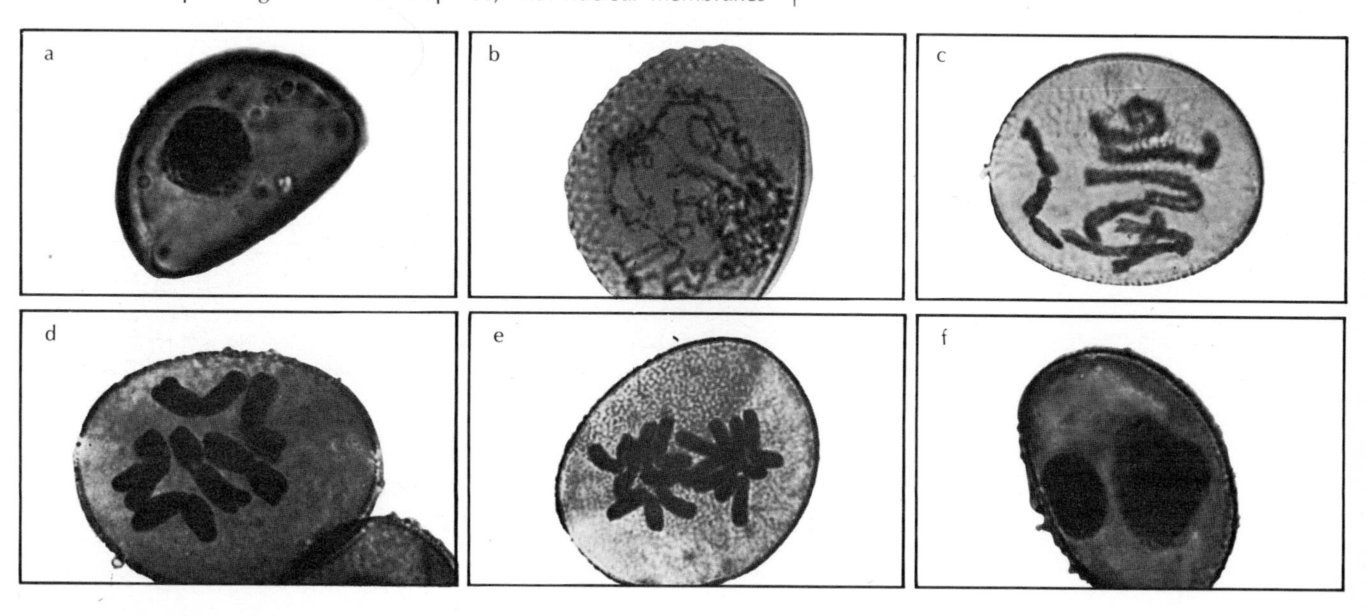

Figure 13.6 (above). Prophase chromosomes of a root tip cell from the May Apple plant (*Podophyllum*). Note the ribbonlike nature of the chromatid pairs.

Figure 13.7 (below). Metaphase chromosomes from a cell treated with colchicine. The spindle has been broken by the drug, and the metaphase position of the chromosomes has been altered somewhat by flattening the cell. Various morphological features of chromosomes can be identified by the colchicine-metaphase technique (see text for explanation).

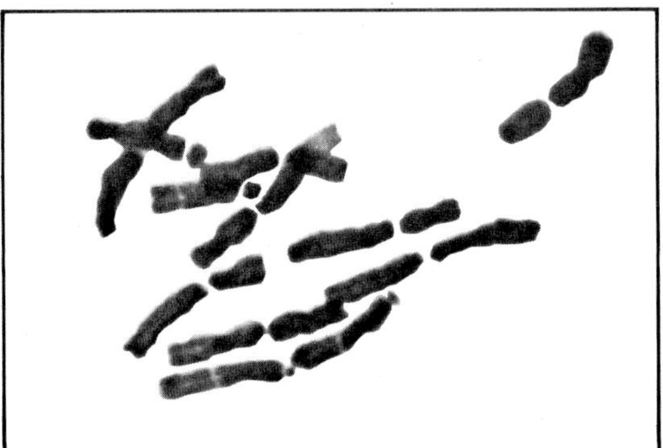

formed around the two sets of daughter chromosomes. The two nuclei contain the same number of chromosomes, and, as far as can be determined, their two sets of chromosomes are identical. As can be seen, however, the two nuclei already show a difference in appearance and will perform different functions. The larger of the two nuclei will never divide again, but it has metabolic functions in programming the further development of the pollen grain. The smaller nucleus is in a separate cell, and it will later divide again and form the two sperm nuclei that can function in fertilization of an appropriate egg cell.

Figure 13.6 shows the prophase chromosomes of a cell from a root tip of the May apple plant (*Podophyllum*). The pairs of chromatids, which become more distinct in later phases of mitosis, can be seen in this micrograph. The group of chromosomes in Figure 13.7 are in a cell that has been treated with the drug *colchicine*, an alkaloid that blocks mitosis at metaphase. The chromosomes are in the metaphase position but freed from the spindle fibers and oriented by the cells having been flattened before the photograph was taken. This group of chromosomes at the colchicine-metaphase illustrates some of the morphological features of chromosomes.

Each chromosome has at least one constriction, which appears as a light spot in the chromosome. This narrow region is the centromere, or spindle attachment. The two parts on each side of the centromere are called the *arms*. The overall length of a chromosome and the ratio of its arm lengths are identifying features. In a diploid cell, such as this root cell, there are two chromosomes of each type. The pair of similar chromosomes are called homologous chromosomes, or simply *homologs*. In bisexual organisms, one homolog is inherited from each parent.

In some cells, each kind of chromosome can be distinguished from the others by its size and shape. Various species of mammals have from 17 to more than 40 kinds of chromosomes, and some of these may appear identical under the light microscope. On the other hand, the six homologs of the May apple cell are readily distinguishable. Four of the six different homologs have a single constriction, or nonstaining gap. The other two homologous pairs (four chromosomes, each with two chromatids) have two constrictions each. In this species, the two shortest chromosomes have two constrictions. One constriction is the centromere; the other is the nucleolar organizer region. The May apple nucleus may have four nucleoli, one at each organizer region, or in some cells two or more of the nucleoli may fuse; therefore, the number of nucleoli may vary from one to four. The nucleoli disappear during prophase and are re-formed during telophase. During metaphase and anaphase, the nucleolar organizer regions appear as nonstaining gaps in the chromosomes. Determination of which gap represents the centromere and which represents the nucleolar organizer can be made by observing the cells in true metaphase to see where the spindle fibers attach. In the two chromosomes at the lower right of Figure 13.7, the constriction near the middle is the nucleolar organizer, and the constriction near one end is the centromere.

STUDIES OF THE CELL CYCLE

A cell spends most of its time in interphase. Because few events associated with division can be observed with the light microscope in interphase cells, this stage was long thought to be a period of mitotic inactivity. However, the interphase stage now is known to be the active, metabolic stage when

Figure 13.8. The diagram illustrates the life cycle of a "typical" cell. Note the great disparity between interphase and the other cell division stages. The total time required to complete one cycle varies from one cell type to another and with varying environmental conditions, such as temperature.

most or all of the components of the cell are synthesized. In a cell preparing for division, nearly every component may be doubled in amount before prophase. The DNA molecules of the chromosomes are replicated during interphase, and the mitotic process may be regarded as an intricate mechanism that precisely divides the two sets of DNA between the daughter cells. The other parts of the cell are divided less precisely, and the two daughter cells may differ in size and in cytoplasmic components.

A technique called autoradiography provided the first tool for the study of DNA synthesis during interphase (Interleaf 13.1). This technique reveals that there is an interval, or gap (called the *G_1 period*), of several hours after the end of telophase during which no DNA but much RNA and protein is synthesized. In the bean root cells used in the original studies, the G_1 period lasts about six hours after the start of interphase. The G_1 period is followed by an *S-phase* of DNA synthesis (chromatin replication), which lasts for about six to eight hours. After the S-phase, there is another gap, the *G_2 period*, of about six hours during which no DNA synthesis occurs. The G_2 period ends with the beginning of prophase.

Studies of DNA synthesis in many kinds of plant and animal cells have shown the same general cycle of activity that was observed in the bean root cells (Figure 13.8). In procaryotes, however, similar studies show that DNA synthesis proceeds almost continuously during the short life of the cell.

Autoradiographic studies have revealed a great deal about the events that occur during interphase, when few changes can be observed directly with the light miscroscope. It is now clear that "resting stage" is an extremely inaccurate description for interphase.

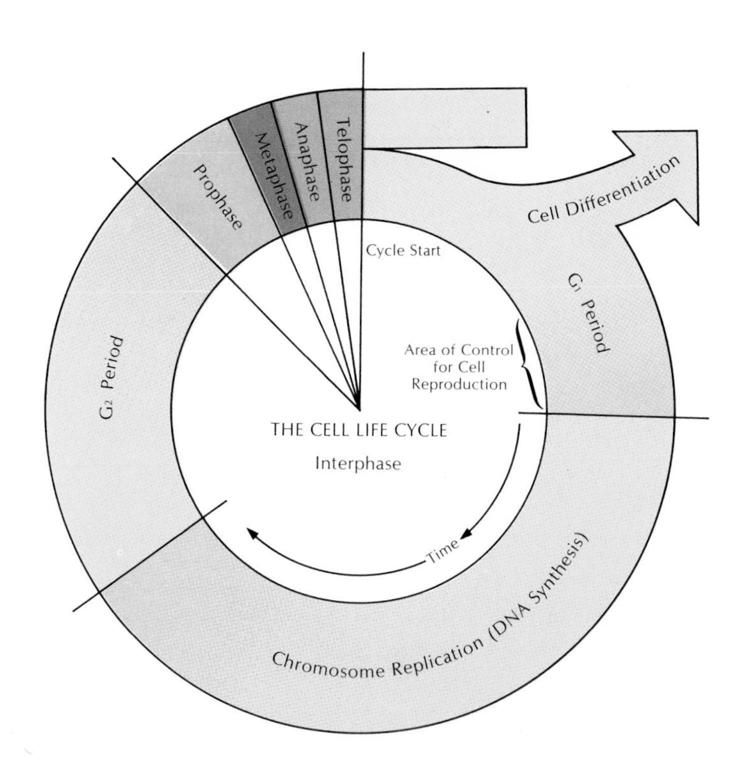

Figure 13.9. Autoradiographs of a cell labeled with H^3-cytidine, which is utilized in the synthesis of both DNA and RNA. In the autoradiograph at left, note that only interphase nuclei are labeled, as cells in other stages of division do not incorporate cytidine into the nucleic acid structure. On the right, the cytoplasm is labeled.

Interleaf 13.1
AUTORADIOGRAPHY

In the technique of autoradiography, a cell is "labeled" by radioactive material that is incorporated into particular cellular structures. The cell is then placed in close contact with photographic film, which is exposed by the radiation. When the film is developed, it shows a picture of the cellular structures into which the radioactive material was incorporated.

The first available radioactive labeling material was phosphorus-32. P^{32}-phosphate supplied to a cell is incorporated into nucleotides, the building blocks of DNA. If the cell is synthesizing DNA, some of the radioactive nucleotides will be built into the new DNA molecules as they form. When the photographic film (which has been placed in close contact with the cell for an appropriate time) is developed, black silver grains will appear wherever radiation (beta rays) from decaying atoms of P^{32} struck the film. If the radioactive thymidine has been built into DNA molecules, the dark grains will lie in long lines. If no DNA is being synthesized, the dark grains will be scattered randomly through the cell.

The technique of autoradiography was further improved through use of carbon-14 and hydrogen-3 (tritium) as labels. The lower energy radiations from these isotopes penetrate the film for shorter distances, permitting more exact location of the radioactive structures. J. H. Taylor and his colleagues first prepared thymidine labeled with H^3 and used this method to study DNA synthesis in cells. Since then, many other precursor compounds labeled with H^3 have been used to study the synthesis of DNA, RNA, proteins, and other large molecules in the cell.

The simplicity and elegance of autoradiography have made this technique very popular among cell biologists. Many new features of cellular reproduction and metabolism have been revealed through autoradiographic studies. Figure 13.9 is an autoradiograph of a cell labeled with H^3-cytidine, which is utilized in the synthesis of both DNA and RNA. The photograph actually is taken through a light microscope, looking at the developed film that had been in contact with the cell. The radioactive atoms that produced the black spots (silver grains) over this nucleus were incorporat-

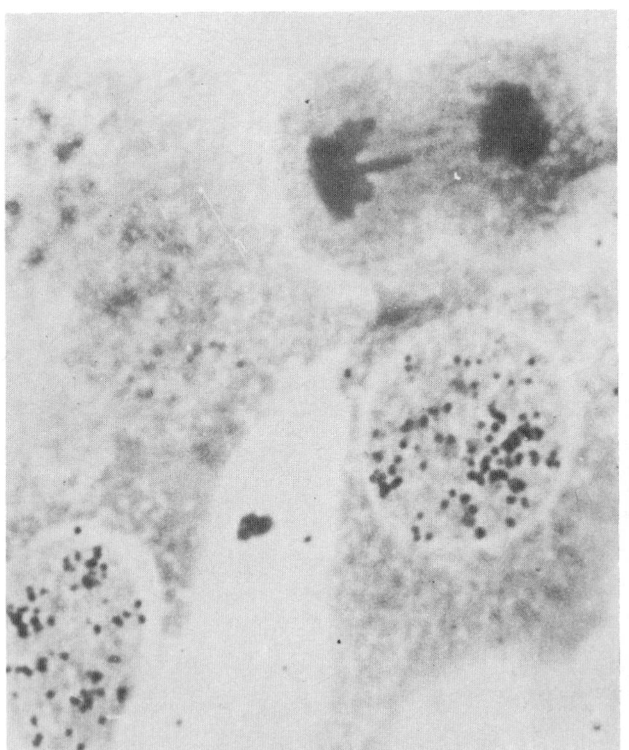

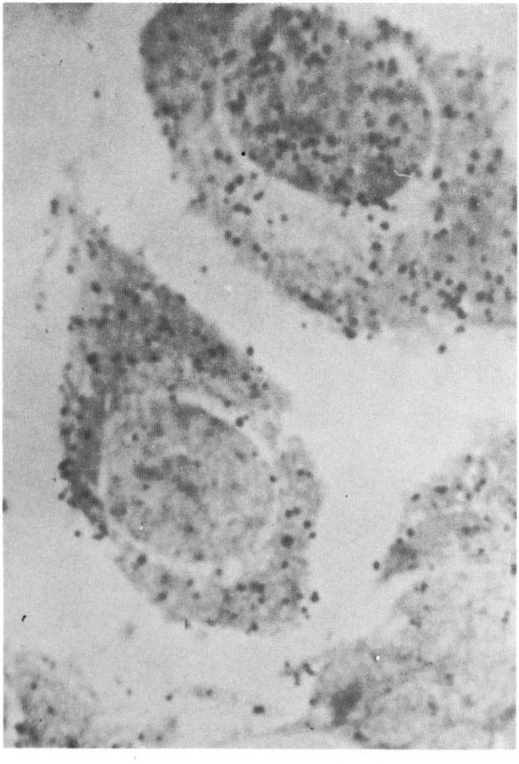

Figure 13.10. Living cancer cells (HeLa strain) attached to a glass surface and growing in a monolayer culture. The ovoid cell in the center is in mitosis (metaphase). Note the condensed chromosomes aligned on the equatorial plate. The bright halo effect surrounding the mitotic cell is characteristic of rounded objects viewed by phase contrast, which is the type of microscopy used here.

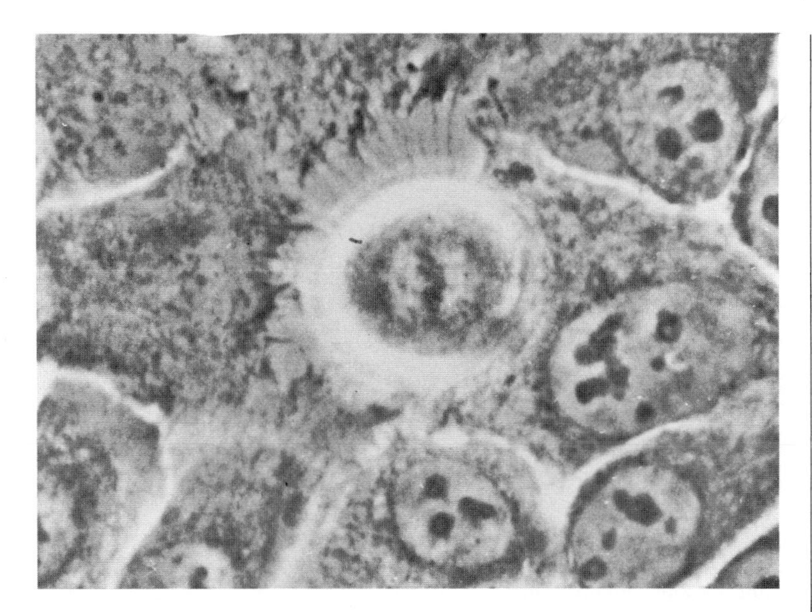

ed primarily in RNA molecules. This fact is not apparent from observation of the photograph but is determined from a knowledge of the experimental procedures and the intermediate substances from which the cell originally received the isotope.

TISSUE CULTURE

Cells can often be isolated from tissues and grown in cultures, much as bacteria are grown in the laboratory. Through repeated mitotic divisions, a single cell may produce a population of cells very similar in both appearance and behavior. Such cells grow independently in the culture, much as microorganisms do. Because all of the cells in the culture behave similarly, it is possible to remove samples of cells at intervals and examine them with various techniques to determine what the entire population is doing at each stage.

The same technique can be used with some populations of cells taken from a particular organ or tissue—for example, cells from the root of a plant or from the liver or kidney of an animal. However, the populations of cells from such organs are not as uniform as cloned cells—that is, cells derived by division from a single cell.

Vertebrate animal cells grown in cultures have many uses in addition to the study of cell division. For example, they are useful in the study of the growth and reproduction of viruses. Such cells have been used as hosts for growing viruses for the production of vaccines. Many cancer cells have been grown in culture, and studies of such cells may eventually contribute the necessary links for understanding the nature of cancer.

Isolated cells in culture typically grow best when they can attach to glass or certain plastic surfaces (Figure 13.10). However, experimenters have isolated a few kinds of cells that can grow suspended in the appropriate fluids. By using either attached or free cells, the experimenter can vary the cell's environment in ways that are impossible within the whole organism. Isotopic labels can be supplied or removed. Samples of the population can be harvested for experimental study, leaving a supply of similar or identical cells for future studies or for the production of new cultures. The cells can be frozen by appropriate methods and stored for years at low temperatures; when thawed, they will grow and initiate new cultures.

Over the past few decades, cell and tissue culture studies have complemented the studies of cells taken from living organisms. Each kind of experimentation contributes some information to the growing knowledge of cell biology.

Figure 13.11 (left). Untagged chromosomes of *Bellevalia* in metaphase. Note how large and distinct the eight chromosome pairs are in this specimen.

Figure 13.12 (right). Radioactively tagged chromosomes of *Bellevalia* in metaphase. These chromosomes have duplicated once in the radioactive solution, and both members of each pair are labeled.

In most studies of fixed and stained tissues, about two-thirds of the cells are found to be in interphase even in tissues where rapid division is known to occur. Therefore, early investigators assumed that a cell spends about two-thirds of the cell cycle in interphase. Studies with autoradiographic techniques and time-lapse cinematography have confirmed this picture of the cell cycle. The time required for a complete cell cycle and the amount of time required for each stage in the cycle vary greatly among different kinds of cells but are extremely constant among different individual cells of the same kind. The length of the cell cycle does vary with physiological conditions and temperature. Motion pictures of cell division and experiments dealing with the effects of temperature on cell division have been made possible by techniques of tissue culture (Interleaf 13.2).

The length of the cell cycle varies from less than an hour to more than 20 hours in various kinds of cells. In most cases, interphase occupies 60 to 95 percent of the cell cycle. The other phases of the cycle are accomplished fairly rapidly, usually within 10 to 60 minutes. About two-thirds of this time is devoted to prophase and about one-third to telophase. Metaphase and anaphase occupy small periods of time within the mitotic cycle.

Chromosomal Replication

Synthesis of the major components of the chromosomes occurs during S-phase, although some events essential to chromosomal reproduction may occur in G_2 or in prophase. In addition to DNA replication, the synthesis of the basic proteins (histones) associated with chromosomal DNA occurs during S-phase. RNA and other protein components of the nucleus are synthesized throughout interphase.

Studies in which selective labeling of RNA was used show that nearly all of the RNA synthesized in the nucleus is either transported out to the cytoplasm or broken down in the nucleus within a few hours. The histone molecules are as stable as the DNA with which they are associated. Molecules of DNA and histone remain intact for many cell generations, although additional new molecules are synthesized in each cycle of cell division.

Chromosome replication has been extensively studied through the use of H^3-thymidine labeling. This substance is known to be used almost exclusively in DNA synthesis. Root cells of several plants were the first test subjects for H^3-thymidine labeling studies. Because the plant *Bellevalia* has only eight large chromosomes in each cell, particularly clear results were obtained from studies of its root cells (Figure 13.11). Bulbs and roots of the plant were immersed in a solution containing H^3-thymidine for one to two

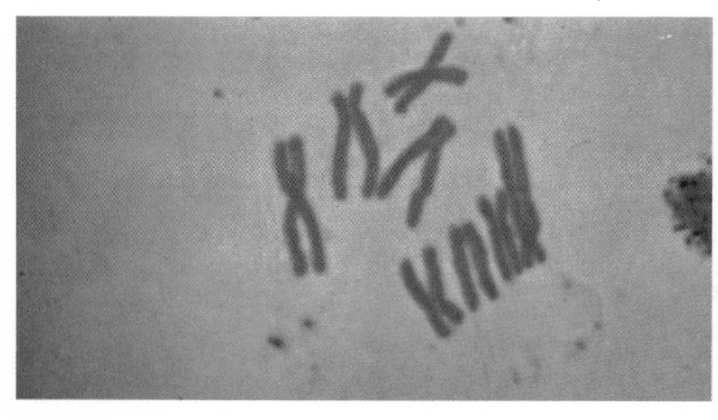

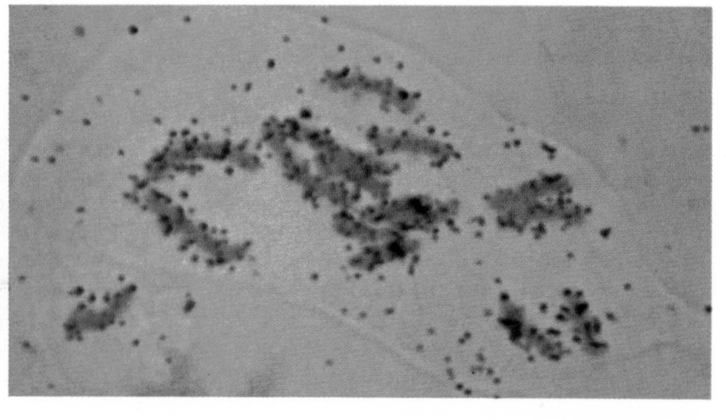

Figure 13.13 (above). Radioactively tagged chromosomes of *Bellevalia* in metaphase that have been allowed to duplicate once in the radioactive solution and once after the cells were removed. Only one member of each chromatid pair is tagged, except where segments have broken and have been exchanged between sister chromatids.

Figure 13.14 (below). Proposed diagrammatic representation of chromosome organization and replication that interprets the autoradiographic results. Solid lines represent nonlabeled units, and those units in dashed lines are labeled. The dots represent grains in the autoradiographs.

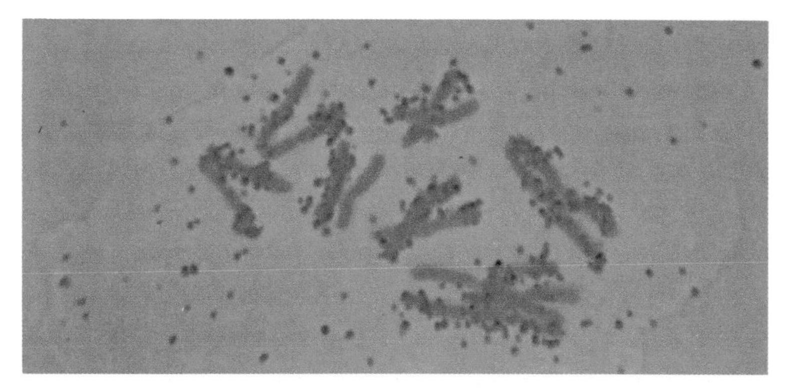

hours. Some of the roots then were fixed and squashed on glass slides, while the others were put into a nonradioactive solution to continue growing. The autoradiographs showed that about one-third of the cells had incorporated the labeled thymidine into their DNA. This observation was consistent with the expectation that about one-third of the cells would be in S-phase at any given time.

Autoradiographs of labeled cells reaching metaphase showed that all of the chromosomes were about equally labeled and that both chromatids of each pair had incorporated the H^3-thymidine into their DNA (Figure 13.12). This observation confirmed the model of DNA replication suggested by Watson and Crick (Chapters 4 and 15). Each chromosome entering S-phase contains a double strand of DNA. During DNA replication, this strand separates and a new complementary strand is built on each of the original strands. Thus, each chromatid would be expected to contain one original unlabeled strand and one labeled strand built during immersion in the H^3-thymidine solution.

Next, samples were taken from roots that had grown in nonradioactive solutions for 36 hours after immersion in H^3-thymidine solution. Because the complete cell cycle was known to take about 24 hours, these cells would be expected to have passed through one S-phase in the unlabeled solution. The chromosomes of each labeled cell would have entered that S-phase with a double strand of DNA in which one strand was labeled and the other unlabeled. When the strands separated, each would acquire an unlabeled complementary strand. Thus, cells in metaphase after 36 hours would be expected to have chromosomes in which one chromatid was labeled and its sister chromatid was unlabeled. Some cells did show the expected pattern (Figure 13.13), but there were significant exceptions.

Some chromatids had only a portion of their length labeled. However, when both sister chromatids were examined, enough labeled segments were available in each metaphase chromosome to account for one fully labeled chromatid in each chromosome. The most reasonable way to account for these observations is to assume that the whole unit of DNA in a chromosome does not necessarily remain intact through an entire S-phase. Some breaks and exchanges between sister chromatids must occur (Figure 13.14). If each chromatid is composed of one long DNA molecule or of many shorter molecules joined end to end, the observations are explainable.

The arrangement of DNA in chromatids is still a controversial matter, but it seems clear that very long chains are folded to form nucleoprotein fibrils about 250 A in diameter. The DNA molecules, which are only about

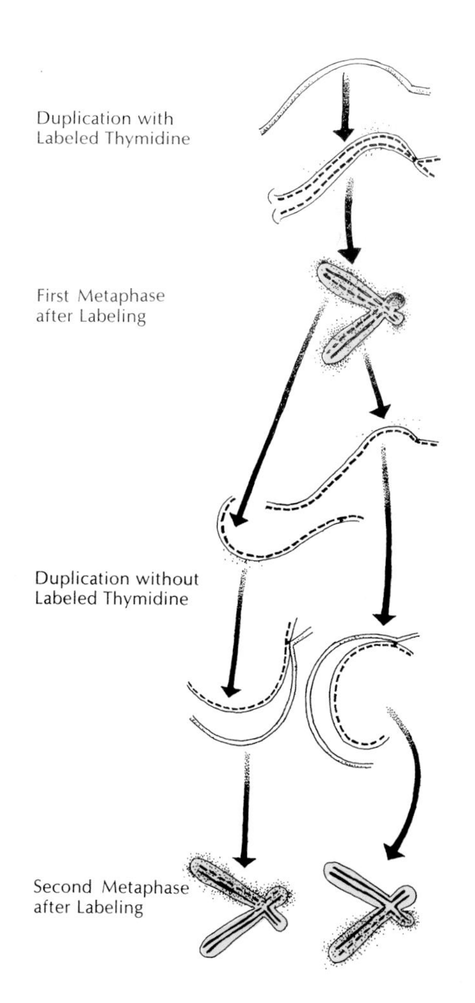

Figure 13.15. Electron micrograph of a thin section of nucleus. The predominant fibers are 250 A, but finer fibers suggest a possible unwinding of the chromatin.

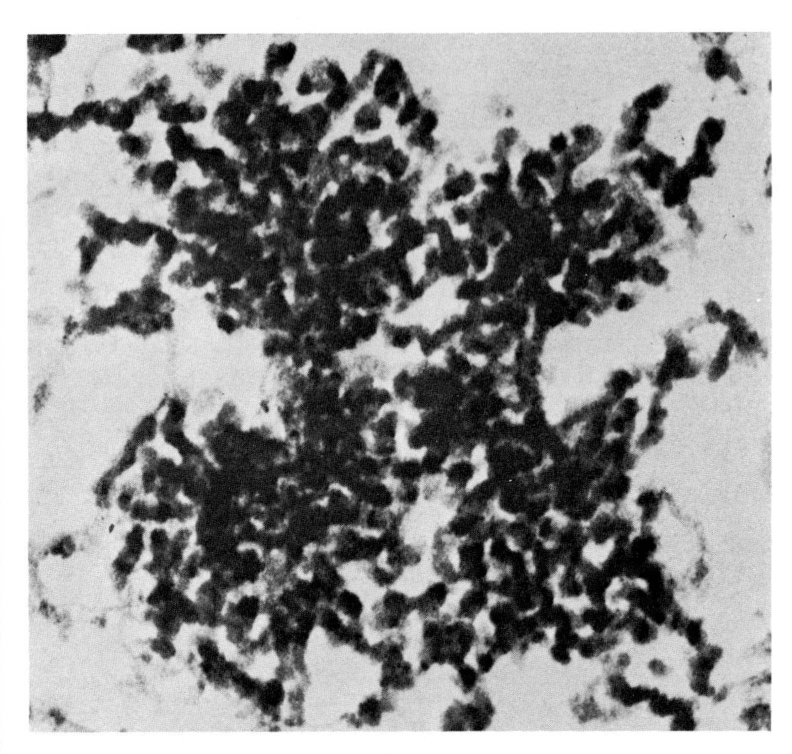

20 A in diameter, are folded or coiled in some unknown way in the larger fibrils. The fibrils, in turn, appear to extend as rather loose loops from the central axis of the chromatid (Figure 13.15).

In certain tissues, such as the salivary glands of the larvae of flies and mosquitoes, giant chromosomes are formed. These giant chromosomes are more than 1,000 times the volume of the corresponding normal chromosomes. They are thought to be formed by repeated replications of the DNA molecule, forming perhaps 1,000 copies of the DNA molecule within a single giant chromosome. Because the giant chromosome is about the same length as the corresponding normal, uncoiled interphase chromosome, the giant chromosome is thought to be made up of DNA strands lying side by side with little or no coiling. For this reason, the giant chromosomes are called *polytene* (many-stranded) chromosomes.

Regular patterns of light and dark bands appear on the polytene chromosomes. These patterns have been mapped and have been shown to be consistently the same for a given kind of chromosome from a given species. Mutations that are expressed in various physiological abnormalities in the insects in some cases cause alterations in the band pattern of the polytene chromosomes. Studies of these changes made it possible to map the location of particular genes on the chromosomes of organisms such as the fruitfly (Chapter 14).

Centrioles

All eucaryotic cells except those of higher plants have centrioles, and these structures appear to play a major role in the formation of the mitotic spindle. Two pairs of centrioles exist in the interphase cell, with the centrioles of each pair at right angles to each other. As prophase begins, the pairs of centrioles separate. Each pair lies at the center of an aster, and the spindle fibers appear to be spun from the centrioles as they separate. Electron micrographs, however, show that the spindle fibers do not actually contact the centrioles. Each centriole is made up of nine groups of microtubules, usually with three microtubules in each group. These microtubules are similar in appearance to the microtubules that make up the spindle fibers. The centrioles also appear to play a role in the formation of cilia and flagella. In the cilia and flagella, patterns of microtubules show a clear relationship to the microtubules of the centriole, but the microtubules of the structure apparently formed by the centriole do not appear to be attached to the centriole.

During telophase, as the two daughter nuclei form, each pair of centrioles replicates to give two pairs of centrioles to each daughter cell. The replication of centrioles thus may be regarded as the first step in preparation for the next cycle of mitosis, occurring even before the previous cycle is completed. The mechanism by which the centrioles replicate is unknown. The role of the centrioles in construction of spindles and asters also is unknown. Although the centrioles appear to play a major role in this process in most eucaryotic cells, the higher plant cells that lack centrioles are able to construct a mitotic apparatus almost identical to that of the cells having centrioles.

Cytokinesis

The final step in cell division, occurring at the end of telophase, is cytokinesis, or cell cleavage. In this process, the nuclei and cytoplasm of the daughter cells are separated by the plasma membrane to form two complete and independent cells. In some cells, such as those of striated muscle tissues, mitosis may occur without cytokinesis, thus forming multinucleate cells.

The mechanism of cytokinesis is markedly different in animal and in plant cells. In plant cells, a structure called the *cell plate* forms along the equatorial plate of the spindle during telophase. This cell plate appears to be composed of membranes from the Golgi complexes, which gradually fuse to form new plasma membranes that separate the two daughter cells. The growth of the new membranes probably proceeds from near the center of the cell outward, until they fuse with the membrane of the parent cell.

In animal cells, the cell membrane constricts or furrows, gradually closing in until the two daughter cells are separated. The mechanism by which this furrowing is carried out is unknown but probably involves contractile microfilaments. The fact that cell cleavage has been observed in cells from which the mitotic apparatus has been removed rules out the hypothesis that the cell membrane is pulled inward by the spindle or aster fibers. Hydrolysis of ATP is apparently involved in the process of furrowing, and there is some evidence that a contractile protein is involved in the process, obtaining energy for its contraction from ATP.

MEIOSIS

Mitosis maintains the constancy of chromosome number from one cell generation to the next. Mitosis alone, however, does not make sexual reproduction possible, for fusion of sex cells with the full chromosome number would double the number of chromosomes in the resulting cell. The possibility of sexual recombination is provided by the process of meiosis, which produces haploid cells from diploid ones, thereby making subsequent cell fusion feasible (Figure 13.16). The genetic reshuffling made possible by

Figure 13.16. Sequence of meiotic stages in the division of *Lilium michiganense*. (Courtesy Carolina Biological Supply Company)

1. Early prophase. The chromosomes are becoming visible.
2. Late prophase. The thick double threads (chromatids) lie close to the periphery of the nucleus.
3. Metaphase I. The chromosomal pairs are arranging themselves along the equatorial plate.
4. Anaphase I. The individual chromosomes of each pair separate and move toward the poles, possibly pulled by the contracting spindle fibers.
5. Telophase I. The chromosomes have reached the poles and are bunched up tightly. They are separating into the chromatin network of the daughter nuclei, and a new nuclear membrane is forming.

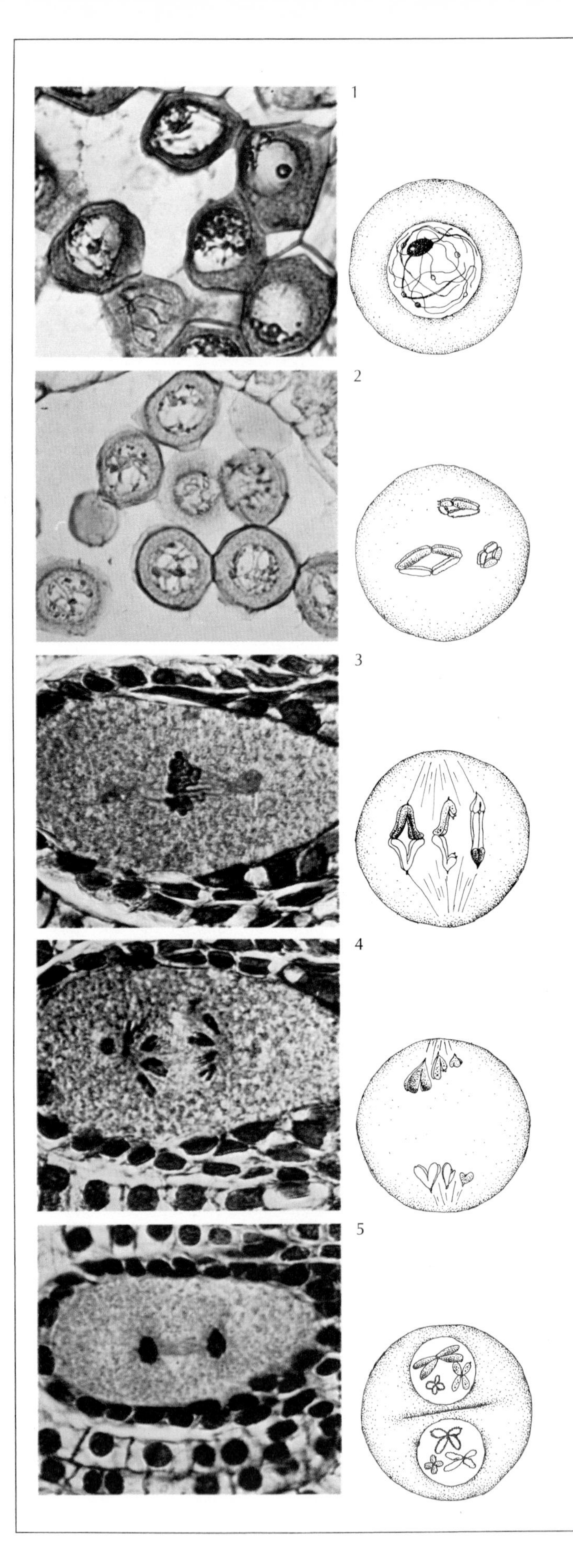

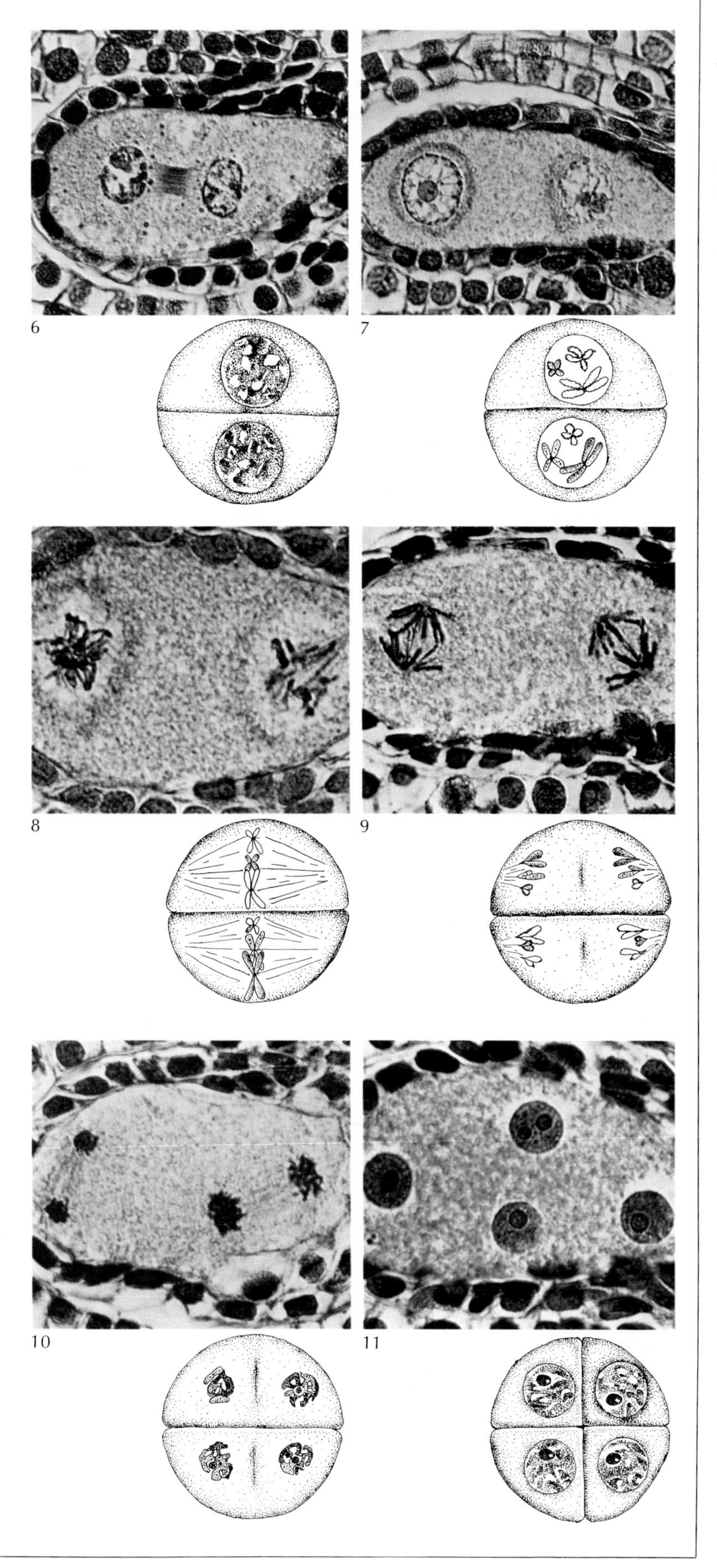

6. Interphase. Two nuclei are visible. Note the remnants of the spindle fibers.
7. Prophase II. Chromosomes are condensing, and the spindle is forming as the nuclei prepare for the second division.
8. Metaphase II. Spindle fibers are appearing, and the chromosomes are lining up along the equatorial plate.
9. Anaphase II. The centromeres have separated, and the chromosomes are moving to the opposite poles of the spindle.
10. Telophase II. The chromosomes have separated, and nuclear membranes are forming around the haploid sets of chromosomes. The plasma membranes will separate the daughter cells.
11. Spores. The diploid cell has become four haploid spores.

sexual reproduction almost certainly accounts for the explosive evolution of eucaryotic organisms into the spectacular array of higher animals and plants that now populate the earth.

Sexual reproduction is identified almost exclusively with diploid organisms, occurring only as exceptional processes among haploid organisms. However, sexual reproduction probably first arose as a means of genetic reshuffling among haploid organisms. Its introduction required the development of meiosis as a method of returning the fused cell to its normal haploid condition. Thus, diploidy, which provides enormous adaptive advantages for an organism (Chapter 36), was at first an intervening condition existing briefly after sexual reproduction. The condition was reversed in the evolution of higher organisms, with diploidy becoming the dominant phase and haploidy the intermediate one.

Meiosis in diploid organisms involves two successive cell divisions accompanied or preceded by a single chromosomal replication (Figure 13.17). The first meiotic division (meiosis I) is similar to mitosis in some respects but involves an exceptionally long and complex prophase. In this division, pairs of homologous chromosomes are separated, but each chromosome retains both chromatids. In the second meiotic division (meiosis II), which follows a brief interphase, the chromatids are separated. Mitosis produces two diploid daughter cells from a single diploid parent, and meiosis produces four haploid daughter cells from a single diploid parent.

The movements of chromosomes and other cell structures during meiosis, as seen in the light microscope, were described before 1900, but the molecular events that initiate the process, that lead to pairing and segregation of the chromosomes, and that result in reciprocal changes between homologs are only now beginning to be understood.

Meiosis I

Meiosis I is characterized by a very long and complex prophase. As in other prophases, the first visible change is a condensation of chromatin into long, tangled, threadlike chromosomes less than a micron in diameter. Because prophase I of meiosis is so complex, it has been subdivided into several stages. The first stage is called the *leptotene* stage. The leptotene chromosomes appear longer and thinner than the chromosomes of mitotic prophase and are distinguished by the presence of a series of dark granules, or chromomeres, along the chromosomes. Although studies of DNA synthesis indicate that each chromosome must already contain two double strands of DNA at the beginning of the leptotene stage, the chromosomes are not visibly double as they are in mitosis.

In the *zygotene* stage, each chromosome pairs with its homolog in a very regular fashion. Each chromomere can be seen to match up with the corresponding chromomere on the homolog. During this stage, the chromosomes continue to shorten and thicken. The joining of homologous chromosome pairs is called *synapsis*. Homologous chromosomes always synapse in pairs even in polyploid nuclei, where several copies of each homolog are present. At the close of the zygotene stage, the nucleus appears to have only the haploid number of chromosomes, but each apparent chromosome is actually a pair of homologs closely bound together. These pairs are called *bivalents*.

In the *pachytene* stage, further shortening and thickening of the chromosomes occurs. During this stage, segments are interchanged between pairs

Figure 13.17. Comparative diagram of meiosis and mitosis. *Printed by permission from J. D. Watson,* Molecular Biology of the Gene, *© 1965, J. D. Watson.*

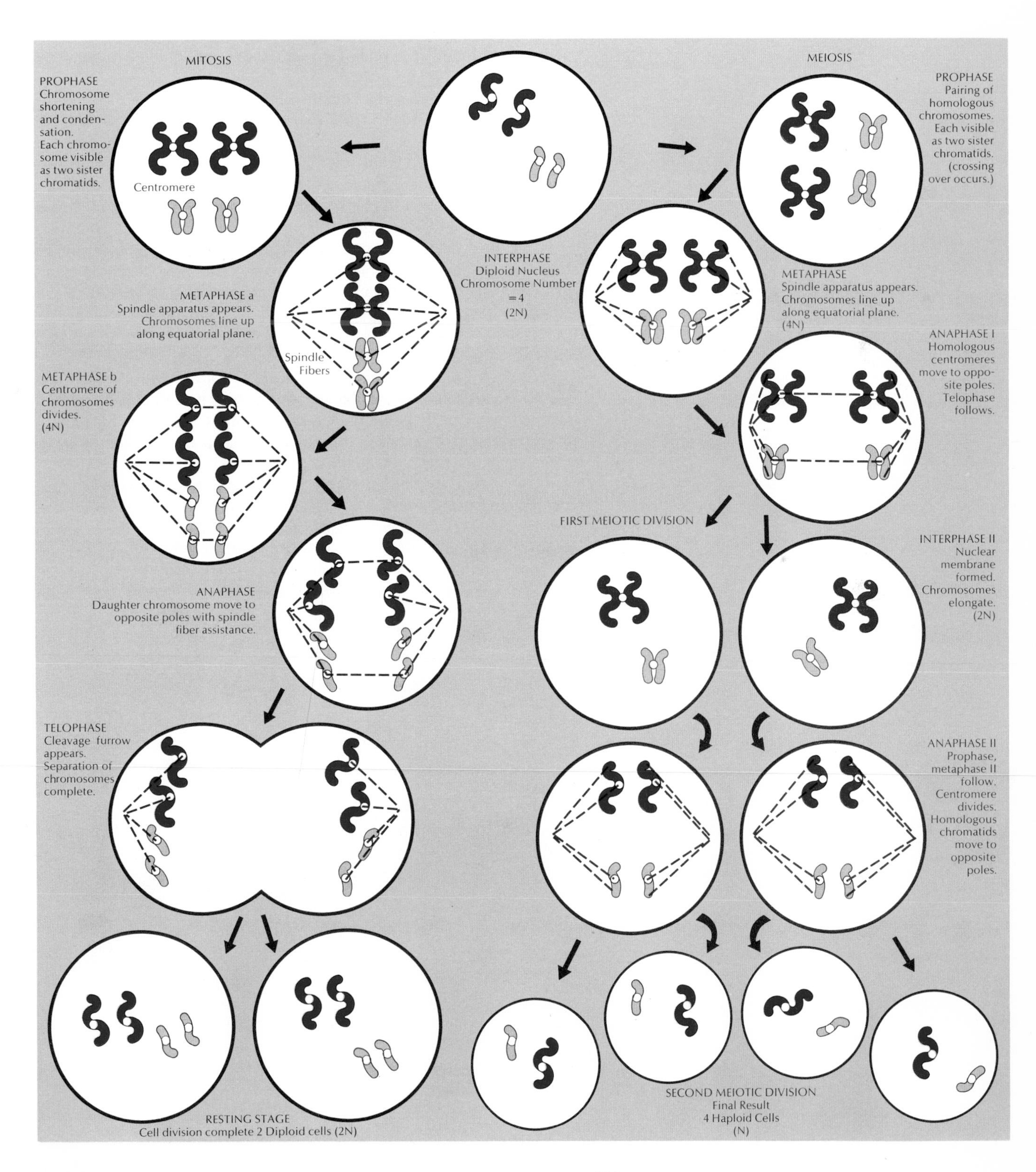

MITOSIS
PROPHASE
Chromosome shortening and condensation.
Each chromosome visible as two sister chromatids.
Centromere
METAPHASE a
Spindle apparatus appears. Chromosomes line up along equatorial plane.
Spindle Fibers
METAPHASE b
Centromere of chromosomes divides.
(4N)
ANAPHASE
Daughter chromosome move to opposite poles with spindle fiber assistance.
TELOPHASE
Cleavage furrow appears. Separation of chromosomes complete.
RESTING STAGE
Cell division complete 2 Diploid cells (2N)
INTERPHASE
Diploid Nucleus
Chromosome Number =4
(2N)
MEIOSIS
PROPHASE
Pairing of homologous chromosomes. Each visible as two sister chromatids. (crossing over occurs.)
METAPHASE
Spindle apparatus appears. Chromosomes line up along equatorial plane.
(4N)
ANAPHASE I
Homologous centromeres move to opposite poles. Telophase follows.
FIRST MEIOTIC DIVISION
INTERPHASE II
Nuclear membrane formed. Chromosomes elongate.
(2N)
ANAPHASE II
Prophase, metaphase II follow. Centromere divides. Homologous chromatids move to opposite poles.
SECOND MEIOTIC DIVISION
Final Result
4 Haploid Cells
(N)

Figure 13.18. Micrograph of a single bivalent. The chromatids cross each other at the points of the two chiasmata.

of homologous chromatids. This process apparently involves breakage of the chromatids at corresponding points, interchange of the two segments, and rejoining of the chromatids.

In the *diplotene* stage, the homologous chromosomes begin to separate, and the pair of chromatids making up each chromosome is visible. The chromosomes do not separate entirely but are joined together at their ends and cross each other at points called *chiasmata* (Figure 13.18). Each chiasma represents the approximate point at which an exchange of chromatid segments between the two chromosomes occurred. One chiasma is formed for each bivalent, and longer bivalents generally form several chiasmata.

In the next stage, *diakinesis*, the chromosomes become maximally shortened and thickened. The chiasmata appear to move toward the ends of the bivalents until each bivalent is held together only by chiasmata at its ends. Diakinesis is the final stage of prophase I, and it is accompanied by disintegration of the nucleolus and nuclear membrane and the formation of a spindle.

Metaphase I is similar to the metaphase of mitosis, with the fully condensed chromosomes becoming aligned on the equatorial plate and spindle fibers attaching to the centromeres. However, in metaphase I of meiosis, the centromeres of each bivalent usually are separated and lie somewhat off the equatorial plate toward the poles.

In *anaphase I*, the chromosomes move toward the poles of the spindle. Each chromosome is made up of a pair of chromatids, joined by the centromere. As the two centromeres of the pairs of bivalents move toward opposite poles, any remaining chiasmata slide apart, and the pairs of homologous chromosomes are separated from each other. Anaphase I ends with a complete haploid set of chromosomes clustered at each of the poles, but these chromosomes are not identical to the chromosomes that existed

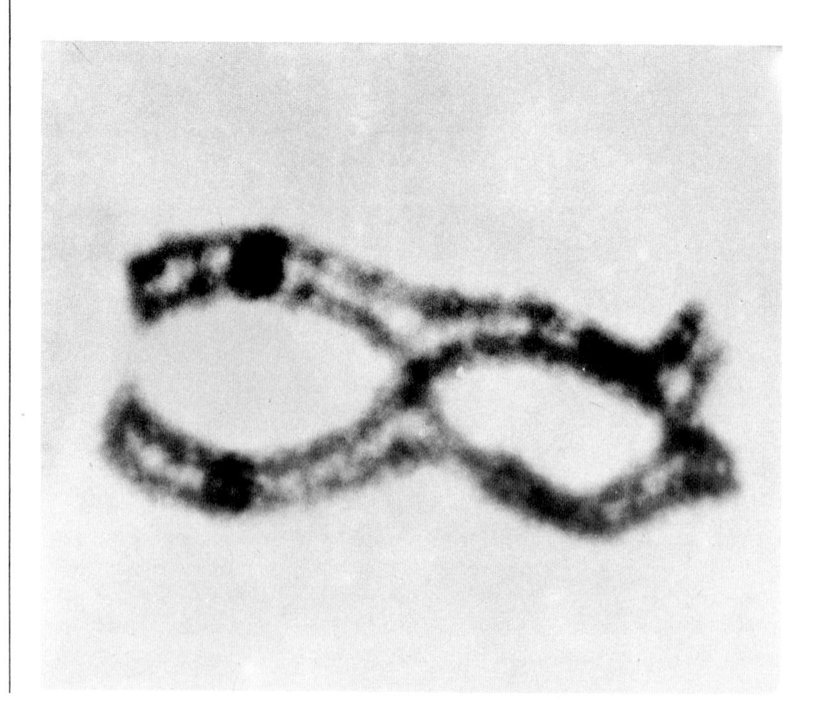

at the beginning of prophase I. Each chromosome contains one of its original chromatids and one chromatid that is a mixture of segments from its own original chromatid and from a chromatid of its homolog.

In *telophase I*, nuclear membranes form around the two sets of chromosomes, the chromosomes uncoil, and a cell membrane is formed between the two nuclei.

After a relatively brief *interphase*, the two haploid cells enter meiosis II. In some species, there is no noticeable interphase, and telophase I may proceed without any formation of nuclear or cell membranes and without uncoiling of the chromosomes. In any case, the chromosomes pass from meiosis I to meiosis II without further chromosomal replication.

Meiosis II

Prophase II is relatively brief, particularly in those species with minimal telophase I and interphase. It is marked by condensation of the chromosomes and formation of the spindle. In *metaphase II*, the chromosomes become aligned on the equatorial plate of the spindle, and the chromosomes replicate and become attached to the spindle fibers. In *anaphase II*, the daughter centromeres separate, and the individual chromatids move to opposite poles. During *telophase II*, the nuclei are formed around the resulting haploid sets of chromatids, which now may be called chromosomes, and plasma membranes separate the daughter cells.

The second meiotic division is similar to mitosis, but begins with a haploid number of chromosomes and ends with the formation of daughter cells that contain a haploid set of chromatids. Each of the haploid cells (gametes) produced by meiosis could fuse with a haploid cell from another individual to form a normal diploid cell. This diploid cell will contain a complete set of chromosomes from each of the parent organisms.

Germ Cells

The process of meiosis occurs only in the germ cells of sexually reproducing plants and animals. In animals, these cells are found in organs called gonads.

In the male animal, the germ cell gives rise by repeated mitotic divisions to a large number of cells called *gonocytes*, which become specialized to *spermatogonia*. Each spermatogonium divides twice mitotically to form four *spermatocytes*. Each of the spermatocytes then undergoes meiosis to form four haploid *spermatids*. These cells are then transformed into the specialized structures called *spermatozoa*, or sperm.

In the female animal, the gonocytes become specialized to form *oogonia*. Each oogonium divides twice mitotically to form four *oocytes*. Each oocyte undergoes meiosis to form four haploid *ootids*. One of the ootids develops into an *ovum*, or egg. The other three become *polar cells*, which eventually disappear. The function of the polar cells is unknown.

In plants, meiosis occurs at various times during the life cycle. In some cases, male and female sex cells similar to the sperm and ova of animals are formed. In other cases, the products of meiosis are asexual spores. In some plants, the haploid stage involves a significant portion of the life cycle of the organism.

The result of meiosis and sexual reproduction is to give each diploid cell a reshuffled combination of genetic information. Each zygote obtains a complete haploid set of chromosomes from each of its parents. Both sets are duplicated in each of the body cells, which arise by mitotic division of the zygote. When gametes are formed by meiosis, the assortment of the members of each chromosome pair occurs in an independent fashion, so that each gamete is likely to contain a mixture of maternal and paternal chromosomes in its complete haploid set. Thus, in sexually reproducing organisms, each mutation that happens to arise is shuffled and recombined with many other possible mutations to increase astronomically the combinations of different genetic messages that can be found within a population.

Research on Meiosis

Many questions about meiosis remain unanswered today, and research is under way in many laboratories to learn more about this vital process. The pairing of homologous chromosomes during the zygotene stage of meiosis I is an extremely important and little-understood process.

One of the basic problems is that of explaining the mechanism by which the homologous pairs of chromosomes locate each other. Because the homologs contain the same sequences of genes, their DNA nucleotide sequences are similar. It is known that single-stranded DNA molecules of similar nucleotide sequences will locate one another in a solution and become attached to form a double strand. There is no evidence that a similar chemical attraction will occur between double-stranded DNA molecules. Furthermore, the DNA of the chromosomes is thought to be enclosed by protein molecules that protect it from hydrolyzing enzymes.

The process of synapsis, in which the homologs become attached to form a bivalent, is also poorly understood. Recent electron-microscope studies have suggested that, in at least some species, the ends of all chromosomes are attached to the nuclear membrane during prophase. The ends of homologous chromosomes are located close to each other on the nuclear membrane, suggesting that synapsis might begin at the ends of the chromosomes and proceed toward the middle. This suggestion has been confirmed in light-microscope studies of *Tradescantia* and of *Agriotis mancus* (a species of beetle). On the other hand, studies of synapsis in *Drosophila* indicate that synapsis begins at a few apparently random points on the chromosomes. The points of synapsis increase as time elapses until eventually the two homologous chromosomes are joined all along their length.

Electron-microscope studies have shown that a structure called the *synaptinemal complex* exists between homologs in all bivalents. Cycloheximide, a potent inhibitor of protein synthesis, also acts to inhibit synapsis and maintenance of the synaptinemal complex. Apparently, the continuous synthesis of protein is required to maintain synapsis. The chemical structure of the synaptinemal complex is not yet known, but preliminary studies suggest that it contains protein and lipid. Other important studies are under way in an attempt to discover the nature of the mechanism that forms chiasmata and interchanges segments of homologous chromatids.

CELL DIVISION AND HEREDITY

By 1900 the general sequence of events in mitosis and meiosis had been described. Scientists had not discovered that chromosomes exist in pairs of different kinds, but they did know that meiosis results in a halving of the chromosome number. Most biologists, acting on what was known about cell division, assumed that hereditary information is carried in the nucleus.

By observing the regular assortment of chromosomes between daughter cells, several biologists were able to speculate that the chromosomes are involved in transmission of hereditary information.

Thus, the stage was set for rapid progress with the rediscovery of Mendel's concept of hereditary factors. Biologists at the turn of the century knew that a zygote is formed by the fusion of sex cells from male and female parents and that the fusion of nuclei plays an important role in this process. They knew that all cells of the mature organism are derived through mitotic divisions of the zygote and that the sex cells are formed by meiotic divisions of certain specialized body cells. These processes provided obvious parallels to the reassortment of factors hypothesized by Mendel. Thus, twentieth-century geneticists moved rapidly into a search for the physical carriers of heredity within the cell nucleus.

FURTHER READING

More detailed information about mitosis and meiosis will be found in books by Brachet and Mirsky (1961), Harris (1963), Hughes (1952), and White (1961). More general discussions are found in articles by Mazia (1953, 1955, 1961).

14
Classical Genetics

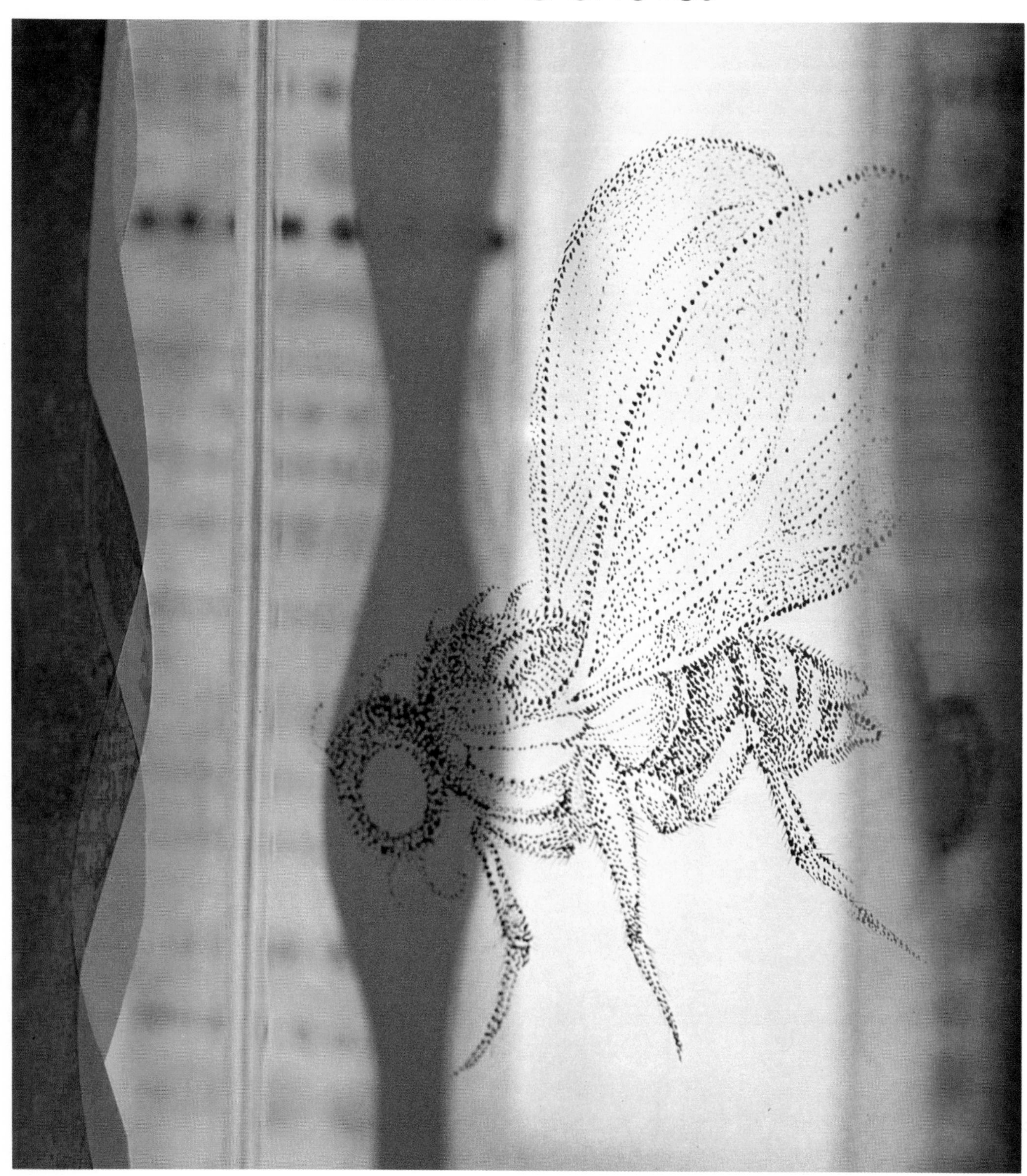

The first decade of this century was a period of great activity in biology, and the studies of heredity must have seemed only one minor sidelight of progress that included discovery of hormones, proof of the existence of viruses, discovery of human blood types, development of techniques for tissue culture, Pavlov's studies of conditioned reflexes, discovery that enzymes act as catalysts, and clarification of the anatomy and organization of the nervous system. In the physical sciences, subatomic particles were being described, and quantum theory and relativity were upsetting old ideas.

In retrospect, it is possible to pick out certain key observations and theories that seem to mark a steady progress toward understanding of inheritance. Such historical selection always tends to make things much simpler than they appeared at the time.

For the first two decades of this century, students of heredity extended Mendel's principles to explain the many apparent exceptions they had discovered and attempted to discover the physical basis for Mendel's hypothetical hereditary factors. William Bateson played a leading role in this work. He introduced such terms as "homozygote," "heterozygote," and "F_1 and F_2 generations" to clarify discussions of experiments and theories. He coined the word "genetics" to describe the study of heredity. The word "gene" was introduced by the Danish botanist Wilhelm Ludwig Johannsen as a term for the specific unit within the cell that corresponds to one of Mendel's hereditary factors.

Most investigators had assumed that all chromosomes in a cell are essentially equivalent. The mechanisms of mitosis and meiosis were viewed simply as devices that put half of the chromosomes into each daughter cell. The American cytologist T. H. Montgomery, Jr. showed in 1901 that the cell contains pairs of distinctive kinds of chromosomes. Each sperm or egg cell contains a complete haploid set—one of each chromosome kind. Nuclear fusion during fertilization gives the diploid zygote (and each body cell that forms from it) two complete chromosome sets—one of maternal origin and the other paternal. Montgomery assumed that meiosis again separates these sets, giving half of the gametes the maternal set and the other half, the paternal set. He worked with cells from grasshoppers because the chromosomes in many insect cells are unusually visible and few in number. Another American working with grasshoppers suggested that the chromosomes play a role in determining the sex of an individual. C. E. McClung described the existence of two kinds of sperm cells—one with an extra, or accessory, chromosome that is not present in the other kind of sperm cell. McClung concluded that eggs fertilized by sperm having the accessory chromosome develop into males, whereas eggs fertilized by sperm lacking the accessory chromosome develop into females (the reverse of the actual situation, as it later turned out).

Another study of grasshopper cells was made by W. S. Sutton. In the species he studied, Sutton found that the 11 chromosomes in a sperm or egg cell can be distinguished by size and that a fertilized egg cell or body cell contains a pair of each kind of chromosome. In meiosis, one member of each pair goes into each reproductive cell, as Montgomery had observed. At first, Sutton thought—as Montgomery had—that the maternal chromosome set is separated from the paternal set in meiosis. However, Sutton strongly suspected that the chromosomes are the carriers of the hereditary factors. If so, the chromosomes must be assorted independently

Figure 14.1. Germ plasm continuity model. Only the germ cells (black) carrying the hereditary information are sustained through the generations. The body, or somatic, cells (yellow) comprising the individual (encircled) are destined for momentary existence, then death.

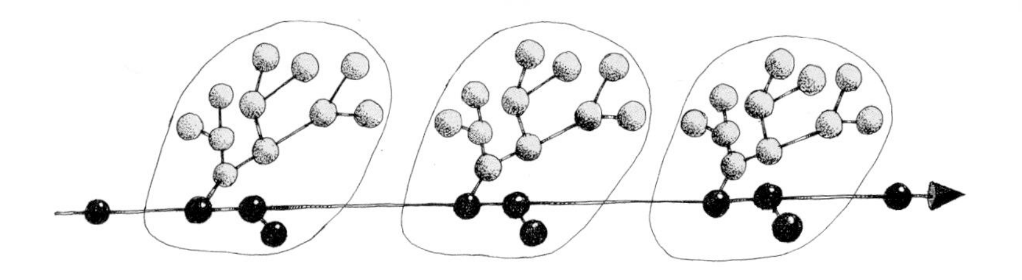

among the reproductive cells, as are the hereditary factors in Mendelian theory. Although Sutton was unable to prove that the gametes do receive a mixture of maternal and paternal chromosomes, his microscopic observations convinced him that his suspicions were probably true.

Sutton concluded that the hereditary factors are closely associated with the chromosomes. He pointed out that there are far more hereditary factors than there are chromosomes in any species, so that a number of different factors must be associated with a single chromosome. Thus, a variety of recessive and dominant factors might be expected to remain together on a single chromosome, rather than assorting independently as assumed by Mendel.

By 1905 one of Sutton's professors, Edmund B. Wilson, was able to clarify the nature of sex determination in insects. In some species, the cells of males contain an unpaired accessory chromosome, whereas the cells of females contain a pair of this chromosome. Half of the sperm cells contain the accessory chromosome and half do not. An egg fertilized by a sperm cell containing the accessory chromosome forms a zygote with a pair of this kind of chromosome and develops into a female. An egg fertilized by a sperm cell without the accessory chromosome forms a zygote with an unpaired extra chromosome and develops into a male.

In a second group of insect species, all sperm cells contain the same number of chromosomes, but there are nevertheless two kinds of sperm. The sperm may contain either a large or a small form of one of the chromosomes, which Wilson called the idiochromosome. Every egg cell contains a large idiochromosome. Thus, fusion of egg and sperm can lead either to a zygote with a pair of large idiochromosomes (a female) or to a zygote with one large and one small idiochromosome (a male).

Wilson observed that those species with an accessory chromosome may be regarded as special cases in which the small idiochromosome has disappeared. Later research has confirmed most of Wilson's conclusions and has shown that the same pattern of sex determination exists in most sexually reproducing plants and animals. However, there are many species—for example, butterflies, moths, birds, amphibians, and some fishes—in which the zygote with identical idiochromosomes, now called *sex chromosomes*, develops into a male and the zygote with dissimilar sex chromosomes or an unpaired sex chromosome develops into a female. In these species, there are two kinds of eggs and only a single kind of sperm.

Wilson's report provided one clear example of chromosomal control of a hereditary character and thus helped to support the theory of chromosomal inheritance. In the meantime, the Mendelists were struggling to explain the results of the ever-increasing number of hybridization experiments being performed. They attempted to fit some of these results to various odd ratios such as 7:1:1:7 or 15:1:1:15. In some cases, they were forced to assume that a single factor can exist in more than two forms, that a single character is controlled by two or more independently assorted factors, or that dominance can be incomplete in certain factors (Bateson, 1913; Bateson, *et al.*, 1902–1909).

THOMAS HUNT MORGAN AND THE FRUITFLY

Thomas Hunt Morgan, who had specialized in embryology before becoming interested in genetics, was convinced that sudden changes, or mutations, play a more important role in evolution than do the hypothetical recombinations of hereditary factors described by the Mendelists. Like many

Figure 14.2 (above). The fruitfly *Drosophila melanogaster*.

Figure 14.3 (lower left). Life cycle of *Drosophila*.

Figure 14.4 (lower right). Chromosomes of *Drosophila*. There are four pairs of chromosomes in each cell. Note the differences in the sex determiner chromosomes of males and females.

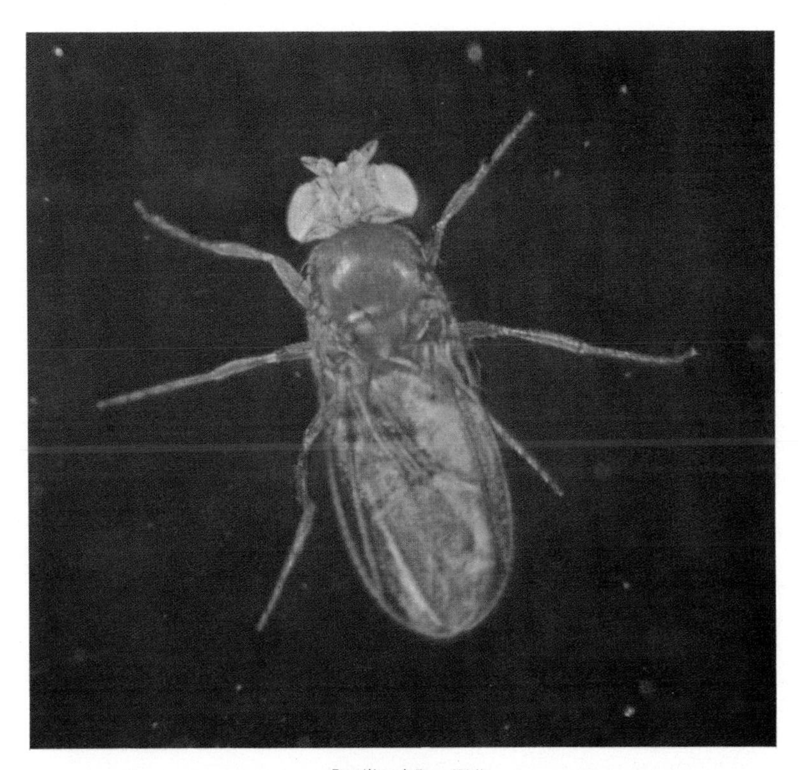

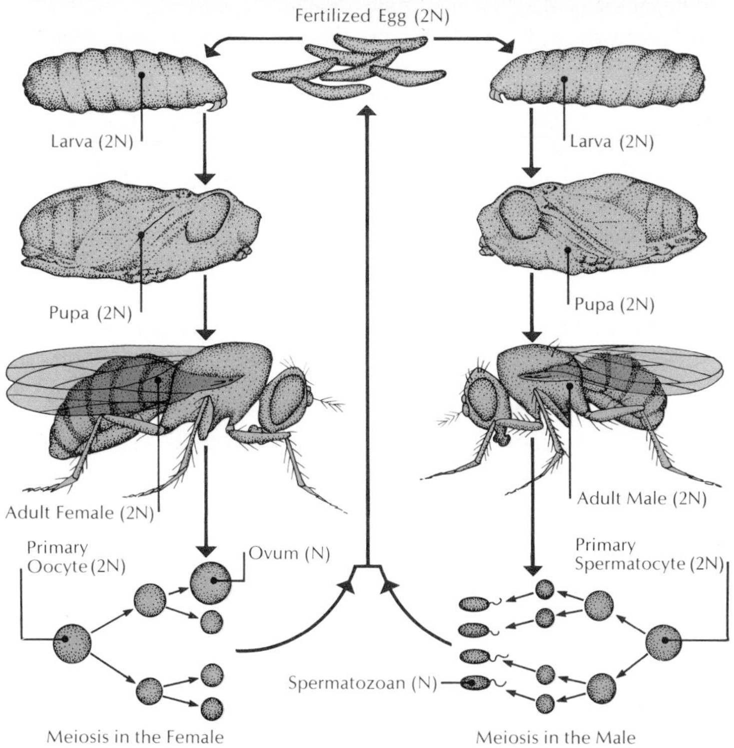

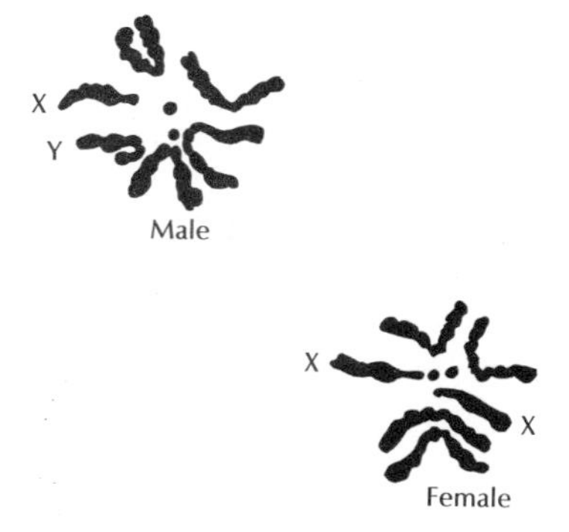

other researchers exploring heredity and the chromosomes, Morgan turned to insects as experimental subjects. It is easy to maintain large populations of small insects; they reproduce and grow rapidly, and their cells and chromosomes can be viewed easily under the microscope. Morgan had heard of some experiments being done with *Drosophila melanogaster*, a small fruitfly. These little flies thrive on a diet of mashed fruit or yeast, can be kept by the hundreds in half-pint milk bottles, and require only about 12 days to reach maturity—thus providing some 30 generations each year for genetic studies. Furthermore, each *Drosophila* cell has only four chromosome pairs, making this an ideal organism for study in the search for simple relationships between heredity and chromosomes.

Morgan subjected his fruitflies to heat, cold, x-rays, radioactivity, and various chemicals, but he was unable to detect any mutations produced by these treatments. Then, in April 1910, Morgan discovered a single, white-eye male fly in a bottle of normal, red-eye flies. No other white-eye flies had appeared in the dozens of generations through which Morgan had observed this population, so he was sure that the white-eye male represented a mutation.

Morgan mated the white-eye male with wild-type, red-eye females from the same generation. The F_1 generation produced by this cross was entirely red-eye, as Mendelian principles would predict if the white eyes are produced by a recessive gene. Next, Morgan allowed the F_1 generation to interbreed. The resulting F_2 generation contained 3,470 red-eye and 782 white-eye flies. This result is rather far from the 3:1 ratio predicted by Mendel's laws, but the reappearance of the white-eye character showed that Morgan had indeed located a new heritable character.

One property of the F_2 generation was incompatible with Mendel's laws. All 782 white-eye F_2 flies were males. The red-eye F_2 flies included 2,459 females and 1,011 males. At first, Morgan thought that the white-eye character might somehow be impossible in females, but when he crossed the original white-eye male with some of his red-eye daughters from the F_1 generation, he obtained 129 red-eye females, 132 red-eye males, 86 white-eye males, and 88 white-eye females. In the results of this cross, white-eye females were as common as white-eye males.

Clearly, the inheritance of the white-eye character is somehow related to the hereditary determination of sex. In his first report on the white-eye mutant, Morgan (1910) used Mendelian principles to explain the results of his crosses by assuming that the eye-color factor and the sex-determining factor are linked together rather than assorting independently. Morgan was well aware of the sex chromosomes in insects and knew that one of *Drosophila*'s four chromosome pairs is responsible for sex determination. In the cells of females, this pair is made up of two rod-shape chromosomes (now called X chromosomes), whereas in the male the pair consists of one rod-shape X chromosome and one J-shape Y chromosome. Later, Morgan showed that the results of his experiments with red-eye and white-eye flies are consistent with the assumption that the eye-color gene is carried only on the X chromosome—that is, eye color is a *sex-linked* trait.

Because the white-eye factor is recessive, the female will have white eyes only if she carries the white-eye factor on both X chromosomes. In a male, on the other hand, the presence of the white-eye factor on the single X chromosome will lead to development of white eyes. This fact explains why only males of Morgan's original F_2 generation showed white eyes.

Figure 14.5. The inheritance of white-eye color, a sex-linked recessive trait, in *Drosophila*. (A) A wild-type (red-eye) crossed with a white-eye produces an F_1 generation in which all are red-eye, but the females are "carriers" (W^+W). The F_1 flies are crossed among themselves to produce an F_2, three-quarters of which are red-eye males and females and one-quarter of which are white-eye males. (B) A homozygous recessive white-eye crossed with a wild-type (red-eye) produces an F_1 in which all females are red-eye and all males are white-eye. The F_1 flies are crossed among themselves to produce an F_2 of one-half red-eye males and females and one-half white-eye males and females.

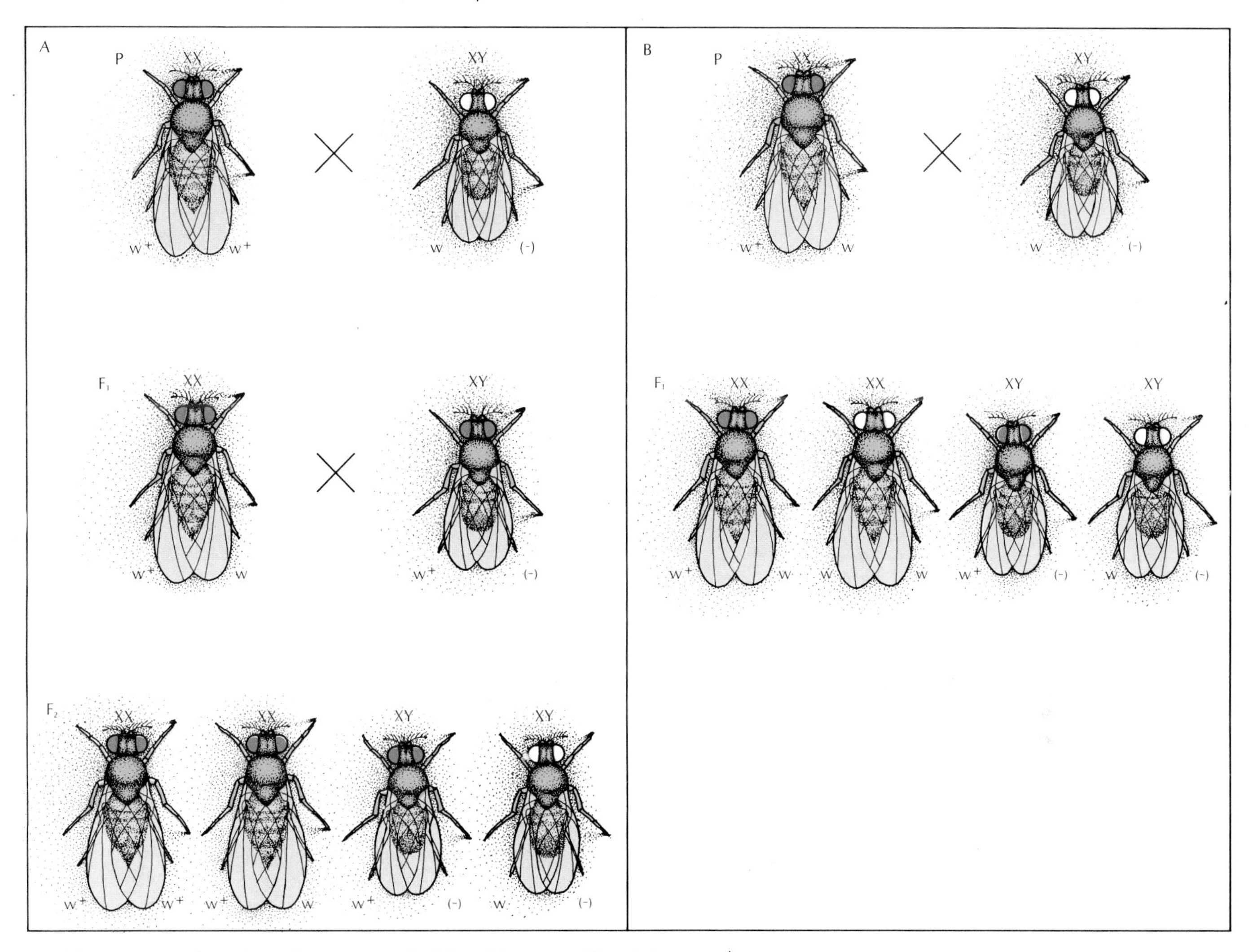

Morgan soon found another mutant fruitfly, this one with miniature wings—a characteristic that also proved to be sex-linked. Before the end of 1910, Morgan had discovered 40 different mutant characters in *Drosophila*, a number of which are sex-linked. Other characters that are not sex-linked proved to be linked to one another. For example, if a fly with purple eyes and a black body is crossed with a wild-type fly (red eyes and "gray" body—actually yellowish gray with dark bands), the characters of purple eyes and black bodies tend to appear together in offspring generations, rather than assorting independently.

Every bit of shelf space in Morgan's laboratory was crowded with bottles of flies, each bottle with a label describing the genetic background of the flies inside. Morgan and his students devoted themselves to the laborious task of anesthetizing each bottle of flies with ether and carefully examining each fly under the microscope to classify its characters. Some of the mutant characters involved subtle alterations of the shape of minute bristles on the fly body or slight changes in the shade of red in the eyes. It was not unusual for the ether to begin to wear off, and the weary researcher in the midst of a

Figure 14.6. Morgan's crosses of the white-eye mutant. The results of the original cross are shown in the 2 top figures, and back cross results are shown below.

Interleaf 14.1

DIAGRAMS OF MORGAN'S CROSSES

Morgan's crosses with his original white-eye mutant fruitfly can be represented with checkerboard diagrams. The symbols for male (♂) and female (♀) used by biologists are derived from the astrological symbols for Mars (a shield and spear) and Venus (a looking glass). The symbols ♂ ♂ and ♀ ♀ represent the plurals, "males" and "females." Figure 14.6 outlines the crosses that Morgan performed.

In the original cross of the white-eye male with wild-type females, the females produce only w^+ gametes, whereas the male produces both w and $(-)$ gametes. Thus, the F_1 generation contains equal numbers of w^+ $(-)$ and w^+w genotypes, or equal numbers of red-eye males and red-eye females. (Actually, in his F_1 generation, Morgan obtained 1,237 red-eye males and females and 3 white-eye males. He assumed that these three males represented the appearance of new mutations, and he did not include them in his statistics.)

When the members of the F_1 generation are interbred, each sex produces two kinds of gametes. (The possible combinations are shown in Figure 14.6a.) The predicted composition of the F_2 generation is 25 percent red-eye males, 25 percent white-eye males, and 50 percent red-eye females. Actually, Morgan obtained about 24 percent red-eye males, 18 percent white-eye males, and 58 percent red-eye females.

Morgan also crossed the white-eye male with heterozygous red-eye females from the F_1 generation of the previous cross. Again, each sex produces two kinds of gametes, but this time equal numbers of four different phenotypes are predicted (Figure 14.6b). Morgan actually obtained 30 percent red-eye females, 30 percent red-eye males, 20 percent white-eye females, and 20 percent white-eye males.

A cross such as this one—in which an individual from the F_1 generation is crossed with its parent or with an individual of a genotype identical to the parent—is called a *back cross*. Another useful cross is the *test cross*, in which a heterozygote is crossed with an individual homozygous for the recessive alleles of the gene or genes being studied. Morgan performed a number of back crosses and test crosses to be sure that the white gene does behave in the predicted fashion in every case.

Morgan's experimental results are much further from the predicted ratios than were Mendel's results. These results are not surprising in view of the relatively small numbers of flies in Morgan's populations. In fact, even with his large populations of pea seeds, Mendel's results show less random deviation from the predicted ratios than would be expected. In the crosses cited above, white-eye flies consistently appear in smaller numbers than predicted. This result might be merely chance, but if it is consistently found in a great many experiments, it will have to be explained. Perhaps another gene is involved, or perhaps the white-eye flies are more likely than red-eye flies to die before the zygote develops into an adult. A great many test crosses and the counting of a very large number of flies may be necessary to resolve such uncertainties.

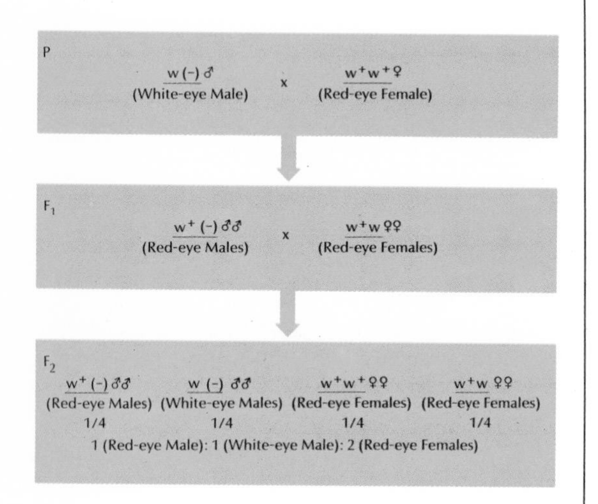

♀ Gametes \ ♂ Gametes	w^+	(-)
w^+	w^+w^+ (Red-eye ♀)	w^+ (-) (Red-eye ♂)
w	w^+w (Red-eye ♀)	w (-) (White-eye ♂)

♀ Gametes \ ♂ Gametes	w	(-)
w^+	w^+w (Red-eye ♀)	w^+ (-) (Red-eye ♂)
w	ww (White-eye ♀)	w (-) (White-eye ♂)

P

w (-) ♂ (White-eye Male) x w^+w ♀ (Red-eye Female)

F_1

w^+ (-) ♂♂ (Red-eye Males) 1/4 — w (-) ♂♂ (White-eye Males) 1/4 — w^+w ♂♂ (Red-eye Females) 1/4 — ww ♀♀ (White-eye Females) 1/4

1 (Red-eye Male): 1 (White-eye Male): 1 (Red-eye Female): 1 (White-eye Female)

count of hundreds of flies would find his research population beginning to take wing and disappear into the corridors.

By 1915 Morgan and his students had identified nearly 100 different mutant characters in *D. melanogaster.* More than 20 of these characters are sex-linked and are controlled by factors carried on the X chromosome. The remaining characters fall into three groups, with the characters of each group tending to remain linked together. The four linkage groups correspond nicely with the four chromosome pairs of *Drosophila*, and one of the groups contains only a few characters, as might be expected from the fact that one of *Drosophila*'s chromosome types is little more than a small dot.

Further evidence that genes are carried by the chromosomes was soon to come from Morgan's laboratory. His important work with *Drosophila* led many other geneticists to begin experimenting with this insect, and it became the most common organism for genetic research. In fact, because this inconspicuous little fly is of minor importance to man as a pest or otherwise, someone once remarked that God must have created *Drosophila* just for Morgan.

LINKAGE AND CROSSING OVER

Geneticists working with *Drosophila* developed a standard set of symbols to represent the various genes. This symbolism has been used with some other organisms, but it is not universally adopted. The various forms of the same gene are now called *alleles*. For example, the eye-color gene first discovered by Morgan and carried on the X chromosome has a recessive mutant allele (w), which tends to produce white eyes, and a dominant wild-type allele (w^+), which tends to produce red eyes. In genetics, the symbol + always indicates a wild-type allele. The letter used for a particular gene is an abbreviation of the name given to the mutant character (in this case, white). A capital letter is used if the mutant allele is dominant; a lower-case letter is used if the wild-type allele is dominant. Thus, B represents a mutant allele for a condition called bar-eye, which is dominant to its wild-type allele, B^+.

The white gene is carried only on the X chromosome, so the white-eye phenotype is produced by a ww genotype in females or by a $w(-)$ genotype in males. The symbol $(-)$ is used to represent the Y chromosome, which carries no allele for this gene. The red-eye phenotype is produced by a $w^+(-)$ genotype in males, but the genotype of a red-eye female may be either the wild-type w^+w^+ or the hybrid w^+w.

The bar-eye character is also sex-linked. In the pure bar-eye character, the form of the eye is a narrow red bar (Figure 14.7). An intermediate form called wide-bar-eye is observed in some females produced in cross-breeding experiments. Males may be either wild-type or bar-eye in phenotype, corresponding to the genotypes $B^+(-)$ or $B(-)$. Three genotypes are possible in the female: B^+B^+, B^+B, or BB. Females with the heterozygous B^+B genotype show the intermediate wide-bar-eye phenotype—an example of *incomplete dominance.* The presence of the recessive, mutant allele does have some effect on the phenotype of the heterozygote.

In his early work with *Drosophila* mutants, Morgan found some mutant characters that appear only as a result of the combined effects of two recessives. For example, certain wing defects do not occur if the individual has a dominant wild-type allele for either of two different genes. Thus, Morgan found himself postulating multiple genes as he had once criticized the

Figure 14.7a (above). Variations in eye structure and color in the fruitfly *Drosophila melanogaster*.

Figure 14.7b (below). Inheritance of the sex-linked recessive trait bar-eye in the fruitfly. In a male that has the recessive gene and a female that is homozygous recessive for the trait, the eye form is a narrow bar. A female heterozygous for the trait has a wide-bar phenotype—an example of incomplete dominance.

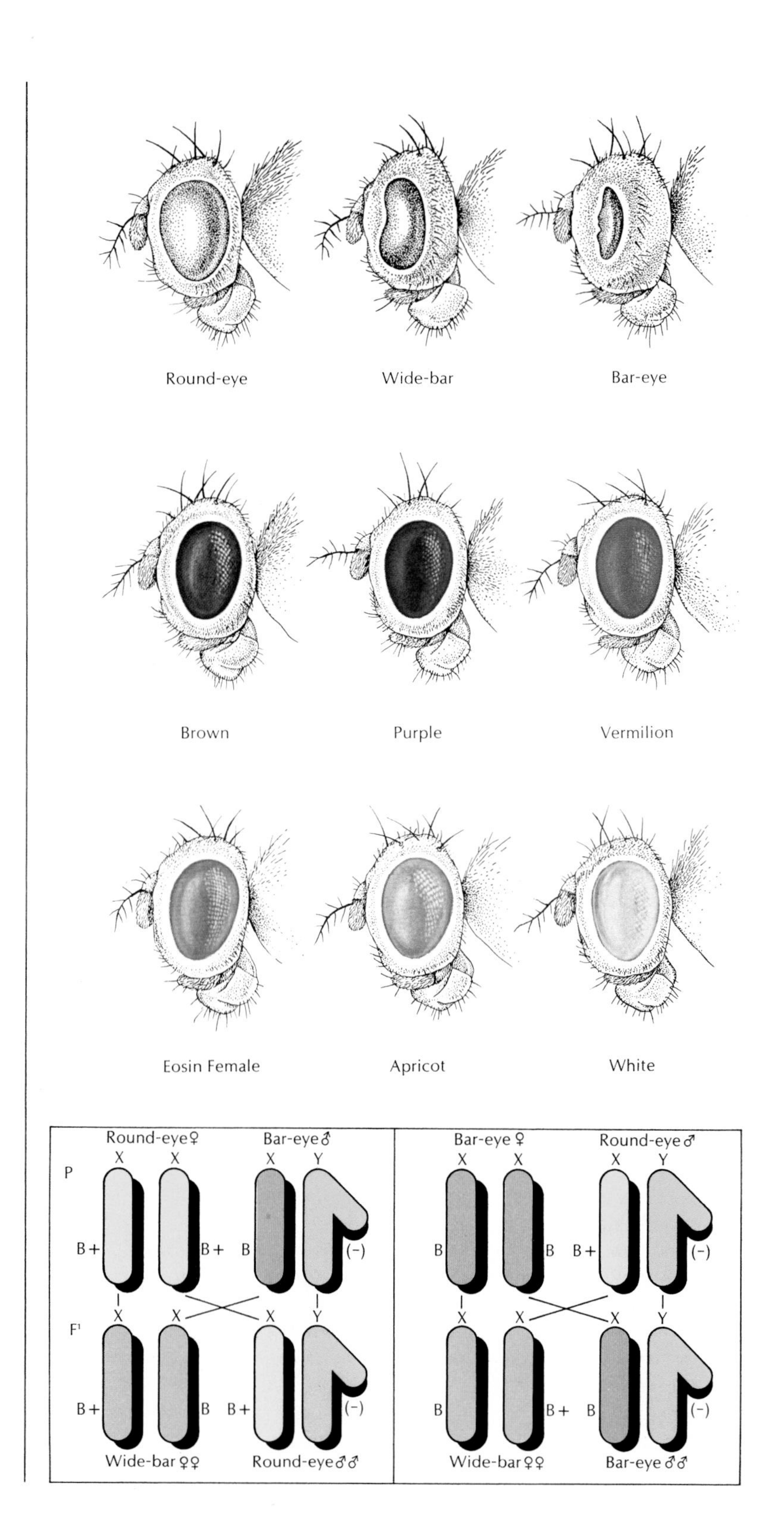

Figure 14.8. Sturtevant's cross utilizing two sex-linked recessive genes. Sturtevant crossed a long-wing, vermilion-eye female (m^+v/m^+v) with a miniature-wing, red-eye male ($-/mv^+$). These F_1 offspring were interbred to produce an F_2 generation. The gametes and predicted genotypes are shown in the Punnett square.

Mendelists for doing, but Morgan's postulates were based upon following hundreds or thousands of individuals through many generations of various test crosses. Incomplete dominance can be explained by a relatively minor modification of Mendel's original laws, merely recognizing that in some cases the heterozygous phenotype may be intermediate between the homozygous phenotypes. Gene linkage can be explained by adopting Sutton's hypothesis that the genes are linked together on chromosomes and thus cannot assort independently.

Among the early mutant genes Morgan discovered in *Drosophila* are those for miniature wings (*m*) and vermilion eye color (*v*). Both are sex-linked genes carried by the X chromosome, and therefore the two genes may be expected to be linked to each other. In both cases, the mutant allele is recessive to the wild-type allele. In an early experiment, one of Morgan's students, A. H. Sturtevant, crossed a long-wing, vermilion-eye female with a miniature-wing, red-eye male. The female was homozygous for both genes, so her genotype may be represented as m^+v/m^+v. In cases of linked genes, the slant is used to separate the alleles found on the two chromosomes of the pair, making it easier to keep track of the linked alleles in later assortment. The genotype of the male was $(-)\ (-)/mv^+$.

All of the gametes produced by the female must be of the genotype m^+v. The male will produce equal numbers of gametes containing an X chromosome with genotype mv^+ and of gametes containing a Y chromosome with genotype $(-)\ (-)$. Therefore, the offspring, or F_1 generation, of this cross should consist of equal numbers of the two genotypes, mv^+/m^+v and $(-)\ (-)/m^+v$. These genotypes correspond to the two phenotypes: long-wing, red-eye females and long-wing, vermilion-eye males. As expected, Sturtevant obtained an F_1 generation with approximately equal numbers of these two phenotypes.

An F_2 generation was then produced by interbreeding members of the F_1 generation. Because the alleles appearing on the same chromosome should remain linked together, the F_1 females should be able to produce only two

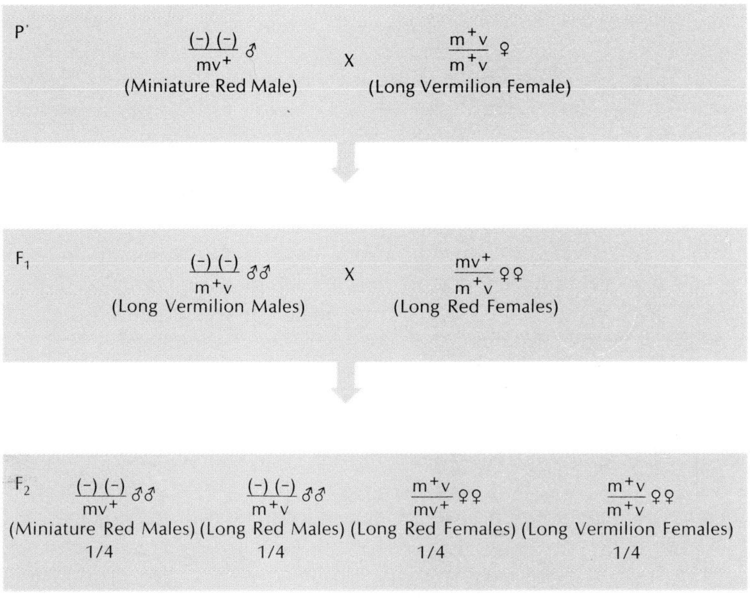

P: $\frac{(-)\ (-)}{mv^+}$ ♂ (Miniature Red Male) X $\frac{m^+v}{m^+v}$ ♀ (Long Vermilion Female)

F_1: $\frac{(-)\ (-)}{m^+v}$ ♂♂ (Long Vermilion Males) X $\frac{mv^+}{m^+v}$ ♀♀ (Long Red Females)

F_2: $\frac{(-)\ (-)}{mv^+}$ ♂♂ (Miniature Red Males) 1/4; $\frac{(-)\ (-)}{m^+v}$ ♂♂ (Long Red Males) 1/4; $\frac{m^+v}{mv^+}$ ♀♀ (Long Red Females) 1/4; $\frac{m^+v}{m^+v}$ ♀♀ (Long Vermilion Females) 1/4

♀ Gametes	♂ Gametes: $(-)\ (-)$	♂ Gametes: m^+v
mv^+	$\frac{(-)\ (-)}{mv^+}$ (Miniature-wing Red-eye ♂)	$\frac{m^+v}{mv^+}$ (Long-wing Red-eye ♀)
m^+v	$\frac{(-)\ (-)}{m^+v}$ (Long-wing Vermilion-eye ♂)	$\frac{m^+v}{m^+v}$ (Long-wing Vermilion-eye ♀)

Figure 14.9. Schematic model of chromosome crossover. (1) Two homologous double-stranded chromosomes, one bearing the linked alleles A and B and the other bearing the linked alleles a and b are joined in synapsis. (2) Corresponding breaks occur in one chromatid of each pair and the fragments are exchanged. (3) After crossing over, one chromatid of the first chromosome bears alleles A and b and one chromatid of the second chromosome bears alleles a and B.

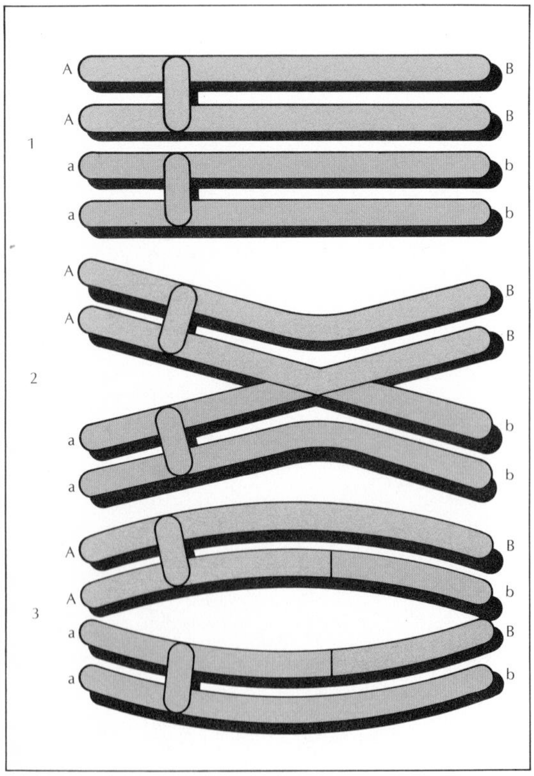

kinds of gametes: mv^+ and m^+v. The male gametes should also be of two kinds: *(−) (−)* and m^+v. Recombination of these gametes yields four genotypes, each of which corresponds to a different phenotype. Thus, the expected F_2 generation would contain equal numbers of long-wing, red-eye females; long-wing, vermilion-eye males; long-wing, vermilion-eye females; and miniature-wing, red-eye males. Table 14.1 compares the predicted and observed phenotypes of the F_2 generation.

Table 14.1
Observed and Predicted Phenotypes of F_2 Generation

PHENOTYPE	Observed		Predicted
	Number	Percentage	Percentage
Long-wing, red-eye males	8	1.7%	0.0%
Long-wing, red-eye females	138	29.4	25.0
Long-wing, vermilion-eye males	117	24.9	25.0
Long-wing, vermilion-eye females	110	23.4	25.0
Miniature-wing, red-eye males	97	20.5	25.0
Miniature-wing, red-eye females	0	0.0	0.0
Miniature-wing, vermilion-eye males	1	0.2	0.0
Miniature-wing, vermilion-eye females	0	0.0	0.0

Source: A. H. Sturtevant, "The linear arrangement of six sex-linked factors in *Drosophila*, as shown by their mode of association," *Journal of Experimental Zoology 14* (1913): 43–59.

There are two unexpected results. First, although three of the four expected phenotypes are present in approximately the expected proportions, there are fewer miniature-wing, red-eye males than expected. Morgan and Sturtevant found that miniature-wing phenotypes always occur in much smaller numbers than predicted in any cross. They concluded that this phenotype has a low viability. That is, zygotes that would develop into miniature-wing adults tend to die before reaching maturity and thus are not counted by the researcher.

The second unexpected result is the presence in the F_2 generation of small numbers of two phenotypes that were not expected to appear. Morgan suggested that the appearance of such phenotypes might be due to an exchange, or *crossing over*, of alleles between the homologous chromosomes in the female during the meiotic division that forms the gametes. Such crossing over cannot occur in the male because there is no second X chromosome with which alleles can be exchanged. (It was later discovered that crossing over never occurs on any of the chromosomes of the male *Drosophila*, but this species is unusual in this respect.)

Crossing over, or exchange of alleles, would make it possible for some female gametes to have the genotypes *mv* and m^+v. Combination of these female gametes with the two kinds of male gametes could produce the additional genotypes (and phenotypes): *(−) (−)/mv* (miniature-wing, vermilion-eye male), $(-)\,(-)/m^+v^+$ (long-wing, red-eye male), m^+v/mv (long-wing, vermilion-eye female), and m^+v/m^+v^+ (long-wing, red-eye female). The new female genotypes produce phenotypes indistinguishable from those expected by linked assortment, but the two new male phenotypes are those that Sturtevant actually observed. The very small number of miniature-wing, vermilion-eye males is consistent with the assumed low viability of the miniature-wing phenotype. Among the F_2 males, there are 214 of the expected phenotypes and 9 of the cross-over phenotypes. Therefore, crossing over occurred during formation of 9/223, or 4 percent of the gametes. The frequency of crossing over in the females cannot be determined from this cross.

Morgan immediately began to search for a physical explanation of the phenomenon of crossing over and found a clue in the observations of the Belgian cytologist F. A. Janssens, who described the process of chiasmata formation at the beginning of meiosis. When homologous pairs of chromosomes come together in synapsis, it appears that some material is interchanged between chromatids of the two chromosomes. Janssens suggested that the chromatids break at corresponding places and interchange equal segments. Thus, when the chromosomes are pulled apart in anaphase, each of the separating chromatids contains some segments derived from the other chromosome of the pair.

Morgan and his colleagues quickly realized that the physical crossing over described by Janssens provides exactly the mechanism needed to account for the crossing over of genes indicated by their experiments with *Drosophila.* Chiasmata do not form between the X and Y chromosomes during meiosis in the male. Furthermore, the chromatids always seem to exchange exactly corresponding segments. These observations are consistent with the genetic observations that crossing over does not occur with sex-linked genes in the male and that genes are exchanged but never lost during crossing over.

Morgan recognized another important implication of this mechanism for crossing over. He had found that the percentage of crossing over remains roughly constant for any given pair of genes but varies greatly among different pairs of genes. Some pairs of genes are completely linked so that crossing over is never observed between them. With other pairs, the percentage of crossing over reaches about 50 percent. Morgan suggested that the genes fall at particular locations along the chromosomes. If two genes happen to be located close together, they will rarely be separated during chiasmata formation; if crossing over occurs, they will cross over together, and the crossing over will not be detected by breeding experiments. On the other hand, if the two genes are located near opposite ends of the chromosome, crossing over would be expected almost every time that chiasmata are formed. The fact that cross-over frequency seldom rises above 50 percent is probably due to multiple crossing over, which could not always be detected in breeding experiments. Morgan pointed out that cross-over frequencies should provide a means of mapping the relative locations of genes on the chromosomes.

By observing an unusual chromosomal abnormality in *Drosophila,* Curt Stern (1931) was able to provide definite proof of the physical reality of crossing over. Chromosomal fragments attached to the ends of the X chromosome made it possible to show microscopically when segments of the X chromosomes were interchanged. In every case, a physical interchange of the chromosome segments was accompanied by genetic crossing over (determined through breeding experiments) between genes located near the ends of the chromosome.

CHROMOSOME MAPPING

Morgan and Sturtevant soon were able to show the relative positions of a number of genes on the X chromosome of *D. melanogaster.* Within a few years, they prepared *chromosome maps* (Interleaf 14.2) for each of the four

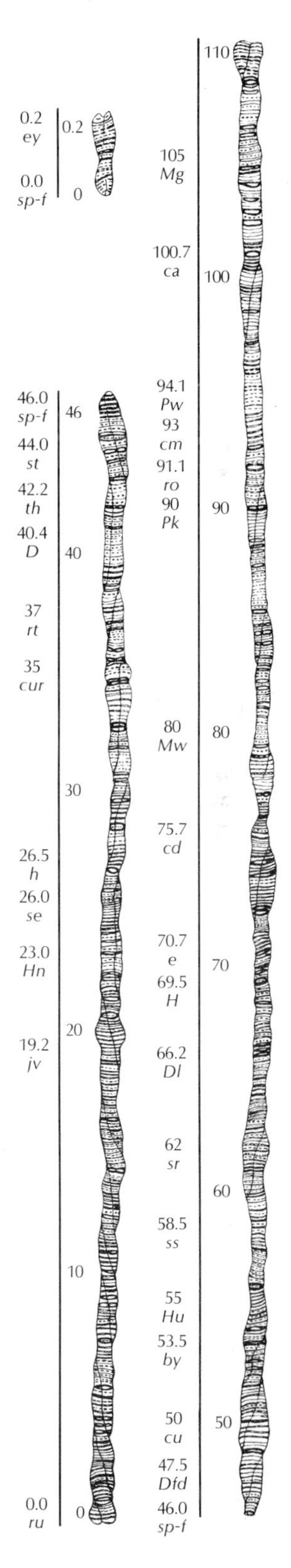

Figure 14.10. Map of the salivary gland chromosomes of *Drosophila melanogaster.*

Interleaf 14.2

CHROMOSOME MAPS

Sturtevant (1913) prepared the first chromosome map of *Drosophila melanogaster*, using Morgan's suggestion that the cross-over frequency provides an index of the distance between any two genes. He used six sex-linked genes studied in the early work in Morgan's laboratory. Two of the genes were found to be at the same location and have subsequently been found to represent different mutant alleles of the same gene. In the following discussion, modern terminology is used for the genes, but Sturtevant's original experimental results are used as data.

The genes used in this mapping are *w* (white eyes), *v* (vermilion eyes), *m* (miniature wings), *r* (rudimentary wings), and *y* (yellow body color). Sturtevant determined the cross-over frequencies for various pairings of these genes (Table 14.2).

Table 14.2
Cross-Over Frequencies for Sex-Linked Gene Pairs

GENES CONCERNED	Cross-Over Frequency
w and *v*	29.7%
w and *m*	33.7
w and *r*	45.2
w and *y*	1.0
v and *m*	3.0
v and *r*	26.9
v and *y*	32.2
m and *y*	35.5
r and *y*	37.6

Source: Adapted from A. H. Sturtevant, "The linear arrangements of six sex-linked factors in *Drosophila*, as shown by their mode of association," *Journal of Experimental Zoology* 14 (1913): 43–59.

Sturtevant realized that distance between genes might not be the only factor affecting cross-over frequency. For example, some parts of the gene might break more easily than others, thus giving a higher cross-over frequency for that part of the gene. However, the maps were prepared to represent only the "statistical distance" between genes based on the assumption that distance is proportional to cross-over frequency. Later work did show that various parts of the gene are more susceptible to crossing over, and therefore the cross-over maps are now known to represent somewhat distorted pictures of the physical locations of the genes on the chromosomes. However, independent techniques confirm the sequence of genes determined by cross-over studies.

As a unit of distance, Sturtevant chose a length of the chromosome such that, on the average, 1 crossing over will occur in that length for every 100 gametes formed. In other words, cross-over frequency expressed as a percentage is used as an index of distance. For example, the distance between genes *w* and *v* is 29.7 units, between genes *w* and *m* is 33.7 units, and so on.

On a line representing the X chromosome, *w* may be placed at an arbitrary position, with *v* 29.7 units away from it. The distance between *w* and *m* is 33.7 units, and the distance between *v* and *m* is only 3.0 units. The only position for *m* that is compatible with these distances is 3.0 units beyond *v*. (A position on the other side of *w* would be much too far from *v*.) The distances do not quite add up—the distance from *w* to *m* (33.7 units) should be equal to the sum of the distance from *w* to *v* (29.7 units) and the distance from *v* to *m* (3.0 units). However, the discrepancy is quite small and could be due to inaccuracies in determining the cross-over frequencies.

Next, *r* can be added to the map. It is 45.2 units from *w* and only 26.9 units from *v*, so it must be located beyond *v* and *m*. (Sturtevant had not yet measured the cross-over frequency between *m* and *r* when he prepared this first map.) This time the discrepancy is more serious; the sum of cross-over frequencies between *w* and *v* and between *v* and *r* is 56.6 percent, whereas the cross-over frequency between *w* and *r* is only 45.2 percent.

Finally, *y* can be placed on the map. It is only 1.0 unit from *w*, but which side should it go on? The data are inconsistent, because *y* is farther from *v* and *m* than is *w*, but it is closer to *r* than is *w*. Sturtevant chose to trust the figures obtained for the shorter distances and to put *y* on the map to the left of *w*. Using *y* as the arbitrary beginning of the chromosome and using the shortest distance measurements available to place the chromosome distances between genes, Sturtevant obtained his first simple map of the X chromosome.

He found that cross-over frequencies over long distances on the map are always smaller than the value predicted by adding the cross-over frequencies for intermediate distances. He explained this discrepancy by pointing out that "double cross-overs" within the long distance would leave the genes being studied on their original chromosomes and thus would not be detected in the breeding experiments. Therefore, any cross-over frequency measured by the outcome of crossing experiments will be lower than the true frequency of crossing over. The discrepancy will be largest for the longest distances. For this reason, Sturtevant used measurements between adjacent genes to construct his map (Figure 14.11a).

Crossing experiments involving three or more genes can be used to check the hypothesis of double cross-overs. Figure 14.11b summarizes an experiment of this

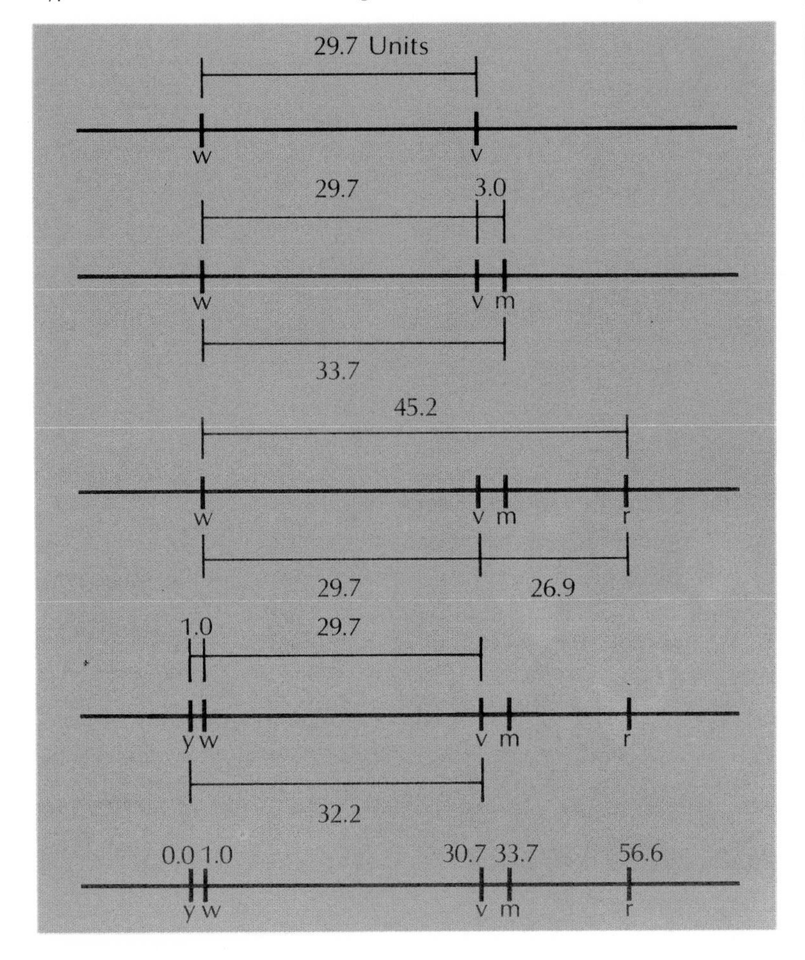

Figure 14.11a. Maps of Sturtevant's experiment involving *w*, *v*, *m*, *r*, and *y* genes.

Figure 14.11b. Summary of an experiment used to check the hypothesis of double cross-overs.

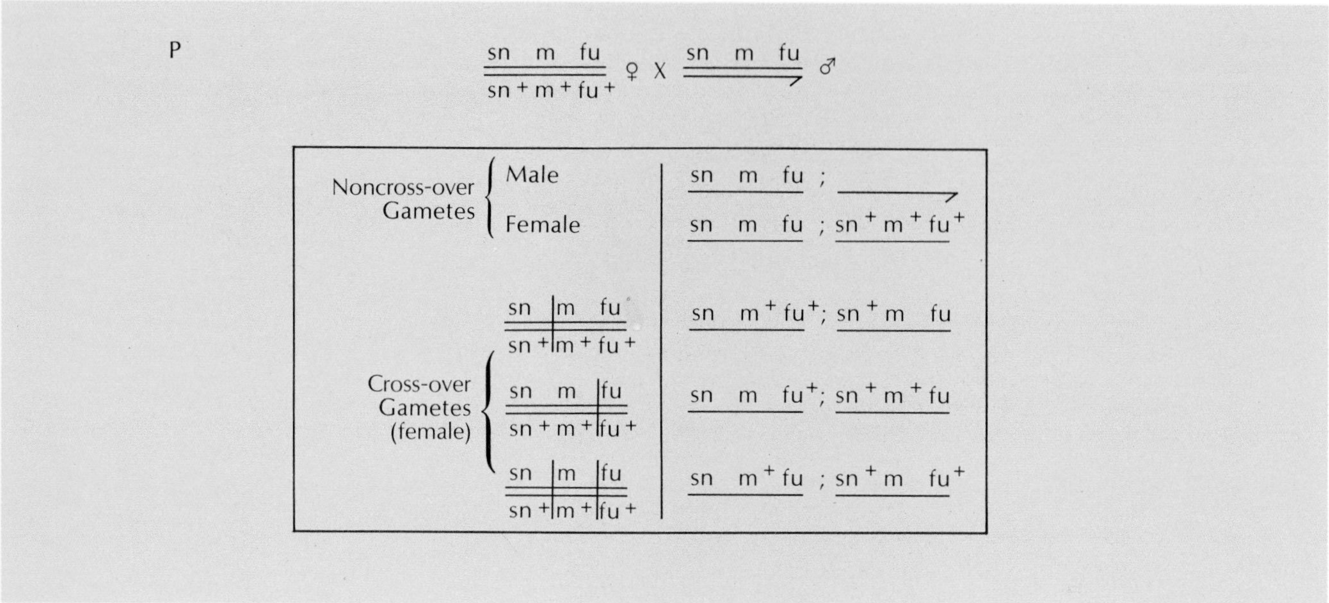

P

$\frac{sn \; m \; fu}{sn^+ \; m^+ \; fu^+}$ ♀ X $\frac{sn \; m \; fu}{\rightharpoonup}$ ♂

Noncross-over Gametes	Male	sn m fu ; ———
	Female	sn m fu ; $sn^+ \; m^+ \; fu^+$
Cross-over Gametes (female)	$\frac{sn \mid m \; fu}{sn^+ \mid m^+ \; fu^+}$	sn $m^+ \; fu^+$; sn^+ m fu
	$\frac{sn \; m \mid fu}{sn^+ \; m^+ \mid fu^+}$	sn m fu^+; $sn^+ \; m^+$ fu
	$\frac{sn \mid m \mid fu}{sn^+ \mid m^+ \mid fu^+}$	sn m^+ fu ; sn^+ m fu^+

F$_1$

	Genotypes	Phenotype	Number	Percentage
Noncross-over	$\frac{sn \; m \; fu}{sn \; m \; fu}$ ♀ ; $\frac{sn \; m \; fu}{\rightharpoonup}$ ♂	Singed-bristle Miniature-wing Fused-wing	3,661	66.3
	$\frac{sn \; m \; fu}{sn^+ \; m^+ \; fu^+}$ ♀ ; $\frac{sn^+ \; m^+ \; fu^+}{\rightharpoonup}$ ♂	Wild-type	3,672	
sn/m Cross-over	$\frac{sn \; m^+ \; fu^+}{sn \; m \; fu}$ ♀ ; $\frac{sn \; m^+ \; fu^+}{\rightharpoonup}$ ♂	Singed-bristle Normal-wing	665	12.1
	$\frac{sn^+ \; m \; fu}{sn \; m \; fu}$ ♀ ; $\frac{sn^+ \; m \; fu}{\rightharpoonup}$ ♂	Normal-bristle Miniature-wing Fused-wing	676	
m/fu Cross-over	$\frac{sn \; m \; fu^+}{sn \; m \; fu}$ ♀ ; $\frac{sn \; m \; fu^+}{\rightharpoonup}$ ♂	Singed-bristle Miniature-wing	1,041	18.5
	$\frac{sn^+ \; m^+ \; fu}{sn \; m \; fu}$ ♀ ; $\frac{sn^+ \; m^+ \; fu}{\rightharpoonup}$ ♂	Normal-bristle Fused-wing	1,003	
Double Cross-over	$\frac{sn \; m^+ \; fu}{sn \; m \; fu}$ ♀ ; $\frac{sn \; m^+ \; fu}{\rightharpoonup}$ ♂	Singed-bristle Fused-wing	165	3.1
	$\frac{sn \; m^+ \; fu}{sn \; m \; fu}$ ♀ ; $\frac{sn^+ \; m \; fu^+}{\rightharpoonup}$ ♂	Normal-bristle Miniature-wing	173	

kind. Three genes located on the X chromosome are studied. The *miniature* gene (small wings) has already been discussed. The other two genes are *singed* (curled and twisted bristles) and *fused* (certain wing veins joined together). A female heterozygous for all three characters is crossed with a male showing the recessive mutant traits.

In addition to the four kinds of gametes that can be produced by linked reassortment, there are six kinds of female gametes that can be produced by crossing over. A hooked line has been used to represent the Y chromosome to emphasize the fact that crossing over cannot occur between X and Y chromosomes. In this kind of test cross, both males and females can be used in estimating cross-over frequency. Both the Y chromosome and the male's completely recessive X chromosome allow the alleles on the chromosome from the female to be detected in the phenotype.

Double crossing over did occur in 3.1 percent of the gametes. Because these cases represent crossing over both between *sn* and *m* and between *m* and *fu*, these cases should be counted in determining cross-over frequencies for both regions. In other words, the map distance between *sn* and *m* is 15.2 units (12.1 + 3.1), and the map distance between *m* and *fu* is 21.6 units (18.5 + 3.1). In a two-gene crossing experiment using only *sn* and *fu*, the double cross-overs would have been indistinguishable from the noncross-overs. Such an experiment would have indicated a distance of 30.6 units between *sn* and *fu* instead of the 36.8 units obtained in this experiment. It is possible that some double cross-overs have occurred in the regions between genes and not been detected or that some of the gametes recorded as single cross-overs actually represent triple cross-overs. Therefore, cross-over frequencies must always be regarded as minimum values for the true map distance being determined.

Even if the sequence of the three genes had not been determined earlier from two-gene crossings, it could be deduced from the results of this experiment by assuming that the smallest frequency phenotypes in the F_1 generation represent double cross-overs. Because the *m* alleles are exchanged in that group, *m* must lie between the other two genes.

chromosomes of this species, showing the relative locations of each of the 85 mutant genes then known (Morgan, *et al.*, 1915).

The consistent results of the mapping convinced Morgan and most other biologists that the genes are indeed located in linear order along the chromosomes. They found no evidence of branching or parallel chains of genes on a single chromosome. Every known gene could be assigned a consistent location on one of the chromosomes. Occasionally, two genes were found to have the same apparent location. In some cases, further experiments with larger numbers of flies revealed a very low frequency of crossing over, so that the two genes could be assigned locations very near to each other. In other cases, no crossing over could be found. For example, at a particular point near one end of the X chromosome, several mutant genes for different eye colors have been assigned the same location. It appears that all of these are different mutant alleles of the same gene, for all are recessive to the wild-type character.

Another of Morgan's graduate students, Calvin B. Bridges, discovered something that put the chromosomal theory of heredity almost beyond doubt. Bridges (1916) noticed that a vermilion-eye female occasionally turns up among the offspring of a cross between vermilion-eye females and red-eye males. Because the allele for vermilion eyes is recessive, a vermilion-eye female must have two of these alleles. Yet, in this cross, the male cannot contribute a mutant allele. Bridges guessed that these two alleles might occur if the X chromosomes of the mother fly failed to separate during meiosis, thus giving the egg a pair of X chromosomes, each carrying the mutant allele for vermilion eyes. If this egg were fertilized by a sperm carrying a Y chromosome, the resulting zygote would have the abnormal genotype *vv*(−). When Bridges examined the cells of the unexpected vermilion-eye females, he found exactly what he had predicted: two X chromosomes and a Y chromosome. This experiment provided dramatic and convincing support for the theory that genes are carried on the chromosomes.

Thus, Morgan—who had begun his career in genetics as an opponent of the Mendelists—became one of the outstanding proponents of a modified, chromosomal version of Mendelian genetics. He and his colleagues became the recognized leaders of genetic research and theory.

CHROMOSOMAL ABNORMALITIES AND MUTATIONS

The extra X chromosome in some vermilion-eye females was not the only chromosomal abnormality that Bridges discovered. If some eggs received an extra X chromosome during meiosis, it might be equally possible for both X chromosomes to go into the polar body, leaving the egg with no sex chromosome. Fertilization of such an egg by a sperm with an X chromosome would produce a zygote with a single unpaired X chromosome. Bridges found this condition in certain male flies that produce immotile sperm and thus do not reproduce. Apparently, the Y chromosome plays some role in the normal development of sperm. Fertilization of the egg without a sex chromosome by a sperm with a Y chromosome would produce a zygote with an unpaired Y chromosome. Bridges was unable to locate any flies with this condition, so he concluded that such zygotes die at a very early stage of development. Apparently, this condition is a lethal chromosomal abnormality. A fourth possibility would involve fertilization of an egg with two X chromosomes by a sperm with a third X chromosome.

Figure 14.12a (upper left). The normal male karyotype. This photomicrograph of a chromosome smear preparation was taken from a white blood cell of a normal male. The 23 chromosome sets have been repositioned in rank order to facilitate examination. Note the size disparity between the X and Y chromosomes.

Figure 14.12b (upper right). The normal female karyotype. This photomicrograph of a chromosome smear preparation was taken from a white blood cell of a normal female. These cytological preparations are made during the metaphase stage of the cells' mitotic division. Note the X chromosomes.

Figure 14.13 (lower left). Photomicrograph of the chromosomes found in the white blood cell of a human female with Turner's syndrome. Note the single X chromosome. Phenotypic expression of this chromosome abnormality results in retarded sexual development and sterility.

Figure 14.14 (lower right). Photomicrograph of the chromosomes found in the white blood cells of a Down's syndrome male. Note the additional chromosome to set number 21. This chromosome abnormality is phenotypically expressed as a form of Mongolian idiocy.

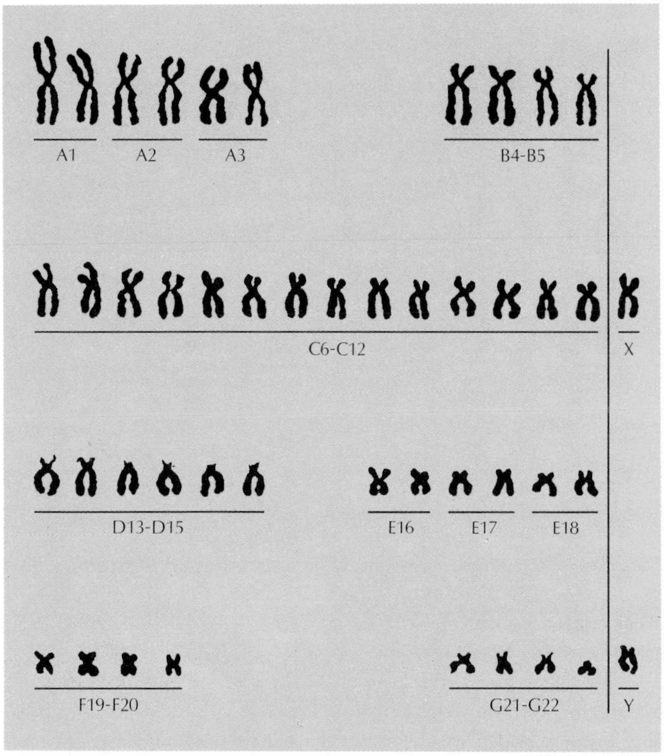

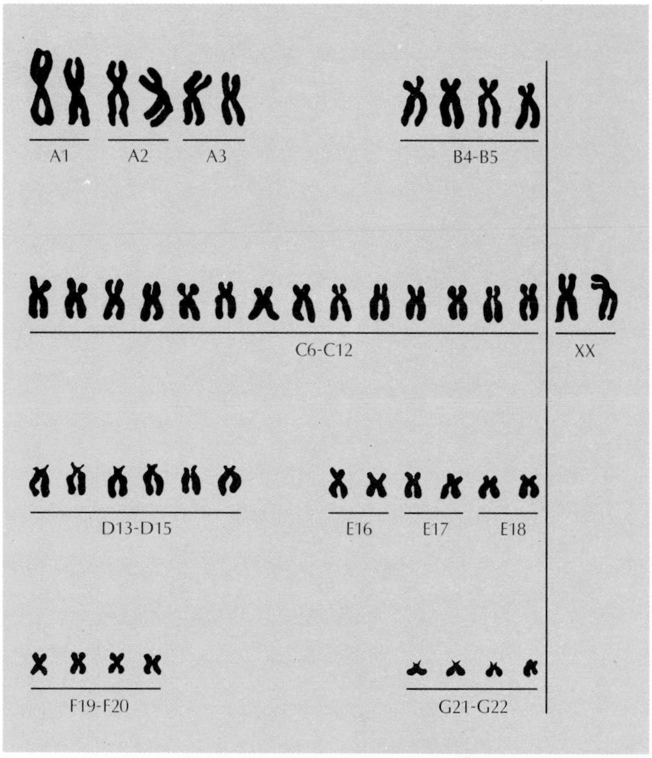

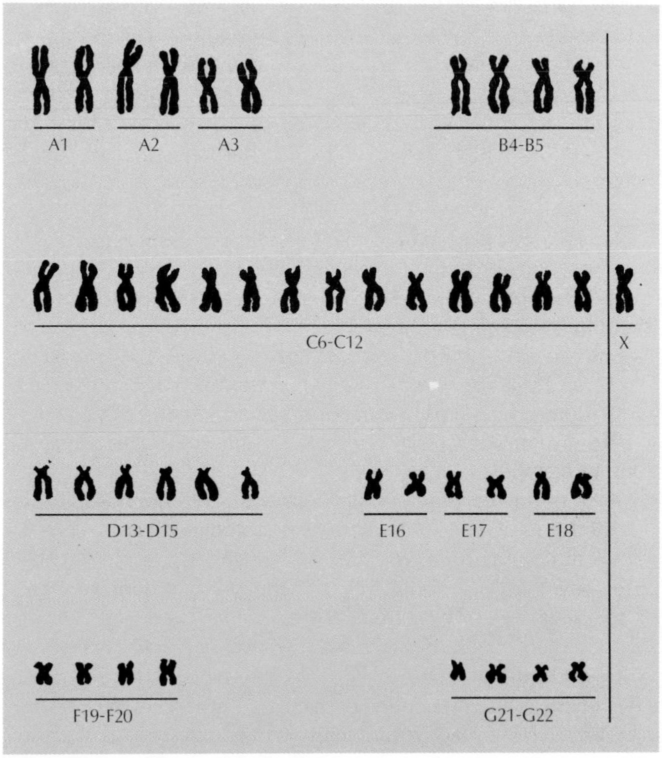

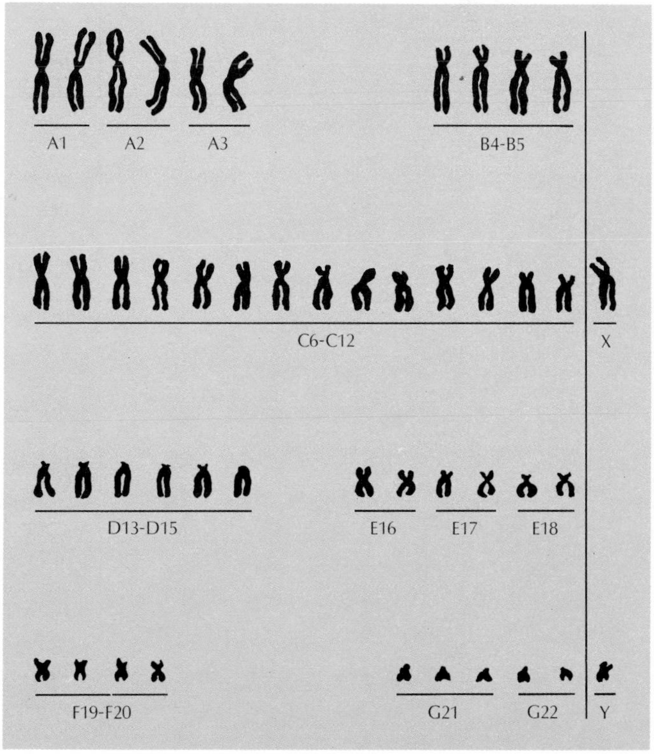

Figure 14.15. Photomicrograph of the chromosomes found in the white blood cell of a human male with Klinefelter's syndrome. Note the XXY genotype, which phenotypically results in severe mental retardation and may result in either a lack of male secondary sex characteristics or in the development of some female secondary sex characteristics.

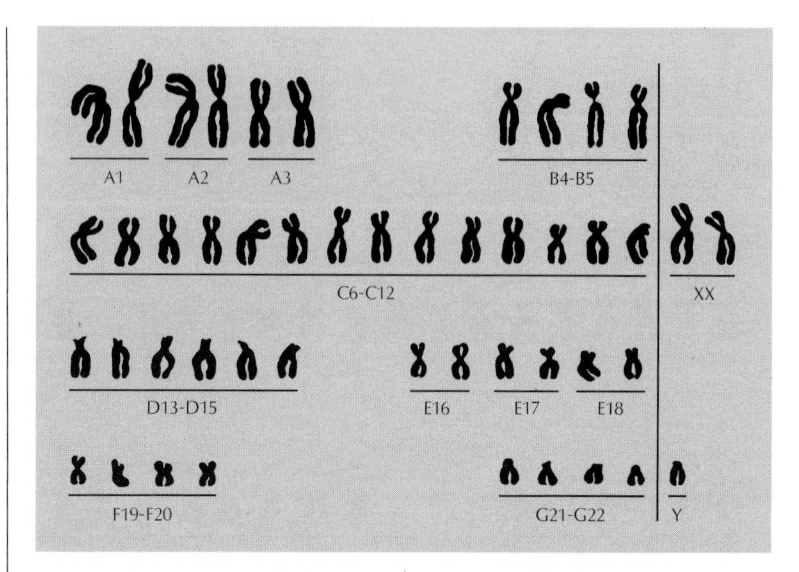

Bridges found cells containing three X chromosomes in certain female flies that die before reaching maturity.

Such variations in chromosomal number have subsequently been shown to cause many hereditary abnormalities. The absence of one chromosome is now called *monosomy*. An example of monosomy in humans is Turner's syndrome, which results from the lack of one of the X chromosomes and is expressed phenotypically in retarded sexual development. The presence of an extra chromosome is called *trisomy*. The presence of two X chromosomes in a human male causes a condition called Klinefelter's syndrome. Such a male is severely retarded mentally and either fails to develop normal secondary sexual characteristics of a male or develops some female characteristics such as enlarged breasts or broad hips. The XXY human male is invariably sterile. Down's syndrome—caused by the presence of three copies of one of the other human chromosomes—results in a form of Mongolian idiocy.

Most cases of trisomy in humans result in highly abnormal embryonic or fetal development and are lethal. The general term for the absence or duplication of part of the normal diploid chromosome set is *aneuploidy*. In most cases of aneuploidy in animals, the zygote fails to develop into an adult. In cases where adulthood is reached, the aneuploid individual is usually unable to reproduce. Thus, aneuploidy is seldom transmitted from one generation to the next. In some plants, a number of varieties with extra copies of particular chromosomes do exist, and the aneuploid condition may be passed along to offspring.

The presence of one or more complete extra sets of chromosomes is a condition called *polyploidy*. Cells with three complete sets of chromosomes are called triploid, those with four sets are called tetraploid, and so on. Because polyploidy interferes with normal synapsis during meiosis, polyploids are characterized by extremely low fertility.

Polyploidy in plants often results in exaggeration of certain characteristics of the flowers or of the fruit—for example, the size and number of petals or the fleshiness and sugar content of the fruit. Because such characteristics may be prized by farmers or gardeners, many domesticated plants are

Figure 14.16. The frequency of sex-linked lethal recessive mutations in *Drosophila* is a function of the x-ray dosage. The relationship between dosage and lethal mutations is a linear one.

polyploid mutants. Many of these mutants (particularly triploids) do not reproduce effectively by sexual means but must be reproduced through grafting or slipping techniques. Those polyploid species that can reproduce sexually (such as many crop plants that are tetraploid) pass the polyploidy along to their descendants. Polyploids, however, are usually not viable in animals.

Some hybrid plant species are formed by a combination of complete chromosome sets from the two parental species. For example, suppose that one species has the diploid set of chromosomes AABBCC and a fairly similar species has a different set XXYYZZ. In most cases, fusion of an ABC gamete from one species with an XYZ gamete from the other will produce a sterile hybrid with chromosomes ABCXYZ. Because the chromosomes do not pair properly at synapsis, the gametes of the hybrid will have incomplete chromosome sets and will be inviable. If each species becomes polyploid and thus produces diploid gametes, the fusion of an AABBCC gamete with an XXYYZZ gamete can produce a zygote with a complete diploid set of six chromosomes: AABBCCXXYYZZ. This hybrid can produce gametes with full chromosome sets because all of its chromosomes have homologous pairs. Furthermore, the hybrid will produce relatively infertile triploid zygotes if it crosses with one of the parental species. Thus, the hybrid becomes reproductively isolated and is apt to evolve rapidly into a species of quite different characteristics from the parental species. It appears that this process has been of great importance in the evolution of new plant species.

Other chromosomal abnormalities result from the breakage and rejoining of chromosomes. When a chromosome breaks, the broken ends act as if they are sticky and tend to rejoin. In most cases, the rejoining results in restoration of the original chromosomal configuration, but other things can happen. Segments may be duplicated or deleted, a part of the chromosome may be inverted, or a part of one chromosome may be joined onto another. Many of these changes are lethal to the cell in which they occur or result in severe abnormalities of zygote development. If the alteration does not prevent mitosis or meiosis, the altered chromosome may be replicated and passed on to future generations. For example, the bar-eye mutation of *Drosophila* is now known to involve duplication of a short segment of the X chromosome.

The genetic research in Morgan's laboratory was hampered by the extreme rarity with which new mutations appear naturally. Hundreds of thousands of flies were examined in order to find a few hundred mutant genes. Yet another of Morgan's students, Herman J. Muller, hit upon a new idea. He reasoned that most mutations involve only a single gene and that he needed to attack the chromosome with some agent that would affect only a tiny part of its length. High-energy radiation seemed to be the only thing that could accomplish the necessary microscopic damage. Muller (1927) subjected a group of *Drosophila* to a dose of x-rays so strong that some of the flies became sterile. He then crossed the remaining fertile flies with wild-type flies. When the offspring matured, Muller found more mutants than he could have hoped for. Comparison with a control group of untreated flies showed that the radiation had increased the mutation rate by 15,000 percent.

A large number of the new mutant genes proved to be lethal. Some were "dominant lethals," detected by the decreased reproduction rate of the treated flies. Others were "recessive lethals," which allowed the heterozy-

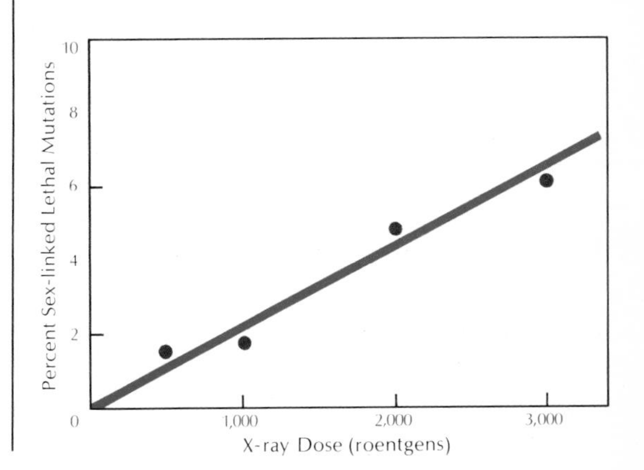

Figure 14.17. Photomicrograph depicting the giant salivary cell chromosomes of *Drosophila*. Note the distinct banding—dárk bands containing DNA and light bands containing RNA. Note the two distinct chromosome puffs. These puffed-out regions may be areas of gene activity where RNA transcription is taking place.

gous first-generation flies to mature but proved fatal to their offspring that were homozygous for the mutant allele. Many of the mutant characters already known were produced among the radiated flies, as well as a number of new mutant characters of similar nature—splotched wings, sex-combless, and so on. Under the microscope, many cases of chromosome breakage, inversion, and exchange could be seen. Chromosome mapping of some of the mutants confirmed the alteration of gene locations in a number of cases.

Other investigators soon showed that x-rays are equally effective in producing mutations in other organisms. With this new research tool to produce large numbers of mutations, genetic research could proceed much more rapidly. The nature of mutations could be studied far more easily. And perhaps of most importance, the fact that radiation produces mutations provided an explanation of the source of naturally occurring mutations. Cosmic rays, ultraviolet light, and natural radioactivity provide a constant source of low-level radiation that strikes all organisms. This natural radiation accounts for the natural appearance of mutations at a low rate in all organisms. These mutations provide the variability of characters that can be reassorted through Mendelian mechanisms to provide a variety of phenotypes within a population. This variety makes possible evolution through natural selection.

GIANT CHROMOSOMES

Cytologists had known since 1881 that the chromosomes in the salivary glands of *Drosophila* larvae are about 100 to 200 times as long as the chromosomes of other cells and have a banded structure. This fact did not come to the attention of geneticists, however, until T. S. Painter (1933) made a careful study of these cells using staining to make the banded patterns more visible. Painter showed that the pattern of banding is very regular for homologous chromosomes within individuals or between individuals of the same species. Using the banding pattern, he identified even small cases of inversion, duplication, deletion, exchange of chromosomal fragments, and so on. By comparing characteristics or organisms and the banding patterns of their salivary gland chromosomes, Painter identified the particular bands that correspond to particular genes. Thus, he prepared chromosome maps by a method entirely independent from the crossing over method used by Morgan's group. Comparison of Painter's physical maps with Morgan's sta-

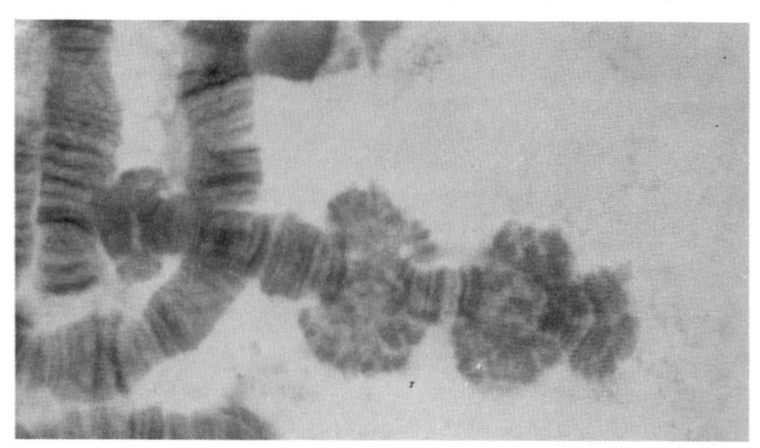

tistical maps revealed exactly the same sequence of genes, although the spacing between genes was different. Apparently, the frequency of crossing over does not depend solely upon distance between genes; some parts of the chromosome are more liable to cross over than are others. However, Painter's work dramatically confirmed the validity of the indirect inferences about gene order made by Morgan and Sturtevant.

Use of the giant chromosomes – which are produced by several consecutive duplications of the chromosomes to produce a total of more than 1,000 chromosomes lying side by side – made possible much more detailed chromosome maps (Bridges and Brehme, 1944). Studies of such detailed modifications of the chromosome revealed facts that forced a revision of prevailing ideas about genes. Most geneticists had assumed that genes are independent units that are merely carried by the chromosomes. If so, the effects of a gene should be the same regardless of its location. However, studies with the giant chromosomes showed that the same two genes may have different effects if they are on the same chromosome instead of on two different chromosomes, or if their sequence is inverted. Such *position effects* indicate that the genes are more closely related to one another on the chromosomes than had earlier been thought (Dobzhansky, 1936; Lewis, 1950).

GENES AND PHENOTYPIC CHARACTERS

As Morgan's group discovered and mapped genes in *Drosophila*, they found that almost every character of the adult organism is affected by a number of different genes on different chromosomes. For example, the color and shape of the eye are controlled by more than a dozen different genes on each of the three large chromosomes and by a few other genes on the small fourth chromosome. In addition, some genes have several different recessive alleles, each leading to development of a slightly different phenotypic character. To complicate matters further, there are many cases of incomplete dominance.

While Morgan's group and many other geneticists concentrated upon locating the genes and working out their relationships to various characters in the adult organism, other geneticists began to seek an understanding of the way in which the gene controls the development of these characters. How can a change in a minute segment of a chromosome affect a character such as eye color or wing shape in an adult *Drosophila?*

A brilliant suggestion was made early in the history of genetics by a British biochemist, Archibald Garrod (1909). A number of human diseases had long been known to be most common among the descendants of persons who also had the disease – that is, to be hereditary. Many of these diseases were known to be associated with the presence of unusual substances or the absence of normal substances in the urine or in the affected parts of the body. Combining ideas from the latest discoveries in genetics and in biochemistry, Garrod suggested that these diseases are "inborn errors of metabolism." An inherited, defective gene leads to the body's failure to produce a needed enzyme. Because of the absence of this enzyme, some portion of the normal metabolic processes goes astray. Chemicals that should have been broken down accumulate, or other necessary chemicals are not manufactured. Garrod showed that the same explanation can be applied to many of the mutant characters in *Drosophila* and other organisms. Furthermore, some human diseases show sex linkage similar to that

Figure 14.18 (above). Dr. George Wells Beadle, Nobel Prize-winning geneticist noted for his work on *Neurospora* leading to the one gene–one enzyme hypothesis.

Figure 14.19 (below). Photograph of the red bread mold *Neurospora crassa*.

observed in *Drosophila*. Wherever enough data were available to study the appearance of these diseases in successive generations, the ratios of phenotypes found proved quite similar to the results of experiments with recessive mutant genes in *Drosophila* and other organisms.

Some geneticists attempted to discover the relationships between genes and enzymes. Most of these attempts were relatively unsuccessful because the genetics and development of organisms are so complex that it is almost impossible to find simple relationships.

Again, the key to further discoveries was the use of a new organism particularly suited to the problems being investigated, and again the advance was the work of one of Morgan's colleagues. George Wells Beadle worked with Sturtevant on crossing experiments with *Drosophila* and corn, but Beadle felt that this sort of work was a dead end—there were many details to be resolved but little hope of any major new discoveries.

Beadle later joined forces with Edward L. Tatum, a biochemist at Stanford University. They decided to abandon research with *Drosophila*, an organism whose genetic properties had been thoroughly mapped but whose biochemistry was complex and poorly understood. Instead, they needed an organism with simple biochemical properties. In 1940 they began to work with a red bread mold, *Neurospora crassa*. The haploid cells and relatively short life cycle (ten days between sexual generations) of this mold make it suitable for genetic research. Its ability to reproduce asexually and rapidly enables a researcher to create a sizable sample of any genotype in quantities suitable for biochemical analysis. *Neurospora* thrives in a simple culture medium containing mineral salts, sugar, and the vitamin biotin.

When *Neurospora* reproduces sexually, haploid cells from two individuals fuse to form a zygote. Meiotic division of the zygote produces four haploid cells, each of which then divides by mitosis. The eight spore cells that result are neatly lined up in a spore sac. Because of the orderly process of cell division within a confining structure, the sequence of the spores in the sac can be directly correlated with the separation of chromosomes in the various divisions of the zygote. With a microscope, a trained laboratory assistant can isolate one spore sac, remove the eight spores in sequence, and place each into a tube of culture medium. Rapid asexual reproduction

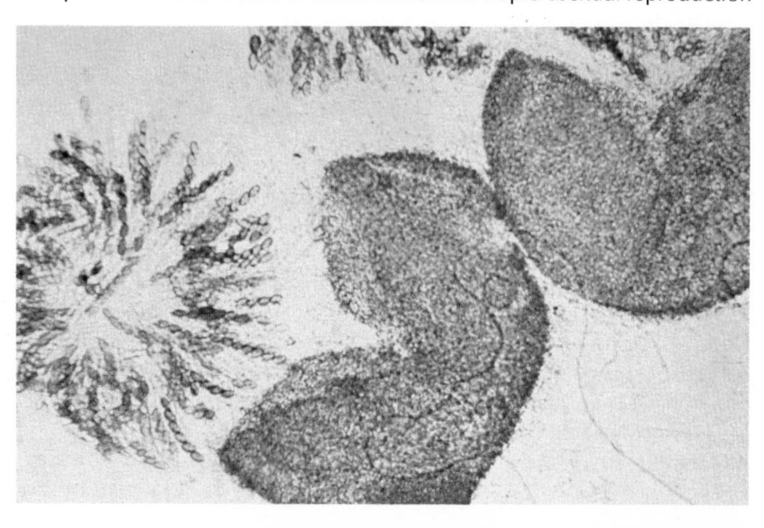

Figure 14.20. The life cycle of *Neurospora crassa*.

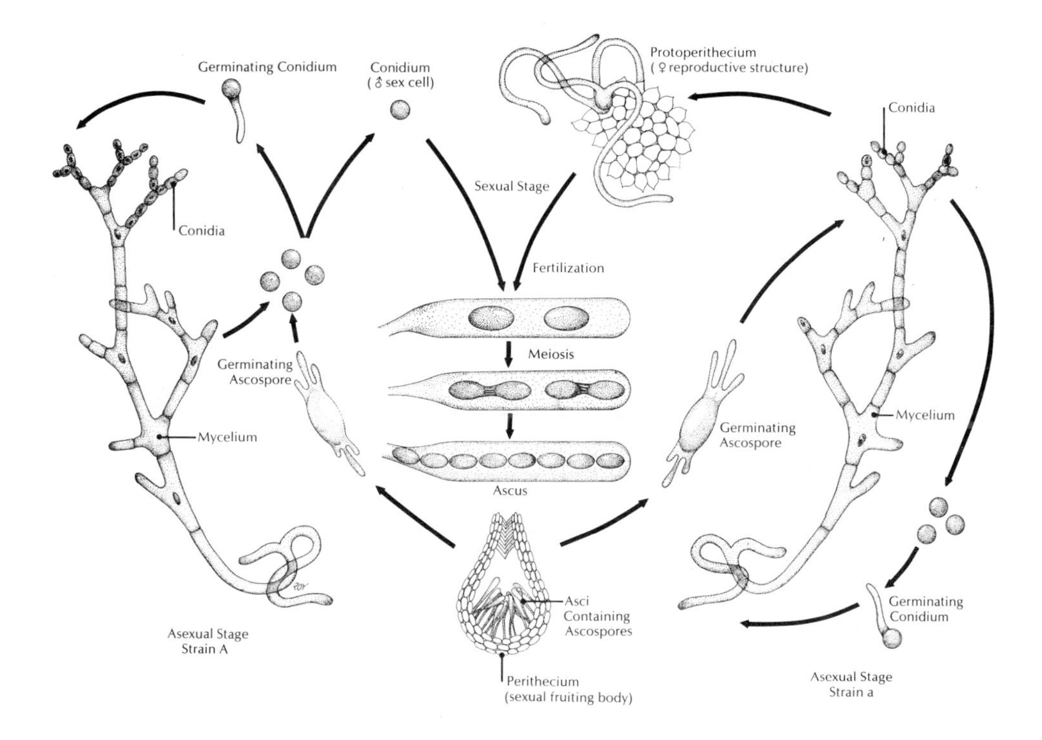

leads to a sizable population derived from a single chromosome set. Thus, the complex crossing experiments needed to analyze the genetic composition of *Drosophila* can be avoided for the most part.

Neurospora synthesizes all the substances it needs from the mineral salts, sugar, and biotin in the minimal culture medium. Beadle and Tatum reasoned that this complex metabolic machinery must involve a great many enzymes. If the presence of a mutant gene results in the absence of a particular enzyme (as they strongly suspected from their earlier work with *Drosophila*), the mutant mold should fail to synthesize some needed substance. It will then fail to survive on the minimal medium. By finding out what substance it needs for survival, they should be able to correlate a particular mutant gene with the failure of a particular synthetic step in metabolism. Biochemical studies could then reveal the enzyme needed for that particular synthesis. Thus, Beadle and Tatum hoped to establish a one-to-one correlation between the genes of *Neurospora* and the enzymes that it normally manufactures.

They began by exposing cultures of *Neurospora* to x-rays in hopes of producing mutations. From the irradiated *Neurospora*, they obtained 2,000 spores, each of which was planted in a tube containing a complete medium made of yeast and malt extracts in addition to mineral salts and sugar. This complete medium contains most of the substances that the mold normally synthesizes for itself, so that even mutant strains will thrive in it. After sizable populations had been established in each tube by asexual reproduction, a small sample of each culture was transferred to a tube containing the minimal medium. Three of the samples failed to grow on the minimal medium. For each of these strains, samples were tested on a sequence of media, each consisting of the minimal media plus one vitamin,

Figure 14.21 (A) A portion of the pathway of arginine biosynthesis in *Neurospora*. An essential enzyme is required in each step of the process in order to catalyze the reactions culminating in arginine synthesis. (B) Growth requirements of three groups of arginine-requiring mutants in *Neurospora*.

amino acid, or other organic chemical present in the complete medium. In this way, Beadle and Tatum (1941) discovered exactly what substance the mutant strains require for growth. One proved unable to synthesize pyridoxine (vitamin B_6), another, thiamine (vitamin B_1), and the third, para-aminobenzoic acid. The thiamine molecule is composed of two parts, a pyrimidine half and a thiazole half. The second mutant strain was not only able to grow with thiamine added to the minimal medium but also was able to grow with only thiazole added. Apparently, it could synthesize the pyrimidine part of the thiamine molecule but not thiazole.

In later studies, Beadle and Tatum worked out various pathways of synthesis in the normal organism by study of mutant strains. For example, one mutant strain grew only if the amino acid arginine was added to its minimal medium. Another mutant strain grew on either arginine or citrulline. A third strain accepted arginine, citrulline, or ornithine. Presumably, arginine is normally synthesized in *Neurospora* by the reaction pathway: ornithine → citrulline → arginine. The ornithine is synthesized from simpler precursors. Each step in the pathway is catalyzed by a specific enzyme.

A

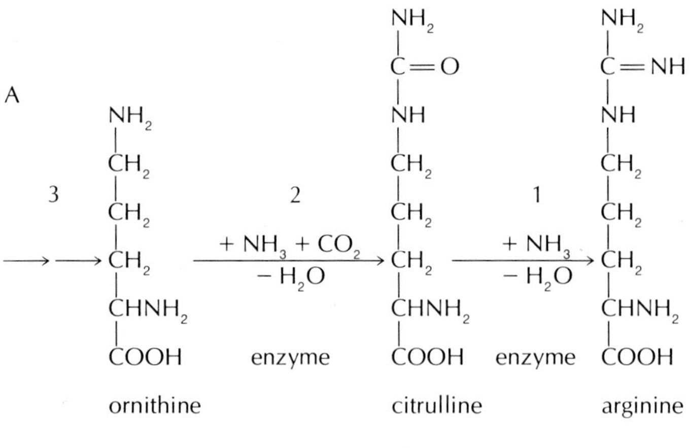

ornithine: NH_2–CH_2–CH_2–CH_2–$CHNH_2$–COOH

$\xrightarrow{3}$ ornithine $\xrightarrow[-H_2O]{+NH_3+CO_2}$ (2, enzyme) citrulline $\xrightarrow[-H_2O]{+NH_3}$ (1, enzyme) arginine

citrulline: NH_2–C(=O)–NH–CH_2–CH_2–CH_2–$CHNH_2$–COOH

arginine: NH_2–C(=NH)–NH–CH_2–CH_2–CH_2–$CHNH_2$–COOH

B

Arginine-requiring mutant	growth on minimal	ornithine	citrulline	arginine	reaction blocked
1	–	–	–	+	1
2	–	–	+	+	2
3	–	+	+	+	3

The first mutant strain lacked the enzyme that converts citrulline to arginine. Neither ornithine nor citrulline did any good for this strain; it needed arginine to survive. The second strain lacked the enzyme that converts ornithine to citrulline. If supplied with citrulline, it could synthesize arginine. The third strain lacked an enzyme for an earlier step in the pathway and thus was able to convert either ornithine or citrulline into arginine.

In every case, Beadle and Tatum found that a particular mutant strain was deficient in only a single step in a reaction pathway and therefore presumably in a single enzyme. They concluded that each enzyme is produced under the direction of a single gene. This postulate has come to be known as the *one gene–one enzyme hypothesis*. They demonstrated that each synthetic deficiency is in fact inherited as a single gene in crosses of the mutant strain with normal strains (Beadle, 1945a, 1946).

Analogous experiments with many different organisms have subsequently revealed mutations that affect single steps in many different biosynthetic pathways. The one gene–one enzyme hypothesis was modified slightly when it was later discovered that many enzymes are composed of two or more separate polypeptide chains, each of which is produced under the direction of a separate gene. This discovery led to the restatement of the postulate as the *one gene–one polypeptide hypothesis*, and it now has been extensively confirmed (Wagner and Mitchell, 1964).

The work of Beadle and Tatum established *Neurospora* as a genetic research subject second in popularity only to *Drosophila*. Even more important was the dramatic evidence that biochemical studies could carry understanding of heredity a great deal further. From the 1940s onward, genetics and biochemistry became ever more closely related fields. Now that the general mechanism by which the gene directs cell metabolism through enzymes was understood, the next step was clearly an understanding of the mechanism by which the gene controls synthesis of enzymes. What is the mechanism of gene action? Solution of this problem was the work of molecular geneticists in the last two decades.

FURTHER READING

Many of the important theoretical and research reports mentioned in this chapter and the next are reprinted in the collection edited by Peters (1959). Historical accounts and further discussion of the material in this chapter and the next are given by Beadle and Beadle (1966), Muller (1951), and Sullivan (1967). Burdette (1962) describes many of the experimental techniques commonly used in genetics research.

A full understanding of Mendelian and classical genetics principles can best be developed by working through a number of examples. For more detailed information and exercises, see introductory genetics texts such as those by King (1965) and Srb, Owen, and Edgar (1965).

For further information about *Drosophila* see Bridges and Brehme (1944), Demerec (1950), Demerec and Kaufman (1961), Herskowitz (1952, 1958, 1964), Morgan (1926), Morgan and Bridges (1916), Muller (1939), Sang (1956), and Strickberger (1962). Further information about *Neurospora* is given by Beadle (1945b), Bonner (1946), Horowitz (1950), and Lindegren (1932, 1933).

15
Molecular Genetics

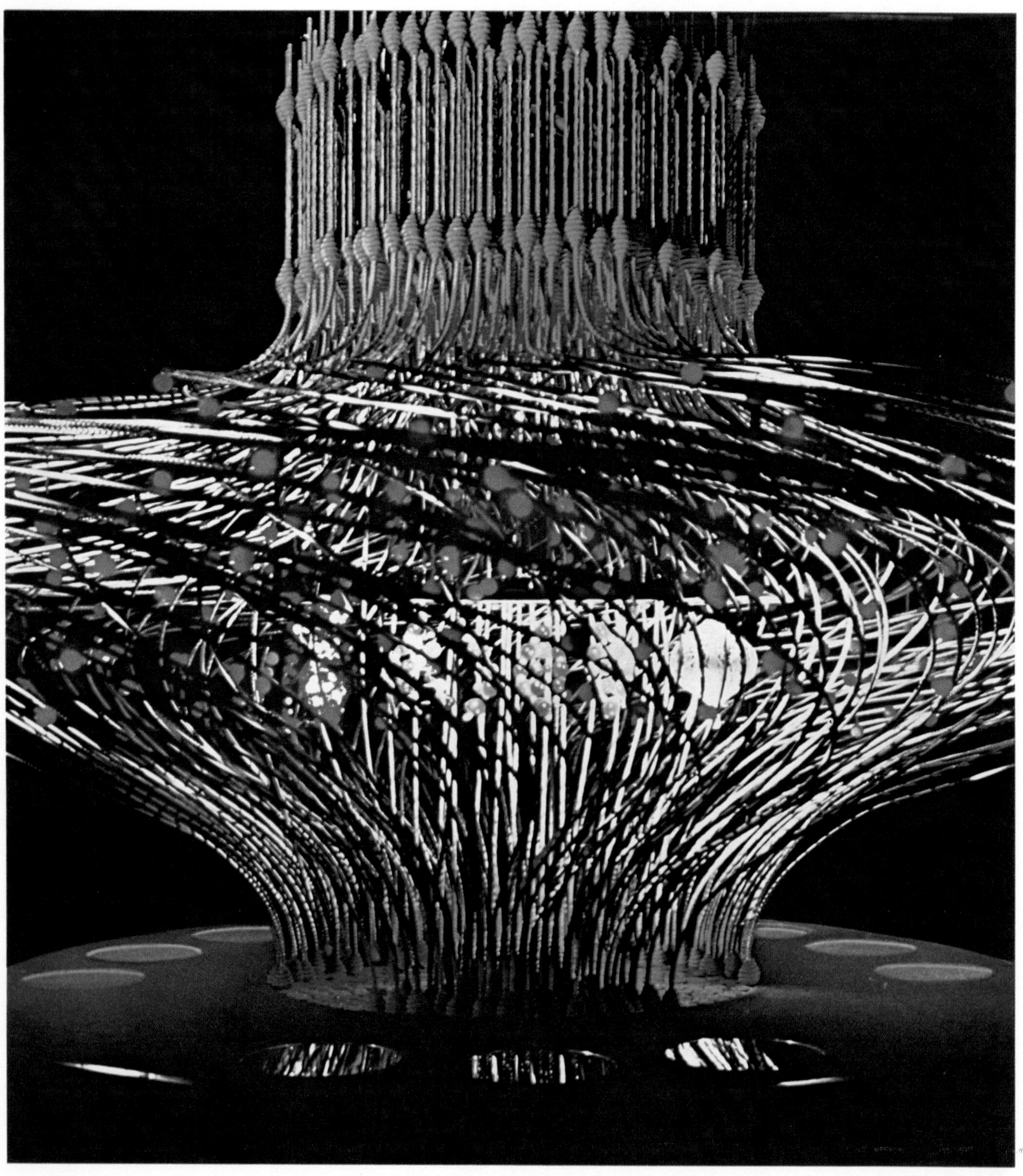

The greatest achievement in biology in this century has been the development of the field of molecular genetics. Ingenious and imaginative experimentation—combined with powerful tools of chemical analysis—has produced a remarkable series of discoveries leading to a precise description of the chemical nature of the gene and a deep understanding of gene function and gene mutation. The central principle by which genes replicate and provide genetic information to the cells of an organism is now understood in detailed molecular terms. These discoveries are now being extended by an enormous and intensive research program on a worldwide scale. The effort is directed toward a full understanding of the biochemical basis of structure and function of all living systems. The potential value of this knowledge is beyond estimation.

Already the accomplishments of molecular geneticists have led to new triumphs of medical technique. The possibility of a prevention or cure for cancer and hereditary disease gives new public significance to experiments and theories, and it is not unusual to find accounts of esoteric research results in the newspapers or the newsmagazines. Some biologists have begun to worry about the wisdom of further progress toward the ability to manipulate human genes. With the example of atomic power always in mind, one wonders whether the human race is ready to use such power in a beneficial fashion. For the time being, the techniques of genetic engineering remain a futuristic speculation, and most geneticists are convinced that the almost limitless beneficial results of further knowledge of genetic processes outweigh the danger of its misuse.

NUCLEIC ACIDS AND CHROMOSOMES

Nucleic acids were discovered at almost the same time that Mendel was experimenting with pea plants. By 1881 it was established that chromatin either is composed of or is closely associated with nucleic acids, but its importance went unnoticed.

At the turn of the century, a number of discoveries forced a reassessment of the role of nucleic acids. Careful study of chromosomes showed that chromatin varies in amount during different parts of the cell cycle and even seems to disappear entirely from many cells during interphase. Because chromatin and nucleic acids were believed to be the same substance, biochemists considered the disappearance of chromatin observed in the staining of many cells as evidence that nucleic acids are broken down during certain parts of the cell cycle. An obvious requirement of the genetic material is that it must be stable, and nucleic acids were not. At the same time, proteins were shown to be polymers that acted as enzymes controlling most of the chemical machinery of the cell. In addition, they were thought to be stable. Biochemists came to view proteins as the primary chemicals of life. Obviously, they reasoned, the persistent and extremely complex proteins—not the impermanent nucleic acids—must be the genes. The relatively simple molecules of nucleic acids, consisting of only four bases, merely provided structural support for the chromosomes; the associated proteins, with their great variety of structures and chemical properties, were the carriers of heredity information.

Suggestive Evidence

During the first half of the twentieth century, much evidence suggested that the nucleic acids, not the proteins, were the genetic material. Ultraviolet

Figure 15.1. Absorption spectrum of DNA and protein in the ultraviolet range. Note that the number of mutations at any wavelength follows the pattern of the DNA absorption spectrum and not that of the protein.

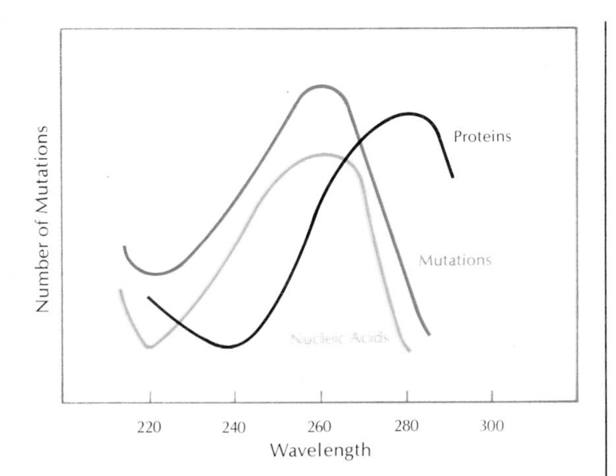

light was known to cause mutations in cells, presumably because its absorption by the cell produced chemical modifications of the genetic material. Therefore, the wavelengths of ultraviolet light that were absorbed most strongly would be expected to produce the most mutations. Because the absorption spectra (that is, the amount of absorption versus different wavelengths) of nucleic acids and proteins are significantly different, a spectrum plotting the number of mutants produced versus wavelength might indicate the identity of the genetic material. The results coincided with the spectrum of nucleic acids but not with that of proteins. In the wavelengths where nucleic acids strongly absorbed ultraviolet light, there were many mutations; where absorption was small, there were few mutations.

In any given species of animal, the amount of DNA is the same in all cells except sperm or egg cells. Even during the phases of the cell cycle when the chromatin seems to disappear, the DNA content remains constant. The sperm and egg cells contain just half as much DNA as do the other body cells, as would be expected if the offspring received one-half of their genetic material from each parent. Other studies of this nature revealed a much greater variation in the amount of protein or other chemicals present from cell to cell. Many similar experiments on a variety of organisms and cell types gave consistent results: although the amount of DNA in each cell may vary from one type of organism to another, each cell of a particular organism has the same amount of DNA. In short, DNA is present in constant quantity proportional to the number of chromosomes present, just as would be expected for the hereditary material, whereas proteins and other substances fail to meet this criterion of constancy.

Another suggestion of DNA's suitability as a genetic molecule came from studies of its stability. The breakdown of genetic molecules would represent loss of genetic information with no means of rebuilding the destroyed genes. Therefore, the macromolecule that stores genetic information must be highly stable. The stability of DNA was confirmed by use of radioactive labels. In most experiments, DNA is labeled by supplying H^3-thymidine or C^{14}-thymidine during interphase, when DNA is being synthesized. These experiments reveal that radioactivity, once incorporated into the DNA, is not freed from the DNA molecule so long as the cell remains alive. DNA, once made, does not undergo breakdown and resynthesis in the living cell but shows the high degree of stability expected of the genetic material. On the other hand, similar experiments reveal that all other types of macromolecules within the cell—proteins, carbohydrates, and RNA—undergo breakdown and resynthesis with considerable reshuffling of atoms.

The scientific discovery that established the groundwork for the subsequent era of molecular genetics was made in the 1920s by Fred Griffith. Griffith was studying the bacterium *Diplococcus pneumoniae* (also called pneumococcus), an organism that causes pneumonia. He was trying to develop a method for immunization against the disease, but he failed to achieve his objective. Instead, he obtained puzzling experimental results that could not be explained by the knowledge of the time.

Griffith worked with two strains of bacteria. On a nutritive plate, one strain formed colonies that had a smooth surface; the other strain formed colonies with a rough surface. The smooth (S) strain was highly virulent (infectious and damaging to the host organism), whereas the rough (R)

Figure 15.2. Diagram of Griffith's experiment. The genetic composition of the bacteria *Diplococcus pneumoniae* is transformed by the addition of heat-killed cells of a different strain. This experiment set the stage for the proof that DNA is the genetic material of the cell. *Printed by permission from J. D.* Watson, Molecular Biology of the Gene, © *1965, J. D. Watson*

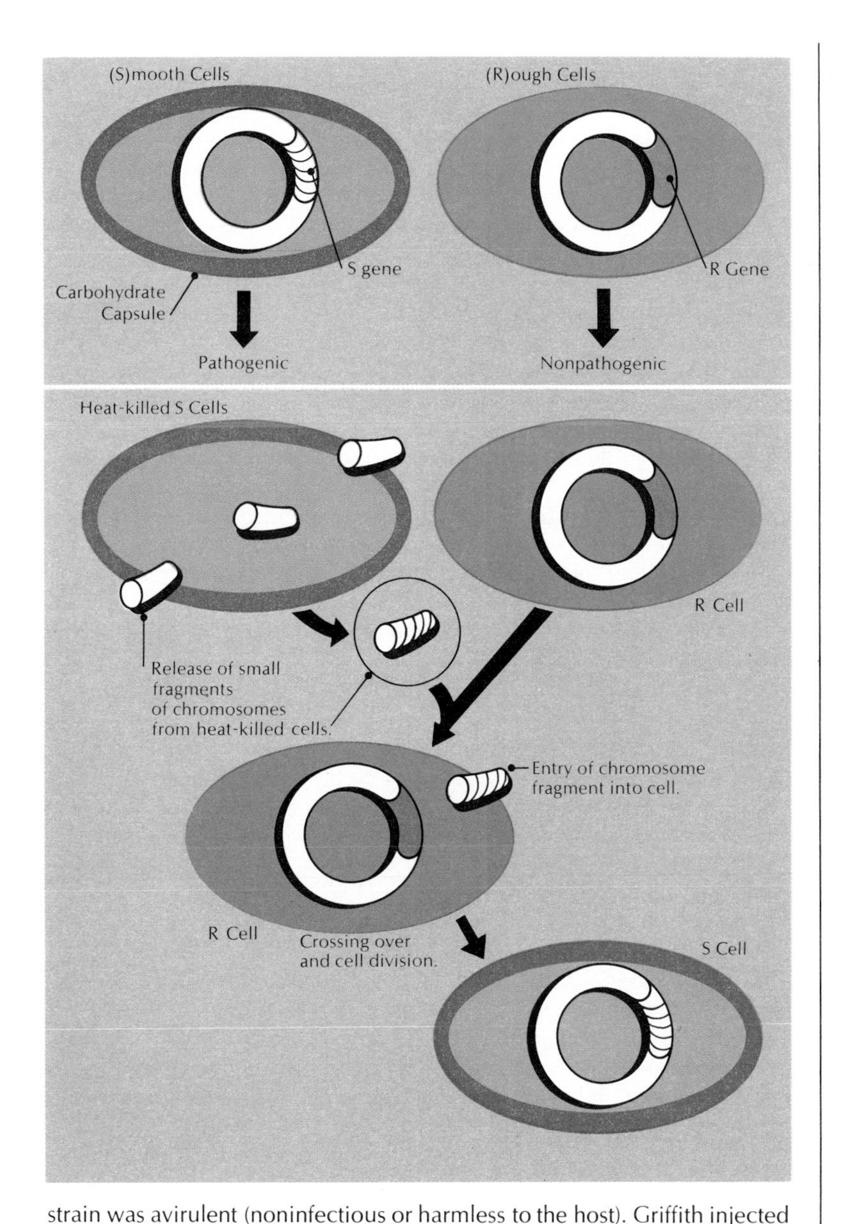

strain was avirulent (noninfectious or harmless to the host). Griffith injected S bacteria into mice and found, as he expected, that the bacteria multiplied rapidly and the mice developed pneumonia and died within a few days. The injection of R bacteria into mice led to no signs of illness, and the bacteria disappeared from the host animals within a short time. The explanation for these results was well known. The S bacterium has a polysaccharide cell wall, or capsule, that inhibits the attack of leucocytes (white blood cells). No such capsule protects the R bacterium, and the leucocytes readily ingest and destroy it.

The ability of a bacterium to produce a capsule is a hereditary trait. When S bacteria undergo binary fission, the offspring are also S bacteria.

Similarly, the offspring of R bacteria possess the avirulent R traits. It is not the capsule of the virulent strain that damages the host organism. Large populations of R bacteria would be equally damaging, but the body's defense mechanisms eliminate defenseless R invaders before they can establish large populations.

In a series of routine experiments designed to confirm these theories about the pneumonia bacteria, Griffith prepared a vaccine consisting of S bacteria that had been killed by heating to 60°C. Although the capsules still surrounded these dead bacteria, their presence in the bloodstream produced no ill effects in the mice. Next, Griffith prepared a vaccine containing a mixture of live R bacteria and dead S bacteria, a mixture that also should have been harmless to the mice. To his surprise, mice injected with this vaccine developed pneumonia and died. Microscopic examination showed that the sick mice contained large populations of living bacteria with polysaccharide capsules. When isolated on nutrient plates, these bacteria continued to produce offspring with virulent S traits. Somehow the presence of dead S bacteria in the mouse transformed the live R bacteria into S bacteria—a transformation that involved permanent change of a hereditary trait. How could the presence of dead bacteria transform the genetic information of live bacteria? The only explanation that Griffith could suggest was that live R bacteria ingested the dead S bacteria and somehow were transformed by this diet.

Griffith's experiments were later duplicated outside the living cell. A mixture of dead virulent bacteria and living avirulent bacteria in a suitable medium led to the genetic transformation of the living bacteria. Once again, this transformation occurred only in the presence of dead virulent cells. Later experiments showed that a fluid extracted from the dead virulent bacteria can carry out the genetic transformation of the avirulent strain. This material came to be known as the *transforming factor*. It was Griffith's work that set the stage for the proof that DNA is the genetic material.

The Proof

In 1944 O. T. Avery, C. M. MacLeod, and M. McCarty succeeded in purifying the transforming factor. This material proved to be DNA, as evidenced by its susceptibility to DNase, an enzyme that specifically destroys DNA molecules; an extract treated with DNase will not cause genetic transformation. Other substances such as proteins proved ineffective in transforming avirulent bacteria.

The work of Avery, MacLeod, and McCarty represented the first direct proof that a molecule of DNA can carry genetic information.

Further confirmation of the genetic role of DNA soon came from studies of the viruses that infect the bacterium *Escherichia coli*. These viruses were known to consist only of protein and DNA, but it had been assumed that the protein plays the active role in the virus' takeover of a bacterial cell for its own reproduction. However, early electron micrographs using shadowing techniques showed that the phage attach to the bacterial cell wall and become flat, empty shells, suggesting that only part of the phage enters the cell.

The assumption that phage DNA is the active agent in takeover of the bacterial cell was confirmed in an experiment performed by Alfred Hershey and Martha Chase (1952). They grew phage-infected bacteria in media containing either radioactive sulfur (sulfur-35) or radioactive phosphorus

The following series of electron micrographs and autoradiographs describes the progressive sequence of the "central dogma," which can be schematically shown as:

DNA —*transcription*→ *RNA* —*translation*→ *PROTEIN*
replication

Figure 1 is an autoradiograph of DNA replicating in *E. coli.* Figure 2 shows a tracer experiment indicating the role of RNA as an intermediate in protein synthesis. Radioactive uridine (found only in RNA) is first seen in the nucleus and later in the cytoplasm. Figure 3 is an electron micrograph of a polyribosome with a thin strand of linking mRNA—the level of translation. The final autoradiograph (Figure 4), completing the sequence, shows the production of protein with tritium-labeled leucine (an amino acid) in guinea pig pancreatic cells.

Interleaf 15.1

THE "CENTRAL DOGMA"

The process of life is intimately connected to molecules. Nucleic acid and proteins are two of the main containers of that life. The progressive panels of this interleaf reveal the mechanisms by which information is transmitted from the nuclear DNA out to the cytoplasm, where proteins are formed. Panels B, C, D, and part of E show the events that take place in the nucleus, whereas panels E and F show those occurring in the cytoplasm.

The "central dogma" is the key to understanding this molecular basis of life. Panels B and C depict DNA in the various conventions that biochemists use to define molecules. DNA is progressively shown by its chemical formula, stick-ball form, softened stick-ball form, space-filling—or effective size—model (based on electron-cloud size), and finally in symbolic form in panel D. Panel D shows the two strands separating and self-replication occurring. It is this self-replication mechanism that ensures that genetic information is preserved and transmitted from one generation to the next.

In order to present a clear view of events, the perspective of these illustrations has been distorted somewhat and the individual strands separated so that the three-dimensional quality of the molecule can be observed. Enzyme-mediated reactions throughout this interleaf are depicted symbolically by yellow-orange "clouds" and the hydrogen bonding between the bases by yellow shading. The student will find that the rendition of this complex process has been portrayed with accuracy, and in exploring its intricacy, he will be able to note strand polarity, synthesis direction, hydrogen bonding, base pairing, and so on.

When panels B and E are brought together, the formation of messenger RNA on DNA templates, or transcription, is shown. The final panels E and F are interpretations of the frozen reality of electron micrographs. The panels show an illustrative view of a protein being formed on the mRNA template—the process of translation. Ribosomal subunits are shown on the endoplasmic reticulum (ER) linking up with the mRNA. They form one long polyribosome. Transfer RNAs, with the individual amino acid that each one carries for a specific codon on the mRNA, then participate in the protein building. As the polyribosome recedes into the ER, the final product, a complicated protein, is released. A certain degree of liberty has been taken in sacrificing scale proportions so that the student can begin to appreciate the more dynamic and architectural realities of the inner space of the cell world.

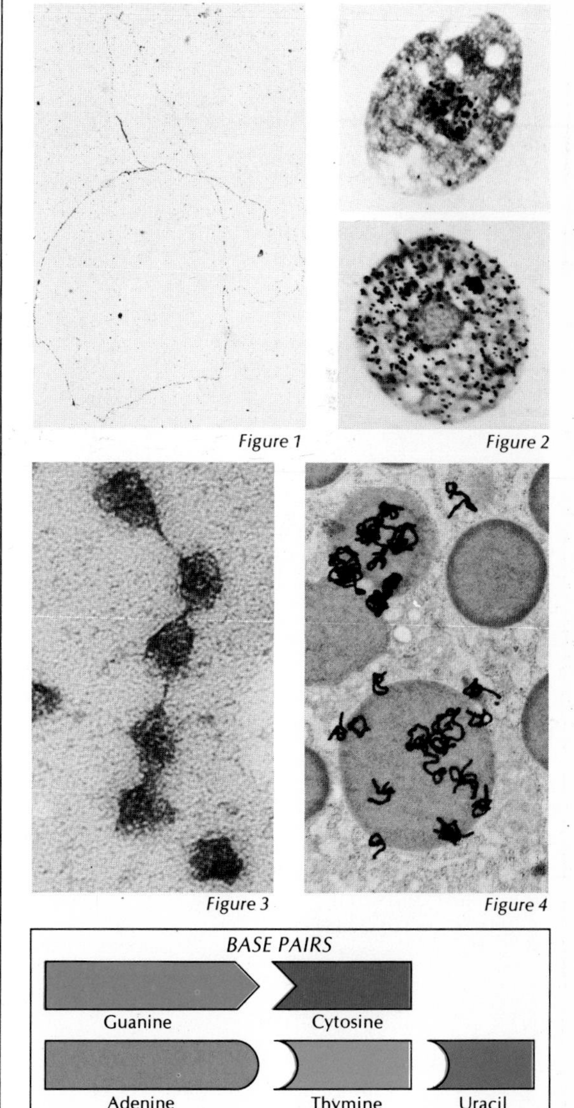

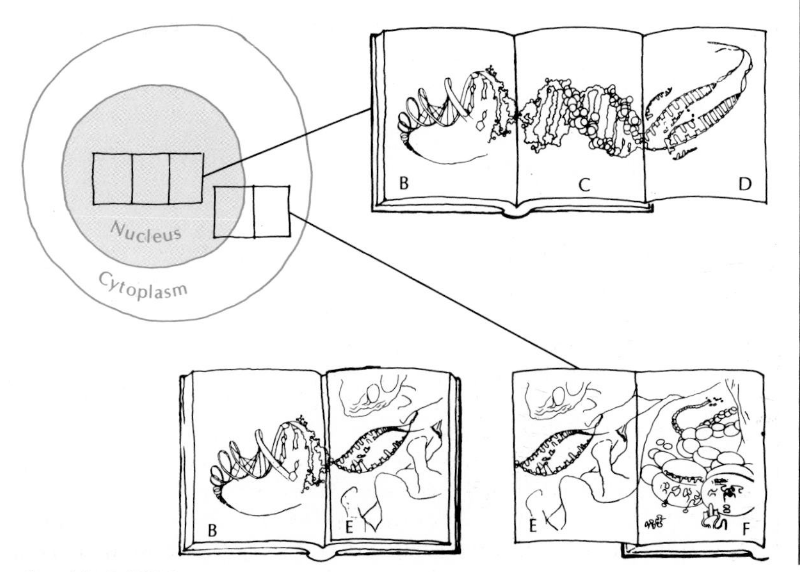

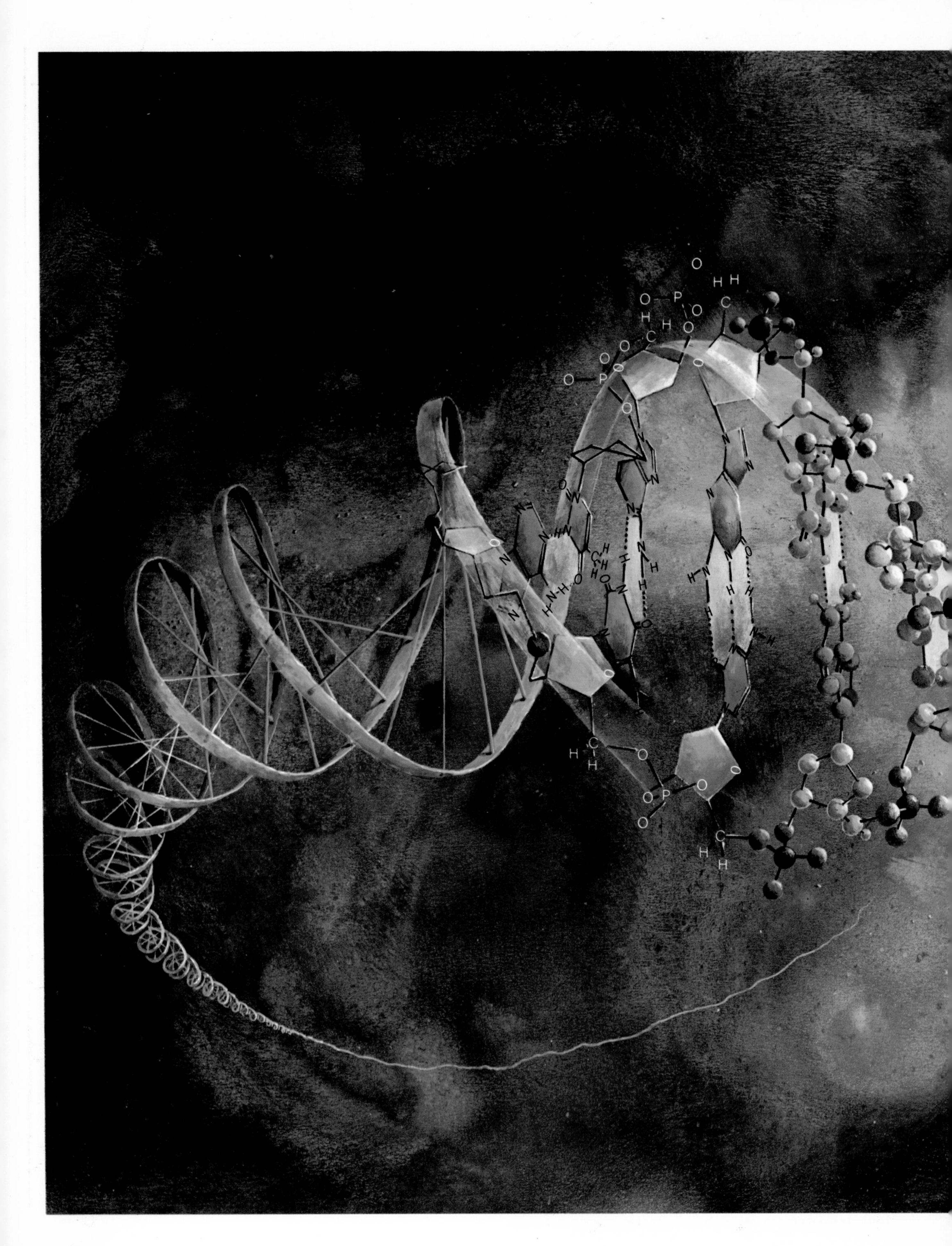

B

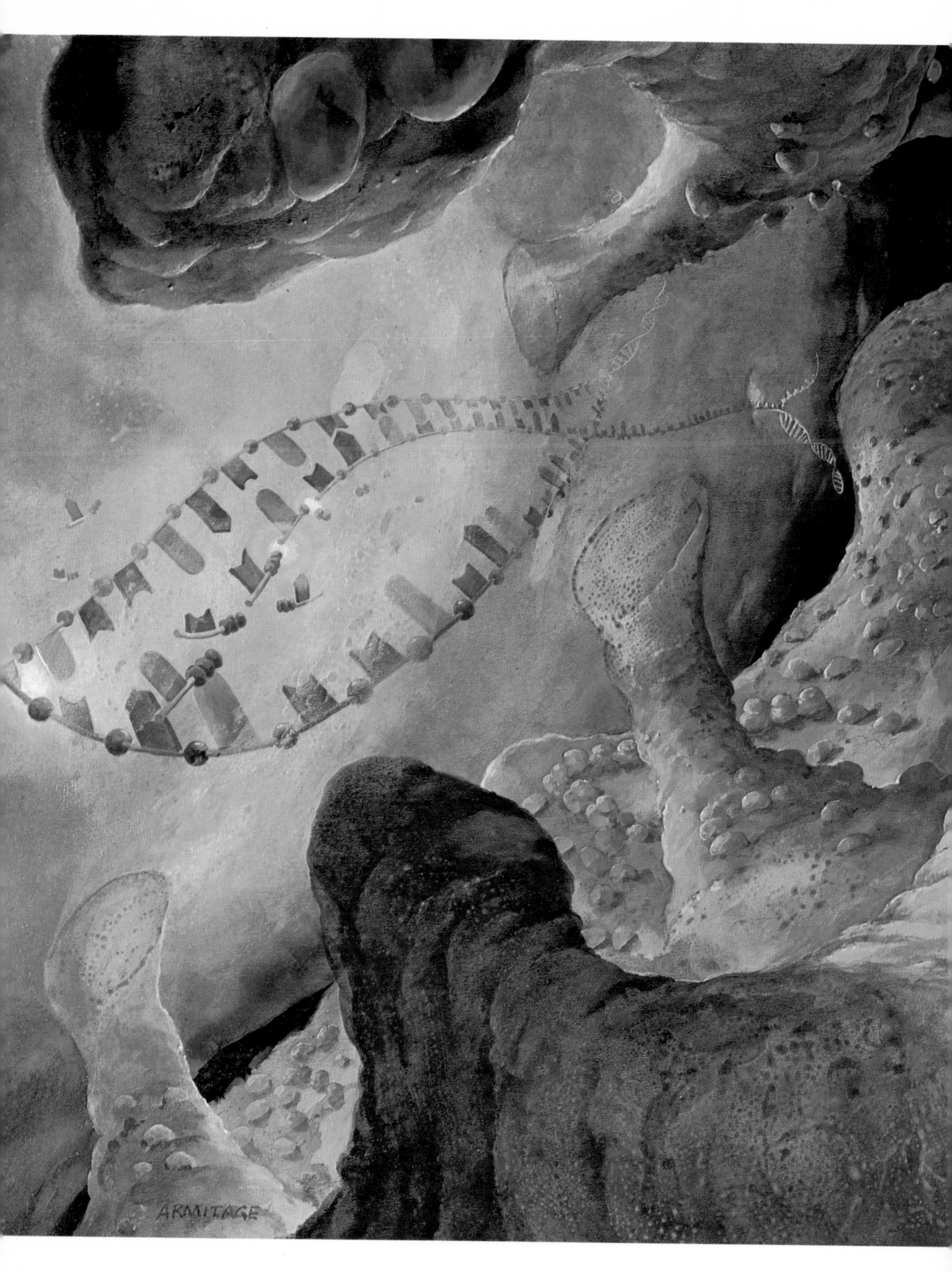

E

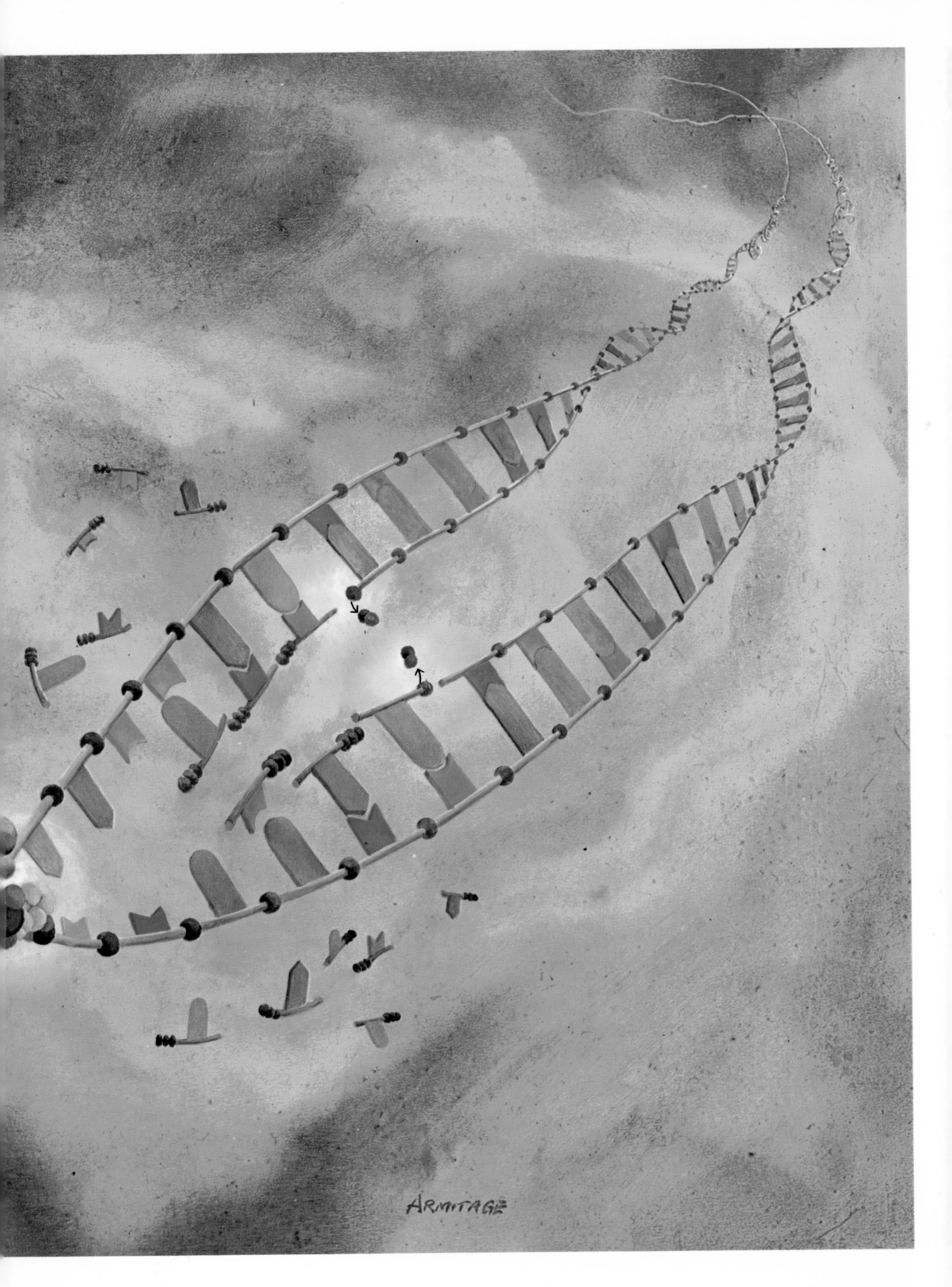

D

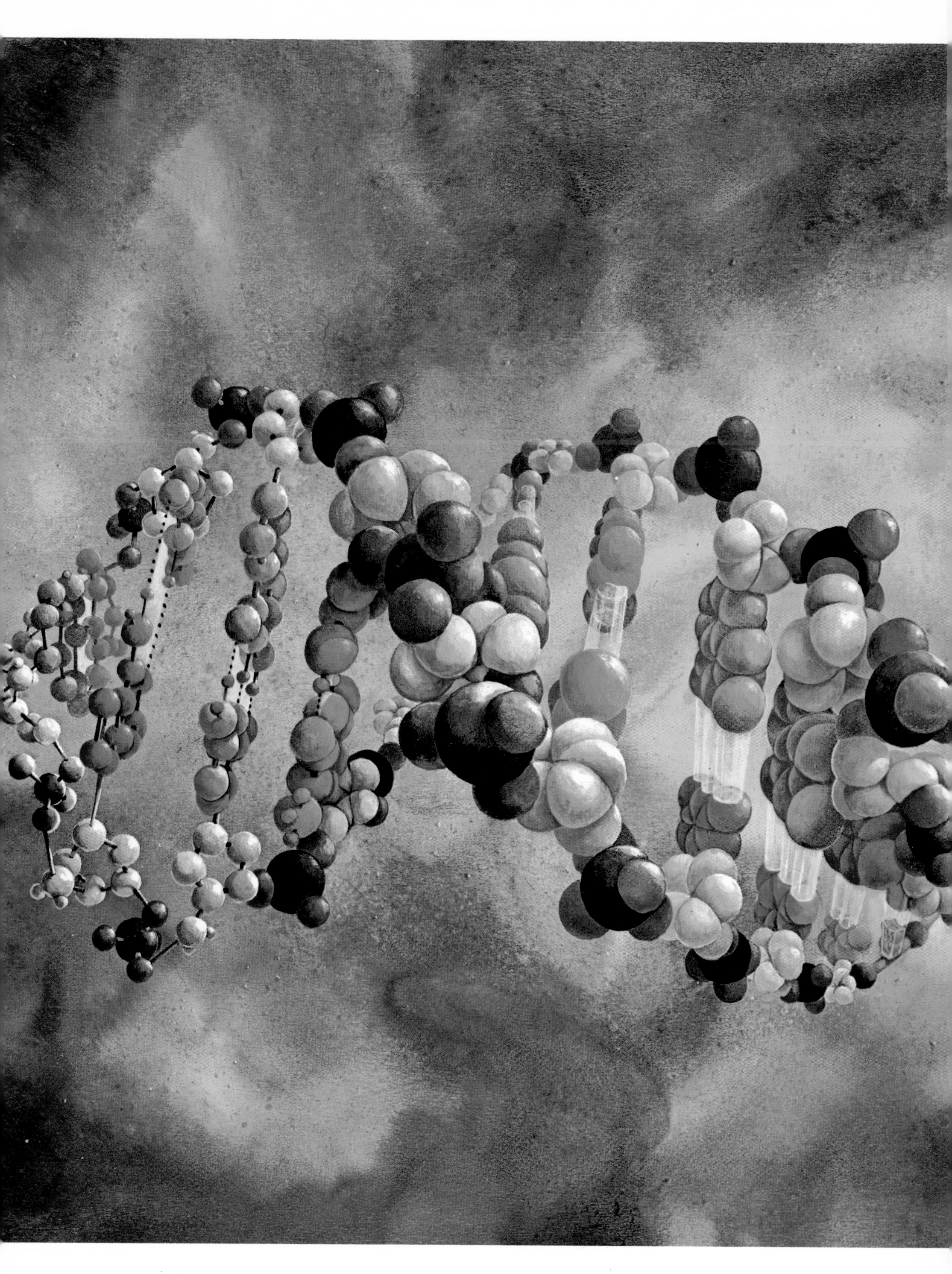

C

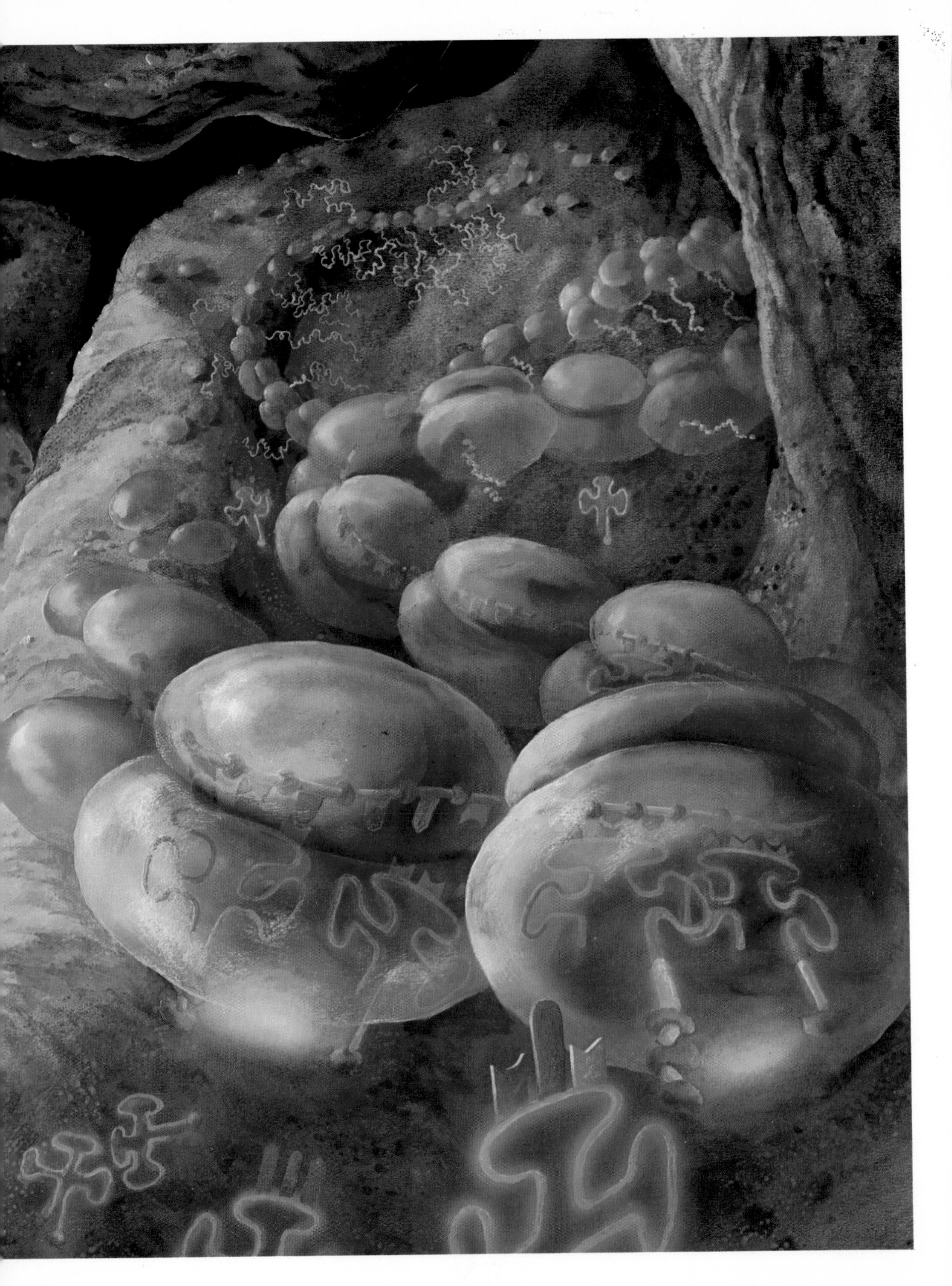

F

Figure 15.3. Diagram of the Hershey-Chase experiment with radioactive labeled bacteriophage. The radioactive nucleic acid enters the bacterial cells, and its entry alone (without protein) is sufficient for faithful reproduction of more viruses.

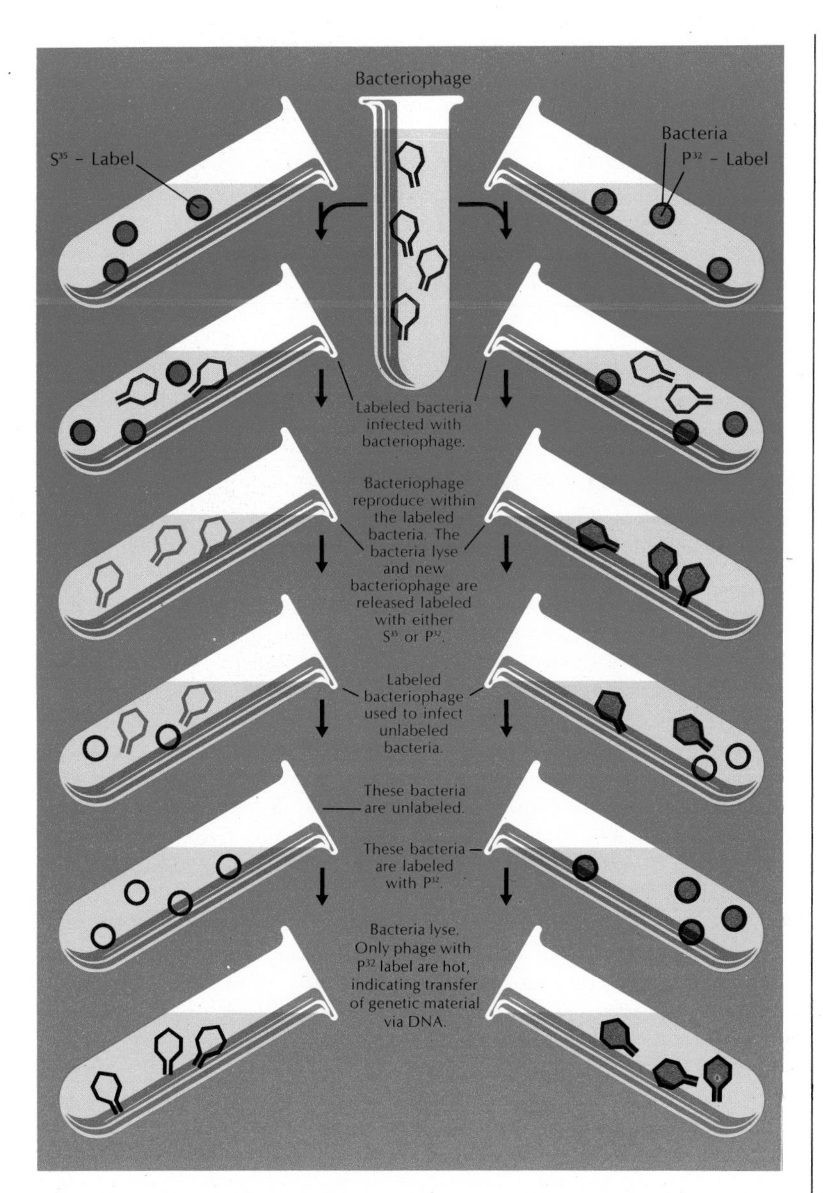

(phosphorus-32). Sulfur atoms were incorporated into proteins but not into DNA. Phosphorus atoms were built into nucleic acids but not into proteins. Thus, phage that were produced within bacteria in the S^{35} medium incorporated the radioactive atoms into their proteins, whereas phage grown in the P^{32} medium incorporated radioactive atoms into their DNA (Figure 15.3).

When phage grown in the S^{35} medium are used to infect nonradioactive bacteria, almost no radioactivity is found inside the bacterial cells after infection, showing that little if any of the phage protein enters the cells. When the bacteria are infected with phage containing P^{32}, most of the radioactivity is found inside the bacterial cell after infection, showing that most of the phage DNA does enter the cell. This experiment revealed that only the phage DNA is needed in order for the phage to take over the bacterial cell and reproduce itself. The DNA must carry all of the genetic instructions needed to put together large numbers of complete phages inside the bacterial cell.

Other viruses, such as the tobacco mosaic virus (TMV), were known to contain only RNA and proteins. Heinz Fraenkel-Conrat showed that he could cause mutations in TMV by chemical modification of the RNA but not of the protein. Thus, he proved that RNA is the carrier of genetic information for TMV. Fraenkel-Conrat later succeeded in infecting tobacco leaves with pure RNA extracted from TMV. Thus, RNA could also serve as genetic material.

From observations such as those described above and many others, it has been established beyond any doubt that DNA is the genetic substance in all cells, as well as in some viruses. Genes must be composed of DNA.

THE STRUCTURE OF DNA

The powerful evidence—obtained in the 1940s and early 1950s—that DNA served the genetic role stimulated an intense effort to determine the chemical structure of the DNA molecule. It had long been known that DNA is composed of the four deoxynucleotides—thymidylic acid, adenylic acid, cytidylic acid, and guanylic acid. The molecular structures of these four nucleotides were known, but the structure and size of the polymer remained to be determined.

Erwin Chargaff analyzed the relative amounts of the four bases in DNA from various kinds of organisms. Using the technique of paper chromatography for separation of the bases, Chargaff found considerable variation in the base ratios of DNA from different species, but his results showed remarkable constancy in the base ratios of DNA from different cells of a single species. The DNA of each kind of organism has its own specific ratio of different nucleotides, a ratio that might be the result of a specific sequence of nucleotides that could serve as a means of encoding genetic information. Chargaff noted that in all the DNA samples he studied the ratios of adenine to thymine, of guanine to cytosine, and therefore of total purines to total pyrimidines, "were not far from one."

William Astbury had made an attempt at x-ray diffraction study of DNA in 1938. From the diffraction patterns, he concluded that the DNA molecule is a long chain of repeating units, with the flat planes of the organic bases oriented at right angles to the long axis of the molecule, rather like "a pile of pennies." Pauling and Corey later concluded that DNA has a helical structure, and the British biochemist J. M. Gulland argued from chemical considerations that DNA is probably a double chain in which hydrogen bonds between bases hold the two chains together. Another group headed by Maurice Wilkins had begun a crystallographic analysis of DNA structure in 1950.

Enter Watson and Crick

When James D. Watson arrived at Cambridge University to work in the laboratories of Francis H. C. Crick, he was convinced that the structure of DNA would be found through x-ray crystallography. Crick shared his enthusiasm, and both set about to solve the riddle of the DNA structure. Together they brilliantly synthesized the results of many workers into a

Figure 15.4. The base pairs of DNA. Adenine is bonded to thymine by two hydrogen bonds, and guanine is bonded to cytosine by three hydrogen bonds.

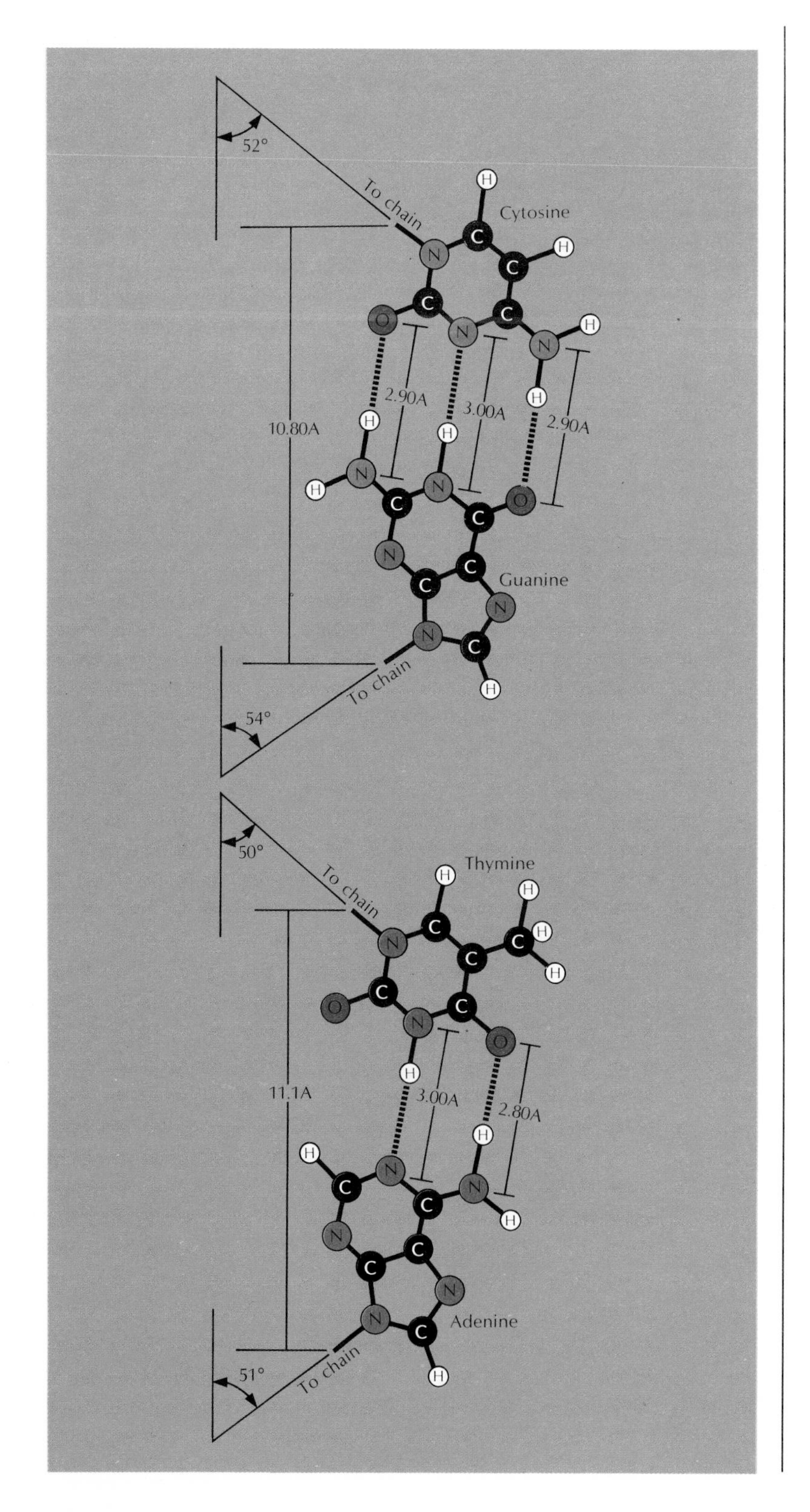

coherent model consistent with the known data on DNA. Most importantly, their model could account for Chargaff's data as well as offer a plausible mechanism for DNA replication.

Watson and Crick decided that there was good evidence that DNA was a two-chain, helical molecule with a constant width of 20 A. The twin phosphate chains ran up the outside, and the base pairs in the center were held together by hydrogen bonds. However, the manner in which the bases were arranged was still a mystery.

Because purines are much larger than pyrimidines, pairing of like bases along the chains would cause the width of the helix to vary considerably from one base pair to the next. After Watson had built unsuccessful models using purine-purine or pyrimidine-pyrimidine base pairing, a friend pointed out to him that he was using the forms of the bases commonly shown in textbook diagrams, and these were almost certainly the wrong stereochemical isomers for the base groups in DNA. It was this clue that enabled Watson to discover the true nature of the base pairing. With this structure in mind, he cut scale models of the base groups out of cardboard and began trying to fit them together in various ways. Suddenly, he realized that an adenine-thymine pair held together by two hydrogen bonds is identical in size to a guanine-cytosine pair joined by two or three hydrogen bonds. Such pairing would explain Chargaff's base ratios showing equal amounts of purines and pyrimidines, and it would also give the double helix a very regular structure, meaning it would have constant width. Furthermore, it would provide an intriguing means of DNA replication. The molecule could split and each strand could then build upon itself a complementary strand to restore the original molecule. As Watson and Crick remarked in their publication on DNA structure: "It has not escaped our notice that the specific pairing we have postulated immediately suggests a possible copying mechanism for the genetic material." Soon afterward, the model structure was found to be fully consistent with the x-ray data of Wilkins. Almost immediately, the Watson-Crick model of DNA structure was acclaimed by biologists and geneticists as one of the major biological discoveries of the century.

Replication of DNA

The most exciting feature of the Watson-Crick model of DNA structure is the obvious possibility for such a structure to act as a self-replicating molecule. The mechanism of DNA replication causes at least a partial separation of the strands of the duplex at the replicating point, so that the hydrogen-bonded bases are exposed. The free nucleotides then "recognize" the complementary base via hydrogen bonding, and the sequence of bases in the parental strand orders them in the daughter strand. Once again, the important point is base pairing. These monomers are then polymerized to form the daughter strand. Because the two initial strands are themselves complementary, the result is two double-stranded molecules. In this way, DNA molecules replicate without loss of information. It is thus useful to think of parental strands as templates for the synthesis of daughters.

The process described above is called *semiconservative replication* because each of the daughter molecules contains one of the two strands of the parent DNA molecule. Proof that DNA replication is semiconservative was soon provided by Matthew Meselson and Franklin W. Stahl in an experiment with *Escherichia coli* (Figure 15.5). Bacteria were grown for many

Figure 15.5. Diagram of the Meselson-Stahl experiment, a demonstration of complementary strand separation during DNA replication by the use of the cesium chloride (CsCl) density gradient technique. *Printed by permission from J. D. Watson,* Molecular Biology of the Gene, *© 1965, J. D. Watson*

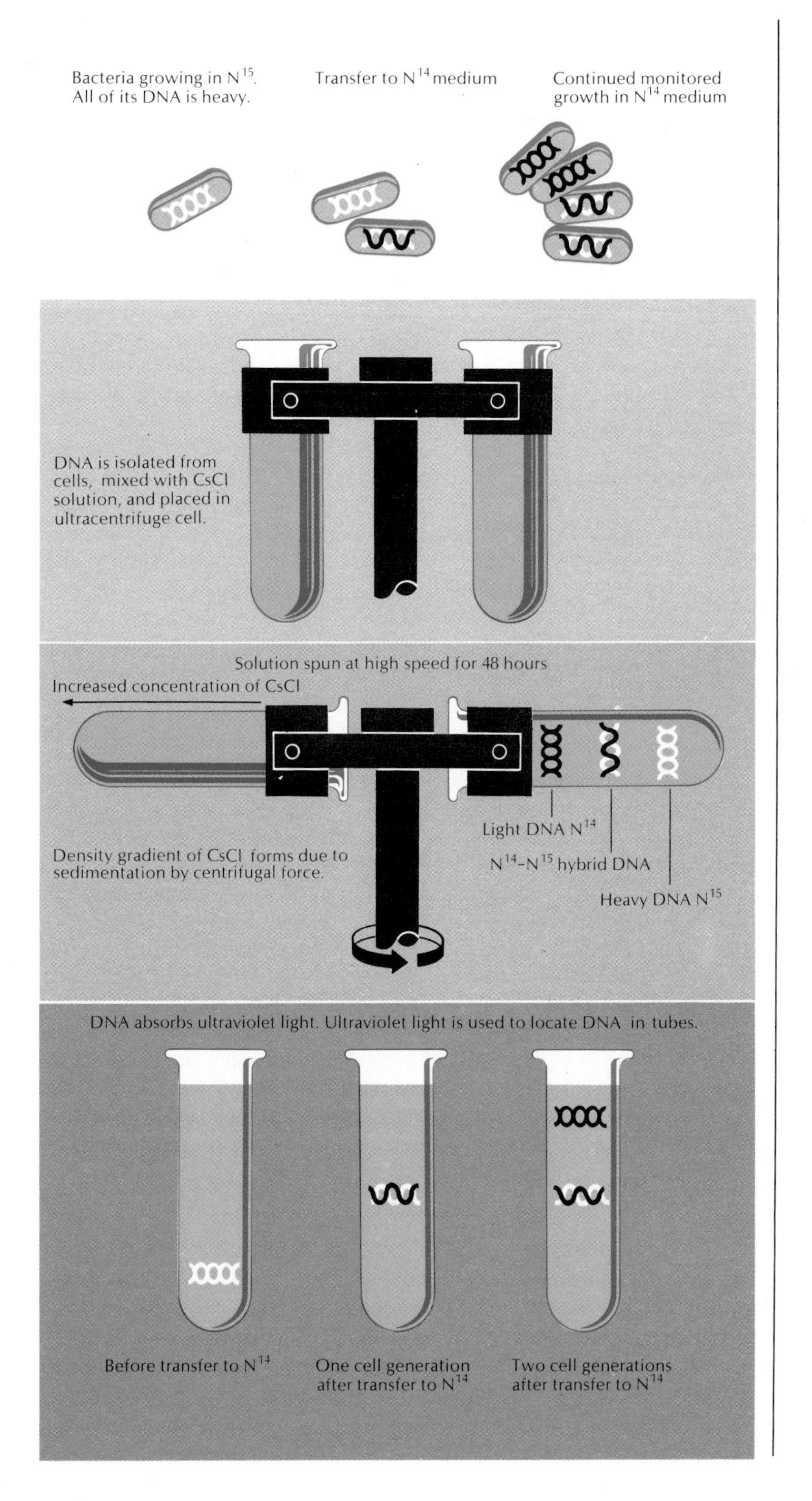

Bacteria growing in N^{15}. All of its DNA is heavy.
Transfer to N^{14} medium
Continued monitored growth in N^{14} medium
DNA is isolated from cells, mixed with CsCl solution, and placed in ultracentrifuge cell.
Solution spun at high speed for 48 hours
Increased concentration of CsCl
Light DNA N^{14}
N^{14}-N^{15} hybrid DNA
Heavy DNA N^{15}
Density gradient of CsCl forms due to sedimentation by centrifugal force.
DNA absorbs ultraviolet light. Ultraviolet light is used to locate DNA in tubes.
Before transfer to N^{14}
One cell generation after transfer to N^{14}
Two cell generations after transfer to N^{14}

Figure 15.6. Autoradiograph of the replicating *E. coli* K12 Hfr chromosome. The DNA of the chromosome was labeled with H^3-thymidine for two generations of DNA replication and extracted with lysozyme. The same structure is diagrammed at the upper right. It is divided into three sections (A, B, and C) that arise at the two forks (X and Y). X shows the growing point where replication is occurring, and Y the finishing point. The chromosome is about two-thirds replicated.

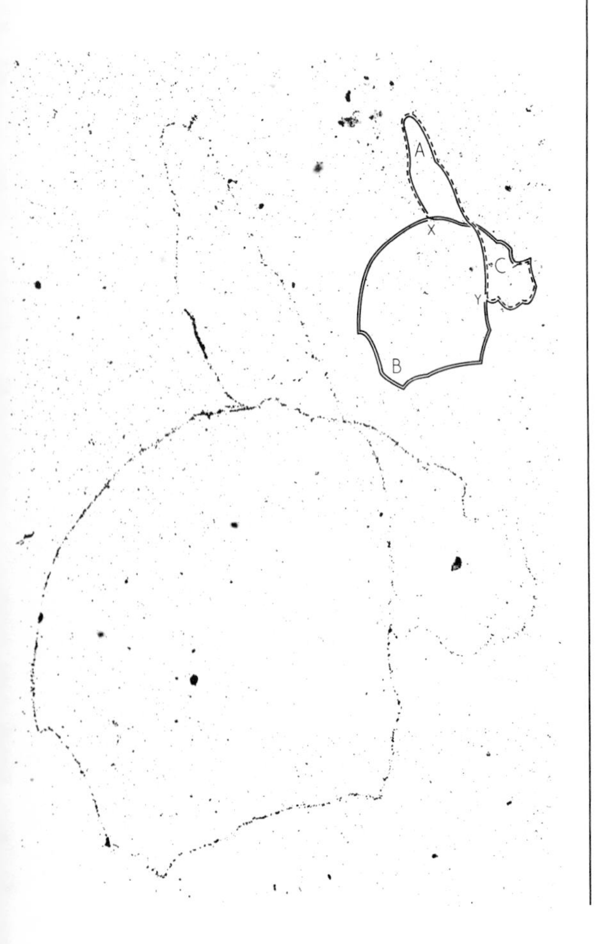

generations on a nutrient medium containing the heavy, nonradioactive isotope nitrogen-15. During the DNA synthesis that precedes binary fission, nucleotides containing N^{15} were incorporated into the bacterial DNA. After several cycles of cell division, most of the normal N^{14} in the DNA had been replaced by the heavier N^{15}, making the molecules about 1 percent heavier and denser than normal.

Very small differences in the density of macromolecules can be detected by a technique called density-gradient centrifugation. With this process, Meselson and Stahl (1958) separated DNA containing N^{15} in one or both chains from DNA containing N^{14} in both chains. When DNA from the bacteria grown in the N^{15} medium (heavy DNA) was mixed with DNA from bacteria grown in a normal N^{14} medium (normal DNA) and the resulting mixture centrifuged, the two kinds of DNA formed distinct layers. The bacteria with heavy DNA were then transferred into a normal medium.

After one cycle of cell division, each DNA double strand, or duplex, will have replicated, forming two daughter duplexes. According to the semiconservative replication scheme, each of the daughter duplexes should contain one of the original heavy chains and one light chain constructed from the precursors available in the normal N^{14} medium. As predicted, DNA extracted from these bacteria after a single division cycle in the normal medium formed an intermediate layer—halfway between the positions of the heavy and the light DNA duplex layers in the first test.

Cells allowed to undergo two replications in the N^{14} nutrient should produce a population of cells whose DNA would consist of an equal mixture of light ($N^{14}-N^{14}$) duplexes and intermediate ($N^{14}-N^{15}$) duplexes. After three replications of the DNA, 75 percent should be of light intensity and 25 percent of intermediate density. These predictions also were confirmed in this experiment.

The work by Meselson and Stahl showed clearly that each DNA chain remains intact during replication but that the double strand separates and each strand forms a new complementary strand as predicted by the hypothesis of semiconservative replication. Experiments of similar nature have been subsequently performed with eucaryotic cells undergoing mitosis. These experiments also confirmed the hypothesis of semiconservative replication. Other possible replication mechanisms have been proposed, but no evidence has been obtained to support these alternative schemes. It thus appears that DNA replication in all cells occurs through a semiconservative mechanism.

The general requirements for DNA synthesis are now known. A DNA chain must be present to act as a template, nucleotide precursors must be available, and DNA polymerase must be present to catalyze the formation of bonds between nucleotides as they become aligned on the template. There are important details of DNA replication, however, that are not yet clear. Autoradiographs (Figure 15.6) are among the strong evidence that synthesis of complementary strands occurs at the same time as the separation of the two strands of the parent molecule—that is, the chains do not completely separate before synthesis of the daughter strands begins.

One point not yet understood is the process that causes the two chains of the duplex to begin separating from one another so that replication can occur. Presumably, some molecule—possibly a protein—is able to rupture the hydrogen bonds that hold the two chains together. Once the separation has been initiated at some site on the DNA molecule, the processes of DNA replication may be sufficient to cause the continuing unzippering of the two chains.

Replication is not an easy task. For example, an *E. coli* is able to undergo binary fission about 20 minutes after it has been formed as a daughter cell of a similar fission. The separation of the two strands of the molecule requires them to untwist because the strands are intertwined in the helical coil. To untwist the 360,000 turns of the *E. coli* DNA within a few minutes would require a rate of rotation that is almost incredible. There must be 3.6 million nucleotides of just the right kinds available for each of the strands to build its complementary strand. Within another few minutes, these 7.2 million precursor molecules must be moved to the proper locations, fitted into place, and joined together to form the new double strands. Self-replication of DNA must involve an astounding speed and precision of molecular movements.

Another major problem yet to be solved is the control of initiation of DNA replication. Apparently, some mechanism exists to control the onset of the S-phase when DNA is synthesized. This process is of extreme importance because the health and proper development of an organism depend heavily upon proper control of cell regulation. In cancer, for example, a defect in this regulatory process causes cells to reproduce far too rapidly.

THE ROLE OF RNA

The one gene–one polypeptide hypothesis implies that the genetic information implicit in the sequence of base pairs in the DNA molecule must somehow serve to direct the assembly of a polypeptide chain with a particular sequence of amino acids. Even before the Watson-Crick model was proposed, it had become clear that RNA plays a key role in the process of protein synthesis.

RNA is the intermediate responsible for the transfer and decoding of the genetic information. Here, too, base pairing provides the mechanism. In virtually every case, RNA molecules are synthesized on a DNA template exactly as DNA synthesis is guided by the parental template. However, the monomers are ribonucleotides and the enzymology is somewhat different. Figure 15.4 illustrates synthesis of RNA on a DNA template. It is important to note that the base-pairing rules are exactly the same and that the RNA is complementary to the DNA template. The process of DNA-directed RNA synthesis is known as *transcription*, because the information contained in the DNA sequence is copied into RNA using the same alphabet (remember that uracil pairs exactly as thymine). Again, it must be emphasized that the rules of secondary structure, namely base pairing, govern the expression of genetic information.

The synthesis of RNA requires that the hydrogen bonds joining the two DNA chains in the double helix be temporarily disrupted to allow one chain to act as a template for RNA synthesis. After the newly synthesized RNA molecule is dissociated from the DNA, the DNA duplex might re-form by regenerating the hydrogen bonds, or the process of RNA synthesis might be repeated many times. In fact, it would be possible for many RNA molecules to be synthesized at the same time if a new RNA chain began to form as each partially completed chain began to peel away. In 1969 O. L. Miller, Jr. and Barbara R. Beatty at Oak Ridge National Laboratory succeeded in

Figure 15.7 (left). High-resolution electron micrograph of nucleolar genes from an amphibian oocyte. These genes code for ribosomal RNA. Each cluster of fibrils represents the simultaneous transcription by about 100 polymerase molecules. Each fibril is an rRNA precursor molecule. The triangular shape of each cluster results from each fibril being in a different state of completion. Those near the apex of the triangle are just commencing synthesis. (× 25,000)

Figure 15.8a (middle). Autoradiograph of a cell fed on H^3-uridine for 5 minutes and then killed. All the RNA is clustered in the nucleus—the site of RNA synthesis.

Figure 15.8b (right). Autoradiograph of a cell fed as above, but the radioactive uridine has been replaced by normal uridine. The RNA has moved from the nucleus into the cytoplasm.

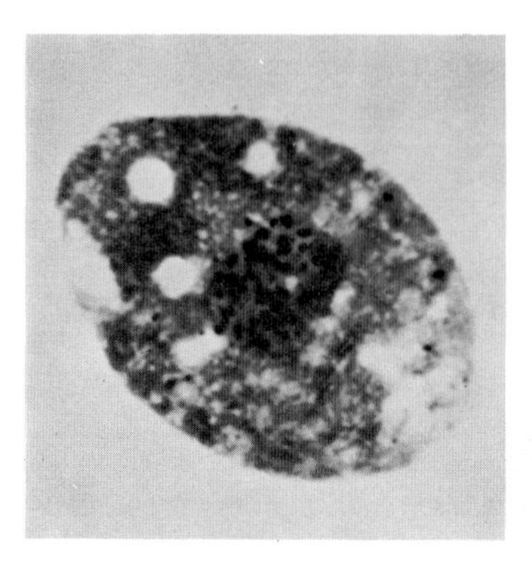

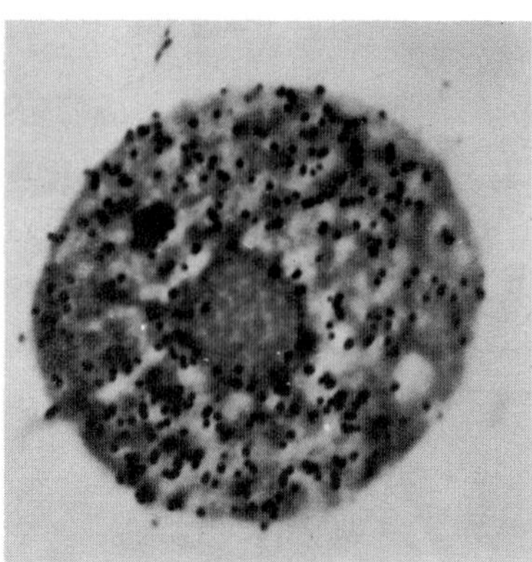

isolating DNA from the nucleoli of amphibian oocytes and preparing high-resolution electron micrographs that show exactly the predicted pattern of RNA synthesis (Figure 15.7).

RNA synthesis can occur in normal cells only in the presence of DNA. Removal of the nucleus from a cell by means of microsurgery results in the immediate cessation of all RNA synthesis in that cell. Furthermore, if a cell is given a radioactive precursor of RNA (such as H^3-uridine), the first RNA molecules to become radioactive are always found in the nucleus (Prescott, 1961).

Figure 15.8a shows an autoradiograph of a cell fed on H^3-uridine for five minutes and then immediately killed to halt synthesis. All of the radioactive RNA is restricted to the nucleus, showing that the nucleus is the site of RNA synthesis. Because there is a great deal of RNA in the cytoplasm of any cell, RNA must move from the nucleus to the cytoplasm. This movement can be confirmed by an extension of the experiment. A cell is fed on H^3-uridine for five minutes as before; then, the radioactive uridine in the nutrient medium is replaced by nonradioactive (normal) uridine, thereby ending the incorporation of radioactivity into RNA. The cell is allowed to live for 60 additional minutes in the nonradioactive medium, synthesizing nonradioactive RNA. When the cell is killed and analyzed by autoradiography, it is evident that a considerable amount of radioactive RNA has moved from the nucleus into the cytoplasm (Figure 15.8b). These experiments were particularly significant in light of earlier observations that suggested a role of RNA in protein synthesis, because proteins are synthesized in the cytoplasm, not in the nucleus.

TRANSLATION

Two processes fundamental to genetic expression have been sketched out: replication of DNA and transcription of DNA into RNA sequences. A third process also employs the rules of base pairing to achieve *translation* of the nucleic acid code (sequence of nucleotides) into the amino acid code (sequence of amino acid residues in a protein). The information coded in DNA base sequences is transcribed into a complementary RNA molecule. This information must then be translated into the information required to produce the primary structure of proteins.

All cells have three major types of RNA molecules. The smallest are transfer RNA (tRNA) molecules, with molecular weights of 2.5 to 3.0 × 10^4. The fact that tRNA molecules are homogenous in size and structure is important for the role they play in protein synthesis. All tRNA molecules have bases that are chemically modified after transcription by specific enzymes. The reason for these modifications is unknown.

Transfer RNA performs the essential step in translating information from nucleic acid into protein; it is the "adapter" between amino acid and its coded counterpart transcribed from DNA. It reads the coded information of the RNA molecule and inserts the corresponding amino acid into a growing polypeptide chain. It must be specifically recognized by an enzyme, which equips the tRNA with the proper amino acid. It must also interact properly with the ribosome, the subcellular organelle in which protein synthesis occurs. Every cell contains at least one tRNA for each amino acid. Often tRNA also regulates transcription of the message itself. Specialized tRNA does not always serve as an amino acid donor but may be responsible for beginning or terminating synthesis of a polypeptide.

Within the ribosomes are *ribosomal RNA* molecules (rRNA). Their functions are not clearly understood, but it is likely that by specifically binding various protein molecules, they are important in organizing a functional ribosome. In addition, interactions between rRNA and messenger RNA or tRNA could be important in the mechanism of protein synthesis. These rRNA and tRNA molecules are generally *stable* gene products—they are not usually degraded after synthesis. Together, these types of RNA form 95 to 98 percent of the total cellular RNA.

The remaining class of RNA molecules is composed of messenger RNA (mRNA). These molecules are transcribed from DNA sequences—"structural" genes—whose information ultimately codes a protein primary structure. Messenger RNA molecules have molecular weights from 100,000 to several million. Only a small percentage of the total cellular RNA is mRNA, but this molecule acts catalytically. A single RNA molecule may be translated many times.

All RNA molecules (except certain viral RNA) are transcribed from a DNA template using the base-pairing rules. The enzymes that catalyze transcription are *RNA polymerases*. In animal cells, there are distinct enzymes, one involved in synthesis of rRNA and another for mRNA. It is possible to isolate the enzyme and use it in a chemically defined system to transcribe isolated or synthetic DNA sequences.

The Code

The information required to specify protein primary structure is contained in the sequence of nucleotides in an mRNA molecule. There are about 20 different amino acids in proteins, but the RNA contains only 4 different types of subunits. If 1 nucleotide residue specified an amino acid, then only 4 amino acids could be coded, using the nucleic acid alphabet. If 2 bases at a time were used, then 16 amino acids could be specified. This number still is not enough, so it is most likely that at least 3 bases are required to specify 1 amino acid; then 64 different combinations of 3 bases are possible, and this number is more than enough to specify the 20 amino acids found in proteins.

It is now known that the specification of nucleic acid to protein is in fact a *triplet* code. In the linear sequence of residues in an RNA molecule, three bases at a time are required to specify one amino acid residue in a protein chain. Each set of three bases is called a *codon*; each codon specifies an amino acid. There are 64 possible codons and only 20 amino acids, so it is clear that some amino acids may be specified by more than one codon (Table 15.1). All 64 possible codons are used in the genetic code. Two triplets (UAA and UAG) appear to act as periods in translation, giving the signal for termination of the polypeptide chain being formed.

Table 15.1
Codons* in the Genetic Code

CODON	Message	Codon	Message	Codon	Message	Codon	Message
UUU	phenylalanine	CUU	leucine	AUU	isoleucine	GUU	valine
UUC	phenylalanine	CUC	leucine	AUC	isoleucine	GUC	valine
UUA	leucine	CUA	leucine	AUA	isoleucine	GUA	valine
UUG	leucine	CUG	leucine	AUG	methionine	GUG	valine
UCU	serine	CCU	proline	ACU	threonine	GCU	alanine
UCC	serine	CCC	proline	ACC	threonine	GCC	alanine
UCA	serine	CCA	proline	ACA	threonine	GCA	alanine
UCG	serine	CCG	proline	ACG	threonine	GCG	alanine
UAU	tyrosine	CAU	histidine	AAU	asparagine	GAU	aspartic acid
UAC	tyrosine	CAC	histidine	AAC	asparagine	GAC	aspartic acid
UAA	STOP	CAA	glutamine	AAA	lysine	GAA	glutamic acid
UAG	STOP	CAG	glutamine	AAG	lysine	GAG	glutamic acid
UGU	cysteine	CGU	arginine	AGU	serine	GGU	glycine
UGC	cysteine	CGC	arginine	AGC	serine	GGC	glycine
UGA	tryptophan	CGA	arginine	AGA	arginine	GGA	glycine
UGG	tryptophan	CGG	arginine	AGG	arginine	GGG	glycine

*Codons are sequences of nucleotides in the RNA. Each is represented by a letter symbolizing its base: U = uracil, C = cytosine, A = adenine, and G = guanine. Each codon causes the addition of a particular amino acid to the protein chain, except UAA and UAG, which indicate the end of a protein chain.

An RNA molecule may be regarded as a string of codons. For each amino acid there exist tRNA molecules that are specific for that amino acid. That is, the tRNA molecules make the connection between codons and amino acids. This connection is achieved by a sequence of three residues in the tRNA primary structure that are complementary to the codon in the mRNA. This part of the tRNA molecule is called the *anticodon*. The codon and the anticodon pair through hydrogen bonds by essentially the same rules of base pairing, and it is by this mechanism that the nucleotide code is translated into the amino acid code. Again, base pairing is the essential mechanism by which genetic information is transmitted.

According to the codon meanings in Table 15.1, mRNA with the base sequence AUGUUUCUCGCGGGG . . . will code for a polypeptide with the amino acid sequence methionine-phenylalanine-leucine-alanine-glycine . . . This translation is based on a nonoverlapping code. Although it was expected that the code would be nonoverlapping, it was necessary to confirm this assumption. The mRNA message above could conceivably be read in a series of overlapping triplets (Figure 15.9), yielding the code sequence AUG-UGU-GUU-UUU-UUC-UCU . . . and the amino acid sequence methionine-cysteine-valine-phenylalanine-phenylalanine-

Figure 15.9 (above). Comparison of the peptide products resulting from the same DNA base sequence being read as triplets in either an overlapping or nonoverlapping fashion. In the overlapping sequence, the reading frame is advanced one case at a time; in the nonoverlapping, three bases at a time. A different sequence of peptides results with an overlapping code than with a nonoverlapping code. The triplet code is known to be nonoverlapping.

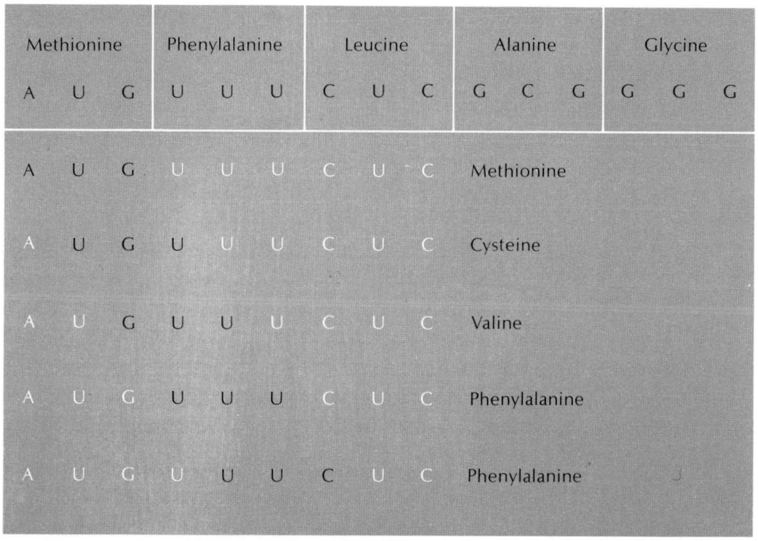

Figure 15.10 (below). ATP activation of an amino acid (AA), which is then transferred to the appropriate end of its specific tRNA. The enzyme amino-acyl synthetase catalyzes the activation of free amino acids by ATP to form an AMP-amino acid intermediate. This intermediate then reacts with a tRNA molecule to form an amino-acid "charged" tRNA molecule, releasing AMP in the process.

Methionine	Phenylalanine	Leucine	Alanine	Glycine
A U G	U U U	C U C	G C G	G G G

A	U	G	U	U	U	C	U	C	Methionine
A	U	G	U	U	U	C	U	C	Cysteine
A	U	G	U	U	U	C	U	C	Valine
A	U	G	U	U	U	C	U	C	Phenylalanine
A	U	G	U	U	U	C	U	C	Phenylalanine

serine . . . Such a reading of the code would occur if the ribosome advanced only one nucleotide each time it added an amino acid to the polypeptide chain. The genetic code was proven to be nonoverlapping by means of rather intricate genetic studies of the bacteriophage T-4.

An important and unexpected property of the genetic code is its universality. The codons represent the same amino acids in all organisms from viruses to multicellular plants and animals. The universality of the genetic code thus implies a common ancestor for all biological systems now in existence.

BIOSYNTHESIS OF PROTEINS

The central element in the process of translation is the mRNA produced by transcription from DNA. The DNA in a chromosome is one continuous molecule (in some cases, at least a few centimeters in length) containing many genes in series. Presumably, a series of nucleotides in the DNA indicates the beginning and end of a gene. Such identification would be essential for the individualized control of the expression of single genes. In bacteria—and perhaps in eucaryotes—certain genes with related functions are transcribed together as a unit with a single control point at one end of the gene series.

Protein synthesis involves the charging of the various tRNA molecules with the proper amino acid sequence. After the mRNA attaches to the ribosomes, the charged tRNA molecules interact by base pairing with the mRNA and donate their amino acids one by one to the growing polypeptide. The tRNA molecules are matched by a series of amino acid activating enzymes that catalyze the addition of a given amino acid to the stem of the appropriate tRNA. For example, the tRNA with the GAA anticodon is recognized by one particular activating enzyme that joins leucine to that tRNA. Each kind of tRNA becomes charged with the appropriate amino acid.

In the next step of translation, tRNA molecules, with their attached amino acids, are brought together with the mRNA molecule (Figure 15.10). The

$$\text{AA} + \text{ATP} \xrightleftharpoons[]{\text{Amino-acyl Synthetase}} \text{AA} \sim \text{AMP} + Ⓟ \sim Ⓟ$$

$$\text{AA} \sim \text{AMP} + \text{tRNA} \xrightleftharpoons[]{\text{Amino-acyl Synthetase}} \text{AA} \sim \text{tRNA} + \text{AMP}$$

Figure 15.11a (left). Ribosomal subunits. A ribosome is made up of a large and a small subunit that each contain RNA molecules and proteins.

Figure 15.11b (right). The *E. coli* ribosome structure. This structure is generally known as the 70S ribosome (S = Svedbergs) — 70S is a measure (sedimentation constant) of the speed of ribosomal sedimentation in a centrifuge. The sedimentation constant of the smaller and larger ribosomal subunits are designated 30S and 50S, and 16S and 23S refer to the same constants of the smaller and larger ribosomal RNA molecules. Bacterial ribosomes are closest in size to the *E. coli* — with 30S and 50S subunits — whereas in larger organisms, the ribosomes are larger (80S) with 40S and 60S subunits. *Printed by permission from J. D. Watson,* Molecular Biology of the Gene, *© 1965, J. D. Watson*

ribosome plays an important role in this process. Each ribosome is composed of a small and a large subunit, each subunit containing an RNA molecule and a number of proteins. The function of ribosomal RNA (rRNA) and ribosomal proteins is not yet fully understood. The first codon of the mRNA binds to the small subunit of the ribosome. This codon is then matched by the corresponding anticodon of the appropriate tRNA. This tRNA molecule, with its attached amino acid, becomes bound to the large subunit of the ribosome in such a way that the codon and the anticodon are hydrogen bonded. The second codon on the mRNA now directs the binding of a second tRNA ($tRNA_2$) by codon-anticodon matching to a second position on the ribosome. The two amino acids on the two bound tRNAs correspond to the first two codons of mRNA.

Because the tRNA molecules are essentially the same size and have the same parallel orientation on the ribosome, the respective amino acids are in proximity to each other. An enzyme on the large subunit is able to link the two amino acids together by a peptide bond. Simultaneously with bond

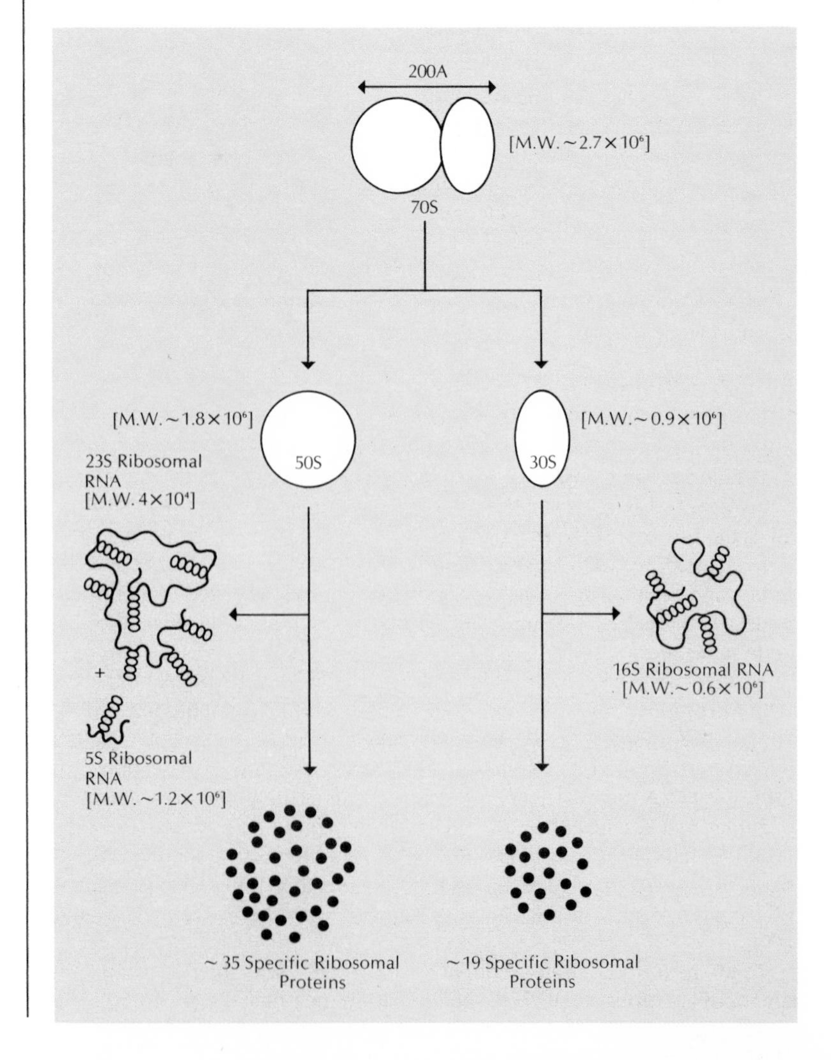

Figure 15.12a (above). Diagram of a polyribosome during protein synthesis, with an mRNA moving from the right to the left. *Printed by permission from J. D. Watson,* Molecular Biology of the Gene, *© 1970, J. D. Watson*

Figure 15.12b (below). Electron micrographs of polyribosomes. The thin strand connecting the ribosomes is the mRNA.

formation, the first amino acid is split from its tRNA, thus freeing the tRNA and leaving the dipeptide bound to $tRNA_2$. Next, the peptide-tRNA complex is shifted to the position originally occupied by $tRNA_1$. It is still hydrogen bonded to the second mRNA codon, so that, in effect, the message has moved along the ribosome a distance of one codon. This move frees the binding site on the ribosome previously occupied by $tRNA_2$. That position may now be occupied by a new charged $tRNA_3$, which corresponds to the third codon of the mRNA. This tRNA is then in a position for its corresponding amino acid to join the growing peptide chain. This process continues polymerizing amino acids until a termination signal is reached, and the completed protein is released from the ribosome and from its tRNA.

There may be a special signal to indicate the beginning as well as the end of a code message in mRNA. Studies of bacterial cells suggest that each code message begins with one of the two codons AUG or GUG. In the starting position on the message, either of these codons directs the placement of a special form of methionine tRNA. In some protein molecules, the

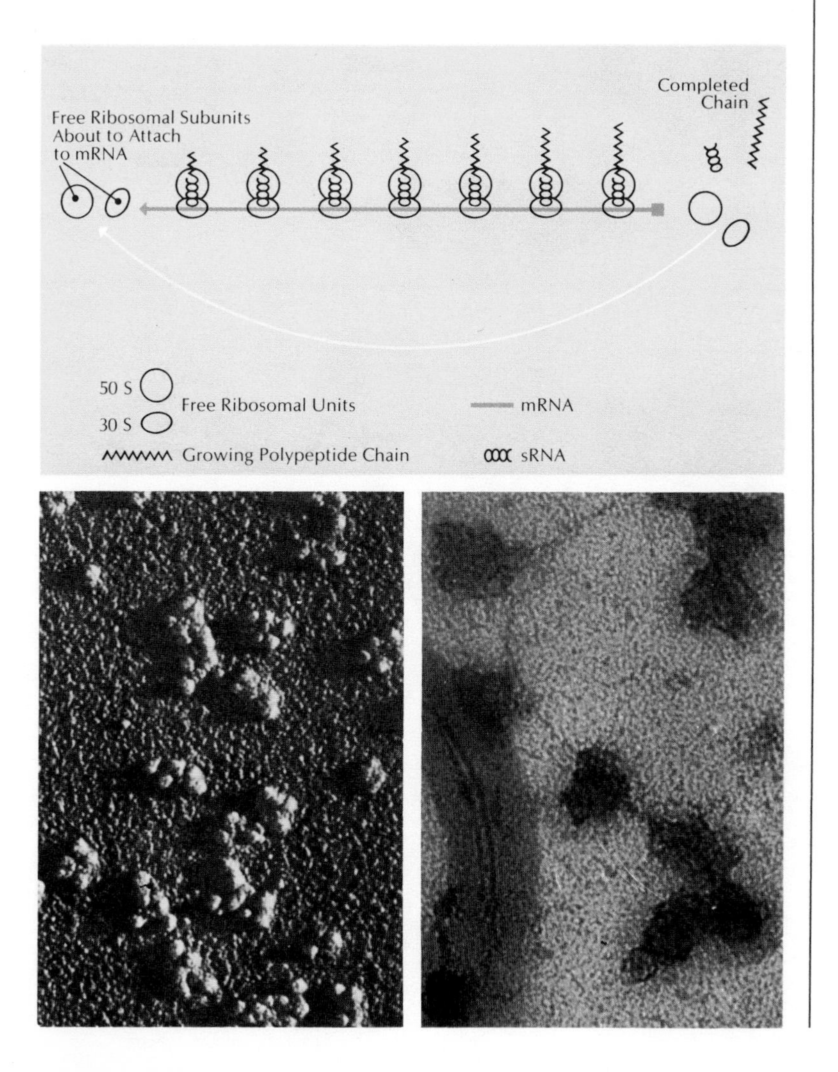

Figure 15.13. Protein synthesis. The left side of the illustration shows the activation of three different tRNAs, with specific anticodons for specific amino acids. ATP provides the energy for this reaction (the ATP's source being the Krebs cycle and oxidative phosphorylation in the mitochondrion). The right side of the illustration depicts the translation process. Each ribosome has two tRNA binding sites. Through enzyme mediation, the polypeptide chain grows by the formation of a peptide bond between, for example, AA-1 and AA-2. After the bond has been formed, the tRNA is ejected from the AA-tRNA binding site. The movement of the mRNA over the ribosomal surface is still unknown, but the growing chain is translocated to a different site on the ribosome, allowing a different tRNA to bind and the transfer process to repeat itself. In this illustration, the ribosomal subunits attach themselves while the mRNA is moving from left to right.

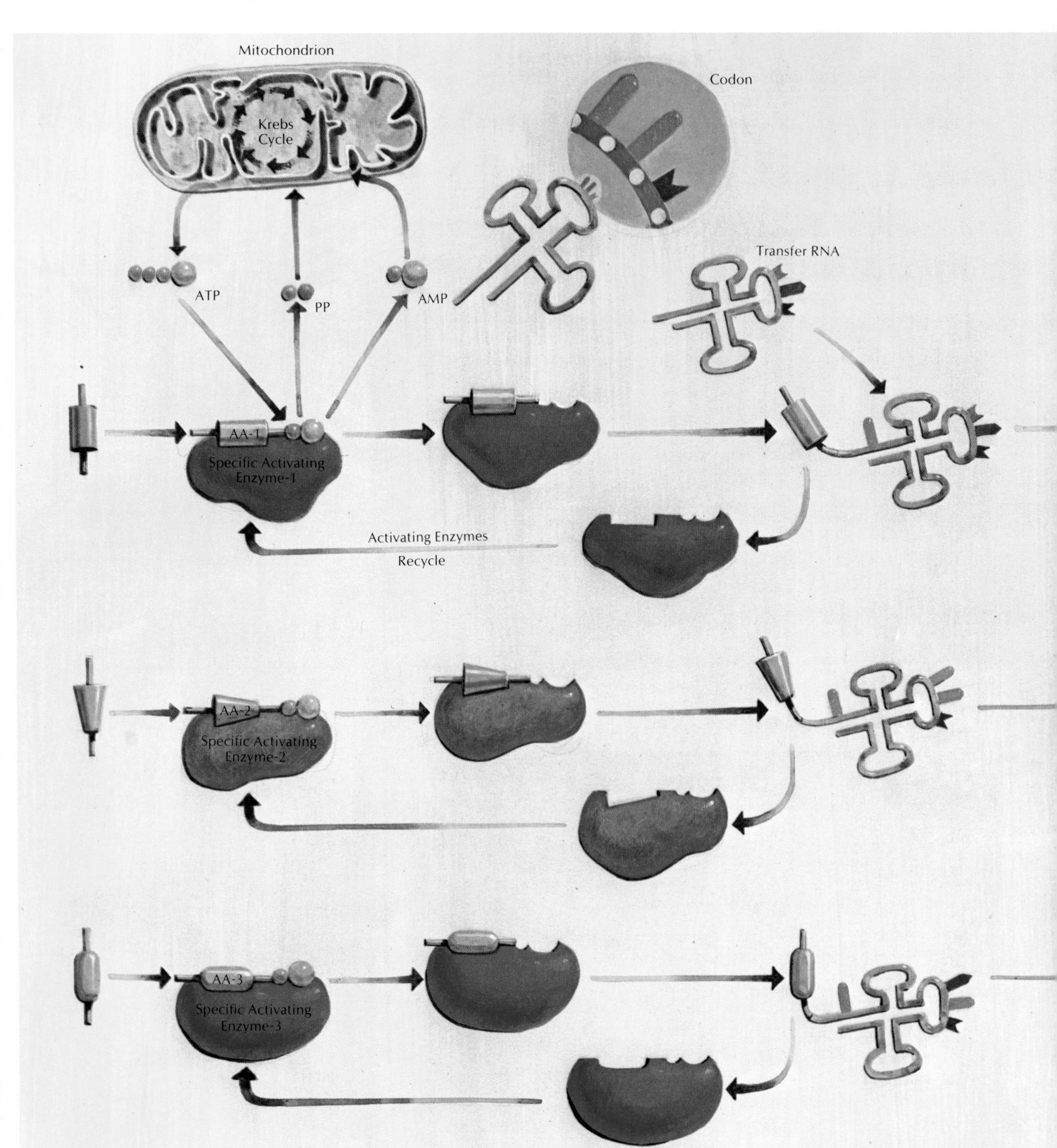

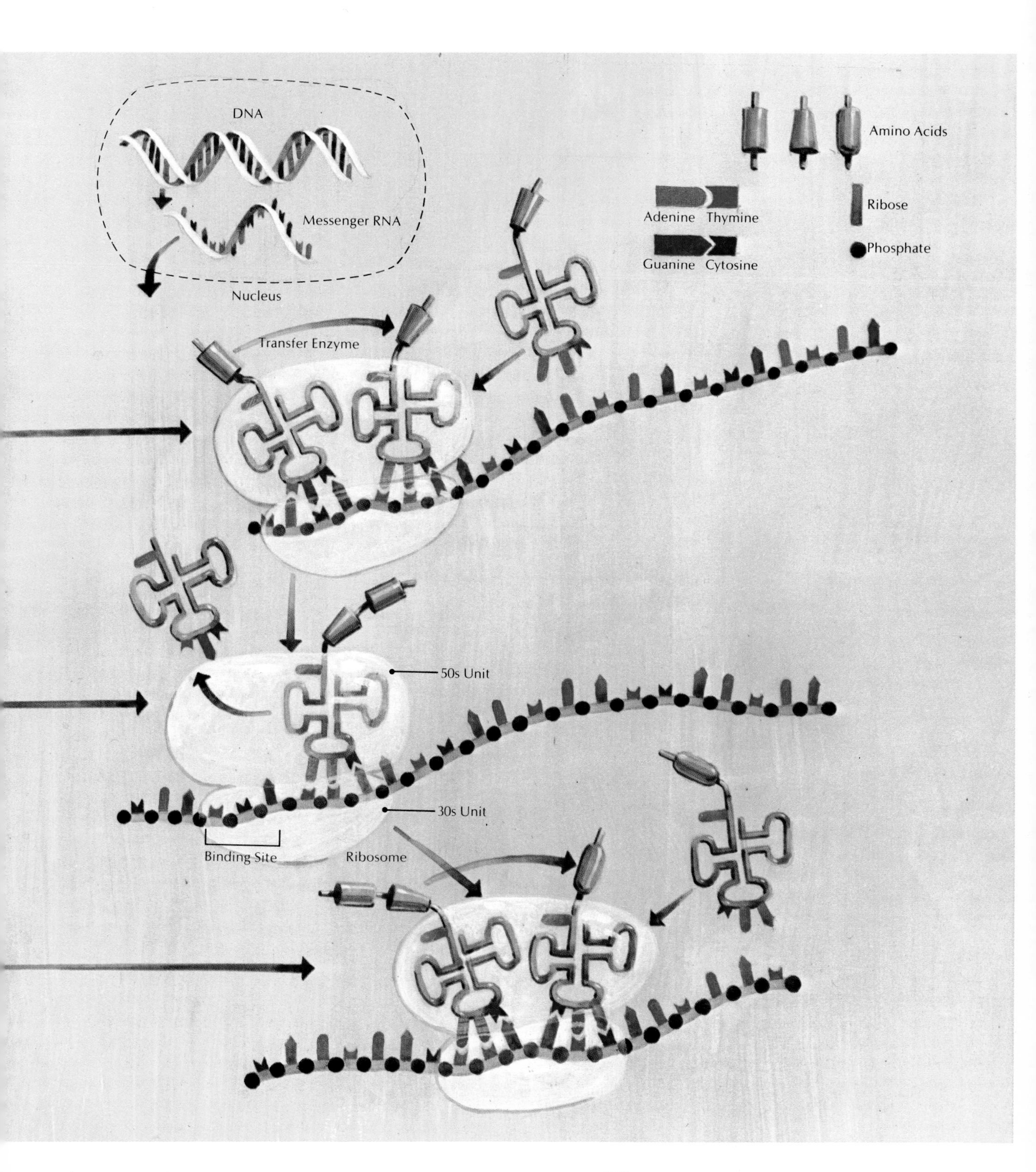
DNA
Messenger RNA
Nucleus
Amino Acids
Adenine
Thymine
Guanine
Cytosine
Ribose
Phosphate
Transfer Enzyme
50s Unit
30s Unit
Binding Site
Ribosome

methionine is later removed from the end of the molecule, presumably by a special enzyme. Recent studies suggest that a similar process involving a special methionine tRNA may also occur in eucaryotic systems. The entire process of synthesizing a polypeptide of 200 amino acids may be accomplished in a fraction of a minute. Several ribosomes may translate a single message at once. Then the ribosomes are strung along the mRNA like beads, forming a structure called a *polyribosome*, or *polysome*. Each ribosome catalyzes synthesis of a polypeptide, so that a single mRNA molecule may direct the synthesis of a number of protein molecules.

The assumption that the linear arrangement of nucleotides in a nucleic acid translates into a matching linear arrangement of amino acids in a protein has been confirmed. The enzyme tryptophan synthetase in *E. coli* is composed of two polypeptide chains that are coded for by two adjacent genes, called the A and B genes. The amino acid sequence of the polypeptide product of the A gene was determined. Several mutant strains exist in which a single amino acid change in the polypeptide chain inactivates the enzyme. Studies using a lysogenic phage made it possible to determine the location of each mutation on the gene. In each case, the position of the mutation is consistent with the position of the altered amino acid in the polypeptide chain. Mutations appear to be caused by a substitution of one nucleotide for another or by deletion of one or more nucleotides in the DNA chain. For example, position 48 in the polypeptide chain is glutamic acid in the normal enzyme, but it is valine in one mutant strain. This alteration is apparently caused by the substitution of the codon GUG for GAG in the corresponding position on the DNA chain of the A gene.

In Vitro SYSTEMS

Much knowledge about molecular mechanisms involved in DNA replication, RNA transcription, and translation has come through the use of *in vitro* systems, which simulate the processes occurring in the cell. These studies involve breaking the cells, then isolating and purifying specific genes, RNA, or special enzymes or complexes. These purified components are then reconstituted under experimental conditions, where specific manipulations of the components often lead to a precise understanding of the individual reactions and their interrelationships.

The enzyme DNA polymerase has been used to synthesize DNA from its constituent nucleotides. At first, there was some doubt as to whether this enzyme made exact copies of the template DNA; recently, however, Kornberg (1960) showed that the enzyme was able to replicate a viral DNA so precisely that it could infect its host just as well as viruses produced in the living cell could have done. Other workers have isolated and purified segments of DNA coding for single sets of genes, including genes for ribosomal RNA, β-galactosidase, and several small viruses. However, these purified genes have not been replicated *in vitro* using DNA polymerase.

RNA polymerase transcribes the DNA base sequence into complementary RNA. It is composed of five subunits plus one additional polypeptide called a sigma factor, which is rather loosely bound to the core enzyme, the five-subunit polymerase. The sigma itself is inactive in transcription, but in combination with the core enzyme it determines which sequences of DNA will be transcribed. Sigma factors are interchangeable so that one sigma can substitute for another to guide the selection of DNA for transcription. Certain viruses are able to direct the synthesis of their own RNA in place of host cell RNA simply by displacing the host's sigma factor and replacing it with their own, which will only transcribe viral genes. Sigma factors may prove to be important in developmental processes where selective gene expression is necessary.

The proof that this polymerase faithfully transcribes DNA into active RNA is shown by the fact that this *in vitro* synthesized RNA can be translated into active enzymes. There are now several *in vitro* systems (called coupled systems) in which purified genes are simultaneously transcribed into RNA and then translated into proteins.

In vitro protein synthesis involves a combination of mRNA, ribosomes, tRNA, activating enzymes, and other enzymes, along with amino acids and an ATP energy source. Such systems have already been invaluable in elucidating the salient features of protein biosynthesis. In addition, they may soon provide answers to problems of translational controls, such as are known to exist in oocytes, seeds, spores, and slime molds (Chapter 17).

GENE TRANSFER

The work of Griffith and Avery proved that DNA could be transferred from one cell to another, producing a stable genetic change in the recipient. Bacteria are now known to have other methods of gene transfer.

In a process called *bacterial conjugation*, certain strains of *E. coli* are able to directly exchange genetic material by forming a type of cytoplasmic bridge between two cells. In any mating pair, only one cell will act as the DNA donor. The donor, however, is not depleted of its genetic material because conjugation will occur with simultaneous DNA replication, with one double strand going to the recipient and the other remaining with the donor. The donor DNA is fully functional in the recipient cell and will be transmitted to that cell's descendants. In this way, bacteria with new combinations of genes can be produced without the slow process of mutation. This process has been appropriately termed a *sexual mating system*; the donor is called the male and the recipient, the female.

The transfer of DNA begins at a particular region on the bacterial chromosome. Thus, the DNA that codes for certain genes enters the recipient before others do. Because it takes about two hours for complete chromosomal transfer, the amount of DNA injected into the recipient cell will be proportional to the time the two cells are in intimate contact. Obviously, genes that are adjacent or close to each other will be transferred together. The order of the genes can then be determined by gently splitting the pairs apart at different times after the onset of mating and then measuring which new genes are present in the recipient cell. This process is called *genetic mapping*. Maps are expressed in terms of minutes after beginning of mating.

Also useful in mapping relative gene positions is phage transduction, in which a few bacterial genes can be incorporated into a replicating virus. After lysing the cell, the virus may then infect another cell, but because it contains a small piece of bacterial DNA in place of some of the viral genes, the virus is unable to kill the cell. Instead, the bacterial DNA transduced into that cell with the phage becomes active. Genes that are transduced together in a single phage are close to each other. By studying a series of various overlapping transductions, biologists can determine the relative order of the genes on the chromosome.

MUTATIONS

Mutations are heritable changes in the genes other than the direct transfer of genetic material between organisms. In general, they represent deletion

Figure 15.14. Illustration of a mutation arising in DNA replication. Substitution of an A for a G results in the production of a normal DNA molecule and a mutant one after replication. Subsequent replications of the mutant DNA will preserve the mutation at that point.

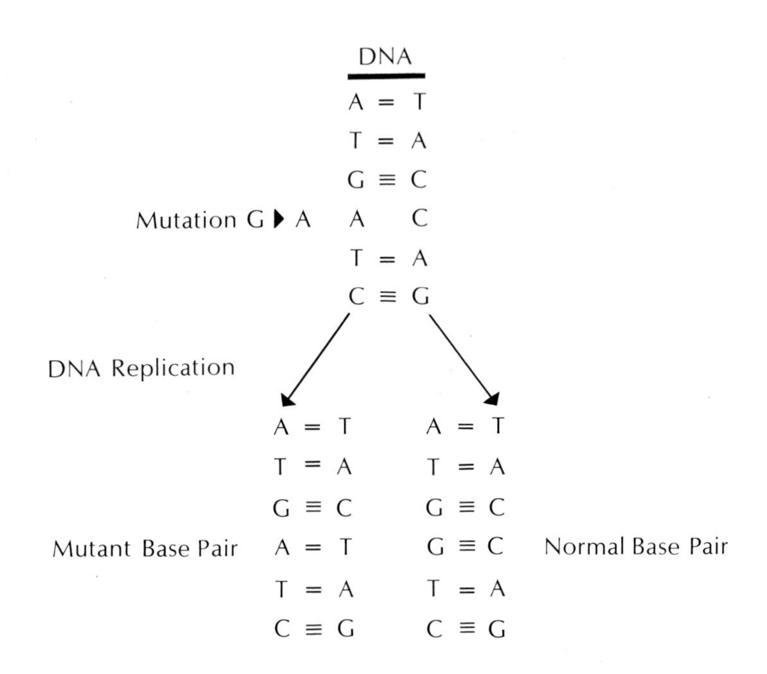

or modification of one or more of the bases in the DNA. In nature, mutation is a rare event, but the rate of occurrence can be increased by certain mutagenic agents, such as x-rays, ultraviolet light, and a variety of substances that chemically react or interact with DNA.

By definition, a mutation is any change in the base sequence of DNA. However, mutations are usually observable only when they affect the phenotype. Many mutations go unnoticed because the change of a single base may not cause a change in the amino acid specified by the code at that position. For instance, a change from GAC → GAU still results in the insertion of aspartic acid into the protein. Occasionally, *suppressor mutations* occur within a gene; they are able to conceal another mutation in the same gene by compensating for the effect of the first mutation.

Point mutations result from the replacement, insertion, or loss of a single nucleotide. Replacement may occur by the substitution of one purine for another or one pyrimidine for another (called a *transition*). The exchange of a purine for a pyrimidine or vice versa is a *transversion*. Some chemicals cause marked changes in the hydrogen bonding properties of the bases so that the proper base pairing no longer occurs. Thus, base-pairing errors are introduced during DNA replication or during RNA transcription. If a transition from G → A occurs on one strand of DNA, then at replication one daughter duplex will have a proper GC base pair (because the complementary C of the unmutated strand will pair with a G), but the mutation will produce an AT pair at the same place.

The insertion or loss of a nucleotide can lead to a *frame shift mutation,* in which the codons are still read three at a time but not in the proper phase. For instance, a DNA base sequence CCAACGGCCGGA would translate into a peptide with the sequence proline, threonine, alanine, and glycine. But if an additional G were inserted at the beginning of the third

Figure 15.15. A frameshift mutation, in which the codons are read out of phase. Addition or deletion of a base still results in the reading of codons as triplets, but now different triplets are read. This alteration causes the insertion of the wrong amino acids in the polypeptide or may even result in the premature termination of the protein. *Printed by permission from J. D. Watson,* Molecular Biology of the Gene, *© 1970, J. D. Watson*

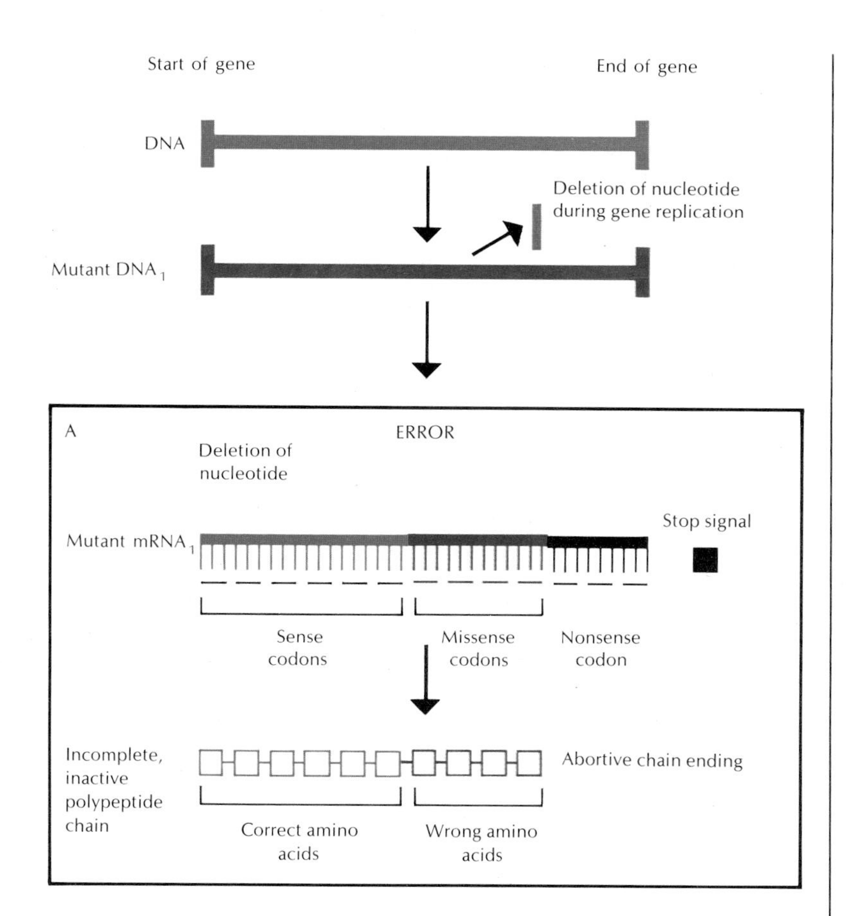

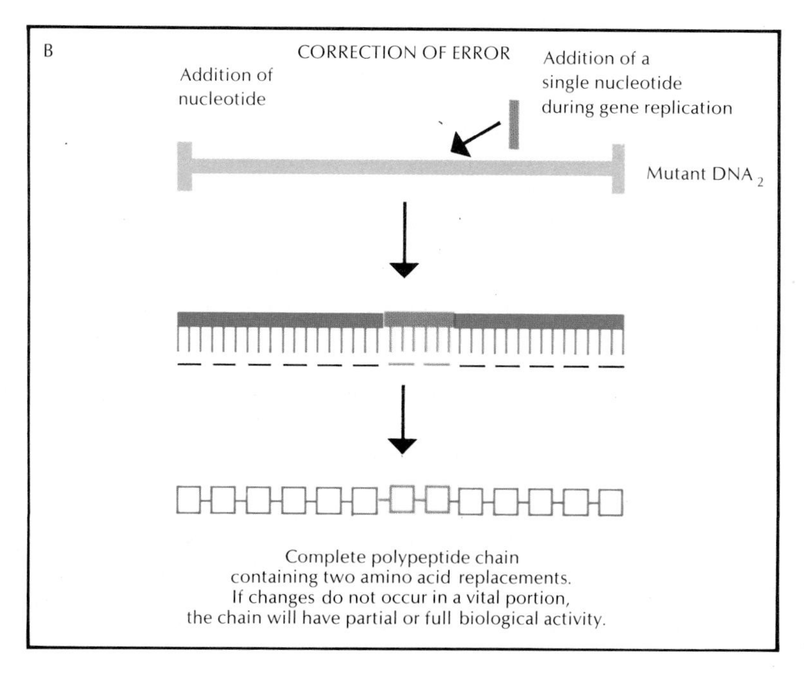

Figure 15.16 (above). Intergenic complementation. In a normal cell, both active gene products 1 and 2 are necessary to produce an active enzyme. A mutation (X) in either gene, which renders either 1 or 2 inactive, leads to the production of an inactive enzyme. If extracts of cells deficient in gene product 1 are mixed with extracts from cells deficient in gene product 2 (A), an active enzyme is produced because the corresponding active gene products may complement each other (B).

Figure 15.17 (below). Intragenic complementation occurs when two DNA molecules in the same cell, each bearing a mutation (X) within the same single gene, are able to give rise to an active product. This product occurs when the two DNA molecules break and reunite in a new combination, making one gene a double mutant (XX) and the other gene normal.

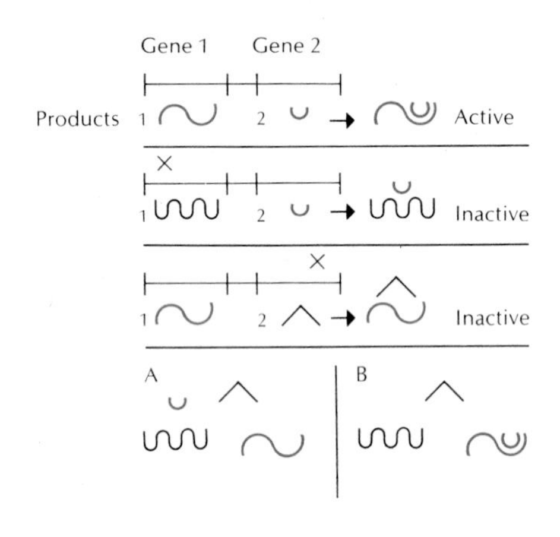

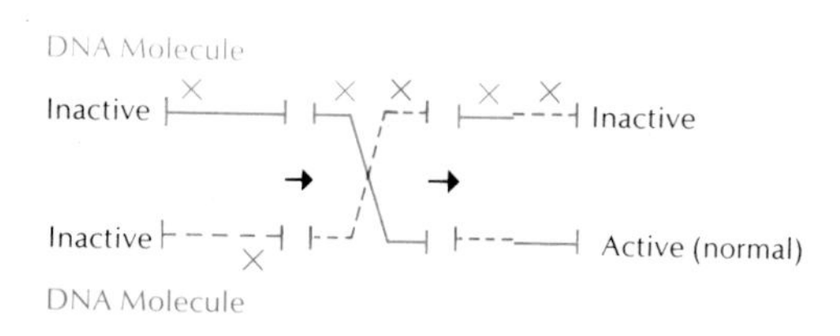

codon, then the third codon sequence would be GGC and would be read as if it were the intended codon. The result would be a peptide sequence of proline, threonine, glycine, and arginine. Often the change of a single amino acid can lead to the production of a totally inactive gene product. The deletion of a base leads to the same kind of reading frame shift. Mutations often arise by the deletion of many bases, and entire genes may be lost.

Mutations often give rise to proteins that are inactive in a catalytic sense. They are still produced by the cell but are nonfunctional. In diploid organisms, recessive alleles can often be thought of as inactive or partially active gene products.

If an enzyme is composed of two different polypeptide chains, which are coded for by separate genes, mutations in one gene do not affect the other gene. Mutations in either gene 1 or gene 2 result in an inactive enzyme. But if an extract is prepared from cells containing a mutation in gene 1 but normal in gene 2, and another extract having a mutation in gene 2 but normal in gene 1 is prepared, and then the two are mixed, enzymatic activity results. The active product of gene 1 is able to complement the active product of gene 2 to give a fully active product. This *intergenic complementation* occurs at the level of the interaction of the gene products. *Intragenic complementation* also occurs, but it involves interaction of the DNA itself within the living cell. If a cell contains two separate copies of the same gene, each one possessing a mutation somewhere within, sometimes progeny that possess a normal gene arise. The two mutations have complemented each other in a way similar to crossing over in meiosis. The homologous regions of each DNA molecule pair precisely, then there is a breakage and reunion of the strands in the new combination. Very fine genetic mapping reveals that the breakage and reunion, or recombination, can occur even within a single codon.

THE FUTURE

For a decade, it was believed that DNA→RNA→protein represented the exclusive direction of flow of genetic information. In 1970 Howard Temin succeeded in demonstrating that cancer-causing RNA viruses contain an enzyme that reverses the flow of genetic information; it catalyzes DNA synthesis from an RNA template. There is evidence that the same enzyme activity is also found in normal cells, and its occurrence raises new questions concerning gene expression.

The transforming viruses may someday be used to introduce healthy genes into cells that lack a required enzyme; such genes might be isolated from healthy cells or synthesized chemically. It would then be possible to actually change an organism's genetic message.

Genetic information is contained in isolated DNA molecules, and this information is inherent in the sequence of bases. These facts imply that genetic information can be synthesized by chemical means. In fact, the chemical synthesis of a gene has already been achieved. Starting with only small molecules and employing straightforward, well-established, organic reaction schemes, scientists have synthesized a gene coding a yeast tRNA.

The implications of genetic knowledge for the future are obvious. Many diseases of metabolism will be cured only when it is possible to change the genetic message of an organism. It must be concluded that, for good or evil, such advances will not be long in coming. It is possible to create genetic information, as polydeoxynucleotide sequences, in the test tube, and it is possible to transform the genotype in certain organisms. To some people, transformation of human genotypes is an attractive proposition, for much human misery could be obviated. To others, it is a disconcerting proposition. Where will it lead? Surely the utmost foresight is required to consider the applications of such knowledge.

FURTHER READING

As an example of the rapid progress in molecular genetics and of the skepticism with which one should regard firm statements of "fact" in textbooks, obtain a copy of a genetics text published in the early 1940s and read the sections dealing with the structure and chemistry of genes and chromosomes.

Probably the best current survey of knowledge about the molecular biology of the gene is that given by Watson (1970). More detailed discussion of experiments and principles will be found in books by Drake (1969), Haggis, *et al.* (1964), Hayes (1968), Herskowitz (1967), Ingram (1963, 1966), and Jacob and Wollman (1961).

For further information about the fine structure of the gene, see articles by Benzer (1962) and Yanofsky (1967).

The many articles published by *Scientific American* over the past few decades provide an accessible and readable review of changing views about nucleic acids and protein synthesis. See particularly the articles by Clark and Marcker (1968), Crick (1954, 1957), Davidson (1965), Deering (1962), Delbrück and Delbrück (1948), Dobzhansky (1950), Doty (1957), Edgar and Epstein (1965), Fraenkel-Conrat (1956), Fruton (1950), Hanawalt and Haynes (1967), Hoagland (1959), Holley (1966), Horowitz (1956), Hurwitz and Furth (1962), Ingram (1958), Isaacs (1963), Jacob and Wollman (1961), Kellenberger (1966), Kendrew (1961), Knight and Fraser (1955), Kornberg (1968), Linderstrom-Lang (1953), Mirsky (1953, 1968), Pauling, Corey, and Hayward (1954), Rich (1963), Sonneborn (1950), and Spiegelman (1964).

16
Reproduction

An adult organism may be logically regarded as a mechanism designed to keep itself functioning until it has contributed to the next generation of its kind. It then deteriorates and dies, removing itself from competition with its offspring for food and space. As the nineteenth-century novelist Samuel Butler put it, "A hen is only an egg's way of making another egg."

Self-reproduction is a basic property of living systems. Because natural selection favors those species that are most successful at reproduction, all other functions of the organism may be regarded as means of enhancing the reproductive function. Reproduction forms new organisms, which embark upon their own careers of growth and development, culminating in their own reproduction. The intertwined processes of reproduction and development may well be considered as the central phenomena of life at the organism level.

TYPES OF ASEXUAL REPRODUCTION

For unicellular organisms, reproduction involves duplication of cellular structures and division of the cell into a pair of daughter organisms. The parent organism does not die but contributes all of its materials to its offspring. There is reason to believe that cell division rejuvenates the unicellular organism, which otherwise would age and die. Even if a cell were immortal, it would still be susceptible to accidental death. The continued survival of a species therefore depends upon its ability to produce new individuals at least as rapidly as old ones die.

For most procaryotic organisms, reproduction is accomplished through the relatively simple process of *binary fission*. During a brief period of growth, all cellular components—including the single, circular chromosome—are duplicated. The cell then splits into two cells, each receiving exactly half of the chromosomal materials and about half of the other materials from the parent cell. The newly formed cells then begin to grow in preparation for the next division. Under suitable conditions, some species of bacteria repeat this fission at 20-minute intervals.

In some bacterial species, daughter cells do not separate fully after division but cling together to form pairs, clusters, or filaments. Many species of blue-green algae form similar clusters or filaments in which the daughter cells are held together after division by a sheath of gelatinous material. In some species, a rapid series of divisions occurs within a single cell, producing a large number of daughter cells, or *endospores*, within the single parental cell wall. The endospores eventually are released from the parental cell, from their own cell walls, and become unicellular adults.

In all these cases, the offspring cell receives a single copy of the parental chromosome. The single chromosome is duplicated again just before each division. Variations in genetic information can arise only through mutations—through spontaneous changes in the nucleotide sequence or through errors in the process of chromosome replication. Variations from this simple pattern are rare among procaryotic organisms. The *actinomycetes*, or funguslike bacteria, form long, threadlike, tubular bodies that probably contain several chromosomes. Presumably, these chromosomes are identical copies of the same chromosome, resulting from successive divisions without the formation of cell walls. In certain bacteria, sexual recombination—the production of a daughter cell with genetic material contributed by more than one parent cell—does occur, although rarely. In a simple form of sexual mating called *conjugation*, one bacterial cell inserts all or part of its chromosome into another cell. The recipient cell later divides to produce daughter cells with single chromosomes created by a recombination of fragments of the parental chromosomes. In another process of genetic transfer called *transduction*, chromosomal fragments are carried from one bacterial cell to another as parts of viral chromosomes. A similar process called *transformation* involves the passage of chromosomal fragments from one bacterium to another, either directly or through the culture medium. These processes of sexual recombination probably are rare in natural populations of bacteria, but they are of importance to geneticists studying the nature of bacterial chromosomes and genes.

Although mechanisms for distributing chromosomes and other parts of the procaryotic cell between the daughter cells probably exist, these mechanisms remain unknown. In eucaryotic unicells, chromosomes are distributed during the process of mitosis, but little is known about the distribution of the other parts of the cell. For more complex unicellular organisms, a great number of organelles either must be duplicated before division and then distributed properly, or they must be produced after division according to instructions carried by the chromosomes.

For many kinds of eucaryotic unicells, reproduction is a matter of mitotic cell division. In many species, however, meiotic division of the diploid adult cell produces haploid gamete cells. A pair of these gametes (often derived from different parental cells) then fuse to form a new diploid individual cell. In many cases, two kinds of gametes are produced by different parental cells: a flagellated gamete that contains little cytoplasmic material and a nonmotile gamete that contains a large amount of cytoplasm. Fusion occurs only when a flagellated gamete happens to run into a nonmotile gamete. There are several species of algae in which daughter cells remain attached to one another after reproduction.

MULTICELLULAR ORGANISMS: ASEXUAL REPRODUCTION

Few multicellular organisms reproduce by a process equivalent to binary fission or cell division. Most multicellular organisms produce single cells or small multicellular fragments that develop into new individuals. The parent, after completing a period of such reproduction, in most cases enters senescence and eventually dies. Asexual reproduction may involve either the formation of multicellular buds or fragments or the production of unicellular spores. Sexual reproduction usually is accomplished by production of unicellular gametes.

Volvox is a green alga that might be regarded as a multicellular plant or as a colony of unicellular organisms (Figure 16.2). Each of the hundreds or thousands of cells in the spherical colony has two flagella. The flagella of the entire sphere beat in a coordinated pattern, moving the sphere through the water with a rolling motion. One particular side of the sphere always is in front during this motion and thus can be called the anterior end of the colony. Asexual reproduction occurs when one of the cells begins to divide, eventually forming a new sphere of cells within the hollow center of the parent colony. Only cells in the posterior half of the colony produce daughter colonies. The daughter colonies remain inside the parent colony until it dies or is broken apart.

In the process of *budding*, one of the cells of the parent organism begins to grow and divide, much as if it were a newly formed zygote. For a time,

Figure 16.1 (upper left). A paramecium dividing. (*Courtesy Carolina Biological Supply Company*)

Figure 16.2 (upper right). *Volvox*, a colonial protistan. Note the daughter colonies within the hollow center of the parent colony.

Figure 16.3a (lower left). The fresh-water coelenterate *Hydra* shown here is reproducing by means of budding. Several buds can be seen attached to the adult's stalk. Each bud will form a miniature adult, like the one to the right, and each will eventually pinch off to assume an independent existence.

Figure 16.3b (lower right). Yeast budding.

this bud continues to grow into a new organism while remaining attached to the parent and drawing nourishment from it. At some stage in its growth, it may become separated from the parent and take up an independent life. In many cases, however, the offspring remain attached to the parent, forming a colony of potential individuals. Each bud is capable of survival if it should be cut off from the colony, but most retain bodily connections and an interchange of nutrients and other materials with the others.

The simple coelenterate *Hydra* is an example of an animal that reproduces by budding. Each bud forms a miniature adult while still attached to the parent organism, then drops off to become independent (Figure 16.3). *Hydra* is of particular interest because the cells of each individual are being continuously and rapidly replaced. Cell division occurs in a zone near the center of the body, and newly formed cells continuously move outward as old cells die and drop off at the base and at the tips of the tentacles. Every few weeks the entire individual is completely replaced. Thus, the individual *Hydra* might be regarded as being in a continuous process of asexual reproduction, which apparently eliminates the process of aging and natural death for the total organism. Other coelenterates, sponges, and flatworms use budding as a means of asexual reproduction.

In some kinds of algae and liverworts, older parts of the plant body may die, leaving the tips to develop into new individuals. Among many kinds of plants and fungi as well as some animals, fragments of the body can

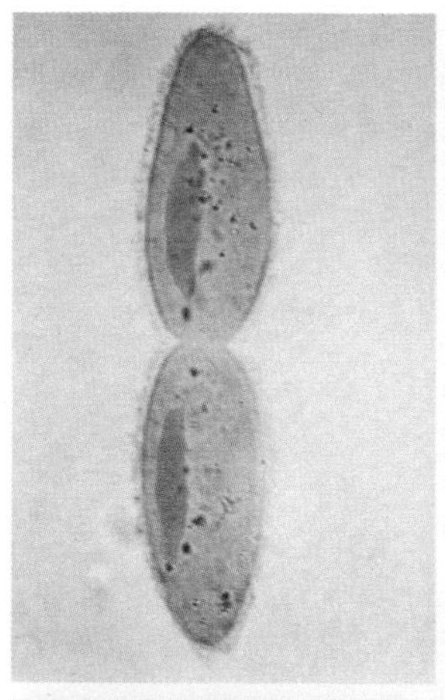

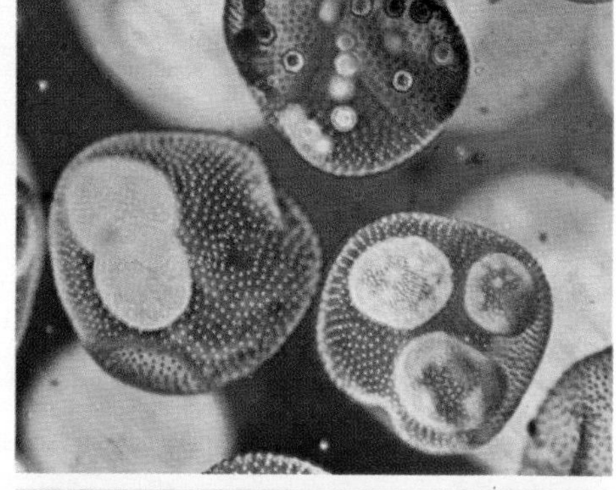

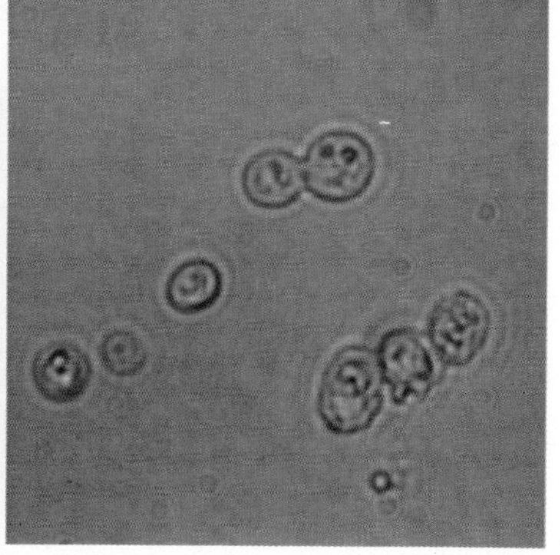

Figure 16.4 (above). Planarian spontaneously dividing. (*Courtesy Carolina Biological Supply Company*)

Figure 16.5 (lower left). Three types of life cycles.

Figure 16.6 (lower right). The alternation of haploid and diploid generations.

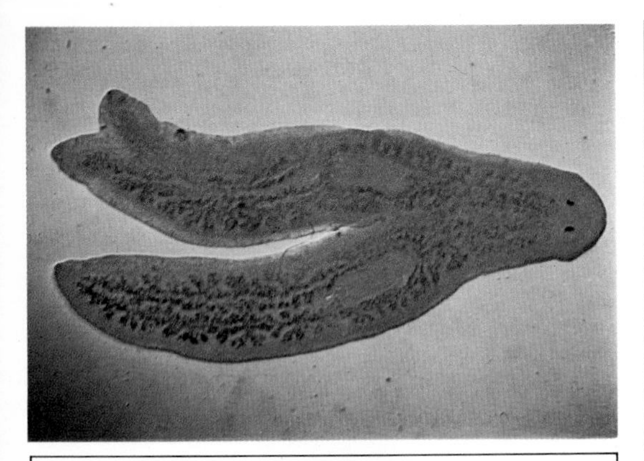

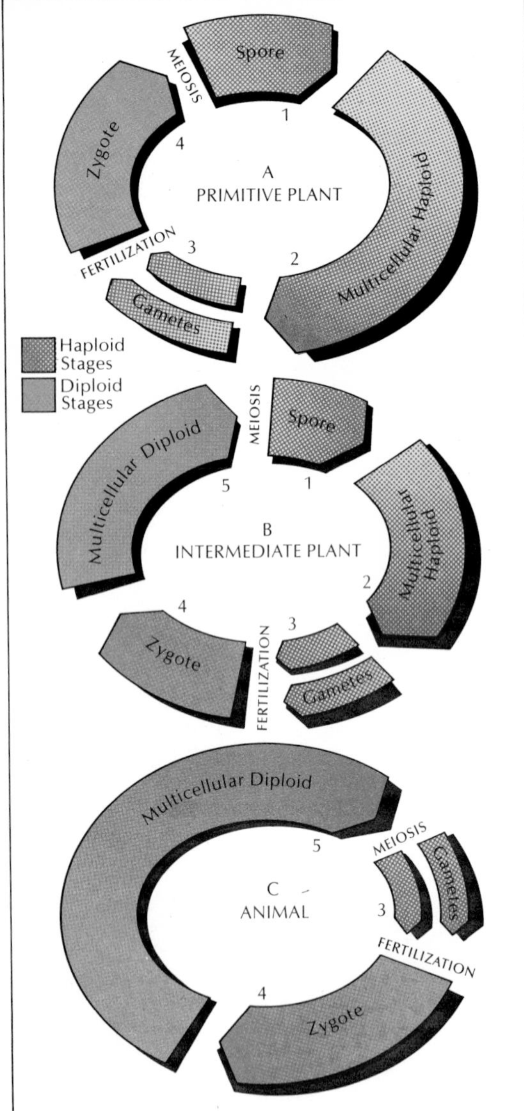

develop into new individuals if some exterior force breaks up the body. Flatworms such as *Planaria* are among the few examples of multicellular organisms that spontaneously divide themselves into more or less equal fragments, each of which develops into a complete new individual.

Asexual reproduction through unicellular spores is found among protists, fungi, and plants, but not among animals. The spores of some species are flagellated; others are immotile and in most cases are transported by wind or water. Thus, the process of reproduction usually involves more-or-less widespread dispersion of the new individuals.

In many species of algae, large numbers of spores are formed by mitotic or meiotic divisions of one or more of the body cells of the parent organism. Each of the *mitospores* formed by mitotic division can develop into an individual similar to the parent plant. A *meiospore* formed by meiotic division will develop into a haploid organism that may be different from the diploid parent in appearance. In such species, the haploid individuals produce gametes, whose sexual fusion creates diploid offspring. Thus, haploid and diploid generations alternate (Figure 16.6). Similar life cycles are found among fungi, ferns, and mosses. In the flowering plants, meiospores are formed within the sexual organs of the adult plant, but they develop within

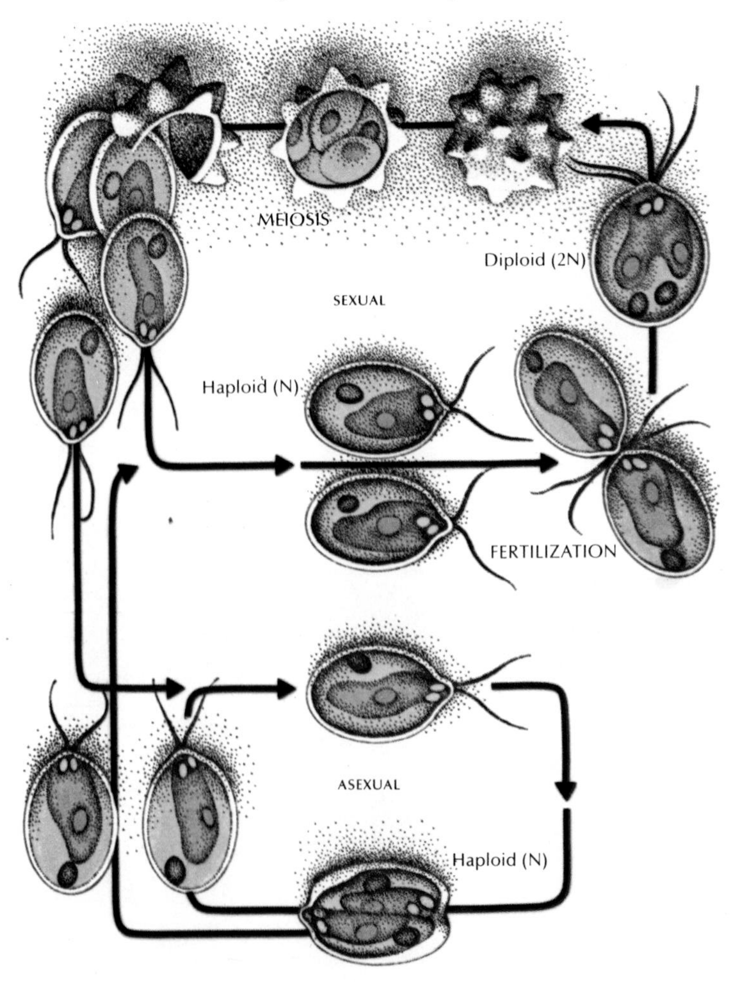

those organs into small haploid individuals. Until they produce gametes, they remain parasitic upon (and hidden within) the diploid parent.

Another form of asexual reproduction is *parthenogenesis*, the development of an unfertilized egg into an adult organism. It occurs spontaneously as an important means of reproduction in some kinds of organisms. Among honeybees, development of an unfertilized, haploid egg gives rise to a drone, a male adult with haploid cells. Fertilized honeybee eggs form diploid, female adults—either queens or workers, depending upon the nutrients supplied during development.

In some cases of parthenogenesis, a diploid egg is formed either through fusion of the egg cell and one of the polar bodies or through a replication of the egg chromosomes without cell division. In other species where parthenogenesis is common (among the aphids, for example), eggs are formed without meiotic division, thus producing diploid eggs that can develop into diploid individuals with the same genetic information as the parents.

MULTICELLULAR ORGANISMS: SEXUAL REPRODUCTION

Almost every kind of multicellular organism is capable of sexual reproduction, although this process may be supplemented by various forms of asex-

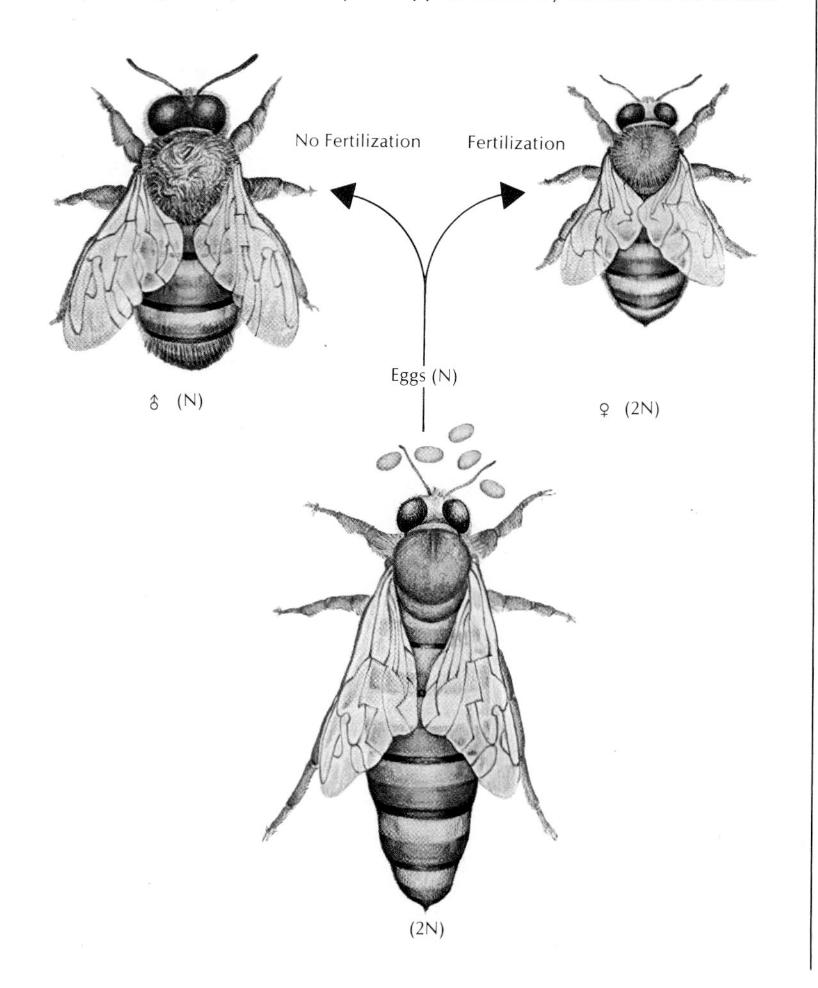

Figure 16.7. An example of parthenogenesis, the development of an unfertilized egg into an adult organism and thus a form of asexual reproduction. In the honeybees shown here, the queen produces eggs that if fertilized become female workers and if not fertilized become male drones.

Figure 16.8a. The alga *Ulothrix* is an organism in which all gametes appear identical.

ual reproduction. Sexual reproduction provides a source of genetic variation in the population. Variations arise in any population through mutations—relatively infrequent, spontaneous changes in the genetic message. For example, suppose that two independent mutations, *A* and *B*, each appear on the average in about 1 of each 1,000 individuals. In a population that reproduces only asexually, about 0.1 percent of the population will show the mutant character *A* and another 0.1 percent will show the mutant character *B*. However, only about 1 individual in each 1,000,000 will happen to undergo both *A* and *B* mutations simultaneously. The only other way that individuals with both *A* and *B* mutations can arise is to wait until one of the descendants of an individual with one mutation happens to undergo the other mutation as well.

On the other hand, in a sexually reproducing population, there is a possibility that an individual with the *A* mutation will mate with an individual with the *B* mutation. Thus, the *A-B* combination can be produced within a single generation by sexual recombination. Similar sexual recombinations can be made of all the different characters that happen to arise by mutation, so that a population that reproduces sexually is likely to possess a much wider variety of genotypes than one that reproduces only asexually. The sexually reproducing population is much more likely to contain individuals suited for survival in changed environmental conditions. Sexually reproducing populations may therefore be expected to survive environmental changes more successfully and to undergo more rapid evolutionary changes. It is not surprising to find that most modern multicellular organisms utilize sexual reproduction. The ability of many unicellular organisms to survive with only asexual reproduction is probably due to their very rapid rate of reproduction, which leads to large populations and frequent chromosome replication, so that favorable mutations are likely to appear in a population within a relatively short time.

In theory, there is no compelling reason why sexual recombination would have to be linked with reproduction. Although the process would be complex for any organism with large numbers of cells, it is possible to imagine an interchange of chromosomes between the cells of two individuals, followed at some later time by asexual reproduction of each individual. In all known cases, however, sexual recombination leads to the formation of a single cell, a zygote, which then develops through repeated divisions into a new adult. The result of this universal process of sexual recombination combined with reproduction is that every cell of a multicellular organism contains the same genetic information. This combination of reproduction and sexual recombination has been favored in the evolution of all multicellular organisms presumably because it ensures that all of the cells of the organism will have the same genetic instructions and therefore will function together smoothly. This combination also underlies the process of natural selection, for it ensures that the reproductive cells of an organism will contain the same genetic information as the body cells that determine the phenotype.

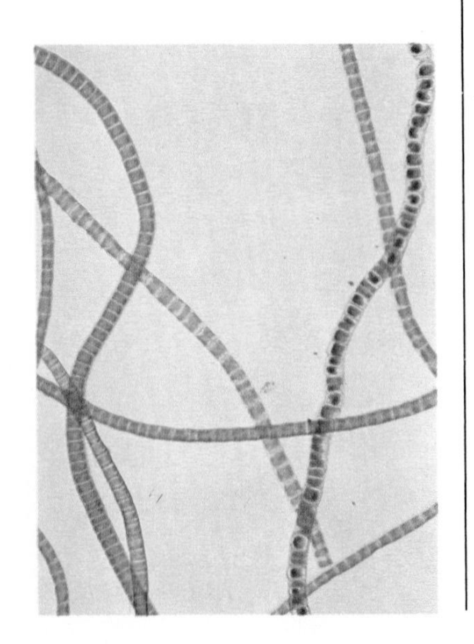

In its simplest form, sexual reproduction involves the formation of haploid gametes that fuse to form a diploid zygote. In some algae, such as *Ulothrix*, all gametes appear identical (Figure 16.8). A gamete fuses with any other gamete that it happens to encounter, although in many species fusion will occur only if the gametes are from different parent individuals. In the case of *Ulothrix*, the adult plant is haploid, and the diploid zygote

Figure 16.8b. Each filament of *Ulothrix* is able to reproduce by sexual and asexual means. The adult plant is haploid and the diploid zygote divides meiotically early in development. Only one of the resulting haploid cells survives to give rise to a new individual.

undergoes meiosis early in its development. Only one of the haploid cells resulting from meiosis survives to give rise to the new individual, thus again ensuring that all cells of the adult plant will have the same genetic information. The process of reproduction through identical gametes is called *isogamy*.

Among most multicellular organisms, sexual reproduction involves two kinds of gametes. One kind tends to be small and highly motile; the other kind is less motile and carries a large supply of nutrients. In this case of *heterogamy*, or *anisogamy*, a particular adult usually produces only one kind of gamete. In the extreme development of heterogamy, one kind of gamete is the sperm—small, active, and equipped with powerful flagella—and the other kind is the egg, or ovum—large, immotile, and packed with nutrients. This extreme form of heterogamy is called *oogamy*; the individuals that produce sperm are males and those that produce eggs are females.

Although the pattern of oogamy is by far the most common form of sexual reproduction among multicellular organisms, it is not universal. In some cases of heterogamy, the two kinds of gametes are very similar in appearance, and the designation of one sex as female and the other as male may be arbitrary. In the green alga *Ulva* (the "sea lettuce"), for example,

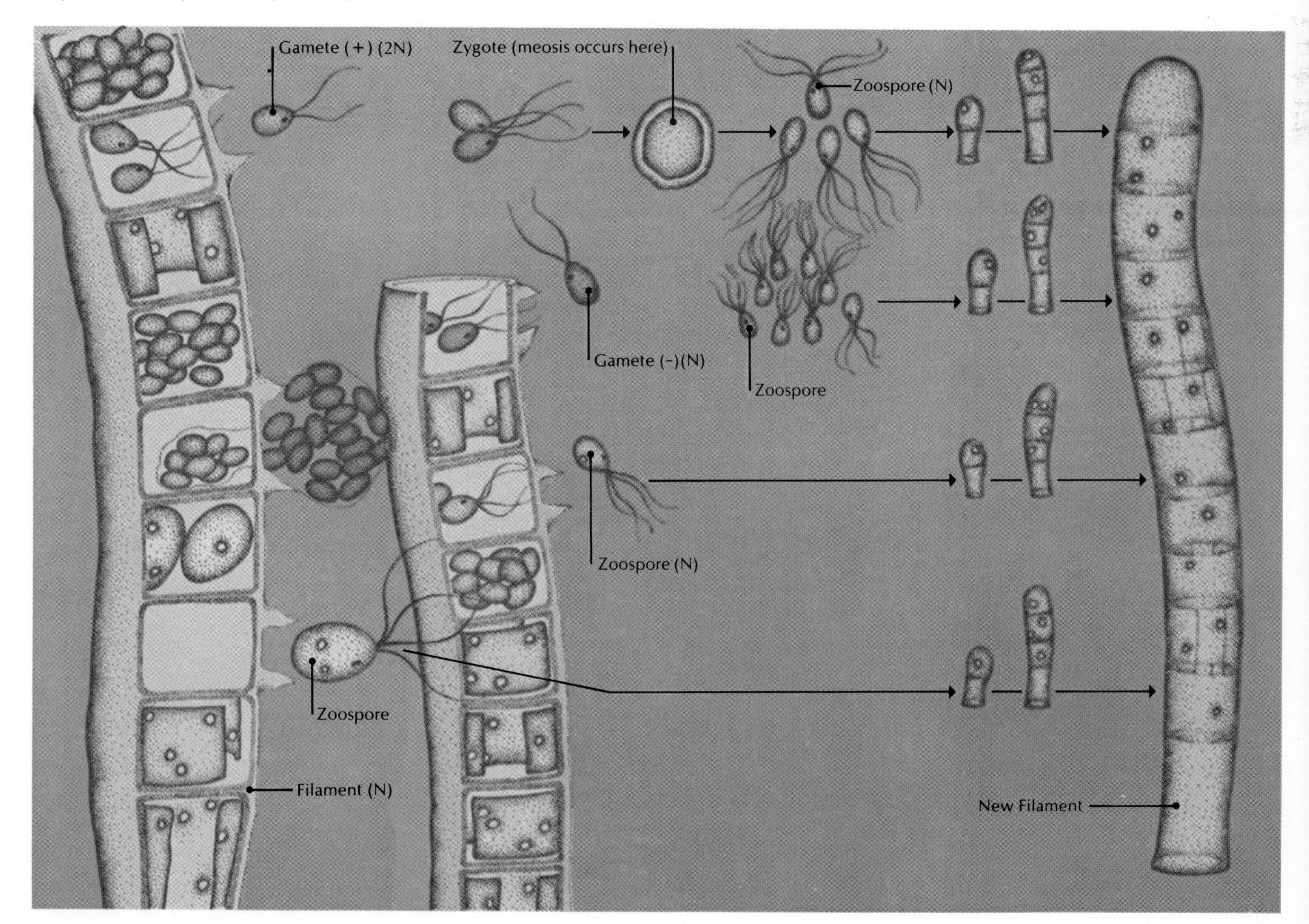

Figure 16.9a (above). The black bread mold *Rhizopus*. *(Courtesy Carolina Biological Supply Company)*

Figure 16.9b (below). Formation of a zygote by two differing strains of *Rhizopus*. This haploid organism normally reproduces asexually through the production of haploid spores. Under certain conditions, however, different individuals will develop filaments whose nuclei fuse and form a zygote.

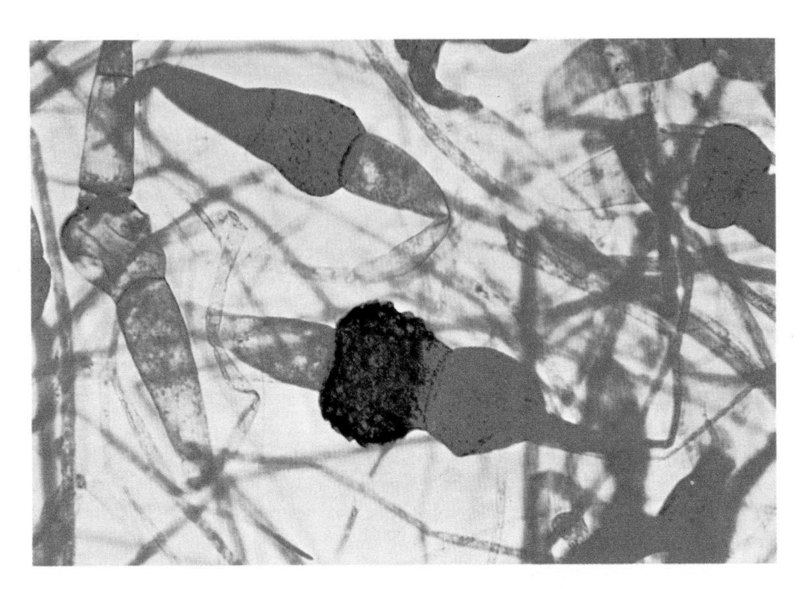

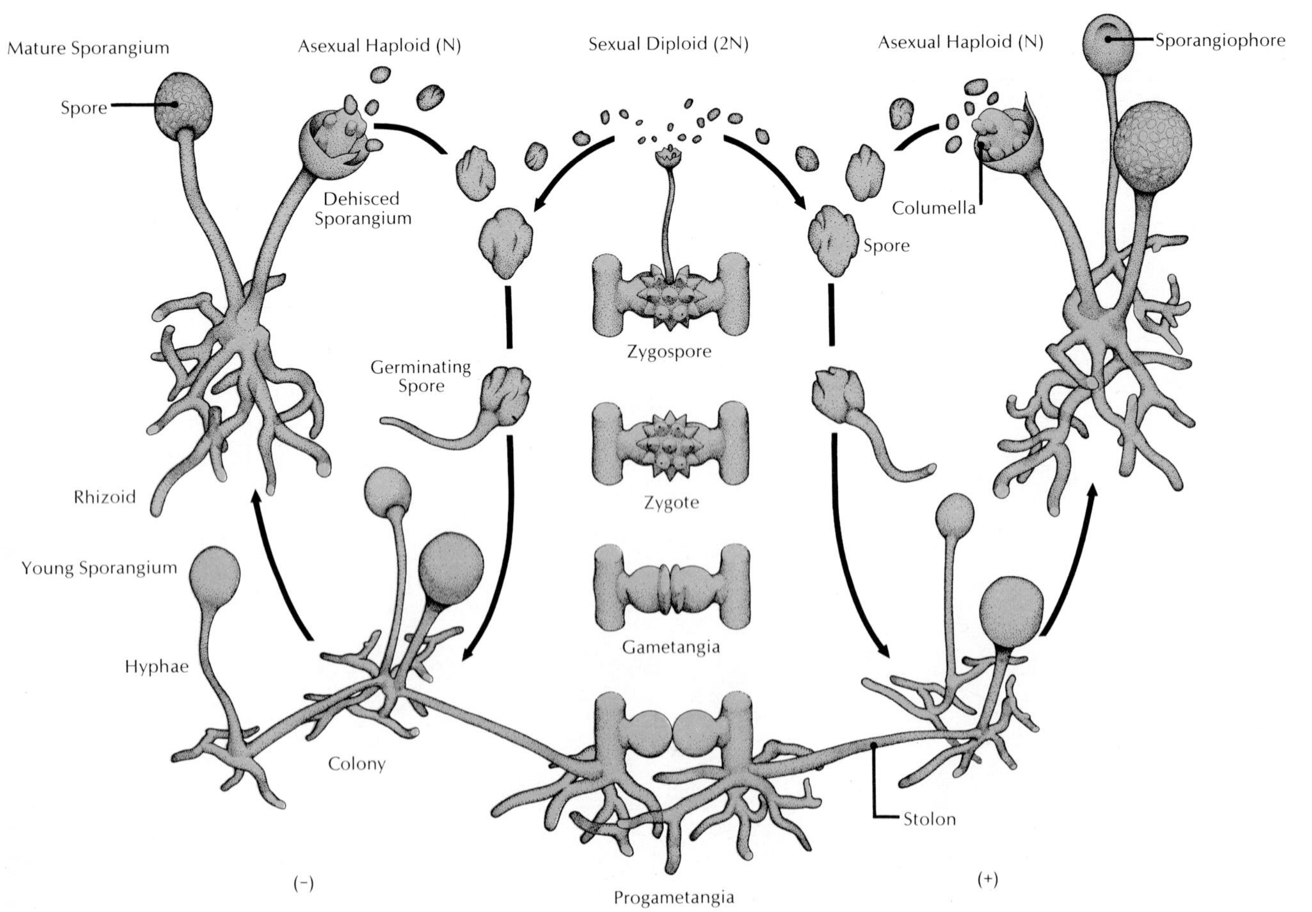

Figure 16.10. Pattern of sexual reproduction in the filamentous green alga *Spirogyra*. Vegetative filament (above). Early conjugation, with protoplasmic bridge being formed between four sets of opposite cells in different filaments (upper middle). Later conjugation, with migration of cellular contents from one cell into the other cell (lower middle). Conjugation is completed (below) with the formation of zygospores (zygotes), each capable of developing a new vegetative filament. (Courtesy Carolina Biological Supply Company)

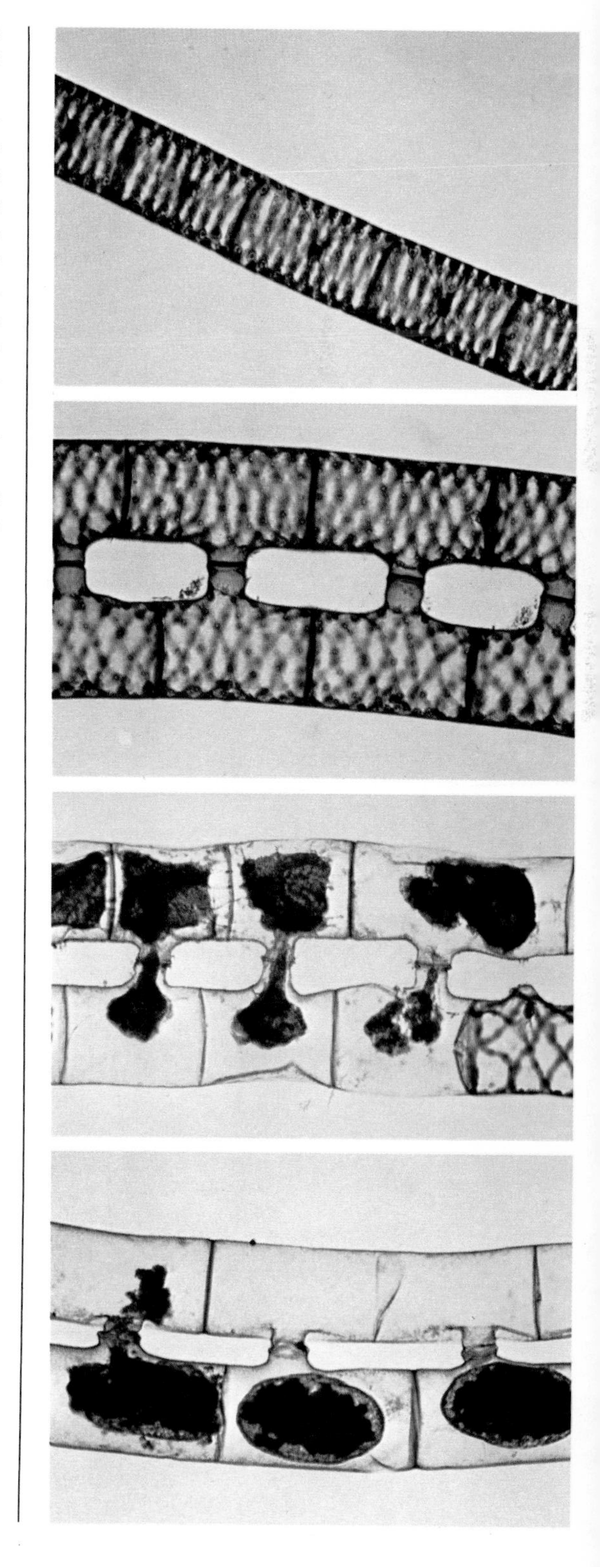

both kinds of gametes are flagellated, although one kind is larger than the other. Among the red algae, on the other hand, neither kind of gamete is flagellated.

In some fungi, the diploid portion of the life cycle has been nearly eliminated. The black bread mold *Rhizopus* is a haploid organism that normally reproduces asexually through the production of haploid spores (Figure 16.9). Under certain conditions, however, sexual reproduction may take place between individuals of two different types. Because individuals of the two types appear physiologically identical, they are called + and − rather than male and female. Projections grow from the filaments of the two individuals and join. At the juncture, nuclei from the two individuals fuse to form a diploid zygote. The zygote is covered by a resistant wall and may remain independent and dormant for some time. Under suitable environmental conditions, the zygote germinates and undergoes meiotic division to produce new haploid spores. Thus, the only diploid part of the life cycle is the dormant zygote.

A similar pattern of reproduction is found in the filamentous green alga *Spirogyra* (Figure 16.10). In this case, the zygote undergoes meiosis upon germinating, but three of the four nuclei formed by meiotic division disintegrate, and the remaining nucleus undergoes mitotic division to form a new filament.

ANIMAL REPRODUCTIVE SYSTEMS

Although asexual reproduction through budding, fission, or fragmentation does occur in some of the invertebrate groups of animals, sexual reproduction is almost universal throughout the animal kingdom. In some of the simpler aquatic organisms, sperms and eggs are released into the water, where fertilization occurs. In most kinds of animals, some form of *copulation* occurs, with the male injecting sperms into the female's body. The fertilized egg cell, or zygote, may develop for some time within the female's body or may be enclosed with nutrient materials in an egg and released from the female's body for further independent development.

The gametes are produced by meiotic division of specialized *germ cells* in the organs called *gonads*. The process of *gametogenesis*, or gamete production, varies somewhat in detail among the different groups of animals, but a generalized description of the process in mammals illustrates the major events.

The reproductive system of a mammal consists of the gonads, the reproductive tract through which the gametes move, and various associated glands. As in most higher animals, each mammalian species has male individuals that produce spermatozoa and female individuals that produce eggs. Hermaphrodism, in which a single individual produces both eggs and sperms, is found among some groups of animals. Fertilization occurs within the reproductive system of the female, and the embryo is retained within the female's body during the early stages of its development. In addition to the differences in the sexual organs of the two sexes (appropriate to their different roles in the formation of the zygote and nurture of the embryo), the sexes can be distinguished by various differences in the form of various other parts of the body — the *secondary sexual characteristics*.

The sex of an individual is determined primarily by genetic inheritance. A mammalian zygote obtaining an X chromosome from each of its parents will normally develop into a female, whereas a zygote obtaining one X and one Y chromosome will develop into a male (Chapter 14). The Y chromosome must carry various genes involved in the development of male characteristics, for an abnormal individual with only a single X chromosome and no Y chromosome (an XO genotype) develops as a female (normal in the mouse, but sterile in the human). The abnormal XXY genotype results in development as a male, but with some female secondary characteristics and severe mental retardation. The determination of sex through sex chromosomes is widespread among animals but is not universal. Among bees, for example, males are haploid and females are diploid. Even among animals having sex chromosomes, the nature of sex determination varies widely. In birds, for example, males have a pair of similar sex chromosomes, whereas females have a pair of different sex chromosomes.

The genetic inheritance of sex is by no means the complete story. The balance of various hormones in the developing embryo—a balance presumably regulated by the genetic information—appears to play the crucial role in controlling the determination of tissues and organs toward male or female forms (Chapter 21).

The determination of the primary germ cells, which will later form the eggs or sperm, occurs at a very early stage in the development of the animal embryo. Cells that form the gonads become determined at a much later stage of development. Thus, the primary germ cells are set aside for that fate very early in development, long before the gonadal structures are visible. Even after the gonads have begun to develop, it is some time before they become determined as either male or female sex organs. The primary germ cells migrate into the developing gonads at a relatively late stage of development. This migration occurs either through amoeboid movement or through transport by the blood (Franchi, *et al.*, 1962).

The male gonads, or *testes*, of a mammal contain a large number of tubules that are made up of germ cells and *Sertoli cells*. The tubules are complexly coiled within the testes and are surrounded by connective tissues and interstitial cells (Figure 16.11). In the mature mammal, the germ cells continue to divide mitotically, producing a continuous new supply. Some of these cells divide meiotically to produce haploid cells that develop into spermatozoa. As the spermatozoa mature, they cluster around the Sertoli cells, whose function is not understood. In some mammals, maturation of spermatozoa occurs only at certain seasons. In others, including man, mature spermatozoa are produced continuously during the reproductive portion of the life cycle. Mature spermatozoa leave the Sertoli cells, move through the tubule, and enter a duct leading to the exterior. The structure of the duct and the nature of the various glands that lubricate the passage of sperm or provide nourishment for sperm cells vary greatly among groups of animals.

The female gonads, or *ovaries*, of vertebrates consist of oval-shape masses of cells, including vascular and connective tissues, as well as the germ cells, *follicle cells*, and *nurse cells*, which provide nutrients for the development of egg cells, or *oocytes* (Figure 16.12). The mammalian female germ cells cease mitotic division before birth, so that the newborn female possesses a complete lifetime supply of immature eggs. The germ cells grow considerably larger than the other kinds of cells that surround them. Gametogenesis begins with meiosis, but the meiotic division is halted at prophase of the first division. The developing oocyte then grows extremely large as it absorbs nutrients from the follicle and nurse cells. The various substances

Figure 16.11 (above). Mammalian testes.

Figure 16.12 (below). Mammalian ovary.

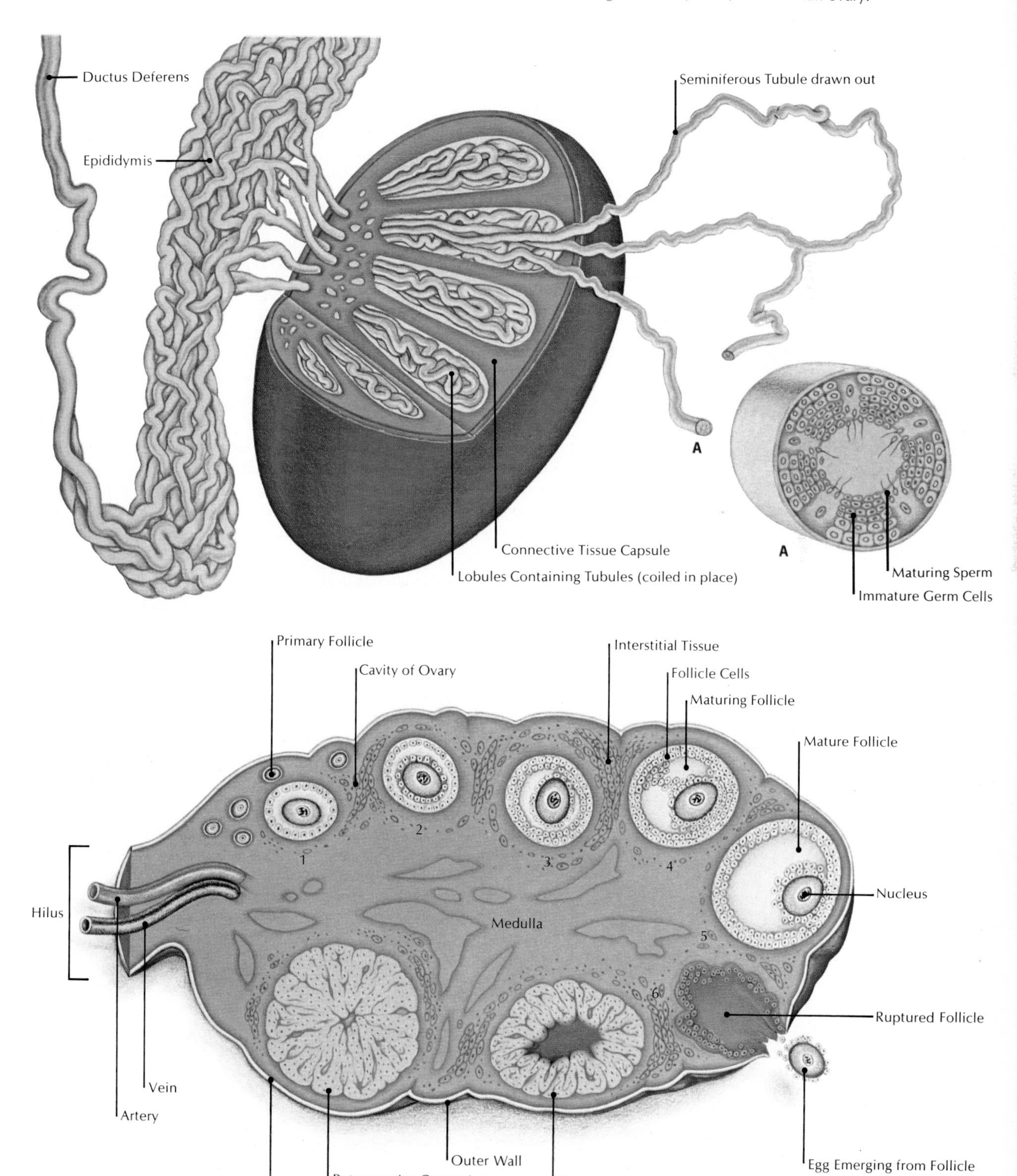

that make up *yolk*—including protein, fat, glycogen, and RNA—are pumped into the rapidly expanding cytoplasm of the oocyte.

Among some organisms, such as birds, the growth of the egg cell continues until it is many thousands of times as large as other cells of the body. The follicle cells form a thin layer over the growing egg cell. The mammalian egg is relatively small and contains only a small amount of yolk, but the layer of follicle cells becomes quite large around the egg cell. When accumulation of yolk has been completed, the oocyte resumes meiotic division. However, the division of cytoplasm during meiosis is unequal. The first division results in the pinching-off from the large egg cell of a very small cell containing little cytoplasm. This *first polar body* eventually disintegrates. In most vertebrates, meiosis is again halted after the first division and is not resumed until sometime after the egg has left the ovary. But whenever the second meiotic division occurs, it also results in the pinching-off of a small cell with very little cytoplasm. This *second polar body* also disintegrates. Thus, a single germ cell in the male gives rise to four spermatozoa, whereas a single germ cell in the female gives rise to only one mature egg.

In various vertebrates, large numbers of mature eggs may be produced at a single time in the life cycle of the female or at seasonal intervals. Eggs of birds may be matured individually at about one-day intervals, but maturation of eggs is normally halted when a batch of eggs has been laid and must be incubated. In the human female, one egg is matured each month during her reproductively active life, but egg maturation is suspended during pregnancy.

When the egg is matured, it moves out of the ovary as a result of the breakdown of some of the follicle cells and connective tissues. Moved by the action of cilia, the flow of fluids, or muscular contractions, the mammalian egg moves through the *oviduct* toward a chamber called the *uterus*. The uterus connects to a passage called the *vagina*, which opens to the exterior and is specialized to accept the male sex organ, or *penis*, permitting the introduction of sperm into the vagina and uterus during copulation. The sperm are carried into the uterus and the oviduct by their own swimming motions and by cilia or muscular contractions of the female reproductive tract. Among mammals, fertilization occurs in the oviduct; among some other groups of vertebrates, it occurs in the uterus. The fertilized mammalian egg becomes implanted in the wall of the uterus, where the early development of the embryo takes place.

Among various groups of animals, there are many variations in the structure of the reproductive organs and tracts. The female insect, for example, has tubules leading directly from the ovaries to genital pores that open to the exterior. In the oviduct of a bird, the egg is successively coated with albumen—a shell membrane—and a hard, limy shell as it moves through the oviduct. Many vertebrates possess paired uteri rather than a single uterus. In most vertebrates and some mammals, the uterus opens into the *cloaca*, which carries excrement and urine to the exterior (Figure 16.13). Only in primates are there separate exterior openings for the vagina, urethra, and anus. Many invertebrates possess an organ called the *spermatheca* (in insects) or *seminal vesicle* (in other groups) in which sperm may be stored within the female reproductive tract so that fertilization of eggs can be carried out over a long period of time after copulation.

The testes of most vertebrates lie inside the body, but in most mammals they are located in a sac called the *scrotum*, which hangs outside the body

Figure 16.13. Diagram of the cloaca of a snake. In reptiles, birds, amphibians, many fishes, and some mammals, the cloaca is the most posterior chamber of the digestive tract; the reproductive, intestinal, and urinary canals empty their products into this cavity.

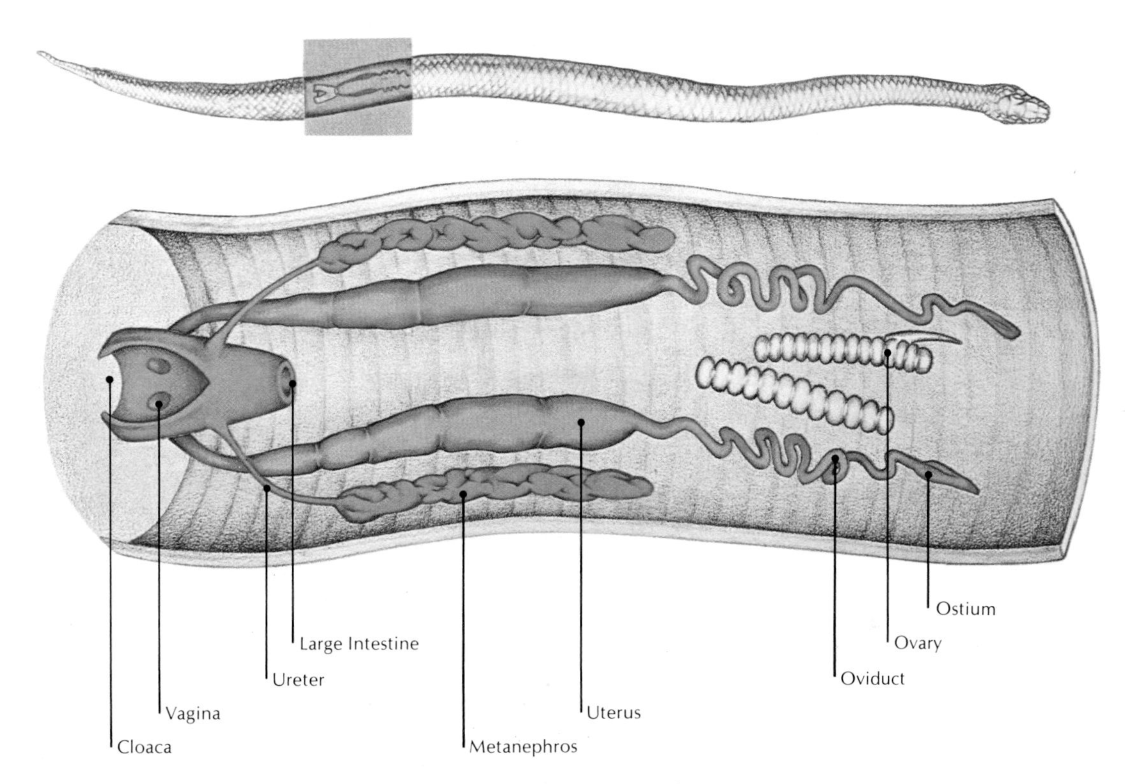

wall. This arrangement keeps the testes at the relatively low temperature needed for spermatogenesis and survival of the sperm. In birds and other vertebrates, various means of internal cooling are used to accomplish the same results. Only among primates and ungulates (hoofed mammals) do the testes remain permanently in the scrotum; in most other groups of mammals, the testes descend into the scrotum only during the mating season.

Among some groups of vertebrates, the sperm are discharged into the cloaca, which then is partially protruded from the male anus and inserted into the cloaca of the female. Insects possess a wide variety of complex copulatory parts, including various kinds of grasping or clasping structures on both males and females, so that mating in many cases is physically impossible with members of slightly different species. When birds mate, the usual pattern is inverted, for the cloacal end of the oviduct is protruded and inserted into the cloaca of the male. Only mammals and turtles have true penes, which are soft structures that become rigid when pumped full of blood. Grooves along the sides of the turtle's penis become ducts for the passage of sperm when the penis is erect. In mammals, the sperm passes through the urethral duct, which extends through the length of the penis. In some mammals, including cattle, a bone adds permanent rigidity to the penile structure.

MATING BEHAVIOR

In the simplest form of sexual reproduction, gametes are shed into the environment, and fertilization occurs whenever two gametes of the proper types happen to meet. Such random fertilization is highly inefficient, with

Figure 16.14a (lower left). The female Surinam toad has "pouches" on her back in which baby toads develop. Just prior to the time of egg-laying, the back of the female toad becomes very spongy. When the eggs are released, the male toad scoops them up and deposits them on her back. He then mounts the female toad and fertilizes the eggs. The pressure of his body forces the eggs down into the spongy layer where they remain until birth.

Figure 16.14b (upper right). Ladybug beetles mating. In many arthropods, sperm from the male is transferred directly to the female by specially modified appendages.

Figure 16.14c (lower right). Mating behavior in amphibians. Shown here are male and female chorus frogs in the mating position (amplexus) and during egg-laying. Fertilization is external; the male elicits egg-laying of the female by grasping her sides and by applying slight pressure. Note the egg mass in the water.

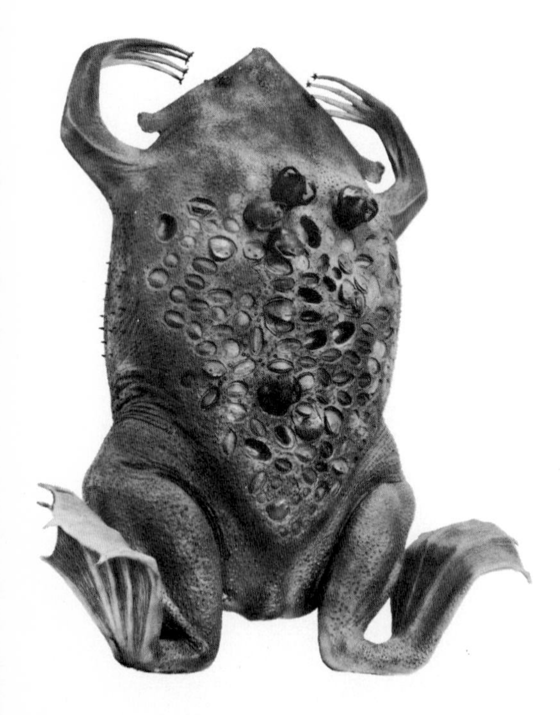

most of the gametes dying before fertilization occurs. A great variety of behavior patterns exist in different species to increase the probability that any given gamete will participate in fertilization. The release of gametes into the environment often is synchronized by chemical means (for example, a substance released with the eggs may trigger other females to release their eggs and males to release sperm) or by synchronization to some external stimulus (such as the tidal cycle). Among many species of fish, the complex behavior pattern involved in mating causes the male to shed the sperm directly over the eggs that have just been laid by the female. The process of copulation achieves the greatest efficiency in guiding spermatozoa to the eggs, but it requires very complex patterns of behavior to synchronize the actions and gametogenesis of the two individuals involved.

Various kinds of courtship behavior help to bring together two individuals of opposite sexes and of the same species, to synchronize the reproductive cycles of the prospective mates, and in some cases to make initial preparations for the care of the offspring. The actual behavior of copulation also involves complex patterns, both instinctive and learned, that ensure that the sperm will successfully be transferred to the reproductive tract of the female and that an egg or eggs will be ready for fertilization. The nature and functions of mating behavior are discussed in more detail in Chapter 31.

CARE OF OFFSPRING

In cases where gametes or spores simply are shed into the environment, the new individual is on its own. It must provide its own protection and nutrition throughout its period of growth and development. In most such cases, vast numbers of new zygotes are produced for each one that manages to survive to maturity. In most species, parents provide some protection or nutrition for the offspring. Various forms of spores, seeds, and eggs provide nutrients and protective coatings for the young individual during the most critical early stage of its growth. Organisms that retain the embryo within the body of the female adult during early growth provide even more protection and nourishment.

Parents (or other adult members of the species) may care for the offspring in a variety of ways that depend upon behavior patterns as well as upon physiological structures. Burrows or nests may be prepared. The adults may carry or cover the young or the unhatched eggs to provide body heat or protection from predators and parasites. Food may be brought to the young, provided from special glands or other sources on the adult body, or stored in the nest or burrow for the use of the young. Such behavior by adults involves complex patterns of reaction to stimuli provided by the presence and the behavior of the young themselves. Behavior patterns that create families and other social groupings provide adult care for the young of many species.

Among some species—particularly mammals—the young remain with the adults for a considerable period of time, learning a variety of behavior patterns that are not genetically determined. This transmission of learned as well as genetic information reaches its peak in some primate species, including the human species.

It is difficult to draw boundaries around the study of reproduction because almost every structure and behavior of a living organism has some relation to the perpetuation of the species through the production of offspring. The process of evolution through natural selection favors the continued

Figure 16.15. Care of offspring. Female scorpion (above) with newly-born young riding on her back. Scorpion courtship consists of a "dance" between partners with the male depositing a germ sac, or spermatheca, on the ground, then maneuvering the female over it. Siamang and baby (middle left). Three-week-old baby cougar cubs with their mother (middle right). As with most mammals, these cougar cubs are dependent on their mother for warmth, protection, food, guidance, and orientation. Gnu and baby (bottom). Giraffe and baby (opposite).

existence of those species most successful in reproducing themselves. In the long run, every characteristic that has been developed and retained through evolution probably contributes to the process of reproduction in one way or another. The picture of the natural world that is consistent with Darwinian theories of natural selection is not so much the often-depicted tooth-and-claw struggle to the death but a continuous effort to produce and protect offspring.

FURTHER READING

More details of reproduction in a wide variety of organisms will be found in many general books on biology, zoology, and botany. The discussions of these matters in books by Hardin (1966), Jessop (1970), and Telfer and Kennedy (1965) are particularly useful. For more extensive discussions of sexual reproduction, see books by Asdell (1964), Berrill (1953), Michelmore (1964), and Van Tienhoven (1968).

Articles of interest in relation to this chapter include those by Jones (1968), Roth and Barth (1967), Rothschild (1956), and Zahl (1949).

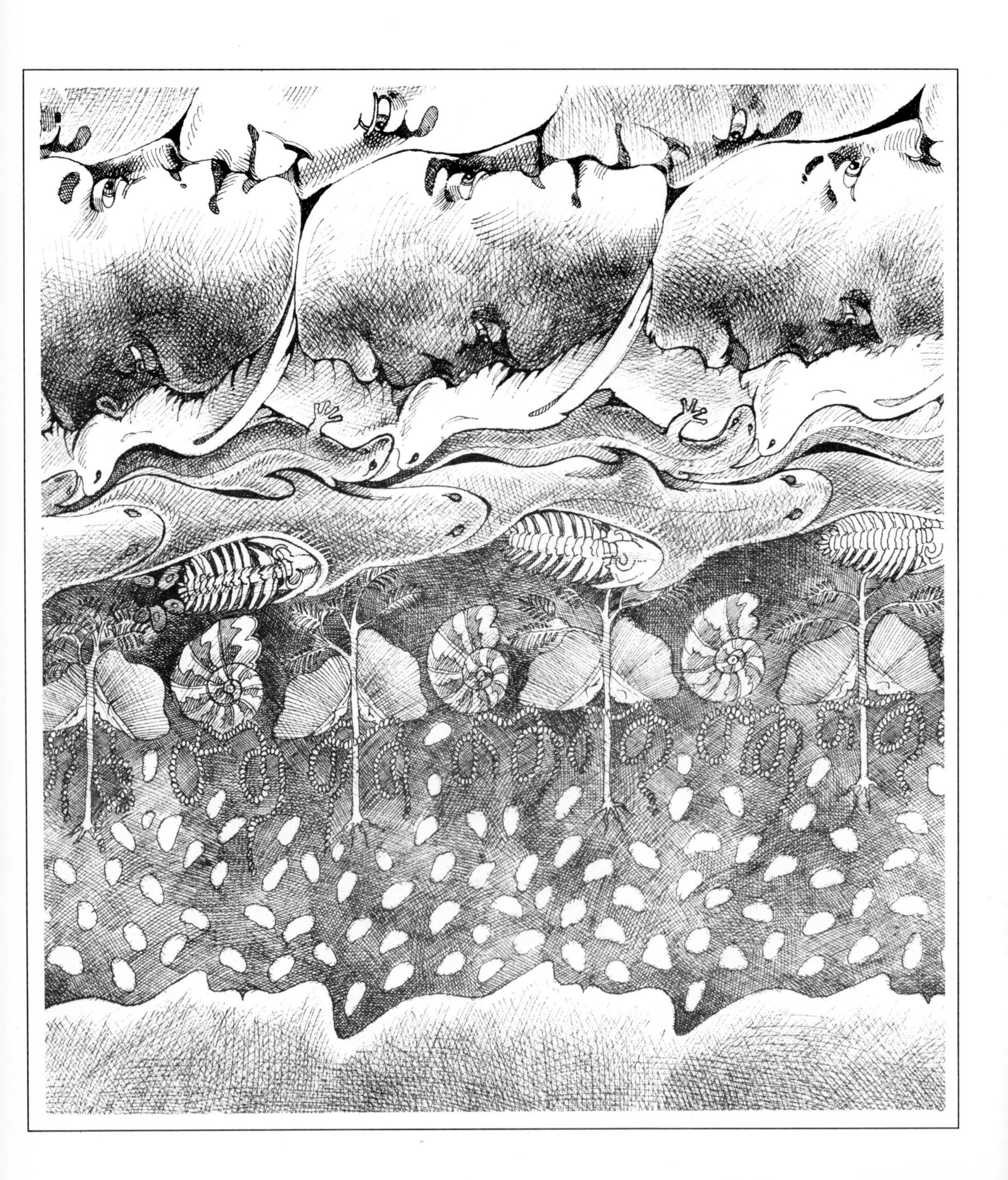

V
Integration

A cell will eventually divide in two, and that is an impressive enough sight, but still all it is doing is to produce another cell like itself. The striking achievement of an egg is to produce things—roots, leaves, legs, eyes, backbones, and so on—which were not in it originally. It does more than merely reproduce itself; it produces something new. Even if you have a certain degree of biological knowledge when you start looking at it—knowing perhaps what everyone seems to know nowadays, that the fundamental characteristics of organisms are determined by the genes inherited from their parents, and that these genes are made of nucleic acid (DNA)—even so, merely to say that the lump of jelly you are looking at contains the right DNA to produce a rabbit leaves an enormous amount unaccounted for. Exactly how does the egg produce legs, head, eyes, intestine, and get up and start running about?

—C. H. Waddington (1966)

17
Cellular Regulation and Control

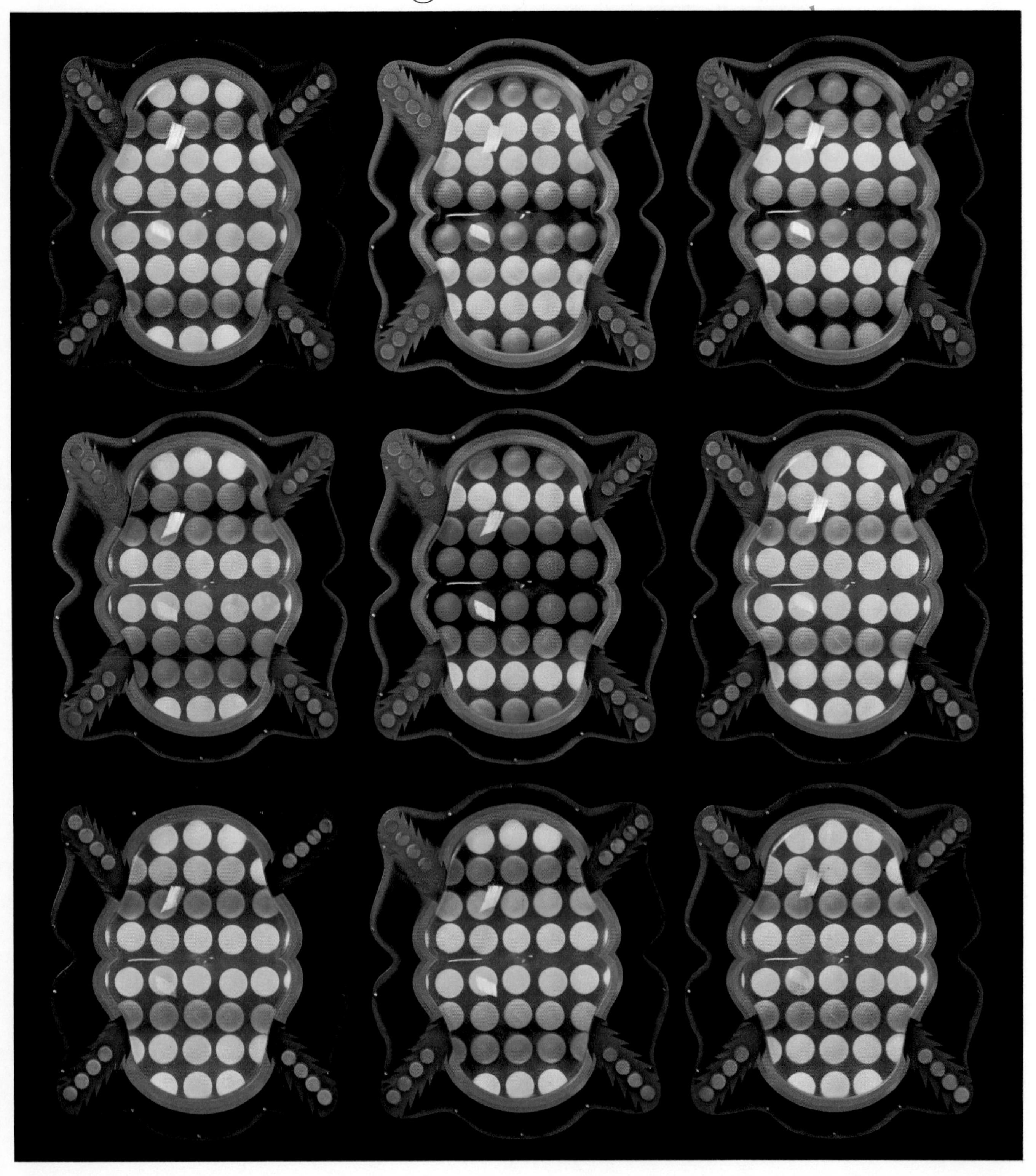

The cell is an exceedingly complex biochemical system. Its life processes require the synthesis and utilization of millions of macromolecules and billions of smaller molecules and ions. Thousands of different kinds of molecules must be produced in the proper numbers and brought together at the appropriate times and places within the cell to serve their functions in metabolic processes. How is this complex system controlled and coordinated?

Part of the controlling mechanism is described by the "central dogma" of molecular genetics. Information specifying the amino acid sequence for a polypeptide chain is encoded in the sequence of base pairs in a gene—a segment of the chromosomal DNA molecule. This information is transcribed into the sequence of base groups in a molecule of messenger RNA, which moves from the gene to a ribosome. At the ribosome, with the aid of transfer RNA, ribosomal RNA, and various enzymes, the polypeptide chain is constructed according to the directions specified by the base sequence. The polypeptide chain thus constructed may act as an enzyme, or it may become part of an enzyme or structural complex.

Because the chromosome contains many genes, each coding for the production of a different polypeptide chain, the cell is able to construct the many different enzymes and structural proteins needed for its life processes. Manufactured enzymes, in turn, can guide the synthesis of necessary nonprotein molecules.

However, the genetic specification of proteins alone is not sufficient to explain the life history of either an individual cell or a multicellular organism. Except for rare cases of mutations, a cell retains the same genes throughout its life. A multicellular organism develops by repeated mitotic divisions of a single zygote, so that each cell of the mature organism possesses at least one complete copy of the zygote's genetic information. If each cell possesses the same set of instructions for protein synthesis, how can different cells synthesize different proteins and exhibit specialized functions? How can a variety of cellular structures and functions develop from a single set of genetic instructions? Obviously, a cell needs a method of selective gene expression. It must be able to control which genes it will express and how much of each gene product it will make.

The cell could exert control at any one of several points along the route from gene to gene product. The duplication of the gene itself could provide many templates for messenger RNA transcription. Direct control over the transcription mechanism could determine which genes would be transcribed and regulate the number of mRNA molecules made from each gene. Likewise, the cell could control the kinds and amounts of mRNA molecules that are translated into proteins and control the assembly of those proteins into functional structures. Even after a protein has been synthesized, it is not beyond the cell's control; there are numerous ways in which a cell could alter and regulate the function of a gene product.

The control of the cellular mechanism is not always a one-way process, with prerecorded instructions in chromosomal DNA blindly guiding the cell along a predestined course. Cellular controls result from interactions of the cell's genetic material with the cytoplasm and, indirectly, with the external environment. Obviously, the living system possesses feedback capability—it reacts to changes in its surroundings and in its own internal status. Although a complete understanding of such genetic control mechanisms lies far in the future, experimental work done in the past few decades has

Figure 17.1. A broad interpretation of the aphorism "What is true of *E. coli* is true of the elephant."

revealed the general nature and some details of the cellular control systems.

Biologists and biochemists have faced many obstacles in their attempts to understand the processes of cellular control in animals. The technical difficulties involved in trying to formulate meaningful answers from the vast array of specialized cells in an adult are staggering. Yet an individual cell removed from an embryo or adult does not always act in the way that it would as a part of the multicellular organism.

THE COLON BACILLUS

Much can be learned from cells grown in tissue-culture media, but such an environment is highly artificial. For this reason, many researchers have turned to simpler organisms for research on cellular control mechanisms, hoping that the fundamental biochemical similarities of all living systems will permit generalization of their findings to more complex organisms. A particularly useful and popular organism for such study has been the colon bacillus *Escherichia coli*, a bacterium normally found in the human intestines (Interleaf 17.1). This bacterium multiplies rapidly, can easily be isolated or grown in large populations, and survives on a simple nutrient medium of glucose, water, and a few inorganic salts. Because a new generation of bacteria can be produced as rapidly as every 20 minutes, *E. coli* are particularly useful for genetic studies. As a result, this bacterium and the bacteriophage that infest it are by far the best-understood organisms in the biosphere.

A surprising number of biochemical mechanisms discovered in the simple cells of *E. coli* have been found in complex eucaryotic cells as well. Some biologists who have studied *E. coli* extensively have humorously overstated the case by coining the aphorism "What is true of *E. coli* is true of the elephant."

Loxodonta africa
Reduction ×100
Escherichia coli
Magnification ×40,000
Elephants are always drawn smaller than life, but an Escherichia coli always larger.

Today, however, a growing number of biologists are advising caution in the uncritical generalization of the results of *E. coli* research to all cells. There is evidence that procaryotes and eucaryotes began following separate evolutionary paths very early in the history of life on earth. Although both kinds of cells may share many mechanisms inherited from their com-

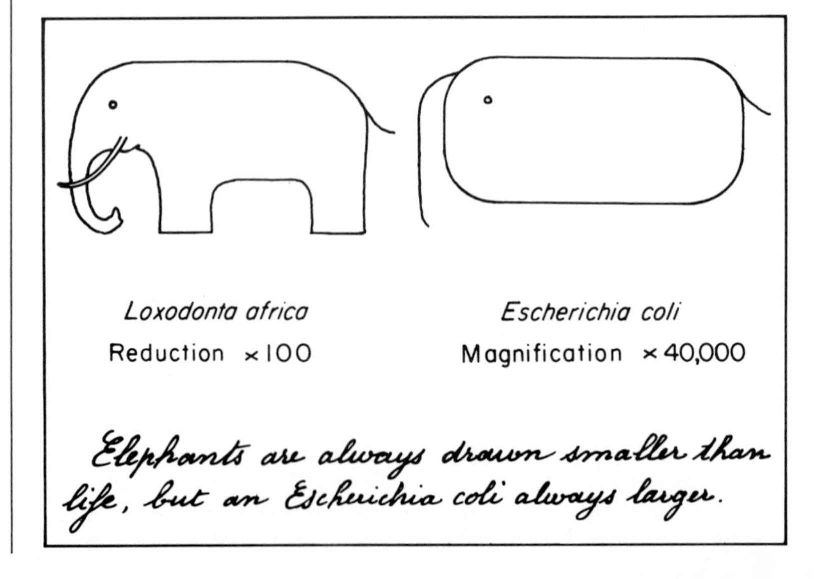

mon ancestors, it is reasonable to expect that evolution has produced many differences between them. Despite the basic differences, there seems to be good reason to hope that the basic principles learned from studies of *E. coli* will point in the right direction for studies of more complex organisms.

Although relatively simple in comparison to eucaryotic organisms, the *E. coli* cell itself is a complex system. Within this microscopic cell are thousands of different kinds of small and large molecules (Table 17.1), many of

Table 17.1
Approximate Chemical Composition of a Rapidly Dividing *E. coli* Cell

COMPONENT	Number of Different Kinds	Average Molecular Weight	Approximate Number of Molecules Per Cell	Percentage of Total Cell Weight
Water (H_2O)	1	18	40,000,000,000	70%
Inorganic ions	20	40	250,000,000	1
Carbohydrates*	200	150	200,000,000	3
Amino acids*	100	120	30,000,000	0.4
Nucleotides*	200	300	12,000,000	0.4
Lipids*	50	750	25,000,000	2
Other small molecules	200	150	15,000,000	0.2
Proteins	2,000–3,000	40,000	1,000,000	15
Nucleic acids				
DNA	1	2,500,000,000	4	1
RNA				6
16s rRNA	1	500,000	30,000	
23s rRNA	1	1,000,000	30,000	
tRNA	40	25,000	400,000	
mRNA	1,000	1,000,000	1,000	

*Including precursors.

Source: James D. Watson, *Molecular Biology of the Gene*, 2nd ed. (New York: Benjamin, 1970) p. 85.

which have not been fully identified. Even with modern techniques, the determination of the three-dimensional structure of a single protein molecule requires many man-years of difficult research. Therefore, the complete molecular structure of the *E. coli* cell may not be known for many decades. Yet enzymes can be recognized in the cell by their catalysis of metabolic reactions, even if their detailed structures remain unknown. Of the approximately 1,000 enzymes needed to catalyze the metabolic reactions known to occur in *E. coli*, many have been isolated and characterized. The *E. coli* chromosome is only large enough to code for about 4,000 proteins, which indicates that 25 percent of the organism's biochemistry is understood.

Because the genetics and biochemistry of *E. coli* are so well known, they have provided most of the available knowledge of cellular regulatory mechanisms. Thus, by necessity, any discussion of regulation must center around bacterial mechanisms. As will be seen, however, bacteria cannot provide the full story.

GENE AMPLIFICATION

One of the most dramatic examples of the regulated expression of genes occurs during the maturation of amphibian oocytes (Chapter 18). The genes

Interleaf 17.1

ECOLOGICAL ASPECTS OF MOLECULAR BIOLOGY

Biologists today estimate that they are aware of about 25 percent of all the specific chemical reactions occurring in *E. coli* and, in many cases, how the cell controls the rates of these reactions. They also have a good understanding of how large molecules such as proteins and nucleic acids are made and, frequently, how these syntheses are controlled. Many of these findings are based on experiments in which the living cells are physically or chemically fractionated into nonliving components. Such results must then be demonstrated in the intact, living cell to prove that they are significant. Similarly, when experiments are performed on whole cells in the laboratory, these results should be interpreted as they apply to the organism in its natural environment, for the demands of this environment restrict and ultimately dictate to the organism what it must do in order to survive. This environment is the mammalian intestine and not the test tube.

E. coli and its ancestors have probably occupied this same relatively static environment for 100 million years, during which time there have been perhaps 10^{30} times as many *E. coli* as mammals. Each of these cells has been subject to the selection pressures of the environment. Because of their rapid growth rate, as compared to mammals, and the fact that they have occupied the same environment for so long, it is reasonable to assume that they are by now optimally adapted—that is, *E. coli* are probably as good at growing in their environment as they could be.

When studied in the laboratory, however, *E. coli* are grown under conditions quite unlike those in the intestine, where food is supplied in intermittent and unpredictable windfalls. The bacteria may experience long periods of near starvation if the host cannot obtain a meal; thus, they normally alternate between periods of feast and famine. In nature they usually divide only once a day, in contrast to once an hour under laboratory conditions where food is both abundant and continuous.

When dividing rapidly in the laboratory, *E. coli* direct their energies very efficiently toward growing and dividing, but when dividing at rates like those in the intestine, they do not appear to allocate their resources as wisely. A cell that wastes any of its limited resources will have a slow growth rate and will soon be outgrown by its more efficient competitors. For example, when *E. coli* are grown under conditions where they divide only once every 24 hours, they contain 7 times as many ribosomes as they need. The large and complex ribosomes are an expensive investment in terms of the cells' materials and energy. Between each division, a cell must synthesize the same total amount of protein regardless of how rapidly it divides. Because a ribosome can assemble the same number of proteins per hour at any growth rate, a cell that must double itself once a day needs only $1/24$ as many ribosomes as a cell doubling once every hour. Therefore, it is surprising to find that slowly growing *E. coli* have such an excess of ribosomes. These extra ribosomes represent about 10 percent of the dry weight of the cells. Each time they divide, they must direct 10 percent of their metabolic raw materials and energy toward the production of these extra ribosomes. Clearly, *E. coli* that did not make the extra ribosomes would be able to divide 10 percent more rapidly under the same culture conditions and would soon outgrow the cells containing the extra ribosomes.

Why hasn't this adaption occurred in nature? There has been ample time. A closer look at the environment provides a probable answer. When the host eats, the *E. coli* are suddenly presented with food for which each cell must compete. Each cell then tries to utilize as much of the food and grow as quickly as possible. The rate at which it can grow is limited by the availability of the synthetic machinery—that is, DNA and RNA polymerases but primarily ribosomes. A cell with excess ribosomes could immediately begin to make protein more rapidly. In contrast, a cell that did not contain excess ribosomes would have to make more ribosomes before it could synthesize protein at the faster rate.

If cells with and without the excess ribosomes were grown with a 24-hour doubling time and were suddenly given unlimited food, calculations show that the *E. coli* with a sevenfold excess will have divided once before the hypothetical cells with no excess have even begun to grow significantly faster. Thus, between meals the bacteria with no excess ribosomes would grow about 10 percent faster, but every time the host eats, the bacteria with excess ribosomes will gain a 100 percent advantage. The host need eat only once every 9 days for the bacteria with excess ribosomes to have a 10 percent selective advantage. Therefore, *E. coli* with extra ribosomes are not inefficient but are very "wise" indeed.

The transport systems of *E. coli* provide a further demonstration that laboratory studies of an organism often need to be correlated to its natural environment to be fully understood. If *E. coli* are grown on lactose at concentrations that occur in the intestine, the growth rate of the cells is limited by the rate at which lactose enters the cell, not by the rate at which it is metabolized. So, why haven't *E. coli* developed a more efficient transport system for lactose? To answer the question, it is necessary to look again at their native environment. The viscosity of intestinal contents is approximately that of lightweight motor oil. In such viscous material, the rate of diffusion of lactose is about 100 times slower than in water. In laboratory media, diffusion is fast relative to the speed at which the "permease" can transport lactose into the cell. However, in the intestine, diffusion is so slow that the cell is able to transport the lactose inside as fast as it can diffuse up to the cell. Thus, the cellular growth rate is limited not by lactose transport from the cell surface to the interior but by the rate of diffusion of lactose through the intestinal contents. Obviously, improvements in the efficiency of the transport system in the intestine would not increase the growth rate and would have had no selective advantage.

An organism must be studied in relation to its natural environment to understand fully and to appreciate its biochemistry, for it is the environment that has shaped the organism's destiny. The phenomena observed at the molecular level are but the organism's methods of coping with the demands placed on it by the environment. In a very real sense, biochemistry is subordinate to environment.

Figure 17.2. The metabolic pathways of *E. coli*.

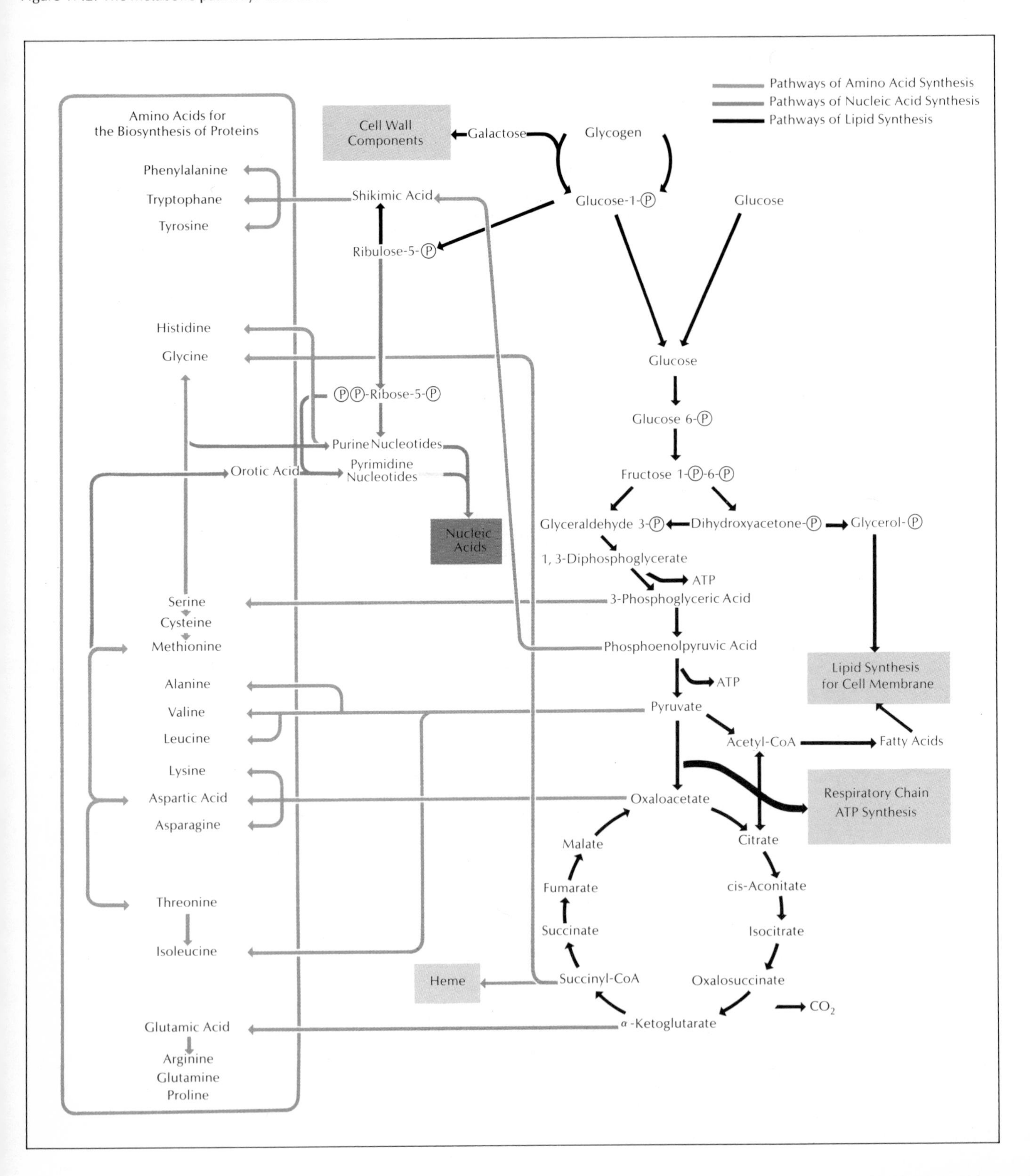

coding for ribosomal RNA are specifically duplicated (as many as 10^3 times), and each copy serves as a template for ribosomal RNA synthesis. Ribosomal RNA synthesis thus proceeds at many times the rate as in other cells, a rate that would be physically impossible with only four sets of ribosomal genes.

Although this increase is the only clearly demonstrated example of gene amplification known at present, there is speculation that it might occur (although less dramatically) in other cell types. Many cells in eucaryotic organisms must produce large quantities of a few gene products, and it is possible that amplification of those genes facilitates the synthesis. In most cases, however, if amplification does occur, the number of extra copies of the gene would probably be small (10 to 100) and, at present, very difficult to detect.

CONTROL OF RNA TRANSCRIPTION

The regulatory mechanism best understood today is the control of transcription of specific messenger RNA in bacteria. Of the number of possible ways in which a cell could control the rate of synthesis of specific mRNA molecules, several have been shown to occur in various organisms. The most extensively studied mechanism controls the synthesis of *inducible* and *repressible* enzymes in *E. coli*.

β-galactosidase, the enzyme that cleaves lactose into glucose and galactose, is the classic example of an inducible enzyme. When no lactose is present, *E. coli* contain only about two molecules of β-galactosidase. When lactose (which must be cleaved to be further metabolized) is added, the bacteria rapidly produce more β-galactosidase—as much as 3,000 molecules per cell. If the lactose concentration is lowered, the cells make intermediate amounts of β-galactosidase. If lactose is subsequently removed from the medium, the cells rapidly cease production of new enzyme molecules. This very sensitive mechanism allows the expression of a gene product only when the cell has need for it and only in the amounts required.

The elucidation of this control mechanism during the last two decades by François Jacob, Jacques Monod, and many others has proved to be one of the major advances in molecular biology. The key component of this regulatory system is a protein known as the *repressor*. When lactose, the *inducer*, is absent, the repressor binds tightly to part of the β-galactosidase gene, physically preventing RNA polymerase from making β-galactosidase mRNA. When lactose (or a synthetic molecule that mimics lactose) is present, it binds to the repressor and changes its conformation so that it can no longer bind to the β-galactosidase gene. The polymerase is then able to make β-galactosidase mRNA, which will serve as the template for many enzyme molecules.

The entire process takes place very rapidly. Within seconds after the inducer enters the cell, the gene releases the repressor and RNA synthesis begins. The first active enzyme molecule appears about two minutes later. When lower concentrations of the inducer are present, the repressor remains bound to the gene part of the time and RNA is made less frequently.

Normal cells contain only about ten molecules of the β-galactosidase repressor, but the recent isolation of mutants that produce a large excess of repressor have made possible its purification and characterization. The binding between the repressor and the inducer is not covalent but is a relatively weak hydrogen bond interaction. Such bonds break and re-form

Figure 17.3. Jacob-Monod model of gene control via a DNA section known as the operon. The operon consists of a regulator gene, an operator gene, and structural genes. Without lactose present in the system, the regulator gene produces a repressor protein. In the absence of an inducer molecule, such as lactose, the repressor binds onto the operator region. This action prevents the activation of the structural genes, and no unnecessary enzyme synthesis results. With lactose present in the system, repressor molecules are synthesized as above, but lactose acts as an inducer for its own enzymatic breakdown. Lactose and repressor combine to form a complex that is unable to bind to the operator region. The "switch" is now turned on, and the structural genes transcribe mRNA, which in turn guides the synthesis of the three enzymes necessary for lactose breakdown.

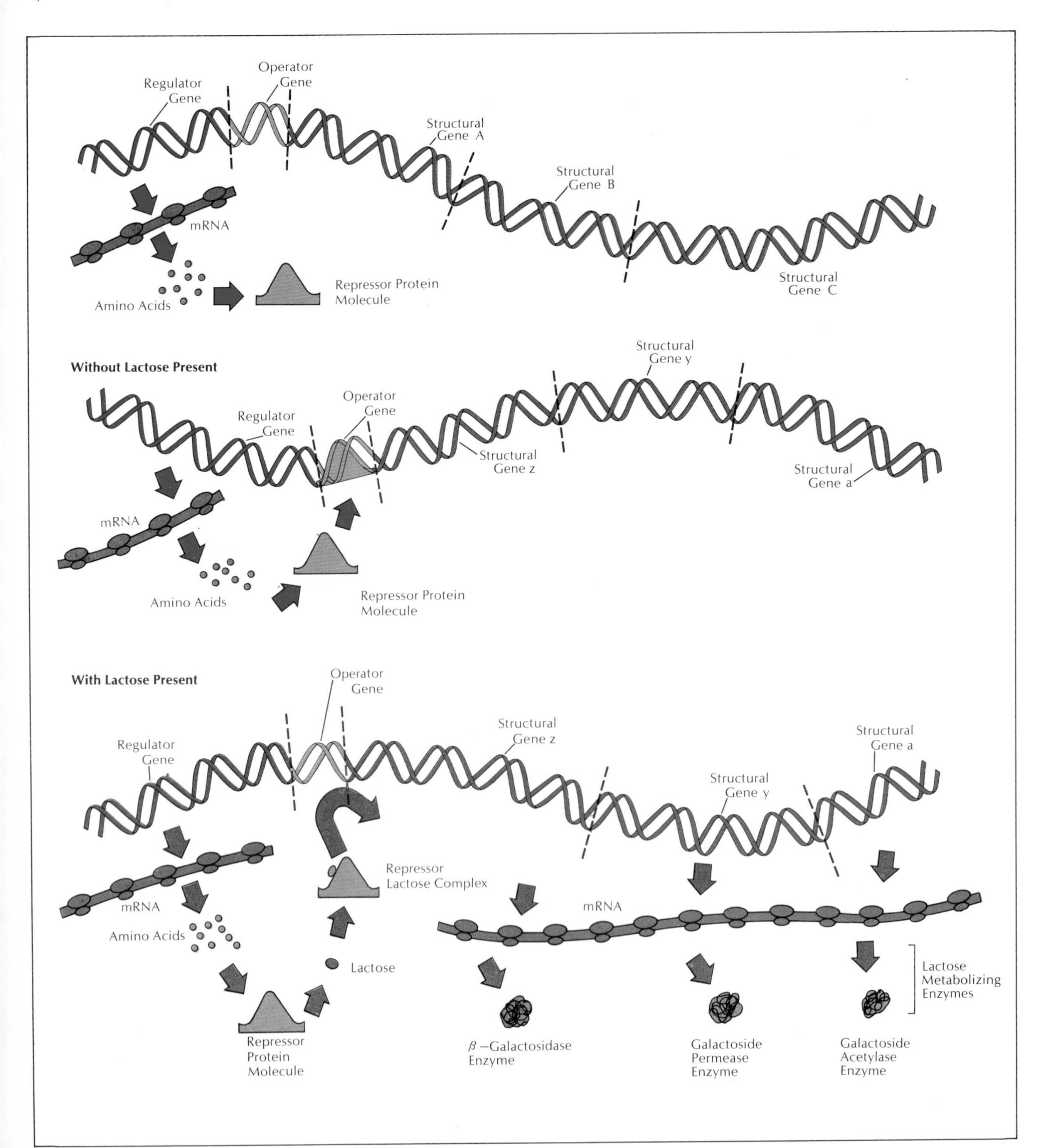

Figure 17.4. Current operon theory. Recent investigations indicate that there is a small region of DNA called the promoter next to the operator gene. The attachment site of the enzyme mRNA polymerase, which is necessary for mRNA transcription, is located at this promoter region. With the repressor molecule in place, the promotor region is blocked, and no transcription takes place.

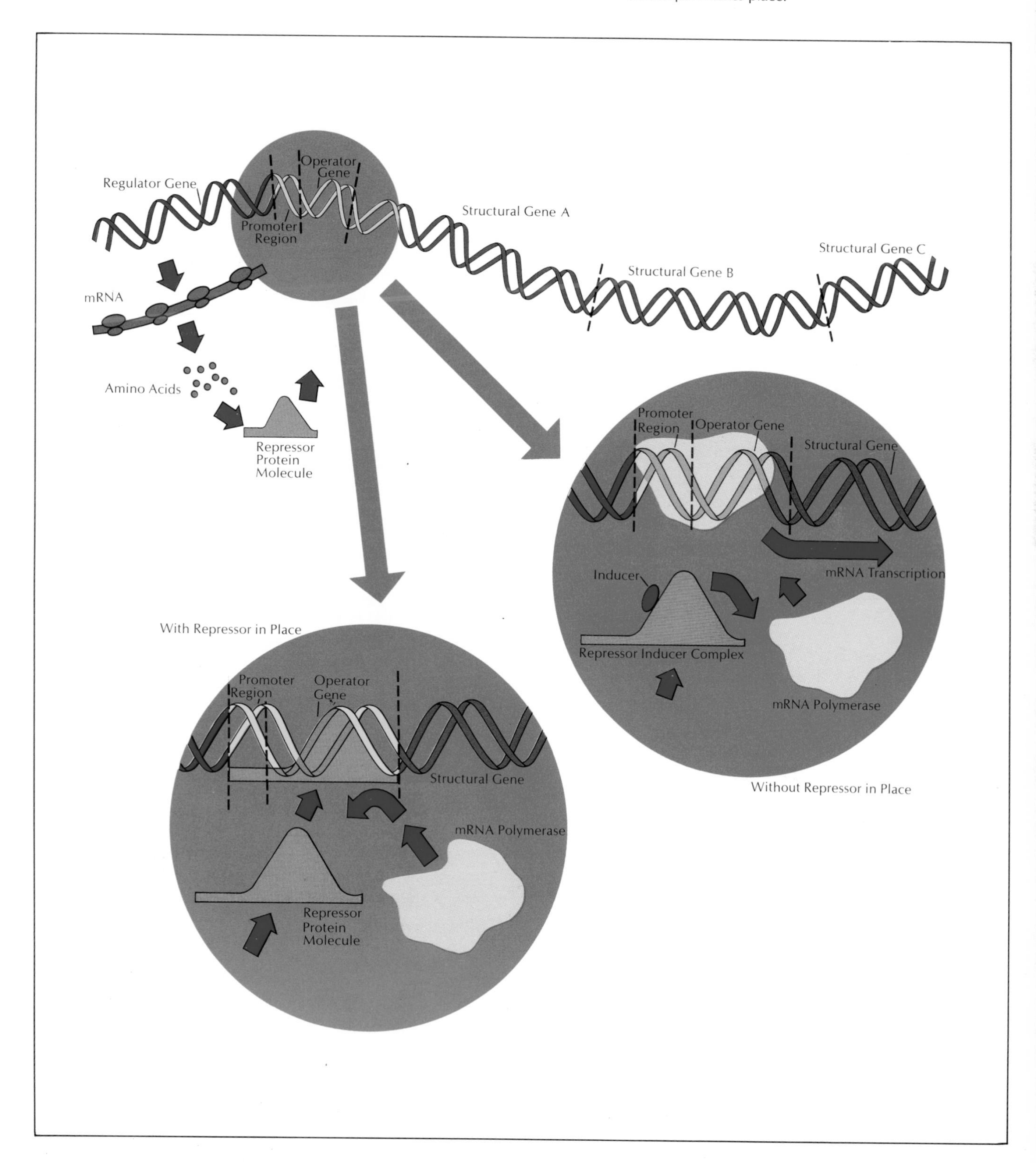

easily, allowing the repressor to adjust quickly to changes in the concentration of the inducer. Enzyme production can thus "turn on" or "turn off" rapidly.

The active repressor (one that is not combined with an inducer) binds to a specific region of DNA called the *operator*. This region, which is about 20 nucleotides long, is near the point at which transcription of the β-galactosidase gene begins. The repressor forms many hydrogen bonds with this nucleotide sequence, which bind it tightly to the operator. Thus, a cell needs only a few active repressor molecules to ensure that at least one molecule is bound to the operator at all times.

A single repressor-operator mechanism can control the expression of more than one gene. The β-galactosidase repressor, for example, also regulates the production of β-galactosidase permease, a protein (located on the cell membrane) that facilitates the passage of lactose into the cell. The two genes are adjacent to each other and the mRNA for both genes is transcribed as one molecule. When the mRNA is translated, separate molecules of the two proteins are produced. A block of two or more adjacent genes, controlled by the same regulatory system, is known as an *operon*. (The operon containing the β-galactosidase gene is known as the lactose operon.) The operon is a very convenient method for coordinating the regulation of enzymes involved, for example, in a single sequence of metabolic reactions. A bacteria such as *E. coli* must regulate quite a few metabolic pathways in order to respond to various nutrients in its environment, and it therefore makes extensive use of operons.

It is also possible for a small molecule to turn off an operon rather than turn it on. The repressors for many operons require the presence of a small molecule called a *corepressor* in order to bind to the operator. In the absence of the corepressor, the repressor is unable to bind and the genes in the operon are expressed. The enzymes involved in the synthesis of many amino acids are regulated in this fashion. In such cases, the corepressor of the operon is an amino acid produced by a set of enzymes. The cell stops synthesis of these enzymes when the amino acid is already present in the medium.

The point at which RNA polymerase attaches to a gene and begins transcribing RNA has recently been shown to have an important regulatory function. This region, known as the *promoter*, does not appear to code for any part of the protein, but it does have a high affinity for the RNA polymerase. The nucleotide sequence of the promoter determines how well the polymerase binds to it and thus how frequently the polymerase transcribes the gene. For example, the promoter controlling the synthesis of mRNA for β-galactosidase repressor has a low affinity for the polymerase, and the gene is transcribed only once or twice every cell generation.

Many genes in bacteria have no repressor mechanism. They are expressed at a constant rate predetermined by the nucleotide sequence of the promoter. Such *constitutive* genes generally code for proteins, which the cell always needs in fairly constant amounts. Inducible or repressible genes also have promoters, however, in which case the promoters control the maximum rate at which the gene can be expressed.

RNA polymerase is a large component consisting of five different subunits, only one of which, the σ (sigma) factor, recognizes and binds to the promoter region. Recent work has shown that different σ factors have different promoter specificities and may play a significant role in regulation.

Figure 17.5. Electron micrograph (left) showing characteristic attachment of RNA polymerase molecules to DNA strands; (right) DNA-dependent polymerase molecules from *E. coli*. (× 400,000)

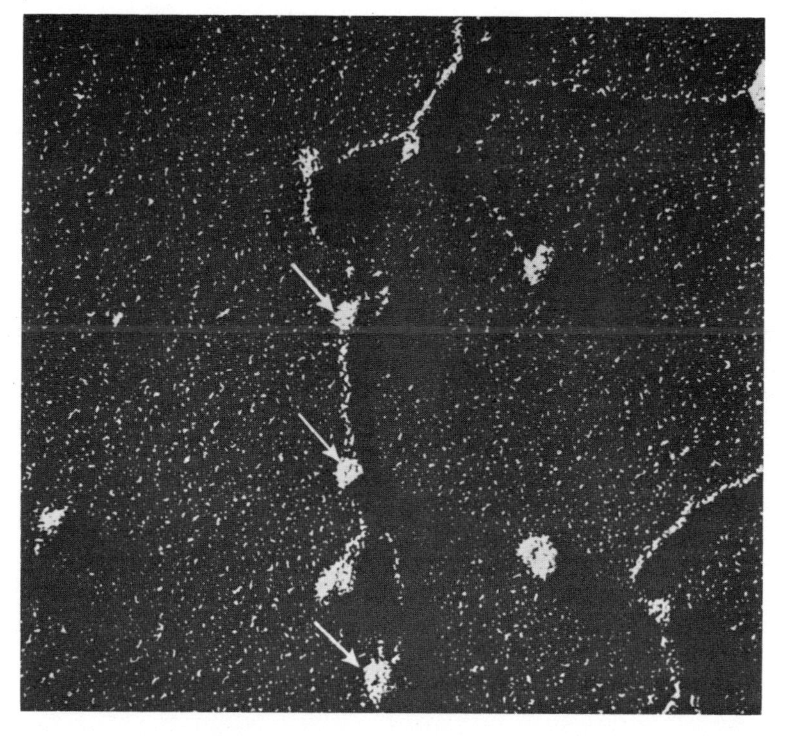

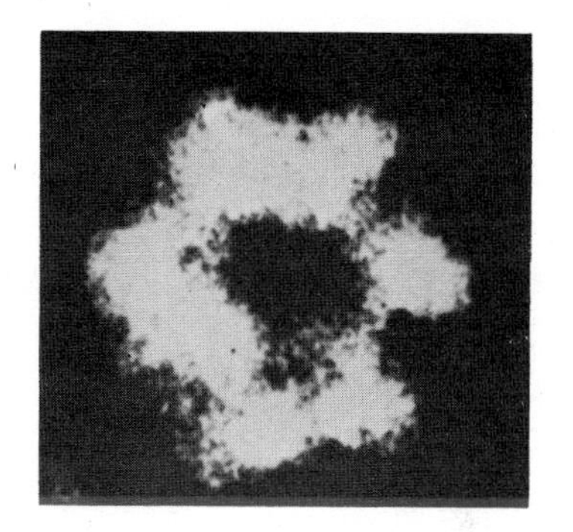

When *Bacillus subtilis* sporulate, for example, a new factor that is specific for genes involved in sporulation appears while at least one of the σ factors specific for the vegetative growth of these cells disappears.

CONTROL OF TRANSLATION

Control of gene expression at the level of protein synthesis is a widespread phenomenon, but the significance and mechanisms of such control are very poorly understood. For example, translational control occurs widely in plant spores and seeds and in animal oocytes. Some of the mRNA for proteins produced early in the development of the organism is present in spores, seeds, and oocytes, but actual protein synthesis by these messages does not begin until germination or fertilization takes place.

Translational control is also important in later developmental sequences, such as in the slime molds. Many new enzymes are made as slime molds aggregate and form fruiting bodies. In some cases, the synthesis of the mRNA for these enzymes begins or is even completed several hours before synthesis of the enzyme begins. Although the mechanism controlling the timing of the translational signals remains a mystery, recent work with viral RNA suggests that the secondary structure (for example, hairpin-type loops formed by base pairing between short nucleotide sequences) can determine whether or not a ribosome recognizes a "start" codon and that changes in secondary structure can affect the translation rate of a protein.

FEEDBACK INHIBITION

In addition to controlling the synthesis of a gene product, it is also possible to control the expression of the product itself. For example, *E. coli* normally synthesize the amino acid isoleucine by a five-step pathway starting with

Figure 17.6. Isoleucine synthesis.

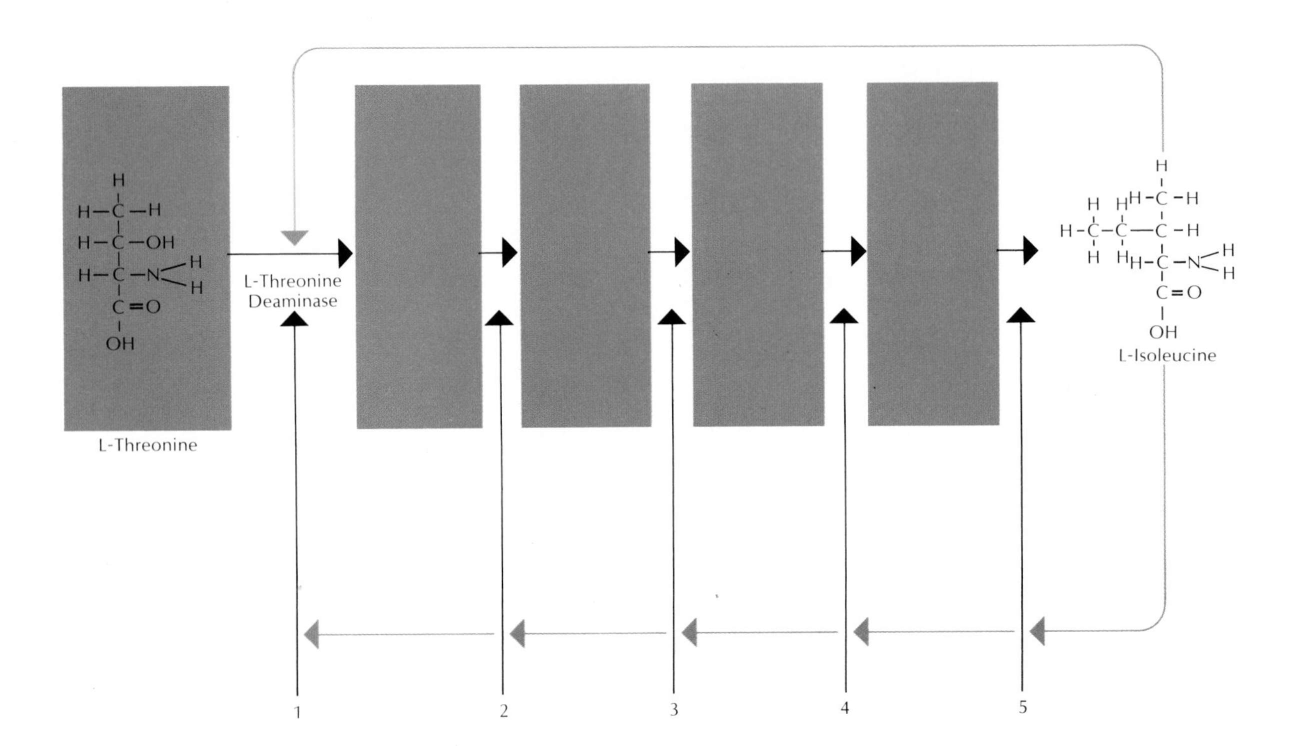

another amino acid, threonine (Figure 17.6). When too much isoleucine is produced or when it is added to the medium, it inactivates the first enzyme of the synthetic pathway, threonine deaminase, thus halting the biosynthesis and accumulation of isoleucine. As excess isoleucine becomes incorporated into proteins and the concentration of the free amino acid drops, threonine deaminase gradually becomes active again. This inactivation, known as *feedback inhibition,* occurs in much the same way as inducer inactivation of a repressor. In this case, isoleucine binds to a specific part of the enzyme molecule and changes its conformation (tertiary structure) so that it is no longer an active enzyme. This freely reversible process enables the enzyme to be active again when isoleucine is removed. Only the first enzyme of a synthetic pathway needs to be inactivated in order to stop synthesis of the amino acid and all of its precursors.

Feedback inhibition allows very rapid, sensitive control over the cells' metabolism. Whenever a product begins to accumulate more rapidly than it is being used, its synthesis is slowed down or halted. As the supply is used up, the synthesis is speeded up again. In contrast, induction or repression provides a somewhat slower adjustment to long-lasting changes in the cells' environment. Many enzymes are subject to feedback inhibition and repression of synthesis by the same end product.

REGULATION IN EUCARYOTIC ORGANISMS

Most of the available knowledge of regulation is virtually limited to bacterial regulation. The big question now is what differences and similarities should be expected in cells of eucaryotic organisms. The requirements a eucaryotic cell places on its regulatory systems are quite different and, in many ways, more complex than the requirements of a bacterial cell.

Most eucaryotic cells undergo a precisely ordered development from an undifferentiated embryonic cell to a highly specialized cell in the adult organism. Many genes must be turned on and off in the right sequence during such differentiations. This situation is quite different than most bacteria experience. On the other hand, most eucaryotic cells live in a rather constant environment. The composition of human blood, for example, remains relatively constant, so that most cells in the body do not need to make large adjustments in their basic metabolic pathways. Eucaryotic cells must be regulated by a wide variety of external factors, not just by small molecules as in bacteria. Many changes that occur during differentiation are triggered by chemical *effectors* that are liberated by one cell or tissue and travel to a *responder cell*, whose metabolism is affected by the chemical signal. In some cases, the biochemical change triggered by the signal is permanent; in others, the change is transitory.

A number of regulatory mechanisms not commonly found in bacteria have already been described in various eucaryotes. Control exerted at the level of protein synthesis seems to play an important role in the timing of enzyme synthesis during differentiation. It now appears that σ factors can also perform this function. For example, a series of genes expressed early in development may include the gene for a new σ factor that is specific for genes to be expressed at a later time. Thus, the second set of genes cannot be transcribed until after the first set. And this second set may include yet another σ factor for a third set of genes, and so on. Another new type of control appears to regulate the degradation of mRNA and proteins. There is now evidence that a cell can vary the amount of an enzyme by varying the rate at which it (or its mRNA) is degraded, keeping the rate of synthesis constant. This mechanism may be more efficient for cells that need to vary the content of an enzyme over only a fivefold or tenfold range.

Almost every observed biochemical differentiation can be explained by variations on basic models such as that diagrammed in Figure 17.4. At the present time, however, such models are merely theoretical constructions.

There has been greater progress in the study of the chemical nature of effectors that pass from one cell to another. From these studies comes evidence to indicate the existence of several different mechanisms for transmitting a signal from one cell to another: (1) exchange of genes; (2) cytoplasmic fusion; (3) exchange of small molecules; (4) control of hormone effectors; and (5) interactions through contact of cell surfaces.

Genetic Exchange

Perhaps the most direct mechanism of cell-cell interaction is the injection of certain genes of one cell into another cell. In this case, DNA acts as the effector, directly transcribing mRNA in the responder cell. Viral transduction, bacterial transformation, and bacterial conjugation are the best-studied examples of this form of interaction.

Direct exchange of genetic information in mammalian systems has been suggested to exist only in a part of the immune response mechanism. There is evidence that when the cell recognizes a foreign particle (antigen), it synthesizes an mRNA molecule coding for an antibody and transmits this

Figure 17.7. The development of a muscle colony from a single myoblast at 3, 6, and 13 days. The multinuclear cells that are formed by the sixth day probably involve cell fusion. Differentiation of myoblasts from a single spindle shape to an amoeboid shape is also shown.

mRNA to secondary cells, where the antibody protein is translated. If subsequent research should support this model, then it is a clear case of cell-cell interaction via nucleic acid.

Cytoplasmic Fusion

Another common form of cell-cell interaction involves the fusion of the cytoplasm of two or more cells to form a multinucleate, *syncytial* tissue. In such a tissue, molecules can travel from one nucleus to another without passing through a cell membrane. Cell fusion occurs in the forest mold *Physarum*, forming a large, yellowish mass containing thousands of nuclei. The amazing property of this organism, or syncytium, is that all the nuclei divide simultaneously.

During chick *embryogenesis* (development of the embryo), large numbers of cells aggregate within the primordial blood vessels and fuse to form syncytial "blood islands." The nuclei in this tissue divide simultaneously about once an hour. As the nuclei multiply, the tissue begins to synthesize hemoglobin rapidly. After a period of hemoglobin accumulation, the blood islands slowly separate into individual cells that make up the original blood cells of the chick embryo. This natural mechanism apparently serves as an

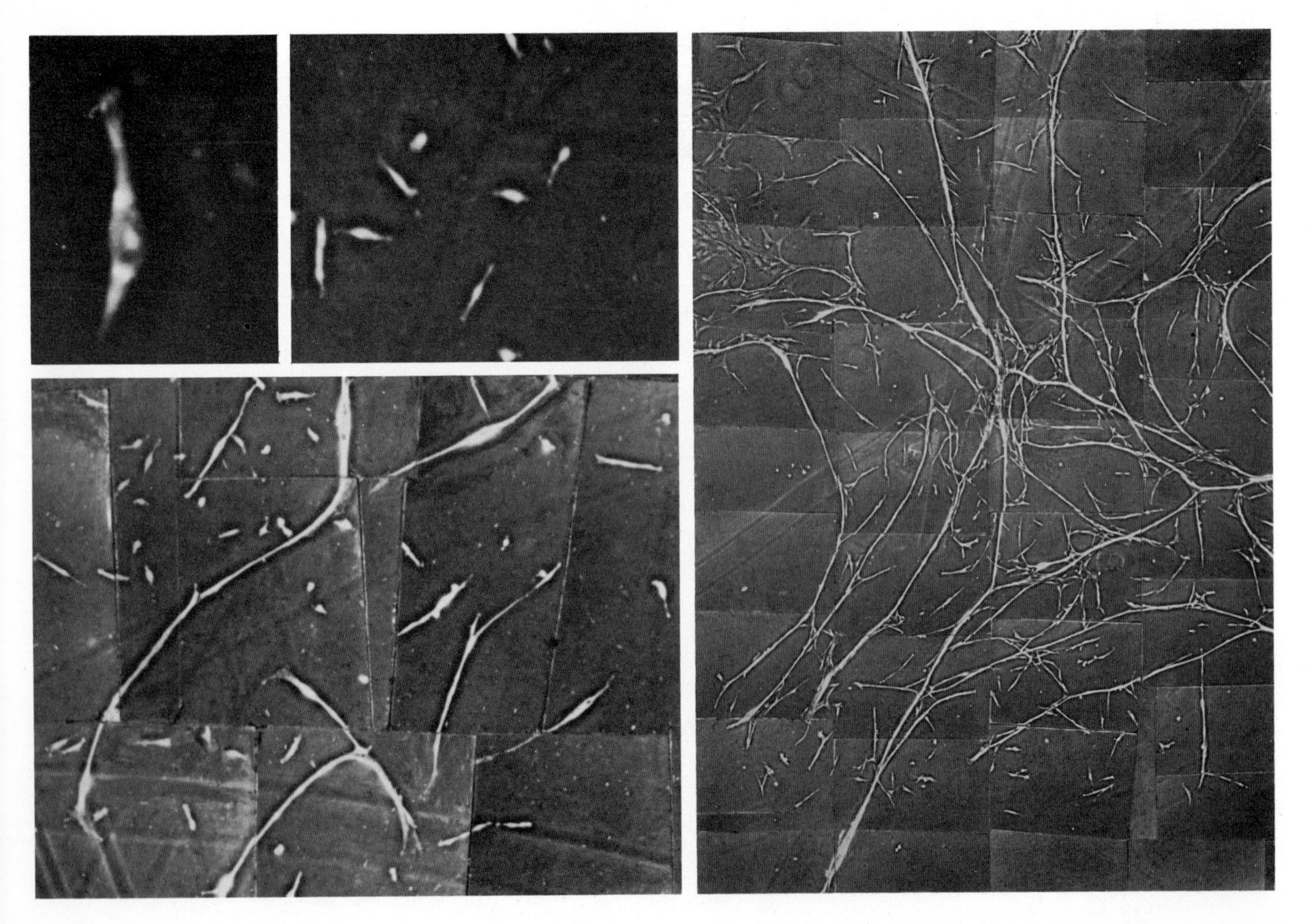

efficient means of initiating hemoglobin synthesis. The genes for hemoglobin synthesis can be turned on simultaneously within the single multinucleate mass, rather than in separate blood cells scattered throughout the embryo.

Cross-Feeding

Most cells, from bacteria to mammalian cells, must communicate with other cells across plasma membranes. A fairly large number of different mechanisms has evolved for accomplishing such communication.

A simple form of cell-cell interaction involving diffusible molecules can be demonstrated with mutant strains of bacteria—for example, with strains that are deficient in enzymes required for the synthesis of arginine from ornithine (→ ornithine → citrulline → arginine). One strain is unable to convert ornithine to citrulline and therefore can survive only in a medium containing citrulline or arginine. A second strain is unable to convert citrulline to arginine and therefore can survive only in a medium containing arginine. However, if the two strains are mixed together, both can survive in a minimal medium. The first strain accumulates ornithine, some of which diffuses out of the cell and into the medium. The second strain converts this

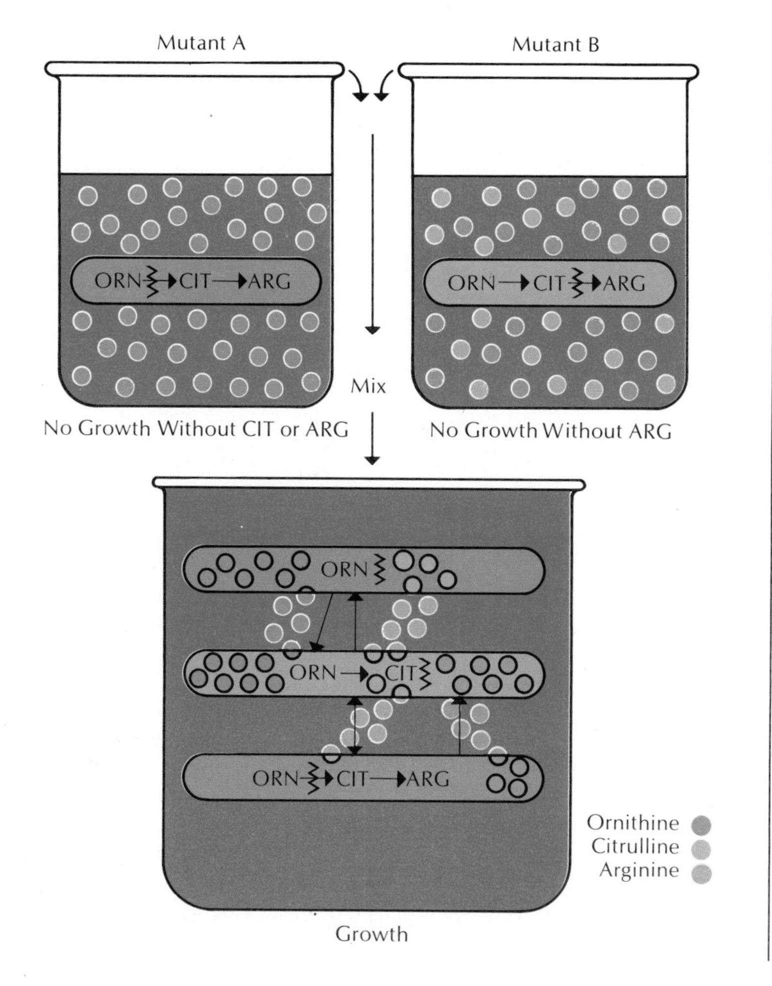

Figure 17.8. Cross-feeding of nutrients in bacteria.

Figure 17.9. The advancing end of this slime mold is moving by amoeboid motion.

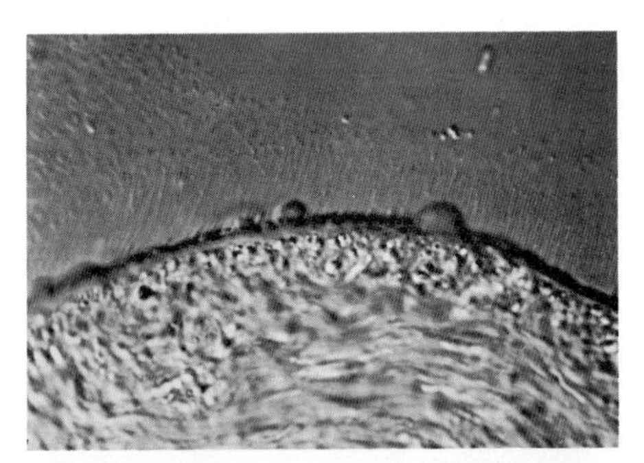

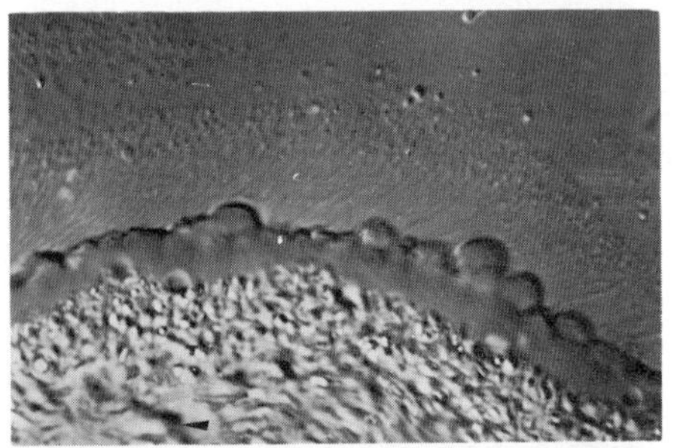

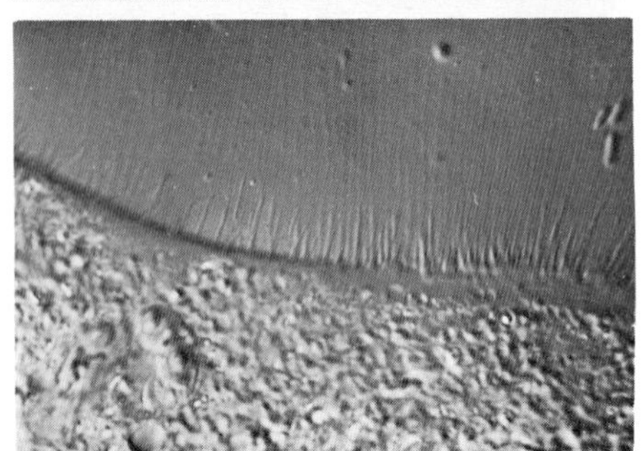

ornithine to citrulline and accumulates citrulline, which diffuses into the medium. The first strain converts the citrulline to arginine, and some arginine diffuses back to feed the second strain (Davis, 1950).

Cross-feeding of nutrients also occurs among mammalian cells in culture. Many necessary nutrients are synthesized by cells and diffuse into the medium. When the population density of cells in a culture is low, the nutrients may be so dilute in the medium that some cells starve. Only when sufficient population density is reached can deficient cells grow without the addition of nutrients. This "mass effect" results from the combined nutrient leakage of many cells, conditioning the medium with a sufficient concentration of nutrients so that all cells may grow. Thus, the growth of one cell depends on the presence of all the others. Interactions of this type in cultures of human cells involve cross-feeding of several different substances (Eagle, 1965).

The cross-feeding of nutrients may be regarded as a simple control signal carrying only the message "grow" or "don't grow," but it is an important control mechanism for tissue development, at least in cultures. In the intact organism, cellular interaction via cross-feeding may be the basis for many mass effects that restrict cellular growth or development until a certain population of cells is achieved. This mechanism guarantees that differentiation of organs will proceed only when there are enough cells to form the required tissue.

Hormone Effectors

Hormonal induction of differentiation occurs during embryogenesis as well as in adult life. Complex organisms have evolved circulatory systems that transport oxygen and food products to the cells and remove CO_2 and waste products. The circulatory system also carries chemical signals among the cells of the organism, thus integrating the functions of various organs and adapting their biochemical properties to changing conditions. Certain specialized cells secrete hormones into the blood, which eventually carries them to responsive "target" cells. When a hormone reaches a target cell, it stimulates a series of biochemical events that result in a specific response. The exact response depends on the specific hormone involved and the nature of the target cell (Chapter 21).

Surface Phenomena

Many cases of cell-cell interaction apparently involve either nondiffusible molecules or direct contact between cells. Such a case is found in the growth and differentiation of myoblasts in tissue culture. Myoblasts differentiate only in the presence of the insoluble protein collagen, which is secreted by fibroblasts. Growth and differentiation of muscle tissue in intact embryos may also depend upon either direct contact with collagen or "conditioning" of parts of the embryo by fibroblast secretions.

As normal cells multiply mitotically in tissue culture, they separate and creep along the surface of the culture chamber by amoeboid motion. This kind of motion is characteristic not only of cultured cells and of the amoeba (from which it derives its name) but also of many embryonic cells and of white blood cells. The cell membrane is distorted by the formation of slender projections, or *pseudopodia*, and by a sort of undulating action. Cytoplasm often streams away from the cell body into the pseudopodia, which then may flare out at their ends or branch to form other pseudopodia.

Figure 17.10. A sequence in the development of the fruiting bodies of the cellular slime mold *Dictyostelium discoideum*.

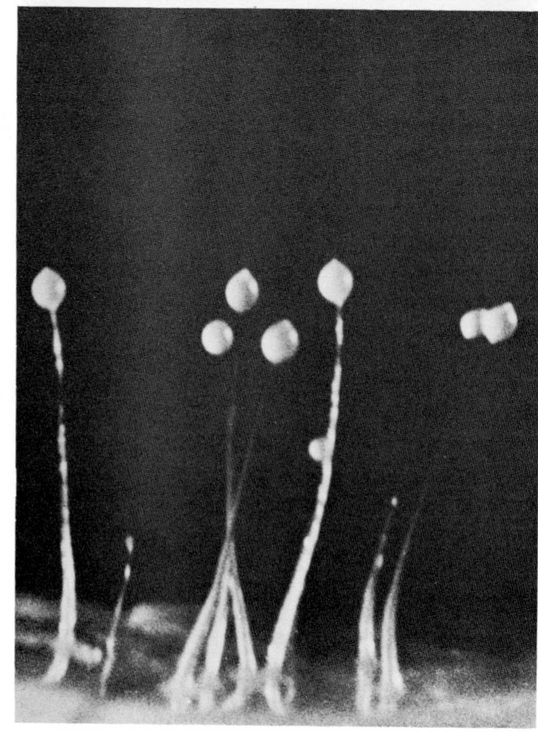

Whichever side of the cell displays the greatest activity of this kind is the "front" of the cell, because this activity tends to pull the cell along with it. If two cells make contact with each other in tissue culture, all amoeboid motion along the surfaces of contact ceases — a phenomenon known as *contact inhibition*. Amoeboid activity on the opposite side of each cell then increases and the cells move apart.

The amoeboid cells in tissue cultures are very active metabolically, synthesizing new DNA, RNA, and proteins. Mitotic division occurs frequently. When the cell population eventually covers all the available surface as a single layer, each cell is necessarily in contact with other cells at all points on its horizontal perimeter. Contact inhibition stops all amoeboid motion, whereupon the rate of synthesis of nucleic acids and proteins decreases. Growth and division are halted dramatically (Abercrombie and Ambrose, 1958).

It appears that contact inhibition plays a major role in normal animal development, causing cells to cease growing and dividing when a tissue has been formed. A tumor may be formed by cancerous cells that continue to grow and divide despite close contact between cells. In some cases, cells become cancerous after infection by an oncogenic, or cancer-producing, virus. In culture, cells infected by such a virus continue to grow and divide even after a complete layer has been formed in the culture dish. Because contact inhibition apparently plays a central role in differentiation and failure of this process may be crucial to the development of cancerous tumors, many biochemical studies of the process are being actively pursued (Egylud and Szent-Györgyi, 1966; Rubin, 1970).

A MODEL SYSTEM

Analysis of cell-cell interaction in embryogenesis is technically difficult because so few cells are involved. In many cases, the interactions of primary interest may occur among fewer than 100 cells. Thus, the biochemist has only a few hundredths of a microgram of material per embryo to work with, and important molecules may be present in such minute amounts that he cannot detect them. A few unusual cases of extreme differentiation — such as the formation of feathers, eye lens, blood, and muscle — have been studied successfully by biochemical techniques, but it has been difficult to relate knowledge about these unusual systems back to the overall development of the organism.

Another approach involves the study of simpler organisms that undergo some sort of developmental processes. One organism that has been extensively studied is the cellular slime mold *Dictyostelium discoideum*. Single cells of this organism grow on decaying forest leaves and are almost identical to amoebae (Raper, 1935). They ingest bacteria and divide mitotically. When the individual cells run out of food, however, an amazing sequence of structural changes transforms a group of the cells into a multicellular organism. The cells gather together to form a fruiting body, within which some cells are transformed into encapsulated spores. The spores can survive extended periods of drought or cold without food. They later germinate to form new, individual, amoeboid cells (Figure 17.10).

The process of differentiation, or specialization, that transforms individual amoeboid cells into the specialized cells of the fruiting body and spores has been extensively studied. Various effectors are exchanged among cells to coordinate the differentiation processes. These effectors act within the

Figure 17.11. The mouse in this photograph was innoculated subcutaneously with a small number of mouse cells transformed by the animal virus SV-40. Within three weeks, these cells grew to produce the large mass visible in the mouse.

cells to exert specific controls over the activation of genes and the timing and rates of translation of various proteins. The precise nature and mode of action of the effectors is still unknown.

CANCER AND CELLULAR CONTROL

Any study of cellular control inevitably comes face to face with its malevolent extreme—the lack of control underlying the wild proliferation of cancer cells. Like normal cells, cancer cells presumably contain a normal complement of DNA. Their abnormal behavior probably is a result of some malfunction of the cellular control mechanism, which removes normal restraints upon growth and division. The cancerous cells then multiply to such an extent that they crowd out and starve other body cells.

Cancer cells may form from almost any kind of living animal cell—cells of the brain, liver, kidney, bone, blood, skin, or other tissues and organs. Most cancer cells formed from specialized cells continue to perform the specialized function of the parent cell. Cancerous hormone-producing cells still produce hormones; cancerous cells of the immune system continue to manufacture antibodies. In almost all respects, the cancer cell is functionally similar to its normal ancestor, but it grows and divides uncontrollably.

Many different influences have been shown to cause a normal cell to lose its restraints upon growth and division and to become cancerous. Among the cancer-causing agents, or *carcinogens*, are atomic radiation, chemicals of many different varieties (including certain combinations of otherwise beneficial drugs), certain hormones, and viruses. The causes vary, but the effects are thought to be essentially the same. In both man and other animals, some individuals apparently inherit a genetic susceptibility to cancer. This hereditary tendency probably is caused by malfunctioning mechanisms that normally counteract the effects of carcinogens.

When a carcinogen acts upon a normal cell, it somehow must disrupt the cell's genetic machinery in such a way that it either alters or destroys the normal checks on growth. The change is permanent. Once the genetic machinery has been altered, each cancer cell produces cancerous daughter cells, and all following generations are cancer cells. Without restraints on growth and division, the cancer cells spread rapidly through the organism and, if they are not controlled, eventually kill it.

Cancer cells clearly differ from normal cells in the way that they interact with other cells, both in the live animal and in tissue culture. Cells in benign tumors are contact inhibited by normal cells but not by other tumor cells. The tumor continues to enlarge because the cells inside it grow without restraint, but it does not spread to other parts of the body because contact with normal cells inhibits cell growth on the surface of the tumor. Malignant cancers, on the other hand, are not contact inhibited by cancer cells or by normal cells. Thus, they grow outwardly from the periphery of the tumor, as well as within it, and can spread to other areas of the body. As mentioned earlier, the loss of contact inhibition is also observed among cancer cells in tissue culture.

Some researchers also suspect that cancer cells are less "sticky" than normal cells. The cancer cells move more freely among one another in culture than do normal cells. This freedom of movement may be related to the failure of contact inhibition and other processes that inhibit growth and division in normal cells. The stickiness of normal cells is very selective. If

Figure 17.12. Cancer cells are the result of some malfunction in the cellular control mechanism. Normal growth restraints are lacking, and the cells multiply in a wild fashion.

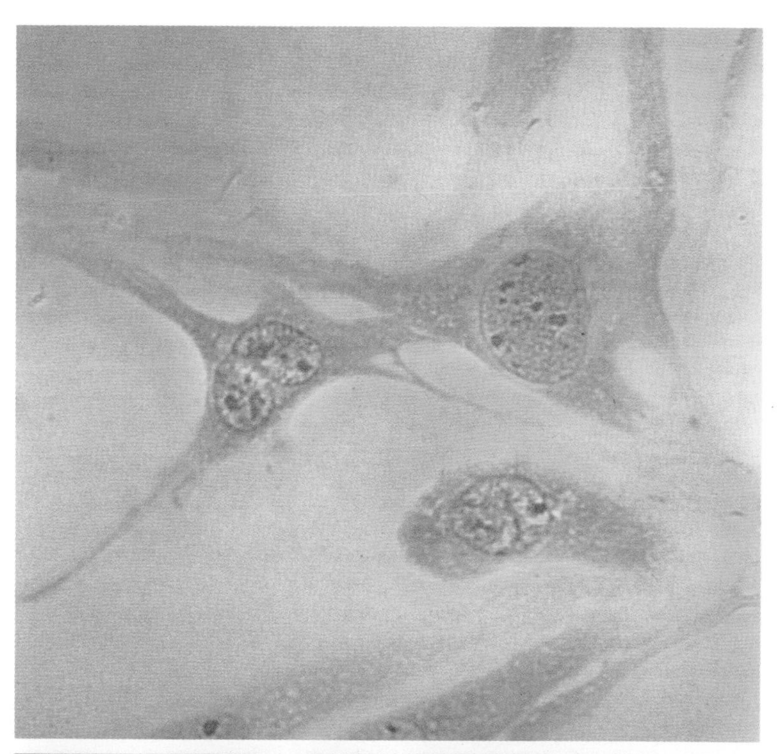

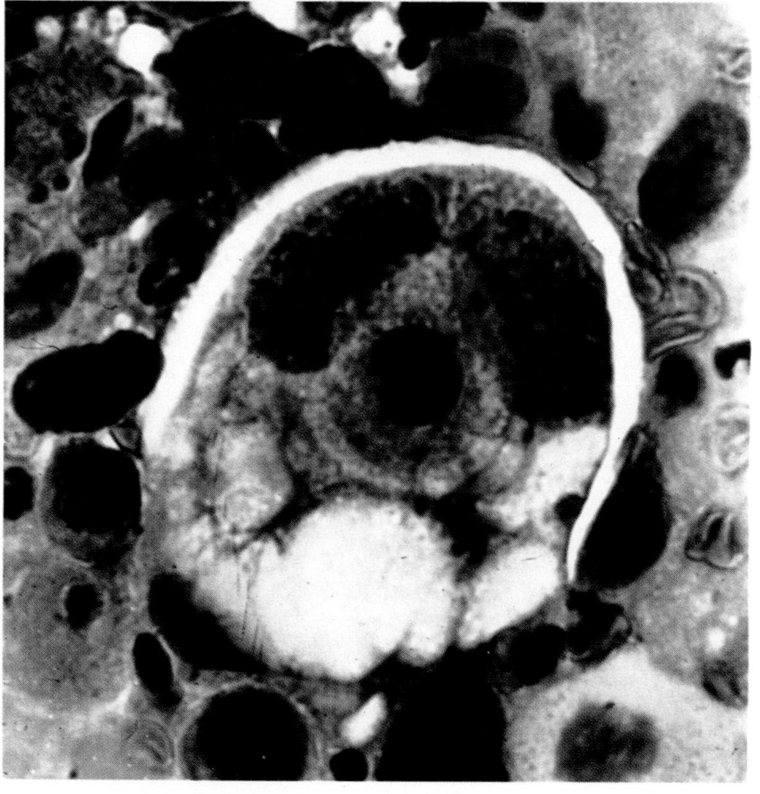

Figure 17.13a (left). Mouse embryo cells. These normal cells grow in monolayers and stop growth when they are contact inhibited.

Figure 17.13b (right). Same cells transformed by a polyoma virus. These cancerous cells are growing wildly because contact inhibition does not occur.

kidney and liver cells are mixed in a culture, the two kinds of cells will separate and form clusters of kidney tissue and clusters of liver tissue. When cancerous liver cells are mixed with cancerous kidney cells, this separation does not occur. These observations support the conclusion that cancer cells have abnormal surfaces that fail to carry out normal interactions with adjacent cells.

VIRUSES AND CANCER

Tissue-culture techniques are used to determine how normal cells differ in their interactions from cancerous cells. Bacterial cells, such as *E. coli*, cannot be used in these studies because they do not form tissues; they separate from one another after division, rather than forming clumps as animal cells do. The cultivation of normal mammalian cells in the laboratory, however, is tedious. Only cells from embryos or from organs such as the kidney and liver—where cell growth is normally relatively unrestrained—can be cultured readily. Cancer cells, on the other hand, can easily be grown for generation after generation in cultures.

Radiation and chemicals cause cancer in many types of cells, but the mechanism by which they do so is difficult to study in cell cultures because these carcinogens affect a number of genes and other cellular components at the same time. Thus, many molecular biologists have come to rely on cancer-causing viruses as a means of exploring the cellular control mechanisms that cancer disrupts.

Viruses are known to act by introducing new genes into a cell, causing the cell machinery to produce new viruses at the expense of its normal structure and functions, and eventually destroying the cell as the new viruses are liberated. A cancer-causing virus also introduces new genes into the cell and causes the cell to produce new viruses, but it does not kill the cell. Instead, it causes the cell to multiply rapidly, with the virus being contained in each new cell that is formed. Experiments carried out in 1968 by Renato Dulbecco and his group showed unequivocally that an animal virus—known as SV-40—introduces a small group of genes into the chromosomes of a normal cell. These genes become irrevocably bound there and cause

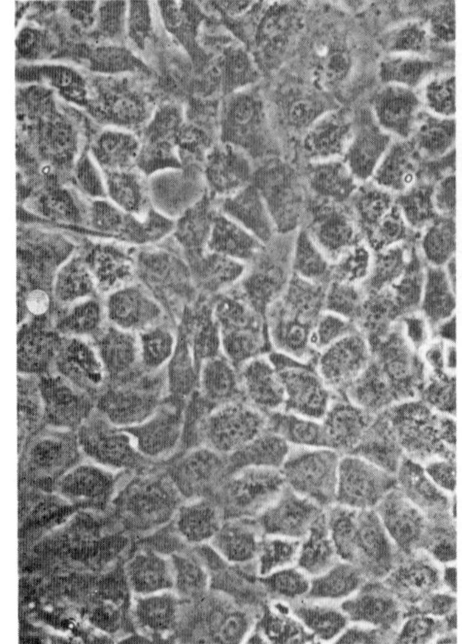

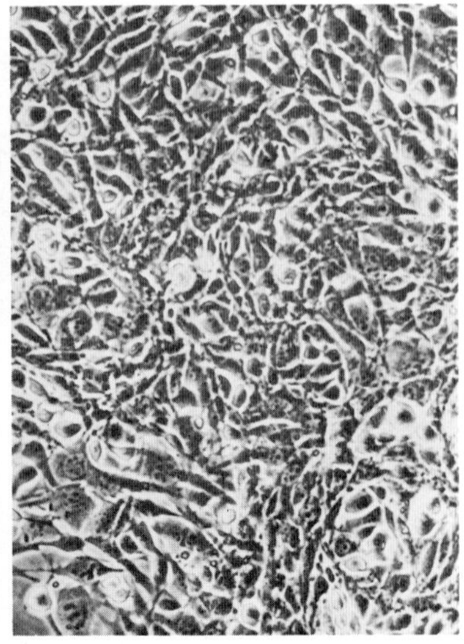

Figure 17.14. RNA type C virus particles, showing mature type C particles in an intercellular space and budding of these particles in the insert at upper right. (× 50,000)

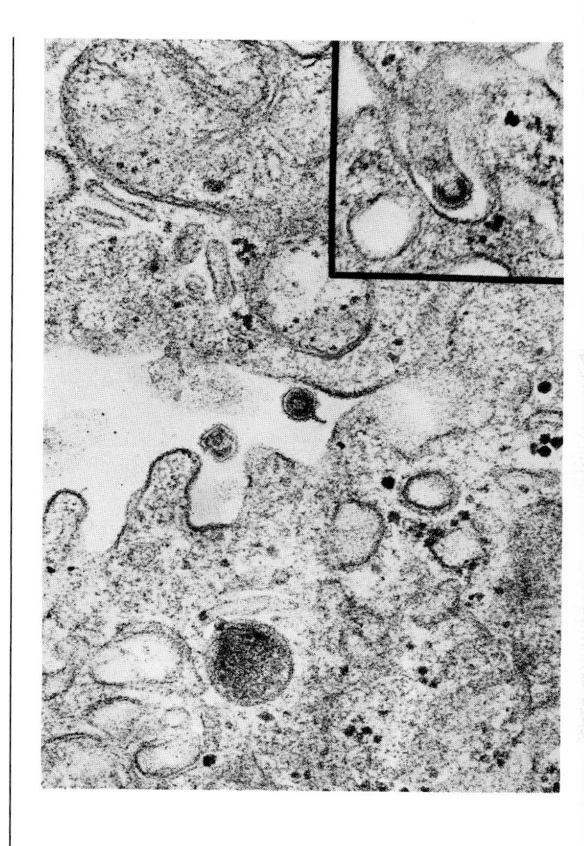

the cell to multiply as a cancer cell, replicating the viral genes in each division. Dulbecco and other researchers chose the SV-40 virus and others very much like it because the number of genes in the viral DNA is very small. Because new research methods make it possible to turn off or delete one gene at a time, it should not be too long before the gene that initiates the cancer process is identified.

The Jacob-Monod theory of cellular control has been used to suggest a model for the cause of cancer. For example, the cancer process might involve production of an inducer that combines with the repressors normally controlling the production of proteins involved in cell growth and division. In late 1969, Robert J. Huebner of the viral carcinogenesis branch of the National Cancer Institute announced his own theory, which is related to the theory of Jacob and Monod. According to the Huebner theory, all normal cells contain a type of RNA virus that stays with the cell from the period of embryonic development to maturity. This RNA virus—called a C-type particle by cancer researchers—contains an *oncogene* that can be triggered into action by built-in genetic defects, cellular aging, or any number of outside factors including carcinogenic chemicals, viruses, or radiation. When the oncogene is activated, suggests Huebner, the substance that it produces releases the inhibitions on the normal DNA of the cell, converting it to a wildly reproducing cancer cell.

Although Huebner's theory has not been definitely proven, RNA C-type viruses are known to cause cancer in a number of different animal species and have been linked, at least by association, with certain human leukemias.

More support for the Huebner theory comes from much earlier work by Henry Kaplan and Ludwik Gross, who showed that irradiation of normal rat cells causes production of carcinogenic RNA viruses. These viruses produce cancer tumors when they are injected into other mice that have not been irradiated. These results suggest that the RNA viruses were present all along, but they were unrecognized until the radiation triggered an inactive gene that caused the cell to become cancerous.

Only a small sampling of the many theories and experimental approaches being used in the study of cancer has been represented here. Cancer research and research on cellular control mechanisms are closely interrelated, and an advance in either area is almost certain to be useful to researchers in the other.

FURTHER READING

The classic paper by Jacob and Monod (1961) sets forth the original version of the operon theory and discusses many of its implications. Further discussions of cellular control processes will be found in articles by Britten and Kohne (1970), D. D. Brown (1967), Changeux (1965), Davidson (1965), Gurdon (1968), Martin and Ames (1964), Miller and Beatty (1969), Moscona (1961), and Ptashne and Gilbert (1970).

18
Development

How does a single cell, the fertilized egg, differentiate into a multicellular organism? It is clear that the fertilized egg of both plants and animals carries in its chromosomes a complete set of genes for all cells in the mature organism. During development, each cell in the organism, excluding the gametes, contains a complete set of these genes. Although the genetic content remains constant, different sets of specialized cells arise during development, producing specific tissues and organs. But how does this process occur?

As indicated in Chapter 17, the process of cellular differentiation is now interpreted in terms of selective gene expression. The fertilized egg of a plant or an animal develops into a multicellular organism by turning on and off specific genes at particular times during development. Although biologists have learned much about this control system, its major mechanisms are far from fully understood, and research in this area of development is being pursued by many biologists today.

Although the basic questions concerning the molecular mechanisms of development remain unanswered, much is known at the cellular level about the sequences of events that occur during the development of a large number of organisms. For many years, developmental biologists have described the orderly and sequential changes occurring in an organism from the time it "begins life" until it dies. In higher plants and animals, this developmental cycle begins with gametogenesis; continues through fertilization, embryogenesis, and maturation into adulthood; and ends with aging and death.

DEVELOPMENTAL PROCESSES

The processes of development are obviously different in plants and animals. Because these processes have been studied by specialists in totally separate fields of research, different concepts and different vocabularies exist for the description of plant development and animal development. Nevertheless, it must be emphasized that both plant and animal developmental biologists are concerned with many of the same questions. In general terms, both attempt to trace the sequence in which various parts of the adult organism are developed and to understand the interactions that cause particular cells to differentiate in particular ways at particular times. More specifically, problems such as wound healing, regeneration, cellular differentiation, cancer, and aging are developmental problems common to both plants and animals.

Plant and animal developmental biologists have been largely concerned with the processes by which a newly formed individual acquires structure, specialized cells, and a net size. These processes fall into three major categories: *morphogenesis, differentiation*, and *growth*. Because the information in each of these areas is voluminous, in this chapter it will be possible to present only a sampling of the available information about development, concentrating chiefly upon the flowering plants and vertebrate animals.

Morphogenesis

Morphogenesis is a general term used to describe processes by which tissues or germ layers are shaped into organs and by which the organism acquires its overall adult shape and form. In plants, morphogenesis is accomplished chiefly through differential growth—that is, through tendencies for cells to elongate or to divide along particular planes and axes. In animal development, movements of cells—either by individual migration from one place to another or as sheets of cells—play a major role in morphogenesis.

One major difference between morphogenesis in plants and animals is that plants contain groups of relatively undifferentiated cells that continue to form new tissues and organs. These embryoniclike cells continue to divide throughout the life of the plant, retaining the capability of adding new tissues and organs continuously and indefinitely in response to environmental changes. On the other hand, morphogenetic movements in animals establish tissue and organ primordia relatively early in development. Once the organ primordia are fixed, the overall shape of the animal is determined, and, except for regeneration or developmental abnormalities, no new organs are produced in the animal.

The ability of a plant to develop new organs throughout its life is called *indeterminate growth.* As a result of this morphogenetic potential, the shape and form of an adult plant varies greatly with the environmental conditions under which the plant grows. Although the individual organs and tissues show forms unique to the species, even the number of organs (such as leaves) may vary greatly from individual to individual. In contrast, because of the way morphogenesis occurs in animals, two individuals of the same species are apt to be quite similar in size and shape and certainly will have the same numbers of various organs.

Differentiation

Adult plants and animals are not simply enlarged copies of the fertilized egg. In a multicellular organism, the thousands or millions of cells produced by divisions of the zygote must become differentiated into many different kinds of specialized cells. Not only must the proper kinds of cells be produced but they must be produced in the proper numbers and assorted into the proper locations in the developing embryo. Each group of cells destined to produce a specific adult tissue passes through a series of biochemical and structural alterations that culminate in the formation of a tissue appropriately specialized for its function. This process is called *differentiation.*

During differentiation in a multicellular organism, cells acquire more and more specific determinations, and the paths open to each cell and its descendants become more and more restricted. In some species, determination occurs at very early stages of embryonic development, apparently as a result of unequal distribution of cytoplasmic components during the early cleavages of the zygote. In other species, determination occurs at a later stage of development, apparently as a result of interactions among neighboring cells. Such control of cell differentiation by influences from neighboring cells is called *induction.*

In most cases, the process of determination cannot be detected by biochemical or structural changes in the cell. The cell in which determination has occurred appears identical to other nonspecialized cells, but observation shows that it now is committed to a particular course of development, which can be modified only partially by outside influences. Specialized structures or chemicals within the cell may not become apparent until many division cycles after determination.

In most cases, once a cell has become determined upon a particular course of differentiation it will pass this tendency on to its descendants. Such cells pass on not only their genetic information but also the regulatory

Figure 18.1a (above). A typical S-shape growth curve. This curve shows the increase in weight of a young corn plant.

Figure 18.1b (below). The growth rate of an insect.

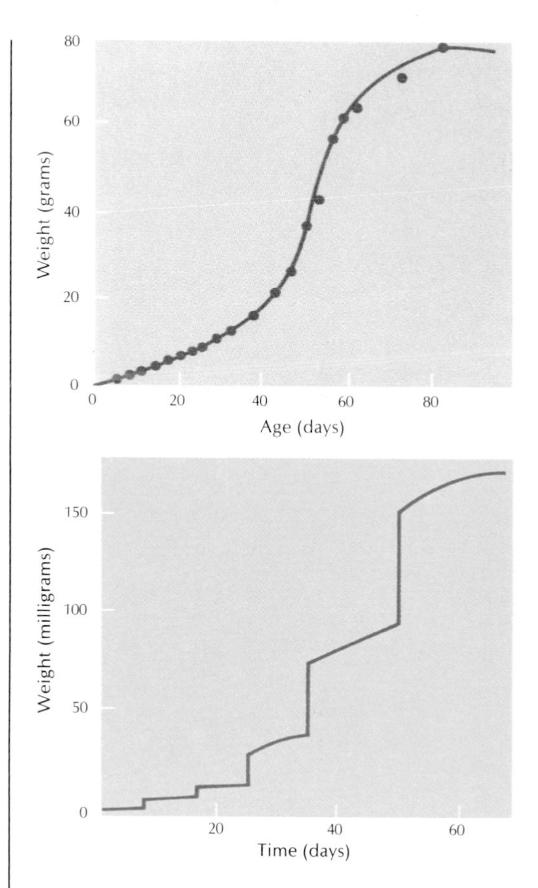

agents that control the use of the genetic information. These instructions might be contained in the portion of the cytoplasm obtained during division, or they might be inherited in the form of certain genes that are more-or-less permanently activated or inhibited.

Although the actual mechanisms of differentiation are not known, it is clear that the course of differentiation in both plants and animals is controlled in time by hereditary influences and by substances entering the cell from its environment.

Growth

Growth is a universal feature of the development of an individual organism. In simple unicellular organisms, binary fission produces daughter cells that are quite similar in structure to the parent cells. After a period of growth in the daughter cells, during which time structures within the cell are duplicated or enlarged, the cell reaches a size at which it is prepared to divide again. In multicellular organisms, far more impressive feats of growth take place in the development from a single cell to the large body of the mature individual. Growth can be measured in terms of length, weight, number of cells, or amounts of various substances. Whatever the measure used, the growth rate varies with time, in most cases reaching a maximum at some point during early life and becoming nearly zero (or even negative) in the mature organism (Figure 18.1). Growth is accomplished both by cell division and by enlargement of cells.

Variations in growth patterns among various organisms are numerous. Many organisms have a simple sigmoid (S-shape) growth curve. Organisms that pass through various abrupt changes in the life cycle—for example, the moltings and metamorphosis of an insect—may have several periods of rapid growth separated by intervals of zero or negative growth. Similarly, a woody plant such as a tree grows each year during favorable seasons.

In most multicellular organisms, the early stages of growth are marked by repeated mitoses of all the cells in the organism. As the number of cells increases, the amount of growth accomplished with each set of cell divisions increases. As development progresses, however, more and more cells become differentiated to serve specialized functions, and they cease to grow and divide. Thus, more and more of the machinery of the organism is diverted from growth to other specialized tasks, and the growth rate decreases. Even in the mature organism, some cells continue to grow and divide, but the rate at which new materials are added is barely sufficient to compensate for the loss of materials through processes of aging.

The source of nutrients providing the energy necessary for growth during embryogenesis varies with different species. In most higher plants, insects, fish, amphibians, and birds, the energy reserves are stored materials such as plant endosperm and animal yolk. In mammals, nutrients from the mother via the placenta nourish the embryo during intrauterine growth. Once the organism begins its development into a mature individual, nutrients for growth are provided from photosynthesis in plants and by active feeding by most animals.

Events that occur during growth are closely interrelated. Changes occurring in one part of the organism often trigger changes in other parts. This complex network of *correlative effects* results in the organized growth of an adult organism. In most cases, these interactions involve chemical messengers, or *hormones*, that can move from cells in one region of the organism to cells in another—for example, growth hormone in man, growth and molting hormones in insects, and the hormone gibberellin in plants.

DEVELOPMENT OF THE FLOWERING PLANT

As higher plants develop, they maintain regions of growth and development throughout the entire life of the organism. Stems and roots are extended and new organs are formed through cell division, elongation, and differentiation in certain growth regions. Growth and development can be regulated by hormones that interact with environmental factors such as heat and light, or that cause specific changes elsewhere in the plant. Although the plant is relatively limited in its ability to respond rapidly to changes in external conditions, it can easily shed old organs and grow new ones as conditions change.

Embryonic Development

Although much less is known about plant development than about animal development, several exciting experiments carried out in recent years have revealed new aspects of the sequences in the embryology of flowering plants that eventually produce the adult.

The embryonic development of the flowering plant occurs in the ovule of the adult. After pollination, a double fertilization occurs, forming a diploid zygote and a triploid endosperm nucleus. The zygote nucleus remains inactive while the endosperm nucleus divides rapidly. During this time, the ovule tissues synthesize and transport material into the zygote to form the surrounding endosperm.

After the endosperm is well developed, the zygote nucleus, which is now surrounded by the endosperm, begins to divide mitotically. The zygote becomes polarized and divides unequally to yield two cells that differ both in size and in contents. In the zygote, vacuoles tend to cluster at one end of the cell, while most of the cytoplasm and organelles move to the other end. The wall formed during the first division of the zygote separates a small, densely cytoplasmic *terminal cell* from a larger, more vacuolate *basal cell.*

The developmental fate of these two cells is quite different. The terminal cell produces cells that differentiate into the embryo itself, whereas the basal cell divides to form a *suspensor cell* and a new basal cell. Further divisions of the suspensor cell produce the stalklike *suspensor*, which attaches the embryo to the rest of the seed.

It is clear that the determination of these two cells is established prior to the completion of the first division. Although there are no experimental data, the obvious unequal division of cytoplasmic components in this case is strikingly similar to the unequal distribution of cytoplasmic components involved in some cases of cell specialization in animals.

The first few divisions of the terminal cell produce a spherical or globular embryo, composed of several cells. As a result of morphogenesis (differential growth) and cellular differentiation, the cells in the embryo become organized into the basic organ primordia of the adult—the shoot apical meristem, the root apical meristem, and, in the case of dicots, the two cotyledons.

The word "meristem" refers to those cells in the plant that continue to divide mitotically. The *apical meristems* are located at the tip (apex) of shoots and roots, and the *lateral meristems* are positioned around the cir-

Figure 18.2. Longitudinal section of a *Capsella* embryo. The darkly stained embryo is surrounded by endosperm. Note the suspensor (tissue strand) connecting the plant embryo to the surrounding wall of the ovule.

cumference of the shoots and roots. These are the embryoniclike cells that have the capability of forming new organs continuously throughout the life of the plant.

Growth regions that will form the cotyledons become established in the globular stage of the embryo. The dicots have two growth regions, producing two cotyledons that grow vertically upward and make the embryo somewhat "heart-shaped." Between the two cotyledons of dicots, or adjacent to the single cotyledon of monocots, is a growth region that forms the apical meristem, or growing tip, of the young shoot. At the other (suspensor) end of the embryo, the apical meristem of the *radicle*, or primary root, becomes established. As the embryo grows, the apical meristems of root and shoot become increasingly separated, pushed apart by the cells they form. Elongation of the axis between the two apical meristems produces the "torpedo-shape" stage of embryo development.

During this torpedo-shape stage, cellular differentiation in the embryo becomes readily apparent. A central core of elongated, densely cytoplasmic cells—the *procambium*—is surrounded by more vacuolate cells of the *ground meristem* (Figure 18.2). Differentiation occurs among the cells that have been formed by divisions of the meristems and left behind in the central part of the embryo as the meristems grow outward.

The extent to which the embryo develops within the ovule varies from species to species, but growth, division, and most metabolic activities are

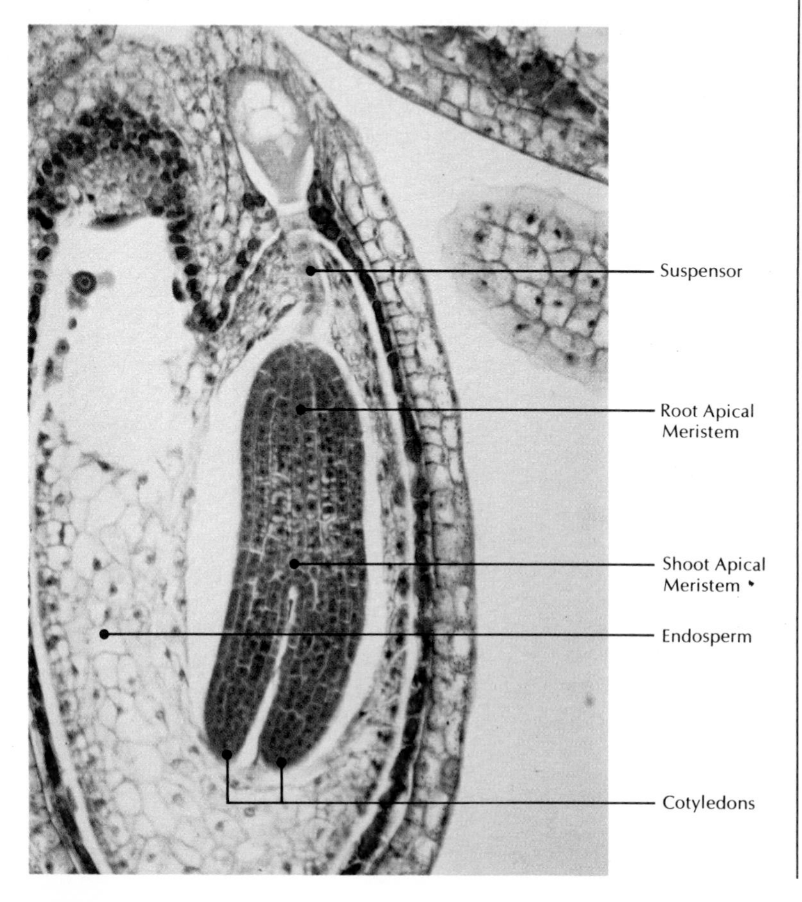

halted when the embryo is enclosed in the seed coat. The embryo remains dormant while the seed is transported away from the parent plant, and in many species dormancy continues for some time after the seed has come to rest. When the seed germinates, growth and development of the embryo resume. The apical meristem of the root divides to form the root tissues. The apical meristem of the shoot gives rise not only to the tissues of the stem but to young leaves, *leaf primordia*, and in some cases, also to *bud primordia*. The procambium develops into the vascular tissues of the plant—the xylem and phloem systems—which carry fluids through the plant body. The ground meristem forms the ground tissue, which may differentiate into various specialized tissues that serve functions of storage, mechanical support, and photosynthesis.

During the 1930s, several experimenters attempted to isolate embryos from ovules and grow them in tissue culture to determine how far the embryo would develop when removed from the adult tissues. Very early embryos would not develop if isolated in a defined liquid culture medium containing minerals, sugars, vitamins, yeast extract, light, carbon dioxide, and oxygen. However, if embryos were not isolated until they had reached the torpedo-shape stage, they would differentiate into normal seedlings. The older the embryo was prior to its isolation, the simpler the culture medium could be to support normal development. For example, if embryos were isolated just after the torpedo-shape stage, they survived and differentiated in medium without vitamins or yeast extract. Embryos isolated at the onset of dormancy develop normally in culture medium containing only minerals.

Johannes van Overbeek and his coworkers (1941) succeeded in culturing very young embryos (in the heart-shape stage) by adding coconut milk, which is actually the liquid endosperm of the coconut seed. Since that time, extracts of the endosperm of other kinds of seeds have been shown to have similar effects in supporting *in vitro* development of very young embryos. An analysis of coconut milk shows that it contains a mixture of basic nutrients such as amino acids and sugars, but more importantly, it contains a complex mixture of hormones.

It is now clear that a mixture of hormones, normally present in the endosperm, is essential for proper development of the early embryo in culture. For example, heart-shape embryos of shepherd's purse (*Capsella*) can be grown in a medium containing various nutrients and a balance of three hormones. In the absence of these hormones, the embryos show little or no growth. An imbalanced mixture of the hormones may cause the embryo to grow into a shapeless, tumorlike mass of cells (Overbeek, *et al.*, 1942; Raghavan and Torrey, 1963).

These experiments demonstrate that hormones present in the endosperm of the zygote are essential for the differentiation of the embryo into the adult plant. Because the endosperm does not begin forming until after fertilization, the influence of the endosperm upon the embryo is not expressed until the early stages of embryogenesis. Therefore, when an early embryo is isolated in a defined medium, without the hormones, it fails to develop.

Single cells taken from carrot *embryos* can be cultured similarly, and under proper conditions these cells form embryolike structures, or *embryoids*, which pass through development stages resembling those of a normal embryo (Figure 18.3). A single cell from an embryo, isolated from

Figure 18.3. Stages in the development of carrot embryos. Shown below are carrot cells.

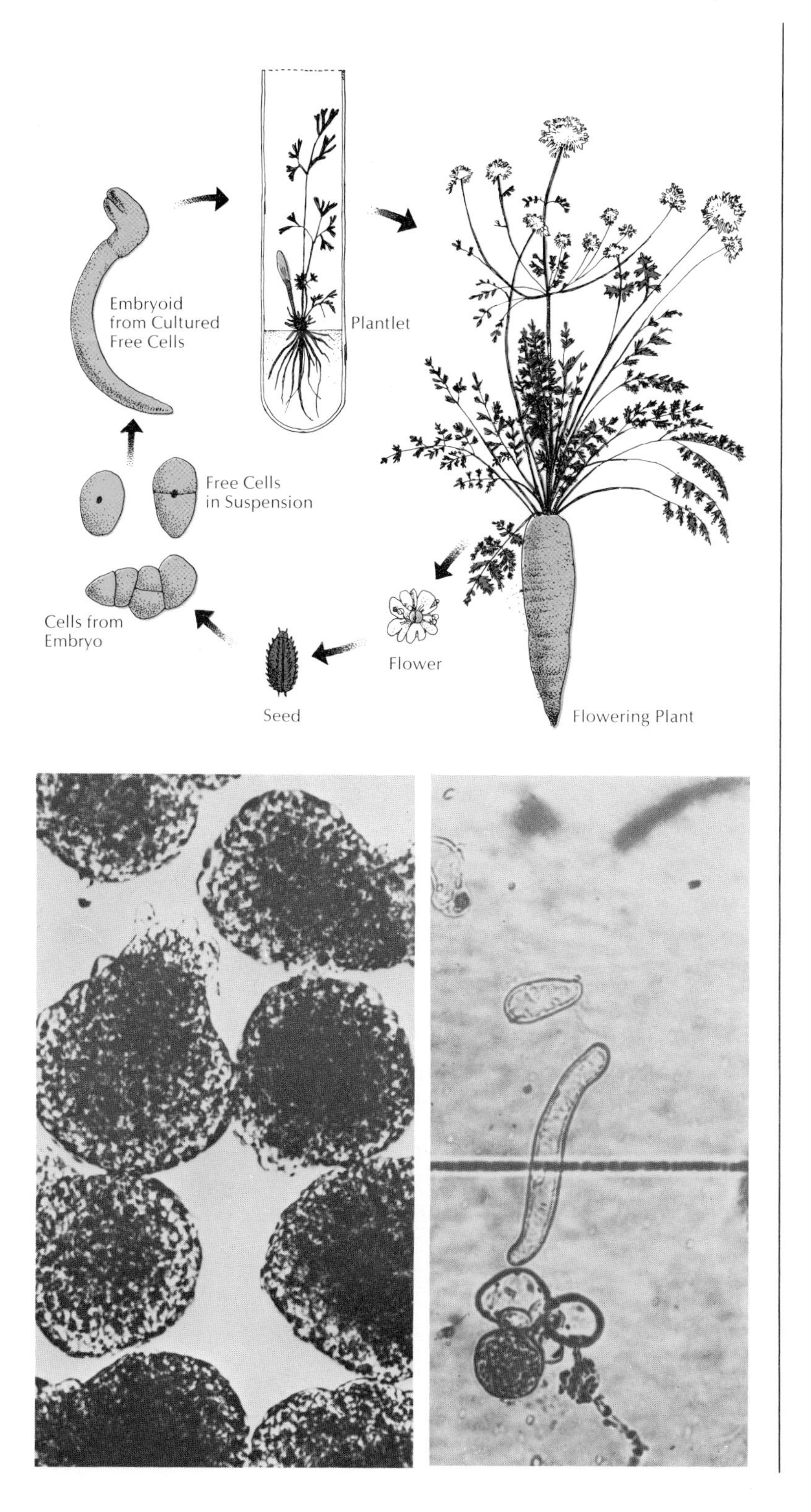

Figure 18.4a (upper left). Surface view of a shoot apex and leaf primordia of the fern *Dryopteris dilatata*. Sixteen leaf primordia can be seen.

Figure 18.4b (lower left). Side view of the shoot apex. Note the compound leaf primordia. A main shoot apex such as this exerts apical dominance over the development of lateral buds.

Figure 18.4c. (right). A scanning electron micrograph of the shoot apex. Primordia of compound leaves are arranged spirally around the apex.

its neighbors and supplied with a medium containing coconut milk, can develop into an apparently normal carrot plant that will flower.

Embryoids have been produced from cells taken from various tissues of many different kinds of plants. Even pollen grains have been cultured to produce embryoids that develop into haploid plants, which are considerably smaller than normal adult plants and which flower but do not produce seeds (Nitsch and Nitsch, 1969).

There has been great interest in the process by which the embryo becomes dormant inside the seed and then resumes growth upon germination (Amen, 1968). The immediate trigger for germination is the absorption of water by the seed under suitable environmental conditions. The tissues of the endosperm swell up as water is taken in, bursting the seed coat and causing hydrolysis of starches and formation of sugars in the endosperm. Growth of the embryo begins as the sugars are transported to it from the endosperm. However, if the embryo end of the seed is cut off before wetting, hydrolysis of starch does not occur in the endosperm. The hydrolysis of starches is catalyzed by the enzyme α-amylase, which is produced by the *aleurone cells*, a group of cells near the base of the embryo. The activity of the aleurone cells is stimulated by a hormone produced by the embryo.

Some seeds—such as wild oats (*Avena fatua*)—that normally remain in the ground ungerminated for years can be germinated quickly by treating the soil with hormones. Seeds that normally germinate only in the light—some varieties of lettuce, for example—will germinate perfectly in the dark if supplied with hormones. Other seeds that normally require darkness for

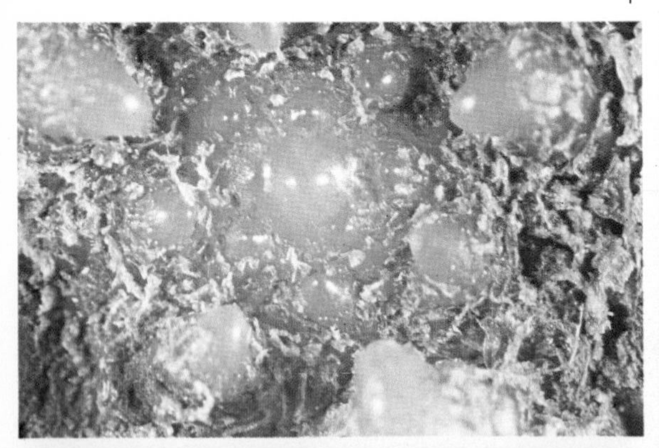

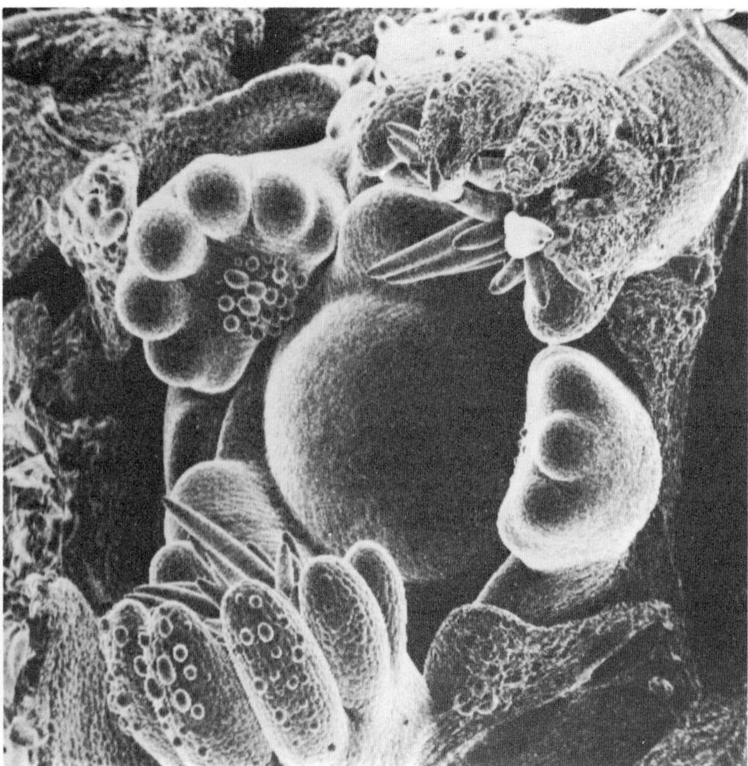

Figure 18.5. A scanning electron micrograph and photograph of the apical dome of *Equisetum* (a horsetail).

germination will germinate in the light if treated with hormones. Seed dormancy in some cases may be nothing more than the failure of the embryo to produce this hormone.

Development of the Shoot Apex

The shoot apex is primarily involved in the initiation of leaf development. The cells in the apical meristem divide mitotically, producing patterns of cells that differentiate into the leaves. Of particular interest to developmental biologists are (1) what initiates leaf development in the daughter cells of the apical meristem and (2) what determines the particular phyllotatic pattern in the plant.

The apical meristem of the shoot is surrounded by leaf primordia that have been formed in a regular sequence by the meristem and are in various stages of development. The location of the primordia on the meristem and the size of the primordia relative to that of the apex vary from species to species. For example, they may be formed singly, in pairs, or in whorls of a greater number.

The tip of the shoot of a flowering plant can be removed and grown in a nutrient medium. If the excised apex bears a few leaf primordia, it will grow in a relatively simple nutrient medium and will eventually give rise to a complete plant with roots and leaves. If the apical dome alone is removed with no leaf primordia, it will only grow in a much more complex medium, indicating that the differentiation of the cells of the apical dome may be dependent on various substances obtained from the cells of the primordia.

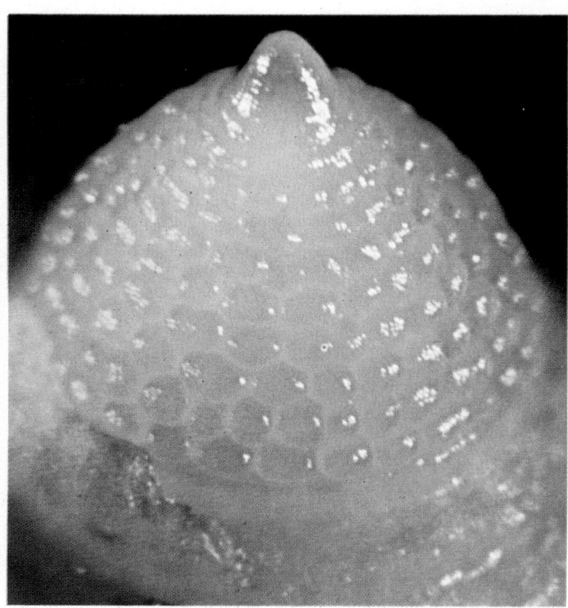

Figure 18.6a (above). Longitudinal view of a shoot apex. Note the formation of a bud primordium (densely stained cells) in the axil of a young leaf.

Figure 18.6b (below). A scanning electron micrograph of young leaf primordium, showing 5 (possibly 7) leaflet primordia.

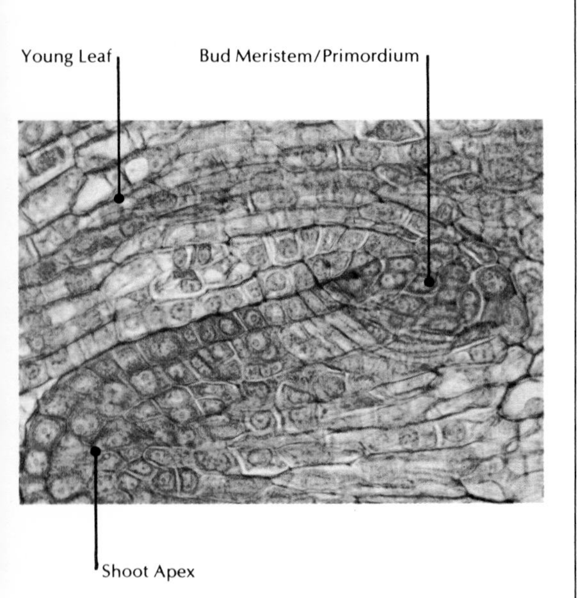

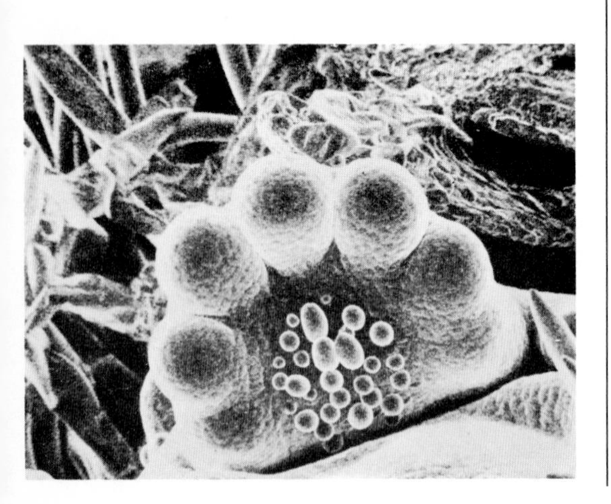

In ferns and other simpler plants, however, the apical cone itself, without leaf primordia, is capable of normal growth and development in a simple medium (Wetmore, 1954).

Many experiments have shown that the pattern of development of primordia can be modified—for example, by incisions in certain places on the growing apex. If the tip of the shoot apex of a flowering plant is bisected, cells along the flanks can develop into new apical meristems, suggesting that the central cells of the apical dome normally exert some influence that inhibits the development of new apices from flank cells. Therefore, it appears that the phyllotatic pattern is affected by the growth of the apex and is regulated by interactions among existing primordia.

Branches of the main stem originate as *axillary buds*, which form from primordia that appear in the axils of the leaf primordia (Figure 18.6). The potential pattern of branching in a plant is closely related to the pattern in which leaf primordia form around the flanks of the apical meristem.

If the site of an as-yet-invisible new leaf primordium of the fern *Dryopteris* is isolated from older primordia by deep cuts, the primordium becomes much larger than normal, indicating that neighboring leaf primordia inhibit the growth of younger ones. If the site of the developing primordium is isolated from the apical cone by a similar cut, the primordium develops into a bud rather than a leaf (Wardlaw, 1949). In fact, even a partially developed primordium will form a bud if it is isolated from the apical cone before it develops the lens-shape apical cell typical of young leaves (Cutter, 1956). This single, apical cell of the leaf develops from one of a group of cells at the surface of very young leaf primordia. However, if the isolating cuts are shallow and only penetrate the surface cell layer of the apical cone, the primordium continues to develop as a leaf rather than a bud.

These experiments suggest that in ferns there is a period in the early development of the perspective leaf primordium in which it can be switched into another path of development. After a certain stage of development is reached, however, this switch can no longer be made. At this point, the cells are determined to differentiate as a leaf. After determination has occurred, a young leaf primordium can be excised completely from the shoot apex and other shoot tissues and grown on a sterile nutrient medium; under these conditions, it will still develop into a recognizable leaf (Feldman and Cutter, 1970). Evidently, the change that takes place in the primordium at the time of its determination is a profound one, but its nature is not yet understood.

The factors that cause particular regions of the apical meristem to begin rapid division and to differentiate into leaf primordia are not understood. It is clear, however, that most of the cells of the meristem possess the potential for becoming primordia and that complex interactions of inhibitory effects from the apical meristem and from older primordia prevent this potential from being expressed except at certain regularly spaced positions.

Branching in the Shoot

In most flowering plants, the lateral buds that may develop into branches are formed in the axil of each leaf primordium. In some species, however, the buds may form only in the axils of certain leaves (for example, every second leaf) or not at all. The degree to which the buds grow out as branches depends largely on a phenomenon called *apical dominance*. Outgrowth of the buds is inhibited or completely prevented by the main shoot apex. If

Figure 18.7a (left). Experiments illustrating the concept of apical dominance. (A) Normal plant growth with lateral buds inhibited by hormones secreted by the apical buds. (B) Plant with apical bud removed, thereby allowing lateral bud growth. (C) Plant with apical bud removed and sealed with a plain agar block—lateral buds develop. (D) Plant with apical bud removed and stump sealed with an agar block containing indoleacetic acid. IAA acts to suppress lateral bud development. (*From* Principles of Plant Physiology *by James Bonner and Arthur W. Galston. W. H. Freeman and Company. © 1952*)

Figure 18.7b (right). Decapitation of shoot apexes results in a significant increase in the growth of lateral branches. Application of a paste of IAA to the severed stump acts to inhibit lateral branch growth, thus replacing the function of the normal plant with an intact apex.

the apex is cut off or damaged, buds along the shoot begin to grow. Gardeners know that removal of the terminal bud on a shoot almost invariably causes the growth of a branch from the next lateral bud below; this knowledge is the basis of pruning. If certain hormones are applied to the cut surface where the terminal bud was removed (in amounts comparable with what the apical bud would have produced), the lateral buds do not grow (Figure 18.7) (Thimann and Skoog, 1933). Similar inhibitions can be observed in a sprouting potato, where a bud that develops first may inhibit all other buds from developing . Although there is still much controversy about the exact mechanism, hormones secreted by the terminal bud play a major role in inhibition of further development of lateral buds on the same shoot (Phillips, 1969).

Stem Elongation

In some plant species, the portions of stem between the leaves (*internodes*) remain much the same length throughout growth, but in most plants the internodes become longer, and an extended stem is formed. Cell divisions responsible for stem elongation occur mainly in a region just below the

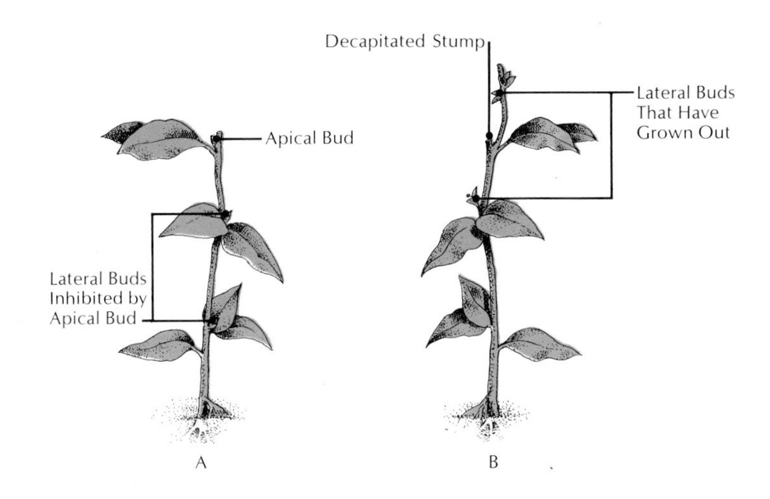

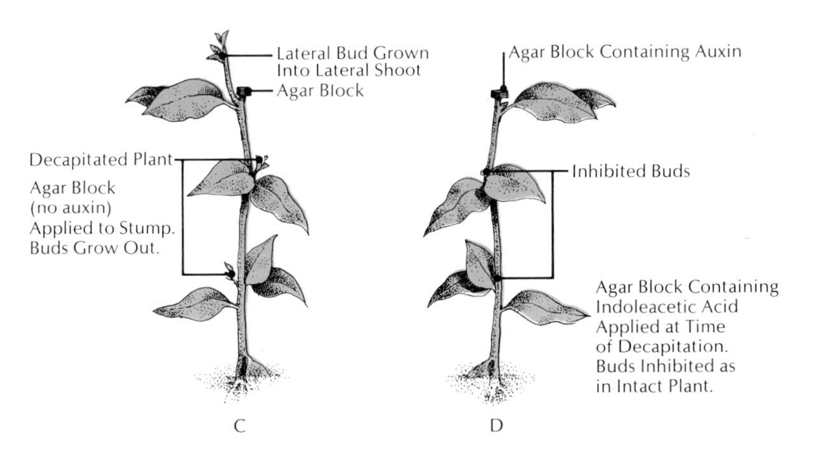

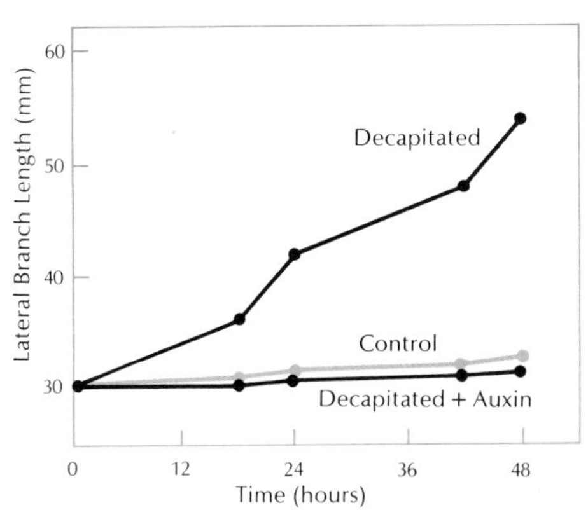

Figure 18.8. Differentiation of vascular tissue. Groups of densely stained orange cells, the procambium, differentiate into xylem (red) and phloem (gray). Only protoxylem and protophloem have been differentiated so far.

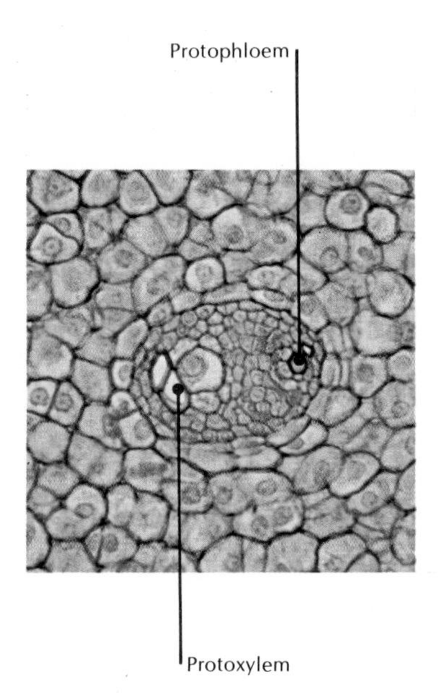

apex—a region sometimes called the *primary elongation meristem* (Sachs, 1965). In at least some cases, it has been shown that hormones stimulate the activity of these meristems by causing elongation and enlargement of these plant cells. If long-stem plants are treated with substances that inhibit the synthesis of hormones, they fail to develop elongated internodes.

Differentiation of Primary Shoot Tissues

The cells formed by division in the apical meristem gradually enlarge and become differentiated to form the various tissues of the shoot. In a longitudinal section through a shoot apex, a region of cellular differentiation is visible just behind the apical meristem. The first sign of differentiation is seen near the meristem, where an outer layer of vacuolate ground tissue cells surrounds a cylinder of more densely staining cells. At a slightly lower level in the shoot, groups of very densely staining cells are visible at positions around the cylinder corresponding to the leaf primordia. The densely staining cells, which are elongated along the axis of the shoot, are *procambium* cells, which will differentiate into conducting, or vascular, tissues. A little farther down the shoot, the procambium cells toward the outside of the stem have differentiated to form the first cells of the phloem, which will conduct nutrients and other substances from the photosynthetic tissues to the rest of the plant. The innermost procambium cells have differentiated as elements of the xylem, which will conduct water and dissolved substances from the roots to the rest of the plant.

Phloem tissue begins to form in older, more mature tissues and grows upward into the developing leaf primordia. Differentiation of xylem tissue begins at the level of attachment of a leaf primordium and proceeds both upward into the young primordium and downward to join the mature elements in the stem.

A transverse section of most dicot stems shows a cylinder of separate vascular bundles or a continuous cylinder of vascular tissue (Figure 18.9). In most monocot stems, the number of vascular bundles is much greater, and they are scattered through the ground tissue (Figure 18.10).

If one of the vascular bundles in the stem is severed, the wound is healed when nonspecialized parenchyma cells begin to differentiate and form a new strand of xylem around the wound. Evidence from many experiments now suggests that parenchyma or procambium cells can only begin to differentiate and form xylem or phloem tissue when they are exposed to sucrose and a hormone that is normally secreted by buds or young growing leaves. The proportion of xylem to phloem is related to the concentration of sucrose in the tissues.

Secondary Growth in the Shoot

In most dicots, a meristematic tissue, the *vascular cambium*, differentiates from procambium remaining between the xylem and phloem—and from parenchyma cells between the vascular bundles—forming a complete cylinder (a lateral meristem) around the xylem portion of the stem. The cambium cells divide predominantly by the formation of walls parallel to the surface of the stem. Tissue formed on the inner side of the cambium differentiates to become xylem, whereas that formed on the outer side becomes phloem. In woody plants that survive over many years, the cambium becomes active in the spring (in temperate climates) and ceases activity in the fall of each year. Because reactivation of the cambium closely follows the

Figure 18.9 (above). Cross section of a typical dicot stem showing the concentric ringlike arrangement of vascular tissue. (*Courtesy Carolina Biological Supply Company*)

Figure 18.10 (below). Cross section of a typical monocot stem showing the well-developed vascular bundles scattered within the ground tissue. (*Courtesy Carolina Biological Supply Company*)

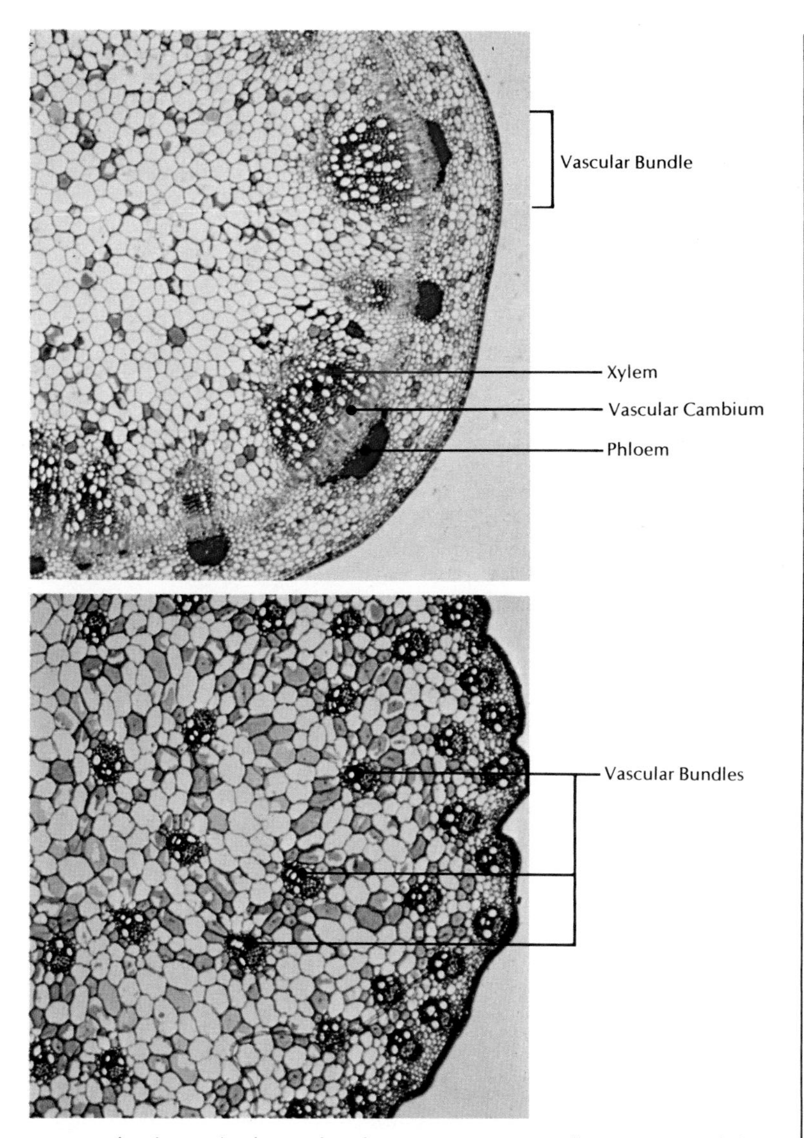

outgrowth of new buds on the shoot system, it was long suspected that some substance secreted by the growing buds might be responsible for activation of the cambium. The activation of the cambium begins near the new buds and progresses along the branches and down the trunk. Removal of the buds before their growth has begun in the spring causes the cambium layer to remain inactive. Relatively normal activity of the cambium layer can be stimulated by applying a mixture of hormones to the sites from which the buds were cut. In nonwoody plants, removal of the growing tip causes failure or cessation of cambial activity. The application of hormones to the cut surface causes the initiation or resumption of cambial activity (Digby and Wareing, 1966).

The effects of light upon shoot growth were demonstrated by Charles and Francis Darwin in 1880. They found that a seedling of the grass family would curve toward the light if illuminated from one side but that curvature would not occur if the extreme tip was covered with an opaque cap. Such patterns of growth toward or away from light sources are *phototropisms*. Because the zone of curvature was well below the tip, the Darwins concluded that some "influence" came down from the tip to the parts below to direct the curvature.

Through a number of ingenious experiments, later researchers demonstrated that the curvature is caused by unequal growth rates on the sides of the stem, that the amount of growth is proportional to the amount of certain hormones received from the tip, and that the amount of hormones produced in various parts of the tip is proportional to the amount of sunlight reaching the part. Some shoots demonstrate *geotropism*—that is, the direction of growth or curvature is determined by gravity. Similar effects of hormones produced in varying amounts by different parts of the tip were shown to account for geotropism.

Flowering

Factors that cause a plant to develop flowers vary from species to species. One important factor is day length—the relative lengths of light and dark periods in each 24-hour day. In temperate regions, the days are longer in the summer and shorter in the winter, and the effects of day length tend to produce seasonal flowering.

It appears that changes in day length lead to production of a hormone in the leaves. This hormone then travels to the shoot apical meristem, where it triggers the beginning of flower production. This hormone, however, has not yet been isolated or identified. Flower production begins with a transformation of the shoot apex. The changes of size and shape vary from species to species. The flowers may develop from flower primordia that form around the flanks of the apex or through a conversion of the whole shoot apex into a floral apex. The floral apex or floral primordium forms the various parts of the flower as lateral appendages.

In many cases, increased cell division in the central apex region is one of the earliest changes observed after induction of flowering; it may occur as soon as 16 hours after treatment to induce flowering has been given. Increased cell division leads to changes in the size and shape of the apex. The eventual differentiation of the flower parts is apparently due to complex interactions of hormones or other factors that alter the environment of individual cells within the floral apex.

Various experiments indicate that the presence of leaves is crucial to the induction of flowering and that the leaves must be exposed to the day-length conditions that trigger flowering. (In some cases, flowering will be induced if as little as 1 square centimeter of a single leaf is exposed to the proper day-length conditions.) Removal of the leaves within a few hours after exposure to the triggering stimulus prevents the induction of flowering, but after a day or two the removal of leaves has little effect upon the further development of the flowers. Stems with leaves that have been exposed to triggering influences can be grafted onto other plants that have not been exposed, and the host plants will flower. These and other experiments strongly support the idea that a substance moves from the leaves to the shoot apex to stimulate flower production.

Although many details of the flowering-induction mechanism remain to be discovered, florists have put the present knowledge to commercial use. By manipulating periods of darkness in their greenhouses, they can make

Figure 18.11. Transverse section (left) and longtitudinal sections (middle and right) of a root tip of the frog-bit *Hychocharis*. The formation of trichoblasts, or hair cells, by unequal divisions of immature epidermal cells is illustrated. The smaller trichoblasts stain more densely.

sure that all their poinsettia plants will be in full bloom for Christmas sales and that their Easter lilies will flower at the proper time. Some plant species respond to particular conditions of temperature for the onset of flowering, rather than demonstrating sensitivity to day length.

The Root

In many dicots, the root system consists of a main, or tap, root with lateral branches. In most monocots, the root system is made up of a number of fibrous roots of fairly similar size. The apical meristem of the root is covered by a *root cap* of parenchyma cells. Cells on the surface of the root cap are worn away as the growing root pushes through the soil but are replaced by new cells added to the inner surface of the root cap through divisions in the apical meristem. At the other side of the meristem, new cells become elongated in a region behind the area of active division. Just in back of this region of elongation, single-cell extensions called root hairs grow out into the soil, and differentiation of the root tissues occurs.

The cells of the root cap control the geotropic response of the root. Removal of the root cap causes a root kept in a horizontal position to grow straight rather than to curve downward. Apparently, the cells of the root cap control the distribution of substances that affect the growth rates of the other root cells (Juniper, *et al.*, 1966).

The hormones that stimulate stem elongation have only a slight elongation effect—or in some cases an inhibiting effect—on root growth. Cell division and elongation occur chiefly in the region relatively near the root apex. The process of root elongation is apparently subject to a less intricate pattern of hormonal and environmental control than is the process of stem elongation.

In the region just below the zone of elongation, extensive differentiation of tissues takes place. Within the epidermal layer of some species, certain cells become more densely cytoplasmic than others, and these *trichoblasts* develop into the root hairs (Figure 18.11). The trichoblasts are

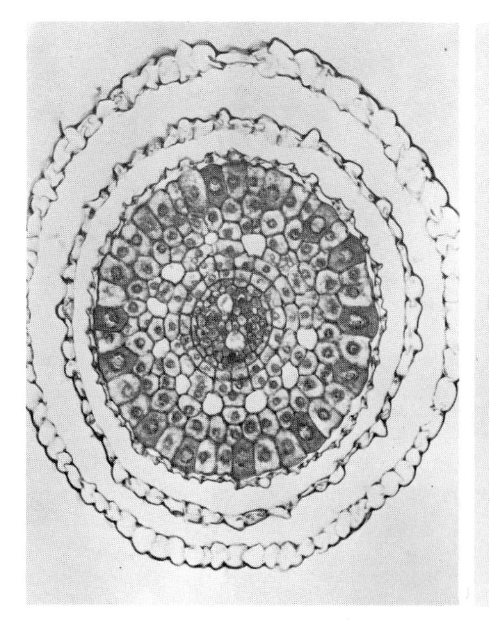

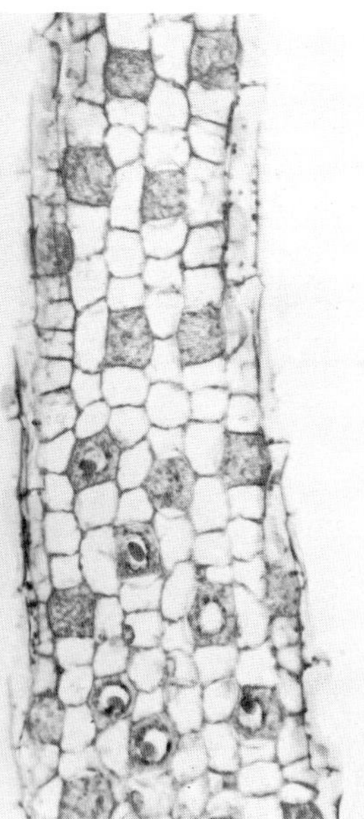

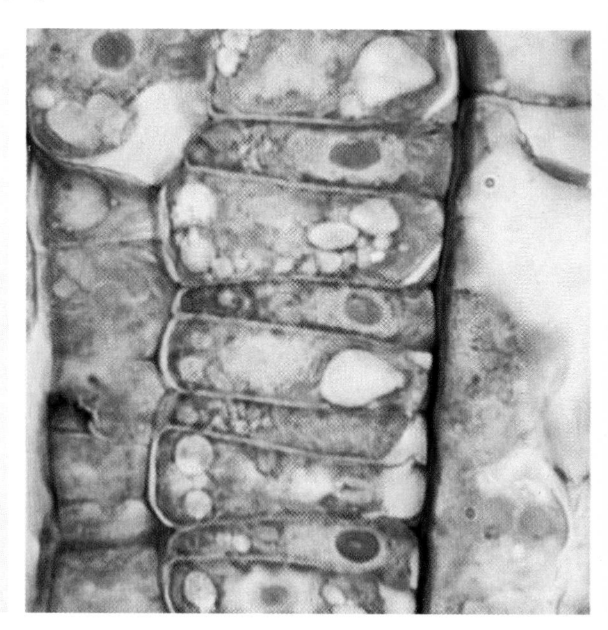

Figure 18.12. The differentiation regions of the root.

characterized by a greater concentration of various enzymes from an early stage of development (Avers, 1958, 1961). Various complex patterns of epidermal differentiation have been observed in several plant species.

As in the shoot, differentiation of ground tissue and a central core of procambium occurs near the apical meristem. Slightly farther back from the tip of the root, the procambium differentiates to form alternating xylem and phloem strands (Figure 18.12). The pattern of the vascular tissues, including the number of strands of xylem, is one of the most characteristic features of the roots of a particular plant species. This pattern is apparently influenced both by the diameter of the root and by concentrations of various hormones.

Cells taken from the outer phloem tissue of adult carrot roots have been cultured in a medium containing coconut milk (Steward, *et al.*, 1958).

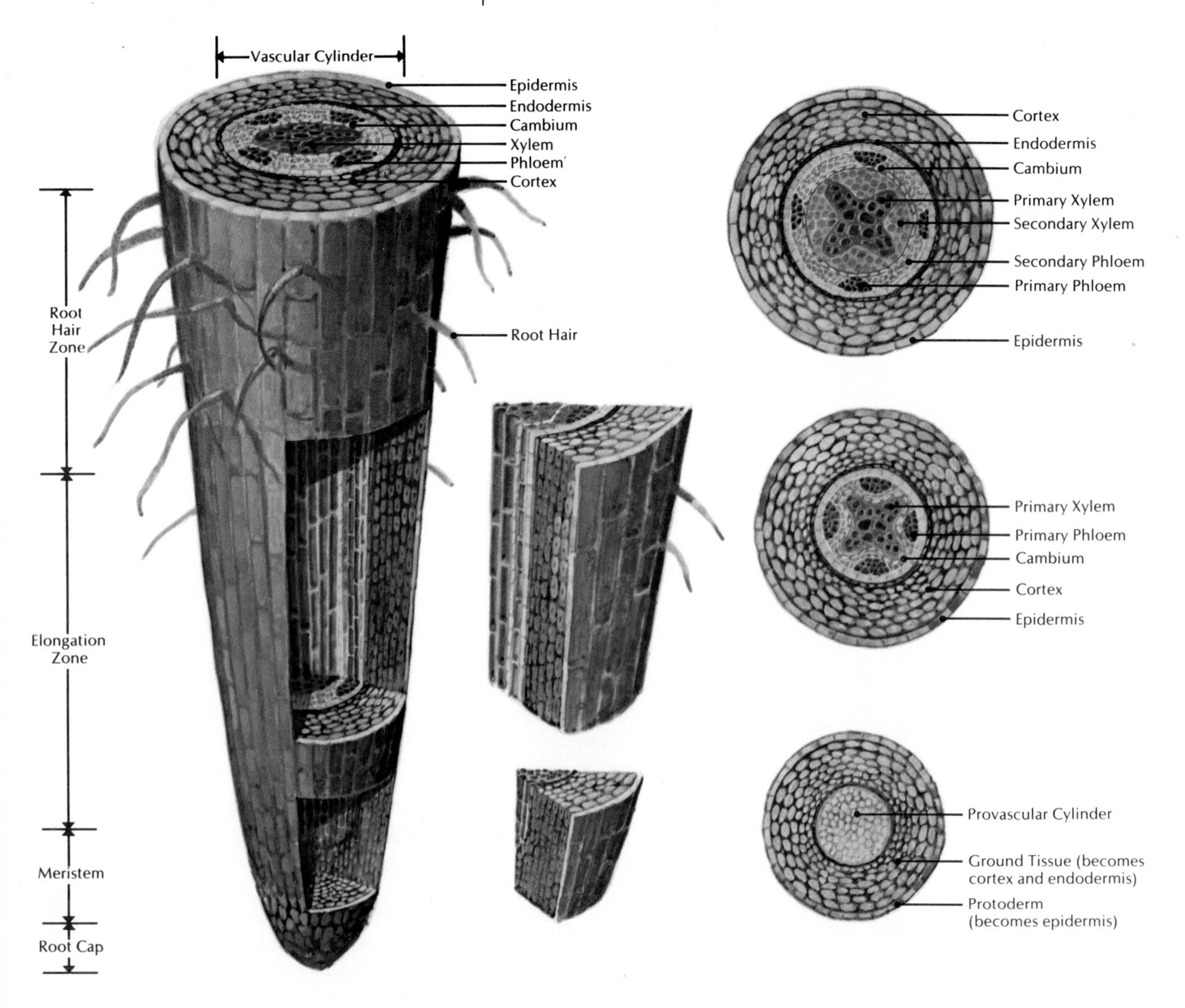

These cells grow, divide, and eventually form structures with small roots. If the structures are then transferred from the liquid medium to a solid medium, they form shoots and eventually whole plants.

This highly significant experiment shows that a single differentiated cell from an adult plant has the capacity to produce through morphogenesis, differentiation, and growth all the cells in an adult plant. Thus, if a fully differentiated cell is isolated from its neighbors and supplied with the appropriate chemical environment, the orderly sequences of selective gene expression necessary for embryogenesis can be reinitiated and maintained. Another important conclusion is that the establishment of the differentiated state in a cell is not an irreversible event.

Lateral organs or branches of the root are not formed at the apex but at some distance back from the root tip. In most flowering plants, the lateral root primordia originate by cell divisions in the pericycle—the outermost layer of the central cylinder. The positions of lateral root formation are related to the positions of vascular tissue in the center of the root. After division of the pericycle, the new cells become organized into a primordium that resembles the main root apex. This structure grows through the cortex to the exterior. The primordium secretes enzymes that dissolve the cortex cells as it makes its way toward the root surface. Relatively little research has been done on the interactions among root primordia, but there is some evidence of interactions similar to those observed in the formation of leaf primordia on the stem.

Major Features of Plant Development

The development of higher plants is characterized by the maintenance of regions of growth and development throughout the life of the organism. Stems and roots are extended, and new organs are initiated as a result of cell division, elongation, and differentiation in the growth regions. Growth and development are regulated by hormonal interactions, some of which are responses to environmental conditions and others of which are responses to development elsewhere in the organism. Although the plant is relatively limited in its ability to respond rapidly to changes in external conditions, it is extremely flexible in its ability to shed old organs or grow new ones as conditions change.

STUDY OF ANIMAL DEVELOPMENT

Animal developmental biologists choose an organism for study because of its convenience and suitability for observation or experimentation on a particular part of the developmental process. As a result, a great deal has been learned about the development of animals such as the sea urchin, frog, fruitfly, bird, and mouse. For example, *oogenesis* (egg formation) is studied more conveniently in frogs, which continue to produce oocytes throughout their reproductive lives, than in mammals, which produce all their oocytes before birth. Fertilization and early development are conveniently studied in the sea urchin, whose eggs can be readily fertilized in a dish of sea water, whereas fertilization in mammals normally occurs within the fallopian tube, where observation is exceedingly difficult. Also, the study of embryonic development is far easier with a frog or an insect, which completes its embryonic life in a few days or less, than with a mammal, which takes months to develop. Although the understanding of mammalian and human development is not complete, enough information is available to make it

Figure 18.13. Successive stages in the embryological development of the frog. From left to right, sequences depicted include: 1 – 6, early cleavage; 7 – 9, blastula; 10 – 11, gastrula; 12 – 15, the formation of the neural tube. (*Courtesy Carolina Biological Supply Company*)

1	2	3	4	5
6	7	8	9	10
11	12	13	14	15

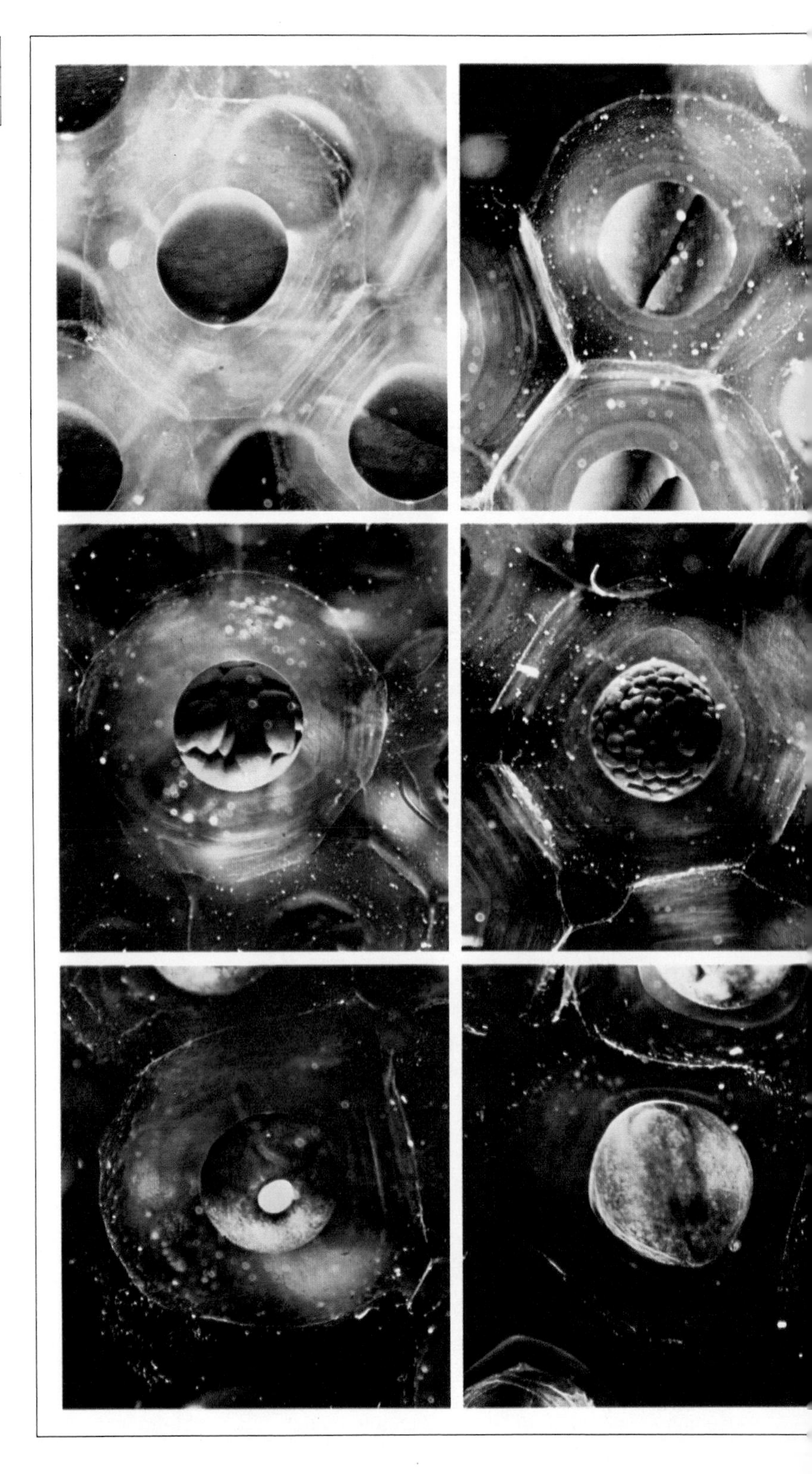

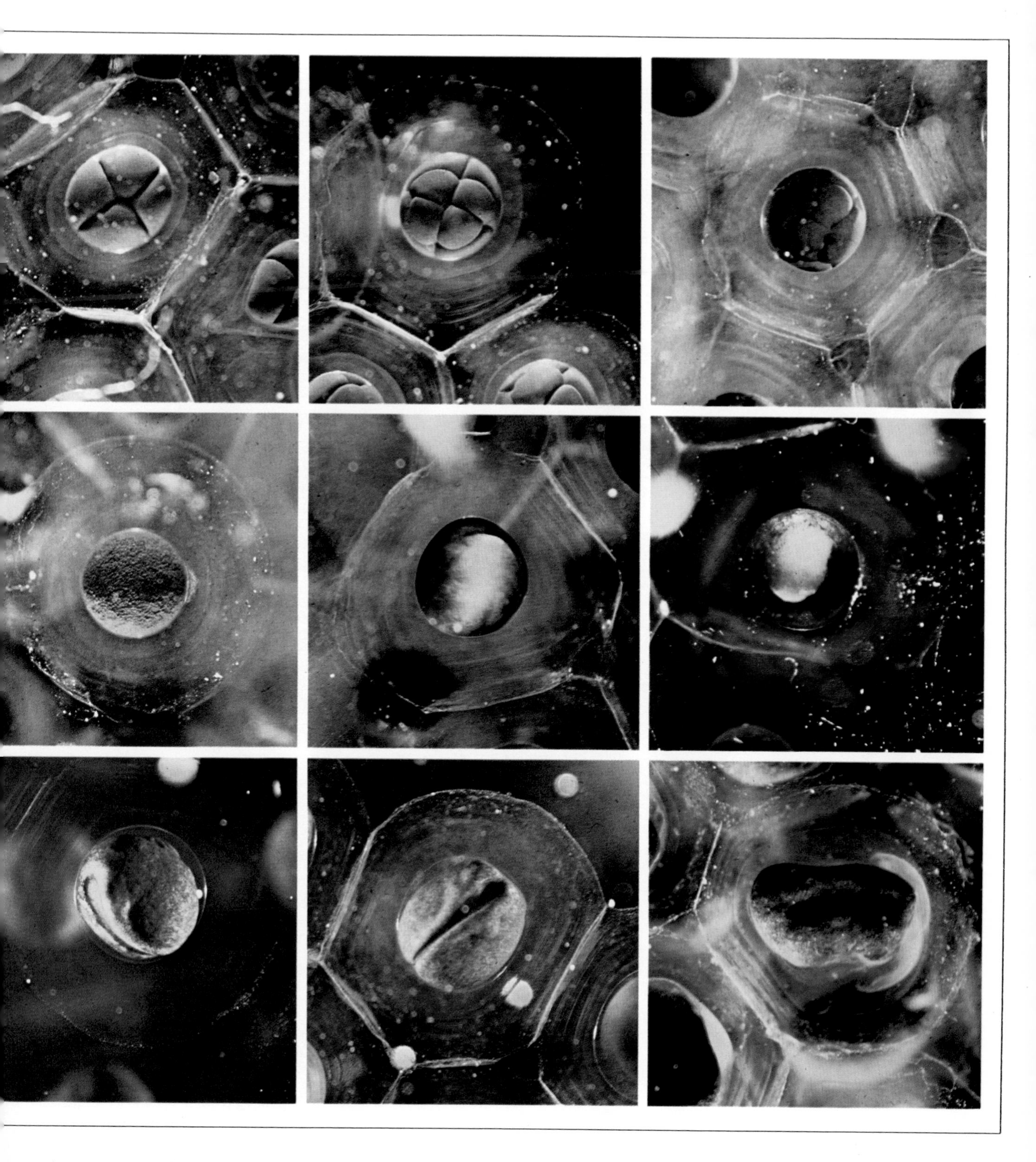

Figure 18.14. Successive stages in the embryological development of the chick. From left to right, sequences depicted include: 1 – 3, primitive streak; 4 – 6, neural tube and somite development; 7, beginning of cephalization, 8 – 11, increased cephalization.

1	2	3	4
5	6	7	8
9	10	11	

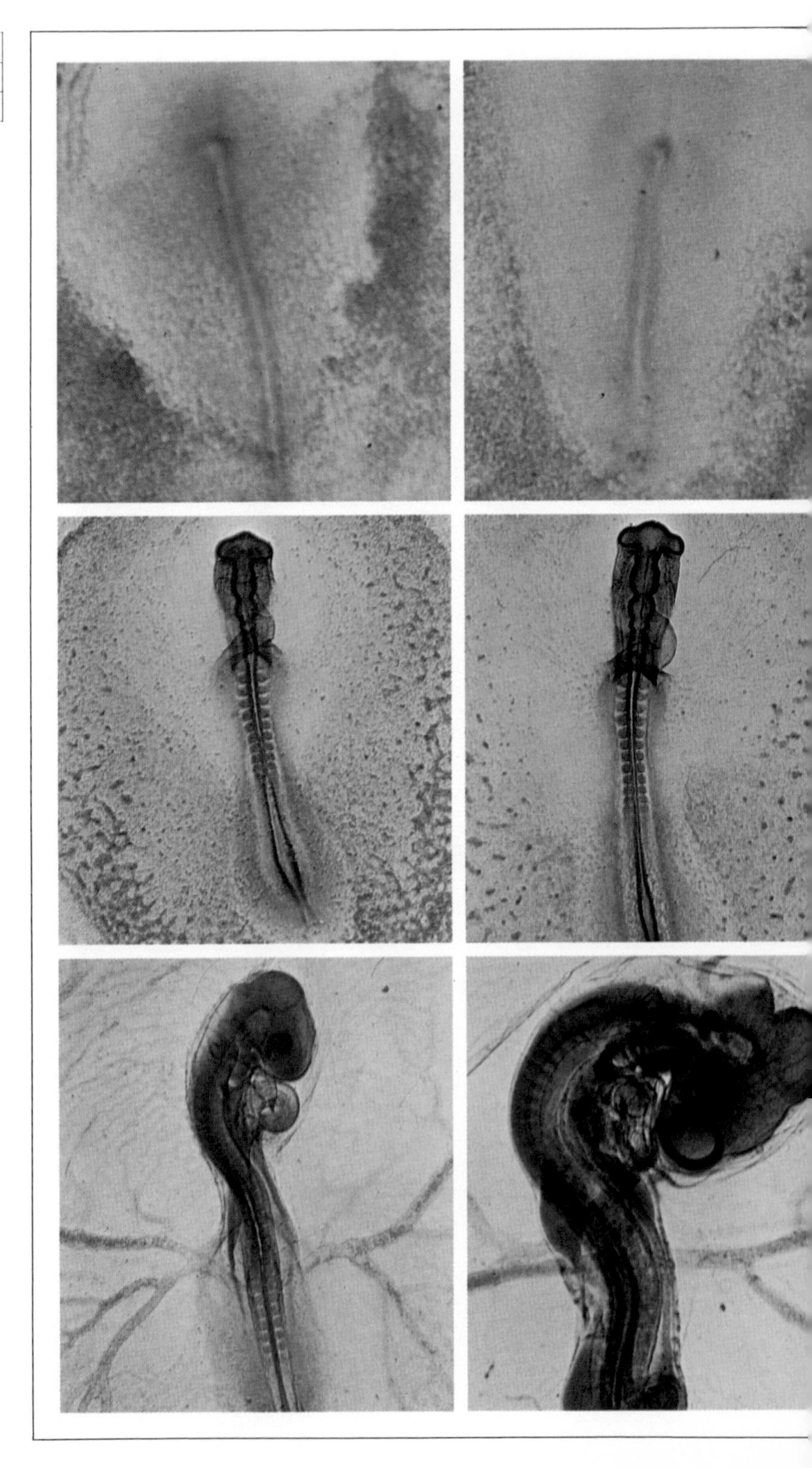

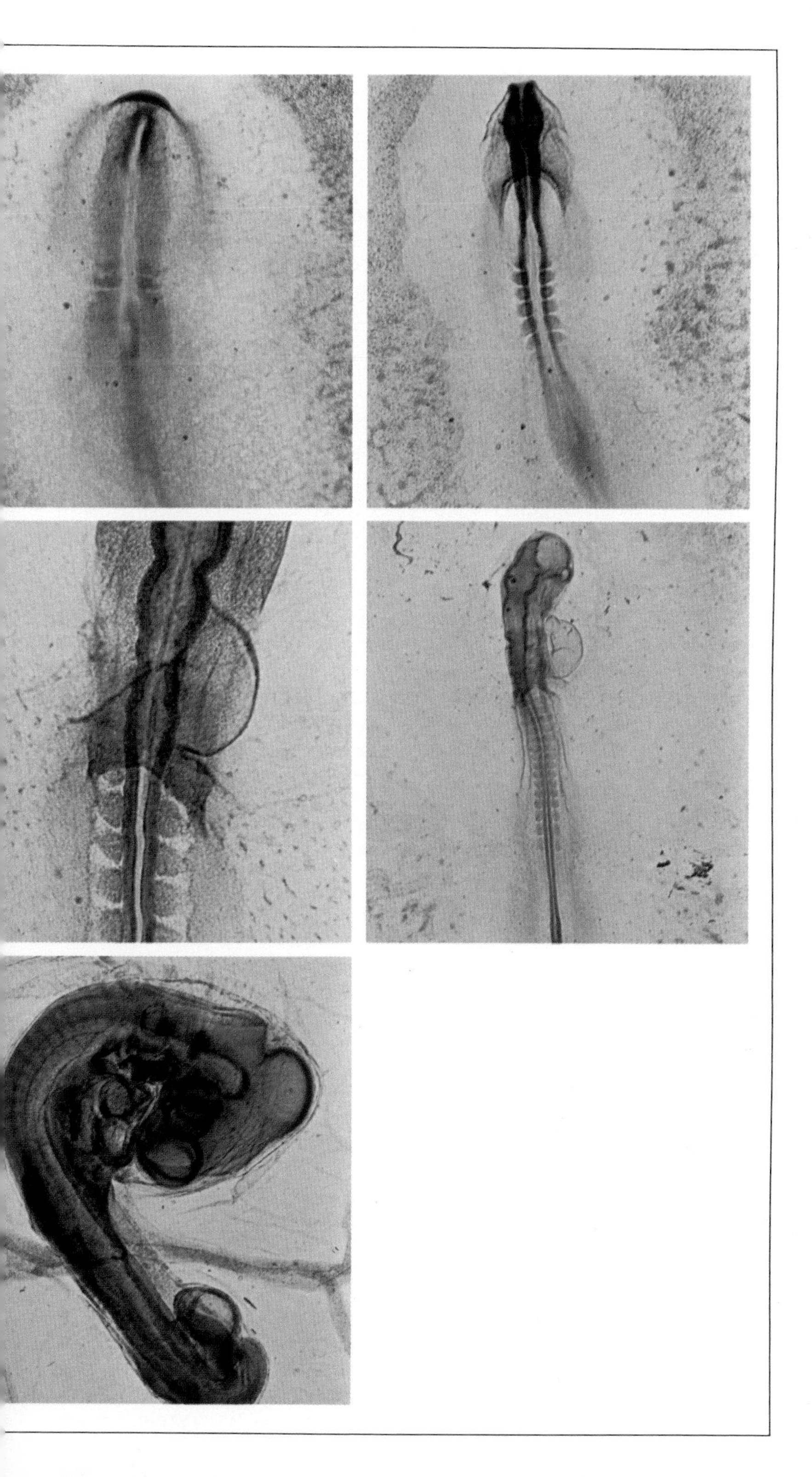

Figure 18.15. Portion of a lampbrush chromosome isolated from a newt oocyte. Note the loops of DNA.

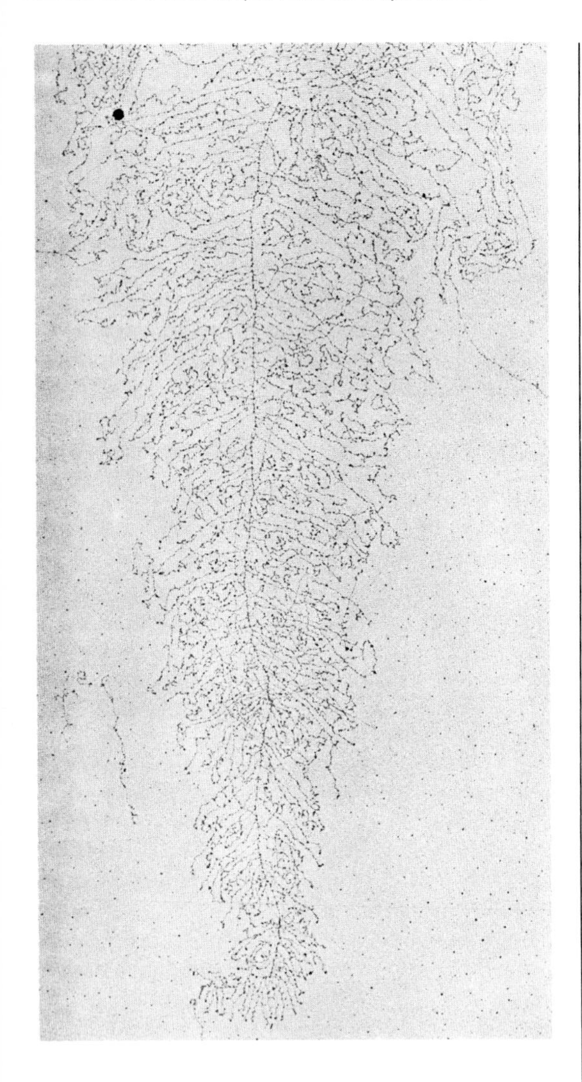

clear that many of the same basic principles involved in the development of lower animals are also valid for mammals.

After many years of observation and experimentation, developmental biologists have described many phenomena of animal development. However, these descriptions—although often quite complex and fascinating—seldom explain the mechanisms involved. Only when the molecular basis of development is thoroughly understood will meaningful control of development become possible. As in plant development, a few of the basic control mechanisms are known, but the unsolved problems remain numerous and extremely important. It is perhaps not incorrect to say that studies by developmental biologists have progressed just far enough for them to begin to discover what they do not know.

Oogenesis

The development of the oocyte plays a crucial role in the determination of the future embryonic development. All eggs contain substances that influence the course of later development (Davidson, 1969). These substances include RNA and a group of molecules called *cytoplasmic determinants*, which are of unknown composition but have a profound influence on development.

The maturation of the frog oocyte takes several months, and for part of this time the oocyte nucleus actively synthesizes various types of RNA. Nearly all the RNA synthesized during this period is ribosomal RNA (rRNA). The great demand for the synthesis of rRNA is enhanced in the oocyte by gene amplification—that is, the genes for two of the three types of rRNA are replicated hundreds of times and exist free in the nucleus as short strands of DNA (Brown and Dawid, 1968). These replicated genes form the cores of many extra nucleoli, in which rRNA is transcribed. Sufficient ribosomes are produced during oogenesis to supply the needs of the embryo through its development up to the time of hatching as a tadpole (Brown and Gurdon, 1964).

In the frog oocyte—and in the oocytes of many other kinds of animals—the chromosomes take on an unusual appearance during oogenesis. The homologous chromosomes have synapsed (paired up) at the beginning of the first meiotic prophase, but further steps in meioses have been suspended. The chromosome strands are thrown into a series of loops of DNA, giving the whole structure a "lampbrush" appearance (Figure 18.15). This lampbrush phase coincides with a period of active messenger RNA (mRNA) transcription, during which about 6 percent of the total genes in the chromosomes are being transcribed. Only about 0.14 percent of the total being transcribed is unique, nonrepetitive DNA (Davidson and Hough, 1969). It is estimated that this relatively small percentage actually represents about 10,000 different genes being transcribed.

The existence of cytoplasmic determinants has been recognized for half a century. When portions of the cytoplasm from a fertilized egg are removed, certain reproducible defects in the developed organism are produced. For example, if portions of the vegetal pole cytoplasm—presumably RNA molecules—are removed or destroyed by ultraviolet irradiation in the frog egg, the resulting frog is normal but sterile (Smith, 1966). Fertility can be reinstated in this case by injecting a small amount of cytoplasm from the vegetal pole region of a normal egg into the irradiated egg. The cytoplasmic substance destroyed by the ultraviolet light suggests that these molecules are RNA, but its identity has not yet been firmly established. No other cytoplasmic factors are capable of producing the germ cells when these cytoplasmic determinants are removed and, in these eggs, at no time in the subsequent development of the animal are the genes for germ cell formation turned on. Although little is known of the chemistry or mode of action of such cytoplasmic determinants, research is being directed toward this problem.

Induction

Cell-cell interactions clearly are important in developmental processes. In the frog embryo, for example, after the major cell movements of gastrulation, cells in a certain region of the embryo begin to fold in and to form the various elements of the nervous system. These cells differentiate into nervous structures only after they make contact with cells that lie directly beneath them, cells that are forming part of the embryonic backbone (Ebert and Sussex, 1970). This process, in which one group of cells influences the development of a second group, is known as *embryonic induction*. Induction takes place even if contact between the two tissues is prevented by a porous filter. Apparently, the induction is caused by a chemical substance passing from the inducing to the induced cells.

The nature of the inducing influence and the mechanism by which it triggers a specific differentiation is unknown. However, once the potentialities of a cell have been restricted to certain differentiations—either by cytoplasmic factors inherited from the egg or by the action of inducers—this specification persists long after the inducing stimulus or conditions have disappeared, and it is passed on by the cell to its descendants. The cells of the embryonic backbone (notochord) in the frog are capable of inducing differentiation of neural structures only during a brief period of their development. During this time, they can even induce development of neural structures from cells that would ordinarily form skin (Ebert and Sussex, 1970). After the period of induction, the affected cells and their descendants remain differentiated and produce the various elements of the nervous system, even though the notochord cells degenerate. The change in the induced cells is so permanent that they can be removed from the embryo and grown in a culture medium, where the induced cells and their descendants continue to form nervous tissue.

Determination

As indicated earlier in this chapter, a cell that has entered upon a particular course of differentiation will pass that tendency on to its descendants; this cell is said to have undergone determination. It may be said to have *firm biases*—a determined cell gives rise to a clone of cells with similar determination (Schneiderman, 1969, and Postlethwait and Schneiderman, 1970). Some reproducing populations of cells with firm biases persist in the adult organism—for example, the cells that form skin, sperm, red blood corpuscles, and so forth.

Some spectacular examples of the stability of firm biases are found among insects. In larvae of fruitflies, tiny nests of cells derived from the embryo give rise to particular adult organs such as legs and genitalia. These nests—known as *imaginal discs*—differentiate into specific adult structures; a leg disc forms a leg, a genital disc forms genitalia, and so on. Fragments of imaginal discs have been cultured in the abdomens of adult flies for more than 1,000 cell generations, and in most cases the resulting cells retain the capacity to form normal adult structures (Hadorn, 1968). Although the program for organ differentiation is normally passed from parent to daughter cells through many generations, such a program may occasionally change spontaneously. For example, cells derived from a leg disc sometimes differentiate into wing structures. When such a change in determination occurs, it usually occurs in a small group of cells at the same time. Each of these cells then transmits the new program to its descendants.

Inheritable stability of determination and differentiation is essential for the long-term coordination of a multicellular organism. However, the ability to adapt to changing circumstances is of equal importance. The control systems that provide the organism with flexible responses have been called *transitory biases* (Schneiderman, 1969). Like firm biases, transitory biases result in specific differentiations. Transitory biases persist only as long as the inducing conditions persist and are not passed on to descendant cells after the inducing conditions disappear. Transitory biases include most hormonal effects in animals and plants, enzymatic induction processes in bacteria, and similar processes. For example, the cells of the thyroid secrete thyroid hormone only when circulating thyrotropic hormone (produced by the pituitary gland) is present. Similarly, the gonads and necessary sex organs remain functional and differentiated only as long as gonadotropin hormones (also secreted by the pituitary gland) are present in their vicinity (Turner, 1966).

Another example of a transitory bias is the action of the juvenile hormone of insects. In insects such as the *Cecropia* silkmoth, the type of cuticle secreted by the epidermal cells varies with the amount of juvenile hormone in the blood (Schneiderman, 1969). When juvenile hormone is present, the epidermal cells secrete a juvenile or larval cuticle. When juvenile hormone decreases in amount, the epidermal cells cease to make larval cuticle and instead make pupal cuticle. When the level of juvenile hormone drops still further, the epidermal cells secrete an adult cuticle. Juvenile hormone apparently imposes a transitory bias on the epidermal cells, causing them to differentiate in a particular fashion. This bias persists only as long as the hormone is present.

Examples of transitory biases are found in both microorganisms and multicellular organisms. However, firm biases are almost exclusively limited to multicellular organisms and some complex acellular, eucaryotic organisms (such as ciliates). A bacterial cell is finely tuned to its changing environment. Within moments after a change in the environment, the rates and the types of synthetic activities are altered to match the new conditions. Life for a cell of a multicellular organism is very different. Such a cell is always influenced by its past history. What it can do depends on the genetic information that it possesses, on what it has done and where it has been in the past, and upon its present position in relation to its neighbors.

In a multicellular organism, the loss of one bias and the assumption of a different one—whether firm or transitory—appears to require replication of the cell's DNA. As larval epidermal cells transform to pupal epidermal cells in an insect, DNA synthesis takes place. Adult epidermal cells secrete larval cuticle in the presence of juvenile hormone, but this transformation is accompanied by DNA replication. If DNA synthesis is prevented, the epidermal cells are unable to undertake a new synthetic program even

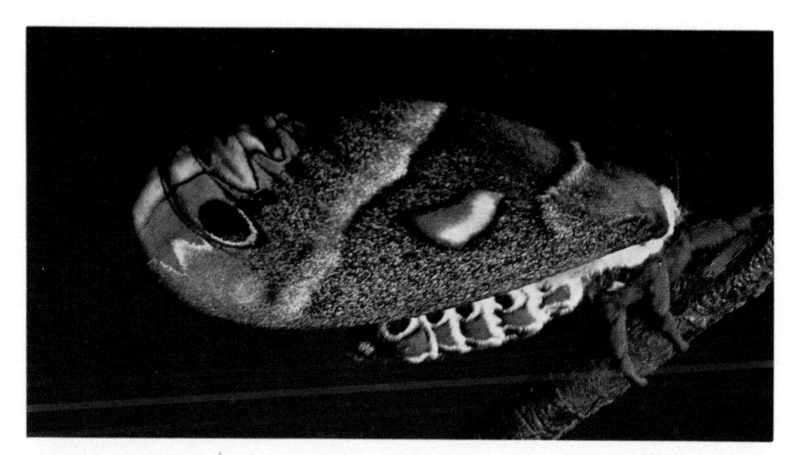

Figure 18.16. The *Cecropia* silkmoth.

though the level of juvenile hormone has changed (Schneiderman, 1969). These findings are consistent with the hypothesis that the reprogramming of cells in multicellular organisms requires some sort of "gene cleaning" that can occur only during DNA replication.

Regulation and Regeneration

Systems that can regulate for, or regenerate, lost parts represent an outstanding case of the loss of stability of cellular determination and differentiation for adaptability. After first or second cleavage in sea urchin embryos, one blastomere can regulate for the loss of the other blastomeres and can produce a complete though small embryo. Similarly, the frog egg can regulate for the loss of almost any part of its cytoplasm.

Regulation also occurs when there are too many rather than too few cells. For example, embryos of a pigmented and of an albino mouse can be mixed together. The mixture of cells re-forms into a single blastula, which can be transplanted into the uterus of a host mother mouse, where it survives and completes its development (Mintz, 1967). The resulting newborn mouse does not have duplicate structures, although its tissues contain cells derived from each of the embryos. Apparently, the early embryo can regulate for the presence of an excess number of cells.

Before an animal has reached the end of its period of embryonic development, most of its tissues take on their mature, differentiated functions. Most of the cells and their descendants will maintain their differentiated conditions within the tissues indefinitely, unless altered by metamorphosis or abnormal circumstances such as disease, uncontrolled cancerous growth, nervous system degeneration, endocrine gland malfunction, and so on. If the equilibrium condition is upset by a wound or amputated appendage, for example, the tissues may respond to restore the balance. A mouse cannot regrow an amputated limb or digit, but the wound caused by amputation will heal. Epidermal cells from the margin of the wound migrate until the wound is covered. In this response to injury, the mouse tissues retain their differentiated characteristics as epidermis, bone, muscle, or connective tissue, although some cells may modify their behavior in response to wounding.

In contrast to the mouse—which responds to amputation by healing the wound—a newt responds by making a new limb in a process called *regeneration*. After a newt's leg is amputated, the wound is covered, as it is in a

Figure 18.17. Regeneration in the starfish.

mouse, by migration of epidermal cells that remain differentiated. However, the other tissues in the vicinity of the wound lose their characteristic differentiated features. Muscle, bone, and connective tissue cells contribute to the formation of a mound of "dedifferentiated cells" beneath the wound epidermis. In this accumulation of cells, the various cells appear identical and cannot be distinguished from each other. Even the electron microscope reveals no distinctions among these cells (Bryant, 1970). These apparently dedifferentiated cells increase in number both by further additions from surrounding tissue and by cell division. The accumulated cells then differentiate and form a new limb to replace the one that was lost.

Regeneration apparently depends upon the ability of the cells first to become dedifferentiated, then to proliferate, and finally to redifferentiate and re-form the limb. There is some evidence that the presence of nervous tissue is in some way necessary for the processes of dedifferentiation and proliferation. Regeneration can be induced in the normally nonregenerating limbs of frogs, lizards, and opossums by increasing the proportion of nervous tissue in the stump (Bryant, 1970).

Nuclear Transplantation and Cell Hybridization

In recent years, the technique of nuclear transplantation has provided information about the mechanisms of cellular determination and differentiation. If the nucleus of a fully differentiated adult frog cell, such as a skin cell or liver cell, is taken out of its own cytoplasm and placed into the egg cytoplasm, it begins behaving as a zygote nucleus and participates in the development of a normal frog. Such experiments, though not as dramatic as the experiments on plant cells, show that the nuclear changes accompanying determination and differentiation in animal cells are not irreversible. Another significant conclusion is that the cytoplasmic environment in which the nucleus resides plays a major role in determining which genes are to be expressed.

Similar findings have come from experiments on somatic cell hybridization. Cells as different from one another as erythrocytes (nucleated red blood cells), fibroblasts, pigment cells, and oocytes can be combined so that there are two nuclei within a fused cytoplasm. This type of "mating" in tissue culture can be performed with cells from organisms as different as chicks and humans. Under normal conditions, the nucleus of the hen erythrocyte does not synthesize RNA or DNA. However, when such a nucleus is transplanted into the cytoplasm of a human cell, it resumes both types of synthesis. These experiments, like the nuclear transplantation experiments in frogs, indicate that differentiation does not cause irreversible changes in the nucleus and that the nucleus is strongly influenced by molecules in the cytoplasm.

The Importance of Position

What clues enable a cell to "know" where it is within the developing organism so that it will differentiate in the fashion suitable to its location? For instance, in the development of chick limb buds, the initial group of limb bud cells may become either hind limbs or wings, depending on their location in the embryo. A limb bud area transplanted early in development will differentiate in a fashion appropriate to its new location. After a certain period of development, hind limb buds transplanted to the wing region continue to differentiate as hind limbs. It has been suggested that cells "estimate" their position within a group by the use of particular markers or reference points (Wolpert, 1969). After determination has occurred, altered positional influences can no longer affect the differentiation of the cells.

The importance of positional influence is illustrated in certain fruitfly mutations that cause various parts of the antenna to transform into leg structures. When such transformations occur, they are extremely precise in terms of the kind of leg structures that arise from particular regions of the antenna. For example, a transformation of the very end of the antenna leads to the production of the end portion of a leg with a claw. A transformation of a central part of the antenna leads to the production of the corresponding central part of a leg. That is, the nature of the leg structure produced by a group of transformed antenna cells depends upon the position of these cells within the antenna. The leg cells respond appropriately to the same set of influences that provide positional information for the antenna cells (Postlethwait and Schneiderman, 1970).

Little is known about the mechanisms through which positional information is specified—the landmarks that cells may use to locate themselves. Among the hypotheses being considered are responses to particular chemical substances produced by neighboring cells or responses to varying gradients of the concentration of such substances across the cell (Wolpert, 1970).

Cellular Communication

There is evidence that adjacent cells can communicate directly with each other, at least during certain critical periods of development. As a fertilized egg becomes partitioned into an increasing number of blastomeres during cleavage, the individual cells continue to remain electrically coupled (Ito and Loewenstein, 1969). If microelectrodes are implanted in adjacent cells, small ions will carry electrical charges directly from one cell to another without passing through the external medium. Such intercellular communication during development might aid in determining a cell's position within a group.

In some well-documented instances, communication between cells involves far more than electrical coupling and ion movements. During oogenesis in some insects, the egg cell has open channels of cytoplasmic communication with surrounding nurse cells (Koch, *et al.*, 1967). Electron micrographs show many organelles in the communication channels, suggesting that free exchange of ribosomes, centrioles, mitochondria, and other cellular constituents may take place. In these insects, RNA is synthesized in the nurse cells and passed into the oocyte, which apparently does not synthesize RNA of its own (Bier, 1963). This transfer of RNA was discovered through autoradiographic studies in which radioactive RNA precursors supplied to the cells were first incorporated in the nurse cells, then moved into the cytoplasmic channels, and finally moved into the cytoplasm of the oocyte.

Shaping Up

As discussed earlier in the chapter, plant morphogenesis is accomplished largely by cell division along particular planes or cell elongation along particular axes. In animals—whose cells lack rigid walls and are more mobile—cell movements play a much more important role, but there are some examples of animal morphogenesis involving oriented cell divisions. The formation of the hollow sphere, or blastula, in sea urchin embryos is accomplished by restriction of the planes of division to radial axes, so that daughter cells are added to the surface of the sphere rather than being pushed to the interior or the exterior (Gustafson and Wolpert, 1967).

Another example is found in the morphogenesis of insects. Clonally derived patches of cells—that is, cells originating from the same ancestral cell—appear as stripes along the length of the legs (Bryant and Schneiderman, 1969), the wings, and the antennae of adult flies. Such patches would be formed if the planes of cell division were consistently oriented at right angles to the long axis of the appendage, forming an elongated line of daughter cells. Patches of similar shape have been shown in other insects—such as bees and butterflies—that undergo complete metamorphosis during development. It has been suggested that oriented cell division is a major factor in morphogenesis of insects (Postlethwait and Schneiderman, 1970).

On the other hand, there is no evidence that oriented cell divisions are responsible for the elongated shape of a chick limb. Differential rates of cell division in different parts of the limb bud may play a part in shaping the limb (Hornbruch and Wolpert, 1970). After the various regions such as upper leg, lower leg, and foot have become determined in the limb, each assumes a different rate of cell division, ultimately producing a properly proportioned limb. Although differential division rates can account for the greater length of the foreleg region as compared to the foot region, it cannot account for the detailed contours of the limb.

Another mechanism involved in animal morphogenesis is controlled cell death. For example, the foot region of a chick's hind limb first develops as a flattened, palette-shape structure. The individual toes become separated from one another by the death of intervening cells (Saunders and Fallon, 1966). A similar mechanism has been discovered in the development of the claws on fly feet (Whitten, 1969). In such cases, individual cells are sacrificed for the good of the entire organism, emphasizing that the multicellular organism—not the individual cells of which it is composed—is the basic unit of multicellular life.

Gastrulation in the sea urchin begins with reorganization of the spherical blastula into a cone shape. The base of this cone then indents into the blastocoele to form the primitive gut. It appears that this indentation is caused by a change in shape of the individual cells that remain attached to

Figure 18.18. Neural formations in a frog embryo. The A series shows very early embryos; B series shows midterm embryos; C series shows late-term embryos with the neural tube almost completely closed. The top series shows the right side of embryos; the middle series shows whole embryos in dorsal view; and the bottom series shows the anterior halves of transversely cut embryos. (See Figure 18.14.)

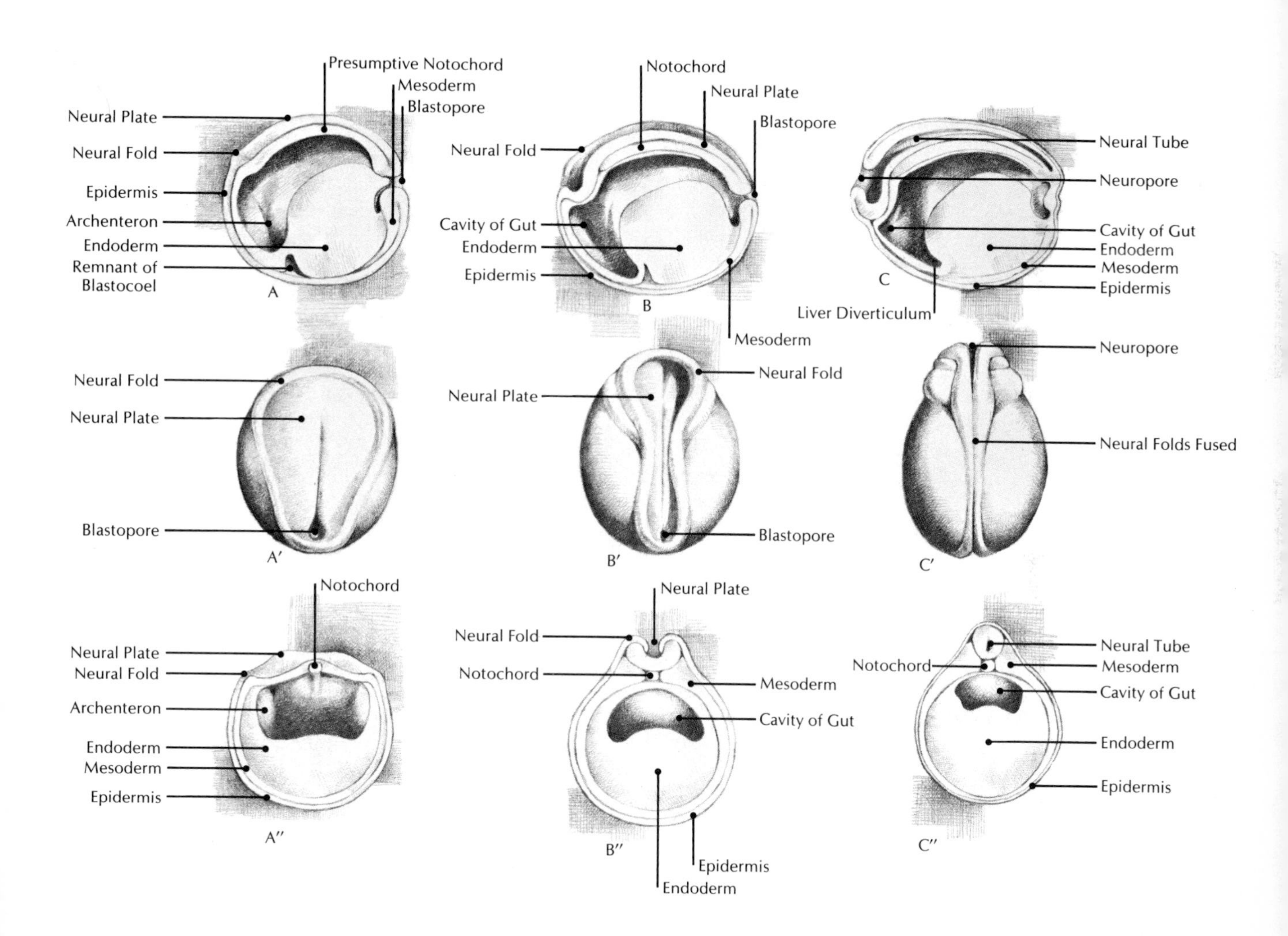

neighboring cells in the sheet. The changes in cell shape may be due to rearrangements of the internal skeletal elements of the cells such as microfilaments. It has also been suggested that the changes in cell shapes result from changes in the degree of adhesion that cells have with their neighbors (Gustafson and Wolpert, 1967). Whatever the mechanism, many important morphogenetic events depend upon changes in cell shape and the resulting movements of sheets of cells.

Examples of individual cells moving to create shapes are less common, but one outstanding example is the migration of neural crest cells in vertebrate embryos (Weston, 1970). After gastrulation, a line of cells along the dorsal side of the embryo begins to divide rapidly, producing a thick *neural plate*, which folds into the embryo to form the *neural groove* (Figure 18.18). The folds along the edges of the neural groove become higher, curve inward over the groove, and eventually fuse, transforming the neural groove into the *neural tube*. This tube later forms the brain and the central canal of the spinal cord. A number of cells from the tips of the neural folds, the *neural crest cells*, migrate away from the neural tube and in new and diverse locations play a profound role in the morphogenesis of a variety of organs and tissues. For example, neural crest cells form spinal ganglia, sympathetic ganglia, head cartilage, pigment cells, and Schwann cells. These cells do not move in random directions away from the neural tube but leave in two well-defined streams – one headed dorsally toward the epidermis and the other ventrally toward the mesenchyme. The mechanism of migration and the positional clues that guide the migrating cells to their destinations in the embryo are as yet unknown.

The individual neurons of the nervous system do not actively migrate, but they do send out extremely long cytoplasmic processes, the axons, into the peripheral tissues. These axons make contact with muscles or sense organs and establish electrochemical communication with them. The factors that guide axons to appropriate target cells at great distances from the cell bodies in the spinal cord are not known. It has been suggested that the target tissues attract nerve axons by production of specific chemicals, but attempts to verify this theory by experiment have been unsuccessful so far (Hughes, 1968). In early development, the organization of the nervous system is quite plastic, for nerves that would innervate (form synaptic connections with) only a small portion of a leg can be made to innervate the whole of a grafted extra leg. Functional organization of nerve connections within the central nervous system occurs after the peripheral connections have been made and is made in a way appropriate to the peripheral connections. Thus, the sensory and motor nerves of an extra leg become coordinated into the nervous system in a way that permits the leg to move in synchrony with the adjacent normal leg (Hughes, 1968).

Further evidence of the functional plasticity of the nervous system comes from experiments in which patches of skin have been interchanged between the belly and the back of tadpoles (Jacobson and Baker, 1969). Just after metamorphosis, a stimulus to the back skin on the belly causes the frog to make wiping motions at the proper location on the belly. A short time later, however, the frog will respond to a similar stimulus by inappropriately attempting to wipe the point on the back where the skin should be located. Apparently, the nerves transmit some kind of information from the transplanted skin to the central nervous system, indicating that the skin is back skin regardless of its real location.

The cellular slime molds are of particular interest because the development of their fruiting bodies represents an extreme example of morphogenesis through cell migration. Under certain conditions such as scarcity of food, the amoeboid cells living as separate individuals migrate toward one another and form a multicellular creature with a characteristic shape. The coordination of the individual cells into a multicellular body is achieved by a chemical attractant called acrasin, which is secreted into the environment by some cells. This chemical overcomes the antagonism that keeps the individual cells apart, and aggregation is then possible (Trinkaus, 1969).

Cells from tissues of multicellular organisms exhibit repulsive tendencies similar to those of the amoeboid cells of the slime molds prior to aggregation. When pieces of embryonic tissues such as heart or cartilage are cul-

Figure 18.19. Photographs and diagrams illustrating the interchange of skin from the back to the belly of the frog.

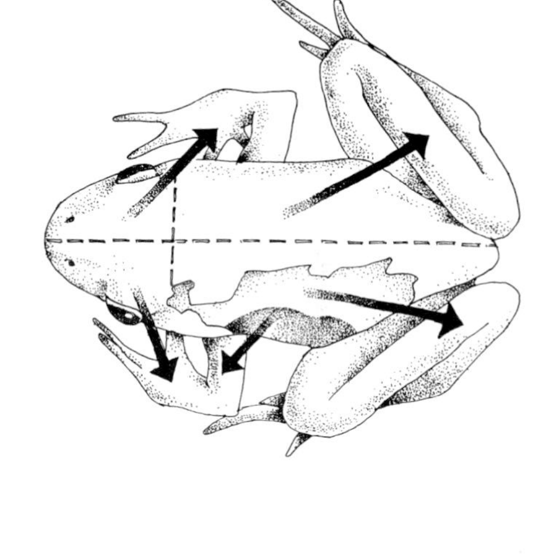

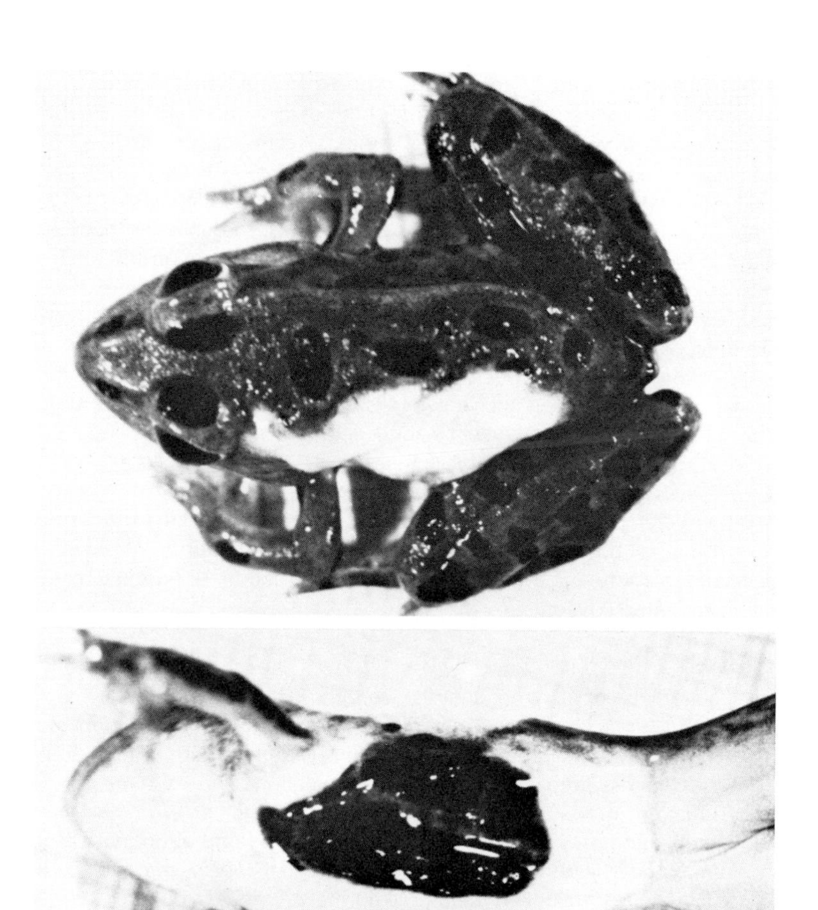

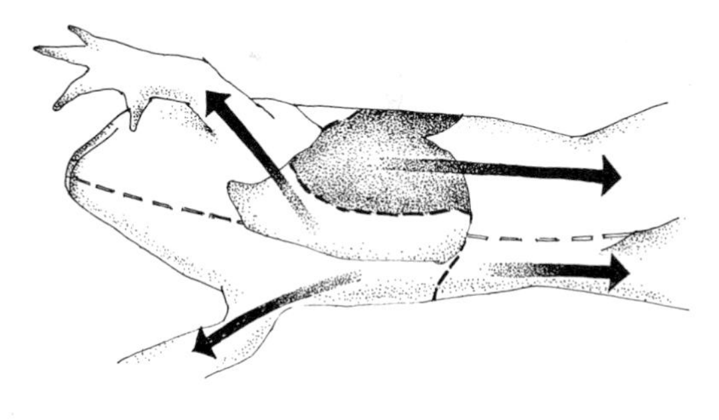

tured on a flat surface, cells leave the tissues and spread out to form a single layer over the surface. Under normal circumstances, they do not crawl over each other. In fact, time-lapse movies show that two cells closely approaching each other stop moving and, if space is available, reverse the direction of their movement so as to move away from each other. Careful examination of the movies shows that the advancing cytoplasmic front of each cell consists of a thin cytoplasmic sheet that is constantly undulating. When this so-called *ruffled membrane* contacts another cell, it stops undulating and the cell becomes temporarily paralyzed, a phenomenon called *contact inhibition*. Soon a ruffled membrane forms on some other part of the cell, and the cell moves away in that direction. Chemicals such as acrasin may suppress the wandering tendencies of cells. The coordinating function of acrasin might also be served by other mechanisms, such as intercellular communication through cell junctions or the secretion of a solid, extracellular matrix that inhibits movement of cells.

Metamorphosis and Maturation

For many animals, the end of embryonic life is the end of the most rapid phase of growth and change. However, for a large number of vertebrates

Figure 18.20. Side of wax cells cut away to expose young honeybees developing inside.

and invertebrates there is an additional phase of rapid change, known as *metamorphosis*, when a juvenile form of the organism is transformed into an adult with a very different mode of life. Organisms such as flies and frogs, which undergo complete metamorphosis prior to sexual maturity, get the best of two worlds by being adapted to one habitat as larvae and to another as adults. In insects such as flies and moths, the changes that take place at metamorphosis are so extreme that the organisms construct many specialized parts of the adult from cells that were set aside in the embryo and never became functional parts of the juvenile organism. These cells are the imaginal discs discussed earlier, which live a virtually parasitic existence in the larva of the insect. At the end of larval life, in response to hormonal changes, many of the larval cells die, whereas the imaginal discs grow to form the pupa and then the adult (Schneiderman, 1969).

In all animals, including those that do not undergo metamorphosis, growth is the most conspicuous developmental phenomenon in the early postembryonic period. This growth appears to be stimulated in vertebrates, as in plants, by hormones and continues until sexual maturity. At the time of sexual maturity, other hormones are secreted, and these promote maturation and inhibit overall growth.

Aging

The phenomena of aging, or senescence, are part of the normal development of an animal. Senescence is the gradual deterioration of function and structure at both cellular and organismic levels. One school of thought is that senescence is primarily caused by accumulation of errors in the genetic information of cells through spontaneous mutations or errors in DNA replication during cell division. Vertebrate cells develop abnormal chromosome numbers relatively rapidly in culture, and it appears that such cells are capable of only a limited number of normal divisions in culture—about 50 in human fibroblasts (Hayflick, 1968). On the other hand, insect imaginal discs proliferated in culture for more than 1,000 cell divisions still exhibit normal differentiation (Hadorn, 1965).

Another possibility is that deterioration of the coordinating systems of the body is responsible for a wide variety of other malfunctions involved in senescence. Age pigments accumulate in neurons, and these and other accumulated waste products might cause general and cumulative deterioration of the nervous system. The human circulatory system also seems particularly susceptible to deterioration with age, and its malfunction could certainly contribute to the senescence of other tissues. Various tissues and organs probably senesce at different rates, and some organized tissues could possibly remain alive indefinitely in an appropriate environment.

At present, the processes of senescence are an inevitable part of the functioning of the mature organism. A great increase in lifespan could be accomplished only by prolonging the period of active growth. In this period, new cells are produced more rapidly than old ones die.

PROBLEMS OF DEVELOPMENTAL BIOLOGY

The problems of developmental biology are many and the answers are few. Fortunately, a number of powerful experimental techniques are available that promise to provide more answers.

The technique of somatic cell hybridization not only holds promise for the elucidation of genetic control mechanisms of development but may make possible the preparation of genetic maps of human chromosomes. With refined surgical instruments such as ultraviolet microbeams and laser beams, biologists can perform fine operations on such tiny structures as chromosomes and mitochondria. By removing microscopic and submicroscopic components of cells, biologists will be better equipped to determine their roles in developmental processes. Drugs that inhibit various parts of the replication-transcription-translation processes and DNA-RNA hybridization techniques are providing new information about the control of gene action and protein synthesis in development (Gall and Pardue, 1969). Improved techniques for culturing cells and tissues and the use of radioactive isotopes as markers are leading to many discoveries about developmental mechanisms (Weston, 1967). Computer modeling of developmental processes is providing a convenient method of testing hypotheses about complex developmental interactions (Waddington and Cowe, 1969, and Ede and Law, 1969).

Because understanding of developmental processes is of major importance to medical progress, there is an active interplay between the researches of developmental biologists and the investigations of medical researchers. Many studies are aimed toward an understanding of malfunctions that cause normal cells to embark upon the uncontrolled division of cancerous tissue. Despite a great deal of interest and many important discoveries, the field of developmental biology has yet to make the major breakthrough to an understanding of major principles that has appeared imminent for the past few years. If that breakthrough is made, it will be of major importance for the general understanding of multicellular organisms and of human medicine.

FURTHER READING

For more specific information about plant development, see books by James Bonner (1966), Esau (1965), Laetsch and Cleland (1967), Leopold (1964), and Steward (1968). A great many books are available as introductions to animal development, including those by Balinsky (1970), Ebert and Sussex (1970), and Waddington (1966).

Articles useful as introductions to general or specialized topics include those by Butler and Downs (1960), Edwards (1966), Frieden (1963), Gurdon (1968), Konigsberg (1964), Moscona (1961), Overbeek (1968), Steward (1963), and Wessells and Rutter (1969).

19
Physiology

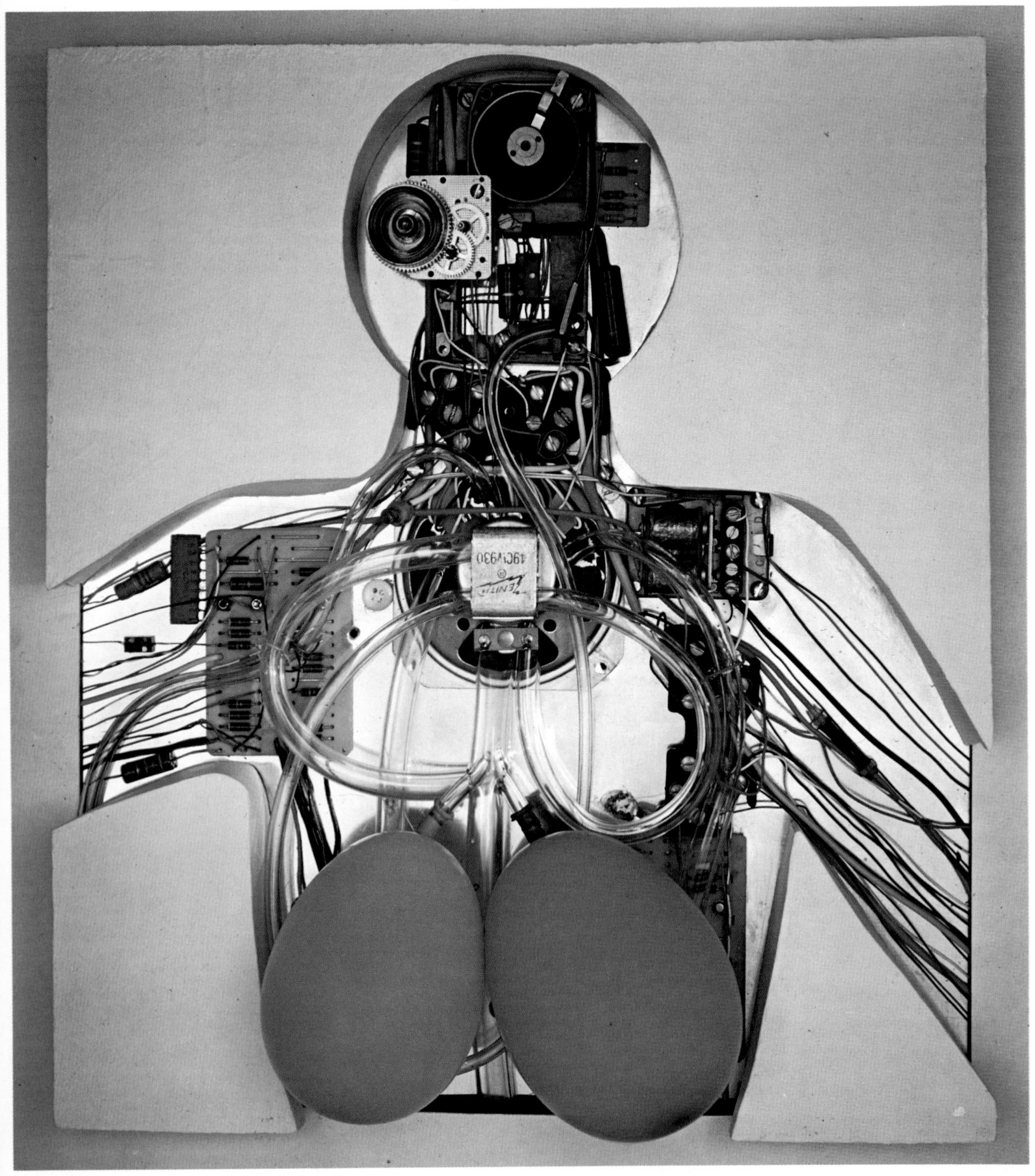

All organisms must cope with essentially the same challenges. Each organism must obtain from the environment the materials needed to build the structures of its body, to repair those structures, and to reproduce. To build and maintain an ordered structure and to carry out the physical and chemical processes of life, every organism must obtain energy. Within the organism, materials must be transported from place to place and made available where needed. Waste products of metabolism must be eliminated selectively from the body. Heat must also be selectively retained or dissipated to keep the internal environment within the rather narrow temperature limits that permit efficient biochemical reactions. Thus, every living thing—from the simplest procaryotic cell to the most complex multicellular organism—maintains *homeostasis*, or balance, by continually adjusting physiological functions for the maintenance of optimal internal conditions.

NUTRITION

The general term "nutrition" refers to the intake by an organism of the materials and energy needed to support life. Every organism must obtain carbon, oxygen, hydrogen, nitrogen, and smaller amounts of other elements in order to synthesize the macromolecules that make up its body.

Nutritional Requirements

Organisms that are able to utilize simple inorganic substances and synthesize all organic molecules needed by their bodies are called *autotrophs*. Organisms that are unable to perform all necessary syntheses and therefore must obtain some organic molecules from other organisms are called *heterotrophs*. There are two types of autotrophs. Photosynthetic autotrophs are able to trap the radiant energy of sunlight and to use this energy to synthesize carbohydrates (Chapter 5). Chemosynthetic autotrophs are bacteria that oxidize various inorganic substances such as sulfur, ammonia, nitrite, iron compounds, and hydrogen and use that energy to synthesize carbohydrates. All other organisms (heterotrophs) can use energy only after it has been trapped by autotrophs and stored in the form of chemical bonds in carbohydrates. The basic nutritional requirement for heterotrophs, then, is carbohydrate as an energy source. Most heterotrophs also require oxygen to burn carbohydrates to carbon dioxide and water—the most efficient method of using the energy in carbohydrates. In addition, most heterotrophs also must obtain certain other ready-made organic molecules. For example, rats and humans cannot synthesize 8 of the 20 amino acids that occur in proteins of all organisms. They must be obtained in the diet. Certain inorganic ions are required by all living things. The diverse but vital functions served by these ions are listed in Table 19.1.

Ingestion and Digestion

The simplest way for a cell to obtain nutrients is to depend upon diffusion to bring nutrients from the environment. The rate of diffusion depends upon the surface area of the cell and upon the difference in concentrations between outside and inside. As the size of a cell increases, the need for nutrients increases roughly in proportion to the volume of the cell (and therefore to the cube of its radius). The surface area of a simple spherical cell increases only in proportion to the square of its radius. The larger an organism is, therefore, the more complexly folded its surface must be in order to

Figure 19.1a. These free-living planarian flatworms have a single opening to the digestive tract, which serves for both ingestion and egestion. This opening is located at the midpoint of the body at the end of a muscular tubular pharynx. To feed, the animal crawls on top of the food, extends the pharynx, and pumps fluids and small bits of food into the digestive tract.

provide sufficient surface area for rapid diffusion of nutrients. Such an organism would also be limited to solutions that contain a relatively high

Table 19.1
Inorganic Ions Required by Living Things

ION*	Some Principal Functions
Na^+	Chief cation in extracellular fluids of most animals; carrier of current in action potentials in most nerves and muscles
Cl^-	Chief anion in extracellular fluids of most organisms
Mg^{++}	Important in maintaining stability of intercellular substances and cell membranes; integral part of the clorophyll molecule; required for muscle contractility; cofactor for many enzymatic reactions
Ca^{++}	Involved in stability of intercellular substances and cell membranes; component of bones and teeth; key factor in initiation of muscle contraction
K^+	Chief cation in intracellular fluids of most organisms; carrier of outward current in action potentials in nerves and muscles
Fe^{++}, Fe^{+++}	Integral part of heme, component of hemoglobin and cytochromes that transport electrons in oxidative metabolism
Co^{++}	Part of vitamin B_{12}, cyanocobalamine
Cu^{++}, Cu^{+++}	Part of hemocyanin molecule that transports oxygen in the blood of some invertebrates; cofactor in mitochondrial electron transport systems
Zn^{++}	Integral part of several enzymes
I^-	Used in amino acids of some structural proteins found in invertebrates; part of thyroxin molecule, the vertebrate thyroid hormone
Mn^{++}	Cofactor for some enzymes
PO_4^{---}	Part of ATP molecule used in energy storage and release; component of bones and teeth; component of nucleic acids
S†	Occurs in several amino acids; S—S bonds contribute to formation of tertiary structure of proteins

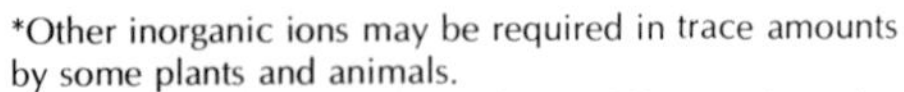

*Other inorganic ions may be required in trace amounts by some plants and animals.
†S is obtained by organisms either as SO_4^{--} or in amino acids; some bacteria can use molecular sulfur or H_2S.

concentration of dissolved nutrients; in very dilute solutions, the rate of diffusion becomes too low to supply the needs of an organism.

Because organic substances are relatively scarce in most environments, heterotrophs must often extract food particles from large volumes of the surrounding fluid or actively pursue and capture their food. Large food par-

Figure 19.1b (left). The coelenterate *Hydra* represents a polyp form that feeds upon small fresh-water zooplankton. The feeding response is elicited by the chemical and tactile stimuli of the prey, which is then captured by nematocyst batteries and pulled into the gastrovascular cavity.

Figure 19.1c (right). This marine jellyfish feeds on small crustacea and fish that it catches with tentacles equipped with nematocyst batteries.

Figure 19.2 (below). Selected features of the digestive tracts of some representative invertebrate animals. From top to bottom are an earthworm, with buccal cavity, crop, gizzard, and intestine; a crayfish, with cardiac and pyloric stomach and intestine; and a mosquito, with a large ventral diverticulum to store the blood meal.

ticles are taken into some sort of internal digestive system, where they are physically and chemically broken down to units that can be absorbed.

Unicellular organisms, using the processes of phagocytosis and pinocytosis (Chapter 10), bring large molecules or small particles of food into the cell by enfolding the food particle in a membrane pocket that is taken into the cytoplasm as a vacuole. Digestive enzymes are secreted into the vacuoles, and the small organic molecules produced by digestion are moved across the vacuole membrane into the cytoplasm by diffusion or active transport.

In the simplest heterotrophic cells, pinocytosis or phagocytosis can occur at any point on the cell surface. In more complex protozoa, a single part of the surface may be specialized for phagocytosis, and cilia are often present to move food particles toward this primitive mouth. Fungi secrete enzymes that break large organic molecules into smaller units, which are then brought into the cell by diffusion or active transport.

Coelenterates such as *Hydra*, jellyfish, or sea anemones feed on small crustacea or fish, which they catch with tentacles equipped with stinging cells (*nematocysts*) that inject a paralyzing poison into their prey. When the prey is immobilized, the tentacles bring it to the mouth, the mouth opens, and the food is stuffed into the digestive cavity. This cavity is a closed pocket lined by cells that secrete digestive enzymes and that absorb nutrients by phagocytosis and active transport. Any undigested residue is ejected back through the mouth.

In all higher animals beginning with the roundworm, the digestive tract is a continuous tube with a mouth at one end and an anus at the other. Food is progressively digested as it moves through the tube, or gut, and processing is not interrupted by each new ingestion. Movement of food through the tube is accomplished by rhythmic contractions, or *peristalsis*, of muscles in the gut walls. In more complex organisms, the length of the digestive system is increased by coiling or folding of the tube; the surface area available for absorption of small organic molecules is increased by foldings or projections of the tube lining; and the tube is differentiated into specialized passageways and cavities within which various parts of the digestive process occur.

The basic chemical process of digestion is one of hydrolysis, breaking down macromolecules to simpler organic molecules such as amino acids, glucose, glycerol, and fatty acids. These smaller molecules are then absorbed across the lining of the digestive tract. Hydrolytic enzymes are released into the digestive tract from specialized secretory cells and organs.

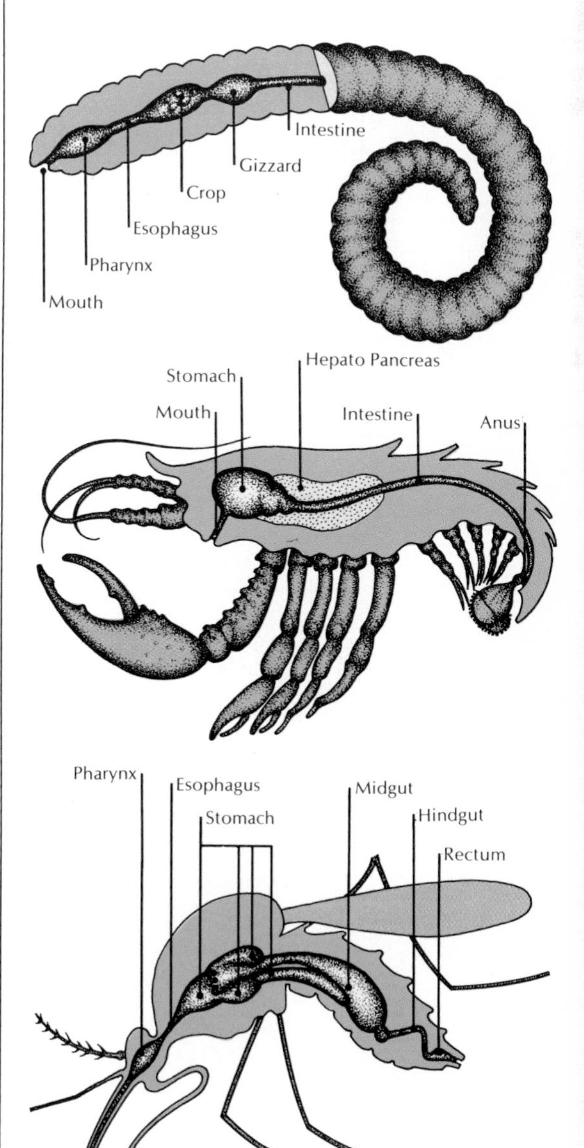

Figure 19.3 (above). Digestive tract of the cow. The cow stomach is divided into four compartments: rumen, reticulum, omasum, and abomasum. In the first two compartments, bacterial and protozoan action break down cellulose to simple sugars; the omasum functions as a water conservation device, whereas the conventional digestive activity of the stomach takes place in the abomasum.

Figure 19.4 (below). Cross section of the gastric mucosa.

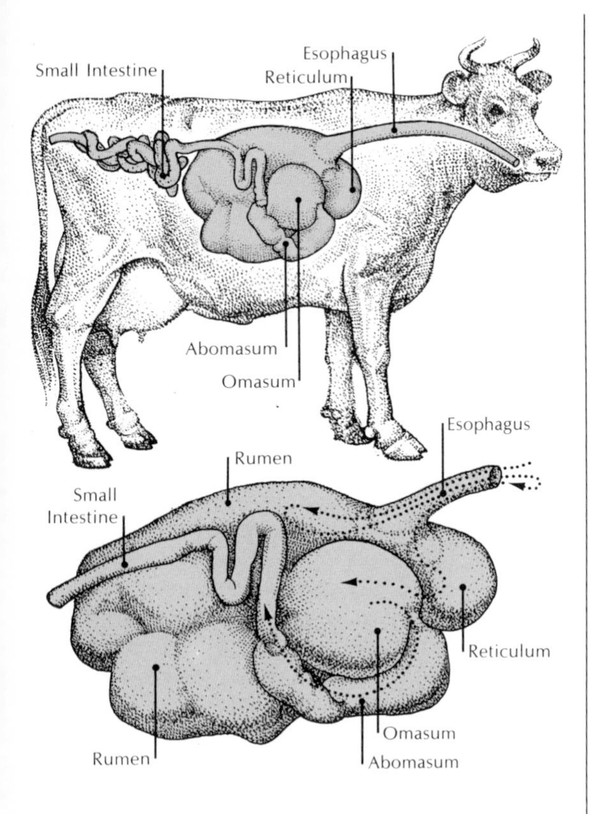

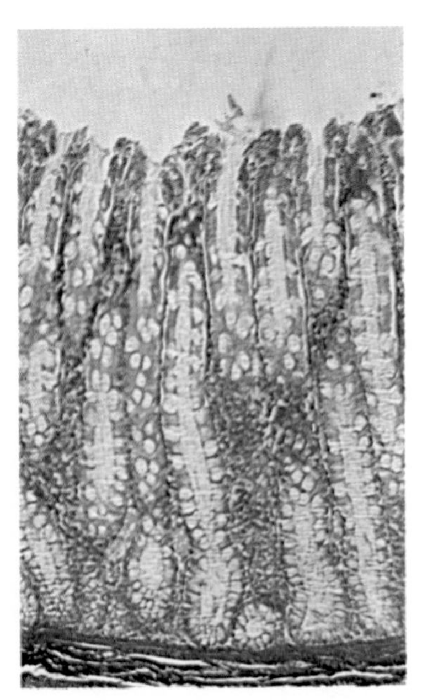

For these chemical reactions to occur, large chunks of food must first be physically broken down into small pieces, allowing access to the digestive enzymes.

The mouth region is specialized for the intake of food into the digestive tract. In many animals, specialized structures such as teeth or beaks assist in the physical disintegration of food. For animals that feed on green plants, the preliminary cutting and crushing of food is particularly important, because the tough cellulose walls of plant cells are extremely resistant to the action of digestive enzymes. Some complex animals use their muscular tongues to bring food into the mouth and to maneuver it within the mouth cavity during chewing and swallowing. In mammals the food mixture is lubricated with *saliva*, which is secreted within the mouth and mixed with the food during chewing and which contains an enzyme that hydrolyzes starches into disaccharides.

Animals that feed intermittently may store food in the *crop*, from which food can be released more slowly as needed into the digestive system. In birds food is crushed internally in the *gizzard*, a muscular sac that contains coarse sand or small stones and has a hard lining. In mammals food moves from the mouth into the gullet, or *esophagus*, which is a tube serving primarily to transport food to the stomach. Muscle contractions move food along the tube, and lubricating fluids are secreted from the glands in the tube walls.

In some grass-eating mammals, such as the cow, the food moves from the esophagus into a large organ called the *rumen*. Within the rumen lives a population of bacteria and protozoa that are capable of breaking down cellulose to simple sugars. In addition, they synthesize many of the vitamins needed by the cow. Food is stored and fermented in the rumen for some time, being regurgitated and rechewed from time to time before it is passed along to the stomach.

The stomach of a mammal is a muscular, baglike organ that can be closed at both ends by *sphincters*, or rings of muscle. Muscle contractions in the stomach wall churn and squeeze the food, thoroughly mixing it with substances secreted by cells and glands in the stomach lining. The epithelial lining of the stomach, or *gastric mucosa*, includes several kinds of cells. The outer parts of folds in the lining are covered mostly with *mucous cells*, which secrête a viscous fluid that coats the stomach lining. This fluid protects the stomach lining from the actions of digestive enzymes and dilutes the food mixture. *Parietal cells*, which secrete hydrochloric acid, and *chief cells*, which secrete the protein-hydrolyzing enzyme *pepsin*, are abundant between folds of the lining. After the food mixture is thoroughly mixed with mucous fluid, acid, pepsin, and other enzymes, the posterior sphincter opens, the stomach contracts, and the food is forced into the small intestine.

Within the long, coiled tube of the small intestine, digestion is completed and the resulting small organic molecules are absorbed into cells that line the tube. The surface of the small intestine is covered with minute, fingerlike projections, or *villi*. The cell surfaces lining the villi are themselves covered with submicroscopic projections, or *microvilli*. These structures produce a great surface area—about 2,000 square feet in the human small intestine—which ensures a complete absorption of nutrients.

The short segment of small intestine closest to the stomach is known as the *duodenum*. Here secretions from the pancreas and liver enter the gut. Pancreatic juice contains enzymes that break down the most common

Figure 19.5. A diagram of the human digestive system.

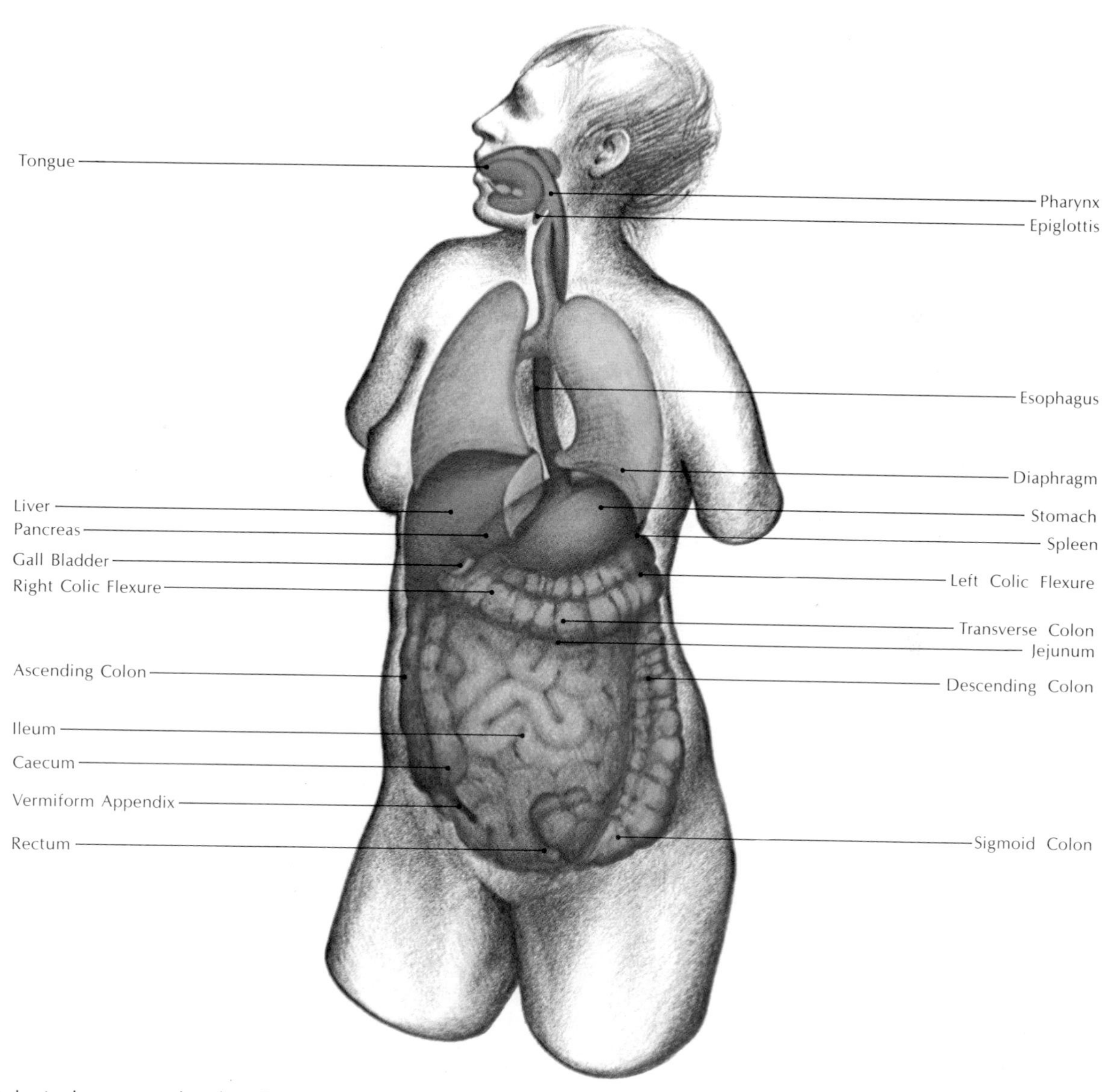

types of biological macromolecules. Pancreatic amylase completes the hydrolysis of starch (begun by salivary amylase) into disaccharides, while other sugar-digesting enzymes act on other sugar chains. Ribonuclease and deoxyribonuclease catalyze the hydrolysis of RNA and DNA into short nucleotide chains. Pancreatic lipase breaks down fats into fatty acids and glycerol. The proteolytic (protein-digesting) enzymes of the pancreas have been intensively studied. Several enzymes with different specificities ensure complete digestion of proteins to amino acids. Trypsin and chymotrypsin cleave long protein chains into short peptides, while aminopeptidase and the carboxypeptidases liberate amino acids from the amino and carboxyl ends of the peptides, respectively. These proteolytic enzymes are synthesized by the pancreas in an inactive form so that they do not destroy themselves or other pancreatic proteins. Upon arrival in the duodenum, these enzymes are activated by residual proteolytic enzymes, which cleave away susceptible sections of the inactive precursor enzyme protein chains, thus unblocking their active sites. In addition to enzymes, pancreatic juice contains sodium bicarbonate, which neutralizes the acid passed down from the stomach and makes the contents of the intestine slightly alkaline. Pancreatic secretion is largely controlled by the hormone secretin.

Another secretion aiding in digestion is produced by the liver in the form of bile. Bile travels from the liver into the gall bladder, where it is stored until it is needed. When food enters the duodenum, bile flows into the duodenum through the bile duct. Bile contains *bile salts*, sodium salts of certain complex organic acids, which function as detergents, breaking up insoluble droplets of fats and making the fat molecules accessible to the action of pancreatic lipase. One of the many functions of the liver is the breakdown of hemoglobin from red blood cells. Heme from hemoglobin is degraded into products that are included in bile as *bile pigments*. Considerable amounts of bile pigments pass out with excrement, giving it a characteristic brown color.

Nutrients are absorbed from the intestine by transport across cell membranes lining the intestine, a process followed by diffusion into the circulatory system. The products of fat digestion – glycerol and fatty acids – cross the membrane passively and diffuse into the *central lacteal*, a lymphatic vessel in each villus. The absorbed nutrients are carried away from the intestine by a capillary network that feeds into the hepatic portal vein. This vein transports dissolved nutrients to another capillary network in the liver. The liver removes excess nutrients for storage and releases nutrients from storage if the blood level of some substance falls too low. In these ways, the liver helps to maintain optimum levels of glucose and amino acids in the blood throughout periods of fasting and gluttony.

Undigested residue moves from the small intestine into the *caecum*, a pouch at the beginning of the large intestine. In many herbivorous animals, such as rabbits, the caecum plays a role similar to that of the rumen. Bacteria living in the caecum of the rabbit convert cellulose to sugar. In carnivorous and omnivorous mammals, the caecum is relatively small or is absent. In man, the short caecum terminates in the appendix. The major part of the large intestine, the *colon*, is a corrugated tube with a smooth lining that has virtually no villi. The colon serves primarily to remove excess water from undigested material. It houses large numbers of bacteria, principally the species *Escherichia coli*, which are also found in lesser numbers elsewhere in the gut. These bacteria synthesize significant amounts of some vitamins, which are subsequently absorbed by the colon and utilized by the body. At the end of the large intestine is a short section, the *rectum*, where undigested waste material is stored until eliminated from the body through the anus.

Absorption in Plants

Fungi and other organisms specializing in absorption of nutrients are restricted to environments that provide a relatively high concentration of nutrient substances. Fungi secrete digestive enzymes that hydrolyze macromolecules outside the organism, breaking them down to simpler organic molecules that can cross cell membranes by diffusion and active transport.

Aquatic plants obtain mineral salts and water by diffusion across the cell membranes, aided in some cases by active transport of certain ions. Land

Figure 19.6. Enlarged photograph of a growing root tip. Note the abundance of root hairs along the margin of the root in the region of cell elongation. These root hairs provide increased surface area for the diffusion and active transport of essential nutrients and water. (*Courtesy Carolina Biological Supply Company*)

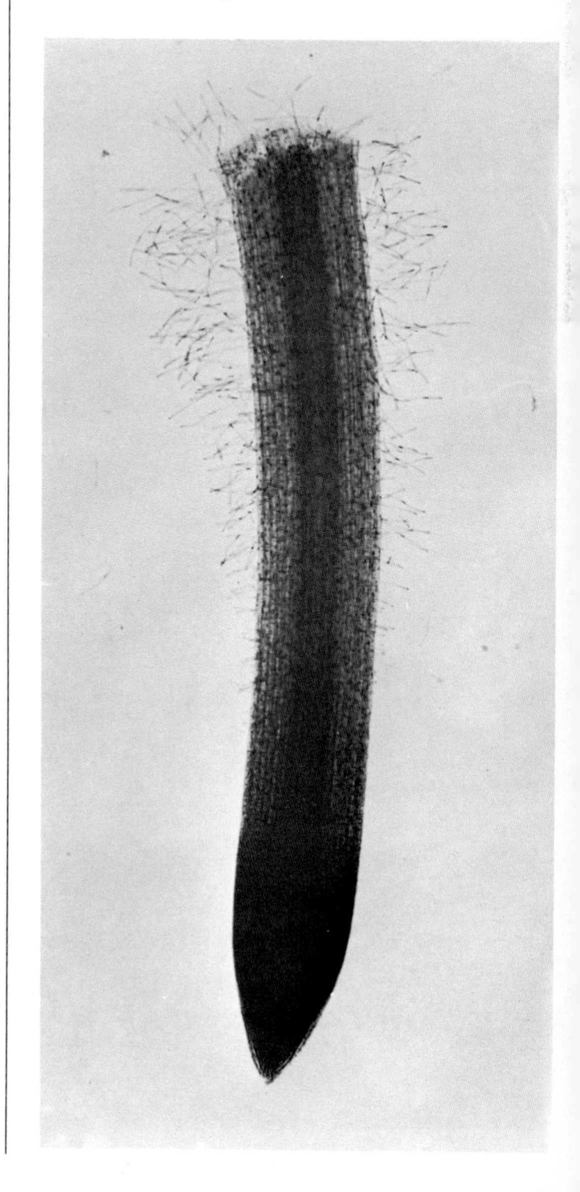

plants have roots specialized to absorb mineral salts and water from the soil. Like villi of the animal intestine, plant root hairs provide a greatly increased surface area across which diffusion and active transport can take place. Most plant root cells are extremely efficient at pumping potassium ions into the organism and pumping sodium ions out. Active transport of other ions has also been demonstrated. The active accumulation of ions produces an osmotic force that drives water into root cells. Because the cells have relatively rigid walls, they do not expand greatly, but a high pressure is built up inside the cells. As water is drawn from inner cells up through the xylem, the pressure in inner cells of the root decreases. High pressure in outer cells forces more water into the xylem, and more water is drawn osmotically from the soil.

GAS EXCHANGE

Small aquatic organisms readily obtain oxygen and carbon dioxide by diffusion from water. In the light, photosynthetic organisms produce more oxygen than respiration concurrently uses up, and the plant must obtain CO_2 from the environment. In the dark, no photosynthesis occurs, and these organisms must acquire oxygen as do all nonphotosynthetic organisms. In order to capture sunlight, the multicellular photosynthetic organism must have cells spread out in relatively thin layers or sheets. Because most of the cells thus have large surface areas in contact with the surrounding water, most aquatic plants are able to obtain the needed gases by simple diffusion directly into the photosynthetic cells and thus require no specialized respiratory systems.

In aquatic organisms that move about to find food, more compact structures for gas exchange appear to be advantageous. In the simplest forms, these structures consist of extensions or indentations of the surface. Most larger aquatic animals possess structures that serve to increase the surface area available for diffusion of oxygen into the organism. With increasing size, however, a means must be provided for delivering dissolved O_2 to cells that may be distant from the site of gas exchange. Therefore, structures specialized for gas exchange bring the dissolved gases in close contact with the animal's blood or body fluid.

In more complex aquatic organisms, extensively convoluted surface regions, or *gills*, are common. Gills are highly specialized to maximize the movement of gases while minimizing the movement of water into cells (Figure 19.7). In most aquatic animals, special mechanisms move a steady stream of water past the gills. Fish generally have sets of muscles in the floor of the mouth and in gill covers that constantly pump water into the mouth and out over the gills. Sharks and mackerel, however, lack the ability to make these breathing movements. They must constantly swim with their mouthes open to pass water over their gills and suffocate if prevented from swimming. The flow of blood within fish gills is in a direction opposite to that in which water flows over the gills. This *countercurrent* system causes the blood leaving the gills to be exposed to oxygen-rich water entering the gills and results in an efficient transfer of oxygen into the organism. Amphibians carry on a substantial fraction of their gas exchange through their moist skins, although they also have lungs or gills.

In one way, obtaining oxygen from air is an easier task than from water. A given volume of air contains 40 times as much oxygen as does the same volume of water. Furthermore, oxygen diffuses 300,000 times more rapidly

Figure 19.7 (right). Structural diagram of a typical teleost (bony fish) gill. The lamella construction of each gill filament serves to increase the surface area and thus the efficiency of gas exchange.

Figure 19.8 (left). Tracheal tube respiratory system as diagrammed in a generalized insect. This system operates by simple diffusion of gases through a complex tube system throughout the body of the organism. The tubes open on the body surface to allow exchange with the external atmosphere.

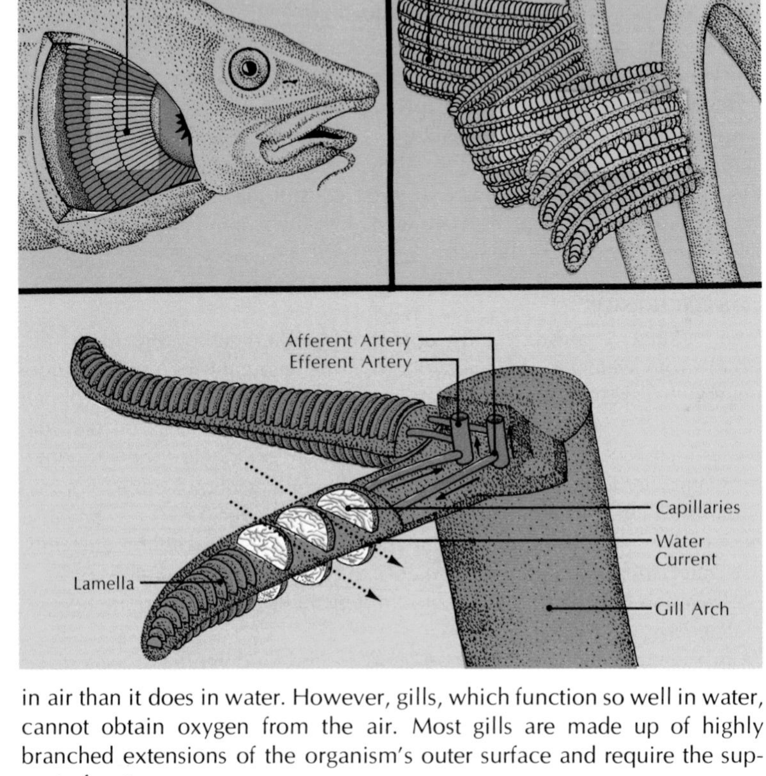

in air than it does in water. However, gills, which function so well in water, cannot obtain oxygen from the air. Most gills are made up of highly branched extensions of the organism's outer surface and require the support of water.

Land animals possess internal respiratory systems that are braced against collapse. The concentration of water vapor within the cavity can remain high, thereby reducing water loss. Small land organisms such as snails and some crustaceans have simple lungs. Insects breathe through a system of *tracheae*, thin tubes that open on the body surface and ramify extensively throughout every tissue, bringing the air supply close to the cells and minimizing the role of the body fluid in the transport of gases.

Most large animals possess lungs, which are alternately emptied and filled with fresh air by active movements, or *ventilation*. Oxygen enters the respiratory system of a land animal in the gaseous state. It then is dissolved in a thin film of water on the lining of the respiratory tract and diffuses across cell membranes in solution.

Air enters the mammalian respiratory system through the nostrils and the nasal chamber, passes through the throat and the windpipe, or *trachea*, and enters the branching tubes, or *bronchi* and *bronchioles* (Figure 19.9). The walls of the trachea, bronchi, and bronchioles are relatively rigid to prevent collapse, but they can be somewhat expanded or contracted to control the flow of air. After branching repeatedly, the tubes, or *alveolar ducts*, end in millions of microscopic pouches, the *alveoli* that make up the lungs. The thin epithelial lining of these pouches, where gas exchange takes place, has a total area of about 750 square feet in man (Comroe, 1966). Each alveolus

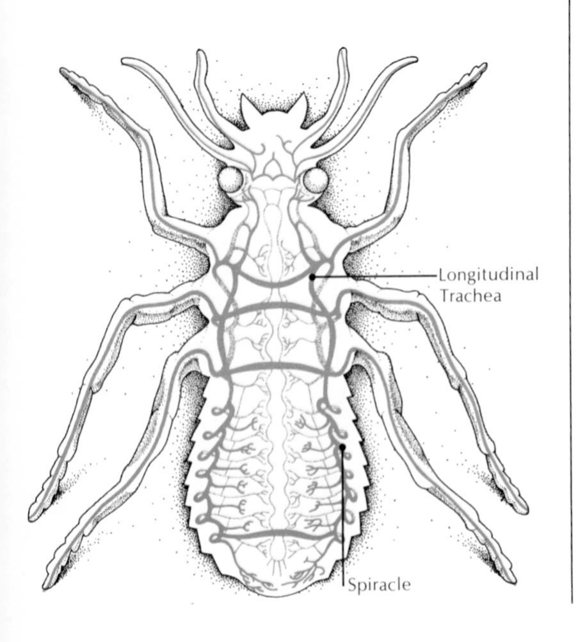

Figure 19.9. The human respiratory system. General view (top); bronchioles terminating in the grapelike clusters of alveoli (lower left); alveoli surrounded by a capillary network (lower right).

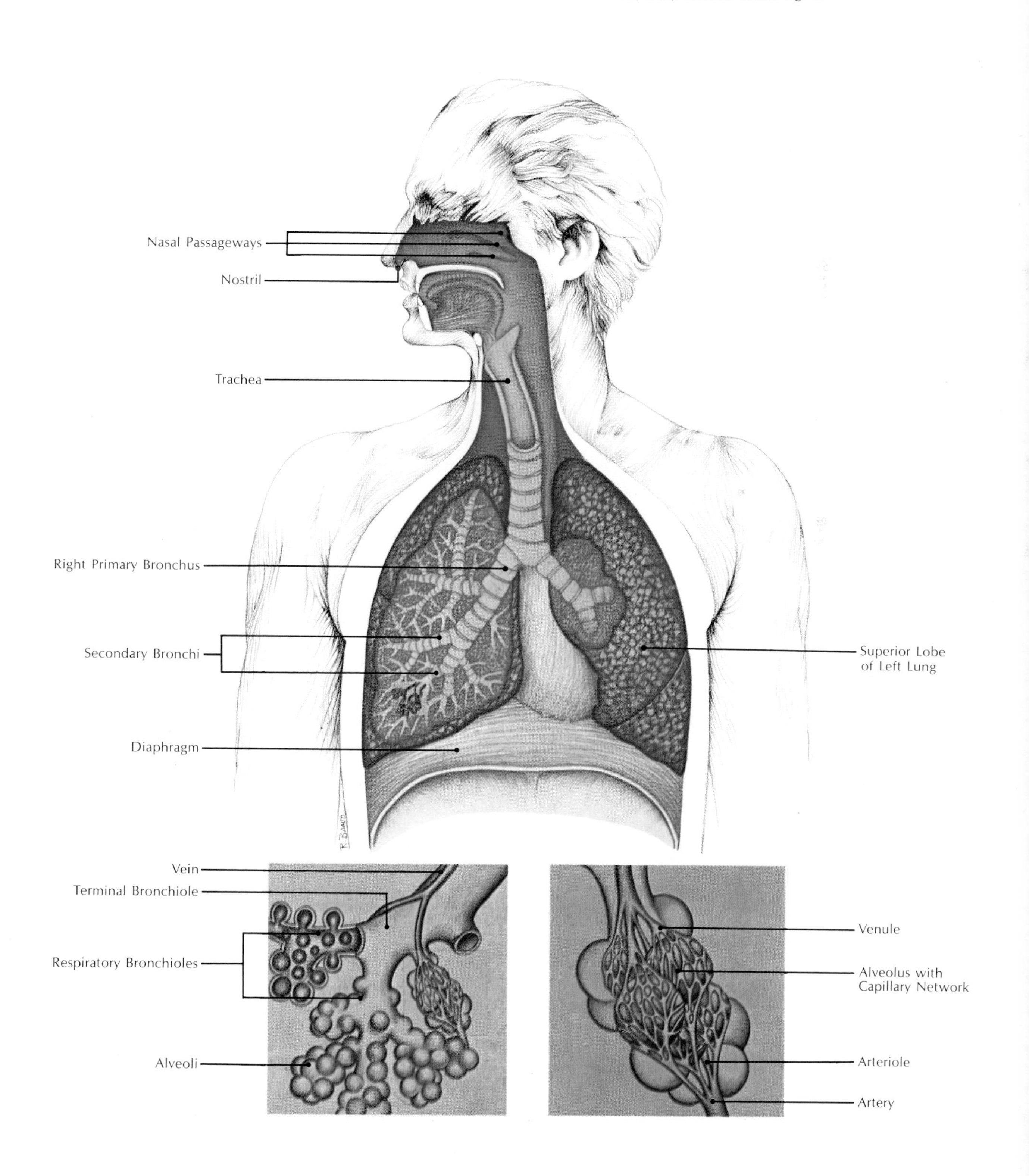

Figure 19.10. Position changes in the diaphragm and rib cage during expiration and inspiration.

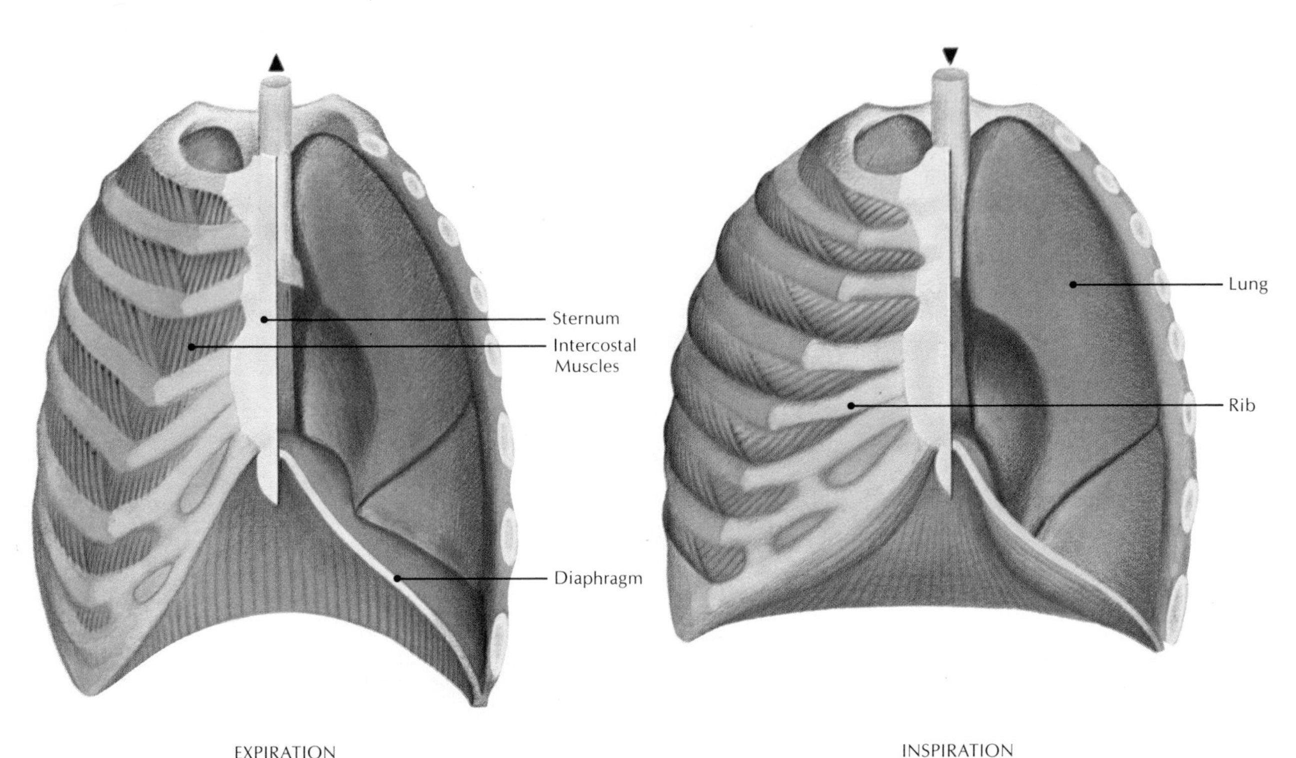

is invested with a capillary network. Blood gives up CO_2 and takes on O_2 as it passes through these capillaries in close contact with the alveolar epithelium.

The lungs occupy most of the chest cavity and are protected by the somewhat flexible rib cage. The floor of the chest cavity is a sheet of muscle called the *diaphragm*. Inhalation is caused by expansion of the rib cage and contraction of the diaphragm. When the diaphragm contracts, it flattens and thus enlarges the volume of the chest cavity. Atmospheric pressure forces air into the lungs so that they expand to fill the enlarged cavity. Exhalation occurs when the diaphragm relaxes and bulges upward, reducing the volume of the chest cavity and forcing air out of the lungs.

The rate of breathing and the extent to which the lungs are refilled on each breath are under partial voluntary control but normally are controlled involuntarily by a respiratory center at the base of the brain. This respiratory center is affected, among other things, by the concentration of CO_2 in the blood.

Like the land animal, the land plant has a surface that is relatively impermeable to water (preventing excessive water loss through evaporation). Therefore, land plants are unable to absorb oxygen or carbon dioxide over their surface, as do aquatic plants. Instead, gases are absorbed at the surfaces of intercellular spaces within the leaves. The site of gas exchange is close to the site of use and release so that the problem of transport is minimized. Gases enter the spaces through microscopic slits in the undersurface of the leaf, the *stomata*. *Guard cells*, which flank the stomata, regulate the flow of gases by opening or closing the slit. The control mechanism of

Figure 19.11a (left). Diagrammatic representation of a leaf section. Note the four distinct tissue layers.

Figure 19.11b (upper right). Stomata-guard cell complexes on the underside of a leaf.

Figure 19.11c (lower right). Photomicrograph of a leaf cross section. Gas exchange takes place through small pores, called stomata, located on the underside of the leaf.

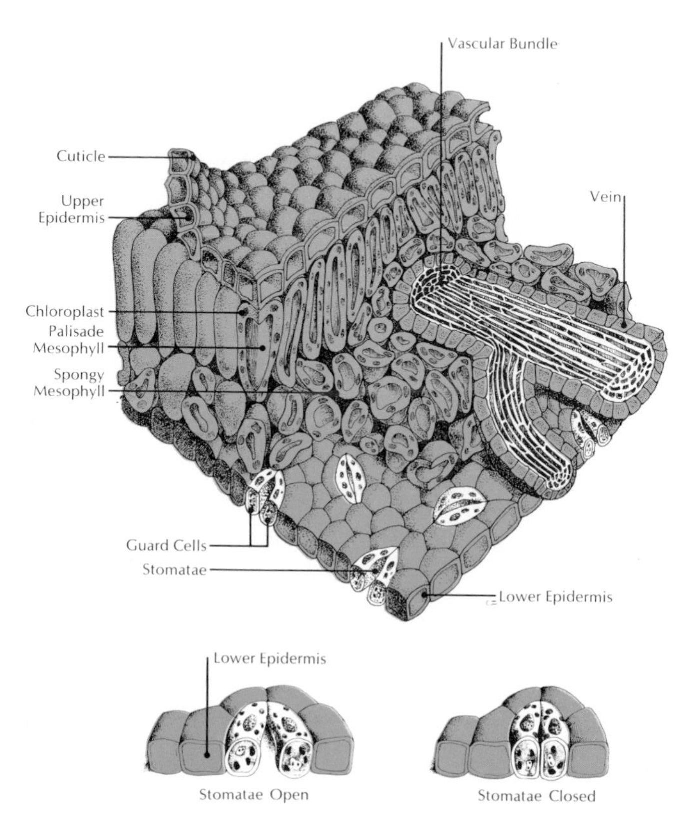

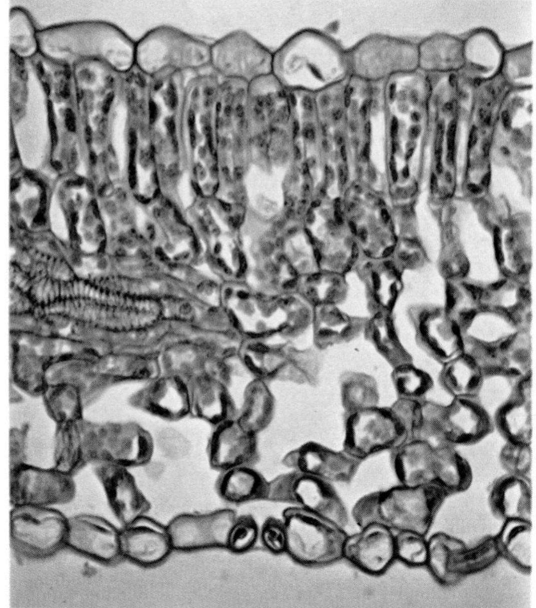

the guard cells is not yet fully understood, but it appears that a decrease in CO_2 concentration within these cells caused by active photosynthesis somehow increases the turgidity of the cells and opens the stomata. If excessive water loss causes wilting, the guard cells become flaccid and the stomata close.

The rate-control mechanisms of the lungs and stomata (as well as a small ring of muscle that opens and closes the tracheae of insects) hold the movement of gases into the respiratory system to the minimum needed for gas supply, thus minimizing the loss of water to the atmosphere.

INTERNAL TRANSPORT

Once nutrient materials have entered the organism, they must be transported to the sites where they can be used in chemical reactions. The respiratory system obtains oxygen from the environment, and this oxygen must be efficiently distributed. Finally, waste products must be transported to the excretory system for elimination. These functions are served by circulatory systems.

Circulatory Systems

In unicellular organisms and small multicellular organisms, diffusion carries materials from parts of the organism where they enter or are formed to parts where they are consumed or expelled. In algae and sponges, where there is relatively little cell specialization, the body organization is such that most cells are exposed to the external sea water or to water circulated through passages in the body. Each cell is able to exchange materials with

Figure 19.12. The closed circulatory system of the earthworm, representing the annelid phylum. A simple tubular heart forces the blood forward along the dorsal vessel and backward along the ventral vessel.

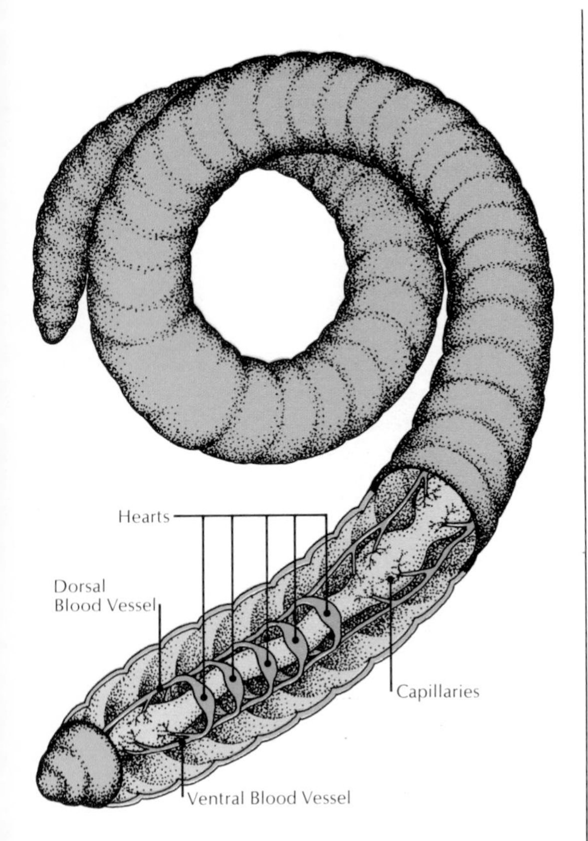

water or to exchange materials with nearby cells that are in contact with water. The flat body structure of algae allows every cell to be near the external environment. In simple animals, water is circulated through internal passages by cilia.

An unusual transport system exists in the coelenterates. Some of the cells in the epithelial layer that lines the body cavity are amoeboid. After engulfing food particles by phagocytosis, these cells move through the organism and distribute the digested nutrients to other cells.

In flatworms, food is taken into the gut and mechanically broken down to small particles. The particles are taken in by cells lining the gut and are digested. Because of the flattened body structure and the small size of flatworms, every body cell is within a few cell widths of some part of the branched gut. Thus, diffusion can carry nutrients to all cells. Animals above the flatworms generally have some sort of specialized circulatory system with one or more pumps to move fluid throughout the system. In molluscs and arthropods, a heart pumps the internal fluid through tubes or vessels that empty into various parts of the body. The fluid bathes the tissues of the body and collects in cavities called blood sinuses, which communicate with a chamber surrounding the heart. When the heart relaxes, blood from the surrounding chamber is sucked into the heart through openings in its walls called *ostia*. When the heart contracts, the ostia close, and blood is forced out through the vessels. This arrangement is called an *open* circulatory system because the blood is not enclosed in vessels during its entire circuit through the body.

The circulatory system of annelids is a closed network of vessels. A simple tubular heart produces waves of muscular contraction that force blood forward along a dorsal vessel and backward along a ventral vessel (Figure 19.12). The movements of blood are somewhat irregular, and fluid may flow either way in various parts of the network.

Echinoderms have well-developed tubular systems extending through the body, but none of these appears to carry nutrients and waste products. The water-vascular system is filled with sea water pushed into the tubes by cilia; the water pressure inside this system helps in movement of body parts. Nutrients are transported within the body by three mechanisms: amoeboid cells, bathing of tissues with sea water in internal cavities, and branching of the gut and digestive system into most parts of the body.

Vertebrates, like annelids, have a closed circulatory system. A heart pumps into large vessels called *arteries*, which then carry the blood to various parts of the body. The arteries branch repeatedly, finally becoming tiny, thin-wall vessels called *capillaries*. As the blood moves through the capillaries, materials such as oxygen and carbon dioxide are exchanged with neighboring cells. The blood then moves into larger, thicker-wall *veins*, which carry it back to the heart.

Blood

The blood of vertebrates is composed of *plasma*, a fluid in which many proteins, ions, nutrients, and waste materials are dissolved. Floating in the plasma are *red blood cells*, several types of *white blood cells*, and a kind of cell fragment called *thrombocytes*, or *platelets* (Figure 19.13).

Materials move between blood and cells by diffusion. Some of these substances are simply dissolved in the blood plasma. However, the red blood cells of vertebrates play a special role in oxygen transport. These cells contain the protein hemoglobin, which gives blood its red color.

Figure 19.13a (upper left). Red blood cells, which function in the transportation and distribution of oxygen in mammals. Note the absence of nuclei.

Figure 19.13b (lower left). Blood platelets. These structures are not usually whole cells but membrane-covered cell fragments, often without nuclei. Platelets are involved in the blood-clotting mechanism.

Figure 19.14 (right). Hemoglobin molecule. The structure of the molecule has been deduced primarily from x-ray diffraction studies. The molecule consists of four closely associated polypeptide chains. Each molecule contains two alpha and two beta chains, each of which binds an oxygen-carrying heme group.

Hemoglobin combines reversibly with oxygen so that mammalian blood is capable of transporting about 60 times more oxygen than would dissolve in plasma. When the oxygen concentration in cells around blood is high (as it is in the capillaries of the lungs or gills), hemoglobin combines readily with oxygen. In body parts where the oxygen concentration is low, hemoglobin readily gives up oxygen and allows it to diffuse out of the blood into the body tissues.

Hemoglobin contains four subunits, each of which consists of a protein chain, globin, and heme (a complex polycyclic ring structure containing iron). Each subunit can bind one oxygen molecule. The four subunits interact in a cooperative or allosteric way so that after one subunit binds an oxygen molecule, the other subunits bind additional oxygen molecules more readily. Hemoglobin can thus perform efficiently in its role of binding oxygen in the lungs and releasing it to the body tissues.

Carbon dioxide, which is produced as a waste product of metabolic respiration, is carried by the blood from the body cells to the lungs or gills, where it is excreted. In the blood, dissolved CO_2 gas is in equilibrium with carbonic acid and bicarbonate ion: $CO_2 + H_2O \rightleftarrows H_2CO_3 \rightleftarrows H^+ + HCO_3^-$. Because the hydration of CO_2 (and the dehydration of carbonic acid) is a relatively slow reaction, red blood cells contain the enzyme *carbonic anhydrase* to catalyze this reaction. This catalysis speeds up the reaction so that bicarbonate can be converted to gaseous CO_2 for exchange during the rapid flow of blood through the capillaries of the lungs.

Dissolved proteins and the platelets are involved in blood clotting, a complex series of reactions that occur in case of injury to the circulatory system. The end result of these reactions is the formation of a clot, which

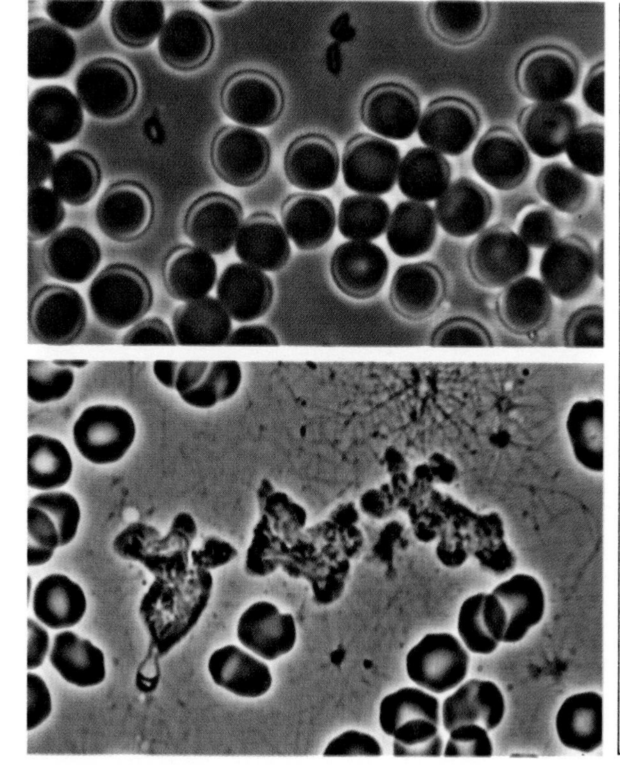

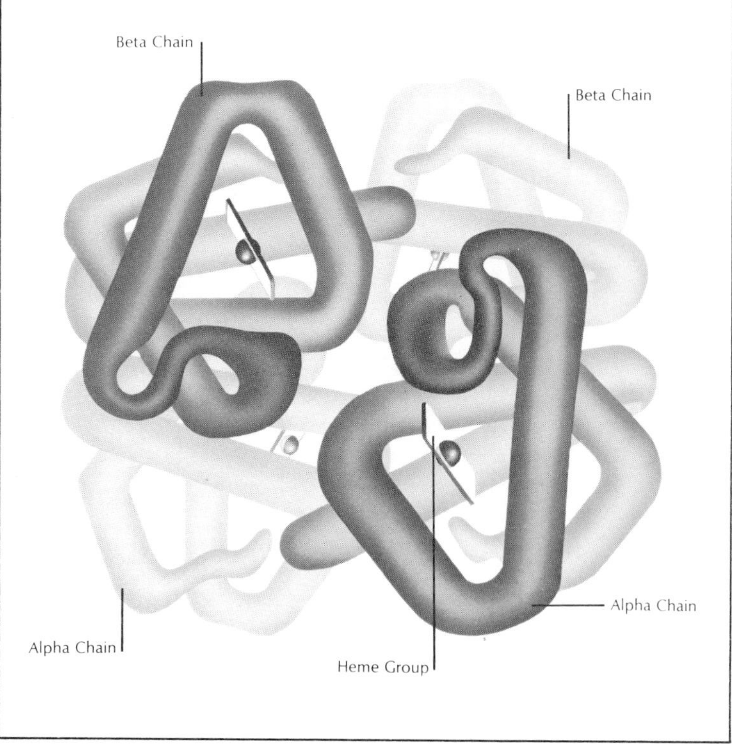

Figure 19.15 (above). The main reactions of the clotting process. Thrombokinase, an enzymatically active substance, is emitted from ruptured platelets and interacts with both prothrombin (inactive) and calcium ions in the blood to produce active thrombin. Thrombin interacts with fibrinogen to form fibrin, an insoluble coagulated protein. Fibrin forms a meshwork of fibers that traps the cellular components of blood, thus forming a clot.

Platelets → Thrombokinase

Prothrombin + Ca^{++} → Thrombin

Fibrinogen → Fibrin

Figure 19.16 (below). The neurogenic heart of a typical arthropod. Note the large central nerve ganglion. Control of heartbeat in these organisms is extrinsic; if nerves to the heart are cut, it stops beating.

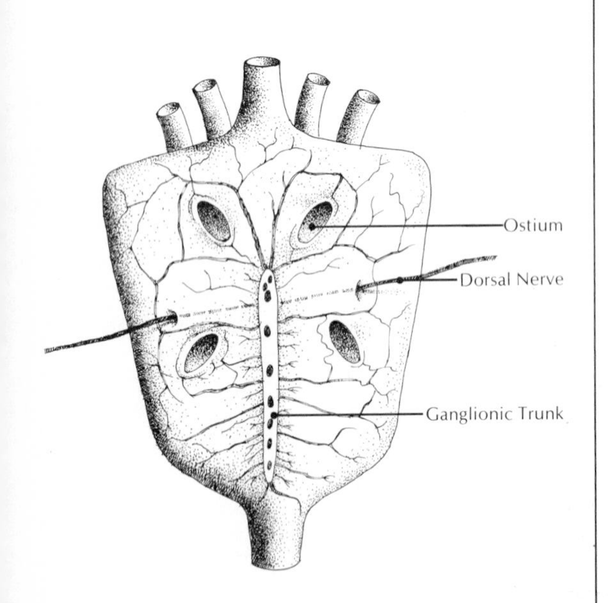

temporarily seals off the injured area until the damage is repaired. An outline of the processes involved in blood clotting is presented in Figure 19.15.

The Heart

Among invertebrates, the form and function of the heart seems to be more directly correlated with habitat and style of life than phylogenetic level. Hearts range from none at all in some small animals to simple spontaneously pulsating blood vessels in annelids and relatively sedentary echinoderms. Molluscs have multichamber hearts with valves to direct the flow of blood.

The muscles that make up animal hearts (except for the hearts of arthropods) are *myogenic*, that is, they contract spontaneously. Nerves from the central nervous system may enter myogenic hearts and affect the strength or frequency of the beat by excitation or inhibition, but if the nerves are cut, a myogenic heart will continue to beat.

The hearts of almost all adult arthropods, on the other hand, are able to contract only on command from the nervous system. Such *neurogenic* hearts generally have a nest of nerve cells, or *ganglion*, that completely control the rate of beat. If the nerve fibers running from these nerve cells to the heart muscle are cut, the heart stops.

The heart of a fish is a two-chamber pump, with a thin-wall *atrium*, which takes in blood from the veins and delivers it to the thicker-wall, more powerful ventricle, which pumps blood under high pressure into the arteries. The arteries carry blood to the gills, where it gains oxygen and gives up CO_2 in a capillary network. Blood then moves into other arteries, which carry it to capillary networks in all parts of the body. Capillaries drain into veins, and blood returns to the heart. The difficulty with this system is that a single pump must provide sufficient pressure to force blood through the resistance of two capillary networks.

The amphibian heart has two atria, which lead into a single ventricle. One atrium receives blood from the lungs; the other receives blood from the general body circulation. The two atria beat simultaneously, and oxygenated and deoxygenated blood are mixed in the single ventricle.

The reptile heart has two atria and two ventricles, although in most species the two ventricles are not completely separate. In the hearts of birds and mammals, oxygenated blood is completely separated from deoxygenated blood. The path of the blood through the two separate circulatory systems of an animal with a four-chamber heart is illustrated in Figure 19.17.

The vertebrate heart is made up of a specialized striated muscle that contracts spontaneously. The mammalian heartbeat is initiated and coordinated in the right atrium by the sinoatrial node, a patch of muscle cells that are specialized for electrical conduction rather than contraction. This *pacemaker* initiates electrical activity that spreads through the cardiac muscle fibers of the atria to another specialized region, the atrioventricular (AV) node, located in the wall between the two atria just above the ventricles. The AV node passes the wave of excitation to a bundle of similar conducting tissue that ramifies throughout the ventricles.

Because the heart muscle is constantly working, it requires a plentiful supply of oxygen to carry out its energy-releasing metabolism. The heart muscle contains its own system of blood vessels. If these vessels become blocked and a portion of the heart muscle becomes short of oxygen, the heart fails to function. One condition that can lead to blockage of the coro-

Figure 19.17. Comparative anatomy and evolution of the vertebrate heart. The four-chamber heart of birds and mammals is actually the result of two subdivided atria and ventricles.

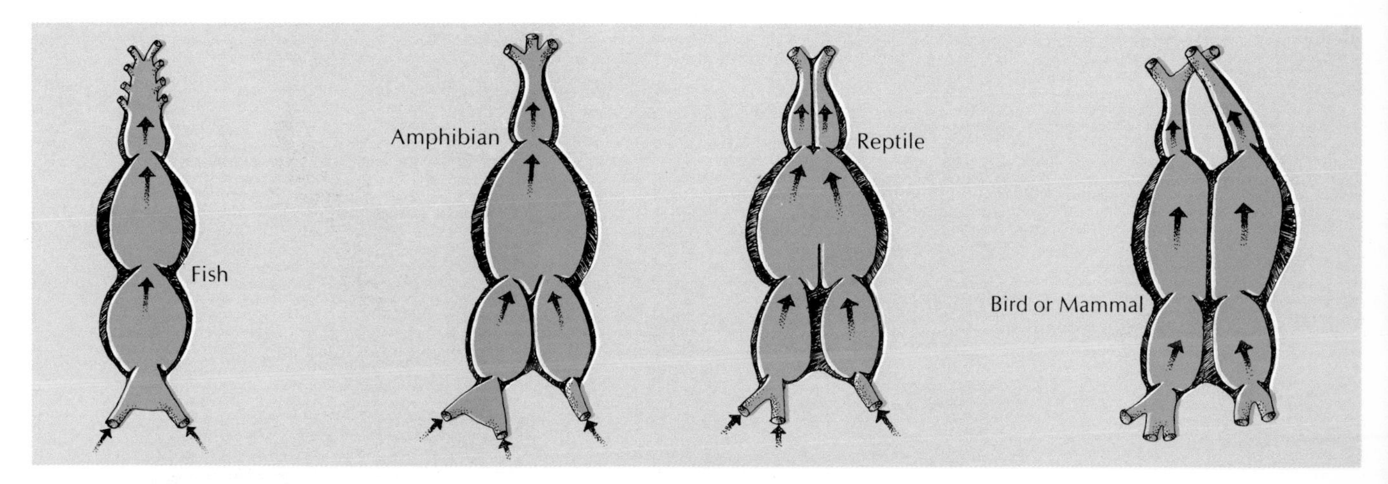

nary circulation is hardening of the arteries, a disease in which there is deposition of the steroid cholesterol in the walls of the arteries. A blood clot traveling in the bloodstream also may interfere with the coronary circulation and cause heart failure. A clot lodged in the heart is called a coronary thrombosis and is a common cause of heart attacks.

Circulation in Mammals

Blood arriving at the heart from the body tissues is high in CO_2 and lower in oxygen. It enters the right atrium, which pumps blood into the right ventricle. The right ventricle sends blood to the lungs via the pulmonary artery. After the blood exchanges its CO_2 for O_2 in the capillaries of the lungs, it returns to the left atrium of the heart via the pulmonary vein. The left atrium pumps blood into the left ventricle, which pumps blood out to all parts of the body through a large artery, the *aorta*. The heart, like all pumps, must have valves in order to prevent backward flow. There are four valves in the hearts of birds and mammals (Figure 19.18).

The arteries have relatively thick walls of muscular tissue, which hold blood under pressure as it moves away from the heart. In a human, the contraction of the ventricle pushes blood into the aorta under a pressure equivalent to about 120 millimeters of mercury. As blood moves away from the heart, the arteries branch to form numerous smaller arteries and eventually lead into the capillary system. Capillary walls consist of a single layer of epithelial cells, permitting ready diffusion of substances from blood to tissue cells and vice versa. The number of capillaries in the body is enormous (a few thousand in a single cubic millimeter of skeletal muscle tissue, for example), and every cell of the body is within a cell or two of a capillary.

Because of the increasing frictional resistance of the arterial system as blood moves away from the heart into an ever-increasing number of smaller vessels, the blood pressure drops to about 35 millimeters of mercury (mm Hg) at the arterial end of a capillary. The concentration of proteins in the plasma produces an osmotic pressure of about 25 mm Hg, which tends to move water into the blood vessel. Thus, near the arterial end of a capillary, there is a net pressure of about 10 mm Hg pushing water out of the capillary into the surrounding cells. At the venous end of the capillary, blood pressure drops to about 15 mm Hg pushing outward, whereas the osmotic pressure remains the same. At the venous end of the capillary,

Figure 19.18. External and internal anatomy of the human heart. Deoxygenated blood from the body flows through the right side of the heart and becomes oxygenated in the lungs. It then returns to the left side of the heart and is pumped to the body. Triangles denote contraction.

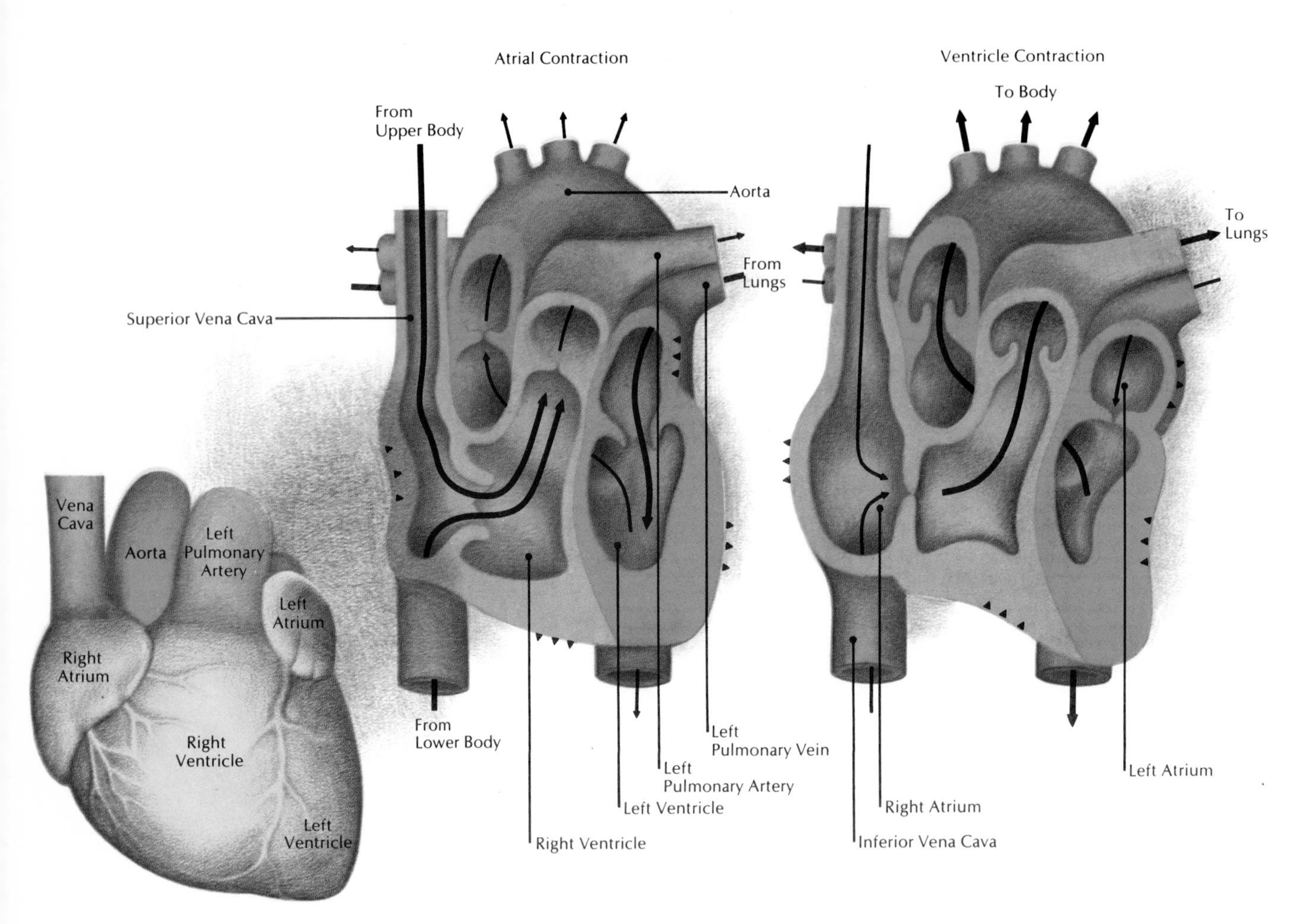

there is a net pressure of about 10 mm Hg pushing water from the cells into the capillary, and the water content of the blood is restored. This movement of water in and out of blood assists in the exchange of dissolved materials with the cells. The balance of pressures is delicate, and any disturbance of blood pressure or of the concentration of proteins in blood plasma can cause bloating or tissue dehydration.

Veins have thinner walls and are less elastic than arteries. A layer of fibrous tissue and a thin layer of muscle surround the epithelial cells that line veins. Veins lack sufficient elasticity to keep blood under pressure. Blood arrives in veins from capillaries under very low pressure, and its movement back through veins to the heart is largely dependent on contraction of skeletal muscles. As the muscles contract, they squeeze blood through the veins that pass through muscle tissue. Valves along the veins keep blood from moving back toward the capillaries. The flow of blood through the circulatory system is controlled both by the rate at which the heart beats and by the relaxation or constriction of muscles in artery walls.

Nerves leading from special centers in the brain can speed up or slow down heartbeat. These centers are activated by various sensory inputs, including receptors that detect unusual stretching of arteries (leading to a

Figure 19.19. The human cardiovascular system. Only the major vessels and organs associated with the circulation system are outlined.

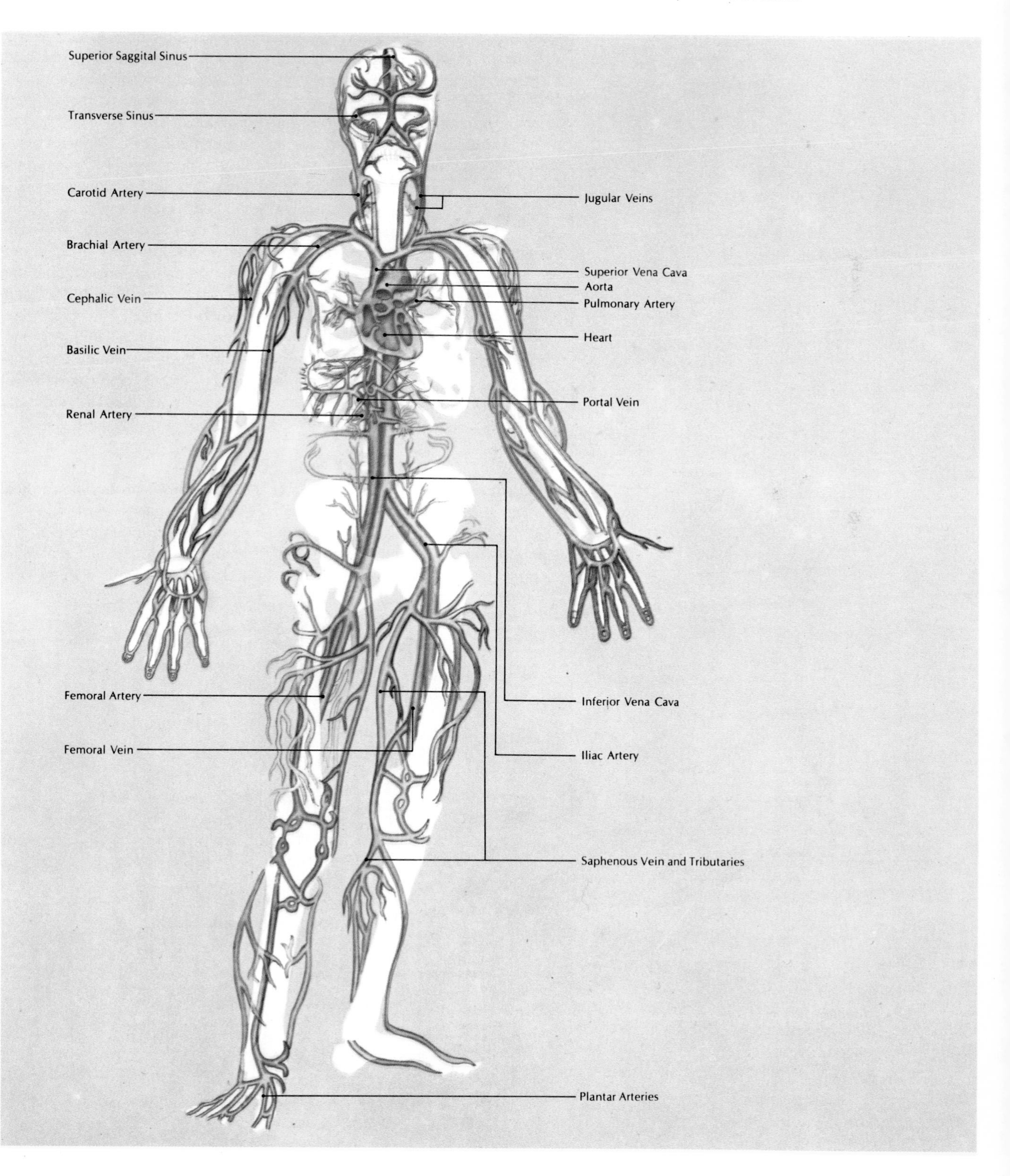

Figure 19.20 (left). The human lymphatic system. Lymphatic vessels function as an auxiliary branch of the main cardiovascular system by returning tissue fluid back to the main bloodstream. Lymph nodes located along the vessels trap foreign matter, including invading bacteria. *(From "The Lymphatic System" by H. S. Mayerson. © 1963 by Scientific American, Inc.)*

Figure 19.21 (right). The vascular tissue of higher plants is a complex of various cell types. Tracheids, vessels, ray cells, fibers, and parenchyma cells all interact both structurally and physiologically as xylem tissue. Sieve tubes, companion cells, ray cells, phloem fibers, and parenchyma cells interact as phloem. Xylem conducts water and dissolved mineral salts and forms supportive tissue (wood). Phloem conducts synthesized products from the leaf canopy to the lower portions of stems and roots.

slowing of heartbeat) or of veins (leading to a speeding of heartbeat) as well as receptors that react to changes in CO_2 or O_2 concentrations of blood. The constriction or dilation of arteries also is directed by nervous impulses. In this way, the flow of blood can be directed toward or away from different parts of the body in response to various conditions.

In vertebrates the lymphatic system, an independent transport system, is involved in movement of tissue fluids outside blood vessels. Some of the fluid that leaves arterial capillaries, together with fluids produced by tissue cells themselves, diffuses into small tubes called lymph capillaries. The tissue fluid, or *lymph*, that enters these vessels moves through the lymph capillaries into larger lymph vessels and eventually into compact, ovoid organs called *lymph nodes*. Foreign particles or microorganisms that wander into tissues are likely to be carried into lymph nodes, for walls of blood capillaries are quite resistant to the passage of such intruders. The small white blood cells known as lymphocytes are found in lymph as well as blood, and the lymphatic system is important to the body's immune responses (Chapter 22). The lymph nodes are drained by other lymph vessels that come together to form two large ducts that empty the filtered lymph into large veins.

Internal Transport in Higher Plants

Fluid movement in higher plants occurs in specialized cells of the vascular system (Chapter 11). Plants do not have fluid-filled spaces between tissues, nor do they have vessels with multicellular walls—both features that play important roles in internal transport in animals. Nutrients and water are conducted in xylem tissue from the roots to body cells through hollow con-

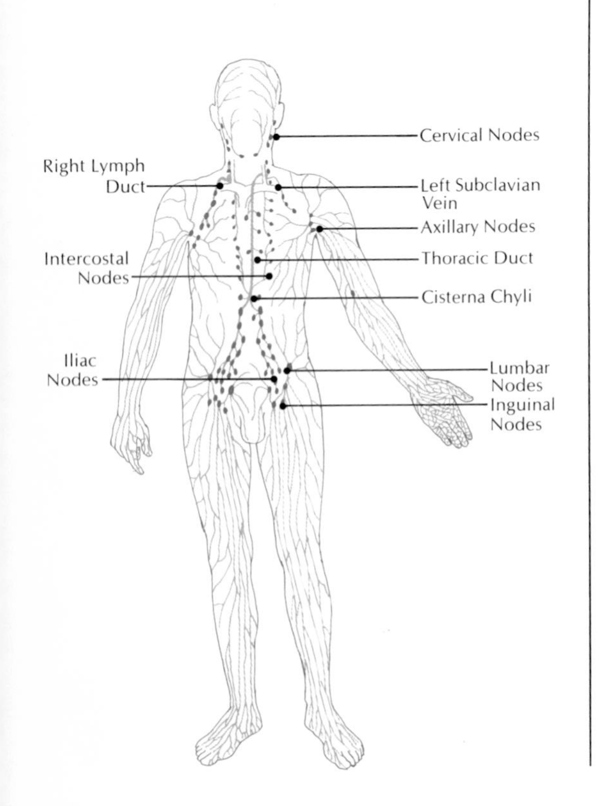

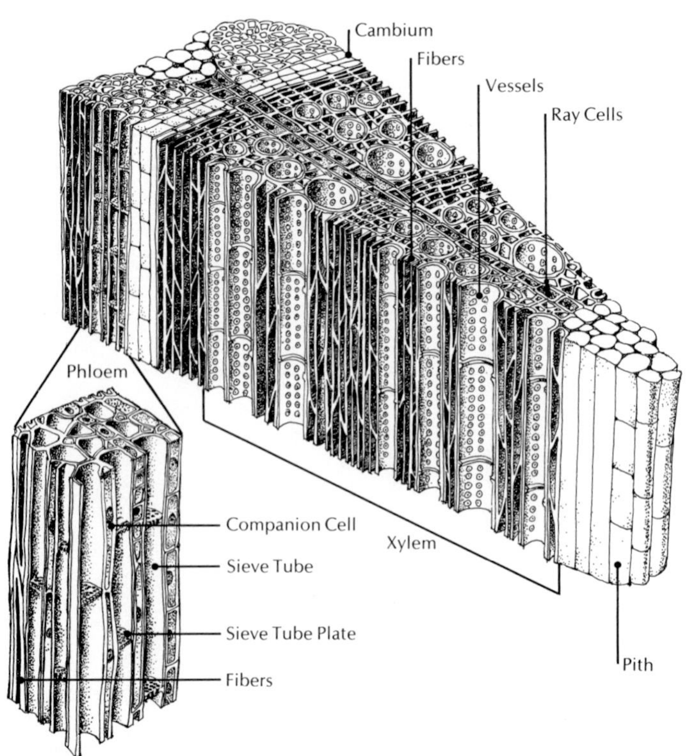

Figure 19.22a (above). Transpiration in the leaf.

Figure 19.22b (below). Water guttation in the strawberry plant. When the humidity is too high to permit sufficient water loss by evaporation, water under pressure is forced out at the leaf ends. This process, in which water droplets are formed, is called guttation. *(Courtesy Carolina Biological Supply Company)*

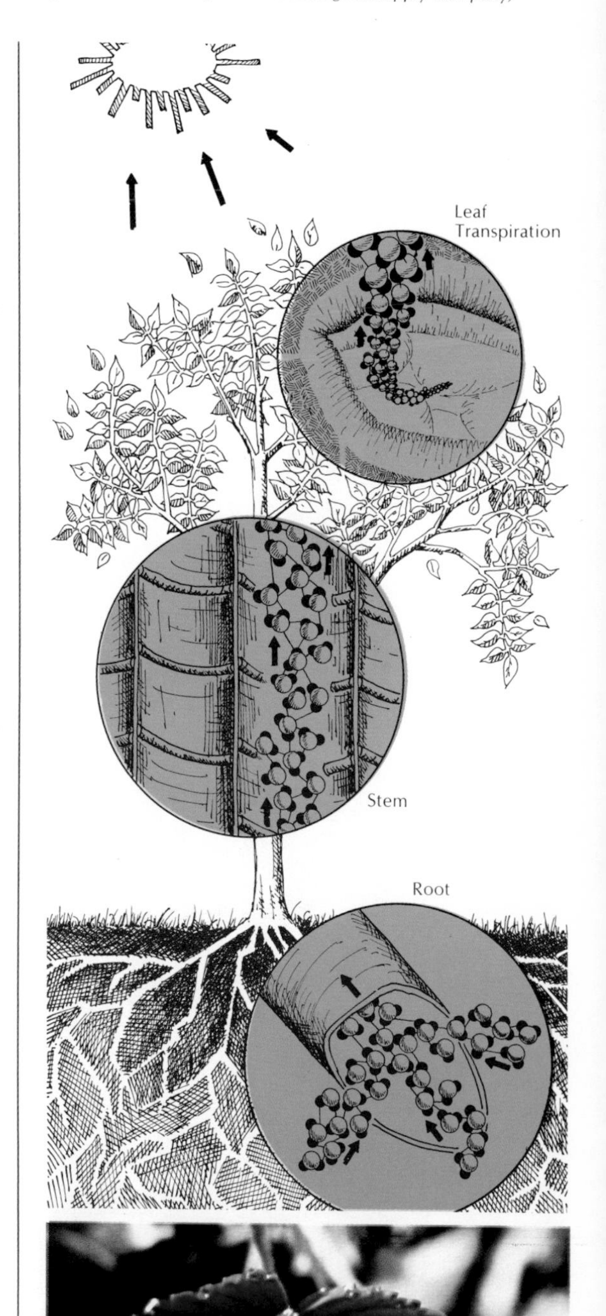

ducting tubes formed by the disappearance of nuclei and cytoplasm from tracheid cells. In many species, tracheids become joined end-to-end to form long vessel elements. The thick cellulose walls of tracheids and vessel elements give strength and rigidity to the plant body and prevent collapse of conducting tissues during periods of desiccation. Organic molecules such as carbohydrates and hormones are transported through the plant body in thin-wall, living cells of the phloem tissue. These living sieve cells may be joined end-to-end to form sieve tubes. The sieve cells retain their cytoplasm and probably their nuclei at maturity, and nutrients are moved from cell to cell through the cytoplasm.

The upward movement of fluids in xylem tissue is related to the loss of water (transpiration) from leaves. Water forms fine columns within the tracheids and vessels, running without interruption up from the roots into the veins of leaves. Hydrogen bonding between water molecules holds this column together, causing it to behave much like a fine wire. The column of water is pulled upward as water is lost through transpiration from the upper end of the column. Experiments show that a column of water inside a thin, airtight tube can withstand a pull of 300 pounds per square inch without breaking—a pull that would be sufficient to lift water to the top of even the highest tree.

Phloem cells are drastically altered by almost any method of observation. Thus, far less is known about the mechanism by which sugars and other materials are moved through the phloem. Transport in the phloem, unlike that in the xylem, depends upon the activities of living cells. The rate of movement through the phloem is far slower than that through the xylem, and sugars move from regions of high concentration toward regions of lower concentration. Radioactive tracers show that different substances move at different rates through the phloem. These studies also show that materials moving from a leaf to the stem divide in the phloem tissue of the stem, with some of the solution moving upward through the stem phloem and most of it moving downward. The upward and downward movements through the phloem occur in separate bundles of vascular tissue. It appears that the upper leaves of a plant supply nutrients chiefly to the apex; the lower leaves supply nutrients chiefly to the roots; and the intermediate leaves supply nutrients in both directions.

Among the mechanisms suggested to account for transport in the phloem are diffusion, cytoplasmic streaming, flow under gravitational pull, osmotic movement across cell membranes, and some form of active transport across sieve cell membranes. None of these hypotheses has proven entirely satisfactory for the explanation of all known facts about phloem transport.

The vascular systems of plants are capable of moving large volumes of fluids, but the rate of movement is much slower than that in animals. Because plants do not expend energy in muscular movements and generally do not require a rapid supply of nutrients to support sudden changes in activity, the slow movements of fluids through the xylem and phloem are sufficient to meet the needs of the organism.

EXCRETION

To maintain proper internal conditions, an organism must not only obtain nutrients but must regulate its internal composition and get rid of waste products. The principal wastes to be handled are carbon dioxide (the major

Figure 19.23. Salt excretion in birds and men. Sea birds such as the gull are capable of a limited sea-water intake due to active secretion of salt by special glands located in the beak. Thus, excess salts can be actively transported out of the body fluids. Man, however, cannot drink sea water. In eliminating the salt, he will lose more water than he has drunk. *(From "Salt Glands" by K. Schmidt-Nielsen. © 1959 by Scientific American, Inc.)*

waste product of cellular respiration), salts, and nitrogenous wastes of protein metabolism. A proper water balance must also be maintained. The functions of maintaining salt and water balance and disposing of nitrogenous wastes are performed by excretory systems.

Ion and Water Balance

Most marine invertebrates are essentially *isosmotic* with sea water—that is, their cells neither shrink nor swell. Because their ionic composition is different from sea water, they must have active transport mechanisms to control their composition. Most marine vertebrates have an osmotic concentration of body fluids far lower than that of sea water. In part, the water exchange is accomplished by making the major portion of the body surface impermeable to water. However, a certain amount of sea water is always swallowed with food, and tissues specialized for gas exchange must come in close contact with sea water. Therefore, there is an osmotic flow of water out of the organism and a danger of dehydration. Organisms that live under these conditions have mechanisms for actively secreting salts and conserving water. For example, higher marine fish excrete small quantities of very concentrated urine and actively transport salt out through the surfaces of the gills.

For organisms living in fresh water, the problem is reversed. Osmotic forces cause water to move into the organism. The fresh-water plants have rigid cell walls that permit the build-up of internal pressure sufficient to counteract the inward movement of water. Animals that live in this environment are covered over the greater part of their surface with a layer im-

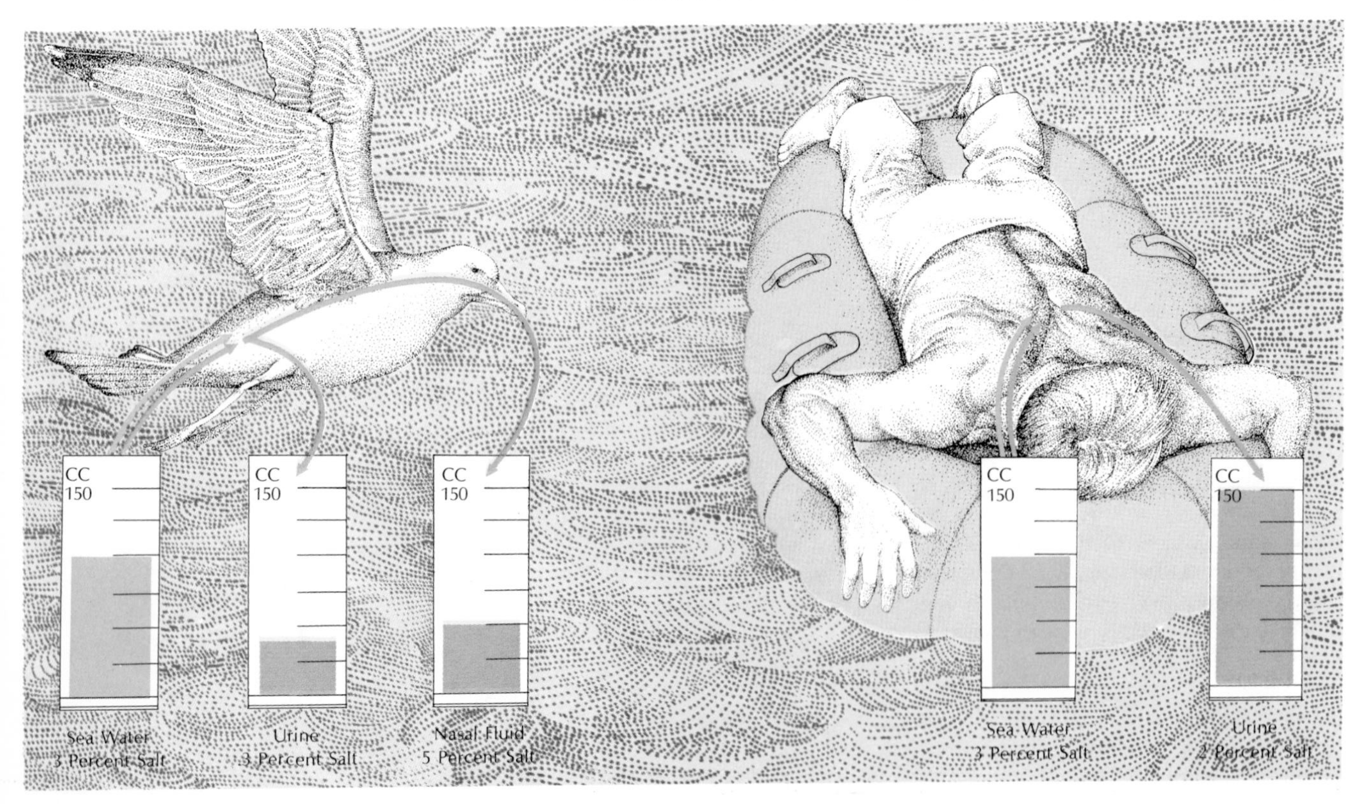

permeable to water. Fresh-water fish and amphibians have kidneys that are specialized for the reabsorption of ions. They excrete large volumes of very dilute urine. Some fresh-water fish also have active transport mechanisms in their gills to take ions out of the water and pump them into their bodies.

Land animals must obtain water by drinking and eating. Because excess salts are frequently included in their food, their excretory systems are specialized for elimination of salts and retention of water. In many land animals, salts are also eliminated through the body surface by sweating.

The kidneys of some desert mammals are so efficient that these animals can survive without drinking water. Water loss from the body is minimized by remaining in closed, humid burrows during the day, by excreting urine and feces with very little water content, and by minimizing loss of water through the skin. (Most of these animals have few sweat glands.) With this rigid program of water conservation, these animals are able to meet their water needs from water obtained during metabolic respiration of carbohydrates. When such animals are limited to a diet of high-protein foods such as soybeans, large amounts of nitrogenous wastes are produced in metabolism. Even with concentrated urine, so much water must be used to get rid of these wastes that the animals become dehydrated without a source of drinking water. However, unlike most other mammals, these desert dwellers can satisfy their thirst with sea water. As shipwrecked sailors have discovered to their dismay, sea water is a hopeless means of meeting water needs. The concentration of salts in sea water is greater than that in human urine, and thus there is a net loss of water from the body in getting rid of the salts. The urine of animals such as the kangaroo rat is more concentrated than sea water, and thus the animal can eliminate the excess salts and still retain some water in its body.

Nitrogenous Waste

Much of the food that an animal takes in consists of proteins that make up the bodies of its prey or of plants. In the digestive process, proteins are hydrolyzed into amino acids, which are absorbed into the cells. Some of these amino acids are reused in synthesizing proteins needed by the organism, but the greater part of this supply is further broken down by the liver as a source of chemical energy. The first step in the metabolism of an amino acid is its conversion into an organic acid by the removal of amino groups. The amino groups are converted into ammonia during this process of deamination.

Ammonia is toxic to the organism if it accumulates in high concentration. In aquatic organisms, most of the ammonia diffuses out of the body in much the same fashion as does carbon dioxide. However, disposal of ammonia would be a major problem for land organisms because ammonia does not diffuse readily into the air as does carbon dioxide. A water solution of ammonia would have to be excreted, and, because only very dilute ammonia solutions can be tolerated, this method of disposal would result in a great deal of water loss from the organism. In adult amphibians and mammals, ammonia is combined with carbon dioxide to form *urea*, a substance that can be tolerated by the organism in concentrations substantially higher than that of ammonia. Even with urea, a substantial amount of water must be used in excretion unless the excretory system is capable of reclaiming water from the urea solution after it is isolated from the general internal fluids in a special excretory system. In egg-laying land animals such as insects, reptiles, and birds, ammonia is converted to *uric acid,* a substance that precipitates out of solution to form a solid and thus is kept out of the internal fluids. The solid or nearly solid uric acid can be expelled from the body with relatively little loss of water.

Excretory Systems

Plants have relatively simple excretory functions. Both the oxygen produced as a waste product of photosynthesis and the carbon dioxide produced as a waste product of respiration serve as a nutrient for the other metabolic function. Excess amounts of either gas are readily diffused into the atmosphere through the same exchange surfaces that bring these materials into the plant body. Stomata regulate this exchange in order to prevent excessive water loss. Membranes of the root epidermal cells selectively move ions in and out of the organism, regulating ion concentrations inside the plant. Because the plant does not take in significant amounts of organic materials, it has relatively few nitrogenous waste products. In some woody plants, nitrogenous compounds such as lignin and alkaloids are deposited in the empty tubes of old xylem tissue, forming the dark and solid heartwood of the stem. It is not clear whether this process serves primarily as a means of disposing of these materials or as a means of strengthening the structure of the stem.

In a simple unicellular organism, waste materials diffuse out of the body. An internal environment with substances present in concentrations different from those in the external environment is maintained through active transport and selective permeability of the cell membrane. Protozoa use active transport to concentrate waste products in vacuoles, which move to the cell membrane and discharge their contents to the exterior.

Most invertebrate excretory systems consist of simple tubes leading from various parts of the body to the exterior. Waste products and water are diffused or actively transported into these tubes by cells that line them, and cilia within the tubes move the solution out of the body. As the fluid moves along the tubes, cells absorb materials from or secrete substances into it, thus closely regulating the composition of the fluid that is finally expelled and the internal composition of the body.

In vertebrates, the kidney adjusts the concentrations of various ions. In addition, the kidney helps to regulate blood concentration of glucose and excretes nitrogenous wastes such as urea, products of hemoglobin breakdown, and creatinine formed as a waste product of muscular activity. Useless materials that find their way into the organism across the intestinal or respiratory epithelia are also excreted by the kidney. These functions are accomplished by a combination of three processes: *ultrafiltration* of blood, *reabsorption* from the filtrate of materials required by the organism, and specific *secretion* of certain materials directly into the filtrate.

The kidney is essentially a collection of many thousands of similar small units called *nephrons*. Each nephron consists of a network of blood capillaries and a *renal tubule.* The wall of the renal tubule is made up of a single layer of epithelial cells resting on a basement membrane. In the other part of the kidney, the *renal cortex*, the tubule ends in a cup-shape structure, *Bowman's capsule*, which surrounds a network of capillaries. A narrow, twisting portion of the tubule (the proximal convoluted tubule) extends from the capsule to a straighter *loop of Henle*, which loops into the inner portion, or medulla, of the kidney and returns to the cortex, where it leads

Figure 19.24a. The flatworm excretory system consists of a long collecting duct along each side of the body. Each of these ducts has input from numerous cul-de-sac flame cells.

Figure 19.24b. Excretion in *Hydra* is accomplished by each individual cell through either passive or active transport. The simple organizational level of this organism does not require sophisticated excretory organs.

Figure 19.24c. The excretory system of a mosquito consists of a group of excretory tubules that collect wastes from the body fluids and empty it into the gut. The malpighian tubules are characteristic of most arthropods.

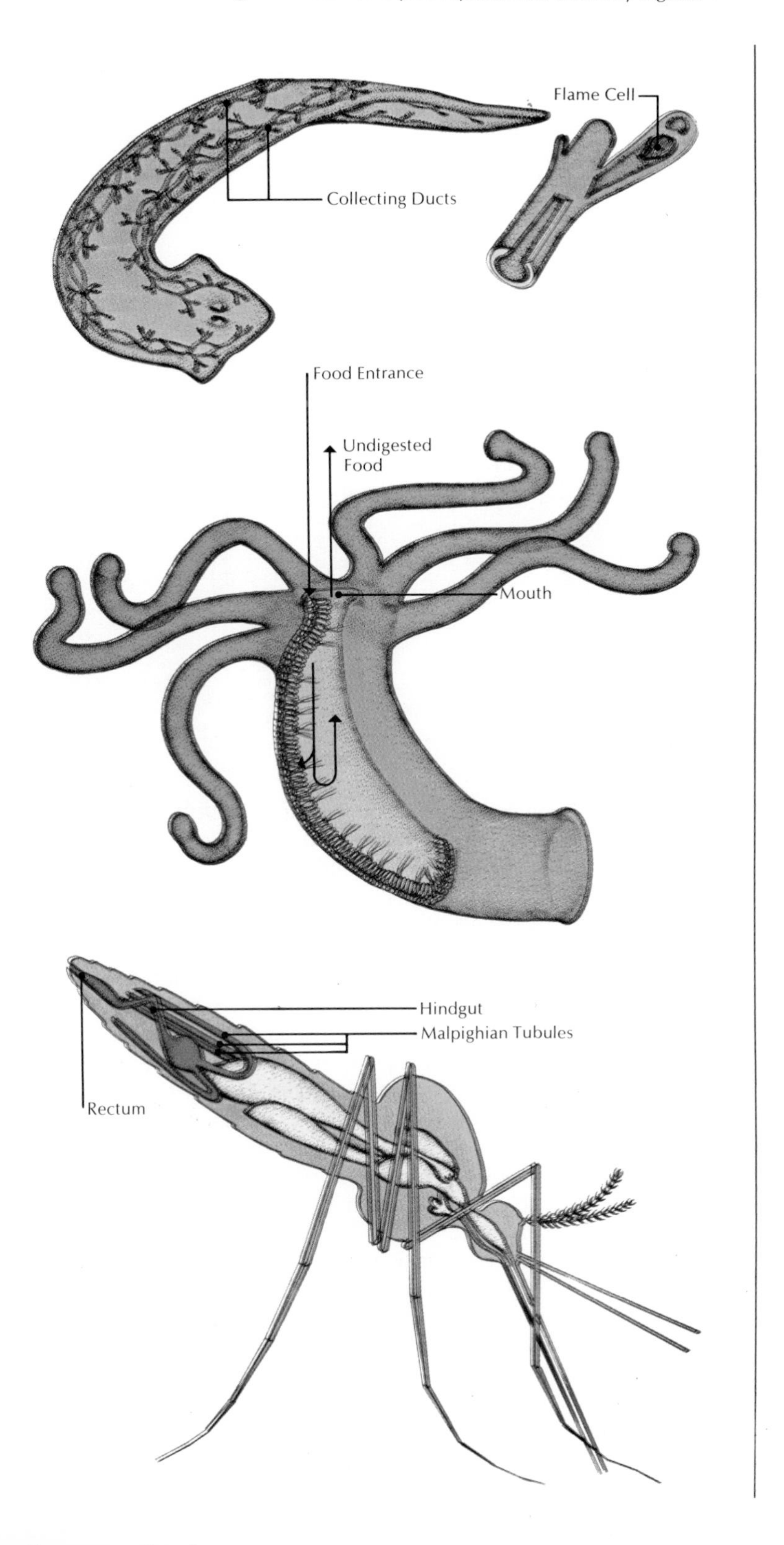

Figure 19.25. The major organ of the human excretory system is the kidney. Three processes, including filtration, reabsorption, and secretion, enable this organ to remove wastes from the blood and simultaneously conserve the useful components of the blood.

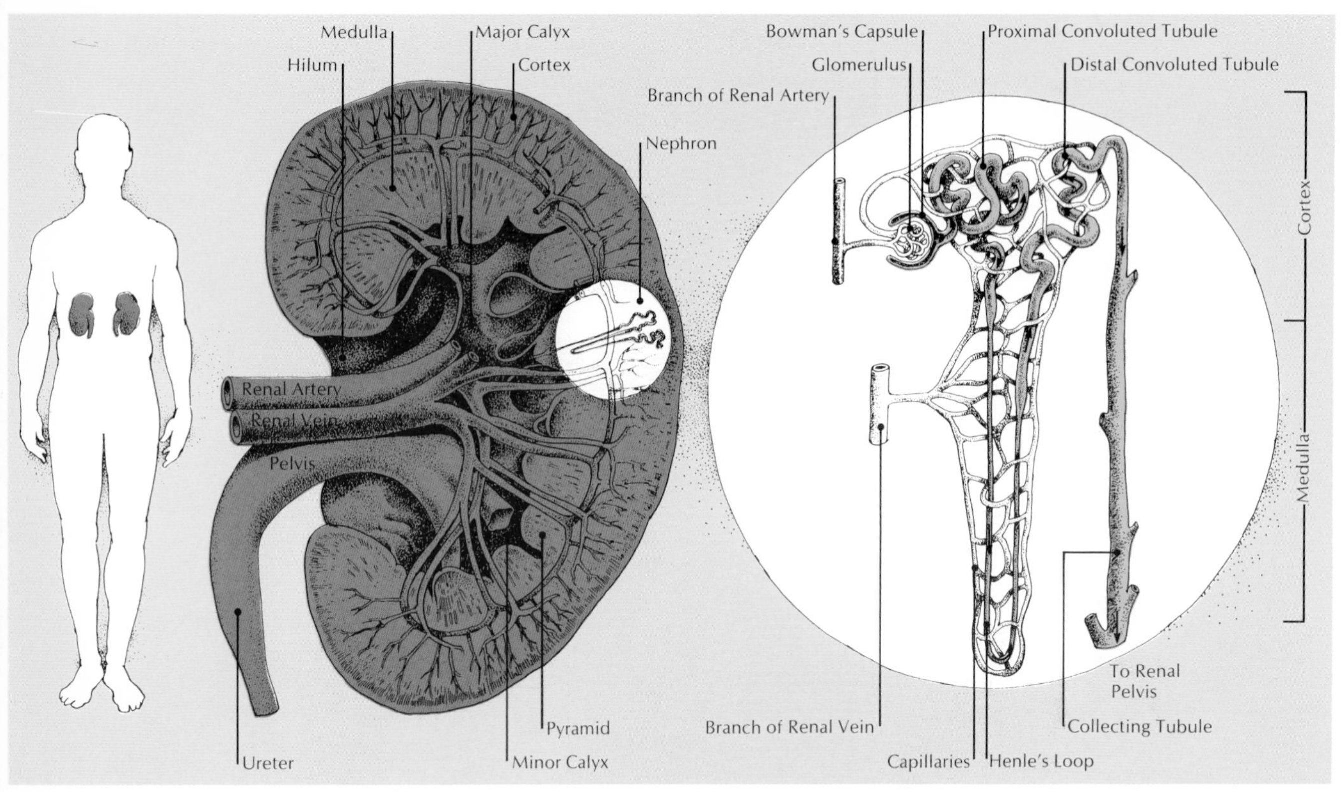

into another narrow, twisted segment of tubule (the distal convoluted tubule). This tubule joins with other tubules to form a larger *collecting tubule*, which leads back through the medulla to the *renal pelvis* in the center of the kidney.

The cluster of capillaries within the cup of Bowman's capsule is called the *glomerulus*. Blood flows into the glomerulus from a branch of the arterial system serving the kidney, and blood flows out of the glomerular capillaries into a smaller arteriole that leads to a second capillary network surrounding the proximal and distal convoluted tubules. From this network, blood collects into the veins leading from the kidney back toward the heart.

Blood filtration takes place in the glomerulus. Because the exit from the glomerular capillaries is smaller than the entrance, considerable blood pressure is built up in these capillaries. Under this high pressure, about one-fifth of the fluid portion of blood is forced through the capillary walls into Bowman's capsule, leaving only blood cells, plasma proteins, and fluid within the capillaries. The ultrafiltrate in the kidney tubule at this stage then contains the same concentration of small molecules as blood, including nutrients and salts in addition to waste products. (The human kidneys filter the blood at such a rapid rate that a volume of plasma equivalent to the total contents of the circulatory system passes through the Bowman's capsules about every 25 minutes.)

The filtrate moves through the renal tubule, where it again comes into proximity with the blood from which it was filtered. Cells lining the proximal and distal convoluted tubules are specialized for carrying on active transport of particular substances. They pump glucose, amino acids, and

Figure 19.26. Diagrammatic representation of osmotic pressure involved in urine formation.

some ions out of the tubule fluid, and these materials then diffuse back into the blood. The high concentration of dissolved materials in blood creates an osmotic pressure that forces water out of the tubule and back into blood.

Excluding water, only substances that are actively transported by the tubule cells are removed from the filtrate. This method of operation makes the nephron a fail-safe system because all waste products and foreign materials not recognized by the tubule cells remain in the filtrate and are passed out in urine. A few substances that are not removed from the blood in Bowman's capsule—hydrogen ions, for example—are actively transported by the tubule cells into tubule fluid.

Each region of the tubule processes a different group of substances. Glucose and amino acids are returned to blood in the twisted region near the capsule, whereas ions move back into the blood in all parts of the tubule except the collecting duct. Under normal conditions, all glucose is returned to the blood, whereas the amount of ions left in the tubule fluid is variable and depends upon the physiological needs of the organism. Sufficient amounts of each ion are actively transported back to the blood to restore the normal homeostatic concentration of each substance in the circulatory fluid. The urine that leaves the collecting duct represents about 1 percent of the volume of fluid that entered the Bowman's capsule.

Most vertebrates excrete urine that is osmotically less concentrated than blood, but mammals and birds produce urine of higher concentration than blood plasma. This concentration is made possible by the structure called Henle's loop, which is present only in the kidneys of these two groups. In the loop, active transport moves sodium ions from tubule fluid into intercellular fluids of the renal medulla. This movement temporarily makes the tubule fluid less concentrated, but it serves to create a strong osmotic pressure that moves water out of the collecting ducts and into the intercellular fluids as the ducts pass through the medulla. This feat of concentration of urine in the collecting duct is accomplished by a countercurrent exchange system that operates on the same principle as that in the gills of fishes.

Urine passes into the ureters—bilateral, thick-wall, muscular tubes that

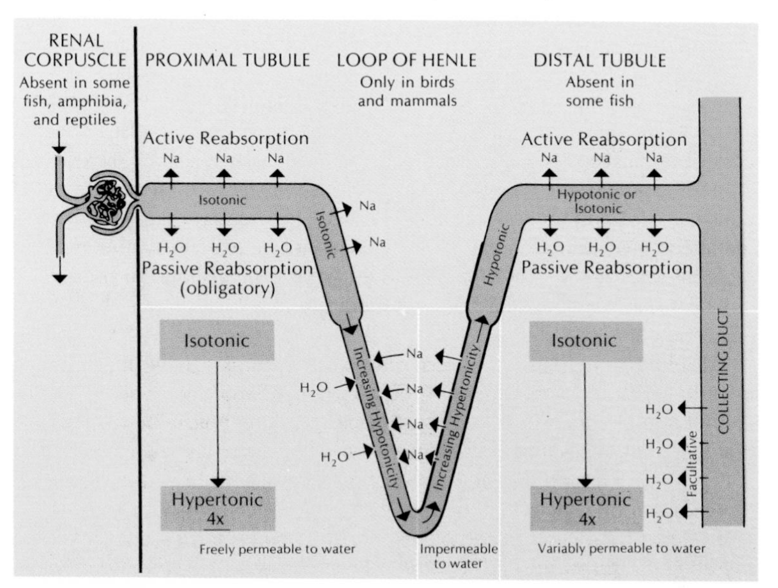

convey fluid from the kidney to the urinary bladder. The bladder serves as a storage vessel for urine. Its epithelial lining is composed of five or six layers of bulging, pear-shape cells. As urine accumulates and distends the bladder, these cells slide past one another and spread to form a thinner membrane of greater surface area. Urine passes from the bladder to the exterior of the organism through the *urethra*, which can be opened and closed by muscles that are under voluntary control.

Urine formation is subject to controls that tend to adjust the urine volume and concentration to counteract changes in the internal environment. These controls are based on two types of information—blood pressure and osmotic concentration of body fluids. When arterial pressure increases, the pressure in the glomerulus rises, forcing more blood through the filter and resulting in an increase in the volume of urine. A corresponding decrease in blood volume reduces the blood pressure and the system returns to normal. The osmotic concentration of body fluids is monitored by sensory receptors in the brain. When the receptors are exposed to an increased concentration of solutes in the fluid that bathes them, they tend to shrink. Osmotic shrinkage excites these receptors and starts a chain of events in the brain that results in the release of a hormone from the posterior pituitary gland. This antidiuretic hormone (ADH) increases the permeability to water of cells lining the collecting tubules in the kidney. The result is an increase in the amount of water returned to blood and the excretion of a smaller volume of more concentrated urine. An increase in the amount of water returned to blood decreases the osmotic concentration of body fluids, again tending to restore normal conditions. Two common substances affect these control mechanisms. Caffeine, found in coffee, stimulates the arteries to contract, raising the blood pressure and producing an increased flow of urine. Alcohol inhibits a component of the system that maintains a steady level of antidiuretic hormone. When the secretion of ADH is cut down, less water is removed from the urine and the volume of urine increases.

TEMPERATURE REGULATION

Enzymes regulate the chemical reactions that make up the life processes of all organisms, but enzymes can function only over a limited range of temperatures. Therefore, in order to survive, organisms must somehow control their internal temperature, either by finding an environment with a suitable temperature range or making some physiological compromise with the environment. The body heat of living organisms comes from the oxidation of foods. Some of the energy obtained in this way is stored in the chemical bonds of ATP, and the rest is released as heat. In one way, heat produced in metabolism is wasted energy because it cannot be used to synthesize new molecules for growth and maintenance. On the other hand, if used well, this heat can maintain the body temperature at a level above that of the environment.

All plants and animals, except birds and mammals, have little control over their body temperatures. Cold-blooded animals, or *poikilotherms*, cannot regulate their body temperatures. Plants and cold-blooded animals that live in climates that are cold for a part of the year have two alternatives—they may live for only one season or go dormant in some fashion during the cold season. Bacteria and fungi can form spores, which are very resistant to cold, and certain seed plants (called annuals) form seeds and die at the end of a growing season. When the environment becomes favorable again, the spores or seeds germinate, producing a new generation. Many plants (perennials) and cold-blooded animals go through a period of inactivity during cold weather. Their metabolic processes slow down, and they can live in a state of dormancy for several months. Nonregulating organisms adapted to cold climates often have higher metabolic rates (as indicated by their oxygen consumption) than similar animals from warmer environments.

Birds and mammals are called *homeotherms*, or warm-blooded, because they can regulate their body temperatures. In these animals, a portion of the brain, the hypothalamus, acts as a thermostat. Alterations in the external temperature are sensed in various ways. There are heat and cold receptors at various points on the body surface. The brain also responds to changes in the temperature of blood and sets various processes in motion to return the body temperature to normal.

If the body temperature rises, several mechanisms are used to dissipate heat. Blood vessels in the skin expand, increasing the amount of blood in a position to be cooled. Sweat glands secrete liquid, which spreads over the skin and evaporates, a physical process that takes up heat. Animals that are covered with fur have few sweat glands; they use evaporation of water from their tongues as a cooling mechanism. Panting moves air over the tongue and speeds evaporation.

If the body temperature falls, the thermoregulatory system tries to minimize heat loss by constricting the blood vessels in the skin. Extra heat is also produced by muscle movements such as shivering and increased activity. If a warm-blooded animal is unprotected in cold environmental temperatures for a prolonged period, it dies because of loss of body heat. Thus, there are many physiological, anatomical, and behavioral adaptations that allow plants and animals to live under all sorts of extreme conditions.

Although the millions of species of organisms represent a diversity of body structures, there is a surprising degree of similarity in the functions performed by organisms. At the biochemical or molecular level, the basic structural components and activities are similar throughout most of the spectrum of living things. Even at the level of tissues and organs, every organism possesses systems specialized to carry out a few vital functions. Although evolution has produced many different mechanisms to perform these functions, the basic similarities are in many ways more striking than the differences.

FURTHER READING

For more detailed discussions of animal physiology, see books by Barrington (1968), D'Amour (1961), Griffin (1962), Hoar (1966), Krogh (1959), Larimer (1968), Prosser and Brown (1961), Scheer (1963), and Knut Schmidt-Nielsen (1964a, 1964b). Further details of plant physiology will be found in books by Galston (1964), Meyer, *et al.* (1960), Ray (1963), Salisbury and Parke (1970), and Steward (1959, 1963, 1964).

Detailed discussions of human physiology will be found in books by Best and Taylor (1961), Guyton (1961), Ruch and Fulton (1960), and Winton and Bayliss (1962).

Among *Scientific American* articles relating to topics of this chapter are those by Adolph (1967), Bartholomew and Hudson (1961), Benzinger (1961), Bogert (1959), Chapman and Mitchell (1965), Clements (1962), Comroe (1966), Fertig and Edmonds (1969), Hock (1970), Irving (1966), Kylstra (1968), Mayerson (1963), Neurath (1964), Knut Schmidt-Nielsen (1959a, 1959b), Knut Schmidt-Nielsen and Bodil Schmidt-Nielsen (1953), Scholander (1957, 1963), Wiggers (1957), Winter and Lowenstein (1969), J. E. Wood (1968), and Zweifach (1959).

20
Plant Hormones

In 1902 British physiologists William M. Bayliss and Ernest H. Starling isolated secretin, a chemical manufactured in the lining of the small intestine when food is present. Secretin moves into the bloodstream and travels throughout the body without apparent effect until it reaches the pancreas. When it reaches the cells of the pancreas, however, that gland begins to manufacture and release the enzymes that assist in digestion. Two years later, Bayliss and Starling concluded that such chemical messengers play many important roles in animal physiology. They coined the term "hormone" (the Greek *'ormōn* means stimulating) to describe substances that are manufactured in one part of the body and distributed to other parts of the body, where they stimulate changes in certain cells, tissues, or organs.

Long before the work of Bayliss and Starling, however, animal and plant physiologists independently had suggested that such chemical messengers exist and play important roles in development and physiology.

PLANT GROWTH SUBSTANCES

Auxins

The first suggestion of chemical messengers in plants was made by Charles and Francis Darwin (1880) in their study of phototropism. Working with plants of the grass family, they found that a seedling illuminated from one side curves toward the light as it grows *unless* its extreme tip is covered with an opaque cap. Because the curved zone is well below the tip, they deduced that an "influence" moves from the tip to the parts below. A clue toward the nature of this influence was obtained in 1911 by a Danish botanist, Peter Boysen-Jensen, who demonstrated that a seedling curves toward the light if the tip is cut off and stuck back on again with gelatin. This experiment strongly suggested that the influence is a chemical substance that can diffuse across the cut through the gelatin. In 1919 a Hungarian researcher, Arpad Paál, showed that the influence accelerates ordinary, noncurving stem growth in the part of the seedling beneath the tip. By this time, it was clear that the influence is a growth-promoting substance that can diffuse through gelatin (Figure 20.1).

In their experiments, Boysen-Jensen and Paál used the *coleoptile*, or first shoot, of the oat (*Avena sativa*). The oat coleoptile, which has become the standard subject for similar studies, consists of a hollow sheath, six to seven cells thick, inside which the first few leaves are tightly rolled up. Two vascular bundles run up the sides, giving the coleoptile an elliptical cross section. In many ways, the coleoptile is like a first leaf sheath without the blade. Both the outer and inner surfaces of the sheath are covered with layers of epidermal cells punctuated by stomata. The coleoptile tip is a domelike or conical structure, solid for a length of about 0.3 millimeter from the top. The uppermost ends of the vascular bundles lie in the lower part of this solid tip. In most experiments, the "tips" removed from coleoptiles consist of considerably more than the minute solid caps.

After Paál's work, several attempts were made to extract the postulated growth substance from ground-up coleoptile tips or other plant substances. P. Stark in 1921 developed a clever technique for detection of the substance. The material to be tested for growth substance is mixed with melted agar, cooled, and cut into small blocks. Coleoptiles are decapitated and the leaves inside are partially pulled out, leaving a small piece of leaf as

Figure 20.1 (above). Boysen-Jensen's experiment demonstrated that an oat seedling coleoptile will still curve toward the light if its tip is cut off and a gelatin layer is placed between the tip and the stump. This evidence strongly suggests that the chemical substance influencing growth can diffuse across the cut through the gelatin.

Figure 20.2 (below). In Went's experiment, an excised coleoptile tip that had been exposed to a unilateral source of light was placed on two agar blocks separated by a razor blade (A). Growth substance from the tip was secreted into each of the blocks, which were then placed in contact with decapitated test plants (B and C). The block that had received growth substance from the shaded side of the tip caused the test plant to curve over twice the amount (16°) than did the plant with growth substance from the lighted side (6°). This experiment showed that light has an inhibiting effect on the amount of growth substance released by the tip or redistributed by the tip tissues.

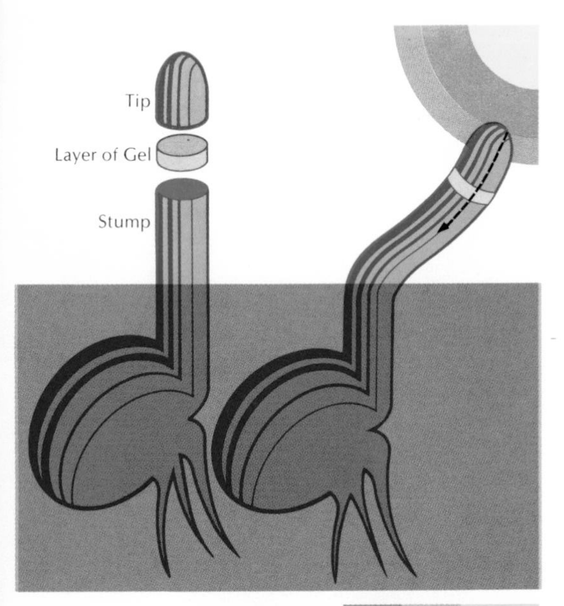

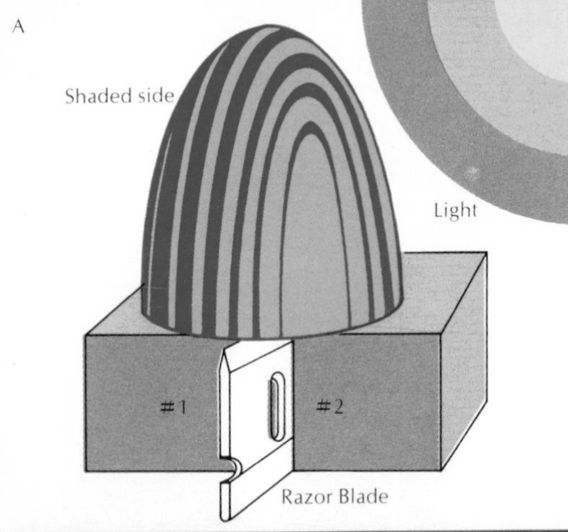

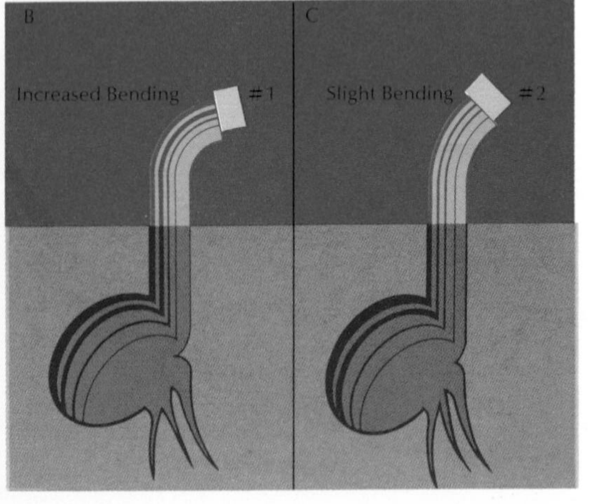

support. Each agar block is applied to one side of a decapitated coleoptile, resting against the remaining piece of leaf. If the agar contains growth substance, the side of the coleoptile that it touches should elongate more than the other side, causing the coleoptile to curve away from the agar block. Unfortunately, none of Stark's coleoptiles curved.

Stark's technique was used by another researcher, however, with greater success. Elizabeth Seubert caused coleoptiles to curve by using materials derived from malt extract and human saliva. If the technique is valid as a means of detecting growth substance, then the substance exists in organisms other than plants. Later, it was detected in several kinds of fungi.

Frits W. Went in 1928 successfully extracted growth substance by letting the living tips secrete it. He placed intact coleoptile tips on agar blocks for a few hours. The agar blocks then produced curved coleoptiles in Stark's test for growth substance. Went demonstrated that the angle of curvature of the coleoptile, measured after the agar block has been applied for a standard time, is proportional to the number of coleoptile tips that had been placed on the agar block and to the time they remained there. Thus, Stark's technique was modified to provide a quantitative measure of the amount of growth substance.

Of Went's many experiments with the plant growth substance, one was of particular importance. Russian botanist Nikolai Cholodny suggested that both phototropism and geotropism in plant shoots and roots are caused by unequal distribution of growth substance (Chapter 18). The effects of light or gravity cause more growth substance to accumulate in one side of the root or shoot tip than in the other. As the substance moves away from the tip, more growth is stimulated on one side of the root or shoot than on the other side, thus causing curvature. Cholodny, working mostly with roots, obtained some indirect evidence to support his hypothesis, but Went was able to confirm it directly. Went illuminated a coleoptile from one side and then cut off its tip. He placed the tip on a line between two small agar blocks separated by a razor blade, so that the previously lighted side of the tip was over one block and the previously shaded side was over the other block. The two blocks were then applied to decapitated test plants in darkness. The block that had received growth substance from the shaded side of the tip produced a curvature of 16°, whereas the block with growth substance from the lighted side of the tip produced a curvature of 6°. In other words, the shaded side of the tip produced more than twice as much growth substance as the lighted side (Figure 20.2). In the same laboratory, H. E. Dolk used similar techniques to demonstrate the mechanism of geotropism. In this case, the tips were laid horizontally and growth substance was collected from their upper and lower sides. The block in contact with the lower side produced about twice as much curvature as that in contact with the upper side.

With a good assay method now available, attempts soon were made to isolate and identify the growth substance. The amount in coleoptile tips, however, was far too small for chemical analysis. Experimenters in Holland found rich sources of growth substance in human urine and in yeast. American researchers obtained large quantities of growth substance from cultures of the fungus *Rhizopus suinus* maintained under certain growth conditions. The growth substance was extracted and analyzed from all three different sources, and in each case it was identified as indole-3-acetic acid, or IAA (Thimann, 1935a). This substance was already known to bio-

Figure 20.3. Three of the most active synthetic auxins—NAA(II), 2,4-D(III), and TCBA(IV)—are shown along with the structural formulae of some of the IAA precursors and derivatives shown in Table 20.1.

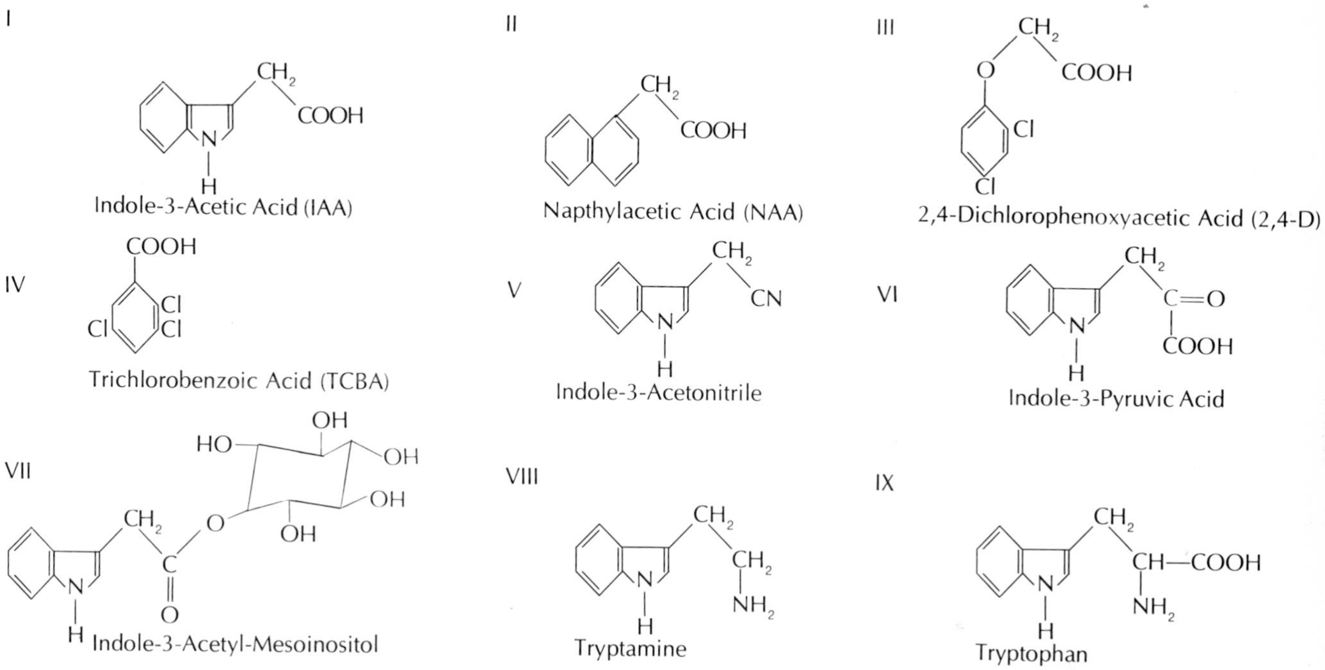

I Indole-3-Acetic Acid (IAA)

II Napthylacetic Acid (NAA)

III 2,4-Dichlorophenoxyacetic Acid (2,4-D)

IV Trichlorobenzoic Acid (TCBA)

V Indole-3-Acetonitrile

VI Indole-3-Pyruvic Acid

VII Indole-3-Acetyl-Mesoinositol

VIII Tryptamine

IX Tryptophan

Table 20.1
Naturally Occurring Derivatives and Precursors of Indole-3-Acetic Acid (IAA)

DERIVATIVE	Plant Source	Process of Conversion to IAA	System Responsible
Indole-3-acetonitrile (V)	cabbage, Brussels sprouts	hydrolysis	nitrilase enzyme (crucifers, cereal leaves)
Indole-3-acetaldehyde	several etiolated seedlings	oxidation or dehydrogenation	aldehyde dehydrogenase (milk, bacteria, and so on)
Ethyl indole-3-acetate	apples (may be artifact of using ethanol)	hydrolysis	esterase
Indole-3-pyruvic acid (VI)	corn seeds (certain cultivars)	oxidative decarboxylation requires 1/2 O_2 & evolves CO_2	spontaneous in warm alkaline solution
N-(3-indolyl) aspartic acid	pea seedlings treated with IAA	hydrolysis to IAA and aspartic acid	heating with alkali
Indole-3-acetyl-mesoinositol and its arabinoside (VII)	corn (certain varieties, each occurring in two modifications)	hydrolysis	spontaneous but hastened by acid or alkali
Tryptamine (VIII)	leaves	oxidative deamination to indole-3-acetaldehyde (q.v.)	monoamine oxidase
Tryptophan (IX)	all plant proteins	oxidative deamination, or transamination to indole-3-pyruvic acid (q.v.) followed by oxidation	not certain that conversion occurs in higher plants free from bacteria
Gluco-brassicin	cabbage family	hydrolysis liberating indole-3-acetonitrile (q.v.), with sulfate and glucose	myrosinase

Figure 20.4 (right). Structural formulae for the gibberellins GA_3 and GA_7.

Figure 20.5 (left). Some dwarf varieties of corn will become almost indistinguishable from naturally occuring tall forms after treatment with gibberellins.

chemists, although it had not been suspected to play a role in plant growth.

In subsequent years, a number of closely related growth compounds have been identified in plant extracts. All seem to owe their activity as growth substances, or *auxins*, to the ability of the plant to convert them into indole-3-acetic acid (Table 20.1). Many additional auxins have been synthesized. These molecules are similar in general structure to IAA (I), but considerable variation is possible. Three of the most active synthetic auxins—NAA(II), 2,4-D(III), and TCBA(IV)—are shown in Figure 20.3 along with structural formulae of the major natural auxins listed in Table 20.1.

Gibberellins

While the auxins were being identified, E. Kurosawa, a Japanese agricultural officer working in Taiwan, was studying a disease that causes rice plants to turn yellow and to grow excessively tall. From diseased plants, he isolated the fungus *Gibberella fujikuroi* and found that healthy rice plants develop the disease symptoms when treated with a medium in which the fungus has grown, although the fungus itself is not transmitted to the plants by this treatment. Apparently, the symptoms are caused by some substance that the fungus secretes into the medium. Chemical work in Japan led to the isolation from the fungus of an extract capable of producing the symptoms. This extract was named gibberellin A, after the fungus. Because of World War II, the Japanese research did not come to the attention of Western biologists for several years. In 1956 John MacMillan in England isolated from the gibberellin A extract a pure compound, which he called gibberellic acid. This compound not only produces the disease symptoms in rice but causes excessive stem elongation in a wide variety of other plants.

Subsequent chemical studies have led to the identification of a whole family of closely related compounds, the gibberellins, which have similar biological effects and varying degrees of activity. On the whole, GA_7 has the highest activity—more than triple that of the original compound, which is now called GA_3 (Figure 20.4). Several of the gibberellins have only very slight activity.

At first, biologists assumed that they had discovered an interesting product of fungal metabolism—a substance that acts as a "plant drug," with effects on plants as unnatural as those of caffeine or opium on man. However, two kinds of experiments first performed about 1956 revealed a very different picture of the role of gibberellins. First, it was shown that gibberellins have their greatest elongating effect on the stems of dwarf plants. Dwarf varieties of peas (such as Little Marvel) grow as tall as the naturally tall varieties (such as Telephone) after treatment with gibberellins. Some dwarf varieties of corn (though not all) become almost indistinguishable from the naturally tall forms after treatment with gibberellins (Figure 20.5). Such experiments suggest that the dwarf varieties are

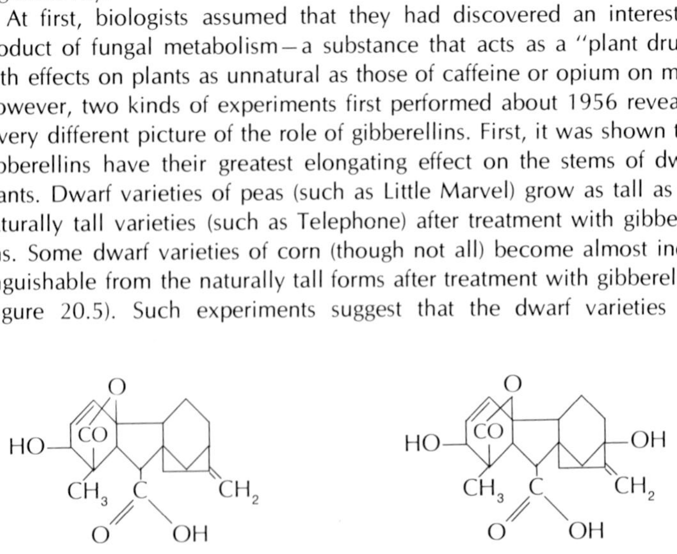

GA_7 GA_3

naturally deficient in gibberellin. Botanists began to suspect that gibberellins, like auxins, are natural plant hormones.

The second kind of experiment provided proof that gibberellins do exist normally in plants. They were found first in the seeds of a very long and straggly desert gourd, *Echinocystis*, and later in many other seeds and in bamboo shoots. Chemically, these natural plant gibberellins are members of the same group of compounds as the substances produced by *Gibberella fujikuroi*. In fact, more than one-third of the 20 or so gibberellins that have been isolated from the fungus subsequently have been found in plants. Thus, the gibberellins represent a second class of plant hormones, chemically and biologically distinct from the auxins.

Cytokinins

The discovery of a third class of plant hormones was made possible by scientific advances that followed discovery of the auxins. The long-sought goal of getting bits of plant tissue to grow in the test tube was achieved through the addition of minute amounts of IAA or synthetic auxins to the nutrient medium. This plant-tissue culture was first accomplished in 1939 in France by R. J. Gautheret with willow cambium tissue and by Pierre Nobécourt with carrot tissue and in the United States by P. R. White with tobacco tumor tissue. Many kinds of plant cells and tissues, including fruits and flowers, subsequently have been grown in culture, and there have been some striking applications—for example, the propagation of orchids has been revolutionized.

The path that led to discovery of cytokinins, the third class of plant hormones, started from the study of the tissue-culture technique itself and from the study of the effects of various additions to the medium. Cultures of tobacco pith cells normally produced masses of callus, or parenchyma tissue.

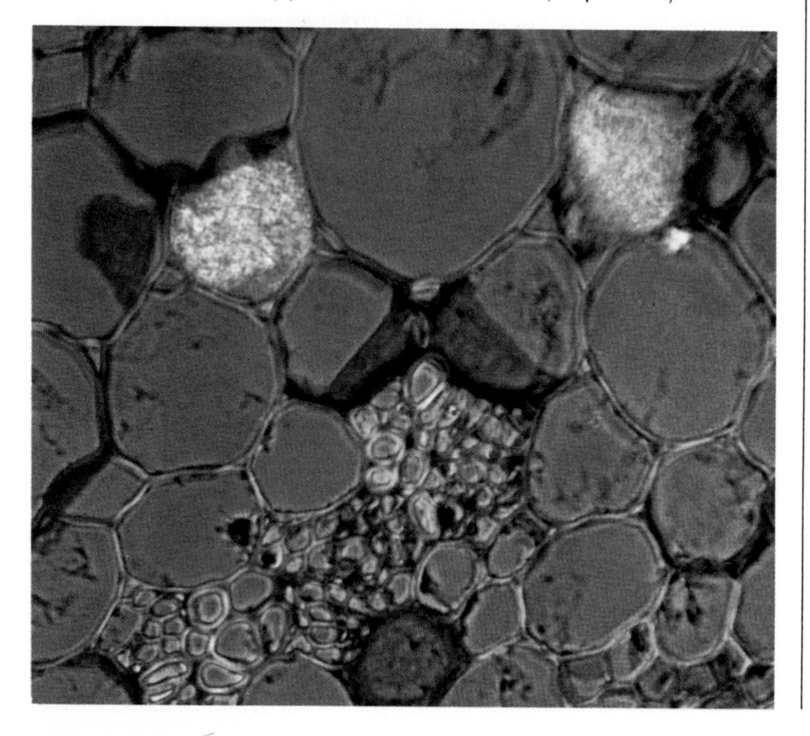

Figure 20.6 (left). Tobacco pith cells. Cultures of these cells were used in experiments that led to the discovery of the bud-forming substance, kinetin.

Figure 20.7(right). The structural formula of cytokinin.

NH—R Hydrophobic Side Chain

Adenine Nucleus

Figure 20.8 (above). The structural formula of kinetin.

Figure 20.9 (below). The structural formula of zeatin.

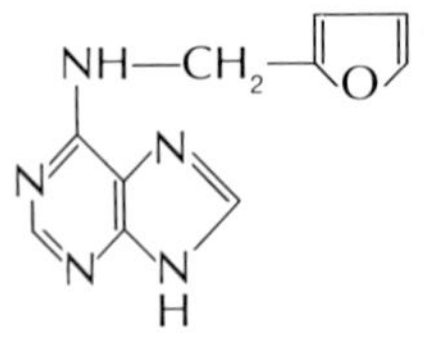

Kinetin

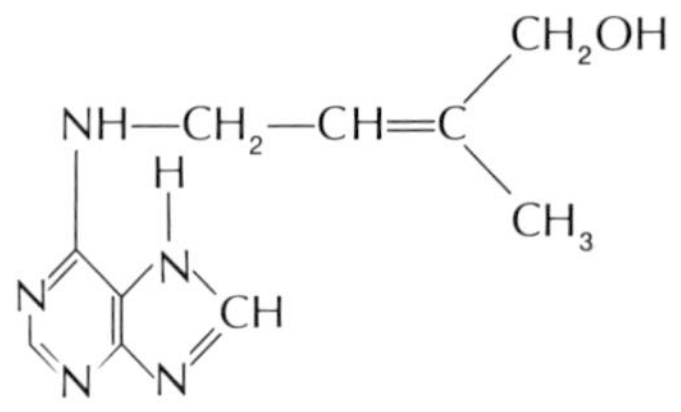

Zeatin

Addition of adenosine and its phosphate to the medium caused an increase in bud formation. Further experimentation revealed that yeast extract is more effective, and yeast nucleic acid is better still. Different samples of yeast nucleic acid extract were found to vary in bud-forming effectiveness, so Folke Skoog and his colleagues at the University of Wisconsin examined some of these samples chemically. The bud-forming activity was traced to a simple constituent, the previously unknown compound 6-furfurylaminopurine, or kinetin (Figure 20.8).

As in the case of gibberellin, kinetin was at first a substance extracted from other organisms and shown to affect the growth of plants. Evidence of a kinetinlike substance in plants was soon obtained, for numerous plant extracts were shown to produce effects on tissue cultures. Particularly effective extracts were obtained from unripe corn and young fruits in general. Preliminary purification studies pointed to a purine—probably an adenine derivative—as the active factor, and researchers in many laboratories set out to isolate and identify the active compound. The search culminated in 1964 with the isolation of zeatin from unripe corn. This compound has a hydroxyallyl side chain instead of the furfuryl group of kinetin (Figure 20.9). Later, a bacterium that causes development of multiple buds in plants was found to produce a compound identical to zeatin except for the absence of the OH group. The family of similar compounds is now called the cytokinins.

Other Plant Hormones

There are probably other plant hormones. There certainly are other compounds that control plant growth and development. A hormone, however, is defined as a chemical messenger—a substance formed in one part of an organism and transported to another part where it acts. Some substances are growth regulators, but they are not readily transported and therefore are not called hormones. Probably in this group lie the many phenolic compounds in plants, which stimulate or inhibit the destruction of auxin by oxidizing enzymes. Phenols with two adjacent OH groups inhibit this process and thus protect the auxin, whereas phenols with only one such group accelerate the destruction. Scores of phenols have been found in plants—the blue, purple, red, and some yellow pigments of flowers, fruits, and autumn leaves belong to this group. Abscisic acid, recently isolated from cotton bolls and from dormant buds (it was at first called dormin), appears to be a widespread inhibiting agent; it is not a phenol but is related to the terpenes and carotenoids. It seems to have not only a general growth-inhibiting effect but also inhibits transpiration; because it is produced in wilting leaves, it helps to protect against drying conditions.

The gas ethylene (C_2H_4) is placed in a special category. In 1901 the plant-damaging effects of coal gas were traced to its ethylene content. Orange growers long had ripened stored oranges by heating them. Biologists demonstrated that the ripening is due to ethylene from the oil heaters used in heating the oranges. One of the damaging effects of ethylene on plants is a stimulation of leaf fall. In the presence of ethylene, the petioles of the leaf blades begin to curve downward on the stem in a characteristic way. In the 1930s, it was noticed that tomato plants suffer such a drooping of the leaves when kept in a closed space with ripe bananas. Development of gas chromatography techniques, which can detect less than one part of ethylene in a billion parts of air, has made possible the proof that all fruits give off ethylene during ripening and that ripening begins when the ethylene content

Figure 20.10a (above). Photograph of etiolated pea seedlings showing the effects of various concentrations of ethylene on their growth during the 48-hour treatment period. The seedling at the far left shows the size of the plants at the beginning of the experiment. The seedling second from the left shows the amount of growth attained during the 48 hours by the untreated control. Remaining seedlings, from left to right, were treated with 10, 20, 40, 80, 160, 220, 640, and 1,280 parts per billion (10^{-9}) of ethylene in a flowing stream of air.

Figure 20.10b (below). Graphs illustrating the effects of various concentrations of ethylene (for 48 hours) on growth (increased fresh weight) of the plumules of 4-day-old pea seedlings in dark and in red light (30 ergs/cm^{-2}/sec).

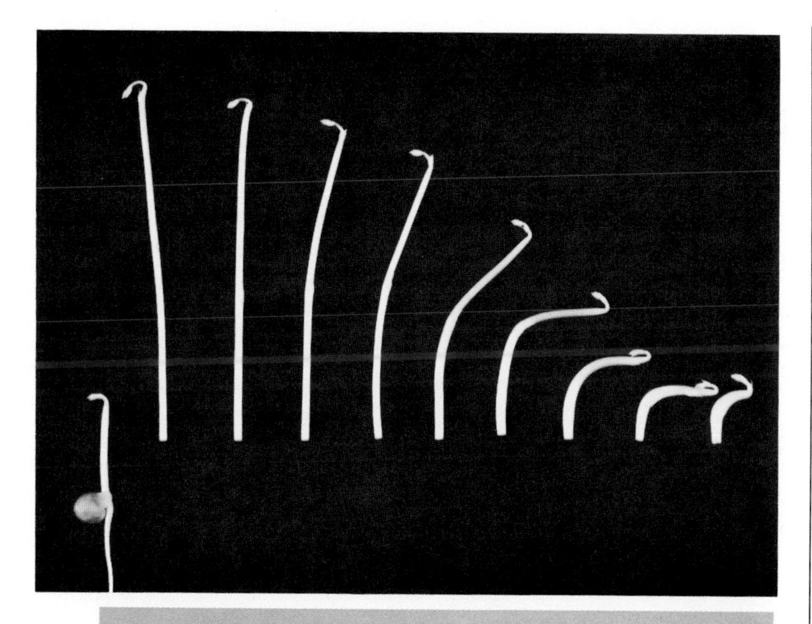

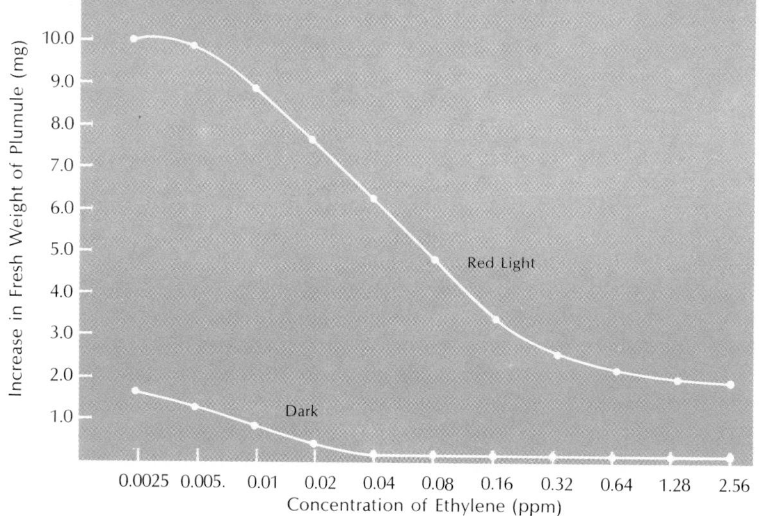

reaches about one part per million of the air in fruit tissues. Pears and cherimoyas have proven to be the most active producers of ethylene during ripening; and citrus fruits the least active.

Ethylene is a ripening substance, but is it a hormone? Application of auxin to many tissues stimulates production of ethylene. Because roots are particularly active ethylene producers under auxin stimulation, some biologists believe that the inhibitory effects of ethylene on root and bud growth are actually caused by increased auxin production. Therefore, ethylene may be regarded as a sort of gaseous hormone.

HORMONAL CONTROL OF ELONGATION PROCESSES

Phototropism in coleoptiles provides the clearest example of hormonal control of elongation processes. Light diverts the transport of auxin across

Figure 20.11. The interaction of gravity and auxin distribution is illustrated when a growing plant is laid horizontally. Auxin coming from the tip is diverted downward across the stem, so that the lower side receives up to 3 times the amount of the upper side. As a result, cell proliferation in the lower side is greater and the shoot curves upward until it regains a vertical position.

the plant and concentrates the growth hormone on the shaded side of the coleoptile tip. Similar effects can be demonstrated with the growing apices of bean, radish, or lupine seedlings. More recently, IAA labeled with C^{14} has been applied to intact tips of seedlings. Light diverts the radioactivity to the shaded side. Thus, there is no doubt that light from one side modifies the distribution of auxin between the two sides of the seedling tip. The side with more auxin (the shaded side) grows more, and therefore the plant curves toward the light.

Gravity acts in a comparable way. When a seedling is laid horizontally, auxin coming from the tip is diverted downward across the tip, so that the lower side receives about two or three times as much as the upper side. In the shoot, the lower side grows more, and the shoot curves upward until it regains the vertical position. In the root, the extra auxin inhibits elongation (perhaps through production of an inhibiting level of ethylene), so the root curves downward (Figure 20.11).

These tropisms show that extremely small changes in the auxin concentration cause significant changes in growth rate. Instead of the normal 50:50 auxin ratio between the two sides of a growing seedling, light or gravity produces a ratio of 67:33 or 75:25, and tropisms result. Because the shoot of a young seedling must find its way to the light and the root must reach the moist lower layers of the soil within the short time that the endosperm can provide food for growth, these tropisms are a matter of life or death for a young plant.

The control of straight growth, or simple elongation, is more complex because both auxin and gibberellin act in the same way. Both hormones

Figure 20.12 (above). A cabbage plant treated with gibberellic acid for some weeks showed amazing growth compared to the control plants (at left) and had to be measure with the aid of a stepladder.

Figure 20.13 (below). Graphs illustrating the relative effects of gibberellic acid (GA) and β-indoly-1-acetic acid (IAA) in stimulating extensive growth of dwarf, tall, and intermediate varieties of garden peas. Stimulation of GA increases progressively as one passes from tall to dwarf varieties while the effect of IAA is relatively constant.

cause elongation, and their effects are additive. In the oat coleoptile, gibberellin has only a small effect; thus, when auxin is applied symmetrically, the resulting increase in growth is a function of the auxin supplied. In peas, beans, and other experimental plants, however, gibberellin plays the major controlling role. A cabbage plant fed gibberellic acid for some weeks had to be measured with the aid of a stepladder (Figure 20.12). The effect of gibberellin is exerted on the internodes or, in monocotyledonous plants such as corn, mainly on the leaf sheaths. In most cases, leaves become longer, thinner, and yellower.

The control of elongation in roots is even more elusive, for here only the lowest concentrations of auxin promote growth. Auxin concentrations high enough to promote the growth of shoots only inhibit that of roots. The level of auxin produced by the root tip is usually slightly inhibitory, so that careful decapitation (without removing the elongating zone just back of the tip) may cause a temporary acceleration of the growth rate. In seedlings, some auxin also reaches the elongating zone from the shoot above. Isolated roots growing in a nutrient medium may show the promotion effect better than roots in complete plants, but it always is small. Gibberellins neither promote nor inhibit root growth. Thus, if any hormone is an important stimulator of root elongation, it remains to be discovered. Even the inhibiting effect of auxin may be indirect and due to its stimulation of ethylene production.

HORMONAL CONTROL OF CELL DIVISION

Growth essentially is enlargement, and for plants (which can sometimes shrink) the best definition of growth is an irreversible increase in volume. Dry weight may increase—for instance, when an old leaf makes starch by photosynthesis—without an increase in volume, whereas volume may increase reversibly in temporary swelling due to osmotic water intake. Nevertheless, many growth processes involve cell division as well, and because cells do not enlarge indefinitely, continued growth usually depends on

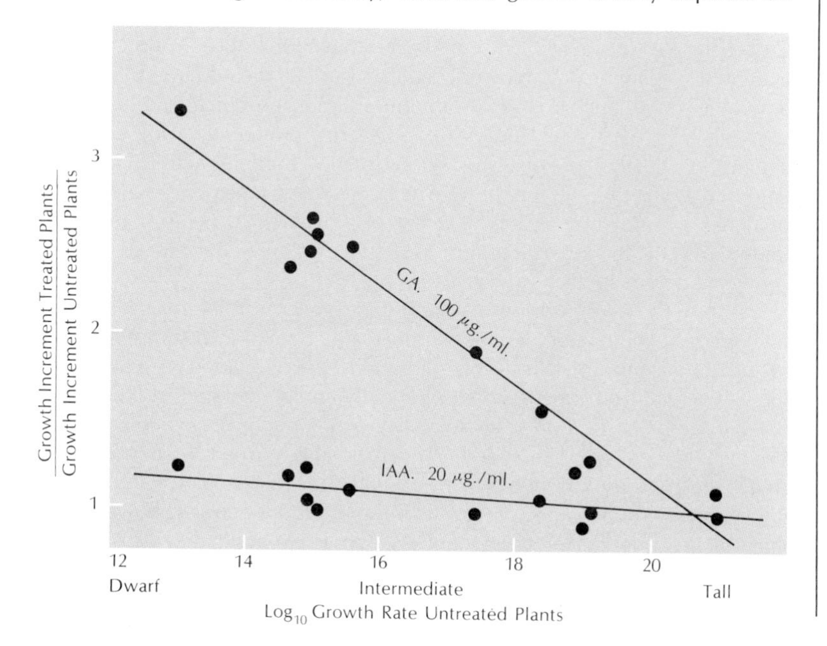

both cell division and cell enlargement. The actions of hormones on cell division are manifold.

Addition of IAA (about one part per million) to a tissue-culture medium makes possible the growth of plant fragments that would otherwise die after a few cell divisions. In the presence of auxin, the cells divide vigorously and repeatedly, making permanent tissue cultures possible. Thus, it is clear that auxin stimulates cell division.

Some tissue cultures show a curious change in auxin requirements after a series of transfers from one auxin-containing medium to another. The tissue changes from firm and solid to crumbly and watery. In its new form, the tissue is able to grow without the addition of auxin to the medium; in fact, its growth is now inhibited by IAA concentrations that formerly were optimal for growth. These *accoutumé*, or adapted, tissues resemble some kinds of tissues derived from plant tumors, which also grow without added auxin. Both tumor and *accoutumé* tissues synthesize auxin, as proved by the fact that more IAA can be extracted from these tissues than they could have accumulated from the nutrient medium. Thus, under some conditions, plant tissues can "learn" to synthesize auxin.

Auxin apparently affects the processes of DNA replication and chromosome doubling during interphase of the cell cycle. Auxin-treated tissues contain many polyploid cells. Cytokinins control synthesis of RNA and cytokinesis, or separation of daughter cells (hence their name). The combination of auxin and cytokinin produces the most actively growing tissue cultures and those with the most normal appearance. The first detectable effect of cytokinin on tobacco pith cells is a drastic increase in the amount of RNA in each cell. Values up to ten times the normal RNA content have been measured. A later effect is multiplication of normal diploid cells, so cytokinin apparently affects mitosis as well as RNA synthesis.

The most striking cell divisions in an intact plant occur in the cambium, where divisions form long new walls in the tangential plane. These divisions and subsequent cell enlargement cause thickening of the stem. At Oxford University in 1933, Robin Snow showed that there is no cambial activity in decapitated sunflower seedlings but that the addition of a small amount of crude auxin extract (from urine) stimulates typical cambial divisions. In later experiments, crystalline IAA has produced strong cambial activity in both herbaceous and woody plants. Cambial activity in trees begins in young twigs in the spring when buds open. It progresses along the branches and down the trunk at a rate of movement close to that determined directly for auxin. Scrapings of the cambium layer taken in the spring are rich in auxin.

Because the cambium normally produces xylem on its inner side and phloem on its outer side, it is not surprising that auxin also stimulates formation of vascular bundles. This formation is very marked in tissue cultures, where many formations are essentially undifferentiated. A local spot of auxin or local insertion of an actively growing bud gives rise to zones of xylem below; if sugar is added, phloem is also formed. In decapitated stems, these zones continue downward until they find their way into existing vascular bundles and join up with them. Oddly enough, however, if that vascular bundle is in connection with an active bud (an auxin source), the newly formed bundle does not fuse with it but is repelled.

A special case of xylem formation occurs when the xylem in a herbaceous stem is cut. New xylem cells gradually differentiate just above the cut and form a C-shape strand of xylem, which finally joins into the old xylem below the cut and thus reestablishes the continuity of the conducting tissue. This process occurs only if buds or young leaves are present on the stem above the cut or if auxin is applied above the cut. The number of new xylem strands so formed is proportional to the auxin concentration. As in tissue cultures, if sugar is added, phloem strands form also. Auxin thus acts to heal wounds in the plant body. The actual types of xylem elements formed, however, appear to depend also on cytokinin.

The formation of roots on stems is a very different phenomenon involving cell division. Roots will form spontaneously on stem cuttings of some plants, particularly if developing buds are present. Botanists long ago observed that roots tend to form directly below a bud and on the same side of the stem. Raymond Bouillenne and Frits W. Went demonstrated in 1933 that extracts from rice grains can mimic the effects of the bud. Application of rice grain extract to the upper end of a cutting stimulates root formation at its base. Apparently, a root-forming hormone is produced in the buds and travels downward through the stem. Auxin moves in this fashion and is a growth stimulator, and it was not long before Kenneth V. Thimann and Joseph Koepfli (1935) in the United States demonstrated that pure IAA applied to the apical end of stem cuttings stimulates root production at the base. This treatment is now used widely by nurserymen to stimulate root growth on cuttings used for plant propagation. Commercial preparations containing synthetic auxins are available for this purpose.

Root development in a stem begins with cell divisions in the layer of cells beneath the epidermis. Other divisions follow rapidly until a conical mass of small cells begins to push outward through the cortex. As this developing root elongates, vascular bundles form behind it and connect with the bundles of the stem. Even on green stems, the root is colorless. It grows out laterally at first and then begins to curve downward, showing a positive geotropism. Thus, differentiation has occurred; the new root shows characteristics typical of root rather than stem tissues. The mechanism that triggers this change in the tissue is completely unknown, except that auxin serves as the initial stimulus to set the changes in motion.

TRANSPORT OF PLANT HORMONES

Throughout the discussion of plant hormone actions, the terms "above" and "below" are prominent. In plants, the directions up and down, as set by gravitation, are very distinct; the properties of the top and bottom—or better, apex and base—of the plant are very different. A cutting can be placed upside down, but roots still form at the basal end of the cutting, although that end is now uppermost. The apex-to-base polarity of the plant is produced by the tendency of auxin to move from plant apices, buds, or young leaves down the stem toward the base. Auxin tends to move from the apex toward the base even when the position of the stem with respect to gravitational forces is altered.

This movement of auxin can be demonstrated readily in a short section cut from an oat coleoptile. An agar block containing auxin is applied to the apical end and a plain agar block to the basal end. After a short time, auxin can be detected in the basal block, and within two or three hours as much as half of the auxin travels through the stem section to the basal block. If the

Figure 20.14. The apex-to-base polarity of auxin movement is illustrated in this series of experiments. A section is cut from a coleoptile (left) and the upper (u) and lower (l) surfaces noted. An agar block containing auxin is applied to the upper surface and a plain agar block to the lower surface. Within 3 hours, the distribution of auxin in both blocks is equal, indicating movement of auxin to the block on the lower surface (both A and B). If the blocks are reversed, and the auxin is applied to the lower surface of the coleoptile, little or no auxin moves through the section into the upper surface agar block (both C and D).

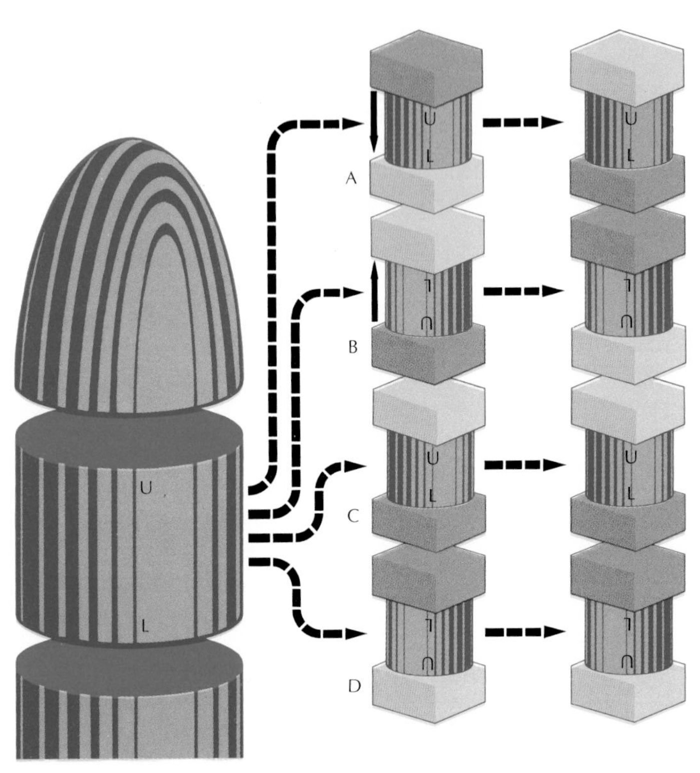

blocks are reversed and the auxin is applied to the basal end, little or no auxin moves through the stem section into the apical agar block. Experimenters can detect auxin by testing the blocks on newly decapitated coleoptiles and measuring the angle of curvature produced. Other researchers confirmed these results by using IAA labeled with radioactive carbon and measuring the level of radioactivity in the agar blocks. The rate of auxin transport in most stem tissues at room temperature is 10 to 12 millimeters per hour. At this rate, a given molecule of auxin would travel from the apex to the base of a 50-foot tree in about 2 months.

The strictness of polarity varies from plant to plant. It is generally high in young cereal seedlings, but in the stems of dicotyledons such as bean and sunflower plants, there is sometimes slight base-to-apex movement. If unnaturally large amounts of auxin are applied, the polarity can be overcome, but it requires 100 to 1,000 times the normal auxin concentrations. Auxins can be used as weed killers because high concentrations of synthetic auxins poured on the soil can be taken up by the roots, as is any other dissolved substance (nitrate, for instance), and drawn up to the leaves in the transpiration stream. Tall trees sometimes can be killed in this way. On the way up through the xylem, some auxin diffuses laterally into the living cortex cells, where it becomes subject to polar transport, which conducts it downward again. Thus, a sort of auxin circulation can occur under artificial conditions.

Transport in roots is not so simply polar. Auxin moves from the base of the shoot down into the root, but auxin also is formed in the root tip and transported from there into the elongating zone behind. Thus, there are two

Figure 20.15. Photograph of a coleus leaf showing the abscission layer at the base of the petiole. Note the small size of the cells comprising the abscission layer. The formation of these cells is dependent upon cessation of auxin flow through the tissues of the leaf petiole.

polarities, neither one very strict, and in the intermediate zone there is very little transport in either direction. Because most of the applied auxin becomes either oxidized or "fixed" in the root and disappears, radioactive auxins are being used in current studies of this transport.

Cytokinin and gibberellin are not subject to such polar transport. Gibberellin moves freely in both directions; it can be applied to the base of a stem, or even to the roots, and causes excessive elongation in the growth zone just behind the stem apex. When applied to the terminal bud, it has the same effects. Cytokinin is transported very poorly in living tissue; it has been shown to move down a petiole only to the extent of 2 percent of the amount applied in 24 hours, which is over 100 times less than IAA. Recently, however, cytokinin has been detected in the bleeding sap that exudes when the stem of a healthy plant is cut off; this sap represents water taken in by the roots and squeezed upward in the xylem by osmotic pressure. The sap is known to contain amino acids synthesized in the roots, and the presence of cytokinin as well may help to explain why roots exert so much effect on the growth rate and greenness of shoots. But this explanation is as yet far from certain.

ABSCISSION AND DEFOLIATION

When leaves become old and when fruits become ripe, they fall off, or *abscise*. The process normally depends on special cells that are formed at the base of the petiole where cell divisions begin to occur as the leaf or fruit gets older. Eventually, the soft cementing material that holds these cell walls together begins to hydrolyze and the cells fall apart. As a result, the leaf or fruit soon is held only by the vascular bundle, which breaks off in the slightest wind. All of this process is inhibited by auxin. The *abscission layer* of special cells does not form while the leaf is young and growing because the young leaf secretes a steady stream of auxin (Figure 20.15). Only when that stream wanes to a trickle and then stops does the abscission process begin. Abscission, like elongation of roots, thus is inhibited by auxin under normal conditions. Surprisingly, if massive amounts of auxin are supplied, abscission is promoted. This opposite action is probably due to ethylene, which is formed in many cells under the influence of excess auxin. Thus, concentrated sprays of synthetic auxins are used to thin crops. By this process, some young fruit fall in the spring, and the remaining apples or pears become bigger. Similarly, the army sprays trees in Vietnam from the air to clear the jungle. Both applications use a fine spray of concentrated synthetic auxin solution.

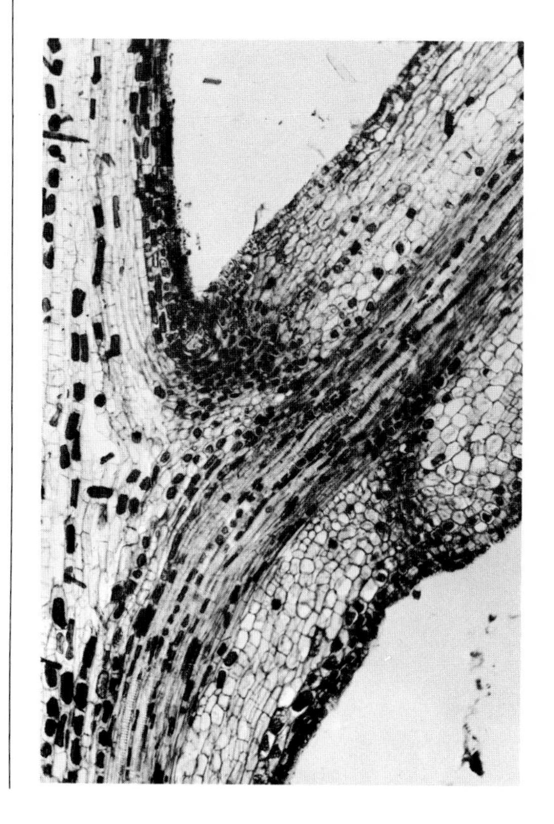

Fruits normally fall in the spring if they have not been fertilized. Without growing seeds, auxin is not produced in the young fruit. It has been shown that the June drop of apples and their fall when ripe in autumn coincide in each case with a minimum in auxin production. The picture is complicated by the fact that gibberellin also promotes abscission somewhat. This hormone seems to reach a maximum at the time of the auxin minimum, so that they work together to promote abscission. Several factors probably interact in this process.

APICAL DOMINANCE AND PLANT INTEGRATION

One aspect of the growth of plants that clearly depends on the interaction between two or more hormones is the influence of buds upon one another.

Figure 20.16. Effects of widespread dissemination of auxins by air on the forests of Vietnam.

Growing buds secrete auxin, which travels down the stem to elicit the formation of roots below. That same auxin inhibits the development of lateral buds on the stem. The growing terminal bud thus prevents other buds from developing (Chapter 18). If the terminal bud is removed and auxin is applied in its place (in amounts comparable with what the bud would have produced), lateral buds remain inhibited. Before auxin was known, such inhibitions were ascribed to the withdrawal of materials for growth by the developing terminal bud, the others thus being starved out. However, auxin applied in such a small concentration that it produces no visible growth of the stem still produces complete inhibition.

The inhibiting action of the terminal bud is incomplete in some plants—larch, gingko (maidenhair), apple, plum, and cherry trees—and as a result they form "short shoots." The lateral buds open and produce a few leaves or a flower, but the lateral shoots do not elongate more than a millimeter or two. They remain short throughout the season and develop into normal, or "long," shoots only if the terminal bud is cut off. Again, the effects of the terminal bud are closely mimicked by the application of auxin.

This action of a growth substance to inhibit a typical growth process has caused much speculation. At least nine theories have been proposed at various times to explain it. It now seems probable that the effect is indirect. Under the influence of auxin, the cells in and around the node begin to produce ethylene, which seems to inhibit the small buds arising from each node. Apparently, internode tissue forms very little ethylene.

Auxin is necessary for growth in tissue cultures, but it produces masses of undifferentiated tissue. When kinetin is added to the medium, the tissue produces numerous buds, suggesting that bud growth (as opposed to inhibition) is favored by kinetin. A simple test system has been devised in which a piece of stem with a single node bearing a bud is floated on sugar solution. After a few days the bud develops and elongates, but if auxin has been added to the solution, the bud remains completely inhibited. If kinetin now is added as well, this inhibition is relieved, and the bud grows just as well as in the controls. About two parts of kinetin to one part of auxin are required for such complete reversal, but partial reversal can be obtained with much smaller amounts. Too much kinetin, however, decreases the bud growth again, so that the phenomenon evidently depends on an exact balance between the two hormones.

Cytokinin also can be applied directly to the bud and cause it to grow out. Because cytokinins are poorly transported, the application must be exactly on the bud and not merely nearby. Auxin, on the other hand, can come from the apex many centimeters away. Application of cytokinin directly to the bud can cause outgrowth not only of the lateral bud itself but also of smaller buds at its base, so that a mass of little buds (called a witch's broom) develops. This mass closely imitates a well-known bacterial plant disease, and it has been shown that cultures of the bacterium on a nutrient medium synthesize a cytokinin. Thus, this particular disease has a rather simple explanation.

The outgrowth of a lateral bud under the influence of cytokinin apparently is due to the formation of a functional vascular bundle leading to the bud. While the bud is inhibited, its vascular connection to the main stem is incomplete, and the units of xylem appear short and not well adapted for conduction. Kinetin causes a connection of normal xylem with long functional units within about 72 hours. A full understanding of hormone action in this or any other function, however, is a long way off.

THE FORMATION AND DESTRUCTION OF AUXIN

Hormones that are effective in such small amounts must be destroyed rapidly, for their accumulation might cause serious abnormalities. Little is known about the destruction of cytokinin and gibberellin, but the destruction of the natural auxin IAA has been well studied. It is brought about by the enzyme *peroxidase*, so called because it normally causes peroxide (H_2O_2) to oxidize organic compounds such as phenols or ascorbic acid, which occur widely in plants. In oxidizing IAA, however, the peroxidase uses oxygen (O_2) instead of H_2O_2; only a trace of H_2O_2 is needed, apparently to keep the enzyme in an active form. The CO_2 of the acid group of IAA is removed and the products rearranged to form a mixture in which 3-methyleneoxindole predominates (Figure 20.17). This compound is totally inactive as a growth hormone and may even have a very weak (and probably unimportant) growth-inhibiting effect.

IAA also can be destroyed by light in the presence of certain activating pigments, such as eosin or riboflavin. Prolonged exposure to bright light is required, and the process is of doubtful biological significance. Ultraviolet light is more effective and can even cause some destruction without an activating pigment, the IAA itself absorbing the ultraviolet light. Among synthetic auxins, 2,4-D is subject to a similar action of bright light in the presence of riboflavin. Its side chain is removed through oxidation, leaving 2,4-dichlorophenol.

The discovery that γ-phenylbutryic acid could suppress the auxin stimulations of growth led to the discovery of chemicals called antiauxins. Since then, various chemicals have been found to be antiauxins. Many of these compounds are synthetic, as naturally occurring ones have not yet been identified chemically.

Strictly speaking, the term "antiauxin" should be used only for those compounds that compete with auxin for the two reaction sites of the sub-

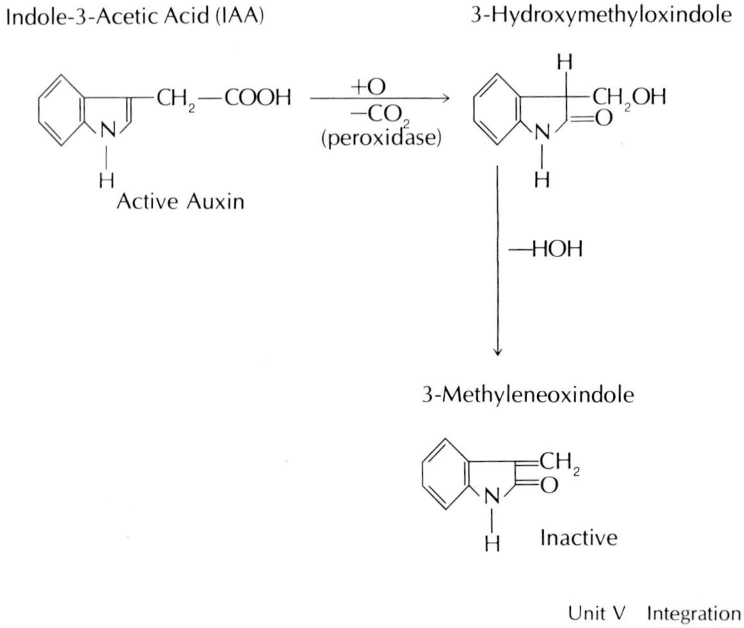

Figure 20.17. Destruction of the natural auxin IAA.

Indole-3-Acetic Acid (IAA)

$-CH_2-COOH$

Active Auxin

$+O$, $-CO_2$ (peroxidase)

3-Hydroxymethyloxindole

CH_2OH

$-HOH$

3-Methyleneoxindole

$=CH_2$

Inactive

strate. An auxin has a benzene ring with the paraposition open and an acid group on the side chain, with specific distance between the two. Antiauxins lack at least one of the requirements of an auxin and so could not fill the two active sites of the substrate.

If only one of the reaction sites is filled, then the product formed is inactive. Antiauxins could fill only one site, thus forming an inactive complex. An excess of auxin could also act as an antiauxin because two molecules could fill up the active sites of one substrate.

THE ROLE OF PLANT HORMONES

Biologists who study animal hormones are accustomed to thinking of the animal body as subject to a multiplicity of hormonal controls—the thyroid hormone controls metabolism, the parathyroid hormone controls calcium deposition, the sex hormones control gonads and secondary sexual characteristics, other hormones control digestive processes, and so on (Chapter 21). These researchers speak of the endocrine system in higher animals and call their science endocrinology. Superimposed on the hormonal system of animals is another control system—faster in action and more localized in its effects—the nervous system.

In contrast, the integration of activity in a higher plant—so far as now is known—is accomplished by a simpler hormonal control system. Relatively few plant hormones are known; these hormones for the most part are not produced in specialized glands. The effects of the plant hormones on growth and metabolism are felt in many parts of the plant rather than being localized in specific target organs as in animals. For these reasons, the study of plant hormones has not become a branch of science as prominent or as independent from general physiology as the science of endocrinology. Nevertheless, hormones are of great importance in plant growth and development, and if more plant hormones are discovered, this field may become a sort of "endocrinology" of plants.

The rather generalized effects of plant hormones suggest a mode of action somewhat different from that of most animal hormones. There is much evidence to suggest that auxin and cytokinin may act through synthesis or modification of specific types of RNA. Cytokinin apparently is incorporated directly into mRNA, which directs protein syntheses in the cell. Gibberellin produces similar effects on preparations of isolated cell nuclei. Thus, the available evidence suggests that plant hormones act directly upon the cellular control mechanisms of the plant cell. This direct effect upon the heart of the cellular mechanism probably accounts for the very general activity of these hormones in plant tissues.

FURTHER READING

Further information about plant hormones will be found in books by Audus (1963), James Bonner (1966), Jensen (1962), Leopold (1955), and Went and Thimann (1937). General books on plant development and physiology listed in Further Readings for Chapters 18 and 19 also contain discussions of plant hormones.

Among the general articles useful as introductions to the study of plant hormones are those by Galston and Davies (1969), Overbeek (1968), and Salisbury (1957). Helgeson (1968) and Letham (1969) deal more specifically with cytokinins. Addicott and Lyon (1969) review research on abscisic acid.

21
Animal Hormones

The idea of animal hormones is not new. Vitalists had recognized that certain substances secreted by glands in various parts of the body have widespread effects throughout the body. For example, it had long been known that castration (removal of the testes) can cause a male to fail to develop normal secondary sexual characteristics. Thus, a mysterious power had long been attributed to the sexual organs. The Austrian physician A. A. Berthold demonstrated that transplantation of a testis into the body cavity of a castrated rooster, or capon, is followed by normal development of male sexual characteristics. Berthold explained this observation in terms of a theory similar to Darwin's theory of pangenesis, suggesting that the sex cells carry particles that travel out to the various parts of the body to direct development. One French physician in 1889 performed a well-publicized series of injections under his own skin of extracts from dog testes. He was 72 years old and claimed that the treatment produced astonishing rejuvenation of his health and sexual prowess.

The control of reproductive functions in the mammalian body provides an excellent example of the complex interweaving of nervous and hormonal controls. This system has been studied extensively because of its importance in animal breeding and in contraception. The control systems affecting development of eggs and sperms are now relatively well understood, although much remains to be learned about the control of the female reproductive system during pregnancy and birth.

MAMMALIAN SEX HORMONES

The discovery of sex chromosomes provided one of the first definite proofs that hereditary information is carried by the chromosomes (Chapter 14). It was demonstrated that the cells of a female mammal contain a pair of X chromosomes, whereas those of a male contain one X and one Y chromosome. The genetic inheritance of sex, however, is not the complete story.

The balance of various hormones in the developing embryo—a balance presumably regulated by genetic information on the sex chromosomes—plays the crucial role in controlling the development of tissues and organs toward male or female forms. Complete reversals of sex have been produced in fishes and amphibia by treating developing eggs with various hormones. Such treatments can lead to the development of apparently normal males or females, regardless of the genetic sexual inheritance of these individuals. For example, appropriate hormonal treatment of an egg with a male genotype can lead to its development as a female with a normal female reproductive system and female secondary characteristics. A sex-reversed amphibian or fish shows mating behavior appropriate to its apparent sex and produces viable eggs or sperm. However, its offspring (if allowed to develop without hormonal treatment) will show abnormal ratios of sexes, for the sexual genotype of the parent is not altered by the hormonal treatments. For example, when a normal individual mates with a sex-reversed individual, two XX genotypes or two XY genotypes may be involved. In the first case, all of the offspring will be of the same sex. In the second case, there will be two offspring of the XY sex for each one of the XX sex and one abnormal YY individual, which probably will not develop beyond the early embryo stage.

Hormonal treatments of newborn mammals do not result in sex reversal because the critical periods of sexual determination occur while the embryo is within the mother's body. If large doses of hormones are injected

Figure 21.1 (right). Sex reversal in a genotypic female medaka (*Oryzias latipes*) as a result of androgen treatments. When mated with normal genetic females, sex-reversed genetic females produced all-female progenies.

Figure 21.2 (left). Mammalian sex hormones, showing the chemical structure of the four interlinking carbon rings.

CH_3 OH

HO

Estradiol

CH_3

C=O

CH_3

CH_3

O

Progesterone

CH_2OH

C=O

HO CH_3 —OH

CH_3

O

Cortisol

CH_2OH

O H

C=O

HO C

CH_3

O

Aldosterone

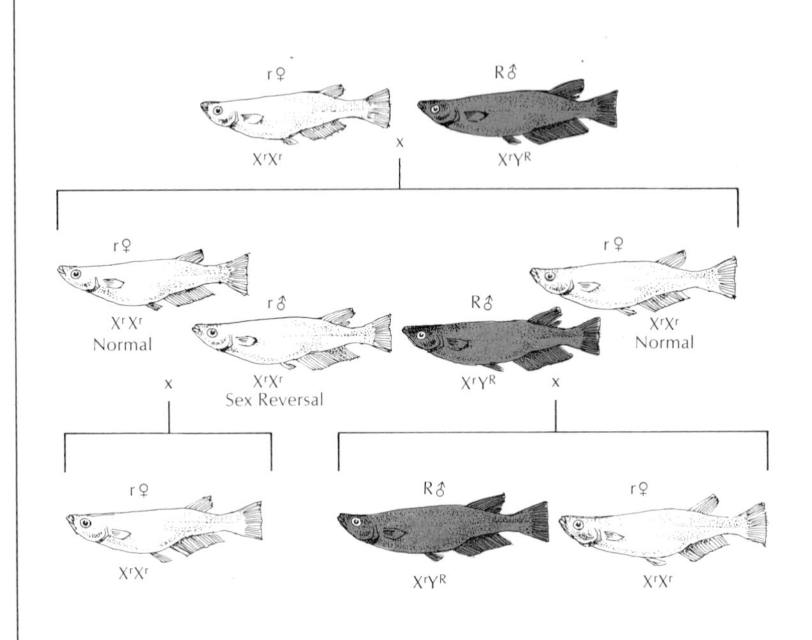

into a pregnant mother, the pregnancy usually results in abortion. However, a condition of partial sex reversal occurs in cattle as a result of the simultaneous development of two embryos (twins) of opposite sex. If the blood supply of the two embryos is intermixed within the mother, the female often develops with undifferentiated sexual organs and with male secondary sexual characteristics. Such an abnormal female is called a *freemartin*, and she is sterile. A freemartin is apparently produced by hormonal influences from the male embryo causing masculinization of the female twin (Lillie, 1917).

Androgens and Estrogens

The hormones involved in the sex-determination phenomena described above involve a group of hormones called the steroid sex hormones, which may be divided into the androgens (typically male hormones) and the estrogens (typically female hormones). Like cholesterol and other sterol lipids, vitamin D, bile acids, and some adrenal hormones, the mammalian sex hormones have a basic chemical structure composed of four interlinking carbon rings (Figure 21.2).

Androgens and estrogens were isolated and identified chemically in the 1930s. About 30 steroid hormones have been discovered in vertebrates, and all are produced by the ovaries, the testes, or the outer portion (cortex) of the adrenal glands, which are located on the kidneys. The cells that secrete these hormones share with the tissues of the kidney a common origin in the embryo.

One important androgen is *testosterone*, which is secreted by some interstitial cells of vertebrate testes during the production period of ma-

Figure 21.3a (upper right). Interstitial cells. A typical arrangement of the interstitial cells in the vertebrate testis is seen at left. At right are shown interstitial cell homologues (boundary cells) in the walls of the testicular lobules of certain teleost fishes. Interstitial cells are responsible for the production of the androgen testosterone.

Figure 21.3b (middle right). The structural formula of testosterone.

Figure 21.4a (lower left). An ovarian follicle of the bat *Myotis lucifugus lucifugus* at two levels of magnification. The micrograph to the right shows the follicle in relation to the whole ovary at lower magnification; the lefthand portion is the same follicle at higher magnification. The follicle cells (surrounding the ova) of the ovary secrete a number of estrogens, which stimulate development of secondary sex characteristics.

Figure 21.4b (lower right). The structural formula of estrogen.

ture sperm. This hormone stimulates development of male secondary characteristics, such as body hair and other male features that appear during puberty in humans. If amphibian or fish eggs are treated with androgens, the eggs will develop into male individuals, regardless of the sexual genotype of the egg. Thus, it appears that initial development of male sexual organs and other male characteristics is determined by the presence of a high level of androgens in the tissues of the embryo at a critical stage of development. Presumably, in mammals the androgens are produced under the guidance of genes on the Y chromosome.

The follicle cells of the ovary secrete a number of estrogens. Like the androgens of a male, these hormones stimulate development of secondary sexual characteristics in other parts of the body. Maturation of female sexual organs in the embryo is stimulated by the presence of a high level of estrogens at the critical period of gonadal sexual determination. The estrogens and a number of other steroid hormones help to regulate the complex cycles of ovulation and pregnancy.

Experiments with mammals show that gonad transplants or gonadal hormone injections can cause changes in reproductive tract structures and in secondary sexual characteristics, but they do not cause transformation of the gonads themselves. Nevertheless, animals subjected to such treatments early in life often are sterile. As in the amphibians, the hormones produce effects in mammals only if present during a critical period of development. In some small rodents, such as mice and rats, this period occurs within the first few days after birth. A female injected with either androgen or estrogen during this period will develop into a sterile adult female. The sterility is caused by the animal's inability to release mature eggs and, in many cases, a failure to exhibit normal behavioral responses to hormones secreted from the ovary at the time an egg is matured and ready for fertilization (Barraclough, 1966).

In mammals with longer gestation periods, such as guinea pigs and monkeys, a similar period of sensitivity to hormones occurs before birth

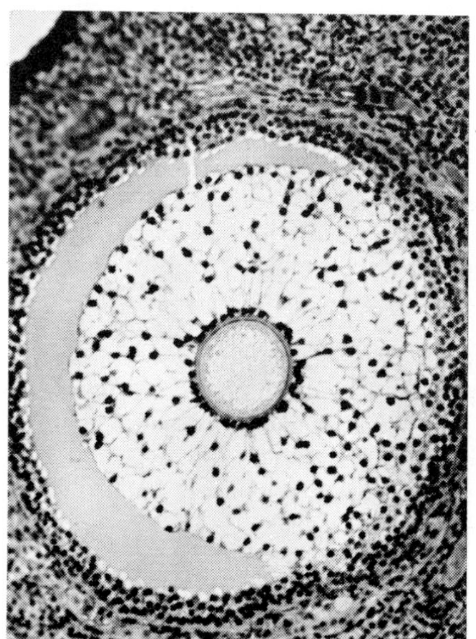

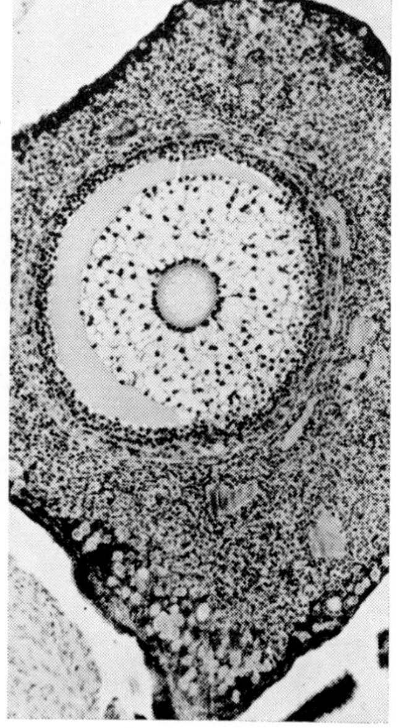

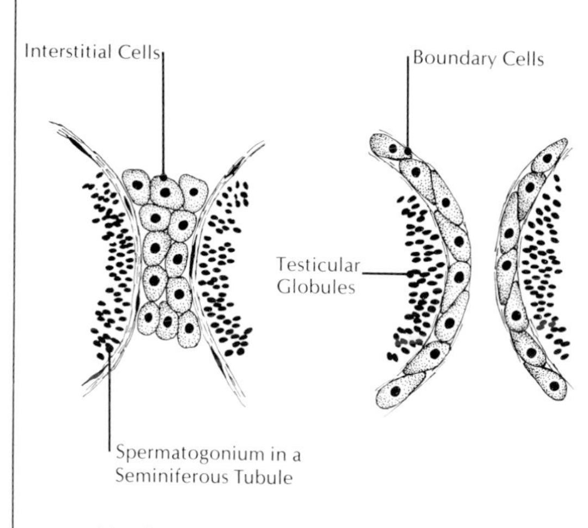

Testosterone

OH
CH_3
CH_3
O

Estrogen

O
CH_3
HO

Figure 21.5. Diagram showing transformation of the genital tracts in mammalian embryos in transition from an undifferentiated stage (above) to the male and the female condition.

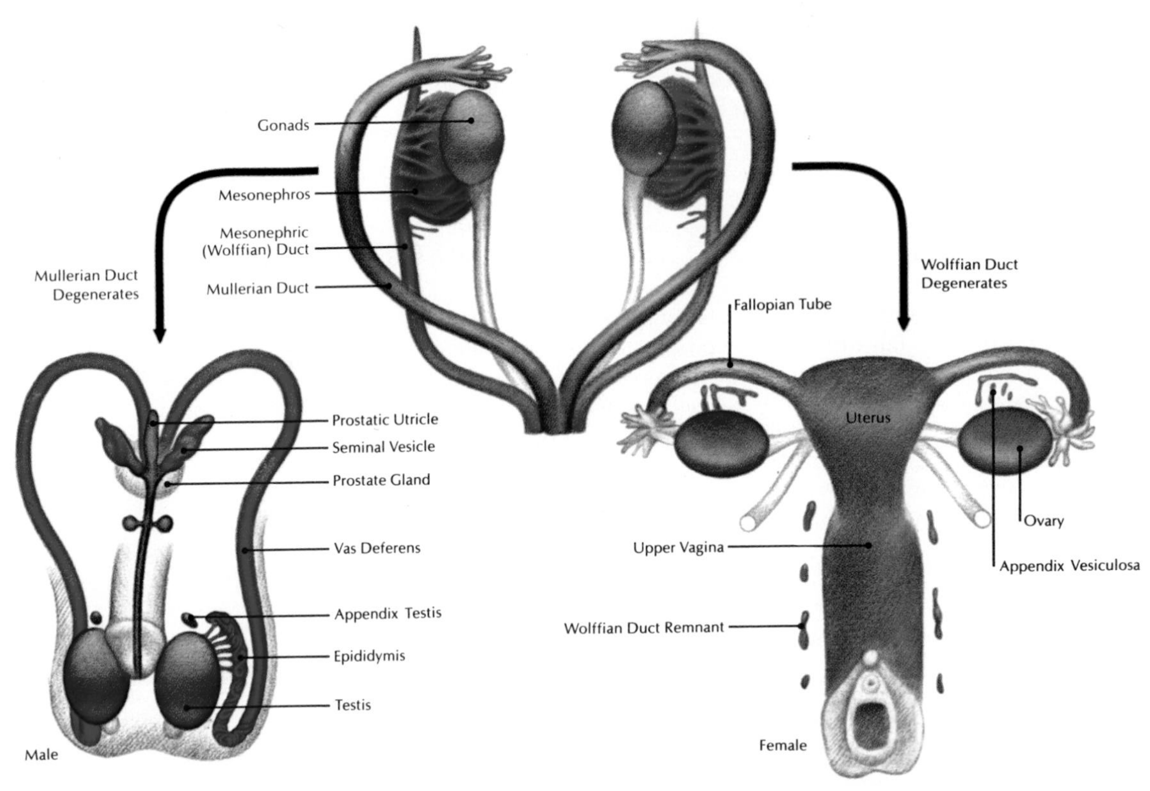

and lasts about five to ten days. A pregnant female injected with androgen during this critical period will produce sterile female offspring.

At an early stage in development of mammalian embryos, immature structures for both the oviducts of the female and the sperm ducts of the male are present. In the normal individual, only one of these embryonic structures develops into the mature reproductive tract; the other degenerates and disappears. If the gonads are removed from the early embryo, both structures remain in the undeveloped form. Thus, secretions from the gonads are apparently responsible both for the development of one reproductive tract and for the breakdown of the primordia of the other (Jost, 1955; Price, 1956). In general, a vertebrate embryo of either sexual genotype contains the primordia for formation of both male and female reproductive systems (Burns, 1961). The development of one or the other system is largely controlled by hormonal concentrations, which presumably are regulated by factors under the control of the genes on the sex chromosomes. It does appear, however, that the sensitivity of tissues to hormonal treatments depends upon the genetic constitution of the individual and that the sexual genotype plays some role in sex determination in addition to the control of hormone levels. In some cases, the primordia for the structures of the sex not genetically indicated are poorly developed and cannot be stimulated into normal growth by hormone treatments.

OVULATION

In most mammals, the release of matured eggs, or ovulation, occurs at regular intervals—for example, every four days in mice, rats, and hamsters. This

Figure 21.6. Estrous cycle of the rat. The hormonal cycles are somewhat idealized.

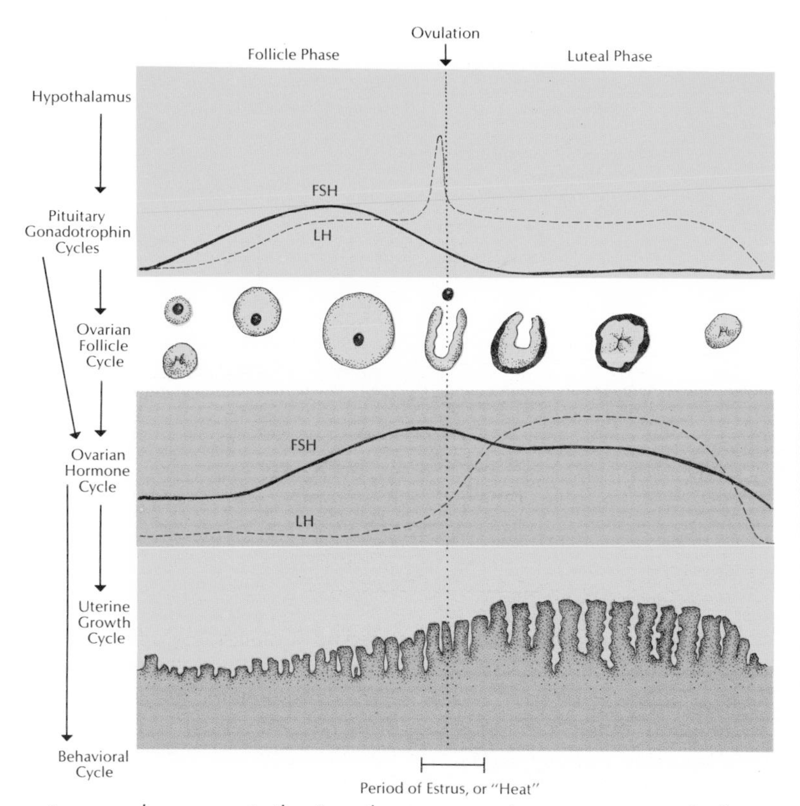

estrous cycle represents the time the ovary requires to mature and release eggs. Because the primary germ cells are determined and cease mitotic division early in development, the ovary has received its entire quota of ova long before the organism reaches sexual maturity. At sexual maturity, when the ovary begins periodically to release the number of eggs characteristic of the particular species, there is a finite population of ova that will be drawn upon throughout the reproductive life of the organism.

Completion of oogenesis and ovulation is dependent upon hormones manufactured by the pituitary gland, which is attached by a stalk to the base of the brain. Although it is a small gland, the pituitary secretes a large number of important hormones involved in the control of many body functions. The activity of the pituitary gland is controlled by chemical "factors" that are produced by neuroendocrine cells in the brain and are passed along to the pituitary through a series of special blood vessels, the pituitary portal system. This system extends from a series of fine capillaries at the top of the pituitary stalk. These capillaries run through a region of the brain projecting from the hypothalamus (which connects the pituitary to the main part of the brain), gather into larger vessels that convey the blood down the pituitary stalk, and then redivide to form capillaries that distribute the blood to the cells of the anterior pituitary. The chemical factors released by neuroendocrine cells in the hypothalamus travel through this blood system to the pituitary and stimulate the release of pituitary hormones into the general blood circulation.

The chemical factors produced in the hypothalamus have been called releasing factors. There is evidence for six different releasing factors, one

Figure 21.7. Schematic diagram illustrating the neural and vascular interrelationships within the pituitary gland. Neuroendocrine cells in the brain secrete chemical factors that pass through the pituitary portal system to the pituitary, where they stimulate the release of pituitary hormones into the general blood circulation.

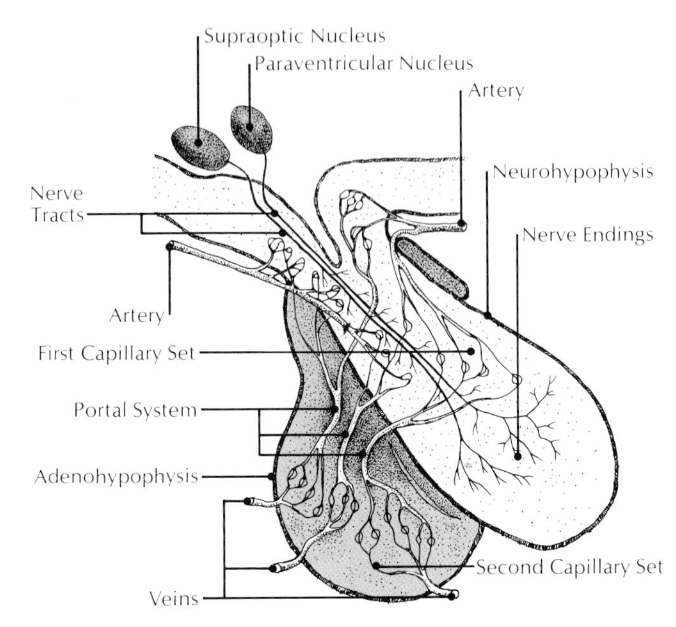

for each of the six hormones manufactured by the anterior pituitary. Two of these hormones, *follicle stimulating hormone* (FSH) and *luteinizing hormone* (LH), are involved in ovulation. The corresponding releasing factors are *follicle stimulating hormone releasing factor* (FSH-RF) and *luteinizing hormone releasing factor* (LH-RF) (McCann and Dharival, 1966).

The maturation of the follicle surrounding the oocyte requires the presence of FSH, which also stimulates the activity of follicle and nurse cells in supplying nutrients to the growing oocyte. These cells are stimulated by FSH and LH to release estrogens into the bloodstream. When the oocyte is mature, its release from the follicle is triggered by a short surge of LH from the pituitary gland. This LH surge may last less than half an hour, but it results in the rupture of the follicle some 10 to 12 hours later.

In some species, a membrane guides the released egg into the reproductive tract. In other species, the ova are released into the body cavity and are carried by ciliary currents into the funnel-shape opening of the oviduct, or *fallopian tube*, which leads into the upper part of the uterus. In mammals the ova remain in the fallopian tube for three to four days, and penetration of the sperm into the ovum (fertilization) occurs within the fallopian tube.

Fertilization is possible for only a short period after the egg is first released into the fallopian tube. Sperm released into the female reproductive tract have a relatively short life because their meager food reserves are soon exhausted by the effort of swimming the length of the uterus and through the fallopian tube toward the ovum.

Blastulation begins immediately after fertilization and is completed by the time the embryo moves from the fallopian tube into the uterus. The tissues of the uterus walls have been prepared for implantation by hormone-stimulated changes that result in thickened walls and increased blood supply. These changes begin under the stimulation of the estrogens released by the follicle cells during oogenesis. When the egg is released from the follicle, the follicle cells form a body called the *corpus luteum*, which secretes the hormone *progesterone*. This hormone stimulates a final thickening of the uterus wall and the development of a rich network of blood capillaries within the wall tissues. By the time the blastula arrives in the uterus, this preparation has been completed, and the blastula becomes embedded in the blood-rich lining, or *endometrium*, of the uterus. The embryo soon develops its own system of blood vessels, including a group of vessels, the *placenta*, that make contact with the capillaries of the endometrium. Through the placenta, nutrients are passed from the mother to the embryo and waste products are removed from the embryo. Normally, there is no direct union of the maternal and embryonic blood supplies; they are separated by the walls of the blood vessels, so that only small molecules are able to pass between mother and embryo.

Ovulation and Sexual Behavior

In all vertebrates, the estrogens act as a signal system to the brain cells, the pituitary, and the reproductive tract. As estrogen production increases in the maturing follicle, the estrogen concentration in the blood reaches a certain critical level and triggers the release of LH-RF in the hypothalamus. The LH-RF, in turn, stimulates production of LH in the pituitary, and LH stimulates ovulation. LH also stimulates the follicle cells to begin immediate production and release of progesterone. About two hours after the beginning of progesterone release, the female rodent stops fending off advances by the male and allows copulation to occur. The egg is not released from the follicle until several hours after the production of progesterone begins.

If the level of estrogens in the blood increases above the critical level, release of LH is inhibited and ovulation is prevented. If high levels of estrogens are maintained, release of both FSH and LH from the pituitary is inhibited, causing the ovarian follicles to remain immature and the reproductive tract to atrophy. The hypothalamus monitors the level of estrogens in the blood, sending appropriate signals to the pituitary to adjust indirectly the rate of estrogen production. There are two regions of estrogen-monitoring cells in the hypothalamus. One is involved in regulation of the pituitary gland, and the other is involved in regulation of sexual behavior (Everett, 1964; Lisk, 1967). Estrogen levels in the blood are increased above the critical level by steroid contraceptive pills, causing the monitoring cells in the hypothalamus to inhibit release of FSH and LH from the pituitary—a process that would normally lead to reduction of blood estrogen levels as maturation of follicles is prevented. The steady intake of estrogens in the pills maintains the high estrogen levels and thus prevents follicle maturation and ovulation as long as the pills are taken.

The corpus luteum is a round mass of cells larger than the follicle and is formed by division and growth of the follicle cells after ovulation. The cells of the corpus luteum respond to LH and FSH stimulation by releasing some estrogens and much progesterone. If fertilization does not occur and an embryo is not implanted in the uterus, the corpus luteum stops producing hormones after a time period characteristic of the species. The estrous cycle can thus be divided into a phase of follicle maturation (follicular phase) followed by a phase of corpus luteum activity (luteal phase).

When the corpus luteum ceases to secrete hormones, the tissues of the endometrium break down and blood is released from the rupturing capillaries. In humans and some other primates, the mass of tissue breakdown causes some of the materials shed from the uterus lining to be released from

Figure 21.8. Diagram showing progressive changes in the endometrial lining of the ovaries, the uterus, and the circulating ovarian hormones.

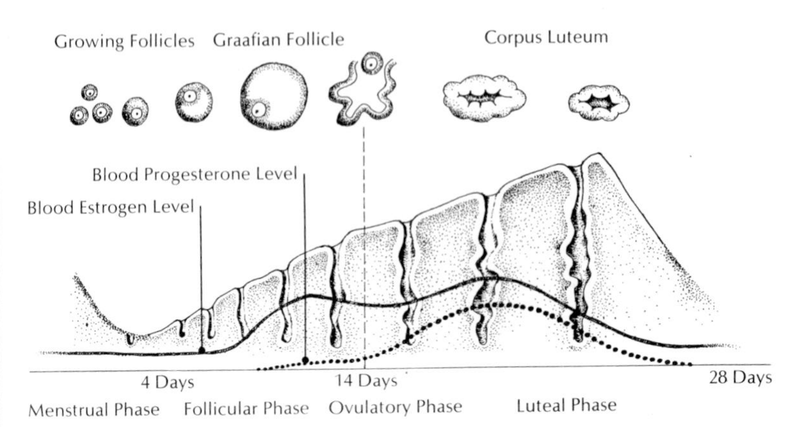

the external opening of the vagina. This process is called *menstruation*, and the estrous cycle in menstruating animals is often called the menstrual cycle.

After the corpus luteum ceases to produce hormones, the estrogen level of the blood normally decreases, causing the hypothalamus to trigger FSH release from the pituitary gland. FSH stimulates the maturation of new follicles, which release estrogens that act upon the uterus lining to halt menstruation and upon the hypothalamus to inhibit further FSH release. With most birth control pills, the intake of estrogens is halted briefly during each cycle, allowing the estrogen level to drop sufficiently to trigger menstruation, but intake of estrogens is resumed before the estrogen level has fallen sufficiently to cause release of FSH from the pituitary. Thus, the enlarged tissues of the uterus lining—whose growth is stimulated by the high estrogen levels—are removed by the normal process of menstruation, but follicle maturation and ovulation are prevented.

Intraspecies Regulatory Mechanisms

Not all species ovulate spontaneously in a regular cycle. Rabbits and members of the cat family produce a set of mature follicles that ovulate only after intense sexual excitement, usually resulting from copulation. In these species, the female becomes sexually receptive, or comes into heat, when follicles are mature. She remains in heat until mating occurs or until the mature ovum breaks down after a few days. If no mating has occurred to cause ovulation, the cat or rabbit comes back into heat after one or two days when another set of follicles has matured. The female thus continues to come into heat throughout the breeding season until mating occurs. In these species, the act of copulation provides the signal for a surge of LH from the pituitary, and ovulation follows some 10 to 12 hours later. These species are *reflex ovulators*, in contrast to the more common *spontaneous ovulators*. The dividing line between spontaneous ovulators and reflex ovulators is not a sharp one, and many outside stimuli can trigger the brain.

When female mice are kept in cages isolated from males, the estrous cycle lengthens from four days to six or seven days. If female mice are kept in large groups—for example, about 30 animals per cage—estrous cycling ceases. If males are introduced into such a cage, a significant number of females mate on the third night after introduction of the males. It is not necessary for the males and females to have contact to cause this readiness in the females but only for odor from the males to reach the females. Appar-

Figure 21.9. Structural formulae of some representative pheromones, or sex attractants. At left are formulae for the following: the pheromone secreted by the honeybee (above); dendrolasin—the pheromone produced by the *Lasius fuliginosus* ant (middle); and gyplure—the pheromone for the gypsy moth. At right is the formula for muskone, the pheromone produced by the musk deer. Most pheromones have a high molecular weight, which accounts for their high potency and narrow specificity.

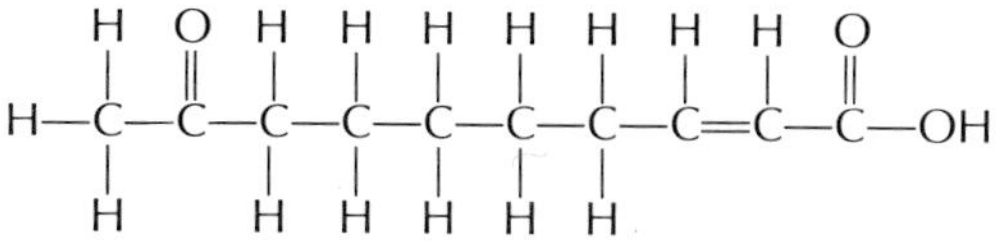

Honeybee Queen Pheromone

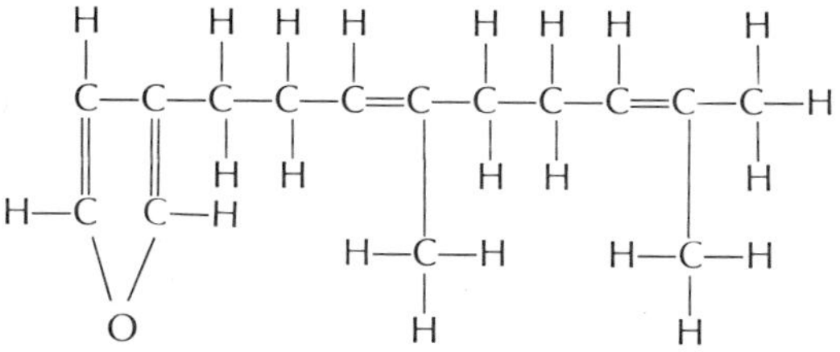

Dendrolasin

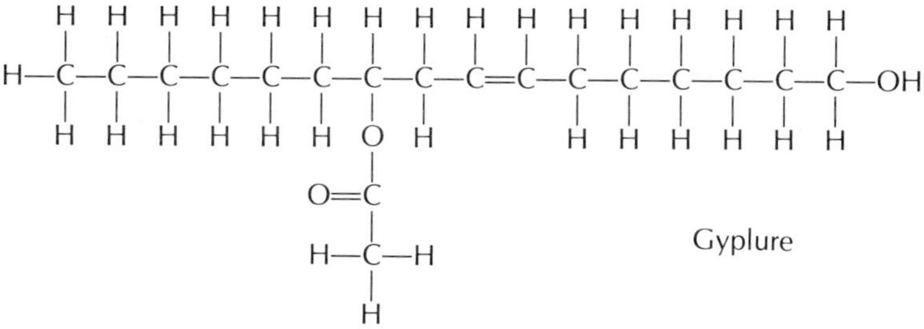

Gyplure

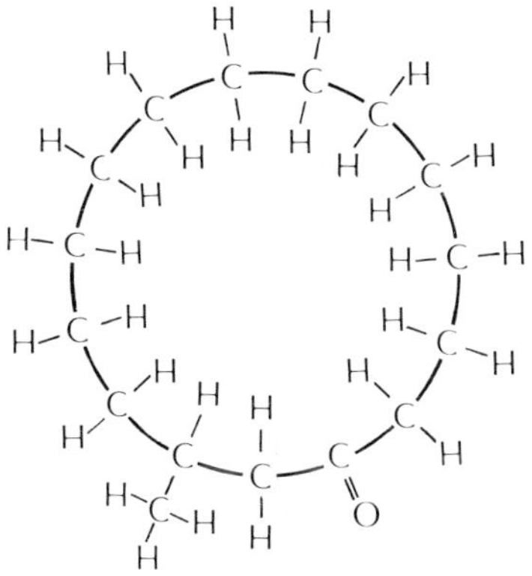

Muskone

ently, some active substance in the odor from the males causes synchronous reinitiation of the estrous cycles in the females. These substances are present in the urine of normal males but not in urine from castrated males.

Chemical substances that regulate sexual activity among members of the same species have been termed *pheromones*. Another example of pheromone action has been observed in mice. If a recently mated female mouse is exposed to a male of a different strain, chances are extremely high that she will not become pregnant but will undergo a new ovulatory cycle and return to heat in three to four days. If the "strange" male remains with her and she mates on returning to heat, she will successfully bear the litter, unless she is again immediately exposed to a different male after mating. This effect can be produced merely by placing the female in a cage containing bedding soiled by the strange male, showing that the signal that results in the so-called pregnancy block is an odor cue (Bruce, 1966).

Such experiments make it clear that ovulation and sexual behavior are highly regulated phenomena whose timing is controlled not only by internal hormonal signals but also by chemical signals between members of the species. Many of the regulatory mechanisms function to increase the likelihood that fertilization will occur just before or after ovulation. In most species, the onset of breeding-season activity in males is triggered by the same environmental signals that act in females (Davidson, 1966). Maturation of

Figure 21.10. The pituitary gland and gonads of a newborn animal can be transplanted into a mature host, where they will rapidly complete development and become capable of normal sexual activity. Here, the testis of an 11-day-old mouse embryo is shown after being grafted into the scrotal testis of an adult host for 30 days. The seminiferous tubules of the graft are approximately the same size as those of the host testis. Interstitial tissue is well developed, and many of the tubules contain spermatids. Mature spermatazoa appear in such grafts after about 35 days.

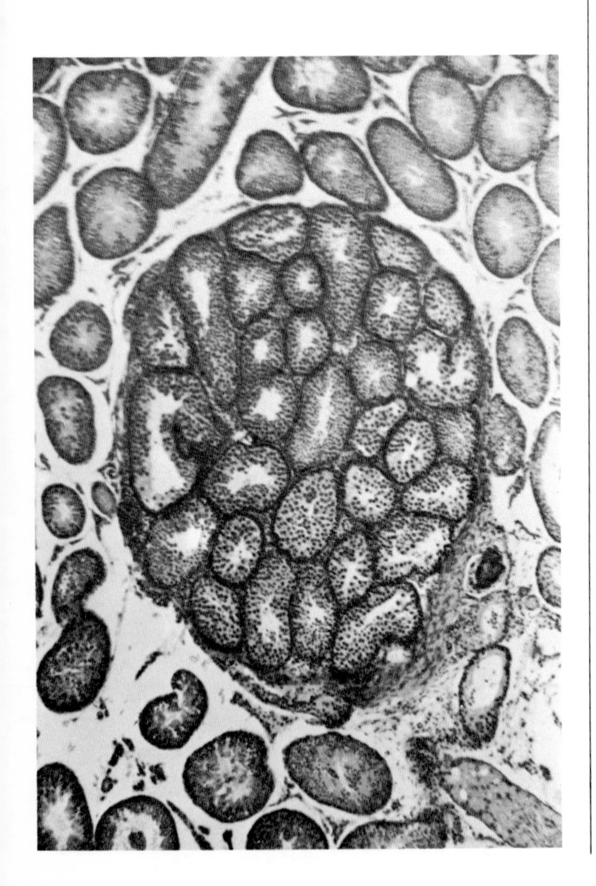

sperm depends upon adequate stimulation of the testes by LH, which also stimulates production of androgens by the interstitial cells. Some FSH may be necessary in the male to complete the maturation of viable sperm.

Mating activity in the male requires an appropriate level of androgens, particularly testosterone, in the circulation. Excessive androgen levels can inhibit LH release from the pituitary, which will result in decreased testosterone production and atrophy of the reproductive tract. As in females, monitoring regions in the hypothalamus control both pituitary activity and sexual behavior, responding to changes in the level of steroid sex hormones in the blood.

PUBERTY

Sexual maturation, often called puberty, is reached at an age that is characteristic for each species. At puberty, the gonads become capable of producing eggs or sperm, and various secondary sexual characteristics appear or are more strongly developed. The immediate cause of these changes is an increased level of androgens or estrogens in the blood. The release of these hormones from the gonadal cells is triggered by release of increased amounts of FSH and LH from the pituitary.

The pituitary gland and gonads of a newborn animal can be transplanted into a mature host, where they will rapidly complete development and become capable of normal sexual activity. Because the maturation of the gonads can occur rapidly in the proper hormonal environment, it appears that this maturation is normally delayed by a lack of releasing factors that would trigger the production of LH and FSH in the pituitary. The hypothalamus-monitoring cells of the immature animal are extremely sensitive to low levels of estrogens or androgens in the blood and inhibit the release of FSH-RF and LH-RF long before the hormone levels in the blood can rise sufficiently to cause gonadal maturation (Critchlow and Bar-Sela, 1967). At puberty, the monitoring cells cease to be inhibited by the low hormone levels, releasing factors are sent to the pituitary, FSH and LH are released by the pituitary, maturation of the gonads and development of secondary sexual characteristics are triggered, and the increased level of estrogen or androgen production from the mature gonads eventually raises the level of these hormones in the blood to the adult critical level for the monitoring cells. The cause of this change in the monitoring cells is not yet known.

Ovaries implanted in male rats that were castrated on the day of birth and then allowed to grow to adult size show periodic ovulations. However, if male rats retain testes until the fifth day after birth, similarly implanted ovaries fail to ovulate. This experiment indicates that the neural mechanism necessary for ovulation can develop in the absence of any hormonal stimuli from the gonads after birth, that a functional ovulatory mechanism can be developed in any animal regardless of its sexual genotype, and that this mechanism normally fails to function in the male as a result of some factor produced by the testes after birth.

Further confirmation comes from experiments with substances called antiandrogens, which prevent androgen from acting upon its target tissues. If antiandrogen is injected into a pregnant female rat throughout pregnancy and treatment to the young is continued for a few weeks after birth, all members of the resulting litter have the external appearance of females (Neuman and Elger, 1965). If an ovary is implanted, ovulation and female

Figure 21.11. The structural formula of cAMP.

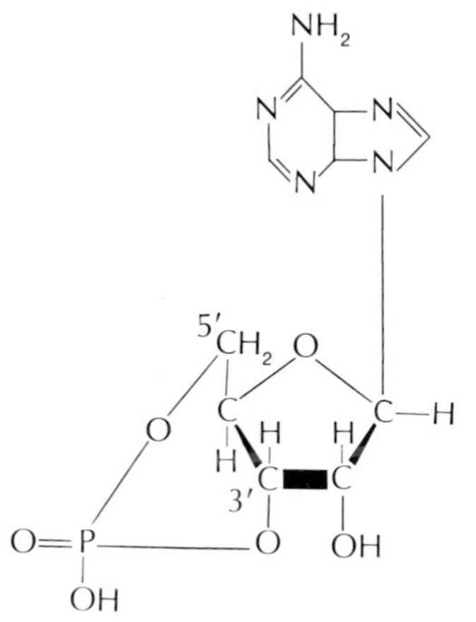

Cyclic Adenosine 3′5′Monophosphate (cAMP)

sexual responses are observed in the genetic males. It seems clear that the condition of maleness is determined by the presence of androgen at certain sensitive periods of early development.

Radioactive hormones can be injected into an animal, and the radioactivity of various tissues from the animal can be measured a few hours later. The radioactivity of the various target tissues—such as uterus, fallopian tube, pituitary gland, and hypothalamus—is many times the level found in the blood or in nontarget tissues. Apparently, the target tissues are able to capture and hold the hormone molecules, presumably as a result of special hormone receptor molecules in the target cells. Adult females that are sterile because of an injection of androgen during early development lack normal responses to injections of female sex hormones. When these animals are tested for hormone receptors by injection of radioactive hormone, little or no retention of the hormone is found in the target tissues. Early treatment with androgen apparently has destroyed the animal's ability to produce receptor molecules for the female sex hormones. This inability may be a major factor in the normal development of maleness.

THE BIOCHEMICAL ACTIVITY OF HORMONES

In man and other vertebrates, many different types of molecules have hormonal activity, and new hormonally active substances continually are being discovered. In addition to the traditional hormones—compounds produced by the endocrine glands—there are other regulatory substances (some not yet chemically identified) that influence the behavior of blood-forming organs, the immune system, and other cellular systems usually regarded as insensitive to hormonal control. Despite this overwhelming diversity, certain generalizations can be made about the chemical basis of hormonal action.

Scientific and medical researchers first became interested in hormones because of certain human diseases associated with malfunctions of the endocrine glands—diabetes, hyperthyroidism, and the often rather spectacular conditions associated with steroid hormone abnormality. It has become clear that hormones play an important part in the functions of virtually every living system, including microorganisms. This universality is fortunate for the biochemist because it offers a possibility for examining the biochemical operation of hormones in simpler organisms, where genetic and environmental factors may be more readily controlled.

Protohormones

The probable evolutionary precursors of hormones have recently been discovered in single, free-living microorganisms. In these small organisms, most metabolic regulation is accomplished by the metabolites themselves. As the concentration of a particular precursor or end product increases, that substance acts either to inhibit the action or synthesis of enzymes involved in its own formation or to induce the synthesis of enzymes that metabolize it. Such compounds have both metabolic and regulatory functions. They may be sugars that are oxidized for energy, amino acids that are incorporated into proteins, and so on. In microorganisms, biochemists have found a few small molecules that have only regulatory functions and that do not play a role in metabolic pathways (Figure 21.11). One of these *protohormones* is adenosine 3′5′ monophosphate, or cyclic AMP (cAMP).

The presence of cAMP in animal cells was detected some time ago, but

Figure 21.12. The *lac* operon mechanism of *E. coli* represented schematically, with lactose present in the system.

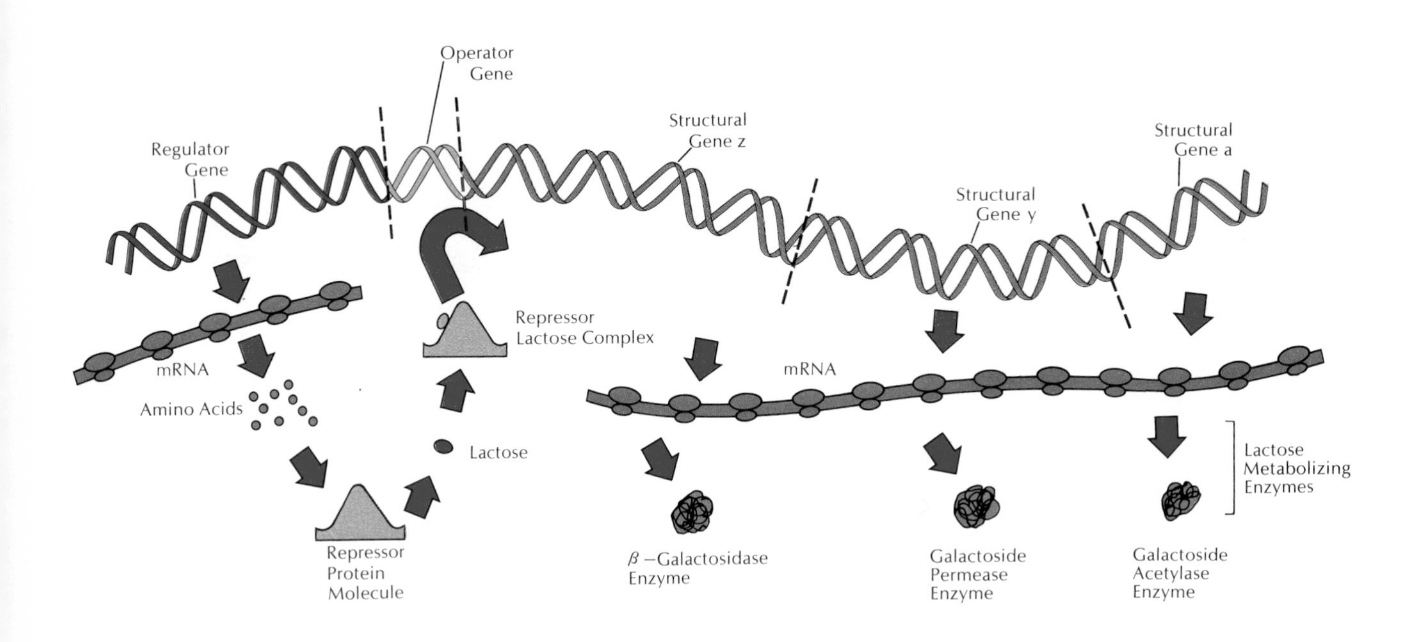

it has been found only recently in bacterial cells. Cyclic AMP is synthesized from ATP in a reaction catalyzed by the enzyme adenyl cyclase, which is found inside bacterial cells. The cyclic nucleotide is degraded in a reaction catalyzed by a specific phosphodiesterase enzyme. Thus, the concentration of cAMP in a bacterial cell depends upon (1) its rate of synthesis, (2) its rate of degradation, and (3) its rate of escape into the medium.

The *lac* operon of *E. coli* provides a good example of the regulatory function of cAMP in the bacterial cell. When lactose is present in the cell, it interacts with the repressor and removes the repressor from the operator site of the operon. With the repressor gone, RNA polymerase can begin transcribing mRNA from the structural genes that code for three enzymes involved in lactose metabolism. Under the influence of these enzymes, lactose is hydrolyzed to a mixture of glucose and galactose. The inducing action of lactose has been understood for some years. It also has been known for some time that *lac* enzymes will not be produced – even in the presence of lactose – if there is sufficient glucose in the medium to sustain bacterial growth. This effect of glucose is called *catabolite repression.* It ensures that the bacterium will not waste energy in degrading lactose so long as sufficient glucose is available for the taking.

The molecular basis of catabolite repression, elucidated only very recently, is relevant to an understanding of hormonal action in higher organisms. When bacteria run out of glucose, the concentration of cAMP within cells increases. In the presence of cAMP, RNA polymerase forms a complex with a specific protein called CR, or CAP. In this complex form, the RNA

Figure 21.13 (above). Schematic diagram illustrating the role of cAMP as a general mediator of carbohydrate metabolism in bacteria.

Figure 21.14 (below). The structural formula of epinephrine.

polymerase attaches more readily to the promoter site of the *lac* operon (Figure 21.13). Thus, cAMP acts as a positive regulator of the *lac* operon, just as the repressor is a negative regulator. Many bacterial operons are controlled by catabolite repression, and cAMP acts as a positive regulator in a similar fashion for each operon. Therefore, cAMP is a general mediator of carbohydrate metabolism in bacteria.

In the case of the *lac* operon, cAMP promotes transcription of mRNA from certain genes. It is also known to stimulate production of the enzyme tryptophanase, which catalyzes the hydrolysis of tryptophan to indole and serine—a step in a pathway that also leads to formation of glucose. Here also, an increase in cAMP concentration leads to an increase in glucose production. Preliminary evidence, however, suggests that cAMP can augment tryptophanase production even when RNA synthesis is prevented. Therefore, cAMP also must stimulate specific protein synthesis at steps in the process other than mRNA transcription. The mechanism of this post-transcriptional effect of cAMP is not yet understood, but it illustrates the fact that increased cAMP concentration mobilizes many different types of cellular responses to meet a deficiency of glucose.

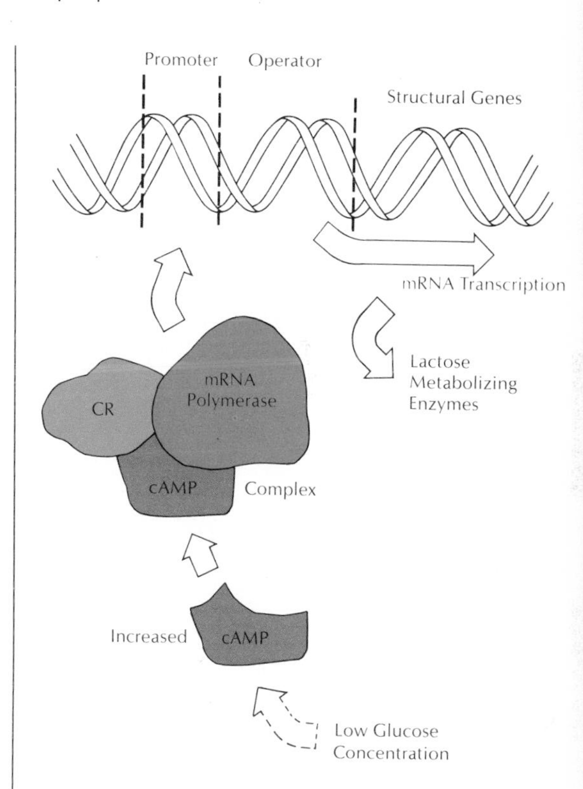

MECHANISMS OF VERTEBRATE HORMONE ACTION

In the cells of vertebrates, adenyl cyclase is found as a component of the plasma membrane rather than free in the cell interior as in bacteria. In this location, cAMP production can be stimulated by various external influences. In fact, cAMP was first detected during studies of the mechanism by which the hormone epinephrine promotes glycogen breakdown in the liver. Earl Sutherland found that epinephrine acts by enhancing cAMP production at the plasma membrane of the liver cells. The cAMP functions within the cell as a "second messenger," triggering the actual metabolic changes associated with the presence of epinephrine at the cell surface. In the liver, as in bacteria, cAMP promotes the production of glucose from a precursor.

In bacteria, cAMP acts within an individual cell, promoting glucose production when the glucose concentration decreases. In cells of vertebrates, however, cAMP functions as the mediator within the cell for other chemical signals arriving at the cell surface. As research has continued, it has become apparent that an astonishing number of vertebrate hormones function simply by stimulating cAMP production in particular target cells. The specificity of response to hormones is therefore determined by the presence of specific hormone receptors at the cell surface and by specific mechanisms within the cell that react to increased cAMP concentration.

For example, the actions of the hormones epinephrine and ACTH (pituitary adrenocorticotrophic hormone) seem very different (Figure 21.14). Epinephrine is produced by the adrenal medulla and stimulates glycogen breakdown in liver cells. ACTH is produced by the pituitary gland and stimulates steroid production in cells of the adrenal cortex. Yet each hormone stimulates adenyl cyclase activity at the surface of its target cells, thus increasing the concentration of cAMP within the target cells. The actions of the two hormones differ because the liver cells have receptors that respond to epinephrine, whereas the adrenal cortex cells have receptors that respond to ACTH. They also differ because the mechanisms within liver cells respond to increased cAMP concentration by more rapid glucose breakdown, whereas the mechanisms within adrenal cortex cells respond to

OH
OH
HOCH
CH_2
NH
CH_3

Epinephrine

increased cAMP concentration by more rapid steroid production. The chemical mechanisms by which hormones act on the cell surface to stimulate cAMP synthesis are not known.

The mechanism by which cAMP acts within the cell is better understood, at least in some cases. In the liver cell, cAMP stimulates the phosphorolysis of glycogen to glucose 1-phosphate. Detailed studies of this process have revealed the general mechanism by which cAMP modifies all such reactions. Cyclic AMP activates a protein kinase. When activated by phosphorylation, the second kinase catalyzes the phosphorylation of the enzyme phosphorylase, which thus becomes activated as the catalyst for glycogen breakdown.

Subsequent research has shown that cAMP is active in virtually every type of mammalian cell, affecting many different types of biochemical reactions. In many cases, cAMP acts to modify enzyme activity, as in the liver cells. In other cases, cAMP stimulates synthesis of specific proteins. In some of the latter cases, inhibitors of RNA synthesis block the stimulating effect of cAMP, indicating that cAMP affects mRNA transcription. In other cases, the stimulating effect of cAMP is unaffected by inhibitors of RNA synthesis, indicating that the cAMP affects posttranscriptional processes.

It is still too early to say whether this unitary hypothesis of cAMP action will be substantiated. Nevertheless, it is an appealing hypothesis to explain the multiple biochemical effects of the cyclic nucleotide.

OTHER POSSIBLE MEDIATORS

The evidence at present suggests strongly that cAMP is not the only mechanism through which hormones act. Steroids, thyroxin, growth hormones, and insulin apparently operate independently of the cAMP concentrations within target cells. It is possible that there are different "second messengers" for these other hormones. In this case, these hormones would react with specific membrane receptors that promote the synthesis of the other intermediate messengers within the target cells. This hypothesis has been rendered more plausible by the discovery of another cyclic purine nucleo-

Guanosine 3′5′ Cyclic Phosphate (cGMP)

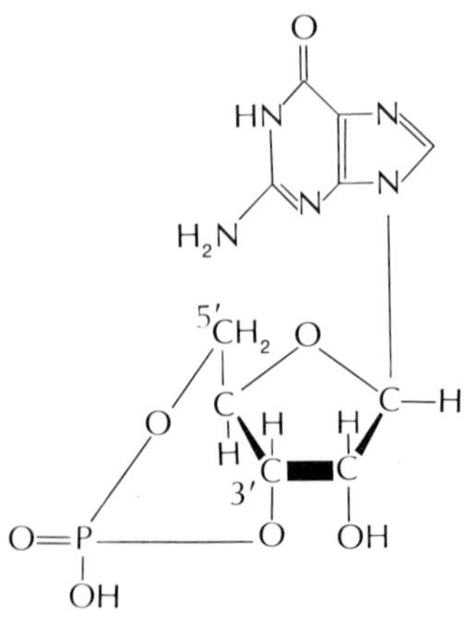

Figure 21.15. The structural formula of cGMP.

Figure 21.16. The structural formula of thyroxin.

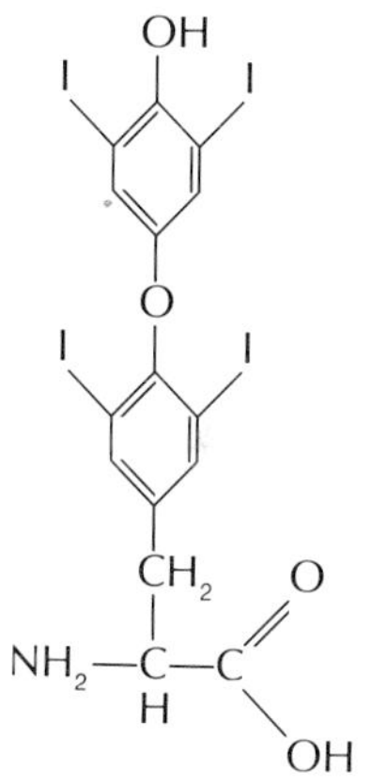

tide, guanosine 3′5′ cyclic phosphate, or cGMP (Figure 21.15). Much less is known about cGMP than about cAMP, but its enzymatic synthesis has been studied. The cGMP concentration within cells is affected by manipulations of hormone concentrations in living animals in a way that strongly suggests a role for cGMP as another mediator of hormone action.

Any other type of molecule that responds to membrane-bound hormones could act in the same way. Recent experimental results suggest that there may be specific membrane receptors for hormones that do not stimulate adenyl cyclase activity. For example, insulin—a relatively large polypeptide—remains hormonally active even when it is bound to an inert polysaccharide carrier so large that the insulin molecule is prevented from entering its target cell (in this case, a fat cell). As with other membrane-active hormones, researchers have investigated the possibility that cAMP is involved in the actions of insulin, but that possibility has been excluded in at least one case. In a line of rat hepatoma cells growing in tissue culture, insulin stimulates synthesis of an internal enzyme. Furthermore, neither cAMP nor its more penetrable dibutyryl derivative affects the rate of enzyme synthesis in these cells. Because insulin stimulates this enzyme even when prevented from penetrating the cells, either the membrane itself regulates the synthesis of the enzyme or, more probably, a molecule other than cAMP acts as an internal mediator for the action of insulin.

Much less is known about the actions of thyroxin and growth hormone. Thyroxin, a small molecule derived from the amino acid tyrosine, has profound effects on metabolism in a number of different vertebrate systems. For example, it is responsible in some way for the metamorphosis process that changes a tadpole into a frog. Thyroxin ultimately influences the synthesis and degradation of many types of macromolecules, but very little is yet known of the mechanisms under direct control of thyroxin.

STEROID HORMONES

Perhaps because of their importance in clinical medicine, the steroid hormones have aroused a great deal of scientific interest. Steroid hormones recently have been shown to have hormonal activity in some organisms simpler than the vertebrates. Apparently, the use of steroid hormones evolved before vertebrate organisms made their appearance, but steroid hormones appear to be a more recent evolutionary development than cAMP. Biosynthesis of steroid compounds has been observed only in eucaryotic cells, whereas cAMP plays an important regulatory role even in procaryotes.

The principal steroid hormones of mammals are the sex hormones (androgens, estrogens, and progesterone) and two types of adrenal hormones (glucocorticoids and mineralocorticoids). A metabolite of the sterol vitamin D has biological activity similar to that of the steroid hormones.

The steroid hormones are generally discussed in terms of their effects on specific target tissues. For example, the reproductive tract is an obvious target tissue for the action of the sex steroids. It is becoming clear that many other tissues—for example, the liver, kidneys, and the nervous system—may also be influenced by sex hormones. Adrenal glucocorticoids usually are regarded as regulators of carbohydrate and protein metabolism in various target tissues. However, it has long been known that glucocorticoids also form the negative part of a feedback control loop by suppressing ACTH release from the pituitary. Glucocorticoids also induce lysis of lymphoid

Figure 21.17. Structural formulae of adrenal hormones.

Glucocorticoids (3 primary hormones)

Cortisol

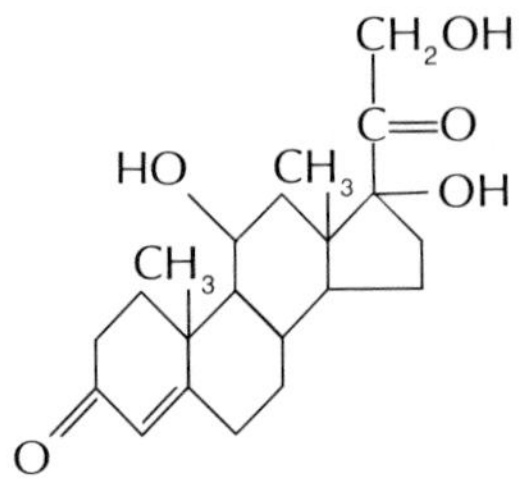

Cortisone

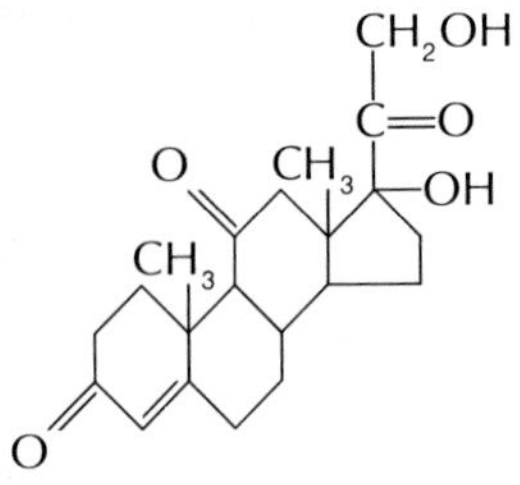

Corticosterone

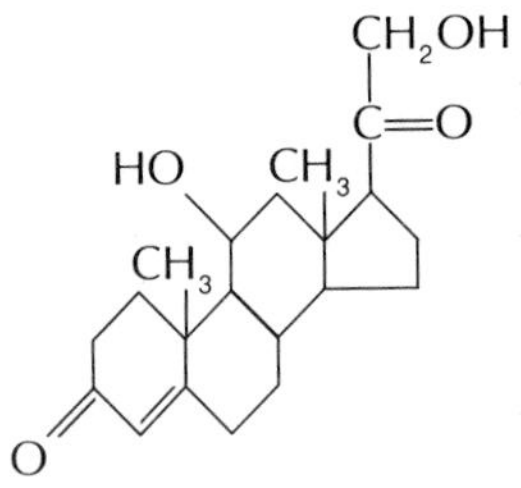

Mineralocorticoids (2 primary hormones)

Adosterone

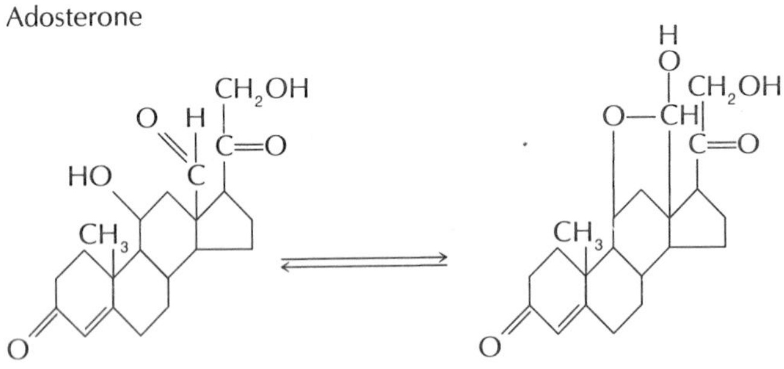

(aldehyde form)

(hemiacetal form)

11-Deoxycorticosterone (DOC)

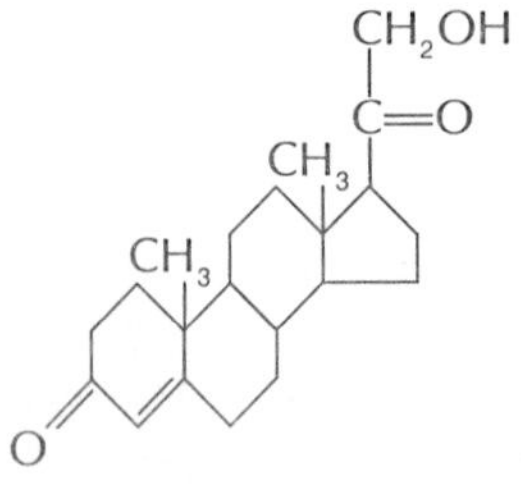

Figure 21.18. Schematic diagram of a theoretical model of hormone-gene interaction. Sex hormones (H) enter the cell and become bound to a receptor molecule (P). The bound hormone can then enter the nucleus and activate specific genes to produce proteins. These proteins in turn bring about the cellular changes triggered by the hormone.

cells, inhibit growth of fibroblasts and of regenerating liver cells, promote development of the pancreas and certain parts of the nervous system during embryonic life, stimulate growth-hormone production by the pituitary, enhance release of free fatty acid from fat tissues, and regulate many other cellular responses as well. Aldosterone—the best studied of the mineralocorticoids—is usually regarded as a regulator of water and ion metabolism in the kidney. It influences sodium transport activity in the toad bladder and probably in other organs as well. In short, the concept that each steroid hormone affects only one or a very few target organs is difficult to defend.

A particular steroid does seem to act primarily as a direct regulator of the synthesis of a few specific macromolecules in cells that are sensitive to the hormone. The precise mechanisms of steroid regulation are unknown, largely because the mechanisms by which protein synthesis is controlled in eucaryotic cells are not yet understood. It is not yet known whether control is normally most prominent at the stage of specific gene transcription, mRNA transport and degradation, or mRNA translation. Until the normal internal control mechanisms of the mammalian cell are better understood, it will be difficult to evaluate the effects of steroids on those mechanisms.

One step in the action of steroids on cells does seem to be common to all systems thus far studied. The first step is the formation by noncovalent bonds of a complex between the hormone and a specific protein receptor molecule, probably in the cytoplasm of the cell (King, 1970). This complex apparently migrates into the nucleus of the cell. The receptors in every case have a number of similar properties, although they respond specifically to different steroid hormones. All receptors yet studied are proteins of about the same size; all form aggregates when studied at low ionic strength in cell-free systems; and all require the presence of free SH groups for the formation of a steroid-receptor complex. Much of the current research on these hormones is directed toward a search for a common underlying mechanism of action for all steroid hormones.

The outstanding question in present steroid research is the exact step at which the steroid-receptor complex affects the process of protein synthesis. Studies of posttranscriptional enzyme induction by steroid hormones in tissue cultures have led to the development of a model of the control mechanism that seems to account for more observations than does a modified

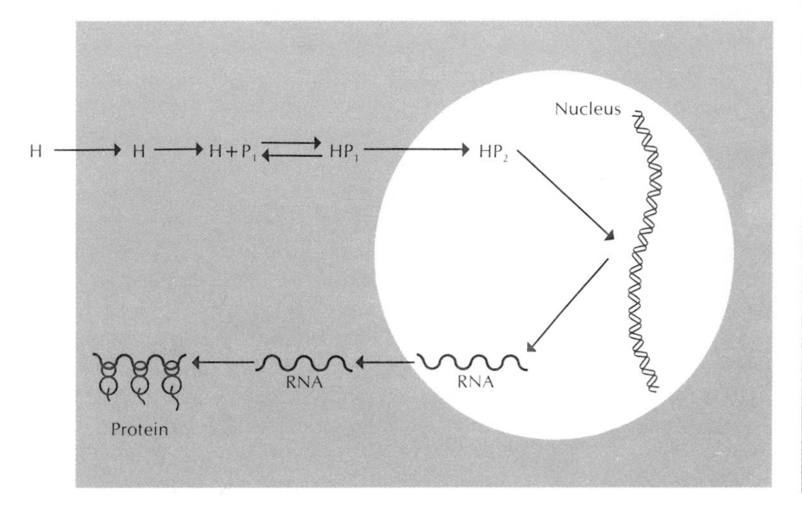

Figure 21.19. Posttranscriptional model. In this model, the hormone receptor alone acts as a repressor for specific mRNA. When the receptor binds to mRNA, it prevents translation of proteins from the mRNA and also makes the mRNA more unstable and subject to more rapid degeneration. Formation of the hormone-receptor complex removes the receptor from mRNA and permits repeated translation of induced protein molecules from the now stable mRNA.

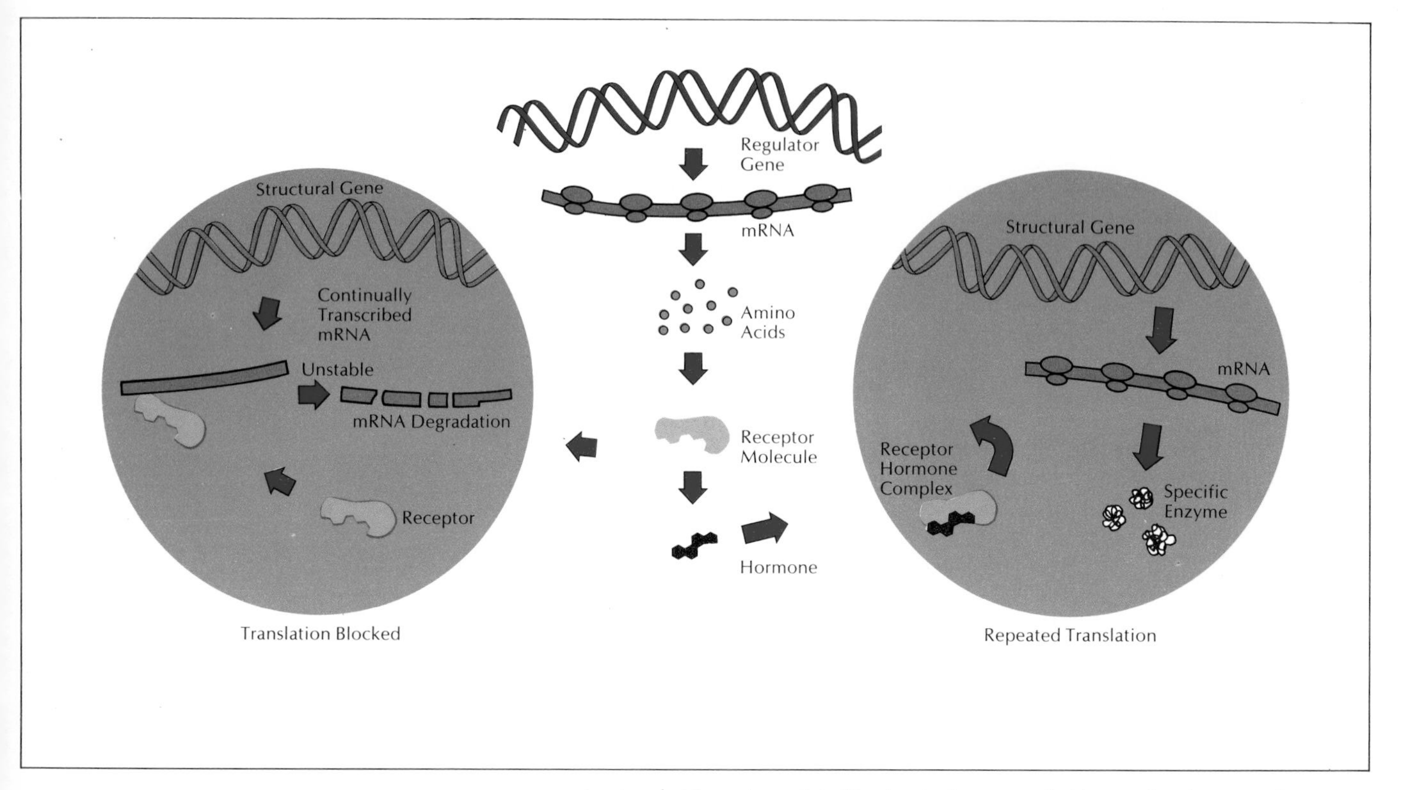

Jacob-Monod model. The basic features of this mechanism are shown in Figure 21.19.

The posttranscriptional model has been proposed in an attempt to account for certain basic facts about the behavior of tissue cultures in response to steroid hormones. The effect of a hormone is to increase the activity of a specific enzyme through an increase in the rate at which that protein is synthesized. However, the hormone does not alter the total levels of protein and RNA synthesis of new RNA, although evidence on this point in various systems is not yet clear. If protein synthesis is inhibited by certain substances, there is no increase in enzyme activity upon addition of hormone but there is an accumulation of RNA (presumably mRNA specific to the induced enzyme), and synthesis of the protein is induced as soon as the inhibiting substance is removed, even if the hormone is no longer present. The hormone must be constantly present in order to maintain the fully induced rate of protein synthesis under normal conditions. However, if specific protein synthesis is first induced by addition of a hormone and then RNA synthesis is inhibited, formation of the induced protein continues at a high rate for some time, even if the hormone is removed.

The posttranscriptional model postulates that the hormone receptor alone acts as a repressor for specific mRNA. When the receptor binds to mRNA, it both prevents translation of proteins from the mRNA and makes the mRNA more unstable and subject to more rapid degradation. Formation of the hormone-receptor complex removes the receptor from mRNA and permits repeated translation of induced protein molecules from the now stable mRNA.

Further research will almost certainly reveal facts that will further modify this model or even precipitate development of entirely new models of hormonal control mechanisms. However, there is good reason to hope that relatively simple control mechanisms will be found responsible for most hormonal actions in eucaryotic cells and that these mechanisms will reveal important features of the normal control mechanisms of eucaryotic cells.

FURTHER READING

General information about animal hormones will be found in books by Barrington (1963), Gorbman and Bern (1959, 1962), Pincus and Thimann (1948–1964), Scharrer and Scharrer (1963), and Turner (1966). More detailed information about sex hormones is given by Van Tienhoven (1968) and W. C. Young (1961). For more information about the cellular slime molds, see the book by J. T. Bonner (1965).

Useful introductory articles on animal hormones include those by Constantinides and Carey (1949), Csapo (1958), Davidson (1965), Fieser (1955), Gray (1950), Jones (1968), Levey (1964), Levine (1966), Rasmussen (1961), Tanner (1968), Williams (1950), Wilson (1963), and Wurtman and Axelrod (1965).

22
Immune Responses

The first great book on immunology, by the Belgian Jules Bordet (1898), opens with the sentence "Life is the maintenance of an equilibrium that is perpetually threatened." The exquisitely balanced system of hormonal and neural regulations involved in the reproductive cycle (Chapter 21) is only one example of the normal mechanisms of self-correcting maintenance and control found in a multicellular animal. The immune responses of vertebrates function primarily to see that no foreign substances or cells disrupt these delicate systems.

In a sense, immunity—as it exists in man and other warm-blooded vertebrates—is a communications system. Foreign material most frequently enters the body as a result of the invasion and multiplication of microorganisms. Other foreign material may enter when a surgeon implants skin or an organ from another individual or when certain types of cancer arise—either spontaneously or in some region of chronic irritation. The first step in the chain of communication is recognition that the invading material *is* foreign—is not-self rather than self. The second step involves multiplication and activation of the cells appropriate to combat the particular alien material present. Finally, these cells or their specialized chemical products must find and destroy the foreign cell, microorganism, or protein. This three-stage process represents the simplest outline of a *primary immune response* to any sort of foreign material.

Immunity, however, is usually thought of in terms of the resistance to further infections that follows an attack of infectious disease. For centuries it has been known that a man with a face marked by smallpox never contracts the disease again. Allowing for an immense variety of details, immunity to smallpox is a prototype of all *secondary immune responses*. Once foreign material has been successfully dealt with, there is an enlarged population of cells able to deal with that particular foreign substance and able to be called into action more rapidly and more effectively.

THE INFLAMMATORY RESPONSE

The presence of a foreign substance in the body triggers a dramatic sequence of events known as the *inflammatory response*. The major participants in this response are the white blood cells, or *leucocytes*, which are amoeboid cells capable of moving through intercellular fluids and tissues of the body as well as the bloodstream. Lymphocytes, granulocytes—neutrophils, eosinophils, and basophils—and monocytes are all types of leucocytes. Most of the leucocytes in the body under normal conditions are *lymphocytes* and *neutrophilic granulocytes* (*neutrophils*). Leucocytes are manufactured in bone marrow, the lymphatic system, and the thymus gland. In general, leucocytes combat foreign materials by engulfing and digesting them, but the details of the inflammatory response are relatively complex.

The process can be analyzed by examining what happens when the tissues of a mouse, for example, are injected with a foreign protein. The injection first causes the breakdown of certain cells in the tissues. These *mast cells* have large granules in their cytoplasm, and the granules appear to be lysosomes filled with hydrolyzing enzymes. The release of enzymes from disintegrating mast cells causes the destruction of other cells in the vicinity, beginning the development of an inflammation (Figure 22.1). Soon a number of neutrophils begin to arrive at the site of inflammation. The neutrophils engulf some of the foreign particles by phagocytosis or endocytosis

Figure 22.1 (above). Primary immune response of inflammatory cells to antigen injected into a mouse (not previously exposed to this antigen). First (upper left), large-granule mast cells in the body tissues come in contact with the antigen and break down. This cellular disintegration releases large numbers of lysosomes, whose enzymes and other substances destroy other cells in the vicinity, initiating inflammation. The first moving defensive cells to arrive at the site are neutrophils, which swallow up some of the foreign particles but soon disintegrate themselves. Next (upper middle), lymphocytes and monocytes reach the area and feed upon the foreign particles and cellular debris. This process causes some of the lymphocytes to enlarge and become macrophages (upper right). Eventually, these cells ingest all the foreign matter and the inflammation subsides. Most of the antigen is broken down into amino acids and sugars by enzymes, but some is maintained in the macrophages by combining with RNA. (*After* Scientific American, *1964*)

Figure 22.2 (below). Secondary immune response of inflammatory cells to antigen injected into a mouse

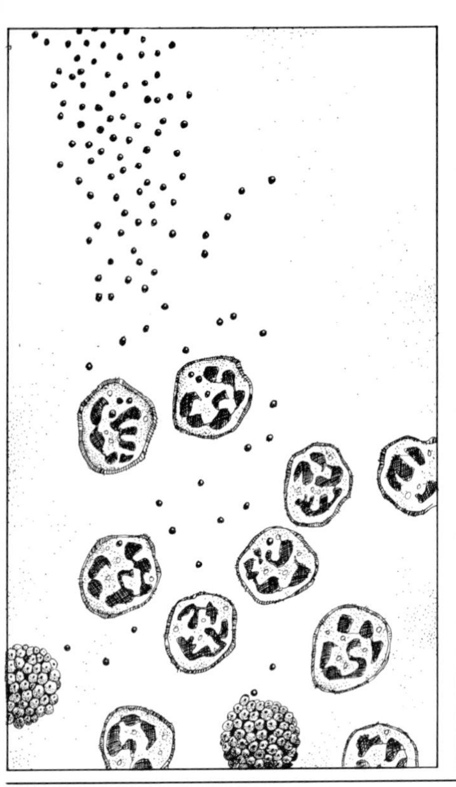

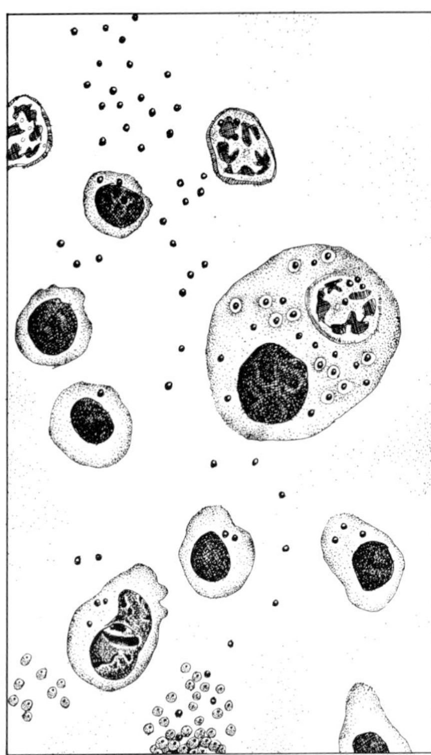

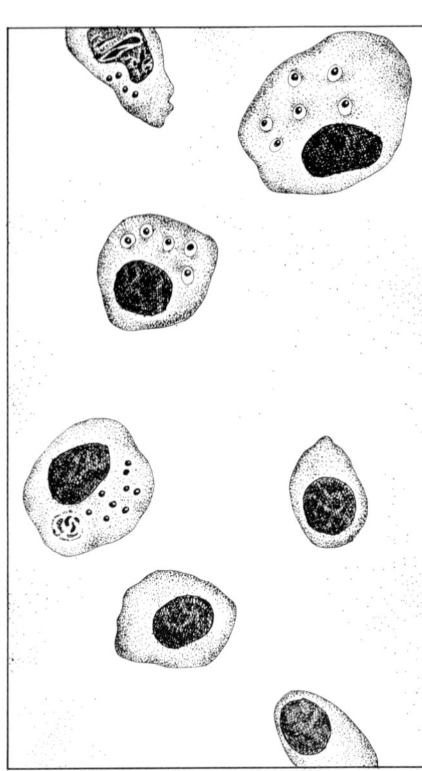

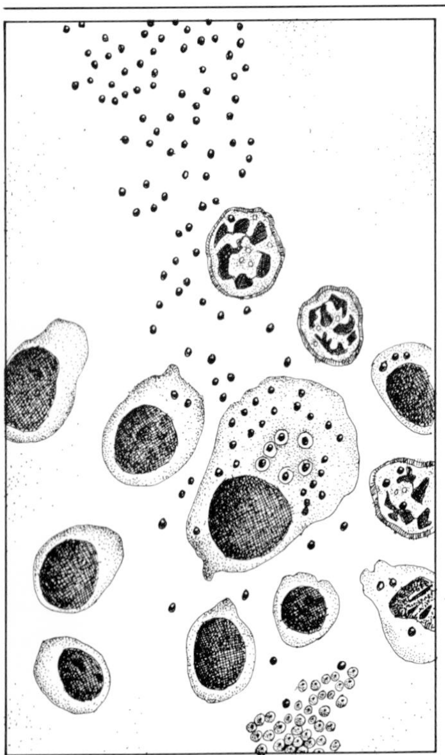

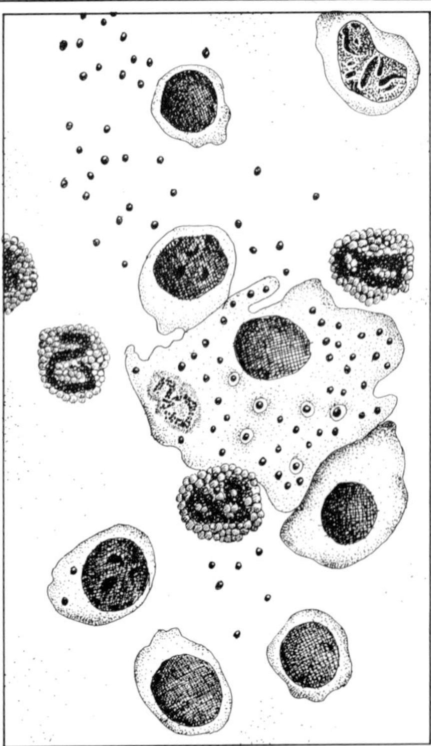

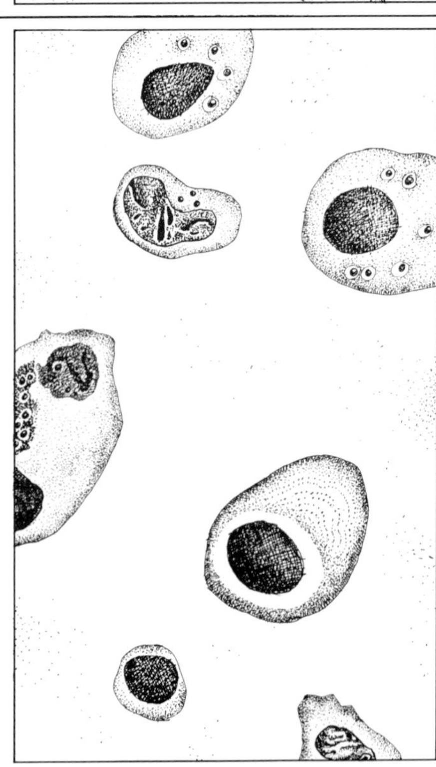

(previously exposed to this antigen). Neutrophils arrive at the site but in fewer numbers, while macrophages arrive in larger numbers (lower left). Some of the macrophages contain antigen in combination with RNA, and these cells interact with esinophils, which cause them to be broken open. More macrophages then move in and engulf pieces of broken cells (lower middle). Some antigen escapes destruction in combination with RNA in the macrophages (lower right). The rapid arrival of macrophages and the ease with which they ingest foreign particles result in a shorter period of secondary inflammation and one that is less severe. (*After* Scientific American, *1964*)

Figure 22.3 (right). Ehrlich's "lock and key" hypothesis explaining the mechanism of immunity. Blood cells with different "locks" will bind to specific antigen-antibody "keys" causing agglutination, or clumping, to occur. Thus, when the key fits the lock, immunity is established.

Figure 22.4 (left). White blood cells.

and internally digest the particles. Within a few hours, however, the neutrophils themselves begin to disintegrate, adding their contents to the growing amount of fluid in the inflamed area.

Among the substances released by the disintegrating mast cells are histamine, serotonin, and heparin. The histamine and serotonin act as local hormones, causing dilation of arterial vessels and constriction of venous vessels near the inflamed site and thereby increasing the supply of blood to the invaded tissues. Heparin prevents clotting, or coagulation, of the blood in the area. The region where the foreign protein was injected becomes swollen, flushed, and suffused with the various products of cell breakdown and secretion that are collectively called pus.

After the neutrophils have begun to break down, lymphocytes and monocytes begin to arrive in the area of inflammation. In their normal condition, these cells have large nuclei surrounded by relatively small amounts of cytoplasm. As they feed upon the foreign particles and cellular debris in the vicinity, many of the monocytes become enlarged to the form called *macrophages*. Some macrophages multiply by cell division, greatly increasing the population of leucocytes in the inflamed area. Eventually, all the foreign material and cellular debris is ingested by macrophages, and the inflammation subsides.

If the mouse is later injected with the same foreign protein, the inflammatory response follows a somewhat different pattern. Again, neutrophils arrive at the invasion site first, but in smaller numbers than responded to the primary infection (Figure 22.2). On the other hand, macrophages arrive sooner and in much larger numbers, and eosinophils also arrive in greater numbers. Some of the macrophages become immobile, swell as they form large fluid vesicles in the cytoplasm, and attract to themselves large numbers of eosinophils, which penetrate and disintegrate the swollen macrophages. The eosinophils and fragments of macrophages are then ingested by other macrophages, which continue to arrive at the site. Other cells that appear in the inflamed area include a large number of *plasma cells*, which contain large amounts of rough endoplasmic reticulum and which secrete proteins rather than ingest materials.

Because of the rapid arrival of macrophages and their greater effectiveness in ingesting foreign particles, the secondary inflammation is less

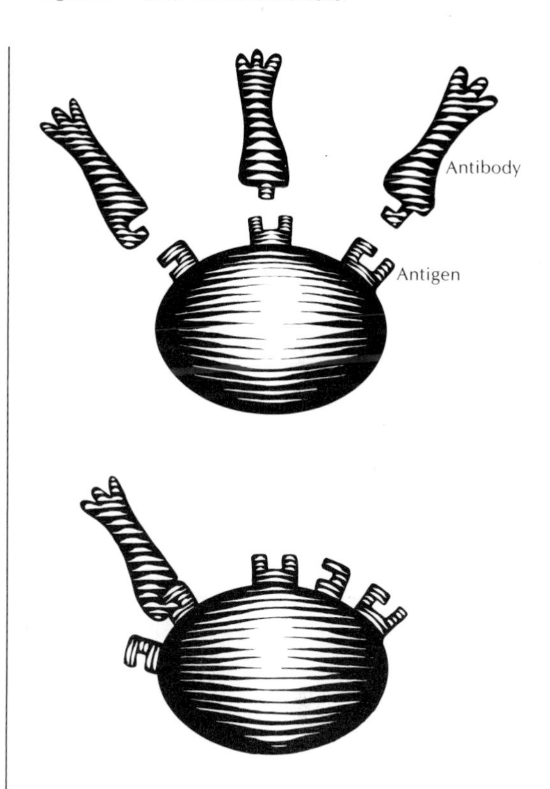

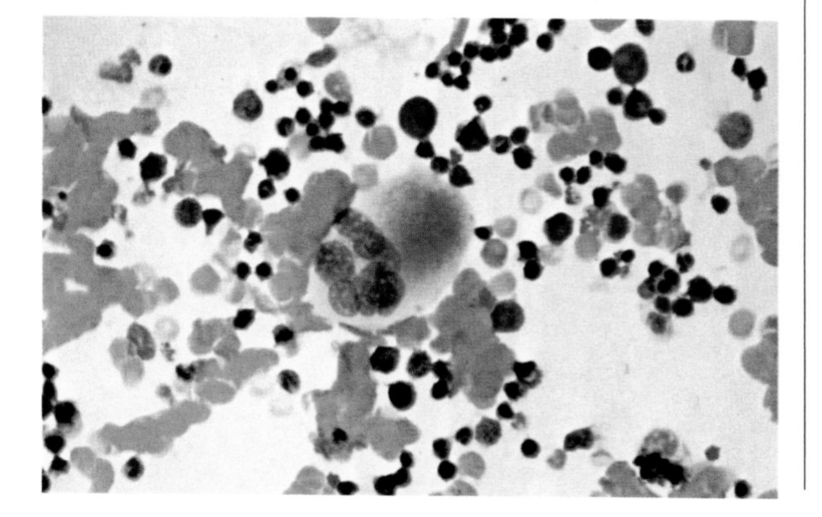

Figure 22.5. The formation of antibodies and the immunization process, according to Ehrlich's theory. Ehrlich suggested that certain cells possess food-capturing receptor side chains. If these receptors capture an antigen and survive, they may be replicated by the cell and released into the blood plasma. These released receptors would then serve as antibodies.

severe and is cleared up more rapidly than was the primary inflammation. As in the case of the primary inflammation, some of the pus and white blood cells may drain out of the wound. The balance is processed through the lymphatic system (Chapter 19). The ultimate fate of the foreign material and cellular debris is either to be ingested and used as food by leucocytes or eventually to be processed through the excretory system of the organism. The damage caused by the inflammation is repaired by the normal processes of wound healing (Chapter 18).

ANTIBODIES

The role of leucocytes in digesting bacteria and other foreign particles was first described by Elie Metchnikoff in the closing decades of the nineteenth century. Metchnikoff regarded the phagocytic action of leucocytes as the body's primary line of defense against foreign materials. At about the same time, however, Emil von Behring and Shibasaburo Kitasato showed that the noncellular, fluid portion, or *plasma*, of the blood contains substances that play a role in the immune response. These substances, which came to be called antibodies, appear in the blood after the primary invasion of a foreign substance, or *antigen*, and are apparently essential to the increased effectiveness of the secondary response. The antibodies were clearly related to the development of immunity against particular diseases or other kinds of invasion, as shown by the success of inoculation as a defense against disease. Therefore, immunologists during the early part of this century turned their attention largely toward the actions of antigens and antibodies.

In the case of bacterial invasions, antibodies immobilize the bacteria or cause them to clump together (agglutinate). Antibodies may immobilize other microorganisms by binding together their cilia or flagella. In the test tube, an antibody may combine with an antigen to form a solid precipitate. In the body, a similar reaction presumably causes neutralization of the antigen, preventing chemical damage to the cells. In each case, however, an antibody acts only against the specific type of antigen that caused its formation in the blood. As more was learned about the action of antibodies, most

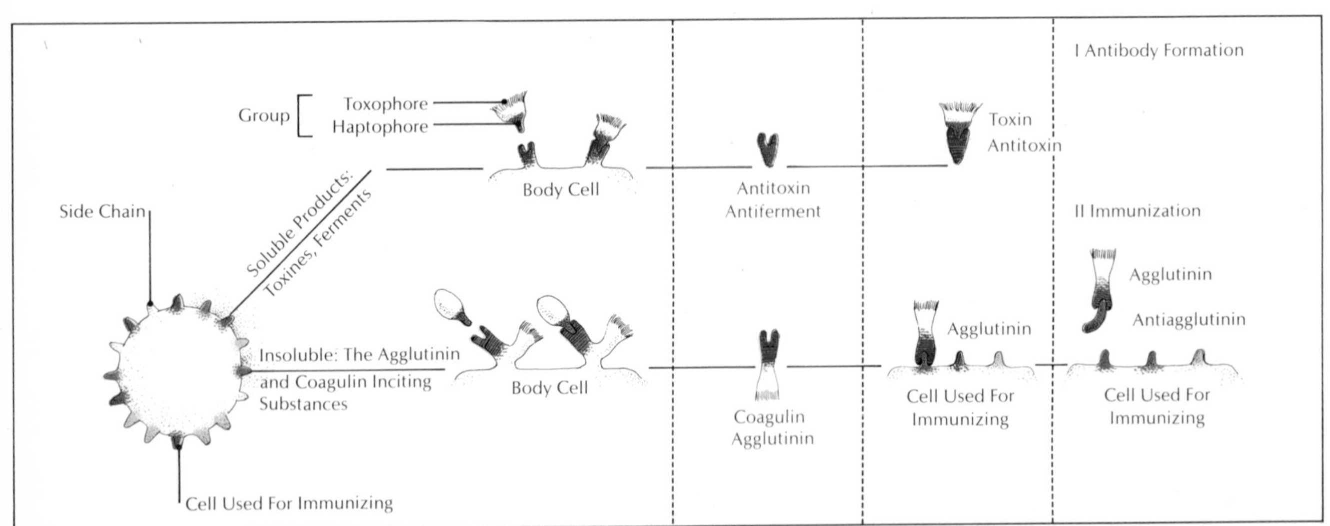

biologists came to regard the ingestive action of the leucocytes as simply a mopping-up operation that followed actual immobilization of the invading materials by antibodies.

One early attempt to explain the formation of antibodies was made by the German bacteriologist Paul Ehrlich in the late 1800s. He suggested that molecules of protoplasm possess side chains that normally serve as receptors in capturing food molecules. If the side chain captures a molecule of bacterial toxin or other antigen—and if the living molecule survives the toxic effects of the antigen—the side chain might be replicated in large numbers by the molecule and released into blood plasma. These released side chains would then serve as antibodies, binding to antigen molecules and immobilizing them.

In 1917 Karl Landsteiner (who had earlier discovered the ABO blood groups) succeeded in causing animals to produce antibodies against artificial antigens, which were prepared by binding small organic molecules called haptens to large carrier proteins. Landsteiner's experiments showed that antibodies can react with antigens other than those against which they were formed only if the new antigens are very similar in molecular shape to the antigen that stimulated antibody production. Furthermore, he showed that a single animal can produce specific antibodies against a great variety of antigens—even artificial antigens that could never have been encountered in nature by this animal or any of its ancestors.

Because so many different antibodies can be produced as needed, the weight of opinion began to shift away from theories that involved formation of antibodies from already existing receptors on molecules or cells. Instead, it began to seem more and more likely that the antibody molecules are somehow created in response to the presence of a specific antigen. By the 1930s, it had become clear that the antibodies are among the proteins of the blood plasma, the so-called *immunoglobulins*. Several biologists independently suggested that antibody proteins might be synthesized in contact with the antigen, taking a complementary shape that would lead to formation of a firm bond between the antibody and any antigen of similar shape.

This "instructive theory" of antibody formation was given a more complete biochemical background by the work of Linus Pauling (1940). At the time, little was known of the process of translation or of the fact that protein shape is determined by amino acid sequence. Pauling suggested that a polypeptide chain was shaped against the antigen template and then held in the new shape by formation of hydrogen bonds. The instructive theory remained the dominant explanation of antibody formation for several years. According to this theory, large globular-protein molecules exist in *immunocytes* (white blood cells involved in antibody formation) in relatively unformed condition. The presence of an antigen causes a cavity (*combining site*) to be shaped in the globular protein, complementing the shape of an active part of the antigen. The release of the specifically shaped protein, now an antibody, then leads to the binding and inactivation of antigen molecules of the same or similar shapes. The shaped antibody molecules can remain within the plasma after the primary infection is eliminated, thus providing a ready reserve of antibodies to combat any further infection of the same antigen without delay.

During the 1940s and early 1950s, extensive experience with blood transfusions and transplantation surgery emphasized the ability of the

Interleaf 22.1

HUMAN BLOOD GROUPS

Many deaths following serious injury or hemorrhaging are due to loss of blood. Why not replace the lost blood with blood from another person or from an animal? This obvious treatment was tried by many physicians over the ages, but the results were often disastrous. In a few cases, the blood transfusion was successful and the patient showed a speedy recovery. In most cases, however, the patient reacted violently to the transfusion and soon died. By the end of the nineteenth century, most European nations had outlawed attempts at blood transfusion. There seemed to be no way to predict in which cases the treatment would prove beneficial.

In the laboratory, experimenters showed that agglutination, or clumping, of red blood cells almost always occurs when blood samples from two animal species are mixed. Similar agglutination occurs often—but not always—when blood samples from two humans are mixed. The Austrian physician Karl Landsteiner began investigating this phenomenon in 1900. He took red cells from the blood of one person and mixed them with blood serum from another person. In some cases the serum caused agglutination of the cells; in other cases it did not. At first, Landsteiner suspected that the blood of some persons lacks the agglutinating factor, either because of illness or because of a hereditary abnormality. To explore the phenomenon further, he obtained blood samples from all the workers in his laboratory and mixed cells and serum in all possible combinations.

Landsteiner soon discovered that serum from a single individual may agglutinate cells from some individuals but not from others. He was able to classify individuals into three groups, each having blood of a type that reacts in particular ways with blood from individuals of other groups. In further studies, Landsteiner found a fourth blood group.

The German bacteriologist Paul Ehrlich recognized this phenomenon as being very similar to antibody-antigen reactions involved in bacterial infections. Ehrlich suggested that agglutination is caused by a "lock-and-key" fitting together of antibodies in the serum with antigens on the cells. He thought that each of Landsteiner's four types of blood contains a different set of antibodies and antigens, with agglutinating combinations possible only between certain pairs of antibodies and antigens.

Landsteiner modified Ehrlich's explanation somewhat, showing that his observations could be explained with only two kinds of antibodies (α and β) and two kinds of antigens (A and B). A antigens combine with α antibodies, and B antigens combine with β antibodies. The blood of a person in group A contains β antibodies in the serum and A antigens on the cells, whereas that of a person in group B contains α antibodies and B antigens. A mixture of these two blood types will always produce agglutination. The blood of a person in group AB contains both A and B antigens but no antibodies. Thus, serum from AB blood can be mixed with either A or B cells without producing agglutination, but the AB cells will be agglutinated by serum from persons of either the A or B groups. The blood of persons in the fourth group (O) contains both kinds of antibodies, but there are no antigens on the corpuscles of O blood.

It was soon demonstrated that blood types are inherited according to simple Mendelian principles (Chapter 12). The production of cell antigens is coded by a particular gene. One allele (I^A) codes for production of A antigens, and the other (I^B) codes for the production of B antigens. Thus, a person with genotype I^AI^A will have type-A blood, and a person with genotype I^BI^B will have type-B blood. The blood of a person with genotype I^AI^B contains both types of antigens and is called group AB. The existence of type-O blood is due to the presence of a third allele for this gene. This allele (i) does not code for either kind of antigen and thus acts as a Mendelian recessive. A person of genotype ii has neither type of antigen on his cells and belongs to group O.

Subsequent research has revealed that the antigens are glycoproteins (called *agglutinogens*) that form part of the surface coating of red blood cells. The combinations of antibodies (called isoagglutinins) and agglutinogens in the blood of the four basic groups are summarized in Table 22.1.

Landsteiner's blood groups made possible the reliable prediction of the outcome of a transfusion. The blood types of donor and recipient can be determined by simple tests with standard serum samples. It is then easy to predict whether the transfusion will cause agglutination. Soon after Landsteiner's research was published, blood transfusion became a standard and indispensable medical treatment.

An understanding of the genetic determination of blood types often has been useful in settling questions of relationships among people. If, for example, a rare mix-up in a hospital nursery leaves some doubt as to which baby belongs to which parents,

Table 22.1.
Genetics of Human Blood Groups

GENOTYPES	Blood Group	Agglutinogens	Isoagglutinins
I^AI^A or I^Ai	A	A	β
I^BI^B or I^Bi	B	B	α
I^AI^B	AB	A and B	none
ii	O	none	α and β

the analysis of blood types may help to resolve the problem. Parents both having type-A blood (genotypes I^AI^A or I^Ai) could not possibly have a child with type-AB or type-B blood. Such genetic analyses also play important roles in many court cases involving disputed paternity. In this case, as in most situations that involve deduction of genotype by observation of phenotype, it should be noted that combined effects of other genes may rarely produce phenotypes that mimic the characteristics produced by the single gene being discussed here. Undoubtedly, several genes are involved in production of proteins associated with blood cells and serum antibodies. The combined effects of changes in such other genes might produce the appearance of type-A blood in a person with a genotype that includes the I^B allele. Thus, there are rare cases where a person whose blood tests as type-B or type-AB may be the offspring of parents with blood that tests as type-A.

Blood-type genetics have been of use also in the study of the races of man. Different populations show different proportions of the three alleles I^A, I^B, and i. For example, the I^B allele appears with a very high frequency in Central Asia and in parts of India, but it becomes less and less frequent in populations farther and farther away from these centers. The corresponding phenotypes (type-B and type-AB blood) show a similar decrease in frequency in populations farther away from Central Asia and India. Among Australian aborigines, nearly 70 percent of the population is of blood type A, and the remainder is of type O; the I^B allele is entirely absent from this population. There are no sharp boundaries between races or population areas. Instead, the frequency of one allele rises and that of another drops as population samples are tested along any particular line.

Landsteiner continued his research on blood types for many years. He found a number of other antigen-antibody pairings that are controlled by other genes. These have less extreme effects in normal transfusions but are of importance in many special cases. Landsteiner discovered the M and N blood groups, and in later work with apes and monkeys he found the *Rh factor*, or Rh antigen, which plays an important role in certain previously unexplained birth difficulties (Chapter 42).

Figure 22.6 (above). Instructive theory of antibody formation (1940). Unformed globular proteins were thought to exist in the immunocyte, which, in the presence of an antigen, conformed to the antigenic determinant (AD). This AD then activated the cell to produce an antibody of a specific shape.

Figure 22.7 (middle). Selective theory (1955). This theory postulated that different antibodies exist in normal blood. The presence of an antigen "selected" the specific antibody whose combining site bound to the AD. Immunocyte activation then leads to specific antibody release.

Figure 22.8 (below). Clonal theory (1959). The most recent theory postulates that there are specific antibody combining sites on the immunocyte cell surface. Antigen contact (via the AD) causes cell stimulation and proliferation. This action produces a clone of similar cells that all produce the same antibody.

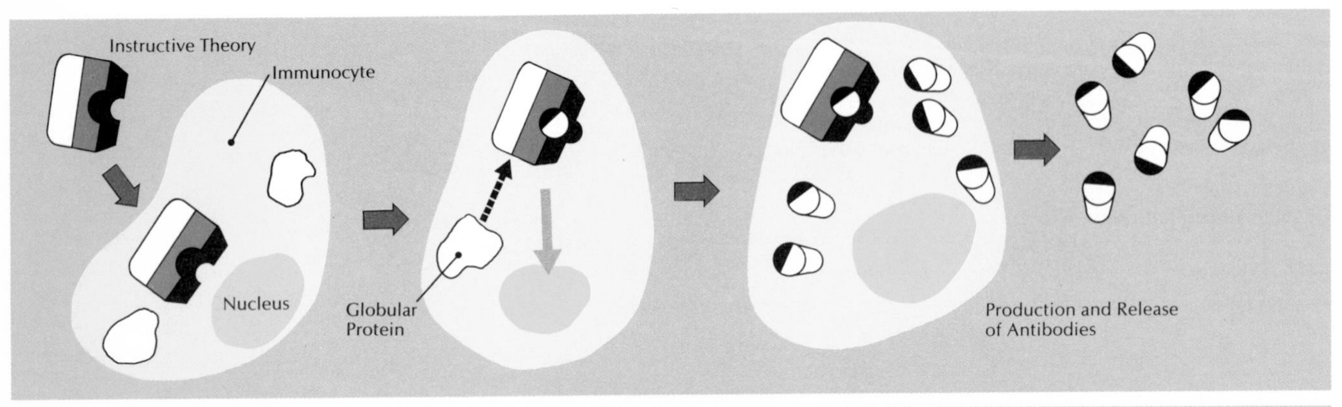

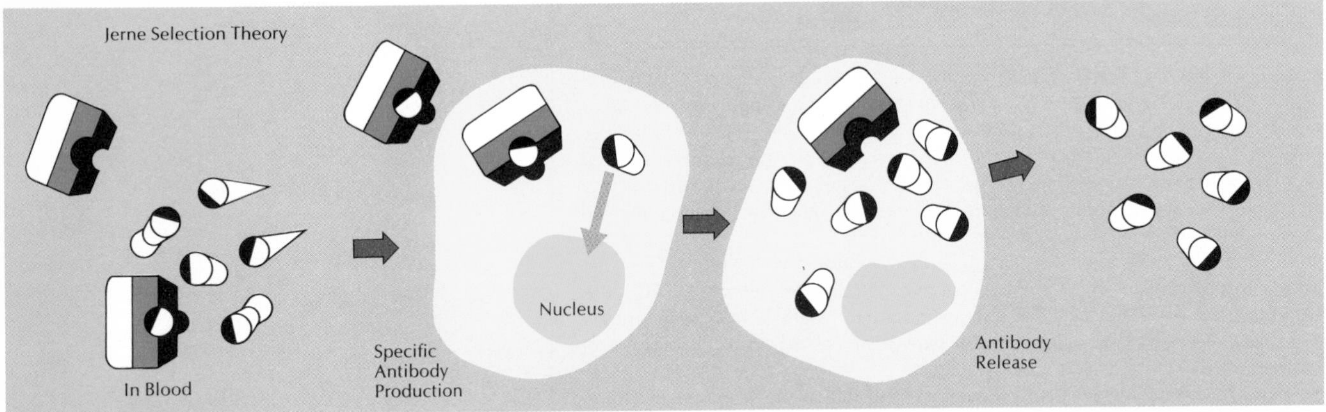

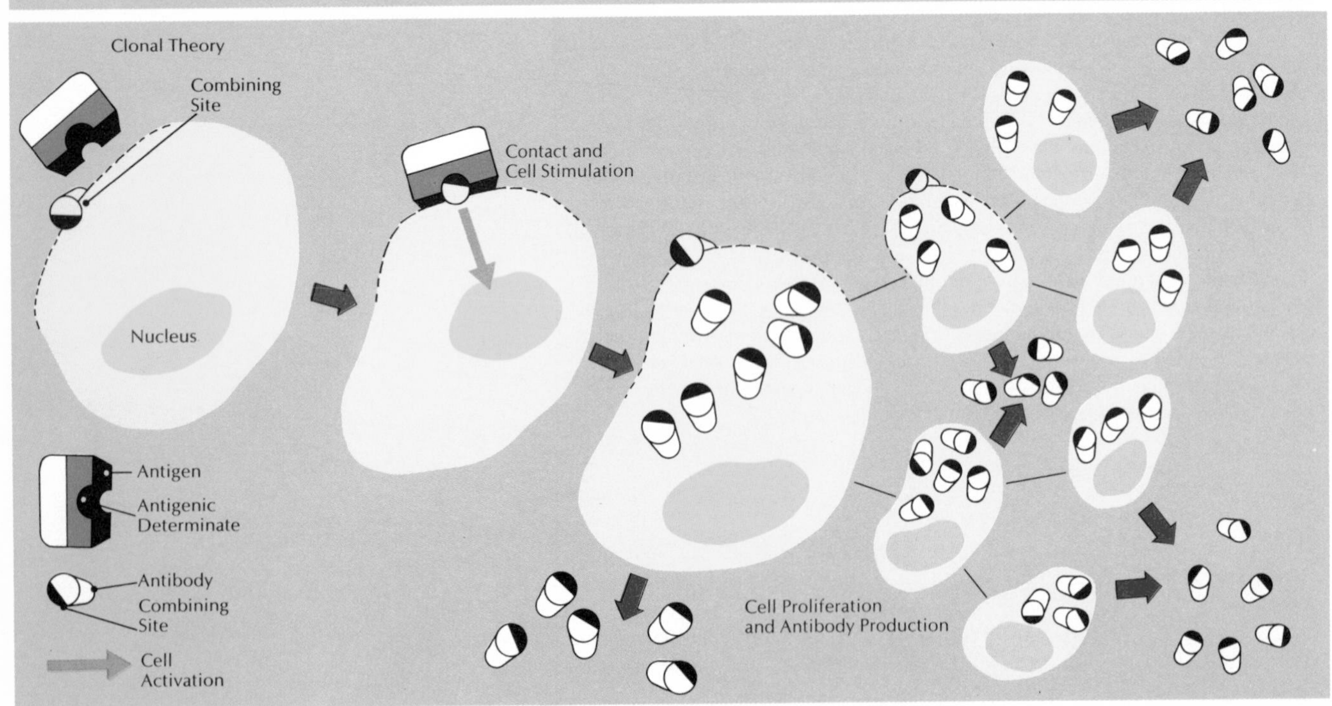

immune system to refrain from attacking materials that are part of the body, even when these materials are transplanted to a new location. A skin graft from another donor is soon attacked and destroyed by leucocytes, whereas a graft of skin from the recipient's own body is not attacked. The instructive theory offered no explanation for this ability of the immmune system to distinguish between self and not-self, between particles formed in its own body and alien particles.

The final blow to Pauling's instructive theory came when studies showed that antibody proteins could be completely denatured with accompanying loss of antigen-binding ability. Renaturation then was carried out in the absence of the antigen, and the refolded antibody was found to have reacquired much of its original antigen-binding activity. Thus, antibody proteins are like other proteins, where the chain folding is determined by the primary structure.

Among the theories developed to replace Pauling's were some additional instructive theories and some selective theories. Instructive theories hold that the antibodies are in some way shaped to complement the structure of the particular antigens present. Selective theories hold that the capability for producing any particular antibody is already present in some immunocyte of the body and that invasion by an antigen causes selective reproduction of that cell and production of antibodies by it and its descendants.

One of the earliest selective theories was offered by N. K. Jerne (1955), who suggested that traces of each kind of antibody exist in normal blood. When an antigen enters the blood, it binds to the specific antibody whose combining site complements some part of its structure (the so-called *antigenic determinant*, or AD). The antigen-antibody complexes are then taken into macrophages, which produce more antibodies of the specific form included in the complex. According to this theory, any natural antibodies capable of reaction with parts of the organism would be combined and removed from the blood early in life, thus accounting for the failure of the immune system to attack tissues transplanted from one part of the body to another.

Jerne's theory, however, did not account for the newly discovered phenomenon of tolerance for foreign antigens that were implanted before a certain stage of development. If an antigen is injected into an embryo or—in some species—a newborn animal, the adult will not show an immune response against this antigen in later injections. A modified selection theory was proposed independently by D. W. Talmage (1959) and Macfarlane Burnet (1959). At present, the *clonal selection theory* proposed by Talmage and Burnet—as subsequently broadened and modified—is generally accepted as the most satisfactory explanation of the immune response (Figure 22.8). This theory and some of the outstanding problems yet unsolved will be described in the following sections.

THE MOLECULAR BASIS OF IMMUNITY

Immunology is one of the biological fields in which research is proceeding at top speed. As in any active biological science, new complexities emerge every few months. If general statements are to stand, they must be carefully chosen. In many cases, it is as well to be honest and say that there are a few observations that do not quite fit but that a particular generalization is so useful in understanding 99 percent of what is observed that it is worth using—at least for the present. There are many precedents for a confidence

Figure 22.9 (left). Lock and key antigen-antibody reaction. Evidence suggests that an antibody molecule has two identical halves, each structured from one large and one small component. A particular antibody reacts with a particular antigen because the configuration of its combining site interlocks with that of the antigen. When the antigen and antibody make contact at the proper angle to bring the complementary patterns together, a union between antigen and antibody is formed, thus immobilizing the antigen.

Figure 22.10 (right). The immunocyte concept of immunity. The surface of an immunocyte contains antibody-type combining sites that can combine with a specific antigen. Binding of an antigen to the cell surface may cause the cell to enlarge and proliferate to produce a clone of similar cells. These cells then become capable of producing large amounts of antibodies as well as multiplying as plasma cells.

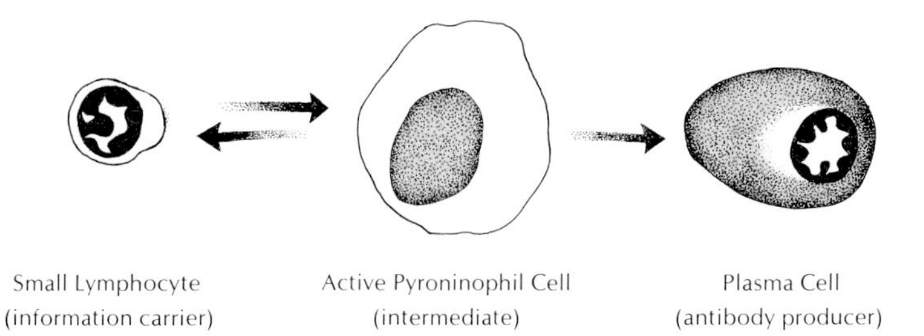

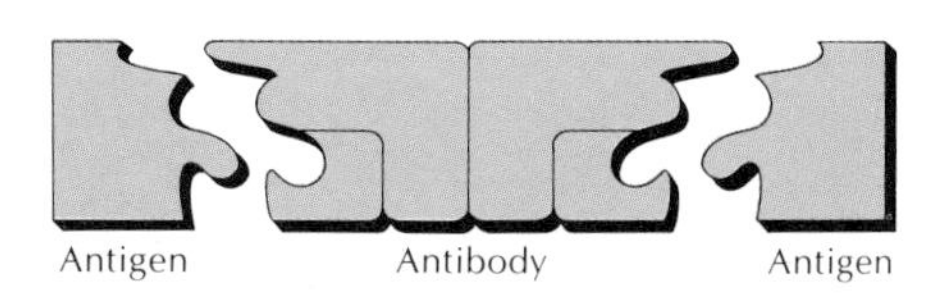

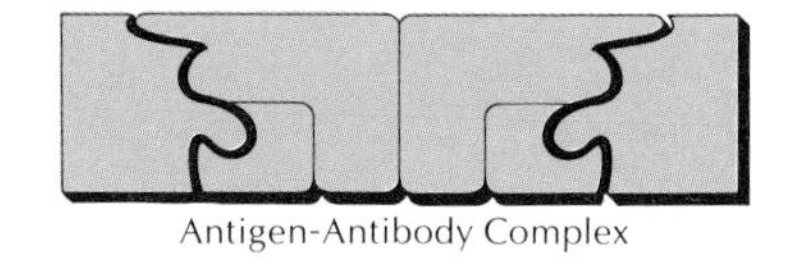

that future work will straighten out the apparent discrepancies. With these minor reservations, it is useful to consider the relationships among antigen, immunocyte, and antibody in terms of *chemical complementarity*.

Each immunoglobulin molecule is a complex protein made up of at least four polypeptide chains. In the most common type of immunoglobulin with antibody activity (immunoglobulin G), there are two parts of the molecule that serve as combining sites. At these sites are two segments of polypeptide chains lying close together and providing a unique three-dimensional pattern—a pattern that, depending upon which amino acids are involved and their sequence, can take on thousands or even millions of different configurations. A particular antibody reacts with a particular antigen simply because the configuration of its combining site interlocks with the antigenic determinant of the antigen, rather like a key fitting into a lock (Figure 22.9). When the antigen and antibody collide at the proper angle to bring the complementary patterns together, a relatively firm union between antigen and antibody is formed. As in the similar relationship between an enzyme and its substrate, the union involves both the interlocking shapes of the molecules and weak bonds such as hydrogen bonds. The antibody will combine only with antigens that possess an AD of very close complementarity to the shape of the combining site.

On the surface of an immunocyte are antibody-type combining sites that, like an antibody itself, can unite with the corresponding AD. On the cell surface, however, this "fixed antibody" acts as a receptor, not only binding the antigen to the cell temporarily but also stimulating the cell to activity. Most commonly, the stimulus causes the cell to enlarge and proliferate to produce a clone of similar cells (Figure 22.10). In some cases, the stimulated cells become capable of producing large amounts of antibody as well as multiplying—these antibody-producing cells are the plasma cells, or plasmacytes.

Recent studies have confirmed that each immunocyte produces only a single kind of antibody (Awdeh, *et al.*, 1970). Perhaps the most important generalization in immunology is that when an immunocyte multiplies, its descendant cells all produce the same type of antibody, whether the antibody is present only as a few surface receptors or is synthesized in large amounts.

The Thymus Gland

One of the most important aspects of recent immunological research has been the recognition of the importance of the thymus gland. The thymus is unlike any other mammalian organ in that it is relatively large in infancy and fades away to a few fibrous shreds in old age. In babies, it is a mass of actively multiplying lymphocytes lying just behind the top of the breastbone (Figure 22.11). The first clues to its importance came in 1960, when Jacques Miller devised a method for surgically removing the thymus from newborn mice. Such mice show serious disturbances in their immune responses, and detailed analysis of their anomalies has been highly fruitful. But for an aid to understanding the function of the thymus, it has proved useful to use what Robert Good calls experiments of nature—genetic diseases that in one way or another involve the immune system. These observations have led to the same conclusion as a wide range of experiments in mice, chickens, rats, and so on. It is clear that there are two major families of immunocytes. One set is thymus-dependent (T-D) cells, whose ancestors

Figure 22.11. The human thymus gland. Immunocytes produced directly by the thymus or the ancestors of cells produced by the thymus appear to be concerned with cell-mediated immune responses.

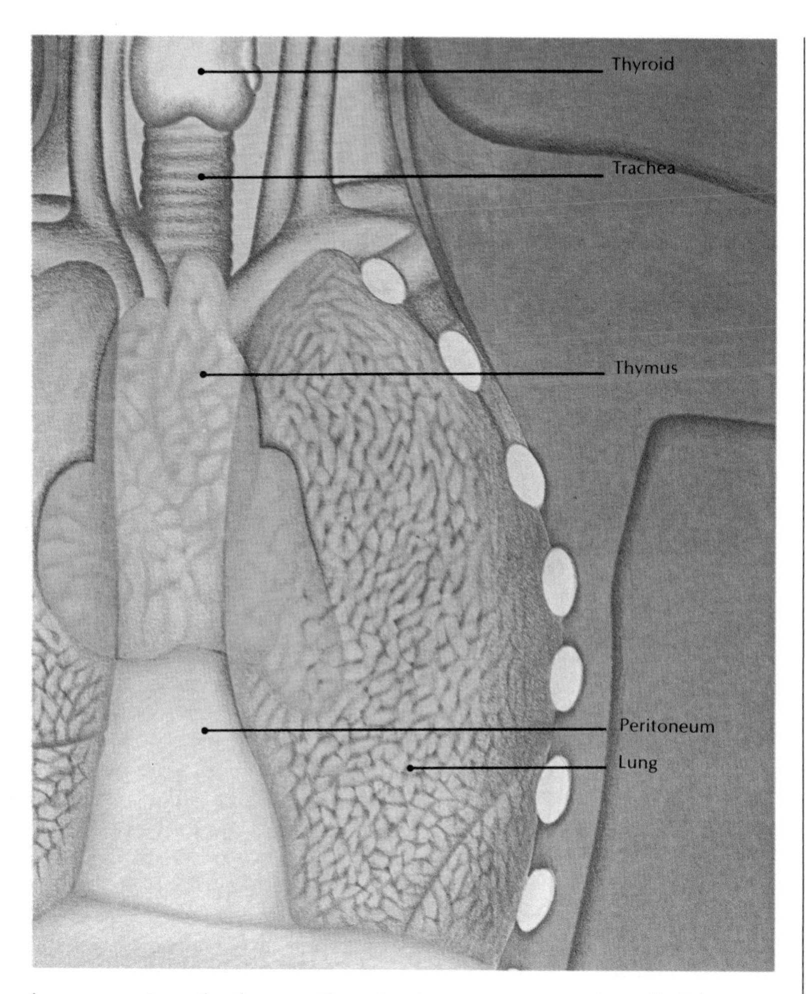

have come from the thymus. The other immunocytes can be called thymus-independent (T-I) for the present, although eventually they will probably be found to be related to some other tissue that plays a role similar to that of the thymus. Some biologists have felt that the evidence justifies calling this analogue of the thymus the *gut-associated lymphoid tissue* (GALT), and this term is often seen in the literature of immunology.

The important feature is that there are two congenital diseases that neatly separate the two groups. In Di George's disease, the embryonic cells that should form the thymus and parathyroid glands fail to develop, and the child is born with no thymus at all. If the effect of the missing parathyroids is remedied by giving the appropriate parathyroid hormone, the remaining symptoms can be assumed to be due to the absence of T-D immunocytes. The second disease, congenital agammaglobulinemia, represents a complete failure of the T-I system with what appears to be a perfectly normal thymus. Both diseases are very rare and have been carefully studied. When they are compared, a clear division of immune function becomes visible.

A child with agammaglobulinemia, whose T-I system is lacking, shows the following differences from a normal child: (1) no antibodies are produced when antigens such as diphtheria toxoid and polio vaccine are injected; (2) there are no plasma cells and only minute amounts of immunoglobulins in the plasma; (3) the child is highly susceptible to pneumonia and other bacterial diseases and—in the days before antibiotic drugs—always died in infancy. In the following types of immune response, however, the child will behave normally; (4) a graft of skin from another child is rejected; (5) the skin shows sensitive reactions to chemical substances, and the child can develop a positive reaction to a tuberculin skin test; (6) some virus infections, such as measles or vaccination against smallpox, run a normal course and are followed by immunity against another attack.

In a case of congenital absence of the thymus, Di George's disease, where T-D cells are absent, a different picture appears: (1) antibodies are produced in response to diphtheria toxin or polio vaccine injections, but in smaller quantities than in normal children; (2) plasma cells are present in normal quantities, and immunoglobulin is in nearly normal amount in the plasma; (3) the child is very "delicate" but is more prone to infection by fungi than by bacteria; (4) skin grafts from another individual are accepted and remain healthy indefinitely; (5) the skin shows no sensitive reactions to foreign substances; (6) measles develops no rash, is often fatal, and—if the child survives—no immunity follows. (The final part of the sixth statement is only a guess based on observations of a related disease.)

The comparison of these two genetic diseases indicates that there are two related but distinct immune systems. The first can be called the antibody-producing system—including the plasma cells—and its main function is dealing with bacterial infection. The second system is concerned with cell-mediated immune responses in which the immunocytes themselves carry out the entire process. Cells derived from thymal ancestors are wholly responsible for the cell-mediated responses and also play an important part in making some types of antibody response possible. In all probability, the thymus-dependent system is of earlier evolutionary origin, and it is in many ways more important than the antibody-producing system. However, because antibodies are more conveniently studied than living immunocytes, much immunological research has focused on the antibody-producing system, and a clear picture of the biological significance of immunity has been obtained from such work. Although the importance of the thymus-dependent system has been increasingly recognized in recent years, any summary of current knowledge and research in immunology must still emphasize the actions of antibodies.

STUDYING ANTIBODIES

Anyone with an interest in biology or medicine has a rough idea of what antibodies are. Before the days of antibiotics, a physician caring for a patient with pneumonia would watch for what was called the crisis, which was a sudden improvement occurring about a week after the onset of symptoms and signifying that adequate supplies of antibodies to deal with the germs—the pneumococci—had been produced. Once that supply of antibodies existed in the blood, the pneumococci would be controlled. There were various ways by which the antibodies could be recognized and their amount measured.

In light of modern knowledge of protein biosynthesis, biochemists have taken a particular interest in the nature of antibodies. Virtually all body proteins are synthesized to a particular, genetically determined pattern to serve some particular function in an organism. Proteins are accurately produced to match patterns indicated in genes through the well-known processes of transcription and translation, involving mRNA, ribosomes, and tRNA. Antibodies, however, appear to exist in an almost infinite variety of forms—such that an antibody is produced to fit almost any sort of foreign organic material. It is not surprising that many biologists have concluded that the antibodies are custom-built to fit the antigens.

Two questions are of particular interest to biochemists: What are antibodies? How are they synthesized in such a vast diversity of patterns? It is not easy to obtain *an* antibody for study. If a rabbit is immunized with some pure protein, enough antibody is produced to react with the protein antigen. It might be reasonable to expect that injection of a pure antigen would lead to production of a pure antibody, but this expectation proves to be far from the case. In fact, what is obtained from the rabbit's blood is a very complex assemblage of immunoglobulin molecules whose one common feature is that they will all unite—in varying degrees of strength—with the antigenic determinants of the protein injected.

Myeloma Proteins

The difficulties of studying mixtures of immunoglobulin molecules were overcome by making use of another experiment of nature. There is a fairly common disease—a mild form of cancer called *myeloma*—in which a single plasma cell undergoes malignant change. It becomes cancerous in the sense that it enlarges and divides into two, the daughter cells continue to divide, and the process continues indefinitely until there are many billions of this sort of plasma cell in the body. The genetic information of the original plasma cell allows and compels it to produce one particular pattern of antibody, and, as the proliferation continues, each descendant acquires the same genetic information. When the disease is fully developed, the blood contains a vast excess of that particular type of antibody. Of special interest to the biochemist is the fact that this antibody is essentially pure. It is a homogeneous population of immunoglobulin molecules, each one made to the same pattern.

Although any random plasma cell may be provoked to cancerous proliferation, the event is so rare that it is most unusual for it to occur twice in a single individual. Each individual myeloma victim has his own unique pattern of antibody, and that pattern differs in some or many ways from the pattern of any other individual's myeloma protein. The antigenic determinant appropriate to unite with the combining site of a myeloma protein has been identified in only a few rare cases. It is relatively certain, however, that if tests were done with any two myeloma proteins, they would react with different determinants.

Application of techniques such as immunoelectrophoresis and comparative studies of normal immunoglobulins and antibodies have yielded a flood of information—so much that only a brief description of the most common and best-studied myeloma protein can be given here. This molecule is of the type called *immunoglobulin G*.

The molecule of immunoglobulin G is made up of four chains bound together with single disulphide linkages. It is a double-end molecule with a combining site at each end (Figure 22.12). Everything is symmetrical; the two light (L) chains are identical, and the two heavy (H) chains also form an

Figure 22.12. Structure of immunoglobulin G.

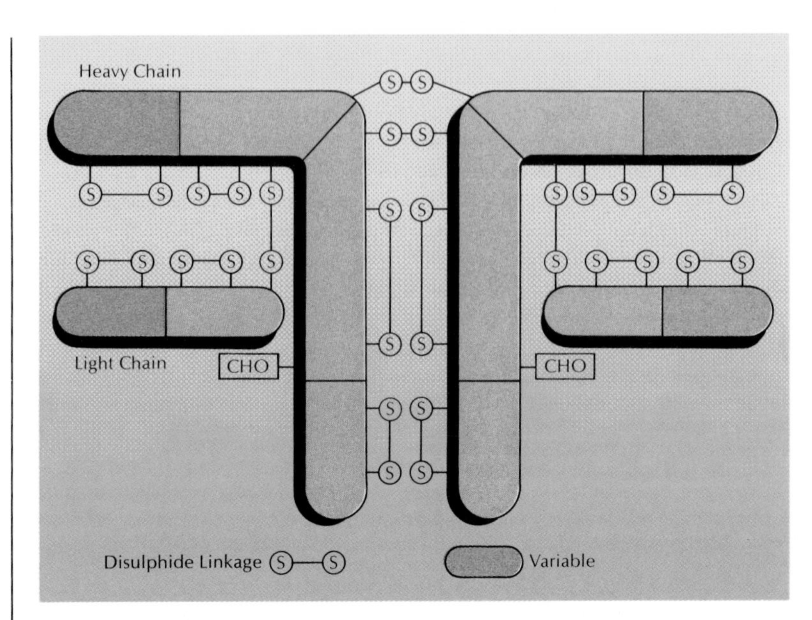

identical pair. The combining sites made by the association of the end segments of one L and one H chain are also equivalent, each reacting with the same antigenic determinants.

When myeloma proteins from a number of different people with this same type of the disease are compared in detail, a new feature becomes evident. Some parts of the molecule, as judged by the sequence of amino acids, are identical in all the proteins or show only one or two discrepancies in amino acids. However, the sections shown in Figure 22.12 as "Variable" differ widely from one protein to another, although they have a basically similar structure. At a position nine places along from the end of this variable segment, for example, the amino acid may be any one of at least six—serine, threonine, glycine, aspartic acid, alanine, or leucine. It is very significant that the combining site is produced by the interaction of the variable segments of the L and H chains. Here, surely, is the chemical basis for the diversity of antibodies.

The Origin of Antibody Diversity

The second and more difficult question remains: How can the plasma cells of the body produce many thousands of different combining-site patterns, each presumably involving a different amino acid segment in the variable segments? How is it that the body can build an appropriate antibody even against some synthetic antigens that neither the individual nor his ancestors could ever have encountered in the past?

According to modern theories of protein biosynthesis, the amino acid sequences of immunoglobulins must be genetically transmitted from each cell to its descendants, coded as nucleotide sequences in genes. How does the body develop a population of plasma cells, each carrying genes that specify a different variable region for its immunoglobulin? One of the biggest puzzles left for molecular biologists after the cracking of the genetic code is the explanation of the source of this genetic diversity. The answer has not been found, but the *sort* of answer that it must be is becoming clear.

According to the clonal selection theory developed by Burnet, the diversity must depend to a large extent on the presence of duplicated genes carried in the germ cells and subject to individual mutations in the course of evolution. The varying evolution of these different (originally identical) genes could account for the presence of different types of immunoglobulins (having different sequences in the "constant" regions) in the same individual. According to Burnet's theory, diversification also arises through somatic mutations – genetic changes that occur in the body cells at various stages of embryonic and later life. These mutations in individual cells produce the variety of possible sequences in the variable regions of the immunoglobulins. These somatic mutations probably involve changes in a single nucleotide in genes specifying the sequence of the variable regions, although there are other possibilities. Finally, Burnet suggests that there must be a process called *phenotypic restriction* through which the cell "chooses" to produce only one of the possible immunoglobulins coded in its genes. Once this selection is made, the type of antibody to be produced by the cell and all of its descendants is fixed.

Various other theories under current consideration differ from Burnet's in certain details, but all postulate the steady production – especially in infancy and childhood – of a great variety of essentially random modifications of the standard antibody patterns inherited by the zygote. This variety of patterns is the raw material from which antigens, when they enter the body, "select" appropriate immunocytes to be stimulated through combination of the AD with the combining site of the immunocyte receptors. The stimulated cells proliferate and make antibodies that will react with the antigen that these cells have "recognized."

TOLERANCE: SELF OR NOT-SELF?

The immune system has to immobilize and eliminate foreign material, but it must have no harmful effect on the proper tissues and components of the body. Thus, there must be a natural immunological tolerance for all accessible components of the body. Tolerance to foreign substances can be induced artificially in various ways – by injecting foreign cells into newborn animals, by using x-rays, or by using drugs. The successful transplantation of organs, for instance, depends upon skillful use of drugs to "trick" the body into tolerating cells and tissues that it would normally reject.

Tolerance is not fully understood, but a simple interpretation, suggested many years ago by Burnet, has not yet been disproved. For a time after its primary differentiation as an immunocyte, the cell is highly sensitive to contact with any antigenic determinant that matches its combining sites. Instead of causing proliferation, the stimulus at this time is so severe that it kills the cell. Such a sensitivity of "newborn" immunocytes would ensure that all cells capable of being stimulated by normal components of the blood and lymph are eliminated from the immunocyte population in this early stage. According to this view, tolerance results simply from the absence of immunocytes with combining sites that match normal components of the body.

Other immunologists have hypothesized the existence of *tolerant immunocytes*, which have the capacity to recognize normal components of the body fluids and to prevent the synthesis of antiself antibodies. However, the existence of such tolerant immunocytes has not yet been demonstrated.

IMMUNE RESPONSES

The complexities of immune responses make it impossible to give a simple summary that includes all known facts about the sequence of events

Figure 22.13. Electron micrograph showing antigens that have become trapped on the branches of dendritic phagocytic cells (DPC) within a lymph node. Also present are the nuclei of small lymphocytes.

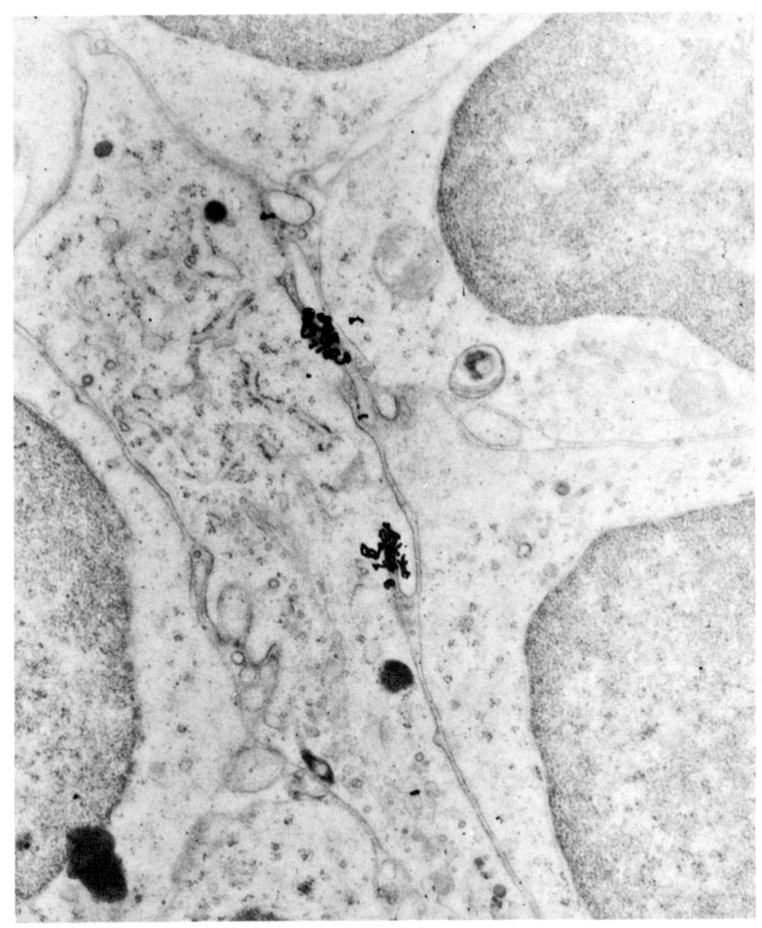

involved. All that can be attempted here are an outline of two well-studied examples and a reasonable—but not unanimously accepted—interpretation of each.

The first example involves the injection into a rat of protein derived from bacterial flagella. This antigen provokes a strong reaction, and the antibody is easily detected and measured in the blood plasma. If this antigen is labeled with radioactive iodine and is injected into the footpad of a rat, the fate of the antigen can be followed in detail by autoradiography. The antigen passes quite rapidly to the lymph node that lies behind the knee-joint and drains the entire foot region. In the lymph node, the antigen is taken up by the phagocytic macrophages, which destroy most of it but may retain some of the antigen. Another part of the antigen is taken up by a different type of phagocytic cell that is present among the masses of lymphocytes in the node. These cells, called *dendritic phagocytic cells*, have long, branching extensions, which are particularly well developed if traces of the antigen are already present in the lymph. Antigen sticks to the surface of these branches and sometimes stays there for weeks (Figure 22.13).

In a living lymph node, lymphocytes are always coming and going and moving about "like a bag of worms." In every lymph node there are ample opportunities for a particular antigen held on the branches of the dendritic phagocytic cells to be contacted by a lymphocyte whose combining sites match its AD. Almost certainly, the stimulation of the immunocyte takes place in this fashion, but immunologists are uncertain of the subsequent processes by which large numbers of antibody-producing plasma cells appear, either in the draining lymph node or elsewhere in the body. A simple—probably too simple—description is that the stimulated lymphocyte enlarges, becomes more mobile, and moves by lymph or blood circulation to any place where the conditions are right for it to settle down and multiply, producing a clone of antibody-synthesizing plasma cells.

The second example—the rejection of a foreign skin graft—demonstrates cell-mediated immunity. Here again, various steps of the process are clear, but the connection between them is still uncertain. It is probably best to give the simplest interpretation that seems to be consistent with the known facts—remembering that future observations may prove this simple explanation inadequate.

When the foreign skin graft is placed in the raw bed from which the animal's own skin was removed (Figure 22.14), there develops a steady traffic of wandering cells from the blood vessels into the general area, some passing into or making frequent contact with the grafted tissue. Fragments of the foreign tissue lodge on these wandering cells, and some are carried to the draining lymph nodes.

In the lymph nodes, the cells carrying foreign antigen appear to meet thymus-dependent immunocytes with complementary combining sites near the central part of the node. There is the appearance of great activity in this area, where immunocytes multiply freely but produce more lymphocytes rather than plasma cells. Many of these active cells do not possess antibodies of the type that matches the antigen, but the stimulation of the "correct" immunocytes has a secondary effect on the others. Both sorts of cells pass into the circulation and, if they reach the site of the foreign graft, tend to lodge there.

The inflammatory response produced by the interaction of these immunocytes with the antigens present in the graft damages the tissues, particularly the newly formed capillaries that are beginning to extend into the graft. Damage to cells accumulates, the blood supply becomes inadequate, and eventually the graft dies and is sloughed off as a scab about two weeks after the implantation.

Figure 22.14. Sequence showing the procedures for experimental skin grafts in mice. (A–B) Dissecting the skin. (C) A fitted graft. (D) Open style grafts.

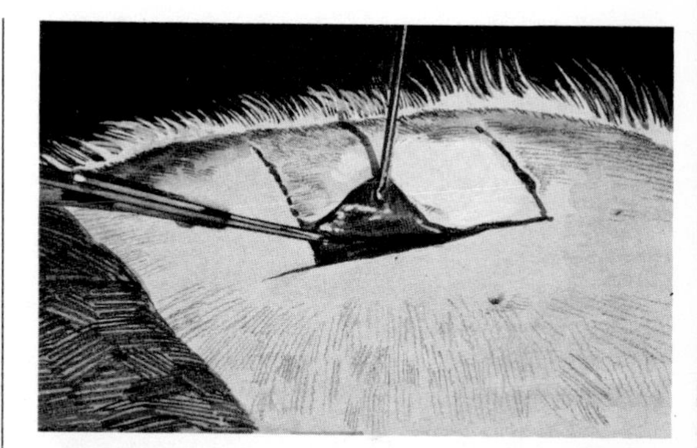

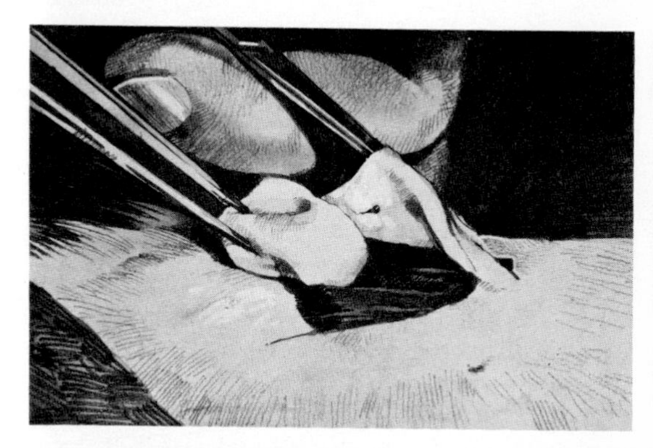

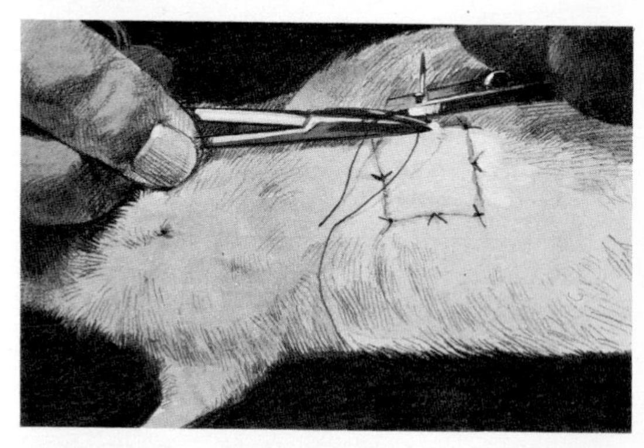

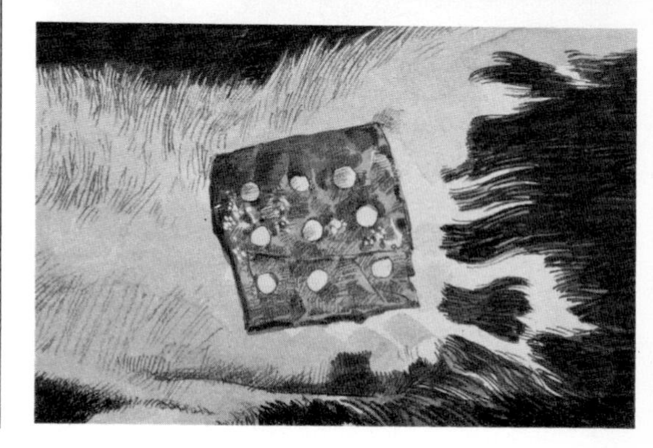

THE SCOPE OF IMMUNOLOGY

Immunology is a science that impinges on virtually every other aspect of biology. It is an area of extremely active research and exciting discoveries, as well as a field with extremely important medical implications.

Much interesting work is being done on the process by which immune systems have evolved and on their comparative forms in different vertebrates. There is growing evidence that the immune system can sometimes—perhaps much more often than yet known—nip an incipient cancer in the bud, and the nature of this immunological surveillance is of obvious relevance to cancer research. The nature of the immunological coexistence of mother and embryo during pregnancy and the occasional breakdown of this mutual tolerance are matters of great interest to many immunological and medical researchers.

The symptoms of measles are due not so much to the actions of the virus that infects the body as to sensitization—the disease is essentially a cell-

Figure 22.15. Bronze statue of Edward Jenner innoculating his son with cowpox germs.

mediated immune response. In fact, pathologists are beginning to suspect that many manifestations of infection are actually unfortunate effects of the immune responses that should deal with them. Of particular interest are the many diseases that appear to be due to an attack by the immune system upon normal body tissues. Among these *autoimmune diseases* are such serious conditions as haemolytic anaemia and lupus, and there is a great deal of evidence that many other conditions such as rheumatism and arthritis involve autoimmune responses. Allergic reactions involve abnormal immune responses; in most cases, tissue damage is caused by an immune response to a substance that would normally be harmless.

The story of immunology began with Edward Jenner's innoculation of a boy with cowpox germs, thus protecting him from smallpox infection. The protection of children against diphtheria, polio, and whooping cough by immunization represents one of the great medical advances of the century

and has resulted in a dramatic decrease in childhood deaths around the world. Paradoxically, this great advance has been a major contribution to the explosive population growth that is a major problem today (Chapter 45). In the field of veterinary diseases, there has been even more effective use of immunization. Much current research is aimed toward development of more effective vaccines and of immunization techniques against diseases not yet controlled.

Immunology provides an enthralling mixture of interests that stretch across the spectrum of biology. Its limits have not been exhausted by any means. Among the problems recently investigated by immunologists—to mention a few of the bizarre but still useful extensions of the field—are the detection of horse meat in hamburgers, the explanation of why certain deodorants produce lumps in ladies' armpits, and the blood grouping of the remains of ancient Egyptians.

FURTHER READING

Burnet (1969) presents a readable summary of current knowledge and theories in immunology, with emphasis upon his own theoretical viewpoint. Kabat (1968) gives a more complete treatment of the molecular aspects of immunology, with thorough bibliographies. Killander (1967) includes reports on much of the research aimed toward an understanding of the synthesis of immunoglobulins. For a general review of current concepts of the nature of antibodies, see Edelman and Gall (1969). Relevant *Scientific American* articles include Allison (1967), Billingham and Silvers (1963), Boyden (1951), Burnet (1954, 1961, 1962), Crowle (1960), Duve (1963), Edelman (1970), Frei and Freireich (1964), Levey (1964), Nossal (1964), Porter (1967), and Williams (1960).

23
Building Blocks of Nervous Systems

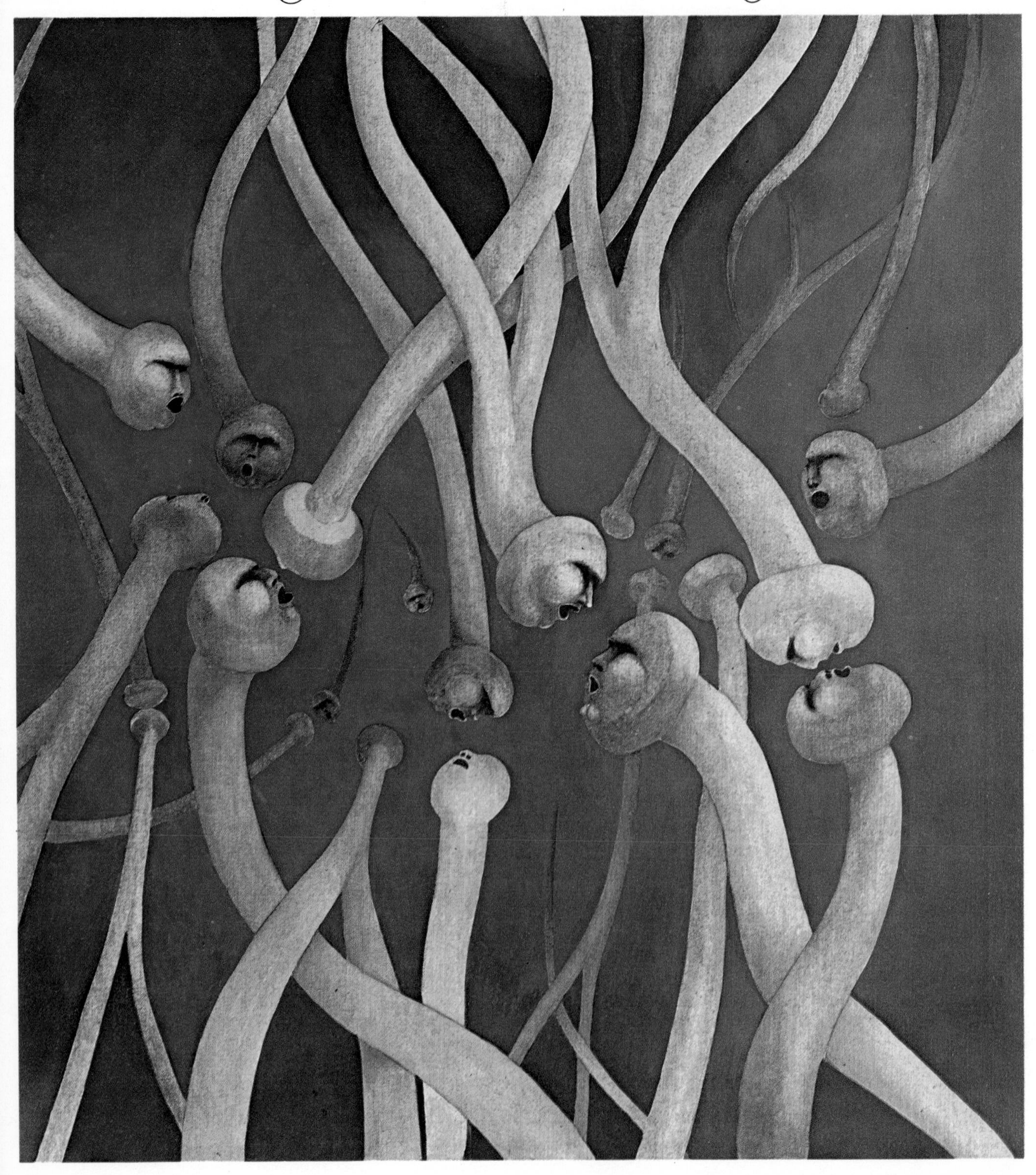

The general anatomy of the nervous system and its function in the control of the animal body were worked out by physiologists over the centuries, culminating in Charles Sherrington's great book, *The Integrative Action of the Nervous System* (1906). But the nature of the nerve impulse, the mechanisms by which it is propagated in the *neuron*, or nerve cell, and the way in which impulses are passed from one neuron to the next have been deciphered only in the past few decades. The study of the nervous system is best begun with an examination of these neurons, the building blocks that make up the fantastically complex nervous systems of higher animals.

An animal's nervous system must gather information about the internal state of the organism and its external environment, evaluate this information, and coordinate activities appropriate to the situation and to the animal's current needs. The gathering of information is performed by *receptors* and *sensory neurons*. The coordinated activities are executed by *effectors*, which may be muscles or glands, and the effectors are controlled by *motor neurons*. The nervous system also contains an immense number of *interneurons*—neither sensory nor motor—whose role is to process the sensory input, evaluate it, and command the motor output. These cells perform an associative function; they form connective links between sensory and motor levels. They may be organized into interconnecting groups and levels with almost unimaginable complexity.

The traditional division of the nervous system into sensory, associative, and motor portions is useful because there are many properties shared by the elements within each division. There are, however, many properties common to particular elements of different divisions, and it is sometimes difficult to say where "sensory" ends and where "associative" begins. Certain cells, for example, originate within the brain but send processes to the peripheral sense organs—such as the auditory receptors—where they presumably modulate the reception of incoming information. Similar difficulties arise with the definition of motor elements. Any division of the complex nervous system into segments, regions, or subsystems must be an arbitrary one. In the living animal, the system functions as a magnificently complex and coordinated whole.

CELLS IN THE NERVOUS SYSTEM

Neurons are the best-studied cells of the nervous system. These highly specialized nerve cells are distinguished by the extremely long processes, or *axons*, that they send through the animal body. The general structure of the neuron is described in Chapter 11.

The vertebrate nervous system includes not only neurons but also closely associated *glial cells*—or *neuroglia*—and the blood vessels, membranes, and other structures closely associated with the nerves. Neuroglia exist in a wide variety of shapes and forms. Although their functions are largely unknown, they traditionally have been thought to function in the support and nourishment of neurons. Neuroglia pack the spaces between nerve cells and closely surround neuron bodies and processes. More than 90 percent of all cells in the vertebrate brain are neuroglia. They are generally smaller than neurons and thus make up only 50 percent of the brain's weight. Neuroglia contain far less RNA than do neurons, and they possess strikingly different kinds of RNA and enzymes.

Two well-studied types of glial cells are *oligodendroglia* and *astrocytes*.

Figure 23.1 (right). Photomicrograph of glial cells, or neuroglia, among the larger nerve cells.

Figure 23.2 (left). Depolarization of a glial cell produced by nerve impulses in the *Necturas* optic nerve (above). The glial membrane potential recorded with an intracellular electrode was 86 mV. In the upper record, depolarization after a single nerve volley peaks in about 150 milliseconds and declines with a half time of about 2 seconds. In the lower record, three stimuli at 1-second intervals produced a summation of glial depolarizations. Glial depolarization effected by four different frequencies of nerve stimulation (below). Trains of electrical stimuli at 0.5, 1, 2, and 5 stimuli per second are applied for about 20 seconds. The amplitude of the glial depolarization, as measured with an intracellular electrode, ranged from about 6 to 7 mV with 0.5/s stimulation to about 17mV with 5/s stimulation.

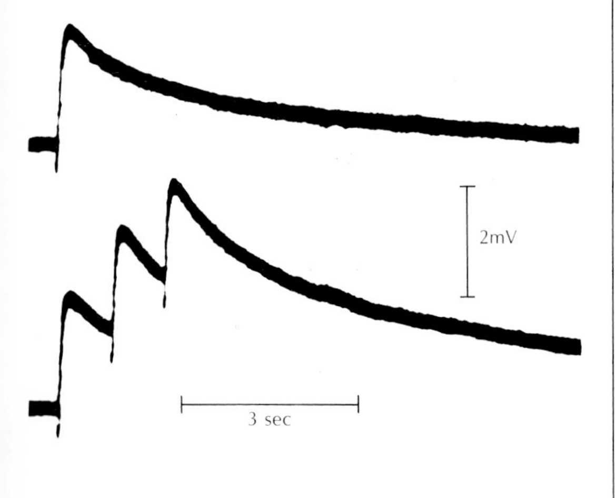

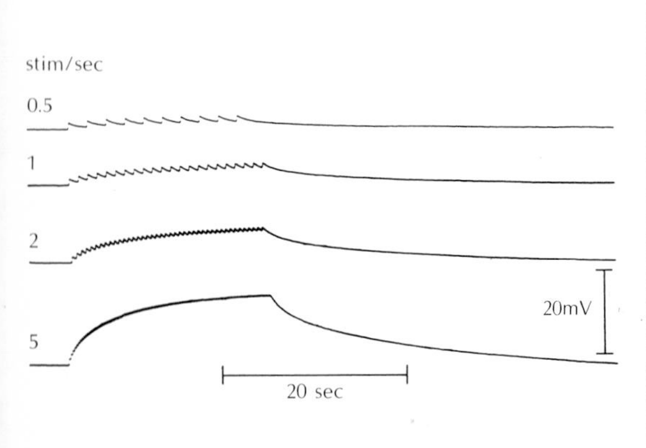

ulation (seconds)

Oligodendroglia are instrumental in the formation of myelin in the central nervous system, as are the Schwann cells in the peripheral parts of the system. The myelin sheaths formed around axons by Schwann cells and oligodendroglia play an important role in speeding movement of impulses along the axons. Oligodendroglia may have other important functions as well, but none has yet been firmly established.

Some of the older suggestions about the functions of astrocytes still seem reasonable today. Undoubtedly, they provide mechanical support for neural tissue. They keep the brain free of sick neurons by phagocytosis—engulfing and digesting them. Astrocytes probably serve in a nutritive capacity because they are intimately associated with capillaries and with neurons. Electron microscope studies indicate that there is always a layer of glial cytoplasm between any neuron and the nearest blood vessel. The astrocytes may mediate transfer of proteins or metabolites between neurons and the bloodstream.

An interesting modern approach to the study of neuroglia has been taken by S. W. Kuffler (1967) and his colleagues at the Harvard Medical School. They used leech nervous systems and optic nerves from *Necturus* (the mud puppy, an amphibian) to demonstrate electrical and ionic relationships of neurons and neuroglia. They found that there is an electrical potential, or *polarization*, between the inside and outside of the glial cell, with the inside about 60 to 90 millivolts (1 millivolt = 1 mV = 0.001 volt) negative with respect to the outside. This *resting potential* depends only on the concentration of potassium ion (K^+) in the extracellular space. The greater the K^+ concentration outside, the less negative the resting potential. (A decrease in the resting potential reduces the polarization and thus is called *depolarization*.)

Kuffler and his colleagues found that the activity of nerve cells induces the glial cell to produce an electrical response, namely the depolarization resulting from the increase in K^+ in the extracellular space. This response is fundamentally different from the nerve impulse of a neuron. There is a

Figure 23.3a (left). Diagram illustrating the variety of types of neurons. Neurons may differ in shape and size of the cell body (soma), in dendrite processes, in length of the axon, and in presence or absence of a myelin sheath.

Figure 23.3b (right). A slice of cerebellar cortex stained to outline neurons.

high electrical continuity between adjacent glial cells, but little electrical connection between the glial cell and the extracellular space or the adjacent neurons. Thus, neuroglia are not well suited for electrical signaling to nerve cells. An older suggestion that glial cells play a vital role in transmission of nerve impulses was disproved in an experiment in which the neurons were washed free of most glial cytoplasm yet maintained the ability to conduct impulses for hours. Thus, neuroglia are not absolutely necessary for conduction of nerve impulses.

The most interesting discovery made by Kuffler's group was that K^+ concentration increases in the region outside a glial cell as nerve impulses pass nearby. The increase of K^+ concentration was detected because it causes a prolonged depolarization of the glial cell membrane. This change in the ionic concentration of the extracellular fluid could have profound effects on the synapses, where one neuron affects the triggering of impulses in another, and possibly on nerve impulses moving through fine nerve fibers. The depolarization may also trigger chemical changes in the glial cells—changes that might ultimately lead to other effects on neurons. The role of glial cells in the nervous system is only beginning to be understood.

ACTIVITY OF NEURONS: THE NERVE IMPULSE

There are many different types of neurons, differing in shape and size of cell body (*soma*), in the presence or absence of specialized dendrite processes, in length of the axon, and in the presence or absence of a fatty myelin sheath covering the axon (Figure 23.3).

An account of the function of neurons must explain how stimuli arising

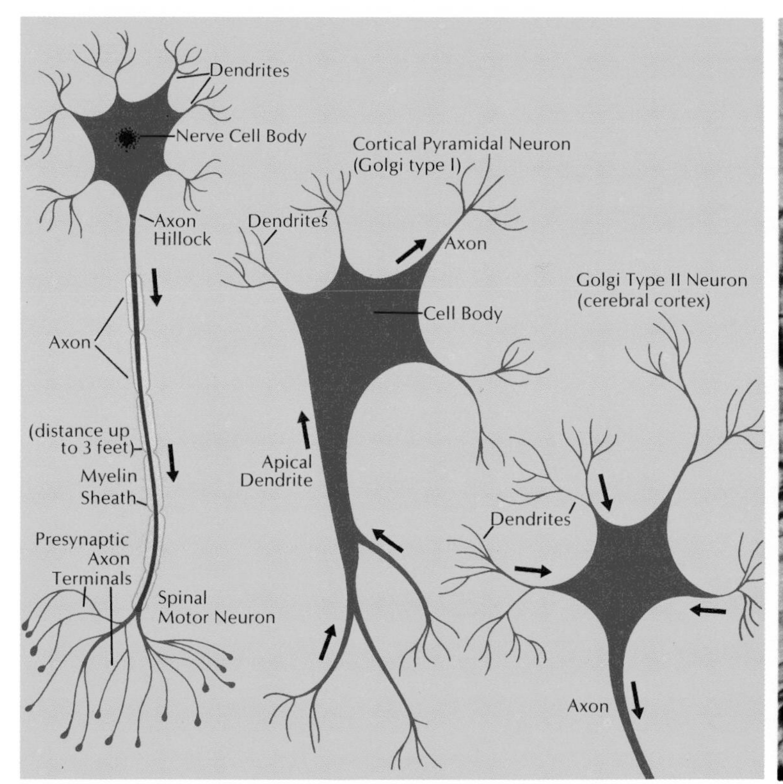

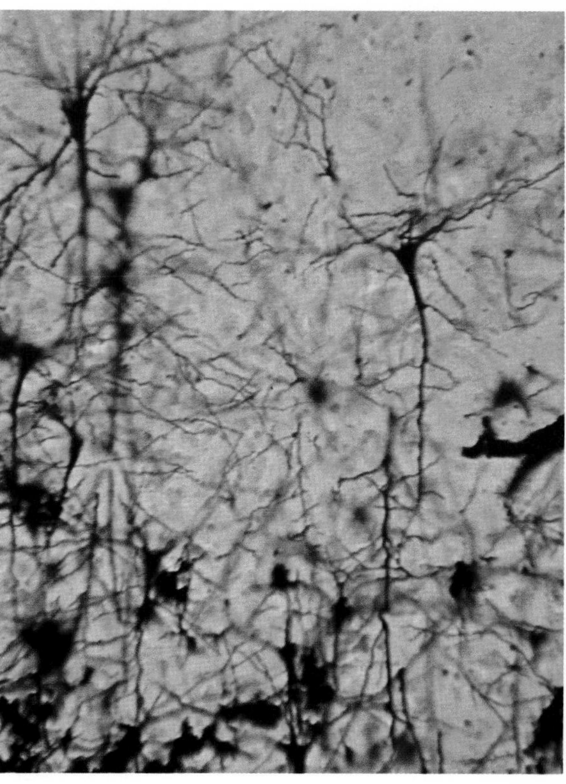

Figure 23.4a (upper left). Micropipette inserted into a very large axon found in the squid. Each scale division equals 33 microns.

Figure 23.4b (lower left). Propagation of action potential by local circuit stimulation. The spatial variation of an action potential at a fixed time is shown in A. Ordinate is transmembrane potential (E_m); abscissa is distance along fibers shown in B and C. The action potential is propagated at a speed of 20 meters per second to the left (arrow). An unmyelinated nerve fiber is shown in B. Transmembrane voltages given in A are presented here by plus and minus signs. Arrows indicate that current flow in a loop is due to the differences in transmembrane potential caused by different ionic permeabilities of the membrane; K^+ is highly permeable at either end, and Na^+ is even more highly permeable in the central region. Propagation is achieved by the depolarization action of local current flow. A myelinated nerve fiber is shown in C. Distance between nodes is shown as twice actual distance. Charge is shown only at nodes because of the very high electrical resistance of the sheath. Local circuit flow is thus from node to node as shown by the arrows.

Figure 23.4c (lower right). Nerve impulse-action potential.

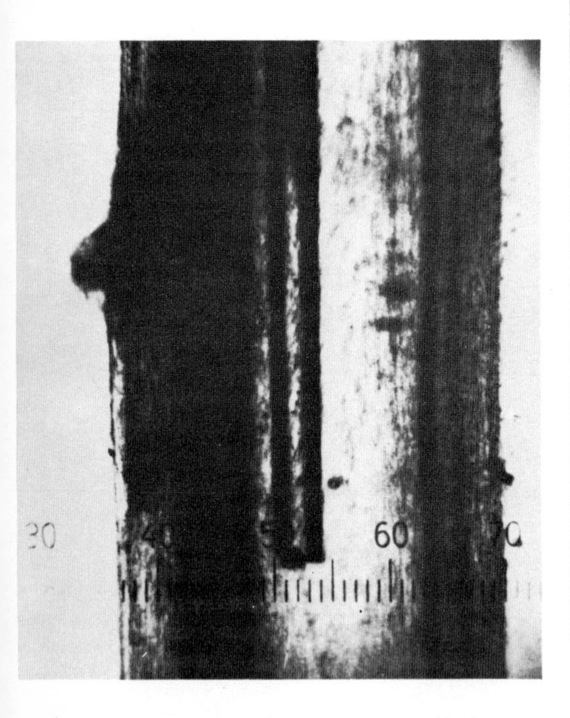

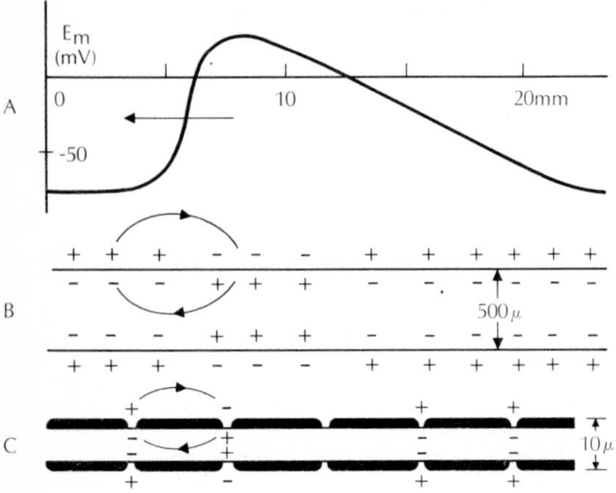

outside the cell are converted into nerve impulses, how the nerve impulse travels along the axon to its destination, and how the impulse stimulates other neurons or effectors at the destination. The electrical and chemical mechanisms that accomplish these functions are similar in most types of neurons. The nature of the nerve impulse is similar in all neurons.

When physiologists first realized that the nerve impulse is electrical in nature, they assumed that nerves act like cables carrying electrical currents through the body. The axon is a cablelike structure, with a core of ion-containing fluid that acts as an electrical conductor, surrounded by a membrane that acts as an insulator. However, the electrical properties of a nerve fiber are extremely poor when compared with standard electrical cable structures. The electrical resistance of the fluid in the axon is about 100 million times greater than that of copper, and the axon membrane leaks electric current about 1 million times more than the sheath of a good cable. Thus, the cablelike performance of an axon is about 100 million million times poorer than that of a typical sheathed-copper cable. If an electric pulse too weak to trigger a nerve impulse is started in an axon, it dies out within a few millimeters' travel along the axon. However, a stronger pulse triggers a unique electrical event, a nerve impulse, that travels along the axon at high speed for an indefinite distance without distortion or loss of strength, something that cannot happen in a normal sheathed-copper cable. The nature of the nerve impulse depends upon the ion-transporting properties of the neuron membrane.

A neuron, like a glial cell, has a resting potential. That is, its membrane normally is electrically polarized, with the inside 50 to 90 mV negative relative to the outside. When a *nerve impulse*, or *action potential*, is triggered, the potential across the axon membrane falls toward zero (depolarizes) and then reverses with the inside positive. The positivity reaches a peak, and then the membrane potential decreases, passes through zero again, and eventually restores the resting potential of −50 to −90 mV (Figure 23.4). At any one spot in the neuron membrane, this complete cycle of polarity changes usually takes about 1 to 5 milliseconds (1 millisecond = 1 ms = 0.001 second). The position of the disturbance of membrane potential travels along the axon at a speed known as the *conduction velocity*. It is this propagation of the nerve impulse along the axon that allows infor-

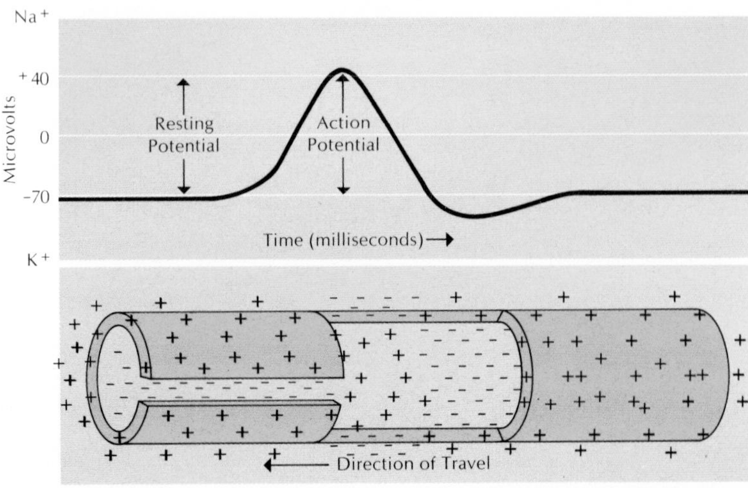

Figure 23.4d. Voltage clamp experiments. Schematic diagram of the apparatus and technique (A); schematic diagram of the feedback amplifier operation (B), and diagram showing component membrane currents (C). Voltage is applied by electrodes on either side of the membrane while an automatic feedback amplifier is used to keep the membrane potential constant at any value selected by the experimenter. Thus, currents resulting from ionic flow across the membrane can be measured when membrane potential is displaced to and held at some fixed value.

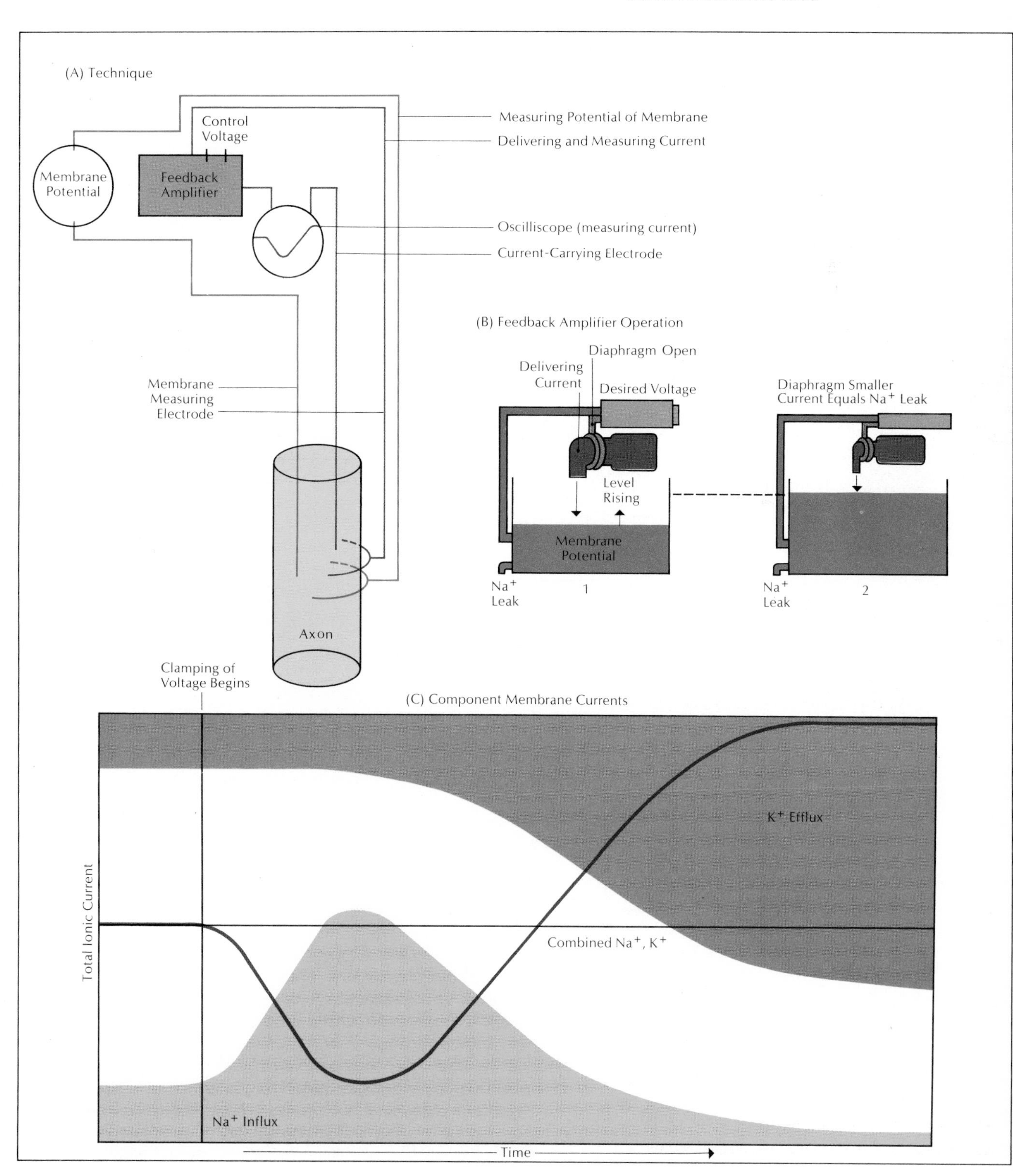

Figure 23.4e. Impulse initiation by local depolarization. Current pulses of fixed duration (indicated below the curves), but of variable size and polarity, cause membrane potential variations represented by the family of curves shown above. Inset diagrams indicate sodium (Na^+) and potassium (K^+) flow through the membrane.

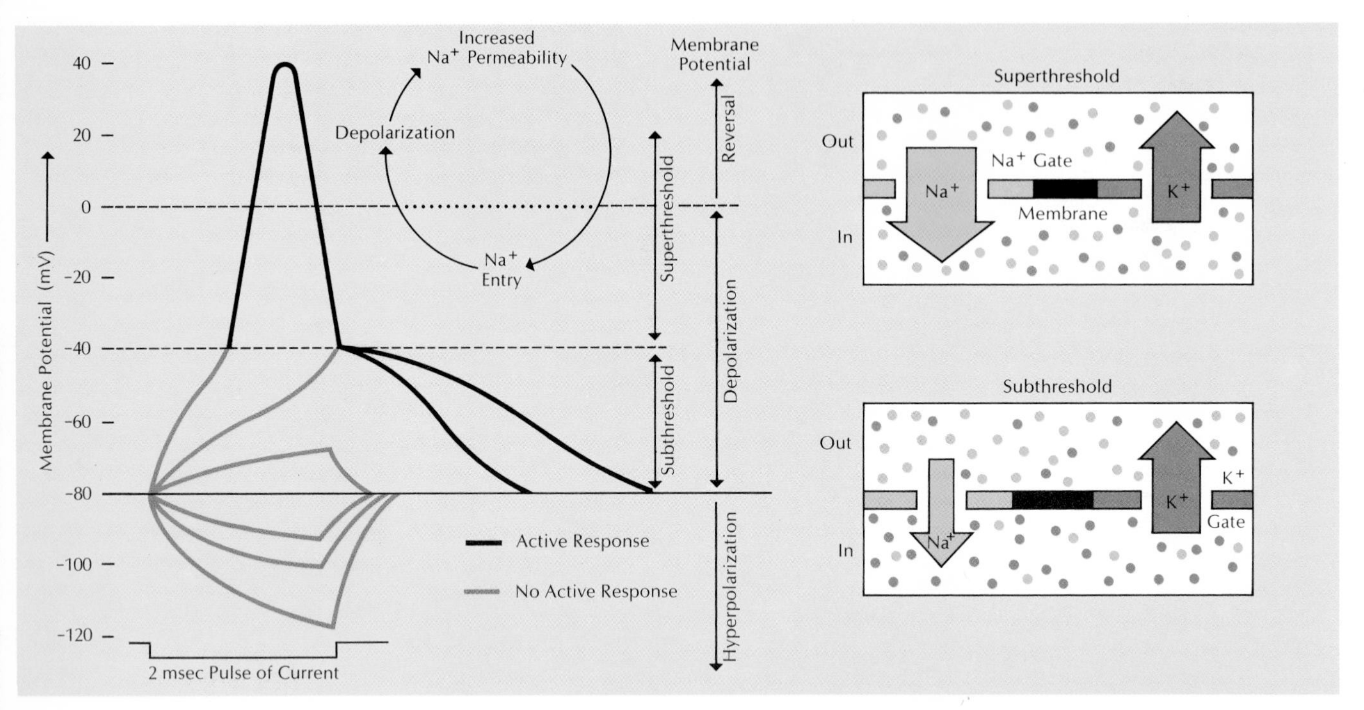

mation to move from one place to another in the nervous system. The nerve impulse is "all-or-none"—it either happens or it does not, and its size, duration, and velocity do not depend on the nature of the stimulus that triggered it. The nerve impulse is *nondecremental*—that is, it does not decrease its size as it travels along the axon from its origin to its destination.

The mechanism by which the impulse is generated and conducted along the axon is now relatively well understood, although there is still much to be learned about its biophysical mechanism. Information about the nerve impulse has come from studies of the giant axon of squid (Hodgkin and Huxley, 1939). Because this axon can be up to 1 mm in diameter, electrical and chemical studies are much easier to make with the squid neuron than with mammalian neurons, which have axons less than 20μ in diameter. Painstaking work with small neurons subsequently has confirmed that the processes of impulse transmission are much the same.

The resting potential is caused by differences in concentrations of certain ions inside and outside the nerve cell. In squid and in most vertebrate neurons, the sodium (Na^+) concentration outside the cell is 10 times greater than the concentration inside, whereas the potassium (K^+) concentration inside is at least 20 times greater than that outside. The Nernst equation, developed by physical chemists, tells what potential must exist across the membrane to keep any particular ionic type in equilibrium when it diffuses freely across the membrane—that is, to have the diffusion inward equal the diffusion outward. This *equilibrium potential* for each ion type is determined by the ratio of its concentrations outside and inside the cell. Another principle of physical chemistry states that the actual membrane potential will depend upon the equilibrium potentials of all the different ion types that can cross the membrane. The relative importance of each ion type in the determination of the membrane potential depends upon the ease with

Figure 23.5. Excitability potential. Ordinate is transmembrane potential; abscissa is time. The absolute refractory period of 0.5 ms follows the peak of an action potential. During this time, the neuron membrane cannot be made permeable to sodium (Na^+) ions. The relative refractory period represents a recovery period where the threshold depolarization necessary to turn on sodium permeability gradually returns to normal.

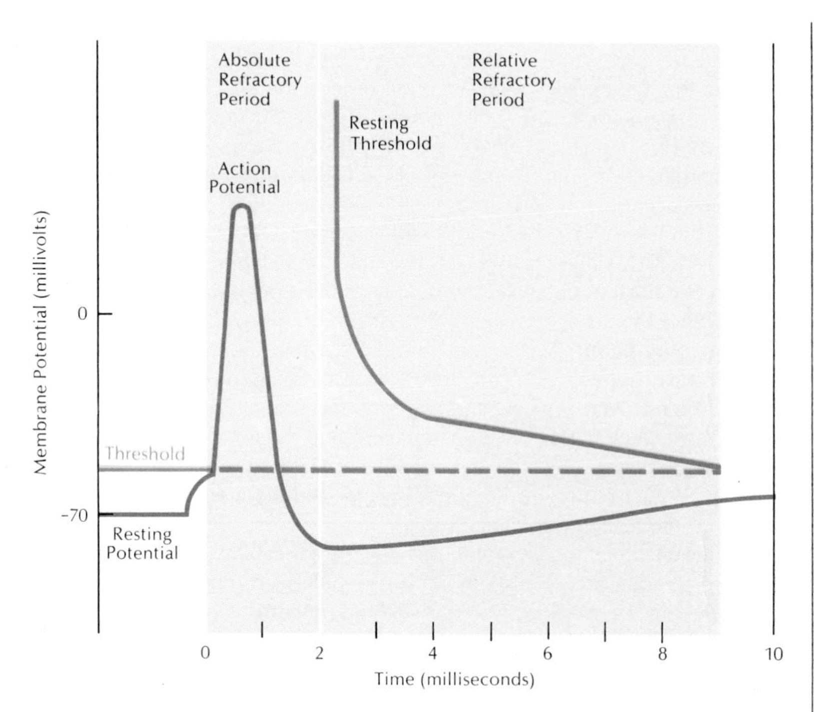

which that ion can cross the membrane—its *ion permeability*. Thus, if K^+ permeability is much higher in the resting axon than Na^+ permeability, the resting potential will be closer to the potassium equilibrium potential (typically about −90 mV) than to the sodium equilibrium potential (typically about +60 mV). The resting potential of the giant squid axon typically is about −70 mV (Hodgkin, 1958).

This explanation of the resting potential is important for an understanding of the action potential. If something caused the axon membrane to become much more permeable to Na^+ than to K^+, then the membrane potential would change toward the +60 mV sodium equilibrium potential. A. L. Hodgkin, A. F. Huxley, and Bernard Katz (1952) performed what is now called a voltage-clamp experiment. In this experimental technique, voltage is applied by electrodes on either side of the membrane. An automatic feedback amplifier is used to keep the potential across the membrane constant at any value desired by the experimenter. Thus, the currents caused by ions moving across the membrane can be measured when the membrane potential is displaced to and held at some fixed value.

In their experiment, Hodgkin, Huxley, and Katz rapidly depolarized the membrane by a certain amount and then held the membrane potential constant at the new level. If the depolarization exceeds a minimum, or *threshold* amount, the Na^+ permeability of the membrane suddenly increases. Sodium ions carrying positive charge rush into the axon. If no equivalent inward current is applied, the membrane potential would rapidly approach the sodium equilibrium potential of about +60 mV. However, with the membrane potential held constant, after a few milliseconds the Na^+ permeability decreases to almost its original level, whereas the K^+ permeability increases greatly. If the membrane voltage were not "clamped," this change would cause the membrane to return to a potential more negative than its resting level — nearly to the K^+ equilibrium potential of about −90 mV.

These findings about changes in ionic conductance enabled Hodgkin and Huxley (1952) to give an accurate description of what happens to a patch of membrane as an action potential passes. Currents from an adjacent patch of membrane depolarize it, causing Na^+ permeability to increase. Sodium ions rapidly move into the cell, further decreasing the voltage difference and further increasing the permeability to sodium ions. This process continues for a fraction of a millisecond; sodium ions moving into the cell make the interior more positive than the exterior, approaching the sodium equilibrium potential. At the end of this brief period of activation, the membrane returns to a normal permeability for sodium ions but has become more permeable to potassium ions, which now move out of the cell toward the lower K^+ concentration of the exterior fluid. As K^+ ions leave the cell, within a few milliseconds the potential difference across the membrane changes, first to a value close to the potassium equilibrium potential, which is more negative than the normal resting potential. The K^+ permeability returns to its normal value, and the membrane gradually returns to the resting potential.

In the voltage-clamp experiment, the Na^+ permeability of the membrane increases gradually as the membrane is depolarized. There is no sudden change in Na^+ permeability as the threshold potential is reached. The sudden development of the action potential is related to the balance between movement of Na^+ and K^+ ions. For depolarizations smaller than the threshold value, the current of K^+ ions flowing outward is greater than the current of Na^+ ions flowing inward, and the net current depolarizes the membrane to its resting potential. When the membrane is depolarized beyond threshold value, the inward current of Na^+ ions exceeds the outward current of K^+ ions, thus further depolarizing the membrane and further increasing the Na^+ permeability. Therefore, once the membrane is depolarized beyond the threshold value, further depolarization occurs at an accelerating rate and the action potential is rapidly developed. The triggering of the impulse is thus a self-regenerative process.

It is easiest to understand propagation of the impulse along the axon in the case of myelinated axons. Because the myelin sheath has very high electrical resistance except at the nodes of Ranvier, most currents must pass through the membrane at these nodes (Tasaki, 1939). As an action potential develops at one node, the sodium influx causes electric current to flow down the axon to the next node of Ranvier. Because of the membrane's electrical capacitance and resistance, the current causes the next node to depolarize beyond the threshold point, increasing Na^+ permeability and triggering an action potential. The action potential jumps from node to node and thus is propagated along the axon.

The propagation process is similar in an unmyelinated nerve. Currents caused by increases in Na^+ permeability at one patch of membrane cause depolarization of adjacent membrane patches, which become active and cause the process to be repeated and the impulse to be propagated. The speed of conduction, or propagation, is influenced by the diameter of the axon (the speed is greater for larger fibers) and by the extent of myelination. The speed ranges from less than 0.1 meter per second to 160 meters per second in different kinds of neurons found in various animal species.

For a period of about 0.5 ms after the peak of an action potential, the neuron membrane cannot be made permeable to sodium ions, no matter

Figure 23.6 (left) Schematic diagram of the ionic pump mechanism. Sodium ions must be pumped out of the neuron and potassium ions pumped in. Because the resting potential is fairly close to the potassium equilibrium potential, but far from the sodium equilibrium potential, the major effect of the ionic pump is the movement of sodium out of the cell. As indicated in the diagram, ATP energy is used to operate the sodium pump active transport mechanism.

Figure 23.7 (right). Compound action potentials. The total action potential of the sciatic nerve of a bullfrog shows several curves, which represent the impulses carried by different size classes of axons. Propagation velocity of the α and β waves is constant in this diagram. S is the stimulating electrode; E_1 and E_2 are recording electrodes, the latter at the killed end of the nerve. Ordinate is the distance from recording electrode E_1; abscissa is time. The starting points of the oscillograph trace show the distances at which the records were taken.

what the depolarization. During this *absolute refractory period*, a new action potential cannot be generated, no matter how intense a stimulus is applied. In the next few milliseconds, there is a gradual recovery in the ability of membrane depolarization to effect an increase in Na^+ permeability. During this recovery period, or *relative refractory period*, the threshold depolarization necessary to turn on the Na^+ permeability (thereby causing another action potential) gradually returns to normal.

The changes during passage of an action potential do not result in large differences in the ionic concentrations inside and outside the neuron. Only about 0.1 to 0.0001 percent of the potassium ions inside axons are lost during the passage of an impulse. However, because neurons are known to be able to carry millions of impulses during their lifetime, it is clear that some mechanism must exist to pump sodium ions out of the neuron and to pump

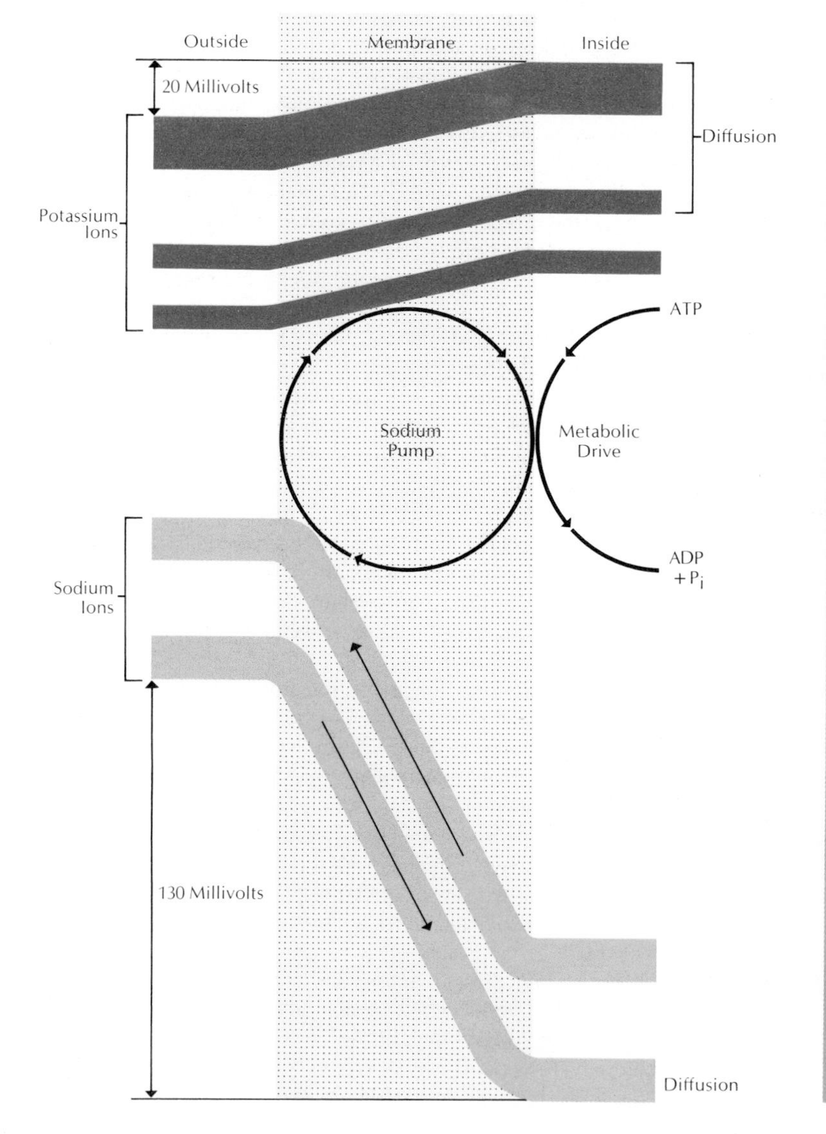

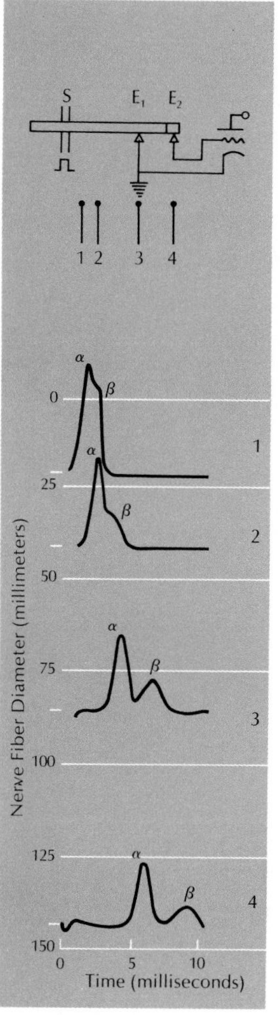

potassium ions in. This mechanism must also counteract the slower diffusions of ions that occur in the resting neuron, as Na^+ leaks in and K^+ leaks out along concentration gradients. Because the ions must be moved against concentration gradients by this mechanism, energy is required to operate the "ion pump."

Because the resting potential is fairly close to the potassium equilibrium potential but far from the sodium equilibrium potential, the major effect of the ion pump must be the movement of sodium ions out of the cell. Studies also suggest that the sodium pump moves potassium ions into the cell. The sodium pump operates slowly but steadily, so that the intermittent entry of sodium ions during passage of impulses is balanced by the continuous expulsion of sodium ions by the pump. Although several transport models have been proposed, the mechanism by which ions are moved across the membrane is not known.

Chemical energy (ATP) is used to operate the sodium pump (Hodgkin and Keynes, 1955; Caldwell, *et al.*, 1960). Because a neuron can continue to generate and transmit impulses for a long time in the absence of oxygen and even when the process of glycolysis is inhibited, it appears that ATP is stored in large quantities near the membrane (Dunham and Glynn, 1961).

The movements of ions by action potentials and by pumps have been confirmed by radioactive isotope studies. Thus, the theory of an ionic basis for the nerve impulse has been generally accepted. The outstanding problem for current research is to discover the mechanism by which the ionic permeabilities of the membrane are changed during the passage of an action potential.

Action Potentials in Nerve Bundles

Axons differ in diameter and degree of myelination, even within a particular sensory nerve bundle in an individual mammal. It is a general property of neurons that conduction velocity increases as axon diameter increases. Often, axon diameters and conduction velocities within a single nerve bundle fall into several distinct classes. If an electrical stimulus is applied extracellularly to a nerve bundle, a larger current is needed to depolarize a small-diameter axon to threshold than is required to trigger an impulse in a large-diameter axon. These facts can explain many observations made about the conduction of impulses in a nerve bundle containing fibers of many different diameters.

First, the stronger the stimulus, the greater the amplitude of the response. Although each axon's action potential is all-or-none, the whole nerve acts in this *graded* manner because as the shock strength is increased, the number of smaller axons or fibers that reaches threshold increases. Second, several humps often are seen on the total action potential. Each hump represents impulses carried by one size class of axons. If the time of arrival of each hump is measured at two distances from the stimulating electrode and the distance between the two sets of recording electrodes is known, then the conduction velocity associated with each size class can be calculated.

These facts explain certain responses in sensory nerves. Different sense organs often have different sizes of axons, and not all sense organs of the same type have the same threshold of excitation or the same diameter of axon. If a person touches a hot stove, he feels the pressure of touching the object long before he feels pain or heat. This experience is largely because touch receptors have thick myelinated fibers, whereas fibers involved in

Figure 23.8 (left) Electron micrograph of the ultrastructure of a synaptic junction. Note the synaptic knobs or boutons.

Figure 23.9 (right). Various types of synaptic junctions.

sensations of pain and heat are smaller in diameter and many are not myelinated. This latter group accounts for the delayed pain felt about one second after the pressure.

ACTIVITY OF NEURONS: SYNAPTIC TRANSMISSION

The action potential explains the conduction of information from one place to another along a single axon. However, there must be a way to pass information to another neuron or to an effector, such as a muscle. Similarly, the neuron must be able to receive information in order to initiate an action potential. Except for pacemaker cells, which generate impulses in the absence of known external stimuli, and some sensory cells, a neuron receives and transmits information at *synapses*. The most common type of synapse in the vertebrate nervous system is the *chemical transmitting synapse*; one neuron (the *presynaptic cell*) excites another (the *postsynaptic cell*) through the release of chemical transmitters from the terminal points of axon filaments. These chemicals diffuse across a small, fluid-filled space (the *synaptic cleft*) between the neurons, act on receptor sites, and thus induce electrical changes in the dendrite or cell body of the neuron with which they come in contact. The unit composed of the presynaptic terminal, the synaptic cleft, and the postsynaptic receptor membrane is known collectively as the synapse. Only since the development of the electron microscope has the structure of the synaptic junction been examined in detail (Figure 23.8).

Each axon filament ends in a presynaptic knob, or *bouton*. Hundreds or thousands of these boutons may lie on the surface of the cell body and dendrites of a single neuron. Electron micrographs reveal that each bouton is separated from the membrane of the postsynaptic cell by a synaptic cleft about 200 A in width. At least three different types of synapses can be distinguished in electron micrographs (Figure 23.9). In *axondendritic* synapses, a small spine often extends from the surface of the dendrite toward the

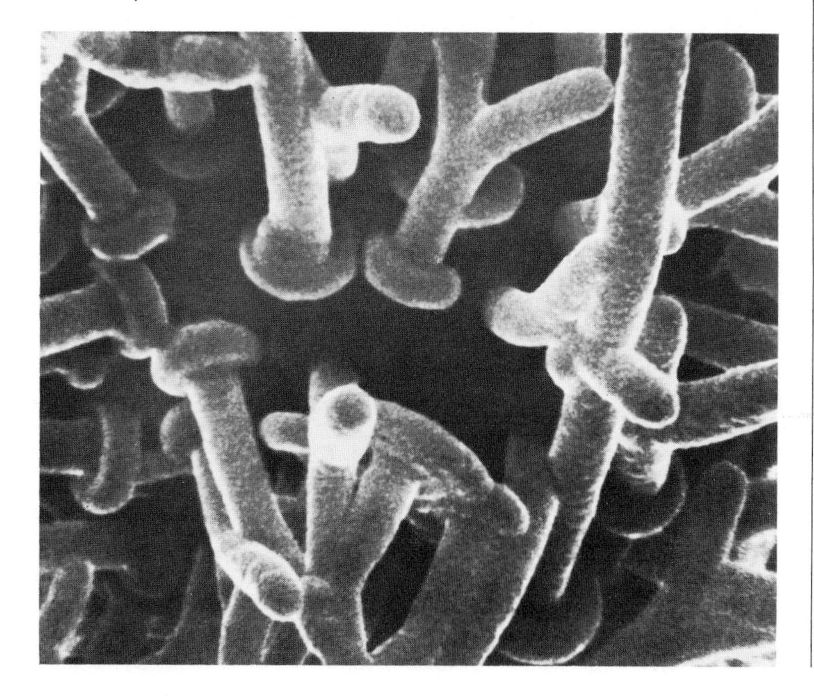

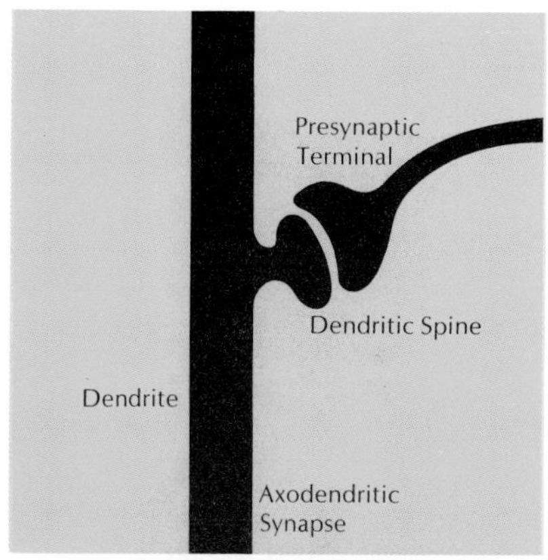

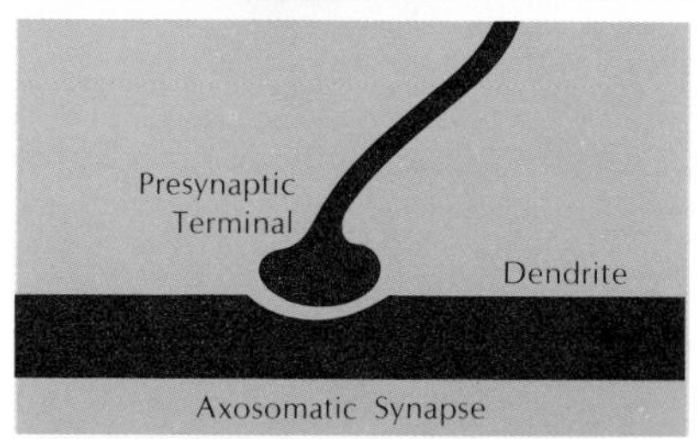

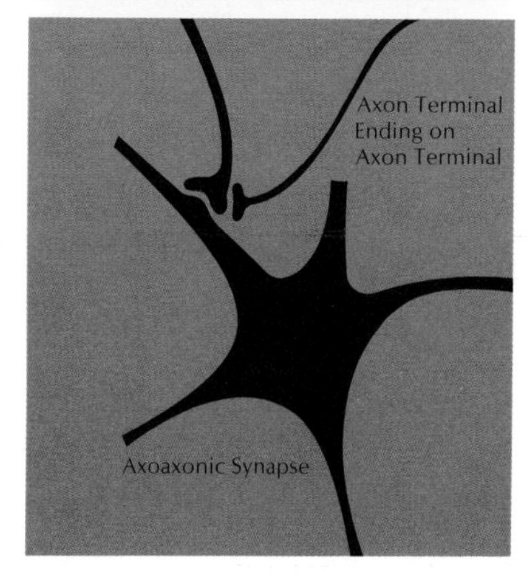

Figure 23.10. Synaptic knobs (above) release short bursts of chemical transmitter substance into the synaptic cleft, where this substance acts on the surface of the nerve cell membrane below. Molecules of the chemical transmitter are stored in vesicles prior to release. The schematic diagram (below) represents a portion of a synaptic cleft with synaptic vesicles in close proximity in the presynaptic terminal. Note the vesicles discharging transmitter molecules into the synaptic cleft. Some transmitter molecules are shown combined with receptor sites on the postsynaptic membrane, thus facilitating membrane pore dilation. At the lower right is an electron micrograph of a synpatic knob containing synaptic vesicles (round bodies).

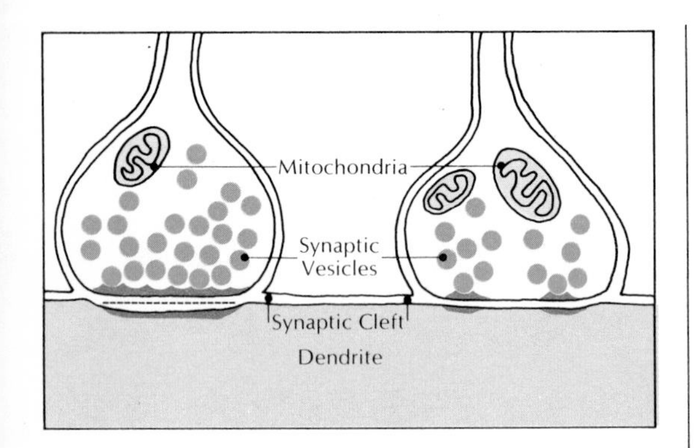

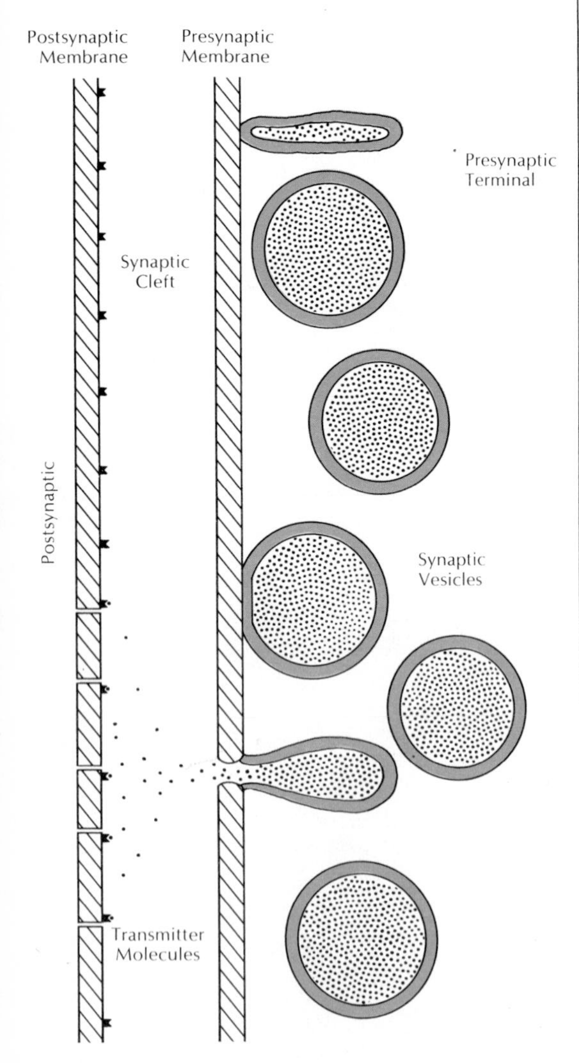

bouton; this type of synapse is found only on dendrites. In *axosomatic* synapses, which are found both on the cell body and on the large stumps of dendrites near the cell body, the bouton lies close to the postsynaptic cell membrane and there is no spine. *Axoaxonic* synapses involve a bouton that synapses on a bouton from another axon, but these synapses are relatively rare and have been observed mainly in sensory pathways.

Release of Transmitter Substance

When an action potential propagates to the axonal arborization of the presynaptic cell, the membranes there depolarize. Depolarization of the axon terminal causes transmitter substance to be released into the synaptic cleft. Although the details of this process are not completely understood, there is enough known to formulate a viable theory. P. Fatt and Bernard Katz (1950, 1951, 1952) studied the synapse between a motor nerve and muscle cell by recording intracellularly from single muscle fibers in the region of the synaptic junction. They discovered that when no action potentials are coming down the nerve, small depolarizations of about 0.5 mV occur spontaneously at random intervals. They called these depolarizations *miniature endplate potentials* (MEPPs). An increase in the concentration of calcium ions in the solution bathing a synapse causes an increase in the mean frequency of MEPPs. Fatt and Katz suggested that each MEPP represents the spontaneous release of a "quantum," or packet, of the chemical transmitter.

When a nerve action potential comes along, they theorized, the quanta are released at a higher rate than normal for a short time. A large number of MEPPs so evoked causes a large depolarization, the excitatory postsynaptic potential. Another experiment proved the accuracy of this hypothesis. Calcium concentration was reduced in the extracellular fluid, causing a drastic decrease in the frequency of MEPPs but no change in the size of each MEPP. When a nerve was stimulated, a small jagged potential occurred, which was clearly the sum of a few MEPPs. As the external Ca^{++} concentration was increased, the number of MEPPs composing the synaptic potential increased until a smooth normal potential was seen. This quantal mechanism of synaptic transmission has been verified for many types of synapses in vertebrate and invertebrate nervous systems. Apparently, calcium attaches to and passes through the presynaptic membrane when the axon terminals are depolarized. When many Ca^{++} ions reach the inner surface of the membrane, a nearby *synaptic vesicle* "fuses" with the

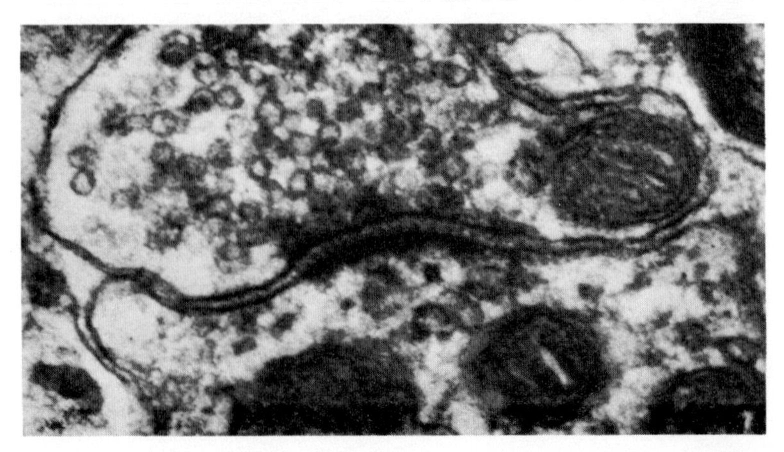

Figure 23.11. Presynaptic inhibition. Cell A has an excitatory synapse on a dendrite of cell C and causes an EPSP in C (as monitored by an intracellular microelectrode) whenever it fires an impulse. If A and B are fired together, a smaller EPSP is seen in C; thus B has an inhibitory effect. Because B fired by itself causes no IPSP in C, it does not have an inhibitory synapse on C but instead must inhibit the effect of B's excitatory synapse.

membrane, dumping its contents (one quantum of transmitter chemical) into the synaptic cleft. Hundreds of quanta may be emptied into the cleft in a single synaptic event at a single synaptic junction.

The release of some chemical transmitter substances into the synaptic junction results in excitation of the postsynaptic cell—a phenomenon called *postsynaptic excitation*—and leads to the production of an action potential in that neuron. However, movement of chemical transmitter substances across some synapses inhibits rather than excites the discharge of the postsynaptic cell. Such an occurrence is known as *postsynaptic inhibition.* In vertebrates, axodendritic synapses usually are excitatory and axosomatic synapses usually are inhibitory, although in a variety of complex synaptic structures in the central nervous system, these generalizations cannot easily be applied (Eccles, 1964).

This picture is complicated further because axoaxonic synapses, in which the axonal terminals of one cell make synapses on the axonal terminals of another cell, affect *presynaptic inhibition.* A special inhibitory transmitter substance from A may decrease the amount of transmitter substance released by an impulse in B, an effect called presynaptic inhibition. Probably there are cases where both excitatory and inhibitory synapses are modulated by these presynaptic interactions.

Throughout the central nervous system, inhibitory synapses counteract the generation of impulses by excitatory synapses. At every synapse of the central nervous system that has been thoroughly investigated, there is a conflict of excitatory and inhibitory action on a single neuron. Apparently, few excitatory synapses have the unchallenged power to excite a nerve impulse in the postsynaptic cell. If there were no inhibitory synapses, a single impulse might cause an explosive spread of excitation throughout the neuronal networks of the nervous system—in other words, convulsions such as those that occur in epilepsy.

In general, an individual neuron is converged upon by many axon terminals, each of which can contribute a quantity of transmitter substance to activate or inhibit the postsynaptic cell. The frequency of discharge for a particular cell at a particular time depends upon (1) the relative quantities of excitatory and inhibitory transmitter substances acting on the membrane of the postsynaptic cell and (2) the stimulus threshold of the cell at the particular time. More than 10,000 synapses may converge upon a single large neuron. The activity of the neuron is influenced by inhibitory and excitatory effects from a very large number of other neurons. Whether or not an impulse is generated in the postsynaptic cell depends upon the summation of the effects of activity in the hundreds of presynaptic cells, and their activity in turn is influenced by many thousands or millions of other neurons. For each synaptic event, only about one millisecond is required. The potentialities of the vertebrate nervous system for the control of complex behavior patterns can readily be appreciated.

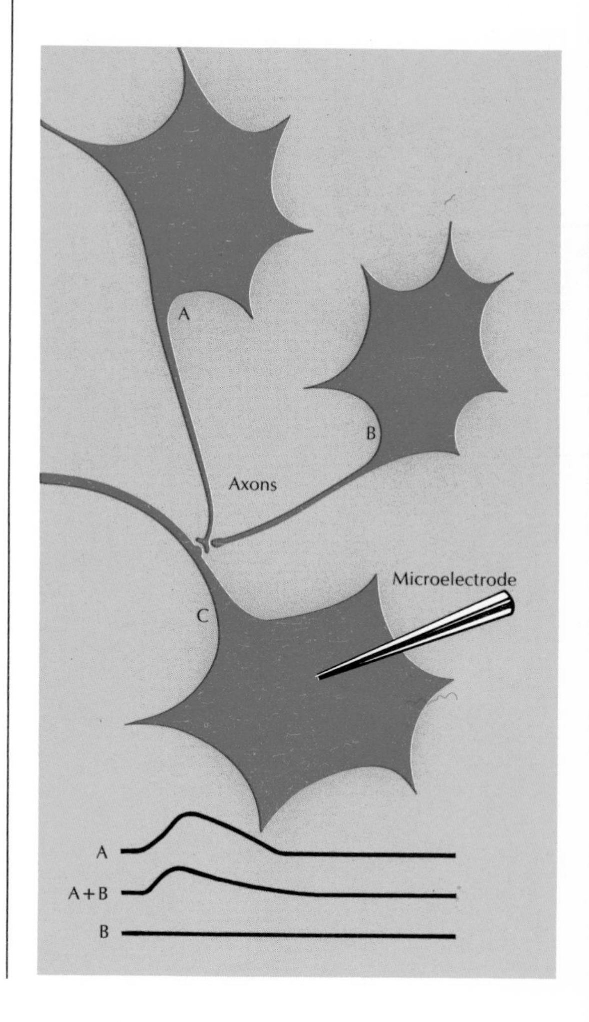

Mechanism of Synaptic Transmission

The nature of synaptic transmission was a subject of debate for many years. The available evidence seemed to support the theory that an action potential in the presynaptic cell creates electrical fields that induce a potential in the postsynaptic cell. There are indeed many examples of *electrical synapses* in invertebrates and in simpler vertebrates. However, today an overwhelming amount of evidence indicates that in mammals the transmission

Figure 23.12. Photomicrograph of a motor neuron with its radiating dendrites and axon. Superimposed on the photograph is a drawing of a microelectrode enlarged 5 times but shown as it would be located for intracellular recording.

across the synaptic junction is accomplished by chemical rather than electrical means. It is interesting to note that John C. Eccles, who was a proponent of the electrical theory in the 1940s, shared the 1963 Nobel Prize largely for his research that provided firm evidence of the chemical mechanism of synaptic transmission.

Great advances in the study of synaptic transmission became possible in 1951 with the development of techniques for recording potentials within a single neuron. A fine glass pipette with a tip diameter of about 0.5μ is filled with a conducting salt solution. If such an electrode is carefully inserted into certain kinds of nerve cells that have been rigidly fixed in place, the cell membrane seals around the glass microelectrode, thus preventing the flow of a short-circuiting current from the inside to the outside of the cell. If this sealing takes place, the nerve cell appears to behave normally for hours. Figure 23.12 shows the dimensions of a microelectrode superimposed on a microphotograph of one of the large neurons that innervates muscle (a motor neuron). Early investigators were fortunate in choosing motor neurons for this research, because intracellular recording has proved to be much easier and more informative in these cells than in any other kinds of neurons. However, in recent years, many other kinds of neurons have been studied in this way with useful results.

Figure 23.13 shows intracellular records of a simple case of excitatory synaptic action. A single set of synaptic excitations causes the potential across the postsynaptic cell membrane to depolarize rapidly and then to return slowly to its resting value. As the number of activated synapses increases, the amplitude of the postsynaptic depolarization becomes larger. In fact, the total magnitude represents a simple summation of the depolarizations produced by each individual synapse. When the depolarization reaches a critical magnitude (in this case, a change of +18 mV, from the resting potential of −70 mV to a potential of −52 mV), an impulse is discharged in the postsynaptic cell. The only effect of further strengthening the synaptic stimulus is to cause a slightly earlier generation of the impulse, which occurs in every case when potential reaches −52 mV.

The depolarizing potential produced in the postsynaptic membrane by the excitatory synapse is called an *excitatory postsynaptic potential*, or EPSP. If the EPSP reaches the threshold potential of the postsynaptic cell, a nerve impulse is discharged. Extensive investigation of a wide variety of nerve cells in the central system indicates that this is the general mechanism of synaptic transmission.

Activation of an inhibitory synapse commonly has just the opposite effect; it causes a further polarization, or hyperpolarization, of the postsynaptic membrane. The effects of individual inhibitory synapses summate in exactly the same way as do excitatory synapses. Because the *inhibitory postsynaptic potential*, or IPSP, opposes the action of the EPSP, the effects of an excitatory and an inhibitory synapse counteract one another.

Excitatory synapses act by increasing the permeability of the postsynaptic membrane to sodium and potassium ions – an effect sometimes called opening the sodium and potassium ionic gates – which results in a net inward current that depolarizes the electrical potential across the membrane. This depolarization can be explained in the same way as the resting potential. In the resting cell, membrane potential is close to the potassium equilibrium potential because the membrane is much more permeable to K^+ than to Na^+. An EPSP increases both K^+ and Na^+ permeabilities, causing

Figure 23.13 (below). Current flow and resulting membrane potential changes in excitatory (red) and inhibitory (blue) synapses. Transmitter substance at an excitatory synapse increases the membrane's permeability to both Na^+ and K^+ ions, resulting in more Na^+ entering the cell than K^+ leaving the cell. The net effect is thus an increase in membrane potential (red curve), which reaches a peak and slowly returns to normal. The inhibitory transmitter in many vertebrate synapses increases the membrane's permeability to K^+ (and sometimes Cl^-) and leads to an efflux of K^+, which generates a decrease in membrane potential (blue curve). If both excitatory and inhibitory synapases are activated in the same neuron, the resulting potential change (purple) will be located between the excitatory and the inhibitory responses. This result is called "spatial summation."

Figure 23.14 (right). IPSP reversal potential. In this diagram, two intracellular microelectrodes are placed in the same neuron. With one microelectrode, positive or negative current is transmitted into the cell, thus setting the membrane potential above (5 – 8) or below (1 – 3) the normal resting potential (4). With the other microelectrode, the additional potential change is recorded in the neuron as an excitatory synapse (red) is first stimulated, then an inhibitory synapse (blue). The height of the excitatory postsynaptic potential (EPSP) increases as the cell's membrane potential (3, 2, 1) is lowered. If the membrane potential (5) is raised, the height of the EPSP decreases until it disappears (6) and finally reverses its direction (7,8). Level 6 is called the equilibrium potential for the EPSP (E_{EPSP}) because the EPSP can cause no further shift in membrane potential. Similarly, the size of the IPSP increases as the membrane potential is raised above the resting potential (5 – 8). The IPSP decreases in size as the membrane potential is lowered, disappears entirely (3), and, on being lowered still further, reverses its direction. Level 3 is therefore called the IPSP equilibrium potential (E_{IPSP}). This fact is explained by the finding that inhibitory and excitatory transmitter substances increase or decrease the synaptic membrane's permeability to different sets of ions.

the membrane potential to rise to a value about halfway between the equilibrium potentials of these two ions. The membrane potential then returns to the resting potential as both Na^+ and K^+ permeabilities return to normal. If the synaptic membrane could be artificially charged to a potential value more positive than the halfway point between Na^+ and K^+ equilibrium potentials, the action of the transmitter substance should cause an inverted synaptic potential, which does happen.

The transmitter substance of an inhibitory synapse causes the postsynaptic membrane to become permeable to potassium and chloride ions. Opening the potassium and chloride ionic gates results in a net outward flow of current, hyperpolarization of the postsynaptic membrane, and development of an IPSP. The generation of the IPSP can be explained in a manner parallel to the explanation of EPSP. Potassium and chloride potentials are more negative than the resting potential, and when these ionic permeabilities increase, the membrane potential decreases toward the lower equilibrium potential. As the transmitter substance is dissipated, the K^+ and Cl^- permeabilities return to this level, and the resting potential is restored. Figure 23.14 shows how IPSPs can be turned upside down by artificially decreasing the membrane potential to a value more negative than the combined K^+ and Cl^- equilibrium potentials.

In the cells of the spinal column, the effect of an IPSP lasts only about 8 ms, slightly less than the 10 ms duration of an EPSP. In the brain, however, the effects of an IPSP last for 100 to 200 ms or more; in the neurons of the brain, a single activation of an inhibitory synapse may counteract many successive activations of excitatory synapses.

Excitatory
Inhibitory
Synaptic Knob
Depolarization
Na^+ Influx
Hyperpolarization
K^+ Efflux
-70 mV
Resting Potential
EPSP
IPSP

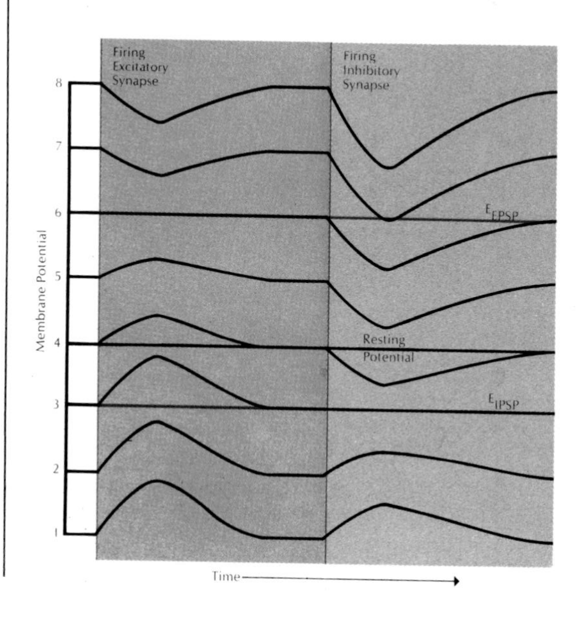

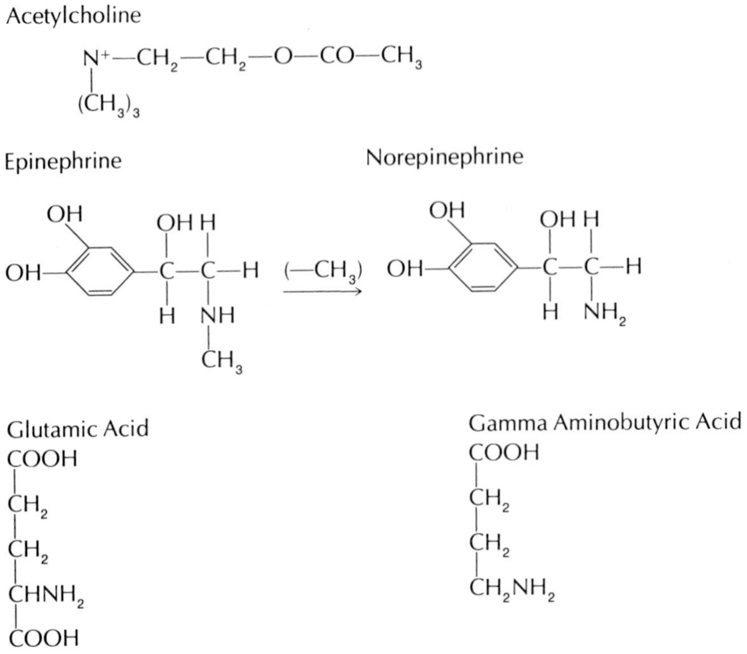

Figure 23.15. Some chemical transmitters. Acetylcholine and norepinephrine are among the transmitters used at excitatory synapses. Glycine and gamma aminobutyric acid have been recently identified as inhibitory transmitters in the mammalian central nervous system.

Acetylcholine

$N^{+}—CH_2—CH_2—O—CO—CH_3$ (with $(CH_3)_3$ on N)

Epinephrine

Norepinephrine

OH, OH — benzene ring — C(OH)(H) — C(H)(NH–CH_3) — H $\xrightarrow{(—CH_3)}$ OH, OH — benzene ring — C(OH)(H) — C(H)(NH_2) — H

Glutamic Acid

COOH — CH_2 — CH_2 — $CHNH_2$ — COOH

Gamma Aminobutyric Acid

COOH — CH_2 — CH_2 — CH_2NH_2

The evidence for chemical transmission of signals across many synaptic junctions must be regarded as conclusive. With few exceptions (among invertebrates), a given presynaptic neuron has only excitatory or only inhibitory synapses with its several postsynaptic cells. Many neurophysiologists believe that a fundamental property of nerve cells is that a given neuron releases only a single transmitter substance from all of its axonal endings. However, it is difficult to establish exactly which transmitter is acting at a given synapse. To prove that a given cell releases a particular transmitter, it must be shown that (1) the cell has the material or machinery necessary to manufacture the transmitter; (2) the transmitter is present at the axon terminal in sufficiently great quantities; (3) stimulation of the neuron causes release of the substance into the synaptic cleft; (4) the postsynaptic cells are sensitive to the substance; (5) a molecular mechanism exists to break down or inactivate the substance after it has done its job, so that it will not accumulate at the synapse and continue to act, thus preventing further transmission of information; and (6) various chemicals that affect synaptic events in the body (such as curare, which prevents EPSPs in skeletal muscle neuromuscular junctions of vertebrates) also have parallel effects on the action of the substance when it is directly applied to the postsynaptic cell.

The advent of the technique of *iontophoresis* has given great impetus to the identification of transmitters. A solution of charged molecules (suspected transmitters) is used to fill a micropipette, which is then placed near a synapse. The micropipette releases a small amount of the substance when an electric current is passed through the pipette. Microchemical techniques, involving chemical assays of very small amounts of synaptic tissues, have also been important, as have histological techniques that show the localization of certain substances. As a result, *acetylcholine* (ACh) and *norepinephrine* were found to be among the transmitters used at excitatory synapses. More recently, *glycine* and *gamma aminobutyric acid* (GABA) were identified as inhibitory transmitters in the mammalian central nervous system.

The identification of transmitter substances at the neuromuscular junction and in a few other kinds of synapses was relatively easy because large numbers of similar nerve cells exist together in these locations. In the spinal cord and brain, however, many different kinds of neurons exist in close proximity, and it is impossible to isolate pure samples large enough for ready chemical analysis. Even in the peripheral nerves of mammalian nervous systems, where ACh and norepinephrine were first identified as transmitters, it is not an easy job. ACh was accepted as the transmitter in one set of synapses, for example, but recently epinephrine was shown to have at least an indirect effect there (R. M. Eccles and Libet, 1961).

In the mammalian central nervous system, ACh was shown to be the transmitter at central synapses made by branches of a motor neuron, which also uses ACh at its peripheral branches where they innervate muscles (Eccles, 1957, 1964). Other candidates for central-nervous-system excitatory transmitters (of which there may be several) are ATP, epinephrine, histamine, glutamic acid, and some other amino acids. Identification of other transmitters may be expected as this research continues.

Longstanding candidates for inhibitory transmitters in the mammalian central nervous system have been 5-hydroxy-tryptamine, gamma aminobutyric acid (GABA), epinephrine, and glycine (an amino acid). Recently, glycine was shown to be an inhibitory transmitter in the spinal cord, and GABA was confirmed to be an inhibitory transmitter in the higher levels of the brain.

Drugs that affect normal synaptic function have been extensively used as tools to learn about transmitters and the processes of chemical transmission. Acetyl cholinesterase is an enzyme that inactivates ACh after it has generated an EPSP. The chemical prostigmine blocks action of this enzyme, thus causing excitatory synapses to produce prolonged synaptic depolarization. If prostigmine is applied to a synapse and has this effect, it provides evidence that ACh is the transmitter. Both botulin toxin (the bacterial toxin involved in some kinds of food poisoning) and curare prevent ACh synapses from functioning. Curare apparently acts by competitive inhibition—the drug reacts with the postsynaptic membrane. It does not depolarize the membrane, but it prevents the ACh from attaching to the same sites. The botulin toxin, on the other hand, apparently acts by preventing the release of ACh from presynaptic terminals. These substances produce muscular paralysis and interfere with involuntary functions of the body by preventing the transmission of impulses across synapses where ACh is the transmitter substance.

Small doses of the poison strychnine act as powerful stimulants of nervous activity by causing uncontrolled contractions of muscles and other symptoms of seizure activity. This effect is the result of the drug's action in blocking synapses involved in inhibitory pathways in the spinal cord. When the drug is present, the interneurons involved in feedback and feedforward inhibition are activated, but their inhibitory synapses fail to exert inhibitory influences on the neurons of the main pathways. Strychnine acts through competitive inhibition of the normal transmitter substance (glycine) in these synapses. Another convulsant drug, bicuculline, antagonizes GABA and thus blocks inhibitory synapses at higher levels of the nervous

Figure 23.16. Diagrams of the two types of inhibitory pathways, recurrent inhibition (A) and afferent collateral inhibition (B). Inhibitory cells are shown in blue.

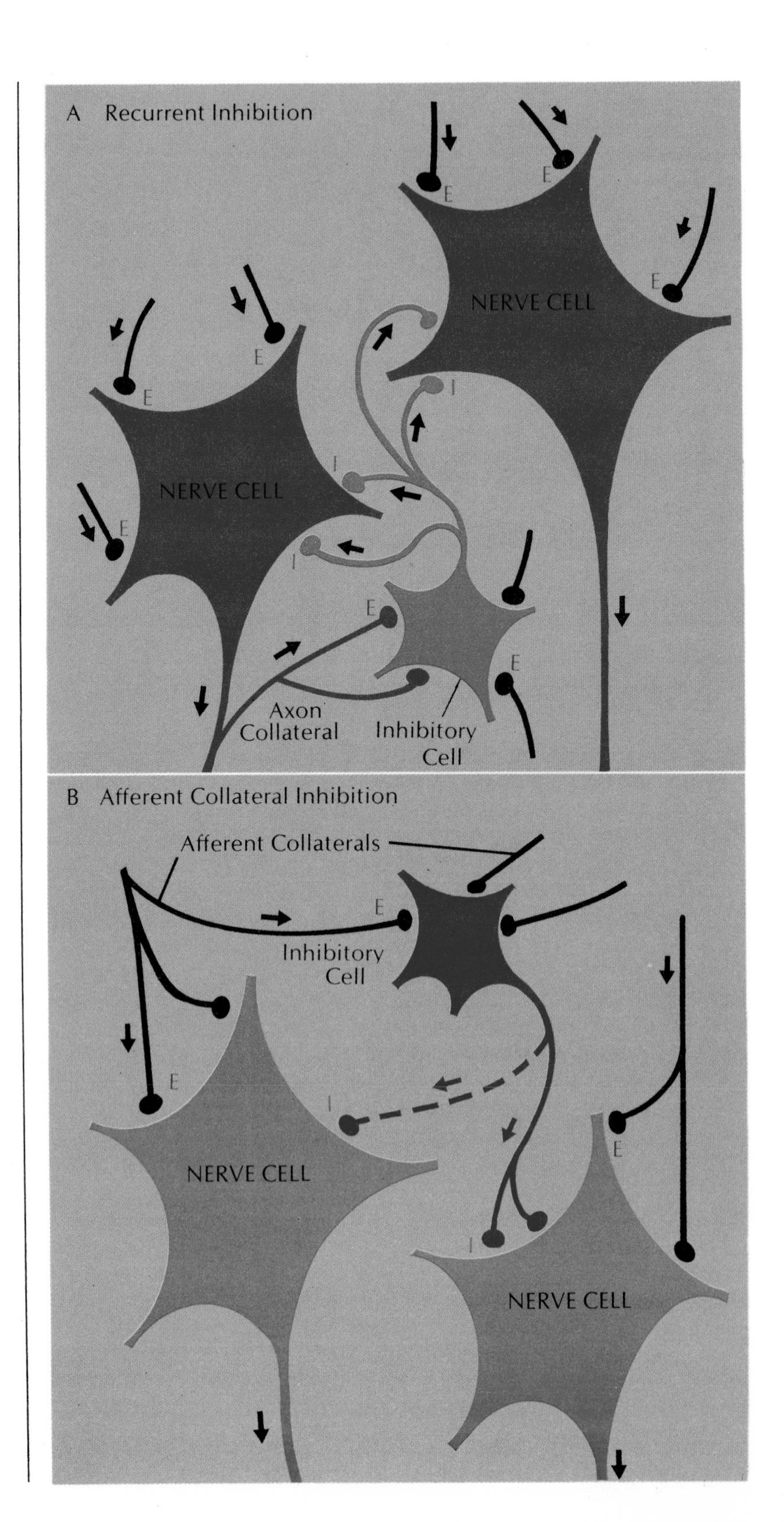

Figure 23.17. Schematic diagram showing pathway to the sensorimotor cortex for cutaneous fibers from the forelimb. Note the inhibitory cells in both the cuneate nucleus and the ventrobasal nucleus of the thalamus. The inhibitory pathway in the cuneate nucleus is of the feedforward type, whereas the inhibitory pathway in the thalamus is of the feedback type.

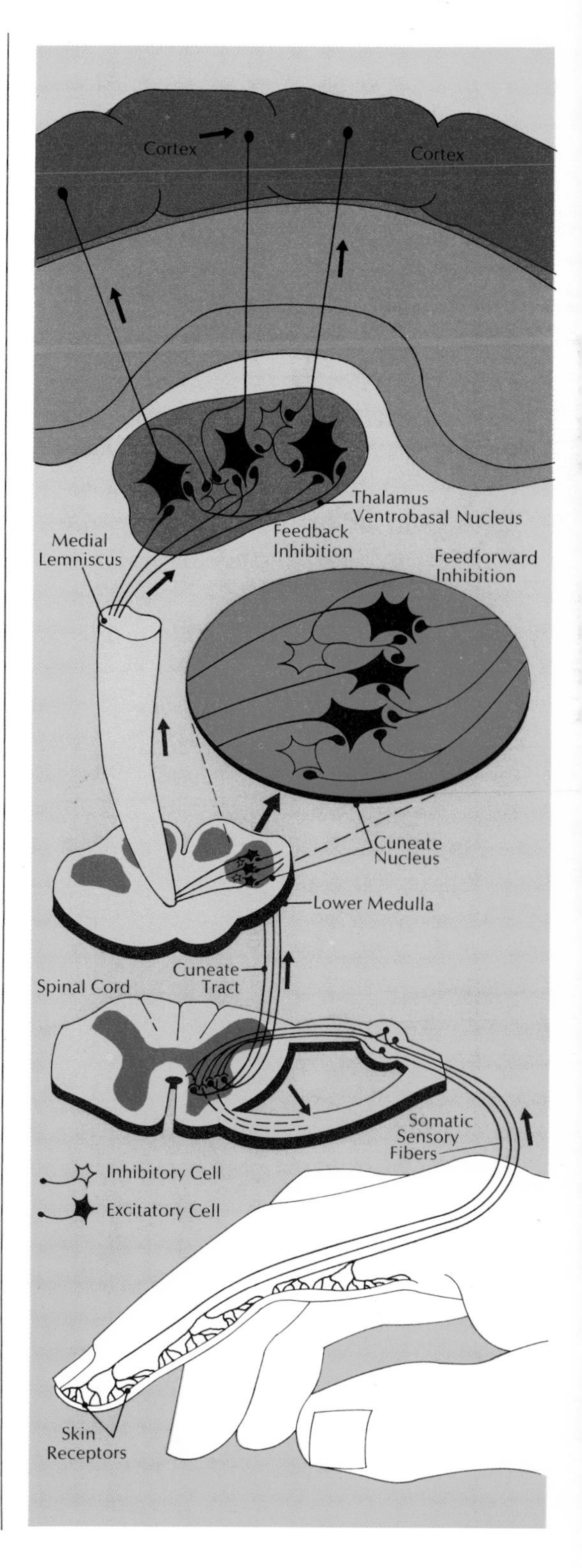

system. Tetanus toxin—which also inactivates inhibitory synapses—is thought to act by preventing release of transmitter substance from the presynaptic terminals of inhibitory synapses.

Many drugs that are commonly used as stimulants or depressants act at the synapses in the central nervous system. The exact nature of their effect cannot be known because not all of the separate connections in the brain nor all of the transmitter substances are known.

Most common nonprescription sleeping pills contain scopolamine, which acts to depress transmission at ACh junctions. Possible side effects of ingesting scopolamine, such as blurred vision and increased heart rate, are the direct results of the drug's action on synapses in parts of the nervous system. The sleep-inducing effects presumably result from some action on the central nervous system at a site or sites as yet unidentified. Barbiturates apparently act by depressing activity at central synapses. Some anesthetics, including ether, evidently act by reducing transmission through interneuron pathways. Still other drugs, such as cyanide, exert their effects upon general processes of cell metabolism, thus affecting the nervous system along with many other systems of the body. It is likely that hallucinogenic drugs also act at the level of the synapse, although little is known about them at this time.

TWO INHIBITORY CONTROL SYSTEMS

As a simple generalization, the two types of postsynaptic inhibitory pathways shown in Figure 23.16 can be regarded as the elementary constituents of all known neural pathways. In *recurrent inhibition*, the axon of a neuron gives off a recurrent collateral, or process, that ends in excitatory synapses on a neuron that has a widespread inhibitory action. Some of the processes from this cell end in inhibitory synapses on the first cell. In this way, whenever the nerve cell discharges an impulse, it automatically activates a recurrent inhibitory pathway, which tends to prevent further discharges from that cell and other similar cells.

In contrast to this recurrent, or feedback, inhibition is *afferent collateral inhibition*, where an afferent nerve fiber (one carrying impulses toward the central nervous system) synaptically exciting a nerve cell gives off a collateral branch that also excites an inhibitory cell. The inhibitory cell, in turn, exerts inhibitory synaptic action on other functionally related neurons—in effect, a feedforward inhibitory action.

After extensive investigations of mammalian central nervous systems, no exception has been found to the rule that nerve cells are always completely inhibitory or completely excitatory in action at all of their synapses; no ambivalent cells with some excitatory and some inhibitory synaptic terminals have ever been found. It is therefore justifiable to label cells as inhibitory or excitatory, according to the nature of their synapses. All primary afferent pathways—the pathways from receptor organs to the spinal cord and brain—are made up of excitatory neurons. These pathways exert inhibitory effects only by synaptic relays through inhibitory neurons (Figure 23.16).

Figure 23.17 illustrates in diagrammatic form the flow of information from skin receptors to the brain. Impulses from the arm travel up the spinal cord in the cuneate nerve tract to relay cells in the cuneate nucleus, where afferent collateral, or feedforward, inhibition may occur. The pathway then crosses to the opposite side of the brain and relays in the thalamus, where recurrent, or feedback, inhibition may occur. The main cells involved in

Figure 23.18 (above). Schematic diagram showing postulated connections of specific afferent fibers to the neocortical pyramidal cells. The inhibitory interneurons with their inhibitory synapses are shown in green. All other stellate cells and the pyramidal cells are assumed to be excitatory. Arrows indicate the directions of impulse propagation. Note that the inhibitory path is through inhibitory cells activated either directly by the afferent fibers or by mediation of excitatory interneurons. Also note the various degrees of complexity of the excitatory pathways to the pyramidal cells.

Figure 23.19 (below). Schematic diagram showing postulated synaptic connections of axon collaterals of pyramidal cells. The single inhibitory interneuron together with its inhibitory synapses on the somata of the pyramidal cells are shown in green. All other stellate cells and the pyramidal cells are assumed to be excitatory and are shown in red. Arrows indicate directions of impulse propagation. Experimental evidence suggests that both excitatory and inhibitory pathways can include interpolated excitatory interneurons.

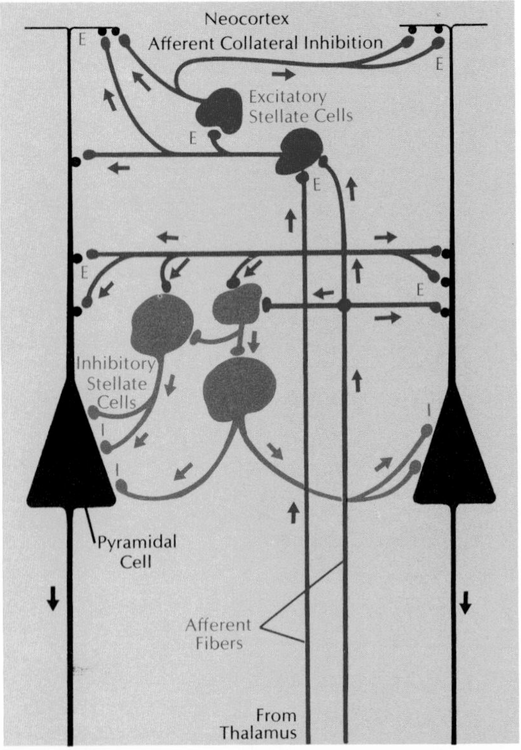

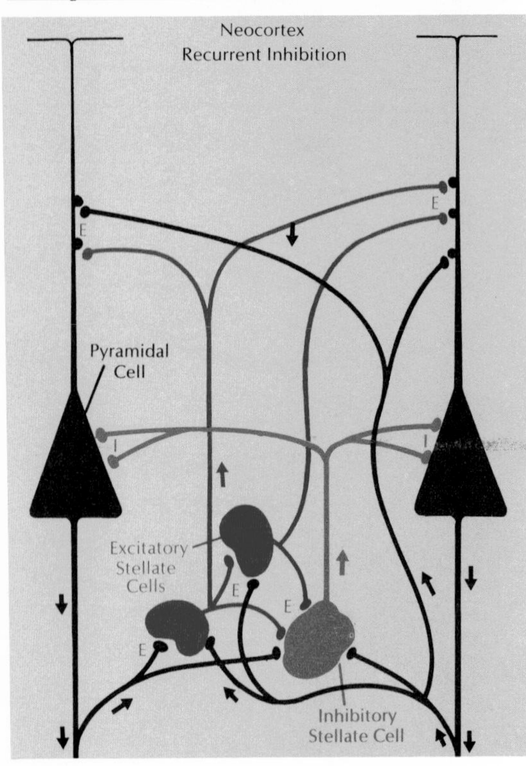

the thalamus have processes leading directly to the cerebral cortex, where neural activity produces effects of conscious experience.

It may seem surprising that the information moving along neural pathways to the brain is subjected to inhibition at each of the synaptic relays. This arrangement serves to separate meaningful signals from "background noise," which originates from random discharges of neurons, and to neutralize the effects of abnormal connections in the nervous system. The nerve fibers grow and interconnect in the developing organism, and there is always the possibility that some disorganization will occur in such an immensely complicated growth process. The inhibition exerted at each synaptic relay tends to eliminate any isolated or random signals, causing the volleys of meaningful impulses from the receptors to stand out sharply.

Figure 23.18 shows schematically the pathways that carry impulses from the thalamus to the cerebral cortex of the brain. Both excitatory and inhibitory neurons lie on the pathways to the pyramidal cells of the cortex. However, this diagram is extremely simplified, and the actual patterns of neuronal activation must be infinitely more complicated. Similar pathways for afferent input to the cortex exist not only for the impulses relayed from various peripheral receptors through thalamic nuclei but also for looping associative fibers. These fibers carry impulses from the pyramidal cells of any area on the cortex to a wide surrounding area of cortex and also to the opposite hemisphere of the brain. Thus, pathways of an almost infinite degree of complexity are built up.

A further complication is illustrated in Figure 23.19, which shows the negative feedback through axon collaterals of the pyramidal cells. In fact, inhibitory feedback usually occurs not directly through an inhibitory interneuron but through one or more excitatory interneurons leading to and from the inhibitory interneuron. These complications make it almost impossible—at least with existing techniques—to trace the complete pathway of any particular afferent input through the brain. Although much has been learned about the general operation of the central nervous system, the preparation of a "wiring diagram" of any major part of the brain is still far in the future.

WHY SO MANY SYNAPSES?

The basic operation of the nervous system depends upon the fact that many almost synchronous excitatory bombardments are necessary to generate an impulse in any neuron and thus to contribute to the further spread of neuronal activity. For an effective spread of activity, each neuron probably must receive synaptic activation from hundreds of neurons and must transmit to hundreds of others. Transmission through the nervous system is therefore more like a wave front than a single impulse (Figure 23.20). There is a kind of multilane traffic in hundreds of neuronal channels. The wave front of activity sweeps over at least 100,000 neurons per second, weaving a pattern in space and time in a way that C. S. Sherrington (1906) likened to the operations of an "enchanted loom." There is a great deal of evidence, for example, that a particular neuron may participate in the patterns of activity developing from many different sensory inputs.

The nervous system may be visualized as a complicated telephone exchange, constructed from some 10 billion unitary components or neurons. Unlike a telephone exchange, however, the nervous system does not merely carry a given message from a single input to a single output. Instead,

Figure 23.20 (above). Schematic diagram of an impulse wave front. The nerve cells of the cortex are laid out as dots on one plane. The multilane traffic in one, evolving a specific neuronal pattern, is shown in black and in another as dark gray. Light gray cells are not activated by either pattern. Note the very dark dots at the crossing of these two lanes, indicating participation of nerve cells in both patterns.

Figure 23.21 (below). Schematic diagram of impulse patterns among active neurons. The arrows show the directions of propagation for two separately evolving patterns (black and dark gray). Note that where the black and gray patterns unite (2 sites), they propagate as one advancing wave.

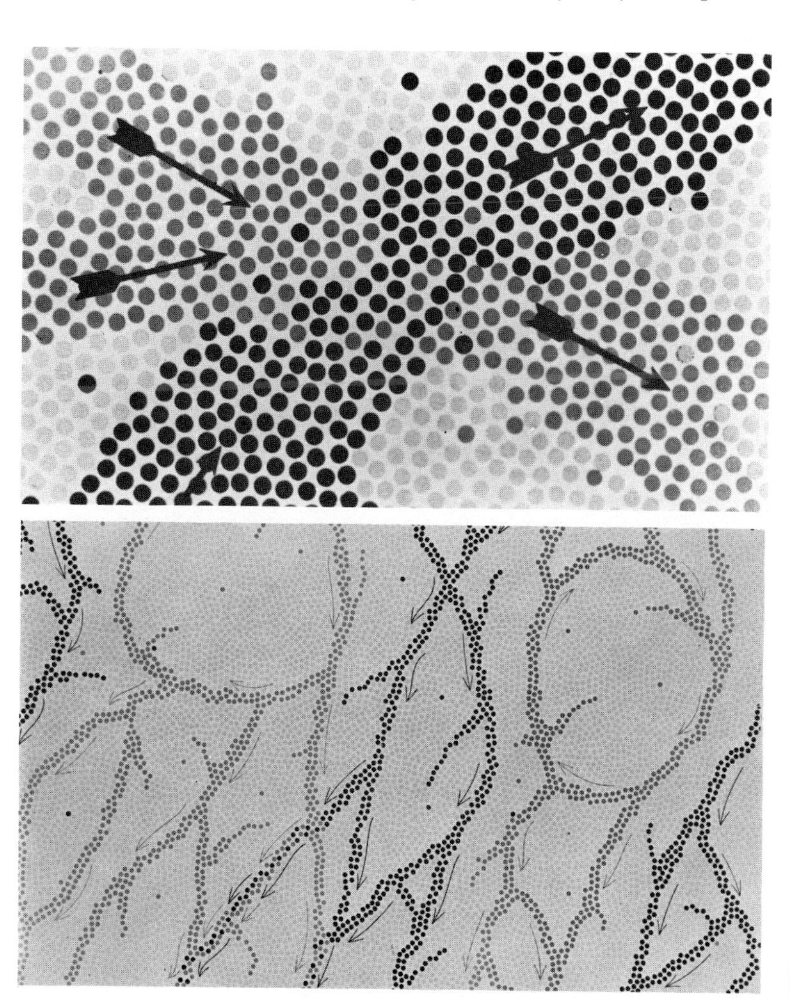

the brain and the rest of the system act to correlate and integrate the incoming signals from an enormous number of sensory channels. For example, there are about 1 million separate nerve fibers leading to the human brain from each eye. The brain not only correlates the input from all of these channels to guide activity in relationship to the visual information but correlates the visual input with other sensory input, as when the eyes and the sense of touch are simultaneously used to guide and control movement.

Within the past decade, enormous advances have been made in the study of the cerebral cortex at very high magnifications in electron microscopy and also in the use of intracellular electrical recording devices to explore the nature of synapses and nerve impulses. However, all of this research has provided only a beginning for the understanding of simple perceptual awareness. The more complex problems of perceptual recognition and judgment have hardly been approached.

There is a great deal of neurophysiological evidence that a conscious experience is always accompanied by some specific activity in the cerebral cortex. For every experience, a specific pattern of impulses in time and space is woven by the meshwork of nerve cells in the brain (Figure 23.21). Any stimulus to a sense organ causes the repetitive discharge of impulses along sensory nerve fibers to the brain. After various synaptic relays in the brain, specific spatiotemporal patterns of impulses are created in the neuronal network of the cerebral cortex. The transmission from sense organ to cerebral cortex utilizes a coded pattern of nerve impulses, rather like a Morse code with dots only in various temporal sequences. Certainly, this coded transmission is quite unlike the original stimulus to the sense organ, and the pattern of neuronal activity evoked in the cerebral cortex is different yet again. As a result of these cerebral activity patterns, people experience sensations that seem to represent events within the body, at its surface, or in the external world.

The investigation of the neuronal mechanism of the cerebral cortex is still at a primitive stage, and therefore it gives only a dim picture of the intricate pattern woven in space and time by the sequential activation of neurons in multilane traffic over the 10 billion components in the cortical slab of cells. It has been estimated that many millions of cells take part in the simplest cortical response. The human cerebral cortex presumably surpasses that of any other animal in its potentiality to develop subtle and complex neuronal patterns of the greatest variety. From this neural complexity must stem the richness of human behavior as compared with that of even the most intelligent of other animals.

The brain events associated with experiences may be caused by local stimulation of the cerebral cortex, by stimulation of some intermediate part of the sensory nervous pathway (as in the "phantom limb" experiences of amputees), or in the usual fashion by activation of sense organs by external stimuli. However, electric stimuli applied directly to the sensory zones of the cerebral cortex usually evoke only chaotic sensations—tingling or numbness in the skin zones, lights and colors in the visual zone, noises in the auditory zone (Penfield and Rasmussen, 1950). Such chaotic responses are to be expected, because electrical stimulation of the cortex must directly excite thousands of neurons, regardless of their functional relationships, thus initiating a spreading field of neuronal activation quite unlike the fine and specific patterns set up by normal input from the sensory organs. A familiar chaotic sensation—involving elements of touch, heat, cold, and pain—arises for a similar reason when a sensory nerve bundle is directly excited, as when the ulnar nerve in the elbow (the "funny bone") is stimulated by a sudden blow.

When gentle repetitive electric excitation is applied to the cerebral cortex, there is a relatively long period between application of the stimulus and the sensory experience. This time lag may be up to 0.5 second with a very weak stimuli, but a more typical lag is about 0.2 second. Each electric stimulus of the repetitive series must excite the discharge of impulses from nerve cells within a few milliseconds. The delay of conscious perception for 0.2 second or longer presumably represents the time needed to elaborate the spatiotemporal pattern that corresponds to the experience. This long time lag is surprising in view of the fact that reactions to stimuli can occur much more swiftly. However, it appears that such rapid reflex reactions—for example, the withdrawal of the hand from a hot object—are carried out before the accompanying sensations such as pain are experienced.

Transmission of an impulse from one neuron to the next takes no longer than 1 ms. Thus, the time lag of 0.2 second preceding a sensation could permit the serial relay of an impulse through as many as 200 synaptic linkages between neurons. Many thousands of neurons are probably activated by the electric stimulus to the cortex, and each neuron in turn activates many nerve cells at each synaptic relay. Thus, it appears that millions of neurons are involved in the pattern that corresponds to a sensation.

This rich tapestry of neuronal activity is required for the perception of even the simplest sensation. Responses involving comparisons, value judgments, correlations with remembered experiences, aesthetic evaluations, and so on must take much longer and involve fantastic complexities in the patterns woven on the "enchanted loom."

FURTHER READING

For more detailed discussions of nerve impulses, see books by Brazier (1960), Eccles (1957), Hodgkin (1964), and Katz (1966). Synaptic transmission is discussed in books by Eccles (1957, 1964), Katz (1966), and McLennan (1969). Thompson (1967) provides a general survey of neurophysiological knowledge related particularly to the study of animal behavior.

Among the *Scientific American* articles particularly related to topics of this chapter are those by Baker (1966), Eccles (1958, 1965), Hubel (1963), Hyden (1961), Kandel (1970), Katz (1952, 1961), Kennedy (1963), Keynes (1958), Luria (1970), and Walter (1954).

24
Nervous Systems

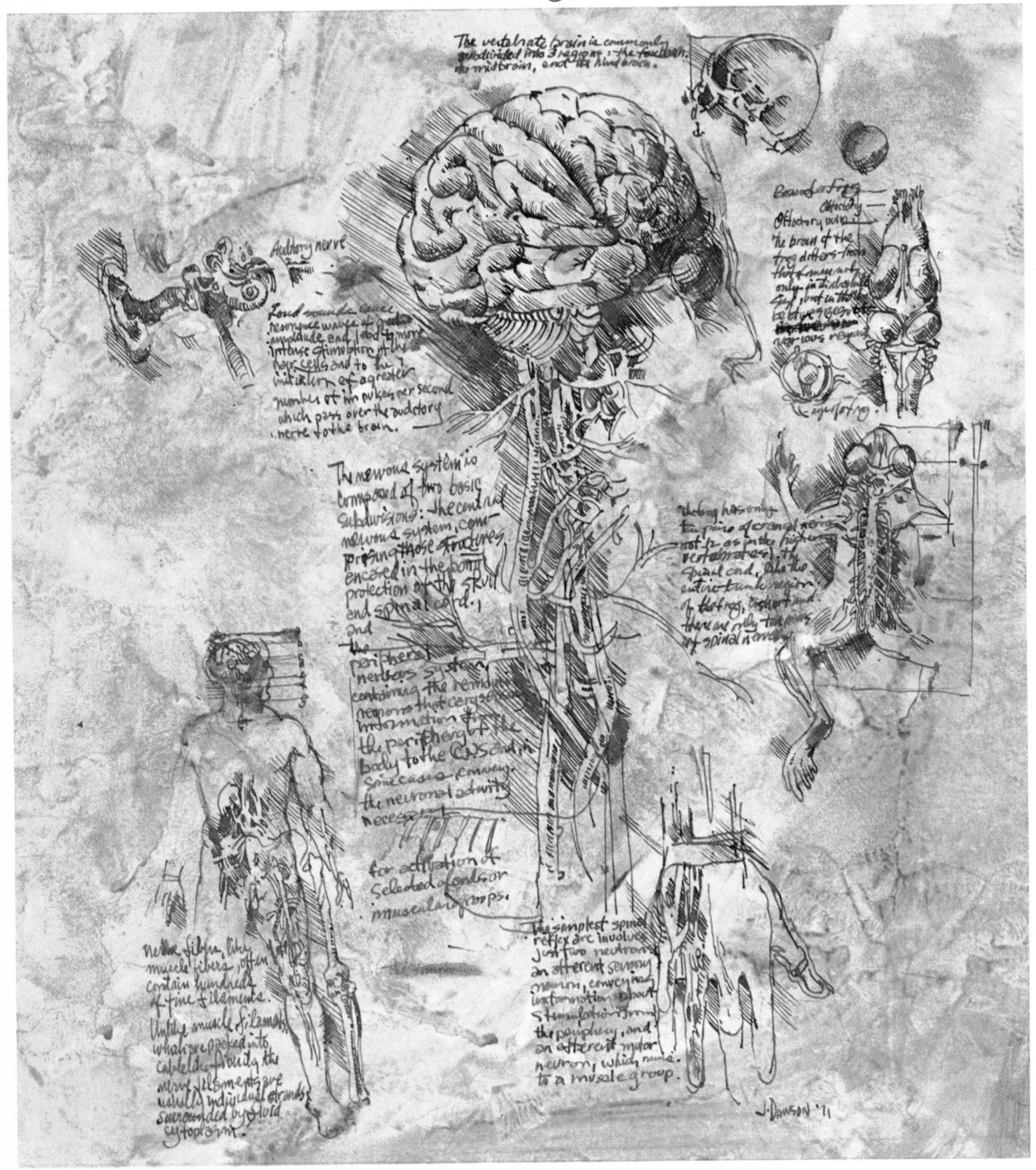

Anyone attempting to describe briefly the structure and function of the nervous system faces much the same dilemma as an electronics engineer attempting to describe the workings of a television set. It is relatively easy—though tedious—to list all the components by name and to describe in precise terms the individual properties of each. It is also possible to describe in more general terms the roles played by various groupings of components in the television set—the tuner, the amplifier, the oscillator, and so on. However, it is far more difficult to show exactly how a given set of components with particular properties interacts to produce the exact effects needed for proper operation.

The scientist studying the nervous system faces a stiffer task. He would find it far more difficult to trace the circuitry and to discover all the components in the circuits than would an engineer studying a TV set. Neurobiologists have developed diverse techniques to explore nervous system function. An experimenter may deliberately damage or electrically stimulate parts of the nervous system. The resulting behavioral changes indicate some of the functions of the affected parts. He may make recordings from single nerve cells, nerve bundles, and large regions of the brain. These recordings provide other clues about the activity of the nervous system. He may also study the anatomy of the major interconnections among parts of the system, but he cannot today make a "wiring diagram" for any vertebrate.

Knowledge of the human nervous system has come from studies of persons with systems damaged by injury or illness. Correlation of behavioral defects with physical damage to parts of the nervous system provides some clues about the functioning of the normal system.

Such studies have revealed much about the functions of major groups of neurons and of major pathways. The properties of individual neurons and small groups of neurons have been investigated in nonhuman mammals, lower vertebrates, and invertebrates. In these animals, and especially in invertebrates, it is possible to see how properties of individual neurons contribute to the general functioning of the nervous system. The general organizational principles discovered in such simple nervous systems have proven useful in understanding the more complex vertebrate nervous systems.

IRRITABILITY WITHOUT NERVES

Even the simplest unicellular organism exhibits some form of *irritability*—the capacity to respond with action to certain kinds of stimuli. Irritability is often listed as one of the basic properties of living systems.

An amoeba is a cell with relatively few organelles in what appears to be a generally formless cytoplasm. Although it lacks apparent specialized structures for sensory reception or response, the amoeba does show a regular behavior pattern. It responds to small food particles or certain chemicals with feeding behavior (phagocytosis) and responds to almost any other stimulus by withdrawing. The nature of its feeding activities varies with the chemical nature of the food stimulus, the degree of activity exhibited by the potential food, and the amount the amoeba has eaten recently. The mechanisms underlying these behaviors are not completely understood, but a stimulus apparently causes local changes in the protein structure of the cytoplasm, and these changes then lead to the general alteration of cytoplasmic properties that produces movement.

Unicellular organisms with organelles such as cilia, flagella, contractile fibrils, and sensory organelles exhibit far more intricate behavior patterns.

Figure 24.1. The giant amoeba *Chaos chaos* capturing and ingesting a paramecium. The nature of the amoeba's activities depends on the integration of such factors as the chemical nature of the food, the degree of activity exhibited by the potential food, and the amount of food the amoeba has eaten recently.

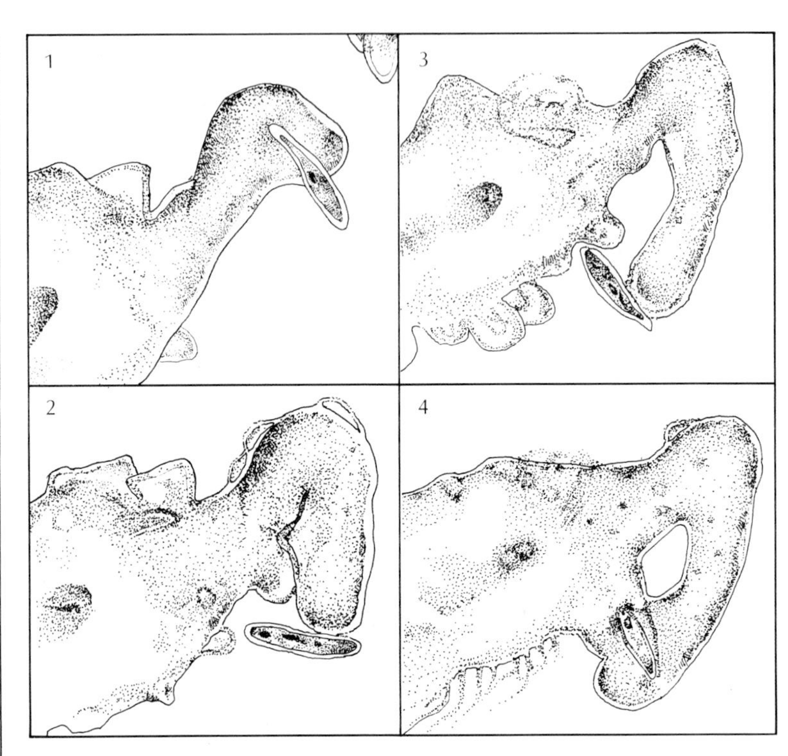

A paramecium that encounters an object or a noxious chemical as it swims will back up by reversing its ciliary beat and then move forward in a different direction. If microelectrodes are attached to these organisms, it is possible to depolarize the anterior end so that the paramecium causes its cilia to beat in reverse. If the posterior end is depolarized, the beat is accelerated. Evidently, some tactual and chemical stimuli cause such depolarizations, thus controlling the cell's behavior.

SIMPLE MULTICELLULAR ANIMALS

The simplest multicellular animals behave as if they are collections of relatively independent cells. Chemical signals spread quite slowly from one cell to another, and, except where contact stimuli occur between touching cells, little communication or coordination takes place. In most multicellular animals, however, specialized cells of the nervous system speed communication among the cells of the body. Neurons may be regarded as cells specialized for response to stimuli (irritability) and for rapid communication of the response through the organism along particular paths (conduction).

Coelenterates are the simplest animals that have well-developed nervous systems. The individual neurons of the coelenterate system are similar to those of higher animals, but the system itself is relatively simple in organization. The coelenterates have radial symmetry; they have neither a head nor anything that could be called a brain. The nervous system is best described as a collection of two-dimensional, interacting nerve nets. A *nerve net* is a group of neurons—in most cases scattered over a surface—whose processes (dendrites and axons) cross and intermingle in a netlike fashion. Synapses occur at many points of contact between processes. Electron micrographs of synapses reveal that there are synaptic vesicles on both sides

Figure 24.2 (left). A "simple" unicellular organism such as *Euglena* is really a complex integrated group of organelles coordinated by chemical signals. A positive phototactic response ensures adequate light to maintain photosynthesis; however, intense bright light will elicit a negative phototaxic response.

Figure 24.3 (right). Nerve net system of the fresh-water coelenterate *Hydra*. Note the absence of a centralized nerve tract and ganglia. The majority of *Hydra* behavior is limited to feeding and defensive contractile responses.

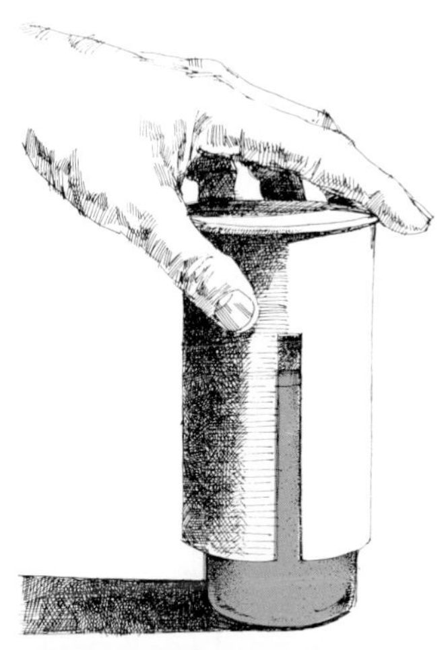

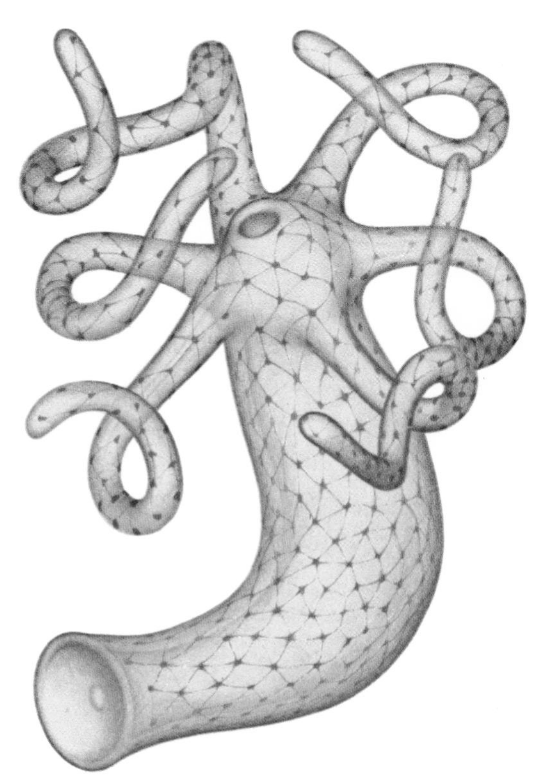

of the synaptic cleft, and the synapse seems to be a two-way connection.

There are two nerve nets spread over the swimming bell of the young medusa (jellyfish stage) of *Aurelia* (Figure 24.4). One nerve is composed of large, bipolar cells with straight processes. This *giant fiber net* lies over the ring of muscle that extends outward from the ring toward the eight rhopalia, or tentaculocysts, which are sensory structures extending from the edge of the bell. The other nerve net is composed of smaller, multipolar cells with shorter processes. This *diffuse net* covers the entire surface of the medusa, and synapses connect the two nets only at the rhopalia. The rhopalia contain equilibrium receptors, or statocysts, and light receptors. They also contain several kinds of neurons whose processes apparently remain within the rhopalia, where they synapse with each other, with receptor cells, and with elements of the two nerve nets. Each of the eight rhopalia serves as a point of interaction among the two nerve nets, the sensory receptors, and the neuronal network of the rhopalium itself. In structure and function, the rhopalia can be considered the prototype of a brainlike structure.

The medusa shows two major kinds of motor activity: (1) a rhythmic, stereotyped, rapid twitch contraction of the bell muscles as it swims and (2) localized, variable, slow contractions associated with feeding or avoidance behaviors and modification of swimming behavior in response to stimuli. Each of the two nerve nets plays a specific role in the total behavior.

If the giant fiber net is cut in such a way that a segment of muscle is isolated from all contact with rhopalia through that net, the isolated segment ceases to show regular swimming contractions. Apparently, then, the signals that drive the coordinated swimming beat originate in the rhopalia and travel to the muscles through the giant fiber net. If cuts are made in such a way that each rhopalium is connected to an area of muscle but is isolated from contact with other rhopalia through the giant fiber net, each segment contracts at an independent rhythm. In the intact animal, the most active or strongest of the eight rhopalia initiates a beat and simultaneously drives the remaining seven via the giant fiber net.

The diffuse net apparently plays no role in the swimming twitch, except perhaps as an indirect inhibitory effect. This net is thought to control the slower contractions involved in feeding or in spasmic avoidance reaction. Swimming contractions are slowed, weakened, or absent when slow contractions are occurring.

The coelenterate nervous system exhibits several principles of organization also found in the more complex nervous systems of higher animals. First, the nervous system is composed of more than one anatomically and functionally separate subsystem. These subsystems interact only at particular centers that also receive sensory information from nearby sense organs. These centers contain pacemakers that initiate rhythmic impulses. External stimuli apparently are not necessary to trigger these impulses. Second, the coordinated activity of the organism involves neural inhibition as well as excitation, and the inhibition may serve to switch the organism from one behavior pattern to another as various subsystems are turned off or on. The qualitative differences in behavior patterns depend largely upon which part of the system is active, not upon changes in the nature of the activity of a single subsystem.

Coelenterates possess complex and specialized receptor organs. Ocelli are cup-shape patches of pigment-containing, light-sensitive cells covered with a transparent, lenslike layer of cells (Figure 24.5). Statocysts contain a

Figure 24.4. Oral view (above) of *Aurelia aurita* showing canal development and location of sensory apparatus (tentaculocysts). Below is shown the life cycle of the scyphozoan medusae (jellyfish) *Aurelia*.

Figure 24.5 (left). Three types of light receptor organs (ocelli) typically found in the tentaculocysts, or rhopalia, of jellyfish.

Figure 24.6 (right). Schematic diagrams of a statocyst receptor organ of a jellyfish. These organs orient the jellyfish with respect to gravity and are housed in the rhopalia. If the bell tilts, the statolith is displaced and touches the hair cells. This action stimulates the nerve fibers and signals the animal to correct its position.

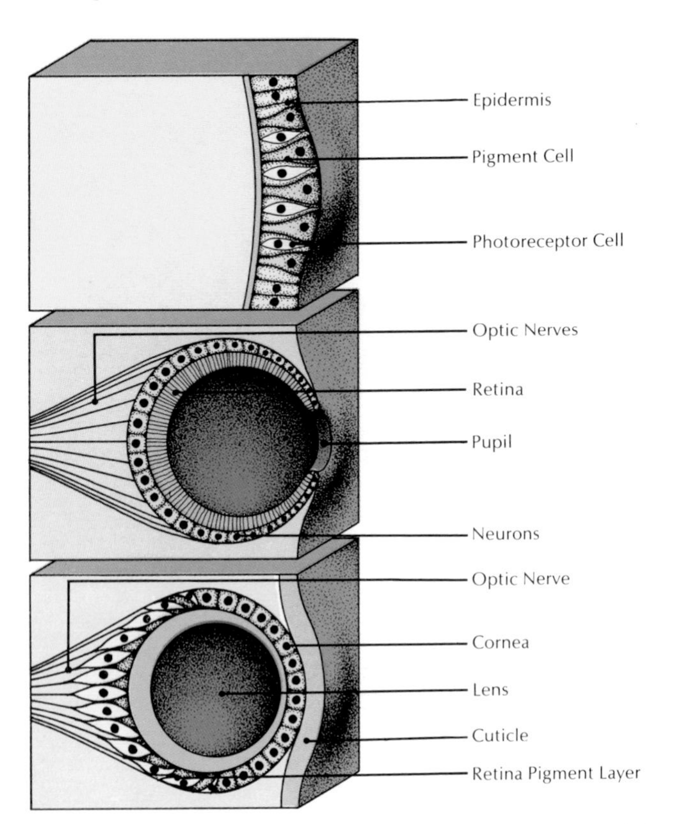

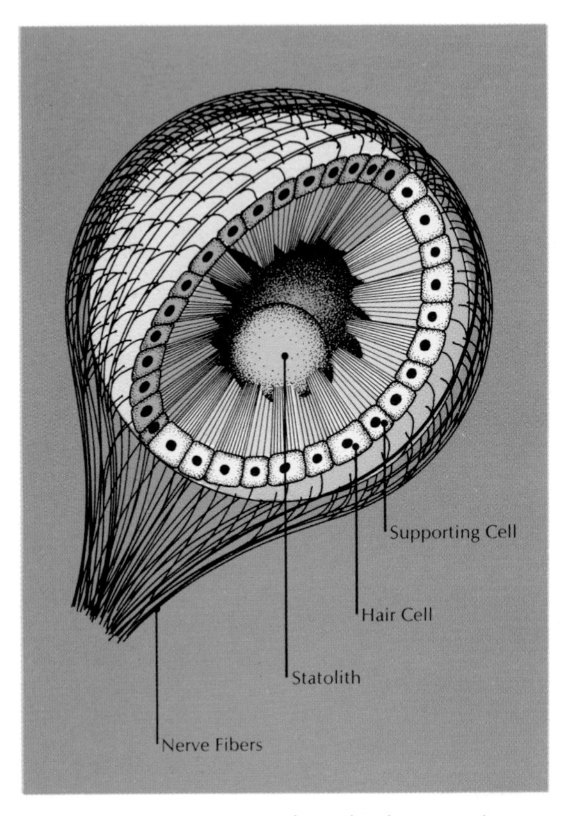

round, hard object made up of organic materials and calcium carbonate (Figure 24.6). This object rolls against various sensory cells as the orientation of the organism changes.

The behavior exhibited by a coelenterate involves a series of interactions among the various sensory, conducting, and effector cells. These interactions permit relatively few variations from a limited repertoire of standardized behavior patterns. A coelenterate's response to a given set of stimuli depends on its internal state and its nervous system. Repeated stimulation by touch, for example, leads to a gradual decrease in the strength of avoidance behaviors elicited by the touch, and repeated feeding leads to a lessening of feeding behaviors, even in the presence of potential food. Overall, this group of animals exhibits a remarkable number of phenomena that higher animals depend on in their nervous systems.

Echinoderms possess more complex nervous systems than those of coelenterates. In addition to nerve nets, bundles of nerves are located well inside the body of the organism. Greater numbers of interneurons increase the number of nerve cells between receptor and effector. Many neurons are closely associated with glial cells (which are absent in coelenterates), and conduction along axons and across synapses occurs only in one direction.

In a starfish, a ring of nerves surrounds the mouth, with radial nerve tracts extending into the arms; a complicated nerve net, or *plexus*, lies beneath the skin. The outer layer of the plexus has a network similar to that of coelenterates, but the inner layer is organized into nerve bundles, or tracts. The nerves of the outer layer form synapses with receptor cells, which are present by the thousands in each square millimeter of skin. The nerve ring and radial nerves are located inside the hard structures of the body wall, as are the muscles. Synaptic connections between the outer and inner nervous systems occur at various thin spots in the body wall.

The interactions of the various parts of the nervous system may be illustrated by the response of a starfish to a light touch at one spot on the lower side of an arm. Nerve impulses spreading out from the stimulated sensory cells through the dorsal plexus cause the spines in the immediate vicinity to bend toward the touched spot. Next, an impulse travels through a chain of nerves within the arm segment and stimulates retraction of the tube foot nearest the touch. Nerve impulses also move along the radial nerve away from the stimulated point, causing a slightly delayed retraction of adjacent tube feet. Finally, the nerve impulses reach the nerve ring and control centers at the base of each arm. The control centers send out impulses along the radial nerve fibers, causing coordinated motions of the arms and tube feet that result in locomotion away from the touched spot. When a starfish is moving, the control center at the base of the leading arm tends to dominate the movement of all tube feet by impulses sent via the nerve ring and radial nerves.

BILATERAL NERVOUS SYSTEMS

Most other animals that have highly differentiated, complex nervous systems are bilaterally symmetrical. Locomotion is usually in a direction such that the head leads, and most major sensory receptors are located on or about the head. Because most of these bilateral organisms are segmented, various elements of the nervous system tend to be repeated in each segment.

In bilaterally symmetrical organisms of increasing complexity, several trends in nervous system organization can be detected. First, there is a trend toward increased cephalization, or development of the size and complexity of the brain. Second, there is an increase in the number of interneurons, or neurons whose processes synapse only with other neurons and not with sensory or effector cells. Third, there is an increased variety of structurally different kinds of neurons and glial cells. Fourth, there is an increase in the variety and differentiation of synaptic regions within the brain. In the more complex animals, the brain has more neurons and a far more complex organization of subsystems than any other part of the nervous system.

There are two major anatomical patterns of organization. In the evolutionary branch containing the molluscs, annelids, and arthropods, the central nervous system is organized into ganglia, or knots of nervous tissue, and each body segment often has one pair of ganglia. Each ganglion contains an outer rind of neuron cell bodies and an interior region of synaptic areas (*neuropil*) and nerve fiber tracts, which connect the various ganglia. In general, each ganglion controls half of one body segment, but in most species the ganglia of several anterior segments are fused to form the brain.

In the chordate branch, the primary organization consists of a continuous tube of nervous tissue extending the length of the organism rather than segmented neuron clumps. Neuron cell bodies, glia, and neuropil (designated *gray matter*) are mixed together in the wall of the tube—not separated as in the ganglia of the arthropod branch. Bundles of axons (or fiber tracts) are separate from the gray matter. They are called the *white matter* because of the glistening appearance of the myelin sheaths formed by Schwann cells that surround each axon.

The nervous system of the earthworm is an example of a simple ganglionated system. In addition to the central nervous system, it has a subepidermal nerve net, but this net apparently functions only in localized and minor responses. Locomotion is accomplished by peristaltic waves of contraction that pass along the length of the body. The contraction of muscles in one segment stimulates receptor cells in the next segment, which trigger the contraction of that segment. An inhibitory mechanism that keeps the wave moving in one direction along the body is apparently built into the nervous system. The cerebral ganglion above the gut in the front segments is well developed and regulates behavior on the basis of input from the sensory organs of the anterior region. The next set of ganglia, located below the gut, is also well developed and appears to initiate many control impulses regulated by the cerebral ganglion. Removal of the cerebral ganglion does not prevent the worm from exhibiting normal burrowing and eating behaviors, but it does limit the worm's ability to modify its behavior in response to changing conditions. The behavior of an earthworm consists of a fairly stereotyped repertoire of responses, which may be modified under varying environmental conditions.

Molluscan nervous systems vary widely in complexity. A chiton has a simple ganglionated and segmented nervous system. In most molluscs, the nervous system has five major ganglia, each of which controls important organs. By studying these simple nerve centers, researchers have made some fundamental discoveries. Many gastropods (slugs, snails, limpets) have giant nerve cell bodies, often over a millimeter in diameter. The pacemaker properties of single neurons and synaptic interactions between neurons have been studied in *Aplysia* (a sea slug). In this animal, simple neural examples of conditioning and training have been studied. Recently, workers have discovered that the large cells in many gastropod ganglia can be named or numbered, and the same cells are present in every individual examined. Each cell seems to have a specific function and anatomy, with even some detail of its complex branching processes repeated in different individuals.

Snails and limpets are capable of limited behavioral modification, but the octopus, a cephalopod mollusc, is better at learning than many vertebrates. Cephalopods have the greatest degree of cephalization of any mollusc, and perhaps of any invertebrate. Their brains are very similar to those of mammals in organization and complexity of structures. They have fiber tracts and layered arrangements of cells and neuropil that are reminiscent of cerebral cortex. It is perhaps because of the complexity of their brains that they are such good learners. They can be taught to discriminate between objects on the basis of touch or sight (Boycott, 1965). Some researchers feel that the mechanism of memory may be more easily discovered in cephalopods than in mammals. The anatomy of their nervous systems has received much attention, but as yet few neurophysiological studies have been done.

Arthropods have highly developed ganglionated nervous systems with advanced cephalization. The nervous systems of crustaceans are distinguished by the relatively small number of neurons (less than 100,000 in the crayfish, for example). Interneurons with numerous and complex processes make possible intricate interconnections despite the small number of cells.

Figure 24.7. Representative invertebrate and vertebrate animals selected to portray the phylogenetic progression in the evolution of nervous systems. The diagram in A shows the bilateral ladderlike nervous system of a triclad flatworm. Note the beginning of cephalization, or brain development, as denoted by the cephalic ganglia. The molluscan nervous system in B, represented by a gastropod, shows placement of ganglia in strategic locations. In C, the ventral solid nerve tract of the earthworm is shown with its segmental arrangement of ganglia. The arthropod nervous system in D also shows the effect of segmentation. Advanced cephalization is also characteristic of this phylum. In E, the primitive vertebrate nervous system is represented by an amphibian, and the advanced vertebrate system is represented by a mammal in F. The central nervous system of all chordates includes a dorsal, hollow, fluid-filled nerve cord.

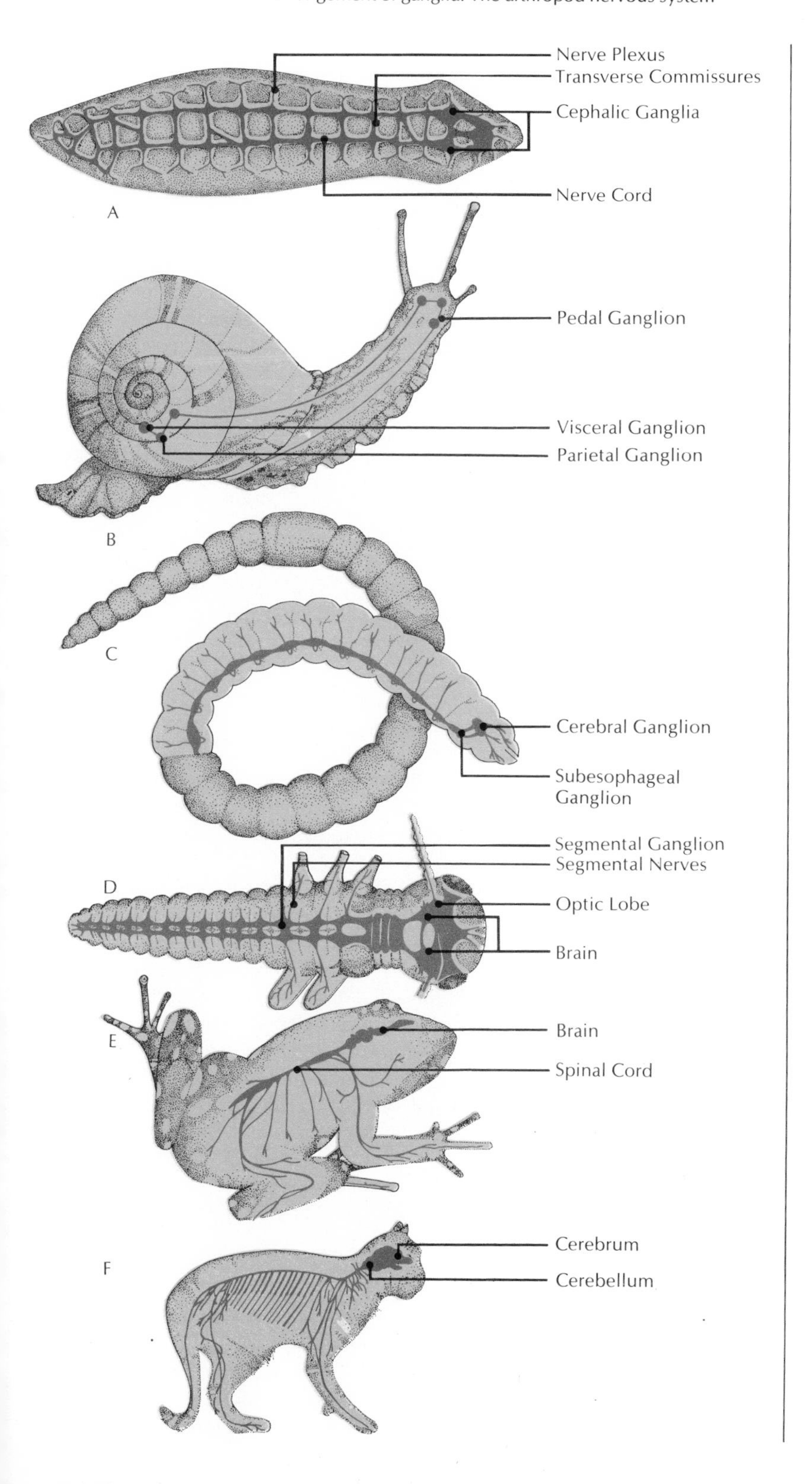

Insects possess extremely specialized sensory organs and complex brains. The study of the genetics and development of nervous systems, a new and exciting field, makes good use of the diversity of experimental subjects presented by the insect world. Many species are capable of complex behavior patterns similar to those of vertebrates. Communication activities in honeybees have been studied extensively and are a good example of complex behavior mediated by a ganglionated nervous system (Frisch, 1967).

VERTEBRATE NERVOUS SYSTEMS

In vertebrates, there is a continuous, hollow, dorsal nerve cord rather than the ganglionated, ventral cord of invertebrates. Centralization of the nervous tissues, decreasing autonomy of outlying ganglia and increasing numbers of neurons, and the complexities of their interconnections characterize the organization of the vertebrate nervous system. The division into sensory, associative, and motor/effector subsystems is probably the most useful way of subdividing any nervous system. But anatomists and physiologists traditionally have divided the vertebrate nervous system into a *central nervous system* (CNS) and a *peripheral nervous system* (PNS). The CNS comprises those neural structures encased in the bony protection of the skull and vertebral column and is composed of interneurons. The PNS includes all nerve processes and neurons that lie outside the CNS. It also forms functional and anatomical links between the sensory receptor system and the CNS and between the CNS and effector organs and glands.

The Somatic Nervous System

Another useful way of dividing the vertebrate nervous system involves consideration of two major parts: a somatic nervous system involved in control of voluntary responses to stimuli and an autonomic nervous system that regulates involuntary activities. Both of these systems include parts of the CNS and PNS, although these divisions are most frequently used in discussions of the PNS.

The central nervous system, together with the elements of the peripheral nervous system that connect it to receptors and effectors, makes up a complete neural network capable of initiating behavioral responses to particular stimuli. The sensory portion of this somatic nervous system includes neural circuits that carry input from each of the sensory systems to specific areas of the brain, where they are processed independently. Other neural circuits respond more generally to arouse or to depress the organism as a result of the interactions of various kinds of stimuli. Similarly, the motor portion of the system includes major nerves that run directly from specific regions of the motor cortex to particular groups of muscles and other nerve systems that function as feedback, stimulation, and inhibition controls over these direct motor responses. The sensory and motor portions of the system are not totally separate but interconnect in spinal reflex arcs as well as within the various parts of the brain.

The Autonomic Nervous System

Control of the involuntary, or vegetative, functions of the body by centers in the CNS is accomplished through the autonomic nervous system (Figure 24.8). This system consists of two divisions: *sympathetic* and *parasympathetic*. In both divisions, the nerve fibers exerting primary control over varied vegetative activities leave the brain stem or spinal cord and synapse

Figure 24.8. The autonomic nervous system with its sympathetic and parasympathetic divisions. Reciprocal innervation of most organs by both divisions is the rule. Regulation of body maintenance activity is the primary function of this system.

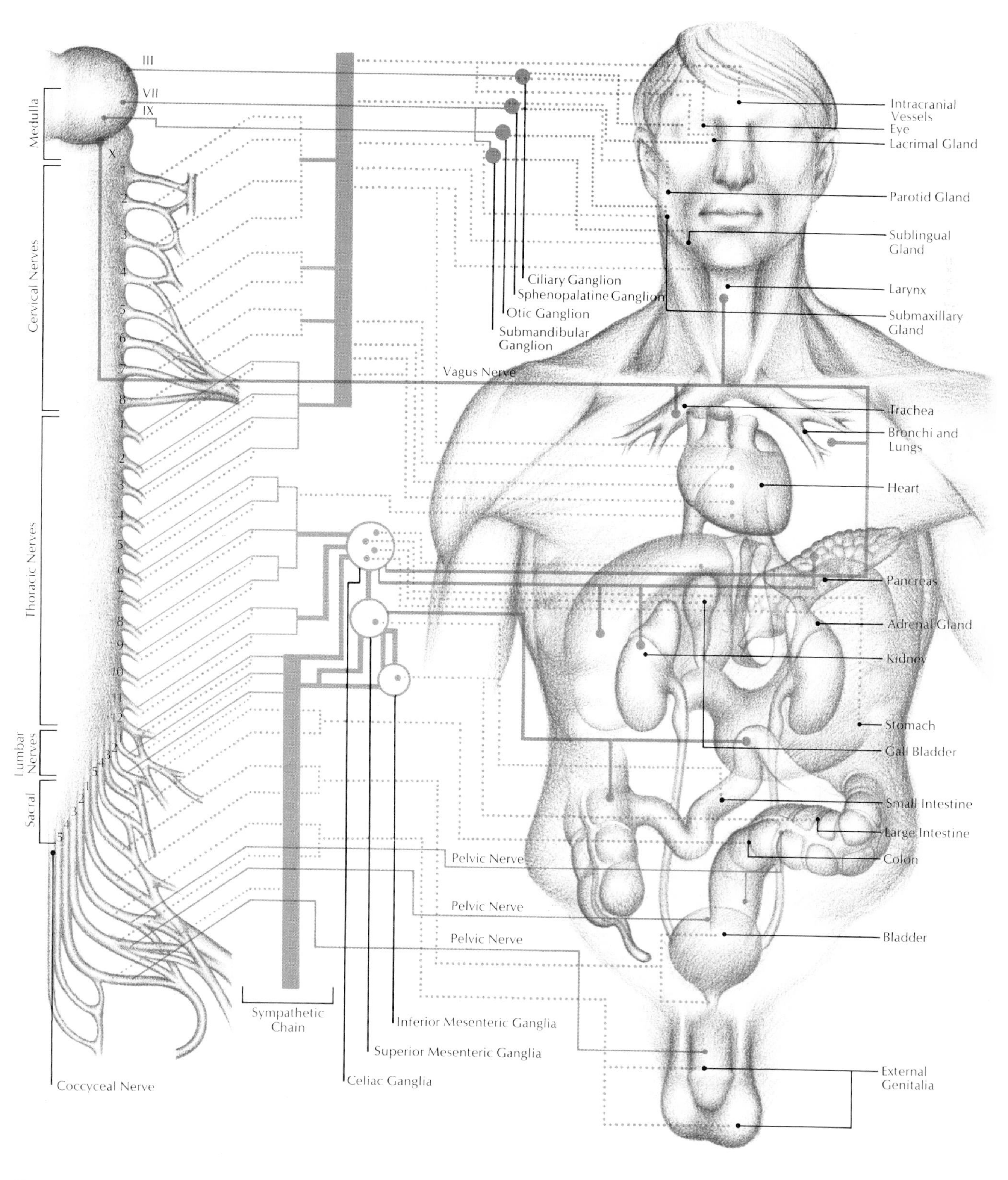

Figure 24.9. An example of a reflex arc as illustrated by the knee jerk reflex.

in ganglia located outside the CNS. In the parasympathetic division, efferent fibers leave either by way of certain cranial nerves or by way of the lower spinal cord, and these fibers synapse in ganglia located near the muscle groups to be innervated. The sympathetic fibers leave the spinal cord in its central region and synapse chiefly in ganglia located near the spinal cord; the fibers that innervate the musculature of the internal organs leave from these ganglia.

In general, the sympathetic division promotes energy expenditure, particularly mobilizing the body resources to meet emergency conditions (increasing blood sugar levels and heart rate, dilating the blood vessels that supply oxygen and nutrients to the skeletal musculature, constricting those blood vessels that supply the skin, and inhibiting digestive processes). The parasympathetic division acts to conserve energy, slowing heart rate and enhancing digestive activity among other things.

The somatic and autonomic nervous systems are not totally independent but are interconnected at various levels, particularly within the brain. The nervous system is also closely interrelated with the hormonal system, and the activities of each affect the other. These complex control mechanisms provide the structural basis that underlies the rich variety of behavior patterns of a vertebrate animal.

The Spinal Cord

The spinal cord is a relatively simple neural system that receives and processes information from sensory receptors and then delivers appropriate impulses to stimulate actions by effectors. It serves as an interface between the brain and the effectors and between sense organs and the brain. The

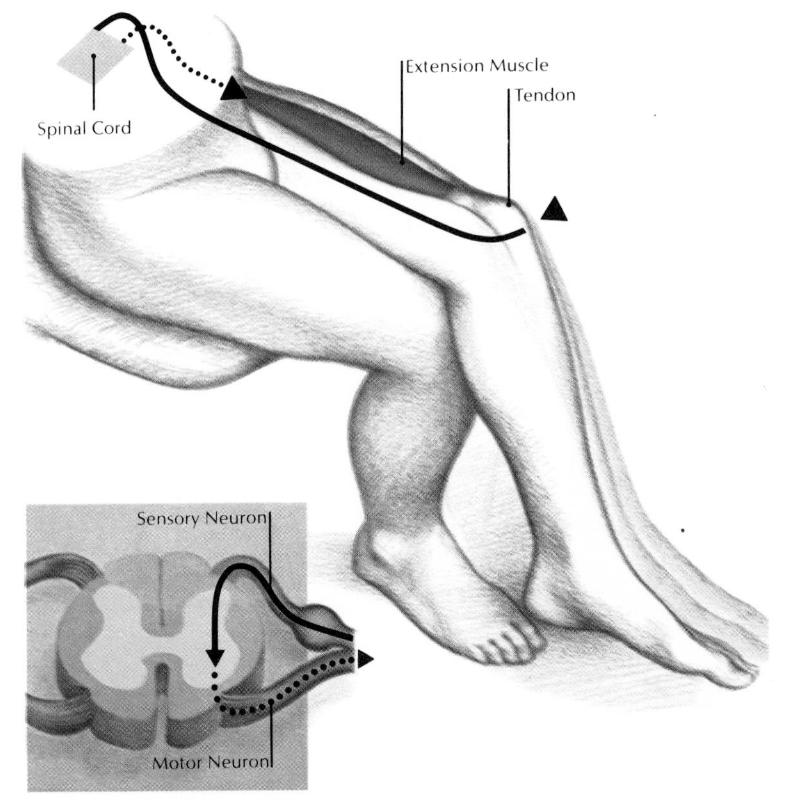

simplest behavioral responses of vertebrates are produced by a spinal *reflex arc* that involves just two neurons—a sensory neuron conveys an impulse *from* a receptor at the periphery to a motor neuron in the cord, which carries an activating impulse *to* a group of muscles. The knee jerk elicited by a tap below the kneecap is the result of the simultaneous activation of many such two-neuron reflex arcs (Figure 24.9). Both the motor neurons (whose cell bodies lie within the spinal cord) and the sensory neurons (whose axons synapse in the spinal cord) are considered parts of the PNS.

In the case of the knee jerk, muscles on the front surface of the thigh are excited and contract, while the motor neurons that lead to muscles on the back of the leg are inhibited from conduction of impulses, causing those muscles to relax. This simultaneous excitation and inhibition of the nerves controlling different muscle groups results in a smooth and continuous motion of the leg. Interneurons are responsible for such coordination of muscle groups; this coordination is a major function of the spinal cord. In a reflex, muscle activation occurs before nerve impulses have traveled up the spinal cord to the brain. Conscious awareness of the stimulus and responses comes only after the reflex action has been triggered. The spinal cord is responsible not only for reflexes but for translation of commands from the brain for movement into coordinated contraction and relaxation of different muscle groups.

The spinal cord is organized into an inner region of gray matter, where the integration of sensory input and motor output occurs, and a marginal region of tracts that carry information between segments of the cord and to or from the brain. Input arrives over axons in the dorsal roots, and output leaves via the ventral roots. There is a pair of dorsal roots and a pair of ventral roots for each spinal segment. Several segments may carry the information to and from each limb. Cell groups, or *nuclei*, of the cord and fiber tracts have been mapped by origin (or input), by destination (or output), and by function. Circuit diagrams for multisegmental reflexes, for some sensory processing, and for some descending control of movement from the brain have been worked out.

As a consequence of both anatomical and physiological studies of the spinal cord, neurosurgeons can detect and localize disease or damage in this organ. A surgeon tests sensory capabilities and reflex function by examining a person's reactions to various stimuli. Some people regard this kind of diagnostic facility as the goal of all nervous system study.

The Brain

The vertebrate brain has traditionally been subdivided into three regions—the forebrain, the midbrain, and the hindbrain. These names are based on the sections of the embryonic neural material from which the brain regions develop, and they do not necessarily correspond to the relative positions of the subdivisions of the adult brain. The hindbrain includes the medulla, pons, and cerebellum; the midbrain contains the colliculi and tegmental nuclei; and the forebrain consists of the cerebral hemispheres and such interior structures as the thalamus and hypothalamus.

THE HINDBRAIN

The *medulla* is similar to the spinal cord in structure and function. Afferent and efferent neurons serving the skin and the head muscles are connected to the medulla, as are the afferent neurons from the sensory organs of balance, taste, and hearing. The medulla controls many automatic or

Figure 24.10. The spinal cord.

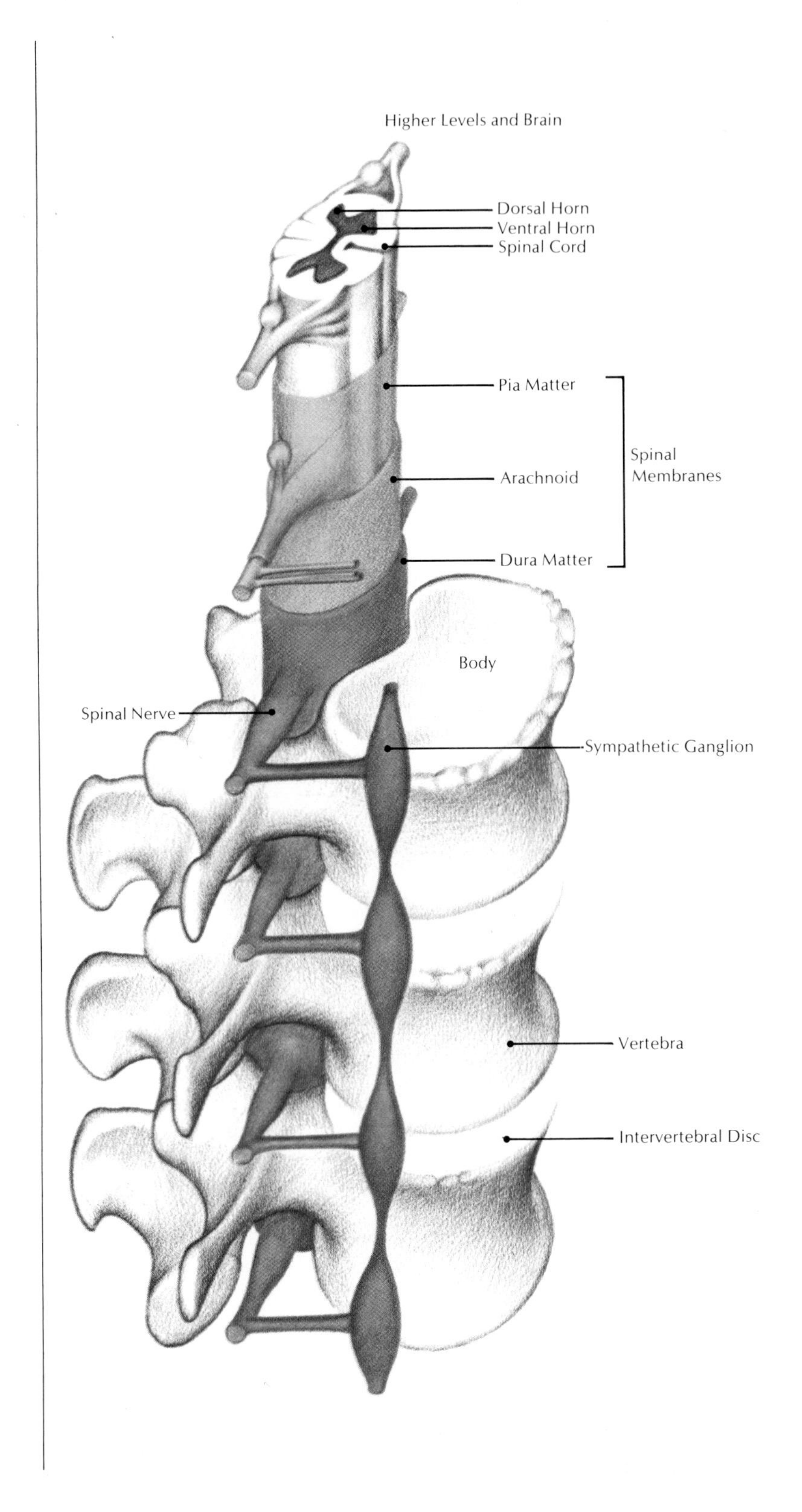

vegetative activities—for instance, those of respiration and circulation.

The *pons* is an extension of the medulla. The pons contains neurons controlling chewing movements, salivation, and facial movements, as well as neurons involved in hearing, respiration, and the inhibition and facilitation of spinal motor nerves. Nuclei of the pons mediate information transfer between the forebrain and the cerebellum, as well as between the cerebellum and the spinal cord.

The *cerebellum* is a lobed structure lying above the pons. The outer layer, or cerebellar cortex, is made up of neuropil and cell bodies, which surround a region of white matter and neuron clusters or nuclei. In primitive vertebrates, the cerebellum is connected to afferent neurons coming from sensory receptors that detect the positions of muscles. In higher vertebrates, the cerebellar cortex is more extensively developed and is connected indirectly to all sensory systems. Medieval anatomists believed the cerebellum to be the seat of the soul, but modern studies indicate that it is involved in the coordination of body movements and the maintenance of posture and equilibrium. If the cerebellum is removed, spasmodic body movements result. The cerebellum apparently contains a number of feedback circuits that continuously modify behaviors on the basis of sensory data about the progress of the actions (Brookhart, 1960).

Recent studies indicate that the cerebellum may play even more central roles in vertebrate behavior. The cerebellum is the only region of the brain other than the cerebral hemispheres and roof of the midbrain in which a cortex layer of gray matter covers a core of white matter. The cerebellar cortex is characterized by Purkinje cells, whose highly branched dendrites gather impulses from a large region of the cortex and send impulses over their axons to relay centers in the central nucleus. Because the organization of the cerebellum is orderly and is much simpler than that of the cerebral cortex, neurophysiologists hope that its function and structure will soon be worked out and will provide clues about the more complex organization of the cerebral cortex. Current speculations are that the cerebellum is involved in time sense, in visualization of continuity and position in time and space, and in modification of learned behavior patterns to suit a wide variety of slightly different conditions.

THE MIDBRAIN

The midbrain, or mesencephalon, is the foremost part of the brain that retains the basic structure of the spinal cord. The midbrain, pons, and medulla together often are called the *brain stem*. The dorsal portion of the midbrain, the tectum, contains neurons involved in the visual and auditory systems. This region is particularly well developed in fishes and amphibians. In mammals, reptiles, and birds, these centers are called the *colliculi*, and they retain their function in sensory information processing. The ventral portion of the midbrain is the tegmentum, which contains some nuclei involved in sensory integration and some involved in oculomotor control (control of eye movements).

THE FOREBRAIN

The rearmost portion of the forebrain consists of the thalamus in a dorsal position and the hypothalamus in front of it. Like other higher parts of the brain, the thalamus and hypothalamus can best be described as sets of nuclei—concentrations or clusters of neurons and glial cells in which neurons

Figure 24.11. Lateral view of the human brain in sagittal section.

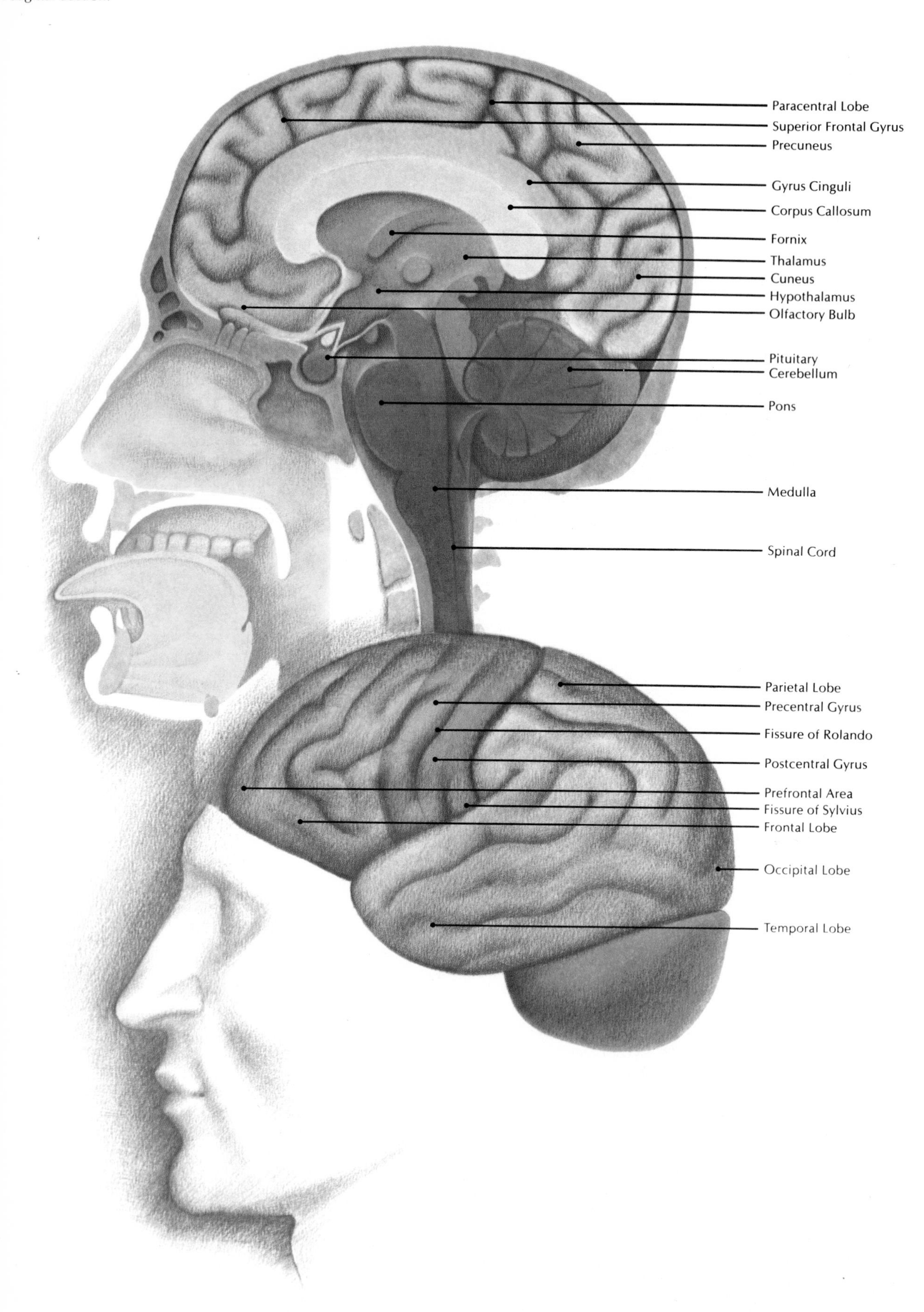

have common connections and consequently common functions.

In mammals, the senses of hearing and vision also send information directly to the sensory nuclei of the thalamus, the *geniculate nuclei.* These nuclei process information and may then send it on to the cortex. In many higher vertebrates, association and discrimination of form and pattern seem to occur in the forebrain, whereas there is some evidence that analysis of spatial relationships (orienting functions) are carried out by the colliculi of the midbrain.

Some nuclei of the thalamus are involved in processing and relaying information from one part of the cerebral cortex to another or from the cerebral cortex to other parts of the brain. Still other nuclei apparently are involved in processing and relaying information between parts of the thalamus and lower regions of the brain.

The hypothalamus is a relatively small set of nuclei, but this structure is of great importance in the regulation of vertebrate behavior. The control of many basic drives (hunger, thirst, and sex), the regulation of pituitary secretion, and the control of the internal environment of the body (blood pressure, heart rate, temperature, and so on) are all mediated by the activity of particular hypothalamic cells.

The foremost portion of the brain, or *cerebrum,* is made up of a group of large nuclei (rather misleadingly called the basal ganglia), surrounded by the layer of neural cells called the cerebral cortex. In lower vertebrates, the dominant structures of the cerebral portion of the brain are the olfactory bulbs, which act as processing centers for sensory input from the organs of smell. In these animals, the cerebral cortex is absent or poorly developed. The cerebral cortex of mammals is highly developed and plays the major role in sensory and motor integration. In most mammals, the olfactory bulbs and olfactory regions of the cerebral cortex are large and complex. In primates, however, the olfactory bulbs are small, and relatively little of the cortex is involved in the olfactory system. Instead, the major portion of the cortex serves to integrate information and regulate the functions of the other portions of the brain. The intricately convoluted outer surface of the human brain is composed of the cerebral hemispheres (Figure 24.11). The degree of convolution is directly proportional to the effective surface area of the cortex, which is thought to be related to information processing capacity. Rodents and primitive mammals have a relatively smooth cortical surface, whereas primates and marine mammals have the most convolutions.

In mammals each sensory system is connected to a specific region of the cortex, and each motor system arises from another specific cortical region. In addition to the sensory and motor regions of the cortex, there are association regions that are not directly connected to either afferent or efferent systems and that have often been assumed to be involved in higher intellectual activities. In higher mammals, the size of the cortex is greater and the amount of association cortex much larger than in lower mammals (Figure 24.12). The relative positions of the various motor and sensory regions are much the same in all mammalian species.

THE HUMAN BRAIN

The human brain is divided by deep fissures into a set of four lobes. The *occipital lobe* is located at the rearmost tip of each hemisphere. The cortex of this lobe is involved in the reception and processing of input from the

Figure 24.12. Comparative mapping of sensory and motor areas of the cerebral cortex in various representative mammals. Note the progressive enlargement of association areas.

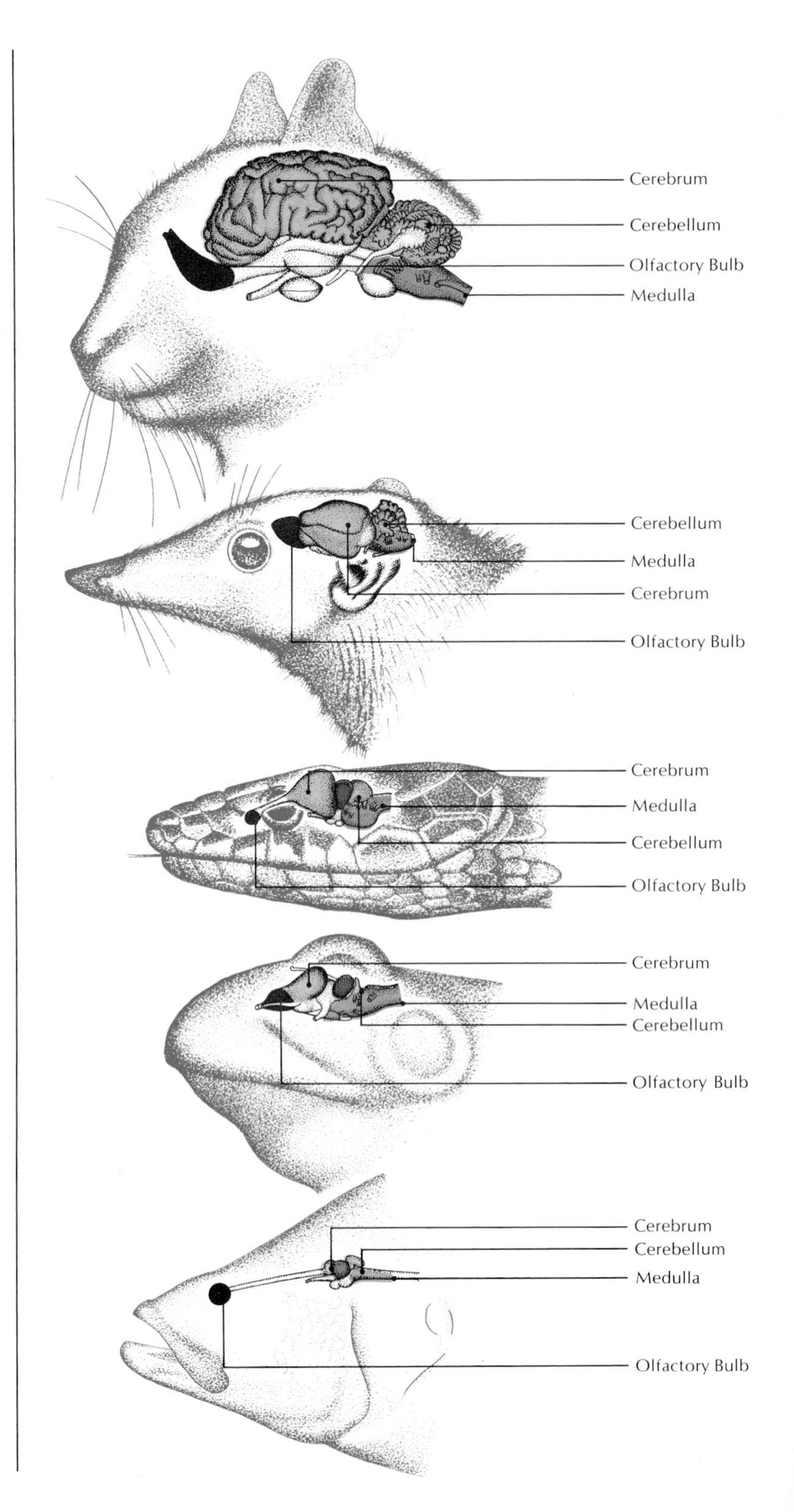

Figure 24.13. The relative proportions of body region representation in the motor cortex (anterior to the central fissure) and in the somesthetic cortex (the post central gyrus).

visual system; injury to parts of the occipital cortex can result in blindness in parts of the visual field. The *temporal lobes* are located on the lower sides of the hemispheres, separated from the remainder of the hemispheres by the fissure of Silvius, or lateral sulcus. The sensory regions involved in the auditory system are located in the temporal lobes, as are certain areas involved in the processing of visual information. Electrical stimulation of parts of the temporal lobe may result in complex auditory or visual illusions and even complete memory sequences. The cortex of this lobe is considered to be associated with the interpretation of sensory data.

The upper, rear portion of each hemisphere—above the lateral sulcus and behind the central sulcus—is the *parietal lobe*. The somesthetic cortex of this lobe contains the sensory regions connected to the receptors involved in detection of bodily position and the primary sensory regions connected with the skin senses. Electrical stimulation of various points along the gyrus just behind the central sulcus reveals that the sensory receptors of the skin are connected to the cortex of this gyrus in a regular pattern (Figure 24.13). The motor region of the human brain is also located along the region near the central sulcus. Motor connections in this cortical region

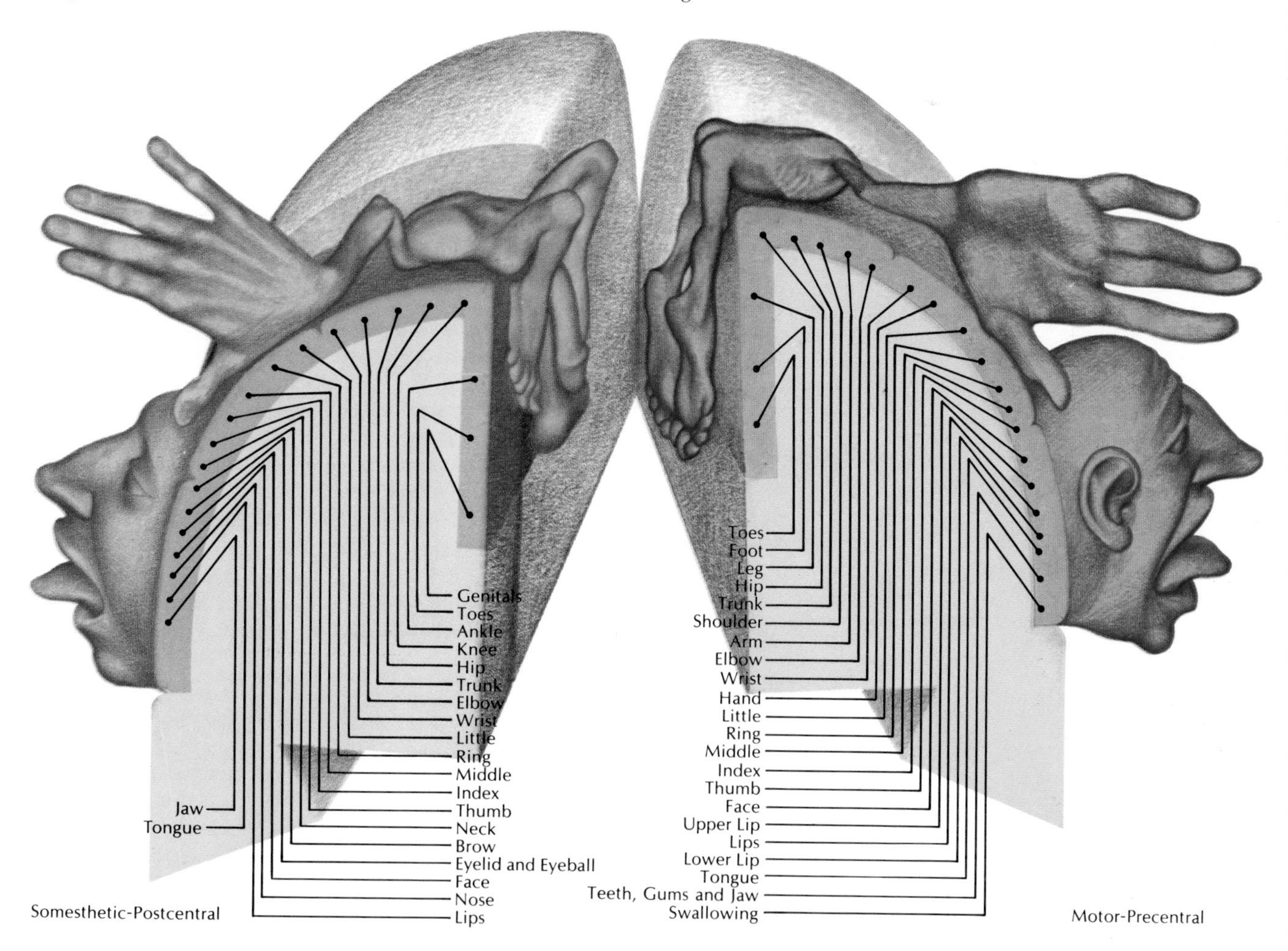

can be mapped by electrical stimulation in a similar fashion and reveal a dramatically different distribution pattern.

For both the motor cortex and the somesthetic cortex there is *contralateral control*; that is, the cortex of the right hemisphere is concerned with activities and sensory reception on the left side of the body, and the left hemisphere is concerned with the right side.

Damage to the somesthetic cortex causes deficits in tactile sensibility, and extensive parietal lobe damage extending beyond the primary sensory area produces unusual difficulty with spatial dimensions, such as misperception of the organization of the environment and distorted self-perceptions of the body.

At the foremost point of each hemisphere is the *frontal lobe*. The region just in front of the central sulcus is primarily concerned with the regulation of fine body movements. On the side of this lobe is a region called Broca's area, which is involved in the use of language. Damage to this area can produce severe language disabilities. An extensive region at the front of the lobe, known as the prefrontal area, is composed of associative cortex, which at various times has been viewed as the seat of intellectual ability or emotional control. The idea that this region is involved in intelligence probably arose from the extensive development of the area in primates and humans, but modern research indicates that intelligence cannot easily be localized in the cortical tissue. Performance on intelligence tests may be relatively unimpaired even after massive removal of frontal lobe tissue. It now appears that behavioral defects resulting from damage to the frontal lobe are of a subtle nature, involving the ability to order stimuli and sort out information.

Although the two hemispheres appear to be mirror images of one another, functional differences between them are apparent. Most people are right-handed, have their primary speech center located in the left hemisphere, and show differing effects in response to damage of the two temporal lobes. Damage to the left temporal lobe impairs intelligence-test performance severely, whereas damage to the right temporal lobe results in a loss of perceptual-test ability. In left-handed persons, these effects are reversed. The chemical or structural differences that must exist between the hemispheres have not yet been discovered.

Although there are many routes of neural communication between the two hemispheres, the primary band of fibers that connects the cortex of one hemisphere with the equivalent region in the opposite hemisphere is known as the *corpus callosum*. This massive group of fibers is involved in the synchronization of the activity of the two hemispheres. When the corpus callosum is surgically severed, responses to visual stimuli learned with one eye may not be evoked when the same stimuli are presented to the other eye.

THE LIMBIC SYSTEM

The limbic system is a group of interconnected structures in the forebrain—including the amygdala, cingulate cortex, hippocampus, and septum—a system that has recently received much attention because of its presumed role in regulating the emotional activity of the organism. Damage to these structures has profound effects on emotional reactivity, turning such highly excitable creatures as monkeys and mountain lions into docile pets or converting tame laboratory rats into ferocious and vicious creatures.

The limbic system is smallest and least developed in the higher animals,

Figure 24.14. The structure of the limbic system.

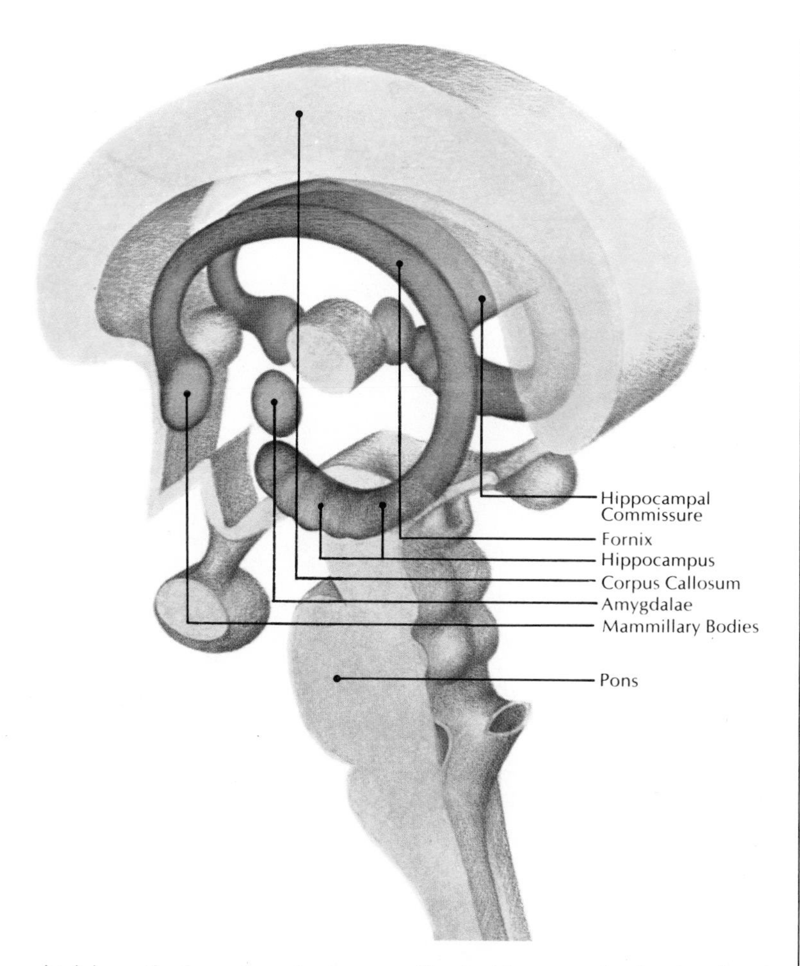

which have the largest cerebral cortex. The limbic system is closely related anatomically and functionally to the hypothalamus, but the exact functions and interactions of its various parts are still a matter of research and debate. Whereas the thalamic nuclei seem to govern expression of emotion, motivation, and so forth, the limbic system seems to play a role in the experiencing of emotions and urges.

THE RETICULAR FORMATION

Within the central core of the midbrain and hindbrain is an intricate latticework of nerve cells called the reticular formation (Figure 24.15). This primitive system, which exists in the brains of all vertebrate species, plays a primary role in regulating the level of alertness of the organism. Damage to the midbrain reticular system produces an animal that sleeps much of the time and exhibits a cortical brainwave pattern characteristic of stages of deep sleep. The system apparently serves to inhibit or to facilitate impulses moving between the brain and spinal cord, thus exercising a general control over the behavioral state of the animal.

ELECTRICAL ACTIVITY OF THE BRAIN

Millions of neurons in the brain are active at any one instant, and their interdependent activity has been likened to an enchanted loom. There is, at

Figure 24.15. The reticular formation serves to inhibit or facilitate the passage of impulses from the sense organs to the cortex. Thus, it acts as a sensory filter and exercises general control over the animal's behavioral state.

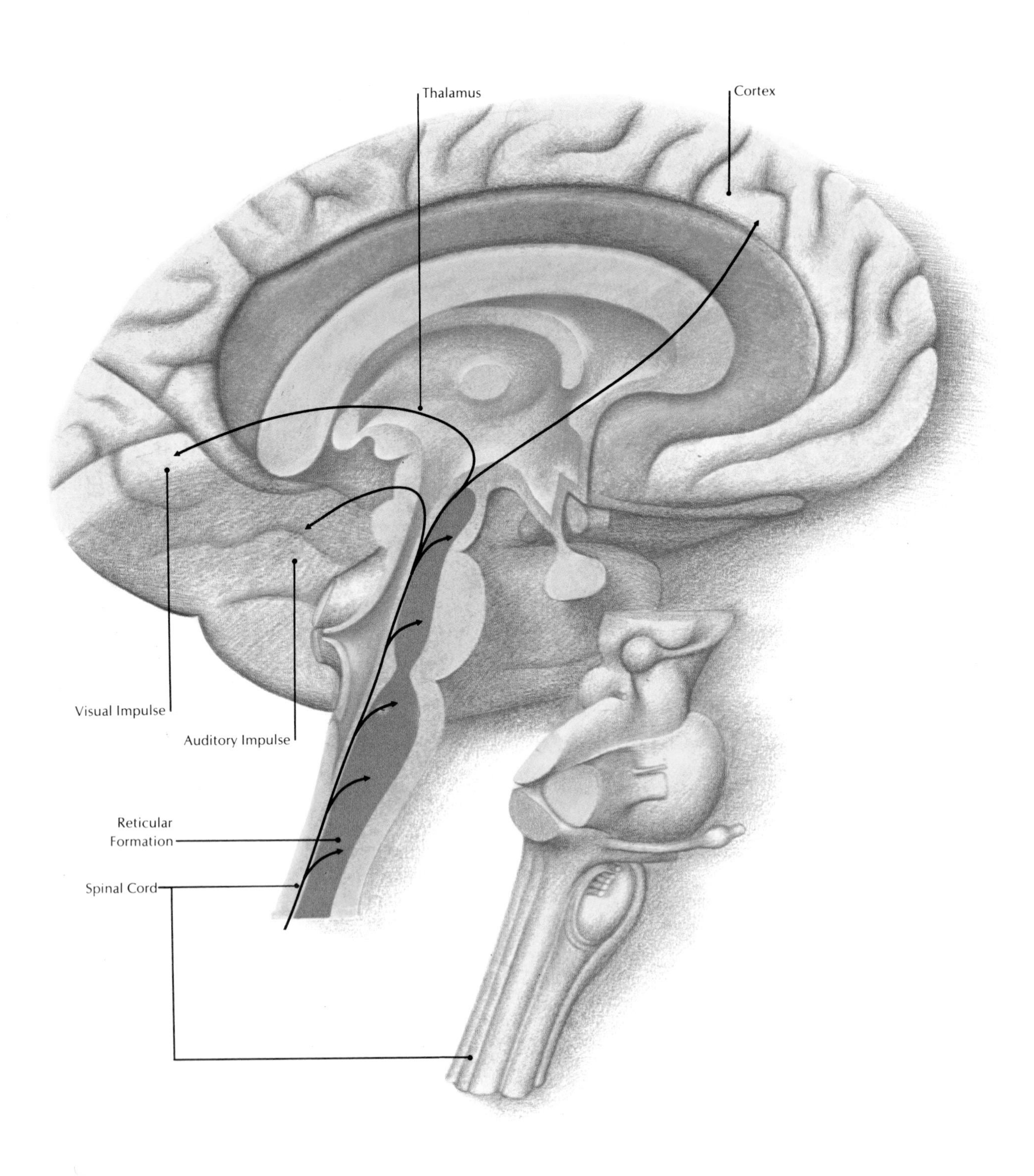

present, little hope that all of the neurons can be monitored simultaneously in order to elucidate brain function. But there are methods for recording from large populations of neurons, and such recordings can give important clues to the interconnections and functions of various regions of the brain.

One such method involves the placing of a large electrode (a millimeter to a centimeter in diameter) on the scalp of an intact animal, on the surface of a brain structure, or deep in the brain tissue. Small, continuous fluctuations of potential can be recorded from such electrodes. The record of the fluctuating potentials, often including simultaneous measurements from many electrodes, is called the *electroencephalogram* (EEG). Much research has been done to elucidate the origin of these potentials. To date, the best guess is that the EEG represents the average activity in millions of axons, cell bodies, and dendrites where action potentials and synaptic potentials are occurring simultaneously.

An alert and mentally active person produces low-level potentials with a random assortment of frequencies of oscillation. A resting person, particularly with his eyes closed, most often exhibits regular waves with a frequency of about 8 to 13 Hertz (*alpha waves*). These waves can sometimes be produced at will by "making the mind blank", and they seem to be present in meditating monks.

The periodic fluctuations exhibited by sleeping subjects have been used to study the phenomenon of sleep. The EEG also has been used to study certain brain malfunctions, such as epilepsy. However, the interpretation of EEG is very difficult, and much remains to be learned. The differences between EEG records from place to place in the brain and from time to time at the same place are greater than the differences between EEGs of different animals.

In records of EEG activity, it is possible to find potentials correlated with sensory stimuli such as flashes of light or sound tones. These potentials are called *sensory evoked potentials* and often require the use of special averaging computers to detect them in the presence of ongoing spontaneous EEG. Sensory evoked potentials measured in various nuclei or brain tracts and potentials evoked by electrical stimulation of other brain structures have contributed greatly to the knowledge of the function of these structures and of their interconnections.

FURTHER READING

Dethier and Stellar (1970) provide a brief introduction to the structure and function of the nervous system. Ochs (1965) and Thompson (1967) provide more detailed accounts of neuroanatomy. For specific information about the vertebrate nervous system, see Ariëns-Kappers, *et al.* (1960), Romer (1970), and Woodburne (1967). Invertebrate behavior is described by Maier and Schneirla (1935) and the invertebrate nervous system is discussed by Bullock and Horridge (1965) and Parker (1919).

For more information about the vertebrate brain, see the books by Butter (1968) and Wooldridge (1963). The autonomic nervous system is described in detail by Kuntz (1953).

Scientific American articles relevant to this chapter include those by Brazier (1962), Eccles (1958), French (1957), Gazzaniga (1967), Held (1965), Holst and Saint Paul (1962), Kennedy (1967), Luria (1970), and Wurtman and Axelrod (1965).

25
Effectors

Figure 25.6 (above). The stimulant effect of acetylcholine on the secretory function of isolated adrenal medullary tissue. Tissue perfused with isosmotic sucrose both with and without calcium.

Figure 25.7 (below). Schematic diagrams with progressive enlargement to show the internal structure of skeletal muscle tissue. Muscle is made up of many interacting muscle fibers (cells), which appear striated under magnification. Individual fibers include myofibrils within which repeated patterns of light and dark bands can be distinguished.

small bundles of neurons enter the adrenal medulla, where they branch, with one axon sending branches to one or more nearby secretory cells. Each secretory cell receives a synapse from one of these neurons. The neurons triggering the release of adrenal hormones originate in the sympathetic branch of the autonomic nervous system and release acetylcholine (ACh) as their neurotransmitter. The results from recordings made by intracellular microelectrodes indicate that the secretory cells have a resting potential, and the cells depolarize if acetylcholine is placed on them. If ACh is placed in the bathing fluid of isolated adrenal medullary tissue, epinephrine and norepinephrine can be detected in the perfusion fluid. The cells have thus been induced to secrete by the addition of ACh. If the calcium is entirely removed from the perfusion fluid and then ACh is added, little secretion can be detected.

MUSCLES

The final general class of effectors is muscle. The skeletal muscles of vertebrates are responsible for postural changes or locomotion, whereas the movements associated with the function of internal organs are performed by smooth muscles (Chapter 11). There is a greater variety of invertebrate muscle types, each suited to the function and habits of the particular organism. In certain fish, for example, there is a highly specialized organ—a descendant of muscle tissue—designed to produce electric discharges that can be sensed by other nearby organisms.

Skeletal Muscle

Figure 25.7 summarizes the levels of organization in a fast-contracting frog

Epinephrine (μg/min)
3
2
1
0
Addition of ACh
Addition of ACh
Ca-free Sucrose
Ca Sucrose
Perfusion Time

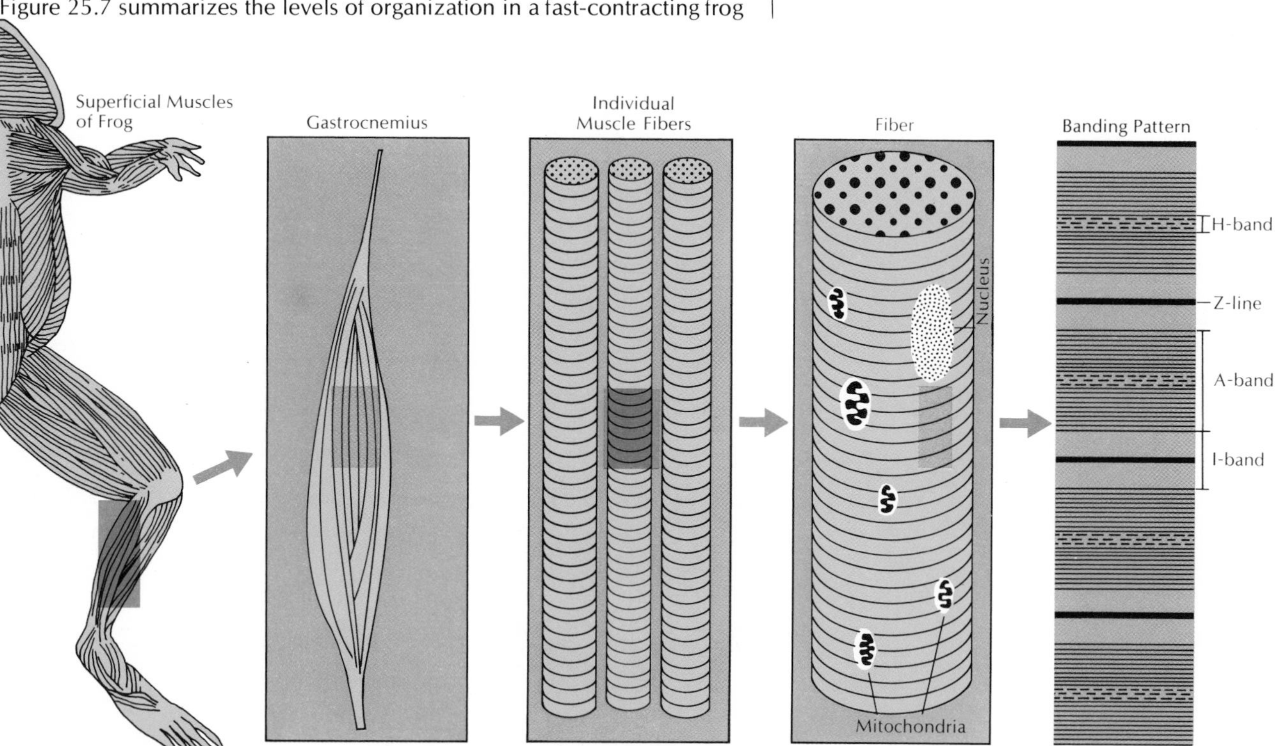

Figure 25.8 (above). Action potentials at the motor end plate in a single skeletal muscle fiber.

Figure 25.9a (below). Electron micrograph of a transverse section through myofibrils in skeletal muscle. Small dots are filaments.

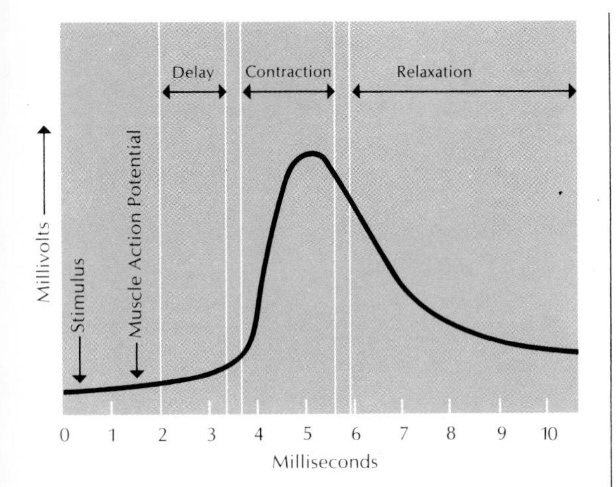

muscle. Much of this discussion of skeletal muscle is based on studies of fast-contracting frog leg muscles, but the principles involved are very similar for other vertebrate skeletal muscles, as well as for certain other types of muscles.

The action of a muscle is initiated when the central nervous system sends a signal via a motor neuron. Motor neurons release acetylcholine onto the end plate, a specialized region of the muscle just under the nerve ending. ACh produces an increase in the membrane's permeability to Na^+ and K^+, depolarizing the muscle membrane and producing an end-plate potential. The end-plate potential normally depolarizes the muscle membrane to a threshold level, and an action potential is fired. The action potential quickly spreads over the cell membrane and lasts a few milliseconds; within 2 to 3 milliseconds after the action potential, the muscle begins to shorten. A simultaneous recording of the membrane potential of the frog muscle and its contraction are shown in Figure 25.8. The sequence of events in muscle action may be summarized as stimulus → muscle action potential → short delay → contraction → relaxation.

The pattern of muscle striation is made up of overlapping thick and thin filaments (Chapter 11). The thin filaments are anchored to the Z-line and interdigitate with thick filaments, which form the darker appearing A-band. For many years, the significance of this pattern was not appreciated, and it was thought that muscle shortening must be caused by the coiling or folding of fibrous protein molecules.

In the early 1950s, however, two groups of researchers in England refuted this hypothesis. A. F. Huxley and R. Niedergerke observed the striation patterns during contraction of single muscle fibers with special arrangements of the light microscope designed to intensify the differences between the dark A-bands and the light I-bands. They found that the length of the

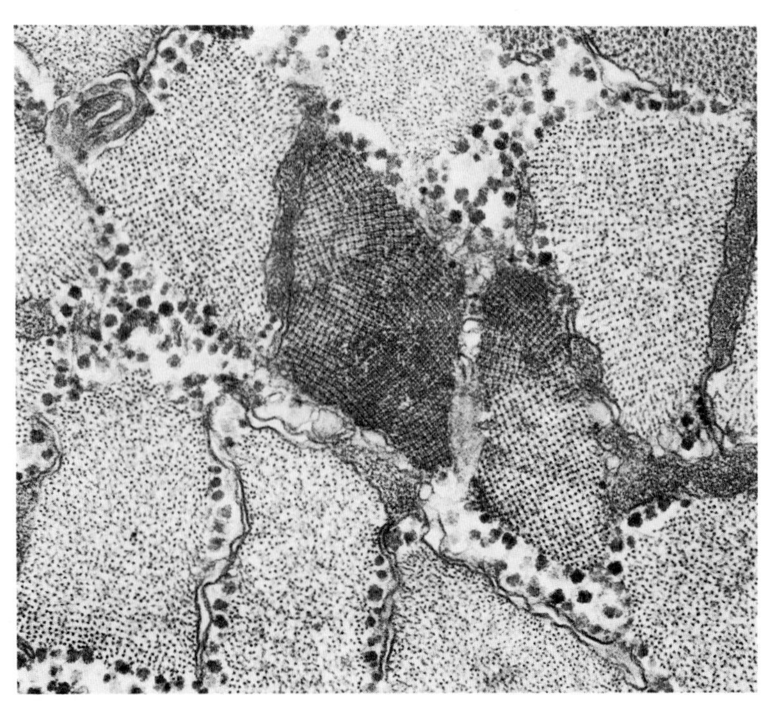

Figure 25.9b. Schematic diagram of a section of striated muscle fiber, showing the relationship of a fiber to the myofibrils that compose it and to the smaller filaments that cause the striations. The pie-shape wedge at the bottom represents a portion of a single fiber. (One fiber is barely visible to the unaided eye.) Each of the circles is a myofibril – one is lifted out and magnified at top, with various components shown in closer detail. The striations in each myofibril are resolved into a repeating pattern of light and dark bands (at right). The sequence of this pattern consists of a Z-line, then an I-band, then an A-band (interrupted by an H-band), another I-band, and finally another Z-line. Electron micrographs have shown that this repeating band pattern is due to overlapping thick and thin filaments (shown here as large and small dots in the myofibril).

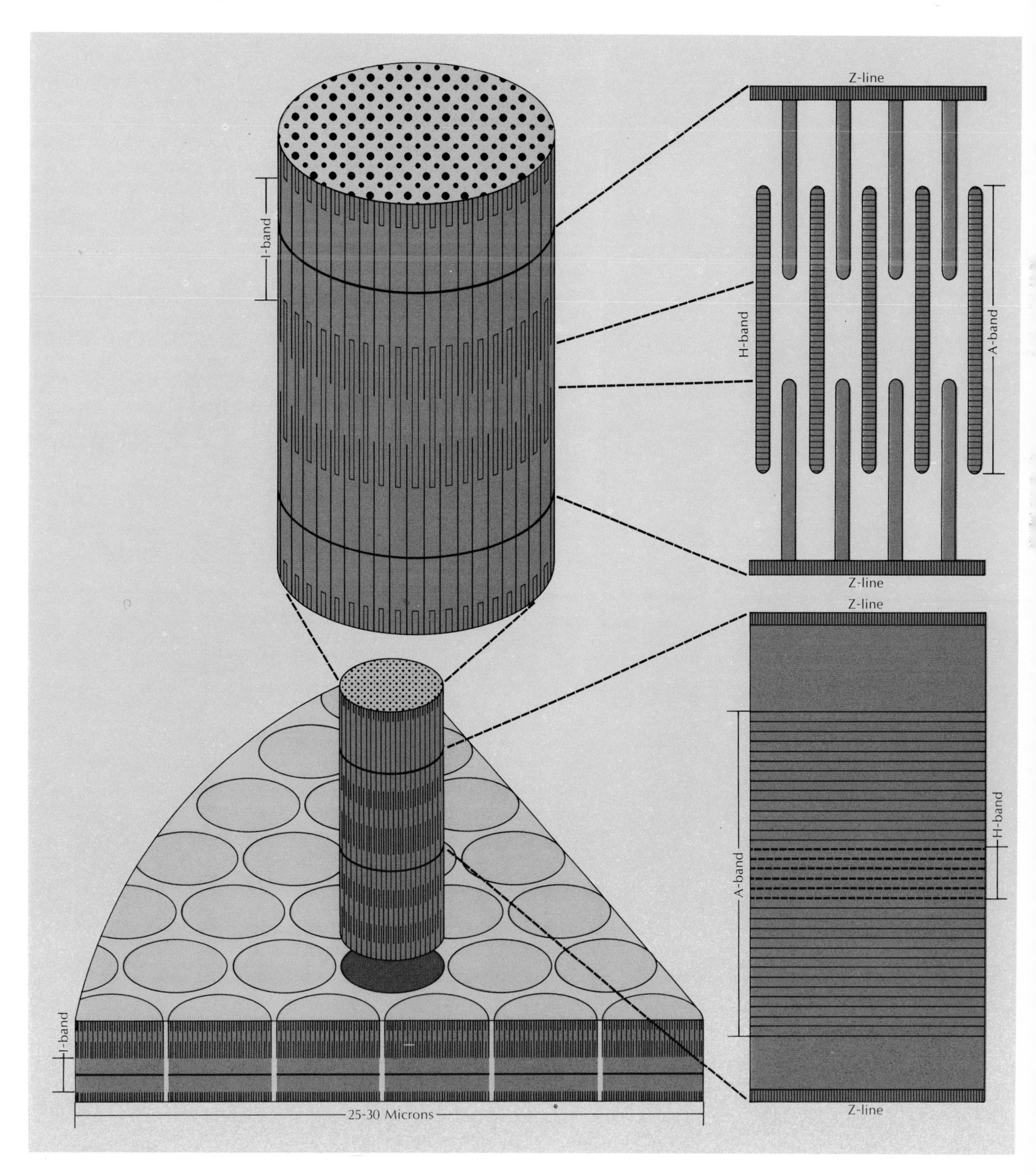

Figure 25.10. The sliding filament model of muscle contraction. Schematic drawing of myofibril filaments in a stretched muscle (A), and schematic drawing showing myofibril filaments in a resting muscle (B).

A-bands remains constant as the muscle contracts, while the I-bands and H-bands become narrower. At the same time, H. E. Huxley and J. Hanson were studying the structure of striated muscle with the electron microscope. They observed muscles fixed in a stretched condition, at rest length, and contracted (Figure 25.10). Each of these teams drew the same conclusion from their studies: the lengths of the filaments in striated muscle do not change during contraction; rather, the two kinds of filaments slide past one another. This mechanism of muscle shortening is called the *sliding-filament model* of muscle contraction. This model does not explain how the force of contraction is generated; the answer to that is not known yet, but there are several good hypotheses. Evidence suggests that the cross-bridges extending from the thick filaments are probably involved.

One hypothesis proposes that the cross-bridges between myosin and actin filaments act as ratchets to pull the filaments past one another (Figure 25.11). A cross-bridge would attach to an active site on an actin filament

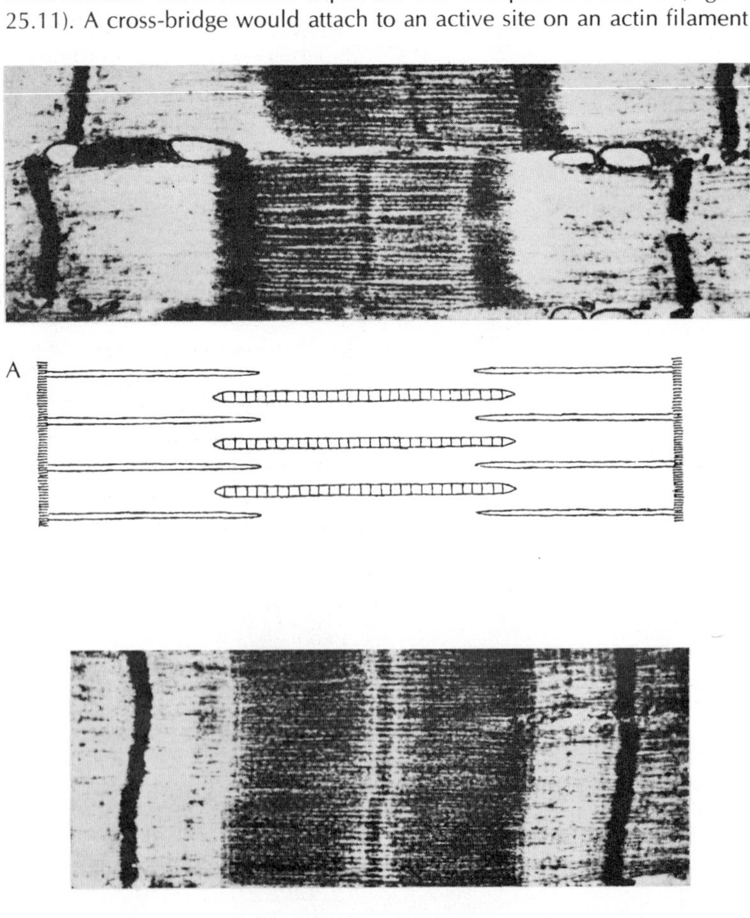

Figure 25.11a. Three-dimensional diagram illustrating how the arrangement of cross-bridges between filaments interact. Thick filaments pull thin filaments along by a type of ratchet action. Each bridge is part of a thick filament but is able to hook onto a thin filament at an active site (dots).

Figure 25.11b. Electron micrograph of filament cross-bridges.

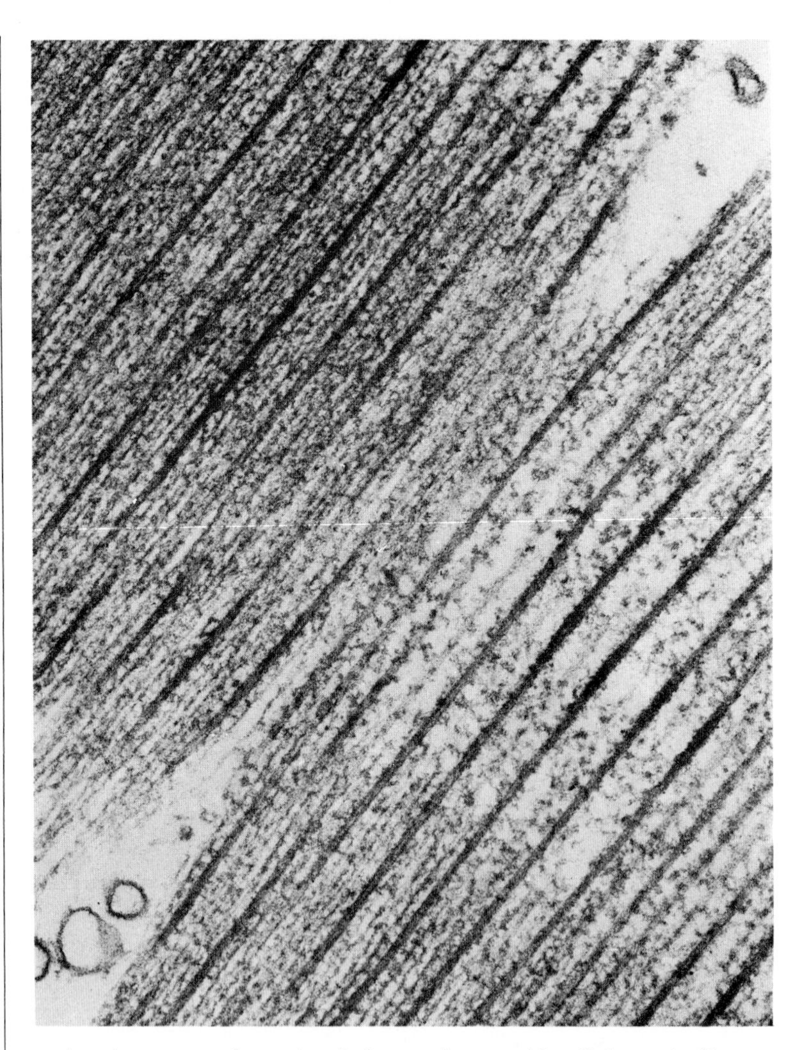

and undergo a conformational change that would pull the actin filament along about 100 A. ATP is hydrolyzed to ADP to provide the energy. Pure myosin isolated from muscles acts as an enzyme for the hydrolysis of ATP and handles the substrate molecules at a rate compatible with the 50 to 100 cycles per second needed to account for the observed speed of muscle contraction. ATP may also be required to release the cross-bridge from the actin filament. This need for ATP to release the cross-bridge would account for the stiffness (*rigor mortis*) of dead muscles; when the cell's supply of ATP is exhausted, the filaments are locked in position by the cross-bridges.

Because the action potential takes place at the surface of the muscle cell, it was reasonable to speculate that some activator substance was produced by the action potential. The activator would then have to diffuse into the contractile apparatus to bring about muscle shortening. In 1949, however, A. V. Hill calculated the time required for anything to diffuse from the surface to the center of a frog muscle 100μ in diameter. He concluded that the diffusion time was much too long to account for the short delay between the action potential and contraction. Some ten years later, electron

Figure 25.12 (above). Schematic three-dimensional diagram of the sarcoplasmic reticulum. This network of microtubules is intimately associated with the myofibrils and may function to conduct activator substances from surface membranes to the myofibrils.

Figure 25.13 (below). Diagram showing electrode experiment described in text. When the electrode is placed over a Z-line, the muscle gives a small local contraction on both sides of that line. If the electrode is placed anywhere else on the surface of the striation pattern, no contraction occurs.

microscope studies of the fine structure of muscle demonstrated that there were invaginations of the surface membrane (t-tubules) and a complex set of internal membranes (sarcoplasmic reticulum) surrounding each myofibril in a specific arrangement with respect to the pattern of thick and thin filaments (Figure 25.12).

A. F. Huxley and R. E. Taylor conducted an experiment in which they used a fine-tip glass micropipette electrode to stimulate localized areas of the muscle cell surface. They found that when the electrode was over a Z-line, the muscle gave a small local contraction on both sides of that Z-line. If the electrode was anywhere else on the surface of the striation pattern, no contraction occurred. In the frog muscle, the Z-line corresponds to the point at which invaginations of the surface membrane, or t-tubules, enter. Along the t-tubule are sites at which the sarcoplasmic reticulum comes in contact with the tubule membrane. When the local stimulation experiment was performed on a lizard muscle instead of a frog muscle, a contraction was produced only at the site opposite the two ends of the A-band. Electron microscopy reveals that the t-tubules enter the lizard muscle fiber at the ends of the A-bands. The results of these local-stimulation experiments implicated the t-tubules in the process of muscle activation.

Meanwhile, evidence from various sources indicated that calcium ions must be involved in muscle contraction. For example, in experiments in which various substances were injected into single muscle fibers, calcium was the only naturally occurring substance that caused a contraction. In the early 1960s, three different laboratories reported that a subcellular fraction of homogenized muscle was capable of accumulating calcium ions against a concentration gradient. This calcium-accumulating material was in the form of small vesicles that appear to have been derived from the sarcoplasmic reticulum. (After the whole muscle is ground up, the membranes of the sarcoplasmic reticulum tend to heal over into small closed vesicles.) When a muscle fiber is soaked in a solution that causes a dense precipitate of calcium salts, the electron microscope reveals that the calcium precipitate has accumulated in the sarcoplasmic reticulum, particularly in that part of the sarcoplasmic reticulum in close contact with the t-tubules (Constantin, Podolsky, and Franzini-Armstrong, 1965). Actin can only interact with myosin to split ATP when a precise concentration of calcium ions is present.

The action potential spreads over the surface of the muscle, and electrical excitation spreads down the t-tubules. In some way, this excitation causes calcium to be released from the sarcoplasmic reticulum, so that actin and myosin can interact and split ATP. This activity somehow results in the sliding of the thick and thin filaments past one another. As long as nerve impulses continue to arrive at the muscle fiber, waves of depolarization continue to move along the sarcolemma and through the t-tubules; Ca^{++} ions continue to be released; and the myofibrils continue to contract with the consumption of energy. When the nerve impulses cease to arrive, the membranes again become polarized, Ca^{++} ions are taken up by the sarcoplasmic reticulum, actin and myosin stop interacting, and the myofibrils relax.

The ATP needed to power muscle contraction is produced by oxidative phosphorylation in the mitochondria (Chapter 6). When the muscle cell is very active, the supply of oxygen may not be sufficient to produce the necessary ATP. The process of glycolysis is then used to obtain energy rapidly. Lactic acid, the end product of glycolysis in animals, accumulates in

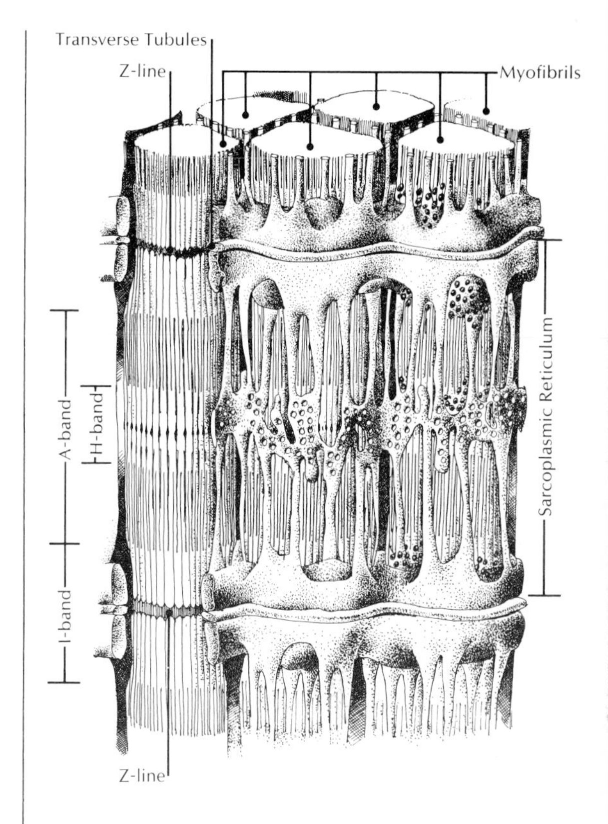

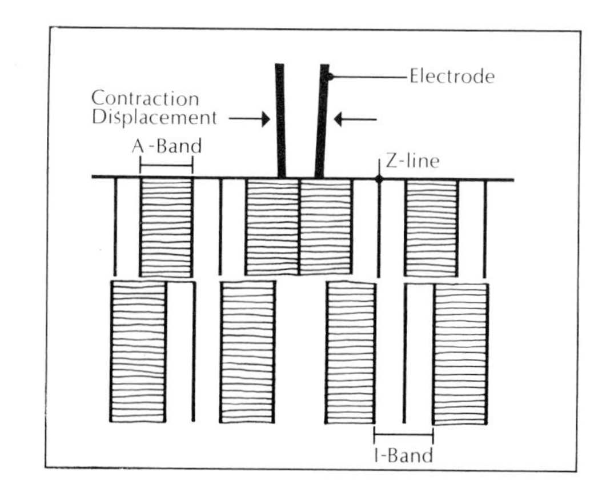

Figure 25.14. At left, the summation of muscular contraction by double stimulation is shown. Records are of the median head of the gastrocnemius muscle responding to two supramaximal stimuli to the motor nerve in succession. Ordinate is relative tension; abscissa is time. At right are recordings of both electrical (e) and mechanical (m) activities of the cat extensor digitorum longus muscle, showing the development of tetanus. Ordinates are relative tension (m) or relative voltage (e); abscissa is time, with 15 milliseconds between successive action potentials in (e). The rate of stimulation is 67 per second.

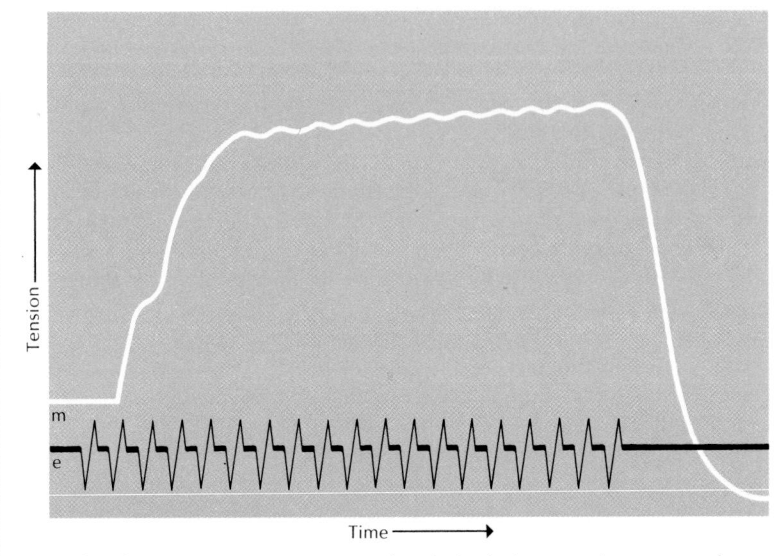

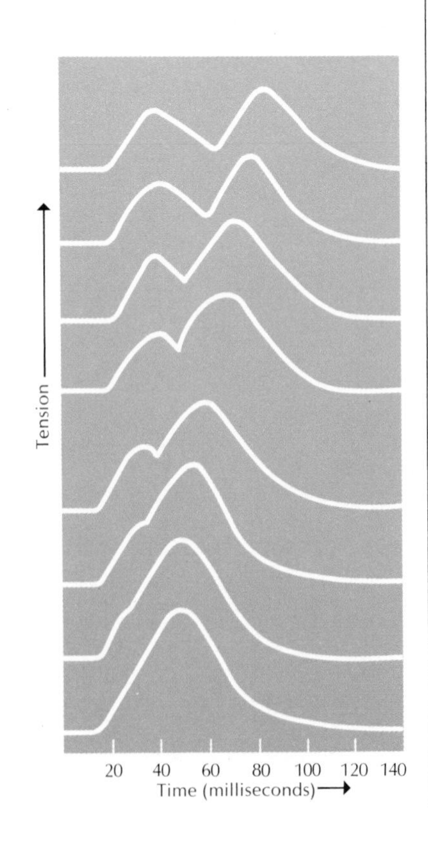

muscles during exertion. By using glycolysis during exertion, a muscle acquires an oxygen debt; eventually, all the lactic acid formed under anaerobic conditions must be reoxidized. Rapid breathing after exertion is associated with "repayment" of the oxygen debt.

An organism must be able to use its muscles in smoothly graded contractions, not just fast twitches. Different animals achieve graded muscular movements by different mechanisms. In the frog muscle, each muscle cell twitches in an all-or-none fashion because a twitch is triggered by an all-or-none action potential. In order to make graded movements, each motor neuron is connected to a small patch of muscle cells. By adjusting the firing pattern of the motor neurons, the central nervous system can produce movements of certain patches of muscle in sequence or together, producing the desired contraction. Some gradation can also be achieved by repetitive firing of the motor neurons. If a motor neuron fires many action potentials in rapid succession, the muscle cell does not have time to relax completely between successive twitches, and the twitches sum. If the frequency of stimulation continues at a high rate, the muscle may not relax much at all between stimuli. When the muscle is unable to relax between stimuli, the muscle is said to be in a state of *tetanus*.

In striated muscles of other animals, the gradation of contraction may be achieved by different patterns of innervation. In arthropods, the innervation patterns of skeletal muscles are very complex. Each muscle fiber may be excited by more than one type of neuron (polyneuronal innervation). Most muscle fibers also receive inhibitory innervation.

Another type of graded contraction is found in some crustacean muscles and in some slowly contracting muscles of lower vertebrates. These muscles do not fire all-or-none action potentials. Instead, they receive many nerve endings from one axon and respond by slow graded contractions to summed end-plate potentials distributed over their surfaces.

Smooth Muscle

The visceral muscles of vertebrates are different in structure from skeletal muscles, and the details of their physiology are not as well understood.

Smooth muscles contract and relax more slowly than striated muscles. Smooth muscles may contract in response to slow, apparently spontaneous depolarization or in response to action potentials. The action potential of smooth muscle differs from that of vertebrate striated muscle in that all the inward current is not carried by Na^+, and an inflow of Ca^{++} can also contribute to depolarization.

In smooth muscle, as in striated muscle, Ca^{++} initiates contraction. Calcium entering from the surface during the action potential may contribute to the initiation of contraction. Because the smooth-muscle cell is very thin (10μ in diameter) compared to the striated-muscle cell and smooth muscle does not contract as rapidly as skeletal muscle, the diffusion of Ca^{++} in from the surface is fast enough to account for the time between excitation and contraction in smooth muscle.

There is much left to be learned about the function of smooth muscle. For example, although the thin filaments of smooth muscle are probably made of actin and the protein myosin can be extracted from smooth muscle, it is not yet known where the myosin is or how it works. Very recent electron microscope studies indicate thick filaments in smooth muscles under some conditions. It may be that smooth muscle also contracts by a sliding-filament mechanism that is triggered by an influx of calcium.

FURTHER READING

Good general coverage of effector systems will be found in the physiology books by Hoar (1966) and Prosser and Brown (1961).

For the original proposals of the sliding-filament model of muscle contraction, see Huxley and Niedergerke (1954), Huxley and Hanson (1954), and the review by Huxley and Hanson (1959).

26
Reception and Action

If an animal is to live and reproduce successfully in a constantly changing world, it must continuously monitor happenings and adjust its relationship to the environment. Animals possess many kinds of sensory systems, which gather important information about the external world and about the animal's internal state. Sensory systems also process this information in such a way as to accentuate important aspects of stimuli and to reduce spurious or unimportant aspects. In a broad sense, the sensory system consists of sets of filters, which abstract specific relevant spatial and temporal patterns from a welter of stimuli, most of which are unimportant to the animal.

Animals also have motor systems, which perform relevant adjustments of the animal relative to the proper environmental stimuli. On the sensory side, patterns of stimulation in space and time must be integrated in the brain and must trigger meaningful responses. These responses also are patterned in space and time, as whole sets of muscles and glands are activated in precise temporal sequences. In a broad sense, the motor system also consists of a set of filters, which select—from an infinite number of possible but meaningless actions—that set of actions appropriate to the animal's current situation.

Although sensory and motor systems are integral parts of the entire nervous system, in this discussion certain parts of the nervous system will be defined rather arbitrarily as a sensory or a motor system. For example, the visual system as discussed here includes the structures of the eye, the visual receptor cells in the retina, and much of the nervous system that is involved in transmission and processing of impulses triggered by visual stimuli. From the examples considered in this chapter, it will be clear that there are parallels between sensory and motor systems, not only in their functions as sets of filters but in the neural networks that organize both input information and output motor commands.

SENSORY SYSTEMS

Sensation and perception—at least in man and other higher animals where these terms are clearly meaningful—are the results of processes occurring in the nervous system, particularly in the brain. A physical *stimulus* in the external or internal environment triggers electrical activity in sensory receptors. The receptor cells in turn stimulate action potentials in the neurons of the afferent pathways of the nervous system. After complex sequences of excitation and inhibition at numerous synapses, a particular pattern of neural activity is stimulated in certain areas of the cerebral cortex. In man, this activity of the cortical cells apparently corresponds to the experience of *sensation*—the awareness of colors, forms, sounds, smells, tastes, and so on. The experience of *perception*—interpretation of the significance of sensations, such as the perception of a particular object in a particular place—is the result of further cortical activity for which exact neural correlates are unknown.

Sense organs are not a specialization limited to animals. Animals have developed a great degree of organ specialization for some types of stimuli, but plants, fungi, and even microorganisms respond very sensitively to light, touch, various chemicals, and gravity. Some photosynthetic bacteria adjust their motions so that, when they are exposed to a spectrum of light projected onto a microscope slide, they congregate at wavelengths that they can utilize for photosynthesis. Bacteria also move toward higher concentrations of oxygen, sugars, and amino acids. Relatively little is known of the mechanisms with which microorganisms detect changes in the environment and adjust their activities accordingly. However, in recent years, some progress has been made in applying the powerful techniques used to study the genetics of bacteria to the study of sensory mechanisms. The biochemical and anatomical analysis of mutant strains deficient in various senses, for example, makes possible the identification of the molecules and structures involved in detection of stimuli. Eventually, these studies may lead to an understanding of the molecular mechanisms of reception, an understanding that will prove as basic and as enlightening as the information about gene action obtained from these simple organisms.

In spite of the greater complexity of sensory systems in more complex organisms, a great many physiologists over the years have chosen to study animal sensory systems. Visual systems of a number of animals, including man, have been studied intensively, and many phenomena of visual sensation and perception have been explained mechanistically (Interleaf 26.1). Because many principles applicable to other sensory systems—insofar as they are known—appear to be much the same as those of the visual system, this chapter describes in some detail present knowledge of the visual system, with a brief summary of knowledge about other sensory systems.

Stimulus, Sensation, and Perception

The sensory process traditionally has been divided into three stages: stimulus, sensation, and perception. The term "stimulus" refers to a physical event, change, or object—a measurable condition or change of condition that exists in the external or internal environment of the organism. The stimulus may be regarded as the physical, measurable, objective event or object that triggers behavior.

Sensation is the subjective experience triggered by a stimulus. Two different persons may experience different sensations arising from the same stimulus. On the other hand, almost identical sensations of flashing lights can be produced by such varied stimuli as a sharp blow to the eyeball, electrical stimulation of the cornea, or certain patterns of light waves in the external environment.

Perception refers to even more subjective mental processes; it involves interpretation of a set of sensations. A particular group of sensations—forms, colors, sounds, and touch sensations—may lead to the perception of a particular person standing in a room. Depending upon the past experience and present mental state of the perceiver, this group of sensations may lead to the perception of a dangerous enemy, a familiar friend, or even a lifeless statue. The earliest field of psychology to develop was sensory *psychophysics*, the study of quantitative relationships between properties of physical stimuli and the subjective experiences reported by human observers. These studies attempted to outline the qualities of stimuli that correspond to differing qualities and strengths of experiences. Later, psychophysical studies used animals as well as humans in an attempt to correlate stimuli and experiences. Psychophysical research continues to provide information about sensory processes that must be explained by the mechanisms proposed by neurophysiologists.

LIGHT

The stimulus for the visual system is light. Objective or measurable properties of the light stimulus are intensity, wavelength, spatial pattern, and

Figure 26.1 (left). A series of time-lapse photographs taken at 4-minute intervals showing growth of a *Phycomyces* sporangiophore in response to a fixed light source.

Figure 26.2 (right). The transparent stalks of the *Phycomyces* sporangiophore act as a lens, focusing light on the side of the stem opposite to the light source. Growth rate is controlled by light intensity; thus, the stalks grow most rapidly where light is most intense. This uneven growth causes the stalk to curve toward the light source. This photograph illustrates a spiral sporangiophore growth pattern in response to a fixed light source; the sporangiophore was placed on a turntable that made 1 revolution every 6 hours and 36 minutes.

Interleaf 26.1

PHYCOMYCES: A SIMPLE SENSORY RECEPTOR

Phycomyces is a funguslike protist that forms a branching mycelium, which will grow on almost anything. From this mycelium, stalklike sporangiophores grow up into the air, each developing at its tip the spherical sporangium that eventually will disperse some 100,000 spores. Each spore is capable of initiating a new mycelium. During its growth, the sporangiophore bends toward any source of blue light (Figure 26.1).

The transparent stalk acts as a lens, focusing light on the side of the stalk opposite to the light source (Figure 26.2). Apparently, the rate of growth in the stalk is controlled by the intensity of light; where light is most concentrated or intense, the stalk grows most rapidly. This uneven growth makes the stalk curve toward the light source.

The elementary processes of this primitive sensory response are more amenable to molecular and particularly to genetic studies than are similar phenomena of phototropism in multicellular oat coleoptiles (Delbrück, 1968). Presumably, the stalk contains pigment molecules that absorb light, become electronically excited, and initiate some biochemical reaction that increases the growth rate. Attempts to identify this pigment molecule have been inconclusive. A likely candidate was a carotene pigment that most strongly absorbs light of just those colors that are most effective in producing the bending effect. However, mutants that lack the carotene pigment also exhibit phototropism, so another pigment must exist. When the pigment is identified, it may be possible to analyze the mechanism by which this simple sensory mechanism operates.

The light-sensing system of *Phycomyces* exhibits several properties similar to those of visual systems of higher animals. Like the vertebrate eye, the *Phycomyces* stalk can respond to light intensities ranging from bright sunlight to dim twilight. A stalk that has been exposed to very bright light is temporarily "blinded" when placed in dim light, but it soon adapts to the less intense stimulus and again bends toward the light source. This action corresponds to the familiar process of dark adaptation, in which a person entering a dimly lit room from bright sunlight requires time to "adjust his eyes" to the room light. Properties such as these, which seem so important in higher animals, occur even in very simple organisms. Genetic and molecular studies of such prototype sensory mechanisms undoubtedly will provide insight into more complex phenomena in animals.

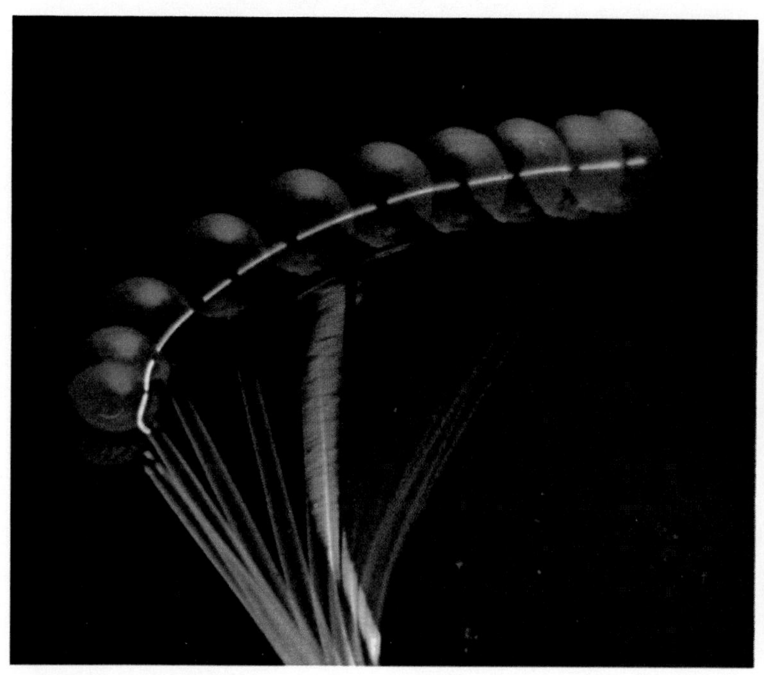

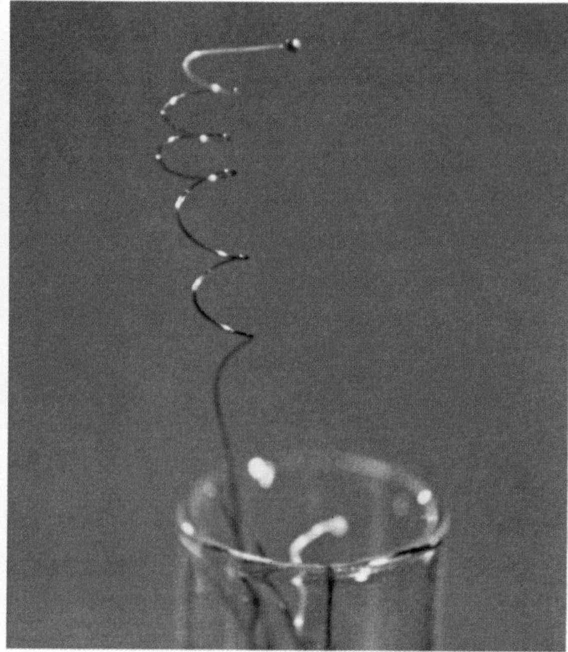

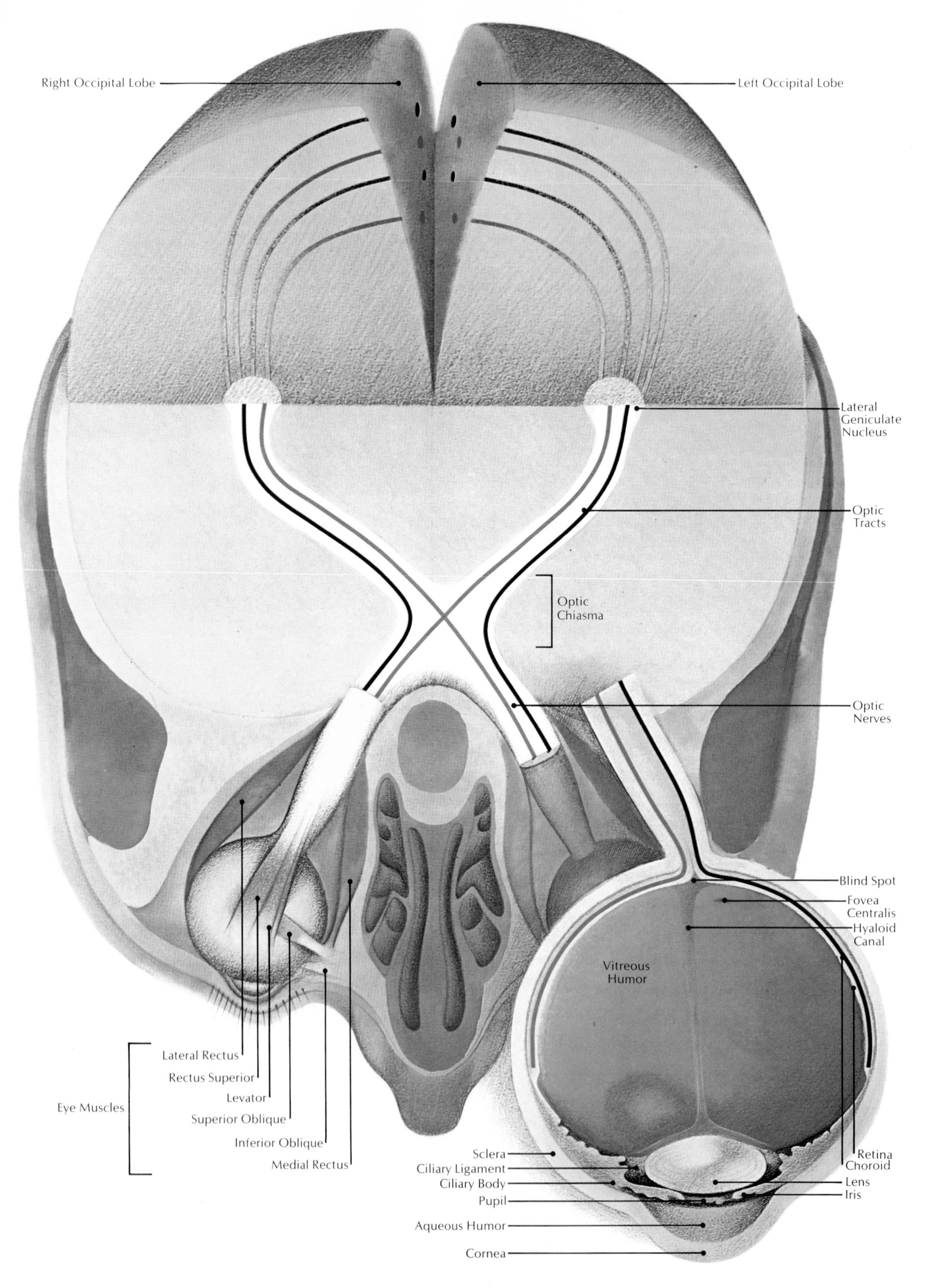
Right Occipital Lobe
Left Occipital Lobe
Lateral Geniculate Nucleus
Optic Tracts
Optic Chiasma
Optic Nerves
Blind Spot
Fovea Centralis
Hyaloid Canal
Vitreous Humor
Retina
Choroid
Lens
Iris
Sclera
Ciliary Ligament
Ciliary Body
Pupil
Aqueous Humor
Cornea
Eye Muscles
Lateral Rectus
Rectus Superior
Levator
Superior Oblique
Inferior Oblique
Medial Rectus

Figure 26.3 (opposite). The eye and its accessory structures as seen *in situ* and in detailed anatomical section.

Figure 26.4 (below). The nearsighted and farsighted eye. Accommodation to near and distant objects in vision. Note the deformation of the lens.

changes in time. The sensations or perceptions corresponding to intensity are those of brightness, and those corresponding to wavelength are colors. Spatial pattern and changes in the spatial pattern over time have more complex counterparts in sensations and perceptions.

THE EYE: ACCESSORY STRUCTURES

The eye is an organ specialized for the collection and detection of light waves and for the initiation of neural activity that ultimately leads to visual sensation. Light reaching the eye is focused on the retina, where the visual receptors are located. The light passes through the cornea, aqueous humor, pupil, lens, and vitreous humor before reaching the retina (Figure 26.3). The lens, by *accommodation*, or by alterations of its shape, plays the major role in focusing the light on the retina. The lens thickens in order to focus near objects and takes a flatter shape to focus far objects (Figure 26.4).

The pupil regulates the amount of light reaching the retina. Vision is sharpest when the diameter of the pupil is smallest because light then passes through only the center of the lens and cornea, where distortions of the image are less than they are toward the periphery. Because the brain does not seem to take into account the pupil size in evaluating brightness, the pupil tends to filter out information about absolute intensity of stimuli.

The human eye is highly mobile. Six muscles control its movement so that it can "pan" or survey over a wide area. Even when the eyes are held aimed in a particular direction, they vibrate constantly, just as the finger does when an attempt is made to hold it motionless at arm's length. This eye tremor was once thought to be a flaw in the control system of the eye, but experiments have established its value. When special optical equipment is used, it is possible to compensate for the tremor and to keep an image focused on one section of the retina. Under this condition, however, the visual sensation disappears within a few seconds as the receptor cells adapt to the unchanging pattern of light on the retina. Thus, tremor is essential to the ability to look at a particular object for any length of time. The movements of the eye are closely controlled by feedback from the visual input, so that the interesting part of the image is kept focused on the most sensitive region of the retina. Clearly, the direction of gaze is very important in determining what area of space is to be examined by the visual system. In fact, man sees well in only a tiny part of the visual field—about 2° of the 360° surrounding his body—which corresponds to a specialized area of receptors in the retina, the *fovea*.

The Eye: Visual Receptors

On reaching the retina, the light passes through layers of nerve fibers and nerve cells before reaching the *rods* and *cones*, the highly specialized receptor cells that lie near the back of the retina (Figure 26.6). These cells contain pigments that, upon the absorption of light, undergo chemical changes resulting in alterations of the electric potential across the cell membrane. The human eye contains 6 or 7 million cones, which are concentrated largely in the fovea, where vision is the sharpest. The rest of the retina, the *periphery*, contains between 75 and 150 million rods, with the highest concentration of rods about 20° from the fovea. (Distance on the retina is measured in degrees because the surface is essentially spherical.)

Rods and cones of vertebrates are long, slender cells with a highly compartmentalized structure. The end of the cell nearest the lens of the eye is

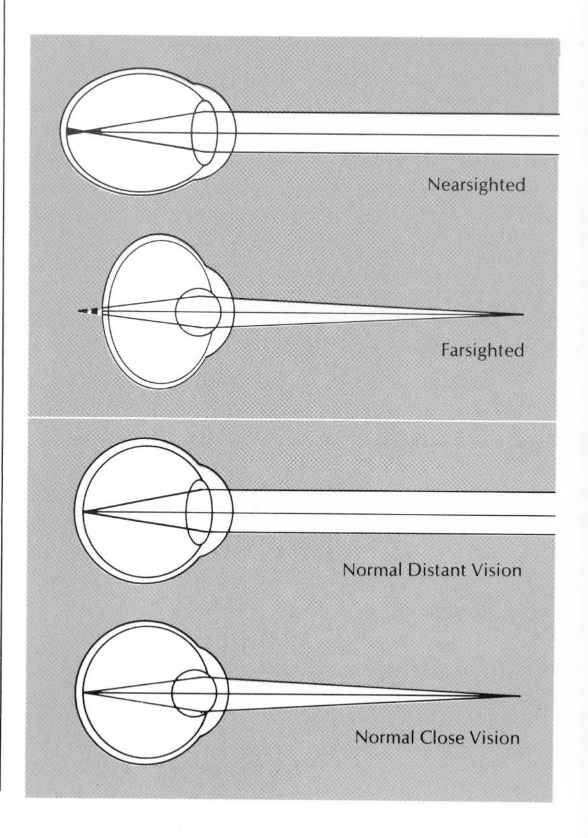

Figure 26.5. Anatomical survey of the four major senses. (A) Structures related to tactile or touch sensation. (B) Photomicrograph of a Pacinian corpuscle (receptor). (C) The organs of taste. (D) The olfactoral apparatus. (E) The auditory organs and accessory structures related to sensations of motion and orientation.

Interleaf 26.2

FOUR SENSES

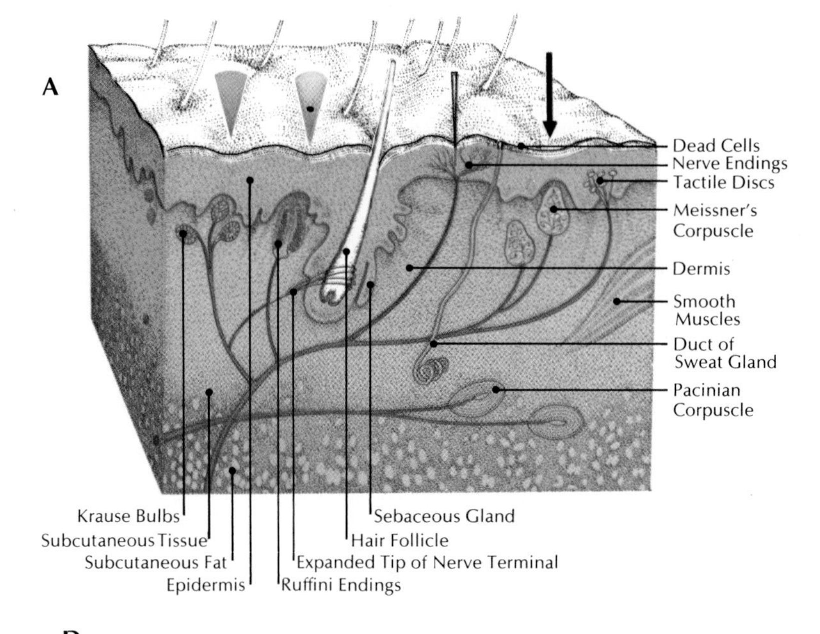

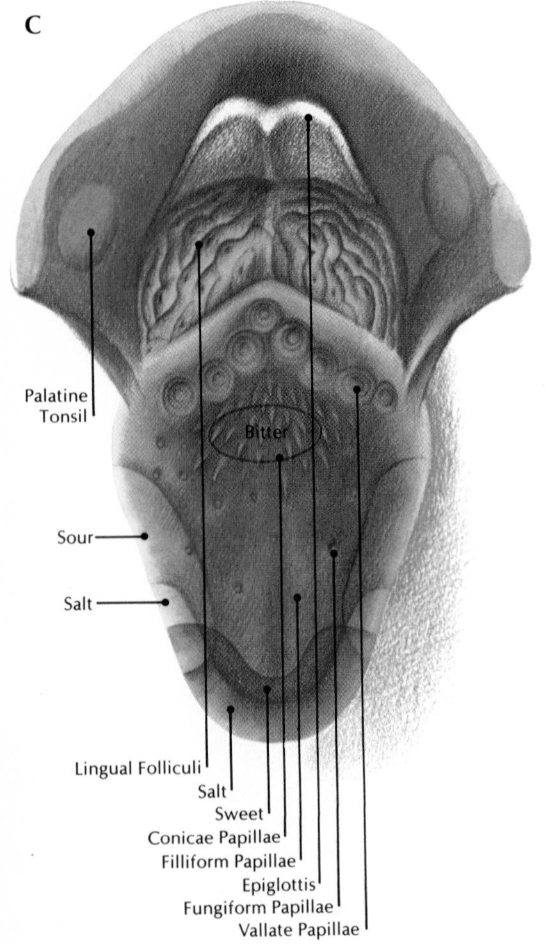

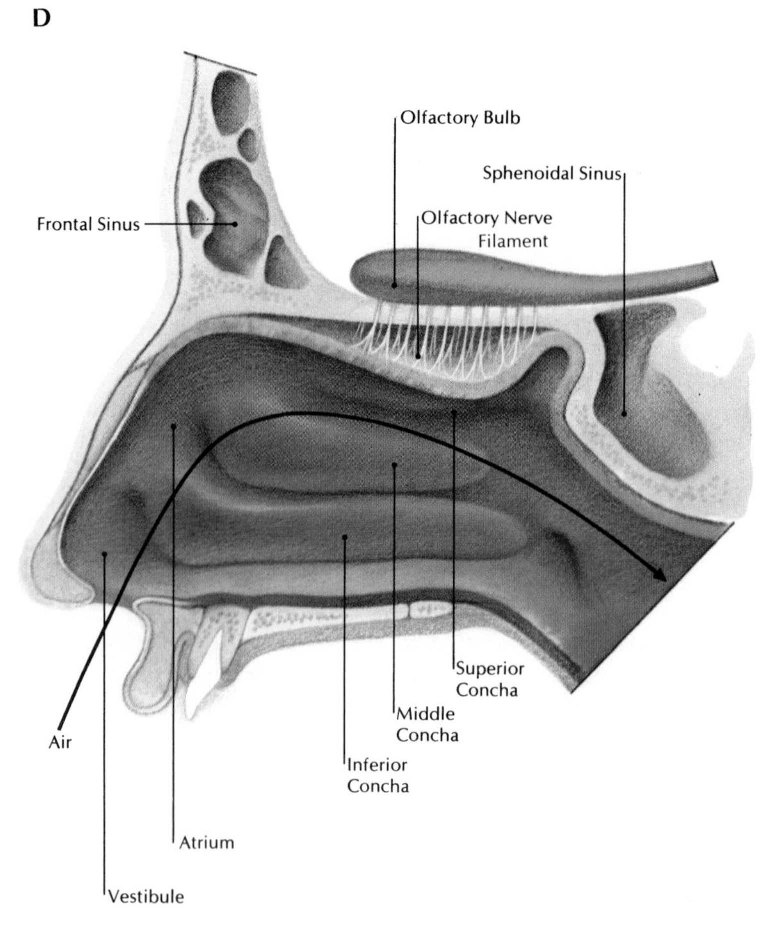

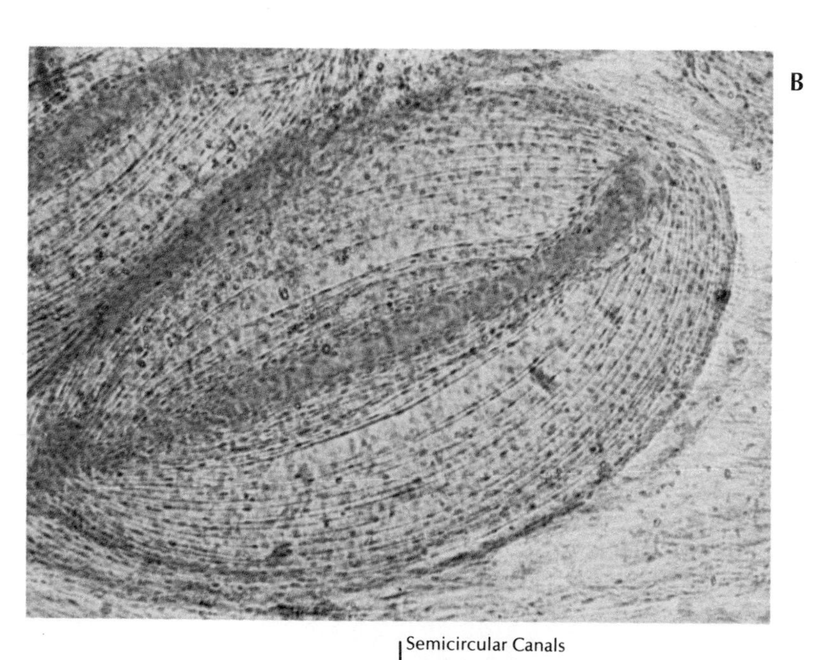
B

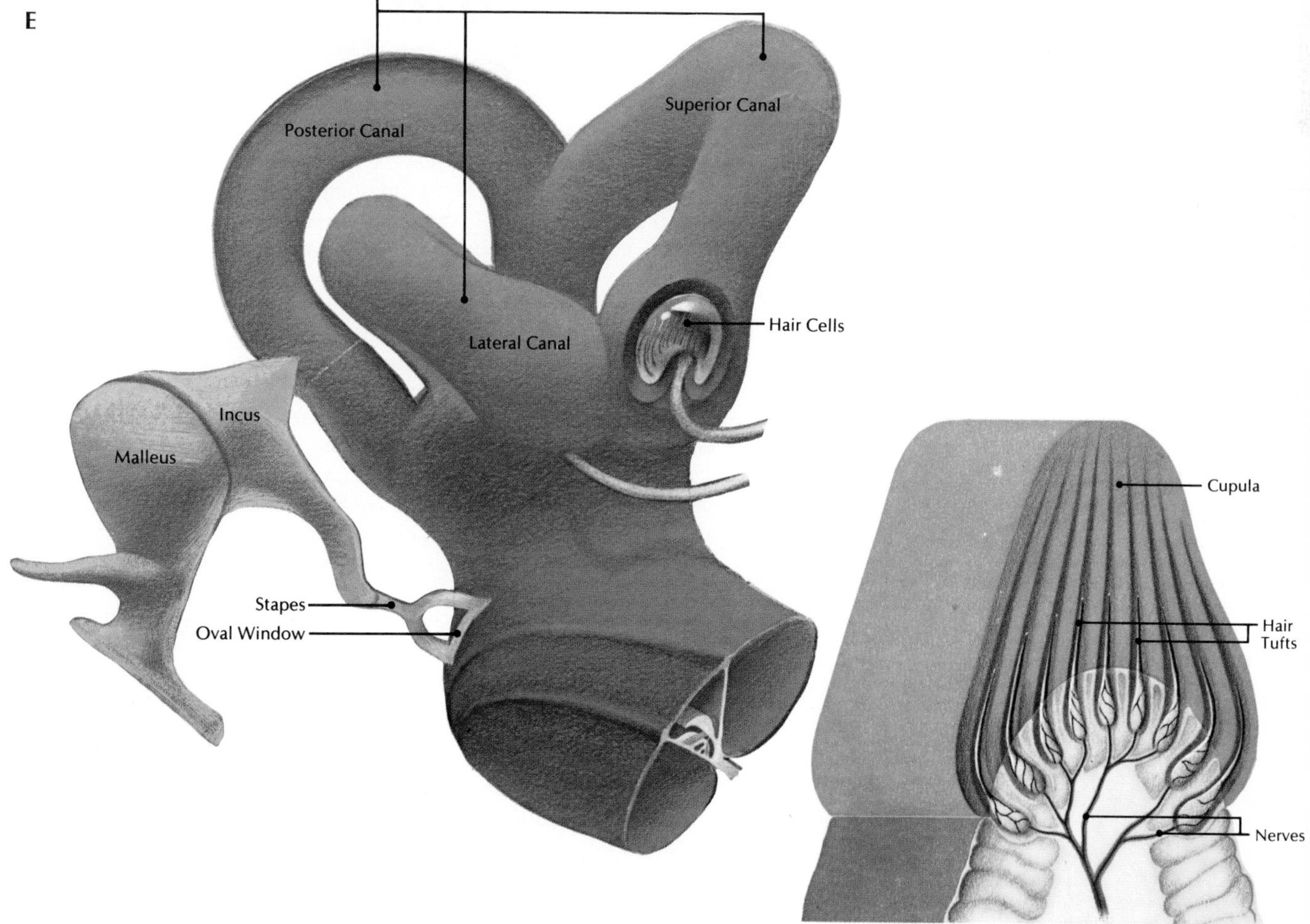
E
Semicircular Canals
Posterior Canal
Superior Canal
Lateral Canal
Hair Cells
Incus
Malleus
Stapes
Oval Window
Cupula
Hair Tufts
Nerves

Figure 26.6. Schematic diagram of the retina in cross section. Note that the receptor cells are not the first structures exposed to incoming light. Note also the decreasing specificity of innervation of the receptor cells away from the macular zone around the fovea.

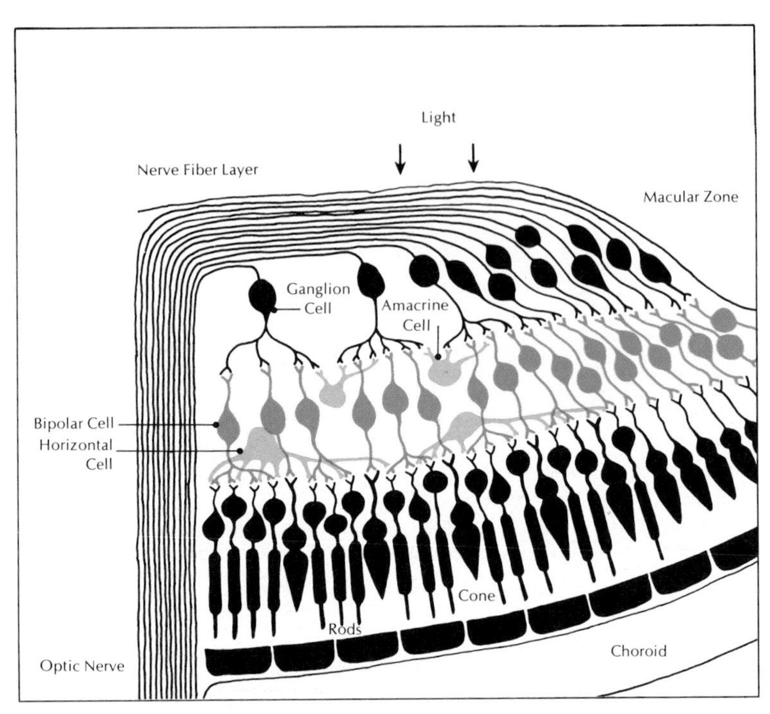

similar to a nerve axon, with a synapse at the end of the process and the cell nucleus near the other end of the axonlike process. In the central body of the cell is a zone containing ribosomes and a Golgi complex. The remainder of the main cell body is densely packed with mitochondria. Connected to the end of the cell by a very thin stalk (a modified cilium) is the *outer segment*, which contains a stack of hundreds of thin, membranal discs. These discs contain the visual pigments that absorb light.

Human visual receptors contain four different pigments, each of which absorbs light maximally at a slightly different wavelength. The rods contain a single pigment, *rhodopsin*, which absorbs maximally at a wavelength of 500 millimicrons. The rods function mainly in dim light, are extraordinarily sensitive, and yield no color sensations. The cone cells function only in relatively bright light and are involved in color vision and in the perception of complex and detailed forms. There are three different kinds of cone receptors, each containing a different pigment. One pigment absorbs maximally in the blue range of the spectrum (450 millimicrons), one in the green range (535 millimicrons), and one in the yellow-red range (575 millimicrons). The rod system and the cone system operate relatively independently. At low light levels—for example, at night—only the rod system functions; in bright light, only the cone system ordinarily functions. However, the two systems appear to operate in a similar fashion.

Other animal species have visual pigments that absorb maximally at quite different wavelengths, and many of these species respond to types of electromagnetic radiation that are imperceptible to humans. However, the visual receptors of all vertebrates and most higher animals show similar structures, with the visual pigments arrayed in membrane layers.

The pigments in rods and cones set severe limits upon the information reported to the brain by the visual system. They limit the region of the electromagnetic spectrum that can be considered by the animal, and they also affect the animal's ability to distinguish colors. This is information filtering in a very strong sense of the word.

TRANSDUCTION

The first step in the excitation of a sensory system is the *transduction* process in which the energy of a specific stimulus is transformed into some other form of energy, most likely by an electrochemical change in the properties of a membrane.

Transduction is usually accomplished by receptor cells, which are highly specialized epithelial cells or neurons. The membrane potential of the receptor cell is altered by effects of stimuli. In most cases, no action potential is produced in the receptor cell itself, but the change in membrane potential varies directly with the intensity of the stimulus, and this change diminishes as it propagates along the membrane of the receptor cell. The graded and decremental alteration in the receptor membrane potential may activate synapses, which in turn excite higher-order neurons of the sensory system. In a few cases, the receptor cell itself produces an action potential in response to the change in membrane potential. This impulse then travels along the axon of the receptor cell and activates synapses in the sensory system. Transduction comprises the events leading from the arrival of stimulus energy to the first detectable change of membrane potential. Recent research indicates that transduction occurs in membranes of the receptor cells.

In the case of vision, the membranes are important because of their relation to the visual pigments. Several investigators have studied the structure of membranes within rod cells by using techniques of x-ray defraction. Their results suggest that pigment molecules are embedded within the surface structure of the membranes that make up the disks.

All known visual pigments are chemically similar. Both the normal visual process and the bleaching process (loss of color upon exposure to light) have been explained in terms of the chemical structure of the pigment molecule (Wald, 1955). The complex molecule of visual pigment is made up of a small molecule, called a *chromatophore*, bound to a larger protein molecule, called an *opsin*. The chromatophore of all mammalian pigments—a substance that is called *retinal*, or retinene—is an aldehyde of vitamin A. The various visual pigments differ only in the structure of the opsin molecule.

The retinal molecule, because of its side chain of alternating single and double bonds, can exist in a number of different isomeric forms, each of which has a slightly different shape (Figure 26.9). Only the *11-cis* form is able to combine with the opsin to form visual pigment. The combination of *11-cis* retinal with opsin is an exothermic reaction that occurs spontaneously.

The only action of light—both in the visual process and in the bleaching process—is to cause the *11-cis* retinal attached to the opsin to be converted to the *all-trans* isomeric form (Hubbard and Kropf, 1958). This isomerization—caused by absorption of a photon—alters the shape of the retinal so that it no longer fits the protein. The retinal then splits off from the opsin; the separate retinal molecule gives the pale yellow color of the first stage in bleaching. The opsin also changes after the retinal has split away; the protein molecule uncoils slightly and reactive groups on the molecule are

Figure 26.7. Comparative anatomical study of the rod and the cone structure of a frog's retina. Light passes through the transparent portion of the cell and is absorbed by pigment molecules in the membranal discs. The visual signal is transmitted back along the cell to the synaptic body and relayed to other retinal cells where transduction of the signal into a neural impulse may be accomplished.

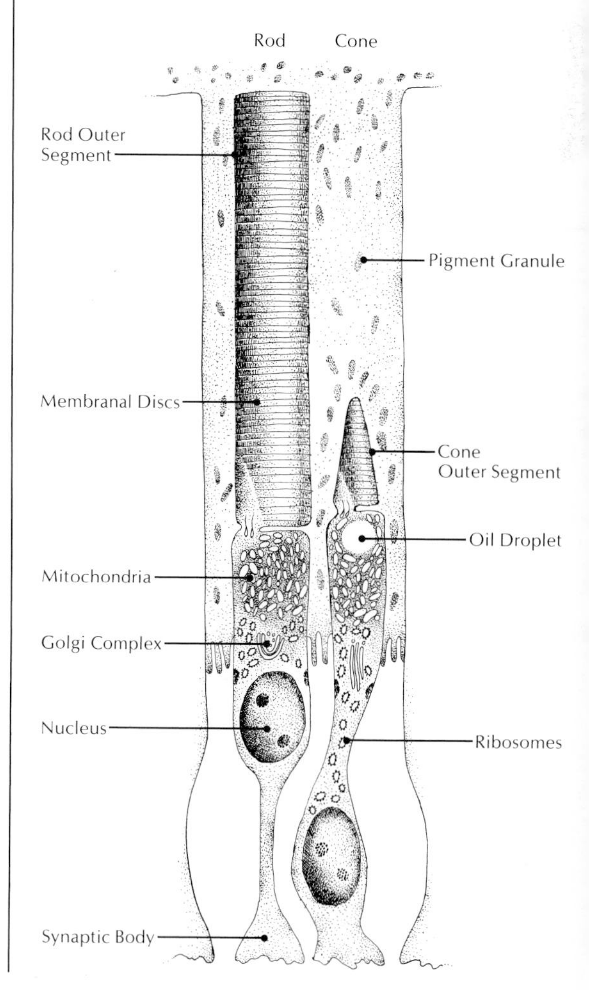

Figure 26.8a (upper left). Schematic diagram of the hair cell-neuromast organs within the lateral line-pressure receptor system (canal organ) of a fish.

Figure 26.8b (lower left). Illustration of monaural acoustical reception. Sound intensity is depicted by color variation with highest intensity corresponding to color intensity.

Figure 26.8c. The auditory organs.

Interleaf 26.3

THE AUDITORY SYSTEM: STIMULUS FILTERING BY THE EAR

The accessory structures of the auditory system—the outer, middle, and inner ear—limit the range of information that comes to the auditory system. The outer ear is directionally selective, both because of its shape and because the head blocks high-frequency sound. The movements of the pinnae by cats and rabbits show that behavioral adaptations have evolved to make use of these direction-filtering properties. In the middle ear, muscles are present that, if contracted, decrease the sensitivity of hearing by decreasing the amplitude of vibration that occurs at the stapes. In many mammals, there is a reflex contraction of these muscles to very loud sounds. This reflex protects the sensory structures against overload. The size and physical properties of the cochlea and basilar membrane limit the frequencies of sound to which an animal is sensitive. Humans hear from about 50 to 18,000 hertz (Hz), whereas cats hear as high as 50,000 Hz, and bats and porpoises hear sounds higher than 100,000 Hz.

The receptor cells of the auditory system are *hair cells*. The bending of the cilia somehow leads to the release of transmitter substance, which triggers nerve impulses in auditory nerve fibers. The hair cells are arranged along the length of the basilar membrane and the cochlea. The physical properties of the basilar membrane—its width, thickness, stiffness, and elasticity—cause it to respond selectively at any given spot to some range of frequencies of sound. Near the stapes, high frequencies cause

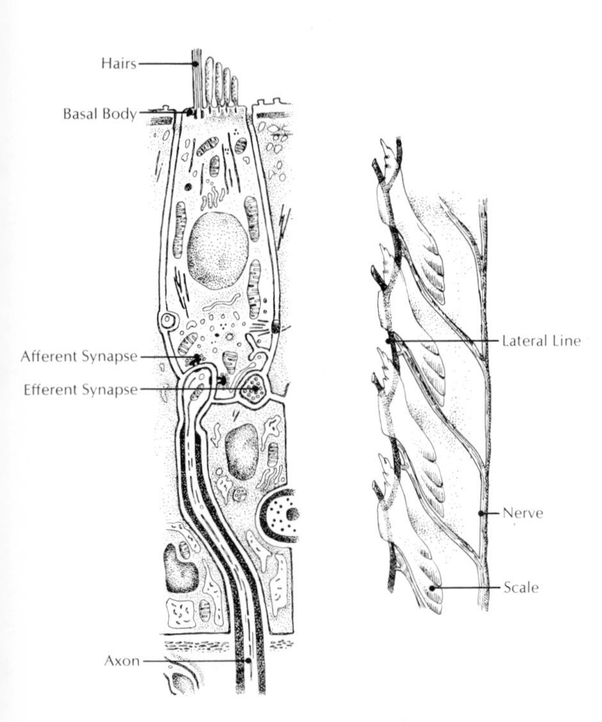

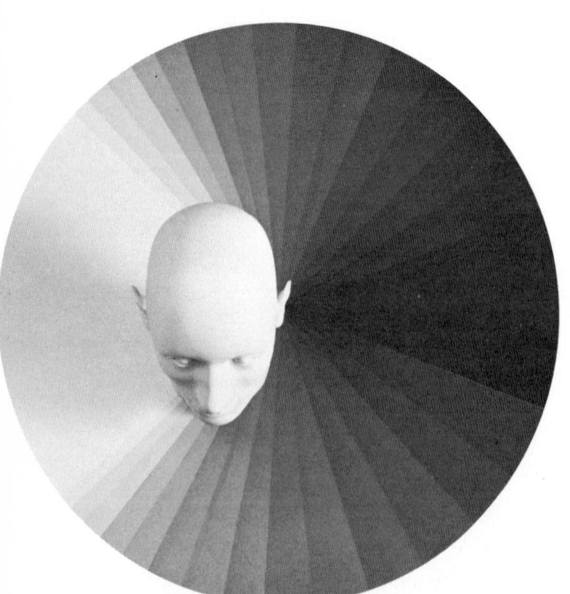

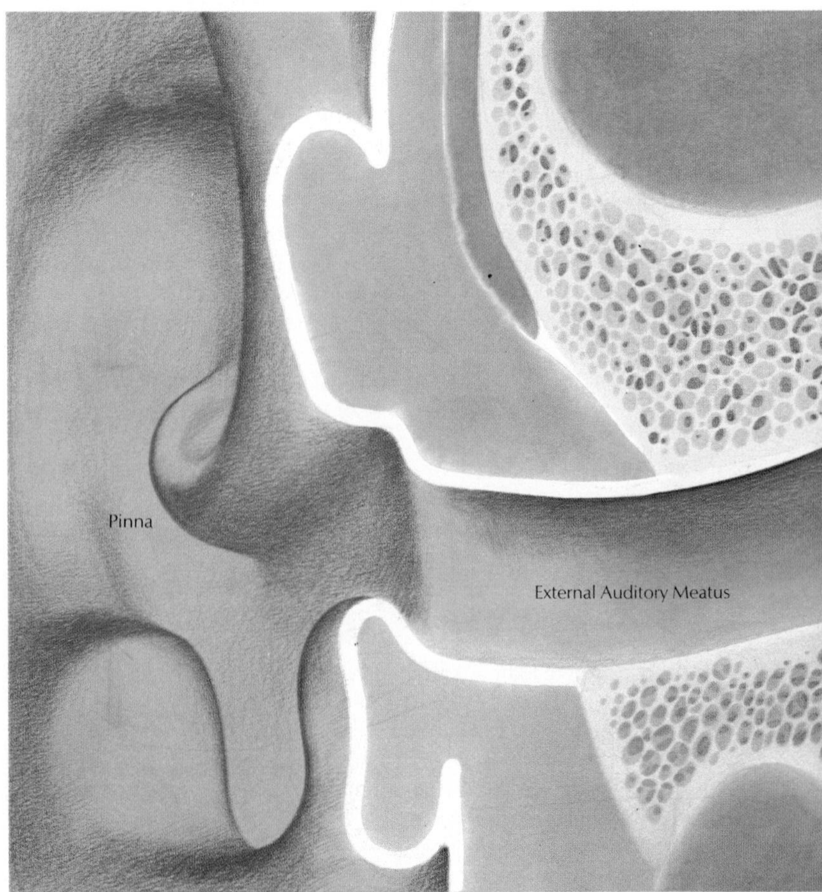

Figure 26.8d (above). Cross section of the cochlear duct showing the tectorial membrane.

Figure 26.8e (middle). Diagram of the cochlea shown coiled and as a straight tapering cylinder.

Figure 26.8f (below). Graphs illustrating the motion of the cochlear basilar membrane in response to high-frequency and low-frequency stimuli.

the basilar membrane to vibrate best, whereas near the helicotrema, low frequencies are most effective. As sounds of various frequencies hit the eardrum and then the oval window (via the stapes), they become fluid waves, which travel around the cochlear spiral and thus send waves of vibration down the basilar membrane. Sounds of any one frequency cause maximum displacements of the membrane at the place determined by width, thickness, stiffness, and elasticity. Auditory nerve fibers attached to the hair cells at that point are stimulated, and they carry the information about the occurrence of the sound to the brain.

The mechanism by which nerve impulses are generated is not certain. The physical arrangement of the *organ of Corti*, in which the hair cells are located, is such that a bending of the basilar membrane causes a shearing force between the hair cells (embedded in the *reticular lamina*) and the *tectorial membrane*. This shearing force bends the cilia, leading to the release of transmitter by the hair cells at their synapses with the auditory nerve fibers. One theory suggests that the cilia, when they are bent, cause a change in the resistance of the hair-cell membrane. The hair cells, normally negatively charged with respect to the endolymph, are then depolarized or hyperpolarized, depending on the direction of the ciliary bending. The depolarization or hyperpolarization then leads to release of transmitter. This theory likens the action of the organ of Corti and the hair cells to a resistance microphone.

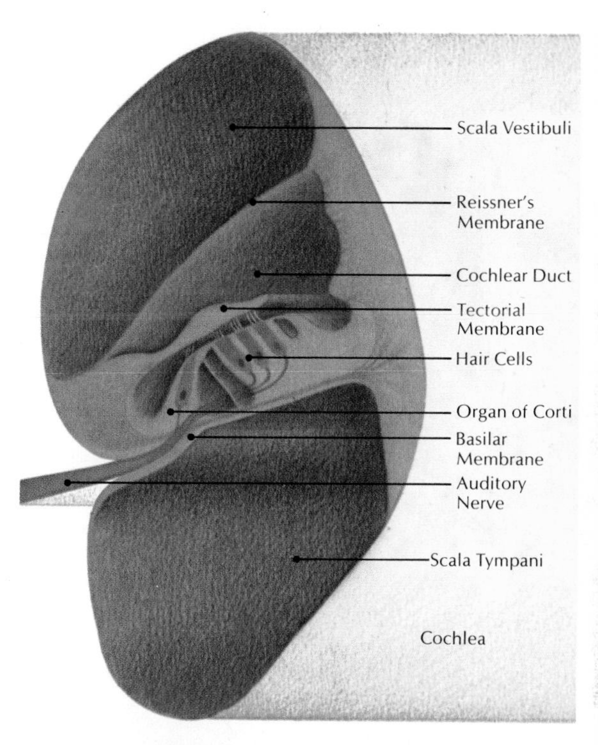

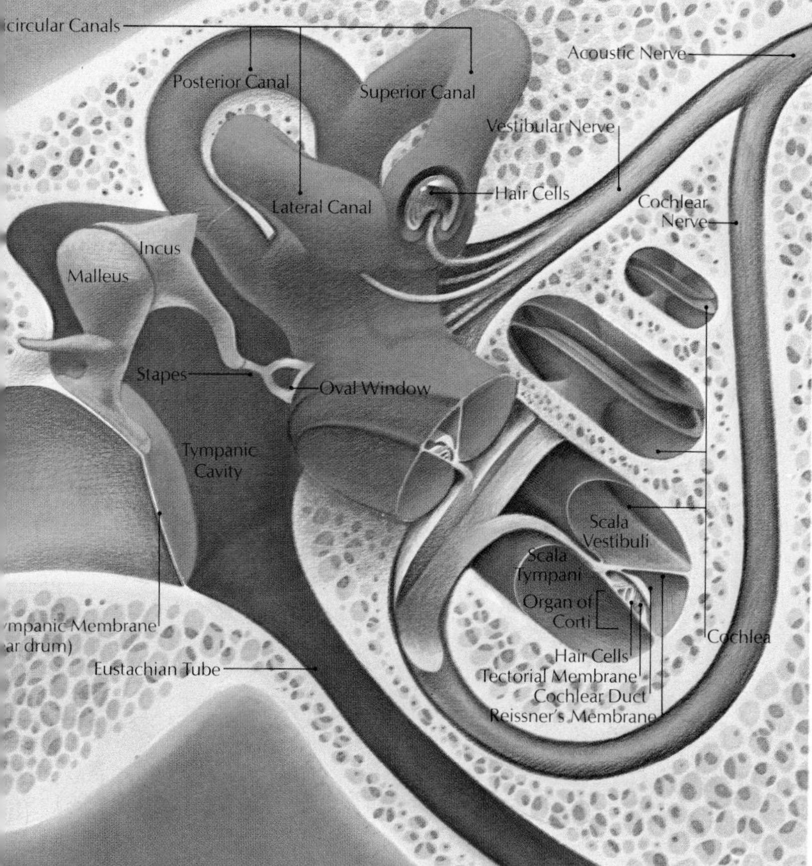

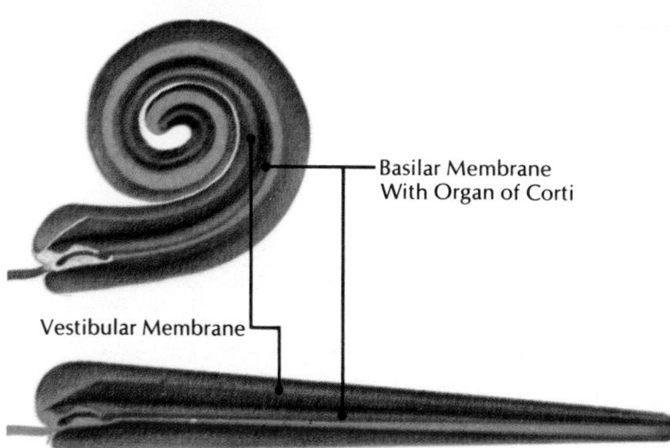

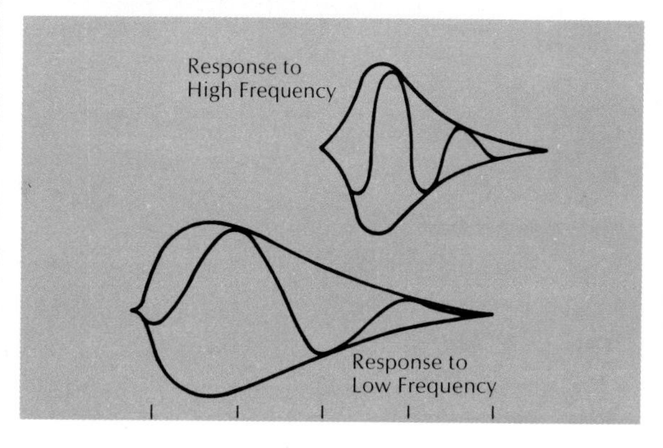

Figure 26.9 (insert). *Cis-trans* isomers of retinal molecule.

Figure 26.10. Recombination in retinal molecules. Light affects the *11-cis* retinal molecule by converting it to *all-trans* isomeric form, which no longer fits the protein opsin. Recombination is accomplished by an enzymatic reaction that converts *all-trans* retinal to the *11-cis* form, which subsequently spontaneously recombines with opsin.

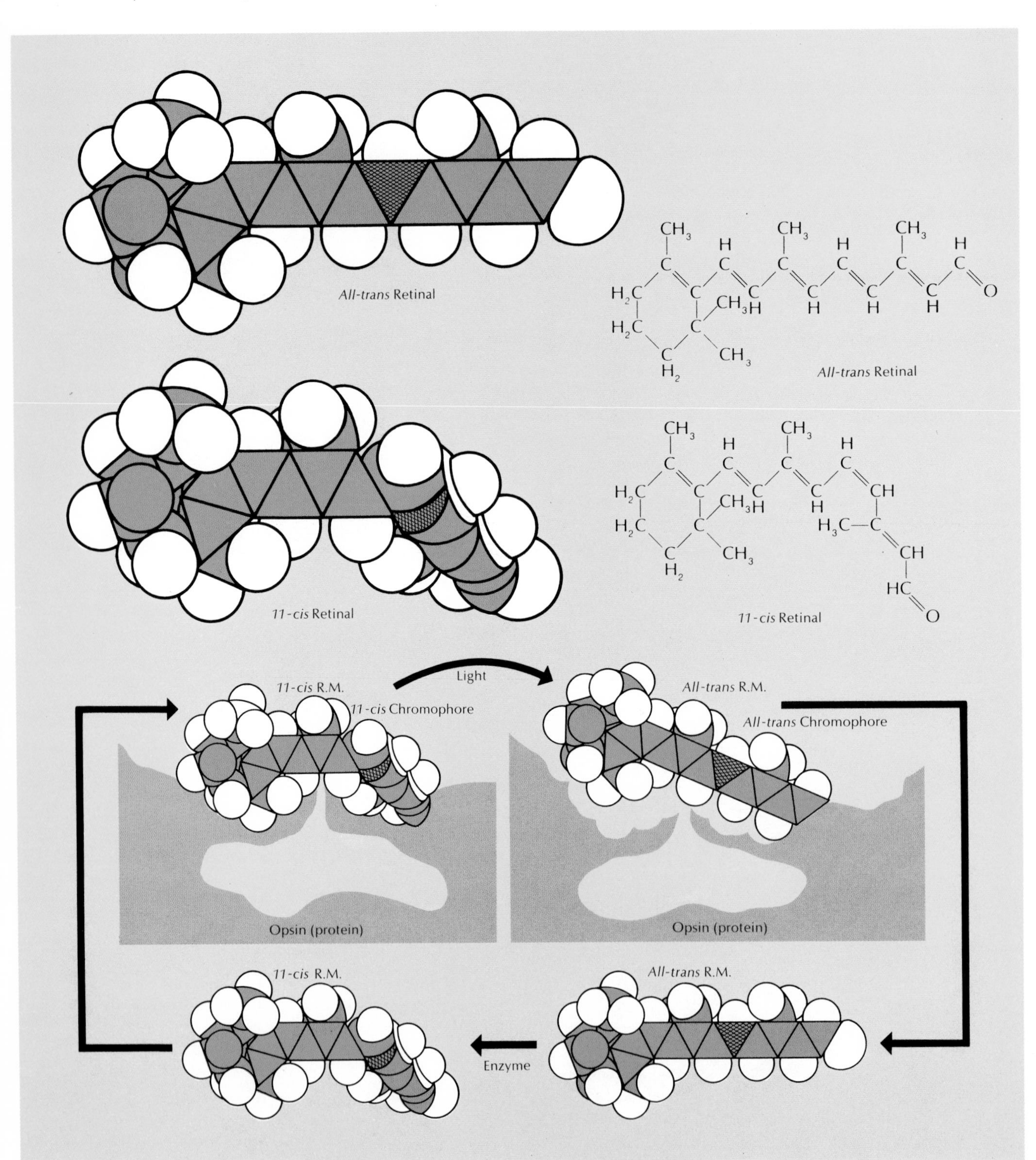

Figure 26.11. Receptor membrane potential changes compared between invertebrate and vertebrate organisms (see text for explanation).

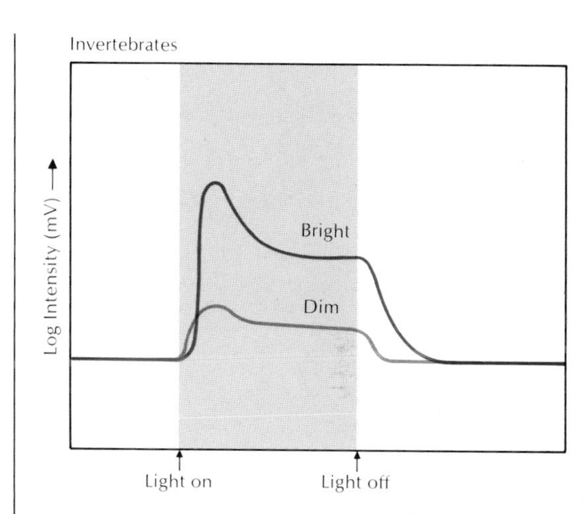

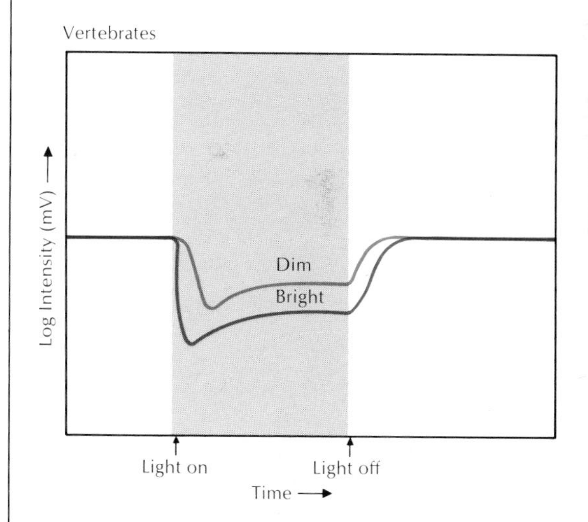

exposed. These changes are believed to be linked to the excitation of the receptor cell. A further chemical reaction may convert the retinal to colorless vitamin A, completing the bleaching process.

In the living receptor, the visual pigment is regenerated by the recombination of retinal and opsin in an enzymatic reaction that converts *all-trans* retinal to the *11-cis* form, which will spontaneously recombine with the opsin (Hubbard and Wald, 1952–1953).

Visual receptor cells, like other cells, possess a resting potential difference across the cell membrane, with the inside of the cell more negatively charged than the outside. Illumination of the receptor cell causes a rapid change in this potential, which is called the generator potential, or receptor potential. The change lasts as long as the stimulus is applied, and it is a graded response—that is, the magnitude of the potential change varies with the strength of the stimulus (Figure 26.11). In most invertebrate receptors, the light-induced change makes the interior of the cell more positive, or depolarizes the membrane. In vertebrate visual receptors, the inside of the cell becomes more negative and the membrane is hyperpolarized (Tomita, 1968). In both cases, the change in potential is apparently caused by alterations of the cell membrane's permeability to ions.

In an invertebrate photoreceptor, light induces a change in the receptor membrane's permeability to positive ions (primarily Na^+), allowing them to flow into the cell (Hagins, *et al.*, 1963). The change in ionic permeability is localized to the illuminated part of the pigmented portion of the receptor. In the vertebrate photoreceptor, however, recent experiments indicate that sodium ions continuously diffuse into the vertebrate photoreceptor under dark conditions and that the effect of light absorption is to decrease the permeability of the membrane to Na^+ (Penn and Hagins, 1969). The process by which splitting of the chromatophore-opsin structure leads to membrane permeability changes is not yet understood, but this is an active area of current research.

In some invertebrate receptors, the potential change induced by light absorption may act to trigger an action potential that moves along the axon of the receptor cell to synapses at its terminal. However, in the photoreceptors of most invertebrates and all vertebrates, an action potential is not generated in the receptor cell. The prolonged potential change induced by light absorption travels along the receptor cell, decreasing in amplitude as it moves (a decremental response). This potential change causes release of transmitter substance at the synaptic junctions of the receptor, which results in excitation of the postsynaptic neurons.

Some receptors generate action potentials; others generate decremental responses that affect synapses. In either case, a stimulus just large enough to produce an action potential is called a *threshold stimulus*. Presumably, the events that occur are similar to what happens when a squid giant axon is electrically depolarized (or when a postsynaptic neuron is activated by a presynaptic neuron). In many receptor systems, a sustained stimulus leads to a sustained receptor potential, which may produce a continuing sequence, or *train*, of spikes (action potentials).

One of the most remarkable properties of animal receptors is their great sensitivity to the appropriate stimulus—in other words, their low threshold. A human rod cell may be excited after the absorption of a single photon. Each receptor cell contains about 30 million molecules of visual pigment, but under the best conditions the absorption of a photon by only one of

Figure 26.12. Rate of impulse generation showing sensory adaptation in a selected group of receptor cells with stimulus applied and maintained at a constant level.

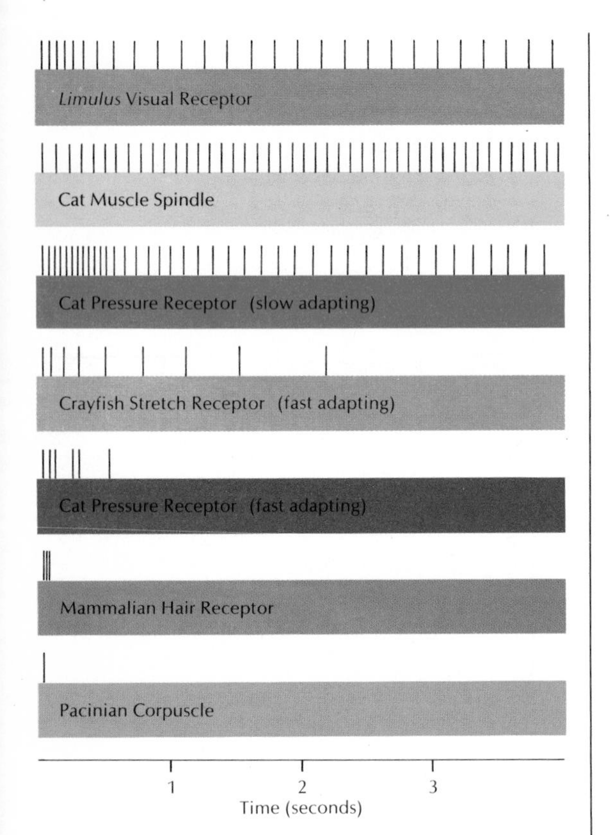

these molecules can trigger the electric response of the receptor cell. But the absorption of a single photon does not lead to a sensation of light. Single receptor cells are continually being excited in random patterns even in the total absence of light. This "noise" in the system is probably caused by spontaneous chemical changes in the pigment molecules. The neural circuits of the visual system are arranged in such a way that the more-or-less simultaneous excitation of five to ten rods is necessary before a person is conscious of a flash of light.

Another important property of many receptors is their capacity for *adaptation*, the adjustment of sensitivity depending on the average level of stimulation. Receptor systems function appropriately over an extraordinarily wide range of stimulus intensities. For example, the human visual system can continue to distinguish forms and shapes through a hundred-million-fold increase in light intensity, and the eye adjusts its sensitivity for each change in the background light. People normally move in environments that vary in light intensity through several orders of magnitude; and the visual system usually requires no more than a few seconds or minutes to adjust to new conditions. The mechanisms of this type of adaptation in sensory systems are not yet well understood.

If a stimulus is applied to a receptor cell, the rate of impulse generation in the receptor or in neurons synapsing with it may decline although the stimulus level remains constant (Figure 26.12). This decline is a form of adaptation and suggests that there is some change in the sensitivity of the receptor, a change that affects a step in the sensory process prior to that of impulse generation.

In the case of vision, several factors may contribute to the overall phenomenon of adaptation. Receptor cells are partially shaded by adjacent pigment-containing cells. Changes in the shape or pigment distribution of these cells have been observed during adaptation. It is not known how the transduction process itself is altered during adaptation. It once was thought that bright light bleaches most of the visual pigments and that this phenomenon was the explanation of visual adaptation. However, it has been shown that much adaptation can occur when only a small percentage of the available pigment molecules are in the bleached state.

In the eye of the horseshoe crab, *Limulus*, adaptation has been studied in the neurons that carry impulses from the eye to the brain. Onset of a stimulus produces a rapid burst of impulses, but the firing rate soon declines although the stimulus remains constant. The nature of the decline in firing rate suggests that impulse generation somehow inhibits further firing of the neuron. If some branches of the axon led back to inhibitory synapses on the membrane of the same neuron, firing of the neuron would produce IPSPs, which would inhibit immediate generation of additional impulses. This model of *self-inhibition* explains the high initial rate of impulse generation (transient "on" response) and the rapid decline to a generally steady level (sustained response) in these neurons. Measurements show that just after an impulse is generated there is a hyperpolarization of the membrane with characteristics identical to those produced by IPSPs in postsynaptic neurons (Chapter 24). However, the feedback branch of the axon has not been observed, and the inhibitory effect may be produced directly within the neuron membrane after firing, rather than by an inhibitory synapse.

Self-inhibition is a form of adaptation. The rate of firing is slowed down so that further increases in strength of light stimuli can be reported by again

Figure 26.13. Photomicrograph of a section of a human retina taken about 1.25 mm from the fovea. Note the top layer consisting of rods, followed by a layer consisting predominantly of bipolar cells with some horizontal cells. The bottom macular zone consists primarily of ganglion and amacrine cells.

increasing the rate temporarily. Self-inhibition occurs much more rapidly than the adaptation of human vision to sudden bright light, but both phenomena have drastic filtering effects on the information passed to the CNS.

NEURAL PROCESSING OF SENSORY INPUT

The excitation of a receptor is only the first step in a sequence of neural activities that may or may not lead to a sensation. Through the complex network of excitatory and inhibitory synapses in the afferent nervous system and brain, the pattern of impulses triggered by receptor cells is transformed into the complex pattern of cortical activity that corresponds to a sensation.

Transduction itself may be regarded as a first step in neural processing. The stimulus excites a graded potential change in the receptor or nerve ending. If this change is sufficiently large, it will trigger one or more all-or-none action potentials in the receptor cell or perhaps in the afferent nerve fibers making synaptic contact with it. Thus, the graded response of the receptor is converted eventually into a pattern of identical nerve impulses.

For many years, neurophysiologists have attempted to discover how the pattern of impulses signals or codes information from the receptor. Two of many possible codes are frequency codes and occurrence codes. In many visual systems, particularly in invertebrates, the number of impulses generated per second increases as the stimulus intensity is increased—an example of a *frequency code*. In cases of *occurrence codes*, a particular receptor generates only a short burst of impulses at the onset of a stimulus (the brightening or dimming of a light). The number of impulses in the burst may signify a relevant stimulus parameter or it may not. Other types of coding have been suggested, especially for nonvisual receptors. Each code imposes limitations on the information available to the nervous system and thus adds to the filtering effect of the sensory system.

A variety of types of processing may occur at higher levels in the nervous system. The outputs of several receptor cells may be arranged in such a way that nearly simultaneous activation of a number of receptors is required to produce a summed EPSP large enough to trigger an action potential in the single afferent nerve serving all of these cells. Other synaptic arrangements may act to inhibit impulses from surrounding cells when a particular receptor has been activated or to trigger impulses in afferent nerves only when certain patterns of receptors are activated. These synaptic arrangements are discussed in the context of the "enchanted loom" in Chapter 24.

In most sensory systems, much of this detailed neural processing occurs deep within the brain and thus has been difficult to study. Moreover, the picture is enormously complicated by lateral connections among the receptors and by afferent nerve fibers leading from the brain to the receptors. However, in the vertebrate visual system, two early stages of neural processing occur within the eye itself. The vertebrate retina is a portion of the visual processing system of the brain that is highly accessible for study. The retina consists of five types of nerve cells (including the receptors), and it has two synaptic layers in which the visual information is initially processed. By the time visual impulses leave the vertebrate eye to move toward the brain, they have already undergone two transformations.

Processing in the Retina

In each of the two synaptic layers of the vertebrate retina, the processes of three kinds of cells interact (Figure 26.13). In the outer synaptic layer, for

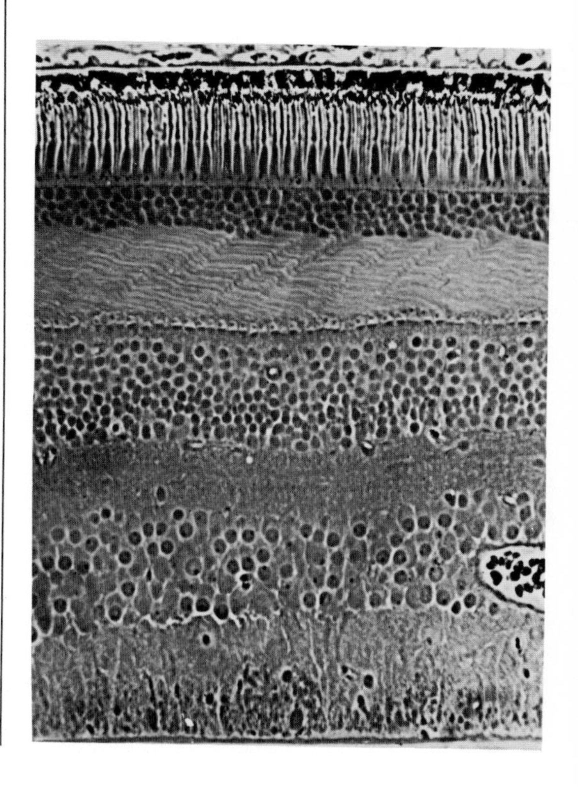

Figure 26.14a (left). The surface of the *Limulus* eye.

Figure 26.14b (right). Photomicrograph of the compound eye of a *Limulus* in cross section. The cornea has been removed; the dark bodies are the sensory parts of the ommatidia.

example, the terminals of the receptors provide input, the dendrites of bipolar cells carry the output, and the horizontal-cell processes provide alternate pathways between receptor terminals and bipolar-cell dendrites. The anatomy of this layer suggests that the receptors excite both the bipolar-cell dendrites and the horizontal-cell processes at specialized contact points. The horizontal cells, in turn, excite bipolar-cell dendrites. Because horizontal-cell processes extend much farther laterally in the retina than do bipolar-cell dendrites, the horizontal cells provide input to bipolar cells that are some distance from the excited receptor cells (Dowling, 1970).

In the second synaptic layer, the axon terminals of the bipolar cells provide the input, the amacrine cells provide lateral connecting pathways, and the ganglion cells, whose axons are the fibers of the optic nerve, carry the output. Fibers from ganglion cells all over the retina lead to the *blind spot*—a small area that contains no receptors—where they gather into the bundles that make up the optic nerve. At the *optic chiasma*, the nerves from the two eyes meet. Each group of fibers is split so that fibers from the left halves of both retinas lead to the lateral geniculate nucleus in the left part of the thalamus, whereas fibers from the right halves of the retinas lead to the right lateral geniculate nucleus (Figure 26.3). From these synaptic areas in the thalamus, nerve fibers called optic radiations carry the visual information to the occipital lobe of the cerebral cortex.

In the search for an explanation of how the parts of the visual system interact to yield clear perceptions of complex objects, scientists have made

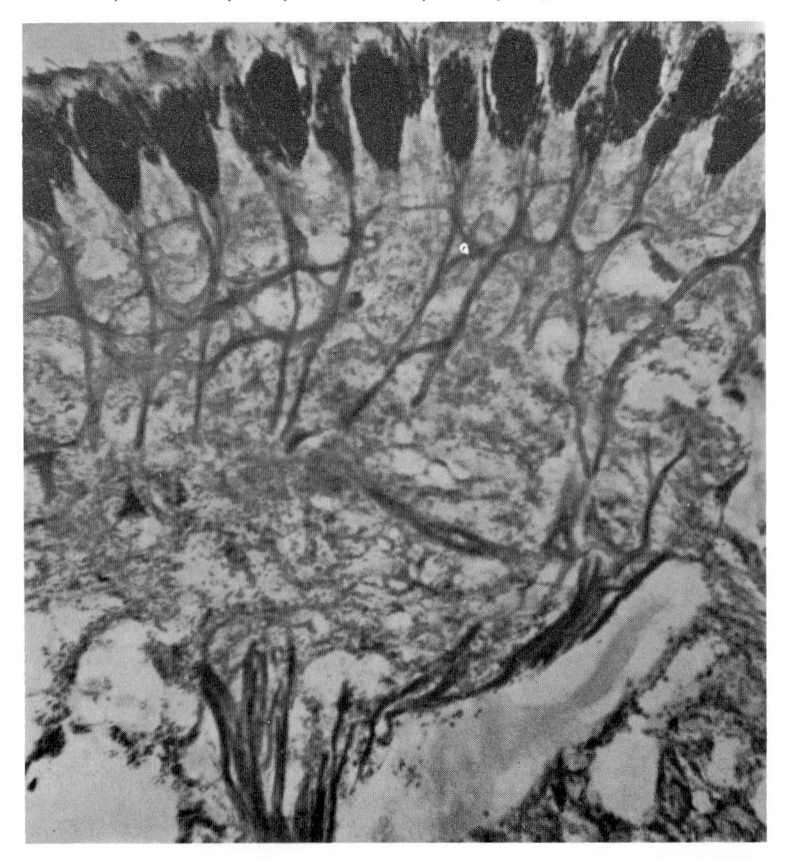

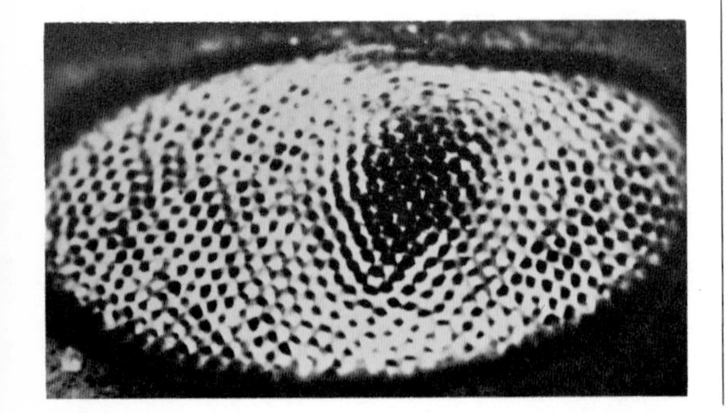

Figure 26.15. A single ommatidium of *Limulus*, greatly enlarged and schematic. The ommatidium consists of 12 pie-shape retinular cells grouped around a central fiber, which is the dendrite of a nerve (eccentric) cell.

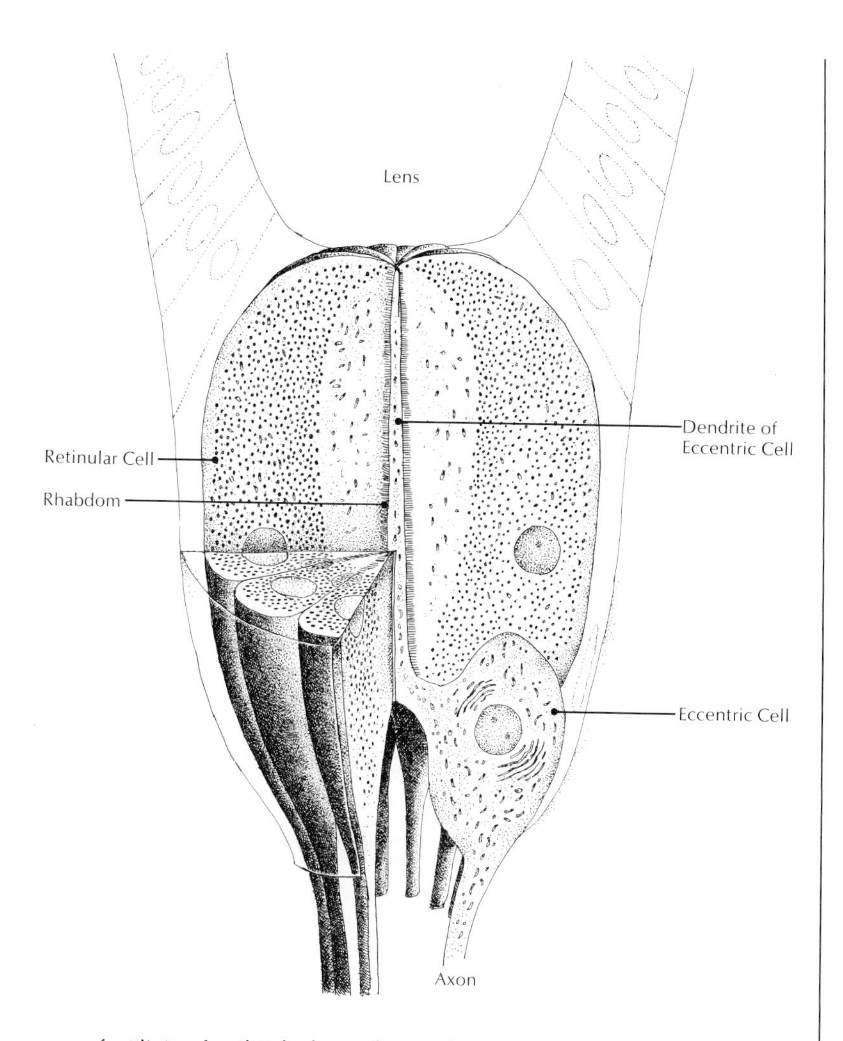

use of a "living fossil," the horseshoe crab (*Limulus*). Not really a crab at all but more closely related to the spiders, this marine animal existed hundreds of millions of years ago in much the same form it has today.

The eye of the horseshoe crab provides scientists with a simple model for the more complex visual system of man and other mammals. The compound eye of *Limulus* is composed of about 1,000 tiny, separate receptors, or *ommatidia* (Figure 26.14). Light falls on the receptors in such a way that neighboring ommatidia receive input from slightly overlapping sections of the visual field. Each ommatidium is composed of about a dozen retinular cells grouped around the dendrite of a neuron called the *eccentric cell.* Retinular cells are apparently the pigment-containing receptors, in which a graded potential is triggered by light stimuli. A sufficient summed potential from retinular cells triggers a nerve impulse in the eccentric cell. The axons of eccentric cells form the optic nerve, but a message traveling through the optic nerve is not simply a set of independent reports from individual ommatidia. Just below the ommatidia, a branching array of nerve fibers make up a plexus that connects each eccentric-cell axon with its neighbors.

When a single ommatidium is stimulated, the synapses of the plexus

Figure 26.16. Mutual inhibition. Two neighboring ommatidia (A and B), illuminated simultaneously, may inhibit the other impulses being generated through interconnecting new fibers.

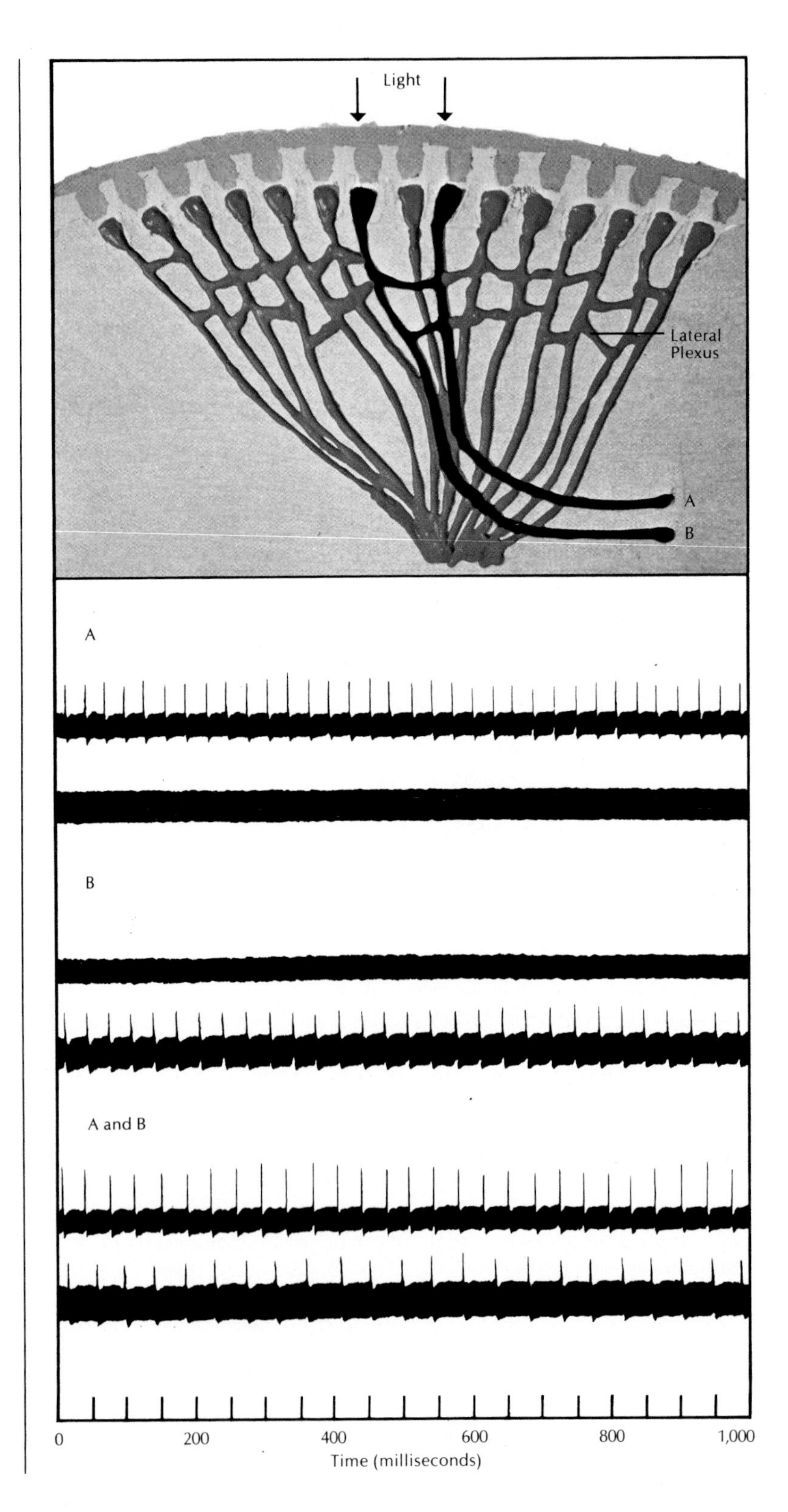

Figure 26.17. Machs' Law. The subjective psychological sensation accentuates a light contour by making a brighter region on the light edge of the border and a darker region on the dark edge of the border.

inhibit impulses in its neighbors. When two adjacent ommatidia are stimulated, each inhibits the other, and the rate of impulses in each nerve is lower than it would be if a single ommatidium were stimulated. The amount of inhibition increases with the frequency of firing of the inhibiting neuron and decreases with distance from the inhibiting neuron. Thus, lighting up the whole compound eye produces a low level of neural activity because all of the cells inhibit each other. However, at a border where light intensity changes in space, some very interesting effects occur. The lighted cells nearest the border are inhibited only by excited cells on one side because the other side is dark and contributes no inhibition. Cells farther from the border in the lighted area are inhibited more because they are surrounded on all sides by excited cells. Thus, the greatest rate of firing occurs just over the bright side of the border. On the dark side, the greatest inhibition is at the other side, so that the darkened cells near the border have depressed firing rates. The effect of the border is emphasized by depressing the firing rate of cells on the dark side and enhancing the firing rate of cells on the bright side.

A similar effect is observed in the human eye. The subjective brightness or sensation reported by a human observer accentuates a light contour by making a brighter region on the light edge of the border and a darker region on the dark edge of the border (Figure 26.17). This effect was first investigated by the physicist Ernst Mach at the end of the nineteenth century, and the apparent light and dark bands are called *Mach bands*. A mechanism of *lateral inhibition* similar to that demonstrated in the eye of *Limulus* occurs in

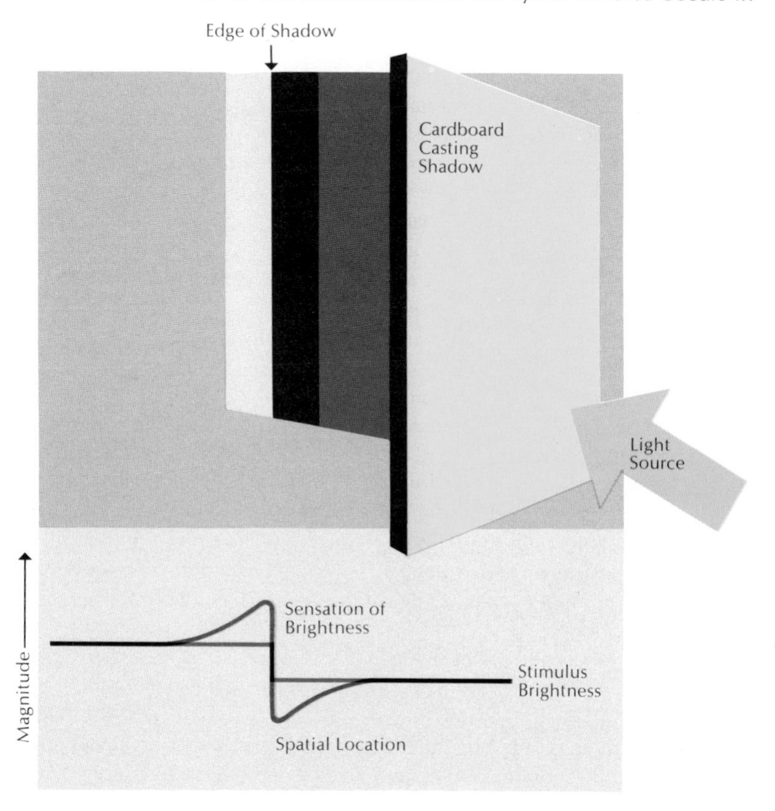

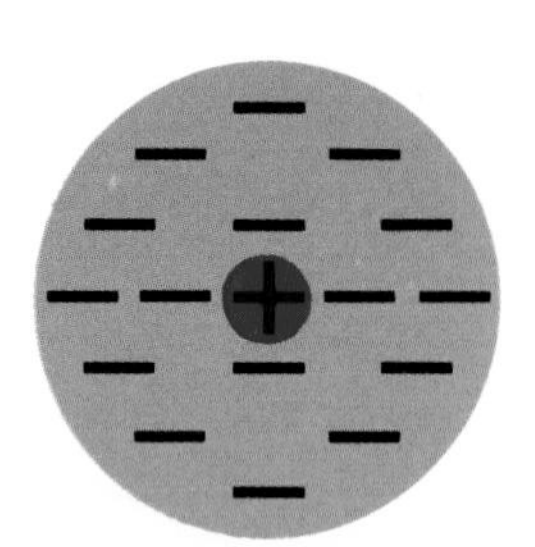

Figure 26.18. Receptive fields in the *Limulus* eye. The minus signs (−) indicate points at which light stimulation reduced cell firing. The plus signs (+) indicate points at which stimulation increased firing.

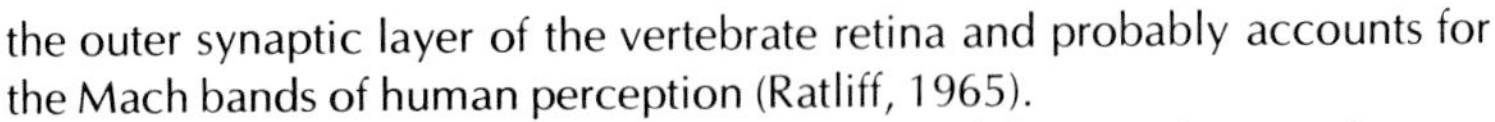

the outer synaptic layer of the vertebrate retina and probably accounts for the Mach bands of human perception (Ratliff, 1965).

Further processing in the more central parts of the visual system has not been analyzed as extensively, but much information has been accumulated about types of processing. In the ganglion-cell layer of the retina in some species, certain cells respond more readily to moving stimuli than to stationary ones (Maturana, *et al.*, 1960). In many cases, these units respond only when the stimulus is moving in a particular direction (Barlow, *et al.*, 1964).

The nature of the synaptic connections in the inner synaptic layer that produce motion or directional sensitivity is not yet worked out, but recent data suggest that the amacrine cells play an important role (Dowling, 1970). Amacrine cells respond with brief, transient potentials at both the onset and cessation of retinal illumination. Such cells select and accentuate dynamic changes in retinal illumination; their response to sustained illumination is very brief. This type of response would be well suited for the detection of movements in the visual field.

Thus, the retinas of some vertebrates process visual information in two distinct ways. The outer synaptic layer performs a spatial analysis of the retinal illumination in relation to visual pattern or form. Contrasts are enhanced through lateral inhibition mediated by the horizontal cells. In the inner synaptic layer, a *temporal* (time-related) analysis is performed. Dynamic qualities of the visual image are accentuated through the action of the amacrine cells. The information moving through the optic nerve fibers to the brain reflects these two stages of processing.

RECEPTIVE FIELDS

From the complicated interconnections of retinal cells, it is easy to see that a ganglion cell—the last in the chain—does not respond simply to events at a single receptor. Its response is influenced by the presence or absence of light stimuli at a large number of receptors. The area of the retina or the portion of visual space within which the stimulus pattern (and changes in the pattern) affect the response of a particular ganglion cell is called the *receptive field* of that cell. Similarly, a receptive field can be defined for any neuron in a sensory system that is influenced by receptors. The set of all stimuli—spatial, temporal, and qualitative dimensions included—that will affect the cell's firing rate under certain conditions is said to constitute the cell's receptive field under those conditions.

In the eye of *Limulus*, the responses of a single eccentric-cell axon are influenced by stimuli striking ommatidia in a receptive field around the particular ommatidium being studied. When the light falls on the ommatidium in which the eccentric cell lies, the cell's rate of firing increases. When the spot of light falls on neighboring ommatidia, the rate decreases. The distance over which some inhibition is effective is about the same in all directions. Thus, the receptive field for a single ommatidium is somewhat circular, with a small excitatory region (*center*) and a larger inhibitory area (*surround*) around it.

Many optic nerve fibers in the visual system of a frog have a receptive field that is oval. The size of the field increases as the intensity of the stimulus is increased. Early research suggested that some fields are excitatory, whereas others are inhibitory—that is, a light stimulus striking anywhere in the field inhibits spontaneous activity in the nerve fiber (Hartline, 1940). Later studies showed that excitatory fields have a larger inhibitory surround; intense stimuli striking in the surround inhibit activity in the fiber (Barlow, 1953). A fiber with an inhibitory receptive field carries a burst of impulses just after light striking the field is turned off. Thus, the inhibitory field can be described as an *off* field, because the fiber carries "off" discharges. The other kind of field can be described as having an *on-center* and an *off-surround.* A spot of light striking the center of the field produces a burst of impulses in the fiber; light striking the surround inhibits impulses in the fiber, but a burst of impulses is produced when the light striking the surround is turned off.

A similar pattern of activity results from interconnections in the mammalian retina but was discovered in recordings made from ganglion cells in the eye of a cat long before the "wiring" of the retina was known (Kuffler, 1953). The receptive fields of cat ganglion cells are about equally divided between those with an on-center and an off-surround and those with the reverse arrangement (Figure 26.20). The center of the field of a ganglion cell is large enough to include input from a number of individual receptor cells, and the surround contributes information from even more. Thus, the response of a ganglion cell does not contain as much information about the location of a stimulus as does that of a receptor. The ganglion cell has lost spatial information, but it has enhanced ability to signal spatial contrast or temporal change of the light that falls within its receptive field.

COLOR PROCESSING IN THE RETINA

The nature of color perception caused many physiologists to suspect that initial color processing takes place in the neurons of the retina. In the nineteenth century, the German physiologist Hermann von Helmholtz concluded from perceptual studies that the eye detects three basic colors and that all other color sensations result from combinations of the three primary colors. This three-color theory received strong corroboration with the discovery of three kinds of cone receptors, each containing one of three major visual pigments. This finding spurred physiologists to even greater efforts in the search for evidence of color information processing in the various stages of the visual system.

Early evidence of cells responsive to color information came from studies of fish retina (Svaetichin, 1956). In these studies, slow and graded potential changes were recorded from cells within the retina, but the particular cells being measured were not identified. Some cells were hyperpolarized by red light and depolarized by green light. Others showed opposing responses to blue and yellow light. These cells give "on" responses to light of certain wavelengths and "off" responses to light of other wavelengths.

Further study of fish revealed that a single retinal ganglion cell may give both an opponent-color response and a center-surround spatial response. In describing the field of such a ganglion cell, it is insufficient to specify merely an off-center and on-surround (or vice versa). Rather, it must be added that the off-center is sensitive to red light, whereas the on-surround is sensitive to blue or green light. It was this sort of finding that forced generalization of the receptive-field concept to include stimulus properties other than purely spatial ones.

Opponent-color responses have been recorded in various cells of the visual systems of a number of organisms that are known to perceive colors. In cats, the cells of the visual center in the thalamus (the *lateral geniculate nucleus*) have receptive fields very similar to those of retinal ganglion cells.

Interleaf 26.4

INFORMATION PROCESSING BY THE AMPHIBIAN VISUAL SYSTEM

Figure 26.19. The frog's retina contains neurons that are more complicated in their responses than are the center-surround units in the retinas of cats or monkeys. For example, the illustrations on this page show the responses of type 1 and type 2 fibers for various stimuli. Type 2 fibers respond to any stimulus in which a sharp border moves within its receptive field, but type 1 fibers are more selective. Neither a moving bar (A) nor a complex picture in which all parts move together (B) evokes a response in the frog. But a small, dark spot (such as a fly) moved against a stationary background elicits a good response (C); a light spot elicits no response (D). The response to a dark spot persists even if general lighting conditions are reduced (E); however, if the dark spot has indistinct edges, this unit will not cause a response (F). Possibly, the type 1 fiber tells the frog that there is a fly (or some other source of food) flying around, and the type 2 fiber simply alerts the frog that something in his visual field is moving.

In the area of prey-catching (opposite page), frog behaviors are not considered to be goal-directed; rather, the actions of frogs seem quite mechanical. In the frog's visual field, a moving spot in a certain place elicits a snap and a tongue flick. But if the spot is too far away, either a jump or a jump-snap action will result (G). If the spot is too far to the side, the frog will "orient" toward it. The decision to jump, orient, or snap is determined in part by the distance between the frog and the prey (H)—it will orient toward a worm that is within the "snap" range but below the glass plate. If a frog is shown a prey object, but is separated from it by a barrier, it may act in a clever manner by trying to climb over the barrier (I) or side-step it (from right to left in J). More detailed experiments have shown that the frog probably does not perceive a worm behind the barrier. Perhaps it has the illusion of seeing a worm at the edge of the barrier in a position requiring a side-step. If an object is placed behind a slotted card, the frog acts as if it perceives a single barrier and orients itself accordingly (K). This orientation causes the frog to lose sight of the object and behave as if it "forgot" the reason for making the original orienting response. It will not pursue the worm further at this point. One explanation for these behaviors is that each moving object that stimulates a spot on the retina causes activity in a particular locus in the optic tectum; the place in the tectum that is excited determines what action the frog will take. If a small incision is made in the appropriate spot in the tectum, the frog will behave as if it had a permanent barrier between itself and any object directly to one side (L). The frog will correctly orient and snap at spots in front of it or behind it, but it will orient incorrectly—toward the edges of an imagined barrier—for an object at the side. This mechanical view of the frog's visual system is somewhat oversimplified. A frog may sit in many different positions (M) and will catch flies in spite of the fact that the position of the fly's image on his retina (and thus the locus of excitation in his tectum) depends upon his posture. Nevertheless, the frog's visual system must not be overestimated. A frog that cleverly snaps at the middle of a mealworm (N) will snap midway between two objects presented simultaneously (O) and will miss them both.

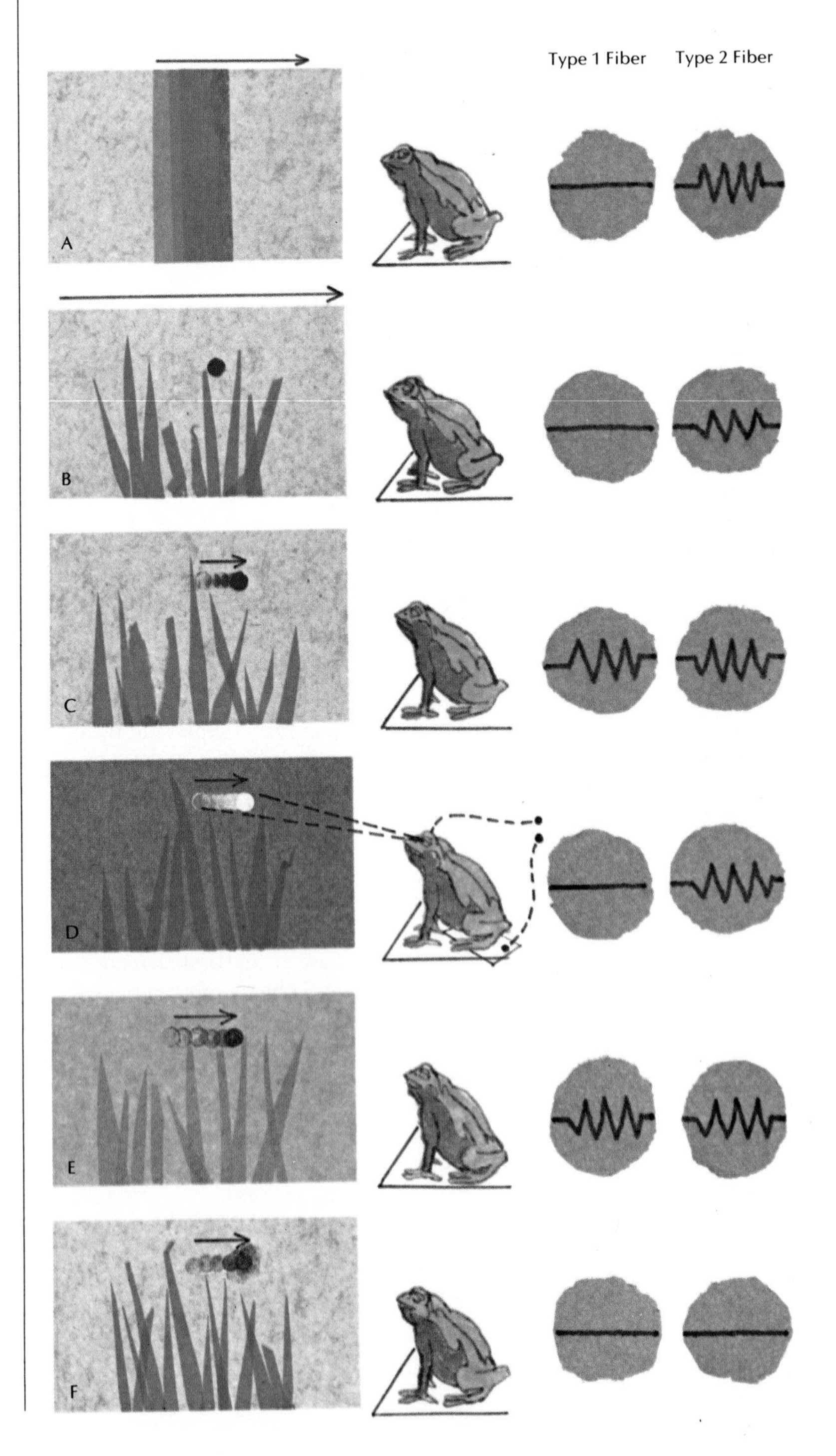

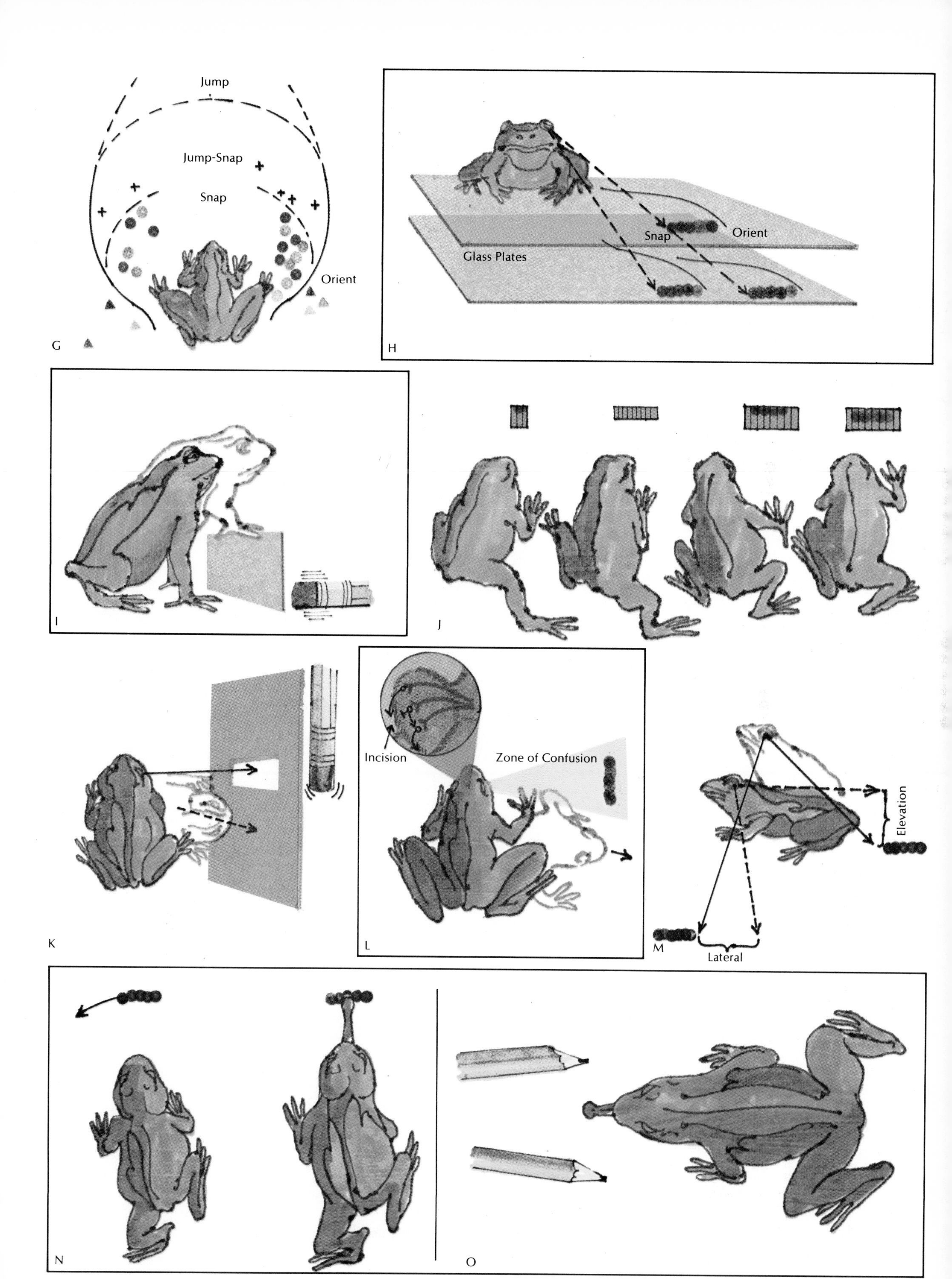
Jump
Jump-Snap
Snap
Orient
G
Snap
Orient
Glass Plates
H
I
J
K
Incision
Zone of Confusion
L
Elevation
Lateral
M
N
O

Figure 26.20. In attempting to understand the processing of visual information, experimenters have mapped the retina by using a moving bar of light. With electrodes, they have been able to trace the effect of light stimulation up through the brain by specific processing cells. When a bar of light hits retinal cells, some are stimulated (red) and others are inhibited (blue). The response of a ganglion cell is the result of the condensation of the information of many receptor cells into an on-off message. (The receptive field is a "map" of a particular grouping of cells within a three-dimensional plane; therefore, many will overlap.) The ability to distinguish contrast is based on this property of either responding to light or its absence.

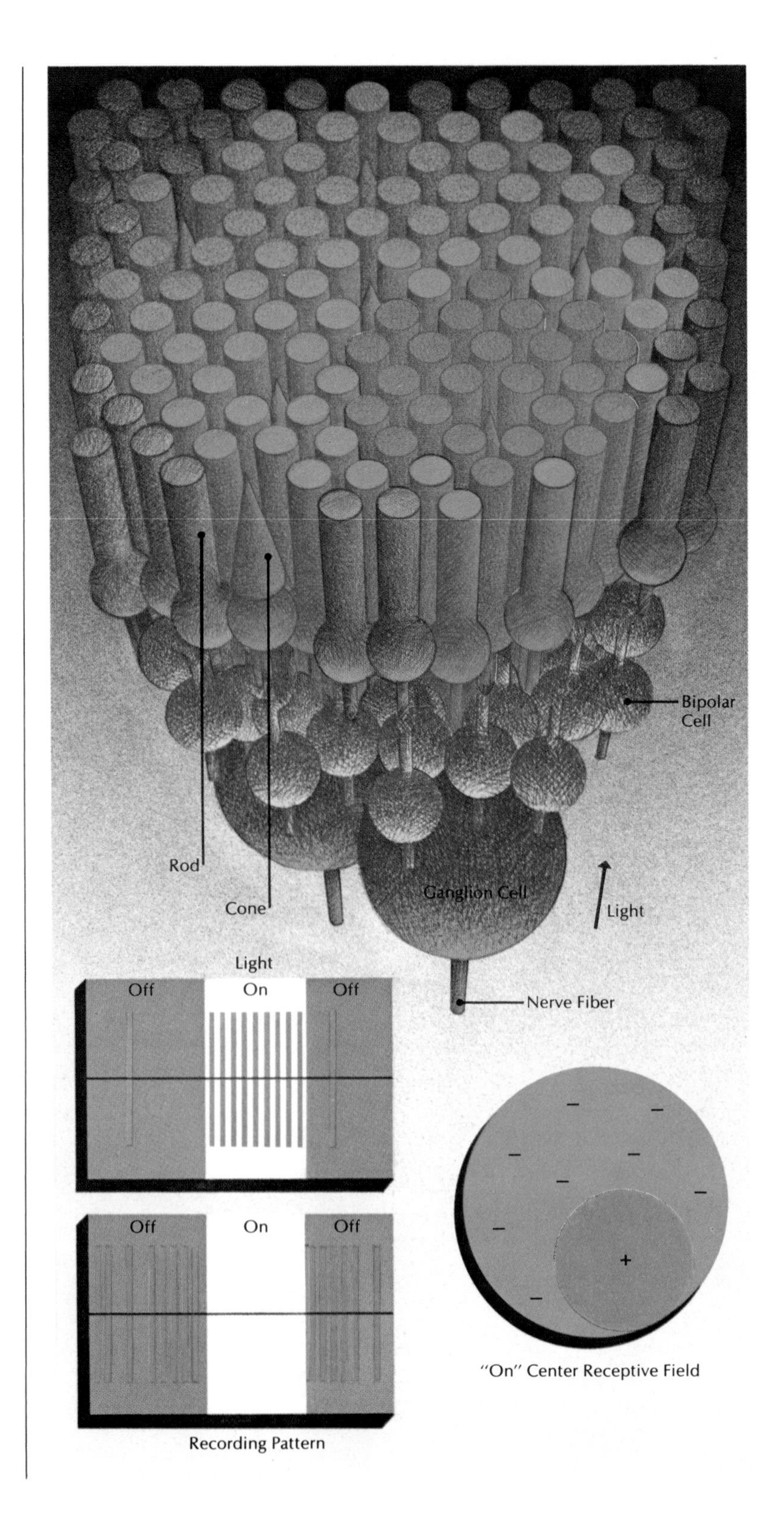

Figure 26.21 (right). Color processing. Cone contribution to a type III on-center cell is shown in A. Three types of cones are represented by colors (red, blue, and green). Receptors are shown on a line through the field center (circle). Cones transmit to the cell through intermediate synapses (not shown). Activation of synapses in the field center produces excitation of the cell, those in the peripheral region to inhibition. In B is shown a schematic representation of a type II cell receiving excitatory input from green-sensitive cones and inhibitory input from blue cones. The schematic representation in C is of a type I cell receiving excitatory input from red-sensitive cones in the field center and inhibitory input from green-sensitive cones in the field periphery.

Figure 26.22 (below). The centers of these two figures are physically identical. The induced colors in such gray areas may come close to that of the surrounding field.

As might be expected from evidence that cats have very poor color vision, cells with opponent-color responses are rare or absent. In monkeys, some cells of the lateral geniculate nucleus do have opponent-color responses (DeValois, *et al.*, 1958).

Subsequent research has revealed several distinct types of receptive fields for cells in the lateral geniculate nucleus of the monkey (DeValois, 1960, 1965). One type of cell has a center-surround opposition, but responds equally well to white light or to light of any color in the visible range, although these cells apparently have no input of information from the rod receptors. There are a great many of the kind of cell first reported in the lateral geniculate nucleus—those with opponent-color responses but no center-surround opposition. There are also a number of cells with responses similar to those of the retinal ganglion cells of fish. These cells have an on-center and an off-surround, or vice versa. Furthermore, either red or green light can be maximally effective for responses of the center if the other color is maximally effective in the surround. A small number of cells show similar opposition between blue and green or between yellow colors. It is possible to diagram neural connections that could produce such response patterns in the lateral geniculate cells (Figure 26.21).

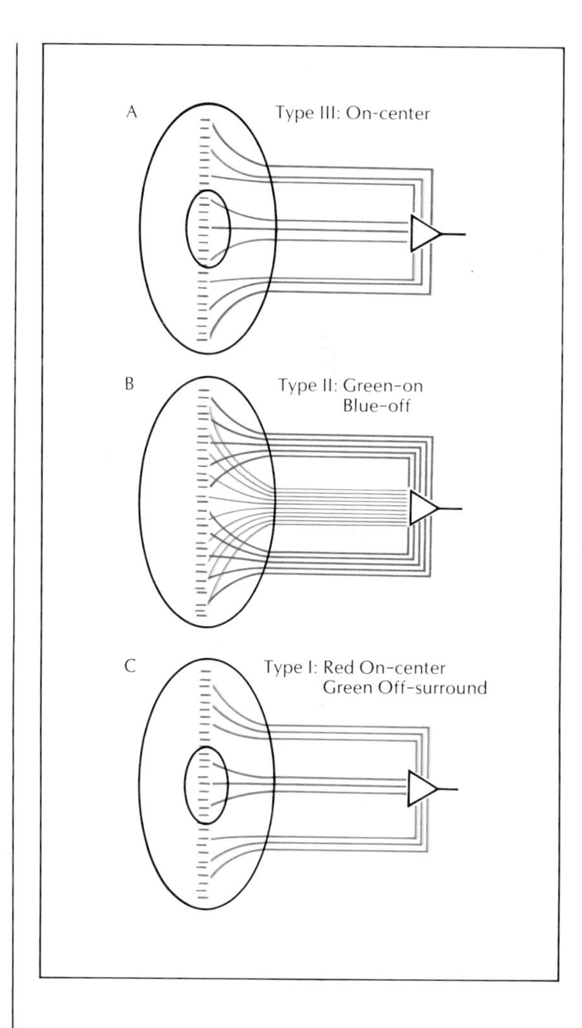

The complex color sensations and perceptions—for example, various illusions and perceptions of colors not "appropriate" to the wavelengths of stimuli—are probably results of excitatory and inhibitory interactions among opponent-color ganglion cells, as well as further interactions in the visual cortex.

The effects of color mixtures are among the perceptual phenomena related to the functions of the opponent cells. Within very broad limits, it is possible to select a set of three monochromatic (single-wavelength) stimuli and, by adjusting their relative intensities, to match any color sensation. There are a few restrictions in the selection of these three so-called primary colors. For example, no two of them can be complementary colors—that is, pairs that when mixed together in correct proportions yield an achromatic, or colorless, sensation. The existence of complementary colors is apparently due to the counteracting effects of opponent cells with opposite excitatory and inhibitory responses. A similar duality of color effects is illustrated in Figure 26.22. The induced colors in such gray areas do not always correspond exactly to the complementaries of the surrounding color field, but the match is quite close. Similarly, if a brief flash of an intense yellow light is presented, a subject is likely to report that shortly after the flash he sees

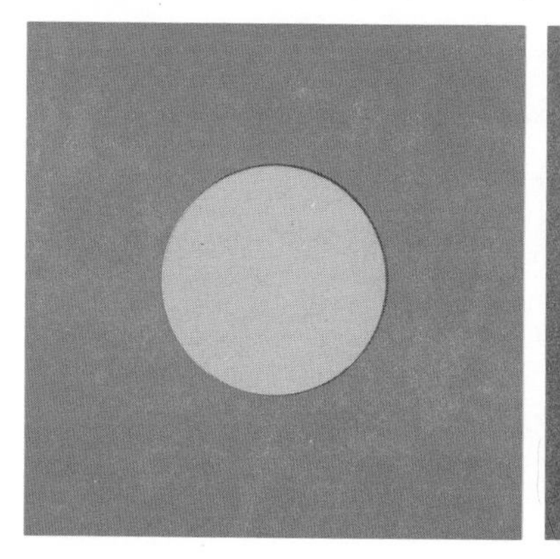

Figure 26.23. Test for color blindness. The discrimination of the numbers in these figures depends entirely on the ability to see colors, as the individual dots are equated for brightness.

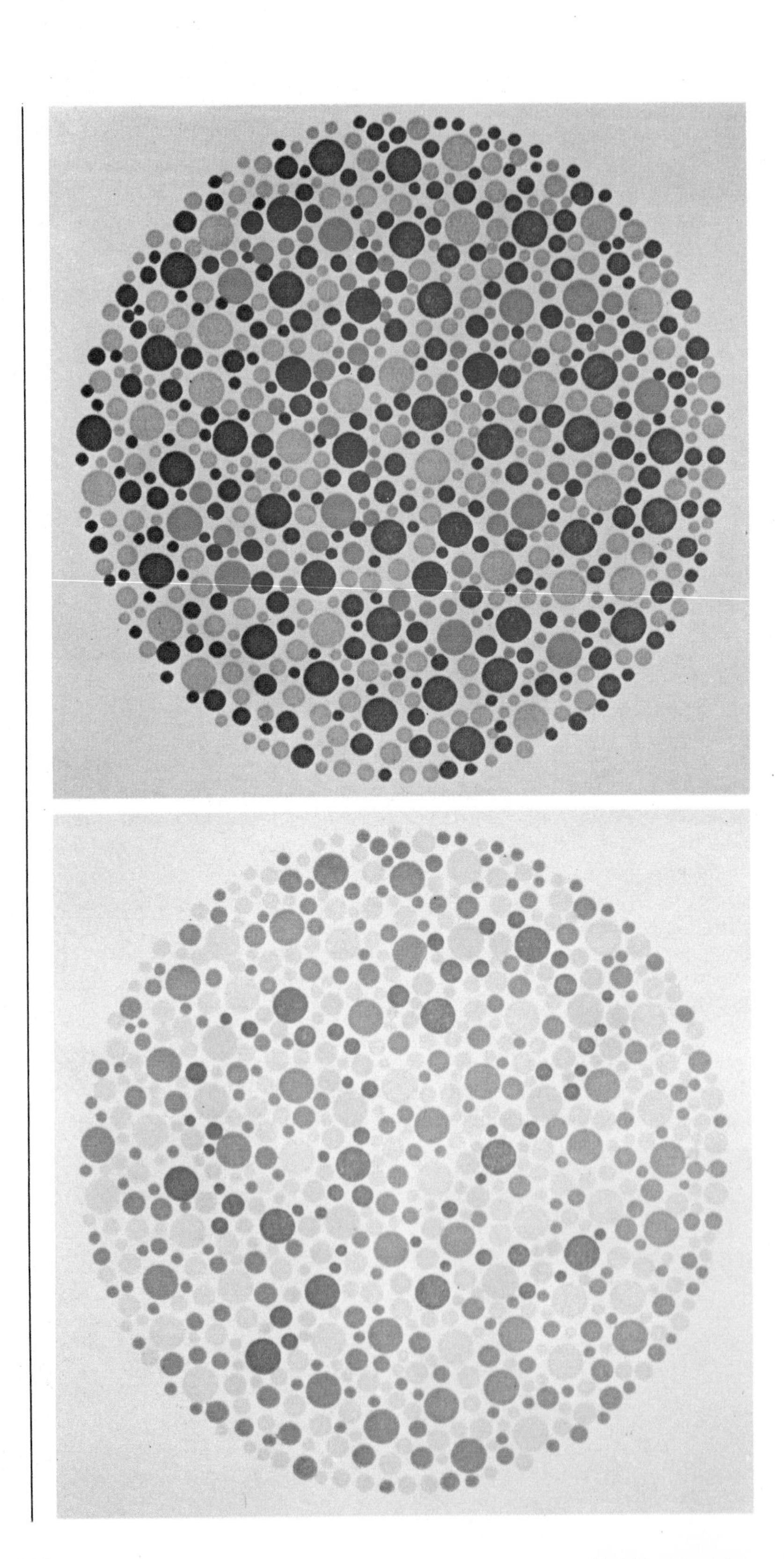

blue light. Alternatively, if a brief intense flash of a green light is presented, the subject may report seeing a red patch after the cessation of the flash.

Anatomical information about the structure of the visual system and experiments such as those described above support the theory that the visual system involves three different opponent systems. The first is the achromatic, light-dark system. The second is a red-green system, in which stimuli of these different wavelengths have opposing effects. The third is a similar yellow-blue system.

The varying forms of human color blindness are due to a lack of one or more types of cones or visual pigments. The most common form of color blindness involves an inability to distinguish reds and greens, apparently because of deficiencies in the cones that activate the red-green opponent system (Figure 26.23).

In the mammalian visual system—particularly in cats and monkeys—the functional organization of the sensory receptors and the first few synaptic layers (in the retina and lateral geniculate) produce drastic changes in the information that the brain receives as compared to the initial pattern of stimuli falling on receptor cells. Spatial information apparently is sacrificed for information about relative intensity differences in center and surround regions. Absolute light levels are not reported (as they could be by steady firing rates with frequencies related to intensity), but onset and offset of stimuli are accentuated. Color information is not reported to the brain in pure form; it is mixed both by opponent-color mechanisms and by spatial color contrasts between center and surround.

This filtering of information occurs in individual neurons and receptors, and in the connections between neurons of the same synaptic layer and between successive layers. Lateral inhibition plays a major role. Many receptors send their information, via bipolar neurons and horizontal cells, to a single ganglion cell. Conversely, each receptor may be connected to many ganglion cells. This combination of convergence, divergence, and lateral inhibition seems to occur in almost every sensory system that has been studied, if the system is followed sufficiently far into the CNS. It seems to be a basic principle of sensory-system construction. In the visual system, it is already evident that some common visual illusions or misperceptions can be explained in terms of these filtering processes.

FURTHER PROCESSING THE VISUAL SYSTEM

The same principles of organization and processing are exemplified in the extension of the visual system to the cerebral cortex, where axons from the lateral geniculate have their endings. Further mechanisms of color processing are not yet fully known, but contour and form detection have been extensively studied at the cortical level. The "simple-cortical" cells, which receive synaptic input directly from the lateral geniculate, have receptive fields radically different from those of their input fibers.

These cells are excited or inhibited when static spots of light are projected onto the retina. But the excitatory and inhibitory zones may be on either side of an edge of particular orientation, or the central excitatory region may be extremely elongated in a particular orientation. These cells respond selectively to the orientations of illuminated patterns on the retina.

Antagonism between an excitatory and an inhibitory region is still present. Diffuse lighting of the entire receptive field produces no response at all in these cells. David H. Hubel and Torsten N. Wiesel (1963), who have

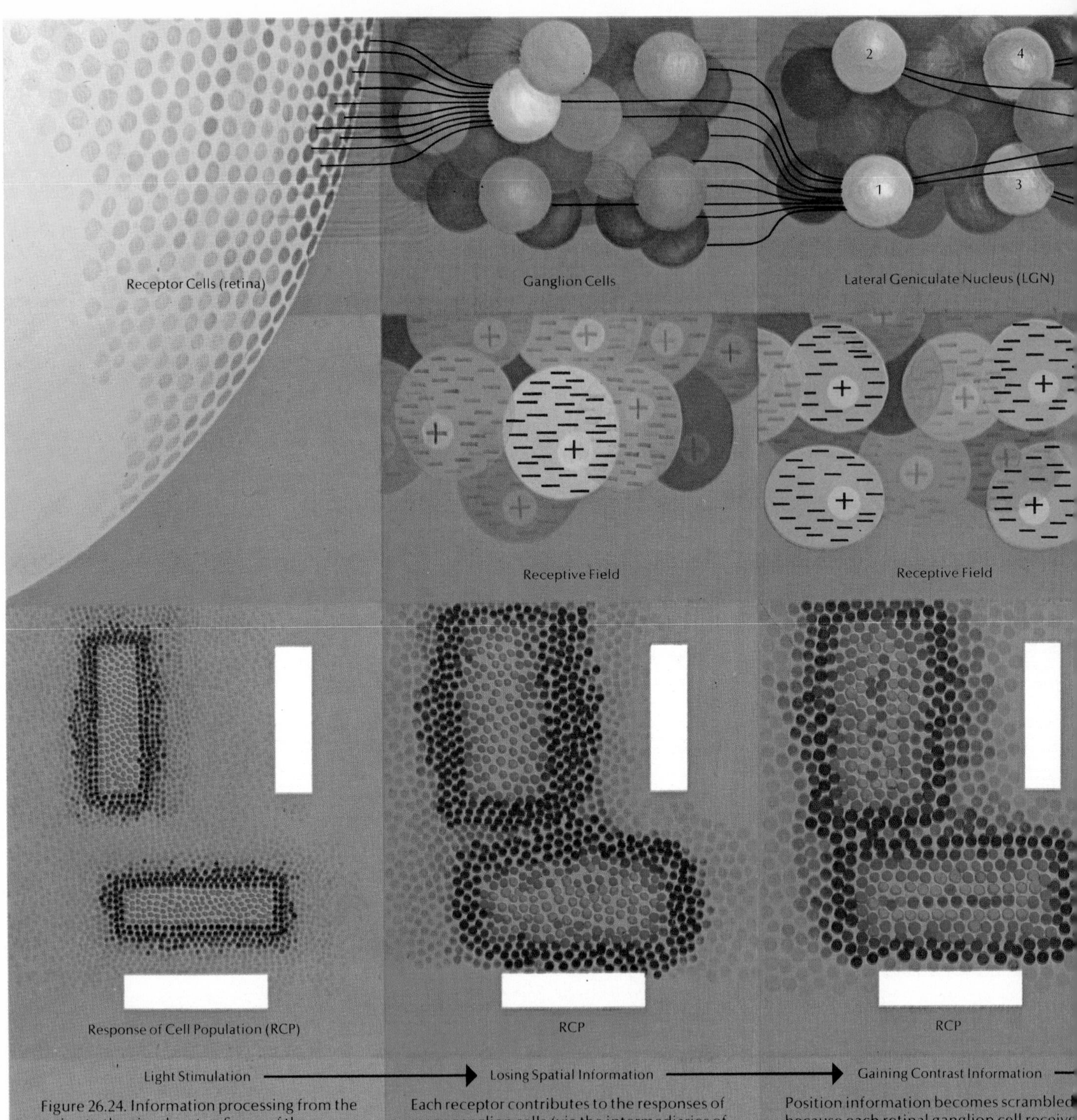

Light Stimulation ⟶ Losing Spatial Information ⟶ Gaining Contrast Information ⟶

Figure 26.24. Information processing from the retina to the visual cortex. Some of the postulated connections between cells are shown in the upper level. Receptive fields are shown in the middle level; excitatory regions are marked with plus signs and inhibitory regions with minus signs (also see Figure 26.20). In the lower level, the responses of cell populations (RCP) to specific stimuli are shown. The degree of red coloration is proportional to the amount of excitation of a cell, the blue to the amount of inhibition, and the brown indicates neither. Receptors are excited only by the light falling on them and are unaffected by the stimulation of neighboring receptors.

Each receptor contributes to the responses of many ganglion cells (via the intermediaries of horizontal cells, bipolar cells, and amacrine cells). Conversely, each ganglion cell is affected by many receptors. A bar of light excites a population of on-center ganglion cells and inhibits a surrounding population of on-center cells. Off-center cells are omitted throughout for clarity. A bar of light will excite or inhibit a ganglion cell depending upon whether it illuminates the center or the surround part of the receptive field; the orientation of the bar has no effect. The ganglion cell no longer "tells" exactly where the stimulus is located.

Position information becomes scrambled because each retinal ganglion cell receive input from many receptors, each ganglior cell sends its output to many cells of the lateral geniculate nucleus (LGN), and each LGN cell receives input from many ganglion cells. A cell whose center and surround are both lighted gives little response. If the whole center but only ha of the surround is lighted, the cell responds vigorously. Thus, at the LGN, border contrast and contours are accentuated. This transformed information is sent to the visual cortex over the optic radiations.

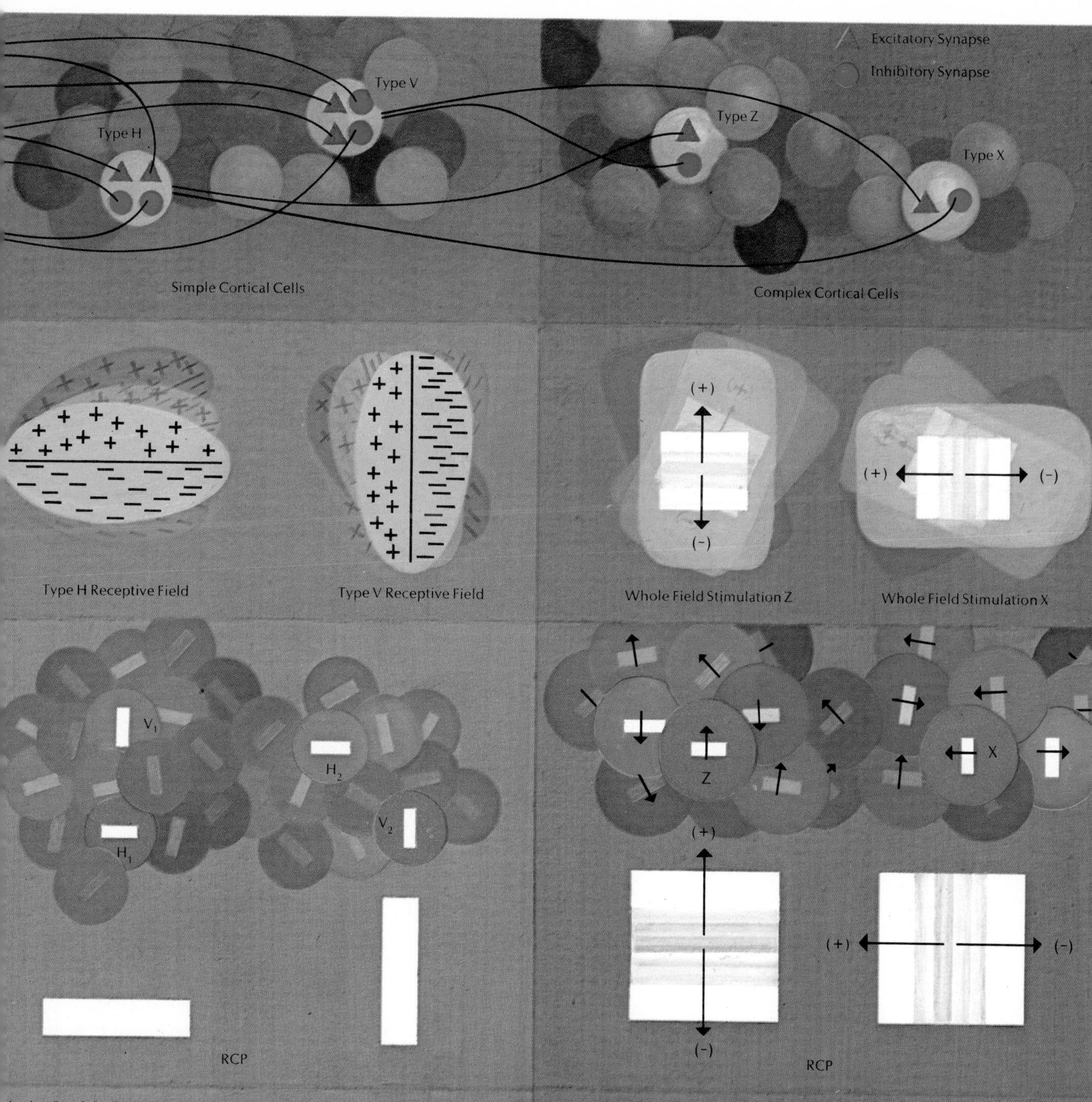

Losing Spatial Information—Gaining Contour Orientation and Contrast → Generalization of Spatial and Motion Information

Each simple cortical cell receives and integrates information from many LGN cells. Often, simple cortical cells are excited by LGN cells whose receptive fields lie on one side of an imaginary line in space and are inhibited by those cells whose receptive fields lie on the other side of that line. This whole process can be seen by following the outputs of the LGN cells 1, 2, 3, and 4 to either type H (horizontal) cells or type V (vertical) cells. The response of a simple cortical cell to stationary or moving stimuli can be predicted from the balance between lighted excitatory and inhibitory areas of its receptive field. Thus, a bar of light that lies entirely within an excitatory area has the same axis as the receptive field and excites the cell strongly. But a bar in a different place with the wrong orientation may inhibit it. Bars in the wrong areas of visual space neither excite nor inhibit.

Complex cortical cells (integrating the information from preceding levels) may receive excitatory input from many simple cortical cells, all of whose receptive fields have the same preferred edge or bar orientation. Each cell may be inhibited by simple cortical cells with a different orientation preference (for example, type Z versus type X). Thus, a complex cortical cell will respond to a bar or border of a certain orientation, regardless of its location in the cell's receptive field. An additional feature of the complex cell is its sensitivity to motion. A properly oriented bar or border, if moved perpendicularly to its axis, will excite sharply for movement in one direction but only slightly, if at all, in the other. (Each input cell is excited as the bar moves from an inhibitory to an excitatory area.) Arrows show movement preference, and bars show orientation preference. The processing described in this illustration probably accounts for recognition of simple contours. Additional processing, presumably along similar lines, permits recognition of real objects.

mapped the receptive fields of many cells in visual systems of cats and monkeys, have proposed a mechanism to account for the fields of these cells. They suggest, for example, that many lateral geniculate cells with on-centers located near a line on the retina all have excitatory synapses with a single cortical cell, whereas lateral geniculate cells with on-centers in a broader area around this line have inhibitory synapses with the same cortical cell. A bar of light falling on the imaginary line on the retina falls on many centers that trigger excitatory inputs to the cortical cell but not upon a proportional area of centers that trigger inhibitory inputs to the cortical cell. A bar of light falling off the imaginary line or at an angle across it stimulates many centers that contribute inhibitory inputs to the cortical cell. The cortical cell is an "oriented bar detector"—it generates an impulse only in response to a bar of light of particular orientation on the retina.

At the next synaptic layer of the cortex, several more complex receptive field types are found. There are cells that fire at a maximum rate if an oriented bar or border is moved across a given area of the retina, always remaining in a particular orientation. Hubel and Wiesel suggested that these cells receive inputs from a series of oriented bar detectors—the orientation and polarity of their excitatory and inhibitory areas determines the kind or direction of motion that excites the higher-level cell. At each synaptic level, there is a progressive generalization of spatial information, with progressive specialization of information about a variable such as orientation, type of border, or movement. This combination, often called an ordered hierarchy of generalization and specialization, continues through higher layers of the processing network. For example, there are cells that generate impulses only if a light bar is present with an end somewhere in the visual field. In most cases, it is possible to hypothesize combinations of excitatory and inhibitory synaptic connections that produce such specialized responses.

Studies of receptive fields of cortical cells in cats and monkeys have provided exciting insights into many aspects of vision other than form perception. Workers in two laboratories recently have discovered cells in the cortex that seem to account for binocular depth perception. Psychophysical studies long ago led to the suggestion that depth is inferred by comparison of the slightly different images cast on the retinas of the two eyes by a three-dimensional object.

The presence of cortical cells that are excited by such "retinal disparities" was demonstrated by focusing a monkey's gaze on a screen and waving a stimulus bar (in this experiment, a yardstick) in front of the screen. A particular cortical cell is found to fire only when the yardstick is, for example, 20 cm in front of the screen—in other words, when the retinal disparity of the images in the two eyes is just right to correspond to a yardstick about 20 cm in front of the screen. If either eye is covered, the responses are greatly reduced. In some cells, if the disparity between the images is altered by as little as 0.1°, the response is completely inhibited.

Other examples of shape-specialized receptive-field complexity could be cited, but the outlines of the processing of visual information seem clear. As the information moves along the sensory pathway, each impulse comes to represent more and more complex and specific data about the light pattern falling on the retina. A neuron deep in the visual cortex will be activated only if a very specific kind of pattern of illumination occurs on the retina.

How the brain integrates and evaluates the visual information coded in the complex pattern of activation in the cortex is not known. Equally obscure is how this sensory information produces or modifies complex behavior patterns of the animal. However, just as the first few steps leading from reception of stimuli to sensation are being unraveled, so are the last few steps that produce the modified behavior animals need in order to survive. The principle of ordered hierarchies applies to motor output as well as to sensory input. The gathering of a pattern of input in space and in time seems in many ways analogous to the generation of outputs organized in space and in time.

MOTOR SYSTEMS

As an animal moves, muscles are contracted and relaxed. These contractions and relaxations must occur in a precise pattern both in space and in time. Different parts of the body must have a particular spatial relationship to one another during any one moment in time, and any one part of the body must move in a particular sequence of related motions over a span of time. This coordination of body movements must result from underlying patterns of muscular contractions.

The contractions may be measured by recording the tension developed by individual muscles during a movement, and this tension proves to be directly related to the electrical activity of the muscle. The more the muscle is electrically excited, the stronger will be the tension produced by that muscle. Electrical activation of muscles is in turn related to the discharge rate of motor neurons that synapse onto the muscles. Each motor neuron has an axon that exits from the CNS and travels in a peripheral nerve to a particular muscle. The motor neurons themselves are activated in the CNS by synapses from other neurons.

Here, then, is another neural design where layers of neurons control succeeding layers of neurons through synaptic connections. The final layer of output neurons controls the effectors themselves. The hierarchical arrangement is analogous to that of the sensory systems already discussed. The critical issue in the organization of patterned behavioral movements is how specific motor neurons are activated in precise sequences so that a particular set of movements can occur at the proper time.

Because each motor neuron innervates a specific muscle and because the contraction of a given muscle causes a specific movement of the body, each motor neuron "codes" for a particular spatial movement. If the CNS properly controls the pattern of firing frequencies of specific motor neurons over time, it will control the proper movements of the body.

Movements of the body can be divided into two broad classes: *postural movements*, which move a body part to a fixed position; and *locomotory movements*, which continuously change the position of body parts. Both types of movements have been studied in detail in the abdomen of the crayfish. These studies illustrate major principles of central motor control. It is hoped that the principles so well illustrated in invertebrates will be applicable—at least in general—to motor systems of vertebrates, much as studies of the simple visual system of *Limulus* have revealed general principles that also apply to vertebrate visual systems.

Crayfish Nervous System

The abdomen, or tail, of the crayfish is segmented. In each segment, there is a basic set of abdominal CNS neurons, peripheral muscles, and exoskeleton. This basic unit of structure is repeated with only minor modifications in

Figure 26.25a (below). Localization of sound in space by binaural comparison of sound intensity. At the top is input into the left ear alone; in the middle is input into the left ear + 300 decibels into the right ear; below is input into the left ear + 50 decibels into the right ear.

Figure 26.25b (opposite left). Neural sharpening of pitch. These three graphs show threshold curves for neurons in the auditory system. The threshold curve shows the lowest intensity of sound that will cause a response at each frequency. At intensities greater than the threshold, there is usually a greater response (unless an inhibitory process is present, as in the bottom graph). The area above the threshold curve is thus the "response" area, and the frequency of greatest sensitivity (lowest threshold) is the "best" frequency. Neurons that receive input from the hair cells (purple) do not show much frequency selectivity (purple curve). Cochlear nuclear units show more selectivity at high intensities but not at lower ones (gray units and curve). Inferior collicular units show good frequency selectivity at all intensities (green units and curve). Lateral inhibition between cells

Interleaf 26.5

INFORMATION PROCESSING BY THE AUDITORY SYSTEM

Auditory nerve fibers receive input from one or many hair cells, each of which is probably excited by a rather broad range of frequencies of sound. If nerve impulses are monitored in an auditory nerve fiber, it is found that this fiber is responsive to a narrower range of frequencies. One possibility that may account for this action is that a nerve fiber may receive excitatory input from some hair cells but inhibitory input from others, which are sensitive to slightly different frequencies. Attempts to show this inhibition have not been completely successful. But it is reasonably certain that lateral inhibition does take place in the *cochlear nucleus*. Frequency discrimination is enhanced in much the same manner that borders are accentuated in the eye of the *Limulus*. In fact, the basilar membrane may be thought of as a spatial array of receptors similar to the retina but where the spatial extent is related to the frequency of sound rather than to the position of the sound source in space. The inhibition from neighbors has the effect of sharpening frequency contrasts. Cells in the cochlear nucleus receive input from many auditory nerve fibers, and the effects of lateral inhibition and efferent collateral inhibition are demonstrable in their threshold curves. If a sound is of appropriate intensity and frequency, it will excite the cell from which the electrophysiologist records. But if it is of a different frequency and intensity, it may inhibit the cell. The *excitatory areas* and *inhibitory areas* of the frequency-intensity plane are defined by such effects of tones on the firing of the cell.

Output from the cochlear nuclei goes to both the thalamus and the midbrain. In the inferior colliculus of the midbrain, cells have narrower excitatory areas; these cells discriminate frequency better than do cochlear nuclear cells. Their inhibitory areas are larger. In bats, which use auditory radar, or *echolocation*, these cells respond well to very closely spaced sounds. Some cells of the inferior colliculus respond well to tones that arrive at the two ears with slightly different intensities or at slightly different times. These cells probably have the function of establishing the direction from which the sound came, because sounds coming from the left side of the head are louder and occur earlier in the left ear than in the right ear. The change of information in the neurons of the auditory system as more central structures of the brain are encountered seem to parallel the changes in visual information.

Complex characteristics of sounds can be recognized by cells in the cortex of bats. Some are specialized to respond to sounds that are frequency modulated—that is, they change in frequency with time. These frequency-modulated tones correspond to motion of the locus of stimulation along the basilar membrane. The mechanisms by which such units are formed are not known, but researchers feel that connections between various neurons found in the medial geniculate body of the thalamus, with appropriate excitatory and inhibitory areas, and cortical neurons, could explain their responses. The elements of complex sounds are pure tones, frequency-modulated tones, and various kinds of noise bursts. Cortical neurons that respond to all of these have been found in some mammals. Perhaps an understanding of the recognition of speech is on the way.

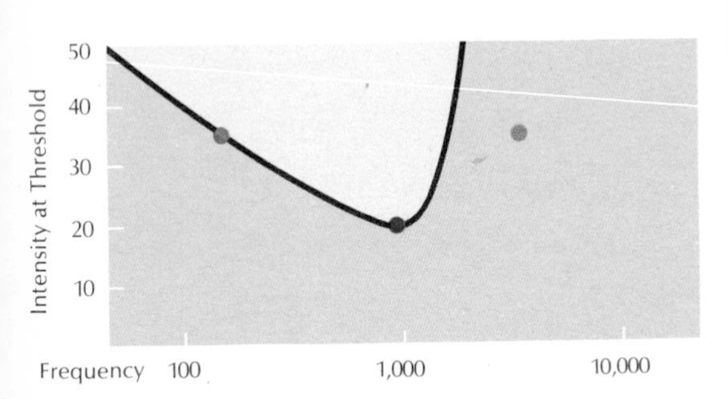

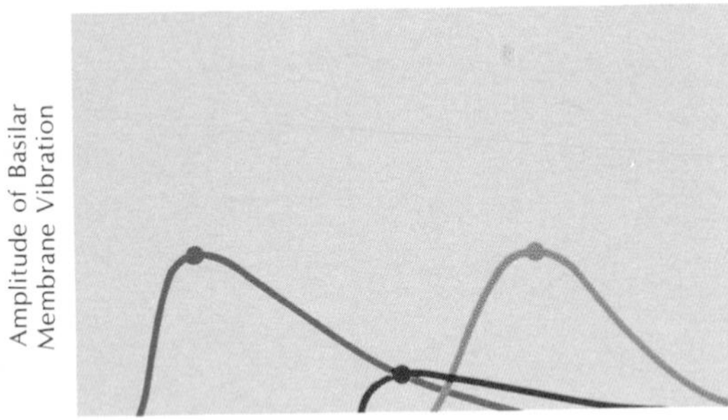

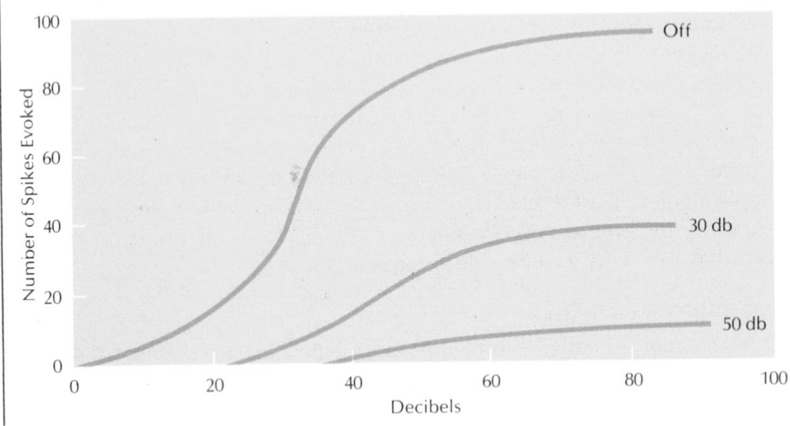

at each of these levels is thought to be responsible for this neural sharpening of pitch. To illustrate this possibility, an inferior collicular unit (green) receives input from the three cochlear nuclear units (light and dark yellow) whose response curves are shown in the middle graph. The stippled response area belongs to the excitatory input cell (light yellow), and the hatched response areas belong to the inhibitory input cells (dark yellow). Therefore, a sound with a frequency and intensity such that it falls in the blue areas of the bottom graph will cause more inhibition than excitation. Only in the central red area will a sound cause more excitation than inhibition. Thus, the inferior collicular cell has a narrow frequency selectivity response area. In practice, this activity does not occur with only three input neurons but with a population of excitatory and inhibitory neurons. Thus, there is a close analogy between the processes that cause neural sharpening of pitch and those that enhance border contrast in the visual system.

Figure 26.25c (right). Neural network describing FM sensitive units, which have no response area for pure tone pulses. When the lower level neurons a and b (having the properties demonstrated by graphs a and b) are connected to neuron c as shown in the upper right diagram, neuron c responds to FM but not to pure tone pulses.

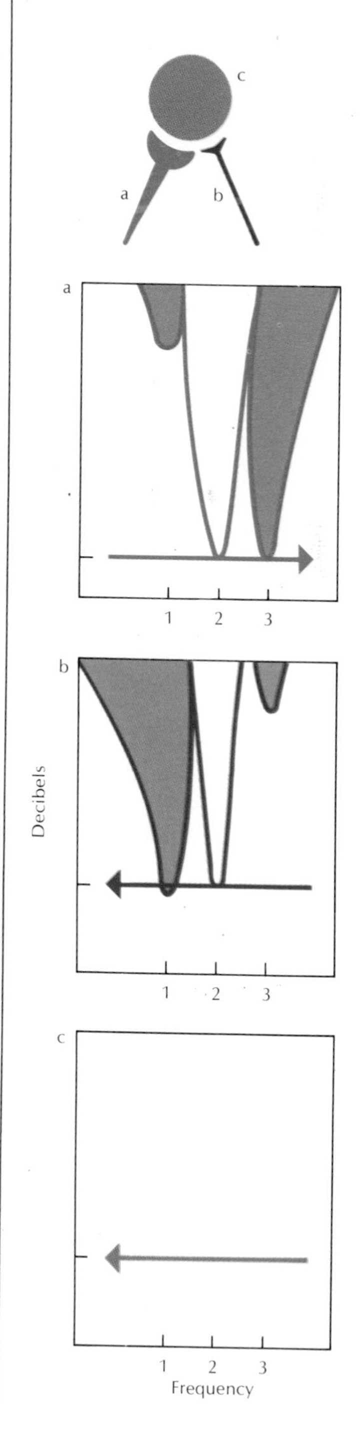

Figure 26.26a (left). Schematic diagram of the third and fourth abdominal segments of a crayfish showing the sensory innervation of the right half of the third ganglion. The first root (shade) from the ganglion innervates the swimmerets. The second root (light) innervates the dorsal muscles (extensors) of the abdomen.

Figure 26.26b (right). Morphology of crayfish flexor motor neurons injected with yellow dye and photographed through a fluorescence microscope.

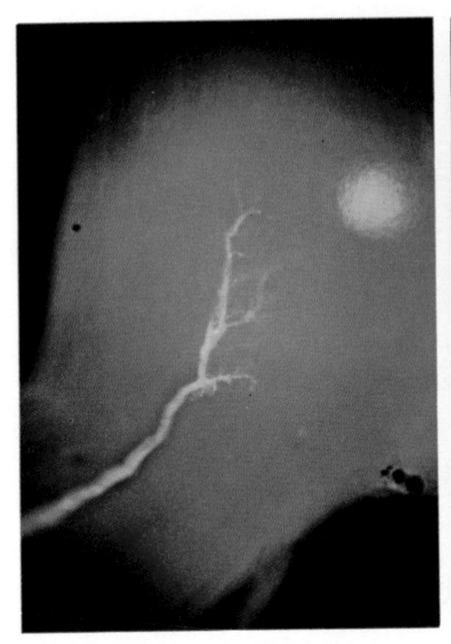

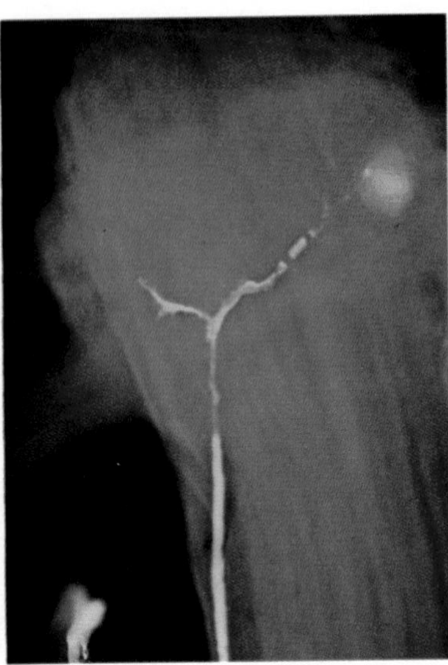

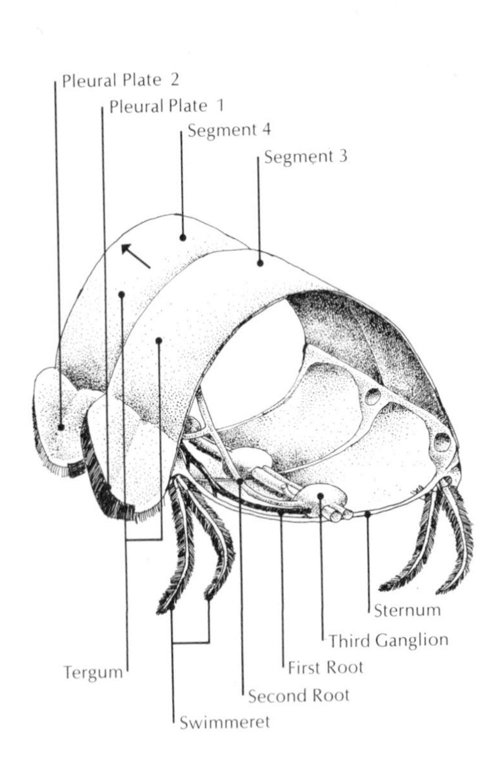

five segments. The sixth segment, the most posterior, is dramatically different from the other five. Most of the work described in the following sections was done with the five anterior segments of the abdomen.

The abdominal nervous system runs along the ventral midline of the abdomen. In each segment, there is a swelling of the nerve cord—the *segmental ganglion*. Within this ganglion are the cell bodies of motor neurons, interneurons, and some sensory neurons. Each ganglion is bilaterally symmetrical; structures of the left half of the ganglion (left hemiganglion) are mirror images of structures in the right hemiganglion.

Three pairs of peripheral nerves exit from each ganglion. The most anterior pair of nerves (the first roots) innervate the swimmerets—small appendages on the lateral edge of the ventral surface that produce a beating motion. (In related crustaceans, this motion aids swimming, but its role is uncertain in crayfish.) Each first root contains the motor axons for one swimmeret and the sensory axons from that swimmeret. The next pair of nerves, the second roots, innervate the dorsal muscles (extensors) of the abdomen. When these muscles are activated, the tail moves into a straight-out, or extended, position. The second roots also carry sensory axons from the dorsal side of the abdomen. The most posterior pair of peripheral nerves in each segment, the third roots, innervate the ventral muscles (flexors) of the abdomen. These muscles when activated cause the tail to curve under the main part of the crayfish body. The third root contains no sensory axons.

Each ganglion is connected with its neighboring anterior segmental ganglion by a bundle of nerves (a connective) and to its neighboring posterior ganglion by another connective. Each connective also is bilaterally symmetrical. The connectives contain no cell bodies and virtually no synapses. They contain only parallel axons transmitting neuronal information among the ganglia.

CRAYFISH POSTURAL CONTROL

The postural control of the crayfish abdomen has been studied in detail by Donald Kennedy (1967) and his collaborators. They observed that the crayfish may hold its tail in any position from one tightly curved underneath the body to one pointing out and slightly upward. Most of the abdominal musculature does not participate in postural control. Only a thin sheet of ventral musculature (the superficial flexors) can perform a postural flexion, and only a thin sheet of dorsal musculature (the superficial extensors) can perform a postural extension. The rest of the abdominal musculature is involved in rapid twitches of the tail used during the escape response and does not contribute to maintenance of a particular posture.

The superficial extensors of each hemisegment (the right or left half of a segment) are excited by 5 motor neurons that leave the CNS in the second root. In addition, there is a single neuron (the peripheral inhibitor) that can reduce the level of excitation in the muscle. Similarly, the superficial flexors in each hemisegment are excited by 5 motor neurons and inhibited by one peripheral inhibitor; all of these axons are located in the third root. Therefore, in each segment, there are 12 neurons that influence postural flexion and 12 neurons that influence postural extension. In the 5 anterior segments of the abdomen, posture is, therefore, completely controlled by a total of 120 neurons. The role of the CNS in postural movements must be to balance properly the activities of these 120 neurons in order to produce a desired posture.

Kennedy hoped to find central interneurons in the connectives that, when stimulated, would produce a coordinated postural response of the abdomen. He dissected fine bundles of axons from the connectives and stimulated these axons while recording from superficial flexor and extensor motor neurons. When he found a bundle containing axons that could influence the postural motor neurons, he further dissected that bundle to obtain only one or two connective axons. With this and other sophisticated techniques, Kennedy was able to stimulate one and only one axon of the connective. He showed that certain central axons, when stimulated above a certain critical impulse frequency, would cause a flexion of the tail, whereas other axons would cause an extension of the tail. Because these central neurons must command motor neurons to fire in a particular fashion that results in a particular posture, they are called *command neurons*.

The postural command neurons are very specific in their motor output. Some command neurons for flexion excite flexor motor neurons in anterior segments more strongly than those in posterior segments, causing the anterior segments of the tail to flex strongly while the posterior segments are only slightly flexed. Command neurons producing slight flexion of anterior segments and strong flexion of posterior segments also exist. Analogous

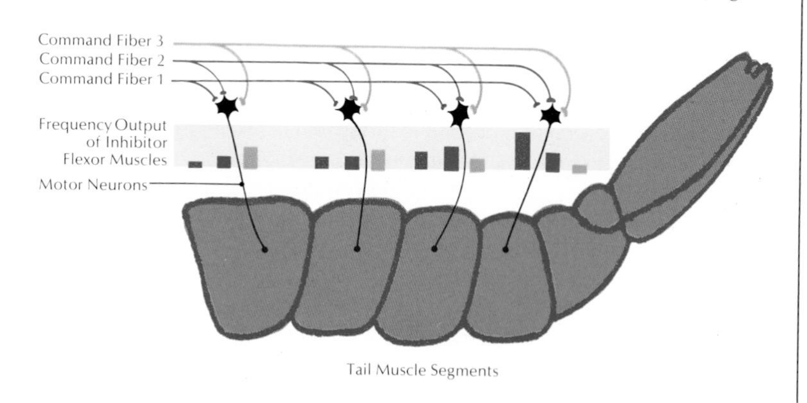

Figure 26.27. Diagram of the pathways of command fibers in the posterior tail muscle segments of the crayfish.

command neurons also control extension of the tail. The effect of any command neuron is stronger or weaker, depending on the frequency at which it is stimulated (Kennedy, *et al.*, 1966).

Because each command neuron, when stimulated, causes a particular postural movement, each command neuron can be said to "code" for a certain position of the body. This body position is the "motor field" of the command neuron. The relationship of a command neuron to its motor field output is the converse of the relationship of a sensory neuron to its receptive field input. A stimulus input causes a response in a higher-level interneuron if the stimulus falls within the receptive field of that interneuron. In many cases, a stronger stimulus produces a greater response in the interneuron. In the motor system, stimulation of the command interneuron causes a motor output in the motor field of that neuron. The greater the stimulation frequency of the command neuron, the stronger the muscular response and the more definite the posture assumed by the crayfish tail. Just as a discharge pattern in a sensory interneuron codes for a particular patterning and position of the stimulus in space and time, the discharge pattern in a command neuron codes for a particular position of the body in space and time.

CRAYFISH LOCOMOTORY CONTROL

The crayfish also has command neurons that can influence dynamic events such as movements of the swimmerets. The command network controlling locomotory muscular output must be more complex than that controlling postural output, because the locomotory movements themselves are more complex than postural movements. The locomotory control system must produce (1) alternate forward motion (returnstroke) and backward motion (powerstroke) of each swimmeret; (2) synchronous forward and backward movements of both swimmerets in a single segment; and (3) matching of the frequencies of swimmeret movement in different segments but with the cycle of each pair slightly delayed in time with respect to the cycle of the next posterior pair.

It was found that for any command neuron to the swimmerets, there is a stimulus *threshold frequency* for the behavior. If the stimulation to the command neuron is below this frequency, no movements of the swimmerets are produced. Above the threshold frequency, stimulation of the command neuron produces a slow oscillation of the swimmerets. As the frequency of stimulation is increased, the oscillation frequency of swimmeret movements increases and the time delay between equivalent movements in neighboring swimmerets on the same side of the body decreases.

The important and startling fact revealed in this experiment is that a single command neuron can control complicated spatial and temporal patterns of muscle contraction needed to produce this behavior. Impulses in a single command neuron cause a number of appendages to carry out coordinated sequences of movements. In the complexity of their coding, these command neurons are analogous to the motion-sensitive cells in the visual systems of mammals.

Organization of Motor Systems

The work of Kennedy and others clearly implies that the output of a single command fiber, through one or many synapses, excites a particular collection of motor neurons. In the case of the locomotory command neu-

Figure 26.28. Influence of visual input on the symmetry of locomotor output in man.

rons, excitation of the command neuron causes excitation of a particular group of motor neurons in a complex temporal pattern. How is the output of the command neuron translated, or decoded, into excitation of particular motor neurons? At present, the answer is not known. It is speculated that the command neuron excites a particular group of interneurons within ganglia and interneurons leading to other ganglia. These neurons in turn excite motor neurons or other interneurons that eventually excite motor neurons. The hierarchical organization of excitatory and inhibitory synapses could lead to generation of a complex output from the simple input, much as if stimulation of a moving-bar detector in a hypothetical inverted visual system could produce a moving-bar pattern on the retina.

In the crayfish, a complex pattern of motor output is obtained from a CNS that is receiving no sensory input. This does not mean that the normal behavior of the crayfish is unaffected by sensory input. On the contrary, sensory information may have a profound influence on output. One of the most profound influences is that of the visual system on the symmetry of output. If blindfolded, most animals (including man) will walk in a circle or spiral rather than in a straight line. This observation implies that the inherent

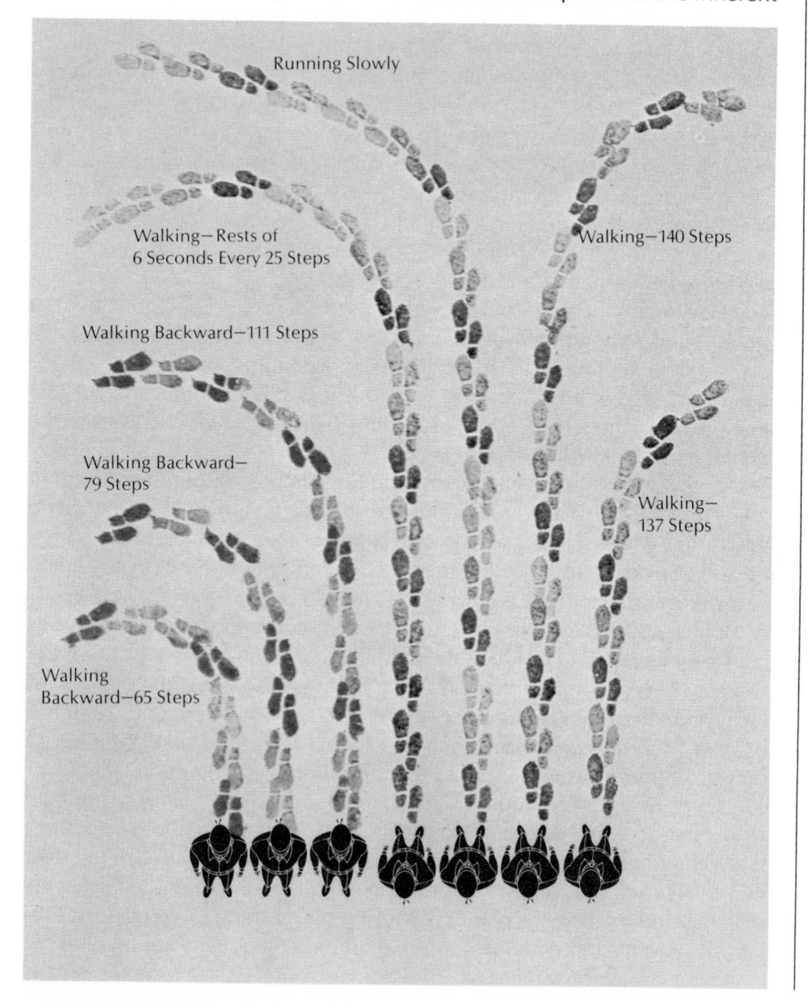

Figure 26.29. Effects of a stable visual image of the environment on the locomotor responses of an insect. The insect is held in place by being glued to the forceps, forcing it to walk and turn the globe. If the rotation is to the left, the insect will walk down the left arm of each Y on the sphere as it comes to it.

motor pattern is asymmetric (or that parts of the body are translating a symmetric pattern into asymmetric movement). Without the blindfold, most animals are able to walk in a straight line. This observation implies that the visual input influences an asymmetric behavior in such a way that the behavior becomes symmetric. The overall pattern of behavior is not changed, but the balance between various parts of the behavior is altered.

The influence of visual input has been explored by Donald M. Wilson (1961, 1966) and others in the walking and flying behaviors of insects. When an animal walks, its visual system reports the movement of the environment past the body. If the animal starts to veer to the left, the environment seems to move to the right. This visual input might provide the clue that causes the animal to adjust the balance of its walking motor output to put itself back on a straight course.

To explore this effect, it is necessary to arrange an experimental situation in which the animal's movements do not influence its visual image of the environment. The animal must be fixed in space; it must walk on a spherical treadmill. No matter which way it walks, its environment will remain fixed with respect to its visual system. Furthermore, the environment can be rotated around the animal, independent of its own movements. In such an experimental situation, most insects show clear evidence of an effect of visual input upon motor output. If the visual world is moved to the left, the insect tries to turn to the left; if the world is moved to the right, the insect attempts to move to the right. Because the insect is securely fixed to a rod, it cannot succeed in changing its own orientation, but the direction in which it would move if free can be determined by noting the movements of the spherical treadmill. These responses to visual input apparently are innate, or "wired into" the nervous system of the insect; they need not be learned at all.

Under normal conditions, these responses help the insect to walk in a straight line. When the visual world starts to drift to the left, it means that the insect is veering to the right. Therefore, an adjustment of the walking behavior to move a little toward the left brings it back into a straight-line motion. Thus, there is a feedback loop in which sensory input adjusts motor output to accomplish a desired behavior.

Another example of sensory input influencing motor output is found in the lobster. Gravity receptors influence the position of swimmerets when the lobster is tilted. The position alterations of the swimmerets produce a force that tends to turn the lobster back into an upright position. As in the case of the visual influence on walking, this response to gravitational sensory input does not alter the basic pattern of motor output. It merely causes the swimmerets on one side of the animal to turn outward during the powerstroke. This slight modification of the basic pattern of swimmeret movement produces the twisting force on the lobster's body and tends to turn it back to a normal position. When the animal is in a normal position, no outward turning is superimposed on the swimmeret motions.

There are also interactions in which motor output alters sensory input. In many sensory organs, motor control is exercised over various structures that gather and focus the energy of physical stimuli. Signals arriving over motor neurons may cause adjustments of these structures—adjustments that alter the intensity or nature of the stimuli reaching the receptors. Obviously, complex interactions of visual and motor systems are involved in watching a moving object as it passes. In some cases, receptor organs

Figure 26.30. Lobster swimmeret position is influenced by gravity receptors. The position alteration of the swimmerets when tilted (top) produces a force that tends to turn the lobster back into an upright position.

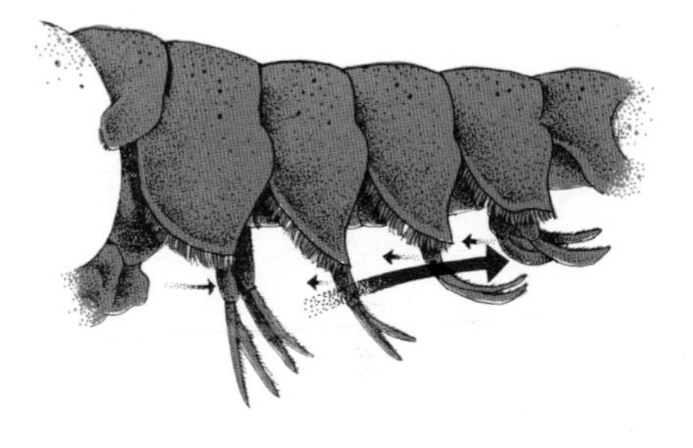

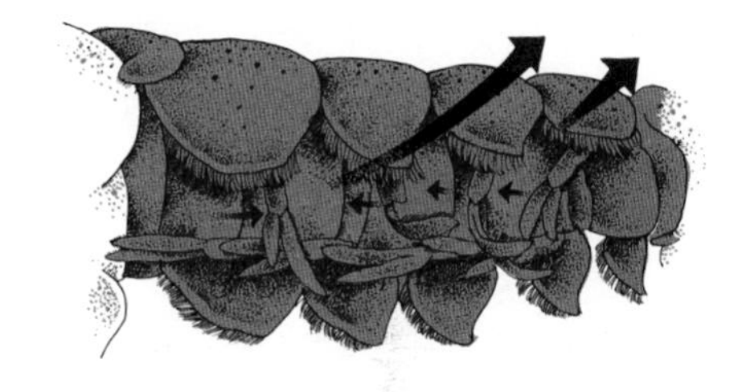

themselves are innervated by efferent neurons from the CNS. Presumably, the activity of the receptor is directly modified by output from the brain.

NEURAL CONTROL OF BEHAVIOR PATTERNS

Studies of sensory and motor systems have produced intriguing clues about the organization of the nervous system. They have revealed a fantastic complexity of built-in processing of information. Receptors and interneurons are connected in such a way that a particular pattern of visual stimuli triggers an impulse in a particular single cell of the mammalian cortex. Command and motor neurons are connected in such a way that stimulation of a single command neuron produces complex swimming patterns in the crayfish.

In some simple cases, it is not too difficult to imagine how sensory input and motor output might be linked. For example, a complex pattern of stimuli in the gravitational receptors of the lobster might eventually trigger a single "tilting-a-little-left-and-forward" detector cell in the CNS. This neuron in turn might excite a single command neuron that would eventually trigger the appropriate pattern of swimming actions to correct the particular tilt. No examples of such complete chains between input and output—reception and action—have yet been demonstrated, however. There is little probability that, in the near future, the human nervous system will be explained as a complex automaton. There are still some 9 billion neurons in the associative areas of the brain with unexplored mechanisms of action. The processes by which learning alters relations between input and output remain almost completely unknown (Chapter 29). What is known thus far merely makes more awesome the complexity and mystery of the "enchanted loom" whose patterns direct behavior.

FURTHER READING

More extensive discussions of the physiological basis of sensory systems are given in books by Dethier (1963), Roeder (1963), Rosenblith (1961), Thompson (1967), and in the *Cold Spring Harbor Symposium of Quantitative Biology* (1965). For more psychological approaches to the study of sensory experiences, see books by Békésy (1960, 1967), CRM Books (1970), Graham (1965), Gregory (1966), Hoch and Zubin (1965), Rock (1966), and Sheppard (1968).

Motor systems are discussed by Roeder (1963) and Wiersma (1967). Also see articles by Bullock (1956, 1961, 1962), Delgado (1965), Hildebrand (1960), Kennedy (1967), Roeder (1955, 1962), Tinbergen (1950), Vowles (1961), Wiersma (1962), and Wilson (1961, 1966, 1968).

Various aspects of sensory systems are discussed in articles by Amoore, *et al.* (1964), Békésy (1956, 1957), Bower (1966), Brindley (1963), Burghardt (1967), Detwiler (1965), Fantz (1961), Gazzaniga (1967), Gregory (1968), Hodgson (1961), Hubbard and Kropf (1967), Hubel (1963), Kaufman and Rock (1962), Kennedy (1963), Kohler (1962), Lissmann (1963), Loewenstein (1960), Luria (1970), Neisser (1968), Pritchard (1961), Rock (1968), Rock and Harris (1967), Roeder (1965), Rosenzweig (1961), Rushton (1962), Simmons (1968), Sperry (1964), and Young (1970).

VI
Biological Behavior

Since animals possess by nature such skills in their voluntary actions which serve the preservation of themselves and their kind, and where many variations are possible; so they possess by nature certain innate skills. . . . A great number of their artistic drives are performed without error at birth without external experience, education, or example and are thus inborn naturally and inherited. . . . One part of these artistic drives is not expressed until a certain age and condition has been reached, or is even performed only once in a lifetime, but even then it is done by all in a similar manner and with complete regularity. For these reasons these skills are not acquired by practice. . . . But not everything is determined completely in the drives of the animals, and frequently they adjust, of their volition, their actions to meet various circumstances in various and extraordinary ways. . . . For if everything and all of their natural powers were to be determined completely, that is, would possess the highest degree of determination, they would be lifeless and mechanical rather than endowed with powers of living animals.

—H. S. Reimarus (1773)

27
Behavior: An Introduction

Individual organisms show an ability to adapt to the environment through morphological, physiological, and behavioral changes. A microorganism moves toward conditions of light and temperature that are best suited for its metabolism. *Escherichia coli* responds to the presence of lactose and the lack of glucose by synthesizing enzymes required for lactose digestion. The stems of a plant curve toward a light source, and the plant may respond to changing day-night ratios by flowering or by dropping its leaves. A female mammal responds to a decreasing concentration of estrogens in her bloodstream by secreting hormones that cause new ovarian follicles to mature, releasing more estrogens into the bloodstream. A newborn human infant responds to a touch on the cheek by turning his head in that direction, and if his lips encounter an object, he responds with sucking and swallowing motions. A trained dolphin leaps out of the water and turns a flip in response to a signal from its trainer. A lecturing professor responds to a brief vocal signal (a question) from one of his students with a long and complex pattern of vocal signals and gestures; this pattern elicits a general response of laughter from the students.

Any of these examples fits within the broadest definition of behavior. Biologists, however, generally use the term "behavior" specifically to refer to the rapid actions of animals—actions that involve contractions and relaxations of muscles under the control of the nervous system. The slower changes, mediated by hormonal control systems or other chemical mechanisms, are studied less frequently by those who call themselves behavioral scientists, although these changes play an important role in the adaptation of an organism to its environment. Both the rapid and the slow behaviors help maintain a steady-state relationship, or *homeostasis*, between an animal and its changing environment.

INNATE AND LEARNED BEHAVIOR

Students of animal behavior have long distinguished between two major kinds of behavior: innate and learned. *Innate behaviors*, or instinctive behaviors, are responses that are exhibited automatically by any individual animal, even when it has been raised without contact with other members of its species. The sucking responses of a newborn human, for example, are invariably made by any normal infant and are apparently incorporated into the nervous system during embryonic development. Many complex patterns of mating and nesting behavior in birds are innate, for they are performed without error by birds that have been reared in isolation with no opportunity to observe the behavior of other birds.

Learned behaviors are patterns that are developed or modified as a result of experiences. They provide a variety of flexible responses that are modified to suit changing conditions. In a sense, innate behaviors represent the genetic learning of a species—individuals with the appropriate behavior patterns tend to survive and reproduce. Through evolution, there is a gradual accumulation of genetically determined behaviors appropriate to the conditions under which the species lives. The capacity to learn is also determined genetically and is subject to natural selection. If conditions alter rapidly, each generation can learn to modify its behavior appropriately and can pass this learning along to the next generation without waiting for the relatively slow process of natural selection to alter the gene pool.

The relative importance of learned and innate behaviors in mammals—and even the existence of the latter—has been the subject of controversy

Figure 27.1. Morphological adaptations in the arrow leaf plant *Sagittaria sagittifolia*. These drawings illustrate the adaptation of the plant to three different habitats: terrestrial, semiaquatic, and fully aquatic. These differences are not hereditary but the result of an interaction between the plant's genotype and its environment. On land, the plant shows extensive root development; under conditions of partial submergence, both broad aerial and ribbonlike aquatic leaves are produced. When the plant is fully submerged, only thin, ribbonlike leaves without supportive tissue are formed.

among students of behavior. The distinction between these behaviors is a useful one, because it has sharpened the concepts of behavioral processes, but most behaviors are a mixture of learned and innate responses.

ANIMATE AND VEGETATIVE BEHAVIOR

The most obvious behaviors are the animate activities, which relate the total organism to its external environment. These activities include orientation and coordination of the body and its parts, movement from one place to another, avoidance of or protection from threats and dangers, cyclic changes of state, communication with other organisms, and specific patterned sequences of behaviors. The vegetative activities—respiration, circulation, and nutrition—maintain the internal environment of the organism within a range that permits continued life. The nervous system controls many vegetative activities.

Vegetative activities regulate such crucial variables of the internal environment as water balance, ionic concentrations, nutrient levels, oxygen and CO_2 levels, acidity, levels of nitrogenous waste products, and temperature. To maintain these variables within an optimal range, the organism must exchange materials and energy with the external environment. These exchanges are accomplished through the respiratory, circulatory, digestive, and excretory systems.

Most physiological systems function continuously throughout the life of the organism, and their activity is rhythmic, as in respiration or heartbeat. In some cases, the rhythm is maintained by nonneural systems, and the nervous system simply modulates the frequency or amplitude of the rhythm. In other cases, the rhythm is maintained by specialized neurons or neuron systems, which are in turn modulated by other neural structures. The study of the neural bases of vegetative activity has provided many insights into the nervous system.

Levels of Animate Behaviors

Animate behaviors range in complexity from the avoidance of an object by a paramecium to social interactions and language in man. It is convenient to organize these behaviors into ascending levels, which suggest easy ways to understand their various mechanisms. Any organizational scheme, however, may occasionally encounter examples that fit only clumsily into its arbitrary categories.

Reflexes are the actions most simply and directly coupled to environmental stimuli. They are elicited by specific stimuli and consist of coordinated effector responses, which are modified only slightly from occurrence to occurrence. Examples include the paramecium's avoidance response and the mammal's eye blink, knee jerk, and sneeze. Reflexes sometimes interact with each other to provide complicated adjustments of the organism relative to its environment. It was once thought that the reflex was the elemental unit of behavior. This view lost favor as it was discovered that many reflexes can be modified through experience.

Most animals maintain a specific *orientation* of the whole body to some aspect or aspects of the external environment. In most cases, this orientation does not result automatically from the shape of the body (as it does in a starfish) but requires energy expenditure and coordinated movements. Neural controls, therefore, must detect any departure from the animal's normal position with respect to some reference point in the environment,

Figure 27.2. Sequential stages in the courtship and mating of the three-spined stickleback fish. The male is attracted by the female's swollen abdomen, and he approaches and chases her. He then maneuvers her into his bubble nest and induces her to lay eggs. After the eggs are laid, he will chase her out of his territory and fertilize the eggs.

Figure 27.3. Euglena (upper left) showing positive phototaxis. Rotifer (lower left) showing kinesis. (*Eric Gravé*)

Figure 27.4. Streak photographs of the avoidance response tracks made by free-flying moths when exposed to a source of ultrasonic pulses mimicking a bat. (A) Track of a moth that did not react to the ultrasonic stimulus (start at arrow). (B) A distant moth turns and flies away from the sound source at left. Turning away from the sound source occurs only when faint sounds reach the moth. At this range, a real bat's sonar could not detect the prey, so this maneuver is a logical one for the slower-flying moth. (C) Turning away as in B. (D) Erratic twisting during power drive in response to intense sounds. The moth eventually stopped flapping and made a free fall to the ground.

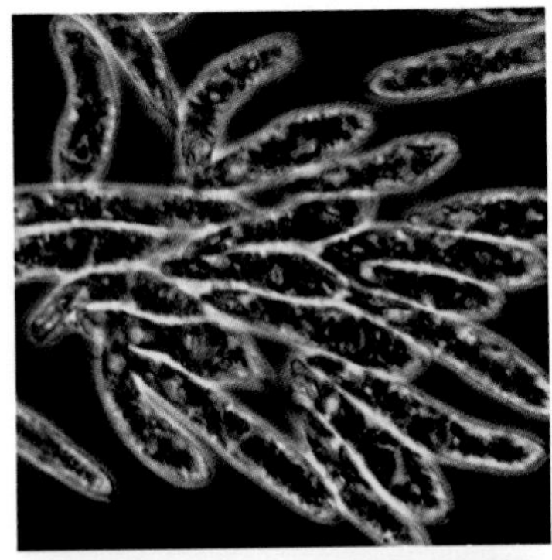

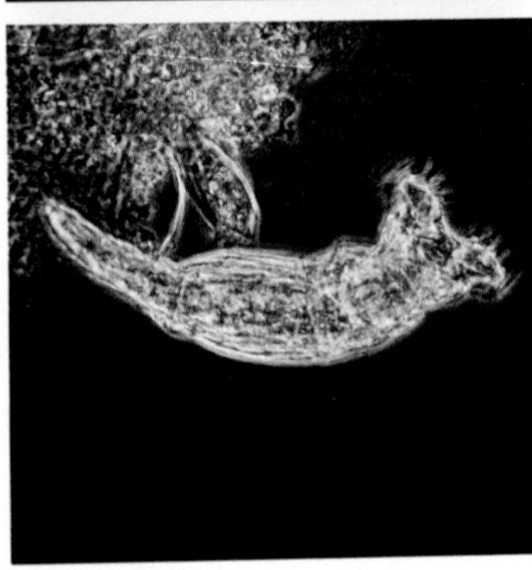

and these controls must initiate muscular movements that will restore the proper orientation to the organism. The most basic and pervasive reference points are the gravitational field (up and down) and the direction of illumination. Contact with the substrate surface, chemical gradients, and wind or water currents can also serve as orienting references for some organisms. There is evidence that some birds use the positions of the stars to orient themselves as they fly over migration routes (Sauer, 1958).

In relatively simple animals, orientation may be accomplished in two different ways. An organism may sense where more favorable conditions lie and move in that direction (*taxis*), or the animal may simply move faster under unfavorable conditions and slow down upon encountering favorable conditions (*kinesis*). Higher animals possess more complicated sense organs that provide the nervous system with enough information to calculate what movements it must make to achieve optimum orientations. Animals usually use appendages to move. Various muscular responses must be coordinated to produce reasonable orientations of appendages (Chapter 23). With the exception of sessile animals, which remain fixed in a single location on a substrate (examples are numerous in the sea), most animals move bodily from one place to another by flying, walking, swimming, or crawling. This locomotion normally is accomplished through specific sequences of coordinated, oriented activities.

Among the most important aspects of an individual's behavior are those activities that serve to protect the animal when it is threatened with danger. In most cases, the *avoidance response* is given a high priority by the ner-

A	B
C	D

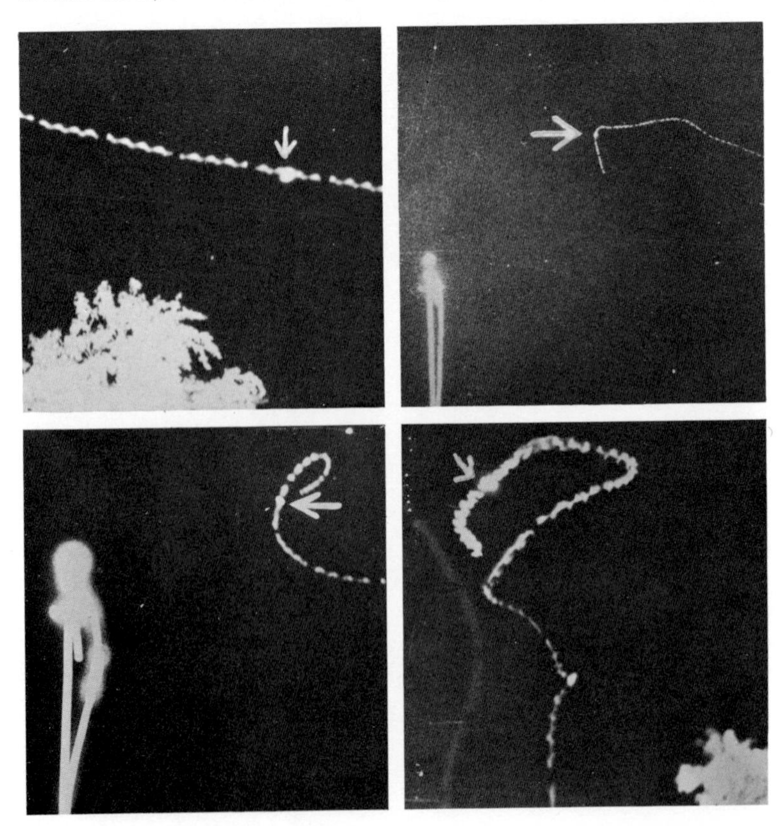

Figure 27.5. Avoidance responses are shown in the group below. The scallop (left) *Chlamys sperculoris* swimming to escape attack by its predator the sea star (*Asterias rubens*). Sow bug (upper and lower right) closing up as an avoidance response. Madagascar day gecko camouflaged (below).

Figure 27.6 (upper right). This figure resembles a hawk if moved to the right and elicits escape responses in young turkeys. If moved to the left, however, it casts a gooselike shadow and fails to elicit avoidance behavior.

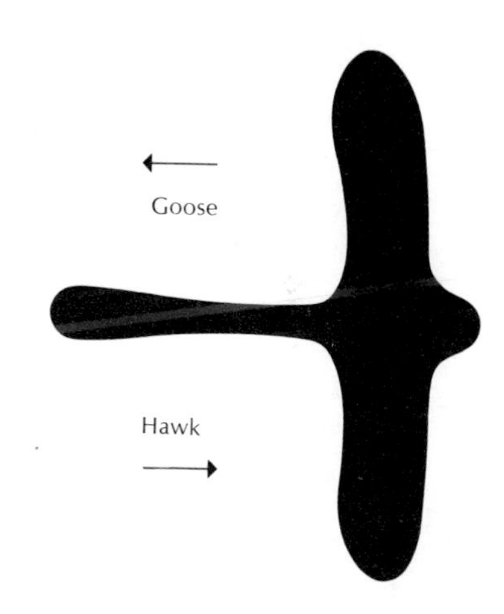

vous system—it tends to take precedence over other possible behaviors. Avoidance behaviors are often quite specific and differ greatly among species. Forest animals flee from an approaching fire. A sudden dimming of illumination—such as might be caused by the shadow of an approaching predator—elicits a sudden and rapid withdrawal of the annelid worm *Sabella* into its tube. The call of a hunting bat may cause a pursued moth to fold its wings and drop toward the earth. Unusual vibration may cause a caterpillar to freeze motionless, looking much like a leafless twig. A normally sessile sea anemone that is touched by a predatory starfish may release its hold on the substrate and swim away violently. Avoidance activities are quite varied; many involve orientation and locomotion. Each activity is elicited by a stimulus that has presumably been associated with danger to an organism during the evolution of a species. The response has survival value for the individual and therefore for the species.

Where avoidance responses help an organism elude danger, *protective responses* minimize damage after danger strikes. Direct injury requires an immediate response from an organism. The first step in any protective response must be an awareness that injury has occurred. Sensory receptors usually provide information about the nature of damage. On the basis of this sensory input, the nervous system generates appropriate coordinated behaviors. Protective responses in mammals, such as blinking and tearing, grooming of wounds, and limping, are familiar to everyone. Examples are found in all groups of animals. A strong pinch to a crab's leg may cause the appendage to break off at a specialized joint. A pinched or damaged earth-

Figure 27.7a (above). An example of protective mimicry. This IO moth has false eyes on its wings to frighten away potential predators.

Figure 27.7b (below). An example of cryptic coloration. The flounder has the ability to change its color pattern to blend in with the background. This camouflage technique is under the animal's neural control.

Figure 27.7c (middle). Three moths camouflaged on the side of a sycamore tree. Many insects have not only undergone morphological adaptations but also behavioral adaptations in perfectly mimicking the form and response of other organisms or inanimate objects.

Figure 27.7d (far right). Protective mimicry in the spicebush larva. This organism also has false eyes.

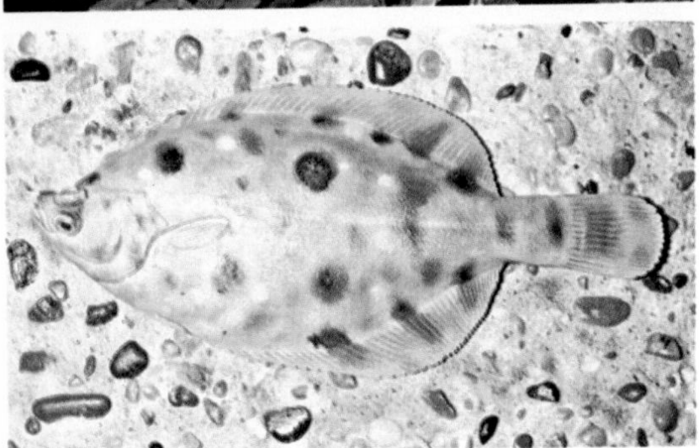

worm may break completely in two, discarding the damaged segments. A lizard may discard its tail if it is grabbed or damaged. These protective behaviors—even those involving self-inflicted damage—increase an animal's chances for survival.

With few exceptions, every organism on earth is subjected to regular, cyclic changes of environmental conditions. Light and temperature vary in a day-night (diurnal) cycle. Currents, water depth, and water temperature vary with the tidal rhythms, and day length, temperature, weather, and sun angle vary with the seasonal cycle. In many cases, environmental changes are so great that optimal behavior during one phase of the cycle is completely inappropriate at another time. Most animals have developed cyclic changes of behavioral state that correspond to these environmental cycles. A diurnal animal sleeps at night and is active during the day; a crepuscular animal is active at dusk or dawn; a nocturnal animal sleeps during the day. Animals living in temperate and subarctic zones show seasonal variations in reproductive activity and may migrate or hibernate at specific phases of the cycle. Such cyclic changes involve profound shifts in the general state of the organism, including but not limited to the nervous system. In many instances, the nervous system is involved because it is the source of neurohormones that trigger general metabolic and behavioral shifts. The precise timing of cyclic activities is usually determined by environmental cues. In many organisms, cyclic activity has become somewhat independent of environmental cues, and the cyclic changes of state may continue regularly in the absence of these cues. This ability to maintain an independent cycle of behavior has been taken by many biologists as evidence of an internal clock or clocks. The evidence suggests that the timing mechanisms of higher animals are located within the nervous system (Chapter 28).

Specific patterned sequences of behaviors are sequences that tend to recur more or less unchanged at appropriate times in the animal's life. Although reflexes may be involved, specific patterned sequences cannot be described simply as a chain of reflexes, each of which triggers the next. A sequence can have both learned and innate components, and it may be modified according to the animal's state, surroundings, and experience. A patterned sequence may be triggered by a specific stimulus or by a broad range of stimuli and circumstances. The latter category includes many directed activities, such as food getting, preparation of or search for shelter, reproduction, and so on. Examples of *stereotyped patterned sequences* include prey captured by the praying mantis and toad and nest building and grooming by birds. These behaviors are relatively unchanged from one occurrence to the next and have been called fixed action patterns. Other patterned sequences, which vary according to conditions or which contain substantial learned components, include the hunting activity of carnivores, the play activity of kittens, and the lovemaking activity of humans.

The behavior of one animal that affects or has the potential of affecting the behavior of another animal is termed *communicative* behavior. This type of behavior allows cooperation between two or more individuals, usually but not always of the same species. Communication usually involves temporal and spatial patterns of movements. These components may be specific patterned sequences, and they may be innate or learned. Human speech, bird song, and courtship in vertebrates and invertebrates are common examples of communicative behavior. The chemical signals of ants and mammals, ritual fights of some fish, birds, and mammals, and the dance of honeybees are less well known.

One of the most complex kinds of behavior that has developed in the animal kingdom is *social behavior*. This category, broadly defined, includes any kind of communicative and cooperative behavior—such cooperation between animals is the essence of social action. Many animals participate in group tasks where the action of one individual is relatively useless without the actions of others. The success of the species rests upon collective effort, rather than upon a few successful individuals. Striking examples of social behavior occur in insects and in mammals—ants, bees, prairie dogs, primates, and so on. Also, school behavior in fish and flock behavior in birds (including some migratory and group nesting behavior) must be regarded as social. Some phases of amphibian and reptilian behavior resemble social behavior. For example, reptiles are notably territorial, and this behavior involves much communication. Invertebrates may group together because there are others of their species present, but it is difficult to tell if the animals are cooperating, or merely forming a desired part of each other's surroundings. There is little doubt, however, that social aspects of group behavior do represent a distinct level of organization.

BEHAVIORAL STATES

A particular set of environmental cues may not evoke the same response in an organism every time it is encountered. On one day, a frog may be interested in black buzzing insects and may spend hours catching and eating them. A few months later, the frog may pay scant attention to insects but spend its time chug-ga-rumming away, preparing for the courtship season. Similarly, a field mouse may, at dawn, start foraging for seeds, whereas at dusk, under similar light and moisture conditions, it may prepare a nest in which to sleep. The sight of a large steak may evoke a complicated set of socially guided behaviors in a hungry college student, but after dinner, the sight of an identical steak may repel him. These examples show a change in the behavioral state of an organism. Numerous conditions in an animal's physiology and nervous system, including the effects of recent and distant past experiences, determine its behavioral state. The concept of behavioral

Figure 27.8 (above). The male stickleback fish may ignore the uncolored top model but readily attack any of the lower ones with a painted red "belly". Note, however, that all patches are on the underside of the model. Apparently, both color and position are important in eliciting the attack response.

Figure 27.9 (below). The feeding response of young herring gulls is most frequently elicited by the model at the bottom. The color patch on the model seems to be a key stimulus in releasing feeding behavior.

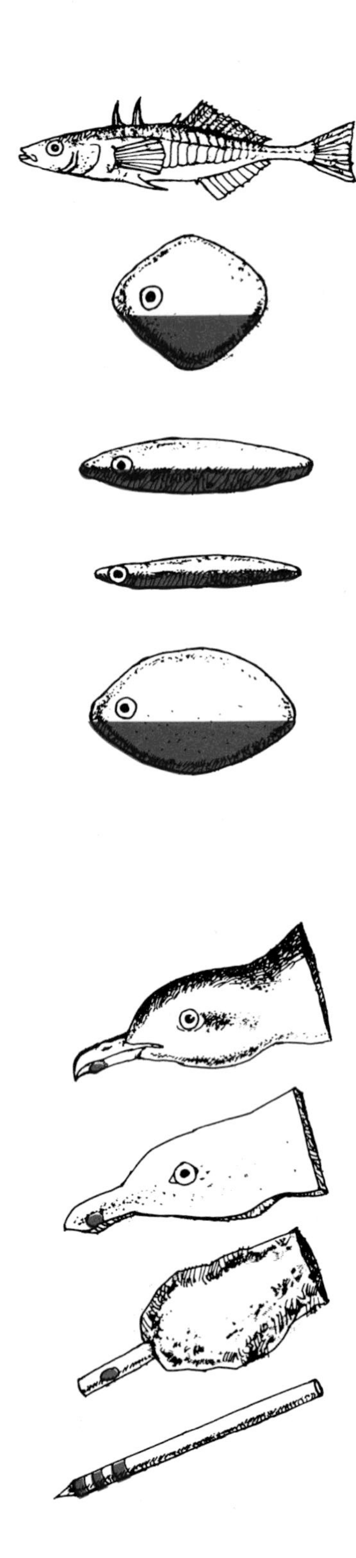

state includes some factors that are relatively well understood, such as hormonal balance, hunger, and so forth. Other factors are included only because the animal's behavior is different now from the way it was then, under similar circumstances, and there must be a reason. Animal experimenters feel that certain human concepts, such as mood, emotion, and whim, have parallels in other animals and that they have bases in the state of the nervous system.

THE SENSORY EXPERIENCE OF ANIMALS

Behavioral acts are triggered by or coordinated with events or stimuli in an animal's environment. The ensemble of environmental signals that can affect an animal's behavior or behavioral state can be called its sensory experience. There is strong evidence that the sensory experiences of other animals differ greatly from those of humans. The sensory world that produces measurable effects on behavior is highly limited. The European naturalist Uexküll described the behavior of a female tick who climbs into the branches of a tree after mating and waits for many weeks until a mammal passes directly under her. Until this time, she appears unresponsive to the barrage of sights, sounds, and odors around her. Only the specific smell of butyric acid generated by mammalian skin glands causes her to respond by dropping from the tree onto her host. In Uexküll's words, "The tick acts like a gourmet who picks the raisins out of a cake."

Students of animal behavior have long recognized that various animal species appear to be aware of different stimuli in the external environment and to respond to various aspects of those stimuli. Although humans are acutely aware of shapes and forms that are perceived visually, many animals show little response to similar changes of their environment. Even animals that do respond to visual stimuli seem to react to aspects of those stimuli quite different from the features that seem obvious to humans.

Ethologists contributed the notion that an *innate release mechanism* exists in an animal's nervous system and that this mechanism recognizes the *releaser* (appropriate environmental stimulus pattern) and triggers a behavioral sequence. The concept is parallel to that of sensory filtering in the nervous system. For example, the male stickleback fish will attack a crude model of another stickleback if the model has a red belly, but it will ignore models that appear much more fishlike to humans if the models lack the red belly (Figure 27.8). Red breast feathers are the most efficient releasers in provoking an attack by a male robin that is setting up its territory. The feeding response of young herring gulls is more affected by the presence or absence of a red patch on the beak of a model of the parent than by distortions that make the model seem very unbirdlike to a person (Figure 27.9).

Studies of these behaviors reveal that the response is not automatically elicited by the presence of a single stimulus. The red patch on a model stickleback is much more effective in eliciting an attack if it is placed on the underside rather than on the top of a model. Both the color and the position of the patch are important in eliciting the attack response. The male stickleback does not always show the same strength of response to a particular model. The probability of a response and its duration and vigor are affected by changes in the internal state of the animal.

A rather remarkable finding is that some artificial stimuli appear to be more effective than natural stimuli in releasing particular behavioral responses. These artificial stimuli are called *superoptimal stimuli*. Oystercatchers and gulls, for example, will attempt to brood a model egg several times larger than their own eggs and will actually choose the giant egg in a preference test, even when the model egg is so large that the bird cannot assume a normal brooding position (Figure 27.10). Thus, quantitative as well as qualitative aspects of an environmental stimulus appear to be important in determining a given response.

The sensory apparatus of any particular species has characteristic limitations. Man, for example, normally cannot detect the polarization of light, but many invertebrates use this property as an important sensory clue for orientation behaviors. Bees are sensitive to ultraviolet light, but man is not: The upper limitation of auditory sensitivity in man is about 20,000 cycles per second, but bats, porpoises, and some rodents can hear up to 100,000 cycles per second. Bats and porpoises navigate by means of echolocation involving frequencies far too high for human hearing. Man can only guess about the nature of the auditory worlds of these animals.

A living fly and a dead fly may appear quite similar to a human, yet these are totally different stimuli to a frog. Some neurons in the frog's brain respond vigorously to a moving black spot, but a dead fly may not even enter the frog's awareness. If the frog does respond, it is an experience totally different from that produced by a moving fly.

To understand behavior, one must understand both the sensory experience by which an animal judges change in its world and the actions that change provokes. In the remaining chapters of this unit, a few examples of behavior will be studied in some detail, with particular emphasis on those cases in which there is some understanding of underlying neural and hormonal mechanisms of sensation and action.

Figure 27.10. An oystercatcher attempting to brood a superoptimal model egg several times larger than its own egg. The bird's preference remains steadfast even though it cannot assume a normal brooding position.

FURTHER READING

General descriptions of modern understanding of animal behavior are given in books by Dethier and Stellar (1970), Hinde (1966), Marler and Hamilton (1966), and McGill (1965).

Books by Lorenz (1952, 1965) and Tinbergen (1951, 1953, 1958) provide enjoyable introductions to ethological studies. Also emphasizing the ethological approach are articles by Dilger (1962), Eibl-Eibesfeldt (1961), Hailman (1969), Lorenz (1958), and Washburn and DeVore (1961).

Typical of the neurophysiological approach to behavior are books by Butter (1968), Ochs (1965), Stevens (1966), and Wells (1962), and articles by Delgado (1970), Isaacson (1970), Melzack (1970), Muntz (1964), and Sperry (1956, 1959).

28
Biological Clocks

Rhythmic events recur everywhere in the environment. The sun rises and sets, the moon moves through its monthly cycle, and the tides and seasons repeat their inexorable rhythms. Their continuity and precision are proverbial: night follows day; and when winter comes, spring cannot be far behind. Temperature, light, humidity, weather, and other environmental characteristics vary along with these cycles.

Animals must adjust their activities to these cycles in order to avoid adverse conditions, to be able to use their senses to best advantage, and to maximize their chances of obtaining adequate food and of producing offspring. They show rhythms in both animate and vegetative behavior. The solar day-night cycle or lunar (tidal) cycle commonly is expressed in activities such as metabolism, locomotion, and food getting. Annual or seasonal behavioral change occurs in relation to reproduction or to periods of adverse weather such as winter or drought. Recently, biologists have realized that direct stimulation by environmental rhythms is not the only factor that guides this rhythmic behavior by animals.

TIMING OF BIOLOGICAL CLOCKS

When organisms are removed from their habitat and placed in conditions of constant light intensity, temperature, pressure, humidity, and chemical composition of the environment, they may continue to exhibit nearly the same rhythmic fluctuations of behavior they showed in their natural lives.

Two categories of hypotheses can be put forward to explain this persistence of biological rhythms. The *exogenous hypothesis* suggests that there are always subtle geophysical factors that affect even isolated organisms. These geophysical factors vary cyclically and *entrain*, or synchronize, the organism's behavior. Geomagnetic fields, cosmic rays, or electrostatic fields have been suggested as entraining stimuli. The *endogenous hypothesis* suggests that organisms possess internal *biological clocks* that determine the timing of changes in behavior and behavioral state. Stimuli from the outside world serve only to reset this clock periodically so that it does not get out of phase with the physical world.

Thus far, it has been impossible to obtain definitive experimental results to indicate that one or the other of these general hypotheses is the correct one. Some researchers suspect that organisms both possess internal timers and respond to subtle environmental clues. Biological timing mechanisms have also been invoked to explain animal homing and direction finding as well as seasonal changes of behavior and physiology. These phenomena seem to require the animal to determine the time of day, or at least the length of the day or night. In the case of homing, many animals navigate by the position of the sun or stars. It seems that the animal must know the time to learn directions from position of celestial bodies. It has been suggested that to synchronize behavior, such as reproduction, to the seasons, an animal needs to measure the length of day or night (days lengthen in the summer) or needs to measure the relation between environmental day and night cycles to an internally generated rhythm. It seems likely that timing phenomena that control these behaviors share some mechanisms with those that control rhythmic activity, but the proof is yet to be discovered.

CIRCADIAN RHYTHMS

Activities that are synchronized to the day-night cycle—that is, have a one-day period—are called *diurnal* cycles, and those that persist with a period

Figure 28.1 (left). Twenty-two day record of the running activity of a small mammal kept in a laboratory darkroom without the daily light-dark cycle. The chart records of each day are placed below one another, making it easy to see that the activity begins earlier each day.

Figure 28.2. Altering the phase of the free-running rhythm of luminescence of *Gonyaulax* cells. All cultures were originally standardized on a 12-hour light cycle and then a 12-hour dark cycle. At the end of a light period, all cultures were placed in the dark. The control remained in the dark, while each experimental culture was subjected to a different 2½-hour light pulse at 1,400 footcandles at 21°C during the time blocks indicated. Light pulses administered during "early night" caused a phase delay, whereas those administered during "late night" produced a phase advance.

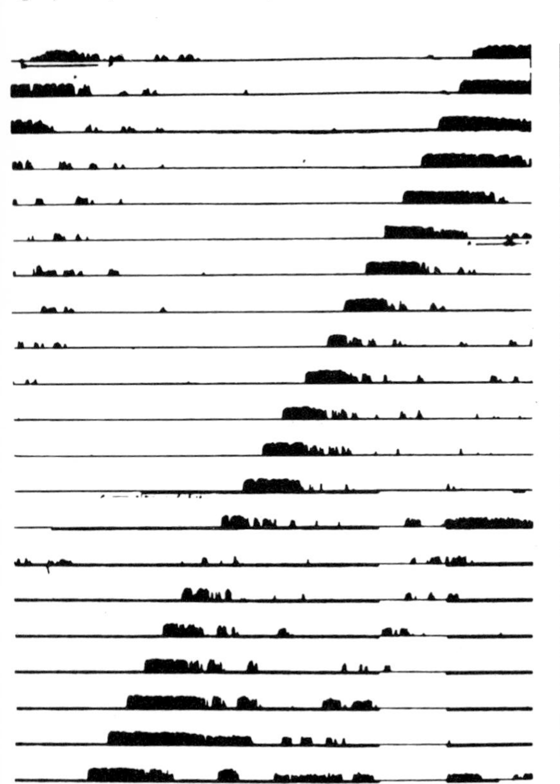

of about one day under constant conditions are called *circadian* rhythms. If there are no external rhythmic stimuli, the *free-running* behavioral cycle is usually a bit different than 24 hours and differs slightly from one individual to the next. Figure 28.1 records the running activity of a small mammal kept in constant darkness in the laboratory. The fact that the activity period drifts in phase relative to real day and night and, more importantly, that it drifts differently in different littermates, poses a problem for proponents of the exogenous theory.

The period of a free-running circadian rhythm shows a slight dependence on temperature. An ambient temperature change of 10°C may speed up or slow down the rhythm by about 10 percent. It is difficult to reconcile this change to the exogenous theory. Such a change also poses problems for endogenous theorists. Chemical reactions, which are usually postulated to be the bases for internal clocks, are also altered by similar temperature changes. As yet, no mechanism for temperature compensation has been proved to account for the small magnitude of temperature effects.

Exogenous theories, of which F. A. Brown, Jr., is an outspoken proponent, suggest that free-running cycles differ from 24 hours and the temperature effects on these cycles are accounted for as follows. The animal does not measure the time of occurrence of peaks in hypothetical cycles but

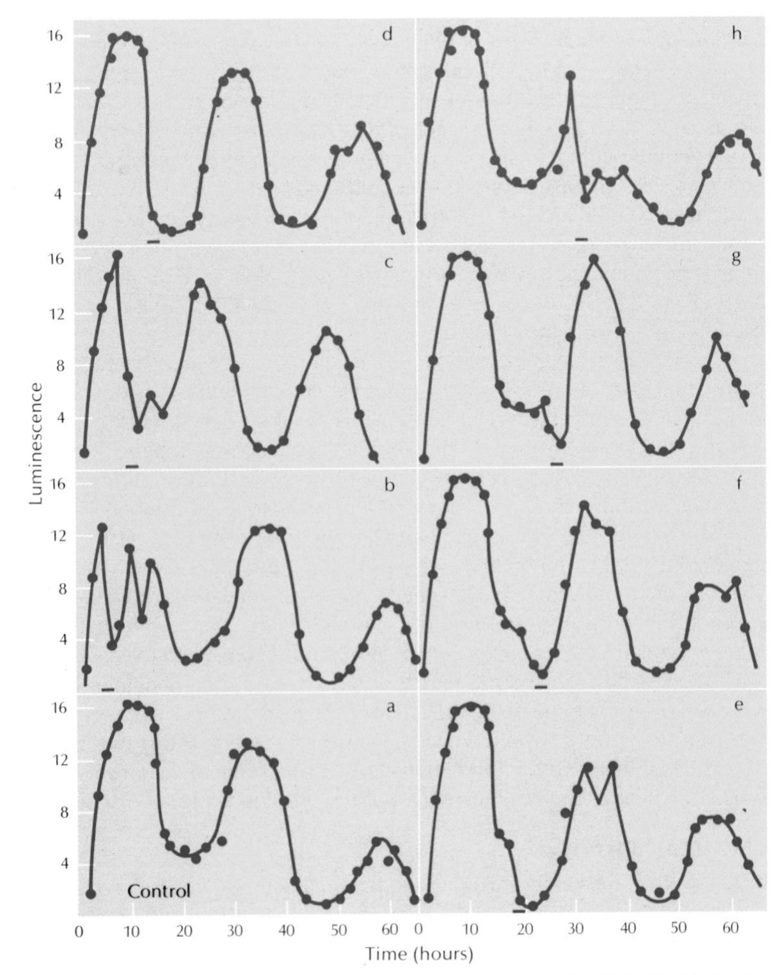

Figure 28.3 (left). Periodicity of spore discharge in the ascomycete fungus *Daldinia concentrica.* The rate of spore discharge can be correlated with time and conditions of illumination. Spore discharge is highly rhythmic within the 24-hour period and is confined chiefly to the dark period, although discharge rate picks up with fading light (dusk) conditions. Day-night periods are indicated by the horizontal strip below the bar graphs. July 6 to July 8 reflects entrainment to a 12,12 light-dark period; July 8 to July 12 reflects entrainment to a 6,6 light-dark period; July 12 to July 16 reflects a free-running rhythm in total darkness. Shown at right is a diagram of the apparatus used for studying the rate of spore discharge.

measures the rate at which they occur. The phase of the animal's behavioral cycle is thus not dependent on the phase of the physical cycle. If an experimental animal expects the conditions of darkness and cool temperature to occur at a certain time, but they do not, then the animal shifts the phase of its activity relative to the basic geophysical cycle. The amount of the phase shift depends on the ambient temperature. The experimenter who reports a change in the rhythm's period is actually measuring a change in the magnitude of a phase shift.

If an animal exhibiting a free-running rhythm is exposed to brief periods of certain stimuli (notably light) at certain times within its activity cycle, the phase of the free-running rhythm may be altered. This alteration is presumably the mechanism by which the endogenous clocks are kept synchronized to the changes in the external world. Proponents of both exogenous and endogenous theories must account for this phasing phenomenon.

Circadian Rhythms in Plants and Animals

As an example of circadian rhythm in a simple life form, consider the ascomycete fungus *Daldinia concentrica*, whose black, ball-like fruiting body lives on dead matter of trees. Its outside is pitted with numerous cavities, each lined with many sacs that ripen and periodically discharge spores. Under natural conditions, most spores are ejected at night; as many as 100 million spores may be liberated overnight by a single fungus. This phenomenon has been investigated in the laboratory in a rather clever and unusual manner. Detached, ripe fruiting bodies, which continue to sporulate over a period of several weeks, are mounted on a model-railroad flatcar and covered with a transparent top with an opening 1 centimeter wide. The loaded flatcar is slowly pulled along a short length of track at a uniform rate by a clock motor and pulley system. On a framework above the rack, 12 glass slides are attached so that the car passes beneath a new slide every 2 hours, traversing the entire set of slides in a 24-hour period. As the sticky spores are discharged through the opening in the covering, they strike and stick to the glass slides. Later, they are washed off and counted. In this manner, the rate of spore discharge can be continuously monitored over the reproductive life of the fruiting body. Spore discharge is highly periodic within the 24-hour period and is confined chiefly to the dark period, although the rate of spore discharge begins to rise before the light ceases (Figure 28.3). Similar observations have been made for the sporulation rhythms of other fungi and for the rhythm of zoospore release in the green alga *Oedogonium cardiacum.*

Some unitary events (which occur only once in an individual's life) are timed by circadian mechanisms. The emergence rhythm of *Drosophila*

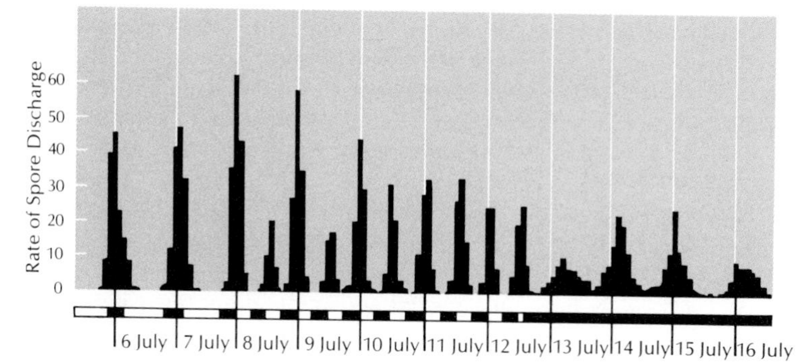

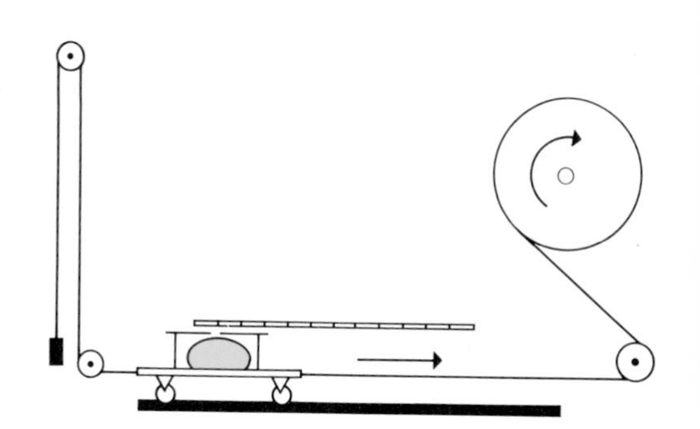

Figure 28.4. The alteration of an innate circadian rhythm controlling the emergence of adult *Drosophila* is illustrated by experimental control of environmental conditions.

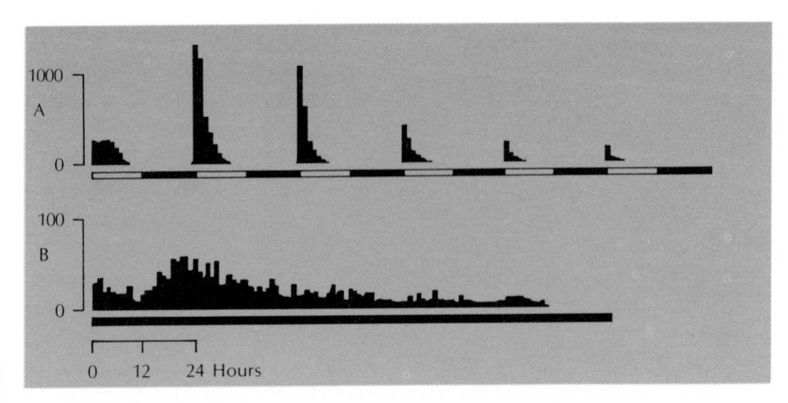

pseudoobscura has been extensively studied using a highly automated system. The so-called bang-box at the heart of the system is furnished with water-jacketed lights, heaters, and clocks and can provide any desired light or temperature sequence. Within the box is a large, square, plastic canister that tapers at the bottom to a small opening. *Drosophila* pupae at various developmental stages are placed in cotton dental plugs, which are inserted in the opening. The canister is suspended from a magnetic device so that the opening is directly above a vial of detergent liquid on a turntable driven by a spring-loaded ratchet device. Every half hour the suspended canister is picked up and dropped ("banged") about 15 to 20 times, thereby knocking into the waiting vial below any adult flies that have emerged during the preceding half hour. The entire turntable tray with its 48 vials is replaced each day. The number of dead flies in each vial is counted, yielding a time distribution of emergence in the population. During a typical experiment, which lasts seven or eight days (corresponding to the total time span of emergence in the population), several thousand flies might emerge.

Figure 28.4 shows the distribution of emergence activity in a population of *Drosophila* maintained at 21°C in an environment where 12 hours of light alternated with 12 hours of darkness. On about day 17 of the 25-day life cycle of the flies, emergence began at about the onset of the light period. Within a few hours emergence stopped, but another peak of activity occurred exactly 24 hours after the first one. The amplitude of the 24-hour rhythm gradually decreased from day to day as the population was "used up," until all flies had emerged by the day 8. Under natural conditions, this species of *Drosophila* is often observed to emerge from the pupa cases a few hours after dawn—a time when humidity tends to be high, a favorable condition for the delicate, emerging fly whose cuticle is still hardening and whose wings are still unfurling as they are pumped up.

Circadian rhythms are present in all mammals, including man, that have been studied. Besides sleeping and waking cycles, isolated men show 24-hour variations in concentrations of chemicals in blood and tissues, in rate of cell division, heart rate, and excretion. More elusive phenomena, such as ability to estimate time or the error rate in problem solving, also exhibit a circadian periodicity.

The most extensive and important studies of human circadian rhythms under constant environmental conditions have been carried out by J. Aschoff and his colleagues at the Max Planck Institute in Germany (Aschoff, 1965, 1967). They have built special underground bunkers with facilities

Figure 28.5. Circadian rhythm of urine excretion (calcium, potassium, and water), body temperature, and wakefulness in a human subject kept under constant environmental conditions in complete isolation.

for human subjects to live in complete isolation for periods of a few weeks.

Measuring devices and controls are outside the bunker. Body (rectal) temperature and the activity pattern are measured continuously; the subject's estimation of time is also recorded. The subject collects his urine at intervals of his choice and stores it in a refrigerator in a small room that serves also for the transfer of mail, food, and one daily bottle of beer. He is asked to lead a "regular" life, to have three meals a day, but not to nap after lunch. Except for carrying out the few experiments (time estimation and so forth), he is free to do whatever he wishes. Subjects are generally not bored or unhappy, and they keep themselves occupied by such activities as studying, reading, and listening to recorded music.

Under such conditions, most human subjects exhibit a circadian day that differs from the normal 24-hour day by only an hour or two. For the great majority of subjects, the period has been observed to be greater than 24 hours (Figure 28.5). After several days, the subject is out of phase with the external world, a fact that he is completely unable to ascertain or sense. In a few individuals, circa-48-hour activity periods occurred without the subjects' suspecting that they were living differently from normal. Subjects can be entrained to light-dark cycles of slightly different periods from their natural one, if these cycles do not differ greatly from 24 hours. One subject was kept in isolation and constant conditions for 19.5 days, during which he experienced 18 "physiological days." At the end of that time, when he was about 12 hours out of phase, the normal light-dark cycle was restored, and the subject quickly shifted phase and became entrained again to the normal cycle. As anyone who has traveled by jet across several time zones knows, such a shift is sometimes an uncomfortable experience whose effects may persist over a period of several days. In the experiment shown, it took a number of days for all of the various rhythmic physiological functions to become fully entrained to the light-dark cycle after the shift.

Such experiments have great importance for the planning of human environmental conditions. In several modern situations, individuals are subjected to artificial activity-rest cycles or schedules. Efficiency as well as accident and error rates of workers vary with time of day, but little is known about the effects of various shift schedules upon these normal circadian

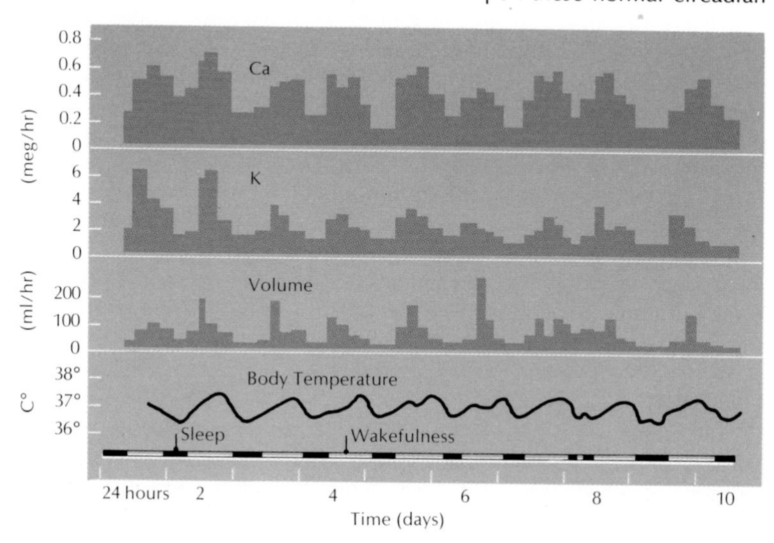

Figure 28.6. Photograph of the fiddler crab *Uca minax*. These intertidal animals exhibit rhythmic change in coloration that can be correlated with light and tidal cycles.

rhythms. Airline pilots who cross time zones, and astronauts on space missions are deprived of normal environmental cycles. Whether this affects them adversely is not known.

TIDAL AND LUNAR RHYTHMS

Although much research in biological rhythms has been concerned with circadian cycles, the study of tidal (12.4-hour), lunar daily (24.8-hour), and lunar monthly (29.5-day) rhythms has raised many fascinating problems (Brown, 1960, 1965). The rhythms of color change and of locomotory activity in the fiddler crab (*Uca minax*) are an extensively investigated example. These crabs, which inhabit the intertidal zone, tend to be darker during the daytime and lighter at night. This color change is accomplished by movements of pigment in chromatophore cells of the epidermis. This color-change rhythm is circadian; it can be entrained by light-dark cycles and will persist in constant darkness for at least two months with little change in period.

However, the color change rhythm also has tidal components. The degree of darkening is greatest when the time of low tide coincides with the period of dark color. Crabs captured on the beach and then kept in continual darkness show the highest degree of darkening at the same clock hour every 15 days, representing the time of coincidence between peaks of the 24.0-hour light-dark cycle and the 12.4-hour tidal cycle. Entrainment with artificial light cycles indicates that both components of the rhythm are phase-shifted by light stimuli. The crabs' activity rhythm is also entrained to the tidal cycle.

It thus appears that a given rhythm in a single organism can simultaneously have both overt tidal and circadian components. Furthermore, the same organism may exhibit still other rhythms that have both tidal and daily periods. Can a single "master timer" control all of these differing rhythms? Or are there separate circadian and lunar clocks underlying the phenomena? Finally, how are these rhythms synchronized and reset? Research is under way to provide answers to these questions.

SEARCH FOR THE BIOCHEMICAL CLOCK

Unicellular organisms are favorable material for the study of the chemical basis of biological rhythms. J. W. Hastings and his coworkers have studied *Gonyaulax*, a dinoflagellate that exhibits circadian rhythms in luminescence, respiration, and photosynthesis. They found that various metabolic poisons can affect the circadian rhythms, but they are not sure if the poisons interfere with the primary clock mechanisms. In particular, actinomycin D (which inhibits DNA transcription into mRNA) stopped the luminescent rhythm. But the rhythm did not halt until after one peak of luminescence had occurred, suggesting that DNA transcription and mRNA translation into protein are required for the luminescent cycles. However, actinomycin D did not affect the photosynthetic periodicity, so the DNA-mRNA-protein oscillation could not be the master clock for *Gonyaulax*.

Working with mammals, R. J. Wurtman and Julius Axelrod (1965) and others have discovered that many organs and glands exhibit circadian rhythms. Particular attention has been given to neuroendocrine organs, such as the adrenal glands, and the pituitary system. Hamster adrenal glands, if maintained *in vitro*, show some metabolic cycles that can be entrained by light-dark cycles. In the intact animal, a hormone secreted

Figure 28.7. Photograph of the dinoflagellate *Gonyaulax polyhedra.* This plankton organism, well known for its production of "red tides," also exhibits circadian rhythms in luminescence, respiration, and photosynthesis.

Figure 28.8. Rhythm of luminescence in the dinoflagellate *Gonyaulax* reflecting entrainment to a 12-hour light-dark period.

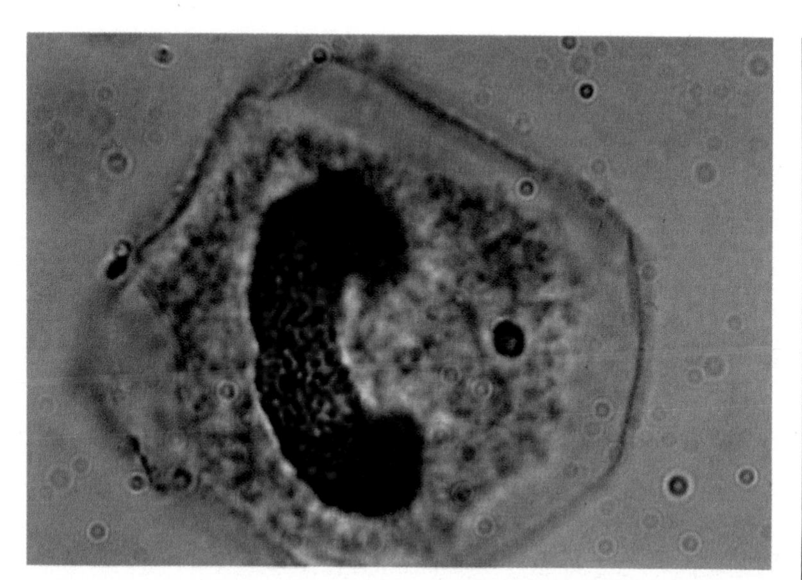

with a circadian rhythm from the adrenal gland controls many body rhythms, such as glycogen in the liver and epidermal mitotic activity. These cycles stop if the adrenal glands are removed, but the hamster's activity rhythm continues. Thus, the adrenal glands are not the seat of the master biochemical clock. Investigators have, in fact, begun to suspect that there are several clocks that tend to be synchronized in healthy animals.

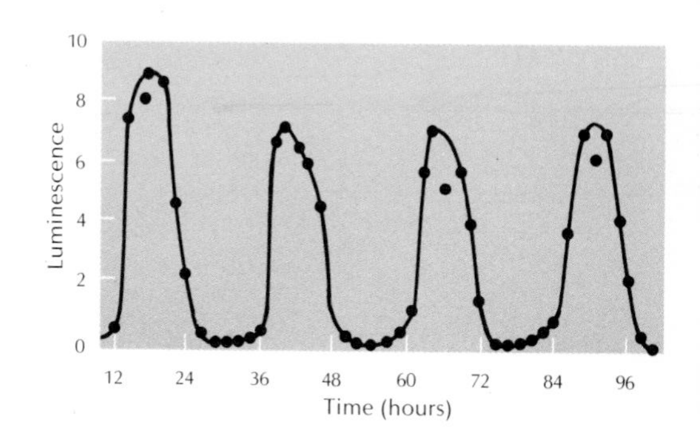

PHOTOPERIODIC PHENOMENA

Many annual and seasonal biological rhythms are photoperiodic phenomena, in that changes are triggered by certain critical timings of the light-dark periods. Among these phenomena are the flowering of plants, the germination of seeds, the hibernation of vertebrates, changes from one stage to another in the life cycle of insects, and the initiation of reproductive cycles and migration in birds and mammals.

As early as 1920, botanists recognized the importance of the length of daylight in initiating the change from vegetative to reproductive phases of plant growth. Tobacco and soybeans, for example, flower only when the day is shorter than a certain critical length, whereas other plants flower only when daylength exceeds a certain critical period. Both hourglass and clock models were proposed to account for photoperiodic phenomena.

According to an hourglass model, daylength is measured through accumulation or disappearance of some cellular component. When the level of this substance reaches a critical value, the photoperiodic phenomenon is triggered. Each day the process is started over.

The search for such a timing mechanism was rewarded with the discovery of *phytochrome*, a protein pigment that has subsequently been found to be involved in a number of plant physiological processes. Phytochrome exists primarily in two forms, red-absorbing P_r and far-red-absorbing P_{fr}. During the day, absorption of red light (wavelength, 660 millimicrons) causes the very rapid conversion of P_r to P_{fr}. During the night, in the absence of red light, the P_{fr} is slowly converted back to P_r. P_{fr} is thought to inhibit the formation of a flowering hormone. In short-day plants, flowering occurs only when the dark period is long enough to decrease the P_{fr}

Figure 28.9a. Testicular enlargement in male white-crowned sparrows is a circadian rhythm entrained to a long-day photoperiod. An experimental group was kept on a schedule of 6 hours of light and 18 hours of dark, with 2-hour blocks of light interrupting the dark period at various selected times. Introduction of the 2-hour light blocks in effect simulated a long-day photoperiod. Note that maximal growth rate occurred after the 2-hour blocks of light administered became equivalent to a long day.

Figure 28.9b. A schematic representation of the functional relationships between neural and hormonal systems in photoperiodically entrained gonadal cycles in birds.

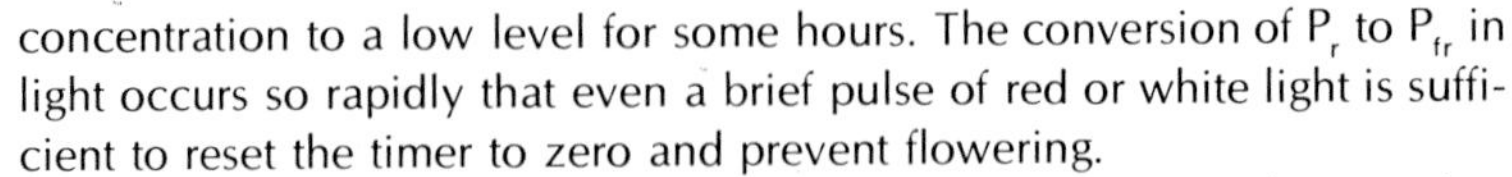

concentration to a low level for some hours. The conversion of P_r to P_{fr} in light occurs so rapidly that even a brief pulse of red or white light is sufficient to reset the timer to zero and prevent flowering.

The phytochrome system was first thought to be an hourglass mechanism that could explain photoperiodic phenomena without any need for other internal timing mechanisms. Soon, however, a number of other observations were made that contradicted this simple model. For instance, the rate of the phytochrome conversion is greatly affected by temperature, but the plants' measurements of daylength appear to be relatively accurate regardless of temperature changes. Biologists began to investigate the possibility that an internal circadian clock might serve as a timer for measurement of nightlength and daylength.

Erwin Bünning in 1936 suggested that the same mechanism responsible for the circadian rhythm of leaf movements in beans might also be involved in photoperiodic timing. He proposed the existence of a circadian clock that causes the alternation of a circa-12-hour period (scotophil) when light inhibits the flowering processes with a circa-12-hour period (photophil) when light enhances the flowering processes. The behavior of short-day plants is explained by assuming that flowering can occur only when the period of daylight falls completely within the photophil. Bünning's hypothesis has been accepted with its broad implications, although subsequent research has refined it in certain details (Pittendrigh, 1966). For example, it has been shown that the phase of the internal cycle of light sensitivity is not necessarily locked to dawn (or the dark-to-light transition) and that the entire scotophil, or dark-requiring phase, is sensitive to light insofar as it affects the entrainment of the rhythm. Light acts both as a photoperiodic inducer (when it falls on the proper parts of the light-sensitive cycle) and as an entraining agent that determines precisely when the phases of the light-sensitive cycle will occur. The phytochrome system probably acts in conjunction with this timing mechanism. It may be part of the mechanism by which light acts to entrain the photophil-scotophil rhythm.

Bünning's hypothesis is appealing in its simplicity and has been applied with equal effectiveness to an explanation of photoperiodic phenomena in

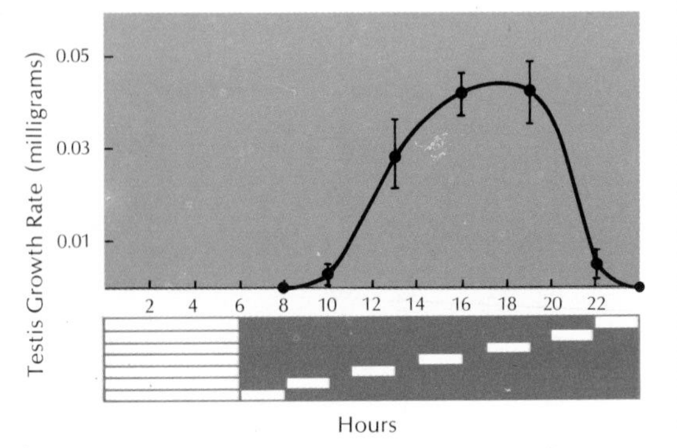

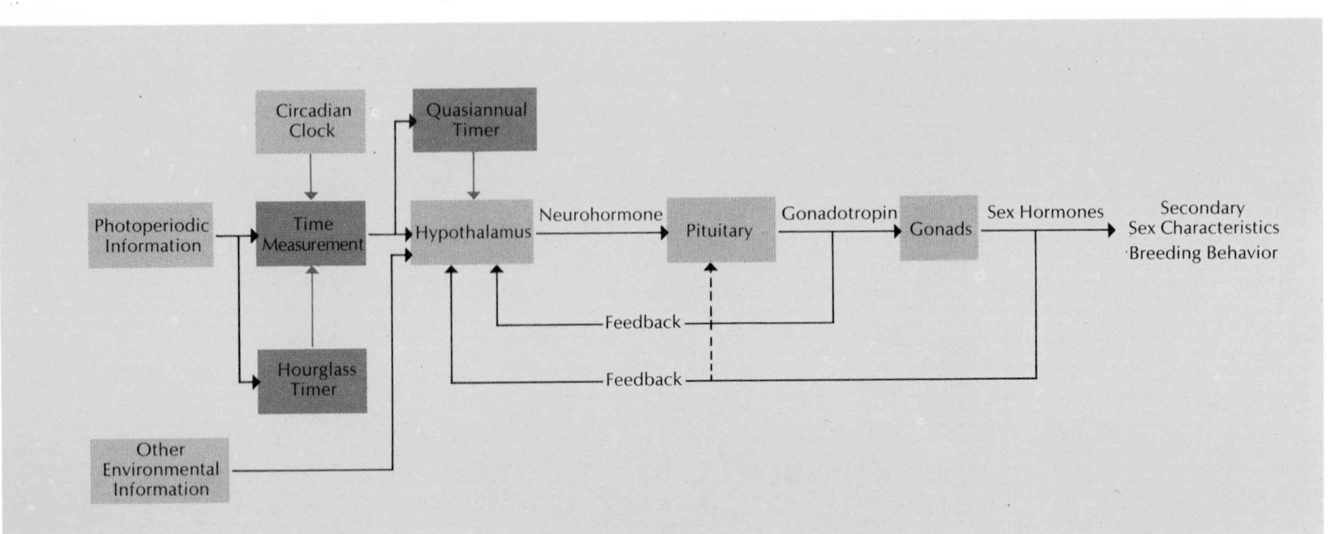

animals, which had long been an obscure area. The common nonmigratory house finch of western North America serves as an example of photoperiodic effects in animals. The testes of the male finch become enlarged in the springtime, a phenomenon common to many bird species. The house finch appears to be a "long-day bird," for the cycle of gonadal enlargement is initiated only in the spring when the light period first exceeds a critical 10- to 12-hour length.

During the early 1960s, a series of experiments was conducted by W. H. Hamner to explore the nature of the finch's biological clock (Aschoff, 1964). Control birds were kept in an environment with a light-dark cycle of 6 hours of light and 18 hours of darkness (written in experimenters' shorthand as *LD: 6,18*). As expected, these birds showed no signs of testicle enlargement, even at the time of year when this phenomenon normally occurs in wild birds. Hamner then exposed experimental birds to light-dark cycles of varying lengths. Each bird experienced 6-hour days, but the intervening dark periods ranged from 6 to 66 hours. Although all the birds experienced short days, some showed testicle enlargement. The cycles that triggered gonadal enlargement were *LD: 6,6, LD: 6,30,* and *LD: 6,54;* cycles of *LD: 6,18, LD: 6,42,* and *LD: 6,66* did not produce testicle enlargement.

These results can be explained by assuming the existence of a circadian rhythm in the photoperiodic response to the onset of a new light period. If the new light period begins 6, 30, or 54 hours after the onset of the dark period, gonadal enlargement is triggered. On the other hand, onset of a new light period 18, 42, or 66 hours after the onset of darkness inhibits testicle enlargement. Because the periods of sensitivity to the light stimulus alternate regularly with periods when the light stimulus has an opposite effect, the mechanism involved appears to be a circadian clock rather than a simple hourglass timer.

ORIENTATION AND CELESTIAL NAVIGATION

Biologists have long been puzzled by the uncanny direction-finding abilities of many animals. For an organism to purposefully move from one place to another, it must have knowledge of the direction toward its goal and a means of orienting itself toward the proper direction as it moves. Many animals can navigate over great distances, moving with great accuracy from a wintering area to a summering area hundreds or thousands of miles away. A newborn animal in some cases is able to navigate this course flawlessly with no previous experience. Similarly, a homing pigeon can return to a home site or breeding ground from remote locations where it has not previously been.

Pioneering studies of starling navigation were made in the early 1950s by G. Kramer, who discovered that the birds apparently use the sun in direction finding. He found that birds in an outdoor aviary show a *migratory restlessness* during the appropriate season; they orient themselves on a perch so that they face in the direction of the normal migratory route (Kramer, 1950). The birds maintain the proper orientation if the sight of the horizon is excluded, but not if their view of the sun is blocked. Furthermore, if the image of the sun is deflected into the cage at various angles by a series of mirrors, the birds alter their orientation by the same angle as the apparent change in the sun's position (Kramer, 1952).

In other experiments, starlings were trained to move in a particular compass direction. Food cups were placed all around a cage, and the birds

Figure 28.10. Sun-compass orientation in birds. The experiment starts at left, with the starlings always trained to feed at the cup in the compass direction south. After training, the birds choose the feeding cups indicated by dots, showing some error in choice of directions. They are placed in an artificial light-dark cycle LD:12,12, 6 hours behind the real time. During the first few days, their hypothetical clocks shift gradually 6 hours behind (black arrow). Consequently, the real time of day when they think it is noon (red arrow) drifts later and later until it reaches 6 P.M. When the birds are briefly exposed to the sun and allowed to look for food, they choose cups west rather than south. During the following days, they are left in constant light; their clocks and thus choices of direction gradually drift. Retrained in the same LD:12,12 schedule, their clocks and choices of direction sharpen up again. They are then shifted to a normal light-dark cycle and, when retested, look for food in the south. To understand why the "sun-compass" theory explains this behavior, imagine the birds' using the rule "When my clock says noon, the sun should be due south of me and that is where I get food."

were fed at the same time each day from a cup in the same true direction from the center of the cage. Tests made at other times of day showed that the birds could compensate for the movement of the sun and still move in the proper true direction (Hoffmann, 1960).

The conclusion from these and other experiments is that the starlings use the *azimuth* of the sun (the angle between the vertical plane through the sun and the observer and the vertical plane lying north-south) in direction finding. Whereas the altitude of the sun above the horizon varies through the year, the azimuth of the sun is always the same at a given clock hour of the day, no matter what the season. If starlings do measure direction by comparison to the sun's azimuth, they must possess some sort of time-compensation mechanism, for the sun's azimuth changes by several degrees each hour. In fact, when Kramer used a stationary artificial "sun" in his laboratory, he found that the birds altered their perch positions during the course of the day by changing their angle to the fixed light source at a rate of approximately 15° per hour. In other words, the birds assumed that the "sun" was moving at its normal rate and adjusted their positions accordingly. Apparently, the birds possess an internal circadian rhythm that enables them to execute time-compensated sun orientation and to solve complex navigational problems.

This hypothesis was tested by K. Hoffmann in an elegant series of experiments. Hoffmann reasoned that a shift in light-dark cycles should reset the internal circadian rhythm and therefore alter the birds' orientation responses. Accordingly, he first trained starlings in natural light-dark cycles to choose a food cup in a particular true direction (Figure 28.10). He then took the birds into the laboratory and placed them under an *LD: 12,12* cycle that was 6 hours behind the local time—that is, the onset of light came 6 hours after true local dawn. After about 12 to 18 days of this artificial day, the birds were tested under natural conditions. When returned to natural sunlight, the birds consistently chose a cup that was about 90° away from the one they had been trained to choose. In other words, the birds assumed that the time was 6 hours earlier than it really was and were adjusting their orientation with respect to the sun accordingly.

Hoffmann then returned the birds to the laboratory, subjected them to constant light and temperature conditions, and periodically tested their orientation under the natural sun. For as long as 28 days, the starlings persisted in choosing the cup 90° from the one to which they had been trained, indicating that the internal rhythm established in the artificial light-dark

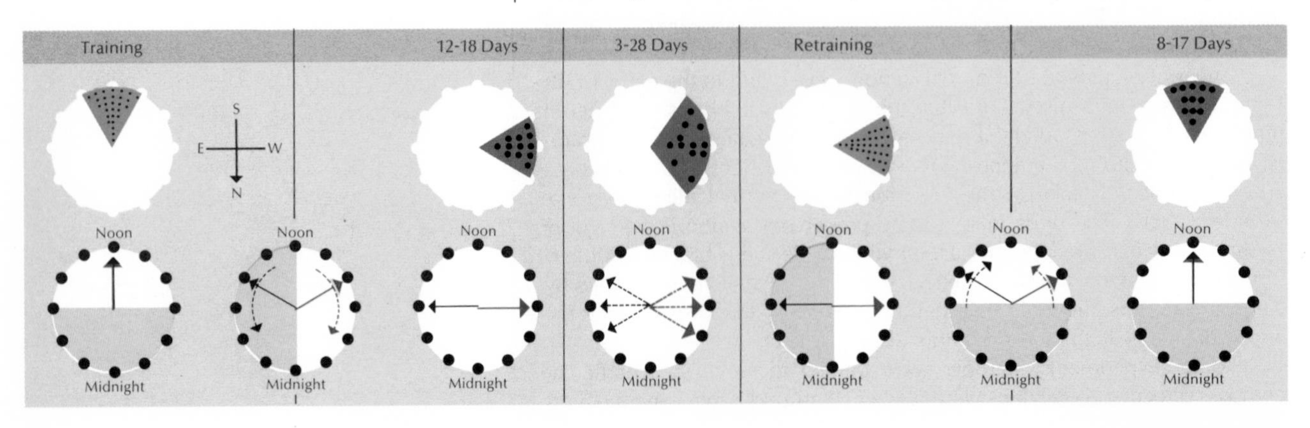

cycle was maintained in the conditions of constant light. Finally, the birds were returned to the shifted light-dark cycle, retested, and then released into the natural light-dark cycle in the outdoor cage. Within 8 to 17 days, the birds again chose the cup to which they had originally been trained, indicating that the internal rhythm had been reentrained to the natural light-dark cycle.

Starlings also show a circadian rhythm of activity, and Hoffmann suggested that the same circadian clock controls both activity and orientation. To test this hypothesis, he first subjected starlings to an *LD: 12,12* cycle and showed that both the onset of activity and the orientation cycle became entrained to this cycle. He then placed the birds in constant light and temperature conditions. The circadian rhythm of activity persisted, but with a period of about 23.5 hours. After about 10 days, Hoffmann tested the birds for orientation under natural sun. The birds tended to choose a cup about 75° from the one to which they had earlier been trained. This result would be expected if the clock used in orientation had been losing half an hour per day, just as the activity rhythm had been doing. Therefore, both the locomotory activity rhythms and the orientation timekeeping appear to be two different functions (or "hands") of the same basic circadian clock (Hoffmann, 1960).

Similar orientation responses have been observed in other species of birds, bees, water striders, fishes, reptiles, mammals, and other organisms, although the involvement of a circadian clock has not been as conclusively demonstrated in these other cases. Certain intertidal crustaceans apparently use the moon for orientation, and there is evidence that European warblers, which migrate at night, may use the stars.

FURTHER READING

An excellent introduction to further reading in biological rhythms is the brief book by Brown, Hastings, and Palmer (1970), which includes presentations of views favorable to geophysical entrainment and favorable to entirely internal mechanisms. Other recent summaries of research and theories on this topic include Aschoff (1964), Bünning (1967), Cloudsley-Thompson (1961), Edmunds (1970), L. Frisch (1960), Mayersbach (1967), Menaker (1970), Moore (1967), Richter (1965), Sollberger (1965), and Sweeney (1969). Surveys of research on photoperiodism can be found in Beck (1968), and in many of the books listed above. Books on navigation and orientation include Autrum (1963), K. Frisch (1950), Griffin (1964), Matthews (1968), and Storm (1967). Introductory and review articles of interest include Aschoff (1963, 1964), Brachmachary (1967), Harker (1958, 1961), Hastings (1959), Hawking (1970), Menaker (1969), Pittendrigh (1961), Richelle (1970), and Sweeney (1963).

29
Learning and Memory

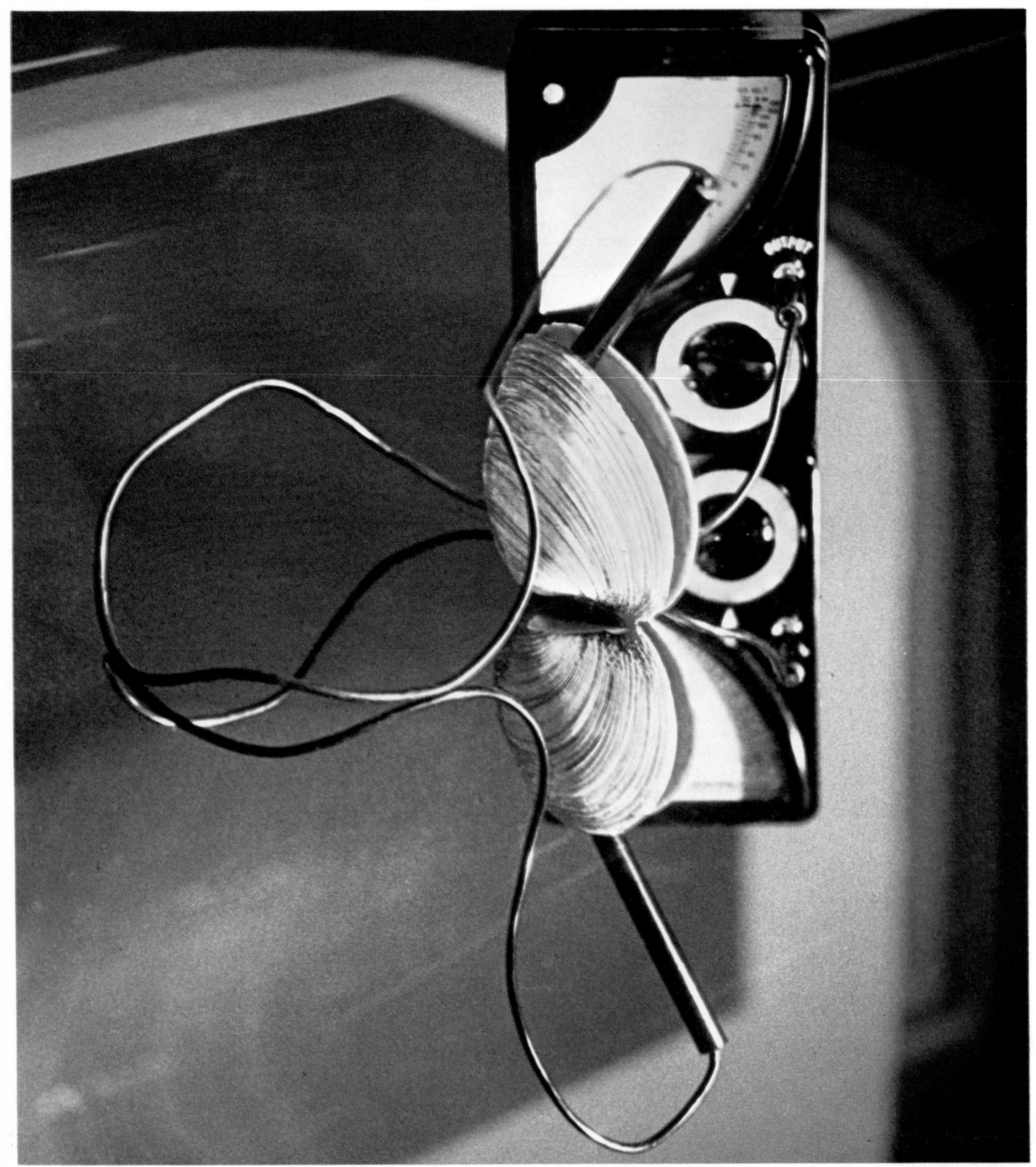

Learning is loosely defined as a change in behavioral responses as a result of experience. *Memory* is the storage within the organism of information about past stimuli and responses. To the behavioral scientist, learning is perhaps the most interesting behavioral phenomenon but one of the hardest to study. Although the process of learning has been studied for nearly a century, very little is known about the actual mechanisms involved.

If an event is to alter behavior at some later time, information must be received and processed through sensory systems, it must be stored and retained, and it must be retrieved and used to alter motor responses to future sensory input. The major problem facing researchers in this area is that none of the processes can easily be studied independently. Most inferences about learning and memory must be made on the basis of overall behavioral performance.

Each organism inherits a set of behavioral responses that are displayed regardless of the experiences of the individual. These are the so-called instinctual, or innate, behaviors that are characteristic of all individuals of the same species. During the life of the individual, the stereotyped behaviors may be modified in certain ways that are not considered true examples of learning. There are some examples of alteration of behavior that are excluded from the "learned" category. First, many species exhibit regular *developmental* changes in stereotyped behavior. A particular pattern of behavior may appear at a given stage of development well after birth, regardless of the experiences of the individual. A human child, for example, appears to go through a process of learning to negotiate stairs. A child who has been restricted entirely to flat surfaces climbs stairs just as skillfully, when he first encounters them, as his identical twin who has practiced stair climbing for several months. The relatively *temporary* effects of sensory adaptation, fatigue, and changes in motivational state are also ruled out as examples of learning.

FORMS OF LEARNING

Habituation is sometimes regarded as the simplest form of learning. For example, an organism may react to a loud noise with a response that involves a convulsive jump, a change in heart rate and other physiological conditions, and an orientation of the head or body toward the source of the noise. If the same loud noise is repeated at intervals, the magnitude of the responses gradually decreases, and they may eventually disappear altogether. What distinguishes habituation from adaptation is that if a novel stimulus is inserted into a train of monotonous ones to which an animal has become habituated, the next repetitions of the monotonous stimuli evoke full responses again. This reaction is called dishabituation. Habituation and dishabituation have been demonstrated in molluscan ganglia and may in this case involve synaptic changes in a single cell.

Another simple form of learning is the phenomenon called *imprinting*. This process has been observed chiefly in birds, but it may exist in some form in other organisms. A young bird "learns" to exhibit certain behaviors toward the first relatively large object it observes after hatching if that object is moving and is emitting regular noises. These behaviors are the ones that young birds normally exhibit toward the mother bird. If a group of ducklings, for example, are exposed immediately after hatching to an experimenter waddling about in a squatting position and making clucking noises,

Figure 29.1. Imprinting baby chicks to a large moving object—a blue ball. The ball is slowly rotated around the runway, eliciting a following response in the young bird. Imprinting normally takes place within a critical time period soon after hatching, when the young birds are highly receptive to environmental stimuli.

the birds will follow him just as ducklings normally follow the mother duck. When he stands up, the ducklings apparently cease to experience him as "mother," and they begin to run about making the noises typical of lost ducklings. As soon as he squats down again and begins clucking, the ducklings again act as if he were the mother duck. Ducklings that have been imprinted in this fashion do not show normal responses to a mother duck (Lorenz, 1937, 1952). In this case of "stereotyped learning," the organism has a limited capacity to modify its behavior on the basis of experience.

Classical Conditioning

In classical conditioning, simple innate stimulus-response patterns of behavior are modified. A behavior that is normally elicited by a certain stimulus (unconditioned stimulus, or UCS) is brought under control of a different stimulus (conditioned stimulus, or CS) by training. Thus, the CS comes to elicit this behavior, whereas before training, it did not.

In classical conditioning, a reflex is often exploited to provide the response (unconditioned response, or UCR) and the CS. The story of Ivan Pavlov's conditioning of the salivation reflex is well known. A dog, if presented with food (UCS), will salivate (UCR). Pavlov paired a light (CS) with the presentation of food (UCS). The light was presented for 5 seconds before

Figure 29.2 (right). Pavlov's apparatus for conditioning salivation in the dog. (*After Pavlov, 1927*)

Figure 29.3 (lower left). Acquisition of the conditioned salivation reflex. (*After Pavlov, 1927*)

Figure 29.4 (lower right). Extinction of the conditioned reflex. (*After Pavlov, 1927*)

food was dropped into the tray. Once every few trials, the light alone was presented without being followed by food. The number of drops of saliva elicited by the light (CR) alone increased as a function of the number of previous trials in which light and food had been paired, eventually reaching a magnitude comparable to that of the UCR (Figure 29.3).

Unconditioned reflexes in most cases persist for life, but conditioned reflexes persist only so long as the unconditioned reflex is occasionally elicited in conjunction with the CS. Thus, if the light (CS) is presented to the organism on a number of successive trials without the UCS (food in the mouth), the amount of saliva elicited by the light gradually declines until it is zero (Figure 29.4). This process Pavlov called *extinction*. The conditioned reflex gradually wanes and is extinguished in the absence of what Pavlov called *reinforcement*—the normal reflex salivation.

Pavlov's experiments showed that the timing of events during conditioning is very critical. If the UCS is presented before the CS, little if any conditioning occurs. On the other hand, if the CS precedes the UCS by too long an interval, the conditioning becomes more difficult to establish. In short, the success of the conditioning is contingent upon the timing of the UCS and the CS.

Pavlov also demonstrated two very important properties of the classical conditioning process: generalization and discrimination. If a particular CR is conditioned to the stimulus of a green light, for example, a similar response will be elicited by a red light or a blue light. The greater the difference between the conditioned stimulus and the test stimulus, the weaker the response that will be elicited. Apparently, classical conditioning does not merely establish a connection between a particular stimulus and a particular response but leads to a process of *generalization*, in which a wide range of stimuli can elicit responses of varying strengths.

However, if the organism is conditioned in appropriate fashion, it can learn to *discriminate* between similar stimuli. For example, if the green light is consistently paired with the UCS and the red and blue lights are consistently presented alone, the organism gradually comes to give the CR only to the green light. The responses to the red and blue lights are extinguished, whereas the response to the green light is reinforced.

Classical conditioning is a very widespread form of learning. It is interesting that human vegetative activity, which cannot be consciously controlled, can often be conditioned. When vegetative activity is conditioned,

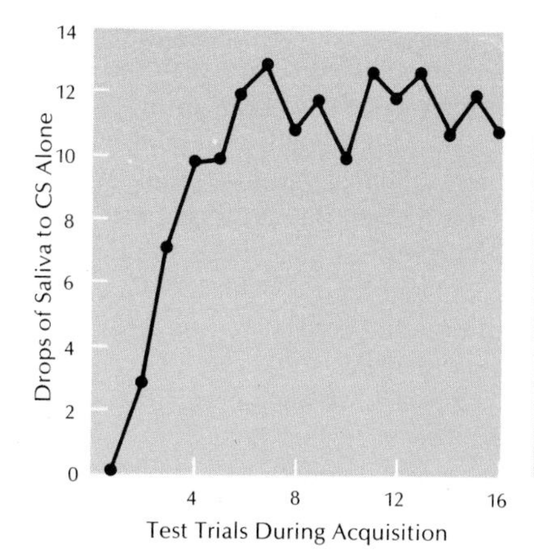

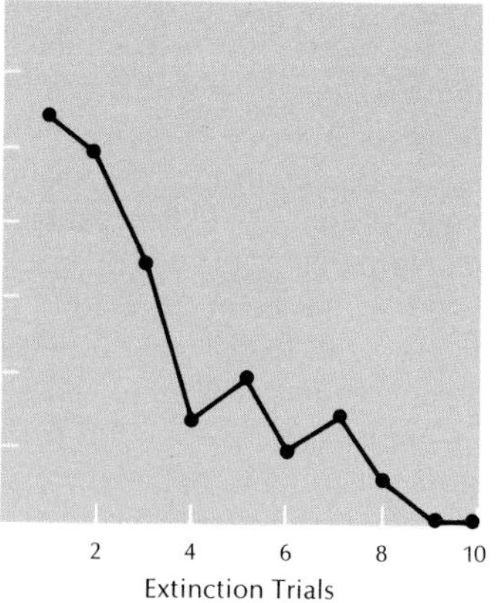

Figure 29.5 (above and middle). Types of training boxes developed by Skinner to test conditioning and learning of operant behavior. Various sensory stimuli can be built into the apparatus for the animal to respond to. Reinforcement (reward or punishment) mechanisms are also included in the apparatus.

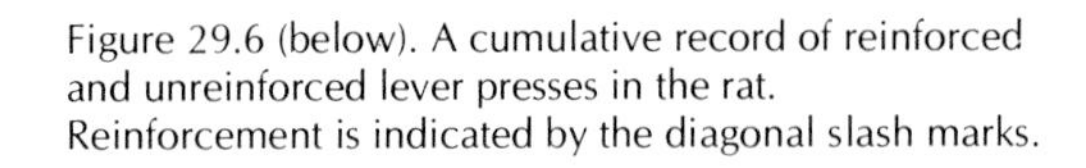

Figure 29.6 (below). A cumulative record of reinforced and unreinforced lever presses in the rat. Reinforcement is indicated by the diagonal slash marks.

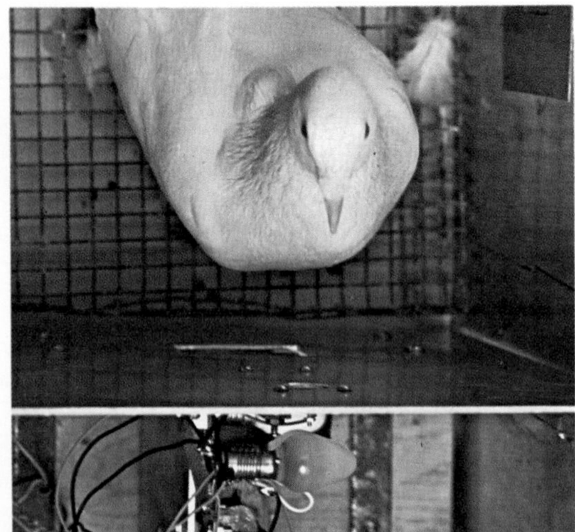

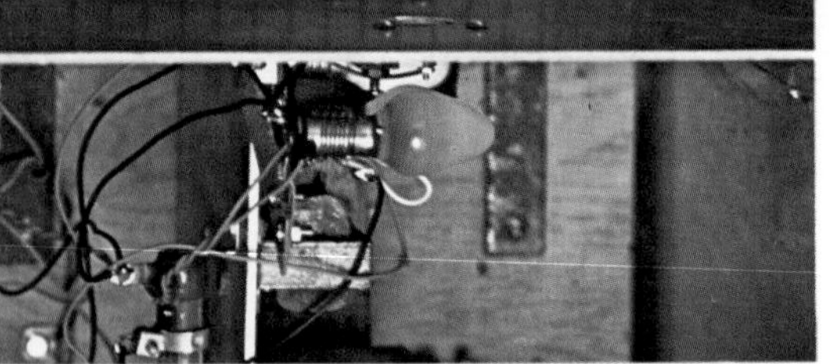

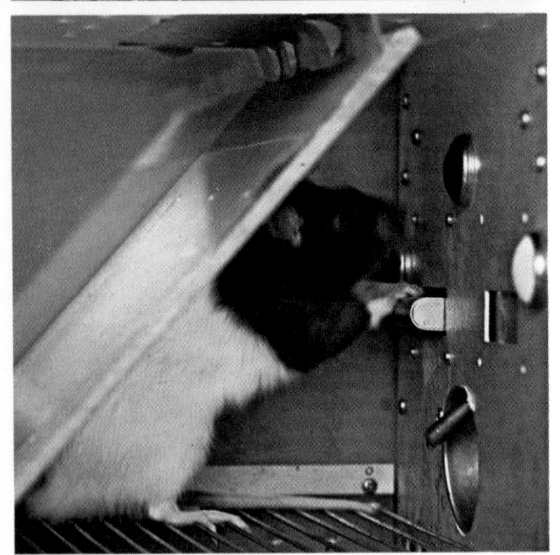

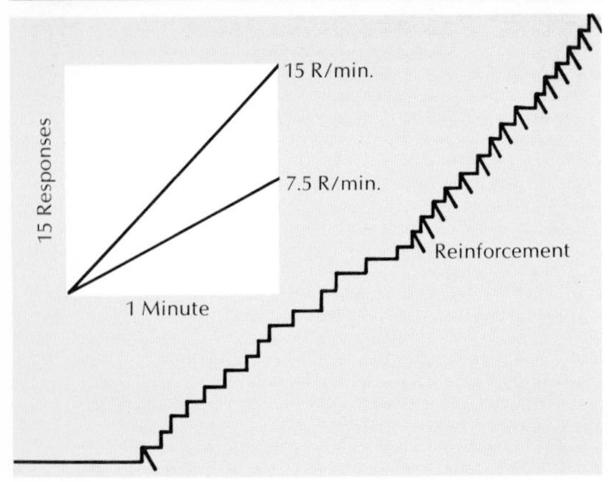

a subject cannot stop himself from giving the conditioned response, unless it is gradually extinguished. Some very simple neural systems exhibit many characteristics of conditioning. An animal whose brain has been severed from its spinal cord can still be conditioned. Single cells in ganglia of some molluscs can exhibit a phenomenon very much like conditioning. There is no doubt that conditioning is a fundamental and even primitive form of learning.

Operant Conditioning

Pavlov observed another kind of learning in his dogs. For example, if he consistently lifted a dog's paw just before putting meat powder in its mouth, the dog would eventually begin to lift the paw itself when it was hungry. In this case, the dog learned to repeat a response that had been followed by a reinforcement. There had been no pairing of an unconditioned stimulus with a conditioned stimulus, and the response was not necessarily one that exists as an unconditioned reflex response.

In general, operant conditioning involves the fact that responses followed by rewards are likely to be repeated, whereas responses followed by punishments are not likely to be repeated. Thus, any normal behavior may become more or less likely to occur again depending on the consequences that follow its occurrence. In operant conditioning, an animal is taught to alter the frequencies or temporal and spatial patterning of behaviors by a system of rewards and punishments. New behavior patterns or alterations in existing stimulus-response relationships can be brought about in this way.

The principle of operant conditioning has always been known in general terms to animal trainers, child raisers, and people manipulators. The trainer compels obedience from an animal by making the comfort of the animal dependent on obeying. The parent controls the behavior of his child by making the child's likes and the avoidance of his dislikes dependent on the behavior that passes for good in the culture to which they belong. The businessman accumulates money by making the spending of money a precondition for either comfort or the relief of discomfort on the part of the buyer. His own desire for money is a result of the fact that the possession of money has usually been followed by pleasant experiences, whereas the lack of money may have led to discomfort in the past.

In the laboratory, the demonstration and explication of operant conditioning has been largely the work of B. F. Skinner. Skinner's training box has become the staple piece of apparatus for the study of the conditioning and learning of operant behavior, not only with animals but also with human beings.

The type of box used for the study of operant behavior is pictured in Figure 29.5. The box contains a lever or key, whose depression by the animal—for instance, a rat—is the operant response under study. Provisions are made within the box for the presentation of various stimuli such as lights or tones. The box contains a mechanism that presents food to the animal, either at the will of the experimenter or as a result of some automatic timing device or the animal's manipulation of the lever. If the animal is hungry, the food acts as a reward, or reinforcement, for the behavior of lever pressing. The result of this reinforcement is that the lever pressing increases in frequency and this behavior persists—the animal presses the lever again and again. Figure 29.6 shows a typical record of a training session. It shows a cumulative record of the number of lever presses as a function of the time elapsed after the rat was placed in the box on a particu-

Figure 29.7. In training programs, reinforcement is used to maintain complex behavior patterns. Here, the behaviors of porpoises and a killer whale were shaped by reinforcing ever closer approximations of the behavior desired by the trainer.

lar day; each reinforcement (operation of the food dispenser) is shown by a diagonal slash mark. After the first lever pressing was reinforced, the rat continued to operate the lever some 15 times without further reinforcement, although the rate of lever pressing began to decrease somewhat. Reinforcement of each lever pressing during the last part of the training session caused a continuing high rate of lever pressing.

An animal in Skinner's box can be trained to operate the lever under certain conditions and not to do so under other conditions. For example, lever pressing might activate the food dispenser only if a blue light is on inside the box. The rate of lever pressing decreases to low levels whenever the blue light is off but quickly reaches high levels whenever the blue light comes on. In this case, the lack of reinforcement has caused the behavior to be extinguished in the absence of the blue light, but reinforcement causes the behavior to be conditioned in the presence of the blue light. The blue light acts as a *contingency factor* for the lever-pressing behavior.

Animals can be trained to discriminate between various stimuli, and phenomena of generalization can be demonstrated in operant conditioning much as they are in classical conditioning. Complex behavior patterns can be *shaped* by reinforcing ever closer approximations to the behavior that is desired. For example, a dolphin may first be reinforced simply for jumping out of the water whenever a particular stimulus is given. Later, the reinforcement may be made only when the dolphin jumps in a particular section of the pool and only when it throws its tail somewhat forward as it reenters the water. By gradually making the standards needed for reinforce-

Figure 29.8. Cumulative records of lever-pressing responses under various schedules of reinforcement. Shown above is a record of responses reinforced on a variable ratio schedule. The subject is rewarded after a preselected number of responses, but the number changes after each reinforcement (slash line). Shown below is a record of responses reinforced on a fixed-ratio schedule. The subject is rewarded after a fixed number of responses, the number of presses remaining constant from one reinforcement to the next. On the opposite page (above) is a record of responses reinforced on a fixed-interval schedule. The subject is rewarded after a fixed period of time has elapsed since the previous reinforced response. The slope of the record, representing the rate of response, increases as the next reinforcement approaches. Most organisms placed on this reinforcement schedule have a response pattern that shows a pause or minimal number of responses following a reinforcement, then they accelerate to a high rate of responses as the reinforcement time approaches. Shown opposite (below) is a record of responses reinforced on a variable-interval schedule. The subject is rewarded after a variable period of time has elapsed since the previous reinforced response. Thus, in the two latter regimens, reinforcement is time-oriented rather than response-oriented.

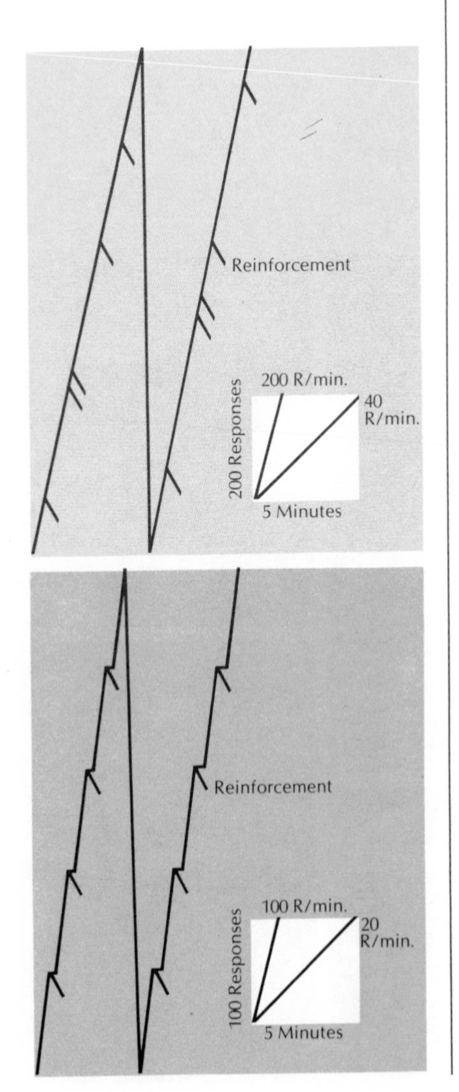

ment stricter, the trainer can teach the dolphin to respond to the stimulus by doing a perfect somersault above the water at a particular spot in the pool—a shaped behavior that the dolphin might never have emitted as a result of random behaviors.

The effects of various schedules of reinforcement on lever-pressing performance are illustrated in Figure 29.8. In many animals, performance is optimum under a variable-ratio schedule in which the lever must be pressed a number of times before the animal is rewarded, but this number is changed after each rewarded response. This kind of reinforcement works very well in gambling houses with slot machines and in college courses with surprise quizzes.

THE MECHANISM OF LEARNING

An explanation of the mechanism underlying the learning process must be expressed in terms of the basic structure of the nervous system and must account for the features of the process observed in psychological experiments. Experience must produce changes of some kind within the nervous system so that a given stimulus results in modified behavior. These changes represent a sort of record of the experience—in other words, a memory, or "memory trace."

For the neurophysiologist interested in learning, several questions are obvious. What is the nature of the change that occurs when a memory is stored? How does a particular sensory input produce a particular memory? How does a stimulus pattern call for playback of a learned behavior? Where is a memory stored?

The theories suggested in answer to these questions have often been regarded as falling into two opposing groups: structural theories and chemical or molecular theories. Structural theories hold that a memory is stored through a change in the anatomy of the nervous system—for example, establishment of new synapses or an increase in the strength of conduction across existing synapses. Chemical theories hold that a memory is stored through changes in the molecules within neurons—for example, synthesis of certain substances that reduce the threshold for activation of particular neurons.

Researchers have asked the question "Where?" for years, for this question affects both structural and chemical theories. Early experiments suggested that damage to a small area of the cortex impairs performance on learned tasks, but no more or less so than similar treatment of any other area of the same size. However, more recent studies have suggested some localization of learning processes in the brain. For example, destruction of the rear part of the cortex impairs the performance of rats in learning tasks that require discrimination between different stimuli but does not impair their ability to learn particular sequences of responses to a single stimulus. On the other hand, destruction of the front part of the cortex impairs performance on sequence learning but does not interfere with sensory discriminations (Gross, *et al.*, 1965). Similarly, particular areas of the monkey's cortex are involved in learning tasks involving different senses or different kinds of problems. However, it is not clear in these experiments whether the brain damage is affecting ability to process sensory information, ability to perform certain kinds of associations, or ability to store and retrieve memories.

Although the associative areas of the cortex do appear to be involved in learning and in memory, it has been impossible to assign a specific function to any particular part of these areas. Most evidence thus far is consistent with the assumption that storage of a memory involves changes occurring over a wide region of the cortex and probably in other parts of the brain as well. This diffuse storage of information has become one of the major characteristics of memory and learning that must be explained by any theory or model.

The two kinds of theories, structural and chemical, are not as distinct as they might seem. It is clear that learning must involve a change in the paths of impulses through the nervous system. Therefore, it must involve changes in the distribution of connections between neurons or changes in the excitabilities of neurons. Any chemical change must be expressed in terms of structural changes within neurons or in synapses between neurons if it is to alter the relationship between sensory input and motor output. Similarly, any structural change can be accomplished only by alterations of the molecules that make up neurons. Thus, the differences among theories lie in the relative emphasis given to the two aspects of the process.

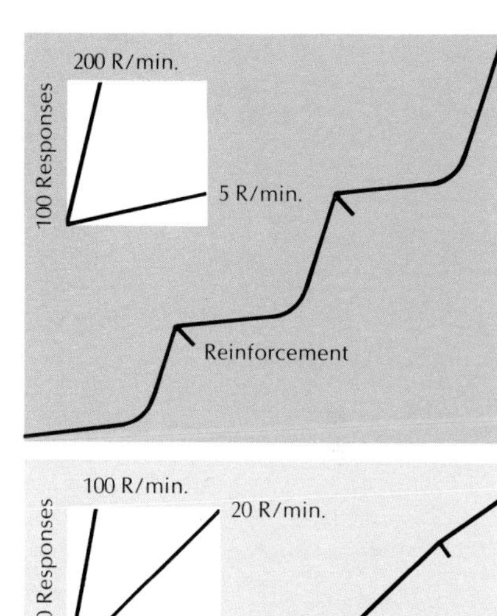

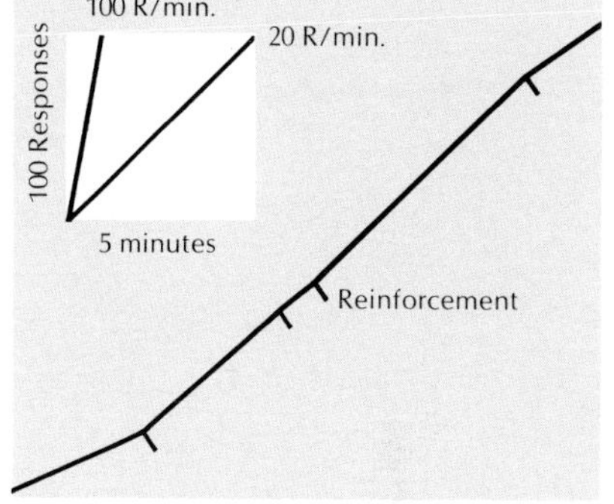

Structural Theories

John C. Eccles (1952, 1958) has argued for a simple structural model of learning processes. He suggests that the use of a neural pathway facilitates later transmissions along it, whereas disuse makes later transmissions more difficult. Certain experiments support this model. For example, if the portion of a spinal nerve peripheral to the dorsal root (sensory) ganglion is cut off, the input of impulses to that nerve is prevented. This lack of input leads to a subsequent shrinkage in the size of the remaining connection to the spinal cord and weakens the monosynaptic reflex evoked by electrical stimulation of the dorsal root. In some cases, the number of synaptic terminals at the axonal ending of the nerve diminish as a result of the cutting. This decrease in number supports the theory that use of a neural pathway increases or maintains the efficiency of that path. This use could serve as a basis for learning.

There is also substantial evidence that use or disuse causes alterations in cortical tissue. Each dendrite of a cortical neuron is covered with hundreds or thousands of minute spines. Recent experiments suggest that the number of spines decreases when input to that area of the cortex is reduced. Presumably, an increase of activity would lead to an increase in the number of spines. Because the spines are apparently involved in the axodendritic synapses, spine growth could be a mechanism that allows greater synaptic contact between two cells. By this mechanism, usage of a particular pathway through the nervous system would facilitate further transmission of impulses along that pathway.

Various kinds of anatomical alterations of the nervous system do result from various kinds of experience, but there is no definite evidence of a particular alteration associated specifically with the storage of a memory. A fascinating study, which also shows the importance of use of neural circuits in order to maintain their normal function, was made in kittens. Normal kittens whose eyes have just opened have neurons in their visual cortex that are sensitive to certain complicated shapes of visual stimulus. Many of the neurons are sensitive to input from either eye. If the kittens are raised with one eye sutured shut for a few critical months, they lose most of the binocular neurons in their visual cortex. The cells are still present, but they are driven by only one eye. Animals deprived of visual form for a few months after they open their eyes appear to have lost most of their cortical neurons that are sensitive to complex shapes or motions. These phenomena show that maintenance of complicated neural circuits requires normal stimulation of some type, at least in one critical period of a kitten's life. Whether this finding can be extrapolated to account for formation of new synapses remains to be seen. If it can, then it will present a concrete hypothesis for memory storage.

Chemical Theories of Learning

Chemical theories of learning propose that memory consists of the manufacture of new molecules or alterations in concentration, location, or availability of existing molecules. Advocates of these theories have interpreted any indications that chemical changes occur along with learning or that chemicals can alter learning as suggestive of an important role for molecules in memory. Among the first and most controversial experiments interpreted as suggesting a chemical theory are a series on transfer of learning in planarians. If a trained planarian is cut in half, the head section regenerates a new tail, the tail section regenerates a new head, and—according to these reports—both of the new individuals show excellent retention of the original learning. It has been reported further that if trained worms are cut in several pieces, all the regenerated sections show retention of the original training. The intriguing fact about these experiments is that all but the head sections must grow an entirely new head ganglion, or "brain," so if they subsequently "remember," the memory cannot be located only in the brain.

These results caused the researchers who did these experiments to explore the possibility that memories might be stored in chemical changes in cells throughout the worm's body. Planarians were trained, cut up, and fed to untrained, cannibalistic planarians. The cannibals were subsequently trained on the same task as that learned by the worms they had eaten. Planarians fed on "educated" worms learned the task faster than did control animals fed on untrained worms (McConnell, 1962). Because RNA is an abundant macromolecule that is clearly involved in major alterations of the cell through protein synthesis, the researchers suspected that RNA might be the "memory molecule." To test this theory, they extracted RNA from trained worms, injected it into untrained worms, and showed that the injected worms performed better at learning than did worms injected with RNA extracted from untrained worms (Jacobson, *et al.*, 1966). However, more recent research has indicated that similar effects may be obtained by feeding planaria with untrained worms that have been subjected to stimulation unrelated to the particular learning task being studied or by injection of RNA from similarly stimulated worms. Moreover, other researchers have failed to confirm the results of the early "memory transfer" experiments in planaria (McGaugh, 1967).

Most recent experiments have used rodents as the subjects for studies of memory transfer. At best, the results have been equivocal. Some sort of memory transfer has been reported in several studies, but none of the researchers has yet been able to specify reliable or optimal procedures for producing such a transfer effect. None of the experiments has provided clear evidence that the effect involved is truly memory transfer rather than simple enhancement of learning ability.

These experiments notwithstanding, it has been thought that memories might be stored and transmitted in the form of nucleotide sequences in

Figure 29.9. Hypothetical structural changes in learning. These diagrams show several possible but unproven ways in which learning could be mediated by anatomical changes in the processes or synapses of neurons. In cases 1 and 2, the increased use of a pathway enhances the efficiency with which its synapses function—the vital involvement of glial cells may or may not be necessary. Case 3 illustrates that some branches of a single axon may be strengthened, whereas others may be weakened as a result of the action at two different synapses. Cases 4 and 5 show structural changes that could account for conditioning. At first, both a CS (1) and a UCS (2) are required in order to fire a UCR (3). If the two stimuli are paired, the synapses of neuron 1 are strengthened so that the CS (1) alone elicits the firing of 3 (which is now the CR). In case 5, the pairing of two stimuli leads to the formation of a new synapse, so that neuron 1 comes to fire both output neurons.

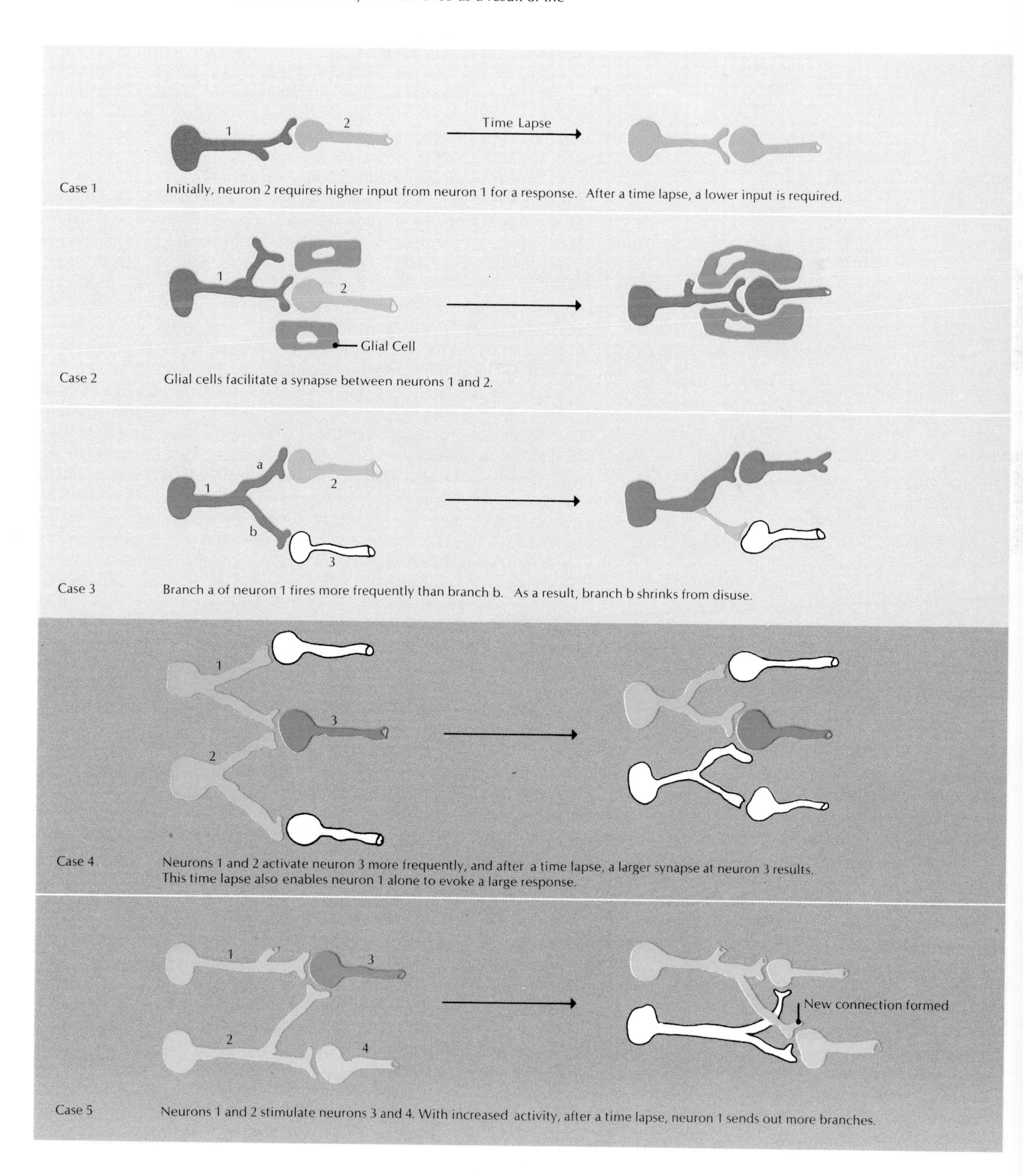

1
2
Time Lapse
Case 1
Initially, neuron 2 requires higher input from neuron 1 for a response. After a time lapse, a lower input is required.
1
2
Glial Cell
Case 2
Glial cells facilitate a synapse between neurons 1 and 2.
a
1
2
b
3
Case 3
Branch a of neuron 1 fires more frequently than branch b. As a result, branch b shrinks from disuse.
1
3
2
Case 4
Neurons 1 and 2 activate neuron 3 more frequently, and after a time lapse, a larger synapse at neuron 3 results.
This time lapse also enables neuron 1 alone to evoke a large response.
1
3
2
4
New connection formed
Case 5
Neurons 1 and 2 stimulate neurons 3 and 4. With increased activity, after a time lapse, neuron 1 sends out more branches.

RNA, much as genetic information is stored and transmitted in the form of DNA nucleotide sequences (Hydén, 1962). The expression of memories supposedly involved translation of RNA into specific proteins. The findings that drugs interfering with protein synthesis disrupt memory storage seemed to support this hypothesis. However, studies with drugs that interfere with RNA synthesis failed to confirm the role of RNA in memory storage.

Nevertheless, several studies have shown that RNA synthesis – although apparently not essential to memory storage – is altered during learning. For example, one series of experiments showed that training of mice enhances RNA synthesis. The effect measured was an increased incorporation of radioactive uridine in trained animals as compared with animals that were stimulated but not trained. The increase in RNA synthesis was observed in brain tissues but not in the liver or kidney, and the enhanced rate persisted for less than an hour (Zemp, *et al.*, 1966). The increased RNA synthesis might not be directly involved in memory storage but might be related to the maintenance or repair of stimulated cells in the nervous system.

Holger Hydén and E. Egyházi (1962, 1963, 1964) developed a sophisticated technique for analysis of the RNA in a single cell. Not only did they find that neurons have very high RNA content and that the amount of RNA increases after stimulation of the neuron, but they found changes in the base ratios of RNA after learning. In one experiment, a rat was taught to balance on a thin wire in order to reach its food. Changes in RNA base ratios were found in nerve cells that are involved in transmission of impulses from the semicircular canals of the ear to the brain. These changes might result from synthesis of new kinds of RNA or, more likely, from changes in the amounts of different kinds of RNA already present. Hydén and Egyházi suggested that changes in RNA reflect changes in the amounts and kinds of transmitter substances being synthesized and thus that the RNA mediates a structural change in the synapses of the neuron. Similar results for glial cells led them to suppose that glial cells are also involved in the chemical processes of memory storage.

More recent research has not cast doubt upon the experimental findings about changes in RNA base ratios, but there is great debate about the proper interpretation of these findings. It does seem clear that memory storage involves changes in RNA content of neurons and protein synthesis within neurons, but the relation of these changes to memory mechanisms is not known.

In several studies with mice and goldfish, drugs that interfere with protein synthesis have been found to impair memory storage. In one series of studies, for example, goldfish were given injections of such a drug either before they were trained in a task or at one of several intervals after the training. Retention tests were then given at various intervals after the injections (Agranoff and Davis, 1968). Animals that were injected prior to training learned normally but "forgot" within a few days (Figure 29.11). Injections after training caused a loss of memory of the conditioned response; the extent of the loss varied directly with the dose of the drug and inversely with the time interval between training and injection. Mice given injections of a similar drug can learn normally at a time when protein synthesis has been reduced to less than 20 percent of its normal level, but the animals "forget" within hours following the training. Injections given after training produce a gradient of memory loss similar to that observed in goldfish (Barondes and Cohen, 1968). Studies with a drug that inhibits RNA synthesis

Figure 29.10. Free-living planarians (flatworms) can be trained in both classical conditioning experiments and in simple maze problems. In the top two photographs, the planarian is in a normal extended position during locomotion. In the lower photographs, the planarian shows a response to an unconditioned stimulus, an electric shock. Worms trained using a variety of experimental regimens have also been used in the study of memory transfer and the chemical theory of learning.

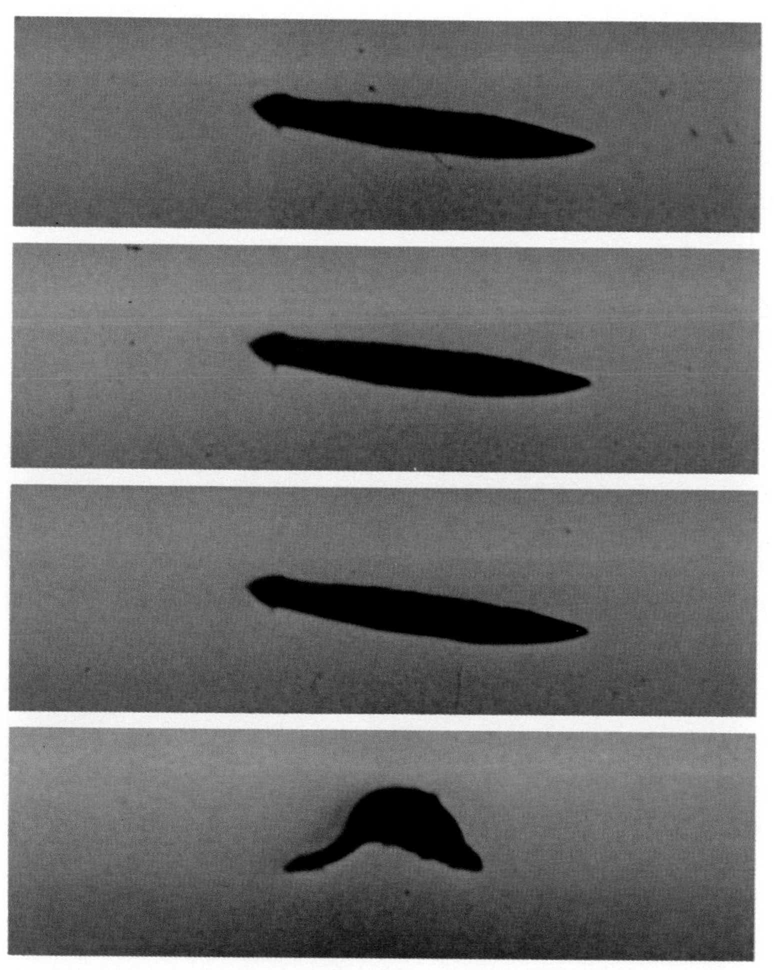

show that memory is not impaired even when RNA synthesis is reduced to 4 percent of its normal level (Barondes and Jarvik, 1964). Unfortunately, the doses used in these studies caused the animals to die within 24 hours after injection, so it is not possible to determine the role of RNA synthesis in long-term memory storage.

A number of conclusions emerged from these and similar drug studies. First, the memory processes involved in retention of recently acquired information appear to be insensitive to the drugs used. Second, the long-term retention of information is impaired by drugs that inhibit protein synthesis, although it is not yet known whether the effects of such drugs on memory are due specifically to their effects on protein synthesis.

Some chemicals have been shown to enhance memory storage. The rate of learning in animals can be enhanced by several central nervous system stimulants, including strychnine, picrotoxin, pentylenetetrazol, and nicotine (McGaugh and Petrinovich, 1965). The enhancement of learning can be obtained by administering the drugs either before or after a training session. The effects of pretraining injections might be due to changes in the sensory or motivational processes of the animals. However, the effects of posttraining injections are almost certainly due to memory-storage

Figure 29.11. The effect of the drug puromycin on memory retention. In the graph above, three groups of animal subjects were used in this experiment: Group A were uninjected controls; Group B were controls injected with a physiological saline solution (having a concentration of 0.15 moles/liter NaCl); and Group C were test animals, injected with 170μg (micrograms) of puromycin. Groups B and C were injected with the solution or the drug within one minute after the training period. Memory retention was essentially normal in both control groups (A and B), whereas the memory retention values for Group C (injected with puromycin) indicate a decline, beginning after 6 hours and becoming virtually complete after 72 hours. In the graph below, two groups of animal subjects (A and B) were tested. Starting on the first day, the subjects were subjected to training and on the fourth day memory retention was measured. In Group A, the effect of 170μg of puromycin injected at various intervals from the moment of training was tested. In Group B, the effect of various doses of puromycin injected within one minute after training was tested. In Group A, subjects showed the most significant memory loss if the drug was injected within the first 30 minutes after training; subjects injected within 60–90 minutes showed a relatively small decline. In Group B, a significant decline in memory was observed for subjects given doses above 90μg; a complete loss of memory was recorded for subjects within the 170–210μg dosage range of puromycin.

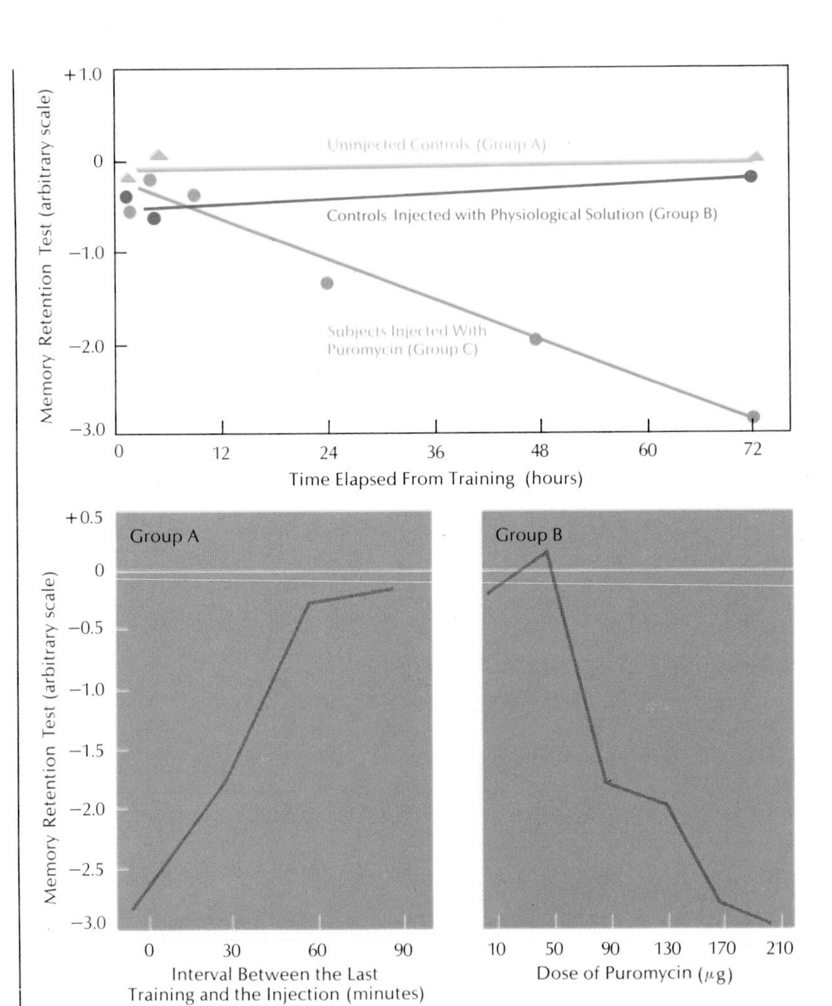

enhancement, for the animals are not drugged during either training or testing. Thus, the drug apparently does not affect the learning process or the retrieval of memory but affects only the consolidation of long-term memory.

Figure 29.12 shows the results of one study in which learning was markedly enhanced by an injection of pentylenetetrazol given each day immediately after the last of three training trials in a maze (Krivanek and McGaugh, 1968). The degree of learning enhancement varied directly with the drug dose. In other experiments using the optimal dosage of this drug, learning enhancement occurred if the drug was injected within one hour before training began or within 15 minutes after the training ended (Figure 29.13). Similar results have been obtained with a variety of drugs and training conditions.

Unfortunately, little is known as yet about the mechanism by which these drugs affect memory storage. Each of the drugs has complex and diverse physiological effects on the nervous system, and which of their effects is related to memory is not known.

THE NATURE OF MEMORY

It is clear that specific memories are not located in any single locus in the brain, and it is clear that some changes in neurons and glia do take place

Figure 29.12 (above). This graph shows how discrimination learning was facilitated by injecting pentylenetetrazol into subjects immediately after daily training. The percentage of correct responses—correlated with the various dosages of the test solution—are plotted. As the drug dosage increases, subjects were able to make a higher number of correct responses. Tests were conducted with groups of mice, and controls were injected with a saline solution.

Figure 29.13 (below). In these graphs, subjects were again given injections of pentylenetetrazol and tested on learning visual discrimination problems. An analysis was made to determine how the timing affected injections of the test solution into subjects both before and after training. The drug was most effective in facilitating the learning process if the subject was injected within the first 15 minutes—either before or after training. Later injections yielded progressively fewer correct responses. Control animals were injected with a physiological saline solution. Groups of 6 mice were used in each problem set.

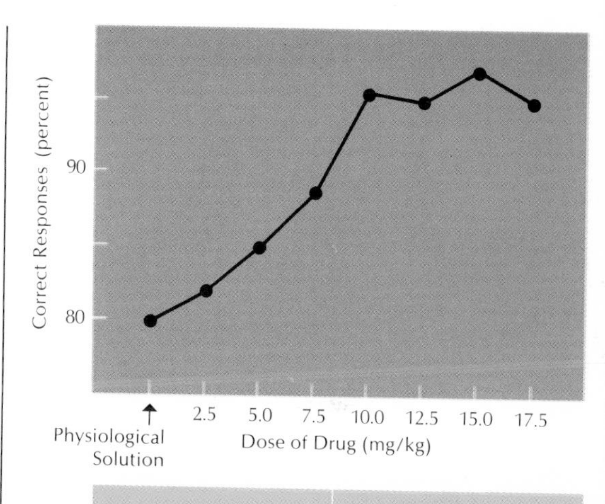

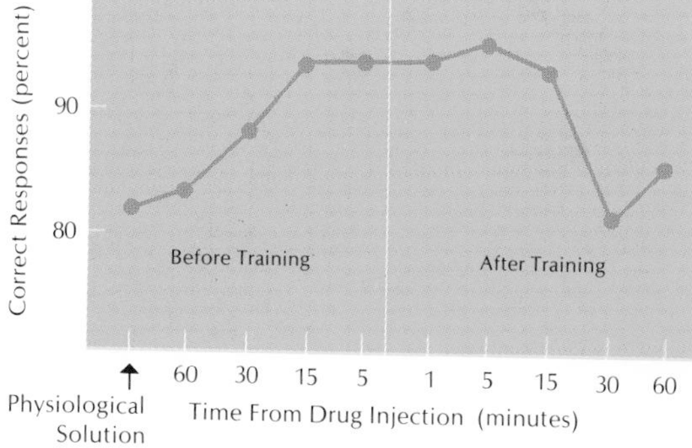

during learning. Neither structural theorists nor chemical theorists have proposed concrete, testable models for the formation, storage, or playback of learned behaviors. But experiments suggest that both chemical and structural changes are important. It seems that, whatever the mechanism, memory must have diffuse (nonlocalized) storage and must be written into and read from this storage in a relatively short time. One intriguing but speculative analogy compares memory storage to the storage of an image in a hologram. The hologram film stores a record of the total wave pattern of light passing through it during exposure. When the image is re-created by passing light through the developed film, the wave pattern is re-created. Damage to any part of the film does not destroy a part of the image but merely makes it less distinct. If two images are superimposed on the same film, both can be seen when the hologram is "played back." However, if a real object similar to one of the images is placed into the appropriate place in the system, the corresponding image is greatly enhanced. These and other analogies to memory processes are highly suggestive, but thus far the hologram analogy has not led to any specific hypotheses about the mechanism of memory storage.

Current knowledge about the mechanisms of memory can be summarized in a few sentences. No particular portion of the brain appears to act as the site for storage of a particular kind of memory or even for memory storage in general. Training does produce electrophysiological, anatomical, and biochemical changes in neural tissue, but the nature of these processes and their relation to memory storage is not yet clear. Chemicals can affect memory but it is not known how. One neurophysiologist, reviewing the history of the search for memory mechanisms, commented: "I sometimes feel . . . that the necessary conclusion is that learning just is not possible" (Lashley, 1950, p. 477). Obviously, learning *is* possible in a wide range of organisms, and the search for an explanation of the mechanisms underlying this process is a major focus of current research in behavior.

FURTHER READING

John (1967) gives an excellent summary of the search for the memory mechanism. Other discussions of memory include books by Adams (1967), Kimble (1965), and Norman (1968). The collection edited by Glassman (1967) includes many of the important research papers describing the search for a biochemical explanation of memory. For more extensive discussions of learning and conditioning, see books by Corning and Ratner (1967), CRM Books (1970), Honig (1966), and Skinner (1938). Holland and Skinner (1961) not only teach about operant conditioning, but do so through a system of programmed learning based upon the theories that they explain.

Magazine articles of interest in relation to this chapter include those by Agranoff (1967), Ceraso (1967), Delgado (1970), Deutsch (1968), DiCara (1970), Gazzaniga (1967), Halstead and Rucker (1968), Lang (1970), Peterson (1966), Scott (1969), and Sperry (1964).

30
Behavioral States

Behavioral state is the ensemble of physiological factors, including the representation of past experience in the nervous system, that may affect the behaviors elicited by various environmental stimuli (Chapter 27). The psychologist's concepts of motivation, set, mood, and emotion are all covered by this designation. What can be said objectively about these subjective phenomena? Psychologists speak of drive and of satiation, but they really mean something about the kinds of behaviors that an animal is currently inclined to exhibit. This chapter will take the view of the *behaviorist* that such phenomena are best described in concrete physiological terms, which convey how the nervous system is provided with the bias to perform some behaviors and not others.

SLEEP

Perhaps the most dramatic difference between two behavioral states is that between waking and sleep. Animals are active and responsive to stimuli in their waking state but usually quiescent, relaxed, and unresponsive to stimuli when they sleep. The rate and depth of breathing are decreased, and there is a corresponding decrease in the oxygen level of the blood and an increase in the level of carbon dioxide and other waste products. The metabolic rate (as measured by oxygen consumption) decreases, and the body temperature drops slightly. But activities of the digestive system are unaffected in the sleep state, and the functions of the kidney and liver are slowed only slightly if at all.

These profound changes in the nervous system are correlated with changes in the electroencephalogram. As measured by the EEG and behavioral assays such as responsiveness to stimuli, there are several levels of sleep and periodic changes from one level to another (Figure 30.1). Four or six times during the night, during this periodic fluctuation of sleep states, a human goes through periods of *paradoxical sleep*, in which the EEG looks like that of an alert, awake person, and *slow wave* sleep (SWS), with a characteristic large-amplitude, low-frequency EEG. During paradoxical sleep, the sleeper reports dreaming and makes rapid eye movements (thus the name REM sleep). This phenomenon has been found in all mammals studied so far. Although its function is not known, it has attracted a great deal of attention, because researchers felt it held a key to the function of sleep itself.

There have been many theories about the origin of sleep. One suggested that something caused sensory input to be cut off from the cortex. A cat whose brainstem has been severed at the base seems to be permanently asleep. The decrease in stimuli from body was implicated as the cause of this sleep state. Later work, however, showed that a diffuse and anatomically poorly defined portion of the brainstem, the *reticular formation*, could be stimulated, and this stimulation would lead to wakefulness even in a cat with no spinal sensory input (French, 1957). Damage to this reticular formation could induce a state of sleep.

It was supposed, then, that strong sensory inputs during sleep could activate the reticular formation, which transmitted a barrage of impulses to the cortex and stimulated it into a state of wakefulness—in effect, breaking up the synchronized patterns of SWS. Researchers then assumed that sleep was caused by a decrease of the impulses from the reticular formation below a certain critical level. This decrease could be caused by a combination of physiological factors, circadian rhythms, and a decrease of external

Figure 30.1. A graph showing the record of a night's sleep (above), and electroencephalograms (EEG) used to identify the stages of sleep shown in graph (below). In REM sleep, or paradoxical sleep, the EEG looks like that of an alert awake person. This sleep state typically includes dreaming and rapid eye movement (REM). Of every 4 hours of sleep, nearly everyone experiences REM periods totaling 1 hour.

sensory stimuli. Further anatomical studies of the reticular formation showed that the system was not quite so simple. Some sensory systems have no direct input to the reticular formation but are connected to the reticular formation only by feedback from the sensory regions of the cortex. Close study revealed that the reticular formation was not the diffuse network of synaptically connected short neurons that it had been thought to be but contained a large number of neuron clusters, or nuclei, and many long axons leading from lower parts of the formation directly to the cortex.

It was also discovered that electrical stimulation of a particular part of the reticular formation in the thalamus induces sleep. Further studies tend to support the theory that certain centers within the reticular formation act to induce synchronized activity in the cortex and thus to induce sleep, whereas other centers act to break up the synchronized patterns of cortical activity and thus to induce wakefulness (Magoun, 1963).

The onset of REM sleep apparently is associated with bursts of impulses that originate in the reticular formation of the pons, travel to the thalamus, and then move along sensory pathways to the cortex. Thus, the transitions between SWS and REM sleep, like those between SWS and wakefulness, are apparently under the control of the reticular formation rather than the sensory systems. The reticular formation has therefore come to be viewed as a system that controls the state of consciousness and activity of the cortex in response to external stimuli and the physiological state of the body.

Studies of the effects of drugs on sleep indicate that certain transmitter substances—serotonin and norepinephrine—may be involved in the changes from one sleep state to another. The drug reserpine causes a gradual decrease in the concentration of these substances. Cats treated with reserpine show a reduction in SWS and distinct changes in the nature of REM sleep. The long-lasting, or *tonic*, phenomena of REM sleep (such as muscle relaxation and the typical REM EEG pattern) disappear entirely, but the short-term, or *phasic*, phenomena of REM sleep (such as the bursts of impulses from the pons, rapid eye movements, and muscular twitches) persist. If a precursor of serotonin is then injected, SWS reappears and the phasic phenomena of REM sleep are suppressed. If a precursor of norepinephrine is given to the cats, the tonic components of REM sleep reappear.

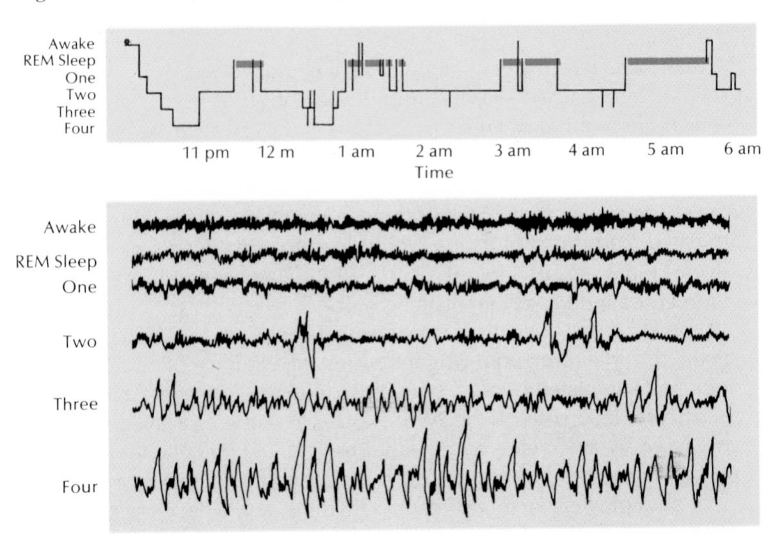

Figure 30.2. An African lioness sleeping during the heat of the midday.

The *raphé system*, a collection of midline cells in the brainstem, contains a high concentration of serotonin. Destruction of the tissues of this system in cats causes a reduction of serotonin in the brain and subsequent insomnia (Jouvet, 1967). Similar results are obtained with a drug that decreases the concentration of serotonin in the brain. The phasic impulses from the pons, which are thought to trigger REM sleep, persist even when the raphé system is completely destroyed. However, the impulses stop when a region in the pons that contains a high concentration of norepinephrine is destroyed.

These observations have led to a theory that SWS is induced by an increase in the amount of serotonin in the raphé system and the diffusion of this substance into the rest of the brain. REM sleep would be induced by norepinephrine secreted from the pons, but this substance could be effective only after the serotonin concentration has reached a high level and SWS has begun (Jouvet, 1969). Such a chemical mechanism could account for the fact that an individual never moves from a state of wakefulness to REM sleep without passing through an intermediate state of SWS.

Although theories of sleep are being modified almost as rapidly as new experimental evidence becomes available, the current belief is that changes from one state to another are controlled neurophysiologically by the reticular formation, and this activity in turn may be controlled by neurochemical changes in the brainstem.

THIRST, OR WATER-SEEKING BEHAVIOR

If a rat is deprived of all water for a day or two, its behavior seems to become oriented toward the goal of finding water. It becomes extremely sensitive to stimuli that have previously been associated with water and tends to ignore or fails to respond to other stimuli that previously evoked responses. There is a temptation to explain this behavior by saying that the rat

Figure 30.3. Rats deprived of water develop a "thirst drive." This motivational state is the result of a multiplicity of physiological changes in the organism culminating in certain predictable water-seeking behaviors.

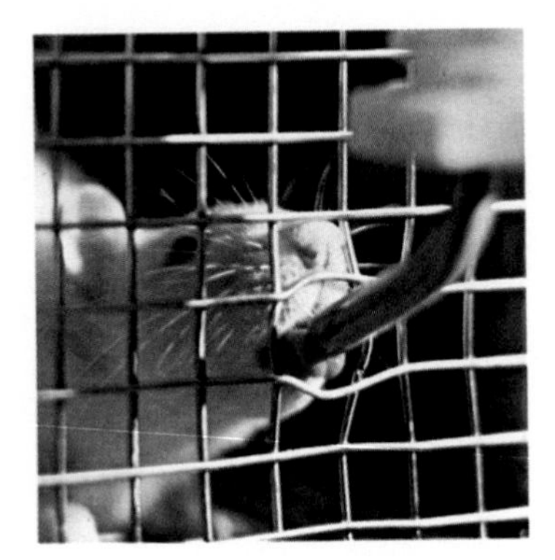

is thirsty, which is another way to say that it tends to perform, or "emit," water-seeking behavior. The concept of thirst adds nothing to this simple observation, unless it can be defined independently in terms of physiological changes within the organism.

What changes does a lack of water cause within the organism? How do these changes affect the nervous system and therefore the behavior of the organism? How does the lack of water specifically alter behavior in such a way that the animal seeks water rather than food or light or sex? And how does drinking water again alter the behavior of the organism to end the water-seeking responses? These questions must be answered if concepts such as "the drive of thirst" or "the need for water" are to be given any real meaning. To continue living, an organism has to obtain a certain amount of water—either too little or too much may be fatal. Behavioral and physiological mechanisms must homeostatically regulate the amount of water in their bodies. When the amount of water in the body falls below the optimal level, certain physiological mechanisms begin to function in order to conserve water. For example, antidiuretic hormone (ADH) is released by neuroendocrine cells in the hypothalamus. This hormone increases the capacity of tubules in the kidneys to reabsorb water; the urine becomes more concentrated and less water is lost in excretion. An excess of water in the body leads to an inhibition of ADH secretion, so that more water is excreted. This hormone regulation as well as other physiological mechanisms help to maintain the homeostatic water balance of the body.

Several behavioral homeostatic mechanisms are also brought to bear. First, the amount that the animal will drink when it does obtain water increases in close proportion to the deficit of water in its body. Second, water comes to serve as a reward, or reinforcer, in conditioning trials. For an animal that has not been deprived of water, this substance acts neither as a reward nor as a punishment.

This rewarding property is manifested by an increase in the strength of behavior patterns that have been reinforced by water. The animal pays more attention to stimuli that have been associated with drinking in the past. It becomes more likely to emit behavior that has been reinforced by water rather than behavior that has been reinforced by food, sex, or other rewards. Punishments, such as electric shock, must be stronger to keep the animal away from water. It becomes more willing to drink bitter water that it rejected before water deprivation began. All these tendencies become stronger as the water deficit in the body becomes larger.

The physiological mechanisms that underlie these behavioral changes have been extensively investigated. Normally, blood and other body fluids contain about 0.9 percent salt. As the body loses water, the percentage of salt—and therefore the osmotic pressure of body fluids (their tendency to attract water across membranes)—increases. It is postulated that the increase in salt concentration is detected by osmoreceptors in the hypothalamus, receptors that trigger the secretion of ADH. Slow injection of a salt solution into the carotid artery that feeds the brain causes an increased secretion of ADH (Verney, 1947).

Normally, an animal stops drinking after it has consumed enough water to restore the normal salt concentration of its blood and body fluids. However, the injection of a hypertonic saline solution (a solution containing more than 0.9 percent salt) into the blood causes water-satiated animals to resume drinking. One experimenter, using goats, implanted tiny tubes lead-

Figure 30.4. Preoptic area in the hypothalamus of the brain where the cells that secrete ADH are located.

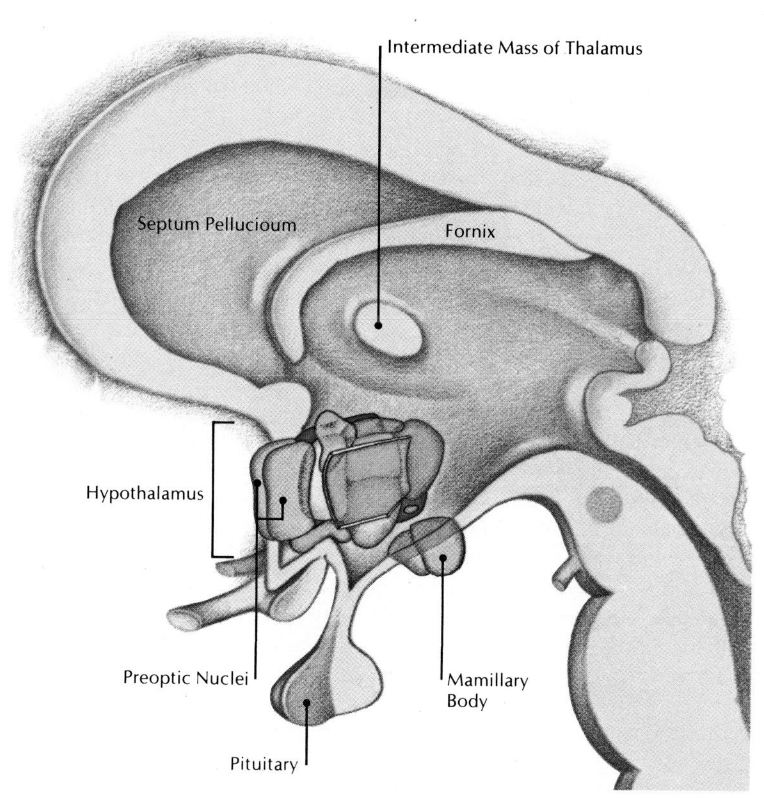

ing down into the area of the brain where the ADH-secreting cells are located. The tubes were cemented to the skull and a small cap placed over the opening of each tube. When the goats recovered from the anesthesia and were behaving normally, fluid could be injected directly into the brain through these tubes. Injection of a normal saline solution (0.9 percent salt) caused no change in drinking behavior, but a minute amount of 2 percent saline solution caused the goats to drink eagerly. Injections of hypertonic saline solution into other parts of the brain had no effect on drinking behavior (Andersson, 1952, 1953).

Further investigations were carried out with cats in an attempt to discover whether injection of salt solution into the hypothalamus simply elicits a drinking reflex or whether it creates the physiological effect that corresponds to normal thirst. In cats that had been deprived of water for relatively short times, injection of 2 percent saline increased the amount of water drunk, whereas injection of pure water decreased the amount drunk. Furthermore, injection of hypertonic saline increased the rate of emission of a response reinforced by water, whereas the injection of pure water decreased the rate of response. The changes in the effectiveness of water as a reinforcer indicate that the saline injections did alter the physiological state of thirst in the organism, not merely elicit an automatic drinking response (Miller, 1957).

In a series of experiments with rats, special electrodes implanted under anesthesia were used to measure the conductivity of brain tissue, which in turn indicated how salty the fluids were. The results confirmed that water deprivation increases the salinity of the brain fluids and that drinking water

Figure 30.5. The effect of water deprivation and injection of salty water on drinking and on tissue conductivity.

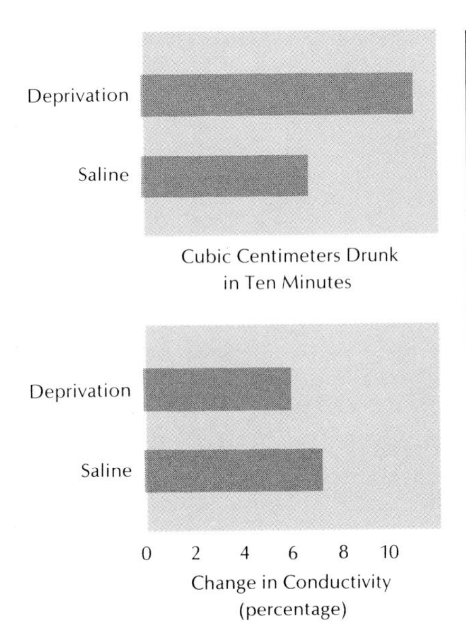

causes the saltiness to decrease. Injection of hypertonic saline into the veins also increases salt concentrations in the brain fluids (Novin, 1962).

Although each veinous injection produced the expected change, a quantitative study of the relative sizes of the changes revealed an unexpected effect (Figure 30.5). The amount of water drunk was not directly proportional to the salt concentration in the brain fluids but was greater for water-deprived organisms than for saline-injected organisms with the same salt concentrations. The experimenter suggested that the slight increase of blood volume resulting from the injections might inhibit drinking, whereas the decrease in blood volume resulting from water deprivation might stimulate drinking behavior.

Further experiments showed that pure water is most effective in satiating the thirst caused by injection of hypertonic saline solution, whereas normal (0.9 percent) saline solution is most effective in satiating thirst caused by reduction of blood volume. These findings indicate that the water-seeking behavior mechanism is capable of discerning the solution that will most rapidly restore the normal salinity and volume of the blood and body fluids. Thus, blood-volume detectors, as well as osmoreceptors, must contribute to the control of water-seeking behavior.

Although water moves relatively rapidly from the digestive tract into the blood, there is a time lag between drinking and the entrance of the water into the blood. However, most animals drink almost exactly the amount of water needed to restore the normal salinity and volume of the blood. What mechanism tells the organism that it has drunk enough before the volume and osmoreceptors in the blood system can be recording normal levels? Experiments with dogs and rats indicate that drinking and swallowing activities and in part the volume of water in the stomach tend to inhibit drinking. Working together, these two mechanisms adjust the amount of water consumed to the amount that is needed. These mechanisms not only act to terminate drinking behavior but also affect the animal's tolerance for bitter water and the effectiveness of water as a reinforcer.

Thus, it is clear that water deprivation leads to a number of changes in the physiological state of the animal. These changes act as stimuli for a number of different receptors whose activation causes hormonal changes as well as physiological changes in the nervous system. As a result of the hormonal and neural changes, the organism exhibits a number of changes in its behavior patterns. All these physiological and behavioral changes may be summarized by saying that the animal is thirsty. Defined in this way, thirst has meaning, for the presence of a state of thirst can be confirmed by observations other than the emission of water-seeking behavior.

TEMPERATURE REGULATION

Mammalian body temperature is controlled by a mechanism that involves receptors in the hypothalamus, neurological and hormonal control systems, and a variety of physiological and behavioral responses that serve to maintain homeostasis. This mechanism is remarkably sensitive and efficient. In environmental temperatures ranging from −20° to 130°F, a human maintains a body temperature of about 97.5° to 99.5°F for a considerable amount of time.

When the body becomes overheated, a variety of responses tend to reduce body temperature. Blood vessels in the skin expand so that more blood comes to the surface and more heat is radiated to the environment.

Figure 30.6a (left). A Masai giraffe feeding at tree top level in an African thorn scrub forest. Utilization of such food sources reduces the giraffes' competitive interactions with other herbivores.

Figure 30.6b (middle). The normal feeding position of the shrub-browsing gerenuk. This graceful African gazelle is noted for feeding on shrubs at intermediate height and thus avoids intense competition with neighboring browsers and grazers who feed on grasses.

Figure 30.6c (upper right). The true African chameleon (*Chamaeleo jacksoni*) in the act of capturing its prey. Note the insect trapped on the sticky end of the lizard's tongue.

Figure 30.6d (lower right). A frog about to snap up a fly The frog's tongue is attached to the anterior portion of the lower mandible and thus has to be "backhanded" out of its mouth to catch its prey.

Sweat glands increase their secretion of fluid, causing greater loss of heat through evaporation. Salivation is increased and panting occurs in many animals, cooling the tongue and mouth. Neuroendocrine cells in the hypothalamus secrete ADH, which is collected and released into the blood by the pituitary gland and which causes the kidneys to increase reabsorption of water, release more concentrated urine, and thus conserve water for use in evaporative cooling. Hunger is inhibited and metabolic processing of food is slowed, decreasing the amount of heat generated within the body. Thirst is stimulated, causing the organism to seek water that can be used in evaporative cooling. This complex pattern of behavioral responses includes functions controlled by the endocrine system as well as the somatic and autonomic nervous systems.

A similar complex of responses occurs when body-temperature becomes too cool. Blood vessels in the skin constrict; secretions from the thyroid gland speed up the metabolic processing of food; shivering releases heat as the muscle cells use up energy reserves; and hunger is stimulated.

If the front portion of the hypothalamus is damaged, a mammal lacks the behavioral and physiological responses to overheating. If the rear portion of the hypothalamus is damaged, a mammal fails to react to cooling of the body. Experiments with electrical stimulation similarly indicate that responses to overheating are controlled by the anterior hypothalamus, whereas responses to cooling are controlled by the posterior hypothalamus. Although heating or cooling of the skin (which contains many temperature

receptors) has some effect in eliciting the temperature-regulating responses, slight heating or cooling of the blood flowing into the brain is far more effective. Direct heating and cooling of the hypothalamus confirms that temperature receptors exist within that portion of the brain and exert primary control over the temperature-regulating responses. A rat whose hypothalamus is cooled exhibits the normal temperature-raising responses and, in fact, suffers from a fever because its body was already at normal temperature.

A lowering of environmental temperature serves as a reinforcer for an overheated animal. Similarly, shaved rats placed in a refrigerator will learn to press a bar in order to turn on a heat lamp. In short, overheating or overcooling produce behavioral states, or drives, very similar to those produced by water or food deprivation.

SEXUAL BEHAVIOR

Although sexual behavior is necessary for the survival of the species, it does not seem to be necessary for the survival of the individual. Nor does it have the same homeostatic functions as the behaviors associated with thirst or control of body temperature. In lower animals, most of the details of sexual behavior seem to be innately determined and under the rigid control of sex hormones and the lower centers of the brain. An appropriate external stimulus is required to trigger the behavior. In the higher mammals, the higher centers of the brain and learning seem to play more important roles.

Sexual behavior certainly has many of the attributes that characterize such drives as hunger and thirst. For example, an adult male rat will learn to run down an alley or to work at some other response in order to obtain access to a receptive female. Although ejaculation is important if a sexual experience is to be maximally reinforcing, there is considerable evidence that other components of sexual activity also can serve as reinforcers. After sufficient sexual activity, the male rat acts as though he were satiated. Sex appears to be a very complex drive and has a greater number of motivational components than thirst or hunger.

Most land animals show sexual behavior only during an annual breeding season. In many species, this season is apparently controlled by a photoperiodic mechanism, but in most cases a combination of internal and external factors determine the time when reproduction can occur. For instance, in species living where rainfall is scarce, reproduction may be triggered by a rainstorm.

Within the broad mating season (which lasts a few months in most species), various factors may determine the more precise cycles of sexual behavior. Some species that mate in the intertidal region show reproductive behavior only at particular phases of the moon. Other short-term cycles are determined by various environmental stimuli (including social behavior patterns) and internal rhythms. In mammals the estrous cycle (a rhythm largely controlled by hormonal mechanisms) causes the female to be sexually receptive at periodic intervals (Chapter 21). In many species, the receptive periods recur regularly throughout the breeding season.

The precise timing of ovulation in birds seems to depend strongly upon such factors as the weather, the building of a nest, the availability of suitable food, and the performance of courtship behaviors with a male of the same species. The intricate (and largely innate) patterns of courtship or precopulatory behavior in many bird species serve several functions. They

Figure 30.7. The three behavioral events involved in the copulatory sequence of the rat. Shown here are the mounting of the male on the female without penetration, intromission with a brief penetration, and ejaculation.

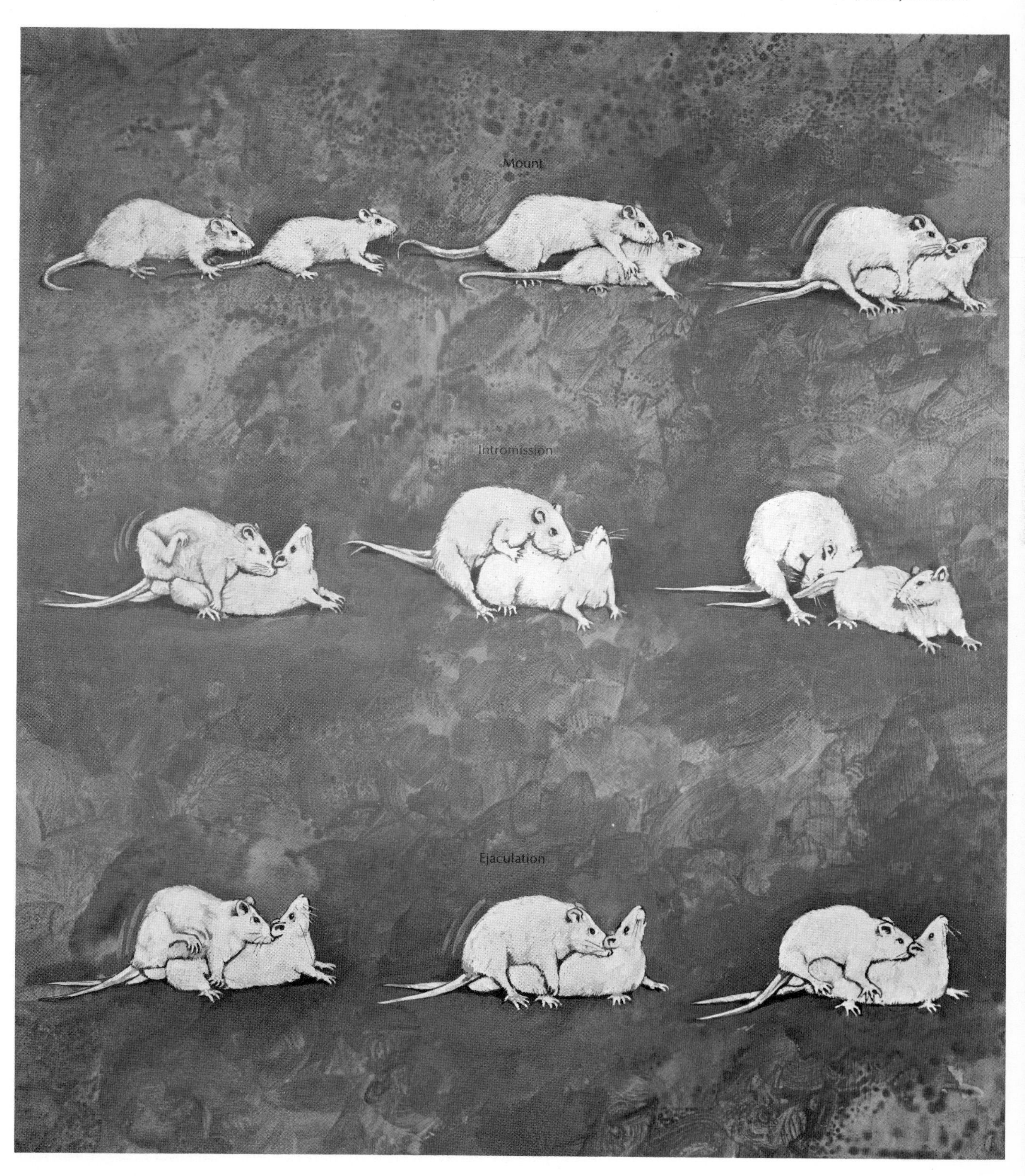

ensure that copulation occurs only between animals of the same species, thus avoiding wasted gametes and time. These behaviors serve as stimuli that bring the reproductive physiological cycles of the male and female into synchronization, thus maximizing the probability of successful fertilization.

Precopulatory behavior in most species serves to bring both partners to a state of sexual excitement, to verify that the female is receptive, and to bring the partners into position for copulation. In most species, copulatory behavior itself is relatively brief. Patterns of precopulatory and copulatory behavior vary incredibly through the animal kingdom, but a brief description of the behavior of the rat will serve as an example.

The female rat is sexually receptive for a period of about 19 hours around the time of ovulation, which recurs at intervals of about 4 or 5 days. If a male rat with previous copulatory experience and a female are placed together in a small cage, the male almost immediately begins following the female and sniffing at her genital region. If the female is not receptive, she continues to flee from the male and he eventually abandons the pursuit. If she is receptive, she responds to his nuzzling of her genital region with an arching of the back that elevates the hindquarters and genital area, a movement of the tail to one side, and a short, jerky forward motion. This response is apparently an involuntary reflex triggered by the male grasping her flanks.

If the female does make the response that indicates her receptivity, the male mounts her and begins a series of rapid thrusting movements of his pelvis. If he fails to penetrate the female after a few thrusts, he dismounts and the sequence begins again. If he does penetrate, he makes a single deep thrust (lasting about 0.2 second) and immediately and vigorously dismounts. He then grooms his genitals or engages in other behavior for 20 to 60 seconds before beginning the cycle again. After about the fifth to eighth penetration of the female, the behavior of the male changes. He does not dismount after the first penetrating thrust but thrusts deeply as many as five times. Ejaculation occurs on the last thrust and is accompanied by an orgasmic spasm of the male's hindquarters. For a few seconds after ejaculation the male holds tightly to the female; then he slowly dismounts.

For a period of about five minutes after this dismounting, the male does not approach the female. Then he begins another cycle of mountings that ends in an ejaculation much sooner than did the first series. If the rats are kept together for several hours, the mounting and ejaculating cycle may be repeated as many as five times, but the interval between ejaculation and the first mounting of the next cycle becomes longer and longer.

Apparently, the initial shallow thrusts that precede penetration serve to orient the male properly for penetration. As soon as genital contact occurs, the male makes the deep thrust and immediate dismount, almost as if it were an involuntary reflex. With each penetration the level of the male's excitement increases, until a penetration provides sufficient further stimulation to trigger the ejaculatory reflex. The reflex itself is largely controlled through the spinal cord, but the inhibition that determines the amount of penile stimulation needed to trigger the reflex is imposed by higher brain centers.

Cessation of copulation is not merely a factor of physical exhaustion, for a male will begin the copulation cycle immediately if placed with a new female as soon as he has stopped copulating with the old one. It appears that the male becomes habituated to the stimuli provided by a particular

Figure 30.8a (upper left). The African lion (*Panthera leo*) during copulation. Prior to mating, lions may exhibit precopulatory play behavior. The male lion often emits a peculiar coughing grunt during intromission.

Figure 30.8b (upper right). The great frigate bird engaged in courtship ritual on a Galápagos Islands hatchery. Note the extended gular pouch (red) of the male, which, when inflated, acts as a sexual stimulant to the female.

Figure 30.8c (lower left). A group of African ringed waterbucks. These antelope are polygamous in nature; the male typically is seen with a group of harem does and fawns.

Figure 30.8d (lower right). After partner selection, mating blue jays exhibit distinctive parental behavior. Here, the male partner is feeding its mate during the incubation period while she is nesting.

mate; these stimuli have less and less effect in triggering his copulatory behavior. If continually supplied with new mates, a male rat may copulate 10 or 15 times before ceasing to emit copulatory behavior in the presence of a new female.

Although the female rat appears to play a very passive role in copulation, she will learn a response that is reinforced by a single mounting from a male and will continue to emit the learned response until she has received about five ejaculations. Clearly, the female rat also has a "sex drive." However, a female will also work in order to avoid sexual contact until a certain preferred interval has passed since the last contact. This interval varies from about 20 seconds after mounting without penetration to about three minutes after an ejaculation (Bermant, 1967).

Similar behavior patterns are observed in guinea pigs, hamsters, and other rodents. Copulatory behavior involves a complex series of responses

Figure 30.9. Most mammals show curiosity or exploratory behavior. Monkeys confined in an environment with little sensory stimulation will quickly learn to press a lever in order to see an event or objects outside their cage. Here, the sight of an electric train is enough to reinforce lever-pressing behavior in this monkey.

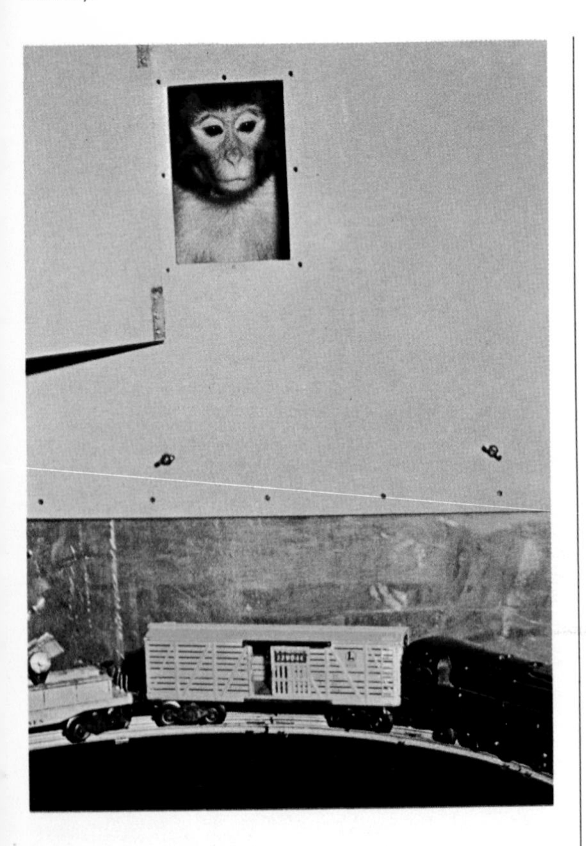

to various visual, olfactory, and touch stimuli. In rodents the mechanism involved in precopulatory behavior and sexual arousal is largely controlled by higher brain centers, whereas the mechanism involved in copulation itself is chiefly under the control of the autonomic nervous system, the spinal cord, and the hypothalamus. In higher mammals, even more of the copulatory behavior is controlled by higher brain centers and is subject to modification through conditioning or learning.

A range of behaviors are involved in the maternal activities of the pregnant female and the activities of both parents in caring for the young after birth, as well as the behaviors of birth itself. All these patterns are similarly controlled by complex drives or internal states, neural and hormonal mechanisms, and external stimuli.

CURIOSITY

Most mammals are curious. Both rats and monkeys will learn new responses for which they are reinforced by the opportunity to explore novel objects. Monkeys confined in a monotonous cubicle learn to press a bar that causes a shutter to be removed from a window so that they can look out and observe activity outside. The monkeys not only learn this new response but continue to work hard to open the shutter (Butler, 1960).

The strength of curiosity as a drive may be illustrated by placing a very hungry rat in a strange box with a supply of familiar food. Until it has completed an examination of the new environment, the rat only nibbles intermittently at the food. The rate of emission of exploratory behavior toward an object decreases with time—the simpler the object, the faster the exploratory behavior disappears. However, even a very familiar object is still explored occasionally.

Little is known about the nature or physiological mechanisms of curiosity. In fact, the general type of activity loosely described as "curiosity" may represent a hodgepodge of quite different behaviors mediated by different mechanisms. One function of curiosity, or exploratory behavior, may be simply the obtaining of adequate sensory stimulation. Humans kept in dark, soundless, motionless quarters (in other words, deprived of most external sensory stimulation) soon show signs of mental and even physiological disturbances. There also seems to be a drive for body activity. Rats learn to press a bar in order to obtain access to an exercise wheel.

AGGRESSION

If a certain part of the hypothalamus of a cat is electrically stimulated, a spectacular burst of aggressive activity is produced. The hair bristles, the pupils open wide, the ears lie back, and the cat hisses and strikes out viciously at a gloved hand or stick inserted into the cage. It will attack a rat in the cage with repeated, slashing paw swipes.

If a slightly different area is stimulated, the cat will make a silent, cold, stalking attack, in which the rat is seized by the neck with a bite that kills it by breaking its neck. A cat stimulated in this way will learn to choose the arm of a Y-maze that leads to a box containing a rat to attack rather than the arm that leads to an empty box.

These experiments demonstrate two rather different types of aggressive behavior, apparently controlled by two different centers in the cat's brain. It is quite possible that there are a number of different types of aggressive

Figure 30.10. Aggression elicited by electric shocks will be directed to another rat, if one is present (above). If no partner is available, however, the rat will attack a doll or other substitute in order to escape pain (below).

behavior and that, like curiosity, this supposed drive actually represents a number of relatively independent neurophysiological control systems.

Aggressive behavior can be elicited by frustration or by painful electric shocks. In the latter case, the tendency to fight can be increased if fighting is rewarded by escape from shock. A rat trained in this way to fight other rats will attack a doll if no rat is present. In many species, groups of individuals living together organize themselves into a dominance hierarchy. After initial aggressive encounters, the more dominant animals control the less dominant ones by mere threats. If a monkey in the middle of a dominance hierarchy is stimulated to aggression by radio-controlled stimulation of the appropriate spot in its brain, it will always attack a less dominant monkey rather than a more dominant one. In the laboratory, an animal shows less aggressive behavior toward an individual that has defeated it in an earlier fight. If it has had a series of successful fights with various individuals, the animal's behavior toward a strange individual becomes more aggressive.

In many animal species, there seems to be a sort of "personal space" around the individual. A novel stimulus may elicit exploratory behavior if it is beyond a certain distance from the animal, but the same stimulus may

Figure 30.11. The puffer fish, or blowfish, with normal body contour (insert at upper right). The same fish is shown below in a defensive pose with body inflated and spines erect. Such a defense mechanism significantly discourages predators.

elicit flight behavior if it is closer than that distance. As the animal becomes familiar with an object through observation from a distance, it slowly approaches. The conflict between curiosity and fear often produces a wavering, back-and-forth movement as the animal seems to approach with curiosity, then retreat in fear, and then approach again somewhat closer. Similarly, another animal may elicit flight or exploratory behavior if it is at a distance but elicit aggressive behavior if it is too close. The dimensions of personal space vary from species to species, from individual to individual, from stimulus to stimulus, and from moment to moment.

Aggression, fear, curiosity, sex, and other nonhomeostatic drives play extremely important roles in the behavior of higher animals. Ethologists have described a great number of fascinating examples of such behavior patterns from a wide range of species. The study of the mechanisms underlying these drives and the elucidation of their nature has barely begun.

FURTHER READING

A large number of books on sleep have been published recently. Among the general discussions of the topic are Foulkes (1966), Luce and Segal (1966), Oswald (1966), and Webb (1968).

For more extensive discussions of behavior from a point of view similar to that of this chapter, see books by Dethier and Stellar (1970), Hinde (1966), Manning (1968), Marler and Hamilton (1966), and Waters, *et al.* (1960). The contributions of ethologists to study of complex behavior have been underrepresented in this chapter; for a more complete picture of their findings, see books by Eibl-Eibesfeldt (1970), Lorenz (1952, 1965), and Tinbergen (1951, 1958). Behaviorist discussions of complex behavior are typified in books by Skinner (1938). Among many introductory articles dealing with complex behavior patterns and behavioral states are those by Dethier (1967), Eibl-Eibesfeldt (1961), Fantino (1968), Gilbert (1962), Harlow and Harlow (1964), Masserman (1967), and Tinbergen (1960).

31
Communication and Social Behavior

Communication and social behavior must involve two parties—one sending the signal and one receiving it—and both parties must derive some advantage from the interaction. In many ways, a flower and a bee communicate with each other. The bee has evolved complex visual recognition capacities that enable it to find a flower. The flower has evolved distinctive shapes and color patterns that the bee can easily recognize. The bee derives food from the flower, and the flower derives the advantages of cross-fertilization from the bee's visits.

Throughout the animal kingdom, there has been a progression of complexity both in mechanisms designed to send signals to other individuals and in mechanisms designed to recognize and interpret signals sent by others. The signals and the recognition systems are often startling in both the intricacy and the efficiency with which they intermesh. Behavior releasers can be chemical, visual, auditory, tactual, or sequences of these and other stimuli, and the sensory systems filter out irrelevant cues.

Social releasers are often marvelously adapted to the properties of the physical world and the modes of communication available. Chemicals with long lifetimes serve for long-range, low-speed communication. Chemicals with shorter lives deliver signals that must act only over short distances and short times. Visual signals give immediate and potentially private communication, and auditory signals give private or general audience, depending upon how they are employed. In animals that must communicate relatively little information, signals tend to be simple, but versatile systems for communicating many different possible messages have evolved as well. Human language is one such system.

SOCIAL RELEASERS

Behavioral researchers have long attempted to identify the components of behavior that trigger responses of other animals. Social releasers are the natural behaviors or signals that influence other animals in a social situation. Those aspects of the natural behavior that are necessary and sufficient to carry the message are called the *trigger features*. Ethologists in particular have analyzed many social situations in an effort to identify such trigger features.

For example, in favorable localities in the southeastern United States, there may be as many as 20 different species of tree crickets singing at the same time. The most prominent contribution to the din of a summer's evening comes from the calling song of the males. Their songs elicit seeking behavior by reproductive females; the songs are simple auditory social releasers. The function of these songs is readily demonstrated experimentally. A cage three feet long is fitted with loudspeakers at each end, and a reproductive female is released in the center. If one loudspeaker transmits recordings of songs of a male of the same species as the female while the other transmits songs of another species, the female moves toward the song of her own species (Walker, 1957).

Electronic analysis of the cricket sounds and tests of the effects of artificially created songs reveal that the males of each species have a particular pulse rate in the calling song. The female responds only to songs with the pulse rate appropriate to her species. The frequencies and quality of the individual chirps, or pulses, in the songs of various species sound somewhat different to human ears. But females ignore those qualities and respond to any artificial song with the correct pulse rate, regardless of the

Figure 31.1 (left). A given motor output (response) may be affected by more than a single input (stimulus), and these various inputs (stimuli) may combine to varying degrees in their effect. A given input typically influences more than one motor output, and these outputs may be affected simultaneously, alternately, or sequentially.

Figure 31.2 (right). The pulse patterns comprising the songs of two species of tree crickets and some artificial songs.

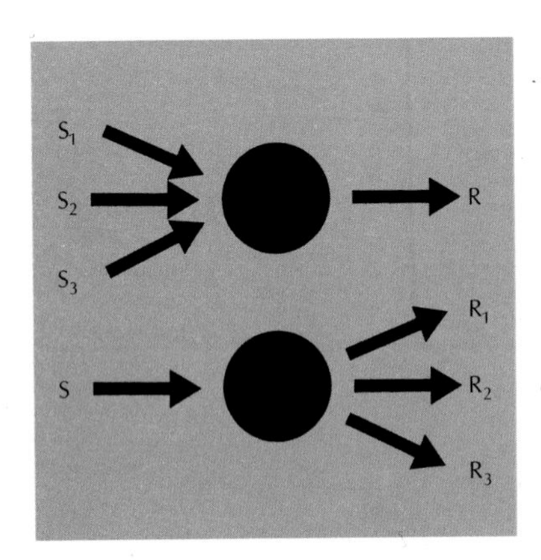

tone structure of the pulses. The irrelevance of song features other than pulse rate has been strikingly demonstrated by use of the temperature dependence of the pulse rate, which speeds up as temperature is raised. Suppose that the male of a given species (A) normally sings with a pulse rate of 50 pulses per second at 70°F. A male of another species (B) that sings with a slower pulse rate can be warmed up until his pulse rate is 50 pulses per second. A recording of this song can then be played to a female of species A at 70°F. She will move toward the source of this song just as if it were the song of a male of species A, even though the pulse tone structure differs from that produced by males of species A.

Anatomical and physiological studies of the auditory system of the tree cricket reveal that this insect is able to detect variations in loudness with great sensitivity, but it has no sensory equipment for detecting variations in frequency of sound. Furthermore, the frequency sensitivity of the female's receptors changes with temperature, so that it precisely compensates for the effect of temperature on the male's pulse rate. The cricket sensory system is designed to measure only one aspect of a seemingly complex song. Pulse rate is the trigger feature of the calling song that elicits mate-seeking behavior. Because the different species that exist together in a given locality

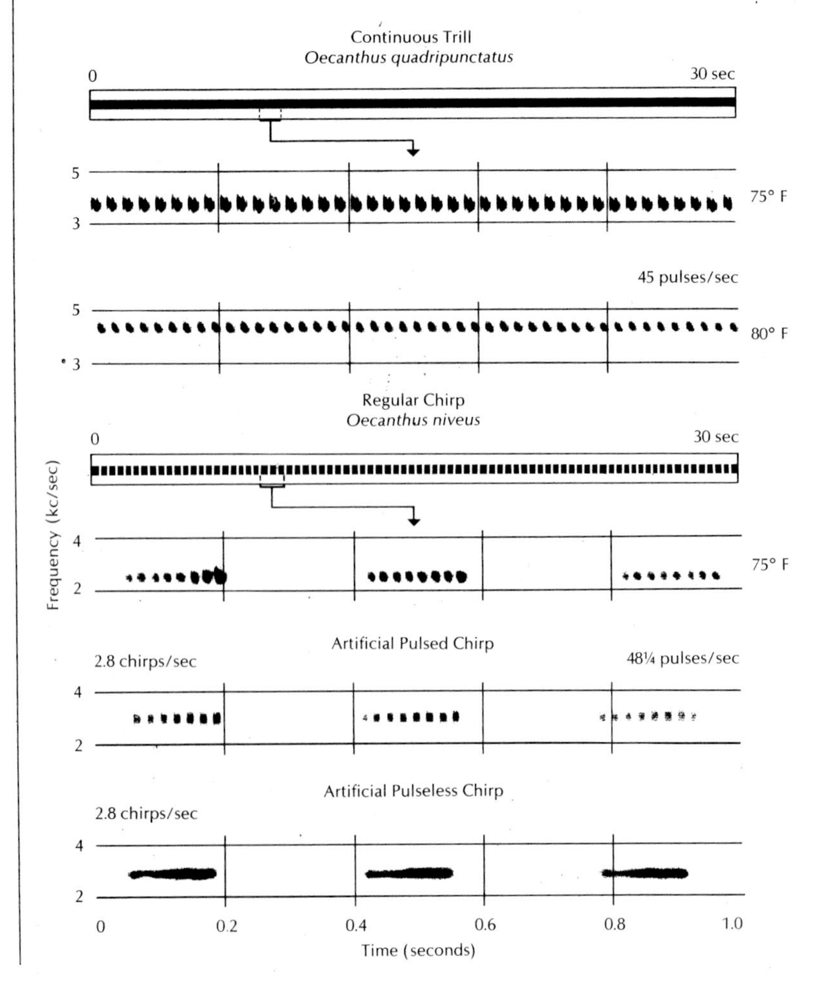

Figure 31.3. Two variations of the apparatus used by Magnus to present different visual stimulus patterns to wild male butterflies (*Argynnis paphia*). The ends of the arms of one apparatus carried a butterfly model with various color patterns and rates of flapping around a circle (above). The ends of the arms of a second apparatus carried a rotating cylinder with alternating black and colored sections, which produced a flicker of color (below).

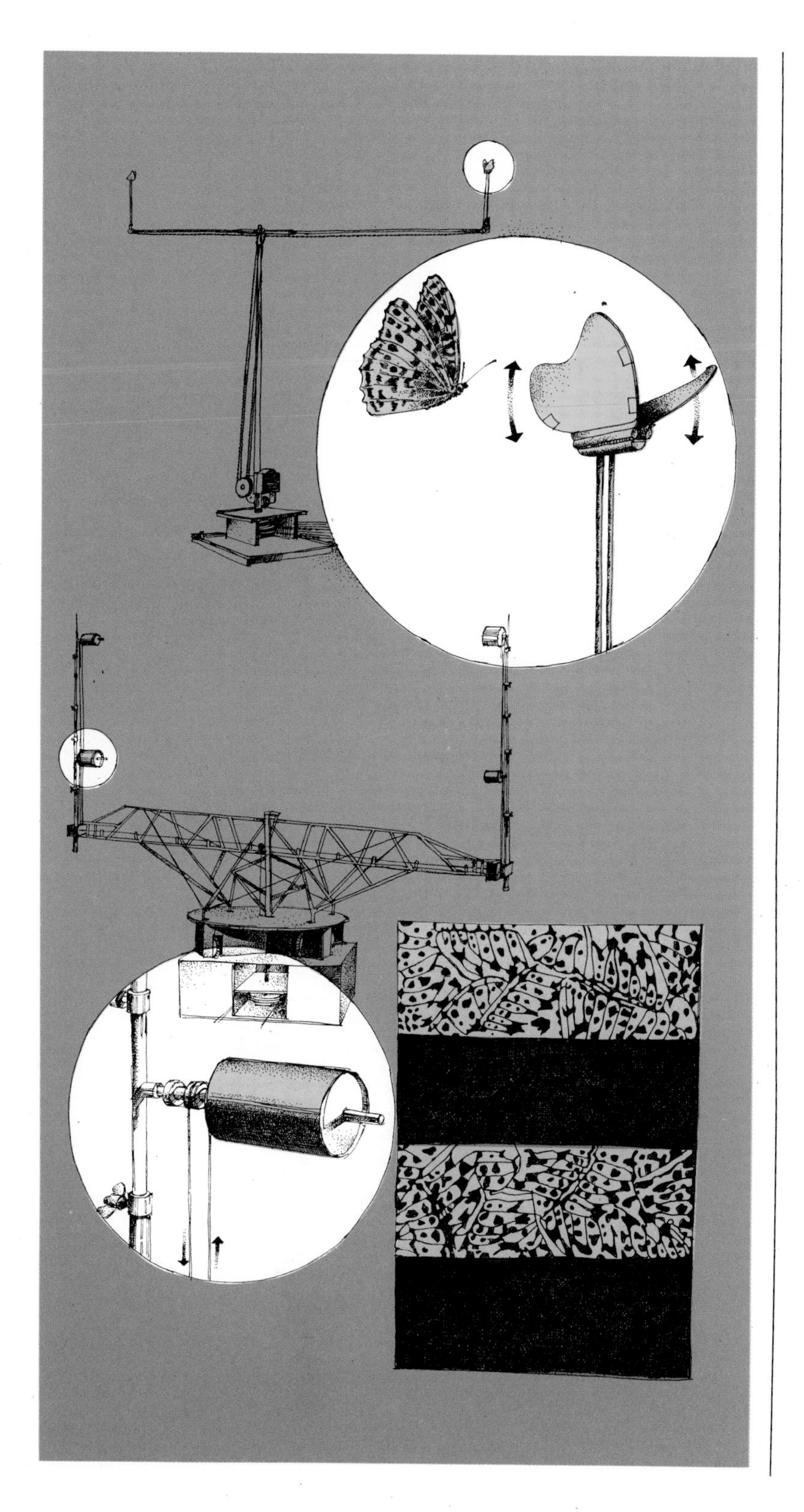

Figure 31.4 (right). Chemical communication in the silkworm moth (*Bambyx mori*). The male can locate a female at distances of less than a meter or two by noting the diffusion gradient in concentration of the pheromone. At greater distances, the male can locate the female only by flying into the wind that brings him the scent. In this illustration, the concentric circles are one meter apart. Tracks of various males are for one-hour periods.

Figure 31.5 (left). Restrictions of chemical communication as illustrated in gypsy moths. The distance and area from which females can recruit males with pheromones is a function of diffusion and air turbulence. At distances of less than 2 meters, males orient themselves by discriminating the diffusion gradient. At distances of greater than 2 meters, they orient themselves by flying into the wind.

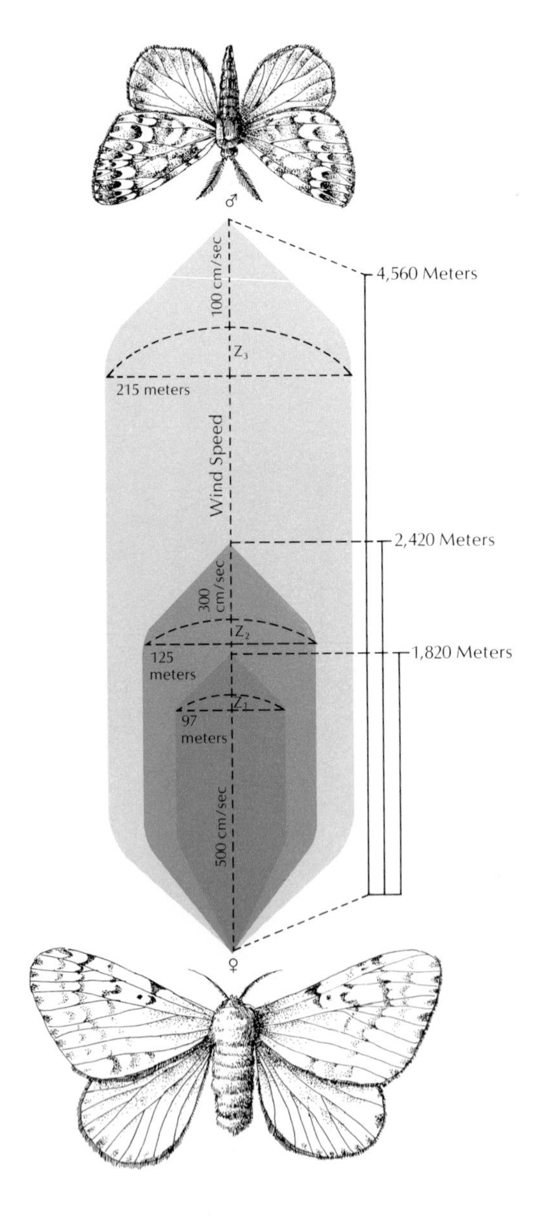

have songs with different pulse rate structure, the cricket calling song evidently serves to promote species isolation during reproductive activities. Communication that promotes such isolation, and prevents abortive interspecies mating attempts, has appeared in most vertebrate animal groups.

Shape, color, and temporal change may all be parts of the trigger features of visual social releasers. Courtship in butterflies illustrates some interesting properties of visual releasers. The first stage of courtship consists of the male's pursuit of the female. D. Magnus devised an apparatus that would move a flapping butterfly model with various color patterns and rates of flapping around a circle. He counted the number of pursuits elicited in a male during a given period of exposure to these models. He discovered that the flapping per se was not important; a rotating cylinder with alternating black and colored sections, which produced a flicker of color, was equally good. In this case, the male paid little attention to shape. The red-orange (the overall color of the female) color was most effective, and the larger the cylinder, the better. When the cylinder was rotated at high speed and the male was offered the choice between this and a more realistic model, he chose the more rapidly flickering one. Only when the cylinder presented a flicker at 140 hertz, considerably faster than a female's flapping, did its effectiveness begin to wane. Thus, a large, rapidly flickering stimulus was a *superoptimal stimulus* — it was a better releaser than is found in real life.

SPECIALIZATION OF SIGNALS

Chemical communication illustrates some of the ways the mode of communication, the signal-carrying vehicle, can be adapted to the nature of the

Figure 31.6. Fire ant worker establishing an odor trail with its extended stinger and the history of a representative odor space of the trail pheromone. The pheromone odor gradually fades with time.

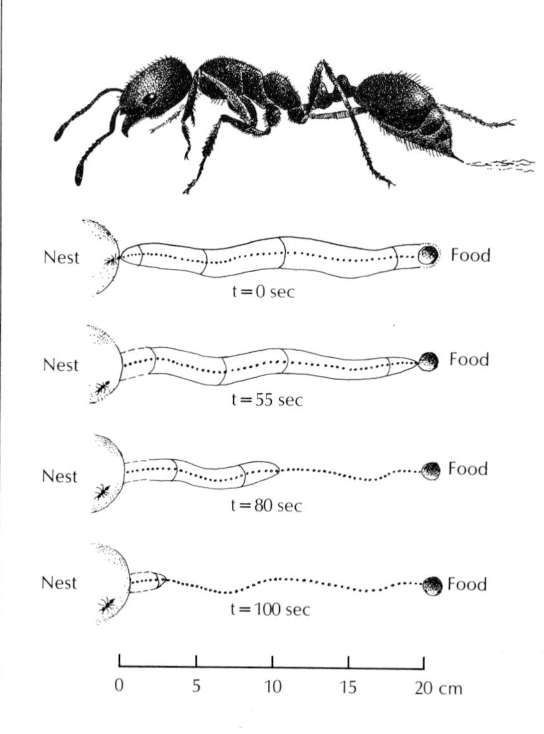

message to be communicated. A species may possess a repertoire of different chemical signals, or *pheromones*, with different diffusion characteristics, produced by various glands on the body, transmitted in different ways and serving such varied functions as marking trails, inducing mating, encouraging aggregations for resting, disseminating alarm, permitting recognition of individuals and groups, and attracting others to a food source. The sensitivity of a male moth to the pheromone emitted by a female is a good example of the species specificity of long-distance communication that can be accomplished through use of pheromones. The male can locate the female at distances of less than a meter or two by noting the diffusion gradient in concentration of the pheromone (Figure 31.4). At greater distances, the male can locate the female only by flying into the wind that brings him the scent (Figure 31.5). Females emit the pheromone only when there is enough wind so that it is likely to be detected by males (Marler, 1969).

The pheromone used for mate location by moths is active over long distances. Ants, on the other hand, have trail-marking substances that fade more rapidly. If the source of food at the end of a trail becomes exhausted, returning foragers who normally must follow scent trails between the nest and food sources do not add their own trail substance to the existing trail. Because the trail gradually fades if it is not renewed by successful foragers, the net effect is that foragers looking for food no longer can find that particular trail, and they spend their effort going to better food sources. Ants also release alarm pheromones, which elicit aggressive behavior either by specialized "soldier" members of their society or by workers. Ants at a short distance from a disturbance are induced to release their own alarm substance, which leads to the spread of the alarm message. The alarm substance must not persist very long, otherwise a minor disturbance at one spot could lead to a chain reaction that would leave the anthill in an uproar for hours after the cause of the original disturbance had passed.

Because of the properties of the auditory systems of vertebrates, a sound source is best localized if it has many different frequency components. Direction cues are furnished by low-frequency sound, which can be compared in phase at the two ears, and by high-frequency sound, which can be compared in intensity at the two ears. Abrupt sounds are easily localized because they arrive at the two ears at slightly different times. Among many animals, especially primates, these characteristics are employed in the construction of auditory signals, whose purpose is to convey the location of the caller. Primates, insects, and frogs all utter series of staccato clicklike sounds that have high- and low-frequency components. These sounds are repeated to ensure that the caller is located. In some monkeys, a lost member of a troop will give a long sequence of such calls.

The converse is the case in many alarm calls, whose purpose is to alert members of the same species to an approaching danger. Of necessity, these calls must be difficult to localize. An acoustic engineer would suggest a single frequency in the intermediate range that did not begin or end abruptly and that was not repeated more than absolutely necessary. The alarm calls given by a number of small birds in the presence of a flying hawk are exactly such sounds. They are simple tones, which fade in and out without an abrupt start or finish (Figure 31.7). In this case, the call apparently carries a warning of danger overhead, and the location of the caller is of little

Figure 31.7 (above). Sound spectrograms of the alarm calls of five species of birds in response to an aerial predator, such as a hawk.

Figure 31.8 (below). The male peacock (*Pavo cristatus*) both before (left) and during (right) courtship display.

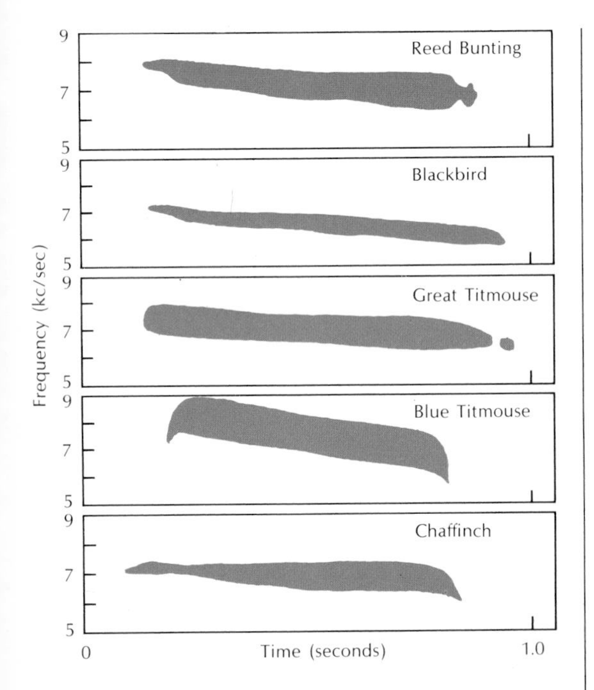

value to others of the same species. On the other hand, there is apparently great survival value for the small bird in minimizing the clues that the hawk can use to locate it. The similarity of the calls of several species is also easily explained. The survival of each species may be enhanced by the fact that it can respond to alarm calls from other species also endangered by hawks.

THE BEGINNINGS OF LANGUAGE

In some species, social releasers are simple sensory cues involving spatial patterns or simple temporal patterns of stimuli. They may be directly related to the activity that is released by the animal perceiving them or to the behavior that the sender is engaged in, such as butterfly flight. But there is a continuum of complexity, especially involving temporal sequences of behaviors, in which sequences serve as releasers. At the same time, there is a continuum of degree of abstraction, or specialization, of signals. Human language is perhaps the ultimate example.

Ritualization of stereotyped behavior sequences may be thought of as one step toward development of the specialized and arbitrary communication signals that form language. Courtship and territorial displays are among the most elaborate examples of ritualized behavior. Many salt- and fresh-water fishes, especially males, stake out territories, which they defend from intruding males of the same species. If an intruder enters the territory of a male fresh-water fish of the family Cichlidae, a ritual territorial encounter occurs. At first, the males turn broadside to each other and spread their fins. If the intruder does not leave, a bout of tail flipping ensues in which the tail is flipped toward the opponent, creating currents that can be felt by the intruder. If this display does not drive him off, a bout of mouth pulling ensues. One fish grabs the other's mouth, and they push and pull each other until one gives up. The vanquished retracts his fins and swims away as the conqueror pursues him with fins erect. Researchers have speculated that the ritualization of the fight, with its emphasis on certain nondestructive displays, minimizes damage to both parties. Often, it is the smaller and younger individuals who would be most badly beaten, and it is these who will provide next year's reproductive material.

Ritual fights usually consist of stylized actions that are similar to real

Figure 31.9. Sound spectrograms illustrating dialects in the songs of 18 male white-crowned sparrows from three areas near San Francisco Bay. The greatest individual variability is seen in the introductory and terminal portions of the call. The middle portions of the call vary little within a population but are consistently different between populations. The time marker indicates 0.5 seconds, and the vertical scale is marked in kHz.

fighting but with applied "rules" that limit damaging behavior. In bird courtship, the displays seem further removed from the acts of reproduction. They involve posturing of the body, dancelike steps or flights, displays of plumage, and often auditory signals. Ethologists believe that many displays, such as that illustrated in Figure 31.8, are derived from the motions a bird makes when it takes flight from the ground or water. Other displays are derived from juvenile behavior, such as food begging. Apparently, during the course of evolution, these actions have become parts of courtship but they have been ritualized until they bear little relation to the current goal of the courtship process. It is speculated that the elaborate displays are needed to ensure that only conspecifics mate and to ensure that the physiological states of the male and female will be right when mating and young rearing occur.

Sequences of communicative behavior may involve both learned and stereotyped components. The bird song is a very elaborate form of temporally patterned communication that is highly arbitrary and bears little if any resemblance to the behavior it attempts to communicate. Vocal apparatus and sensory mechanisms have evolved together with behavior, and they enable generation and recognition of rapid sequences of tones and noise. Birds may have specific songs for territorial displays and courtship invitation, as well as display, alarm, feeding, and many other social functions.

In some bird species, the entire repertoire of signaling behaviors develops normally in birds raised in isolation. In other bird species, however, drastic abnormalities appear under these conditions. An example of such dependence on contact with other members of the species is found in the white-crowned sparrow of central California. The natural song of males of this species varies from place to place, much like local speech dialects in human language (Figure 31.9). Young birds taken from the nest and raised in either individual or group isolation fail to develop the song patterns characteristic of the local dialect. Their songs fall quite outside the class of natural patterns for birds in their area or any other. If the birds are captured a few weeks after the time of fledging and then isolated in soundproof chambers, the song develops normally and the birds display the patterns of the particular local dialect where they were captured.

Isolated nestlings exposed to a recording of normal songs each day (during the period from about 10 days after hatching to roughly 50 days of life) will later develop a normal singing pattern corresponding to the local dialect of the recording to which they were exposed, even though that dialect may be of an area different from the one in which the birds were captured. Thus, the general pattern of the song is apparently inherited, but the details are learned during a certain period of early life. Exposure to normal song outside this period has little or no effect on the song of an individual.

Humans seem to have a similar period of sensitivity in early life during which they must learn some of the details of language behavior. While a child is learning to speak, he conspicuously focuses his attention upon human sounds. Similarly, a white-crowned sparrow confronted during the critical learning period with a choice between white-crowned sparrow songs and recordings of song-sparrow songs (which are also heard on its breeding grounds) learns only the song of its own species. The young bird learns partially by feedback from the sounds that it makes. A sparrow that is deafened after the critical period but before it begins to sing develops a

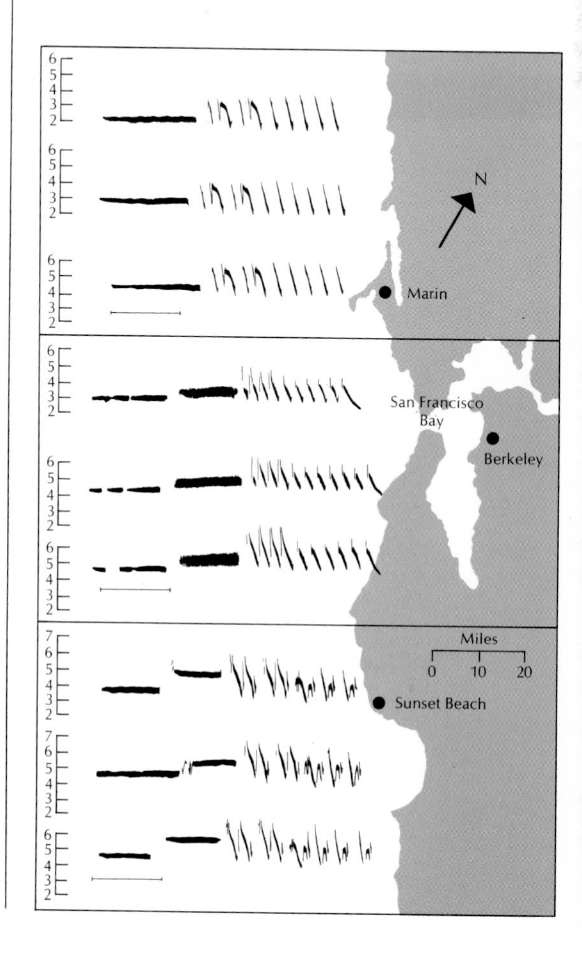

Figure 31.10. Sound spectrograms of the songs of nine male white-crowned sparrows from three areas, raised together in group isolation. Sparrows in Group A_1-A_3 were from Inspiration Point, northeast of Berkeley; those in Group B_1-B_2, were from Sunset Beach; and the remaining birds in Group C_1-C_4 were from Berkeley. Inserts AN, BN, and CN show home dialects from each group.

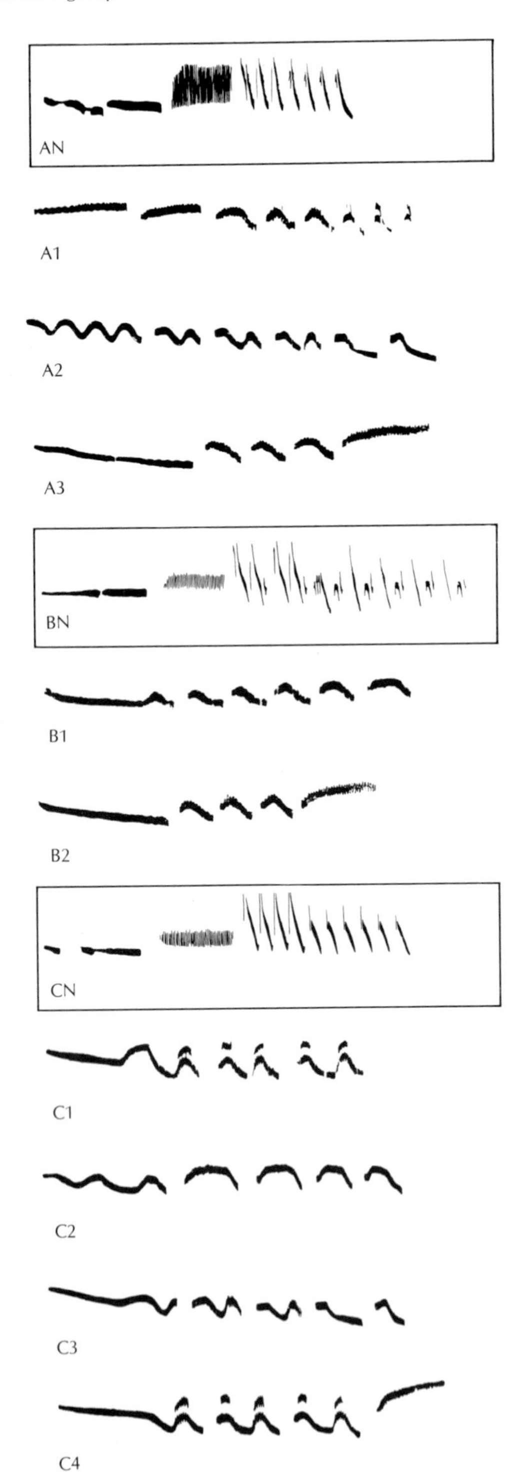

song pattern that is simpler than that of a normal but isolated bird (Figure 31.10). Just as a human deafened late in life can sustain many properties of normal speech, so a sparrow deafened after it has learned to sing can maintain the normal patterns (Figure 31.11).

LANGUAGE

What properties distinguish human language from other forms of animal communication? Linguist Charles Hockett has extensively studied the basic

Table 31.1
Design Features of Human Language

KEY	FEATURE
DF1.	*Vocal-auditory channel.* Sound waves are used for communication.
DF2.	*Broadcast transmission and directional reception.*
DF3.	*Rapid fading.* The sound of speech does not hover in the air.
DF4.	*Interchangeability.* Any adult member of a speech community may act as a transmitter or as a receiver of linguistic signals.
DF5.	*Complete feedback.* The speaker hears everything that is relevant of what he says.
DF6.	*Specialization.* The linguistic signals have no biologically important direct-energetic effects upon other organisms; the signals are important only as triggers for behavioral responses.
DF7.	*Semanticity.* The linguistic signals serve to correlate and organize the life of a community because particular signal elements are associated with features in the world; in other words, some linguistic forms have denotations.
DF8.	*Arbitrariness.* The relation between a meaningful element in the language and its denotation is independent of any physical or geometrical resemblance between the language element and the corresponding feature of the world.
DF9.	*Discreteness.* The possible messages in any language constitute a repertoire of discrete messages rather than a continuously variable repertoire.
DF10.	*Displacement.* It is possible to talk about things that are remote in time, space, or both from the site of the communicative transaction.
DF11.	*Openness.* New linguistic messages are coined freely and easily, and are usually understood in context.
DF12.	*Tradition.* The conventions of any one human language are passed from generation to generation by teaching and learning, not by inheritance.
DF13.	*Duality of patterning.* Every language has a patterning in terms of arbitrary but stable, meaningless signal elements and also a patterning in terms of minimum meaningful arrangements of those elements.
DF14.	*Prevarication.* It is possible to say things that are false or meaningless.
DF15.	*Reflexiveness.* In a human language, it is possible to talk about the very system through which communication is occurring.
DF16.	*Learnability.* A speaker of one human language can learn any other language.

Source: Adapted from C. F. Hockett and S. A. Altmann, "A note on design features," in T. A. Sebeok (ed.), *Animal Communication* (Bloomington: Indiana University Press, 1968), pp. 61–72.

characteristics of human language and he suggests 16 *design features* (DF) shown in Table 31.1, some of which can be found in various other forms of animal communication.

One of the most elementary features of a human language is what Hockett calls *specialization* (DF6). The signals used in human language are the

Figure 31.11. Typical song of an adult male white-crowned sparrow before deafening, one day after the operation, and one year later.

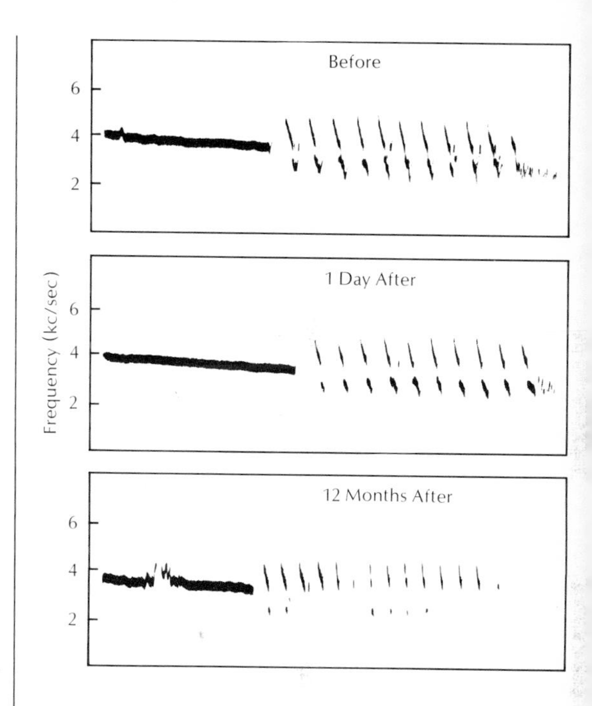

results of specialized actions, not merely incidental outcomes of other ongoing behavior. Their effect on other members of the same species is primarily one of triggering behavioral responses rather than any direct energetic effect. Tree cricket song, bird song and courtship plumage displays, and mammalian vocalizations are examples of such specialization.

Another criterion is *arbitrariness* (DF8). The association between the word "cup" and the object that it denotes is an arbitrary one. Similarly, the difference between species songs in crickets is essentially an arbitrary one. There seems to be no functional reason why one species should have a higher pulse rate at a given temperature than does another species. Many vocal and visual displays of birds, reptiles, and mammals are arbitrary in the same sense. But territorial displays such as tail flips and threat postures are not arbitrary in this sense.

A third criterion is *discreteness* (DF9); distinct words mean distinct things. The several songs and sounds made by crickets, birds, or primates represent a repertoire of discrete messages as do some pheromones. The gradations in a distressed dog's yelping or a cat's courtship vocalizations seem to rule them out of this category.

Tradition (DF12), or dependence on learning from conspecifics, is not a characteristic of cricket communication. The forelegs (on which the hearing organs are placed) can be removed long before the crickets grow old enough to make sounds; deafened crickets still develop normal sound patterns. As in the example of bird song, in many species some or all of the communicative behaviors are learned, but the communicative behaviors of these species lack other design features of human language.

As Hockett has indicated, the major point that emerges from such a survey of animal communication is that most of the design features of human language are known in at least one other animal group. However, in no other species do they all coincide. It is this particular concatenation of features that permitted the explosive development of human language.

The Dance Language of Bees

An important design feature in Hockett's list is *semanticity* (DF7), which refers to the association between particular language signals and objects or events in the external world. One of the best examples of semanticity in nonhuman languages comes from studies of the dancing behavior of the honeybee (K. Frisch, 1967).

The Australian zoologist Karl von Frisch discovered the dance language almost by accident. He was trying to determine the role that colored flowers play in bee feeding and was experimenting with dishes of sugar solution marked with various colors. He noticed that the arrival of a single bee at a feeding dish soon led to the arrival of a stream of worker bees from the hive to which the discoverer had returned. He placed food dishes at various distances from hives and carefully observed the behavior of returning scouts. In this way, he discovered the role of the waggle dance.

Scout bees that discover a rich food source return to the hive and perform a dance within the dark hive on the vertical face of the comb. Other workers within the hive cluster about the dancers, then leave the hive and fly to the food source. The dancing bee moves in a figure-8 pattern, with a straight "waggle run" in the center of the pattern (Figure 31.12). The angle between the waggle run and the vertical is the same as the angle between the direction toward the sun and the direction toward the food source. The

Figure 31.12. The dance language of bees. The round dance (top) is performed only when the nectar source is close at hand and stimulates other workers to search in the general area of the hive. The tail-wagging dance for distant nectar sources indicates to other workers where to fly and how far to fly in relation to the sun's position. The figure 8 (center) is traced at approximately a 120° angle to sun; the abdomen is wagged rapidly, indicating a relatively close location. The figure 8 (bottom) is traced at a 60° angle; the abdomen is wagged slowly, indicating a more distant location.

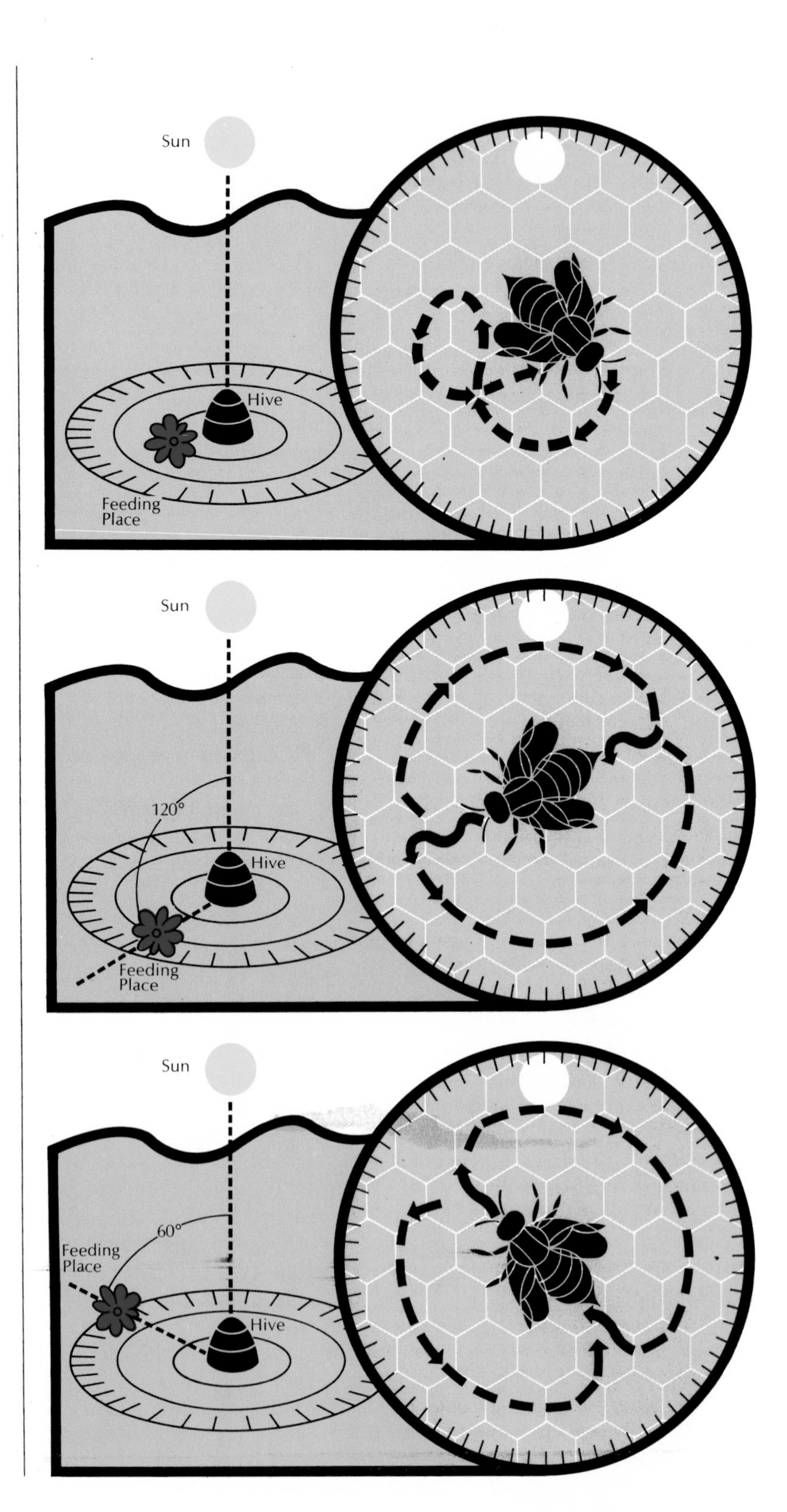

rate of the dancing varies with the distance of the source; a slower dance corresponds to a more distant source. It appears that each feature of the dance has specific semantic denotation in the external environment.

The waggle dance of the honeybee also demonstrates *displacement* (DF10), or the ability to communicate information about events that are separated in time and in space from the communicative behavior itself. There is a necessary delay between the time a scout bee finds a food source and the time the bee communicates about it in the hive. In fact, observations made during the night, when the bees are normally inactive, reveal an occasional insomniac who is still dancing about a food source that it discovered hours before during daylight.

Although a honeybee presumably can communicate a direction or distance that it or its "listener" has never encountered before, there seems to be no way for a bee to communicate other specific information—for example, about a source of danger en route to the food. In this respect, it lacks the *openness* (DF11) that seems to be one of the most distinctive features of human language. The dance language completely lacks other design features such as *tradition* (DF12). Young bees raised in isolation locate food sources and dance normally (Lindauer, 1961). Reception and understanding of information, however, require some practice. Overall, the bee's communication system can carry an impressive amount of information. Although it lacks many important design criteria, it is impressive that it meets so many others, and one is led to wonder what it is about the social structure of bees that caused the evolution of what must be admitted to be a simple language.

Syntax

The capacity to use meaningful syntax seems to be the most important feature that distinguishes human language from other animal languages. Not only does each word in a human language have a denotation, but the words may be combined in various ways to convey complex meanings, subtle distinctions, and new concepts. The rules by which individual symbols or signals can be combined into longer messages are the syntax of a language (Chomsky, 1969). Even among primates—whose language might be expected to be most similar to that of man—evidence for an ability to use syntax is difficult to find.

Primate vocal signals consist of various kinds of roars, grunts, barks, growls, screeches, and screams. Distinct sounds have particular meanings, but most of these sounds represent steps along continuous variations between extreme sounds. For example, T. E. Rowell (1962) distinguishes nine sounds with different meanings in the language of the rhesus monkey, but all of these sounds lie along a continuum between a roar and a squeak. The monkey or ape can convey shades of emotion or meaning by varying the nature of sounds along these continuums, but these is no evidence that sounds are ever combined in meaningful sequences to construct new meanings.

In some primate species, semanticity and discreteness are readily demonstrated. In vervet monkeys, 36 or more distinct auditory signals have been distinguished. These sounds were all uttered in a social context, and each occurred only in certain situations. The situations in which they occurred and the behavior that they seemed to provoke provided a basis for assigning meanings to each sound. Not all of the sounds could be given a sufficiently precise meaning to justify the claim that these monkeys have a 36-word language, but they do have an elaborate set of sound communications.

A particularly striking example of semanticity was provided by the existence of five distinct alarm calls, of which at least three had distinct meanings. Adult females uttered a *chirp* (abrupt, loud, low-frequency calls with higher overtones) whenever a dangerous mammalian predator was sighted. If an avian predator was sighted—for instance, an eagle—the same female would give a *raup* call. This call was also short, but was distinguishable from the *chirp* by a human listener because of its tonal quality. On sighting an eagle, the first adult female (or sometimes juvenile) to see it would emit a *raup* followed by *chirps*. Others of the troop would sometimes join in with *chirps*. Evidently, the initial *raup* was what said "bird," for members of the troop would run for the shelter of undergrowth when they heard it. If the initial call was a *chirp*, the monkeys would head for the trees. The third distinct alarm call was a *chutter* of low amplitude, consisting of repeated sounds with low frequencies. This call was given by females and juveniles, and signaled "snake." On hearing this call, the nearby monkeys would gather around the snake and follow it for a distance. Interestingly, this call was used to warn only of cobras and puff adders. Possibly, only these two were dangerous to the vervets.

In addition to the alarm calls, many differentiable calls were used in situations of interindividual antagonism. Infants had several "lost" calls when they were separated from their mothers, and often these provoked retrieval by the mother. In some calling sequences, the social context seemed important in determining the interpretation of a call. The use of *chirp-raup* for mammals and *raup-raup* for avian predators might be a compelling example of grammatical construction that conveys meaning.

Attempts to teach young chimpanzees to speak have failed almost completely, with the animals mastering only a few simple words and showing no signs of syntactical ability (Kellogg, 1968). However, ethological field studies reveal that chimpanzees have a rich repertoire of visual signals, in which the hands often figure prominently (Van Lawick-Goodall, 1968).

Attempting to duplicate as closely as possible the conditions under which a child normally learns language, Allen and Beatrice Gardner raised a young female chimpanzee, called Washoe, from about 10 months of age in the confines of a fenced yard and a trailer as living quarters shared with humans. All conversation in the presence of Washoe is conducted in American sign language. Washoe has the opportunity to learn the sign language both in special training sessions and by observing people communicating among themselves.

The experiment is still in progress, but Washoe's progress has been dramatic (Gardner and Gardner, 1969). At the age of 3.5 years, Washoe responded appropriately to an estimated several hundred signs. She does not merely apply a sign to the particular object with which she was trained but shows every readiness to generalize freely—not only to other objects of the same class but to photographs of them. The Gardners have movie films of Washoe sitting before a picture book and signing appropriately as the pages are turned: "flower, baby, dog, cat," and so on.

Washoe not only uses signs independently but combines them in sequences that suggest some capacity to employ a meaningful syntax. The most common combinations involve signals the Gardners call emphasizers

Figure 31.13. Hand signs used by the female chimpanzee Washoe within 22 months of the beginning of training. Signs are presented in order of their original appearance in her repertoire.

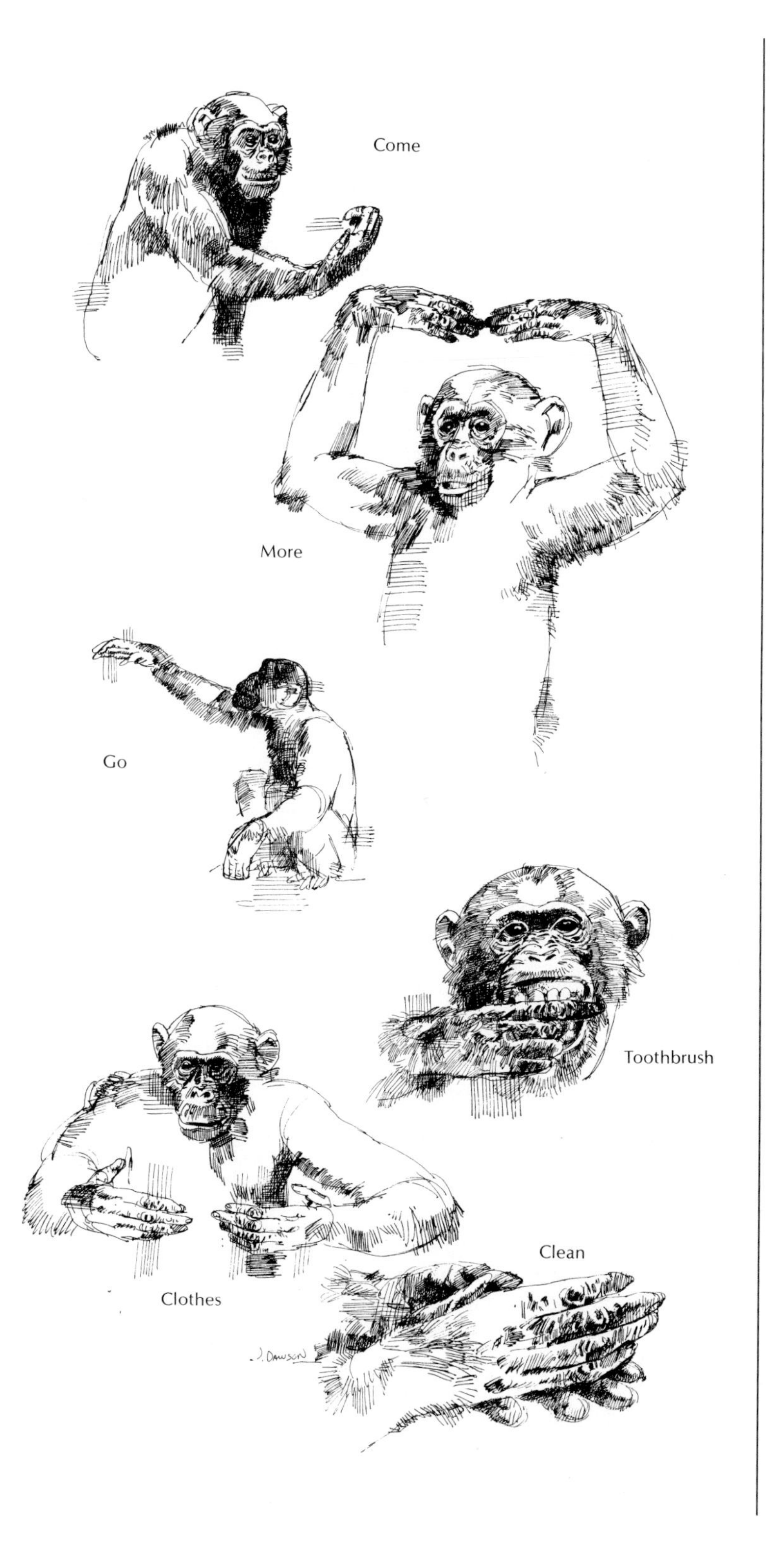

Figure 31.14. Four selected examples of the object-symbols used by Premack and his associates in their training regimen with the female chimpanzee Sarah.

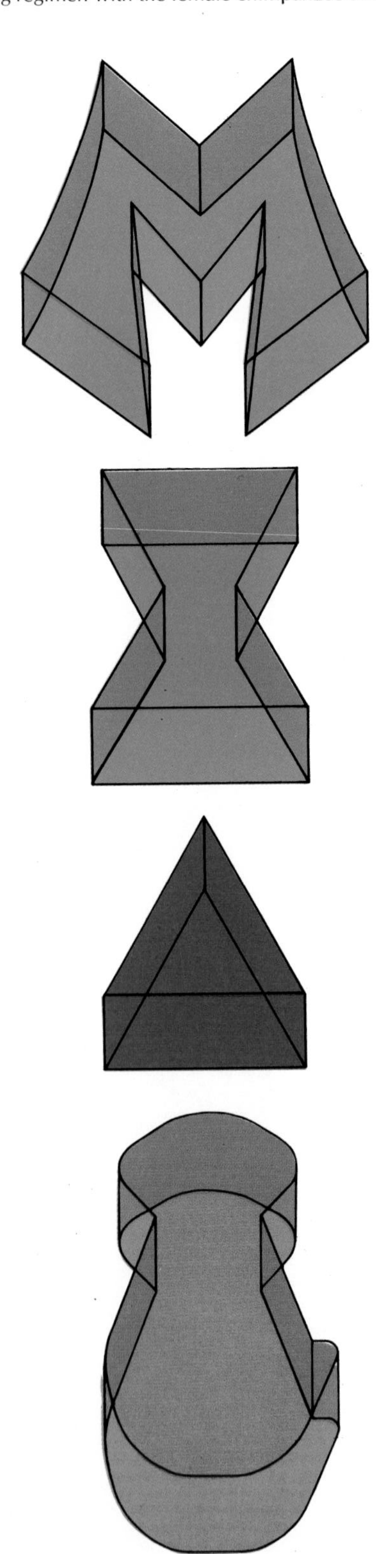

(signals for "please," "come—gimme," "hurry," and "more") with a single other signal. She also uses sequences of more than two signs (such as "you go gimme," "Roger you tickle," or "please Roger come"), all involving names or pronouns. It is difficult to avoid the impression that a very primitive syntax is emerging, although it is possible that Washoe is imitating not only signs but sequences used by others rather than creating new and meaningful combinations of her own.

Even more impressive results are being obtained in experiments conducted by David Premack and his colleagues at the University of California at Santa Barbara. Premack's group began working with a five-year-old chimpanzee named Sarah in 1968, using small plastic pieces of varying shapes and colors as signals. The pieces have metal backings and are arranged in vertical columns on a magnetized slate to form sentences. Sarah seems to be quite proficient at use of syntax. For example, when an experimenter places on the slate the sentence "red on green," Sarah puts a red card on top of a green one. If the experimenter writes "green on red," Sarah responds appropriately. She has learned to carry out such complex instructions as "Sarah insert banana pail apple dish," by putting a banana in a pail and an apple in a dish. She recognizes such abstract words as "shape," "color," "size," "same," "different," and "name." She responds appropriately to a symbol representing a question mark.

During the experiments, Sarah often was required to supply the proper plastic piece to complete a sentence written on the board by an experimenter. Premack (1970) reports that Sarah one day turned the tables on Mary Morgan, one of the experimenters.

Sarah put up a partial statement on one side of the language board (A is on . . .) and then arranged alternate answers on the other side. Sarah would point to each possible answer, and Mary's task (which took her a good while to figure out) was to nod when Sarah pointed to the right one. "The little devil would pass by the solution quickly and try to trick me into a mistake," Mary reported.

Some apes, then, appear to be capable of mastering syntax and other complexities of human language, although they are very limited in vocal skills. These findings raise a new puzzle for behavioral scientists: if some nonhuman primates are capable of mastering the complexities of human language, why have similar languages not evolved in these species?

THE VALUE OF SOCIAL ORGANIZATION

In any species where groups of individuals live in close proximity and engage in social behavior, a fairly complex and regular social organization can be observed. There may be a number of different roles within the group, each with its own pattern of appropriate social behavior. Males, females, old, young, leaders, foragers—the kinds and numbers of roles vary greatly from species to species. In many cases, the assignment to various roles is based upon biological differences (male versus female, old versus young, or, in bees, worker versus drone), but in others the roles result solely from behavioral interactions among biologically similar individuals.

A famous example of social organization among animals is the *pecking order*, first described among hens by T. Schjelderup-Ebbe (1922) and later observed in a wide range of other animals. If hens that have no experience with one another are brought together in a barnyard or a large cage, they show aggressive behavior toward one another. At first, there seems to be no

Figure 31.15. Individual rank in a dominance hierarchy among hens can often be ascertained by observing individual precedence over food or water as is illustrated here.

pattern to the behavior; any two hens are apt to fight upon an encounter. After a period of time, however, a social organization appears within the flock. Encounters no longer result in random aggressive behavior. Each hen exhibits aggression or submission predictably, depending upon which hen she encounters. Thus, a simple pecking order, or dominance hierarchy, is established (Table 31.2). In simple hierarchies, older animals dominate younger ones, larger ones dominate smaller ones, and males dominate females. But there are complications. In breeding seasons, female birds and primates may become dominant. A small group of individuals may cooperate to depose a high-ranking male. In some birds and fishes, a low-ranking male may dominate a higher-ranking one if his own mate is nearby. It is difficult to avoid the impression that human social order is not all new.

Table 31.2
Dominance Hierarchy in a Flock of Hens

Hen 1 pecks	2	3	4	5	6	7	8	9	10	11	12	13
Hen 2 pecks		3	4	5	6	7	8	9	10	11	12	13
Hen 3 pecks			4	5	6	7	8	9	10	11	12	13
Hen 4 pecks				5	6	7	8	9	10	11	12	13
Hen 5 pecks					6	7	8	9	10	11	12	13
Hen 6 pecks						7	8	9	10	11	12	13
Hen 7 pecks							8	9	10	11	12	13
Hen 8 pecks								9		11	12	13
Hen 9 pecks									10	11	12	13
Hen 10 pecks							8			11	12	13
Hen 11 pecks											12	13
Hen 12 pecks												13
Hen 13 pecks none												

Source: Adapted from W. C. Allee, *The Social Life of Animals* (New York: Norton, 1938).

Dominance hierarchies diminish the amount and intensity of fighting and thereby allow more time for pursuits of greater value to the flock or the species. A. M. Guhl in 1956 compared two groups of hens. One group had a stable order, and the other was disrupted by shifting its members. The birds in the disorganized flock "fought more, ate less food, gained less weight, and suffered more wounds." Presumably, in nature, the young of a flock with an established order are more likely to survive and reproduce than if disorder prevails.

Territoriality is also a common form of social organization. Living space and resources such as food and nesting sites are distributed among individuals by this mechanism. In many species, the strongest and most dominant individuals are able to claim territories, whereas weaker individuals fail to do so. They are often unsuccessful in their efforts to obtain a mate and thus fail to sire young.

In migrating populations of song sparrows, the male usually returns to the same area in successive years. In year-around resident populations, the male remains near his territory during the winter. As spring advances, he establishes song posts around the periphery of the territory. At each perch, he sings for several minutes and then moves on, marking the limits of his territory by singing. Intruders of other species usually are ignored, but other male song sparrows are threatened in a ritual manner. If this threat posture fails to dispatch the intruder, more forceful measures are used to drive him away. Female song sparrows are allowed into the territory, and eventually a

Figure 31.16. Social behavior and organization among prairie dogs has placed them among the most successful of the social rodents. The top diagram schematically portrays a 5 acre prairie dog "town" (each square represents 50 square feet). Family territories, or coteries, are shown in brown, with smaller coteries in the process of being established by emigrating adults. Filled in dots indicate large occupied burrows; open dots indicate smaller burrows; small dots indicate small holes. The lower series of drawings depicts a few of the key social interactions that take place between adults. At the top, two individuals of the same coterie meet, exchange an identification "kiss," and go on to groom each other. The bottom series depicts two strangers approaching, exchanging an identification kiss, the tail-raising ceremony, gland sniffing, and departure. The center diorama depicts various aspects of the prairie dogs' life style. Major predators are the hawk, badger, coyote, and weasel; occasional burrow neighbors are the burrowing owl

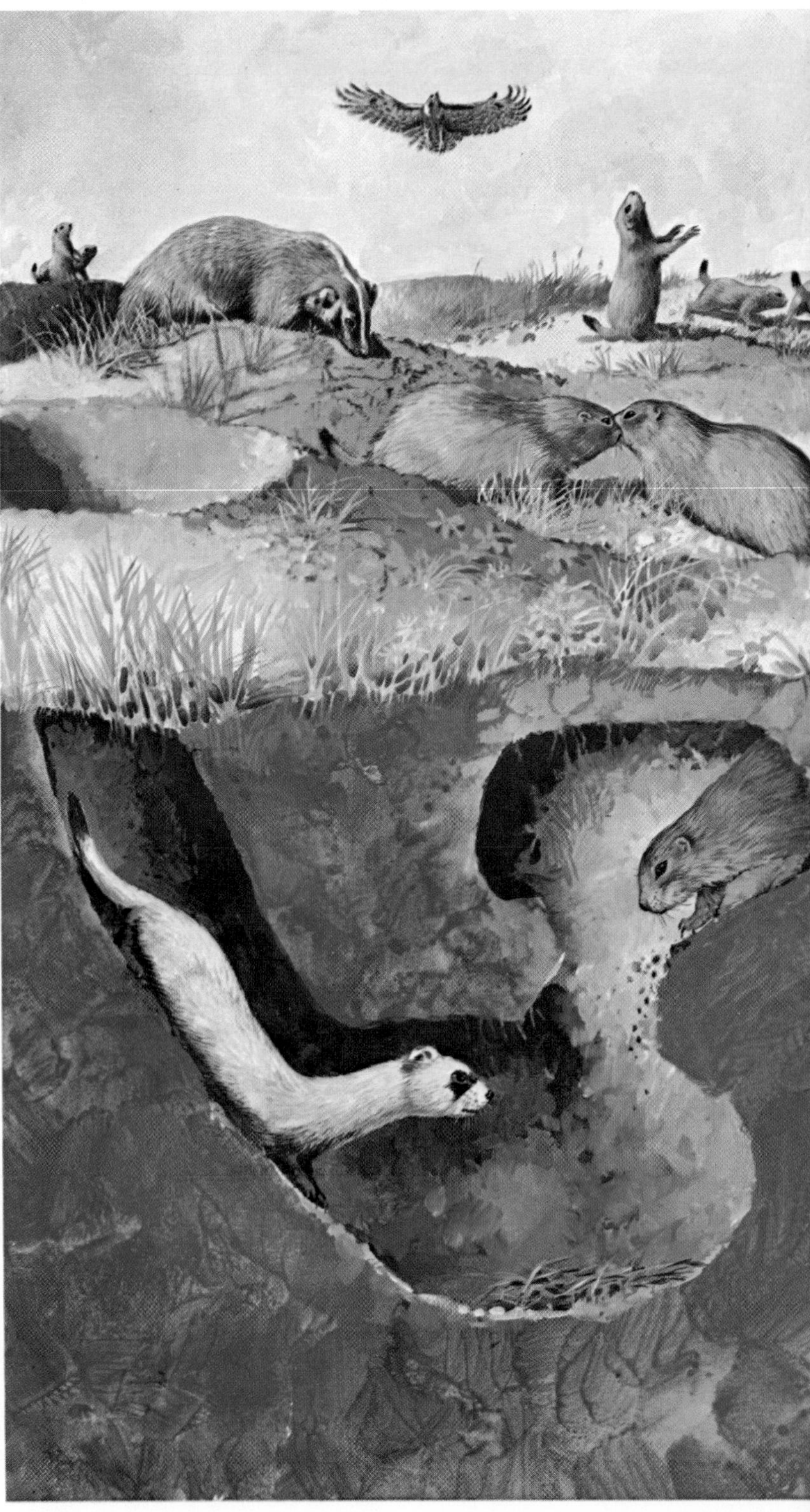

and prairie rattlesnake. The general construction of the burrow system with its raised mound entrance functions to prevent flooding of the burrow and also serves as a lookout point for danger.

Figure 31.17. Photograph of two male impalas in ritualized territorial combat. Participants in these highly stylized battles are rarely injured, as the loser is allowed to withdraw from the scene without reprisal.

pair is formed. The pair join in defense of the territory, where they build a nest, incubate the eggs, and seek food for the young.

The size of territories depends upon both endogenous and external factors. If the size of a population decreases, external pressures are reduced and the size of territories increases—but not without limit. Endogenous factors that influence the behavior of song sparrows keep the territory size from increasing much beyond an acre. Conversely, when the population becomes dense during a series of favorable years, the endogenous behavioral factors prevent the territories from being compressed to less than about half an acre. Thus, innate behavior patterns help regulate population size by imposing an upper limit on the number of successfully breeding adults in any region (Wynne-Edwards, 1962).

Among functions that have been proposed for territorial behaviors (Klopfer, 1969) are (1) to increase the efficiency with which material resources are used by restricting food gathering to a limited area, well known to the animal; (2) to limit the intensity of competition for food, because the minimal territory size limits the total population density; (3) to strengthen pair formation between the sexes; (4) to reduce the effect of predation, because an individual is better able to escape a predator in his well-known territory; and (5) to reduce the level of aggression between members of the species, because in most cases the interaction between two males is reduced to a ritual display of threatening behaviors.

Social organization in birds has been studied for many years. Recently, there have been a number of studies of social behaviors in primates, insects, and fishes. However, there are a great many species for which relatively little ethological or laboratory information is available. Although general patterns such as dominance hierarchies and territoriality have been observed in many different animals, an increasing number of studies, and greater attention to quantitative detail in those studies, have tended to reveal more differences among species than similarities. Even among the primates, there is a wide variety of social organizations and patterns of social behavior. Therefore, most behavioral scientists tend to be skeptical of any attempt at this time to explain human behavior in terms of the few generalizations that have been made about behavior in lower animals.

FURTHER READING

For more extensive general discussions of social behavior, see books by Allee (1951), Etkin (1964), Klopfer (1962, 1969), Klopfer and Hailman (1967), Lorenz (1952), Marler and Hamilton (1966), Scott (1958), Tinbergen (1953, 1958), and Wynne-Edwards (1962). For further discussion of communication among animals, see books by Armstrong (1963), Busnel (1964), Capranica (1965), Haskell (1961), Lanyon and Tavolga (1960), Lenneberg (1967), Lilly (1967), Lindauer (1961), Roslansky (1969), Sebeok (1968), and Thorpe (1961).

Ethological reports of social behavior among various animal species usually are fascinating to read and provide many examples of the behaviors discussed in this chapter. Among the great variety of such reports, good examples are those by Darling (1937), R. Gray (1969), King (1955), Kummer (1968), Morris (1967), and Schaller (1963, 1964).

Useful articles dealing with communication include those by Bennet-Clark and Ewing (1970), Bronowski (1967), Hockett (1963), Jacobson and Beroza (1964), Roeder (1965), Tinbergen (1952), and Wilson (1963). Useful articles dealing with social behavior include those by Barnett (1967), Denenberg and Zarrow (1970), Eibl-Eibesfeldt (1961), Lehrman (1964), Limbaugh (1961), Lorenz (1958), Mykytowycz (1968), Shaw (1962), N. G. Smith (1967), Tinbergen (1960), and Wecker (1964).

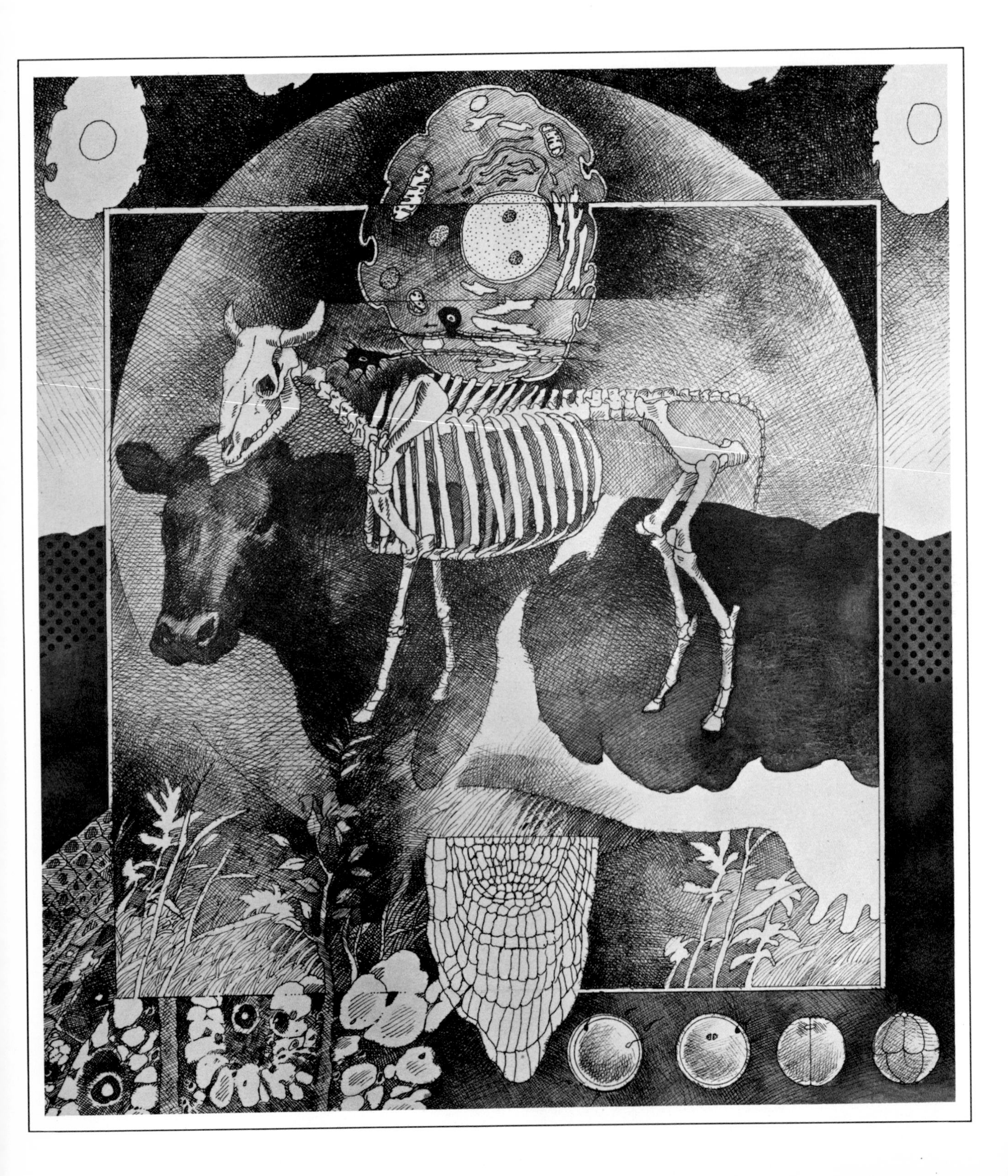

VII Natural History of Organisms

The organism is the middle ground between two levels of biological organization, the cell and the population, in which principles generally relevant to all forms of life are readily identified. As the casting in the mold, the organism has had general properties impressed on it by the die on either side. Every organism is a unit in a population of many organisms of similar structure and behavior; the resulting competitions, which are among the primary concerns of the population biologist, have had profound consequences for the nature of the individual organism. And every organism is itself either a cohesive population of cells or a single cell. This fact also has general consequences for the structure of organisms and how they operate.

—Telfer and Kennedy (1965)

32
Life and the Origin of Species

According to the presently paramount theory, the earth began to form along with the rest of the solar system from a vast cloud of interstellar dust and gas more than 4.7 billion years ago. The oldest terrestrial rocks yet studied—formed about 3.9 billion years ago—contain a great variety of complex organic molecules. From rocks formed about 3.1 billion years ago come fossils of algae and bacteria. The oldest hard-shell animal is thought to date from about 0.7 billion years ago. Thus, the processes of life and the ongoing process of evolution appear to have begun at an early point in the development of the earth.

The original cloud of matter is believed to have coalesced under an increasing gravitational force. As the mass became solid, heat was released and some of the solid constituents were decomposed. Although there is controversy over the exact composition of the early atmosphere, hydrogen, oxygen, carbon, and nitrogen were almost certainly present at an early stage. Chemical properties are presumed to have remained constant, and thus it is possible to develop a model of the early environment.

The atmosphere contained hydrogen, methane, ammonia, and water vapor. Inert gases remained from the initial consolidation. Liquid water was concentrated in terrestrial basins as soon as the surface temperature permitted the liquid phase. Carbon monoxide may have been a constituent of the atmosphere or may have contributed to the formation of such compounds as formaldehyde that would dissolve in the oceans. Hydrogen cyanide may also have been present in the early atmosphere.

SYNTHESIS OF ORGANIC COMPOUNDS

Whatever the exact composition of the atmosphere may have been, it is unlikely that there was free oxygen, as it has been difficult to simulate the synthesis of even simple organic compounds in the presence of strong oxidants. It is possible that some organic compounds were formed before the earth condensed. If such compounds did form in the nebula, much of the carbon captured as the earth formed may have been in simple molecules such as acetylene, ethane, propane, urea, acetone, and acetamide. Most of these compounds would have been destroyed by the heat of condensation, releasing carbon in the form of carbon dioxide or methane.

As concentrations of the various simple molecules built up in the atmosphere and in the seas, more complex molecules became possible. There must have been a source of energy for the endergonic reactions of organic synthesis to have taken place, and this energy may have come from a variety of sources. Sunlight was undoubtedly important. Lightning and perhaps shock waves produced by falling meteors may also have provided useful energy. Once formed, simple organic compounds could have absorbed sunlight of longer wavelengths, providing much greater quantities of energy to carry out further syntheses.

In attempting to simulate conditions as they might have existed on the primitive earth, scientists have synthesized various combinations of molecules in the laboratory. Some of the building blocks of nucleic acids have been formed by heating a concentrated solution of hydrogen cyanide and ammonia for a few days; by bombarding a mixture of methane, ammonia, and water vapor with high-energy electrons; and by heating a mixture of common amino acids. Sugars can be formed in a formaldehyde solution catalyzed by simple chemicals such as calcium hydroxide. Clay can act as a catalyst in the production of sugars in some formaldehyde solutions. The possibility of sugar formation is crucial because sugars act as energy sources as well as components of nucleic acids.

Although the primitive ocean is often described as an "organic soup," the concentrations of simple organic molecules formed by electric discharges, ultraviolet light, and shock waves may have been far too low to permit the more complex syntheses to occur at significant rates. Some mechanism must have concentrated the organic chemicals to permit the formation of polymers.

The simplest concentration mechanism would be evaporation, which would occur in tide pools and lakes. Freezing may have had some concentrating effect, but both evaporation and freezing would also lead to salt formation. A more opportune mechanism of concentration would be the adsorption (clinging) of organic molecules on surfaces of minerals or on the water-air surface. Large droplets of dense organic chemicals may also have served as a similar means of concentrating molecules. It is likely that several mechanisms played important roles in the concentration of the chemicals necessary for the development of living systems.

LIVING SYSTEMS

The precursors of proteins, nucleic acids, and sugars could have formed spontaneously on the primitive earth. The next step—the macromolecules themselves—is crucial, for without these enormous molecules living systems could not have developed.

To what degree of size, organization, and complexity must chemical synthesis proceed before something alive is formed? Could the process of biochemical evolution have consisted simply of the gradual synthesis of more and more complex organic molecules? At what point should it be said that a living system has been formed?

In centuries of meetings, publications, and conversations, biologists have tried to arrive at a simple, specific, concise definition of the word "life." Like most attempts to reduce universal concepts to a phrase, the effort to define life has been only partially successful. Most biologists agree that any distinction between life and nonlife is a matter of process—the collection in one place at one time of numerous events that can be observed singly in nonliving systems but that appear together whenever a system is seen that biologists can agree is living.

In the discussion of complex systems that developed from the original molecular reactions, it must be remembered that the crucial syntheses were reflections of the basic physiochemical and stereochemical properties of matter. In the briefest possible terms—life is an extension of the properties of matter itself.

The central feature of living systems is their ability to reproduce themselves with mutations. The molecule DNA is considered to be the chemical constituent that provides the mechanism for self-reproduction. In addition to directing its own replication, DNA directs the synthesis of RNA and then proteins in the living system. These proteins include the catalysts that make available energy and precursor molecules needed for DNA to replicate. The central problem in describing the origin of life is to account for the origin of such a DNA-protein system.

Somehow, from the aggregation of molecular material available, the molecules were constructed. It may have happened many times before circumstances were right for the next step. Perhaps the first living organism

Figure 32.1 (above). Charles Darwin.

Figure 32.2 (below). Alfred Russel Wallace.

was composed of a nucleic acid associated with a protein that could catalyze the process of duplicating the strands of nucleic acid. Such a system would be self-replicating through the use of preformed organic molecules available in the local environment.

It is also possible, however, that aggregates of molecules rather than individual molecules made up the initial living systems. Perhaps droplets of organic molecules formed in some primitive medium. In these droplets a kind of primitive metabolism might have taken place—accumulating complex molecules from the environment, growing in size, and finally splitting into smaller droplets that would repeat the process.

Whatever the nature of the first living macromolecules, they must have had the ability to replicate and to copy accidental variations (mutations) that occurred during replication. From that point, natural selection could account for the development of more efficient mechanisms of replication, for the development of more effective catalysts, and eventually for the development of a system to capture and transfer energy. Each accidental variation that improved the efficiency of the system or its chances of survival would improve the likelihood of self-replication by that system. In the competition for sources of precursor molecules and of energy as the organic soup was used up, the less efficient systems would be crowded out, while a variety of forms able to utilize different parts of the environment would be likely to develop.

The existence of living things presents scientists with two sorts of problems. On the one hand, scientists can ask about what goes on in an individual organism during its lifetime. Why does a duck egg grow into a duck not a chicken? Why do its developing organs appear in the order they do? How is the adult duck's body temperature kept constant from one minute to the next? Such embryological and physiological questions all concern the individual living system.

But there are other problems raised not by individuals as such but by the ways in which individuals differ from one another. There are many kinds of animals and plants. Why? Why do the various species differ from one another in so many ways? Why are there both ducks and chickens in the world? Most biologists today answer these questions with the Darwinian theory of evolution by means of natural selection.

DARWIN AND WALLACE

The theory of the origin of species by means of natural selection provides a remarkable instance of two scientists arriving independently at the same conclusions. Charles Darwin, 14 years older than Alfred Russel Wallace, was born into a fashionable family of intellectual physicians, successful manufacturers, and country landowners. Charles grew up to be an affable and affluent "young gent," but his thoughtful and industrious side eventually got the upper hand. One scientific achievement followed another, and his careful attention to the stockmarket gradually converted a modest inherited fortune into more than a quarter of a million pounds of capital. By contrast, Wallace grew up in a middle-middle-class household that suffered hard times. Never embittered, Wallace lived stoically through his own investment failures. He shared Darwin's distaste for the social whirl and, when they finally met, the two naturalists quickly became friends. Their great intellectual and personal respect for one another kept any taint of jealousy from their admirable scientific rivalry, and they enjoyed the informal

Figure 32.3. The H. M. S. *Beagle*.

camaraderie of men who have spent their formative years in the tropics.

Darwin had traveled around the world as the naturalist aboard the H.M.S. *Beagle*. He had sailed from England in 1831 with little training in science and with largely biblical views on the history of the earth and its inhabitants. It is impossible to say when he first became convinced that species originate by descent. In July 1837, less than a year after returning to England, he opened his first notebook on the "transmutation of species." By the end of September 1838, his theory of natural selection was complete, and four years later he had composed a manuscript containing virtually all that would appear in 1859 in the book *On the Origin of Species*.

The central claim of that book can be fairly simply stated. According to the Darwinian theory, any natural group of similar species—all the mammal species, for instance—owe their common mammalian characteristics to a common descent from a single ancestral mammal species. The differences among the various mammalian species are due to natural selection, which has brought about gradual changes in the many irregularly branched lines of descent diverging from that common ancestral mammal species. Thus, natural selection is responsible for the distinctive structures, habits, and adaptations of each of the many descendant species.

The theory of natural selection comprises the following argument: (1) The individuals of any particular species are not all alike, and some of these differences are inherited—the fact of hereditary variation. (2) Individuals also differ in their reproductive success. Some individuals do not survive to become parents, and those that do survive contribute different numbers of

Figure 32.4a (above). One of Darwin's finches.

Figure 32.4b (middle). Tuatara, a "living fossil." The history of this organism dates back 200 million years to the age of the dinosaur.

Figure 32.5 (below). Charles Lyell.

offspring to the next generation—that is, individuals compete with each other in the struggle to survive and reproduce. (3) Any individual's chances for reproductive success are partly determined by heredity. Some hereditary characters raise the chances for reproductive success or fitness; others reduce fitness. (4) Therefore, in the long run of successive generations, there is a natural selection of hereditary variants—a selection in favor of the more fit individuals and against the less fit. Wild species are slowly modified by means of natural selection, just as domestic breeds of animals and plants are changed by the practice of selective breeding employed by the farmer and gardener.

Wallace hit on this same theory in 1858. He also believed that new species somehow arise by the modification of earlier ones. When he and his friend Henry Bates traveled to Brazil in 1847, they planned to collect specimens, pay their expenses by selling duplicates through an agent in London, and gather facts, as Wallace wrote Bates at the time, "towards solving the problem of the origin of species," a subject they had already discussed at length. But on Wallace's return in 1852 the problem remained unsolved. In 1854 he was off again, for eight years in the Malay Archipelago. Halfway through these years in Malaysia, after filling his own notebooks on species transmutation, Wallace finally saw how hereditary variation and the struggle for existence together could cause the origin of new species from old. He hastily set down these thoughts in an essay and mailed it to Darwin, with no inkling that Darwin had already reached these conclusions 20 years earlier.

When Darwin heard of Wallace's essay, he put the matter in the hands of two friends, Charles Lyell (a geologist) and Joseph Hooker (a botanist), to whom he had already confided his still unpublished views on species. On July 1, 1858, Lyell and Hooker presented Wallace's essay, together with some extracts from Darwin's writings, as a single communication to the Linnean Society of London. Meanwhile, Darwin quickly abstracted his own vast unfinished manuscript, and within nine months, *On the Origin of Species* was in the shops. The book sold well but, contrary to legend, was not sold out in a day.

LAMARCK: AN INSTRUCTIVE CONTRAST

Much has been written on whether or not the Darwinian theory was original. Inevitably, many historians have concluded that there was little novelty in what Darwin and Wallace were saying. Down through the centuries, from ancient Greek times on, various writers have suggested that new species can arise through the modification of old and that, among all the possible organic types, the world contains only those that can survive the struggle for life. It may seem puzzling, then, that the Darwinian explanation came so late in history. Why were there no Darwinians until the nineteenth century?

Some historians of science will point out that all the Darwinian ideas were familiar to the Greeks, but various prejudices (particularly religious ones) caused these ideas to be suppressed until the age of enlightenment once again allowed free speculation and a proper weighing of the evidence from the study of living organisms. To support their case, these historians invariably talk of various eighteenth-century scientists who, they claim, were on the verge of reaching the Darwinian conclusions. The later writings of Frenchman Jean Lamarck are usually taken to show that the Darwinians

Figure 32.6 (above). Jean Lamarck.

Figure 32.7 (below). An artist's interpretation of increased vital fluids being conserved in reproduction.

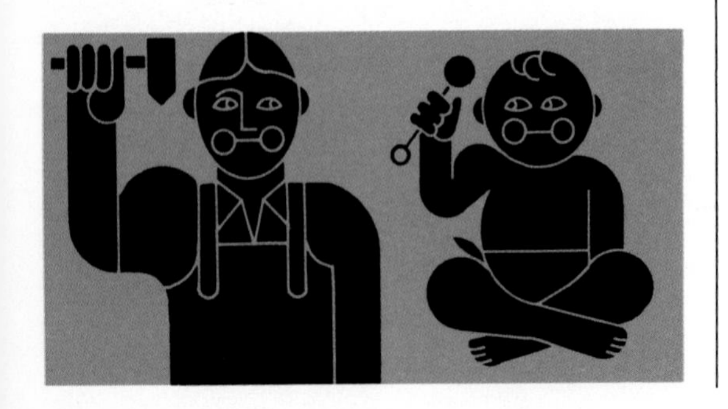

were only refurbishing concepts that had been widely understood and increasingly accepted in the previous century.

In fact, however, Lamarck's writings show exactly the opposite. Lamarck was working out his mature conclusions around 1800, a time well before several crucial developments (especially in geological and geographical theorizing) had begun to raise the particular questions that Darwin and Wallace set out to answer. For Lamarck's generation, there was no need for anything like the Darwinians' theories.

Lamarck never talks of the "origin of species." His belief in spontaneous generation, where the simplest organisms are formed directly from inorganic matter, led him to the conclusion that the more complex plants and animals are produced indirectly by the complexification of these simplest organisms (Chapter 1). This indirect production takes a great many generations and requires eons of time; Lamarck was among the first to hold that the earth had been habitable for many millions of years, not just a few thousand. He argues that the motions of active vital fluids, which are the essential properties of all living bodies, can increase the organization present in a plant or animal, and any gains in organization are conserved in reproduction.

According to Lamarck, however, these progressive increases in the degrees of organization have been modified by subsidiary causes. Vital motions in plants are speeded up by the action of heat and are therefore affected by climatic variations around the world. When responding to changes in their environment, vertebrate animals may adopt new habits, leading to the increased or decreased use of certain organs. The vital fluids in these organs will thus be accelerated, and the development of the organs will be enhanced or suppressed accordingly. Lamarck also believed that all such changes produced by these new habits will be inherited by the descendants. The doctrine that characters due to the environment's influence on an individual will be passed on to its offspring is often called the Lamarckian theory of inheritance, but this view had been a common one since the time of the Greeks. For Lamarck, this claim was simply a special case of his general thesis that the vital fluids of living bodies can produce permanent gains in the degree of organization.

The distinctive features of Lamarck's explanation for organic diversity are most apparent when he makes use of geographical and geological facts about living things. Geography is only briefly mentioned. The presence of different species in different areas of the globe supports the conclusion, Lamarck says, that plant and animal organization are modified by climatic agents, heat in particular.

Fossils enter Lamarck's argument in three ways, each of which was consistent with earlier theories. First, the presence of marine fossils in inland beds testifies, Lamarck says, to past shifts in land and sea distribution. Second, remains of tropical species in now-temperate zones show that climatic changes accompanied those shifts in land and ocean. Third, some fossils do not exactly resemble any living organisms. Some naturalists had thought these fossils to be relics of extinct species. Lamarck denies that species become extinct; fossils are merely additional evidence that organization is affected by altered conditions—that is, fossils are the remains of ancestors of living individuals with modified characters.

Lamarck's theory is not a hypothesis of common descent, which ascribes the common characteristics of a particular species to their common descent from a single species. He claims that mammals are produced by the gradual complexification of reptiles and that this elevation is going on constantly. Although all mammals are descended from reptiles, they are not descended from the same reptiles. Lamarck assumed that all mammals share certain characteristics because they have all reached the same general degree of organization.

GEOLOGY, GEOGRAPHY, AND DEMOGRAPHY

In marked contrast to Lamarck, the Darwinian theory was a theory of the origin of species, a theory of common ancestry worked out in the second third of the nineteenth century as an answer to certain impelling questions about the geographical distribution of species. Moreover, Darwin and Wallace converged on the theory of natural selection because they alone shared a number of theoretical conclusions about species' origins, conclusions that Lamarck had felt no need to consider.

The Darwinians did not need to invent the claim that a common ancestry can explain common characters. That claim was already in use as the standard interpretation of the similarities among the varieties within any individual species. Carolus Linnaeus, along with others, made a rather simple but very important assumption. All the members of any individual species are descended, he said, from a small original stock—perhaps an "Adam and Eve" in each species. Every species has had a history of dramatic multiplication in numbers, dispersion over its geographical range, and limited diversification in character. Linnaeus, like several people before him, had calculated how fast a population would grow, starting with a

Figure 32.8. Carolus Linnaeus.

Figure 32.9 (left). Thomas Malthus.

Figure 32.10 (right). Georges Buffon.

single couple breeding in ideal conditions. Throughout nature, he concluded, there is a balanced "economy," with consumer and producer species supporting and checking each other in a constant struggle for survival.

Thomas Malthus, a British economist who saw little hope for human progress through the Industrial Revolution then under way, supported his pessimistic prophecies with the same simple arithmetic in a famous anonymous publication, *Essay on the Principle of Population*, in 1798. He argued that the size of any population will increase geometrically as a result of normal reproduction. If, for example, each pair of parents produces six offspring, then those offspring could form three pairs, which could in turn contribute 18 children to the next generation. The numbers from the original pair increase generation after generation in a series such as 2, 6, 18, 54, 162 . . . The food supply, on the other hand, cannot continue to increase in this fashion but under the best of conditions could only grow by a roughly fixed amount each generation. It is inevitable, declared Malthus, that the size of the population must be limited by various causes of early death—famine, disease, and war.

Georges Buffon, an eighteenth-century Frenchman, made a generalization about species that had been widely accepted by the early nineteenth century. He declared that the stock of any species would remain recognizably distinct from that of others only as long as no interspecific crossing took place. Species are distinct because they do not cross successfully as a rule; or, as some preferred to say, the inability of two individuals to interbreed successfully is a sign or test that they are two different species descended from separate original stocks.

The most useful application of notions concerning individual species was in the field of geography. Following Buffon's comparison of Old and New World faunas, as well as the observations of scientific travelers such as Alexander von Humboldt, scientists were coming to the conclusion that species had originated at a number of "centers of creation" around the globe rather than dispersing from a single "Garden of Eden" as Linnaeus had held. The origin of species as a class of events was now held to be spread out in space.

Geographical barriers and the assumption of a small original stock in each species seemed to explain the theory that various species had been formed at several centers and not one. For example, there are no tigers in

Figure 32.11. An artist's interpretation of Buffon's "centers of creation."

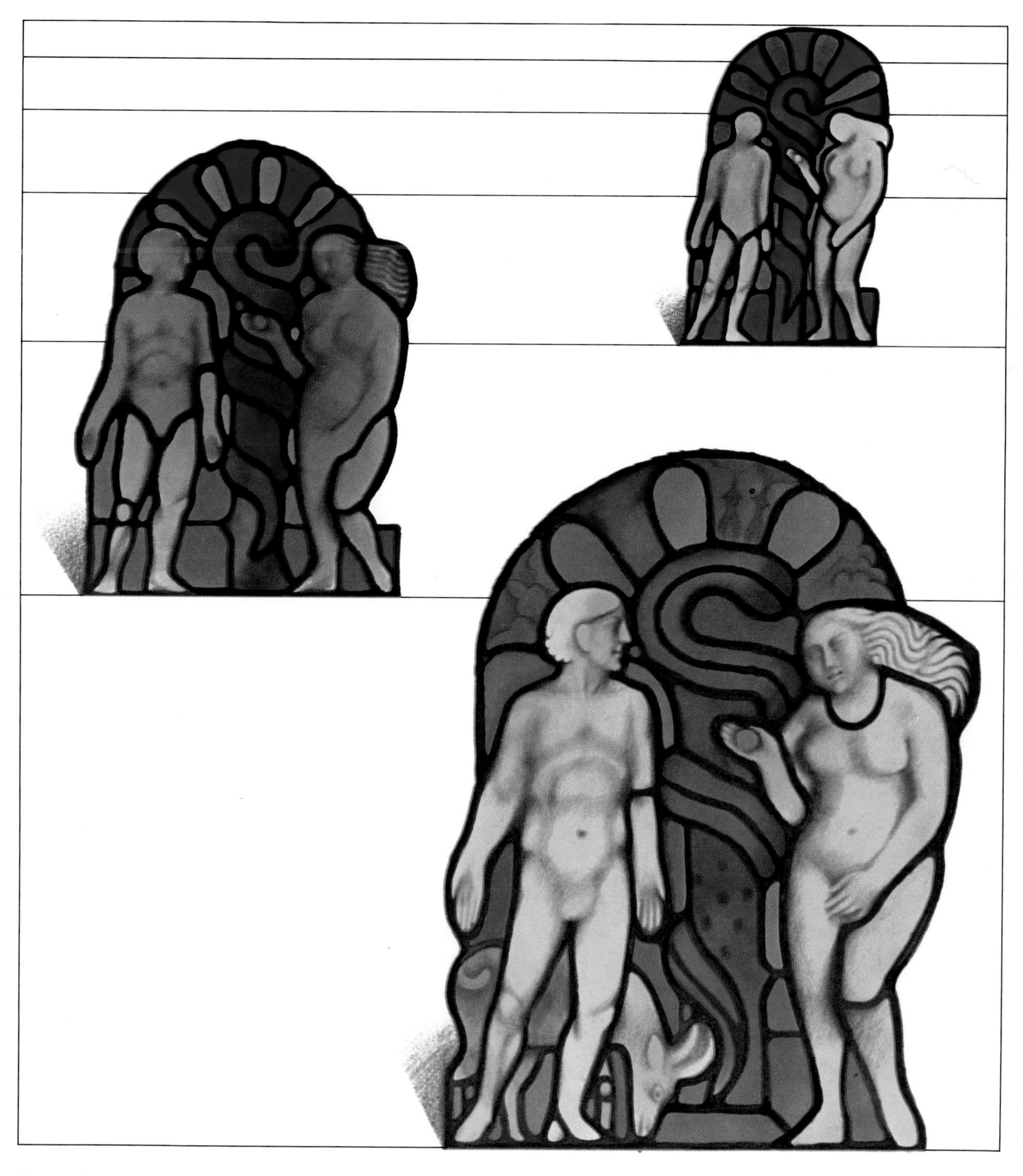

Figure 32.12. The evolution of the horse.

Brazil, but conditions there are ideal for them. According to geographical theorizing around 1800, the tiger species originated at some center of creation outside Brazil and has not been able to get to Brazil. The absence of cougars in India is the same story in reverse, for they were created in the New World and have not reached the Old World.

The next key conclusion on the origin of species came roughly in the decade following 1810. After careful comparisons of fossils from distinct geological formations, naturalists inferred that the earth had witnessed a succession of faunas and floras. In the course of the earth's history as a habitable globe, not only had many species become extinct but they had been replaced by new ones. The origins of species as a class of events had been spread out over millions of years.

IRREGULARLY DIVERGENT DESCENT

The greatest single influence on Darwin and Wallace was Charles Lyell's *Principles of Geology*, published in three volumes from 1830 to 1833. Lyell presented a comprehensive and original picture of new evolutionary conclusions from paleontology and botanical and zoological geography. Lyell opposed the view that throughout history the earth had experienced bursts of activity separated by periods of calm when little happened. In his discussion of the origin and extinction of species, he spreads these evolutionary events evenly in time and in space. The extinction of old species and the origin of new species are events that have occurred slowly but continually throughout the past and all over the globe, with no clumping in time and no crowding in space. Changes in inorganic geography—mountain formation, coast erosion, and so on—also have continued to occur through time.

Lyell's view allowed him to bring geology to the aid of geographical problems in a new way. With a constant exchange of new species for old taking place all over the globe at all times, the *number* of species common to any two areas will be proportional to the age of the barrier between them. A mountain range that dates from before the origin of the oldest extant species will have no living species common to its two sides if it is an effective barrier. In contrast, a very recent mountain range will have a few species peculiar to one side of it.

Darwin and Wallace accepted all these points. They did, however, criticize Lyell on one point: they did not believe that adaptation to conditions could explain the degree of similarity between the species native to opposite sides of geographical barriers. According to Lyell and others, the degree of similarity of those native species—and therefore the number of genera, families, orders, and so on—common to either side of a barrier is strictly determined by the environmental conditions on either side. Darwin and Wallace independently stressed that adaptation to conditions was not the law regulating the geographical representation of supraspecific groups. The similarity between the native species of any two isolated areas is proportional, they argued, to the ease with which species, now probably extinct, have formerly been able to migrate between the two areas.

Placental mammals, a large and important group of the mammalian species, are virtually absent from the eastern Malay Archipelago, although environmental conditions there are precisely like those in the western archipelago. Lyell would have been forced to assume that, when the mammal species now extant were being formed, conditions in the two archipelagos were very different and that a radically different type of mammalian organization was required in the two parts of the archipelago. The Darwinians contended that all or nearly all placental mammals originated outside Australasia and eastern Malaysia, not because placental organization is poorly adapted to Australasian conditions but because they are descended from a single placental ancestor, none—or almost none—of whose descendant species has been able to reach Australasia.

This argument formed the heart of the Darwinian theory, and the most important features of the Darwinian account of organic diversity follow from this reasoning. Darwin and Wallace were concerned with the distinctive adaptations of species to their surroundings, and they were impressed with ecological interactions among species. Lyell had argued that any species expands its population and range only as long as it has access to exactly the right climate and food and as long as it is able to hold its own in competition with other species. Lyell had declined to speculate about the origin of species, but he had held that species become extinct when they lose ground in what he called a struggle for existence.

In arriving at the final clue to the theory of natural selection, both Darwin and Wallace were indebted to Thomas Malthus for his views on populations. Malthus impressed on them the great intensity of competition, not only among different species but also among the individuals of any one species.

The main Darwinian argument for common descent was a simple extension of a familiar geographical thesis. Naturalists had long used the theory of common descent to explain the absence of individual species from suitable dwelling places—ancestors from that species had never managed to reach that region. Darwin and Wallace extrapolated this claim to explain the absence of entire genera, families, and higher orders from areas suitable for them.

The Darwinian theory was not a theory of progress. At this time, some writers claimed there was a law of organic progress—the so-called higher organisms never appeared on a planet until lower types had occupied it for a time. Other writers, confining their attention to the earth, contended that unfavorable conditions prevented higher organisms from originating earlier in the earth's history.

Lyell objected to these proposals, and Darwin and Wallace were careful to show that they were not making progressionist assumptions. If all the allied species of a group are descended from a common ancestor, the Darwinians pointed out, these species could not possibly originate until after that ancestor existed. The origin of mammals, for example, was not delayed by unsatisfactory environmental conditions or by an inherent law of progress. Darwin and Wallace did believe that mammals had descended from fishes. However, they did not base this conclusion on the assumption that mammals are "higher" and fishes are "lower" organisms. Darwin pointed out that the embryos of mammals have gill slits much like those of fishes. He argued that adult mammals have lost their gill slits as a result of natural selection favoring means of air breathing and conservation of water. It would have been equally possible that fishes descended from a mammalian ancestor, but there is no evidence of lung structures in modern fish embryos or adults.

The Darwinians—unlike Lamarck—had no need to postulate any invariable or necessary progress toward higher levels of organization. Before they arrived at the theory of natural selection, both Darwin and Wallace held that the relations between different natural groups can only be represented by an irregularly branched tree. Consider, for example, the first mammalian species—the ancestor of all subsequent mammals. It is most unlikely, Darwin noted, that this species was either the most or the least complex mammalian species that has ever lived. Gains and losses of complexity have occurred during the irregular branching divergence of various lines of descent from the single ancestral species.

In cases such as the Malay Archipelago, this branching descent has taken place in the absence of any dramatic changes in environmental conditions. The mechanism of natural selection offers an explanation of this descent. It explains how various changes arising almost randomly in each species can tend either to be preserved (if they increase chances for survival and reproduction) or to be lost from the species (if they decrease chances for survival and reproduction). This change may involve either an increase or a decrease in complexity. It will become a common feature of the species only if it enhances survival and reproduction (or "fitness"), regardless of whether it seems "higher" or "lower" in a scale of complexity.

The theory of natural selection explained, in Darwin's words, how a variety is converted into a species. It explains how a species comprising several distinguishable varieties (as the human species does) can give rise to a genus or family comprising several distinct species, each distinguished by its own characters and adaptations to a particular ecological niche, and each infertile when crossed with one another.

The principal questions that the Darwinian theory of natural selection attempted to answer were concerned with the origin of species. Like most theories, it created new questions. For example, the Darwinian theory touched upon but offered no solution to the problem of the origin of the first organisms. Likewise, it focused attention on the problem of the origin of hereditary variation. But much of the Darwinian success with the problem of the origin of species was due to the separate treatment of that problem. Far from being a weakness in the Darwinian theory, however, this ability to treat one aspect of the problem was perhaps its greatest contribution.

THE IMPACT OF THE DARWINIAN THEORY

Although the Darwinian theory was a major influence on many areas of thought in the late nineteenth century, its impact on fields other than biology has often been overstated. Many people in other disciplines had adopted broadly evolutionary concepts before the publication of Darwin's book. An obvious example is Karl Marx, who in later years liked to present himself as the Darwin of political and economic science. He even sought unsuccessfully for Darwin's permission to dedicate the English translation of *Das Kapital* to Darwin. However, Marx had arrived at his major theses long before he knew of Darwin and was far more indebted to pre-Darwinian metaphysicians and economists than to any natural scientist.

The importance of Darwin for nineteenth-century religious disputes is also often exaggerated. Most people know of the melodramatic debate between the Bishop of Oxford and Thomas Henry Huxley over the delicate matter of man's ancestry. It is less well known that the two men were unfailingly courteous to one another whenever they met. Many of the issues involved in religious disputes had already been raised in arguments with the geologists who maintained that the earth had existed for more than the few thousand years indicated by a literal interpretation of the Old Testament.

Figure 32.13. Caricatures of Thomas Henry Huxley (above) and the Bishop of Oxford (below).

Figure 32.14. Darwin in caricature (right). In his later life, Darwin's ideas were the subject of cartoons, such as the one shown at left.

SOCIETY FOR THE PREVENTION OF CRUELTY TO ANIMALS. PRES. BERGH
THE ORIGIN OF SPECIES BY
Th. Nast
MR. BERGH TO THE RESCUE.
THE DEFRAUDED GORILLA. "That *Man* wants to claim my Pedigree. He says he is one of my Descendants."
MR. BERGH. "Now, Mr. DARWIN, how could you insult him so?"

But Darwin's work focused attention on the question of human origins.

Although Darwin did not discuss mankind in *On the Origin of Species*, debates on this topic began immediately after the book appeared in 1859. Darwin himself entered this discussion in 1871, with his work *The Descent of Man*. He then insisted that the theory of selection could account quite adequately for the peculiar features that distinguish man from his animal relatives (Chapter 39). Wallace, on the other hand, made an exception of man. For man's bodily characteristics, natural selection was an adequate explanation, but Wallace felt that many of man's mental abilities far exceed what is advantageous for survival. These abilities, Wallace argued without going into further details, are the effects of a "universal mind" presiding over the universe.

Darwin tended to avoid the political and religious debates that involved evolutionary theories. Wallace, however, became increasingly occupied with political and social theory. He was a democratic socialist and, among other things, advocated nationalization of land. Although he supported his political conclusions by arguments from natural selection, his own socialist convictions had originated in his teens before he became interested in the problem of the origin of species. Darwin and Wallace did share an interest in the strictly scientific discussions following the publication of their theory. The geological time scale and the problem of heredity proved to be particularly pressing issues.

THE AGE OF THE EARTH

Lyell had stated that there are no geological traces of a time when the earth was not habitable. He believed that the oldest fossil-bearing rocks are merely the oldest formations that have not yet been melted down or ground up by the earth's inner heat and surface upheavals. These processes have probably destroyed the records of long periods in the history of life on earth.

Some geologists had already opposed Lyell, arguing that the earth is still cooling from a molten state and has been cool enough to support life for only a limited time. In the 1800s, the eminent physicist Lord Kelvin (William Thomson) estimated that the earth could only be about 100 million years old and that life could have existed on the planet for only a small part of this time. Hermann von Helmholtz calculated the rate at which the sun must be contracting in order to convert gravitational energy into light. He concluded that the earth could not be more than 25 million years old, for the sun would have been larger than the earth's orbit before that time.

Wallace worked out his own chronology from various geological, astronomical, and climatological data. His figures agreed reasonably well with the estimates of the physicists, and he remained convinced that natural selection could fully account for the evolution of all organisms except humans. Other biologists—perhaps including Darwin—felt that the theories of the physicists did not allow enough time for natural selection to produce the observed evolutionary effects. Although Darwin never endorsed them, two ways of accounting for rapid evolution became popular with many biologists. Some evolutionists adopted a modified Lamarckian position, suggesting that hereditary variation is not random but is always directed by environmental conditions. Others suggested that new species arise through the sudden birth of dramatically different organisms and that natural selection is active only in determining which new species survive. In the later years of the nineteenth century, few scientists gave unqualified support to natural selection as the major mechanism of evolution.

THE MECHANISM OF HEREDITY

The chronological crisis created by the physicists was only one reason for the development of alternatives to natural selection. An equally rich source of disputes was the total lack of any consensus on the mechanism of heredity. There was no agreement among evolutionists about the mechanism that makes offspring resemble their parents but also allows species to change.

Darwin's own views on this matter have been little studied and are not completely known. He and Wallace always pointed out that, although the theory of natural selection assumes the fact of hereditary variation, the theory makes no assumptions about the causes of this variation. However, Darwin was never without opinions on this matter. He had long been impressed by the great variation among the offspring of domesticated plants and animals, and he felt that the reproduction of parental characteristics is less exact among domesticated organisms than among wild species. He suggested that a change of climate or diet might disturb the hereditary

Figure 32.15. Darwin's pangenesis theory compared with Weismann's germ plasm theory. Weismann theorized that the human body consisted of the somatoplasm (which made up all the body organs except reproductive cells) and the germ plasm (which was the hereditary material isolated early for reproductive purposes only). The germ plasm continued in a stream from generation to generation, producing somatoplasm, but it was completely undifferentiated and not influenced by somatoplasm. On the other hand, Darwin maintained that the germ plasm produced somatoplasm, which in turn produced the pangenes that made up the germ plasm of the following generation.

process and cause an increase in variation, even among wild species. However, Darwin differed from the Lamarckian view in insisting that the nature of the variations would be random. If the species adapts successfully to the changed conditions, it is because natural selection favors the variations that happen to be suited to the new conditions, not because the change in any way directed the production of suitable variations. Darwin never absolutely ruled out the possibility that acquired characteristics could be inherited; however, he never believed it to be the normal mode of inheritance and considered it to play only a minor role in evolution.

There is a persistent but erroneous legend that Darwin suddenly became very Lamarckian in 1867, when he was faced with an objection to natural selection raised by a Scottish physicist and engineer, Fleeming Jenkin. Jenkin pointed out that with many characteristics, such as height in humans, the offspring are usually intermediate between the parents. The character of the offspring appears to be a blend of the parents' characters. Jenkin urged that such "blending inheritance" will cause any hereditary variations to disappear from a species. Because the offspring of the individuals with the new character are most likely to mate with the far more numerous normal individuals, the new character will gradually disappear as it is blended with the normal character. Although the new character might be favored by selection, it will disappear before it can become a majority trait of the species.

Jenkin's objection was not new to Darwin. He had considered this problem himself at least as early as 1842. Over the years, he countered this objection with a number of arguments. Variants with a new character may tend to breed with one another, he suggested, particularly if they become isolated from other individuals by geographical barriers. He also suggested that a new and advantageous character might happen to occur simultaneously among a number of individuals, so that selection can make it a majority trait before blending dilutes its expression in the population.

Darwin's most complete theory of inheritance was his hypothesis of pangenesis, drawn up at least two years before Jenkin raised his objections to natural selection. Darwin suggested that each little part of the structure of an organism reproduces itself by budding off "gemmules," which circulate through the body and end up in the sperm or egg. In fertilization, corre-

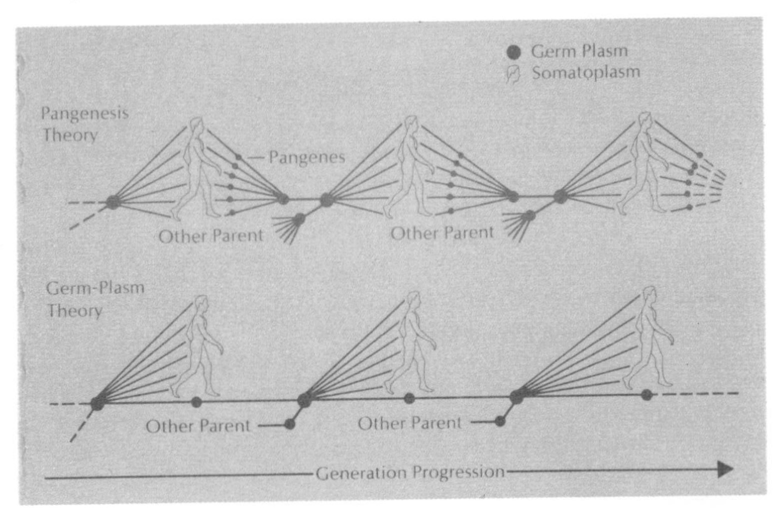

sponding gemmules from the two parents unite. Each combined gemmule then develops into a cell of the type from which the parent gemmules came, showing effects inherited from both parents. The hypothesis of pangenesis was frankly speculative, but Darwin made a great effort to show how it could explain many of the phenomena of inheritance: the normal transmission of the parental characters to offspring; the odd cases that had been reported when mutilations of the parent were apparently inherited by offspring; the reappearance of latent ancestral traits after several generations; a report that the coloring of a mare's offspring had been influenced by earlier matings with different male horses; the effects of domestication upon variation; and the apparent influence of use and disuse upon the development of some organs.

Many of the observations that Darwin attempted to explain were disputed at the time and would not be credited today. Views on inheritance were confused, conflicting, and constantly revised at this time. Wallace rather liked Darwin's pangenesis hypothesis when it was first proposed, but shortly after Darwin's death he warmly embraced a new and quite different theory proposed by August Weismann. Wallace, who lived until 1913, scoffed at the antiselectionist stand taken by many of the early supporters of Mendelian genetics in England. He finally came to deny that Mendel's laws have anything to do with the inheritance of the slight differences that serve as material for natural selection in wild species.

Weismann differed from Darwin's pangenesis hypothesis by suggesting that the hereditary material is set apart quite early in a developing embryo. The hereditary material, or germ plasm, is gathered together in the developing reproductive organs and is unaffected by changes in the rest of the organism through its growth and adulthood. Weismann denied the possibility of Lamarckian inheritance as a part of his general belief that the germ plasm determines the development of the rest of the body but that there are no effects in the reverse direction.

A theory drawing upon both Darwin's and Weismann's hypotheses was proposed in 1889 by Hugo De Vries. His theory of "intracellular pangenesis" incorporated Darwin's idea of persistent hereditary units passed on from parent to offspring. Using Weismann's reasoning, however, De Vries denied that these units, which he called "pangenes," are affected by the other tissues of the body. He suggested that each pangene determines the development of a single character and that different pangenes from the parents can be recombined in various ways in the offspring. By 1900 De Vries' search for the statistical laws governing the transmission of single characters led him to rediscover Mendel's generalizations. He then read and appreciated the forgotten priest's classic paper (Chapter 12). When Johannsen coined the term "gene" in 1909, he did so by dropping part of De Vries' term that had originally conveyed the long-discredited hypothesis of Darwin.

The establishment of the Mendelian theories of particulate inheritance and later discoveries in genetics made it possible to analyze mathematically the ways in which hereditary variations arise and to explore in quantitative detail the effects of natural selection on species. Among other important contributions, the Mendelian account of inheritance showed that characters are not lost through blending, for they can reappear undiluted in later generations.

Early in the twentieth century, new discoveries in physics revealed that much of the earth's present heat comes from radioactivity and that the energy emitted by the sun is obtained from nuclear reactions rather than gravitational collapse. Physicists now began to estimate the age of the earth in billions of years, and the objection that time had not been sufficient to permit evolution through natural selection was removed. By the 1930s, mathematical analysis of genetics had provided a quantitative treatment of evolution through natural selection, and the Darwinian theory has been generally accepted by biologists with very few exceptions since that time.

EVOLUTION: FACT OR THEORY?

Although modern students of evolution have modified Darwin's original theory in many of its details, today the vast majority of scientists accept that species have evolved through natural selection. In fact, some observers have noted a tendency for the theory of evolution through natural selection to become a modern dogma—something that is accepted as a belief rather than being subjected to continual testing through observation and experiment. Biologists tend to pay little attention to arguments against natural selection or evolution, perhaps because many of these arguments are set forth by religious writers with only a superficial knowledge of biological evidence.

However, the process of evolution is not yet fully understood. There seems to be little doubt that natural selection can account for the production of new species, but the details of this process in complex natural situations have not been fully explained mathematically. Modern students of evolution disagree on many of the details of the mechanism by which species originate and on the importance of various factors in this process. Although a number of important points remain to be resolved, there seems to be little doubt that new species have evolved during the history of the earth and that natural selection has played a major role in that evolution.

In retrospect, it can be said that evolution has been a process through which populations or species become better adapted to environmental conditions. However, the term "adaptation" must be used with care, for it often implies a Lamarckian mechanism for evolution. In fact, biologists often tend to discuss evolution in terms that suggest a deliberate effort on the part of individuals or species to modify themselves to meet the demands of the environment. To anyone who has thoroughly studied and absorbed the theory of evolution through natural selection, this way of speaking represents simply a convenient shorthand. If pressed to elaborate his views, the knowledgeable person will explain that the theory of selection implies only that certain organisms are more successful in reproduction and thereby cause certain genes to become more widely represented in the population. Environmental conditions do cause gradual alterations in the genetic characteristics of a species, but this effect operates only through the differential survival and reproduction of the various genotypes produced by sexual recombination and by mutation. For the beginning student in biology, it may be difficult to remember that adaptation of a species to the environment (for example, the invasion of the land by plants and animals) is regarded by most modern biologists not as a progressive change directed by some conscious effort or trend toward "higher" organisms but as the result of a complex modification of the distribution of genes in the population through the effects of breeding and survival. These effects involve not only the influences of environmental conditions but the outcome of apparently random or chance events. Therefore, the detailed and quantitative study of evolutionary mechanisms involves statistical studies of populations (Chapter 34). Brief statements in terms of adaptation or evolutionary trends may provide convenient summaries of complex theories, but they must be regarded with caution, for their implications are often misleading.

FURTHER READING

Darwin's *On the Origin of Species* is available in a number of recent editions. Because the book became longer and less readable in its later editions, the version to read is the facsimile of the first edition (Darwin, 1964). An even better introduction to Darwin's ideas is the paperback anthology by Hyman (1969), which includes not only Darwin's autobiography but his essay of 1844, which contains the material later included in the *Origin*. Many of Darwin's supporters were more graceful and lucid writers than he was. Among their books dealing with debates arising out of Darwin's work are those by Gray (1963), Huxley (1896), and Wallace (1891). Of the many biographies of Darwin, probably the best is that by De Beer (1965). George (1964) gives a biographical study of Wallace and his work.

A more complete discussion of the historical background of Darwinism and the reception of the theory will be found in books by Eiseley (1958) and Greene (1961). Greene (1963) provides a brief introduction to recent Christian thinking on evolution. Among the difficult but important books showing the modern combination of genetic and evolutionary theories are those by Huxley (1942), Mayr (1942, 1963), and Simpson (1949). Ross (1962) shows how evolutionary theories are applied in many different aspects of the study of the universe.

33
Populations

Thomas Malthus (1798) observed that "nature has scattered the seeds of life abroad with the most profuse and liberal hand. She has been comparatively sparing in the room and the nourishment necessary to rear them." Charles Darwin and Alfred Russel Wallace combined this observation with the phenomenon of hereditary variation to account for evolution through a mechanism of natural selection (Chapter 32). An understanding of populations and how they change is important not only for an explanation of evolution but also for an awareness of how energy and materials are utilized by life on earth. The science of *ecology*—the study of interactions among organisms, populations, and the environment—has gained attention in recent decades as biologists have become increasingly aware that the growth of the human population has upset the balance of the biosphere.

A *population* is a group of individual organisms belonging to the same species and living in the same area. All populations living in an area make up the *community*, and the community together with the inorganic environment make up the *ecosystem*. Thus, the largest ecosystem is the earth; the smallest has its limits set by the interest and convenience of the observer.

The most obvious characteristics of a population are its range (the area that it occupies) and the number of individuals that it contains. Also of importance is the *density* of the population, which may vary from place to place within the area. The rate at which new individuals are added to the population through birth and immigration and the rate at which individuals are lost through death and emigration determine the overall *growth rate* of the population. The student of ecology or evolution is interested not only in the growth rate of the population but also in the distribution of individual characteristics such as coloring, size, sex, and age. The ways in which this distribution, or population structure, change with time are significant in discussions of the way species change and give rise to new species.

It is seldom practical to count or measure every individual in a wild population. The population biologist must normally deal with information gathered by sampling the population and thus must work extensively with the mathematical tools and theories of statistics.

POPULATION GROWTH

A major factor in population growth is the rate at which new individuals are produced. As a simple example, consider a population of bacteria in a laboratory culture dish. Under ideal conditions, a bacterium can divide to produce two daughter cells, each of which will divide after 20 minutes. The size of the population doubles each 20 minutes. Table 33.1 shows the calculated growth of such a population, assuming that no individuals are lost through death and that the doubling continues to occur every 20 minutes.

Such an exponential growth pattern cannot in reality continue for long. Even for organisms that reproduce more slowly, the exponential growth pattern would soon lead to populations that could not be supported by the environment. For example, Charles Darwin calculated the theoretical growth of a population of elephants, assuming that each elephant pair begins breeding at age 30, survives until age 100, and produces 6 offspring during its life. He found that within only 750 years a single pair of elephants would give rise to a population of nearly 19 million individuals.

In studies of actual populations, the growth rate begins to decrease after a relatively small number of generations. In a population of yeast cells

Table 33.1
Growth of Hypothetical Bacterial Population*

TIME (*minutes*)	Population Size (*number of individuals*)	Natural Logarithm of Population Size
0	1	0.00
20	2	0.69
40	4	1.39
60	8	2.08
80	16	2.77
100	32	3.47
120	64	4.16
140	128	4.85
160	256	5.55
180	512	6.24
200	1,024	6.93
220	2,048	7.63
240	4,096	8.32
260	8,192	9.01
280	16,384	9.70
300	32,768	10.40
320	65,536	11.09
340	131,072	11.78
360	262,144	12.47

*It is assumed that each individual divides at 20-minute intervals to give rise to two daughter cells and that no individuals are lost from the population through death.

Figure 33.1 (above). The growth curve of yeast cells grown in a laboratory culture. Note the S-shape.

Figure 33.2 (below). The growth curves of yeast cells grown under varying environmental conditions.

grown in laboratory culture, the exponential growth pattern continues for only a few hours. The growth rate of the population does not continue to be proportional to the size of the population (Table 33.2). Instead, the growth rate approaches zero, and the size of the population becomes nearly constant. The growth curve of the population is an S-shape (Figure 33.1). Statisticians call such a curve a *sigmoid curve*.

Table 33.2
Population Growth of Yeast Cells in Culture

TIME (*hours*)	Population Size (*number of individuals*)	Growth Rate (*individuals per hour*)
0	10	0
2	29	9.5
4	71	21
6	175	52
8	351	88
10	513	81
12	594	40.5
14	641	23.5
16	656	7.5
18	662	3

Source: Adapted from Raymond Pearl, *The Biology of Population Growth* (New York: Alfred A. Knopf, 1925).

New individuals continue to be added to the population even after it has reached its equilibrium or maximum size. Although the growth rate has become zero, the birth rate has not become zero. Equilibrium is reached because individuals are dying at the same rate that new individuals are being born. For the time being, only populations where individuals cannot enter or leave the area in which the population lives will be considered. Thus, new individuals are added only through birth and individuals are lost only through death. The equilibrium size of the population changes with environmental conditions. A change in temperature, available space, available food, chemical composition of the environment, or the presence of other populations may alter the size of the equilibrium population (Figure 33.2). It is often said that the environment has a particular *carrying capacity*—a limit to the size of the population that it can support.

Both the rate of births (*natality*) and the rate of deaths (*mortality*) of the population may be affected by environmental conditions as well as by the size and structure of the population itself. If a stable population size is to be achieved, conditions within the population and in the environment must be such that these two rates can become equal.

In most real populations, growth curves approximating the sigmoid curve can be observed, but various deviations from this curve are common. There is often a time delay between a change of conditions and the resulting changes in birth and death rates. Thus, the population may increase beyond the carrying capacity before the factors limiting growth can act to decrease the growth rate to zero. In such a case, the size of the population may fluctuate around the equilibrium size (Figure 33.3). If the time lag is such that the oscillations increase in size rather than damping out, the population may disappear on one of the downward turns when the population reaches zero size. In some extreme cases, the population size increases exponentially far past the carrying capacity and then "crashes" toward a

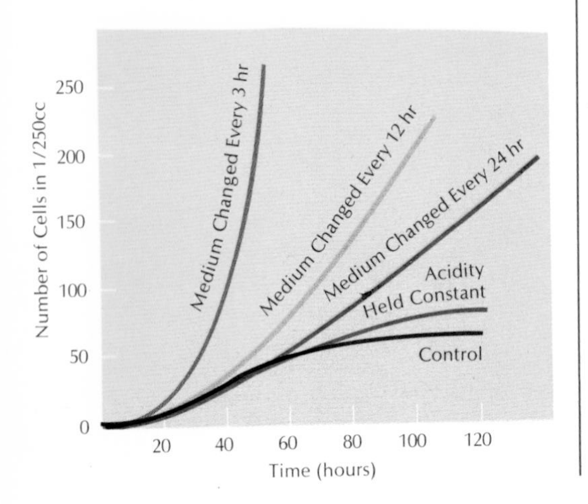

Figure 33.3. The growth curve of sheep following their introduction to an area. Note the initial sigmoid pattern followed by approximate equilibrium.

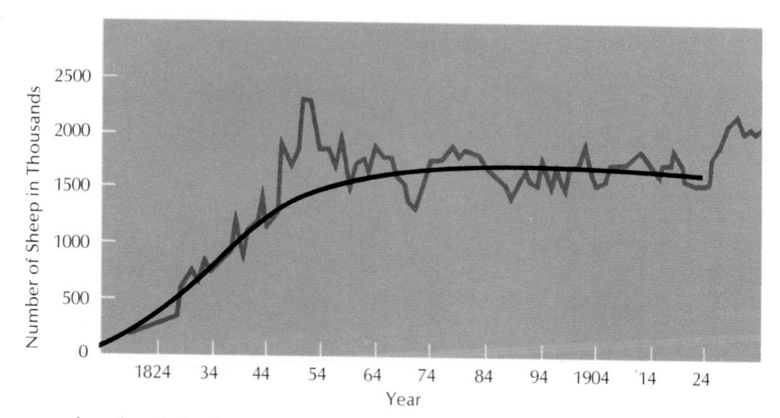

very low level. Such a "J-shape curve" is often observed in the unicellular plant populations that "bloom" in bodies of water in the spring and rapidly exhaust the supplies of nutrients that have accumulated over the winter.

Although the sigmoid curve can be expressed mathematically, this simple mathematical model has little value in predicting the equilibrium size of a population or in understanding deviations from the smooth sigmoid curve. A more detailed understanding of population growth requires a closer look at the opposing factors of natality and mortality.

Natality

The *maximum natality* (sometimes called the physiological natality) of a population represents the maximum rate at which new individuals can be added under optimal conditions. This value is an exponential growth rate in which the rate of births is proportional to the size of the population. In most actual populations, the biologist is more concerned with *realized natality*, the actual rate at which births occur in some specific environment.

Sometimes natality is most easily measured in terms of a *crude birth rate*, such as *n* births per year per 1,000 individuals in the population. However, this method excludes the fact that only certain members of the population produce young. For example, a population in which 90 percent of the individuals are male or one in which 90 percent of the individuals are not yet mature will have birth rates very different from more normally structured populations. To account for the effects of changes in population structure on the birth rate, biologists prefer when possible to work with the *age-specific birth rate*, which is expressed as the number of offspring produced per unit of time by females of a particular age class. Even more useful for most calculations of population growth is the *gross reproductive rate*, which represents the average rate at which females are born to females of each age group.

Theoretical studies rely heavily on the *intrinsic rate of increase*, which takes into account both the age-specific birth rate and the proportion of females that are likely to survive to each age. An accurate estimate of natality can be made only by taking into account the effects of mortality.

Mortality

Mortality rates change with age and other individual characteristics more dramatically than do birth rates. The *crude death rate* represents the number of individuals dying during a period of time for each 1,000 individuals in the population. *Specific death rates* calculated for various age groups are of greater use. Mortality in a population is commonly represented by a *life table*, a form developed by insurance companies. It includes a calculation

Table 33.3
Life Table for Cottontail Rabbits

AGE INTERVAL (*months*)	*of 10,000 rabbits born* Number Living to Beginning of Interval	Number Dying During This Age Interval	Mortality Rate*	Life Expectancy†
0–4	10,000	7,440	0.744	6.5
4–5	2,560	282	0.11	6.6
5–6	2,278	228	0.10	6.5
6–7	2,050	246	0.12	6.5
7–8	1,804	307	0.17	6.4
8–9	1,497	150	0.10	6.4
9–10	1,347	175	0.13	6.3
10–11	1,172	164	0.14	6.3
11–12	1,008	212	0.21	6.3
12–13	796	143	0.18	6.3
13–14	653	98	0.15	6.2
14–15	555	55	0.10	6.0
15–16	500	65	0.13	5.8
16–17	435	31	0.07	5.6
17–18	404	24	0.06	5.3
18–19	380	49	0.13	5.0
19–20	331	36	0.11	4.9
20–21	295	47	0.16	4.6
21–22	248	20	0.08	4.4
22–23	228	39	0.17	4.2
23–24	189	32	0.17	4.0
24–25	157	13	0.08	3.7
25–26	144	7	0.05	3.4
26–27	137	30	0.22	3.1
27–28	107	12	0.11	2.9
28–29	95	13	0.14	2.6
29–30	82	32	0.39	2.4
30–31	50	7	0.14	2.3
31–32	43	9	0.21	2.1
32–33	34	11	0.33	1.9
33–34	23	16	0.70	1.9
34–35	7	3	0.35	2.3
35–36	4	0	0.000	2.0
36–37	4	0	0.00	1.5
37–38	4	0	0.00	1.0
38–39	4	4	1.00	0.5

*Fraction of those alive at the beginning of this interval that die during the interval.
†Average number of months of life remaining at beginning of age interval.

Source: R. D. Lord, Jr., "Mortality rates of cottontail rabbits," *Journal of Wildlife Management* 25 (1961):33–40.

of the *life expectancy* of an individual of a particular age; the life expectancy is the average time that an individual of that age may be expected to remain alive. Table 33.3 is a life table for cottontail rabbits.

From the life table, a statistician can chart a graph showing the percentage of individuals that survive to any particular age. Such survivorship

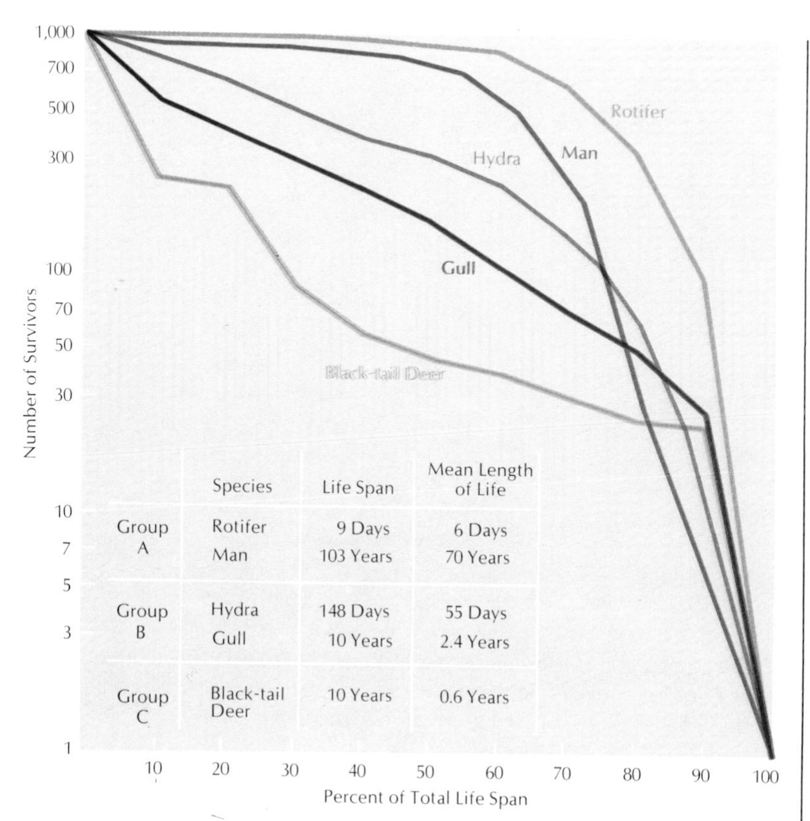

Figure 33.4. Survivorship curves of various general shapes. A majority of large mammals tend to have long life spans.

curves tend to approximate one of three generalized shapes (Figure 33.4). Among cottontail rabbits, nearly three-quarters of the individuals die within the first three months of life; for those that survive, the average life span is about 10.6 months, but some individuals survive for more than 3 years. Thus, the survivorship curve for cottontail rabbits drops rapidly just after birth and then decreases slowly. Similar curves are typical for many small mammals, fishes, and invertebrates. At the other extreme, a majority of large, mammalian individuals tend to live through a relatively long life span, and there is a sudden drop in survivorship at the end of that typical life span. The intermediate case—found among hydra, mice, and many birds—is one in which mortality is roughly constant regardless of age.

POPULATION CONTROL

In any population, natality cannot long exceed mortality. The exponential growth resulting from excess natality soon causes the population to exceed the carrying capacity of the environment, no matter how large that capacity may be. Various factors then cause an increase in mortality, a decrease in natality, or both, ultimately bringing the population to an equilibrium size or causing it to decrease drastically in size. The mechanisms that act to control population size have been studied extensively, but biologists are not yet in agreement on the nature or importance of many of these mechanisms. It is generally agreed that there are two major categories of control mechanisms. *Density-independent* factors are environmental factors that act directly upon individuals in the population; their effects are largely unaffected by the density of the population. *Density-dependent* factors result from interactions within the population; their effects vary with the density

Figure 33.5. Survivorship curves as a function of density.

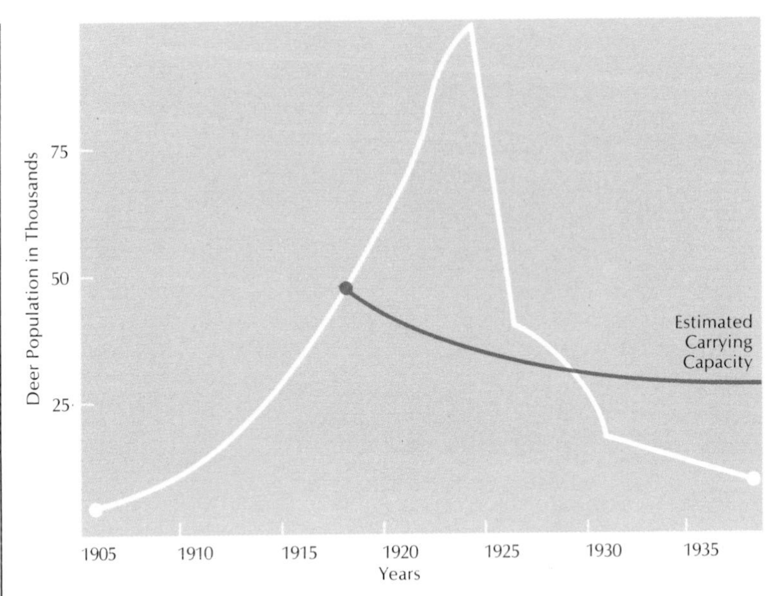

of the population. Most natural populations show cyclic fluctuations of a greater or lesser extent, and in most cases these fluctuations are controlled by both density-independent and density-dependent factors.

Density-Independent Factors

Extreme cases of population control by density-independent events can be seen in many populations that live in harsh conditions. For example, small insects living in desert areas may be able to survive through dry seasons only as eggs or in a dormant state. During rainy seasons, conditions may be such that the population increases exponentially at the maximum possible rate. When the rainy season ends, the environmental conditions are such that nearly all of the population dies. The growth curve for this population consists of a series of J-shape curves occurring at annual or seasonal intervals. The height of each population peak is determined by the number of eggs or dormant individuals left from the last population burst (which, in turn, is a function of the height of that peak) and by the duration of the favorable season.

Some ecologists have argued that such density-independent factors as weather are the major influence on the shapes of population curves that show seasonal peaks (Andrewartha, 1961; Andrewartha and Birch, 1954). Other ecologists have found evidence that, even in extremely seasonal populations, the growth rate is influenced by factors that depend upon the density of the population. For example, the number of black-tail deer that die during a harsh winter may depend not only upon the severity of the weather but also upon the number of deer competing for the available food.

Density-Dependent Factors

Among most populations, the large fluctuations caused by extreme environmental changes are relatively rare. The population size usually does not reach the maximum that could be supported under existing environmental conditions. Instead, as population density increases past some equilibrium value, the growth rate tends to decrease. If the population density drops below this equilibrium value, there is an increase in growth rate. This homeostatic mechanism tends to keep the population at a size somewhat below the maximum theoretical carrying capacity of the environment. Thus, the population is better able to survive extremes of weather or other potential catastrophes without great increases in mortality.

Various forms of social behavior appear to be among the most important homeostatic mechanisms of population control (Chapter 31). Through dominance hierarchies and territorial behavior, the available supply of food and space is shared among the stronger members of the population. These individuals can reproduce normally. If the population is too large for the available resources, extra members cannot secure a territory or an ample supply of food or mates. Thus, the natality of the population is adjusted to match the available resources and size of the existing population (Wynne-Edwards, 1962).

In recent years, there has been much interest in the possibility that physiological changes can be caused by overcrowding. In many species, natality decreases and mortality increases as the population density increases. In some cases, these effects are due to increasing competition for food, increasing opportunities for transmission of disease, or lack of space. However, it has long been known that physiological changes can accompany increasing population densities. During peaks of the population cycle in snowshoe hares, many individuals die from "shock disease," a condition of severe physiological stress characterized by low levels of blood sugar and liver glycogen (MacLulich, 1947). Among many mammalian species, high population densities are associated with various malfunctions of the endocrine system, particularly the adrenal glands. Apparently, the increased population density leads to more frequent aggressive encounters, particularly those involved in establishing territories and dominance hierarchies.

In laboratory experiments with mice and voles, definite physiological changes were found to accompany increases in population density. An increase in the size of a population confined to a constant space led to an increase in the weight of adrenal glands and a decrease in the weight of thymus and reproductive glands. As in the wild populations, these changes were accompanied by decreases in natality and increases in mortality.

Many investigators have stressed the importance of interactions among different species as a mechanism of homeostatic population control. For example, the cyclic fluctuations in lemming population have been explained as the result of interactions between the lemming population and the populations of plants on which it feeds. An increase in lemming population size leads to overgrazing and a deterioration of the soil as it is

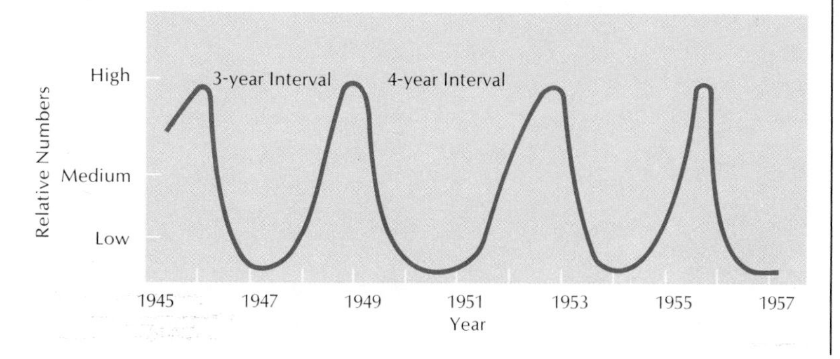

Figure 33.6. Generalized curve of the three- to four-year cycle of the brown lemming population.

Figure 33.7. Populations comprising "simple" ecosystems or those with reduced species diversity may sometimes go awry. As an example, the lemming populations of the Arctic tundra and alpine zones may show a marked increase in response to abundant foods. Lemming predators are unable to check this rapid population explosion and thus disease, eventual lack of food, and a unique behavioral phenomenon known as shock disease act to reduce the lemming population. As the population increases, social contact of an aggressive nature also increases. This constantly increasing level of hostile social encounters affects the lemmings' neurophysiology by causing atrophy and final destruction of the adrenocortical complex. These physiological changes produce behavioral changes. The animal first goes into a state of shock during which it may wander aimlessly (giving some credence to the famous "march to sea"), but eventually this wandering leads to coma and death. (© Walt Disney Productions)

Figure 33.8. The law of tolerance. Each species has a particular range of tolerance within which its life processes can function.

exposed to direct weathering. The overgrazing leads to a food shortage for the lemmings and a lack of cover under which to hide from predators. As the lemming population decreases from the higher death rates, the destruction of the plant populations is lessened. The plant cover is restored in the following growth season, and the stage is set for another population burst among the lemmings (Pitelka, 1957).

HETEROGENEITY OF ENVIRONMENT: HABITATS

No environment is completely homogeneous over a very large area. Even in a desert, plain, or prairie where conditions appear to be monotonously identical, careful study shows differences in soil, temperature, light energy, and moisture from place to place, even within distances of fractions of an inch. These differences are seen on a larger scale in the distribution of mountains, deserts, oceans, and other macroenvironments over the surface of the earth. For each species, it is possible to describe the environmental conditions that make existence possible. For a particular species, there are a maximum and a minimum temperature that can be tolerated. Within the range of tolerance is a smaller range of optimal conditions, within which growth, metabolism, and reproduction occur at optimal rates. Similar ranges of tolerance and optimal conditions can be established for other conditions such as moisture and food supply (Figure 33.8). Tolerance limits may define the broad outlines of the range of a population; different conditions often limit the population at different parts of its range. For example, the distribution of a plant may be limited by cold temperatures in the northern part of its range and by a lack of moisture in the southern part of its range. Within the broad limits of this range, individuals or groups tend to cluster in smaller patches that provide optimal conditions. Often, the details of distribution can be interpreted as the effect of competition—for example, where the local water supply is altered unfavorably by other species.

The general conditions under which populations of a particular species are found are described as the *habitat* of the species. In some cases, the

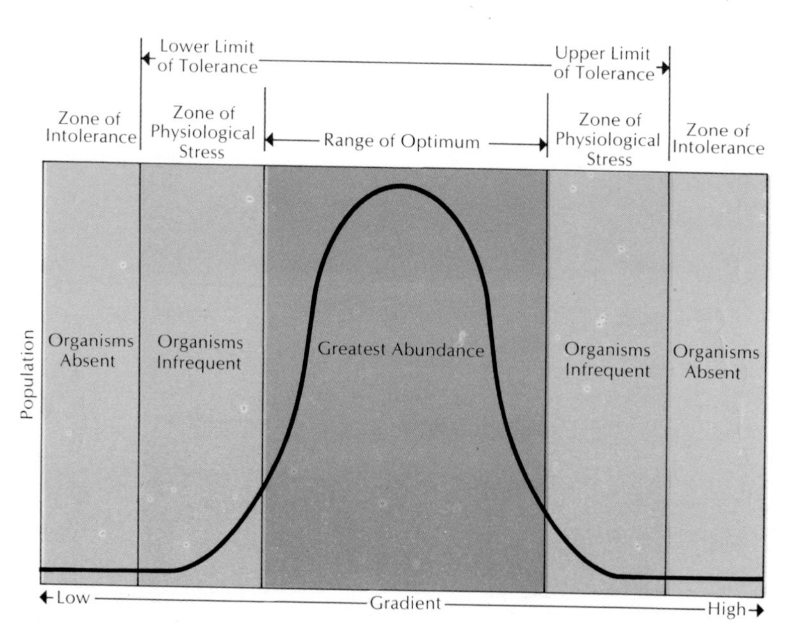

Figure 33.9. Graphs illustrating competitive exclusion.

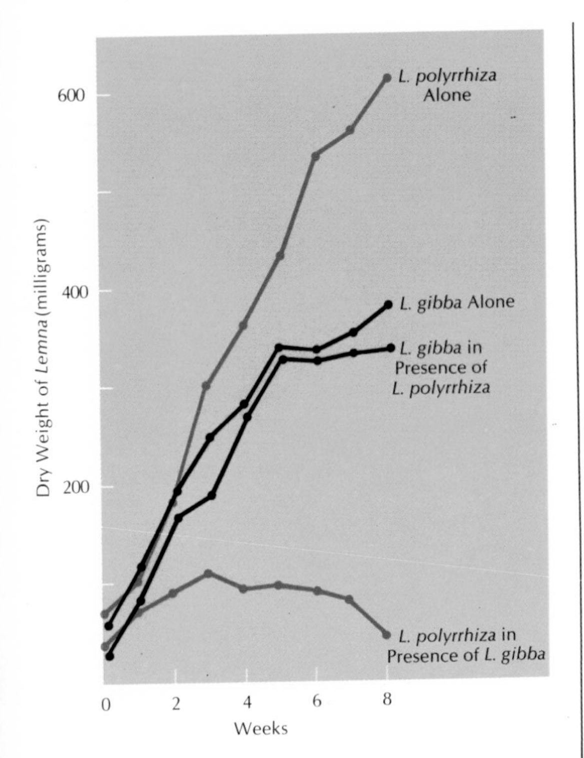

habitat is best described in terms of temperature, moisture, soil or mineral types, or other abiotic factors. In other cases, the habitat is best described in terms of other organisms; for example, the habitat of an insect species may be on the leaves or flowers of a particular plant species. In any case, the description of the habitat actually involves an indication of the environmental factors that most prominently limit the range of the species and of its constituent populations.

If two species are introduced into a uniform, mutually suitable environment and if all resources—food, light, space, water—are available in unlimited supply, both populations will indefinitely continue to grow exponentially. In any real situation, however, the two species are certain to share at least one resource that is in limited supply (space, if nothing else). The two species will have slightly different growth curves. As the size of each population approaches the carrying capacity of the environment (whose conditions now include the presence of the other species, therefore probably giving a lower carrying capacity for each species than either would have alone in the same environment), the growth rates of the two populations decline. Because the two curves are not identical, the growth rate of one population will decline to zero while the other population is still growing, at least slightly. The continued growth of the second population causes it to use still more of the limited common resource, further reducing the growth rate of the first population. Now the first population will begin to decline in size while the second population continues to grow (Figure 33.9). This situation is commonly described by saying that the second species has a *competitive advantage* over the first. The more the second population grows, the lower the carrying capacity of the environment for the first population (because both are competing for a single limited resource), and therefore the second population will eventually crowd out the first population entirely.

The principle of *competitive exclusion* summarizes the above argument by stating that under uniform conditions with at least one resource in limited supply, not more than one species can continue to exist indefinitely.

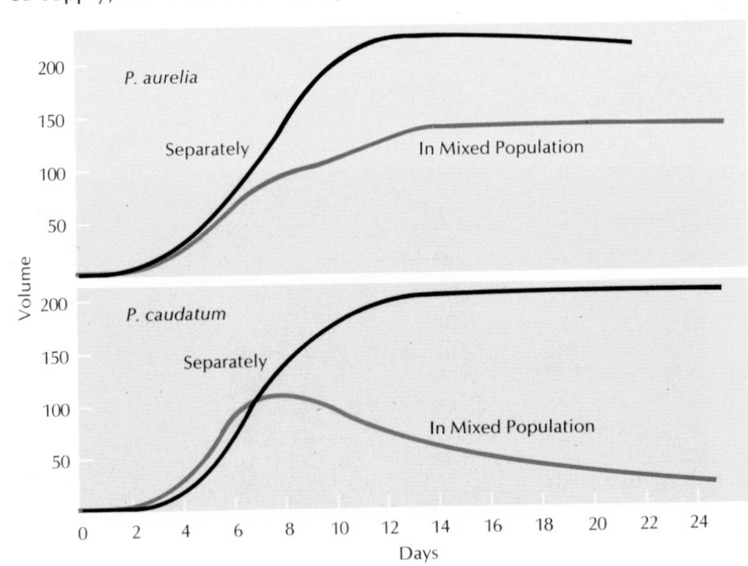

This principle was first stated by A. D. Grinnell in 1904 and has since been repeatedly confirmed by laboratory experiments. The coexistence of numerous species in nature is possible only because of the heterogeneity of the environment from place to place or from time to time.

If the conditions vary slightly from place to place, one species may hold the competitive advantage in some places while another holds the advantage in other places. Environmental changes over time may also permit two species to coexist. For example, suppose that one species holds the competitive advantage under summer conditions, whereas another holds the advantage under winter conditions. The regular alternation of seasons may prevent either species from completely disappearing before it has an opportunity to rebuild its population at the expense of the other.

In natural environments, the principle of competitive exclusion can be expressed in more general terms: no two species can coexist for long if they live in the same habitat and simultaneously exploit without limit a single resource. Careful examination of coexisting species reveals a slight difference in habitat preference or some behavioral or anatomical difference that results in a limitation upon the exploitation of common resources of food, space, and so on. This difference is hardly surprising because species originally were defined on the basis of morphological differences.

To describe the ways in which organisms exploit the resources of their environment, ecologists have developed the concept of the *niche*. If the habitat of a population is simplistically described as the place where the population lives, the niche may be similarly described as the way that the population makes its living. A plant-eating animal and an animal-eating animal occupy different niches, even though they may occupy very similar habitats. The day feeders and the night feeders also occupy different niches in similar habitats. Thus, no two species can long occupy identical niches in identical habitats.

COMMUNITIES AND ECOSYSTEMS

In the heterogeneous natural environment, a single population rarely lives alone in a particular area. In most natural macrohabitats, plant populations utilize solar energy to convert inorganic materials into carbohydrates, animal populations feed upon the plants, other animal populations feed upon the plant eaters, and fungi and bacteria populations convert the remains of plants and animals back into inorganic materials. A group of populations occupying the same area and interchanging materials and energy is called a *biotic community*.

Obviously, a community occupying a desert floor contains species very different from those in a community in a lake or ocean. But some generalizations about the structure of communities can be made. Almost every community contains populations of *producers*, organisms utilizing energy from the nonliving environment and producing complex organic molecules from inorganic substances. Photosynthetic plants and algae are the major producers of most communities, but photosynthetic and chemosynthetic bacteria can also be producers. For a few communities—such as those in caves or deep in the ocean—the major source of energy and organic molecules may consist of organic material that comes from other communities, and producers may be rare or absent.

Decomposers (bacteria and fungi) are also present in nearly every community. If nitrogen, phosphorus, and other crucial elements are to be constantly available to the producers for synthesis of new organic substances,

Figure 33.10. The pyramid of biomass.

these elements must be reclaimed from the remains of dead organisms. Carbon, oxygen, and hydrogen are available in the water and in the atmosphere in relatively large quantities, but even these materials must be recycled on a global scale if life is to continue. Only in an environment with a constant influx of inorganic nutrients and a constant removal of organic debris would it be possible for a community to exist indefinitely without decomposers.

The *consumers* of the community may be present in greater or smaller numbers, depending upon the amount of organic material available from the producers. Although it would be theoretically possible to have a community without consumers, it is very rare to find a community in which some organisms have not taken advantage of the food supply from the producers to establish themselves as consumer populations. *Primary consumers*, or herbivores, obtain their supply of organic nutrients from tissues of the producers. *Secondary consumers*, or carnivores, feed upon the tissues of primary consumers. In communities with dense producer populations there may be *tertiary consumers*, or second-level carnivores, who feed primarily upon the secondary consumers.

A unidirectional flow of energy occurs through the interactions of the community. Energy from the environment is stored in organic molecules by the producers. At each of the other *trophic levels* (primary consumers, secondary consumers, decomposers) of the community, a large portion of that energy is dissipated. The populations of each trophic level consume a portion of the organic material produced in the bodies of the lower level. The amount of energy stored in the organic substances of each level is normally less than 10 percent of the total energy stored in the bodies of the next lower level. The remainder of the energy that is obtained through feeding is expended in the life processes of the organisms; most is eventually reradiated from the earth.

As a result of this inevitable energy loss in the community, the structure of the community may be graphed as a pyramid. Because organisms vary greatly in size, it is normally more useful to examine the pyramid of biomass rather than the pyramid of numbers (Figure 33.10). A pyramid of

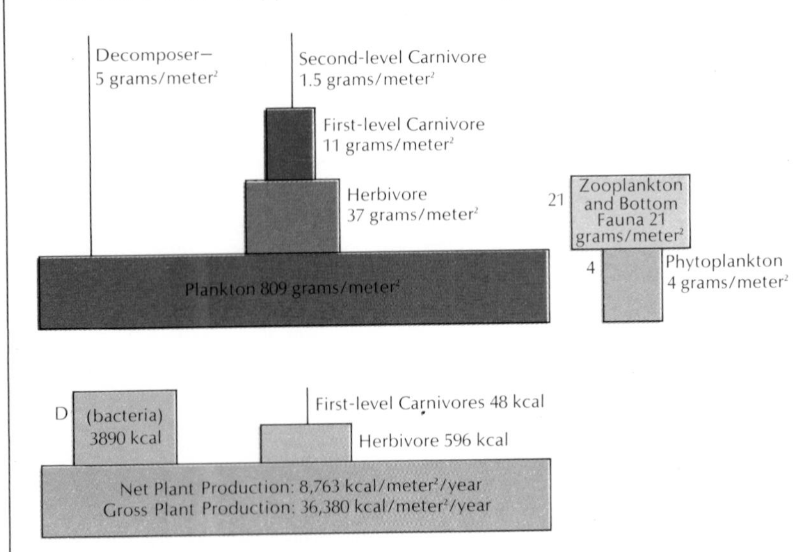

Figure 33.11. The food web of a salt marsh in midwinter. Producer organisms (terrestrial and salt marsh plants) are eaten by herbaceous invertebrates that live on land (grasshopper and snail). Marine plants are consumed by herbivorous marine invertebrates. Fish eat plants from both ecosystems and are in turn eaten by first-level carnivores (great blue heron and common egret). Examples of omnivores that make up a food web are ducks, sparrows, rats, mice and other small rodents, and sandpipers. Another first-level carnivore is the shrew; representative second-level carnivores are the hawk and owl.

Figure 33.12. Flow diagram of energy relationships along a single food chain showing a consecutive pattern of fixation and transfer by the components and considerable respiratory losses at each transfer. P = gross primary production; P_n = net primary production; and P_2, P_3, P_4, and P_5 = secondary production at the indicated levels.

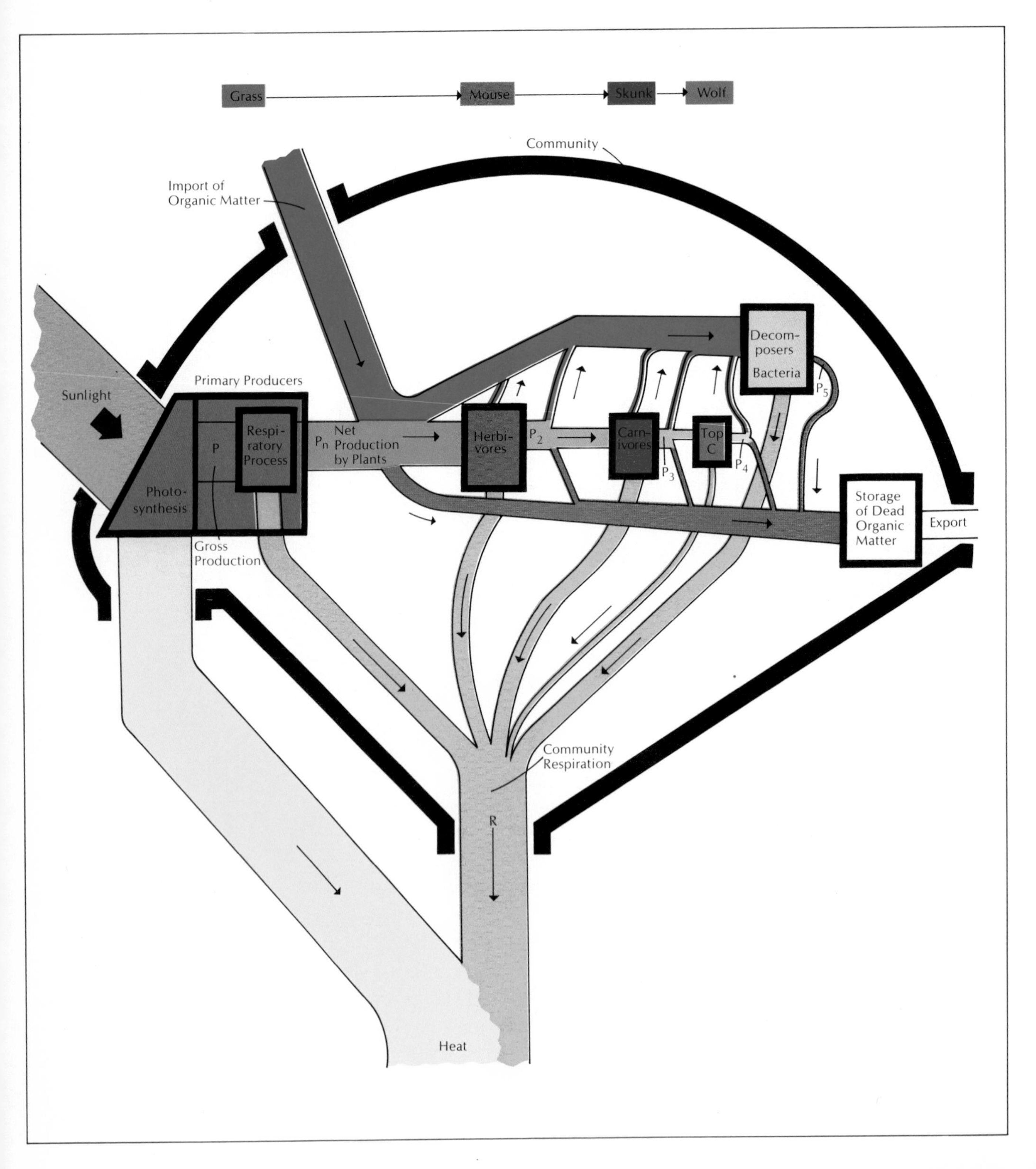

energy is even more informative but normally difficult or impossible to calculate from available information on a natural community.

In most real communities, the actual structure of the trophic levels is quite complicated. Each population feeds upon certain combinations of the other populations. For example, a particular population of secondary consumers may feed primarily upon a single population of primary consumers, but may also occasionally feed upon other primary consumers, other secondary consumers, and even upon the producers themselves. The actual trophic relationships form a complex *food web*, whose structure may vary with seasonal and other changes (Figure 33.11). A simplified picture may be obtained by looking at the energy relationships along a single *food chain* (Figure 33.12).

Few communities have been quantitatively studied in detail. Those that have are relatively small and atypical communities. In a Georgia salt marsh, about 6 percent of the solar energy falling on the surface is utilized by photosynthetic organisms in the production of organic materials. Of the energy converted by the producers, 77 percent is used in respiration within the producers themselves, a figure much larger than that typical of forests or other communities with large plant bodies. Of the energy stored in plant tissues, about 47 percent is dissipated by bacteria (decomposers) and about 8 percent by primary and secondary consumers. The remaining 45 percent of the stored energy is lost to the ecosystem as living and dead organisms

Table 33.4
Energy Utilization in a Salt Marsh Community*

ENERGY USE BREAKDOWN	Kilocalories per Square Meter per Year	Percentage of Total Energy Supply
Total incoming light energy	600,000	100.000
Reflected, heating environment, etc. (not utilized by community)	563,620	93.937
Gross production	36,380	6.063
Producer respiration	28,175	4.696
Building producer tissues (net production)	8,205	1.367
Bacterial respiration	3,890	0.648
Primary consumer respiration	596	0.099
Secondary consumer respiration	48	0.008
Exported from community	3,671	0.612

*Assuming total yearly incoming light energy of 600,000 kilocalories per square meter

Source: Adapted from J. M. Teal, "Energy flow in the salt marsh ecosystem of Georgia," *Ecology* 43 (1962):614–624.

are washed out of the salt marsh. In this community, producers and decomposers play the major role in energy transfer, and a sizable proportion of the energy fixed by the producers is exported from the community (Table 33.4). In a spring community, on the other hand, the primary source of energy is organic debris washed into the community, and little energy is exported (Table 33.5).

An ecosystem chosen to represent a relatively uniform vegetation or animal assemblage is often called a *biome*. Typical biomes include the

Figure 33.13. Community changes as illustrated by wildlife succession in conifer plantations in central New York.

temperate deciduous forest, the coniferous forest, the desert, the tropical rain forest, and the prairie.

The boundaries of communities, ecosystems, and biomes are drawn arbitrarily. Any portion of the environment could be chosen for study as an ecosystem, but such a system is convenient for study only if the flows of

Table 33.5
Energy Relationships in a Cold Spring Community

ENERGY	Kilocalories per Square Meter per Year	Percentage of Available Chemical Energy
Source		
Organic debris entering community	2,350	76.150
Gross photosynthetic production	710	23.006
Immigration of caddisfly larvae	18	0.583
Decrease in producer biomass	8	0.261
Use		
Heat dissipation by community	2,185	70.803
Deposition of organic matter	868	28.127
Emigration of adult insects	33	1.070

Source: Adapted from J. M. Teal, "Community metabolism in a temperate cold spring," *Ecological Monographs 27* (1957):283–302.

energy and materials across its boundaries are minimal or easily measured or estimated. Because each species has its own limits of tolerance for various environmental factors, the ranges of any two populations are seldom identical. Thus, it is normally impossible to draw firm lines around a com-

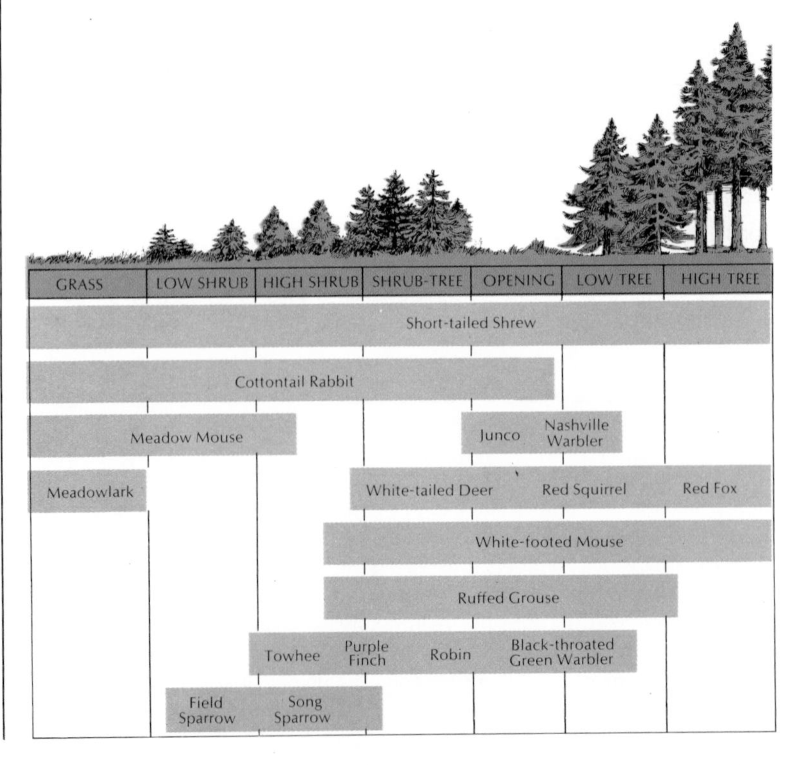

munity that unequivocally separate it from neighboring communities. Biomes similarly grade into one another so that boundaries must be fixed rather arbitrarily.

CHANGES IN COMMUNITIES

Within any community, the sizes of various populations are continually changing. Nearly all populations show sizable fluctuations in size, usually in regular cyclic patterns. As environmental conditions change or new populations immigrate into the community, the balance of competitive advantage may shift and some old populations may dwindle or disappear. Nevertheless, most communities are surprisingly stable. The balance among populations moves now one way, now another, but over a span of a century the total changes are surprisingly small.

The most rapid changes in communities are seen when a forest fire, flood, or other drastic change in the environment suddenly alters the conditions. The barren land produced by a natural catastrophe or human action is first occupied by a relatively simple community of hardy *pioneer organisms*. This community gradually alters the nature of the soil and establishes shade and moisture within which other organisms can grow. As conditions are gradually changed by the organisms, new populations become established and the pioneer populations lose their competitive advantage and disappear. The end product of this process of *succession*—which may pass through a number of distinctive intermediate communities—is a complex and relatively stable community similar to the one that occupied the region before the catastrophe.

The process of succession can occur only where individuals from neighboring regions are able to immigrate into the region of succession. Presumably, such immigration goes on all the time, but only when conditions are just right can individuals become established, compete successfully, and form a growing population.

The detailed and mathematical explanation of how interaction among populations and among individuals within a population can produce new species is among the most important and central ideas of modern biology. This study involves less concern with the overall size and distribution of populations that concerns the ecologist, and a more detailed study of the distribution of individual characteristics within the population. This study of population genetics will be considered in the next chapter.

FURTHER READING

There are a number of good introductory textbooks in the field of ecology, any of which will provide more complete information on populations, communities, and ecosystems. Particularly recommended are those by Kormondy (1969), Odum (1971), and R. L. Smith (1966). More specialized discussions of population growth and interactions among populations will be found in books by Billings (1964), Dasmann (1959, 1964), Elton (1958, 1966), Klopfer (1969), and MacArthur and Connell (1966).

Highly enjoyable books dealing with general ecological principles include those by Bates (1960, 1961), Farb (1963), and Leopold (1949). A selection of articles exemplifying a number of approaches to population and community ecology is included in the collection edited by Hazen (1970). Other useful introductory articles include those by Cooper (1961), Deevey (1970), Evans (1956), Hutchinson (1959, 1970), Powers and Robertson (1966), Wecker (1964), Woodwell (1963, 1967, 1970), and Wynne-Edwards (1964, 1965).

34
Population Genetics

Darwin and Wallace made a convincing general argument for the importance of natural selection in the process of evolution. They lacked, however, a detailed understanding of the mechanism by which inheritance operates and by which hereditary variations appear in a population. With the development of classical genetics in the early years of this century, it became possible to make quantitative estimates of the rates at which the characteristics of a population will change as a result of selection or of other factors. The field of population genetics explores the changes over time in the distribution of particular genetic characteristics in a population. Population geneticists try to understand the evolutionary process quantitatively in terms of the genes, chromosomes, mutations, and other concepts developed by classical and molecular geneticists.

Population genetics is a statistical science concerned with groups of organisms. The individual is important only as a genetic package within the group. For the most part, the population geneticist speaks of the *gene pool*—the total collection of all the allelic forms of all the genes in all the gametes of the population. He is concerned with the proportion of one kind of individual in comparison with other kinds, but he is seldom concerned with the particular combination of genes that occur in a single individual.

The basic unit of study in this field is the *deme*, a population of freely interbreeding individuals. The gene pool of a deme can be described as the *gene frequencies*, which might better be called allele frequencies. The gene frequency is a mathematical expression of the proportion of one allelic form of a gene compared to other allelic forms of the same gene in the gene pool. Also of importance are the *genotype frequencies* and *phenotype frequencies* of the population, both of which are dependent on gene frequencies. Mendel observed the proportions of various phenotypes (such as long-stem and short-stem plants) among his pea plants. By using various controlled breeding experiments, Mendel calculated the proportions of various genotypes among his pea plants. From this information, a geneticist could calculate the gene frequencies of the various alleles involved.

GENETIC EQUILIBRIUM

In order to describe quantitatively the ways in which a gene pool can change from generation to generation, it is necessary to make some simplifying assumptions, many of which may be unrealistic for most natural populations. However, population genetics has built ever more complex models, approximating more and more closely the situations thought to exist in nature. Even computers cannot completely analyze the genetics of a large population. A complete analysis would describe gene frequencies and frequencies of combinations of genes for many thousands of gene loci, most of which have several allelic forms. The interactions of various factors affecting the frequency of these alleles and their combinations are far beyond the capabilities of current programmers and computers. Nevertheless, simplified theoretical models have provided considerable insights into the mechanisms of evolution.

Most examples in this chapter are limited to a single gene locus with only two allelic forms in a bisexual organism, with haploid gametes (sperm and egg) from a male and a female parent combining to form a diploid zygote. The simplest situation for theoretical analysis would be one in which mating occurs in a random manner. That is, any sperm cell is equally likely to be combined with any egg in the population. Such *panmictic populations*

Interleaf 34.1

GENETIC EQUILIBRIUM: THE HARDY-WEINBERG LAW

A population of N individuals makes up a panmictic deme. At a particular gene locus, two allelic forms exist: A and a. For diploid individuals, three genotypes are possible: AA, Aa, and aa. Of the N individuals, let D be the number that are homozygous for the dominant allele (AA), H be the number that are heterozygous (Aa), and R be the number that are homozygous for the recessive allele (aa).

The total number of genes at this locus in the gene pool of the deme is $2N$. The total number of A alleles in the gene pool is $2D+H$, because each AA individual has two A alleles, each Aa individual has one A allele, and the aa individuals have none. The gene frequency (p) of the A allele is the proportion of that allele to the total number of genes at that locus:

$$p = \frac{2D + H}{2N} \qquad (34.1)$$

The gene frequency (q) of the a allele may be determined by similar reasoning:

$$q = \frac{2R + H}{2N} \qquad (34.2)$$

Because $N = D + H + R$, it is clear from adding (34.1) and (34.2) that

$$p + q = 1 = \frac{2D + 2H + 2R}{2N} = \frac{2N}{2N} = 1 \qquad (34.3)$$

If gene frequencies are the same among males and females and if gametes are produced randomly among the population, the proportion of sperm and eggs carrying the A allele will be p, and the proportion carrying the a allele will be q. (It is assumed that N is extremely large, so that random variations from the predicted frequencies are negligibly small.)

Random combination of these gametes to form zygotes yields results that can easily be predicted. The probability of an A sperm combining with an A egg is equal to the products of the proportions of A alleles among the sperms and eggs: $p \times p$, or p^2. Thus, the proportion of AA zygotes is equal to p^2. Similarly, the proportion of aa zygotes is equal to q^2. An Aa zygote can result either from the combination of an a sperm with an A egg or from the combination of an A sperm with an a egg. The probability of either of these events is $p \times q$; therefore, the proportion of Aa zygotes is $pq + pq$, or $2pq$.

To see what happens in a situation of genetic equilibrium, suppose that the initial distribution of genotypes is such that $D = 0.35N$, $H = 0.50N$, and $R = 0.15N$. In this population

$$p = \frac{2D + H}{2N} = \frac{0.70N + 0.50N}{2N} = \frac{1.20N}{2N} = 0.60 \qquad (34.4)$$

$$q = 1 - p = 1 - 0.60 = 0.40$$

Assume that generations are nonoverlapping; each generation grows to adulthood, mates, and then dies without participating in further matings. If mating occurs randomly and all zygotes are equally likely to survive, the numbers of individuals of the various genotypes in the next generation can be calculated:

$$D = p^2N = (0.60)^2N = 0.36N \qquad (34.5)$$

$$H = 2pqN = 2 \times 0.60 \times 0.40 \times N = 0.48N \qquad (34.6)$$

$$R = q^2N = (0.40)^2N = 0.16N \qquad (34.7)$$

In this generation, the genotype frequencies have changed slightly from those of the parent generation. However, the gene frequencies are unchanged:

$$p = \frac{2D + H}{2N} = \frac{0.72N + 0.48N}{2N} = \frac{1.20N}{2N} = 0.60 \qquad (34.8)$$

$$q = 1 - p = 1 - 0.60 = 0.40$$

Because the gene frequencies are unchanged, the genotype frequencies of the following generation will be given by exactly the same calculations as those shown above. Thus, the gene and genotype frequencies of this population will remain unchanged through all future generations. This result can be shown to hold true for any number of alleles and for any gene frequencies.

In a panmictic deme where all individuals are equally likely to survive and reproduce and where there are no disturbing factors such as mutation, gene flow, or genetic drift, the gene frequencies remain unchanged from generation to generation, and the genotype frequencies reach unchanging equilibrium values after at most a single generation of random mating. This generalization is the Hardy-Weinberg law.

The implication of the Hardy-Weinberg law is that a population has genetic inertia that tends to keep its gene frequencies and genotype frequencies constant from generation to generation. The factors that overcome this inertia and alter the gene pool are the factors that produce evolution.

probably are rare or nonexistent in nature. If nothing else, an organism tends to mate with its geographical neighbors. In most natural populations, mating behaviors depend on other nonrandom factors such as a tendency to mate with individuals of similar or different phenotype, with related individuals, or with individuals high in the dominance hierarchy. Nevertheless, in theory, the panmictic population provides a useful model for mathematical analysis.

For the simplest possible model, the following simplifying assumptions are made: (1) mating occurs in random fashion (that is, the population is panmictic); (2) the population is infinitely large, so that random fluctuations from statistically predicted distributions can be ignored; (3) all zygotes are equally likely to survive and to reproduce as mature individuals; (4) the successive generations are nonoverlapping (that is, each generation develops to maturity, mates to produce the next generation, and then either dies or fails to participate in later mating seasons); (5) no individuals enter or leave the population; (6) production of gametes is random, so that the gene frequencies among the gametes produced are the same as the gene frequencies among the parent population; (7) the gene frequencies are the same among the male and female parts of the population; and (8) mutation does not occur.

Under these unrealistic conditions, the gene frequencies remain constant from generation to generation. This deduction was developed independently in 1908 by the English mathematician G. H. Hardy and the German physician W. Weinberg and is known as the *Hardy-Weinberg law* (Interleaf 34.1)

In a population at genetic equilibrium under the conditions of the Hardy-Weinberg law, gene frequencies and, as a consequence, frequencies of various genotypes and phenotypes remain constant from generation to generation. In natural populations, several factors disturb this equilibrium and cause changes in the gene pool. Various behavioral and physiological factors can lead to nonrandom production and combination of gametes during mating. Mutation directly alters the gene pool by substituting one allele for another in a particular gamete; this process is important as a way in which new alleles can be introduced into the gene pool. Selection results from different survival and reproduction rates for various genotypes, thus causing particular genotypes to contribute either more or less heavily to the next generation than they would under random conditions. Gene flow through immigration and emigration can cause significant changes in the gene pool. In small populations, chance effects can lead to changes in the gene pool because random fluctuations make the gene frequencies among gametes or zygotes significantly different from the theoretically expected values. Such random fluctuations lead to *genetic drift* of gene frequencies over time.

Evolutionary changes in a population result from the effects of various factors that systematically or randomly disturb the genetic equilibrium. Selection is only one factor. A quantitative understanding of the role of selection requires the evaluation of the importance and rates of the processes that disturb equilibrium.

GENETIC DRIFT

In deriving the Hardy-Weinberg law, it was assumed that random production of gametes would give the same gene frequencies among the gametes

Figure 34.1 Model illustrating genetic drift. In populations that occur in small numbers, genes may be accidentally lost or fixed in some individuals.

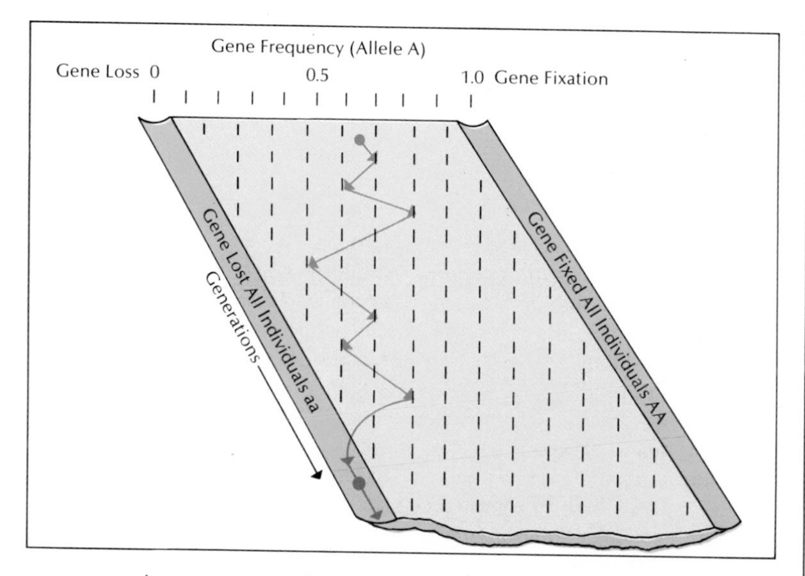

as among the parent population. Similarly, it was assumed that random combination of gametes bearing two alleles in the gene frequencies p and q would yield a population of zygotes with frequencies of p^2, $2\ pq$, and q^2 for the three possible genotypes. In fact, these are the most probable values for the result of such random processes, but the actual results obtained from any particular process can deviate somewhat from the most probable values. For example, if $p = q = 0.50$ in a population of 50 individuals, there is a probability of 0.99994 that p for the next generation will lie somewhere between 0.30 and 0.70 if gametes are produced and combined randomly. As the population becomes larger, the probable fluctuations become smaller. For example, if the population contains 5,000 individuals, there is a probability of 0.99994 that the value of p for the next generation will lie somewhere between 0.48 and 0.52. The Hardy-Weinberg equilibrium assumes a population so large that the probability of a significant fluctuation from the equilibrium gene frequencies is negligibly small. In any real population this large, truly random mating throughout the population is extremely unlikely.

In a large population, the observed gene frequency fluctuates slightly about the value predicted by the Hardy-Weinberg law if no disturbing factors other than genetic drift are operating. In a small population, the fluctuations are much greater. If p should reach zero in some generation, the allele *A* will be permanently lost from the population gene pool; if p should reach 1.00, the allele *a* will be lost. In a population of N individuals where $p = q = 0.50$ initially, the average time before p will reach a value of 0.0 or 1.0 is about $2\ N$ generations. Thus, in a population of only a few hundred individuals, the loss of alleles from the gene pool through genetic drift is quite common.

Genetic drift may be of particular evolutionary importance in small populations whose numbers are occasionally greatly reduced (for example, to less than 100 individuals in a particular generation). Some alleles will be lost from the gene pool in such a population crash, whereas others will be more common among the surviving population than they were in the pre-crash population. Genetic drift is a purely statistical process, and its effects are quite random. The phenomenon of genetic drift can be observed in any relatively small population, even if the particular alleles involved offer no advantages or disadvantages for survival or reproduction. Alleles may be lost or may become more common through genetic drift, regardless of their advantages or disadvantages for the individuals or the population.

MUTATIONS

Mutations are changes in genetic information. Little is known about the exact mechanisms or rates of most mutations. Point mutations, in which one nucleotide in the DNA of a gene is changed to another, are probably the most common and are the most easily analyzed mathematically. More extensive alterations in the sequence of genes or in the size and number of chromosomes probably occur more rarely but may be important in the establishment of new species. Point mutations will be considered in this section as a means by which one allelic form can be substituted for another; other forms of mutation will be discussed later in this chapter.

The rate at which mutations occur in single cells can be readily measured only in unicellular organisms such as bacteria, and even there it is difficult to determine whether a particular phenotypic alteration is the result of mutation of a single gene or of a number of genes. The rate at which spontaneous mutations arise in a particular gene in microorganisms varies between about once in a million and once in a billion divisions.

In cultures of animal cells, mutations of any particular gene are observed about once in a million cell divisions. In a multicellular organism, mutations of most body cells affect only the particular individual. Only mutations of gametes or their precursor cells alter the gene pool of the population. A single mutation can lead to the production of zero, one, or more mutated gametes, depending upon the particular cell affected and the stage of development of gametogenesis in which the mutation occurs.

Suppose a population consists originally of $pA + qa$, where $(p + q) = 1$. If the mutation rate for a particular gene is expressed as a number (u) less than 1.0, ($u \times p$) will represent the proportion of gametes bearing the mutated allele (A changed to a). The observed rates of mutations in animals vary with the particular gene being studied, but the average value of u is about 0.00001. The total number of mutant gametes (up) appearing will vary with the gene frequency (p) of the gene subject to mutation. There will also be a back mutation ($a \rightarrow A$) occurring at a rate of v. If the other conditions of the Hardy-Weinberg equilibrium are met, an equilibrium of gene frequencies will be reached when the gene frequency of a in the gene pool is just large enough to balance the back mutations against the forward mutations (Interleaf 34.2).

Thus, for a population otherwise in genetic equilibrium, mutation shifts the equilibrium frequencies of particular alleles to values that depend upon the rates of mutation and back mutation at the gene locus in question. Mutation may maintain a low frequency of a particular allele against other factors that might tend to make that allele disappear from the gene pool. However, because the observed mutations occur at low rates, mutation cannot account for the rapid rate of evolution in many natural populations.

SELECTION

In a population at Hardy-Weinberg equilibrium, it is assumed that all genotypes are equally likely to survive and to reproduce. In fact, any two

Interleaf 34.2

MUTATION RATES AND GENETIC EQUILIBRIUM

Suppose that a "normal" allele, *A*, mutates to the allele *a* at a rate of u. The back mutation rate ($a \rightarrow A$) is v. The initial gene frequencies of *A* and *a* in the gene pool are p_i and q_i. The change in q each generation (Δq) will be equal to the number of mutations less the number of back mutations:

$$q = up - vq \tag{34.9}$$

An equilibrium will be reached when $\Delta q = 0$, with constant gene frequencies of p_e and q_e, the subscript e signifying the equilibrium frequencies:

$$\begin{aligned} 0 &= up_e - vq_e \\ 0 = u(1 - q_e) - vq_e &= u - uq_e - vq_e \\ q_e(u + v) &= u \\ q_e &= \frac{v}{u + v} \end{aligned} \tag{34.10}$$

$$\begin{aligned} p_e &= 1 - q_e = 1 - \frac{u}{u + v} \\ p_e &= \frac{u + v - u}{u + v} = \frac{v}{u + v} \end{aligned} \tag{34.11}$$

The effect of mutation is to establish a stable polymorphism, with a tendency for gene frequencies to move toward a particular equilibrium set of values. If q is larger than q_e, Δq is negative and the gene frequency of *a* decreases. If q is smaller than q_e, Δq is positive and the gene frequency of *a* increases. Thus, mutation will tend to offset any other factors that tend to move the gene frequencies away from these equilibrium values.

For example, suppose that for a particular gene locus there is a forward mutation rate ($A \rightarrow a$) of $u = 0.0001$ and a backward mutation rate ($a \rightarrow A$) of $v = 0.00005$. In other words, for each 10,000 *A* gametes, one may be expected to mutate to an *a* gamete, and for each 20,000 *a* gametes, one may be expected to mutate to an *A* gamete. In a population not subject to other evolutionary factors, the gene frequencies will eventually reach an equilibrium with the values:

$$\begin{aligned} q_e &= \frac{u}{u + v} = \frac{0.0001}{0.00015} = 0.667 \\ p_e &= \frac{v}{u + v} = \frac{0.00005}{0.00015} = 0.333 \end{aligned} \tag{34.12}$$

Figure 34.2. Graph (with data on opposite page) showing the effects of various selection coefficients upon a population.

differing genotypes are likely to have slightly differing probabilities of producing gametes. Certain alleles are likely to become more common because individuals possessing those alleles have greater success in survival and reproduction. Thus, selection causes gene frequencies to shift away from the constant values expected in a population at genetic equilibrium.

The relative survival value and reproductive capability of a particular genotype vary under different environmental conditions. The *fitness*, or adaptive value, of a particular genotype is defined on a range of values from zero (for a genotype that contributes no gametes to the next generation—for example, a genotype that leads to death before the age of reproduction) to one (for the genotype that proportionately contributes the most gametes to the next generation). Defined in this way, fitness can be determined only through observation of the changes in the gene pool. If one genotype is more successful in reproducing itself than is another, it is more fit than the other. Fitness, then, is a measure of the degree to which a genotype succeeds in reproducing its alleles in the next generation.

A genotype may have high fitness as a result of any combination of a number of different factors—a longer reproductive period, an increased number of offspring, increased efficiency in mating, resistance to disease or environmental stresses at any stage of life, and so on. Figure 34.2 shows the effects of various selection coefficients upon a population initially in equilibrium with $p = q = 0.50$. In Figure 34.3, the same information is recalculated to show genotype frequency of recessive homozygotes. If selection acts equally on all three genotypes (all genotypes are equally fit), the gene frequencies are unchanged, as are the genotype frequencies (line *E*).

In cases of complete dominance, the phenotypes of *AA* and *Aa* genotypes are identical. If these genotypes are more fit than the *aa* genotype, the

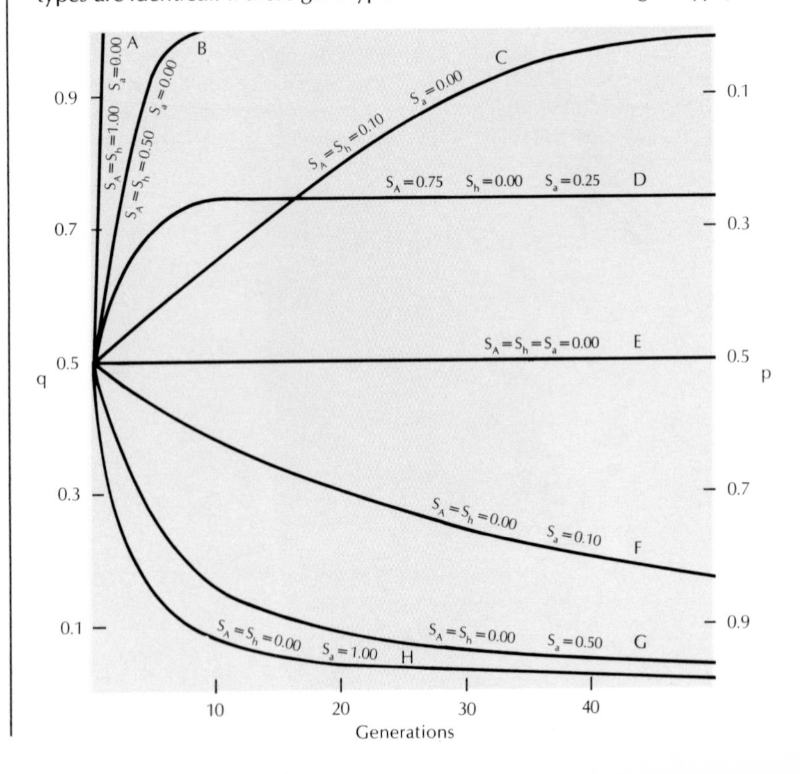

Data for Figure 34.2

Generation	A	B	C	D	E	F	G	H
0	0.500	0.500	0.500	0.500	0.500	0.500	0.500	0.500
1	1.000	0.600	0.514	0.583	0.500	0.487	0.430	0.333
2	1.000	0.706	0.528	0.634	0.500	0.475	0.372	0.250
3	1.000	0.804	0.542	0.669	0.500	0.463	0.324	0.200
4	1.000	0.880	0.556	0.691	0.500	0.451	0.286	0.167
5	1.000	0.932	0.567	0.706	0.500	0.439	0.256	0.143
6	1.000	0.964	0.582	0.716	0.500	0.428	0.231	0.125
7	1.000	0.981	0.592	0.725	0.500	0.418	0.210	0.111
8	1.000	0.990	0.602	0.732	0.500	0.407	0.192	0.100
9	1.000	0.995	0.611	0.736	0.500	0.397	0.177	0.091
10	1.000	0.997	0.626	0.739	0.500	0.388	0.164	0.084
11	1.000	0.999	0.642	0.742	0.500	0.378	0.151	0.078
12	1.000	1.000	0.658	0.744	0.500	0.369	0.141	0.072
13	1.000	1.000	0.674	0.745	0.500	0.361	0.133	0.067
14	1.000	1.000	0.690	0.746	0.500	0.352	0.125	0.063
15	1.000	1.000	0.706	0.748	0.500	0.344	0.118	0.060
16	1.000	1.000	0.721	0.749	0.500	0.336	0.112	0.057
17	1.000	1.000	0.736	0.750	0.500	0.329	0.106	0.054
18	1.000	1.000	0.751	0.750	0.500	0.321	0.101	0.051
19	1.000	1.000	0.766	0.750	0.500	0.313	0.097	0.048
20	1.000	1.000	0.781	0.750	0.500	0.307	0.092	0.046
21	1.000	1.000	0.797	0.750	0.500	0.301	0.089	0.044
22	1.000	1.000	0.810	0.750	0.500	0.295	0.086	0.042
23	1.000	1.000	0.823	0.750	0.500	0.288	0.082	0.040
24	1.000	1.000	0.835	0.750	0.500	0.282	0.079	0.038
25	1.000	1.000	0.847	0.750	0.500	0.277	0.076	0.037
26	1.000	1.000	0.858	0.750	0.500	0.271	0.074	0.036
27	1.000	1.000	0.879	0.750	0.500	0.266	0.071	0.035
28	1.000	1.000	0.889	0.750	0.500	0.260	0.069	0.034
29	1.000	1.000	0.898	0.750	0.500	0.255	0.067	0.033
30	1.000	1.000	0.906	0.750	0.500	0.250	0.064	0.032
31	1.000	1.000	0.914	0.750	0.500	0.245	0.062	0.031
32	1.000	1.000	0.921	0.750	0.500	0.241	0.061	0.030
33	1.000	1.000	0.929	0.750	0.500	0.236	0.059	0.029
34	1.000	1.000	0.935	0.750	0.500	0.232	0.057	0.028
35	1.000	1.000	0.941	0.750	0.500	0.228	0.056	0.027
36	1.000	1.000	0.946	0.750	0.500	0.224	0.054	0.026
37	1.000	1.000	0.951	0.750	0.500	0.220	0.053	0.026
38	1.000	1.000	0.955	0.750	0.500	0.216	0.051	0.025
39	1.000	1.000	0.959	0.750	0.500	0.213	0.050	0.025
40	1.000	1.000	0.963	0.750	0.500	0.209	0.049	0.024
41	1.000	1.000	0.966	0.750	0.500	0.205	0.048	0.023
42	1.000	1.000	0.969	0.750	0.500	0.202	0.047	0.023
43	1.000	1.000	0.972	0.750	0.500	0.199	0.046	0.022
44	1.000	1.000	0.975	0.750	0.500	0.197	0.045	0.022
45	1.000	1.000	0.977	0.750	0.500	0.193	0.044	0.021
46	1.000	1.000	0.979	0.750	0.500	0.190	0.043	0.021
47	1.000	1.000	0.981	0.750	0.500	0.188	0.042	0.020
48	1.000	1.000	0.983	0.750	0.500	0.185	0.041	0.020
49	1.000	1.000	0.985	0.750	0.500	0.182	0.040	0.020
50	1.000	1.000	0.986	0.750	0.500	0.179	0.039	0.019

Values calculated for Hardy-Weinberg conditions except for selection.
Values of selection coefficients for various curves are:

Curve	s_A	s_h	s_a
A	1.00	1.00	0.00
B	0.50	0.50	0.00
C	0.10	0.10	0.00
D	0.75	0.00	0.25
E	0.00	0.00	0.00
F	0.00	0.00	0.10
G	0.00	0.00	0.50
H	0.00	0.00	1.00

Figure 34.3. Graph showing genotype frequency of recessive homozygotes in the same population as shown in Figure 34.2.

frequency of the *a* allele in the gene pool drops rapidly at first, then more slowly as q reaches smaller values (lines *G* and *H*). Even if the recessive homozygotes completely fail to reproduce, the *a* alleles still make up about 2 percent of the gene pool after 50 generations (line *H*). Even the very low mutation rates typically observed may be sufficient to prevent q from dropping below about 1 or 2 percent. If selection against the recessive homozygotes is relatively weak, q decreases very slowly. In the case where recessive homozygotes are 90 percent as fit as the other genotypes, *a* alleles still make up about 18 percent of the gene pool after 50 generations (line *F*).

If the recessive homozygotes are more fit than the other genotypes, the *A* allele is removed from the gene pool relatively rapidly (lines *A*, *B*, and *C*). In the extreme case where *AA* and *Aa* genotypes fail to reproduce, the *A* allele is removed from the gene pool in a single generation (line *A*). Even if the dominant phenotype is 50 percent as fit as the recessive phenotype, the *A* allele will be reduced to less than 1 percent of the gene pool within about 8 generations (line *B*). If the dominant phenotype is 90 percent as fit as the recessive phenotype, removal of the *A* allele is much slower, with about 1 percent of this allele in the gene pool even after 50 generations (line *C*).

In many cases, the heterozygous genotype is more fit than either of the homozygous genotypes. This result often occurs in cases of incomplete dominance, when the heterozygous phenotype is intermediate in characteristics between the homozygous phenotypes. In such a case, the gene frequencies tend toward an equilibrium value in which both alleles are retained in the gene pool (line *D*). Unless one homozygous genotype is much more fit than the other, both alleles will be retained at rather high frequen-

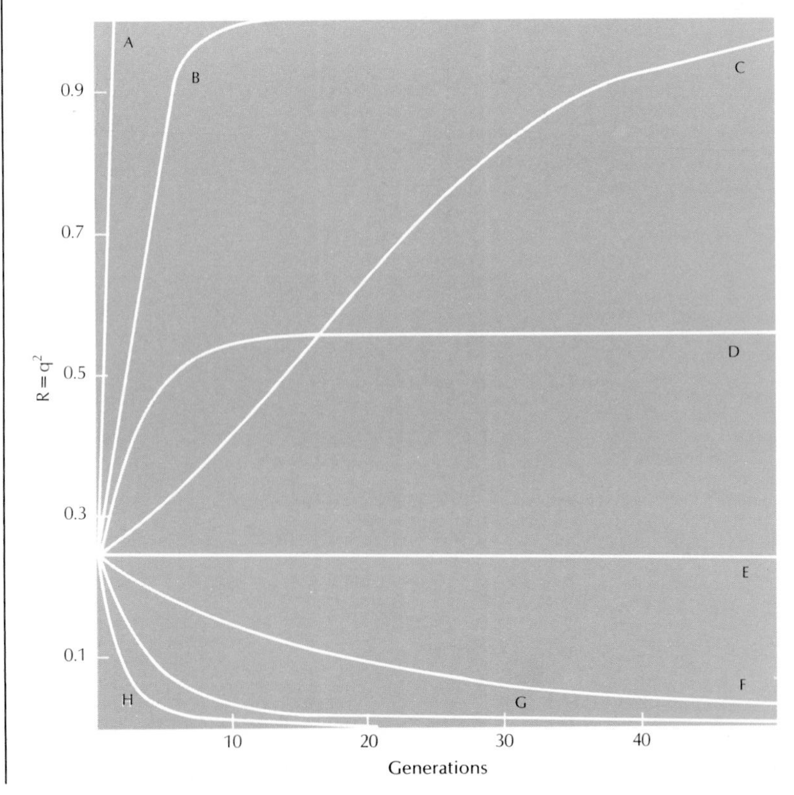

cies. Thus, such a population is able to adapt rapidly to a new equilibrium if changed conditions alter the selection coefficients of the genotypes.

If the homozygotes are more fit than the heterozygotes, an unstable equilibrium exists with $p = q = 0.50$. However, the slightest increase in the frequency of either allele (for example, due to genetic drift) will lead to a steady elimination of the less common allele from the gene pool. In this case, it is a matter of chance whether the allele retained in the gene pool is that with the more fit homozygous phenotype.

MULTIPLE FACTORS

In any real population, the gene frequencies at any time are likely to result from the combined effects of mutation, gene flow, genetic drift, and selection. Furthermore, in real organisms genes are not assorted independently but are linked together on chromosomes. Thus, the frequency of alleles of one gene can be influenced by the factors affecting the frequencies of alleles of a linked gene. The effects of recombination through crossovers must then be considered. For example, selection may favor a particular chromosome containing a number of relatively fit linked alleles, even though individual alleles in this group may not be the most fit possible for their particular genes.

Attempts to develop complete mathematical representations of genetic changes in realistic populations are still in a primitive state. However, combinations of simplified models make it possible to draw some general conclusions about the factors acting together on natural populations.

Mutation is a significant influence on the gene pool only when the resulting mutant allele is rare in the gene pool, because mutation rates are too low to alter significantly large gene frequencies. The effects of selection are strongest on relatively common alleles in the gene pool. Selection acts very slowly upon rare alleles—particularly if they are recessive—either to eliminate an unfavorable allele or to increase the frequency of an advantageous one. Genetic drift is a significant factor only in a very small population, where it may account for the random loss of alleles from the gene pool. Gene flow is of significance in establishing equilibria among semi-isolated populations, but its effects upon a particular gene pool become significant only if the migrants make up a relatively large part of the population or if the gene pool of the migrants is very different from that of the population.

The combined effects of mutation, gene flow, selection, and genetic drift can significantly alter the gene pool of a population and can therefore account for evolutionary changes in species. Under most conditions, however, selection is apt to be the most important factor. Can these effects account for the evolutionary changes that are known to have occurred in nature? Is this model of the evolutionary mechanism adequate to account for the evolution of life on earth? These questions have not been answered to the satisfaction of all. Although many population geneticists now feel that selection is the major factor in evolution, some put more emphasis on random events. All agree, however, that known mechanisms of population genetics can account for the evolution of modern life in the time that has passed since the origin of the earth.

EVOLUTION THROUGH NATURAL SELECTION

The evolutionary theory of Darwin and Wallace, somewhat restated in the terminology of modern genetics, holds that adaptation of populations, or evolutionary change, occurs through accumulation of small genotypic changes. New genotypes are introduced through mutation. Genotypes that improve the fitness of the individual increase in frequency within the population as a result of selection. Because selection coefficients are different in different environments, any two populations prevented from interbreeding will gradually develop differing gene pools.

Selection, however, operates upon phenotypes and only indirectly upon genotypes. In some cases—particularly where the phenotypic character is strongly influenced by environmental as well as genetic factors—this distinction becomes quite important. Selection favors the survival and reproduction of certain phenotypes. To the extent that these phenotypes share similar or identical genotypes, selection alters the gene pool.

A number of examples of evolutionary change through selection have been studied. These observations show that the effects predicted by population geneticists do occur in natural populations. Study of these relatively simple examples of rapid evolutionary change has given evolutionists greater confidence in their theories about the long-term evolutionary changes in species.

DIRECTIONAL SELECTION: CHANGE TOWARD EXTREMES

Directional selection occurs when an extreme phenotype enjoys maximal fitness, and the gene frequencies are shifted toward high proportions of particular alleles. A classic example of natural directional selection is seen in a variety of species of moths and butterflies that live in industrial areas of Britain. As the terrain and vegetation have become darkened by the effects of smoke and other industrial pollution, dark-colored phenotypes have become predominant in many species of moths and butterflies where light-colored phenotypes formerly predominated.

One of the earliest and best-studied examples of this phenomenon is the moth species *Biston betularia*, known as the peppered moth. The typical phenotype of this species was light-colored with dark speckling. About 1845 a dark, or melanic, form of this species was first reported near Manchester, England. Because the melanic form was seldom seen by collectors, it is estimated that this phenotype made up less than 1 percent of the population at that time. By 1895 the population of *B. betularia* near Manchester was about 99 percent melanic individuals. In this population, the frequency of the melanic phenotype shifted from less than 1 percent to about 99 percent in only 50 generations. (This species breeds only once each year, forming annual, nonoverlapping generations.)

In most cases studied, melanism, or dark coloring, is due to the effect of a dominant allele at a single gene locus. In populations made up entirely of typical individuals (recessive homozygotes), melanic individuals do appear occasionally, apparently as a result of mutations. However, the mutation rate is far too low to account for the rapid increase in the proportion of melanic individuals in industrialized areas. Presumably, the melanic genotype is introduced into a typical population by mutation, but some form of selection must act in industrialized areas to rapidly increase the proportion of melanic phenotypes in the population.

If the melanism is due to the effect of a single, dominant allele, only individuals of the recessive homozygous genotype have the typical phenotype. Both heterozygotes and dominant homozygotes have the melanic phenotype. Presumably, the typical phenotype enjoyed a strong selective advantage under nonindustrialized conditions, so that the frequency of the

Figure 34.4 (left). Graph showing the changes in melanic phenotype frequencies expected with various selection coefficients.

Figure 34.5a (upper right). *Biston betularia*, the peppered moth, and its black form (*carbonaria*) at rest on lichened tree trunk in the countryside.

Figure 34.5b (lower right). The same two moths at rest on a soot-covered oak trunk near the city of Birmingham, England.

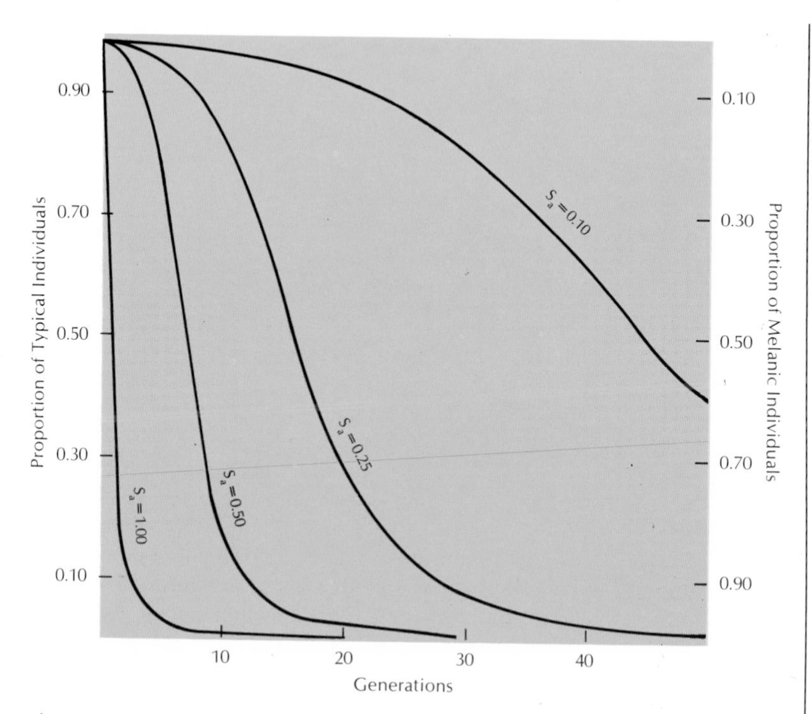

dominant allele in the gene pool remained at a very low level determined by an equilibrium between mutation and selection. When environmental conditions changed as a result of industrialization, the melanic phenotype gained the selective advantage. Figure 34.4 shows the changes in frequency of the phenotypes expected with various selection coefficients. A change from 1 percent to 99 percent melanic phenotypes in 50 generations could be accomplished with a fitness of about 0.70 for the typical phenotype and 1.00 for the melanic phenotype (Haldane, 1924). In fact, conditions did not change suddenly, so the fitness of the typical form was probably higher at first, gradually increasing to a value below 0.70 in the fully industrialized environment.

The selection in the moth population was explained by the hypothesis that predators notice and kill moths that are colored conspicuously differently from the surface on which they rest. In nonindustrialized areas, typical individuals are almost invisible on lichen-covered tree trunks, whereas melanic individuals are very conspicuous. Industrial pollution leads to a reduction of the amount of lichen on tree trunks and a deposit of dark soot on the surface. On these polluted trunks, the melanic form is inconspicuous and the typical form is highly visible (Figure 34.5). Thus, in nonindustrialized areas, predators that visually locate their prey destroy many more melanic moths than typical ones, whereas in industrialized areas, the selection coefficients due to predation are reversed. Confirmation of this hypothesis was found in the fact that trends toward melanism were noted only in species that spend daylight hours resting motionless on trunks or branches. No trend toward melanism was noted in species that spend most of the day flying or in species that are hidden from predators by their resemblance to green or dead leaves (Kettlewell, 1958).

When the phenomenon of industrial melanism among moths and butterflies was first noted about 1890, it was assumed to be some direct effect

Figure 34.6. Graph showing that a 0.50 change to a selection coefficient for the melanic phenotype will cause a 99 percent melanic population to return to 99 percent typical in about 20 generations.

of pollution upon individual organisms. Because the changes in the population were so rapid, Darwinian explanations were greeted with considerable skepticism. After the theories of population genetics were developed, this skepticism seemed to be justified. Selection coefficients in natural populations were believed to be about 0.01 or slightly larger. Few biologists were willing to accept the existence of a selection coefficient as high as 0.30, and a lower coefficient could not account for the rapid changes in the moth population (Figure 34.4). The most widely accepted explanation was a direct effect of the pollution upon the mutation rate. Some investigators claimed to have produced mutation rates as high as 0.08 by chemical treatment of moth adults, larvae, and eggs (Fisher, 1933; Harrison, 1928). Other investigators failed in attempts to confirm these observations.

In the 1950s, new direct evidence confirmed the existence of a high degree of selection. Melanic and typical moths, marked so that they could be identified upon recapture, were released in a nonindustrialized area where the natural population was almost entirely made up of typical individuals. The percentage of typical individuals recaptured was three times greater than the percentage of melanic individuals recaptured (Kettlewell, 1956). When a similar experiment was carried out in an industrialized area with a native population consisting of nearly 90 percent melanic individuals, the percentage of melanic individuals recaptured was three times greater than the percentage of typical individuals recaptured (Kettlewell, 1955). These experiments indicated a selection coefficient of about 0.5 for the melanic form in a nonindustrialized region and a selection coefficient of about 0.5 for the typical form in a nonindustrialized region. This evidence confirmed selection rates high enough to account for the rapid change in the frequencies of melanic phenotypes among the natural moth populations.

The hypothesized mechanism of selection also was confirmed experimentally by H. B. D. Kettlewell and Niko Tinbergen. Moths were released

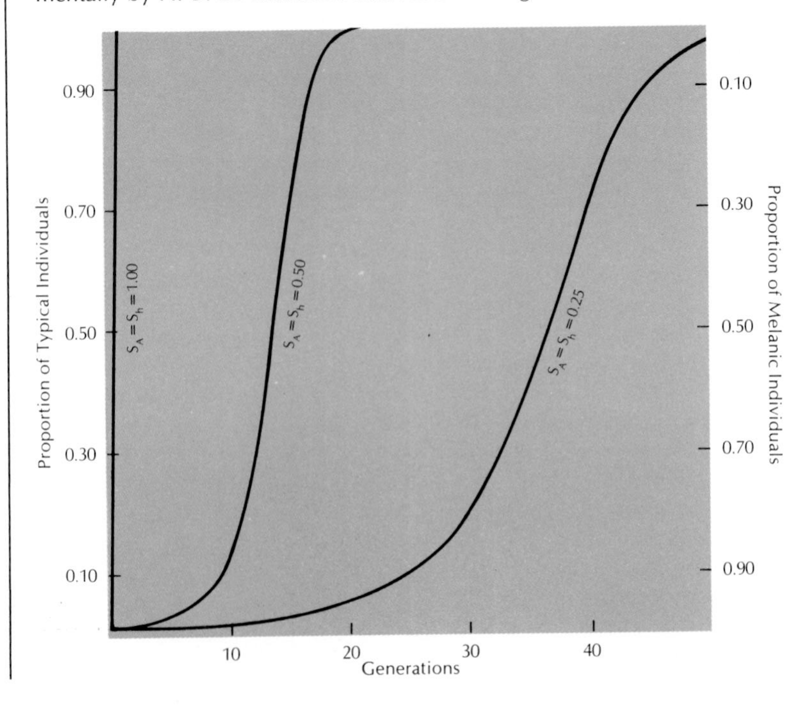

in various areas and predation was observed and photographed from a blind. In nonindustrialized areas, various birds ate about six times as many melanic moths as typical moths, although both kinds were released in equal numbers. In industrialized areas, birds ate about three times as many typical moths as melanic ones. In many cases, birds were observed to eat conspicuous moths while overlooking nearby individuals whose coloring matched the background surface.

In recent years, pollution in England has been sharply reduced by the use of smokeless fuels. As the countryside returns to its preindustrial condition, the proportions of typical moths are again increasing. Moth species adapt through changes in gene frequencies, changing the phenotype frequencies within relatively few generations so that most of the population is made up of individuals that are most fit for survival and reproduction under the prevailing conditions. If conditions change immediately to give a selection coefficient of 0.50 for the melanic phenotype, a population now of 99 percent melanic individuals will return to a condition of 99 percent typical individuals within little more than 20 generations (Figure 34.6).

ADAPTIVE RADIATION

Fossil records show that the first appearance of a particular major kind of organism in an area is usually followed by a relatively rapid process of speciation and divergence. That is, many distinct populations become established, and interbreeding among the populations decreases. The gene pools become sufficiently distinct that the various populations may be regarded as separate species. Each species becomes adapted to a different niche or ecological habitat in the region.

This process of *adaptive radiation* has been best studied among the birds, probably because of the vast numbers of persons who have gathered information about the distribution and characteristics of bird populations. One classic example of adaptive radiation was first observed by Charles Darwin among the finches of the Galápagos Islands. In fact, this observation was very important in helping Darwin develop his ideas about species and their origin. The Galápagos Archipelago consists of five relatively large islands and a number of nearby smaller islands (Figure 34.7). In general, the smaller islands are both farther from one another and farther from the center of the archipelago than are the larger islands. Whatever animals now live on the islands must be descended from individuals that crossed the 600 miles of ocean separating the islands from South America during the million or more years since the islands were formed by volcanoes pushing up from the sea bottom. Only two kinds of mammals, five kinds of reptiles, six kinds of songbirds, and five kinds of other land birds are found on the islands. Some species, such as the famed giant tortoises, are very different from species on the mainland and presumably have been evolving independently on the islands for a long time. Other species are identical or quite similar to mainland species and therefore are assumed to have reached the islands relatively recently.

On the islands, 13 different species of finches have been identified (Figure 34.9). They differ in size, in beak form, in feeding habits and other behavioral patterns, and in many minor characteristics. Although the species differ sufficiently that they have been grouped in six different genera, they are sufficiently similar in major characteristics that they are grouped as a single subfamily, Geospizinae, of the finch family (Fringillidae). The

Figure 34.7 (above). The Galápagos Archipelago.

Figure 34.8. Two inhabitants of the Galápagos Islands. The giant tortoise (middle) and the marine iguana (below) are very different from mainland species.

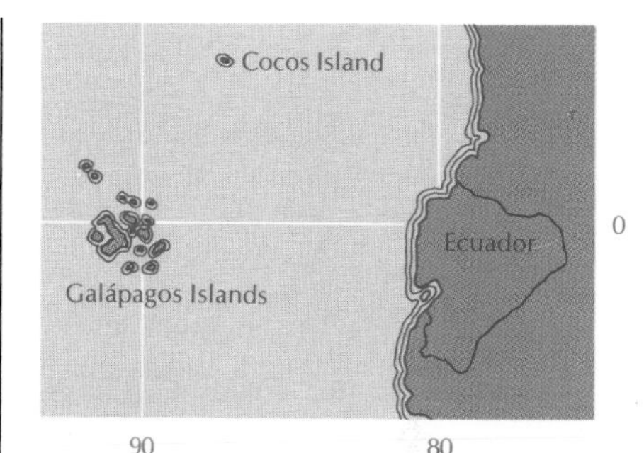

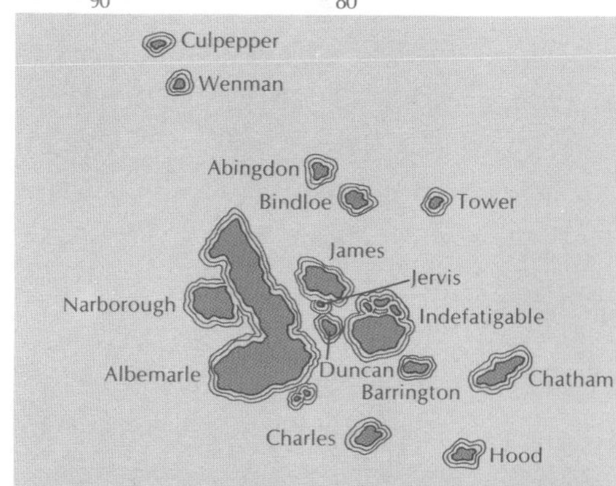

Figure 34.9. Darwin's finches.

Figure 34.10. Various species of birds with different beak structures. Like Darwin's finches, these Hawaiian honeycreepers have specialized in different foods depending upon their beak structure.

subfamily, often known as Darwin's finches, is sufficiently different from other finches of the world that only one species found elsewhere than the Galápagos Islands is included in the subfamily. The other species is found only on Cocos Island, about 400 miles northwest of the Galápagos.

Each of the larger islands has nine or ten species of finches living on it, except the easternmost island (Chatham), which has only seven species. Most of the smaller islands, particularly those far from the center of the archipelago, have only a few species of finches. The various species have specialized in the use of different kinds of food, and the differences in beak structure are closely related to the kind of food eaten (Figure 34.10).

Because the various species differ in their choice of habitats and in the food they seek, there is relatively little competition among them. There are six species of ground finches (*Geospiza*) that feed from seeds lying on the ground, and these species tend to live in dry coastal areas. Four of the ground-finch species are found together on most of the islands. Three of these live in similar habitats and feed from seeds, but they do not compete because their beaks are of different sizes and they tend to choose different sizes of seeds. The fourth species has a longer beak and feeds mainly on

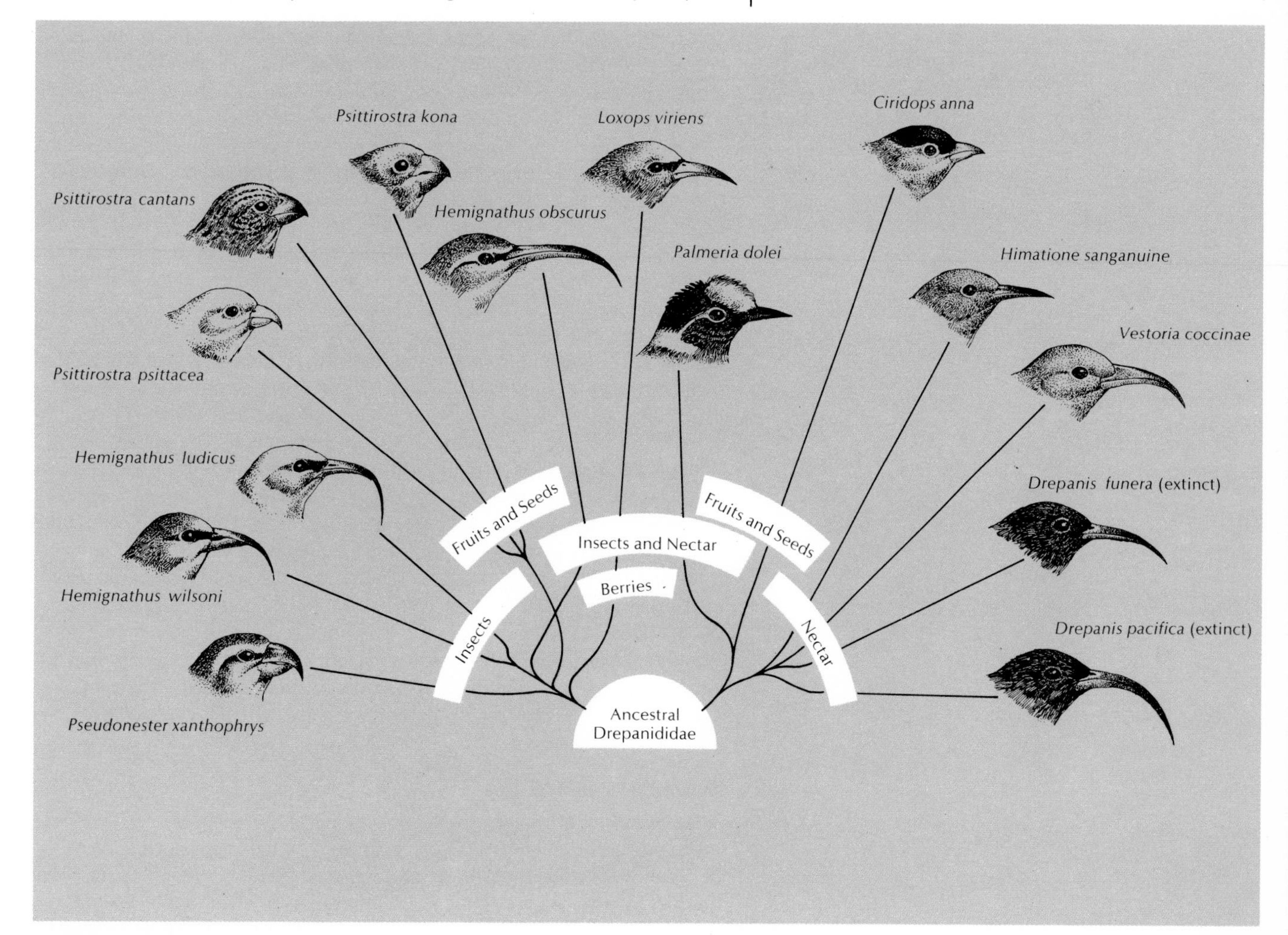

prickly pear cactus. The other two species of ground finch are found mostly on the outer islands, where the more common ground-finch species are rare or missing.

Six other species are tree finches, living mostly in the moister forests of the central part of the islands and feeding chiefly on insects. One remarkable species has a chisel-shape bill much like that of a woodpecker and feeds upon insects that it finds in the crannies of tree bark. Lacking the long tongue of a woodpecker, this finch carries with it a cactus spine or twig to pry insects out of cracks and holes—one of the few known examples of the use of a tool by an animal other than a primate. Another species specializes in capturing insects from mangrove swamps. Still another has a parrotlike bill and lives mainly on a diet of fruits and buds. The remaining three species of tree finches are similar, but they differ somewhat in size and bill shape and presumably specialize in the capture of different kinds of insects.

The remaining two species of finches are quite different from the ground and tree finches. One species (*Certhidea olivacea*) is similar to a warbler in appearance and habits. This species lives in both coastal and forest regions and feeds mainly on insects in bushes. The lone species of Cocos Island lives in the forest and feeds on insects, nectar, and fruits.

Modern evolutionists agree with Darwin's speculation that all the different species of finches on the Galápagos Islands have descended from a single, original species that reached the islands a long time ago (Lack, 1947). The original population probably spread out over all the islands and gradually developed some variation of characteristics in the slightly differing environments available. The relatively isolated populations on the small, outer islands probably tended to change their gene frequencies most drastically as a result of genetic drift in small populations and of their isolation from gene flow. Eventually, some of these isolated populations probably became sufficiently different from the interbreeding populations on the main islands that they were no longer capable of interbreeding with members of the main population. Thus, when members of the isolated populations happened to reach the main islands, they formed the nucleus of new populations there. Individuals in these new populations would be most likely to survive if they were able to utilize food sources or habitats overlooked by the more numerous main population. The effects of competition between populations on the main islands would force the new species to diverge still more in their characteristics, taking advantage of the large area and wide range of habitats and food sources on the large islands. Eventually, members of these new species would find their way to the outer islands, where further divergence due to isolation could occur.

Thus, adaptive radiation of the finches on the Galápagos Islands probably involved a cycle with new species arising through isolation on the outer islands, dispersing to inner islands, where further divergence would be caused through selection in competition with well-established species, and eventually dispersing back to the outer islands, where the cycle could begin again.

EVOLUTION OF SPECIES

Adaptive radiation provides one explanation of the origin of species. Other mechanisms have been proposed to account for the origin of certain groups of organisms or as mechanisms of general importance in evolution. Nonrandom mating behaviors, in which an organism tends to select a mate of similar appearance or behavior, may be as important as geographical isolation in establishing populations free from gene flow where evolutionary changes can lead to establishment of a new species. The establishment of new species of plants through polyploidy is relatively common. Through aberrations in the process of cell division, new individuals may be formed with multiples of the normal complement of chromosomes. These polyploids—which can be regarded as a form of mutations—reproduce themselves in future cell divisions, so that a new species of tetraploid individuals may result from a single abnormal division in a diploid species.

Similarly, it has been argued that major mutational changes in chromosomes—inversions of long sequences of genes, changes in chromosome number, or translocations in which sequences of genes are interchanged between chromosomes—may have been the source of major new groups of organisms among animals. Although most such mutations result in organisms of drastically unfavorable fitness, the extremely rare case in which the mutant was dramatically different but of superior fitness could lead to rapid adaptive radiation to form a new group of organisms competing in many different niches. Thus far, there is little evidence to support this hypothesis in the evolution of animal species.

In general, the theory that new species arise and change through evolutionary processes in which selection is a major factor is now almost universally accepted. Changes within a species through directional selection, the formation of new species through geographical isolation and adaptive radiation, and the formation of new plant species through sudden changes in chromosome number have been demonstrated convincingly in limited numbers of particular cases. Population genetics has provided quantitative models to account for evolutionary changes under very unrealistic simplified conditions. There is great reason to hope that extension of these simplified models to more complex and realistic conditions will explain most of the evolutionary changes that have occurred among life on earth.

On the other hand, many major problems in evolutionary theory remain to be solved. No fully satisfactory and detailed explanation of the conditions under which species arise and evolve has been developed. There is no generally accepted explanation for the apparent bursts of evolutionary change indicated by the fossil record of life on earth. The detailed evolutionary history of most modern species has yet to be worked out.

FURTHER READING

Among the many useful general discussions of evolution and natural selection are the books by Dobzhansky (1951), Ehrlich and Holm (1963), Grant (1963), Mayr (1942, 1963), Moody (1962), Simpson (1949, 1953), Stebbins (1950, 1966), Tax (1960), and Wallace and Srb (1970). More specific discussions of population genetics are given by Ford (1964), Hamilton (1967), and Li (1955) and in the *Cold Spring Harbor Symposia of Quantitative Biology* (1955, 1959).

Smith (1966) includes an excellent section on evolution and a good bibliography. Lack (1947) discusses the finches of the Galápagos in detail and the implications of this example for a general theory of evolution. Hardin (1959, 1969) and Lerner (1968) discuss many of these principles with special emphasis on their application to the human species.

Among many general articles dealing with evolution, useful introductions to particular aspects of the field will be found in those by Allison (1956), Camin and Ehrlich (1958), Dobzhansky (1950, 1959), Grant (1951), Mangelsdorf (1950, 1953), Sheppard (1961), Stebbins (1960), and Wright (1932).

35
Population Interactions and Communities

A survey of the organisms to be found at any one place at any particular time reveals a sample in time and space of the biotic community. The sample can be described in terms of numbers of individuals and species, distribution in time and space of the individuals and populations, and observed activities and interactions. The characteristics of the sample can be explained in terms of interactions among populations and among individuals within populations.

As an example, consider a sample of a tropical forest community, taken on one morning during the dry season. Some characteristics of the sample can be explained in terms of the present interactions among organisms. For example, the sample includes an unusually large number of solitary bees—3,482 individuals of 31 different species. The bees are found on the flowers of a large adult tree, *Andira inermis*, in the sample. Each morning during its month-long flowering period, this tree produces about 10,000 new flowers, each of which contains about 0.001 cubic centimeter of nectar. Like most other tropical trees, *Andira* grows during the rainy season and flowers during the dry season. The female bees gather nectar for food and to provision their nest cells; the male bees collect nectar for "fuel" to keep themselves alive while they search for female bees.

The characteristics of the sample can be explained further through a consideration of events in the recent past—within a few life spans of the longest-lived individuals in the community. The *Andira* tree has such a large crop of flowers because it has spread its canopy to the maximum possible size—shading out its competing neighbors rather than being shaded out by them. Its extensive growth was possible because during the last rainy season it used all of its energy reserves for growth of branches and leaves rather than flowering a second time. When the growing season began with the first rains, it was prepared with an extensive canopy to take maximum advantage of the photosynthetic energy available for further growth. Furthermore, while its neighbors may have been hampered in their growth by attacks of plant-eating insects, the *Andira* tree remained relatively free of such pests. Part of the reason for its freedom from insect pests is the great distance of this tree from other *Andira* trees. The caterpillars and leafhoppers that feed upon new *Andira* leaves did not travel the great distance from infected *Andira* trees to this lone individual.

This *Andira* tree happens to be far from other individuals of its species because long ago a bat carried an *Andira* fruit far from the parent tree before eating the fruit and dropping its seed. This tree has not become surrounded by its own offspring because most of every seed crop has been destroyed by the weevils that maintain a local population under each *Andira* tree.

There are so many bees because last year the *Andira* and its neighbors of other species flowered at this time, allowing the female solitary bees to build and provision the maximum possible number of nest cells. Furthermore, during the past year, many of the old flowering trees in the vicinity were cut down by a lumbering operation, greatly reducing the number of food sources for bees and thus increasing the number of bees feeding on each tree, including the *Andira*. Yet another reason for the large number of bees is that the previous dry season was unusually severe. It is likely that few of the parasites that normally infest the bees' nests survived the long, dry, and hot search between individual nests, and the mortality due to parasites among this year's crop of young bees was consequently quite low.

These examples are only a few of the interactions among organisms in

Figure 35.1. The *Andira* tree with details of its leaves.

the present and the recent past that have produced the characteristics of the community sample. The list could be extended almost indefinitely, involving interactions among all of the species present and the nonliving environment. Still another level of explanation can be reached by considering interactions stretching over the distant past—a perspective often termed an evolutionary time scale. In effect, these evolutionary explanations represent the summation over a long time of many interactions of the sort described above.

The behavior of the bees and the *Andira* tree in any interaction is greatly influenced by their genetic programming, the result of natural selection acting over many generations, influenced largely by the multitude of ecological interactions that confronted the ancestral populations of bees and *Andira* trees. The relationship between present-day phenotype and those past interactions can only be inferred, for no human observer was present to record what happened. However, an examination of the present and recent past events in the context of the entire community can lend support to these inferences.

The *Andira* tree flowers in the dry season because individuals of past *Andira* populations that flowered during the rainy (growth) season probably were shaded out by individuals that devoted all their resources to rapid growth of leaves and branches. These individuals did not produce as many seeds as did the individuals that flowered during the dry season, when no species was expanding its canopy. Furthermore, the trees that did not produce a large number of flowers per day were visited less frequently by bees. Their flowers were less likely to be pollinated at all, particularly by pollen

grains carried from other trees. Thus, the individuals that flowered out of phase with the other trees of the forest or that produced small numbers of flowers for any reason became genetically isolated from the general population and, because of their selective disadvantage, these genotypes were eventually eliminated.

The isolated *Andira* tree of this sample suffered little insect damage primarily because most leaf-eating insects in the tropical forest community are relatively host-specific. In their evolutionary past, the insects have become specialized to overcome the particular biochemical, morphological, and behavioral defenses of particular tree species. Insects that have become adapted to overcome the defenses of some other tree species are a minor threat to *Andira* trees. On the other hand, the *Andira* population has survived because it has been able to change its biochemical, morphological, and behavioral genotype sufficiently to keep one step ahead of the insects that are adapting specifically to attack it.

The ability of the *Andira* population to make these changes of genotype heavily depends upon the extensive genetic recombination occurring during cross-pollination. The asexual reproduction of self-pollination contributes little to the evolutionary change of the population gene pool (Chapter 34). In this light, the role of the bees, the large flower crop, and the time of flowering take on added significance.

Andira trees with large seeds enjoyed a selective advantage because the greater supply of nutrients enhanced the survival chances of the seedling. The larger fruits were more attractive to bats and thus more likely to be transported a sufficient distance from the parent tree to minimize competition between parent and offspring for light and nutrients and to minimize intertree exchange of pest populations.

The evolution of bee populations has been influenced strongly by interactions with the tree populations. Some bees responded to cues in the physical environment that were synchronized closely with the cues used by *Andira* and other trees to time flowering. These bees emerged from their underground nest cells a year after the construction of the cells and found an ample supply of food. Those bees responding to other cues emerged at times when flowers were scarce or unavailable and produced few offspring. Female bees that concentrated their foraging on the few large trees in full bloom were able to produce a large number of nest cells in the short time that these trees bloomed. Thus, these females produced more offspring than females that expended flying time and energy searching for less conspicuous and smaller food sources during this blooming season. On the other hand, bees with the searching behavior pattern were able to specialize in feeding on other species of plants with less conspicuous flowers and longer flowering periods and thus to avoid competition with the successful bees specializing in trees such as *Andira*. Female bees that nested in isolated patches of ground produced offspring with a lower rate of mortality due to nest parasites than did bees that nested in large aggregations.

The listing of population interactions over evolutionary time scales could also be expanded indefinitely. Such inferences can lead to an incredible set of fairy tales about the possible significance of this or that interaction between pairs of populations. The nearly infinite complexity produces a range of possible speculation that has led many biologists to throw up their hands in despair, abandoning hope of finding any general and reliable principles of evolutionary interaction and concentrating instead upon the

Figure 35.2. An artist's interpretation of the interactions between bee populations and *Andira* trees.

Figure 35.3. Bushes of *Leucaena glauca* on an island off Puerto Rico. The brown seed pods can be seen emerging above the vegetation.

mechanisms of interaction that can be observed and experimentally studied within short time spans. Such principles can be generated, but they must be based upon close observation and experimentation with existing systems. Pure description of interactions in natural systems – no matter how fascinating and complex (and therefore seemingly intellectually sophisticated) – is of little use unless it makes possible the development of some general statements or principles that can be used to predict the characteristics of new systems.

The general principles that will be discussed in this chapter have already been suggested by the brief examination of the interactions among *Andira inermis* and the other populations in its community. Many other general principles will undoubtedly be discovered as further research is done in the relatively new area of population and community ecology.

The most basic feature of any community is the struggle to obtain and to retain energy and the basic building blocks that allow an organism to obtain and to retain energy. The success of this manipulation of energy is ultimately evaluated by the success of the organism in having its genotype represented among the members of future generations. This success is not necessarily proportional to the energy harvested, retained, or expended. The expenditure of a small amount of energy to place a toxic compound in a small number of seeds and thus permit all of the seeds to survive without predator damage may be just as successful evolutionarily as an immense amount of energy expended on seed production (so that seed numbers are great enough to satiate seed predators and ensure survival of a few seeds). In a habitat where large amounts of energy are needed for other life processes, survival may require the development of a means of seed production that uses little energy. On the other hand, in a habitat where abundant energy is available for seed production, there may be some advantage in the production and dispersal of a large number of seeds even though few survive. The competition for energy underlies each of the general principles that will be explored through various examples in the following sections.

CONSPECIFIC SEPARATION IN SPACE AND TIME

The distance in time and space between individual organisms of the same species (*conspecific* individuals) is determined both by interactions among the individuals of this species and by interactions with other populations. Consideration of extreme examples will help to clarify the kinds of interactions involved.

Suppose that a tree seed is carried a long distance by water, wind, or bird and happens to land on an oceanic island unpopulated by its species. A consideration of its fate reveals the importance of interactions within the tree population of the mainland community. The isolated individual has left far behind the entire complex of host-specific predators and parasites that caused high mortality among each generation of new members in the tree population. Evolutionary pressures have resulted in a genotype that produces a large number of seeds in each crop in order to ensure survival of a few seedlings. On the island, many thousands of seeds in a viable condition may be dispersed by the single tree.

For example, *Leucaena glauca*, a small and shrubby legume of the Central American mainland, made it to the island of Puerto Rico but left behind at least three species of pea weevils, or bruchid beetles, that destroy more than 90 percent of each seed crop on the mainland (Janzen, 1969a).

Although this plant is relatively rare on the mainland, it is extremely common in Puerto Rico, partly because each island plant disperses as many as 100 times more viable seeds than does each mainland plant.

The plant that immigrates to an island also leaves behind an array of species that carry pollen from one plant to another or that help to disperse its seeds. Most of these species subsist in part or in whole on the flowers and fruits of the mainland population. The single island plant must produce its first generation of offspring through self-pollination. If this species has evolved mechanisms to reduce self-pollination, the production of viable seeds may be so small that the island population is unable to get started, even in the absence of host-specific predators. Even as the population density of reproductive adults begins to increase, the absence of species that serve as pollinators on the mainland may lead to a heavy reliance on self-pollination. In addition to the lack of genetic variability, because all individuals are descended from the single migrant individual, the limitation of sexual recombination can lead to a low rate of adaptation in the island population for the first several generations at least.

The lack of species that disperse fruits or seeds on the mainland also may hamper the spread of the island tree population. Most fruit-eating vertebrates utilize a wide variety of different kinds of fruits, but such animals are rare on most islands, probably because they cannot survive without fruit-bearing trees. But fruit-bearing trees may not become well established without the animals to disperse their seeds. Thus, the nearly simultaneous arrival of immigrants from both populations may be necessary for the establishment of a stable community.

In the first few generations of the island tree population, each individual will have little competition from other members of the same species. Because of the absence of dispersal agents, a seed is more likely to germinate beneath its parent and near its siblings than is the case in mainland communities (Carlquist, 1966). A similar proximity of conspecific individuals is found in populations that rely upon wind or other inanimate agents for seed dispersal, as is common among the tree species of temperate forests.

At the other extreme, close proximity of conspecific adults has important consequences of a quite different sort for the interactions within the population. For example, wind pollination is a highly effective out-crossing mechanism over the short intertree distances in coniferous forests but is very ineffective in tropical forests, where conspecific individuals are widely spaced (Whitehead, 1969). In the spring season of temperate zones, wind pollination can occur much earlier and more reliably than can pollination that relies upon insects. Conspecific individuals are so closely spaced that the tree of the temperate zone can bear male flowers at a different time than female flowers. The tree thus avoids self-pollination, but there is a high probability that some nearby tree will be producing flowers of the opposite sex at the appropriate time (Sharp and Sprague, 1967).

On the other hand, proximity of conspecific adults can be disadvantageous to the population because it maximizes the ease of movement from plant to plant of host-specific seed predators and parasites. In the case of wind-carried diseases, even wide spacing between conspecific individuals provides little protection. The evolution of mechanisms for resisting attack is the only hope for survival of the host species. On the other hand, natural selection will tend to eliminate parasite species that are too successful in destroying their host populations, so that evolutionary trends in both the

Figure 35.4. An artist's interpretation of the web of dependency as described in the text.

Figure 35.5. A swarm of migratory cicadas, commonly known as locusts, attacking corn.

parasite and host populations tend toward a balance in which a number of host individuals are able to survive parasite attack. In the case of diseases carried by insects—such as the Dutch elm disease, which is caused by a fungus carried from tree to tree by a bark beetle—moderate spacing of conspecific adults may slow or stop the spread of infection sufficiently that the death of diseased adults will be matched by production of new adults. The population density of adults will decrease after the introduction of such a disease until this moderate spacing is attained or a mechanism of resistance to infection is evolved. The latter situation has been observed in a number of crop plants (Painter, 1967).

Separation of conspecific individuals in time or in space may have an advantage in reducing the damage caused by less mobile or less lethal predators and parasites. For example, an acorn buried by a squirrel escapes in space from an immense population of insects (dominated by acorn weevils of the genus *Curculio*). This escape is accomplished close to the parent tree and by a movement of only a few inches downward. Predation by squirrels is the price that the adult tree pays for protection of new nuts buried before insects can find them.

In the coniferous forests of the Pacific northwest, escape from predators and parasites is accomplished by separation in time (Smith, 1968). Many trees of the temperate zone share the trait of producing seeds only during years of a "mast crop." Mast crops are separated by several years of nearly sterile vegetative growth. Populations of insects and squirrels that feed upon the seeds decline during the "off years," despite the high density of adult trees in the area. When the mast crop is produced, the small populations of predators are soon saturated with food and many seeds escape to germinate. In the Pacific northwest, as many as six species of conifers in the community may synchronize the timing of their mast crops. This synchronization allows a much greater proximity in space of conspecific adult trees for a given amount of seed predation, because the trees have escaped the predators by spacing their reproductive activities in time.

A high density of conspecific individuals in space may be accomplished despite high predation if there is sufficient separation in time. An extreme example is provided by the periodical cicadas, or "locusts," which spend a period of 13 to 17 years underground as larvae. After this long larval period, the entire population of highly predator-susceptible adults emerges simultaneously. This sudden appearance of large numbers of insects is not anticipated by the predator populations. The stomachs of all the local vertebrate predators are soon stuffed, and most of the female cicadas oviposit successfully. The few cicadas that emerge in years before and after the general emergence stand little chance of survival because they lack the usual predator-avoidance behavior patterns of cicada species that emerge annually. Thus, natural selection in this population tends to maintain the extreme synchrony of emergence (Lloyd and Dybas, 1966).

If conspecific adults are closely packed in both space and time, they usually have very effective morphological defenses (such as the hard shells of oysters and barnacles) or chemical defenses against predators and parasites. Among the few exceptions to this generalization are populations on islands that have not developed or cannot sustain predator populations (MacArthur and Wilson, 1967).

One of the best examples of chemical defense in dense populations is seen in the communities of mangrove swamps, which are found all over the

Figure 35.6 (left). Gum exudate from the trunk of a large legume tree in a Costa Rican lowland tropical forest. The gum has been produced as a response to a boring moth larva that feeds on the cambium of the tree.

Figure 35.7. The deciduous Central American tree *Hymenaea courbaril* (upper right) and a close-up of one of its flowers (lower right). The flower is bat pollinated but may also be moth pollinated.

world with one to five tree species in each community. Although most of the species of mangrove trees have evolved from different families, they are all characterized by high tannin content in their bark and foliage. Their wood is so insect resistant that it is eagerly sought for fence posts, and these trees supplied the raw material of the tannin industry for many years. When an insect eats the foliage, the high concentration of tannins makes it almost impossible for the insect to utilize any of the proteins in the leaf (Feeny, 1968; Feeny and Bostock, 1968). This same effect apparently protects the mangrove seeds from insect damage, and the trees produce great numbers of seeds nearly all year around.

In a similar fashion, conifers are protected from insect attack by systems of oleo-resin ducts, which not only produce chemicals that discourage insect feeding but also drown or mechanically push out bark beetles that attempt to bore into the living tree (Stark, 1965). When, for some reason, such as drought stress or senescence, the oleo-resin pressure of the tree decreases, the increased success of insect attack often leads to a population explosion among the tree's parasites. The large number of insects attacking healthy trees may then cause the failure of the oleo-resin defense system in much of the tree population.

Cases of intermediate proximity of conspecific adults are found in the tropics, where virtually any combination of the defense mechanisms mentioned in this section may be found within a single forest. A single plant species may employ mast years, chemical defenses, and a relatively low conspecific density. One such plant, *Hymenaea courbaril*, is common in the deciduous forests of Central America. It is most famous for its produc-

tion of the resin that, when fossilized, forms much of the New World amber (Langenheim, 1969). The adult plants, usually spaced at distances of 50 to many hundreds of meters, produce fruit only at intervals of three to five years. Because the individual trees are not synchronized in fruit production, the fruit crop is only one-third to one-fifth as dense for the seed-eating weevil (*Rhinochenus stigma*) as might be thought from a simple count of the adult trees. The weevil must move between the relatively widely spaced trees, and it will find relatively few of those trees to be suitable hosts for the next year. In northern Central America and in Puerto Rico, where this weevil is not present, trees of this species bear fruit every year (Janzen, 1970).

Among vertebrates, a high conspecific density is commonly accompanied by some form of intraspecific territorial behavior. As the probability increases that one member of the species will find the resource that another member is seeking, there is greater value in expending time and energy to expel that other member from the area. This kind of interaction has been demonstrated in the case of blackbirds nesting on marshes, where the size of nesting territories is directly related to the productivity of the marsh plants (Orians and Horn, 1969). Similarly, meadow mice (voles) show a dramatic increase in the intensity of territorial defense as the density of the mouse population rises. In fact, the mice may become so aggressive that they never mate (Krebs, *et al.*, 1969).

The spacing of conspecific individuals is influenced by many kinds of interactions, both within the population and with other populations in the community. Whereas a high population density provides advantages for maximal use of resources and ease of mating, other factors such as intraspecific competition and predation tend to give an advantage to populations of low density. The balance that is achieved in any particular case can be explained only by a thorough examination of the interactions experienced by the population on both the evolutionary and the ecological time scales.

INTERSPECIFIC SEPARATION IN SPACE AND TIME

The separation in space and time between organisms of different species is affected by all of the interspecific interactions within the community. The population densities of other species greatly influence the distance that an organism will maintain from its neighbors of other species.

Between a simple community made up of a few populations (for example, the community of an arctic, alpine, desert, or seashore splash-zone region) and a complex community of a lowland wet tropical forest, there is an increase in the number of close relationships between pairs of species—mutualisms, parasitisms, predations, and so on. For each species in the complex community, population density is likely to depend heavily upon the densities of a number of other populations, as well as upon the weather and other parameters of the physical environment. Some of these interdependencies have been illustrated in preceding sections, but a few more examples will help to underscore their importance in the community structure.

Numerous cases of mutualism between plants and pollinators or dispersal agents exist in most communities (Faegri and van der Pijl, 1965; van der Pijl, 1968). The frequency of such mutualisms increases as the daily and seasonal fluctuations of the physical environment become more regular and as the availability of sunlight and water becomes greater. Two species can become dependent upon one another for various resources only if both species have been present in the habitat at the necessary times

Figure 35.8 (above). Bat pollinating a flower. The bat's wings are folded and partly wrapped around the inflorescence. In this way, pollen is transferred to the bat's body and subsequently may be deposited on another inflorescence.

Figure 35.9 (below). Male orchid bees pollinating an orchid. The relationship of these bees to the orchids has evolved into a form of mutualism that is obligatory for both parties.

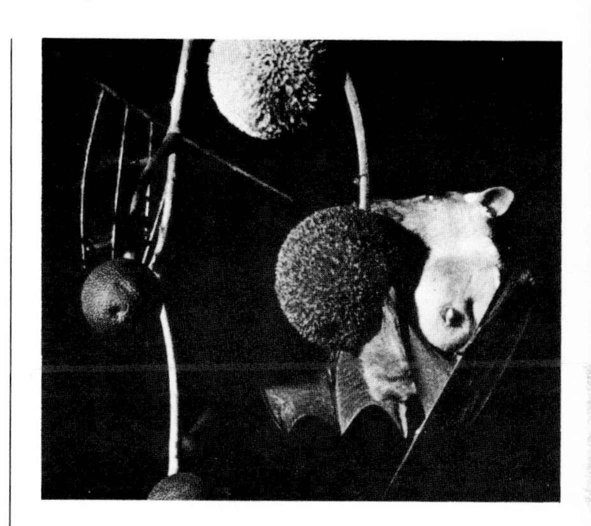

in the necessary numbers and in the necessary age classes over the evolutionary time scale. Along a north-south line from the Arctic to Panama, the number of species of mammals per unit area increases regularly (Simpson, 1964). However, the great increase in the number of mammal species from southern Mexico to Panama is accounted for primarily by an increase in bat species. If bats are excluded from the sample, the increase in species density stops in southern Mexico, and the diversity of mammals is approximately constant from there south to the Equator. Furthermore, the number of insect-eating bat species is about the same from California to Panama; the increase is in bat species that feed on fruits, flowers, fish, birds, small mammals, and other bats. The greatest increase is among the bats that visit fruits and flowers. There are many bat-pollinated flowers (such as *Hymenaea courbaril*) and bat-dispersed fruits (such as *Andira inermis*) that begin to appear in plant communities in lowland southern Mexico, and their numbers rise steadily toward the Equator (Baker and Harris, 1957). The key to this relationship lies in the fact that fruits and flowers are relatively nutrient-poor food sources, particularly for reproduction and for the storage of winter or dry-season fat reserves. A bat specialized on fruits and flowers must have them available throughout the year.

Several closer mutualisms are immediately apparent to any researcher working on tropical community structure. Botanists investigating the extreme diversity of tropical orchids have described a number of the intricate interspecific relationships that support this diversity (Dodson, *et al.*, 1969). One of the most complex mutualisms is that between male "orchid bees" (*Euglossa, Euplusia, Eulaema*) and orchids such as *Catasetum, Gongora,* and others that are pollinated only by these bees. The male bees appear to require highly volatile chemicals that are produced by the flowers. The bees collect the compounds from modified parts of the plants with small brushes on the front feet and then transfer the material to special grooves in the hind legs, where the substance is absorbed into the bee's body. At times, the bee gets enough of these chemicals to alter his flight behavior dramatically. In the process of obtaining the compound, the bee is guided by a series of complex floral structures to a position where the pollinia (a pair of sacs containing pollen) are glued to some predictable part of his body. Upon entering a female flower at a later date, the bee is again guided by peculiar flower morphology to a position where the pollinia are removed by the stigma of the female flower. The bees cannot survive without the chemicals produced by the orchids, and the orchids can be pollinated only by bees with these specific behavior patterns. The coevolutionary steps

Figure 35.10 (left). Sequence illustrating a bee pollinating a flower. Note how the structure of the flower accommodates the bee.

Figure 35.11 (right). The solitary bee (*Anthophoridae*) in the act of pollination. Note the abundance of pollen on the hairs of the body and rear legs.

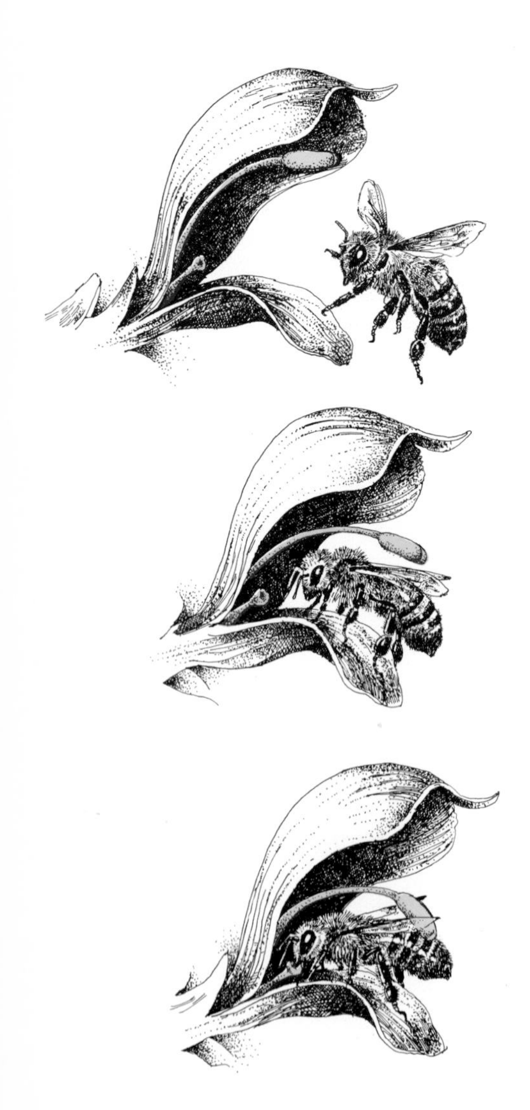

leading to the development of this obligatory mutualism are a subject of speculation. Further coevolution of the two species produced the very specialized behaviors and morphologies observed in the modern populations. Both the orchid bees and the orchids they pollinate are absent from most Caribbean islands—an emphatic example of the difficulty of simultaneously establishing two populations that are heavily dependent upon each other. Because the male bees cannot survive without the orchids and the bee population cannot survive without males, most of the large number of plant species pollinated by females of these bee species on the Central American mainland are also missing from the islands.

Another close mutualism is displayed by dry- and moist-habitat shrubs of the genus *Acacia*, which appear in early stages of succession in the plant communities of Central America. Ants of the genus *Pseudomyrmex* live in the swollen thorns of the plant, gain their sugar from nectaries on the leaves, feed their larvae with modified leaflet tips that are rich in proteins and steroids, and have a nearly continuous food supply because these species of *Acacia* remain green during the dry season (in contrast to other *Acacia* species not associated with ants). The ants, in turn, drive away plant-eating insects and prune back vines and shrubbery that might crowd out the *Acacia*. This activity is of immediate benefit to the ants because it keeps the *Acacia* strong and healthy and ensures a more continuous and abundant food supply. The larger the ant colony, the more effective the continuous protection that it provides for the plant; thus, both ants and *Acacia* can maximize their growth through this close mutualism. The coevolution of this interaction has continued to a point that the leaves and shoot tips of these *Acacia* species lack the insect-repelling compounds found in most plant leaves and therefore cannot survive without the ants.

In drier or cooler habitats, tropical ant-plant interactions disappear from the community. In areas with long dry seasons, the *Acacia* cannot hold its leaves throughout the year, and the ant colony cannot survive through the dry season to protect the plant during following rainy seasons. In cooler regions, the ants spend more and more time inactive inside the thorns and less and less time protecting the leaves of the host. In both cases, the interspecific interaction disappears from the community (leading to a simpler

Figure 35.12. Ants of the genus *Pseudomyrmex* shown moving about on the host *Acacia* plant. The ants derive their nourishment from the plant and in turn protect it from plant-eating insects and encroachment from close growing vines and shrubbery.

community) because environmental conditions made the interaction impossible, although the conditions were not directly harmful to either participating population (Janzen, 1966, 1967, 1969b, 1969c).

On the other hand, the ant-*Acacia* interaction has enabled this mutualistic pair to invade much wetter (and more complex) communities than have other *Acacia* species. The interaction enables the ants to avoid almost all direct competition with other ant species. The mutualism, which is possible only in the tropical habitat, allows the "stacking" of more species into the community than is possible in less favorable environments where ants and plants must be more generalized in their life styles for survival.

In lowland Mexico at the Tropic of Cancer, the northern boundary of the range of the ant-*Acacia* mutualism, both seedling *Acacia* plants and *Pseudomyrmex* queens capable of founding new colonies are found in the community. However, they are present only because of continual immigration from adjacent habitats where the ant colonies and plants can reach reproductive maturity. The environmental conditions are not quite suitable for the maintenance of the mutualism. This example emphasizes the importance of the presence of individuals of the proper age class in the habitat if a mutualism is to persist.

Predator-prey interactions have effects opposite those of mutualisms. For maximal survival, the prey must maximize the distance in time and space between himself and the predators, whereas the predator must minimize this distance. Each time that a predator is successful in capturing a prey individual, he slightly increases the average distance between predator and prey and thus improves the survival chances of the remaining prey. As a result, predator-prey interactions are extremely "density-dependent." The success (or failure) of the predator is dependent upon the density of the prey population and vice versa. The percentage of the remaining prey population that will be captured by predators declines as the predators are successful in reducing the prey population. Thus, it becomes increasingly profitable for the predator to switch over to hunting some other prey population, a situation that may even cause effective immunity from predation until a prey population can build itself back up again. A homeostatic balance tends to exist in which both predator and prey populations fluctuate around an equilibrium density.

The community, then, can be viewed as a collection of populations, some of which are declining and some of which are rising at any given instant of time. The rates of rise and decline depend upon such factors as the efficiency of the predators in using energy and materials from the prey to reproduce, the reproduction rate of predators and prey, the ability of the predator to shift from one prey species to another, the costs to the predator in energy and materials of maintaining generalized metabolism and behavior patterns that enable it to utilize several prey species, the number of secondary predators feeding on the primary predators, the amount of resources available to the prey, and so on.

One of the most important effects of predator-prey interactions is the reduction of competition between prey species that share a common predator. For example, the starfish *Piaster* is a major predator on sessile molluscs and barnacles of the intertidal zone. If the starfish is excluded from the community, one or two of the sessile species soon crowd or starve out the other sessile species because of their competitive advantage in feeding and reproduction. However, if the starfish is allowed access to the simplified community, it removes many individuals in these successful sessile populations, leaving space for immigration of individuals of several other species. In other words, the addition of a single predator species leads to a large increase in the total number of prey species (Paine, 1966).

The same principle applies to forest trees (Janzen, 1970). As in the intertidal zone, members of the prey species compete for space, which represents the opportunity to gather energy (sunlight) and nutrients (inorganic ions from the soil). The predator species include a large array of generalist and host-specific insects, fungi, and vertebrates that eat seeds and seedlings. Beginning with predation on the developing embryo in the green seed pod, the predators cause a steady attrition of the plant population size in each generation up through dispersal and germination of the seed. In the absence of such predators, one tree species would tend to dominate any particular habitat. The predator's ability to depress the density of any one tree species to a point where it does not competitively displace the other tree species is primarily a function of how effectively the predators on juvenile plants can move across space and time between successive crops of seeds and seedlings. In habitats where an occasional late spring frost kills most members of a major seed predator population, there is a large population of tree seedlings that year. Which species of tree will be represented among the new adult trees produced from these seedlings in later years is determined by the outcome of competition among the seedlings of various species. Even if it is a relatively poor seedling competitor in a predator-free environment, the species that has the most effective defenses against predation is most likely eventually to dominate the community.

Along the gradient from temperate zone habitats to wet lowland tropical habitats, the probability declines that any one species of tree will escape predators either through sudden environmental fluctuations or through weather conditions that are generally harmful to the predator. The result is a situation that has been recognized for many years: tropical lowland wet forests are characterized by a large number of species of trees, each present in a population density much lower than those of the relatively few tree species found in temperate zone communities (Ashton, 1969).

The same phenomenon of density-dependent mortality stated previously in general terms applies to the populations of trees and their predators. As predation on juveniles of a particular tree species increases, the probability of an embryo surviving to form an adult tree decreases. As the population of this tree species declines, the predators have to move farther between host seed crops, fewer predators reproduce, and fewer juvenile plants are destroyed. As fewer juvenile plants are destroyed, the population of the tree species begins to rise and the entire cycle repeats itself. Thus, the density of adult trees of each species tends to approach some equilibrium value, and the actual number of trees—if observed over a long period of time—fluctuates around this value. Any factor that decreases the efficiency of the predators will lead to an increase in the equilibrium density of tree populations, but the number of species and/or the density of other species in the habitat must decline (because there is only a finite amount of space in the habitat for the tree crowns).

PROXIMATE AND ULTIMATE FACTORS

For any particular trait of an individual, a population, or a community, two kinds of explanatory causes may be distinguished: immediate physiological stimuli (the "proximate factors") and long-range selective pressures (the "ultimate factors").

When the physical environment undergoes regular and major fluctuations, many species in the habitat show periodic changes in behavior correlated with those environmental cycles. For example, in some relatively dry tropical habitats, many tree species flower shortly after the end of the rainy season. This regular time of flowering is of great importance to the large complex of bee species that emerge at this time to harvest pollen and nectar from these trees. The fruits that are produced from these flowers during the dry season may be of special importance to many mammals and birds because fewer leaves and insects are available as food than during the rainy season.

In such a case, it is possible to search for the direct physiological stimuli that cause each tree to become reproductive and to explore the physiological mechanisms by which these stimuli lead to changes in the physiology and morphology of the trees. Such an investigation leads to one kind of explanation for the flowering of many trees at a particular time of the year, an explanation in terms of proximate factors.

In general, it is more profitable to explore the benefits that accrue to a tree flowering at this time. The assumption can then be made that this characteristic has provided a selective advantage over the evolutionary time scale and that selection has developed some physiological mechanism that enables the tree to flower regularly at the appropriate time. From this point of view, the important causes of the synchronized flowering are the ultimate factors that provide a selective advantage for trees flowering at this time. It is not surprising that different species have evolved different mechanisms responding to different environmental cues for advantageous timing.

A number of selective pressures may lead to the flowering of many trees during the early part of the dry season (Janzen, 1967). First, and probably most important, the tree that uses all available energy for vegetative growth during the growing season is most likely to maintain an optimal location in the general vegetative canopy. It can then store any excess energy obtained throughout the growing season and use these reserves for reproduction after dry-season leaf-fall when, for the most part, no vegetative competition is occurring. The reserves can always be used for vegetative growth if an emergency does arise during the growing season. Second, a tree that flowers out of phase with the remainder of the population will have to rely on generalist pollinators (because the specialized pollinators of the species are not present or engaging in pollinating behaviors) and on self-pollination. So long as the tree is not self-pollinating, it will not contribute to the gene pool of succeeding generations, and the tendency to flower out of phase will be eliminated from the population. This effect is increased as more and more of the trees in the population become better and better synchronized, because the bees coevolve a peak in emergence to match the cycle of the plants. Third, flowering during the rainy season may lead to rain damage of the flowers and pollen and makes it more difficult for bees to arrive at the plant at a regular time of day to find newly opened flowers. Fourth, by maturing its fruits before the end of the dry season, the tree maximizes the probability that new seeds will be on the ground at the beginning of the next rainy season.

In short, any new genotype that is able to utilize an environmental change as a cue and to flower at the beginning of the rainy season is likely

Figure 35.13 (above). Leaves and fruit of tropical forest *Ficus* tree.

Figure 35.14 (middle). Gall wasps feeding on the pods of *Asclepias linaria*. A similar interaction occurs between certain highly host-specific gall wasps and fig populations. The fig wasps pollinate the trees and in turn are dependent on them as a microhabitat and as a source of food.

Figure 35.15 (below). Latex from a superficial wound of *Ficus*. Many fig species are noted for producing copious amounts of milky latex that contains toxic compounds. These compounds are probably an effective defense against insects.

to have a selective advantage. As such genotypes become more common in the population and as bee populations coevolve matching cycles of emergence, the selective advantage of flowering at this time increases.

In some cases, the tree may not respond to the environmental cue until it has stored up enough energy to produce a full crop of fruits and flowers. This is probably the case with the *Hymenaea courbaril* trees, which set fruit only every three to five years.

As is often the case, however, there are organisms in the community that have evolved radically different strategies to handle very similar problems. For example, in the tropical forests there are some tree populations such as figs (*Ficus*) in which some individual trees are setting fruit at any time of the year. The small seeds are dispersed by bats and birds and are dropped along watercourses, where germination can occur throughout the year (removing one of the selective advantages of seasonal flowering). Because their roots are in wet soil along the river and they remain evergreen in most habitats, figs are in vegetative competition with trees nearly all year (removing another selective advantage of seasonal flowering). Because the figs are pollinated by tiny, highly host-specific gall wasps and are not damaged by rainfall, the other selective advantages of seasonal flowering have little effect on the fig population. In addition, there is a strong selective advantage in this population for continuous fruiting. Because the wasps can develop only within the figs of a particular host species and die very soon after emerging from the fig, there must be another fig tree with fruits ready for pollination nearby at the time the fig wasps emerge, or the population of fig wasps will decline and disappear. Because the fig trees are pollinated only by the wasps, the fig population will also disappear if the distance in time and space between fruiting trees becomes very large.

In view of the small distances between conspecific individuals in the fig populations, it is not surprising that *Ficus* species are noted for copious quantities of milky latex that contains toxic compounds, probably an effective defense against many insects. The direct physiological stimuli that cause a particular tree to begin fruiting at a particular time are only a minor part of the data needed to understand the *Ficus* population as a part of the interactions of the community.

ENERGY COMPETITION WITHIN THE ORGANISM

Each organism must carry out at least three major functions with the finite amount of energy and materials that it can obtain from the habitat per unit time: maintenance, reproduction, and defense. The allotment among these three activities is a function both of the physical environment and of the traits of the other organisms in the habitat.

It is apparent from the study of physiology and biochemistry that there are many opportunities for competition between different cells, tissues, and organ systems for the finite amount of energy and materials taken in by an individual plant or animal. A particular environmental challenge may strongly influence the balance among these competing systems and therefore influence the allotment of resources among the functions of maintenance, reproduction, and defense. Often, the environmental challenge takes the form of an interaction with some other species in the habitat.

For example, the "secondary substances" produced by plants have long been regarded as waste products. These substances include terpenes, alkaloids, free amino acids, cyanogenic glycosides, resins, and a variety of other chemicals, but for the most part the particular molecules found among the secondary substances of a given species are unique to that species and closely related species. In fact, the secondary substances often provide a "chemical fingerprint" that permits unambiguous species identification. It is no coincidence, however, that these substances are also generally or specifically toxic to an immense array of insects, fungi, and vertebrates, many of which are potential predators or parasites of the plant. A growing body of evidence indicates that the particular substances present in a plant species (or in a particular part of each plant) is a coevolved result of the presence of particular animal species in the community and the resources available to the plant. For example, plants of the bean family (Leguminosae) – whose root nodules bear nitrogen-fixing bacteria – use primarily nitrogen-rich defensive compounds (alkaloids and free amino acids, such as canavanine and L-Dopa). Depending upon the extent of predator and parasite threat, the plant may devote more or less of its metabolic resources to the production of defensive compounds. Normally, the plant produces compounds that require the least expenditure of energy and materials. It tolerates a certain amount of damage in order to minimize the resources devoted to defense and to maximize the resources available for other activities.

In effect, the *Pseudomyrmex* ants of the swollen-thorn *Acacia* plants are a multipurpose defensive device for the plants. The plant pays for its defense with the resources needed to produce various adaptations that support the ants.

ENERGY AND ITS IMPORTANCE IN COMMUNITY STRUCTURE

Because the flow of energy through the community is obviously a major component of the community structure, it has been traditional to think of those organisms that store or process large amounts of energy as the most significant members of the community. The deceptiveness of this assumption can be appreciated easily by noting the effects of the disappearance of some inconspicuous insect populations.

When the herbaceous weed St. John's wort (*Hypericum*) was accidentally introduced into pastures of the West Coast, it quickly spread and became one of the most common members of the plant community. A small and inconspicuous leaf beetle (*Chrysolina geminata*) was introduced from Australia and quickly caused near extinction of the plant. The beetle browses on the leaf rosettes at the base of the plant until the plant is so weakened that it loses out in competition for light and space with other plant species. In the sunny parts of pastures, where the beetle grows rapidly, the plant is now locally extinct. It survives, however, in the shade of fencerows, where the beetle does poorly. Both beetle and plant are now very rare in the pasture community. Even at the peak of its population growth after introduction, the direct contribution of the beetle to the total energy flow of the community was trivial (Huffaker, 1959).

A very small daily drain on the resources of an organism may have a very large cumulative effect. A caterpillar eating off 1 centimeter from the tip of a growing tree branch removes only an insignificant fraction of the tree's stored energy. However, the loss of the growing apex means the loss of a number of mature leaves that would have appeared later and would have carried out a great deal of photosynthetic activity for the plant. Not only does the delayed leaf production mean loss of energy gathered, but it

Figure 35.16. Moth caterpillars feeding on a leaf. Many such caterpillars feed on the new foliage of their host plants but cannot feed on mature leaves that may contain higher levels of secondary toxic substances. Thus, certain plants may regulate both the quality and quantity of the toll exacted by their insect enemies.

may also mean loss in height or status within the canopy. By the time a new axillary shoot has elongated to replace the original branch end, the branches of neighboring plants may have grown considerably higher.

Similarly, a few ounces of succulent plant matter can make the difference between survival and death for a deer in the late winter, when fat reserves are low and the threat of predation or starvation is high. Those few ounces, however, are a very trivial part of the deer's annual intake.

In short, the temporal and spatial distribution of resources can be as important as their nutritive value. A few calories invested in production of nectar may lead to heavy and accurate cross-pollination by bees over long distances; the same amount of energy put into a few more seeds might have a much smaller impact on later representation of the parental genotype in later tree populations. Natural selection provides the only test in which the value of various energy allocations are judged by their success in reproduction.

PREDATORS DO NOT EAT LATIN BINOMIALS

In thinking about community structure, it is easy to fall into the trap of regarding species as natural units made up of individuals with uniform morphological and behavioral characteristics. The name of an organism is no more than a convenient pigeon-holing device; predators do not select their prey on the basis of its Latin binomial label. The members of a species come in many ages, sexes, behaviors, sizes, and so on.

If a species of lygaeid bug is observed to feed on many kinds of seeds, dead insects, fecal matter, and leaf parts, it is tempting to regard this species as polyphagous—able to eat many kinds of food. Such a classification, however, tends to obscure the fact that the female bug can obtain the nutrients and other materials necessary for egg production only from one or two of the seed species. The other foods are adequate only for fuel to keep the female going until she locates the appropriate seeds and can reproduce (Sweet, 1960).

Many caterpillars feed freely on new foliage of their host plants, but they cannot feed on matured leaves, which contain various toxic secondary substances. Thus, the insect may be sitting in the middle of a pure stand of its "host plant" and yet be starving to death. Tropical trees produce new leaves in periodic bursts of activity, separated by long periods when no new leaves are produced, thus separating leaf crops in time and helping to keep the caterpillar population low. Selection strongly favors such genotypes in relatively warm climates of moderate rainfall, where plant-eating insects would otherwise thrive.

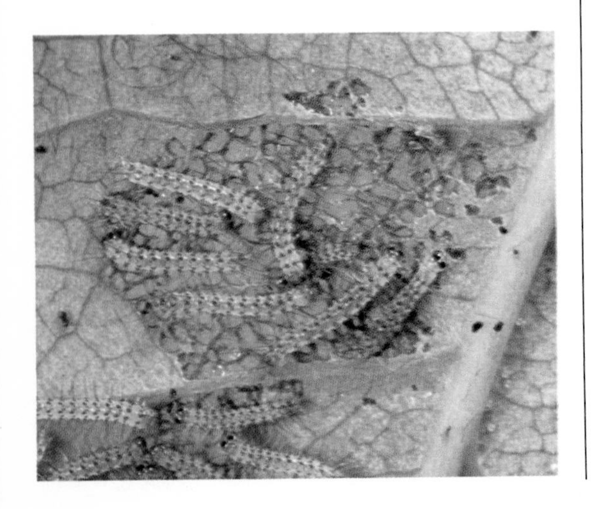

It is often noted that the bigger the animal, the longer the list of plant species on which it can feed. There are two components of this observation that are of interest in the study of community structure. First, these plants generally can be arrayed in a rough order of preference, which is probably determined by such things as content of toxic substances, ease of finding, speed of digestion, energy content, work to process, and exposure to predators in finding and feeding. For larger animals, many of these factors are not very important. Second, the mere presence of all of these plants in the habitat does not ensure survival and reproduction. For example, it is commonly noted that big game animals eat small amounts of many species of plants. It is very tempting to speculate that if the animal were forced to feed on only a single plant species, it might die of poisoning from the secondary substances of the leaves. In short, in addition to passing its food through the processing of the stomach acids and of the bacterial culture in the intestines, the large browsing mammal also dilutes the secondary substances from one plant species with those of many others. Thus, it avoids ingesting a harmful amount of any one toxic substance. In a system such as this one, the animal could starve if half of its potential food species disappeared from the habitat, even though each of those remaining became twice as common.

Although an acorn and its parent oak tree are both individuals of the species *Quercus alba*, the removal of one by a predator has a very different effect on the community from the removal of the other. Similarly, if ants of the species *Solenopsis geminata* are consistently found in the gut of a bird, it makes an immense difference to the eventual structure of the community whether they are queen, male, or worker ants.

Generally, what an organism is doing or would have done is more important than its species name or classification in the significance of an interaction within the community.

THE STUDY OF COMMUNITIES

Although a great deal has been learned about population interactions in many biotic communities, only the first steps have been made toward a general understanding of the basic principles of community structure. The basic concepts are easily lost within the maze of complex and fascinating interactions that are revealed by any thorough ecological study of a natural community.

Increasing attention is being focused upon all aspects of ecology because of the growing realization that the large and technological human population is drastically altering the interactions and equilibria of the ecosystem. Much of the current interest in ecology is directed toward problems of immediate economic and political significance: oil spills in the oceans, phosphates in sewage, mercury in streams, and so on. In the heat of debate on such matters, it is often difficult to separate impartial scientific observation and prediction from deeply held opinions about values and goals for society. Whether the current public interest in ecology proves to be another "fad" or not, the scientific field of ecology will continue to grow and change rapidly as new information and techniques from other sciences are brought to bear on the study of ecological interactions. Clearly, the findings of ecologists will have great significance for almost every realm of human planning and endeavor.

FURTHER READING

Among many books dealing with community structures and population interactions, the following are useful introductions with helpful bibliographies: Cox (1969), Dasmann (1968), Elton (1966), Hutchinson (1965), Klopfer (1969), MacArthur and Connell (1966), Odum (1971), Smith (1966), Tubbs (1969), and Whittaker (1970).

Articles of interest in relation to this chapter include those by Batra and Batra (1967), Cole (1958), Connell and Orias (1964), Ehrlich and Raven (1967), Gilliard (1963), Grant (1951), Klopfer and MacArthur (1961), Limbaugh (1961), MacArthur (1965), Petrunkevitch (1952), and Went (1955).

36
Unicellular Organisms

Until the seventeenth century, it was easy to regard the world of life as a realm divided into two kingdoms: plants and animals. However, the world of microorganisms discovered by Anton van Leeuwenhoek proved to be a complex one, and biologists eventually were forced to recognize that attempts to classify all unicellular organisms as animals or plants are not very useful (Chapter 2). The importance and variety of microorganisms were recognized only gradually. To the modern biologist, microorganisms are of importance for at least three reasons. First, some microorganisms play essential roles in the cycling of elements and the flow of energy through the biosphere. Second, single-cell organisms provide an opportunity for observation of life processes in relatively simple and easily isolated forms. Third, there are many reasons to believe that multicellular organisms—plants, fungi, and animals—evolved from unicellular ancestors.

Today, it is clear that even the smallest microorganism is a relatively complex molecular system (Chapters 7 and 8). Even a bacterial cell only a few microns in diameter is capable of carrying out a rich variety of metabolic reactions, as well as the fundamental genetic processes of DNA replication, RNA transcription, and protein synthesis.

Most microorganisms are classified in the kingdoms Monera and Protista (Chapter 2). However, a few organisms of microscopic size are multicellular. These microorganisms, along with some single-cell organisms that are very similar in structure to the cells of certain multicellular organisms, are classified in the kingdom Plantae, Fungi, or Animalia.

MONERA: BACTERIA

The procaryotic cells of the Monera are the smallest and simplest systems known that unquestionably can be classified as living organisms. However, even the tiny cell of a mycoplasma is a complex molecular mechanism, capable of carrying out a wide range of metabolic reactions and passing through a life cycle of several stages. Biologists are not yet agreed on the proper classification of mycoplasmas, but the kingdom Monera is generally agreed to include the bacteria and the blue-green algae.

A bacterium possesses a cell wall composed of a complex polymer of amino acids, amino sugars, and different amounts of carbohydrates, lipids, and other macromolecules. In some species, this rigid cell wall is surrounded by a capsule of polysaccharides or polypeptides. Within the cell wall is a plasma membrane similar in structure to the membranes of eucaryotic cells. Ribosomes are present in large numbers, but mitochondria and plastids are absent. The bacterial chromosome is twisted through the cytoplasm or clumped into nuclear areas, but there is no nuclear membrane. In those species that possess flagella, the flagella appear as simple fibers in electron micrographs, lacking the (9 + 2) arrangement of filaments characteristic of flagella in eucaryotic cells. In the photosynthetic bacteria, pigments are arranged in minute, ovoid bodies called *chromatophores*, which show no sign of the parallel layers typical in the chloroplasts of eucaryotic cells.

As a bacterium grows, it increases in size and duplicates each part of its structure, including the chromosome. It then undergoes binary fission, splitting itself in half to form two new daughter cells. In the presence of adequate nutrients and under suitable environmental conditions, this process can be very rapid. This simple and brief life cycle ensures that a bacterium encountering favorable conditions can rapidly produce a thriving,

Figure 36.1. The nitrogen cycle. The rate of cycling as well as the relative amounts of each compound are controlled by various populations of microorganisms. Nitrogen-fixing microorganisms include bacteria, fungi, and blue-green algae.

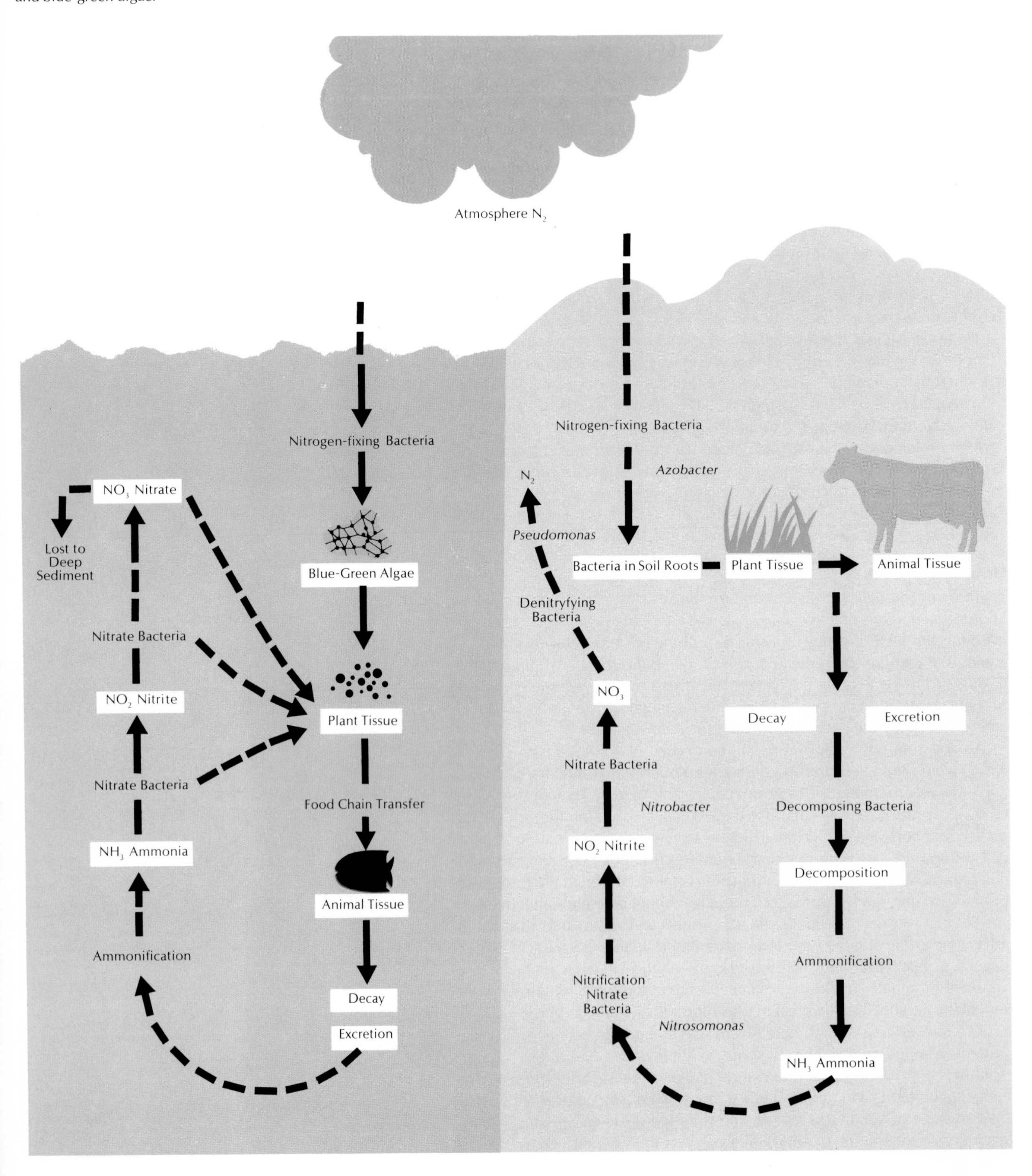

Figure 36.2. Certain nitrogen-fixing bacteria live in a symbiotic state within the root systems of leguminous plants. The bacterial colonies form root nodules from which to direct their chemical processes. The concept of crop rotation was predicated on alternating leguminous crops, which replenished soil nitrogen, with crops that heavily utilized nitrogen. *(Courtesy Carolina Biological Supply Company)*

sizable population of bacteria. When the supply of nutrients is exhausted or the conditions become unfavorable, if only a few bacteria can survive to reach a new source of nutrients, the population will soon grow again.

It appears that many bacteria are able to suspend much of their metabolic activity and to survive for some time in the absence of a supply of nutrients. Other species form a wall of polypeptides around the nucleus and certain other parts of the cytoplasm to create a resistant and nonmetabolic structure called a *spore*. Such a spore becomes a normal bacterium when it encounters suitable conditions, even after a long period of desiccation or a short period of high temperatures.

The ability to survive unfavorable conditions for a considerable period of time and then to reproduce rapidly when conditions become suitable has apparently been a major factor in the success of bacteria. In addition, the flagellated bacteria are able to move toward more favorable environments. For example, the flagellated, photosynthetic bacteria tend to move away from dark areas and toward light.

The autotrophic bacteria require only inorganic nutrients, using chemosynthesis or photosynthesis to produce their own organic molecules from carbon dioxide and water. Most heterotrophic bacteria are able to survive on a relatively simple diet. Many species can be grown in the laboratory with a nutrient medium containing only amino acids and inorganic substances. Some species (including *Escherichia coli*) can either utilize amino acids, producing ammonium salts as waste products, or utilize sugars and ammonium salts to synthesize their own amino acids.

Many species of bacteria play important roles in the cycling of nitrogen through the biosphere (Figure 36.1). *Nitrosomonas*, a chemosynthetic bacterium living in the soil, can oxidize ammonia to nitrite, thus obtaining the energy to synthesize organic molecules from CO_2. *Nitrobacter* obtains its chemosynthetic energy by oxidizing nitrite to nitrate. Several species of soil bacteria, including *Pseudomonas*, reduce nitrate to gaseous nitrogen. Still other species, such as *Azotobacter*, can utilize gaseous nitrogen in producing amino acids. In various combinations, such soil bacteria play important roles in freeing nitrogen from organic macromolecules or obtaining it from the atmosphere and making it available in suitable form to be used by plants in constructing new macromolecules. Because they possess such a wide range of nutritional capabilities, the bacteria play very important roles in cycling many elements rapidly.

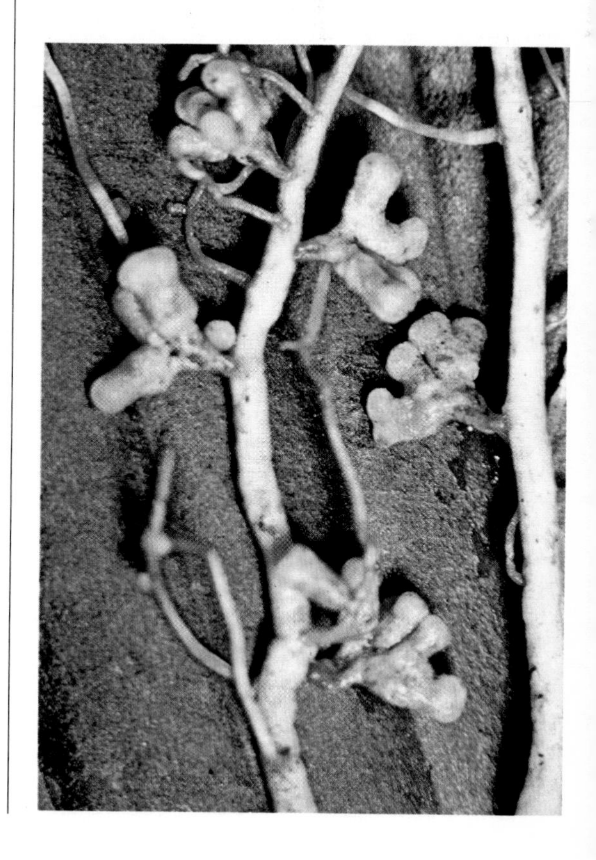

The rate at which mutations appear in bacteria is similar to that in other organisms. A mutation of a particular gene may be expected to occur about once in each 100,000 to 1 billion individuals. However, because bacteria can reproduce so rapidly, the chances are good that a favorable mutation may occur within a population threatened by environmental change. For example, if a population is being destroyed by the drug penicillin, it is likely that a few individuals possessing a mutation that makes them resistant to penicillin could arise before the population is completely destroyed. If this resistance occurs, the mutants will have little competition for available nutrients and will reproduce rapidly to produce a large population of penicillin-resistant bacteria.

Most bacteria, particularly the actinomycetes, secrete substances that inhibit the growth of other organisms. Many antibiotics, used to destroy disease-causing bacteria in plants and animals and to prevent bacterial decay of food, have been obtained by purification of these substances.

Figure 36.3a (above). The common blue-green alga *Anabaena*, noted for its "chain of beads" aggregation.

Figure 36.3b (middle). Large masses of blue-green algae often form scumlike masses on the surface of polluted ponds and lakes.

Figure 36.4 (below). A blue-green alga of the genus *Gloeotrichia*. Note the clusters of basal cells toward the center of the photograph. *(Walter Dawn)*

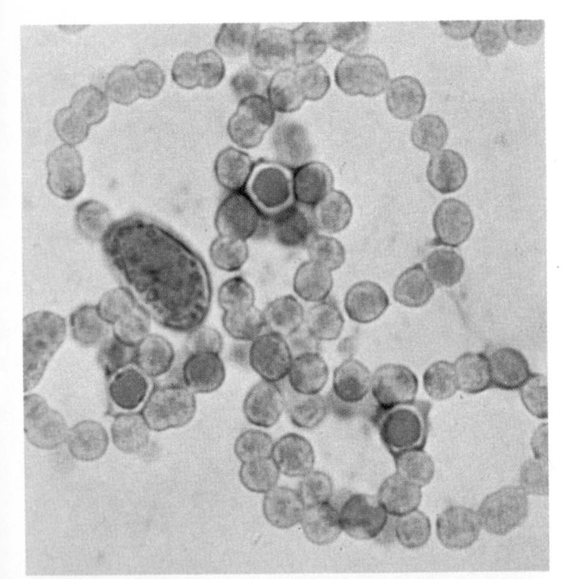

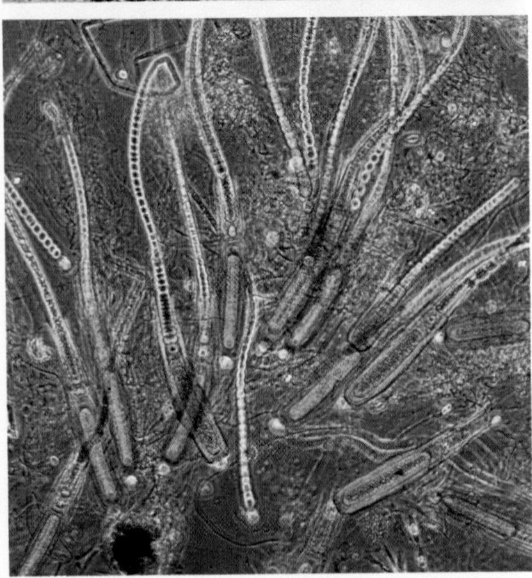

Actinomycin – an antibiotic obtained from the actinomycete *Actinomyces* – acts by interfering with RNA synthesis. Bacitracin – obtained from the bacterium *Bacillus* – interferes with the process of cell wall formation in certain kinds of bacteria, thus preventing cell division.

Antibiotics have been of great importance in medicine, and hundreds of them have been obtained from various species of bacteria and fungi. In nature, these substances apparently serve to protect the bacterium from competition of other species in the search for nutrients.

MONERA: BLUE-GREEN ALGAE

Like the bacteria, the blue-green algae are procaryotic, lacking mitochondria, plastids, centrioles, (9+2) flagella or cilia, and nuclear membranes. The cell walls are composed of hemicellulose, cellulose, and pectin. In most species, the plasma membrane is complexly folded to form parallel membrane layers within the cytoplasm, on which the photosynthetic pigments and probably other enzymes involved in metabolic reactions are located.

All blue-green algae are photosynthetic autotrophs. Some utilize gaseous nitrogen in forming amino acids, a property shared only with the nitrogen-fixing species of bacteria. The blue-green algae, or cyanophytes, are very widely distributed. At least a few species are apt to be present in any location where there is moisture and light. Some species live in snow, whereas others are found in hot springs at temperatures up to at least 75°C.

All cyanophytes contain chlorophyll *a* as the major photosynthetic pigment; an accessory pigment, phycocyanin, gives the cells a blue tinge. The cyanophytes differ in pigments both from the photosynthetic bacteria – which have only bacteriochlorophyll – and from the eucaryotic algae – which have at least one other chlorophyll in addition to chlorophyll *a*.

The cyanophytes are divided into two classes: the Chroococcales and the Hormogonales. The species of Chroococcales form spherical cells that may exist independently or may be held together in colonies by the layers of gelatinous material surrounding the cells. Most colonial species form spherical aggregates of cells, but some form flat plates or hollow spheres. There is no evidence of cell differentiation in the Chroococcales, and it appears that colonies are formed because daughter cells are held together by the gelatinous material around the parent cell.

The Hormogonales have elongated cells that cling together to form long filaments, also apparently held together by the gelatinous sheath. In some species, all of the cells are similar, and the filament elongates as the cells anywhere along the chain divide and grow. In others the filament has a large basal cell, with cells diminishing in size toward the other end of the filament. In such species, most cell division occurs near the middle of the filament. In the genus *Gloeotrichia*, symmetrical clusters of filaments are formed with the basal cells joined together in the center (Figure 36.4). Many species appear to have branched filaments, but close examination shows that the apparent branches are actually formed by independent chains joined together.

In many of the Hormogonales species, large cells called *heterocysts* are found at intervals along the filaments. These cells lack photosynthetic pigments, but staining shows that they contain large amounts of DNA. A small pore appears to penetrate the cell wall at one or both ends of the heterocyst.

In some cases, heterocysts separated from the filament have been observed to form pigments and then begin to divide to form a new filament. It has been suggested that the heterocysts represent a simple form of a reproductive cell, but the normal function of these specialized cells has not yet been definitely determined.

In some unicellular species, a rapid series of divisions occurs within a single cell, producing a large number of daughter cells called *endospores* within the single parental cell wall. The endospores eventually are released from the parental cell wall, form their own cell walls, and become adult individuals. Under unfavorable conditions, some species form thick, resistant cell walls and appear to become nonmetabolic spores. The spores of some Hormogonales species germinate to form a short thread of spherical cells. The thread eventually develops into a normal filament. This thread is capable of movement, but the mechanism is unknown.

Origin of the Monera

The bacteria and the blue-green algae are similar in many ways. Both contain a cytoplasm that lacks vacuoles, appears more rigid than the cytoplasm of eucaryotic cells, and is far more resistant to damage by heating, desiccation, and chemicals than is the cytoplasm of eucaryotes. Both have cell walls containing the macromolecules known as mucopeptides. For these reasons, it has been suggested by some scientists that both the ancient bacteria and the ancient blue-green algae evolved from an unknown procaryotic ancestor.

On the other hand, the two groups differ in many important ways. The cyanophytes are not heterotrophic and do not possess flagella. The photosynthetic pigments of the two groups are very different. The cyanophytes use water as an electron donor in photosynthesis and release free oxygen, whereas the photosynthetic bacteria use molecules such as hydrogen sulfide as electron donors and do not release free oxygen. For these reasons, some biologists believe that the two groups evolved independently from different procaryotic ancestors.

The fossil record has not helped scientists to decide between these two hypotheses. Fossils that appear to be remains of bacteria and cyanophytes have been found in South Africa in rocks formed about 3.1 billion years ago. These fossils are the oldest known remains of organisms and seem to indicate that the Monera were among the earliest organisms present on the earth. Thus, there is still some doubt about the evolution of the Monera from simpler ancestors or from nonliving systems.

PROTISTA: ALGAE (PROTOPHYTES)

The kingdom Protista includes some algae (plantlike organisms) and the protozoans (animallike organisms). All of the Protista are eucaryotic organisms. Most are unicellular, but some species form multicellular colonies with different degrees of cell differentiation. The distinction between plantlike and animallike forms does not seem to be important in this kingdom. In some species, a single cell is capable of obtaining organic molecules through photosynthesis, absorption, or ingestion—depending upon environmental conditions.

The phylum Euglenophyta contains both autotrophic and heterotrophic species. The euglenophytes do not have cell walls, and most species are flagellated unicells. *Euglena* is a photosynthetic autotroph that lives in damp mud or in fresh water rich in organic substances. The outer portion of the cytoplasm is rigid, holding the cell in its elongated shape (Figure 36.5). Electron microscopy shows that the rigidity is caused by a bandlike structure that winds helically around the cell. The nucleus, mitochondria, and plastids are found in the inner, more fluid cytoplasm. Small plastids serve as storage for food reserves. Large chloroplasts containing chlorophylls *a* and *b* and accessory pigments are scattered through the cell. At the anterior, or front end, of the cell is a small indentation, the gullet. Vacuoles that form in the cytoplasm migrate to the gullet and apparently discharge their contents to the exterior. The flagellum arises from a basal body in the floor of the gullet. *Euglena* is pulled forward through the water by whipping motions of the flagellum. A small granular body, the *stigma* (located near the gullet), contains carotenoid pigments and is thought to play a role in detecting light and enabling the cell to swim toward brighter areas.

Astasia is another euglenophyte, very similar in structure to *Euglena* but lacking chloroplasts. It lives as a heterotroph, ingesting small particles of organic matter through the gullet. The ancestors of the Euglenophyta probably were photosynthetic, and the heterotrophic species have probably evolved through loss of chloroplasts.

The phylum Chrysophyta includes two classes: the Chrysophyceae (golden algae) and the Bacillariophyceae (diatoms). The golden algae are mostly flagellate unicells, forming an important part of the population of small organisms in the oceans and in fresh water. They have cell walls of hemicellulose and silica. The chloroplasts contain chlorophyll *a* and accessory pigments including carotenoids that give the cell a golden color. Food is stored in plastids in the form of leucosin (a polysaccharide) and fats. The coccolithophores are a group of golden algae that form tiny plates of calcium carbonate (coccoliths) in their cell walls. These elaborately sculptured plates make up a large part of the thick chalk layers found in Cretaceous rocks, formed about 125 million years ago. Like the Euglenophyta, the golden algae reproduce chiefly by mitotic division, although sexual recombination has been observed in rare instances.

The diatoms form intricate cell walls of hemicellulose impregnated with silica. These walls have beautiful markings or sculptures on the outer surface (Figure 36.7). The diatoms are abundant in fresh and marine waters, and their cell walls make up thick deposits of "diatomaceous earth," particularly in rocks formed during the Tertiary period, about 10 to 70 million years ago. The cytoplasm of the diatom cell forms a thin layer around a large central vacuole. Some species are capable of small, jerky movements, but the mechanism of motion is unknown—for most species lack flagella.

The diatoms reproduce through cell division, with each daughter cell retaining half of the old cell wall and forming a new complementary half. The new half is formed to fit inside the old half, and thus the cells tend to become smaller with each generation. Eventually, a small cell sheds the cell wall, existing as an *auxospore* during a growth period and then forming a new cell wall. In some species, gametes are produced through meiotic division, and the fusion of two haploid gametes creates a diploid auxospore. Some species have been observed to form thick-wall spores under unfavorable conditions. The chloroplasts of diatoms contain chlorophylls *a* and *c*, carotenoids, and other accessory pigments. Food is stored in plastids in the form of leucosin and fat. The diatoms are grouped with the golden algae

Figure 36.5 (upper left). The flagellated unicell *Euglena*. The euglenophytes do not have cell walls and may be functional autotrophs and heterotrophs, depending on environmental factors. (*Eric Gravé*)

Figure 36.6 (upper middle). *Vorticella*, a sessile ciliate protistan creates a current of water with its cilia and directs food particles into its gullet. The recoil action of its contractile stalk is illustrated. (*Courtesy Carolina Biological Supply Company*)

Figure 36.7 (upper right). Diatoms from intricate cell walls (shells) of cellulose are impregnated with silica. Thus, they literally live in a "glass house." A sample of the varied shapes of these shells is illustrated.

Figure 36.8 (below). Close-up photographs of cell walls or shells of diatoms.

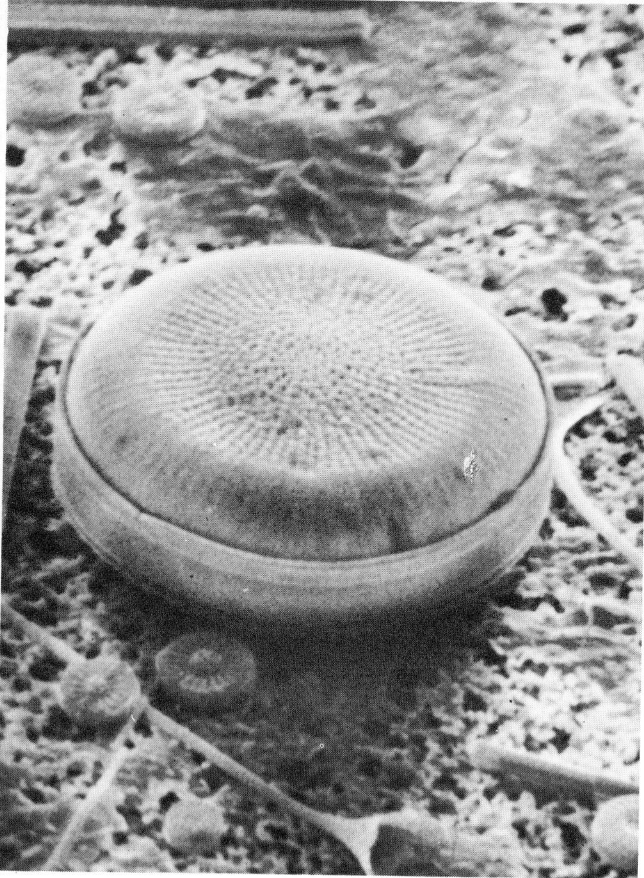

Figure 36.9 (right). A slime-moldlike form of yellow-green alga, representing the group *Xanthophyta*.

Figure 36.10 (above). Zoospores.

Figure 36.11 (below). Germinating aplanospores.

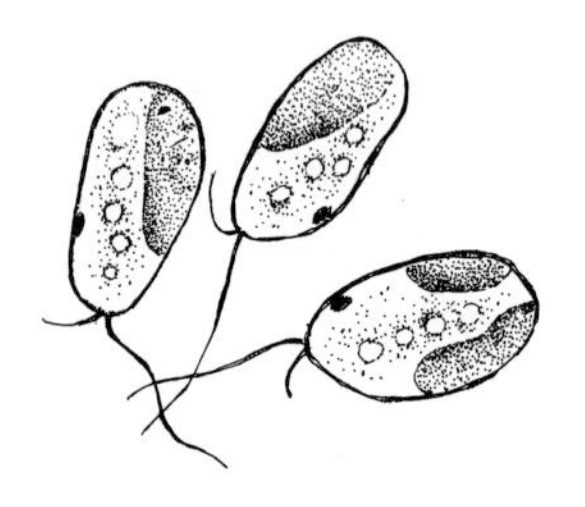

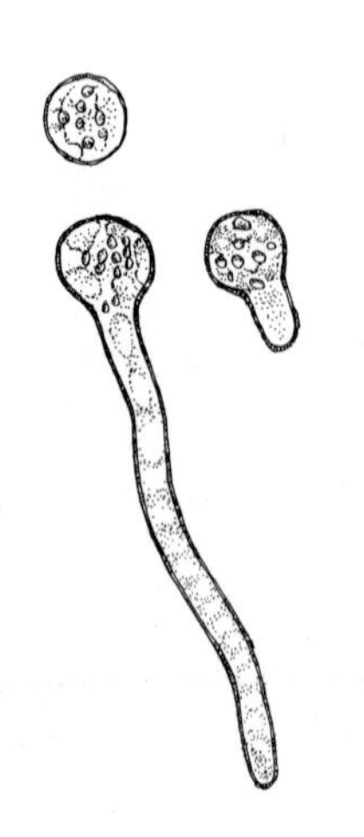

because of biochemical similarities and because both groups produce two-part cell walls of hemicellulose and silica.

Many taxonomists group the Xanthophyta, or yellow-green algae, in the phylum Chrysophyta. Like the chrysophytes, the xanthophytes form cell walls of hemicellulose and silica; in most species, the cell wall is composed of two overlapping segments. The chloroplasts contain chlorophylls *a* and *e*, carotenoids, and possibly other accessory pigments. Food is stored in plastids in the form of leucosin and fats.

Although there are several different species of xanthophytes, they are relatively rare and are found mainly in fresh water. Some species are flagellated, some form small colonies held together by an outer layer of mucilage, some form nonflagellated unicells, and a few form short filament-shape colonies. One filamentous species produces both flagellated *zoospores* and thick-wall *aplanospores*. A few species of xanthophytes form multinucleate organisms with rootlike extensions, or *rhizoids*, that attach to the substrate. In these species, a portion of the cell may become separated by a membrane, and its nuclei develop into zoospores or aplanospores that are shed into the water.

The phylum Pyrrophyta includes three classes—the Desmophyceae, the Dinophyceae, and the Cryptophyceae. The pyrrophytes possess chlorophylls *a* and *c* and various accessory pigments. They store food reserves in the form of starch or fats and produce cell walls of cellulose and hemicellulose. Some of the Desmophyceae are bioluminescent. The Dinophyceae, or dinoflagellates, have two grooves on the cell wall, one running along the body and the other around it. One flagellum lies within each of these grooves (Figure 36.12). A few species of this class lack flagella, and a few species form filamentous colonies. The nonflagellated and filamentous species produce flagellated zoospores, whereas the flagellated unicells reproduce by simple cell division. Sexual reproduction through fusion of haploid gametes may exist in a few species. The Cryptophyceae include a

Figure 36.12 (left). A fossil dinoflagellate. Many of these organisms are noted for their bioluminescence. Most dinoflagellates have two flagella for locomotion. A few are noted for tremendous population growths known as "blooms." One such bloom is called the red tide. (*Eric Gravé*)

Figure 36.13 (right). Zooplankton and phytoplankton.

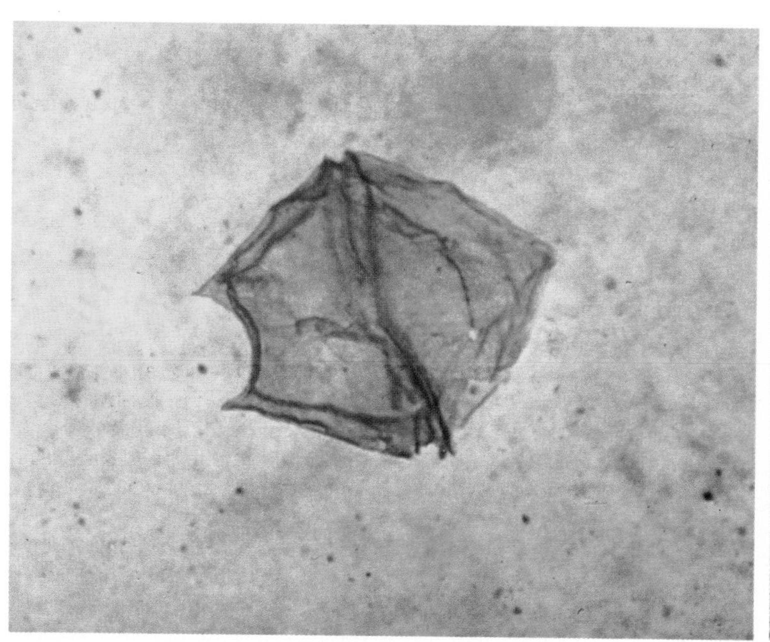

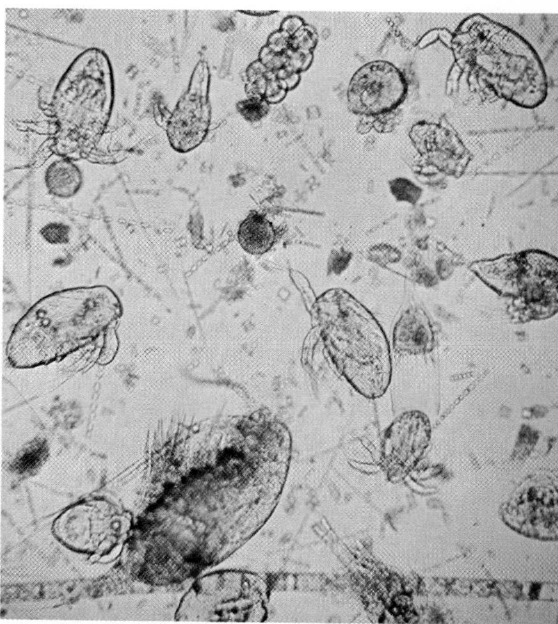

few species of flagellated and nonflagellated unicells that reproduce by cell division or production of zoospores.

The protistan algae are particularly abundant as members of the *plankton*, the community of small, floating organisms in the upper layers of oceans and of fresh waters (Figure 36.13). Within the upper layers of water, ample sunlight is available for photosynthesis. Mineral nutrients dissolved in the water from the sediments on the bottom are brought to the surface by turbulence or upwelling currents. The flagellated algae are able to swim through the water, bringing themselves constantly into contact with a fresh supply of nutrients. The nonflagellated forms are slightly more dense than the water and tend to sink slowly, thus moving into new surroundings as they deplete the nutrients around them. A large number of unicellular algae sink into the dark depths, where they cannot photosynthesize but serve as food for deep-water organisms. However, the plankton population is continually replenished through reproduction and through the return of sinking organisms to the surface by turbulence. Many of the ornate surfaces found on the cell walls of protistan algae are thought to play a role in adjusting the rate at which the organism sinks or is lifted by currents.

Multicellular plants are rare among the plankton, and thus the protistan algae are responsible for most of the photosynthesis that occurs in the oceans. The heterotrophic organisms of the oceans ultimately must depend upon the algae of the plankton for their supply of organic nutrients. Because the oceans cover the greatest part of the earth's surface, the planktonic algae are thought to be responsible for the greater part of the photosynthesis that occurs on earth and therefore for the greater part of the free oxygen that is continually supplied to the atmosphere. Recently, some biologists have become alarmed about the effects of DDT and other pesticides washed from the land into the oceans. These chemicals tend to accumulate in the upper few meters of the ocean water and have already reached significant concentrations even very far from land. There is some evidence

Figure 36.14. A plasmodiophore. These parasitic organisms feed on algae, fungi, and higher plants. Spores penetrate the host organism and develop into a multinucleate plasmodium, which absorbs nutrients from the host.

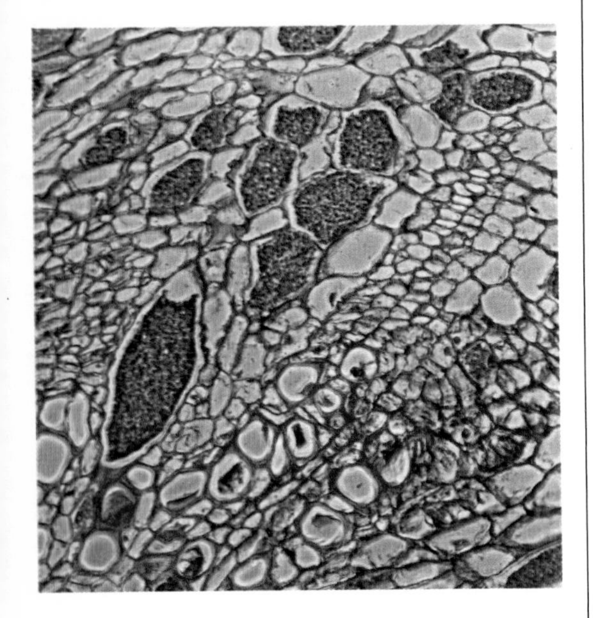

that the pesticides may interfere with the reproduction of at least some species of algae, but the effects of the present or expected levels of pesticides on the planktonic algae are as yet unmeasured.

The protistan algae are found in almost any environment that provides moisture and light, although they have not been as successful as the blue-green algae in penetrating extremely hot or cold environments. The "moss" that forms a green coating on moist soil or the shaded sides of trees is usually composed of algae. The slimy green appearance of most stagnant water is caused by various species of algae.

FUNGUSLIKE PROTISTANS

The phyla Hyphochytridiomycota and Plasmodiophoromycota include organisms that specialize in obtaining organic nutrients through absorption. The hyphochytrids are mostly unicellular organisms that exist in sea water that is rich in organic materials. A series of rhizoids holds the cell to the substrate. Reproduction occurs through the production of flagellated zoospores. A few of the species of hyphochytrids live as parasites on algae and fungi.

The plasmodiophores are parasitic organisms that feed on algae, fungi, and plants. Spores germinate to form flagellated gametes that swim through the water (through films of rainwater or dew on land) and penetrate the host organism. These gametes may fuse to form a zygote, or they may create a new organism without sexual fusion. (The details of this portion of the life cycle are not yet well understood for the various species.) Cell division without formation of new cell walls leads to the formation of a multinucleate *plasmodium*. After a certain period, during which the plasmodium feeds by absorbing organic substances from the host organism, thin walls divide the plasmodium into multinucleate structures called *sporangia*. Spores are formed within the sporangia and released to the environment.

Like the fungi, the funguslike protistans appear to exist as haploid cells during most of the life cycle. Formation of a zygote is immediately followed by meiosis, restoring the haploid condition. The funguslike protistans are relatively rare and appear to be of little importance to man, except for two species of plasmodiophores that cause destructive plant diseases in cabbage and potatoes.

PROTISTA: PROTOZOANS

The animallike Protista, or protozoans, include a number of phyla that obtain their principal organic nutrients through ingestion of other organisms or particles of organic matter. Most protozoa are motile, using flagella, cilia, or amoeboid movements to move through the water or over a substrate. Few protozoans have been grown successfully in a chemical medium, but those that have been grown require many amino acids and dozens of other organic compounds for normal growth and reproduction. Because these heterotrophic organisms are unable to synthesize many of the macromolecules that they need, they depend upon the bodies or waste products of other organisms as a source of nutrients. Smaller protozoans feed chiefly upon bacteria, and larger protozoans feed chiefly upon algae and smaller protozoans.

The phylum Sporozoa includes a number of parasitic species, most of which live inside the cells of other organisms during some part of their life cycle. In the mature stage of the life cycle, most sporozoans are incapable

Figure 36.15. Life cycle of the malarial organism *Plasmodium vivax*. This sporozoan life cycle involves the *Anopheles* mosquito and man.

of motion, and they feed by absorbing nutrients from the cell in which they live. Some species produce an immature stage that is capable of amoeboid movement and of feeding by phagocytosis. The life cycles of most sporozoans are quite complex, often involving stages that live in different organisms. The life cycle of *Plasmodium*, which causes malaria in humans, may be taken as an example (Figure 36.15). Amoeboid *sporozoites* live in the salivary glands of the mosquito genus *Anopheles*. When the mosquito bites a human, it injects saliva, which contains a chemical that prevents coagulation (clotting) of the human blood and permits the insect to suck up a supply of blood on which to feed. The sporozoites enter the human bloodstream with the mosquito saliva and move into the red blood cells. In the blood cells, the sporozoite undergoes a period of growth and development, followed by a series of mitotic divisions that produce large numbers of *merozoites*. When the blood-cell wall disintegrates, the merozoites move passively through the blood and enter other blood cells, where the process of asexual reproduction is repeated to produce still more merozoites. The liberation of merozoites from blood cells is synchronized throughout the bloodstream, occurring every 24 hours. Toxic substances released along with the merozoites cause the recurrent fevers that typify malarial attacks.

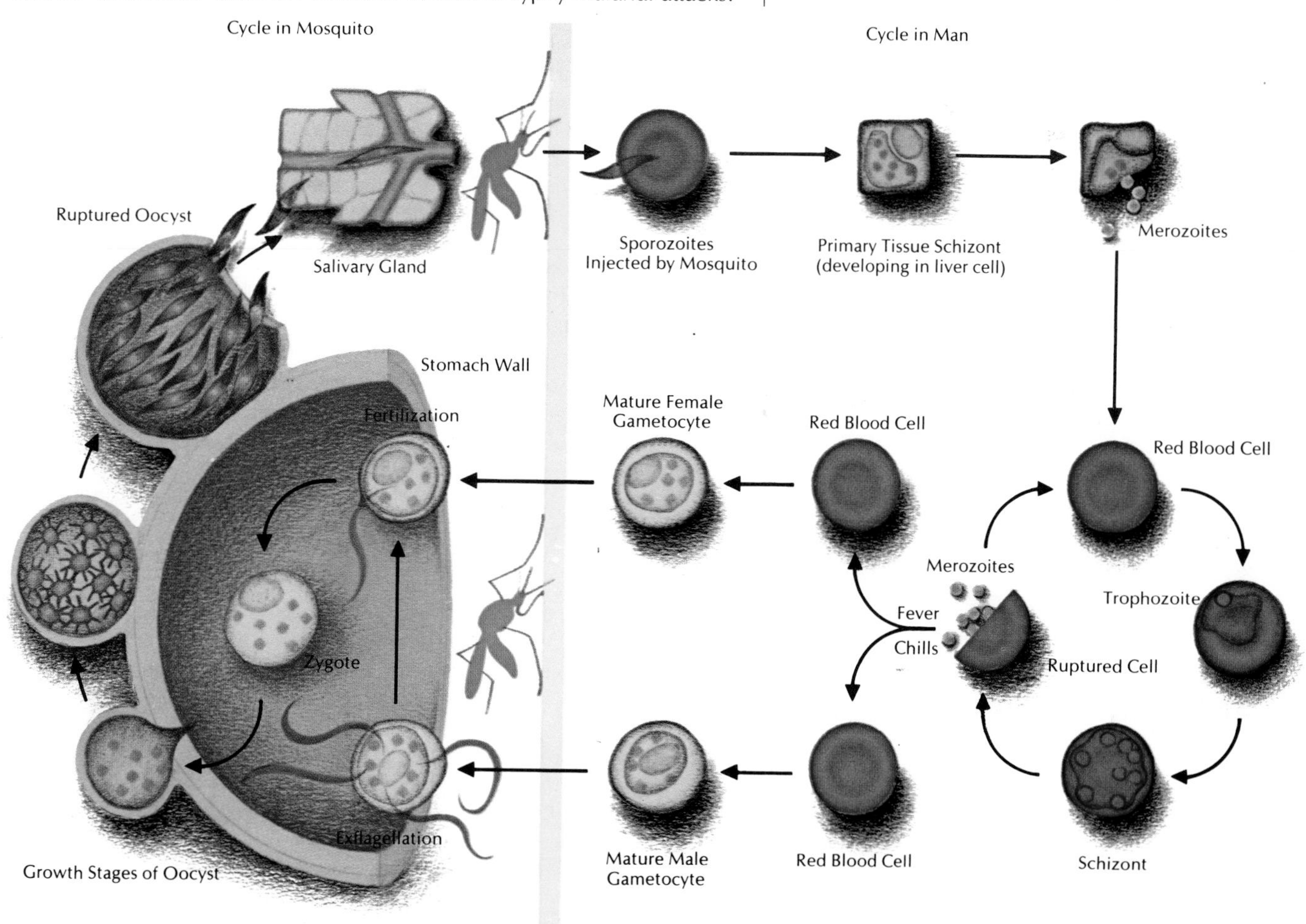

Figure 36.16 (right). Various parasitic flagellates of trypanosome groups live in the blood of various vertebrates. One of the most important forms, *Trypanosoma gambiense* (pictured here), causes African sleeping sickness. *(Courtesy Carolina Biological Supply Company)*

Figure 36.17 (left). A fossil radiolarian. Radiolaria are amoebas with shells, or tests, of chitin and siliceous spines. They are common oceanic plankton and feed on microorganisms and suspended debris. *(Eric Gravé)*

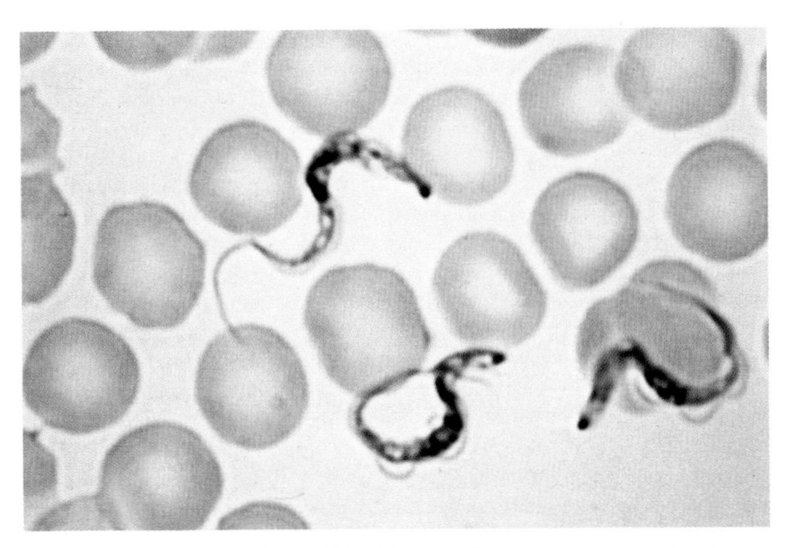

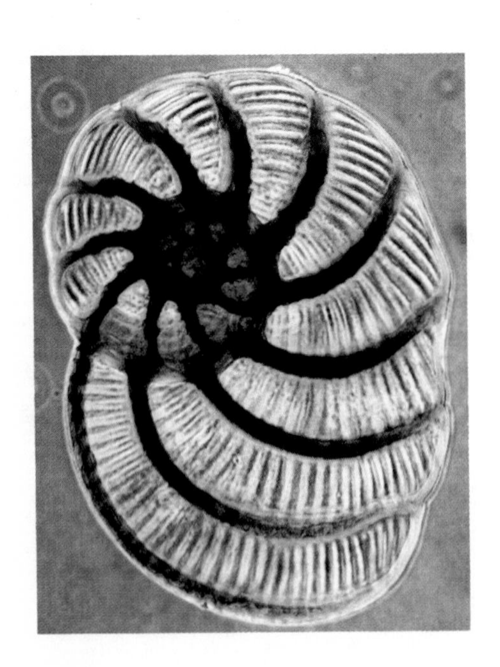

Some merozoites develop within blood cells into *macrogametocytes* or *microgametocytes*. If a mosquito biting a human already infected with *Plasmodium* takes these gametocytes into its stomach with the blood that it sucks, the gametocytes undergo meiotic division to form *macrogametes* and *microgametes*. In the mosquito's stomach, the amoeboid microgametes fuse with immotile macrogametes to form diploid zygotes. The zygote moves by amoeboid motion into the intestinal wall of the mosquito, where it forms a thick-wall *oocyst*. Inside the oocyst, large numbers of sporozoites are formed by mitotic divisions. When the oocyst bursts, the sporozoites move through the body fluids of the insect to the salivary glands, thus completing the life sycle.

The phylum Cnidosporidia includes parasitic organisms similar to the sporozoans but differing in certain details of the life cycle. The cnidosporozoans spend most of their life in a single host, form gametes soon after the formation of spores, and have a complex spore that is developed from a small number of cells. (The oocyst is considered to be the spore of the sporozoans.) The cnidosporozoans are parasites of invertebrates and fishes and, in a few cases, of reptiles and amphibians.

The phylum Zoomastigina includes all of the flagellated heterotrophs not grouped in the Euglenophyta or Pyrrophyta (Figure 36.16). Unlike the heterotrophs grouped with the algae, the zoomastigophores do not store food in the form of starchlike compounds, although a few species produce glycogen. Most species of zoomastigophores live as parasites in plants and animals, some feeding by absorption and some by ingestion. There is no cell wall, but there is a rigid pellicle within the cell membrane. In many species, the pellicle is flexible enough to permit some degree of amoeboid movement.

The phylum Sarcodina includes the heterotrophic protists lacking flagella or cilia during the major part of their life cycle and moving by changing the shape of the cell through flow of cytoplasm (Figure 36.17). The actinopods have slender, radiating pseudopods, or armlike extensions. The cytoplasm in these pseudopods is more rigid toward the core; some species have a rigid filament along the axis. The pseudopods act as sticky traps for food, which may be ingested on the pseudopod or drawn toward the body

Figure 36.18 (left). Shells of various marine radiolaria. Note the winglike extensions of the shells, which aid in flotation. *(Eric Gravé)*

Figure 36.19 (middle). The protozoan class Sarcodina includes both naked and shelled amoebas. Pictured here under the phase contrast microscope is an unshelled amoeba. *(Eric Gravé)*

Figure 36.20 (right). The stalked ciliate *Stentor*. This organism is noted for its complex gullet and elaborate ciliation pattern. Note the conspicuous light-dark spotted pattern denoting membranelles that ring the gullet area. *(Eric Gravé)*

Figure 36.21 (below). The ciliate protozoan *Paramecium*. This mobile unicellular organism is an active predator in the fresh-water pond. *(Eric Gravé)*

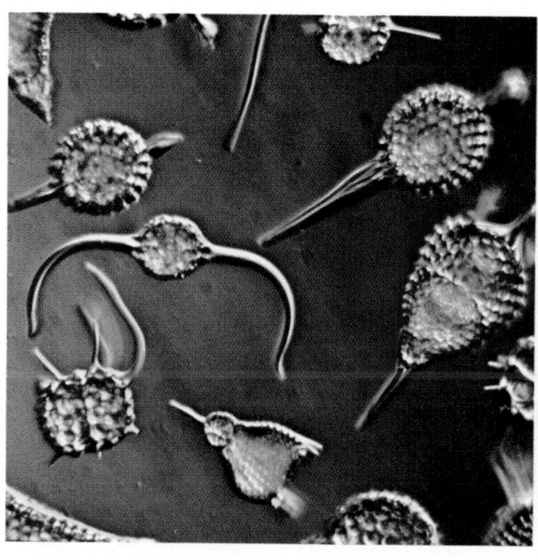

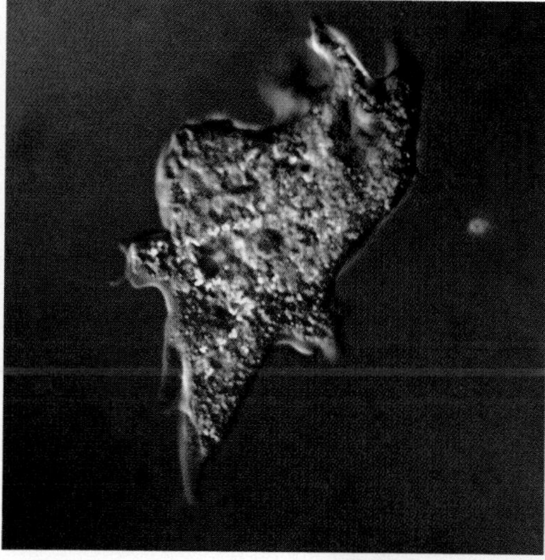

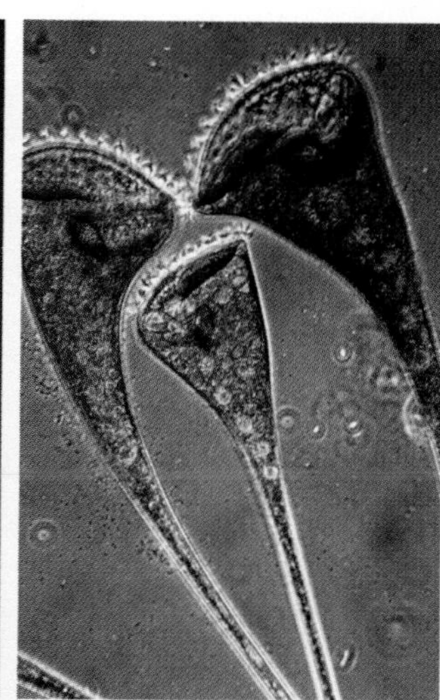

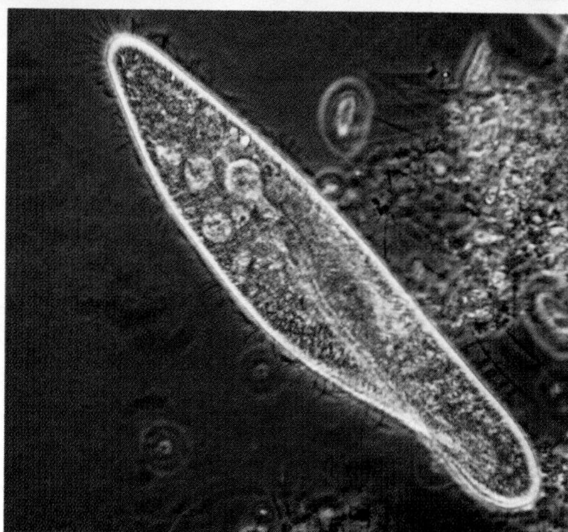

for ingestion. Among the most beautiful actinopods are the radiolarians, with cell capsules composed of chitin or similar materials and skeletal spines lying within the cell or passing through the capsule into the pseudopods. Reproduction in some species is by simple division, whereas in some species a small portion of cytoplasm containing a nucleus is separated from the mature organism and later builds its own cell capsule and skeleton. Sexual reproduction through flagellated gametes has been observed in some species. Radiolaria are common in the oceanic plankton, where they feed on other microorganisms, and their skeletons have been found in rocks formed more than 425 million years ago.

The phylum Sarcodina also includes the rhizopods, which move with the extension of pseudopods. Among the rhizopods are the amoebas, which lack a cell wall and include both free-living and parasitic species. Another group of rhizopods, the foraminiferans, form shells, or *tests*, of siliceous material. In most species, the test lies within the cytoplasm. Filamentous, branched pseudopods extend outside the cell to capture and digest food. Most foraminiferans live on the ocean bottom, and the tests of species such as *Globigerina* form thick layers among sedimentary rocks.

The phylum Ciliophora includes organisms that possess cilia during some stage of the life cycle (Figure 36.20). Most of the ciliophores possess both macronuclei and micronuclei, and most reproduce asexually. Some ciliates show highly specialized cell surfaces, including structures similar to the mouths, anuses, and holdfasts of multicellular organisms. In many species, the cilia on the surface sweep food particles toward the mouth. The suctorians have hollow tentacles that capture prey. The food is ingested on the tentacle and flows down the center of the tentacle to a food vacuole at its base.

Origins of the Protista

Both theoretical considerations and fossil evidence suggest that the eucaryotic Protista evolved from earlier procaryotic cells. Considerable evidence has now been gathered to support the hypothesis that the earliest eucaryotes originated as a combination of two or more procaryotic cells. According to this hypothesis, a blue-green alga became established inside the cell of a large heterotrophic procaryote (perhaps a bacterium), possibly as a result of surviving a process of ingestion. Through mutation and natural selection, the alga and the host cell came to divide simultaneously and each came to depend on the other for certain nutrients. The alga evolved into a chloroplast, retaining its own DNA, ribosomes, and photosynthetic apparatus but losing most of its other structures. The host cell came to depend upon the chloroplast as a source of organic nutrients and developed structures and behavioral patterns that would facilitate photosynthesis in the chloroplast. Through a similar process, another species of bacterium became established within the cell, specialized in respiratory metabolism, and evolved into the mitochondrion. Some biologists believe that flagella, cilia, and centrioles originated through a similar establishment of a spirochaete bacterium within the host cell. Finally, it is thought that the nuclear membrane and endoplasmic reticulum evolved through a complex infolding of the plasma membrane of the host cell.

Although more and more biologists are beginning to support this general picture of the evolution of the first eucaryote, there are many different hypotheses about the sequence in which these events occurred and the nature of the early Protista. The fossil record offers little clarification of this question. Structures that appear similar to modern unicellular algae recently have been found in rocks formed about 1.3 billion years ago. If these fossils do prove to contain nuclei and other eucaryotic organelles, the origin of eucaryotes must have occurred more than twice as long ago as had been previously thought. The oldest undoubted fossils of eucaryotic algae are found in rocks formed about 500 million years ago.

One recent theory about the origin of the eucaryotes suggests that the first eucaryote was a unicellular alga, similar to some of the modern green algae (Klein and Cronquist, 1967). From this primitive eucaryote evolved the green algae (Chlorophyta of the kingdom Plantae), the Euglenophyta, the Pyrrophyta, and the Chrysophyta. According to this theory, the funguslike Protista and the protozoans evolved from various species of Euglenophyta and Pyrrophyta that specialized in absorption or ingestion and eventually lost their chloroplasts.

Another recent theory holds that the first step toward the eucaryotes occurred when aerobic bacteria became established in a simple procaryotic cell, forming mitochondria and converting the host cell to an organism similar to an amoeba. At a later time, a spirochaete bacterium became established in the cell, forming a flagellated amoeboid, and the nuclear membrane and mitotic apparatus were developed. The various phyla of protozoans then evolved from this primitive protozoan. The funguslike protozoans evolved from certain protozoans specializing in absorption rather than ingestion. The algae, according to this hypothesis, appeared relatively late in evolution, when blue-green algae became established in some protozoans to form chloroplasts (Margulis, 1968).

At the present time, there is little evidence to permit a choice between these two theories or to rule out other possibilities. However, careful studies of the combinations of organelles existing in living species and comparisons of the DNA in various organelles and cell nuclei, as well as new fossil discoveries, may make possible a more confident outline of the evolutionary history of the Protista in the near future.

Whatever their origin, it is clear that the Protista proved remarkably successful at life in the planktonic regions of the oceans and deep lakes. The existing species of Protista include almost every imaginable variation of structure and metabolism. No multicellular organism has been able to displace the Protista from their dominance of the upper waters of the oceans. Although multicellularity is apparently of little value to planktonic organisms, those organisms that evolved toward multicellularity proved to be far more suited for life on the land.

FURTHER READING

General introductions to the study of microorganisms are given by Carpenter (1967), Pelczar and Reid (1965), and Stanier, *et al.* (1970). DeKruif (1926) records the fascinating history of "the microbe hunters." Relationships between man and his beneficial and harmful microparasites are discussed by Simon (1963) and Zinsser (1935)—and in particularly entertaining fashion by Rosebury (1969). Wedberg (1963) discusses microorganisms of sewage, soil, and the atmosphere and their interactions with man. Sussman (1964) describes a number of simple laboratory investigations of microorganisms.

The protistan algae are discussed by Chapman (1962), Fritsch (1945), and Smith (1955). Simple and entertaining introductions to protozoans are given by Curtis (1968), Headstrom (1968), Jahn and Jahn (1949), and Manwell (1968). More detailed and technical information on protozoans is given by Corliss (1961), Kudo (1954), and Mackinnon and Hawes (1961).

37
Plants and Fungi

Most multicellular organisms that have specialized in a photosynthetic, autotrophic way of life are grouped in the kingdom Plantae. Some unicellular and heterotrophic organisms also are included in this kingdom because of close structural or biochemical resemblances to the plants. There are three phyla of algae, which incorporate the familiar seaweeds of the nearshore ocean and some species that inhabit moist areas on land. The phylum Bryophyta includes the mosses and liverworts, organisms that have specialized somewhat for existence in moist terrestrial habitats. The remaining phyla of the kingdom are the vascular plants, organisms that are well adapted to life on land.

The taxonomic organization arranges the plants in order of increasingly complex and specialized tissues and organs, but it is difficult to resist reading into this sequence the story of the gradual conquest of the land by plants. In fact, available fossil evidence suggests that a gradual evolution took place, beginning with simple unicellular algae, progressing through colonies of cells and simple multicellular organisms, and eventually leading to the development of flowering land plants. However, in this chapter the progression presented will be for the most part an arbitrary sequence of modern organisms, selected to show one possible model of the way in which complex plants could have evolved from unicellular algae through a series of small changes. All available evidence suggests that a similar progression of increasing complexity occurred through time as the plants evolved, but there is a great deal of dispute about details of the actual evolutionary sequence.

Many clues about the evolutionary origins of complex structures come from studies of their development during the life of an individual organism. Every multicellular organism begins its life as a single cell. In the process of cell division and differentiation that follows, it is often possible to see how a complex structure is developed from a precursor that has structures similar to those that exist in simpler organisms. Although the major emphasis of this and the following chapter will be the structural and functional organization of multicellular organisms, some details of developmental patterns will be mentioned where they can help to elucidate the nature and origin of structural features.

CHLOROPHYTA

Because the green algae (phylum Chlorophyta) are similar in many ways to complex land plants (they have the same photosynthetic pigments and utilize starch for food storage), they have been of particular interest to botanists. Within this phylum can be found a series of organisms composed of nearly identical cells, but ranging in complexity from unicellular organisms to relatively complex multicellular plants.

The unicellular green alga *Chlamydomonas* is thought to be similar to the early eucaryotic organisms. Its biflagellate cell is less than 25 microns long and contains a single, large, cup-shape chloroplast (Figure 37.1). The flagella end of the cell is the front, or *anterior*, end, with the cell being pulled through the water by the synchronized beating of the flagella. A nucleus lies near the center of the cell, often concealed within the cup of the chloroplast. Near the anterior end are two contractile vacuoles that discharge their contents to the exterior of the cell at brief intervals. Starch granules are formed in a particle called the *pyrenoid*, which is located within the chloroplast near the back, or *posterior*, end of the cell. Near the

Figure 37.1a (upper left). The biflagellate, unicellular green alga *Chlamydomonas*. Different species of *Chlamydomonas* occur in fresh-water ponds and streams and in various marine habitats. (*Eric Gravé*)

Figure 37.1b (right). Life cycle of *Chlamydomonas*.

Figure 37.2 (lower left). *Volvox*, a flagellated, colonial green alga. The colony is a hollow sphere. Note the dark green daughter colonies forming within the parent colony.

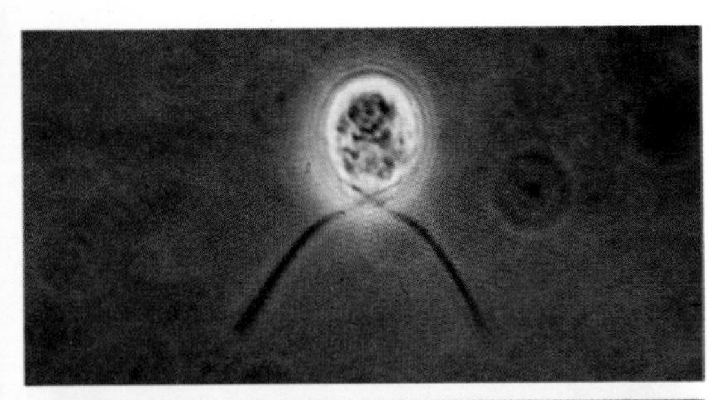

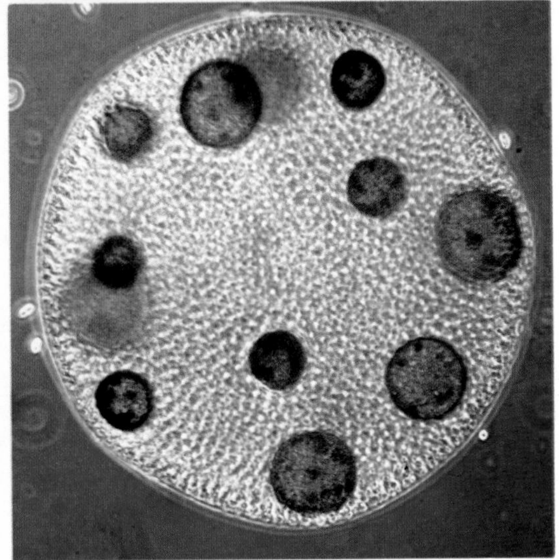

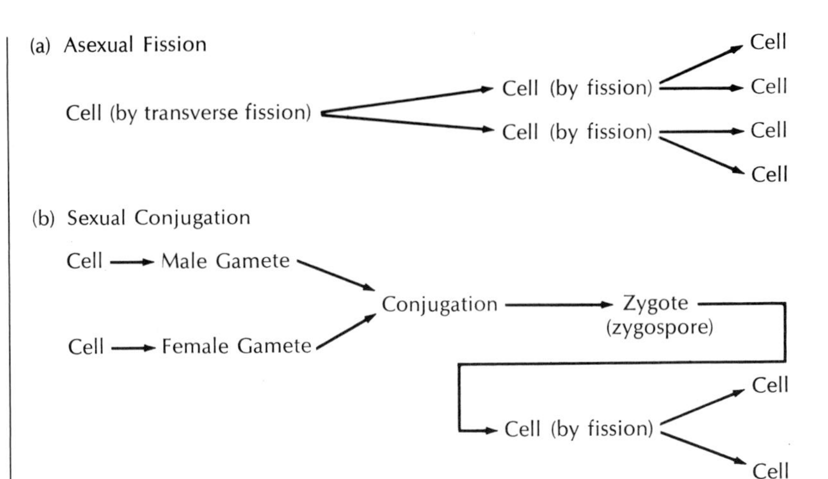

rim of the chloroplast is an eye-spot containing carotenoid pigments. Although the eye-spot is believed to play a role in the ability of this alga to swim toward brighter light, mutants lacking an eye-spot are also able to orient toward light. The cell membrane is surrounded by a cellulose wall. The chloroplast pigments include chlorophylls *a* and *b*, carotenoids, and xanthophylls.

Chlamydomonas reproduces asexually when conditions are favorable. The flagella disappear and mitotic division produces two to eight daughter cells (depending upon the species). The daughter cells develop flagella and then break out of the parental cell wall. Sexual reproduction also occurs in *Chlamydomonas*. The flagella of the parent cell are withdrawn, and cell division produces up to 64 daughter gametes, which develop cell walls and flagella before breaking out of the parental cell wall.

PHAEOPHYTA

Most species of brown algae (phylum Phaeophyta) live in the oceans, where most are seaweeds of the nearshore zone. Only three rare genera of brown algae are found in fresh water. Abundant xanthophyll pigments give these algae a brownish or almost black color. The simplest brown algae are formed of branching filaments. (No unicellular or colonial species are known.) Most species show oogamous sexual reproduction, and in many species there are alternating haploid and diploid generations with somewhat different forms.

The largest and most complex of the brown algae are the kelps such as *Laminaria*. The body of the diploid generation is composed of a holdfast, a stemlike *stipe*, and a blade (Figure 37.3). There is a definite differentiation of cells into various tissues within the *stipe*. The algal body is covered with an outer layer of mucilage. Underneath is a layer of cells that continues to divide throughout the life of the plant. Beneath these cells is a region of less pigmented, elongated cells. In the stipe center is a core of branched and intertwined filaments, and the inner portion contains columns of elongated cells, whose end walls are perforated by groups of small pits. These cells appear to be specialized to play a role in transport of fluids through the plant body.

At certain times of the year, cells in some areas of the blade develop into sporangia. Haploid zoospores are produced by meiosis within the

Figure 37.3a (right). *Laminaria andersonii*, a common cosmopolitan kelp found on Whidby Island, Washington.

Figure 37.3b (upper left). Generalized life cycle of a kelp.

Figure 37.4 (lower left). The brown alga *Fucus*. Note the reproductive receptacles with their wartlike covering. (*Walter Dawn*)

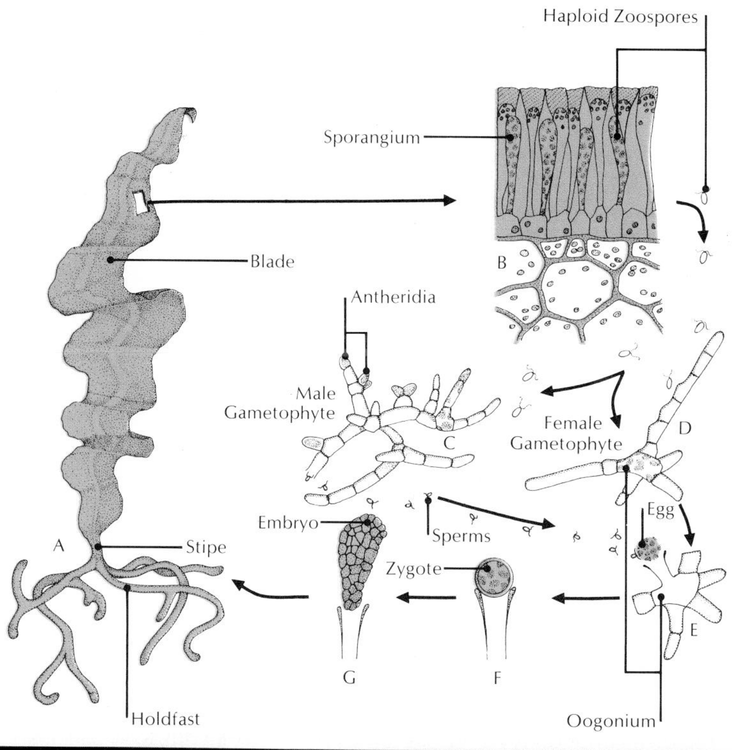

Figure 37.5a (above). Generalized life cycle of a typical red alga.

Figure 37.5b (below). Photomicrograph of the cystocarp of the red alga *Polysiphonia*. (Courtesy Carolina Biological Supply Company)

sporangia. The zoospores develop into small haploid plants called *gametophytes*, which are quite different in appearance from the diploid *sporophyte* plants. Motile male gametes and nonmotile female gametes are produced in separate gametophytes. Fusion of the gametes produces a zygote, which remains attached to the female gametophyte during the early stages of its development.

Because unicellular and colonial brown algae have not been discovered, the evolutionary origin of this phylum is a matter of some dispute. Klein and Cronquist (1967) hypothesized that the Phaeophyta evolved from ancient green algae. Although the kelps are the largest and most complex of all algae, there is no evidence that any land plants evolved from this phylum.

RHODOPHYTA

Like the brown algae, the red algae (phylum Rhodophyta) grow almost exclusively in the ocean, predominantly near the shore. Many species live in the deep waters of warm seas, and some form calcareous external skeletons that aid in building coral reefs. The bright pink color of most Rhodophyta is due to biloprotein pigments. Fossil skeletons of red algae have been found in rocks formed about 500 million years ago.

The red algae differ from other algae in many important ways. Flagella are not present at any stage of the life cycle; even the male gametes are immotile. The range of body types present in this phylum is greater than that of any other algal phylum. These facts suggest that the Rhodophyta became an independent evolutionary line very early in the history of cellular life. The Rhodophyta may have evolved independently to the eucaryotic condition before the evolution of eucaryotic cells in the green algal line

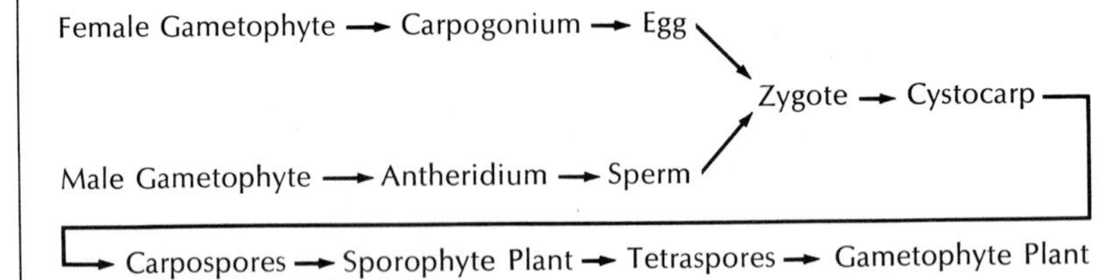

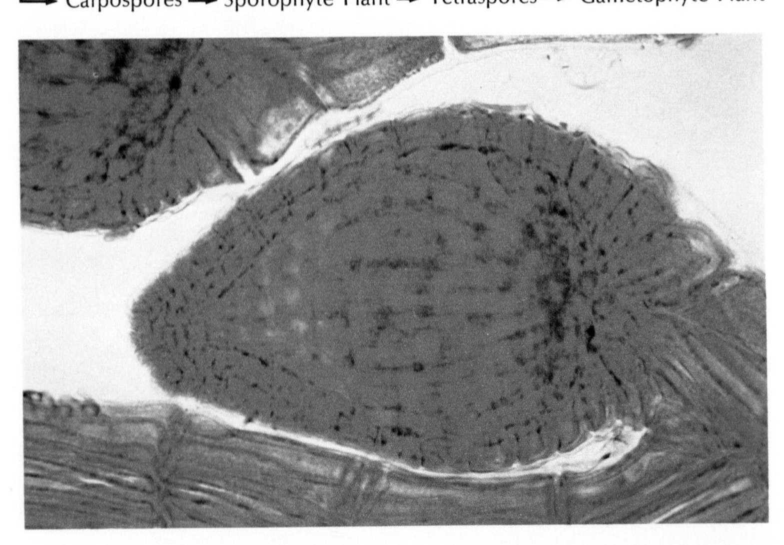

that is theorized to have produced all other eucaryotic organisms. However, the Rhodophyta may have diverged from the other algae soon after the evolution of the first eucaryotes.

The bodies of many red algae show relatively complex differentiation similar to that described for kelp, and the sexual structures and reproductive cells are highly specialized. The life cycles of many species are very complex. Some include a haploid gametophyte, a diploid sporophyte, and a haploid sporophyte. Others include one haploid and two diploid generations. Like the brown algae, the success of the red algae appears to have been largely limited to the oceans, and there is no evidence that land plants evolved from the Rhodophyta.

Figure 37.6. A species of red algae. Note the delicately branched body of the plant.

EVOLUTIONARY TRENDS AMONG THE ALGAE

Although some caution must be used in theorizing evolutionary sequences based upon a logical grouping of modern organisms, the available fossil evidence seems to support the hypothesis that colonial and multicellular forms have evolved independently in several lines of algae. It is possible to guess about the advantages that multicellularity would give to an organism. In the nearshore waters, an organism attached to the substrate is bathed in nutrients by rushing currents while holding itself in a location of suitable sunlight and other conditions. For such an organism, the development of a large multicellular body greatly increases the amount of photosynthesis that can be carried out. In the deeper parts of the ocean, there seems to be less value to multicellularity and the unicellular algae continue to form the dominant part of the plankton population.

For those algae entering moist land environments, multicellularity and differentiation provide very clear advantages. Parts of the structure can be lifted into the air and specialized for photosynthesis, whereas other parts can be specialized to obtain water and mineral nutrients from the soil and to hold the plant in place.

Another apparent trend leads from asexual reproduction in the simplest algae through isogamy to oogamy in most complex algae. For large organisms that must reproduce slowly, sexual reproduction provides a clear advantage in increasing the number of new genetic combinations that appear in each generation. The mutation rate alone is rapid enough to provide considerable variation among bacteria, but sexual recombination assures a supply of differing forms among the offspring of more slowly reproducing organisms. If environmental conditions change, a species having a variety of individual characteristics is more likely to have some of its members survive to carry on the species. The advantage of oogamy appears to lie in specialization of the motile gamete for rapid motion over a long distance and in specialization of the immotile gamete for storage of nutrients and other materials to support the early development of the zygote. Development of structures and chemical substances that guide the motile gametes to the egg, a condition observed in many of the more complex algae, makes oogamy a relatively efficient means of sexual reproduction.

BRYOPHYTA

The bryophytes are small, relatively simple land plants that grow chiefly in damp, shaded locations. Mosses (class Musci) are widespread on the continents. In some bogs, high mountains, and polar regions, they form a dominant part of the vegetation. Liverworts (class Hepaticae) and hornworts

Figure 37.7a. Photograph of the common liverwort *Marchantia*. Note the fingerlike clusters of sporophytes growing from the flat, leaflike gametophyte plant.

(class Anthocerotae) are less abundant and less able to withstand brief periods of desiccation. A bryophyte spore develops into a small haploid filament, the *protonema*, that closely resembles the filamentous green algae from which this phylum is thought to have evolved. After a period of growth and development, the protonema forms buds that give rise to erect leafy plants in the mosses (Figure 37.7). The leafy plant and the protonema with its multicellular rhizoids make up the haploid gametophyte generation. Multicellular sexual organs are formed at the apex of the leafy structures. A nonmotile egg cell is formed in an *archegonium*, and large numbers of flagellate sperm cells are formed in an *antheridium*.

In many species of mosses, each gametophyte bears only one kind of sexual organ. Release of sperm from the antheridium is triggered by water, through which the sperm can move to the archegonium. In some species, splashing raindrops apparently serve to carry sperm from one plant to another. The final movement of a sperm cell into an archegonium is apparently guided by a chemical substance secreted by the egg. Fusion with the egg produces a diploid zygote, which develops within the archegonium to form a diploid sporophyte plant that is permanently attached to the gametophyte, receiving its nutrients from the gametophyte tissues. A spore capsule, or *sporangium*, forms at the top of the sporophyte. Haploid spores are produced within the capsule. When the spores are mature, the capsule opens and the spores are carried to new locations by wind. Those landing in favorable locations germinate to form new protonemata. The liverworts and hornworts lack multicellular rhizoids and differ from the mosses in details of the structure of the sporophyte and gametophyte plants.

A few species of mosses reach heights of a few feet, but most species are a few inches or less in height. Apparently, size is limited by the slow rate at which nutrients can move from the rhizoids to the photosynthetic structures. Most bryophytes have a layer of *epidermis*, a tissue that forms a thin covering over the parts of the plant exposed to the atmosphere. The epidermal cells have a thin coating of waxlike *cutin* over the outer surfaces. Although the epidermis helps to reduce water loss through evaporation, few of the mosses can endure prolonged dry conditions. The plants can grow only where moisture is frequently present, because sexual reproduction depends upon movement of sperm cells through water. Spores are covered with a thick cutin layer and can endure relatively prolonged dry periods. In many species, cells of the protonema can survive underground during unfavorable conditions and germinate to create a new plant when conditions again become favorable. Fossils of complex mosses and liverworts have been found in rocks formed more than 350 million years ago. It is thought that the bryophytes evolved independently from the green algae, but at about the same time as the more complex land plants.

VASCULAR PLANTS

All green plants, other than algae and bryophytes, possess specialized conducting, or vascular, tissues. Many taxonomists group all these plants in a single phylum, the Tracheophyta. Vascular plants have two distinct types of vascular tissue. *Xylem* carries water and mineral nutrients from roots to photosynthetic cells, and *phloem* tissue carries organic products of photosynthesis from photosynthetic cells to the rest of the plant body. Development of vascular tissues has permitted these plants to live in many habitats

Figure 37.7b (above). Generalized life cycle of a moss.

Figure 37.8a (middle). Generalized life cycle of a liverwort.

Figure 37.8b (below). Various moss species growing on rocks in the rapids zone of a stream. Big Spring State Park, Missouri.

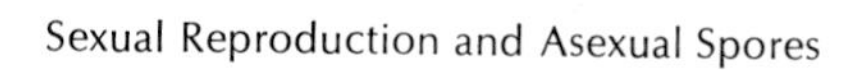

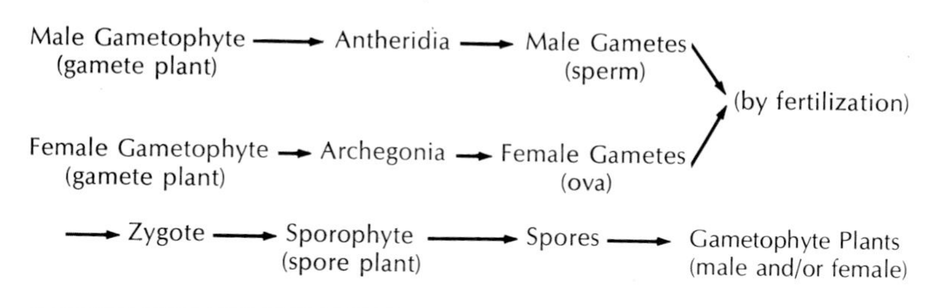

(a) Asexual Fragmentation

Plant Body (by fragmentation) → Each Fragment Develops a Plant Body

(b) Asexual Cupules and Gemmae

Plant Body → Cupules → Gemmae → Each Gemmae Develops a Plant Body

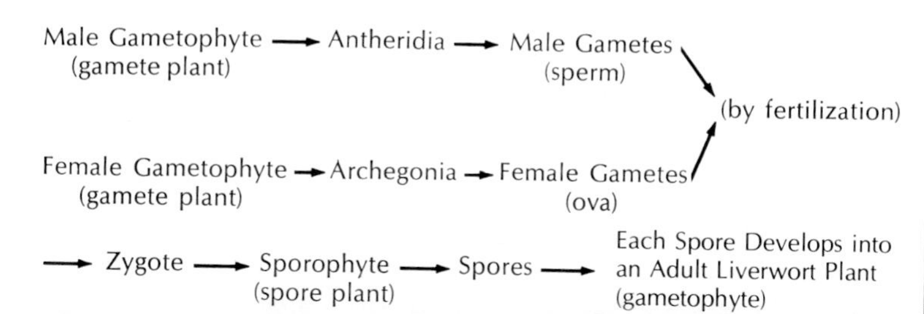

Figure 37.9. Photomicrograph of a pine needle cross section. Note the lobed parenchyma cells just under the epidermal layer. (*Courtesy Carolina Biological Supply Company*)

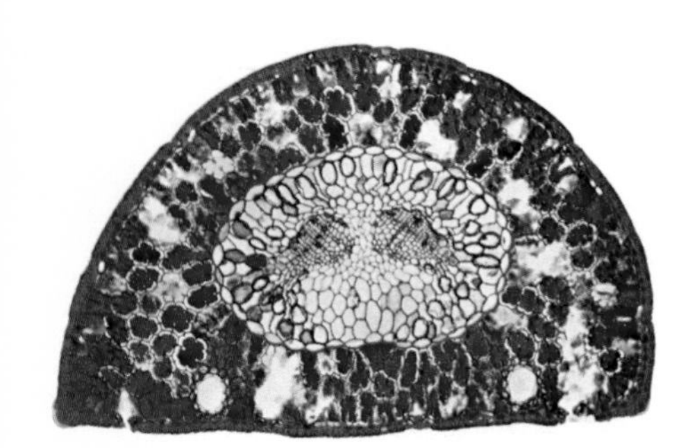

on land, with roots deep into the soil to trap moisture and leaves high in the air to capture sunlight.

Most of the cells of the vascular tissues are essentially small tubes oriented along the axis of the stem or root. Often, the ends of the cells lie at a sharp angle to the axis of the tube, increasing the surface area between adjacent conducting cells. In the flowering plants (and a few species of other vascular plants), individual conducting cells become joined to form continuous tubes extending throughout nearly all of the plant body.

The most abundant conducting tissue in the vascular plant body is xylem. During their maturation, thick-wall *tracheid* cells of the xylem become hollow conducting tubes due to the disappearance of the nucleus and cytoplasm. In most flowering plants, another conducting cell, the *vessel element*, loses its contents and its end walls during maturation, becoming joined end-to-end with other vessel elements to form long, tubelike *vessels*. The thick cellulose walls of the tracheids and vessel elements give strength and rigidity to the plant body and prevent collapse of the conducting tissues during periods of desiccation. Small, thin places called *pits* form in the side walls of the tracheids and vessel elements. The pits of adjacent cells are aligned, and these pits (which are covered by a pit membrane) provide a passage for solutions to move from one cell to another within the xylem.

In addition to the dead cells that form conducting tubes, the xylem contains thin-wall, living cells that make up a generalized kind of tissue called *parenchyma*, which is capable of cell division. In some species, the parenchyma cells are scattered throughout the xylem tissue. In other species, the parenchyma cells form rays extending radially between aggregations of tracheids and vessel elements. The parenchyma cells play a role in lateral movements of fluid through the xylem and in carbohydrate storage. Fiber cells also are formed within the xylem, adding further strength to the tissue and, in some cases, also serving for storage of organic molecules.

The other major tissue of the vascular system is phloem, made up of living *sieve cells* that may be joined end-to-end to form *sieve tubes*. Fiber and parenchyma cells are also present in the phloem. Recent observations suggest that some, if not all, sieve cells may retain their nucleus; earlier, it had been thought that all sieve cells lost their nuclei at maturity but retained the cytoplasm and other organelles.

Between the vascular tissues and the epidermis in roots and stems is a tissue called the *cortex*. It is composed largely of parenchyma cells, but fibers are also present in some species. The cells of the cortex serve many functions, helping to transport various molecules from the vascular tissues to the plant surface and vice versa, storing organic molecules and mineral nutrients, and carrying out photosynthesis in the aerial portions of the plant. The epidermis forms a protective covering over the outside of the plant outside the cortex. On the aerial portions of the plant, the epidermal cells produce a coating of waxy cutin, which, together with cellulose, forms the light gray coating, or *cuticle*, on certain leaves and fruits. The waxy substance on the cuticle can easily be wiped off.

Entry of gases into the body of the plant to reach the photosynthetic cells of the cortex (or mesophyll of the leaf) is accomplished through openings called *stomata*. Each stoma is enclosed by a pair of distinctive, bean-shape *guard cells*. These cells can change in size and shape, varying the size of the opening. When the stomata are open, carbon dioxide and oxygen dif-

Figure 37.10a (above). Photograph of a living insect-trapping pitcher plant.

Figure 37.10b (below). Photograph of a living sundew plant with a trapped insect. Note the conspicuous secretory hairs.

fuse from the air into the spaces between cortex cells, and water vapor moves out of the plant. The diffusion of water vapor into the atmosphere is called transpiration.

The epidermis of the root lacks a cuticle, permitting solutions to move readily from the soil into the root cells. Fine *root hairs*—formed by extension of walls of epidermal cells near the root tip—extend into the soil. On the aerial portions of some plants, epidermal cells become elongated to form hairs or scales over the plant surface. In some species, these structures apparently shade the plant or reduce transpiration; in other species, they protect the plant from being eaten by animals.

Most photosynthetic activity is carried on in the *leaves*. Although leaves are highly variable in size and form, most are composed of a green, flat, relatively thin, and broad *blade* attached to a *petiole*, or stalk. Most leaves are placed horizontally, with one flat surface turned upward toward the sunlight and the other one shaded. The blade—consisting of a few layers of green, photosynthetic *mesophyll* cells—is largely supported by the pressure of water in the vacuoles of the mesophyll cells. On a very hot day, the excessive loss of water by transpiration may cause the leaf to become limp and flabby. The leaf blade is also supported by a branching framework of *veins*, or vascular strands, that ramify through the blade. The soft, delicate mesophyll tissue of the leaf is enclosed above and below by the transparent, firm epidermis with its waxy cuticle coating.

The internal structure of the leaf in complex plants shows several features that appear to contribute to photosynthetic efficiency (Chapter 19). Just beneath the upper epidermis, the *palisade mesophyll* is composed of several layers of cylindrical, photosynthetic cells arranged at right angles to the surface. These cells perform most of the photosynthesis of the leaf. In the lower portion of the leaf beneath the palisade cells is a system of irregularly shaped cells called the *spongy mesophyll*. These cells are loosely interconnected and are separated by conspicuous intercellular spaces, which are connected with numerous stomata in the lower epidermis, allowing water vapor, carbon dioxide, and oxygen gases to move between the photosynthetic cells and the atmosphere under the control of the stomata. The leaf possesses an internal aeration system of impressive proportions. In the catalpa tree, for example, the internal surface of the leaf is about 12 times greater in area than the external surface.

In some plants, the leaves are highly modified, specialized structures. For example, insectivorous plants, which supplement their photosynthetic production of organic molecules by catching and digesting insects, have leaves that bear various types of secretory hairs and glands. Some are modified in shape to form pitcherlike receptacles in which the insects become trapped (Figure 37.10).

Some leaves are *simple* and undivided; others are *compound* leaves, composed of many small leaflets. In addition to the *foliage leaves* that carry out photosynthesis, vascular plants may have *scale leaves*, which form protective wrappings around buds and underground stems; *bracts*, which protect the basal parts of flowers; and *cotyledons*, or seed leaves. The cotyledons are the first leaves formed in the developing embryo of a seed-producing vascular plant and are specialized to serve as food-storage organs.

The leaves play an important role in the movement of water through the xylem. Water diffuses from the soil through the membranes of the root epidermal cells, particularly in the root hairs. Within the root, the water moves toward the vascular tissues, both by diffusion along cell walls and by osmosis across cell membranes. When the water enters the tracheids and vessels of the xylem, it is actually pulled upward through the plant under the control of forces acting in the leaves.

Tremendous quantities of water are transpired from the leaves, particularly during the day, when photosynthesis is occurring and the stomata are open. One corn plant, for example, may transpire as much as 54 gallons of water during its life span of about 14 weeks. If transpiration occurs more rapidly than water can be supplied to the shoot system through the xylem, the leaves may wilt and the plant may die. Although transpiration may be detrimental to the plant, it cannot be avoided entirely. The plant must obtain carbon dioxide and oxygen from the atmosphere and therefore must exchange gases with the air. Although the stomata with their guard cells help to control this exchange and to limit the loss of water vapor, the plant cannot prevent water vapor from leaving its body whenever it is bringing in other gases from the air.

However, the loss of water through transpiration does have certain effects that are useful to the plant. It has been shown experimentally that transpiration cools the leaves, helping to keep the cells at a temperature where their metabolism can function efficiently. In wilting leaves, the guard cells close the stomata, and transpiration (as well as most photosynthesis) is essentially halted. The temperatures of these leaves may rise as much as 6°C above the air temperature. Transpiration also serves to set in motion the upward movement of water within the xylem cells (Chapter 19).

Psilophyta

The phylum Psilophyta includes fossil species that lived during the early Devonian period some 350 to 500 million years ago. Like the more complex algae, these simple vascular plants have a slender, dichotomously branching body and show little tissue and organ differentiation. They lacked leaves and had no clear distinction between aerial and underground portions. (The more familiar root-shoot vascular plant body organization did not evolve until middle Devonian or later.) Internally, these plants had a slender cylinder of xylem surrounded by a narrow band of elongate cells that may have been phloem.

Spores were produced by meiosis on the aerial portions in multicellular

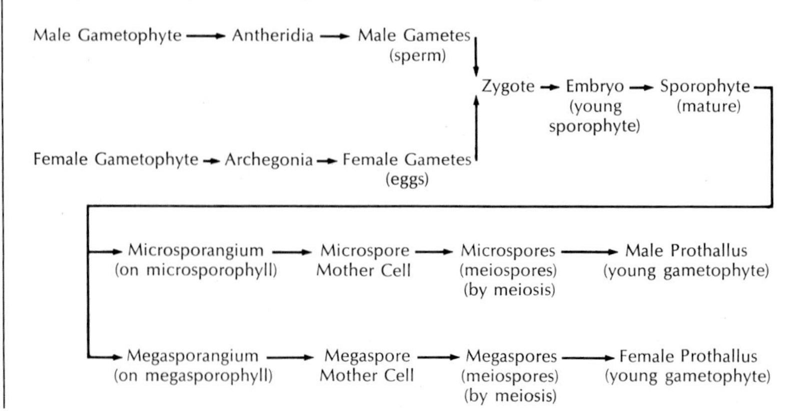

Figure 37.11a. Life cycle of a club moss.

Figure 37.11b. Photograph of the club moss *Lycopodium*. Note the spirally arranged leaves and the terminal sporangia.

sporangia. The spores, protected by a cutin layer, were presumably scattered by the wind. They may well have germinated to form a gametophyte. No fossils, however, of the gametophyte phase have yet been discovered.

The psilophytes were once believed to be the ancestors of all land plants, but it is now thought that several different lines of land plants may have evolved—probably from the green algae—at about the same time. In any case, the psilophytes do not closely resemble any known algae, so their evolution probably had progressed for some time before the formation of the earliest fossils yet known. The rocks formed during the period just preceding the appearance of fossil land plants indicate a period of frequent and extensive droughts. Such intermittent dry periods would have given considerable survival advantage to algae possessing mutations and gene recombinations for more waterproof coverings, for conducting tissues, and for means of reproduction that could take advantage of intermittent moisture. Many evolutionists believe that the drought-resistant algae produced by natural selection during this period later began to move into the moister land environments, where further environmental pressures led to the natural selection of the bryophytes, psilophytes, and other simple land plants.

Lycopodophyta

The lycopods, or club mosses, include more than a thousand living species grouped in five genera. The lycopods appeared about 370 million years ago. They play a relatively minor role in the modern plant world, but fossils show that they were abundant about 300 million years ago, even forming large trees. Today, however, only small plants a few inches high represent this phylum.

In lycopods, there is a clear differentiation between the structures of the root and of the aerial part of the plant, or *shoot*. The small, simple leaves contain a single vascular strand, or *vein*. Sporangia are produced in or near the leaf *axils*, the points where leaves join the stem. In many species, the leaves with sporangia at their axils are grouped into clublike cones at the tips of branches. The life cycle includes a gametophyte generation consisting in most species of a plant very different in appearance from the sporophyte. In most species, the gametophyte is a relatively simple organism with little tissue differentiation. In some relatively modern species, the gametophyte develops within the spore and produces gametes very soon after germination.

Arthrophyta

The arthrophytes, or horsetails, are represented today by about 25 species grouped in the genus *Equisetum*. Like the lycopods, the arthrophytes appear in the fossil record of a little more than 345 million years ago and were abundant as trees and herbs for some time. The arthrophytes have differentiated roots and shoots. In the living species, the leaves are small and have single veins, forming *whorls* at intervals along the stem. The sporangia are produced on highly modified umbrella-shape structures that cluster into a conelike organ at the tip of a branch. The life cycle is similar to that of the lycopods, with a small, relatively undifferentiated gametophyte and flagellated sperm cells that move through water to reach the egg cells.

Many botanists believe that the psilophytes, lycopods, and arthrophytes represent three distinct lines of simple vascular plants that independently

Figure 37.12a (left). Photograph of the field horsetail *Equisetum*. Note the fertile stalks with terminal sporangia and the infertile vegetative stalks.

Figure 37.12b (right). Life cycle of the common horsetail (*Equisetum arvense*).

(a) Asexual spores

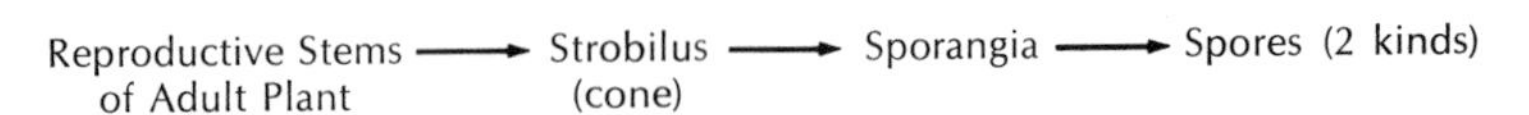

(b) Sexual Reproduction

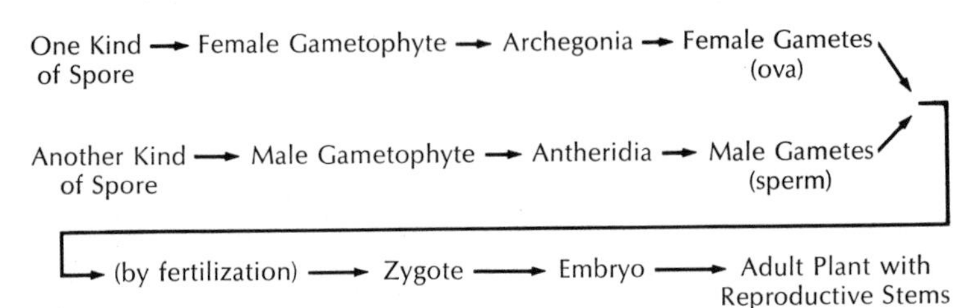

evolved from the green algae and for a time occupied dominant positions among the land vegetation. Later, they were crowded out by the more successful and more complex land plants, leaving only a few surviving species of lycopods and arthrophytes.

Pterophyta

Like the vascular plants, the pterophytes, or ferns, appear in the fossil record of about 345 million years ago and were abundant about 300 million years ago. The ferns, however, survived and remain an important part of the modern vegetation, with some 9,000 living species.

The most obvious feature differentiating the ferns is the large and complex leaf of the sporophyte, with its branching pattern of veins. Living ferns range from large, treelike species that grow in tropical rain forests to small plants that float in water (Figure 37.13). Most ferns grow in moist, shady locations, for they require a film of water through which sperm cells can move from antheridia to archegonia.

The sporophyte consists of large leaves arising from a stem, or *rhizome*, that is underground in many species. Often, the leaves are compound, with many small leaflets. Roots grow from the stem just beneath the leaf bases.

In most species, the sporangia are clustered in compact groups (*sori*) on the underside of the leaves. Spores falling on moist soil germinate to produce small, simple gametophytes. The haploid gametophyte is an independent, photosynthetic plant, living on the forest floor and absorbing nutrients from the soil through small rhizoids. Archegonia and antheridia are produced on the gametophyte, and sperm cells from antheridia swim to archeogonia, where they fuse with egg cells to form zygotes. The diploid zygote develops within the antheridium, absorbing nutrients from the gametophyte body during the early stages of its growth. As the diploid sporophyte becomes established, the gametophyte dies and disintegrates.

The ferns are thought to have evolved from the psilophytes. The earliest known fossil ferns are from rocks formed some 345 million years ago. It is difficult to account for the success of the ferns in surviving into the present while the psilophytes became extinct. Perhaps the structure of the ferns permitted the plants to survive occasional dry periods more readily and to take advantage of less abundant moisture for sexual reproduction. The ability of the underground rhizome to survive through an unfavorable season

Figure 37.13a (above). Photograph of a group of living temperate zone ferns in a woodland habitat.

Figure 37.13b (below left). Generalized fern life cycle.

Figure 37.14 (right). Close-up photograph of the underside of a fern frond showing developing sori, or spore cases.

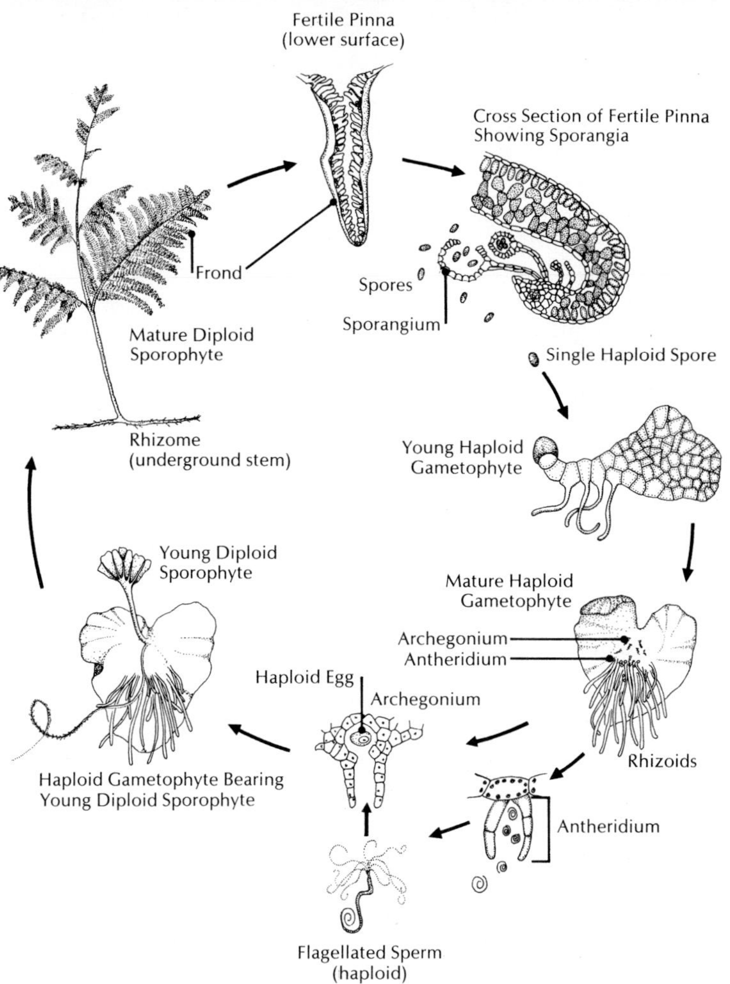

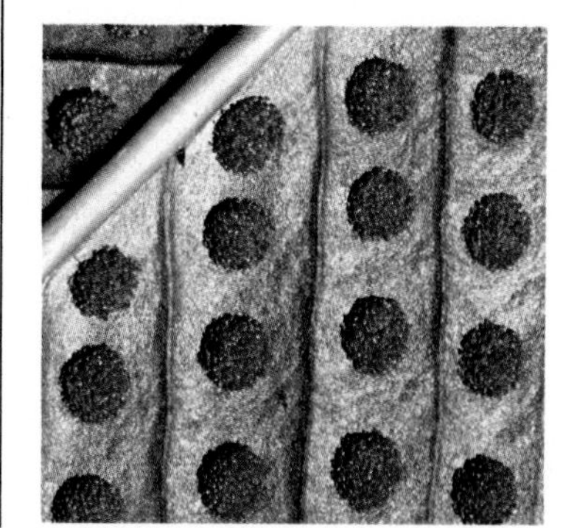

Figure 37.15. Photograph of ferns growing in a temperate rain forest habitat. Olympic National Park, Washington.

and then regenerate the leaves may be an important adaptation for survival on land, permitting the species to survive even if conditions do not permit successful sexual reproduction during a particular season.

Seed Plants

The nonvascular plants and certain vascular plants (such as ferns) are able to survive on land, but they are limited to relatively moist and shady environments. At certain crucial points in the life cycle, harsh environmental conditions easily can destroy the plant. Perhaps the most important limitation is the need for a film of water through which sperm cells can move from the antheridium to the archegonium. Even when water is available, this journey is a relatively difficult one, and only a small percentage of the sperm cells released succeed in fertilizing egg cells. The young embryo that develops from the zygote is a delicate and easily destroyed structure. In such plants as the ferns, it is given some initial nourishment and protection by the tissues of the gametophyte, but it still must assume an independent autotrophic existence relatively soon and must develop at the location of the female gametophyte within a short time after fertilization. The gametophyte itself is a delicate organism in plants such as the ferns, usually lacking complex vascular tissues and being able to survive only in quite moist conditions.

The seed plants (phyla: Cycadophyta, Coniferophyta, and Anthophyta) possess adaptations that protect the plant against these dangers. These adaptations have been so successful that the seed plants are by far the most conspicuous members of the modern land vegetation, and they apparently have crowded out the simpler vascular plants that dominated the landscapes some 300 million years ago.

CYCADOPHYTA

The modern cycads consist of only about 90 living species that grow mostly in the tropics. The sporophyte is a relatively large plant with a thick, treelike stem and a massive central *taproot.* A rosette of large, compound leaves forms at the apex of the stem, which is covered with the leaf bases of leaves shed in earlier years. The stem grows very slowly, and plants only a few feet high are often hundreds of years old.

The sporangia are produced on the scales of cones. *Macrospores* are produced in large cones on some trees; *microspores* are produced in smaller cones on other trees. Each microspore develops into a multicellular *pollen grain,* and these grains are scattered from the cones by wind. Each macrospore develops into a female gametophyte with an archegonium containing a large egg cell. The female gametophytes remain within the cones, surrounded by a structure called the *integument,* which forms from tissues of the parent sporophyte. A pollen grain is caught on a small drop of fluid at the lip of an opening in the integument and is drawn into a pollen chamber within the integument, where the male gametophyte within the pollen grain germinates and grows a tube, extending into fluids near the female gametophyte. Two ciliated sperm cells are produced in the male gametophyte and are released from the tube. One enters the archegonium and fuses with the egg cell to form a zygote.

Several months may elapse between the time the pollen grain is trapped in the pollen chamber and the time that fertilization of the egg cell occurs. Immediately after fertilization, the embryo begins to develop, absorbing nutrients from the tissues of the female gametophyte. The cells of the embryo divide several times and begin to differentiate, forming a stemlike structure with two or more cotyledons, or seed leaves. The nutrient materials obtained from the female gametophyte are stored in the cotyledons. The integument forms a hard seed coat around the embryo. The seed coat with its enclosed embryo forms a *seed.*

The seed may be freed from the cone and carried some distance by wind, water, or animals. The embryo remains dormant inside the seed until suitable conditions of temperature and moisture cause it to resume growth. The root and stem then break through the seed coat and grow by using the nutrients stored in the cotyledons, which remain partially enclosed in the seed. The stem soon develops leaves, and the new sporophyte begins its independent existence as a photosynthetic autotroph.

In the life cycle of the cycad, there is no need for a film of water to bring the gametes together for fertilization. The delicate gametophytes are well protected; the female gametophyte remains hidden inside the cone produced by the sporophyte, and the male gametophyte is exposed only in the form of the dormant and resistant pollen grain during its journey from a

Figure 37.16a (above). Close-up photograph of cycad cones.

Figure 37.16b (below). Photograph of a large cycad, the sago palm.

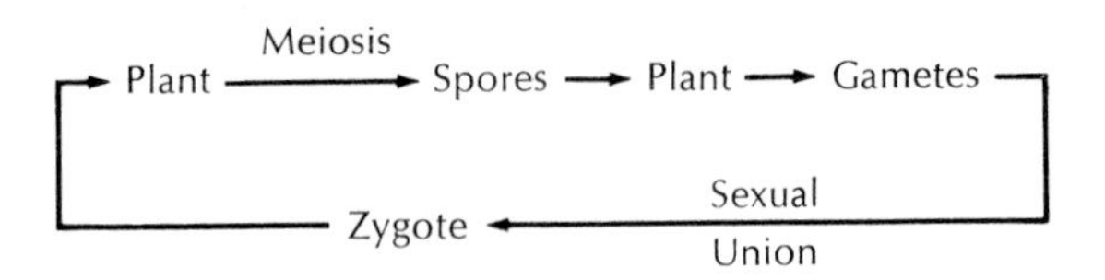

Figure 37.17 (opposite). Generalized life cycle of a cycad.

Figure 37.18 (below). Bishop pines at Iverness Ridge in Point Reyes National Park.

male tree to a female tree. The embryo carries out its early growth while still within the protecting tissues of the parent sporophyte. It is protected by the tough seed coat during its dispersal from the parent tree and begins growth with a store of nutrients derived from the female gametophyte.

Some 300 million years ago, a group of cycadophytes known as *seed ferns* flourished in the forests of that time. The seed ferns resembled the modern tree ferns, with an unbranched stem and a crown of large, fernlike, leaves. Unlike the ferns, however, these ancient cycadophytes produced seeds. Some 250 million years ago, the seed ferns are hypothesized to have given rise to the ancestors of the modern cycadophytes.

CONIFEROPHYTA

The conifers (Coniferophyta) are prominent members of the present land vegetation, especially at high altitudes and in colder regions. In most conifers, the leaves are relatively small and simple. Lateral branches are formed from the main stem, and the form of the mature tree is pyramidal in most species. Elongation of stems and roots occurs through cell division in apical meristems. A *cork cambium* near the outer surface of the cortex divides to form new cortex cells and thick-wall *cork cells.* The cork cells lose their cytoplasm and nucleus and form cork, the porous tissue that makes up the protective bark of the tree. A *vascular cambium* between the xylem and phloem layers produces new vascular cells. As the tree grows, the woody xylem layer becomes thicker and thicker. Because the central stem, or trunk, is long and straight in most conifers, these trees are a valuable source of lumber.

The life cycle of conifers is similar to that of cycads, but with a few major differences. The pollen and seed cones are usually produced in different regions of the same plant. When the pollen grain germinates within the

Figure 37.19. Generalized life cycle of a pine.

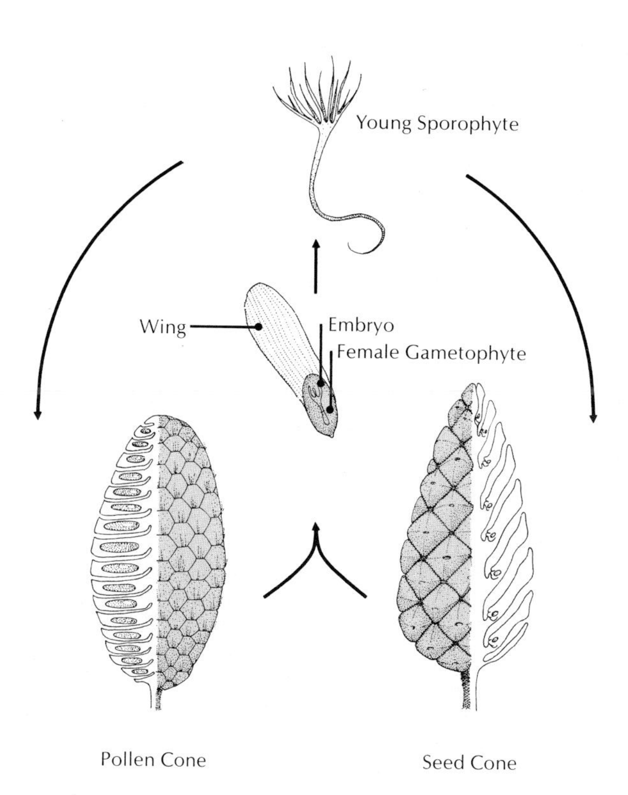

female cone, it extends a tube into the archegonium. A nucleus from the male gametophyte moves directly into the egg cell without formation of a separate male gamete. The apical meristems of both the stem and the root are well developed in the embryo before it ceases growth to enter the dormant seed stage.

Fossil conifers are found in rocks formed about 350 million years ago. The elimination of the need for a fluid-filled space through which the sperm cell could move from the male gametophyte to the female gametophyte may have been the slight advantage that permitted the conifers to thrive into modern times while the cycads largely disappeared. Presumably, the conifers were even more prominent in the land vegetation before they were crowded out of the warmer regions by the flowering plants.

ANTHOPHYTA

Although the anthophytes, or flowering plants, were relative latecomers in evolution, they are by far the most abundant of modern land plants. More than 250,000 species have been identified, and they exist in a great variety of body forms. The oldest undoubted fossils of flowering plants are found in rocks formed about 130 million years ago, and by 80 million years ago the anthophytes had become the dominant members of the land vegetation. Some fossils that might represent early flowering plants have been found in rocks formed as long as 200 million years ago, but there is no firm evidence as yet that these plants did produce flowers.

Like the cones of cycadophytes and conifers, the flowers of anthophytes are highly specialized reproductive organs produced at the tips of stems.

Figure 37.20a (above). Generalized angiosperm life cycle.

Figure 37.20b (below). Photograph of a flowering eucalyptus tree (*Eucalyptus sideroxylon*).

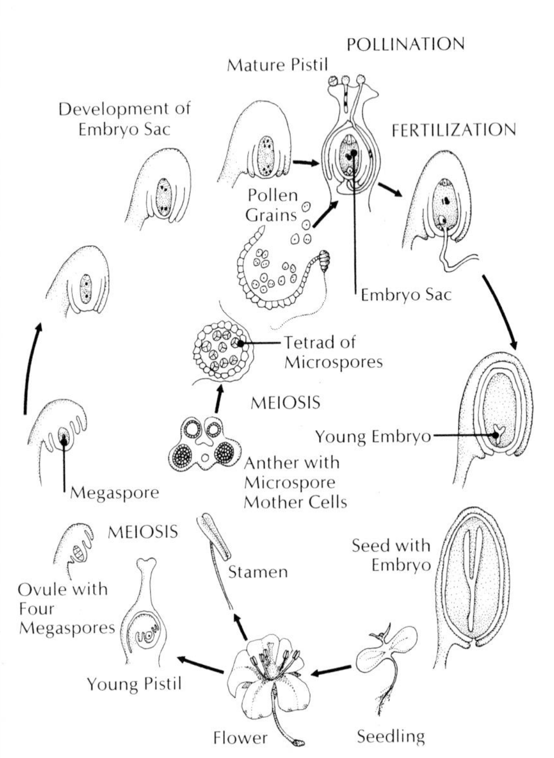

Although there are wide variations among the multitudinous species of anthophytes, most flowers show the general features outlined in Figure 37.21. The prominent parts are a group of *petals* forming the *corolla*, which is set within a group of *sepals* forming the *calyx*. The petals are white or colored in most flowers, whereas the sepals are green and leaflike. The corolla and calyx together constitute the *perianth*.

Within the perianth are the *stamens*, the pollen-producing organs, and the *pistils*, the seed-producing organs. In many flowers, a group of pistils may be fused together. The enlarged base of the pistil is the *ovary*, within which are several *ovules*. Megaspores, or haploid spores, are formed within the ovules. The slender portion of the pistil above the ovary is the *style*. The tip of the style, the *stigma*, is a structure modified to capture pollen grains. Each stamen is composed of a stalklike *filament* tipped by a pollen-containing *anther*.

The *meiospores*, or pollen grains, are produced within the anther. Each meiospore consists of a single haploid cell enclosed by a distinctively sculptured wall. While still within the anther, the meiospore cell divides to produce two cells of unequal size. A further division of the smaller cell to produce two sperm nuclei may occur before the pollen grain leaves the anther or, in some species, after pollination. In either case, the pollen grain may be regarded as an immature male gametophyte when it leaves the anther.

Each ovule consists of a large central *megasporocyte* (also called a megaspore mother cell) enclosed by one or two layers of integuments. A small opening, the *micropyle*, occurs in the integument layer of the ovule. By meiotic division, the megasporocyte produces four haploid megaspores, three of which disintegrate. One of the megaspores divides mitotically to produce a saclike female gametophyte, which in most species contains eight cells. Because of its shape, the female gametophyte was long ago called the *embryo sac*, a term still in use.

The pollen grains are transferred from the anthers of one plant to the stigmas of another by wind, water, or animals (usually insects). In most species, various mechanisms of structure or timing prevent self-pollination of a plant. When it reaches the stigma, the pollen grain produces a tube that grows down through the style, enters the ovary, penetrates the micropyle of an ovule, and pushes into the embryo sac. The tip of the pollen tube then releases the two sperm nuclei into the embryo sac. One of the sperm nuclei fuses with the egg nucleus (in most cases, the nucleus nearest the micropyle) of the embryo sac, forming the diploid zygote. The other sperm nucleus fuses with two nuclei near the center of the embryo sac to form a triploid nucleus called the *primary endosperm nucleus*. This process of *double fertilization* occurs only in the flowering plants.

The remaining five nuclei of the embryo sac disintegrate. The zygote divides mitotically to form the embryo. The primary endosperm nucleus undergoes rapid mitotic division to form a tissue called the *endosperm*, whose cells become filled with starches and other organic materials supplied by the parent sporophyte. During this early stage of development, the embryo becomes differentiated into an axis with a root apical meristem at one end and a shoot apical meristem at the other. One or two cotyledons form near the shoot apex. In a few species, such as orchids, the embryo enters a dormant state after only a few cell divisions of the zygote have been completed.

During the development of the endosperm and embryo, the ovule expands. The integuments are transformed into a hard seed coat that encloses

Figure 37.21a (upper left). Close-up photograph of blue flax showing the stamen and pollen.

Figure 37.21b (right). Generalized structure of a flower. The entire male portion is called the stamen; the entire female portion is called the pistil.

Figure 37.22 (lower left). Photograph of a Frangipanti flower. These small tropical trees are grown for their showy and fragrant flowers, which are popular for leis in Hawaii.

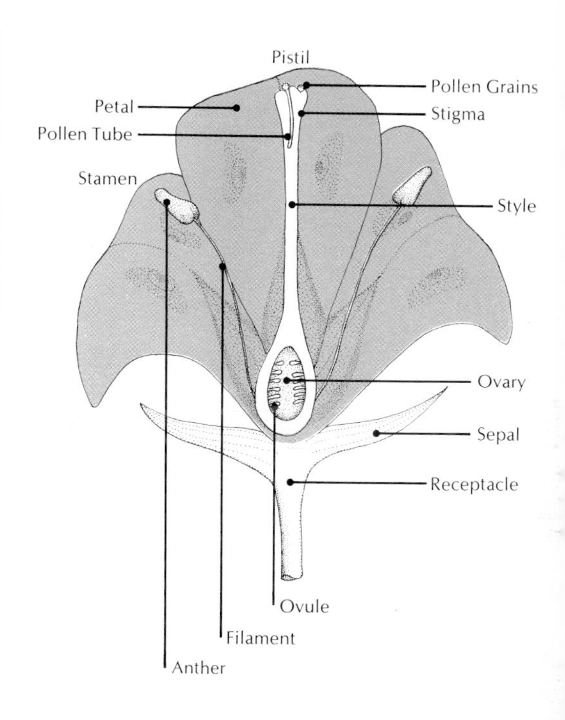

Figure 37.23. Photomicrograph of pine pollen grains.

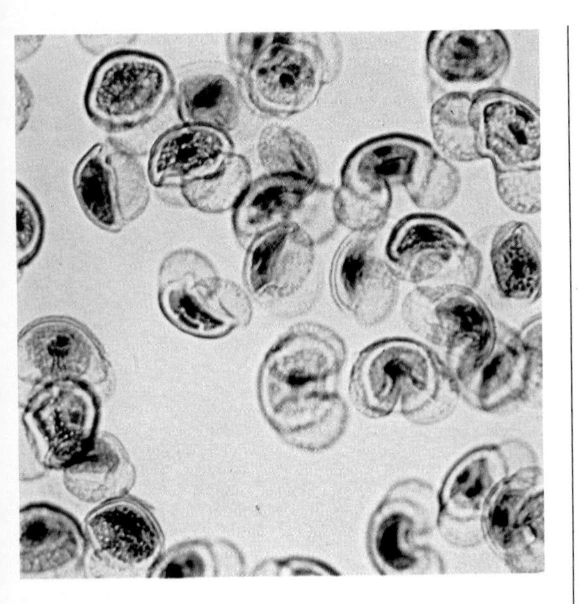

the endosperm and embryo, which is now in a dormant state. After the seed is transported to a new location away from the parent plant, the embryo resumes growth when conditions are favorable, supported in its early stages by the nutrients of the endosperm.

A great diversity of mechanisms for seed dispersal are present among the flowering plants. The seeds of orchids are minute, dustlike objects readily transported by wind. The cells of the cotton seed coat are differentiated into very long hairs (cotton fibers) that aid in dispersal. The seed coat of flax becomes sticky so that the seed tends to become attached to the coat of passing animals. The seeds of some violets and euphorbeas have an appendage, the *caruncle*, that is rich in protein and fat. Ants gather the seeds in order to feed on the caruncle, carrying the undamaged seed coat with its enclosed embryo to new locations.

In many flowering plants, the seeds are dispersed with the aid of a structure called the *fruit*, which is essentially a container for the seeds. While the seeds are developing from the ovules, the ovary (and, in some species, other parts of the flower) develops into the fruit. At maturity, the fruits of many species split open, releasing the seeds. In peas and other legume plants, the fruit splits open explosively when dry, throwing the seeds several feet. In poppies, the fruit has a circular row of holes at its apex through which the seeds sift out as the fruit is shaken by wind.

In many other species, the seeds remain enclosed within the fruit during dispersal. The fruits of maple and ash trees possess flattened wings, enabling the fruit with its enclosed seeds to travel some distance from the parent tree by a whirling motion similar to the rotation of helicopter blades. The fruits of cocklebur, sandspur, and many other plants have hooks or barbs that facilitate attachment to the fur of animals or the clothing of humans. In many species, the fruit becomes fleshy, juicy, and edible. Birds and other animals eat these fleshy fruits, and the seeds pass through the digestive system unharmed and, in some cases, even better prepared for germination.

Flowers occur in a great variety of shapes, sizes, and colors, and the appearance is largely determined by the nature of the perianth. Much of this variety undoubtedly serves to facilitate pollination of the plants by insects, birds, and other animals. Various species of animals search out particular shapes, colors, or scents of flowers. The animal obtains nectar or pollen from the flower for food, and, in the process, some pollen from the anthers is attached to the animal. When the animal visits a second flower for food, some pollen is transferred to the stigmas of the second flower.

In many cases, there is an amazing correspondence between the flower form and the habits of the pollinating insect or bird. The flowers of one member of the phlox family (*Ipomopsis tenuituba*) have petals united to form a tube. This tube is closely similar in length to the proboscis of the hawk moths that pollinate the flower. Furthermore, the flowers open about sundown, just as the moths begin to search for food, and they produce a greater fragrance and flow of nectar at night, when the moths are active. It seems clear that the evolution of the flowering plants and of the animals that pollinate them has been a mutual interaction in which changes in each kind of organism have affected the selection of the other.

The success of the flowering plants appears to be due to a combination of several factors—development of highly differentiated vascular tissues, exploitation of the ability to survive unfavorable seasons in the seed stage

Figure 37.24 (above). Photograph showing seeds in the open seed pods of the marsh marigold.

Figure 37.25 (below). Photograph showing seed dispersal in the dandelion *Taraxacum officinale*.

or to drop leaves and greatly reduce the rate of metabolism until favorable conditions return, and effective utilization of animals as agents of pollination and seed dispersal. Because of their complexity and their present dominance of the land vegetation, the flowering plants are regarded as the most advanced members of the plant kingdom.

FUNGI

The kingdom Fungi includes those multicellular organisms that specialize in absorptive heterotrophy, obtaining a supply of organic nutrients by absorbing them from the environment. In one way, they might be regarded as

Figure 37.26. Life cycle of the black bread mold *Rhizopus nigricans*.

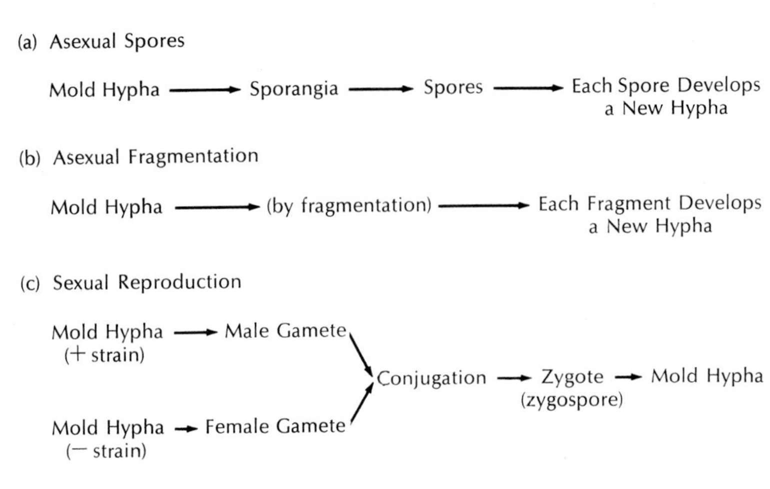

animals that simply absorb food passively rather than ingesting it actively. However, because most fungi are immotile and structurally resemble algae, they have traditionally been classified in the plant kingdom.

Some unicellular organisms are included in the fungi because of their close similarity to more complex organisms of this kingdom. The more complex fungi have a body composed of filaments (called *hyphae*), which in most species are made up of multinucleate cells or of long multinucleate tubes without cross-walls to separate individual cells. The entire body is called the *mycelium*. The absorbing body of most species shows little differentiation, but the more complex species have reproductive bodies made up of differentiated tissues. Although some fungi have life stages that move in an amoeboid fashion, most species live embedded in a source of nutrients and are immotile. The life cycles of most species include both sexual and asexual reproduction.

In the classification system used in this book, the slime molds (Myxomycophyta) are classed as a phylum of the kingdom Fungi, although there is considerable disagreement among taxonomists about the proper classification of this group of unusual organisms. The true slime molds (class Myxomycetes) form a multinucleate body (*plasmodium*) without cell walls and travel by amoeboid movement over damp soil or on the undersides of logs and rocks. The plasmodium feeds both by absorbing nutrients and by ingesting small food particles. Under favorable conditions, the plasmodium may grow large enough to cover an area of several square feet, although most are much smaller. When food becomes scarce or the plasmodium is threatened with desiccation, the cytoplasm of the plasmodium gathers into clumps or "blebs," forms cellulose cell walls, and builds multicellular sporangia (Figure 37.28). Within the sporangia, haploid spores are produced by meiotic division. The spores, scattered by wind, germinate upon suitable moist substrates to form flagellated gametes. Fusion of two gametes produces a flagellated diploid zygote, which eventually loses its flagella and becomes amoeboid. As this cell feeds, it undergoes nuclear division and eventually forms a plasmodium.

The cellular slime molds (class Acrasieae) show a similar life cycle, except that the zygotes remain as independent, uninucleate amoeboid organisms during the feeding stage and come together to form a multicellular *pseudoplasmodium*, which then develops into a multicellular sporangium.

Figure 37.27 (above). Photomicrograph of live budding yeast cells.

Figure 37.28a (middle). Close-up photograph of the plasmodium of the slime mold *Physarum*.

Figure 37.28b (below). Photograph of the sporangia, or fruiting bodies, of the slime mold *Physarum*.

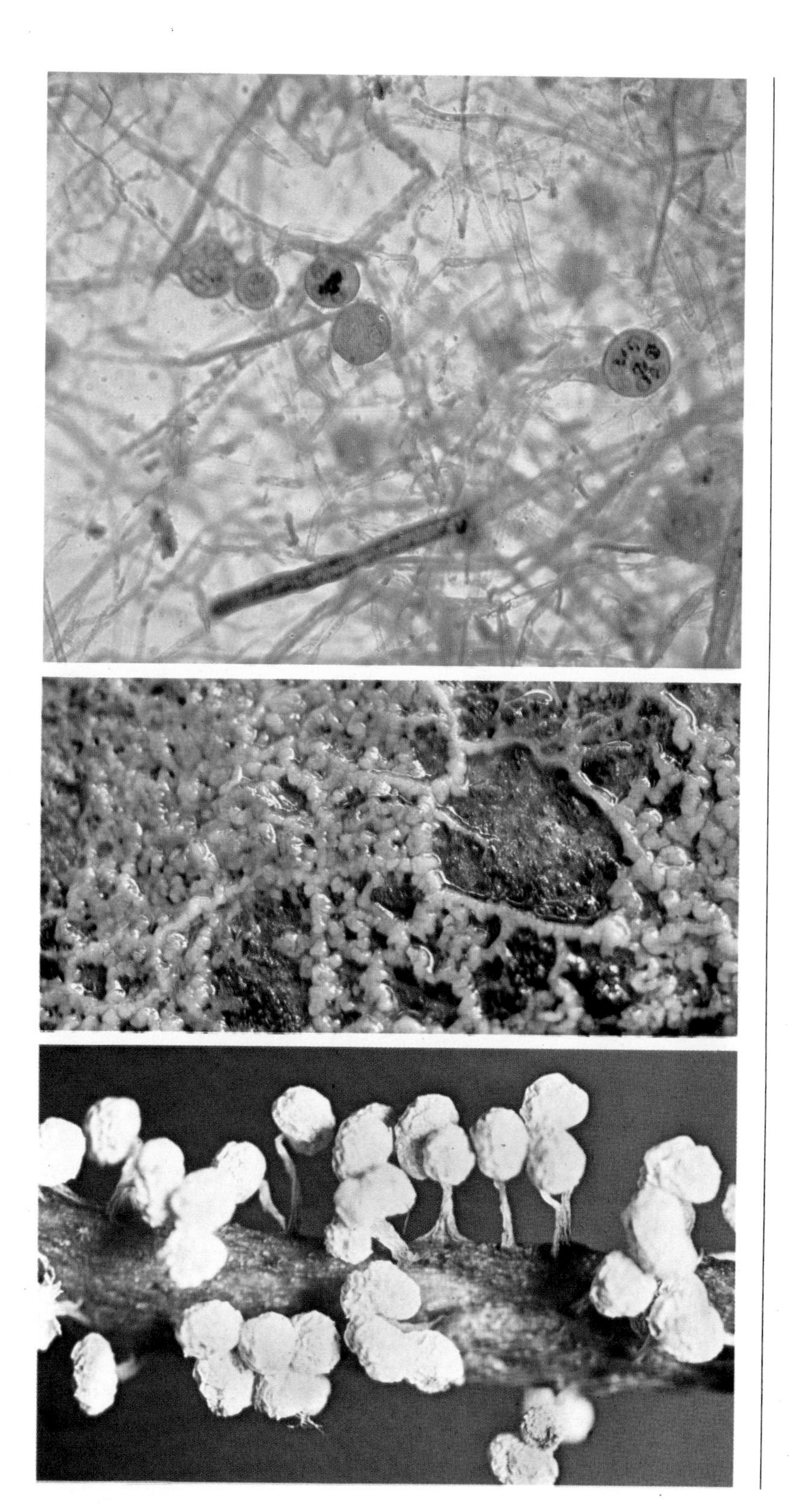

Interleaf 37.1

STUDIES OF DIFFERENTIATION IN *Dictyostelium discoideum*

The cellular slime mold has proven useful in studies of differentiation processes. In the laboratory, the free-living amoeboid cells of *Dictyostelium discoideum* can be grown in a broth medium. Thus, billions of identical healthy cells can easily be collected for biochemical study. When the cells are washed free of the medium and placed on wet filter paper, subsequent stages in development occur synchronously in all the cells. Thus, every stage in multicellular differentiation is open to convenient biochemical analysis.

Because the cells grow as individual amoebae, their genetic processes can be studied by the same powerful techniques that have proved so suitable with bacteria and other microorganisms. Amoebae can be treated with mutagenic chemicals and spread out in a dish with bacteria as food. Cells that survive under the conditions being studied grow and eat their way through the lawn of bacteria, making a clear spot, or *plaque*. Because each strain of amoebae forms a distinctively shaped plaque, rare mutants can easily be detected, isolated, and analyzed by noting unusual plaques and collecting cells from them. This technique has yielded mutant strains with alterations of a large variety of normal functions and differentiations.

Almost every stage of development of *D. discoideum* raises interesting questions of cellular interactions that are similar to those found in vertebrate embryogenesis and metabolism. The first thing the individual cells must do when they are induced to develop is to get together. The cells move toward a higher concentration of cAMP (Bonner, *et al.*, 1969). As a few cells happen to come together, they secrete more cAMP into the area, and additional numbers of cells are attracted toward the spot. In this way, large numbers of dispersed cells gather into a compact mass. Because cAMP diffuses more rapidly than the amoebae move, the entire area tends to become saturated with cAMP. The cells solve this problem by excreting a slowly diffusing enzyme that breaks down cAMP to a compound that does not attract the cells.

The alteration of amoebae that causes them to move toward higher cAMP concentration may be the first step in differentiation. In the amoeboid state, the cells normally are attracted toward some unknown substance excreted by bacteria, thus tending to move toward food. When aggregation begins, the cells cease to move toward the bacteria and begin to move toward each other along the cAMP concentration gradient. This alteration is one of the clearest examples known of a sharp biochemical switch of a nature that can be called differentiation (Bonner, *et al.*, 1970).

After the cells have aggregated, they become mutually adhesive and remain as a single unit throughout the rest of development. The aggregate now begins to move as a worm-shape *pseudoplasmodium*, or slug, containing up to 100,000 cells. At this stage, the front cells begin to undergo the biochemical differentiations that will transform them into stalk cells, while the posterior cells begin to differentiate into spore cells. The proportion of prespore cells to prestalk cells is always 2:1, no matter how many cells have aggregated in the pseudoplasmodium. This observation clearly points to a mechanism that allows the cells to interact throughout the organism in order to set up the correct ratio of spore to stalk cells in the final fruiting bodies. Somehow the cells at the rear must be affected by the number of cells in front of them.

Several theories have been proposed to account for this interaction. One of the simplest theories suggests that interaction is mediated by the extracellular polysaccharide that is secreted by the cells and covers the whole organism with a layer of slime. The slime remains stationary on the supporting surface, and the cells move through it. Because the cells at the rear must follow on the slime laid down by cells at the front, the tip cells are the only ones free to control the direction of movement. Thus, the motion of thousands of previously independent cells is integrated into organismic behavior. Moreover, because the cells at the front secrete the slime and then move out of it, they are surrounded by a thinner layer of slime than are cells at the rear. It has been suggested that it is the amount of this polysaccharide coating that determines the course of differentiation. Cells at the front, surrounded by less polysaccharide, transform into stalk cells, whereas those at the rear, embedded in far more polysaccharide, transform into spore cells. This theory accounts for correct proportioning in a pseudoplasmodium containing any number of cells if the rate of movement of the pseudoplasmodium is directly proportional to its length and if the rate of slime secretion is proportional to the surface area. There is some evidence to indicate that these relationships do exist (Bonner, *et al.*, 1953, 1955). Mechanisms involving specific cellular adhesions and cell migration leading to differentiation in various ratios occur in the embryogenesis of almost all complex organisms.

All of the biochemical steps that result in spore and stalk cell differentiation in *D. discoideum* occur after the amoebae have ceased feeding and occur without growth or division of the cells. Amino acids needed to make new proteins are derived from breakdown of old proteins. Carbohydrates for the synthesis of the slime layer and stalk polysaccharides are derived from stored glycogen and the metabolism of other molecules already in the cells. The biochemical conversions require the synthesis of many new enzymes. At present, about 15 enzymes are known to accumulate during development, and 11 have been studied in detail. Some of these enzymes are involved in the breakdown of carbohydrates and amino acids, some catalyze steps in the synthesis of new polysaccharides, and others are formed only in spore cells, where they direct the formation of specialized polysaccharides (Newell, *et al.*, 1969). Although the studies of these enzymes have greatly increased understanding of the spatial and temporal pattern of differentiation, they have not led to any explanation of the nature of cell-cell interactions that cause the synthesis of these enzymes at specific times in specific cells. Nevertheless, such interactions must occur, and further studies are under way to attempt to elucidate them.

One of the clearest demonstrations of cell-cell interaction can be seen between pairs of mutant strains. For instance, two mutant strains that fail to form pseudoplasmodia can go on to form normal fruiting bodies with well-differentiated spore and stalk cells if cells of the two strains are mixed and incubated together. Alone, neither strain undergoes the biochemical differentiations required for fruiting body formation, but together they synthesize the enzymes required for normal differentiations (Yanagisawa, Loomis, and Sussman, 1967). This interaction may result from simple cross-feeding, as in the example of bacterial mutants (Chapter 17).

A slightly more complex cell-cell interaction that controls specific biochemical differentiations can be demonstrated during aggregation of normal cells. If formation of pseudoplasmodia is prevented, the synthesis of several enzymes in the amoebae is inhibited. However, if the cells are allowed to aggregate for eight hours and then prevented from normal pseudoplasmodium formation, the enzymes accumulate normally. It appears that the first steps in the induction of these enzymes occur during the first eight hours and that they require intimate contact of the amoebae. Because this period occurs long before the accumulation of mRNA for the enzymes, it has been called commitment (Loomis and Sussman, 1966). Experiments designed to isolate a "committer" that is transferred between the cells have not yet been successful. The interaction appears to be very subtle and may involve modification of a surface polysaccharide or of the cell membranes.

Some of the individual cells differentiate to form the stalk, others form the spore-bearing capsule, and still others undergo meiosis to form the spores. These organisms are of particular interest to biologists studying the mechanisms of cell differentiation and coordination (Interleaf 37.1).

Eumycophyta

Although they are relatively inconspicuous members of the biosphere, the true fungi, or eumycophytes, play a very important role in the cycling of elements. Along with the heterotrophic bacteria, they are responsible for most of the breakdown of organic molecules in the bodies of dead organisms, releasing the various elements in the form of simple inorganic molecules that can be used by plants as nutrients. Because many of the fungi are specialized to absorb nutrients from living plants and animals, they are also important to humans as the cause of many plant and animal diseases.

The true fungi, being completely specialized for absorptive nutrition, lack the ingestive-feeding plasmodial stage that characterizes the slime molds. Most species secrete digestive enzymes into the environment. These enzymes catalyze the initial breakdown of complex organic molecules into substances that can be absorbed through the membrane of the fungal body. The fungi lack the cuticle or corky bark that protects plants from desiccation on land. Because they are nonphotosynthetic, the fungi are able to avoid desiccation by living in the soil or inside trees and fallen logs. An aerial fruiting body is created for the dispersal of spores during reproduction.

The hyphae of most fungi are enclosed in a firm cell wall of chitin. If cross-walls or septa are present between cells, they have small openings that permit movement of cytoplasm and nuclei between cells.

The great variety of fungi (more than 200,000 species are known, and about 1,000 new species are discovered each year) rivals that of the plant kingdom, although most fungi are relatively small organisms. There is some evidence that the kingdom Fungi includes several independent evolutionary lines that emerged separately from the Protista.

The chytrid fungi (class Chytridiomycetes) are mostly microscopic organisms, ranging from unicells that live as parasites in plant cells to tubular forms that anchor themselves to a food source with threadlike rhizoids. The chytrids produce motile reproductive cells, each with a posterior flagellum.

The oomycetes (water molds, white rusts, and downy mildews) range from biflagellate unicells to species with complexly branched hyphae. The oomycetes produce a thick-wall *oospore* that develops from a large, nonmotile female gamete inside a specialized *oogonium*, or egg-producing cell. The oomycetes are of importance to man because they include species that cause plant diseases such as the late blight of potatoes and downy mildew of grapes.

The zygomycetes are a small but interesting group that includes bread molds, parasites of insects, and species that capture animals. Spores are produced in stalked sporangia that rise from the mycelium. Sexual reproduction may occur by the union of two adjacent hyphae. Fusion produces a thick-wall zygote, or *zygospore*, that remains dormant for some time and later germinates to form spores. Some species of this class are able to capture small worms with specialized loops formed of hyphae that tighten around the animal as it attempts to crawl through. Specialized hyphae then grow into the animal and it is digested and absorbed. Some species live as

Figure 37.29. Photograph of bracket fungi growing on the bark of a tree. Certain species of bracket fungi can rot the heartwood and sapwood of living trees, causing severe damage.

parasites in insect bodies, forming reproductive structures on the surface of the insect body after it has died.

The ascomycetes, or sac fungi, include unicellular and mycelial forms. The hyphae are divided into segments by perforated septa. Haploid spores are formed by meiotic division within a special, elongated cell (*ascus*) that forms at the end of a hypha. Each of the four spores divides mitotically, producing eight *ascospores* in each ascus. The ascospores are ejected through an opening in the end of the ascus and are scattered by the wind. Among the unicellular ascomycetes are the yeasts. The parasitic powdery mildews are mycelial ascomycetes. In the cup fungi, the hyphae intertwine to form a rather complex fruiting body, or *ascocarp*.

The basidiomycetes, or club fungi, are similar to the ascomycetes in many ways, but haploid spores are produced on minute stalks that form on the outside of a specialized terminal cell called a basidium. In most species, only four basidiospores are formed on each basidium, and these are scattered by the wind when mature. Many species produce umbrella-shape fruiting bodies, which are composed of closely intertwined hyphae that form pseudotissues. The complex structures of the fruiting body, or *basidiocarp*, can be separated into individual hyphae simply by pulling them

Figure 37.30 (opposite). Generalized life cycle of the common field mushroom (*Agaricus campestris*).

Figure 37.31 (left). Photograph of the puffball *Lycoperdon perlatum*. This species is noted for its large, pear-shape basidium.

Figure 37.32 (below). A large leafy type of lichen. Lichens are combinations of algae and fungi in a mutualistic relationship. The alga provides nutrients through photosynthesis and the fungus provides moisture and a slightly acidic environment, in which the alga thrives.

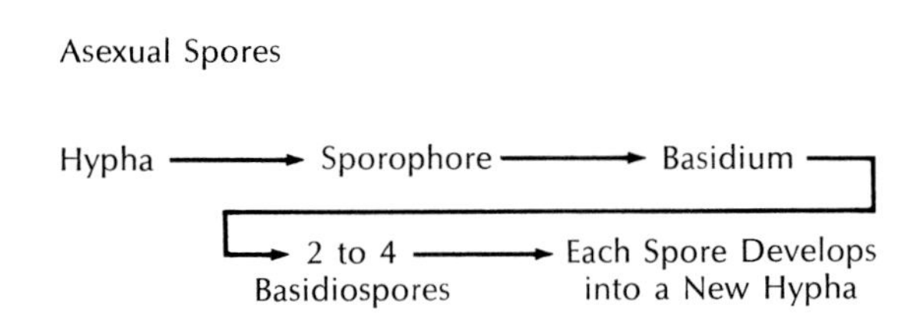

apart with a needle. Therefore, these structures are not considered to be made of true tissues. In many species, when the basidiospore germinates, it forms a uninucleate hypha. If this hypha contacts another hypha, they may fuse and the nuclei from one hypha move into the other. The nuclei remain side by side without fusing, dividing simultaneously as the hypha grows. Fusion of the nuclei to produce a true diploid nucleus does not occur until the formation of the basidium, just before meiosis. The basidiomycetes are thought to have evolved from the ascomycetes, with both groups probably invading the land at about the same time as the green plants.

Lichens

Lichens occur in many forms, including powdery layers on rocks or tree bark, thin crusts that look something like paint splotches on rocks or bark, leaflike scales protruding somewhat from the substrate surface, and tiny shrublike structures growing upward from surfaces or hanging from trees. Until the 1860s, the lichens were thought to be a group of plants, but in 1867 the Swiss botanist Simon Schwendener discovered that a lichen is actually made up of a species of blue-green or green algae and a species of fungi living closely together. The alga can be a chlorophyte, a cyanophyte, or, in a very few lichens, a xanthophyte. In most lichens, the fungus is an ascomycete, but basidiomycetes are present in a few lichens. The algal cells may be scattered among the fungal hyphae, or they may form distinct layers within the pseudotissues of the fungus. In most lichens, rhizoids from the hyphae penetrate into the algal cells. The interaction of the two species produces features in each that are different from those of free-living relatives. In other words, the lichen has properties of its own that make it a distinctive organism, not simply a mixture of two kinds of organisms living in close proximity.

The fungus most likely obtains organic nutrients from the alga. In turn,

the alga is shaded and protected from desiccation by the fungus. Experiments with radioactive tracers have confirmed that organic substances move rapidly from the algal cells into the fungus. Some biologists believe that the fungus should be considered a parasite, providing little benefit to the alga but absorbing nutrients from it. Among the wide variety of lichens, examples can be found to support either view.

Because in most cases the alga and the fungus can be recognized and classified as members of a genus of free-living organisms, most taxonomists do not create a separate category for the lichens but regard them as a case of *symbiosis*—the close association of two species. However, the lichens are a most interesting example of a situation in which something very much like a new kind of organism has been formed by a combination of two other kinds of organism. It is a situation not greatly unlike the origin postulated for the first eucaryotic cells.

FURTHER READING

Adams, *et al.* (1970) give a full discussion of the topics treated in this chapter. Sire (1969) provides a beautiful collection of color photographs of various plant structures. For further details on the structure and reproduction of various plant groups, see books by Bold (1964), Eames (1961), Foster and Gifford (1959), and Smith (1955). The evolutionary history of the plant kingdom is discussed in books by Andrews (1947, 1961), Delevoryas (1962), Good (1956), and Scagel, *et al.* (1965). Classification of plants is treated by Lawrence (1951), Porter (1967), Sokal and Sneath (1963), and Solbrig (1970).

For further information on the fungi, see books by Alexopoulos (1962), Alexopoulos and Bold (1967), Christensen (1965), and Smith (1955, Vol. 1).

Among many *Scientific American* articles relevant to this chapter are those by Arditti (1966), Batra and Batra (1967), Beadle (1948), Biale (1954), Clevenger (1964), Jacobs (1955), Lamb (1959), Naylor (1952), Overbeek (1968), Preston (1968), Salisbury (1958), and Scott (1962).

38
Animals

The evolutionary theme of the animal kingdom is ingestion as a way of feeding. Like fungi, animals are heterotrophic, obtaining their food from other organisms. Animals, however, are multicellular organisms that ingest food—taking tissues of other organisms in through their mouths for digestion within their bodies, rather than absorbing organic substances through the outer body surface as fungi do.

In the function of living communities, plants are producers that create food, animals are consumers that use some of this food, and fungi and bacteria are decomposers that break dead remains of both plants and animals down into inorganic substances. There are varied ways in which animals act as consumers. Herbivores eat living plant tissues; carnivores eat other animals, obtaining food energy from plants secondhand (or thirdhand or fourthhand); fungivores feed on the tissues of fungi; scavengers feed on the dead remains of other organisms; omnivores (man, for example) feed on some combination of living plants or animals or other foods; and parasites live on or in other organisms and take their food from their hosts. Many aquatic animals are filter feeders, straining from the water a mixture of different kinds of small organisms and dead particles. Many filter feeders are omnivores, but some specialize in particular kinds of food drawn from the water.

Ingestion as a way of life has led to the evolution of some of the structures considered to be characteristic of animals. First, most animals possess a mouth and a digestive tract that has glands or glandular cells, which secrete enzymes and other chemicals that assist in the digestion of food. Most animals have other organ systems that supplement the essential function of ingestion and digestion: a system of external respiration to bring in sufficient oxygen for the metabolism of the food, an excretory system to dispose of the waste products of metabolism, and a circulatory system to transport food, oxygen, and wastes within the organism. In addition, the great majority of animals have organ systems that make possible movement to seek out suitable food, to pursue food organisms, or to escape being eaten by carnivores. These organ systems include the muscular system; the skeletal system, to which the muscles are attached; the sensory system, by which food or danger is perceived; and the nervous system, which coordinates the sensory and muscular systems and directs appropriate behavior. Active movement, which once was considered characteristic of all animals, is an important adaptation for the ingestive way of life, but it is not needed by some filter feeders and parasites.

There are more than 1 million species of animals. Only 13 of the major phyla of the animal kingdom will be described in this chapter, of the 20 or so phyla recognized by most zoologists.

PORIFERA

A rock picked up at the seashore is coated with an orange crust a fraction of an inch thick, soft to the touch, apparently motionless. This crust is a typical sponge (Porifera). Why is this organism classified as an animal? Microscopic examination reveals a complex, peculiar organization (Figure 38.2). Water canals lead from openings on the surface of the sponge, through chambers and passageways to exits. Lining the chambers are distinctive cells with flagella that are surrounded by cytoplasm-filled collars. These collar cells are the feeding cells of the sponge, and they ingest food particles from the water as the flagella push a current through the canals. The

Figure 38.1. Characteristics of the animal phyla.

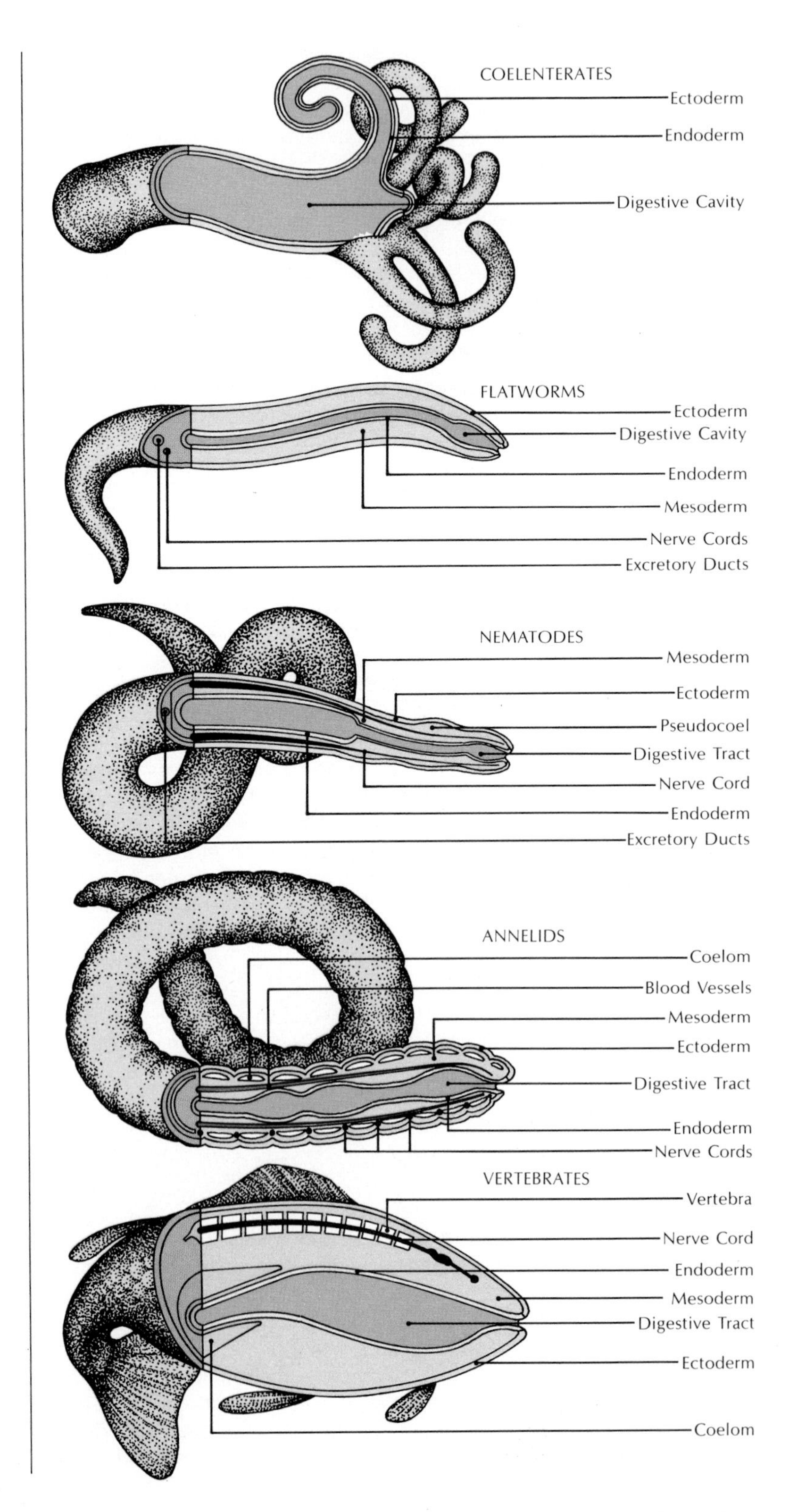

Figure 38.2a (right). A large sponge from the Carribean area.

Figure 38.2b (left). The internal structure of a sponge gives a clue to its filter-feeding life style. The body wall is perforated with canals lined with flagellated feeding cells (choanocytes), which keep a current of water moving through the animal and trap food particles.

sponge is thus, like many other aquatic animals, a filter feeder.

The body of the sponge is made up of a number of kinds of specialized cells, held in their places by a skeletal structure. The collar cells have been observed to pass food particles on to amoeboid cells, which may digest the food particles and then move through the tissue, passing the digested food on to other kinds of cells. Some cells, with long protoplasmic processes that unite with processes from other similar cells, form a nervous network through the body of many species of sponges. Muscle cells contain contractile fibers whose contraction can reduce the size of the canal openings on the surface of the sponge. Other cells secrete the materials of the skeleton.

In different groups of sponges, the skeleton is made of protein fibers, sharp spicules of calcium carbonate or of silica (glass), or mixtures of these materials. The familiar bath sponge (if it is not made of plastic) comes from a sponge with a skeleton of protein fibers. The great majority of sponges live in, and strain their food from, sea water, but some sponge species have become adapted to fresh water.

Almost any small piece broken or cut from an adult sponge can undergo cell division and produce a complete organism. This regeneration is a form of asexual reproduction. Sponges also reproduce sexually; sperms and eggs are formed from other cells, and the fertilized egg develops into a minute swimming larva. The sponges are thought to have evolved from colonial forms of flagellated protozoans. The oldest fossil sponges are found in rocks formed about 600 million years ago, and structures that may be the remains

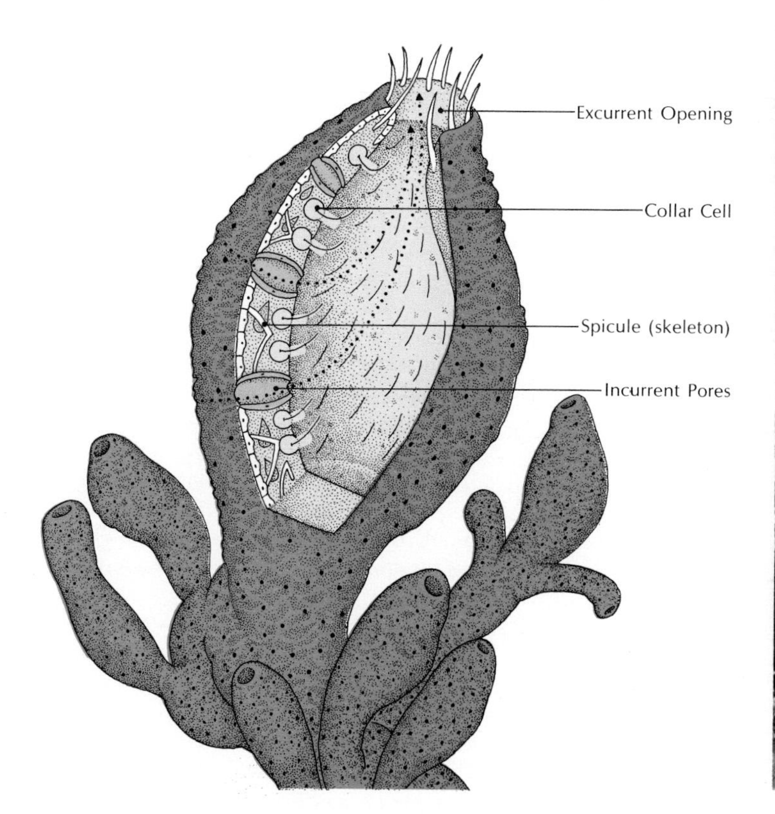

Figure 38.3a (above). Sea anemone from the Cayman Islands, British West Indies.

Figure 38.3b (below). A close-up photograph of star coral on the Cayman Islands, British West Indies.

of spicules from the skeletons of calcareous sponges have been found in even older rocks.

CNIDARIA

The phylum Cnidaria, or Coelenterata, includes such organisms as the sea anemones, corals, jellyfishes, and fresh-water hydras (Figure 38.3). The body organization of the coelenterates is simple in comparison with that of most other animals, but it shows a more definite tissue development than that of the sponges. Most of the coelenterate body is composed of two layers of cells: an outer layer that includes stinging cells, and an inner layer lining the digestive cavity. In many species, gelatinous material occurs between the two cell layers, and in some species there are other cells between the layers (Figure 38.4). In *Hydra*, there are no muscle tissues—movement is made possible by contractile fibers in the bases of some cells. However, some advanced coelenterates have real contractile tissues. The digestive system is a sac with only one opening that is lined by glandular and digestive cells. The nervous system consists of a simple network of cells with processes contacting one another.

The coelenterate body organization permits only simple feeding (and, for unattached forms, swimming) behavior. However, coelenterates are carnivores that catch and kill their food. The body surfaces—particularly on the tentacles—bear stinging cells with inverted threads. When an animal touches the tentacle, the threads evert, penetrating the tissues of the animal and injecting a poisonous substance. Swimmers who have blundered into the tentacles of a large jellyfish are aware of the consequences of being

Figure 38.4. Comparative cutaway views of the fresh-water coelenterate *Hydra* and the jellyfish *Aurelia*.

Figure 38.5a. Diagrammatic view of the internal anatomy of a planarian flatworm and a cross-sectional view of a planarian through the pharyngeal region of the body.

penetrated by thousands of these poison-coated threads. The great majority of coelenterates live in the sea, but the hydra and its relatives and a small number of jellyfishes are adapted to fresh water.

All coelenterates reproduce sexually. In most species, sperms and eggs are released into the water by male and female individuals, and fusion of the gametes occurs outside the adult organisms. The zygote develops into a free-swimming, ciliated larva. In different coelenterates, the larva may develop into a jellyfishlike adult (*medusa*) or may become attached to the substrate and develop into a hydralike adult (*polyp*). In some species, a polyp is formed that later buds off small medusae, which are the sexually mature organisms.

The evolutionary origin of the coelenterates is largely unknown. An ancestry among the colonial protozoans has been suggested. Fossils that appear to be remains of medusae and polyps have been found in rocks formed about a billion years ago, and the hard skeletons formed by corals are widespread in rocks formed during the last 500 million years.

PLATYHELMINTHES

Flatworms (Platyhelminthes) are the simplest animals that have well-defined bilateral (or two-sided) symmetry instead of the radial symmetry of the

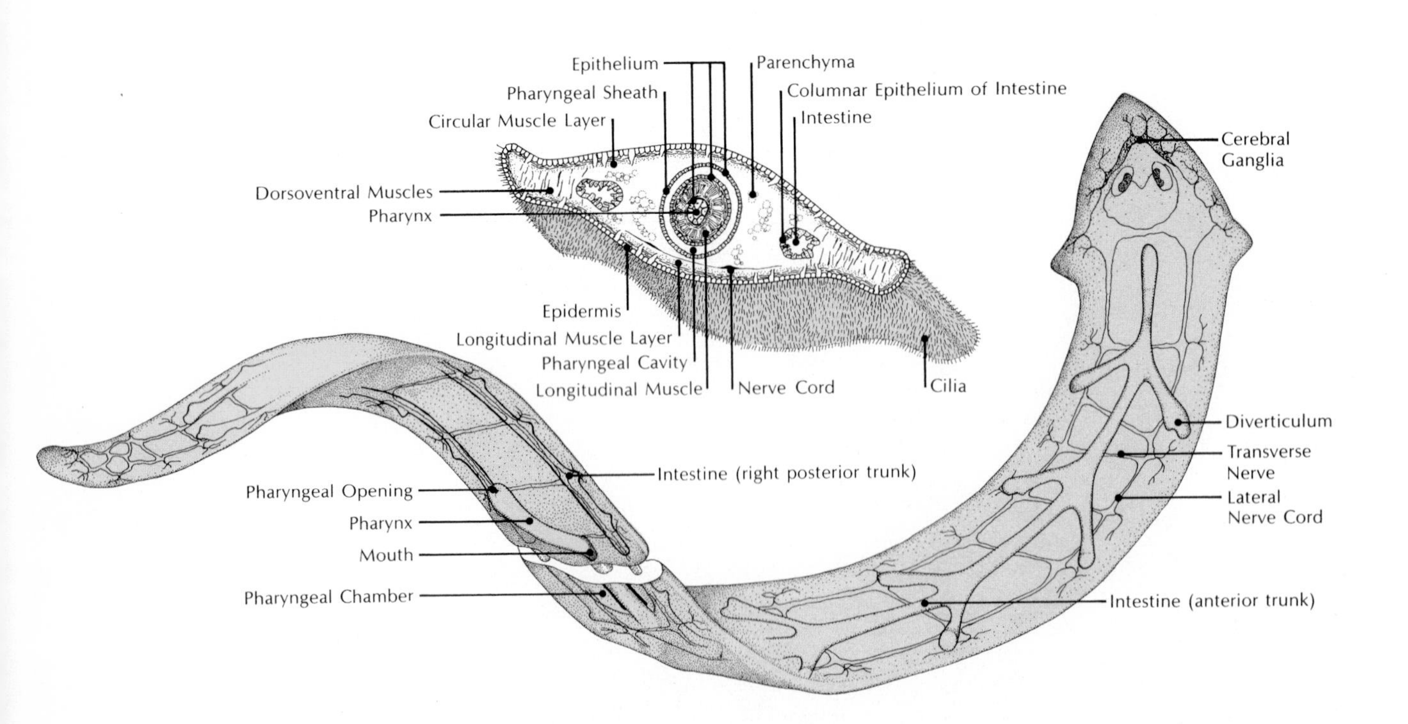

Figure 38.5b. The triclad flatworm *Stylochus*.

Figure 38.6a (upper right). The liver fluke *Fasciola hepatica*. *(Courtesy Carolina Biological Supply Company)*

Figure 38.6b (lower left). Diagram of the liver fluke to show general structural features.

Figure 38.6c (lower right). Copulating liver flukes.

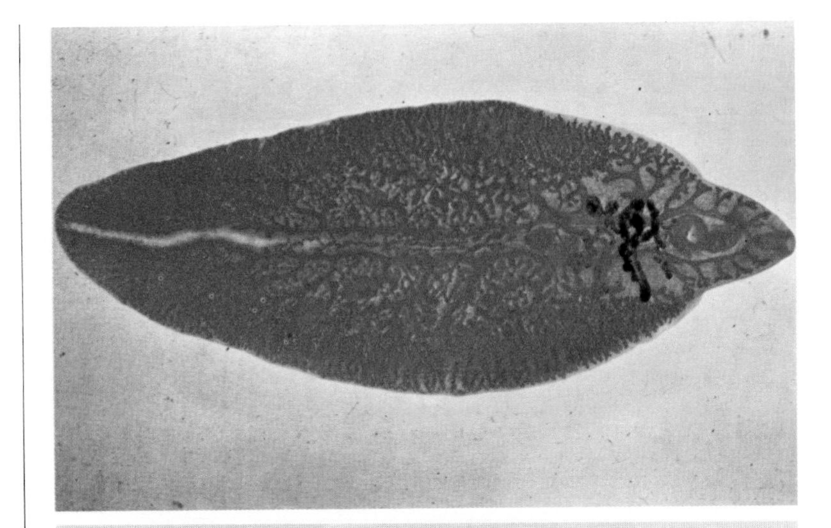

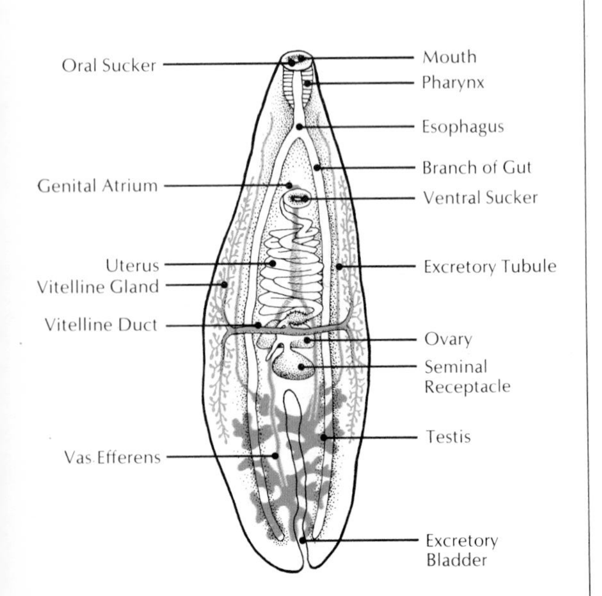

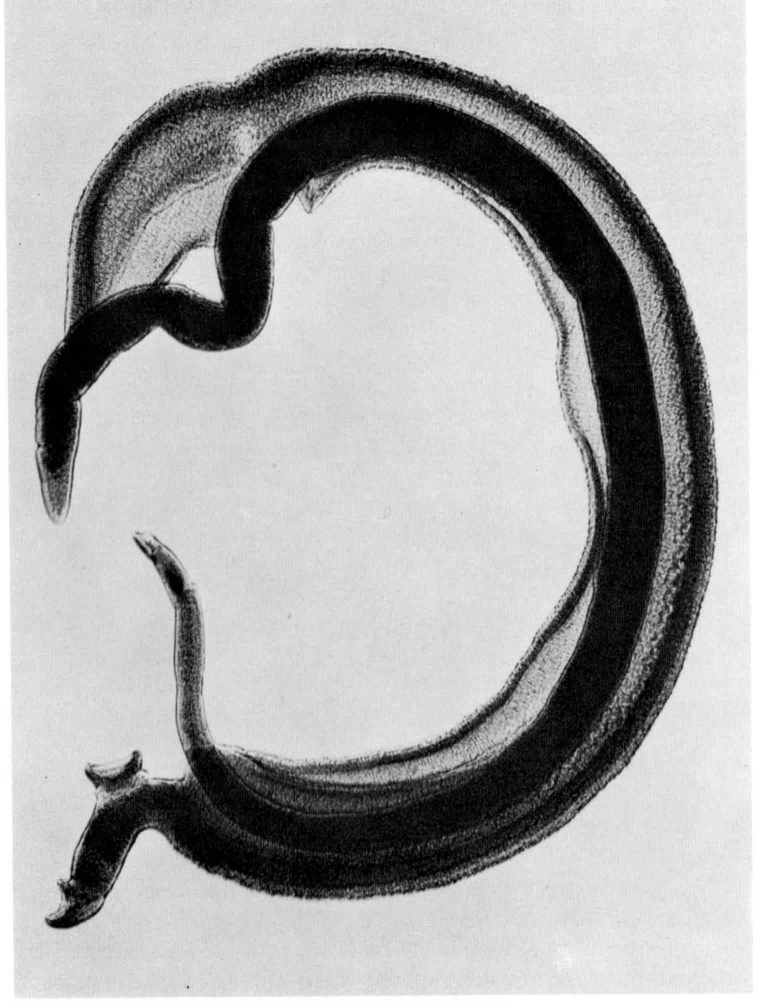

Figure 38.6d. Life cycle of the blood liver fluke *Schistosoma mansoni*. Adults reside in the liver, whereas copulation of the worms typically takes place in the hepatic portal system. Fertilized ova pass through the alimentary canal and begin larval development in water, then the snail host. Metacercana larva ultimately penetrate the human host while the person is in water, possibly bathing or washing clothes.

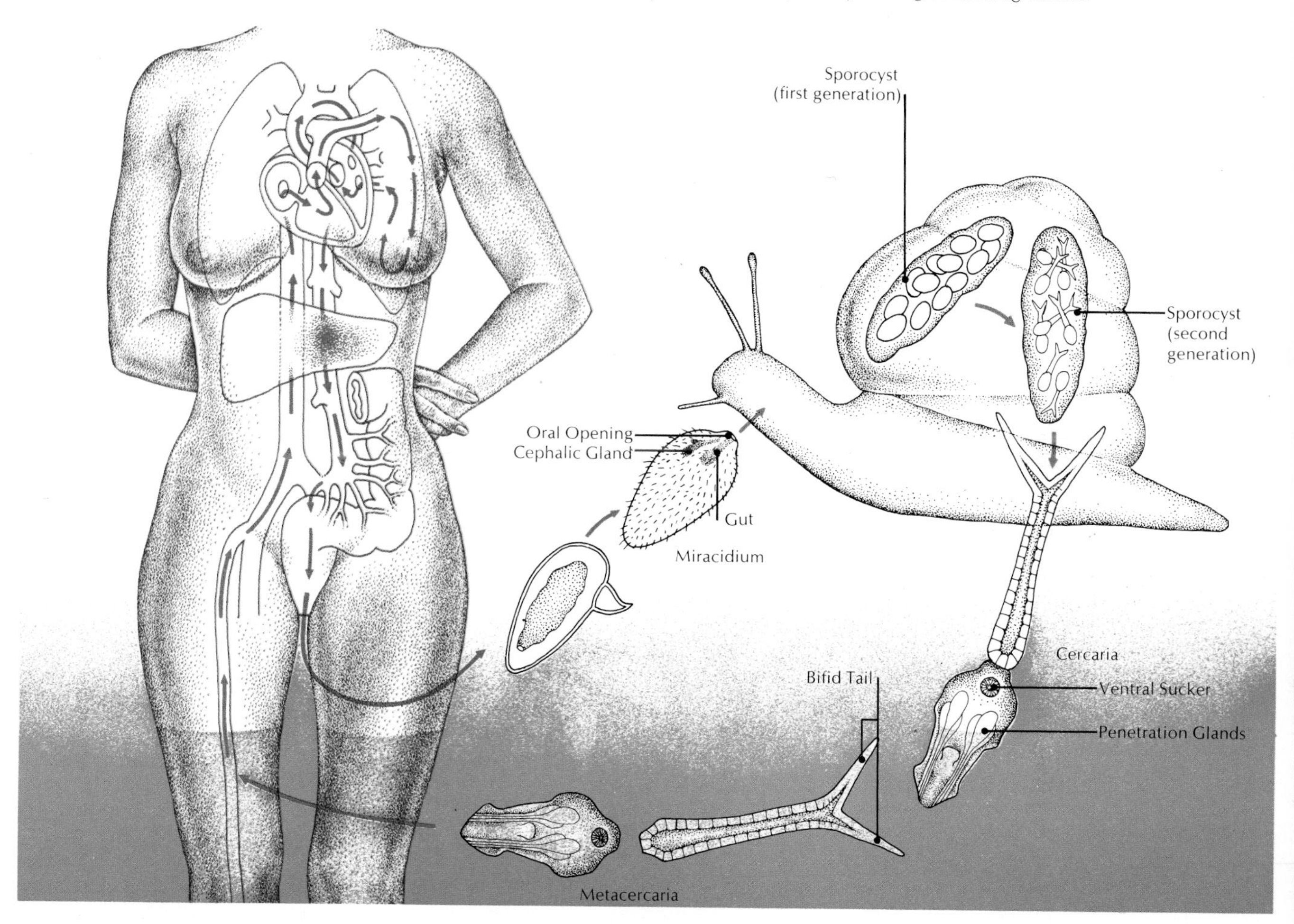

sponges and coelenterates (Figure 38.5). In the flatworm, as in the coelenterate, the digestive tract is a blind sac. However, in the body of the flatworm, extensive muscular and other tissues occupy the space between the inner and outer cell layers, and there is the beginning of a central nervous system—a pair of *ganglia* (clusters of nerve cells) at the head end of the worm. Many of the free-living flatworms are scavengers, sucking dead organic matter into the digestive cavity through the mouth and *pharynx* (throat); others are predators.

The Platyhelminthes include a diversity of marine forms, a good number of fresh-water species (among them the planarians), and a few terrestrial species. Among some flatworms, however, parasitism is a way of life—a phenomenon that is almost absent among sponges and coelenterates. Two large groups of flatworms—the flukes and the tapeworms—are wholly parasitic and include many parasites of importance to man. Schistosomiasis, one of the major diseases afflicting man, is caused by a fluke living in the blood vessels of the human body (Figure 38.6). Tapeworms, which have less severe effects on human health, are common in temperate climates such as that of the United States. Different species of tapeworms may enter the human body when uncooked beef, pork, or fish containing larval forms (in minute capsules, or cysts) is eaten. Although the flukes are similar to free-living members of the phylum, the tapeworms are evolutionary deviants. In adaptation to parasitism, they have evolved a kind of segmented

Figure 38.7. A mature tapeworm proglottid showing the fully developed male and female reproductive systems. At right are scolices of five species of tapeworms. Note the variety in holdfast structures: suckers, hooks, grooves, and lobes.

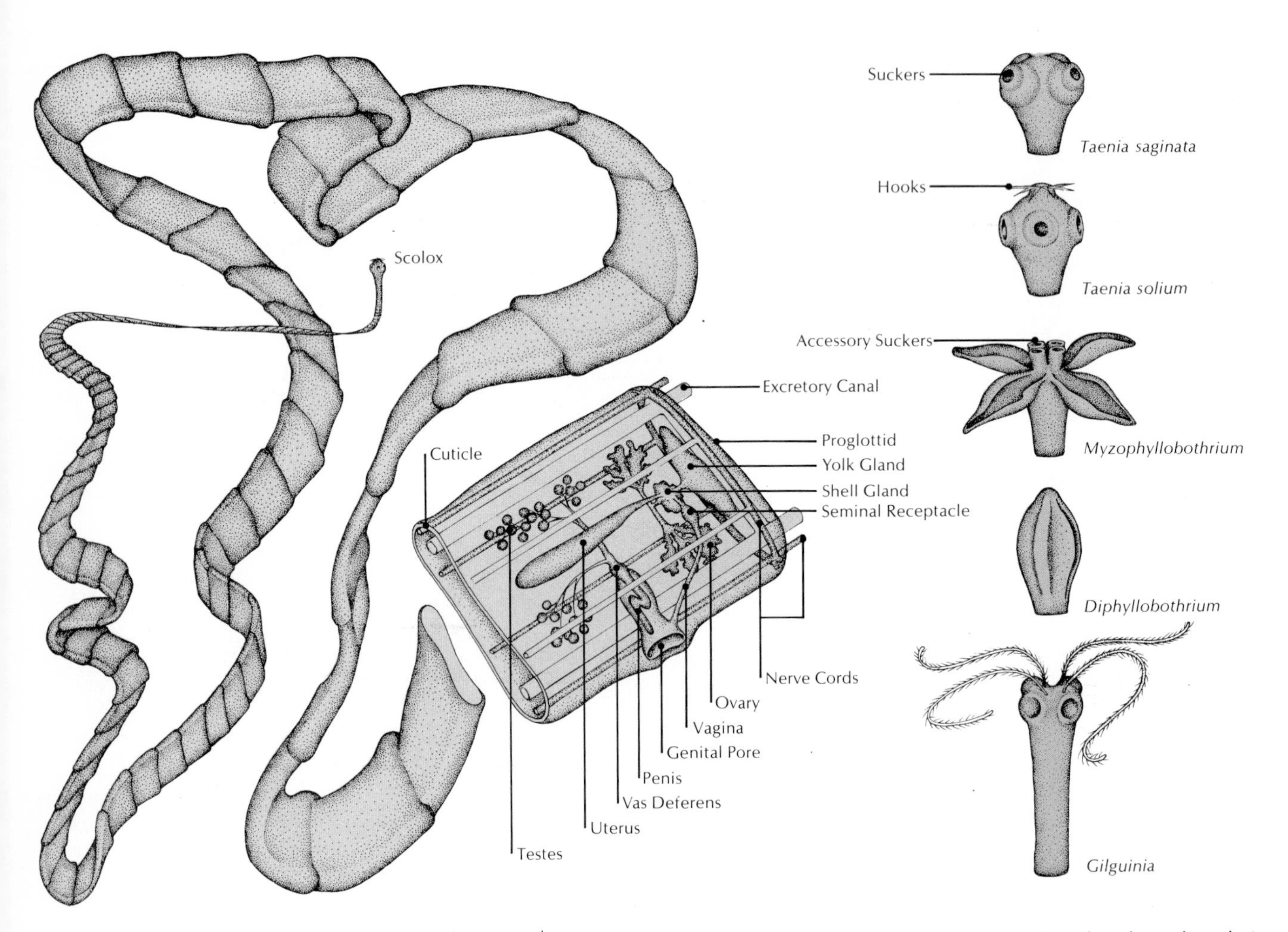

structure and an elaborate reproductive system, but they have lost their digestive tract and feed by absorption of nutrients from the fluids of the host's body (Figure 38.7).

Almost all flatworms are *hermaphroditic*—that is, both sperm and egg cells are produced by each individual. In most species, an egg capsule is formed around the zygote and a supply of nutrient material (*yolk*), and this egg capsule is released into the environment. In many forms, a free-swimming larva develops from the zygote; in others, the swimming larval stage is omitted from the life cycle.

The flukes and tapeworms have specialized and, in some cases, elaborate life cycles that may include production of millions of eggs, development during the cycle of distinctive larval stages, and infection of more than one kind of host. Thus, the fluke of schistosomiasis has a free-swimming larva that bores into the tissues of snails, where it develops into other larval stages, from which comes a different kind of free-swimming larva that penetrates the skin of man and then develops into an adult in the human body.

Because the bodies of flatworms contain no hard parts, this phylum has left no trace in the fossil record except for some markings that might be fos-

Figure 38.8 (above). A free-living nematode roundworm.

Figure 38.9 (below). The trichina worm (*Trichinella spiralis*) responsible for causing trichinosis.

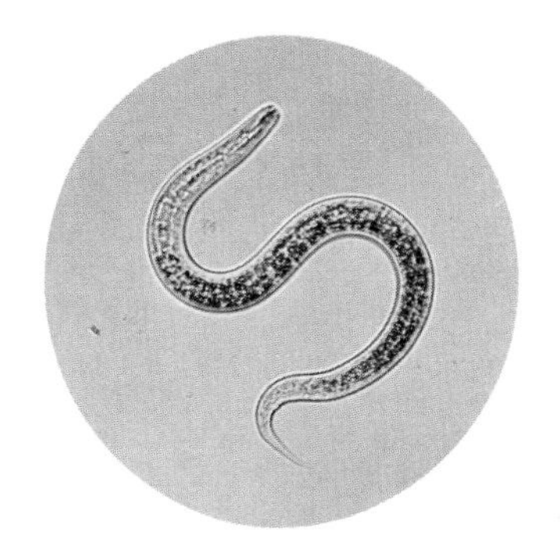

silized burrows or tracks. The structure of the flatworm suggests that it probably evolved relatively early in the history of the animal kingdom and that the higher animal phyla may have evolved from primitive flatworms. Certain similarities between coelenterates and flatworms have led to various theories that one group evolved from the other, but the origin of these two phyla is still very much a matter of speculation.

ASCHELMINTHES

A variety of other kinds of worms, most of them with cylindrical rather than flattened bodies, are grouped in the phylum Aschelminthes. (Many taxonomists prefer to divide the Aschelminthes into several phyla.) The body organization of aschelminths shows several evolutionary advances over that of flatworms. The saclike gut has been replaced by a true digestive tract leading from the mouth through several differentiated passages and chambers in which the food is processed (pharynx, stomach or intestine, and cloaca) to an *anus*, or outlet. Another advance is the appearance of a *false coelom*, a liquid-filled space (not lined by special cells) between the body wall, with muscles on the outside and the digestive tract on the inside. *Dorsal* (upper) and *ventral* (lower) nerve chords run the length of the body from a nerve ring with ganglia around the pharynx. Ducts for waste excretion also run the length of the body, and in many aschelminths (as in the flatworms) these ducts connect to special excretory cells called *flame cells*.

The largest group of aschelminths is the *nematode worms*, or roundworms. These actively moving worms have cylindrical bodies tapered at each end. They are white or transparent, with a body design based upon tubular organs within a tubular body wall (Figure 38.8). Nematodes are abundant in the soil, in water, and as parasites in plants and animals. The most dangerous parasitic worm in the United States—the pork worm *Trichinella*—enters the human body as larval cysts in pork that has been inadequately cooked. The larva of the hookworm lives in the soil of the south-

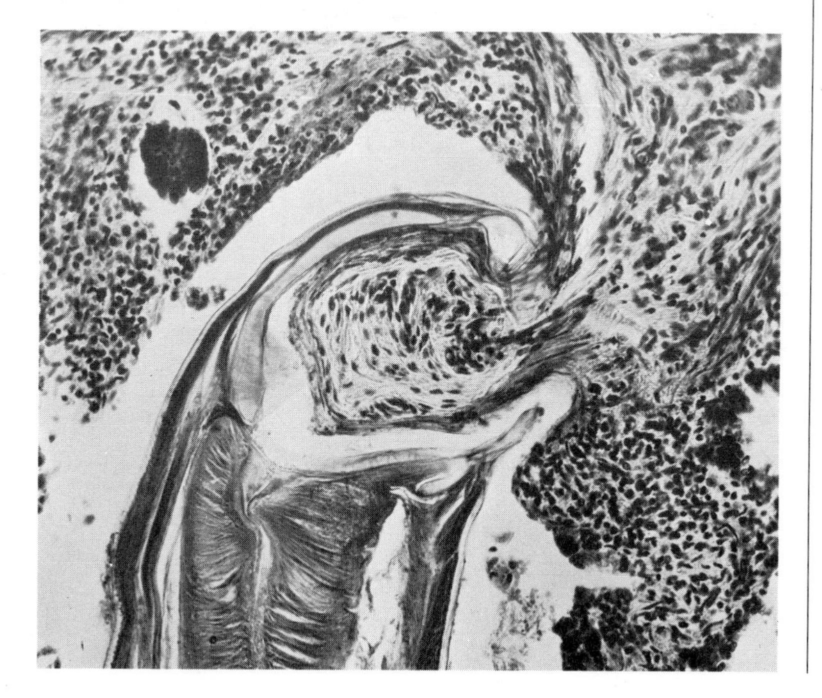

Figure 38.10. Life cycle of the trichina worm (*Trichinella spiralis*).

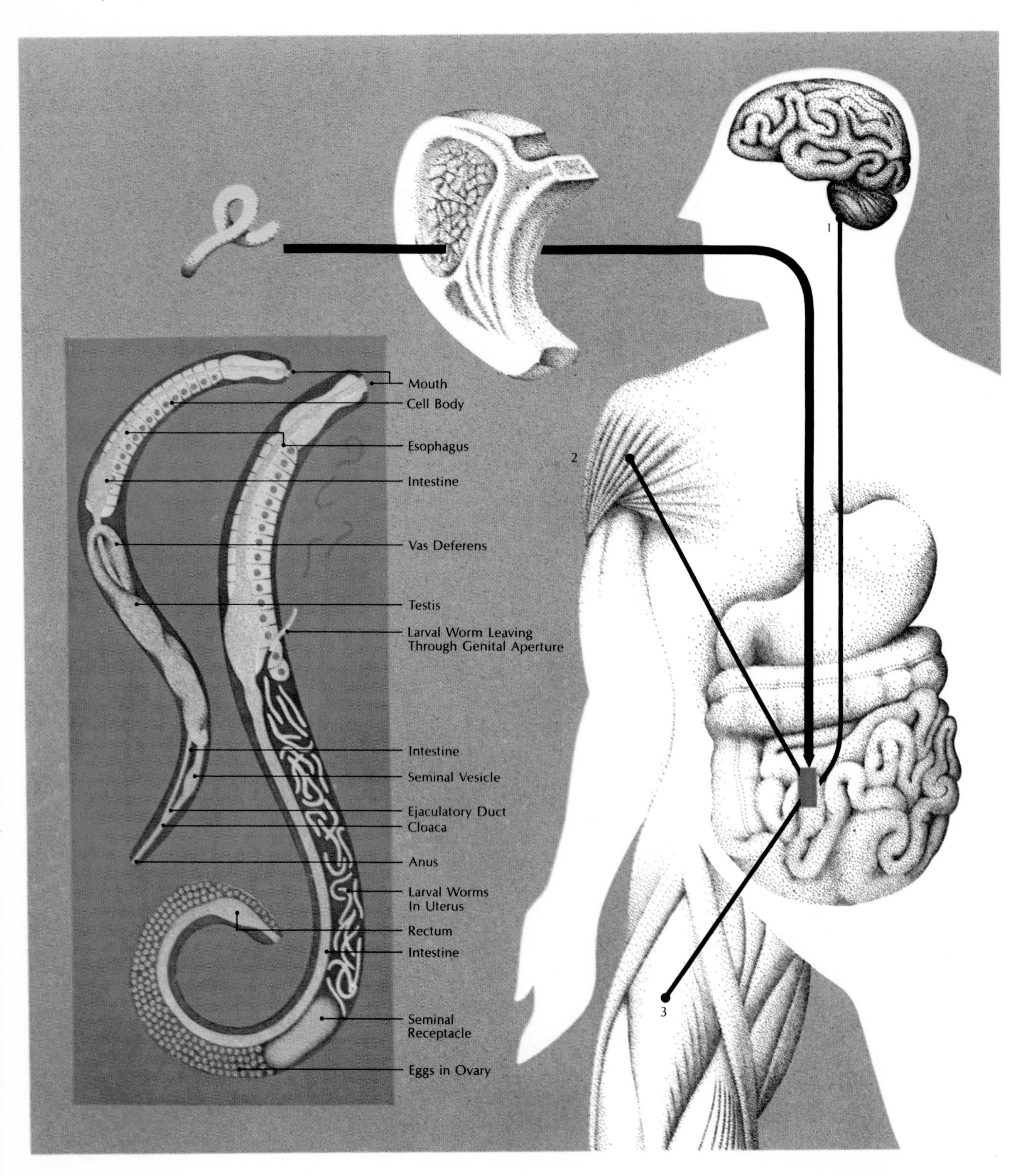

eastern United States and enters the human body by penetrating the skin of a bare foot. Several other roundworm parasites of man live in temperate climates, and a still greater number are found in the tropics.

Another important group of aschelminths is the *rotifers*—microscopic, transparent, swimming, and creeping animals with crowns of cytoplasmic hairs or of cilia that appear to rotate like wheels and that bring currents of water (and particles of food) in toward the mouth. Rotifers are among the most abundant animals in ponds and other aquatic environments.

Most aschelminths have separate male and female individuals. In these individuals, sexual reproduction involves copulation, a process in which the male injects sperm into the female's body, where a zygote is formed by fusion of a sperm with an egg cell. Among the rotifers, however, another form of reproduction—*parthenogenesis*—is common: the female produces diploid eggs that develop into adult individuals without fertilization. In one peculiar nematode (*Trichosomoides crassicauda*, a parasite that lives in the urinary tract of the rat), the minute male lives within the reproductive organs of the larger female. The zygote is released from the female body enclosed with nutrient materials in an egg.

Like flatworms, aschelminths have left no fossil record. Some biologists believe that aschelminths evolved from primitive flatworms. Others suggest that rotifers were the original aschelminths and that they evolved from animals related to primitive annelids.

TENTACULATA

Tentaculates (phyla Entoprocta, Bryozoa, Brachiopoda, and Phoronida—sometimes grouped together as phylum or superphylum Tentaculata)

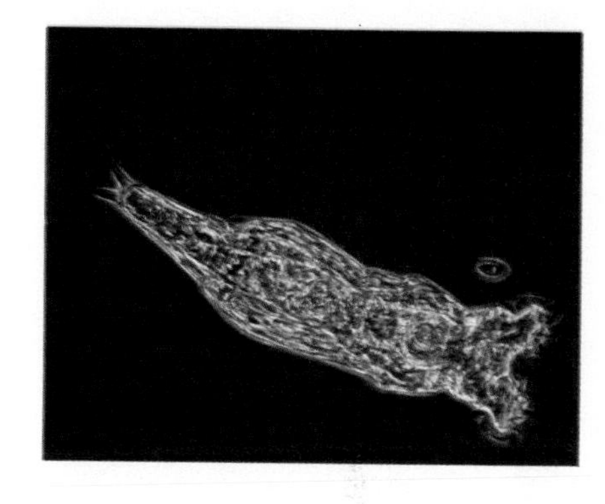

Figure 38.11a. A rotifer. These microscopic multicellular organisms show complex system development. (*Eric Gravé*)

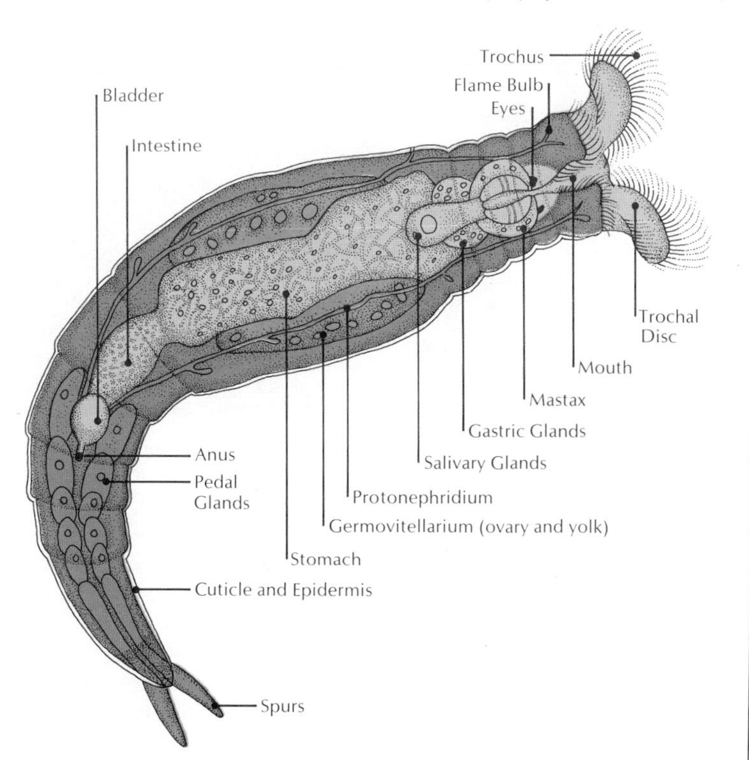

Figure 38.11b. Rotifer structure. The body of a rotifer can be divided into head, trunk, and tapered foot. The foot has cement glands for attachment to a substrate.

Figure 38.12a (above). Photograph of a bryozoan. Note the tentacles, which unlike the entroproct are encased in a sheath.

Figure 38.12b (lower left). Photograph of a living entoproct. Note the ring of ciliated tentacles, the lophophore, that surrounds the mouth.

Figure 38.12c (lower right). Brachiopod, or lampshell. All internal structures except for a portion of the lophophore are hidden by the partially closed valves.

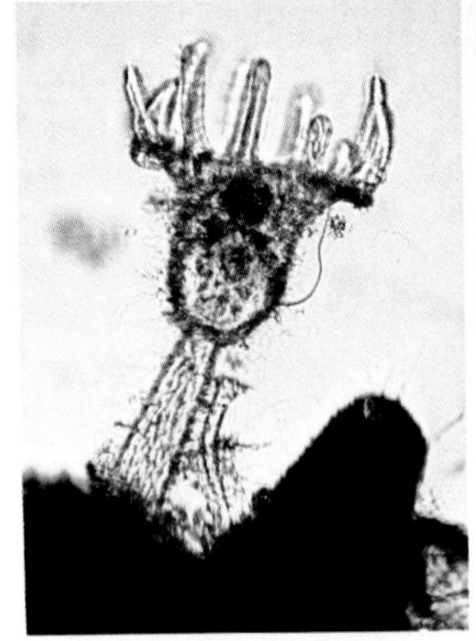

are animals with a horseshoe-shape, ciliated crown of tentacles. As in the rotifers, the cilia beat in such a way as to draw a current of water to the animal, which then extracts food particles from the current. A similar ciliated crown appears as a means of filter feeding in certain protozoans, rotifers, and annelids as well as in the tentaculates. The digestive tract of a tentaculate is U-shaped, looping from the mouth through the digestive chamber, or stomach, back to an anus located somewhere near the mouth.

The entoprocts have the simplest body organization of the tentaculates. Like the aschelminths, the entoprocts have a false coelom. These "moss animalcules" are small, colonial animals that form thin crusts on rocks and other surfaces or form branching structures that look something like the plants of mosses. Reproduction may be sexual or by budding. Sexual reproduction leads to the formation of ciliated larvae that leave the female adult and swim about for a time before attaching to the substrate and developing into an adult. No fossil entoprocts have been found.

Among the other three groups of tentaculates, the body cavity is a true coelom, lined with a special layer of cells. The bryozoans are "moss animalcules" similar to the entoprocts in appearance and reproduction. They are most readily distinguished from the entoprocts by the fact that the anus lies outside the ring of tentacles rather than within it as in the entoprocts. Many species of bryozoans form stony, leathery, or fibrous structures within which the individual animals of the colony are embedded. Fossils of these structures have been found in rocks formed as long as 500 million years ago.

The brachiopods, or lampshells, are tentaculate animals that live within a hinged pair of shells much like the shell of a clam. Brachiopods are abundant in the fossil record of the past 600 million years, but relatively few species have survived into the present. The halves of the shell, or *valves*, are of unequal size and lie above and below the body of the brachiopod. Male and female animals are distinct individuals, and sperm and egg cells are released into the water and either fuse within the shell of the female or fuse free in the water, forming free-swimming larvae that develop into new adults.

The phoronid worms live in tubes in mudflats, with the tentacles extended above the mud surface. Some species are hermaphroditic; others have individuals of separate sexes. Fertilization takes place in the body of the female in some species and free in the water in other species. Free-swimming, ciliated larvae are developed by all phoronids. Although some species build tubes of cemented sand grains around their bodies, no fossils have been found that can be assigned with certainty to worms of this group.

Although it is suggested that the tentaculates may have evolved from a common ancestor, little can be said with certainty about the origin of this group.

ANNELIDA

The elongated, cylindrical body organization of the animals commonly called worms is the most common body form in the animal kingdom. The Annelida also are worms but with a more complex body organization than that of the aschelminths. Whereas the body of a nematode is smooth and continuous, the body wall of an annelid is marked off into a series of disclike or ringlike segments. Internally, the annelids have a true coelom, and in many annelids this cavity is divided into fluid-filled compartments by partitions between the segments. Excretory organs—including ciliated funnels in the body cavity and tubes leading from them to the outside of the body—appear in each segment for much of the length of the body. As in the aschelminths and all higher animals, the digestive system is essentially a tube differentiated into a number of organs with different functions. The nervous system includes a pair of ganglia that are connected to a nerve loop around the pharynx and a nerve cord that runs down the ventral side of the body and connects ganglia in each segment.

The annelids possess an important system not present in simpler animals—a circulatory system. A ventral blood vessel runs parallel to the

Figure 38.13. Sagittal schematic diagram through an entoproct.

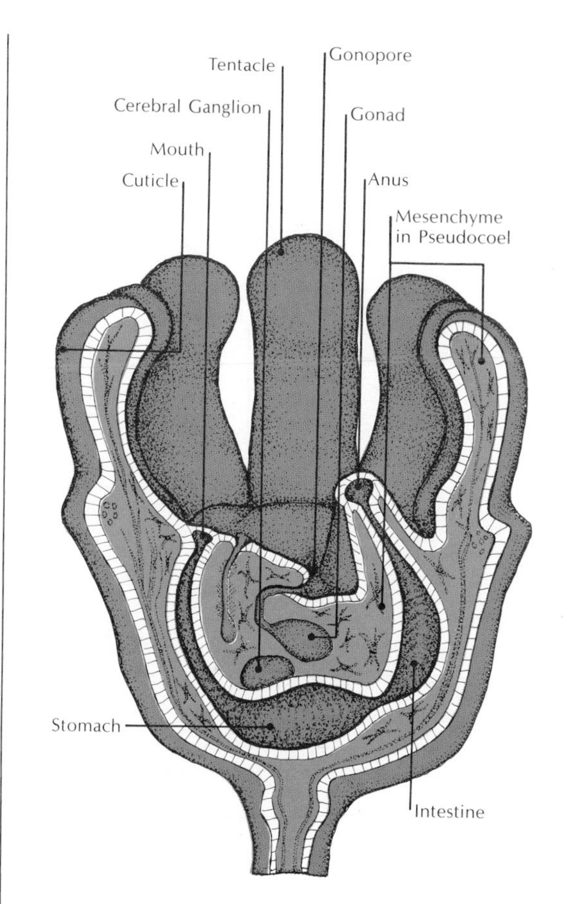

Figure 38.14. Earthworm anatomy. External features and internal anatomy as revealed by sectioning.

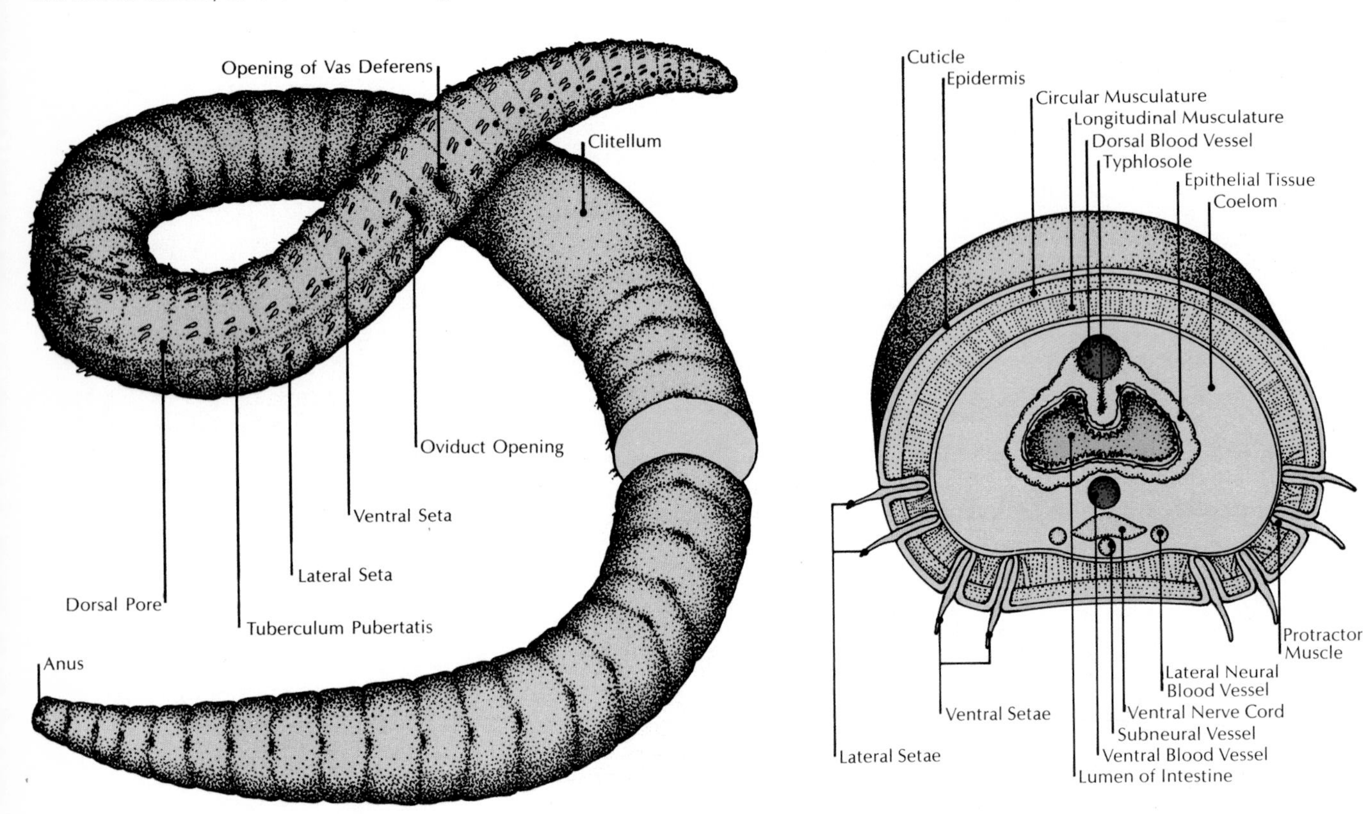

nerve cord along the lower side of the body, and a second, dorsal blood vessel runs along the upper side of the digestive tract. In the earthworm, five muscular tubes, or hearts, on each side of the esophagus (the passage between pharynx and stomach) connect the two vessels. They are connected also by branch vessels and capillaries carrying blood to different parts of the body. Presumably, the annelids have reached a degree of size and complexity at which a special system is needed for transport of food from the digestive tract and oxygen from the body surface to the various organs and tissues.

The phylum includes three major classes. The Polychaeta include a diversity of marine worms, many with appendages on each segment or concentrated at the head end. These appendages variously serve different polychaetes for locomotion, for food collection, and as gills. The Oligochaeta include the familiar earthworms and others found primarily in soil and fresh water. The Hirudinea, or leeches, possess suckers with which they attach themselves to a vertebrate host. Using tiny teeth in one of the suckers, the leech painlessly cuts a small hole in the skin of the host animal and—aided by a chemical substance that prevents blood clotting—sucks up enough blood within a few hours to keep it supplied with nutrients for many months. It then drops off the host and resumes its existence in water or moist soil.

Most polychaetes are bisexual (males and females are separate animals), but most oligochaetes and leeches are hermaphroditic. In most polychaetes, gametes are produced by temporary specialization of mesodermal cells (the cells between the outer and inner tissue layers). The gametes are

Figure 38.15a (upper left). Polychaete tubeworm attached to star coral from Cayman Islands, British West Indies.

Figure 38.15b (lower left). Two leeches attached to the fin of a fish host.

Figure 38.16 (right). Copulating earthworms. Each worm is transferring sperm to its partner that will later be used to fertilize the eggs when they are laid in the cocoon. Development of terrestrial earthworms is direct—that is, without larval stages.

released into the water through ducts in the segments (or by rupture of the adult body wall in some species). Fertilization occurs free in the water, and the zygote develops into a distinctive, ciliated, free-swimming larva. The larva later goes through a metamorphosis and becomes a segmented adult worm. Oligochaetes and leeches have specialized reproductive systems of a design that, in most species, permits reciprocal copulation between two hermaphroditic individuals. In these groups and in some polychaetes, eggs are laid enclosed in cocoons and the free-swimming larval stage does not develop. Instead, the eggs hatch to release immature worms.

Because of the absence of hard parts, the fossil record of annelids is sparse, but the chitinous jaws of polychaetes have been found in rocks formed more than 500 million years ago. There were no doubt other more primitive annelids, of which no recognizable fossil record exists.

MOLLUSCA

The name Mollusca means "soft-bodied," but the most familiar molluscs are known for the hard shells that provide skeletal support and protection for the soft body. This phylum is very diverse, with more than 50,000 known living species and numerous extinct species known from fossil records. The molluscs are unsegmented and are characterized by the development of a massive muscular organ, or foot, behind the mouth. Most species possess a complex digestive system, a circulatory system with a heart, a respiratory system with complex gills, an excretory system with a pair of "kidneys," a nervous system with well-developed ganglia and with eyes and other sense organs and muscular and reproductive systems.

Figure 38.17. A fossil nautiloid cephalopod.

In most molluscs, the sexes are separate, and the zygote develops into a free-swimming larva. The close resemblance of this larva to that of the polychaete annelids suggests an early evolutionary connection between these two phyla.

There are six major classes of living molluscs. The Monoplacophora are rare and simple marine molluscs with some internal segmentation of the body. The Amphineura, or chitons, also are quite simple; these marine animals have a flattened body and an eight-part, segmented shell. They live firmly attached to a rock by suction generated by the foot, moving very little and slowly, and feeding on algal film that coats the rock. The Scaphopoda, or marine toothshells, have curved, tubular shells and live buried in sediment on the ocean floor with only the tip of the shell protruding into the water. Water is circulated in and out of siphons in the tip of the shell.

The class Gastropoda includes snails (with a single, coiled shell), limpets (with an uncoiled shell), nudibranchs and slugs (with little or no shell), and pteropods (distinctive gastropods that float in the surface waters of the ocean). Gastropods, like chitons, feed by means of a hard rasp, the *radula*, with which many forms file surface material from plants or the surface film of microscopic organisms from rocks. Most snails are herbivores or scavengers, but a few species are predators that use the radula to penetrate the shell of another mollusc on which they feed. The snails are the most widely distributed of the mollusc classes, occurring in the ocean, in fresh waters, and on land. Land snails have lost their gills and have developed a lunglike cavity in the *mantle* (the fleshy covering of the mollusc body on which the shell is formed) that makes it possible for them to extract oxygen from the air.

The Pelecypoda, or Bivalvia, include the clams, scallops, oysters, mussels, and their relatives, characterized by a shell composed of two halves. Unlike brachiopods, pelecypods have shells composed of two nearly equal halves located on the sides of the body. Most pelecypods are filter feeders that have evolved a remarkably efficient use of the gills for feeding, as well as for external respiration. The beating of the cilia draws water through small pores and tubes of the gills, where food particles are caught in mucus on the gill surfaces. The food-bearing mucus is carried forward to the mouth and ingested. Most pelecypods are marine, but a few live in fresh water (none live on land). Many burrow in sand, rocks, wood, or coral reefs.

The Cephalopoda are large molluscs with tentacles (evolved from the foot), well-developed eyes, and highly developed nervous systems and behavior. The nautiloids, of which *Nautilus* is the only living genus, had large, spiral, chambered shells. Many of the cephalopods have largely or wholly lost the external shell or have replaced it by an internal skeleton as in the squid and octopus. The cephalopods are carnivores of the oceans, and some are large, powerful, and quick hunters. The squid moves by jet propulsion, ejecting water through a funnel-shape tube from its muscular mantle cavity.

Fossil gastropods are found in rocks formed about 600 million years ago; nautiloids make their appearance in rocks about 550 million years old; chitons, pelecypods, and scaphopods first appear in the fossil record between 450 and 500 million years ago; and the various cephalopods other than nautiloids appear as fossils in rocks formed in the past 400 million years. Cephalopods were abundant until about 70 million years ago—with such now-extinct forms as the ammonoids being common fossils—but

Figure 38.18. Structural diagrams comparing the internal organization of (from top to bottom) a chiton, a scaphopod (or toothshell), a snail, a clam, and a squid.

Figure 38.19a (above). Limpet and other marine organisms associated on a rock substrate.

Figure 38.19b (middle left). Scaphopod, or toothshell, with mantle extended over the shell.

Figure 38.19c (middle right). A marine nudibranch laying a whorl of eggs.

Figure 38.19d. A sea slug laying large gelatinous eggs. Many eggs must be produced in order for a few to survive predation.

Figure 38.19e (below). A squid, representing the cephalopod class of molluscs. Squids are noted for their ability to undergo rapid color change and therefore escape detection by their enemies.

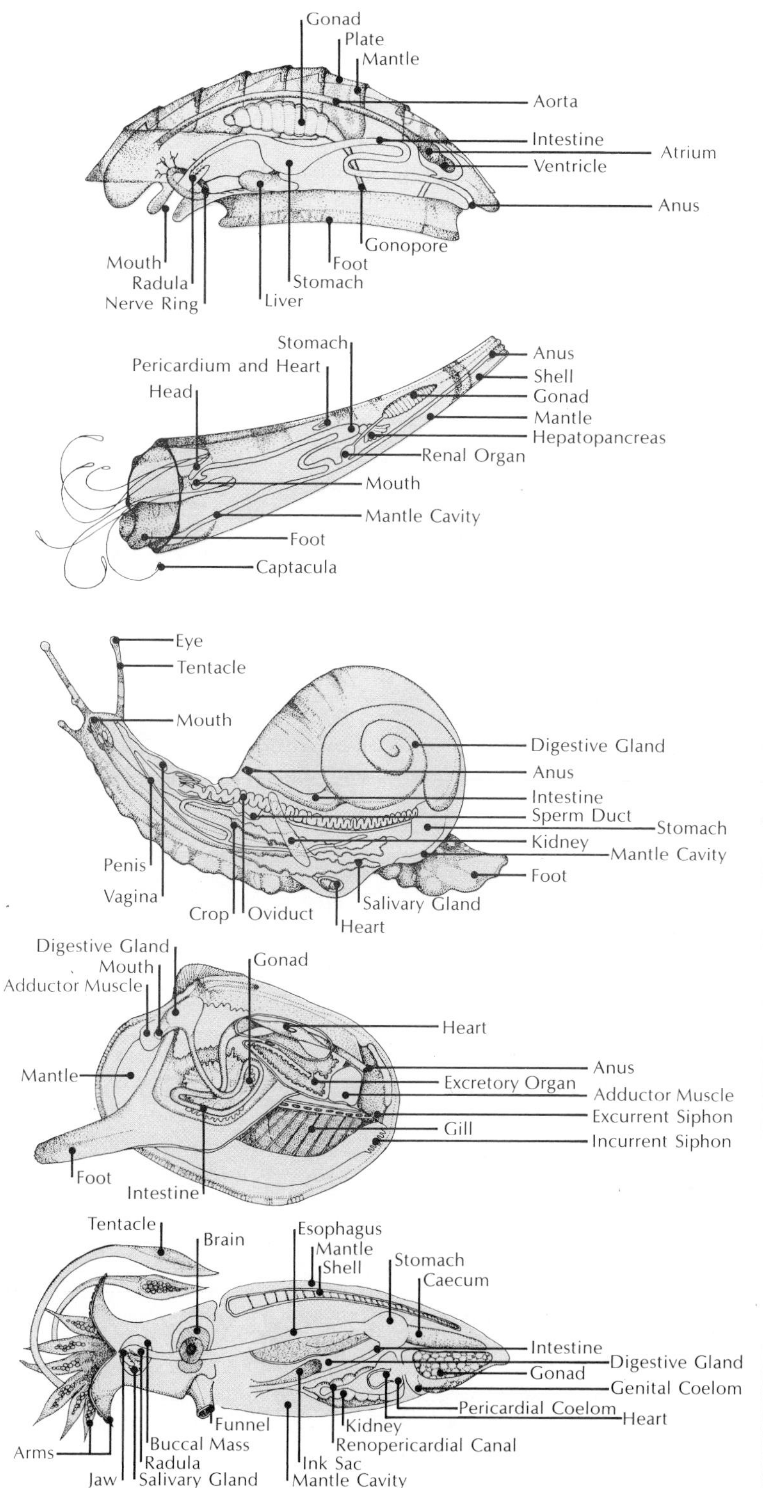

Figure 38.20a (above far left). A honeybee, representing the arthropod class Insecta.

Figure 38.20b (above middle). A garden spider, representative of class Arachnida.

Figure 38.20c (above far right). A centipede, representing class Chilopoda.

Figure 38.20d (middle right). A millipede, representing class Diplopoda.

Figure 38.20e (below left). *Daphnia*, the water flea, is a common fresh-water crustacean. *(Courtesy Carolina Biological Supply Company)*

Figure 38.20f (below right). Hermit crab living in a snail shell.

have somewhat decreased in abundance and numbers of species. On the other hand, the gastropods and pelecypods have become somewhat more abundant and diverse within the past 150 million years.

ARTHROPODA

In number of species and diversity of forms, the Arthropoda must rank first among the phyla of the living world. More than a million species of arthropods are known (some 850,000 of them insects), and estimates of the total number of arthropod species yet to be discovered and classified run as high as several million.

The arthropods are distinguished by the possession of jointed appendages (the appendages of the Polychaeta and other simpler animals are not jointed) and by the possession of a segmented, chitinous, external skeleton covering the whole body. Like annelids, arthropods have a segmented

Figure 38.21a (above). A fossil trilobite. This organism is thought to have been the basic prototype giving rise to contemporary arthropod lines.

Figure 38.21b (below). A fossil eurypterid. This ancient group of Arachnids was abundant in the ancient seas about 400 million years ago. Some species reached gigantic proportions of over 6 feet in length.

body, a coelom, a circulatory system, and a ventral nerve cord running back from a paired brain in the head. However, particularly in the more complex arthropods, the segments are differentiated and have become parts of distinct regions of the body—for example, the head, thorax, and abdomen of the insects. As part of this differentiation, the appendages of these segments have been variously lost or modified—those of the head segments, for example, have evolved into mouth parts. Some simpler arthropods have legs on many of the body segments. In more advanced arthropods, the number of legs is reduced to ten (crabs and lobsters), eight (spiders), or six (insects). In insects, all six legs are attached (with the wings) to the thorax, or midpart, of the body.

In many arthropods, the head is much more highly developed than in the annelids. It bears complex and variously adapted mouth parts, a more advanced brain, glands releasing chemical regulators (hormones), and advanced sense organs that may include both single eyes and compound eyes with many lenses as well as highly sensitive chemical receptors. As in the other more complex animals, the highly evolved sensory, nervous, and muscular-skeletal systems make possible complex and highly adaptive behavior. The interrelation and integration of parts of the arthropod is made possible both by the nervous system and by the circulatory and hormonal systems. In many respects, the arthropods represent an extension of evolutionary trends from annelids and molluscs. Yet it is clear that the arthropods illustrate one of the peaks of evolution in the living world. Only a few of the many classes and subphyla of arthropods can be mentioned here.

The Onychophora are wormlike, simple, terrestrial arthropods of the tropics and are of interest because of certain similarities to annelids. Onychophores may be descendants of the primitive arthropods, which probably evolved from annelids. Fossil onychophores and other primitive arthropods have been found in rocks formed about 550 million years ago.

The Trilobita, or trilobites, were the dominant animals of the seas from about 600 million until about 350 million years ago, but they became extinct about 230 million years ago. The flattened body was divided into a head, thorax, and tail. The head and tail were covered by solid shields and the thorax by a jointed exoskeleton. Two grooves running the length of the body gave trilobites their distinctive, three-lobed appearance. A pair of legs appeared on each segment, with each leg divided into a jointed lower branch used for crawling and a featherlike upper branch that may have acted as a gill. Trilobites probably swam, crawled, and burrowed in the ocean bottom.

The Crustacea are numerous and predominantly marine, although there are some fresh-water forms and a few land dwellers (pillbugs and tropical land crabs). The crustacean body is organized into a *cephalothorax* (combined head and thorax) and abdomen. The number of legs is far greater in simpler forms than in higher forms. Copepods—small, shrimplike crustaceans—are among the most abundant animals in the world, forming a prominent part of the plankton in oceans and fresh waters. Barnacles, lobsters, crabs, shrimps, crayfish, and prawns are among the other familiar crustaceans. Fossils of small crustaceans appear in rocks formed about 550 million years ago, but the more familiar large, ten-legged forms make their first appearance in rocks about 300 million years old.

The Arachnida include spiders, ticks, mites, and scorpions. Most of the members of this class live on land. In spiders, ticks, and mites, the body

Figure 38.22. Schematic diagrams comparing the external organization of body structures in the various arthropod classes.

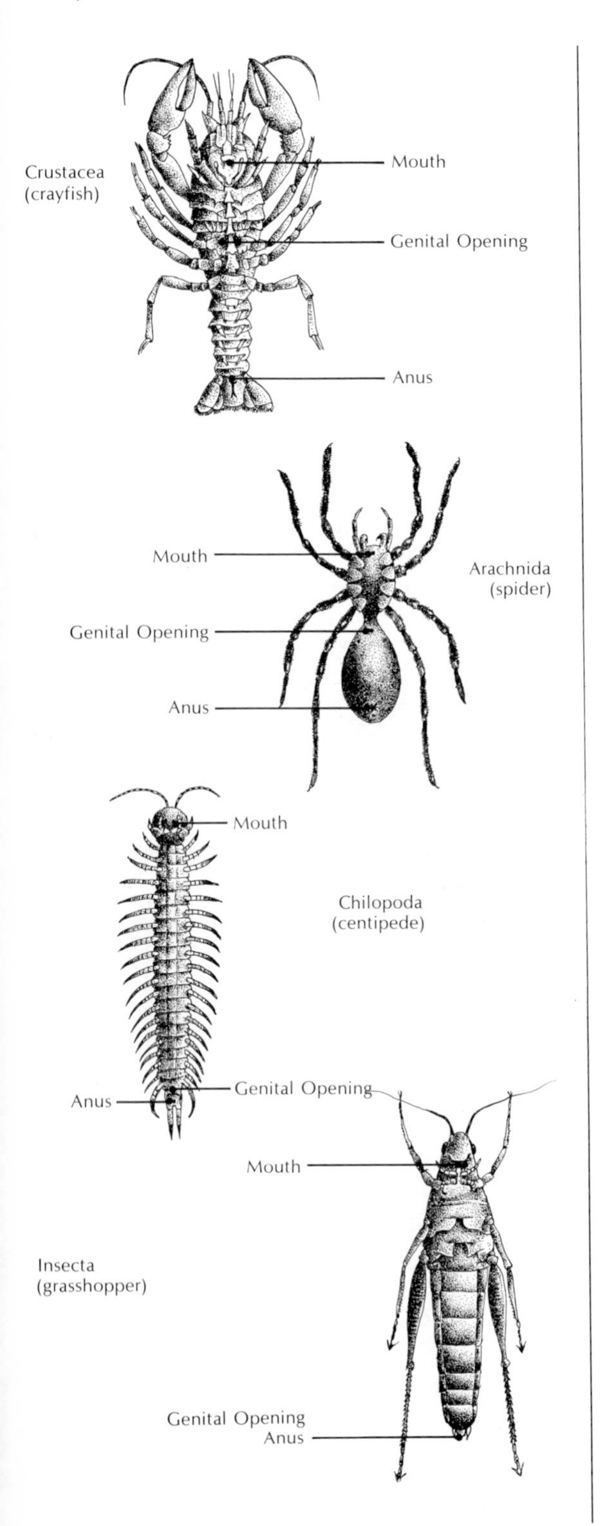

segments are almost unrecognizable, and the body is organized into a cephalothorax and a saclike abdomen. The horseshoe crab, or king crab, is an interesting member of this class. The organization of its body shows clearly that it is more closely related to spiders than to crabs, and of particular interest is the fact that this genus appears to have been so successful that natural selection has led to few recognizable changes in the past 200 million years. An extinct group of arachnids, the eurypterids, or "sea scorpions," were abundant in the oceans from about 500 million to about 225 million years ago. Some species were giants among arthropods—more than six feet in length.

The Myriapoda include two distinct classes: millipedes (Diplopoda) and centipedes (Chilopoda). Both millipedes and centipedes are terrestrial animals with clearly segmented, elongated bodies and many legs. Although thousands of living species are known, fossil myriapods are rare, and the oldest came from rocks formed about 300 million years ago.

The Insecta include more species than all other groups of animals combined. Most insects are terrestrial, although many are found in fresh water and a few in the ocean. Numerous insects live as parasites, many of them on other insects. Like arachnids, myriapods, and onychophores, insects have a tracheal respiratory system. Tubes, or *trachea*, extend from openings on the body surface, branching into more minute tubes that lead to all parts of the body and carry oxygen to the tissues. This type of respiratory system is very different from the lungs of the land vertebrates as an adaptation to the problem of air breathing. The higher insects have larval stages that differ widely from the adult stages of the same species in structure and often in way of life and food supply. Unlike the small and short-lived larval stages of most other animals, insect larvae generally grow to the full size of the adult and actually live longer than insect adults. Adaptation of larvae to diverse and distinctive food supplies—along with the advantages of small size, advanced organization, and complex behavior—has contributed much to the spectacular success of the insects in occupying the land environments of the world with great numbers of individuals and species. Wingless insects are found in rocks formed about 375 million years ago, with winged insects making their appearance in the fossil record about 300 million years ago. Many insects depend on flowering plants for their food supply, and, in turn, many of the flowering plants depend on insects to fertilize their eggs by carrying pollen grains from plant to plant.

ECHINODERMATA

Among the Echinodermata are starfishes, sea urchins, sand dollars, brittle stars, sea cucumbers, and sea lilies. Echinoderms are quite unusual in their body organization. Although they develop from larvae with bilateral symmetry, the adults appear radially symmetrical or nearly so. Echinoderms lack segmentation, heads, and distinct brains. A true coelom and a circulatory system are present, but most forms have no special systems for excretion or external respiration. There is a unique water-vascular system with water tubes connected to numerous tube-feet that can be expanded and contracted by water pressure. The tube-feet are used for attachment to the substrate, for movement, and for holding food. All echinoderms live in the oceans, and there is a great variety of motile and attached forms. Most echinoderms possess hard, calcareous plates in the skin layer, and in

Figure 38.23a (left). Brittle or serpent stars. These echinoderms have a reduced number of tube feet and move by lateral undulations of their arms.

Figure 38.23b (upper middle). Sea urchin. These echinoderms have an endoskeleton formed of welded calcium carbonate plates.

Figure 38.23c (lower middle). The echinoderm sea cucumber. These organisms have tube feet restricted to two parallel rows and move slowly about the bottom, functioning as detritus feeders.

Figure 38.23d (right). Crinoids, or sea lilies. These filter-feeding echinoderms, with their stalk and branching arms, resemble plants rather than animals.

Figure 38.24 (below). Schematic diagrams showing sea star anatomy, external-aboral view. A cross section of an arm shows details of the water vascular system and tube feet.

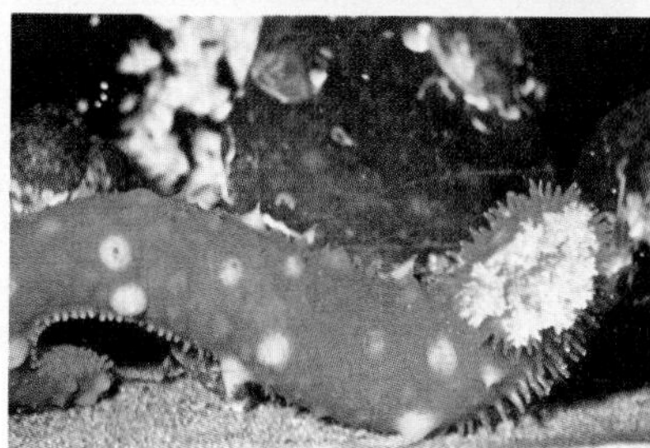

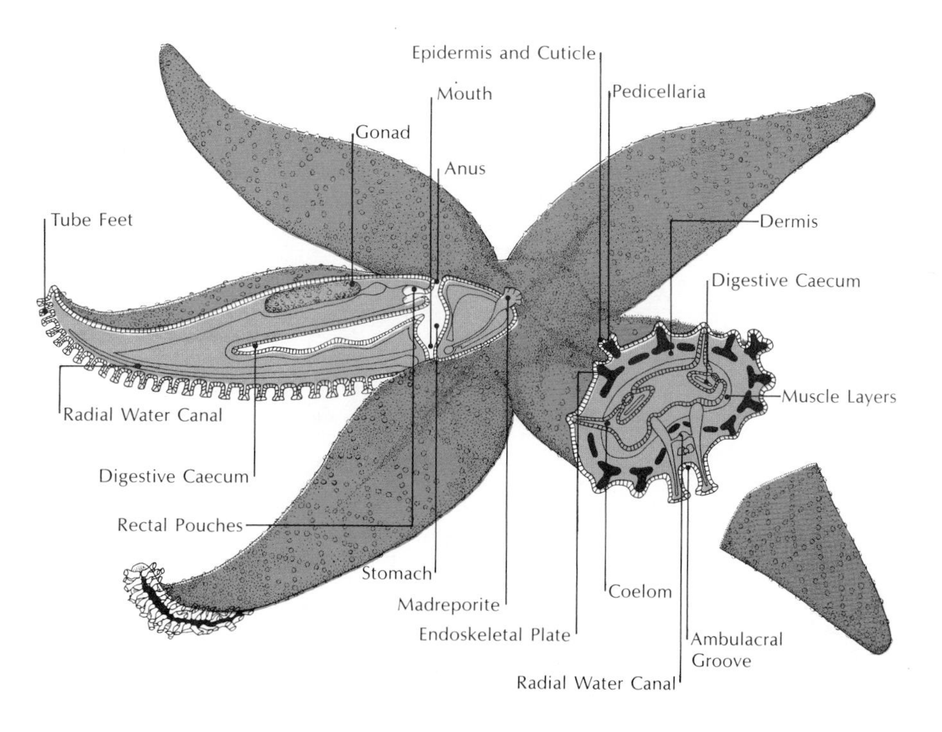

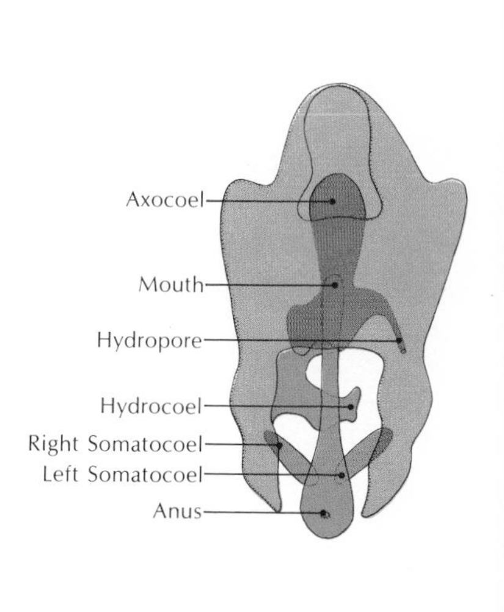

Figure 38.25a (left). Photograph of the sea grape tunicate. Note the incurrent and excurrent siphons.

Figure 38.25b (right). Sagittal schematic diagram of an adult tunicate. The animal feeds by drawing water through the incurrent siphon into the basketlike pharynx.

the sea urchins the plates are welded together to form a solid exoskeleton.

Sea lilies, or crinoids, are filter feeders that live attached to the sea bottom by a stalk. Branching, flexible arms surround the mouth at the top of a rigid exoskeleton, but the crinoids have no tube-feet. Starfishes are stiff but flexible and are able to crawl slowly. They capture shellfish and other prey with their strong tube-feet, and some species feed by everting the stomach around the prey. Brittle stars are similar to starfish but have longer, more slender arms that are capable of more rapid motion. Sea urchins move slowly by means of tube-feet and long spines, and they feed chiefly on dead organic matter and seaweed. Sea cucumbers are elongate animals with leathery skins and tentacles around the mouth. They feed on small animals or dead organic matter in mud or sand.

Attached echinoderms similar to modern crinoids appear as fossils in rocks formed almost 600 million years ago, and both starfishes and sea urchins make their appearance within the next 100 million years. The fossil record suggests that the attached echinoderms appeared before the free-living forms; perhaps, in fact, their peculiar radial structure evolved originally as part of their adaptation to attached life. The sexes are separate in echinoderms. Gametes are released into the water for external fertilization, and the zygote develops into a bilaterally symmetrical, swimming larva. Sea urchins and other echinoderms have proven useful experimental subjects for studies of many basic biological processes, including gamete production, fertilization, embryonic development, and nervous function.

CHORDATA

The Chordata include the vertebrate animals (of which man is one) and some other less familiar organisms. Most chordates have segmented bodies. The segments are difficult to recognize in the higher forms, except for the backbone (or vertebral column) and the nerves, muscles, and blood vessels closely associated with it, which do show segmented structure. Chordates

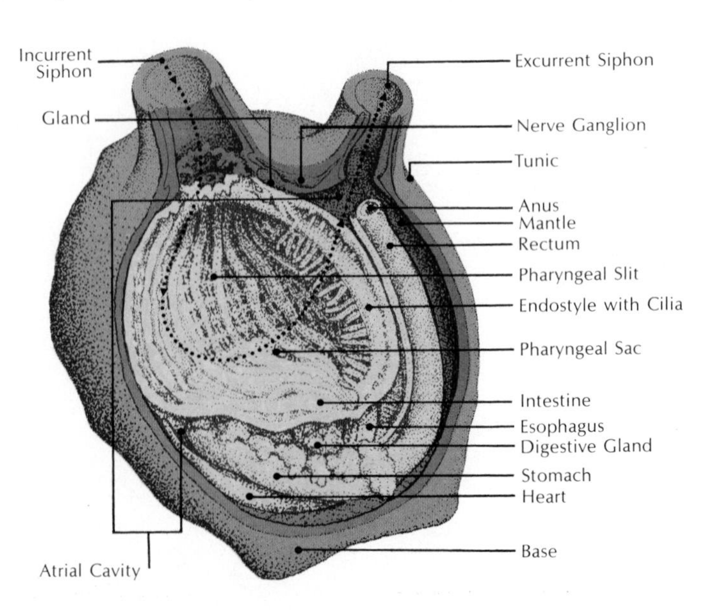

differ from annelids and arthropods in that they have a dorsal, tubular nerve cord (the spinal cord and brain) rather than a ventral, solid nerve cord. Chordates are also distinguished by the possession of an internal skeleton. In the most primitive forms, the skeleton consists of a longitudinal stiffening rod—the *notochord* (for which the phylum is named). In the advanced forms, however, a bony vertebral column replaces the notochord, and other bones or cartilaginous structures are present. Aquatic chordates have gill slits for external respiration in the pharynx; in terrestrial chordates, such gill slits appear in the embryo but are not present in adults. Chordates have true coeloms, advanced excretory and circulatory systems, and the most advanced nervous and hormonal systems of the animal kingdom.

The primitive chordates—including tunicates (or sea squirts) and lancelets—are small marine animals. They are of great interest for their bearing on early evolution of the chordate phylum, but study of the primitive chordates has not yet provided clear evidence of the ancestors from which the chordates evolved. The phylum Hemichordata (sometimes included in the Chordata) includes the acorn worms and their relatives. These animals have ciliated, swimming larvae strikingly similar to those of the echinoderms and also have a structure corresponding to that of the notochord. The hemichordates thus suggest what at first glance seems one of the least likely relationships among animal groups—a distant, early evolutionary connection between the echinoderms and the chordates.

The subphylum Vertebrata includes animals with vertebral columns, of which there are seven classes with living members. The class Agnatha includes primitive, round-mouthed (jawless) vertebrates—which appear in the fossil record about 470 million years ago—and are now extinct except for certain eellike parasites (lampreys and hagfish).

Members of the class Chondrichthyes—including sharks, skates, and rays—are relatively primitive fishes that have skeletons of cartilage rather than bone. These fishes, most of which live in the oceans, are unusual in that they retain large amounts of urea in the blood and body fluids. The concentration of these dissolved materials in the fluids within the body is approximately equal to that of the sea water outside. The urea in the blood thus is an adaptation that prevents loss of body water by osmosis—a loss that would occur if the concentration of dissolved substances were lower within the body than in the surrounding sea water. Fertilization is accomplished through copulation, and the female of most species lays eggs enclosed in leathery shells. Fossilized teeth indicate that sharks made their appearance in the oceans at least 400 million years ago.

The more advanced fishes with bony skeletons, including many marine and many fresh-water species, are grouped in the class Osteichthyes. Fossils of primitive bony fishes have been found in rocks formed as much as 450 million years ago. These early vertebrates were covered with an armor of bony scales and had lungs as well as gills. A few species of living osteichthyans retain lungs, but in most species the lung is no longer used for respiration. Instead, it has become modified to serve as a swim bladder that helps in adjusting the buoyancy of the fish. The concentration of solutes in the body fluids of osteichthyans is much lower than that in sea water. Marine species drink large amounts of sea water to replace the fluids constantly being drawn out of their bodies by osmosis and must expend metabolic energy in getting rid of the salt that they acquire. Today, the osteichthyans

Figure 38.26a (left). Blue cromies, representing the true bony fishes.

Figure 38.26b (right). Diagram of the general anatomy of a teleost (bony) fish.

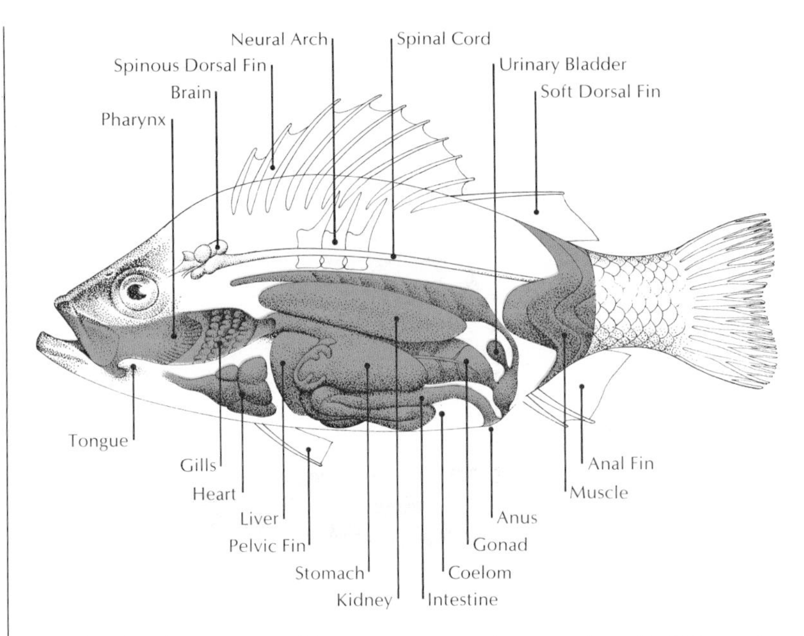

are the dominant group of large water animals; more than 20,000 species are known, with a great variety of forms and ways of life.

It is thought that both the chondrichthyans and the osteichthyans evolved from the jawless fishes at about the same time. The different osmotic adaptations to sea water take on special interest in this connection. It is likely that both the bony fishes and the sharks (and their relatives) evolved from jawless ancestors in fresh water. The two groups later invaded the seas, using different adaptations to overcome the osmosis problem. Evolutionary divergence produced many marine species in each group, while (for unknown reasons) the sharks and their relatives virtually disappeared from fresh waters. Of particular interest is a group of primitive osteichthyans, the crossopterygians, which were among the early marine forms of the bony fishes and which were long thought to have become extinct about 75 million years ago. In 1939, however, a living crossopterygian was captured from the deep ocean off Africa, and since then several more specimens have been caught.

The class Amphibia includes frogs, toads, and salamanders, the most primitive terrestrial vertebrates. Most amphibians spend the early stages of life in fresh water and the later stages on land fairly close to water. The larval stages of most species have gills that are lost in the adult stages. Similarities of biochemistry and structure suggest that the primitive amphibians evolved from primitive crossopterygians whose lungs became more highly developed for life on land and whose stubby fins became modified to form legs. Fossils of primitive amphibians are found in rocks formed about 360 million years ago. Like the primitive land plants that were becoming abundant at about the same time, these early land animals remained dependent upon water for reproduction and were unable to thrive in dry locations.

Included in the class Reptilia are lizards, alligators, turtles, snakes, and the extinct dinosaurs. Unlike the amphibians, the reptiles are truly terrestrial. They lay large eggs that can develop on land and thus can be completely independent of water bodies throughout their life cycles. The reptilian egg

Figure 38.27a. Representative amphibians. Green tree frog (above). Salamander (middle).

Figure 38.27b (below). Schematic diagram of the major internal organs of a male frog.

Figure 38.28a. Representative reptiles. Mangrove snake (upper left). Box turtle (lower left). Chameleon (right).

Figure 38.28b (below). Reptilian anatomy as represented by a schematic cut-away diagram of a lizard.

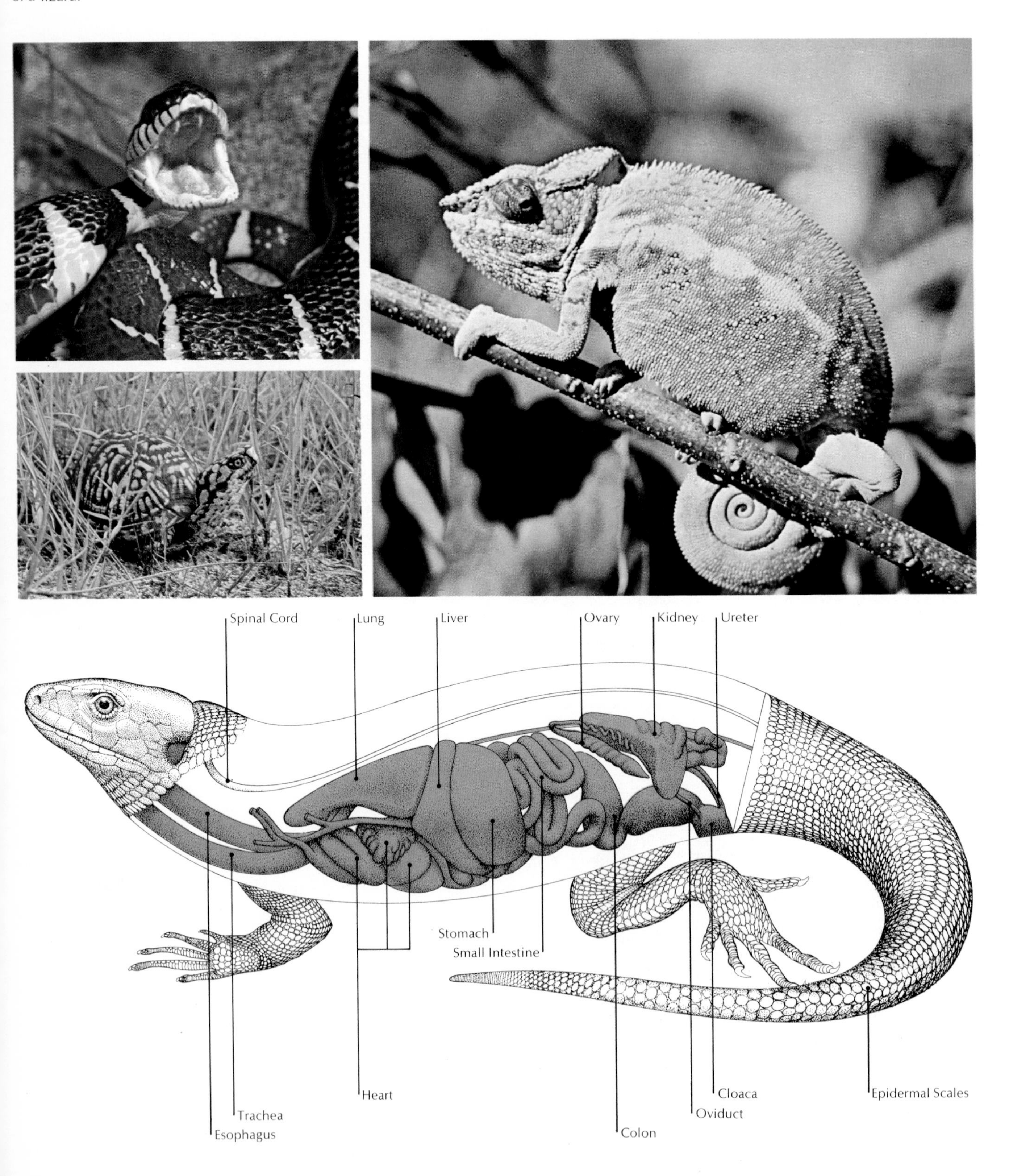

Figure 38.29a (right). White mynah bird (India).

Figure 38.29b (left). General internal anatomy of the pigeon.

is covered with leathery or calcareous substances that protect the embryo from desiccation, and a body of food for the early development of the embryo is stored in the yolk of the egg. Copulation is a well-developed mechanism of fertilization, so a body of water is unnecessary for reproduction. The young hatch from the eggs as tiny replicas of the adults, ready to feed and avoid danger. Although they grow and in most cases change in proportions after hatching, they do not pass through a larval stage. The reptiles possess more highly developed respiratory systems, mechanisms of locomotion, circulatory systems, and means of preserving body water while eliminating waste products than do the amphibians. Primitive reptiles appear in the fossil record about 290 million years ago and probably evolved from primitive amphibians. During the period from about 250 million to about 70 million years ago, the reptiles were the dominant large animals on the land, in the oceans, and in the air. Today, birds and mammals occupy many of the niches in the biosphere once occupied by reptiles, and many groups of reptiles such as the dinosaurs have become extinct.

Class Aves, the birds, includes animals characterized by wings, feathers forming protective coverings and wing surfaces, very efficient systems of lungs and air chambers for external respiration, and highly developed sensory and nervous systems. Fossils of primitive birds are found in rocks formed about 150 million years ago. Except for their feathers, these early birds are reptilelike in body organization, and in that respect, so are modern birds. There are more than 8,500 species of living birds, with a great variety of forms and ways of life. Like mammals, birds maintain a constant temperature within the body and a high rate of metabolism, regardless of external conditions.

The class Mammalia is thought to have evolved from a group of reptiles different from those that gave rise to birds. Like birds, the mammals developed an efficient circulatory system and a complex nervous system, as well as the ability to maintain a steady internal temperature. Hair protects mammals in the same manner that feathers protect birds. In most mammals, the embryo is nourished within the mother's body rather than being

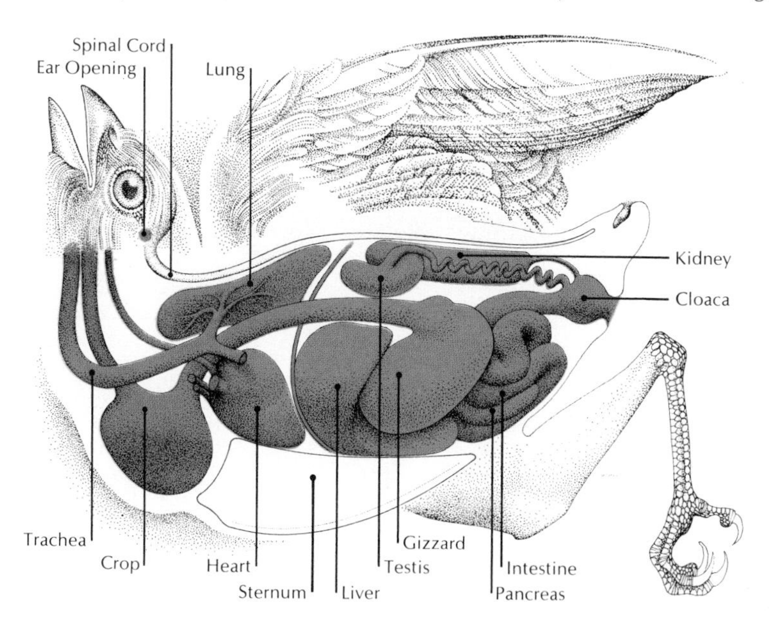

Figure 38.30a. Representative placental mammals marmosets (upper left). Sea lion cub (lower left). Cheetahs (right).

Figure 38.30b (below). General internal structure of a mammal.

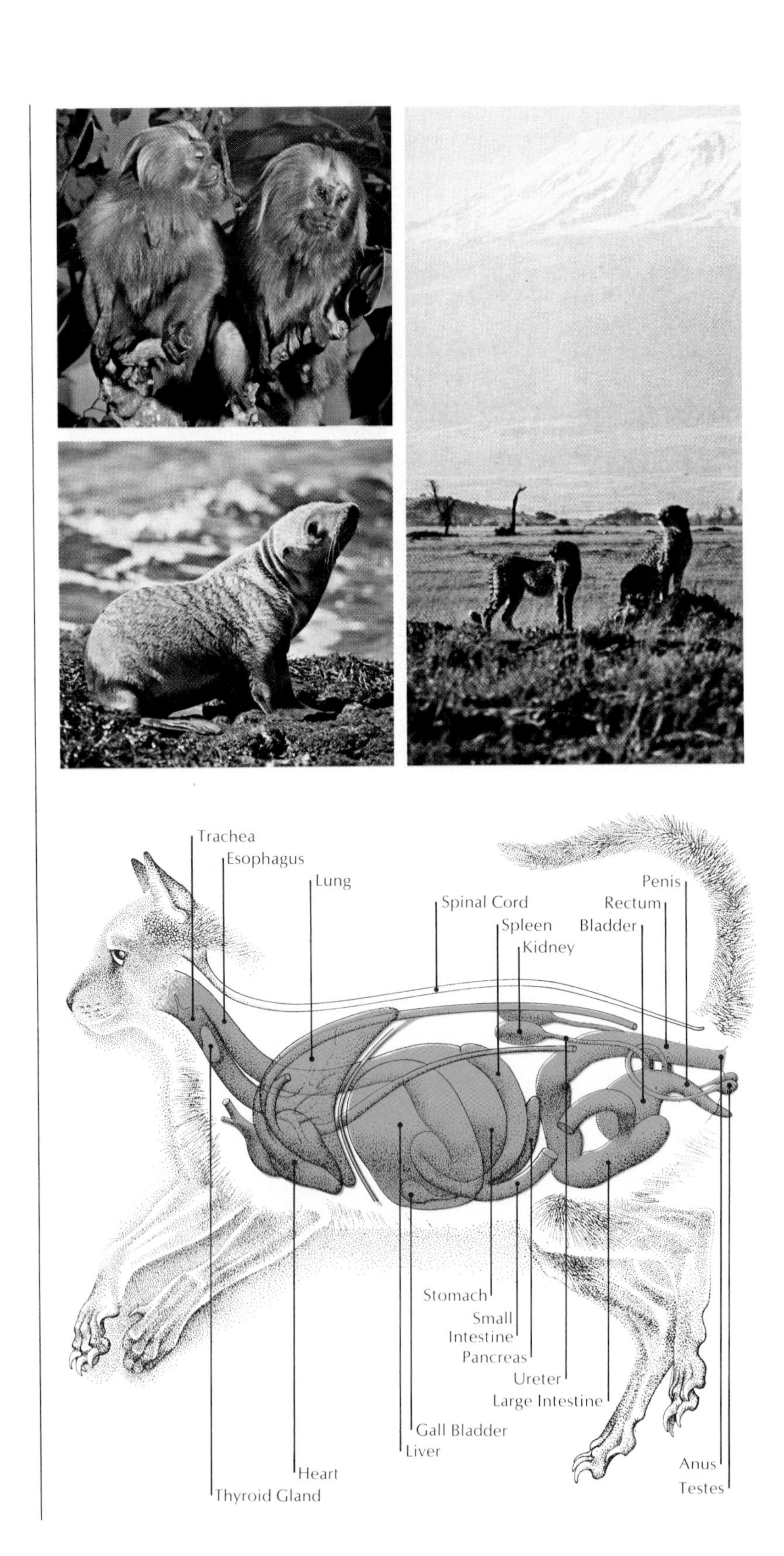

enclosed in an egg. After birth, the young mammal is relatively helpless for some time and is generally fed on milk from the mother's mammary glands. In general, mammals seem to have fewer behavioral patterns that are genetically determined and more that are learned during the periods of infancy or adulthood.

Fossils of primitive mammals have been found in rocks formed about 180 million years ago. These early mammals were very small creatures that apparently remained rather rare and inconspicuous in a landscape dominated by reptiles. About 70 million years ago, within a relatively short period of time, many groups of reptiles and other animals became extinct or greatly diminished while the numbers of mammals greatly increased. The reasons for the ending of the "Age of Reptiles" and the beginning of the "Age of Mammals" are still very much a matter of debate.

EVOLUTION OF ANIMALS

It would be most interesting to know how each group of animals evolved from earlier groups. The complete story of evolution will never be known, for too much of that evolution occurred in the remote past and involved soft-bodied organisms that were not preserved as fossils. The hypotheses of evolutionary relationships given in this chapter have been derived from three kinds of evidence. First, there are some cases in which a series of fossils shows the gradual evolutionary development of one kind of organism into another kind. For example, in the vertebrates, good fossil evidence is available to support the development of reptiles from amphibia and the development of birds and mammals from reptiles. However, no comparable series of fossils has been found to show how the Cyclostomata might have evolved from earlier organisms.

A second kind of evidence involves the use of careful studies of structures in living organisms. Often, it is possible to place these organisms in a sequence that forms a possible evolutionary scale, showing the development of more complex structures from simpler ones. Thus, it is possible to suggest that the Monoplacophora are relatively simple molluscs and that the snails, clams, and cephalopods are more complex molluscs with features that could have developed from certain structures in the Monoplacophora. Such studies must be applied with great caution, for more complex structures can evolve into simpler ones under certain kinds of selective pressures, and it must be remembered that the Monoplacophora living today have also been subject to evolutionary changes for millions of years and may be quite different from the most primitive molluscs.

A third source of evidence about evolution comes from the study of embryonic development. In many cases, structures that have been lost in the adult stages during evolution still appear briefly in some form during the embryonic stages. Great caution should also be exercised in the interpretation of this evidence, for it is by no means true that "ontogeny recapitulates phylogeny." Evolution also affects the embryonic stages of organisms, and in many cases the embryonic structures are different from ancestral structures. The highly adapted larvae of most insects, for instance, differ greatly from ancestral organisms. On the other hand, the appearance of gill slits and a notochord in all higher vertebrates (including man) during embryonic development is thought to be an indication of the evolution of the higher vertebrates from primitive chordates.

A major feature of evolution is the conquest of different environments. Life originated in the sea and went through its earliest evolutionary stages

Figure 38.31. Life on a coral reef as envisioned 330 million years ago. Note the predominance of cephalopods.

there. The chemistry of marine organisms and the osmotic concentration of their cytoplasm and body fluids are closely adapted to the chemistry of sea water. A marine animal placed in fresh water suffers osmotic flooding—water enters its tissues in great quantities and bloats the animal. To invade fresh water, marine animals had to evolve a series of structures including, in most cases, impermeable protective coatings, excretory organs to pump out excess water, and a means of absorbing salts from fresh water. According to present theories of evolution, these structures are not developed in any purposeful fashion. Rather, those mutations that happen to produce useful structures lead to greater opportunities of reproduction for the organisms that possess them. Thus, the proportion of the mutant gene to the former gene in the population gradually increases, and over many generations the average characteristics of the species change. In many cases, the population remaining in the old environment remains almost unchanged, while a population invading a new environment undergoes sufficient alteration to become a new and distinct species.

Only certain groups in some of the marine phyla produced forms that were capable of the transition to fresh water. Invasion of the land involved still more drastic alterations for avoidance of desiccation. Land animals must solve problems of controlling water balance in the body, obtaining oxygen without losing too much water, and bringing male and female gametes together in the absence of a body of water. These problems are less severe for organisms that live in the soil than for those that live above it. From the phyla that succeeded in fresh water, some invaded the land as soil animals, and a small number of phyla succeeded above the soil as true land organisms.

Terrestrial animal life is overwhelmingly dominated by the chordates and arthropods. Among other phyla, only the molluscs (represented by the snails) have more than a trivial place among land animals.

Wherever a place to live and a source of food is available, it is likely that some group of organisms eventually will become adapted to exist there. The opportunity to survive with relatively little competition places great survival value on those mutants that invade a new habitat or utilize a new food supply. Land life may be rigorous in some respects, but it has offered the chordates and arthropods great rewards in terms of possibilities for such evolutionary diversification. Among insects and vertebrates, evolution on land has led to diverse ways of life and complex, highly adapted behavior patterns. Insects and spiders have tended to evolve toward complex but fixed inherited behavior patterns, or instincts. Vertebrates have tended to evolve toward learned and therefore more flexible behavior as an addition to or replacement of instinctual behavior. The intelligence of man may be regarded as the extreme product of this direction of evolution among the vertebrates.

Man's intelligence has enabled him to develop a means of communication and organization of social groups and a high degree of aggressiveness and success in modifying his environment to meet his desires. Unfortunately, man has not proven very successful in limiting his own population to match available resources or in controlling aggressiveness against members of his own species. The study of life reveals many examples of evolutionary trends that led to highly specialized forms—forms that proved to be so strongly developed in one direction that they were unable to solve new problems and to survive for long. Can the human intelligence that has built high civilizations supporting dense urban populations also develop the means to limit and govern those populations? Can man restrain population growth to match resources, halt the degradation of the environment, and control the destruction of members of his own species? The next century may well reveal whether man is really the most successful species in the long story of life's evolution.

FURTHER READING

General discussions of the animal kingdom are given by Hanson (1964), Storer and Usinger (1965), Villee, *et al.* (1963), and Weisz (1966). Biosystematics of the animal kingdom are discussed by Mayr (1963), Mayr, *et al.* (1953), Rothschild (1961), and Simpson (1961). For further information on animal reproduction, development, physiology, and hormones, see Chapters 16, 18, 19, and 21 and their Further Readings. Nervous systems and animal behavior are discussed in Chapters 24 through 31.

For further information about invertebrates, see the books by Barnes (1963), Borradaile and Potts (1961), Brown (1950), Buchsbaum (1948), Buchsbaum and Milne (1960), Easton (1960), and Hyman (1940–1955). More detailed information about insects is given by Farb (1962), Hutchins (1966), Imms (1964), Klots and Klots (1959), and Snodgrass (1935).

General discussions of vertebrates are given by Romer (1966, 1970) and Young (1962). More specific discussions of amphibians are given by Cochran (1961); of reptiles by Carr (1963) and Schmidt and Inger (1957); of fishes by Herald (1961), Lagler, *et al.* (1962), and Ommanney (1963); of birds by Austin (1961), Gilliard (1958), Peterson (1963), Van Tyne and Berger (1959), and Welty (1963); of mammals by Sanderson (1955), Walker (1964), and Young (1957).

Among the many *Scientific American* articles dealing with animals are those by Barnett (1967), Bartholomew and Hudson (1961), Cockrill (1967), Fertig and Edmonds (1969), Irving (1966), Isaacs (1969), Jensen (1966), Kooyman (1969), McVay (1966), Ruud (1965), Savory (1962, 1966, 1968), Southern (1955), and Wecker (1964).

VIII
The Human Organism

In the past, brave attempts have been made to clarify the uniqueness of man and they continue but . . . we have not gone much beyond the first reassuring platitudes. We have no deep knowledge. We need also to have an understanding of the gulf between human and other life forms as it has developed in varying cultural traditions, and the implications of these findings for the preservation of the natural world, for the prevention of extinctions of threatened plants and animals even if human affairs retain—as indeed they inevitably will—a central position in the preoccupation of man.

—C. J. Glacken (1970)

39
Human Origins

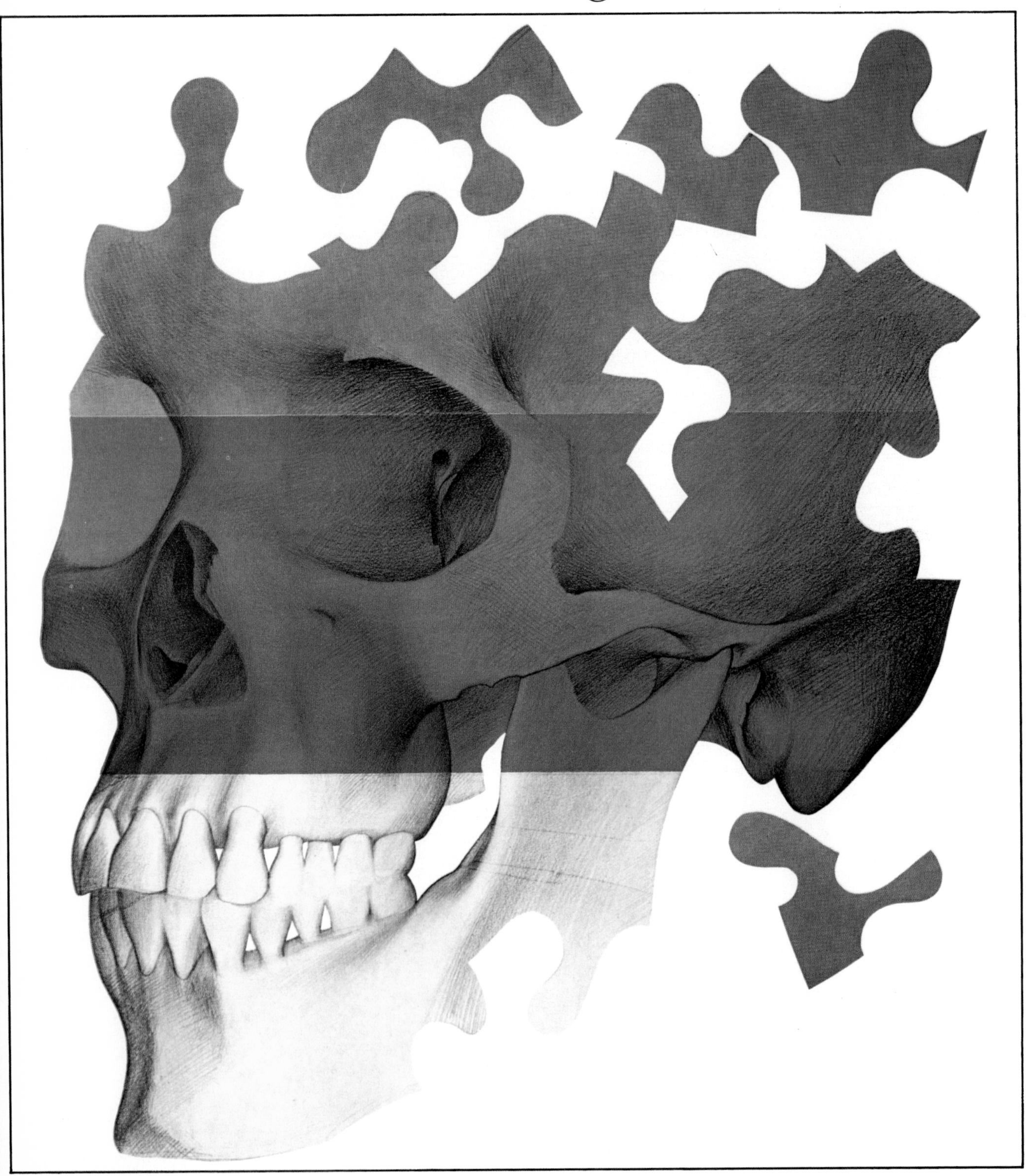

One of the most controversial aspects of Darwin's theory of evolution was its implication that the human species might have evolved from lower animals. Darwin himself avoided the touchy issue in *The Origin of Species*, but the opponents of his theory were quick to point out this "insulting" aspect. Darwin's strongest defender in scientific and public debates was Thomas Henry Huxley, who was deeply interested in evolutionary problems. He is reported to have reacted to Darwin's book with the age-old line: "Now why didn't *I* think of that?"

In a famous debate in 1860, Huxley faced Bishop Samuel Wilberforce of Oxford at a meeting of the British Association for the Advancement of Science. Wilberforce attacked Darwin's theory from several angles, using scientific arguments that had been supplied to him by the zoologist Richard Owen, who was a major opponent of Darwinism. To cap his argument, the Bishop turned to Huxley and inquired sarcastically whether he traced his descent from the apes on his father's side or his mother's side of the family. In a letter written of few months later, Huxley recalled his reply:

> When I got up I spoke pretty much of the effect—that I had listened with great attention to the Lord Bishop's speech but had been unable to discover either a new fact or a new argument in it—except indeed the question raised as to my personal predilections in the matter of ancestry—That it would not have occurred to me to bring forward such a topic as that for discussion myself, but that I was quite ready to meet the Right Rev. prelate even on that ground. If then, said I, the question is put to me would I rather have a miserable ape for a grandfather or a man highly endowed by nature and possessing great means and influence and yet who employs those faculties and that influence for the mere purpose of introducing ridicule into a grave scientific discussion—I unhesitatingly affirm my preference for the ape.
>
> Whereupon there was unextinguishable laughter among the people, and they listened to the rest of my argument with the greatest attention . . . (Foskett, 1953).

In 1863 Huxley published his classic essays in *Man's Place in Nature*. He argued that man is related much more closely to the apes (family Pongidae—chimpanzees, gorillas, orangutans, and gibbons) than to the quadrupedal monkeys (family Cercopithecidae). His views were widely publicized at the time, and for over a century many scientists have believed that man's closest living relatives are the African apes. However, this view did not go unchallenged, and almost every conceivable theory of human origins has been suggested. Modern theories range from a separation of the human lineage 50 million years ago to as little as 5 million years ago.

THEORIES OF HUMAN ORIGIN

Theories of human origin can be grouped into three major categories. First, some theories postulate a separation of the human lineage before the separation of monkeys from apes. Such theories place the origin of the human species more than 40 million years in the past. This view has seemed reasonable to only a few scientists. A more widely accepted variant has been the tarsioid hypothesis, which maintains that the modern tarsier (a small, nocturnal, jumping, insect-eating creature) is more closely related to man than any other living primate and that all similarities between man and apes are a result of parallel evolution of distinct lineages.

Figure 39.1. Tarsier (*Tarsius syrichta*). This nocturnal, insect-eating prosimian is found only on a few islands in Southeast Asia.

The second group of theories holds that lines leading to man, apes, and Old World monkeys all became distinct at about the same time—perhaps 20 to 30 million years ago. This hypothesis of a separation in Oligocene or early Miocene time is the most common modern viewpoint (Leakey, 1960). According to this view, if there was an ape in human ancestry at all, it was a very early and unspecialized form. Man is regarded as no more closely related to the apes than to baboons.

The third group of theories cling to the classic position of Huxley—that man is most closely related to the apes and particularly to the African apes. According to these theories, the human lineage became separated in the late Miocene or in the Pliocene, about 5 to 15 million years ago. These theories have gained support recently from studies comparing DNA of various primate species, but this viewpoint is regarded with some skepticism by the majority of anthropologists today.

Obviously, there are intermediate positions, and scientists change their minds. Everyone who has been concerned with the study of human evolution knows that the fossil record is sufficiently fragmentary to permit considerable latitude in interpretation. Even so, especially considering the massive efforts in comparative anatomy, it is surprising that after more than a hundred years of effort there is very little agreement among experts. Even if the more extreme positions are ignored, the comparative anatomical and paleontological information can be interpreted to support an origin of the human lineage anywhere between 30 million (Oligocene) and 10 million (late Miocene or early Pliocene) years ago. This difference of reasonable opinion is so great as almost to preclude any discussion of events determining the course of human evolution. In spite of all the effort—all the study of anatomy, all the fragmentary fossils—there are still a few anthropologists who agree with Huxley and many more who hold a variety of divergent points of view.

UNITARY ORIGINS AND VARIABILITY

No well-informed person today believes that modern man evolved from a single, "pure," homogeneous, ancestral stock. Nor can it be argued that the various races of modern man have evolved separately as distinct, homogeneous lineages. The inescapable fact is that variation, or heterogeneity, typifies every living population. Heterogeneity provides the material for evolutionary change and adaptation; it is perpetuated by the mechanism of chromosomal recombination. At every stage of hominid evolution, the hominid family has consisted of a complex group of populations, displaying a wide range of diversity.

One broad means of approaching the problem is through biochemical evidence. It is possible to estimate genetic affinity between *Homo sapiens* and other primates by comparing the detailed organization of genetic material. Evidence from such studies suggests that there is more similarity between the nucleotide sequences of man and the chimpanzee than between man and the rhesus monkey. The implication is that man and the chimpanzee are more closely related evolutionarily than are man and the rhesus monkey.

It is also possible to compare degrees of reaction to injections of human serum—the more profound the reaction, the greater the evolutionary separation is presumed to be. Once again, it appears that man is more closely related to the apes than to the monkeys.

Both of these lines of evidence make use of evolutionary changes in proteins. In attempting to establish a time scale, scientists must assume that mutations occur at a regular rate. Thus, the 42 differences in the amino acid sequences between horse and man are assumed to represent a minimum of 52 mutations over the 150 million years that the two lines have been evolving independently (75 million years each since the split into two lines).

This assumption leads to the further one that the smaller number of differences between the monkeys and man would have, on the same time scale, indicated separation of those lines some 20 to 25 million years ago—and a much shorter time for the separation of the lines for man and apes.

The problem in determining the time of separation of human lineage from that of the apes is that the separation is so recent that statistically reliable numbers of differences have not yet been accumulated. The fossil record of the genus *Australopithecus*—clearly belonging to the hominid family—makes a separation of less than 5 million years impossible. The biochemical data apparently make a separation of more than 10 million years almost equally impossible. Additional study of proteins probably will narrow these limits and may provide a generally accepted evolutionary clock.

There is no danger of confusing any individual of the modern human species with other animals. He shares many structural features with the apes that are his closest living relatives and with other primates and mammals, but *Homo sapiens* is beyond doubt a distinct species. Any two individual

Figure 39.2. An artist's interpretation of human origins.

Figure 39.3 (above). The evolution of the skull from ape to modern man. From top to bottom, note the increasing size of the brain case (area above eye sockets) and decreasing size of the face (area below eye sockets). In addition, the canine teeth of the ape are considerably larger than those of man.

Figure 39.4 (below). Neanderthal skulls collected from around the world show as much variation in physical features as do those in contemporary populations. At left is the skull of a classic Neanderthaler, Monte Circeo man, taken from an Italian cave over 40,000 years old. The skull of the La Ferrassie man (middle) was found in southwestern France—the front teeth are extremely worn, perhaps from chewing animal skins. On the right is the skull of a Tabūn woman found on Israel's Mount Carmel. Unlike the Europeans, Neanderthalers of the Middle East displayed less classic Neanderthal features, including higher foreheads.

humans are more like one another than they are like other kinds of primates. For example, a human skull is readily distinguished from an ape skull by its large cranium, prominent forehead, pointed chin, smooth surfaces, large mastoids, and lack of prominent ridges above the eyes and across the back of the skull (Figure 39.3). The human spinal cord attaches almost beneath the skull's center of gravity, reflecting the upright and bipedal human posture, whereas that of the ape attaches near the rear of the skull. On the other hand, within the human species, there exists a wide range of variations on the basic human pattern. The term "modern man" embraces such diverse morphological types as the various pygmy and "giant" groups of Asia and Africa. The enormous geographical spread of the human species since the end of the Ice Ages—into America, Australia, Oceania, and the circumpolar regions—implies a corresponding increase in genetic heterogeneity.

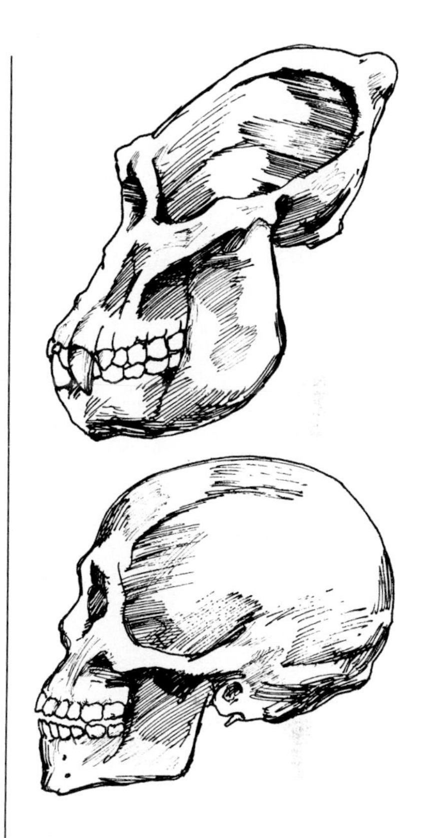

What was the human species like in the past? Consider the human population during Upper Pleistocene times—say, about 50,000 years ago, near the close of the Ice Ages. A number of fossil remains dated at about this time have been found at various sites around the Old World. The first and most famous remains were found in 1856 in the Neanderthal valley of western Germany, but remains with similar characteristics have subsequently been found in other parts of Europe, Africa, the Middle East, Russia, Java, and China. Those that can be dated with some accuracy appear to have been deposited between 110,000 and 35,000 years ago. These remains are similar enough to modern man to be classified as *Homo sapiens* but are distinct enough to be called Neanderthal man rather than modern man. The skull of Neanderthal man differs from that of modern man in its more massive bones, larger eye sockets, flatter top, less prominent forehead and chin, and more prominent ridges above the eyes and across the back of the skull. In short, it is slightly more apelike than the skull of modern man, but still far more like a modern man than like an ape.

Neanderthal skulls collected from around the world show that the human population of this time varied as widely in characteristics as does the modern population (Figure 39.4). Furthermore, skulls found at Swanscombe

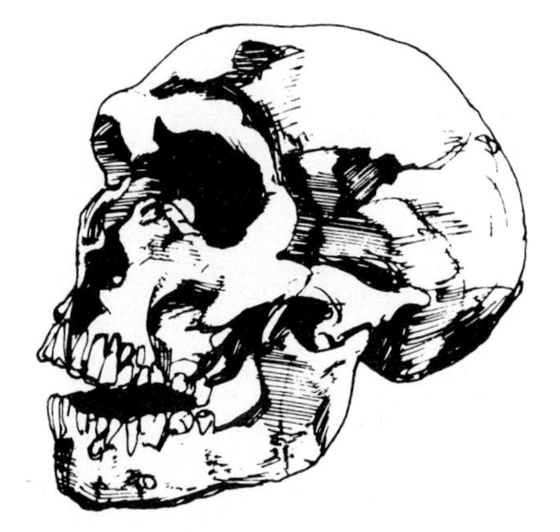

Figure 39.5. Map showing the locations of major excavation sites around the Old World.

SOME FOSSIL RECORDS OF HOMINID EVOLUTION

Site	Approximate Age (millions of years before present)	Date Found
Homo sapiens sapiens		
1 Czechoslovakia	0.03	1894
2 France	0.02–0.03	1868
3 Sarawak	0.04	1959
4 South Africa	0.04	1933
5 Kenya	0.06	1942
6 Sudan	0.02	1924
7 China	?	1934
8 Java	?	1890
Homo sapiens neanderthalensis		
9 Belgium	0.40–0.07	1886
10 Britain	0.15–0.25	1935–1955
11 France	0.07–0.15	1947
12 France	0.04–0.05	1908
13 France	0.04–0.06	1909–1921
14 France	0.04–0.06	1908–1921
15 France	0.15	1949
16 Germany	0.06–0.12	1914–1925
17 Germany	0.04–0.07	1856
18 Germany	0.15–0.20	1933
19 Gibraltar	0.04–0.07	1848
20 Greece	0.04–0.07	?
21 Italy	0.04–0.07	1939
22 Italy	0.06–0.10	1929, 1935
23 Yugoslavia	0.03–0.05	1895–1905
24 Israel	0.04–0.05	?
25 Israel	0.07	1924
26 Israel	0.04–0.05	1929-1934
27 Iraq	0.05–0.07	1949, 1954
28 Java	0.02	1931–1933
29 South Africa	0.06	1953
30 Morocco	0.04	1960
31 Zambia	0.04	1921
Homo erectus		
32 Germany	0.50	1907
33 Hungary	0.35	1965
34 China	0.60	1963
35 China	0.30–0.40	1927
36 Java	0.50	1937–1938
37 Java	0.70	1936, 1941
38 Java	0.50	1891
39 Java	0.70	1936
40 Algeria	0.35	1954
41 South Africa	0.50	1949
42 Tanzania	0.50	1960
43 Tanzania	1.0	1964
Australopithecus africanus		
44 Chad (Yayo)	0.5	1960
45 South Africa	1.0?	1947–1962
46 South Africa	1.0?	1936–1957
47 South Africa	1.0?	1924
48 Tanzania	1.5–1.8	1959–present
Australopithecus robustus		
49 South Africa	0.5	1938–1941
50 South Africa	?	1948–1952
51 Tanzania	1.7	1959
52 Tanzania	0.7	1964

(England) and Steinheim (Germany) and believed to be about 150,000 to 200,000 years old have been claimed to be more similar to modern man than to Neanderthal man (Figure 39.5). These skulls raise the possibility that populations with characteristics similar to those of modern man have existed continuously for at least 150,000 years and therefore existed at the same time as Neanderthal man. Until about 30 years ago, this view was widely held, but most of the finds once believed to represent very old remains of modern man have turned out to be either deliberate hoaxes or misinterpretations of relatively recent remains. Thus, there has been a trend toward a belief that modern man evolved within the past 50,000 years or so from some populations of Neanderthal man.

Repeated investigations of the Swanscombe site have confirmed its age, but sophisticated analyses of the skull fragments recently have cast doubt upon their similarity to bones of modern man (Weiner and Campbell, 1964). The Steinheim skull is badly fragmented and distorted, and recent studies not only discredit its resemblance to modern man but question whether it should be assigned to *Homo sapiens* or to *Homo erectus*. Other skull fragments of about the same age found at Fontéchevade in France are sometimes claimed to be modern in appearance, but most anthropologists regard these pieces as too fragmentary to permit any conclusions. In short, although the evidence is incomplete, the present view is that the diverse populations of modern man evolved from equally diverse populations of Neanderthal man.

Fossils of essentially modern man are found at sites dated at about 35,000 to 45,000 years ago in France (the famous Cro-Magnon man), Africa, and southeastern Asia. It is not yet clear whether modern man evolved in certain isolated Neanderthal populations and then spread out over the Old World, competing with and displacing Neanderthal, or whether there was a general evolution of Neanderthal populations toward modern characteristics as the use of tools placed greater survival value on agility and less value on massive jaws and teeth as all-purpose weapons and tools. Present evidence seems to suggest a general trend in all human populations from Neanderthal toward modern characteristics, and cultural remains (which are far more abundant than fossil bones) seem to show a continuous sequence of modifications in technological features from the Neanderthal populations through historical times.

Tracing hominid evolution still further back in time, a similar pattern emerges—insofar as it can be deduced from fragmentary fossil remains. About 100,000 to 200,000 years ago, there existed a diversity of forms intermediate between *Homo erectus* and *Homo sapiens*, with Steinheim and Montmaurin man probably near the Neanderthaloid end of the continuum and Fontéchevade man probably near the more "modern" end. Yet further back in time—perhaps 500,000 to 1,000,000 years ago—there is evidence of populations of *Homo erectus* scattered over the Old World. These remains are sufficiently different from Neanderthal and modern man to be classified as a different species, showing even more pronounced apelike features than those of Neanderthal man (Figure 39.7). Fossils from Java and China demonstrate that there was wide variation within the human population of this time also (Figure 39.8).

There seems to be a break of a few hundred thousand years in the record between the most recent remains of *Homo erectus* and the oldest remains of Neanderthal man, but this gap may be due to the relatively imprecise

Figure 39.6. Comparative study of ape and primitive human skulls showing chimpanzee skull (above), Swanscombe skull (middle left), Steinheim skull (middle right), Neanderthal skull of Tabūn man (lower left), and *Homo erectus* skull (lower right).

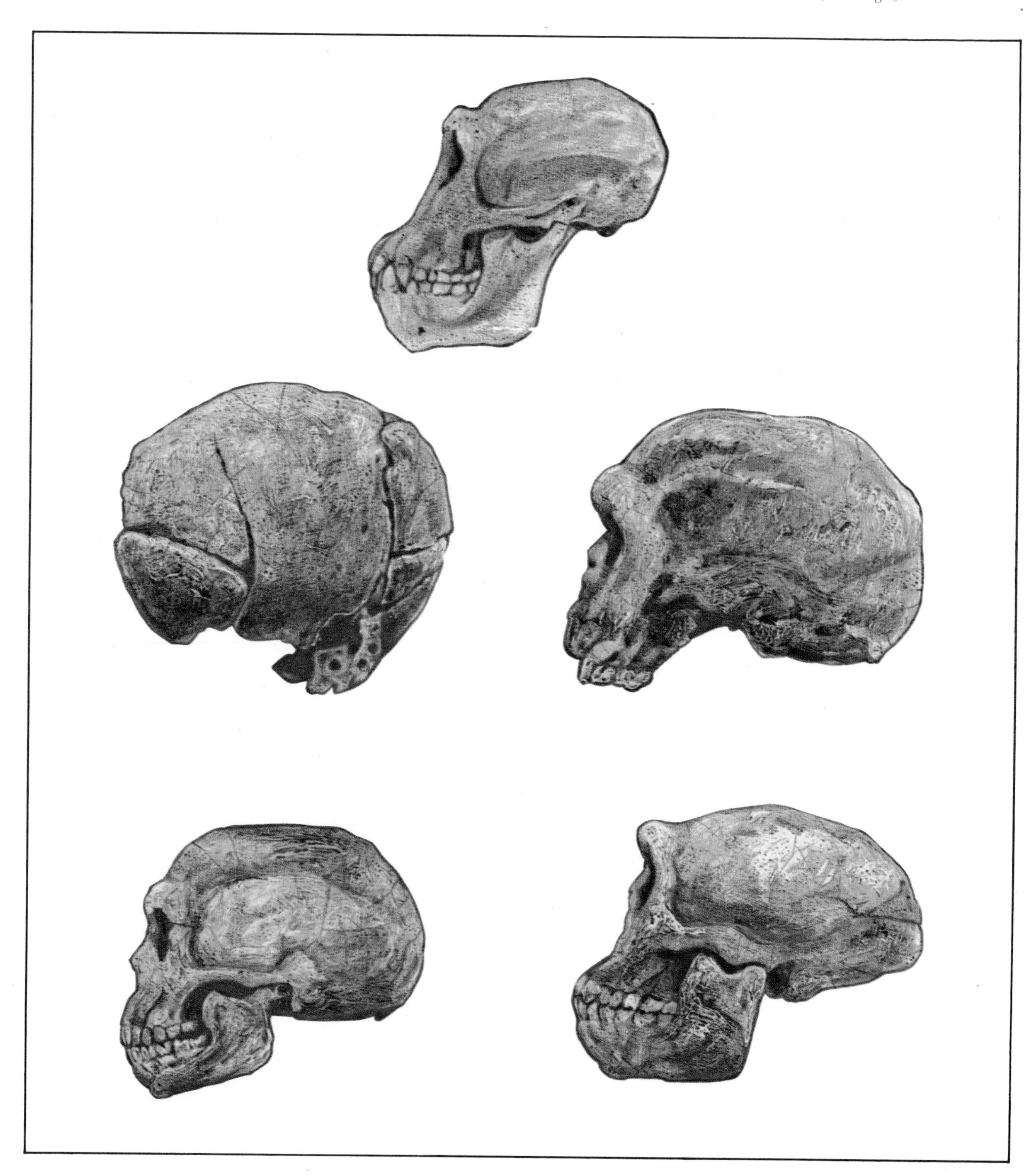

Figure 39.7. Phylogenetic scheme indicating the relationships within the family Hominidae.

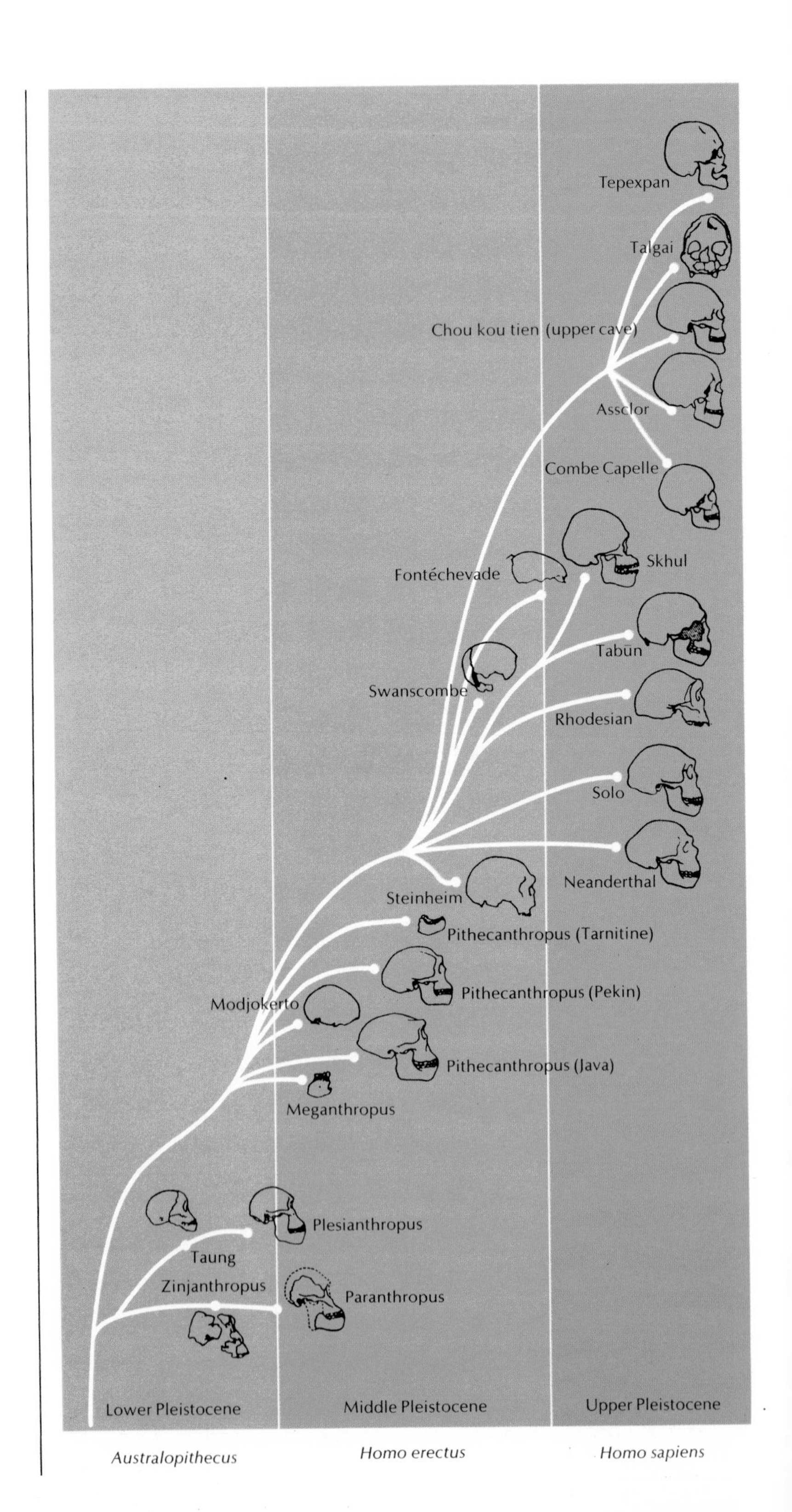

Figure 39.8. Comparative study of selected hominid skulls: Java man (upper left); Pekin man (upper right); *Australopithecus africanus* (lower left); and *Paranthropus* (lower right).

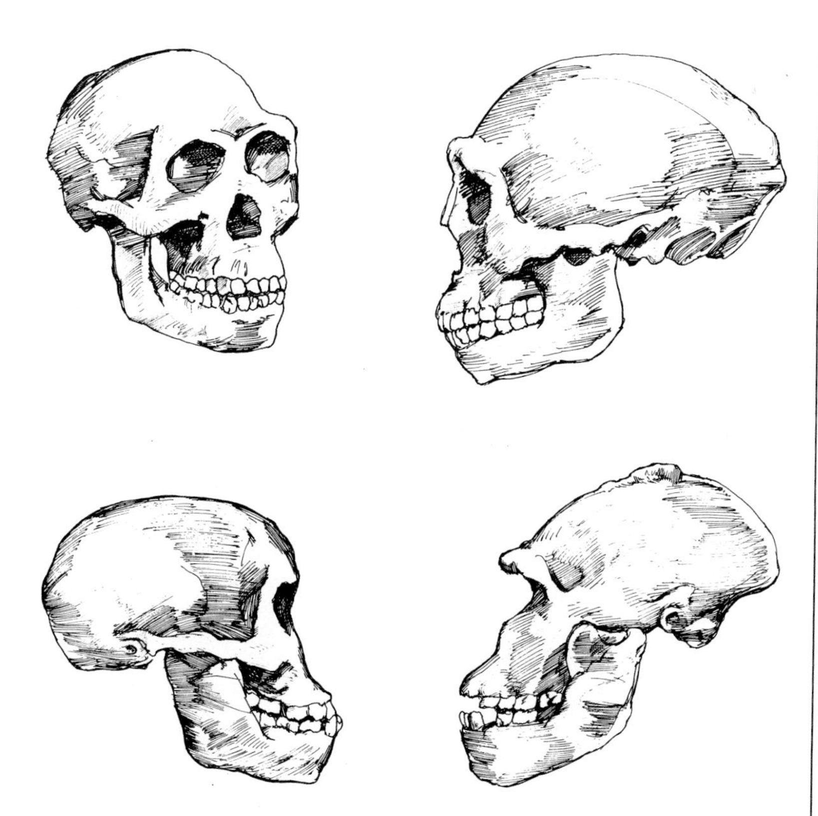

means of dating used on most fossil finds. The transitional nature of the oldest Neanderthal fossils and the continuous record of stone tools and other cultural remains suggest that there was a continuous evolutionary sequence from diverse populations of *Homo erectus* to diverse populations of Neanderthal man.

Yet further back in time, about 1 to 2 million years ago, fossil evidence from East Africa and South Africa shows the existence of populations of *Australopithecus* with sufficient variety that two distinct species have been named. In fact, many anthropologists classify *A. robustus* as a distinct genus, *Paranthropus*, and consider it to be a manlike ape that branched off from the main line of human evolution, perhaps coexisting for a time with the more human *A. africanus* but eventually becoming extinct.

At Olduvai Gorge in Tanzania, Louis S. B. Leakey (1965) believes he has found evidence of three distinct hominid species—*Australopithecus robustus, Homo erectus*, and what he calls *Homo habilis*—living in the same area at about the same time. However, other anthropologists regard his *Homo habilis* remains as a variety of *Australopithecus africanus* and dispute the identification of the *Homo erectus* remains (Clark, 1967).

Obviously, much work needs to be done to sort out the taxonomic and temporal-spatial relationships among these early hominids, but the evidence does seem to indicate that even in these early times the hominid family was characterized by diversity as much as by unity. The sum of all the fossil evidence seems to support a view that human evolution should be regarded as a continuum, with a wide variety of forms existing at any time and a continual and gradual shift of genetic and phenotypic characteristics

Figure 39.9 (left). Schematic diagram illustrating the "spectrum" or "continuum" in the evolutionary development of the hominids.

Figure 39.10 (right). The most common tools of the Oldowan culture are crude choppers, made by removing flakes along one side of a pebble to form an irregular cutting edge.

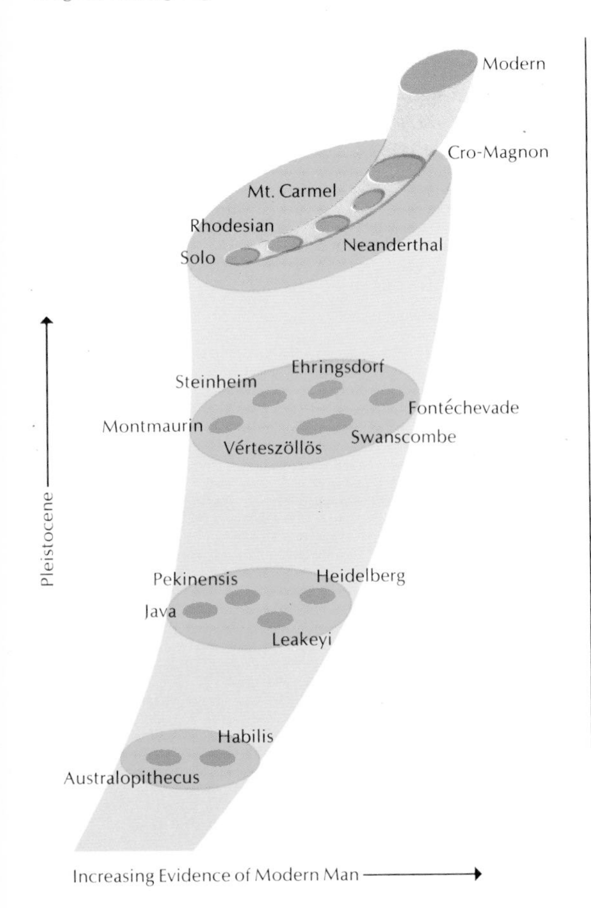

through time (Figure 39.9). This shift has resulted mainly from the influence of natural selection on the interconnected network of partially isolated populations. Certain general trends have characterized the evolution of all hominid populations over time, but these trends have been more marked in some regional populations, whereas other regional populations have been eliminated along the way.

MAN

The fossil record seems to lead back in time to an origin of the hominid stock—particularly of the ancestors of the genus *Homo*—somewhere in the region of East and South Africa. There is no evidence that *Australopithecus* moved north of equatorial regions. It seems safe to conclude that the human species developed in a hot climate. Evidence to support this statement comes from five fields: archaeology, paleontology, climatology, anatomy, and physiology.

Archaeological evidence deals with the stone tools that apparently represent the first clear evidence of manlike behaviors on the part of early hominids. For a century or so, claims were put forward that the earliest stone tools were made in various parts of Europe, including regions that were glaciated during early hominid development. Over and over again, these claims have been disproved. If the tools proved to be authentic, they were found to be more recent than had been claimed. If they proved to be as old as claimed, they were found to be natural rather than man-made. It now seems relatively clear that the earliest and simplest implements were the pebble tools of the Oldowan culture (Figure 39.10), and the oldest known Oldowan artifacts are from beds formed about 2 million years ago at Olduvai Gorge in East Africa.

Similarly, paleontological evidence has discounted claims that man originated in Europe or Java, but the oldest known fossil hominids now are the *Australopithecus* remains found by Leakey in the same deposits as the oldest tools.

A variety of anatomical and physiological evidence suggests that man originated in a hot climate. One of man's interesting characteristics is his

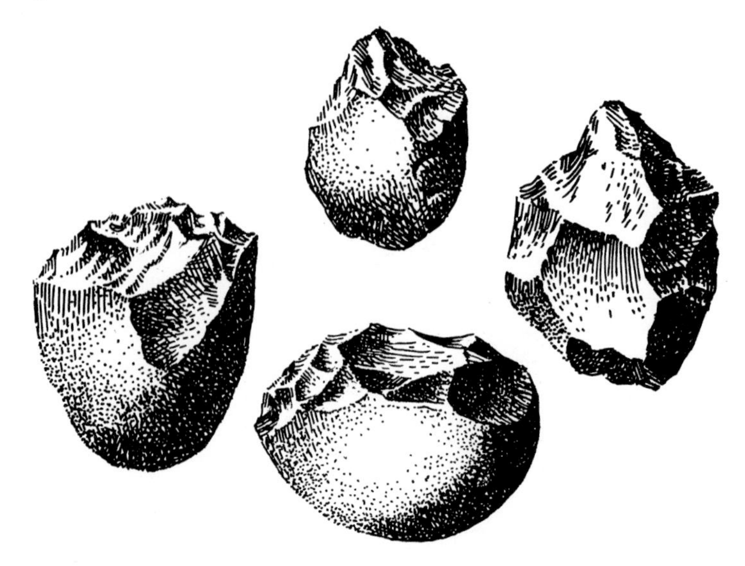

relative hairlessness in comparison to other primates. This hairlessness can most plausibly be explained as an adaptation to permit activity in the heat of a tropical day. In the skin of most animals, including the apes, there are a large number of *apocrine glands* in close association with the hairs. The hair and the glands develop together from the same embryonic structure. Apocrine glands produce a slight but very odorous secretion, which probably serves as a social or sexual signal to other members of the species. In man, both hairs and apocrine glands are scattered sparsely over the skin, but *eccrine glands* are found in high density over the entire skin surface (Figure 39.11). These glands produce sweat in great abundance, making possible rapid heat loss through evaporation. A thick hairy coat traps a layer of still air, greatly reducing evaporation from the skin surface even if there is a strong air movement over the body. Thus, both the abundance of eccrine glands and the sparseness of hair are adaptations suited to increasing the rate of heat loss through evaporation.

Man is the only predator capable of continuing vigorous hunting activity during the tropical day, and his habit of sleeping at night and being active during the day (the reverse of the pattern of most carnivores) suggests that he has long specialized in midday activity. There are about 2 million sweat glands scattered over the skin of an individual human, and there is little variation in this number among modern men living in a wide range of climates. Some population groups in very cold climates show more development of body hair than those now living near the Equator, but the differences are relatively slight. It appears that evolution has modified man's heat-loss equipment only very slightly since he left the tropical regions where he developed.

Physiological tests of heat tolerance have been performed on samples drawn from many racial and regional varieties of mankind. In general, they reveal a universal ability to cope with heat load through a phenomenon called heat acclimatization. With continued exposure to heat, there is a marked improvement in the body's ability to control its temperature. At first, skin and rectal temperatures climb, and the heart rate increases as blood is pumped through the surface regions for cooling. Within a few

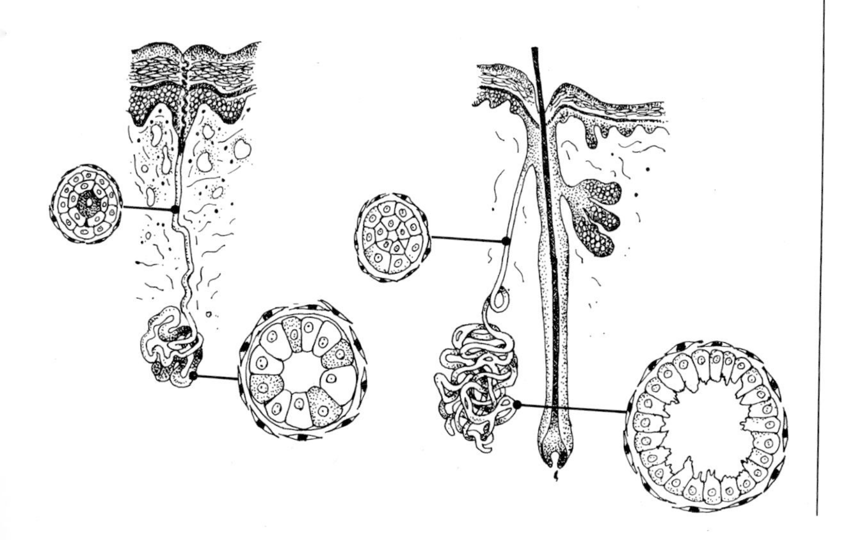

Figure 39.11. Schematic diagrams of the human eccrine (left) and apocrine (right) sweat glands, showing the coil and duct in cross section.

Figure 39.12. Diagram showing changes in selected human physiological parameters during acclimatization to heat.

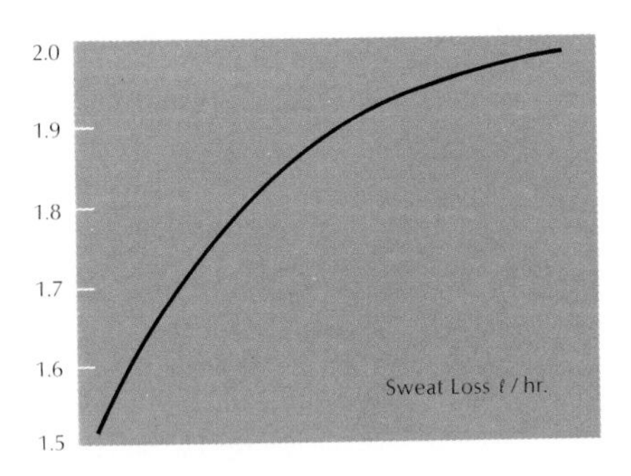

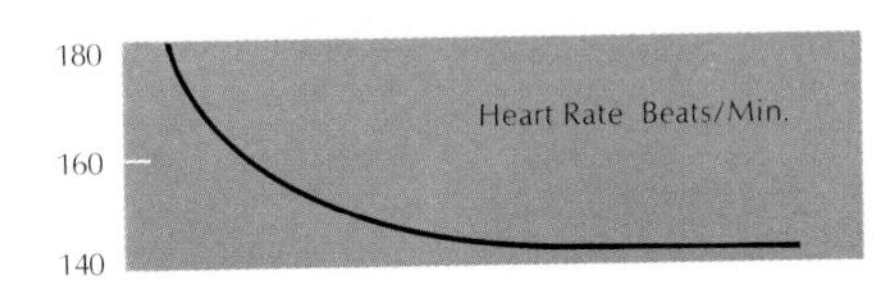

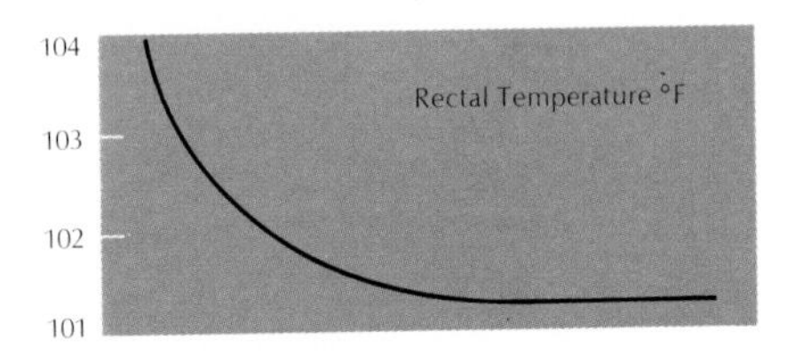

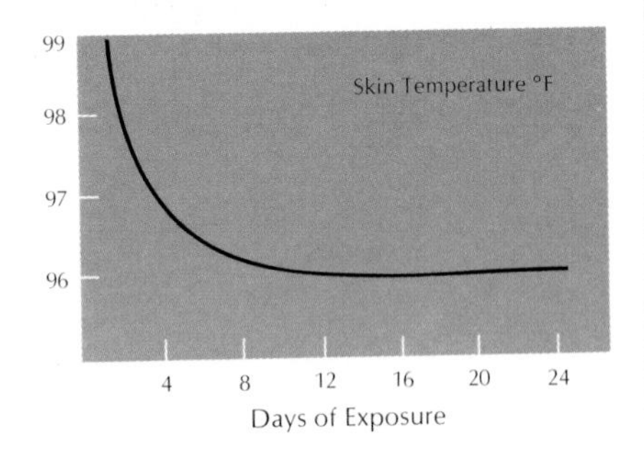

days, the rate of heat loss through sweating climbs sharply, and heart rate and temperatures drop back to more normal levels (Figure 39.12). Even persons from very cold climates are able to adapt relatively quickly to conditions of sustained heat. Again, the evidence suggests that man evolved efficient means of dealing with a hot environment and that this mechanism has been very slightly altered by evolution of groups living in cold climates.

The bipedal locomotion of man is not suited to sustained speed, but it is well adapted to long-distance travel with minimum expenditure of energy. The hunting style of early man probably was much like that of surviving hunting and gathering groups in tropical regions. A group of men simply trot along after some large animal, keeping it moving continuously for days at a time. Eventually, exhausted by its lack of food and continued activity through the midday heat, the animal becomes easy prey for spears or clubs. Man's biological heritage has made him a very efficient agricultural laborer, able to cope with hot, humid conditions quite as well as with the hot, dry, sunny conditions of his original habitat. In fact, he is able to work even harder in the cooler habitats to which he has migrated.

Other physiological evidence of man's tropical origin involves the "critical temperature," the temperature at which an animal begins to increase its heat production, mainly by shivering. The critical temperature of a tropical animal is much higher than that of an arctic animal, which is insulated from external cold by hair and fat. The critical temperature for a naked man is about 28°C, about the same as that of most tropical animals. Man is severely limited in his ability to withstand cold without artificial protection. His heritage as a hunter makes him able to maintain a high level of activity and heat production, and human populations in temperate regions do have a slightly thicker layer of fat. However, few if any human populations rely predominantly on physiological or morphological adaptations for protection from cold. Rather, man produces an artificial and fairly stable microclimate within his clothing and dwellings, reproducing the conditions of his original tropical habitat. Eskimos have brought this technique of climatic bioengineering to a high degree of effectiveness.

Humans show well-established diurnal rhythms of temperature and other physiological functions. These rhythms seem related to the regular daily cycle of the tropical environment, and many of them are modified slightly or not at all even in populations living under conditions of continuous light or darkness in the polar regions. The temperature rhythm does show some adaptability to changes in daily routine.

Skin color shows much more modification in groups living in cooler regions. Dark skin color, produced by a high concentration of melanin, is an essential adaptation for a hairless animal subjected to the strong ultraviolet radiation of tropical regions. It protects against sunburn and some forms of skin cancer. Those populations that live in regions of lower radiation intensity do show some modification toward lighter skin color.

The characteristic of skin color—perhaps because it is one of the few obvious characteristics that has been modified by evolution since man spread out over the globe—seems to many people an obvious justification for the division of mankind into distinct races. This conclusion, however, cannot be supported by the evidence.

Sunburn not only is painful but disables the sweat gland mechanism and thus interferes with ability to withstand external heat. The genetically dark-skin individual—or a lighter-skin individual with an acquired tan—is able

Figure 39.13 (left). Reflectance spectrophotometry is a process used to measure the varying intensities of skin color in different populations throughout the world. The amount of skin reflectance is measured by a reflectometer, which is used in conjunction with a spectrophotometer, an instrument that analyzes skin luminosity, or brightness.

Figure 39.14 (right). The primate line may have evolved from arboreal, placental mammals such as the one depicted here, the prosimian (*Plesiadapis*.).

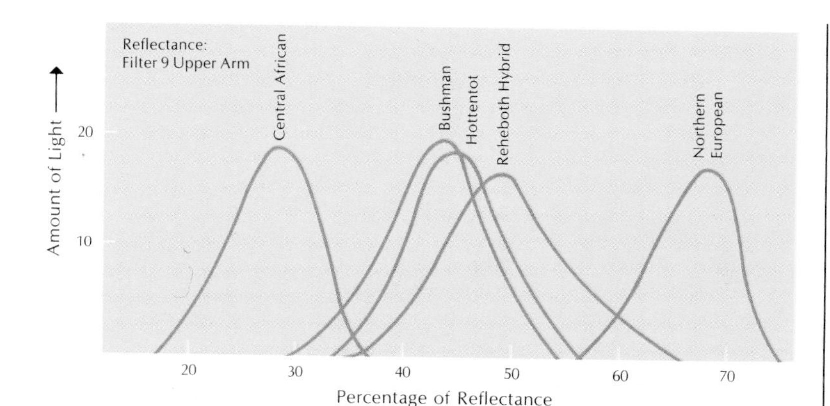

to resist this short-term damage to his skin. He also is less likely to develop skin cancer as a result of ultraviolet radiation. A quite moderate degree of pigmentation is sufficient to ensure almost complete protection (Figure 39.13). It is not surprising, then, that degree of skin pigmentation is closely correlated with ultraviolet radiation in the environment and is relatively independent of other "racial" characteristics. Among the Caucasoid populations, the Mediterranean, South Indian, and Yeminite groups are examples of darker pigmentation. Among Mongoloids, there are the dark-skin peoples of southeastern Asia. Even the Amazonians—relatively recent arrivals in the tropics—have a skin color comparable to that of some Negroids. The skin color of Negroids shows a very wide range, from the relatively light Hottentot and Bushman groups (who are genetically similar to their darker neighbors in other respects) to the very dark Sudanese and West African groups. Australians and Melanesians also show a variety of pigmentations.

Thus, the evidence of physiology and anatomy strongly suggests that man is adapted for life in a tropical climate—an adaptation that has been modified only slightly during the 25,000 years or so that he has moved out of the equatorial regions. If man evolved in a tropical region a few million years ago, it seems almost certain that he did so in Africa. At about the time early hominids were emerging, the northern hemisphere had cooled substantially and was moving into a period of successive glaciations. The fossil evidence indicates that varieties of arboreal, apelike primates had been living in northeastern Africa for many millions of years. Some of these forms seem quite plausible ancestors for the first hominids.

With some rather risky extrapolations from the available evidence, it is possible to suggest a plausible explanation of human origins in East Africa. During the Pleistocene epoch and glaciations of the northern regions, beginning a few million years ago, Africa was becoming increasingly arid. One effect of this climatic change was a decrease in the extent of thick forests and an increase in the extent of scattered woodlands and savannas. Some apelike primates probably lived in the forests much like modern apes, spending a good deal of time in the trees for feeding, sleeping, and escaping from predators. Like most primates, they fed chiefly on fruits, leaves, nuts, berries, and so on, but were perfectly willing to eat meat when a small bird or animal happened to be captured.

As the forests dwindled, some groups of these primates became adapted to life in the relatively open woodlands. They were large enough to be

Figure 39.15. Selected racial varieties in the contemporary global population. Such racial distinctions are based on a variety of features including skin color, eyes and hair type, and shape of the skull.

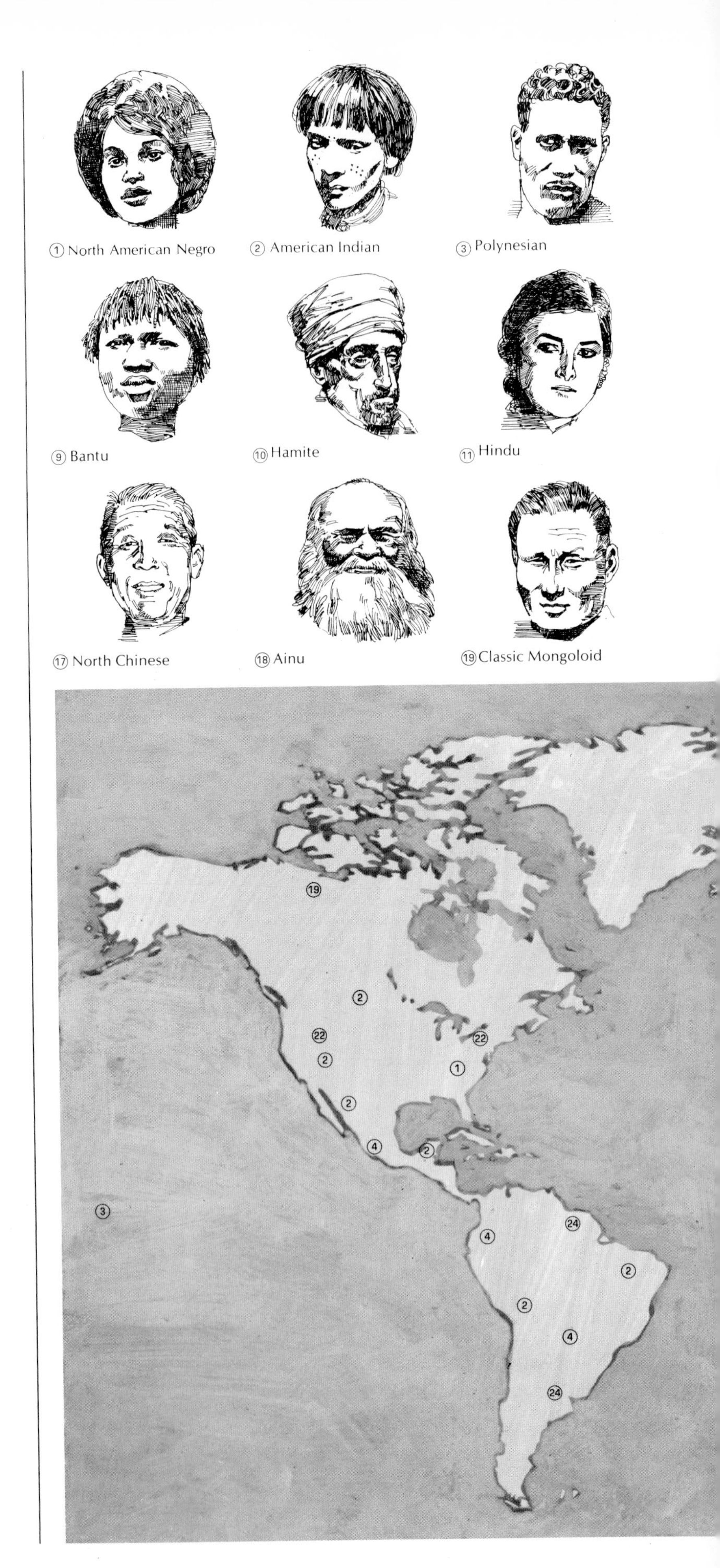

④ Ladino
⑤ Sudanese
⑥ Negrito
⑦ Bushman
⑧ South African Negro
⑫ Tibeto-Indonesian Mongoloid
⑬ Murrayian
⑭ Carpentarian
⑮ Melanesian
⑯ Southeast Asiatic
⑳ Turkic
㉑ Northeast European
㉒ Northwest European
㉓ Nordic
㉔ Mediterranean

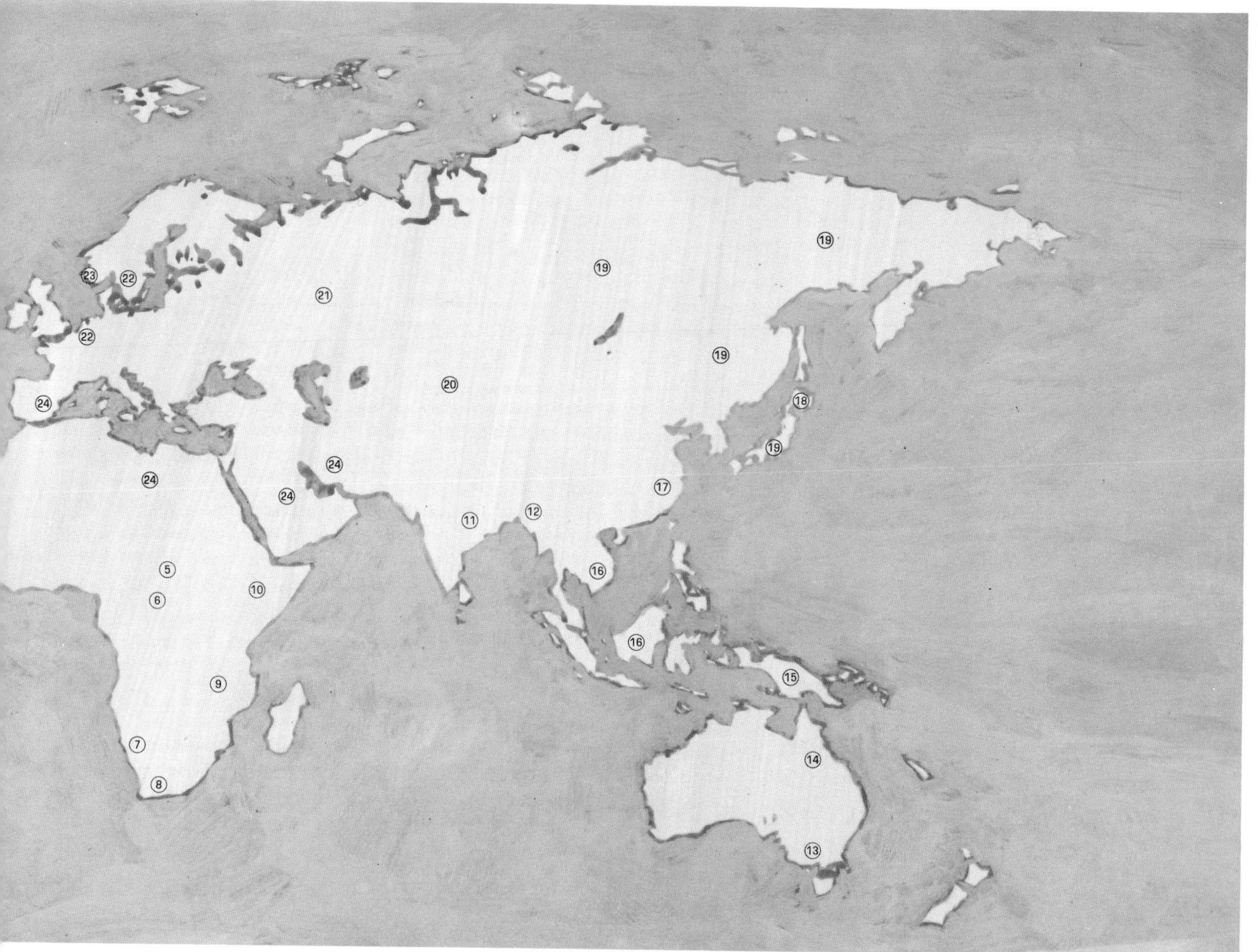

Figure 39.16. A reasonable interpretation of man's progression up the evolutionary scale.

somewhat free of attacks by predators, and they developed the knuckle-walk as a means of locomotion on the ground. They seldom ventured out on the drier and more open savannas and still spent much of their time in the trees. They retained the herbivorous habits of their forest ancestors. This picture is consistent with what can be deduced of the way of life of *Australopithecus robustus* — apparently a very apelike form.

However, other groups adapted to life on the savannas — perhaps because their home woodlands were dwindling and they were forced to forage on the savannas surrounding them, or perhaps because increasingly long and severe dry seasons limited their dependence upon their herbivorous diet. These groups probably came to depend more and more upon catching small animals for food. Such a diet would lead to selection for agility, a lithe body, a more bipedal posture, and improved intelligence. Thus, some groups of *A. robustus* evolved into *A. africanus*, and the two species probably existed simultaneously for a long time — one group inhabiting the drier grassier regions and the other living in the moister woodland regions.

While *A. robustus* remained much the same for millions of years, probably living much as modern chimpanzees do, the selective pressures of the harsh savanna life were producing rapid evolutionary changes in *A. africanus*. He became more and more efficient at bipedal locomotion and his teeth developed into a relatively modern form, suited to an omnivorous diet. He developed characteristics similar to those of carnivores, losing the need to nibble plant food continuously and becoming adapted to occasional meals. He adopted a home base in a cave or rock shelter, where a group lived, raised their young, and shared the food they captured. He became able to control his defecation, so as not to foul this home site. Most importantly, at some point he began to extend the primate ability to pick up sticks and use them as simple tools. He learned to use bones, sticks, and rocks as weapons and as tools and eventually began to shape these objects for better tools. Once this process had begun, selection would favor further increases

Figure 39.17. Neanderthal man, *Homo sapiens Neanderthalensis*. Skeletal remains of Neanderthal people dating back 80,000 years have been found in western Europe. Here, a man is depicted building a fire in his cave shelter.

in upright posture (to free the hands), agility, and intelligence. Thus, *A. africanus* began the relatively rapid evolution into humanity, becoming the active and relentless hunter of the hot, midday savanna, while *A. robustus* continued his apelike existence in the woodlands.

GENETIC AFFINITY AND GENETIC VARIABILITY

The fundamental uniformity and variability of the human population can be clearly illustrated by direct data about the genetic constitution of *Homo sapiens*. The 46 chromosomes of the zygote and the configuration of individual chromosomes are universal characteristics of the species. Modifications in minor features of chromosome shape may exist in various populations, and abnormalities of chromosome number are the basis for congenital malformations of various kinds. However, the human chromosomal composition is sufficiently universal that fertile matings are known for almost every conceivable combination of racial or geographical groups of modern man.

If interbreeding does occur to some extent among all human populations, then every individual—potentially, if not actually—shares in the gene pool of the species. The total population can be divided into large subgroups such that the members of each subgroup mate much more often with each other than with members of any other subgroup. These subgroups can be further subdivided, and so on. This hierarchy of semi-isolated populations arises because of various factors, including geographical separation, cultural restrictions, and religious beliefs. Thus, the gene pool of the species consists of various partially isolated gene pools, or *isolates*.

By studying blood samples, serologists can identify and characterize at least 25 different genetic systems, including the well-known blood groups, variations in hemoglobin and plasma proteins, and various systems of red cell isoenzymes. Each of these genetic characters is controlled by an array of alleles at particular gene loci, and the alleles are represented to a greater or lesser extent in every human population. Only rarely is a particular allele

Figure 39.18. The distribution in aboriginal populations of the *M* allele of the MNSs blood group system. The frequency of *N* in each area can be calculated as 1.0 minus the frequency of *M*.

found to be completely absent from a local group or to exist only in limited populations.

World maps showing distribution of the various genetic markers reveal that divisions between populations drawn on the basis of one genetic character usually differ from those drawn on the basis of another character. For example, the blood-group allele *M* is world-wide in occurrence, but its gene frequency varies from about 5 percent in New Guinea to more than 90 percent in parts of America (Figure 39.18). The Duffy gene (Fy^a) is similarly widespread, but shows a very different distribution pattern, ranging from 100 percent in Australasia to about 5 percent in Africa.

Even on the basis of serological characters, each controlled by a single gene locus, it is impossible to distinguish discrete populations within the human species. There is a great deal of overlapping and intergrading between populations, which are even more obvious for characters controlled by many gene loci—for example, skin color, body size and shape, and facial features. It is this fact that makes a simple classification of mankind into races a rather arbitrary exercise and one that simultaneously obscures the similarities between races and the variety within them.

No particular genotype can be best suited to every kind of environment, yet the environment varies continuously, both from place to place and from time to time at the same place. Under these varying conditions, with gene flow among populations living in different environments, most alleles are maintained in the population. Furthermore, the fitness of a particular allele varies with the nature of the other alleles in the same genotype. Some genetic changes arising by mutation do seem to be harmful in any external or genotypic situation. Such variability may be regarded as an unavoidable evil. But most variability has value to the population. The existence of diverse genotypes enables the population to exploit varied local conditions within its range and to respond to changes in the environment over time.

In short, there is no ideal genotype toward which the population tends. There is only a range of diversity, composed of the array of genotypes that yield a suitable degree of adaptiveness to environments that the population commonly encounters. As the environment changes from time to time or place to place, the advantages of various genotypes change also. The gene frequencies for alleles at each particular locus are most sensitive to certain kinds of environmental conditions. Because all of the environmental conditions do not change together from one place to another, neither do the gene frequencies for all of the loci.

Within the entire human species, large groups can easily be distinguished on the basis of one characteristic or another. These characteristics may correspond to major variations in environmental conditions or to limitations upon gene flow between partially isolated populations. However, it appears that gene flow among all human populations is sufficiently great to cause gradations in most phenotypic characters between any one population group and its neighbors.

For the species as a whole or for any population group within it, maintenance of variability ensures maximal use of the range of conditions in the present environment and a maximal opportunity to adapt rapidly to changing conditions over time. Biologically and culturally, *Homo sapiens* as a species and as individuals shows a high degree of variability and a correspondingly high degree of adaptability.

ADAPTIVE BEHAVIOR

The human species is distinguished from other animals most prominently on the basis of its behavioral or technological adaptability. Although the biological characteristics of the species have changed relatively little in the past 25,000 years, there has been a dramatic change in behavioral characteristics that has completely altered the relationship between the human species and the rest of the ecosystem.

As in biological characteristics, the human species has always been typified by both unity and diversity in its technological characteristics. The Oldowan pebble-tool culture developed into the Acheulian hand-axe culture (Figure 39.19). With elegant hand axes and a variety of other flaked-stone tools, simple hunting bands of the Acheulian culture roamed over the African plains and far beyond—across Morocco into Europe (as far as the North Somme valley and Britain) and eastward over Asia into the Punjab. There are differences in local styles of tool making, but the similarities are great enough to suggest a high level of cultural interchange throughout the population during this period.

In the Far East, a different culture developed independently from a preceding pebble-tool technology. The Eastern culture produced choppers and chopping tools with a single flaked cutting edge (Figure 39.20). In India the two cultures met and in some places fused.

Thus, the major characteristic of human cultural development—the coexistence of cultural innovation and cultural diffusion—is clearly displayed in the Middle Pleistocene epoch. Innovation may take the form of major separate developments—as in the Acheulian hand-axe culture and

Figure 39.19. Representative Acheulian tools.

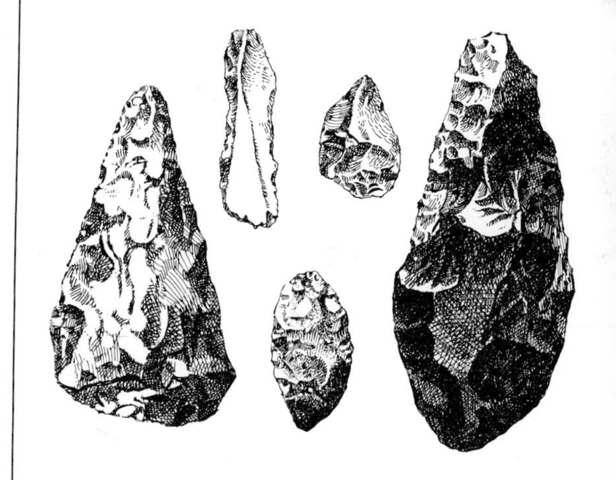

Figure 39.20. Chopping tools from China and India.

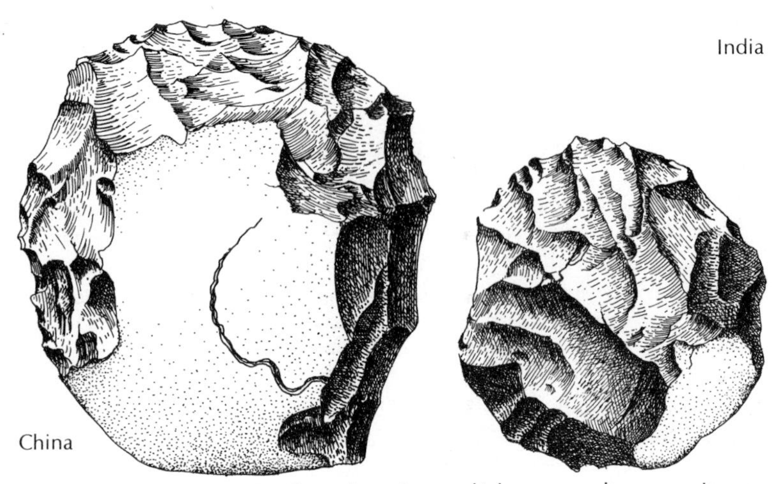

the contemporaneous Far East chopping-tool/chopper culture—or it may appear merely as small local modifications within each tradition. Diffusion and exchange may proceed over an enormous area, as in the case of the Acheulian culture. On the other hand, it may remain quite regional and circumscribed.

It is possible to regard every cultural response in this early period as compounded in different proportions of borrowing from older or neighboring traditions and modifying to meet local and present needs. Although the two Middle Pleistocene technologies are quite distinct, both emerged from similar pebble-tool predecessors. Present evidence suggests that the single species *Homo erectus* was responsible for all these developments. The boundaries between different cultures may or may not correspond with significant changes in biological characteristics.

The two traditions showed a steady development toward more advanced tools, and in later stages both traditions adopted similar techniques. For example, the Levalloisan core-shaping technique involves striking triangular points from cores with prepared faces and faceted striking platforms. This technique was adopted in cultures that are otherwise developmentally distinct over the whole area inhabited by the early *Homo sapiens* population, roughly from 80,000 to 20,000 years ago.

There is a clear technological advance from the Acheulian and chopping-tool/chopper complexes of *Homo erectus* times to the industries developed by early *Homo sapiens*. There is also a continuity of tradition and a steady diffusion of technique from stage to stage. The technological progress of the later *Homo sapiens* hunting peoples beginning about 25,000 years ago is even more striking in its rapidity and elaborateness. Stone artifacts, blades, and small flints of beautiful workmanship evidence steady improvement in techniques. Even more striking are the skillfully fashioned objects of antler, ivory, and bone and cave paintings and sculptures. Such advances occurred in all human populations, although the extent of the advance varies from one local population to another and earlier traditions persisted in many populations.

Throughout human history, populations differing conspicuously in one or several biological characteristics have successfully exploited the same geographical region. India, southwestern Africa, and the Malayan Archipelago are examples of areas now inhabited by a variety of different peoples.

Figure 39.21. European tool types are the basis for classifying Paleolithic cultures. The crude pebble tools contrast with the more sophisticated blade and core tools.

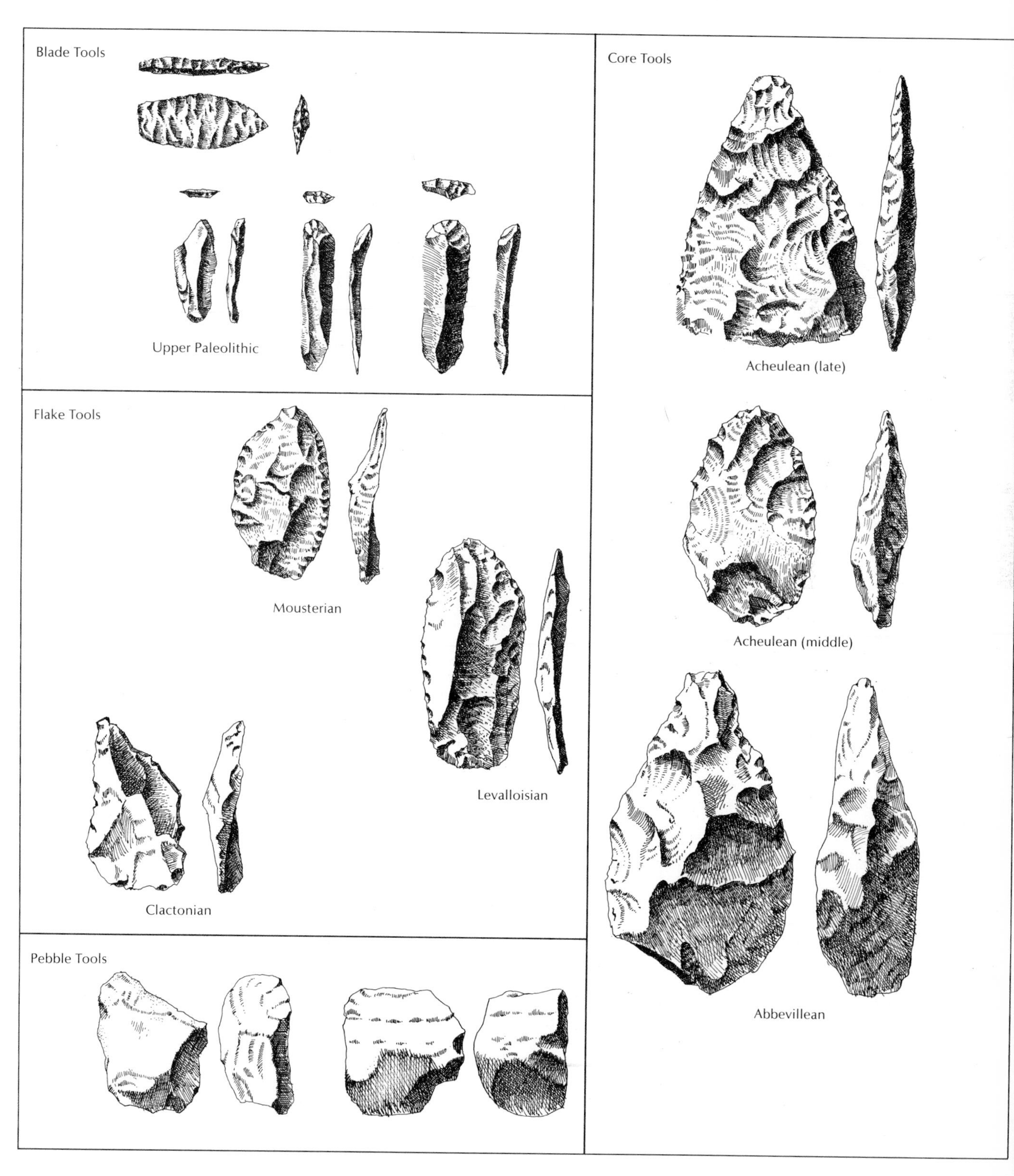

By far the most spectacular examples are the spread of European groups into Australia, Africa, and America; of Japanese and Chinese into the Pacific and America; of Indians into Africa, America, Asia, and Europe; and of Africans and Asians into America. If the native genotypes possess an inherent biological advantage for local conditions, it is easily overcome by the technological and biological adaptiveness of the "invading" humans.

Some habitats are more favorable than others for the development of certain technologies. An industry based on fishing, for example, could hardly develop in or be successfully imported into a desert region. On the whole, however, the habitat has not greatly restricted the appearance of economies of all stages. Most habitats can be exploited in a variety of ways. The resulting ecosystems may vary greatly, particularly in population densities, but the technology of advanced cultivators facilitates settlements of high density in almost every habitat.

There is little correlation between human "races" and the occupation of particular habitats or the practice of particular economies (Table 39.1). In particular habitats, one and the same group has established, often in historical sequence, a succession of ecosystems.

Table 39.1
"Economies" Practiced by Various "Races"

ECONOMIES	Negroid	Caucasoid	Mongoloid	Australo-Veddoid
Food gatherers	Bushmen	"Mesolithic" communities	Fuegians, Californians	Australians
Higher hunters and fishermen	Pygmies, Hadza	(Ainu) Mesolithic communities	Eskimos, American Indians	Veddas, Australians
Pastoralists	Bantu, Masai, Hottentots	Lapps Iron-Age Nordics	S. Manchurians	— — —
Nomads	Nilotes, Sahara Negroids	Bedouins, Tunreg	Mongols, Turkestan, Chuckchi	— — —
Simple cultivators	Bantu	Indo-Dravidians	S. Mongols, Malays	Melanesians
Advanced cultivators	W. Africans	Mediterranean peoples	Hopi, Maya, Indonesians, N. Mongols	Polynesians

With increasing technological control over both the external environment and the human organism itself, the biological significance of diversity as a mechanism for species survival has decreased. Medical technology makes possible the survival of genotypes that were once lethal or highly disadvantageous. Thus, modern societies can tolerate an increase in the "load" of deleterious alleles in the gene pool, some of these alleles being carried in a "masked" condition by every normal individual.

It is not surprising that in modern conditions of controlled environment, variations in biological characteristics are of decreasing significance. Europeans moving to the tropics may be more prone to malaria or to heat disorders than are the indigenous populations. Americans of Negroid ancestry tended to suffer more frostbite during the Korean War than did those of European ancestry. Light-skin individuals in South Africa or Australia are

Figure 39.22. World-wide pattern of distribution of the gene for sickle cell anemia.

more prone to skin cancer than are the darker native inhabitants. Island peoples are more susceptible to poliomyelitis or measles than are mainland inhabitants. In each of these cases, the effects of modern medical technology tend to offset any effect that these biological differences might have on survival rates and selection.

With increased environmental control and cultural diffusion rates, the geographical variations in external conditions for human individuals are becoming less and less significant. In short, there have been very significant alterations in the conditions that have produced the biological and cultural evolution of the human species over the past few million years. The effects of these changes upon the course of present and future human evolution are discussed in Chapter 44.

FURTHER READING

General discussions of factors influencing man's evolution are given in books by Baker and Weiner (1966), Barnett (1953), Brace (1967), CRM (1971), Claiborne (1970), Hooton (1947), Laughlin and Osborne (1967), Steward (1955), Tax (1960), and Weiner (1970), and in articles by Baker (1960), Brace (1964), Brues (1959), Coon (1955), Emiliani (1968), Meggers (1954), Schreider (1964), and Weiner (1971).

For further information about the physical and cultural characteristics of early man, see books by Augusta (1957), Bordes (1968), Braidwood (1967), G. Clark (1967), W. E. Le Gros Clark (1967), Lee and DeVore (1968), Macgowan and Hester (1962), Oakley (1966), Pfeiffer and Coon (1963), Pfeiffer (1969), and Ucko and Dimbleby (1969).

Discussions of diversity in modern man will be found in books by Mourant (1954) and Race and Sanger (1962). Also of interest in regard to this topic are articles by Allison (1955, 1964), Barnicot (1957), Garn (1957), Hulse (1955, 1962), Newman (1961), Newman and Munro (1955), and Roberts (1953).

40
Human Reproductive Physiology

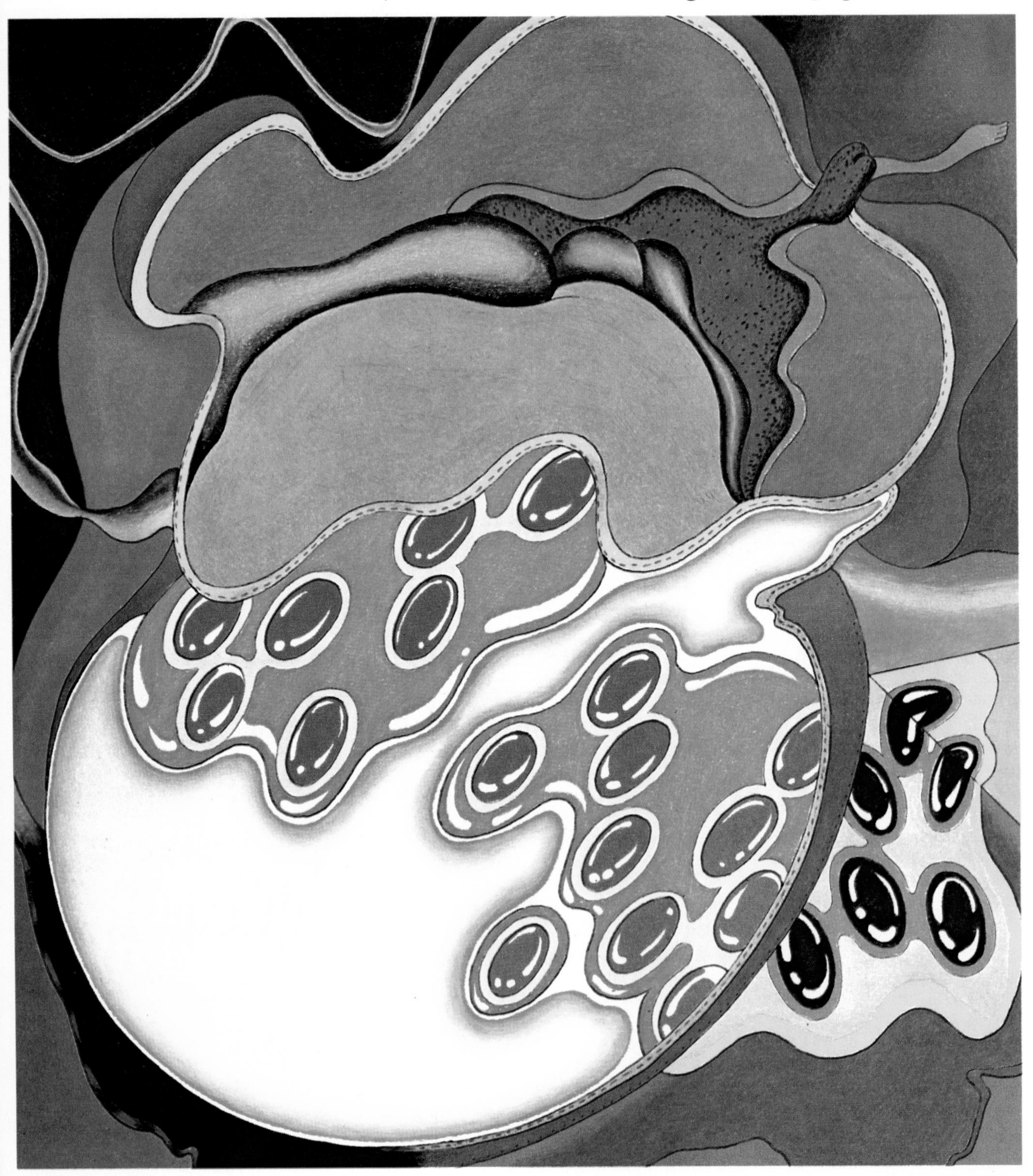

There are significant differences between man and other primates in sexual behavior and physiology. Little is known of the evolutionary development of these differences, but a possible pattern can be described.

Precursors of *Australopithecus* are presumed to have had a social organization similar to that of present ground-dwelling primates. Because of the more strenuous duties they performed, males were generally larger than females. Females underwent a periodic estrous or heat cycle, corresponding to ovulation, during which they actively sought copulation. Groups were lead by mature males, who prevented the younger males from breeding until the younger males could force the older from dominance.

The estrous cycle began to diminish in survival importance as the family group developed as a basic unit. The female vagina became longer—perhaps as a defense against infection—and the male penis became longer and lost the usual mammalian stiffening cartilage. These changes and diminished clitoral size are thought to indicate an adoption of the ventral position in copulation.

FEMALE REPRODUCTIVE SYSTEM

The reproductive system of the human female consists of the ovaries, fallopian tubes (oviducts), uterus, vagina, and external genitals (Figure 40.1). The *ovaries* are flattened glands located at the sides of the pelvic cavity and attached by ligaments to the uterus and fallopian tubes. The *fallopian tubes* extend from ovaries to uterus, forming a passageway for ova, or egg cells. The ovarian end of the tube expands into a funnellike structure with a fringed border; this funnel fits closely over part of the ovary surface. The tube is lined with cilia that move the ovum toward the uterus, where the fallopian tube ends in a minute opening in the uterine wall.

The *uterus*, or womb, is a pear-shape organ lying between the bladder and the rectum. The fallopian tubes open into its upper end, and the lower end of the uterus opens into the vagina. Ovaries, fallopian tubes, and uterus are all attached by ligaments to the walls of the pelvic cavity. The thick uterine walls are made of smooth muscle. The inner lining, or *endometrium*, consists of glandular tissue and a supporting framework. Normally, the uterine cavity is almost closed, but during pregnancy it expands enormously. The lower part of the uterus, near the opening into the vagina, is called the *cervix*.

The *vagina* is a dilatable, muscular tube about 3 or 4 inches long, leading from the exterior to the opening of the cervix. The vagina lies almost at right angles to the uterus, and the muscular rim of the cervix extends into the vagina. The vaginal muscle layers are much thinner than those of the uterus, and the interior of the vagina is lined with a thin layer of nonglandular epithelial cells.

The external genitals, or *vulva*, include a number of structures. The *mons veneris* is a mound of fatty tissue over the pubic bone at the front of the genitals; it becomes covered with pubic hair at puberty. The *labia majora* are prominent, thick folds of skin and fatty tissue, lying on the sides of the genitals and covered with pubic hair on their outer surfaces. The labia majora develop from the same embryonic structures as the scrotum of the male. The *labia minora* are smaller folds of skin and fatty tissue, lying inside the labia majora and forming the rim of a depression called the *vestibule*. Where the labia minora meet at the top of the vestibule is a small erectile

Figure 40.1a. Anterior view of the uterus and ovaries.

organ, the *clitoris*, that corresponds to the penis of the male. It is composed of spongy tissues that are expanded and stiffened by blood engorgement during sexual excitement, and its tip is highly sensitive.

Within the vestibule are the openings of the urethra (urinary tube) and vagina. Stretched across the front of the vaginal opening is a fold of mucous membrane called the *hymen*. In the young female, it commonly exists as a partial closure of the vaginal opening and usually is torn during childhood or during the first sexual intercourse. The *glands of Bartholin* are located on either side of the vagina just beneath the skin surface and open through very small ducts into the vestibule in the grooves between the hymen and the labia minora. Fluid from these glands lubricates the vaginal opening during intercourse. The anus—the external opening of the rectum—is located well behind the genitals.

Menstrual Cycle

The menstrual cycle prepares, renews, and refreshes the reproductive system for 30 to 40 years of a woman's life. This renewal is the only case in human physiology where blood loss is a sign of good health rather than injury. *Menstruation* occurs as blood-rich, glandular tissue is shed from the endometrium (lining of the uterus) if conception fails to take place during

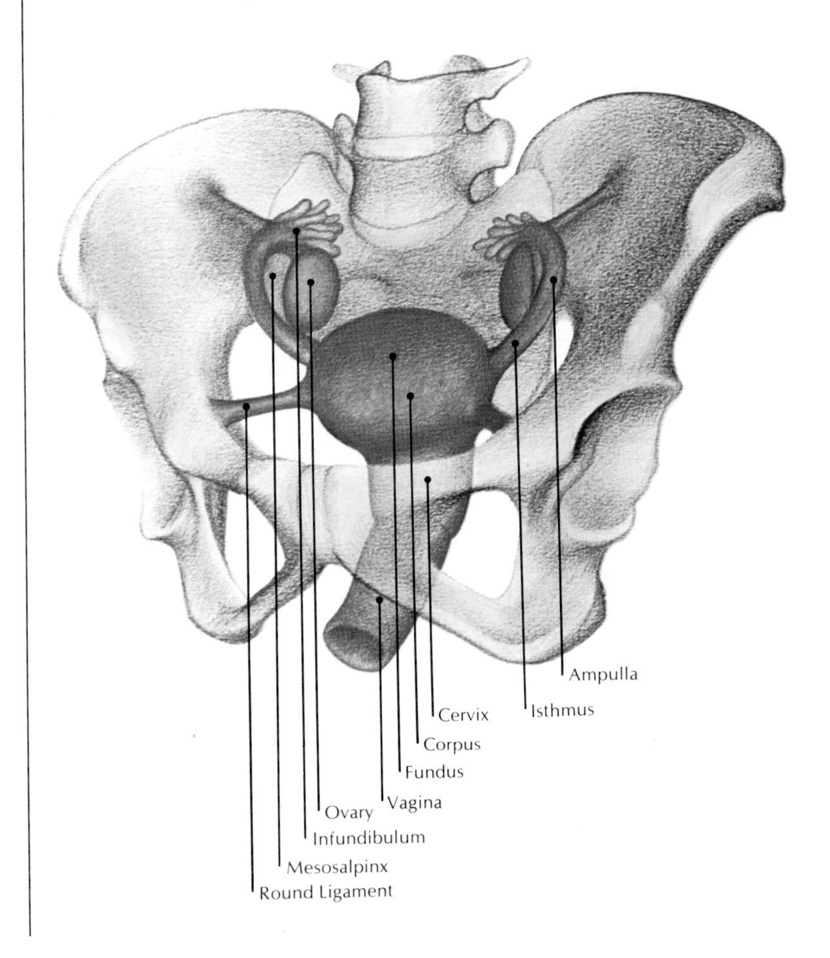

the monthly cycle. Menstruation occurs 14 to 16 days after ovulation and is triggered by changes in blood levels of hormones secreted by the ovaries.

The average menstrual cycle lasts 27 to 30 days, and the *menstrual phase* of the cycle, during which blood flows to the exterior, usually lasts from 3 to 6 days. Many variations are considered normal if they are repeated during each cycle.

As endometrial tissue is being sloughed from the uterine lining, follicle-stimulating hormone (FSH) is released by certain cells of the pituitary gland in increasing amounts (Chapter 21). This secretion apparently is controlled primarily by FSH releasing factor (FSH-RF) formed in a specific area of the hypothalamus by neurosecretory cells. The increasing level of FSH in the blood stimulates further development of an undetermined number of young follicles in the ovaries. As the follicles grow, they release increasing amounts of estrogen into the blood. FSH alone does not stimulate estrogen secretion by the follicles, but secretion is stimulated by a high level of FSH in combination with a small amount of luteinizing hormone (LH), which is secreted by another group of cells in the pituitary gland.

As bleeding stops at the end of the menstrual phase, the endometrium is thin, with a few blood vessels and only the base of the endometrial glands remaining. As the level of estrogens in the blood increases, the *proliferative*

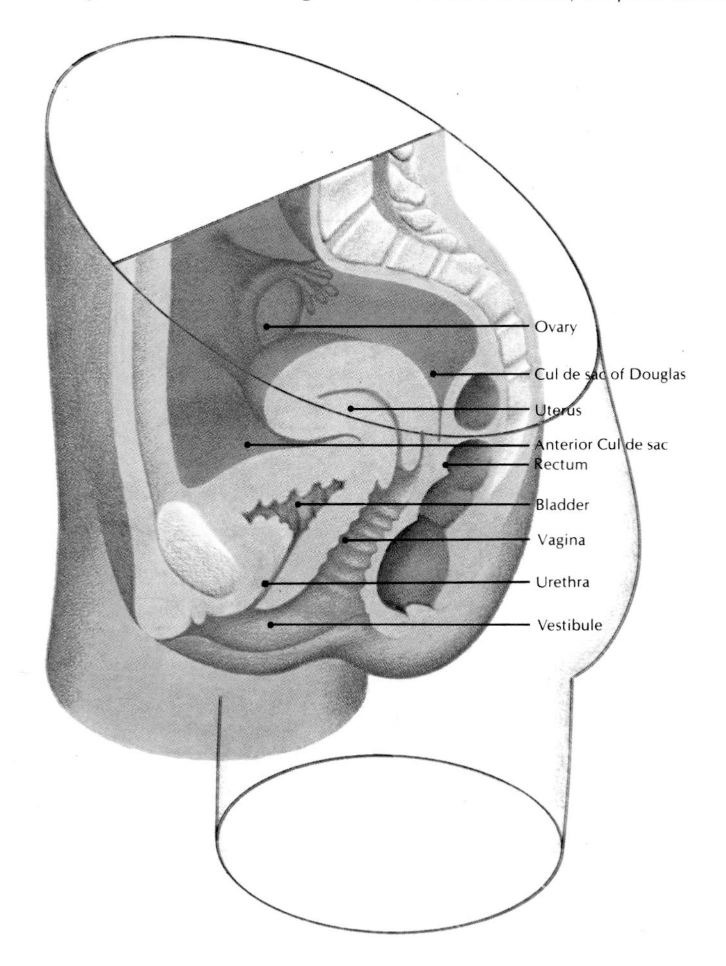

Figure 40.1b. The normal female pelvis, in sagittal section.

Figure 40.2 (above). The female external genitals.

Figure 40.3. (below). Menstruation phases. Note the interaction of pituitary and gonadal hormones and their effect in the endometrium.

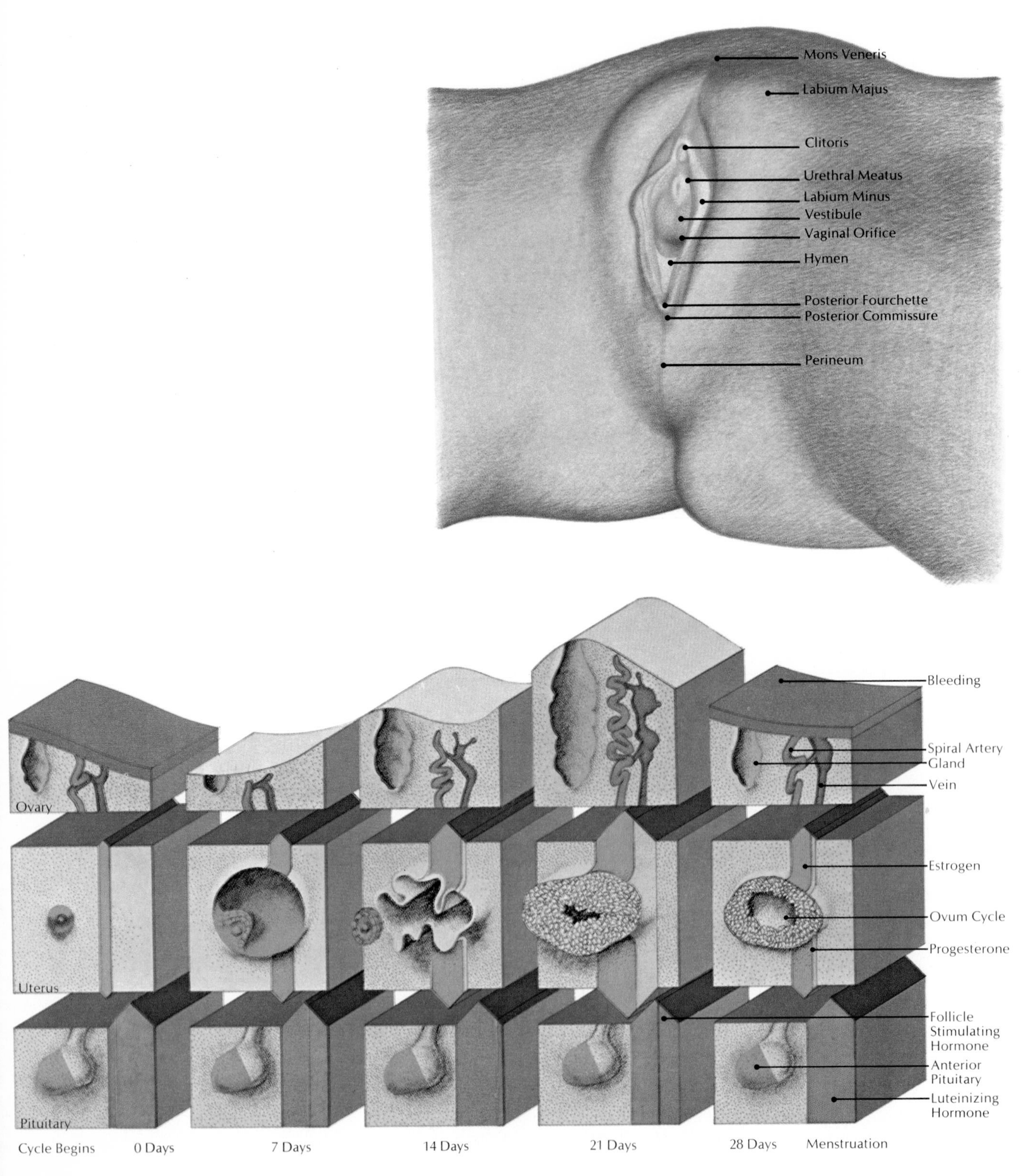

Figure 40.4. Luteinizing hormone (LH), follicle-stimulating hormone (FSH), and basal body temperatures throughout a menstrual cycle in an ovulating woman. Note the surge of LH at midcycle.

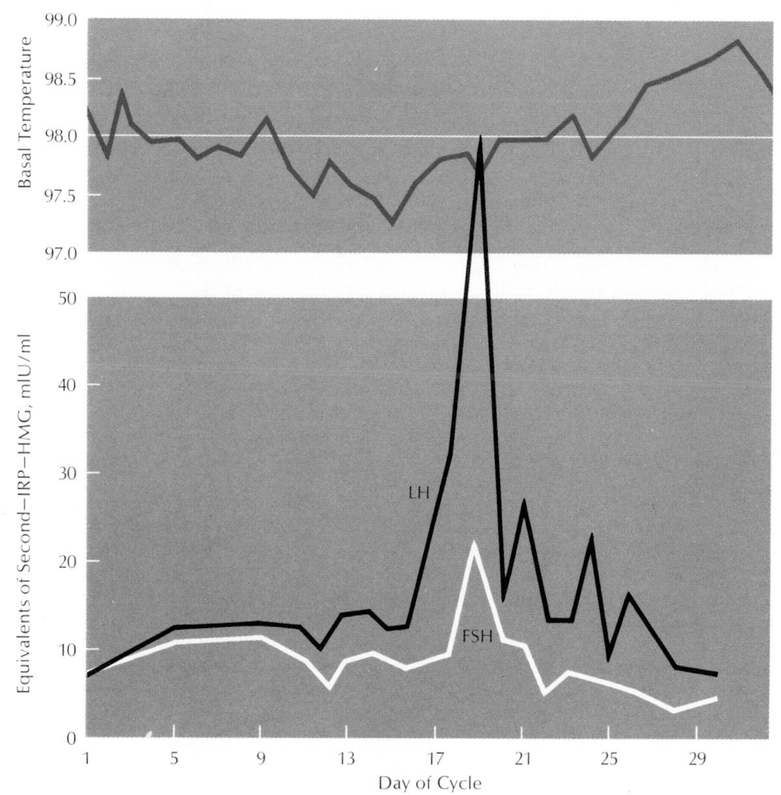

phase of the menstrual cycle begins. During this phase, the glands, blood vessels, and surface epithelial cells of the endometrium are regenerated. Branching capillaries extend from arteries in each endometrial segment. These capillaries and the associated network of venous capillaries will make maternal blood available to the fertilized ovum if it implants.

As estrogen level rises during the proliferative phase, production of FSH-RF in the hypothalamus is suppressed, and the secretion of FSH from the pituitary gradually declines. At the same time, production of LH-RF in the hypothalamus is stimulated, and there is a gradual increase in secretion of LH from the pituitary. Throughout the follicular phase, the follicles in the ovary have been developing under the influence of FSH. They move to the surface of the ovary and form blisterlike elevations on its surface. The entire follicular structure swells as its cells grow and the ovum within it matures.

The *ovulatory phase* begins with ovulation, or release of the ovum from the follicle, and usually occurs about 12 to 16 days before the beginning of the next menstrual phase. Ovulation is triggered by a surge of the LH level in the blood, which results from stimulation of the pituitary by a burst of LH-RF from the hypothalamus. The neurosecretory cells in the hypothalamus apparently respond to a certain critical level of estrogens in the blood, setting off the chain of hormonal signals that leads to ovulation. The ovum becomes suspended in fluid within the follicle, and the translucent surface of the follicle protrudes above the ovarian surface and ruptures. Follicular fluid and the ovum are extruded, or exploded, into the pelvic cavity, and the ovum is picked up by the fringed extensions of the fallopian tube.

Immediately after ovulation, there is a slight decrease of estrogen, which probably causes the slight bleeding that occurs in some women at this time.

Figure 40.5. The normal male pelvis, in sagittal section.

The secretion of another hormone, progesterone, after ovulation causes a distinct increase in basal body temperature of about 0.4 to 1.0°F. This temperature elevation is maintained until the onset of menstruation. Ovulation probably occurs within the 48 hours immediately preceding the maintained temperature rise. The endometrium continues to thicken and blood vessels increase during the ovulatory phase.

The ovum undergoes rapid degeneration within about 24 hours after ovulation if it is not fertilized. Fertilization is thought to occur in or near the fringed portion of the fallopian tube, and transfer of the fertilized ovum along the tube usually takes about 3 days. The tiny embryo implants in the endometrium about 6 or 7 days after fertilization.

After ovulation, the *corpus luteum* (yellow body) is formed from the remaining cells of the ruptured follicle. The cells enlarge, differentiate, and migrate to form a structure quite different from the follicle. Some cells secrete progesterone, and the level of this hormone in the blood increases as the *luteal phase* of the menstrual cycle begins. The mature corpus luteum secretes both estrogen and progesterone. These hormones cause further changes in the endometrium to prepare it for embryo implantation. The network of capillaries and the superficial layer of the endometrium thicken.

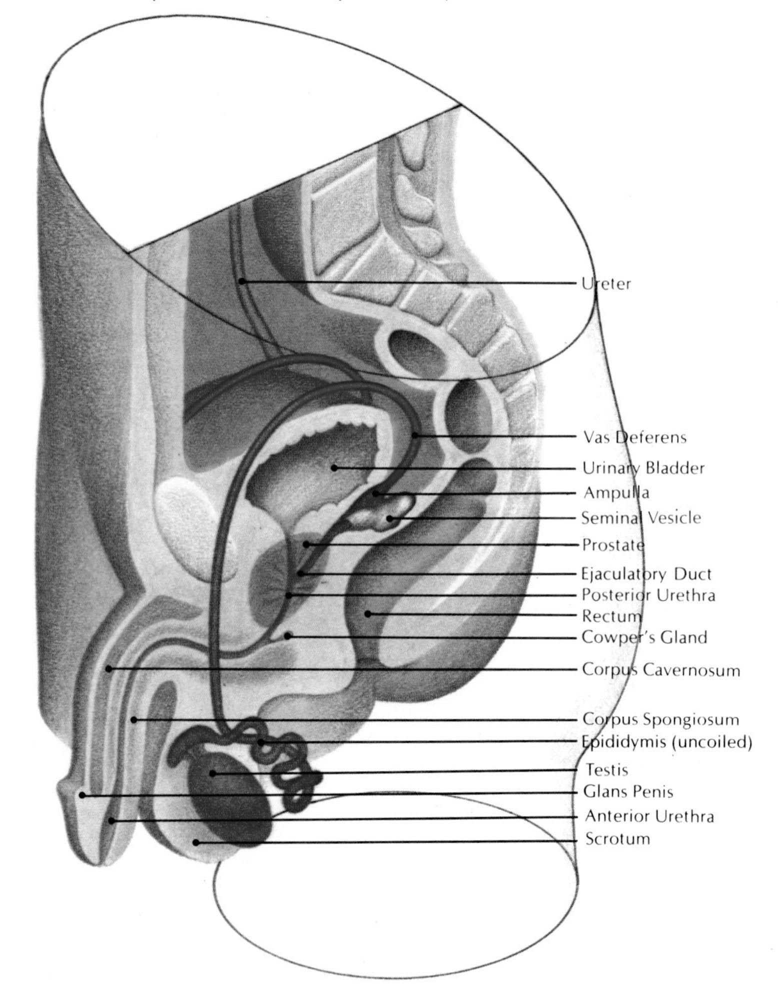

Figure 40.6. The male reproductive system in an anterior view.

The elaborate build-up of the endometrium is much more pronounced in humans than in other primates. It is this extensive preparation for the embryo that causes humans to menstruate so much more profusely than do other mammals.

Increased blood progesterone levels during the luteal phase apparently are responsible for deterioration of the corpus luteum, which begins 8 or 9 days after ovulation, or 5 or 6 days before the beginning of the next menstrual phase. As the corpus luteum disintegrates, levels of estrogen and progesterone in the blood decrease, and within a few days menstruation begins as the endometrium disintegrates. If the ovum is fertilized, however, certain cells secrete hormones that maintain the corpus luteum in an active state. Thus, if fertilization occurs, the estrogen and progesterone levels remain high and breakdown of the endometrium does not occur.

MALE REPRODUCTIVE SYSTEM

The reproductive system of the human male consists of the testes, scrotum, seminal ducts, seminal vesicles, ejaculatory ducts, accessory glands, and the penis (Figure 40.5). The *testes* are the glandular organs in which spermatozoa are produced. Before birth, the testes move from the abdominal

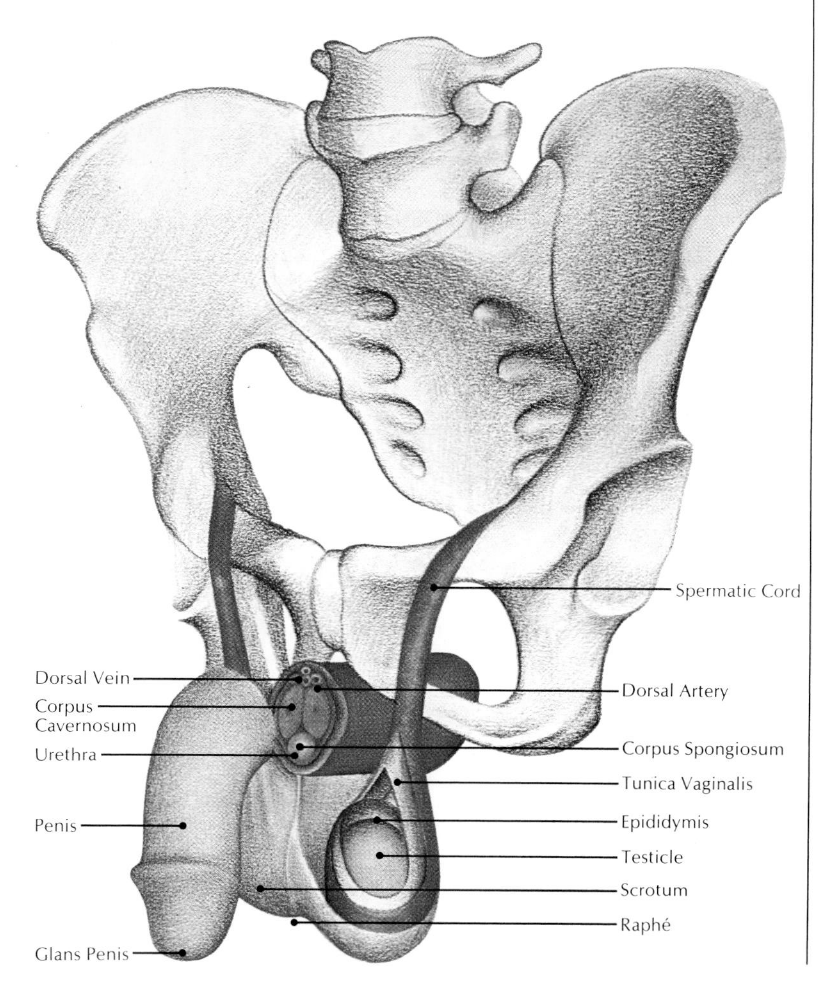

cavity down into the *scrotum*, a pouch of skin that hangs near the base of the penis. Each testis is an oval organ containing interstitial, or supporting, cells and the hundreds of coiled tubules within which sperm are formed. The testis is covered by several layers of membrane, as well as the thin muscle and skin layers of the scrotum.

All of the tubules within the testis lead into about a dozen small ducts that leave the upper surface of the testis and drain into the *epididymis*, a single thin tube about 20 feet long but coiled and closely packed into the appearance of a thicker tube that rests on the top and back of the testis, held there by some of the surrounding membranes.

The epididymis functions in several ways to help move spermatozoa from the testis toward the penis. Muscle and ciliated cells propel the sperm along the duct system. The sperm enters the epididymis against considerable pressure. The epididymal tubules remove fluids, cell debris, and pigment accompanying the sperm. They also digest and absorb sperm if they are produced in greater numbers than needed. The epididymis also serves to increase the fertilizing capacity of sperm, for sperm removed from the ducts near the testis have poorer fertilizing ability than those taken from the other end of the epididymis.

Near the bottom of the testis, the duct of the epididymis connects with the *seminal duct*, or vas deferens. This duct travels up the side of the testis next to the epididymis as part of the spermatic cord, up to a canal passing through the abdominal wall above the pubic bone, across the pelvic cavity beside the bladder, over the ureter, and down to the back of the bladder, where it meets the seminal duct from the other testis. Alongside the end of the duct, between the bladder and the rectum, is an oblong pouch called the *seminal vesicle*, which contains a single fine tube, packed in close coils.

The seminal duct and the tube from the seminal vesicle join to form the *ejaculatory ducts*. The ejaculatory ducts lie side by side and open into the urethra just below the bladder. The urethra, which extends from the bladder to the tip of the penis, carries both urine and seminal fluid to the exterior. Surrounding the urethra, just below the bladder, is an organ composed of both glandular and muscular tissue, the *prostate gland*. This gland secretes most of the seminal fluid. Fluid from the prostate gland and seminal vesicles are involved in the maintenance and activity of the spermatozoa. The fluid from the prostate enters the urethra through a small opening between the openings of the ejaculatory ducts. Near the base of the penis, two small glands empty into the urethra. These *bulbourethral glands*, or Cowper's glands, also contribute to the seminal fluid.

The *penis* is a cylindrical organ composed of three long bodies of spongy tissue, bound together with fibrous tissue. During sexual excitation, blood is pumped into the spongy tissues, making the entire structure larger and rigid. Two of the spongy, or *cavernous*, bodies are attached to the pubic bone at the sides of the opening in the bottom of the pelvis. They run along the sides of the penis and are fused together to form the main part of the penis shaft. Within the groove between these two bodies on the lower side of the penis is another cavernous body that surrounds the urethra. This body is expanded at the base of the penis to form a bulblike structure beneath the bulbourethral glands. It also expands at the tip of the penis to form a caplike structure, the *glans penis*, that is wrapped around the end of the penis. The urethra ends in a slitlike opening at the summit of the glans. The entire penis is covered by a thin, loosely attached layer of skin that is

Figure 40.7. Schematic diagram of spermatogenesis within the seminiferous tubules of an adult male.

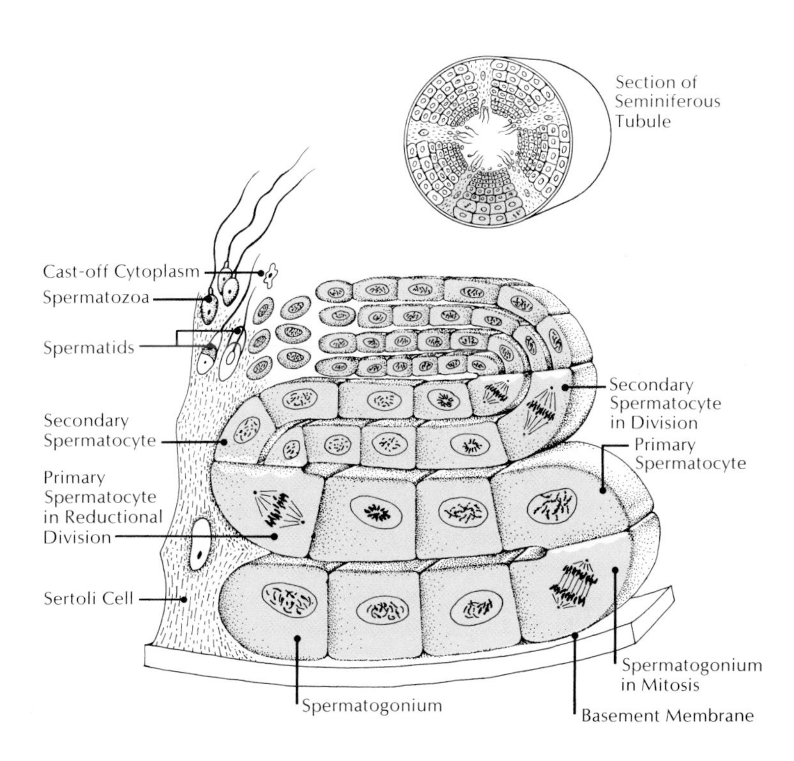

continuous with the scrotum and with the skin surrounding the genitals. Near the tip of the penis, the skin is folded back on itself to form the *foreskin*, or prepuce. This fold of skin may be removed soon after birth by circumcision. Unlike other primates, humans have no bone or cartilage within the penis.

Spermatogenesis

Spermatogenesis in mammals is described in Chapter 16. As in other mammals, human spermatogenesis normally is continuous throughout the male's lifetime, although it may be interrupted by a variety of hormonal and other agents. Spermatozoa are continually being formed in the normal testes, although some decrease of mitotic activity in old age may slow the further differentiation of primary spermatocytes. Both spermatogenesis and the secretion of the testicular hormone *testosterone* are regulated by the steroid hormones secreted by the pituitary. FSH affects some critical step in the development of the primary spermatocyte, and it appears that the maturation of early spermatids into mature spermatozoa also is dependent upon FSH. Spermatogenesis up to the primary spermatocyte stage apparently occurs automatically without the influence of pituitary hormones.

Interstitial cells of the testes are differentiated during the prepubertal period from certain interstitial connective tissue cells. Spermatogenic activity begins at the same time, but complete development of the interstitial cells lags behind that of germinal cells. The development of interstitial cells is dependent upon *interstitial cell stimulating hormone* (ICSH), a pituitary hormone that plays a role analogous to that of LH in the female. ICSH

Figure 40.8. Histological study of the adult human testis. Photomicrograph of a cross section from a normal testis of a 30-year-old man. Seminiferous tubules that are fully developed contain all the cells of the spermatogenic series.

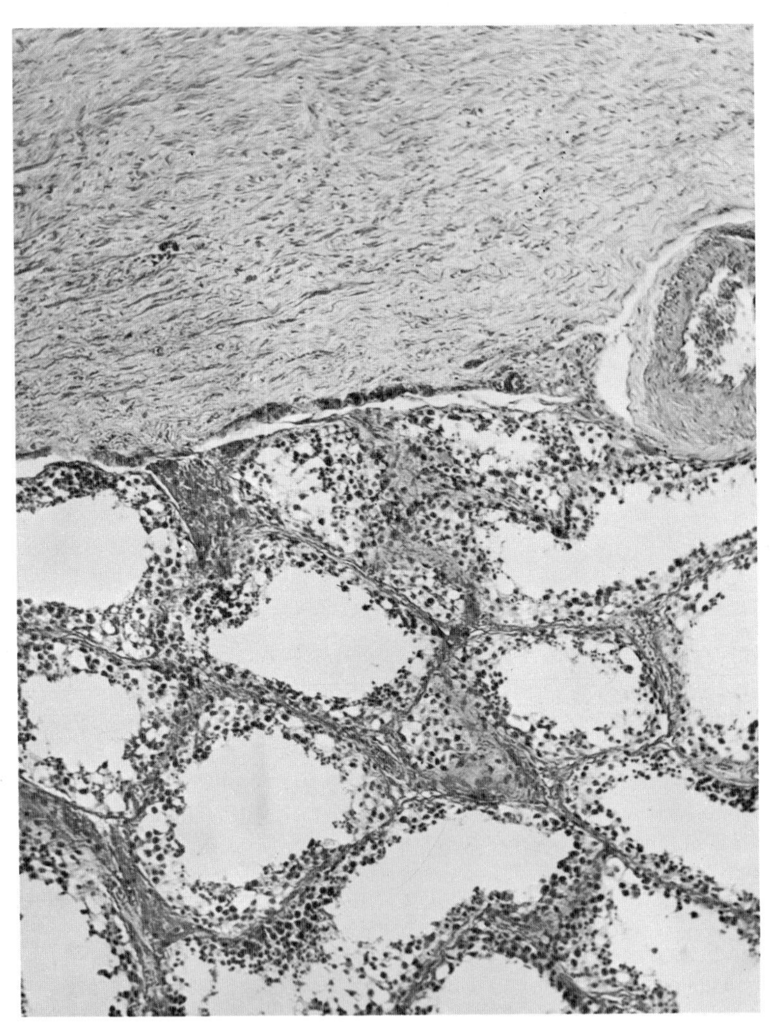

stimulates the growth of interstitial cells as well as mitotic activity among spermatogonia. FSH and ICSH stimulate spermatid formation. The androgens secreted by the interstitial cells maintain a favorable environment within the tubules for sperm cell development.

The hormones produced by the interstitial cells also affect the accessory reproductive organs. They stimulate the prostate gland and the seminal vesicles to produce the major portion of the seminal fluid, or *semen*. Testosterone inhibits spermatogenesis by suppressing release of FSH from the pituitary. Thus, the male does have a complex interbalancing system of sexual hormones but no prominent cycle of hormone levels.

Ejaculatory Mechanism

During intercourse, the erect penis is inserted into the vagina of the female, and semen is ejected forcefully from the urethra in the process called *ejaculation*. The ejaculatory process occurs in two stages. The first stage involves contractions of the entire genital tract—from epididymis through seminal ducts to seminal vesicles. Fluid from the prostate gland is pumped into the upper urethra, where it is joined by fluid from the seminal ducts and seminal vesicles. During the course of ejaculation, prostatic fluid repeatedly is added to the seminal fluid content by regularly recurring contractions of the prostate gland. As seminal fluid collects in the urethra, the urethral bulb at the base of the penis involuntarily expands to twice or three times its normal size, in preparation for the second stage of ejaculation. As the ejaculatory process begins, the sphincter (muscle ring) at the bladder opening closes, preventing seminal fluid from being forced back into the bladder and also preventing urine from being forced into the seminal fluid.

The second stage of ejaculation begins with a relaxation of the urethral sphincter at the base of the prostate, allowing the seminal fluid to flow into the distended urethral bulb and penile urethra. Contractions of muscles in the perineum (the region between testicles and anus) and the penis propel the fluid along the penile urethra. Regular contractions of the urethral bulb also aid in propulsion.

The first third of the seminal fluid expelled contains about 75 percent of all sperm in the ejaculation. The later part of the emission consists largely of fluids from the seminal vesicles and prostate gland. The seminal vesicles convert glucose into fructose, which is present in the semen in concentrations about 3 or 4 times as great as glucose concentration in the blood. This sugar apparently supplies an energy source for spermatozoa.

MOVEMENT OF SPERM THROUGH THE FEMALE TRACT

Sperm ascend the cervical canal very rapidly and have been found in the uterus within minutes after ejaculation. Because sperm swim at only about 0.5 cm per minute, forces other than their own motility must transport them through the female genital tract. After the sperm have entered the cervix, a complex system of contractions, fluid currents, ciliary action, and "anatomic barriers" help to influence the rate at which sperm move toward the site of fertilization. Some evidence now suggests that substances in the seminal fluid cause relaxation of the uterine walls, aiding the flow of cervical secretions and sperm into the cavity of the uterus.

In most cases, sperm reach the cervical mucus within 1.5 to 3 minutes after ejaculation. After sperm get through the cervix, contractions of the uterus probably play a significant role in transporting the sperm to the fallopian tubes. Sperm may reach the fallopian tubes within 30 minutes after ejaculation, and this rapid ascent cannot be explained solely on the basis of sperm cell motility. The uterine contractions may be stimulated by oxytocin released from the pituitary during orgasm or by substances in the seminal fluid. Deposition of certain components of semen in the vagina is known to affect the state of muscles in the uterus within 7 to 10 minutes after the deposition.

Normal fertility obviously depends upon various anatomical, physical, and chemical changes in the cervix and in the mucus that it secretes. Cervical mucus, normally alkaline, reaches a peak of alkalinity at the time of ovulation. This alkalinity helps to overcome or minimize the detrimental effects on sperm motility and survival of the slightly acidic vaginal mucus. If cervical mucus is plentiful, sperm may be able to move directly from the alkaline semen into the alkaline cervical mucus and thus avoid the unfavorable vaginal acidity, which is sufficient to halt sperm motility.

As the uterus contracts rhythmically, a small amount of mucoid fluid moves up and down through the genital tract. The velocity of flow

Figure 40.9. Composite schematic diagram of the human spermatozoan greatly enlarged.

Figure 40.10 (opposite). Diagram illustrating the major physiological events that comprise ovulation, fertilization, and implantation.

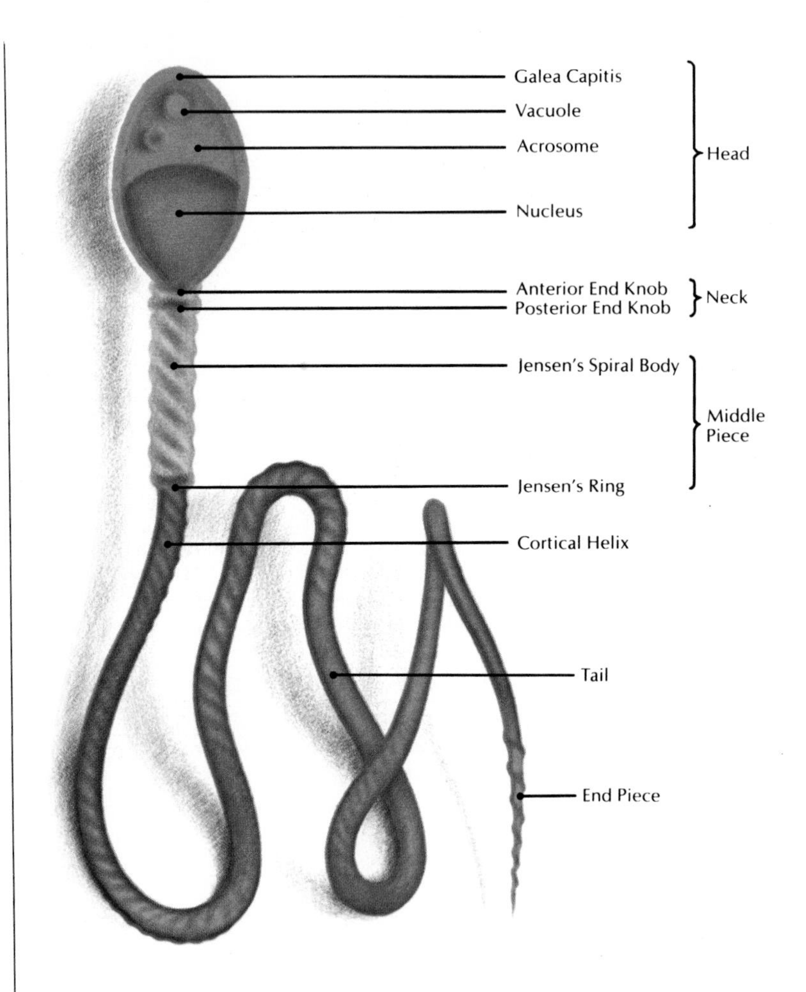

approaches zero at the surface of the uterine lining and is greatest at the center of the cavity. The overall movement of the fluid is toward the cervix. It is not yet known with certainty whether sperm motility is essential for movement within the uterus and fallopian tubes. There is a heavy loss of sperm within the female tract. Secretions from the female organs, unfavorable conditions of acidity or alkalinity, and phagocytosis by leucocytes—as well as physiological changes in the sperm cells themselves—contribute to the loss of active sperm.

One mystery not yet solved is the question of how the tiny sperm, swimming at random through the relatively huge passages of the fallopian tubes, find the ovum at all. A mechanism of chemotaxis has been suggested, but evidence to support this suggestion is lacking. A flow of fluid in the tubes toward the ovary, vigorous contractility of the muscles at the ovarian end of the tubes, and the relatively large size of the ovum and its attached cells may be sufficient to account for the high probability of sperm meeting ovum.

In most species, including man, several sperm contact each ovum, but normally only one penetrates the innermost layer, the vitelline membrane,

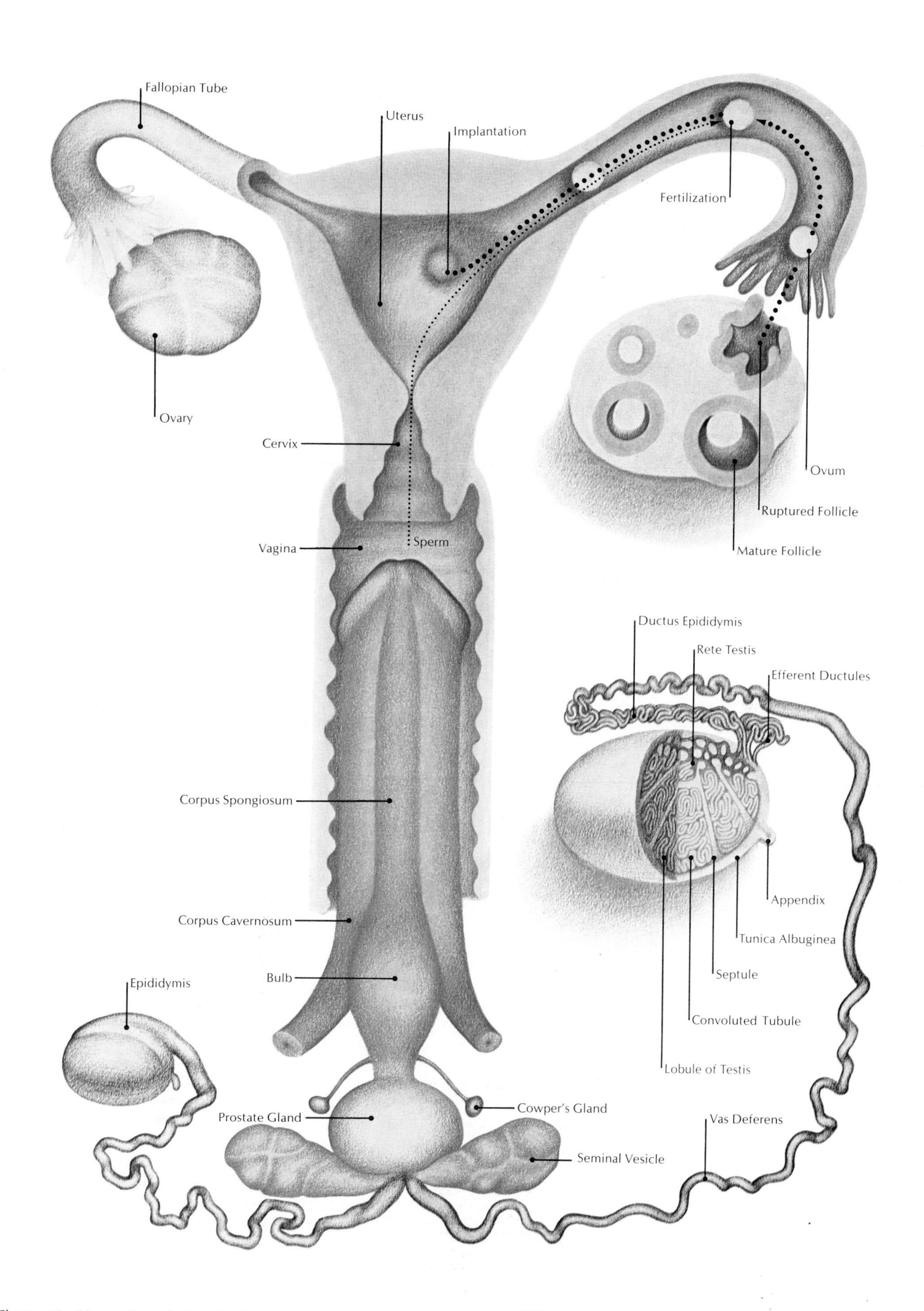
Fallopian Tube
Uterus
Implantation
Fertilization
Ovary
Cervix
Ovum
Ruptured Follicle
Mature Follicle
Vagina
Sperm
Ductus Epididymis
Rete Testis
Efferent Ductules
Corpus Spongiosum
Appendix
Tunica Albuginea
Corpus Cavernosum
Septule
Epididymis
Bulb
Convoluted Tubule
Lobule of Testis
Prostate Gland
Cowper's Gland
Vas Deferens
Seminal Vesicle

Figure 40.11. Photomicrograph of a living human egg at the instant of fertilization. Note the first polar body (left) indicating that the ova still has one more meiotic division to complete. Clearly visible are numerous spermatozoa in contact with the ovum, only one of which will reach the egg nucleus.

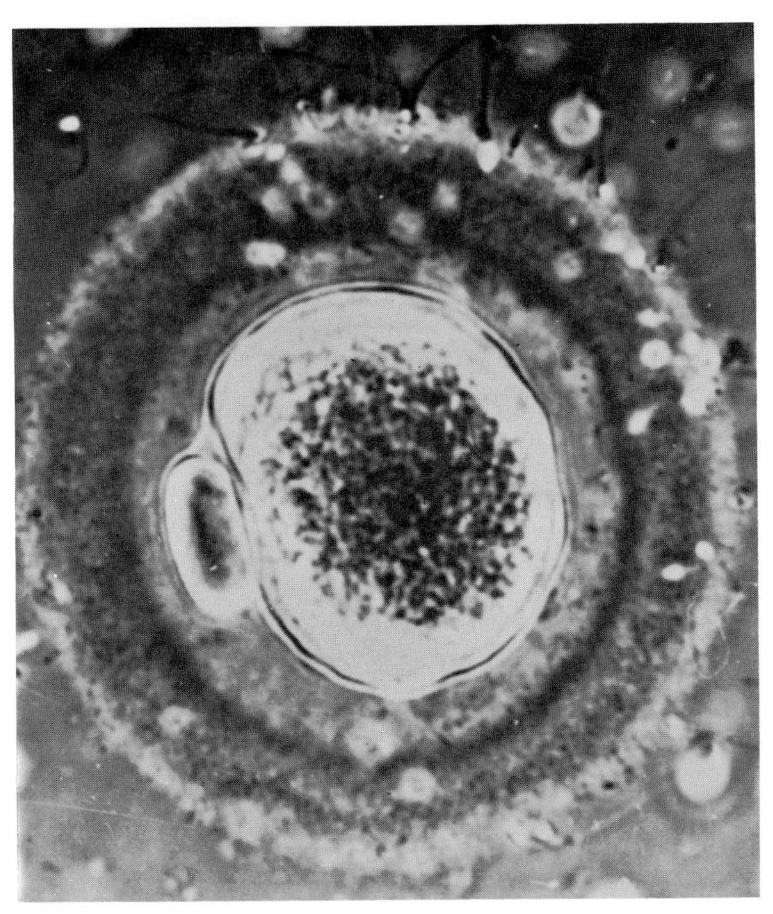

and fertilizes the ovum. The nature of this block to multiple fertilization is being studied chemically and ultramicroscopically, but its mechanism is not yet known.

IMPLANTATION

The endometrium has been prepared for implantation of the embryo by the effects of progesterone during the menstrual cycle. The uterus is almost completely collapsed in the normal state, so that the cavity consists of little more than a slit between roughly triangular anterior and posterior walls. At this stage in the menstrual cycle, the coiled arteries reach nearly to the endometrial surface, and capillaries form a dense network just below the surface. Thus, there is likely to be an almost immediate blood supply at the point where implantation occurs.

The ovum, up to this time, has been leading a relatively anaerobic existence, but it probably has begun to reach the limits of its ability to exist and grow as a free-floating organism. Implantation provides an immediate source of further nutrients.

The placenta forms a rich network of small blood vessels from the embryonic circulatory system. These vessels penetrate the endometrial tissues and blood vessels, which then break down to form a spongy, blood-filled tissue around the placental vessels. Thus, the fetal blood is separated from maternal blood only by the walls of the fetal capillaries, and exchange of substances between the two circulatory systems is rapid and efficient.

In most mammals, the placental vessels lie next to maternal vessels rather than penetrating them. Humans share the more efficient form of placenta with apes, monkeys, insectivores, and some rodents. In apes, the placenta develops more rapidly than in other primates. In humans, the formation of the placenta occurs most rapidly of all, and the preparation of the endometrium for implantation and placentation is far more extensive than in other mammals. All these human specializations are related to the rapid growth required of the fetus (Chapter 42).

FURTHER READING

For further comparative information about sexual physiology and behavior in man, other primates, and other animals, see books by Beach (1965), Ford (1964), Ford and Beach (1951), Harrison, *et al.* (1964), Hill (1953 – 1962), Jeanniere (1967), Kinsey, *et al.* (1948, 1953), Masters and Johnson (1966), Morris and Morris (1966), Washburn (1962), and Rowlands (1966). Morris (1967) offers some interesting speculations about the evolution of human patterns in relation to those of animals.

Demarest and Sciarra (1969) give an excellent series of illustrations of human reproduction, with accompanying explanations. More detailed descriptions of human reproductive anatomy are given by Dickinson (1949) and Goss (1966). Good general discussions of reproduction are found in books by Gallien (1963), C. F. Lloyd (1964), MacGillivray (1963), and Nalbandov (1964). The roles of hormones in the reproductive processes are treated in detail by Corner (1963), Turner (1966), and W. C. Young (1961).

Also see Further Readings for Chapters 16 and 21.

41
Human Sexual Behavior

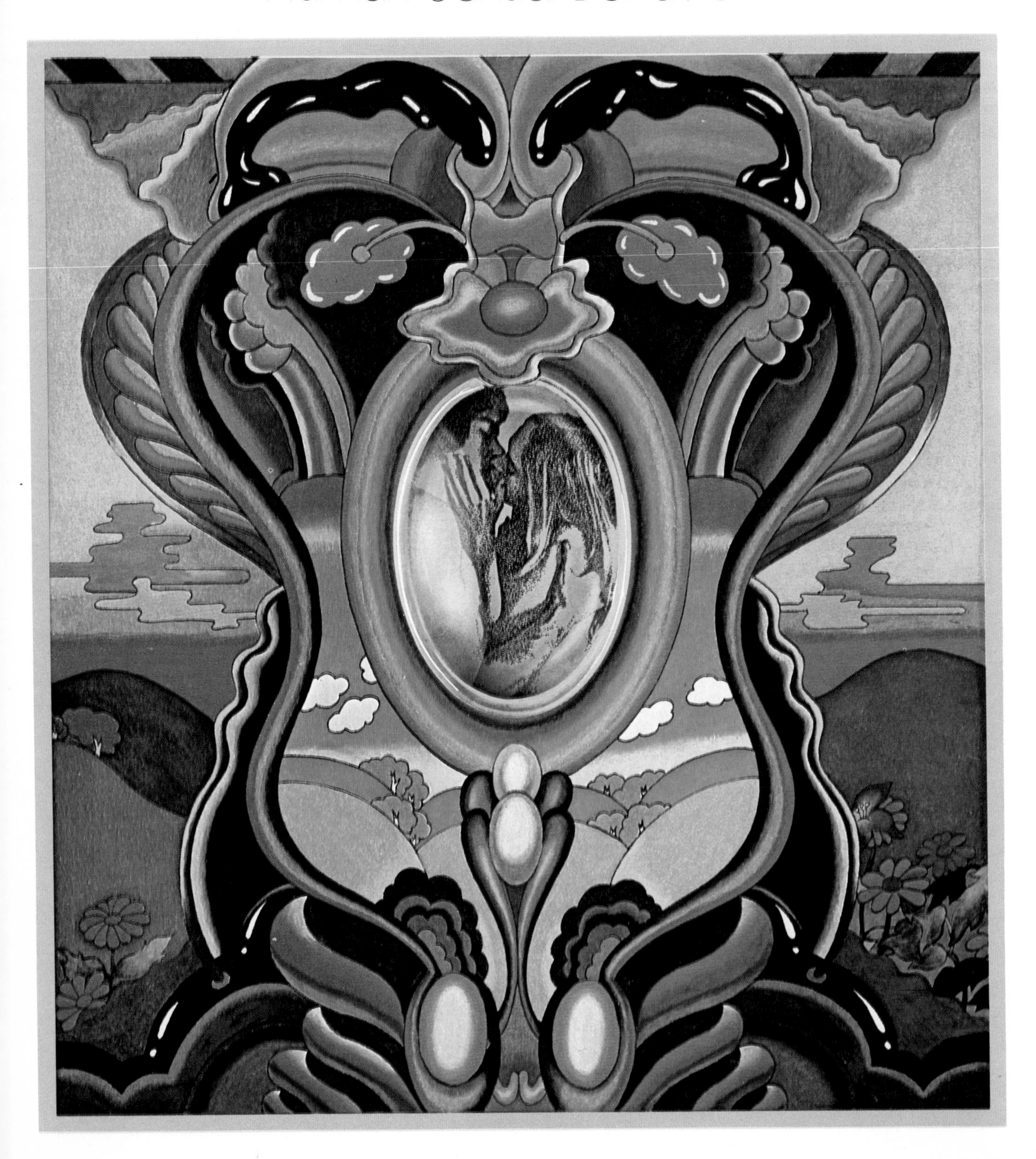

Physiologists and behavioral scientists have devoted great efforts to the study of basic animal needs, or drives. Millions of scientific man-hours have been spent in answering basic questions about eating, drinking, breathing, and sleeping. But what about sex? Although an individual can survive without sex, a species cannot.

Physiologists have investigated the embryological origin of spermatozoa and ova, and anatomists have mapped the genital tracts. The growth of the embryo after fertilization has been explored in great detail. Behavioral scientists have devoted much effort to exploration of courting behaviors, and anthropologists have speculated about the influences of mating patterns on social structure. Yet—until very recent years—both physiologists and behavioral scientists avoided study of the sexual act itself, particularly in the human species. The scientist who chose to investigate this aspect of sex could expect ridicule from his colleagues at the very least.

In 1938 Alfred C. Kinsey began to collect information about human sexual behavior on a widespread, statistically reliable basis, and he established the Institute for Sex Research in 1947. The publication of the "Kinsey reports" on sexual behavior in the human male and female (Kinsey, *et al.*, 1948, 1953) was greeted with controversy, but for the first time it became possible to speak with some confidence about the sexual practices of a reasonably large sample of the human species. The data obtained in the Kinsey studies is limited to the behavior that interviewees would admit, and no one would claim that a cross section of mid-twentieth-century Americans necessarily are typical of all humans in all cultures. Yet the Kinsey studies did provide a factual basis for beginning a study of human sexual behavior, and some sociologists, psychologists, and anthropologists were encouraged to begin collecting information to extend this knowledge into other areas.

When William Masters and Virginia Johnson established the Reproductive Biology Research Foundation at Washington University, they hoped to answer two basic questions: (1) what physiological changes occur in men and women as they respond to sexual stimulation and (2) what controls the response patterns? Until their study was published, little was known about the physiological and behavioral nature of the female sexual response.

COMPARATIVE SEXUAL BEHAVIOR

Although man is to a large extent unique in his cultural adaptations, he is also an animal with responses and behaviors partially determined by his mammalian inheritance. In a number of respects, human sexual behavior is similar to that of other mammals, and these similarities probably reflect man's genetic background derived from millions of years of mammalian evolution. There seem to be at least ten traits of sexual behavior that are characteristic of most mammalian species. To a greater or lesser extent, man shares most of these traits.

1. There are two neurologically determined patterns of sexual behavior—one for "masculine" mounting behavior and the other for "feminine" mounted behavior. The animal that mounts the other is more active and exhibits rhythmic thrusting motions of the pelvic area. The mounted animal tends to be relatively immobile. In most species, mounting behavior is most common among males and mounted behavior among females, but both patterns do occur in either gender.

2. Suitable physical stimulation increases sexual excitement and produces various physiological symptoms, such as increased heart rate, more

Figure 41.1. "Noah's Ark," by Kerstin Apelman Öberg.

rapid breathing, muscular tension, and swelling (tumescence) of the external soft parts of the body. In humans, tumescence is most noticeable in the genitalia and in the female breasts.

3. Continued stimulation culminates in a sexual climax (orgasm). The male orgasm involves spasms of the pelvic musculature, which cause ejaculation of semen. The existence of a female orgasm in mammals has been the subject of great debate, but the physiological data of Masters and Johnson (1966) make it clear that orgasm in the human female is analogous to that in the male. Some investigators have claimed that the female orgasm is unique to the human species. Very little information is available about physiological changes in nonhuman females during intercourse. The few data available do reveal changes in blood pressure and heart rate similar to those that accompany the human sexual arousal.

4. Patterns of sexual behavior appear in the "play" behavior of young mammals before they reach the age of sexual maturity. Mounting behaviors are commonly exhibited by immature males (and sometimes females) of most mammalian species, but typical female sexual behavior seems to be

Figure 41.2. Heart rate in human female (upper left) and human male (lower left) before, during, and after coitus. Also shown is the heart rate in a male dog during sexual activity (lower right). An increased heart rate is one of the most common mammalian reactions to tactile or erotic stimulation, and invariably it results from sexual arousal. (*After Kinsey, 1953*)

more dependent upon the hormones released during puberty. In more permissive human cultures, young children often go through all the motions of sexual intercourse (except ejaculation).

5. Sexual intercourse (coitus) is preceded by sexual stimulation of varying duration. Most—but certainly not all—of this precoital stimulation is done by the male. Such intense activity heightens the sexual responses of both male and female.

6. Homosexual behavior—that is, males mounting males or females mounting females—is common in mammalian sexual behavior. Individuals who limit their sexual behavior to such homosexual contacts have not been reported in studies of wild nonhuman species, but such individuals can be produced under laboratory conditions.

7. Considerable oral stimulation precedes or accompanies coitus in many mammalian species. Nuzzling, licking, and gentle biting are common. Oral stimulation of the genitalia by the male has been reported both in chimpanzees and in gorillas (Ford and Beach, 1951; Schaller, 1963). Oral stimulation is commonplace in human sexual behavior: kisses, "love bites," and mouth-genital contact. In most of the United States, mouth-genital stimulation is a crime—ironically, this behavior is often grouped with "crimes against nature."

8. Rapid erection and equally rapid loss of erection of the penis are typical mammalian traits shared by humans.

9. Periodicity of sexual activity is characteristic of females in all nonhuman mammalian species and of males in some species. In addition to the

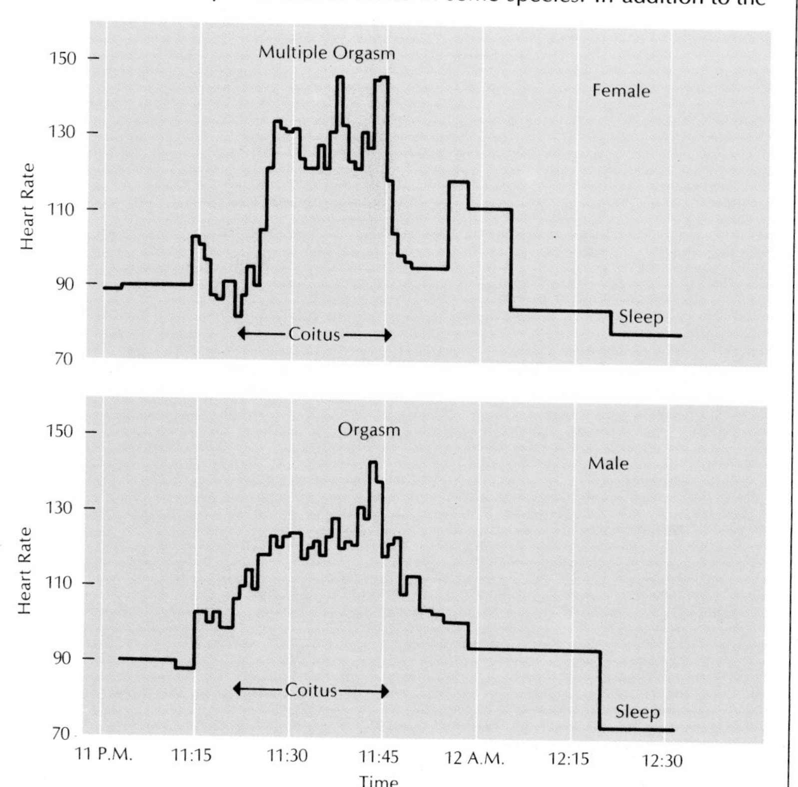

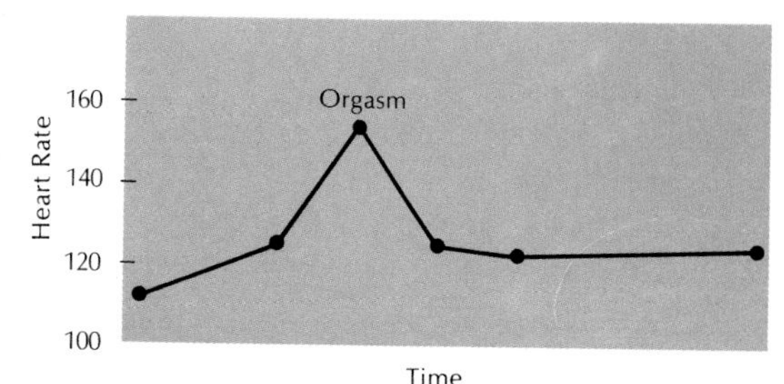

Figure 41.3. Comparative study of the cycles of erotic desire described by 906 American women and cycles of coital behavior observed in 9 female chimpanzees. The data pertaining to apes indicate that the highest number of coital acts took place when the female was in the stage of maximal genital swelling and when ovulation was imminent. In contrast, the cycles of human females indicate that maximal sexual desire most commonly occurred either a day or two preceding or following the period of menstrual flow. Thus, the period of maximum desire is least likely to occur in the phase of the cycle when ovulation is impending. Although these curves for two primate species are almost directly opposed, it must be remembered that each is a measurement of a different parameter— namely, *desire* not coitus in the human, and *coitus* not desire in the chimpanzee.

periodic exhibition of sexual behavior by the female during estrus, many species exhibit sexual behavior only during certain breeding seasons of the year. The almost complete lack of sexual periodicity among humans seems to be unique among mammalian species. A minority of women do seem to show greater desire for sexual behavior at certain stages of the menstrual cycle. Presumably, lack of sexual periodicity in humans is related to the extensive cerebral control of sexual behavior, largely removing it from the direct control of hormonal cycles and periodic external stimuli.

10. Sexual behavior in most mammalian species is typified by a wide range of individual variations. In nonmammalian species, individuals are more uniform and stereotyped in sexual interest and activity. The trend toward individual variation also is related to increasing cerebral control of sexual behavior, and this trend reaches its peak in the human species, where some individuals desire daily sexual activity and others are satisfied with little or no sexual outlet.

Anthropological Perspective

Comparative studies of mammalian sexual behavior shed considerable light upon universalities of the human behavior patterns. Further insight comes from comparative studies of various human cultures made by anthropologists and historians. Because most sex research has been done in

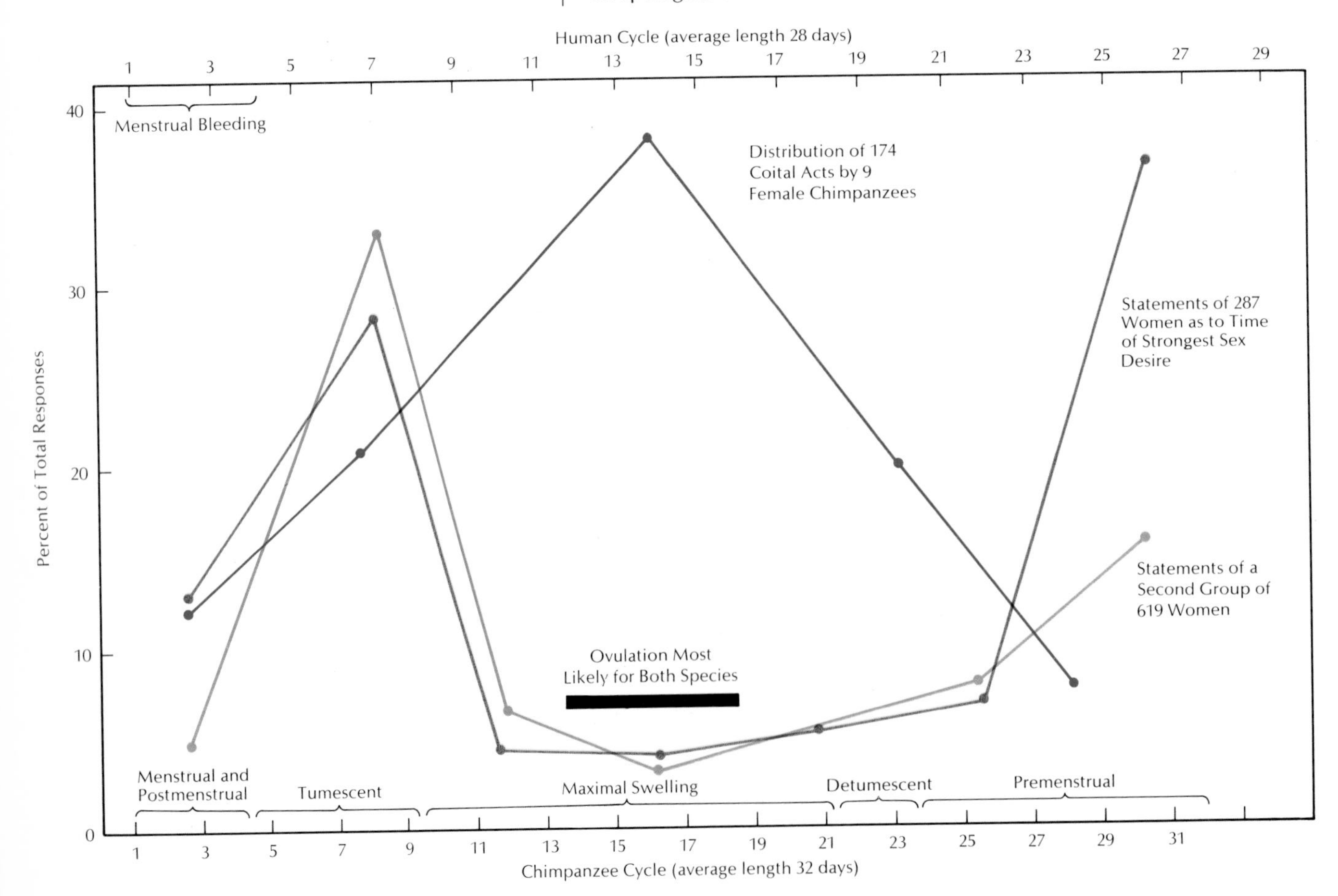

Figure 41.4. Shown here (in bar-graph form) is the seasonal duration of mating activity for some representative nonhuman mammals. Southern hemisphere seasons have been converted to northern hemisphere equivalents, and several locations are included. Within the animal kingdom, reproduction may occur at any time within the year and may change from year to year in duration and intensity.

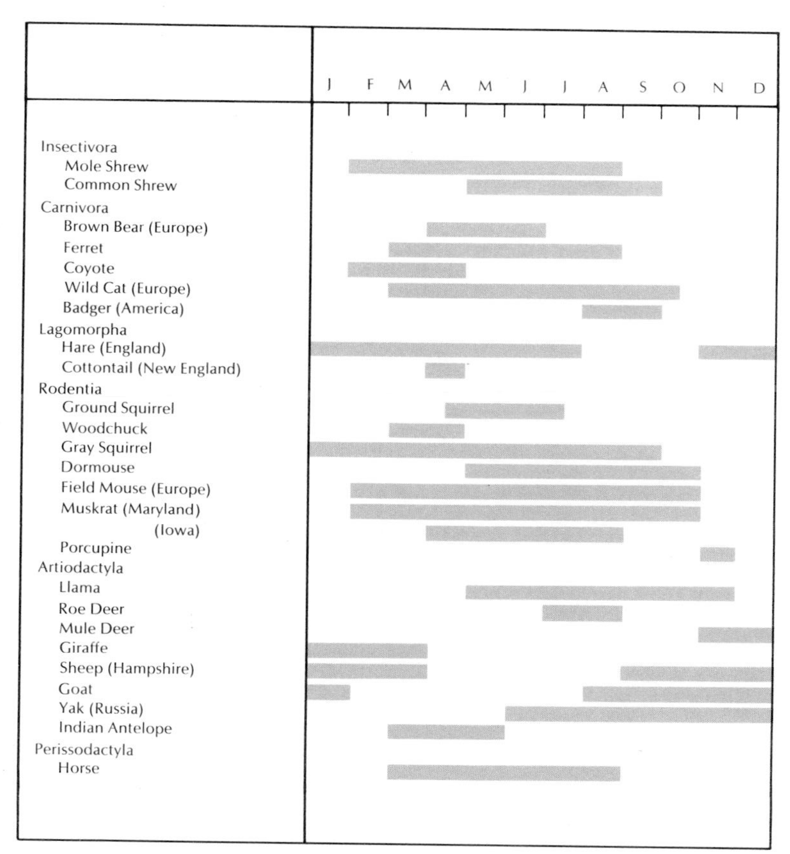

recent Western civilizations, many findings and inferences drawn from them are valid only within the framework of these civilizations. A broader and less culture-bound viewpoint is necessary for an adequate understanding of human sexuality.

All human societies must to some degree govern the sexual behavior of their members in order to curb disruptive competition and to establish a social stability that allows a society to define the rights and obligations upon which social functioning depends. The most common and successful method of managing sexuality has been the institution of marriage. Every known human society has had some form of marriage that entails complex rights and duties and fixes responsibility for the care of children.

Societies also attempt to govern various forms of sexual activity by systems of reward and punishment. Societies vary enormously, however, in what they regard as undesirable and desirable. A behavior that is constructive in one society may be socially disruptive in another.

The variation in sexual moralities among different societies is so great that it is dangerous to speak of any behavior as being normal for all societies or to assume that any behavior is a part of "human nature." Sexual norms, values, goals, and restrictions are determined culturally.

Most modern Western societies have been unusually repressive, but remarkably inconsistent. An almost endless list of examples could be cited, but a few will suffice. Sexual attractiveness is held up as a major goal, but sexual activity outside of marriage is forbidden. Behaviors forbidden to the

Figure 41.5. People around the world differ in their celebration of the marriage ceremony. Even within the same culture, there may be a wide variation in the type or setting of the ceremony. Marriage ceremonies of the United States and the Belgian Congo are shown at the upper left and right. Also included are Grecian and Indian couples (upper middle, left and right); members of Russian and Austrian wedding parties (lower middle, left and right); and ceremonies in Rumania and Japan (lower left and right).

Figure 41.6. Two aging prostitutes on the streets of New York.

young are condoned in the adult. Different socioeconomic classes have different ideas as to what is right or wrong in sexual behavior. Virginity is regarded as a virtue in the young but as evidence of emotional or social maladjustment in the adult. This painful inconsistency in beliefs and values may account for the sensitive and emotional reactions to discussion of sexual behavior in these societies.

Anthropological studies indicate that there is no positive or negative social value inherent in any particular sexual act. The context in which the act takes place determines its value and meaning. Even acts as seemingly disruptive as rape or incest can be tolerated or usefully employed under special circumstances. Indeed, in many preliterate societies, behavior ordinarily forbidden is allowed on particular ceremonial occasions or under unusual conditions. Such socially accepted exceptions to the rules seem to serve a useful function as a safety valve, a temporary relief from repression that might otherwise prove intolerable. Traces of such "ceremonial license" are visible in American society: for example, kissing another man's wife is ordinarily taboo but is permitted at weddings and on New Year's Eve under the mistletoe.

A second major lesson to be learned from anthropology is that humans are enormously plastic. Human behavior can be conditioned to a remarkable extent, and even inherent behavior patterns can be suppressed or

Figure 41.7. Photographic interpretation reflecting the ambiguous and often conflicting nature of many sexual mores and values upheld in Western societies.

modified to a great degree. Female orgasm provides an excellent illustration. In some societies, females are conditioned to regard coitus as a distasteful duty to be avoided if possible, and females in these societies achieve orgasm only rarely. In sexually permissive societies, such as those in parts of Polynesia, both females and males are taught to regard coitus as a desirable activity and as one of life's greatest pleasures. Females in these societies almost always reach orgasm and sometimes experience several orgasms in one act of coitus. A normal physiological response is being inhibited in one case and facilitated in the other. The power of cultural conditioning cannot be overstated.

The complex interrelationship between inherent traits and cultural conditioning is poorly understood, and the "nature versus nurture" argument still rages. In human behavior, however, cultural factors clearly override genetic and neurophysiological factors in all but extreme cases. Persons with genetic abnormalities usually behave according to the sexual patterns in which they were trained as children rather than according to their genetic, "real" gender. Castrated females (females whose ovaries have been removed) continue to function quite adequately sexually. Even many castrated males retain their sexual interest and potency for many years. Homosexual behavior seems quite independent of hormonal balance. Cultural conditioning is most powerful during childhood and youth, when people are more easily molded. With increasing age humans become more resistant to change, and established sexual patterns are difficult to modify after the second decade of life.

Masturbation and Petting

Discussion of specific forms of sexual behavior is hampered by a lack of sufficient anthropological data and must rely heavily upon sociological studies conducted in Western civilizations. Nevertheless, some useful generalizations are possible.

Self-masturbation is common among humans and sometimes begins before puberty. It appears to be more common among males than among females. In the United States, approximately nine out of ten males have masturbated, but only one-half to two-thirds of the females report having masturbated. The typical male masturbates far more frequently than most females who do masturbate. Masturbation is replaced largely by sexual activity with other persons and is relegated to the status of a supplementary activity or one to be used when sexual activity with others is unavailable.

Manual stimulation of the genitalia is the most common technique of masturbation for both males and females. Insertion of an object into the vagina or insertion of the penis into some object is probably less common. Many females reach orgasm more quickly and easily in masturbation than in coitus, because the speed and intensity of stimulation are wholly under the female's control in masturbation, whereas in coitus she must adjust to the partner's behavior.

Orgasm during sleep very rarely has been reported in anthropological accounts. In the United States, roughly three-quarters to nine-tenths of males and about two-fifths of females report this experience. Frequencies of orgasm in sleep generally are low. Curiously enough, orgasm in sleep is most common among adolescent males and tends to disappear later in life, whereas in females it is uncommon in adolescence and gradually increases with age, reaching its maximum in the third and fourth decades of life.

Figure 41.8 (above). Accumulative incidence of masturbation for the total U.S. male population. Graph shows the percentage of the total male population with masturbatory experience by each of the indicated ages. As depicted in the graph, over 90 percent of the total population has engaged in masturbation at some time to achieve orgasm. All data is based on total population, irrespective of marital status. (*After Kinsey, 1948*)

Figure 41.9 (below). "The Lovers" by Pablo Picasso.

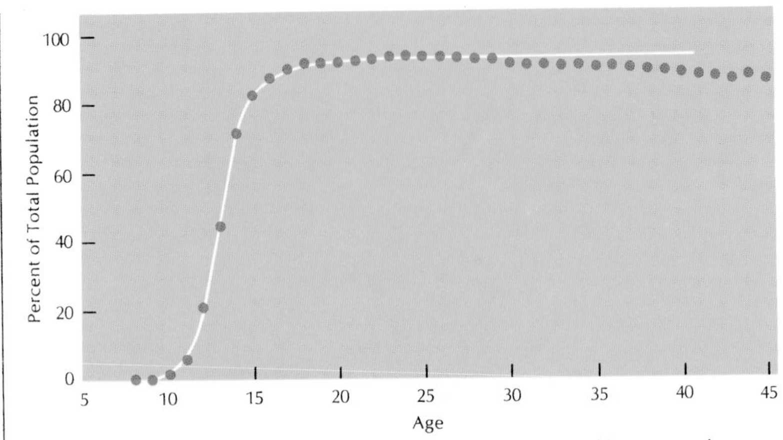

Orgasm in sleep usually, but not always, is accompanied by erotic dreams. Contrary to popular belief, "wet dreams" are not entirely a compensatory phenomenon for sexual deprivation. Occurrence of orgasm during sleep in most cases seems to be unrelated to the presence or absence of other sexual activity.

Heterosexual petting, which is best defined as physical stimulation not involving insertion of the penis into the vagina, exists in three forms. First, there is petting as a goal in itself, a pleasurable activity that may result in orgasm. In this form, it is often used as a substitute for coitus by young persons wishing to avoid coitus. Second, it may be used by a male with the hope of sexually arousing the female to the point where she will permit coitus. Third, when coitus has already been agreed upon by both individuals, petting is used as a preparation for coitus, and in this form petting is known as foreplay. Foreplay seems to be a universal characteristic of human coitus.

Petting techniques vary considerably with individual wishes and degrees of sexual sophistication. Kissing on the mouth is usual in Western civilization, but it is rare or absent in some societies. Manual stimulation of the female breast by the male is nearly universal, and oral breast stimulation is rather common. However, in a few societies this technique is regarded as too similar to infant suckling and therefore is considered inappropriate. Interestingly enough, females rarely stimulate the male breast—both females and males generally are unaware that this area is a potential source of male sexual arousal. Manual stimulation of the genitalia is very common among sexually experienced individuals in virtually all societies.

Oral stimulation by the male of the female genitalia (cunnilingus) is a matter on which societies and subgroups within a given society differ markedly. In some, it is common; in others, it is rare or absent and is regarded as unnatural or perverse. Although cunnilingus is punishable as a felony offense in most of the United States, such mouth-genital contact is commonplace, occurring in over half of the marriages. Oral stimulation by the female of the male genitalia (fellatio) is equally common in the United States. The scant anthropological data suggest that fellatio generally is more acceptable than cunnilingus, and a similar bias can be found in the United States. Nevertheless, it is also a felony in most states. Other techniques of stimulation exist but are less common.

Premarital petting is important as an opportunity to learn about sexual responses of the opposite gender and to obtain sexual and emotional gratification without the more serious commitments of marriage. In most Western societies, premarital petting tends to follow a stereotyped sequence with each new step initiated by the male—hugging and kissing, followed by breast and leg caressing, and proceeding to genital stimulation. This behavior pattern in Western societies seems to be a basic step toward a mature heterosexual adjustment and an integral part of courtship as a prelude to marriage.

Coitus

In every human society, coitus is regarded as the ultimate and natural culmination of a serious heterosexual relationship. However, societal attitudes toward coitus vary with its context. Coitus in marriage is universally regarded as a right or even a duty. There is a great variety of societal attitudes toward coitus under other circumstances. Although premarital coitus is officially discouraged in many societies, nearly all societies permit or excuse male premarital coitus. In only about half of human societies are females given as much freedom as males in regard to premarital coitus. Actual practice may vary greatly from publicly expressed attitudes. In the United States, where premarital coitus is—at least, in theory—disapproved and in most states is against the law, at least two-thirds of males and one-half of females of upper and middle socioeconomic classes engage in premarital coitus. Males tend to have premarital coitus with more partners than do females; females generally begin to have coitus only after a fairly strong, affectionate relationship has developed. Most females who have premarital coitus in the United States do so with males whom they hope to marry.

The social and psychological significance of premarital coitus is linked to social attitudes. In many societies, it is expected behavior and serves a useful role in the selection of a spouse; in such societies, there seldom are negative psychological consequences. In societies that condemn the activity, there are possible negative consequences—social sanctions, guilt reactions, and a greater possibility of exploitation of the female. Even so, it is interesting to note that the great majority of American females who have had premarital coitus report that they do not regret it.

The frequency of marital coitus varies greatly not only in different societies but among individual couples within the societies. Early marriage is characterized by a high frequency of coitus. The American male usually sets the minimum limit of frequency, whereas the female sets the maximum, because the male ordinarily is the partner desiring more frequent coitus. However, later in life as the male interest declines, male and female may become more equal in their sexual drives. The wife who in early marriage complained of too frequent coitus may in middle age complain of its infrequency. Although age does reduce sexual activity, as it does all physical activity, it need not eradicate it. If individuals are in reasonably good health, coitus can continue into extreme old age. In addition to its physiological benefit and its value as one of life's pleasures, marital coitus can be an affirmation of affection and desirability that holds a marriage together.

Extramarital coitus—that is, coitus between a married person and someone other than the spouse—often carries with it the potential of marital and social disruption: jealousy, neglect of duties, new social bonds, disputed paternity, new emotional allegiances, and so on. Consequently, most societies either prohibit extramarital coitus or else permit it only under well-defined circumstances. Sometimes it is allowed only with specified persons,

Figure 41.10. Comparison of aging patterns with frequency of coitus among married males and females. The graph illustrates how active median frequencies of orgasm attained in sociosexual contacts in the female are affected by the patterns of male activity. (*After Kinsey, 1953*)

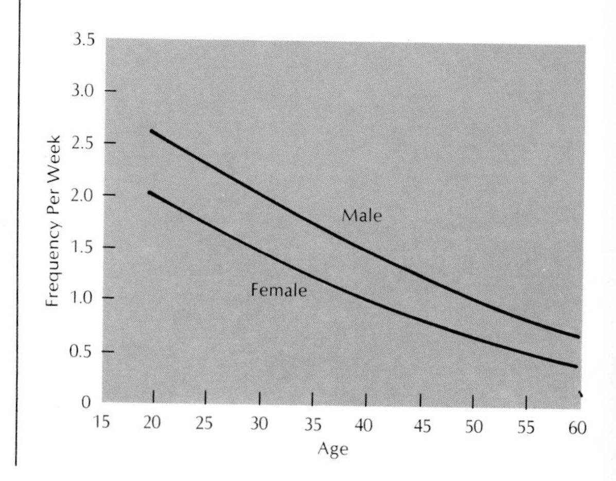

only during certain ceremonies or festivities, only with the permission of the spouse, or only when extenuating circumstances exist. Despite official disapproval, at least one-half of American husbands and one-quarter of American wives experience extramarital coitus.

Postmarital coitus—occurring after a marriage has terminated but before a new one has begun—has received very little attention from societies and from students of human behavior. Restrictions tend to be minimal; even societies opposed to premarital and extramarital coitus tend to close their eyes to postmarital coitus, realizing that persons accustomed to regular coitus in marriage hardly can be expected to refrain after marital status changes. This realization is well founded; the great majority of widowed or divorced males and females continue to have coitus.

Homosexuality

Homosexuality, chiefly among males, was not uncommon in many preliterate or ancient societies. Most preliterate societies accept homosexual activity by certain individuals who are considered to be different from ordinary beings. Only a small minority of societies have treated homosexuality as normative and expected behavior, and in these societies it does not interfere with heterosexuality but coexists with it in noncompetitive fashion. Indeed, homosexuality and heterosexuality should not be viewed as antagonistic, because these tendencies exist, in various ratios, in most individuals. Exclusively homosexual persons are rather rare—about 4 percent of the adult males and 1 percent of the adult females in the United States.

There seems to be no single cause for exclusive homosexuality, but there are a number of predisposing factors, none of which alone is suffi-

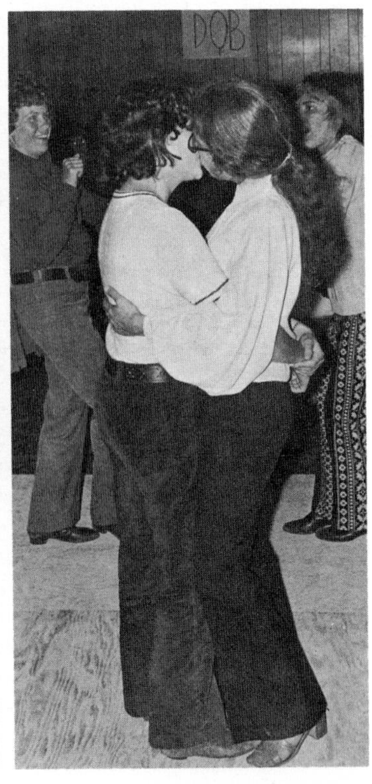

Figure 41.11. Some members of the female homosexual organization Daughters of Bilitis at a social function.

Figure 41.12. Graph depicting the development of heterosexuality and homosexuality in U.S. males by age periods. The number of males with no sociosexual response rapidly decreases between the ages of 5 and 20. Males with predominately heterosexual responses rapidly increase in number until they account for 90 percent of the population. Males who are homosexual in response are most abundant in preadolescence and adolescence, gradually declining with increased age. (*After Kinsey, 1948*)

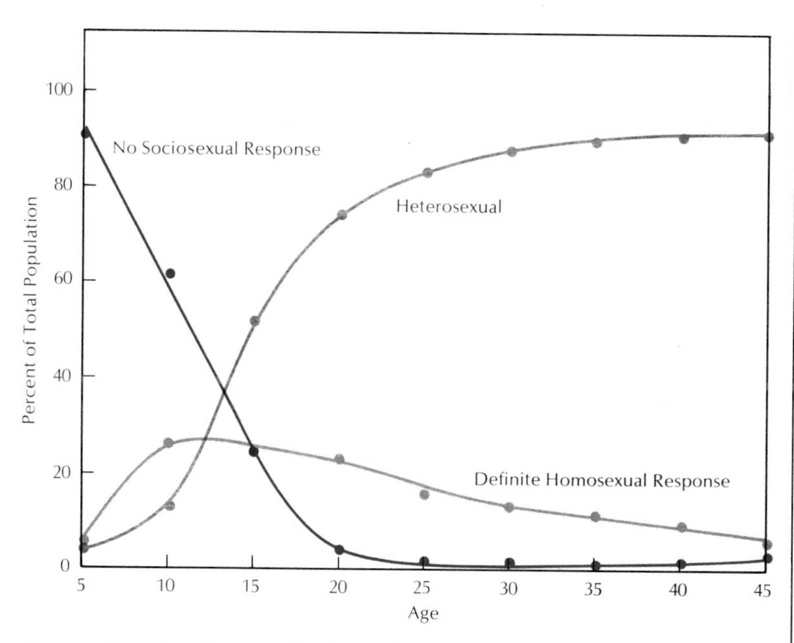

cient explanation. Distorted relationships between the child and parents seem an important factor in some cases, but, in general, anything that interferes with heterosexual expression correspondingly predisposes an individual toward homosexuality. Thus, societal restrictions on heterosexual behavior push some individuals toward homosexuality. Incidental, temporary homosexual activity, such as that occurring during adolescent experimentation, is rather common. About one-third of American males and one-fifth of females have had such experiences.

Extensive or even exclusive homosexuality is not in itself proof of mental or emotional illness. It is not a disease but a preference. If they live in a society that condemns and punishes their behavior, homosexuals are more prone to neurosis and stress symptoms. The great majority of individuals who are chiefly or wholly homosexual cannot be identified by dress or customary behavior, and they lead lives that (except for the sexual sphere) are basically ordinary and normal. Persons who are predominately homosexual tend to gather together, generally in larger cities, to form homosexual communities and a homosexual subculture. A similar tendency to congregate for mutual support is seen in other minority groups.

There are other sexual manifestations that are statistically abnormal and viewed by most societies as pathological, criminal, or both. These include sadomasochism (sexual pleasure in giving or receiving pain or humiliation), pedophilia (a sexual preference for children), and fetishism (sexual attraction to inanimate objects). Although these manifestations have been the subject of much psychological study and writing, they are extremely uncommon.

PHYSIOLOGY OF COITUS

Until a few years ago, little more could be said about human sexual behavior than the general statements of statistical frequency of certain acts summarized above. The research of William Masters and Virginia Johnson (1966) has made possible much more specific description of the typical

Figure 41.13. Graphs summarizing the female sexual response cycle (left) and the male sexual response cycle (right).

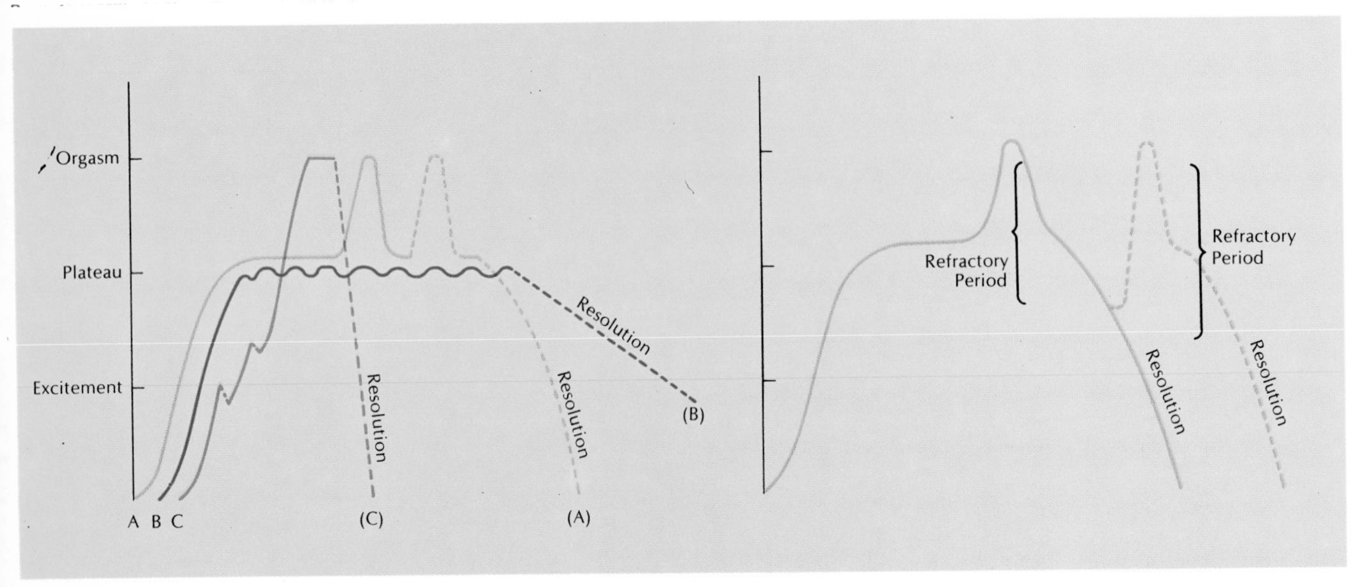

physiological processes during coitus. Because Masters and Johnson were among the first to study physiological changes during sexual stimulation, they had to develop new methods and equipment, particularly for examining the female response, which is more difficult to observe because it is internal. As the investigation proceeded, their techniques included simple measurement of anatomical parts during different phases of stimulation, intrauterine electrodes to measure uterine contractions, electrocardiograph tracing to monitor rate and level of cardiac contractions, as well as equipment to measure respiration and blood pressure. Data were collected during masturbation as well as during natural and artificial coitus. The optical properties of plastic penises used for artificial coitus allowed observation and cinematographic recording of intravaginal physiological responses. The reaction patterns of both female and male were found to be essentially independent of the method of stimulation employed.

The sexual response cycle of both women and men may be divided into four separate phases: the excitement phase, the plateau phase, the orgasmic phase, and the resolution phase (Figure 41.13). The following descriptions of cycles are a composite of individual cycles. Individual variations are marked and usually concern duration not intensity of the response. The initial and final phases are of significantly longer duration than the other two.

Female Cycle

EXCITEMENT PHASE

Many parts of the body in addition to the genitalia respond to sexual stimulation. Heart rate and blood pressure elevation begin in this phase, and the breasts show nipple erection and a slight increase in size. Size alteration is a result of vasocongestion, the concentration of blood in certain blood vessels. Vasocongestion is a primary response to sexual stimulation and is found in a number of different body parts, including the labia minora and the clitoris. It is instrumental in the development of the sex flush, which appears late in this phase. The sex flush is a superficial vasocongestive

Figure 41.14. Diagrammatic study of the anatomical and physiological changes within the female pelvic organs accompanying the sexual response cycle: (1) normal female pelvis, lateral view; (2) female pelvis in the excitement phase; (3) plateau phase; and (4) orgasmic phase.

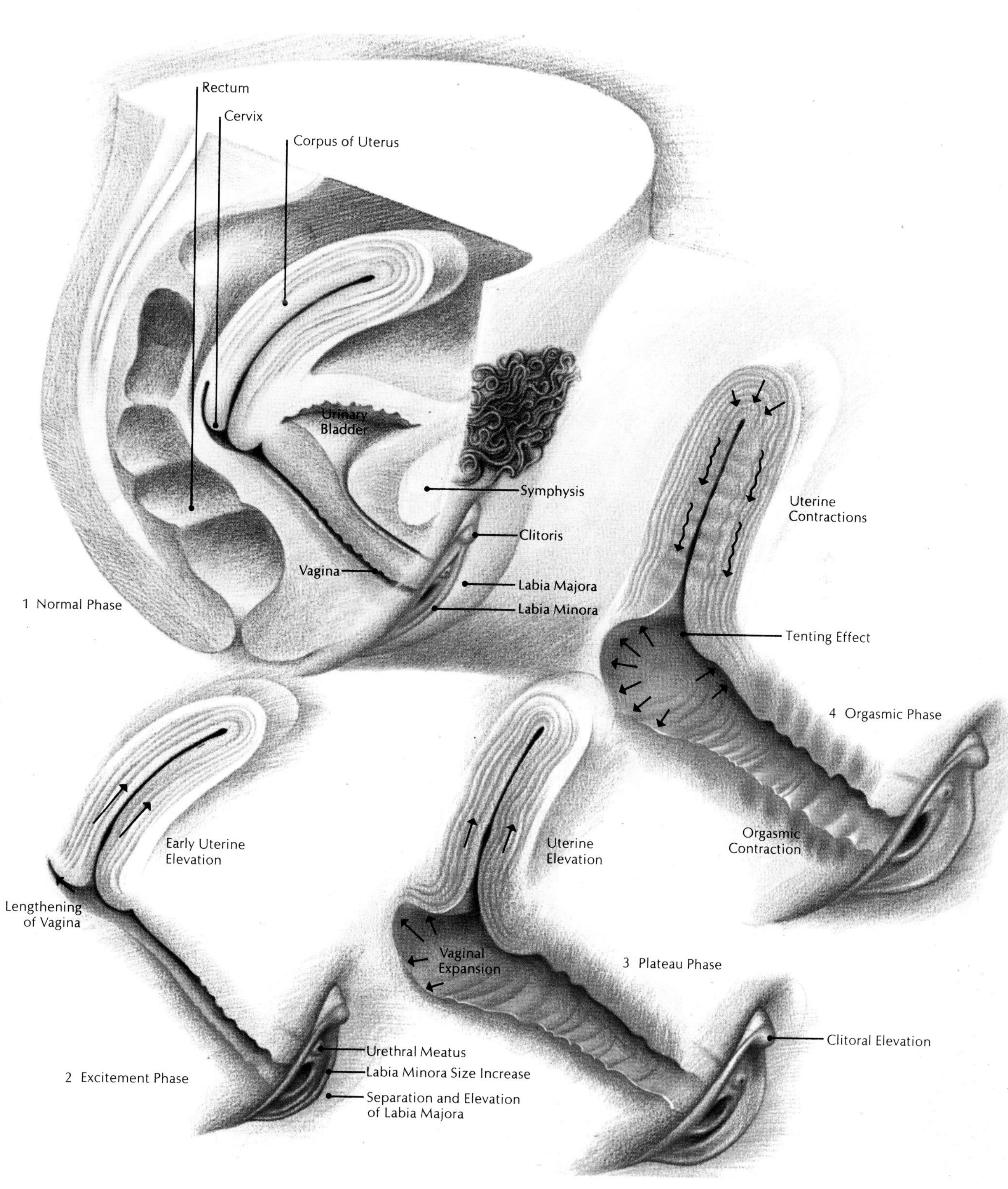

reaction that causes a rash to spread from the breasts over the lower abdomen and shoulders as sexual tension increases. It was observed in 75 percent of the women on at least one occasion.

Myotonia, an increase in muscular tension, is another generalized sexual response involving both voluntary and involuntary muscles. Coital position has a role in determining which muscle groups will exhibit this reaction during a particular sexual response cycle.

Genital changes during this initial phase include clitoral glans (head) expansion and vaginal lubrication. The tumescence (swelling) of the glans is a result of vasocongestion and always accompanies sexual tensions. Because of the nature of the response, it was observable to the unaided eye in only about half of the women studied. Vaginal lubrication is not a result of glandular activity in the vagina, for there are no glands in the vaginal walls. The lubricating fluid passes through the pores of the vaginal tissue, like sweat, in response to significant dilation of a number of blood vessels and to the consequent massive vasocongestive condition. Like lubrication, vaginal expansion is involuntary.

PLATEAU PHASE

The first part of this phase is mainly a continuation of reactions that originated during the preceding phase. If effective sexual stimulation is maintained, the sex flush spreads, myotonia and vasocongestion increase, and breast expansion advances. The clitoral glans and shaft now withdraw. The more effective the stimulation, the more marked this withdrawal is.

The expansion that affected the inner two-thirds of the vagina during the excitement phase now extends to the outer one-third. This tissue, even more than the inner area, becomes grossly engorged with venous blood. Both the outer third of the vagina and the labia minora are involved in this major vasocongestive reaction. Together they form the anatomical basis for the physiological expression of orgasm and are called the orgasmic platform.

The uterus showed partial elevation in the preceding phase. It now attains complete elevation and moves backward toward the spine. It remains in that position until the onset of the resolution phase. The physiological reaction that leads to uterine elevation is unknown. As sexual tensions increase and orgasm becomes imminent, further increases in respiration, heartbeat, and blood pressure are seen.

ORGASMIC PHASE

Immediately before orgasm the physiological responses described for earlier phases are at their most intense level. The breasts are enlarged, with nipples erect. Sex flush and vasocongestion are widespread. The clitoris is completely withdrawn. The vagina is lubricated and extended with the orgasmic platform well developed. The uterus is elevated. Only myotonic tension becomes even more intense with orgasm.

The physiological onset of orgasm is indicated by contractions in certain organs. The orgasmic platform is first. It displays a long contraction, lasting from 2 to 4 seconds and then a series of shorter contractions about 0.8 second apart. The factors that initiate this response have not been identified. They may be neural, hormonal, muscular, or some combination of factors. The platform contractions are accompanied by uterine contractions of a less definite pattern and often by contractions of the external rectal sphinc-

Figure 41.15. Auguste Rodin's sculpture, "The Kiss."

ter. The frequency of these rectal contractions usually is a function of the intensity of the particular orgasmic response. No particular reaction to the orgasmic condition has been observed in the breasts or the clitoris. As indicated in the phase graphs, some women exhibit the ability to experience several orgasms in a relatively short period of time.

Both women and physicians long have believed that orgasm resulting from vaginal stimulation (as in normal intercourse) is different from that resulting from clitoral stimulation alone (as in self-masturbation or manual stimulation by a partner). Freudian theory supported this view and maintained that, in normal sexual development, women move from maximal response to clitoral stimulation to maximal response to vaginal stimulation. Masters and Johnson's data yield no support for this view. They found the physiological and anatomical responses from any form of effective stimulation to be essentially the same.

RESOLUTION PHASE

This phase is characterized by return to the normal, or precoital, condition. The return is fairly rapid if orgasm has occurred. The average time is about 5 to 10 minutes. Myotonia, sex flush, and the orgasmic platform subside quickly. The clitoris regains normal size and position. Changes resulting from vasocongestion disappear throughout the body in an irregular flow of abatement. This process is much slower if plateau levels of tension were reached but orgasm did not occur. In this case, 30 minutes or more may be required. Complete resolution does not occur between cycles if more than one orgasm is attained.

Again, it should be stressed that the cycle described here is what Masters and Johnson see as an average cycle. Many elements may be missing in any normal cycle, but this composite picture is representative of the more than 10,000 cycles they observed.

One of the principal findings of these studies was the great extent to which the woman's entire body is involved in her response to effective sexual stimulation. Myotonia, vasocongestion, sex flush, and other specific reactions were observed in numerous parts of the body. Orgasm often was accompanied by a general reduction in perceptual acuity and the appearance of perspiration. The importance of these peripheral states in the subjective experience of orgasm is unexplored.

The newly found properties of the clitoris are of equal interest. It is the focal point of sensual response in the woman's pelvic area. It is unique among all human organs because its only known function is that of receiving, responding to, and transmitting neural messages of sexual stimulation. It contains both afferent and efferent neural pathways. Its ability to respond to stimulation received at other points exposes the fallacy involved in typical "marriage manual" instructions to maintain direct clitoral stimulation throughout coitus. This contact is both difficult and unnecessary, because the clitoris reacts to the sensory input of other areas, such as the vagina. In most cases, this secondary stimulation is sufficient to cause orgasm. No correlation between intensity of orgasm and clitoral size or positioning was observed.

Another fallacy is the common belief that coitus during menstruation causes physical distress to the woman. About 10 percent of the women who were menstruating regularly during their involvement with this research objected to sexual activity during menstruation on religious or aesthetic grounds. About 40 percent expressed no particular like or dislike for sexual activity during menstruation. Approximately 50 percent voiced a desire for sexual activity during the menses. Of particular interest were the 10 percent who reported frequent use of masturbation to relieve discomfort accompanying the onset of menstruation.

Male Cycle

A remarkable finding of the Masters and Johnson studies was the existence of numerous similarities between male and female sexual response patterns. The similarities are much more pronounced than the differences, as is apparent in the description of the male cycle.

EXCITEMENT PHASE

Breast enlargement is not part of the male cycle. Nipple erection is seen less frequently in men than in women. When it does occur, it usually is initiated late in this phase. Other extragenital responses include increased heart rate and blood pressure. Elevated tension (myotonia) in both voluntary and involuntary muscles appears early in this phase.

Early genital reactions are most visible. Penile erection results from massive vasocongestion in the three cavernous bodies that run the length of the penis. Some myotonia also is involved. Unlike female vasocongestion, mainly arterial blood is collected in the penis. Vasocongestion in this and other areas is the man's primary physiological response to sexual stimulation. Vasocongestion and thickening of the scrotal skin are involved in the elevation of the scrotal sac. However, testicular lifting is accomplished through shortening of the spermatic cord by an involuntary contraction of the muscles near the cord. This response begins in this period and is completed during the next.

PLATEAU PHASE

The penis, which reached full erection during the excitement phase, shows a slight increase in circumference during this phase. This late development is restricted to the head, or glans penis. As testicular elevation progresses, an increase in testicular size becomes apparent. Testes show about a 50 percent size increase, and if plateau levels of sexual tension are prolonged without orgasm, up to 100 percent increase may occur.

As orgasm is approached, a preejaculatory fluid originating in Cowper's gland may extrude from the head of the penis. Frequently, it contains active sperm. Orgasmic approach also occasionally is accompanied by the sex flush observed in women. In men, it usually covers the chest, neck, face, forehead, and sometimes shoulders and thighs. Impending ejaculation also is signaled by increased heart rate, breathing rate, blood pressure, and myotonia.

ORGASMIC PHASE

Although ejaculation and orgasm conveniently are discussed as two parts of a single occurrence, it should be remembered that they are not inseparable. Orgasm without ejaculation – that is, without the expulsion of seminal fluids – is experienced frequently by prepubertal boys.

With the exception of fluid discharge, the physiological responses that form the basis for the male's and female's experience of orgasm are quite similar. Both involve the loss of general muscular control as well as massive

Figure 41.16. Diagrammatic study of the anatomical and physiological changes accompanying the male sexual response cycle: (1) normal male penis, lateral view; (2) excitement phase; (3) plateau phase; (4) orgasmic phase; and (5) resolution phase.

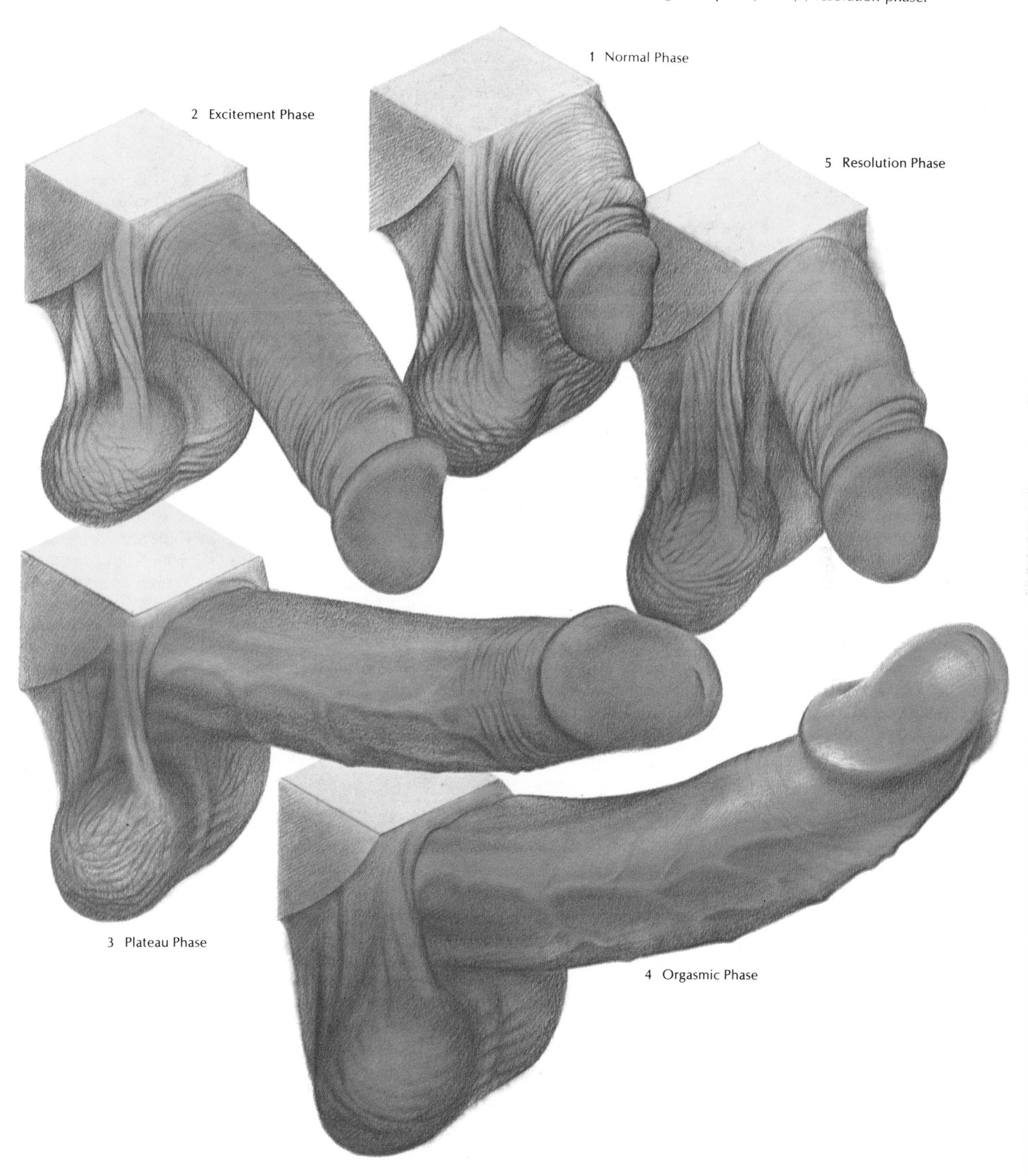

contractions in the genital area, including the rectal sphincter. The contractions of the orgasmic platform and of the penile urethra both occur at intervals of 0.8 second during part of the orgasmic experience. Heart rate, breathing, and blood pressure are at their highest during orgasm in both sexes. The contractions of the accessory reproductive organs (prostate, seminal vesicles, ejaculatory duct) and the resulting seminal emission distinguish the male response from the female response. The whole body participation described for women is also the case for men, as myotonia and vasocongestion are observed throughout the body.

RESOLUTION PHASE

Following orgasm, the return to the unstimulated state is usually quite rapid. The sex flush fades, and perspiration sometimes is seen in the immediate postorgasmic period. Detumescence (loss of swelling) of the penis occurs with the loss of localized vasocongestion. This loss also leads to the descent of the testes into the relaxed scrotum. A general decrease of muscle tension then appears, usually no more than 5 minutes after the beginning of this phase. Nipples that showed erection in the excitement phase display no additional reactions during plateau or orgasmic phases and may require as much as an hour to return to their normal position.

Myths about the male sexual response are even more numerous than those about the female, perhaps because the external genitalia of men makes their major reactions visible and consequently more a topic of discussion than the invisible and little-known female response. Part of Masters and Johnson's research was an investigation of the validity of commonly held beliefs about the male sexual response. For the most part, they found these beliefs to be without basis in fact.

Contrary to popular belief, circumcision has no apparent influence on sexual behavior. A comparison of circumcised and uncircumcised subjects revealed no differences in degree of ejaculatory control. A study of men of all ages showed that, although some erectile difficulty may appear with age, impotence is in no sense an inevitable result of aging. In addition, this type of impotence is a reversible condition in men of all ages.

Penis size has long been believed to be predictive of a number of other factors. In some ways, it was supposed to be correlated with general body build, sexual adequacy, and the degree of female satisfaction. A careful survey of these factors revealed no such correlations. This study also failed to support the belief that the large penis exhibits a greater erectile ability than the small penis. Masters and Johnson found the exact opposite to be true. The small penis usually doubles its length during full erection; the large one increases to 1.5 times its flaccid length. Within very wide limits, penis size is not a significant factor in either partner's sexual response.

CURRENT AND FUTURE RESEARCH

As it becomes increasingly more respectable in the eyes of grant-giving foundations and universities, sex research will increase. This trend has been quite visible in recent decades. In the 1950s, the number of people devoting most of their time to research on human sexual behavior could be counted on the fingers of both hands. Following Kinsey's famous publications, which served to legitimatize such work, the number grew rapidly.

Although continued growth of research on human sexual behavior is probable, the direction of growth can only be guessed because it tends to follow public and government interests. Whatever is viewed as a social problem will receive attention and funds. Thus, the recent emphasis has been on homosexuality and on the effects, if any, of pornography, and these problems may well continue to engender research. The relationship of sex and violence is a likely candidate for future work, and the increasing youthfulness of the population may focus attention on heterosexual problems common in early marriage. More physiological studies following in the footsteps of Masters and Johnson are likely, and a moderate amount of experimentation can be anticipated. In addition, there are many basic questions connected with the origins of behavior and the interrelationships between sexuality and the other aspects of life that desperately need investigation.

FURTHER READING

The encyclopedia edited by Ellis and Abarbanel (1969) is a useful general reference for information about sexual behavior. Beach (1965) and Ford and Beach (1951) summarize most available information about sexual behavior in nonhuman species. Interesting anthropological views of sexual roles in various societies are given by Mead (1949) and Seward and Williamson (1970). The legal aspects of sexual behavior are discussed in a collection edited by Slovenko (1965).

The "Kinsey reports" (Kinsey, *et al.*, 1948, 1953; Gebhard, *et al.*, 1958, 1965) remain the major source of sociological information about American sexual behavior. Masters and Johnson (1970) also have published a study of inadequate sexual responses and their causes. Gebhard, *et al.* (1970) discuss female sexuality.

Societal attitudes toward premarital sex are discussed by Reiss (1967). There are a great number of books on homosexuality, with that by West (1968) being one of the best. Other interesting viewpoints are given by Bieber (1962), Churchill (1967), Ellis (1965), Lindner (1961), Marmor (1965), Ovesey (1969), Ruitenbeek (1963, 1968), Schofield (1966), and Storr (1964).

Among many books offering general information and advice about sexual behavior from varying points of view, the following are particularly interesting and useful: Ellis (1958), O'Relly (1968), Schoenfeld (1968), and Young (1966).

42
Human Embryology

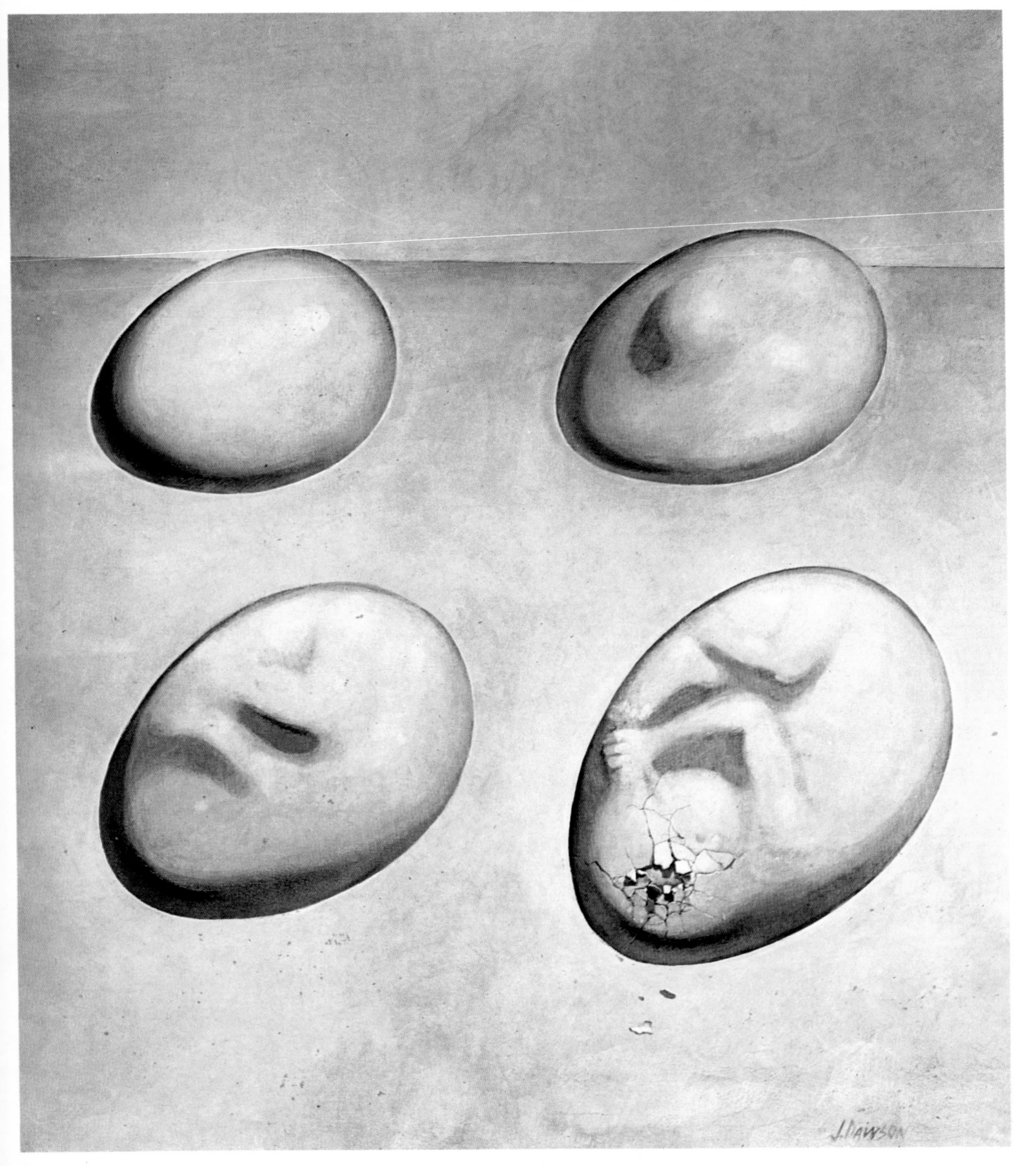

When does life begin? Philosophers, theologians, doctors, and biologists have argued this question for centuries. Some say life begins at fertilization; others consider the crucial time to be when implantation occurs, when the embryonic heart begins beating, or when the embryo becomes a fetus with recognizably human characteristics. Many have argued that life begins with the first independent kicks or squirms of the fetus ("quickening"), but others claim that life does not begin until the newborn takes his first breath and becomes relatively independent of his mother's physiological systems. To most modern biologists—particularly to embryologists—these arguments are irrelevant. Life began only once, billions of years ago, and every modern living thing has a direct connection to that primitive life and to all of life.

Each fertilized human ovum contains 23 pairs of chromosomes, and each chromosome is believed to carry at least 1,000 genes. Every gene (or a precursor of every gene) has been handed down through generations and comes together in the composite genetic heritage of each new individual. Yet the number of possible genetic combinations is so great that each individual is of necessity a new and unique combination of those units, not identical to any other combination ever before assembled.

It is convenient to describe events of prenatal development in terms of a time scale that begins with fertilization on day 0. By day 21, about 100 progenitors of germ cells arise in a vestigial (ancestral) structure called the yolk sac. The cells of the germ plasm multiply rapidly, and by day 27 there are 1,000 or more of them. They begin to migrate by amoeboid movement toward the region where the ovary or testes later will form. It is one of the many marvels of the developmental process that these germ cells arise at the right time, move in the right direction, and arrive at their permanent location just as it is properly prepared to receive them.

In a female embryo, there are an estimated 600,000 of these primitive germ cells, or oocytes, by week 6. By week 20, they have increased to about 7 million, but by birth they are reduced to about 2 million. At puberty, they have been further reduced to about 400,000, and at the height of sexual maturity (age 22 years), they may number about 300,000. Of these 300,000 oocytes, only about 400 will mature to ova—an average of 1 each 28 days for about 40 years. Theoretically , a maximum of 40 ova could be fertilized (the maximum reported for any particular woman is about 25). Thus, from about 7 million potential ova in the embryo, only 25 to 40 could possibly contribute to the next generation.

In the embryo destined to be a male, a similar sequence of events occurs. The germ cells migrate to the genital ridges, which will later develop into testes. Final maturation of spermatozoa does not begin until puberty, but this process then continues throughout the life of the male. Although it takes more than 2 months for a primitive germ cell to mature into a spermatozoan, this process goes on at such a rate that millions of mature sperm may be produced each day, for as long as the normal male survives after puberty. Although the typical female loses reproductive ability at about age 45 or 50, the male can fertilize mature ova as long as he lives, under normal conditions.

Because of chromosomal reassortment and crossing-over during meiosis, no two germ cells produced by the same individual are likely to have exactly the same set of genetic messages. If each woman is capable of producing about 400 ova during her lifetime, and each male produces about

Figure 42.1. (above). Living human ovum enlarged about 200 times. The human egg is approximately the size of one of the periods on this page. Unlike many other vertebrate ova, the human egg lacks large amounts of yolk and instead is dependent on nourishment from the mother's blood via the placenta. (*From Rugh and Shettles*, From Conception to Birth: The Drama of Life's Beginnings, *Harper & Row, 1971*)

Figure 42.2 (below). Living, active human sperm highly magnified. Although the sperm has 85,000 times less volume than the human egg, it carries one-half the genetic material required for the complete development of the human infant. (*From Rugh and Shettles, 1971*)

Figure 42.3 (lower right). A fertilized human egg, 12 hours after fertilization. Note the two discarded polar bodies, which have been produced during meiotic division of the egg nucleus. The first cleavage of the egg typically takes place 30 hours after fertilization. (*From Rugh and Shettles, 1971*)

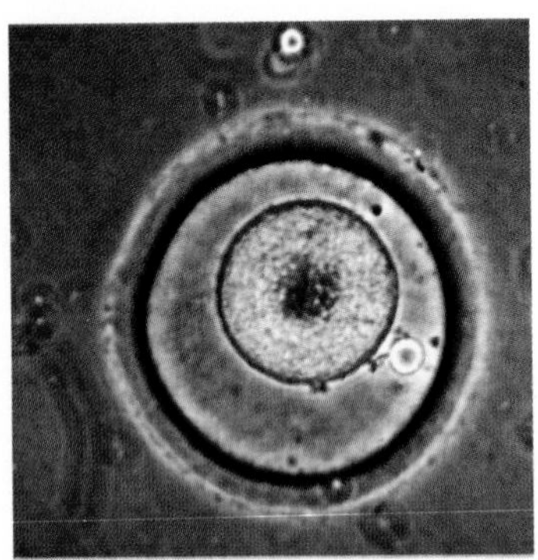

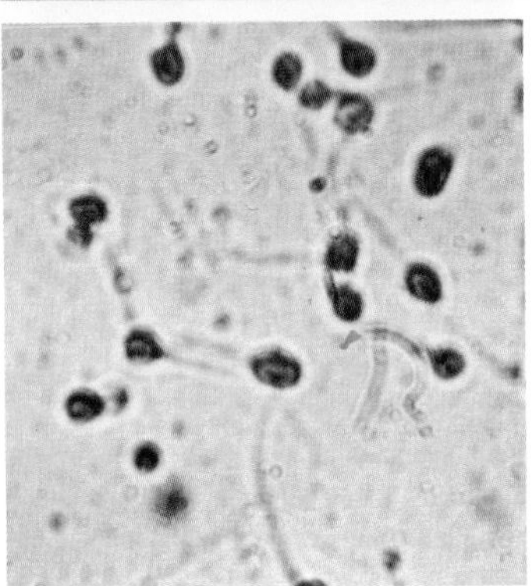

360 million mature sperm at any given time, more than 14 trillion different combinations of ovum and sperm are possible in the mating of a particular human pair. Thus, each child is a unique and very rare combination of genetic possibilities.

Every mature ovum carries one X chromosome, but half of the spermatozoa carry X chromosomes and half carry Y chromosomes. If an X-bearing sperm fertilizes the ovum, an XX zygote results, and this zygote normally will develop into a female. If a Y-bearing sperm fertilizes the ovum, an XY zygote results and normally will develop into a male. Sex therefore is determined genetically at the moment of fertilization, depending on which of the two types of sperm (occurring in equal numbers) reaches the ovum first. For some reason—possibly a minute difference in weight or activity favoring the Y-bearing sperm—there are 106 boys for every 100 girls born in the overall population.

If active sperm cells are in the female genital tract within the period from 48 hours before ovulation to about 12 hours after ovulation, fertilization is very likely to occur. Within seconds after the head of the sperm touches the egg, the fertilized ovum reacts with violent undulating movements that bring the sperm and ovum nuclei close together, thus restoring the full diploid set of 23 chromosome pairs.

Fertilization occurs without the knowledge of the woman who will house the embryo and fetus, but this moment sets in action the sequence of processes that will lead 266 days later (plus or minus a week) to delivery of a child composed of some trillions of cells that have arisen from the single fertilized ovum, about 1/175 of an inch in diameter. If a woman misses a menstrual period, she may hope (or fear) that she is pregnant, but at this time pregnancy can be detected with certainty only by various tests that detect the effects of the hormone secreted from the chorion of the embryo. Human female urine is injected into a female mouse, rat, or rabbit, and its ovaries are examined at a certain interval thereafter. If the woman is at least 4 weeks pregnant, her urine will contain the hormone and its breakdown products, and these will cause rupture of ovarian follicles of the test animal

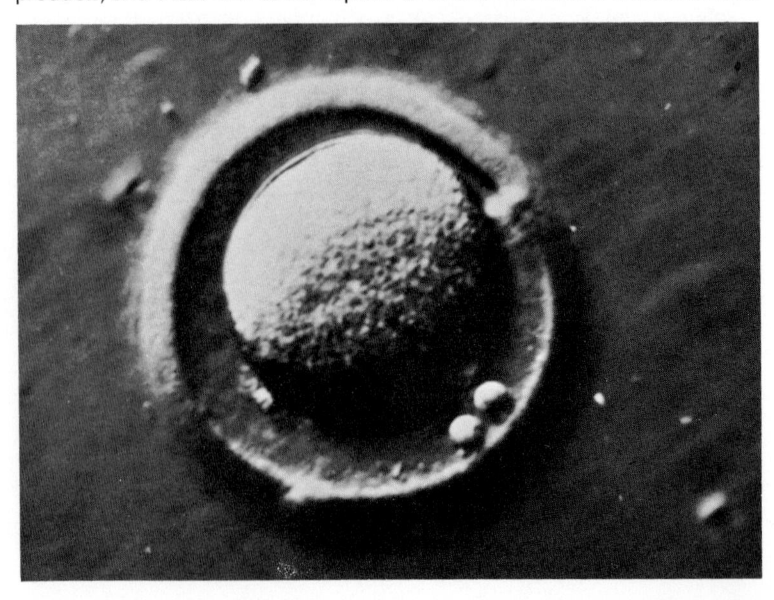

that are seen as blood clots in the ovaries. There are also the so-called frog tests and some recently developed chemical tests that determine in a matter of minutes whether the hormone is present in the urine. Pregnancy may cause the woman to experience a slightly elevated body temperature, some morning nausea, fatigue, or tingling breasts shortly after the first missed menstrual period.

THE FIRST TRIMESTER

The 9 months, or 266 days, of pregnancy conveniently are divided into trimesters, each consisting of 3 months. The first trimester involves basic organization from an apparently unorganized zygote into a fetus with recognizable human features. It is during this period of organ formation that drugs, viral diseases, or radiation may cause *congenital anomalies*, defects that are not heritable. These defects are brought about by certain traumatic events, including alcoholism, smoking to excess, contraction of German measles, influenza, polio, and other diseases, or direct exposure of the developing embryo and fetus to more than 10 roentgens of radiation.

There is a growing tendency to eliminate drug prescriptions for the pregnant woman because the drugs most often prescribed are tranquilizers, sedatives, and so on, all of which can and do affect the embryo and fetus. The so-called hallucinatory drugs are believed to be capable of some damage to the chromosomes of the embryo. Recent experience with the tranquilizer thalidomide (which causes such grotesque fetal anomalies as shortening or loss of limbs, mental retardation, and so on) has greatly increased awareness of dangers in taking any drugs during pregnancy. Alcohol, in contrast to most drugs, has its greatest effect during the third trimester, when alcohol concentrations are 10 times as great in the fetus as in the mother, especially in the fetal brain and liver. After the organs have formed, they cannot be damaged so readily, but more subtle and functional defects can be caused in the second and third trimesters, affecting behavior and intelligence.

During the first month, the embryo reaches a total length of about 1/8 of an inch. It is composed of millions of cells, all derived from the single-cell zygote. During the first month, the embryo increases its weight about 500 times, a greater change than at any later month. Some 30 hours after fertilization, the zygote divides into 2 equal cells, and 10 hours later into 4 cells. By 72 hours, there are 16 cells—each identical in size and appearance to the original zygote. With each cleavage, the cells become smaller and smaller. They all are enclosed in a translucent membrane (the *zona pellucida*), which holds them together like parts of a mulberry. This stage is called the *morula*. While cleavages are occurring, the embryo is propelled downward through the oviduct toward the uterus, largely by means of hairlike cilia lining these cavities, but also by muscular contractions in their walls.

During the next 3 days, the morula is free within the uterine cavity, but then it finds a place to implant, usually in the posterior uterine wall. The endometrium has been prepared by the mother's hormones (Chapter 40). At the time of implantation, the embryo is a hollow sphere. One side of the sphere (the *trophoblast*) is thin and will give rise to the membranes that surround the embryo. The other side of the sphere is thick with cells that make up the embryo proper (known as the *inner cell mass*). The total sphere is called the *blastocyst*, and the blastocyst cavity in its center is filled with a fluid. The zona pellucida disappears, and then the trophoblast secretes a

Figure 42.4a (above). Human egg at the two-cell stage of division. (*Courtesy Carnegie Institution of Washington*)

Figure 42.4b (below). Human embryo after 4 divisions, which result in 16 cells. These 16 cells are similar because differentiation has not begun. (*Courtesy Dr. L. B. Shettles*)

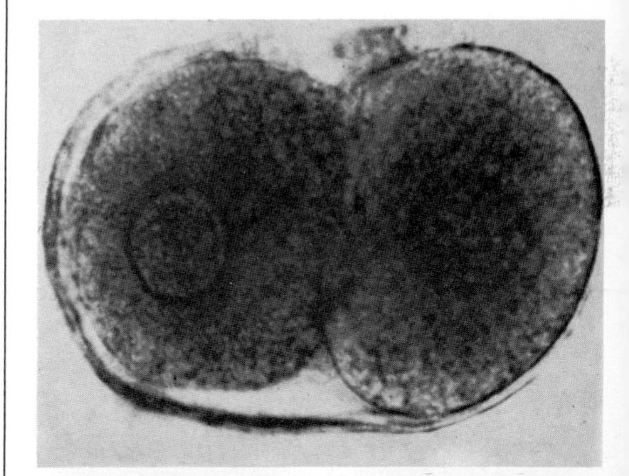

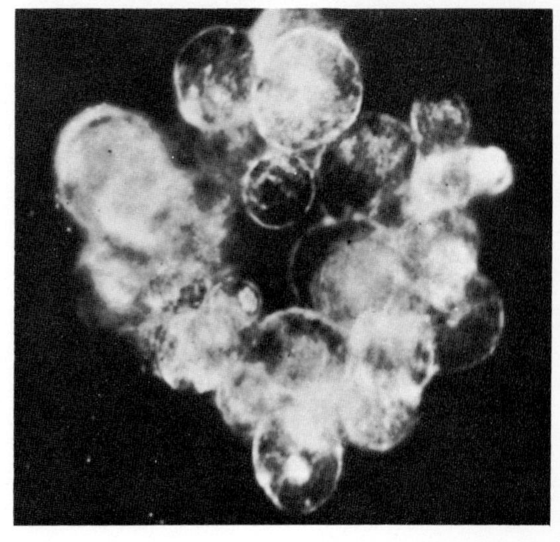

Figure 42.5. Size chart comparing embryos of varying ages.

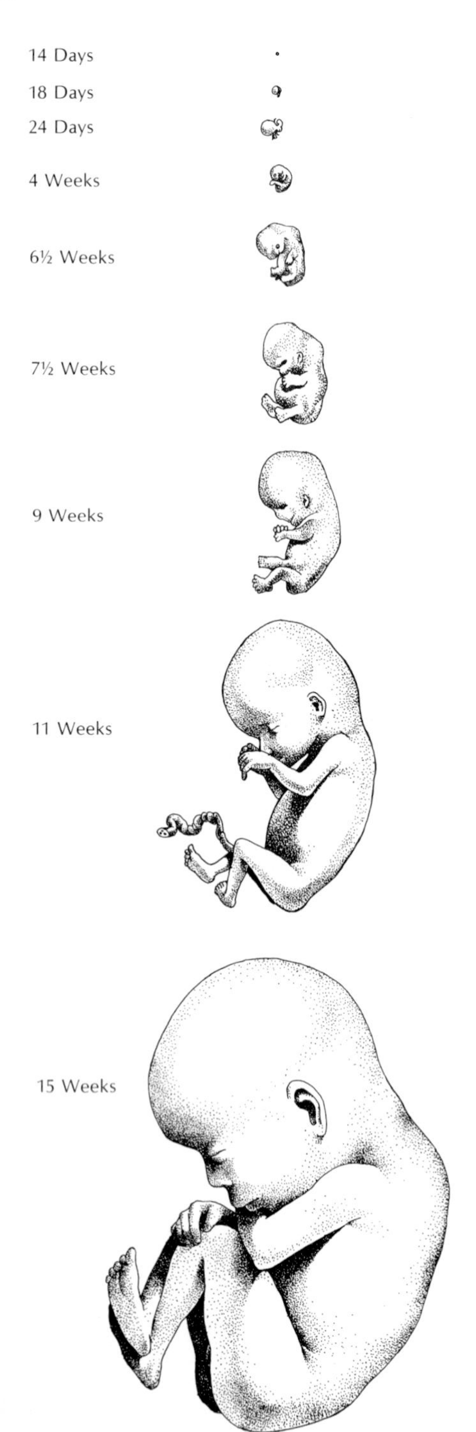

proteolytic enzyme that helps it to invade and penetrate the lining of the uterus. As it does so, the blastocyst digests away some of the blood vessels in its path, so that the embryo comes to lie in a minute pool of the mother's blood. By day 8, the embryo has bored its way beneath the surface lining of the uterus and is surrounded by bloody fluid rich in glycogen, a food needed by the embryo at this time. By day 12, implantation is completed with the aid of the uterine tissues, and the endometrium closes over the injured surface of the uterus.

Because fertilization generally occurs 14 days after the onset of the last menstruation, implantation is completed at about the time of the next expected menstruation. Sometimes implantation releases a small amount of blood into the uterine cavity, giving the woman the mistaken impression of a much reduced but timely menstrual flow. Outgrowths from the trophoblast, known as *chorionic villi*, penetrate the maternal tissues and begin formation of the *placenta*, an organ mutually formed by the embryo and mother. The placenta is a transfer area for wastes and carbon dioxide from the fetus and for nutrients and oxygen from the mother to the fetus. The placenta at birth weighs about 1 pound and is discarded, along with the umbilical cord, as the *afterbirth*.

The human embryo makes great progress in basic development before the end of the first month. It forms a *primitive streak*, which is its main body axis, and the trophoblast begins to form membranes that enclose the embryo and protect it physically. Of the various systems, the nervous system is the first to start development. It appears as a neural plate, neural fold, and then neural tube (day 24), which will become the brain and spinal cord and give rise to many of the cranial and spinal nerves. The forward part of the neural tube walls thickens and the tube expands, soon (day 26) to be pinched to form four primary brain vesicles, each of which will form a specific part of the brain. Only the germ cells take precedence over brain and nervous system development. Associated with the brain are the sense organs—eyes, nose, and ears—which begin to appear at this time. Even the lens and retina of the eyes start to form.

On each side of this neural axis the embryo acquires, during this first month, 32 pairs of muscle blocks, or *somites*, which will give rise to most of the voluntary muscles of the body, much of the skeleton, and the skin. By day 28, there are 3 pairs of visceral arches on each side of the pharynx. These arches are vestiges of distant ancestry, where they were related to functioning gills, but the human embryo never develops gills. A membrane over the mouth ruptures in the first month, even though the embryo will not take in any food for a long time.

Elements of the lower jaw appear, followed shortly by those of the upper jaw. The lung primordia (day 27), and the beginnings of trachea and bronchi (day 31) are formed. Blood cells appear in the yolk sac and in the chorionic membrane, and simultaneously (day 21) the tubular heart appears. By day 24, the heart begins to pulsate, even before it is completely developed. Its first beats are irregular and slow, but they increase over a period of weeks until they may occur as rapidly as 180 per minute. Once the heart starts beating (day 24), it continues without any interruption for as long as the individual lives—perhaps 80 or 90 years—averaging about 100,000 beats per day. At the end of the first month, the valve that prevents backflow of blood between the auricle and ventricle begins to form, even though the heart is beating regularly. By day 27, the embryo forms a pair of primitive

Figure 42.6. In the embryonic period, the embryo is markedly curved and shows a distinct head fold, which contains swellings and markings that will soon become the jaw, eyes, and ears of the individual. *(Courtesy Carnegie Institution of Washington)*

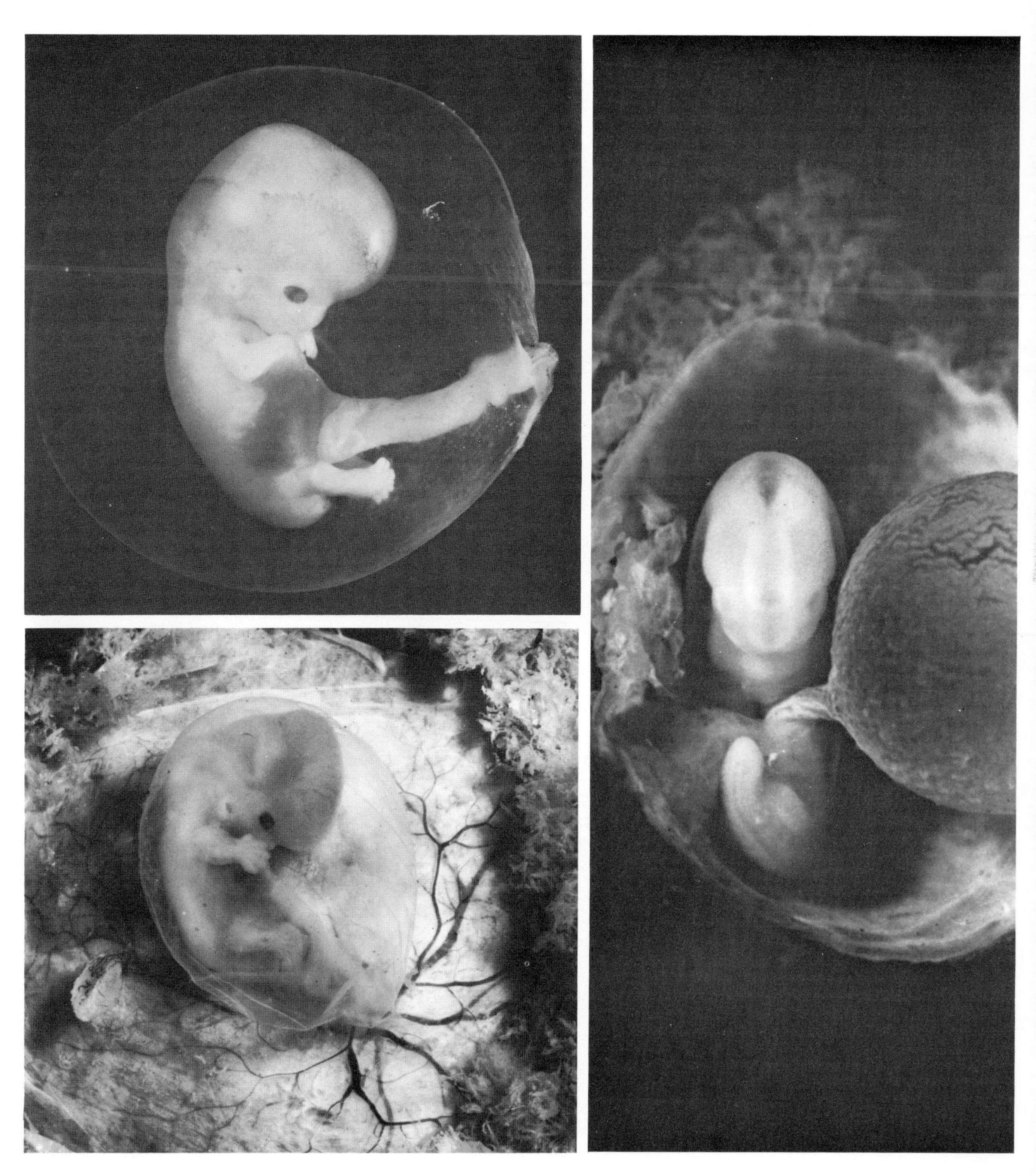

Figure 42.7 (above). Human fetus between 4 and 5 months. Note the recognizable ears, nose, eyes, and mouth. The fetus is entirely enclosed in a watery amniotic sac, which serves as a protective water cushion that absorbs jolts, equalizes pressures, and facilitates change of fetal posture. (*Courtesy Dr. Roberts Rugh*)

Figure 42.8 (below). Human fetus at 40 days, lateral view. Note the deep cleft, or isthmus, in the brain, separating the midbrain from the hindbrain.

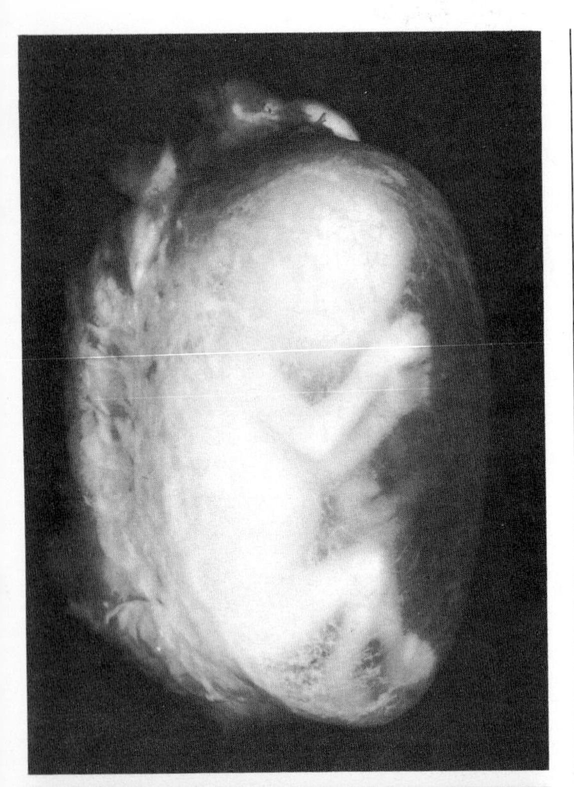

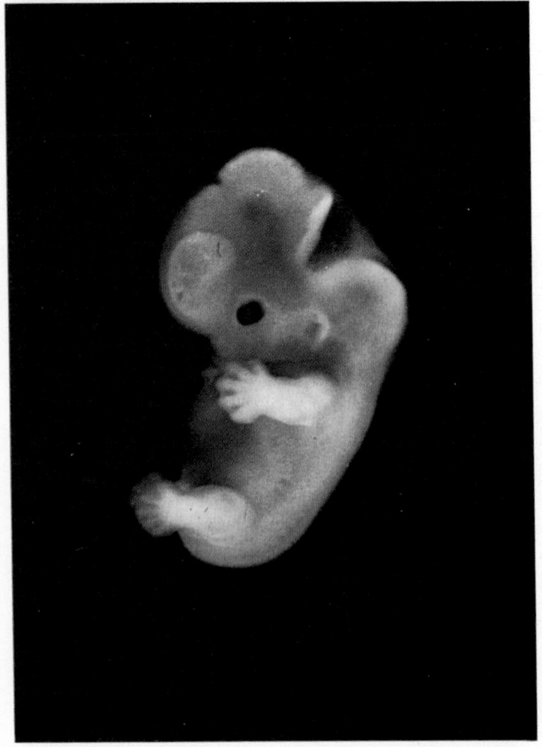

kidneys known as the *pronephroi*. By about day 30, a second pair of kidneys—the *mesonephroi*—form, which will function briefly for the developing fetus. Both mesonephric kidneys are joined posteriorly to the cloaca (an excretory chamber) by nephric ducts, which are embryonic ureters. Budlike thickenings develop (day 28) that will become arms. The primordia of the important endocrine organs, such as the thyroid, anterior pituitary, dorsal and ventral pancreas, all make their appearance just before the end of the first month.

The human zygote loses no time in becoming a multicellular structure. Before the first month is over, certain groups of cells are determined to become very specific and necessary but as yet undeveloped organ systems. All major systems begin to form, although none is functional. Before the woman is certain that she is pregnant her embryo has become a very complex and intricately associated group of organ primordia, although the whole structure is smaller than the mother's smallest fingernail.

The events of the second month become vastly more complex. By the end of the second month, the growing organism measures about 1.25 inches in total length and has small but definite limbs, a body with a distinct head, recognizable ears, nose, open but lidless eyes, and open mouth. It is entirely enclosed in a watery *amniotic sac*, which affords some physical protection against contusions, adhesions, and temperature changes. The amniotic fluid is gradually increased but is kept clean by constant exchange with the mother's circulatory system so that it is completely replaced about every 3 hours. The embryo can swallow or inhale this water without harm and does so. The body stalk, or umbilical cord, now is evident and joins the circulatory system of the embryo to the placenta, where gaseous, nutritious, and waste exchanges are made with the mother's circulatory system, putting an increasing burden on her kidneys and lungs.

During the second month, somites are added to make a total of 49 chromosome pairs. The germ cells reach the fetal kidney region (mesonephros). Arm and leg buds are distinct by day 35. The vestigial tail of earlier stages begins to be resorbed. The eyes are forming with lens and retina, the ear canals arise, and jaws develop. By day 46, the reproductive organs begin to form ovaries or testes, although secondary sex characteristics are not yet developed. Before the end of the second month, fingers and toes are distinguishable on the webbed paddlelike appendages. Ossification centers appear in the jaws and clavicle regions, indicating the beginning of skeletal formation. The first 2 months generally are considered to be the period of the embryo. By the end of the second month, the growing organism resembles a miniature human and is called a *fetus*.

By the end of the third month, the fetus is about 3 inches in body length (half of that length is head) and weighs about 0.5 ounce. Its thumb and forefinger are opposed to each other, an evolutionary advantage over other vertebrates. The muscles and their nervous connections are so much better developed that the breathing, eating, and general body movements become more purposeful. Variations between fetuses begin to appear—behavior becomes individualized, particularly with regard to facial expressions. Numerous taste buds have developed—theoretically, the fetus has a keener sense of taste than it will at any future time, because many of these buds are lost during later development. Probably the most pronounced changes are seen in the reproductive and excretory systems, which arise in close proximity to each other and retain some mutual functions throughout

Figure 42.9 (left). The development of hands and feet. In the fifth week, hands are a "molding plate" with finger ridges. In the sixth week, finger buds have formed. In the seventh and eighth weeks, the fingers, thumbs, and fingerprints form; note the prominent touch pads. In the third month, the pads regress, and the hands are well formed. The feet begin to form in the sixth week and 48 hours later have large toe ridges. The heel appears by the end of the sixth week and grows out in the next five days. In the third month, feet are well formed. *(Courtesy Carnegie Institution of Washington)*

Figure 42.10 (upper right). A fetus in a man's hand. *(From Rugh and Shettles, 1971)*

Figure 42.11 (lower right). Photograph of an x-ray of a 2-month fetus. Note the extensive development of the skeleton, with centers of ossification appearing as darkened regions. The limb bones and digits are almost completely ossified, although the rib cage remains mostly cartilage. *(From Rugh and Shettles, 1971)*

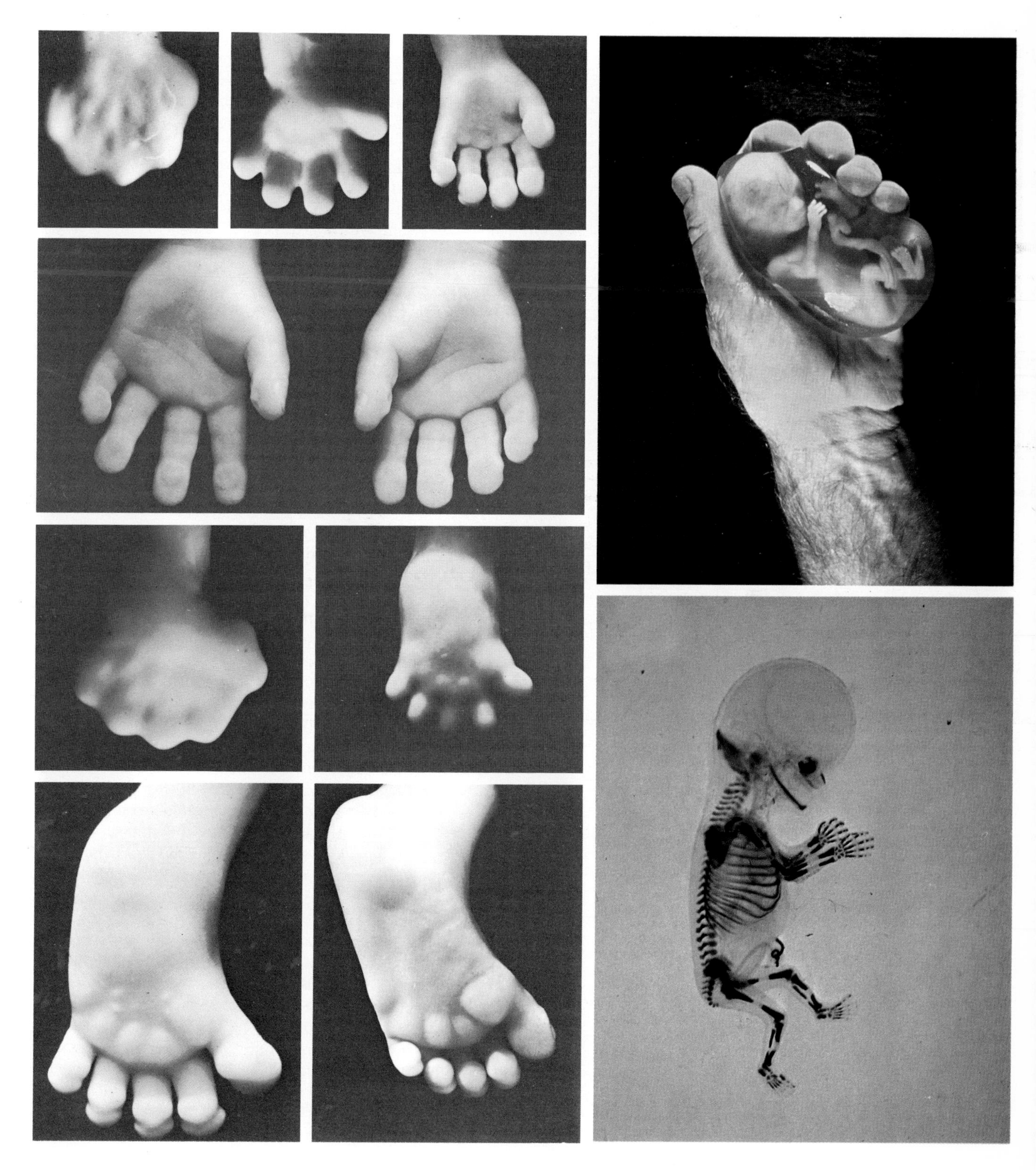

Figure 42.12. Human fetus at 3 months with placenta attached. The placenta provides for physiological exchange between the developing fetus and the mother by permitting the passage of gases, nutrients, and metabolic wastes. By the end of the third month, the fetus is about 3 inches long; the head continues to be relatively large, with a high prominent forehead, a prominent nose, external ears level with the lower jaw, and eyelids that are sealed shut. *(From Rugh and Shettles, 1971)*

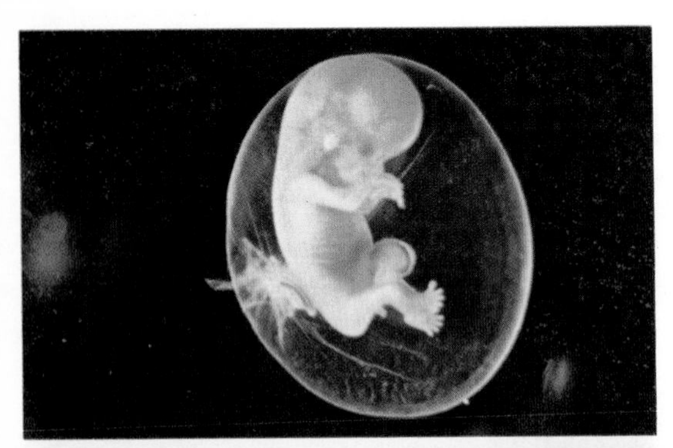

life. The fetus excretes urine into the amniotic fluid, and its waste products diffuse to the circulatory system of the mother. X-rays show that many of the cartilage centers are changing into bone; the skeleton is forming extensively. Even the digits of the hands and then the feet become ossified. The skin has numerous sweat glands, some of which now become the mammary glands of either sex. Facial hair appears by day 70.

The third month completes the first trimester of pregnancy, and the human fetus has acquired all of its major organ systems. Although it is a functioning organism, it still is unable to survive if removed from the mother. During the second trimester, there will be refinements in all newly developing systems.

THE SECOND TRIMESTER

Although the brain starts to form at an early stage, the rest of the body grows faster than the head after the first trimester. Eventually, the head will be only 10 percent of the total body length. By the end of the fourth month, the fetus weighs 4 to 6 ounces and is 7 to 8 inches long. His back becomes straight as the internal viscera enlarge and are enclosed by the abdominal wall, and he can hold his head erect. The skin is so well formed that the palms of the hands and soles of the feet have distinct patterns of lines that distinguish this fetus from all other humans. The fetus reacts to stimuli at first with total body response and then gradually with typical reflexes. He spontaneously stretches his arms and legs, movements that the mother may be able to feel. His heart now pumps regularly, some 25 quarts of fluid per day (mostly recirculated blood), and this pumping can be heard with a stethoscope.

The blood is now formed by bone marrow rather than by the liver or the spleen. The brain is beginning to form convolutions to add to its surface area and complexity. The eyes are beginning to be light sensitive, but the ears are not yet sensitive to sound. The placenta is now disc-shaped and covers a large portion of the inner surface of the uterus.

By the end of the fifth month, the fetus is 9 to 11 inches long and weighs about half a pound. He is now freely mobile within his watery amniotic sac, surrounded by a quart of amniotic fluid, which is changed at the rate of 6 gallons per day. This watery environment protects him from physical injury and temperature fluctuations. Although the 5-month fetus would appear to be fully formed, if severed from the placenta, he would be unable to survive outside of the uterus. The youngest known surviving premature infant was born at 25 weeks, weighed 650 grams, and required almost complete artificial sustenance by temperature regulation, intravenous feeding, oxygen supplementation, and so on. The fetus does have complete lungs, but the alveoli are not yet functional, even though the fetus does take amniotic fluid into his lungs. The fetus' skin is covered by a cheesy, pastelike substance (*vernix caseosa*) composed of loose cells and secretions from the sebaceous and sweat glands. This substance is protective, but the skin is not ready for the colder atmosphere. The digestive organs are well formed but are not ready to take in and digest food.

All embryos of "higher" animals go through some embryonic stages having structures that resemble embryonic structures of "lower" animals. The human embryo and fetus are no exception. Every human embryo develops structures similar to those associated with the development of gills, a tail, and primitive kidneys in embryos of other animals. Vestiges of some

Figure 42.13. Human fetus at 4 months. During this period of development, the fetus doubles in length to almost 6 inches and weighs about 4 ounces. Note the extensive number of superficial blood vessels in the head region. *(From Rugh and Shettles, 1971)*

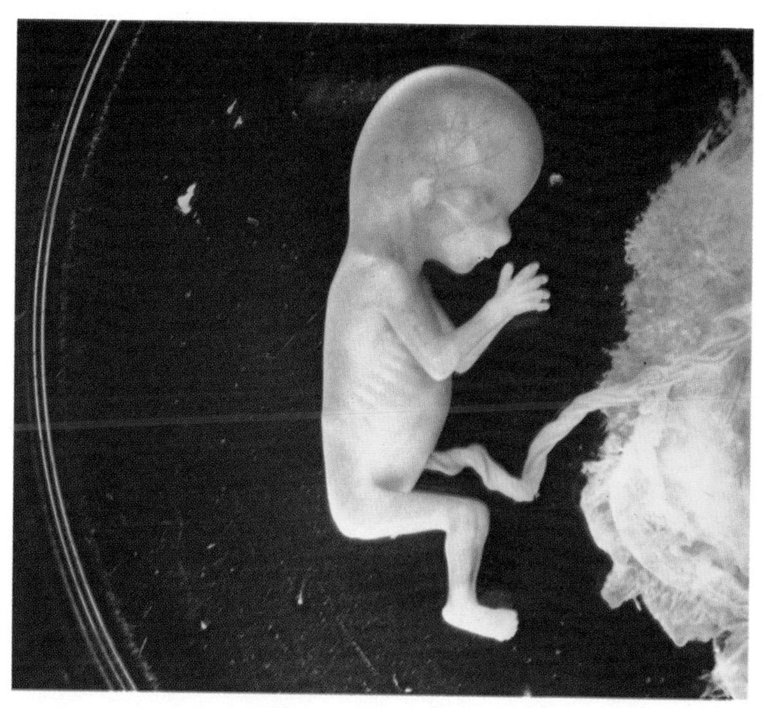

structures without apparent functions persist in the adult human, presumably remnants of the heritage of distant animal ancestors. Some are retained because other associated organs are absolutely necessary to survival. The appendix is one structure that is useful to the rabbit but apparently not to man. About 6 percent of all babies are born without the tail being completely resorbed, so the tail is surgically removed. Reduction or loss of ancestral and vestigial structures occurs extensively during the fifth month. But other features are retained, such as the firm reflex grip of the hand that was so important to prehuman primates. Fingernails and toenails begin to form—in some babies, the nails must be cut at birth to keep them from scratching themselves. Up to this time, growth and expansion have been the most prominent changes, but now some organs (particularly the skin) begin to slough off outer layers of cells and replace these with new cells. This loss and regeneration of cells goes on throughout life, and at adulthood each person makes a complete exchange of most body cells approximately every seven years. This loss probably does not apply to the nervous system, brain, or skeleton but only to the so-called soft tissues.

Most exciting for the parents is the fact that kicking and turning of the fetus become almost constant during the fifth month, except during periods when he is known to be sleeping. The elbows, knees, and buttocks of the fetus can be identified through the stretched abdominal wall of the mother. Occasionally, the fetus will hiccough every few seconds, but such spells do not last long. Some mothers schedule their periods of sleep and rest to coincide with those of the fetus to avoid interruption of their own sleep. There is an increasing burden on the mother's lungs, heart, and kidneys, as she deals with increasing loads of fetal oxygen, nutrition, and excretion.

At the sixth month (end of the second trimester), the fetus measures about 12 to 14 inches and weighs about 1.5 pounds. Now the fetus is a

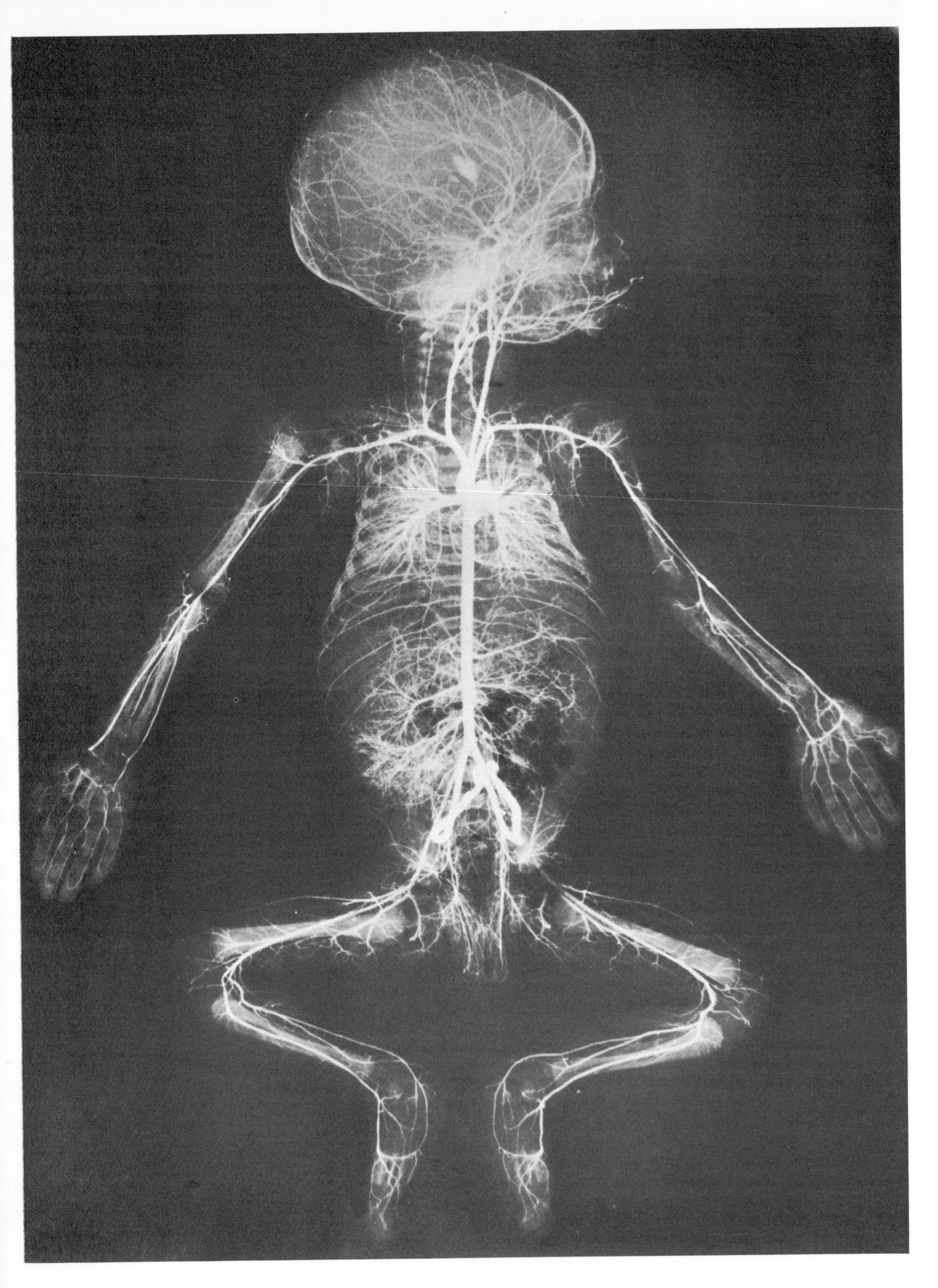

Figure 42.14. A radiograph of the circulatory system of a 7-month-old fetus, which has been injected via the umbilical artery with a microopaque solution.

well-proportioned miniature human being and may be able to survive if delivered prematurely by Caesarean section. His skin is red and wrinkled, and he looks rather like an old person. The skin of a 6-month-old fetus begins to have hair follicles, of which 500,000 will eventually form. In addition, each square inch of skin is developing 700 sweat glands, 100 sebaceous (or oil) glands, and 21,000 cells sensitive to heat, pressure, and pain. The skin is still covered by a relatively thick (1/8 inch) vernix caseosa, which protects the active fetus from bruising. The digestive tract (gut) becomes filled with dead cells from its lining, plus a greenish secretion from the gall bladder (bile). This material remains in the gut until shortly after birth. Reflex activity extends to the foot, where the plantar reflex develops (a downward flexion following tickling).

While the skeleton has been developing, some of the 222 bones develop directly, without going through a cartilage phase. Skeletal formation is very active during the sixth month, and the mother must take in more calcium than she needs for herself in order to provide adequately for the mineralization of the fetal skeleton. The fetal bones are only 12 percent calcium, but this amount will increase to 90 percent in the adult.

THE THIRD TRIMESTER

The third trimester is the hardest for the pregnant woman, because of the added weight of the fetus, the increased pressure on her own body organs, and the increased demands the fast-growing baby makes on her system. She must breathe, digest, excrete, and circulate blood for two individuals. The mother's blood volume at 7 months is 30 percent greater than normal, and her blood pressure and circulatory rate also are above normal. About 16 percent of the mother's blood is in her uterus and the placenta. The fetal heart beats vigorously now, but the mother's heart works harder than ever to keep her circulation and that of the placenta flowing. Breathing may become somewhat difficult because of pressure against the lungs, until the process of "lightening" occurs several weeks before delivery. This process involves a change of position by the fetus, moving head downward and low in the pelvis. Lightening is a great relief to the mother, particularly through relaxation of pressure of her diaphragm.

The third-trimester fetus may survive outside the uterus if it is removed by Caesarean section. Although only 10 percent of those delivered during the seventh month survive, the survival probability increases to 70 percent in the eighth month and 95 percent in the ninth month. In the last trimester, the fetus must obtain large amounts of calcium, iron, and nitrogen through the foods that the mother eats. During this period, 84 percent of the calcium that the mother consumes goes into the fetal skeleton. About 85 percent of the iron that she takes in goes into the hemoglobin of the fetal bloodstream. Nitrogen is needed as a major constituent of the many proteins being synthesized as the fetal nervous system and brain complete the final, rapid stages of their growth. The great importance of a proper maternal diet during the final trimester has been recognized only recently. It now appears that the tendency of persons from low socioeconomic classes to have lower than average scores on various kinds of intelligence tests may be more closely related to maternal diet during the final trimester of pregnancy than to either genetic inheritance or postnatal education.

Antibodies of the immune systems can be transmitted through the placenta between the maternal and fetal blood systems during the last trimester. Almost any infection or disease contracted by the mother—measles, mumps, whooping cough, scarlet fever, colds, or influenza—will cause her to develop antibodies against the toxins of the disease. The antibodies pass to the fetal bloodstream, giving the child the same immunity to those diseases for about 6 months after birth. This exchange of antibodies can have undesirable effects in rare cases. If the mother has Rh^- blood and her husband has Rh^+ blood, the baby may inherit the Rh^+ factor from his father. Some of the Rh^+ antigens may enter the mother's bloodstream during birth (they normally cannot get across the placenta during pregnancy). The mother then may form antibodies against the Rh^+ factor. If later fetuses have Rh^+ blood, the antibodies from the maternal bloodstream may enter the fetal bloodstream and begin to immobilize the red blood cells of the fetus, causing erthroblastosis fetalis. Thus, second or later children of such a marriage have been stillborn in about 35 percent of the cases. A recently developed technique for transfusing blood of a certain type into the fetus during weeks 30 through 37 has saved most of the "Rh babies" that would otherwise have died. An even more recent discovery is a drug called Rhogam, which can be used to prevent the mother's antibodies from harming future fetuses.

During the ninth month, the fetus often seems less active than in earlier months, because his own growth has caused him to become confined in a space that permits very little movement. By this time, the uterus may have stretched to 60 times its original size, reaching a volume of about 4 liters. Yet the smooth muscles of the uterus wall, after the baby is born, are able to contract and to return the uterus to its prepregnancy volume, about the size of an orange.

LABOR AND DELIVERY

It is possible that the placenta has a major role in determining the date of final delivery of the child. As early as the seventh month, the placenta begins to check its growth and activity. Before birth, it shows areas of degeneration and breakdown of its circulatory bed. This change seems to force the child to seek a new environment, usually occurring 256 to 276 days from the time of conception, or about 280 days from the beginning of the last menstrual period. Although 75 percent of babies are born within this period, prematurity is not abnormal, particularly for third or later births. In a first-born, prematurity may be related to alcoholism, smoking, or failure of the cervix to retain the baby in the uterus.

During the ninth month, the rate of growth and development of the fetus slows. If he continued at the early growth rate after birth, he would weigh 200 pounds at his first birthday. Fortunately, there is some regulating mechanism that limits growth to certain heights and weights.

The birth of the child is anticipated by certain events collectively known as labor and delivery, which is probably the single most critical period in the life of the child, as he emerges from a warm and watery chamber into a cold, airy, and unlimited environment.

Labor is divided into three steps. *Dilatation* (lasting 2 to 16 hours) begins with the onset of labor contractions and ends with the full expansion of the cervical canal. The contractions usually last for 25 to 30 seconds and come at intervals of about 15 to 20 minutes. Toward the end of this stage is the period of "crowning," when the baby's head makes its appearance through the cervix and vaginal opening. The second or *expulsive stage* lasts from 2

Figure 42.15. The birth of a child. *(From Rugh and Shettles, 1971)*

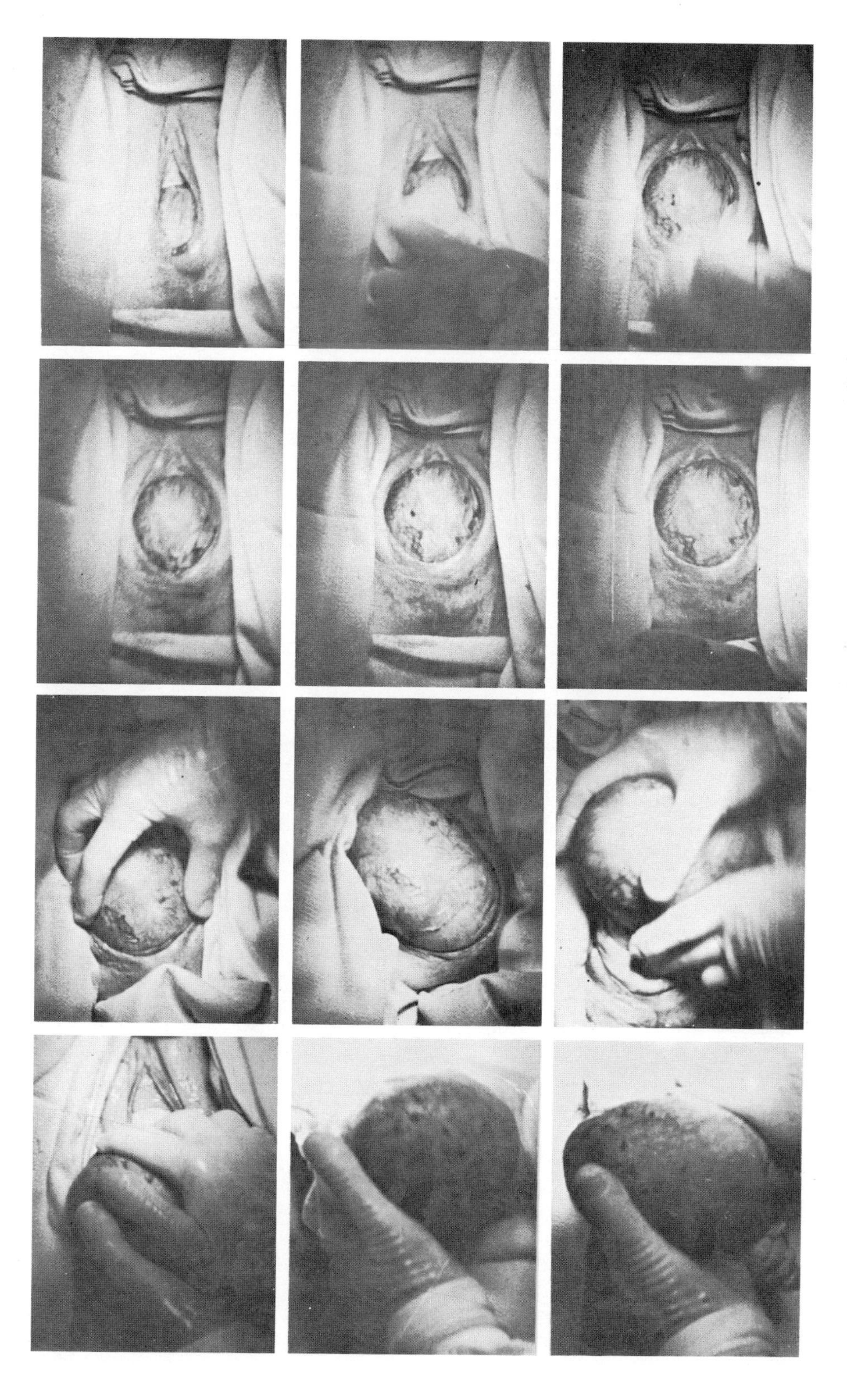

Figure 42.16. Fraternal twin fetuses. When double **ovulation** occurs and these ova are fertilized by two **spermatozoa**, the two fetuses share the uterus during **pregnancy**. Because fraternal twins are independent of **each other**, each develops a placenta. *(From Rugh and Shettles, 1971)*

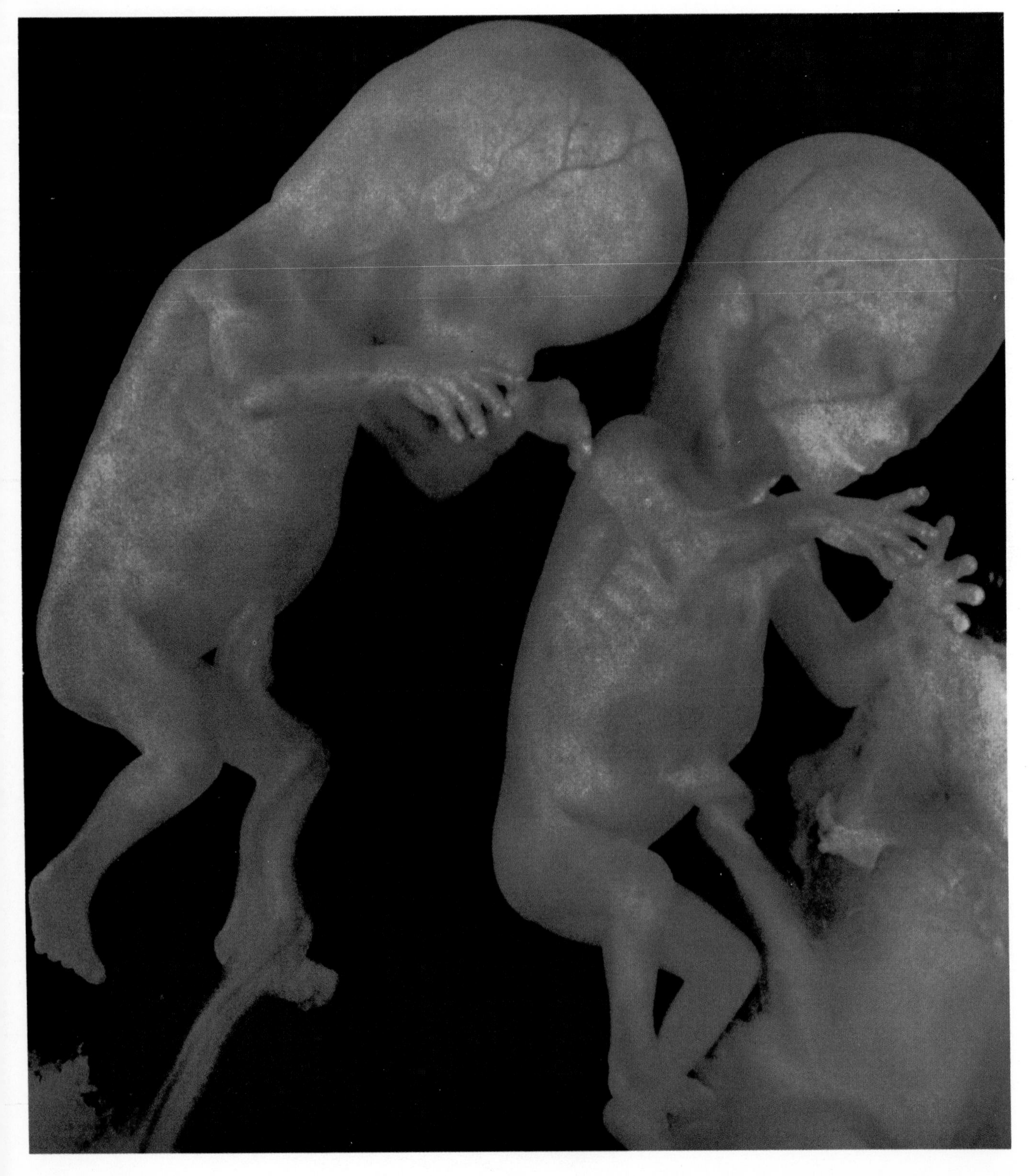

Figure 42.18 (upper left). Representative samples of various contraceptive devices presently available.

Figure 42.19 (upper right). An IUD in place.

Figure 42.20 (lower left). IUD percentage rates. Data from 27,000 women confirms that expulsion is the major drawback of intrauterine devices, particularly in the first two years.

Figure 42.21 (lower right). Oral contraceptive percentage rates compared with other means of contraception.

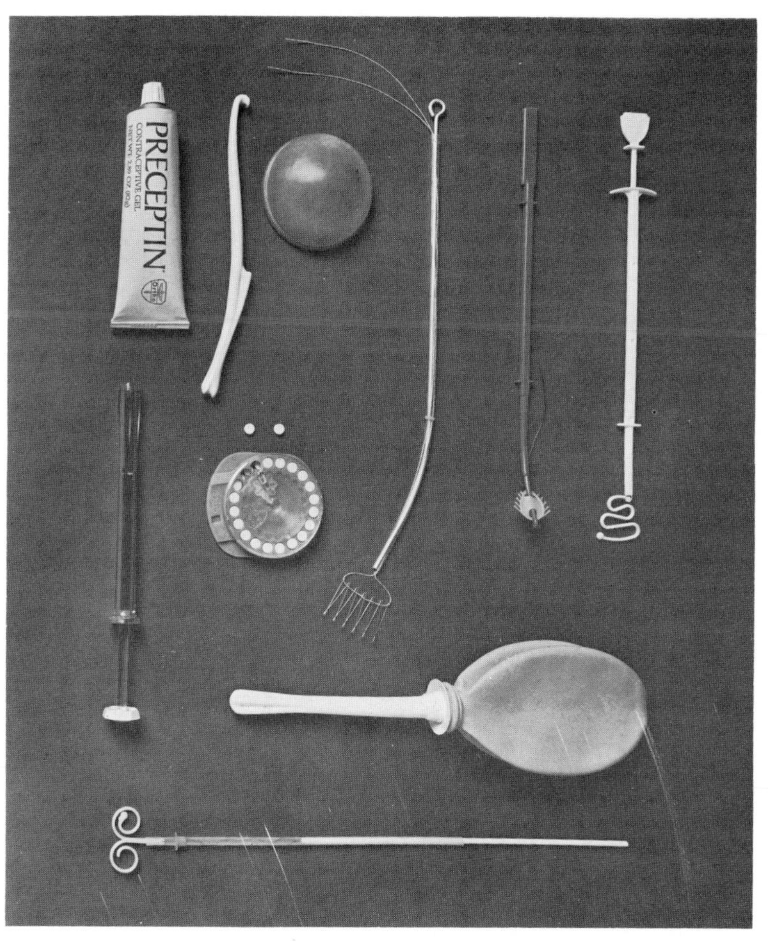

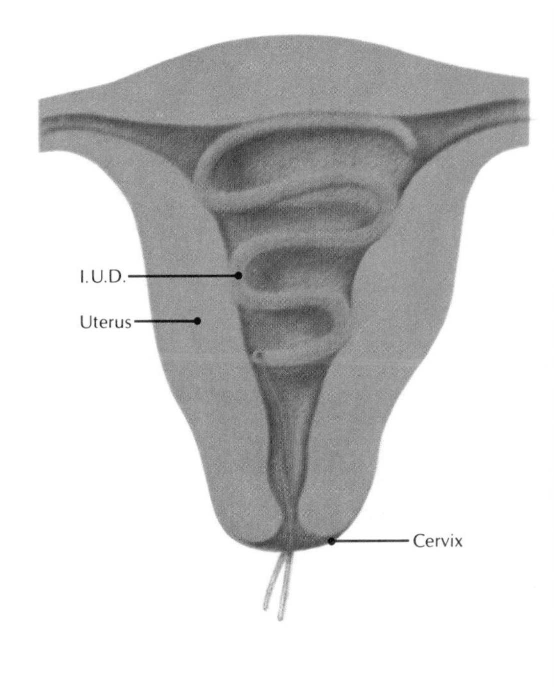

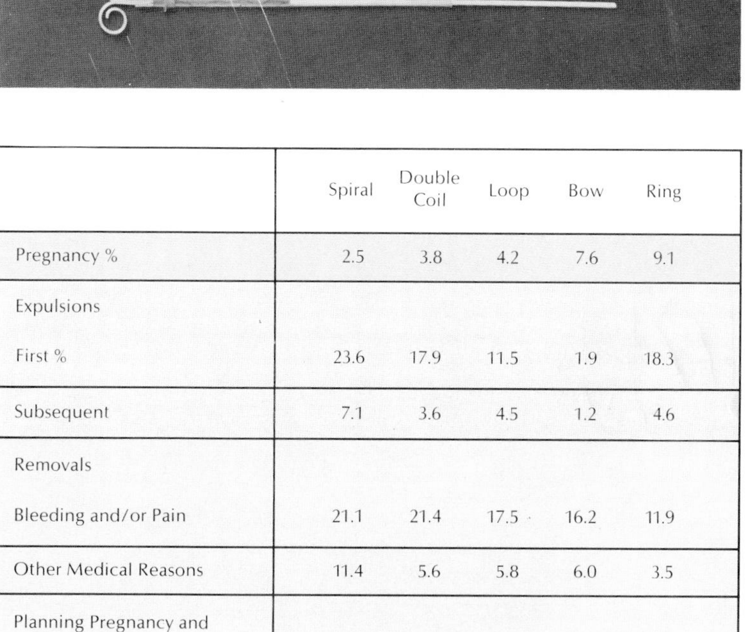

	Spiral	Double Coil	Loop	Bow	Ring
Pregnancy %	2.5	3.8	4.2	7.6	9.1
Expulsions					
First %	23.6	17.9	11.5	1.9	18.3
Subsequent	7.1	3.6	4.5	1.2	4.6
Removals					
Bleeding and/or Pain	21.1	21.4	17.5	16.2	11.9
Other Medical Reasons	11.4	5.6	5.8	6.0	3.5
Planning Pregnancy and Other Personal Reasons	8.9	4.8	6.7	6.0	7.6

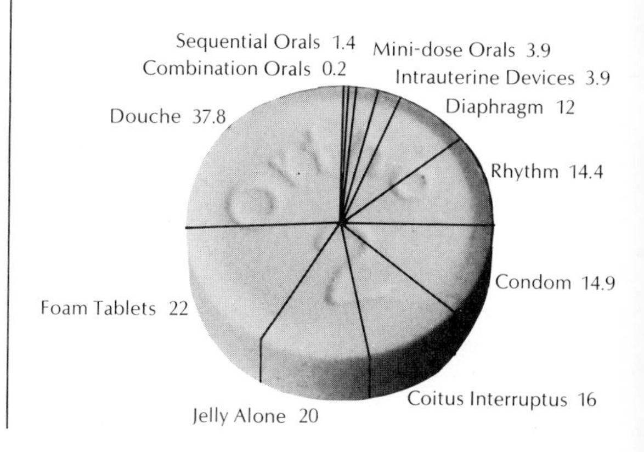

Figure 42.22. Diagrams showing insertion and placement of the diaphragm.

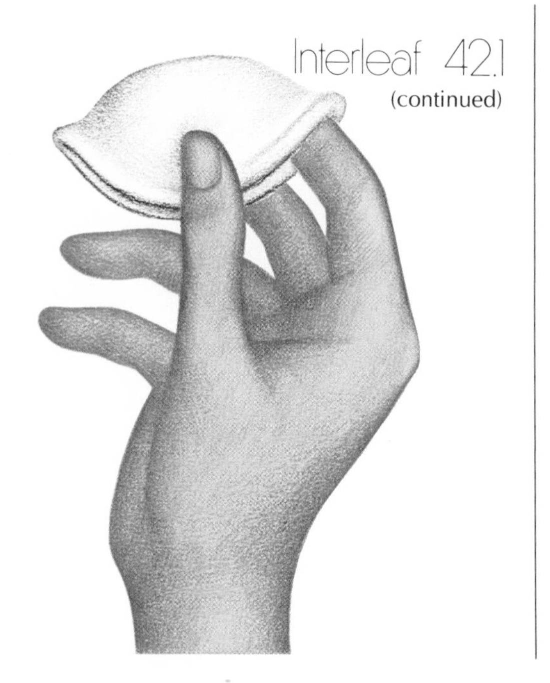

Birth-control pills contain estrogen and progesterone in relative and absolute amounts that vary from one kind of pill to another. In most forms of pills, the woman takes one pill each day, starting on the fifth day after the beginning of her menstrual period and continuing for 20 days. She then stops taking the pills until menstruation begins again. The pills alter the hormonal balance of the blood, preventing ovulation from occurring. However, if the woman is to remain healthy, it is necessary to allow menstruation to occur each month. One major disadvantage of the pills is that the woman must remember to take one each day during the proper part of her cycle. If the pills are taken properly, the risk of pregnancy is only a small percentage or less.

There has been controversy recently about evidence that use of birth-control pills increases the chance of blood clotting, which may lead to death in some cases. The extent of this risk is not yet determined to the satisfaction of all investigators, but it appears quite small. About 22 of every 1,000 pregnancies result in death of the mother, so the risk of using pills may be much smaller than the risk of getting pregnant.

The pills also cause other unpleasant symptoms in some women, such as nausea, swelling of the breasts or other parts of the body, headaches, allergic rashes, and depression or irritability.

Until development of the IUD and pills, the most common contraceptive devices were the condom and the diaphragm. The condom, or rubber, is simply a thin, balloonlike rubber sheath that fits over the penis and traps the semen. Its disadvantages include a tendency to develop leaks or tears (particularly if lubricated with petroleum jelly, which dissolves the rubber), a dulling of pleasurable sensations for both partners, and the need to stop foreplay to put on the device after erection is achieved. The best condoms offer only about 85 percent reliability. The diaphragm is a dome-

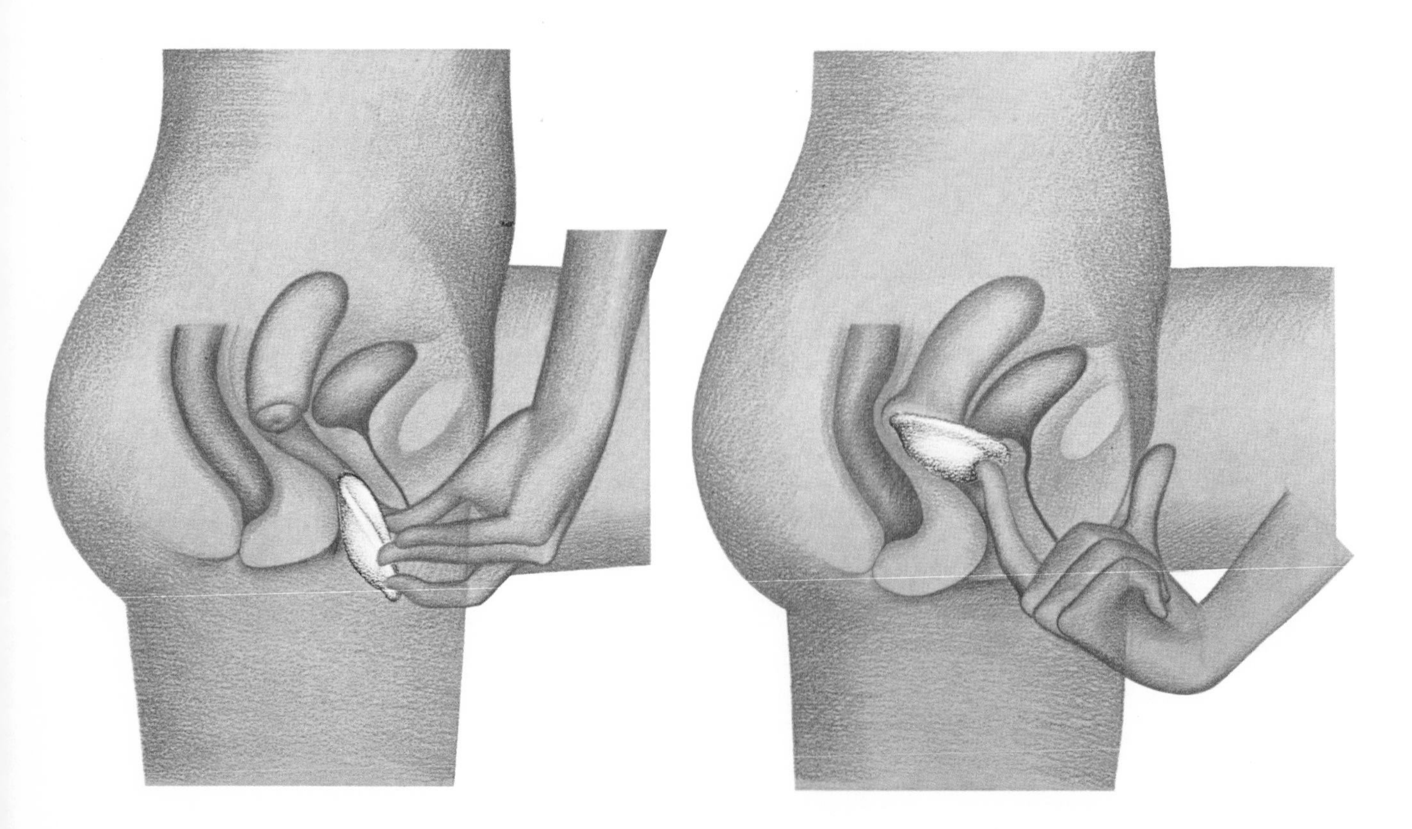

shape membrane of rubber with a metal ring around the edge, shaped to fit over the cervix and block the passage of sperm. A spermicidal jelly is used to block any leaks around the diaphragm and to help hold it in place. It is about 90 percent effective and almost completely safe. It also has the advantage of being inserted before intercourse if desired. On the other hand, the diaphragm can slip out of place or develop tiny leaks.

Other methods of contraception are rather ineffective, but probably slightly better than nothing. Spermicidal suppositories or aerosol foams may be inserted in the vagina before coitus, or a fluid may be used to wash out the vagina after coitus. The rhythm method is based on abstinence during the period when ovulation is most likely to occur. Unfortunately, ovulation can occur at any time during the menstrual cycle (even during menstruation) on some occasions, so that the rhythm method usually proves little better than luck. Withdrawal just before ejaculation is often used to avoid conception but is also very unreliable. Not only does it require good timing and strong willpower, but many active sperm may be released in drops of fluid that ooze from the penis long before ejaculation.

Many new methods of birth control are under experimental study at the present time, including a pill that can be taken after fertilization and will prevent implantation (the "morning-after" pill), a hormone that can be administered in monthly injections, a form of hormonal contraceptive that can be implanted under the skin and will remain effective for months or years, and various forms of semipermanent but reversible sterilization. A number of forms of male contraceptives are also being developed, but men seem to have little motivation for regular use of contraceptive devices or medications. Probably none of these new contraceptive methods will be generally available for several years.

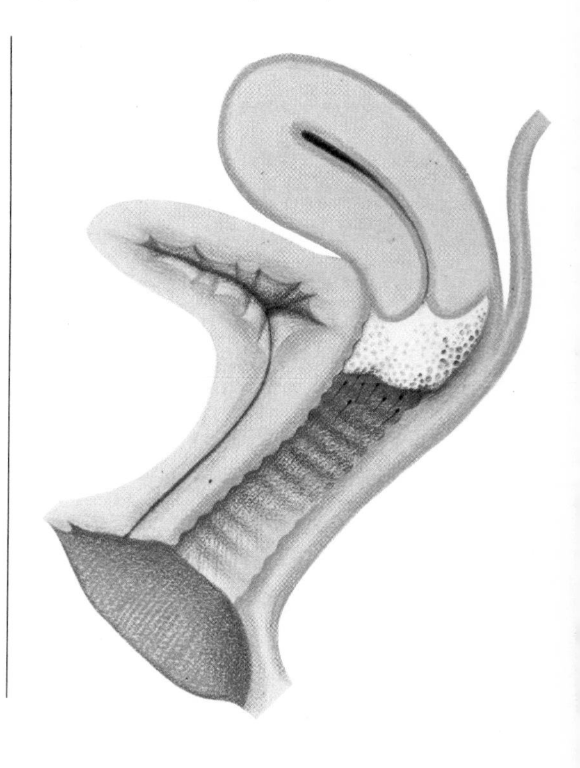

Figure 42.23 (upper right). Diagram illustrating the mechanical barrier of vaginal foam placed over the cervix.

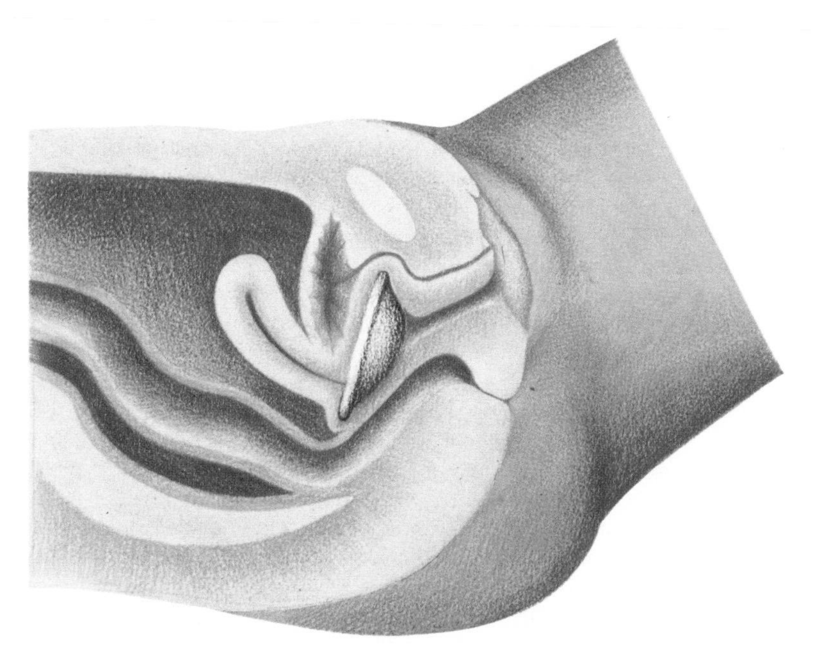

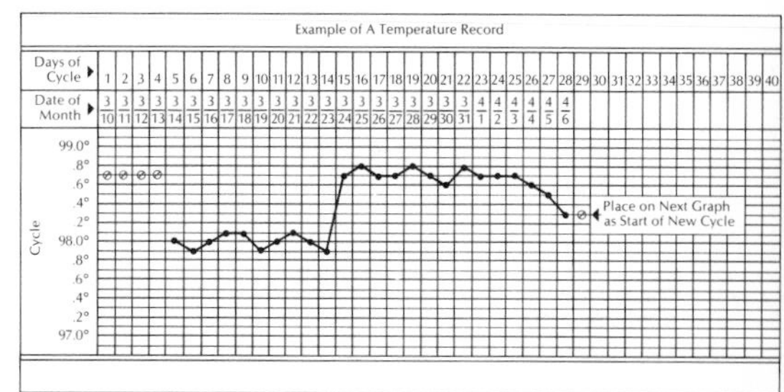

Figure 42.24 (lower right). Basal temperature record in graph form showing a typical 28-day menstrual cycle. The rise in temperature following ovulation (day 15 in this graph) is due to the thermogenic effect of progesterone secreted by the corpus luteum.

embryo or fetus that is developing abnormally for any reason—genetic or other defects—and about half of all miscarriages involve embryos with serious defects in structure or physiology. Some miscarriages in later stages of pregnancy occur because the cervix is unable to hold the fetus within the uterus. A weak cervix now can be detected long before problems occur and can be corrected surgically. There is approximately 1 miscarriage for each 4 live births in the United States.

Intentional abortions are far more common than miscarriages. Therapeutic abortions are performed in cases where the fetus is believed to be abnormal or where the birth of a child is believed to represent a serious risk to the health—mental or physical—of the mother. Laws specifying conditions under which therapeutic abortions can be performed vary from country to country and from state to state within the United States. Many states are in the process of revising their abortion laws, in most cases widening the range of circumstances under which abortions can be performed. Abortions performed illegally also are extremely common—estimates run as high as 1 illegal abortion for each 2 live births.

Most therapeutic abortions are performed during the first trimester, using a surgical procedure called dilatation and curettage, or D & C—the same procedure normally used to obtain samples of the uterine lining or to remove unwanted substances or growths from the uterus. The vagina and the cervix are distended, or dilated, by use of special instruments that can be inserted and then expanded, creating an opening through which the surgeon can work. The interior of the uterus is then scraped with an instrument called a curette—a wide, smooth loop of steel attached to a special handle. The scraping dislodges the embryo, along with a certain amount of endometrial tissues and blood.

If the operation is done by a skilled surgeon under sterile conditions, complications develop in fewer than 1 of each 100 patients. With suitable pain-killing drugs, there is little discomfort associated with the operation. Bleeding soon stops and the woman recovers completely within a few days. In illegal abortions, however, there is a very serious risk of complications and of death. The abortionist may not be a skilled surgeon, the equipment used may be inappropriate, and conditions may be unsanitary. In such cases, the uterus may be punctured or infected, and serious injury or death is not uncommon.

Other techniques of abortion have been developed. One method gaining favor for use in very early stages of pregnancy involves the use of a mild vacuum or suction device to pull the early embryo from the uterus. Of the many methods for "do-it-yourself" abortion that are tried by desperate women, none is very effective and most are extremely dangerous. The risk of infection or injury to the genital tract is very high for any method that involves insertion of objects into the vagina and through the cervix. Without the special equipment of the surgeon, it is essentially impossible to insert any object into the uterus without severe injury to the woman. Any drug capable of inducing an abortion may possibly kill the mother. Falls, heavy exercise, hot baths, or lifting weights will harm the mother long before they affect the embryo. In short—no matter how desperate a woman may be—she almost always will be better off to have a baby than to try an illegal abortion.

The questions of the legality and morality of deliberate abortions pose serious problems for modern societies. Is an abortion equivalent to the murder of a totally helpless individual? Or is it a basic right of a woman to rid her body of a parasitic organism that she does not choose to support? Between such extreme views are a variety of ethical, legal, and religious positions. Similar problems are posed by techniques of birth control (Interleaf 42.1) – some of which actually involve abortion at a very early stage of pregnancy (for example, by preventing implantation). All of these problems have gained significance for modern man in view of the threat of overpopulation (Chapter 45). Although many biologists have strong opinions about these topics, the science of biology cannot contribute directly to a solution of such moral or legal issues. However, biology can reveal basic facts about the physiological processes involved and about comparable behaviors in other animal species. These facts at least provide a sound factual background upon which moral and legal decisions can be based.

FURTHER READING

For detailed information on the physiological development from the zygote through the fetus and birth, see books on human embryology by Hamilton, *et al.* (1962) and Patten (1968). Rugh and Shettles (1971) describe and illustrate development from conception through birth. There also are a variety of books designed for pregnant women that summarize facts about prenatal development. For further information about premature births, see Babson and Benson (1966) and Drillien and Ellis (1964).

Many reports and comments concerned with the continuing debate over the effects of "hallucinatory" drugs such as LSD on the embryo or fetus have appeared in *Science* from 1969 through the present. Effects of maternal nutrition on prenatal development are discussed by Collaborative Perinatal Research Project (1964).

Further information about birth is given by American Academy of Pediatrics (1964), Davis and Middleton (1968), Schwartz (1961), and Winchester (1968). For various views and facts about abortion, see the books and articles by Copeland, *et al.* (1969), Hardin (1968), Lader (1969), Roemer (1967), and Tietze and Lewit (1969).

43
Drugs and Human Behavior

The variety and quantity of drugs used in American society have increased rapidly in recent years. Although drug use often is regarded as a problem of adolescents and youth, increased drug consumption is one phenomenon that is widespread on both sides of the generation gap. Commercials in the mass media condition all age groups to regard drugs as quick and easy solutions for insomnia, depression, nervousness, and worry, as well as a wide range of more physical ailments. One recent survey of American families found an average of 30 different drugs per household, most of them being used primarily by adults (Louria, 1968). Almost every modern American probably takes some kinds of drugs in a deliberate attempt to alter his experiences chemically.

The following sections discuss some of the drugs more commonly used in the United States to modify or alter human behavior, ignoring the even greater variety of drugs used to treat various physical complaints, most of which also have effects on the nervous system. The term "drug" is used here to mean any substance that can produce a temporary change in neurophysiological functions, thoughts, feelings, or behavior when it is physically taken into the body.

DRUG EFFECTS

Behavior alteration associated with drug use results from the interplay of both drug and nondrug factors. Drug factors include pharmacological properties, dosages, methods of use (ingestion, injection, or others), frequency of use, and long-term cumulative effects. Besides biological factors, nondrug factors, which include various psychosocial interactions, also are important determinants of drug effects. The user's expectations of what will happen when he consumes a particular drug are crucial elements in these psychosocial interactions. His expectations are determined by his own personality, recent events in his life, and the physical, interpersonal, and social environment in which he uses the drug.

As dosages are increased, drug factors predominate, and psychosocial factors become relatively less important. For example, with high doses of alcohol, the user becomes progressively more sedated and eventually comatose, regardless of his personality and the social setting. On the other hand, with less potent or lower dosages of drugs, nondrug factors may predominate. For example, the effects of a marijuana "high" are much easier to obtain in a social group of marijuana users than in an austere laboratory setting.

OPIATES

Heroin, morphine, methadone, and opium are members of a group of drugs called opiate alkaloids. To drug users, the opiates are known as hard stuff or heavy drugs. These drugs usually are classified as narcotics, and they are used widely in both therapeutic and illicit settings. Heroin commonly is injected or sniffed, opium smoked or ingested, and morphine injected. Among many slang names, the most common are black (opium), white stuff (morphine), and H, horse, junk, smack, or stuff (heroin). A user may purchase drugs from a dealer or be tricked into an attempted purchase from an undercover detective (narc). He may purchase a small packet of drugs (a balloon, bundle, deck, piece, or nickel or dime bag) or capsules (caps). He may "get burned" if the dealer takes his money and delivers fake drugs or nothing at all. He may begin by sniffing or taking an occasional injection,

Figure 43.1. Heroin advertised as safe medicine in 1900. In 1874, a chemical modification of morphine was developed, but there was little interest in the new compound called heroin until 1890, when it was proposed as a cure for narcotic addiction. Even more amazing is the fact that the Bayer Company of Germany was widely marketing heroin as a sedative for coughs, placing it in the same category as aspirin. This advertisement appeared in U.S. pharmacy journals as late as 1900.

BAYER
PHARMACEUTICAL PRODUCTS.
We are now sending to Physicians throughout the United States literature and samples of
ASPIRIN
The substitute for the Salicylates, agreeable of taste, free from unpleasant after-effects.
HEROIN
The Sedative for Coughs,
HEROIN HYDROCHLORIDE
Its water-soluble salt,
You will have call for them. Order a supply from your jobber.
Write for literature to
FARBENFABRIKEN OF ELBERFELD CO.
40 Stone Street, New York,
SELLING AGENTS.

but he usually soon graduates to subcutaneous injections (skin popping) to support a small, irregular habit. Tolerance effects soon lead to intravenous injection (mainlining) and heavy addiction (being hooked). The heavy addict usually is forced to steal in order to obtain the $50 or more per day needed to buy drugs. He worries about hiding his supply of drugs (stash), hypodermic equipment (outfit), and the needle marks (tracks) on his arms or legs. Injection of the drug (shooting up) with a hypodermic needle usually leads to a rush of pleasurable sensation (flash), but may cause a severe reaction or death (overdose, or OD). Most of these slang terms originated in black ghettos several decades ago and have now spread through the youth subculture.

Acute pharmacological effects of the opiates include relief of tension and anxiety, decrease in physical drive, increased drowsiness, and relief from pain. Some users experience a sense of well-being and euphoria, but other individuals—particularly if they are not anxious or not in pain—experience distinct unpleasantness after taking opiates (Beecher, 1959). In a therapeutic setting, opiates can be effective in pain relief, sedation, diarrhea control, and cough suppression. The biochemical mechanisms by which opiates induce these effects are unknown.

Chronic use of opiates often leads to the classical addiction syndrome. *Tolerance* is a common feature of this syndrome—in other words, a given dose no longer produces the desired effect, so that higher doses are taken to renew this effect. This pharmacological principle is utilized by addicts in their often futile attempts at "chipping"—that is, using low doses infrequently to avoid becoming "hooked." Too often, however, the intervals shorten, the dosages increase, and the increments of tolerance accumulate.

When an individual has become accustomed to opiates, he must maintain a certain blood level or suffer withdrawal symptoms. These unpleasant symptoms include weakness, muscular aches, vomiting, diarrhea, cold sweats, chills, and goosebumps. While the addict is on his habit, he is strongly motivated to secure a supply of opiates to avoid these unpleasant symptoms.

Compulsive drug seeking, or *psychological dependence*, another facet of the addictive syndrome, is pronounced in most opium addicts. Opiates are extremely effective in reducing distress. Because of this pharmacological effectiveness, an individual quickly learns through operant and classical conditioning to associate opiate use with replacement of distress by pleasant feelings. Even after years of enforced abstinence, the former user—when confronted by distressing conditions similar to those in which he used drugs in the past—is impelled to repeat the pattern of drug use that is associated with abrupt change from distress to well-being.

In the last few years, there has been a significant increase in heroin use among middle-class young people. Most of them commonly use a variety of mind-altering substances, regardless of legal restraints. These people know the hazards of the opiates, but they think they are invulnerable. Future studies will be necessary to determine if these youthful opiate addicts from privileged backgrounds will experience the difficulty in staying off the habit that traditionally has been part of opiate addiction.

Recently, some encouraging treatments for opiate addicts have been developed, including long-term maintenance of the addict on methadone. Methadone, a long-acting opiate, decreases the user's craving for heroin, while allowing him to function almost normally in personal, social, and

Figure 43.2 (left). Nineteenth-century ad for Mrs. Winslow's Soothing Syrup. Influences that helped to increase the magnitude of the addiction problem were the unreserved medical use of opium and the increase of self-medication with opium preparations. Tucked away in many of the most widely promoted and most respected patent medicines were opium and morphine salts. Nostrums such as Ayer's Cherry Pectoral, Jayne's Expectorant, Pierce's Golden Medical Discovery, and Mrs. Winslow's Soothing Syrup are advertised here on a trade card.

Figure 43.3 (right). Opium addiction problem in nineteenth-century China. Opium addicts were not numerous until an epidemic of opium smoking spread throughout China. The emperors of China issued edict after edict prohibiting the use and importation of opium, but they were unable to enforce these measures. This Daumier caricature of 1859 depicts the British, who were the largest importers of opium, pouring it down the throats of the Chinese.

Figure 43.4 (upper left). Early hypodermic syringe, 1860. Morphine, the most active ingredient of opium, was discovered in 1805 by a German apothecary named Frederich Serturner. With the invention of the hypodermic syringe in the 1850s, a new method of administration was employed, increasing the addiction problem. Shown here is one of the earliest hypodermic syringes known.

Figure 43.5 (below left). Opium pipes. The first, widespread illicit use of opium began when people started smoking it in a pipe (as seen here). The gum opium is rolled into small pills and placed in the bowl of the pipe. As the pill burns, it gives off a sickening, sweet odor. The fumes are drawn into the lungs of the smoker through the stem of the pipe. Once used widely by the Chinese, the smoking of opium in this fashion has nearly disappeared today.

Figure 43.6 (upper right). Heroin administration. "Shooting up" has become the most common method of taking heroin.

Figure 43.7 (lower right). Opium smoking.

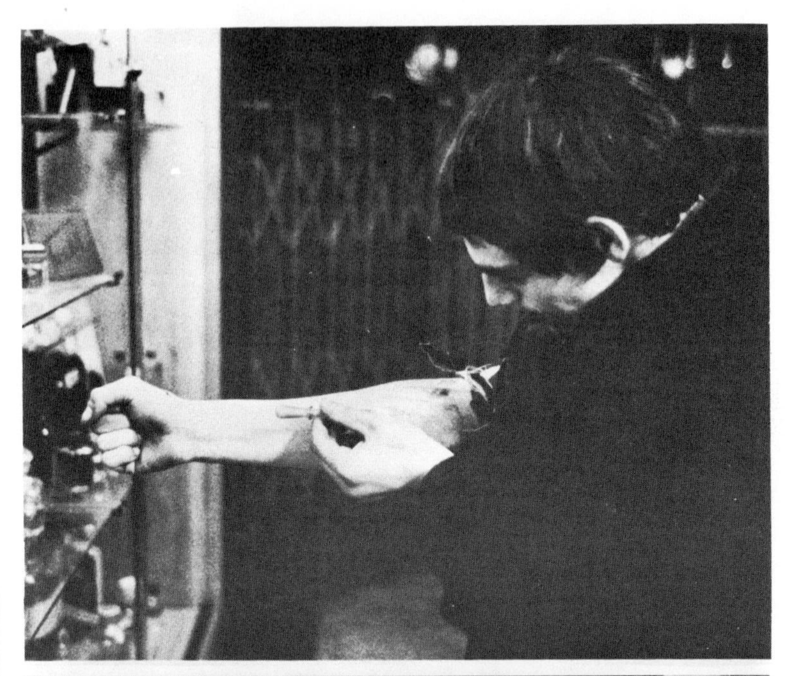

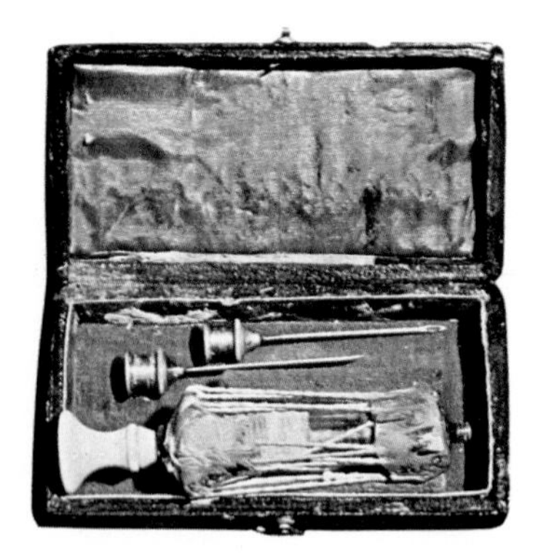

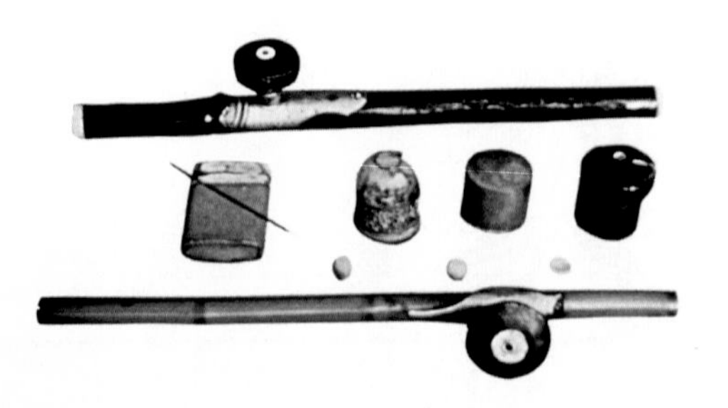

vocational spheres. These new developments for the treatment of opiate addiction are being greeted with cautious enthusiasm because of the known time lag between use of new drugs and recognition of adverse effects.

AMPHETAMINES

The amphetamines are a group of synthetic chemicals classified pharmacologically as stimulants of the central nervous system. Their acute effects have been extensively investigated in a number of well-designed studies. Because amphetamines have been used widely in both clinical and illegal settings since the 1930s, certain long-term effects also are known.

Amphetamines include several different drugs. The names of the drugs can be confusing because each drug has several "proper" names and because slang terminology changes far too rapidly to be recorded accurately in any book. Each particular drug has a full chemical name (such as α-methylphenethylamine), a simplified chemical name (such as amphetamine sulfate), and a variety of brand names (such as Benzedrine), as well as the slang names that vary from place to place and from time to time.

The amphetamines include amphetamine sulfate (Benzedrine, bennies, beans, cartwheels), dextroamphetamine sulfate (Dexedrine, dexies), and methamphetamine sulfate (Methedrine, Desoxyn, crank, crystal, meth). Amphetamines generally are known as speed, uppers, pep pills, whites, or crank. Their chemical structure is similar to that of epinephrine (Adrenalin), the hormone released by the adrenal gland at times of fright. Either amphetamine or epinephrine in the bloodstream gives a sudden burst of energy, overcoming effects of fatigue. The heart pounds rapidly, there is pallor and sweating, and the body in general is mobilized for rapid muscular action (the "fight or flight response").

Although the behavioral effects of the different amphetamines are not identical, they are sufficiently similar to permit discussion of them as a single kind of drug. In general, they produce a more sustained and less extreme version of the behavioral effects of epinephrine. Amphetamines usually are taken orally as pills, but the powdered form or its solution may be injected into veins or muscle tissue, sniffed through the nasal membranes, or absorbed through the skin.

In low dosages, amphetamines produce a general increase in alertness, a sense of well-being, wakefulness, and decreased feelings of fatigue. They can produce significant improvement in activities requiring extreme physical effort and endurance, such as athletic or military maneuvers. Amphetamines tend to shorten reaction time, improve motor coordination, and enhance the ability to make simple responses to data displays. These effects seem to represent actual improvement above normal levels in certain tasks that do not require complex cognitive skills. On the other hand, normal subjects on low amphetamine dosages show no modification in intellectual functions that involve comprehension, problem solving, and judgment (Weiss and Laties, 1962). Similar findings have been made for a variety of drugs. Most drugs seem ineffective for the enhancement of performance on more complex cognitive and creative tasks. However, amphetamines definitely can counteract the inhibiting effects of fatigue, boredom, and other factors on performance of a wide variety of tasks.

Other well-documented effects of amphetamines in low dosages include suppressed appetite, increased heart rate and blood pressure, and altered sleep patterns. The exact mechanisms of amphetamine action are unknown, but there is definite stimulation of the cerebrum, particularly the sensory cortex. Part of the effect is probably indirect—amphetamines apparently induce the release of norepinephrine and possibly other stimulating chemicals within the body.

The amphetamines have demonstrated their effectiveness in the control of certain pathologically hyperactive, usually brain-damaged children. Unfortunately, at this time there is also indiscriminate use of the amphetamines for many other types of childhood disturbances despite the lack of evidence supporting their efficacy. Amphetamines are also used for a number of other clinical problems including treatment of narcolepsy (a rare condition that includes the compulsive tendency to fall asleep), obesity,

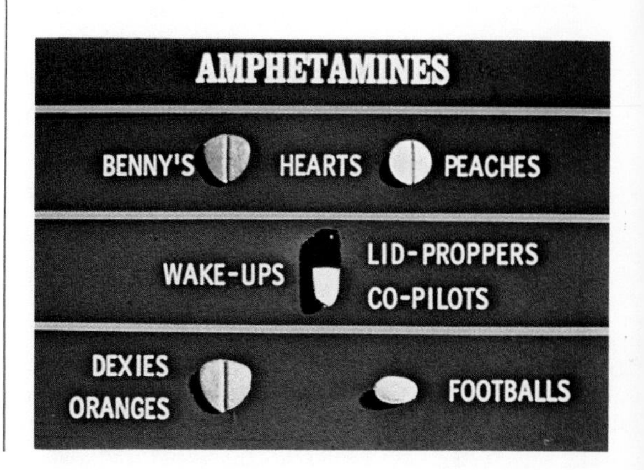

Figure 43.8. Amphetamines.

Figure 43.9. Stimulants. The most powerful central nervous system stimulants are the amphetamines.

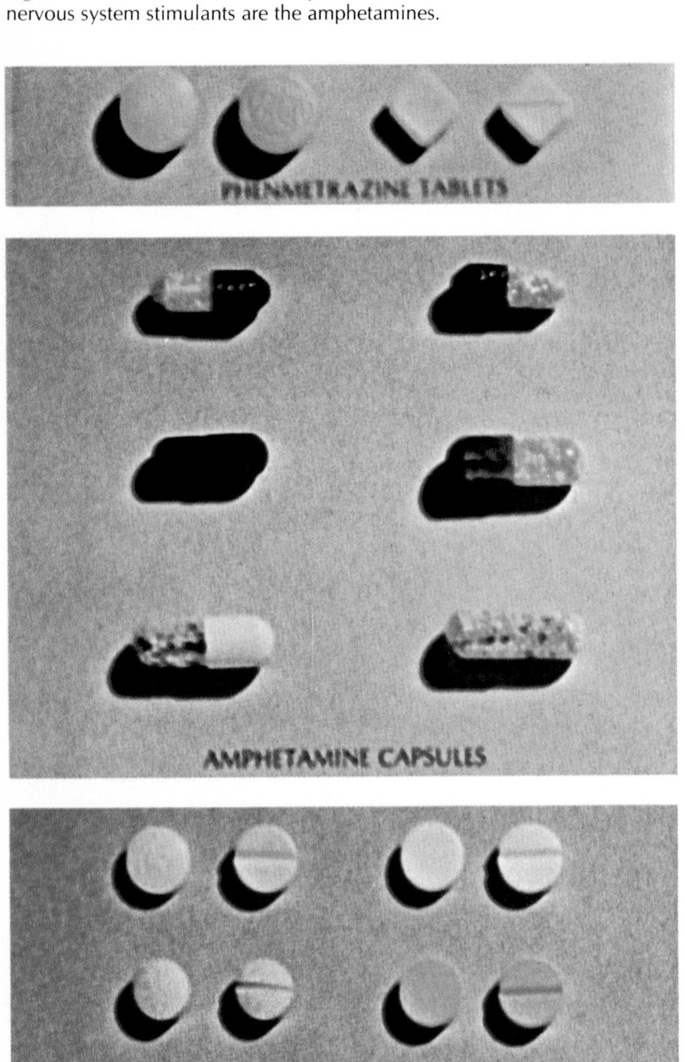

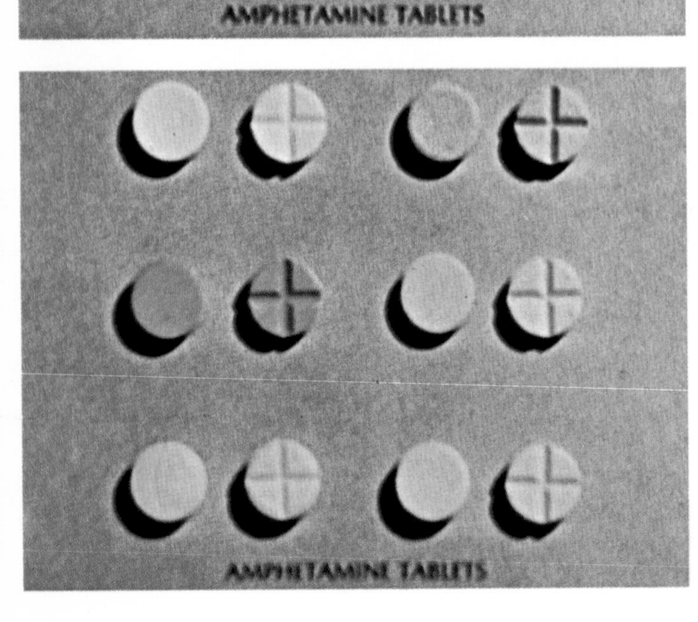

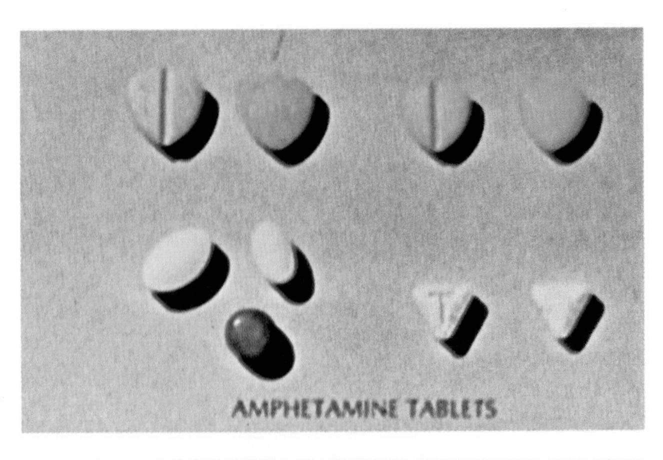

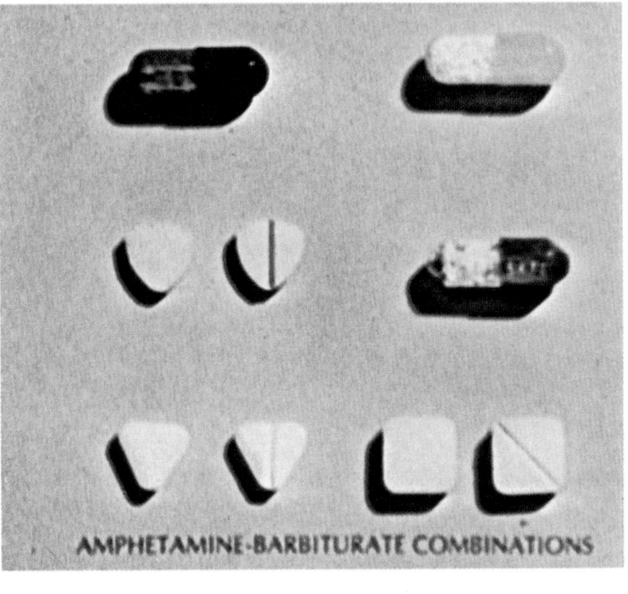

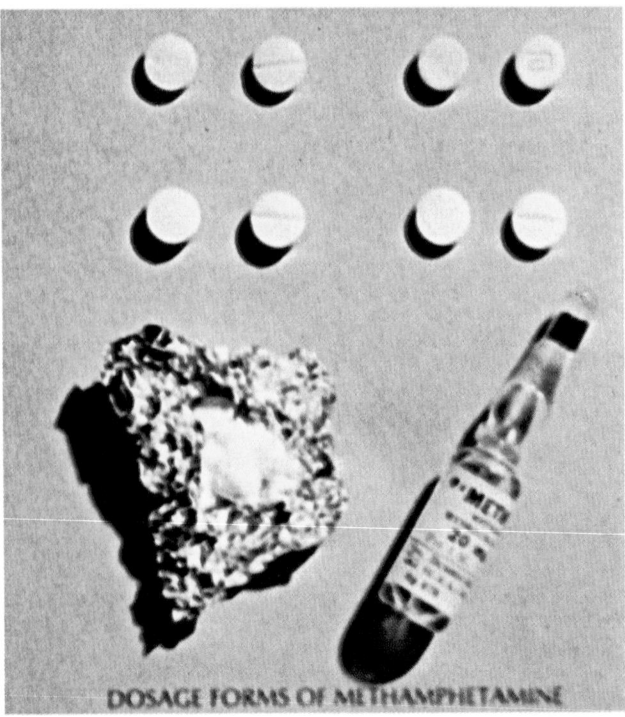

enuresis (bed wetting), fatigue, and minor depression. For these conditions, medical researchers have increasing doubts about the long-term efficacy of amphetamines. Other drugs may be more beneficial with fewer risks of toxicity.

One common pattern of amphetamine use in nonmedical settings involves taking low oral doses for limited time periods. This pattern is exemplified by students cramming for examinations, overtired businessmen, military personnel on extended missions, and long-distance truck drivers. In general, this pattern seems relatively benign, although instances of unfavorable effects involving bad judgment have occurred.

A second pattern of amphetamine use involves oral ingestion of moderate doses over prolonged periods of time. This pattern is followed by people who are trying to control their weight and by individuals who are chronically fatigued or depressed and are attracted by the activating, euphoric properties of the amphetamines. However, with time, tolerance develops and the same amount of amphetamine no longer induces the desired effects. Thus, the dosage may be gradually increased. This pattern of amphetamine use can lead to severe paranoid psychoses, manifested by unfounded suspiciousness, hostility, hallucinations, and persecutory delusions (Connell, 1958). These paranoid reactions result from the interplay of preexisting personality trends toward paranoid behavior, reactions of significant individuals in the user's environment, and other psychosocial forces, but recent studies have confirmed the crucial contributions of drug factors in inducing psychoses (Griffith, *et al.*, 1970).

A third pattern of amphetamine use is that of intravenous injection of large doses. This pattern may involve a few injections on irregular occasions, or it may follow the now common sequence of a "run": self-administered injections every few hours for periods of several days to a week or more (Kramer, *et al.*, 1967). Immediately after each injection, the user experiences a generalized pleasurable feeling—a "flash" or "rush." He feels invigorated and perhaps euphoric for several hours. However, he then

Figure 43.10 (right). The chemical structure of amphetamines.

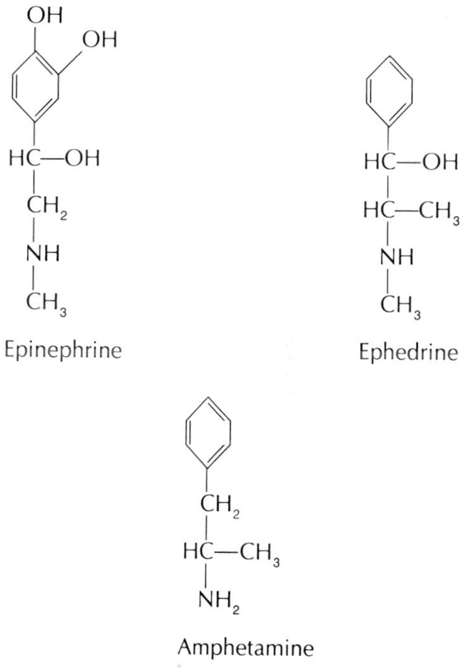

OH
OH
HC—OH
CH_2
NH
CH_3
Epinephrine

HC—OH
HC—CH_3
NH
CH_3
Ephedrine

CH_2
HC—CH_3
NH_2
Amphetamine

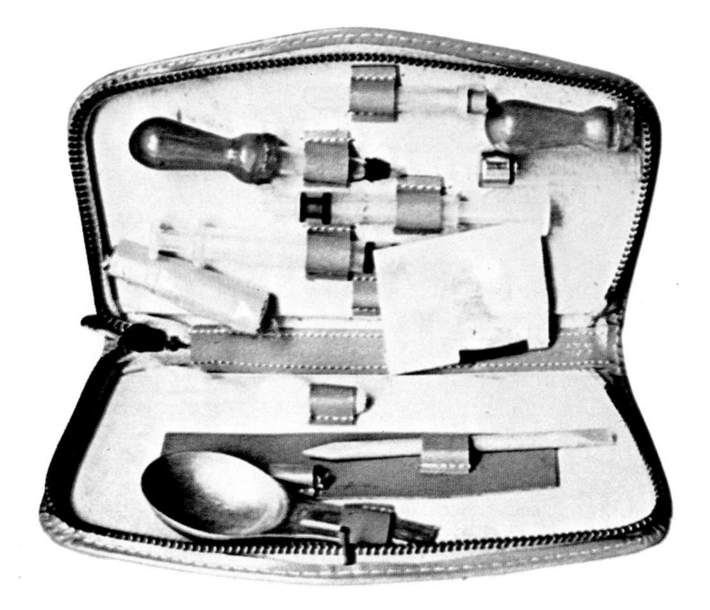

Figure 43.11 (below). Speed kit. Methamphetamine, or "speed," is most commonly "mainlined" (taken intravenously).

experiences the gradual onset of irritability, a vague uneasiness, and uncomfortable aching sensations. To avoid these discomforts and to recapture the pleasure of the initial rush, the user is compelled to take another injection. During this "speed binge," the individual often goes without sleep, adequate food, or liquid and may gradually develop paranoid behavior. When this prolonged run is finally over, there is a considerable period of discomfort, several days of hypersomnia, and an extended period of lethargy with variable amounts of depression.

In contrast to the first two patterns of illicit amphetamine use, where increased tendencies toward violence are not apparent, the chronic, compulsive, intravenous use of amphetamines is associated with inclinations toward aggression (Tinklenberg and Stillman, 1970). The pharmacological effects of amphetamines that can contribute to assaultive behavior include greater tendencies to take physical action quickly, decreased inclinations for a mediating pause to consider future consequences, exaggerated self-confidence, and a high incidence of paranoid reactions with irritability, suspiciousness, and hostility. In addition to these drug effects, certain psychosocial factors that increase the probability of violence commonly operate among intravenous amphetamine users. These factors include prevalence of concealed weapons, dominant group attitudes of distrust and suspiciousness, peer pressures that demand physical retaliation for any perceived wrongs, and the lack of social or other constraints against violence in this subculture.

Although amphetamines may enhance aggressiveness, there are well-recognized clinical uses of amphetamines that paradoxically decrease tendencies toward violence in individuals with certain brain disorders. As described earlier, amphetamines occasionally are helpful in reducing destructive outbursts in brain-damaged and hyperactive children. Thus, one pattern of amphetamine use is associated with violent, assaultive behavior, yet the very same drug given under medical supervision, usually in lower doses, can be helpful in reducing violent behavior.

Psychosocial factors also help determine sexual behavior with amphetamines. Although general stimulating properties make amphetamines feasible drugs for enhancing sexual drive, individual predispositions are more important. In general, people who are vigorously inclined toward sex in a nondrugged state are undaunted by the usual doses of amphetamines. Conversely, people who do not have strong sexual predispositions without drugs are unlikely to change with amphetamines or other psychoactive drugs. However, chronic use of amphetamines (and other drugs) often is associated with impairment of sexual behavior.

Reports indicate that in the last few years, amphetamine abuse has increased dramatically among young people in England, with no indications of any downward trend. The illicit drug-use patterns in the Haight-Ashbury area of San Francisco began before the summer of 1967, predominantly with LSD and marijuana use, with an insignificant amount of intravenous amphetamine use. By 1969 this pattern had changed to one of extensive high-dose intravenous amphetamine use. Subsequently, there is evidence that amphetamine use is declining, and many users are switching to heroin (Smith and Meyers, 1971).

Within these changing trends are a number of complex psychological and social forces that influence the development of amphetamine users. Of crucial importance is the ready availability of the drug and of people who

Figure 43.12. A photographic impression of a speed trip.

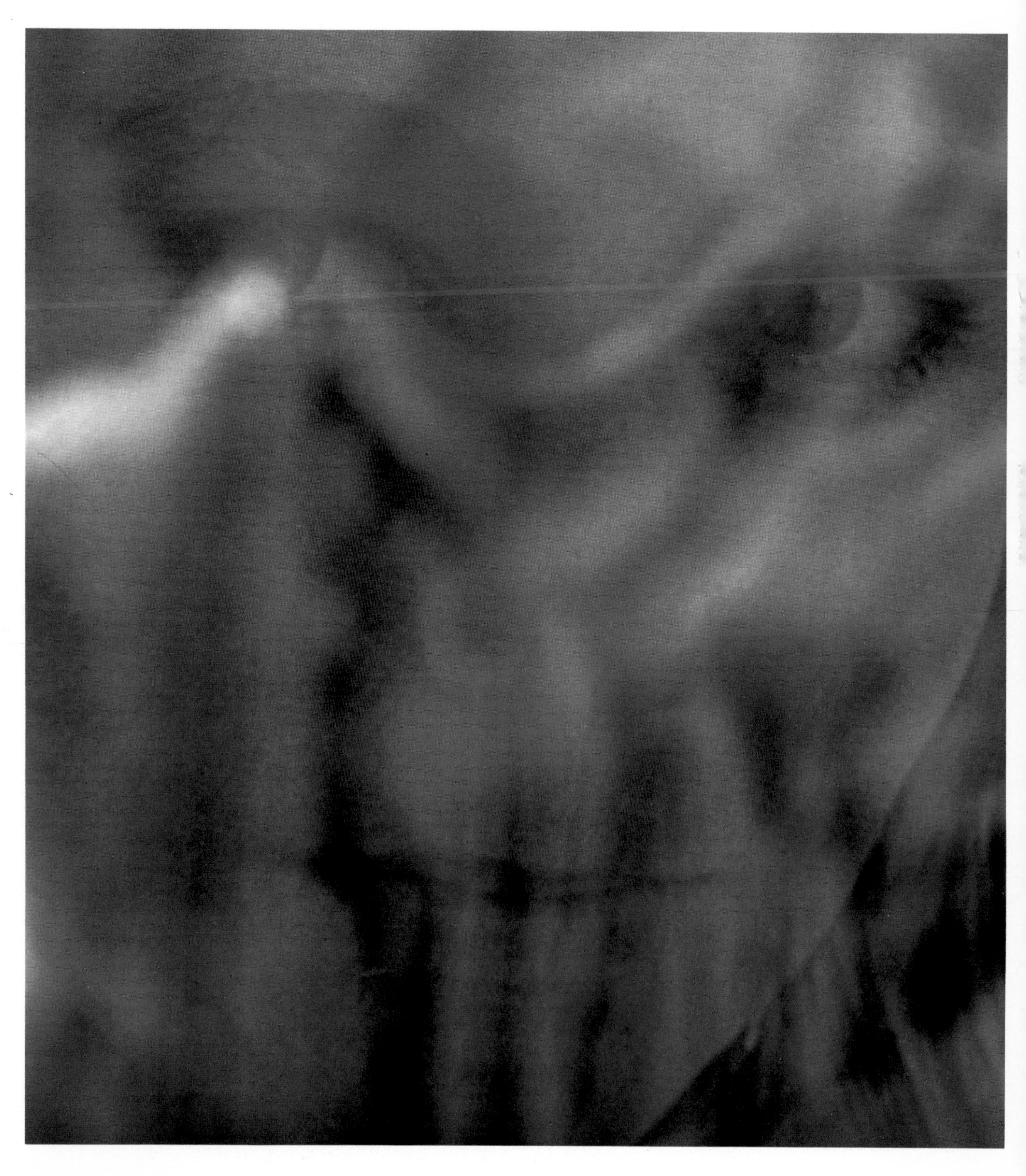

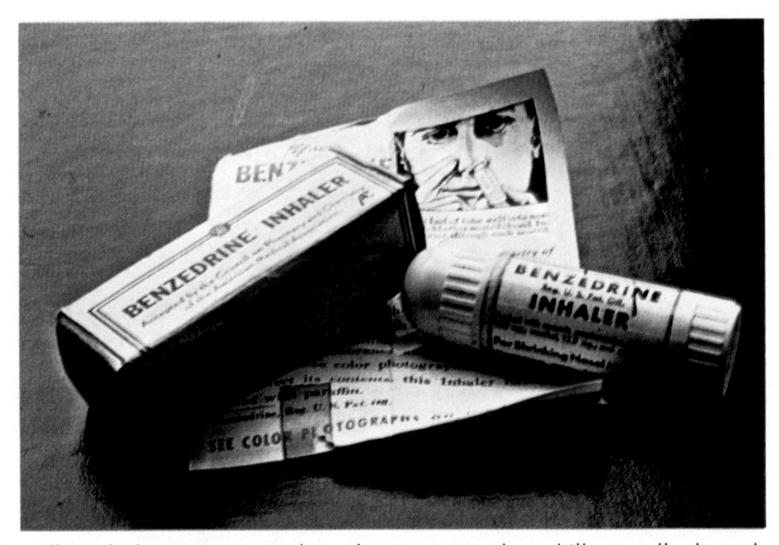

Figure 43.13. Benzedrine inhaler, 1932. Amphetamine was synthesized in 1927 by a California pharmacologist, who turned over his patent rights to Smith Kline and French Laboratories. Because the compound was found to be effective in relieving nasal congestion, SK&F introduced the benzedrine inhaler in 1932. It was soon noted that amphetamine was a stimulant, and newspaper publicity led to the abuse of amphetamine. Misuse brought amphetamine under legal control, so "bennie" users started buying up the benzedrine inhalers (which were available without a prescription). They cracked open the tube, removed the benzedrine-impregnated paper and soaked it in hot liquid to extract the amphetamine. SK&F placed an emetic in the Benzedrine impregnated paper to stop its misuse, but the "bennie-busters" soon found a new way to extract the amphetamine.

will teach the curious neophyte the necessary drug skills, usually through direct modeling. If these factors exist, a variety of individuals with disparate personality characteristics seem vulnerable to becoming chronic, compulsive amphetamine users. The belief that a particular set of personality traits predisposes one to use a particular drug does not seem altogether valid. However, after the pattern of compulsive, intravenous amphetamine use has become firmly established, regardless of predrug personality characteristics, abusers tend to decline in school or work performance and, not uncommonly, may engage in criminal activity to support their habit.

Although the term "addiction" raises formidable semantic problems, some facts about amphetamine addiction are generally accepted. Amphetamine users can develop tolerance. Although withdrawal symptoms are less pronounced than with the opiates, there is usually some physical discomfort associated with abrupt cessation of high-dose amphetamine use. Despite enforced abstinence of months or years beyond the point where physiological drug factors are likely to be operative, many users are compelled to return to amphetamines by a force comparable to that in heroin users (Kramer, *et al.*, 1967). This phenomenon illustrates that, as in the case of the "hard narcotics," physical tolerance and withdrawal are not as important in addictive patterns of behavior as are psychosocial forces, such as conditioning and response generalization. Getting an amphetamine (or heroin) addict to "kick" his habit is relatively easy; preventing him from returning to it is extremely difficult. Finally, chronic amphetamine use is associated with the most important aspect of addiction—harm to the individual and his associates. Research has established that compulsive amphetamine use gradually can destroy the personal, occupational, and social structures of the user's life to leave an existence focused only on procurement and use of the drug. This progressive exclusion of formerly meaningful activities closely resembles the classical addiction syndrome ascribed to the use of heroin and other opiates.

There is clinical evidence that intravenous amphetamine use encourages experimentation with other intravenously administered drugs including the opiates. An individual who uses amphetamines may not develop a biological deficit that can be satisfied only with drugs, but he has learned

Figure 43.14. Coca and cocaine advocated as safe drugs, 1896. Aside from caffeine, the first potent stimulant came from the land of the Incas. South American Indians used coca leaves as a stimulant by chewing little balls made from the leaf of the plant. *Erythroxylon coca* was introduced into American medicine in 1886. The alkaloid cocaine was discovered by Koller of Austria in 1884 and suggested as a local anesthetic. Coca and cocaine were added to many of the patent medicines and beverages. As late as 1896, coca leaves and cocaine were being widely advertised as safe products in ads like the one shown here.

certain drug-using patterns of behavior, which may generalize to include other drugs. The importance of classical and operant conditioning in the drug-using process is repeatedly evident. Thus, the act of using a needle to inject something into the body can acquire pleasurable features—the act of injection itself is reinforced by the pleasurable sensations that follow immediately. Any unpleasant consequences of drug taking are likely to follow much later and thus have little effect on the conditioning of the drug-taking behavior.

Another process whereby intravenous amphetamine use can lead to other drugs involves the amphetamine run. People who follow this drug-use pattern frequently try a variety of sedating agents to reduce discomfort associated with cessation of the run. They often discover, for example, that heroin not only is extremely effective in reducing the distress of "coming down" but also can produce direct euphoric effects, particularly in individuals experiencing the stress that is integral to most illicit drug use. With time, heroin use may become more frequent, and "secondary" addiction may ensue. In the first six months of 1969, over 50 percent of the cases referred to the drug-treatment program in San Francisco were former intravenous users of amphetamine who had switched to heroin.

COCAINE

Legally, cocaine is classified as a narcotic, but it only partially fits the medical definition of a narcotic in that it relieves pain but does not induce sleep and stupor. Its pharmacological effects of profound central-nervous-system stimulation and euphoria are similar to those of amphetamines. Cocaine is derived from the coca shrub of South America and was the first local anesthetic to be introduced into medical and dental practice. For centuries, this drug has been ingested by some Indian tribes of the Andes Mountains to improve existence under extremely harsh conditions. Cocaine is rarely used medically for injection because its intense stimulation of the central nervous system can lead in large doses to acute paranoia and in some cases to sudden death from heart or respiratory failure. It also has been abandoned as a local anesthetic because of its dangerous side effects, which include damage to the tissues that it contacts, severe headaches, anxiety, and dizziness.

Among illicit drug users, cocaine is known as coke, snow, she, or girl. Traditionally, it was sniffed into the nostrils so that the drug could diffuse through the mucous membranes of the nose. In chronic cocaine sniffers, local tissue damage often destroys the partition between the nostrils. In the United States today, cocaine is used as a needle drug. Recently, its use has spread from the inner city to include more affluent, middle-class drug users. Tolerance and withdrawal symptoms are minimal, but some people chronically abuse cocaine. As with amphetamine users, some of these individuals develop assaultive tendencies and paranoid reactions.

BARBITURATES

Barbiturates are the most common of a group of sedative-hypnotic drugs that are classified as central-nervous-system depressants. Drug users refer to barbiturates as downers, and particular drugs are known by a variety of names that generally refer to capsules of common trade brands. Such names include yellows or yellow jackets (Nembutal), reds or red birds (Seconal), blue heavens (Amytal), and rainbows or red devils (Tuinal).

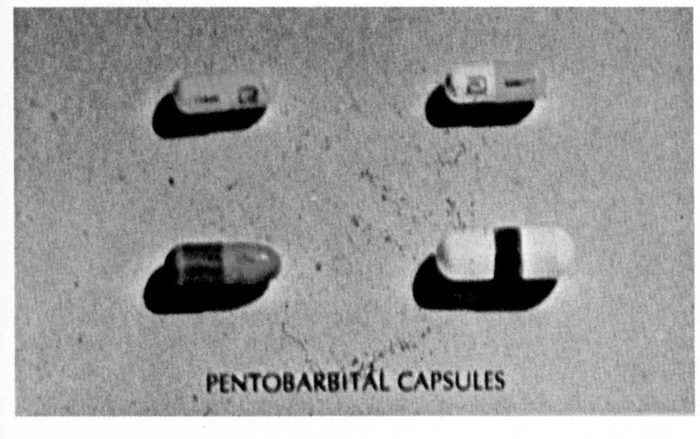
PENTOBARBITAL CAPSULES

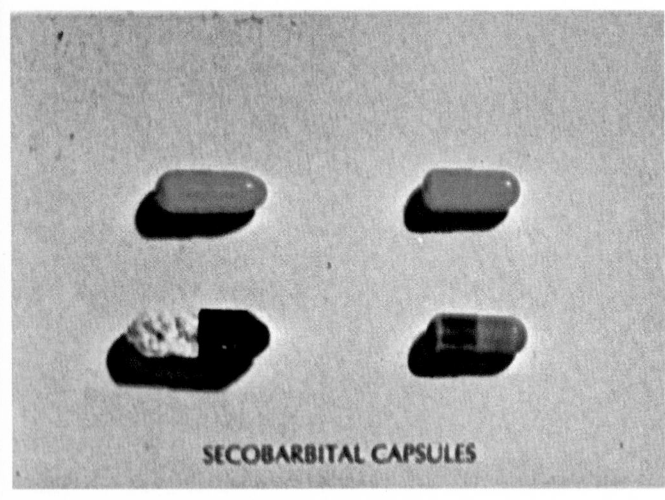
SECOBARBITAL CAPSULES

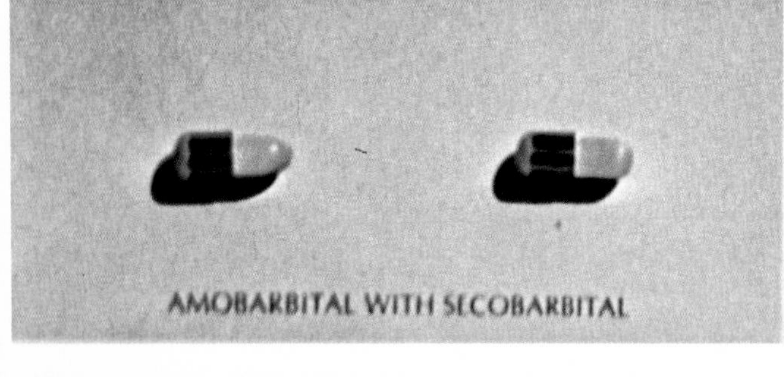
AMOBARBITAL WITH SECOBARBITAL

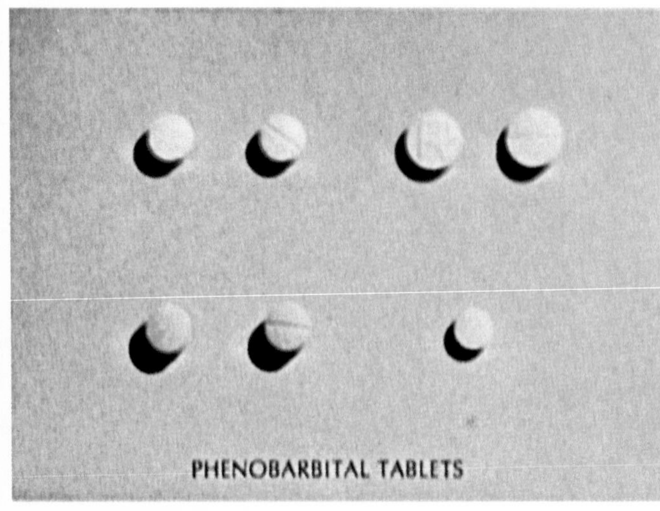
PHENOBARBITAL TABLETS

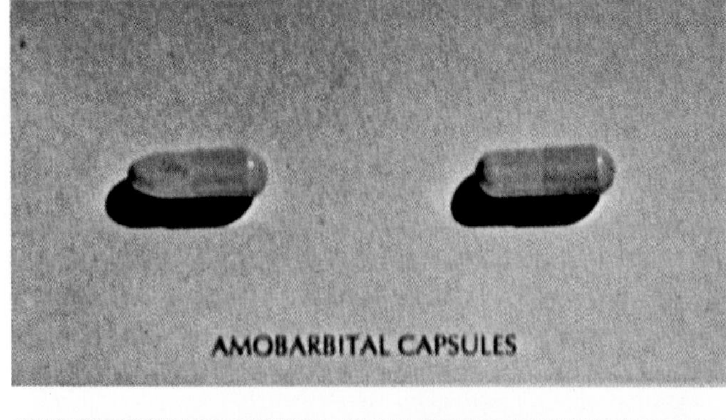
AMOBARBITAL CAPSULES

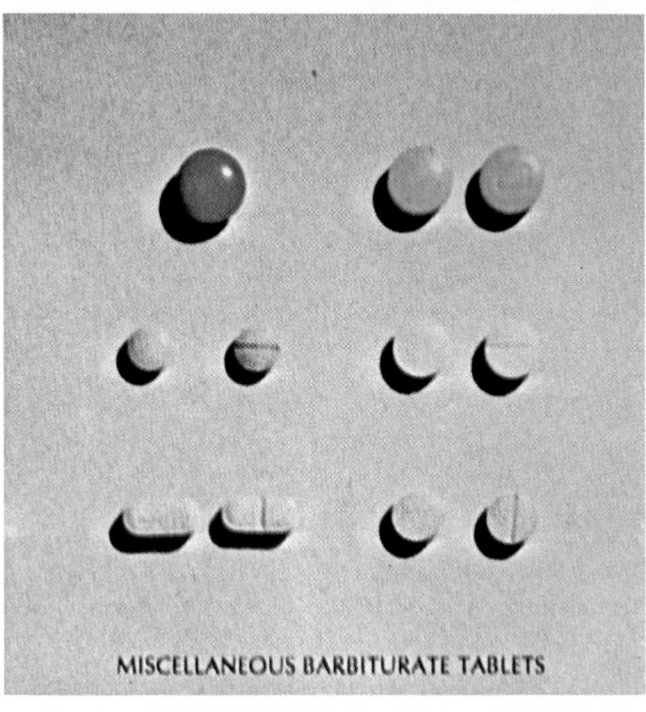
MISCELLANEOUS BARBITURATE TABLETS

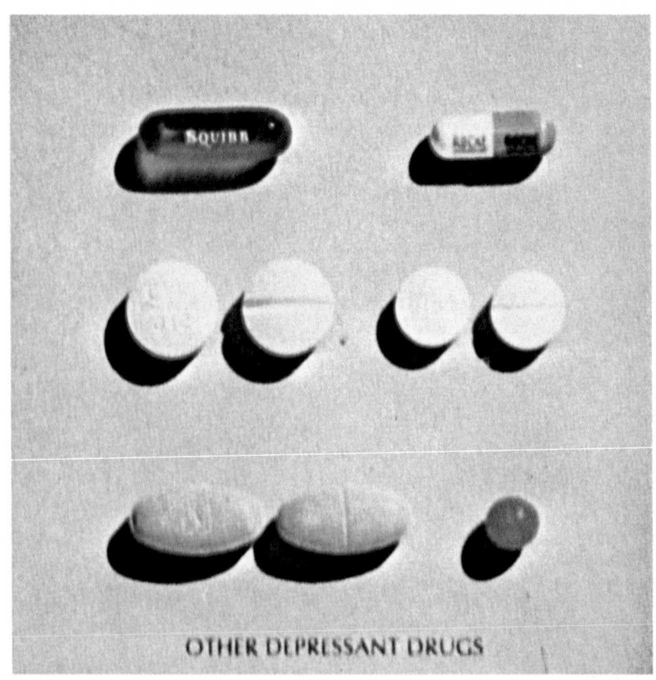

OTHER DEPRESSANT DRUGS

Figure 43.15 (opposite). Depressants, or sedatives, include alcohol, bromides, chloral hydrate, and paraldehyde. However, the most potent and most commonly abused of the sedatives are the barbiturates. There are many different barbiturates on the market, and most of them have been given brand names by the manufacturers. They are medically prescribed in the form of tablets, capsules, syrups, and injectables.

Recently, a number of nonbarbiturate sedative-hypnotics—such as Valmid, Thalidomide, and Doriden—have been introduced. Most of these newer drugs initially have been marketed as "safe" and nonaddictive and subsequently have been discovered to produce addiction and other ill effects. For example, Thalidomide was used for several years before its adverse effects on prenatal development were discovered. Sedative-hypnotics act primarily on the higher cortical structures, although they have some effects on midbrain and brainstem functions. Behavioral effects that occur are similar to those of alcohol and include variable amounts of anxiety reduction, changes in mood, slurring of speech, muscular incoordination, and eventual drowsiness or sleep. As with alcohol, there is great variation in the impairment of cognitive functions, depending on the task and individual (Evans and Davis, 1969).

Barbiturates and other sedative-hypnotics are used medically for treatment of insomnia, seizure control, daytime sedation, and in combination with other drugs, such as stimulants and analgesics. These drugs usually are ingested but may be injected. They are also used illicitly to alter awareness and to reduce discomfort associated with the effects of other drugs.

Almost all sedative-hypnotics can lead to the classical addictive syndrome. Tolerance can develop rapidly, sometimes up to 15 times the usual sedating dose. An individual accustomed to high doses of these drugs often experiences more severe withdrawal symptoms than those associated with alcohol or opiate withdrawal. Like the alcoholic, the barbiturate abuser usually is unable to function adequately in his daily activities. The chronic barbiturate user often demonstrates mental confusion and frequently is obstinate, irritable, and abusive. These characteristics are in marked contrast to the opiate addict, who usually remains passive and oriented to his environment (Wikler, 1970). Certain barbiturate users become violent, particularly those who combine barbiturates with intravenous injection of a stimulant, such as methamphetamine or cocaine (Tinklenberg and Stillman, 1970).

As a means for self-destructive behavior, barbiturates and other sedatives clearly pose a major health problem. Barbiturate overdose is currently the most frequent method of suicide among American women and accounts for over 3,000 known deaths each year in the United States. In addition, there are accidental deaths resulting from the combined use of

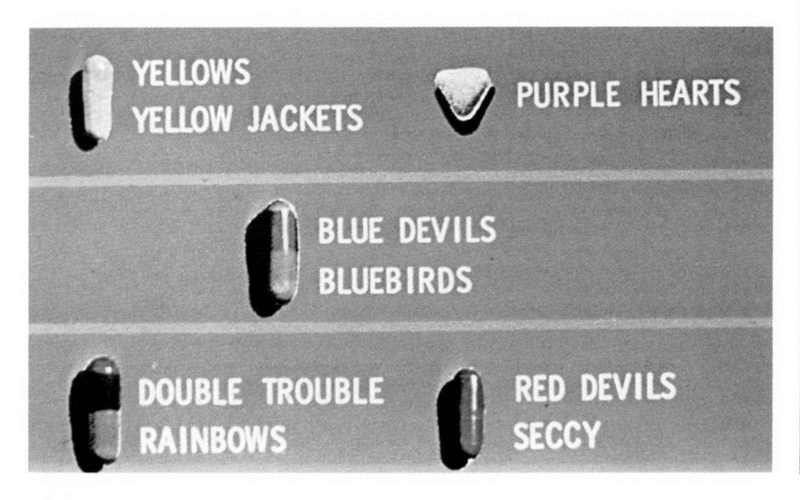

Figure 43.16. Barbiturates.

barbiturates and alcohol, neither of which would have proved fatal by itself. Both drugs have depressant effects that can combine to produce cessation of breathing.

PSYCHEDELIC OR PSYCHOTOMIMETIC DRUGS

The controversial nature of LSD and other drugs with similar effects is reflected in the lack of a commonly accepted name for this group of drugs. The various names emphasize different aspects of the drug effects. Those who regard these drugs as aids for expansion of awareness call them *psychedelic* (mind-manifesting or mind-expanding) drugs. Those who emphasize their sensory effects call them *hallucinogenic* (hallucination-producing) drugs. And those who emphasize the similarity between symptoms produced by these drugs and symptoms produced by severe personality disturbances call them *psychotomimetic* (psychosis-mimicking) drugs. The term "psychedelics" will be used in the following discussion because it is the shortest and probably the most common term for this group of drugs, but these drugs do possess some properties implied by each of the names.

The psychedelic drugs are distinguished by their marked effects on normal patterns of thought and perception and moods. Intellectual impairment, sedation, stimulation, or addictive potential are minimal in the usual dose ranges. These drugs include lysergic acid derivatives, such as LSD; phenylethylamine derivatives, such as mescaline; indole derivatives, such as psilocybin; piperidine drugs, such as Ditran; and phenylcyclidine (Sernyl). These drugs possess unknown mechanisms of action, but these mechanisms seem to involve changes in the turnover rates of brain amines, particularly serotonin. Of this group, LSD is the most common and will be discussed as a representative example.

The ingestion of one-millionth of an ounce of LSD, barely visible to the naked eye, will produce noticeable effects in most people. These effects include somatic changes (dizziness, weakness, nausea, pupillary dilatation, and hyperreflexia); perceptual alterations (distorted time sense, visual aberrations, heightened auditory acuity, and synesthesia—the blending of two sensory modalities such as hearing colors and seeing sounds); and psychic symptoms (rapid changes in mood, depersonalization, distortions in body image, and dissociation of the self from external reality). These effects often occur in a sequential pattern. Somatic changes come first, then perceptual alterations, and finally psychic changes, although there is considerable overlap among these three phases (Hollister, 1968).

These effects, particularly in lower doses, are significantly determined by the personality and expectation of the user, recent events in his life, the setting, and other psychosocial forces (Blum and associates, 1964). Many effects are difficult to describe in verbal terms because they are along different dimensions of usual awareness. Some users interpret these sensations in transcendental and religious terms, although the effects are not always pleasant. Severe panic reactions ("bummers" or "bad trips") and other effects are more common with these potent drugs than with milder marijuana. Although the probability of adverse reactions is reduced by having a serene mood without anger or hostility, secure surroundings, and trusted companions, the guarantee of a tranquil experience is a myth. LSD users report having hundreds of good trips, with pleasurable and ecstatic experiences, before inexplicably having a bummer, replete with monstrous perceptions and delusions of being trapped in a drugged state of madness.

Figure 43.17 (upper left). LSD is one of the most potent drugs known, producing effects in microgram quantities. Illicit LSD samples (as shown here) are found in capsule and tablet forms and on sugar cubes.

Figure 43.18 (lower left). A synthetic compound called dimethyltryptamine (or DMT) is found in the seeds of various West Indian and South American plants. It produces hallucinations similar to mescaline but of a shorter duration.

Figure 43.19 (right). Psilocybin is a substance extracted from a Mexican mushroom. It was used by primitive societies for "divination and communion with supernatural powers." Psilocin also is extracted from the mushroom, and both compounds produce an effect similar to mescaline.

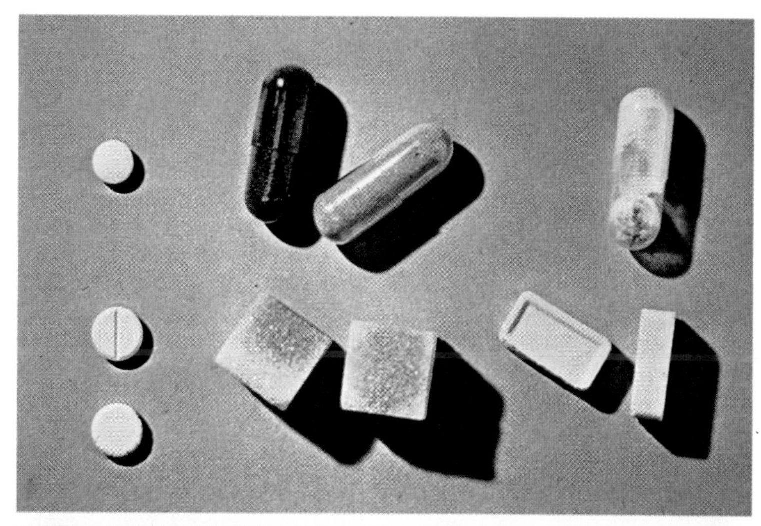

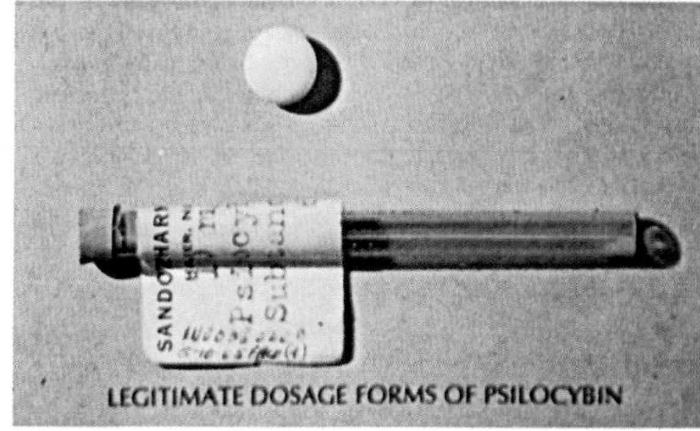

These unexplainable bummers are more common with high doses. Unfortunately, accidental high doses do occur because the purity of LSD varies widely, and the extremely small amounts in each dose require skillful measurement.

There are other unfavorable effects with LSD. Prolonged psychotic reactions and "flashback" phenomena may occur in both occasional and chronic users. Flashbacks are sudden, unexpected perceptual distortions and bizarre thoughts of an LSD trip that occur after the pharmacological effects of the drug have worn off (perhaps months or years after the last ingestion). The brain mechanisms involved in this phenomenon are unknown. Apparently, certain cues, either from the environment or from within the individual, trigger a perceptual-feeling process similar to that "imprinted" during the LSD trip. In a cybernetic fashion, this LSD-associated process generates additional cues that trigger more perceptual alterations.

Although the controversy surrounding LSD and cellular damage is not adequately resolved, the available evidence indicates that the ingestion of

Figure 43.20. Peyote, known by the ancient Aztecs, has been used for centuries to produce euphoria. The "mescal buttons" obtained from the peyote cactus contain mescaline, which acts as a hallucinogen.

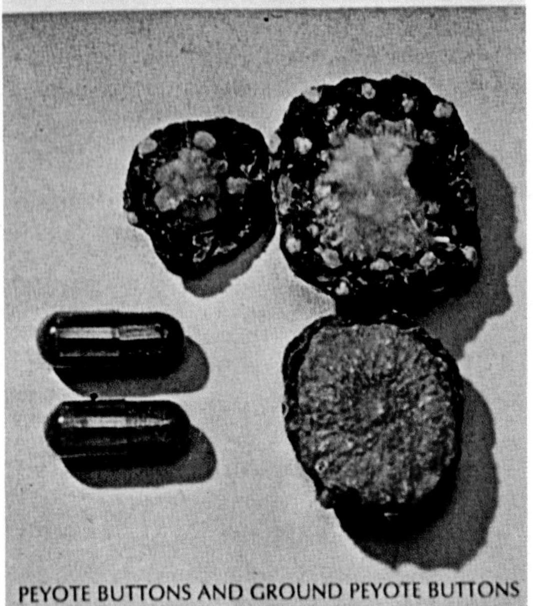

moderate doses of *pure* LSD does not damage chromosomes *in vivo*, does not cause detectable genetic damage, and is not a teratogen or a carcinogen in man (Dishotsky, *et al.*, 1971). However, illicit LSD does increase the rate of chromosomal damage and the rate of spontaneous abortions. These changes may be attributable to the impurities that are virtually always found in illicit LSD.

Although some people claim that LSD and other psychedelics enhance creativity, more controlled studies fail to support these assertions. LSD may increase both the user's subjective appraisal of his creativity and the frequency of actual production of original (infrequent) responses. However, LSD impairs the critical synthetic operations of the creative process—the sequential coordination of original responses into a pattern that is relevant to the problem or endeavor at hand. LSD sometimes is claimed to improve interpersonal sensitivity and to facilitate various kinds of psychotherapy. Most controlled studies indicate that, although the LSD user may claim new insights, objective changes in his behavior usually are not significant.

The use and effects of drugs similar to LSD vary along certain dimensions. The effects, especially with low doses, depend to a large extent on nondrug factors such as the expectations of the individual (set) and the physical and social environment (setting). Mescaline, in the form of peyote buttons from the Mexican cactus, is the ceremonial drug of the (Indian) Native American Church. Ditran and related drugs have profound central anticholinergic effects, resulting in greater intellectual impairment than with most other psychedelics, as well as considerable confusion and disorientation. Sernyl differs from the other psychedelics by markedly reducing sensory input and creating various degrees of sensory deprivation. One of the common street derivatives of Sernyl is appropriately dubbed the "peace pill." Dimethyltryptamine (DMT) and diethyltryptamine (DET) have many characteristics similar to LSD, but their effects may occur more rapidly and may last less than an hour (Efron, 1967). Because of their intense yet brief effect, these drugs go by the name of "the businessman's trip." In contrast, dimethoxy-methylamphetamine (DOM or STP) can induce psychedelic effects for longer periods of time. Psychedelic effects also can be obtained from morning glory seeds, nutmeg, the datura plant (jimson weed), and many other botanical products. Most of these plant products induce unpleasant side effects such as nausea, vomiting, and diarrhea. Their long-term effects are largely unknown, but scattered reports indicate that these "natural" sources of psychedelic experiences may be at least as dangerous as synthetic drugs.

The major danger of illicit drug use—particularly use of psychedelic drugs, which usually are manufactured by amateur chemists—is inherent in the ingestion of a potent substance of unknown composition in unknown dosage. Several studies involving analysis of drugs purchased illegally revealed that street drugs—particularly those sold as LSD—seldom contain what they are claimed to contain. Various mixtures of dangerous drugs such as strychnine, belladonna, and amphetamines commonly are added to or sold as LSD.

Newer psychedelics probably will continue to be developed and most will be used briefly until experience shows them to be less effective than older drugs or more fraught with unpleasant side effects. However, a few undoubtedly will demonstrate specific features that make them useful among certain groups of people. In addition, the study of LSD and other

Figure 43.21. Woman smoking marijuana, eighteenth century. In mythology, the hemp plant was produced as a nectar by the gods. Island physicians were prescribing the hemp plant for a variety of ailments, and traveling vendors were selling concoctions of marijuana as an aphrodisiac. Marco Polo wrote of the famed Assassins of Arabia who enlisted recruits by giving them brief visits to "paradise" with the help of hashish. This eighteenth-century miniature shows a woman using a hookah to smoke hashish or marijuana.

psychedelics is providing useful information toward an understanding of the biology of the brain and its relation to dimensions of the mind.

MARIJUANA

Marijuana is a crude preparation of resins from Indian hemp (a plant named *Cannabis sativa* by Linnaeus in 1753). The resins contain behavior-altering chemicals called tetrahydrocannabinols (THC). The concentration of THC in any given sample of cannabis varies widely according to such factors as where the plant was grown, what parts of the plant are being used, and how it was cultivated, harvested, and cured. Thus, there will be a wide range of potencies in different cannabis preparations. At one extreme is hashish (hash), which is a potent cannabis preparation, usually grown in the Middle East, Africa, or India. At the other extreme are the relatively weak preparations obtained from cannabis plants found throughout the United States. Recently, THC has been produced synthetically, but the difficulty and expense of the procedure limit this source.

The cannabis plant is found throughout the world, from the tropics to the Arctic. Cannabis preparations usually are smoked, but may be consumed as "tea" or mixed into confections and other foods. In the United States, synonyms for marijuana are myriad and continuously changing: pot, weed, grass, tea, hay, mary jane, and others. Despite the variety of terms for cannabis preparations and the different parts of the world from which these preparations can be obtained, the predominant behavior-altering components are just one group of chemicals: the tetrahydrocannabinols. The terms "cannabis" and "marijuana" can be used almost interchangeably, with "marijuana" usually referring to lower-potency preparations.

Marijuana does not fit neatly into any of the established drug categories because its effects vary considerably depending on dose and the time course of drug action. In small doses, marijuana initially induces mild euphoria and some central-nervous-system stimulation; later, the effect of sedation becomes more prominent. In larger doses, marijuana effects more closely resemble those of the psychotomimetics than those of any other drugs. Although in the United States cannabis preparations are usually regulated under the narcotics laws, their pharmacology differs significantly from that of opium, heroin, and other drugs controlled by these laws.

The biochemical and neurological effects of marijuana are not well understood, although there appear to be alterations in functions of the cerebral cortex, thalamus, limbic system, and reticular activating system. In addition to directly exerting their effects, the tetrahydrocannabinols are partially metabolized into compounds that may have biological activity.

The usual effects from low doses of marijuana include altered perceptions of sounds, colors, spatial configurations, and other sensory phenomena. As with most mind-altering drugs, there is wide variability of these perceptual changes, and there are marked differences among individuals in their awareness and interpretation of these alterations. These differences are determined largely by the user's expectations of how marijuana will affect him and by what he has learned from his past experiences, if any, with the drug. The chronic user may interpret his experience with low-potency marijuana as a "high," whereas the neophyte may not describe significant perceptual changes from the same dosage.

The effects of cannabis preparations on mood are variable. The usual pattern is initial slight apprehension, then a sense of pleasant weariness

Figure 43.22. Loose marijuana and some rolled "joints," or "numbers."

intermittently interspersed with pronounced euphoria. Later, sedation becomes apparent, and sleep is common (Hollister, 1971b).

The cognitive changes induced by marijuana differ widely, depending on the individual, the dose, and the cognitive task being measured. Marijuana seems to affect short-term memory (coherent retention of events from the preceding few seconds), ability to sustain attention or shift attention appropriately from one focus to another, and the ability temporarily to integrate events of the past with present input toward future goals (Tinklenberg, *et al.*, 1970; Melges, *et al.*, 1970). Marijuana can slow performance on certain information-processing tasks in a fashion similar to alcohol (Jones and Stone, 1970). However, performances on a range of other cognitive tasks either are not measurably impaired by low doses of marijuana or the individual is able to compensate effectively for any drug effects.

The behavioral changes induced by marijuana seem to follow a cyclic pattern of waxing and waning, rather than a smooth "dose effect" curve. These "ups and downs" have important implications for social regulations, because the individual may be relatively clear at one moment and impaired the next (Clark, *et al.*, 1968).

The initial physical effects of marijuana are minimal. Most pronounced are increases in heart rate, dilatation of the conjunctival blood vessels (which create characteristic blood shot eyes), and reduced physical strength (Hollister, 1971a). Marijuana can increase appetite in some people. Although certain brain-wave changes after cannabis ingestion have been identified, the exact alterations in brain function are unknown (Jones and Stone, 1970).

Although marijuana once was used to treat a variety of human illnesses, at present there are no definite medical indications for prescribing marijuana. The use of marijuana for its sedative or euphoric effects is constrained by the marked individual variation in response to the drug, as well as by the cognitive and perceptual alterations.

With higher doses of THC (as with most psychoactive drugs), behavioral effects are determined more by drug properties themselves and less by psychosocial factors such as individual expectations. With these higher

Figure 43.23. Marijuana leaves. The psychoactive agents are located in a sticky resin that is produced by the glandular hairs, which are particularly abundant on the leaves and female flowers.

doses (and occasionally with low doses in particularly susceptible individuals), the perceptual and cognitive changes previously described may be markedly exaggerated. Subtle alterations of sensory input become gross distortions, and the user may feel inundated by myriad sensory cues from his environment. Extensive cognitive disintegration may prohibit him from coherently directing his thought patterns. In contrast to LSD-induced bad trips (where perceptual disorders often predominate), thought disorders (particularly of a paranoid nature) seem to be more frequent with marijuana.

A frequent interpretation of these alterations, especially by the neophyte, is that he is losing control over his mind and will be permanently "crazy." The user's mood changes from the usual pleasant lassitude to various degrees of anxiety, including rare incidents of panic and aggressiveness. These adverse reactions are similar in some respects to bad trips with LSD and other potent psychotomimetics. However, marijuana bad trips occur much less frequently, are less severe, and are usually less prolonged than with LSD.

The long-term effects of cannabis consumption are inadequately understood at this time. However, on the basis of experience with other drugs, one can reasonably predict that the body systems and functions most affected by the drug are most vulnerable to unfavorable long-term consequences. Thus, in a percentage of susceptible, chronic, long-term cannabis users, it may be expected that there eventually will be persistent impairment in certain cognitive task performances involving immediate memory, temporal integration of past and present events with future goals, and sustained attention. Similarly, a few vulnerable long-term users of marijuana are likely to show persistent changes in mood and in their usual sensory abilities, especially regarding time perception. A percentage of the chronic smokers of cannabis preparations will manifest disorders of the lungs and bronchi. However, again drawing from experience with other drugs, it is unlikely that all chronic cannabis users will manifest these or other harmful effects. There is great individual variability in susceptibility to long-term effects of drugs, just as there is great variability in the acute manifestations of drug effects. Certain individuals are biologically susceptible to long-term effects of drug usage, whereas others are relatively immune to the same cumulative doses. Finally, there is an inevitable time lag between the widespread use of a drug and the recognition of any long-term deleterious effects. Tobacco, with its long history of use as a "harmless" drug, provides a salient example.

Is marijuana addictive? Although some tolerance can develop, the need for increasing doses to produce the same effects usually is minimal with low-potency preparations. The crucial issue—what percentage of users progressively increase their dosage—is unresolved. Clinical impressions suggest that these tendencies do occur among some marijuana users, but not to the marked degrees so apparent with users of opiates or intravenous amphetamines. Although unpleasant symptoms can occur upon sudden withdrawal of the drug in chronic marijuana users, these symptoms are usually mild in contrast to the extreme distress experienced during opiate withdrawal. Thus, marijuana users are not driven to continued use of the drug to avoid withdrawal symptoms, as opiate users so commonly are.

However, as discussed with the amphetamines, addictive patterns of behavior are determined not only by physical tolerance and withdrawal but

Figure 43.24. French "hippies" advocating use of marijuana, 1849. It was not until the early nineteenth century that physicians in western Europe began exploring the possible medical uses of cannabis. About the same time, a group of intellectuals began self-experimentation with the plant. Theophile Gautier organized the famed Club des Haschischins in Paris, and in 1849, this French almanac shows a wizard peering into the future through his telescope while a line of nineteenth-century "hippies" parade past carrying banners proclaiming the use of hashish and ether.

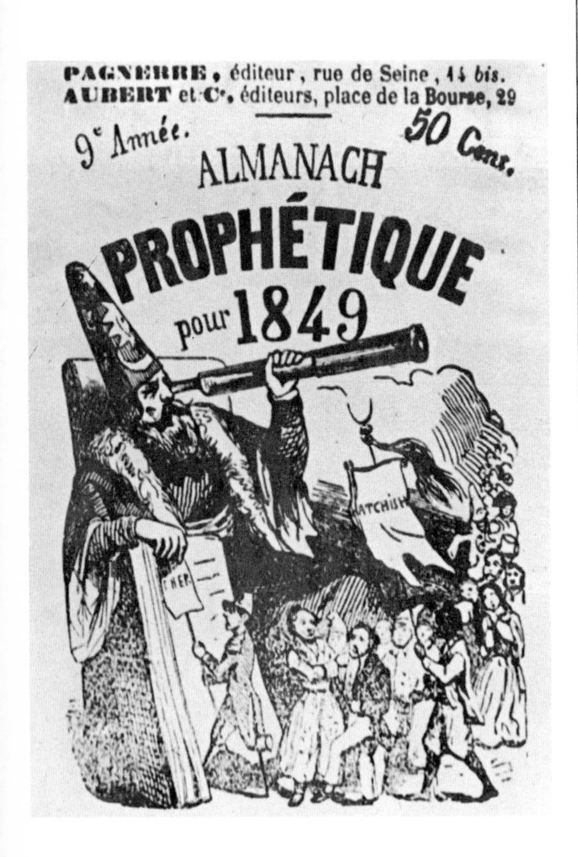

also by a variety of psychosocial forces (Wikler, 1970). Both the direct effects of marijuana and features intrinsic to the use of marijuana are pleasurable to many people; thus, the use of marijuana is rewarded, and patterns of subsequent use are influenced by these reinforcing factors. Perhaps most people who try marijuana in contemporary social settings experiment a few times with the drug but then for a variety of reasons cease using it. Another group of people continue to use marijuana on a regular basis but are able to abstain voluntarily without undue difficulty. However, as with virtually all other drugs with pleasurable effects, a certain percentage (probably low) of chronic marijuana users increasingly center their lives around marijuana, gradually excluding other meaningful and satisfying activities. Another small percentage of users will begin to abuse other drugs, including alcohol. If more potent cannabis preparations become available, the percentages in these latter groups will probably increase.

Although many of the severe legal sanctions against marijuana in the United States were initially established because of alleged tendencies for marijuana to induce violence, most available evidence indicates that limited marijuana use reduces violent activities (Tinklenberg and Stillman, 1970). The disinclination for aggression seems to stem from the interplay of the drug's significant action-reducing effects and the psychosocial attitudes of passivity or nonviolence among many current users of marijuana in the United States. However, there are a few individuals who demonstrate idiosyncratic responses, including assaultive behavior. Likewise, because many of the effects of marijuana can be suppressed with effort, a delinquent gang or individual with aggressive plans is unlikely to be deterred by marijuana. Finally, the chronic use of marijuana is associated in some people with altered attitudes and distorted judgment so that probabilities for violence are increased. Again, the importance of considering psychosocial contributions to drug-affected behavior is evident.

The effects of marijuana on driving are poorly understood. Not only is very little known about the effects of marijuana, but very little is known about what factors generally contribute to good or poor driving performance (Kaplan, 1970). Crucial factors in driving may include not only effects of marijuana on attention and psychomotor skills but also the effect of drugs in altering general aspects of personality function, such as risk taking and assertiveness. The intermittent "waxing and waning" effects of marijuana must be considered in regard to driving and other complex activities.

Cannabis preparations have been used widely in many different countries for centuries. At present, on a world-wide basis, marijuana is an intoxicant second in popularity only to alcohol. In the United States, there is a current trend toward the increased use of marijuana, particularly among young people. The immediate trigger for drug use comes through imitation and modeling; most marijuana users first learned about the drug from an older, respected friend who was willing to teach drug use. The few available systematic studies of long-term marijuana users in the United States suggest that the chronic user is more likely than the average person to use other drugs, including excessive amounts of alcohol, and that he shows greater instability in work, residence, and marriage (McGlothlin, *et al.*, 1970). Only future research will tell whether these generalizations will apply to individuals now beginning to use marijuana.

The final question regarding marijuana is that of future social regulation. The history of tobacco, caffeine, alcohol, and other drugs is one of repeated failure in religious and political attempts to prohibit use of a drug already accepted by a significant portion of the population (Blum and associates, 1969b). These precedents suggest that if marijuana use continues to gain popularity, government involvement with marijuana eventually will become directed toward taxation, quality control, and regulation of distribution.

ALCOHOL

Alcohol refers to a group of beverages—including distilled spirits, wine, and beer—that contain ethyl alcohol along with other ingredients. Alcohol usually is classified pharmacologically as a central-nervous-system depressant, with particular depressant effects on the cerebral cortex and reticular activating system. The apparent excitatory effects of alcohol are due primarily to release of inhibitory controls, although there may be some direct stimulating effects.

The acute effects of alcohol on human behavior are well known. They usually include variable amounts of anxiety reduction and increases in verbal and assertive behavior. The use of alcohol is associated with many kinds of violence (Tinklenberg, 1971). Alcohol causes a gradual reduction in many motor and cognitive skills, although there is considerable variation, depending on task and dosage. Maximum behavioral effects occur when blood alcohol concentration is increasing—that is, when the central nervous system is becoming acclimatized to the alcohol. An individual quickly develops tolerance, so that the same dose no longer produces the same behavioral changes.

There is wide individual variability of response to any given dose of alcohol. Low doses of alcohol usually produce slight changes in mood and level of tension. The same low doses in a few people precipitate the poorly understood clinical entity of "pathological intoxication," with grossly impaired memory function, disorientation, marked perceptual changes, and exaggerated, impulsive, aggressive behavior.

Behavioral responses also vary with different doses of alcohol. Although it is tempting to observe effects of alcohol or other psychoactive drugs at a

Figure 43.25. Ancient distillation scene, sixteenth century. Alcohol is the oldest sedative known to mankind, and during the Middle Ages when the process of distillation was perfected, extremely potent alcoholic beverages became available.

Figure 43.26. Graph showing the relationship between blood alcohol level and the probability of causing an accident. Subjects with a blood alcohol level of just under 0.04% were about as likely to cause accidents as drivers with no alcohol in their system. As the blood alcohol level rises, however, the probability of having an accident rises sharply. A driver with 0.06% had an accident probability of double that of one with 0.00–0.04%; with 0.10%, 6 or 7 times as high; and with 0.15%, over 25 times as high.

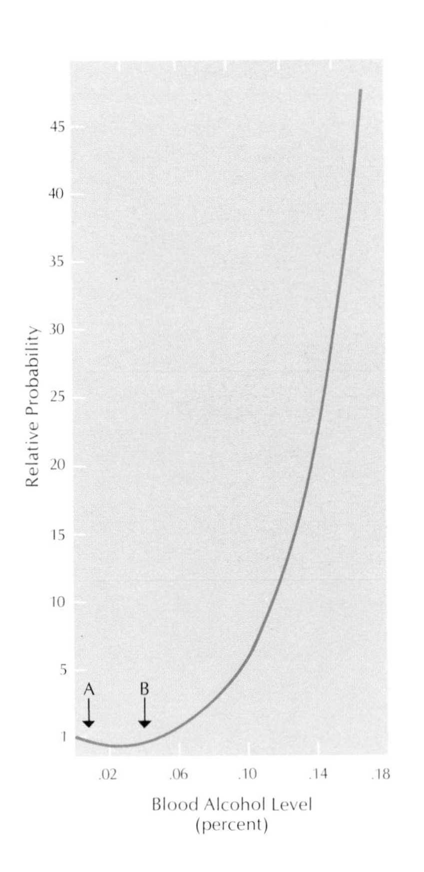

particular dosage and to infer similar behavior at all dose levels, such inferences are likely to be inaccurate. Instead, as with most drugs, there is a dose-response curve, with different behavioral probabilities at different dose levels. For example, one systematic study on drinking and driving behavior suggests that different doses of alcohol are associated with opposite behavioral probabilities (Zylman, 1968). Figure 43.26 is partially derived from that study. The estimated number of drinks is a rough guide in which 1 drink refers to 1 cocktail, 5 ounces of table wine, or 1 bottle of beer. The cited number of drinks is a minimum for an average-size person (150 pounds). More drinks would be required to achieve a given blood level if the individual were heavier, if absorption were delayed by food in the stomach, or if the drinking took place over a longer period of time. The entry labeled "relative probability" represents the increased chances that a drinking driver would have an accident as compared to a nondrinking driver.

At the dose levels of arrow *A*, small additional amounts of alcohol *decrease* the chances of an accident, when all drivers are considered. But when blood alcohol is at the level indicated by arrow *B* or higher, additional drinks markedly *increase* the probability of an accident.

The effects of different doses of alcohol on behavior also involve a condition called state dependency. In this condition, information acquired while the central nervous system is in a certain physiological state—for example, under the influence of a given dose of alcohol—is most efficiently recalled when that state is closely replicated (Goodwin, *et al.*, 1969). In practical terms, if a student has a certain blood alcohol level while he is studying for exams, subsequent recall of learned material usually will depend upon how closely the original physiological state (blood alcohol level) is replicated. Thus, state dependency limits the use of drugs and alcohol for enhancement of human learning and creativity. State dependency is not limited to drugs but is important in sleep deprivation, hypnosis, and other altered physiological conditions.

The chronic use of alcohol definitely is associated with tolerance and withdrawal symptoms, most vividly demonstrated as delirium tremens. Excessive use of alcohol may be accompanied by nutritional disorders involving the central and peripheral nervous systems, the gastrointestinal system, and organ systems such as the liver. Among chronic users, there may be very different disorders. One individual may show primarily central-nervous-system effects, with profound memory impairment (Korsakoff's syndrome) and no liver damage; another individual, with a very similar drinking and nutritional history, may have advanced liver disease but no memory deficits.

The treatment of chronic alcoholism universally is regarded as difficult. The best treatment programs entail the judicious use of a variety of therapeutic techniques, such as Alcoholics Anonymous; individual, group, and especially couple psychotherapy; and chemical therapies, such as disulfiram (Antabuse). The success of these and other regimens for chronic alcoholism depends crucially on mutual agreement on treatment goals between patient and therapist.

The significance of alcohol usage must be considered in the context of a changing society. Until a few decades ago, excessive drinking was unlikely to affect more than the drinker and perhaps a few of his close associates. However, in a crowded, technological society—with automobiles, air-

Figure 43.27. The plight of the alcoholic. This man from the Bowery area of New York City lives in an abandoned car or a shack made of crates. *(Alan Mercer)*

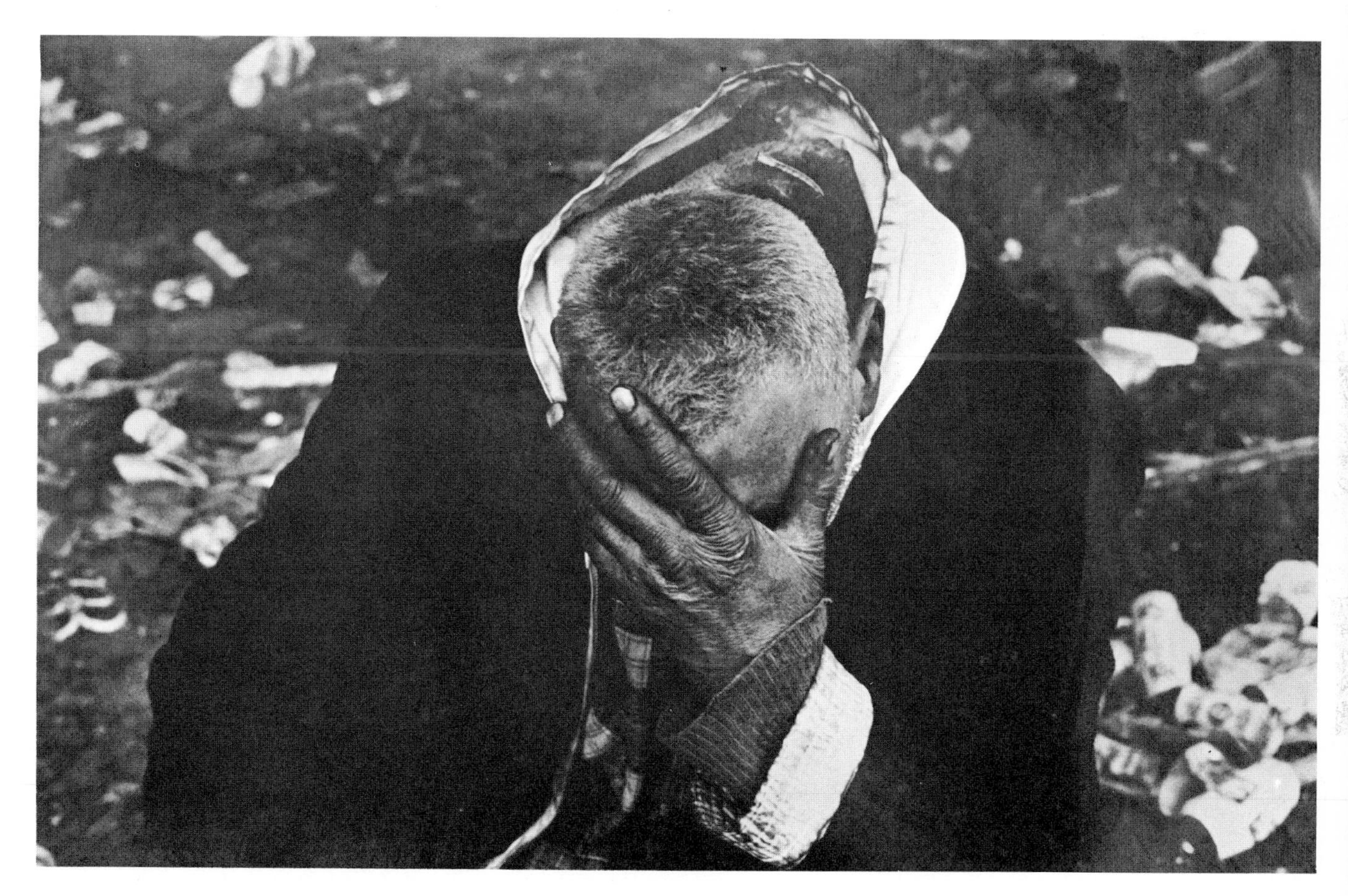

planes, dense city populations, and so on—behavior impaired by alcohol or other drugs becomes increasingly harmful as it impinges on an increasing number of other individuals.

CAFFEINE

Caffeine is a central-nervous-system stimulant that can be made synthetically but generally is obtained from tea, coffee, and other plants. The use of tea and coffee has a long history, demonstrating that the drinks most consistently used through human history are those that affect the central nervous system. Both substances initially were used in religious settings, later were touted for medicinal virtues, and then came to be used widely for a variety of noninstitutionalized purposes. For a time, coffee users of the Near East were subject to criminal penalties, including death. In spite of these extreme sanctions, coffee drinking spread in the Near East as well as in most other parts of the world.

The mild stimulation of the central nervous system resulting from ordinary doses of caffeine is manifested in increased alertness, a sense of well-being, and decreased feelings of fatigue or boredom. Caffeine can prolong the amount of time during which the individual can perform physically exhausting work, and it can sustain and perhaps improve performance on a variety of simple repetitive tasks. However, more complex intellectual

Figure 43.28. America Indian with pipe. The Indians were the first to cultivate the tobacco plant.

functions, such as reading comprehension, are not altered appreciably except when normal performances already have been lowered by fatigue, boredom, alcohol, or other factors. To some extent, caffeine can counteract these decrements in performance on a variety of tasks (Weiss and Laties, 1962).

Other effects of caffeine include increased muscular tremor, increased gastric secretion, constriction of cerebral vessels, and some mild cardiac and respiratory stimulation. Higher doses may cause nervousness, irritability, headache, and disturbed sleep. Although caffeine is used clinically to produce mild cerebral stimulation, to combat respiratory depression, and as an ingredient in headache remedies, other drugs are usually more effective.

Different countries have favorite drinks that contain caffeine. People of the United States prefer coffee; tea is most popular in England; and maté is the favorite drink containing caffeine in South American countries. These beverages, as well as the caffeine-containing cola drinks, are used in a wide variety of settings with only rare, adverse effects in healthy individuals. Although some tolerance develops to caffeine, and there may be increased irritability and headaches upon abrupt cessation of the drug, physical consequences of long-term use of the drug are nil. As with other psychoactive drugs, psychological habituation to caffeine takes place, but this behavior does not intrude significantly on other aspects of the individual's life.

TOBACCO

Tobacco is composed of dried leaves of *Nocotiana tabacum*. This plant originally was cultivated for smoking by American Indians. In the early 1600s, European explorers and colonists quickly acquired the habit and transported tobacco smoking back to the old country. Despite admonitions by the Establishment of that day against the "foul weede," the wide use of tobacco spread quickly. According to one authority, the spread of tobacco qualifies as the most dramatic "epidemic" of drug use in history (Blum and

associates, 1969b). Again, it is apparent that when a drug has obtained widespread public acceptance and does not have immediate harmful effects, religious and political prohibitions are ineffective in curbing its use.

Initially, tobacco was promoted for a variety of medicinal purposes, and for a time it was called *herba panacea* – a cure for all ills. Later, tobacco smoking was considered a costly but harmless habit. The long-term effects have only been discovered in the last two decades – a higher incidence of lung cancer, emphysema, coronary artery disease, and fetal abnormalities. More than 300 years elapsed between widespread use of tobacco and recognition of its adverse effects.

Tobacco smoke produces a mixture of semisolid and gaseous materials, including nicotine, hydrocarbons, so-called tobacco tars and resins, and other constituents. Nicotine is responsible for most of the immediate behavioral and pharmacological effects. These effects are complex and highly dependent on dosage. Depending on the dosage, the effects of nicotine on the peripheral nervous system can be either excitatory or tranquilizing. Nicotine doses from smoking produce only moderate effects on the central nervous system, including some stimulation, but these doses do not appreciably affect complex functions such as learning.

Other activities can be pursued simultaneously with smoking, and if these activities are reinforced, smoking also is reinforced. Thus, for one person, smoking becomes necessary for social tranquilization, and for another it becomes important for relief of boredom. A person who smokes while working, studying, or talking soon finds that smoking has become an integral part of his habit patterns. The smoking habit is difficult to break, because it has been reinforced by so many different consequences and because smoking behavior is triggered by so many different stimuli.

DRUGS AND DELIRIUM

Many drugs – including the psychedelics – sometimes are associated with an abnormal behavioral condition called delirium, which results from general impairment of brain function. This condition is characterized by changing levels of awareness to surrounding events, by decreased ability to maintain attention on any specific task, and by variable amounts of mental confusion, including difficulties in maintaining time-space orientation. These changes often are associated with dizziness, lightheadedness, and euphoria. Hallucinatory experiences, especially those of a visual nature, often are described. These impairments in mental functioning may be associated with temporary changes in formal neurological functioning – for example, incoordination and changes in reflexes. Although delirious states resemble the usual effects of psychedelics, they differ in that the psychedelic drugs in usual doses seldom induce disorientation.

Inhalants are a common group of drugs that induce delirium. These inhalants include glue, various aerosols, volatile solvents, kerosene, gasoline, anesthetic agents, such as nitrous oxide (laughing gas), and many others. Most inhalants can impair brain function directly, but some act indirectly by interfering with oxygen exchange in the lungs, and sudden death by asphyxiation or irreversible tissue damage can occur.

Excessive amounts of bromides, belladonna alkaloids, and most other drugs used for tranquilization may produce delirium. These drugs are readily available to most people, either as over-the-counter proprietary preparations or through illicit channels. Because many of these agents are cheap

Figure 43.29. Common inhalants. This group includes glue, gasoline, and paint thinner, and several items commonly used to inhale the fumes (rags, paper bags, and so on).

Table 43.1
Major Substances Used for Mind Alteration

OFFICIAL NAME OF DRUG OR CHEMICAL	Slang Name(s)	Legitimate Medical Uses (present and projected)	Potential Psychological Dependence	Potential for Tolerance (leading to increased dosage)
Narcotics (opiates, analgesics)		Treatment of severe pain, diarrhea, and cough	High	Yes
Opium	Op			
Heroin	Horse, H			
Morphine				
Codeine				
Percodan ®				
Demerol ®				
Cough syrup (Cheracol ®, Hycodan ®)				
Amphetamines		Treatment of obesity, narcolepsy, fatigue, depression; anesthesia of the eye and throat	High	Yes
Benzedrine ®	Bennies			
Methedrine ®	Crystal, Speed			
Dexedrine ®	Dexies or Christmas trees (spansules)			
Cocaine	Coke, Snow			
Barbiturates		Treatment of insomnia and tension; induction of anesthesia	High	Yes
Nembutal ®	Yellow Jackets			
Seconal ®	Red Devils			
Phenobarbital	Phennies			
Doriden ® (glutethimide)	Goofers			
Chloral hydrate				
Miltown ®, Equanil ® (meprobamate)				
LSD	Acid, Sugar	Experimental study of mind and brain function; enhancement of creativity and problem solving; treatment of alcoholism, mental illness, and the dying person (chemical warfare)	Minimal	Yes (rare)
Psilocybin				
Mescaline (peyote)	Cactus			
Cannabis (marijuana)	Pot, Grass, Tea, Weed, Stuff	Treatment of depression, tension, loss of appetite, sexual maladjustment, and narcotics addiction	Moderate	No
Hashish				
Alcohol	Booze	Rare	High	Yes
Whiskey, gin, beer, wine	Hooch	Sometimes used as a sedative (for tension)		
	Juice			
Caffeine	Java	Mild stimulant; treatment of some forms of coma	Moderate	Yes
Coffee, tea,				
Coca-Cola ®				
No-Doz ®, APC				
Nicotine (and coal tar)	Fag	None (used as an insecticide)	High	Yes
Cigarettes, cigars				
Glue		None (except for antihistamines used for allergy and amyl nitrite for some episodes of fainting)	Minimal to moderate	Not known
Gasoline				
Amyl nitrite				
Antihistamines				
Nutmeg				
Nonprescription "sedatives"				
Tranquilizers		Treatment of anxiety, tension, alcoholism, neurosis, psychosis, psychosomatic disorders and vomiting	Minimal	No
Librium ® (chlordiazepoxide)				
Phenothiazines				
Thorazine ®				
Compazine ®				
Stelazine ®				
Reserpine (rauwolfia)				
Antidepressants		Treatment of moderate to severe depression	Minimal	No
Ritalin ®				
Dibenzapines (Tofranil ®, Elavil ®)				
MAO inhibitors (Nardil ®, Parnate ®)				

Source: Adapted from Joel Fort, M.D., *The Pleasure Seekers: The Drug Crisis, Youth and Society* (Indianapolis: Bobbs-Merrill, 1969), pp. 236–244.

Potential for Physical Dependence	Usual Short-Term Effects (psychological, pharmacological, social)	Usual Long-Term Effects (psychological, pharmacological, social)
Yes	CNS depressants; sedation, euphoria, relief of pain, impaired intellectual functioning and coordination	Constipation, loss of appetite and weight, temporary impotency or sterility; habituation, addiction with unpleasant and painful withdrawal illness
No	CNS stimulants; increased alertness, reduction of fatigue, loss of appetite, insomnia, often euphoria	Restlessness, irritability, weight loss, toxic psychosis (mainly paranoid); diversion of energy and money; habituation
Yes	CNS depressants; sleep induction; relaxation (sedation); sometimes euphoria; drowsiness; impaired judgment, reaction time, coordination, and emotional control; relief of anxiety-tension; muscle relaxation	Irritability, weight loss, addiction with severe withdrawal illness (like DT), diversion of energy and money; habituation
No	Production of visual imagery, increased sensory awareness, anxiety, nausea, impaired coordination; sometimes consciousness-expansion	Usually none; sometimes precipitates or intensifies an already existing psychosis; more commonly can produce a panic reaction when person is improperly prepared
No	Relaxation, euphoria, increased appetite, some alteration of time perception, possible impairment of judgment and coordination (probable CNS depressant)	Usually none; possible diversion of energy and money
Yes	CNS depressant; relaxation (sedation); sometimes euphoria; drowsiness; impaired judgment, reaction time, coordination and emotional control; frequent aggressive behavior and driving accidents	Diversion of energy and money from more creative and productive pursuits; habituation; possible obesity with chronic excessive use; irreversible damage to brain and liver, addiction with severe withdrawal illness (DT)
No	CNS stimulant; increased alertness; reduction of fatigue	Sometimes insomnia or restlessness; habituation
No	CNS stimulant; relaxation (or distraction) from the process of smoking	Lung (and other) cancer, heart and blood vessel disease, cough, and so on; habituation; diversion of energy and money; air pollution; fire
No	When used for mind alteration, generally produces a "high" (euphoria) with impaired coordination and judgment	Variable—some of the substances can seriously damage the liver or kidney
No	Selective CNS depressants; relaxation (relief of anxiety-tension); suppression of hallucinations or delusions; improved functioning	Sometimes drowsiness, dryness of mouth, blurring of vision, skin rash, tremor; occasionally jaundice, agranulocytosis
No	Relief of depression (elevation of mood), stimulation	Basically the same effect as tranquilizers, above

Figure 43.30. This James Gillray caricature of 1802 portrays one of the humorous results of the chemical lectures demonstrating "new discoveries in pneumaticks or an experimental lecture on the powers of air." These more formal lectures subsequently became traveling road shows where lecturers demonstrated the exhilarating effects of ether vapor and laughing gas. Sometimes the demonstrations turned into laughing gas parties or ether frolics and became of considerable concern to community leaders.

and available and have definite psychoactive properties, they frequently are used to adulterate more expensive psychedelics. Unfortunately, these agents can increase the hazard of bad trips by adding the dimension of delirium and complicating treatment of the bad trip.

PSYCHIATRIC DRUGS

The phenothiazines, butyrophenones, and other related drugs have properties that make them useful for treating psychoses. Their exact mechanisms of action are unknown, but there is ample evidence of their effectiveness in psychotic individuals for decrease of responsiveness to external and internal stimuli, decrease of excessive motor activity, and dampening of hallucinatory and delusional behavior. In psychotic or otherwise severely disturbed individuals, these marked effects can be accomplished without sedation, thus allowing the individual to continue other activities. The antipsychotic drugs pose a number of interesting questions for future investigation. One enigma centers around the fact that psychotic individuals can take high doses of these agents and continue daily activities, whereas nonpsychotic people on the same doses become sleepy, confused, or otherwise disabled. Another intriguing phenomenon of the antipsychotic drugs is their apparent lack of addictive potential. Virtually all other potent psychoactive drugs claim a few habitués, but no addicts have been reported with the phenothiazines.

The antianxiety drugs include meprobamates, benzodiazepine derivatives, and the hydroxyzine group. These drugs reduce anxiety, muscular tension, and irritability with dosages not markedly hypnotic (sleep-inducing) for most people. The antianxiety drugs are used extensively for treatment of anxiety, as well as treatment of psychoneurotic reactions, muscular spasms, and many other conditions. These drugs are not effective, however, in treating psychoses.

Recently, a number of important advances have been made in the development of psychiatric drugs that influence mood. *Antidepressants*, or mood elevators, include the monoamine oxidase (MAO) inhibitors and the tricyclics. The MAO inhibitors have a number of dangerous side effects and largely have been replaced by the less toxic tricyclics. In certain individuals, these drugs induce significant elevations in mood and increased activity and drive. Their exact mechanism of action is unknown. The precise therapeutic indications for these drugs are less well defined than the indications for antipsychotic medication, but systematic studies show that they are more effective in individuals who demonstrate the "pure" severe depressive mood with slowed body movements, reaction time, and stream of thought and speech. The more the individual's depressed behavior is mixed with other components such as anxiety, hostility, and hyperactivity, the less effective the antidepressants seem to be.

The *lithium salts* are among the newest drugs for the treatment of disorders in mood. Lithium is effective in altering mania—with its exaggerated moods of unrealistic euphoria, inappropriate self-confidence, and poorly controlled hyperactivity—toward more nearly normal conditions. In addition, prophylactic lithium is useful in treating manic-depressive disorders, where the individual oscillates between extremes of mania and severe depression. As with the antidepressants, lithium seems most effective in those individuals whose mood disorders are "pure" and not mixed with disturbances of thinking or perception.

The recent development of psychiatric drugs that are very specific in their behavioral effects indicates that further developments in psychopharmacology will be made during the next few decades. Many new agents will be introduced to treat behavioral disorders. It will be difficult to develop drugs that will do more than dampen behavioral alterations toward the norm and that will actually enhance normal human behavior along specific dimensions.

There can be no doubt that man will continue to use chemicals to alter his behavior. This practice has existed since the beginning of recorded history, and cultural anthropologists find evidence of far earlier drug use in legends and languages. Further biological research may make it possible to avoid many of the dangers of drug use and to obtain exactly the behavioral effects that are desired in a particular instance.

FURTHER READING

For general information about drugs and their effects, see books and articles by Blum and associates (1969b), Claridge (1970), Cohen (1969), Louria (1968), and Tinklenberg and Stillman (1970). Useful handbooks for information about names of drugs and their effects are those by Burak (1970), Lingeman (1969), and Modell (1970). Problems of drug abuse, addiction, and treatment of addicts are discussed from various viewpoints by Eldridge (1967), Leech and Jordan (1967), McLean and Bowen (1970), Pollock (1969), Redlich and Freedman (1966a), Wikler (1970), and C. W. M. Wilson (1968).

For further information about amphetamines, see works by Kalant (1966) and Russo (1968). For information about marijuana, see works by Bloomquist (1971), Kaplan (1970), and McGlothlin and West (1968). Alcohol and alcoholism are discussed by Blum and Blum (1967), Catanzaro (1968), Fox and Bourne (1971), and Redlich and Freedman (1966b). For further information about opiates, see works by Kramer (1969) and Wikler (1970).

Psychedelic drugs are discussed by Blum and associates (1964) and Cohen (1964, 1967, 1969).

44
Human Possibilities

Now is a particularly interesting time in human evolution to attempt to look ahead at the future of the human species. During most of his existence on earth, man has been numerically insignificant, has been preyed upon by carnivores, and has supported himself by hunting and gathering; therefore, he has been greatly influenced by variations in the ecosystem of which he is a part. During this long period of evolution, man's ancestors have been responding to environmental forces and social interactions only dimly perceived at the time.

Until the agricultural revolution about 10,000 years ago, the change in circumstances of living was slow indeed. Since then, the rate of change has accelerated. Within the last century, man himself has changed the circumstances of his daily living more than in all preceding human existence. Man now dominates – even if he does not control – the ecosystem (Chapter 45). Today, he finds himself in a powerful position to alter the world and himself, and he is currently and painfully realizing the awesome implications of this responsibility. Whether the human species continues to survive no longer depends upon predators and parasites or on changes of the environment – it now depends on man.

HUMAN GENOTYPES AND SELECTION

The central tenet of evolutionary theory is the principle of natural selection. Operationally, this tenet reduces to the concept that, because of genotypically based advantages, certain individuals have a higher probability of survival and reproduction than others. The genetic basis for human survival and reproduction probably is changing, with an associated change in the nature of natural selection. But in precise terms of gene frequencies, human biologists are almost totally ignorant of exactly what is happening. The story of the resistance to *P. falciparum* malaria conferred by the gene responsible for the sickle cell trait is well known, and it is possible to predict with some accuracy what the disappearance of malaria from the world would mean for the frequency of that particular gene. There are several other genes whose selective implications are reasonably well understood, but beyond that, there is a lack of knowledge.

The potential complexity of the human genotype is staggering. The amount of DNA in a human gamete is sufficient to supply the code for approximately 10 million polypeptides. In contrast, the bacteriophage T-4 has sufficient DNA to code for approximately 160 polypeptides. The proportion of the genetic material that is actually utilized, however, is unknown. In T-4, some 100 different genetic loci associated with specific functions have been identified. In the very simple bacteriophage ϕx174, the coding potential, on the basis of DNA present, is probably less than 10 polypeptides, and 8 different genetic functions have been identified. Thus, for ϕx174, there are specific functions for all or almost all of the genetic material; for T-4, there are specific functions for a minimum of approximately 60 percent of the coding potential.

How very different is the situation for man, for whom almost 2,000 distinct inherited traits have been recognized. These traits include such malformations as achondroplasia (a kind of dwarfism) and lobster claw hand (a skeletal deformity). However, knowledge of such malformations does not reveal what the normal, nonmutant form of the responsible gene does. Intelligence and stature are known to have a genetic component, based on many genes; the way that these individual genes function is unknown.

Figure 44.1. Chromosome map of the bacteriophage virus T-4. The locations of gene loci controlling a variety of functions is shown. Mat. Def. = maturation defective; DNA Neg. = no synthesis of phage DNA under nonpermissive conditions.

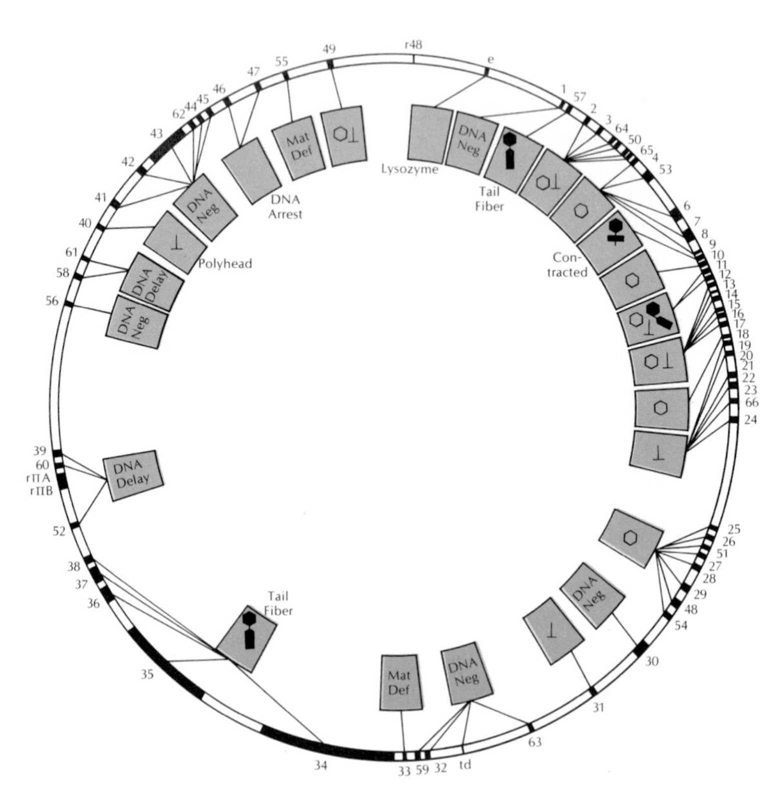

Furthermore, as a result of biochemical screening, dozens of inherited variations are now recognized in the complex proteins found in human blood serum or red blood cells. In most cases, however, nothing is known about the biomedical implications of possessing a different kind of serum protein or red cell enzyme. The inability to define the effects of an abnormal gene implies a considerable gap in knowledge of the role of its normal counterpart in the total economy of the organism. For only a few dozen genes can it be stated how changing them in a specified manner would influence human functioning. Genetically, man is an unknown proposition.

Furthermore, it is impossible at this critical juncture in the world's history to visualize the future environment in which the human genotype will be called upon to function. If overpopulation is not brought under control, not only will the standard of living fall, but the widespread famines that could result from crop failure will threaten the cultural integrity of the countries concerned. On the other hand, if population control is achieved, theoretically man could move into a plateau of relative plenty. These diverse developments might well result in very different environments, in which different kinds of human beings would survive, reproduce, and be judged "successful."

In recent years, it has been fashionable to think that man has somehow placed himself beyond the reach of natural selection. It is true that he has altered its base, but selection still exists. Death during infancy and childhood is less common, and the causes of mortality have changed, but even today in the United States, some 15 percent of all liveborn infants do not survive to complete the reproductive period. Differential fertility, the other

Figure 44.2. An artist's interpretation of the human condition, contrasting the circumstances that could arise from overpopulation with those that might be realized with population control.

Figure 44.3. A cemetery in the borough of Queens, New York. This picture emphasizes the conditions of overcrowded urban centers.

mechanism that determines the contribution of each person to the gene pool of the next generation, is today for women (but not for men) perhaps even greater than in prehistoric times.

Faced with these uncertainties, it would be foolhardy to predict the biological future of man. That future depends to a large extent on conscious human decisions rather than on the inexorable operation of natural laws. However, first it is possible to discuss some feasible and probable developments well within human power and their likely consequences and then to glance briefly at some of the far-out possibilities.

THE IMMEDIATE FUTURE

In the immediate biological future of man, there are several possible developments, well within human choice, about which conscious decisions must be reached and the probable decisions seem clear. Although these decisions are not spectacular, nevertheless they have enormous implications for the human species.

Zero Population Growth

It is clear that the world's population cannot continue to expand at the increasing rate of the past several centuries. Some experts feel that the number of people on the earth already exceeds the planet's long-range capacity to support such a number comfortably. They predict that—because population growth cannot possibly be brought under control within the next decade or two—the numbers of people in certain large areas soon will so far exceed the carrying capacity of those areas that large-scale famine will ensue with the first crop failures. Other experts feel that there is room for modest expansion in the near future, depending on how the earth's resources are utilized. A clear decision between the various viewpoints is impossible. Surely, the humanitarian point of view demands the more cautious approach. The principle of zero population growth emphasizes that everything possible must be done to stabilize population numbers at once, although this decision does not preclude further growth should future resources prove much greater than they seem at present. At present, the average United States family includes about 3.3 children. In the future, if this number were reduced to 2.5 children per family, the rate of growth in the United States would eventually decrease to zero (Ehrlich and Ehrlich, 1970).

Any decision to curb population growth, to be effective, requires an unprecedented type of concerted action on the part of governments and religious authorities. It is clear that the first question in any effort to reach a world-wide policy is by what guidelines the reproductive performance of an individual couple should be constrained. More specifically, should an effort be made to promulgate world-wide standards for parenthood?

Such an effort would probably stall almost immediately because of differing opinions. One acceptable approach—and even this one will find some dissenters—is a simple quota system, with a maximum of three children per couple. Because some people fail to marry, are infertile, or limit family size to less than three, the average number of children per couple would be approximately two. After attaining the quota, a couple could be sterilized if they so desire.

It is important to emphasize the genetic consequences of such a program. With present low infant and childhood mortalities, differential fertility is the principal mechanism whereby the composition of the gene pool is changed. Two children will not between them have a representative of each and every one of the genes present in their parents. Because of the laws of genetic segregation and recombination, in such a replacement population there may be some change in the gene pool from one generation to the next. On the other hand, if this method of population control were adopted, there would be a much greater stabilization of the gene pool than in a "state of nature." Because, in the final analysis, biological evolution consists of changes in the gene pool, the rate of evolution would be slowed down at a time when some feel it should be accelerated. Unfortunately, neither the knowledge nor the wisdom for choices based on individual merit are at hand. It is far more important to bring population growth under control quickly than it is to make provision for the small amount of human evolution that might occur in the next 50 years.

A program to control population growth will succeed only with a total commitment from many sources of authority, such as governments, religions, labor unions, and agricultural cooperatives. It is doubtful that this commitment—and a concurrent deceleration of the birth rate—can materialize rapidly enough to avert the famines prophesied by some. The program can succeed only if it threatens the status of no one. A future development in the direction of a stabilized population seems necessary and highly probable.

Protected Gene Pool

Much has been written about the threat to the gene pool posed by man's increasing exposure to ionizing radiation. More recently, the problem of chemical mutagenesis—from potent insecticides, herbicides, air pollutants, artificial sweeteners, and so on—has attracted increasing attention. A reasonable current estimate of the genetic effects of human exposures to man-made sources of radiation determines that such exposures produce about $1/30$ to $1/60$ as many additional mutations as those that occur spontaneously. This figure is small, based on animal experimentation rather than human data, but translated into millions of births, it implies some increase in the number of defective children (although no one can make a reasonable guess as to how many). The effect of chemicals is even less predictable. Extensive studies are urgently needed.

If it is proven beyond doubt that current exposures to radiation and chemicals are increasing the mutation rate, with the likelihood of deleterious consequences to some future child, then man faces a frustrating dilemma. Radiation is a by-product both of medical diagnostic procedures designed to alleviate human suffering and of atomic energy installations designed to produce the energy that the culture demands to support its standard of living. Exposure to pesticides results from efforts to ensure bountiful crops, and exposure to air pollutants results from modern transportation methods. The total elimination of these noxious influences at the moment is incompatible with the functioning of the culture. It is desirable to minimize these exposures—but rationally—so that the best possible balance between gain and loss is reached in the end.

Like stabilizing population, reducing the mutation rate is a conservation measure. Supporters of such measures visualize a slight increase in mutation rates but only a minor increase over the course of the next 100 years,

Figure 44.4. An artist's interpretation of the equilibrium or balance between cultural gains and environmental losses that must be achieved if the quality of life is to be maintained.

which is as far ahead as it is feasible to look. This effort to minimize exposures to mutagens will contribute to a stabilization of the gene pool.

Improvements in the Quality of Life

The near future is likely to bring three kinds of developments related to improvement of the quality of human life. These are (1) improved treatment of genetically determined disease, (2) decreases in proportions of genetically defective children through genetic counseling and prenatal diagnosis, and (3) improvements in the environmental factors that limit realization of genetic potentials.

Man already has the theoretical capability of eliminating diseases due to nutritional deficiencies and most infectious agents. What is left for future control are the genetically determined diseases and those diseases—such as gout, diabetes, and high blood pressure—that result from an interaction between genotype and external environment. Most people are unaware of the spectrum of approaches presently available for the treatment of genetic disease (Scriver, 1967).

Some genetic diseases result from a block in one step of a chemical reaction. An example is phenylketonuria (PKU), a disease caused by a great decrease in the body's ability to convert the amino acid phenylalanine to tyrosine. Because of this block, large quantities of phenylalanine and certain related compounds accumulate. For reasons not yet clear, this accumulation has serious consequences for the developing central nervous system that produce mental retardation. Special diets now have been developed—diets that are very low in this essential amino acid but that still contain enough phenylalanine to meet physiological requirements. If the diagnosis is made and therapy instituted within the first few weeks of life, mental development appears to be normal.

Other genetic diseases result from the inability of the body to synthesize essential compounds. For example, normal development depends upon adequate amounts of the thyroid hormone. If there is a deficiency, a severe form of physical and mental retardation termed cretinism results. The biosynthetic pathway by which the thyroid hormone, thyroxin, is synthesized from the amino acid tyrosine is now well known. Each of the successive steps in this pathway is under genetic control, and genetically determined failures for most of these steps are known. Any of these failures would cause a child to become a cretin. If the diagnosis is made early in life and the end product of this chain, thyroxin, is administered, development is normal.

Some biochemical reactions also require a low-molecular-weight compound termed a coenzyme or cofactor for the optimal function of the enzyme. Such a coenzyme may actually bind to the enzyme in order to facilitate its proper functioning. Vitamin D is known to act as a coenzyme for certain reactions. One dominantly inherited disease, known as vitamin D-resistant rickets, is characterized by all the skeletal deformities usually seen in rickets due to vitamin D deficiency. This disease can be prevented if a child is given approximately 100 times the normal requirement for vitamin D. Although the mode of action is not entirely clear, one possibility is that a mutation has altered the binding of the coenzyme (vitamin D) to the enzyme concerned. This inefficient binding can be corrected by greatly increasing the availability of the coenzyme.

Currently, there are some new possible approaches to the study of genetic disease, none of which has yet found application. Some human diseases may be due to failure of a normal operator or regulator gene (Chapter 17). One such disease is the very rare, recessively inherited disease oroticaciduria, characterized by a severe deficiency of two enzymes, orotidine-5′-monophosphate pyrophosphorylase and orotidine-5′-monophosphate decarboxylase. Certain critical evidence is lacking to prove that an operator or regulator gene mutation is involved, but the fact that two consecutive enzymes in a biosynthetic pathway are affected simultaneously renders this hypothesis quite possible. The result of the missing enzymes is a deficiency of the pyrimidine uridine; affected children are characterized by retarded development and by anemia. When cells from affected persons are maintained in tissue culture, the metabolic defects persist. If barbituric acid (a close chemical relative of the barbiturates used in tranquilizers) is introduced into the culture medium, the cells develop greatly increased levels of both enzymes affected by the gene, from less than 1 percent of normal activity to 50 percent. Although this level is far short of normal, it does illustrate the partial ability of a chemical to reverse the basic enzymatic defect in a genetic disease. The feasibility of this approach for the treatment of the disease is not yet known.

The specifications for the synthesis of a particular protein are taken to ribosomes by messenger RNA, and the proper amino acid building blocks for each protein are assembled at the ribosome by transfer RNA. It is possible that some human diseases result from genetically determined defects in these two essential components of the protein synthesizing machinery. If such diseases do exist, the possibility of replacement therapy is problematical. However, there may be other approaches to defects in such systems. In a tissue culture system utilizing cells derived from a tumor of human liver, the process of translation of the mRNA for the enzyme tyrosine transaminase can be influenced by a steroid hormone of the adrenal cortex. Thus, the potential for influencing cell function at this level does exist.

One obvious practical disadvantage of the therapeutic approaches discussed thus far is that treatment usually must be continued throughout life. A fanciful example of the therapeutic possibilities of the future can be based on sickle cell anemia, a severe disease known to be caused by a single pyrimidine substitution in the structural gene that codes for one of the two protein chains of normal hemoglobin. Based on analogy with bacterial experiments, it may someday be possible to remove bone marrow from an affected person, to alter the region of the chromosome in which that gene is found (through transformation or transduction), and then to reimplant the treated cells, now capable of making normal hemoglobin. At present, this possibility is extremely remote. The practical difficulties and dangers in the development of such an approach are enormous. When placed in tissue culture, bone marrow cells quickly stop making hemoglobin. Thus, by the time the genetic change could be effected, the cell might have lost its specialized functions. Therefore, the success of the transformation could not be tested, nor (if it could) would the transformed cells make hemoglobin after their return to the body. However, the real danger is that far more than the gene in question might be altered, and the replaced cells might also have the potentialities for cancerous growth. This development as applied to man now seems remote, but it is an example of the possibilities stemming from genetic advances of recent years. It is important to remember that persons so treated would still transmit the defective gene and, with their health improved, might transmit the gene to more children than would otherwise be the case.

By virtue of their genetic constitution, some parents run a much greater risk of producing defective children than others. Genetic counseling is a service aimed at providing such parents with exact risk figures, so that they can plan their families more intelligently. This service is relatively new but soon will be available at all major medical centers.

Until recently, a set of "high-risk" parents could take positive action only by a decision to limit family size—perhaps accompanied by adoption. Now, however, there are developments that significantly alter the strategy of counseling. Using the technique of amniocentesis, a physician can remove a small quantity of amniotic fluid from the uterus during the early months of a pregnancy. He can then immediately examine the cells present in the fluid to determine the sex of the child or grow the cells in culture. Two examples will illustrate the potential diagnostic significance of this development. First, imagine a woman, known to be a carrier of the sex-linked gene responsible for severe (type A) hemophilia, whose husband is normal. Her sons have a 50 percent chance of hemophilia, but her daughters will be normal (although half of them will carry the gene). Such a woman would probably prefer to have only daughters. The fact that she is carrying a son now can be determined at an early stage of pregnancy, allowing the option of an abortion. The second example involves Down's syndrome, a severe mental defect, often accompanied by other abnormalities, due to an extra chromosome 21. In some families, there occurs a particular chromosomal abnormality (translocation) that predisposes to an abnormality in meiosis, resulting in this syndrome. The carriers of this particular abnormality have about a 10 to 30 percent chance of producing an affected child. By monitoring the pregnancies of such women, doctors can detect this defect at an early stage of pregnancy.

In each case, the diagnosis becomes a means of controlling disease only if followed by an abortion. This limiting factor places a premium on early diagnosis and raises moral questions concerning the indications for an

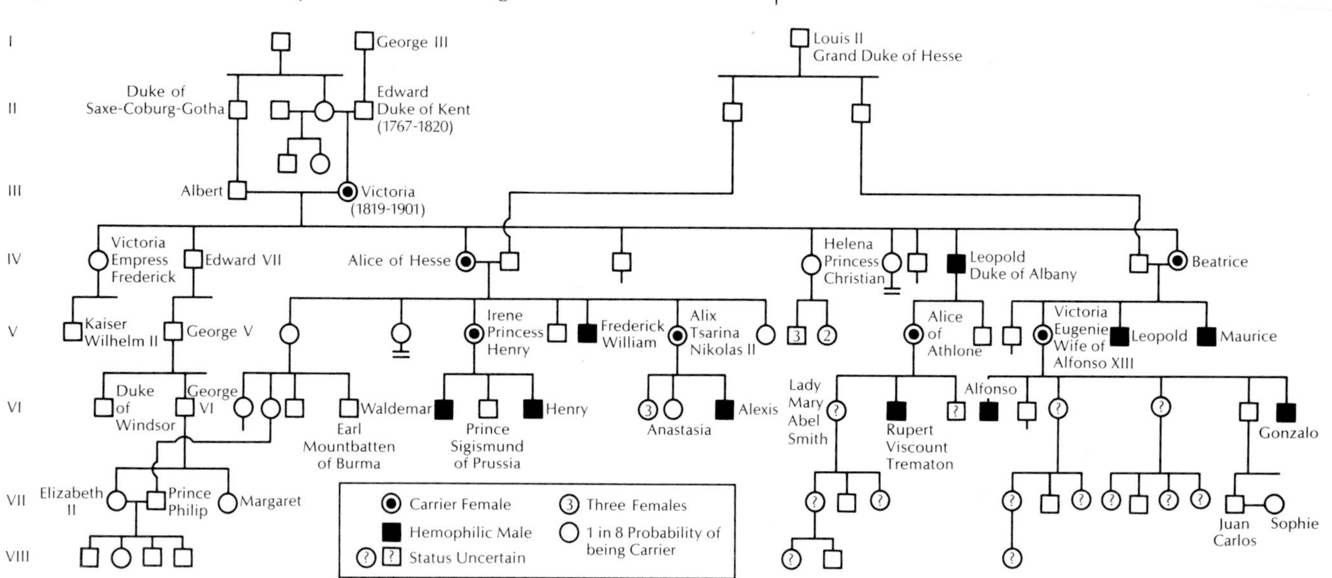

Figure 44.5. Pedigree chart tracing the inheritance of hemophilia, a sex-linked recessive gene, through the royal families of Europe. Queen Victoria can be traced as the original carrier of the hemophilic gene.

abortion. Because there is a growing tendency to justify a severe defect as a legal indication for abortion, the issue now is more religious than legal. And because a couple whose religious beliefs do not permit abortion are unlikely to seek prenatal diagnosis, there is not likely to be a conflict.

The number of diseases that can be diagnosed prenatally is very small, and the high-risk couple may not be identifiable until they already have borne one defective child. In some instances, there are tests for carrier states in prospective parents, but because the cost of universal administration of such tests might be so great as to far outweigh the cost of caring for the affected children, there is a problem in medical economics. Thus, it would be unreal at present to expect a dramatic impact on the frequency of genetic disease from this development.

Each human individual in some way falls short of reaching his full genetic potential. For some people, born into exceptionally favorable circumstances, the discrepancy between the possible and the attained is slight. But for others, who are born to an ill-cared-for mother, living on a diet so deficient as to impair the development of the central nervous system, and given limited schooling and job opportunities, the gulf between the innate potential and the realized performance may be wide. The complexity of the human genotype is such that each individual probably differs from every other and requires a tailor-made environment for maximum realization of his potentialities. Obviously, a tailor-made environment is out of the question. But it is not out of the question to develop distribution of material goods, of educational opportunities, of medical services, and of the physical setting for living that ensures each person essentially the same running start on life.

What is involved here is "cultural engineering." Man now so dominates his environment that he consciously must assume responsibility for it in all of its aspects. The development of a "cure" for one of the genetic diseases mentioned above is dramatic. A child so cured gains a useful life. But probably no more than 1 or 2 percent of all children ultimately will be found to have clear-cut, genetically determined, metabolic defects subject to simple manipulation. What can be done for the others, who face an increasingly complex environment? The manipulation of that environment so as to add an average of 10 IQ points to each individual's intellect, accompanied by a proper social orientation, conceivably could determine whether man will cope successfully with his increasingly complex world.

It may seem naïve to hope that man might consciously strive toward this optimal environment. But what are the alternatives? What greater challenge does society face? Surely the production of more material goods will not better man's condition, nor will further improvement of means of self-destruction.

Environmental engineering embraces both positive and negative aspects of life, with the negative features posing greater long-term problems to man. There are major, man-made changes in the environment that probably should be decelerated or stopped: increasing atmospheric CO_2; increasing particulate content of the atmosphere; accumulation of diverse chemicals in water; soil erosion and destruction; replacement of fertile farms and woodlands by highways and towns; rising noise levels; and thermal pollution. But what of the positives?

As man manipulates his environment to further develop minds and bodies, his cultural milieu will impose a sense of values. The range of cul-

Figure 44.6. In order to improve the quality of life, man must first redefine his physical goals. The present value system, which tends to equate a "better" life with the consumption of more and more material goods, creates serious problems of trash accumulation and disposal.

tures that have been encountered on the earth leave no doubt of the great plasticity of the human animal. Compassion, sensitivity, and thoughtfulness are largely acquired rather than transmitted traits. Although the range of human adaptability is great, there are limits. One limitation involves population numbers and the disposition of people. The endrocrine functions of rats and mice are in part a function of population density. If the same is true for man, then perhaps aspects of present living arrangements must be altered to give cultural engineering a full chance.

Because of the genetic diversity of man, there can be no one best environment. In the foreseeable future, society probably will not be able to provide each child with a special environment. There may be an intermediate course whereby, at a fairly early age, children get started on special "tracks" according to their special needs, either nutritional or educational. Although most of the concern about tracks has thus far been directed toward the disadvantaged, now equal emphasis must be directed toward other groups. It will be difficult to establish such tracks and still retain the integrity of the nuclear family, considered to be so important for normal development.

A small, possible opportunity to tailor the microenvironment to the individual is provided by the newly recognized condition of $alpha_1$ antitrypsin deficiency. The severe form of this entity, characterized by very low levels of this enzyme inhibitor in the blood serum, is caused by homozygosity for a single gene. There is growing circumstantial evidence that people with this disorder who are subjected to air pollution, especially those who smoke heavily, are unusually prone to chronic obstructive lung disease

Figure 44.7. Incidence of emphysema in two cities with contrasting levels of air pollution. In a 1960–1966 study, post-mortem examinations were performed on the lungs of 300 residents of heavily industrialized St. Louis, Missouri, and an equal number from relatively unpolluted Winnipeg, Canada. The subjects were matched by sex, occupation, socioeconomic status, length of residence, smoking habits, and age at death. These data clearly suggest a link between air pollution and pulmonary emphysema.

(emphysema) and die prematurely from respiratory failure. Although the evidence seems conclusive that all cigarette smokers are unusually prone to lung cancer, persons with this defect have an added hazard. Good environmental engineering (preventive medicine, in this case) would establish the identity of such persons at an early age.

CONSEQUENCES FOR THE GENE POOL

The developments discussed above would improve the quality of life for the next generation. Some of them have thought-provoking implications for future generations. The humanitarian orientation of the present generation actually may degrade the gene pool of the future. In the past, a child with phenylketonuria or genetically determined cretinism seldom reproduced offspring. If that child is rendered phenotypically normal, the rate of transmission of the defective gene to the next generation will be increased.

Prenatal diagnosis and abortion of a genetically defective child may also (paradoxically) result in an increased rate of transmission of defective genes to the next generation. For example, many recessively inherited entities toward which prenatal diagnosis is or will be directed are compatible with prolonged existence but a very low rate of reproduction. Under past conditions, an affected child usually would remain in the home for a variable period of time, absorbing parental love and attention. Given any kind of family planning, it seems a reasonable hypothesis that such a child often would tend to inhibit the parents from having further children. The advent of prenatal diagnosis might alter the dynamics of the situation, for example, in the case of hemophilia A.

If limited reproduction on the part of hemophiliacs is assumed—certainly true until recently—in the past, approximately one-third of existing genes for hemophilia have been lost in each generation, with the gene frequency being maintained by mutation. Even today the reproduction of hemophiliacs is greatly impaired. Suppose that female carriers of the gene for hemophilia can be identified and that, given the prenatal diagnosis of sex, all such carriers will elect to abort male fetuses (with a 50 percent probability of hemophilia) and carry to term only females. Further, assume that in the world of tomorrow each couple has only two liveborn children who survive and reproduce. Without prenatal diagnosis, each carrier mother will have an average of 0.5 hemophilic sons, whose genes seldom will be transmitted. In other words, she will on the average have one daughter, who will have a 50 percent chance of possessing the gene for hemophilia. But, if instead she now has two daughters, both with a 50 percent chance of possess-

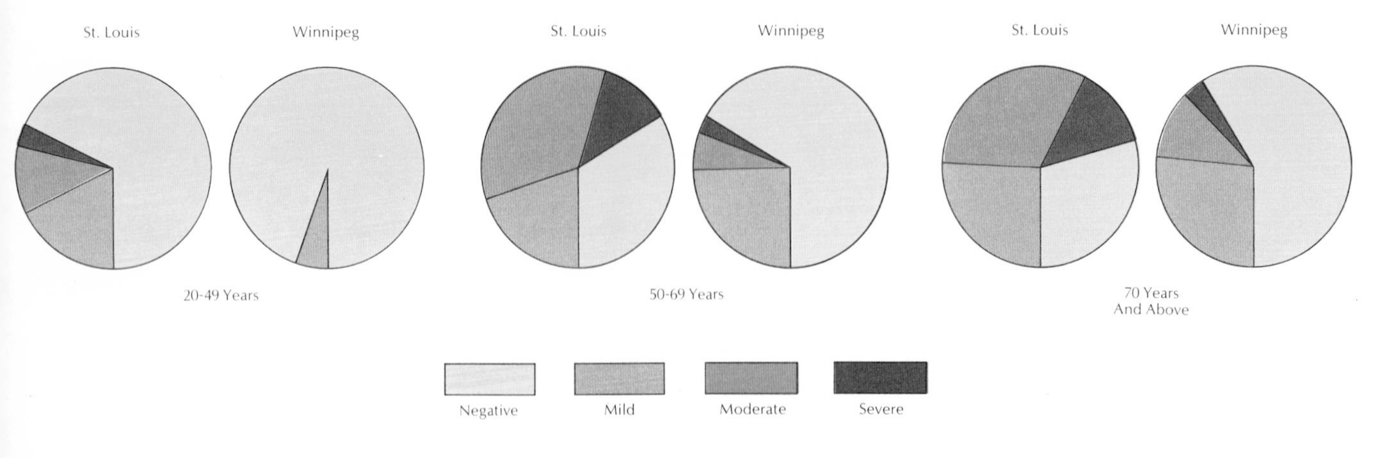

Figure 44.8. Tables showing the incidence of hemophilic victims and carriers (A) and assuming normal genetic segregation (B). In A, assuming the special conditions described in the text statistics are given for five generations of offspring produced by a normal father and carrier mother (4 offspring per generation). In B, the statistics relate to the the conditions in the text (2 offspring per generation).

Potential Number of Hemophiliacs and Carriers Produced by One Carrier Mother and One Normal Father

A Assuming Normal Segregation

GENERATION	Normal Offspring Produced (♀ and ♂)	Carrier Offspring Produced (♀)	Hemophilic Offspring Produced (♂)	Total Offspring Produced (♀ and ♂)	Fraction and Percentage of Carriers to Total Population Produced
1	2	1	1	4	$1/4 = 25$
2	12	3	1	16	$3/16 = 18.7$
3	56	5	3	64	$5/64 = 7.8$
4	240	11	5	256	$11/256 = 4.3$
5	992	21	11	1,024	$21/1064 = 2.0$
Total	1,302	41	21	1,364	$41/1364 = 3$

B Assuming Special Conditions (described in text)

GENERATION	Normal Offspring Produced (♀ and ♂)	Carrier Offspring Produced (♀)	Hemophilic Offspring Produced (♂)	Total Offspring Produced (♀ and ♂)	Fraction and Percentage of Carriers to Total Population Produced
1	1	1	0	2	$1/2 = 50$
2	3	1	0	4	$1/4 = 25$
3	7	1	0	8	$1/8 = 12.5$
4	15	1	0	16	$1/16 = 6.2$
5	31	1	0	32	$1/32 = 3.1$
Total	56	5	0	62	$5/62 = 8$

ing the gene for hemophilia, then the probability of transmitting the gene to the next generation is doubled. Now the gene no longer is subject to negative selection and, because of continuing mutation pressure, in three generations its frequency in the population will approximately double.

That developments in medicine and sanitation might have adverse implications for the gene pool is scarcely a new thought. It has been voiced repeatedly in recent generations. But only recently has detailed knowledge of human genetics permitted some quantification of the possibilities, although much is yet to be learned. However, it seems safe to say that, given the low frequency of the simply inherited genetic diseases and the relatively slow rate of gene frequency change when selective conditions are altered, there is little prospect of a runaway process. The present ethical framework really gives no choice but to alleviate suffering by every means available. The challenge is to alleviate suffering and to develop a social structure within which continuing biological adaptation to changing conditions is possible.

THE DISTANT FUTURE

The potentiality for genetic change within the human species is staggering. An analogy commonly drawn is the history of the dog since domestication. From the starting point of the wolf (with possible admixture from several other wild canines), breeds have been developed as disparate as the St. Bernard, Chihuahua, bulldog, and dachshund. There seems little doubt that selective breeding of the human species could lead to the same diversity. Even if such a possibility was ethical, by what standards should it be guid-

Figure 44.9. An artist's interpretation of selective breeding of the human species—diversity or stereotype?

ed? In the case of the dog, bred primarily for hunting or other specialized purposes, the standards are simple. Standards are also simple in developing the egg-laying machine that is the White Leghorn chicken or the machine for converting plant food to animal protein or milk that is seen in the Charolais and Holstein breeds of cattle. But for what qualities would man be bred? There are perhaps only two qualities on which there would be near agreement: intelligence and nonaggression. And agreement even on these two perhaps is reached from a negative approach: stupidity and unbridled aggression can scarcely be seen as assets in this complicated world. Furthermore, depending on how the world goes, the desirability of even these two traits must be carefully scrutinized. The complicated world into which man is moving calls for leadership; to what extent can this trait be divorced from aggression? And if automation becomes the order of the day, what of high intellects condemned to operate a console of buttons?

No brief chapter on the future of man could be complete without some reference to genetic engineering. This term is used in a variety of ways; some geneticists prefer to restrict it to directed changes in the transmissible genetic material (genetic surgery in the germ line), but others would extend it to include some of the possible therapeutic developments discussed earlier. The detailed structure of the gene is now understood, and significant steps have been taken toward the synthesis of the gene. Genes can be transferred from one (bacterial) cell to another by the techniques of transformation and transduction. The theoretical possibilities for altering the human genotype are obvious. Experimentation already is under way with human cells growing in tissue culture and with experimental mammals.

The opening of this chapter stressed how little is known of genetic man. Elsewhere in this book are examples of how man's ingenuity has outstripped his wisdom. The ultimate now left for him to violate is that mysterious strand of DNA, some 3 billion years in evolving, that makes each human what he is. In a sense, that strand is altered by developments such as those discussed above that alter the gene pool. These are nonpurposive alterations stemming from a strange mix of ethics and ignorance. But now to alter the strand consciously—even in the obvious case of correcting the genetic defect responsible for phenylketonuria—tampers with a situation of such complexity that clearly the time is not yet at hand. Moreover, dramatic and intellectually appealing as this development is to some, it is quite certain that the simpler developments discussed earlier will have a far greater numerical impact on the gene pool.

Man can see his own future only very dimly. The world today is in what can only be called an unstable state. The human species is in a rapid transition, from man the hunter to, perhaps, man the master. The success of that transition is by no means assured. There are those who feel the time to make that transition is rapidly being exhausted. If the next 100 years are the critical ones, there is no prospect in so short a time of developing the superman who will solve these pressing problems: we must make do with what we are.

FURTHER READING

The future evolution of the human species is discussed by Dobzhansky (1960, 1962), Dubos (1965), Medawar (1960), Stebbins (1970), and Wolstenholme (1963). Further information about genetic engineering and other deliberate biological efforts to alter human evolution is given by Handler (1970) and Scriver (1967).

45
The Environment

During the past decade, increasing concern about the human "population explosion" and its effects upon the environment has been expressed by biologists, conservationists, and even politicians. It seems clear that man must deliberately choose to limit his increase in numbers. If he fails to do so, his prolific rate of population increase will be countered inevitably by a rising death rate from factors such as famine, plague, and war.

On the one hand, it is argued that pessimists have always proclaimed imminent disaster for the human species because its population is about to outrace its food supply but that agriculture and technology have always met the challenge and the population has continued to increase. On the other hand, it is argued that the human population now is fast approaching absolute limits to its growth and that a decline in the growth rate soon is physically inevitable. Both sides of this argument can present convincing evidence to back their views. In order to obtain a balanced view of human population phenomena, it is necessary to look at the situation in the perspective provided by the study of human origins.

BEHAVIORAL LEGACY OF THE SAVANNA

The sources of both human greatness and human tragedy were stamped into the genetic program of the species by events and forces thousands and even millions of years ago. The single event with the greatest impact was perhaps the Pliocene drought that persisted through some 10 million years in Africa. This drought pushed back the ancient, protective forest that had cloaked Africa through the preceding 12 million years of the Miocene epoch, replacing it with the parklike habitat called the savanna. A new habitat calls for a new life style, and one group of semierect apes was quick to respond. They soon became a dominant species—scavenging, hunting, and eventually warring. The new life style developed by these man-apes to exploit the savanna was similar to that of the wolf. How did the peaceful herbivore of the forests evolve into the cunning strategist of the open plains?

Except for fossils, direct evidence of evolutionary events is nonexistent, but the comparative approach suggests some explanations. All three of the great apes now living (the chimpanzee, orangutan, and gorilla) are forest dwellers. Assuming that early Pliocene human ancestors behaved somewhat as these apes do today, scientists can make some educated guesses about human behavioral background. First, if not tool makers, the human ancestors probably were tool users. All modern apes use tools. Chimpanzees and gorillas crush leaves and sop up water with these makeshift sponges. Thin twigs or leaf ribs are placed into termite colonies, later to be removed and stripped of the attacking soldier termites that have clamped on with their mandibles. Enraged apes throw rocks and attack with clubs.

Just as important as tools, though, is the troop organization of semiterrestrial apes. A dominance hierarchy of adult males is the "spinal cord" of nearly all primate troops. This social pattern provides order, stability, leadership, and predictable behavior when danger threatens. Groups of males also are adventurous. Adult male bands of chimpanzees roam widely on scouting expeditions, calling and drumming trees loudly when a good food source is found.

If it were not for intelligent weapon using and the cooperative adventurousness of male man-apes, they hardly could have become successful savanna animals. Exploratory male groups and the use of tools would have

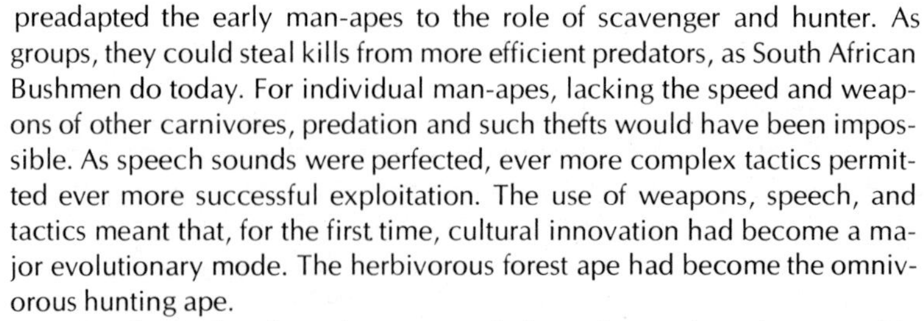

preadapted the early man-apes to the role of scavenger and hunter. As groups, they could steal kills from more efficient predators, as South African Bushmen do today. For individual man-apes, lacking the speed and weapons of other carnivores, predation and such thefts would have been impossible. As speech sounds were perfected, ever more complex tactics permitted ever more successful exploitation. The use of weapons, speech, and tactics meant that, for the first time, cultural innovation had become a major evolutionary mode. The herbivorous forest ape had become the omnivorous hunting ape.

Most physical anthropologists now believe that such a change in life style of some Pliocene apes set the stage for the origin of true men millions of years later. Great reproductive advantage must have accrued to those groups and individuals making the best use of culture. As a consequence, growth and improvements in the brain regions responsible for dexterity, memory, learning, and thought (particularly the cerebral cortex) were at a premium. Increase in brain size was slow at first, but by the end of the Pliocene it was virtually explosive, increasing at between one and two cubic centimeters per 1,000 years.

Cultural evolution seems to have lagged a few thousand years behind the physical evolution of the brain. The hunting-gathering life style of the Pliocene and Pleistocene persisted until the last recession of the glaciers. Then, with the invention of agriculture and animal domestication about 11,000 years ago, culture began to evolve in an exponential fashion. Metals soon replaced stone for tools and weapons. The pace then quickened with the industrial revolution. With the coming of the technological, biomedical, nuclear, and space ages, man now changes more behavioral patterns in a generation than he did in a millennium before the invention of agriculture. The biological evolution of the large human brain made possible civilization and mass communication. Now the natural selection of ideas and technology (rather than the natural selection of genotypes) shapes human behavior and does so in a way that seems to shrink time. For example, man has learned in 60 years to fly faster and higher than birds did in 60 million years.

During the millions of years on the savanna, hunting and killing became ends in themselves. Food had to be pursued, subdued, and transported to the waiting females and infants. Natural selection must have favored genotypes (individuals) with a passion for the hunt. Man evolved as a hunter and remained a hunter through more than 99 percent of his history. As long as evolution proceeded slowly, the ability to kill presented no serious problems. All social predatory species have evolved behavior patterns that buffer against excesses of violence and prevent individuals from killing each other (Lorenz, 1966). Students of animal behavior recognize these as submissive displays, or appeasement gestures. Baboons "present" their hindquarters to calm an angry superior. A wolf seemingly bent on murder during a squabble will be pacified when the weaker animal turns away its head and exposes the vulnerable and arched side of the neck. Humans are in trouble today because appeasement postures function only in proximity. In hand-to-hand combat with only fists or clubs, there is time for the beseechings and pleas of the vanquished to elicit the mercy of the strong. But in rapid succession came the arrow, the bullet, and the thermonuclear ICBM. These cultural advances simply outpaced and outdistanced man's capacity for pity. It is probable, but not proven, that massacres of women

Figure 45.1. The man-ape *Australopithecus*, hunter and gatherer of the African savanna.

Figure 45.2. Whereas cultural evolution formerly lagged behind the physical evolution of the human brain, today the natural selection of ideas and technology shapes human behavior in a way that seems to shrink time.

Table 45.1
Birth, Death, and Growth Rates of Selected Modern Nations (1969)

NATION	Birth Rate*	Death Rate†	Growth Rate‡ (*percentage*)	Doubling Time (*years*)
Costa Rica	45	7	3.8	18
Philippines	50	15	3.5	20
Honduras	49	17	3.4	21
Mexico	43	9	3.4	21
Pakistan	52	19	3.3	21
Zambia	51	20	3.1	23
Iran	50	20	3.1	23
Mongolia	40	10	3.0	23
United Arab Republic	43	15	2.9	24
Brazil	38	10	2.8	25
Albania	34	8.6	2.7	26
Laos	47	23	2.6	27
Nigeria	50	25	2.5	28
Turkey	46	18	2.5	28
India	43	18	2.5	28
Indonesia	43	21	2.4	29
Chile	33	10	2.3	31
Cambodia	41	20	2.2	32
Guinea	55	35	2.0	35
Nepal	41	21	2.0	35
Canada	18	7.3	2.0	35
WORLD AVERAGE	34	15	1.9	37
Australia	19.4	8.7	1.8	39
Argentina	23	9	1.5	47
Mainland China	34	11	1.4	50
Uruguay	21	9	1.2	58
Greece	18.5	8.3	1.2	58
Puerto Rico	26	6	1.1	63
Portugal	21.1	10	1.1	63
Japan	19	6.8	1.1	63
United States	17.4	9.6	1.0	70
France	16.9	10.9	1.0	70
Denmark	18.4	10.3	0.9	78
Switzerland	17.7	9.0	0.9	78
Spain	21.1	8.7	0.8	88
Norway	18	9.2	0.8	88
Poland	16.3	7.7	0.8	88
Italy	18.1	9.7	0.7	100
Finland	16.5	9.4	0.6	117
Ireland	21.1	10.7	0.5	140
Czechoslovakia	15.1	10.1	0.5	140
West Germany	17.3	11.2	0.4	175
Hungary	14.6	10.7	0.3	233
Belgium	15.2	12.2	0.1	700
East Germany	14.8	13.2	0.1	700

*Rates in number of individuals per 1,000 total population.

†Low death rates in rapidly growing countries are largely a result of the small proportion of older individuals in those populations.

‡Calculated growth rates include allowance for immigration and emigration, which is very significant in some countries.

Source: Population Reference Bureau, *1969 World Population Data Sheet* (Washington, D.C.: Population Reference Bureau, Inc., 1969).

and children were at least more taxing emotionally when repeated clubbings instead of automatic rifles were involved. Violence is often erotic, and rape may have saved many an unfortunate female from a worse fate. Modern weapons combined with Pliocene appeasement signals have led man to the very brink of species suicide.

HUMAN POPULATION GROWTH

If the "appeasement gap" somehow is overcome for the next few decades and humans manage either through disarmament or through luck to survive to the year 2000, the species still will face the challenge of another biological anachronism—fecundity appropriate to the Pleistocene circumstances. Understanding of this problem requires a short detour into concepts of demography.

The size of a population, like the size of an individual, depends on the ratio between inputs and outputs. For an individual, the relevant input is energy in the form of food, and the relevant output is energy in the form of heat. The inevitable consequence of a sustained input-to-output ratio larger than one is obesity; a ratio smaller than one leads to starvation. For a population, natality (births) is the input, and mortality (deaths) is the output. The population grows if more individuals are added through births than are removed through deaths. Human natality and mortality statistics normally are given in the form of whole numbers. For example, the 1970 United States birth rate of about 18 means that about 18 births occurred during the year for each 1,000 persons in the average total population for that year. The 1970 death rate was about 10. The growth rate (r) of the population is the difference between birth and death rates: $(18 - 10) = 8$ per 1,000, or 0.8 percent. If the United States population continues to grow at this rate, the size of the population will double about each 70 years. Birth rates, death rates, growth rates, and doubling times for various modern societies are given in Table 45.1.

The human species existed for hundreds of thousands of years before its population reached 0.5 billion around the year 1650. By 1850 it had doubled to 1 billion. The next doubling occurred in 80 years (by 1930), and the population is about to double again in 46 years to 4 billion by 1976. The present doubling time is 35 to 37 years. In the year 2000, there will be 4 humans on the planet for every 1 that existed in 1900, or 28 for every person living at the time of Christ. As shown in Figure 45.3, the human population *and the growth rate of the human population* both are increasing at an accelerating pace. Projected population sizes are given in Table 45.2.

The forces behind the present explosive growth of humanity are not

Table 45.2
World and Regional Population Projections

YEAR	Projected Population *(millions of persons)*							
	World	Africa	Asia	North America	Latin America	Europe	Oceania	USSR
Mid-1969	3,551	344	1,990	225	276	456	19	241
2000 (UN estimate based on constant fertility)	7,552	860	4,513	388	756	571	33	402
2000 (UN medium estimate)	6,130	768	3,458	354	638	527	32	353

Source: Population Reference Bureau, *1969 World Population Data Sheet* (Washington, D.C.: Population Reference Bureau, Inc., 1969).

Figure 45.3. Growth rate of the human population for the past one-half million years. If Old Stone Age were in scale, its base line would extend about 18 feet to the left.

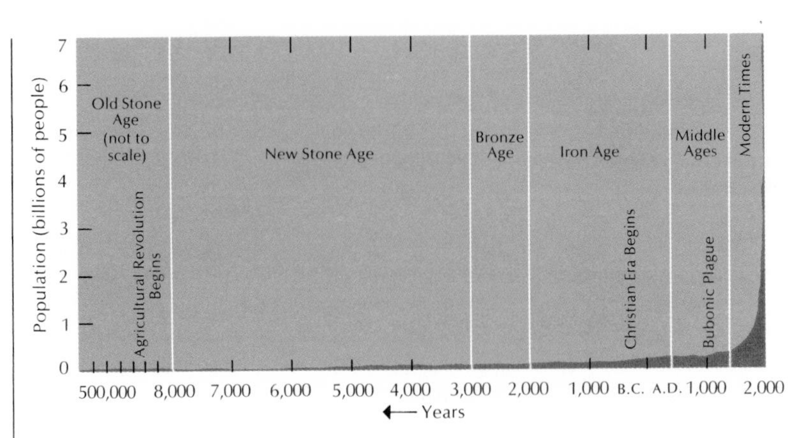

difficult to find. What is happening today is a precipitous decline in the world-wide death rate. Deaths, especially among infants and children, are occurring at lower rates, largely because of the discovery that microorganisms cause infectious diseases. A relatively small economic investment in antibiotics, immunization programs, and sanitation is bringing about a major demographic revolution in the underdeveloped countries.

Before medical science intervened, man's natural birth rate was just another example of what Malthus described as nature's "profuse and liberal hand." Man was no exception to the biological law of surplus. All organisms are geared to produce a surplus of offspring to compensate for the everpresent mortality factors: disease, famine, predation (or war), and genetic failure. In parts of Africa, Latin America, and Asia, Pliocene and Pleistocene mortality still persist, and 50 percent of children die before the age of 6. As late as 1952, 34 percent of all babies died within a year of birth in northeastern Brazil. When the mortality burden is lifted suddenly, man's genetically programmed, Stone Age fecundity races on oblivious to the change.

Birth rate can change. It also is under environmental control. Lactation, for instance, inhibits ovulation and helps to space babies at two- to three-year intervals if the mother nurses. Behavioral and mechanical controls can be effective. Births can be prevented with different efficiencies by birth-control methods and by abstention. Abortion is still the most widely used birth-control procedure. A recent study in Chile showed that 30 percent of pregnancies are terminated by abortion, most of them clandestine and dangerous. Several studies in Europe indicate that about half of pregnancies in cities are terminated by illegal abortion. Many are self-inflicted or are performed by untrained amateurs. It has been estimated that 4 percent of Italian abortions result in maternal death. Abortions are legal and inexpensive in Japan, China, and some eastern European countries. When performed by an experienced physician, abortion is safer than completing the pregnancy. As the pill, the intrauterine device, and other methods become more freely available, abortion will be used less. Until then, it will continue to be the only effective brake on uncontrolled population growth.

If birth rate is controllable, why does its decrease lag so far behind that of the death rate, especially in nonindustrialized nations? The answer has to do with motivation.

THE VICIOUS SPIRAL

In general, parents raise about as many children as they want, and they want the number that will maximize their social and economic status. Tra-

Figure 45.4. Changes in the rate of mortality among selected Asian nations. The average rates of 1945–1949 are compared with those of 1960–1961. The decrease in death rate is due largely to "death control" through world health organizations.

ditional, conservative, space-limited societies may have zero or even negative growth rates. On a circumscribed atoll in the South Pacific or on tiny farms in land-hungry eastern European countries, severe controls over family size long have been practiced. Relatively low growth rates are characteristic of all industrialized countries. Doubling times for some industrialized and nonindustrialized countries were given in Table 45.1. Why the disparity of growth rates between the two kinds of societies?

The reasons for low growth rates in the currently industrialized countries are historical. While industrialization was occurring, infant and child mortality was high, and explosive population growth was prevented. Even as late as 1900, European death rates were 18 to 20 per thousand, and in the United States infant mortality (death during the first year of life) was 16 percent. In addition, rising population densities, mechanization of agriculture, and a shift from rural-farming families to urban-worker families made smaller families more desirable. In preindustrial agriculture, children—even rather young ones—contribute to the productivity of the farm. In cities, children are strictly consumers until they become employable. Therefore, as long as the nuclear (immediate) family is a stable institution, city parents are wise to have only a few children. By the time medicine and sanitation made it possible for nearly all children to survive to adulthood in the industrialized countries, small families were seen to be beneficial. Birth rates fell accordingly.

But the tables have been turned on the large majority (about 90 percent) of the world's people. Most are subsistence farmers. Large families are generally desirable, not only because they mean more working hands but because surviving sons are the only form of old-age insurance. Into these relatively stable societies have recently been injected two perturbing factors: death control and rapid change in traditional social patterns.

At the close of World War II, humanitarian public health programs began employing broad-spectrum biocides such as DDT to control insects that carry malaria. Insect-control, sanitation, and immunization programs triggered a sharp decrease in death rates. Just prior to the widespread control of mosquitoes with DDT in 1946, Ceylon had a death rate of 22 per thousand per year. Within 8 years, the death rate had dropped to 10 per thousand.

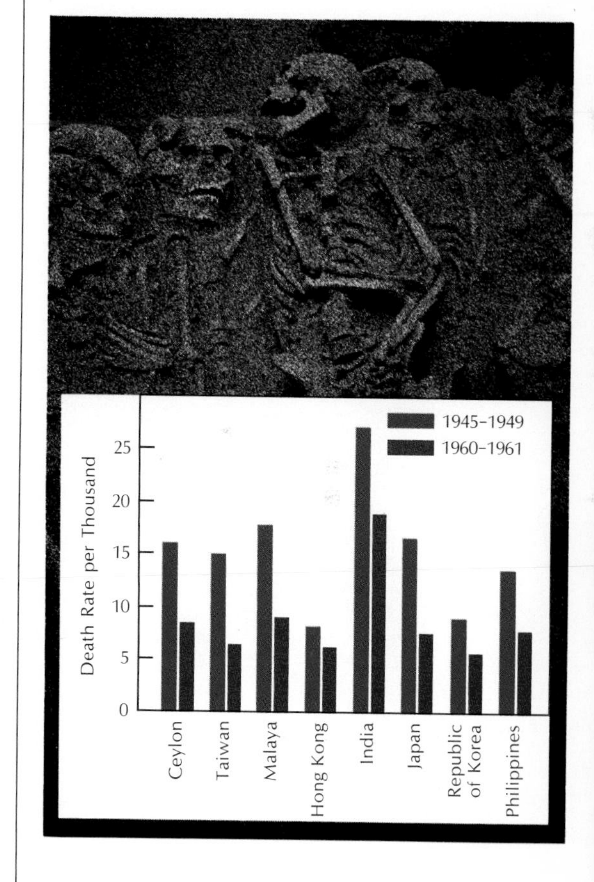

Increasing numbers of children are placing severe strains on schools, hospitals, and food, just when impoverished governments are hoping to divert their limited resources to support of industrial development and improvement of the standard of living. To make things worse, the rich industrialized nations—with only about 15 percent of the world's population—are using over 50 percent of the world's annual production of nonrenewable fossil fuels and ores, much of which originates in the poor countries. The United States alone—6 percent of the world population—gobbles up about 35 percent of these raw materials. Out of 77 strategic raw materials, the United States must import 66 of them—40 of these from unstable areas (Murphy, 1969). The implications of Western consumption are serious. The required amounts of many metals, including copper and zinc, exceed known and projected reserves. If, miraculously, all the world's people were raised to the level of affluent consumption enjoyed by Americans, the earth would be depleted of all ores virtually overnight (Brown, 1967).

Many modern countries were not well endowed to begin with. In fact, many of the so-called undeveloped countries of Asia, North Africa, the Middle East, and Central America exist on the ruins of the most ancient

Figure 45.5. Man's short-sighted exploitation of natural resources has left billions of blighted acres unfit for all but marginal subsistence.

civilizations. More often than not, their high-grade ores are depleted, their soils eroded, and forests cut by the preindustrial civilizations. Not only are their peoples off to a late start in the modern race to the good life, but they are handicapped by inferior track conditions as a result of the thoughtless exploitation of the environment by their predecessors.

The accelerating pace of social change aggravates the problem further. With mass transport of people, cheap magazines, and transistor radios, images of well-nourished, well-sheltered, well-amused Westerners flow in a one-way stream from rich to poor, West to East, and urban to rural. Expectations rise, and untutored, unprepared hordes migrate to the misery of cities, hoping for a better life while traditional values and customs are cast aside. In less chaotic times, changes like these take centuries not years, and there is time to accommodate and compensate. Poverty and impatience are an explosive mixture.

Birth rates decline when people desire smaller families. Small families become desirable when children are expensive to feed, clothe, and educate and when parents need not live with the terror that child mortality will leave them alone and without support in old age. In short, the precondition for lowered birth rates is security—a rare commodity. Furthermore, parental anxiety is aggravated by the absence of visible improvement in quality of life. So there is a vicious spiral of babies → poverty → more babies → more poverty.

FOOD

The essence of tragedy is the remorseless working of destiny. The tragedy of the human population explosion is the ultimate finiteness of the planet earth and the absolute authority of the laws of thermodynamics. Whether the carrying capacity may be a human population of 3 billion or 300 billion, there is a limit. It can be and has been argued just which resource actually will be limiting first. On the input side, it might be food, water, energy, oxygen, or resistance to disease that sets the limit. On the output side, it might be chemical pollution or heat. Whatever the practical limit, the logic

Figure 45.6. A bread sculpture showing the geographical distribution of hunger in the United States. Malnutrition (darkened areas) and deficiency diseases debilitate as many as 20 million people in the United States. Retarded mental development and stunted growth are just a few of the consequences that might result from a diet deficient in requirements such as protein.

is inescapable that those who espouse continued population growth for economic, religious, or other reasons are short-sighted at best or irresponsible and callous at worst. Estimates of the largest human population that could be supported at the United States standard of living and consumption usually center around 1 billion—less than one-third of today's population (Hulett, 1970).

Moreover, to discuss the ultimate carrying capacity of the planet is an academic exercise when a majority of the humans alive at this moment are hungry and ignorant. As a species, man presently furnishes the materials and facilities to allow about 400 million individuals to reach their full genetic potential. For the other 90 percent, one or more of the following essentials are lacking: good nutrition, quality formal education, and sufficient medical care. The country with the highest per capita income, the United States, still has a long way to go in these three areas. A recent study in the United States pointed to the widespread occurrence of malnutrition and deficiency diseases (Figure 45.6). Chronic hunger in the United States debilitates as many as 20 million people. Children in depressed regions typically are stunted. One study showed that 6-year-olds in Appalachia are 1 to 2 inches shorter than national norms. A fair standard of health-care systems is infant mortality; the United States' rate is 22 per 1,000 live births, compared to 13 in Sweden and 19 in Britain. The humane goal that every citizen be educated so as to maximize his contribution to society is far from realized, even in the United States.

Food certainly is the most critical shortage at this point. People may not riot over substandard or nonexistent schools or over a lack of doctors and

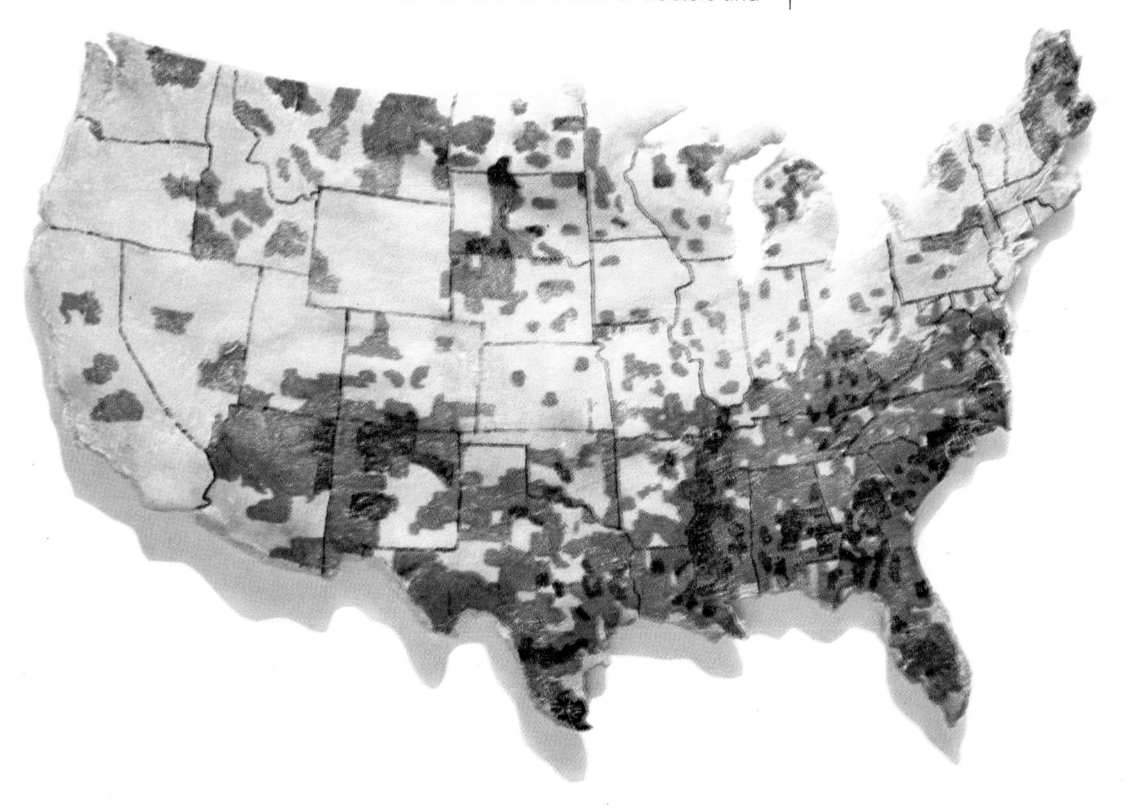

Figure 45.7 (above). Potential farmland on a global scale based on the nature of the soil.

Figure 45.8 (below). Changes in world land utilization (1882–1952). The net addition of tilled acreage amounts to 210 million hectares during a period when the world population grew by almost 2 billion people. Thus, the gain is reduced to only 0.1 hectare (1/4 acre) per individual.

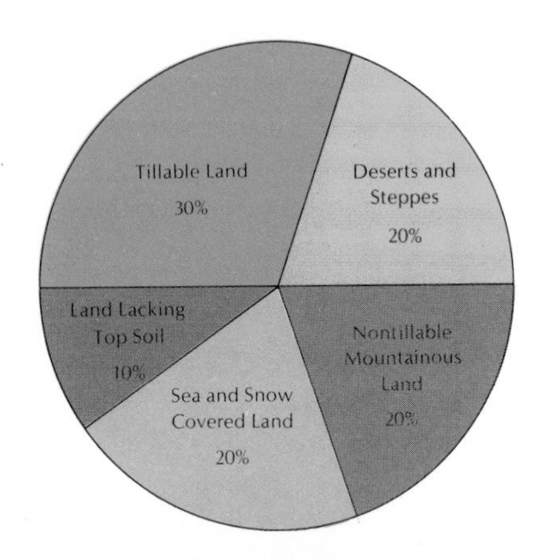

	1882	Percent	1952	Percent	CHANGE 1888–1952	Percent
Forest	5.2	45.4	3.3	29.6	-1.9	-36.8
Desert and wasteland	1.1	9.4	2.6	23.3	+1.5	+140.6
Built-on land	0.87	7.7	1.6	14.6	+0.73	+85.8
Pastures	1.5	13.4	2.2	19.5	+0.7	+41.9
Tilled land	0.86	7.6	1.1	9.2	+0.24	+24.5
	9.53	83.5	10.8	96.2	+1.27	+12.9
Area not especially utilized	1.81	16.5	0.27	3.8	-1.54	-79.9
Total	11.34	100	11.07	100	-0.27	-2.4

Changes in Land Utilization 1882–1952
(billion hectares)

hospitals, but they object strenuously to starving, especially when others are eating more than they need. Social instability is the stuff of which wars are made.

Today, at least half of the world's people do not meet the minimal nutritional standards. What are the prospects for feeding another 3 billion people in 30 years? Broad methods for increasing food production include: (1) breaking new lands, (2) enhancing productivity of lands already exploited, (3) preventing storage losses, and (4) agricultural and technical innovation of various sorts. Biologists, ecologists, and agricultural experts continue to fill many books with detailed discussions of the improvements that can be expected in each of these areas (Pirie, 1967). Attempts to forecast future developments always are risky, and the "experts" by no means agree on what to expect. However, a few things seem fairly certain.

Excluding the Antarctic, 80 percent of the earth's surface is ocean. These waters will continue to produce important protein supplementation, but seafood is not a significant source of the basic food requirement—calories. Optimistic estimates for the year 2000 predict that the oceans can furnish 30 percent of the minimal protein requirements but only 3 percent of food energy needs for the world (Ricker, 1969). Figure 45.7 shows that the land itself is not all potential Iowa farmland. Experts agree that between 60 and 70 percent is nontillable desert, tundra, or mountains or is snow-covered or lacking topsoil. The rest, between 30 and 40 percent, is potentially useful for farming and grazing. Virtually all of the best farmlands now are under the plow and produce more than 80 percent of the world's food (Hendricks, 1969). Most tillable land not already being farmed (less than half of the available farmland) is used for grazing, or it is in the tropics. Virtually all of these lands are classed as marginal because of low fertility or unpredictable rainfall. Experiments in farming these lands with modern intensive agricultural practices usually terminate in disaster. Food optimists often ignore the collapse of the "virgin lands" scheme in Russia and of the ground-nut scheme in East Africa. Attempts to farm in tropical rainforests, using iron-rich land in West Africa and Brazil, ended disastrously with laterization, a process by which tropical soils quickly harden to pavement. Only 2.8 percent (950 million acres) of dependable land remains uncultivated. Only 7 percent of the total land on the planet has the right combination of rainfall, nutrients, temperature, and topography to permit normal agriculture (Paddock and Paddock, 1968). The large empty spaces that delude the naïve into complacency lack essential resources, usually minerals or water. It is no accident, for example, that Nevada is sparsely populated. Water is so expensive to move that it is not economical even for rich nations to farm such dry lands. Poor nations simply cannot afford large-scale irrigation, even if water were available; it usually is not.

Even more discouraging is the fact that good lands are being lost to urbanization, erosion, and drying (becoming deserts) much faster than new land is being broken. Figure 45.8 shows changes in land utilization between 1882 and 1952. The pace of these changes has been accelerating since then. The 950 million acres left is a pitifully small bit of land—less than 10 years' supply at the current rate of population growth and farming productivity.

Other supposed panaceas also prove hollow upon examination. In theory, the productivity of present agricultural lands can be doubled or even tripled in many regions. All that is needed is increased input of water and nutrients (fertilizers). It sounds simple, but it costs money. Why does India with a population of 550 million use the same amount of fertilizers as Sweden with 8 million? Why does the entire continent of South America use no more than Holland? The answer to this question is the universal answer to all questions that involve technology. The "technological fix" is expensive. Agricultural extension workers must be trained and sent to remote villages. The mining of minerals, the building of factories, and the construction of dams, roads, railroads, and aqueducts to move and process the necessary materials all require enormous outlays of nonexistent capital. And now that the economic aid contributions of the most charitable country, the United States, have dried up to less than 1 percent of its budget, the prospects are even gloomier.

Most United States foreign aid offered to the underdeveloped nations since World War II has been in the form of long-term, low-interest loans. However, the interest due on these loans has snowballed over the years. The annual interest payments that Latin American nations must make to the United States on old loans now exceed the current foreign aid assistance from this country—in other words, the Latin American nations are currently losing money on American aid (Paddock and Paddock, 1968).

Newspapers and magazines encourage optimism about solutions such as high-yield "miracle" wheat and rice varieties (Curtis and Johnston, 1969). Less publicity is given to the fact that these grains require expensive and skilled application of fertilizer. The fertilizer can be obtained only by purchasing it from industrialized nations—thus further increasing the flow of capital out of the country—or by diverting badly needed investment funds into construction of fertilizer plants. Further investment is needed to transport the fertilizer to the farms and to educate the farmers in its proper use. Furthermore, the new grains require abundant water supplies—further investment is needed for irrigation systems. To get the most from the supergrains, additional investments for pesticide programs and mechanized harvesting are needed.

In a few cases, all of these problems have been met. For example, the 1968 wheat harvest in India and Pakistan was about 35 percent above previous records—although excellent weather probably accounted for most of the increased production. On the other hand, the people of the underdeveloped nation have very little extra money to spend for the additional grain. When the same food expenditure is spread over a larger amount of food, the result is a decrease in the price the farmer can get for his grain. Thus, the farmer who has invested heavily in the new technology finds himself rewarded by a dwindling market. The low rate of fertilizer use in India in 1969 suggests that farmers have decided to stick with older farming methods rather than invest in the new technology with its attendant risks.

Similar difficulties beset the other promises of the technocrats. The possibilities will not be realized unless poor countries can acquire large amounts of capital for long-term investment or unless investors in rich countries foresee a high profit from investments in the industries of the poor countries. Such panaceas as synthetic foods, edible bacteria grown on petroleum, nuclear reactor-based agricultural-industrial complexes, and algae farming all turn out to be economically impractical or unprofitable. With 2 billion hungry people today and between 10 and 20 million (mostly children) dying as a result of malnutrition every year, the species can hardly wait very long for the drawingboard miracles to become reality. One potential panacea is largely ignored by cosmic dreamers—better food storage. Most people are shocked to learn that between 20 and 50 percent of the food produced is lost through spoilage. Unfortunately, governments prefer dramatic, technological *tours de force,* including hydroelectric projects, to practical but mundane programs to improve storage techniques.

The quality of food is diminishing as well. Even in the United States, diets have deteriorated since 1955, according to a Department of Agriculture study (Kelsay, 1969). Protein (or its component amino acids) is an essential dietary constituent. To increase productivity, plant breeders have created larger, more starchy grains. The protein percentage has steadily declined. At least 12 percent of an individual's caloric intake should be protein. Most of the high-yield grains provide only 5 to 10 percent protein. Western countries import huge amounts of protein in the form of fish meal and soya bean cakes, largely to supplement the diets of their livestock. Corn of the modern hybrid varieties has so little protein that it must be mixed with protein additives to be fit for hogs. Grains once supplied a balanced, protein-sufficient diet. In many places today, grain must be supplemented with expensive fish, beans, milk, eggs, or meat. Again, quality is being sacrificed for quantity. A small amount of wholesome food is being replaced by a larger amount of unwholesome food.

The human cost is staggering. There is increasing evidence that the short stature, slow minds, and physical deformities of many of the world's people result in large part from malnutrition. In East Pakistan alone, 50,000 people a year become permanently blind from vitamin A deficiency. About 80 percent of brain growth occurs in the first 3 years after birth, although only 20 percent of body growth occurs in these years. Protein deficiency during the brain's rapid growth phase stunts its growth and now is widely believed to result in permanent mental retardation, decreasing IQ scores by about 5 to 15 points.

A South African study compared a group of grossly undernourished children raised under atrocious conditions to a group that was well fed and well cared for. The groups seemed very similar genetically. By the age of 8 years, the undernourished group averaged 3.5 inches shorter, 5 pounds lighter, and 1 inch less in head circumference (which is directly proportional to brain size) than the well-nourished group. Other studies have shown that enrichment of diet after dwarfing has occurred does not reverse the effect on brain size, although it partially compensates for stunting of body size. Scientists in Latin America have investigated the relation between nutritional stunting and mental retardation. Significant differences in intersensory organization (a major variable in IQ testing) were found in Guatemalan Indian children and were correlated with their heights. Short stature was largely the effect of malnutrition (Cravioto, *et al.*, 1966). According to a recent calculation, today there are 300 million people ($^{1}/_{12}$ of the total population, but a much higher proportion in parts of Africa, Asia, and Latin America) who are the silent victims of dietary retardation (Scrimshaw, 1968).

If all the world's protein were distributed equally, there would today be just enough to go around. Actually, the wealthier nations harvest or buy the lion's share, and their consumption of high-quality animal protein contrasts vividly with that of the poorer nations (Table 45.3). Denmark, for instance, imports 240 pounds of high-protein foods per person each year, much of it as supplemental feed for animals. The United States now imports four-fifths of its seafood, one-half of which is used as protein supplement in animal foods. Much of this food is purchased from countries where large parts of the population are undernourished. Americans can afford to pay more to feed dogs and cats than Egyptians or Peruvians can to feed their babies.

The future holds no miracles. The ocean's production of fish, now about 60 million metric tons a year (40 percent of the world's protein), can be increased to about 100 million metric tons by the year 2000. Because of the projected population increase, this amounts to a per capita *loss* in protein. One recent development that could be very beneficial is the breeding of grain varieties that have higher than normal amounts of the essential amino acids lysine and tryptophane. This breeding will substantially increase the protein quality of grains, assuming they can be developed for each crop and climate. Because the lead time for an agricultural innovation is usually a decade or less, it is unlikely that these grains will have much impact on nutrition before "the time of famines."

Table 45.3
Protein Availability in Different Regions (1961–1962)

AREA	Food per Person	
	Total Protein (*grams/day*)	Animal Protein (*grams/day*)
North America	93	66
Western Europe	83	39
Mexico	68	20
South Asia	50	7
Far East (including China)	56	8

Source: K. E. Watt, *Ecology and Resource Management: A Quantitative Approach* (New York: McGraw-Hill), 1968.

The hopeless cycle can be stopped. The forces that can stop it are population control, agricultural research, and medical and economic aid.

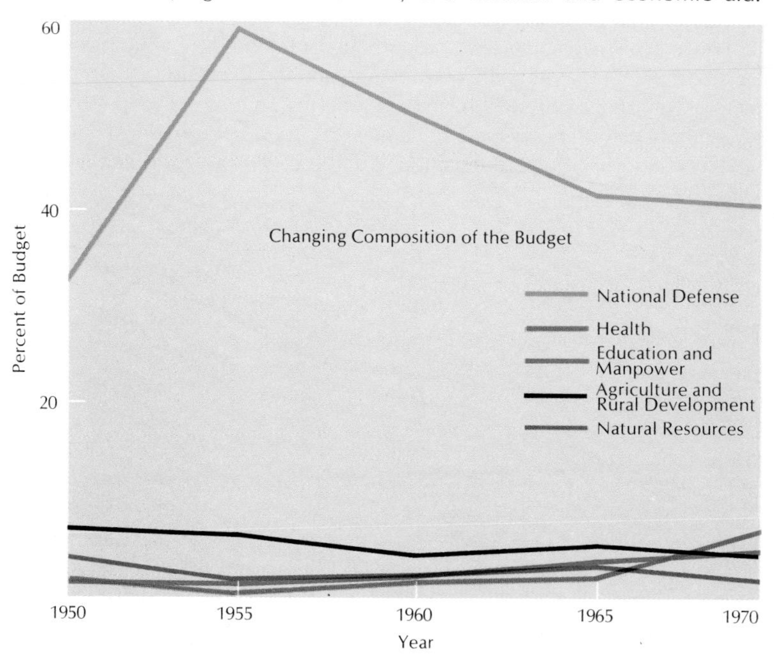

Figure 45.9. Graph showing percentage of U.S. Budget spent on the military versus domestic welfare.

Figure 45.10. World food production contrasted with consumption. Food projection for developing regions (left). Grain production and consumption pattern (right).

Population-control programs alone will succeed only when people see the economic desirability of small families, as they have in Western industrialized countries. Cheap, simple, and reliable contraception would be very helpful, but birth-control pills, intrauterine devices, and other existing methods plus abortion will do until superior methods are developed.

Very probably, the only way to motivate people toward smaller families is to eliminate both malnutrition and infant and child mortality. Parents then can plan for the future without the need to overbreed as a hedge against disease and death. Once this plan is achieved, population-control propaganda and programs may fall on receptive ears.

Even this plan may be academic. Such a massive program in public health has few adherents. The sad reality is that the rich nations are importing more protein than they are exporting, spending 40 to 50 times more on their military establishments than on economic aid, and doing almost nothing to increase agricultural efficiency in the developing countries. Only 1 percent of the United States' research and development budget now goes to agriculture. The efforts of international organizations are woefully underfinanced, and the progress made by the Rockefeller and Ford Foundations to develop better grains for Asia and Latin America is a noble drop in the bucket. Even if more-productive, high-protein grains could be developed for every climate, it would only be a first step. Much more water, fertilizers, pesticides, and skills are needed to exploit the new grains. Roads must be built to supply the farms and to deliver the surplus to markets and storage facilities. Farmers must be guaranteed a market and profits, or they will not invest in an expensive program of intensive agriculture. This endeavor would require more money by orders of magnitude than is presently expended.

The food shortage now is chronic for over half the world's people—the proportion that is undernourished or short of calories. In 10 to 20 years, famine may have reached crisis proportions for hundreds of millions. About 10 to 20 million die annually of starvation and its effects now. In 30 years, the present food supply will have to be doubled, merely to keep the average amount of food per person the same as it is today. To provide a minimal healthy diet for the entire population, it will be necessary to triple the pres-

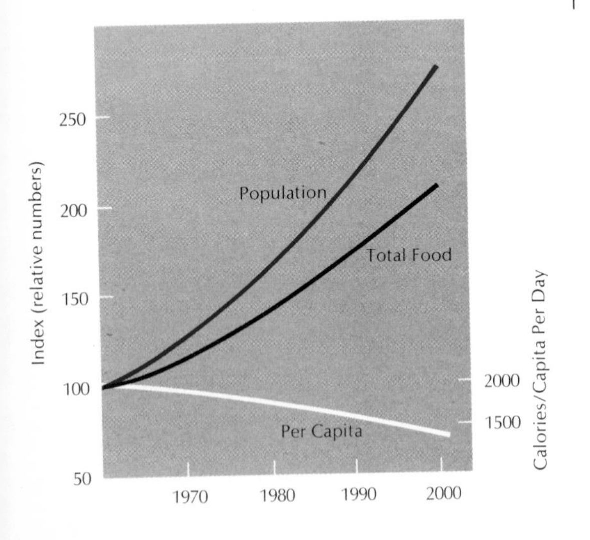

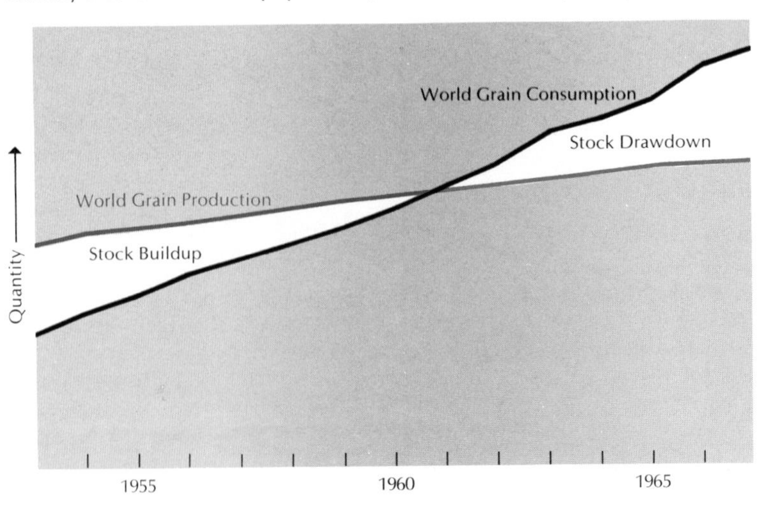

ent food supply. Even if population-control programs were to become effective in the near future, the population growth would not decrease notably for some time. Rapidly growing populations have a very high proportion of children, who will continue to mature and produce children of their own in ever greater numbers—even if each individual has fewer children. And effects of modern medicine will enable these people to live into old age while their children and grandchildren are adding to the population.

Table 45.4

Projected Increases by Age in Five Countries—1960 or 1961 to 1975 or 1976—on Assumptions of Continuity in Fertility and Declines in Mortality

VARIABLE AND COUNTRY		Total	Selected Age Groups 5-14	15-44	45-60
Population (millions)					
India:	1961	423.6	100.1	187.6	45.7
	1976	600.6	150.3	252.4	64.2
Pakistan:	1961	95.4	24.3	41.0	9.1
	1976	145.6	38.8	60.4	13.6
Indonesia:	1960	93.3	21.4	41.5	10.0
	1975	137.4	34.7	57.2	15.1
Mexico:	1960	34.1	9.2	13.9	3.4
	1975	54.5	14.0	22.5	4.5
Brazil:	1960	66.1	16.7	28.6	6.5
	1975	98.3	25.8	41.5	9.4
Amount of Change (millions)					
India		177.0	50.2	64.8	18.5
Pakistan		50.2	14.4	19.4	4.5
Indonesia		44.0	13.3	15.7	5.0
Mexico		20.3	5.5	8.7	1.2
Brazil		32.2	9.1	12.9	2.9
Change (percent)					
India		41.8	50.1	34.5	40.5
Pakistan		52.7	59.3	47.2	49.9
Indonesia		47.2	62.3	37.9	50.2
Mexico		59.6	59.4	62.7	35.6
Brazil		48.7	54.5	45.0	45.3

Source: Adapted from United Nations, Department of Economic and Social Affairs, Population Division, *Future population estimates by sex and age:* Mexico, Report I, *The population of Central America (including Mexico). 1950–1980,* Table 2, p. 42 (New York, 1954), 84 pp.; Brazil, Report II, *The population of South America, 1950–1980,* Table 4, p. 73 (New York, 1955), 139 pp.; Indonesia, Report III, *The population of Southeast Asia (including Ceylon and China: Taiwan), 1950–1980,* Table 6, pp. 138–139 (New York, 1958), 1966 pp.; Pakistan, India, Mainland China, Report IV, *The population of Asia and the Far East, 1950–1980* [Pakistan, Table xvii, p. 109; India, Table viii, p. 100 (high fertility projection)].

Today, shipments of free wheat from the United States are the only thing that prevents famine for many nations. Even with its highly mechanized and efficient agriculture, the United States cannot long continue to offset the population growth of the world. The U.S. Department of Agriculture predicts that Americans will be unable to meet the food needs of the world by 1984; other experts predict that the food will run short as early as 1975. It must be remembered that many millions are starving to death even now.

These dates refer to the time when a major famine will strike some country, causing starvation of a major portion of its population, and the United States will have no surplus food to send in aid.

THE DYING PLANET

In its frantic attempts to provide itself food, shelter, and other comforts, the human species is ravaging the earth's thin skin of life. Lumbering has nearly shaved the continents of their forest mantles. North America, Russia, and a few small nations still retain forest reserves, but demand for pulp, building materials, and wood annually shrinks the planet's remaining acreage of trees. Reforestation in the United States is not keeping up with deforestation. If forests were only vertical columns of wood, the problems would not be so great, but they are much more. Forests harbor most of the planet's species of animals and plants, they protect and produce topsoil, and they conserve water by preventing rapid runoff. Man, in his frenzy of breeding and feeding, is destroying his life-support systems. Deforestation has lowered water tables throughout the world and may be the reason for the drying out of many parts of the tropics.

Rational land-use programs rarely are instigated. Instead, the best is made of a worsening situation. Industry has exploited rapid runoff from deforested, eroding lands with hydroelectric dams. Farmers have learned to exploit the silt-laden flood waters from deforested lands hundreds and even thousands of miles away. Often both groups vehemently resist reforestation because it would jeopardize their supply of runoff waters for turbines and irrigation. In the meantime, topsoil is lost at alarming rates. Thousands of years may be needed to accumulate an inch of topsoil. Due to poor agricultural practices, the loss in some places today can be measured in inches per year. About 1 percent per year disappears from the *best* farmland.

Complex ecosystems such as forests and savannas are homeostatic systems. A change in the numbers of a few species or a gradual shift in climate can be compensated. These ecosystems can survive for eons. Man destroys these ecosystems with their shock-absorbing feedback properties. In their place he puts a single-species ecosystem—a crop. The crop needs constant protection and care. It has none of the homeostasis of a natural ecosystem. It can be devoured by an invasion of a single insect species. Winds can knock down the feebly rooted, quickly grown plants. Irrigation may be needed to supplement precipitation. Early rains may wash away the seedlings. Late rains may cause the plants to rot. If the crop is successfully harvested and stored, up to one-third or even one-half will be consumed by insects and rodents. And, finally, nutrients removed from the soil by harvesting must be replenished. No wonder an Iowa farmer expends more calories in petroleum fuel than he produces in grain. Only a rich nation can afford a gasoline-based agriculture.

The farmer is never more than a stride ahead of calamity; neither is the world. The loss of a single summer's crop in the northern hemisphere would exhaust humanity's slim margin of survival. Only North America has enough food stored to withstand such a disaster. This possibility is not as remote as it sounds. In 1815 the volcano Tombura on Sumbawa threw 150 cubic kilometers of ash into the atmosphere. There was no summer in much of the northern hemisphere the following year.

Attempts to feed today's population and the 72 million mouths added annually result in the addition of more tiers to the shaky tower of the simplified ecosystem. Modern agriculture now is totally dependent on artificial additives—pesticides and inorganic fertilizers. In turn, these create other stresses.

Pesticides are powerful agents of natural selection. When they are applied, the most susceptible insects die, leaving the least susceptible to reproduce. Because selection promotes the evolution of resistance in pesticide targets, it becomes necessary to increase the dose and frequency of spraying, thus intensifying the selection on the target organisms and increasing environmental contamination. Resistance to the three major groups of insecticides (chlorinated hydrocarbons, organic phosphates, and carbamates) now is so widespread that crops of many kinds are on the verge of collapse. In some areas, withdrawal of pesticides is the best solution. Natural pest control can reassert itself—as happened, for example, in the cocoa crop in the state of Sabah, Malaysia (Conway, 1969). Sprayings were not preventing the defoliation and death of many trees. Because the worst insect outbreaks began *after* the introduction of a heavy spraying program, it was decided to stop application of pest species.

Unfortunately, such experiments are uncommon, and farmers everywhere are increasingly becoming slaves of pesticides. The evolution of resistance by pests necessitates heavier and heavier sprayings. Frequent substitutions of new, more deadly pesticides also are encouraged. It is a vicious cycle—a cycle promoted by the ecologically disastrous recommendations of the pesticide industry.

But these are only the direct effects. The indirect effects of pesticides are more in the public consciousness. One group of pesticides, the chlorinated

Figure 45.11. Schematic representation of the path of DDT residues through the food chain. Note the increasing concentration of the residue as one proceeds up the food chain to the carnivore levels.

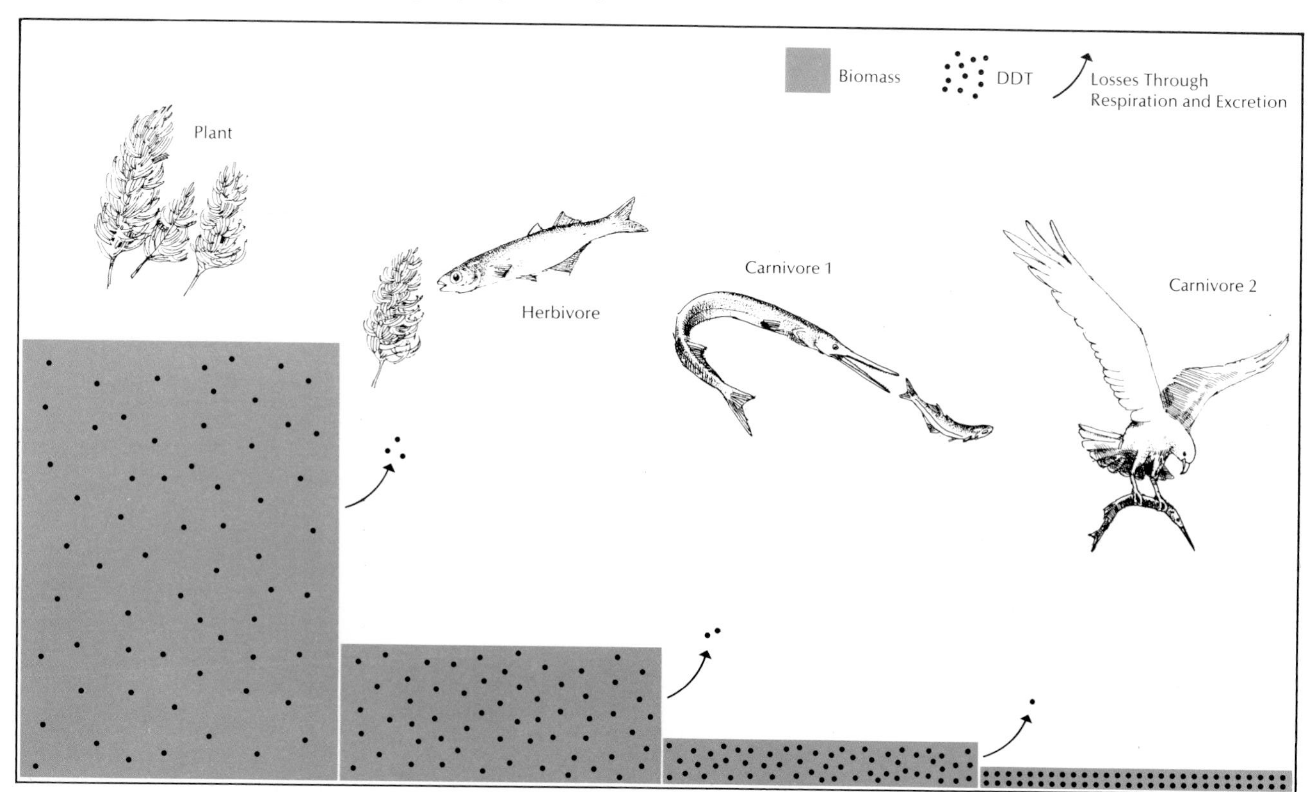

Figure 45.12. DDT residues, including the derivatives DDD and DDE, have been reported in most ecosystem food webs. Note the especially high concentrations in the eggs of the carnivorous birds.

LOCATION	ORGANISM	TISSUE	CONCENTRATION (parts per million)
U.S. (average)	Man	Fat	11
Alaska (Eskimo)			2.8
England			2.2
West Germany			2.3
France			5.2
Canada			5.3
Hungary			12.4
Israel			19.2
India			12.8–31.0
U.S. – California	Plankton		5.3
U.S. – California	Bass	Edible Flesh	4–138
U.S. – California	Grebes	Visceral Fat	Up to 1,600
U.S. – Montana	Robin	Whole Body	6.8–13.9
U.S. – Wisconsin	Crustacea		0.41
U.S. – Wisconsin	Chub	Whole Body	4.52
U.S. – Wisconsin	Gull	Brain	20.8
U.S. – Missouri	Bald Eagle	Eggs	1.1–5.6
U.S. – Connecticut	Osprey	Eggs	6.5
U.S. – Florida	Dolphin	Blubber	About 220
Canada	Woodcock	Whole Body	1.7
Antarctica	Penguin	Fat	0.015–0.18
Antarctica	Seal	Fat	0.042–0.12
Scotland	Eagle	Eggs	1.18
New Zealand	Trout	Whole Body	0.6–0.8

hydrocarbons (including DDT, DDD, dieldren, and lindane), is highly resistant to oxidation and enzymatic attack. This resistance, plus relative insolubility in water, results in the accumulation of these substances in the tissues of exposed organisms. With some important exceptions, including insect strains with evolved resistance, organisms cannot dispose of chlorinated hydrocarbons as they do natural wastes and toxins. Predators and filter feeders at the top of food chains naturally accumulate the most. Many predatory and oceanic bird populations now are on the verge of extinction. It is still too early to predict what effects will result from the high pesticide levels in whales, porpoises, sea birds, fishes, crabs, shellfishes, and men. Further, no one knows what will be the effect of eliminating top predators from the ocean. In 1971 the United States government, despite lip service to environmental quality and research, is planning to pump more into research on diseases primarily affecting the aged (cancer and heart disease), while insignificant funds are available to assess to ecological deterioration.

Agricultural collapse and widespread famine immediately would follow a sudden termination of pesticide use. Insect-borne disease also would increase significantly. According to statistics of the World Health Organization, about 10 million people have been saved from fatal effects of malaria by the use of DDT in antimosquito campaigns. To escape from the horns of this ecological dilemma will require time, intelligence, and—above all—research. The latter requires the support of an alert public.

Inorganic nitrate and phosphate fertilizers are responsible for a growing tide of ecological disasters. First, inorganic fertilizers "loosen" the nitrogen cycle in the soil by shortcutting certain steps. As with pesticides, the more they are used, the more they are needed. Second, like pesticides, these fertilizers do not remain where they are applied but are dispersed and concentrated in unexpected places. Lakes are the most seriously affected habitats, because they are nutrient traps. Great blooms of algae are promoted by the

Figure 45.13 (above). Inorganic nitrates and phosphates often fertilize monstrous growths of algae in fresh-water ponds and lakes. When these algae "blooms" subside, the subsequent bacterial decomposition of their bodies depletes the surface water of oxygen and a "kill" of other organisms occurs.

Figure 45.14 (below). The accumulation of large masses of organic matter has accelerated the eutrophication process to the extent that Lake Erie has "aged" the equivalent of 1,500 years in the last 50.

nitrates and phosphates. Oxygen is depleted by bacterial decomposition of the masses of algae, and die-offs of fish foul the shores. Only the hardiest, "rough" fish, such as carp, can withstand much oxygen depletion.

The lakes age prematurely under these conditions. In the absence of oxygen, the decomposer organisms cannot keep up with the rain of algal corpses. Organic sludge accumulates rapidly on the bottom. The "eutrophication" of Lake Erie has caused it to age the equivalent of 1,500 years in the last 50 (Powers and Robertson, 1966). Biologists used to console themselves with the belief that once the lakes were no longer fertilized by entering nutrients, the nutrients would be forever locked up in the sludge at the bottom and eutrophication would cease. Now there is evidence that anaerobic conditions promote recycling of nutrients in such a lake. Thus, even if nutrient pollution were to cease, eutrophication would continue. The problem is world-wide and will become more serious as intensive agriculture spreads to meet increasing food needs.

Reliance on nitrates already is affecting man directly. High nitrate levels are showing up in wells in agricultural areas throughout the United States. Nitrate itself is relatively innocuous, but bacteria in the intestines of infants convert it to nitrite. In the blood, nitrite combines with hemoglobin to produce methemoglobin. Methemoglobin does not have the affinity for oxygen of hemoglobin, and its presence can lead to respiratory distress and suffocation. The disease is becoming a serious public health problem in the agricultural valleys of California as well as in other states.

Can man and other higher animals evolve resistance to pesticides and other chemical pollutants? Perhaps, but the penalties are forbidding. Insects have evolved resistance by virtue of natural selection. Usually, only a fraction of a percentage of the population survives a heavy application, and many generations of massacres precede the appearance of significant numbers of resistant individuals. The cost of quickly increasing the frequency of the genes that might confer resistance in man is beyond anything our species has suffered yet, at least if resistance is to be acquired in a few generations.

Technology in the developed countries is growing even faster than the population. Per capita energy consumption is doubling about every 10

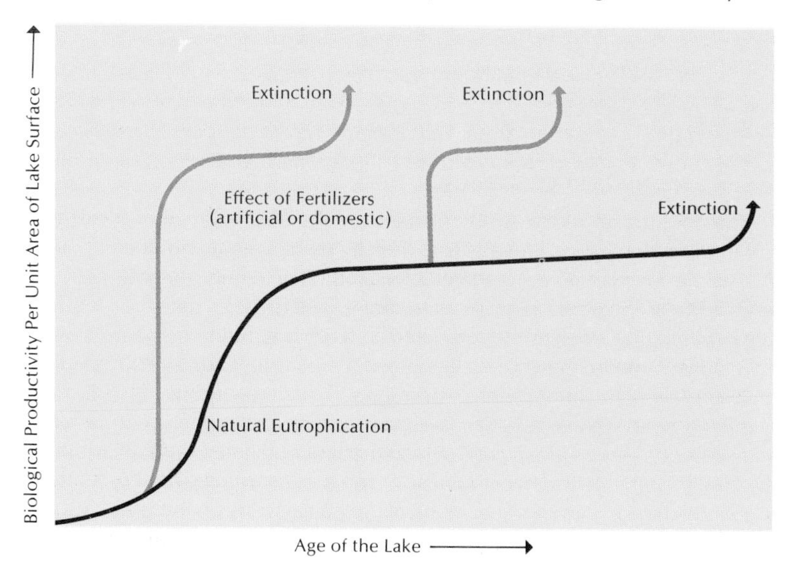

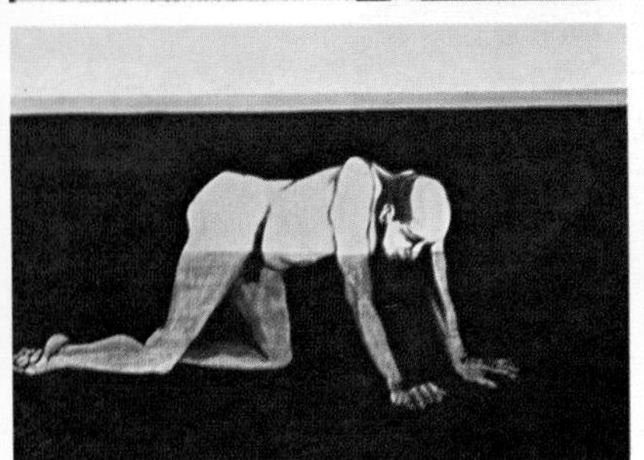

years in the United States, and only 20 percent of this increase is attributable to the growing population. This profligate use of resources is devastating the earth's fossil fuels and other nonrenewable resources (Committee on Resources and Man, 1969). Present Western affluence is possible only by living off accumulated capital. Forests, petroleum, and high-grade ores will be depleted in about 50 years. The earth's exhausted crust never again will be able to be midwife to the birth of an industrial civilization (Drucker, 1969).

Notwithstanding technology's chemical pollution of the water and air, the most serious long-term threat may be heat. The atmosphere is a complex machine. Its temperature and patterns of circulation determine climate, and climate determines the distribution and productivity of ecosystems, including crops.

Technology and agriculture are altering the heat balance of the atmosphere in two ways: (1) locally and directly by heating it with the exhausts of factories, power plants, and vehicles; and (2) generally and indirectly by increasing its load of CO_2 and dust particles. The particles tend to shield the atmosphere from the sun's energy and so tend to cool the planet, whereas CO_2 adds to the "greenhouse effect" and so tends to warm the planet. The CO_2 content of the atmosphere has increased 12 percent since 1880. At the moment, neither trend is dominant. In the long run, though, with energy consumption growing exponentially and pollution abatement a possibility, heating is probably the eventual result. Some meteorologists are predicting a few degrees' rise in average atmospheric temperature by the year 2000. A very likely consequence would be the melting of the ice pack in the Arctic

Ocean, which, in turn, probably would result in a northward shift of the winter storm belts, perhaps followed by another ice age—an end to productive agriculture in the northern hemisphere. However, present knowledge is insufficient to predict with certainty the degree or even the direction of the eventual climatic changes (Claiborne, 1970).

In evolutionary or historical perspective, men rarely have been physically secure. With the exception of the last few decades in a few countries, half the children born have died before reaching maturity. Famine, drought, and disease always have been lurking near. The disease- and danger-ridden eons of human existence provide the context for the present crises. Those who read this book are recent escapees of this dangerous past—part of a small human elite for whom life is not always a battle for mere survival. Man's future will resemble the past more than the present. Once again, survival is everyone's issue.

Perhaps humans are psychologically programmed to be optimistic. Man the hunter certainly could not have survived without hope. Even though the population-environment crisis leaves little room for rejoicing, it would clearly be nonadaptive, if not abnormal, to despair. Does the situation justify any rational optimism? Yes, if man acts soon. But it must be stated emphatically that no action short of a revolution in human values will suffice.

With the power to kill the planet must come a compelling commitment to protect it. From an exploitive environmental ethic, man must shift to a therapeutic environmental ethic. This beautiful but sick blue and white spaceship needs stewards, not gremlins—especially not fecund gremlins.

Old cultural patterns and values are disintegrating today as never before. It is a chaotic, painful time. Transitions always are. But into the vacuum left by the deterioration of conservative ways there is emerging a young generation unafraid of change and disgusted by the wasteful, exploitive, and often impersonal societies bequeathed to them. Will quality replace quantity? Will aesthetics replace greed? Will a brave and free new world rise phoenixlike from the dying planet? Only if you do something about it.

FURTHER READING

Probably the best sources of information about the human population and its effects on the environment are books by Ehrlich and Ehrlich (1970) and Young (1968). Also of great value are books by Bresler (1968), Harrison Brown (1954, 1967), Committee on Resources and Man (1969), Cox (1969), Heer (1968), Rudd (1964), Shepard and McKinley (1969), and the publications of the Population Reference Bureau, Inc.: *Population Bulletin, PRB Selections*, and the *World Population Data Sheet.*

Books speculating about human biological inheritance and its contribution to present crises include those by Ardrey (1961, 1966, 1970), Lorenz (1966), Morris (1967, 1969), and Tiger (1969). Useful articles dealing with topics related to this chapter include those by Edwards (1969), Holt (1969), Julian Huxley (1956), McVay (1966), Pequegnat (1958), Plass (1959), and Woodwell (1967).

Among the great many books expressing various views on the current ecological crisis, the following are particularly recommended: Blake (1964), Borgstrom (1967, 1969), Boulding (1964), Bronson (1968), Commoner (1967), Dasmann (1965, 1968), Day and Day (1965), Drucker (1969), Ehrlich (1968), Freeman (1968), Hardin (1969), Hauser (1963), Hopcraft (1968), Jarrett (1969), McHarg (1969), Paddock and Paddock (1968), Rienow and Rienow (1967), Udall (1963, 1968), and Whyte (1968).

Contributing Consultants

Preston Adams received his doctorate from Harvard University in 1959 and is associate professor of botany at DePauw University. He has published several research papers on the taxonomy of flowering plants based upon extensive field study in the southeastern United States, Cuba, Mexico, and Central America. In his teaching, he has sought to combine traditionally separate botanical specialties—for example, the merging of plant morphology with paleobotany into an inquiry into the evolutionary development of plants. This interest in innovative teaching led him to coauthor an inquiry-oriented textbook, *The Study of Botany*. He is currently interested in the ecological approach to plant science, viewing botany as the science through which man is related (through the influence of his activities) to the world ecosystem.

Thomas Peter Bennett is professor and chairman of biological sciences at Florida State University. He received his Ph.D. from Rockefeller University, where he studied under Fritz Lipmann. His research interests include basic biochemical studies, the biochemical basis of cellular control, and the regulation of protein synthesis. He is the author of two textbooks in biochemistry and many teaching articles in biology and biochemistry, including *Elements of Protein Synthesis*, an instructional model and text for teaching protein synthesis.

Konrad E. Bloch, the winner of the Nobel Prize in Medicine and Physiology in 1964, is chairman of the Department of Chemistry at Harvard University. He received his Ph.D. in biochemistry from Columbia University and went on to teach there and at the University of Chicago. His research interests center on the intermediary metabolism of amino acids and lipids, the mechanism of peptide bond formation, and the biosynthesis of steroids. Dr. Bloch is a fellow of the National Academy of Sciences.

John Tyler Bonner is chairman of the Department of Biology at Princeton University. He received the Ph.D. from Harvard University and has taught at the Marine Biological Laboratory, University of London, and Brooklyn College. Dr. Bonner has held the Guggenheim Fellowship and the National Science Foundation Senior Postdoctor Fellowship, and he was a Sheldon Traveling Fellow and a Rockefeller Traveling Fellow. He is the author of *The Cellular Slime Molds, The Evolution of Development, Size and Cycle, The Ideas of Biology, The Scale of Nature*, and numerous other publications.

Susan Bryant is an assistant professor in the Department of Developmental and Cell Biology at the University of California at Irvine. She received her early training and her Ph.D. in biology from the University of London. At Case Western Reserve University, she spent two years as a postdoctoral fellow in the laboratory of Professor Marcus Singer, a renowned student of regeneration. Dr. Bryant's research interests lie in the problems involved in the regeneration of vertebrate appendages.

Frank Macfarlane Burnet was born in Australia and completed his medical course at Melbourne University. His first research work on the agglutinin reactions in typhoid fever was begun in the Walter and Eliza Hall Institute of the Melbourne Hospital in 1923, and he was Director of the Hall Institute from 1944 to 1965. Burnet's work has covered several fields, but he has had a continuing interest in immunology since his early work (1931) on staphylococcal toxin and antitoxin. In 1960 he shared the Nobel Prize for Physiology and Medicine with Professor Peter B. Medawar for the "discovery of immunological tolerance." His clonal selection theory of immunity has acted as a stimulus to experimentation over the last ten years and has led to a considerable change in the biological approach to the phenomena of immunity. Since retirement, his interests have broadened to include a concern for the social implications of science as well as a continuing interest in immunology. During his presidency of the Australian Academy of Science, he was primarily interested in the Academy's Science and Industry Forum, and as Chairman of the Commonwealth Foundation, he has been directly concerned with the problems of developing countries. He was knighted in 1951, received the Order of Merit in 1958, and became a K.B.E. (Knight Commander of the British Empire) in 1969.

Michael Crichton was graduated summa cum laude from Harvard College and received his doctorate from Harvard Medical School. He is the author of several books including *The Andromeda Strain* and *Five Patients*. He is currently on leave of absence from the Salk Institute for Biological Studies in La Jolla, California.

Elizabeth G. Cutter, professor of botany at the University of California at Davis, received her early training and the Doctor of Science degree from the University of St. Andrews in Scotland. In 1964 she earned her Ph.D. at the University of Manchester, England. Her research interests center on plant morphogenesis and anatomy, with an emphasis on the study of cell and organ differentiation in plants, and experiments on the growth and development of ferns and flowering plants (including aquatic plants). In addition to numerous research papers, she is the author of *Plant Anatomy: Experiment and Interpretation* (1969 – 1971) and she edited *Trends in Plant Morphogenesis* (1966).

Max Delbrück, a Nobel Prize recipient, is a professor of biology at the California Institute of Technology. After receiving his doctorate in theoretical physics at the University of Göttingen, he spent three years in postdoctoral studies in England, Switzerland, and Denmark. In Denmark, he became interested through Niels Bohr in the relations between physics and biology—an interest that he pursued first in Berlin, then at the California Institute of Technology, and later at Vanderbilt University as a phage researcher. After World War II, he accepted a professorship at the California Institute of Technology, where his research interests have shifted from molecular genetics to sensory physiology. Dr. Delbrück is a fellow of the National Academy of Sciences. At present, he is studying the primary transducer processes of sense organs using the sporangiophores of *Phycomyces* as a model system.

Joyce A. F. Diener, articles editor for *Psychology Today* magazine, graduated from Ohio Wesleyan University with honors in psychology in 1965. She received her doctorate from the University of California at Berkeley in 1970. Her graduate research centered around reproductive behavior. She completed studies of maternal behavior in beagles and copulatory behavior in rats under the direction of Frank A. Beach.

John E. Dowling is professor of biology at Harvard University. After graduation from Harvard College in 1957, he spent two years at the Harvard Medical School. He then switched to biology and obtained a Ph.D. from Harvard in 1961. He taught in the Harvard Department of Biology until 1964, when he went to Johns Hopkins. He remained at Johns Hopkins until 1971. His major interests are in the structure and function of the vertebrate retina, in problems of visual adaptation, in the synaptic organization of the retina as revealed by electron microscopy, and in intracellular recording.

John C. Eccles, professor of physiology and biophysics at the State University of New York at Buffalo, received his Ph.D. from Oxford University. He was awarded the Nobel Prize in Physiology and Medicine in 1963 and was knighted in 1958. He has taught at universities in Australia, New Zealand, England, and the United States. Dr. Eccles has served as President of the Australian Academy of Science and is a foreign associate of the National Academy of Sciences. His expertise in neurophysiology is reflected in his publications, which include *Physiology of Nerve Cells, The Physiology of Synapses, Brain and Conscious Experience, The Cerebellum as a Neuronal Machine,* and *Facing Reality.*

Leland N. Edmunds, Jr., is an associate professor of biology at the State University of New York at Stony Brook. After obtaining his Ph.D. from Princeton University in 1964, he spent a summer at the University of Costa Rica. His research interests include biological clocks, cell biology, and control of the cell cycle. Dr. Edmunds belongs to several professional organizations, including the American Society for Microbiology and the American Society of Plant Physiologists.

J. S. Finlayson is a research chemist at the National Institutes of Health, where since 1958 he has been engaged in studying the biochemistry of fibrinogen and other proteins of the blood plasma. He holds M.S. and Ph.D. degrees from the University of Wisconsin and teaches introductory biochemistry and mathematical preparation for biochemistry at the Foundation for Advanced Education in the Sciences, Bethesda, Maryland.

William Fishbein is an assistant professor of psychology at the City College of the City University of New York. He followed his Ph.D. from the University of Colorado with a postdoctoral fellowship in psychobiology at the University of California at Irvine with James L. McGaugh. His general interests are in the neurobiological bases of behavior. His research efforts have included pharmacological, biochemical, electrophysiological, and behavioral studies of sleep. His recent concern has been with the processes occurring during sleep that may influence the mechanisms underlying long-term memory storage.

Paul Gebhard is a professor of anthropology and the director of the Institute for Sex Research at Indiana University. Educated at Harvard University in the field of anthropology, he received a Ph.D. in 1947. Dr. Gebhard was a coauthor with Alfred Kinsey of *Sexual Behavior in the Human Female* and was the senior author of *Pregnancy, Birth and Abortion*, *Sex Offenders: An Analysis of Types*, and *The Sexuality of Women*.

Terrell H. Hamilton is a professor of zoology at the University of Texas. He took his Ph.D. at Harvard University in 1960 and was a recipient of the Guggenheim Fellowship in biochemistry in 1964. His research interests include the areas of evolution theory, hormonal and reproductive physiology, and adaptive aspects of sexual dimorphism and species-area phenomena.

Peter H. Hartline received a B.A. from Swarthmore College in physics and later pursued his interest in animal behavior through an M.A. at Harvard University. Because he believed the base of behavior to be the nervous system, he did his doctoral work in the field of neuroscience and received the Ph.D. from the University of California at San Diego in 1969. His main research and scientific interests are in the processing of information in sensory systems and its relation to the perception and behavior of animals.

J. Woodland Hastings, professor of biology at Harvard University, is engaged in studies on the molecular mechanism of bioluminescence and its biological role in energy metabolism. He received his Ph.D. from Princeton University in 1951. He has served as managing editor of the *Journal of General Physiology*, as a member of the Commission on Undergraduate Education in the Biological Sciences, as a member of the National Science Foundation's Panel on Molecular Biology, and as a member of the Space Biology Subcommittee. Dr. Hastings is currently on advisory committees for the National Science Foundation, the Red Sea Marine Research Station, and the Marine Biological Laboratory at Woods Hole, Massachusetts.

Jonathan Hodge received his doctorate in the history of science from Harvard University in 1970. He has taught at the University of Toronto and has been a visitor for two years at the University of California at Berkeley. An historian of biology and philosophy, he has been especially concerned with the origins of the Darwinian theory. He has papers on Jean Lamarck and Robert Chambers forthcoming and is working on a book analyzing different traditions in the treatment of origins and species from Plato and Aristotle to the present.

John Holland, professor of biology at the University of California at San Diego, took the Ph.D. from the University of California at Los Angeles in 1957. He worked at the University of Washington, at the University of California at Irvine, and was a visiting scientist at the Institute of Molecular Biology at the University of Geneva in Switzerland. His current research focuses on the biochemistry of animal and human viral infections, and he has published numerous papers in this field.

Yashuo Hotta received the Ph.D. in biology from Nagoya University in Japan. He was a postdoctoral fellow at the Plant Research Institute of the Canadian Department of Agriculture and a research associate at the University of Illinois. Currently, Dr. Hotta is an associate research biologist at the University of California at San Diego. His interest in the chemistry of nucleic acids is reflected in his numerous publications.

Tom D. Humphreys II earned his Ph.D. in zoology at the University of Chicago in 1962. Before taking his present position as an assistant professor of biology at the University of California at San Diego, Dr. Humphreys taught at the Massachusetts Institute of Technology and at the Marine Biological Laboratory. In addition to his research activities in the biochemistry of specific cell association and the molecular biology of development, he is a member of the Planning Committee for the University Community.

Daniel H. Janzen, associate professor of biology at the University of Chicago, is well known for his studies in the field of population and community ecology of plants and animals in natural habitats. After receiving his Ph.D. from the University of California at Berkeley, Dr. Janzen taught at the University of Kansas and at the Smithsonian's Summer Institute in Systematics. He is the author of numerous publications and is on the editorial committee of the *Annual Review of Ecology and Systematics.*

William A. Jensen, chairman of the Botany Department at the University of California at Berkeley, received both his M.S. and Ph.D. from the University of Chicago. He has done considerable research abroad, and his primary botanical interests lie in the areas of histochemistry, cytology, and cell development.

Robert W. Kistner received his M.D. degree from the University of Cincinnati, where he also completed his internship and a residency in obstetrics. He received his training in gynecology at Johns Hopkins Hospital, Baltimore, Kings County Hospital, Brooklyn, and the Boston Hospital for Women. He is now senior gynecologist and obstetrician at the Boston Hospital for Women. He has written over 150 articles in the medical literature and is the author of *Principles and Practice of Gynecology, Progress in Infertility*, and *The Pill—Fact and Fallacy.*

Hans Adolf Krebs, recipient of the Nobel Prize in Medicine in 1953, took his early professional training in Germany. In addition to receiving an M.D. degree from the University of Hamburg, Germany, and an M.A. from the University of Cambridge, England, he has been awarded numerous honorary doctorate degrees from universities throughout the world. He has been a professor of biochemistry at Oxford University, England, since 1954, and has contributed many articles on biochemistry to professional journals. Dr. Krebs is a fellow of the National Academy of Sciences.

Lee H. Kronenberg is currently a predoctoral fellow in molecular biology and biochemistry at the University of California at San Diego. He received his undergraduate training in biophysical chemistry at the University of California at Berkeley. His present research interests include the regulation of gene expression, ribonucleic acid metabolism in eucaryotic cells, biological aspects of the secondary structure of nucleic acids, and interferons and related problems in virology. He has taught introductory physics and chemistry at Berkeley and biochemistry, molecular biology, and chemical evolution at San Diego. He participated in the development of the Contemporary Natural Sciences course at Berkeley and remains deeply committed to communication of science to nonscientists and novices.

Richard C. Lewontin took his Ph.D. in zoology from Columbia University in 1954 and received a Fulbright Fellowship in 1961. He is now a professor of biology at the University of Chicago. His primary research interests lie in the areas of population genetics, ecology, and evolution.

Robert D. Lisk has contributed various review chapters and numerous research articles on neural regulation of reproduction to professional journals. He received his Ph.D. from Harvard and is now a professor of biology at Princeton University. He is a member of the Society for Study of Reproduction, the International Brain Research Organization, the Endocrine Society, and is on the editorial boards of several journals.

William F. Loomis, Jr., a recipient of a Ph.D. from the Massachusetts Institute of Technology, is an assistant professor of biology at the University of California at San Diego. His main research interests lie in the chemical basis of the control of gene activity, and he is studying this phenomenon in the cellular slime mold *Dictyostelium discoideum.*

Vincent T. Marchesi received his Ph.D. from Oxford University in 1961 and his M.D. from Yale University in 1963. After an internship and residency in pathology at Washington University, Dr. Marchesi was a research associate at Rockefeller University. He is presently chief of the Section on Chemical Pathology at the National Institute of Arthritis and Metabolic Diseases. His research interests include cell interactions in inflammatory reactions, the chemistry and structure of cell membranes, and the properties of tumor cells.

Peter Marler first earned a Ph.D. in botany at the University of London in 1952, then pursued studies at the University of Cambridge, England, where he later took a Ph.D. in zoology. He is presently a professor at Rockefeller University and maintains extensive research interests in the field of animal behavior—particularly relative to the function and evolution of animal communication systems. He has a special interest in the vocalizations of birds and the ontogenetic basis of bird song, and he is a member of several professional ornithological societies.

Donald M. Maynard, professor of biology at the University of Oregon, earned his A.B. degree in biology at Harvard College and received his Ph.D. in zoology from the University of California at Los Angeles in 1955. He was a recipient of a postdoctoral fellowship from the National Science Foundation and received the Guggenheim Fellowship in 1964. His primary research interests include comparative neurophysiology, invertebrate physiology, and the physiology of behavior, and he has contributed numerous papers in these areas.

James L. McGaugh earned his Ph.D. from the University of California at Berkeley in 1959. Before taking his present position as chairman of the Department of Psychobiology at the University of California at Irvine, he taught at San Jose State College and at the University of Oregon and did postdoctoral work in Rome. He is the author of numerous publications concerning the neurobiology of learning and memory. He is on the advisory boards of many journals in the field and is editor of *Communications in Behavioral Biology.*

Stanley L. Miller received his early training in chemistry and his Ph.D. at the University of California. Now a professor of chemistry at the University of California at San Diego, he focuses his research on the origin of life, enzyme mechanisms, and mechanisms of general anesthesia.

James V. Neel, author of the chapter entitled "Human Possibilities," is chairman of the Department of Human Genetics at the University of Michigan Medical School. After receiving his Ph.D. and M.D. degrees from the University of Rochester, he did field work in Japan and then organized the Department of Human Genetics at the University of Michigan. Dr. Neel is a member of the National Academy of Sciences and the American Academy of Arts and Sciences, is past president of the American Society of Human Genetics, and is on the editorial boards of numerous journals.

David M. Phillips is an assistant professor of biology at Washington University in St. Louis, Missouri. He received his Ph.D. in zoology at the University of Chicago and was a postdoctorate fellow at Harvard Medical School. His research interests include the mechanisms of cellular motility and some aspects of cellular nucleic acid metabolism. His studies on motility involve cinematographic analysis of the swimming movement of spermatozoa from various species of insects and mammals and the correlation of these swimming movements with differences in structure.

David M. Prescott, who has been an exchange scientist to the USSR for a one-month lecture tour, is a professor in the Department of Molecular, Cellular, and Developmental Biology at the University of Colorado. After receiving his Ph.D. from the University of California at Berkeley, Dr. Prescott was an American Cancer Society Fellow at Carlsberg Laboratory in Copenhagen and a Markle Scholar in Medical Science. His research focuses on mechanisms of chromosome replication and function, chromatid exchanges during the cell cycle, exchange of proteins between the nucleus and cytoplasm, and the factors regulating the initial synthesis of DNA during cellular reproduction.

Eugene Rabinowitch was born in Russia and took his Ph.D. in chemistry at the University of Berlin. Throughout his distinguished career, he has been singled out for many honors, including the Guggenheim Fellowship, the honorary Doctor of Science from Dartmouth College and Brandeis University, and the 1965 Kalinga Prize from UNESCO. He has contributed numerous articles to professional journals and has authored several publications, including a multivolume work on photosynthesis (1945–1956). He is currently a professor at the State University of New York, with research interests in photobiology and photochemistry.

Roberts Rugh has taught and conducted research in embryology for 44 years. He received his Ph.D. from Columbia University, where he was, until his recent retirement, a professor of radiology in the College of Physicians and Surgeons. His research for the last 23 years has focused on the effects of radiation on the embryo and fetus, and the 221 titles and 6 books he has had published are almost exclusively concerned with embryonic development. Dr. Rugh's most recent publication, *From Conception to Birth: The Drama of Life's Beginnings*, which he coauthored with Dr. L. B. Shettles, is illustrated with color pictures he has taken of human fetuses.

Howard A. Schneiderman is dean of the School of Biological Sciences, director of the Center for Pathobiology, and professor of biological sciences at the University of California at Irvine. Dr. Schneiderman received his Ph.D. in physiology from Harvard University in 1952. He was assistant and then associate professor of zoology at Cornell University between 1953 and 1961. In 1961 he joined Case Western Reserve University as professor and chairman of the Department of Biology and director of the Developmental Biology Center. He gave up the chairmanship in 1966 to accept the Jared Potter Kirtland Distinguished Professorship of Biology at Case Western Reserve University; this post was held until leaving for UCI in 1969. His principle research interests have been the physiology and development of insects. He has published more than 150 papers, particularly in the field of insect endocrinology and developmental biology. His recent research has been on mechanisms of determination, pattern formation and intercellular communication in imaginal discs and embryos of *Drosophila*, and mode of action of juvenile hormones and molting hormones.

Michael Soulé, an assistant professor of biology at the University of California at San Diego, received his Ph.D. from Stanford University in 1964. He was a research associate in population biology at Stanford University and studied under Paul Ehrlich. Dr. Soulé is primarily interested in reptilian thermoregulation, insular biogeography, and ecological and evolutionary theory with emphasis on the significance of intraspecific variation.

Paul S. G. Stein is an assistant professor of biology at Washington University. After receiving his Ph.D. in neurological sciences from Stanford University in 1970, he was a postdoctoral fellow at the University of California at San Diego.

Payson R. Stevens has studied molecular and biological science at the City University of New York and oceanography at Scripps Institution of Oceanography. He also took classes at the School of Visual Arts and worked with the Bread and Puppet Theater in New York. As a graphics consultant, he has used his skills in these fields to try to open the paths of curiosity to the poetry and music of biology. His DNA and Structure-Function interleaves are attempts to break away from the often alienating presentation of scientific material.

Albert Szent-Györgyi was born in Budapest, Hungary, and received his M.D. degree from the University of Budapest. In 1927 he took a Ph.D. at Cambridge University. He started research as a medical student in histology, then turned to physiology, pharmacology, bacteriology, and chemistry. In 1937 he was awarded the Nobel Prize for the elucidation and discovery of the catalytic functions of C_4-dicarboxylic acids and the isolation of vitamin C. Dr. Szent-Györgyi has taught at universities in Holland, Hungary, and England. He is also a fellow of the National Academy of Sciences. He is currently director of the Institute for Muscle Research at the Marine Biological Laboratory in Woods Hole, Massachusetts.

J. Herbert Taylor received his Ph.D. from the University of Virginia. He has been a consultant for the Oak Ridge National Laboratory and the Brookhaven National Laboratory and has taught biology courses at Columbia University, University of Tennessee, and University of Oklahoma. He is now a professor of biological sciences at the Institute of Molecular Biophysics at Florida State University. Dr. Taylor's research interests include chromosome structure and behavior, DNA replication, autoradiographic studies of macromolecular synthesis in cells, and characterization of the molecular units of chromosomes.

Kenneth V. Thimann, professor of biology and provost of Crown College, University of California at Santa Cruz, received his Ph.D. from Imperial College, University of London, and holds an honorary doctorate from Universität Basel in Switzerland. A recipient of several prizes and fellowships, he has served on advisory committees for the President's Medal for Science and to the NASA Council on Biosciences for Manned Orbiting Missions. He is a fellow of the American Academy of Arts and Sciences and the National Academy of Sciences as well as numerous foreign academies. Dr. Thimann has also served on the editorial boards of *American Naturalist, Annual Review of Plant Physiology, Biological Abstracts*, and on the governing board for the American Association for the Advancement of Science. His main research interest is in the field of plant biochemistry and physiology.

Jared R. Tinklenberg, an assistant professor of psychiatry at Stanford University, received his M.D. at the University of Iowa, his medical training at Yale, and his psychiatric training at Stanford. In addition to his teaching post, Dr. Tinklenberg is the assistant director of the Stanford University In-Patient Psychiatric Service and a research associate at the Palo Alto Veterans' Administration Hospital. His research interests lie in the field of clinical psychopharmacology, especially the effects of tetrahydrocannabinol (the active ingredient of marijuana) on the mind. In accord with this interest, Dr. Tinklenberg serves as a consultant on drug abuse to numerous public and private agencies. His other research interests include the study of thought disorders in psychotic individuals, the relations between drug use and violence, and the long-term effects of drug use.

Gordon M. Tomkins received his M.D. from Harvard and his Ph.D. from the University of California at Berkeley. Before becoming a professor of biochemistry at the University of California at San Francisco, Dr. Tomkins served as chief of the Laboratory of Molecular Biology at the National Institute of Arthritis and Metabolic Diseases. He was awarded the Prize in Molecular Biology in 1967 and has held the Mider lectureship at the National Institutes of Health and the Jesup lectureship at Columbia University. Over 120 of his research papers have been published to date.

Harold C. Urey took the Ph.D. from the University of California at Berkeley and went on to do postdoctorate work as an American-Scandinavian Foundation Fellow at the Niels Bohr Institute for Theoretical Physics in Copenhagen. He was awarded the Nobel Prize in Chemistry in 1934. In addition to being professor emeritus at Chicago University and university professor emeritus at the University of California at San Diego, he is a consultant to NASA and an advisory member of the Space Science Board of the National Academy of Sciences. His fields of interest include the entropy of gases, chemical problems of the origin of the solar system, the origin of life, the separation of isotopes, and the temperatures of the ancient seas of the world.

James D. Watson achieved international fame for his elucidation of the double helical structure of DNA. This research, performed in collaboration with Francis Crick from 1951 to 1953 at Cambridge University, England, earned for him the Nobel Prize in Medicine and Physiology in 1962. He has worked on the structure of RNA at the California Institute of Technology and on the mechanism of protein synthesis and the replication of viruses at Harvard University, where he is professor of biology. He is also affiliated with the Cold Spring Harbor Laboratory of Quantitative Biology and is a fellow of the National Academy of Sciences. Dr. Watson's textbook, *Molecular Biology of the Gene*, is based on a series of ten lectures he has given to undergraduate biology students at Harvard and is widely used as a supplementary biology text throughout America.

J. S. Weiner, convener of the Human Adaptability Section of the International Biological Programme, is a professor of Environmental Physiology of the University of London and director of the Medical Research Council Unit of Environmental Physiology at the London School of Hygiene and Tropical Medicine. Dr. Weiner took his medical degree at St. George's Hospital, London, and his Ph.D. at London University. He is known for his leading role in the exposure of the Piltdown forgery and his definitive study of the Swanscombe fossil. He has also made field studies on Bushmen, Hottentots, and other African peoples. For these studies, he received the Rivers Memorial Medal. He is the author of *The Piltdown Forgery* and *Man in Natural History*, and he has published many papers on various aspects of climatic physiology and human evolution.

Robert H. Whittaker obtained his Ph.D. in ecology at the University of Illinois. He has been a weather observer and forecaster for the Army Air Forces, a senior scientist in the Radiological Sciences Department at Hanford Laboratories in Richland, Washington, a visiting scientist at Brookhaven National Laboratory, and is currently a professor of biology in the Section of Ecology and Systematics at Cornell University. He is the author of *Communities and Ecosystems* and *Classification of Natural Communities* and of articles on vegetation analysis, forest productivity, and other research problems. His interests range from the structure and function of natural communities to the evolution and broad classification of organisms. He is also vice president of the Ecological Society of America.

Special Consultation

BIOLOGY TODAY could not have been developed without the assistance, counsel, and encouragement of many individuals whose names are not listed above. In particular, we would like to express our gratitude to the following persons, whose help has been greatly appreciated:

Robert Coffman
Greg Erickson
Hudson Freeze
Ernest Habicht
Gail Habicht
Tom Hahn
William Hazen
Maryanna Henkart
Donald McQuarrie
Don Patt
Gail Patt
Ray Salemme
Bernard Weinstein

Bibliography

Abercrombie, M., and E. J. Ambrose. 1958. "Interference microscope studies of cell contacts in tissue culture," *Experimental Cell Research*, 15:332–345.

Adams, J. A. 1967. *Human Memory*. New York: McGraw-Hill.

Adams, Preston, J. J. W. Baker, and G. E. Allen. 1970. *The Study of Botany*. Reading, Mass.: Addison-Wesley.

Addicott, F. T., and J. L. Lyon. 1969. "Physiology of abscisic acid and related substances," *Annual Review of Plant Physiology*, 20:139–164.

Adolph, E. F. 1967. "The heart's pacemaker," *Scientific American*, 216 (March):32–37. Also Offprint No. 1067.

Agarwal, K. L., *et al.* 1970. "Total synthesis of the gene for an alanine transfer ribonucleic acid from yeast," *Nature*, 227: 27–34.

Agranoff, B. W. 1967. "Memory and protein synthesis," *Scientific American*, 216 (June):115–122. Also Offprint No. 1077.

———, and R. E. Davis. 1968. "Evidence for stages in the development of memory," in F. D. Carlson (ed.), *Physiological and Biochemical Aspects of Nervous Integration*. Englewood Cliffs, N.J.: Prentice-Hall.

Alexopoulos, C. J. 1962. *Introductory Mycology*. New York: Wiley.

———, and H. C. Bold. 1967. *Algae and Fungi*. New York: Macmillan.

Allee, W. C. 1938. *The Social Life of Animals*. New York: Norton.

———. 1951. *Cooperation Among Animals*. Rev. ed. New York: Schuman.

Allen, R. D. 1962. "Amoeboid movement," *Scientific American*, 206 (February):112–122. Also Offprint No. 182.

Allison, A. C. 1955. "Aspects of polymorphism in man," *Cold Spring Harbor Symposia on Quantitative Biology*, 20:239–255.

———. 1956. "Sickle cells and evolution," *Scientific American*, 195 (August):87–94. Also Offprint No. 1065.

———. 1964. "Polymorphism and natural selection in human populations," *Cold Spring Harbor Symposia on Quantitative Biology*, 29:137–149

———. 1967. "Lysosomes and disease," *Scientific American*, 217 (November):62–72. Also Offprint No. 1085.

Alston, R. E., and B. L. Turner. 1963. *Biochemical Systematics*. Englewood Cliffs, N.J.: Prentice-Hall.

Amen, R. D. 1968. "A model of seed dormancy," *The Botanical Review*, 34: 1–31.

American Academy of Pediatrics. 1964. *Standards and Recommendation on the Hospital Care of New Born Infants*. Evanston, Ill.: American Academy of Pediatrics.

Amoore, J. E., J. W. Johnston, Jr., and Martin Rubin. 1964. "The stereochemical theory of odor," *Scientific American*, 210 (February):42–49. Also Offprint No. 297.

Andersson, Bengt. 1952. "Polydipsia caused by intrahypothalamic injections of hypertonic NaCl solutions," *Experientia*, 8:157–158.

———. 1953. "The effect of injections of hypertonic NaCl solutions into different parts of the hypothalamus of goats," *Acta Physiologica Scandinavia*, 28:188–201.

Andrewartha, H. G. 1961. *Introduction to the Study of Animal Populations*. Chicago: University of Chicago Press.

———, and L. C. Birch. 1954. *The Distribution and Abundance of Animals*. Chicago: University of Chicago Press.

Andrews, H. N. 1947. *Ancient Plants and the World They Lived In*. Ithaca, N.Y.: Comstock.

———. 1961. *Studies in Paleobotany*. New York: Wiley.

Arditti, Joseph. 1966. "Orchids," *Scientific American*, 214 (January):70–78. Also Offprint No. 1031.

Ardrey, Robert. 1961. *African Genesis*. New York: Atheneum.

———. 1966. *The Territorial Imperative*. New York: Atheneum.

———. 1970. *The Social Contract*. New York: Atheneum.

Arey, L. B. 1963. *Human Histology*. Philadelphia: Saunders.

Ariëns-Kappers, C. U., G. C. Huber, and E. C. Crosby. 1960. *Comparative Anatomy of the Nervous System of Vertebrates*. 3 vols. New York: Hafner.

Armstrong, E. A. 1963. *A Study of Bird Song*. London: Oxford University Press.

Arnon, D. I. 1960. "The role of light in photosynthesis," *Scientific American*, 203 (November):104–118. Also Offprint No. 75.

Aschoff, Juergen. 1963. "Comparative physiology: diurnal rhythms," *Annual Review of Physiology*, 25:581–600.

———. 1964. "Survival value of circadian rhythms," *Symposia of the Zoological Society of London*, 13:79–98.

——— (ed.). 1965. *Circadian Clocks*. Amsterdam: North-Holland.

———. 1965. "Circadian rhythms in man," *Science*, 148:1427–1432.

———. 1967. "Human circadian rhythms in activity, body temperature, and other functions," in A. H. Brown and F. G. Favorite (eds.), *Life Sciences and Space Research*. Vol. 5. Amsterdam: North-Holland, pp. 159–173.

Asdell, S. A. 1964. *Patterns of Mammalian Reproduction*. 2nd ed. Ithaca, New York: Cornell University Press.

Ashton, P. S. 1969. "Speciation among tropical forest trees: some deductions in the light of recent evidence," *Biological Journal of the Linnean Society*, 1: 155–196.

Asimov, Isaac. 1962. *The World of Carbon*. New York: Collier.

———. 1962. *The World of Nitrogen*. New York: Collier.

———. 1964. *A Short History of Biology*. New York: Natural History Press.

Audus, L. J. 1963. *Plant Growth Substances*. Rev. ed. New York: Interscience (Wiley).

Augusta, Josef. 1957. *Prehistoric Man*. London: Spring Brooks.

Austin, O. L. 1961. *Birds of the World*. New York: Golden Press.

Auturm, H. (ed.). 1963. *Animal Orientation*. Berlin: Springer-Verlag.

Avers, C. J. 1958. "Histochemical localization of enzyme activity in the root epidermis of *Phleum pratense*," *American Journal of Botany*, 45:609–613.

———. 1961. "Histochemical localization of enzyme activities in root meristem cells," *American Journal of Botany*, 48:137–143.

Awdeh, Z. L., A. R. Williamson, and Brigitte Askonas. 1970. "One cell—one immunoglobulin: origin of limited heterogeneity of myeloma proteins," *Biochemical Journal*, 116 (January): 241–248.

Babson, S. G., and R. C. Benson. 1966. *Primer on Prematurity and High-Risk Pregnancy*. St. Louis, Mo.: Mosby.

Baker, H. G., and B. J. Harris. 1957. "The pollination of *Parkia* by bats and its attendant evolutionary problems," *Evolution*, 11:449–460.

Baker, J. J. W., and G. E. Allen. 1965. *Matter, Energy, and Life*. Reading, Mass.: Addison-Wesley.

Baker, P. F. 1966. "The nerve axon," *Scientific American*, 214 (March):74–82. Also Offprint No. 1038.

Baker, P. T. 1960. "Climate, culture and evolution," *Human Biology*, 32:3–16.

———, and J. S. Weiner (eds.). 1966. *The Biology of Human Adaptability*. London: Oxford University Press.

Balinsky, B. K. 1970. *An Introduction to Embryology*. 3rd ed. Philadelphia: Saunders.

Bangham, Alec. 1971. "Model for biological membranes," *New Scientist*, 49: 63–64.

Barlow, H. B. 1953. "Summation and inhibition in the frog's retina," *Journal of Physiology*, 119:69–88.

———, **R. M. Hill, and W. R. Levick.** 1964. "Retinal ganglion cells responding selectively to direction and speed of image motion in the rabbit," *Journal of Physiology*, 173:377–407.

Barnes, R. D. 1963. *Invertebrate Zoology*. Philadelphia: Saunders.

Barnett, H. G. 1953. *Innovation: The Basis of Culture Change*. New York: McGraw-Hill.

Barnett, S. A. 1967. "Rats," *Scientific American*, 216 (January):78–85. Also Offprint No. 1060.

Barnicot, N. A. 1957. "Human pigmentation," *Man*, 144:1–7.

Barondes, S. H., and H. D. Cohen. 1968. "Memory impairment after subcutaneous injection of acetoxycycloheximide," *Science*, 160:556–557.

———, **and M. E. Jarvik.** 1964. "The influence of actinomycin-D on brain RNA synthesis and on memory," *Journal of Neurochemistry*, 11:187–195.

Barraclough, C. A. 1966. "Modifications in the CNS regulation of reproduction after exposure of prepubertal rats to steroid hormone," *Recent Progress in Hormone Research*, 22:503–528.

Barrington, E. J. W. 1963. *Introduction to General and Comparative Endocrinology*. Oxford: Clarendon Press.

———. 1968. *The Chemical Basis of Physiological Regulation*. Glenview, Ill.: Scott, Foresman.

Bartholomew, G. A., and J. W. Hudson. 1961. "Desert ground squirrels," *Scientific American*, 204 (November):107–116. Also Offprint No. 1120.

Bassham, J. A. 1962. "The path of carbon in photosynthesis," *Scientific American*, 206 (June):88–100. Also Offprint No. 122.

———, **and Melvin Calvin.** 1957. *The Path of Carbon in Photosynthesis*. Englewood Cliffs, N.J.: Prentice-Hall.

Bates, Marston. 1960. *The Forest and the Sea*. New York: Random House.

———. 1961. *The Nature of Natural History*. Rev. ed. New York: Scribner.

Bateson, P. P. G. 1966. "The characteristics and context of imprinting," *Biological Review*, 41:177–220.

Bateson, William. 1913. *Mendel's Principles of Heredity*. 3rd ed. Cambridge: Cambridge University Press.

———, *et al.* 1902–1909. *Reports to the Evolution Committee of the Royal Society*. 5 vols. London: Harrison & Sons.

Batra, S. W. T., and L. R. Batra. 1967. "The fungus gardens of insects," *Scientific American*, 217 (November):112–120. Also Offprint No. 1086.

Beach, F. A. (ed.). 1965. *Sex and Behaviour*. New York: Wiley.

Beadle, G. W. 1945a. "Biochemical genetics," *Chemical Review*, 37:15–96.

———. 1945b. "Genetics and metabolism in *Neurospora*," *Physiological Reviews*, 25:643–663.

———. 1946. "Genes and the chemistry of the organism," *American Scientist*, 34: 31–53.

———. 1948. "The genes of men and molds," *Scientific American*, 179 (September):30–39. Also Offprint No. 1.

———, **and Muriel Beadle.** 1966. *The Language of Life*. New York: Doubleday.

———, **and E. L. Tatum.** 1941. "Genetic control of biochemical reactions in *Neurospora*," *Proceedings of the National Academy of Sciences*, 27:499–506.

Beals, R. L., and Harry Hoijer. 1965. *An Introduction to Anthropology*. 3rd ed. New York: Macmillan.

Beck, S. D. 1968. *Insect Photoperiodism*. New York: Academic Press.

Beecher, H. K. 1959. *Measurement of Subject Response*. New York: Oxford University Press.

Békésy, George von. 1956. "Current status of theories of hearing," *Science*, 123: 779–783.

———. 1957. "The ear," *Scientific American*, 197 (August):66–78.

———. 1960. *Experiments in Hearing*. E. G. Wever (tr. and ed.). New York: McGraw-Hill.

———. 1967. *Sensory Inhibition*. Princeton, N.J.: Princeton University Press.

Bell, C. R. 1967. *Plant Variation and Classification*. Belmont, Calif.: Wadsworth.

———, **and C. L. F. Woodcock.** 1968. *The Diversity of Green Plants*. Reading, Mass.: Addison-Wesley.

Bennet-Clark, H. C., and A. W. Ewing. 1970. "The love song of the fruit fly," *Scientific American*, 223 (July):84–92.

Benzer, Seymour. 1962. "The fine structure of the gene," *Scientific American*, 206 (January):70–84. Also Offprint No. 120.

Benzinger, T. H. 1961. "The human thermostat," *Scientific American*, 204 (January):134–147. Also Offprint No. 129.

Berelson, B. R., *et al.* (eds.). 1966. *Family Planning and Population Programs*. Chicago: University of Chicago Press.

Bermant, Gordon. 1967. "Copulation in rats," *Psychology Today*, 1 (July):52–60.

Berrill, N. J. 1953. *Sex and the Nature of Things*. New York: Apollo.

Best, C. H., and N. B. Taylor. 1961. *The Physiological Basis of Medical Practice*. 7th ed. Baltimore: Williams & Wilkins.

Biale, J. B. 1954. "The ripening of fruit," *Scientific American*, 190 (May):40–44. Also Offprint No. 118.

Biddulph, Susann, and Orlin Biddulph. 1967. *Morphology of Plants*. 2nd ed. New York: Harper & Row.

Bieber, Irving (ed.). 1962. *Homosexuality: A Psychoanalytic Study of Male Homosexuals*. New York: Basic Books.

Bier, Karlheinz. 1963. "Synthese, interzellulärer Transport, und Abbau von Ribonukleinsäure im Ovar der Stubenfliege *Musca domestica*," *Journal of Cell Biology*, 16:436–440.

Billingham, R. E., and W. K. Silvers. 1963. "Skin transplants and the hamster," *Scientific American*, 208 (January): 118–127. Also Offprint No. 148.

Billings, W. D. 1964. *Plants, Man, and the Ecosystem*. Belmont, Calif.: Wadsworth.

Blackwelder, R. E. 1963. *Classification of the Animal Kingdom*. Carbondale: Southern Illinois University.

———. 1967. *Taxonomy and the Diversity of Animals*. New York: Wiley.

Blake, Peter. 1964. *God's Own Junkyard: The Planned Deterioration of America's Landscape*. New York: Holt, Rinehart and Winston.

Bloom, William, and D. W. Fawcett. 1962. *A Textbook of Histology*. 8th ed. Philadelphia: Saunders.

Bloomquist, E. R. 1971. *Marihuana, the Second Trip*. Beverly Hills: Glencoe.

Blum, E. M., and R. H. Blum. 1967. *Alcoholism*. San Francisco: Jossey-Bass.

Blum, R. H., and Associates. 1964. *Utopiates*. New York: Atherton.

———. 1969a. *Students and Drugs*. San Francisco: Jossey-Bass.

———. 1969b. *Society and Drugs*. San Francisco: Jossey-Bass.

Bogert, C. B. 1959. "How reptiles regulate their body temperature," *Scientific American*, 200 (April):105–120. Also Offprint No. 1119.

Bold, H. C. 1964. *The Plant Kingdom*. 2nd ed. Englewood Cliffs, N.J.: Prentice-Hall.

Bonner, David. 1946. "Biochemical mutations in *Neurospora*," *Cold Spring Harbor Symposia on Quantitative Biology*, 11:14–24.

Bonner, J. T. 1965. *The Molecular Biology of Development*. New York: Oxford University Press.

———. 1966. *The Cellular Slime Molds*. 2nd ed. Princeton, N. J.: Princeton University Press.

———, **A. Chiquoine, and M. Kolderie.** 1955. "A histochemical study of differentiation in the cellular slime molds," *Journal of Experimental Zoology*, 130: 133–157.

———, **P. Koontz, and D. Paton.** 1953. "Size in relation to the rate of migration in the slime mold *Dictyostelium discoideum*," *Mycologia*, 45:235–240.

———, *et al.* 1969. "Acrasin, acrasinase and the sensitivity to acrasin in *Dictyostelium discoideum*," *Developmental Biology*, 20:72–87.

———, *et al.* 1970. "Evidence for a second chemotactic system in the cellular slime mold, '*Dictyostelium discoideum*'," *Journal of Bacteriology*, 102:682–687.

Bordes, François. 1968. *The Old Stone Age*. New York: McGraw-Hill.

Bordet, Jules. 1898. "Sur l'agglutination et la dissolution des globules rouges par le sérum d'animax injectés de sang defibriné," *Annales de l'Institut Pasteur*.

Borgstrom, Georg. 1967. *The Hungry Planet*. New York: Collier.

———. 1969. *Too Many: A Study of Earth's Biological Limitations*. New York: Macmillan.

Borradaile, L. A., and F. A. Potts. 1961. *The Invertebrata*. 4th ed. New York: Cambridge University Press.

Boulding, K. E. 1964. *The Meaning of the 20th Century: The Great Transition*. New York: Harper & Row.

Bower, T. G. R. 1966. "The visual world of infants," *Scientific American*, 215 (December):80–92. Also Offprint No. 502.

Boycott, B. B. 1965. "Learning in the octopus," *Scientific American*, 212 (March):42–50. Also Offprint No. 1006.

Boyden, A. A. 1951. "The blood relationships of animals," *Scientific American*, 185 (July):59–63.

Brace, C. L. 1964. "The fate of the 'classic' Neanderthals: a consideration of hominid catastrophism," *Current Anthropology*, 5: 3–43.

———. 1967. *The Stages of Human Evolution: Human and Cultural Origins*. Englewood Cliffs, N.J.: Prentice-Hall.

Brachet, Jean, and A. E. Mirsky (eds.). 1959–1964. *The Cell*. 6 vols. New York: Academic Press.

Brachmachary, R. L. 1967. "Physiological clocks," *International Review of Cytology*, 21:65–90.

Braidwood, R. J. 1967. *Prehistoric Men*. Glenview, Ill.: Scott, Foresman.

Braude, A. I. 1964. "Bacterial endotoxins," *Scientific American*, 210 (March):36–45. Also Offprint No. 177.

Brazier, M. A. B. 1960. *The Electrical Activity of the Nervous System*. New York: Macmillan.

———. 1962. "The analysis of brain waves," *Scientific American*, 206 (June): 142–153. Also Offprint No. 125.

Bresler, J. B. (ed.). 1968. *Environments of Man*. Reading, Mass.: Addison-Wesley.

Bridges, C. B. 1916. "Non-disjunction as proof of the chromosome theory of heredity," *Genetics*, 1:1–52, 107-163.

———, **and K. S. Brehme.** 1944. "The mutants of *Drosophila melanogaster*," *Carnegie Institution of Washington Publication No. 552*.

Brindley, G. S. 1963. "Afterimages," *Scientific American*, 209 (October): 84–93. Also Offprint No. 1089.

Britten, R. J., and D. E. Kohne. 1970. "Repeated segments of DNA," *Scientific American*, 222 (April):24–31.

Bronowski, J. 1967. "Human and animal languages," in *To Honor Roman Jakobson* (Paris: Mouton), pp. 374–394.

Bronson, William. 1968. *How to Kill a Golden State*. Garden City, N.Y.: Doubleday.

Brookhart, J. M. 1960. "The cerebellum," in J. Field, *et al.* (eds.), *Handbook of Physiology and Neurophysiology*. Vol. 2. Washington, D. C.: American Physiological Society, pp. 1245–1280.

Brown, D. D. 1967. "The genes for ribosomal RNA and their transcription during amphibian development," *Current Topics in Developmental Biology*, 2: 47–73.

———, **and I. B. Dawid.** 1968. "Specific gene amplification in oocytes," *Science*, 160:272–280.

———, **and J. B. Gurdon.** 1964. "Absence of ribosomal RNA synthesis in the anucleolate mutant of *Xenopus laevis*," *Proceedings of the National Academy of Sciences*, 51:139–146.

Brown, F. A. 1950. *Selected Invertebrate Types*. New York: Wiley.

Brown, F. A., Jr. 1960. "Response to pervasive geophysical factors and the biological clock problem," *Cold Spring Harbor Symposia on Quantitative Biology*, 25:57–71.

———. 1965. "A unified theory for biological rhythms," in Juergen Aschoff (ed.), *Circadian Clocks*. Amsterdam: North-Holland, pp. 231–261.

———, **J. W. Hastings, and J. D. Palmer.** 1970. *The Biological Clock: Two Views*. New York: Academic Press.

Brown, Harrison. 1954. *The Challenge of Man's Future*. New York: Viking.

——— (ed.). 1967. *The Next Ninety Years*. Pasadena: California Institute of Technology.

Brown, Robert. 1833. "The organs and mode of fecundation in Orchideae and Asclepiadeae," *Transactions of the Linnean Society of London*, 16:685–746.

Bruce, H. N. 1966. "Smell as an exteroceptive factor," *Journal of Animal Science*, 25 (Supplement):83–87.

Brues, Alice. 1959. "The spearman and the archer—an essay on selection in body build," *American Anthropologist*, 61: 457–469.

Bryant, P. J., and H. A. Schneiderman. 1969. "Cell lineage, growth, and determination in the imaginal leg discs of *Drosophila melanogaster*," *Developmental Biology*, 20:263–290.

Bryant, S. V. 1970. "Regeneration in amphibians and reptiles," *Endeavour*, 29: 12–17.

Buchsbaum, R. M. 1948. *Animals Without Backbones*. 2nd ed. Chicago: University of Chicago Press.

Buchsbaum, R. M., and L. J. Milne. 1960. *The Lower Animals: Living Invertebrates of the World*. Garden City, N.Y.: Doubleday.

Bullock, T. H. (ed.). 1956. *Physiological Triggers and Discontinuous Rate Processes*. Washington, D.C.: American Physiological Society.

———. 1961. "The origins of patterned nervous discharge," *Behaviour*, 17:48–59.

———. 1962. "Integration and rhythmicity in neural systems," *American Zoologist*, 2: 97–104.

———, **and A. Horridge.** 1965. *The Structure and Function of the Nervous System in Invertebrates*. San Francisco: Freeman.

Bünning, Erwin. 1967. *The Physiological Clock*. Revised 2nd ed. New York: Springer-Verlag.

Burak, Richard. 1970. *The New Handbook of Prescription Drugs*. New York: Ballantine.

Burdette, W. J. (ed.). 1962. *Methodology in Basic Genetics*. San Francisco: Holden-Day.

Burghardt, G. M. 1967. "Chemical perception in newborn snakes," *Psychology Today*, 1 (August):50–59.

Burnet, Sir Macfarlane. 1954. "How antibodies are made," *Scientific American*, 191 (November):74–78. Also Offprint No. 3.

———. 1959. *The Clonal Selection Theory of Acquired Immunity*. Nashville, Tenn.: Vanderbilt University Press.

———. 1961. "The mechanism of immunity," *Scientific American, 204* (January):58–67. Also Offprint No. 78.

———. 1962. "The thymus gland," *Scientific American, 207* (November): 50–57. Also Offprint No. 138.

———. 1969. *Self and Not-Self*. London: Cambridge University Press.

———, **and W. M. Stanley** (eds.). 1956–1959. *The Viruses*. 3 vols. New York: Academic Press.

Burns, R. K. 1961. "Role of hormones in the differentiation of sex," in W. C. Young (ed.), *Sex and Internal Secretions*. Vol. 1. Baltimore: Williams & Wilkins, pp. 76–158.

Busnel, R. G. (ed.). 1964. *Acoustic Behavior of Animals*. Amsterdam: Elsevier.

Butler, R. A. 1960. "Acquired drives and the curiosity-investigative motives," in R. H. Waters, *et al.* (eds.), *Principles of Comparative Psychology*. New York: McGraw-Hill, pp. 144–176.

Butler, W. L., and R. J. Downs. 1960. "Light and plant development," *Scientific American*, 203 (December):56–63. Also Offprint No. 107.

Butter, C. M. 1968. *Neuropsychology: The Study of Brain and Behavior*. Belmont, Calif.: Brooks/Cole.

Butterfield, Herbert. 1957. *The Origins of Modern Science: 1300–1800*. London: Bell.

Buvant, Roger. 1969. *Plant Cells*. New York: McGraw-Hill.

Cairns, John. 1966. "The bacterial chromosome," *Scientific American*, 214 (January):36–44. Also Offprint No. 1030.

Caldwell, P. C., *et al.* 1960. "The effects of injecting 'energy-rich' phosphate compounds on the active transport of ions in the giant axons of *Loligo*," *Journal of Physiology*, 152:561–590.

Camin, J. H., and P. R. Ehrlich. 1958. "Natural selection in water snakes (*Natrix sipedon* L.) on islands in Lake Erie," *Evolution*, 12:504–511.

Campbell, B. G. 1966. *Human Evolution: An Introduction to Man's Adaptations*. Chicago: Aldine.

Capranica, R. R. 1965. *The Evoked Vocal Response of the Bullfrog*. Cambridge, Mass.: M.I.T. Press.

Carlquist, Sherwin. 1966. "The biota of long-distance dispersal. III. Loss of dispersibility in the Hawaiian flora," *Brittonia*, 18:310–335.

Carpenter, P. L. 1967. *Microbiology*. 2nd ed. Philadelphia: Saunders.

Carr, Archie. 1963. *The Reptiles*. New York: Time-Life.

Catanzaro, R. J. 1968. *Alcoholism: The Total Treatment Approach*. Springfield, Ill.: Thomas.

Ceraso, John. 1967. "The interference theory of forgetting," *Scientific American*, 217 (October):117–124. Also Offprint No. 509.

Changeux, Jean-Pierre. 1965. "The control of biochemical reactions," *Scientific American*, 212 (April):36–45. Also Offprint No. 1008.

Chapman, C. B., and J. H. Mitchell. 1965. "The physiology of exercise," *Scientific American*, 212 (May):88–96. Also Offprint No. 1011.

Chapman, V. J. 1962. *The Algae*. New York: St. Martin's Press.

Chargaff, E. 1955. "Isolation and composition of the deoxypentose nucleic acids and of the corresponding nucleoproteins," in E. Chargaff and J. N. Davidson (eds.), *The Nucleic Acids*. Vol. 1. New York: Academic Press, pp. 307–371.

Chemical Education Material Study. 1963. *Chemistry: An Experimental Science*. San Francisco: Freeman.

Chomsky, Noam. 1969. *Aspects of the Theory of Syntax*. Cambridge, Mass: M.I.T. Press.

Christensen, C. M. 1965. *The Molds and Man*. 3rd ed. Minneapolis: University of Minnesota Press.

Churchill, Wainwright. 1967. *Homosexual Behavior Among Males*. New York: Hawthorn (Arco).

Claiborne, Robert. 1970. *Climate, Man, and History*. New York: Norton.

Claridge, Gordon. 1970. *Drugs and Human Behavior*. New York: Praeger.

Clark, B. F. C., and K. A. Marcker. 1968. "How proteins start," *Scientific American*, 218 (January):36–42. Also Offprint No. 1092.

Clark, Grahame. 1967. *The Stone Age Hunters*. New York: McGraw-Hill.

Clark, L. D., B. Lincoln, and E. N. Nakashima. 1968. "Experimental studies of marihuana," *American Journal of Psychiatry*, 125:379–384.

Clark, W. E. Le Gros. 1967. *Man-Apes or Ape-Men?: The Story of Discoveries in Africa*. New York: Holt, Rinehart and Winston.

Clements, J. A. 1962. "Surface tension in the lungs," *Scientific American*, 207 (December):120–130. Also Offprint No. 142.

Clevenger, Sarah. 1964. "Flower pigments," *Scientific American*, 210 (June): 84–92. Also Offprint No. 186.

Cloudsley-Thompson, J. L. 1961. *Rhythmic Activity in Animal Physiology and Behavior*. New York: Academic Press.

Clowes, F. A. L. 1959. "Apical meristems of roots," *Biological Review*, 34:501–529.

———. 1961 *Apical Meristems*. Oxford: Blackwell.

———. 1965. "Meristems and the effect of radiation on cells," *Endeavour*, 24:8–12.

Cochran, D. M. 1961. *Living Amphibians of the World*. Garden City, N.Y.: Doubleday.

Cockrill, W. R. 1967. "The water buffalo," *Scientific American*, 217 (December): 118–125. Also Offprint No. 1088.

Coghill, G. E. 1929. *Anatomy and the Problems of Behavior*. New York: Hafner.

Cohen, Sidney. 1964. *Drugs of Hallucination*. London: Secker & Warburg.

———. 1967. *The Beyond Within: The LSD Story*. New York: Atheneum.

———. 1969. *The Drug Dilemma*. New York: McGraw-Hill.

Cold Spring Harbor Symposia on Quantitative Biology. 1955. *Population Genetics: The Nature and Causes of Genetic Variability in Populations*. Cold Spring Harbor, N.Y.: Cold Spring Harbor Laboratory of Quantitative Biology.

———. 1959. *Genetics and 20th Century Darwinism*. Cold Spring Harbor, N.Y.: Cold Spring Harbor Laboratory of Quantitative Biology.

———. 1965. *Sensory Receptors*. Cold Spring Harbor, N.Y.: Cold Spring Harbor Laboratory of Quantitative Biology.

Cole, L. C. 1958. "The ecosphere," *Scientific American*, 198 (April):83–92. Also Offprint No. 144.

Collaborative Perinatal Research Project. 1964. *Five Years of Progress*. Bethesda, Md.: National Institute of Neurological Diseases and Blindness, National Institutes of Health.

Committee on Resources and Man. National Research Council. 1969. *Resources and Man: A Study and Recommendations*. San Francisco: Freeman.

Commoner, Barry. 1967. *Science and Survival*. New York: Viking.

Comroe, J. H., Jr. 1966. "The lung," *Scientific American*, 214 (February): 56–68. Also Offprint No. 1034.

Conant, J. B. 1951. *Science and Common Sense*. New Haven, Conn.: Yale University Press.

Conn, E. E., and P. K. Stumpf. 1966. *Outlines of Biochemistry*. 2nd ed. New York: Wiley.

Connell, J. H., and Eduardo Orias. 1964. "The ecological regulation of species diversity," *American Naturalist*, 98: 399–414.

Connell, P. H. 1958. *Amphetamine Psychoses*. Maudsley Monograph 5. London: Chapman & Hall.

Constantin, L. L., R. J. Podolsky, and Clara Franzini-Armstrong. 1965. "Localization of calcium-accumulating structures in striated muscle fibers," *Science*, 147: 158–160.

Constantinides, P. C., and Niall Carey. 1949. "The alarm reaction," *Scientific American*, 180 (March):20–23. Also Offprint No. 4.

Conway, G. R. 1969. "Pests follow the chemicals in the cocoa of Malaysia," *Natural History*, 78 (February):46–51.

Coon, C. S. 1955. "Some problems of human variability and natural selection in climate and culture," *American Naturalist*, 89:257–280.

Cooper, C. F. 1961. "The ecology of fire," *Scientific American*, 204 (April):150–160. Also Offprint No. 1099.

Copeland, H. F. 1956. *The Classification of Lower Organisms*. Palo Alto, Calif.: Pacific Books.

Copeland, W. E., J. C. Ullery, and G. F. Essig. 1969. "Therapeutic abortion," *Journal of the American Medical Association*, 207:713–715.

Corliss, J. O. 1961 *The Ciliated Protozoa*. Oxford: Pergamon.

Corner, G. W. 1963. *Hormones in Human Reproduction*. New York: Atheneum.

Corning, W. C., and S. C. Ratner (eds.). 1967. *The Chemistry of Learning*. New York: Plenum Press.

Cox, G. W. (ed.). 1969. *Readings in Conservation Ecology*. New York: Appleton-Century-Crofts.

CRM Books. 1970. *Psychology Today: An Introduction*. Del Mar, Calif.: CRM Books.

———. 1971. *Anthropology Today*. Del Mar, Calif.: CRM Books.

Cravioto, Joaquin, E. R. De Licardic, and H. G. Birch. 1966. "Nutrition, growth and neurointegrative development: an experimental and ecologic study," *Pediatrics*, 38:319–372.

Crick, F. H. C. 1954. "The structure of the hereditary material," *Scientific American*, 191 (October):54–61. Also Offprint No. 5.

———. 1957. "Nucleic acids," *Scientific American*, 197 (September):188–200. Also Offprint No. 54.

———. 1962. "The genetic code," *Scientific American*, 207 (October):66–74. Also Offprint No. 123.

———. 1966. "The genetic code: III," *Scientific American*, 215 (October):55–62. Also Offprint No. 1052.

Critchlow, B. V., and M. E. Bar-Sela. 1967. "Control of the onset of puberty," in Luciano Martini and W. F. Ganong (eds.), *Neuroendocrinology*. Vol. 2. New York: Academic Press, pp. 101–162.

Crombie, A. C. 1959. *Medieval and Early Modern Science*. Cambridge, Mass.: Harvard University Press.

Crowle, A. J. 1960. "Delayed hypersensitivity," *Scientific American*, 202 (April):129–138.

Csapo, Arpad. 1958. "Progesterone," *Scientific American*, 198 (April):40–46. Also Offprint No. 163.

Curtis, B. C., and D. R. Johnston. 1969. "Hybrid wheat," *Scientific American*, 220 (May). Also Offprint No. 1140.

Curtis, Helena. 1968. *The Marvelous Animals: An Introduction to the Protozoa*. Garden City, N.Y.: Natural History Press.

Cutter, E. G. 1956. "Experimental and analytical studies of pteridophytes. XXXIII. The experimental induction of buds from leaf primordia in *Dryopteris aristata Druce*," *Annals of Botany*, N. S. 20:143–165.

D'Amour, F. E. 1961. *Basic Physiology*. Chicago: University of Chicago Press.

Dampier, W. C. 1958. *A History of Science*. New York: Cambridge University Press.

Danielli, J. F., and Hugh Davson. 1935. "A contribution to the theory of permeability of thin films," *Journal of Cellular and Comparative Physiology*, 5:495–508.

Darling, F. F. 1937. *A Herd of Red Deer*. London: Oxford University Press.

Darwin, Charles. 1859. *On the Origin of Species by Means of Natural Selection, or the Preservation of Favoured Races in the Struggle for Life.* London: John Murray.

———. 1871. *The Descent of Man, and Selection in Relation to Sex.* 2 vols. London: John Murray.

———. 1964. *On the Origin of Species by Charles Darwin: A Facsimile of the First Edition with an Introduction by Ernst Mayr.* Cambridge, Mass.: Harvard University Press.

———, **and Francis Darwin.** 1880. "Sensitiveness of plants to light: its transmitted effects," reprinted in M. L. Gabriel and Seymour Fogel (eds.), *Great Experiments in Biology.* Englewood Cliffs, N.J.: Prentice-Hall, 1955, pp. 142–146.

Dasmann, R. F. 1959. *Environmental Conservation.* New York: Wiley.

———. 1964. *Wildlife Biology.* New York: Wiley.

———. 1965. *The Destruction of California.* New York: Collier.

———. 1968. *A Different Kind of Country.* New York: Macmillan.

Davidson, E. H. 1965. "Hormones and genes," *Scientific American,* 212 (June): 36–45. Also Offprint No. 1013.

———. 1969. *Gene Activity in Early Development.* New York: Academic Press.

———, **and B. R. Hough.** 1969. "High sequence diversity in the RNA synthesized at the lampbrush stage of oogenesis," *Proceedings of the National Academy of Sciences,* 63:342–349.

Davidson, J. M. 1966. "Control of gonadotropin secretion in the male," in Luciano Martini and W. F. Ganong (eds.), *Neuroendocrinology.* Vol. 1. New York: Academic Press, pp. 565–611.

Davis, B. D. 1950. "Studies on nutritionally deficient bacterial mutants isolated by means of penicillin," *Experientia,* 6:41–50.

Davis, D. C., and LeMar Middleton. 1968. "Rebirth of the midwife," *Today's Health,* 46:32–37.

Davis, P. H., and V. H. Heywood. 1963. *Principles of Angiosperm Taxonomy.* Princeton, N.J.: Van Nostrand.

Davson, Hugh, and J. F. Danielli. 1943. *The Permeability of Natural Membranes.* London: Cambridge University Press.

Day, L. H., and A. T. Day. 1965. *Too Many Americans.* New York: Delta.

De Beer, Gavin. 1965. *Charles Darwin: A Scientific Biography.* Garden City, N.Y.: Doubleday.

Deering, R. A. 1962. "Ultraviolet radiation and nucleic acid," *Scientific American,* 207 (December):135–144. Also Offprint No. 143.

Deevey, E. S., Jr. 1970. "Mineral cycles," *Scientific American,* 223 (September): 148–158.

DeKruif, Paul. 1926. *Microbe Hunters.* New York: Harcourt Brace Jovanovich.

Delbrück, Max. 1968. "Molecular biology—the next phase," *Engineering and Science,* (November):36–40.

———, **and M. B. Delbrück.** 1948. "Bacterial viruses and sex," *Scientific American,* 179 (November):46–51. Also Offprint No. 1104.

Delevoryas, Theodore. 1962. *Morphology and Evolution of Fossil Plants.* New York: Holt, Rinehart and Winston.

Delgado, J. M. R. 1965. "Sequential behavior induced repeatedly by stimulation of the red nucleus in free monkeys," *Science,* 148:1361–1363.

———. 1970. "ESB" (Electrical Stimulation of the Brain), *Psychology Today,* 3 (May):48–53.

Demarest, R. J., and J. J. Sciarra. 1969. *Conception, Birth and Contraception: A Visual Presentation.* New York: Blakiston.

Demerec, Milislav (ed.). 1950. *Biology of Drosophila.* New York: Wiley.

———, **and B. P. Kaufman.** 1961. *Drosophila Guide.* Washington, D.C.: Carnegie Institution of Washington.

Denenberg, V. H., and M. X. Zarrow. 1970. "Rat pax," *Psychology Today,* 3 (May):33–35, 58–60.

DeRobertis, E. D. P., W. W. Nowinski, and F. A. Saez. 1965. *Cell Biology.* 4th ed. Philadelphia: Saunders.

Dethier, V. G. 1963. *The Physiology of Insect Senses.* London: Methuen.

———. 1967. "The hungry fly," *Psychology Today,* 1 (June):64–72.

———, **and Eliot Stellar.** 1970. *Animal Behavior.* 3rd ed. Englewood Cliffs, N.J.: Prentice-Hall.

Detwiler, S. R. 1965. "The eye and its structural adaptations," *American Scientist,* 53:327–346.

Deutsch, J. A. 1968. "Neural basis of memory," *Psychology Today,* 1 (May): 56–61.

DeValois, R. L. 1960. "Color vision mechanisms in the monkey," *Journal of General Physiology,* 43 (Part 2):115–128.

———. 1965. "Analysis and coding of color vision in the primate visual system," in *Sensory Receptors.* Cold Spring Harbor, N.Y.: Cold Spring Harbor Laboratory of Quantitative Biology, pp. 567–580.

———, *et al.* 1958. "Responses of single cells in monkey lateral geniculate nucleus to monochromatic light," *Science,* 127: 238–239.

DiCara, L. V. 1970. "Learning in the autonomic nervous system," *Scientific American,* 222 (January):30–39.

Dickerson, R. E., and Irving Geis. 1970. *The Structure and Action of Proteins.* New York: Harper & Row.

Dickinson, Alice. 1967. *Carl Linnaeus: Pioneer of Modern Botany.* New York: Watts.

Dickinson, R. L. 1949. *Human Sex Anatomy.* 2nd ed. Baltimore: Williams & Wilkins.

Digby, J., and P. F. Wareing. 1966. "The effect of applied growth hormones on cambial division and the differentiation of the cambial derivatives," *Annals of Botany,* N. S. 30:539–548.

Dilger, W. C. 1962. "The behavior of lovebirds," *Scientific American,* 206 (January):88–98. Also Offprint No. 1049.

Dippell, R. V. 1962. "Ultrastructure of cells in relation to function," in W. H. Johnson and W. C. Steere (eds.), *This Is Life.* New York: Holt, Rinehart and Winston.

Dishotsky, N. I., *et al.* 1971. "LSD use and genetic damage," *Science,* 172. (in press).

Dobzhansky, Theodosius. 1936. "Position effects on genes," *Biological Review,* 11:364–384.

———. 1937. *Genetics and the Origin of Species.* New York: Columbia University Press.

———. 1950. "The genetic basis of evolution," *Scientific American,* 182 (January):32–41. Also Offprint No. 6.

———. 1951. *Genetics and the Origin of Species.* 3rd ed. New York: Columbia University Press.

———. 1959. "Variation and evolution," *Proceedings of the American Philosophical Society,* 103:252–263.

———. 1960. "The present evolution of man," *Scientific American,* 203 (September):206–217. Also Offprint No. 609.

———. 1962. *Mankind Evolving: The Evolution of the Human Species.* New Haven, Conn.: Yale University Press.

Dodson, C. H., *et al.* 1969. "Biologically active compounds in orchid fragrances," *Science,* 164:1243–1249.

Doty, Paul. 1957. "Proteins," *Scientific American,* 197 (September):173–184. Also Offprint No. 7.

Dowling, J. E. 1970. "Organization of vertebrate retinas," *Investigations in Opthamology.* (in press).

Drake, J. W. 1969. *The Molecular Basis of Mutation.* San Francisco: Holden-Day.

Drillien, C. M., and R. W. B. Ellis. 1964. *The Growth and Development of the Prematurely Born Infant.* Baltimore: Williams & Wilkins.

Drucker, P. F. 1969. *Age of Discontinuity: Guidelines to Our Changing Society.* New York: Harper & Row.

Dubos, Rene. 1965. *Man Adapting.* New Haven, Conn.: Yale University Press.

Dulbecco, Renato. 1967. "The induction of cancer by viruses," *Scientific American,* 216 (April):28–37. Also Offprint No. 1069.

Dunham, E. T., and I. M. Glynn. 1961. "Adenosinetriphosphate activity and the active movements of alkali metal ions," *Journal of Physiology,* 156:274–293.

Dupraw, E. J. 1968. *Cell and Molecular Biology.* New York: Academic Press.

Durham, Anthony. 1971. "How a virus assembles itself," *New Scientist and Science Journal,* 49:200–203.

Duve, Christian de. 1963. "The lysosome," *Scientific American,* 208 (May):64–72. Also Offprint No. 156.

Eagle, Harry. 1965. "Metabolic controls in cultured mammalian cells," *Science,* 148: 42–51.

Eames, A. J. 1961. *Morphology of the Angiosperms.* New York: McGraw-Hill.

———, **and L. H. MacDaniels.** 1947. *An Introduction to Plant Anatomy.* New York: McGraw-Hill.

Easton, W. H. 1960. *Invertebrate Paleontology*. New York: Harper & Row.

Ebert, J. D., and I. M. Sussex. 1970. *Interacting Systems in Development*. 2nd ed. New York: Holt, Rinehart and Winston.

Eccles, J. C. 1952. *The Neurophysiological Basis of Mind*. Oxford: Clarendon Press.

———. 1957. *The Physiology of Nerve Cells*. Baltimore: Johns Hopkins Press.

———. 1958. "The physiology of imagination," *Scientific American*, 199 (September):135–146. Also Offprint No. 65.

———. 1964. *The Physiology of Synapses*. New York: Springer-Verlag.

———. 1965. "The synapse," *Scientific American*, 212 (January):56–66. Also Offprint No. 1001.

Eccles, R. M., and B. Libet. 1961. "Origin and blockade of the synaptic responses of curarized sympathetic ganglia," *Journal of Physiology*, 157:484–503.

Echlin, Patrick. 1966. "The blue-green algae," *Scientific American*, 214 (June): 74–81. Also Offprint No. 1044.

Ede, D. A., and J. T. Law. 1969. "Computer simulation of vertebrate limb morphogenesis," *Nature*, 221:244–248.

Edelman, G. M. 1970. "The structure and function of antibodies," *Scientific American*, 223 (August):34–42.

———, **and W. E. Gall.** 1969. "The antibody problem," in E. E. Snell (ed.), *Annual Reviews of Biochemistry*. Vol. 38. New York: Annual Reviews, Inc.

Edgar, R. S., and R. H. Epstein. 1965. "The genetics of a bacterial virus," *Scientific American*, 212 (February):70–78. Also Offprint No. 1004.

Edmunds, L. N., Jr. 1970. *Biological Clocks: Temporal Organization in Living Systems*. New York: Benjamin.

Edwards, C. A. 1969. "Soil pollutants and soil animals," *Scientific American*, 220 (April):88–98. Also Offprint No. 1138.

Edwards, R. G. 1966. "Mammalian eggs in the laboratory," *Scientific American*, 215 (August):72–81. Also Offprint No. 1047.

Efron, D. H. 1967. "Ethnopharmacologic search for psychoactive drugs," *Public Health Service Publication No. 1645*, pp. 374–382.

Együd, L. G., and Albert Szent-Györgyi. 1966. "Cell division, SH, ketoaldehydes, and cancer," *Proceedings of the National Academy of Sciences*, 55:388–393.

Ehrlich, P. R. 1968. *The Population Bomb*. New York: Ballantine.

———, **and A. H. Ehrlich.** 1970. *Population, Resources, Environment: Issues in Human Ecology*. San Francisco: Freeman.

———, **and R. W. Holm.** 1963. *The Process of Evolution*. New York: McGraw-Hill.

———, **and P. H. Raven.** 1967. "Butterflies and plants," *Scientific American*, 216 (June):104–113. Also Offprint No. 1076.

Eibl-Eibesfeldt, Irenäus. 1961. "The fighting behavior of animals," *Scientific American*, 205 (December):112–122. Also Offprint No. 470.

———. 1970. *Ethology: The Biology of Behavior*. New York: Holt, Rinehart and Winston.

Eiseley, Loren. 1958. *Darwin's Century: Evolution and the Men Who Discovered It*. New York: Doubleday.

———. 1969. *The Unexpected Universe*. New York: Harcourt Brace Jovanovich.

Eldridge, W. B. 1967. *Narcotics and the Law: A Critique of the American Experiment in Narcotic Drug Control*. 2nd ed. Chicago: University of Chicago Press.

Ellis, Albert. 1958. *Sex Without Guilt*. Rev. ed. New York: Lyle Stuart.

———. 1965. *Homosexuality: Its Causes and Cure*. New York: Lyle Stuart.

———, **and Albert Abarbanel** (eds). 1969. *Encyclopedia of Sexual Behavior*. 2nd ed. 2 vols. New York: Ace Books.

Elton, C. S. 1958. *The Ecology of Invasions by Animals and Plants*. London: Methuen.

———. 1966. *The Pattern of Animal Communities*. London: Methuen.

Emiliani, Cesare. 1968. "The Pleistocene epoch and the evolution of man," *Current Anthropology*, 9:27–47.

Esau, Katherine. 1965. *Plant Anatomy*. 2nd ed. New York: Wiley.

Etkin, William (ed.). 1964. *Social Behavior and Organization Among Vertebrates*. Chicago: University of Chicago Press.

Evans, F. C. 1956. "Ecosystem as the basic unit in ecology," *Science*, 123: 1127–1128.

Evans, W. O., and K. E. Davis. 1969. "Dose-response effects of Secobarbitol on human memory," *Psychopharmacologia*, (Berlin) 14:46–61.

Everett, J. W. 1964. "Central neural control of reproductive functions of the adenohypophysis," *Physiological Reviews*, 44:373–431.

Ewing, J. A. 1967. "Non-narcotic addictive agents," in A. M. Freedman and H. I. Kaplan (eds.), *Comprehensive Textbook of Psychiatry*. Baltimore: Williams & Wilkins.

Faegri, Knut, and Londert van der Pijl. 1965. *The Principles of Pollination Ecology*. New York: Pergamon.

Fantino, Edmund. 1968. "Of mice and misers," *Psychology Today*, 2 (July): 40–43, 62.

Fantz, R. L. 1961. "The origin of form perception," *Scientific American*, 204 (May):66–72. Also Offprint No. 459.

Farb, Peter. 1962. *The Insects*. New York: Time-Life.

———. 1963. *Ecology*. New York: Time-Life.

Fawcett, D. W. 1958. "Structural specialization of the cell surface," in S. L. Palay (ed.), *Frontiers in Cytology*. New Haven, Conn.: Yale University Press.

———. 1966. *The Cell: Its Organelles and Inclusions*. Philadelphia: Saunders.

Fatt, P., and Bernard Katz. 1950. "Some observations on biological noise," *Nature*, 166:597–598.

———. 1951. "An analysis of the end-plate potential recorded with an intracellular electrode," *Journal of Physiology*, 115:320–370.

———. 1952. "Spontaneous subthreshold activity at motor nerve endings," *Journal of Physiology*, 117:109–128.

Feeny, P. P. 1968. "Effect of oak leaf tannins on larval growth of the winter moth, *Operophtera brumata*," *Journal of Insect Physiology*, 14:805–817.

———, **and H. Bostock.** 1968. "Seasonal changes in the tannin content of oak leaves," *Phytochemistry*, 7:871–880.

Feldman, L. K., and E. G. Cutter. 1970. "Regulation of leaf form in *Centaurea solstitialis* L. II. The developmental potentialities of excised leaf primordia in sterile culture," *Botanical Gazette*, 131: 39–49.

Fender, D. H. 1964. "Control mechanisms of the eye," *Scientific American*, 211 (July): 24–33. Also Offprint No. 187.

Fenner, F. J. 1968. *The Biology of Animal Viruses*. 2 vols. New York: Academic Press.

Fertig, D. S., and V. W. Edmonds. 1969. "The physiology of the house mouse," *Scientific American*, 221 (October): 103–110.

Fieser, L. F. 1955. "Steroids," *Scientific American*, 192 (January):52–60. Also Offprint No. 8.

Fisher, R. A. 1933. "On the evidence against the chemical induction of melanism in Lepidoptera," *Proceedings of the Royal Society* (London), Series B, 112: 407–416.

Fogg, G. E. 1968. *Photosynthesis*. New York: American Elsevier.

Ford, C. S. 1964. *Comparative Study of Human Reproduction*. New York: HRAFP.

———, **and F. A. Beach.** 1951. *Patterns of Sexual Behaviour*. New York: Harper & Row.

Ford, E. B. 1964. *Ecological Genetics*. London: Methuen.

Foskett, D. J. 1953. "Wilberforce and Huxley on evolution," *Nature*, 172:920.

Foster, A. F., and E. M. Gifford. 1959. *Comparative Morphology of Vascular Plants*. San Francisco: Freeman.

Foulkes, David. 1966. *The Psychology of Sleep*. New York: Scribner.

Fox, Ruth, and Peter Bourne (eds.). 1971. *Alcoholism: Progress in Research and Treatment*. New York: Academic Press.

Fraenkel-Conrat, Heinz. 1956. "Rebuilding a virus," *Scientific American*, 194 (June):42–47. Also Offprint No. 9.

———. 1962. *Design and Function at the Threshold of Life: The Viruses*. New York: Academic Press.

——— (ed.). 1968. *Molecular Basis of Virology*. New York: Van Nostrand Reinhold.

———, **and R. C. Williams.** 1955. "Reconstitution of active tobacco mosaic virus from its inactive protein and nucleic acid components," *Proceedings of the National Academy of Sciences*, 41: 690–698.

Franchi, L. L., A. M. Mandl, and Solly Zuckerman. 1962. "The development of the ovary and the process of oogenesis," in Solly Zuckerman (ed.), *The Ovary*. Vol. 1. New York: Academic Press, pp. 1–88.

Freedman, A. M., and H. I. Kaplan (eds). 1967. *Comprehensive Textbook of Psychiatry*. Baltimore: Williams & Wilkins.

Freeman, O. L. 1968. *World Without Hunger*. New York: Praeger.

Frei, Emil III, and E. J. Freireich. 1964. "Leukemia," *Scientific American*, 210 (May):88–96.

French, J. D. 1957. "The reticular formation," *Scientific American*, 196 (May):54–60. Also Offprint No. 66.

Frey-Wyssling, Albert, and Kurt Muhlethaler. 1965. *Ultrastructural Plant Cytology*. New York: American Elsevier.

Frieden, Earl. 1963. "The chemistry of amphibian metamorphosis," *Scientific American*, 209 (November):110–118. Also Offprint No. 170.

Frisch, Karl von. 1950. *Bees: Their Vision, Chemical Senses, and Languages*. Ithaca, N.Y.: Cornell University Press.

———. 1967. *The Dance Language and Orientation of Bees*. Cambridge, Mass.: Harvard University Press.

Frisch, Leonora (ed.). 1960. *Biological Clocks*. Cold Spring Harbor, N.Y.: The Biological Laboratory.

Fritsch, F. E. 1945. *The Structure and Reproduction of the Algae*. 2 vols. New York: Cambridge University Press.

Fruton, J. S. 1950. "Proteins," *Scientific American*, 182 (June):32–41. Also Offprint No. 10.

Gall, J. G., and M. L. Pardue. 1969. "Formation and detection of RNA-DNA hybrid molecules in cytological preparations," *Proceedings of the National Academy of Sciences*, 63:378–383.

Gallien, Louis. 1963. *Sexual Reproduction*. New York: Walker.

Galston, A. W. 1964. *The Life of the Green Plant*. 2nd ed. Englewood Cliffs, N.J.: Prentice-Hall.

———, and P. J. Davies. 1969. "Hormonal regulation in higher plants, *Science*, 163: 1288–1297.

Gardner, R. A., and B. T. Gardner. 1969. "Teaching sign language to a chimpanzee," *Science*, 165:664–672.

Garn, S. M. 1957. "Race and evolution," *American Anthropologist*, 59:218–224.

Garrod, A. E. 1909. *Inborn Errors of Metabolism*. London: Oxford University Press.

Gazzaniga, M. S. 1967. "The split brain in man," *Scientific American*, 217 (August): 24–29. Also Offprint No. 508.

Gebhard, P. H., *et al.* 1958. *Pregnancy, Birth and Abortion*. New York: Harper & Row.

———, *et al.* 1965. *Sex Offenders*. New York: Harper & Row.

———, Jan Raboch, and Hans Giese. 1970. *The Sexuality of Women*. New York: Stein & Day.

George, W. B. 1964. *Biologist Philosopher: A Study of the Life and Writings of Alfred Russel Wallace*. London: Abelard-Schuman.

Gerard, R. W. 1961. *Unresting Cells*. New York: Harper & Row.

Gilbert, P. W. 1962. "The behavior of sharks," *Scientific American*, 207 (July): 60–68. Also Offprint No. 127.

Gilliard, E. T. 1958. *Living Birds of the World*. Garden City, N.Y.: Doubleday.

———. 1963. "The evolution of bowerbirds," *Scientific American*, 209 (August):38–46. Also Offprint No. 1098.

Glassman, Edward (ed.). 1967. *Molecular Approaches to Psychobiology*. Belmont, Calif.: Dickenson.

Good, Ronald. 1956. *Features of Evolution in the Flowering Plants*. London: Longmans, Green.

Goodwin, D. W., *et al.* 1969. "Alcohol and recall: state-dependent effects in man," *Science*, 163:1358–1360.

Gorbman, Aubrey, and H. A. Bern (eds.). 1959. *Comparative Endocrinology*.New York: Wiley.

———. 1962. *A Textbook of Comparative Endocrinology*. New York: Wiley.

Goss, C. M. (ed). 1966. *Gray's Anatomy of the Human Body*. 28th ed. Philadelphia: Lea & Febiger.

Graham, C. H. (ed.). 1965. *Vision and Visual Perception*. New York: Wiley.

Grant, Verne. 1951. "The fertilization of flowers," *Scientific American*, 184 (June): 52–56. Also Offprint No. 12.

———. 1963. *The Origin of Adaptations*. New York: Columbia University Press.

Gray, Asa. 1963. *Darwiniana: Essays and Reviews Pertaining to Darwinism*. (Edited by A. H. Dupree). Cambridge, Mass.: Harvard University Press.

Gray, G. W. 1950. "Cortisone and ACTH," *Scientific American*, 182 (March):30–36. Also Offprint No. 14.

Gray, Robert. 1969. *The Great Apes: The Natural Life of Chimpanzees, Gorillas, Orangutans and Gibbons*. New York: Norton.

Greene, J. C. 1961. *The Death of Adam: Evolution and Its Impact on Western Thought*. New York: New American Library.

———. 1963. *Darwin and the Modern World View*. New York: New American Library.

Gregory, R. L. 1966. *Eye and Brain: The Psychology of Seeing*. New York: McGraw-Hill.

———. 1968. "Visual illusions," *Scientific American*, 219 (November):66–76. Also Offprint No. 517.

Grew, Nehemiah. 1682. *Anatomy of Plants*. London.

Griffin, D. R. 1962. *Animal Structure and Function*. New York: Holt, Rinehart and Winston.

———. 1964. *Bird Migration*. Garden City, N.Y.: Doubleday.

Griffith, John D., *et al.* 1970. "Experimental psychosis induced by administration of D-amphetamine," in E. Costa and S. Garattini (eds.), *Amphetamines and Related Compounds*. New York: Raven.

Gross, C. G., S. L. Chorover, and S. M. Cohen. 1965. "Caudate, cortical, hippocampal, and dorsal thalamic lesions in rats: alternation and Hebb-Williams maze performance," *Neuropsychology*, 3: 53–68.

Gross, Ludwik. 1961. *Oncogenic Viruses*. New York: Pergamon.

Gunsalus, I. C., and R. Y. Stanier (eds.). 1960–1964. *The Bacteria: A Treatise on Structure and Function*. 4 vols. New York: Academic Press.

Gurdon, J. B. 1968. "Transplanted nuclei and cell differentiation," *Scientific American*, 219 (December): 24–35. Also Offprint No. 1128.

Gustafson, T., and L. Wolpert. 1967. "Cellular movement and contraction in sea urchin morphogenesis," *Bological Review*, 42:442–498.

Guyton, A. C. 1961. *Textbook of Medical Physiology*. 2nd ed. Philadelphia: Saunders.

Hadorn, Ernst. 1965. "Problems of determination and transdetermination," in *Genetic Control of Differentiation*, Brookhaven Symposium in Biology No. 18. Upton, N.Y.: Brookhaven National Laboratory.

———. 1968. "Transdetermination in cells," *Scientific American*, 219 (November):110–120. Also Offprint No. 1127.

Haggis, G. H., *et al.* 1964. *Introduction to Molecular Biology*. New York: Wiley.

Hagins, W. A., H. V. Zonana, and R. G. Adams. 1963. "Local membrane current in the outer segments of squid photoreceptors," *Nature*, 194:844–848.

Hailman, J. P. 1969. "How an instinct is learned," *Scientific American*, 221 (December):98–106.

Haldane, J. B. X. 1924. "A mathematical theory of natural and artificial selection," *Proceedings of the Cambridge Philosophical Society*, 1–2:158–163.

Hall, A. R. 1954. *The Scientific Revolution, 1500–1800: The Formation of the Modern Scientific Attitude*. London: Longmans, Green.

Halstead, W. C., and W. B. Rucker. 1968. "Memory: a molecular maze," *Psychology Today*, 2 (June):38–41, 66–67.

Hamilton, T. H. 1967. *Process and Pattern in Evolution*. New York: Macmillan.

Hamilton, W. J., *et al.* 1962. *Human Embryology*. 3rd ed. Baltimore: Williams & Wilkins.

Hanawalt, P. C., and R. H. Haynes. 1967. "The repair of DNA," *Scientific American*, 216 (February):36–43. Also Offprint No. 1061.

Handler, Philip (ed.). 1970. *Biology and the Future of Man*. New York: Oxford University Press.

Hanson, E. D. 1964. *Animal Diversity*. 2nd ed. Englewood Cliffs, N.J.: Prentice-Hall.

Hardin, Garrett. 1959. *Nature and Man's Fate*. New York: Holt, Rinehart and Winston.

———. 1966. *Biology: Its Principles and Implications*. 2nd ed. San Francisco: Freeman.

———. 1968. "Abortion—or compulsory pregnancy," *Journal of Marriage and the Family*, 30:246–251.

——— (ed.). 1969. *Population, Evolution, and Birth Control: A Collage of Controversial Ideas*. 2nd ed. San Francisco: Freeman.

Hardy, G. H. 1908. "Mendelian proportions in a mixed population," *Science*, 28:49–50.

Harker, J. E. 1958. "Diurnal rhythms in the animal kingdom," *Biological Review*, 33: 1–52.

———. 1961. "Diurnal rhythms," *Annual Review of Entomology*, 6:131–146.

Harlow, Harry, and Margaret Harlow. 1967. "The young monkeys," *Psychology Today*, 1 (September):40–47.

Harris, R. J. C. (ed.). 1963. *Cell Growth and Cell Division* (Symposia of the International Society for Cell Biology, Vol. 2). New York: Academic Press.

Harris, M. 1964. *Cell Culture and Somatic Variation.* New York: Holt, Rinehart and Winston.

Harrison, G. A., *et al.* 1964. *Human Biology.* New York: Oxford University Press.

Harrison, J. W. H. 1928. "Induced changes in the pigmentation of the pupae of the butterfly *'Pieris napis'* L., and their inheritance," *Proceedings of the Royal Society*, Series B. 102:347–353.

Hartline, H. K. 1940. "The receptive field of the optic nerve fibers," *American Journal of Physiology*, 130:690–699.

Haskell, P. T. 1961. *Insect Sounds.* London: Witherby.

Hastings, J. W. 1959. "Unicellular clocks," *Annual Review of Microbiology*, 13: 297–312.

Hauser, P. M. 1963. *The Population Dilemma.* Englewood Cliffs, N.J.: Prentice-Hall.

Hawking, Frank. 1970. "The clock of the malaria parasite," *Scientific American*, 222 (June):123–131.

Hayes, William. 1968. *The Genetics of Bacteria and Their Viruses: Studies in Basic Genetics and Molecular Biology.* 2nd ed. New York: Wiley.

Hayflick, Leonard. 1968. "Human cells and aging," *Scientific American*, 218 (March):32–37. Also Offprint No. 1103.

Hazen, W. E. (ed.). 1970. *Readings in Population and Community Ecology.* Philadelphia: Saunders.

Headstrom, Richard. 1968. *Nature in Miniature.* New York: Knopf.

Heer, D. M. (ed.). 1968. *Readings on Population.* Englewood Cliffs, N.J.: Prentice-Hall.

Held, Richard. 1965. "Plasticity in sensory-motor systems," *Scientific American*, 213 (November):84–94. Also Offprint No. 494.

Helgeson, J. P. 1968. "The cytokinins," *Science*, 161:974–981.

Hendricks, S. B. 1969. "Food from the land," in Committee on Resources and Man, *Resources and Man.* San Francisco: Freeman, pp. 65–85.

Herald, E. S. 1961. *Living Fishes of the World.* Garden City, N.Y.: Doubleday.

Hershey, Alfred, and Martha Chase. 1952. "Independent functions of viral protein and nucleic acid in growth of bacteriophage," *Journal of General Physiology*, 36(1): 39–56.

Herskowitz, I. H. 1952. *Bibliography on the Genetics of Drosophila, Part 2.* Farnham Royal, Slough, Buckshire, England: Commonwealth Agricultural Bureau.

———. 1958. *Bibliography on the Genetics of Drosophila, Part 3.* Bloomington: Indiana University Press.

———. 1964. *Bibliography on the Genetics of Drosophila, Part 4.* New York: McGraw-Hill.

———. 1967. *Basic Principles of Molecular Genetics.* Boston: Little, Brown.

Herz, Werner. 1963. *The Shape of Carbon Compounds.* New York: Benjamin.

Hildebrand, Milton. 1960. "How animals run," *Scientific American*, 202 (May): 148–157. Also Offprint No. 1114.

Hill, W. C. O. 1953–1962. *Primates: Comparative Anatomy and Taxonomy.* 5 vols. New York: Wiley.

Hinde, R. A. 1966. *Animal Behavior.* New York: McGraw-Hill.

Hirschmann, Ralph, *et al.* 1969. "Studies on the total synthesis of an enzyme, V. The preparation of enzymatically active material," *Journal of the American Chemical Society*, 91:507–508.

Hoagland, M. B. 1959. "Nucleic acids and proteins," *Scientific American*, 201 (December):55–61. Also Offprint No. 68.

Hoar, W. S. 1966. *General and Comparative Physiology.* Englewood Cliffs, N.J.: Prentice-Hall.

Hoch, P. H., and J. Zubin (eds.). 1965. *Psychopathology of Perception.* New York: Grune & Stratton.

Hock, R. J. 1970. "The physiology of high altitude," *Scientific American*, 222 (February):52–62.

Hockett, C. F. 1963. "The problem of universals in language," in J. H. Greenberg (ed.), *Universals of Language.* Cambridge, Mass.: M.I.T. Press, pp. 1–29.

———**, and S. A. Altmann.** 1968. "A note on design features," in T. A. Sebeok (ed.), *Animal Communication.* Bloomington: Indiana University Press, pp. 61–72.

Hodgkin, A. L. 1958. "Ionic movements and electrical activity in giant nerve fibers," *Proceedings of the Royal Society*, Series B. 148:1–37.

———. 1964. *The Conduction of the Nervous Impulse.* Springfield, Ill.: Thomas.

———**, and A. F. Huxley.** 1939. "Action potentials recorded from inside nerve fiber," *Nature*, 144:710–711.

———**, A. F. Huxley, and Bernhard Katz.** 1949. "Ionic currents underlying the activity in the giant axon of the squid," *Archives des Sciences Physiologiques*, 3:129–150.

———**, and A. F. Huxley.** 1952. "Properties of nerve axons (I): movement of sodium and potassium ions during nervous activity," *Cold Spring Harbor Symposia on Quantitative Biology*, 17: 43–52.

———**, A. F. Huxley, and Bernard Katz.** 1952. "Measurement of current-voltage relations in the membrane of the giant axon of *Loligo*," *Journal of Physiology*, 116:424–448.

———**, and R. D. Keynes.** 1955. "Active transport of cations in giant axons from *Sepia* and *Loligo*," *Journal of Physiology*, 128:28–60.

Hodgson, E. X. 1961. "Taste receptors," *Scientific American*, 204 (May):135–144. Also Offprint No. 1048.

Hoffman, Klaus. 1960. "Experimental manipulation of the biological clock in birds," *Cold Spring Harbor Symposia on Quantitative Biology*, 25:379–387.

Hokin, L. E., and M. R. Hokin. 1965. "The chemistry of cell membranes," *Scientific American*, 213 (October):78–86. Also Offprint No. 1022.

Holland, J. G., and B. F. Skinner. 1961. *The Analysis of Behavior: A Program For Self-Instruction.* New York: McGraw-Hill.

Holley, R. W. 1966. "The nucleotide sequence of a nucleic acid," *Scientific American*, 214 (February):30–39. Also Offprint No. 1033.

Hollister, L. E. 1968. *Chemical Psychoses, LSD and Related Drugs.* Springfield, Ill.: Thomas.

———. 1971a. "Hunger and appetite after single doses of marihuana, alcohol and dextroamphetamine," *Clinical Pharmacology and Therapeutics*, 12: 44–49.

———. 1971b. "Marihuana in man: three years later," *Science*, 172:21–29.

Holst, Erich von, and Ursula von Saint Paul. 1962. "Electrically controlled behavior," *Scientific American*, 206 (March):50–59. Also Offprint No. 464.

Holt, S. J. 1969. "The food resources of the ocean," *Scientific American*, 221 (September):178–194. Also Offprint No. 886.

Holter, Heinz. 1961. "How things get into cells," *Scientific American*, 205 (September):167–180. Also Offprint No. 96.

Hooke, Robert. 1665. *Micrographia.* London.

Honig, W. H. (ed.). 1966. *Operant Behavior: Areas of Research and Application.* New York: Appleton-Century-Crofts.

Hooton, E. A. 1947. *Up from the Ape.* Rev. ed. New York: Macmillan.

Hopcraft, Arthur. 1968. *Born to Hunger.* Boston: Houghton Mifflin.

Hornbruch, Amata, and L. Wolpert. 1970. "Cell division in the early growth and morphogenesis of the chick limb," *Nature*, 226:764–766.

Horowitz, N. H. 1950. "Biochemical genetics of *Neurospora*," *Advances in Genetics*, 3:33–71.

———. 1956. "The gene," *Scientific American*, 195 (October):78–90. Also Offprint No. 17.

Hotchkiss, R. D., and Esther Weiss. 1956. "Transformed bacteria," *Scientific American*, 195 (November):48–53. Also Offprint No. 18.

Hubbard, Ruth, and Allen Kropf. 1958. "The action of light on rhodopsin," *Proceedings of the National Academy of Sciences*, 44:130–139.

———**, and George Wald.** 1952–1953. "*Cis-trans* isomers of vitamin A and retinene in rhodopsin synthesis," *Journal of General Physiology*, 36:269–315.

Hubel, D. H. 1963. "The visual cortex of the brain," *Scientific American*, 209 (November):54–62. Also Offprint No. 168.

———, **and T.N. Wiesel.** 1963. "Receptive fields of cells in striate cortex of very young, visually inexperienced kittens," *Journal of Neurophysiology*, 26: 994–1002.

Huffaker, C. B. 1959. "Biological control of weeds with insects," *Annual Review of Entomology*, 4:251–276.

Hughes, Arthur. 1952. *The Mitotic Cycle.* New York: Academic Press.

———. 1959. *A History of Cytology*. New York: Abelard-Schuman.

Hughes, A. F. W. 1968. *Aspects of Neural Ontogeny*. New York: Logos Press.

Hulett, H. R. 1970. "Optimum world population," *BioScience*, 20:160–161.

Hulse, F. S. 1955. "Technological advance and major racial stocks," *Human Biology*, 27:184–192.

———. 1962. "Race as an evolutionary episode," *American Anthropologist*, 64: 929–945.

Hurwitz, Jerard, and J. J. Furth. 1962. "Messenger RNA," *Scientific American*, 206 (February):41–49. Also Offprint No. 119.

Hutchins, R. E. 1966. *Insects*. Englewood Cliffs, N.J.: Prentice-Hall.

Hutchinson, G. E. 1959. "Homage to Santa Rosalia, or why are there so many kinds of animals?" *American Naturalist*, 93: 145–159.

———. 1965. *The Ecological Theater and the Evolutionary Play*. New Haven, Conn.: Yale University Press.

———. 1970. "The biosphere," *Scientific American*, 223 (September):44–53.

Huxley, A. F., and R. Niedergerke. 1954. "Structural changes in muscle during contraction," *Nature*, 173:971–973.

Huxley, H. E. 1958. "The contraction of muscle," *Scientific American*, 199 (November):66–82. Also Offprint No. 19.

———. 1965. "The mechanism of muscular contraction," *Scientific American*, 213 (December):18–27. Also Offprint No. 1026.

———, **and Jean Hanson.** 1954. "Changes in the cross-striations of muscles during contraction and stretch and their structural interpretation," *Nature*, 173:973–976.

———. 1959. "The structural basis of the contraction mechanism in striated muscle," *Annals of the New York Academy of Sciences*, 81:403–408.

Huxley, Julian. 1942. *Evolution: the Modern Synthesis*. New York: Harper & Row.

———. 1956. "World population," *Scientific American*, 194 (March):64–76. Also Offprint No. 616.

Huxley, T. H. 1896. *Darwiniana: Essays*. New York: Appleton-Century-Crofts.

Hydén, Holger. 1961. "Satellite cells in the nervous system," *Scientific American*, 205 (December):62–70. Also Offprint No. 134.

———. 1962. "The neuron and its glia—a biochemical and functional unit," *Endeavour*, 144–145.

———, **and Endre Egyházi.** 1962. "Nuclear RNA changes of nerve cells during a learning experiment in rats," *Proceedings of the National Academy of Sciences*, 48: 1366–1373.

———, **and Endre Egyházi.** 1963. "Glial RNA changes during a learning experiment in rats," *Proceedings of the National Academy of Sciences*, 49:618–624.

———, **and Endre Egyházi.** 1964. "Changes in RNA content and base composition in cortical neurons of rats in a learning experiment involving transfer of handedness," *Proceedings of the National Academy of Sciences*, 52: 1030–1035.

Hyman, L. H. 1940–1955. *The Invertebrates*. 5 vols. New York: McGraw-Hill.

Hyman, S. E. (ed.). 1969. *Darwin for Today*. New York: Viking Press.

Imms, A. D. 1964. *A General Textbook of Entomology*. 9th ed. London: Methuen.

Ingram, V. M. 1958. "How do genes act?" *Scientific American*, 198 (January):68–74. Also Offprint No. 104.

———. 1963. *The Hemoglobins in Genetics and Evolution*. New York: Columbia University Press.

———. 1965. *The Biosynthesis of Macromolecules*. New York: Benjamin.

Irving, Laurence. 1966. "Adaptations to cold," *Scientific American*, 214 (January): 94–101. Also Offprint No. 1032.

Isaacs, Alick. 1963. "Foreign nucleic acids," *Scientific American*, 209 (October): 46–50. Also Offprint No. 166.

Isaacs, J. D. 1969. "The nature of oceanic life," *Scientific American*, 221 (September):146–162. Also Offprint No. 884.

Isaacson, R. L. 1970. "When brains are damaged," *Psychology Today*, 3 (January): 38–42.

Ito, S., and W. R. Loewenstein. 1969. "Ionic communication between early embryonic cells," *Developmental Biology*, 19:228–243.

Jacob, François, and Jacques Monod. 1961. "Genetic regulatory mechanisms in the synthesis of proteins," *Journal of Molecular Biology*, 3:318–356.

———, **and E. L. Wollman.** 1961. *Sexuality and the Genetics of Bacteria*. New York: Academic Press.

———, **and E. L. Wollman.** 1961. "Viruses and genes," *Scientific American*, 204 (June):92–107. Also Offprint No. 89.

Jacobs, W. P. 1955. "What makes leaves fall?" *Scientific American*, 193 (November):82–89. Also Offprint No. 116.

Jacobson, A. L., Clifford Fried, and S. D. Horowitz. 1966. "Planarians and memory," *Nature*, 209:599–601.

Jacobson, Marcus, and R. E. Baker. 1969. "Development of neuronal connections with skin grafts in frogs: behavioral and electrophysiological studies," *Journal of Comparative Neurology*, 137:121–142.

Jacobson, Martin, and Morton Beroza. 1964. "Insect attractants," *Scientific American*, 211 (August):20–27. Also Offprint No. 189.

Jahn, T. L., and F. F. Jahn. 1949. *How to Know the Protozoa*. Dubuque, Iowa: Brown.

Janzen, D. H. 1966. "Coevolution of mutualism between ants and acacias in Central America," *Evolution*, 20:249–275.

———. 1967. "Synchronization of sexual reproduction of trees within the dry season in Central America," *Evolution*, 21: 620–637.

———. 1969a. "Seed-eaters versus seed size, number toxicity and dispersal," *Evolution*, 23:1–27.

———. 1969b. "Birds and the ant X *Acacia* interaction in Central America, with notes on birds and other myrmecophytes," *Condor*, 71:240–256.

———. 1969c. "Allelopathy by myrmecophytes: the ant Azteca as an allelopathic agent of Cecropia," *Ecology*, 50:147–153.

———. 1970. "Herbivores and the number of tree species in tropical forests," *American Naturalist*, 104:501–528.

Jarrett, Henry (ed.). 1969. *Environmental Quality in a Growing Economy*. Baltimore: Johns Hopkins Press.

Jeanniere, Abel. 1967. *Anthropology of Sex*. New York: Harper & Row.

Jensen, David. 1966. "The hagfish," *Scientific American*, 214 (February): 82–90. Also Offprint No. 1035.

Jensen, W. A. 1962. *Botanical Histochemistry: Principles and Practice*. San Francisco: Freeman.

———. 1964. *The Plant Cell*. Belmont, Calif.: Wadsworth.

———, **and R. B. Park.** 1967. *Cell Ultrastructure*. Belmont, Calif.: Wadsworth.

Jerne, N. K. 1955. "The natural selection theory of antibody formation," *Proceedings of the National Academy of Sciences*, 41: 849–857.

Jessop, N. M. 1970. *Biosphere: A Study of Life*. Englewood Cliffs, N.J.: Prentice-Hall.

John, E. R. 1967. *Mechanisms of Memory*. New York: Academic Press.

Jones, J. C. 1968. "The sexual life of a mosquito," *Scientific American*, 218 (April):108–116. Also Offprint No. 1106.

Jones, R. T., and G. C. Stone. 1970. "Psychological studies of marihuana and alcohol in man," *Psychopharmacologia*, 18:108–112.

Jost, A. 1955. "Modalities in the action of gonadal and gonad stimulating hormones in the fetus," in *Memoirs of the Society of Endocrinology*, No. 4. London: Cambridge University Press, pp. 237–248.

Jouvet, Michel. 1967. "Mechanisms of the state of sleep: a neuropharmacological approach," in S. S. Kety, *et al.* (eds.), *Sleep and Altered States of Consciousness*. Baltimore: Williams & Wilkins, pp. 86–126.

———. 1969. "Biogenic amines and the states of sleep," *Science*, 163:32–40.

Juniper, B. E., *et al.* 1966. "Root cap and the perception of gravity," *Nature*, 209: 93–94.

Kabat, E. A. 1968. *Structural Concepts in Immunology and Immunochemistry*. New York: Holt, Rinehart and Winston.

Kalant, Harold, and R. D. Hawkins. 1969. *Experimental Approaches to the Study of Drug Dependence*. Toronto: University of Toronto Press.

Kamen, M. E. 1963. *Primary Processes in Photosynthesis*. New York: Academic Press.

Kandel, E. R. 1970. "Nerve cells and behavior," *Scientific American*, 223 (July): 57–70.

Kaplan, John. 1970. *Marijuana: The New Prohibition*. New York: World.

Katz, Bernhard. 1952. "The nerve impulse," *Scientific American*, 187 (November):55–64. Also Offprint No. 20.

———. 1961. "How cells communicate," *Scientific American*, 205 (September): 209–220. Also Offprint No. 98.

———. 1966. *Nerve, Muscle, and Synapse*. New York: McGraw-Hill.

Kaufman, Lloyd, and Irvin Rock. 1962. "The moon illusion," *Scientific American*, 207(July):120–130. Also Offprint No. 462.

Kellenberger, Edouard. 1966. "The genetic control of the shape of a virus," *Scientific American*, 215 (December):32–39. Also Offprint No. 1058.

Kellogg, W. N. 1968. "Communication and language in the home-raised chimpanzee," *Science*, 162:423–427.

Kelsay, J. L. 1969. "A compendium of nutritional status studies and dietary evaluation studies conducted in the U.S. 1957–1967," *Journal of Nutrition*, 99 (Supplement 1, Part 2):123–166.

Kelso, A. J. 1970. *Physical Anthropology: An Introduction*. Philadelphia: Lippincott.

Kendrew, J. C. 1961. "The three-dimensional structure of a protein molecule," *Scientific American*, 205 (December):96–110. Also Offprint No. 121.

Kennedy, Donald. 1963. "Inhibition in visual systems," *Scientific American*, 209 (July):122–130. Also Offprint No. 162.

——— (ed.). 1965. *The Living Cell: Readings from Scientific American*. San Francisco: Freeman.

———. 1967. "Small systems of nerve cells," *Scientific American*, 216 (May): 44–52. Also Offprint No. 1073.

———, W. H. Evoy, and J. T. Hanawalt. 1966. "Release of coordinated behavior in crayfish by single central neurons," *Science*, 154:917–919.

Kettlewell, H. B. D. 1955. "Selection experiments on industrial melanism in Lepidoptera," *Heredity*, 9:323–342.

———. 1956. "Further selection experiments on industrial melanism in Lepidoptera," *Heredity*, 10:287–301.

———. 1958. "Industrial melanism in the Lepidoptera and its contribution to our knowledge of evolution," *Proceedings of the Tenth International Congress of Entomology (1956)*, 2:831–841.

Keynes, R. D. 1958. "The nerve impulse and the squid," *Scientific American*, 199 (December):83–90. Also Offprint No. 58.

Killander, Johan (ed.). 1967. *Nobel Symposium 3: Gamma Globulins—Structure and Control of Biosynthesis*. New York: Interscience (Wiley).

Kimble, D. P. (ed.). 1965. *The Anatomy of Memory*. Palo Alto, Calif.: Science & Behavior Books.

King, J. A. 1955. *Social Behavior, Social Organization, and Population Dynamics in a Black-tailed Prairie Dog Town in the Black Hills of South Dakota*. Ann Arbor: University of Michigan Press.

King, R. C. 1965. *Genetics*. 2nd ed. New York: Oxford University Press.

King, Roger. 1970. "Reception of a sex hormone," *New Scientist*, 46(704): 472–473.

Kinsey, A. C., W. B. Pomeroy, and C. E. Martin. 1948. *Sexual Behavior in the Human Male*. Philadelphia: Saunders.

———, *et al.* 1953. *Sexual Behavior in the Human Female*. Philadelphia: Saunders.

Klein, R. M., and Arthur Cronquist. 1967. "A consideration of the evolutionary and taxonomic significance of some biochemical, micromorphological, and physiological characters in the thallophytes," *The Quarterly Review of Biology*, 42:105–296.

Klopfer, P. H. 1962. *Behavioral Aspects of Ecology*. Englewood Cliffs, N.J.: Prentice-Hall.

———. 1969. *Habitats and Territories: A Study of the Use of Space by Animals*. New York: Basic Books.

———, and J. P. Hailman. 1967. *An Introduction to Animal Behavior: Ethology's First Century*. Englewood Cliffs, N.J.: Prentice-Hall.

———, and R. H. MacArthur. 1961. "On the causes of tropical species diversity: niche overlap," *American Naturalist*, 95: 223–226.

Klots, A. B., and E. B. Klots. 1959. *Living Insects of the World*. Garden City, N.Y.: Doubleday.

Knight, C. A., and Dean Fraser. 1955. "The mutation of viruses," *Scientific American*, 193 (July):74–78. Also Offprint No. 59.

Koch, E. A., P. A. Smith, and R. C. King. 1967. "The division and differentiation of *Drosophila* cystocytes," *Journal of Morphology*, 121:55–70.

Kohler, Ivo. 1962. "Experiments with goggles," *Scientific American*, 206 (May): 62–86. Also Offprint No. 465.

Konigsberg, I. R. 1964. "The embryological origin of muscle," *Scientific American*, 211 (August):61–66. Also Offprint No. 191.

Kooyman, G. L. 1969. "The Weddell seal," *Scientific American*, 221 (August): 100–106. Also Offprint No. 1156.

Kormondy, E. J. 1969. *Concepts of Ecology*. Englewood Cliffs, N.J.: Prentice-Hall.

Korn, E. D. 1966. "Structure of biological membranes," *Science*, 153:1491–1498.

Kornberg, Arthur. 1960. "Biologic synthesis of DNA," *Science*, 131: 1503–1508.

———. 1968. "The synthesis of DNA," *Scientific American*, 219(October):64–78. Also Offprint No. 1124.

Kramer, Gustav. 1950. "Orientierte Zugaktivität gekafighter Singvogel," *Naturwissenschaften*, 37:188.

———. 1952. "Experiments on bird orientation," *Ibis*, 94:265–285.

Kramer, J. C. 1969. "New directions in the management of opiate dependence," *New Physician*, 18:203–209.

———, V. S. Fischman, and D. C. Littlefield. 1967. "Amphetamine abuse," *Journal of the American Medical Association*, 201:305–309.

Krebs, C. J., B. L. Keller, and R. H. Tamarin. 1969. "Microtus population biology: demographic changes in fluctuating populations of *M. ochrogaster* and *M. pennsylvanicus* in southern Indiana," *Ecology*, 50:587–607.

Krebs, H. A. 1950. "The tricarboxylic acid cycle," *Harvey Lectures*, 44 (1950): 165–199.

———, and H. L. Kornberg. 1957. *Energy Transformations in Living Matter: A Survey*. Berlin: Springer-Verlag.

Krivanek, Jara, and J. L. McGaugh. 1968. "Effects of pentylenetetrazol on memory storage in mice," *Psychopharmacologia*, 12:303–321.

Křiženecký, Jaroslav (ed.) 1965. *Fundamenta Genetica: The Revised Edition of Mendel's Classic Paper with a Collection of 27 Original Papers Published during the Rediscovery Era*. Brno, Czechoslovakia: Moravian Museum. Prague: Publishing House of the Czechoslovak Academy of Sciences. Oosterhout (N.B.) The Netherlands: Anthropological Publications.

Krogh, A. M. 1959. *The Comparative Physiology of Respiratory Mechanism*. Rev. ed. Philadelphia: University of Pennsylvania Press.

Kudo, Roksabro. 1954. *Protozoology*. 4th ed. Springfield, Ill.: Thomas.

Kuffler, S. W. 1953. "Discharge patterns and functional organization of mammalian retina," *Journal of Neurophysiology*, 16: 37–68.

———. 1967. "Neuroglial cells: physiological properties and a potassium mediated effect of neuronal activity on the glial membrane potential," *Proceedings of the Royal Society (London)*, Series B, 168: 1–21.

Kummer, Hans. 1968. *Social Organization of Hamadryas Baboons: A Field Study*. Chicago: University of Chicago Press.

Kuntz, Albert. 1953. *The Autonomic Nervous System*. 4th ed. Philadelphia: Lea & Febiger.

Kylstra, J. A. 1968. "Experiments in water-breathing," *Scientific American*, 219 (August):66–74. Also Offprint No. 1123.

Lack, David. 1947. *Darwin's Finches: An Essay on the General Biological Theory of Evolution*. London: Cambridge University Press.

———. 1953. "Darwin's Finches," *Scientific American*, 188 (April):66–72. Also Offprint No. 22.

Lader, Lawrence. 1969. "First exclusive survey of non-hospital abortions," *Look*, (January 1, 1969):63–65.

Laetsch, W. M., and R. E. Cleland (eds.). 1967. *Papers on Plant Growth and Development*. Boston: Little, Brown.

Lagler, K. F., J. E. Bardach, and R. R. Miller. 1962. *Ichthyology*. New York: Wiley.

Lamb, I. M. 1959. "Lichens," *Scientific American*, 201 (October): 144–156. Also Offprint No. 111.

Lang, P. J. 1970. "Autonomic control," *Psychology Today*, 4 (October):37–41, 86.

Langenheim, J. H. 1969. "Amber: A botanical inquiry," *Science*, 163: 1157–1169.

Lanyon, W. E., and W. N. Tavolga (eds.). 1960. *Animal Sounds and Communication.* Washington, D.C.: American Institute of Biological Sciences.

Larimer, James. 1968. *Introduction to Animal Physiology*. Dubuque, Iowa: William C. Brown.

Lashley, K. S. 1950. "In search of the engram," *Symposia of the Society for Experimental Biology*, 4:454–482.

Laughlin, W. S., and R. H. Osborne (eds.). 1967. *Human Variation and Origins: Readings from Scientific American*. San Francisco: Freeman.

Lawrence, G. H. M. 1951. *Taxonomy of Vascular Plants*. New York: Macmillan.

Leakey, L. S. B. 1960. "Recent discoveries at Olduvai Gorge," *Nature*, 188: 1050–1052.

———. 1965. *Olduvai Gorge, 1951–1961*. London: Cambridge University Press.

Ledbetter, M. C., and K. R. Porter. 1964. "Morphology of microtubular of plant cells," *Science*, 144:872–874.

———. 1970. *Introduction to Fine Structure of Plant Cells*. New York: Springer-Verlag.

Lee, R. B., and Irven DeVore (eds.). 1968. *Man the Hunter*. Chicago: Aldine.

Leech, Kenneth, and Brenda Jordan. 1967. *Drugs for Young People: Their Use and Misuse*. Oxford: Religious Education Press (Pergamon).

Lehninger, A. L. 1961. "How cells transform energy," *Scientific American*, 205(September):62–73. Also Offprint No. 91.

———. 1964. *The Mitochondrion: Molecular Basis of Structure and Function*. New York: Benjamin.

———. 1970. *Bioenergetics: The Molecular Basis of Biological Energy Transformations*. 2nd ed. New York: Benjamin.

Lehrman, D. S. 1964. "The reproductive behavior of ring doves," *Scientific American*, 211(November):48–54. Also Offprint No. 488.

Lenneberg, E. H. 1967. *Biological Foundations of Language*. New York: Wiley.

Leopold, A. C. 1949. *A Sand County Almanac*. New York: Oxford University Press.

———. 1955. *Auxins and Plant Growth*. Berkeley: University of California Press.

———. 1964. *Plant Growth and Development*. New York: McGraw-Hill.

Lerner, I. M. 1968. *Heredity, Evolution and Society*. San Francisco: Freeman.

Letham, D. S. 1969. "Cytokinins and their relation to other phytohormones," *BioScience*, 19:309–316.

Levey, R. H. 1964. "The thymus hormone," *Scientific American*, 211(July): 66–77. Also Offprint No. 188.

Levine, R. P. 1969. "The mechanism of photosynthesis," *Scientific American*, 221 (December):58–70. Also Offprint No. 1163.

Levine, Seymour. 1966. "Sex differences in the brain," *Scientific American*, 214 (April):84–90. Also Offprint No. 498.

Lewis, E. B. 1950. "The phenomenon of position effect," *Advances in Genetics*, 3: 73–115.

Li, C. C. 1955. *Population Genetics*. Chicago: University of Chicago Press.

Lillie, F. R. 1917. "The free-martin: a study of the action of sex hormones in the fetal life of cattle," *Journal of Experimental Zoology*, 23:371–452.

Lilly, J. C. 1967. *The Mind of the Dolphin: A Nonhuman Intelligence*. Garden City, N.Y.: Doubleday.

Limbaugh, Conrad. 1961. "Cleaning symbiosis," *Scientific American*, 205 (August):42–49. Also Offprint No. 135.

Lindauer, Martin. 1961. *Communication Among Social Bees*. Cambridge, Mass.: Harvard University Press.

Lindegren, C. C. 1932. "The genetics of *Neurospora*—II. Segregation of the sex factors in the asci of *N. crassa, N. sitophila, and N. tetrasperma*," *Bulletin of the Torrey Botanical Club*, 59:119–138.

———. 1933. "The genetics of *Neurospora*—III. Pure bred stocks and crossing-over in *N. Crassa*," *Bulletin of the Torrey Botanical Club*, 60:133–154.

Linderstrom-Lang, K. U. 1953. "How is a protein made?" *Scientific American*, 189 (September):100–106. Also Offprint No. 23.

Lindner, Robert. 1961. *Must You Conform?* New York: Grove.

Lingeman, R. R. 1969. *Drugs from A–Z: A Dictionary*. New York: McGraw-Hill.

Lisk, R. D. 1967. "Sexual behavior: hormonal control," in Luciano Martini and W. E. Ganong (eds.), *Neuroendocrinology*. Vol. 1. New York: Academic Press, pp. 197–239.

Lissmann, W. H. 1963. "Electric location by fishes," *Scientific American*, 208(March):50–59. Also Offprint No. 152.

Lloyd, C. F. 1964. *Human Reproduction and Sex Behavior*. Philadelphia: Lea & Febiger.

Lloyd, Monte, and H. S. Dybas. 1966. "The periodical cicada problem, II," *Evolution*, 20:466–505.

Loewenstein, W. R. 1960. "Biological transducers," *Scientific American*, 203 (August):98–108. Also Offprint No. 70.

Loewy, A. G., and Philip Siekevitz. 1963. *Cell Structure and Function*. New York: Holt, Rinehart and Winston.

Loomis, W. F., Jr., and Maurice Sussman. 1966. "Commitment to the synthesis of a specific enzyme during cellular slime mold development," *Journal of Molecular Biology*, 22:401–404.

Lord, R. D., Jr. 1961. "Mortality rates of cottontail rabbits," *Journal of Wildlife Management*, 25:33–40.

Lorenz, K. Z. 1937. "The companion in the bird's world," *Auk*, 54:245–273.

———. 1952. *King Solomon's Ring: New Light on Animal Ways*. New York: Crowell.

———. 1958. "The evolution of behavior," *Scientific American*, 199 (December):67–78. Also Offprint No. 412.

———. 1965. *Evolution and Modification of Behavior*. Chicago: University of Chicago Press.

———. 1966. *On Aggression*. New York: Harcourt Brace Jovanovich.

Louria, D. B. 1968. *The Drug Scene*. New York: McGraw-Hill.

Luce, G. G., and Julius Segal. 1966. *Sleep*. New York: Coward McCann.

Luria, A. R. 1970. "The functional organization of the brain," *Scientific American*, 222 (March):66–78.

Luria, S. E., and J. E. Darnell. 1968. *General Virology*. 2nd ed. New York: Wiley.

Lyell, Charles. 1830–1833. *The Principles of Geology*. 3 vols. London: John Murray.

MacArthur, R. H. 1965. "Patterns of species diversity," *Biological Reviews*, 40: 510–533.

———**, and J. H. Connell.** 1966. *The Biology of Populations*. New York: Wiley.

———**, and E. O. Wilson.** 1967. *The Theory of Island Biogeography*. Princeton, N.J.: Princeton University Press (Monographs in Population Biology).

MacGillivray, Ian. 1963. *Outline of Human Reproduction*. Baltimore: Williams & Wilkins.

Macgowan, Kenneth, and J. A. Hester, Jr. 1962. *Early Man in the New World*. Rev. ed. Garden City, N.Y.: Doubleday.

Mackinnon, D. L., and R. S. J. Hawes. 1961. *An Introduction to the Study of Protozoa*. London: Oxford.

MacLulich, D. A. 1947. "Fluctuations in the numbers of varying hare (*Lepus americanus*)," *University of Toronto Biological Series No. 43*.

Magoun, H. W. 1963. "Central neural inhibition," in M. R. Jones (ed.), *Nebraska Symposium on Motivation*. Vol. 2. Lincoln: University of Nebraska Press, pp. 161–193.

Mahan, B. H. 1969. *University Chemistry*. 2nd ed. Reading, Mass.: Addison-Wesley.

Maier, N. R. F., and T. C. Schneirla. 1935. *Principles of Animal Psychology*. New York: McGraw-Hill.

Malthus, Thomas. 1798. *An Essay on the Principle of Population*. London: Johnson.

Mangelsdorf, P. C. 1950. "The mystery of corn," *Scientific American*, 183 (July): 20–24. Also Offprint No. 26.

———. 1953. "Wheat," *Scientific American*, 189 (July): 50–59. Also Offprint No. 25.

Manning, Aubrey. 1968. *An Introduction to Animal Behavior*. Reading, Mass.: Addison-Wesley.

Manwell, R. D. 1968. Introduction to Protozoology. New York: Dover.

Margulis, Lynn. 1968. "Evolutionary criteria in thallophytes: a radical alternative," *Science*, 161:1020–1022.

———. 1970. *Origin of Eukaryotic Cells*. New Haven, Conn.: Yale University Press.

Marler, Peter. 1969. "Animals and man: communication and its development," in J. D. Roslansky (ed.), *Communication*. Amsterdam: North-Holland, pp. 25–62.

———**, and W. J. Hamilton III.** 1966. *Mechanisms of Animal Behavior*. New York: Wiley.

Marmor, Judd (ed.). 1965. *Sexual Inversion: The Multiple Roots of Homosexualism.* New York: Basic Books.

Martin, R. G., and B. N. Ames. 1964. "Biochemical aspects of genetics: the operon," *Annual Reviews of Biochemistry*, 33:235–258.

Masserman, J. H. 1967. "The neurotic cat," *Psychology Today*, 1 (October): 36–39, 56–57.

Masters, W. H., and V. E. Johnson. 1966. *Human Sexual Response.* Boston: Little, Brown.

———. 1970. *Human Sexual Inadequacy.* Boston: Little, Brown.

Matthews, G. V. T. 1968. *Bird Navigation.* 3rd ed. New York: Cambridge University Press.

Maturana, H. R., *et al.* 1960. "Anatomy and physiology of vision in the frog (*Rana pipiens*)," *Journal of General Physiology*, 43:129–175.

Mayersbach, H. von (ed.). 1967. *The Cellular Aspects of Biorhythms.* New York: Springer-Verlag.

Mayerson, H. S. 1963. "The lymphatic system," *Scientific American*, 208 (June): 80–90. Also Offprint No. 185.

Mayr, Ernst. 1942. *Systematics and the Origin of Species: From the Viewpoint of a Zoologist.* New York: Dover.

——— (ed.). 1957. *The Species Problem.* Washington, D.C.: American Association for the Advancement of Science.

———. 1963. *Animal Species and Evolution.* Cambridge, Mass.: Harvard University Press.

———. 1969. *Principles of Systematic Zoology.* New York: McGraw-Hill.

———, E. G. Linsley, and R. L. Usinger. 1953. *Methods and Principles of Systematic Zoology.* New York: McGraw-Hill.

Mazia, Daniel. 1953. "Cell division," *Scientific American*, 189 (August):53–63. Also Offprint No. 27.

———. 1955. "The organization of the mitotic apparatus," *Symposia of the Society for Experimental Biology*, 9: 335–357.

———. 1961. "How cells divide," *Scientific American*, 205 (September): 100–120. Also Offprint No. 93.

McCann, S. M., and A. P. S. Dharival. 1966. "Hypothalamic releasing factors and the neurovascular link between the brain and the anterior pituitary," in Luciano Martini and W. F. Ganong (eds.), *Neuroendocrinology*, Vol. 1. New York: Academic Press, pp. 261–296.

McConnell, J. V. 1962. "Memory transfer through cannibalism in planarians," *Journal of Neuropsychiatry*, 3 (Supplement 1):42–48.

McElroy, W. D., and H. H. Seliger. 1962. "Biological luminescence," *Scientific American*, 207 (December):76–89. Also Offprint No. 141.

McGaugh, J. L. 1967. "Analysis of memory transfer and enhancement," *Proceedings of the American Philosophical Society*, 111: 347–351.

———, and L. F. Petrinovich. 1965. "Effects of drugs on learning and memory," *International Review of Neurobiology*, 8: 139–196.

McGill, T. E. (ed.). 1965. *Readings in Animal Behavior.* New York: Holt, Rinehart and Winston.

McGlothlin, W. H., D. J. Arnold, and P. K. Rowan. 1970. "Marihuana use among adults," *Psychiatry*, 33:433–443.

———, and L. J. West. 1968. "The marihuana problem: an overview," *American Journal of Psychiatry*, 125: 370–378.

McHarg, Ian. 1969. *Design with Nature.* New York: Natural History Press.

McLennan, Hugh. 1969. *Synaptic Transmission.* Philadelphia: Saunders.

McLean, G. R., and Haskell Bowen. 1970. *High on the Campus: Student Drug Abuse: Is There an Answer?* Wheaton, Ill.: Tyndale House.

McVay, Scott. 1966. "The last of the great whales," *Scientific American*, 215 (August):13–21. Also Offprint No. 1046.

Mead, Margaret. 1949. *Male and Female.* New York: William Morrow.

Medawar, P. B. 1960. *The Future of Man.* London: Methuen.

Meggers, B. J. 1954. "Environmental limitation to the development of culture," *American Anthropologist*, 56:801–824.

Melges, F. T., *et al.* 1970. "Marihuana and temporal disintegration," *Science*, 168: 1118–1120.

Melzack, Ronald. 1970. "Phantom limbs," *Psychology Today*, 4 (October):63–68.

Menaker, Michael. 1969. "Biological clocks," *BioScience*, 19:681–689.

——— (ed.). 1970 (in press). *Symposium on Biochronometry.* Washington, D.C.: National Academy of Sciences.

Mendel, J. G. 1866. "Versuche Uber Pflanzen-Hybriden," *Verhandlungen Des Naturforschenden Vereines in Brunn*, 4: 3–47. [Reprinted in facsimile in *Journal of Heredity*, 42 (1951):1–47.]

Mercer, E. H. 1962. *Cells: Their Structure and Function.* Garden City, N.Y.: Doubleday.

Meselson, M. S., and F. W. Stahl. 1958. "The replication of DNA in *Escherichia coli*," *Proceedings of the National Academy of Sciences*, 44:671–682.

Meyer, B. S., D. B. Anderson, and R. H. Bohning. 1960. *Introduction to Plant Physiology.* New York: Van Nostrand Reinhold.

Michelmore, Susan. 1964. *Sexual Reproduction.* Garden City, N.Y.: Natural History Press.

Miller, N. E. 1957. "Experiments on motivation: studies combining psychological, physiological, and pharmacological techniques," *Science*, 126:1271–1278.

Miller, O. L., Jr., and B. R. Beatty. 1969. "Visualization of nucleolar genes," *Science*, 164:955–957.

Miller, W. H., Floyd Ratliff, and H. K. Hartline. 1961. "How cells receive stimuli," *Scientific American*, 205 (September): 222–238. Also Offprint No. 99.

Mintz, Beatrice. 1967. "Gene control of mammalian pigmentary differentiation. I. Clonal origin of melanocytes," *Proceedings of the National Academy of Sciences*, 58: 344–351.

Mirsky, A. E. 1953. "The chemistry of heredity," *Scientific American*, 188 (February): 47–57. Also Offprint No. 28.

———. 1968. "The discovery of DNA," *Scientific American*, 218 (June):78–88. Also Offprint No. 1109.

Modell, Walter (ed.). 1970. *Drugs in Current Use and New Drugs.* 16th rev. ed. New York: Springer.

Montagna, William. 1965. "The skin," *Scientific American*, 212 (February): 56–66. Also Offprint No. 1003.

Moody, P. A. 1962. *Introduction to Evolution.* 2nd ed. New York: Harper & Row.

Moore, J. A. 1963. *Heredity and Development.* New York: Oxford University Press.

Moore, Shirley. 1967. *Biological Clocks and Patterns.* New York: Criteron Books.

Morgan, T. H. 1910. "Sex-limited inheritance in *Drosophila*," *Science*, 32: 120–122.

———. 1926. *The Theory of the Gene* (2nd ed. 1928). Baltimore: Williams & Wilkins. (Reprinted in 1964 by Hafner, New York.)

———, and C. B. Bridges. 1916. "Sex-linked inheritance in *Drosophila*," *Carnegie Institution of Washington*, No. 237.

———, *et al.* 1915. *The Mechanism of Mendelian Heredity* (2nd ed. 1923). New York: Holt, Rinehart and Winston.

Morowitz, H. J., and M. E. Tourtellotte. 1962. "The smallest living cells," *Scientific American*, 206 (March):117–126. Also Offprint No. 1005.

Morris, Desmond. 1967. *The Naked Ape.* New York: McGraw-Hill.

——— (ed.). 1967. *Primate Ethology.* Chicago: Aldine.

———. 1969. *The Human Zoo.* New York: McGraw-Hill.

Morris, Ramona, and Desmond Morris. 1966. *Men and Apes.* New York: McGraw-Hill.

Moscona, A. A. 1961. "How cells associate," *Scientific American*, 205 (September):142–162. Also Offprint No. 95.

Mourant, A. E. 1954. *The Distribution of the Human Blood Groups.* Springfield, Ill.: Thomas.

Muller, H. J. 1927. "Artificial transmutation of the gene," *Science*, 66: 84–87.

———. 1939. *Bibliography on the Genetics of Drosophila.* Edinburgh: Oliver & Boyd.

———. 1951. "The development of the gene theory," in L. C. Dunn (ed.), *Genetics in the 20th Century.* New York: Macmillan, pp. 77–99.

Muntz, W. R. A. 1964. "Vision in frogs," *Scientific American*, 210 (March): 110–119. Also Offprint No. 179.

Murphy, George. 1969. "Introduction of a bill to amend the agricultural marketing agreement act of 1937," *Congressional Record* (May 20, 1969): 13072.

Mykytowycz, Roman. 1968. "Territorial marking by rabbits," *Scientific American*, 218 (May):116–126. Also Offprint No. 1108.

Nagel, Ernest. 1961. *The Structure of Science: Problems in the Logic of Scientific Explanation.* New York: Harcourt Brace Jovanovich.

Nalbandov, A. V. 1964. *Reproductive Physiology.* 2nd ed. San Francisco: Freeman.

Nash, L. K. 1963. *The Nature of the Natural Sciences.* Boston: Little, Brown.

Naylor, A. W. 1952. "The control of flowering," *Scientific American*, 186 (May):49–56. Also Offprint No. 113.

Neisser, Ulric. 1968. "The processes of vision," *Scientific American*, 219 (September):204–214. Also Offprint No. 519.

Neumann, F., and W. Elger. 1965. "Proof of the activity of androgenic agents on the differentiation of the external genitalia, the mammary gland, and the hypothalmic-pituitary system in rats," in A. Vermeulen and D. Exley (eds.), *Androgens in Normal and Pathological Conditions: Proceedings of the Second Symposium on Steroid Hormones.* New York: Excerpta Medica Foundation, pp. 168–185.

Neurath, Hans. 1964. "Protein-digesting enzymes," *Scientific American*, 211 (December):68–79. Also Offprint No. 198.

Newell, P. C., J. S. Ellingson, and M. Sussman. 1969. "Synchrony of enzyme accumulation in a population of differentiating slime mold cells," *Biochemica et Biophysica Acta*, 177: 610–614.

Newman, M. T. 1961. "Biological adaptation of man to his environment: heat, cold, altitude, and nutrition," *Annals of the New York Academy of Sciences*, 91: 617–633.

Newman, R. W., and E. H. Munro. 1955. "The relation of climate and body size in U. S. males," *American Journal of Physical Anthropology*, 13:1–17.

Nitsch, J. P., and C. Nitsch. 1969. "Haploid plants from pollen grains," *Science*, 163:85–87.

Nordenskiöld, Erik. 1960. *The History of Biology.* New York: Tudor.

Norman, D. A. 1968. *Memory and Attention: An Introduction to Human Information Processing.* New York: Wiley.

Nossal, G. J. V. 1964. "How cells make antibodies," *Scientific American*, 211 (December):106–115. Also Offprint No. 199.

Novikoff, A. B., Edward Essner, and Nelson Quintana. 1964. "Golgi apparatus and lysosomes," *Federation Proceedings*, 23:1010–1022.

Novin, Donald. 1962. "The relation between electrical conductivity of brain tissue and thirst in the rat," *Journal of Comparative and Physiological Psychology*, 55:145–154.

Oakley, K. P. 1966. *Frameworks for Dating Fossil Man.* 2nd ed. Chicago: Aldine.

Ochs, Sidney. 1965. *Elements of Neurophysiology.* New York: Wiley.

Odum, E. P. 1971. *Fundamentals of Ecology.* 3rd ed. Philadelphia: Saunders.

Olby, R. C. 1966. *Origins of Mendelism.* London: Constable.

Ommanney, F. D. 1963. *The Fishes.* New York: Time-Life.

O'Relly, Edward. 1968. *Sexercises—Isometric and Isotonic.* New York: Pocket Books.

Orians, G. H., and H. S. Horn. 1969. "Overlap in foods and foraging of four species of blackbirds in the potholes of central Washington," *Ecology*, 50: 930–938.

Oswald, Ian. 1966. *Sleep.* Baltimore: Penguin.

Overbeek, Johannes van. 1968. "The control of plant growth," *Scientific American*, 219 (July): 78–81. Also Offprint No. 1111.

———, M. E. Conklin, and A. F. Blakeslee. 1941. "Factors in coconut milk essential for growth and development of very young *Datura* embryos," *Science*, 94:350–351.

———. 1942. "Cultivation *in vitro* of small *Datura* embryos," *American Journal of Botany*, 29:472–477.

Ovesey, Lionel. 1969. *Homosexuality and Pseudohomosexuality.* New York: Science House.

Paddock, William, and Paul Paddock. 1968. *Famine—1975! America's Decision: Who Will Survive?* Boston: Little, Brown.

Paine, R. T. 1966. "Food web complexity and species diversity," *American Naturalist*, 100:65–75.

Painter, R. H. 1967. "Plant resistance to insects as applied to breeding vegetable crops," *Proceedings of the 17th International Horticultural Congress*, 3: 259–273.

Painter, T. S. 1933. "A new method for the study of chromosome arrangements and plotting of chromosome maps," *Science*, 78:585–586.

Parker, G. H. 1919. *The Elementary Nervous System.* Philadelphia: Lippincott.

Pasteur, Louis. 1861. "Mémoire sur les corpuscles organisés qui existent dans l'atmosphere, examen de la doctrine des générations spontanées," *Annales des sciences naturelles*, 16:5–98.

———. 1866. *Etudes sur le vin.* Paris.

Patt, D. I., and G. R. Patt. 1969. *Comparative Vertebrate Histology.* New York: Harper & Row.

Patten, B. M. 1968. *Human Embryology.* 3rd ed. New York: McGraw-Hill.

Pauling, Linus. 1940. "Theory of the structure and process of the formation of antibodies," *Journal of the American Chemical Society*, 62:2643–2657.

———. 1960. *The Nature of the Chemical Bond.* 3rd ed. Ithaca, N.Y.: Cornell University Press.

———, R. B. Corey, and Roger Hayward. 1954. "The structure of protein molecules," *Scientific American*, 191 (July):51–59. Also Offprint No. 31.

Pearl, Raymond. 1925. *The Biology of Population Growth.* New York: Knopf.

Pelczar, M. J., and R. D. Reid. 1965. *Microbiology.* 2nd ed. New York: McGraw-Hill.

Penfield, Wilder, and Theodore Rasmussen. 1950. *The Cerebral Cortex of Man.* New York: Macmillan.

Penn, R. D., and W. A. Hagins. 1969. "Signal transmission along retinal rods and the origin of the electroretinographic α-wave," *Nature*, 223:201–205.

Pequegnat, W. E. 1958. "Whales, plankton and man," *Scientific American*, 198 (January):84–90. Also Offprint No. 853.

Peters, J. A. (ed.). 1959. *Classic Papers in Genetics.* Englewood Cliffs, N.J.: Prentice-Hall.

Peterson, L. R. 1966. "Short-term memory," *Scientific American*, 215 (July): 90–95. Also Offprint No. 499.

Peterson, R. T. 1963. *The Birds.* New York: Time-Life.

Petrunkevitch, Alexander. 1952. "The spider and the wasp," *Scientific American*, 187 (August):20–23. Also Offprint No. 1097.

Pfeiffer, J. E. 1969. *The Emergence of Man.* New York: Harper & Row.

———, and C. S. Coon. 1963. *The Search for Early Man.* New York: American Heritage.

Phillips, D. C. 1966. "The three-dimensional structure of an enzyme molecule," *Scientific American*, 215 (November): 78–90.

Phillips, I. D. J. 1969. "Apical dominance," in M. B. Wilkins (ed.), *Physiology of Plant Growth and Development.* London: McGraw-Hill, pp. 165–202.

Pincus, Gregory, and K. V. Thimann (eds.). 1948–1964. *The Hormones.* New York: Academic Press.

Pirie, N. W. 1967. "Orthodox and unorthodox methods of meeting world food needs," *Scientific American*, 216 (February): 27–35. Also Offprint No. 1068.

Pitelka, F. A. 1957. "Some aspects of population structure in the short term cycle of the brown lemming in northern Alaska," *Cold Spring Harbor Symposia on Quantitative Biology*, 22:237–251

Pittendrigh, C. S. 1961. "On temporal organization in living systems," *Harvey Lectures*, 56:93–125.

———. 1966. "The circadian oscillation in *Drosophila pseudoobscura* pupae: a model for the photoperiodic clock," *Zeitschrift fuer Pflanzenphysiologie*, 54:275–307.

Plass, G. N. 1959. "Carbon dioxide and climate," *Scientific American*, 201 (July): 41–47. Also Offprint No. 823.

Pollock, M. B. 1969. "The drug abuse problem: some implications for health education," *Journal of the American College Health Association*, 17:403–411.

Population Reference Bureau. 1969. *1969 World Population Data Sheet.* Washington, D.C.: Population Reference Bureau, Inc.

Porter, C. L. 1967. *Taxonomy of Flowering Plants.* 2nd ed. San Francisco: Freeman.

Porter, K. R., and Clara Franzini-Armstrong. 1965. "The sarcoplasmic reticulum," *Scientific American*, 212 (March):72–80. Also Offprint No. 1007.

Porter, R. R. 1967. "The structure of antibodies," *Scientific American*, 217 (October):81–90. Also Offprint No. 1083.

Postlethwait, J. H., and H. A. Schneiderman. 1970. "A clonal analysis of development in *Drosophila melanogaster:* morphogenesis, determination, and growth in the wild type antenna," *Developmental Biology* (in press).

Powers, C. F., and Andrew Robertson. 1966. "The aging Great Lakes," *Scientific American*, 215 (November):94–104. Also Offprint No. 1056.

Premack, David. 1970. "The education of S*A*R*A*H: a chimp learns the language," *Psychology Today*, 4 (September):54–58.

Prescott, D. M. 1961. "RNA and protein replacement in the nucleus during growth and division and the conservation of components in the chromosome," in International Society for Cell Biology, *Cell Growth and Cell Division*, Vol. II. New York: Academic Press, pp. 111–128.

Preston, R. D. 1968. "Plants without cellulose," *Scientific American*, 218 (June): 102–108. Also Offprint No. 1110.

Price, Dorothy. 1956. "Influence of hormones on sex differentiation in explanted fetal reproductive tracts," *Josiah Macy, Jr. Foundation Conferences on Gestation*, 3:173–185.

Pritchard, R. M. 1961. "Stabilized images on the retina," *Scientific American*, 204 (June):72–78. Also Offprint No. 466.

Prosser, C. L., and F. A. Brown, Jr. 1961. *Comparative Animal Physiology*. 2nd ed. Philadelphia: Saunders.

Ptashne, Mark, and Walter Gilbert. 1970. "Genetic repressors," *Scientific American*, 222 (June): 36–44.

Rabinowitch, Eugene. 1948. "Photosynthesis," *Scientific American*, 179 (August):24–35. Also Offprint No. 34.

———**, and Govindjee.** 1965. "The role of chlorophyll in photosynthesis," *Scientific American*, 213 (July):74–83. Also Offprint No. 1016.

———. 1969. *Photosynthesis*. New York: Wiley.

Race, R. R., and Ruth Sanger. 1962. *Blood Groups in Man*. 4th ed. Oxford: Blackwell.

Raghavan, V., and J. G. Torrey. 1963. "Growth and morphogenesis of globular and older embryos of Capsella in culture," *American Journal of Botany*, 50(6):540–551.

Raper, K. B. 1935. "*Dictyostelium discoideum*, a new species of slime mold from decaying forest leaves," *Journal of Agricultural Research*, 50:135–147.

Rasmussen, Howard. 1961. "The parathyroid hormone," *Scientific American*, 204 (April):56–63. Also Offprint No. 86.

Ratliff, Floyd. 1965. *Mach Bands*. San Francisco: Holden-Day.

Ray, P. M. 1963. *The Living Plant*. New York: Holt, Rinehart and Winston.

Raylor, R. L., J. I. Maruer, and J. R. Tinklenberg. 1970. "The management of 'bad trips' in an evolving drug scene," *Journal of the American Medical Association*, 213:422–425.

Redi, Francesco. 1668. *Experienze intorno alla generazione degl' inettis*. Florence.

———. 1668. *Experiments on the Generation of Insects*. (Translated from the Italian edition of 1688 by Mab Bigelow.) Chicago: Chicago Open Court, 1909.

Redlich, F. C., and D. X. Freedman. 1966a. "The drug addictions," in F. C. Redlich and D. X. Freedman (eds.), *The Theory and Practice of Psychiatry*. New York: Basic Books, pp. 725–747.

———. 1966b. "Alcoholism," in F. C. Redlich and D. X. Freedman (eds.), *The Theory and Practice of Psychiatry*. New York: Basic Books, pp. 748–777.

Reiss, Ira. 1967. *The Social Context of Premarital Permissiveness*. New York: Holt, Rinehart and Winston.

Rich, Alexander. 1963. "Polyribosomes," *Scientific American*, 209 (December): 44–53. Also Offprint No. 171.

Richelle, Marc. 1970. "Biological clocks," *Psychology Today*, 3 (May):33–35, 58–60.

Richter, C. P. 1965. *Biological Clocks in Medicine and Psychiatry*. Springfield, Ill.: Thomas.

Ricker, W. E. 1969. "Food from the sea," in Committee on Resources and Man, *Resources and Man*. San Francisco: Freeman, pp. 87–108.

Rienow, Robert, and L. T. Rienow. 1967. *Moment in the Sun*. New York: Dial.

Roberts, D. F. 1953. "Body weight, race and climate," *American Journal of Physical Anthropology*, 11:533–558.

Robertson, J. D. 1959. "The ultrastructure of cell membranes and their derivatives," *Biochemical Society Symposia*, 16:3–43.

———. 1960. "The molecular structure and contact relationships of cell membranes," *Progress in Biophysics and Biophysical Chemistry*, 10:344–418.

———. 1962. "The membrane of the living cell," *Scientific American*, 206 (April):64–72.

Robinson, Robert. 1966. "Origins of petroleum," *Nature*, 212:1291–1295.

Rock, Irwin. 1966. *Nature of Perceptual Adaptation*. New York: Basic Books.

———. 1968. "When the world is tilt: distortion—how we adapt," *Psychology Today*, 2 (July):24–31.

———**, and C. S. Harris.** 1967. "Vision and touch," *Scientific American*, 216 (May):96–104. Also Offprint No. 507.

Roeder, K. D. 1955. "Spontaneous activity and behavior," *Scientific Monthly*, 80: 361–370.

———. 1962. "Neural mechanisms of animal behavior," *American Zoologist*, 2: 105–115.

———. 1963. *Nerve Cells and Insect Behavior*. Cambridge, Mass.: Harvard University Press.

———. 1965. "Moths and ultrasound," *Scientific American*, 212 (April):94–102. Also Offprint No. 1009.

Roemer, Ruth. 1967. "Abortion law: the approaches of different nations," *American Journal of Public Health*, 57:1906–1922.

Romer, A. S. 1966. *Vertebrate Paleontology*. 3rd ed. Chicago: University of Chicago Press.

———. 1968. *The Procession of Life*. New York: Universe.

———. 1970. *The Vertebrate Body*. 4th ed. Philadelphia: Saunders.

Rook, Arthur (ed.). 1963. *The Origins and Growth of Biology*. Baltimore: Penguin.

Rosebury, Theodor. 1969. *Life on Man*. Berkeley: Medallion.

Rosenblith, W. A. (ed.). 1961. *Sensory Communication*. Cambridge, Mass.: M.I.T. Press.

Rosenzweig, M. R. 1961. "Auditory localization," *Scientific American*, 205 (October):132–142. Also Offprint No. 501.

Roslansky, J. D. (ed.). 1969. *Communication*. New York: Fleet.

Ross, H. H. 1962. *A Synthesis of Evolutionary Theory*. Englewood Cliffs, N.J.: Prentice-Hall.

Roth, L. M., and R. H. Barth, Jr. 1967. "The sense organs employed by cockroaches in mating behavior," *Behaviour*, 28:58–94.

Rothschild, Lord. 1956. "Unorthodox methods of sperm transfer," *Scientific American*, 195 (November):121–132.

Rothschild, N. M. V. 1961. *A Classification of Living Animals*. London: Longmans.

Rowell, T. E. 1962. "Agonistic noises of the rhesus monkey (*Macaca mulatta*)," *Symposia of the Zoological Society of London*, 8:91–96.

Rowlands, I. W. (ed.). 1966. *Comparative Biology of Reproduction in Mammals: Symposia of the Zoological Society of London*. No. 15. New York: Academic Press.

Rubin, H. 1970. "Overgrowth stimulating factor released from Rous sarcoma cells," *Science*, 167:1271–1272.

Ruch, T. C., and J. F. Fulton (eds.). 1960. *Medical Physiology and Biophysics*. 18th ed. Philadelphia: Saunders.

Rudd, R. L. 1964. *Pesticides and the Living Landscape*. Madison: University of Wisconsin Press.

Rugh, Roberts, and Landrum Shettles. 1971. *From Conception to Birth: The Drama of Life's Beginnings*. New York: Harper & Row.

Ruitenbeek, H. M. (ed.). 1963. *The Problem of Homosexuality in Modern Society*. New York: Dutton.

———. 1968. *Homosexuality and Creative Genius*. New York: Astor-Honor.

Rushton, W. A. H. 1962. "Visual pigments in man," *Scientific American*, 207 (November):120–132. Also Offprint No. 139.

Russo, J. R. 1968. *Amphetamine Abuse*. Springfield, Ill.: Thomas.

Ruud, J. T. 1965. "The ice fish," *Scientific American*, 213 (November):108–114. Also Offprint No. 1025.

Sachs, R. M. 1965. "Stem elongation," *Annual Review of Plant Physiology*, 16: 73–96.

Salisbury, F. B. 1957. "Plant growth substances," *Scientific American*, 196 (April):125–134. Also Offprint No. 110.

———. 1958. "The flowering process," *Scientific American*, 198 (April):108–117. Also Offprint No. 112.

———**, and R. V. Parke.** 1970. *Vascular Plants: Form and Function*. 2nd ed. Belmont, Calif.: Wadsworth.

Sanderson, I. T. 1955. *Living Mammals of the World*. Garden City, N.Y.: Doubleday.

Sang, J. H. 1956. "The quantitative nutritional requirements of *Drosophila melanogaster*," *Journal of Experimental Biology*, 33:45–72.

Sauer, E. G. F. 1958. "Celestial navigation by birds," *Scientific American*, 199 (August):42–47. Also Offprint No. 133.

Saunders, J. W., and J. F. Fallon. 1966. "Cell death in morphogenesis," in Michael Loche (ed.), *Major Problems in Developmental Biology* (25th Symposium of the Society for Developmental Biology). New York: Academic Press.

Savory, Theodore. 1963. *Naming the Living World: An Introduction to the Principles of Biological Nomenclature*. New York: Wiley.

Savory, T. H. 1962. "Daddy longlegs," *Scientific American*, 207 (October): 119–128. Also Offprint No. 137.

———. 1966. "False scorpions," *Scientific American*, 214 (March):95–100. Also Offprint No. 1039.

———. 1968. "Hidden lives," *Scientific American*, 219 (July):108–114. Also Offprint No. 1112.

Scagel, R. F., *et al.* 1965. *An Evolutionary Survey of the Plant Kingdom*. Belmont, Calif.: Wadsworth.

Schaller, G. B. 1963. *The Mountain Gorilla: Ecology and Behavior*. Chicago: University of Chicago Press.

———. 1964. *The Year of the Gorilla*. Chicago: University of Chicago Press.

Scharrer, Ernest, and Berta Scharrer. 1963. *Neuroendocrinology*. New York: Columbia University Press.

Scheer, B. T. 1963. *Animal Physiology*. New York: Wiley.

Schjelderup-Ebbe, T. 1922. "Beiträge zur Sozialpsychologie des Haushuhns," *Zeitschrift fuer Psychologie*, 88:225–252.

Schleiden, M. J. 1838. "Beitrag zur Phytogenesis," *Archiv fur Anatomie, Physiologie, und Wissenschaftliche Medicin*, Part II. Vol. 5, pp. 137–176.

Schmidt, K. P., and R. F. Inger. 1957. *Living Reptiles of the World*. Garden City, N.Y.: Doubleday.

Schmidt-Nielsen, Knut. 1959. "The physiology of the camel," *Scientific American*, 201 (December):140–151. Also Offprint No. 1096.

———. 1959. "Salt glands," *Scientific American*, 200 (January):109–116. Also Offprint No. 1118.

———. 1964a. *Animal Physiology*. 2nd ed. Englewood Cliffs, N.J.: Prentice-Hall.

———. 1964b. *Desert Animals: Physiological Problems of Heat and Water*. Oxford: Clarendon Press.

———**, and Bodil Schmidt-Nielsen.** 1953. "The desert rat," *Scientific American*, 189 (July):73–78. Also Offprint No. 1050.

Schneiderman, H. A. 1969. "Control systems in insect development," in Samuel Devons (ed.), *Biology and the Physical Sciences*. New York: Columbia University Press, pp. 186–208.

Schoenfeld, Eugene. 1968. *Dear Doctor HIPpocrates: Advice Your Family Doctor Never Gave You*. New York: Grove.

Schofield, Michael. 1966. *Sociological Aspects of Homosexuality*. Boston: Little, Brown.

Scholander, P. F. 1957. "The wonderful net," *Scientific American*, 196 (April): 96–107. Also Offprint No. 1117.

———. 1963. "The master switch of life," *Scientific American*, 209 (December): 92–106. Also Offprint No. 172.

Schreider, Eugene. 1964. "Ecological rules, body-heat regulations, and human evolution," *Evolution*, 18:1–9.

Schwann, Theodor. 1839. *Mikroskopische Untersuchungen uber die Ubereinstimmung in der Struktur und dem Wachsthum der Thiere und Pflanzen*. Berlin. (English translation by Henry Smith, published in 1847 by the Sydenham Society, London).

Schwartz, Philip. 1961. *Birth Injuries of the Newborn*. New York: Hafner.

Scolnick, E. M., *et al.* 1971. "RNA dependent DNA polymerase activity in mammalian cells," *Nature*, 229:318–321.

Scott, B. I. H. 1962. "Electricity in plants," *Scientific American*, 207 (October): 107–117. Also Offprint No. 136.

Scott, J. P. 1958. *Aggression*. Chicago: University of Chicago Press.

———. 1969. "A time to learn," *Psychology Today*, 2 (March):46–48, 66–67.

Scrimshaw, N. B. 1968. "Infant malnutrition and adult learning," *Saturday Review*, (March 16, 1968):64.

Scriver, C. R. 1967. "Treatment in medical genetics," in J. F. Crow and J. V. Neel (eds.), *Proceedings of the Third International Congress of Human Genetics*. Baltimore: Johns Hopkins Press, pp. 45–56.

Sebeok, T. A. (ed.). 1968. *Animal Communication*. Bloomington: Indiana University Press.

Segal, S. J., and C. Tietze. 1969. "Contraceptive technology: current and prospective methods," *Reports on Population/Family Planning*. New York: The Population Council.

Seward, G. H., and R. C. Williamson (eds.). 1970. *Sex Roles in Changing Society*. New York: Random House.

Sharp, W. M., and V. G. Sprague. 1967. "Flowering and fruiting in the white oaks. Pistillate flowering, acorn development, weather, and yields," *Ecology*, 48: 243–251.

Shaw, Evelyn. 1962. "The schooling of fishes," *Scientific American*, 206 (June): 128–138. Also Offprint No. 124.

Shepard, Paul, and Daniel McKinley (eds.). 1969. *The Subversive Science: Essays Toward an Ecology of Man*. Boston: Houghton Mifflin.

Sheppard, J. J. 1968. *Human Color Perception*. New York: American Elsevier.

Sheppard, P. M. 1961. "Some contributions to population genetics resulting from the study of the Lepidoptera," *Advances in Genetics*, 10: 165–216.

Sherrington, C. S. 1906. *The Integrative Action of the Nervous System*. London: Scribner's.

Siebold, C. T. E. von. 1845. "Bericht uber die Leistungen im Gabiete der Anatomie und Physiologie der wirbellosen Thiere," *Archiv, fur Anatomie, Physiologie, und wissenschaftliche Medicin*, 12:1–120.

Simmons, J. A. 1968. "The sonar sight of bats," *Psychology Today*, 2 (November): 50–52, 54–57.

Simmons, J. L. (ed.). 1967. *Marihuana: Myths and Realities*. North Hollywood, Calif.: Brandon House.

Simon, H. J. 1963. *Microbes and Men*. New York: McGraw-Hill.

Simpson, G. G. 1949. *The Meaning of Evolution*. New Haven, Conn.: Yale University Press.

———. 1951. "The species concept," *Evolution*, 5:285–298.

———. 1953. *The Major Features of Evolution*. New York: Columbia University Press.

———. 1961. *Principles of Animal Taxonomy*. New York: Columbia University Press.

———. 1964. "Species density of North American recent mammals," *Systematic Zoology*, 13:57–73.

———. 1969. *Biology and Man*. New York: Harcourt Brace Jovanovich.

Singer, Charles. 1959. *A History of Biology to About the Year 1900: A General Introduction to the Study of Living Things*. 3rd ed. New York: Abelard-Schuman.

Sire, Marcel. 1969. *Secrets of Plant Life*. New York: Viking.

Sistrom, W. R. S. 1969. *Microbial Life*. 2nd ed. New York: Holt, Rinehart and Winston.

Skinner, B. F. 1938. *The Behavior of Organisms*. New York: Appleton-Century-Crofts.

Slovenko, Ralph (ed.). 1965. *Sexual Behavior and the Law*. Springfield, Ill.: Thomas.

Smith, C. C. 1968. "The adaptive nature of social organization in the genus of tree squirrels *Tamiascuirus*," *Ecological Monographs*, 38:31–63.

Smith, D. S. 1965. "The flight muscles of insects," *Scientific American*, 212 (June): 76–88. Also Offprint No. 1014.

Smith, G. M. 1955. *Cryptogamic Botany*. 2nd ed. Volume 1: *Algae and Fungi*. Volume 2: *Bryophytes and Pteridophytes*. New York: McGraw-Hill.

Smith, H. M., and Olwen Williams. 1970. "The salient provisions of the International Code of Zoological Nomenclature: a summary for nontaxonomists," *BioScience*, 20:553–557.

Smith, H. W. 1953. "The kidney," *Scientific American*, 188 (January):40–48. Also Offprint No. 37.

Smith, L. D. 1966. "The role of a 'germinal plasm' in the formation of primordial germ cells in *Rana pipiens*," *Developmental Biology*, 14:330–337.

Smith, N. G. 1967. "Visual isolation in gulls," *Scientific American*, 217 (October): 94–102. Also Offprint No. 1084.

Smith, R., and F. Meyers. 1971. *The Market Place of Speed*. Berkeley: University of California Press.

Smith, R. L. 1966. *Ecology and Field Biology*. New York: Harper & Row.

Snodgrass, R. E. 1935. *Principles of Insect Morphology*. New York: McGraw-Hill.

Sokal, R. R. 1966. "Numerical taxonomy," *Scientific American*, 215 (December): 106–116. Also Offprint No. 1059.

———, **and P. H. A. Sneath.** 1963. *Principles of Numerical Taxonomy*. San Francisco: Freeman.

Solbrig, O. T. 1970. *Principles and Methods of Plant Biosystematics*. New York: Macmillan.

Sollberger, Arne. 1965. *Biological Rhythm Research*. New York: American Elsevier.

Solomon, A. K. 1960. "Pores in the cell membrane," *Scientific American*, 203 (December):146–156. Also Offprint No. 76.

Sonneborn, T. M. 1950. "Partner of the genes," *Scientific American*, 183 (November):30–39. Also Offprint No. 39.

Southern, H. N. 1955. "Nocturnal animals," *Scientific American*, 193 (October):88–98. Also Offprint No. 1095.

Sperry, R. W. 1956. "The eye and the brain," *Scientific American*, 194 (May): 48–52. Also Offprint No. 1090.

———. 1959. "The growth of nerve circuits," *Scientific American*, 201 (November):68–75. Also Offprint No. 72.

———. 1964. "The great cerebral commissure," *Scientific American*, 210 (January):42–52. Also Offprint No. 174.

Speirs, R. S. 1964. "How cells attack antigens," *Scientific American*, 210 (February):58–64. Also Offprint No. 176.

Spiegelman, S. 1964. "Hybrid nucleic acids," *Scientific American*, 210 (May): 48–56. Also Offprint No. 183.

Srb, A. M., R. D. Owen, and R. S. Edgar. 1965. *General Genetics*. 2nd ed. San Francisco: Freeman.

Stanier, R. Y., Michael Doudoroff, and E. A. Adelberg. 1970. *The Microbial World*. 3rd ed. Englewood Cliffs, N.J.: Prentice-Hall.

Stanley, Wendell, and E. G. Valens. 1961. *Viruses and the Nature of Life*. New York: Dutton.

Stark, R. W. 1965. "Recent trends in forest entomology," *Annual Review of Entomology*, 10:303–324.

Stebbins, G. L. 1950. *Variation and Evolution in Plants*. New York: Columbia University Press.

———. 1960. "The comparative evolution of genetic systems," in Sol Tax (ed.), *Evolution After Darwin*, Vol. 1. Chicago: University of Chicago Press, pp. 197–226.

———. 1966. *Processes of Organic Evolution*. Englewood Cliffs, N.J.: Prentice-Hall.

———. 1970. "The natural history and evolutionary future of mankind," *American Naturalist*, 104:111–126.

Stent, G. S. 1963. *Molecular Biology of Bacterial Viruses*. San Francisco: Freeman.

Stern, Curt. 1931. "Zytologisch-genetische Untersuchungen als Beweise fur die Morgansche Theorie des Faktorenaustauschs," *Biologisches Zentralblatt*, 51:547–587.

———, **and E. R. Sherwood** (eds.). 1966. *The Origin of Genetics: A Mendel Source Book*. San Francisco: Freeman.

Stern, Herbert, and D. L. Nanney. 1965. *The Biology of Cells*. New York: Wiley.

Stevens, C. F. 1966. *Neurophysiology: A Primer*. New York: Wiley.

Steward, F. C. (ed.). 1959. *Plant Physiology*. Vol. 2. New York: Academic Press.

———. 1963. "The control of growth in plant cells," *Scientific American*, 209 (October):104–113. Also Offprint No. 167.

——— (ed.). 1963. *Plant Physiology: A Treatise*. Vol. 3. New York: Academic Press.

——— (ed). 1964. *Plants at Work*. Reading, Mass.: Addison-Wesley.

———. 1968. *Growth and Organization in Plants*. Reading, Mass.: Addison-Wesley.

———, **M. O. Mapes, and K. Mears.** 1958. "Growth and organized development of cultured cells. II. Organization in cultures grown from freely suspended cells," *American Journal of Botany*, 45:704–708.

Steward, J. H. 1955. *Theory of Culture Change*. Urbana: University of Illinois Press.

Stoll, N. R., *et al.* 1964. *International Code of Zoological Nomenclature Adopted by the XV International Congress of Zoology*. London: International Trust for Zoological Nomenclature.

Storer, N. W. 1966. *The Social System of Science*. New York: Holt, Rinehart and Winston.

Storer, T. I., and R. L. Usinger. 1965. *General Zoology*. 4th ed. New York: McGraw-Hill.

Storm, R. M. (ed.). 1967. *Animal Orientation and Navigation*. Corvallis: Oregon State University Press.

Storr, Anthony. 1964. *Sexual Deviation*. Baltimore: Penguin.

Strickberger, M. W. 1962. *Experiments in Genetics with Drosophila*. New York: Wiley.

Sturtevant, A. H. 1913. "The linear arrangement of six sex-linked factors in *Drosophila*, as shown by their mode of association," *Journal of Experimental Zoology*, 14:43–59.

Sullivan, Navin. 1967. *The Message of the Genes*. New York: Basic Books.

Sussman, A. S. 1964. *Microbes—Their Growth, Nutrition and Interaction*. Boston: D. C. Heath.

Svaetichin, Gunnar. 1956. "Spectral response curves from single cones," *Acta Physiologica Scandinavia*, 39 (Supplement 134):19–46.

Swanson, C. P. 1969. *The Cell*. 3rd ed. Englewood Cliffs, N.J.: Prentice-Hall.

Sweeney, B. M. 1963. "Biological clocks in plants," *Annual Review of Plant Physiology*, 14:411–440.

———. 1969. *Rhythmic Phenomena in Plants*. New York: Academic Press.

Sweet, M. H. 1960. "The seed bugs: a contribution to the feeding habits of the Lygaeidae (Hemiptera: Heteroptera)," *Annals of the Entomological Society of America*, 53:317–321.

Swift, Hewson. 1965. "Nucleic acids of mitochondria and chloroplasts," *American Naturalist*, 99:201–227.

Talmage, D. W. 1959. "Mechanism of the antibody responses," in Raymond Zirkle (ed.), *A Symposium on Molecular Biology*. Chicago: University of Chicago Press, pp. 91–101.

Tanner, J. M. 1968. "Earlier maturation in man," *Scientific American*, 218 (January): 21–27. Also Offprint No. 1091.

Tasaki, I. 1939. "The electro-saltatory transmission of the nerve impulse and the effect of narcosis upon the nerve fiber," *American Journal of Physiology*, 127: 211–227.

Tatum, E. L., and Joshua Lederbert. 1947. "Gene recombination in the bacterium *Escherichia coli*," *Journal of Bacteriology*, 53:673–684.

Tax, Sol (ed.). 1960. *Evolution After Darwin*. 3 vols. Chicago: University of Chicago Press.

——— (ed.). 1960. *Evolution after Darwin*. Vol. 2: *The Evolution of Man: Man, Culture, and Society*. Chicago: University of Chicago Press.

Taylor, G. G. 1963. *The Science of Life: A Picture History of Biology*. New York: McGraw-Hill.

Taylor, J. H. 1958. "The duplication of chromosomes," *Scientific American*, 198 (June):36–42. Also Offprint No. 60.

Teal, J. M. 1957. "Community metabolism in a temperate cold spring," *Ecological Monographs*, 27:283–302.

———. 1962. "Energy flow in the salt marsh ecosystem of Georgia," *Ecology*, 43: 614–624.

Telfer, William, and Donald Kennedy. 1965. *The Biology of Organisms*. New York: Wiley.

Thimann, K. V. 1935. "Growth substances in plants," *Annual Review of Biochemistry*, 4:545–568.

———. 1935a. "On the plant growth hormone produced by *Rhizopus suinus*," *Journal of Biological Chemistry*, 109: 279–291.

———. 1963. *The Life of Bacteria*. 2nd ed. New York: Macmillan.

———, **and J. B. Koepfli.** 1935. "Identity of the growth-promoting and root-forming substances of plants," *Nature*, 135: 101–102.

———, **and Folke Skoog.** 1933. "Studies on the growth hormone of plants. III. The inhibiting action of the growth substance on bud development," *Proceedings of the National Academy of Sciences*, 19(7): 714–716.

Thompson, R. F. 1967. *Foundations of Physiological Psychology*. New York: Harper & Row.

Thorpe, W. H. 1961. *Bird Song: The Biology of Vocal Communication and Expression in Birds*. Cambridge: Cambridge University Press.

Tietze, Christopher. 1969. "New intrauterine devices," *Studies in Family Planning*. No. 47. New York: The Population Council.

———, **and Sarah Lewit.** 1969. "Abortion," *Scientific American*, 220 (January):21–27.

Tiger, Lionel. 1969. *Men in Groups*. New York: Random House.

Tinbergen, Niko. 1950. "The hierarchical organization of nervous mechanisms underlying instinctive behaviour," *Symposia of the Society for Experimental Biology*, 4:305–312.

———. 1951. *The Study of Instinct*. Fair Lawn, N.J.: Oxford University Press.

———. 1952. "The curious behavior of the stickleback," *Scientific American*, 187 (December). Also Offprint No. 414.

———. 1953. *Social Behaviour in Animals*. London: Methuen.

———. 1958. *Curious Naturalists*. New York: Basic Books.

———. 1960. "The evolution of behavior in gulls," *Scientific American*, 203 (December):48–130. Also Offprint No. 456.

Tinklenberg, J. R. 1971. "Alcohol and violence," in R. Fox and P. Bourne (eds.), *Alcoholism: Progress in Research and Treatment*. New York: Academic Press.

———**, and R. C. Stillman.** 1970. "Drug use and violence," in D. N. Daniels, M. F. Gilula, and F. M. Ochberg (eds.), *Violence and the Struggle for Existence*. Boston: Little, Brown, pp. 327–365.

———, *et al*. 1970. "Marihuana and immediate memory," *Nature*, 226: 1171–1172.

Tomita, Tsuneo. 1968. "Electrical responses of single photoreceptors," *Proceedings of the IEEE*, 56:1015–1024.

Trinkaus, J. P. 1969. *Cells into Organs: The Forces that Shape the Embryo*. Englewood Cliffs N.J.: Prentice-Hall.

Tubbs, C. R. 1969. *The New Forest: An Ecological History*. New York: Transatlantic Arts.

Turner, C. D. 1966. *General Endocrinology*. 4th ed. Philadelphia: Saunders.

Ucko, P. J., and G. W. Dimbleby. 1969. *The Domestication and Exploitation of Plants and Animals*. Chicago: Aldine.

Udall, Stewart. 1963. *The Quiet Crisis*. New York: Holt, Rinehart and Winston.

———. 1968. *1976: Agenda for Tomorrow*. New York: Harcourt Brace Jovanovich.

van der Pijl, Londert. 1968. *Principles of Dispersal in Higher Plants*. New York: Springer-Verlag.

Van Lawick-Goddall, Jane. 1968. "A preliminary report on expressive movements and communication in the Gombe Stream chimpanzees," in P. C. Jay (ed.), *Primates*. New York: Holt, Rinehart and Winston, pp. 313–374.

Van Tienhoven, Ari. 1968. *Reproductive Physiology of Vertebrates*. Philadelphia: Saunders.

Van Tyne, Josselyn, and A. J. Berger. 1959. *Fundamentals of Ornithology*. New York: Wiley.

Verney, E. B. 1947. "The antidiuretic hormone and the factors which determine its release," *Proceedings of the Royal Society*, Series B, 135:25–106.

Villee, C. A., W. F. Walker, and F. E. Smith. 1963. *General Zoology*. 2nd ed. Philadelphia: Saunders.

Virchow, Rudolf. 1858. *Cellular Pathology*. Frank Chance (tr.). New York: Dover.

Voeller, B. R. (ed.) 1968. *The Chromosome Theory of Inheritance: Classic Papers in Development and Heredity*. New York: Appleton-Century-Crofts.

Vowles, D. M. 1961. "Neural mechanisms in insect behavior," in W. H. Thorpe and O. L. Zangwill (eds.), *Current Problems in Animal Behavior*. Cambridge: Cambridge University Press, pp. 5–29.

Waddington, C. H. 1966. *Principles of Development and Differentiation*. New York: Macmillan.

———**, and R. J. Cowe.** 1969. "Computer simulation of a molluscan pigmentation pattern," *Journal of Theoretical Biology*, 25:219–225.

Wagner, R. P., and H. K. Mitchell. 1964. *Genetics and Metabolism*. 2nd ed. New York: Wiley.

Wald, George. 1954. "The origin of life," *Scientific American*, 190 (August):44–53. Also Offprint No. 47.

———. 1955. "The photoreceptor process in vision," *American Journal of Ophthalmology*, 40(Part II):18–41.

———. 1959. "Life and light," *Scientific American*, 201 (October):92–108. Also Offprint No. 61.

Walker, E. P. 1964. *Mammals of the World*. 3 vols. Baltimore: Johns Hopkins Press.

Walker, T. J. 1957. "Specificity in the response of female tree crickets (Orthoptera, Gryllidae, Oceanthinae) to calling songs of the males," *Annals of the Entomological Society of America*, 50: 626–636.

Wallace, A. R. 1891. *Natural Selection and Tropical Nature: Essays on Descriptive and Theoretical Biology*. London: Macmillan. (Reprinted by Gregg International Publishers, 1969.)

Wallace, Bruce, and A. M. Srb. 1970. *Adaptation*. 3rd ed. Englewood Cliffs, N.J.: Prentice-Hall.

Walter, W. G. 1954. "The electrical activity of the brain," *Scientific American*, 190 (June):54–63. Also Offprint No. 73.

Wardlaw, C. W. 1949. "Experiments on organogenesis in ferns," *Growth*, 13(supplement):93–131.

Washburn, S. L. (ed.). 1962. *Social Life of Early Man*. Chicago: Aldine.

———**, and I. DeVore.** 1961. "The social life of baboons," *Scientific American*, 204 (June):62–71.

Waters, R. H., D. A. Rethlingshafer, and W. E. Caldwell. 1960. *Principles of Comparative Psychology*. New York: McGraw-Hill.

Watson, J. D. 1968. *The Double Helix*. New York: Atheneum.

———. 1970. *Molecular Biology of the Gene*. 2nd ed. New York: Benjamin.

Watson, J. D., and F. H. C. Crick. 1953. "A structure for deoxyribose nucleic acid," *Nature*, 171:737–738.

Webb, W. B. 1968. *Sleep: An Experimental Approach*. New York: Macmillan.

Wecker, S. C. 1964. "Habitat selection," *Scientific American*, 211 (October): 109–116. Also Offprint No. 195.

Wedberg, S. E. 1963. *Paramedical Microbiology*. New York: Van Nostrand Reinhold.

Weinberg, W. 1908. "Über den Nachweis der Verebung beim Menschen," *Jahresheft des Vereins für vaterländische Naturkunde, in Württemberg*, 64: 368–382.

Weiner, J. S. 1967. "Physical anthropology: an appraisal," *American Scientist*, 45:79–87.

———. 1971. *Natural History of Man*. New York: Universe.

———**, and B. G. Campbell.** 1964. "The taxonomic status of the Swanscombe skull," in C. D. Ovey (ed.), *The Swanscombe Skull*. London: Royal Anthropological Institute.

Weiss, Bernard, and V. G. Laties. 1962. "Enhancement of human performance by caffeine and amphetamines," *Pharmacological Review*, 14:1–36.

Weisz, P. B. 1966. *The Science of Zoology*. New York: McGraw-Hill.

Wells, M. J. 1962. *Brain and Behavior in Cephalopods*. Stanford, Calif.: Stanford University Press.

Welty, J. C. 1963. *The Life of Birds*. New York: Knopf.

Went, F. W. 1955. "The ecology of desert plants," *Scientific American*, 192 (April): 68–75. Also Offprint No. 114.

———**, and K. V. Thimann.** 1937. *Phytohormones*. New York: Macmillan.

Wessells, N. K., and W. J. Rutter. 1969. "Phases in cell determination," *Scientific American*, 220 (March):36–44.

West, D. J. 1968. *Homosexuality*. Chicago: Aldine.

Weston, J. A. 1967. "Cell marking," in F. H. Wilt and N. K. Wessells (eds.), *Methods in Developmental Biology*. New York: Crowell, pp. 723–736.

———. 1970. "The migration and differentiation of neural crest cells," *Advances in Morphogenesis*, 8:41–114.

Wetmore, R. H. 1954. "The use of 'in vitro' cultures in the investigation of growth and differentiation in vascular plants," *Brookhaven Symposia in Biology*, 6: 22–40.

White, M. J. D. 1961. *The Chromosomes*. 5th ed. London: Methuen.

Whitehead, D. R. 1969. "Wind pollination in the angiosperms: evolutionary and environmental considerations," *Evolution*, 23:28–35.

Whittaker. R. H. 1969. "New concepts of kingdoms of organisms," *Science*, 163: 150–160.

———. 1970. *Communities and Ecosystems*. New York: Macmillan.

Whitten, J. M. 1969. "Cell death during early morphogenesis: parallels between insect limbs and vertebrate limb development," *Science*, 163:1456–1457.

Whyte, W. H. 1968. *The Last Landscape*. Garden City, N.Y.: Doubleday.

Wiersma, C. A. G. 1962. "The organization of the arthropod central nervous system," *American Zoologist*, 2: 67–78.

——— (ed.). 1967. *Invertebrate Nervous Systems: Their Significance for Mammalian Neurophysiology*. Chicago: University of Chicago Press.

Wiggers, C. J. 1957. "The heart," *Scientific American*, 196 (May):74–87. Also Offprint No. 62.

Wikler, Abraham. 1970. "Some implications of conditioning theory for problems of drug abuse," in P. H. Blachly (ed.), *Drug Abuse: Data and Debate*. Springfield, Ill.: Thomas.

Wilkins, M. H. F., A. R. Stokes, and H. R. Wilson. 1953. "Molecular structure of deoxypentose nucleic acids," *Nature*, 171:738–740.

Williams, C. A., Jr. 1960. "Immunoelectrophoresis," *Scientific American*, 202 (March):130–140. Also Offprint No. 84.

Williams, C. M. 1950. "The metamorphosis of insects," *Scientific American*, 182 (April):24–28. Also Offprint No. 49.

Wilson, C. W. M. 1968. *Adolescent Drug Dependence*. New York: Pergamon.

Wilson, D. M. 1961. "The central nervous control of flight in a locust," *Journal of Experimental Biology*, 38:471–490.

———. 1966. "Insect walking," *Annual Review of Entomology*, 11:103–122.

———. 1968. "The flight-control system of the locust," *Scientific American*, 218 (May):83–90. Also Offprint No. 1107.

Wilson, E. O. 1963. "Pheromones," *Scientific American*, 208 (May):100–114. Also Offprint No. 157.

Winchester, J. H. 1968. "Rescuing newborns with minisurgery," *Today's Health*, 46:52–56.

Winter, P. M., and Edward Lowenstein. 1969. "Acute respiratory failure," *Scientific American*, 221 (November):23–29.

Winton, F. R., and L. E. Bayliss. 1962. *Human Physiology*. 5th ed. Boston: Little, Brown.

Wollman, E. L., and François Jacob. 1956. "Sexuality in bacteria," *Scientific American*, 195 (July):109–118. Also Offprint No. 50.

Wolpert, L. 1969. "Positional information and the spatial pattern of cellular differentiation," *Journal of Theoretical Biology*, 25:1–47.

———. 1970. "Developing cells know their place," *New Scientist* (May 14, 1970):322–325.

Wolstenholme, Gordon (ed.). 1963. *Man and His Future*. Boston. Little, Brown.

Wood, J. E. 1968. "The venous system," *Scientific American*, 218 (January):86–96. Also Offprint No. 1093.

Wood, W. B., Jr. 1951. "White blood cells vs. bacteria," *Scientific American*, 185 (February):48–52. Also Offprint No. 51.

Wood, W. B., and R. S. Edgar. 1967. "Building a bacterial virus," *Scientific American*, 217 (July):60–74. Also Offprint No. 1079.

Woodburne, L. S. 1967. *The Neural Basis of Behavior*. Columbus, Ohio: Merrill.

Woodwell, G. M. 1963. "The ecological effects of radiation," *Scientific American*, 208 (June):40–49. Also Offprint No. 159.

———. 1967. "Toxic substances and ecological cycles," *Scientific American*, 216 (March):24–31. Also Offprint No. 1066.

———. 1970. "The energy cycle of the biosphere," *Scientific American*, 223 (September):64–74.

Wooldridge, D. E. 1963. *The Machinery of the Brain*. New York: McGraw-Hill.

Wright, Sewall. 1932. "The roles of mutation, inbreeding, crossbreeding, and selection in evolution," *Proceedings of the Sixth International Congress on Genetics*, 1:356–366.

Wurtman, R. J., and Julius Axelrod. 1965. "The pineal gland," *Scientific American*, 213 (July):50–60. Also Offprint No. 1015.

Wynne-Edwards, V. C. 1962. *Animal Dispersion in Relation to Social Behavior*. New York: Hafner.

———. 1964. "Population control in animals," *Scientific American*, 211 (August):68–74. Also Offprint No. 192.

———. 1965. "Self-regulating systems in populations of animals," *Science*, 147: 1543–1548.

Yanagisawa, K., W. F. Loomis, Jr., and M. Sussman. 1967. "Developmental regulation of the enzyme UDP-galactose polysaccharide transferase," *Experimental Cell Research*, 46:328–334.

Yanofsky, Charles. 1967. "Gene structure and protein structure," *Scientific American*, 216 (May):80–94. Also Offprint No. 1074.

Young, J. Z. 1957. *The Life of Mammals*. New York: Oxford University Press.

———. 1962. *The Life of Vertebrates*. 2nd ed. New York: Oxford University Press.

Young, L. B. (ed.). 1968. *Population in Perspective*. New York: Oxford University Press.

Young, R. W. 1970. "Visual cells," *Scientific American*, 223 (October): 80–91.

Young, Wayland. 1966. *Eros Denied: Sex in Western Society*. New York: Zebra Books (Grove Press).

Young, W. C. (ed.). 1961. *Sex and Internal Secretions*. 3rd ed. 2 vols. Baltimore: Williams & Wilkins.

Zahl, P. A. 1949. "The evolution of sex," *Scientific American*, 180 (April):52–55.

Zemp, J. W., *et al.* 1966. "Brain function and macromolecules, I. Incorporation of uridine RNA of mouse brain during short-term training experience," *Proceedings of the National Academy of Sciences*, 55:1423–1431.

Zinsser, Hans 1935. *Rats, Lice and History*. Boston: Little, Brown.

Zweifach, B. W. 1959. "The microcirculation of the blood," *Scientific American*, 200 (January):54–60. Also Offprint No. 64.

Zylman, Richard. 1968. "Accidents, alcohol, and single cause explanations: lessons from the Grand Rapids study," *Quarterly Journal of Studies on Alcohol*, (Supplement 4):212–233.

Glossary

a

abortion (ah-bor′shun): premature expulsion of a fetus, especially before it is independently viable, either by miscarriage or by surgery.

abscission (ab-sish′un): in botany, the normal separation of flowers, fruit, and leaves from plants.

absolute refractory period: see refractory period.

absorption spectrum: a graph of the amount of electromagnetic radiation absorbed by a medium, as it varies with the wavelength of radiation passing through the medium.

acceptor (ak-sep′tor): a substance that unites with hydrogen or oxygen in an oxidation-reduction reaction, and so enables the reaction to proceed.

acetone (as′e-tōn): a colorless, volatile organic solvent that dissolves fats but is miscible with water.

acetylcholine (as′e-til-kō′lēn): a reversible acetic acid ester of choline; one of the chemical transmitters that function at the excitatory synapse.

acid (a′sid): a substance that releases hydrogen ions when dissolved in water or can donate protons in chemical reactions; it has a sour taste, turns blue litmus red, and unites with bases to form salts.

actin (ak′tin): a protein found in myofibrils; acting with myosin, it is responsible for contraction and relaxation of muscles.

actinomycin D (ak′ti-nō-mī′sin): an antibiotic that binds to DNA and blocks subsequent synthesis of mRNA.

action potential: a nerve impulse or progressive change in polarity along a nerve fiber. (See nerve impulse.)

activated complex: a transitional molecule formed when two reactant molecules possessing the necessary energy of activation come together; it decomposes to yield the products of the reaction.

active site: the region on a protein to which the substrate of an enzyme attaches so that it is activated to react with another molecule.

active transport: movements of molecules through the cell membrane against the natural direction of diffusion; it requires the expenditure of chemical energy.

actomyosin (ak′to-mī′ō-sin): a complex protein consisting of actin and myosin; it is thought to interact with ATP to cause muscle contraction.

adaptive radiation: the diversity that arises among species as each adapts to a unique set of environmental conditions.

adenine (ad′e-nēn): a nitrogenous base and a constituent of nucleic acids. (See purine.)

adenosine diphosphate (ah-den′o-sēn dī-fos′fāt): a product formed in the hydrolysis of ATP with the concurrent release of organic phosphate and energy. (Abbreviated: ADP)

adenosine triphosphate (ah-den′o-sēn trī-fos′fāt): a compound occurring in all cells and serving as a source of energy for physiological reactions such as muscle contraction. (Abbreviated: ATP)

adenylic acid (ad′e-nil′ik): one of the four deoxyribonucleotides that make up DNA.

ADP: see adenosine diphosphate.

aerobe (e′rōb): an organism that uses oxygen in carrying out respiratory processes.

afferent collateral inhibition: a type of inhibition that occurs when an afferent nerve fiber, which synaptically excites a nerve cell, gives off a collateral branch that also excites an inhibitory cell.

afferent nerve fiber: a nerve fiber carrying impulses toward the central nervous system.

afterbirth (af′ter-berth): the mass of placenta and allied membranes expelled from the uterus after the birth of a child.

agammaglobulinemia (a-gam′mah-glob′ū-linē′mē-ah): deficiency or absence of gamma globulin in the blood despite the presence of a thymus; a rare congenital disease significant in the study of immune responses.

agglutination (ah-glū-ti-nā′shun): A joining together; the aggregation or clumping of suspended particles, cells, or molecules.

alkali metal: any of the group of univalent metals including potassium, sodium, lithium, rubidium, cesium, and francium, whose hydroxides are alkalis.

alkaline-earth metal: any of a group of bivalent, chemically-active metals including beryllium, magnesium, calcium, strontium, barium, and radium; the hydroxides are alkalis but are less soluble than those of the alkali metals.

alcohol (al′ko-hol): a hydrocarbon molecule in which one or more hydrogen atoms are replaced by hydroxyl (—OH) groups.

aleurone (al′yah-rōn): protein granules found in a single layer of cells in the outermost portion of the endosperm in some plant seeds.

allele (ah-lēl′): one of two or more contrasting genes that determine alternative characters in inheritance because they are situated at the same locus in homologous chromosomes.

alpha helix: a polypeptide chain folded into a spiral structure and held together by hydrogen bonds. (Abbreviated: α-helix)

alveolar duct (al-vē′ō-lar): a small tubule leading to an alveolus or air sac in the lungs.

alveolus (al-vē′ō-lus): a small saclike dilation; for example, the microscopic air pouches that occur in the lung tissue. (Plural: alveoli.)

amino acid (ah-mē′nō): a type of carboxylic acid containing the amino group $—NH_2$; a constituent of all proteins, which are amino acid polymers.

amniocentesis (am′nē-ō-sen-tē′sis): removal of a small quantity of amniotic fluid from the uterus during pregnancy.

amniotic sac (am′nē-ot′ik): the thin, transparent but tough membrane that encloses the embryo and produces the amniotic fluid.

amphetamine (am-fet′ah-mēn): a group of synthetic chemicals used to stimulate the central nervous system, suppress appetite, increase heart rate and blood pressure, and alter sleep patterns.

amyloplast (ā′mil-ō-plast): a starch-storing, nonpigmented plastid of plant cells; a type of leucoplast.

anabolism (ah-nab′ō-lizm): constructive metabolism; any constructive process by which simple substances are converted by living cells into more complex compounds.

anaerobe (an′ah-rōb): an organism that does not require oxygen for respiration but makes use of processes such as fermentation and glycolysis to obtain its energy.

anaphase (an′ah-fāz): the stage of mitosis between the metaphase and telophase, in which the daughter chromosomes move apart toward the poles of the spindle.

anaphase I: in meiosis, the stage in which the pairs of homologous chromosomes are separated from each other; the stage ends with a complete set of chromosomes, each containing two chromatids, clustered at each pole.

anaphase II: in meiosis, the stage in which the daughter chromosomes, formerly chromatids, separate and move to opposite poles; the stage results in four haploid sets of chromosomes.

androgen (an'dro-jen): a typically male hormone; one of the hormones involved in sex determination.

aneuploidy (an'ū-ploi'dē): the state of having more or fewer than the normal diploid number of chromosomes.

angstrom (ang'strem): one hundred-millionth of a centimeter; a unit used in measuring the length of light waves. (Abbreviated: A or Å)

anion (an' ī-on): a negatively charged ion.

anisogamy (an-ī-sog'ah-mē): reproduction through the union of gametes that are different in appearance.

annulus (an'ū-lus): a ring-shape or circular structure; for example, a small round pore in the nuclear membrane of a eucaryotic cell; it permits the passage of macromolecules. (Plural: annuli.)

anomaly (ah-nom'ah-lē): any marked deviation from the normal or usual.

anther (an'ther): the male sex organ in a flowering plant.

antheridium (an'thah-rid'i-um): a structure producing the male gamete in lower plants, such as ferns and mosses. (Plural: antheridia.)

anthrax (an'thraks): a carbuncle or other infection caused by the anthrax bacillus. Malignant anthrax is a disease of cattle and sheep but may also occur in man.

antianxiety drugs: psychotherapeutic drugs used chiefly to treat anxiety and neurotic conditions and to reduce stress.

antibiotic (an'ti-bī-ot'ik): any chemical substance, produced usually by microorganisms, with the capacity to inhibit or destroy bacteria and other organisms.

antibody (an'ti-bod'ē): a globulin type of protein that combines with and renders harmless foreign substances introduced into an animal by infectious processes.

anticholinergic (an'ti-kō'lin-er'jik): blocking the passage of impulses through the parasympathetic nerves.

anticodon (an'ti-kō'don): a set of three nucleotides in the tRNA primary structure that is complementary to the codon in the mRNA.

antidepressant (an'ti-dē-pres'ant): an agent that prevents or relieves mental depression; for example, the so-called "mood elevator" drugs.

antidiuretic hormone (an'ti-dī'u-ret'ik): a hormone released by neuroendocrine cells in the hypothalamus; it increases the capacity of tubules in the kidneys to reabsorb water. (Abbreviated: ADH)

antigen (an'ti-jen): a protein that stimulates the production of antibodies or reacts with them.

antigenic determinant: a structural characteristic of an antigen; its shape enables it to bind to an antibody by fitting into the cavity that is the combining site of the antibody. (Abbreviated: AD)

antipsychotic drugs: psychotherapeutic drugs used in the treatment of psychotics to alter response, decrease excessive motor activity, and dampen delusional behavior without inducing sleep.

anus (ā'nus): the opening through which the solid refuse of digestion is excreted.

aorta (ā-or'tah): the main trunk of the arterial system; it conveys blood from the heart to all of the body except the lungs.

apical dominance: in plants, inhibition of the branching of axillary buds by hormones secreted by the terminal bud.

apical meristem: the cells that occur at the tip of a plant root or shoot; they initiate growth in length.

aplanospore (ā-plan'ah-spōr): a thick- wall, nonmotile, asexual spore formed within a cell in certain algae and fungi.

apocrine gland (ap'o-krēn): a secretory organ, the cells of which partly disintegrate to form the secretion but then are quickly restored.

archegonium (ar'kah-gō'nē-um): the female reproductive organ in mosses, ferns, and some higher plants.

arginine (ar'ji-nēn): one of the essential amino acids that make up plant and animal proteins.

arteriosclerosis (ar-te're-ō-skle-rō'sis): a condition involving loss of elasticity, thickening, and hardening of the arteries.

artery (ar'ter-ē): a relatively small, thick-wall vessel through which blood passes from the heart to the tissues. (See vein.)

ascocarp (as'kah-karp): the fruiting body of the cup fungi.

ascospore (as'kah-spōr): a spore formed within an ascus, an enlarged cell of certain fungi.

ascus (as'kus): the sac in which the sexual spores are formed in sac fungi.

asexual reproduction: reproduction without the fusion of sexual cells; for example, fission or budding.

aster (as'ter): the radiating structure surrounding the centrosome of some cells; it is seen at the beginning of mitosis.

astrocyte (as'trō-sīt): a star-shape glial cell with many functions, including mechanical support for the neural tissue.

atomic number: the number of protons in the nucleus of the atom of a particular element; the number is different for each element.

atomic weight: the average weight of an atom of an element; units of weight are based on 1/12 the mass of the carbon-12 atom.

ATP: see adenosine triphosphate.

atrium (ā'trē-um): a chamber; for example, one of the chambers of the heart.

autoimmune diseases: diseases that are thought to be due to attacks by the immune system upon normal body tissues.

autonomic nervous system: the part of the central nervous system that controls the involuntary or vegetative functions of the body.

autotroph (aw'to-trōf): an organism that uses an external source of energy to produce organic nutrients from simple inorganic chemicals.

auxin (awk'sin): a general name for a class of hormonelike substances formed in the actively growing parts of plants and having the power to affect subsequent growth and development.

auxospore (awk'so-spōr): a fused pair of gametes (or a zygote) of unicellular algae, such as diatoms.

avirulent (ā-vir'u-lent): not lethal or not causing fatal disease; often applied to microorganisms.

avoidance response: any form of behavior that helps an organism to elude a source of danger or an unfamiliar stimulus; for example, running or crouching.

axillary bud: a bud growing within the angle formed by the upper side of a leaf or stem and the supporting stem or branch.

axoaxonic synapse (ak'sō-ax-on'ik): an axon-to-axon synapse; a type observed mainly in sensory pathways.

axodentritic synapse (ak'sō-den-drit'ik): an axon-to-dendrite synapse, a type found only on dendrites.

axon (ak'son): the fiberlike extension of a neuron that conducts and transmits impulses away from the nerve cell body; an efferent nerve fiber.

axonal arborization: a number of fine branches extending from the transmitting end of the axon of a nerve cell.

axosomatic synapse (ak'sō-sō-mat'ik): an axon-to-soma synapse found both on the cell body and on the large stumps of dendrites near the body.

bacillus (ba-sil'us): a rod-shape bacterium.

bacitracin (bas'i-trā'sin): a proteinoid substance produced by bacteria and used as an antibiotic.

back cross: the cross of a hybrid with its parent or with one of the parental genotypes.

bacterial transformation: the conversion of bacteria from one form to another by treatment with foreign DNA.

bacteriochlorophyll (bak-te'rē-ō-klo'ro-fil): a green pigment, similar to chlorophyll, produced by purple bacteria and instrumental in bacterial photosynthesis.

bacteriophage (bak-te'rē-o-fāj): a virus that infects bacteria.

balanced polymorphism: an equilibrium ratio of homozygous phenotypes within a population.

barbiturate (bar-bit'u-rāt): a salt of barbituric acid; a group of sedative drugs that depress the central nervous system.

bar-eye (bar'ī): a sex-linked character carried by a gene on the X chromosome in fruitflies; the form of the eye is a narrow red bar.

basal body: basal corpuscle; a small thickening at the base of each cilium of ciliated cells.

basal cell: the larger of two cells formed in the first mitotic division of a plant zygote; it divides to form the suspensor cell and a new basal cell.

base: a substance that releases hydroxyl ions when dissolved in water or that can accept protons in chemical reactions; bases have a bitter taste, turn red litmus blue, and unite with acids to form salts.

basidiocarp (bah-sid'ē-o-karp): the fruiting body of the club fungi.

basidium (bah-sid'ē-um): the structure on which the spores are borne, usually at the end of slender projections, in the club fungi.

behaviorism: the school of psychology based upon objective observation and analysis of human and animal behavior without reference to consciousness.

benzene (ben'zēn): a clear, inflammable, volatile distillate from petroleum; it is used as a solvent for fats, resin, India rubber, and certain alkaloids.

β-galactosidase (bā'tah-gah-lak'to-sī'dās): an inducible enzyme that cleaves lactose (milk sugar) into the component sugars glucose and galactose.

bilayer (bī'lā-er): a bimolecular layer; a layer two molecules in thickness.

bile (bīl): a secretion from the liver that aids in the digestion of fats.

bile salt: any of the salts produced by the conversion of cholesterol in the digestive tract; important in the digestion of fats.

bimolecular layer: a bilayer; a layer two molecules in thickness.

binary fission: a form of asexual reproduction in bacteria and other lower forms of life; the cell divides itself into two approximately equal parts.

biological clock: a hypothetical mechanism explaining rhythmic fluctuations of behavior in animals.

biome (bī'ōm): an ecosystem chosen to represent a relatively uniform vegetation or animal assemblage; for example, a grassland or a tropical rain forest.

biosphere (bī'o-sfēr): the part of the earth's crust, waters, and atmosphere where living organisms can subsist.

biosystematics (bī'o-sis-te-mat'iks): biological classification and nomenclature.

biotic community (bī-ot'ik): a group of populations occupying the same area and interchanging materials and energy.

biotin (bī'o-tin): a B-complex vitamin widely distributed in plant and animal tissues.

bladder (blad'der): a sac, situated to the rear of the pelvic cavity, that serves as a reservoir for urine.

blastocyst (blas'tah-sist): a blastula; the early developmental stage of a metazoan following the morula stage and consisting of a layer of cells forming a hollow sphere.

blastulation (blas'tu-lā'shun): the transformation of a fertilized egg into a blastula, a single layer of cells surrounding a fluid-filled cavity

blind spot: a small area in the retina of the eye that contains no receptors; the area where fibers from the ganglion cells of the retina gather into a bundle comprising the optic nerve.

bond energy: the approximate amount of energy stored in a chemical bond; the same or greater energy is required to break the bond.

bouton (bo͞o'ton): the presynaptic knob at the end of the axon filament.

Bowman's capsule: a double-wall, cup-shape structure surrounding a glomerulus in the kidney.

bract (brakt): a leaf that protects the basal part of a flower; the bract itself may resemble a flower.

brain stem: the portion of the brain including the midbrain, the pons, and the medulla.

Broca's area: an area of the left cerebral hemisphere thought to be responsible for coordinating the use of language.

bronchiole (brong'kē-ōl): one of the smaller air passages in the lungs of mammals.

bronchus (bron'kus): either of the two branches of the trachea or air passage leading to the lungs.

budding: a form of asexual reproduction in which a cell divides itself into two unequal parts; the larger part is considered the parent and the smaller part, the bud.

bulbourethral gland (bul'bo-yah-rē'thral): either of two small glands near the base of the penis that contribute to the seminal fluid.

C

caecum (sē'kum): a pouch between the small and the large intestine, into which the small intestine empties residue.

calorie (kal'o-rē): as a unit of heat, the amount of heat required to raise the temperature of one gram of water 1°C; a dietary calorie consists of 1,000 of these units.

Calvin cycle: a cyclic pathway of sugar biosynthesis from carbon dioxide that occurs during the dark phase of photosynthesis.

calyx (kā'liks): the sepals or outer group of floral parts, usually green but may resemble part of the flower.

cambium (kam'bi-um): a layer of meristemic tissue between the inner bark (phloem) and the wood (xylem) in the stems and roots of plants; it produces all secondary growth.

cancer (kan'ser): a tumorous growth resulting from the abnormal and uncontrolled growth of cells in plants or animals.

cannabis (kan'ah-bis): a plant of the genus *Cannabis;* the hemp plant from which marijuana and hashish are derived.

capillary (kap'i-lar-e): a microscopic network of vessels through which the blood moves from the arteries into the veins.

capsid (cap'sid): the protein coat of the tobacco mosaic virus.

capsomere (cap'so-mēr): a subunit of the protein coat surrounding the tobacco mosaic virus.

capsule: a fibrous, fatty, cartilaginous, or membranous structure surrounding another structure, organ, or part; for example, the protective sheath surrounding the cell wall in certain bacteria.

carbohydrate (kar'bō-hī'drāt): an organic substance belonging to a class of compounds represented by sugars, dextrins, starches, and celluloses; it contains carbon, hydrogen, and oxygen.

carbonic anhydrase: an enzyme in the red blood cells that catalyzes the hydration of carbon dioxide and the dehydration of carbonic acid; involved in carbon dioxide transport.

carcinogen (kar'si-no-jen): any cancer-producing substance.

carnivore (kar'nah-vōr): a flesh-eating animal or an insect-eating plant.

carotenoid (kar-rot'e-noid): the large class of yellow to orange fat-soluble pigments found in nearly all plants and in some animals.

carrier molecule: a hypothetical "revolving" molecule providing the mechanism of active transport; thought to exist in the cell membrane.

carrying capacity: a limit to the size of the population that a particular environment can support.

cartilage (kar'ti-lij): a rubbery connective tissue, usually part of the skeleton, which is composed of proteins (collagen), polysaccharides, and occasional cells.

caruncle (kar'ung-k'l): in botany, a protuberance of certain flowering plant seeds that is rich in protein and fat.

castrate (kas'trāt): to render an individual incapable of reproduction by removal of the gonads.

catabolism (kah-tab'ol-lizm): destructive metabolism; an energy-yielding process by which complex organic substances are converted by living cells into simpler compounds.

catalyst (kat'ah-list): a substance that stimulates a chemical reaction but is not used up in the reaction and does not form part of the final product.

cation (kat'ī-on): a positively charged ion.

cell (sel): the basic subunit of any living system; the simplest unit that can exist as an independent living system.

cell-free extract: a fluid containing most of the soluble molecules of a cell; produced by rupturing the membranes of living cells and centrifuging the suspension.

cell lysis: destruction of the cell that occurs when the cell wall or membrane is broken down by physical, chemical, or viral agents.

cell plate: a structure that forms along the equatorial plate during telophase in plant cells; it forms the new plasma membranes that separate the two daughter cells.

central dogma: the basic relationship of DNA, RNA, and protein: DNA serves as a template for both its own duplication and the synthesis of RNA; and RNA, in turn, is the template in protein synthesis.

central nervous system: one of the two traditional divisions of the vertebrate nervous system; it includes the neural structures (interneurons) encased in the skull and in the vertebral column. (Abbreviated: CNS)

centriole (sen′trē-ōl): an organelle in the cytoplasm of the cell; it forms the spindle pole during mitosis and meiosis.

centromere (sen′tro-mēr): a central point on the chromosome to which the spindle fiber is attached during mitosis and meiosis.

centrosome (sen′tro-sōm): a region of differentiated cytoplasm containing the centriole in a cell and usually located near the nucleus.

cephalization (sef′al-i-zā′shun): in the evolutionary development of animals, a trend toward enlargement of the anterior part of the nervous system to form the brain.

cephalothorax (sef′ah-lo-thor′aks): the combined head and thorax of certain crustaceans and arachnids (arthropods).

cerebellum (ser′a-bel′um): a lobed region of the vertebrate brain; it lies above the pons and is involved in the coordination of body movements and the maintenance of equilibrium.

cerebral cortex: the external layer of neural cells or "gray matter" of the forebrain (cerebrum); it functions in sensory and motor integration and is highly developed in humans.

cerebrum (ser′e-brum): the largest portion of the brain; it controls voluntary movements and coordinates mental actions.

cervix (ser′viks): the lower and narrower section of the uterus opening into the vagina.

chemical energy: the potential energy incorporated in chemical bonds.

chemical synapse: the most common type of synapse in the vertebrate nervous system; chemical transmitters released from terminal points of the axon cross the synaptic cleft and excite the adjacent cell. (See electrical synapse.)

chemosynthesis (ke′mo-sin′the-sis): the biosynthesis of carbohydrate from carbon dioxide and water by means of the energy derived from chemical oxidations, rather than from absorbed light; carried out by certain bacteria and algae.

chiasma (kī-as′mah): an X-shape crossing such as the crossing of chromosomes in the diplotene stage of meiosis. (Plural: chiasmata.)

chlorophyll (klo′ro-fil): any of several green, light-absorbing pigments essential as electron donors in photosynthesis; found in green plants and certain bacteria.

chlorophyll *a:* a yellowish-green pigment found in all plants.

chlorophyll *b:* a blue-green pigment found in green land plants and green algae.

chlorophyll *c:* a brownish-green pigment found in some plants, such as algae and diatoms.

chloroplast (klo′ro-plast): the chlorophyll-bearing body in cells; the plastid in which photosynthesis takes place.

cholesterol (kō-les′ter-ol): a complex alcohol occurring in animal fats and oils, bile, blood, brain tissue, milk, egg yolk, the liver, kidneys, adrenal glands, and the sheaths of nerve fibers. It is found mostly in higher animals (mammals).

chorion (ko′re-on): the protective and nutritive envelope of cells surrounding a fertilized ovum.

chorionic villus: a threadlike outgrowth from a trophoblast that penetrates the uterus and begins formation of the placenta.

chromatic aberration: imperfect refraction of light rays of different wavelengths passing through a lens; it results in fringes of color around the image.

chromatid (krō′mah-tid): one of two coiled filaments making up a chromosome; these filaments separate in mitosis, each going to a different pole of the dividing cell.

chromatin (krō′mah-tin): a granular, readily-stainable portion of the nucleus of plant and animal cells; the carrier of the genes in inheritance.

chromatophore (krō-ma′to-for): an elastic, pigmented cell responsible for rapid color changes in some animals; a pigmented body in some plant cells.

chromomere (krō′mo-mēr): one of the beadlike granules of chromatin comprising a chromosome.

chromophore (krō′mo-for): the part of a pigment molecule that absorbs visible light (is colored); for example, the chromophore of the visual pigment rhodopsin is retinene.

chromoplast (krō′mo-plast): any pigment-producing plastid other than a chloroplast.

chromosomal arm: either of two parts of a chromosome divided by the centromere.

chromosome (krō′mo-sōm): a microscopic, threadlike body that is composed of chromatin and that appears in the nucleus of a cell at the time of cell division. Chromosomes contain the genes and normally are constant in number within a species.

chromosome mapping: plotting of the relative positions of certain genes with respect to the positions of the other genes in the same linkage group.

cilium (sil′ē-um): a short, hairlike process in some types of animal cells; cilia are capable of beating in unison to produce locomotion or to move particles along the surface of the cell. (Plural: cilia.)

circadian (ser′kah-dē′an): pertaining to rhythmic biological cycles recurring at approximately 24-hour intervals, even under constant conditions.

class: the usual major subdivision of a phylum in the classification of plants or animals; it usually consists of several orders.

classical conditioning: a training method in which a conditioned stimulus (CS) and an unconditioned stimulus (UCS) are paired to effect a change in stimuli; for example, Pavlov's dog salivated in response to a light (CS) after learning to associate it with food (UCS).

cline (klīn): minor variations in genetic traits within a species population; it results from individual adaptations to slightly different environments.

clitoris (kli′to-ris): one of the female genital organs; a small, erectile body located at the top of the vaginal vestibule where the labia minora meet.

cloaca (klo-a′kah): a cavity into which the intestinal, urinary, and reproductive ducts open in most vertebrates and some mammals.

clonal selection theory: the theory that all antibody-forming cells are genetically identical but occur as different varieties (clones), each making one type of antibody protein with a characteristic amino acid sequence.

clone (klōn): a strain of cells descending from a single cell.

coccus (kok′us): a spherical-shape bacterium.

cochlea (kok′lē-ah): the coiled cavity of the inner ear containing the receptors for hearing.

codon (kō′don): a set of three nucleoside bases (in DNA) required to specify one amino acid residue in a protein chain synthesized on the DNA template.

coelom (sē′lom): the cavity between the body wall and the digestive tract of higher metazoan animals.

coenocyte (sē′nah-sīt): the fusion of two or more cells, forming a single, large, multinucleate cell, as in some fungi or algae.

cofactor (kō′fak-tor): any organic or inorganic substance necessary to the function of an enzyme.

coitus (kō′i-tus): the act of sexual intercourse.

colchicine (kol′chi-sēn): an organic alkaloid that speeds up the evolutionary processes in plants by doubling chromosome numbers.

coleoptile (kō′lē-op′til): the first leaf to appear above the ground in a grass seedling; it forms a sheath around the stem tip.

collar cell: the flagellated cell of sponges; each cell consists of a flagellum that creates a current in the water and a cytoplasmic collar that ingests the food particles filtered from the water.

collenchyma (ko-leng′ki-mah): an elongated plant cell with thickened walls; it is frequently present as support in maturing plant tissues.

colliculus (ko-lik′u-lus): any of the sensory control centers in the midbrain of mammals, reptiles, and birds. (Plural: colliculi.)

collision theory: the assumption that chemical reactions involve collisions between the molecules, atoms, or ions of the reactants.

colloid (kol′oid): any finely divided substance in permanent suspension in a medium; small groups of organic molecules bearing electrical charges and dispersed in a liquid.

colon (kō′lon): the major part of the large intestine; it removes excess water from undigested material.

command neurons: neurons in the central nervous system that command motor neurons to fire in a particular fashion, resulting in a particular posture or motion of the body.

communicative behavior: any behavior of an organism that affects or has the potential of affecting the behavior of another organism.

competitive advantage: the advantage that a growing population has over a declining population when both are competing for the same limited resources.

competitive exclusion: the principle that, under uniform conditions with at least one resource in limited supply, not more than one species can continue to exist indefinitely.

competitive inhibition: a reduction in the rate of an enzymatic reaction that is due to the inhibiting molecule and the substrate molecule competing on equal terms for the active site on the enzyme.

conditioned response: the response given to a conditioned or learned stimulus.

conditioned stimulus: a stimulus that elicits a response as a result of training.

conduction velocity: the speed at which the disturbance of membrane potential (change in polarity) travels along the axon or nerve fiber.

congenital (kon-jen′i-tal): pertaining to any condition present at birth.

conjugation (con′ju-gā′shun): a form of sexual reproduction found in certain bacteria; a bridge is formed between the mating cells and some DNA from the donor moves through this bridge to the recipient cell, which then reproduces by dividing itself in half.

conjunctival (kon′junk-ti′val): pertaining to the delicate membrane that lines the eyelid and covers the exposed surface of the eyeball.

conspecific (kon′spi-sif′ik): pertaining to individuals or populations of the same species.

constitutive gene: a bacterial gene having no repressor mechanism; it is expressed at a constant rate predetermined by the nucleotide sequence of the promoter.

consumer (kon-su′mer): an organism that feeds upon other organisms or their products.

contact inhibition: prevention of amoeboid motion caused by collision with another cell.

contractile vacuole: a small cavity containing watery fluid, seen in the protoplasm of certain one-cell organisms; it gradually increases in size, then contracts to expel the fluid through a pore in the cell membrane.

contralateral control: control of the opposite side; the right hemisphere of the brain is concerned with activities of the left side of the body, and the left hemisphere is concerned with the right side.

convergent evolution: a similarity of form or structure among different organisms resulting from their adaptation to similar habitats rather than from similar heredities.

copulation (kop′yah-la′shun): sexual union in which the male injects sperm into the body of the female.

corepressor (ko′re-pres′er): a small molecule that is required to bind certain gene repressors to the operator gene.

cork cambium: the cambium or growing tissue that forms the bark of a tree.

corolla (kah-rol′ah): collective term for the petals of a flower.

corpus callosum (kor′pus kah-lō′sum): a band of nerve fibers that connects the two cerebral hemispheres and coordinates activities in both.

corpus luteum (kor′pus lo͞o′tē um): the body formed by the ovarian follicle cells after the egg is released; it secretes the hormone progesterone.

correlative effects: changes occurring in one part of an organism that trigger changes in other parts.

cortex (kor′teks): in botany, a tissue between the epidermis and the vascular tissue in roots and stems; an outer layer or rind of an organ such as the kidney or brain.

cotyledon (kot′i-lē′don): the seed leaf of the embryo of a flowering plant.

covalent bond: a bond formed by the sharing of electrons by two atoms.

creatine (krē′ah-tēn): an amino acid found in muscle, where it functions as a source of phosphate essential in muscle contraction.

crepuscular animal (kre-pus′ku-lar): an animal that is active mainly at dusk or at dawn.

cretinism (krē′tin-izm): a congenital disease caused by a deficiency of thyroid hormone and resulting in physical and mental retardation.

crista (kris′ta): a ridge; a projection or projecting structure; for example, the inner membrane of the mitochondrion folded into sheets or tubules called cristae. (Plural: cristae.)

crop: a pouchlike enlargement of the gullet, in which food is stored in birds; the food is released to the digestive system as needed.

cross fertilization: the fertilization of one flower by the pollen of another.

crossing over: an exchange of material between adjacent chromatids of an homologous pair of chromosomes; it is caused by a chromosome break and produces genetic combinations different from those of the parents.

cunnilingus (kun′ah-ling′gus): oral stimulation of the female genitalia.

cuticle (kyo͞o′tik′l): in botany, a light gray film composed of cutin and cellulose that covers certain leaves and fruits.

cutin (kyo͞o′tin): a transparent, waxy substance that coats the outer surface of plants.

cytidylic acid (sī′ti-dil′ik): one of the four deoxyribonucleotides that make up DNA.

cytochrome (sī′tō-krōm): an iron-containing proteinoid compound that functions in the transport of electrons during metabolic oxidation reactions.

cytokinesis (sī′tō-ki-nē′sis): biochemical and structural changes in the cytoplasm occurring during cell division and fertilization.

cytokinin (sī′tō-kin′in): a class of hormonelike substances occurring in yeast, bacteria, and green plants; the cytokinins encourage bud formation.

cytoplasm (sī′to-plazm): the nonparticulate living matter in a cell, exclusive of that in the nucleus. (See karyoplasm.)

cytosine (sī′to-sēn): a nitrogenous base and a constituent of nucleic acids. (See pyrimidine.)

***D* amino acid:** one of the two forms (isomers) of every amino acid; it can be synthesized in the laboratory but is not found in most living systems.

dark stage: one of the two stages of photosynthesis; the stage in which the high-energy, chemically reactive molecules produced during the light-limited stage are converted into oxygen and sugar through reactions not involving light energy. (Also called the enzymatic stage.)

daughter cell: any cell formed by division of a parent cell.

decomposer (dē′kom-pō′ser): an organism (bacterium or fungus) that converts dead organic materials into inorganic materials.

dehydrogenase (dē-hī′dro-jen-ās): an enzyme that releases hydrogen from an organic substrate so that it can pass to a hydrogen acceptor during catabolism; an oxidizing enzyme.

delirium (di-lēr′ē-um): a mental disturbance often accompanied by hallucination and usually characterized by changing levels of awareness, decreased attention span, mental confusion, and altered time-space orientation.

deme (dēm): a population of freely interbreeding individuals; it usually corresponds to a species or a group of similar species.

dendrite (den′drīt): a short, branched extension from the body of a nerve cell; it conducts impulses toward the cell body.

deoxyribonuclease (dē-ok′si-rī′bōn-ū′klē-ās): an enzyme that specifically destroys DNA molecules. (Abbreviated: DNase)

deoxyribonucleic acid (dē-ok′si-rī′bō-nū′-klē′ik): a nucleic acid containing deoxyribose that is found chiefly in the nucleus of cells; it functions in the transfer of genetic characteristics and in protein synthesis. (Abbreviated: DNA)

deoxyribonucleotide (dē-ok′si-rī′bō-nū′-klē-ō-tīd): a compound consisting of the sugar 2-deoxyribose, a purine or pyrimidine base, and a phosphate group; a constituent of DNA.

depolarization (de′pō-lar-i-zā′shun): reduced polarization (electrical difference) resulting from a decrease in resting potential.

design feature: in linguistics, one of the basic characteristics distinguishing human language from other forms of animal communication.

desmosome (des′mo-sōm): a bridge corpuscle; a fold in the plasma membrane forming a tight connection between adjacent cells.

detritus (de-trī′tus): particulate material resulting from the degeneration and decay of organisms or inorganic substances in nature.

diakinesis (dī′ah-ki-nē′sis): in meiosis, the stage in which the nucleolus and nuclear membrane disintegrate preparatory to prophase I.

diaphragm (dī′ah-fram): a sheet of muscle that separates the chest cavity from the abdominal cavity of higher animals.

dicot (dī′kot): any of the higher plants that produce two seed leaves; a dicotyledon.

dictyosome (dik′tē-ō-sōm): a common form of Golgi apparatus found in both plant and animal cells; it consists of cisternae surrounded by a halo of tubules and small spherical vesicles.

differentiation (dif′er-en′shē-ā′shun): the series of biochemical and structural changes that groups of cells undergo in order to form a tissue appropriate for its specialized function.

Di George's disease: congenital absence of the thymus and the immunity it produces; a rare disease significant in the study of immune responses.

dihybrid (dī-hī′brid): the offspring of parents differing in two characters.

dinoflagellate (dī′nō-fla′gel-lāt): a type of unicellular, marine alga usually having two flagella and a cellulose wall.

diploidy (dip′loi-dē): the state of having two full sets of chromosomes.

diplotene (dip′lo-tēn): in the prophase of meiosis, the stage in which homologous chromosomes begin to pair and in which their division into chromatids becomes visible.

discrimination training: training an organism to respond differently to each of two similar but distinct stimuli.

disulfide bridge: a covalent bond formed between the sulfur atoms of two molecules.

diurnal animal (dī-er′nal): an animal that sleeps at night and is active during the day.

DNA: see deoxyribonucleic acid.

DNA polymerase (po-lim′er-ās): the enzyme that catalyzes the formation of DNA from deoxyribonucleoside triphosphates.

DNase: see deoxyribonuclease.

dominance (dom′i-nans): a genetic principle; when parents differ in one characteristic, their offspring usually resemble one of the parents rather than a blend of the two characters.

dominance hierarchy: the organization of social animals in a natural situation in which the more dominant animals control the less dominant ones by threat and aggressive action.

dominant character: a Mendelian character that develops when transmitted by a single gene.

dorsal (dor′sal): pertaining to the back or denoting a position more toward the back surface than some other object of reference. (See ventral.)

double bond: a bond in which two pairs of electrons are shared by two atoms.

double helix: a double spiral; for example, the structure of the DNA molecule in which two polynucleotide chains are entwined.

Down's syndrome: mongolism; usually caused by the presence of an extra sex chromosome.

duodenum (dū′o-dē′num): the short segment of small intestine closest to the stomach.

eccrine gland (ek′rin): an organ that pours its secretions into an external cavity; for example, a sweat gland.

echolocation (ek′ō-lō-kā′shun): an auditory system of "vision" used by some animals, such as bats and whales, to locate objects in their surroundings; high frequency sounds emitted by the animal bounce back from objects, revealing their location and distance.

ecosystem (ēk′ō-sis-tem): a system formed by the interaction of a community of organisms with its environment.

ectoderm (ek′to-derm): the outer tissue layer of an animal embryo; also the surface layer of coelenterates, such as jellyfish.

effector (ef-fek′tor): a muscle or gland that receives the nerve impulses that activate muscle contraction and gland secretion.

efferent nerve fiber: a nerve fiber carrying impulses away from the central nervous system.

ejaculation (e-jak′u-lā′shun): the ejection of semen from the penis.

ejaculatory duct: the tube that connects the seminal vesicle and urethra of the male and serves as a passage for sperm during copulation.

electrical synapse: a type of synapse in which transmission across the synaptic cleft is accomplished by electrical rather than chemical means; occurs in invertebrate animals and simpler vertebrates. (See chemical synapse.)

electrolyte (e-lek′tro-līt): a solution that conducts electricity by means of ions.

electromagnetic radiation: radiation consisting of electromagnetic waves, including radio waves, light, x-ray, and gamma rays.

electron (e-lek′tron): an elementary particle that is a fundamental constituent of matter, has a negative charge, and exists independently or as a component outside the nucleus of an atom; its mass is only 1/1840 of the mass of a proton or neutron.

electronegativity (e-lek′trō-neg-a-tiv′i-tē): the electron-attracting ability of a nucleus involved in a covalent bond.

electron resonance: a process by which molecules exchange energy.

electron shell: a group of orbitals surrounding the nucleus of an atom.

electrostatic (e-lek′trō-stat′ik): pertaining to static electricity or to electricity contained or produced by charged bodies.

element: in chemistry, a simple substance that cannot be decomposed by chemical means; it is made up of atoms that are alike in their chemical properties but may differ in atomic weight and radioactive properties.

Embden-Meyerhof pathway: the sequence of reactions involved in the metabolism of glucose; it occurs in both fermentation and glycolysis.

embryo (em′brē-ō): the early or developing stage of any organism, especially the developing product of a fertilized egg.

embryogenesis (em′brē-o-jen′e-sis): the development of an embryo.

embryonic induction: the process in which one group of embryo cells influences the development of another group.

emphysema (em′fi-sē′mah): abnormal swelling or inflation of an organ, especially the lung, with air or gas.

endergonic (end′er-gon′ik): characterized by the absorption of energy.

endoderm (en′do-derm): the inner tissue layer of an embryo; also the inner lining of body cavities in coelenterates, such as corals and jellyfish.

endogenous hypothesis (en-doj′e-nus): any concept that accounts for biological rhythms by assuming the presence of internal biological clocks.

endometrium (en-dō-mē′trē-um): the glandular lining of the uterus; the tissue in which the ovum lodges and develops.

endoplasmic reticulum (en′do-plas′mik re-tik′u-lum): membrane sheets folded through the cytoplasm to form a complex system of tubules, vesicles, and sacs; it is studded with ribosomes.

endosperm (en′do-spurm): a multicellular tissue formed inside a developing seed and serving in the nutrition of the embryo.

endospore (en′do-spor): a resistant daughter cell produced within the parental cell wall in some microorganisms; it becomes dormant during adverse conditions.

endotoxin (en′do-tok′sin): a toxic substance released when bacteria die and disintegrate; it causes animal cells to produce interferon. (Also called bacterial pyrogen.)

endocrine gland (en′do-krin): any gland that pours its secretion into the blood, which carries the secretion to the target organ.

entropy (en′tro-pē): that portion of the energy of a substance not available for the performance of useful work.

enzyme (en′zīm): any of various complex organic compounds originating from living cells and capable of producing certain chemical changes in organic substances by catalytic action.

epidermal tissue (ep′i-der′mal): the outermost surface of an organism; the outer layer and protective covering of roots, stems, and leaves in most plants and of the skin in most animals.

epididymis (ep′i-did′i-mis): a cordlike structure along the top and back of the testes in which spermatozoa are stored.

epinephrine (ep′i-nef′rin): a hormone secreted by the adrenal medulla; used to stimulate the heart, to constrict the blood vessels, and to relax the bronchi in asthma.

epithelial tissue (ep′i-thē′lē-al): animal tissue covering internal and external surfaces; it consists of tightly packed cells with little intercellular space.

equatorial plate: a hypothetical plane upon which the chromosomes arrange themselves as the cell approaches the metaphase in mitosis.

equilibrium (e-kwa-lib′ri-um): the condition existing when a chemical reaction and its reverse reaction proceed at equal rates.

erythroblastosis fetalis (e-rith′ro-blas-tō′sis fe-ta′lis): a hemolytic anemia of the fetus or newborn infant caused by transmission of an antibody through the placenta from mother to offspring.

esophagus (e-sof′ah-gus): the gullet; the tube that transports food from the mouth to the stomach.

essential amino acid: a necessary amino acid, which an animal organism must obtain by breaking apart protein molecules in food.

ester (es′ter): any compound formed from an alcohol and an acid by removal of water.

estrogen (es′tro-jen): a typically female hormone; one of the hormones involved in sex determination.

estrous cycle: a recurrent series of physiological changes in the organs of female mammals between periods of maximum sexual receptivity.

eucaryotic (yū-kar′ē-o′tic): having a true nucleus; said of cells surrounded by a membrane and containing a nucleus surrounded by a nuclear membrane.

euploidy (yū-ploi′dē): the state of having one or more complete extra sets of chromosomes.

eutrophication (yū-trof′i-kā′shun): enrichment of a body of water by an accumulation of nutrients that support a dense growth of plant and animal life; the decay of this organic matter depletes the water of oxygen.

evolution (ev′o-lū′shun): the continuous genetic adaptation of organisms or species to the environment by selection, hybridization, inbreeding, and mutation.

excitation (ek′sī-tā′shun): in physics, a process by which an electron absorbs energy and moves to a higher energy level; in biology, the inducement of a nerve impulse by stimulation.

excitatory postsynaptic potential: the depolarization that the excitatory synapse produces in the membrane of a postsynaptic cell; the magnitude of depolarization depends upon the number of contributing synapses. (Abbreviated: EPSP)

exergonic (ek′ser-gon′ik): characterized by the release of energy.

exocrine gland (ek′so-krin): any gland that secretes its product into a duct leading directly to the target organ.

exogenous hypothesis (eks-oj′e-nus): any concept that accounts for biological rhythms by assuming that external geophysical factors, which vary cyclically, influence the behavior of organisms.

exothermic reaction (ek′so-thur′mik): a chemical change in which heat is released.

f

fallopian tube (fal-lō′pē-an): the oviduct; a tapering duct connecting the ovary and uterus of mammals; the tube in which fertilization occurs.

family: the major subdivision of an order or suborder in the classification of plants or animals; it usually consists of several genera.

fatty acid: a class of weak organic acids containing only carbon, hydrogen, and oxygen; they are present as glycerides (esters) in all animal and vegetable fats.

feedback inhibition: inhibition of the enzymatic activity of an enzyme in a metabolic pathway by the end product of that pathway.

fellatio (fel-ā′shē-o): oral stimulation of the male genitalia.

fermentation (fur′men-tā′shun): anaerobic respiration; the partial enzymatic breakdown (catabolism) of organic fuels by an organism in the absence of oxygen. The usual end product is alcohol or lactic acid.

fertilization (fer′ti-li-zā′shun): the union of male and female gametes, the spermatozoa and ova.

fetishism (fe′tish-izm): an attraction to an object that, although not of a sexual nature, causes an erotic response.

fetus (fē′tus): the unborn offspring of any live-bearing animal; the developing young in the human uterus after the end of the second month. (Before eight weeks, it is called an embryo.)

fibroblast (fī′bro-blast): a connective tissue cell that synthesizes the extracellular fibers binding tissues together.

fibrous protein: a protein molecule with a linear structure, such as collagen.

filter feeder: an animal that obtains its food, usually in small particles, by filtering it from water.

first filial generation: all of the offspring produced by the mating of two individuals. (Symbol: F_1)

first ionization energy: the amount of energy needed to remove one electron from an atom. (Symbol: I_1)

first polar body: a very small cell pinched off a large egg cell in the first meiotic division of the oocyte; it disintegrates.

fitness: a measure of the degree to which a genotype succeeds in reproducing its alleles in the next generation.

flagellum (flah-jel′um): a whiplike appendage; the organ of locomotion of sperm cells and of certain bacteria and protozoa. (Plural: flagella.)

flame cell: a hollow, bulb-shape structure formed by one cell or several small cells and terminating the branches of the excretory tubules of certain lower invertebrates.

follicle cell: a cell, found in the ovary, that helps to nourish the developing egg cell; a different kind of follicle cell produces animal hair.

follicle stimulating hormone: one of the hormones manufactured by the pituitary; it is essential to the maturation of the follicle surrounding the oocyte and also stimulates the supply of nutrients. (Abbreviated: FSH)

food chain: the natural transfer of food energy from producers (green plants) to successive consumers (herbivores and carnivores); certain species characterize a given food chain.

food web: energy interrelationships in any community with special reference to feeding habits; a typical food web includes green plants, herbivores, carnivores, omnivores, and detritus feeders.

forebrain (for′brān): one of the three subdivisions of the vertebrate brain; it consists of the cerebral hemispheres and interior structures such as the thalamus; also called prosencephalon.

foreskin (for′skin): a skin fold at the end of the penis; the fold that is removed by circumcision.

founder effect: rapid changes in gene frequencies among emigrants from an established population.

fovea (fō′vē-ah): a specialized area of receptors in the retina of the eye; the area of clearest vision where most of the cones are located.

free energy: the energy that is actually available to do useful work. (Symbol: ΔG)

freemartin (frē′mar-tin): a female calf born as a twin to a male calf; it is usually sterile as a result of hormonal influences from the male embryo.

frontal lobe: one of the four lobes dividing the hemispheres of the human brain; it is involved in the regulation of fine body movements.

fundamental tissue: tissue forming the base or foundation of the plant body; it is composed largely of a single kind of specialized cell.

fungivore (fun′ji-vōr): an organism that feeds on fungi.

g

G_1 period: an interval of several hours after telophase during which no DNA is synthesized in a cell; it is followed by S-phase.

G_2 period: second of two periods in which no DNA is synthesized in a cell; it follows S-phase and precedes prophase.

gamete (gam′ēt): either of two cells, male or female, whose union is necessary in sexual reproduction to initiate the development of a new individual; gametes contain half the full complement of hereditary material.

gametogenesis (gam′e-tō-jen′e-sis): the formation in the body of gametes, or male and female reproductive cells (sperm and eggs).

gametophyte (gah-mē′tah-fīt): the sexual, haploid generation of a plant that produces gametes; it alternates with an asexual, diploid, sporophyte generation.

gamma aminobutyric acid (ah-mē′nō-byū-tir′ik): a fatty acid derivative that functions as one of the inhibitory transmitters at the synapse in the central nervous system. (Abbreviated: GABA)

gamma globulin: a protein component of blood plasma that contains antibodies effective against certain microorganisms, such as those causing measles and infectious hepatitis.

ganglion (gang′glē-on): a group of nerve cells whose cell bodies are located outside the central nervous system; in some invertebrates, separate ganglia control segments of the body.

gastric mucosa: the membranous stomach lining.

gastrulation (gas′trū-lā′shun): the process of cell division and differentiation in which the young embryo acquires its three germ layers.

gene (jēn): the unit of heredity transmitted in the chromosome; interacting with other genes, it controls the development of hereditary character.

gene flow: gradual changes in the genetic make-up of a population as a result of emigration and immigration of organisms.

gene frequency: a mathematical expression of the proportion of one allelic form of a gene compared to other allelic forms of the same gene in the gene pool.

gene pool: the total aggregate of allelic forms of all the genes in a population of organisms.

genetic code: the genetic information of genes, which determines hereditary traits.

genetic drift: random fluctuations in the composition of the gene pool caused by chance mortality of individuals.

genetic mapping: the plotting of relative gene positions on a chromosome.

geniculate nuclei (je-nik′yah-lāt): the sensory nuclei of the thalamus that process information concerning hearing and vision.

genotype (jen′o-tīp): the assortment of genes that comprise the fundamental hereditary constitution of an individual.

genotype frequency: a mathematical expression of the proportion of one genotype to other genotypes in the same population.

genus (jē′nus): the major subdivision of a family or subfamily of plants or animals; it usually consists of more than one species. (Plural: genera.)

geotropism (jē′o-trō′pizm): the influence of gravity upon growth; for example, the tendency of many plants to grow vertically.

germ cell: the sexual reproductive cell produced in the gonad.

germ layer: any of three embryonic tissue layers – ectoderm, mesoderm, and endoderm – that differentiate into various organs during development.

ghost corpuscle: a red blood corpuscle from which the hemoglobin has been removed.

gibberellin (jib-e-rel′in): a class of hormonelike substances occurring naturally in plants; the gibberellins encourage stem growth.

gill: a respiratory organ of various aquatic organisms; it enables them to obtain dissolved oxygen from the water and eliminate carbon dioxide.

gizzard (giz′ard): a muscular sac containing coarse sand or small stones, in which food is crushed in the digestive system of herbivorous birds and fishes.

gland: a cellular organ specialized for secretion.

gland of Bartholin: one of the glands located on either side of the vagina just beneath the skin that secrete fluid to lubricate the vaginal opening during intercourse.

glans penis: a caplike structure at the end of the penis.

glial cell (gli′al): one of the small but numerous cells that fill the spaces between nerve cells and are thought to function in their support and nourishment. (Also called neuroglia.)

globular protein: a protein molecule whose structure is roughly spherical.

globulin (glob′yah-lin): a class of proteins characterized by their solubility in dilute salt water rather than in fresh water; present in plant and animal tissues.

glomerulus (glo-mer′yu-lus): a cluster; for example, the cluster of capillaries within the cup of Bowman's capsule in the kidney.

glucose (glū′kōs): a six-carbon sugar occurring in many fruits, animal tissues, and fluids; the common source of metabolic energy resulting from breakdown of starch or glycogen.

glycerol (glis′er-ol): an oily three-carbon alcohol that combines with fatty acids (as an ester) to form the glycerides of natural fats and oils. (Also called glycerin.)

glycine (glī′sēn): aminoacetic acid; one of the inhibitory transmitters at the synapse in the central nervous system.

glycogen (glī′ko-jen): a complex polymer of glucose; the chief carbohydrate storage material in animals.

glycolipid (glī′ko-lip′id): a lipid combined with carbohydrate substances (sugars).

glycolysis (glī-kol′i-sis): the enzymatic breakdown of sugars into simpler compounds.

glycoprotein (glī′ko-prō′tēn): any conjugated protein with an attached carbohydrate group.

Golgi (gol′jē): a complex system of folded membranes containing substances to be excreted from a cell; sometimes called Golgi apparatus, Golgi bodies, or Golgi complex.

gonad (gōn′ad): animal reproductive organ; an ovary or testis.

gonocyte (gon′o-sīt): a gamete-producing cell formed by repeated meiotic divisions of the germ cell in a male animal.

guanine (gwah′nēn): a nitrogenous base and a constituent of nucleic acids. (See purine.)

guanylic acid (gwa-nil′ik): one of the four deoxyribonucleotides that make up DNA; its structure includes guanine.

guard cell: one of two cells that regulate the flow of gases by opening and closing the stoma in a leaf or a young stem.

gullet (gul′it): the esophagus or any structure resembling a throat; for example, a cell indentation found in certain protozoans and thought to be the opening through which food is ingested.

gut-associated lymphoid tissue: a tissue that is thought to produce the family of lymphocytes (immunocytes) that do not arise from the thymus gland. (Abbreviated: GALT)

gyrus (jī′rus): a ridge in the cerebral cortex of the brain. (Plural: gyri.)

habitat (hab′i-tat): the specific place or environmental situation in which an organism lives.

habituation (hah-bich′ū-ā′shun): the simplest form of learning; the magnitude of response gradually decreases as the organism grows accustomed to a particular stimulus, such as a sudden loud noise.

hair cells: epithelial cells with hairlike processes; for example, the receptor cells of the auditory system.

hallucinogenic (hah-lū'si-no-jen'ik): producing hallucinations.

halogen (hal'o-jen): any of the electronegative elements—fluorine, chlorine, bromine, and iodine, all of which combine with metals to form saltlike compounds.

haploid (hap'loid): having a single set of chromosomes per individual or cell, as in gametes. (See diploid.)

haploid number: the number of chromosomes contained in a haploid cell of a particular species. (Symbol: *n*)

Hardy-Weinberg law: a generalization describing the equilibrium established between gene frequencies in a population after a generation of random mating and interbreeding.

heat content: energy content; a thermodynamic quantity. (Symbol: *H*)

heat of reaction: the heat content of the products minus the heat content of the reactants. (Symbol: ΔH)

hemisphere (hem'i-sfēr): in anatomy, either of the two symmetrical halves of the brain; it includes the frontal, temporal, parietal, and occipital lobes.

hemoglobin (hē'mo-glō'bin): the reddish protein-iron complex of the red blood corpuscles; it binds oxygen for transport to the tissues.

hemophilia (hē'mo-fil'ē-ah): an abnormal condition of males that is inherited through the mother; characterized by a tendency to bleed immoderately because of improper coagulation of the blood.

herbivore (her'bah-vōr): an animal that feeds on plants.

heredity (he-red'it-tē): the genetic transmission of a particular quality or trait from parent to offspring.

hermaphrodite (her-maf'ro-dīt): a single individual that possesses both male and female reproductive organs.

heterocyst (het'ah-ro-sist): one of the large cells found at intervals in filaments of blue-green algae; it lacks photosynthetic pigments but contains a large amount of DNA.

heterogamy (het'er-og'ah-mē): reproduction resulting from the union of two sex cells (gametes) that differ in size and structure.

heterotroph (het'er-o-trōf): an organism such as an animal or chlorophyll-free plant or bacterium that depends on an external source of organic substances for its food and energy.

heterozygous (het'er-o-zī'gus): having different members (alleles) on a particular gene; not true-breeding with respect to a given characteristic. (See homozygous.)

high-energy bond: a bond that releases at least 5 kcal/mole when broken; for example, the phosphate bonds in ATP that are used for storing and transferring chemical energy in living systems.

Hill reaction: a partial photosynthetic process in which a substitute oxidant, such as a ferric compound, is used in place of carbon dioxide as an acceptor of electrons from chlorophyll.

hindbrain (hīnd'brān): one of the three subdivisions of the vertebrate brain; it includes the medulla, pons, and cerebellum.

holdfast: a rootlike organ or part serving for attachment of algae, such as kelp.

homeostasis (hō'mē-ah-stā'sis): a tendency toward stability and constancy in metabolic conditions; also, a steady-state relationship between an animal and its environment.

hominoid (hom'i-noid): a member of the superfamily that includes the great apes and man.

homolog (hom'o-log): one of a pair of similar chromosomes in a diploid cell; in bisexual organisms, one homolog is inherited from the male parent and one from the female parent.

homozygous (hō'mo-zī'gus): having identical members (alleles) on a particular gene; true-breeding with respect to a given characteristic. (See heterozygous.)

hormone (hor'mōn): a chemical substance, produced in the body, that has a specific regulatory effect on the activity of a receptive organ.

hybrid (hī'brid): an organism produced from parents of two different species or genotypes.

hydration (hī-drā'shun): the binding of water molecules to a substance by weak hydrogen bonds.

hydrocarbon (hī'dro-kar'bon): an organic compound that contains only carbon and hydrogen.

hydrogenation (hī'dro-jen-ā'shun): chemical reduction; combination with hydrogen.

hydrogen bond: the weak electrostatic attraction between a hydrogen atom and a more negative atom of another molecule.

hydrolysis (hī-drol'i-sis): the splitting of a compound into fragments by the chemical addition of water.

hydrophilic (hī'dro-fil'ik): having an attraction for water; for example, polar (charged) substances.

hydrophobic (hī'dro-fo'bik): repellent to or having no attraction for water; for example, nonpolar substances.

hymen (hī'men): a membranous fold that partially blocks the opening of the vagina.

hypha (hī'fah): one of the threadlike elements of the mycelium of a fungus. (Plural: hyphae.)

hypnotic drugs: drugs used to treat insomnia and anxiety by inducing depression and sleep.

hypothalamus (hī'po-thal'ah-mus): a cluster of neurons lying in front of the thalamus in the forebrain; it affects control of basic drives, regulation of pituitary secretion, and control of the body's internal environment.

I

IAA: see indoleacetic acid.

imaginal disc: a tiny nest of cells in the embryo that gives rise to particular organs in the adult insect.

immunity (i-myū'ni-tē): the power of living cells to resist or overcome an infection.

immunocyte (i-myū'no-sīt): a type of white blood cell involved in antibody formation.

immunoglobulin (im'yū-no-glob'yah-lin): a protein of the blood plasma that acts as an antibody.

imprinting: behavior attached to stimuli very early in life and generally not reversible; it occurs at critical stages of development.

inclusion (in-klū'zhun): in cells, bodies of inactive materials, such as droplets or crystals, enclosed in the cytoplasm.

independent assortment: the random behavior of genes on separate chromosomes during meiosis; the result is recombination, or a new mixture of genetic material in the offspring. (See recombination.)

independent effector: any cellular organelle that functions without neural stimulation.

indeterminate growth: the proliferation of unspecialized cells that have the potential to form new organs throughout the organism's life.

indoleacetic acid (in'dōl-ah-sē'tik): a hormonal substance obtained from yeast, fungus, and human urine and used to affect the growth of plants. (Abbreviated: IAA)

inducer (in-dūs'er): in virology, a chemical substance, such as viral RNA, that stimulates animal cells to release the protein interferon; in biosynthesis, a substrate that increases the production of enzymes by blocking gene repressors.

inducible enzyme: an enzyme whose rate of production can be increased by the presence of inducers in the cell.

induction (in-duk'shun): control of cell differentiation by chemical substances from neighboring cells.

inert gas: a gaseous element distinguished by its lack of chemical activity.

inflammatory response: the process by which white blood cells combat foreign materials by engulfing and digesting them.

inhalant (in-hā'lant): a substance taken into the body by way of the nose and trachea.

inhibition (in'hi-bish'un): reduction of the rate of enzymatic reactions when the substrate and other substances compete for an active site on the enzyme.

inhibitory postsynaptic potential: the hyperpolarization potential produced in the postsynaptic membrane by the inhibitory synapses. (Abbreviated: IPSP)

innate release mechanism: in behavior, the means by which an organism's nervous system "recognizes" a certain stimulus and produces an appropriate response.

inorganic (in′or-gan′ik): pertaining to chemical substances that do not contain carbon and are mostly not derived from living organisms.

instructive theory: any of several theories in immunology based on the assumption that antibodies are in some way shaped to complement the structure of the particular antigens present in the body.

intercalated disc (in-ter′kah-lāt-ed): a cellular junction that interconnects cardiac muscle fibers.

interferon (in′ter-fer′on): a protein released by animal cells in response to infection or chemical agents (inducers); viral RNA is a potent inducer of interferon.

interneuron (in′ter-nū′ron): nerve cells that form a connective link between the sensory cells and the motor cells; they process the sensory input and command the motor output.

internode (in′ter-nōd): the portion of a plant stem between two leaf nodes.

interphase (in′ter-fāz): the interval during which a cell is not dividing and the chromatin material is dispersed.

interstitial cell (in′ter-stish′al): a cell that supports and forms the matrix of a living structure; for example, the glial cell of the nervous system that binds, insulates, and nourishes the neuron.

interstitial cell stimulating hormone: a pituitary hormone that stimulates growth of gonadal interstitial cells, mitotic activity among spermatogonia, and production of semen. (Abbreviated: ICSH)

intrauterine device (in′trah-yū′ter-in): a coil or loop of plastic inserted into the uterus as a means of contraception. (Abreviated: IUD)

ion (ī′on): an electrically charged atom or group of atoms formed by the loss or gain of one or more electrons.

ionic bond: a bond formed by the electrical attraction between the cation of one atom and the anion of another.

ionization (ī′on-i-zā′shun): the separation of a substance in solution into its constituent ions.

isoagglutinin (ī′sō-ah-glū′ti-nin): a type of protein in red blood cells that forms antigens.

isogamy (ī-sog′ah-mē): reproduction through the union of gametes that are identical in appearance.

isosmotic (ī′sos-mot′ik): having the same osmotic pressure; for example, the isosmotic properties of marine invertebrate tissues and of sea water.

isotope (ī′so-tōp): any of two or more forms of a chemical element having the same number of protons in the nucleus but a different number of neutrons. Isotopes may be stable or unstable (radioactive).

k

karyoplasm (kar′ē-o-plazm): the nonparticulate living matter in the nucleus of a cell. (See cytoplasm.)

kidney (kid′nē): in vertebrates, the organ that secretes the urine (soluble waste products).

kinesis (ki-nē′sis): behavior of an organism that is elicited, accelerated, or decelerated by certain stimuli but is not directly guided by these stimuli.

kingdom: any of five categories into which organisms are usually classified: Monera, Protista, Plantae, Fungi, and Animalia; each consists of two or more phyla.

Klinefelter's syndrome: A condition characterized by mental retardation and sexual abnormalities and attributed to the presence of two X chromosomes in a male; an example of trisomy.

Krebs citric acid cycle: a biochemical reaction sequence in aerobic organisms in which the pyruvate produced as the end product of glycolysis is degraded to carbon dioxide and water, producing energy.

l

L **amino acid:** one of two forms (isomers) of amino acid; it is found in the proteins of living organisms.

labia majora (lā′bē-ah mah-jor′ah): the large, outer folds of skin covering the vulva.

labia minora (lā′bē-ah mi-nor′ah): the inner folds of skin bordering the vagina and covered by the labia majora.

lac **operon** (lak op′er-on): the group of three genes coding for the formation of three proteins involved in β-galactoside hydrolysis and transport.

lacteal (lac′tē-al): a lymph vessel in a villus of the intestinal wall of mammals.

lateral meristem: a layer of growing cells along the lateral surface of plant organs; responsible for organ growth.

leptotene (lep′to-tēn): in meiosis, the stage in which the chromosomes are slender and threadlike.

leucine (lū′sēn): an amino acid essential for growth in infants and for nitrogen equilibrium in human adults.

leucocyte (lū′ko-sīt): a white blood cell that functions in combating infections.

leucoplast (lū′ko-plast): a colorless plastid of plant cells in which starch, lipids, or proteins are stored.

leucosin (lū′kah-sin): a polysaccharide; one of the forms (along with fats) in which food is stored in the plastids of diatoms.

life table: a table showing the number of individuals who die at any given age; compiled from death statistics for selected population groups.

light-limited stage: one of the two stages of photosynthesis; the stage in which light energy is converted by chlorophyll to chemical energy and stored as NADPH and ATP. (Also called the photochemical stage.)

limbic system (lim′bik): an area of the forebrain, including the hypothalamus, amygdala, septal area, and cingulate gyrus; it is related to the activities of the visceral organs and to emotional behavior.

lipase (līp′ās): an enzyme, secreted by the pancreas, that aids in the digestion of fats.

lipid (lip′id): any of a group of fatty organic compounds that constitute, along with proteins and carbohydrates, the chief organic components of living cells.

lithium (lith′ē-um): an alkaline-earth metal, the salts of which are used as drugs for the treatment of mood disorders, especially manic-depression.

liver (liv′er): a large, glandular organ in higher animals that functions in producing bile, storing glycogen, and carrying out other metabolic processes essential to life.

loop of Henle: a U-shape turn in the renal tubule of the mammalian kidney.

luteal phase (lū′tē-al): the period of the menstrual cycle during which hormones secreted by the corpus luteum prepare the endometrium for embryo implantation.

luteinizing hormone (lū′tē-in-ī′zing): one of the hormones manufactured by the pituitary; it stimulates the release of estrogen into the bloodstream and the release of the mature oocyte from the ovarian follicle.

lymph (limf): a clear, yellowish fluid, containing white blood cells, that is derived from the tissues of the body and conveyed to the bloodstream by the lymphatic vessels.

lymph node: an organ of the lymphatic system that produces lymphocytes and purifies the lymph.

lymphocyte (lim′fo-sīt): a variety of white blood cell derived from the lymph glands.

lysogen (lī′so-jen): an agent or condition causing the dissolution (lysis) of cells; a state of bacterial cells in which viral and bacterial DNA are joined in an integrated chromosome. (See prophage.)

lysogenic bacteria: bacteria containing viral DNA as prophages.

lysogenic phage: a viral cell capable of causing cell lysis in bacteria. (See prophage.)

m

Mach bands (mäk): a phenomenon of vision in which the eye accentuates a light gradient by perceiving a darker band along the dark border and a lighter band along the light border.

macrogamete (mak′rō-gam′ēt): the female gamete of the malarial parasite.

macrogametocyte (mak′rō-gah-mē′to-sīt): the female form of the malarial parasite; one of the gametocytes formed in the red blood corpuscles of a human infected with malaria.

macronucleus (mak′rō-nū′klē-us): the larger of two types of nuclei in a cell.

macrophage (mak'rō-fāj): an enlarged, transitional monocyte that helps to relieve inflammation by ingesting foreign material and cellular debris.

mantle (man't'l): fleshy fold that usually secretes the shell of the mollusc.

mass number: the total number of protons and neutrons in the nucleus of an atom.

mast cell: a cell with a small, spheroid nucleus and large granules in its cytoplasm; it is widely distributed in connective tissue.

mast crop: a seed crop; in certain plants, seeds are produced at intervals following several years of nearly sterile vegetative growth.

matrix (mā'triks): the basic material or medium of biological structures in which organized bodies develop or are suspended; for example, the substance within the inner membrane of mitochondria.

mechanism (mek'ah-nizm): the theory that the phenomena of life are based on the same chemical and physical laws that apply to inorganic substances. (Opposed to vitalism.)

medulla (me-dul'lah): a region of the hindbrain that is like the spinal cord in structure and function; it controls certain autonomic processes, such as respiration and circulation.

megasporocyte (meg'ah-spor'ah-sīt): the central part of the ovule; by meiotic division it produces four haploid megaspores. (Also called the megaspore mother cell.)

meiosis (mī-ō'sis): the process of cell division by which chromosome number is halved during formation of sperm and egg cells.

meiospore (mī'o-spor): a spore produced by meiosis; it is always haploid.

melanin (mel'ah-nin): a pigment found in the skin and hair of man and responsible for yellowish to blackish coloration in many animals and plants.

menstrual phase: the period of the menstrual cycle during which blood is discharged; in humans, it occurs 14 to 16 days after ovulation and usually lasts from three to six days.

menstruation (men'strū-ā'shun): the periodic discharge of blood and mucosal tissue from the uterus—in humans and in certain other primates—when fertilization does not occur.

meristem (mer'i-stem): the undifferentiated embryonic tissue from which permanent tissues are derived in higher plants; the site of active cell division.

merozoite (mer'o-zō'īt): one of the stages in the life cycle of certain sporozoan organisms; in malaria, a sporozoite undergoes mitotic divisions to form merozoites in the red corpuscles of human blood.

mesonephros (mes'o-nef'ros): one of the embryonic excretory organs of vertebrates; it becomes the functional kidney of fishes and amphibians and becomes part of the epididymis in higher vertebrates. (Plural: mesonephroi.)

mesophyll (mez'ah-fil): the parenchyma tissue that forms the interior of a leaf blade.

messenger RNA: a ribonucleic acid, produced in the cell nucleus, that transcribes the genetic code of DNA in chromosomes; when attached to ribosomes, it serves as the template for determining amino acid sequence in protein synthesis. (Abbreviated: mRNA)

metabolism (me-tab'o-lizm): all of the physical and chemical processes in an organism by which protoplasm is produced, maintained, and destroyed and by which energy is made available for its functions.

metamorphosis (met'a-mor'fo-sis): a change in shape or structure; for example, the transformation of a larva into an adult insect.

metaphase (met'ah-fāz): an intermediate stage of mitosis, between the prophase and anaphase, when the chromosomes lie in a plane at the equator of the spindle.

metaphase I: in meiosis, the stage in which pairs of homologous chromosomes become aligned on the equatorial plate with spindle fibers attached to the centromeres.

metaphase II: in meiosis, the stage in which daughter chromosomes (paired chromatids) become aligned on the equatorial plate and become attached to the spindle fibers.

methemoglobin (met'hē'mo-glō'bin): a brownish compound of oxygen and hemoglobin formed in the blood by the action of various drugs and oxidizing agents.

methionine (me-thī'o-nēn): a sulfur-bearing amino acid essential for growth in infants and for nitrogen equilibrium in human adults.

micelle (mī'sel): a microscopic particle or unit of cell structure composed of polar molecules, such as phospholipids; a colloidal particle.

microfibril (mī'kro-fī'bril): a very small fiber; for example, the walls of plant cells are layers of such fibers embedded within a matrix.

microgamete (mī'krō-gam'ēt): the male gamete of the malarial parasite.

microgametocyte (mī'krō-gah-mē'to-sīt): the male form of the malarial parasite; one of the gametocytes formed in the red blood corpuscles of a human infected with malaria.

micron (mī'kron): one-thousandth of a millimeter; a unit of linear measure. (Abbreviated: μ)

micronucleus (mī'krō-nū'klē-us): the smaller of two types of nuclei in unicellular organisms, such as protozoa.

micropyle (mī'krah-pīl): an opening at the apex of an ovule or the corresponding opening in the testa of a seed; the pollen tube penetrates the ovule through the micropyle.

microsurgery (mī'krō-ser'jer-ē): dissection of minute living structures under a microscope.

microtubule (mī'krō-tū'byūl): a microscopic tube or hollow cyclinder commonly found in the cytoplasm of most eucaryotic cells; it is made up of 13 filaments arranged in a circular pattern.

microvillus (mī'krō-vil'lus): a tiny fold projecting from the membrane of a cell; microvilli increase the surface area and absorptive capability of the cell. (Plural: microvilli.)

midbrain (mid'brān): one of the three subdivisions of the vertebrate brain; it contains the colliculi and tegmental nuclei; also called mesencephalon.

millimicron (mil'li-mī'kron): one-thousandth of a micron: a unit of linear measure. (Abbreviated: $m\mu$)

miniature endplate potentials: small, spontaneous depolarizations that occur at random intervals in the resting nerve fiber. (Abbreviated: MEPP)

mitochondrion (mī'to-kon'drē-on): an organelle in the cytoplasm of a cell; it contains enzymes and functions in metabolism of the cell.

mitosis (mī-tō'sis): the usual method of cell division characterized by complex chromosome movements and exact duplication of chromosomes.

mitospore (mī'to-spōr): a spore produced by mitosis; it may be haploid or diploid, depending upon the chromosome complement of the parent cell.

mitotic apparatus: cell parts involved in the process of mitosis: the asters, the spindle, and, in some cells, the centrioles.

mole (mōl): gram molecule; the molecular weight of a substance expressed in grams.

molecular weight: the weight of a molecule of a substance as compared with the weight of hydrogen; it is equal to the sum of the weight of the constituent atoms.

monocot (mon'o-kot): any of those higher plants that produce one seed leaf; a monocotyledon.

monocyte (mon'o-sīt): a large, mononuclear leucocyte or white blood cell.

monolayer (mon'o-lā'er): a monomolecular layer; a layer only one molecule in thickness.

monomer (mon'o-mer): a single molecule of a compound of relatively low molecular weight. (See polymer.)

monosomy (mon'o-sō'mē): the absence of one chromosome from the complement of an otherwise diploid cell.

mons veneris (monz ven'er-is): the mound of fatty tissue over the pubic bone of the female.

morphogenesis (mor'fo-jen'e-sis): the processes by which tissues or germ layers of organisms are shaped into organs with adult shape and form.

morula (mor'yah-lah): a solid mass of dividing cells resulting from cleavage of the ovum; it develops into the blastocyst.

motor neuron: one of the cells that control the effectors in the nervous system by conveying motor impulses to them.

mucopeptide (myū′ko-pep′tīd): a general term describing any substance composed of proteins and polysaccharides (carbohydrates).

mucous (myū′kus): of or pertaining to mucus; mucoid.

mucus (myū′kus): the slime or viscous secretion of mucous cells or membranes; it consists of proteins and/or polysaccharides (carbohydrates).

mutation (myū-tā′shun): an abrupt departure from the parent type, as when an individual differs from its parents in one or more inheritable characteristics as the result of a change in a gene or chromosome.

mutualism (myū′chū-al-izm): a relationship or symbiosis in which two or more organisms benefit from their association and may be unable to survive without it.

mycelium (mī-sē′lē-um): the vegetative body of certain complex fungi; it consists of filaments or hyphae.

mycoplasma (mī′kō-plaz′mah): pleuropneumonialike organisms, the smallest known cells; these have been identified as the cause of respiratory disease in man and animals but are also found as free-living organisms. (Also called PPLO.)

myelin sheath (mī′e-lin): the fatty cell membranes surrounding the axon of some nerve cells.

myeloma (mī′e-lō′mah): a common but comparatively mild form of cancer.

myofibril (mī′o-fī′bril): any of the slender, protein threads in the skeletal muscle cell that are responsible for its contractibility.

myofilament (mī′o-fil′ah-ment): any of the filaments that make up the myofibril in the skeletal muscle cell.

myogenic (mī′o-jen′ik): of muscular origin; also used to describe spontaneous muscular activity in living tissues.

myosin (mī′o-sin): a globular protein in muscle; along with actin, responsible for the contraction and relaxation of muscle.

myotonia (mī′o-tō′nē-ah): an increase in muscle tension.

n

NADP or NADPH: see nicotinamide adenine dinucleotide phosphate.

narcotic (nar-kot′ik): in medicine, any agent that produces narcosis (stupor and insensibility); legally, other painkilling drugs such as cocaine are also considered narcotics.

natality (nā-tal′i-tē): the rate at which births occur in a population.

natural selection: a process of evolution resulting in the continuation of only those forms of plant and animal life that are best adapted to reproduction and survival in the environment.

nematocyst (ne-mat′o-sist): a minute stinging structure with which certain coelenterates, such as jellyfish, inject a paralyzing poison into their prey or enemies.

nephron (nef′ron): one of many microscopic, cellular units that make up the mammalian kidney; includes a network of blood capillaries and a renal tubule.

nerve impulse: an electrochemical signal created in the neurons and transmitted to other cells.

nerve net: a group of neurons, usually scattered over a surface, whose dendrites and axons cross and intermingle in a netlike fashion; for example, the nervous system of coelenterates.

nerve plexus: a system of intermingled nerve fibers (nerve net) and bundles or tracts; occurs in some higher invertebrates, such as the starfish.

neural crest cell: one of the cells that migrate from the neural tube to form a variety of organs and tissues in the developing vertebrate embryo.

neural groove: the ingrowth of the neural plate of a vertebrate embryo; its edges eventually fuse to form the neural tube.

neural plate: a thick plate, formed by rapid cell division, on the dorsal side of the vertebrate embryo; it folds into the embryo to form the neural groove.

neural tube: the embryonic tube that forms the brain and spinal cord of a vertebrate.

neurofibril (nū′ro-fī′bril): one of the delicate threads or microtubules running in every direction through the cytoplasm of a nerve cell.

neurogenic (nū′ro-gen-ik): of nervous origin; also used to describe a muscular contraction stimulated by the nervous system.

neuroglia (nū-rog′lē-ah): see glial cell.

neuron (nū′ron): a nerve cell, the structural unit of the nervous system.

neuropil (nū′ro-pīl): a synaptic area in the interior of a nerve ganglion.

neurosecretion (nū′ro-se-krē′shun): secretion by nerve cells.

neurosecretory cells: neurons that act as effectors by secreting hormones into the bloodstream.

neutron (nū′tron): an elementary particle that is a fundamental constituent of the nucleus in all atoms except hydrogen; it has nearly the same mass as the proton but carries no electrical charge.

neutrophilic granulocyte (nū′tro-fil′ik gran′yah-lō-sīt): a variety of white blood cell containing granules that are neutrophilic, or readily stainable by neutral dyes. (Also called neutrophil.)

niche (nich): the distinctive way in which organisms of a species use the resources of their habitat or environment; where resources are limited, no two species can long occupy the same niche in the same habitat.

nicotinamide adenine dinucleotide phosphate: a coenzyme that functions in oxidation and reduction reactions (hydrogen transfer). (Abbreviated: NADP or NADPH)

Nissl body (nis′′l): any of the nucleoprotein granules occurring in the cytoplasm of nerve cells.

nitrogenous base: a nitrogen-containing molecule having the properties of a base (the tendency to acquire a hydrogen atom); for example, the purines and pyrimidines.

nocturnal animal (nok-ter′nal): an animal that sleeps during the day and is active at night.

node of Ranvier (rahn′-ve-ā′): a depression occurring between each pair of successive Schwann cells comprising the myelin sheath of neurons.

noncompetitive inhibition: reduction in the rate of enzyme reaction occurring when the inhibiting molecule becomes tightly attached to the enzyme's active site or permanently alters it.

nonpolar molecule: a molecule in which the electrical potential is symmetrically distributed; for example, a neutral fat.

norepinephrine (nor′ep-i-nef′rin): a hormone secreted by the adrenal medulla; one of the chemical transmitters functioning at the excitatory synapse.

notochord (nō′tah-kord): a rodlike cord of cells forming the chief axial structure of the chordate body; in mammals, it is present only in the embryonic stage and is supplanted by the vertebral column.

nuclear membrane: the outer membrane of the cell nucleus.

nuclear pore: an apparent opening in the nuclear membrane of a eucaryotic cell.

nuclease (nū′klē-ās): any of several enzymes that split nucleic acids (DNA or RNA) into basic units (nucleotides).

nucleic acid (nū-klē′ik): any of a group of acids that carry genetic information; the two kinds are RNA and DNA.

nucleolar organizer: a small piece of chromosomal DNA that lies within the nucleolus and contributes to the formation of the nucleolus itself and of ribosomal RNA.

nucleolonema (nū′klē-ō′lo-nē′mah): a network of fibers formed by the organization of granules in some nucleoli. (Plural: nucleolonemata.)

nucleolus (nū-klē′o-lus): an organelle within the nucleus of a cell; it contains large amounts of RNA and protein and is the site of ribosomal RNA synthesis.

nucleoside (nū′klē-o-sīd): a type of compound consisting of the sugars ribose or deoxyribose and the nitrogenous bases adenine, guanine, cytosine, and thymine or uracil.

nucleotide (nū′klē-o-tīd): a molecule consisting of phosphate, a 5-carbon sugar, and a purine or pyrimidine base; the chemical unit comprising DNA and RNA.

nucleus (nū′klē-us): a central body or core; in a cell, an organelle consisting of a number of characteristic structures surrounded by a nuclear membrane. In an atom, the positively charged mass composed of neutrons and protons occupying only a small fraction of the volume but possessing most of the mass.

nurse cell: a cell, found in the ovary, that helps to nourish the developing egg cell.

O

occipital lobe (ok-sip′i-tal): one of the four lobes dividing each hemisphere of the human brain; it is involved in the reception and processing of input from the visual system.

oleo-resin (ō′lē-ō-rez′in): a mixture of an essential oil and a resin; it occurs in nature, for example, as a substance secreted by conifers to repel insect pests.

oligodendrocyte (ol′i-gō-den′dro-sīt): one of the cells forming the sheath around axons in the central nervous system.

oligodendroglia (ol′i-gō-den-drog′lē-ah): cells instrumental in the formation of myelin in the central nervous system; these cells are present in the myelin sheath and help to convey impulses along the axons.

omnivore (om′nah-vōr): an animal that eats both plants and other animals.

oncogenic virus (on′ko-jen′ik vī′rus): a tumor-forming virus; instead of killing the animal cell, it transforms the cell that it enters.

one gene–one enzyme hypothesis: the postulate that each enzyme is produced under the direction of a single gene.

one gene–one polypeptide hypothesis: the postulate that the sequence of base pairs in the DNA molecule directs the assembly of a polypeptide chain with a particular sequence of amino acids.

oocyte (ō′o-sīt): a plant cell, formed by division of the oogonium, that undergoes meiosis to form four haploid ootids.

oogamy (ō-og′ah-mē): reproduction through the union of nonidentical gametes–that of the male being small and motile, and the female gamete being larger and nonmotile.

oogenesis (ō′o-jen′e-sis): the process of egg formation from sex cells in plants or animals.

oogonium (ō′o-gō′nē-um): the primary plant cell from which the ovarian egg arises; it divides twice to form four oocytes. (Plural: oogonia.)

oospore (ō′ah-spōr): a fertilized egg within an oogonium in algae and fungi.

ootid (ō′o-tid): one of four plant cells formed from the oocyte by meiosis; one of the four develops into an ovum, while the other three, the polar cells, disappear.

operant conditioning: a form of behavior training in which an animal learns to adopt responses that are suited to the events that follow, whether reward or punishment.

operator (op′ah-rā′ter): a chromosomal region capable of interacting with a specific repressor to control the function of an adjacent operon.

operon (op′er-on): a block of two or more adjacent genes controlled by the same regulatory system.

opsin (op′sin): the protein portion of the visual pigment rhodopsin found in the retina; the protein is released from the chromophore retinene during the bleaching process in the eye.

optic chiasma (kī-as′mah): the junction of the two optic nerves leading from the retina of the eye to the brain.

orbital (or′bi-tal): the approximate path on which an electron moves around the nucleus.

order: the major subdivision of a class or subclass in the classification of plants or animals; it usually consists of several families.

organ of Corti: the organ that contains special sensory receptors, the hair cells, that function in hearing.

organelle (or′gan-el′): a specialized structure having a definite function in a cell; for example, the nucleus, the mitochondrion, the Golgi apparatus, or the ribosome.

organic (or-gan′ik): pertaining to carbon-containing substances that may be derived from living organisms.

organic acid: an organic compound that contains a carboxyl group (—COOH).

orgasm (or′gazm): the physical and emotional sensations experienced at the climax of a sexual act; in the male, it results in ejaculation.

orientation (or′ē-en-tā′shun): the voluntary location of an organism with respect to some reference point in its environment.

oroticaciduria (or-ō′tik-a′si-du′rē-ah): a rare congenital disease caused by enzyme deficiency and characterized by retarded growth and anemia.

osmoreceptor (oz′mō-rē-cep′tor): a specialized sensory nerve ending that is stimulated by changes in osmotic pressure of the surrounding medium.

osmosis (os-mō′sis): the passage of pure solvent from a lesser to a greater concentration of solute through a membrane separating two solutions; the membrane is permeable to the solvent but impermeable to the solute.

ossification (os′i-fi-kā′shun): the formation of bone.

ostium (os′tē-um): an orifice; in anatomy, the opening of a tubular structure or passage between two cavities. (Plural: ostia.)

ovary (ō′vah-rē): the female sexual gland in which the ova are formed.

oviduct (ō′vi-dukt): the tube connecting the ovary and the uterus of higher animals.

ovulation (ōv′u-lā′shun): the discharge of a mature egg from the ovary.

ovulatory phase: the period of the menstrual cycle during which the ovum is released from the follicle (ovulation).

ovule (ō′vyūl): the egg cell of a plant.

ovum (ō′vum): the female reproductive cell. (Plural: ova.)

oxidation (ok′si-dā′shun): the loss of one or more electrons in a chemical reaction; combination with elemental oxygen.

oxidation number: a number equal to the net charge on an atom or an ion; in molecules, the charge that the atom would bear if the electrons were completely held by the nucleus that attracts them most strongly.

oxidation-reduction reaction: a chemical reaction in which oxidation numbers of atoms or molecules are changed. (Also called redox reaction.)

oxidizing agent: a molecule or an atom, such as oxygen, that accepts electrons from an element in the process of oxidation. (Also called the oxidant.)

pacemaker (pās′māk-er): a small spot of neuromuscular tissue that generates a sequence of impulses to govern heartbeat.

pachytene (pak′e-tēn): in the prophase of meiosis, the stage following synapsis in which the homologous chromosome threads shorten, thicken, and exchange segments.

palisade mesophyll: several layers of photosynthetic cells lying just beneath the upper epidermis of the leaf in higher plants.

pancreas (pan′krē-as): a gland, located behind the mammalian stomach, that secretes a variety of digestive enzymes into the duodenum.

pandemic (pan-dem′ik): a widespread epidemic disease.

pangenesis (pan-jen′e-sis): Darwin's comprehensive concept of heredity; the hypothesis that every atom or unit of the whole organism reproduces itself by means of pangenes or hypothetical units of heredity.

panmictic population (pan-mic′tic): a population in which mating occurs in a random manner; any sperm cell is likely to be combined with any egg cell in the population.

paradoxical sleep: the level of sleep in which dreams and rapid eye movements (REM) occur; it recurs four to six times during the night.

parasite (par′ah-sīt): an organism that lives on or in another organism and depends upon the host for its food.

parasympathetic nervous system: one of the two divisions of the autonomic nervous system; it acts to conserve energy by slowing down the body's processes and counteracts the sympathetic system.

parenchyma (par-eng′ki-mah): the fundamental tissue of plants; it is composed of nonspecialized, thin-wall cells that support the plant body and store water and nutrients.

parietal lobe (pah-rī'e-tal): one of the four lobes dividing the hemispheres of the human brain; it is involved with skin senses and detection of bodily position (proprioception).

parthenogenesis (par'the-nō-jen'e-sis): unisexual reproduction, as occurs in certain lower animals; reproduction by the development of an egg not fertilized by a spermatozoan.

passive transport: movement of molecules through the cell membrane by natural diffusion—from the side of higher concentration toward the side of lower concentration.

Pavlovian conditioning: see classical conditioning.

pecking order: see dominance hierarchy.

pedophilia (pē'do-fil'ē-ah): sexual perversion toward children.

pellicle (pel'i-k'l): a thin skin or film.

pelvis (pel'vis): a cavity or region in the lower trunk of vertebrates, or the bony girdle enclosing the cavity.

penis (pē'nis): the male sex organ.

pentylenetetrazol (pen'ti-lēn-tet'rah-zol): a drug that stimulates the central nervous system and has been used to enhance learning.

pepsin (pep'sin): an enzyme that aids in the digestion of proteins; secreted by cells in the stomach lining.

peptide (pep'tīd): a compound containing two or more amino acids in which the carboxyl group of one acid is linked to the amino group of the other.

peptide linkage: a covalent bond in which the nitrogen in the amino group of one amino acid is attached directly to the carbon in the carboxyl group of another amino acid.

perianth (per'ē-anth): the corolla and the calyx of a flower.

periderm (per'i-durm): the outermost surface or bark of adult trees; it is composed of cork cells.

periodic table: a chart on which the elements are listed in order of increasing atomic number; each vertical column includes elements of similar chemical properties.

peripheral nervous system: one of the two traditional divisions of the vertebrate nervous system; it includes all nerve processes and neurons outside the central nervous system. (Abbreviated: PNS)

peristalsis (per'i-stal'sis): rhythmic contractions of muscles in the walls of the gut; causes movement of food through the digestive tract.

peroxidase (per-ok'si-dās): an enzyme that causes peroxide to oxidize certain organic compounds in plants; it prevents the accumulation of excess indoleacetic acid.

petiole (pet'ē-ōl): a leafstalk; the slender stalk by which a leaf is attached to a stem.

phage (fāj): see bacteriophage.

phagocyte (fag'o-sīt): any amoeboid cell that ingests microorganisms or other cells and foreign particles.

phagocytosis (fag'o-sī-tō'sis): the process by which a cell ingests microorganisms, other cells, or foreign particles.

pharynx (far'inks): the part of the alimentary tract between the mouth cavity and the esophagus; in mammals, it is also part of the air channel from nose to larynx.

phenotype (fē'no-tīp): the observable characteristics that depend on genetic and environmental factors of an organism.

phenotype frequency: a mathematical expression of the proportion of one phenotype to other phenotypes in the same population.

phenotypic restriction: a theoretical process in which a cell "chooses" to produce only one of the possible immunoglobulins coded in its genes.

phenylalanine (fen'il-al'ah-nēn): a water-soluble amino acid obtained from egg white or skim milk; it is essential for growth in infants and for nitrogen equilibrium in human adults.

phenylketonuria (fen'il-kē'to-nū'rē-ah): a congenital disease involving faulty metabolism of phenylalanine producing mental retardation; a naturally occurring amino acid. (Abbreviated: PKU)

pheromone (fer'ah-mōn): any substance secreted by an organism that stimulates a physiological or behavioral response in another individual of the same species.

phloem (flō'em): one of the vascular tissues in plants; it consists of sieve tubes and companion cells and transports nutrients both up and down.

phosphocreatine (fos'fō-krē'ah-tēn): a compound, found chiefly in muscle, formed by the enzymatic interaction of an organic phosphate and creatine; it breaks down to provide energy for muscle contraction.

phospholipid (fos'fō-lip'id): a lipid containing phosphorus in addition to fatty acids, glycerin, and a nitrogenous base; an important constituent of the cell membrane.

phosphorylation (fos'for-i-lā'shun): introduction of the trivalent phosphate group into an organic molecule.

photolysis (fō-tol'i-sis): chemical decomposition by the action of light.

photon (fō'ton): a unit or quantum of electromagnetic radiation.

photoperiodism (fō'tō-pe'rē-ah-dizm): any rhythmic response of plants or animals to varying periods of light and darkness.

photophil (fō'to-fil): a period of about 12 hours during which light enhances the flowering process in plants.

photosynthesis (fō'to-sin'the-sis): the biological process in which plants transform carbon dioxide and water into carbohydrates by using solar energy absorbed by chlorophyll.

phototropism (fō'to-trō'pizm): the influence of light upon growth; for example, the tendency of many organisms to grow toward or away from light.

pH scale: a scale, varying from 1 to 14, used to specify the acidic or basic character of a solution.

phycobilin (fī'kō-bi'lin): the group of compounds that includes the orange-red and blue pigments in red and blue-green algae.

phycocyanin (fī'kō-sī'ah-nin): a blue pigment found in blue-green algae; with chlorophyll, it contributes to the process of photosynthesis.

phyllotactic (fil'ah-tak'tic): pertaining to the arrangement of leaves on a stem.

phylum (fī'lum): the major primary subdivision of the plant or animal kingdom; it consists of one or more related classes. (Plural: phyla.)

phytochrome (fī'tō-krōm): a plant pigment consisting of protein and a chromophore; its two forms function mainly in the influence of light on flowering.

pigment (pig'ment): any colored substance; a compound that absorbs some wavelengths of visible light and reflects others.

pinocytosis (pi'no-si-tō'sis): a process by which cells engulf liquids.

pistil (pis'til): the seed-bearing organ of a flower; when complete, it consists of ovary, style, and stigma.

pituitary gland (pi-tū'i-tar'ē): a small, oval endrocrine gland attached to the base of the brain; it secretes a large number of important hormones.

placenta (plah-sen'tah): a mass of blood vessels and membranes in the uterus of pregnant mammals; it functions in the exchange of nutrients and waste products between mother and embryo (or fetus).

plankton (plangk'ton): a collective term for the organisms, large and microscopic, that float or drift at random in a body of water.

plasma (plaz'mah): the fluid portion of the blood in which blood cells are suspended.

plasma cell: one of the cells, derived from lymphocytes, found in chronically inflamed connective tissue; it secretes protein.

plasma membrane: the outer layer or membrane of a cell.

plasmodium (plaz-mō'dē-um): a multinucleate mass of protoplasm formed by cell division without the formation of new cell walls in slime molds.

plastid (plas'tid): a specialized organelle of a plant cell; various plastids determine coloration, synthesize carbohydrates, or convert glucose to starch, lipids, or proteins.

platelet (plāt'let): a thrombocyte or disclike cell fragment occurring in large numbers in the blood of mammals; it aids in coagulation.

polar cell: a small, impotent plant cell formed, along with the ovum, from division of the oocyte.

polarity (pō-lar'i-tē): the exhibition of opposite or unequal effects at different points in a body or system; for example, the unequal distribution of electrical charges on a molecule.

polarization (pō'lar-i-zā'shun): an electrical potential or uneven distribution of chemical ions or charges; for example, the inside of the glial cell has a negative potential with respect to the outside. (See resting potential.)

polar molecule: a molecule in which the electrical potential is not symmetrically distributed.

pollen (pol'en): the microspores, or male fertilizing elements, of flowering plants.

polymer (pol'i-mer): a compound, usually of high molecular weight, formed by the linking of simpler molecules or monomers.

polynucleotide (pol'ē-nū'klē-o-tīd): a linear sequence of nucleotides in which the sugars are linked through phosphate groups.

polyp (pol'ip): the attached, hydralike form of various coelenterates consisting of a fleshy stalk with tentacles; it occurs singly or as part of a colony.

polypeptide (pol'ē-pep'tīd): a polymer of two or more amino acids (peptides); a protein of low molecular weight or portion of a protein.

polyphagous (pah-lif'ah-gus): pertaining to an organism that eats many kinds of food.

polyploidy (pol'ē-ploi'dē): see euploidy.

polysaccharide (pol'ē-sak'ah-rīd): a complex carbohydrate molecule; a polymer of sugars, including starches, glycogen, and cellulose.

polysome (pol'ē-sōm): clusters or rows of several ribosomes attached to a single mRNA molecule and found within the cytoplasm of a cell.

polytene (pol'ē-tēn): a giant chromosome containing many strands of chromatin.

pons (ponz): an extension of the medulla; it contains certain neurons controlling facial movements, hearing, and respiration, and others relating to the spinal motor nerves.

postsynaptic cell: the neuron that receives a nerve impulse after it crosses the synapse.

postsynaptic excitation: a stimulation of the postsynaptic cell caused by release of chemical transmitter substances into the synaptic cleft; it leads to the production of an action potential in the postsynaptic neuron.

postsynaptic inhibition: a suppression of the postsynaptic cell that occurs when the movement of chemical transmitter substances across the synapse inhibits rather than stimulates the discharge of the postsynaptic cell.

PPLO: see mycoplasma.

presynaptic cell: the neuron that transmits a nerve impulse across the synapse.

presynaptic inhibition: a condition existing in axoaxonic synapses when a substance released by one axon inhibits the release of transmitter substances by the other axon.

presynaptic vesicle: a microscopic sac in the terminal branches of the axon; it discharges a chemical substance into the synaptic cleft in response to a nerve impulse.

primary consumers: animals (herbivores) that feed upon the tissues of producers (green plants).

primary elongation meristem: an active region just below the apex of the plant stem where cell divisions responsible for stem elongation occur.

primary endosperm nucleus: a triploid nucleus formed in the flowering plant embryo sac by fusion of one sperm nucleus with two egg nuclei; it divides to form the nuclei of the endosperm.

primary immune response: the natural reaction of living tissues to foreign organisms or materials; it involves mobilization of special cells and chemical substances for the purpose of destroying the foreign agents.

primitive streak: an elongated mass of cells forming the main body axis in early embryos.

primordium (prī-mor'dē-um): the first discernible indication of an organ or structure in embryonic development. (Plural: primordia.)

procambium (pro-kam'bē-um): the meristem from which the vascular tissues of a plant develop.

procaryotic (prō-kar'ē-o'tic): lacking a true nucleus; said of cells in which the nuclear materials are found within the cytoplasm.

producer (prō-dū'ser): an organism utilizing energy from the nonliving environment and producing complex organic molecules from inorganic substances; for example, photosynthetic plants.

progesterone (prō-jes'ter-ōn): the hormone produced by the corpus luteum; it prepares the uterus for the reception and development of the fertilized ovum.

proliferative phase: the period of the menstrual cycle during which glands, blood vessels, and surface epithelial cells of the endometrium are regenerated.

proline (prō'lēn): an alcohol-soluble amino acid found in all proteins.

promoter (pro-mō'ter): the region of a gene (DNA strand) to which RNA polymerase attaches; it regulates the frequency with which the polymerase transcribes the genetic code in RNA synthesis.

pronephros (prō-nef'ros): the first embryonic kidney, nonfunctional in mammals after birth. (Plural: pronephroi.)

prophage (prō'fāj): the transitory union of viral and bacterial DNA that precedes cell lysis.

prophase (prō'fāz): the first stage of mitosis, including all the processes preceding the separation of chromosomes.

prophase I: in meiosis, the stage in which the chromosomes are shortened, the nucleoli disappear, and the nuclear membrane disintegrates.

prophase II: in meiosis, the relatively brief period marked by condensation of the two sets of homologous chromosomes (resulting from the first meiotic division) and formation of the second spindle.

proplastid (prō-plas'tid): a simple plastid thought to be the parent structure from which all different plastid types develop; also, a body present in the cells of plants grown in the dark that changes to a chloroplast if the plant is placed in the light.

prostate gland (pros'tāt): a gland surrounding the neck of the bladder and the urethra in the male; it secretes most of the seminal fluid.

protective response: any form of behavior that tends to minimize damage from an injury; for example, blinking or limping.

protein (prō'tēn): any one of a group of complex organic nitrogenous compounds widely distributed in plants and animals and required for all life processes; a polymer of amino acids.

protein conformation: the three-dimensional arrangement of certain protein molecules.

proteolytic (pro'tē-o-lit'ik): pertaining to proteolysis; the hydrolysis of proteins into simpler compounds by means of enzymes.

protohormone (prō'to-hor'mōn): a small molecule, found in microorganisms, thought to be the evolutionary precursor of the hormone; these molecules have regulatory functions but do not play a role in metabolic pathways.

proton (prō'ton): an elementary particle that is a fundamental constituent of all atomic nuclei; it has a positive charge equal in magnitude to that of the negative electron.

protonema (prō'tah-nē'mah): a small haploid filament produced by germination of the spore in mosses and related plants; the leafy plant arises from this structure.

protoplasm (prō'to-plazm): living matter; the essential material of all plant and animal cells.

pseudoplasmodium (sū'dō-plaz-mō'dē-um): a multicellular structure formed by the union of zygotes of slime molds; it develops into a multicellular sporangium.

pseudopodium (sū'dō-pō'dē-um): a temporary projection from the membrane of a cell; it is involved in both feeding and amoeboid motion. (Plural: pseudopodia.)

psychedelic (sī'ki-del'ik): characterized by relaxation, freedom from anxiety, and highly imaginative thought patterns.

psychotomimetic (sī-kot'ō-mi-met'ik): producing manifestations resembling those of psychosis.

purine (pyur'ēn): a type of bicyclic (two ring) nitrogenous base, such as adenine and guanine; constituent of DNA and RNA.

pus (pus): a yellowish semifluid composed of the products of cell breakdown and secretion resulting from inflammation.

pyrenoid (pi're-noid): a proteinoid body within the pigmented cells of algae; it functions in starch formation.

pyrimidine (pih-rim'e-dēn): a type of monocyclic (one ring) nitrogenous base, such as thymine, uracil, and cytosine; constituent of DNA and RNA.

pyruvate (pī'rū-vāt): a salt or ester of pyruvic acid; an intermediate stage in the glycolysis (catabolism) of sugars.

quantum (kwon'tum): a fundamental unit of radiant energy as defined by the quantum theory. (Plural: quanta.)

quantum mechanics: the branch of mechanics that is applicable to systems at the atomic and nuclear level.

r

radiant energy: energy transmitted by electromagnetic waves; also called light energy.

radicle (rad'ik'l): the primary root of a plant.

radioactive label: a radioactive atom introduced into a molecule to facilitate observation of its metabolic transformations.

radioactivity (rā'dē-ō-ak-tiv'i-tē): the phenomenon, exhibited by certain elements, of emitting corpuscular or electromagnetic radiations as a result of changes in the nuclei of atoms.

radula (raj'ū-lah): a flexible, tongue or filelike structure in the anterior part of the digestive tract of most molluscs; it is used to chew or, in some groups, is protruded to file surface material from rocks or plants.

rapid eye movements: eye movements that usually occur when a sleeper is dreaming. (Abbreviated: REM)

reactive group: a grouping of atoms that appears in many different kinds of molecules and has characteristic structure and activity; for example, the hydroxyl group (—OH) or the carboxyl group (—COOH).

receptive field: the area of the retina, with its variety of receptor cells, that affects the response of a ganglion cell; in general, the complete set of stimuli that affect the firing rate of a nerve cell under any specified condition.

receptor (re-sep'tor): a sensory nerve ending that responds to the internal state and external stimuli of an organism.

recessive character: a genetic trait that develops in an organism only when both genes determining the trait are present.

recombination (rē-kom-bi-nā'shun): the result of independent segregation and assortment of genes on separate chromosomes during meiosis. (See independent assortment.)

rectum (rek'tum): a short section at the end of the large intestine where waste material is stored prior to elimination.

recurrent inhibition: a type of "feedback" inhibition in which a nerve cell activates, and is then inhibited by, a second neuron via a return pathway.

red blood corpuscle: an anucleate blood cell containing hemoglobin and adapted for the transport of oxygen.

redox reaction: see oxidation-reduction reaction.

reducing agent: a molecule that donates electrons to another molecule (the oxidant) in a chemical reaction; also called the reductant.

reduction (rē-duk'shun): in chemistry, the gain of electrons by a compound or atom in a chemical reaction.

reductionism (rē-duk'shun-izm): the theory that living and nonliving systems obey the same chemical and physical principles; it implies that all biological processes can be reduced to terms of chemical and physical processes.

reflex (rē'fleks): an involuntary response to a stimulus; for example, the eye blink, knee jerk, or sneeze.

reflex arc: a two-cell pathway in the nervous system that produces the simplest behavioral responses of vertebrates; a sensory neuron conveys an impulse from a receptor at the periphery to a motor neuron in the cord.

reflex ovulator: any species in which ovulation does not normally occur until provoked by the act of coitus or by some intense psychic stimulus. (See spontaneous ovulator.)

refractory period: a brief recovery period following a nerve impulse, during which the nerve cannot generate a new impulse (absolute refractory period) and then gradually returns to normal (relative refractory period).

regeneration (rē-jen'er-ā'shun): the natural replacement of a lost tissue or part; it occurs to a limited extent among higher animals.

reinforcement (rē'in-fors'ment): in behavior conditioning, the experimental procedure of immediately following a response with a reinforcer, such as a food reward.

relative refractory period: see refractory period.

renal (rē'nal): pertaining to the kidney.

replication (re'pli-kā'shun): in genetics, the duplication of hereditary material (complementary chromosomes) that takes place in meiosis.

repressible enzyme: an enzyme whose rate of activity is decreased when the intracellular concentration of certain metabolites increases.

repressor (rē-pres'or): a compound that inhibits the synthesis of a specific enzyme by a gene.

respiration (res'pi-rā'shun): the process of breathing or taking in oxygen; also, the liberation of metabolically useful energy from fuel molecules within cells.

respondent conditioning: see classical conditioning.

responder cell: a cell whose metabolism is affected by a chemical effector.

response generalization: the act or process of responding to a stimulus similar to but distinct from the conditioned stimulus.

resting potential: a polarization or electrical difference between the inside and outside of inactive cells in excitable tissues such as nerve or muscle; potential depends on the relative concentrations of potassium ions.

resting spore: a resistant reproductive body or cell that remains dormant during adverse environmental conditions; it germinates when conditions are favorable.

reticular formation (re-tik'u-lar): a network of cells extending from the upper spinal cord, through the medulla and pons; it is believed to be an important center for the regulation of activation and alertness.

reticular lamina: a thin layer of tissue in which the hair cells are embedded within the organ of Corti.

retina (ret'i-nah): a layer of light-sensitive cells, lining the rear of the eyeball, on which images are focused; it contains the rods and cones.

retinal cones: pigmented receptor cells in the eye near the back of the retina; cones function in bright light and color vision.

retinal rods: pigmented receptor cells in the eye near the back of the retina; rods function in dim light vision but not in color discrimination.

retinene (ret'i-nēn): a yellow derivative of vitamin A and the chromophore of visual pigments; for example, rhodopsin consists of retinene (or retinal) combined with the colorless protein opsin.

Rh factor: a type of antigen (protein) in the blood.

rhizoid (rī'zoid): a rootlike extension that attaches certain plants, including yellow-green algae, to the substrate.

rhizome (rī'zōm): a rootlike underground stem that produces shoots above ground and roots below.

rhodopsin (rō-dop'sin): one of the four pigments contained in human visual receptors; the sole pigment of retinal rods.

ribonucleic acid (rī'bō-nū-klē ik): any of the nucleic acids containing ribose; found chiefly in the cytoplasm of cells. (Abbreviated: RNA)

ribonucleotide (rī'bō-nū'klē-ō-tīd): a compound consisting of the sugar ribose, a purine or pyrimidine base, and a phosphate group; a constituent of RNA.

ribosomal RNA (rī'bo-sōm'al): a type of RNA molecule found in the ribosomes that functions in protein synthesis. (Abbreviated: rRNA)

ribosome (ri'bo-sōm): any of several minute granular particles composed of protein and RNA and found in the cytoplasm of a cell.

ribulose (rī'byū-lōs): a five-carbon sugar found in green plants; important in photosynthesis. (See ribulose diphosphate.)

ribulose diphosphate: a diphosphate ester and activated form of ribulose that functions as the carbon dioxide acceptor in photosynthesis.

RNA: see ribonucleic acid.

RNA polymerase (po-lim'er-as): one of the enzymes that catalyze transcription of the genetic code from DNA to RNA.

root cap: a hollow cap of cells covering the growth tip of a root and protecting its meristematic cells from damage as the tip pushes through the soil.

ruffled membrane: the undulating "front" of a moving amoeboid cell.

rumen (rū'min): a large organ in grass-eating animals where food is stored and fermented and from which food is regurgitated and rechewed before being passed along to the stomach.

S

saliva (sah-lī'vah): a secretion from the glands of the mouth; it moistens and softens food and contains an enzyme that starts the digestion of starches.

salt: the ionic product of the reaction between a base and an acid.

sarcolemma (sar'kō-lem'ah): the membrane that surrounds every striated muscle fiber.

sarcoplasmic reticulum: a complex membranous structure found between the myofibrils in skeletal muscle.

satellite virus: a very small virus that can infect only a cell also infected by a larger virus.

scale leaf: a leaf that forms a protective wrapping around a bud or an underground stem.

Schwann cell: one of the cells forming the sheath around peripheral nerve fibers.

scotophil (skō'tō-fil): a period of about 12 hours during which light inhibits the flowering process in plants.

scrotum (skrō'tum): the sac that contains the testes.

sebaceous (se-bā'shus): of or pertaining to fatty or greasy substances; for example, sebaceous glands in the skin produce oil.

secondary consumers: animals (carnivores) that feed upon the tissues of primary consumers (herbivores).

secondary immune response: a resistance to further infection acquired by tissues following an infectious disease or exposure to foreign agents.

secondary sexual characteristic: physical development that occurs at puberty and distinguishes one sex from the other but is not essential to reproduction.

second filial generation: all of the offspring produced by the mating of two individuals of the first filial generation. (Symbol: F_2)

second ionization energy: the amount of energy needed to remove a second electron from an atom. (Symbol: I_2)

second polar body: a very small cell pinched off a large egg cell in the second meiotic division of the oocyte; it disintegrates.

secretion (se-krē'shun): a substance that is elaborated and released by a cell or gland; for example, any substance other than waste expelled by a cell.

segregation (seg're-gā'shun): a genetic principle; when a hybrid reproduces, half its cells transmit the dominant character of one parent and the other half, the recessive character of the other parent.

selective breeding: the improvement or development of breeds by selective mating and hybridization.

selection coefficient: a numerical method of representing the fitness of a genotype.

selective theory: any of several theories holding that the capability for producing a particular antibody is naturally present in some immunocyte in the body and that invasion by an antigen causes selective reproduction of that cell.

self-fertilization (self' fer-til-i-zā'shun): the process by which pollen of a given plant fertilizes the ovules of the same plant.

semen (sē'men): the thick, whitish, sperm-bearing secretion of the male reproductive organs.

semiconservative replication: the process by which one of the two strands of DNA in a given gene is copied by each daughter molecule.

seminal duct (sem'i-nal): the passage for conveyance of semen and spermatozoa from the epididymus to the ejaculatory duct. (Also called vas deferens.)

seminal vesicle: an enlarged portion of the male reproductive duct in which sperm are stored prior to copulation.

senescence (se-nes'ens): gradual deterioration of function and structure caused by aging.

sensory neuron: any neuron that receives and conveys sensory nerve impulses.

sepal (sē'pal): one of the individual leaves of the calyx of a flower.

serotonin (ser'o-tō'nin): a compound found in the brain, intestines, and blood platelets; it is used as an experimental agent to alter the states of sleep.

Sertoli cell: an elongated nurse cell in the tubules of the testes; maturing spermatozoa cluster around these cells.

sessile (ses'il): pertaining to organisms that are permanently attached; for example, corals or higher plants.

sex chromosomes: the chromosomes associated with the determination of sex; in mammals, the pairs of chromosomes designated XY (male) and XX (female).

shaped behavior: a pattern of response instilled by training that reinforces successive steps leading to the desired final response.

sickle cell anemia: a hereditary disease caused by a defective form of hemoglobin and characterized by sickle-shape red blood cells.

sieve cell: a specialized, thin-wall cell through which fluids move in the phloem of the plant body.

sieve tube: a tube formed in the phloem of the plant body by the joining end-to-end of the sieve cells.

sigma factor: a control substance that causes the enzyme polymerase to attach to certain sites on the DNA of a virus or a bacterium.

single bond: a chemical bond in which one pair of electrons is shared by two atoms.

slow wave sleep: a level of sleep devoid of rapid eye movements (REM) and dreams; such periods alternate with periods of paradoxical sleep. (Abbreviated: SWS)

social behavior: any kind of behavior that involves communication or cooperation between organisms.

social releaser: the natural behavior of an organism that influences other animals in a social situation.

solar energy: radiant energy derived from the sun and consisting of visible and invisible rays.

solute (sol'yūt): a substance dissolved in a solvent to form a solution.

solvent (sol'vent): a fluid that dissolves or is capable of dissolving solutes.

soma (sō'mah): a body; for example, the body of a cell. (Plural: somata.)

somite (sō'mīt): one of the paired muscle blocks arranged along the neural tube of the embryo and giving rise to voluntary muscles, most of the skeleton, and the skin.

sorus (sōr'us): a cluster of sporangia on the back of a fern leaf. (Plural: sori.)

species (spē'shēz): the major subdivision of a genus or subgenus; it is composed of related individuals that resemble each other and are able to breed among themselves but usually not with other species.

specific patterned sequences: sequences of behavior that tend to recur at appropriate times in the animal's life; the pattern can include both learned and innate behavior and may be modified under varying circumstances.

spermatheca (spur-mah-thē'ka): the seminal vesicle of certain arthropods.

spermatid (sper'mah-tid): a cell derived from a spermatocyte by meiosis; it develops into a spermatozoan.

spermatocyte (sper'mah-to-sīt): a cell of the next to the last generation of cells that divide to form sperm.

spermatogenesis (sper-ma'to-jen'e-sis): the process of sperm formation from sex cells in plants or animals.

spermatogonium (sper'mah-to-gō'nē-um): a male germ cell originating in a seminal tubule and dividing twice to form four spermatocytes. (Plural: spermatogonia.)

spermatozoan (sper'mah-to-zō'on): a motile sperm cell. (Plural: spermatozoa.)

S-phase: period of DNA synthesis in a cell; it follows G_1 period and procedes G_2 period.

sphincter (sfingk'ter): a ringlike band of muscle that opens and closes a passage.

spinal cord: the bundle of neurons running through the spinal column of vertebrates; a relatively simple neural system that mediates between receptors and effectors (or the brain).

spindle (spin'd'l): a structure formed in a cell nucleus during mitosis and meiosis; it consists of fine "threads" radiating out from the centrosomes and connecting them.

spirillum (spi-ril'um): a corkscrew-shape bacterium. (Plural: spirilla.)

spleen (splēn): a highly vascular, glandular organ, situated in man at the cardiac end of the stomach; it functions in the formation of lymphocytes, in the destruction of worn-out erythrocytes, and as a reservoir for blood.

spongy mesophyll: a layer of cells forming the lower portion of a leaf (below the palisade cells) and comprising the leaf's internal aeration system.

spontaneous generation: the early belief that living organisms are generated spontaneously from nonliving matter.

spontaneous ovulator: any species in which ovulation occurs periodically as a result of an internal stimulus. (See reflex ovulator.)

sporangium (spor-ran'ji-um): a spore-producing structure in plants or microorganisms. (Plural: sporangia.)

spore (spōr): the asexual reproductive element of lower organisms, such as bacteria, protozoans, or ferns and mosses; capable of developing directly into an adult.

spore coat: a firm, virtually impermeable membrane that surrounds a resting spore; it encloses the nucleic acids and a small part of the cytoplasm of the original cell.

sporophyte (spōr'ah-fīt): the asexual, diploid generation of a plant that produces spores; it alternates with a sexual, haploid gametophyte generation.

sporozoan (spōr'ro-zō'an): any of a class of parasitic protozoans that reproduce by forming spores and have no organs of locomotion; for example, the malarial parasite.

sporozoite (spō'ro-zō'it): one of the stages in the life cycle of many sporozoan organisms; sporozoites from the salivary glands of mosquitoes cause malaria in man.

stabilizing selection: a natural process removing from a population (by mortality) those individuals with extreme physical characteristics.

stamen (stā'men): the pollen-bearing or male organ of a flower.

starch: a white, tasteless polymer of glucose (sugar) found in most plants, where it functions as stored nutrient.

state-dependency: a condition affecting memory; information acquired while the central nervous system is in a certain physiological state is most efficiently recalled when the state is closely replicated.

stereotyped patterned sequence: any type of behavior that is relatively unchanged from one occurrence to the next—also called fixed action pattern; for example, nest building by birds.

sterol lipid: a type of lipid molecule consisting of fatty acids attached to complex alcohols (sterols).

stigma (stig'mah): the female or pollen-receiving part of a flower; also, a small granular body containing carotenoid pigments and found in certain protozoans.

stipe (stīp): a stalk or a stemlike part of a plant or an animal.

stoma (stō'mah): a small opening; for example, a microscopic slit on a leaf or young stem that permits the exchange of gases. (Plural: stomata.)

stroma (strō'mah): a framework; for example, the material surrounding grana and thylakoids in the inner membrane of the chloroplast in higher plants. (Plural: stromata.)

style (stīl): in botany, the stalklike part of a pistil; it connects the stigma with the ovary.

subphylum (sub-fī'lum): a category of related classes within a phylum.

substrate (sub'strāt): in biochemistry, any substance acted upon by an enzyme; in biology, the medium upon which an organism grows.

sulcus (sul'kus): a groove; for example, a shallow fissure in the cerebral cortex. (Plural: sulci.)

superclass (sū'per-klas): a category of related classes within a subphylum.

superinfection (sū'per-in-fek'shun): a second or subsequent infection by the same virus or a different virus.

superoptimal stimuli: artificial stimuli that are more effective than natural stimuli in releasing particular behavioral responses.

surface antigen: a particular chemical configuration in the outer surface of a cell; it affects the cell's reaction to other cells bearing the same or different antigens.

surface tension: a cohesive force that acts to preserve the integrity of a surface; for example, the resistance to rupture possessed by the surface film of a liquid.

suspensor (sus-pen'ser): the stalklike structure that attaches the plant embryo to the rest of the seed.

symbiosis (sim'bī-ō'sis): the close association of two organisms of different species for mutual or one-sided benefit.

sympathetic nervous system: one of the two divisions of the autonomic nervous system; it promotes energy expenditure by mobilizing the body's resources and counteracts the parasympathetic system.

synapse (sin'aps): the microscopic junction between neurons or between a neuron and a receptor or effector cell; the neuron receives or transmits information across this junction.

synapsis (si-nap'sis): the pairing off of homologous chromosomes in the zygotene stage of meiosis.

synaptic cleft: the fluid-filled space within the synapse through which a nerve impulse is transmitted by the diffusion of chemical substances. (See chemical synapse.)

synaptic vesicle: one of many small, rounded bodies on an axon that empty fluids into the synaptic cleft.

synaptinemal complex (sin-ap'ti-nē'mal): a structure found between paired homologous chromosomes during prophase of meiosis; it is thought to be composed of protein.

syncytial tissue (sin-sish'al): a multinucleate tissue formed by the fusion of the cytoplasm of two or more cells and by the loss of cell membranes.

systole (sis'to-lē): a contraction or drawing together; for example, the contraction of the heart.

t

tannin (tan′in): an astringent compound found especially in the bark and foliage of certain trees; it functions in repelling insects and parasites.

taxis (tak′sis): behavior of an organism that involves movement in a specific direction relative to the source of stimulation.

taxon (tak′son): a taxonomic category such as genus or species. (Plural: taxa.)

tectorial membrane (tek-to′rē-al): a membrane located in the cochlea of the ear and in contact with the cilia of hair cells; in sound reception, the membrane stimulates hair cells by its movement.

tectum (tek′tum): the dorsal (anatomically, the upper) portion of the midbrain containing neurons involved in the visual and auditory systems.

tegmentum (teg-men′tum): the ventral (anatomically, the lower) part of the midbrain; it contains nuclei involved in sensory integration and in the control of eye movements.

telophase (tel′o-fāz): the final stage of mitosis in which the chromosomes reorganize to form daughter nuclei, and the cytoplasm divides to form two complete daughter cells.

telophase I: in meiosis, the stage in which nuclear membranes form around the two sets of homologous chromosomes, the chromosomes uncoil, and the spindles disappear.

telophase II: in meiosis, the stage in which nuclei are formed around four haploid sets of chromosomes, and plasma membranes separate the four daughter cells.

template (tem′plāt): a pattern or mold; for example, the model of coded information stored in the DNA molecule from which other DNA or RNA molecules are replicated.

temporal lobe: one of the four lobes dividing the hemispheres of the human brain; it is involved in auditory and visual functions.

teratogen (ter′ah-to-jen): an agent that causes physical defects in the developing embryo.

terminal cell: the smaller, denser of two cells formed in the first division of a plant zygote; it contains most of the cytoplasm and organelles. (See basal cell.)

test cross: a back cross in which a heterozygote is crossed with an individual homozygous for the recessive alleles of the gene or genes being studied.

testis (tes′tis): the male gonad; the glandular organ in which spermatozoa are produced. (Plural: testes.)

testosterone (tes-tos′ter-ōn): the hormone produced by the testes; it stimulates development of male secondary sex characteristics.

tetanus (tet′ah-nus): the state of a muscle that is unable to relax between stimuli; continuous contraction.

T-even phage: a virulent strain of bacteriophage; it infects the bacterium *Escherichia coli* and kills the host cell.

thalamus (thal′ah-mus): a cluster of neurons and glial cells in the rearmost part of the forebrain; it is involved in hearing and vision.

thermal energy: the part of the energy of a system that is due to the motion of the molecules; a function of temperature.

thiamine (thī′a-mēn): a B-complex vitamin and an essential cofactor in some biochemical reactions.

threshold stimulus: the specific stimulus intensity below which a given irritable tissue exhibits no response.

thrombocyte: see platelet.

thylakoid (thī′la-koid): one of the flattened sacs that form a complex system in the inner membrane of the chloroplast.

thymidine (thī′mi-dēn): a pyrimidine deoxyribonucleoside isolated from liver and a component of DNA.

thymidylic acid (thī′mi-dil′ik): one of the four deoxyribonucleotides that make up DNA.

thymus gland (thī′mus): an organ, lying behind the top of the breast bone, that is large during infancy but atrophies in adults; it produces cells (immunocytes) that function in immune responses.

thyroxin (thī-rok′sin): a thyroid hormone synthesized from the amino acid tyrosine; used in the treatment of hypothyroidism.

tobacco mosaic virus: the virus causing the mosaic disease of tobacco; it is frequently used in virus studies. (Abbreviated: TMV)

trachea (trā′kē-ah): in humans, the tube descending from the larynx to the bronchi; in insects, a complex system of branched, air-filled respiratory tubules. (Plural: tracheae.)

tracheid (trā′kē-id): an elongated, tapering xylem cell adapted for conduction and support.

transcription (tran-skrip′shun): the enzymatic process by which the base sequence of chromosomal DNA is transferred to messenger RNA, which forms a complementary copy.

transduction: in genetics, the transfer of genetic information from one cell to another by a virus; in sensory response, the process in which the energy of a stimulus, such as light or sound, is converted into another form of energy, such as a nerve impulse.

transfer RNA: relatively small molecules of nucleic acid that act as carriers of specific amino acids during protein synthesis on ribosomes. (Abbreviated: tRNA)

transformation (trans′for-mā′shun): a change in form or structure; for example, the change in an animal cell caused by the entrance of an oncogenic virus.

transient polymorphism: the state of a population that temporarily consists of two or more different phenotypes.

translation (trans-lā′shun): the biosynthetic process by which amino acid sequences in proteins are determined by base sequences in the mRNA template on the ribosome.

triad (trī′ad): a membranous structure that crosses the myofibril, connects adjacent myofibrils, and extends to the sarcolemma in skeletal muscle cells.

trichoblast (trik′o-blast): a densely cytoplasmic cell that occurs in the epidermal layer of some plants; these develop into the root hairs.

trichocyst (trik′o-sist): an effector organelle embedded in the body surface of ciliated protozoa and containing a fine, hairlike filament capable of being ejected; possibly a defense mechanism.

triglyceride (trī-glis′er-īd): a lipid molecule consisting of glycerol with three fatty acids attached as esters; it serves as a form of long-term energy storage in animal cells.

triose (trī′ōs): a three-carbon sugar.

trisomy (trī′sō-mē): the presence of an extra (third) chromosome of one type in an otherwise diploid cell.

trophic levels (trof′ik): the successive levels of assimilation in a food chain; plant producers constitute the lowest level and dominant carnivores, the highest level.

trophoblast (trof′o-blast): the thin-wall side of a blastocyst; it develops into the fetal membrane with nutritive functions.

tropism (trō′pizm): a growth response to an external stimulus.

tropomyosin (trō′po-mī′o-sin): a protein found in the myofibrils of skeletal muscle cells.

tubal ligation (tū′bal li-gā′shun): an operation in which the fallopian tubes are tied or severed to prevent pregnancy.

tumescence (tu-mes′ens): a swelling or the condition of being swollen.

Turner's syndrome: an abnormality characterized by retarded growth and sexual development and attributed to the lack of one of the X chromosomes; an example of monosomy.

u

ultrastructure (ul′trah-struk′chur): the arrangement of the smallest elements making up a body or cell; small details visible only by electron microscopy.

umbilical cord (um-bil′i-k′l): a cord connecting the fetus with the placenta and transmitting oxygen and nourishment (via the fetal blood) from the mother to the fetus.

tell us what you think...

Students today are taking an active role in determining the curricula and materials that shape their education. Because we want to be sure **Biology Today** is meeting student needs and concerns, we would like your opinion of it. We invite you to tell us what you like about the text—as well as where you think improvements can be made. Your opinions will be taken into consideration in the preparation of future editions. Thank you for your help.

Your Name ____________________ School ____________________

City and State ____________________ Course Title ____________________

How does this text compare with texts you are currently using in other courses?

☐ Excellent ☐ Average
☐ Very Good ☐ Poor
☐ Good

Name other texts you consider good and why. ____________________

Do you plan to sell the text back to the bookstore or keep it for your library? ☐ Sell it ______ ☐ Keep it ______

Circle the number of each chapter you read because it was covered by your instructor.

Unit I. Chapters: 1 2
Unit II. Chapters: 3 4 5 6
Unit III. Chapters: 7 8 9 10 11
Unit IV. Chapters: 12 13 14 15 16
Unit V. Chapters: 17 18 19 20 21 22 23 24 25 26
Unit VI. Chapters: 27 28 29 30 31
Unit VII. Chapters: 32 33 34 35 36 37 38
Unit VIII. Chapters: 39 40 41 42 43 44 45

What chapters did you read that were not assigned by your instructor? (Give chapter number.) ____________________

Please tell us your overall impression of the text.

	Excellent	Very Good	Good	Average	Poo
1. Did you find the text to be logically organized?	____	____	____	____	____
2. Was it written in a clear and understandable style?	____	____	____	____	____
3. Did the graphics enhance readability and understanding of topics?	____	____	____	____	____
4. Did the captions contribute to your understanding of the material?	____	____	____	____	____
5. Were difficult concepts well explained?	____	____	____	____	____

Can you cite examples that illustrate any of your above comments? ____________________

Which chapters did you particularly like and why? (Give chapter number.) ____________________

Which chapters did you dislike and why? ____________________

After taking this course, are you now interested in taking more courses in this field? ☐ Yes ☐ No

Did this text have any influence on your decision? ☐ Yes ☐ No

General Comments: ____________________

at topics would you like included in the text that are not covered now?__________

you find the interleaves understandable and useful? ☐ Yes ☐ No

Unit II, "The Physical Basis of Life," provide you with dequate explanation of the chemistry necessary for an nderstanding of contemporary biology? ☐ Yes ☐ No

you use the involvement manual? ☐ Yes ☐ No

es, how useful did you find it?__________

ve you seen the *Biology Today* films? ☐ Yes ☐ No

es, how helpful were they?__________

uld you like a study guide to accompany the text? ☐ Yes ☐ No

FIRST CLASS
Permit No. 59
Del Mar, Calif.

BUSINESS REPLY MAIL
No Postage Stamp Necessary if Mailed in the United States

Postage will be paid by

CRM BOOKS

Del Mar, California 92014

Just fold, staple and mail. No stamp necessary.

unconditioned response: the response given naturally to an unfamiliar stimulus.

unconditioned stimulus: an unfamiliar stimulus that elicits a response.

unit membrane: description of the basic triple-layer structure characteristic of all cell membranes.

uracil (yu'rah-sil): a nitrogenous base and a constituent of nucleic acids. (See pyrimidine.)

ureter (yu-rē'ter): the tube that conveys the urine from the kidney to the bladder.

urethra (yū-re'thrah): the tube through which urine is carried from the bladder and discharged from the body.

uterus (yū'ter-us): the chamber in which the fetus is contained and nourished in female animals.

V

vaccination (vak'si-nā'shun): the injection of vaccine in order to induce immunity.

vagina (vah-jī'nah): the canal in the female that receives the penis in copulation.

valence (vā'lens): a number representing the number of electron pairs that can be shared in covalent bonds or the number of electrons that are gained or lost in the formation of ions.

valence electron: an electron in the outer shell of an atom and one that can be shared with or lost to other atoms.

valine (val'ēn): a white, water-soluble amino acid derived from plant and animal proteins; it is essential to growth in infants and to nitrogen equilibrium in human adults.

vascular cambium: a group of cells that forms a continuous sheath between the xylem and phloem in the stems and roots of plants.

vascular tissue: a tissue containing vessels through which fluids are conducted; in plants, the xylem and phloem.

vasectomy (vas-ek'to-mē): surgical removal of the seminal duct in males, usually as a means of contraception.

vasocongestion (vas'o-kon-jest'shun): the concentration of blood in certain blood vessels.

vein (vān): a relatively large, thin-wall vessel through which blood passes from the tissues back to the heart. (See artery.)

ventral (ven'tral): pertaining to the under surface of an organism or denoting a position more toward the under surface than some other object of reference. (See dorsal.)

vernix caseosa (ver'niks kas'e-ō'sah): a pastelike substance that covers the skin of the fetus; it is composed of loose cells and secretions from sebaceous and sweat glands.

vessel element: a conducting cell in the xylem of most flowering plants; such cells lose their end walls during maturation and become joined to form long, tubelike vessels.

villus (vil'us): a small protrusion, especially a protrusion from the surface of a membrane. (Plural: villi.)

virion (vir'i-on): a particle of the tobacco mosaic virus; it includes a coiled molecule of RNA surrounded by a protein coat.

virulent (vir'yu-lent): capable of causing disease.

virus (vī'rus): a submicroscopic, noncellular particle that consists of either DNA or RNA and a protein coat; it is parasitic and reproduces by means of synthetic processes in the host cell.

viscera (vis'er-ah): a collective term for the internal abdominal organs of an animal.

vitalism (vī'tal-izm): a concept of life that ascribes the unique properties of a living organism to a vital principle distinct from chemical and physical laws. (Opposed to mechanism.)

vitamin (vī'tah-min): any of a group of unrelated organic substances essential in small quantities for normal metabolism.

vulva (vul'vah): the external portions of the female genital organs.

W

waggle dance: the "dance" performed by a scout bee to communicate the location of a food source to other bees in the hive.

X

xanthophyll (zan'tho-fil): any of the yellow, orange, or red pigments that occur in all green plants and in some animals; it resembles carotene.

X chromosome: in humans, the differential sex chromosome carried by half the male and all the female gametes.

xylem (zī'lem): one of the vascular tissues in plants; it conducts water from the roots upward and represents the wood of the plant.

Y

Y chromosome: in humans, the differential sex chromosome carried by half the male gametes.

yolk (yōk): a part of the ovum that serves as nutrient for the developing embryo; mostly fat and protein.

yolk sac: a sac containing yolk used for the nourishment of embryonic and immature fishes, reptiles, and birds; in most mammalian embryos, there is a similar organ, but it does not contain yolk.

Z

zona pellucida (zō'nah pel-lū'sid-ah): a transparent, noncellular membrane enclosing the morula stage of the mammalian embryo.

zoospore (zō'ah-spōr): an asexual, flagellated spore produced by certain algae and fungi.

zygospore (zī'gah-spōr): a cell formed by fusion of two similar gametes, as in certain algae and fungi.

zygote (zī'got): a fertilized ovum in plants or animals; the diploid cell resulting from fusion of male and female gametes.

zygotene (zī'go-tēn): in the prophase of meiosis, the stage in which each chromosome pairs with its homolog and the chromosomes shorten and thicken.

Index

a

b

d

e

f

I

n

O

q

r

S

t

U

V

W

X

Y

Z

Credits and Acknowledgments

Introduction

xx—Terry Lamb; xxi—(left) John Dawson after "The Lymphatic System" by H. S. Mayerson, © 1963 by Scientific American, Inc. All rights reserved, (right) Richard Oden, xxii—(left) Richard Oden, (center) Roberts Rugh and Landrum B. Shettles, (right) Douglas Armstrong; xxiii—(left) Thomas Anderson, (center) Richard Oden, (right) The Bettmann Archive, Inc.; xxiv—(right) Carolina Biological Supply Company; xxv—(top left) John Dawson, (bottom left) Grant Heilman, (center) John Dawson, (right) Richard Oden; xxvi—(top left) The Bettmann Archive, Inc., (bottom left) Steve McCarroll, (center) Carolina Biological Supply Company, (right) John Dawson; xxvii—(left) The Granger Collection, (right) Richard Oden; xxviii—(left) Wesley Parke, (top center) Ward's Natural Science Establishment, (bottom center) Carolina Biological Supply Company, (right) John Dawson; xxix—(left) The Granger Collection, (right) Richard Oden; xxx—(left) Carolina Biological Supply Company, (center) Douglas Armstrong, (right) John Dawson; xxxi—(left) United Press International, (center) John Brenneis.

Unit I/Chapter 1

2—Terry Lamb; 4—David MacDermott; 6—(left) The Granger Collection, (right) courtesy The Upjohn Company, Kalamazoo, Michigan; 7—(bottom) The Bettmann Archive, Inc., (top) John Dawson; 8-11—The Bettmann Archive, Inc.; 12, 13—The Granger Collection; 14—Wide World Photos; 15—NASA, Manned Spacecraft Center; 16—Wide World Photos; 17—Charles Carter.

Chapter 2

18—William MacDonald/IBOL; 21, 22—Douglas Armstrong; 24—(top) M. Woodbridge Williams, (bottom) Ward's Natural Science Establishment; 25—(top left) Carl Zeiss, Inc., (top right) Jeroboam/Smith; 26—(top) Hugh Spencer, (center) Grant Haist, (bottom left) Allan Roberts, (bottom right) Jack Dermid; 27 (left to right, top to bottom) Eric Gravé (2), Dennis Brokaw, Grant Heilman, Jack Dermid, San Diego Zoo Photo/Ron Garrison and F. D. Schmidt, Philip Hyde, Jack Dermid, Dr. Franklin Kosdon; 28—(left to right, top to bottom) James W. Dutcher, Gordon Robilliard, Jeroboam/Smith, Photo by Runk/Schoenberger (Grant Heilman), M. Woodbridge Williams, William H. Amos, Jeroboam, Douglas P. Wilson; 29—Runk/Schoenberger, Grant Heilman; 30—(left to right, top to bottom) James W. Dutcher, Paul Villiard/Saugerties, New York, Dr. Franklin Kosdon, Jack Dermid, Richard F. Trump/Biophotography, Grant Haist, James W. Dutcher, San Diego Zoo Photo/Ron Garrison and F. D. Schmidt; 31—(left to right, top to bottom) Corson Hirschfeld, Grant Heilman/Lititz, Pa., San Diego Zoo Photo/Ron Garrison and F. D. Schmidt, James W. Dutcher, (bottom row) San Diego Zoo Photo/Ron Garrison and F. D. Schmidt; 32—San Diego Zoo Photo/Ron Garrison and F. D. Schmidt; 33—William MacDonald/IBOL.

Unit II/Chapter 3

34—Terry Lamb; 36—John Brenneis for *Scientific American;* 38—(left) Culver Pictures, (right) The Granger Collection; 39—(left) Tom Lewis, (right) Douglas Armstrong; 40—Tom Lewis; 42-46—Douglas Armstrong; 47—Tom Lewis; 48-49—Douglas Armstrong; 50—Tom Lewis; 51-52—Douglas Armstrong; 53-56—photographs by Ray Salemme; 56 (top) and 57—Douglas Armstrong; 58—Grant Heilman/Lititz, Pa.; 59-60—Douglas Armstrong.

Chapter 4

64—Charles Carter; 66—(top) Douglas Armstrong, (bottom) Tom Lewis; 67—(top) Tom Lewis, (bottom) Douglas Armstrong; 68-69—Douglas Armstrong; 70—(top) The Granger Collection, (bottom) Douglas Armstrong; 71—John Dawson; 72, 73, 77, 78—Douglas Armstrong; 79—Tom Lewis; 80-82—Douglas Armstrong; 83—Douglas Armstrong, redrawn by permission from Watson, *Molecular Biology of the Gene* © 1970, J. D. Watson.

Chapter 5

84—construction by Terry Lamb, photography by Werner Kalber/PPS; 86—William MacDonald/IBOL; 87—(left and right) The Bettmann Archive, Inc., (center) The Granger Collection; 88-90—Douglas Armstrong; 92—The Art Works; 94—The Art Works, (top) after Lehninger, *Biochemistry*, p. 495, © 1970, Worth Publishers, Inc.; 96—(top) The Art Works, (bottom) Tom Lewis; 98—The Art Works; 99—Douglas Armstrong; 100—construction by Donald Fujimoto, photographv by Werner Kalber/PPS; 101—courtesy Dr. R. S. Wolfe; 102—fossils courtesy M. Lawrence Stevens, photography by Werner Kalber/PPS.

Chapter 6

104—construction by Joyce Fitzgerald, photography by Werner Kalber/PPS; 106—(left) John Dawson, (right) Douglas Armstrong after Shoemaker and Garland, *Experiments in Physical Chemistry*, 1967, with permission of McGraw-Hill Book Company; 107—Douglas Armstrong; 108—Tom Lewis; 110—(top) Douglas Armstrong, (bottom) Tom Lewis after Payson Stevens; 111—Douglas Armstrong, redrawn by permission from Lehninger, *Bioenergetics*, © 1965, W. A. Benjamin; 112—from U.S. Naval Medical School, *Color Atlas of Pathology*, Vol. 2, 1956, courtesy J. B. Lippincott Co.; INTERLEAF: all illustrations by Robert Kinyon/Millsap and Kinyon after Payson Stevens; B—(electron micrograph top) Dr. K. R. Porter, (electron micrographs center and bottom) Dr. H. Fernandez-Moran, Department of Biophysics, University of Chicago; C—(electron micrographs top to bottom) William A. Jensen, Myron Ledbetter, Garth Nicholson/Salk Institute, R. B. Park; 113, 115—Douglas Armstrong.

Unit III/Chapter 7

118—Terry Lamb; 120—Diane MacDermott; 122—(bottom left) The Granger Collection, (bottom right) The Bettmann Archive, Inc., (top left) John Dawson, (top right) The Bettmann Archive, Inc., 123—(left) The Bettmann Archive, Inc., (center and right) The Granger Collection; 124—courtesy The Sheffield City Museum, England; 125—Tom Lewis after Swanson, *The Cell*, 3rd edition, Prentice-Hall, 1967; 126—William A. Jensen; 127—(top) Douglas Armstrong after Jensen and Park, *Cell Structure*, Wadsworth Publishing Co., (bottom) H. E. Hinton; 128-129—Robert Kinyon/Millsap and Kinyon; 130—(top) Douglas Armstrong, (bottom) David M. Phillips; 131—(top) Daniel Branton, from *Proceedings of the National Academy of Sciences*, 55:1048, 1966, (center and bottom) David M. Phillips, (bottom right) Douglas Armstrong; 132—courtesy Dr. Elliott Robbins; 133—K. R. Porter; 134—David M. Phillips; 135—(left) David M. Phillips, (right) Douglas Armstrong; 136—K. R. Porter; 137—William A. Jensen; 138—(top) William A. Jensen, (center) David M. Phillips, (bottom) Robert Kinyon/Millsap and Kinyon; 139—(top) Douglas Armstrong, (bottom) David M. Phillips; 140—David M. Phillips; 141—(top) David M. Phillips, (bottom) William A. Jensen; 142—H. E. Hinton; 143—(left) William A. Jensen, (right) D. and H. Adamson, courtesy Professor R. D. Preston.

Chapter 8

146—Diane MacDermott; 148—(left) John Dawson, (right) John B. Watterbury, Department of Bacteriology and Immunology, University of California, Berkeley; 149—(top and bottom right) John Dawson, (bottom left) Dr. Norma Lang, University of California, Davis; 150—Philip Kirkland; 151—(left) Dr. George B. Chapman, (right) John Dawson; 152—T. F. Anderson, E. L. Wollman, and F. Jacob; 153—(top) The Granger Collection, (center) courtesy Bronx Botanical Gardens, (bottom) R. A. Packer, Iowa State University; 154—from Smith and Holdeman, *The Pathogenic Anaerobic Bacteria* (1968), courtesy of Charles C. Thomas, Publisher, Springfield, Illinois, and Center for Disease Control, HSMHA, USDHEW; 155—(bottom) Patrick Echlin; 156—(top left) John Dawson, (top right) William T. Hall, Electro-Nucleonics Laboratories, Inc., (bottom) Douglas Armstrong; 157—John Dawson after Swanson, *The Cell*, 3rd edition, Prentice-Hall, 1967, p. 76.

Chapter 9

158—Philip Kirkland; 160—(top) *Biochim. Biophys. Acta*, 88 (1964), p. 146 (Fig. 2), courtesy A. K. Kleinschmidt, (center left) Jeroboam, (center right and bottom) courtesy the Virus Laboratory, University of California, Berkeley; 161—Robert Kinyon/Millsap and Kinyon; 162—(top) courtesy the Virus Laboratory, University of California, Berkeley, (bottom left) John Dawson, (bottom right) H. Fernandez-Moran; 163—(top) Dr. L. D. Simon and Dr. T. F. Anderson, Institute for Cancer Research, Philadelphia, (bottom) F. A. Eiserling; 164—Tom Lewis, redrawn by permission from Watson, *Molecular Biology of the Gene*, © 1965, J. D. Watson; 167—(left and center) from *The Journal of General Virology*, Vol. 7, p. 77, reprinted by permission of Cambridge University Press, (right) Douglas Armstrong; 169—Ron Estrine; 171—Tom Lewis redrawn by permission from

Watson, *Molecular Biology of the Gene*, © 1965, J. D. Watson; 172—Tom Lewis after William Keeton, *Biological Science*, W. W. Norton & Co., 1967.

Chapter 10

174—David MacDermott; 176—Tom Lewis; 177—(left) Tom Lewis, (right) Dr. A. D. Bangham; 178—Tom Lewis; 179—Vincent T. Marchesi; 180–182—Tom Lewis; 183—(top left and bottom right) Tom Lewis, (top right and bottom left) Vincent T. Marchesi; 184–187—Tom Lewis; 188—(bottom) Vincent T. Marchesi.

Chapter 11

190—Karl Nicholason; 192—(bottom left) James W. Dutcher, (right) James Endicott; 193—(top right) John Dawson, (bottom left) George Lower from National Audubon Society; 194—Franklin Photo; 195—(right) Dr. E. G. Cutter; 197—Dr. E. G. Cutter, from the *American Journal of Botany*, Vol. 51, p. 320; 199—Robert Kinyon/Millsap and Kinyon; 200—(bottom left) Douglas Armstrong, (bottom right) Clara Franzini-Armstrong; 202—Dr. E. G. Cutter, from *Plant Anatomy*, Part 2, p. 26, Fig. 2.14b, Addison-Wesley and Edward Arnold (Publishers) Ltd.

Unit IV/Chapter 12

204—Terry Lamb; 206—construction by Joyce Fitzgerald, photography by Werner Kalber/PPS; 209 and 210—courtesy Moravian Museum; 211—Douglas Armstrong; 213—Douglas Armstrong; 216—Wide World Photos.

Chapter 13

218—Masami Teraoka; 220—(top) after G. R. Taylor, *The Science of Life*, Thames and Hudson, p. 317, (bottom) Elton Stubblefield; 222–223—drawings by John Dawson; 224—Andrew S. Bajer; 225, 226—J. Herbert Taylor, Institute of Molecular Biophysics, Florida State University, Talahassee; 227—Tom Lewis; 228—J. Herbert Taylor; 229—courtesy Dr. Elliott Robbins; 230—J. Herbert Taylor; 231—(left) J. Herbert Taylor, (right) John Dawson; 232—Elton Stubblefield; 234–235—drawings by John Dawson; 237—Tom Lewis after Watson, *Molecular Biology of the Gene*, © 1965, J. D. Watson; 238—Dr. James Kezer, Department of Biology, University of Oregon.

Chapter 14

242—construction by Joyce Fitzgerald, photography by Werner Kalber/PPS; 244—John Dawson after Singer, *A History of Biology*, 1959, by permission of Murnat Publications, Inc.; 245—(top) photo by Runk/Schoenberger, Grant Heilman, (bottom left) Roseann Litzinger, (bottom right) John Dawson; 247—John Pierce; 248—Douglas Armstrong; 250—(top) Ray Bravo, (bottom) Tom Lewis; 251—Douglas Armstrong; 252—Tom Lewis after William Keeton, *Biological Science*, W. W. Norton & Co., 1967; 253—after Bridges, *Journal of Heredity*, 26:60–64, 1935; 255, 256—Douglas Armstrong; 259, 260—William D. Loughman, Donner Laboratory, University of California; 261—from Levine, *Genetics*, Holt, Rinehart and Winston, 1968; 262—Ulrich Clever; 264—(top) Wide World Photos, (bottom) Gabriele Wunderlich; 265—Douglas Roy.

Chapter 15

268—courtesy of The Upjohn Company, Kalamazoo, Michigan; 270—Neill Cate; 271—Tom Lewis redrawn by permission from Watson, *Molecular Biology of the Gene*, © 1965, J. D. Watson; 273—Tom Lewis after Payson Stevens; 275—Douglas Armstrong; INTERLEAF: color illustrations by Frank Armitage after Payson Stevens; A—(left) Tom Suzuki, (electron micrograph top left) John Cairns, (electron micrograph top right) David M. Prescott, (electron micrograph bottom) Henry S. Slayter, 277—Tom Lewis redrawn by permission from Watson, *Molecular Biology of the Gene*, © 1965, J. D. Watson; 278—John Cairns, Cold Spring Laboratory; 280—(left) Oscar Miller, Oak Ridge National Laboratory, (center and right) Dr. David M. Prescott; 283—Neill Cate; 284—(left) John Dawson, (right) Douglas Armstrong redrawn by permission from Watson, *Molecular Biology of the Gene*, © 1970, J. D. Watson; 285—(top) Douglas Armstrong redrawn by permission from Watson, *Molecular Biology of the Gene*, © 1970, W. A. Benjamin, Inc., (bottom) Alexander Rich, MIT; 286, 287—John Dawson; 290–292—Neill Cate; 291—redrawn by permission from Watson, *Molecular Biology of the Gene*, © 1970, J. D. Watson.

Chapter 16

294—construction by Sheila Ordean, photography by Werner Kalber/PPS; 297—(top right) Jeroboam/Smith, (bottom left) Ward's Natural Science Establishment, (bottom right) Runk/Schoenberger (Grant Heilman); 298—(bottom left) Tom Lewis after Keeton, *Biological Science*, W. W. Norton & Co., 1967, (right) Richard Oden; 299—James Endicott; 300—Gabriele Wunderlich; 301—Roseann Litzinger;302—(bottom) Larry Sharp; 305—Ray Bravo; 307—James Endicott; 308—(left) American Museum of Natural History, (top right) Grant Haist, (bottom right) Allan Roberts; 310—(top and center right) Allan Roberts, (center left and bottom) San Diego Zoo Photo/Ron Garrison and F. D. Schmidt; 311—San Diego Zoo Photo/Ron Garrison and F. D. Schmidt.

Unit V/Chapter 17

312—Terry Lamb; 314—construction by Masami Teraoka, photography by William MacDonald/IBOL; 316—courtesy François Jacob; 320—Douglas Armstrong; 322–323—Douglas Armstrong after Payson Stevens; 325—H. Fernandez-Moran; 326—Douglas Armstrong; 328—Irvin R. Konigsberg, from "Clonal Analysis of Myogenesis," *Science*, 140:1273–1284, June 21, 1963; 329—Douglas Armstrong; 330—Dr. Earl Ettienne, Oakland University, Rochester, Michigan; 331—J. T. Bonner; 332—Dr. Renato Dulbecco; 333—(top) Jeroboam/Smith, (bottom) Gabriele Wunderlich; 334—Dr. Renato Dulbecco; 335—Leon Domchowski, Department of Virology, M. D. Anderson Hospital, Houston, Texas.

Chapter 18

336—Karl Nicholason; 339—Douglas Armstrong; 341—Dr. E. G. Cutter; 343—(top) John Dawson, (bottom) Dr. Walter Halperin; 344—(left) Dr. E. G. Cutter, (right) Dr. R. H. Falk; 345—(left) Dr. R. H. Falk, (right) Dr. E. G. Cutter; 346—(top) Dr. E. G. Cutter, from *Phytomorphology*, Vol. 17, 1967, p. 440, fig. 4, (bottom) Dr. R. H. Falk; 347—(left) John Dawson, (right) Douglas Armstrong; 348—Dr. E. G. Cutter; 351—(left and center) reprinted by special permission from Cutter, *Plant Anatomy: Experiment and Interpretation*, Part I, Addison-Wesley and Edward Arnold (Publishers) Ltd., (right) Dr. E. G. Cutter; 352—John Dawson; 356–357—Ward's Natural Science Establishment; 358—O. L. Miller and Barbara R. Beatty, Biology Division, Oak Ridge National Laboratory; 361—Allan Roberts; 362—M. Woodbridge Williams; 365—John Dawson after Balinsky, *Introduction to Embryology*, W. B. Saunders, 1970; 367—(left) Marcus Jacobson, (left) John Dawson; 368—Lynwood M. Chace.

Chapter 19

370—construction by Earnie and Helga Kollar, photography by Werner Kalber/PPS; 372—Ward's Natural Science Establishment; 373—(top left) Allan Roberts, (top right) Robert Hermes, National Audubon Society, (bottom right) John Pierce; 374—(top) James Endicott; 375—John Pierce; 378—(left) Larry Sharp, (right) James Endicott; 379—Ray Bravo; 380—Richard Oden; 381—(left) John Dawson, (bottom right) Runk/Schoenberger (Grant Heilman); 382—John Pierce; 383—(left) Walter Dawn; 384—(top) Neill Cate, (bottom) John Dawson; 385—John Dawson; 386—Richard Oden; 387—Peter Mendez; 388—John Dawson; 389—(top) Joseph Garcia; 390—James Endicott; 393—Roseann Litzinger; 394—John Dawson; 395—Douglas Armstrong after Pitts, *The Physiological Basis of Diuretic Therapy*, C. C. Thomas, 1959.

Chapter 20

398—Richard Oden; 400—(top) Tom Lewis after Keeton, *Biological Science*, W. W. Norton & Co., 1967, (bottom) Tom Lewis; 402—photograph courtesy Bernard Phinney; 403—Walter Dawn; 405—(top) courtesy John D. Goeschl, (bottom) Neill Cate after J. D. Goeschl and H. K. Pratt, from F. W. Wightman and G. Setterfield (eds.), *Biochemistry and Physiology of Plant Growth Substances*, Runge Press, Ltd, Ottawa, 1968; 406—Tom Lewis; 407—(left) Neill Cate after Brian and Hemming, *Plant Growth Substances*, Interscience Publishers, Inc., 1963, with permission of Physiologia Plantarum, (right) courtesy of S. H. Wittwer, Michigan State University and Michigan Agricultural Experiment Station; 410—Tom Lewis after Keeton, *Biological Science*, W. W. Norton & Co., 1967; 412—A. H. Westing.

Chapter 21

416—Richard Oden; 418—John Dawson after Yamamoto, *Journal of Experimental Zoology*, 137, 1958; 419—(top) John Dawson, (bottom) from Wimsatt and Kallen, *Anatomical Record*, 129, 1957; 420—Peter Mendez; 421–424—John Dawson; 426—Hiroshi Asakawa from Turner and Bagnara, *General Endocrinology*, W. B. Saunders, 1971; 428, 429, 433—Douglas Armstrong; 434—Douglas Armstrong after Payson Stevens.

Chapter 22

436—construction by Suzan Anson, photography by Werner Kalber/PPS; 438—John Dawson after R. S. Speirs, "How Cells Attack Antigens," © 1964 by Scientific American, Inc. All rights reserved; 439—(top) Roseann Litzinger, (bottom) Philip Feinberg; 440—John Dawson; 444—Tom Lewis after Payson Stevens; 446—(left) Douglas Armstrong, (right) John Dawson; 447—Ron Estrine; 450—Tom Lewis; 452—F. M. Burnet; 453—John Dawson; 454—The Granger Collection.

Chapter 23

456—Paul Slick; 458—(left) Dr. Richard K. Orkand, (right) Lester Bergman and Associates; 459—(left) Tom Lewis, (right) Lester Bergman and Associates; 460—(top left) A. L. Hodgkin, (bottom left) The Art Works, after Hodgkin and Huxley, *Cold Spring Harbor Symposia on Quantitative Biology*, 25:379–387, 1952, (bottom right) Douglas Armstrong; 461—Douglas Armstrong after Payson Stevens; 462, 463—Douglas Armstrong; 465—Douglas Armstrong after "How Cells Communicate" by Bernhard Katz, © 1961 by Scientific American, Inc. All rights reserved; 467—(left) United Press International, (right) Tom Lewis; 468—(left) Douglas Armstrong (bottom left) after J. C. Eccles, "Modes of Transmission Within Nerve Cells and Between Nerve Cells,"

Nova Acta Leopoldina, 28:33–57, 1964, (bottom right) S. L. Palay; 469—Tom Lewis; 470—Sir John C. Eccles; 471—Douglas Armstrong, (right) after J. C. Eccles; 474—Tom Lewis; 475—Tom Lewis after Payson Stevens; 476—Tom Lewis; 477—Sir John C. Eccles.

Chapter 24

480—John Dawson; 482—John Dawson; 483—(top) John Dawson, (center and bottom left) John Born, (right) Richard Oden; 485—Larry Sharp; 486—John Pierce; 489—Douglas Roy; 491–496—Richard Oden; 498—Douglas Roy; 499–502—Richard Oden.

Chapter 25

504—construction by Joyce Fitzgerald and Edward Douglas, photography by John Oldenkamp/IBOL; 506—(left) John Dawson; 507—John Dawson; 508—John Dawson; 509—(top) Douglas Armstrong, (bottom) Tom Lewis; 510—(top) Douglas Armstrong, (bottom) Clara Franzini-Armstrong; 511—Tom Lewis; 512—courtesy Hugh E. Huxley; 513—Tom Lewis; 514—Clara Franzini-Armstrong; 515—(top) John Dawson, (bottom) Douglas Armstrong; 516—Douglas Armstrong.

Chapter 26

518—construction by Paul Slick, photography by Robert Van Doren; 521—David S. Dennison, courtesy Max Delbrück; 522—Richard Oden; 523—Neill Cate; 524–525—illustrations, Richard Oden; 526—after C. H. Graham, *et al., Vision and Visual Perception*, John Wiley & Sons, Inc., 1966; 527—John Dawson; 528—(top left) John Dawson, (bottom left) Werner Kalber/PPS; 528 and 529—Richard Oden; 529—(top and center right) Richard Oden, (bottom right) The Art Works; 530—Douglas Armstrong; 531—The Art Works; 533—John Dowling, from Boycott and Dowling, *Philosophical Transactions B.*, 255:109, 1969; 534—Floyd Ratliff, from *Mach Bands: Quantitative Studies on Neural Networks*, Holden-Day, Inc.; 535—John Dawson; 536—(top) illustration, Tom Lewis, photography, Werner Kalber/PPS; 537—Douglas Armstrong after Payson Stevens; 538—after de Valois in *Frontiers in Physiological Psychology*, Academic Press, 1966; 540–541—Tom Lewis; 542—John Pierce; 543—(right) Douglas Armstrong; 544—courtesy Dr. Israel Dvorine; 546–547—Tom Suzuki after Payson Stevens; 550–551—Douglas Armstrong; 552—(top) Dr. Allen J. Selverston; (bottom left) John Dawson; 553—Tom Lewis; 555—Tom Lewis; 556, 557—John Dawson.

Unit VI/Chapter 27

558—Terry Lamb; 560—David MacDermott; 562—John Dawson after Wallace and Srb, *Adaptation*, Prentice-Hall, 1964; 563—John Dawson; 564—(bottom right) from *Animal Behavior*, 10:300–304, 1962, courtesy Kenneth D. Roeder; 565—(top left) Douglas P. Wilson, (top right, 2) photography by Grant Heilman/Lititz, Pa., (bottom) Dr. Franklin Kosdon; 566—(top left) Lynwood M. Chace, (center) Allan Roberts, (right) Lynwood M. Chace, (bottom) Allan Roberts; 568—John Dawson after Tinbergen, *The Study of Instinct*, Clarendon Press, Oxford, 1951; 569—Thomas McAvoy, *Life* Magazine, © Time, Inc.

Chapter 28

570—Al Kahn Design; 572—(left) from K. S. Rawson, *Photoperiodism and Related Phenomena in Plants and Animals*, No. 55, pp. 791–800, © 1959, American Association for the Advancement of Science, (right) from J. W. Hastings, *Photophysiology*, Vol. 1, pp. 333–361; 573—from Ingold and Cox, 1955, *Annals of Botany*, 19:201–209; 574—Douglas Armstrong; 575—The Art Works; 576—courtesy Frank A. Brown, Jr.; 577—(top) J. Woodland Hastings, (bottom) Douglas Armstrong; 578—Douglas Armstrong; 580—Neill Cate.

Chapter 29

582—David MacDermott; 584—Steve McCarroll/IBOL; 585—(top) John Dawson, (bottom) Douglas Armstrong; 586—(top and center) Steve McCarroll/IBOL, (bottom) Douglas Armstrong; 587—Steve McCarroll/IBOL; 588–589—Douglas Armstrong; 591—Douglas Armstrong; 593—courtesy Dr. James McConnell; 594–595—Neill Cate.

Chapter 30

596—Richard Oden; 599—Dr. Franklin Kosdon; 600—Steve McCarroll/IBOL; 601—James Endicott; 602—Neill Cate after Novin, 1962; 603—(left and center) Leonard Lee Rue III, (top right) Corson Hirschfeld, (bottom right) Runk/Schoenberger (Grant Heilman); 605—John Dawson; 607—(top left) Grant Heilman/Lititz, Pa., (top right) Grant Haist, (bottom left) Leonard Lee Rue III, (bottom right) Richard F. Trump/Biophotography; 608—photo by Valentin Scheglowski, courtesy of Dr. R. A. Butler; 609—courtesy Neal Miller, *Journal of Abnormal and Social Psychology*, 43:155–178, 1948; 610—James W. Dutcher.

Chapter 31

612—Tom Lewis; 614—(top) Douglas Armstrong, (bottom) T. J. Walker, Department of Entomology, University of Florida; 615—John Dawson; 616—(left) John Dawson, (right) Tom Lewis; 617—John Dawson, adapted from E. O. Wilson, "Chemical Communication Among Workers of the Fire Ant, *Solenopsis saevissima, Animal Behavior*, Vol. 19, 1962, pp. 134–164; 618—(left) Peter Marler, (right) San Diego Zoo Photo/Ron Garrison and F. D. Schmidt; 619–621—Peter Marler; 622—Tom Lewis; 625—John Dawson; 626—after Dr. David Premack; 627—A. M. Guhl; 628–629—John Dawson; 630—Grant Heilman/Lititz, Pa.

Unit VII/Chapter 32

632—Terry Lamb; 634—Karl Nicholason; 637—(top) The Bettmann Archive, Inc., (bottom) The Granger Collection; 638—The National Maritime Museum, England; 639—(top) George H. Harrison/Grant Heilman, (center) Allan Roberts, (bottom) The Granger Collection; 640—(top) The Bettmann Archive, (bottom) Tom Lewis; 641—The Granger Collection; 642—The Granger Collection; 643—Richard Oden; 644—(all but bottom right) The American Museum of Natural History, (bottom right) David W. Miller; 646—John Dawson; 647—The Granger Collection; 648—The Granger Collection; 650—Douglas Armstrong.

Chapter 33

654—Patricia Peck; 656–661—Douglas Armstrong; 659—after Kormondy, *Concepts of Ecology*, Prentice-Hall, 1969; 660—after Allee, *et al., Principles of Animal Ecology*, Saunders, 1949; 663–666—Douglas Armstrong; 664—after Gause, *The Struggle for Existence*, Williams & Wilkins, 1934; 666—after Smith, *Ecology and Field Biology*, Harper & Row, 1966; 667—John Dawson; 668—Douglas Armstrong; 670—John Dawson.

Chapter 34

672—Joseph Garcia; 677—Douglas Armstrong after Ehrlich and Holm, *The Process of Evolution*, McGraw-Hill, 1963; 680, 682, 685—Neill Cate; 685—photography from the experiments of H. B. D. Kettlewell, Department of Zoology, Oxford University; 686—Neill Cate; 687— (top) Douglas Armstrong; (center and bottom) photo by George H. Harrison, from Grant Heilman; 688—John Dawson; 689—John Dawson.

Chapter 35

692—Richard Oden; 694—John Dawson; 696—John Dawson; 697—Daniel H. Janzen; 699—Howard Saunders; 700—P. W. Hay, National Audubon Society; 701—Daniel H. Janzen; 703—(top) photograph by H. G. Baker and B. J. Harris, *Evolution*, II:499–460, 1957, (bottom) Daniel H. Janzen; 704—(left) John Dawson; (right) Alexander B. Klots; 705—Daniel H. Janzen; 708—(top) Herbert C. Weihrich, (center and bottom) Daniel H. Janzen; 710—Treat Davidson, National Audubon Society.

Chapter 36

712—construction by John Dawson, photography by Werner Kalber/PPS; 714—Tom Lewis after Payson Stevens; 715—(top) Ward's Natural Science Establishment, (center) Lynwood Chace, National Audubon Society, (bottom) Walter Dawn; 719—(top right) Douglas P. Wilson, (bottom left and right) Jeroboam/Stevens; 720—(left) John Dawson, (right) Grant Haist; 721—(right) Grant Heilman/Lititz, Pa.; 722—Walter Dawn; 723—John Pierce.

Chapter 37

728—construction by Joyce Fitzgerald, photography by Werner Kalber/PPS; 731—(top left) Douglas Roy, (bottom left) Walter Dawn, (top right) F. T. Haxo; 733—Douglas P. Wilson; 734—Grant Haist; 735—Jack Dermid; 737—(top) Grant Haist, (bottom) Hugh Spencer; 739—Grant Haist; 740—Grant Haist; 741—(top) Dennis Brokaw, (bottom left) Douglas Roy, (bottom right) Grant Haist; 742—Philip Hyde; 743—William H. Amos; 744—Philip Hyde; 745—Douglas Roy; 746—(top) Douglas Roy, (bottom) San Diego Zoo Photo/Ron Garrison and F. D. Schmidt; 747—(top left) Grant Heilman/Lititz, Pa., (top right) Douglas Roy, (bottom left) San Diego Zoo Photo/Ron Garrison and F. D. Schmidt; 749—(top) Grant Haist, (bottom) Grant Heilman/Lititz, Pa.; 751—(top) Runk/Schoenberger (Grant Heilman), (center) Hugh Spencer, National Audubon Society, (bottom) Hugh Spencer; 755—Philip Hyde; 756—(top left) Grant Haist, (bottom right) Dennis Brokaw.

Chapter 38

758—Richard Oden; 760—Douglas Roy; 761—(left) James Endicott, (right) James W. Dutcher; 762—James W. Dutcher; 763—Richard Oden; 764—Douglas Roy; 765—William H. Amos; 766—(left) Ray Bravo, (bottom right) Armed Forces Institute of Pathology; 767—Ray Bravo; 768—Roseann Litzinger; 769—(top) Runk/Schoenberger (Grant Heilman), (bottom) Armed Forces Institute of Pathology, Neg. No. 33819; 770—Larry Sharp; 771—(bottom) James Endicott; 772—(top and bottom left) Runk/Schoenberger (Grant Heilman), (bottom right) Allan Roberts; 773—Betsy Kluga; 774—John Pierce; 775—(top left) James W. Dutcher, (bottom left) William H. Amos, (right) Allan Roberts; 776—American Museum of Natural History; 777—(illustration) Douglas Roy, (photos left to right, top to bottom) Dennis Brokaw, James W. Dutcher, M. Woodbridge Williams, Allan Roberts, James W. Dutcher; 778—(left to right, top to bottom) Grant Haist, Dennis Brokaw, Allan Roberts (2), Carolina Biological Supply Company, James W. Dutcher; 779—American Museum of Natural History; 780—John Dawson; 781—(left) M. Woodbridge Williams, (top center) William H. Amos, (bottom center) Jeroboam, (right) Gordon A. Robilliard, (illustration bottom) Douglas Roy; 782—(left) Runk/Schoenberger (Grant Heilman), (right)

Betsy Kluga; 784 – (left) James W. Dutcher, (right) Richard Oden; 785 – (top) San Diego Zoo Photo/Ron Garrison and F. D. Schmidt, (center) Jack Dermid, (bottom) Richard Oden; 786 – (top left) San Diego Zoo Photo/Ron Garrison and F. D. Schmidt, (top right) Dr. Franklin Kosdon, (center left) Jack Dermid, (bottom) Richard Oden; 787 – (left) Richard Oden, (right) San Diego Zoo Photo/Ron Garrison and F. D. Schmidt; 788 – (top left) San Diego Zoo Photo/Ron Garrison and F. D. Schmidt, (top right) Dr. Franklin Kosdon, (center left) James W. Dutcher, (bottom) Richard Oden; 790 – Allan Roberts.

Unit VIII/Chapter 39

792 – Terry Lamb; 794 – Richard Oden; 796 – San Diego Zoo Photo/Ron Garrison and F. D. Schmidt; 798 – Terry Lamb; 799 – Karl Nicholason; 800 – 801 – illustration by Paul Slick, photography by Werner Kalber/PPS; 803 – Karl Nicholason; 804 – Douglas Armstrong; 805 – Karl Nicholason; 806 – (top) John Dawson, (bottom) Douglas Armstrong; 807 – John Dawson; 808 – Neill Cate; 809 – (top) Neill Cate, (bottom) Karl Nicholason; 810 – 811 – John Dawson; 812 – Robert Kinyon/Millsap and Kinyon; 813 – Allan Roberts; 814 – Douglas Armstrong adapted from A. E. Mourant, *The Distribution of Human Blood Groups*, Blackwell, 1954; 815 – John Dawson; 816 – 817 – John Dawson, 819 – Douglas Armstrong.

Chapter 40

820 – Masami Teraoka; 822, 823 – Richard Oden; 824 – (top) Richard Oden, (bottom) Roseann Litzinger; 825 – Douglas Armstrong; 826, 827 – Richard Oden; 829 – Ray Bravo; 832, 833 – Richard Oden; 834 – from L. B. Shettles, *Ovum Humanum*, Urban and Schwarzenberg, 1960.

Chapter 41

836 – construction by Masami Teraoka, photography by Werner Kalber/PPS; 838 – Kerstin Apelman Öberg, *Erotic Art*, Vol. 2, courtesy Eberhard and Phyllis Kronhausen; 839 – The Art Works after Kinsey, Pomeroy, and Martin, *Sexual Behavior in the Human Female*, W. B. Saunders, 1953; 840 – The Art Works after Ford and Beach, *Patterns of Sexual Behavior*, Harper & Row, 1951; 841 – after J. W. Everett, "The Mammalian Female Reproductive Cycle and Its Controlling Mechanisms," in W. C. Young (Ed.), *Sex and Internal Secretions*, Vol. 1, Williams and Wilkins, 1961; 842 – (left to right, top to bottom) Joe Loper, Otto Lang/Photophile, Costa Manos/Magnum Photos, Marilyn Silverstone/Magnum Photos, Elliott Erwitt/Magnum Photos, Kirk Breedlove/Photophile, Inge Morath/Magnum Photos, Dennis Stock/Magnum Photos; 843 – Franklynn Peterson/Black Star; 844 – William MacDonald/IBOL; 846 – (top) The Art Works after Kinsey, Pomeroy, and Martin, *Sexual Behavior in the Human Male*, W. B. Saunders, 1948, (bottom) The National Gallery of Art, Washington, D.C., Chester Dale Collection; 847 – The Art Works after Kinsey, Pomeroy, and Martin, *Sexual Behavior in the Human Female*, W. B. Saunders, 1953; 848 – Jenny Inge/Daughters of Bilitis; 849 – The Art Works after Kinsey, Pomeroy, and Martin, *Sexual Behavior in the Human Male*, W. B. Saunders, 1948; 850 – The Art Works after Masters and Johnson, *Human Sexual Response*, Little, Brown, 1966; 851 – Richard Oden; 853 – The National Gallery of Art, Washington, D.C., Gift of Mrs. John W. Simpson; 855 – Richard Oden.

Chapter 42

858 – John Dawson; 860 – from L. B. Shettles, *Ovum Humanun*, Urban and Schwarzenberg, 1960; 861 – (bottom) from L. B. Shettles, *Ovum Humanum*, Urban and Schwarzenberg, 1960; 862 – John Dawson; 864 – (bottom) courtesy Landrum B. Shettles; 865 – (left column) courtesy The Carnegie Institution of Washington; 868 – Wesley W. Parke, courtesy University Park Press, Baltimore; 874 – Peter Mendez; 875 – (photos top left and bottom right) William MacDonald/IBOL, (top right) Richard Oden; 876 – 877 – Richard Oden.

Chapter 43

880 – Paul Slick; 882 – 883 – National Audiovisual Center; 884 – (left top and bottom) National Audiovisual Center, (top right) Ian Berry/Magnum Photos, (bottom right) Marilyn Silverstone/Magnum Photos; 885 – 887 – National Audiovisual Center; 889 – William MacDonald/IBOL; 890 – 901 – National Audiovisual Center; 902 – Neill Cate; 903 – Alan Mercer; 904 – The Bettmann Archive, Inc.; 905 – Werner Kalber/PPS; 908 – National Audiovisual Center.

Chapter 44

910 – Masami Teraoka; 912 – Douglas Armstrong after Luria and Darnell, *General Virology*, Wiley, 1966; 913 – Karl Nicholason; 914 – Bernard Wolff/Photo Researchers; 916 – Karl Nicholason; 919 – The Art Works after Dr. Victor A. McKusick; 921 – (left) Dan Morrill, (right) Allan Roberts; 922 – Douglas Armstrong; 924 – construction by Karl Nicholason, photography by Werner Kalber/PPS.

Chapter 45

926 – Payson Stevens; 928 – Karl Nicholason; 929 – Werner Kalber/PPS, William MacDonald/IBOL; 932 – Neill Cate after *Population Bulletin*, Vol. 18, No. 1; 933 – photography by Dan Morrill; 934 – Dan Morrill; 935 – Tom Lewis; 936 – Neill Cate, (bottom) from R. R. Doane, *World Balance Sheet*, Harper & Row, 1957; 939 – Douglas Armstrong; 940 – (left) Douglas Armstrong after J. Bonner, "The Next Ninety Years," from R. P. Schuster (ed.), *The Next 90 Years*, California Institute of Technology, p. 40; 943 – John Dawson; 944 – Neill Cate; 945 – (top) photograph by Allan Roberts, (bottom) Douglas Armstrong; 946 – (left to right, top to bottom) Ken Heyman, Roger J. Adams, Jack K. Clark, Dave Newcomb, Ronald Slusarenko, Greg Brockmeyer; 947 – John Avery.
Cover design by Don Fujimoto.
Photography by Jeroboam/Smith
and William MacDonald/IBOL.

Film Supplement for Biology Today Films

The Biology Today Film Series deals with three of the most active fields of research in the biological sciences: cell structure and function, muscle function, and the origin of life and evolution. The medium of film has been used to great advantage to develop concepts in these areas that are often difficult to grasp through the written word. The following are synopses of topics covered in the films.

THE ORIGIN OF LIFE AND EVOLUTION

The process of evolution serves as the primary unifying concept for the many fields of study in biology. Evidence that life forms are not static but are constantly changing in response to the demands of their environments comes from such diverse disciplines as molecular biology, physiology, and ecology.

In essence, evolution is the process by which an entity develops from a simpler to a more complex state. In a broad sense, biological evolution is only a specialized manifestation of the evolution of the universe. The film *The Origin of Life and Evolution*, by using metaphorical images, makes a bold attempt to synthesize today's knowledge and speculation concerning the beginnings of life and the developments leading to life's current state of complexity.

As developed in the film, the basics of evolution are quite simple. That all organisms exhibit variability in characteristics and that all organisms reproduce more offspring than survive are readily observable traits. Two conclusions can be drawn from these observations. First, those individuals that are well suited to survival in the prevailing environment will be more likely to live and reproduce than less-suited individuals. Second, assuming that there is a mechanism for passing characteristics from an individual to its offspring, the characteristics of the better-suited individuals are more likely to be passed to the next generation.

By using the Galápagos Island finches and the bacterium *Escherichia coli* as examples, the film shows that the process of evolution selects the individuals best adapted to the environment. Differential migration and genetic drift are shown to modify the process.

The opening sequences of *The Origin of Life and Evolution* examine the beginning of life. Later sequences show that evolution has occurred throughout the history of life and is occurring today.

The film speculates that at one point in the past the universe existed as a massive, solitary body in the void of space. The "big-bang" theory of the origin of the universe hypothesizes that the explosion of this primordial atom resulted in its fragmentation. Over the succeeding billions of years, these fragments cooled and condensed into galaxies. Stars, planets, and other solitary units of mass formed within the galaxies.

In its primitive stages (approximately 5½ billion years ago), the earth was a molten ball laced with boiling upheavals and fiery explosions. There was no atmosphere as we know it today, only a gaseous storm sweeping around the earth and out into space. Eventually, the earth began to cool, and the debris of volcanic activity resulted in the accumulation of an atmosphere consisting of carbides, hydrogen, methane, ammonia, and water. The atmosphere contained no free oxygen; the oxygen that was released from terrestrial activity was instantly combined with other elements.

Scientists have attempted to duplicate this early chemical and energetic environment with laboratory apparatus. In one experiment, they found that the interaction of an electronic discharge with a mixture of ammonia, methane, hydrogen, and water resulted in the formation of amino acids and sugars, the basic building blocks of living systems. Thus, there is experimental evidence that the constituents of life arose inevitably because of the prevailing conditions at the surface of the earth.

The existence of the building blocks of living systems does not imply the concurrent existence of life. Life, however, can be functionally defined as a system that can mutate, reproduce, and metabolize. It is hypothesized that random chemical interactions on the surface of the earth resulted in such a system. The supposed time of the origin of life on earth is 3½ billion years ago. Because all life contains genetic material in the form of DNA, science speculates that DNA or a similar molecule was produced at about this time.

Many biologists believe that life began when the first cells formed. It has been theorized that the first cells were bacterialike organisms that existed in a nonoxygen atmosphere. Among the products of evolution were organisms containing the pigment molecule chlorophyll, which has the ability to trap and utilize energy from the sun to synthesize food materials. This process, termed photosynthesis, was vital in increasing the cells' abilities to survive, for the first photosynthetic cells no longer had to rely solely on externally produced nutrients for their nourishment.

The first bacteria, in addition to giving rise to photosynthetic cells, were probably the predecessors of flagellated cells. Because of their capacity for mobility, flagellated cells were able to move through their surroundings to gather food and thus did not need the photosynthetic process to survive. These cells were most likely the first animal cells on earth.

Life first came into existense in an environment of water and remained there for a long time. The land was devoid of life. It is speculated that the first organisms to survive out of water were primitive plants. Later, amphibians became the first vertebrates to leave the water and inhabit solid ground. They were still dependent on water, however, and had to return there to lay their jellylike eggs. Over millions of years, the amphibians gave rise to reptiles, which eventually completely severed dependence on water as an environment.

At about the same time that evolution of vertebrates was giving rise to an abundance of species of amphibians and reptiles, evolution among the invertebrates included the development of a diversity of insect species. Their structures embodied many extremely advantageous adaptations, including the ability to fly, an armored exterior skeleton for protection, and a segmented body that was easily modified.

Mammals evolved later in evolutionary history. They carried their eggs within their bodies and had the ability to maintain their body temperature and survive in extreme weather conditions. They proved relatively successful in adapting to a broad spectrum of environments.

One line of evolution of the mammals gave rise to man. Approximately 25,000 years ago the world population of *Homo sapiens* was about 3,300,000 individuals. Today, the world's human population is over

1,000 times as large and is rapidly increasing.

To what do we owe our amazing evolutionary success? Culture lies at the root of man's unusual adaptability and, hence, his survivability. Man has created symbolic communication in the forms of language and the written word, produced food by efficient methods of agriculture, created the method of scientific investigation, developed industrial technology, and grouped his kind into huge communicating populations.

Today mankind exists as a single (albeit internally antagonistic) world population. Man's development of culture has enabled him to not only adapt to the environment but to extensively alter the environment to suit his needs and desires. Hopefully, the future will bring the development of a culture that will enable us to live at peace with ourselves and our environment.

THE CELL: A FUNCTIONING STRUCTURE PART I

The first part of this film concerns the characteristics of the living cell — the basic unit of life. Emphasis is placed on the cell as a dynamic, functioning entity as well as a stable, reproducible structure. Furthermore, the size of the cell is charted relative to atomic, molecular, and macroscopic levels.

Because most cells can be examined only on a microscopic level, the elucidation of cell anatomy has closely paralleled the development of the microscope. The film presents a visual survey of the images characteristically produced by the simple, compound, dark field, phase-contrast, Nomarksian Interference, polarizing, fluorescent, electron, and scanning electron microscopes. Although the electron microscope can provide an exquisitely detailed picture of the anatomy of a cell, because the specimen must be killed and chemically treated, the sight of the animate characteristics of a living cell is lost. Through the technique of cinemicroscopy (movies taken through a microscope), the film shows the incredible variety of one-celled animal life found in a drop of pond water.

All living creatures need nourishment to sustain themselves. Cells are no different. For example, the didinium, one of the creatures seen in the drop of pond water, spears its prey with a tiny "harpoon." Another one-celled organism, the amoeba, is shown extending its jellylike mass around the other creatures that make up its diet.

The white blood cells of the human body ingest nourishment in the same way as does the amoeba. A sequence in the film shows how this ingestion, called phagocytosis, can also serve as a scavenging process for foreign matter and waste in the bloodstream.

What happens to the large particles of food and dissolved molecular nutrients taken into cells? A look at the interrelationship of vacuoles, which are warehouses of cellular components, and lysosomes, which contain enzymes capable of degrading large molecules into reusable fragments, demonstrates the cell's digestive capabilities.

Plants obtain nourishment in an entirely different way than do animals. Cinemicroscopy of the water lily, for example, allows an examination of some of the activities and structures in plant cells that enable them to synthesize their own food from small molecules using energy captured from the sun.

One of the most intricately beautiful activities of any cell is its mode of reproduction. The process of mitosis, the means by which a cell ensures the faithful replication of its component structure, is outlined in detail. More emphasis is placed upon the dynamics of the phenomenon than upon the series of stages. The film follows the coiling of the chromosomes, their pairing and migration to the "equator" of the cell, and their eventual separation into two daughter cells. In addition, the role of the spindle fibers and centrioles in pulling the chromosomes to opposite ends of the cell is illustrated. The film sequence of mitosis also introduces the nuclear membrane, a structure that is one distinguishing characteristic between the "higher" organisms, or eucaryotes, and the "lower" organisms, or procaryotes.

In allowing simultaneous viewing of both types of cells, the film makes some of the structural distinctions very clear. Looking more closely at a procaryotic cell, such as a bacterium, the viewer begins to see a subtle organization of the cell that has just recently been recognized and that has allowed these organisms to remain fairly constant in structure through eons of natural selection.

From a description of how cells divide, the film examines the question of how often they divide. Some procaryotic cells, such as bacteria, may divide every twenty minutes if conditions are optimal. In contrast, human skin cells divide every ten hours; and there are some cells, such as those in a human liver, that divide only when the organ has been injured. Cinemicroscopy also allows one to view human brain cells, which cease to divide shortly after birth and, in fact, decrease in number with age.

Returning to an electron micrograph of a typical eucaryotic cell, the film depicts some of the intricate organization of the cell's cytoplasm—the region between nucleus and membrane that, under light microscopes, appears to be merely unorganized liquid. Dr. Richard McIntosh of the University of Colorado, Boulder, discusses microtubules, the organelles of the cytoplasm that are associated with changes in the shape of a cell, with the motion of cell particles, and with motion of the cell itself.

One aspect of cell anatomy that receives much attention today is the membrane. The film includes a discussion of the membrane as a living boundary. It is a fluid, chemically functioning surface that allows the cell to communicate with its neighbors and interact with its environment. One of the most important features of the cell membrane is its participation in the process of transport of particles into and out of cells. Through animation the film presents the current theory of membrane structure and the way in which a protein component of the membrane might transport particles from the surface to the cytoplasm.

Membranes also form boundaries for the many subcompartments of a eucaryotic cell. In this case enzyme molecules are integral components of the membrane, and animation demonstrates how a single enzyme molecule might function to regulate a repetition that breaks down 100,000 other molecules in one second.

The intense activity of an enzyme leads into the final discussion of the film, which revolves around the question of the eventual "aging" of molecules as part of the process of life and death at the molecular level.

THE CELL: A FUNCTIONING STRUCTURE PART II

In Part II, the viewer can take a closer look at some of the important chemical events that take place within the cell.

Proteins function as both structural units and chemical catalysts. Protein catalysts, or enzymes, regulate nearly all of the cell's anabolic (synthetic) and catabolic (degradative) activities. An enzyme carries out its catalytic activity by using the building blocks (amino acids) that make up the chain-like backbone of the molecule and the unique three-dimensional folding of the chain itself. The composition of the 20 amino acids commonly found in proteins is covered in the film, as is the nature of the protein that they form. The final folding of the peptide chain is discussed in terms of the structural characteristics of the "active site" and the flexibility of protein molecules.

The film takes a brief look at some of the functions that proteins perform in addition to enzymatic catalysis and demonstrates the critical role that this class of macromolecules plays in controlling bodily functions.

A fascinating phenomenon unfolds as the function of the genetic code is traced from DNA (deoxyribonucleic acid) to RNA (ribonucleic acid) and finally to the bonding together of the peptide chain from separately coded amino acids on the ribosomes. The structures of the nucleotides that make up the two chains of DNA are described through the use of animation. The film examines the complex three-dimensional configuration of the whole DNA molecule, commenting on how this structure enables it to perform such a vital role in the cell. The disease sickle cell anemia is used to illustrate how a single mistake in the sequence of nucleotides that make up the DNA template for hemoglobin can result in a serious malfunctioning in the oxygen-carrying capacity of blood.

The film emphasizes that the formation of proteins requires an energy source. The cell's energy is supplied by the mitochondrion. Upon close examination, the viewer discovers a fascinating correspondence between structure and function in this organelle, which facilitates the transfer of nutrient bond energy into ATP (adenosine triphosphate), the energy-storage molecule of the cell. In the mitochondrial membrane system, the arrangement of groups of enzymes involved in a series of sequential reactions provides an excellent example of structure-function relationships within the cell.

The film then explores an equally intriguing structure—the chloroplast of the plant cell—in which a series of chemical reactions called photosynthesis harnesses photo energy from the sun for use in the synthesis of nutrient materials. Here, as in the mitochondrion, the viewer sees the way in which the structure of the organelle enhances the efficiency of its function.

All the cells in the human body contain identical genetic material, at least in type if not in quantity. How, then, can different cells perform unique functions or a single cell regulate its enzyme level in response to environmental changes? A film look at the Jacob-Monod model of gene repression in the bacterium *Escherichia coli* illustrates one way in which cytoplasmic nutrient levels can affect the early steps of protein synthesis in the nucleus. The components involved and the functioning of this system are developed in detail.

The film concludes with a look at the phenomenon of cellular communication as exemplified by the development of a fertilized egg and the rhythmic, synchronous beating of heart cells in tissue culture. The well-characterized interactions of blastula cells in developing invertebrate eggs lead to a discussion of the mesenchymal-epithelial interactions that occur during the differention of some of human organ systems.

MUSCLE

One of the most striking differences between plants and animals is the mobility of animals as opposed to the almost total immobility of plants. The only movement that most plants can achieve is the slow process of growth.

Because plants synthesize their own food by the process of photosynthesis, they do not need to move for their nourishment. But immobility can be a disadvantage for a plant; a flower next to a lake may wilt and die for lack of water unless its roots grow to reach it. An animal, however, does not have to wait for growth. It can move through the environment to reach food or water. The effectors of this superior mobility are an animal's muscles, groupings of specialized cells that contract in response to stimuli. The film *Muscle* summarizes the knowledge concerning the function of muscle tissue and illustrates the amazing diversity of movements effected by muscles. The chemical basis of muscular contraction is also discussed.

The diversity of movement of which man is capable is provided by over 600 different muscles acting in combination. In the human body, as well as other vertebrates, there are three different types of muscles. These are extensively demonstrated with animated drawings based on the anatomical sketchbooks of Leonardo Da Vinci.

The first class is striated muscle, which enables movement. Striated muscles (also termed skeletal muscles) are attached to the bones by tendons. The second type is smooth muscle, which functions mainly in the digestion of food. Smooth muscles make up most of the digestive tract, including the swallowing mechanism, the esophagus, and the stomach. Smooth muscle also provides contraction in the bladder, arteries, veins, intestines, and the uterus in women.

Smooth muscles are involuntary—that is, they are not consciously controlled but are controlled by the autonomic nervous system. Nerve impulses originating in the control centers of the brain and spinal cord travel down nerve cells to their ends, where the impulses innervate the muscle. The smooth muscles receive two different kinds of impulses—one that stimulates the muscles to contract and one that relaxes them. Smooth muscles characteristically contract slowly. The body systems that utilize smooth muscle (such as the digestive system) do not require fast action to work effectively.

The third type of muscle is cardiac, or heart, muscle, which is responsible for pumping blood. Blood brings nourishment to all parts of the body (including all the muscle systems) in the form of food materials and oxygen. Blood is also responsible for removing waste products. Transfers of materials into and out of the blood occur across the walls of the capillaries, the smallest of blood vessels. Cardiac muscle combines characteristics of both smooth and striated muscles. Outwardly, the cells of cardiac muscle look much like striated muscle cells. However, as with smooth muscle, their contraction is involuntary.

Cardiac muscle has a unique characteristic. Unlike most smooth and striated muscle, it does not need to be externally innervated by the nervous system to act. It has a self-contained, or myogenic, origin of contraction. The nerves that innervate the heart do not initiate contraction; they only modify it.

The heartbeat originates in a specialized area of muscle cells called the pacemaker. Its pace is transmitted by electrical couplings to adjacent cells, causing them to contract rhythmically. The contraction of the cells flows in this manner across the heart in a wavelike motion.

If a single cardiac cell is isolated, it will beat or contract by itself. The contact of the two cardiac cells together gives rise to a coupling, with the fastest cell governing the rate of both. From the first contraction of the heart in the human embryo, the heart steadily and automatically contracts more than once per second, or over 100,000 times every day throughout life.

Animation in the style of Da Vinci continues with an elucidation of the chemical processes by which muscle contraction occurs. The roles of membrane potential and active transport, as well as the "sliding filament" theory of muscle contraction, are shown in graphic detail.

The film concludes with two researchers from the University of Colorado, Dr. Thomas Budzynski and Dr. Johann Stoyva. Their work is involved with monitoring muscle action potentials and feeding back this information to patients with severe muscle contractions or muscle tension headaches. This feedback loop trains patients to control muscles, such as the frontalis muscle of the forehead, thereby relieving tension headaches. A patient is shown being trained. Similar treatments are being developed for insomnia, psychosomatic illnesses, and hypertension.

BIOLOGY TODAY Book Team

John H. Painter, Jr. · *Publisher*
Susan Harter · *Publishing Coordinator*
Cynthia MacDonald · *Acting Editor*
Carolyne Hultgren · *Editorial Assistant*
Donald Fujimoto · *Senior Designer*
Dale Phillips · *Associate Designer*
Linda Higgins · *Assistant Designer*
Richard Carter · *Design Assistant*
Sheridan Hughes · *Production Supervisor*

John Ochse · *Sales Manager*
William Bryden · *Science Marketing Manager*
Jan Lindsey, Georgene Martina, Jaque McLoughlin,
Nancy Taylor · *Sales Coordinators*
Nancy Hutchison Sjöberg · *Rights and Permissions Supervisor*

CRM BOOKS

Richard Holme · *President and Publisher*
John H. Painter, Jr. · *Publisher, Life and Physical Sciences*
Roger E. Emblen · *Publisher, Social Sciences*
Tom Suzuki · *Director of Design*
Charles Jackson · *Managing Editor*
Henry Ratz · *Director of Production*

Officers of Communications Research Machines, Inc.

Charles C. Tillinghast III · *President*
Richard Holme · *Vice-President*
James B. Horton · *Vice-President*
Paul N. Lazarus III · *Vice-President*
Wayne E. Sheppard · *Vice-President*